World Tribology Congress

UK Organizing Committee

Dr Bill Roberts (Chairman)
Dr Alex Alliston-Greiner, Plint & Partners Limited
Professor Tom Bell, University of Birmingham
Professor Brian Briscoe, Imperial College of Science, Technology & Medicine
Mr David Carnell, NSK-RHP European Technology Centre
Dr Les Hampson, National Centre of Tribology
Dr Ian Hutchings, University of Cambridge
Dr H Peter Jost, President, International Tribology Council (Ex-Officio)
Mr Stephen Maw, Neale Consulting Engineers Limited
Dr David Parkins, Cranfield Institute of Technology
Mr Julian Reed, Neale Consulting Engineers Limited
Dr Rob Rowntree, European Space Tribology Laboratory, Chairman, IMechE Tribology Group,
Dr Brian Roylance, University of Wales
Professor Hugh Spikes, Imperial College of Science, Technology & Medicine
Professor Chris Taylor, University of Leeds
Mr Bob Wood, Consultant to the Tribology Group, IMechE
Dr David Yardley, David Yardley & Associates

Patron of the Congress
HRH The Prince Philip, Duke of Edinburgh KG KT

Abstracts of papers from:

I MECH E
150th Anniversary
1847 - 1997

World Tribology Congress

8–12 September 1997

Organized by the Tribology Group of
the Institution of Mechanical Engineers (IMechE)

Sponsored by

NSK-RHP
AEA Technology plc
GKN
Climax Molybdenum Company
Pall Industrial Hydraulics

Published by Mechanical Engineering Publications Limited for
the Institution of Mechanical Engineers, Bury St Edmunds and London.

First Published 1997

This publication is copyright under the Berne Convention and the International Copyright Convention. All rights reserved. Apart from any fair dealing for the purpose of private study, research, criticism or review, as permitted under the Copyright, Designs and Patents Act, 1988, no part may be reproduced, stored in a retrieval system, or transmitted in any form or by any means, electronic, electrical, chemical, mechanical, photocopying, recording or otherwise, without the prior permission of the copyright owners. Reprographic reproduction is permitted only in accordance with the terms of licences issued by the Copyright Licensing Agency, 90 Tottenham Court Road, London W1P 9HE. *Unlicensed multiple copying of the contents of this publication is illegal.* Inquiries should be addressed to: The Publishing Editor, Mechanical Engineering Publications Limited, Northgate Avenue, Bury St. Edmunds, Suffolk, IP32 6BW, UK. Fax: 01284 704006.

© With authors

ISBN 1 86058 109 9

A CIP catalogue record for this book is available from the British Library.

Printed by Bookcraft Limited, Bath, UK

The Publishers are not responsible for any statement made in this publication. Data, discussion, and conclusions developed by authors are for information only and are not intended for use without independent substantiating investigation on the part of potential users. Opinions expressed are those of the Author and are not necessarily those of the Institution of Mechanical Engineers or its Publishers.

Acknowledgements

Associated Bodies

The First World Tribology Congress is organized with the full support of:

Egyptian Society of Tribology*
Finnish Society for Tribology*
Gesellschaft für Tribologie*
Institute of Materials
Institute of Physics Tribology Group
Institution of Chemical Engineers
Italian Tribology Group of AIMETA*
Japanese Society of Tribologists*
Japan Society of Mechanical Engineers
Leeds-Lyon Symposium
Lithuanian Scientific Society Tribologija*
NORDTRIB
Polish Tribological Society*
Romanian Tribology Association*
Royal Society of Chemistry
Russian Association of Tribology Engineers*
Russian National Tribology Committee*
Scientific Society of Mechanical Engineers

Sectie Tribologie BvMK*
Slovak Society of Tribology*
Slovenian Society of Tribology*
Société Tribologique de France*
Society of Bulgarian Tribologists*
Society of Chemical Industry
Society of Tribologists and Lubrication Engineers*
Society of Tribologists of Belarus*
South African Institute of Tribology*
Tribology Division of the American Society of Mechanical Engineers*
Tribology Institution of the Chinese Mechanical Engineering Society*
Tribology Society of India*
Tribology Society of Nigeria*
Ukrainian National Tribology Society*
Wear of Materials
Yugoslav Tribology Society*

*Member of the International Tribology Council

Related Titles

Title	Author/Editor	ISBN
New Directions in Tribology	Edited by Ian Hutchings	1 86058 099 8
An Introductory Guide to Industrial Tribology	J D Summers-Smith	0 85298 896 6
A Tribology Casebook	J D Summers-Smith	1 86058 041 6
History of Tribology	D Dowson	1 86058 070 X
Lubrication and Lubricant Selection	A R Lansdown	1 86058 029 7
Lubricants in Operation	U J Moller & U Boor	0 85298 830 3
Lubrication of Gearing	W J Bartz	0 85298 831 1
Analysis of Rolling Element Bearings	Wan Changsen	0 85298 745 5

For the full range of titles published by MEP contact:

Sales Department
Mechanical Engineering Publications Limited
Northgate Avenue
Bury St Edmunds
Suffolk
IP32 6BW
UK

Tel: 01284 724384
Fax: 01284 718692

800 varieties of lubricants.
There are still some mechanisms
that work without us.

With more than 800 types of lubricants sold in 70 countries, TOTAL can be satisfied: lubricating almost everything, everywhere. But TOTAL doesn't rely on this success and carries on innovating unswervingly and testing its lubricants in the most extreme conditions. This is the reason why you may find TOTAL lubricants in car-races, in the desert sands of long-distance car rallies and in the heat of Formula 1 circuits. Nevertheless, a few mechanisms are so perfect they will never need us.

TOTAL

Gearboxes

Gears

Spiradrive

100 A CENTURY OF GEAR DESIGN AND MANUFACTURE

The name Davall has become synonymous with high quality gears and gearboxes. With our design and prototyping departments we have been able to satisfy the needs of such diverse gearbox projects as a clockwork radio through to Main Battle Tank. In addition, we design and manufacture SPIRADRIVE® a unique gear system able to offer zero backlash and a high torque transmission capability in a small envelope.

To find out more call us on 01707-265432.

DAVALL
The Davall Gear Company Ltd

Welham Green, Hatfield, Hertfordshire AL9 7JB
Telephone: 01707-265432 Fax: 01707-268536
Email: davall@dial.pipex.com

Driven by Quality

Contents

Abstracts of Oral Sessions

MONDAY

M1	Thick film bearings – I	3
M2	Lubricant chemistry and rheology – I	21
M3	Wear mechanisms and contact mechanics	39
M4	Surface engineering – I	55
M5	Solid lubrication	73
	Measurement and simulation – I	81

TUESDAY

TU1	Thick film bearings – II	91
	EHL and boundary lubrication – I	99
TU2	Lubricant chemistry and rheology – II	109
TU3	Wear in metallic systems	127
TU4	Surface engineering – II	143
TU5	Magnetic storage systems	161

WEDNESDAY

W1	EHL and boundary lubrication – II	179
W2	Automotive and other applications	195
W3	Wear by hard particles – I	211
W4	Wear in ceramic systems	227
W5	Bio-tribology	243

THURSDAY

TH1	Rolling bearings	263
TH2	Industrial problems and test methods	281
TH3	Wear by hard particles – II	297
TH4	Wear in polymer systems	313
TH5	Manufacturing and maintenance	327

FRIDAY

F1	Environmental issues	345
	Seals	349
F2	Tribology in extreme environments	353
F3	Micro- and nano-tribology	361
F4	Thermal effects and tribo-chemistry	369
F5	Measurement and simulation – II	377

Contents

Abstracts of Poster Sessions

Corresonding oral session

MONDAY

M1/TU1	Thick film bearings	387
M2/TU2	Lubricant chemistry and rheology	443
M5	Solid lubrication	473
M4/TU4	Surface engineering	479

TUESDAY

M3/TU3	Wear and friction in sliding systems	517
TU5	Magnetic storage systems	579
F1	Rotordynamics	593
TH5	Tribology in manufacturing and maintenance	597
TU1	Tribology education and training	633

WEDNESDAY

TU1/W1	EHL and boundary lubrication	645
W3/TH3	Wear by hard particles	665
W4	Tribology of ceramic materials	675
M5/F5	Measurement and simulation	691
W5	Bio-tribology	731
W2	Practical applications of friction and wear	749

THURSDAY/FRIDAY

F3	Micro- and nano-scale tribology	773
TH1	Rolling bearings	795
F4	Thermal effects in tribosystems	815
F4	Tribochemistry and corrosion wear	823
TH4	Tribology of polymers	831
TH2/TH5	Industrial problems and solutions	843
F2	Tribology in extreme environments	861
W2	Practical applications of friction and wear	869
F1	Environmental issues	901

Foreword

On this unique occasion a number of major international and national tribology meetings scheduled for 1997 have been integrated into the Congress, including ASME and STLE events, NORDTRIB, EUROTRIB '97 and the Diamond Jubilee Conference commemorating the first International Conference on Lubrication and Wear, held in London in 1937, at the very same venue.

This prestigious Congress is a true World focus for tribology. It brings together the principal engineers, scientists and practitioners from around the world working on *the science and technology of interacting surfaces in relative motion* in all its diverse aspects. The meeting will review the state of the art, define current research, explore future opportunities and consolidate progress in tribology at a world level. We have five Plenary Lectures and thirty-four Invited keynote presentations by renowned international experts, which are backed by high-quality oral and poster presentations covering all aspects of tribology. The opportunities for participants to obtain an overview of recent advances in all areas of tribology, and to make contacts on a truly international scale, are unparalleled and I trust that delegates will grasp this unique opportunity to make new contacts with fellow workers from around the world.

The Organising Committee was extremely gratified by the tremendous international response to the call for papers. This volume contains the abstracts of over 900 contributions, from some 45 countries, which will be presented over the five days of the Congress.

We hope that delegates will enjoy a stimulating, interesting and memorable meeting.

Finally, I take this opportunity to thank all the Tribology Societies, world-wide, which have given the Congress their support. And last, but by no means least, our sincere thanks go to our Sponsors for their generous financial contributions which have been crucial to underpinning all our efforts towards making a success of this major event, the largest of its kind ever organized by the Institution of Mechanical Engineers.

Bill Roberts

Chairman, Organizing Committee
First World Tribology Congress

MONDAY
8
SEPTEMBER

Page number

M1	Thick film bearings –I	3
M2	Lubricant chemistry and rheology – I	21
M3	Wear mechanisms and contact mechanics	39
M4	Surface engineering – I	55
M5	Solid lubrication	73
	Measurement and simulation	81

AN EXPERIMENTAL STUDY OF THRUST PAD FLUTTER

WILLIS W. GARDNER
Waukesha Bearings Corporation, Waukesha, WI, USA

ABSTRACT

Tilting pad type thrust bearings are commonly used on both the active and inactive (loaded and unloaded) sides of thrust collars of turbomachinery rotors. The thrust pads on the unloaded side operate with relatively large "film thicknesses" due to the axial end clearance used. This clearance prevents the "unloaded" bearing from operating with a significant self-generated hydrodynamic load, which in turn adds to the load on the "loaded" bearing, in addition to generating additional heat and power loss,

Fatigue failure of temperature sensor leadwires to unloaded thrust pads, and abnormal wear at contact points of these pads, in several field applications, prompted an investigation into the cause of these phenomena. Since this occurred only on "unloaded" thrust bearings, flutter (vibration) of these thrust pads was suspected. Laboratory tests were conducted to study this phenomena.

A 267mm (10.5 inch) double thrust bearing was used for these tests. Non-contact proximity probes were used to "view" the back faces of two pads in the unloaded bearing and one pad in the loaded bearing. Vibration amplitude was of primary interest but frequency was also monitored. Pressure probes were mounted in the oil discharge annulus of the housing to monitor both static and dynamic pressure data. Means to vary the restriction in the tangential discharge were used, including an "open" discharge.

Unloaded side thrust pad flutter is not an unknown phenomenon. Thrust bearing tests comparing lubricant supply methods produced this action and were reported in (1). Pad vibration was evident from the fatigue failure of instrumentation wiring to the unloaded pads, and also was manifested as an audible clicking sound emanating from the housing. Those problems were eliminated by the use of an oil control ring around the thrust collar, generating a backpressure in the discharge annulus. It was noted, however, that this oil control ring (discharge restriction) adversely affected the thrust bearing power loss.

Figure 1 is a plot of pad vibration amplitude vs oil flow rate over a range of flows (from the tests reported here), showing the large vibration amplitudes found at the higher flows, in addition to those that develop at the low end of the flow range. The "low flow" vibrations are consistent with the formation of incomplete oil films. The "high flow" vibrations had the nature of resonant vibrations resulting from an exciting force in tune with a natural oscillating frequency of the pad/oil film system.

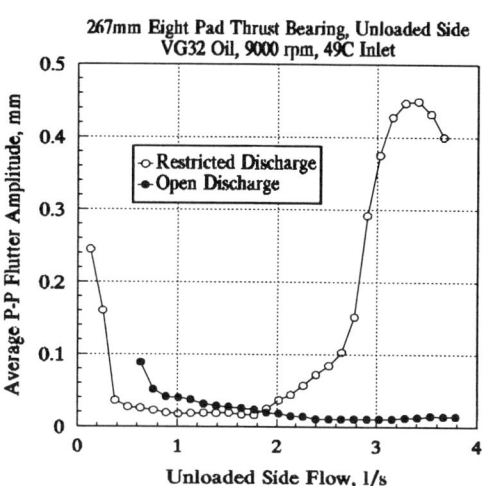

Fig. 1: Flutter Amplitudes

A review of fluid flow technology, and specifically (2), indicated that vortices were possibly being formed, within the oil flowing in the annular chamber surrounding the thrust bearing, by the throat (cutwater) where this annular chamber connects to the tangential discharge. These vortices would be reflected as pressure pulsations and would represent an oscillating force on the thrust pads as these vortices traveled around the annular chamber. Pressure oscillations were found in the oil discharge annulus at the same frequency as the pad vibrations.

These tests studied the effects of a number of geometry and operating variables on this pad flutter and the results of these are included in the paper. The primary findings were:

1. An unrestricted, open discharge essentially eliminated the "high flow" pad flutter, indicating that this phenomenon is related to the action of the oil in the discharge paths.
2. This unrestricted, open discharge did not eliminate the "low flow" pad flutter. Increased restriction of the discharge required lower flows for this "low flow" flutter to develop, consistent with pad starvation as the cause.

REFERENCES

(1) Mikula, A. M., and Gregory, R. S., "A Comparison of Tilting Pad Thrust Bearing Lubricant Supply Methods," ASME JOURNAL OF LUBRICATION TECHNOLOGY, Vol.105, No. 1, Jan. 1983, pp. 39-47.

(2) Blevins, R. D., *Flow-Induced Vibration*, 2nd Edition, Krieger Publishing Company, Malabar, Florida, USA, 1994.

SPRING-SUPPORTED THRUST BEARINGS USED IN HYDROELECTRIC GENERATORS: LABORATORY TEST FACILITY

J.H. YUAN and J.B. MEDLEY
Department of Mechanical Engineering, University of Waterloo, Waterloo, Ontario, Canada N2L 3G1
J.H. FERGUSON
General Electric Canada Inc., Peterborough, Ontario, Canada K9J 7B5

ABSTRACT

A test facility was developed for the experimental investigation of large spring-supported thrust bearings that are used in hydroelectric generators. The test bearing had an outer diameter (OD) of 1.168m thus making it large enough to simulate many of the features of the thrust bearings in the field, yet small enough to be placed conveniently in an engineering laboratory. The load is imposed by a hydraulic piston through a hydrostatic bearing. Instrumentation included a distribution of thermocouples and pressure taps throughout the facility and particularly in the thrust bearing pads. In addition, eddy current probes were located in the rotor to measure oil film thickness. This facility is unique in its ability to study relatively large spring-supported thrust bearings in the laboratory rather than being forced to rely on field measurements (1).

The test bearing had 12 pads of approximately sector shape (1.168m OD, 0.711m ID, 41.1mm overall thickness) and the groove between pads had a constant width of 52.3mm. Each pad was supported by 15 coil springs and had a ring groove in the center into which oil was injected to provide a hydrostatic oil film at start-up and shut-down (Fig. 1). These dimensions are typical of many full-scale bearings in service and therefore, heat conduction through the pad thickness and convective heat transfer in the groove should be representative.

The test facility permitted the study of many features of spring-supported thrust bearings, which previously have only been examined theoretically (2). For example, using Terresso 46 oil at 70°C and increasing loads, pad temperature and film pressures increased while film thicknesses decreased, as expected. However, for a given load, decreasing speed led to a drop in measured values of both temperature, film thickness and, surprisingly, pressure. This shift in the pressure distribution may have been caused by cavitation or thermal crowning of the pad.

In general, the film thickness measurements appeared plausible but were not without difficulties. Under the load of 177kN per pad, probe number 3 indicated a film thickness in the vicinity of zero (Fig. 2) and thus the absolute film thickness values were judged to be inaccurate. The source of this inaccuracy might have been the procedure for determining the voltage level for zero film thickness by stopping the rotor motion which allow the warmer pad surface to shift the probe temperature.

The present study illustrated that a useful test facility was now available. In particular, the oil film thickness could be measured and, although the absolute magnitudes were not accurate as yet, the pad shapes did seem plausible. Future work on comparisons with the theory of Ettles (2) is in progress.

Fig. 1: Spring-supported thrust bearing pads

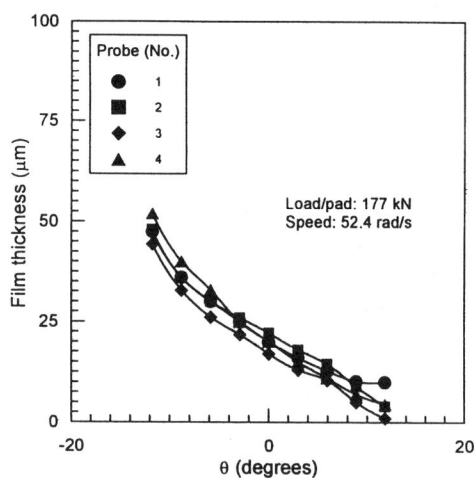

Fig. 2: Oil film thickness distribution

REFERENCES

(1) J.H. Vohr, ASME, J of Lubrication Technology, Vol. 103, 1981.
(2) C.M. Ettles, ASME, J of Tribology, 1991

Phenomena of Instability in the Fluid Film of thrust Bearings

Masami HARADA and Jyutaro TSUKAZAKI
Depertment of Mechanical Engineering, Saitama University, Urawa, Japan
Mitsuaki SEKI
Production Engineerinng Centre, Canon Inc., Ami, Japan

ABSTRACT

Experimental investigation of the instability in the fluid Film of thrust bearings are presented.

Flow between a stational and a rotational disk of 180 mm and 80 mm diameter are used as a model of fluid film of thrust bearings. The gaps between two disks are varied from 0.15 mm to 4.0 mm.

Stational disk is made of glass to visualize the flow. Water is used as a working fluid. In order to detect the onset of instability, two types of fine particles are used. One is the powder of barium sulfate. The average diameter of the particles is less than 5 μm. The other is aluminum powder. The average diameter is about 50 μm.

The Shaft rotational speed is changed from 0 to 1400 r/min. Within these shaft rotational speeds and gap ranges, the reciprocal Ekman number $1/E_k$ ($=\omega h^2/\nu$) changes from 1.0 to 100. Where ω, h and ν is an angular velocity of the shaft (rad/sec), a gap and a kinematic viscosity respectively. Under appropriate lighting, the flows are observed and taken a photo by using a conventional 35mm camera.

Photographs show that the four types of flow patterns are observed. The first one is the instability generating in high Reynold's number region (Fig. 1a), and the vortices are nearly concentric (this type of instability is called type I instability). Fig. 1b shows the second type of instability. It is observed in rather low Reynold's number region and called type II instability.

The third one is the cross patterns observed on the rotating disk (Fig. 2) and it is observed only in narrow gaps and low Reynold's number regime. We call this type III instability. In the just outside of the cross pattern, we can see spiral pattern. We call this type IV 0 instability.

Relations between Critical Reynold's number Re_δ ($=\omega r\delta/\nu$) and reciprocal Ekman number $1/E_k$ are showen in Fig. 3. Where r is the location of the inner boundary of each instability and $\delta = (\nu/\omega)^{1/2}$. Subscript I, II, III, IV 0 express the criteria for type I, II, III, IV 0 instability respetively and "0" means the value at the outer boundary. The criteria for the type I instability is agree well with other experimental and analytical results. Silivat's experimental results agree with the critria for type II and type IV 0.

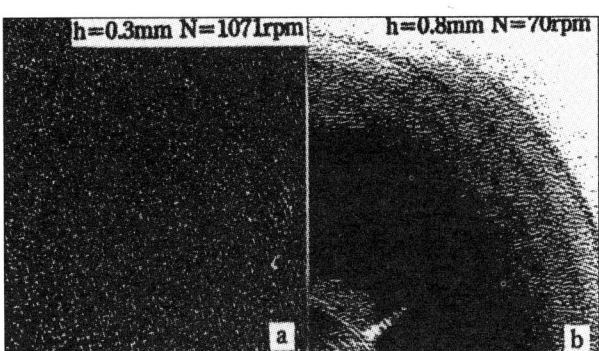

Fig.1 Photographs of voltices and Instability

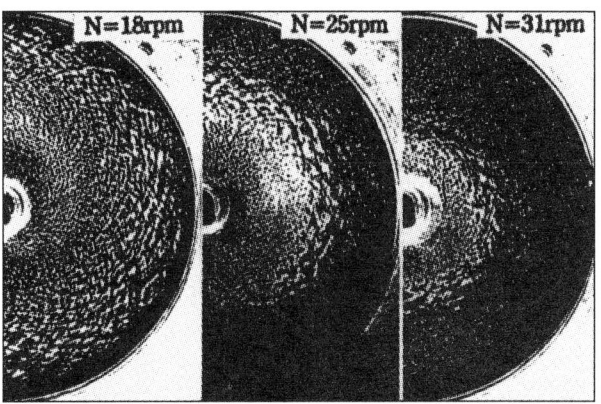

Fig.2 Photgraphs of cross patterns for h=0.7mm

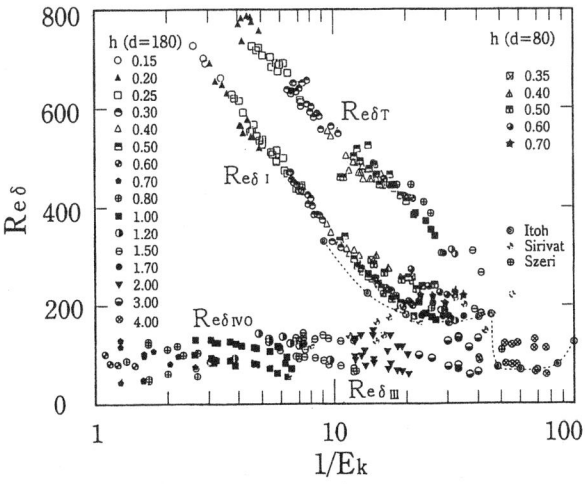

Fig.3 Relation between critical Reynold's number Re_δ and reciprocal Ekman number $1/E_k$

POWER LOSSES IN THE PIVOTED SHOE JOURNAL BEARING

K R BROCKWELL and W DMOCHOWSKI
National Research Council, U-89, Alert Road, Uplands, Ottawa, K1A 0R6, Canada
S DECAMILLO
Kingsbury, Inc, 10385 Drummond Road, Philadelphia, PA, 19154, USA

INTRODUCTION

The pivoted-shoe journal (PSJ) bearing is today's industry standard for rotordynamically sensitive rotating machinery and is widely used in a variety of high power-density machinery. Recently, attention has focused on the steady-state behaviour of the PSJ bearing, particularly in regard to reducing oil flowrate requirement and power loss, and increasing load carrying capacity. Ideally, these objectives should be reached without raising bearing operating temperatures.

This paper presents results from an experimental and theoretical study of the power losses in a 150 mm diameter PSJ bearing with an L/d ratio of 0.46. Offset and centre pivoted shoes were tested for both "on-pad" and "between-pads" loading conditions.

EXPERIMENTAL STUDY

The experimental study was performed on a new test facility that measures the static operating characteristics of radial type bearings, with particular reference to bearing power loss. Load from a pneumatic loading system is applied to the test bearing through two hydrostatic bearing systems. The upper spherical hydrostatic bearing ensures good alignment between test bearing and shaft, and also permits bearing torque reaction to be measured even under heavily loaded conditions. The second lower hydrostatic bearing provides lateral freedom for the test bearing. The test facility is driven by a 112 kW variable speed DC motor and has maximum load and speed capabilities of 25 kN and 15,500 rpm respectively.

THEORETICAL STUDY

A quasi 3D thermohydrodynamic model of the PSJ bearing was used to calculate the viscous shearing losses at the surface of each shoe (1). This model accounts for heat conduction across the oil film, heat convection in the circumferential direction, turbulence, oil mixing and pad deflection. Besides shearing losses, the model also calculates parasitic losses in the bearing. These include churning losses which result from oil exposed to the moving surfaces, and kinetic losses which are associated with the delivery of oil into the bearing housing.

RESULTS AND DISCUSSION

Measured and calculated bearing power losses for the offset pivot bearing with a load of 14 kN are shown in Figure 1 below.

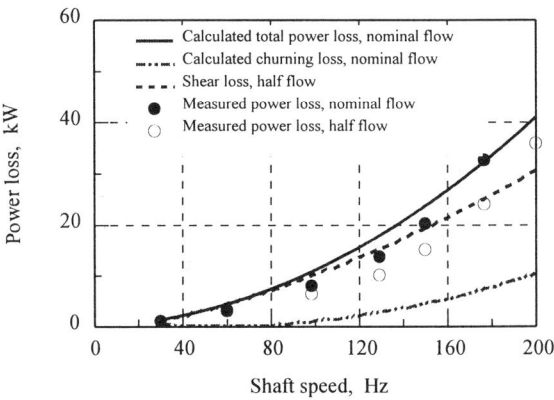

Fig. 1: **Measured and calculated power losses**

The experimental results confirm that reductions in oil flowrate will lead to reductions in bearing power losses. As Figure 1 shows, this seems to be the result of reducing the churning losses in the bearing cavity. This conclusion is reached on the basis that, for all shaft speeds, the calculated churning losses are similar to the difference between the measured power losses at nominal and half flowrate conditions (in this analysis, the half oil flowrate is similar to the calculated side flow of the bearing). Thus, it is justified to assume that, for the half flowrate condition, the bearing cavity contains only a limited amount of oil. Furthermore, because bearing metal temperatures rise only slightly with a 50% reduction in flowrate, it appears that bearing efficiency (power loss) can be improved significantly without compromising operational efficiency (bearing metal temperature).

REFERENCES

(1) W. Dmochowski et al, ASME Journal of Tribology, Vol. 115, 219-226 pp, 1993.

AN OPTIMUM DESIGN METHOD FOR TILTING PAD JOURNAL BEARINGS

MASAYUKI KURITA
Mechanical Engineering Research Lab., Hitachi Ltd., Kandatsu, Tsuchiura, Ibaraki 300 Japan
MASATO TANAKA*
Department of Mechanical Engineering, University of Tokyo, Hongo, Bunkyo-ku, Tokyo 113 Japan

ABSTRACT

This paper proposes a computer-aided optimum design for the hydrodynamic bearings used for turbomachinery, gives a computer programme coded by the authors, shows how it works in the case of tilting-pad journal bearings, and confirms the solution obtained is optimum.

First, the authors studied practical design procedures of tilting-pad journal bearings widely used in industries, and they clarified the framework and the flow chart for bearing design. Designing a bearing was found to involve obtaining a solution of the bearing specification (design variables) such as bearing clearance, preload factor, and bearing width, under given operating conditions. This solution must satisfy the required bearing performance (state variables) such as minimum film thickness, maximum pressure in the oil film, frictional torque, and oil flow rate required.

Next, behaviours of the state variables were analyzed using the iso-viscous hydrodynamic lubrication theory. It was discovered that the combination of the gradient-projection method and the direct-search method is the most suitable for this field of design.

The design procedures, the suitable optimizing method, and the results of sensitivity analysis were rearranged into an "optimum design template", consisting of three parts (Fig.1): a design worksheet, filling rule of the worksheet, and a solver programme. With the help of the filling rule, a design engineer fills the blanks in the worksheet. Then, the solver programme repeats calculation automatically and obtains an optimum solution.

Table 1 shows an optimum design example and its obtained solution. Load, rotational speed, and journal diameter are entered as constants. Both upper and lower limits are given to the design variables. One appropriate state variable has to be selected as an objective function, while constraints are imposed on the other variables. By dividing the design space by the mesh and calculating objective function on each point, the solution was verified to be truly optimal (Fig. 2).

The optimum design template presented here enables design engineers to obtain a satisfactory solution of bearing specification easily, directly, and rapidly by entering operating conditions and required bearing performance. Furthermore, the obtained solution, guaranteed to be optimum, has higher quality than that obtained by means of conventional design methods.

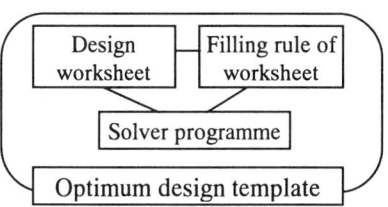

Fig. 1 Optimum design template

Table 1: An optimization example

CONSTANTS	
Load (kgf)	= 1000
Rotational speed (rpm)	= 4000
Journal diameter (m)	= 0.1
DESIGN VARIABLES	
Pad L/D	[0.4, 0.5] 0.46
Pad pivot position	[0.5, 0.6] 0.51
Bearing radial clearance (mm)	[0.075, 0.085] 0.075
Preload factor	[0.4, 0.8] 0.4
Angular extent of pad (deg.)	[48, 60] 60
Effective viscosity of oil (Pa s)	[0.021, 0.026] 0.026
Load direction	= LOP
Number of pads	= 5
STATE VARIABLES	
Minimum film thickness (mm)	> 0.02
Frictional loss (kW)	
Required oil flow rate (l/min)	Objective function
Maximum pressure (MPa)	< 10
Spring coefficient (kgf/mm)	
Damping coefficient (kgf s/mm)	

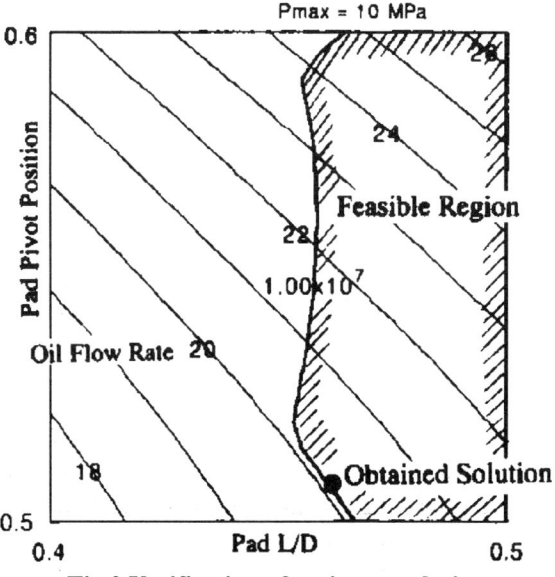

Fig.2 Verification of optimum solution

TRANSIENT THERMOELASTOHYDRODYNAMIC STUDY OF TILTING-PAD JOURNAL BEARINGS - APPLICATION TO BEARING SEIZURE

P. MONMOUSSEAU, M. FILLON and J. FRENE,
Université de Poitiers, UMR CNRS 6610, Laboratoire de Mécanique des Solides,
S.P.2M.I., BP 179, 86960 Futuroscope cedex, FRANCE.

ABSTRACT

Nowadays, the tilting-pad journal bearings are used under more and more severely demanding operating conditions. During start-up, the rapid increase of the temperature in the bearing solids leads to the thermal expansion of both the pads and the shaft. In this case, the operating bearing clearance decreases and, when it tends toward zero, seizure occurs. The evolution of the main characteristics (temperature, pressure, film thickness and displacements) versus time is analyzed in the case when a seizure occurs.

The transient behavior of the tilting-pad journal bearings is very important during acceleration : both the temperature in the film due to the viscous dissipation and the temperature in the bearing elements (shaft and pads) increase quickly, especially when the acceleration of the shaft is significant.

Few studies about lubricated mechanisms have been published. In 1975, Conway-Jones and Leopard (1), studied the seizure of a tilting-pad journal bearing experimentally. They concluded that nominal rotational speed and initial temperature were the main causes of bearing seizure. Recently, Pascovici and al. (2) performed a theoretical investigation about plain bearing seizure and, Kucinschi (3) completed a comparison between theoretical results (using a simple lumped model) and experimental data : a strong agreement has been established. But, these studies treated only the case of an unloaded plain journal bearing.

A transient thermal analysis of a tilting-pad journal bearing submitted to an acceleration while operating in safe conditions has been performed. Once all the thermal phenomena (viscous dissipation in the fluid and heat transfer in the pads. in the shaft and in the housing) and the thermomechanical displacements have been taken into account, theoretical results are confirmed with experimental data (4).

When the bearing operates under severe operating conditions, the thermal displacements of the bearing elements induce a decrease in the radial bearing clearance and can lead to the bearing seizure. For example, when the tested bearing (4 pads, L/D = 0.7, C_b = 50.10^{-6} m) without external load is accelerated from 0 to 10000 rpm in 5 s, the theoretical seizure occurs after approximately 50 s (Fig. 1). Just before the seizure occurs, when the minimum film thickness is equal to few micrometers, the maximum pressure is high and the maximum temperature reaches 150 °C.

Most of the solid displacements are taken into account in this study and it is shown that the thermal displacements are more dominant than the mechanical ones. While the shaft diameter (0.1 m) and pad thickness (0.02 m) variations lead to a decrease in the radial bearing clearance (negative values), the housing thickness (0.03 m) and the pivot deflection lead to an increase in the radial bearing clearance (positive values). The seizure phenomenon is rapid and thus, the thermal expansion of the housing is not very significant (17 % of the total displacements). The seizure is partially due to the thermal expansion of the pad (20 %) and mainly to the thermal expansion of the shaft (55 %). The pivot deflection represents only 8 % of the total displacements.

During the transient regime, the locations of the maximum temperature, the minimum film thickness and the maximum pressure are moving toward the pivot zone because of the pad rotation and the shaft's thermal expansion.

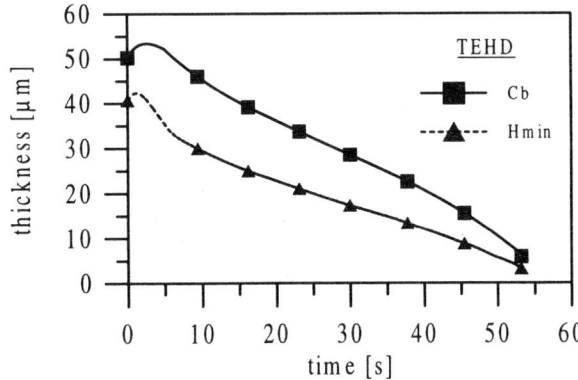

Fig. 1: Transient decrease of the radial bearing clearance and the minimum film thickness

REFERENCES

(1) Conway-Jones J.M. and Leopard A.J., Proc. 4th Turbomachinery Symposium, Texas A & M University, 55-63 pp., 1975.
(2) Pascovici M.D., Khonsari M.M. and Jang J.Y., ASME Journal of Tribology, Vol. 117, 744-747 pp., 1995.
(3) Kucinschi B., Proc. of the 7th International Conference on Tribology, ROTRIB'96, Bucharest, Vol. 18, 103-109 pp., 1996.
(4) Monmousseau P., Fillon M. and Frêne J., ASME/STLE Tribology Conf., San Fransisco, October 13-17, paper 96-TRIB-29, 1996.

TEMPERATURE PREDICTION IN JOURNAL BEARING USING A NEW CAVITATION MODEL

K.N. MISTRY
S.V. Regional College of Engg & Tech., Surat-395 007, India
S. BISWAS and K ATHRE
Indian Institute of Technology, New Delhi - 1 IOOl6 India

ABSTRACT

The analytical model assuming the cavitation region as an axial contraction of uncavitated film predicts rising temperature which is contradictory to the experimental evidence. The 'weak bond' between the stationary bearing inner surface and the lubricant film and the nearly constant velocity profile of the lubricant across film in the cavitation zone reveal that the viscous shear in the cavitation zone is almost negligible. Hence in the absence of viscous force and the pressure force in the cavitation zone, it is suggested in this paper that, the lubricant film sustained due to surface tension and centrifugal force only. Using this approach an elemental heat balance incorporating the width of streamer is established by the authors to predict the temperature in the cavitation region of journal bearing.

Mistry et al. (1) have noted that by incorporating the effective length (EL) model proposed by Fillon et al. (2), the bush inner surface temperature predicted especially for the low or moderate eccentricity ratio exhibits more or less rising trend in the cavitation zone. The modified form of the energy equation proposed by Mistry et. al. (1) by neglecting the viscous shear effect and applying the weightage factors in terms of the eccentricity ratio, provided technique to predict the bush inner surface temperature much closer to the experimental values.

The reason for reduction of energy dissipation in cavitation region was explained by Heshmat and Pinkus (3). In this paper, the width of streamer is estimated based on the balance of surface tension and the centrifugal force keeping the lubricant film in equilibrium in the cavitation zone The energy equation is modified for the cavitation region by incorporating the width of streamer and by neglecting the viscous shear term. The analysis includes the consideration of the mixing at the groove between the cold oil and the recirculating hot oil. The temperature profile of the midplane along the circumference is plotted in Fig. 1

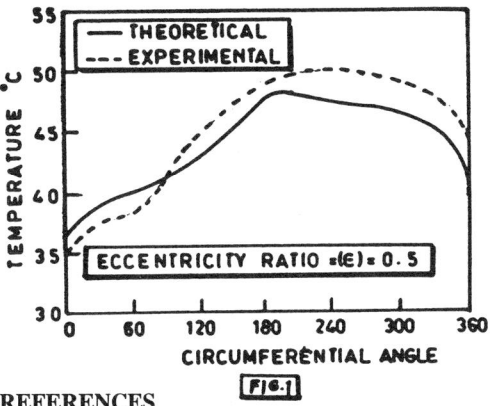

REFERENCES

(1) Mistry K., Biswas S., and Athre, K., WEAR, Vol. 159, pp. 79-87, 1992.
(2) Fillon, M., Frene, J. and Boncompain, R., Proceeding of 13th Leeds-Lyon Symposium, pp. 27-47, 1986.
(3) Heshmat, H. and Pinkus, O., ASME, Journal of Tribology, Vol 108, pp. 231-243, 1986.

START-UP TRANSIENT TEMPERATURE DISTRIBUTION IN A MECHANICAL FACE SEAL

Traian CICONE
Polytechnic University of Bucharest, Spl.Independentei 313, Bucharest 77206, RO
Bernard TOURNERIE, David REUNGOAT and Jean Christophe DANOS
Laboratoire des Mecanique des Solides, Universite de Poitiers, Futuroscope, SP2MI, 86960 Futuroscope, FR

ABSTRACT

Acceleration from stand-by conditions to operating speed occurs very quickly, causing high temperature gradients to develop along the rotor of a mechanical face seal (MFS). These temperature gradients can be large enough to cause undesired thermal deflection of the rotor, a possible cause of seal damage.

Few experiments are reported for transient regimes in a MFS. Doust and Parmar (1), and later Parmar (2) measured temperature variation in the stator of a noncontacting MFS, in correlation with film thickness. Although excellent instrumented and carefully conducted, the experiments do not offer an image of the temperature distribution in the rotor. Tournerie et al. (3) demonstrated that infrared thermography can be successfully used for transient temperature measurements in the seal interface.

The aim of the present experimental work was to obtain a more detailed information on the temperature distribution in the components of a MFS. The experiments were made on a model of an inside pressurized MFS (3) having a specially designed rotor with variable conicity (RCV) allowing to control the initial conicity of the active surface

The field data are obtained by infrared thermography, using an original technique (Fig.1). It offers simultaneously, the radial temperature distribution in the seal dam, the axial temperature distribution on the lateral surface of the rotor, and the temperature on the back of the stator. The power dissipation in the seal was evaluated by measuring the power consumption of the driving d.c. motor, with and without the seal. The partition of the heat flux between the rotor and the stator was estimated using the temperature measured on the back of the stator.

The experiments shown that during start-up the axial temperature gradient in the rotor is greater than during steady-state conditions. As conicity due to thermoelastic deformation is a function of this gradient, it is found a possible cause of the initiation of thermoelastic instabilities.

In a preceding work, Cicone et al. (4) presented an analytical model to calculate the transient temperature distribution in the rotor of a MFS, based on a fin model with variable input heat flux. A first stage of investigation permitted a partial comparison between theory and experiments. The results shown a good agreement for temperature distribution along the rotor. However, the fin model underestimates the time delay to steady-state conditions (Fig. 2). Two problems need further investigations: the more accurate estimation of the convection coefficients and the partition of the heat flux at the interface.

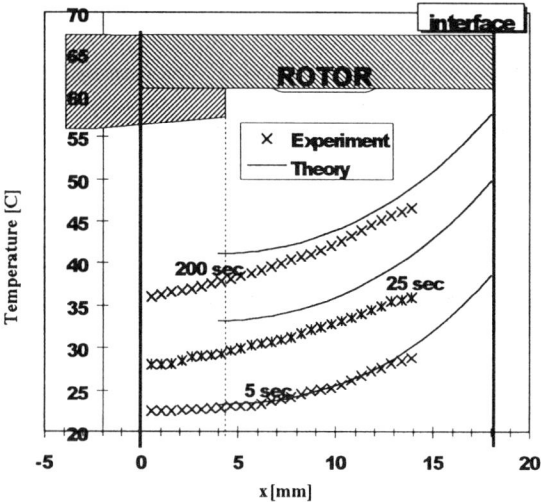

Fig. 2. Temperature variation along the rotor

REFERENCES

(1) T.G. Doust, A. Parmar, 11-th Int. Conf. on Fluid Sealing (BHRA), Cannes, Paper F4, 1987.
(2) A. Parmar, Proc. of 12-th Int. Conf. on Fluid Sealing (BHRA), pp 576-595, 1993.
(3) B. Tournerie, D. Reungoat, J. Frene, Trans. ASME J. of Tribology, Vol 113, No 3, pp 571-576, 1991.
(4) T. Cicone, D. Reungoat, B. Tournerie, Proc. of the 7th Int. Conf. of Tribology ROTRIB'96, Vol 3, pp 26-36 Bucharest, Romania, 1996.

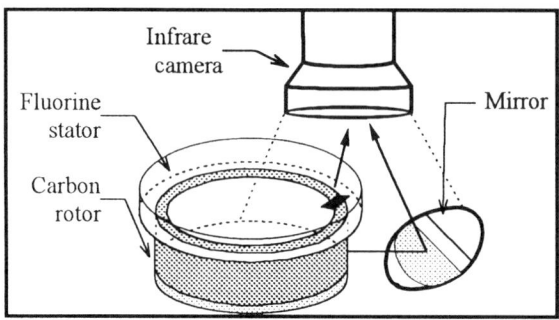

Fig. 1: Infrared camera arrangement

THD ASPECTS IN ALIGNED FACE SEALS

Véronique PERSON, Bernard TOURNERIE, Jean FRENE.
Laboratoire de Mécanique des Solides, UMR CNRS 6610, Université de Poitiers
SP2MI, boulevard 3, Téléport 2, BP 179, 86960 FUTUROSCOPE CEDEX - FRANCE

ABSTRACT

Very few thermohydrodynamic (THD) analyses relative to seals have been performed. Etsion and Pascovici (1), (2), developed an analytical model based upon simplifying hypotheses which allows for the determination of the temperature distributions in the sealing gap and on the faces. Assuming steady-state conditions, they have shown that the misaligned case can be derived from the aligned one.

The aim of this study is to solve the THD problem for an aligned seal with smooth faces operating under steady-state conditions without requiring too many restrictive assumptions. The model shown in figure 1 does not required axial symmetry.

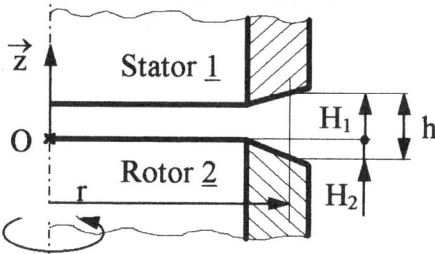

Figure 1 : Aligned face seal model

By using the classic hypothesis concerning the thin viscous films one obtains:

- the velocity components in the fluid film

$$v_r = \frac{\partial p}{\partial r}(I - \frac{I_1}{J_1}J) \quad , \quad v_{\theta'} = \omega r(1 - \frac{J}{J_1})$$

$$v_z = -\frac{1}{r}\int_{H_2}^{z} \frac{\partial (rv_r)}{\partial r} d\xi + cste \quad , \text{where:}$$

$$I(z) = \int_{H_2}^{z} \frac{\xi d\xi}{\mu(r,\theta',\xi,t)} \quad , \quad I_1 = I(H_1)$$

$$J(z) = \int_{H_2}^{z} \frac{d\xi}{\mu(r,\theta',\xi,t)} \quad , \quad J_1 = J(H_1)$$

- the generalised Reynolds equation

$$\frac{\partial}{\partial r}\left(rG_1\frac{\partial p}{\partial r}\right) = 0, \text{ with } G_1 = \rho\int_{H_2}^{H_1}\frac{(z-H_2)}{\mu}\left(z-\frac{I_1}{J_1}\right)dz$$

- the energy equation

$$\rho C_p\left\{v_r\frac{\partial T}{\partial r} + v_z\frac{\partial T}{\partial z}\right\} = \mu\left\{\left(\frac{\partial v_r}{\partial z}\right)^2 + \left(\frac{\partial v_{\theta'}}{\partial z}\right)^2\right\} + K\frac{\partial^2 T}{\partial z^2}$$

The equation of conduction in the rings takes the form:

$$\frac{1}{r}\frac{\partial T}{\partial r} + \frac{\partial^2 T}{\partial r^2} + \frac{\partial^2 T}{\partial z^2} = 0$$

These equations are solved using finite difference techniques. They are coupled by the temperature and, therefore, are solved simultaneously using an iterative procedure. Different types of conditions may be imposed on the boundary surfaces defined in figure 2.

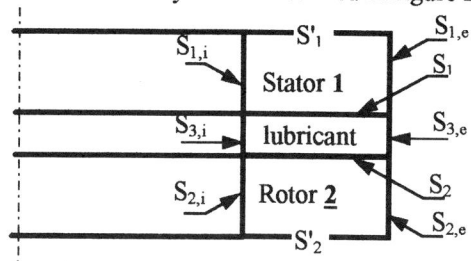

Figure 2: Boundary surfaces

In figure 3, the isothermal lines in the oil film and in the rotor are shown for one typical example. This result was obtained for an externally pressurised seal (1 MPa), with tapered faces and a minimum film thickness of 5 µm. The stator is assumed to be insulated, then S_1 is an adiabatic surface. At the film-rotor interface S_2, the heat flux continuity is applied. On the surfaces $S_{2,i}$ and $S_{2,e}$, heat is transferred by convection. On the surface S'_2, the temperature is imposed. In this case, the heat dissipated in the film is mainly evacuated by conduction to the shaft.

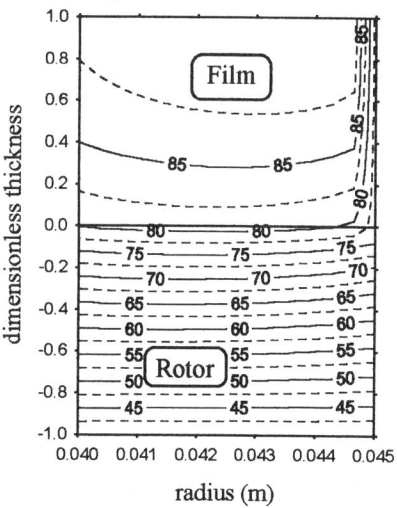

Figure 3 : Isothermal lines (°C).

REFERENCES

(1) PASCOVICI M.D., ETSION I., "A thermo-hydrodynamic analysis of a mechanical face seal", Journal of Tribology, ASME Trans., 114, 4, pp 639-645, 1992.

(2) ETSION I., PASCOVICI M.D., "A thermohydrodynamic analysis of a misaligned mechanical face seal", Tribology Trans., Vol. 36, 4, pp 589-596, 1993.

Submitted to *IMechE Journal of Engineering Tribology*

STABILITY ANALYSIS OF MECHANICAL SEALS WITH TWO FLEXIBLY MOUNTED ROTORS

J. WILEMAN
Laboratoire de Mécanique des Solides, Université de Poitiers, 86960 Futuroscope CEDEX, FRANCE
I. GREEN
Woodruff School of Mechanical Engineering, Georgia Institute of Technology, Atlanta, Georgia 30332-0405 USA

INTRODUCTION

This work investigates the dynamic stability of a seal configuration in which both seal elements are flexibly mounted to independently rotating shafts. Face seals normally consist of a rotor attached to a rotating shaft and a stator attached to a stationary housing. However, sealing between two shafts requires replacing the stator with a second rotor. Such a seal could be used for applications such as interstage seals in gas turbine engines with counterrotating stages.

Dynamic analyses have shown that in the typical rotor/stator configuration, the seal will be more stable when the rotor is flexibly mounted rather than the stator. This work derives stability criteria for the two-rotor seal in which both rotors are flexibly mounted.

A kinematic model for this configuration, denoted FMRR, was described in (1), and rotor dynamic coefficients were derived to model the incompressible fluid film which couples the two rotors. These equivalent stiffness and damping coefficients allow the fluid film properties to be introduced directly into the equations of motion (2). The stability analysis is based upon this set of linear equations.

STABILITY ANALYSIS

The four equations of motion, two for each rotor, represent a moment balance between the dynamic moments and the moments applied by both the flexible supports and the fluid film. The equations are expressed in an inertial reference frame, and the independent variables are the components of the rotor tilts about the inertial axes. If support damping is neglected, the homogeneous equations of motion can be solved in closed form. In the general case, however, the stability limits must be obtained using a Routh-Hurwitz analysis.

To obtain the Hurwitz polynomial, a harmonic solution is assumed, and the two equations of motion for each rotor are combined using complex notation. The resulting system of two coupled complex linear equations is expressed in matrix form, and the Hurwitz polynomial is obtained by taking the determinant and multiplying it by its complex conjugate, yielding an eighth degree polynomial equation of the form

$$\sum_{k=0}^{8} a_k \lambda^k = 0$$

whose coefficients a_k are real functions of the design parameters of the seal. The system is stable if all eight of the Hurwitz determinants formed using these coefficients are positive. For a completely defined seal design, the values of all of the design parameters are substituted to obtain the values of the determinants.

PARAMETRIC STUDY

In this work, values are substituted for all of the design parameters except a single shaft speed, and requiring all of the Hurwitz polynomials to be positive establishes the critical speeds at which the system makes the transition from stable to unstable operation. This technique was first applied to a reference seal design; then, one parameter at a time in the design was changed to determine its effect upon stability.

The speeds of shafts 1 and 2 are denoted ω_1 and ω_2, respectively, and their signs are consistent with the right-hand rule. For a fixed ω_1 the seal will be stable for a range of ω_2 roughly centered about ω_1 and divided between the counterrotating and corotating regimes. The effects of changing the various design parameters are either to reduce this operating range, to shift it relative to ω_1, or both.

Ten different design parameters were tested: the inertial ratio of element 2; the moments of inertia, support stiffness, and support damping of both elements; the pressure drop; the coning angle; and the design clearance. The stability range is always larger for the case in which both rotors are short, indicating that the gyroscopic moments stabilize the FMRR seal with two short rotors. Lengthening one of the rotors both reduces the stable speed range and skews it toward the counterrotating regime, indicating that the increased fluid film stiffness in this regime transfers restoring moments from the short rotor to the long.

The stability range of the system is almost independent of the support stiffness of either element, but is more dependent upon the support damping coefficient, particularly as the damping approaches zero. The influence of the pressure drop, coning angle and design clearance depends upon whether one of the rotors is long and is much more pronounced for short rotors.

REFERENCES

(1) J Wileman, I Green, *ASME Journal of Tribology*, 1991, Vol. 113, No. 4, pp 795-804.
(2) J Wileman, I Green, *ASME Journal of Tribology*, 1997, Vol. 119, No. 1, pp. 200-204.

STATIC AND DYNAMIC CHARACTERISTICS OF PERMEABLE AND POROUS SQUEEZE FILMS - AN EXPERIMENTAL INVESTIGATION

A.ALMAKHLAFY, M.M.MEGAT AHMED, D.T.GETHIN AND B.J.ROYLANCE
Department of Mechanical Engineering, University of Wales, Swansea, SA2 8PP, U.K.

ABSTRACT

Squeeze films are to be found functioning in all manner of applications, ranging from the performance of, for instance, connecting rod 'big-end' bearings in automobile engines, screen printing operations, and the human knee joint. The main aim of the experiments undertaken was to examine the response of the squeeze film to three different forms of loading: steadily applied, impact and dynamically varying. The purpose of performing the experiments is to provide the means for establishing the response characteristics of the squeeze film and thereby to compare its behaviour, notably in terms of its dynamic response, for different solid boundary conditions and fluid film properties. It is intended that by also achieving a good correlation with relevant theoretical models, a basis will have been established for developing suitable design procedures and criteria for a wide range of operating conditions in practise.

INTRODUCTION

A single apparatus was designed and built to perform tests in which the measured pressure was used as the common means for comparing the performance under the three different operating conditions. Tests were conducted using rigid, and also 'compliant', steel surfaces, as well as porous media, for Newtonian and non- Newtonian fluid film conditions. A circular thrust bearing geometry was utilised throughout the entire test programme.

Theoretical models were developed and compared with the experimental data for each condition to permit the determination of performance characteristics beyond those attainable with the present experimental apparatus.

RESULTS

The results in Figure 1 shows the effect, as a function of time on film thickness, velocity of approach, and central pressure, respectively, to the application of a steadily applied load; first, between two opposing rigid steel surfaces, and then between a rigid steel surface and a thin steel diaphragm. The response characteristics of the compliant surface as the applied load is varied is shown in Figure2.

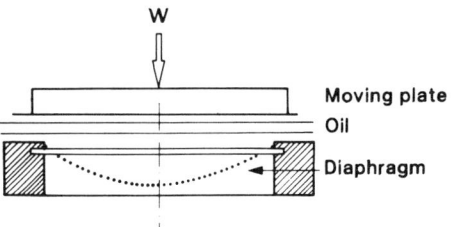

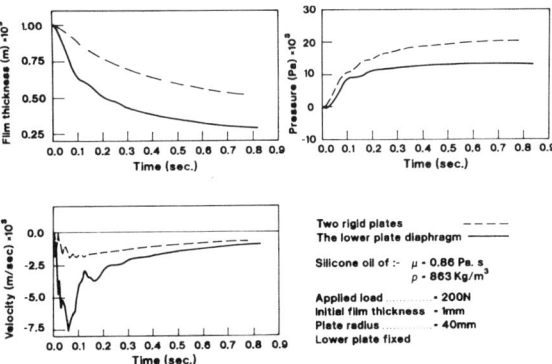

Fig. 1: Rigid and compliant surface squeeze film bearings - steadily loaded

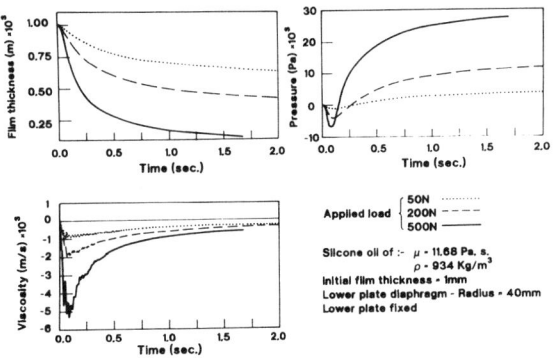

Fig. 2: Effect of applied load - compliant bearing

CLOSURE

The experimental results show good agreement with theory. The test apparatus developed can be utilised to compare the performance of different materials and lubricants under static and dynamic operating conditions.

A MIXED LUBRICATION MODEL FOR COLD STRIP ROLLING

W. R. D WILSON and H-S. LIN
Center for Surface Engineering and Tribology, Northwestern University, Evanston, IL 60208-3111, USA.

N. MARSAULT
Center de Mise en Forme des Materiaux, Ecole Nationale Superieure des Mines de Paris, Sophia Antipolis, France.

ABSTRACT

Wilson and Sheu's model (1) for the flattening of workpiece asperities under conditions of bulk plastic strain and the average Reynolds equation proposed by Wilson and Marsault (2) for conditions of large fractional contact area are combined to model mixed lubrication of strip rolling. The rolls are assumed to be relatively smooth while the strip is assumed to have a Christensen height distribution and arbitrary lay characterized by the Peklenik surface pattern parameter γ_s. Analyses for the inlet zone, work zone and outlet zone and allowance for the influence of pressure on viscosity are included in the model. Some approximations and special numerical techniques are used to obtain solutions over a wide range of conditions.

Figure 1 shows the typical variation of film thickness with speed predicted by the model. The results are plotted using the non-dimensional film thickness H_t and speed S defined by

$$H_t = h_t / R_q$$

and

$$S = a\ \mu_o\ U_r\ /\ k\ R_q\ x_1$$

respectively, where h_t is the mean film thickness, R_q the original strip roughness, a the roll radius, μ_o the roll radius, U_r the roll speed, k the strip shear strength and x_1 the length of the work zone.

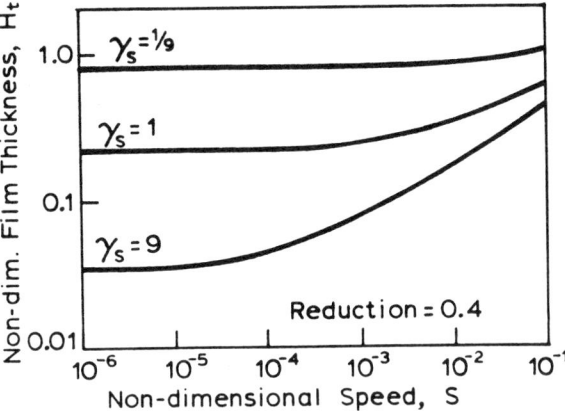

Fig. 1: Theoretical variation of film thickness with speed

As previously found by Wilson and Chang (3), and Chang et al. (4), mixed lubrication effects persist over a very wide range of speeds, perhaps as much as six orders of magnitude. Under high speed conditions, the generation of hydrodynamic pressures is dominated by wedge action in the inlet zone. Under low speed conditions, pressures are generated by hydrodynamically by wedge action in the converging channels created by asperity flattening as the strip passes through the roll bite. With high reductions, the percolation threshold may be reached in the roll bite, particularly with strip surfaces with isotropic and transverse lays. This leads to a "hydrostatic lubrication" condition in which the lubricant is trapped in isolated pockets in the strip. In general, strip surfaces with longitudinal lays yielded the smallest film thicknesses and the highest frictional conditions while transverse strip surfaces yielded the highest film thicknesses and lowest frictional conditions. Isotropic strips yielded intermediate levels of film thickness and friction.

The theoretical model has been validated by using "oil drop" film-thickness measurements and surface roughness measurements in rolling 5052-0 aluminum alloy. Strip surfaces with longitudinal and transverse lays, produced by abrasive paper, and isotropic surfaces, produced by sand blasting, were used with a variety of liquid lubricants. In general, the experimental film thickness data showed good quantitative agreement with the predictions of the theory. In the case of the isotropic surfaces, it was necessary to increase the effective hardness of the asperities in the model to allow for the physical and geometric hardening of the surfaces relative to the Wilson and Sheu model. The roughnesses measured in the experiments could also be predicted from the theoretical film thickness using a simple analysis.

REFERENCES

(1) W. R. D. Wilson and S. Sheu, "Real Area of Contact and Boundary Friction in Metal Forming," Int. J. Mech. Sci., Vol. 30, pp. 475-489, 1988.

(2) W. R. D. Wilson and N. Marsault, "Partial Hydrodynamic Lubrication with Large Fractional Contact Areas, in 'Manufacturing Science and Engineering-1995", Kannitey-Asibu, E., ed., ASME, New York, pp. 1187-1192, 1995, also to be published in J of Tribology.

(3) W. R. D. Wilson, and D-F. Chang, "Low Speed Mixed Lubrication of Bulk Metal Forming Processes," J. of Tribology_, Vol. 118 , pp. 83-89, 1996.

(4) D-F Chang, W. R. D. Wilson and N. Marsault, "Lubrication of Strip Rolling in the Low-Speed Mixed Regime" Tribology Trans., Vol. 39, pp. 407-415, 1996.

CALCULATION OF FRICTION FOR STEADY-STATE LOADED JOURNAL BEARINGS UNDER MIXED LUBRICATION CONDITIONS

L. DETERS, G. FLEISCHER and D. BARTEL
Machine Elements and Tribology, University of Magdeburg, Universitaetsplatz 2, D-39106 Magdeburg, Germany

ABSTRACT

Mixed lubrication of a steady-state loaded journal bearing is represented in the well-known Stribeck curve on the left of the minimum of the coefficient of friction. It means the simultaneous existence of two states of friction, the friction of the hydrodynamic lubrication and the solid friction. The load-carrying capacitiy of the bearing under hydrodynamic lubrication conditions results from the surface geometry and the motion of the fluid, and that under solid friction conditions is produced by the micromechanic deformation of the softer bearing surface. The adhesion between the surfaces of the shaft and the bearing can be neglected, as these are covered with a thin film of lubricant.

Mixed lubrication will occur, when the external bearing load F cannot be supported only by the lubricant. As the criterion for the transition from hydrodynamic to mixed lubrication the decreasing of the film thickness under the critical transition film thickness h_{0tr} is used. Coming below the transition film thickness at low sliding speeds or high loads, the asperities of the shaft start to deform the bearing surface right through the present lubricant (1), (2). Now the macroscopic hydrodynamic resp. elasto-hydrodynamic analysis has to be expanded by the micro-soliddynamic. In the region of mixed lubrication both parts of the load-carrying capacity also have to appear in the calculation of friction. While the parts of the load-carrying capacity F_h (hydrodynamic) and F_s (solid) and also the parts of the friction force F_{fh} and F_{fs} can be added simply, the arising coefficients of friction represent functions of the parts of the load-carrying capacity. Last named make only comparative contributions to the coefficient of friction of mixed lubrication according to the following equation of Vogelpohl (4) (Fig. 1).

$$f_m = f_s \frac{F_s}{F} + f_h \frac{F_h}{F}$$

The work for the micromechanic deformation of the softer bearing material carried out by the asperities of the shaft is an essential point of the calculation. The highly variable coefficient of friction of solid friction is obtained as a function of the lubricant film thickness resp. of the number of revolutions of the shaft out of the direct registration and mathematical treatment of the surface topography of the shaft resp. by using relevant empirical data about the microgeometry and by the aid of the energy for the mainly plastic deformation of the near-surface bearing material elements. This procedure requires the modelling of a representative micro deformation process, which should consider the multilayer design of a modern high performance bearing (3). This task was theoretically largly solved and compared with results out of experiments.

In summary it may be said:
1. The coefficient of friction of the hydrodynamic lubrication aim like the film thickness at a finite value caused by the support of the asperities, when the sliding speed resp. the number of revolutions tends to zero.
2. The coefficient of friction of solid friction must not be assumed as constant within the mixed lubrication region.
3. Caused by the presence of a very thin lubricant film the coefficient of friction of solid friction is substantially smaller than the coefficient of friction of dry friction, as a pushing aside of the lubricant film and adhesion don't occur.

Therefore even under mixed lubrication conditions a stable resp. safe operation of lubricated bearings with minimum wear is possible, what is also confirmed in practical applications.

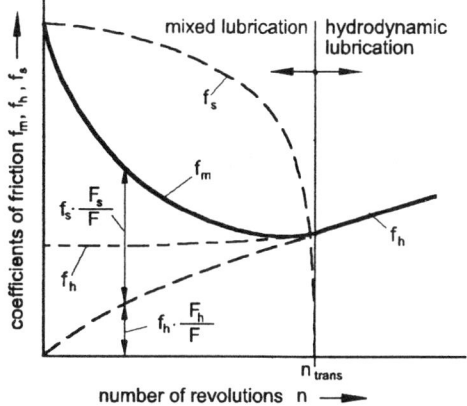

Fig. 1: Stribeck curve with parts of friction

REFERENCES

(1) G. Fleischer, Schmierungstechnik, Vol.13, No.12, p.356-360, 1982.
(2) O. Bodenstein, Schmierungstechnik, Vol.19, No.7, p.196-201, 1988.
(3) J. Hadler, Dissertation, Otto-von-Guericke-Universität Magdeburg, 1994.
(4) G. Vogelpohl, Z.VDI, Vol.96, p.261-268, 1954

THE INFLUENCE OF FILM SHAPE ON HYDRODYNAMIC PRESSURE GENERATION IN A SLIDING LINE CONTACT

D S MEHENNY and C M TAYLOR
Department of Mechanical Engineering, University of Leeds, Leeds LS2 9JT, UK

ABSTRACT

The aim of this paper is to explore the influence of transverse surface waviness on the generation of hydrodynamic pressures in an infinitely wide sliding contact. The case of a rigid inclined slider and plane separated by an iso-thermal Newtonian lubricant is analysed by solving the Reynolds equation numerically. The transverse ridges are modelled by superimposing sinusoidal waves onto the smooth surfaces of the slider and plane. Examples are presented for stationary as well as moving transverse waviness. For the case of stationary waviness, a modified form of the Reynolds equation for one-dimensional steady flow is employed to describe comprehensively the pressure profile in terms of the shape of the lubricant film. This new approach is in contrast to that of previous work (1) which attempts to interpret the influence of transverse ridges in terms of their size and position in the contact. In its integrated form, the Reynolds equation may be written,

$$\frac{dP}{dX} = 12\left(\frac{H - H_m}{H^3}\right)$$

where H_m is the film thickness at maximum pressure. Differentiation yields the new "H_m form" of the Reynolds equation,

$$\frac{d}{dX}\left(\frac{dP}{dX}\right) = 12\left(\frac{3H_m - 2H}{H^4}\right)\frac{dH}{dX}$$

Fig. 1 shows an example of an inclined slider with two bumps and the resulting pressure profile. The pressure profile is divided into regions according to film slope (converging or diverging) and film thickness (H).

The film shape approach is also employed to interpret the resulting pressure and pressure gradient profiles in the case of moving waviness. The influence of waviness height and wavelength on load carrying capacity, maximum pressure, lubricant flow rate and friciton coefficient is examined.

PUBLICATION

The paper will be submitted for publication in the Proceedings of the Institution of Mechanical Engineers, Part C: Journal of Mechanical Engineering Science.

Converging film ($dH/dX < 0$)		
Region	Film thickness	Pressure profile
C1	$\frac{3}{2}H_m < H$	pressure rises at an increasing rate
C2	$H = \frac{3}{2}H_m$	pressure gradient inflection point
C3	$H_m < H < \frac{3}{2}H_m$	pressure rises at a decreasing rate
C4	$H = H_m$	pressure gradient is zero
C5	$H < H_m$	pressure falls at an increasing rate
Pressure cannot fall at a decreasing rate in a converging film		
Diverging film ($dH/dX > 0$)		
Region	Film thickness	Pressure profile
D1	$H < H_m$	pressure falls at a decreasing rate
D2	$H = H_m$	pressure gradient is zero
D3	$H_m < H < \frac{3}{2}H_m$	pressure rises at an increasing rate
D4	$H = \frac{3}{2}H_m$	pressure gradient inflection point
D5	$\frac{3}{2}H_m < H$	pressure rises at a decreasing rate
Pressure cannot fall at an increasing rate in a diverging film		

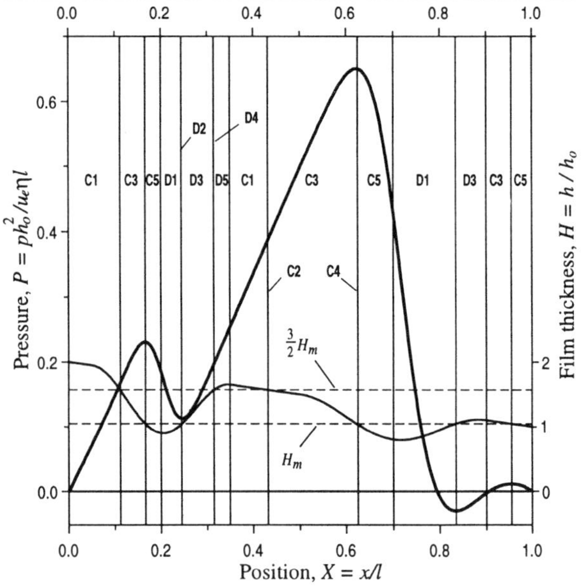

Fig. 1: Pressure profile regions in an iso-viscous film

REFERENCES

(1) E. S. Song, PhD thesis, University of Leeds, 1992.

A NEW FOIL AIR BEARING TEST RIG FOR USE TO 700°C AND 70,000 RPM

DR. C. DELLACORTE
NASA Lewis Research Center, 21000 Brookpark Rd., MS 23-2, Cleveland, Ohio 44135

ABSTRACT

A new test rig has been developed for evaluating foil air bearings at high temperatures and speeds. These bearings are self acting hydrodynamic air bearings which have been successfully applied to a variety of turbomachinery operating up to 350°C. This unique test rig is capable of measuring bearing torque during start-up, shut-down and high speed operation. Load capacity and general performance characteristics, such as durability, can be measured at temperatures to 700°C and speeds to 70,000 rpm. This paper describes the new test rig and demonstrates its capabilities through the preliminary characterization of several bearings. The bearing performance data from this facility can be used to develop advanced turbomachinery incorporating high temperature oil-free air bearing technology.

PARTIAL CONTACT AIR BEARING CHARACTERISTICS OF TRIPAD SLIDERS FOR PROXIMITY RECORDING

YONG HU
Quinta Corporation, 1415 Koll Circle #101, San Jose, CA95112, USA
PAUL M. JONES and PAUL T. CHANG
Iomega Corporation, 800 Tasman Drive, Milpitas, CA95035, USA
DAVID B. BOGY
Department of Mechanical Engineering, University of California, Berkeley, CA94720, USA

ABSTRACT

Proximity recording introduces some new head/disk interface characteristics as compared to the conventional non-contact recording. Different design philosophy has led to several distinct partial contact tripad air bearing designs including the shaped rail negative pressure and straight rail positive pressure tripad sliders. The impact of these new characteristics and design differences on the head/disk reliability needs to be thoroughly understood in order to ensure a successful proximity recording. This paper investigates the partial contact air bearing characteristics of three representative tripad slider designs.

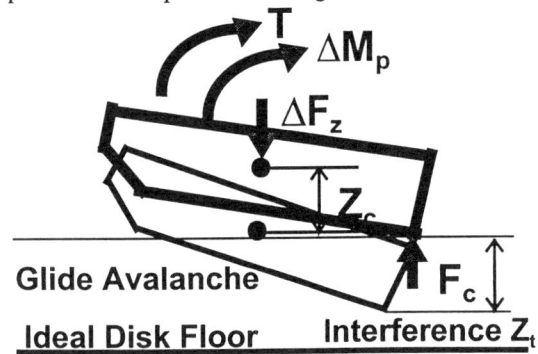

Fig.1: Schematics of force equilibrium of a proximity recording tripad slider

The contact stiffness is developed to measure the air bearing compliance. Figure 1 shows the schematics of proximity recording, in which the minimum fly height is designed at or close to glide avalanche. The force equilibrium requirement of a slider flying at the glide avalanche yields the following expression for the contact stiffness

$$\text{Contact Stiffness} = \frac{\partial F_c}{\partial z_t} = \frac{K_p}{b^2 + \frac{K_p}{K_z}}$$

where K_p and K_z are air bearing pitch and vertical stiffnesses, respectively. b is the distance from the pivot point to the trailing edge.

The Fukui-Kaneko linearized Boltzmann equation is used to model the nonlinear rarefaction effects in the modified Reynolds equation for the ultra-low fly height application. The Greenwood-Williamson asperity-based contact model is employed to model the slider/disk contact. The suspension dynamics are integrated into the air bearing simulation using modal analysis.

A positive pressure tripad slider and two negative pressure tripad sliders were studied. The contact stiffness has been shown to be able to provide a reasonable indication of the partial contact air bearing compliance. The positive pressure tripad slider has a smaller contact stiffness, while the two negative pressure tripad sliders exhibit flatter profiles of the gap fly height and the contact force across the disk surface. The contact force and contact stiffness increase with altitude. The increase in contact stiffness is dominated by the pitch drop with the altitude. The contact take-off simulations show that the two negative pressure tripad sliders take off faster than the positive pressure tripad slider. Smaller partial contact air bearing pitch stiffness, though reducing the contact stiffness, renders the ability for a slider to follow the disk waviness in proximity thereby increasing the fly height and contact force modulations. The simulations of both track seeking and crash stop dynamics indicate that the contact force modulations during these events strongly depend on the partial contact air bearing roll stiffness. A higher roll stiffness produces smoother track accessing and crash stop performances. Overall, we conclude that it is feasible to design a negative pressure tripad slider that combines the fast take-off and uniform fly height and contact force profile of the negative pressure sliders and compliant air bearing of the positive pressure tripad slider. However, we do have to design the correct partial contact air bearing pitch stiffness that is a compromise between a small contact stiffness and smooth disk waviness following performance.

REFERENCES

(1) Hu, Y., "Head-Disk-Suspension Dynamics," Ph.D. Dissertation, University of California at Berkeley, 1996.

TWO-LAYER POROUS CERAMIC AEROSTATIC GUIDEWAY BEARINGS

C J ROACH, J R ALCOCK, J CORBETT and D J STEPHENSON
School of Industrial and Manufacturing Science, Cranfield University, Cranfield, Bedford, MK43 0AL, UK

ABSTRACT

Recent technological advances, particularly in the fields of microelectronics, ultra precision machining and measurement, have put substantial demands on the accuracy and performance of manufacturing equipment. Aerostatic guideways and bearing systems are being increasingly used in such critical applications. The advantages they offer include a) low heat generation resulting in low thermal distortion, b) ultra precision positioning capability, c) precise relative motion at high speeds, d) excellent cleanroom capability, and e) excellent wear characteristics.

The first paper noted on the theory of porous aerostatic bearings came from Sheinberg and Shuster in 1960 (1). An up to date review has been carried out in the content of this work, which includes more recent publications in Germany and Japan (2, 3).

This research programme is developing a range of novel porous ceramic aerostatic bearings which meet the continuing need for improved ultra precision guideway bearings. The research has the potential for commercial application in a wide range of nano fabrication processes, for example the next generation of diamond grinding machines, lithography and other nanotechnology processes, including measurements, which require ultra precision positioning and alignment systems.

The paper discusses an entirely novel 2-layer structure to enhance the performance and manufacturability when compared to bearings produced by other means. It has resulted in bearing structures which contain a multitude of open cells in order that the ceramic structure can provide the function of a gas diffusing material. These act as minute restrictions uniformly distributed over the entire bearing surface which provide the best pressure distribution, and damping conditions, combined with the highest load capacity and stiffness. The theory of porous aerostatic bearings has been widely published but two major problems, to date, have been
a) the availability of porous materials with predictable flow properties before and after machining, and
b) pneumatic instability due to a "dead" volume of air entrapped amongst the pores.

The first problem originates from the production and processing of the porous material, and requires remedial action in the form of an additional adjustment process, in which the pores are partially sealed with successive layers of a blocking agent such as shellac solution or electroless nickel plating, until the permeability of the material reaches the design value. The second problem occurs when using bearing materials which have relatively coarse particle size (approaching the designed working clearance of the bearing). The extra dead volume of air, between the peaks of the porous surface and the narrowest part of the porous passage, is sufficient to cause pneumatic instability.

The recent research in Japan and Germany suggests the use of a 2-layered porous medium as a solution to the problems indicated above. This involves a material with a bulk core of very coarse, highly permeable structure which has a 'thin' surface layer with a relatively high restriction to fluid flow. The core essentially provides structural strength to resist deflection under fluid pressure, in order to maintain the functionally sensitive geometry of the bearing, while the flow restriction is provided almost entirely by the thin fine pored surface layer. The use of a fine and pored layer at the bearing surface is essential to eliminate the instability which causes a dead air volume in the bearing gap. The economic production of this structured material in batch production volumes has, however, yet to be achieved. Researchers in Germany, for example (2), have resorted to laser drilling to create the fine pores of the required size and distribution density on the surface of a porous bronze core material, with the surface pores closed up by 'over grinding' prior to drilling. These problems can be overcome by using a closely controlled 2-layerd porous ceramic material (4).

Methods of producing appropriate 2-layer ceramic porous structures which overcomes these problems are described together with data comparing their performance with more conventional bearing theories.

REFERENCES

(1) S A Sheinberg, V G. Shuster, Machine and Tooling, 31(11), 1960, 24-29pp.
(2) B Schulz, M Muth, Proceedings of the 8th International Precision Engineering Seminar, Compiègne, France, May 1995, Elsevier Science Inc. 533-536 pp.
(3) M Okano, Researches of the Electrotechnical Laboratory, No. 952, April 1993.
(4) Y B P Kwan, PhD Thesis, Cranfield University, March 1997.

THE ELASTOHYDRODYNAMIC FRICTION AND FILM FORMING PROPERTIES OF LUBRICANT BASE OILS

SELDA GUNSEL
Pennzoil Products Company, P O Box 7569, The Woodlands, TX 77387, USA
STEFAN KORCEK
Ford Motor Company, P O Box 2053, Dearborn, MI 48121-2053, USA
MATTHEW SMEETH and HUGH SPIKES
Tribology Section, Department of Mechanical Engineering, Imperial College, London SW7 2BX

INTRODUCTION

In recent years, lubricant manufacturers have become increasingly preoccupied with the development of low friction and thus energy-efficient lubricants. The initial requirement was to produce highly fuel efficient, automotive engine oils but there is now also interest in minimizing energy losses in transmissions and other lubricated components.

One class of system where friction control is particularly complex is in high pressure, elasto-hydrodynamic (EHD) contacts as found in cams and gears. Here, three lubricant properties play a role. Viscosity and pressure viscosity coefficient (α-value) combine to determine EHD film thickness and thus the extent to which the contact operates in full film rather than boundary lubrication, while within full film lubrication, friction is determined by the limiting EHD friction coefficient. The ideal EHD oil would thus be one with high α-value (to form a thick EHD film and thus avoid any boundary friction) and low EHD friction coefficient, to minimize friction within the full film regime. Unfortunately there are suggestions in the literature of a positive correlation between α-value and limiting shear stress, although this has been disputed by other authors. This would imply that the selection of a lubricant with a high α-value in order to yield a high film thickness might carry the penalty of a high limiting shear stress and thus high friction.

This paper describes EHD film thickness and friction measurements in a rolling/sliding ball on flat contact using a range of base fluids typical of those employed in engine oil formulation. These measurements have been used to determine both effective α-value and limiting shear stress and thus explore how these vary with fluid type and also to test a possible correlation between these two properties.

RESULTS

Fig. 1 shows α-value plotted against limiting EHD friction coefficient for all fluids at all test temperatures. A straight line plot through this data has $R^2 = 0.85$ showing that, for the set of fluids tested, there is a strong correlation between α-value and limiting EHD friction coefficient.

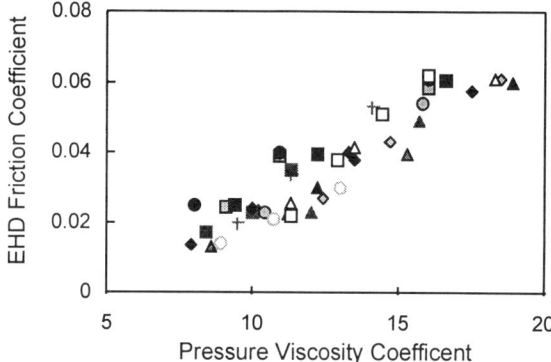

Figure 1. Correlation between EHD friction coefficient and α-value

Fig. 2 shows EHD friction coefficient plotted against the dynamic viscosity of the fluid at the prevailing test temperature for all fluids tested at low temperature (40 and 50°C). There is no significant dependence of EHD friction coefficient upon viscosity but there is a distinct banding of the test fluids into two groups. Base fluids, C, F, G, P, R and, to a lesser extent, J and S show considerably lower friction coefficient than the other fluids tested. These are either synthetic hydrocarbons or non-conventionally refined (i.e. wax-isomerized, hydrocracked and isodewaxed) fluids and all belong to API base fluid groups II, III and IV.

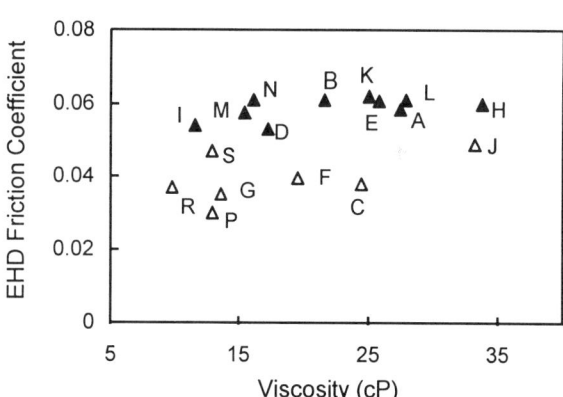

Figure 2. Range of measured EHD friction coefficients

Submitted to *Tribology Transactions*

FRICTION PROPERTIES OF BASE OILS AND ANTI-WEAR AGENTS FOR WET CLUTCH

TORU MATSUOKA

Lubricants Research Laboratory, Mitsubishi Oil Co Ltd., 4-1, Ohgimachi, Kawasaki-ku, Kawasaki 210, Japan

ABSTRACT

There are various requirements for automatic transmission fluids (ATF) used to control the torque capacity and the anti-shudder performance. The optimized friction coefficient and its durability are particularly important among other factors. Friction modifiers have been the only target of investigation to improve the friction property, while high-performance automatic transmission is demanding a new breakthrough other than friction modifiers to extend durability of friction properties.

Under such circumstances, a decreased friction coefficient at lower velocities as well as torque capacity, and thus a specific level of friction coefficient, are required to improve the anti-shudder performance. In this paper, we used modified SAE No.2 friction tester, and effects of the type of base oil as well as the molecular structure and the content level of anti-wear agents on friction properties are reported.

First, evaluated effects of base oils on friction properties have revealed that an increased paraffin content in mineral oil decreases friction coefficients. In particular, poly-α-olefin oil, a synthetic oil (paraffinic), has a friction coefficient lower than those of normal mineral oils. On the other hand, polybutene, a synthetic oil, has a higher friction coefficient especially at lower velocities. This may be attributed to the molecular structure of polybutene. However, a synthetic naphthene base oil, which has a traction coefficient as high as that of polybutene, has a friction coefficient similar to those of mineral oils. This has revealed that the traction coefficient in the EHL (elastohydrodynamic lubrication) region is not related to the friction-generating mechanism in wet clutches.

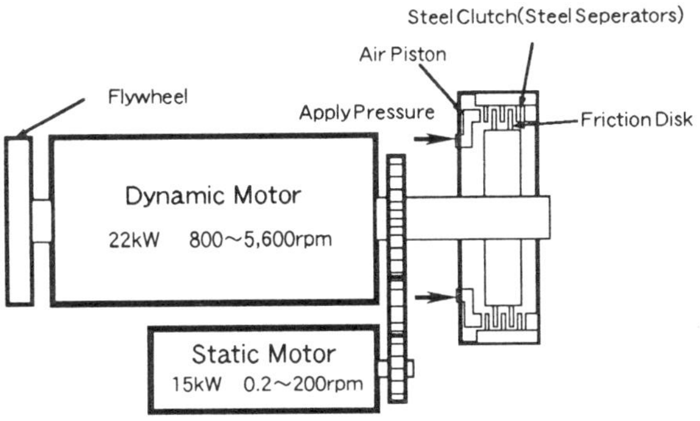

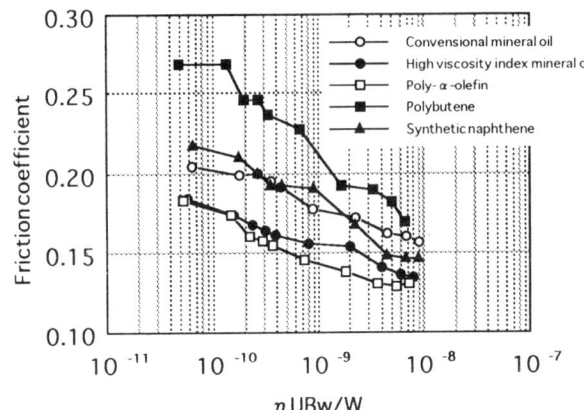

Fig.1: Modified SAE No.2 Friction Tester

Fig.2: Friction Properties of Base Oils

Second, the effects of molecular structure and addition level on friction properties have been investigated by adding phosphoric-ester anti-wear agents to mineral oils. Longer hydrophobic alkyl groups have higher abilities to reduce the level of friction coefficient at lower velocities, and could reduce it even at higher velocities. On the other hand, 0.2% or more of a phosphoric ester has caused its adsorption all over the surfaces of the friction plate and the steel plate, making the level of friction coefficient to saturate.

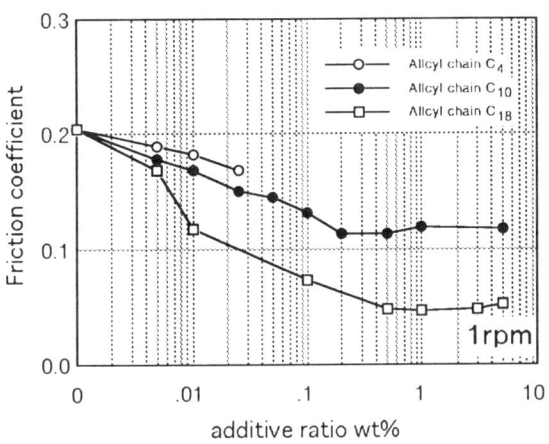

Fig.3: Friction properties of phosphoric-ester anti-wear agents

GIBBS ENERGY OF FLOW UNDER EHD CONDITIONS

E N DIACONESCU

Mechanical Engineering Department, University of Suceava, 1 University Street, Suceava, 5800, Romania

ABSTRACT

The theories for viscosity based on both activation energy and free volume assume that the jump of a molecule in a vacancy, the elementary act of flow, takes place when this possesses a certain activation energy and it is adjacent to a sufficiently large free volume. Macedo and Litovitz (1) derived the viscosity by simply superposing these two effects. Based on quantum mechanics, Diaconescu (2 to 4) expressed the viscosity in terms of two molecular Gibbs free energies, Δg_h for vacancy formation and Δg_j, for jump activation. The sum $\Delta g_h + \Delta g_j$ represents the parameter Eyring (5) measured to be the molecular Gibbs energy of flow $\Delta g = \varepsilon$. He found experimentally that, at ambient pressure, Δg is about 0.408 of the vaporisation heat. This heat is a good estimation of lattice energy in the liquid state, in its turn defined as half of the sum of all potential interactions between a molecule of the liquid and all the others.

This work aims to express these energies in terms of pressure and temperature dependent lattice energy.

A vacancy is generated when a lattice element jumps from the first sub-superficial layer on the free surface. The vacancy Gibbs energy, equal to the work done against intermolecular forces by the molecule performing a jump, is the difference between the potential energies this molecule possesses in these two layers. This is estimated by using a Lennard-Jones-London intermolecular potential. The equilibrium inter-molecular distance under pressure is found as the root of the equation of balance between the pressure forces and the potential interactions.

To avoid the use of non-harmonic lattice oscillations, an increase of temperature is visualised as an equivalent negative pressure variation producing a similar volume change by means of compressibility and thermal expansion coefficient of the liquid.

The dependence of intermolecular distances on pressure and temperature alters the intermolecular potential and therefore the lattice energy and Gibbs energy for a vacancy formation. Numerical estimations of these changes show that the lattice energy increases nearly linearly with pressure and decreases slightly in a similar fashion with temperature. The ratio of vacancy to lattice energy increases with pressure tending towards 0.35. At high pressures, typical for EHL, the temperature has little effect.

To derive Gibbs free energy for a molecular jump, two particular positions of the observed molecule are considered. In the first, its centre of interaction is placed in a lattice point neighbouring a vacancy. The second corresponds to a maximum of interaction potential with neighbouring molecules. The difference between the potential energies the involved molecules possess in the two positions is defined as the Gibbs free energy for a molecular jump. The numerical results indicate that this energy decreases asymptotically with increasing pressure towards 0.18 of lattice energy. Again, at EHD pressures, the temperature has an insignificant effect.

The molecular Gibbs energy for flow activation, defined as the sum of vacancy and jump free energies and shown in Fig. 1, does not depend practically on temperature at higher pressures and tends towards a fraction of 0.533 of the lattice energy. As the pressure lowers, the effect of temperature becomes important. At ambient pressure, this energy decreases with increasing temperature, and reaches a value of 0.412 at the standard temperature, in good agreement with Eyring experimental value.

This energy allows now to assess the variation of viscosity with pressure and temperature.

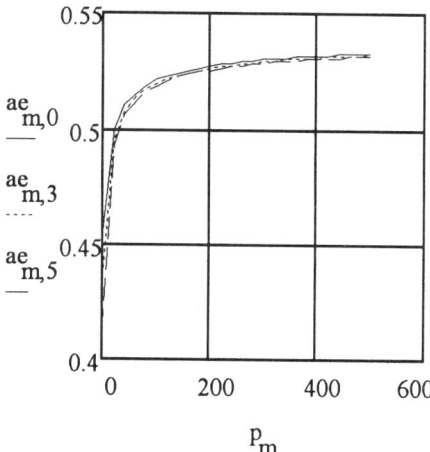

Fig 4: Gibbs free energy for flow activation

REFERENCES

(1) P.B. Macedo and T.A. Litovitz, Journal of Chemical Physics, Vol 42, No1, 1965, 245-256pp.
(2) E.N. Diaconescu, Acta Tribologica, Vol. 1, No1, 1992, 49-61pp.
(3) E.N. Diaconescu, Solidlike Behaviour of EHD Oil Films, EEC Sc. Report ERB-CIPA-CT-92-0526, 1993.
(4) E.N. Diaconescu, Leeds-Lyon Symp., 1993, 18p.
(5) S. Glasstone, K.J. Laidler and H. Eyring, The Theory of Rate Processes, McGrow-Hill Book Co., 1941.

PREDICTING EHD FILM THICKNESS OF LUBRICANT POLYMER SOLUTIONS

H MITSUI
Showa Shell Sekiyu K.K., 123-1 Shimokawairi, Atsugi-shi, Kanagawa 243-02, JAPAN
HUGH SPIKES
Tribology Section, Department of Mechanical Engineering, Imperial College, London SW7 2BX

A systematic study has been made of the elastohydrodynamic (EHD) film-forming properties of a range of well-characterized, low dispersivity polymer solution. Ultrathin film interferometry was employed, both to enable EHD film thicknesses to be measured very accurately and also so that film formation in the very thin film, boundary regime could be explored.

In the thick, EHD regime, above about 50 nm where boundary lubrication can be neglected, it is found that, as expected from previous work, polymer solutions form films lower than those predicted from the viscosities of their bulk solutions although still higher than those obtained from the base oil. This behaviour is shown in figure 1, where measured film thickness for a solution of polyisoprene (PIP) in linear dialkylbenzene (LDAB) is compared to that predicted from the polymer solution viscosity and also that measured using the polymer-free LDAB. It was observed that film thickness falls progressively further and further below the theoretical value as both polymer concentration and polymer molecular weight were raised, presumably because of increasing shear thinning.

determined by measuring the viscosity of very low polymer concentration solutions. The effective relative viscosity η_r (viscosity of solution divided by viscosity of base oil) is then given by;

$$\eta_r = 1 + [\eta]c \qquad (1)$$

where c is the polymer concentration.

Figures 2 compares three plots of relative viscosity versus polymer concentration. One is η_r of the polymer solution measured using low shear rate viscometry. This rises rapidly with polymer concentration due to polymer entanglement. One shows the dilute solution η_r predicted from equation 1. The third is the effective relative viscosity from EHD film thickness measurements. It can be seen that the EHD film thickness and predicted dilute solution values correspond closely. Similar results were found with other polymers systems. This implies that EHD film thicknesses of polymer solutions may be predicted from dilute solution viscosities based on intrinsic viscosity measurements.

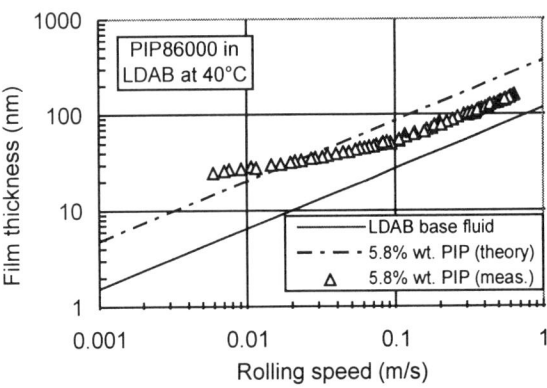

Figure 1. EHD film thickness for polymer solution (PIP MWt 86000 in LDAB)

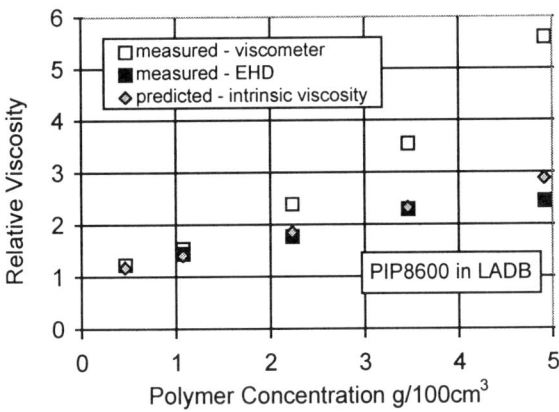

Figure 2 Relative viscosities of PIP solution in LDAB determined by (i) capillary viscometry, (ii) intrinsic viscosity method (equation 1), (iii) EHD film thickness measurements

In this study, a means was sought for predicting EHD film thickness using conventional, low shear rate viscosity measurements. It was hypothesized that, within the inlet of an EHD contact, the shear stresses are so high that a polymer solution shear thins fully down to its "dilute solution" state, where polymer-polymer interactions are lost and the fluid behaves as a solution of entirely independent polymer coils. In such a state, the effective viscosity can be predicted using the intrinsic viscosity of the solution, $[\eta]$,

This work also examined polymer boundary film formation and showed that, whilst some polymers such as PIP give an enhancement in film thickness in the thin film regime due to polymer adsorption, (as shown in figure 1), others, such as polymethylmethacrylate, show a reduction in film thickness in this regime due to polymer depletion.

MODELING OF MIXED HYDRODYNAMIC AND BOUNDARY LUBRICATION: LAYERED STRUCTURES AND SHEAR THINNING

SIYOUL JANG
Lab. Manuf. & Productivity, Massachusetts Institute of Technology, Cambridge MA 02139 USA
JOHN A. TICHY
Dept. Mech. Eng., Aero. Eng. & Mech., Rensselaer Polytechnic Institute, Troy NY 12180-3590 USA

ABSTRACT

Lubricants for modern engines have various kinds of additives. In particular, viscosity index improvers of high molecular weight polymer content are added to base oils to reduce temperature sensitivity. In this paper, modeling of boundary and hydrodynamic lubrication is undertaken in the case where such polymer additives are used.

Shear Thinning

A VI improver reduces the viscosity at high shear rate while stabilizing viscosity at high temperature. The polymeric additive within the lubricant thickens at high temperature, giving the oil its multigrade characteristics. However, these polymeric VI improvers produce non-Newtonian behavior during flow. When the shear rate is high, the multigrade lubricant shows a shear thinning effect. At high shear rate conditions, above about 10^6 s^{-1}, which is normal driving condition in engine components, the effective thickening provided by the polymer decreases significantly. This may lead to the reduction of friction and oil film thickness in the critical contacts of the engine. Therefore, we seek to investigate the potential to balance frictional efficiency against wear protection by computational analysis.

Solid-like Layers on Surfaces

We consider an intermediate regime between boundary and hydrodynamic lubrication. With high-molecular-weight polymers, it is found that polymeric molecules stick to the boundary surfaces in the form of immobile solid-like layers, the thickness being approximately 50 nm. Such layers may comprise about half of the minimum film thickness (~0.1 µm) under the severe sliding conditions of engine bearings or other moving parts. Shear thinning occurs in a central region of the film, away from the thickened layers.

Across this kind of polymer layer, an anomalous increase of apparent viscous resistance occurs. Following other studies, we choose to investigate this kind of molecular microstructure by regarding it as an immobile porous medium layer attached to a contact surface. In the region away from this solid-like polymeric layer, the bulk lubricant is likely to show the common non-Newtonian shear thinning effect.

Thus, we consider the film as consisting of three regions: two immobile solid-like layers and a shear thinning bulk fluid film. Earlier experimental studies by Chan and Horn have shown that such an interpretation can correctly predict squeezing forces down to two nanometers (about three or four molecular lengths).

Methodology

From the concept described above, a modified Reynolds' equation is developed. In our model, how the molecular structure reacts to the contiguous solid surface is implicitly described with three parameters: layer thickness δ, effective porosity of the layer α (as in prior studies), and relative viscosity μ^* (the viscosity of the fluid in the layer divided by the low shear viscosity of the fluid in bulk). Fluid film pressures are calculated for various values of porous layer thickness and porosity. These two factors have the greatest influence on contact load capacity. The thickness magnitude depicts polymeric additive activity depend-ing on the molecular structure and how the molecules react to the presence of the solid surface.

Results and Summary

The layer immobility is expressed by an *effective* viscosity of the porous layer, which is the resistance to motion of the layer (as opposed to the viscosity of the fluid in the pores). The effective viscosity of the porous layer varies from 2 to 100 times that produced by a the Newtonian viscosity in our computations. The Newtonian fluid case means that δ and α are equal to zero, and shear rate does not induce viscosity changes. If the porous layer thickness increases while keeping other parameters unchanged, as $\mu^* = 10$ and $\alpha = 0.1$, the load capacity increase significantly.

Compared to Newtonian case, all the molecular microstructure cases cause higher pressure levels. This result is consistent with the physical idea that boundary lubrication kinematic conditions (film thickness and surface speeds) cannot produce sufficient viscous forces by the classical hydrodynamic theory. The present analysis shows that other mechanical forces put in play may contribute to the boundary lubrication action.

Submitted to *ASME Journal of Tribology*

ON THE RHEOLOGICAL BAHAVIOR OF SYNTHETIC FLUIDS THICKENED BY SOLID ADDITIVES

P. VERGNE and P. PRAT

Laboratoire de Mécanique des Contacts, Institut Européen de Tribologie, UMR CNRS 5514
INSA, bâtiment 113, 20 avenue Einstein, 69621 Villeurbanne cedex, France

INTRODUCTION

The paper deals with wet lubrication for space mechanisms, for which noise and vibrations reduction and lifetime increase are now required. These objectives could be met if the lubricant localization is well controlled and if its different losses are reduced.

Lubricating suspensions seem to be potential lubricants for these applications where a good compromise between liquid properties (low viscosity, mobility) and grease properties (yield stress, adherence) must be found. The aim of the study is to show how the suspensions rheological behavior varies with their composition, and to find the consequences for the contact replenishment point of view.

LUBRICANTS

A set of suspensions has been elaborated from two low vapor-pressure base oils, a linear PFPE (noted Z) and a synthetic hydrocarbon (noted P). Three thickeners have been selected :
- PTFE A is a dry powder (mean diameter of 4 µm),
- PTFE V is made of low molecular weight flat particles in suspension, with a mean size of 9 µm,
- MoS_2 is a dry powder (mean particle size about 7µm).

The nomenclature includes the base oil nature (Z or P), the PTFE type (A or V), its volumic concentration (in %) and finally the addition of 1% vol. of MoS_2 (-1) or not (0).

YIELD STRESS RESULTS

It is firstly observed (figure 1) that compared to organic greases, a similar thickener content leads to similar materials which show significant yield stresses, indicating a solid-like structure before flow.

It also appears that the thickening effect of PFTE V is more efficient than the PTFE A. At similar concentration, PV and ZV lubricants show higher values than those measured with PA and ZA. This observation cannot only be explained by the peculiar nature of PTFE V. We note, at constant concentration, that PV suspensions lead to higher yield stresses than ZV ones, whereas PA show lower values than ZA. This inversion suggests that a strong physico-chemical interaction between oil P and PTFE V occurs and is superposed to the natural thickening effect of PTFE V.

The influence of MoS_2 on the yield stress is discussed in the paper : it is analyzed following the competition between geometrical and physico-chemical effects when the PFTE concentration is increased.

The study of the temperature influence has shown that the suspensions yield stress is also very stable.

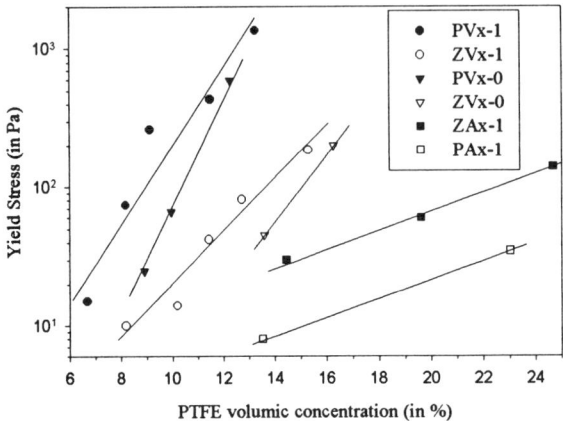

Figure 1 : Yield stress as a function of the PTFE concentration.

APPARENT VISCOSITY

The suspensions behave as non Newtonian fluids. The apparent viscosity continuously decreases when the shear rate is increased. At high shear rate, it tends toward an asymptotic value, function of the base oil viscosity, particle concentration and size. Concerning the influence of the different components and their interactions, the tendencies observed on the yield stress values are still valid and apply for the viscosity results at low and medium shear rates.

Between -20 and 60°C and at a given shear rate, it is found that the apparent viscosity variation of the suspensions is much lower than the base oil one.

CONCLUSIONS

For a set of lubricating suspensions, the influence of the nature and concentration of the base materials on the rheological parameters has been shown and discussed according to the possible interactions occurring between them. This can be used to formulate suspensions from rheological criteria.

For their potential use in space mechanisms, the following points must be underlined :
- their localization will be helped by the yield stress which also limits their migration,
- low running torque are expected due to the low apparent viscosity measured at high shear rate,
- their rheological properties are weakly temperature dependent, which facilitates cold starts and limits the running torques at low temperature.

EFFECTS OF UREA GREASE COMPOSITION ON SEIZURE OF BALL BEARINGS

MICHIHARU NAKA, HIROYUKI ITO, HIDEKI KOIZUMI and YOICHIRO SUGIMORI
NSK Ltd.,R&D Center, 1-5-50 Kugenuma Shinmei, Fujisawa, Kanagawa, 251 JAPAN

ABSTRACT

In recent years, urea greases containing a urea compound as a thickening agent have been increasingly used in bearings for automotive electrical devices and bearings operated at high speed running conditions. This trend is attributed to the superior heat resistance, shear stability and water resistance of urea greases compared to the properties of lithium soap greases that were dominantly used as lubricant for rolling bearings formerly.

The authors previously presented urea greases that were considered as being practical and suitable specific applications(1), and reported(2)on the superior seizure life of synthetic oil-urea greases to that of other greases .

Endo et al. uniformly applied three urea greases, each containing a urea compound of different composition, to steel plates and heated them for 1000hrs at 150℃, and reported that their deterioration rates were different(3).

However, no systematic study has been reported on the relation between the composition of urea compounds and the seizure life of rolling bearings. With attention paid to the composition of urea compounds, the authors tested the rotation of ball bearings containing various urea greases to examine their grease leakage rates, oil separation rates, degradation by oxidation, wear, and seizure life.

Deep-groove ball bearings with non-contact rubber seal (having an inside diameter of 25mm and an outside diameter of 62mm) were used as the test bearings. The bearing were filled with 3.4g (35% of the bearing space volume) of test grease. Test conditions are under a temperature 140℃, running speed 10000rpm, radial load 98N, and axial load 98N.

Fig. 1 shows the grease life test results of Greases A, B, C, F, G. Among the Greases A, B, and C whose base oil was poly α-olefin, the aromatic diurea Grease C exhibited the longest life, while the life of the aliphatic diurea Grease A was shortest. The heat stability and shear stability of the urea cpmpounds were reflected in the behavior of the greases in the bearings to show a good correlation with the seizure life. Among the Greases B, F and G containing alcyclic diurea as a thickener, the life of the diphenyl ether based Grease G or polyol ester-based Grease F was longer than the life of the Grease B based on poly α-olefin. The grease life correlated with the heat resistance and evaporation of the base oils.

Ball bearing operation tests were performed with different synthetic oil-diurea grease compositions. Findings from the tests are as follows:

Differences in the heat resistance and shear stability of diurea compounds used as thickeners resulted in differences in grease leakage rate and oil separation rate, and the results were directly reflected in the seizure-resisting performance of the greases. The aromatic diurea greases showed the longest life. Those that showed the next longest life were alcyclic greases and aliphatic greases,in this order. In the aliphatic-alcyclic diurea greases, a slight change in the amine compounding ratio caused a notable difference in the grease life.

Greases containing a diurea compound and a base oil showed a difference in performance when compared with those containing the same diurea compound but a different type of base oil. The heat resistance and evaporation characteristic of the base oils were reflected in the seizure performance of the greases.

This paper were submitted to STLE ,and will be published in Tribology Transactions or Lubrication Engineering.

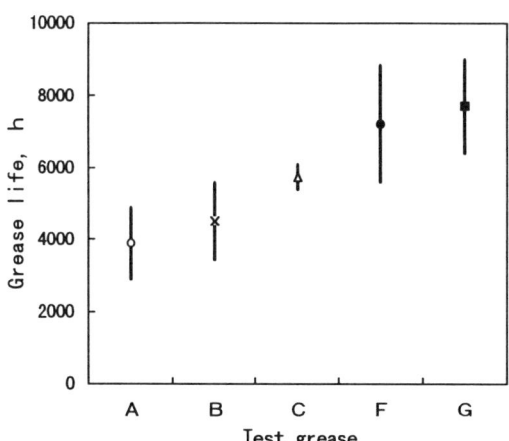

Fig. 1: Grease life test

REFERENCES

(1) M.Naka, NSK Technical Journal, No.650, page12-20, 1989.
(2) H.Ito, H.Koizumi, M.Naka, Proc. Int. Tribo. Conf. YOKOHAMA, page931-936, 1995.
(3) T.Endo, NLGI Spokesman, Vol.57, No.1, page532-541,1993.

Submitted to *Tribology Transactions*

THE DESIGN OF LUBRICIOUS OIL-IN-WATER EMULSIONS

MONICA RATOI and HUGH SPIKES
Tribology Section, Department of Mechanical Engineering, Imperial College, London SW7 2BX
RONALD HOOGENDOORN
Unichema International, Gouda PO 2800, The Netherlands

INTRODUCTION

Oil-in-water (O/W) emulsions are widely used both in metal rolling and also as fire resistant hydraulic fluids. In both applications they need to possess an adequate degree of lubricating ability to protect rubbing parts from high friction, seizure and excessive wear. In practice many commercial O/W emulsions show very poor lubricating ability and this can give rise to problems in service. A key aspect of this poor lubricating performance is that some O/W emulsions are unable to form adequate elastohydrodynamic films in high pressure contacts. Recent work has shown that this weakness results from inability of the oil phase in some emulsions to wet rubbing metal surfaces and thereby produce fully-flooded lubricated contacts (1).

A fundamental study has been carried out on the influence of O/W emulsion composition on surface wetting and thus on lubricant film formation in rolling, concentrated contacts. Two important surface chemical parameters are identified, the displacement energy and the adhesion tension. These enable the role of the emulsifier and the lubricant base fluid on wetting to be quantified and thus point the way to designing O/W emulsions with optimal film-forming properties.

THEORETICAL MODEL

One very important factor controlling wetting is emulsifier concentration. By measuring EHD film thickness, it has been shown that there is an optimum emulsifier concentration to ensure maximum wettability and thus maximum film thickness (1). This concentration has a value of about 0.2 to 0.6 of the critical micelle concentration; a level which can be explained using surface chemical principles to quantify how easily oil phase displaces water phase from a solid surface. This concept has been tested using both an anionic and, in the current paper, a nonionic emulsifier system.

A second variable which should play an important role in determining oil wetting from O/W emulsions is the nature of the base fluid. Base oils have a range of polarities, varying from species with permanent dipoles such as esters and polyglycols, through mineral oils where some of the components are easily polarizable to the synthetic oligomers such as polyalphaolefins with only very weak polarizability. The polarity of a base oil is likely to significantly influence its tendency to wet metal surfaces, so that a consideration of the wetting and corresponding film-forming ability of base oils of different polar types and viscosities is essential in optimising emulsion performance.

To quantitatively compare the wetting ability of base oils of different polarity it is desirable to carry out experiments in the absence of surfactants in order that the results be generally applicable. In this case, a useful measure of wetting ability of oil is the "adhesion tension". When the two fluids are oil and water respectively, adhesion tension is given by:

$$\tau_{SOW} = \gamma_{SW(O)} - \gamma_{SO(W)} = \gamma_{OW} \cos\theta_{SOW}$$

where γ refers to appropriate surface tension and θ to the contact angle. For oil to displace water from the solid surface, τ_{SOW} must be positive and the more positive its value, the more readily adhesion of the oil on the steel surface will occur. Adhesion tension can thus be determined by measuring the contact angle at the oil/water/steel boundary. Because the adhesion tension determined in this way takes account of the mutual saturation effects of the two phases in contact, its use is especially suitable for wettability studies of different types of oil.

Figure 1 shows the influence of base oil type on EHD film formation for five O/W emulsions of otherwise identical composition. In this study, these and similar results are used to confirm the validity of the adhesion tension criterion.

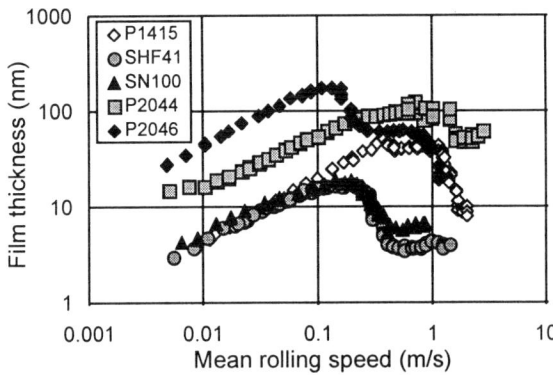

Figure 1. Influence of base oil on EHD film thickness

REFERENCES

1. Ratoi-Salagean, M., Spikes, H.A. and Rieffe, H.L. "Optimizing Film Formation by Oil in Water Emulsions". Accepted for publication, STLE Trans.

LUBRICATING PROPERTIES OF THE RAPESEED BASED OILS

A. ARNŠEK, J. VIŽINTIN
University of Ljubljana, Faculty of Mechanical Engineering, Centre of Tribology and Technical Diagnostics,
Bogišičeva ul. 8, 1000 Ljubljana, Slovenia

ABSTRACT

Biodegradable, vegetable oils have been used as lubricants for a long time. However, since the introduction of mineral oils, they have dominated the lubricant market. Today, due to growing environmental concerns, vegetable oils are finding again their way into lubricants for industrial and transportation applications. Vegetable oils can offer significant environmental advantages with respect to resource renewability, biodegradability, as well as offering satisfactory performance in a variety of applications. But not all types of vegetable oils are suitable for lubricants. For good performance, special requirements on lubricant must be fulfilled i.e. good lubricity, good corrosion protection, compatibility with other materials, and fair oxidative and hydrolytic stability and low temperature behaviour associated with the triglyceride. For reasons of quality and of course economic reasons, rapeseed oil is mostly used for lubricants.

For some commercially available hydraulic/transmission lubricants from the market, based on rapeseed oil the lubricating characteristics i.e. coefficient of friction and wear characteristics are investigated in a wide range of operating conditions in an oscillating contact. Since their efficiency in gear contacts, scuffing load capacity and pitting resistance are also unknown, additional tests on the FZG test machine were performed. The results are compared with corresponding commercial mineral based oils.

Sliding friction at the point of contact causes shear forces in contact which influence pitting failure. Another result of the high sliding friction is load-dependent power loss, and consequently oil and gear heating. Basic friction and wear characteristics were investigated in oscillating point contact under boundary lubricating conditions. Amplitude, frequency, temperature and test duration were kept constant for all tests (1000 μm, 50 Hz, 80 °C, 120 min), only the contact pressure was varied (1, 2, 2.5 and 3.17 GPa).

Scuffing and pitting are typical gear failures. Scuffing is a severe form of adhesive wear which occurs when there is a sudden and severe mode of lubrication breakdown. Scuffing can be prevented with a thick oil film, sufficient to separate the surfaces almost completely, or with AW and EP additives. In this case reaction films having developed on the flank surfaces also resist scuffing. Scuffing load capacity for oils was investigated using standard FZG procedure.

Pitting is to be regarded as a type of fatigue failure characterised by the fact that particles of material will break off the surfaces of the gear teeth after a certain number of meshing cycles has been reached. The main reason for pitting is that the critical contact pressure for a specific material has been exceeded. There is a number of effects influencing pitting, but using a FZG test procedure in our case, only the influence of the lubricant was considered. It is known that the lubricant viscosity, the lubricant type and the lubricant temperature have a rather strong influence on pitting resistance while the additive type and percentage have only a minor influence on the endurance level. Tested oils were so chosen that at the test temperature (90 °C) the viscosities were almost the same for all tested oils. Only the lubricant type and perhaps the additive type was considered in the pitting test.

The results obtained show, with one exception, significant reduction in average coefficient of friction for rapeseed oils as compared to mineral oils. It was expected that rapeseed oils exhibit a very good lubricity, because they contain organic straight chain compounds with polar end groups. Although, the lowest friction occurs with rapeseed oils, their anti-wear characteristics are worse than by mineral oils, especially at high contact stresses. This can be explained by the fact that due to base oil high polarity, the AW/EP additives are not efficient enough.

The scuffing load capacity for tested oils is in general similar for all tested oils, showing comparable quality of the rapeseed oils with the mineral oils. Interesting is that the measured oil temperature in the gear box at the end of the each load stage was approximately 8-13 °C lower when rapeseed oils were tested than when mineral oils were tested. This can be explained by lower coefficient of friction and on the other hand by better separation of the meshing gears by the lubricating film when rapeseed oils were used. Lower viscosity-temperature dependence of these oils (VI 210) compared to mineral oils (VI 90) shows that, although viscosity at the starting temperature of each load stage (90 °C) were the same for all oils, with increasing temperature during the test, the thicker film and better separation of the contact surfaces was obtained with rapeseed oils.

The tests have shown a better pitting resistance of the rapeseed oils compared to that of the mineral oils. Differences can be explained by lower sliding friction at the point of contact. This leads to smaller tangential stresses at the surface which can efficiently prevent fatigue failure associated with surface-initiated cracks.

All results with detailed explanation will be presented at the conference.

ELECTRIFICATION OF OIL AND FILTER ELEMENT - PART I
CRACKING OF OIL MOLECULES BY SPARK DISCHARGES

AKIRA SASAKI and SHINJI UCHIYAMA
KLEENTEK Industrial Co., Ltd. 2-7-7 Higashi-ohi, Shinagawa-ku, Tokyo 140, Japan
TAKASHI YAMAMOTO
Dept. of Mech. Eng., Tokyo University of Agri. & Tech., 2-24-16 Nakamachi, Koganei, Tokyo 184, Japan

ABSTRACT

It is known that contamination of oil is harmful to hydraulic and lubricating systems. Therefore use of filters is recommended to protect hydraulic and lubricating systems (1). It has been made clear that polymerized oil oxidation products called sludge are also harmful to hydraulic and lubricating systems (2). We know that oils are oxidized in 1 or 2 year use and that polymerized and oil insoluble varnishes are produced in oils, even though machines are operated at oil temperature of around 45°C. It has not been clear why oils with oxidation inhibitors are oxidized at such mild oil temperature and in such a short time.

Although it is widely known that static electricity is generated by friction of two dielectric materials like filter element and oil (3), it has been overlooked that oil is damaged by spark dischargs of the static electricity accumulated on the filters, as the filters are connected to the grounded machines by metal pipes and spark discharges of electric charge cannot be seen easily.

Authors have several times noticed that static electricity is generated when oils are cleaned with mechanical filters. In order to demonstrate electrification of oil and filter element in the process of filtration, the filter test stand was insulated and the potentials of both the oil and the filter housing were measured by an electrometer. With a dry depth filter element, the potential of the oil increased from 8.45kV to 21.1kV and that of the insulated filter housing from -25kV to -50kV, when the pump flow rate was increased from 0.75 l/min. to 4.0 l/min. With a humid depth filter element, the potential of the oil increased from -1.2kV to 3.3kV and that of the insulated filter housing from -25kV to -40kV in proportion to increase of pump flow rate from 0.75 l/min to 4.0 l/min.

Very interesting finding was that the potentials of the oil were higher with grounded filter housing than with the insulated filter housing. This suggests that a kind of electric condenser is formed between the filter housing and the filter element. The potentials of the filter hosing and the test stand frame were measured at five different points. The results of the measurement suggest that there is high possibility of discharging the static electricity accumulated in the oil at the sharp cut edge of the grounded pipes.

The static electricity accumulated on the filter element was taken out of a filter element by induction through the insulated filter housing and introduced to a spark discharge test device which was made of a pair of electrodes fixed on a Teflon frame with 1mm gap. A flat electrode was connected to the filter housing by a high voltage electric cable and a needle electrode to the ground. The wired spark discharge test device was placed in oil in a beaker. In a minute after the pump operated, spark discharges started between the pair of electrodes in oil at every 1~10 second depending on the oil flow rate. Gas bubbles and a trace of carbon particles were ejected at every spark discharge. When the filter housing was grounded, the spark discharges disappeared, because the electric charge on the filter element was grounded.

A depth type filter and a pleated type one which were available on the market were tested and the potentials of the filtered oil samples were measured in a Faraday cage with an electrometer. Within the tested flow rates, it was confirmed that the oil was electrified by filtration and that the potentials of the filtered oil were proportional to the oil flow rate.

Filter elements used in factories were examined. Carbon particles were found on the filter meshes. The pleated filters used were cut and examined. The pleated filter element made of unwoven fibers had a center core with punched holes. The sharp edges of the punched holes which are adjacent to the pleated lines of the filter fiber are carbonized. These facts indicate possibility of spark discharges of static electricity in the filters.

The results of the study demonstrate that static electricity will generated by friction of oil with a filter element and that spark discharges of the accumulated static electricity on a filter element will crack the oil molecules.

REFERENCE

(1) AHEM, "Guide Lines to Contamination Control in Hydraulic Fluid Power Systems " P5, 1985
(2) Sasaki, A, et al, Lubrication Engineering, 45,3, 1989, 140-146pp.
(3) Lauer, J. L. & Antal, P. G., J. Colloid Interface Sci. 32, 3, 1970, 407-423pp.

AUTOXIDATION OF MODEL ESTER LUBRICANTS

EIJI NAGATOMI[†], JOHN R LINDSAY SMITH and DAVID J WADDINGTON
Department of Chemistry, University of York, Heslington, York YO1 5DD, UK
ANDREW J HOLMES
Shell Research and Technology Centre, Thornton, PO Box 1, Chester CH1 3SH, UK

It is desirable to understand the mechanism of liquid phase oxidation of lubricant oils, since autoxidation is usually the main process leading to their degradation in engines. While the mechanisms of autoxidation of hydrocarbon lubricant base fluids are well defined, those of synthetic oils, such as the pentaerythrityl esters which are increasingly being used for lubricants under more extreme conditions, are poorly understood (1). The overall aim of this work has been to elucidate the autoxidation mechanisms of these synthetic lubricants using a selection of simpler compounds to model the key structural features of the esters and their oxidation intermediates. Here we report the results we have obtained from studies of the autoxidation of the neopentyl esters of butyric (**1**), pivalic (**2**), 3,3-dimethylbutyric (**3**) and 2,2-dimethylbutyric (**4**) acids.

```
         CH3    O
          |     ||
   CH3-C-CH2-O-C-CH2-CH2-CH3
          |
         CH3
      neopentyl butyrate (1)

         CH3    O   CH3
          |     ||   |
   CH3-C-CH2-O-C-C-CH3
          |         |
         CH3       CH3
      neopentyl pivalate (2)

         CH3    O   CH3
          |     ||   |
   CH3-C-CH2-O-C-CH2-C-CH3
          |         |
         CH3       CH3
      neopentyl 3,3-dimethylbutyrate (3)

         CH3    O   CH3
          |     ||   |
   CH3-C-CH2-O-C-C-CH2-CH3
          |         |
         CH3       CH3
      neopentyl 2,2-dimethylbutyrate (4)
```

The ester oxidations were carried out at 165° C in a glass reactor using a constant flow of dry oxygen. The volatile products were collected in two traps. Analysis of liquid and gaseous products was by gas chromatography and the identity of the compounds was confirmed by comparison with authentic materials, using gc, ms and nmr.

The main products from the autoxidation of esters **1**, **3** and **4** are hydroperoxides, the parent carboxylic acid and neopentyl alcohol, whereas with **2**, the parent carboxylic acid, pivalic acid, is the only major product and neopentyl alcohol is a minor product.

The hydroperoxide concentrations from the autoxidation of **1**, **3** and **4** initially increased with time to a maximum value, after which the acids appeared and became the main products. Similar behaviour has been observed in the autoxidation of hydrocarbons, and is explained in terms of the formation of hydroperoxide, by rapid free radical decomposition of the hydroperoxide; their decompositions lead to the formation of secondary products (2). In the case of **2**, the hydroperoxide concentration, during the period of study did not reach a maximum and the yields of the oxidation products were significantly smaller than those from the other esters.

It was felt that it was possible that the ester was being hydrolysed to the parent alcohol and acid. Results from a series of hydrolysis experiments of **1** under nitrogen showed that hydrolysis does occur when both water and carboxylic acid are present. However, the amount of both butyric acid and neopentyl alcohol formed under nitrogen with these additives is considerably smaller than that under oxygen in their absence. This result suggests that both the parent carboxylic acid and alcohol are mainly formed by oxidation and comparatively small amounts by hydrolysis.

The substrates were chosen so that they contained three different types of *sec*-hydrogen atoms namely, α-alcohol (**1, 2, 3** and **4**), α-acyl (**1** and **3**) and β-acyl (**1** and **4**). It is possible to use the relative rates of reaction of the esters to calculate the relative reactivity of these hydrogens in each substrate. Thus using the detailed products studies from the autoxidation of the four types of neopentyl esters, the reactivaties of the different *sec*-hydrogen atoms in the esters are calculated. The mechanisms of oxidation will be discussed.

REFERENCES
(1) V N Bakunin and O P Parenago, *J. Synth. Lub.*, 1992, **9**, 3073
(2) For example, S Baine and P E Savage, *Ind. Eng. Chem. Res.*, 1991, **30**, 2185

[†]Present address: Showa Shell Sekiyu KK, 4052-2, Nakatsu, Aiko-gun, Kanagawa, Japan

HIGH TEMPERATURE AMINE ANTIOXIDANT FOR SYNTHETIC LUBRICANT

JOHN T. LAI
BFGoodrich Company, 9921 Brecksville Rd, Brecksville, OH 44141
STEVEN G. DONNELLY
R.T.Vanderbilt Company, 30 Winfield St, P.O.Box 5150, Norwalk, CT 06856

ABSTRACT

Dialkyl diphenylamine (DADPA) and Alkyl N-phenyl-1-naphthylamine (APANA) are well-known lubricant antioxidants(1).

For high temperature applications in synthetic ester lubricant, the oligomers of DADPA and/or APANA, through oxidative couplings are better antioxidants (2) (3). We found that when a mixture of DADPA and APANA, with greater than 1:1 molar ratio, was subjected to oxidative coupling, the product invariably consists of the dimer of DADPA and a cross-coupled oligomers of DADPA and APANA, with more DADPA than APANA in the same molecule(4):

$$(DADPA)_2 + (DADPA)_x (APANA)_y$$
$$x > y;\ x \geq 2$$

The reason for the result is because DADPA has only one position for coupling, while APANA has more than two positions for the coupling.

This led to the finding that monoalkyl diphenylamine(MADPA) will homo-oligomerize upon oxidation:

MADPA can also replace APANA in the previously mentioned oxidative coupling with DADPA, resulting in the similar types of oligomer mixtures, namely, the dimer of DADPA and cross-coupled oligomers of DADPA and MADPA, where more DADPA than MADPA are in the same molecule(5):

$R^1, R^2, R^3 =$ *t*-Butyl, *t*-Octyl, Styryl

$$(DADPA)_2 + (DADPA)_x (MADPA))_y$$
$$x > y;\ x \geq 2$$

Many commercial lubricant antioxidants are mixtures of DADPA and MADPA, and they are much cheaper than APANA.

All the reported oligomers are excellent antioxidants, providing better protection for the ester lubricant at high temperature than any of the monomers, or their combinations.

References:
(1) D. R. Randell, U.S. Patent 3,696,851 (1972).
(2) M. Brald and D. Law, U.S. Patent 3,573,206 (1970).
(3) Chem. Abstrt. 97, 165681p, (1982).
(4) J. Lai and D. Filla, Patent pending.
(5) J. Lai, U.S. Patent 5,489,711 (1996).

EVALUATION OF TRIBOLOGICAL PROPERTIES OF SYNTHETIC LUBRICANTS UNDER HIGH TEMPERATURE AND HIGH SPEED CONDITIONS

KOICHI HACHIYA, HIDETO YUI, YOSHIO SHODA, SHINYA YOKOI
Research Institute of Advanced Material Gas-Generator, 1-5-50 Kugenuma Shinmei, Fujisawa, Kanagawa, 251 JAPAN
MICHIHARU NAKA
NSK Ltd.,R&D Center, 1-5-50 Kugenuma Shinmei, Fujisawa, Kanagawa, 251 JAPAN

ABSTRACT

Tribological properties of synthetic lubricants of a dipentaerythritol ester (DiPE), a tripentaerythritol ester (TriPE), and a 5-ring polyphenyl ether (5P4E) were evaluated using a unique oxidation stability tester and a high-temperature, high-speed ball/disc tribotester. The test conditions simulated a high-temperature (300 °C) high-speed (60 m/s) bearing.

The oxidative stability of lubricants was evaluated using a unique oxidative stability tester. A bearing incorporated in an aluminum housing was attached to the reactor tube and the bearing temperature was controlled to 315 °C through a high-frequency inductive heater. Air (50 cm^3/min) and lubricant (1.0 cm^3/min) were fed to the heated bearing. The bearing was rotated by means of a rotation actuator mounted on top of the reactor tube. The bearing was placed under an axial load of 4 N (Maximum hertzian stress: 0.5 GPa). Lubricant properties of the fluids following the 40 hour oxidation test were compared to the brand-new fluids to determine the level of viscosity increase, the change in total acid number, the molecular weight distribution (Mw/Mn) by gel permeation chromatography (GPC), the level of weight loss, and the quantity of deposit on the bearing. An oil specified by MIL-L-23699C (MIL) was also tested to facilitate comparison with other lubricants.

As for oxidative stability, 5P4E showed a superior oxidative stability for all parameters evaluated except for deposition. The level of viscosity increase, Mw/Mn and TAN change were equal in TriPE and DiPE while deposit forming tendency was lower in TriPE than in the other lubricants. However, Wt Loss was lower in DiPE than in TriPE. When comparing with MIL, the oxidative stability of DiPE and TriPE seems to be adequate.

Dynamic test was performed using a high-temperature (200-300 °C), high-speed (25-60 m/s) tribotester which provides rolling contact between a rolling Si3N4 ball specimen and a flat M50 disc specimen. A ball and a flat disc were arranged to be in contact with each other at a pitch circle diameter (PCD) of approximately 100 mm and both were directly driven independently by motors.

No seizure nor surface damage occurred during all the test conditions even though the tests were performed up to 85% of slide/roll ratio where the PV value (P:Maximum hertzian stress, V:Sliding velocity) was 48 GPa·m/s. It is therefore expected that all the traction tests would be conducted under full EHL conditions.

As examples of test results, Fig. 1 shows the maximum traction coefficient as a function of temperature and the slide/roll ratio at the maximum traction. In some cases the maximum traction force occurred at slide/roll ratios above 10%, thus traction values at 10% were used instead in these cases. The values of the maximum traction coefficient shown in the figure are normalized on the basis of the maximum traction coefficient of MIL at 200 °C, 40 m/s, 1.0 GPa, being taken as 1.0.

DiPE and 5P4E showed a large traction coefficient even at 300 °C, and its dependency on temperature was high. In contrast, TriPE showed less traction coefficient under many of the test conditions, and the traction properties of TriPE are less dependent on external factors such as temperature, and rolling velocity.

Overall oxidative stability and traction property results suggests TriPE may be a more suitable high-temperature and high-speed bearings lubricant than either DiPE or 5P4E.

This paper were submitted to STLE, and will be published in Tribology Transactions or Lubrication Engineering.

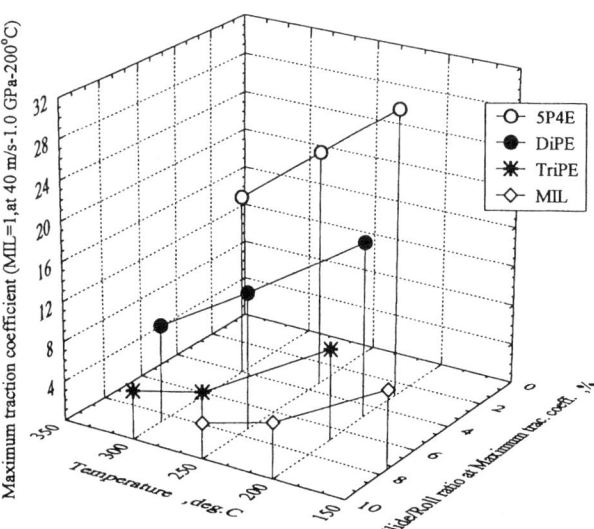

Fig. 1: Influence of temperature on maximum traction coefficient.

Submitted to *Tribology Transactions*

EVALUATION OF DEGRADATION INHIBITORS IN POLY(HEXAFLUOROPROPENE OXIDE) FLUIDS

WILLIAM R. JONES, JR.
NASA Lewis Research Center, Cleveland, OH 44135, USA
KAZIMIERA J.L. PACIOREK, WEN-HUEY LIN, STEVEN R. MASUDA
Lubricating Specialities Co., Vernon, CA 90058, USA

ABSTRACT

Perfluoropolyalkylethers (PFPAE), as represented by a series of commercial fluids (1)-(3) and new experimental compositions (4), are a family of high viscosity index and an exceptional thermal and oxidative stability. A number of investigations have assessed the thermal oxidative stability of these materials in the presence of metals, metal oxides and metal halides (5)-(8). These studies have shown that all these materials in the presence of metals undergo degradation and cause metal corrosion, in particular at elevated temperatures and in oxidizing atmospheres. Corrosion and fluid degradation were also observed at ambient temperatures in boundary lubrication situations (9).

It is believed that under boundary lubrication conditions, high temperatures are reached at the molecular level leading to thermodynamically favored fluoride formation. These metal fluorides, in turn, promote scissions of the PFPAE chains, as originally postulated by Gumprecht (1) and later substantiated by Carré (9) and others (5).

Two families of materials, phosphines and phosphorus containing heterocyclics, such as phospha-s-triazines and diphosphatetraazacyclooctatetraenes, have been identified as effective inhibitors of the degradation processes (5). The initially synthesized phospha-s-triazines exhibited hydrolytic instability due to the presence of a -OCF(CF$_3$) linkage adjacent to the ring. Replacement of this group with a -OC(CF$_3$)$_2$ moiety in monophospha-s-triazines, was accomplished (10) which alleviated this shortcoming. Corresponding diphospha-s-triazines and diphosphatetraazaocta-tetraenes were also synthesized (11). Recently, new phosphate ester additives were developed and shown to represent a very effective family of degradation inhibitors (12) Thus, the objective of the work reported here was to evaluate the action of various alloys: 440C, M-50 steel, Pyrowear 675, Cronidur 30 and Ti (4Al, 4Mn) and the effect of different phosphorus based inhibitors on two poly(hexafluoropropene oxide) fluids (143AC and 16256) at elevated temperatures in the presence of oxygen.

The degradation promoting action of the ferrous alloys in the 16256 fluid were comparable; Ti (4Al, 4Mn) alloy was significantly more detrimental. The overall rating of the additives was: phosphates > phosphate/diester mixture > phosphine ≥ phospha-s-triazines. The 16256 fluid was less responsive to additive inhibition than 143 AC. Phosphate esters were fully effective over 24 hour exposure in the 16256/440C steel and 16256/Ti (4Al, 4Mn) systems at 330°C. In general, the phosphine was less effective in the presence of ferrous alloys than the phosphates and phospha-s-triazines.

REFERENCES

(1) W. H. Gumprecht, ASLE Trans., 9, pp 24-30, 1966.
(2) D. Sianesi et al, Wear, 18, pp 85-100, 1971.
(3) Y. Ohsaka, Petrotech, 8 (9), pp 840-843, 1985.
(4) W. R. Jones, Jr. et al, NASA TM 106299, 1993.
(5) K. J. L. Paciorek and R. H. Kratzer, J. Fluorine Chem., 67, pp 169-179, 1994.
(6) W. R. Jones, Jr. et al, Ind. Eng. Chem. Prod. Res. Dev.., 24, pp 417-420, 1985.
(7) D. J. Carré and J. A. Markowitz, ASLE Trans., 28, 1, pp 40-46, 1985.
(8) P. H. Kasai, Macromolecules, 25, pp 6791-6799, 1992.
(9) D. J. Carré, ASLE Trans., 29, 2, pp 121-125, 1986.
(10) K. J. L. Paciorek et al, U. S. Patent 5, 326, 910, 1994
(11) K. J. L. Paciorek, SBIR Phase II Report, NAS3-26976, CR-198524, NASA Lewis Res Ctr, Cleveland, OH, 1996
(12) K. J. L. Paciorek et al, U. S. Patent 5, 550, 277, 1996.

SPECTROSCOPIC ANALYSIS OF PERFLUOROPOLYETHER LUBRICANT DEGRADATION DURING BOUNDARY LUBRICATION

PILAR HERRERA-FIERRO
Ohio Aerospace Institute, Brookpark, OH 44142, USA
BRADLEY A. SHOGRIN
Case Western Reserve University, Cleveland, OH 44106, USA
WILLIAM R. JONES, JR.
NASA Lewis Research Center, Cleveland, OH 44135, USA

ABSTRACT

The class of liquid lubricants, known as the perfluoropolyethers (PFPEs), have been used for space applications for many years (1) and more recently as lubricants for magnetic recording media (2). These fluids are also candidates for advanced gas turbine engine applications (3). Although these materials are quite stable compared to conventional lubricants, they do degrade at high temperatures in contact with catalytic surfaces and at room temperature in tribological contacts (4). In fact, these tribochemical degradation products allow unformulated PFPEs to survive in boundary lubricated contacts by forming low shear boundary films, such as FeF_3 (1), (5).

A number of investigators have studied the fate of fluorinated materials in boundary lubricated contacts using a variety of surface analytical techniques. Sugimoto and Miyake (6) studied the progression of degradation on polychlorotrifluoroethylene and PTFE sliding against 440C steel. They concluded that these polymers are progressively degraded into a fluorine deficient material and finally into an amorphous carbon network and metallic carbides.

Novotny et al (7) and Karis (8) reported on the degradation of PFPE lubricants on magnetic media and in ball mill experiments. They concluded that PFPEs are removed from the contact area by both physical displacement and degradation. The dominant process was tribochemical scission resulting in lower molecular weight fragments that can desorb and carboxylic acid fragments that can adsorb on the surfaces. Carré and Markovitz (9) reported the formation of fluorinated carboxylic acids with a branched PFPE in the presence of Lewis acids at elevated temperatures.

The objective of this paper was to further investigate the progressive degradation of a branched PFPE during sliding experiments with a ball-on-disc apparatus. Micro-Raman and micro-FTIR were used to analyze the surfaces. Test conditions included: a sliding speed of 0.05 m/s, a 3N load, 440C steel specimens, 25 °C, and a dry nitrogen atmosphere.

Discs were coated with thin (600 Å), uniform films of the PFPE. When the friction coefficient surpassed the value obtained with an unlubricated control, the film had either been physically displaced or partially transformed into a "friction polymer." Infrared analysis of this friction polymer indicated the presence of a polymeric fluorinated acid species (R_fCOOH). Raman spectroscopy indicated the presence of amorphous carbon in the wear track and in the friction polymer. Some reaction mechanisms are suggested to explain the results.

REFERENCES

(1) W. R. Jones, Jr., STLE Trans., 38, 3, pp 557-564, 1995.
(2) J. F. Moulder et al, Appl. Surf. Sci., 25, pp 445-454, 1986.
(3) C. E. Snyder, Jr. and R. E. Dolle, ASLE Trans., 19, pp 171-180, 1976.
(4) K. J. L. Paciorek and R. H. Kratzer, J. Fluorine Chem., 67, pp 169-175, 1994.
(5) P. Herrera-Fierro et al, NASA TM 106548, 1994.
(6) I. Sugimoto and S. Miyake, J. Appl. Phys., 65, 2, pp 767-774, 1989.
(7) V. J. Novotny et al, J. Vac. Sci. Technol., A12, pp 2879-2886, 1994.
(8) T. E. Karis et al, J. Appl. Poly. Sci., 50, pp 1357-1368, 1993.
(9) D. J. Carré and J. A. Markovitz, STLE Trans, 28, pp 40-46, 1985.

COMPARISON OF REACTED FILM FORMATION OF THREE DIFFERENT TYPES OF PERFLUOROPOLYETHERS WITH A HEATED STEEL SURFACE

MASABUMI MASUKO and NORIKATSU KAMIO[*]
Tokyo Institute of Technology, Department of Chemical Engineering, O-okayama, Meguro-ku, Tokyo 152, Japan
[*] Graduate school student. Currently at Tonen Corporation.

ABSTRACT

Perfluoropolyether (PFPE) lubricants are used for space applications and for high-temperature machinery, and a substantial body of research on their chemistry already exists. With regard to space mechanisms, because future space machinery will require better tribological performance of lubricants, PFPEs may no longer be as successfully applicable as before. Delaying degradation is a major requirement for extending the lubricant life. Accordingly, recent efforts have been made to improve the performance of PFPEs, most of them focusing on PFPE degradation stability under contact with solid surfaces. Because some space mechanisms will be operated under a boundary or mixed lubrication regime for long periods, antiwear performance of lubricants will be also critical for ensuring a long service life.

It has already been reported that PFPEs can successfully lubricate metal surfaces without seizure under high or ultra-high vacuum with no antiwear additives. This is due to the formation of a tribological reaction surface layer that is believed to consist of a metal fluoride. As often observed with conventional EP additives, excess reaction to produce the EP surface film causes excess wear, known as chemical wear or corrosive wear. Therefore, the reactivity of metal fluoride formation is likely to be an important factor in explaining PFPE performance.

In this study, the production of an inorganic layer by thermal reaction of PFPEs with a steel surface is investigated.

Three different kinds of PFPE, types Fz, D, and B were used for the sample oil. The reaction experiments were carried out using a "Hot-wire" method in air and nitrogen atmospheres to investigate the effect of oxygen. The thickness of the reaction surface film was derived by calculation between the electrical resistance of the test wire before and after reaction. The rate constant K of film formation was obtained by applying the parabolic law of corrosion.

As shown in Fig. 1, the value of K in the air environment was largest for type Fz, followed by type D, and smallest for type B. As shown in Fig. 2, it is also found that the presence of oxygen increased the rate constant K both for types Fz and D, but did not increase the value of K in the case of type B. It is theorized that the formation of a reactive fluorine intermediate, such as acyl fluoride, controls the reactivity of PFPE, and that this reactivity is accelerated by the presence of oxygen.

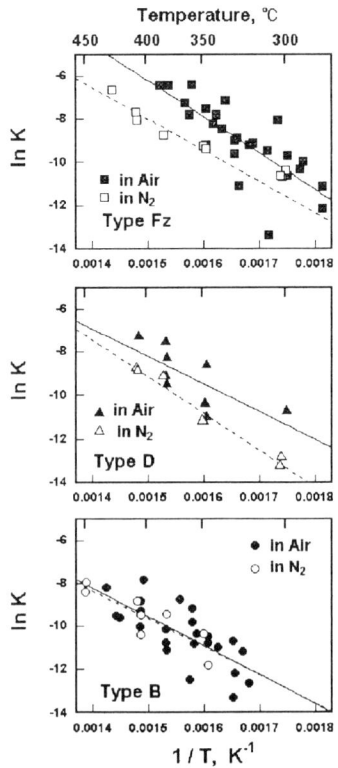

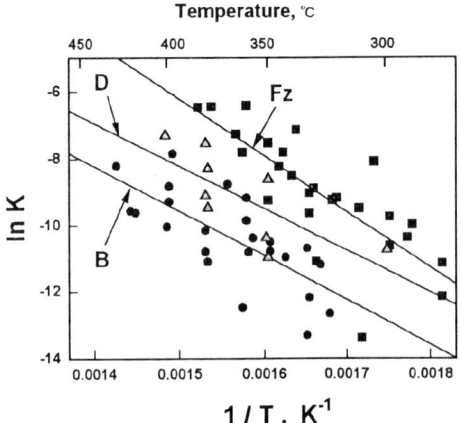

Fig. 1 Arrenius plot in air

Fig. 2 Comparison of results

EVALUATION OF BOUNDARY-ENHANCEMENT ADDITIVES FOR PERFLUOROPOLYETHERS

BRADLEY A. SHOGRIN
Case Western Reserve University, Cleveland, OH 44106, USA
WILLIAM R. JONES, JR.
NASA Lewis Research Center, Cleveland, OH 44135, USA
PILAR HERRERA-FIERRO
Ohio Aerospace Institute, Brookpark, OH 44142, USA
TZUHN-YUAN LIN AND HAJIMU KAWA
Exfluor Research Corp., Round Rock, TX 78664, USA

ABSTRACT

Perfluoropolyethers (PFPE) are widely employed as lubricants for space applications because of their excellent thermal and chemical stability and are particularly effective in the elastohydrodynamic range (1). However, when used as a boundary lubricant, PFPE performance is less predictable (2). The most significant problem encountered when using PFPEs in the boundary lubrication regime is the unavailability of soluble additives. Some soluble phosphorus based additives (phosphines and phosphatriazines) have been developed for anti-corrosion and anti-degradation (3)-(4).

Masuko et al (5) studied a series of PFPE derivatives (acids, alcohols and phosphate esters) in a linear PFPE basestock. A vacuum four-ball apparatus was used. These additives yielded some antiwear activity, with the PFPE acid being the most effective. Sharma et al (6) reported antiwear activity for a soluble ketone and alcohol in a linear PFPE based on tetrafluoroethylene. Nakayama et al (7) reported wear behavior in vacuum and various atmospheres for two PFPE additives: a carboxylic acid and an aminophenylsulfone. The acid reduced wear in vacuum, while the sulfone accelerated wear under the same conditions.

Recently, NASA has developed a vacuum four-ball apparatus and a test protocol for the evaluation of liquid lubricants and greases for space applications (8). Therefore, the objective of this work was to evaluate a series of newly synthesized soluble additives for their antiwear activity in a poly (hexafluoropropene oxide) basestock. Types of additives included a phosphate, a thiophosphate, a β-diketone, a benzothiazole, an amide and a sulfite. Additives were evaluated at a one weight per cent concentration by measuring steady state wear rates. Test conditions included: vacuum ($< 5.0 \times 10^{-6}$ Torr), 200 N load, a speed of 100 rpm, room temperature (~ 23 °C), 440C stainless steel bearing balls and a total test duration of 4 hours.

Wear rates for each formulation were determined from the slope of wear volume as a function of sliding distance. These results are summarized in Table 1. All additives yielded reductions in mean wear rates of at least 55 percent, with the exception of the benzothiazole which had no effect. Two of the additives, an amide and a sulfite, reduced the mean wear rate by at least 80 percent. IR and Raman analysis indicated the severity of wear can be correlated to the amount of surface fluorinated polymeric acid species and amorphous carbon, in and around the wear scar.

Table 1 - Wear Rate Summary

Additive Name	Mean Wear Rate ($\times 10^{-10}$ mm^3/mm)
Basestock	6.34 ± 3
Phosphate	2.16 ± 1.4
Thiophosphate	2.92 ± 1.6
β-Diketone	1.88 ± 0.23
Benzothiazole	6.09 ± 3.8
Amide	1.01 ± 0.42
Sulfite	1.20 ± 0.15

REFERENCES

(1) W. R. Jones, Jr., Trib. Trans., 38, 3, p 557, 1995.
(2) P. L. Conley and J. J. Bohner, 24th Aerospace Mech. Symp., NASA CP 3062, p 213, 1990.
(3) C. Tamborski et al, U. S. Patent 4, 454, 349, 1984.
(4) R. H. Kratzer, J. Fluorine Chem., 10, p 231, 1977.
(5) M. Masuko et al, Wear, 159, p 249, 1992.
(6) S. K. Sharma et al, J. Synthetic Lub., 7, p15, 1990.
(7) K. Nakayama et al, Wear, 192, p 178, 1996.
(8) M. Masuko et al, Lubr. Eng., 5, 11, p 871, 1994.

Submitted to *Society Tribologists and Lubrication Engineering*

INFLUENCE OF A HARD SURFACE LAYER ON THE LIMIT OF ELASTIC CONTACT: ANALYSIS USING A MODIFIED GW MODEL

T NOGI and T KATO
Department of Mechanical Engineering, the University of Tokyo,
7-3-1 Hongo, Bunkyo-ku, Tokyo, 113 Japan

ABSTRACT

A hard surface layer, such as ceramic coatings, is considered to be effective to realize elastic contact and to reduce wear of materials. In some cases, however, rupture or removal of the hard layer occurs mainly because of plastic deformation of the substrate (1). Thus the limit of elastic contact for layered surfaces must be predicted to specify such parameters as the layer thickness, substrate hardness and surface roughness.

In this paper, the contact of an elastic halfspace with an elastic layer against a rigid rough surface is considered and the Greenwood-Williamson model (2) is modified to predict the limit of elastic contact for layered surfaces. The summit height distribution is assumed to be Gaussian. All asperities are viewed as spheres with a same radius independent of their neighbours. Numerical solution developed by Chen and Engel (3)(4) for the contact of a layered halfspace against a rigid spherical asperity is used to calculate the contact area, pressure and internal stresses as a function of the penetration.

Fig. 1 shows the limit of elastic contact defined to be when $A_p/A = 0.02$ where A and A_p are the real and plastic contact area respectively. The equivalent layer thickness h^*, equivalent nominal pressure p^* and plasticity index ψ^* are defined as follows

$$h^* = \frac{h}{\sqrt{\beta \sigma_s}}, \quad p^* = \frac{1}{D_s \beta \sigma_s} \frac{p}{A_n k_1}, \quad \psi^* = \frac{\mu_1}{k_1} \sqrt{\frac{\sigma_s}{\beta}} \quad (1)$$

where h, p, D_s, β and σ_s are the layer thickness, nominal pressure, asperity density, asperity radius, RMS summit heights respectively. The results correspond to a TiN layer and a steel substrate (shear modulus $\mu_1 = 100$Gpa, $\mu_2 = 79$Gpa; shear strength $k_1 = 4.1$Gpa, $k_2 = 0.93$Gpa; Poisson ratio $\nu_1 = \nu_2 = 0.3$ where suffixes 1 and 2 denote the layer and the substrate respectively). It can be seen that the critical values of p^* increase as h^* increases.

Fig. 2 shows the limit of elastic contact defined to be when $N_{sp}/N_s = 0.02$ for $p^* = 0.5$ where N_s and N_{sp} are the number of contacting and plastically contacting summits respectively. Results of the numerical simulation by a more rigorous real surface model (5) are also depicted by ● (plastic) and ○ (elastic). Agreement of these two models is fairly good and it is suggested that the present model is a useful tool for the design of a hard surface layer realizing elastic contact and reducing wear.

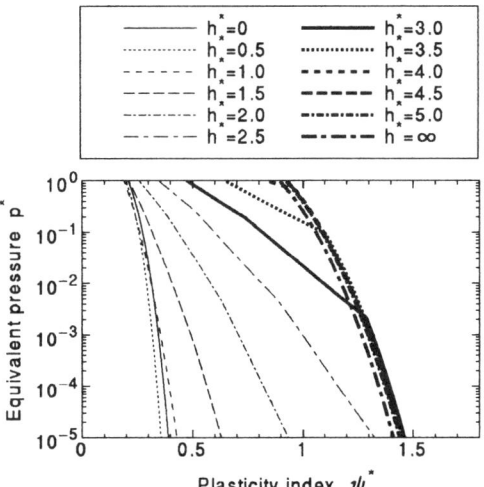

Fig. 1: Elastic/plastic contact map for various equivalent layer thicknesses

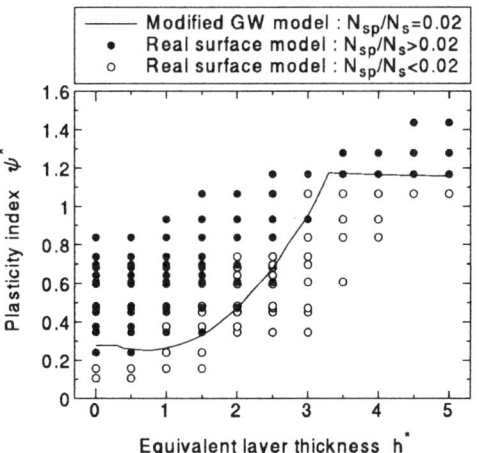

Fig. 2: Elastic/plastic contact map for $p^* = 0.5$

REFERENCES

(1) K. Komvopoulos, N. Saka, N. P. Suh, ASME J. Trib., Vol. 109, 1987, pp. 223-231.
(2) J. A. Greenwood, J. B. P. Williamson, Proc. Roy. Soc. Lond., Vol. A295, 1966, pp. 30-319.
(3) W. T. Chen, Int. J. Engng. Sci., Vol. 9, 1971, pp. 75-800.
(4) W. T. Chen, P. A. Engel, Int. J. Solids Struct., Vol. 8, 1972, pp. 1257-1281.
(5) T. Nogi, T. Kato, ASME J. Trib., Vol. 119, 1997, to be published.

Submitted to *ASME Journal of Tribology*

ELECTRICAL CONTACTS: THE USE OF LUBRICANTS AGAINST FRETTING

J.SWINGLER
Electrical Engineering, University of Southampton.
J.A. HAYES
Electronic & Electrical Engineering, Loughborough University, Loughborough.

INTRODUCTION

The development of electronic/electrical systems and devices over recent years has brought with it more demands on the electrical contact and its reliability (1)(2). It is important that the contact resistance remains low and stable throughout operation (3). This is achieved by maintaining a good surface condition void of excessive corrosion and debris.

FRETTING IN CONTACTS

The full paper of this title (4) highlights one of the most important electrical connector failure mechanisms: fretting, which results in high contact resistance and intermittent resistive faults. Causes of this failure mechanism are **i) External Vibrations, ii) Thermal Differential Expansion** arising from heating externally or internally, and **iii) Alternating Magnetic Fields**. Microscopic physical and chemical processes are described in detail and a model is presented which aids the description of various features involved in the fretting of connector contacts. This is illustrated in Figure 1.

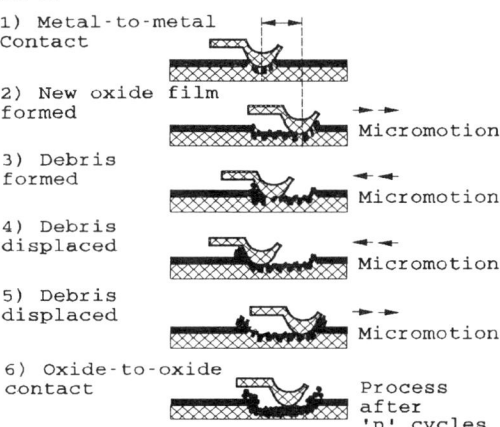

Figure 1 Fretting Corrosion

LUBRICANTS

Lubricants are becoming common place in electrical connectors as they ensure better performance and reliability of the contact surfaces. This is achieved by a number of processes: **i) Reduction in Corrosion Film Formation** as the lubricant acts as a barrier between the contact surface and the environment, **ii) Reduction in Wear** as the lubricant reduces the mechanical area of contact between the two surfaces, & **iii) Debris Suspension and Transport** as the lubricant aids in moving debris away from the contact area under the fretting action. The choice of lubricant for connectors depends upon the application and operating environment. Usually a tenfold improvement can be observed. However, the lubricant if not selected casrefully, may offer little performance enhancement and can even be detrimental to the connector. The lubricant must posses a number of properties to be of any use to the connector system and these are discussed in the full paper along with some experimental results which demonstrate improved performance and detrimental behaviour. Figure 2 shows a number of tests at different operating temperatures of contact resistance measurements against fretting cycles.

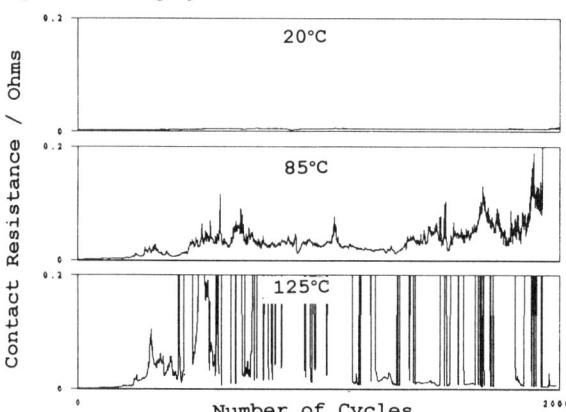

Figure 2 Contact Resistance of Tin/lead Connector Contact with a Lubricant During Fretting

REFERENCES

(1) Whitley, J.H., "Connector Requirements and Technology.", "Convergence '74", International Colloquium on Automotive Electronic Technology, 1974.

(2) Stennett, N.A. & Hayes, J.A., "Connector Reliability - A 'No Fault Found Problem'?", Proc. Electric Contact in Automotive, Aeronautical and Space Applications, pp145-150, 1991.

(3) Holm, R., "Electric Contacts.", 4th ed., Springer-Verlag, Berlin, 1968.

(4) Swingler, J. & Hayes, J.A., "Electrical Contacts: The Use of Lubricants Against Fretting." Submitted to WEAR, 1997.

(5) Antler, M., "Electrical Effects of Fretting Connector Contact Material: A Review.", Wear 106, No.1-3, pp5-33, Nov. 1985.

(6) Savage, D.W., "An Overview of Connector Lubrication.", Proc. 33rd Electronic Components Conf., pp400-403, 1983.

A MODEL FOR MICRO-SLIP BETWEEN FLAT SURFACES BASED ON DEFORMATION OF ELLIPSOIDAL ELASTIC ASPERITIES - PARAMETRIC STUDY AND EXPERIMENTAL INVESTIGATION

LARS HAGMAN and ULF OLOFSSON
Machine Elements Department of Machine Design KTH S-100 44 Stockholm Sweden

ABSTRACT

Micro-slip is a phenomenon that occurs between contacting surfaces when a frictional load, less than that necessary to produce macro-slip, is applied to the contacting surfaces. A model is presented for micro-slip between a flat smooth surface and a flat rough surface. The rough surface is covered with uniformly distributed ellipsoidal elastic asperities (see figure 1). A numerical 2^4 factorial design was used to evaluate the model. An experimental investigation was also performed to verify the model.

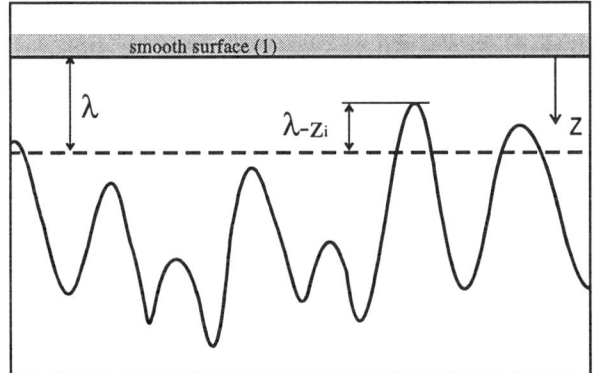

Figure 1. *Flat surface in contact with rough surface.*

The model is based on deformation of ellipsoidal elastic asperities and the total frictional load F becomes

$$F = F_{spring} + F_{slip} = \int_0^{z_{Li}} F_i CA dz + \int_{z_{Li}}^{\lambda} \mu P_i CA dz$$

where F_{spring} is the frictional load from the active asperities which have not reached their limiting tangential deflection, F_{slip} is the contribution from asperities which have reached their limiting tangential deflection, C is the number of contacts and the approach of the surfaces, A is apparent area of contact and P is the normal load. In figure 2, numerical results are presented for two numerical test cases. Figure 3 shows results from the experimental investigation.

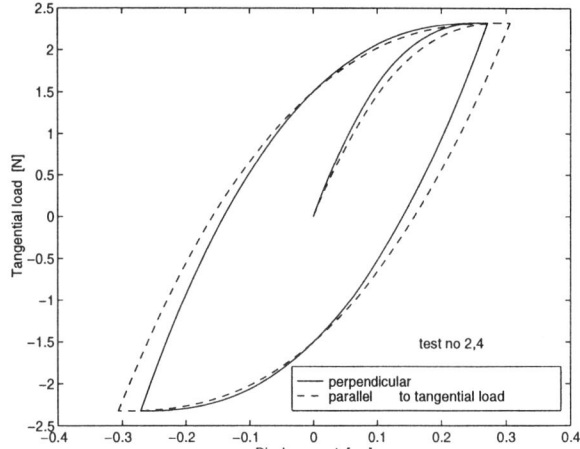

Figure 2. *Micro-slip curve. The major semi-axes of the ellipsoidal asperities are perpendicular and parallel to the tangential load respectively.*

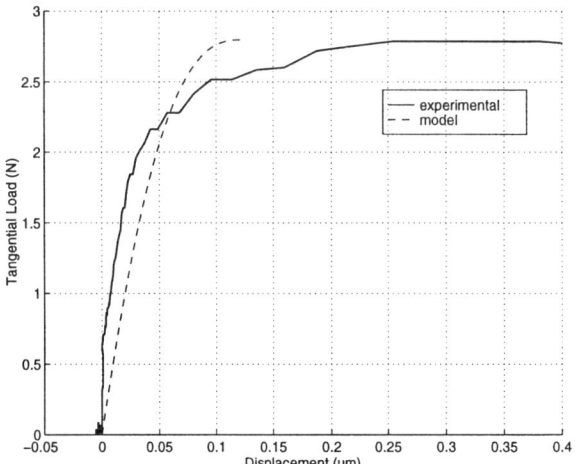

Figure 3. *Tangential load versus displacement. Continuous line -experimental result. Broken line -result calculated by the model. Contact pressure is 100MPa*

The results show that
- the anisotropy
- the material combination
- number of contacts of the contacting surfaces

influence
- the tangential stiffness at zero displacement
- the length of the micro-slip zone
- the energy dissipated in the contact.

Submitted to *Tribology International*

EFFECT OF LATERAL RESOLUTION ON TOPOGRAPHICAL IMAGES AND 3D FUNCTIONAL PARAMETERS.

H. ZAHOUANI, R. VARGIOLU, JL. LOUBET, PH. KAPSA, TG. MATHIA

Laboratoire de Tribology et Dynamique des Systèmes, UMR CNRS 5513.
Ecole Centrale de Lyon, 36 Avenue Guy de Collongue, 69131 Ecully cedex, France
& Ecole Nationale d'Ingénieurs de St Etienne, 42000 France.
Institut Eurorpéen de Tribologie.

ABSTRACT

Traditionally, surface topography analysis has consisted in assuming that the measurement points rest on a hypothetical surface whose mathematical properties, like stationarity, continuity and differentiability, are good. Thus all methods of mathematical analysis can be fully used. Unfortunately this hypothetical surface is only a figment of our imagination, for not only does the measuring instrument interfere with our view of the surface, but also, even on the atomic scale, the matter is discontinuous.

This imaginary surface derived from our measurements is therefore only a partial vision which helps us to perceive reality. This problem of scale which we have tried to show here can be found whenever we try to quantify a physical magnitude in tribology, as all the magnitudes plotted from a rough surface area (the real contact area, watertightness, bearing area, stress, notion of hardness, friction force, wear volume...) will necessarily depend on the observation scale.

The influence of the analysis scale on the characterisation of surface roughness is set out in this work. First the effect of lateral resolution on the local morphology of a random engineered surface is shown. When the geometry of the tip of a tactile profilometer, or the sampling step of a laser profilometer is varied. The lateral resolution influences not only the roughness amplitude parameters, but also, spectacularly, bearing area, developed area, the void or material volume, **figure (1)**. In the second half of this work the contribution of fractal geometry is examined. This is because, through its roughness index, called the Hölder index, it may be independent of the measurement.

Fractal geometry has allowed us to simulate the behaviour of various roughness parameters as a function of their degree of fractality, that is, as a function of a parameter which is independent of the measuring process, and which reproduces the various roughness scales through the Hölder coefficient or the fractal dimension, which play the role of indicators of the degree of perturbation of the roughness.

There is now a lot of work to be done, using these indicators as roughness scale coefficients in a considerable number of magnitudes which depend on the measurement scale, so as to be free of the apparatus' resolution.

By using the Weierstrass Mandelbrot function as a three-dimensional multi-scale model of the topography [1], the evolution of the bearing area **figure (2)**, the developed area and the volume as a function of the Hölder roughness index can be shown [2,3].

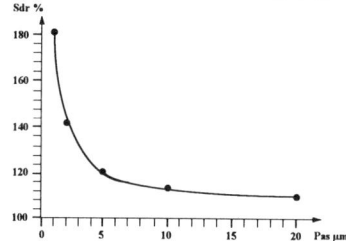

Figure (1): Incidence of lateral resolution on developed area.

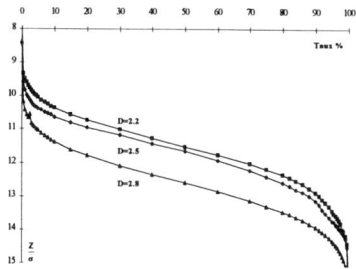

Figure (2): Bearing area curve of Weierstrass Mandelbrot type surfaces with various fractal dimensions.

REFERENCES

[1] A.Majunder and C.L Tien, "Fractal characterisation and simulation of rough surfaces.", 1990, Wear, Vol. 136, pp. 313-327.

[2] H.Zahouani, R.Vargiolu, JL.Loubet "Fractal models of surface topography and contact mechanics", à paraître dans Mathematical and Computer Modeling, special issue in Contact Mechanics 1996.

[3] A.Majunder and B.Bhushan ,"Role of fractal geometry in roughness characterisation and contact mechanics of surfaces", 1990, ASME J.Tribology, Vol. 112, pp.205-216.

THE USE OF ULTRASOUND TO STUDY ROUGH SURFACE CONTACT

R S DWYER-JOYCE,
Department of Mechanical Engineering, University of Sheffield, Sheffield, S1 3JD, England.
B W DRINKWATER
Department of Mechanical Engineering, University of Bristol, Bristol, BS8 1TR, England

ABSTRACT

The measurement of ultrasonic reflection has been used to study aspects of the contact between two rough surfaces. An incomplete contact will reflect a proportion of an incident ultrasonic wave. This proportion is known as the reflection coefficient. If the ultrasound wavelength is large compared with the size of the gaps in the plane of the interface, then the reflection can be modelled by considering the contacting bodies and the interface as an array of springs in series (1). The reflection of the wave is then a function of the stiffness of the interface and the frequency of the wave. The authors in a previous work (2) demonstrated the applicability of this approach.

This experimental principle has been employed in two configurations. An ultrasonic transducer has been used to scan a range of interfaces to study the extent and distribution of reflections within a bulk geometric contact. Figure 1 shows the ultrasonic reflection signal from the contact between a rubber cylinder pressed onto a rough steel flat. The lighter coloured areas show regions of low reflection and hence higher conformity.

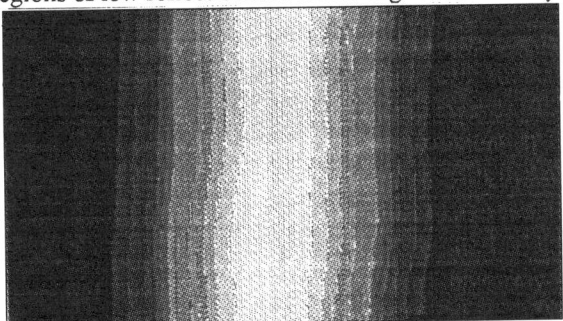

Fig.1: Reflection map of the contact between a rubber cylinder and a flat surface (5mm width contact)

In other experiments the reflection has been measured over a small region of an interface for a range of contact conditions. The reflection is then related to the interface stiffness using the spring model approach.

Figure 2 shows the stiffness of the contact between two grit blasted aluminium surfaces during three loading cycles. Plastic deformation takes place during the first loading. Subsequent cycles are predominantly elastic although some hysteresis and further plasticity occurs. The source of this hysteresis is believed to be irreversible adhesion.

This method has been used to study the effect of contact element material, surface roughness, and the loading path.

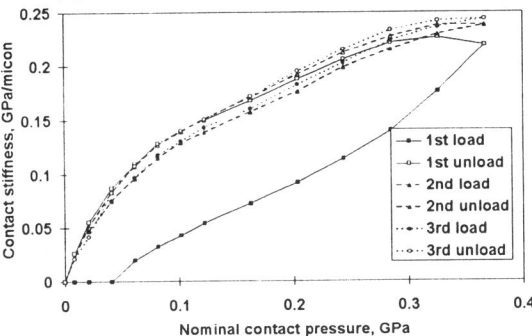

Fig.2: The stiffness of an aluminium interface during three loading cycles.

The stiffness of an interface is a function of the number, size, and proximity of the asperity micro-contacts. It is not possible to uniquely predict real contact area from a stiffness measurement.

Statistical (3) and numerical (4) models of elastic rough surface contact have been adapted to predict contact stiffness. The results have been compared to reflection measurements from the elastic unloading of a range of contact cases. The models under predict the stiffness because they do not consider the enhanced conformity of two surfaces which have been previously plastically pressed together.

REFERENCES

(1) Kendall, K. and Tabor, D., 1971, Proc. Royal Soc. A, Vol. 323, pp. 321-340.
(2) Drinkwater, B. W., Dwyer-Joyce, R. S., & Cawley, P., 1995, Proc. Royal Soc. A, Vol. 452, No. 1955.
(3) Greenwood, J. A. and Williamson, J. B. P., 1966, Proc. Royal Soc. A, Vol. 295, pp. 300-319.
(4) Webster, M. N. and Sayles, R. S., 1986, Trans. ASME, J. Trib., Vol. 108, No.3, pp. 314-320.

NUMERICAL MODELLING OF PARTICLE-DETACHMENT

A. ELEŐD

Technical University Budapest, Faculty of Transportation Engineering, Bertalan L. u. 2., Budapest, 1111 H

ABSTRACT

Based on the results of the tests that are given by modern testmachines, the "third body" theory queries from a lot of points of view the usability and validity of the classical wear theories or wear mechanisms.

The wear is a result of a complex processing, but the stipulation of the wear-beginning is, in any case, the getting moving of the particle-detachment.

The particle-detachment can happen - except the extreme case of micro-cutting - practically in three different ways:
- by cold fracture (on elastic contact field in consequence of a specific fracture-work that is reduced to zero under influences of state-factors),
- by fatigue fracture (as a consequence of the cumulated damage, under elastic contact conditions),
- by plastic fracture (due to the exhausted of the deformation capacity in the plastic contact area).

All the three fracture processes have a common characteristics of: the fracture will happen, if the local deformation-capacity of the material exhausts for some reason.

Former studies (1) brought in connection the deformation capacity to the specific fracture work of the material. It is widely known that the specific fracture work depends considerably on the state-factors, especially on the stress state of the material. The specific fracture work can not be used directly as a material law in case of the numerical modelling of particle detachment.

According to the Ziaja's assumption verified by theoretical and experimental ways (2), the small environment of the crack turns into locally instability previously of the plastic fracture, namely the 3D stress state will be converted in this small environment of the crack into plane stress state, and inside of this, the strain state will become a plane strain state. In this case the material needs a minimum energy to the fracture.

In a common plane stress and plane strain state, the plastic fracture begins, if the plastic yield condition and the plane state characterising stress-ratio will be satisfied:
- plane stress state is if: $\sigma_3 = 0$
- inside of this a plane strain state is if: $\sigma_2 = \sigma_1 / 2$
- in this case the yield condition is: $k_f(\bar{\varepsilon}) = \sigma_1 \sqrt{3}/2$

The stress causing the plastic fracture ($\bar{\sigma} = k_f$) can be determined by a special tensile test (2). This value can be used to the determination of the plastic material law.

In the contact model (see the Fig.1.), the measured residual-stress-state of the surfaces developing during the working of them, as well as the influence of the residual stresses on the deformation-capacity of the particles was also considered. We defined the material-laws for the different surface-layers from the yield curve of the basic material with consideration of the residual-stress-state of the layers as an actual yield point. The deformation-capacity of the particles of the surface was defined by the stress causing the plastic fracture of the basic material. The charge of the surfaces was on the one hand a structural face load, on the other hand a rigid body-like displacement of one of the surfaces in relative motion.

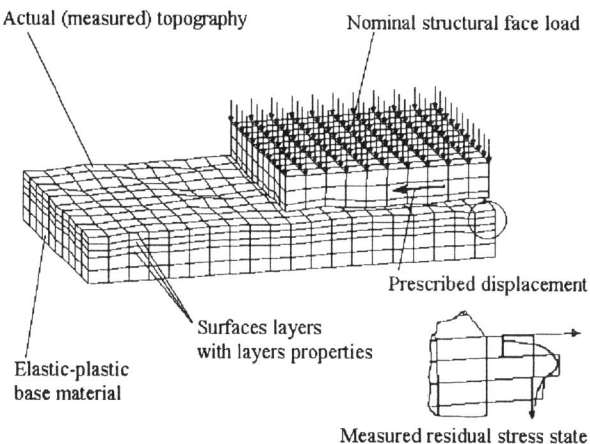

Fig. 1: The composition of the model

The calculated results were evaluated from point of view of the stress-components, such as it is shown in Fig. 2.:

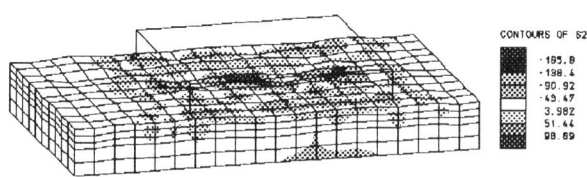

Fig. 2: The distribution of the 2nd stress component

REFERENCES

(1) G. Fleischer: Zur Energetik der Reibung. Wissenschaftliche Zeitschrift der Technischen Universität "Otto von Guericke" Magdeburg 34 (1990) Heft 8.

(2) Gy. Ziaja: Alakítási folyamatok határállapotai. Akadémiai doktori értekezés, Budapest, 1994.

Finite-Element Modeling of Plastic Deformation, Crack Growth and Wear Particle Formation in Sliding Wear

Ko, P.L.,
National Research Council of Canada, Vancouver, B.C., Canada V6T 1W5
IYER, S.S.
Babcock & Wilcox, Cambridge, Ontario, Canada
VAUGHAN, H., and GADALA, M.
Dept. of Mech. Eng., UBC, Vancouver, B.C., Canada V6T 1Z4

ABSTRACT

Two surfaces in sliding contact are examined. Roughness is modeled by surface asperities having the form of randomly spaced circular corrugations, which ideally represent plane strain conditions perpendicular to the direction of sliding. Contact stresses between corresponding corrugations on the sliding surfaces are evaluated using ANSYS 5.1.

INTRODUCTION

The existence of microscopic surface and subsurface cracks in metal components is well established. During sliding between two metals the forced contact between the two sets of surface asperities produces local elastic-plastic deformation together with extension of the microscopic cracks. Repeated cyclic loading can cause cumulative crack extension leading eventually to the formation of slivers, platelets and particles some of which are removed from the surfaces and become wear particles. A mathematical model which can predict the production rate of wear particles is fundamental in the design of machine components for a specified wear-life. There have been many attempts to use finite element methods to study the responses of contacting bodies. Ham et al. (1) examined rolling-sliding contact across an elastic-perfectly-plastic half-space, Ohame and Tsukizoe (2) predicted wear particle formation on a level surface based on crack propagation concepts. In this analysis rough surfaces are characterized by asperities discretised into cylindrical-type corrugations or low-profile spherical irregularities (3).

WEAR PARTICLE DETACHMENT MODELING

The interface of this contact problem is first modeled using general purpose contact elements. The contact stresses reveal the inherent nature of the ratchetting phenomenon in a cyclic contact loading situation. The zone of maximum stress intensity is identified and the information used for modeling wear particle formation.

Next, a finite-element fracture mechanics model is developed incorporating an assumed crack of known length and orientation in the smooth half space which is loaded by a cylindrical asperity in the normal direction. The stress intensity factor is calculated and related to crack extension using Paris Law for fatigue crack growth extension. A new approach, which significantly differs from the conventional approach of assuming the normal and tangential loads are shared equally by every contacting asperities, is used to model crack growth. This approach assumes that some asperities are subjected to a larger share of the tangential load and a smaller share of the normal load than other asperities. The direction of crack growth is modeled as being perpendicular to the direction of the first principal stress, essentially creating a Mode I fracture. The process is repeated until the crack extends and turns toward the surface and eventually forming a single wear particle. Later, wear volume is calculated by applying surface statistics.

NUMERICAL SIMULATIONS

Figure 1 shows the path of crack growth for a simulation using 25 μm incremental crack length with the initial crack at a shallow angle to the loading surface. Some estimated wear volumes from the finite-element model are given in Table 1, they are in fairly good agreement with some earlier experimental results (4) even though the model being two dimensional has an idealized wear particle geometry. Future models will take into consideration non-linear and random effects.

Table 1 Computation of wear particle characteristics

Crack length μm	Crack Orientation deg.	No. of cycles	Max. Depth μm	Aspect ratio length : dep.	Wear volume mm³
30	10	2998	6.54	12:1	0.008
30	0	4935	10.28	5.7:1	0.019
25*	10	2430	5.0	8.3:1	0.0074
25	0	4521	9.86	6.08:1	0.017

Typical experimental mass loss of 410 s.s. flat (Rc30) under similar loading conditions: 2000-5000 cycles, 0.04-0.12 mg (Ref. 4).

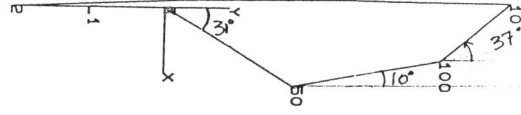

Fig. 1. *Path of crack growth for example #3 in Table 1.

REFERENCES

(1) G. Ham, C.A. Rubin, G.T. Hahn and V. Bhargava, J. of Tribology, Vol. 110, pp 44-49, 1988.
(2) N.Ohame and T.Tsukizoe, Wear, 61(2), pp 333-339, 1980.
(3) S.S. Iyer and P.L. Ko, ASME Publication PVP-Vol. 328, pp 369-377, 1996.
(4) P.L. Ko, G, Knowles and H. Vaughan, Proceedings, Int. Tribology Conference, Vol.1, pp 181-186, 1995.

Submitted to *Wear*

DYNAMIC INSTABILITIES IN THE SLIDING OF TWO LAYERED ELASTIC HALF-SPACES

GEORGE G. ADAMS
Department of Mechanical Engineering, Northeastern University, Boston, MA, 02115, USA

ABSTRACT

The sliding of two surfaces upon each other is such a common occurrence. Yet this phenomenon remains one which is not well-understood. A better understanding of the nature of sliding friction can lead to methods which reduce friction and wear. It may be possible to design *surfaces* in such a way as to significantly diminish both friction and wear. Such "designer surfaces" could have surface topographies and surface layering selected in order to optimize sliding properties. Clearly there are enormous economic incentives to reduce friction and wear. It has been estimated that the annual losses due to friction and wear in the U.S.A. alone are over $200 billion.

Martins, Guimarães, and Faria (1) investigated the sliding of elastic and viscoelastic half-spaces against a rigid surface. Dynamic instabilities were found for cases in which the friction coefficient and the Poisson's ratio were both large. These instabilities were thought to play a role in Schallamach waves. In another investigation Adams (2) showed that the steady sliding of two elastic half-spaces is dynamically unstable. The instability mechanism is essentially one of slip wave destabilization. Steady-state sliding is shown to give rise to a dynamic instability in the form of self-excited motion. These self-excited oscillations are generally confined to a region near the sliding interface and can eventually lead to either partial loss-of-contact or to regions of stick-slip motion.

It is the purpose of the current investigation (3) to determine the effect of surface layers on dynamic instabilities in the sliding of two elastic half-spaces. Consider two perfectly flat layered elastic half-spaces sliding with respect to each other with constant velocity and constant coefficient of sliding friction. Each body consists of a single uniform layer of elastic material perfectly bonded to a semi-infinite elastic solid of differing material properties. The sliding speed is assumed to be much less than the wave speeds of any of the materials. The solution of the plane strain equations of dynamic elasticity is decomposed into two parts, *i.e.* the nominally steady-state solution which is invariant in time and space, and the unsteady part, the stability of which is investigated here. The unsteady solution is taken in the form of a spatially periodic propagating motion with a time-varying *amplitude* of oscillation.

Solutions of the plain strain equations of elasticity, subject to appropriate boundary conditions, have been obtained. These solutions are also subject to the inequality constraints that the contact pressure and the slip velocity both be nonnegative. The former condition is necessary to prevent loss-of-contact and the latter inequality ensures that slip is uni-directional. Finally, it is of interest to determine the normal and shearing stresses at the interface between each layer and its corresponding half-space, as these stresses can be expected to play an important role in delamination.

The results show that self-excited oscillations exist for a wide range of material combinations. The magnitude of these dynamic instabilities depends on the shear moduli ratios and on the mismatches in the shear wave speeds of the sliding materials. Their existence is due to destabilization of waves at the sliding interface. Depending upon the phase relation between the unsteady part of the contact pressure and the slip velocity, the energy dissipated at the sliding interface may be greater (stable motion) or less (unstable motion) than the work done in moving the bodies. These dynamic instabilities can eventually give rise either to partial separation or to regions of stick-slip. The greater the mismatch in shear wave speeds of the sliding materials, the less likely is stick-slip as opposed to loss-of-contact. Furthermore for a given material pair, high contact pressure and slow sliding speeds make stick-slip more apt to occur.

The normal and shearing stresses at the bonded interfaces have also been determined. These dynamic stresses can be greater in magnitude than the nominally steady-state stresses. Furthermore the dynamic stresses fluctuate rapidly as they travel with the slip wave velocity of the materials, rather than at the much slower sliding speed. Thus these dynamic stresses are apt to play an important role in delamination of surface layers.

ACKNOWLEDGEMENT
This material is based upon work supported by the National Science Foundation under Grant No. CMS-9622196.

REFERENCES
(1) J.A.C. Martins, J. Guimarães, L.O. Faria, ASME Journal of Vibration and Acoustics, Vol. 117, pp. 445-451, 1995.
(2) G.G. Adams, ASME Journal of Applied Mechanics, Vol. 62, pp. 867-872, 1995.
(3) G.G. Adams, ASME Journal of Tribology, to appear.

AN ASPERITY MICRO-EXPLOSION MODEL FOR BRITTLE MATERIAL SLIDING WEAR:
I. ANALYTICAL MODEL

T. N. YING and S. M. HSU

National Institute of Standards and Technology, Gaithersburg, MD 20899, USA

ABSTRACT

For brittle materials, fracture process dominates in the deformation and wear. Tensile cracks have been found to be a predominant mechanism. We have monitored the interface contact phenomena of a one pass experiments of a silicon nitride ball on silicon nitride ball using a high speed video camera. Micro-explosion of wear particles flying from the contact junction was captured on the video. Further experiments using quartz and other transparent material pairs revealed the presence of shear cracks.

Our experimental evidence suggests that the elastic energy stored from the first point of contact continues to build while tensile cracks begin to propagate. However, the direction of the maximum shear stress changes with the contact position. The first principle stress, σ_1, reaches a maximum value of $2\mu P_o$ and the direction of the stress coincides with σ_x when x equal to (-a). An array of tensile cracks occurs at the trailing edge position during the collision. The video records show that the fully extended cone crack length at a coefficient of friction 0.7 is much longer than that at 0.2. This is consistent with theory. As the tensile crack continues to propagate, the contact point continues to move forward with the tensile crack at its wake. There comes a point when the tensile crack length is larger than the contact area, then the new surfaces of the tensile crack constitute a new boundary condition for the contacting area. We used a finite element program to calculate the stresses in this situation. The results of the analysis showed that beneath the contact area there is no significant tensile stress due to the formation fo the new crack boundary. The principal shear stress is

The largest shear stress is located at the center of the contact area. Because there is no tensile stress ($\sigma < 0$) to compete with, the shear stress is the only driving force to initiate and propagate a new crack. When the shear stress is larger than a critical value τ_c, a forward propagating crack occurs. This crack, therefore, is the shear crack. The direction of the crack is along the maximum shear stress trajectory. Although the analysis is qualitative, the calculation agrees with the observation.

It is noted that the driving force of crack initiation is a function of $\sigma_1 - \sigma_3$, but the hydrostatic stress is a function of $\sigma_1 + \sigma_3$. For brittle materials in an elastic region, Poisson's ratios are usually less than 0.3. The hydrostatic pressure, σ_o in a plain strain case will be

$$\sigma_o = \frac{\nu+1}{3}(\sigma_1 + \sigma_3) \qquad (1)$$

The built-in hydrostatic pressure then is a function of stress ratio λ and stress intensity factor K_{IIc}. a large K_{IIc} value also results in a high hydrostatic pressure. In a three dimensional compression state, before a shear stress reaches a critical value to initiate a crack, a hydrostatic pressure term must coexist. Once a shear crack propagates, the built-in hydrostatic pressure must be released at the same time. This sudden release of the hydrostatic pressure is the main driving force behind the explosion.

THREE-DIMENSIONAL CONTACT STRESS ANALYSIS USING THE BOUNDARY ELEMENT METHOD

J G LEAHY and A A BECKER
Department of Mechanical Engineering, University of Nottingham, U.K.

ABSTRACT

Numerical methods such as the Finite Element Method (FEM) and the Boundary Element Method (BEM), are increasingly being used to solve engineering problems. The BEM is particularly well suited to contact problems since the contact variables are coupled directly without the need for special purpose elements, and local mesh refinement is relatively easy since only the boundary needs to be modelled. The present authors are involved in the development of solution algorithms for contact problems using the BEM in a multi-domain BE formulation.

Practical contact problems involve frictional forces which are generated when two or more bodies come into contact. If the ratio of normal to tangential forces exceeds the prescribed coefficient of friction, μ, then frictional slipping occurs. However, if this ratio of forces is less than μ then a no-slip condition is imposed, and there is no relative displacement between the two contacting bodies at the interface.

Sub-division into sticking and slipping zones and the extent of the contact region is usually unknown prior to the analysis and large changes in the contact area are possible. For this reason, an iterative approach has been adopted (1) (2) whereby the area is estimated and the correct contact conditions are determined on a trial and error basis. Any nodes with normal tensile stresses in the contact region are deleted in the next iteration resulting in a smaller contact region. Similarly, detection of overlap suggests that the contact area is too small for the given load and the associated nodes are included in the next iteration.

For three-dimensional contact, displacements and tractions have been coupled in matrices based on a local co-ordinate system, enabling no-slip, frictionless and frictional contact analysis. Such development has necessitated the generation of a number of benchmark cases for the evaluation of accuracy (3). The commercial FE package, ABAQUS, has been used for comparison purposes.

Normal stress, shear stress and relative tangential displacements are obtained for the case of a punch in contact with a foundation (3). Increasing μ only lowers the normal stress by a small percentage, Fig. 1. For each value of μ there is a distinct change in shear stress gradient with increasing distance along the contact interface, Fig. 2. This signifies the change from a sticking region to a frictional slipping region. This is also reflected in the results for relative tangential displacement, Fig. 3. As expected, increasing the value of μ results in a larger stick region.

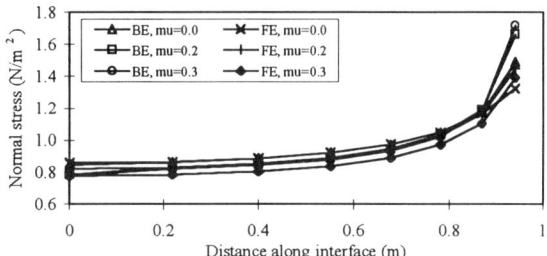

Fig. 1: Normal stress distribution for a punch in contact with a foundation

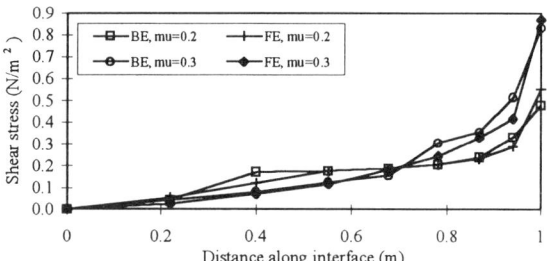

Fig. 2: Shear stress distribution for a punch in contact with a foundation

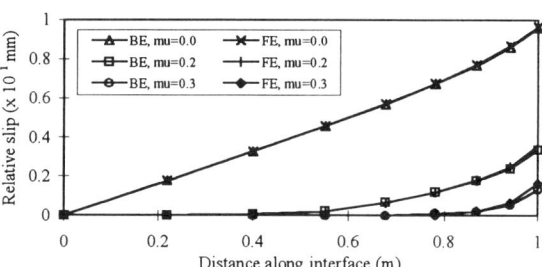

Fig. 3: Relative slip distribution for a punch in contact with a foundation

REFERENCES

(1) O A Olukoko, A A Becker, and R T Fenner, Journal of Strain Analysis, Vol. 28, pp. 293-301, 1993.

(2) O A Olukoko, A A Becker, and R T Fenner, International Journal for Numerical Methods in Engineering, Vol. 36, pp. 2625-2642, 1993.

(3) J G Leahy and A A Becker, Proc. NAFEMS World Congress on Design, Simulation & Optimisation, pp. 335-346, NAFEMS, 1997.

NEW METHOD OF MEASURING REAL CONTACT AREAS BY USING A THIN PET FILM
(EFFECT OF SURFACE ROUGHNESS ON PET FILM METHOD)

ISAMI NITTA
Graduate School, Science and Technology, Niigata University, Niigata 950-21, JAPAN
AKIRA MOROHASHI
Kokusai Electric Co.,Ltd, Toyama Works2-1, Yasuuchi, Yatsuo-machi, Nei-gun, Toyama, 939-23, JAPAN
CHIKASHI OTANI
Precision Engineering, Chiba Institute of Technology, Chiba, 275, JAPAN

ABSTRACT

A new method of measuring the distribution of real contact areas between two solid surfaces was developed in a previous paper (1). In this paper, an effect of surface roughness on this method has been investigated.

One of two solids in contact has to be transparent since a contact microscope must be used. Thus, the real contact areas between a turned surface of aluminum alloy and a right-angled glass prism were measured by the PET-film method and the contact microscope.

The PET (polyethyleneterephthalate) film, 0.9 μm thick, was inserted at the contact interface and then pressed at a given pressure for 1 min. It was indented by surface asperities at real contact points, as shown in Fig.1. After unloading, the PET film was removed from the contact interface and the indented areas where the plastic deformation occurred were measured automatically with an image processing through an optical microscope that was modified so that its stage can be controlled by a micro computer.

The contact microscope that can observe the entire nominal contact area, annular area of 3 mm in inner diameter and 5 mm in outer diameter, was used to examine the distributions of real contact areas measured by the PET-film method. A method of total reflection of polarized light was used to improve accuracy of the conventional contact microscope. A polarized light beam was projected on the hypotenuse of the glass prism which was in contact with the turned aluminum surface, and the reflecting light from the real contact points of the aluminum surface was intercepted by an analyzer. Thus real contact areas were much more clearly observed than without the polarized light, as a dark image in bright background.

The distributions of real contact areas measured by two different methods were relatively in good agreement with each other. However, the PET-film method depended upon the surface roughnesses of the specimens.

To examine the effect of surface roughness, the aluminum specimens with various surface roughness from 2 μm to 14 μm as maximum surface roughness, Ry, were prepared. The real contact areas could be measured by the PET film method if the Ry was larger than about 5 μm, as shown in Fig. 2. A shape of a tip of surface asperity on the aluminum specimen also affected the PET film method. It was observed that blunt tips of the surface asperities did not give sufficient indentations on the PET films even if the Ry was over 5 μm.

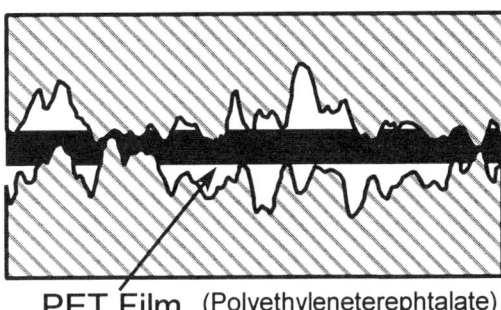

Fig. 1: Principles of the PET film method

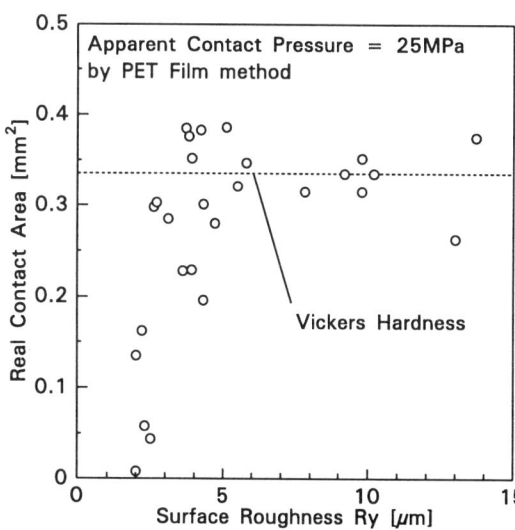

Fig.2: Measured real contact areas with Ry

REFERENCES
(1) Isami NITTA, Wear Vol.181-183, (1995) 844-849.

CONTACT, STRESS AND THERMAL ANALYSIS OF SLIDING COMPOSITE-STEEL REAL SURFACES

KLAUS FRIEDRICH - JOACHIM FLÖCK
Institute for Composite Materials, University of Kaiserslautern, 67663 Kaiserslautern, Germany
KÁROLY VÁRADI - ZOLTÁN NÉDER
Institute of Machine Design, Technical University of Budapest, H-1521 Budapest, Hungary

ABSTRACT

Between two sliding bodies, under load transmission, there are real contacts over only small areas of the nominal contact area due to the surface waviness and surface roughness. This is also true for a composite-steel sliding contact. In the pin-on-disk system, especially at the beginning of the running-in period there are only asperity contacts. After running-in the real contact areas are much bigger and the real contact pressures therefore become smaller.

The aim of this investigation is to study the contact, stress and thermal behaviour of sliding composite-metal surfaces by both numerical and experimental ways during the wear process.

The composite material is a unidirectional continuous carbon fiber/polyether etherketone system. Sliding on the composite structure is investigated in three different orientations (normal, parallel and antiparallel) relative to the fiber direction (Figs. 1a, 1b and 1c).

The contact algorithm (1), (2) "works" on discretised surfaces. Due to the nature of the composite materials, anisotropic material properties are used in the 3D "half-space" FE model to obtain the elements of the influence matrix for the contact analysis.

To avoid the very high peaks of elastic contact pressure, an approximate elastic-plastic contact analysis is used introducing the plastic limit pressure as the possible highest pressure, as specified for the anisotropic structure studied.

In the thermal analysis the frictional energy is converted into a heat source that is partitioned between the contacting bodies. Steady state heat conduction is assumed in the contact temperature calculation.

The numerical results are used as "input data" for a connecting FE stress and thermal analysis of the real matrix/fibre micro structure. This FE models (Figs. 1a, 1b and 1c) contain a very small "micro-segment" of the composite material structure, assuming a regular arrangement of fibres. By this FE model different failure modes will be simulated like fibre debonding, buckling, and fracture.

The wear process is followed by repeated experimental measurements of the surface roughness of both components for the contact, stress and temperature calculations.

According to the results of the anisotropic contact analysis, if dry friction is assumed, the real contact area, being formed by many small contact spots, is very small compared to the nominal contact area.

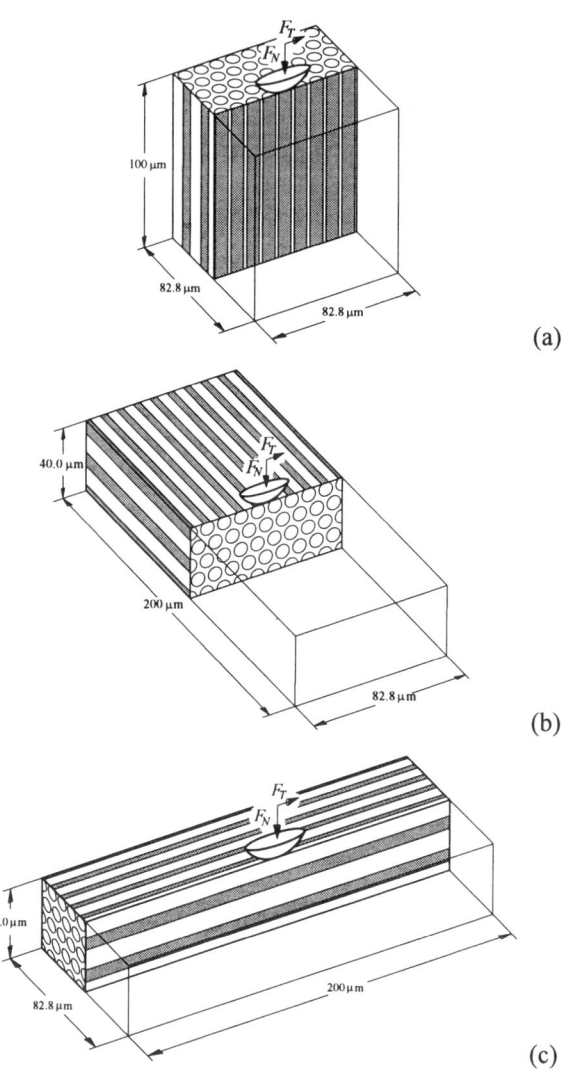

Fig. 1: The composite micro models for N, P and AP fibre orientations

ACKNOWLEDGEMENT

The presented research task was supported by the Deutsche Forschungsgemeinschaft (DFG FR675/19-1) as part of the German-Hungarian research co-operation on contact mechanics of different materials.

REFERENCES

(1) K. Váradi, Z. Néder, Tribologia, 3-1996, 237-261
(2) K. Váradi, Z. Néder, K. Friedrich, Wear 200 (1996), 52-62

SPATIALLY RESOLVED MECHANICAL PROPERTIES OF A "TPO" USING A FREQUENCY SPECIFIC DEPTH-SENSING INDENTATION TECHNIQUE

A. C. Ramamurthy, Ford Motor Company
B. N. Lucas, Nano Instruments, Inc.
W. C. Oliver, Nano Instruments, Inc.

ABSTRACT

Thermoplastic olefins (TPO's) are now the materials of choice for .automotive fascia applications. An area of concern to the automobile manufacturer is the vulnerability of painted facias to "tribological forces" encountered in normal service life. Several scenarios involving mild physical contacts encountered in parking lots and with other stationary objects may bring about removal of paint and in some cases subsurface delamination within the TPO substrate.

TPO's for automotive fascia applications are typically processed by injection molding techniques. Molded TPO fascias are usually sprayed with an adhesion promoter (2-4 microns) followed by automotive flexible topcoats (60 to 80 microns). This step is followed by a baking process involving exposure of facias to 250^0 F temperature for twenty five minutes. The resulting structure, morphology, as related to OEM paint finishes, play a significant role in governing service life . It is our belief that micro mechanics of the surface and near surface regions play a dominant role in determining service life performance. The measurement of these near surface mechanical properties is therefore of vital importance.

One of the most advanced techniques available for spatially resolved mechanical properties' measurements is depth-sensing indentation. It is now possible to measure many mechanical properties, both static and dynamic, using depth-sensing indentation techniques. Recently, using an extension of a frequency specific indentation technique from Pethica and Oliver[1] and Loubet et al[2], it was shown that it is possible to measure the linear viscoelastic properties of a solid polymer[3]. It was found that both the contact stiffness (S) and the contact damping coefficient (C) vary linearly with the indentation depth. The slopes of these straight lines were respectively found to be proportional to the storage modulus, G', and the loss modulus, G" through the geometry of the contact.

Figure (1) is a plot of the elastic storage modulus versus the depth of indentation obtained from the indentation of the TPO surface. The result represents the average of ten indentations performed in a 200 μm x 50 μm array. The results suggest the presence of a very stiff or high modulus layer in the near surface region. The measured modulus is then observed to decrease rapidly as a function of depth before starting to rise toward a bulk value at deeper depths.

The second set of experiments were performed on a TPO cross-section. These results (figure not shown here), support the finding of the indentations from the surface in that a very compliant layer exists in the 20 - 100 μm depth range. The modulus is then again seen to rise toward a constant value through the bulk before showing mirror symmetry across the centerline toward the other surface. The nearly constant value of the modulus obtained through the bulk of the cross-section is known to be in good agreement with values obtained from bulk testing techniques.

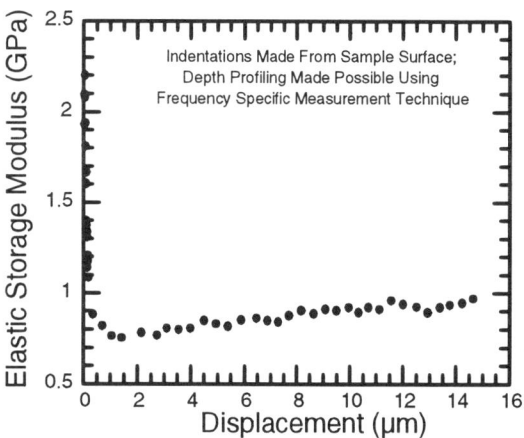

Figure1. A plot of the elastic storage modulus at 45 Hz versus the depth of the indentation into the TPO surface.

REFERENCES

1. J. B. Pethica and W. C. Oliver in *Thin Films: Stresses and Mechanical Properties* , eds. J. C. Bravman, et al. (Mater. Res. Soc. Symp. Proc. **130**, 1989) 13.

2. J.-L. Loubet, M. Bauer, A. Tonck, S. Bec., and B. Gauthier-Manuel in *Mechanical Properties and Deformation Behavior of Materials Having Ultra-Fine Microstructure* , eds. M. Nastasi, D. M. Parkin, and H. Gleiter (Kluwer Academic Publishers 1993) 429.

3. J.-L. Loubet, B. N. Lucas, and W. C. Oliver in *NIST Special Publication 896, Conference Proceedings: International Workshop on Instrumented Indentation.* , eds. D. T. Smith (NIST 1995) 31.

SUBSURFACE CONTACT STRESS AS AFFECTED BY ASPERITY ROUGHNESS LAY ORIENTATION AND FRICTION

SI C. LEE

The Ohio State University, 206 West 18th Avenue, Columbus, Ohio 43210-1107

ABSTRACT

Surface topography has a significant effect on the tribological performances of the contacting systems. The key parameters responsible for initiating surface pitting and wear are the distributions of the asperity contact pressure and the associated near surface stress field. A small change in the distribution of the heights, widths, and curvatures of the asperity peaks may have a noticeable effect on the micro-contact behaviour of the rough surfaces. In this paper, a series of numerical simulations were performed in order to study the effects of surface topography on the subsurface contact stress. The nominal contact geometry for the simulations was that of a ball-on-flat contact. The rough surfaces were generated numerically to the pre-set specifications of the statistical roughness parameters as required in the simulations. These surface topographies were then superimposed on top of the nominal contact geometries. Five different surface lay orientations were examined; the aspect ratios corresponded to $\gamma = 1/6$ (a near transverse surface), $\gamma = 1/3$, $\gamma = 1$ (an isotropic surface), $\gamma = 3$, and $\gamma = 6$ (a near longitudinal surface). Three different values of the coefficient of friction were used in the simulations for a given load: $\mu = 0.0$, 0.1, and 0.25.

Histograms for the Mises stress distribution and the principal tensile stress distribution were generated for each combination set of the simulation variables g and m. Each histogram represents the averaged results of five different surfaces which have similar statistical roughness parameters. The square simulation domain, having the length of 2.4 times the hertzian width (a), was divided into 151 x 151 grids for the numerical calculations. The stresses were evaluated for 31 different layers of depth into the solid.

The highest values of the Mises stress occurred approximately at the depth of 0.01a below the surface for the simulation range investigated. The stress histogram data also showed that the isotropic surface ($\gamma = 1$) had the least favourable stress distribution, having the highest stress values. The stress distribution for the $\gamma = 3$ surface was significantly lower than that of the isotropic case. However, when the $\gamma=3$ results were compared with those of the $\gamma = 6$ surface, the stresses were only slightly higher. This suggests that the $\gamma = 6$ surface would have a stress distribution very similar to the pure longitudinal surface. Likewise, comparisons of the transverse surfaces resulted in a similar conclusion where $\gamma = 1/6$ would behave like a pure transverse surface. Finally, of the five surface topography types examined, the longitudinal surface gave the most favourable stress distribution.

MICRO-SLIP OF ENGINEERING SURFACES

S ANDERSSON
Machine Elements, Department of Machine Design,
Royal Institute of Technology, KTH, S-100 44 Stockholm, Sweden

ABSTRACT

Micro-slip is, according to the ASM Handbook, Volume 18, defined as "small relative tangential displacement in a contacting area at an interface, when the remainder of the interface in the contacting area is not relatively displaced tangentially". The phenomenon is important in many applications, for example; in press fit joints, road and wheel contacts, and railway track and wheel contacts. Micro-slip is also important in other rolling and sliding contacts such as in rolling bearings and gears.

Surface topography and elastic deformation of the interacting surfaces play an important role in the process. Friction variation and plastic deformation have an influence as well. Micro-slip can therefore be observed in engineering contacts to different degree depending on the boundary properties, form, waviness and roughness of the surfaces.

The micro-slip effect has mainly been studied for pure sliding, and sliding and rolling nonconformal contact surfaces. Figure 1 shows the results of numerical simulations for an elastic sphere, with different surface roughness, sliding against an elastic flat surface (1). The presented relationships between friction force and tangential displacement for smooth and rough surfaces are typical for a nonconformal contact during initial motion.

Tangential loading of conformal engineering contact surfaces also produces an apparent elastic and plastic displacement, similar to the curve shown in Figure 1. The "plastic" part contains mainly local slip in the contact. Experimental studies have been made with annular specimen. For engineering steel surfaces the displacements for such contacts can amount to 5 - 20 µm before complete sliding occurs in the contact. This displacement is mainly due to waviness and roughness of the surfaces.

The creep and micro-displacement in rolling and sliding contacts has been extensively studied due to its importance in railway technology (2). Most research has treated the friction and creep of such contacts. The form of the contact surfaces has a strong influence on the slip distribution in the contact. The local wear of a surface is often assumed to be directly proportional to the local slip and thus also strongly dependent on the slip distribution in the contact. The wear will change the form and topography of the contact surfaces and also the slip and pressure distribution in the contact. For example this occurs in a boundary lubricated cam and follower contact in a combustion engine, or in a spherical roller thrust bearing contact. The interrelation between the wear distribution, motion of the interacting bodies, and the contact pressure distribution, has a crucial significance on the performance and life of such contacts (3). Similar effects of micro-displacements in contacts can be seen in gears and other high performance rolling and sliding contacts.

The friction properties during initial motion and oscillating motion of highly elastic contacts, for example, the contact between seal and cylinder in a hydraulic or pneumatic cylinder, is strongly dependent on the micro-displacement in the contact. It is found that; with the aid of micro-slip relations, good frictional models can be formulated and used successfully in computer control of such machine elements.

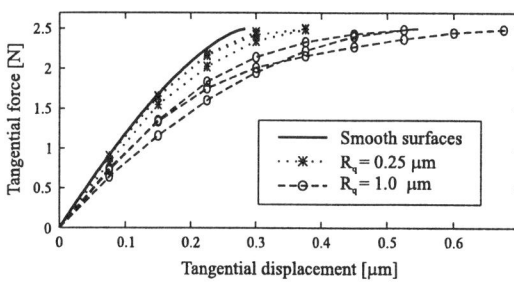

Fig.1: Micro-slip of an elastic sphere (R=5mm) on an elastic flat surface.

REFERENCES

(1) S. Björklund, S. Andersson, Wear 179 (1994) 117-122.
(2) J.J. Kalker, Kluwer Academic Publishers, The Netherlands.1990.
(3) A.B-J. Hugnell, S. Björklund, S. Andersson, Wear 199 (1996) 202-210.

NOVEL HIGH WEAR RESISTANT DLC COATINGS DEPOSITED BY MAGNETRON SPUTTERING OF CARBON TARGETS.

A.H.S. JONES, D. CAMINO, D.G. TEER, J. JIANG*.

Teer Coatings Ltd., 290 Hartlebury Trading Estate, Hartlebury, Worcestershire, DY10 4JB. UK.
*Dept. of Aeronautical and Mechanical Engineering, University of Salford, Salford, M6 4WT. UK.

ABSTRACT.

Diamond-like carbon (DLC) is a hard, low friction coating of great interest in terms of its tribological properties. Originally produced from the break down of hydrocarbon gas using r.f. power (1), the films were highly stressed. The inclusion of a small amount of metal by simultaneously sputtering from metal targets was found to reduce the internal stress (2). Following this work, amorphous hydrogenated DLC films were produced at Teer Coatings using a closed field unbalanced magnetron system (3). These films provided low friction, low wear rates and the ability to protect the counterpart making them suitable for a variety of wear resistant applications such as cutting tools and automobile parts.

The performance and suitability of DLC coatings in automobile applications has been studied. Conventional hydrogenated DLCs gave good results but attempts made to improve the tribological properties by optimising deposition parameters produced only small improvements.

Alternative deposition methods were investigated and a new coating was developed where graphite targets were sputtered in a pure argon atmosphere, again utilising the closed field unbalanced magnetron system. This versatile process allowed the variation of deposition parameters to produce a series of hydrogen free pure carbon films ranging from transparent, electrically insulating films to opaque, conductive films.

The dark conductive films were of most tribological interest producing very low friction and wear rates. This paper describes a series of tribological tests comparing the conventional hydrogenated DLC with the new sputtered carbon films. Reciprocating and uni-directional pin on disc friction and wear tests have shown that the new pure carbon films have a lower coefficient of friction and a specific wear rate more than one order of magnitude better than the conventional hydrogenated DLC when tested against an unlubricated 5mm diameter WC-6%Co ball. Figure 1 shows the graph of friction versus the number of cycles for both types of film in a reciprocating wear test at 100N load, the conventional DLC fails at 6500 cycles but the new DLC survives with a wear rate of ~2.4×10^{-17} $m^3(Nm)^{-1}$. The effect of load on wear rate has been studied and also the film's performance under lubrication.

Taper sections of the film were produced using a ball crater, and applying these to wear tracks allowed wear analysis and interface examination.

Attempts have been made to study the structure the sputtered carbon film using Raman spectroscopy, glancing angle XRD and EELS, and some of the early results are reported briefly.

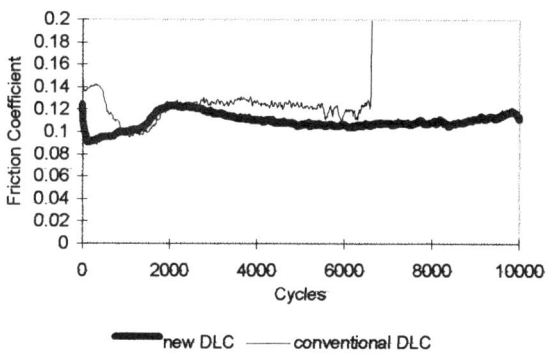

Figure 1. Graph showing friction vs. number of cycles for conventional hydrogenated DLC and new pure carbon DLC.

REFERENCES.

1. Holland, S. M. Ojiha, *Thin Solid Films*, 58 (1979)107.
2. H. Dimigen and H. Hubsch, *Philips Tech. Rev., 41 (1983-4) 186.*
3. D.P. Monaghan, D.G. Teer. P.A. Logan, I. Efeoglu, R.D. Arnell, *Surf. Coat. Tech., 60 (1993) 525-530.*

Surface Engineering of Ti Alloys for Low Wear Load Bearing Surfaces

K. MAO, A. BLOYCE and H. DONG
The Wolfson Institute for Surface Engineering, School of Metallurgy and Materials
The University of Birmingham, Birmingham B15 2TT, UK

ABSTRACT

The wider use of titanium and its alloys in engineering components is frequently restricted by its poor tribological properties. A titanium surface rubbing against steel even when well lubricated, galls and wears rapidly under quite low loads. Conventional surface treatments have had limited success in reducing sliding wear, and the deep case-hardening processes used on steel surfaces are not possible for titanium. Surface engineering of titanium alloys, i.e. duplex surface treatments, have therefore been developed to overcome this problem (1).

These surface engineering techniques are still in undergoing development, and in order to accelerate this process, parallel research is being conducted to develop a contact mechanics model for simulation of the service behaviour of surface treated machine elements, such as gears and cams. Conventional Hertz theory is restricted to frictionless smooth surfaces and perfectly elastic solids. Significant progress in the field of non-layer surface contact over the last few decades has been associated largely with the removal of these restrictions. A proper treatment of friction at the interface of bodies in contact has enabled the elastic theory to be extended to both slipping and rolling contact in a realistic way. However, in the case of layered surface contact, the understanding of their contact mechanics is still in an early stage. The modelling of layered surface contact is of great analytical and practical importance in many engineering applications where the load bearing capacity, friction and wear characteristics can be significantly improved by depositing thin layers of hard wear resistant materials. The present model (Fig.1) is based on modern theories of multi-layered surface contact, taking account of multi-layered structures, real surface roughness and friction effects (2).

Amsler wear tests have been carried out where a hardened steel disc is rotated against a titanium alloy counterface with and without Titanium Oxygen treatments (3) respectively under rolling/sliding contact to simulate the titanium gear performance. Table 1 shows the steel against the untreated titanium alloy Amsler wear test results and it can be seen from it that the untreated titanium alloy weight loss is high, as well as titanium transferred to its counterface steel surface (with negative weight loss). However, the Titanium Oxygen treated titanium alloy Amsler wear lightly as well as its counterface steel surface (Table.2). Furthermore, some titanium gear tests have been carried out and safe running results have been achieved. Good agreements between theoretical predictions and experimental results have been achieved. Further high performance can be achieved by combining an oxygen diffusion treatment with a DLC coating. For the same load and geometry, the titanium alloy combining an oxygen diffusion treatment with a DLC coating shows much higher load carrying capacity comparing to the TO treated titanium alloy.

Table.1 Amsler Test 1

Specimen	weight loss (grams)	cycles
Ti6Al4V	1.0329	3.6×10^4
Steel	-0.0011	3.6×10^4

Table.2 Amsler Test 2

Specimen	weight loss (grams)	cycles
TO treated Ti6Al4V	0.0027	4.3×10^4
Steel	0.0002	4.3×10^4

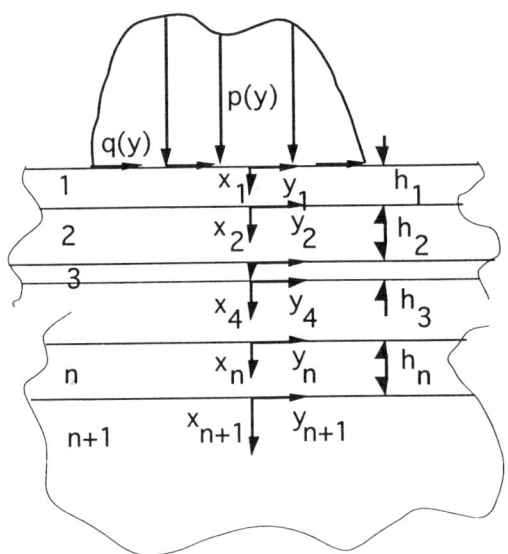

Fig.1: Multi-layer surface contact model

REFERENCES
(1) A. Bloyce, P. H. Morton and T. Bell, Vol. 5, ASM Handbook, 1994, 835-851pp
(2) K. Mao, T. Bell and Y. Sun, Journal of Tribology, ASME, 119 (2) 1997, In Press.
(3) H. Dong, A. Bloyce and T. Bell, 10th Congress of the IFHT, Brighton, 1996.

MODIFICATIONS OF α-MoO₃ THIN FILMS TOWARDS CONDUCTING, TEMPERATURE RESISTANT AND SOLID LUBRICANT COATINGS.

H. KLEIN, Y. MATHEY, D. PAILHAREY
GPEC UMR 6631 CNRS, Université de la Méditerranée 13288 Marseille CEDEX 9 FRANCE
F. TORREGROSSA
IBS, 101 Z.A. Les Pradeaux 13850 Gréasque FRANCE

Considering on the one hand the growing demand for solid lubricant materials, and on the other hand the limitations of classic solid lubricants such as MoS₂ for high temperature use (>450°C), we have undertaken the preparation under thin film form of a material combining high temperature resistance, solid lubricant properties and electrical conductivity.

Due to their bidimensional structure which induces anisotropy in all their properties, lamellar materials are of great interest for our goal. Moreover their properties can be modified and even tuned under certain conditions by intercalation of chemical species in the Van der Waals gap.

For these reasons, we choose to start from a lamellar oxide, α-MoO₃, which is thermally stable up to 850°C and appears to be an interesting host for our project, although it is insulating and has poor lubricant properties.

We have realized α-MoO₃ synthesis under thin film form by DC magnetron sputtering, and performed frictional study of this material on the nanoscopic scale using an Atomic Force Microscope (AFM) in order to compare such data to those obtained on the macroscopic scale. We have modified these films by post-treatments such as ion implantation, or by co-sputtering of other metals such as Re in order to achieve electrical conductivity.

AFM is known as a convenient tool for imaging surfaces up to atomic resolution. Thanks to its conception AFM is also an excellent tool for probing fundamental forces, Van der Waals and electrostatic forces for instance, and more particularly friction forces. Recently AFM has brought a new understanding in fundamental mechanisms of friction and adhesion on the nanometer scale [1]. But it could be also powerful for the comprehension of friction phenomena (wear and degradation) on the nanometer scale [2], which are of great interest for industry.

By scanning the AFM probe (a sharp tip mounted on a flexible cantilever) over a surface, the interactions between the tip and the surface induce cantilever deflections followed by the deviation of a laser beam. These deflections contain both topographic and frictional informations about the surface. According to several models (Amontons' law for instance) the friction force F_f is dependent on the normal load F_n. In addition, on the nanometer scale the friction force intensity depends also on further parameters such as surface pollution which modifies the characteristics of the tip/surface, surface roughness which causes normal load variations or tilt of the probed surface out of the AFM scan plane and leads to variations of the signal level recorded on the detector.

In this study we present friction measurements on α-MoO₃ (0k0) or (0k0)+(00l) oriented films compared to those obtained on a sample without specific orientation and on a single crystal cleaved along its highest density plane (0k0 plane).

For our tests, samples are investigated with AFM in air. All the experiments presented are performed with the same tip (a "V shape" silicon nitride, 193 μm length, normal spring 0.03N.m⁻¹) on a 1μm² area at 1μm.s⁻¹ scan speed. During the tests, we make F_n variable, firstly increasing to a maximum value, and then decreasing to the initial value in order to put in evidence reversible and irreversible phenomena due to friction (wear for instance). For all the samples, during the load/unload cycle of F_n, the relationship between F_f and F_n is linear according to Amontons' law. Furthermore, there is no wear on our samples during measurements.

We have also performed such measurements on MoS₂ and NbSe₂ single crystals [3]. On these materials it appears surface deformation (MoS₂) or local wear (NbSe₂) under the tip stress. In these cases there is no linear relation F_f vs F_n, and we cannot extract nanoscopic μ values.

From these experiments it appears that when surfaces are unaltered during measurements, friction behaviour on the nanometer scale can be compared to classical behaviour on the macroscopic scale.

Our results indicate that it is possible to distinguish frictional properties of material such as α-MoO₃ under different orientations, and to investigate local variations of the friction coefficient by scanning particular crystals of a polyoriented film. At the present step of this research, ion implantation and Re to Mo substitution appear to improve the conducting properties of our films but no significant modifications of their frictional behaviour have been obtained yet. We are actually correlating these nanoscopic results to macroscopic ones obtained with an alumina ball sliding against the same samples.

REFERENCES
[1] J. Krim, *Scientific American*, 10, 48, 1995.
[2] B. Bhushan, J.N. Israelachvili, U. Landman, *Nature*, 374(13), 607, 1995.
[3] H. Klein, Y. Mathey; D. Pailharey, *Surf. Sci*, accepted for publication, april 97.

Submitted to *Tribology Letters*

BOUNDARY ELEMENT MODELLING OF COATED MATERIALS IN BALL/FLAT CONTACT.

R. KOUITAT NJIWA; R. CONSIGLIO; J. von STEBUT
Laboratoire de Science et Génie des Surfaces, URA CNRS 1402 INPL-Ecole des Mines,
Parc de Saurupt, F-54042 Nancy, France

ABSTRACT

Thin coatings are increasingly adopted for optimisation of various surface and bulk material properties. A low cycle friction fatigue laboratory test (under overall elastic contact pressure) is used for assessment of the coating's intrinsic mechanical strength as well as the rheological properties of the coating/substrate composite (1, 2). A judicious selection of the experimental parameters (radius of the indenter, total load) such that the most constrained points are known in advance requires modelling.

Consider homogeneous materials, satisfying the half space approximation (HSA). Analytical solutions of the type given by Hamilton (3) are used. In the case of composites, such useful expressions do not exist. Therefore numerical methods for solving engineering problems are required. We use the boundary element method adequate for the solution of overall elastic contact problems valid for many tribo-systems.

First, we study the *static elastic indentation of a homogeneous body by a ball*. The effect of loading is described by a hertzian type pressure field. The evolution of the calculated von Mises stress (σ_{VM}) along the symmetry axis is compared to the analytical one. We introduce the notion of relative sizes which correspond to the ratio of true sizes (specimen width, L and and thickness, t) over the Hertz contact radius, a_H. Then the HSA is valid if the relative sizes L/a_H and t/a_H of the indented medium are greater than roughly 60. Below this value, the maximum of σ_{VM} is higher than analytical prediction based on the HSA.

Next, take a *coating/substrate composite*. We show that, the evolution of σ_{VM} depends on the ratio of coating and the substrate Young's moduli, E_c and E_s. The relative coating thickness t/a_H becomes the key parameter. For certain experimental conditions there are two maxima of σ_{VM}. If $E_c/E_s=0.41$ this holds for $t/a_H \sim 0.5$.

For $E_c > E_s$, σ_{VM} presents a maximum in the coating and another one at the interface (Fig. 1). When friction is included in the analysis, the state of stress is mainly affected in the coating leading to higher values of σ_{VM}. Moreover its maximum approaches the surface. When the friction coefficient is high enough, the maximum is at the loaded surface.

Consider that, under load, the value of the overall mean pressure, P_m, and the yield stress σ_{eq} of the material under test are available from experiment. Then $P_m = C\sigma_{eq}$. Suppose $E_c < E_s$, the factor $C = C_0$ for the substrate, and $C<C_0$. For the same experimental conditions (load and indenter radius) our numerical analysis shows that, the maximum value of σ_{VM} is localiseded in the coating. When $C>C_0$, this maximum is localiseded in the substrate.

This work proves the reliability of the boundary element modelling of coated materials. We show that the state of stress in the composite under spherical indentation depends on Young's moduli of the coating and the substrate. This state is also strongly influenced by the relative dimensions of the sample (in particular t/a_H), which are closely related to the applied load and the indenter radius.

The concept of a thin coating depends on contact conditions. Once they are specified, a thin coating can be conveniently defined as having a relative thickness such that the numerical and the analytical HSA solutions deviate by more than a given fraction, itself depending on the precision of the desired approximation.

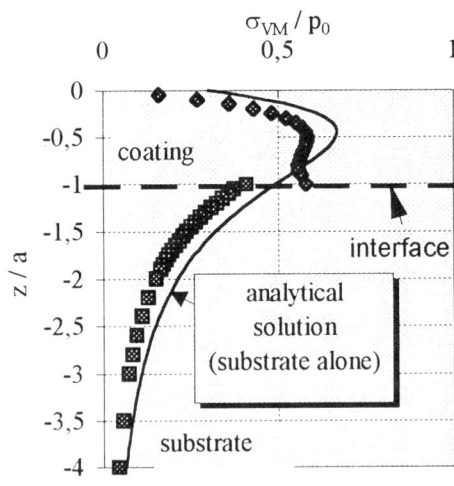

Figure 1: Evolution of σ_{VM} along the axis of symmetry of the sample. $t/a_H = 1$; $E_c/E_s = 2.44$.

REFERENCES

(1) A. Darbeida, J. von Stebut, M. Assoul, J. Mignot, J. Mech. Tools Manufact. 35, (1995), 177

(2) A.Darbeida, J.von Stebut, in «Tribologie et Ingénierie des Surfaces». Ed. J.von Stebut, SIRPE, Paris, 1996, p.79

(3) G.M.Hamilton, Proc. Inst. Mech Eng. 197C, (1983), 53

SUBSTITUTION OF ZINC PHOSPHATED COATING IN COLD BULK FORMING BY DEVELOPMENT OF APPROPRIATE TOOL COATINGS

D. SCHMOECKEL and M. RUPP
Institute for Production Technology and Forming Machines, Technical University Darmstadt,
Petersenstr. 30, 64287 Darmstadt, Germany

ABSTRACT

In the area of cold bulk forming, the extremely high tribological loads involved require the use of lubricant carrier coatings (zinc phosphat) such as reactive soaps or molybdenum disulphide. In certain applications, this lubricating system is supplemented by cold pressing oils. Application und removal are performed by chemical means in various baths and associated with substantial environmental, cost and personel burdens.

In the scope of a research project, the possibility of substituting the currently used carrier coatings and solid lubricants by means of appropriate hard coatings on tools was studied.

$3\times10^{17}C^+/cm^2$ $3\times10^{17}N^+/cm^2$

Fig. 1: TiN-Coating modified by Ion Implantation (1)

For basic studies of hard material coating system an analogy tests based on free cupping were developed for wear testing in cold bulk forming. With the help of wear studies, CVD on Ti basis (TiC, TiN, Ti(C,N)) as well as PVD with TiN and CrN were analysed. The Ti-based coating seemed to offer adequate protection against adhesion. It has also shown that the CVD layers demonstrate the best results with regard to chipping and wear resistent. In addition, hard material coatings with fringe layers modified by ion implantation were studied for the first time. (Fig. 1) The tool life could be increased by ion implantation.

Based on the basic studies, the coatings were applied under production conditions. The product studied was a commercial M 16 bolt with different lengths. In the conventionell manufacturing process, hot-rolled and phosphated wire is used.

On these studies, no lubricant was used on the wire material. Bright steel was processed, with a minimum of oil. Such minimum of oil application, allows tool to performe to life testing, using various coatings. For these tests, CVD TiC/TiN, CVD TiC/Ti(C,N), PVD TiN and PVD CrN coatings were used and studied under otherwise identical production conditions.

After short production use, the two PVD coating variants showed chipping in the reducing edge zone, resulting in local failure of the coating and thus of the tools. As a consequence of this there was an increase of roughness in a small area of the bolt circumference. In contrast with this, the CVD coating variants showed uniform wear of the coating along the entire circumference. After coating was abraded, the roughness values increased continuously over the entire circumference of the shank.

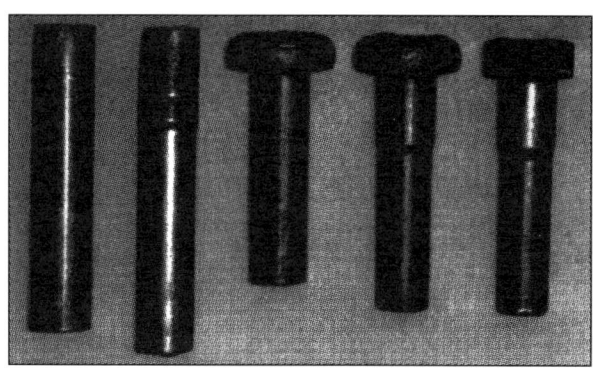

**Fig.2: Stage Plan of a Hexagonal Bolt M16
(1.Stage: Reducing, 2. Stage: Upsetting, 3. Stage: Reducing, 4 Stage: Burring)**

When comparing the coating variants with respect to their life behaviour, the CVD coatings frequently showed longer life than the PVD coatings. In the following diagram (Fig. 3) different coating variants are compared with regard to the number of parts produced. Using minimum lubrication, CVD coatings permitted 300,000 to 400,000 parts (or 25 km wire, 36 t of wire) to be produced in the first stage. After that, the tools had to be replaced. Assuming optimum conditions for minimum lubrication this will allow production to be performed at slightly reduced costs already.

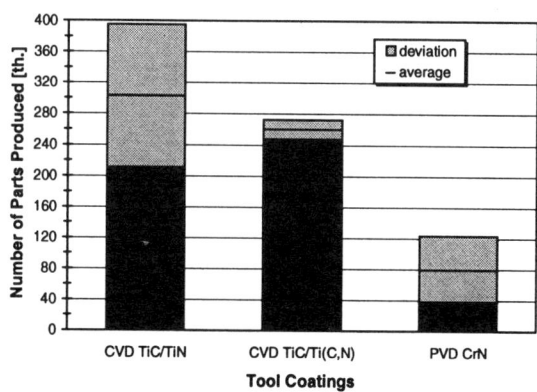

Fig. 3: Life Behaviour of the Various Coating Variants with Minimum Lubrication

REFERENCES

(1) Seidel, F.: Final Report, Project BMBF 13N6214, Stiftung Institut für Werkstofftechnik (IWT)

THE SCRATCH TEST: A SENSITIVITY AND REPRODUCIBILITY STUDY

J. MENEVE and K. VERCAMMEN
Materials Technology Centre, Vlaamse Instelling voor Technologisch Onderzoek, Boeretang 200, 2400 Mol, Belgium
P. ANDERSSON and S. VARJUS
VTT Manufacturing Technology, P.O. Box 1702, FIN-02044 VTT, Finland
D. CAMINO and D.G. TEER
Teer Coatings Ltd., 290 Hartlebury Trading Estate, Worcs. DY10 4JB, United Kingdom
N.M. JENNETT, J.P. BANKS and S.R.J. SAUNDERS
Centre for Materials Measurement and Technology, National Physical Laboratory, Teddington, TW11 0LW, UK
J. von STEBUT
Laboratoire de Science et de Génie des Surfaces, Ecole des Mines, Parc de Saurupt, 54042 Nancy Cedex, France

ABSTRACT

The scratch test is extensively used, in industry and research laboratories, to assess the adhesion strength of coated surfaces. In the scratch test, a diamond Rockwell C stylus is drawn over the sample surface under an increasing normal load. Coating adhesion is assessed by the determination of a critical normal load at which the coating is detached from the substrate. In a wider scope, the scratch test can be, and should be, considered as a repeatable method for determining the mechanical integrity of surfaces, coated surfaces as well as any other bulk or composite surfaces.

As with other engineering test methods, however, scratch test results are influenced by test specific factors such as the scratching velocity and indenter geometry, which may lead to large scatter in scratch test results even when the tests are performed on a single homogeneous test sample. For example, a round robin experiment on scratch testing conducted in 1988 showed that the reproducibility of scratch test results for different stylus-instrument combinations was poor: variations in the critical load values up to 70% were observed (1). Whereas the influence of normal loading rate, specimen displacement rate and indenter geometry is well documented in literature, the possible effects of varying friction between the scratch stylus and the specimen are to date not well understood (2).

In the present work, three sets of geometrically and crystallographically characterised scratch styli were produced out of two different natural Ia type diamonds as well as a synthetic Ib type. Reference TiN coated AISI M2 steel specimens were used, with controlled interfacial properties using in situ Auger electron spectroscopy (AES) during physical vapour deposition (PVD) of the coatings. The aim of the study was to determine the possible influence of different diamond impurity levels on the frictional behaviour of the stylus. In addition, the effects of the diamond crystal orientation with respect to the scratching direction were assessed, as well as the influence of transfer layers originating from preceding scratches. It is well known that ambient conditions may strongly influence friction and wear during tribological testing, therefore the sensitivity of the scratch test to varying relative humidities and temperatures was studied, and optimised cleaning procedures for the scratch styli and the specimen were established. To determine the influence of inherent instrument properties such as the instrument frame compliance, a reproducibility experiment including three different instrument types was organised. During this experiment, each of the test specific parameters were controlled and kept constant, after traceable calibration of each instrument involved.

One of the main conclusions of this study is that the possible influence of diamond stylus material impurities or crystallographic orientation on the friction between that stylus and the specimen surface, and hence the critical load values generated by the scratch test, is completely obliterated by uncertainties in the stylus tip shape. For commercial indenters deviations up to 100 µm from the specified 200 µm tip radius were found, as well as other surface defects such as cavities and cracks, which in turn resulted in critical load uncertainties in the order of 50%. Clearly there is a strong case for the availability of a certified reference material (CRM) for this well recognised test method.

The results of this work will be used to produce an improved European Standard on the scratch test, based on earlier work within the European Standards Committee CEN TC 184/WG 5 'Test Methods for Ceramic Coatings' (3).

REFERENCES

(1) A.J. Perry, J. Valli, P.A. Steinmann, Surface and Coatings Technology, 36 (1988) 559.
(2) P.A. Steinmann, Y. Tardy, H.E. Hintermann, Thin Solid Films, 154 (1987) 333.
(3) ENV 1071-3:1994 E: Determination of Adhesion by the Scratch Test, CEN Central Secretariat, Stassartstraat 36, 1000 Brussels, Belgium.

This work is supported by the European Commission under Standards, Measurements and Testing contract MAT1-CT94/0045.

A COMPARISON OF WEAR AND CORROSION-WEAR OF PVD TIN COATINGS UNDER UNI- AND BI-DIRECTIONAL SLIDING

P. Q. WU, D. DREES, L. STALS*, J. P. CELIS
Department of Metallurgy and Materials Engineering
Katholieke Universiteit Leuven, de Croylaan 2, B-3001 Leuven, Belgium
* IMO, Limburgs Universitair Centrum, B-3590 Diepenbeek, Belgium

ABSTRACT

Unidirectional and bidirectional tests are two of the most popular laboratory methods for evaluating the friction and wear behaviour of materials. The choice of a test method depends on the type of information searched for and the field application expected. It was reported that wear debris have great influence on the friction and wear behaviour of contacting materials, and that the influence varies with different test methods (1-3). The objective of this article is to compare friction and wear behaviour of PVD TiN coatings (4 μm on ASP23) sliding vs corundum in three different aqueous environments (50% RH air, demineralised water, 0.02M H_3PO_4) under unidirectional and bidirectional test conditions.

Unidirectional tests were done in two different pin-on-disk geometries (unidir.I and unidir.II) with the same test parameters (5N, 0.2m/s). Bidirectional test parameters were chosen to enforce gross-slip (5N, 10Hz, 100μm). These test set-ups are schematically shown in Fig. 1.

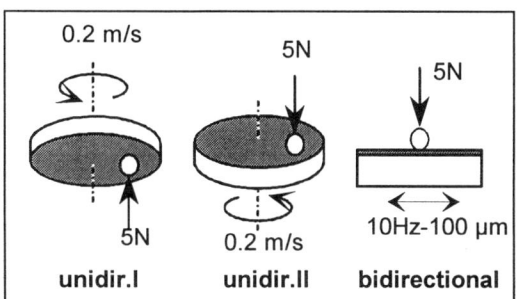

Fig. 1 : Schematical test set-ups

Results show that the coefficient of friction is lower in bidirectional sliding (Fig. 2) but wear of TiN is more aggressive (Fig. 3) in air as well as in water. The friction evolution is the same for unidirectional I and II set-ups when tested in air. However, when tested in water and phosporic acid solution, the evolution demonstrates a significant difference between the two unidirectional set-ups. Different wear mechanisms are operative under both test conditions. In bidirectional tests, a tribo-oxidation mechanism is identified in the air and water, and a lubricating triboactivated phosphation mechanism dominates in H_3PO_4. The wear mechanism in unidirectional tests in water and air is mild oxidational wear (4). In unidirectional II and using H_3PO_4, the wear debris can have a lubricating action.

From these results, H_3PO_4 appears as a candidate for oil-free lubrication in bidirectional and some unidirectional wear contacts.

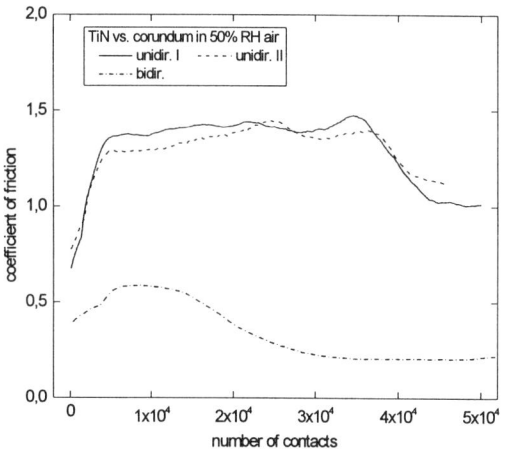

Fig. 2 : Friction evolution of TiN vs corundum

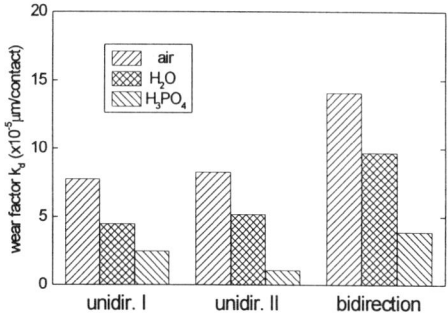

Fig. 3 : Wear factor expressed per contact

REFERENCES

(1) H. Mohrbacher, B. Blanpain, and J. P. Celis, ASTM-STP 1278, (1996), 76-93.
(2) I. L. Singer, S. Fayeulle and P. D. Ehni, Wear, 149(1991), 375-394.
(3) G. M. zu Kocher, T. Gross and E. Santner, Wear 179 (1994), 5-10.
(4) D. Drees, Gecombineerd slijtage-corrosie onderzoek van dunne harde deklagen, Ph.D. thesis (in Dutch), K.U.Leuven, May 1997.

Submitted to *Surface and Coatings Technology*

EFFECT OF THERMOCHEMICAL TREATMENTS ON THE SLIDING WEAR MECHANISMS OF STEELS UNDER BOUNDARY LUBRICATION

JISHENG E* and D.T. GAWNE**
Department of Materials Engineering, Brunel University, Uxbridge, Middlesex, UB8 3PH, UK

ABSTRACT

The effect of thermochemical treatments on the material removal mechanisms and wear behaviour of steels under boundary lubricated conditions has been studied. A controlled procedure using mild abrasives under a light load was adopted in order to avoid long-term effects from severe wear in the running-in stages of pin-on-disc testing. The results indicate that the friction and wear varies with the different thermochemical treatments. The wear increases with an increase in the hardness of steels.

Electron microscopy of the sliding surfaces showed that wear of the untreated steel took place by abrasive and adhesive mechanisms, Figures 1a and 1b. Heat treatment by austenitizing, quenching and tempering (through-hardening) reduced the wear rate and carburizing produced a further reduction in wear rate. The principal wear mechanisms in the both through-hardened and carburized steels became abrasive and delamination wear with adhesive wear becoming unimportant, Figures 1c-1f. Electron microscopy on cross-sections through the sliding surfaces of the carburized and through-hardened steels showed extensive plasticity, cracks, and delaminated wear phenomena. The elimination of adhesive wear as a major wear mechanism is attributed to the influence of hardness on junction growth and the emergence of delamination wear to the effect of nano-crystalline carbides on fatigue life[1].

Plasma nitriding resulted in a further reduction in wear rate and the effective elimination of delamination wear, which was attributed to the very fine and coherent precipitates existing both in the grain boundaries and within the grains[2]. Electron microscopy on the worn surfaces indicated that the principal material removal mechanisms were micro-pitting and abrasive wear, Figures 1g-1h.

Microstructural examination of the worn surfaces of the plasma-nitrided steel revealed the presence of white layer regions resulting from plastic deformation[3]. They play two roles. On the one hand, the formation of the white layer regions on the sliding surfaces increases their hardness, which acts to reduce the wear rate. On the other hand, metallurgical examination of the worn surfaces showed that micro-pitting occurred at white layer regions to the detriment of the wear performance.

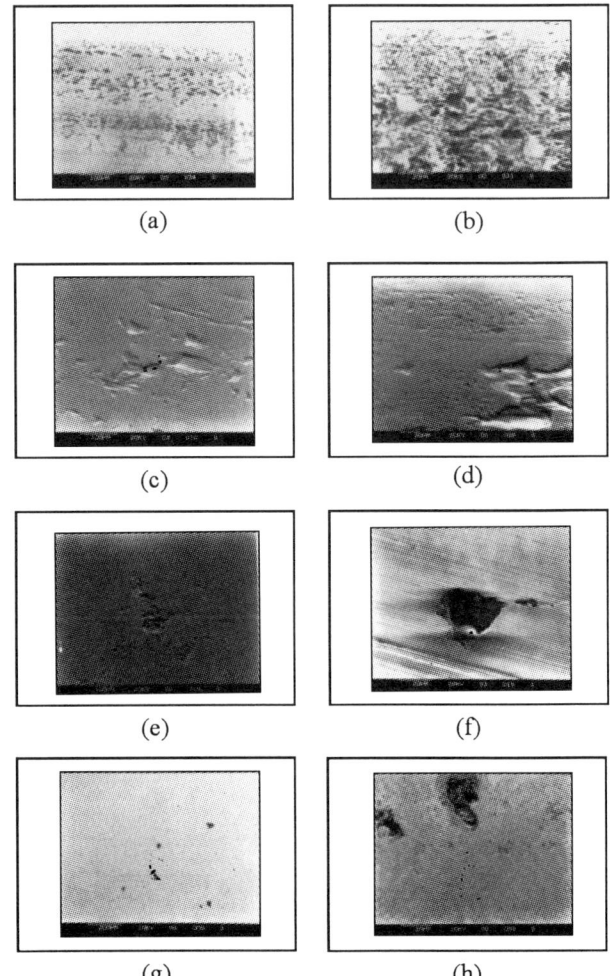

Fig. 1 SEM micrographs of the worn surfaces of (a and b) the untreated, (c and d) the through hardened, (e and f) the carburized and (g and h) the plasma-nitrided steels after (a-f) step-load tests up to the 400 N applied load and (g and h) long term tests at 900 N applied load under boundary lubricated conditions

REFERENCES

(1) Krauss,G., ASM Handbook, Vol. 3, Edt by Henry,S.D et al, ASM International, USA, 1992, 363.
(2) O'Brien,J.M., ibid , 420.
(3) Bulpett,R., Eyre,T.S. and Ralph,B., Wear, **162-164**(1993), 1059.

*Present address: GKN Technology Limited, Birmingham New Road, Wolverhampton, WV4 6BW, UK

**Present address: South Bank University, 103 Borough Road, London, SE1 0AA, UK

ENDURANCE OF IBAD MoS$_2$/TiN COATINGS: CONTRIBUTIONS OF ORIENTATION AND BONDING

L E SEITZMAN and I L SINGER
Naval Research Laboratory, Code 6176, Washington DC, USA.

ABSTRACT

It is the objective of this study to evaluate the role of coating orientation and adhesion on MoS$_2$ endurance. Previous work has suggested that these parameters are critical in controlling MoS$_2$ endurance, (1) (2), but attempts to isolate these factors are usually complicated by changes in other parameters such as density or stoichiometry. Recently, it was shown that the orientation of MoS$_2$ crystallites could be reproducibly controlled without introducing additional changes in coating structure(3)(4) by using ion beam assisted deposition (IBAD). Also, the IBAD process is easily modified to produce changes in interface chemistry that should affect coating adhesion without altering coating microstructure.

IBAD MoS$_2$/TiN coatings were grown on hardened steel substrates at temperatures between 315K and 573K using ion/atom ratios between 0.02 and 0.10. Coating orientation was evaluated using X-ray diffraction. Endurance was measured using a steel ball sliding against the coated steel in dry air at a mean pressure of 1.4 GPa and a sliding speed around 0.6 m/s. Stoichiometry and density were evaluated using Rutherford backscattering. In one experiment, the effect of the sputter cleaning on coating adhesion and life was investigated. One of a pair of substrates was masked during the sputter cleaning that precedes coating deposition, then uncovered so both pairs received the same coating. Additional tests run on this pair of coated steel were Brale indentation, scratch adhesion testing, and Auger spectroscopy.

Coatings had a density about 90% of bulk MoS$_2$ and a S/Mo ratio very close to 2. In general, the C and O content in the film was below the detection limit of 2%. The coatings exhibited a mixture of basal and edge orientation, with edge orientation generally greater as temperature and ion/atom ratio increased. On average, coating endurance increased with increasing deposition temperature. For coatings deposited between 315K and 373K, endurance was reduced by the presence of edge oriented crystallites. In contrast, endurance on coatings grown at 473K and 573K with and without substantial edge orientation almost always exhibited high endurance. No influence of coating thickness on endurance was observed.

Sputter cleaning was found to increase endurance, indentation critical load, and scratch critical load. Auger analysis near the Brale indents indicated two loci of failure: primarily the MoS$_2$-TiN interface and secondarily the TiN-steel interface.

The results suggest that bonding of the coating is the dominant factor that influences MoS$_2$ endurance on steel, while crystallite orientation is much less important. These findings are consistent with recent evidence that IBAD MoS$_2$ wears rapidly in the first 10% of coating life, leaving a thin (<100 nm) remnant of the original coating to accommodate sliding for the majority of the wear test(5). The life of the remnant lubricant film should depend strongly on coating adhesion.

REFERENCES

(1) P.D. Fleischauer, ASLE Transactions, Vol. 27, No. 1, page 82, 1984.
(2) T. Spalvins, J. Materials Engineering and Performance., Vol. 1, No. 3, page 347, 1992.
(3) L.E Seitzman, R.N. Bolster, I.L. Singer, J. C. Wegand, Tribology Transactions, Vol. 38, No. 2, page 445, 1995.
(4) L.E. Seitzman, R.N. Bolster, I.L. Singer, Thin Solid Films, Vol. 260, page 143, 1995.
(5) K.J. Wahl and I.L. Singer, Tribology Letters, Vol. 1, page 59, 1995.

TRIBOLOGICAL PROPERTIES OF SPUTTERED MoSx FILMS IMPROVED BY HIGH ENERGY Ga and C ION IMPLANTATION

N.WATANABE, N.ABE, K.YASUMOTO and M.NISHIMURA
Department of Engineering, Hosei University, 3-7-2 Kajinocho, Koganei, Tokyo 184, Japan

ABSTRACT

It has been reported that tribological properties of MoSx films are improved by inert gas ion implantation 1,2). In this study, we used high energy gallium and carbon ions as ion source, expecting to elongate film wear life.

Substrate was a SUS440C disk, roughness of which was 0.05 μ m. MoSx films to the thickness of 1 to 1.7 μ m were deposited on the substrates by RF magnetron sputtering. Gallium and carbon ions were implanted on the films using high energy, high current tandem accelerator. Mating surface was SUS 440C steel ball with a diameter of 7.938 mm. Friction tests were conducted in air and 10^{-5} Pa at a load of 9.8N and a sliding speed of 0.5 m/s.

Figure 1 compares the friction coefficient and the wear life of the untreated film with those of the films implanted with Ga ions at a dose of 1.0×10^{15}, 1.0×10^{16} and 2.0×10^{17} ions/cm², respectively. Accelerating voltage for the ion implantation was set to maximize the concentration of implanted ions at the interface.

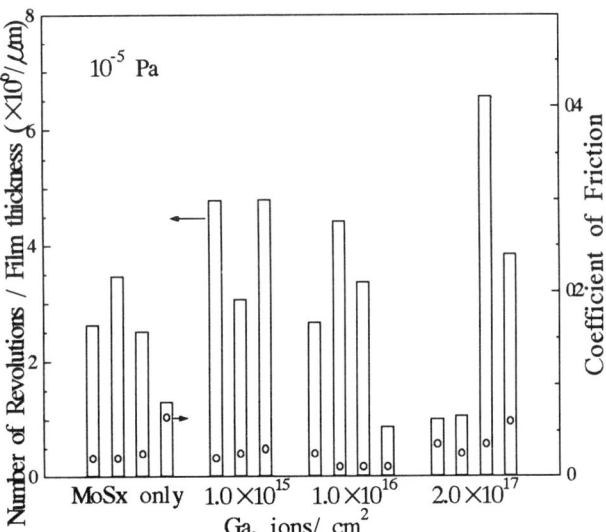

Figure 1. Comparison of test-result(1)

In fig. 1, Ga ion implantation shows little effect in improving film wear life. Undesired variation of the wear life is observed when the dose of ions is 2.0×10^{17} ions/cm², notwithstanding that the wear life is improved in some cases.

From this result, we changed the accelerating voltage so that the maximum concentration be at the center of the film thickness.

We tested carbon ion implantation as well. In this case we selected the accelerating voltage so as to maximize the interface concentration. At the same time we changed the duration of the ion bombardment from previous 5 minutes to 1 minute, considering that this process might be applied to solid lubricated ball bearings.

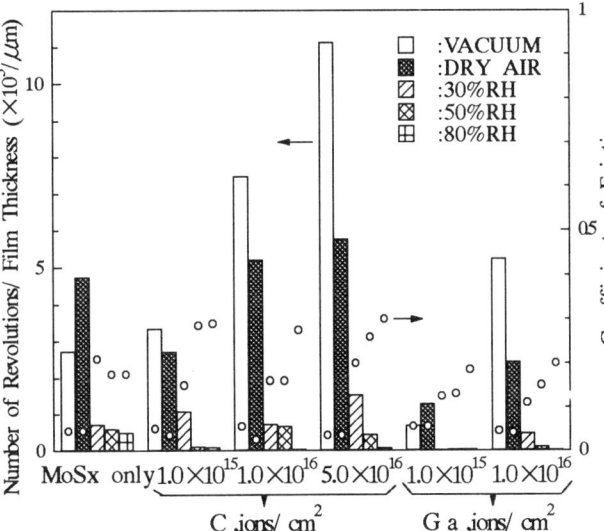

Figure 2. Comparison of test-result(2)

The results given in Fig. 2 indicate the carbon ion implantation is effective in elongating the wear life in dry air and vacuum but not in high humid air. In contrast, gallium ion implantation is again disappointing.

The scale of lifetime in Fig.2 is an order of magnitude shorter than the one in figure 1. This is attributed to the shortened prebombardment. Longer pretreatment causes cleaner surface and more suitable surface roughness.

All things considered, the optimum ion dose appears to be between 5.0×10^{16} and 2.0×10^{17} ions/cm² for both ion implantation.

REFERENCES

1) N.Mikkelesen, J.Chevallier, & G.Sorensen Appl.Phys.Lett.Vol.52,No.14 (1988) 1130.
2) J.Chevallier, S.Olessen,& G.Sorensen, Appl. Phys. Lett.Vol.48, No.13 (1986) 876.

WEAR RESISTANT LAYERS ON AL AND MG ALLOYS BY LASER TREATMENT

ANDREAS WEISHEIT, ROLF GALUN and BARRY L. MORDIKE
Institut für Werkstoffkunde und Werkstofftechnik, TU Clausthal, Agricolastr. 6, 38678 Clausthal-Zellerfeld, Germany

Introduction

Al and Mg alloys find more and more applications in industry because weight saving is nowadays a major demand in many areas. Especial-ly the car manufacturers are interested in replacing steel by light metals for certain engine components like the cam follower or the valve. This will only be successful if the wear resistance is improved. A technology which has a promising potential to achieve this requirement is laser surface melting and alloying.

Cast Aluminium Alloys

Remelting of cast Al-Si alloys leads to a very fine distribution of the Si-phase accompanied by a hardness increase of 20-40 % related to the as cast condition. An increased hardness is achieved by alloying. Ni and Ni-based alloys have proved to be the best choice /1/. Depending on the amount of the precipitation of the intermetallic compound Al3Ni the hardness can be 2-10 times that of the as-cast alloy.

A pin-on-roller wear test was conducted for remelted and alloyed layers on the hypereutectic alloy AlSi18CuMgNi. The test simulates sliding under an alternating load. The main wear mechanisms are adhesion and surface fatigue. As shown in fig. 1 even remelting leads to a significant decrease of the wear rate. This can be easily understood considering the increase in hardness and surface area of the fine Si-precipitations and the homogeneous microstructure. Alloying leads to a further reduction of the wear rate only if the hardening is moderate. The brittleness of very hard layers causes a change in wear particle formation from micro cutting to micro fracturing. The later generates bigger debris and the wear rate increases again.

A potential application for this treatment is the cam follower. Until today it is not possible to manufacture it completely out of Al because of its poor wear behaviour. Motor tests with laser alloyed cam followers will soon be conducted to assess the improved surface properties.

Magnesium Alloys

First attempts to improve the wear resistance of Mg alloys were made by laser alloying with Al, Ni, Cu and Si /2/, resulting in a microstructure of Mg solid solution and intermetallics. Scratch tests showed an improved wear resistance (fig. 2), but the corrosion resistance (tested in 3 % NaCl solution) decreased because of the different corrosion potential of Mg and its intermetallics. This negative effect can be avoided when the layer consists of 100 % intermetallics. However, this leads to an undesired embrittlement. A more sophisticated solution is the addition of Al which improves the corrosion behaviour of the Mg solid solution significantly. Additionally, an even decreased wear rate is observed attributed to the formation of Al-Ni and Mg-Al-Cu intermetallics.

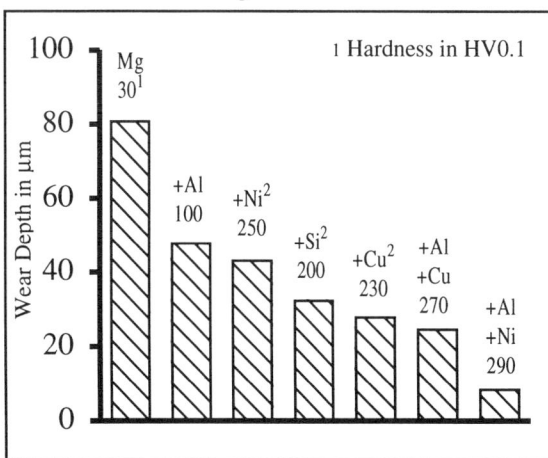

Fig. 2: Wear depth of cp Mg and laseralloyed samples of cp Mg and Mg0.8 Al(2) in a scratch test; counter part 100 Cr 6 steel ball

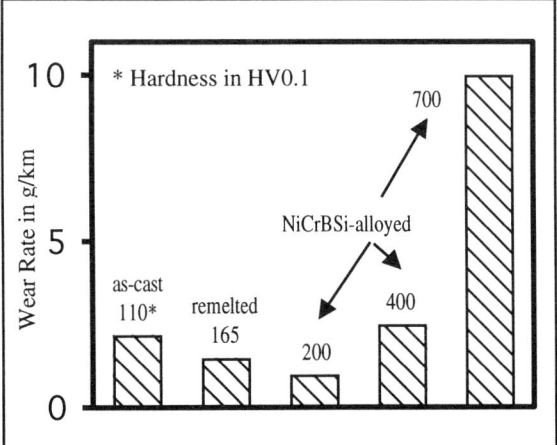

Fig. 1: Wear rate of as-cast and laser treated samples of AlSi18CuMgNi in pin-on-roller test; counter part 100 Cr 6

References

1 Weisheit, A., et al.: Aluminium (1996) 7/8, 522

2 Galun, R., et al.: J. of Las. Appl. (1996) 8, 299

Nitrogen Plasma Immersion Ion Implantation for Surface Treatment and Wear Protection of Stainless Steels

Carsten Blawert and Barry Leslie Mordike
Institut für Werkstoffkunde und Werkstoffkunde, Technische Universität Clausthal,
Agricolastr. 6, 38678 Clausthal-Zellerfeld, Germany

Abstract

Plasma immersion ion implantation is a suitable surface treatment for stainless steels. At moderate temperatures nitrogen remains in solid solution without forming nitrides. This increases the surface hardness and the wear resistance while the passivation of the steels remains possible.

Introduction

Stainless steels have a relatively low hardness and therefore mostly a poor resistance to wear. Without surface treatment, the strength and hardness can be only increased by cold forming or age-hardening, but these have almost no effect on the wear behaviour. Thermochemical surface treatments such as conventional nitriding or coatings have limited application due to the high temperatures used during these processes or due to a lack of adhesion to the substrate. Hardening by nitrogen is quite effective but the precipitation of CrN becomes a problem above 673 K and the steels lose their good corrosion resistance. PI^3 can be considered as a suitable surface treatment because it offers nitriding treatments below this critical temperature.

The PI^3 Process

In the PI^3 process the workpiece is immersed in a nitrogen plasma. On applying high negative voltage pulses to the workpiece the nitrogen ions of the plasma are accelerated towards it. The energy of the ions is used to heat the workpiece. By regulating the repetition rate of the high voltage pulses the treatment temperature can be controlled from 150°C to 550°C.

At elevated temperatures (above 300°C) the PI3 becomes a hybrid process (Fig.1). Nitrogen uptake occurs not only by implantation but also by a nitriding type of surface reaction when no high voltage is applied.

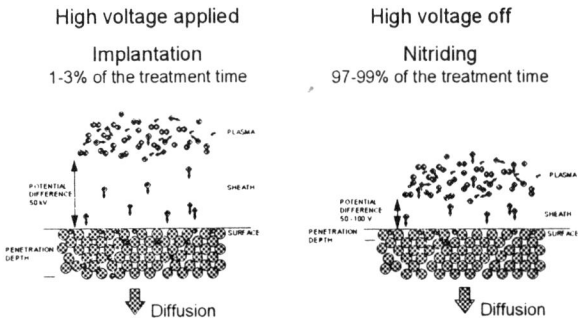

Fig. 1: Nitrogen uptake during PI^3 at elevated temperatures

Structure and Nitrogen Depth Profile

In austenitic and austenitic-ferritic stainless steels a phase called expanded austenite or S-phase (Fig. 2) was detected while the phases found in ferritic stainless steels are still unknown.

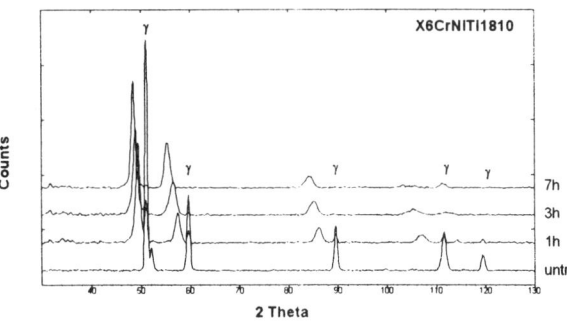

Fig. 2: Diffraction patterns of expanded austenite

Since nitrogen remains in solid solution diffusion of nitrogen into the material can occur, forming expanded austenite layers of several micrometer thickness in a few hours.

Hardness, Wear and Corrosion

The increase in surface hardness (Fig. 3) leads to a change in the dominant wear mechanismen. For untreated stainless steel severe adhesive and abrasive wear was observed while treated samples revealed only very mild abrasive wear. Best results were obtained when the specimen and the counterpart were treated.

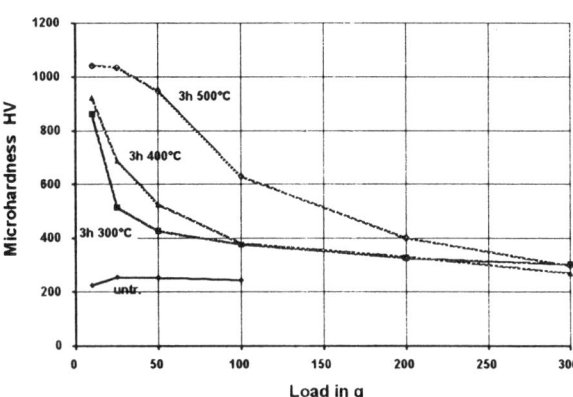

Fig. 3: Surface hardness of X6CrNiTi1810 after different PI^3 treatment temperatures

A controlled treatment is possible, keeping the nitrogen in solid solution with only a minor effect on the passivation property of the chromium and an overall good corrosion resistance is maintained while the wear resistance is increased.

WEAR OF PHYSICAL VAPOR DEPOSITION TiN COATINGS SLIDING AGAINST Cr-STEEL AND WC COUNTERBODIES

KEE-RONG WU
Department of Marine Engineering, National Kaohsiung Institute of Technology, Kaohsiung, Taiwan, ROC
RAYMOND G. BAYER
Consultant in Tribology, Vestal, NY, USA
PETER A. ENGEL and D. C. SUN
Department of Mechanical Engineering, State University of New York at Binghamton, Binghamton, NY, USA

ABSTRACT

Titanium nitride (TiN) coatings are widely used because of their good wear resistance. The method of coating by physical vapor deposition (PVD) has many superior features over other methods of deposition. Among the parameters affecting the wear resistance of PVD TiN coatings, the most crucial ones are the deposition temperature and the coating thickness. The paper presents an experimental investigation of the wear characteristics of PVD TiN coatings produced at two deposition temperatures (around 220°C and 371°C) and for a range of coating thicknesses (0.48 μm to 3.25 μm).

A ball-on-flat configuration was used in the study. The flat specimens, made of M2 tool steel with TiN coatings, were slid against either a 52100 chromium steel (Cr-steel) or a tungsten carbide (WC) counterbody ball in unlubricated condition. These two counterbody materials were chosen because of their different properties. Cr-steel was softer and more closely simulated ordinary workpiece surfaces, while WC was harder and had a greater thermal conductivity. The normal load used was 1.8 N. The wear volume was measured and wear scar examined with a 3-D optical profilometer. The transferred layers on the worn surfaces were analyzed with the scanning Auger microscopy (SAM) and an electron probe microanalyzer (EPMA).

Quantitative data were obtained for the wear volume and coefficient of friction as functions of the sliding cycles. The effects of the deposition temperature and coating thickness were determined, and different wear mechanisms for different material pairs were identified.

In the case of TiN versus Cr-steel, wear debris were generated in and around the sliding track as the test proceeded. The central part of the wear track, in parallel with the sliding direction, was initially undamaged. The profiles of the wear scars are shown in Fig. 1, where the X/radius scan is along the sliding stroke and the Y/circumference scan is perpendicular to the sliding stroke. Wear of the coating occurred preferentially in the outer regions of the wear track. A shiny golden yellow color covered these regions, while a shiny rusty brown color appeared in the central undamaged region. This undamaged part of the coating surface was covered by a thin layer of transferred materials. It was shown by EPMA that the transferred materials were iron and iron oxide originated from the Cr-steel ball. The undamaged central region, however, collapsed after more sliding cycles. After that wear rapidly proceeded and a regular shaped wear scar was formed. The presence of transferred wear debris in the wear track of the coating surface as well as on the flattened ball surface was a major factor controlling the wear behavior.

In the case of TiN versus WC, no transferred layers were observed. The wear track was covered with shiny flakes of mixed pink and golden colors. It was found by SAM that the loose debris wiped off the wear track were titanium and titanium oxide. Some amount of tungsten was also present in the debris. The predominant wear mechanism on the TiN coating was determined to be mechanical rather than chemical.

It was concluded that both the lower and higher deposition temperature coatings could significantly reduce wear in the case of TiN versus Cr-steel. The data showed that the optimal thickness for the lower deposition temperature coatings was much smaller than that for the higher deposition temperature coatings.

It is believed that the presented wear data, from a ball-on-flat reciprocate sliding configuration and at such a small normal load, were obtained for the first time. Details and more results of the study are contained in the paper of this title, which has been submitted to the ASME Journal of Tribology for publication.

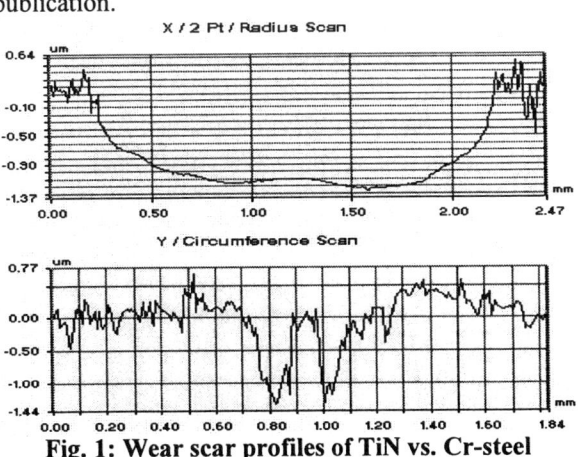

Fig. 1: Wear scar profiles of TiN vs. Cr-steel

CORROSION AND WEAR RESPONSE OF S-PHASE COATINGS

P. A. DEARNLEY
Department of Mechanical Engineering, University of Leeds, Leeds LS2 9JT, UK.
K. L. DAHM*, A. J. BETTS[+] and G. A. WRIGHT[+]
*Department of Chemical & Materials Engineering, [+]Department of Chemistry,
University of Auckland, Private Bag 92019, Auckland, New Zealand

ABSTRACT

Recent research by the authors has focused on the beneficial use of metastable coating materials for improving the aqueous corrosion-wear resistance of stainless steels (1,2). Amongst various possibilities a new coating material, essentially nitrogen super-saturated austenite and known sometimes as "S-phase" (1), appears well suited to providing protection from the conjoint action of corrosion and wear. A series of such coatings, typically 5µm thick, have been produced using unbalanced magnetron sputter deposition. Coatings were applied to polished (Ra~0.05µm) 316L austenitic stainless steel substrates previously ultrasonically degreased in isopropyl alcohol and sputter cleaned, for 90 minutes, using a radio frequency glow discharge plasma with 500 watts of forward power. During the deposition cycle a DC bias of -50 volts was used; substrate temperatures ~200°C were achieved. Total gas flow (Ar + N_2) was maintained at 10 sccm and the throughput of nitrogen was controlled in the range of 0 to 40 vol.%. A constant chamber pressure of 2 mTorr was used for all deposition cycles.

flow rate approaching maxima in both cases for those coatings processed with ~40 vol.% N_2. A maximum hardness of 20 GPa (determined by nanoindentation) was achieved.

Anodic polarisation in 1M HCl showed that *all* coatings were more passive than untreated 316L austenitic stainless steels. Coating passivity increased with increasing nitrogen flow, attaining equal levels in those coatings produced using ≥20 vol.%N_2, Fig 1. Accelerated immersion corrosion tests in 10wt% $FeCl_3$ at 20°C showed coatings produced with ≥20 vol.% N_2 to be corroded by a factor of at least 225 times less than untreated 316L.

Sliding wear resistance was similarly improved, Fig 2, but there appeared little advantage in exceeding 20 vol% nitrogen during the deposition cycle even though hardness of the coatings could be further increased. *All* coatings reduced the overall rate of wear despite significant plastic flow of the substrate: the Hertzian sub-surface shear stress (τ_{max}~398 MPa) far exceeded the shear yield strength (k~250MPa) of the substrate.

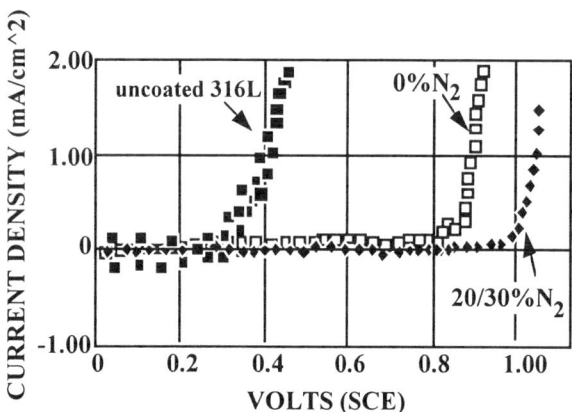

Fig 1: Anodic polarisation data obtained in IM HCl at 25°C (forward diretion only)

The elucidation of coating crystal structure was made possible by the use of glancing incidence X-ray diffraction ($\alpha=2°$). Stainless steel coatings deposited with 0% N_2 (Ar only) were body centred cubic. All other coatings, using nitrogen flow rates ≥5 vol.%, exhibited face centred cubic symmetry. The latter also showed a strong {200} preferred orientation. Unit cell dimensions and hardness increased with increasing N_2

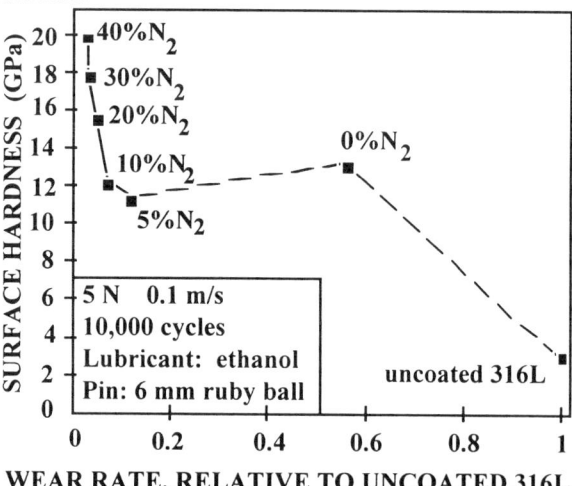

Fig 2: Sliding wear test results

REFERENCES

(1) K. L. Dahm and P. A. Dearnley, Surface Engineering, Vol. 12, No.1, 61-67pp. 1996.
(2) P. A. Dearnley, N. Dortmann, K. L. Dahm and H. Weiss, Surface Engineering, Vol. 13, No.2, in press, 1997.

Comparison of Tribological Characteristics of Sputtered MoS$_2$ Films Coated with Different Apparatus

MINEO SUZUKI

Space Technology Research Group, National Aerospace Laboratory, Chofu, Tokyo, JAPAN

ABSTRACT

Tribological characteristics of sputtered MoS$_2$ films have been investigated by many researchers. However, the reported tribological data, especially for wear life, are sometimes very different and/or contradictory. The cause of this is partly due to the use of different type of friction tester and test procedure, however, it seems that the nature of the MoS$_2$ film was different because the film was deposited using different apparatus, deposition procedure and conditions.

In this study, tribological characteristics of sputtered MoS$_2$ films, coated with 6 different apparatus by 5 different organizations, are evaluated and compared. Tested MoS$_2$ films are shown in Table 1. All the films were sputter-deposited on 440C stainless steel disks with a thickness of 1 μ m. The deposition procedure and conditions were optimized at each organization. Ball-on-disk friction tests were carried out in vacuum, nitrogen gas and dry air environments at a sliding speed of 0.5 m/s and an applied load of 10 N. Some films were also evaluated in disk/disk (edge loading) configuration with slip ratios of 1% and 10% at a rotating speed of 0.5 m/s and a load of 150N. Maximum contact stresses are 1.16 GPa for the ball-on-disk tests and 0.75 GPa for the disk/disk tests, when calculated assuming there exists no film.

Wear life of the sputtered MoS$_2$ film obtained in a series of the tests are shown in Fig.1. Repeatability of wear life in ball-on-disk tests was examined in vacuum. Although the number of tests was limited, data scatter for wear life of the same film was within a factor of 2, except for Film F. However, durability of the MoS$_2$ film was much different even when the sputtering apparatus was the same magnetron type. In addition, the effect of test configuration on wear life depended on the type of the films. In roll+slide configuration, life was much longer than in pure sliding for Film C, essentially the same for Film B and E, and a little shorter for Film A. As for the effect of test environment, all the tested films showed similar tendency, and the life was in the order of in nitrogen gas > in dry air > in vacuum.

These results suggest that difference in tribological behavior observed in this study was caused by the difference in nature of the MoS$_2$ film. An EPMA analysis showed that S/Mo ratios were 1.7-1.8 for Film A, B, C and E, 1.4 for Film D, and 1.0-1.3 for Film F. Thus factors other than S/Mo ratio is responsible to the different tribological performance.

Table 1 Tested MoS$_2$ Films

Film	Sputtering Apparatus	Target * (MoS2 source)	Pressure during Sputtering
A	RF	HP	5 E-2 Torr
B		NA	2 E-2 Torr
C	Magnetron	HP	6 E-3 Torr
D		CIP	2 E-3 Torr
E		HP	5 E-3 Torr
F	ECR Ion Gun	CIP	1.5E-4 Torr

* HP:Hot-Press   CIP:Cold Isostatic Press
NA:Data Not Available

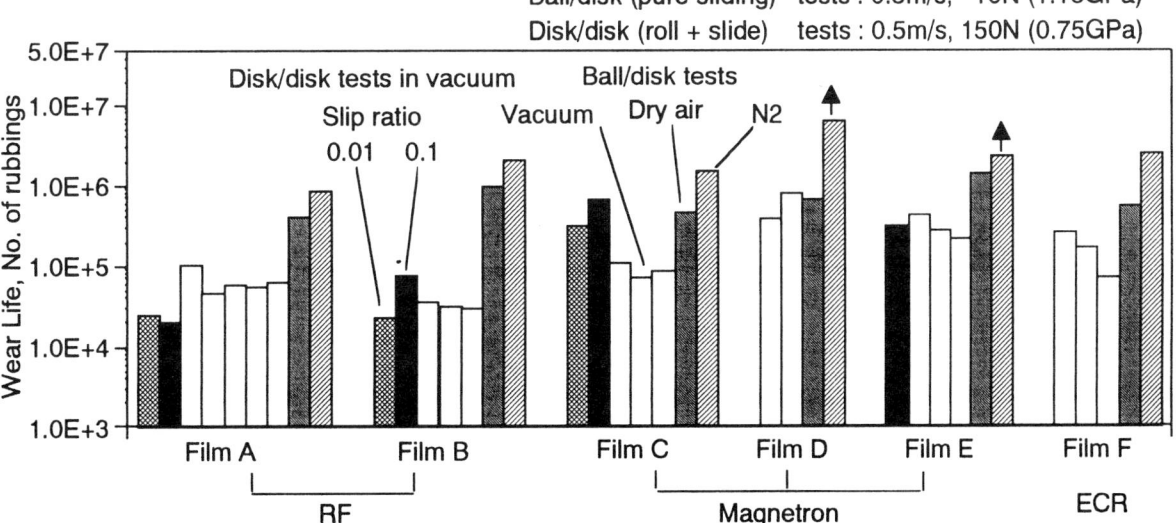

Fig.1 Wear Life of MoS$_2$ Films

HIGH-CURRENT-DENSITY ION IMPLANTATION OF NITROGEN AND TRIBOLOGICAL PROPERTIES OF FERROUS MATERIALS.

A.V. BYELI, V.A. KUKAREKO, O.V. LOBODAEVA, S.K. SHYKH
Physical-Technical Institute, 4 Zhodinskaya Str., Minsk, Republic of Belarus.

Ion implantation is a promising technique for improving the tribological properties of surfaces. Its main drawbacks have generally considered to be a small depth of ion penetration and comparatively high cost of processing. The present work examines high-current-density, elevated-temperature ion implantation of nitrogen in different ferrous materials that have successfully addressed these drawbacks and enabled rapid saturation of comparatively deep layers.

Data have been presented that demonstrate solid solutions, precipitates of new phases and sometimes amorphization of surface layers with thicknesses up to several μm and more are induced by high-current-density ion implantation. As the role of diffusion for elevated-temperature implantation exceeds that of simple ballistic delivery of ions, the importance of the microstructure and chemical composition of the surface layer dramatically increase. The model for fast diffusivity of implanted atoms is presented, which confirms rapid saturation of a deep surface layer.

Ion implantation dramatically improves the wear resistance of investigated materials. Tribological properties, phases formed, and depth distribution of implanted atoms are controlled by the implantation parameters. Particularly, ion implantation at ~620 K results in a rather homogeneous hardening of the surface layers which enhances wear resistance. The main hardening mechanism in this case is solid solution strengthening. Higher implantation temperatures (750-770 K) are beneficial because they yield new phases and more effective hardening of the surface, but they result in the development of a heterogeneous structure and a decrease in the corrosion resistance of the surface. Both of these factors can be unfavourable for a number of tribological applications and should be taken into account to determine optimum parameters of ion beam treatment.

TRIBOLOGICAL EVALUATION OF LASER TREATED SURFACES BY MEANS OF CAVITATION

MICHAEL BOHLING
NU*TECH*GmbH, Ilsahl 5, 24536 Neumünster, Germany
ALFONS FISCHER
Universität GH Essen, Werkstofftechnik, Universitätsstr. 15, 45117 Essen, Germany
ASTRID ZIEGE
Fachhochschule Lübeck, Fachbereich AN, Stephensonstr. 3, 23562 Lübeck, Germany

ABSTRACT

Cavitation may appear at ship propellers, water turbines, fluid pumps, as well as on pipes. Microjets (diameter: 1 µm) hit the surface with a speed of about 200 to 1000 m/s and lead to local impact stresses of several 100 MPa to GPa. This brings about a loss of material, deterioration of performance, and destruction of the entire machine part. The main wear mechanism is surface fatigue. Thus, one measure to counter wear by cavitation is change of materials. Beside others this can be brought about by surface techniques gaining fine microstructures with sufficient hardness and toughness in order to achieve tribological stability.

The major advantage of laser cladding lies within the high energy density, which can be varied within a wide range and governs the heat input. This might bring about fine homogenous microstructures depending on the chemical composition of the involved materials and cladding parameters (2).

The aim of this work is the characterization of wear behavior of technical laser claddings under cavitation.

Testing Materials: Laser cladded Fe-, Ni-, Co-base materials have microstructures consisting of a metal-matrix (MM) with fine eutectic (fHP) and/or coarse primary (cHP) hard phases.

	MM	MM+fHP	MM+fHP+cHP
Fe	FeC	FeW6Mo5V2C	
Ni		NiCr20Fe18Nb5Ti	NiCr40Mo4C
Co		CoCr29W13	

Table 1: Groups of Materials Tested

The metal matrices of FeC and FeW6Mo5V2C are martensitic, while those of the Ni- and Co-base materials have an austenitic structure. FeW6Mo5V2C has fine hard phases of M_6C-type (M=W, Mo), while eutectic Ni_3Nb as well as fine primary TiC solidify from the melt of NiCr20Fe18Nb5Ti. Eutectic M_7C_3 (M = Cr, Co) appear within CoCr29W13 and NiCr40Mo4C, while the latter has coarse primary Cr-carbides of the same type as well. These microstructures render hardness values between 280HV10 (FeC) and 1022HV10 (FeW6Mo5V2C).

Testing and Results: Tests were carried out in accordance with ASTM G32 (3). The incubation period is limited by the first measurable weight loss. Cavitation rates were determined by weight loss related to the testing time. The set up was calibrated using annealed pure Ni. The reproducibility is better than 10 %. The worn surfaces were investigated using an SEM in order to determine the wear appearances.

	Incubation period [min]	Erosion Rate [µg/min]
FeC	130	80
FeW6Mo5V2C	240	3
NiCr20Fe18Nb5Ti	150	100
CoCr29W13	240	3
NiCr40Mo4C	180	22

Table 2: Results of Cavitation Tests

Discussion: During incubation period worn surfaces appear like being chemically deep etched. Metal matrix as well as eutectic hard phases are worn at the same wear rate bringing about a loss of material being equally distributed over the entire microstructure. The differences between the materials are related to the stability of the metal matrix, only, which is governed by its strength (FeW6Mo5V2C) and/or capability to work harden (CoCr29W13). The highest corrosion rate shows the austenitic NiCr20Fe18Nb5Ti cladding, because of solidification cracks at the interfaces between MM and fHP. Even the softer martensitic FeC cladding shows a lower erosion rate. It should be mentioned, that pores, which normally appear within laser claddings, deteriorate the cavitation behavior in a similar way. NiCr40Mo4C has a distinctly lower erosion rate. The loss of material concentrates on the metal matrix and the eutectic. Thus, the cHP protrude from the surface until they loose their support and are worn by microcracking.

THE WEAR BEHAVIOR OF A THIN MoS$_2$ COATING, AS STUDIED BY TRIBOSCOPIC MEASUREMENTS IN FRICTION AND ELECTRICAL CONTACT RESISTANCE

BELIN M.*, WAHL K.J.** and SINGER I.L.**
* Laboratoire de Tribologie et Dynamique des Systèmes, UMR 5513
École Centrale de Lyon, B.P. 163 - 69 131 Écully Cedex, France.
** Naval Research Laboratory, Tribology Division
Washington D.C., 20375, USA

INTRODUCTION

The wear resistance of thin MoS$_2$ layers has been extensively studied in literature. The occurrence of material transfer, mechanical and tribochemical wear has been recently evidenced. Nevertheless, the high durability of this type of solid lubricant is still a matter of controversy. One explanation for this phenomenon is that friction and contact resistance are both controlled by material transfer processes (1). This paper deals with this question, bringing an original contribution.

EXPERIMENTAL

To explore the wear and failure of MoS$_2$ further, we use spatially-resolved friction f and electrical contact resistance Rc measurements, coupled with *ex situ* optical and surface analysis - Raman spectroscopy, AES and EDS. We have performed some wear experiments in reciprocating configuration, on duplex layers of 40 nm MoS$_2$ on 35 nm of TiN elaborated by ion-beam assisted deposition onto a steel substrate. Experiments have been performed at low speed, high contact pressure (Po=1 GPa), in different environments: dry air (RH < 1%), ambient air (RH = 40%) and dry nitrogen.

RESULTS

We confirm that the friction level strongly depends on the environment. As an example, the friction coefficient value is typically 0.10 in ambient air, and 0.02 in dry air. In addition, the durability of the layer is strongly affected by humidity: catastrophic breakdown is occurring at N<1500 cycles in the case where RH > 40%, see Fig. 1 and 2. Spatially-resolved data maps are presented, called triboscopic diagrams. The friction maps indicate that friction coefficient is very homogeneous along the wear track, over most of the life of the test before the catastrophic breakdown. On contrast, the contact resistance exhibits local variations along the wear track, much earlier in sliding.

The results show that the electrical contact resistance is very sensitive to incipient failure. They are discussed in terms of the role of material transfer processes in durability of this solid lubricant. The combined triboscopic and analytical methodology allows Rc and f evolution to be correlated to variation in the active interfacial film thickness (coating + transfer), in the case of ambient air. A simple model of contact is settled, based on assumptions concerning the real contact area, the shearing process and the ohmic electrical conduction process When Rh = 40%, the approximate value of resistivity is found to be close to 0.54 Ohm.m. In the case of dry air, we get evidence that the ohmic model is no more valid.

The effect of tribochemistry and orientation of MoS$_2$ is then discussed. The influence of atmosphere and the potential for using Rc measurements for *"early failure warning"* in solid lubricants are then highlighted.

FIGURES

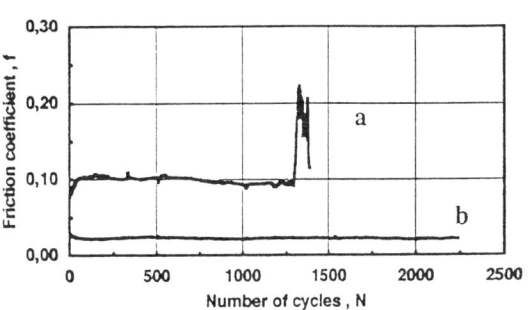

Fig.1 : Evolution of the friction coefficient f of a 40 nm thin IBAD MoS$_2$ layer, in ambient air (a) and dry air (b) (Rh = 40% and <1 % resp .)

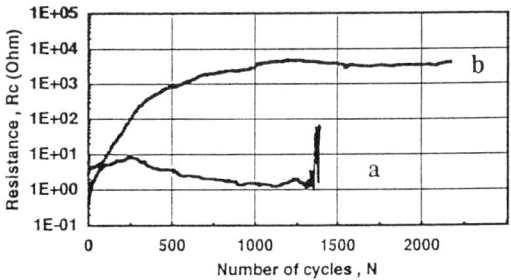

Fig.2 : Evolution of the electrical contact resistance Rc of the same sample, in ambient air (a) and dry air (b) (Rh = 40% and <1 % resp.)

REFERENCE
(1) K.J. Wahl, Belin M. and I.L. Singer, "A tribosopic investigation of the wear and friction of MoS$_2$ in a reciprocating sliding contact", submitted to Wear, 1997.

THE INFLUENCE OF ATMOSPHERE ON THE LIFE OF SOLID LUBRICATED BALL BEARINGS

YASUO YOSHII, NAOFUMI HIRAOKA
Energy and Mechanical Research Laboratories, TOSHIBA Corporation, 4-1 Ukishima-cho, Kawasaki, Japan
AKIRA SASAKI, YOSHIFUMI NODOMI, NOBUO KENMOCHI
Toshiba Komukai Works, TOSHIBA Corporation, 1 Toshiba-cho, Saiwai-ku, Kawasaki, Japan

INTRODUCTION

This paper presents a study on the effects of atmosphere on the durability of solid lubricated ball bearings. Solid lubricated ball bearings have already been successfully used for applications in solar array paddle drive mechanisms, antenna pointing mechanisms, etc. for spacecraft (1)(2). Considering the need for functional testing before launch, it is desirable for the bearing to be durable both on earth and in space.

RESULTS AND DISCUSSION
BALL BEARING TEST

Several life tests of ball bearings under different radial load conditions were carried out in laboratory air and in vacuum. The balls and races of the test bearings were 440C stainless steel coated with sputtered molybdenum disulfide (MoS_2), and the retainers were made of PTFE composite containing glass fibers and molybdenum particles. The test bearings were 12 mm in internal diameter, 28 mm in external diameter, and 8 mm in width. Figure 1 shows the bearing life with respect to atmosphere as a parameter. The endurance life of ball bearings in air was reduced to less than 1/50 that in vacuum. After the running tests, the ball surfaces were analyzed by XPS. Although the ball surfaces in both in-air and in-vacuum tests were transfer-coated with PTFE from the retainer, the ball surfaces in the in-air tests were worn, while the ball surfaces in the in-vacuum tests remained intact.

Furthermore, it was found that the transferred film on the ball surfaces included the ingredients of glass fibers after the running tests. This result indicates that the shorter bearing life in air is not due only to the poor lubrication behavior of sputtered MoS_2 in air, but also to the transferred film.

BALL-ON-DISK TEST

Ball-on-disk friction and wear tests were carried out to evaluate the PTFE transferred film properties in air and in vacuum. PTFE composite disks and 440C bearing steel balls were used. The diameter of the balls was 9.525 mm (3/8 in). The sliding speed was 300 rpm at a track radius of 20 mm, and the applied load was 9.8 N (dead weight). Figure 2 shows the PTFE composite disk surface after the test. It was observed to be smooth under in-air conditions compared with in-vacuum conditions. Since the wear of glass fibers in air was much greater than that in vacuum, glass debris was scattered over the surface of PTFE resin in air.

From these results, we suggest that the transferred film including wear debris of glass fibers abrades solid lubricated ball bearings in air.

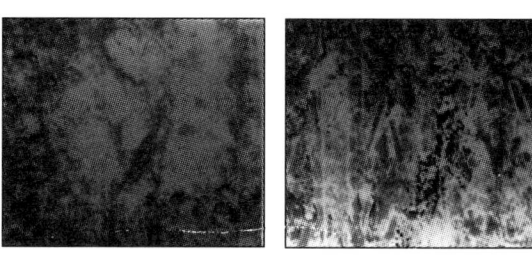

(AIR)   (VACUUM)
Fig. 2: SEM images of wear scars on disks

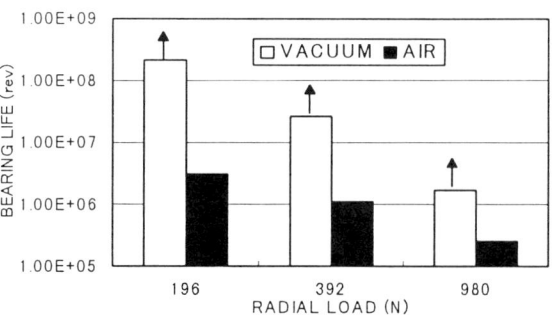

Fig. 1: Life of solid lubricated ball bearings

REFERENCES
(1) T. Kawamura et al.; Proceedings of Fourth European Symposium on Space Mechanisms and Tribology, 1989, pp.101-107.
(2) Y. Yoshii et al.; Proceedings of the Japan International Tribology Conference, 1990, pp.1839-1844.

MATERIAL TRANSFER IN JOURNAL BEARING IN UNLUBRICATED SLIDING

ADAM POLAK
Cracow University of Technolgy, ul. Warszawska 24, 31-155 Krakow, Poland

ABSTRACT

The paper presents investigation of the phenomenon of material transfer in journal bearing plastic - steel and carbon material - steel operating in dry friction conditions.

The aim of the investigation was to describe the mechanism of transfer film formation as a macro-coating in a journal bearing. Majority of authors(1) describe material transfer in tribological pair plane - plane and first of all describe the first layer of the transfer film formation on metal surface. However, as shown in the presented paper, the process of transfer film formation as a whole, and the mechanism of the film formation on the surface of the bearing bush are interesting and require explanation.

The mechanism of transfer film formation due to specific friction conditions in a journal bearing has been described i.e.; pressures in the journal - bush contact zone have different values at the beginning and end of the contact zone and at the point of perpendicular force load, wear particles are kept in the bearing and participate in the process of friction again.

METHODS AND MATERIALS

The tribological pair used was steel journal and bearing bush made from: carbon material WG of three different degrees of graphitization: 35%, 50%, 95%; polyamide PA6; polyoxymethylene POM; polyethylene HDPE. Also carbon materials WG impregnated with bearing alloy (SnSb11Cu3) or lead alloy (PbSb9) were tested, and the plastics mentioned above were graphite-modified. The bearing bush materials were selected in such a way that they differed in structure and properties.

The tests were carried out at rotation of the journal and motionless bearing bush and with motionless journal and motion of the bush. The tests were run with no hard abrasive particles and in the presence of those (quartz grains) in the friction zone. The pressure range was p = 0.1 to 1.5 MPa and friction velocity v = 0.1 to 1.2 m/s.

RESULTS

On the basis of the tests:
' some phenomena of transfer film formation mechanism on the bearing bush in various friction conditions have been presented;
' it has been demonstrated that some phenomena connected with material transfer are common for materials significantly different in structure and physical and chemical properties;
' different structure of the transfer film at the beginning, in the middle zone and at the end of journal-bearing bush contact zone has been shown;
' the effect of hard abrasive particles in the bearing (quartz dust) on transfer film formation has been shown;
' the following models of material transfer have been proposed: at rotation of journal and motionless bush (Fig.1.a), in the presence of abrasive particles at rotation of journal and motionless bush (Fig.2.a), with motionless journal and bush motion (Fig.1.b), in the presence of abrasive particles with motionless journal and bush motion (Fig.2.b).

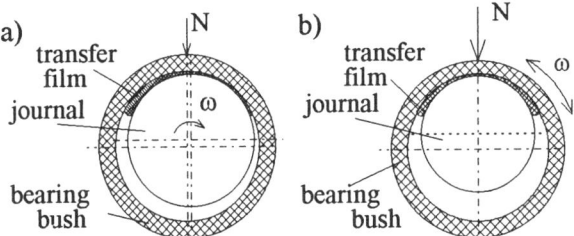

Fig.1. Transfer film on bearing bush in case of
a) motionless bearing bush and journal motion
b) motionless journal and bearing bush motion

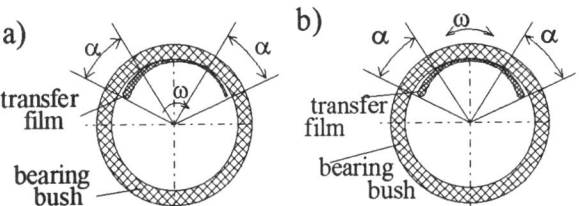

Fig.2. Transfer film on bearing bush in the presence of quartz dust. α - zone of loosened transfer film

REFERENCES

(1) Abarou S., Bahadur S., Bellow D.G., Brayant P.J., Braza J.F., Brendle M., Briscoe B.J., Cooper J.R., Dowson D., Eiss N.S., Evans P.D., Feyuelle S., Fisher J., Furst R.E., Fusaro R.L., Golmore R., Gong D., Gutshell P.Z., Herold J.A., Jain V.K., Kennedy F.E., Lamouri S., Lancaster J.K., Langlade C., Lin Heng Yao, Oliver R., Play D., Pooley C.M., Pytko S., Roberts J.C., Sebastian S., Smyth K.A., Stolarski T.A, Stuart B.H., Sviridyonok A.I., Tabor D., Tanaka K., Taylor L.H., Turgis P., Tweedale P.J., Viswanath N.P., Wood K.C.

MULTILAYERED ADAPTIVE LUBRICANTS FOR HIGH TEMPERATURE APPLICATIONS

J. S. ZABINSKI AND S.D. WALCK
Materials Directorate, Wright Laboratory, Wright-Patterson Air Force Base, OH 45433-7750, USA
J.E. BULTMAN
University of Dayton Research institute, Dayton, OH 45469

ABSTRACT

Solid lubricants that perform well in relatively modest temperature ranges from subambient to above 800°C are available. For example, transition metal dichalcogenides are useful to about 400°C, PbO from 450 to 650°C, and CaF_2/BaF_2, above about 700°C. However, there are few candidate coatings showing potential for use over the entire temperature range (i.e., subambient to > 800°C). As part of an effort to produce durable solid lubricant coatings for extreme environments, adaptive lubricants were investigated. Adaptive lubricants are materials designed to maintain low friction and provide wear protection by undergoing chemical reactions with elements in the environment as temperature increases (1). The as-deposited coating provides lubrication up to a conversion temperature where high temperature lubricant forms. Using this scheme, it is possible to maintain low friction to the thermal limit of the high temperature solid lubricant. A critical problem with candidate adaptive lubricants is their limited ability to withstand thermal cycling. The reaction of low temperature lubricant to high temperature lubricant is irreversible. Once the coating is completely converted to high temperature lubricant, friction increases as temperature decreases. One promising technology to improve thermal cycling is based on using functional multilayers. The multilayer architecture investigated was designed to preserve adaptive components by separating them with diffusion barriers.

The objective of this study was to determine the chemistry and transport properties of oxygen and adaptive components in thin film coatings with and without diffusion barriers. $ZnO-WS_2$ adaptive lubricant coatings were grown from composite targets by pulsed laser deposition using 248nm light. These coatings provide low friction at room temperature and react with O_2 above 500°C to form $ZnWO_4$ which has good lubrication properties at high temperature. The single layer coatings were heated from 400-800°C in 100°C intervals for different durations. After heating, cross-section TEM, selected area diffraction, and XPS sputter depth profiling were used to determine chemical and microstructural changes. It was demonstrated that Zn diffused toward the surface and oxygen towards the substrate. The topmost layer of the coating reacted with the O_2 and formed $ZnWO_4$. Having the high temperature lubricant form on the surface is desirable because it lubricates the contact region. The immediate subsurface region had a significant fraction of WO_3 present which is an intermediate in the $ZnWO_4$ reaction process. Below the zone penetrated by O_2, WS_2 began to crystallize. To prevent depletion of Zn in the bulk and to prevent oxygen penetration of the entire coating, several diffusion barriers including Ti, TiC, Al_2O_3, and Au were evaluated. These barriers were deposited between two adaptive lubricant layers and heated from 400-800°C as for the single layer coatings. Titanium and TiC were eventually oxidized as the temperature increased to 600°C, but they were effective in slowing the diffusion process. Titanium carbide was effective in preventing the diffusion of Zn. Gold was not effective because it nucleated and provided diffusion pathways. Alumina stopped the diffusion of both oxygen and zinc. The wear life of multilayered adaptive lubricant/alumina coatings was up to two orders of magnitude longer than coatings without alumina interlayers at 500°C. Wear life was proportional to the number of $ZnO-WS_2$ layers. Friction coefficients before failure remained below 0.1. The relative ratio of ZnO/WS_2 had a significant effect on coating wear life. Coatings with lower ZnO concentration had the longest wear lives below 300°C, while coatings with higher ratios lasted longer at 500°C. The effectiveness of the different barriers as a function of their microstructure and thermochemistry is discussed.

Adaptive lubricants show promise for providing low friction and long wear life over a broad temperature range.

References

(1) J. S. Zabinski, S. V. Prasad and N. T. McDevitt, "Advanced Solid Lubricant Coatings for Aerospace Systems", *Proceedings of the NATO/Agard Conference, Sesimbra*, Portugal, AGARD Conference Proceedings 589 (AGARD-CP-589), May 6-10, 1996; *or Tribology Letters*, To be published.

E:mail: zabinsj@ml.wpafb.af.mil
Phone: +1 (937) 255-8544

FRICTION CHARACTERISTICS OF TUNGSTEN DISULFIDE (WS_2)-ZINC OXIDE (ZNO) NANOCOMPOSITE FILMS

S. V. Prasad
University of Dayton Research Institute, Dayton, OH45469-0168
N. T. McDevitt
Ramspec Research, 2941 E. Mohave Dr., Dayton OH 45424
J. S. Zabinski
Wright Laboratory, Materials Directorate (WL/MLBT)
Wright Patterson Air Force Base, Ohio 45433-7750

ABSTRACT

As part of an ongoing effort to synthesize new generations of lubricants that are effective in extreme environments, we previously reported the development of nanocrystalline zinc oxide (ZnO) films by pulsed laser deposition (1). The study showed that the nanocrystalline ZnO films could deform plastically at room temperature and their coefficient of friction (COF) could be significantly less than that of a conventional hot pressed ZnO disk (2). Tungsten disulfide (WS_2) is a well known solid lubricant belonging to the family of transition metal dichalcogenides. It provides an ultralow friction surface and long wear life in dry environments and in vacuum. However, the presence of condensable matter in the environment, notably moisture, causes a rise in its friction coefficient. Similarly, at elevates temperatures, the tribological performance of WS_2 deteriorates as a result of oxidation.

The objective of the current study is to develop WS_2-nanocrystalline ZnO composite films that are effective over an extended range of operating conditions (e.g., strain rates and stresses) and environments. Nanocrystalline ZnO powder obtained from a commercial source (average particle sIze: 25 nm) and the WS2 powder (particle size: 1-2 µm) were mixed using a V-cone blender. The WS_2-ZnO composite films with varying fractions of ZnO were burnished on metallographically polished substrates with a lint-free cloth. For comparison, pure ZnO and pure WS_2 films were also burnished under simialr conditions. A few WS_2-ZnO composite films were grown using an excimer laser. These films have shown promise as adaptive lubricants (i.e., lubricants that can undergo designed chemical changes to continually maintain low friction) (1), and are evaluated here at elevated temperatures for comparison to burnished films. Friction measurements were made using a ball-on-flat tribometer. Room temperature tests were run in dry nitrogen and in humid air with 85% relative humidity. Elevated temperature tests were run in air at 300°C and at 500°C. Wear scars and trasfer films on the counterface were analyzed using Raman spectroscopy and scanning electron microscopy (SEM).

Results showed that at ambient temperatures and in dry environments, the composite films provided a low friction friction coefficient ($\mu = 0.05$) similar to that of pure WS_2. In humid air, the friction coefficient of the composite films was significantly less than that of pure WS_2. At elevated temperatures, the composite films were able to provide low friction ($\mu = 0.04-0.06$) for prolonged lenghts time where as the pure WS_2 films were found to oxidize within a short span of time. The tribochemical reactions and the synergistic roles played by nanocrystalline oxides and transition metal dichalcogenides in imparting solid lubrication in extreme environments are discussed.

REFERENCES

(1) J. S. Zabinski, S. V. Prasad and N. T. McDevitt, "Advanced Solid Lubricant Coatings for Aerospace Systems", Proceedings of the NATO/Agard Conference, Sesimbra, Portugal, May 6-10, 1996, 3.1-3.12pp; or Tribology Letters, To be published.

(2) S. V. Prasad and J. S. Zabinski, "Tribological Behavior of Nanocrystalline Zinc Oxide Films", Wear, Vol. 203-204, 1997, 408-506pp.

Submitted to *Tribology Transactions*

SOLVING TRIBOLOGICAL PROBLEMS WITH SOLID LUBRICANTS

T J RISDON
Climax Molybdenum Marketing Corporation, 124 Pearl Street, Ypsilanti, MI 48198-0407, USA
RÜDIGER HOLINSKI
Dow Corning GmbH, Rheingaustraße 53, D-65201, Wiesbaden, Germany

ABSTRACT

Since the middle of this century, new machine designs have emerged which are not always successfully lubricated by conventional lubricants. Solid lubricants, originally utilized in the aerospace industry, have become tribological problem solvers particularly in dry lubricating applications. Today, solid lubrication is a new dimension within the realm of available lubrication techniques.

Only a few solids have dry lubricating properties. The most commonly used solid lubricants include molybdenum disulfide and graphite. These materials are characterized tribologically by their extremely good adhesion on most metal substrates and by their excellent film forming properties. In some applications, the solid lubricant(s) may be mixed with other solid additives which, in themselves are not considered effective lubricants, to impart improved performance over that afforded by the solid lubricant(s) alone.

The forms in which solid lubricants can be used include burnished films, bonded lubricant coatings, composites, and sputtered films. In addition, solid lubricants are used in greases, pastes (high solids content greases) and as dispersions in various types of lubricating oils.

This paper discusses a number of industrial applications where solid lubricants are successfully being used due to the prevailing tribological conditions where conventional lubricants could not provide the desired performance. The examples cited in the paper focus on dry lubricating film applications where solid lubricants are employed in one of three forms: burnished or tumble-coated films, composites, or bonded lubricant coatings.

In a production facility for the manufacture of semiconductors, some anti-friction bearings must operate in extremely high vacuum for extended periods. Any lubricant used must not contaminate the vacuum environment. Bearing elements were coated with molybdenum disulfide using a tumbling method which produced a thin adherent film on the rolling elements and raceways, giving the desired low friction and wear.

Another way of utilizing solid lubricants in anti-friction bearings is dry lubricating composites. These materials contain a high solid lubricant content in a polymeric matrix which is in rubbing contact with the rolling elements. Bearings operating in high vacuum or at elevated temperatures are able to give many times longer life or can provide lifetime lubrication for the component. Examples of the successful use of bearings lubricated by this approach include vacuum chambers for sputtering processes and tile plant kiln truck bearings which can operate without attention for two years at temperatures ranging up to more than 200 C.

Bonded lubricant coatings, which are sometimes described as lubricating paint films, have been found to be effective in the dry lubrication of threaded parts. In nuclear reactors used in electrical power generating stations, studs used to secure hoods on pressure vessels are prone to galling and seizing during assembly and disassembly. Special anti-friction coatings, free of halogens which may cause corrosion, have been developed which can tolerate wide temperature ranges while preventing thread damage.

Bonded coatings are also being increasingly utilized in automotive applications. Specially formulated bonded lubricant coatings have successfully replaced tin-plate on piston skirts as means of preventing scuffing of the aluminum piston rubbing against the cast iron cylinder bore in the engine. The new coating provides longer wear protection and reduced engine noise.

FRICTION AND WEAR PROPERTIES OF PLASMA SPRAY COATING FILMS FOR HIGH-TEMPERATURE SEAL UP TO 1000°C

SHINYA SASAKI

Mechanical Engineering Laboratory, 1-2 Namiki, Tsukuba-shi, Ibaraki 305 JAPAN

ABSTRACT

A study has been carried out to develop a high temperature sliding seal for the ceramic gas turbine engine (CGT)'s regenerator. The solid-lubricanting coatings, which consisted of the combinations of NiO, Ni/Cr, Cr_3C_2/NiCr, ZrO_2/CaO, CaF_2, BaF_2, Ag_2O and graphite, were synthesized by low pressure plasma spraying.

Friction and wear characteristics were evaluated at temperatures from 50°C to 1000°C by rubbing the coatings against a cordierite disk. After the sliding tests, the worn surfaces were analyzed by SEM-EDX and XRD. The analytical results showed that the tribological properties of coatings were greatly related to the surface layers. Formation of CaF_2/BaF_2 rich layer led a low friction and a low wear property. On the other hands, formation of solid-state reaction product ($Mg_{0.4}Ni_{0.6}O$) between the coating and the cordierite increased friction coefficients, and the transfer of cordierite debris caused a seizure and/or a severe wear.

Among the sprayed coatings, the Cr_3C_2/NiCr based coating containing CaF_2/BaF_2 and Ag_2O showed the excellent tribological properties, which should satisfy the specifications for the basically designed CGT engine.

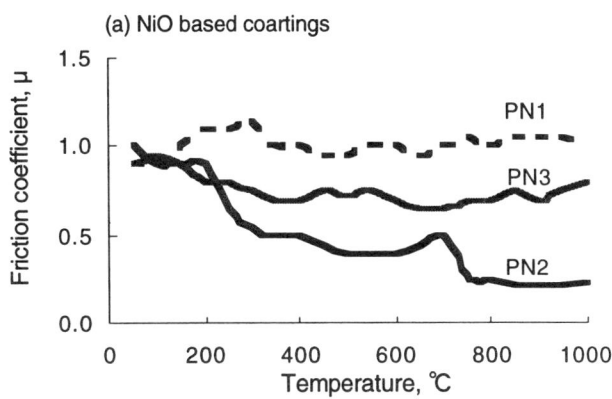

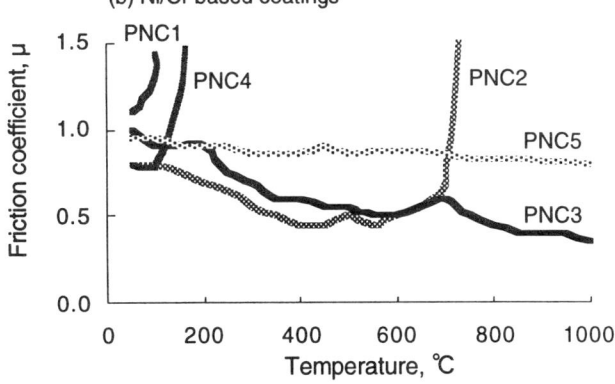

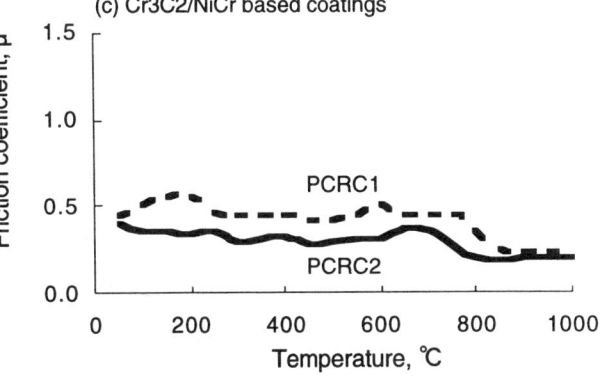

Fig.1 Fricrion behavior of each films with increasing temperature.

Table 1 Composition ratio of each coating films (wt%)

Coatings	NiO	Ni/Cr	ZrO_2/CaO	Cr_3C_2/NiCr	BaF_2/CaF_2	Ag_2O	Graphite
PN1	80				20		
PN2	65				20	15	
PN3	85						15
PNC1		100					
PNC2		70				30	
PNC3		65			20	15	
PNC4		70	30				
PNC5		30	25	30	15		
PCRC1				70	30		
PCRC2				65	20	15	

ASSESSMENT OF DATA REPEATABILITY & REPRODUCIBILITY IN TRIBOMETER TESTS FOR SPACE APPLICATIONS

J M CUNNINGHAM
STL, AEA Technology plc, Risley, Warrington WA3 6AT, UK.

ABSTRACT

Practical data on friction and wear is obtained from simple testing devices known as *"Tribometers"*. However, it is common for different laboratories using similar test machines to produce contradictory results. The Round-Robin was implemented to assess the data repeatability & reproducibility in friction and wear data for space applications. All tests were conducted following the recommendations made in a Guide-Line compiled by ESTL (1).

The test program was split into three phases. In **Phase (i)** the repeatability & reproducibility in friction & wear data from three material pairs was assessed using five nominally identical **pin-on-disc (POD)** tribometers (point contact & pure sliding motion) located at five different test laboratories. In **Phase (ii) & (iii)**, the ability of two additional tribometers employing different contact geometries to rank the performance of the same set of tribo-materials was assessed. These tests were conducted on a **two-disc {edge loaded} tribometer (2DE)** (line/elliptical contact & combined roll/slide motion) and **a two-disc {face loaded} (2DF)** tribometer (flat contact & pure sliding motion), located at two additional test laboratories.

The three material selected for investigation were: **A** - 52100 Bearing Steel vs Sputtered Molybdenum Disulphide (MoS_2) coated 52100 Steel. **B** - Leaded Bronze (LB9) vs 52100 Bearing Steel. **C** - Vespel SP-3 vs 52100 Bearing Steel. Each material combination was tested three times. All tests were conducted in vacuum $< 5 \times 10^{-6}$ torr. The main test parameters are listed in **Table 1**.

Table 1 - Test Parameters

Phase	Test Parameters	Units	A	B	C
(i) POD	Load	N	10	5	5
	Contact Stress	MPa	821 (mean Hertzian)	124 (mean Hertzian)	22 (mean Hertzian)
	Motion Type	-		Pure Sliding	
	Motion Speed	rpm	300	60	60
	Sliding Speed	m/sec	0.5	0.1	0.1
	Test Completion Criteria	-	Friction > 0.1 (0.15 during run-in).	10,000 revs (1000m) or friction > 0.5.	
(ii) 2DE	Load	N	100	5	5
	Contact Stress	MPa	578 (mean Hertzian)	125 (mean Hertzian)	22 (mean Hertzian)
	Motion Type	-		Rolling & sliding - roll:slide ratio 10:1	
	Motion Speed	rpm	Motor 1: 240 Motor 2: 216	Motor 1:50 Motor 2:45	
	Sliding Speed	m/sec	0.5	0.1	
	Test Completion Criteria	-	Friction > 0.1 (0.15 during run-in).	8,000 revs (1000m) or friction > 0.5.	
(iii) 2DF	Load	N	50		5
	Contact Stress	MPa		0.22(load/apparent area of contact)	
	Motion Type	-		Pure Sliding	
	Motion Speed	rpm	280	60	60
	Sliding Speed	m/sec	0.5	0.1	0.1
	Test Completion Criteria	-	Friction > 0.1 (0.15 during run-in).	10,000 revs (1000m) or friction > 0.5.	

Seven laboratories participated in the Round-Robin: Austrian Research Centre, B. Verkin Institute (Ukraine), CSEM (Switzerland), Hosei University (Japan), Sulzer Innotec (Switzerland), National Aerospace Laboratory (Japan), National Physical Laboratory (UK).

In general, friction and wear data were found to be repeatable within each laboratory, however the inter-laboratory reproducibility was more varied. A comprehensive presentation of the test results is outside the scope of the article (refer to Ref 2), however, in order to highlight one anomaly in an otherwise excellent set of friction data, the pin-on-disc tests results from Test Group C - Vespel SP-3 vs Steel from five different test laboratories are shown in **Figures 1.**

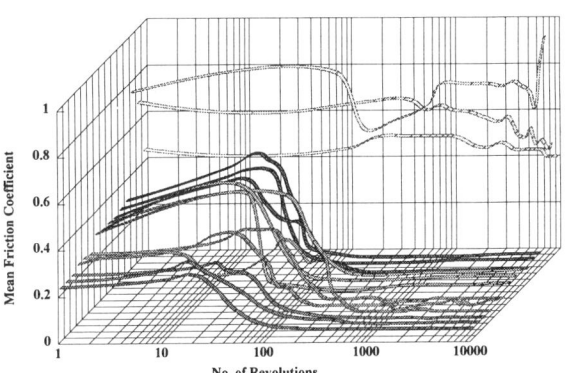

Figure 1 - POD Test Results - Vespel SP-3

In the POD study, four of the five test laboratories produced identical results. The 2DE & 2DF machines also produced comparative data. The anomalous results from one laboratory were attributed to the high partial pressures of water vapour in the test chamber inhibiting the generation of transfer films. Comparative results between tribometer types were not obtained for Leaded Bronze as friction coefficients were higher in the POD tests (0.3-0.5) than the 2DE & 2DF tests (0.2).

The Round-Robin test results showed that repeatable and reproducible friction and wear data could be generated by pin-on-disc tribometers using standardised test procedures provided the correct environment is achieved. Comparable friction data can be achieved using different tribometer types, however, the ability to produce comparable data depends on the susceptibility of the material combination tested to be influenced by contact geometry and motion type.

REFERENCES

(1) J. M. Cunningham. Tribometer User's Guide Lines for Space Applications. ESTL/TM/139. Dec. 1994.
(2) J. M. Cunningham. Tribometer Round-Robin: A Practical Assessment of Data Repeatability & Reproducibility in Tribometer Tests for Space Applications. ESTL/TM/170. Oct 1995.

PREDICTION OF THE LUBRICATION AND WEAR OF PISTON RINGS – THEORETICAL MODEL

M. PRIEST, D. DOWSON and C.M. TAYLOR

Department of Mechanical Engineering, The University of Leeds, Leeds, LS2 9JT, UK

ABSTRACT

It is proposed that a complete understanding of the tribological performance of piston rings in reciprocating internal combustion engines can only be achieved when both lubrication and wear are considered in combination. The running profile of the piston ring that slides against the cylinder wall wears significantly in service, even with wear resistant materials and coatings, such that the ring profile after only a short period of running in the engine differs greatly from that of the component as new.

Modification of the ring profile by wear has a large effect on lubrication, friction and oil transport at the interface between the piston ring and cylinder wall, which then in turn modifies the wear conditions (1). This interaction between lubrication and wear has important implications for fuel consumption, oil consumption, hydrocarbon exhaust emissions and durability of internal combustion engines in which the piston ring pack is widely recognised as having a pivotal role.

Figure 1 shows the measured running profile of a new top compression ring from a Caterpillar 1Y73 single cylinder diesel engine. Details of the engine geometry and operating conditions can be found in (2). The design of the ring is typical of a top compression ring with a symmetric, barreled convex profile and a wear resistant coating of electroplated chromium.

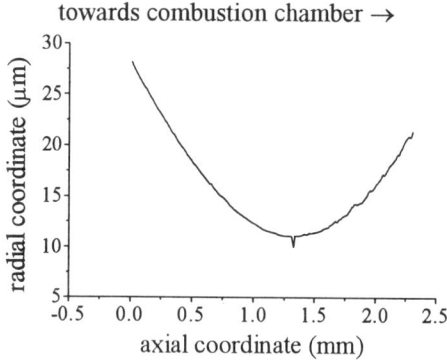

Figure 1: New piston ring face profile

Figure 2 shows the same ring measured after 120 hours running in the engine at a constant load and speed with an SAE 10W30 lubricant. The profile is characterised by a well defined worn region, with much less curvature than the new profile, and a profile peak displaced towards the crankcase flank of the ring. Theoretical performance predictions (1) indicate that the new ring operates for 41% of the engine cycle in the mixed or boundary lubrication regimes. With the worn ring the situation is much improved, with mixed or boundary lubrication for only 7% of the cycle.

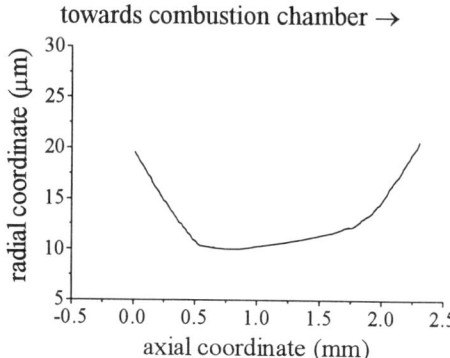

Figure 2: Piston ring face profile after 120 hours running in a diesel engine

This paper describes a numerical model that predicts the dynamics, lubrication and wear of piston rings interactively. The motion of the piston ring within its groove in the piston is characterised by the torsional twisting of the ring section about its centre of mass and the axial movement of the ring within the groove, often referred to as ring lift. The lubrication problem is complicated as piston rings in moving through a single stroke in an engine may experience boundary, mixed and full fluid film lubrication. Detailed account is therefore taken of the effect of the surface topography on hydrodynamic pressure generation and of any contact between the surfaces. The model predicts the inter-ring gas pressures, twist angle, ring lift, oil film thickness, friction and oil transport for each ring of the piston ring pack throughout the engine cycle. Wear of the piston ring is computed from the predicted dynamics and lubrication performance plus specific wear rates determined from bench test rig experiments.

Examples of the predictions of the model are presented alongside experimental measurements for the same operating conditions. Detailed validation of the model will be dealt with in further publications.

REFERENCES

(1) Priest M., Ph.D. thesis, University of Leeds, UK, October, 1996
(2) Taylor R.I., Brown M.A., Thompson D.M. and Bell J.C., SAE Paper 941981, 1994

EXPERIMENTAL STUDY BY MEANS OF THIN LAYER ACTIVATION OF THE HUMIDITY INFLUENCE ON THE FRETTING WEAR OF STEEL SURFACES

P. DE BAETS, F. VAN DE VELDE
Department of Mechanical and Thermal Engineering, University of Gent, Sint-Pietersnieuwstraat 41, B9000 Gent, Belgium
K. STRIJCKMANS
Institute of Nuclear Sciences, University of Gent, Proeftuinstraat 86, B9000 Gent, Belgium
G. KALACSKA
Department of Mechanical Engineering Technology, University of Gödöllö, Pater K. u. 1, H2103 Gödöllö, Hungary

ABSTRACT

Fretting wear occurs when two normally loaded surfaces suffer oscillatory relative motion with small amplitude. The corresponding microscopic wear volumes are very difficult to measure with classical methods. For this reason Thin Layer Activation (TLA) has been developed for accurate and sensitive wear and material transfer measurements (1).

In the present research fretting experiments in air with different humidity levels are conducted between a bearing steel ball and flat steel specimen vibrating against each other under gross-slip regime (slip amplitude 35 µm). The wear is continuously measured by means of the Normal Displacement technique and at the end of the experiment by means of the Spherical Cap Modelling (SCM) and Thin Layer Activation (TLA).

From friction results it is concluded that during the very first cycles of the fretting process the initial oxide layer is scraped off from the surface and metal-metal contact occurs. The adhesion accompanying this metal-metal contact is not influenced by the ambient humidity. After about 500 to 2000 cycles (independently of the humidity content) a drop in the friction coefficient can be seen, related to the protective action of the freshly formed oxide layers and the increase of the real contact surface. At this stage of the fretting process the influence of the humidity is mainly the boundary lubrication of the contacting surfaces, rather than oxidation inhibition.

It is found that during the first 5000 fretting cycles the normal displacement rapidly becomes negative due to oxidation of the rubbing surfaces and formation of wear particles staying trapped in the contact surface. The larger 'negative wear' at lower humidity level indicates greater wear particle formation.

Figure 1 shows the wear and material transfer as a function of different ambient humidities after 33 000 fretting cycles. The SCM-results are consistent with literature reports (3)(4) explaining that at high humidity the water adsorption prevents oxygen from reacting with the steel surface producing a passivating oxide layer. But, it can be seen that above R.H. 60 % a serious discrepancy between TLA and SCM-results exists, indicating that at high humidity the wear scar area increases, in contrary to the wear scar volume. This means that the above explanation does not hold and that a high humidity is beneficial for fretting wear. The decreasing possibility of forming protective oxide layers at high humidity is thus overruled by some mechanism leading to a decrease of the fretting wear with increasing humidity. The material transfer form the flat onto the spherical specimen decreases with increasing humidity suggesting that a high ambient humidity prevents adhesion between the contacting surfaces and consequently inhibits adhesive wear. This conclusion is consistent with the lower friction under high humidity.

After 100 000 fretting cycles a wear curve similar to figure 1 is found. As from previous research it was already found that at this number of cycles abrasion is the predominant wear mechanism, it seems that a high humidity also protects the contact surfaces against abrasive wear.

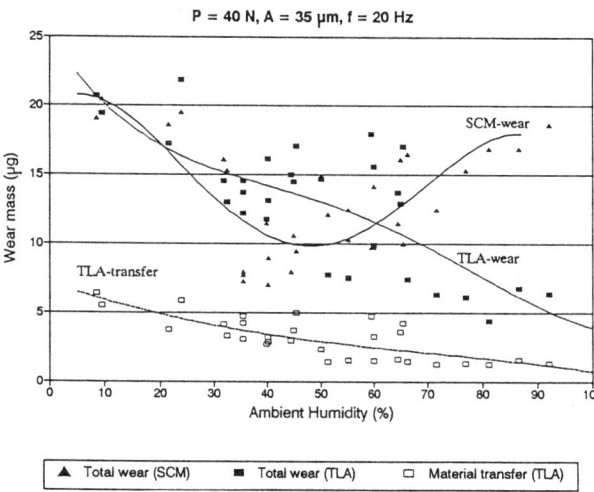

Fig 1. Fretting wear results as a function of humidity

REFERENCES

(1) De Baets P. Strijckmans K., Europ. Journ. Mech. Eng., Vol. 41, 2 (1996), 67
(2) Wright K.H.R, Proc. Instn. Mech. Engrs. (1952), Part 1B, 556
(3) Godfrey D., Bailey J.M., Lub. Eng. (1954), June, 155
(4) Goto H., Buckley D.H., Trib. Int. (1985), Vol. 18, 237

SIMULTANEOUS MEASUREMENT OF A PISTON OIL FILM THICKNESS AND A TEMPERATURE USING A DUAL FLUORESCENCE METHOD

SHUZOU SANDA, HIDETO INAGAKI and TAKASHI NODA
Toyota Central R & D Labs. Inc., Nagakute, Aichi, 480-11, JAPAN

ABSTRACT

We have been developed a method which enables to measure a thickness and a temperature of a lubricant oil film simultaneously with high response by extending a laser induced fluorescence (LIF) technique(1)(2). The method was applied to evaluate an instantaneous thickness and a temperature of an oil film between a piston and a cylinder bore under a real operating condition of an internal combustion engine.

Measurement System

The effect was made use of that fluorescence intensities of dyes change with a temperature (3). Fluorescence spectrum induced by He-Cd laser (442nm) in a lubricating oil containing two dyes, Coumarin 6 (C6) and Rhodamine B (RB) has two peaks in wavelengths of around 530nm and 580nm, respectively. The peak intensity of RB, $I2(580nm)$ was found to decrease in a higher rate than that of C6, $I1(530nm)$, with a temperature rise, as shown in Fig.1(a). An oil film temperature can be determined uniquely from the ratio $I2/I1$, Fig.1(b). An oil film thickness, which is in propotion to an absolute intensity $I1$ under an isothermal condition, can also be determined by using a calibration curve including a correction of a temperature effect, Fig.1(c).

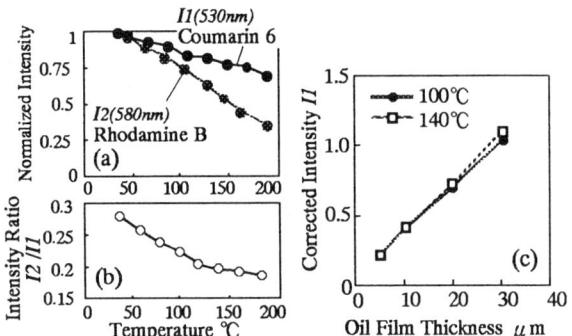

Fig. 1 Calibration Curves
(Coumarin 6:0.25g/ℓ , Rhodamine B:2 g/ℓ)

An LIF system was set up as shown in Fig.2, in which induced fluorescence intensities in wavelengths of 530 ± 10nm (C6--Ch.1) and 580 ± 10nm(RB--Ch.2) are measured independently by using two dichroic mirrors and two PMTs.

An oil film thickness and a temperature were evaluated from outputs Ch.1 and Ch.2 by using relations obtained formerly by a precise calibration. A capability of temperature measurement was confirmed by the calibration in the range from room temperature to 180°C within an error of ± 10°C.

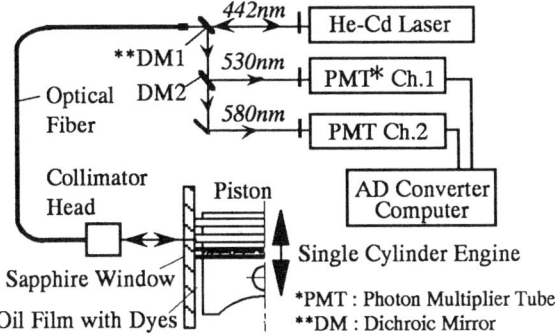

Fig. 2 Measurement System

Results on the Piston Oil Film

The system was applied to the measurement on a piston oil film under real operating conditions. The oil film temperature on the piston top land region was evaluated higher under a high speed firing condition than under a motoring condition, as shown in Fig.3, which is reasonable.

A theoretical analysis of mixed lubrication was also carried out considering viscosity change with temperature. The measured oil film thickness on a first compression ring was in better agreement with the prediction using the measured temperature than with the isothermal analysis.

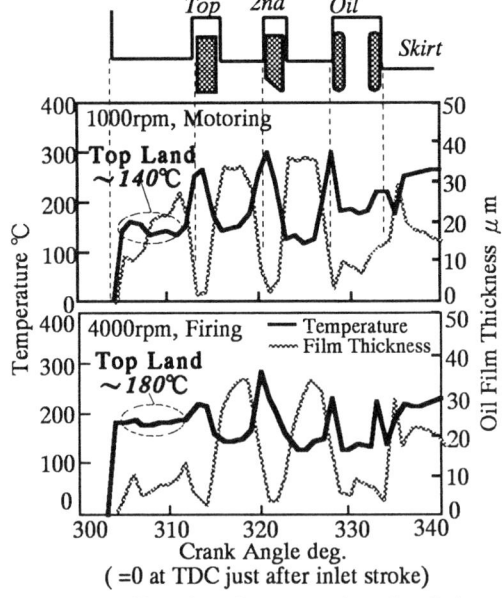

Fig. 3 Results (Compression Stroke)

REFERENCES

(1) D.P.Hoult et.al., SAE Paper No.881587, 1988
(2) S.Sanda et al., Proc.IMechE C465/014, p155, 1993
(3) A.Azetsu et al., Proc.JSME, 940-30,p518, 1994 *(In Jap.)*

THE APPLICATION OF RAMAN MICROSCOPY TO HIGH TEMPERATURE FRETTING WEAR

M MARSH, I R McCOLL and B NOBLE
Materials Engineering and Materials Design, University of Nottingham, University Park, Nottingham, NG7 2RD, UK
J SKINNER
Nuclear Electric plc, Barnett Way, Barnwood, Gloucester, GL4 7RS, UK
S WEBSTER
Department of Physics and Astronomy, University of Leeds, Woodhouse Lane, Leeds LS2 9JT, UK

ABSTRACT

The fretting wear behaviour of metals at elevated temperatures involves interactions between wear and corrosion processes which can result in the transfer of debris across the fretting interface and the formation of 'oxide glaze pads'. These pads can adhere tenaciously to one or both contacting surfaces and markedly change the fretting wear behaviour. Under some circumstances, debris transfer may occur preferentially in one direction across the fretting interface, resulting in the formation of mounds and complementary pits.

The fretting scars formed on metals at high temperatures are normally small and generally exhibit a number of distinct regions, which makes identification of the debris/corrosion products difficult, without resorting to transmission electron microscopy. However, Raman microscopy enables oxide areas as small as 2 μm in diameter to be characterised, and its use in the study of oxidation products formed on metals at high temperatures has been reported (1,2).

In this paper we report on the fretting wear behaviour of three iron-chromium-nickel alloys and the use of Raman microscopy to identify the fretting wear-corrosion products. Tests were carried out on: 9Cr iron (9% Cr, 1% Mo, 90% Fe), Inconel 600 (16% Cr, 76% Ni, 8% Fe) and 310 stainless steel (25% Cr, 20% Ni, 55% Fe). A 'crossed cylinder on cylinder' specimen arrangement was used. Both 'like-on-like' and 'mixed-metal' specimen pairs were investigated. Tests were carried out in a mildly oxidising atmosphere, CO_2 at a pressure of 3.5 Bar, at a mean specimen temperature of 450°C, under a normal force of 10 N. A fretting stroke of 200 μm, applied at a frequency of 66 Hz, and a fretting distance of 7 km were normally used. Raman measurements were undertaken on a Renishaw Raman microscope.

Figs. 1 and 2 show example Raman spectra from 9Cr iron and Inconel 600 specimens, after a 'mixed-metal' test. There are marked differences between the spectra obtained from the worn (glazed) and unworn areas on each specimen, showing a clear interaction between the fretting wear and corrosion processes. The spectrum from the glazed area on the 9Cr iron specimen (Fig. 1) indicates predominantly the presence of Fe_3O_4 (≈ 660 cm^{-1}), whereas that from the unworn surface suggests Fe_2O_3 (≈ 200 to 600 and ≈ 1300 cm^{-1}).

The spectrum from the glazed area on the Inconel 600 specimen (Fig. 2) similarly reveals Fe_3O_4, whereas that from the unworn surface suggests NiO (≈ 530 cm^{-1}) as well as Fe_3O_4 and traces of other oxides.

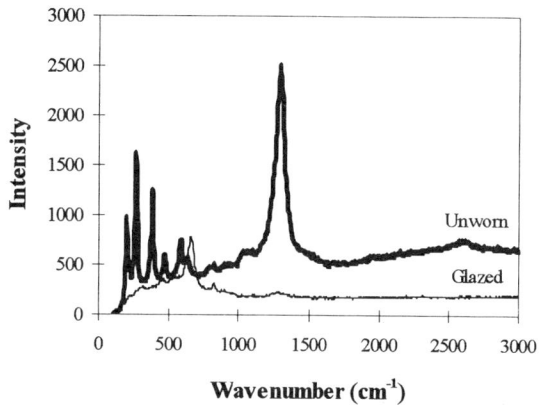

Fig. 1: Raman spectrum from a 9Cr iron specimen.

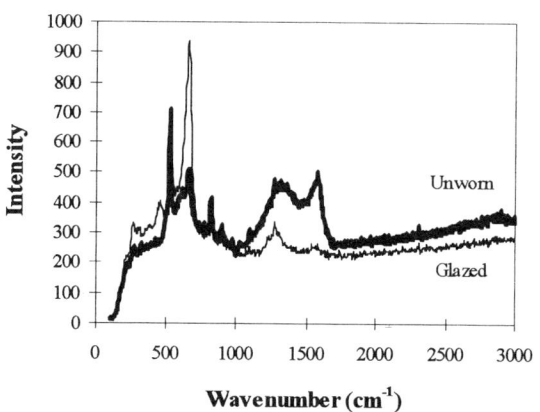

Fig. 2: Raman spectrum from an Inconel 600 specimen.

REFERENCES

(1) D J Gardiner, C J Littleton, K M Thomas and K N Strafford, *Oxid. Met.* **27** 1-2 (1987) 57-72.

(2) D J Gardiner, *Proceedings of the 2nd International Conference on the Microscopy of Oxidation, March 1993, Cambridge, England*, The Institute of Materials (1993) 36-43.

SIMULATION OF THE POLISHING WEAR OF DIESEL SOOTED OIL WITH AN ELASTOHYDRODYNAMIC TRIBOMETER

E. VARENNE, D. MAZUYER, J.M. GEORGES,
Ecole Centrale de Lyon, Laboratoire de Tribologie et Dynamique des Systèmes UMR CNRS 5513, B.P. 163, F-69131 Ecully Cedex, France,
B. CONSTANS
Centre de Recherche Elf Solaize, Chemin du Canal, B.P. 22, F-69360 Saint Symphorien d'Ozon, France

ABSTRACT

Among the hypothesis that could explain the role of soots in the wear of diesel engines, their aggregation in the inlet of the contact creating an oil starvation is often advanced (1-2). Then, the knowledge of the interparticle interactions (steric, electrical, van der Waals forces) whose equilibrium governs the stability of the particles in the lubricant is very important to understand the wear mechanisms. Nevertheless, the proposed wear processes in the literature (3-4) do not completely explain the wear behaviour of different lubricants and the role played by the metallurgy of the solid surfaces used in the diesel automotive engines. Besides the particles interactions, the dwell times of the contact point and the cinematic lengths of sliding the surfaces must be considered (5).

To simulate the tribological and the cinematic conditions of the cam/tappet contact, a tribometer realising an elastohydrodynamic lubricated contact between a ball and a disc has been developed. The ball and the plane can be moved independently and simultaneously from 500 µm/s to 5m/s. The contact can work by controlling the ratio rolling/sliding. The disc is made of glass coated with a 50 nm chromium layer and makes in contact with the steel ball an optical resonant cavity that allows us to precisely measure the oil thickness (6). The dynamic behaviour of the soots in a high pressure contact is studied according to the cinematic conditions applied to the contact. These experiments show that when the conditions of speed in the EHD tribometer are close to the real kinematics of the cam/tappet contact (see Fig. 1), the chromium coating is worn with an used oil in a mild wear process due to the polishing of surfaces by the soots while the same new formulation let the metallic surface unworn.

It has been proved that a new lubricant with the same viscosity as an used oil causing an important wear of the cam in a diesel engine does not wear of the chromium coating which shows that the wear is not related to the viscous properties of the oil. Finally, the mean wear rate deduced from this simulation has the same order of magnitude as the wear rate measured after a standard engine test and is equal to 0.005 nm/cycle. This low value indicates that the wear process of the cam/tappet contact is statistic and is governed by the probability of presence of soot particles in the contact.

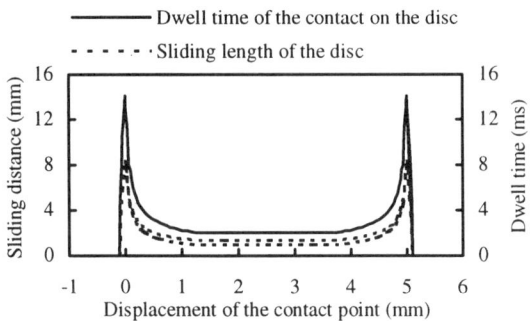

Fig. 1 : Sliding length and dwell timeof the contact point of the disc in the EHL tribometer

From these results, we proposed a model of abrasive wear by the soots partcles that takes into account their real displacement relative to both the sliding surfaces. These displacements strongly depend onto the interactions between the particles and the boundary films adsorbed on the solid surfaces. Four main parameters govern the wear process by soots particles :

- The real kinematic of the cam/tappet contact related to the conception of the Diesel engine,
- The interactions between the soots ans the surfaces depending on the formulation of the oil and the metallurgy of the cam s and the tappets,
- The volume fraction of the soots in the oil that varies according to the interparticles forces,
- The wear rate due to one partcicle or to an aggregate of particles which also depends on the physico-chemisrty of the surfaces and the oil.

REFERENCES

(1) K. Yoshida, T. Sakurai, Journal of JSLE, International Edition n°10, pp. 133-138, 1989,
(2) Ph. Colacicco, D. Mazuyer, STLE Transactions, 38-4, pp. 959-965, 1995,
(3) R.G. Rounds, SAE Technical Paper Series n°770829, 1977,
(4) K. Narita, K. Kakugawa, T. Miyaji, International Tribology Conference, Yokohama, Japan, 1995,
(5) J.C. Bell, T.A. Colgan, Tribology International, 2, 24, pp. 77-84, 1991,
(6) C.A. Foord, W.C. Hammann, A. Cameron, ASLE Transactions, 11, pp. 31-43, 1968.

STUDY OF THE TRIBOLOGICALLY TRANSFORMED STRUCTURE CREATED DURING FRETTING TESTS

E. SAUGER and L. VINCENT
Laboratoire IFoS, UMR 5621, Ecole Centrale de Lyon, 69131 Ecully cedex, FRANCE
L. PONSONNET and J.M. MARTIN
Laboratoire LTDS, URA CNRS 855, Ecole Centrale de Lyon, 69131 Ecully cedex, FRANCE

INTRODUCTION

Fretting tests carried out on various materials (iron, nickel, aluminium and titanium alloys) showed that during the first cycles debris particles were always detached from a new structure formed within the first few hundred cycles inside the bulk material. This structure is called the Tribologically Transformed Structure (TTS) (1). Understanding both the nature and formation of this structure is the aim of this study.

ABSTRACT

Different techniques were used in this study :

- Optical micrographs revealed several distinct areas in each case. Hardly detected before etching, one of these areas was between 20 and 100 µm thick, and not identified as compacted debris : it was indeed the TTS.

- Microhardness measurements on various materials showed that the TTS was much harder (about 1100 $Hv_{0.025}$) and appeared brittle. Its brittleness proved to lead to wear particle formation.

- SEM analysis detected neither traces of transfered material from the counterface nor oxides: TTS composition thus seemed identical to the one of the bulk material. Thence TTS does not originate from a phenomenon close to mechanical alloying (MA).

- TEM has been performed on Ti6Al4V ($\alpha+\beta$ structure). Detailed crystallographical analysis showed very fine grained microstructures with no preferential orientation of the α-phase. Energy Dispersive X-ray analysis carried out on thin foils confirmed SEM analysis. Key-issue arised from the localization of all the vanadium which could not be contained in the α-structure of the TTS according to equilibrium diagrams. Moreover, the TTS of iron alloys (called white phase) is made of ferrite with the same chemical composition as the bulk material, i.e. with in some cases apparently over 1 % Carbon in the BCC α-structure. Taking these results into account, Blanchard modelled TTS formation. According to him, large plastic deformations are responsible for the nucleation of α-nuclei with segregation of the vanadium at their boundaries. Nuclei are developed until sufficient vanadium is contained in the grain boundaries to stop their growth. In this case, alloying elements ensure a finer and a more stabilized structure. This model may explain the extreme brittelness of the pure titanium TTS which could be formed by coarse grains of α-phase in the absence of alloying elements. On the other hand, Rigney and al (2) assumed that oxygen from the atmosphere plays the main role in the process of TTS formation. But TEM observations made on TTS of pure titanium showed that grain size was in the same range as the Ti6Al4V. Brittleness of pure titanium TTS was hence not due to coarse grain size.

In order to detect oxygen and localize vanadium, Electron Energy-Loss Spectroscopy (EELS) was performed because of its very interesting spatial resolution and its great sensitivity to oxygen. EDX analysis was indeed not able neither to localize precisely the vanadium, nor to quantify oxygen. EELS spectra were recorded on the TTS, on one debris zone, on the initial α and β grains and on the crystalline and amorphous TiO_2. The latter were used as reference materials for this study. Oxygen was first detected in TTS but in much lower concentrations than in debris zone whose spectra were very close to the ones of amorphous TiO_2. Shifts in the Energy Loss Near Edge Structures (ELNES) were next observed growing from titanium to TiO_2 through TTS and debris zone. No significative shift was detected in titanium ELNES, except for the wear particles which are closer to the ones of titanium oxide. Identical remarks were made on the vanadium.

Oxygen detected in the TTS probably came either from the initial chemical composition or natural diffusion as its structure was far from the equilibrium one.

Work under completion now focuses on titanium, oxygen and vanadium mappings. To achieve this goal, a new experimental technique, called "Imaging-Spectrum" (IS) is being performed on the debris and TTS. IS consists in acquiring energy-filtered images. Mappings of titanium and oxygen were recorded Work is now in progress as to the vanadium location. No conclusion could be made at present because of the small amounts of vanadium and of the thin foil thicknesses.

CONCLUSION

Because of differences between TTS and debris, EELS study clearly shows that TTS does not originate from a phenomenon close to MA. Their origin probably comes from large plastic deformation occured in the bulk material during fretting tests.

Thanks to EELS analysis, we detected oxygen in the TTS of Ti6Al4V but no evidence was given on its part played in TTS formation.

Further work is needed on the alloying element location to explain structure and characteristics (great hardness, brittleness...) which lead to TTS degradation.

REFERENCES

(1) P. BLANCHARD, PhD Thesis, 1991, Ecole Centrale de Lyon, Ecully, France.
(2) D. RIGNEY, Wear, No. 100, 195-219 pp, 1984.

STUDY ON THE CRITICAL AMPLITUDE OF RELATIVE DISPLACEMENT FOR THE FRETTING WEAR OF ENGINEERING METALLIC MATERIALS IN AIR

HOZUMI GOTO and SHUNJI OMORI

Fukuoka Institute of Technology, 3-30-1, Wajirohigashi, Higashi-ku, Fukuoka, 811-02, JAPAN

ABSTRACT

Fretting wear tests for a carbon steel, an aluminum alloy, a brass, and stainless steels were carried out in air at various levels of relative displacement S and contact load P. Mean wear rates after a given number of fretting cycles were measured. The existence of critical relative displacement S_{cw} for a rapid increase in wear was confirmed for all of the metals tested. Therefore, the correlation between S_{cw} and mechanical properties of the metals such as fatigue strength was investigated while evaluating the results by the present authors and other researchers.

A driving unit of fatigue testing machine was used for an oscillation source of fretting motion. Flat-on-cylinder type contact i.e. line contact was employed as a contact mode. A repeated reciprocating circular motion of the cylindrical specimen by the driving unit caused small relative displacement or slip between the cylindrical and flat specimens in contact, resulting in the occurrence of fretting damage between the contact surfaces.

Same metal combinations were employed for both the specimens. The frequency was 20 Hz and the load P was kept at 67, 122, and 171 N. The peak-to-peak amplitude of S was changed in the range of 0 to 450 μm. The specimens were fretted during 5×10^5 cycles. The temperature and relative humidity of the environment were kept at 30 ± 3 ℃ and 60 ± 5 %.

The wear rates of the metals as a function of relative displacement S show the following facts. Only surface irregularities due to plastic deformation are observed in a range of very small relative displacement. As S is increased, the minimum wear becomes detectable with the present measurement method. A fairly large amount of wear is measured above a critical relative displacement S_{cw}. The mean wear rate drastically increases linearly with the relative displacement at $S \geqq S_{cw}$. The straight line can be determined with a method of least squares by using experimental points of the portion where wear is linearly increased. A point of intersection between the horizontal axis and the straight line can be determined as the critical relative displacement S_{cw} with an extrapolation method. The S_{cw} value is equal to 35 for the steel, 65 μm for the 18Cr-8Ni stainless steel, 75 μm for the 12Cr stainless steel, 90 μm for the aluminum alloy, and 70 μm for the 60/40 brass at a load of 67 N.

Asperities in contact on rubbing surfaces are subjected to repeated stresses by fretting motion. When S is very small, the asperities are subjected to repeated stresses lower than fatigue strength in the stick region where the contact surfaces undergo no microslip. Since it is supposed that the asperities do not cause fatigue fracture, wear particles will not be generated. When S is increased, the asperities in a microslip portion are subjected to repeated stresses slightly higher than fatigue strength in the mixed region of stick and microslip on the contact surfaces. Hence, very small quantities of wear particles are generated. When S reaches S_{cw}, the whole area of the contact surfaces is in macroslip. All asperities in contact are subjected to repeated stresses higher than fatigue strength. High and low cycle fatigue fracture and shear fracture accompanied by adhesion generates a lot of wear particles. Thus, it can be demonstrated that the wear rate drastically increases above S_{cw}.

A close relationship should be present between fatigue strength of the metals and S_{cw}. Therefore, from the fretting data of the metals including copper by the present authors and other researchers, the relationship between S_{cw}/P and repeated strain σ_w/E is obtained as shown in Fig. 1, where E is the modulus of elasticity and σ_w the fatigue strength at 10^7 cycles. A strong correlation is found between S_{cw}/P and σ_w/E. Since the coefficient of correlation for a straight line is about 95 %, the correlativity is concluded to be very high. Thus, the generation mechanism of wear particles by surface fatigue contributes greatly to fretting wear.

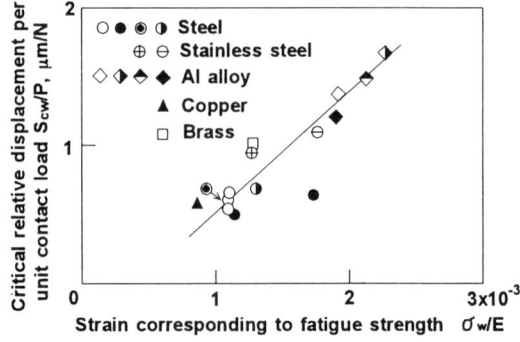

Fig. 1: Relationship between S_{cw}/P and σ_w/E

TUESDAY 9 SEPTEMBER

SESSION		Page number
TU1	Thick film bearings – II	91
	EHL and boundary lubrication – I	99
TU2	Lubricant chemistry and rheology – II	109
TU3	Wear in metallic systems	127
TU4	Surface engineering – II	143
TU5	Magnetic storage systems	161

MOBILITY ANALYSIS FOR ENGINE BEARINGS: ADAPTATION TO INCORPORATE SHEAR THINNING LUBRICANTS

D Han and C M Taylor

Department of Mechanical Engineering, The University, Leeds LS2 9JT, UK.

ABSTRACT

The Mobility Method for analysis of dynamically loaded bearings was introduced about thirty years ago (1). Since that time it has remained the foremost design technique for the study of the performance of automotive engine bearings. It has a remarkable robustness and has mainly been used for the prediction of the cyclic minimum film thickness in main, big end and gudgeon pin bearings. it has been extended to enable consideration of the maximum pressure and lubricant flow rate predictions. Application of the design technique is made in the hope that the simple model will reflect sufficient reality such that the trends in variations with parametric changes will be correct. As with the analyses of all the major friction devices in the internal combustion engine, the predicted magnitudes can only be benchmark and not absolute. This is reflected by the range of assumptions normally inherent in its use, that is: the use of the short bearing analysis (but not necessarily); circumferential symmetry; no consideration of oil film history; no consideration of elastic and thermal distortions; a well aligned journal and bush; Newtonian lubricant behaviour; the neglect of shaft intertia; and isothermal analysis. Various authors have addressed specific limitations but inevitably such considerations complicate the analysis and hinder the effectiveness of the design approach offered by the mobility method.

In the present paper the non-Newtonian lubricant behaviour is addressed. Such characteristics are indicated in Figure 1 according to various models, the Cross model reflecting reality most closely.

Analytical studies of shear thinning will be reviewed and simple and more detailed models for investigating the influence of shear thinning associated with the high molecular weight polymer additives in engine lubricants will be detailed. A method of using the existing mobility method of analysis but incorporating mobility vectors reflecting shear thinning behaviour will be presented. The development retains the simplicity and robustness of the existing mobility method of analysis and will enable current computer analyses to be modified with extreme simplicity in order to investigate shear thinning effects.

Figure 2 shows predicted results for minimum film thickness and maximum film pressure for the big end of a modern four cylinder sutomotive engine.

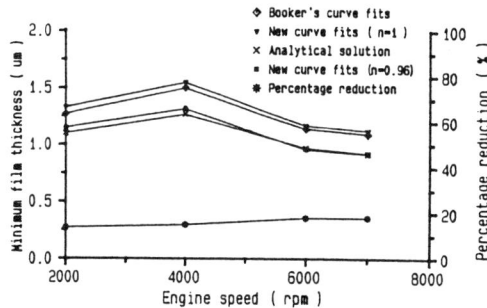

Fig. 2(a) Minimum film thickness of a big-end bearing at different speeds obtained by different solutions.

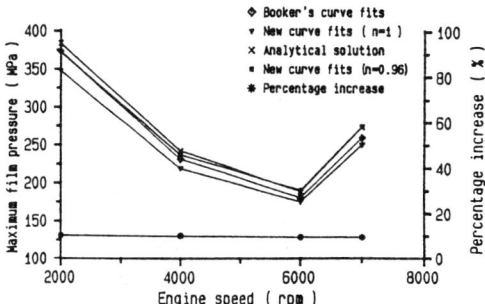

Fig. 2(b) Maximum film pressure of a big-end bearing at different speeds obtained by different solutions

REFERENCE

(1) J F Booker, Dynamically Loaded Journal Bearings: Mobility Method of Solution, Trans. ASME, Jrl. Basic Engineering, Sept. 1965, pp 537-546.

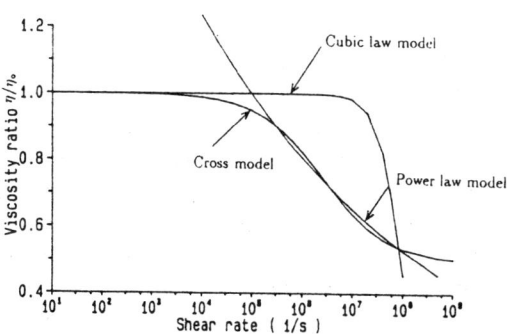

Fig. 1: Viscosity variation with shear rate predicted by different shear thinning lubricant models.

Elastic influences on the hydrodynamic of the connecting rod big end bearing

Dr. THEO MOSSMANN and Prof. Dr. RUDOLF HALLER
Institute of Machine Design and Automotive Engineering, University of Karlsruhe,
Kaiserstr. 12, D-76128 Karlsruhe, Germany

Abstract:

With increasing engine power and reduced stiffness of the bearing bush, it becomes evident to include the elastic deformations into the calculations. A finite element program was developed for steady and transient loaded journal bearings with taking the deformation of the bearing housing into account.

The independent pressure built-up in the lubrication gap filled with oil, guarantees the separation of the two sliding partners, journal and bearing bush. The mathematical-technical description of the hydrodynamic load process in the bearing is based on the Reynolds differential equation. As the relative film thickness H is expanded to an elastic part H_V the enlarged examination of the hydrodynamic process leads to a coupled system of differential equations.

For the study a 4-stroke otto-cycle-engine with an engine capacity of 1,3 l and an engine performance of 40 KW has been selected. The results refer to a number of revolutions n = 4500 1/min. The big end connecting rod bearing experiences a transient load during the working cycle due to the operational processes of the piston engine. The course of the load, as well as the amount the direction, can be well illustrated in a polar diagram. For the later considerations the lower part is of uttermost interest; in this case the force is directed to the gap of the connecting rod.

In a first step the housing of the connecting rod big end was modelled as a simple cylinder. The main dimensions are chosen in agreement to the contour of the connecting rod. The only change we made, was the reduction of the boundary restrictions on the outer border. The first model A has only local elasticity as all nodes on the outer diameter are fixed. The sequence of the following FE models, beginning with model B and ending with model F, can be characterised by the kind of boundary condition: with running letters the area of fixed nodes on the border is reduced, beginning with B (180°) over C (150°), D (120°), E (90°) up to F (64°).

Figure 1 shows the pressure curve p in the lubrication gap over the circumferential direction φ. The pressure curves for the different model A - F correlate to the same point of time α = 0°. α is the actual angle of the crank shaft and is used as a reference mark of time. In this case intaking of air in the cylinder is starting and the bearing load is nearly vertically directed to the bottom of the connecting rod. The pressure curve A has still a typical hydrodynamic profile: a plain ascent and a steep peak. With increasing elasticity the pressure curve changes: the amount is reduced and the curve becomes flatter like a plateau. Finally the pressure curve is proceeded to the contour of a double peak. In

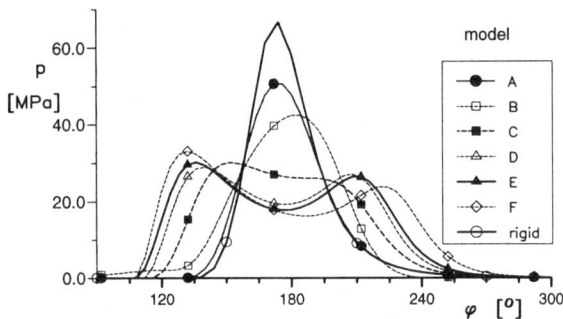

Fig.1: Pressure curve p for different calculation models comparison to the rigid case (p_{max} = 65 MPa.) the pressure maximum of model D/E/F is reduced to p_{max} = 32 MPa. These changes in the pressure curve correlate well with the elastic deformation of the bearing housing respectively with the lubrication gap influenced by this deformation.

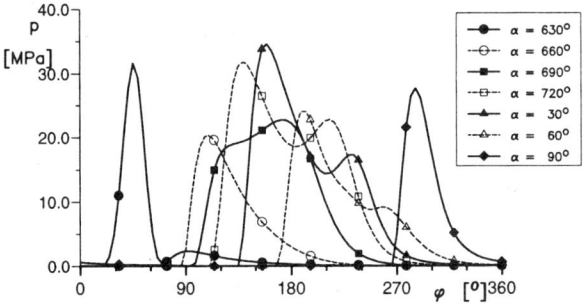

Fig. 2: Pressure curve p at different moments

In a second step a more realistic and detailed finite element model of the connecting rod big end was created. The next figure 2 shows the pressure curves in the middle plane on the bearing surface at certain distinguished, chronological moments. At the beginning (α = 630°) and at the end (α = 90°) the pressure curves show the typical hydrodynamic profile. In the next moment (α = 690°) the pressure peak changes to a plateau on a high level. And in the following point of time (α = 720°, 30°) one can find a pressure curve with a double peak, a big one and a reduced one. In comparison to the so called rigid case, the pressure maximum is reduced to a considerable amount and the pressurised area has increased.

The study shows that the maximum pressure values obtained by rigid calculation are considerably reduced by the "self-curing effect", e.g. by the deformation of the margin and the simultaneous supporting effect of adjacent bearing sections. The altered distribution of pressure in the lubrication gap is mainly effective in the region of the lower half of the bearing brass.

MEASUREMENTS OF PRESSURE IN A SQUEEZE FILM DAMPER WITH AN AIR/OIL BUBBLY MIXTURE*

SERGIO E. DIAZ and LUIS A. SAN ANDRÉS
Mechanical Engineering Department, Texas A&M University, College Station, TX 77843-3123, USA

ABSTRACT

High performance turbomachine designs demand larger power to weight ratios, with ever increasing speeds and using light-flexible rotors. These systems, however, are more susceptible to excessive synchronous vibration and rotordynamic instabilities.

Squeeze film dampers (SFDs) are virtually the only mean to provide damping to rotating machinery, and have been successfully used in industrial applications and jet engines to reduce excessive vibration amplitudes and instabilities (1,2). However, large amplitude SFD journal motions while traversing critical speeds enhance the ingestion of air and lead to operation with a bubbly mixture of lubricant and air, a condition called gaseous cavitation (1,2,3). This phenomena, though widely reported to occur, lacks firm analytical understanding (1) and controlled tests are presented here showing its effect on SFD performance.

The experiments report measurements of the dynamic pressure field in a SFD apparatus operating with a controlled bubbly mixture of oil and air. The journal describes circular centered orbits at a fixed whirl frequency and the damper is fully submerged in an oil bath. The dynamic squeeze film pressures become less reproducible for consecutive cycles of journal orbital motion and an unstable zone of null squeeze pressure generation in the dynamic pressure fields, related to gas cavitation, becomes prevalent for mixtures with a large enough content of air. In this zone of constant pressure the work done by the journal motion relates to a change in the lubricant volume rather than a change on its pressure.

Fig. 1 shows the dynamic pressure at two different axial locations along with the local film thickness for a lubricant with 10% of air in volume. This mixture composition lays in a range in which the zone of null pressure generation (between points a and b) appears only in some of the journal motion cycles.

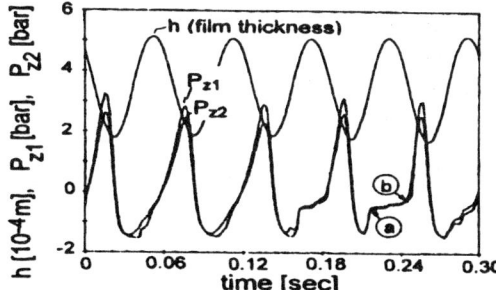

Fig. 1: Incipient gaseous cavitation regime

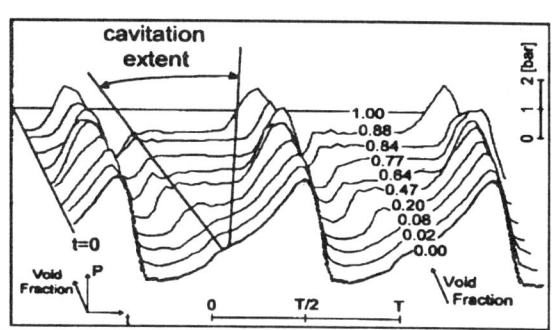

Fig. 2: Development of the gaseous cavitation zone

The transition condition between the condition in which the constant pressure zone never appears and the one in which it appears for every motion cycle of period "T" is hereby called incipient gaseous cavitation regime.

Fig. 2 shows the development of the gaseous cavitation on the pressure field with the mixture void fraction. The extension of the constant pressure zone grows with the amount of air present in the lubricant. This also results in decreasing peak-to-peak squeeze film pressures as the mixture void fraction (air/oil volume ratio) increases (see Fig. 3). The power required to drive the SFD test rig decreases with increments in the air/oil volume ratio, and thus, provides evidence of a reduction in the mixture viscosity and damping capability of the test SFD.

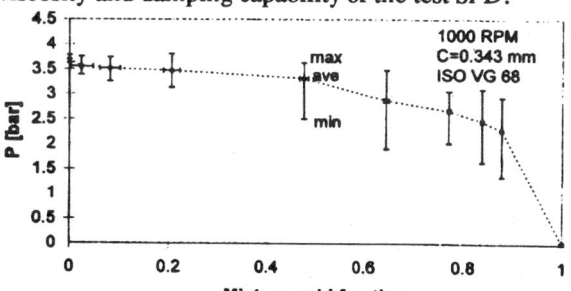

Fig. 3: Peak-to-peak dynamic pressure vs. mixture void fraction

REFERENCES

(1) Walton, J., Walowit, E., Zorzi, E., Schrand, J., ASME J. of Lub. Tech., vol. 109, pp. 290-295, (1987).
(2) Zeidan, F. Y., Vance, J. M., Rotating Machinery Dynamics, DE-vol. 18-1, ASME Conference, pp. 237-242, (1989).
(3) Zeidan, F. Y., Vance, J. M., Tribology Trans., vol. 32 1, pp. 100-104, (1989).

* Accepted for publication in the STLE Tribology Transactions.

75 YEAR REVIEW: PAST, PRESENT AND FUTURE OF WATER-LUBRICATED ELASTOMER BEARINGS

ROY L. ORNDORFF, JR.
BFGoodrich Aerospace, Birchwood Drive, Kent Ohio, 44240 USA

ABSTRACT

Water-lubricated elastomer bearings have evolved from an inauspicious start early in this century when a piece of rubber hose was used as an emergency replacement for a failed bearing in a flooding California mine. The development process can be traced through a number of sources. Progress was slow because no one has been able to extend Reynolds' equations to elastomer bearings.

TIME LINE

The engineer who started it all was Charles Sherwood. His first patent (1) dated 1922, shows multiple protrusions on the bearing surface. He later patented a practical spiral groove design. A research report by Annis (2) compared sandy-water testing results for a spiral groove and a new fluted (flat land) bearing. The spiral groove bearing had more bearing wear, caused more shaft wear, started squeaking sooner and absorbed twelve horsepower compared to four. Busse and Denton (3) in 1931 concluded that the angle of approach of the shaft to the flat land automatically changed with load similar to the tilting pads in thrust bearings and that "the classical laws of bearing lubrication do not apply ." Brazier and Holland-Bowyer (4) and Fogg and Hunwick (5) reported on progress and discussed some friction comparisons. Their results were, like those of Busse and Denton, somewhat flawed because no one then realized the effects on friction and wear of rubber hardness, thickness and surface finish. Shannikov (6) in 1939 described the first successful attempt to experimentally prove the existence of hydrodynamic lubrication in rubber bearings, contrary to the results of Busse and Denton.

Beatty and Cornell (7) reported the change from natural rubber to nitrile as the standard elastomer because of reversion (liquefied blistering), caused by combat damage to U.S. Navy ships at the Battle of Midway in June 1942. The severity had never been envisioned in lab testing and proved difficult to duplicate. They succeeded, helped by a fortuitously out-of-balance test machine.

Gusman, et al (8) wrote in 1959, the only book exclusively devoted to water-lubricated rubber bearings.

Smith and Schneider (9) were experienced U.S. Navy research engineers who reported in 1963 on rubber and other bearings from a naval users perspective. They discussed special importance of low static and running friction, minimum noise, long life, and the use of a new abrasive wear test machine as a screening tool.

Fuller (10) in 1984 discussed a number of water-lubricated bearing designs. Orndorff (11) discussed significant reductions in friction and wear achieved by reducing the rubber thickness and by molding the rubber surface extremely smooth. A pocket is formed in the surface, trapping lubricant, and is called Plastoelastohydrodynamic lubrication. This bearing is widely used in all classes of U. S. Navy ships and those of many foreign navies.

Orndorff and Finck (12) reported in 1996 on the development and testing of a slippery polymer alloy (SPA). This is the first water-lubricated material that appears to solve the total system wear problem. Also discussed was the theory and mechanism of bearing noise generation.

This review points the way to new, heavy duty water-lubricated environmentally firendly machinery applications.

REFERENCES

(1) C. Sherwood, U.S. patent 1,416,988; 1922.
(2) B. Annis; BFGoodrich Co. Akron, OH, 1925.
(3) W. Busse and W. Denton, ASME Trans., 1932.
(4) S.A. Brazier, et al, Proc.I Mech E, 1937.
(5) A. Fogg, et al, Proc. IMechE,1937.
(6) V. Shannikov, Kauchuk I Rezina, No. 12, pp. 18-22, 1939.
(7) R. Beatty, et al, RUBBER WORLD, Nov- Dec. 1949.
(8) M. Gusman, et al, State Scientific-Technical Publishing of the Petroleum and Mining Literature, Moscow, 1959.
(9) W. Smith, et al, Naval Engs. pp. 841-854, 1963.
(10) D. Fuller, "Theory and Practice of Lubrication for Engineers," John Wiley & Sons, pp. 439-469, 1984.
(11) R. Orndorff, Jr. Naval Engs. Vol. 97, No. 7, 1985.
(12) R. Orndorff, Jr., et al, The Royal Institution of Naval Architects, 1996.

Comments	Ring	Wear/year Rotor	Total
1. PTFE	23.6	2.46	26.1
2. Polyurethane	6.69	1.03	7.72
3. PEEK	4.13	1.55	5.68
4. BFG SPA	1	0	1
5. 30% Mica-Filled PTFE	8.74	2.06	10.8
6. Polyurethane	2.26	2.26	4.52

(Thickness loss, ratio to SPA)
(Stainless steel rotors)

Table: Pump Seal Abrasive Slurry Wear

THE DEVELOPMENT OF POROUS CERAMIC WATER HYDROSTATIC BEARINGS FOR ULTRA PRECISION APPLICATIONS

R J ALMOND, J CORBETT and D J STEPHENSON
School of Industrial and Manufacturing Science, Cranfield University, Cranfield, Bedford, MK43 0AL, UK

ABSTRACT

This paper presents the outcome of preliminary research aimed at improving the rotational accuracy and performance of spindle bearing assemblies. Specific areas sought for improvement include, an increase in power efficiency, higher rotational speed and higher levels of stiffness. Improvements in the performance envelope under these strategic headings are expected to yield a new generation of ultra precision machine tool spindles.

An increasing demand for higher precision in manufacturing, necessitates machine structures and spindle assemblies of much higher static and dynamic stiffness than currently achieved. Additionally, error motions in the sub micron or even nanometre range, demand very low levels of seismic and acoustic vibration and very low levels of thermal drift. These requirements are central to the provision of an economic ultra precision machining capability.

The medium term accuracy of oil hydrostatic bearings has traditionally been limited by the thermal effects caused by heat generated in viscous shear. Aerostatic bearings have improved thermal characteristics, but suffer from relatively low stiffness and load capacity due to the compressibility of air. New concepts are therefore required, and the current programme is using fundamental principles to develop a new generation of porous bearings.

The current investigation has shown that it is possible to use porous hydrostatic principles where, effectively, a multitude of integral restrictors and pockets are dispersed over the entire bearing shell. Earlier difficulties with this method regarding permeable consistency (1) have been overcome by the ceramic material choice and bearing manufacturing technique. This arrangement results in a significant increase in bearing stiffness, as a maximum pressure is maintained over, almost, the complete bearing length.

Alternative approaches are being investigated to produce the journal bearings. For example, the choice of ceramic is governed by both mechanical and physical properties, and ideally the material should provide the highest possible strength, stiffness, dimensional stability and ease of processing to ensure optimised pore morphology. Hot Isostatic Pressing (HIPing) has been employed to produce the controlled porous structure, from carefully graded ceramic powder.

The use of water as a hydrostatic fluid offers many potential advantages. In particular its superior thermal characteristics have improved the traditional thermal limitations on hydrostatic bearing performance. Various mathematical theories, largely based on the Reynolds Equations and Darcey have been evaluated in order to predict bearing performance (2). Optimum substrate permeability values for various bearing design parameters were calculated. By this method the viscous permeability target for the materials development phase of this project was set at $10^{-14} m^2$. Additionally, an estimate of substrate density and average pore size, was made by using Carmen modified Kozeny Theory (3).

The paper discusses:
a) How the manufacturing process has been derived to produce consistent ceramic hydrostatic aqua-porous (HA) bearing shells.
b) Optimisation of both processing parameters and bearing performance parameters, regarding bearing shell production.
c) The evaluation of bearing performance when compared to different published bearing theories.
d) A direct experimental comparison of bearing performance with an equivalent oil hydrostatic bearing.

This research will have potential for commercial exploitation over a wide range of ultra precision and high speed machining applications. For example, the next generation of diamond turning and grinding machines for damage free direct grinding of brittle and hard materials will require very high spindle stiffness. Manufacturing in metal, plastics and semiconductor materials will exploit the high speed machining benefits. Further, the economical and power auditing trend of modern manufacturing industry, will welcome the economy and cleanliness of this new hydrostatic system.

REFERENCES

(1) A Kumar, N S Rao, Tribology International, October 1994, Butterworth Heinemann Ltd, Vol. 27, No. 5, 299-305pp.
(2) S Biringen, J. Fluid Mech. Vol. 148, 413-442 pp, 1984.
(3) A E Scheidegger, The physics of flow through porous media, University of Toronto Press, 1993.

Submitted to *IMechE Journal of Engineering Tribology*

Fluid Whirl Phase Relationships in Rotordynamic Instability

WILLIAM D. MARSCHER
Mechanical Solutions, Inc., 9 Sylvan Way, Ste. 360, Parsippany, New Jersey 07054 USA

ABSTRACT

Rotordynamic instability in rotating machinery was first recognized in "shaft half speed whirl" and "shaft whip" problems encountered by steam turbine manufacturers during the 1920's, as they increased operating speeds. Since then, there has been significant research into the influence on this phenomenon of the whirl of the fluid surrounding the rotor. Today it is understood that, typically, rotor instability problems are the result of the cross-coupled stiffness forces induced by rotor motion within whirling fluid flow fields. Such forces can become stronger than the opposing viscous damping forces, particularly within fluid film bearings and annular seal passages, such as compressor and steam turbine labyrinth seals, and pump wear rings.

Unfortunately, predicting whether or not cross-coupling-induced rotor instability is likely to occur for a given design relies upon "black box" damped eigenvalue computer programs. Although these programs are theoretically precise, their predictions are dependent upon linearized bearing, seal, and impeller passage dynamic coefficients. These coefficients usually are based, in turn, upon either simplified bulk flow analytical predictions, 3-D computational fluid dynamic analysis without benefit of fluid/ structure interaction, or component testing using small circular orbits and with over-constrained (by test rig design) but potentially dominant fluid/ rotor interactions.

This paper proposes an alternate approach to determine 1) why most confirmed cases of rotordynamic instability occur at about half running speed; 2) whether some rotor systems are resistant to rotordynamic instability, regardless of cross-coupling; and 3) whether or not bearing and seal coefficients critical to stability can be modeled independently of the rotor orbital motion.

The proposed model's predictions are based upon the whirl force versus response (i.e. input/ output) phase angle, which is a function of the ratio of the rotor/ bearing system's natural frequency to the frequency of net fluid whirl. Sources of fluid whirl which are considered include impeller exit flow angle, leakage flow through rotor wheel side passages as well as close clearance sealing cavities and bearing films, fluid entrainment associated with rotor spin, "stirring" caused by mechanical whirl, conservation of angular momentum within radial flow fields, and Couette flow fluid shear effects. First principles models in each of these cases indicate that it is an unusual situation for swirl within turbomachinery to be at other than somewhat less than half running speed. This implies that unstable rotor whirl at frequencies other than 40 to 50 % speed is unlikely.

A basic premise of the paper is that instability is, by definition, a response that is bi-laterally coupled to the force which is creating it. The initiating force may be quite nominal, but the response which results causes an increase in the force that is greater than the entire original force, through a distinct physical feedback mechanism. In this manner force causes response, which causes force, in a ever-increasing spiral, until the boundary conditions are changed enough by the eventually large motion that the feedback reaches a maximum, or the machine fails.

In a rotor system, in order for such feedback to occur in practice, a plausible physical mechanism for the feedback must be present. A cross-coupled vector (caused for example by a skewed static pressure field induced by swirling flow) can supply that mechanism, but only under certain special circumstances. These include that 1) the rotor whirl be at the same frequency as the fluid whirl (in terms of the static pressure distribution if not the fluid particles); 2) the whirl frequency be at a rotor or support structure natural frequency; 3) the affected natural frequency must have low modal damping times rotational frequency as compared to the cross-coupled stiffness; and 4) the rotor whirl must be in forward precession and/ or the stator whirl be in backward precession. The rotor/ fluid whirl coincidence is required to provide a forced response operating mechanism that has potential for phase-lag feedback. The natural frequency/ whirl frequency coincidence is needed for a 90 degree phase shift between the cross-coupled force's instantaneous direction and the direction of the eventual (phase-lagged) response to that force, to change neutral feedback into negative feedback. The relatively weak damping versus cross-coupling is needed to allow the cross-coupled force to vectorially dominate, ensuring unstable negative rather than stabilizing positive feedback from the net vector of cross-coupling vs. damping forces. Likewise, the correct precessional sense is needed to result in negative rather than positive feedback

Currently-used damped eigenvalue or time-transient computer programs predict instability potential by estimating modal damping, as calculated from linearized bearing, seal, and impeller coefficients. Such analyses, which do not calculate the coefficients on a motion-interactive basis, may provide misleading predictions concerning the potential for rotordynamic instability. For reliable predictions, a new approach is needed that accounts for all significant mechanisms of fluid whirl, and calculates the nonlinear dependence of the rotordynamic coefficients upon the rotor motion as well as on the fluid whirl physics.

HEAT FLOW INDUCED ALTERATION OF ROTOR DYNAMIC PROPERTIES

B LARSSON
Department of Mechanical Engineering, Linköping University, Linköping, Sweden

ABSTRACT

Rotors, in particular large machines such as steam turbines, sometimes experience a behavior referred to as "spiralling" see (1) which means that they constantly change their synchronous vibrations. This can lead to problems such as difficulties to perform a proper balancing and in serious cases to a situation where it is impossible to operate. The phenomenon is associated with asymmetric temperature distribution in the shaft and feedback coupling from this to vibration. Contacts between rotating and stationary part in e. g. seals may be one explanation. Another is that vibrations in the bearing cause a skew symmetric heating of the journal and a consequent bow.

Dimarogonas (2) studied the dynamics of the thermal bow, Keogh and Morton (3) the whole closed loop, from bearing vibration to asymmetric heating on to thermal shaft bow back to bearing vibration. This is represented in figure 1. As a result of the feedback instability may occur. Unbalance sensitivity will be altered and small residual unbalance after balancing can result in a shaft bow which augments vibrations considerably. The time coefficient is long, it can take hours before a bow reaches steady state (provided that it is stable).

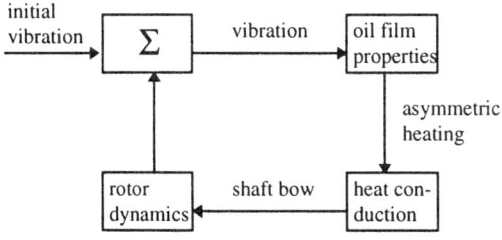

Fig. 1: Rotor dynamic couplings

Experiments (4) indicate a linear relation between vibration and asymmetric journal temperature. A theoretical solution of the three-dimensional, time transient temperature in a dynamic oil film has so far not been accomplished. Most resent work (5) deals with a special case where the journal performs a circular orbit around the bearing center. Based upon the results so far available, a linear relation between vibration and skew temperature has been assumed in the present work.

In the next step asymmetric heat is conducted in the shaft, resulting in a temperature field. In turn a volume integral of this gives a shaft bow. Here dynamic aspects must be regarded. The three dimensional transient heat conduction is quite evolved to solve (here a finite difference approximation is used), and

questions about stability and robustness to model errors leads to multivariate feedback control theory (multiple bearings are coupled through the rotor dynamics). Effort has been spent on modeling an accurate state space representation with as low model order as possible. A guide line for this is to study the Bode diagram (figure 2) of the heat conduction dynamics. Earlier work (2) has treated a similar problem and the model is here extended to catch the full solution with absolute errors (abs(correct - approximation)) under certain levels. A goal function for error minimizing has been evaluated and optimization of different models has been done. With a rational transfer function of order five (polynomial degree five in numerator and denominator in the frequency) maximum absolute error below 1 % of steady state solution, and less than 1e-5 in low frequency region has been achieved.

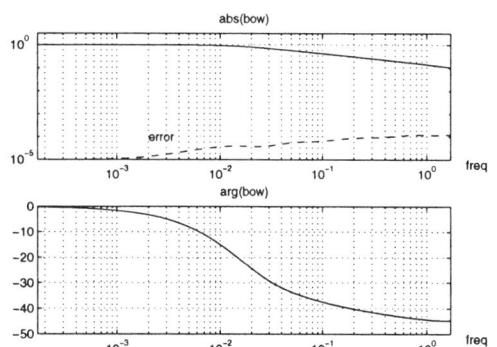

Fig. 2: Bode plot of full solution and model error

Characteristic time constants for the shaft bow are measured in minutes, i. e. much longer than the time constants for the rotor vibrations. Consequently, the steady state solution for the dynamics of the rotor can be used. Robust stability (model errors included) and time dependent unbalance response are evaluated.

REFERENCES

(1) W. Kellenberger, Journal of Mechanical Design, Vol. 102, 1980, pp. 177-184.
(2) A. D. Dimarogonas, Dissertation, Rensselaer Polytechnic Institute, Troy, New York, 1970.
(3) P. S. Keogh, P. G. Morton, Proceedings of the Royal Society of London, series A, Vol. 445, 1994, pp. 273-290.
(4) B. Hesseborn, Measurements of temperature unsymmetries in bearing journal due to vibration, Internal report ABB Stal, 1978.
(5) P. G. Tucker, P. S. Keogh, ASME Journal of Tribology, Vol. 118, 1996, pp. 356-363.

EHD FILM THICKNESS IN NON-STEADY STATE CONTACTS

JOICHI SUGIMURA
Department of Energy and Mechanical Engineering, Kyushu University, Fukuoka 812-81, Japan
WILLIAM R JONES JR
NASA Lewis Research Center, Cleveland, OH 44135, USA
HUGH A SPIKES
Tribology Section, Imperial College, Exhibition Road, London SW7 2BX, UK

ABSTRACT

This paper describes a study of EHD film thickness in non-steady state contact conditions. A modification of ultrathin film interferometry is employed which is able to measure accurately both central film thickness and film thickness profiles 50 times a second (1). Film thickness with two perfluoropolyethers (PFPE) and two mineral base oils are investigated in a number of different types of non-steady state motion, including acceleration/deceleration, stop/start and reciprocation. The results demonstrate a range of transient behaviors of EHD film where thicknesses deviate from those found in steady state conditions.

Contacts undergoing acceleration or deceleration form film thicknesses which differ from those produced at constant velocity. Accelerating films are thinner and decelerating contacts thicker than expected, as shown in Fig. 1. The measured film thicknesses normalized by those at constant velocity are plotted against acceleration in Fig. 2, which demonstrates that the deviation from steady state thickness is proportional to the rate of acceleration.

In unidirectional stop/start motion, fluid entrapments form in the contact upon halting. The subsequent leakage of oil from the contact is fluid specific. For the mineral oil studied, the minimum film thickness at the perimeter of the contact fell within one tenth of a second to a very low value whilst a much thicker minimum film thickness was maintained for PFPEs, as shown in Fig. 3.

Reciprocating, nominally constant speed motion gave film behavior similar to unidirectional stop/start except that the minimum film thickness did not have time to fully collapse during reversal and the build up of film thickness after reversal was slower than in stop/start, possibly because of starvation.

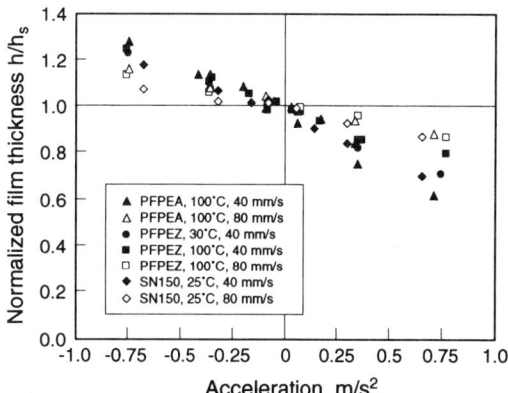

Fig. 2 Dependence of normalized film thickness h/h_s on acceleration, h_s being film thickness at constant velocity

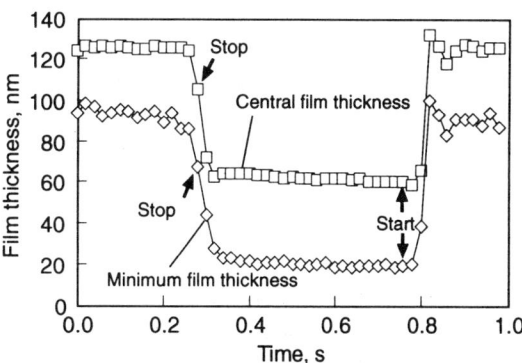

Fig. 3 Minimum and central film thickness variation during a stop/start cycle; PFPEZ, 30°C, 1Hz, 90mm/s

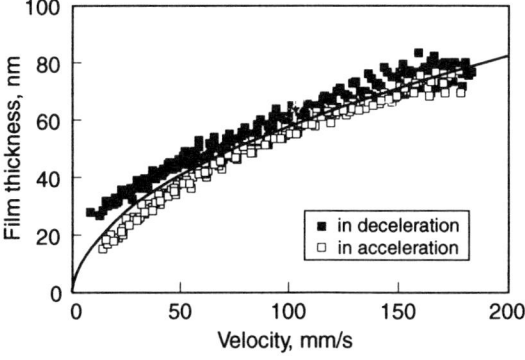

Fig. 1 Changes in central film thickness with speed in an acceleration/deceleration test at 1 Hz; PFPEZ, 100°C; solid line shows steady state film thickness

REFERENCE

(1) J. Sugimura and H. A. Spikes, presented at the 23rd Leeds-Lyon Symposium on Triblology, Leeds, 1996, to be published in the Proceedings, Elsevier.

Submitted to *ASME Journal of Tribology*

A COMPREHENSIVE EHL SOLUTION APPLIED TO SPUR GEAR DYNAMICS

J. M. WANG

Analogoy Inc - A TDK Group Company, Portland, OR 97223, USA

ABSTRACT

It has been recognized that, for high speed gears, the predominant mode of failure is critically influenced by film thickness and contact temperature between gear teeth. In most gear drives, the actual number of teeth engaged at one time varies. Hence, the number of teeth supporting the total load varies as gears travel through the engagement cycle. As a result, the thickness of film thickness changes along the line of action, even though the external applied load may be constant.

In searching for a better approximation, Adkins and Redzimovsky (1) used actual gear tooth profiles and tried to solve the lubricant film thickness variation governed by the time-dependent Reynolds equation. However, their studies suffered from the drawback of neglecting the effects of surface deformation as well as pressure and temperature dependent viscosity of the lubricant. Dowson and Higginson (2) applied their steady state isothermal elastohydrodynamic lubrication theory to predict the film thickness between gear teeth at the pitch point. Nevertheless, the film thickness predicted by steady state EHL theory is questionable. Lin and Medley (3) recognized the limitation and a modification was introduced to account for rapidly changing in curvature. Their results showed that considering velocities as a function of the local coordinate at an instant time had a significant effect on the film thickness in some cases. Unfortunately, their analyses have not been extended to the entire gear contact zone. Moreover, the model was established under isothermal condition with a constant static load (unity contact ratio) on a single pair of teeth.

In this study, the detailed expressions for the radius of curvature and the velocities of a pair of spur gear in contact are derived as the function of both instant time and the local coordinate based on gear kinematics. The film shape function consists of the actual tooth contact configuration and elastic deformation under pressure. The squeezing (or transient) effect is analyzed by taking in account the dynamic load due to mesh cycle motion along the line of action. The thermal effect is also incoporated in the analysis along with a nonlinear limiting shear stress model.

The established dynamic model is then solved along with the previously developed comprehensive lubrication model (4). The important factors such as: non-Newtonian properties of lubricants, surface roughness, elastic deformation, temperature, squeezing effect and dynamic load are studied in an integrated way to see various interactions among different influences.

The results have shown that the pitch point EHL film thickness does not reliably represent the minimum EHL film thickness. Isothermal EHL is in error both in magnitude and in trend.

The dynamic load sharing profile has a dominant impact on the actual location of the minimum film thickness along the line of action. The squeezing effect has an important influence on both the minimum film thickness and its location, while the kinematics of gear has less effects under a given rotation speed.

The contact ratio determines the dynamic load sharing profile. A larger contact ratio results in a higher film thickness with relative higher pressure peak. Furthermore, a higher contact ratio reduces total flash temperature on the contacting tooth surfaces.

The pressure peak varies along the line of action. Its maximum value appears near the tip and/or the root of the teeth. Therefore, the peak contact stress may be significantly underestimated if it is calculated at the pitch point.

Surface roughness has a moderate effect on the film thickness. The effects become more significant when the roughness amplitudes approach the nominal film thickness.

Non-Newtonian behavior may significantly alter the level of the minimum film thickness and temperature distribution, but it has only small effect on pressure peak.

REFERENCES

(1) R.W. Adkins, E.I. Radzimovsky, J. of Basic Engineering, ASME Trans., page 655-663, 1965.
(2) D. Dowson, G.R. Higginson, Elastohydrodynamic Lubrication - The Fundamentals of Roller and Gear Lubrication, Pergamon Press, London, 1966.
(3) Z.G. Lin, J.B. Medly, Wear, page 143-150, 1984.
(4) J.M. Wang, V. Aronov, 96-TRIB-26, ASME/STLE Joint Tribology Conference, San Francisco, 1996.

A STUDY OF TEMPERATURE RISE OF OIL DUE TO COMPRESSION

KEIJI IMADO, YUKIHITO KIDO and HIROOMI MIYAGAWA
Faculty of Engineering, Oita University, 700 Dannoharu, Oita-shi, 870-11, JAPAN
FUJIO HIRANO
Professor Emeritus, Kyushu University, 4-10-12 Takamiya, Minami-ku Fukuoka-shi, 815, JAPAN

ABSTRACT

In analysis of oil film condition in EHD, estimation of viscosity change with pressure and temperature is one of the most important but difficult problem for theoretical treatment (1). As the temperature rise in operating condition negatively affects the viscosity value, correct estimation of temperature rise is required for practical purposes. In most studies, temperature rise is calculated by assuming heat is generated mainly from the shearing of an oil film (2). This simple assumption seems to be not so appropriate for high pressure condition such as in EHD contact under pure rolling or at the center of a squeeze film, where adiabatic compression plays a major role in heat generation.

In this study, two kinds of experiments were carried out to estimate temperature change of oil under compression. They were impact test and quasi-adiabatic compression test. Impact was applied with a steel ball mounted at the end of a leaf spring impinging against an oiled sapphire glass. Temperature was measured by means of an infrared radiation thermometer. Impact velocity was measured with the Laser Doppler vibrometer. Temperature rise up to 45 °C was found in the impact tests. Furthermore as shown in Fig. 1, a close relation between temperature rise and $\alpha \cdot A_{max}$ was found for oils of relatively low viscosity, where α was the pressure-viscosity coefficient and A_{max} was the maximum acceleration of the hammer during impact. As far as temperature rise is concerned under a given impact condition, there was a certain optimum combination of the pressure-viscosity coefficient and the initial viscosity.

In order to account for these facts, quasi-adiabatic compression tests were carried out using a high pressure vessel. Test oil was pressurized almost adiabatically in a cylinder by a plunger which was quickly pushed into the cylinder by hydraulic power. Temperature was measured by a thermocouple exposed in the oil. Figure 2 shows the major results. Temperature rise increases linearly with an increment of volumetric strain regardless of oil kind.

As for the order of temperature rise of a corresponding oil in each experiment, it was reversed with each other. Namely, the higher temperature rise was recognized for an oil of smaller pressure-viscosity coefficient in the compression test. In other words, for an oil of large free volume, temperature rises easily due to adiabatic compression such as gas. On the contrary, for an oil of small free volume, temperature rise was also small. On the other hand, larger temperature rise was recognized for an oil of larger pressure-viscosity coefficient in the impact test. As far as temperature rise is concerned for entrapped oil in the impact test, it can be concluded from these experiments that the pressure-viscosity coefficient has two effects. One is increasing the volume of oil to be entrapped by increasing the viscosity and prevents the oil from squeezing out. The other is it essentially decides the temperature rise under adiabatic compression.

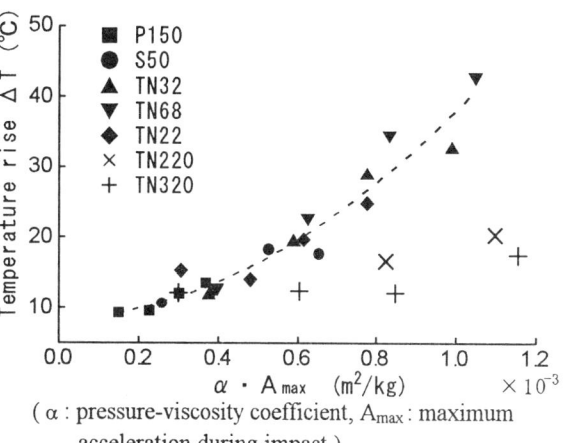

(α : pressure-viscosity coefficient, A_{max} : maximum acceleration during impact)

Fig. 1: Relation between $\alpha \cdot A_{max}$ and ΔT

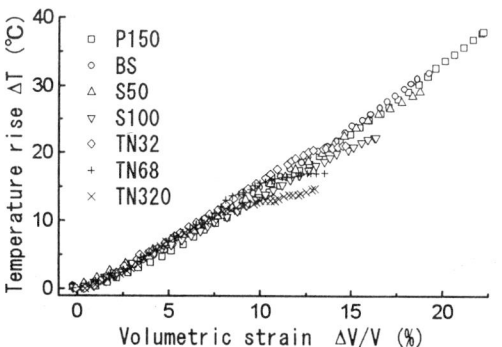

Fig. 2: Relation between volumetric strain and ΔT

REFERENCES

(1) J. Sorab, W.E. Vanarsdale, Tribology Transaction, vol.34, No.4, 1991, 604-610pp.
(2) B.J. Hamrock, D. Dowson, Ball bearing lubrication, John Wiley & Sons (1981) 151-152pp.

ELECTROSTATIC CHARGING PRECURSOR TO SCUFFING IN LUBRICATED CONTACTS

R J K WOOD, M BROWNE and M T THEW

Department of Mechanical Engineering, University of Southampton, Southampton, SO17 1BJ, UK

ABSTRACT

Brief industrial tests found that electrostatic charge was generated during the early stages of adhesive wear or scuffing in an oil lubricated FZG gear scuffing rig. A ONR funded fundamental research programme into this phenomena has been established at the University of Southampton. If the electrostatic charge generation process can be well understood it has potential to lead to a powerful predictive maintenance tool for oil lubricated systems. This paper covers aspects of the work evaluating the electrostatic signatures of adhesive wear generated within oil lubricated metal to metal sliding contacts. Previous published work in this area is scarce with only Nakayama (1) detecting charged particles and photons emitted when scratching ceramics in a hydrocarbon atmosphere using an active sensor system. By varying the potential on the sensor both positive and negative charges emanating from the surface could be identified.

Adhesive wear has been studied under controlled lubrication conditions using a pin-on-disc wear machine incorporating a fixed earthed 6mm diameter En31 steel ball loaded onto an earthed rotating disc of the same material. This sliding point contact was subjected to a range of contact loads (10-150N) with the initial Hertzian contact pressures between 0.5 and 1.5 GPa and sliding speeds between 2 and 7 m/s. Specially designed passive split ring primary sensors were used to monitor electrostatic charge levels around the ball on disc contact. Button and line array electrodes were used as secondary sensors positioned over the disc wear track between 30° and 120° around from the contact, in the direction of sliding, to evaluate charge transportation. The tests used Shell Vitrea ISO 32 oil which is a blend of base oils with no additives, to eliminate the complex effects of such additives and was delivered to the point contact uncharged.

Experiments at high speed (7 m/s) and moderate load (50N) induced the thin oil film formed by elastohydrodynamic lubrication within the point contact to breakdown, allowing strong metal-metal adhesion or scuffing. Background charge levels of up to 0.03pC were observed during testing while scuffing resulted in charge levels of between 1 and 4 pC, with individual charging events lasting approximately one millisecond and generally monopolar.

Analysis of the electrostatic charge signals prior to scuffing revealed a unique charge generating mechanism lasting approximately 10ms, consisting of several monopolar events, usually occurring once per revolution and on occasion up to three times per revolution. These 'precursor' signals increased in magnitude with time and led directly into the scuff. The duration between precursor identification and scuffing was random and on one occasion lasted three minutes.

Optical microscopy and surface profilometry revealed a unique morphology associated with the suspected 'precursor' sites on the disc wear track. A 25% increase in wear depth was observed at these sites. Electron microscopy of etched precursor generating ball scars revealed the presence of white layers which have a much higher carbon concentration than the bulk material. The increase in carbon content is thought to be by diffusion. Similar studies on scuffed wear scars revealed much larger areas of white layers. Vickers hardness tests of the En31 steel balls gave the order of untested En31 (640 Hv) < precursor En31 (720-780Hv) < scuffed En31 (810 Hv). This increase in hardness is thought to be associated with an increasing white layer presence associated with thermomechanical transformation at the surface to high carbon austenite or martensite. Thus, it appears that it is the onset of phase transformation or the subsequent effects of this transformation on the adhesive wear process which causes measurable charge generation and hence a precursor signal. One such effect is the generation of wear debris which could be charged.

Continuing studies are outlined to further understand the mechanisms of electrostatic charge generation and to define the conditions under which precursor charging can be expected. Electrochemical techniques are discussed employing microelectrodes to generate a known type and amount of charge in oil which will enable the electrostatic sensors to be calibrated and for their response characteristics to be evaluated. Sensor geometry and location will also be examined.

REFERENCES

(1) K Nakayama, H Hashimoto, Wear, Vol. 185, 1995, pp 183-188.

ELASTOHYDRODYNAMIC FILM FORMING WITH SHEAR THINNING LIQUIDS

SCOTT BAIR

George W. Woodruff School of Mechanical Engineering, Georgia Institute of Technology, Atlanta, Georgia 30332 USA

ABSTRACT

The prediction of lubricant film thickness in concentrated contact is well-developed for Newtonian liquids. The machine designer, for example, may specify a lubricant viscosity which will ensure a complete separation of machine elements of known surface finish. It is now understood that all liquids used as bases oils must shear thin at some critical shear stress but that this shear stress may or may not be attained in the film forming (pressure boosting) zone of contact. Some success has been obtained in prediction of film thickness for some special reological models using a generalized Reynolds equation similar to the one developed by Dowson (1). These formulations require a functional form for the constitutive law which is readily integrable and have not been compared with experiment. There have been recent advances in experimental technique for measuring the non-linear shear response of lubricants under pressure (2-4). Data generated using these techniques have been fitted to empirical rheological models which have been useful in predicting traction for concentrated contacts with sliding. It should be natural to apply the same rheological measurements to film thickness prediction.

In a recent note (5), we have introduced a simple numerical method for obtaining the incompressible film thickness solution for one-dimensional contact and a liquid with a second Newtonian. In this paper the simple numerical method is extended to include sliding. Solutions are generated and compared with experimental film thickness for several liquids which display interesting rheological behavior.

An example is given in the figure for a solution of polyisobutylene in decalin. The parameters for the rheological model were obtained from the data of Lodge (6) using time/pressure shifting and from a high-pressure falling body viscometer. Film thickness measured in rolling contact is shown as the two points. The Newtonian prediction using the low shear viscosity of the blend (μ_1) is marked as $\eta = \mu_1$ and using the viscosity of decalin as $\eta = \mu_2$. The ordinary Newtonian prediction overestimates film thickness by two orders of magnitude. Improved predictions are obtained by making the viscosity a function of shear stress, $\eta(\tau)$.

In addition, useful insights are provided concerning the effects of pressure-viscosity behavior for Newtonian liquids, sliding and starvation for non-Newtonian liquids and the relevant shear stress for film forming.

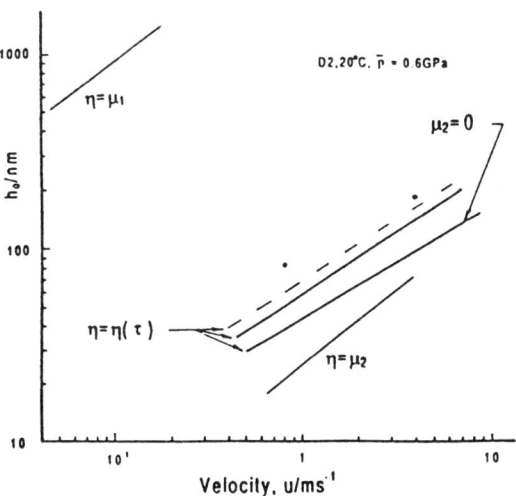

Fig. 1: Film thickness solutions for the highly elastic liquid, D2, and measured thickness.

REFERENCES

(1) Dowson, D., International Journal of Mech. Scien., 1962, pp. 159-170.
(2) Bair, S. and Winer, W.O., STLE Tribology Transactions, 1993, p. 721.
(3) Wong, P.L., Lingard, S. and Cameron, A., Proc. Leeds-Lyon Symp. Tribology.
(4) Zhang, Y. and Ramesh, K. T., ASME Journal of Tribology, 1995.
(5) Bair, D. and Khonsari, M., ASME Journal of Tribology, 1996, pp. 341-343.
(6) Lodge, A.S., Al-Hadithi, T. and Walters, K., Rheologica Acta, 1987, pp. 29-32.

Transient Solutions with Interfacial Slip in EHL Point Contacts

P. Ehret, D. Dowson, C.M Taylor
Institute of Tribology, Department of Mechanical Engineering, University of Leeds, Leeds LS2 9JT, UK

ABSTRACT

Recent experiments [1,2,3,4] of point contact elastohydrodynamic lubrication (EHL) have led to intriguing interferometric observations of film thickness which contradict the results predicted by traditional EHL theory. With smooth surfaces, pure sliding conditions and for a steady-state regime, a large dimple is visualised at the centre of the conjunction instead of a flat plateau. In a previous study [5], the authors have shown that this effect can be attributed to slip conditions across the film, so that the mean velocity of the lubricant no longer corresponds to the entrainment velocity of the surfaces. As suggested by Oldroyd [6], it was shown that the slip varies along the conjunction in the same manner as the shear stress at the wall. At high shear rates, the slip is simply proportional to the limiting shear stress. Of particular importance in the modelling of the problem is the fact that a modified Reynolds' equation can still be derived to accommodate the slip conditions. Thus the Multigrid Multi-Integration techniques previously developed can still be used to accelerate the convergence of the problem.

In physical terms, slip conditions are capable of embodying various effects that traditional EHL theory cannot account for, since this is constrained by the assumptions of continuum mechanics and constant pressure and linear shear-stress across the film. A plausible explanation is to refer to the slip as an additional kinematic condition at the interface wall/lubricant. The underlying concept is based upon the assumption that, in the high pressure region of the contact, the lubricant is in a glassy state and surrounded by thin layers of lubricant in the liquid state. Slip conditions allow rapid velocity changes in these thin layers to be modelled as a velocity discontinuity between the wall and the core of the glassy lubricant.

In the present paper, the previous study is extended to investigate time dependent effects. Experimental transient conditions [1] are modelled in which a thin layer of lubricant is entrapped by impacting a steel ball on a glass disk. The surfaces are subsequently subjected to rolling or sliding motions to release the volume of lubricant enclosed. Conditions of slip, and lubricant rheology are explored in order to match the unusual observations obtained by Kaneta et al [1]. The figure attached represents the changes in film thickness when pure sliding conditions are imposed after impact. The direction of the flow is from left to right.

PUBLICATION

The paper will be submitted for publication in ASME Journal of Tribology

REFERENCES

(1) **Kaneta, M., Kamzaki, Y., Kameishi, K., Nishikawa, H.** Non-Newtonian Response of Elastohydrodynamic Oil Films. *Proc. Japan Int. Tribology Conf. Nagoya (1990)*, 1195-1700.
(2) **Kaneta, M., Nishikawa, H., Kameishi, K., Sakai, T., and Ohno, N.** Effects of Elastic Moduli of Contact Surfaces in Elastohydrodynamic Lubrication. *J. Trib. (Trans. ASME F)*, 114:75–80, 1992.
(3) **Kaneta, M.** Necessity of Reconstruction of EHL Theory. *Japanese Journal of Tribology*, 38:860–868, 1993.
(4) **Kaneta, M., Nishikawa, H., Kanada, T., Matsuda, K.** Abnormal Phenomena Appearing in EHL contacts. *J. Trib. (Trans. ASME F)*, 118:886–892, 1996.
(5) **Ehret, P., Dowson, D., and Taylor, C.M.** On Interfacial Slip in Elastohydrodynamic Conjunctions. *Submitted for publication Proc. Roy. Soc. London*, pgs. 45, 1996.
(6) **Oldroyd, J.G.** The Interpretation of Observed Pressure Gradients in Laminar Flow of Non-Newtonian Liquids Through Tubes. *Colloid Sci*, 4:333–342, 1949.

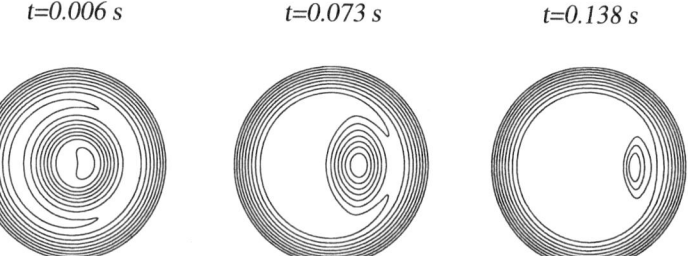

Film thickness contours at different time steps for a rapid sliding start with interfacial slip after impacting a thin layer of lubricant between a steel ball and a glass disk

PREDICTION OF SLIDING TRACTION IN ELASTOHYDRODYNAMIC LUBRICATION

A V OLVER and H A SPIKES
Department of Mechanical Engineering, Imperial College, London SW7 2BX

ABSTRACT

Prediction of traction (friction) in lubricated rolling sliding contacts, such as are encountered in gears and cam and tappet systems, remains a challenging problem despite the development of the realistic Maxwell-Eyring-limiting shear stress model by Johnson and co-workers in the 1980's. This is largely because there is a strong coupling between the elastohydrodynamic traction and the film temperature. An added complication is that the heat conducted into the rubbing surfaces, as well as influencing traction directly, also determines the temperature in the *inlet* to the contact and hence the thickness of the elastohydrodynamic film.

In the present paper, the traction model of Johnson et al is combined with heat transfer analysis of the contacting bodies as well as the film thickness regression equation. In addition, the variation of lubricant rheological properties with temperature and pressure based upon the measurements of Muraki et al [2] has been included. The traction equation is expressed in dimensionless form and is solved using a simple iterative scheme, which in many cases allows estimation of the traction without the use of a computer. Closed form equations for the friction are given for each of the traction regimes reported in [1].

The figure shows the predicted friction coefficient for two identical steel discs as a function of the slide roll ratio. The lubricant was assumed to have the viscosity temperature characteristics of ISO VG 220 together with the rheology of the mineral oil reported in [2]. The results lie in the viscoelastic shear thinning regime except for those having a slide to roll ratio of between 2 and 6 percent which have reached the limiting shear stress.

A striking feature of the results is the high temperature reached by the lubricant in sliding elastohydrodynamic films under quite mild conditions. This means that the available lubricant rheological data, which has mostly been obtained at lower temperatures, is of limited use.

REFERENCES

[1] Evans, C R and Johnson, K L, Regimes of Traction in Elastohydrodynamic Lubrication, Proc I Mech E, 200 (1986) 313-324

[2] M Muraki, M, Matsuoka, T and Kimura, Y, Influence of Temperature Rise on Non Newtonian Behaviour of Fluids in EHD Conditions, in Holmberg, K and Nieminen, I, (eds) Proceedings of the 5th International Congress on Tribology, Volume 4, Espoo, Finland, 1989, pp 226-231.

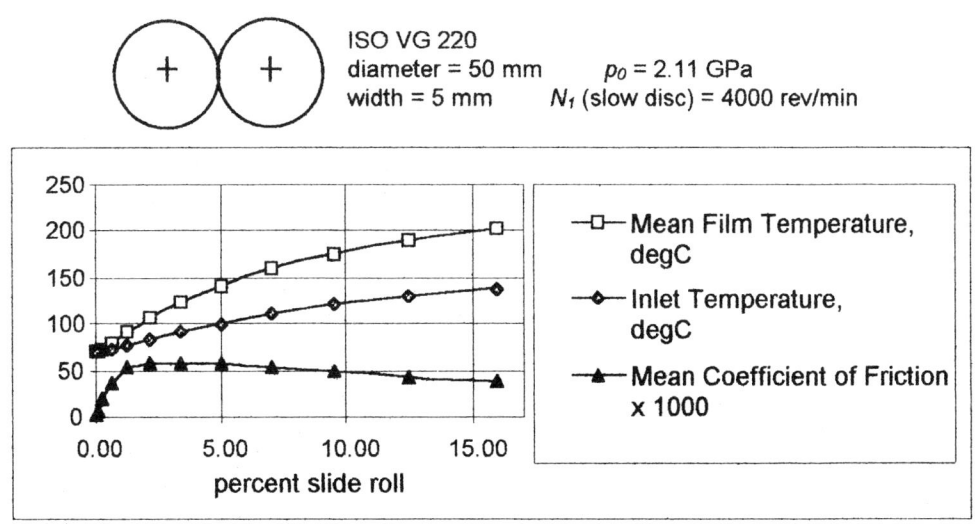

THE STUDY OF ELASTOHYDRODYNAMIC LUBRICATION OF ROLLING/SLIDING POINT CONTACTS BY COMPUTER DIFFERENTIAL COLORIMETRY

I KŘUPKA, M HARTL and M LIŠKA

Faculty of Mechanical Engineering, Technical University of Brno, Technická 2, 616 69 Brno, The Czech Republic

ABSTRACT

Nowadays, the theory of elastohydrodynamic (EHD) lubrication predicts lubricant film thickness distribution in highly loaded contacts with very good accuracy. Many experimental works have been carried out to validate it and good agreement has been found. However, some interesting phenomena waiting for explanation have been observed. The dissimilarity between EHD lubricant film thickness profiles under pure rolling and sliding conditions is one of them. Sanborn et al (1) found the local film thickening just before the EHD exit constriction under pure sliding. The similar one was described by Kaneta et al (2). But they did not find this phenomenon under pure rolling as it was difficult to measure such slight changes in lubricant film thickness by conventional optical interferometry.

This paper describes the use of computer differential colorimetry for the mapping of EHD lubricant film thickness in rolling/sliding point contacts. This newly developed experimental technique (3) enables accurate, quick and automatic evaluation of an EHD film thickness distribution with high resolution from chromatic interferograms obtained from a conventional optical test rig (a steel ball and transparent disc).

room temperature (25±0.5°C) and max. Hertz pressure 0.425 GPa. It can be seen that the central film thickness values are almost identical with those predicted by Hamrock and Dowson equation. As to the minimum film thickness, at higher rolling speeds experimental film thicknesses are above theoretical ones, while with decreasing rolling speed experimental values go under these predicted by the theory.

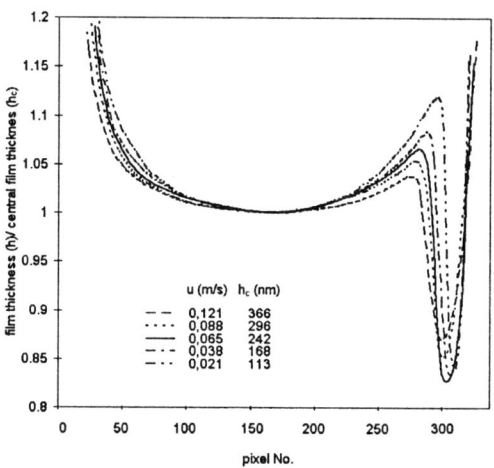

Fig. 2: Relative film profiles along the centre line in the direction of motion

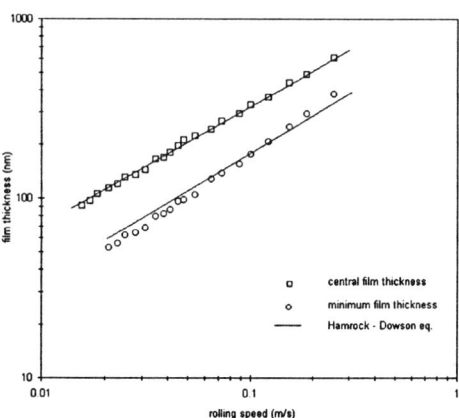

Fig. 1: Log central and minimum film thickness vs. log speed obtained by differential colorimetry

Fig. 1 depicts the central and minimal film thickness values measured by differential colorimetry versus rolling speed (for naphthenic base oil) and their comparison with the Hamrock and Dowson film thickness equations. Experiments were carried out at

From experimental results in Fig. 2 the local film thickening is evident and reaches the value of about 15 nm which is well under the resolution of conventional optical interferometry. It can be seen that this thickening changes slightly with increasing rolling speed.

Similar experiments have been carried out for wider range of loads, speeds, slide to roll ratios and material parameters.

REFERENCES

(1) D. M. Sanborn, W. O. Winer, ASME Trans., Jour. of Lubr. Tech., 93, 2, pp. 262-271, 1971.
(2) M. Kaneta, H. Nishikawa, K. Kameishi, T. Sakai, N. Ohno, ASME Trans., Jour. of Tribology, 114, 1, pp. 75-80, 1992.
(3) M. Hartl, I. Křupka, M. Liška, Elastohydrodynamic Film Thickness Mapping by Computer Differential Colorimetry, World Tribology Congress '97.

DENT INITIATED SPALL FORMATION IN EHL ROLLING/SLIDING CONTACTS

GANG XU and FARSHID SADEGHI
School of Mechanical Engineering, Purdue University, West Lafayette, Indiana 47907, USA
MICHAEL R. HOEPRICH
The Timken Company, Canton, Ohio 44706, USA

INTRODUCTION

Surface initiated failure due to debris denting effects has been recognized as one of the major failure mechanisms in EHL contacts. Debris will cause indentations on the contacting surfaces resulting in high stress concentrations on the surface, leading to surface crack initiation, spall propagation and fatigue failure.

Experimental results are provided from a ball-on-rod rolling contact fatigue tester with the rod pre-dented with a single large dent. The results indicate that the spall usually initiated at the trailing edge of the dent on the driving surface. These cracks and spalls can also be created in the absence of lubricant.

SPALL INITIATION MODEL

Based on the accumulated plastic strain and damage mechanics concept, a line contact spall initiation model was developed to investigate the dent effects on spall initiation and propagation, as illustrated in Fig. 1. The model was consisted of three parts: the EHL solution with dent effect; the internal stress and plastic strain accumulation calculation; and the damage law and spall initiation and propagation.

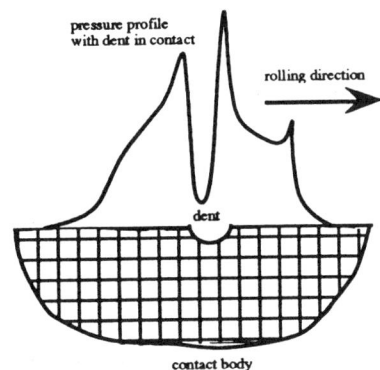

Fig. 1: EHL contact crack initiation model

The near surface volume of the contact solid was divided into many small metal cells and a damage variable D was associated with each cell. For each cell a damage law was applied to determine whether the cell is undergoing damage or not. The damage law was expressed as:

$$\frac{dD}{dN} = \frac{Y}{S_m} \cdot \Delta\varepsilon_p \cdot H(\Delta\varepsilon_p - \Delta\varepsilon_D)$$

where S_m is the material constant, $\Delta\varepsilon_p$ is the accumulated cyclic plastic strain in one cycle and Y is the strain energy density release rate. If the cell on the surface is damaged, then it is removed from the surface and a spall will be formed. If the damaged cell occurs below the surface, then a sub-surface void is generated, this void could grow to the surface depending on the EHL running conditions. The spall will further modify the surface geometry and initiate a new spall, hence, the spall will propagate.

RESULTS

The results indicate that the location of spall initiation depends on the EHL and dent condition. Spalls can initiate at either the leading or trailing edge of the dent depending on the surface traction, as illustrated in Fig. 2.

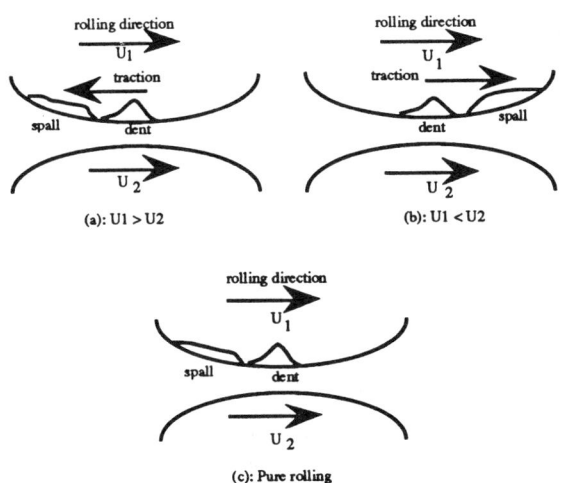

Fig. 2 : Spall location vs. EHL condition

The results on crack initiation life and the effects of material properties were also provided.

STREAM FUNCTION AND STREAMLINES FOR VISUALIZING SIDE FLOWS IN EHL ELLIPTICAL CONTACTS

HSING-SEN S. HSIAO and BERNARD J. HAMROCK
Dept. of Mechanical Engineering, The Ohio State University, Columbus, OH 43210, USA
JOHN H. TRIPP
SKF Engineering and Research Centre B.V., The Netherlands

ABSTRACT

Streamlines of the flow in an elastohydrodynamically lubricated elliptical contact will clearly reveal the trajectories of the lubricant particles that pass by, pass through, and flow back from the Hertzian contact zone, or circulate somewhere in the flow passage. Unlike those for cylindrical contacts given in Shieh and Hamrock (1)(2) and Hsiao and Hamrock (3)(4), the stream function (which is used to construct the streamlines) for elliptical contacts becomes three-dimensional. Solving for the complete field of a three-dimensional stream function is found very difficult and needs to rely on a suitable numerical method. However, under some special conditions, the stream function on some particular planes becomes two-dimensional and closed form solutions are readily available. On other planes, pseudo-two-dimensional stream functions may exist and they may be solved by using a modified fourth-order Runge-Kutta numerical method. The first part of this abstract, which includes Figs. 1, 2, and 3, is to demonstrate some of the streamlines that vivify the flow paths formed due to entraining and blocking the lubricant by the bounding solid surfaces. Figures (1) and (2) show the typical streamlines on the central plane of a flow under pure rolling and simple sliding, respectively. Figure (3) shows the streamlines on the midfilm plane of a flow which is bounded by two geometrically symmetric solid surfaces and under pure rolling. In these streamlines, reversed and passing flows are revealed. In Fig. (3) the side flow and the deflected (by-passing) flow are also depicted.

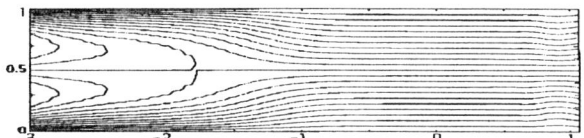

Fig. 1 Typical central plane streamlines for bisymmetric flow under pure rolling. Note that the horizontal axis is the entrainment direction normalized with Hertzian half-width; the vertical axis is the film thickness direction normalized with the film thickness itself.

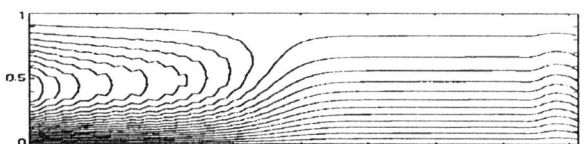

Fig. 2 Typical central plane streamlines for symmetric flow under simple sliding. See note in Fig. 1. Note also that, here, the lower surface moves in the positive entrainment direction while the upper surface is stationary.

In the second part of this abstract, introduced are the conceptual column stream function and column streamlines which are derived by using the column continuity equation with the column mass fluxes given in Hsiao et al. (5). A plot of the column streamlines becomes a very convenient tool to visualize and measure the rate of a compressible flow which passes through an EHL elliptical contact. Figure (4) shows a typical plot of column streamlines for a symmetric flow. It indicates that only a very small part of the entrained lubricant actually passes through the Hertzian zone, the flattened part of the flow passage.

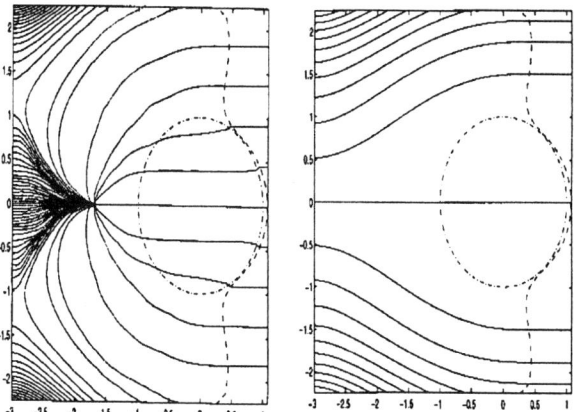

Fig. 3 Typical midfilm plane streamlines for bisymmetric flow.

Fig. 4 Typical column streamlines for symmetric flow.

Note that for both Figs. (3) and (4) the horizontal axis is the entrainment direction normalized with Hertzian half-width; the vertical direction is the side flow direction normalized with the Hertzian half-width in the side flow direction.

REFERENCES

(1) J A Shieh, B J Hamrock, *Proceedings of US/Taiwan Joint Symposium on Tribology*, pp. 69-78, 1989.
(2) J A Shieh, B J Hamrock, ASME *Journal of Tribology*, Vol. 113, No. 2, pp. 372-377, 1991.
(3) H-S S Hsiao, B J Hamrock, ASME *Journal of Tribology*, Vol. 114, No. 3, pp. 540-552, 1992.
(4) H-S S Hsiao, B J Hamrock, ASME *Journal of Tribology*, Vol. 116, No. 4, pp. 794-803, 1994.
(5) H-S S Hsiao, B J Hamrock, J H Tripp, "FEM System Approach to EHL Elliptical Contacts," to be published in ASME Journal of Tribology.

SYNERGISTIC EFFECTS OF ALKYLAMINES ON FRICTIONAL PROPERTIES OF OILINESS AGENTS

YOSHIHARU BABA, KIYOSHI HANYUDA, SUNAO IKEDA and RYUJI MARUYAMA
Showa Shell Sekiyu K.K., Central R&D Laboratory, P.O. Box 5, Aikawa
4052-2, Nakatsu, Aikawa-cho, Aikoh-gun, Kanagawa-ken, Japan

INTRODUCTION

Since modern slideway lubricants should possess excellent frictional properties to prevent stick-slip motion and to perform accurate positioning, various types of oiliness agents like sulphurised fatty oils, phosphoric esters and fatty acids are generally included. However, these components often have maximum concentration limits because some have disadvantages in terms of corrosion, solubility, cost increase, etc.

In this report, it was found that the frictional properties of commonly used oiliness agents can be significantly improved by the additions of small amounts of certain types of alkylamines. The synergistic effects of alkylamines were investigated in detail through positioning accuracy tests using a numerically controlled machine tool. The mechanisms of the synergism were studied using Langmuir-Blodgett (LB) mono-molecular layer techniques.

EXPERIMENTS

Reference oils (68 cSt kinematic viscosity at 40 °C) based on solvent refined paraffinic mineral base oils and commonly used oiliness agents were available. Test oils containing various types of alkylamines in the reference oils were prepared.

The frictional properties of test and reference oils were measured with a vertical type machining centre which was a positioning accuracy test based on JIS-B6338-2.8(1). Actual positions of the slideway against commanded positions were measured using a laser length measurement apparatus. The positioning accuracy of each oil was obtained in the form of lost-motion which is the number of pulse counts needed for the start of reverse travelling of the slideway. The lost-motion of the reference oils containing various types oiliness agents are compared in Fig.1, with these oils containing a small amount of alkylamine as an additional amine, to identify which oiliness agents show synergistic frictional improvements when combined with oleylamine. In addition, to investigate which type of amines are effective in terms of synergism, the positioning accuracy tests were performed for oils additionally formulated with different types of alkylamines in the reference oil, containing amine salts of di- and mono-octylphosphoric acids (see Fig.2).

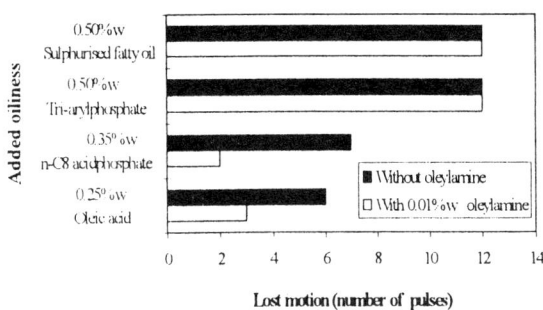

Fig.1 Effects of oleylamine on the lost-motions of various oiliness agents

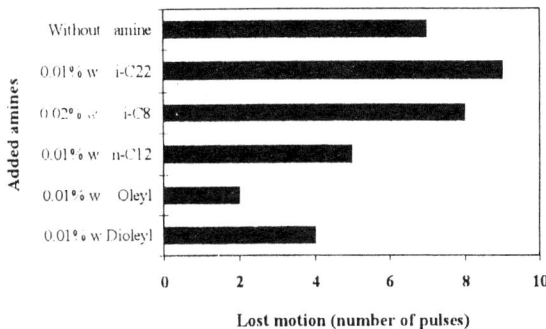

Fig.2 Effects of various types of alkylamines on the frictional property of an alkylamine salt of n-octylphosphoric acid

CONCLUSIONS

The positioning accuracy tests showed that oiliness agents with high polarity showed synergistic frictional improvements when included with alkylamines, and that amines with linear alkyl groups were most effective.

The mechanism of this synergism is discussed in terms of the LB technique which considers mono-layers of polar oiliness agents mixed with alkylamines at the air/water interface.

REFERENCES

(1) JIS B6338 2.8 Minimum set unit feed test for machining centres (vertical type), 1985

FRICTION REDUCING AND ANTIOXIDANT CAPABILITIES OF ENGINE OIL ADDITIVE SYSTEMS UNDER OXIDATIVE CONDITIONS

STEFAN KORCEK, RONALD K. JENSEN, MILTON D. JOHNSON, ARUP K. GANGOPADHYAY, and MICHAEL J. ROKOSZ
Ford Motor Company, Ford Research Laboratory, MD 2629/SRL, Dearborn, MI 48121-2053, USA

ABSTRACT

Engine oil additives play a very important role in reducing friction and wear in lubricated engine contacts operating under boundary or mixed lubrication regimes. A very effective additive combination used for this purpose in recently introduced advanced fuel efficient engine oils contains molybdenum dialkyldithiocarbamate, $Mo(dtc)_2$, and zinc dialkyldithiophosphate, $Zn(dtp)_2$. This combination could provide superior friction reducing capability and good antiwear properties. It has been shown by Igarashi (1) that the above two types of additives, when present together, undergo ligand exchange reactions leading to formation of mixed ligand products, $Mo(dtc)(dtp)$ and $Zn(dtp)(dtc)$, and full ligand exchange products, $Mo(dtp)_2$ and $Zn(dtc)_2$ (Fig. 1). However, it has also

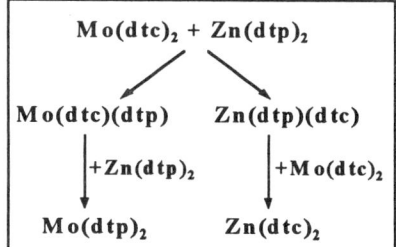

Figure 1 - Ligand Exchange Reactions

been shown that during oil aging both original additives, $Mo(dtc)_2$ and $Zn(dtp)_2$, are gradually consumed due to oxidation since they act as antioxidants (2). In order to develop an understanding of these processes and their effects on retention of friction reducing behavior, we have investigated the occurrence of the above ligand exchange reactions in various base oils under oxidative conditions simulating the engine environment (NO_2 initiated oxidation at 160°C).

The results of our investigations show that both the ligand exchange reactions and the friction reducing capabilities of these additives are significantly affected by oxidation (Fig. 2) and, consequently, also by antioxidants either added or naturally present in base oils or formed during oxidation. Based on our studies we have concluded that:

- the formation of molybdenum ligand exchange products under oxidative conditions is accelerated by formation of hydroperoxides which preferentially react with zinc exchange products; this selectively removes zinc exchange products from the system and shifts the equilibrium towards increased formation of molybdenum exchange products,
- the presence of peroxy radical trapping antioxidants substantially reduces formation of hydroperoxides thus slowing down formation of molybdenum exchange products; this is also observed in base oils containing aromatics, but no sulfur compounds where radical trapping antioxidants are formed by oxidation of aromatic components of oil, and
- the presence aromatics and sulfur compounds in the solvent refined base oils accelerates consumption of $Zn(dtp)_2$ and suppresses formation of molybdenum exchange products; this could possibly be explained by an increased rate of initiation of oxidation in these systems.

Results of our investigations suggest that interactions between $Mo(dtc)_2$ and $Zn(dtp)_2$ during engine oil aging and consequently also retention of friction reducing capabilities of additive systems containing these compounds are determined not only by these additives themselves but also by base oil composition and by the presence of other peroxide decomposing and peroxy radical trapping additives. This finding provides important directions for formulating engine oils with improved retention of fuel efficiency throughout extended service intervals.

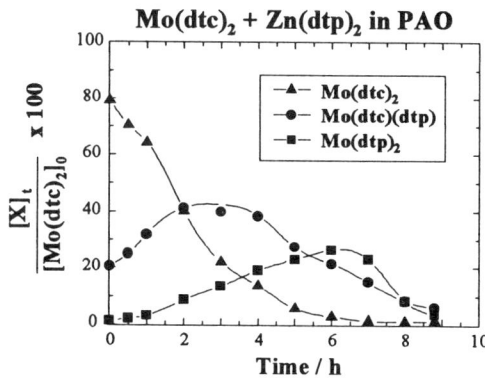

Figure 2 - Ligand Exchange During Oxidation

REFERENCES

(1) I. K. Yagishita and J. Igarashi, *Prepr. JAST Tribology Meeting*, Fukuoka, October, 673-6 (1991).
(2) S. Korcek, R. K. Jensen, M. D. Johnson and E. M. Clausing, *Intenational Tribology Conference, Yokohama 1995*, 733-737 (1996).

STUDY OF THE INFLUENCE OF ORGANIC MOLYBDENUM COMPOUND ON VALVE TRAIN WEAR CONTROL PERFORMANCE OF GASOLINE ENGINE OIL

MITSUHIRO NAGAKARI, SOUICHI YAMAMOTO, HIROYUKI SAKURAI and KOUICHI KUBO
Central R&D Laboratory, Showa Shell Sekiyu K. K., Aikawa P.O. BOX 50, Kanagawa-ken, Japan

ABSTRACT

In order to meet current and future fuel economy targets the application of friction modifiers (FMs) to gasoline engine oils is receiving considerable attention. It is well known that Organic molybdenum compounds are effective FMs for automotive engine lubricants under high temperature and low speed operating conditions. However, the effect of such compounds on lubricant wear control performance is not well understood.

API SJ class SAE 5W-20 test oils containing primary and secondary zinc dithio phosphate (ZnDTPs) individually and in combination with molybdenum dithiocarbamate (MoDTC) were prepared (Table 1).

The friction coefficients of each test oil were measured with a SRV high frequency friction tester. Addition of MoDTC to these oils gave friction reduction (Table 1).

The valve wear control performance has been studied using a motored gasoline engine which is a current Japanese 1.5 liter, 4 cylinder OHC design with a pivoted follower valve train system (1). The test oil containing primary ZnDTP (Oil C) gave slightly higher cam nose height loss and severe scuffing on the rocker pad after 100 hours test time. However, addition of MoDTC to the test oil C (Oil D) showed excellent valve train wear control. In contrary, the test oil containing secondary ZnDTP (Oil A) showed excellent valve train wear control performance but addition of MoDTC to the oil A (Oil B) gave severe scuffing on the tappet surface (Table 1). These results indicate a strong interaction between the MoDTC and ZnDTP type in terms of generating a scuffing protection film.

Table 1: Test Oils and Results of Friction Reduction and Scuffing

	Oil A	Oil B	Oil C	Oil D
SAE Vis. Grade	5W-20			
Type of ZnDTP	Secondary		Primary	
Elements P mass%	0.095			
FMs	none	MoDTC	none	MoDTC
Elements Mo mass%	-	0.04	-	0.04
Friction Coefficient	High	Low	High	Low
Scuffing	Good	Poor	Poor	Good

To understand the wear mechanism in this valve train wear test, samples of the worn tappet surface after 10, 25, 50 and 100 hour test duration were analyzed for surface elements by electron probe microanalysis (EPMA) and X-ray photo electron spectroscopy (XPS).

Examination of the worn surface by EPMA where no scuffing was present showed there were some differences between each oil in terms of intensity of P, S, Zn atoms (Fig.2). For oil A (sec-ZnDTP without MoDTC), there was a consistent reduction of P, S and Zn intensity with time, but oil C (pri-ZnDTP) showed that an increase of Zn and S intensity with time in spite of decreasing P intensity. Analysis of the Zn compound by XPS showed there were both Zn-O and Zn-S materials on the worn surface with each oil. The proportion of Zn-O for oil A was higher than that for oil C. A glassy phosphorus wear protection film generated by ZnDTP is generally composed of mainly Zn, P and O (2). In the case of test oils without the Mo compound, it is possible that secondary ZnDTP tends to generate a Zn-O rich layer which may be effective in preventing scuffing.

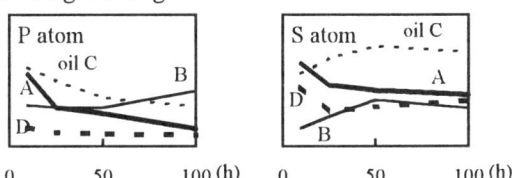

Fig. 2: Results of EPMA analysis of the worn surface on P and S atom

Results of XPS analysis of that specimens run with Mo (oil B and D) showed the presence of MoS_2 in the reaction film. Oil D (pri-ZnDTP and MoDTC) also showed the presence of MoO_3 whose proportion increased with time. However MoO_3 was not present on the wear surface of the components run on oil B (sec-ZnDTP and MoDTC) (Fig.3). It may be speculated that the formation of MoO_3 reaction film gave excellent wear protection in this test.

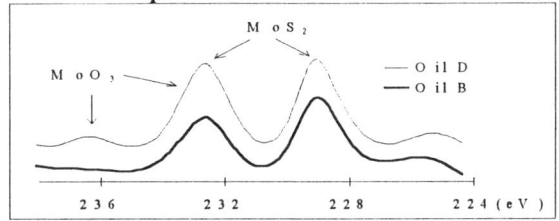

Fig. 3: XPS Chart of the worn surface of Oil B and D (100hr)

REFERENCES
(1) Takasi Kikuchi et.al., J-SAE Paper 924063, (1992)
(2) C.J.Bell et.al., Wear Particles, Elisevier Science Publishers, 387, (1992)

SURFACE CHEMICAL CHARACTERISATION OF BOUNDARY LAYER LUBRICANT ADDITIVES ON STEEL

C. A. SOTO, K. MATSUMOTO, A. ROSSI AND N. D. SPENCER
Laboratory for Surface Science & Technology, Department of Materials, Swiss Federal Institute of Technology, ETHZ, CH-8092 Zurich
M. BINGGELI
Blaser Swisslube Ltd. Hasle Ruegsau Switzerland

ABSTRACT

In boundary lubrication, additives react with surfaces under tribological conditions resulting in the formation of tribo-films. These films have generally been poorly characterised and thus the mechanism of film growth and removal (wear) is largely unknown. This is particularly true for metal-free additives, which are increasingly being used as an environmentally friendlier alternative to heavy metal containing ones.

The focus of this work is to investigate the chemical composition of tribo-films formed on 52100 steel (U.S.A. notation) by commercially available metal-free oil additives under a variety of tribo-conditions.

The tribological measurements were performed with a flat-on-disk tribometer (CSEM "pin-on-disk") using steel/steel tribo-pairs. The load was kept constant at 10N while the velocity was varied stepwise from 1 to 10,000 mm/min. This procedure resulted in Stribeck-like curves which can then be compared so as to evaluate the relative performance of the oils. Pure polyalpha olefin (viscosity of 31cS) was used as the base oil and the additive used was an alkyl phosphoric acid neutralised with a dialkylammine (Irgalube 349 Ciba-Geigy). The steel was mechanically polished and the final R_a roughness was 0.6 μm for the flat and 0.01 μm for the disk. The steel pair was immersed in lubricant during the tribo-test. The experiments were performed in air at 22°C. The relative humidity was measured during each run. The X-ray photoelectron spectrometer was a PHI 5600, the high resolution spectra being acquired in FAT mode with a pass energy of 23.5eV. The analysed area was 0.4 mm in diameter.

A plot of the friction coefficient versus velocity is shown in fig. 1. The additive reduces the friction coefficient at low velocities and prevents the occurrence of stick-slip. The error bars do not represent the amplitude of stick-slip events but rather the standard deviation of independent tribo-measurements under the same conditions. XPS spectra were collected for disks after tribo-testing at various velocities. For comparison, XPS results from the pure additive (frozen liquid, deposited on sputter-cleaned gold), model compounds and steel samples immersed in pure additive following a variety of mechanical treatments, are presented. Small-area XPS was performed on different regions of the sample in order to characterise the contacted and non-contacted regions. In table 1, the intensity ratios (corrected for the sensitivity factors from the spectrometer manufacturer) for selected treatments are given. The steel in the "as received" state (cleaned with cyclohexane in an ultrasonic bath) was found to be coated with an insoluble organic layer. However, this layer is sufficiently thin to allow detection of an underlying oxide layer, mainly constituted of iron (III) oxide and iron oxy-hydroxide. The results show that even in the absence of tribo-stress and at room temperature, the additive reacts dissociatively with oxidised samples (sputtered and reoxidised), while almost no additive was adsorbed on the as received sample. The results also show that further reaction takes place upon tribo-stress and wear under boundary lubrication conditions (20mm/min).

After tribological testing the film was found to contain phosphorus and nitrogen. The identification of the chemical state of phosphorus in the additive and on the steel surface has been carried out with a two dimensional (Wagner) chemical state plot, representing the measured Auger electron signals (PKLL) versus the photoelectron binding energy (P2p). This method reveals chemical information that cannot be obtained from photoelectron or Auger lines alone. The presence and the different chemical states of phosphorus following tribo-testing in different regions of the disk will be discussed and correlated with the tribological results.

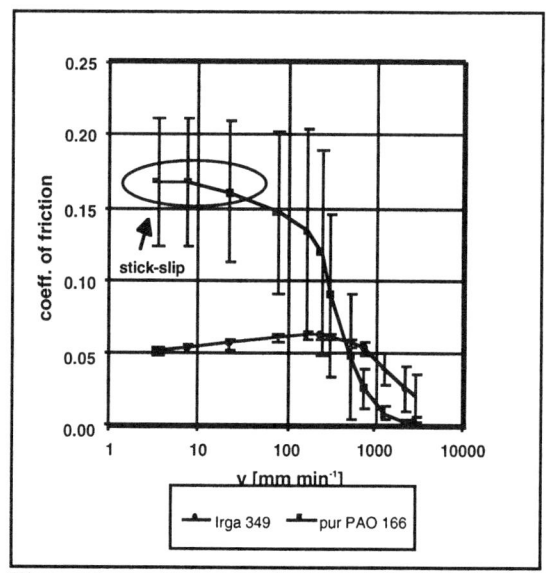

Figure 1. Friction-velocity curves for selected oils. Load 10N, steel on steel system

	C/P	C/Fe	P/Fe
Frozen Irga 349	36.65	--	--
As Received*	23.72	1.06	0.05
Sputtered and Reoxidised*	5.11	0.84	0.17
Tribo 2% Irga 349	2.85	1.08	0.38

Table 1. Ratios of Key Elements Irga 349 Tests.
* Samples immersed for 3 days in the pure additive

THE CHEMISTRY OF ANTIWEAR FILMS GENERATED BY THE COMBINATION OF ZDDP AND MODTC EXAMINED BY X-RAY ABSORPTION SPECTROSCOPY

MASOUD KASRAI, JEFF N CUTLER, KATHRYN GORE, GREG W CANNING and G MICHAEL BANCROFT
Department of Chemistry, University of Western Ontario London, Ontario, N6A 5B7, Canada,
KIM H TAN
Canadian Synchrotron Radiation Facility, University of Wisconsin, WI 53589, USA

ABSTRACT

Oil-soluble Mo-S compounds are well established as antiwear and extreme pressure additives in lubricating oils and greases (1). Molybdenum dialkyldithiocarbamates (MoDTC) have been investigated by several groups in recent years as a friction modifier in automobile engines (2,3). The effect of MoDTC compounds on friction reduction in most investigations has been attributed to surface films containing primarily MoS_2 and other molybdenum oxides. In recent years we have shown that X-ray absorption near edge structure (XANES) is a powerful technique for antiwear film characterization (4-6).

In order to obtain a better understanding of the decomposition mechanisms of antiwear and antioxidant engine oil additives, such as zinc dialkyldithiophosphate (ZDDP) and molybdenum dithiocarbamate (MoDTC), tribochemical and thermally deposited films were prepared from oil solutions containing varying concentrations of these two additives. The chemistry of these films was studied by using X-ray absorption (XANES) spectroscopy and X-ray photoelectron. The antiwear properties were examined on a Plint rubbing machine by measuring the amount of wear as a function of time and the concentrations of the additives.

The combined results of S L-edge and K-edge XANES of tribofilms generated from neat MoDTC indicates that the film is composed of a MoS_2 like film and a sulphate. The sulphate is mostly on the surface. On the other hand, films generated from a mixture of MoDTC and ZDDP do not show the sulphate form and mostly contain MoS_2-like and ZnS-like species (see Fig. 1). The presence of as low as 100 ppm ZDDP is sufficient to prevent the oxidation of sulphur to sulphate and Mo (IV) to Mo (VI). ZDDP also facilitates the formation of a better MoS2-like film.

Results obtained from the P L-edge and K-edge XANES show that there is some interaction between ZDDP and MoDTC at low concentration of ZDDP (< 500 ppm). At the low concentration, the nature of the polyphosphates film resemble a short chain polyphosphate that clearly affects the wear performance of the film. At high concentrations the effect is not noticeable. The S K-edge data clearly show that the counter ion in the phosphates is zinc and not iron.

The S XANES spectra of the thermal films suggest that major part of sulphur in the films undergoes oxidation to sulphate. The phosphate form is also affected by the presence of MoDTC. Generally, short chain polyphosphates were found for the thermal films.

Using mixtures of ZDDP and MoDTC, the friction is reduced, and wear is comparable to using ZDDP alone. The chemistry of the tribofilms is related to wear and friction measurements.

REFERENCES

(1) P.C.H. Mitchell, Wear, (1984), 100, 281-300.
(2) E.R. Braithwaite and A.B. Greene, Wear, 1978, 26, 405-432.
(3) M. Muraki and H. Wada, *Lubricants and Lubrication, Tribology Series*, Vol. 30, Dowson, D et al. Ed., Elsevier, (1995), pp. 409-422.
(4) Z. Yin, M. Kasrai, M. Fuller, G.M. Bancroft, K. Fyfe, and K.H. Tan, Wear (1997) 202, 172-191.
(5) Z. Yin, M. Kasrai, G.M. Bancroft, K. Fyfe, and K.H. Tan, Wear (1997) 202, 192-201.
(6) M. Fuller, Z. Yin, M. Kasrai, G.M. Bancroft, E.S. Yamaguchi, P.R. Ryason, P.A. Willermet, and K.H. Tan, Tribology International, (1997), 30, 305-315.

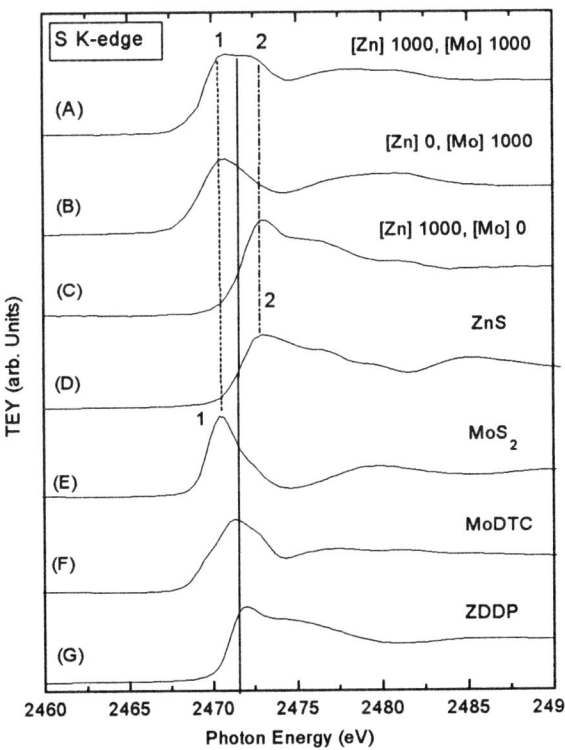

Fig. 1: S K-edge XANES (TEY) spectra of tribofilms generated from combinations of ZDDP and MoDTC, as compared with model compounds.

BOUNDARY FILM FORMATION BY ZnDTPs AND DETERGENTS USING ECR

E S YAMAGUCHI
Oronite Global Technology, Chevron Chemical Company, Richmond, California 94802, U.S.A.
P R RYASON and T P HANSEN
Chevron Products Company, Global Lubricants, Richmond, California 94802, U.S.A.
S W YEH
Chevron Research and Technology Company, Analytical Sciences, Richmond, California 94802, U.S.A.

ABSTRACT

Both overbased detergents and zinc dithiophosphates (ZnDTPs) form electrically insulating boundary films. Prior work at this laboratory has shown that the rates of resistive film formation can be investigated with electrical contact resistance (ECR) measurements (1). A study was undertaken to compare resistive film formation rates of separate commercial overbased detergents, a commercial secondary ZnDTP, and combinations of those additives dissolved in a light neutral oil. As tribocouple, a hard 52100 steel ball was slid upon a hard 52100 steel disk at 100°C. Additionally, XPS analysis was performed upon the wear scar surfaces generated by sliding wear in the base oil, in the various base oil-additive mixtures, and of a new (unused) ball.

Overbased calcium sulfonate, overbased magnesium sulfonate, and overbased calcium phenate all formed resistive films rapidly when base oil-detergent mixtures were used to lubricate the tribocouple in sliding wear. Under the same conditions, the secondary ZnDTP-base oil mixture formed a resistive film more slowly than did any of the detergent-base oil mixtures. When binary combinations of detergent with ZnDTP were used in admixture with the base oil, resistive film formation was inhibited. Of the detergents, overbased calcium phenate, in combination with ZnDTP, took the greatest length of time to form a resistive film. Typically, film resistance proceeds with an initial induction period, with little change in resistance, followed by a relatively rapid, but sometimes sporadic, increase in film resistance. Sheasby and Caughlin have made similar observations for calcium sulfonate and magnesium sulfonate in combination with ZnDTP in a direct observation wear machine (2).

XPS analysis was undertaken on the wear scar surfaces to obtain some information on the surface composition of the wear scar surface. Perhaps the greatest difference between the results of the present work and that previously reported (3)(4) is that for overbased calcium sulfonates. Two groups have reported that, in a rubbing contact, calcite is formed in the contact from the amorphous calcium carbonate micelles of the overbased calcium sulfonate suspension. In the work reported here, no carbonate ion was evident in the XPS spectra of the wear scar surfaces from overbased calcium sulfonate (OBCS), overbased magnesium sulfonate (OBMS), nor from the OBCS-ZnDTP or OBMS-ZnDTP combinations. However, carbonate was observed on the wear scar surface from the overbased calcium phenate (OBCP)-ZnDTP combination.

Clearly, profound additive interactions occur. Presumably, counterparts of the observations made in the tribometer experiments described here are to be found in practical sliding contacts, e.g., engines. As such, ECR measurements may help to guide the work of synthetic and oil formulation chemists.

REFERENCES

(1) E S Yamaguchi, P R Ryason, and T P Hansen Tribology Letters Vol. 3, No. 1, 1997, pp 27-33.
(2) J S Sheasby and T A Caughlin, Lubricants and Lubrication (Leeds-Lyon 21), 1995, pp 399-408.
(3) J M Georges and D Mazuyer, J. Phys. Condens. Matter, Vol. 3, 1991 pp 9545-9550.
(4) T Palermo, S Giasson, T Buffeteau, B Desbat, and J M Turlet, Lubr. Sci., Vol. 8, No. 2, 1996, pp 119-127.

WEAR BEHAVIOUR OF NEUTRAL, BASIC ZINC DIISOBUTYLDITHIOPHOSPHATES AND BIS(O,O-DIISOBUTOXYPHOSPHINOTHIOYL)-DISULPHIDE IN THE PRESENCE OF SOOT.

L. MARGIELEWSKI and S. PŁAZA
Department of Chemical Engineering and Environmental Protection, University of Łódź,
Pomorska 163, 90-236 Łódź, Poland

ABSTRACT

Combustion-generated soot in the crankcase of diesel engines in the presence of zinc dialkyl-(aryl)dithiophosphate (ZDTP) can significantly increase engine wear.

The prowear action of soot and carbon black in the presence of ZDTP was investigated using different tribometers; four-ball (1-4), plane-on-plane (5), reciprocating piston-cylinder (6), ball-on-flat-disk (7) and cross-pin (8) and also engine test (9).

The investigators have presented different hypotheses. The hypotheses of adsorption of ZDTP have been mentioned most often. According to Rounds the effect of soot on wear in the presence of ZDTP is that soot preferentially adsorbs the acid antiwear species formed from ZDTP before they have a chance to form the antiwear coatings on the rubbing surfaces. Based on adsorption studies and engine tests Hosonuma and co-workers reported that diesel soot adsorbs the zinc-containing ZDTP decomposition by-products but phosphorus containing compounds are adsorbed far less.

According to other authors carbon black and soot act as an abrasive.

Since the major degradation pathway of ZDTP in rubbing contact is oxidation which produces the disulphide $[(RO)_2PS]_2$ and a basic salt $[(RO)_2PS]_6Zn_2O$, these compounds were used in our investigations.

Our adsorption studies (10) of neutral and basic ZDTPs from n-hexadecane solution on carbon black have demonstrated that these compounds are strongly oxidized to bis(O,O-dialkoxyphosphinothioyl)disulphide (DS) at 80°C. Investigation of DS adsorption on carbon black at this temperature has revealed that DS is oxidized to elemental sulphur (ES) which may be due to corrosive wear of rubbing surface.

Wear tests were made on a four-ball wear tester (Load = 300 N, Speed = 1400 r/min, Oil Temp.= 80°C, Time = 15, 30, 60 and 90 min)

The antiwear (AW) properties of both ZDTP salts and DS without soot are little influenced by time of friction. Zinc salts have significantly better AW behaviour then DS. Wear volumes in case of DS are about 5 fold bigger. Presence of dispersed soot in samples caused increase in volume in comparison to the samples without soot by factor ranging from 2.5, 3.9 and 1.9 for 15 min to 13.3, 9.4 and 2.8 for 90 min wear tests in case of neutral salt of ZDTP, basic salt of ZDTP and DS respectively.

The antiwear (AW) properties of both ZDTP salts in the presence of soot, in contrast to the AW activity of DS in the same condition, are strongly dependent of time of friction. Wear volume after 15 minutes wear tests is 3.5 times smaller for DS in comparison to both ZDTPs and falls down to 1.1 after 90 minutes.

Additionally the wear scars were tested using SEM/EDAX method and the soot was tested using X-Ray fluorescence analysis.

The analysis of the wear scars after tests using SEM/EDAX technique show change in the concentrations of the sulphur, zinc and phosphorus in the AW layer. The ratios of zinc and sulphur to phosphorus are comparable to the ratios of this elements corresponding to the zinc polythiophosphates. After tests done in the presence of soot on the scars zinc was not present, and zinc polythiophosphate layer was unable to be formed.

Analysis of the soot using X-Ray fluorescence technique showed increase in the ratios of concentration of zinc to phosphorus (3:1 onto the soot) in comparison to the original compound (0.5:1). this indicated that ZDTPs are oxidised onto the soot surface.

Oxidation of ZDTP salts to the DS onto soot is the major reason for the antagonistic effect of this form of carbon on the AW performance of well known zinc dialkildithiophosphates.

REFERENCES

(1) Rounds F. G., SAE Papers 770829.
(2) Rounds F. G., SAE Papers 810499.
(3) Rounds F. G., Lubr. Eng., Vol. 40, 7, 394, 1984.
(4) Kawamura, Ishiguro T., Lubr. Eng., Vol. 43, 572, 1987.
(5) Hosonuma K., Yoshida K., Matsunaga A., Wear, Vol. 103, page 297, 1985.
(6) Berbezier J., Martin J.M., Kapsa P., Tribol. Int., Vol. 19, No. 3, page 115, 1986.
(7) Ryason P.R., Chang I.Y., Gilmore J.T., Wear, Vol. 137, page 15, 1990.
(8) Nagai, Endo H., Nakamura H., Yano H., SAE Papers 831757.
(9) Yahagi Y., Tribol. Trans., Vol. 20, page 365, 1987.
(10) Płaza S., Margielewski L., Tribology Trans., Vol. 36, No. 2, page 207, 1993.

Submitted to *Tribology International*

IDENTIFICATION AND CHARACTERISATION OF SURFACE CHEMICAL ASPECTS IN SLIDING FRICTION

M. BINGGELI,
Blaser Swisslube AG, 3415 Hasle-Rüegsau, Switzerland
C. SOTO, N.D. SPENCER,
LSST, ETH Zürich, Switzerland

INTRODUCTION

The chemical state and the chemical properties of surfaces greatly influence the sliding behaviour of two touching bodies. One of the most spectacular phenomena is the so called stick-slip motion. This is a well known phenomenon from daily life (squeaking door, violin string, etc.). Technically, it occurs in the boundary lubrication regime, at low speeds or under poor lubrication conditions. Stick-Slip has been studied in numerous attempts, but can not yet be entirely explained or modelled, at least not for real, non-idealised systems. The complexity of the involved chemical and physical processes is overwhelming.

It is empirically known, that polar components such as fatty acids and esters, do a good job in stick-slip prevention. Traditionally, it is believed that a high static friction coefficient is responsible for the occurrence of stick-slip. The emerge of very high resolution surface analytical techniques, such as XPS, scanning TOF-SIMS and AFM/STM in the last decade stimulate new efforts in the field.

The first goal of the presented study is to show that real machine guideways can be simulated with a flat on disk tribometer, the second, to address some first results on the way to correlate the surface chemistry of (steel) surfaces with their actual tendency to undergo stick-slip motion.

For the tribosystem, we chose to look at guideways of tool machines. This represents an industrially relevant system where stick-slip is a major concern. The lubrication of such guideways is an underrated problem in the opinion of the authors. To achieve the ever narrower tolerances and to avoid the risk of tool fracture the efficient suppression of stick-slip motion is a key factor. This is a serious condition to meet, especially with respect to the other duties of the guideway oil, as e.g. corrosion protection and a good compatibility with the coolant.

EXPERIMENT/RESULTS

Pure base fluids and blends thereof with friction modifying additives were systematically evaluated in regard to stick-slip. The first set of experiments was conducted on a pilot guideway of the engineering school of Darmstadt (Germany). This guideway allows to accurately measure friction forces under realistic circumstances at extremely low speeds down to 0.1 mm/min.

The very interesting results readily evidenced that stick-slip tendency can not be derived only based on static friction force considerations. Furthermore, it became clear, that the definition of the static friction coefficient is a very problematic one at this moment. The static friction force found for a macroscopic system seems to be a question of the time scale or the speed control rather than of actual underlying physical processes. This is opposed to the findings with e.g. friction force microscopy, where real interatomic 'hopping' with discrete static friction force values can be observed.

The results of the pilot guideway served as the baseline for measurements on a flat on disk system. By creating adequate experimental conditions it became possible to obtain Stribeck (friction-velocity) curves in very good agreement with those formerly measured on the pilot guideway. Thus, we concluded it legitimate to carry on studying fundamentals of stick-slip motion on the bench test apparatus.

Several base stocks and blends were tribo-stressed on the pin/disk machine. Then, the samples were investigated with surface analytical methods. We find that EP/AW-additives undergo several chemical transitions before reaching their final state. Particularly, we looked at a derivative of phosphorous acid exhibiting extraordinary friction modifying properties. It turns out that a phosphate rich layer is being built up under tribo-stress conditions. The higher the pressure, the thicker the amorphous layer becomes. It was found to be up to 1000 nm thick.

CONCLUSIONS

It was possible to show the comparability of a bench top tribometer with a pilot guideway, thus allowing to carry out the evaluation of guideway oils in a fast and cost effective way. Further, the combination of surface analytics and tribology is successfully started and has brought about very encouraging results for future work. With the presented approach we hope to contribute to a more profound understanding of the processes that govern the stick-slip phenomenon on real surfaces.

MICELLAR CALCIUM BORATE AS AN ANTI-WEAR ADDITIVE

VIRGINIE NORMAND and JEAN MICHEL MARTIN
Laboratoire de Tribologie et dynamique des systèmes UMR 55 13, Ecole Centrale de Lyon, F-69131 Ecully, France
KIYOSHI INOUE
Central Technical Research Laboratory, Nippon Oil Co., Ltd., Yokohama 231 Japan

ABSTRACT

In this study, a new generation of lubricant anti wear additives are described.

The anti wear action of two types of micelles have been studied by TEM analysis of wear particles generated during a friction test between two ferrous substrates (boundary lubrication conditions) (1).

Here are described the two types of micelles :
- The Calcium Carbonate Overbased Salicylates (or CC micelles) are made of a calcium carbonate core ($CaCO_3$) and an organic shell (called salicylates).
- The Calcium Borate Overbased Salicylates (or CB micelles), are micelles in which calcium carbonate have been replaced by calcium borate $Ca(BO_2)_2$ (2).

All wear debris, directly deposited on a carbon grid, were investigated by analytical electron microscopy (Diffraction electron and Electron Energy Loss Spectroscopy techniques).

Conclusions have been made, regarding to the efficiency of the CC and CB micelles. CC micelles seem not to be very efficient during the friction process ; wear debris stemming from CC micelles tests (shown figure 1) contain a high rate of iron (as a crystalline iron oxyde Fe_2O_3). Their antiwear action is limited as far as they are not preventing the formation of Fe_2O_3 known as a long term abrasive compound.

Fig. 1
Crystalline wear particle (from CC micelles)

Unlike CC used micelles, CB ones lead to the formation of an amorphous film, poorer in iron than crystalline debris observed at the same time (made of iron oxyde Fe_2O_3) (See an example of amorphous wear particle figure 2).

We assume, by analysis of EELS Spectra (shown on figure 3), that the amorphous matrix is made of calcium and iron borate ; this matrix appears to contain a large number of small calcite grains (5-10 nm).

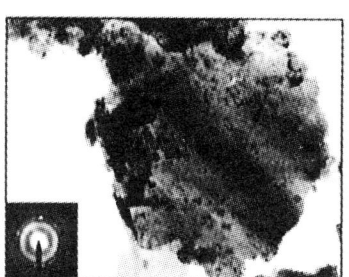

Fig. 2
amorphous wear particle (from CB micelles)

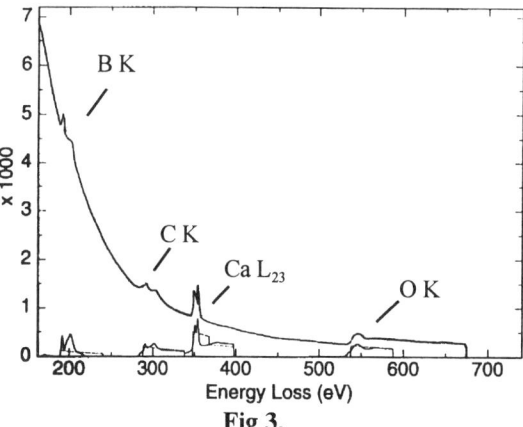

Fig 3.
EELS spectra of an amorphous wear particle

CONCLUSION

CB micelles, thanks to the presence of boron (2), lead to the formation of an amorphous film (at the microscopic level), preserving the contact metallic surfaces from wear due to shearing forces.

This film acts by substituting film / film contacts for metal / metal ones and by limiting the formation process of iron oxyde Fe_2O_3. : boron (Z = 5) behaves as an oxidation inhibitor and a glass former, like phosphorus for zinc dithiophosphates.

REFERENCES

(1) Mansot J.L., Hallouis M., Martin J.M., 'Colloïdal antiwear additives' 2- 'Tribological behaviour of colloïdal additives in mild wear regime' Colloids and surfaces, vol 75, p 25-31, 1993.

(2) Inoue K. 'Calcium borate overbased detergents' Lub. Eng., vol 49, No.4, p 263-268, 1992.

Submitted to *JAST*

THE UTILIZATION OF NOVEL BENCH SCREENING TECHNIQUES IN THE DEVELOPMENT OF ANTIWEAR ADDITIVES FOR LUBRICANTS

CYRIL A. MIGDAL, ROBERT G. ROWLAND, and JOHN R. BARANSKI
Uniroyal Chemical Company, Inc., Benson Road, Middlebury, CT 06749

ABSTRACT

We have investigated a modification to two bench tests that are widely used in the evaluation of potential antiwear additives for gasoline and diesel engine oils. Improving the reliability of economical bench tests is important, as antiwear agents are ultimately selected based on performance in expensive engine tests. Both industry and government have great interest in finding improved antiwear agents, particularly those that do not include zinc or phosphorus.

Addition of cumene hydroperoxide (CHP) to the test formulations is the key parameter investigated. The CHP is reported (1) to serve as a pro-oxidant, catalyzing sample degradation, and better simulating conditions encountered in actual lubricant usage. The purpose of the present study was to determine if the pro-wear effect of CHP observed in engine tests (1) and in the four-ball test (2) could also be observed in a reciprocating rig.

Test results are obtained using a Falex Four-Ball Friction and Wear Machine, following ASTM D 4172, and a reciprocating Cameron-Plint High Frequency Friction Machine (TE77 model). The reciprocating rig employs a 100 Newton load, 30 Hertz frequency, and a 2.35 mm stroke length. The temperature of the Cameron-Plint test chamber is increased from ambient to 150 °C during the run. Using the two test rigs therefore allows us to look at both rotational and reciprocating motions, and simulate low and high temperature wear conditions.

Data was collected in both a passenger car motor oil and a heavy duty diesel engine oil. Wear scar diameter measurements are reported for both formulations in each test. Into each blend tested is 1 wt % of CHP (80%, technical grade). The data in Figure 1shows the effect of CHP on the wear scar diameter in fully formulated oils in ASTM D 4172, with and without ZDDP present.

The data shows that with No CHP present there is very little difference in wear scar diameter between oils with and without ZDDP present. Without CHP present the gasoline formulation does show a slight difference between blends with and without ZDDP, but the the diesel formulation actually shows the oil without ZDDP to be have a smaller wear scar. In contrast, with CHP present both formulations show a significant difference in wear scar diameters between blends with and without ZDDP.

Four-Ball Wear Results of Engine Oils (Effect of CHP)

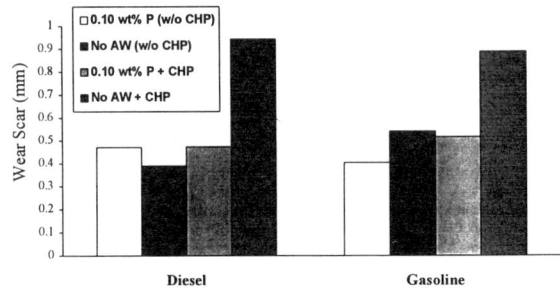

Figure 1

The data in Figure 2 shows the effect of CHP on the wear scar diameter in fully formulated oils in the Plint test described above, with and without ZDDP present.

Cameron-Plint Wear Results of Engine Oils (Effect of CHP)

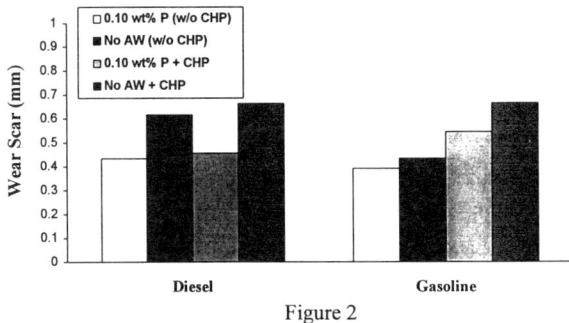

Figure 2

The data in Figure 2 shows that with No CHP present there is some difference in wear scar diameter between diesel oils with and without ZDDP present. The difference in the diesel oil becomes slightly greater when CHP is present. Without CHP present the gasoline formulation only shows a slight difference between blends with and without ZDDP, but with CHP present the effect is more pronounced. The effect of adding CHP to promote wear is greater in the rotational motion four-ball test than in the reciprocating motion Plint test.

REFERENCES

(1) J.J. Habeeb, W.H. Stover, ASLE Trans, 30, 4, 419 (1987).
(2) F. Rounds, Tribology Trans, 36, 2, 297 (1993).

Tribological Investigation of Environmentally Compatible Extreme-Pressure Additives

PIERRE VOUMARD, ERIC JEANPETIT AND EBERHARD PFLÜGER
Centre Suisse d'Electronique et de Microtechnique S.A., Tribocoatings, Rue Jaquet-Droz 1, CH-2007 Neuchâtel

INTRODUCTION

Ashfree compounds, avoiding the presence of metallic atoms, are synthesized by the major additive suppliers to replace ZnDTP derivatives as extreme-pressure and anti-wear additives. Further efforts are made to eliminate heteroatoms like phosphorus or sulfur. Nevertheless, even if it is recognised that these new additives may be as efficient as ZnDTP, there are comparatively fewer data available. We have undertaken a systematic evaluation of ashfree EP additives. Their performances are compared with those of classical ZnDTP. For environmental and economical consideration, it is important to reduce the additive concentration to the minimum needed to keep the desired performance.

EXPERIMENTS

All the tests have been performed on CSEM pin-on-disk tribometers at room temperature and at a relative humidity maintained at 50 %. The disks and the 6 mm ø balls used are made out of AISI 52100 steel. The tribological tests are run systematically by changing the concentration of the additives.

RESULTS

In some cases it can clearly be observed that additives used at a high concentration have negative influence on the anti-wear properties (see figure 1).

Interestingly Hitec E655 has anti-wear properties that are strongly dependent on the base oil. In a diester an optimum occurs at low concentration (about 0.5%). On the contrary in poly-alpha-olefin (PAO 6) the anti-wear properties are expressed only at concentrations higher than 2%. Finely HITEC E655 has no anti-wear properties when mixed in a triester. It should be noticed that this base oil already has very good tribological properties. It can be assumed that there is competition between the base oil, that is a very polar compound, and the additive on the surface.

In order to minimize the effect of the base oil and of its interactions with the additives, the focus was put on the PAO that is a very unreactive solvent. Figure 2 shows that in those test conditions the Hitec E655 has very little effect and that one percent of the additive does not reduce the wear rate in comparison with the pure base oil. Irgalube 211 behaves well in moderate stress as an antiwear. On the contrary at higher load, the extreme pressure capabilities are poor. On the other side the Irgalube 349 has anti-wear and extreme-pressure capabilties with reduction of the wear rates of about three orders of magnitude.

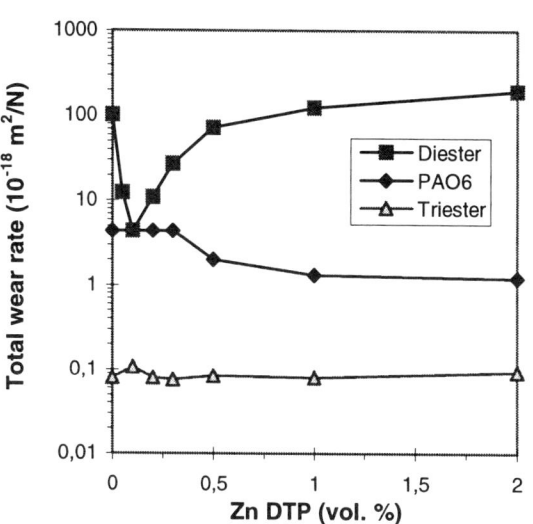

Fig. 1: Wear rate vs. Hitec E655 concentration in different base oils. Test conditions: load: 5 N, sliding speed: 10 cm/s

CONCLUSIONS

The ashfree additives tested exhibit interesting anti-wear and extreme-pressure properties. They are linked to the formation of phosphorus rich layers that can grow already at room temperature. This capability makes this type of additive very suitable for applications where heat dissipation is negligible, like fine mechanics.

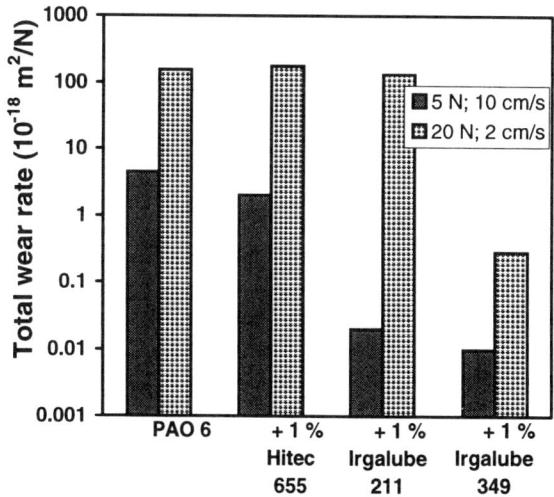

Fig. 2: Wear rate without and with 1% of different additives in a PAO base oil.

Submitted to *Lubrication Science*

TRIBOLOGICAL BEHAVIOR OF SOME ANTIWEAR ADDITIVES IN VEGETABLE OILS

U.S.CHOI, B.G.AHN and O.K.KWON
Tribology Research Centre, KIST, PO BOX 131, Cheongryang, Seoul, Korea
Y.J.Chun
Department of Ind. Chem. Eng., Chungnam Polytch. Univ., Hongsung, Chungnam-do, 350-800, Korea

ABSTRACT

Recently due to growing the worldwide interest in environmental issues, there have been a growing concern for the use of vegetable oils having excellent biodegradability and also improved limitation stability by additive formulation as more environmentally acceptable base fluids (1)(2).

Antiwear additives are used for wear reduction in tribological systems. However, under severe condition, thermal activation energy of free hydrogen released by exothermic decomposition of the hydrocarbons contribute to the thermal instability at the frictional contact junction, resulting in film rupture and failure (3)(4).

To solve this problem, the new additive, 3.5 di-t-butyl 4-hydroxy benzyl phosphonate(DBP) has synthesized and its antiwear performance investigated and compared with that of the conventional additive, TCP in vegetable oils using Shell Four-Ball wear tester. The new additive provides excellent antiwear performance under high temperature and sliding velocity conditions. To investigate the wear mechanism of DBP additive, thermal degradation tests were conducted for idenfication of exothermic and endothermic reactions of hydrogen free radicals in heated vegetables oils using differential scanning calorimeter, and also the surface analysis of worn balls was carried out using EDAX and optical microscope. This study is to describe the effect of the tribological behavior of antiwear additives in vegetable oils and to establish the wear mechanism of the new additive, DBP.

The effect of sliding velocity on the antiwear performance is shown by the data in Figure 1. The results were obtained at a load of 392N, a test duration of 60 min, bulk oil temperatures of 75 and 150°C and sliding velocities of 30.7, 40.1 and 69.1 cm/sec. As show in Figure 1, wear rose with increasing sliding velocity. Without the additive function, the transition from wear to severe wear with sliding velocity was inevitable in such high stress and temperature conditions due to the accumulation of heat and the subsequent thermal instability of frictional contact junction.

When TCP added, there was influenced the phenomenon little in spite of the improvement on antiwear performance. But the new additive DBP showed relatively constant wear ratio in spite of increasing sliding velocity, and higher antiwear performance compared with TCP under severe conditions. This can be proved by surface analysis and thermal degradation test.

On the basis of the experimental results, it is deduced that the wear mechanism of TCP is protective film formation, and its antiwear capability depends upon the shear strength of the film formed. On the other hand, with DBP, the thermal energy dissipation was much reduced and thermal instability was reduced by the hydrogen scavenging reaction which is an endothermic reaction process.

In conclusion, the new additive, DBP has been shown the dual function of hydrogen scavenging and protective film formation and to be a highly effective antiwear additive in vegetable oils.

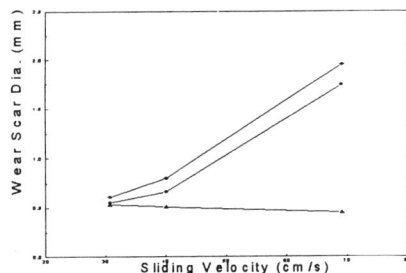

Fig.1: Effect of sliding velocity on antiwear performance under olive oil : ◆ nonadditive, ● TCP, ▲ DBP.

REFERENCES

(1) O. O. Steven, Lub. Eng., 1989, pp 685-692
(2) K Lal and V Carrick, Ecological and Economical Aspects of Tribology, 1991, 2-9.1-14.
(3) F.T. Barwell and O.K.Kwon, Proc. Plenary Session of 3rd Int. Tribology Congr., Vol.1, 1981.
(4) U.S.Choi and O.K.Kwon, KSTLE, Vol.7, No.1, 40-45pp, 1991.

LUBRICATION PERFORMANCE OF MODEL ORGANIC COMPOUNDS IN HIGH OLEIC SUNFLOWER OIL

ICHIRO MINAMI*
Kochi University of Technology, Tosayamada Kochi, 782, JAPAN
(*was a visiting scientist at Lubrizol from '95-'96)
HYUN-SOO HONG, and NARESH C MATHUR
The Lubrizol Corporation, Wickliffe, OH 44092, USA

ABSTRACT

The effect of additives on the lubricity of high oleic sunflower oil was studied. Linear alkyl compounds with a polar functional group at the terminal carbon were selected as model friction modifiers. Antiwear and antifriction properties depend on the functional group in the order of carboxylic acid > amine > amide. Thiol exhibits negative effect on lubricity.

INTRODUCTION

The demand for environmentally friendly base oils in the lubricants industry has increased due to growing concern on environment protection. Synthetic esters have been studied as an alternative base oils to mineral oil. However, these synthetic ester lubricants are typically more expensive than mineral-based lubricants.

Recently, high oleic sunflower oil has received substantial interest due to its good oxidative stability and improved low temperature properties (1). It also has good lubricating properties. However, the lubricating properties of high oleic sunflower oil can be improved with appropriate additive technology (2).

In this study, four linear alkyl compounds having a polar functional group at the terminal carbon were examined as model vegetable oil additives.

RESULTS AND DISCUSSION

Four model compounds were used in this study. Their code and structure are; ACID (RCO2H), AMINE (RCH2NH2), AMIDE (RCONH2), and THIOL (RCH2SH). Pin (8620 steel) on disk (8620 steel) tests were performed at 50 or 75°C under the contact pressure of 0.5-0.9 GPa (corresponds to normal load of 34.3-200N).

Anti-wear properties: Very low wear was observed at 50°C in the load range of 34.3-141 N. Almost no additive effect was observed since base sunflower oil itself lubricated efficiently under these conditions. At the load of 200 N, the effect of additives was observed. ACID, AMINE, and AMIDE prevented wear to some extents, whereas THIOL promoted wear. At 75°C the amount of wear volume with THIOL was higher than that with base fluid under all conditions examined in this study. ACID, AMINE, and AMIDE exhibit antiwear properties in the order of ACID > AMINE > AMIDE.

Antifriction properties: Four friction types were found by monitoring friction torque during the test:
Type A: constant friction coefficient at steady state.
Type B: friction gradually increases during the test.
Type C: friction gradually decreases during the test.
Type D: friction is fluctuating throughout the test.

The Type A represents a desirable lubricant while the Type D represents a poor lubricant. Type B and Type C may be acceptable. Duplicated test result of friction trace under each conditions indicate that ACID, AMINE, and AMIDE are generally good or acceptable lubricants. Improvement in lubrication was observed with ACID and AMINE under higher temperature and higher load conditions.

THIOL and ACID are reported to exhibit antiwear properties in mineral oil (3). However, combination of the two additives decreases antiwear effect. There might be some interaction between the additives. The reaction of carboxylic acids or esters with thiols to afford thioacids is known (4). Since thioacids are more acidic than original carboxylic acids, they are potentially corrosive wear causing.

CONCLUSION

Taking wear volume and friction type into account, lubricity can be described in the order of ACID > AMINE > AMIDE >> THIOL. Except for THIOL, the tendency of antifriction is closely related to the results in the mineral oil (5).

REFERENCES

(1) Lawate, S.S., "CRC Handbook of Tribology," in press.
(2) Asadauskas, S., Perez, J.M., Duda, J.L., Lubrication Engineering, 52, 12 877-882 (1996).
(3) Forbes, E.S., Reid, A.J.D., ASLE Trans., 16, 1 50-60 (1973).
(4) Janssen, M.J., "The chemistry of carboxylic acids and esters" edited by S.Patai, p705-764 (Chapter 15), John Wiley & Sons Ltd., London (1969).
(5) Jahanmir, S., Beltzer, M., ASME J. Tribol., 108, 1 109-116 (1986).

THE SURFACE CHEMISTRY OF EXTREME PRESSURE LUBRICATION

WILFRED T. TYSOE

Department of Chemistry and Laboratory for Surface Studies, University of Wisconsin-Milwaukee, Milwaukee, WI 53211, USA

ABSTRACT

Chlorinated hydrocarbons are commonly added to the base fluid to synthesize lubricants used under extreme-pressure (EP) conditions. It is shown that the seizure load measured using a pin and v-block apparatus increases with the addition of chlorinated hydrocarbons to a base fluid consisting of a poly α-olefin. Two types of behavior are observed; one in which the seizure load initially increases but then reaches a plateau after the addition of a limiting amount of additive. This behavior is exemplified by chloroform and methylene chloride. In a second type of behavior, demonstrated when using carbon tetrachloride as an additive, the seizure load appears to increase continually and very rapidly with increasing chlorine concentration without showing any sign of reaching a plateau.

It has been demonstrated that the interfacial temperature in the EP regime during lubrication with the pin and v-block apparatus varies linearly with the applied load and that temperatures in excess of 1000 K can be attained (1). A temperature calibration is obtained by using pins of known melting point which revealed that the temperature rise is ~2.5 K/kg applied load in this apparatus. Both methylene chloride and chloroform thermally decompose at these temperatures forming a film that consists of a layer of iron chloride and which also incorporates small (~50 Å diameter) carbon particles. This is consistent with the nature of the surfaces of the pins and v-blocks which are also found to contain iron halides.

It is proposed that this halide layer provides a solid, anti-seizure lubricant which is removed by abrasion and continually replenished by reaction with the additive. Both the film wear and growth rates can be independently measured and the film thickness during a pin and v-block experiment (X) calculated. The results show that films of ~microns in thickness can be deposited. Since the reactively formed halide film prevents seizure, it is assumed that seizure takes place when the film thickness $X \to 0$. This idea successfully predicts the variation in seizure load with additive concentration when using methylene chloride as additive and, with appropriate modification, also that of chloroform.

It is further demonstrated that, at sufficiently high temperatures, carbon tetrachloride can thermally decompose on iron to form a carbide layer (2). It is shown, by careful measurements of the wear rate for different applied loads, that iron chloride still provides the anti-seizure layer at low loads but that iron carbide (Fe_3C) fulfills this role at much higher applied loads (and interfacial temperature).

The chemistry of chlorinated hydrocarbons was studied on clean iron in ultrahigh vacuum using molecular beams. This approach is particularly useful for measuring gas-phase products. These experiments show that both methylene chloride and chloroform completely thermally decompose to yield only hydrogen as:

$$CH_2Cl_2 + Fe \to FeCl_2 + C + H_2$$

$$\tfrac{2}{3}CHCl_3 + Fe \to FeCl_2 + \tfrac{2}{3}C + \tfrac{1}{3}H_2$$

The carbon is present as small (~50 Å diameter) carbonaceous particles and the number of particles depends on the stoichiometry of the chlorinated hydrocarbon and this variation may affect the tribological behavior.

Similar experiments using carbon tetrachloride shows that this thermally decomposes to yield only carbon and iron chloride according to:

$$\tfrac{1}{2}CCl_4 + Fe \to FeCl_2 + \tfrac{1}{2}C$$

although a small amount of C_2Cl_4 is detected.

Finally, heating co-adsorbed carbon+chlorine films, formed from either carbon tetrachloride or methylene chloride in ultrahigh vacuum, reveals that the carbon diffuses much more rapidly when co-adsorbed with four chorines (from CCl_4) than when co-adsorbed with two (methylene chloride). This effect may explain the increased tendency for carbon tetrachloride to form a carbide and therefore rationalizes its superior anti-seizure behavior. That is, the activation energy to carbon diffusion into the bulk is decreased by the addition of electronegative chlorine to the surface.

REFERENCES

(1) T.J. Blunt, P.V. Kotvis and W.T. Tysoe, Tribology Letters Vol. 2, page 221, 1996

(2) P.V. Kotvis, J. Lara, K. Surerus and W. T. Tysoe, Wear, Vol. 201, page 10, 1996

$FeCl_3 \cdot 6H_2O$ - A potential new Lubricant for Cold Forging of Stainless Steel

T. STEENBERG, E. CHRISTENSEN AND N. J. BJERRUM

Materials Science Group, Department of Chemistry, Technical University of Denmark, Build. 207, 2800 Lyngby, Denmark

ABSTRACT

Cold forging of stainless steel is a potentially important process but there is a need for better lubricants. Currently iron oxalate with MoS_2 is used, but this system is insufficient for more severe cold forging processes (i.e. backward can extrusion) and undesirable environmental aspects in the coating process. The idea of using $FeCl_3$ as a lubricant for cold forging of stainless steel (1) presented here originated from chlorinated organic EP additives (i.e. chlorinated paraffin oil). The EP mechanism of these compounds has been investigated by various authors and it is believed that decomposition to $FeCl_2$ at the steel surface plays a very important role (2). $FeCl_3$ would be expected to react with the steel surface to form $FeCl_2$ during the metal forming process:

$$2FeCl_3 + Fe \Leftrightarrow 3FeCl_2$$

We have applied the lubricant film by dipping the test specimens in a melt of $FeCl_3 \cdot 6H_2O$ at 50°C. Chemical analysis (visible absorption) of an aqueous solution of the lubricant film showed an increasing amount of $FeCl_2$ after surface expansion. This confirms that the lubricant film consist of two layers: $FeCl_2$ at the steel surface and $FeCl_3$ on top. During the metal working process a considerable surface expansion occurs, and $FeCl_3$ reacts with the formed nascent surfaces and form $FeCl_2$. In this case the lubricant film "regenerate" itself during the metal working process.

The proposed lubricating mechanism has also been confirmed by mechanical tests. The capabilities of this new lubricant has been evaluated by using a backward can (or cup) extrusion test.

The surface expansion and pressure was measured during the test in order to compare and evaluate the friction. This was also done to see whether there was formation of a "dead zone" in front of the punch nose. This did not occur and the surface expansion was similar in all experiments. The lubricant was also mixed with MoS_2 and graphite.

The results show that MoS_2 and graphite had no effect on the lubrication and friction during the backward can extrusion (p_2 constant) but improved the removability of the cup from the container afterwards.

We believe the explanation for this is that $FeCl_3$ is an excellent lubricant during the "dynamic" part of the process (surface expansion) but perform worse under more "static" conditions.

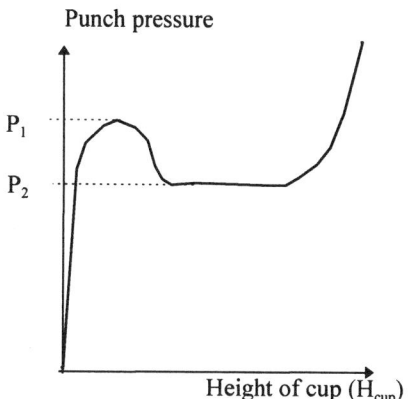

Fig. 1

This idea is also in agreement with the observations that additions of MoS_2 or graphite lowered p_1. A H_{cup}/D_{cup}-ratio of more than two could be obtained without pick-up, sticking or seizure (D_{cup} is the diameter of the cup).

REFERENCES

(1) Steenberg, T., Christensen, Bjerrum, N. J., Kønig, B. G. - *Use of Transition metal Halogenides, especially $FeCl_3$, as a Lubricant in Cold Forming processing of Stainless Steel*, Patent Appl. PTC/DK95/00422, 1995.

(2) Yates, J. T., Smentkowski, V. S., Linsebigler, A. L. - *Surface Science and Extreme Pressure Lubrication - CCl_4 Chemistry on Fe(110)*, NATO ASI Series E: Applied Sciences, Vol. 220, pp. 313-322, 1992.

THE ROLE OF OIL ADDITIVES IN THE FAILURE OF BALL BEARINGS

A. A. TORRANCE and N. C. BRENNAN
Department of Mechanical and Manufacturing Engineering, Trinity College, Dublin 2, Ireland
J. E. MORGAN
Department of Mechanical Engineering, University of Bristol, University Walk, Bristol, BS8 1TR, UK
G. T. Y. WAN
SKF Engineering & Research BV, 3430 DT Nieuwegein, The Netherlands

ABSTRACT

This paper describes current research aimed at identifying the effects of additive interaction on bearing life. The research is divided into two sections, (a) dynamic rolling contact fatigue (RCF) tests using a novel advanced test machine, and (b) static hot-wire reactivity tests using a variation of the "Barcroft" apparatus system. A feasibility study into the deterioration of deep grooved ball-bearings, by Torrance et al (1) showed that the L_{10} life of bearings was reduced using additives under certain conditions. This L_{10} reduction was shown to be due to the chemical reactivity of the oil additive.

The RCF test machines, at the University of Bristol, were designed fifteen years ago to test the life of bearing steels machined under varying grinding conditions. The machines use the lubricant both to apply the contact stress and to provide lubrication, as described by Stokes (2). This type of loading has advantages over more conventional methods of testing. The machine consists of six balls loaded hydrostatically against a test specimen, with the lubricant in the system being be filtered during the tests. Under conditions of relatively low oil flow, the pressure behind each ball is equal and constant.

The machines were designed to test cylindrical specimens with inner diameters of 25 (mm), using balls of 11.112 (mm) diameter. The machines, have been shown to produce max. Hertzian stress of 2.7 to 3.0 (GPa). These stresses are somewhat lower than standard "4 ball machines" which can operate at 5 (GPa). and higher. Using lower stresses for set periods of time can however establish how surface deterioration develops in real use. SKF 6305 inner raceways were used under the following conditions. The lubricating oil used was Shell Vitrea M100, with Sulphur/Phosphorus and Bismuth additives added in concentrations of up to 5% (wt). The temperature of the oil was raised to 70°C to promote chemical activity. Surface roughnesses of the raceways were 0.5µm. and 0.03µm. (cla.), and all tests were run at speeds of 1000 rpm for periods of up to 700 hours. Surfaces and sections of the races were examined using optical and scanning electron microscopy. Profilometry was used to quantify changes to surface topography.

The second part of this research used a variation of the well known "Barcroft" hot wire reactivity apparatus (3). Using the principle of the Wheatstone bridge to investigate reactivity, four lengths of thin iron wire were immersed in identical oil baths, all of which contained the Shell Vitrea M100 base oil and one of which contained the same base oil with a concentration of additive. The metal wires act as resistors in the circuit, as seen in Fig 1. A current was passed through the bridge which caused the temperature of the wires to increase. The wires reacted with the oil and this chemical reaction produced a reduction in the diameter of the wire. The temperature of reaction was found using the temperature/resistance characteristics of iron wire. The wires were allowed to react for a set period of time during which the voltage across the bridge was measured. The reaction films were assumed to be non-conducting and the reactions were assumed uniform over the wires. If the temperature of reaction of the wires are assumed equal then the potential drop across the bridge is caused by the relative reactivity of the single wire immersed in the bath containing additive. The relative reaction rates found using the hot-wire rig was used to predict the extent of reaction on the raceway surfaces of the RCF tests and hence its effect on bearing life.

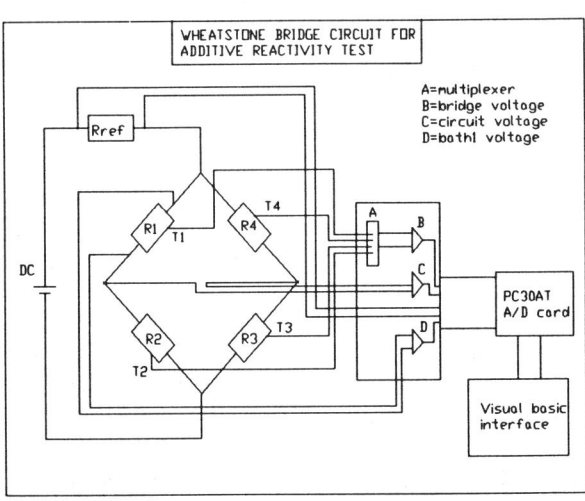

Fig. 1: Four Wire Hot Wire Circuit

REFERENCES

(1) A A Torrance, J E Morgan and G T Y Wan, Wear, 192, (1996) pp. 66-73.
(2) R J Stokes et al, Proceedings of the Institute of Petroleum, Number 1, (1981).
(3) F T Barcroft, Wear, 3, (1960), pp. 440-453.

INFLUENCE OF EXTREME PRESSURE ADDITIVES ON THE PERFORMANCE OF GRINDING FLUIDS IN CREEP FEED GRINDING

TOSHIAKI WAKABAYASHI
Faculty of Engineering, Kagawa University, 1-1 Saiwai-cho, Takamatsu, Kagawa 760, Japan
HIDEO YOKOTA, TATSUYA SHINADA AND SEIJI TAKAHASHI
Central Technical Research Laboratory, Nippon Oil Company Ltd., 8 Chidori-cho, Naka-ku, Yokohama, Kanagawa 231, Japan

ABSTRACT

Creep feed (CF) grinding has preferably been used in the machining process of such hard materials as nickel-base alloys and high speed steels because of its superior efficiency. In CF grinding, since the contact between the wheel and the workpiece is under considerably severe lubricating conditions, grinding fluids should possess not only sufficient cooling capacity but also excellent lubricating capability. Nevertheless, regarding studies on their performance in CF grinding, much attention has so far been attracted to the cooling effects (1)(2)(3), resulting in still unclear significance of their lubricating action.

Under the circumstances, extreme pressure (EP) additives can generally act as an effective lubricant. Organo-chlorine, organo-sulphur and organo-phosphorus compounds are typical EP additives and, recently, the combinations of organo-sulphur compounds and overbased sulphonates are also of great interest because they have demonstrated synergistic lubricating capability in several metal cutting operations (4)(5). This study therefore investigates the influence of these EP additives on the performance of grinding fluids in CF grinding.

Using a surface grinder with a vitrified cubic boron nitride (CBN) wheel, 4mm wide and 0.5mm deep grooving of a high speed steel (HRC60) is carried out. Wheel surface speed is 25m/s and work feed speed is 0.33m/s. During each grinding, the maximum electric power is measured; the lower this power, the better the performance of a grinding fluid tested. Test fluids are prepared by blending EP additive into mineral oil. EP additives chosen are chlorinated paraffin, polysulphide, amine phosphate and overbased calcium sulphonate; test fluids are designated as sample oils CL, S, P and CA, respectively. In addition, a test fluid containing the combination of polysulphide and overbased calcium sulphonate, designated as sample oil SCA, is included.

The evaluation by grinding power presents the following order of the grinding performance of the sample oils examined: (excellent) CL>SCA>S>P>CA (poor). Chlorinate paraffin can effectively decrease grinding resistance and, in contrast to several cases of ordinary metal cutting, the combination of polysulphide and calcium sulphonate is not as much excellent as chlorinated paraffin in CF grinding. Further, results of a static heating test, where a piece of CBN wheel is boiled in a sample oil, suggest that high reactivity of the combined system of polysulphide and calcium sulphonate degrades the bond of the CBN wheel and thus promotes the detachment of abrasive grains. One evidence of this detrimental interaction is shown in Fig.1 demonstrating that the bond material (mainly silicon carbide) is largely reduced by the influence of polysulphide and calcium sulphonate.

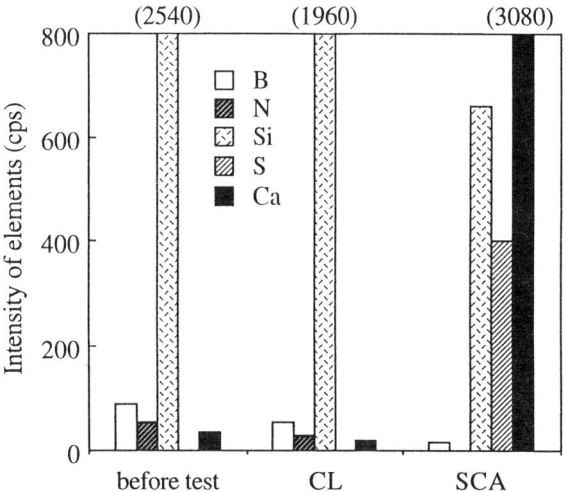

Fig. 1: Results of elemental analysis (EPMA) on the wheel surfaces after the static heating test

REFERENCES

(1) J. W. Powell, T. D. Howes, Proc. 16th Int. MTDR Conf., page 629, 1978.
(2) S. Ohishi, Y. Furukawa, J. JSME (Part C), Vol.50, No.460, page 2399, 1984.
(3) J. Shibata, T. Goto, T. Akiyama, J. JSME (Part C), Vol. 56, No. 527, page 1940, 1990.
(4) H. Hong, A. T. Riga, J. M. Cahoon, J. N. Vinci, Lub. Eng., Vol.49, No.1, page 19, 1993.
(5) T. Wakabayashi, H. Yokota, M. Okajima, S. Ogura, J. JAST, Vol.39, No.9, page 784, 1994.

GROSS PLASTIC FRETTING OF METALS

ÅSA KASSMAN RUDOLPHI and STAFFAN JACOBSON
Uppsala University, Materials Science, Box 534, S-751 21 Uppsala, Sweden

ABSTRACT

Fretting under circumstances of massive plastic deformation, *gross plastic fretting*, can be considered a separate mechanism for wear and surface damage. Gross plastic fretting is of importance in e.g. electrical contacts, which require plastic deformation to fulfil their main function (1,2).

As established in the field of fretting under near elastic conditions, three fretting regimes are distinguished also in gross plastic fretting. The regimes have been given the denominations: *Gross weld regime, Temporary weld regime* and *Gross slip regime*. The regime prevailing is primarily determined by the combination of fretting amplitude and normal load. All three regimes involve plastic deformation which have distinct effects on the surface damage, the friction and the adhesion to the counter face.

Under conditions of loading above the point of yield and further exposure to high friction forces, the fretting mechanisms deviate substantially from classical elastic "Mindlin fretting" (3). This is obvious from Fig. 1, which illustrates the contact pressure distribution before and after fretting in the gross weld regime (small fretting amplitudes). Due to contact area growth during fretting, the pressure distribution becomes very different from that estimated from Hertzian theory.

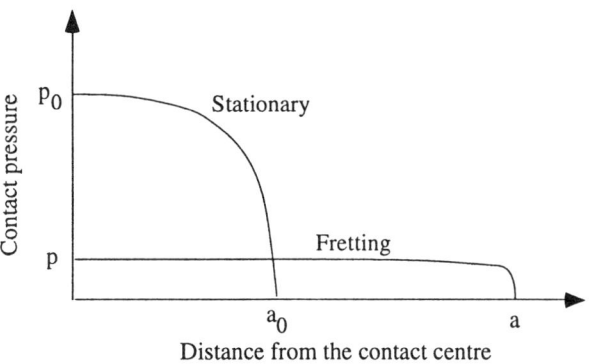

Fig. 1: Schematic of the pressure distribution over the contact area before and after fretting in the gross weld regime.

In many respects the gross weld regime constitutes a unique tribological contact situation. It is characterised by substantial contact area growth and large scale welding. The contact area may grow to very large sizes, see Fig. 2. The figure demonstrates the contact area growth of a 35 μm silver coating contact as estimated from contact resistance readings and as measured in the SEM. The growth is caused by substrate deformation in combination with thinning of the silver coating within the welded area. For thick silver coatings the latter mechanism may result in long silver strings leaving the contact area, see Fig. 3.

In this work the gross plastic fretting has been studied for a number of homogeneous metals (Cu, Ag, Al, brass and steel). The influence of normal force, displacement amplitude and vibration frequency on the surface damage and contact behaviour is discussed.

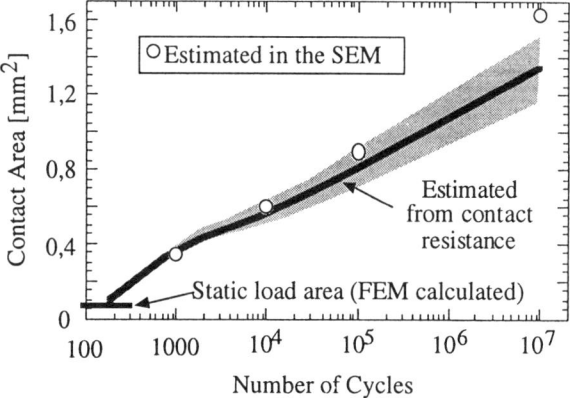

Fig. 2: Contact area development during fretting in the gross weld regime. 35 μm silver coated copper. Crossed cylinders geometry (Ø 10 mm). Normal load 50 N, fretting amplitude 20 μm (p-p).

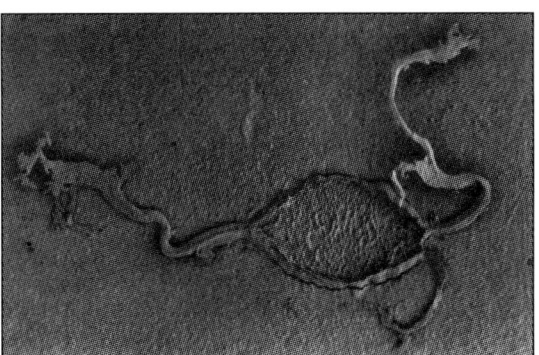

Fig. 3: Typical appearance of the contact area after 10^5 fretting cycles in the gross weld regime (SEM). 35 μm silver coated copper. The scar diamater is about 1 mm.

REFERENCES

(1) Å. Kassman Rudolphi, S. Jacobson, Wear 201, 1996, 244-254pp.
(2) Å. Kassman Rudolphi, S. Jacobson, Wear 201, 1996, 255-264pp.
(3) R.D. Mindlin, J. Applied Mechanics 16, 1949, 259-268pp.

CONTRIBUTION OF THE SOLID THIRD BODY TO THE TRIBOLOGY OF A RADIAL-FACE SEAL

PASCALE JACQUEMARD, YVES BERTHIER, MARIE-HELENE MEURISSE
Laboratoire de Mécanique des Contacts, IET, INSA de Lyon, F69621 Villeurbanne
JACQUES MAURY
Laboratoire Etanchéité, DTE/SLC CEA VALRHO, F26702 Pierrelatte
MICHEL MOREAU
Département Spécialités et Matériaux Avancés, Carbone Lorraine, F92231 Gennevilliers

ABSTRACT

Radial-face seals are mechanisms which fulfil the functions of velocity accommodation and leakproofing between a rotating shaft and its fixed housing. They do this by the contact of two annular faces, or 1st bodies. In the case of the seal studied the sealed fluid is water, one 1st body is in carbon-graphite and the other in silicon carbide. Radial-face seals usually operate in the hydrodynamic or mixed lubrication depending on the thickness of the sealed fluid film separating the 1st bodies. The mixed lubrication implies interactions between the 1st bodies causing particles detachment. In the literature relative to radial-face seals, the contribution of the sealed fluid or fluid 3rd body to the load capacity and the friction torque is well modelized by the Reynolds equation (1). On the opposite side, the authors do not generally take into account the contribution of the particles detached from the 1st bodies or solid 3rd body.

In order to identify the contribution of the solid 3rd body, we used an approach which consists in identifying all influent parameters of the contact with the right sequence of causes and consequences (2). This approach relies on the evaluation of the flows of the 3rd body, fluid or solid (Fig. 1). We distinguish :
- the *source flows* (Q_s) which correspond to the supply of fluid (Q_s^f) and the particle detachment or solid flow (Q_s^s);
- the *internal flows* (Q_i) which correspond to the circulation of 3rd body in the contact (Q_i^s, Q_i^f);
- the *ejection flows* (Q_e) which correspond to the ejection of 3rd body from the contact (Q_e^s, Q_e^f).

The ejection flow of the fluid 3rd body is calculated with the Reynolds equation. On the opposite side, we are not able to measure the flow and rheology of the solid 3rd body with classical methods. This is due to its nature and thickness ranging about a few tenths of a micrometer. We get round this difficulty by reconstructing the flows from the morphologies of 1st and 3rd bodies after various running times and visualisation tests. With this approach, we identified first the contributions of the mechanism and the 3rd body fixing the materials of the 1st bodies (3). Then, we identified the influence of the carbon-graphite's constituents which is the purpose of this article.

The source flow of the solid 3rd body is first activated near the inner circumference of the carbon-graphite 1st body because of thermoelastic deformations. Then, the solid 3rd body contributes locally to the velocity accommodation and re-activates the source flow towards the outer circumference of the 1st bodies. We achieved a qualitative and relative classification of the 3rd body flows depending on the carbon-graphite's constituents. The internal flow of the solid 3rd body is evaluated by means of an appreciation of cohesion, adherence and "ductility" of the 3rd body. For example, we observe that an augmentation of the proportion of carbon black results in a decrease in the 3rd body internal flow. Consequently, the 3rd body keeps tracking in the contact and causes the friction increase (Fig. 1). From the 3rd body flows, the functions of the seal can be modelized while integrating the influence of the materials parameters.

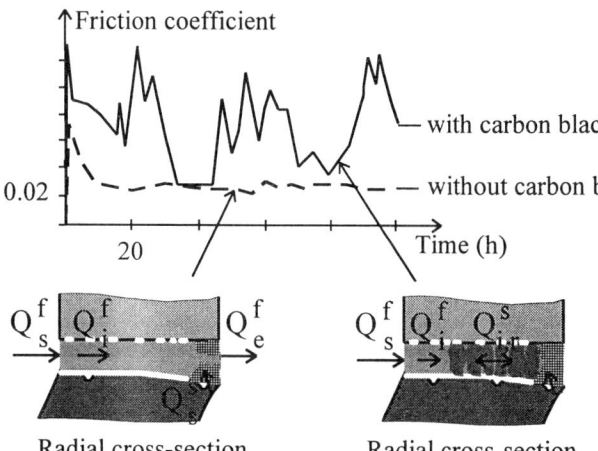

Fig. 1 : **Influence of carbon black**

REFERENCES

(1) V. Person, B. Tournerie, J. Frene, Proceedings of the A.S.M.E./S.T.L.E. Lubrication Conference, San Francisco-U.S.A., October 1996, pre-print 27 p.
(2) Y. Berthier, Proceedings of the 22nd Leeds-Lyon Symposium on Tribology, Lyon-France, September, 1995. Elsevier 1996, 21-30pp.
(3) P. Jacquemard, Proceedings of the 22nd Leeds-Lyon Symposium on Tribology, Lyon-France, September, 1995. Elsevier 1996, 91-102 pp.

THE EFFECT OF MICA CONTENT ON THE WEAR OF MICA-ALUMINIUM COMPOSITES

V.K.Srivastava* and G.H.Borahni
Department of Mechanical Engineering, Institute of Technology, B.H.U., Varanasi- 221 005, INDIA

ABSTRACT

Aluminium is highly deformable having a thin oxide cover, which is easily ruptured during sliding under pressure. Thus it tends towards cold working and powder will stick to the die. Deonath et al [1] have worked in wide range of mica dispersion in aluminium alloys by conventional casting technique. They observed that the weakness of bonding is inferred from the presence of voids around the mica particles. Mica is a superior solid lubricant than graphite under dry friction conditions because of its high oxidation resistance and chemical inertness. Several studies on the application of wear and friction of metal-matrix composites are available in the literature. However, the wear of solid surface is complex process, which under many conditions includes both chemical attack and physical change.

The present work has investigated the wear and friction properties of mica-aluminium composites prepared by a PM route, with the variation of sliding speed and load. The commercial purity aluminium powder (-200 to +300 mesh) and mica powder (-200 Mesh) were used for the preparation of mica-aluminium composites. They were mixed by mechanical mixing for 1 h in a cylindrical plastic blender. Green cylindrical pellets 11 mm in diameter and 10 mm in height were compacted from both ends at a pressure of 472 Mpa and a nominal rate of $\approx 10^{-3}$ s^{-1} in a rigid steel die on a single acting hydraulic press. The green compacts were sintered in a tubular furnace at 883 $\pm$ for 1 h under vacuum (≈ 10 Mpa). Linear wear rates for the 0 to 4% mica specimens sintered in vacuum were measured with the variation of load and sliding speed. The results indicate that the wear rate increases with increase of load and sliding speed. In the 4 % wear rate at low loads could be the result of removal of mica particles from the compacts. However, under more severe loading conditions, the mica particles are smeared over the disc where they act as a lubricant, thus reducing friction and lowering wear rate. The addition of mica to commercial purity aluminium powder mixes resulted in enhanced densification and also gave the better wear and friction values.

REFERENCE

(1) D.NATH, S.K.BISWAS AND P.K.ROHATGI, WEAR, 60, 1980, 60-68PP.

*Visiting Fellow, Department of Materials, QMW, University of London, E1 4NS, U.K.

EFFECTS OF MICRO (ROCKING) VIBRATIONS AND SURFACE WAVINESS ON WEAR AND WEAR DEBRIS

M. D. BRYANT
Mechanical Engineering, The University of Texas at Austin, Austin, Texas 78712-1063
ATUL TEWARI
Sterling Information Group, 1717 W 6th St, Austin, Texas 78703
DAVID YORK
McDonnell Douglas Aerospace, 13100 Space Center Blvd, Houston, Texas 77059

ABSTRACT

Recent work [1]-[6] showed micro-vibrations (10-100 µm, 10 to 100 Hz) can reduce wear 50%. We slid carbon brushes against copper rotors with smooth (heights < 50 µm) and wavy surfaces. Waves induced rigid body vibrations of the brush near the natural frequency (≈ 160 Hz) of the brush-spring system, including translations normal to the sliding surface, and rocking motions with rotation vectors parallel to the sliding surface.

Measured (Fig. 1a) were wear rates (µg/s) vs. speed (rpm) for brushes of different geometry and load. Wear rates on the wavy rotor WR_w (open squares), material lost per 24 hours, were always less than WR_s (solid squares) on the smooth rotor, and almost half WR_s near 600 and 1000 rpm. Curves were repeatable: points represented at least three trials. Later trials [2]-[5] with different rotors, surfaces, materials, frequencies, and current [4], [5] gave similar reductions. Wear of cathode brushes, generally much higher than wear of anode brushes, was verified [5] on a conventional smooth slipring (Fig. 1b: compare -40 A solid square, dashed curve to +40 A solid triangle, solid curve). Micro vibrations reduced cathode wear to levels of anode wear [5], (Fig. 1b: compare - 40 A open square, dashed curve to + 40 A open triangle, solid curve). Contact voltage drop [3]-[5] showed rocking vibrations never disconnected the brush from the rotor; 20 kHz bandwidth measurements [2], [3] showed normal and tangential contact forces on the both rotors similar.

Brush and holder clearances were reduced while sliding over a slightly wavy (8 to 20 µm) steel surface. Tighter fits restricted rocking, looser fits permitted. Plotted (Fig. 2) were wear rate vs. speed with clearance a parameter. Normal and rocking motions were measured.

We found: a) Micro-vibrations reduced brush wear on steel. b) No rocking (too tight, trial 0) gave "smooth" wear. c) An optimal fit (200 µm clearance, trial 2) gave least wear. Kinematics permitted optimum rocking, 10^{-3} to 10^{-1} degrees. d) Fits too loose increased wear beyond smooth (trial 8), allowing impacts of brush and holder or rotor. e) Rocking with rotation vectors parallel or perpendicular to sliding gave similar wear reduction, 50% or more. f) Rocking with a rotation vector perpendicular to sliding generated "chatter". g) Rocking with a rotation vector parallel to sliding was quiet. h) Wear reduction can occur at low frequency (20-70 Hz), amplitude (rocking ≈ 0.001°, normal ≈ 20 to 70 µm), and/or waviness (8 µm)

Wear particles were SEM inspected. At lower speeds particles from wavy and smooth surfaces were similar. At higher speeds smooth surface particles were larger than wavy: they were often snowball like compactions of sub-particles similar to those shed from the wavy surface. Hypothesis: small wear particles shed from a slider running over a wavy escape the sliding interface through gaps opened by vibrations; without gaps, smooth surface particles become entrapped and compacted. Finally, displacements of rocking vibrations correlated to the size of the gaps required for particles to escape.

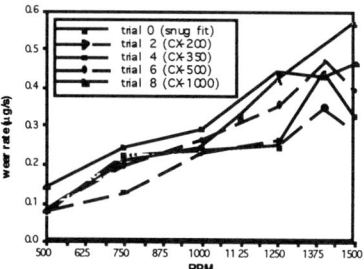

Fig. 2: Wear rate vs. speed, clearance (µm) a parameter.

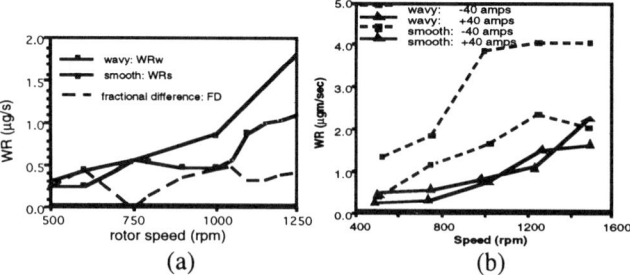

Fig. 1: Wear rate (µg/s) vs. speed (rpm) for brushes on smooth and wavy rotors, (a) without (b) with ±40.

REFERENCES

[1] M. Bryant, J. Lin, Wear, Vol. 170, 1993.
[2] M. Bryant, J. Lin, ASTM STP 1247, 1995.
[3] J. Lin, M.. Bryant, ASME J. Trib., Vol. 118, 1996.
[4] M. Bryant, A. Tewari, J. Lin, IEEE Trans. Comp, Hybrids, & Manuf. Technol.-Part A, Vol. 18, 1995.
[5] M. Bryant, A. Tewari, Proc. of the 49th Meeting of the Soc. for Machinery Failure Prevention Techn, 1995.
[6] M. Bryant, J. Lin, U.S. Patent 5,466,979, 1995.

EFFECTS OF MATERIAL PAIR PROPERTIES AND SLIDING HISTORY ON FRICTON AND WEAR BEHAVIOR OF METALS

DAE-EUN KIM, DONG-HWAN HWANG and SANG-JO LEE
Department of Mechanical Engineering, Yonsei University, Shin-chon Dong 134, Seoul, Korea
IN-HA SUNG
Multimedia Business Division, LG Electronics Inc., Pyung Taek, Korea

ABSTRACT

The purpose of this work is to investigate the factors that dictate the friction and wear behavior of metals, particularly from the view point of material properties of the sliding pair and wear particle dynamics. In order to identify the relative importance of material compatibility (1) versus mechanical interaction of the material pair on the friction and wear behavior, tribological tests were performed using pure metal specimens which were selected based on their degrees of compatibility and hardness ratio. Hardness ratio was chosen as the parameter that represent the mechanical aspect of the frictional interaction.

Pure metals such as Mo, Ti, Ag, Zn, Sn, Al, Cu, Pb, etc. with purity better than 99% were used in the experiment. The metal pairs were strategically selected based on their levels compatibility and hardness ratio. To assure that the state of the specimens at the onset of the sliding tests remain relatively constant, similar surface roughness as well as pin geometry were maintained for all sliding pairs.

Friction and wear experiments in dry condition were conducted using both pin-on-disk and pin-on-reciprocator type tribotesters. Normal force of 200gf was used at a linear sliding speed of about 0.1~0.2m/s for the tests and all experiments were conducted in ordinary laboratory environment. The frictional force and coefficients were measured by a strain gage sensor and the data were recorded by a PC based data acquisition system.

Experimental results show that hardness ratio affect the initial friction coefficient values (Fig.1) but the compatibility parameter does not seem to influence the initial stage of sliding significantly. When the friction coefficient reaches steady state, neither compatibility (Fig.2) nor the hardness ratio dominate the frictional behavior for the given conditions. Rather, the magnitude is found to be affected more by the sliding history. As shown in Fig.2 the magnitude of the pin-on-disk tribotester, which represents the uni-directional motion, is found to be higher than the results of the pin-on-reciprocator (bi-directional motion) tribotester. Also, wear rates for the pin-on-disk tests were found to be generally higher than those of pin-on-reciprocator tests. Thus, it may concluded that uni-directional sliding causes more severe damage to the surface compared to bi-directional sliding. It is presumed that the sliding motion of the pin and plate affects the wear particle dynamics, which in turn influences the friction and wear behavior.

The findings of this research suggest that frictional interaction of the materials undergoing sliding contact cannot be simply characterized by material pair properties like compatibility or hardness ratio. Rather, factors such as sliding history and particle dynamics may be more critical in practical tribosystems.

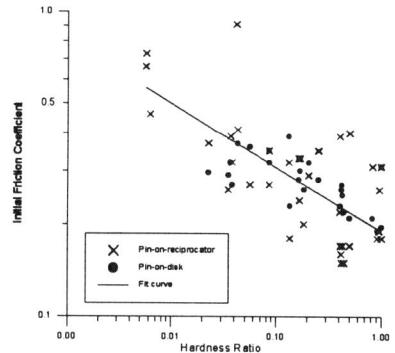

Fig. 1 Initial friction coefficient vs. hardness ratio

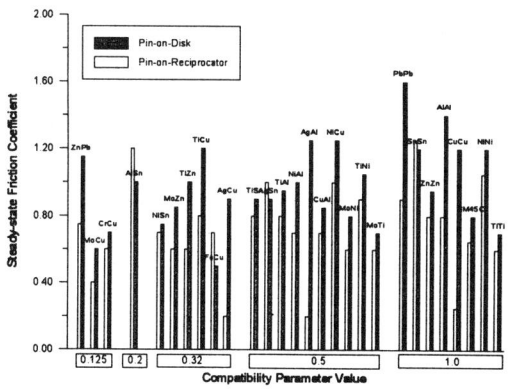

Fig. 2 Steady-state friction coefficient vs. compatibility parameter

ACKNOWLEDGMENT

This work was supported by the Korea Science and Engineering Foundation under the grant number 951-1009-055-2.

REFERENCE

(1) E. Rabinowicz, ASLE Transactions, Vol. 14, 1971, pp. 198-205.

Submitted to *ASME Journal of Tribology*

WEAR PROPERTIES OF Al₂O₃ FIBER AND PARTICLE REINFORCED ALUMINUM COMPOSITES

YOSHIRO IWAI and TOMOMI HONDA
Dept.of Mechanical Engineering, Fukui University, Bunkyo 3-9-1, Fukui, 910, Japan
YOSHIHUMI IWASAKI
Mitsubishi Heavy Industries, LTD, Takasago, Hyogo, 676, Japan

INTRODUCTION

One of the major incentives for the technological development of the aluminum matrix composite materials(MMC) reinforced with ceramic whiskers, fibers or particles has been the potential of these composites as new wear resistant materials for a number of tribological applications.

In this study, the effects of fiber and particle on wear process and wear rate were clarified and the wear mechanisms of MMC were discussed.

MATERIALS AND TEST PROCEDURE

The matrix material and MMC were produced by a high pressure and low speed die casting technique. The matrix material was ADC12 aluminum alloy (composition(wt%) : Cu, 2.5%; Si, 10.8%; Al, bal) and the reinforcements were of several combinations of Al₂O₃ fibers and particles of various sizes. One is Al₂O₃ fiber reinforced aluminum composites (designated as MMCf), where the fibers were 4 μm in diameter with fiber volume fraction Vf ranging from 3% to 26%, and length ranging from 200 μm to 40 μm. The other is Al₂O₃ fiber plus particle reinforced aluminum composites (designated as hybrid MMC), where the particle diameter d were ranged from 0.6 μm to 20 μm with particle volume fraction Vp=20%. The counterface material was nitrided SUS440B, whose Vickers hardness was 1200 and nitrided depth was about 70 μm.

Wear tests were carried out by rubbing a disk of the matrix material or MMC against a pin of nitrided SUS440B in room air. The sliding velocity was 0.1 m/s and the contact load was 10 N. The MMC disk test piece was 30 mm in diameter and 5 mm in thickness. The pin test piece had a flat surface with a diameter of 4 mm. The diameter of the wear track was 23 mm. The surfaces of both test pieces were polished using grade #1200 emery paper. The mass loss was measured with a precision balance.

EXPERIMENTAL RESULTS AND DISCUSSION

The severe wear occurred for matrix material, and MMCf with low volume fraction, but didn't occur for MMCf with high volume fraction and hybrid MMC. This is because reinforcements prevented wear crack propagation and plastic flow in the initial severe wear. Fig.1 shows the variations in the steady-state wear rate of MMC as a function of Vf and d. The wear rate of MMC decreased with an increase of Vf and d. The wear rate was minimum for the MMCf with Vf=9%, and increased slightly with increasing Vf. The wear rate rapidly decreased with increasing d and nearly became equal to zero. From the wear rate and the appearance of the worn surfaces, the wear mechanism models were classified three types as shown in Fig.2, namely; I. Adhesive wear dominated by matrix material, and MMCf with low volume fraction, II. Three body abrasive wear for MMCf with high volume fraction, and hybrid MMC with small particles, III. Two body abrasive wear for hybrid MMC with large particle diameter.

CONCLUSION

Aluminum reinforced with fibers and particles of Al₂O₃ can improve dry sliding wear resistance.

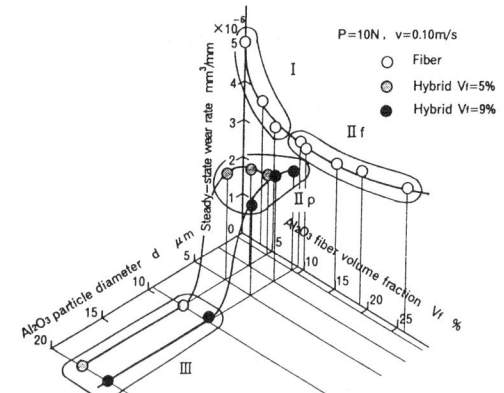

Fig.1 Variations in the wear rate against Vf and d

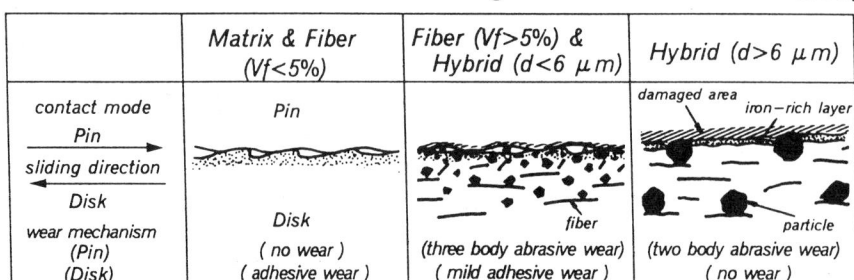

Fig.2 Wear mechanism models

BEHAVIOUR OF SINTERED HIGH SPEED STEEL MATRIX COMPOSITES UNDER FRETTING CONDITIONS: QUANTIFICATION OF WEAR

B. MARTIN - L. VINCENT
Ecole Centrale de Lyon IFoS/MMP UMR 5621 BP 163 69131 Ecully Cedex - France
H. ZAHOUANI - P. KAPSA - R. VARGIOLU
Ecole Centrale de Lyon LTDS UMR 5513 BP 163 69131 Ecully Cedex - France

Developing wear resistance of materials used in automotive parts, such as cam lobes, appears to be an essential challenge for the coming years.

In this way, new composites were created, allowed by the Powder Metallurgy route. In this study, several sintered high speed steel matrix composites were processed: they consist in a M3/2 matrix, in which ceramic particles (TiC) and solid lubricant (MnS) are included, in order to respectively reduce wear and minimise friction.

These composites were tribologically tested using the fretting method which consists in applying a small amplitude oscillatory motion between two solid surfaces in contact (Fig 1).

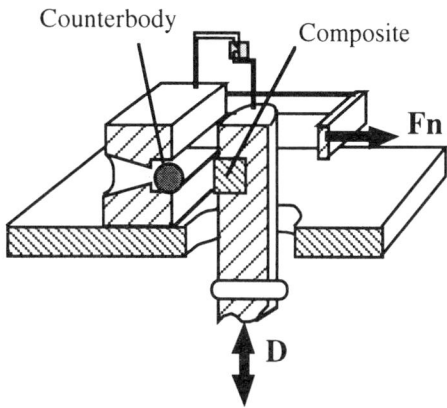

Figure 1: Fretting machine

The samples submitted to fretting tests classically present two kinds of damage, depending on the running conditions: Wear and Cracking. This study was only focussed on the wear damage, with the formation of the tribologically transformed structure (TTS) and the quantification of the wear volume.

Wear analysis was conducted using a 3-Dimensional profilometry which enables the determination of the wear volume (Vw) for each test. This study shows how this quantification may be related to the cumulated dissipative energy (Edc) during the fretting test and to the Coefficient of Friction (μ), using the equation:
$$Vw = \alpha' + \beta' Edc \quad (eq\ 1)$$

Therefore, the variation of the wear volume as a function of the cumulated dissipative energy allows the understanding of the wear mechanisms, in which the nature of the additives plays an essential role, considering the formation of the TTS and the wear kinetic.

On the other hand, a second equation links Edc to the imposed displacement D:
$$Edc = \alpha + \beta\,\mu D \quad (eq\ 2) \quad with\ \beta = 4NFn$$
and Eq 1 and 2 can be combined:
$$Vw = \alpha'' + \beta''D \quad (eq\ 3) \quad (Fig\ 2).$$

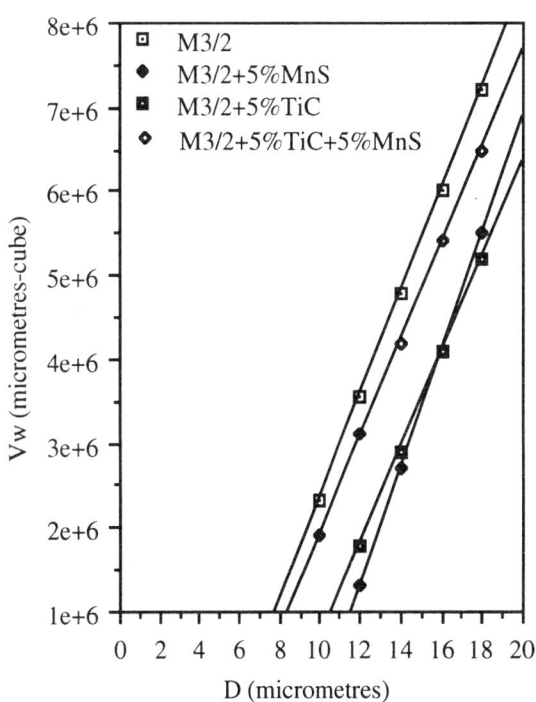

Figure 2: $Vw = \alpha'' + \beta''D$ for the matrix and the composites

In conclusion, the wear induced by fretting can be minimised using a high speed steel matrix, in which MnS and/or TiC particles are included.

A study on the quantitative descriptions of seizure behaviors of steels at low temperature in a vacuum

K OKADA and Q OUYANG

Dept of Mechanical system Engineering, Faulty of Engineering, Yamanashi University, Kofu, JAPAN

ABSTRACT

Seizure, a kind of special friction phenomenon, usually results in the dangerous failure of sudden stop in mechanical engineering systems[1]. Up to now it is almost impossible to predict and to prevent the occurrence of seizure for industrial applications due to its unexpectedness. In this paper, a systematic study on the seizure properties of several common steels was carried out by friction tests at low temperature in a vacuum.

Friction experiments were performed on a pin-on-disc friction tester with a cryogenic system. The temperature in the sample chamber can be controlled in the range of 300K to 16L in a vacuum of 10^{-3} to 10^{-5} Pa. SUJ1 steel bearing ball was utilized as the hemispherical pin with a diameter of 16mm. The fixed disc was made of S45C carbon steel or SUS304L stainless steel. The rotation speed of the pin was 10rpm corresponding to the sliding speed of 5mm/sec.

The influence of temperature on the seizure properties was conducted on a SUJ1/S45C pair at a constant applied load of 50N. It is found that at a certain temperature, there exists a definitive time of seizure occurrence. After seizure occurred, the friction coefficient increased about more than ten times. There exists a critical magnitude of temperature. Below this magnitude, the seizure will not occur even after several hours of steady sliding. The relation between the friction time before seizure occurs and surface temperature T can be written as

$$t = K/(T-T_c) \quad (1)$$

Where T_c is the critical temperature and K is a constant for a certain friction condition.

The macro adhesive areas were measured at various stages of seizure when surface temperature was kept at 200K (Fig.1). In order to produce seizure, macro adhesive area must grow to a critical size, and after seizure occurs the adhesive area stays at this saturation value, S_s and no longer changes during the seizure process.

If we plot the Ln of S_s versus the reciprocal of surface temperature 1/T as shown in Fig.2, we find that their relationship as logarithmic

$$S_s = S_0(p,v)\exp[-\alpha/(T-T_c)] \quad (2)$$

where T_c is the critical temperature that we have defined before. α quantitatively expresses the material properties connected with the tendency to seizure S_0 represents the maximum adhesive area at high temperature.

Using these two formulas, we can describe the process of seizure in following way: When two metals are brought into contact and rubbed together, at first both adhesive area and temperature on their interface increase with friction time, while the friction during this period is approximately steady. The surface temperature increases to Tc after sliding for some time t which is determined by formula (1), the micro adhesive regions grow into a saturation value Ss as can be seen in formula(2). At this instance friction and wear suddenly changes to the severe form of seizure.

REFERENCES
(1) D.Barnes, F.Stottm and G.Wood, wear, Vol.45, No.2, 199-209pp.

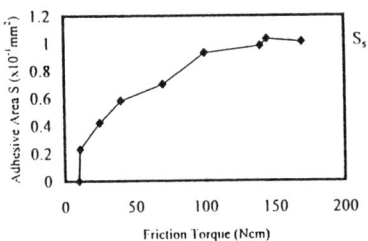

Fig.1. Adhesive area versus friction torque.

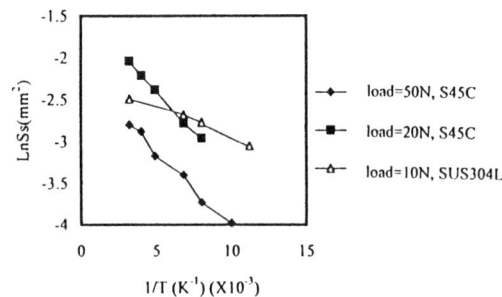

Fig.2. Saturation adhesive area Ss versus inverse temperature.

THE ROLE OF TRIBOPARTICULATES IN DRY SLIDING WEAR

JIAREN JIANG
Research Institute for Design, Manufacture and Marketing, Salford University, Salford M5 4WT, UK
F.H. STOTT and M.M. STACK
Corrosion and Protection Centre, UMIST, Manchester M60 1QD, UK

ABSTRACT

During sliding of solid components, the presence of triboparticulates, either accumulated wear debris particles or artificially-supplied fine particles, considerably decreases wear. Under some other conditions, the entrapment of hard wear debris particles within the rubbing interface can be detrimental, leading to three-body abrasion. The main purpose of the research reported in this paper was to investigate the roles of wear debris particles in dry sliding wear transitions and to determine the mechanisms for the development of wear-protective load-bearing layers via the compaction of wear debris particles.

In this study, a nickel-base alloy, N80A, was used. Wear tests were carried out on a pin-on-disk reciprocating rig in dry oxygen at temperatures from 20^0 to 250^0C. The average sliding speed and load were 83 mm s^{-1} and 15 N respectively.

At the various temperatures, apparent wear transitions from a high rate to a low rate were observed. Corresponding to each wear rate transition, the contact resistance between the rubbing surfaces sharply increased from near zero to a more positive, much higher value. The high resistance contact was more stable, while the wear rate was less, at the higher temperatures than at 20^0C.

SEM observations showed that, in the very early stages of sliding at the various temperatures, contact between the rubbing surfaces was essentially metal-to-metal. After the transition in contact resistance, compact wear debris particle layers were developed; these layers acted as load-bearing areas. Particles within the layers were oxygen-rich. At 20^0C, the surfaces of the compact layers contained only distinguishable very fine wear debris particles. At 250^0C, smooth, 'glaze' layers formed on top of the compact particle layers. At an intermediate temperature, 150^0C, the load-bearing layers consisted of the types observed at both 20^0C and 250^0C. At all temperatures, the compact particle layers overlaid the metallic wear surfaces developed in the earlier stages of sliding. The high resistance compact particle layers had a surface coverage of 20 to 50% of the wear scars and were mainly distributed along approximately two thirds of the central part of the wear track.

According to experimental observations, the following model for wear transitions observed at the various temperatures is presented.

In the initial stages of sliding, metal-to-metal contact occurs and metallic wear debris particles are generated. While very large wear debris particles are easily removed from the rubbing interface to cause direct material loss, some of the wear debris particles are entrapped within the rubbing interface, undergoing further deformation, fragmentation and comminution and becoming smaller in size. The fine particles are agglomerated somewhere on the wear surfaces, often in troughs previously formed during the metal-to-metal contact. As sliding continues, the agglomerated clusters of particles are compressed to form compacted load-bearing layers. During these processes, the fine particles are oxidized and sintered to some extent. At elevated temperatures, the rates of sintering and oxidation are faster than at 20^0C. As a result, more solid layers are formed at elevated temperatures and the wear surfaces are better protected from further wear. The compact particle layers are wear-protective. The wear transition occurs when the area covered by such layers reaches a critical value.

According to the JKR adhesion model (1), the strength of adhesion, σ_a, between two spherical particles has been derived; this strength is in the range of 746 to 1608 MPa for metallic particles with diameters from 0.1 to 1 μm. Even when attenuation of the adhesion due to particle surface roughness is considered, this value is still very high. The adhesion strength increases with decreasing particle size and with increasing temperature.

The experimental results obtained in this study and many observations reported in the literature can be explained according to the presented model.

It is finally concluded that triboparticulates play an important role in sliding wear. Fine particles can agglomerate and adhere together to form compact layers. When a sufficient area of such layers has developed on the contacting surfaces, the latter are protected from further severe wear; as a result, there is a transition from a high to a lower rate of wear.

REFERENCE

(1) K.L. Johnson, K. Kendall and A.D. Roberts, Proc. Roy. Soc. Lond., A324 (1971) 301-313.

Submitted to *Wear*

SLIDING WEAR BEHAVIOUR OF ALUMINIUM-BASED METAL MATRIX COMPOSITES PRODUCED BY A NOVEL LIQUID ROUTE

P. H. SHIPWAY, A.R. KENNEDY and A.J. WILKES
Department of Materials Engineering and Materials Design, University of Nottingham,
University Park, Nottingham, NG7 2RD.

ABSTRACT

Discontinuously reinforced aluminium matrix composites have emerged from the need for light weight, high stiffness materials. Significant increases in stiffness and strength can be conferred even with small reinforcement volume fractions. Many of the applications for which composites are advantageous also require enhanced tribological performance. The sliding wear behaviour of aluminium based MMCs has received much attention in the literature, and whilst the behaviour is complex, it is generally demonstrated that reinforcement of aluminium alloys results in a reduction in sliding wear rate and in an increase in load at which the transition from mild to severe wear occurs. However, wear rates of counterfaces increase with reinforcement and when considering the whole system, the benefits of reinforcement are less clear.

The reinforcement type is significant in determining the wear rate of the composite. Most work has examined composites reinforced with SiC or Al_2O_3 particles but others reinforcement types such as TiC have also shown to give favourable results. The method of composite manufacture influences the mechanical properties of the material, and thus will also affect the tribological performance.

In this work, 10μm TiC particles were added to three aluminium alloys, namely a casting-type alloy (A356), a wrought-type alloy (Al-4Cu) and a commercial purity aluminium (CP). The composites were manufactured by a modified liquid metallurgy technique (1) where the dispersion of particles in the liquid metal is achieved not by vigourous stirring but by spontaneous incorporation as a result of wetting of the particles by the liquid alloy. The method, and choice of particle-matrix combination ensure uniform distribution of the reinforcing particles within the alloy and strong interfacial bonds between the reinforcement and the metal.

The TiC reinforced composites were compared with a *Duralcan* composite; its alloy type was close to A356 and it contained 15 vol% SiC particles with average particle size of 13μm. All the composites were extruded at 400°C with an extrusion ratio of 13:1. The sliding wear behaviour of the extruded composites was examined with a pin-on-disc wear test, with composite pins sliding against a mild steel disc at 1 m s^{-1} under a range of applied loads.

The hardnesses of the composites were greater than those of the unreinforced materials; TiC is a known grain refiner in aluminium alloys and this also resulted in a hardness increase. Wear data from the steady state periods for all the alloys were obtained; those for the CP alloy and for CP reinforced with both 10 vol% and 20 vol% TiC reinforcement are shown in Figure 1 (values of wear rates are shown above the bars, along with values of the wear coefficient, K).

For all the alloys, particulate reinforcement caused a reduction in wear rate, and an increase in the load for the transition from mild to severe wear. In the mild wear regime, the wear coefficients of all the alloys and composites were very similar; A356 reinforced with 10 vol% TiC was the hardest and thus most wear resistant alloy. Figure 1 shows that for the CP alloy, high wear coefficients were observed at all loads for the unreinforced material, but reinforcement with 20 vol% TiC resulted in low wear coefficients.

Wear of the mild steel counterfaces was observed when slid against the composite materials; at a given reinforcement volume fraction, increase in wear of the composite due to increased applied load resulted in increased counterface wear. When both composite and counterface wear are considered, an optimum volume fraction for lowest system wear exists.

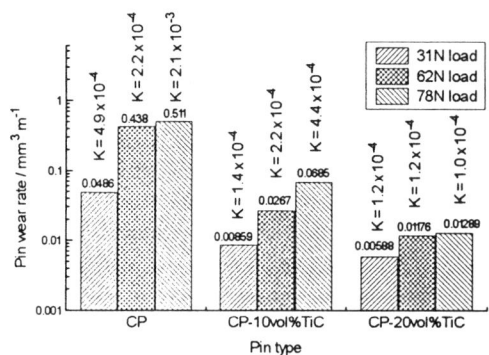

Fig. 1: Wear rates and coefficients of CP alloys and composites

REFERENCE

(1) A.R. Kennedy, D.G. McCartney, J.V. Wood in F.H. Froes, C. Suryanarayan, C.M. Ward-Close (eds.), Synthesis/Processing of lightweight metallic materials, TMS, Warrendale, Pa, 1995, pp 261 - 273.

FRICTION AND WEAR PROPERTIES OF OIL-RETAINING POROUS SILICON CARBIDE IN SLIDING CONTACT

Tadayuki JIMBO and Seiichiro HIRONAKA

Department of Inorganic Materials, Faculty of Engineering, Tokyo Institute of Technology, Ookayama, Meguro-ku, Tokyo 152, Japan

INTRODUCTION

Many studies have been reported which indicate that SiC is one of the best friction materials among ceramics. However, desirable friction and wear properties of SiC cannot always be obtained wiyhout lubrication. This study tried to improve the friction and wear properties of SiC and develop nonoiling friction materials using oil-retaining porous SiC.

EXPERIMENTAL

The friction and wear properties of dense silicon carbide (SiC-A) and oil-retaining porous silicon carbide (SiC-B) in air were studied by using a pin-on-disk friction machine. SiC-B was prepared by retaining perfluoropolyether in porous SiC having the porosity of 15-20 vol% under reduced pressure. The friction tests were carried out under the conditions of various sliding velocities and loads and various combinations of SiC-A and SiC-B (pin/disk), as shown in Table 1.

RESULTS AND DISCUSSION

The friction and wear properties and morphologies of frictional surfaces of these SiC greatly changed dependent on the test conditions. SiC-A pin showed a lower and more stable friction coeffi-cient than those of self-mated SiC-A independent on sliding velocities and loads, when it was slid against the SiC-B disk, as shown in Figs.1 and 2. The least wear amount and smoothest frictional surfaces was also obtained in the SiC-A/SiC-B system. This is due to the lubrication effect of the oil supplied from SiC-B disk at the friction interface.

The friction and wear mechanisms of these SiC were classified into four modes depending on the relation between friction coefficient and specific wear amount and SEM observation of frictional surfaces.

Table.1. Properties of SiC-A and SiC-B Specimens

Properties	Sample		
	SiC-A	SiC-B	
Bulk density, g/cm³	3.1	2.8	2.7*
Porosity, volume%	<4	<2	15-20*
Bending strength**, MPa	813.4	519.4	519.4*
Young's modulus, GPa	392	343	294*
Poisson's ratio	0.13	0.17	0.15*
K_{IC}***, MPa m$^{-1/2}$	3.5	3.5	-*

*: Before oil-retaining
**: JIS R 1601 3-point bending test (at room temparature)
***: K_{IC}: Fracture toughness

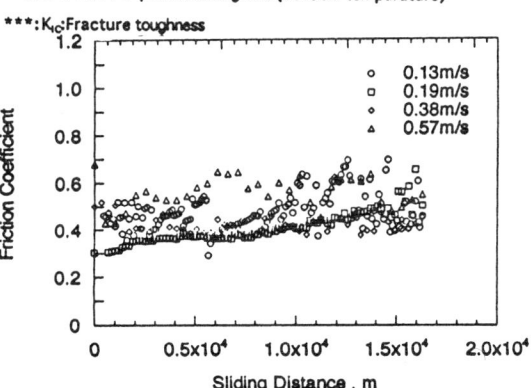

Fig.1. The friction coefficient as a function of sliding distance in SiC-A/SiC-A system under 9.8N at various sliding velocities.

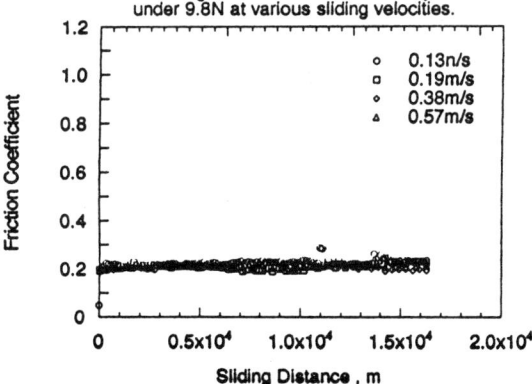

Fig.2. The friction coefficient as a function of sliding distance in SiC-A/SiC-B system under 9.8N at various sliding velocities.

NON-STATIONARITY IN THE SLIDING FRICTION OF UNLUBRICATED AND BOUNDARY LUBRICATED POLYETHYLENETEREPHTHALATE

F HOLLWAY, P J DOYLE, O J GALLAGHER and M J ADAMS
Unilever Research Port Sunlight Laboratory, Quarry Road East
Bebington, Wirral, L63 3JW

ABSTRACT

An experimental study of polyethyleneterephthalate (PET) is described in which the friction of planar specimens against a smooth steel sphere was monitored during multi-pass sliding. The friction reduced with the sliding distance and the extent of this non-stationarity increased with the applied normal load and the degree of pre-damage resulting from abrasive cleaning.

INTRODUCTION

In a study of polyethyleneterephthalate (PET) from different sources, we observed that the friction of both nominally clean and boundary lubricated specimens reduced by significantly varying extents during each pass of multi-pass sliding measurements. Such non-stationarity has been reported previously for fibrillar (1) and planar (2) specimens. As in the case of other organic polymers, it is known to arise from the entrapment of wear debris or the development of a transfer film on the counter-surface. A detailed study of the wear mechanisms for PET in contact with polished steel has been described by Yamada and Tanaka (2). However, the factors controlling frictional non-stationarity have not been studied specifically. This is important for quantifying the effectiveness of lubricants which is complicated if the frictional force in the clean state is ill-defined.

EXPERIMENTAL

Planar specimens of PET (3.2mm sheet supplied by Eastman Kodak) were cleaned by overnight soaking in aqueous 0.01M sodium dodecyl sulphate with some specimens being further cleaned by gentle abrasion with silicon carbide paper. The frictional measurements were carried out by multi-pass sliding of a steel sphere (Insley Industrial Ltd; radius 9.5mm polished to an Ra of 0.304 ± 0.036μm) at a velocity of 0.4 mm s^{-1} and applied loads in the range 10-100g. Contact region conditions included both dry state and fully flooded with lubricant. Surface topography of the sliding tracks was examined using laser profilometry and interference microscopy.

RESULTS AND DISCUSSION

Figure 1 shows some typical friction data for a fully lubricated contact as a function of sliding distance (taken over 30 cycles). Whilst the dry data is somewhat erratic it may be clearly seen that abrasive pre-cleaning induces non-stationarity which increases with the normal load. The sliding on the abraded surface was accompanied by visible polishing which was confirmed by surface topography measurements.

That surface mechanical damage induces frictional non-stationarity is an important consideration in the use of mild abrasion for cleaning PET. This is a relatively common practise since an equilibrium state is not readily achieved using chemical methods and prolonged exposure may cause environmental stress cracking. Although the influence of the surface topography of the counter-surface on polymer wear is well established (3), little is known about the effect of pre-damage on the friction. The current work is consistent with the so-called 'flip-flop' wear mechanism in which the tips of asperities are subjected to plastic fatigue (4). While this process acts to smooth the wear track, sufficient debris remains entrapped to cause the non-stationarity. Current work is aimed at spectroscopic analysis of the steel countersurfaces to investigate the possibility of the formation of transfer films.

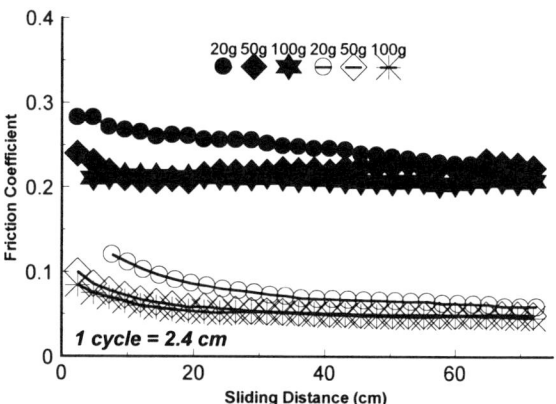

Fig. 1: Effect of applied load and cleaning regime for PET lubricated with 0.02M oleylamine in aqueous acetic acid at pH 6; (filled symbols = surfactant cleaned, unfilled symbols = abraded with grade 1000 'Wet'N'Dry' paper)

REFERENCES

(1) B. J. Briscoe, T. K. Wee. A. Winkler and M. J. Adams, 'Polymer Wear and its Control' ed. L-M Lee (ACS, Washington) 1985, 375-387 pp.

(2) Y. Yamada and K. Tanaka, ib. 363-374 pp.

(3) A. E. Hollander and J. K. Lancaster, Wear, Vol. 25, page 155, 1973.

(4) B. J. Briscoe and P. D. Evans 'Tribology in Particulate Technology' ed. B. J. Briscoe and M. J. Adams (Adam Hilger, Bristol) 1987, 335-350 pp.

THE WEAR PROPERTIES OF METALLIZED CARBON SLIDER IN UNLUBRICATED SLIDING AGAINST COPPER TROLLEY UNDER ELECTRIC CURRENT

SHUNICHI KUBO
Railway Technical Research Institute, Kokubunji, Tokyo, 185, JAPAN
KOJI KATO
Tohoku University, Sendai, 980-77, JAPAN

INTRODUCTION

The wear of contact strip on a pantograph of electric railway vehicle is mainly governed by arc discharge occurring simultaneously with contact break between strip and trolley wire[1]. This paper shows the effect of arc discharge on wear rate of contact strip.

EXPERIMENTAL PROCEDURE

Materials are Cu-impregnated baked carbon for strip and Cu for trolley wire. Strip is pressed against a 5 mm wide Cu band set on a rotating disk of 1000 mm diameter, representing the trolley wire. Sliding velocity is 27.8 m/s (100km/h), sliding time 30 minutes and normal load 49 N. DC electric current of 0 to 200 A, 100V flows from trolley to strip. Trolley surface is made rough by abrasion with a Cu-based sintered metal alloy.

Electric potential between trolley and strip and circuit current are measured all through sliding test to evaluate energy of arc discharge due to contact break. Wear rate of strip is calculated from its weight loss. Wear profile on strip is also measured.

RESULTS

Fig.1 shows electric potential and current during one revolution of disk, about 110 msec. Rectangular pulses indicating the beginning of each revolution of disk are also shown. In Fig.1 pulses having a potential higher than approximately 10 V correspond to contact break with arc discharge. Maximum duration of a contact break is 7 to 8 msec.

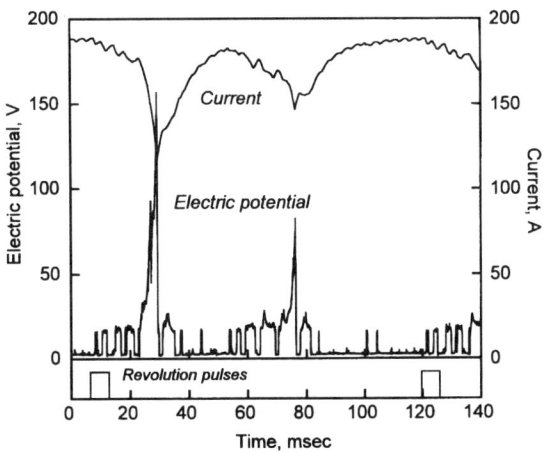

Fig.1: Electric potential and circuit current

Fig.2 shows a cross sectional profile of wear track on strip after sliding test. In the portion of strip where arc discharge occurrence concentrates, surface roughness increases.

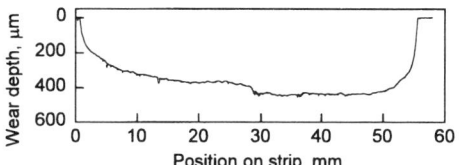

Fig.2: Wear profile on a strip

Arc discharge energy of every contact break is calculated by summation of products of potential and current during break. Accumulated energy generated by arc discharge during sliding test is evaluated for each tested strip. Wear rate of strip is plotted against the accumulated energy in Fig.3, which shows a proportional relationship between wear rate and accumulated discharge energy. Accumulated energy is normalized by a unit sliding distance.

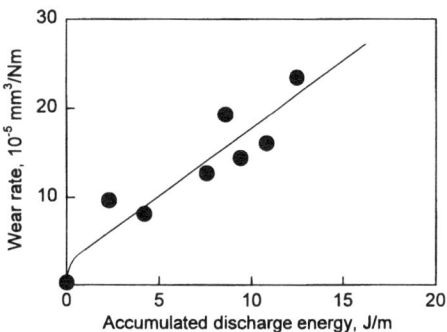

Fig.3: Wear rate of Cu-impregnated carbon strip with accumulated energy

CONCLUSIONS

In sliding wear test under electric current for the combination of Cu-impregnated baked carbon strip and Cu trolley, wear rate of strip is in proportion to the accumulated arc discharge energy due to contact break.

REFERENCE

(1) S.Kubo, H.Tsuchiya and J.Ikeuchi: Quarterly Report of RTRI, **38**, 1, 1997, 25-30pp.

WEAR BEHAVIOURS OF Al-Si ALLOY AGAINST STEEL UNDER THE LUBRICATION OF ORGANIC COMPOUNDS

W M LIU and Q J XUE
Laboratory of Solid Lubrication, Lanzhou Institute of Chemical Physics,
The Chinese Academy of Sciences, Lanzhou 730000, P. R. CHINA

ABSTRACT

Al-Si alloys are extensively used in the tribological applications such as internal combustion engines, plain bearings, compressor and refrigerator. Although Al-Si alloys meet many of the requirements for such applications, their poor resistance to seizure and difficult lubrication make them vulnerable in the sliding process. The lubricity of alcohol, partial ester and diol for aluminum was investigated, and the antiwear mechanism of formation of 5 or 6 rings complex of diol and aluminum was suggested (1, 2). In this work, the wear behaviours of Al-Si alloy against steel under the lubrication of liquid paraffin (as base oil) with 3 % wt additive of N and O containing compounds such as ethyleneglycol, ethylenediamine, ethanolamine, 8-hydroquinoline and N, N-dibutylethanolamine were investigated used a Timken Tester at a sliding velocity of 2.05 ms^{-1} (800 rpm) at room temperature. The chemical composition of the boundary film formed on the surface of Al-Si alloy during the sliding process was evaluated using XPS (PHI-5702).

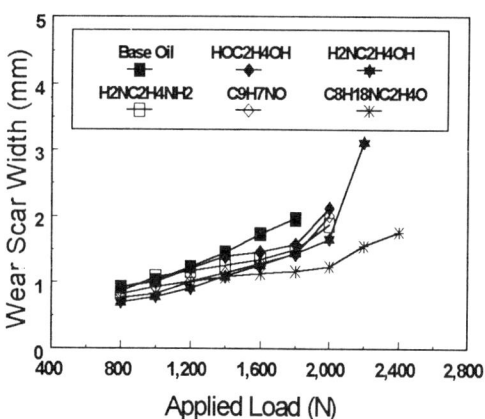

Fig. 1: Wear scar width of Al-Si block as a function of applied load under the lubrication of liquid paraffin containing 3 % wt additive
(Timken, 2.05 m/s, 20 min, 25 ℃)

Results in Fig. 1 show that at a load higher than 1200 N, the five tested organic compounds exhibit quite different antiwear performance for the Al-Si alloy.

With the lubrication of ethanolamine and N, N-dibutylethanolamine, smaller wear scar width and higher load-carrying capacity (referred as the failure load) were observed, as compared to the wear results of Al-Si alloy lubricated by ethylenediamine or ethyleneglycol.

XPS analyses were conducted for the elements in the boundary film of Al-Si alloy after tests under the load of 1600 N. From the binding energies of Al$_{2P}$ at 74.6 eV, it is concluded that the formation of aluminum oxides, while from the binding energies of Al$_{2P}$ at 73 to 73.8 eV and the binding energies of N$_{1S}$ at 401.5 eV of N, N-dibutylethanolamine and 400.5 eV of 8-hydroquinoline, it is concluded that the formation of complex of aluminum and the N, O-containing compounds as illustrated in Fig.2.

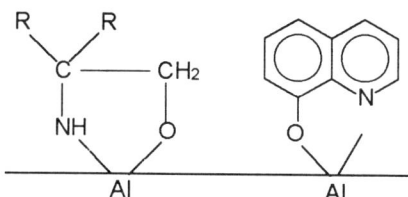

Fig. 2: Interaction of ethanolamine and 8-hydroquinoline with aluminum

The following conclusions were drawn:
(1) Diamine, 8-hydroquinoline and ethanolamine as additives in liquid paraffin reduce wear of aluminum-on-steel system, and a longer chain length N, N-dibutylethanolamine exhibits better antiwear performance and higher load-carrying capacity as compared to ethanolamine.
(2) Surface analyses of the rubbed surface of Al-Si alloy suggest the interaction between Al-Si alloy and 8-hydroquinoline or ethanolamine, which probably results in the formation of complexes of aluminum and ethanolamine or aluminum and 8-hydroquinoline.

REFERENCES
(1) B.W. Hotten, Lubrication Engineering, 30(1974) 398-403.
(2) Yong Wan, Weimin Liu and Qunji Xue, Wear, 193(1996)99-104.

FRETTING WEAR CHARACTERISTICS OF ZIRCALOY-4 TUBE

TAE-HYUNG KIM
Graduate school, Department of Mechanical Engineering, Kyungpook National University, Taegu, Korea
KWANG-HEE CHO
Korea Atomic Energy Research Institute, Taejon, Korea
SEOCK-SAM KIM
Department of Mechanical Engineering, Kyungpook National University, Taegu, Korea

ABSTRACT

In general fretting wear means the surface damages between interfaces of materials to move relatively within a few of hundreds of micrometers(1). Almost all of situations to cause the fretting wear damage are due to the unexpected accidental small amplitude vibration or the change of stresses. For typical example, the tube system vibrates unexpectedly in consequence of the fluid flow with high temperature and high pressure in the reactor and the heat exchangers of nuclear power plant and the fretting damage can be brought about owing to the flow-induced vibration(2).

However, as the fretting wear is occurred by the combined process of a lot of wear mechanisms, it is very difficult to analyze in spite of its importance.

Therefore, to examine the fretting wear characteristics of the fuel rod material, Zircaloy-4 tube, the experiment was carried out at various conditions.

This study was concentrated on the influences of normal load, slip amplitude and number of cycles as the main factors of fretting.

A fretting wear tester was designed for this experiment and cylinder to cylinder contact at right angle was used for a wear test method.

The results of the study showed that the wear volume increases abruptly with the increase of normal load at given cycles but the change of wear volume under the slip amplitude of 100μm is hardly shown even though the normal load increases. It seems that the change of wear volume is more dependent on normal load when the slip amplitude is longer than a certain slip amplitude.

The slip amplitude showing the transition is defined as the critical slip amplitude. It is known that the critical slip amplitude varies with the materials but exists in the range of 50 and 100μm for almost all of materials.

The critical slip amplitude of Zircaloy-4 tube used in this study is about 100μm. When the materials deformed by normal load form the contact area and move relatively within very small distance, only the elastic deformation is repeated on the central section of the contact area and the surface damage is not observed because slip is hardly occurred. And, the wear volume becomes so small. This contact region is called stick-slip region.

It was found that the existence of the critical slip amplitude was caused by the stick-slip region in which materials deform elastically.

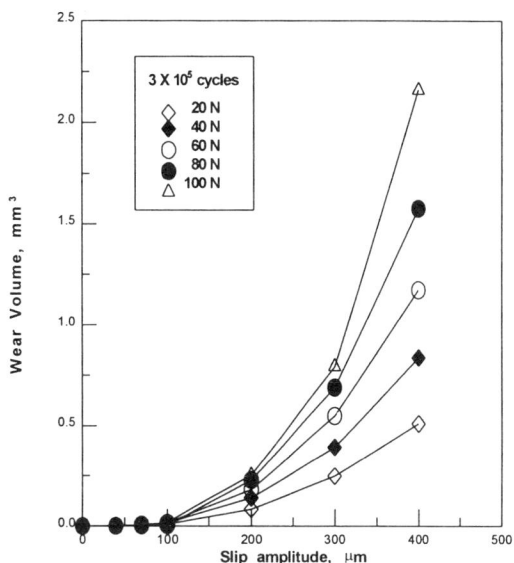

Fig. 1 Wear volume as a function of slip amplitude at 3×10^5 cycles

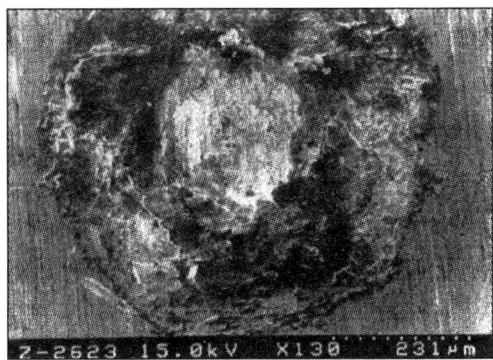

Fig. 2 SEM photograph showing stick-slip (70μm, 80N, 3×10^5 cycles)

REFERENCES

(1) R.B. Waterhouse, Fretting Corrosion, p.36, Pergamon Press, 1975.
(2) P. L. Ko, Journal of Tribology, 107, 1985.

Improvement of Tribological Performance of Carbon Brushes in Electrical Motors

DR. RÜDIGER HOLINSKI and FRANZ HEIDENFELDER
Dow Corning GmbH, Rheingaustr. 34 Wiesbaden, Germany

ABSTRACT

In an electrical motor energy has to be transferred to a rotating collector. This is accomplished by carbon brushes, which are in sliding contact to the copper commutator. Electrical energy is transferred through the surface of the dry sliding contacts. Life of electric motors depends on wear life of both sliding components.

A carbon brush for automotive applications consists primarily of graphite, copper powder, a binder, abrasives and a solid lubricant which reduces friction and wear of the sliding components. Surface investigations revealed that a solid lubricant film is formed at the sliding surface of the carbon brush. A transfer film of the solid lubricant is also formed on the surface of the copper commutator. These solid lubricant films protect the substrates from wear and reduce friction.

It was found that a certain concentration of molybdenum disulphide powder in a brush formulation leads to a formation of rather thick MoS_2 films on surfaces of the copper collector. After a certain sliding distance the transfer film blisters off and a new film forms on the commutator surface from the reservoir of the brush formulation.

Tested was an alternative solid lubricant which consisted primarily of a mixture of sulphides of bismuth, zinc and tin. This solid lubricant package in a carbon brush also lead to a formation of a transfer layer on the surface of a copper commutator. However, these films were found to be much thinner than MoS_2 films and had much longer wear life. This means adhesion of additive films on the copper substrate and cohesion within the film were comparably strong.

This effect resulted in better protection of the copper substrate or lead to a dramatic reduction of commutator wear and a substantial reduction of brush wear (fig. 1). These thinner transfer films also reduced resistance of electrical current.

Investigation of contact areas of brush and commutator revealed, that surface roughness was reduced by 50% by the new additive, which obviously was distributed more uniformly in the brush matrix. Also friction and frictional temperature was reduced by the mixed sulfide additive.

In sliding experiments of both types of carbon brushes against copper commutators noise level was determined. It was found that the new additive generated 50% less sliding noise than MoS_2 which can be contributed to the reduced surface roughness of brush and commutator.

It was demonstrated that different carbon brush additives lead to different transfer films of commutators of different adhesion and film thickness. This has a substantial effect on wear of copper commutators, wear of carbon brushes and surface roughness.

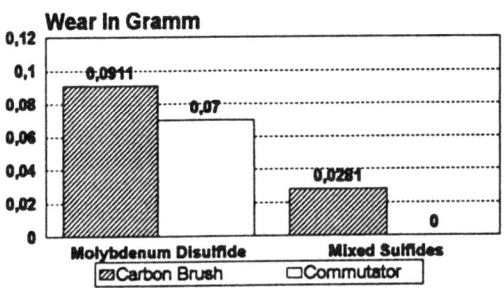

Fig. 1: Wear of carbon brushes and commutator

EFFECT OF STRESS STATE OF DIFFERENT TiN FILMS ON THEIR TRIBOLOGICAL BEHAVIOR

ZHUANG Daming, LIU Jiajun, ZHU Baoliang
(Tsinghua University, 100084 Beijing, P.R. China)
ZHOU Zhongrong
(Southeast Jiaotong University, 610031 Chengdu, P.R. China)
Leo VINCENT, Philipe KAPSA
(Ecole Centrale de Lyon, 69131 Ecully Cedex, France)

ABSTRACT

The deposited ceramic films on a suitable substrate can dramatically improve its wear-resistance. But not every method is successful in commercial applications. One of the key reasons is the poor films-substrate adhesion, which is closely related with the internal stress state at both sides of interface (1). The deposited films can peel off if their strain energy exceeds the interface energy (2).

In this paper the TiN films were deposited by the methods of ion beam enhanced deposition (IBED), plasma chemical vapour deposition (PCVD) and ion plating (IP) on a 52100 steel substrate respectively. A MSF-2M model X-ray stress diffractomter was employed for measuring the internal stress state. The friction and wear tests of different TiN films were carried out on a SRV testing machine at different load and frequency (velocity) conditions. Fig.1 shows the stress distribution at both sides of film-substrate interface for different TiN films. Based on the measurement result of internal stress, its effect on microhardness, bonding strength with substrate and tribological behavior of different TiN films was analysed systematically.

The main conclusions of this research can be drawn as follows:

(1) The internal stress state in all IBED, PCVD and IP TiN films is compressive. Whereas that of 52100 steel substrate adjacent to the interface is varied from compressive to tensive with different deposition methods.

(2) The internal stress state shows obvious effect on the hardness and bonding strength of TiN films, which are increased in the order of IP-PCVD-IBED with the decreased internal stress in the same order.

(3) IBED TiN film shows better tribological performances than PCVD and IP TiN films due to its appropriate stress state.

(4) The wear mechanism of TiN films is also determined by their stress state. The IP and PCVD TiN films with larger internal stress are easily failed by carcking and peeling, while the IBED films shows much better peeling-resistance, its main wear mechanism is abrasive wear.

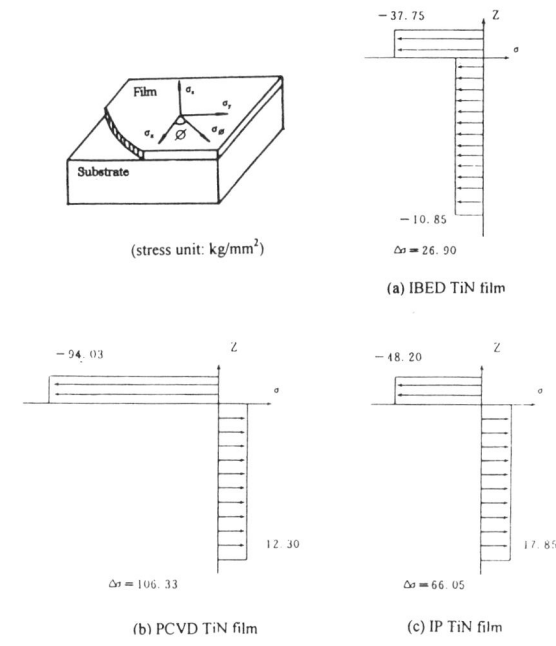

Fig.1 Stress distribution at both sides of film-substrate interface for different TiN films.

REFERENCES

(1) B.Kramer Thin Solid Films 108(1983)117-175
(2) Tian Minbo, Liu Deling Handbook of film science and technology, Beijing Machinary Industry Press 1991, 8-9

LUBRICATION BY DIAMOND AND DIAMONDLIKE CARBON COATINGS

KAZUHISA MIYOSHI
National Aeronautics and Space Administration, Lewis Research Center, Cleveland, Ohio 44135, USA

ABSTRACT

This investigation examined the lubrication, friction, and wear properties of fine-grain CVD (chemical-vapor-deposited) diamonds and DLC (diamondlike carbon) films in sliding contact with CVD diamond pins. Four types of relatively smooth surfaces (6 to 37 nm in R_{rms}) of CVD diamond films were produced: as-deposited, fine-grain diamond (1); as-polished, coarse-grain diamond (2); polished and then fluorinated, coarse-grain diamond (3); and polished and then nitrogen-ion-implanted, coarse-grain diamond. Two types of ion-beam-deposited DLC smooth surfaces were produced at ion energies of 1500 and 700 eV (4).

The CVD diamond and DLC films were characterized by scanning and transmission electron microscopy, Rutherford backscattering spectroscopy, hydrogen forward scattering, Raman spectroscopy, Fourier transform infrared spectroscopy, x-ray photoelectron spectroscopy, x-ray diffraction, surface profilometry, and scanning probe microscopy.

Unidirectional, rotating, sliding friction experiments were performed at room temperature in humid air (relative humidity, 40 %), in dry nitrogen (relative humidity, <1 %), in an ultrahigh vacuum (10^{-7} Pa), and in distilled water.

Results indicated that the as-deposited, fine-grain CVD diamond, the polished, coarse-grain CVD diamond, and the fluorinated, coarse-grain CVD diamond had a low coefficient of friction (<0.1) and wear rate ($\leq 10^{-6}$ mm^3/N•m) in humid air, dry nitrogen, and water. However, they had a high coefficient of friction (>0.1) and wear rate (order of 10^{-4} mm^3/N•m) in ultrahigh vacuum where removing some contaminant surface film from the contact area of diamond film results in a stronger interfacial adhesion between the diamond pin and the diamond film, raising the coefficient of friction and wear rate (5).

The nitrogen-ion-implanted diamond and the DLC had a low coefficient of friction (<0.1) and wear rate ($\leq 10^{-6}$ mm^3/N•m) regardless of environment (Fig.1). A thin surficial layer (<0.1 µm thick) of amorphous, nondiamond carbon produced on the CVD diamond by ion implantation and the hydrogenated nondiamond carbon surface of the DLC had a low shear strength and a low surface energy at the contact area. In general, the combination of low shear strength or low surface energy of the thin, nondiamond carbon surface layer and the small contact area gives low coefficients of friction (6).

In conclusion, regardless of environment, ion-beam-deposited DLC and nitrogen-ion-implanted CVD diamond films can be used as effective wear-resistant, self-lubricating coatings. However, as-deposited, fine-grain CVD diamond, polished, coarse-grain CVD diamond, and fluorinated, coarse-grain CVD diamond can be used as effective wear-resistant, self-lubricating coatings in humid air, in dry nitrogen, and in water, but they are not acceptable for solid-film lubrication applications in ultrahigh vacuum. The polished, coarse-grain CVD diamond film had an extremely low wear rate of far less than 10^{-10} mm^3/N•m in water.

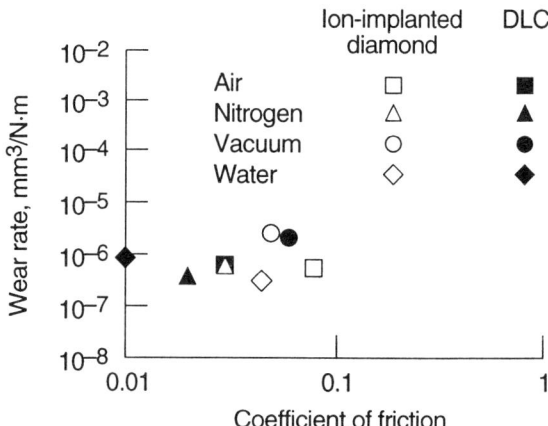

Fig. 1.—Coefficients of friction and wear rates of nitrogen-ion-implanted CVD diamond film and DLC film in sliding contact with CVD diamond pins.

REFERENCES

(1) R. L. C. Wu, et. al., J. Appl. Phys., Vol. 72, No. 1, July 1992, pp. 110–116.
(2) M. Murakawa, S. Takeuchi, Mat. Sci. Eng., Vol. A140, 1991, pp. 759–763.
(3) S. Miyake, Appl. Phys. Lett., Vol. 65, No. 9, 1994, pp. 1109–1111.
(4) R. L. C. Wu, et. al., Mat. Res. Soc. Symp. Proc., Vol. 354, 1995, pp. 63–68.
(5) F. P. Bowden, A. E. Hanwell, Proc. R. Soc. London Ser. A, Vol. 295, 1966, pp. 233–243.
(6) I. L. Singer, Proc. of NATO Advanced Study Institute on Fundamentals of Friction, Kluwer Academic Publishers, Dordrecht, 1992, pp. 237–261.

Tribology of diamond coated face seals

P. HOLLMAN, H. BJÖRKMAN, A. ALAHELISTEN and S. HOGMARK
Department of Materials Science, Uppsala University, P.O Box 534, SE-751 21 Uppsala, Sweden
G. ANDERBERG
HUHNSEAL AB, P.O Box 288, SE-726 04 Landskrona, Sweden

ABSTRACT

Mechanical face seals are used in many applications such as pumps and engines. Due to high demands on sliding speed, load carrying capacity, wear resistance and capability to resist elevated temperatures, high performing seals are generally made of cemented carbide or ceramics (1). However, these materials display a high friction coefficient - normally around 0.5, - which causes energy loss and heat formation. Extensive heat generation is a problem which could result in thermal fatigue of the surfaces (cracks) and of course seizure which could end the life of the component.

One way to decrease the friction is to deposit a diamond coating on one or both faces of the seal. Diamond is known to generate low friction, high wear resistance and high thermal conductivity. However, as-deposited diamond coatings generally have a high surface roughness which results in a high friction coefficient and extensive wear of the counter material in sliding contact. This problem is avoided by polishing the coating or by using two diamond coatings sliding against each other, they would then polish each other during a short running-in period.

In the present study, face seals with inner/outer diameters of 18/22 mm, made of cemented carbide (CC) (96 wt% WC and 4 wt% Ni) were coated with diamond using the combustion flame technique. The equipment consists of a special flame nozzle with 8 holes in a circle and a rotating sample holder.

Acetylene (99.5%) and oxygen (99.95%) with a flow of 3.3 and 3 SLM, respectively, are used as source gases. The flow is controlled by mass flow controllers. An IR camera is used to monitor the temperature of the sample and a PID regulator controls the temperature of the sample to 800 °. Before deposition the samples were etched for 5 minutes in 50% HNO_3 to remove Ni from the surface. The diamond deposition time of each sample was 20 minutes.

Friction tests were performed according to Fig. 1. During these tests the surface pressure was 0.5 MPa and the sliding speed 1.5 m/s.

Furthermore, self mated diamond coated seals were also tested in a component test rig using pressurised water (about 5 MPa) as a medium and a sliding speed of 1.5 m/s. SiC - Graphite and Al_2O_3 - Graphite were used as reference.

Raman spectroscopy, scanning electron microscopy (SEM), light optical microscopy (LOM) and atomic force microscopy (AFM) were used to characterise the diamond coatings before and after the friction tests.

After deposition the coatings reveal a rough surface with facets (normal CVD diamond coating morphology) with a grain size of 4 μm. Raman spectroscopy shows a sharp peak at 1332 cm^{-1} and a broad peak around 1550 cm^{-1}. The coating thickness was 3.5 μm except for the inner edge where the thickness was 4.5 μm as measured by SEM.

In dry sliding, self mated diamond shows a very low start value of the friction, 0.05. Slowly, the friction increases during 500 seconds and stabilise at 0.2. When using a polished diamond coating against CC the friction starts at 0.35 and decreases down to 0.25. During both these tests there were no noise and the friction was stable and showed no measurable fluctuations. A much higher friction, 0.8, was obtained in the test between self mated CC. Depending on extensive vibrations and heat generation this test was interrupted after 900 seconds. During the test there was also a continuos sound indicative of seizure.

The wear process of the diamond coatings begins with smoothening of the asperities. Thereafter the wear rate is extremely low. When studying the worn diamond in SEM and LOM it displays an extremely smooth surface with a roughness of $R_a \approx 0.6$ nm as measured with AFM. Raman spectra from the worn diamond surface showed no changes compared to as-deposited diamond.

In the component test no leak was seen and the friction diamond - diamond, SiC - Graphite and Al_2O_3 - Graphite was 0.004, 0.014 and 0.012, respectively. After the test, SEM revealed that some areas of the diamond coating and the CC had spalled at the inner rim of the seal.

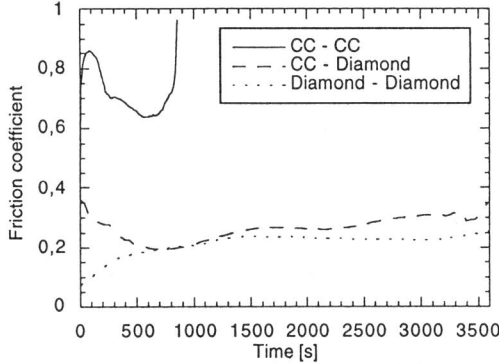

Fig. 1: Friction coefficient in dry sliding

ACKNOWLEDGEMENT

NUTEK/NFR and the Ångström Consortium are gratefully acknowledged for the financial support.

REFERENCES

(1) S. Hogmark, M. Olsson, International Symposium on Advanced Ceramics for Structural and Tribological Applications, Vancouver, B.C., Canada August 19-23 1995.

Processing-Dependent Microstructure, Properties and Wear of Plasma Sprayed Alumina Coatings

L.C. ERICKSON and T. TROCZYNSKI
Metals & Materials Engineering, University of BC, Vancouver, BC, Canada
H. TAI and D. ROSS
Northwest Mettech Corporation, Richmond, BC, Canada
H.M. HAWTHORNE
National Research Council Canada, Vancouver, BC, Canada

INTRODUCTION

The influence of the processing conditions on the coating microstructural integrity and the resultant impact on the coating's wear resistance are evaluated. Plasma sprayed coatings were prepared from pure α-Al_2O_3 powders of different average particle size (5, 10 and 18 µm) to investigate the influence of splat size, phase composition and indentation resistance on the wear properties.

EFFECTS OF VARYING PROCESS PARAMETERS

Two process parameters were varied between high and low values: percent hydrogen as a secondary plasma gas and orifice diameter of the plasma torch nozzle. Deposition of an ideal coating results from exposing the powder particles to an appropriate temperature for an adequate period of time. Tabulated below is the series of spray runs for each powder size with the deposition parameters that were varied and their effects.

Table 1 - Processing Conditions

Spray Run #	%H_2	Effect on Heat Input	Torch Nozzle Size [in]	Effect on Residence Time [s]
APS	NA	Low	NA	Moderate
1	10	Low	1/2	Less
2	10	Low	9/16	More
3	20	High	1/2	Less
4	20	High	9/16	More

Runs #1-4 were sprayed using an axial feed plasma spray torch, while Run APS was sprayed using a conventional radial feed torch.

MICROSTRUCTURE AND WEAR PROPERTIES

The coating microstructures were analyzed with respect to crystallographic phase composition (XRD), microhardness (HV_{300}), splat thickness and splat diameter (SEM). Abrasion and erosion tests using large particles were performed. These results can be used as a statistical measure of microstructural integrity, i.e. inter-splat cohesion, in a plasma sprayed coating (1,2).

Due to rapid solidification, a well melted α-Al_2O_3 particle will transform to γ phase. This produces a dense, uniform coating with good wear resistance. Figure 1 shows the abrasive wear rate of coatings as a function of both spray conditions and powder size.

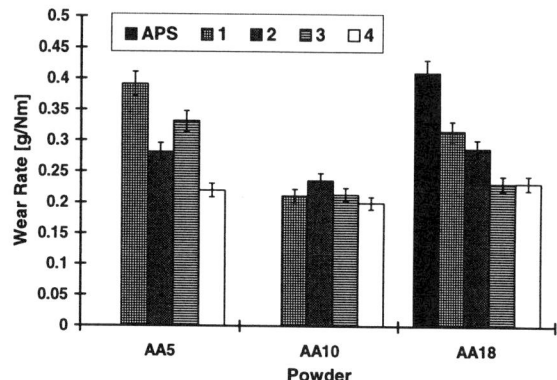

Fig. 1 - Differences in abrasive wear rates vs. coating powder size and processing conditions.

There is a trend in which coatings deposited under higher heat input and more residence time have lower wear rates, higher hardnesses, and less retained α phase (AA5-4, AA10-4, AA18-4). In addition, splat thickness decreases and splat diameter increases due to greater flattening of the droplet.

Larger particles require higher heat input to produce more wear resistant coatings, but are not significantly affected by changes in residence time (AA18-3 and AA18-4). Smaller particles also require the higher heat input, however they are greatly affected by changes in residence time (AA5-X). The coating AA5-3 has a high wear rate possibly because spraying with a low powder residence time corresponds to a high particle velocity which results in the droplet "splashing" on impacting the substrate. The resultant volume loss is indicated by reduced splat diameter and thickness. Microhardness is high due to recrystallization of the splashed particles and re-entrainment in the coating. Optimum powder particles result in coatings with superior wear resistance and relative insensitivity to spray parameters (AA10-X).

ACKNOWLEDGEMENTS

Thanks to R. Westergard, Uppsala University, for help with abrasion tests.

REFERENCES

(1) H.M. Hawthorne, L.C. Erickson, D.Ross, H.Tai and T. Troczynski, Wear, Vol. 203-204, pp 709-714, 1997.
(2) L.C. Erickson, R. Westergard, U. Wiklund, N. Axen, H.M. Hawthorne and S. Hogmark, Submitted to Wear.

FRICTION AND WEAR BEHAVIOUR OF CARBON-BASED COATINGS (a-C, a-C:H, Me-C:H AND DIAMOND) IN VIBRATING TRIBO CONTACTS

DIETER KLAFFKE AND ANDRÉ SKOPP

Bundesanstalt für Materialforschung und -prüfung (BAM) VIII.12 "Fretting, Microtribology, Cryotribology"
Unter den Eichen 44 - 46, D-12200 Berlin, Germany

ABSTRACT

Thin carbon-based coatings (a-C, Me-C:H, a-C:H, diamond) were tribologically characterised under vibrating sliding conditions in a ball-on-disk configuration against balls of bearing steel (100Cr6, AISI52100) or alumina. The friction and wear behaviour of flat coated specimens has been investigated in the gross-slip regime in a broad range of strokes (25 µm to 10 mm), frequencies (1 Hz to 20 Hz) and loads (1 N to 20 N). Tests were performed at room temperature in different test rigs, allowing the variation of relative humidity of the ambient air. Tests were carried out in dry air, normal air and moist air with relative humidities (R.H.) of 3 %, 50 % and 100 %, respectively. Beside standard tests with 10^5 cycles, test series were performed with number of sliding cycles in the range 10^3 to $1.2 \cdot 10^6$ in order to investigate initiation and evolution of wear.

In all tests the coefficient of friction and the linear wear of the system were measured continuously. Volumetric wear was calculated separately for both specimens using the size and profiles of the wear scar.

The tribological behaviour of carbon based coatings was analysed and compared, using a data base. This tool allows to study the tribological effects of common and new created parameters. Selection criteria can be defined easily. Exemplary results will be given.

In order to analyse wear processes and to correlate friction and wear processes with tribochemical changes different surface analytical tools were used as optical micrography (OM), atomic force microscopy (AFM), scanning electron microscopy (SEM) with EDX, electron probe micro analysis (EPMA), micro laser Raman spectroscopy (LRS), and small spot ESCA.

Friction and wear as well as transitions of friction and wear depend strongly on the type of carbon-based coating and on relative humidity. For a given tribocouple and at a given constant relative humidity, friction and wear depend most significantly on a so called power parameter, P, given as the product of stroke, frequency and load.

Systematically performed test series, storage in a data base and plotting the results in dependence of varied parameters give important hints for standardisation of testing methods as well as hints on application ranges of coatings and their limits. To reduce the number of tests for the characterisation of the tribological behaviour of carbon based coatings the use of the power-parameter-concept was found to be helpful. The tendency of increasing or decreasing friction and wear in dependence of P depends on relative humidity as well as on tribopairings (used coating and counterbody).

The development of testing procedures, standardisation activities, and the interpretation of test results of thin hard coatings have to consider the effects of test parameters. For carbon-based coatings very low friction coefficients below 0.05 are observed. For special test conditions extremely low wear rates are found that are below the limit of resolution (10^{-8} mm^3/Nm).

A data base was developed in order to store all relevant attributes and results of vibrating tests. Some hundred data sets with test results of carbon based coatings are included.

Keywords: Carbon based coatings, vibrating sliding tests, friction, wear, effects of operational parameters and relative humidity, triboreaction layers

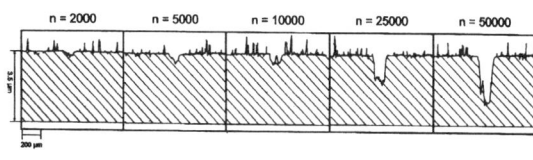

Fig. 1: Evolution of wear of Me-C:H coated disk against Al$_2$O$_3$ tested in air of 50 % relative humidity under vibrating sliding conditions (coating thickness: 3.5 µm)

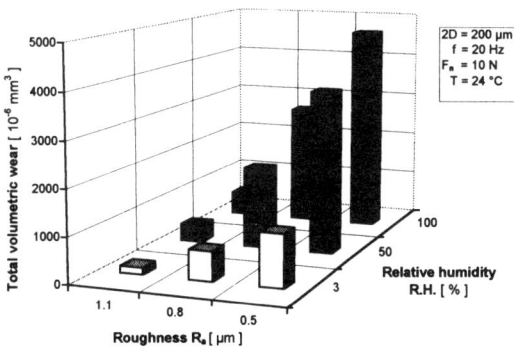

Fig. 2: Volumetric wear of alumina ball, running against diamond coatings, versus relative humidity and coating roughness

THE EFFECT OF SURFACE TEMPERATURE ON FRICTION PROPERTY OF CARBON NITRIDE COATING BY ION BEAM ASSISTED DEPOSITION

NORITSUGU UMEHARA, KOJI KATO and TAKAHIRO SATO
School of Mechanical Engineering, Tohoku University, Sendai, 980-77, Japan

ABSTRACT

Carbon nitride coating is expected as hard overcoat of magnetic rigid disk with long life (1-2). Because it was predicted theoretically that carbon nitride could be harder than diamond, if its structure has similar structure to β-Si_3N_4. On the other hand, it was reported that the wear of overcoat of magnetic rigid disk is influenced of surface temperature strongly (3). Therefore it is important to know the effect of surface temperature on the friction properties of carbon nitride coating. In order to clarify tribological properties of carbon nitride, amorphous carbon nitride coatings were produced with ion beam assisted deposition, and evaluated tribological properties. In this paper, the effects of surface temperature on friction and wear properties were investigated.

Carbon nitride coatings were deposited on silicon wafer disks with the diameter of 50mm and the thickness of 0.07mm. After Ar ion bombardment for cleaning disk, Ar ion beam sputtering of carbon and implantation of nitrogen ion by N ion beam were done simultaneously. The acceleration energy and electric current of N ion beam was 500 eV and 40 mA/cm^2, respectively. The thickness of coating deposited was about 100nm. According to ESCA analysis, carbon nitride coating contained 90 at% carbon and 10 at% nitrogen. No oxygen was found in the bulk material of the coating.

Pin-on-disk type friction test was conducted for carbon nitride coating against Si_3N_4 ball in vacuum chamber. During friction, surface temperature was controlled within the range from 20°C to 270°C by the electric heater from the back side of the substrate. Pressure in the vacuum chamber was 2.6x10^{-6} Torr. Normal load and sliding speed were 10 mN and 0.007m/s, respectively. Normal load and friction were measured by the strain gage system during friction.

When surface temperature was lower than 18°C, friction coefficient was constant during sliding. But the surface temperature is higher than 18°C, friction coefficient gradually decreased with sliding distance and reached to a certain stable value. The relationship between the reached values of friction coefficient and surface temperature is shown in Fig.1. Friction coefficient decreased with surface temperature of substrate. At surface temperature of 270°C, 0.026 of friction coefficient was obtained. Same dependency of friction property against temperature in vacuum was reported for DLC (4). The reason of this low friction can be considered that some low shear strength material was generated on the surface at high temperature. But we could not detect any new material by SIMS. Temperature dependency of friction coefficient of Si wafer also was investigated. For Si wafer, friction coefficient rapidly increased with surface temperature as shown in Fig.1. Regarding on wear property, no wear scar was observed with optical microscope.

From all of these results, it can be concluded that carbon nitride coating produced ion beam assisted deposition is expected as promising coating at high temperature in high vacuum.

REFERENCES
(1) E.C.Cutiongco, D. Li, Y.-W. Chung and C.S Bhatia, Trans. ASME, J. Tribol., Vol.118, (1996),543.
(2) A. Khurshudov, K. Kato, D. Sawada, Proc. of Int. Conf. of Tribology, Yokohama, (1996),1931.
(3) A. Khurshudov, M. Olsson, K. Kato, Proc. of Int. Conf. of Tribology, Yokohama, (1996)1937-1942.
(4) K. Miyoshi, Adv. Info. Storage Syst., 3(1991)147.

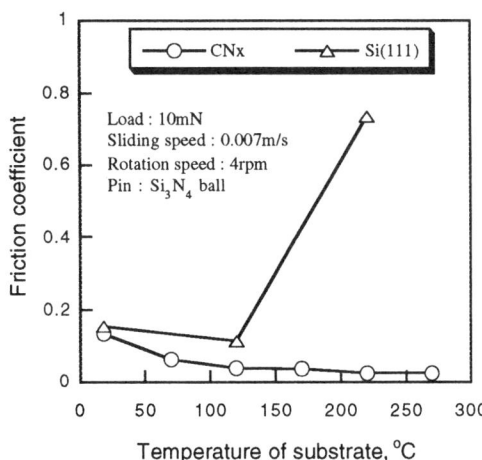

Fig. 1 The effect of surface temperature on friction coefficient of amorphous carbon nitride (CNx) and Silicon (Si)

CONTRIBUTION TO INVESTIGATION OF TRIBOLOGYCAL PHENOMENON AT DIFFERENT STEELS WITH PLASMA SURFACE COATINGS

D.KAKAŠ, Lj.MAŽIBRADA and B.ŠKORIĆ
Faculty for Technical Sciences, University of Novi Sad, 21000 Novi Sad, Trg D.Obradovića 6, YU

ABSTRACT

Property of plasma surface layers influence significantly on wear resistance. Our previous investigation shows that the base material has also great influence on wear resistance |1|. Combined layers consist of plasma nitrided and plasma deposited TiN or TiAlN could improve this property especially in the case when the base material has to be relatively softer |2,3|. Optimal hardness of high speed steel is about 65 HRC but in the case of hot working steel (H11) optimal hardness is 44-46 HRC.

In this paper some experiences connected with investigation of temperature flow at contact zone of tribologycal pair was presented. At the same time the wear resistance was measured using criterion of dimensional exchanges during the wear. Our results show that temperature in wear zone increased continually according to some exponential function. At the same time the dimensional exchanges of wear zone exchanged also according to the other type of exponential function. That was the reason for our investigation of correlation between these two parameters.

This phenomenon was investigated at two different steel (M2 and H11) on uncovered samples, samples with single TiN hard layer and samples covered with combined layers. The firs combined layer was produced by plasma nitriding and consequently plasma deposition of TiN. The second combined layer was produced with combination of plasma nitriding and plasma deposition of TiAlN. Depth of plasma nitriding for M2 high speed steel was about 50 μm and consist only of diffusion zone. On the other tool steel (H11) the depth of plasma nitriding was about 100 μm. For the both steel the roughness of surface was varied. One group of samples was grinded and the other was polished.

Wear resistance was investigated by universal Amsler equipment with simulation of sliding without lubrication. Countermaterial was produced like ring with contact at the point with the flat surface of the samples. High specific contact stresses at the contact zone was produced by load 50 N and 200 N. Compared results of correlation between temperature increasing and wear zone dimension it was concluded that significant difference exist relating the type of surface layer. Figure 1 shows this phenomenon for M2 steel loaded with 50 N during the wear. Uncovered sample has correlation $\Delta T = 24.753$ WZ with standard deviation 0.984 (WZ – Dimension of wear zone). Sample with combined layer PN+TiN has correlation $\Delta T = 0.4687 - 27.7424$ WZ $+ 125.3$ WZ^2 with standard deviation 0.999. Temperature increase (ΔT) were higher for conbined layer than for uncoated sample what could be explained with fact that the contact area was significantly smaller.

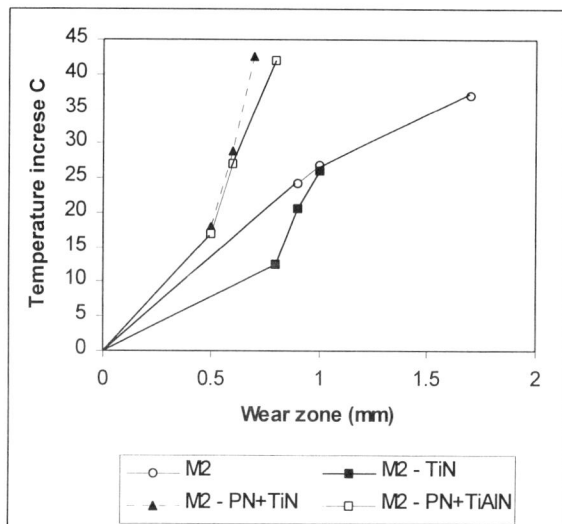

Fig.1. Temperature increase in relation to wear zone

Type of line for combined layers M2-PN+TiN and M2-PN+TiAlN is very similar like the wear resistance under created conditions. Shape of line for single TiN layer is very similar to combined layer but characterized with more intensive wear. This result was explained by investigation of exchanges in wear zone morphology by scanning electron microscopy.

The results presented in this paper could give some contribution to better understanding of wear processes at contact surface of tribologycal pair.

REFERENCES

(1) B. Škorić, D. Kakaš, Materials and Manufacturing Processes, Vol. 10, No. 1, 1995, 133-138 pp.
(2) D. Kakaš, Lj. Mažibrada, Z Kolumbić, Proc. of 10th Int. Colloquium – Tribology, Eslingen 1996, Vol. 3, 1975-1982 pp.
(3) D. Kakaš, Lj. Mažibrada, Materials and Manufac. Processes, Vol. 10, No. 3, 1995, 565-570 pp.

SURFACE ENGINEERING BY POWDER CO-INJECTION MOULDING

J R ALCOCK, P M LOGAN and D J STEPHENSON
School of Industrial and Manufacturing Science, Cranfield University, Cranfield, Beds., MK43 0AL, UK

ABSTRACT

A development of the powder injection moulding process, powder co-injection moulding (PCM), has recently been introduced (1). The new process allows surface engineering for wear-resistance during powder injection moulding rather than as a post-production step. This paper presents results on the production of wear-resistant surface layers in metallic, ceramic and mixed systems by PCM.

Powder injection moulding is a technique for the mass production of intricately-shaped components. Polymeric binders are mixed with metal or ceramic powders and the mixture heated and then forced under pressure into a die cavity. The particle assembly takes on the shape of the mould and retains that shape through the subsequent debinding (polymer removal) and sintering stages.

For PCM, a twin-barrelled injection moulding machine was used, in which the barrels were connected to a single injection nozzle by a hydraulically controlled valve (2). Binder and powder were injected from the barrels in three stages:

In stage (A1) material from barrel A was injected into a mould to form the skin of the compact.

In stage (B) a shut-off valve was repositioned to allow injection of the core material from barrel B. This secondary injection forced the skin material to the outer parts of the mould cavity.

In stage (A2) the valve was opened to A for a short shot of material to complete the skin at the injection point. The mould was then packed under pressure.

A binder system, containing combinations of carnuba and paraffin wax, stearic acid and polypropylene, was utilised.

Feed-stocks were initially blended in a shaker-mixer for 30 minutes at room temperature. Each was subsequently compounded in a twin-screw co-rotating extruder, at a screw speed of 200-300 rpm and temperature of 95-200 °C. The resulting feed-stock was granulated.

The fabrication of a metallic system with a skin of 316L stainless steel (Osprey Metals), and core of OM grade carbonyl iron (BASF) was investigated (3). The skin mix contained approximately 5 vol%, 25 µm alumina particles as a secondary, wear-resistant phase.

The apparent viscosity of the stainless-steel powder-binder mix was greater than that of the iron powder-binder mix, owing to the higher powder loading. This appeared to prevent break through of core material to the skin of the compact.

Prevention of delamination of the core and skin during sintering was found to be critical to the success of the PCM process. Hence, control of the sintering rate of the skin and core powders was required. For the metallic system delamination was prevented by reducing the volume fraction of powder in the core to 12.5% less than in the skin. The increased 'differential densification' of the carbonyl iron retarded its sintering rate.

Following moulding, the wax binder constituents were removed by solvent debinding in heptane at 85 °C. Thermal debinding and sintering were performed as a two-step process. Initially, the compacts were debound and part sintered in a hydrogen atmosphere. The compacts were then sintered under vacuum to 1300 °C, followed by furnace cooling. No delamination was visible. The core showed some un-reduced porosity of size less than 10 µm. The density of the sintered compacts measured by a liquid immersion method was 90%. Figure 1 shows the interface between the stainless steel surface layer (left) and the core of an injection moulded compact.

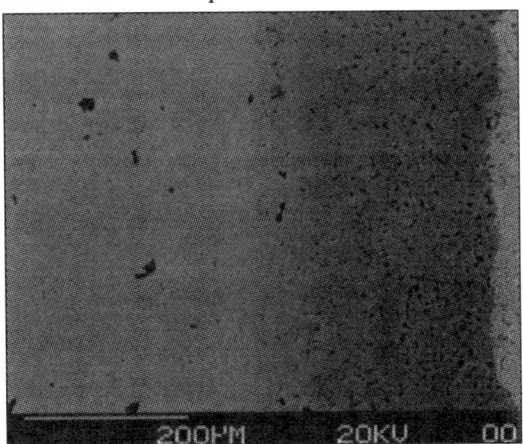

Fig. 1.: A surface-engineered stainless steel - carbonyl iron compact produced by powder co-injection moulding

A surface engineered ceramic system of alumina has also been investigated. Compacts were fabricated with a skin of 0.5 µm and core of 1 µm alumina.

REFERENCES

(1) UK Patent Application 9607 718.5, 1996.
(2) D F Oxley, D J H Sandiford, Plastics and Polymers, Vol. 39, page 288, 1971.
(3) J R. Alcock, P M Logan, D. J. Stephenson, Journal of Materials Science Letters, Vol. 15, pp 2033-2035, 1996.

WEAR CHARACTERISTICS OF THERMALLY SPRAYED TiC-Ni BASED COATINGS

P VUORISTO and T MÄNTYLÄ
Institute of Materials Science, Tampere University of Technology, P.O. Box 589, FI-33101 Tampere, Finland
L-M BERGER and M NEBELUNG
Fraunhofer Institute for Ceramic Technologies and Sintered Materials, Winterbergstr. 28, D-01277 Dresden, Germany

ABSTRACT

Thermally sprayed WC-Co and Cr_3C_2-NiCr coatings are widely used as wear and corrosion resistant coatings in many industrial applications. Another hardmetal system on the base of TiC-Ni has found recently a lot of interest, because of its expected good properties, such as high wear and corrosion resistance, low friction properties, high temperature stability, etc. In collaboration between the authors, novel TiC-Ni based spray dryed and sintered powders have been developed for thermal spray processes (1-3). The powders are characterized by a nearly spherical morphology and well-developed hardmetal-like microstructure. These powder properties provide that coatings with a dense microstructure, high wear resistance and without oxidation of TiC can be prepared by atmospheric plasma spray (APS), detonation gun spray (DGS) and high-velocity oxyfuel (HVOF) spray processes.

Table 1 and 2 present the results from the rubber wheel abrasion wear test (abrasive 0.1-0.6 mm quartz sand, 23 N contact force) and solid particle erosion wear tests (erosive 0.1-0.6 mm quartz sand, impact angle 30°, velocity 80 m/s) for WC-Co, Cr_3C_2-NiCr and TiC-Ni based coatings. Plasma spraying is a well-suitable process for TiC-Ni based powders; TiC-Ni based coatings show better abrasion and erosion wear properties in comparison to corresponding WC-Co and Cr_3C_2-NiCr coatings. In case of DGS and HVOF sprayed coatings, TiC-Ni based coatings do not reach the same level of abrasion wear resistance as the WC-Co coatings, but show similar or better properties than the Cr_3C_2-NiCr coatings, depending on the composition of the TiC-Ni based coatings. An explanation to this is that in these high-velocity spray processes, in contrast to the APS process, the amount of retained WC in the coatings is high enough to provide good wear resistance.

The abrasion wear test results clearly show also that alloying the plain TiC-Ni system in the hard phase with Mo and in the binder phase with Co improves the wear properties. This is believed to be due to improved bonding between the hard phase and the metallic binder. SEM and XRD studies have shown that in the (Ti,Mo)C-NiCo coatings the hard phase consists of a core/rim-type microstructure typical of corresponding sintered materials. Alloying of TiC with N seems also to improve the wear resistance, possibly due to improvements in the coating microstructures. In the erosion wear test, the TiC-Ni base coatings show high wear resistance. The results clearly show the high potential of the novel TiC-Ni coatings for use as protective coatings. Further studies will be focused on the optimization of the powder compositions for different industrial applications.

Table 1: Abrasion wear resistance of thermally sprayed WC-Co, Cr_3C_2-NiCr and TiC-Ni coatings.

Coating	Spray method	Volume loss (mm^3)
WC-12%Co	APS Ar/H_2	19.8
Cr_3C_2-25%NiCr	APS Ar/H_2	26.1
(Ti,Mo)C-28%NiCo	APS Ar/H_2	16.7
WC-12%Co	DGS	7.0
Cr_3C_2-25%NiCr	DGS	15.9
TiC-30%Ni	DGS	18.7
TiC-40%Ni	DGS	19.1
(Ti,Mo)C-28%NiCo	DGS	13.9
(Ti,Mo)CN-28%NiCo	DGS	11.7
(Ti,Mo)C-37%NiCo	DGS	14.6
(Ti,Mo)CN-36%NiCo	DGS	10.4
WC-12%Co	HVOF	7.5
Cr_3C_2-25%NiCr	HVOF	15.0
(Ti,Mo)C-28%NiCo	HVOF	15.7
(Ti,Mo)CN-36%NiCo	HVOF	13.6

Table 2: Erosion wear resistance of thermally sprayed WC-Co, Cr_3C_2-NiCr and TiC-Ni coatings.

Coating	Spray method	Volume loss at 30° (mm^3)
WC-12%Co	APS Ar/H_2	7.3
Cr_3C_2-25%NiCr	APS Ar/H_2	13.1
(Ti,Mo)C-28%NiCo	APS Ar/H_2	4.3
WC-12%Co	DGS	3.6
Cr_3C_2-25%NiCr	DGS	6.1
(Ti,Mo)C-28%NiCo	DGS	3.7

REFERENCES

(1) P. Vuoristo, K. Niemi, T. Mäntylä, L.-M. Berger, M. Nebelung, pp. 309-315 in Proc. 8th National Thermal Spray Conf., Sept. 1995, Houston, Texas, USA.
(2) P. Vuoristo, T. Stenberg, T. Mäntylä, L.-M. Berger, M. Nebelung, pp. 729-734 in Proc. 9th National Thermal Spray Conf., Oct. 1996, Cincinnati, Ohio, USA.
(3) P. Vuoristo, T. Mäntylä, L.-M. Berger, M. Nebelung, in Proc. 10th United Thermal Spray Conf., Sept. 1997, Indianapolis, Indiana, USA, to be published.

Submitted to *Surface and Coatings Technology*

NANOLAYERED AND NANOCOMPOSITE TRIBOLOGICAL COATINGS COMBINING AMORPHOUS AND CRYSTALLINE MATERIALS: A BREAK-THROUGH IN SURFACE ENGINEERING?

A. A. VOEVODIN and J. S. ZABINSKI

Materials Directorate, Wright Laboratory, Wright-Patterson Air Force Base, OH 45433-7750, USA

ABSTRACT

The improvement of tribological characteristics of surfaces by the application of thin coatings is one of the major focuses of surface engineering. Intensive research in this direction recently resulted in the development of composite coatings combining amorphous and crystalline materials. A number of materials schemes could be considered, among which the combination of ceramic and diamond related materials could provide a most unique combination of hardness, toughness, and low friction coefficient (1).

In our research, particular attention was given to amorphous and crystalline materials deposited at low (below 100 °C) substrate temperatures to protect steel surfaces from wear and friction losses at high contact loads. This included amorphous diamond-like carbon (DLC), crystalline carbides, nitrides, and metals.

Design concepts were formulated for tough nanolayered coatings made of crystalline and amorphous materials (2). Amorphous materials were used as dislocation barriers to achieve desirable hardness, while crystalline materials were used for controlled plastic deformation to dissipate crack energy and improve coating toughness. A functionally gradient concept was applied to build up a sufficient load support and reduce substrate compliance (3). These concepts allowed the usage of super-hard (60-70 GPa) DLC materials for protection of moderately hard (12 GPa) steel surfaces against sliding wear at above 1 GPa Hertzian contact pressures.

Another conceptual developments were load-adaptive nanocrystalline carbide / DLC composites (4). Their nanostructure was engineered to achieve a plastic respond of the hard composites to loads exceeding their elastic limit. This change of behavior from hard to plastic provided a self-adjustment of the coating surface at above critical loading, preventing brittle failure and achieving a composite super-toughness. An example could be a TiC/DLC composite consisted of 10-50 nm TiC grains embedded with about 5-10 nm separation into sp^3 hydrogen-free DLC matrix (4). Such composite coatings had a hardness of hard ceramics (30-35 GPa), but also could deform plastically up to 40% at loads above their elastic limit without cracking and/or delamination. This load-adaptation was combined with a low friction of the DLC phase.

Friction and wear mechanisms of engineered coatings were considered with an attention to the influence of the DLC phase. One finding was that a friction induced $sp^3 \rightarrow sp^2$ carbon phase transition of hydrogen-free DLC promotes self-lubrication (5) of engineered composite coatings in humid environments. A result of this self-lubrication was a low friction coefficient below 0.1. The wear losses in nanolayered and nanocomposite coatings were decreased by several orders of magnitude in comparison to single layer ceramic TiN and TiCN coatings. In unlubricated sliding tests, steel disks with engineered coatings of about 2 µm thickness successfully operated against sapphire balls for above 1 mln cycles at initial contact pressures about 2 GPa, maintaining a 0.07 friction coefficient in a humid air environment.

Thus, the new developments in nanocrystalline / amorphous composites opened a possibility to engineer unique wear protective coatings, which are not only very hard, tough, and have a low friction coefficient, but also load-adaptive and self-lubricated. This could be a new break-through in surface engineering.

REFERENCES

(1) A. A. Voevodin, M. S. Donley and J. S. Zabinski, Surface Coat. Technol., 1997, in press.
(2) A. A. Voevodin, S.D. Walck and J. S. Zabinski, Wear, 1997, in press.
(3) A. A. Voevodin, M. A. Capano, S. J. P. Laube, M. S. Donley and J. S. Zabinski, Thin Solid Films, 1997, in press.
(4) A. A. Voevodin, S. V. Prasad and J. S. Zabinski, J. Appl. Phys., 1997, in press.
(5) A. A. Voevodin, A. W. Phelps, M. S. Donley and J. S. Zabinski, Diamond Relat. Mater., Vol. 5, 1996, pp. 1264-1269.

E-mail: voevodaa@ml.wpafb.af.mil
Phone: +1 (937) 255-9001

MODELING TRIBO-CONTACTS WITH MICRO SLIP IN COATED SOLIDS

LANSHI ZHENG and S. RAMALINGAM
University of Minnesota, Minneapolis, MN USA 55455

ABSTRACT

When hard coats are used with light alloys to improve their wear properties, the elastic constant mismatch at the film-substrate interface can lead to high local stresses and, possibly, failure of the coating under static or dynamic in-service loading. Tribo-contact modeling with displacement formulation has been used to calculate the stresses and displacements at the film-substrate interface (1)(2). Film thickness, the film-substrate elastic modulus ratio and the ratio of contact width to film thickness were varied.

Reported results are based on calculations with prescribed normal and tangential surface tractions. But, in a typical tribo-contact, the contact is subjected to a load with a prescribed tribo-pair contact geometry. Contact length and/or pressure distribution are unknown at the outset. In such contacts, there may be partial or complete slip at the contact surface. This problem is addressed in this paper.

Normal and shear tractions at the contact are first determined for a tribo-pair subject to particular micro slip contact conditions. Stresses within the film and substrate are then calculated and presented as a function of friction, slip, film thickness and elastic properties to facilitate coating design.

The tribo-contact analysis of interest here is a mixed boundary value problem. Traction distribution at the contact will have to be determined from the problem geometry, properties of the coating, and those of the substrate. To do so, the coating and substrate are assumed to be isotropic and homogeneous with elastic constants E_1 and E_2 and Poisson's ratios μ_1 and μ_2. The coated solid is considered to be semi-infinite, occupying the half-plane $-\infty < x < +\infty$; $y \geq 0$, with contact loading along $y = 0$. The counter face, with a prescribed geometry or profile $f(x) = 0$, is rigid and the contact load is $-P$.

In this problem, the displacements in film u_1, v_1 and in the substrate u_2, v_2 must satisfy the biharmonic equations $\nabla^4 u_j = 0$ and $\nabla^4 v_j = 0$; $j = 1, 2$. To fulfill this, the displacements in the film and the substrate of the coated solid are written in terms of U_j and V_j, the Fourier transforms of u_j and v_j. Once U_j and V_j with six unknown constants are determined, the stresses in film ($j = 1$) and substrate ($j = 2$) are determined with the Mitchell-Beltrami equations.

The six unknowns in the four displacement equations are determined from displacement continuity at the film-substrate interface $y = h$ and continuity of two stress components there. Two more equations are obtained from the boundary conditions at $y = 0$. These allow U_j and V_j to be determined explicitly.

The normal pressure $p(x)$ and shear traction $q(x)$ acting at the contact region, and the contact length (for incomplete contact) are determined using numerical methods. Distributions $p(x)$ and $q(x)$ are approximated with two series expansions in terms of even Chebyshev polynomials T_{2n} and coefficients s_n. Coefficients s_n are determined by noting that the counter face forces the surface of film to deform and conform with the shape of rigid 'indentor' over some contact length. A displacement constraint with contact loading $p(x)$ and $q(x)$ then prevails locally. Normal and shear tractions are zero outside the contact.

For a complete contact, contact length is known, and in an incomplete contact, the stress at contact extremities are zero. Contact length and normal pressure $p(x)$ hence become determinate from the total load applied - P and the indentor geometry.

When partial slip prevails, as in a cylindrical contact over a contact length $2a$ and 'no slip' over a stick zone of length $2c$, shear traction $q(x)$ is determined from $\dfrac{\partial u_1|_{y=0}}{\partial x} = 0$ for $|x| \leq c$ and $q(x) = f|p(x)|$ for $c \leq |x| \leq a$.

Normal tractions have been calculated using this approach for a rigid cylinder (R = 100 µm; contact load P = 0.1 N/µm) on solid aluminum and TiN-coated Aluminum with 1.2 µm TiN film. Higher contact stiffness for the coated solid is found to be reflected in the reduced contact length and in the non-Hertzian contact stress distribution under otherwise identical contact conditions.

Corresponding shear traction distributions for zero slip (c/a = 0) and slip over 25%, 50% and 75% of the contact lengths have also been determined (solid Al and TiN-coated Al with 1.2 µm TiN film; contact load = 0.1 N/µm; friction coefficient = 0.2). Calculated results show significant changes in traction distributions due to the presence of the thin hard coat.

Once $p(x)$ and $q(x)$ have been determined, film-matrix interface stresses as well as von Mises stress everywhere in the film and the substrate can be readily calculated. These calculated quantities are presented in this paper (submitted to: J. Tribology) as a function of contact geometry and loading for a number of film-matrix combinations of interest for the wear protection of light alloys with thin hard coats.

REFERENCES

(1) S. Ramalingam, and L. Zheng, Tribology International, Vol. 28, 1995, 145-161 pp.
(2) L. Zheng and S. Ramalingam, Surface & Coatings Technology, Vol. 81, 1996, 52-71 pp.

Submitted to *ASME Journal of Tribology*

THE INVESTIGATION OF FRICTION AND WEAR PROPERTIES OF Al_2O_3 - TiO_2 PLASMA-SPRAYED COATINGS AT ROOM AND AT ELEVATED TEMPERATURES

ANNA CSORDÁS-TÓTH, GYULA KISS, JÁNOS TAKÁCS, LAJOS TÓTH and PÉTER GÁL
Dept. of Mechanical Eng. Technology TU Budapest, Budapest 1111 Bertalan u. 2
FRIEDRICH FRANEK and ANDREAS PAUSCHITZ
Institute fur Feinwerktechnik TU Wien, A-1040 Wien Floragasse 7/2

ABSTRACT

In this paper we describe the preparation and investigation of the tribological properties of plasma-sprayed coatings made from Al_2O_3-TiO_2 powder mixture, prepared by conventional and mechanofusion methods. For the determination of these properties two different types of laboratory rig were used, one of these modeled sliding friction, and as an alternative the other one rotation motion. The coefficient of friction, wear volume and the reliability of coatings were measured at room and at temperatures of 250 and at 500 °C.

The material of the substrata of both specimen types is annealed, unalloyed steel with 0,45 % carbon content. After machining, cleaning and activation the plasma-spraying was done with M-500 T APS Plasma-Technik AG equipment. The titania content of the coatings was 0, 3, 13 and 40 %. The thickness of alumina-titania coatings was ≈ 0,5 mm. The average roughness of the ground surface was ≈ 0,5 μm. The measured Hanemann hardness of the ground surface was 850 - 1100 HV.

The counterpart, the "pin" in the pin and plane rig was a ball-bearing steel ball (100 Cr6), in the case of the pin and disc machine it was a sintered Si_3N_4 (SSN) ball. The hardness of the steel ball was 840 HV, ⌀ = 10mm. The hardness of the SSN ball was ≈ 2200 HV, ⌀ = 11,2mm. The SSN ball has an outstanding heat-resistivity, chemical stability and retains it hardness up to 1400 °C (1).

The pin and plane investigation is suitable for the determination of the coefficient of friction at room temperature, when only steady mild wear, with a slight abrasive wear is taking place. In this way the fiction process without severe abrasive wear, and with out deformation or destruction of surfaces was modeled. Such types of sliding friction take place - for instance - in mud pumps, where plasma-sprayed surfaces work together with quenched steel. Using data from preliminary experiments we determined the loading condition so that during the fiction test only mild steady wear occurred.

According to our measurements the coefficient of friction of the mechanofusion powder mixture is the same or a little smaller, than of the conventional mixture. The result of the test also underlines that the addition of TiO_2 decreases the coefficient of friction.

this result is congruent with data found in the literature (2).

During the pin and disc investigation the coefficient of friction, the wear volumes of the plasma-sprayed and of the SSN balls were measured. The measurements of the wear rate were possible due to the significant difference in hardness between the Si_3N_4 balls and the average hardness of the coating. During the steady mild wear the Si_3N_4 balls and the plasma-sprayed coating scratched, ploughed and wore each other. At the start of the frictional process there is a point of contact, but with the progress of the wearing process the contact area will continuously increase. At 3000 m sliding distance the diameter of the contact area was about 1 mm. The width of the plasma-sprayed surface was too about 1 mm.

The result of the cold and warm test show that the increase of the titania content decreases the coefficient of friction and the wear volume of coatings as well as of the Si_3N_4 balls. It means that the titania content increases the wear stability of coatings and even decreases the wear volume of SSN balls. The measured wear volumes and the coefficient of friction value of the Al_2O_3, TiO_2 coatings underline the data found in the literature (3).

The result of the pin and disc measurements are further proof that the integrity of coatings made from mechanofusion or conventional powder mixture are almost the same. We did not find the scaling or destruction of the coatings. We might state, that the reliability of the coatings made from both powder mixtures are adequate and satisfactory. Both powder mixture types are well sprayed with the selected technology. The coefficient of friction value, the wear volume of mechanofusion coatings and even the wear volume of the Si_3N_4 balls are somewhat less in the case of mechanofusion powders in comparison with conventional powders.

REFERENCES

(1) Arthur F. McLean, Dale L. Hartsock, Ceramics and Glasses ASM International 1991, 667- 689 pp.
(2) Valesek I. , Tribologiai Kezikonyv, Tribotechnik KFT Budapest, 1996, 41-70 pp.
(3) K. - H. Habig, M. Woydt , VDI Berichte Nr. 670, 1988, 683-697 pp.

SLIDING WEAR OF PLASMA-SPRAYED CHROMIUM OXIDE-SILICA COATING

H S AHN and S K LEE

Tribology Research Center, Korea Institute of Science and Technology, 39-1, Hawolgok-dong, Songbuk-Gu, Seoul, 136-791, Korea

ABSTRACT

Cr_2O_3-based coatings have been received a considerable attention for the use in combustion engines as wear-resistant coatings[1][2]. The main wear mechanisms of plasma-sprayed Cr_2O_3-based coatings were reported to be plastic deformation, crack formation and spalling due to fatigue, brittle fracture and material transfer[3][4]. The contact configuration, however, used in these works were either ring-on-block or roller-on-block with a uni-directional sliding.

The overall goal of this research is to evaluate the possibility of applying plasma-sprayed chromium oxide-silica coating to both a piston ring and a cylinder liner and identify their friction and wear behaviour.

In the present study, we employed a bench test technique which simulates the reciprocating motion of a piston ring against the cylinder liner. Both the plate and disc specimens were spray coated. Dry sliding tests were conducted with 80 N load at an oscillating frequency of 20 Hz for 90 minutes and at a temperature of 450 °C to simulate the high top ring reversal temperature. For comparison, tests at room temperature were also conducted.

Both the coated plates and discs tested at 450 °C showed a marked decrease in the wear rate in the initial period of sliding and reached a steady-state level of extremely low wear rate ($<10^{-9}$ g/Nm) after around 4 minutes sliding whereas the coatings tested at room temperature showed the similar level of steady state wear rate but much lower initial wear. Fig.1 shows the total weight loss of the plates and discs due to wear. It is seen that the wear loss of the plates tested at 450 °C is much higher than that tested at room temperature.

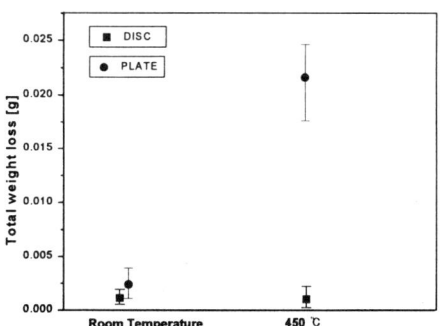

Fig. 1: Total weight loss of plasma spray coated plates and discs after wear test.

By contrast to the wear behaviour of the coatings, the coatings tested at 450 °C exhibited a lower coefficient of friction (~0.4) than the coatings tested at room temperature which was in the range of 0.5-0.6.

A SEM analysis revealed that dispersed smooth surface layers were developed on the worn surface for both test conditions. These smooth patches in the wear track were formed by plastic deformation of adhered and compacted debris particles to the surface. The smooth layer seems to work as a wear-protective, load-bearing area. The greater quantity of the smooth layer formed on the wear track of the specimen tested at room temperature than that tested at 450 °C may be attributed to the lower wear loss of coatings tested at room temperature than tested at 450°C.

The smooth layers and the as-ground surface were analyzed by X-ray photoelectron spectroscopy(XPS). XPS reveals that the chemical composition of the smooth layer formed on the wear track varies depending on the test temperature. Considerable quantity of CrO_3 composition was detected at room temperature test whereas CrO_2 composition was detected at 450 °C test.

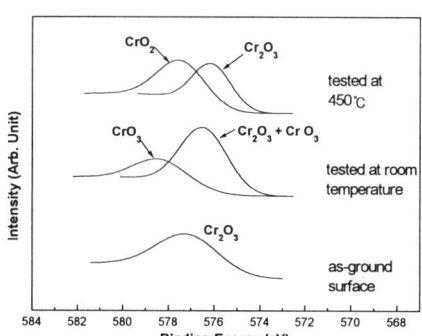

Fig. 2: Cr 2p X-ray photoelectron spectra of the coating surface.

CrO_2 composition in the smooth layer seems more favorable than CrO_3 in reducing the wear loss. The plasma-sprayed chromium oxide-silica coating can be a promising candidate as a wear-resistant coating in internal combustion engines.

REFERENCES

(1) M.E. Woods, W. Bryzik and E. Schwarz, SAE Technical Paper 940951, 1994.
(2) F. Rasteger and A.E. Craft, Surface and Coatings Technology, 61 (1993) 36-42.
(3) J.M. Cuetos, E. Fernandez, R. Vijande, A. Rincon and M.C. Perez, Wear, 169 (1993) 173-179.
(4) Y. Wang, Y. Jin and S. Wen, Wear, 128 (1988) 265-276.

Submitted to *Tribology International*

ROLLING CONTACT FATIGUE PERFORMANCE OF PLASMA SPRAYED COATINGS

R. AHMED & M. HADFIELD

Brunel University, Department of Mechanical Engineering, Uxbridge, Middlesex,
UB8 3PH (United Kingdom)

ABSTRACT

This experimental study describes the rolling contact fatigue (RCF) performance and the failure mechanisms of plasma sprayed WC-12%Co coatings. WC-Co coatings are known to provide a high resistance to sliding, abrasive and hammer wear in hostile environments (1). The advancements in plasma spray coatings due to higher velocity and temperature of the impacting lamella call for investigations into new applications (2). One possible application is the rolling element bearing. The complex lamella and anisotropic microstructure of plasma spray coatings means the analytical approaches to evaluate the fatigue resistance are expensive to solve. Hence, an experimental approach was adopted in which a modified four ball machine was used to investigate the performance of plasma sprayed rolling element cones.

The modified four ball machine (figure 1) which models the configuration of a deep groove rolling element ball bearing was used as an accelerated method to compare the rolling contact fatigue resistance of materials. In this case a coated rolling element cone replaced the upper drive ball. RCF tests were conducted in conventional rolling element steel ball bearing (steel lower balls) and hybrid ceramic bearing (ceramic lower balls) configurations. Plasma sprayed coatings were deposited on a bearing steel (440-C) substrate at different thicknesses. The substrate material was selected because of its high RCF resistance, good adherence to the coating material and high hardness to support the coating microstructure. The coated rolling element cones were ground and polished to achieve a good surface finish on the rolling elements. RCF tests were conducted under the various tribological conditions of contact stress, lubricant and test configurations. The frictional torque in the cup assembly was measured during the RCF tests with the aid of a force transducer. The speed of the planetary balls was monitored using an accelerometer and fast fourier transformation of the vibration signal. This enabled the experimental measurement of the slip between the coated cone and the driven balls. Analytical approach was also adopted to calculate the micro-slip between the contacting rolling elements.

The failed rolling elements, coating debris and the lower planetary balls were analyzed for surface observations under the Scanning Electron Microscope (SEM), Electro Probe Microscope Analysis (EPMA) and fluorescent die investigations. The RCF test results are discussed with the aid of coating microstructure, talysurf analysis of the wear track, microhardness measurements and micro-slip investigations. The test results indicate that the coating performance was dependent upon the tribological conditions during the test. The failure mode was observed to be coating delamination.

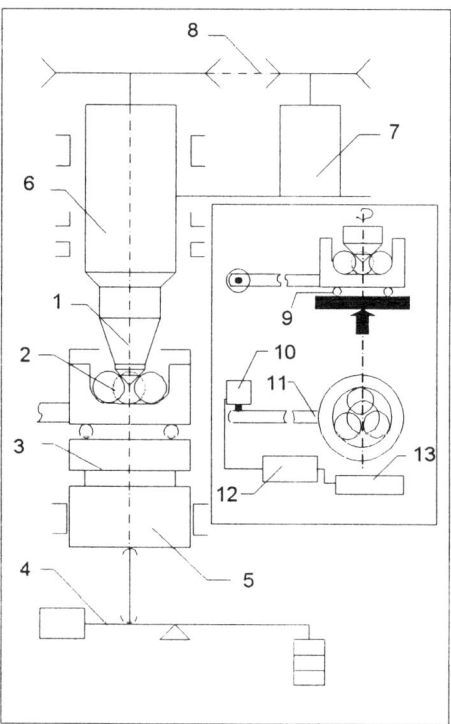

Fig. 1: Schematic of the modified four ball machine
(1, coated cone and collet; 2, Planetary balls; 3, heater plate; 4, loading lever; 5, loading piston; 6, spindle; 7, driving motor; 8, belt drive; 9, thrust bearing; 10, force transducer; 11, torque arm; 12, digital readout; 13, printer)

REFERENCES

(1) P. Vouristo, K. Niemi, A. Makela & T. Mantyla, Proceedings of the National Thermal Spray Conference, Anaheim, CA, ASM International, 1993, 173-178pp.

(2) R. Ahmed & M. Hadfield, Tribology International, Vol. 30, No. 2, 1997, 129-137pp.

THE EFFECT OF MECHANICAL STRESSES IN THE COATING FATIGUE FAILURE OF PVD COATED CEMENTED CARBIDE INSERTS IN MILLING

K.-D. BOUZAKIS, K. EFSTATHIOU AND N. VIDAKIS

Laboratory for Machine Tools and Machine Dynamics Aristoteles University of Thessaloniki, GR-54006, Greece

ABSTRACT

The performance of PVD coated hardmetal inserts is significantly influenced by the mechanical properties of the cemented carbides as well as the fatigue behaviour of their coatings. During interrupted cutting processes, such as milling, the cutting tool is subjected to impacting cutting loads that usually initiate the fatigue fracture of the compound, decreasing herewith the insert performance. The coating fatigue lifetime is predictable, with the aid of developed procedures (1), (2) as well as the static strength of hardmetals (3). In this paper, the cutting behaviour of coated inserts is examined experimentally and analytically.

The coated inserts are tested under constant cutting conditions per experiment, and the number of cuts that leads to coating or substrate failure is determined. The failure is depicted through SEM and microanalyses investigations of the examined cutting edges. Furthermore, through a Finite Element Method (FEM) simulation of the contact between the tool and the workpiece, the distributions of critical stress components that are associated with cohesive coating fatigue failure mode are determined.

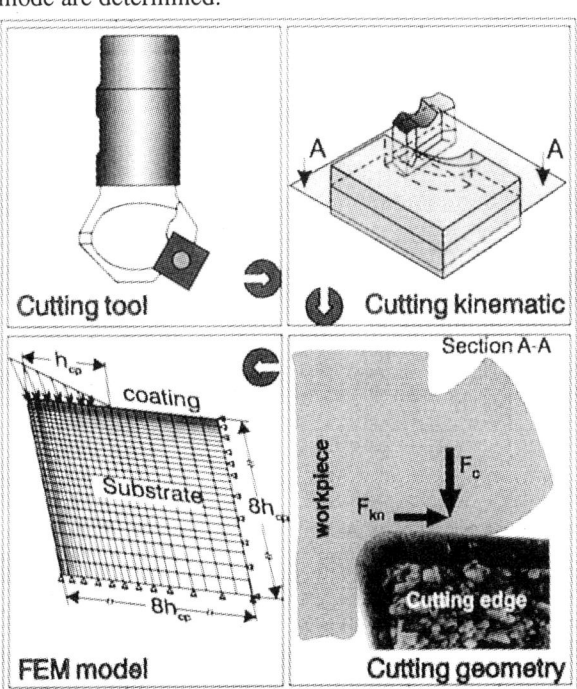

Fig. 1: FEM modelling procedure of the cutting procedure

The computational procedure considers every geometric and technological specification of the milling procedure as well as the material properties for the coating and the tool (see figure 1). Essential parameters for the FEM simulation, such as cutting forces, chip compression ratios etc., are experimentally determined. The stress results are associated to an already developed coating fatigue lifetime prediction model. The calculated maximum number of impacts that the cutting edge is able to withstand, are in a good agreement with the experimental results as it is presented in the Woehler diagram of figure 2. At the top of this figure, SEM pictures of the failed cutting edge tips, due to static overloads, as well as of the coatings without and with failure can be observed.

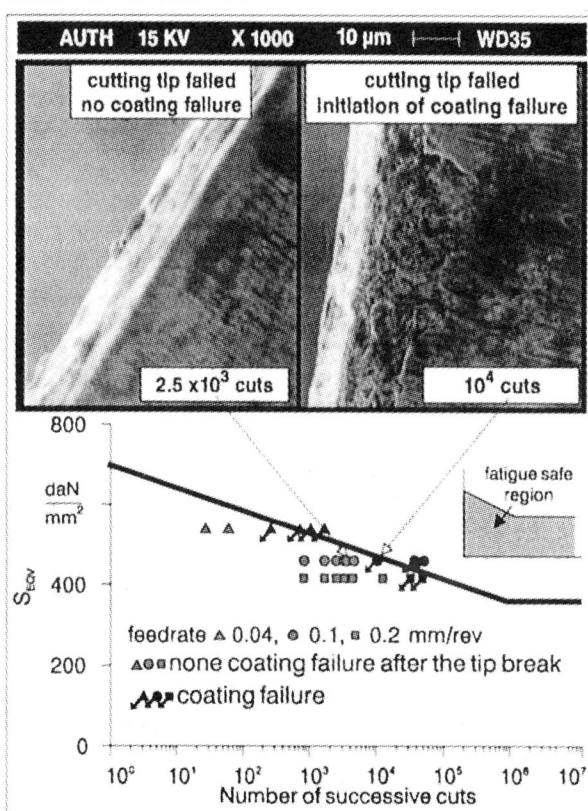

Fig. 2: Typical experimental and FEM results

REFERENCES

(1) Bouzakis K.-D., et.al., Surface and Coatings Technology 86-87 (1996):549-556.
(2) Bouzakis K.-D., Vidakis N., 1996, Effect of the Mechanical Stresses Developed during Gear Hobbing on the Fatigue Failure of Tool Coatings. In press in Int. Journal for Manufacturing Science and Production
(3) K. Brooks, 1987, Word directory and handbook of hardmetals, 4th edition, international carbide data.

Submitted to *Wear*

THE RESPECTIVE ROLE OF OXYGEN AND WATER VAPOR ON THE TRIBOLOGY OF HYDROGENATED DIAMOND-LIKE CARBON COATINGS

C. DONNET, T. LE MOGNE, L. PONSONNET, M. BELIN
Laboratoire de Tribologie et Dynamique des Systèmes, UMR 5513, École Centrale de Lyon,
B.P. 163 - 69 131 Écully Cedex, France.
A. GRILL, V. PATEL,
IBM Research Division, T.J. Watson Research Center, Yorktown Heights, NY 10598, U.S.A.

INTRODUCTION

The tribological behavior of diamond-like carbon coatings (DLC) strongly depends both on the nature of the film and the chemical nature of the environment during friction. In particular, oxygen and/or water are known to increase friction and wear in ambient air, compared to inert atmospheres such as dry nitrogen (1).

The present study proposes to discriminate between the respective effects of water vapor and oxygen on the friction behavior of an hydrogenated DLC coating exhibiting ultralow friction (friction coefficient below 0.02) in ultrahigh vacuum.

EXPERIMENTAL

Using a UHV reciprocating pin-on-flat tribometer, friction tests have been performed in progressively increasing partial pressures of pure oxygen and pure water vapor, at a maximum Hertzian pressure of 1 GPa. Friction tests have also been performed at fixed gas pressures. The pressure has been varied from less than 10^{-10} hPa to maximums of 60 hPa and 25 hPa, for oxygen and water vapor respectively. The maximum water vapor pressure corresponded to a relative humidity RH=100% at room temperature.

The structure and chemical composition of the initial film and wear particles have been investigated by electron energy loss spectroscopy in a transmission electron microscope, in order to elucidate the friction mechanisms responsible for the tribological behaviors observed in the two different gaseous environments

RESULTS

It was found that the progressive increase or decrease of oxygen pressure (in the investigated pressure range) during friction does not affect the ultralow friction behavior observed in UHV. Conversely, a change of the friction coefficient from 0.02 to more than 0.1 has been observed when the water vapor pressure increased during friction above 4 hPa (RH=17%). Friction tests performed at fixed pressures of water vapor showed a lower pressure threshold, in the range of 0.1 - 1 hPa, for the same change of the friction coefficient.

The friction mechanisms are discussed by comparing the structure and composition of the film and the wear particles produced in UHV, oxygen and water vapor.

FIGURES

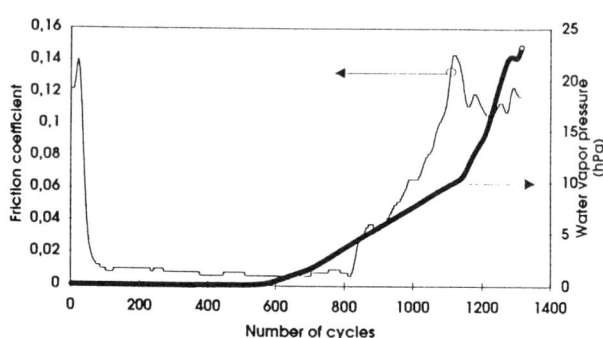

Fig.1 : Friction coefficient of a-C:H *vs* number of reciprocating cycles, at different water partial pressures

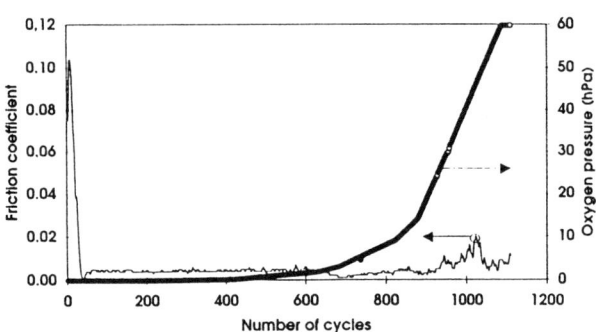

Fig.2 : Friction coefficient of a-C:H *vs* number of reciprocating cycles, at different oxygen partial pressures.

REFERENCE

(1) C. Donnet, M. Belin, T. Le Mogne, A. Grill and V. Patel, in Protective Coatings and Thin Films, Y. Pauleau and P.B. Barna (eds.), Kluwer Academic Publishers, 1997, p. 229.

SURFACE DURABILITY OF THERMALLY SPRAYED WC CERMET COATING IN LUBRICATED ROLLING/SLIDING CONTACT

AKIRA NAKAJIMA and TOSHIFUMI MAWATARI
Department of Mechanical Engineering, Saga University, 1, Honjo-machi, Saga-shi, Saga 840, JAPAN

ABSTRACT

Thermal spraying has been applied in various fields of industry as one of the most important material processing or surface modification technologies(1)(2). Among the rest, the sprayed coatings with high hardness material like WC cermet generally offer superior tribological properties such as wear resistance and sliding performance, so that a wide variety of applications to the contact surfaces of machine elements are anticipated. However, available research data on surface durability or tribological properties of sprayed coating in rolling/sliding contact conditions are extremely few except some simple evaluation tests such as indentation hardness test, scratch test, and wear test under sliding conditions.

In the present investigation, the authors have conducted rolling/sliding contact tests of rollers with a WC-Cr-Ni cermet coating(3) formed by HVOF (high velocity oxygen-fuel flame spraying)(1). Experiments were carried out using a two-roller testing machine having a center distance of 60mm. A pair of rollers D (with coating) and F(without coating) were rotated under nearly pure rolling (slip ratio $s \fallingdotseq 0\%$) or rolling/sliding ($s=-14.8\%$ or $s=+12.9\%$) conditions. The outside diameter of rollers was 60mm and the effective track width was 10mm. The coatings of about $25\mu m$, $40\mu m$, $90\mu m$ in thickness were formed onto the D rollers. As D roller materials, a carburized and hardened alloy steel (Hv$\fallingdotseq$760) or a thermally refined carbon steel (Hv$\fallingdotseq$320) were used. The F roller material was a carburized steel. After spraying, the coated D roller surfaces were finished smooth to a mirror-like condition with a roughness of $0.2\mu m$Ry by grinding and polishing (the hardness of coating Hv$\fallingdotseq$950). The mating carburized steel F rollers without coating were finished by precision cylindrical grinding(Ry$\fallingdotseq 0.1\mu m$, the surface hardness Hv$\fallingdotseq$800). The rotational speed was about 3600rpm on the driving side and a maximum Hertzian stress in the range of $P_H=1.0\sim 1.4$GPa was applied in line contact. A mineral gear oil without EP additives was supplied at a flow rate of 15cm^3/s. The oil temperature was kept at 45℃ and the corresponding oil viscosity ν was 52.5mm^2/s.

Each test was continued up to $N=2\times 10^7$ cycles unless any serious damages occurred. Fig.1 shows some results obtained using carburized steel roller with coating. In the case of pure rolling conditions, neither flaking nor surface distress occurred. In the case of

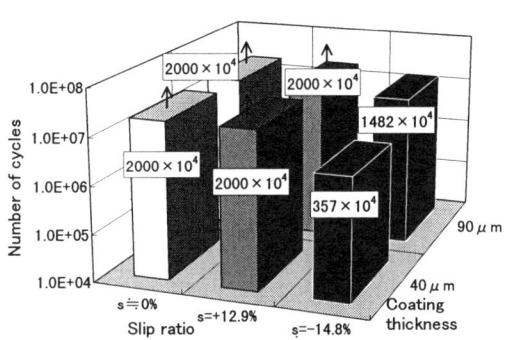

Fig. 1: Effects of slip ratio and coating thickness (P_H=1.4GPa, carburized steel).

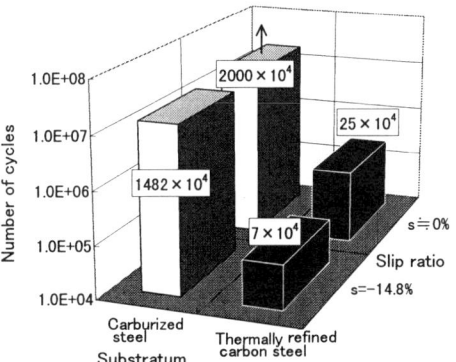

Fig. 2: Effect of substratum hardness (P_H=1.4GPa, coating thickness$\fallingdotseq 90\mu m$).

rolling/sliding conditions, flaking was apt to occur when the coated roller was placed on the slower side (s=-14.8%), while the occurrence of flaking was restrained when the coated roller was placed on the faster side(s=+12.9%). Fig.2 shows the effect of substratum hardness of the coated roller upon the life to flaking. As shown in the thermally refined carbon steel roller, the life became very short when the substratum hardness lowered. On the whole, the life to flaking had a tendency to be long as the thickness of coating increases, though significant differences in the friction and in the behavior of oil-film formation were not recognized.

REFERENCES
(1)Y.Harada, Bull Jpn. Inst. Metals, 31, page 413, 1992.
(2)K.Tani, Thermal Spraying Technique, 9[4], page 57, 1990.
(3)K.Tani, H.Nakahira, K.Miyajima, and Y.Harada, Mater. Trans. JIM, 33, page 618, 1992.

FRACTAL ANALYSIS OF HARD DISK SURFACE ROUGHNESS AND CORRELATION WITH START/STOP FRICTION

F.E. KENNEDY*, C.A. BROWN**, J. KOLODNY*, & B.M. SHELDON*
*Thayer School of Engineering, Dartmouth College, Hanover, NH 03755 USA
**Dept. of Mechanical Engineering, Worcester Polytechnic Institute, Worcester, MA 01609 USA

ABSTRACT

The decrease in roughness of computer hard disks to meet increasing storage density requirements can lead to excessive friction, termed 'stiction', during start-up of the disk. Stiction forces can be particularly high after a period when the slider has been resting on the disk. Stiction has been attributed to the formation of meniscus bridges between the disk and slider, owing to the presence of thin layer of lubricant and/or condensed water vapor on the disk surface (1, 2). The magnetic recording industry is trying to optimize the surface roughness of the disks, especially in so-called 'landing zones', to limit stiction without causing wear during contact start/stop operation.

Scale-independent fractal methods have been tried for characterization of surface roughness (3), but many engineered surfaces are neither self-similar or self-affine over significant ranges in scale. To overcome this limitation, new fractal analysis techniques have recently been developed to deal effectiively with roughness scales (4). The purpose of this work was to use new scale-sensitive fractal parameters to characterize the surface topography of hard disks, and to correlate the fractal parameters with measured stiction behavior of textured hard disks.

Two different sets of hard disks were tested in this work. Glass disk substrates had been subjected to two different chemical texturing treatments. The glass disks were uncoated and unlubricated. Lubricated aluminum disks were also tested. Half of the aluminum disks had been polished, while the rest had been subjected to a mechanical texturing process.

The topographies of all disks were measured at eight locations using an atomic force microscope. The glass disks were scanned in the non-contact mode to avoid damage to the tip of the AFM cantilever, whereas the aluminum disks were scanned in the contact mode. The surface topography data were analyzed by the patchwork method, which is an area-scale fractal analysis technique involving a virtual tiling of the surface with triangular patches (4). Two parameters were chosen to characterize the surface topography: smooth-rough crossover (SRC) and area-scale fractal complexity (ASFC). SRC, which is most influenced by the large scale surface features, is the scale, or triangular patch size with units of area, above which the surface appears smooth and below which it appears rough. ASFC, which is primarily influenced by small scale surface features, is a unitless parameter related to the fractal dimension over two decades of areal scale.

Glass or aluminum disks which had been textured by the same process and in the same batch had similar values of ASFC and SRC, but those values differed for disks subjected to a different texturing process. Glass disks which had been subjected to a similar chemical texturing process, but in a different batch, had statistically different values of SRC. ASFC was also found to give a good indication of wear of the AFM tip.

Stiction behavior of the disks was determined in a series of 1-rpm drag tests. Two friction values were measured: an initial stiction value measured upon initiation of motion after a 30 second rest time, and a stick-slip friction value measured by taking the standard deviation of friction variations in a 15 second interval during the contact sliding tests. Ten sliding tests were run on each disk, each on a different track. Tests of the unlubricated glass disks were run at three different relative humidity values, 21%, 31% and 48%. All tests of the lubricated aluminum disks were run at 20% relative humidity.

Figure 1 shows a summary of the stiction data for tests of the unlubricated glass disk. The relationship between stiction force and surface roughness, as characterized by the SRC parameter, depended upon the thickness of the liquid layer present on the disk surface, which is determined by the relative humidity. Results for the lubricated aluminum disks followed a similar trend. The results provide further evidence that stiction is strongly affected by menisci which link the slider and disk surfaces. Currently available meniscus models rely on a variety of traditional roughness parameters to characterize the surface topography (1, 2). The scale-dependent fractal parameters used in this work appear to be more appropriate for use in a meniscus model, since they include information about the area of surface features as well as their height.

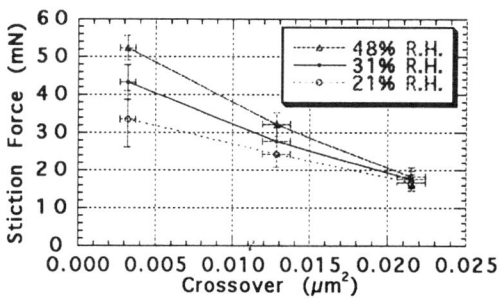

Figure 1. Stiction force as a function of Smooth-Rough Crossover for textured glass disks at different relative humdity.

REFERENCES

(1) H. Tian & T. Matsudaira, ASME J. Tribology, v.115 (1993), pp. 28-35.
(2) C. Gao, X. Tian & B. Bhushan, Tribology Trans., v.38 (1995), pp. 201-212.
(3) A. Majumdar & B. Bhushan, ASME J. Tribology, v.112 (1990), pp. 205-216.
(4) C.A. Brown, P.D. Charles, W.A. Johnsen & S. Chesters, Wear, v.161 (1993), pp. 61-67.

Submitted to *ASME Journal of Tribology*

MEASUREMENT TECHNIQUES FOR IN-SITU STUDIES OF WEAR MECHANISMS IN MAGNETIC THIN-FILM DISKS

SCOTT M. FOREHAND and BHARAT BHUSHAN
Computer Microtribology and Contamination Laboratory
Department of Mechanical Engineering
The Ohio State University, Columbus, Ohio 43210-1107, U.S.A.

ABSTRACT

Rigid disks are the principal on-line magnetic storage media in computers, and the tribological study of disk durability is of significant interest to the computer industry. Specifically, as the demand for larger storage density grows, the need for lower head flying height increases. Lower flying height, however, increases the potential for asperity contacts at the head-disk interface, which results in higher wear rates.

Durability tests in the past have focused primarily on in-situ measurement of interface friction and post-failure evaluation of wear tracks. These approaches are limited in that friction and wear are related but independent, and post-failure measurements give only end results. To design machine elements with low friction and wear, "precursors to failure" need to be understood. One approach is a test apparatus that provides simultaneous in-situ localized information about specific wear phenomena, such as interface friction, acoustic emission, formation of wear debris and wear track progression, during a continuous drag test.

The present work utilizes the test apparatus which is the same as that used by Bhushan and Forehand (1) (2), with the addition of an acoustic emission counter. The counter registers a count every time the acoustic emission signal from the head-disk interface exceeds a certain threshold. This additional data gives further insight into the nature of the contacts at the head-disk interface. The test apparatus of Bhushan and Forehand is further modified to perform a CSS test. This is accomplished by using a microchip to control the power supply to the disk drive motor. The microchip causes the motor to ramp up to top speed, hold there for a specific length of time, ramp down to rest, and hold there for a specific length of time before starting the cycle again. In this setup, disk surface reflectance is measured in addition to coefficient of static friction in order to observe wear track progression.

Contact start/stop (CSS) tests and continuous drag tests are used for detailed in-situ analysis of wear mechanisms in magnetic thin-film rigid disks. Coefficient of static friction and disk surface reflectance are measured during the contact start/stop tests, and coefficient of kinetic friction, acoustic emission rms, acoustic emission count, and disk surface reflectance are all measured during the continuous drag tests. It is the first time that friction, acoustic emission and optical reflectance sensors are used simultaneously for wear studies. In-situ measurements are used to determine precursors to failure of magnetic thin-film disks, to compare wear processes in CSS and drag tests, and to correlate CSS and drag tests. Disk wear is seen to follow a pattern of lubricant depletion and/or localized polishing of overcoat asperities, followed by significant sliding contact between the head slider and overcoat asperities. The sliding distance associated with CSS tests is less damaging to disks than continuous drag sliding distances. A combination of sensors used in this study, in particular an optical sensor used to measure changes in disk reflectance, has been demonstrated to be valuable in understanding the failure mechanisms.

Several physical phenomena contribute to the wear process in magnetic thin-film disks. The first phase of wear is believed to be lubricant depletion and/or localized polishing of overcoat asperities, followed by more significant sliding wear between the head slider and disk overcoat asperities. Lubricant depletion and localized polishing of asperities reveal themselves as an increase in coefficient of static friction during a contact start/stop test and an increase in coefficient of kinetic friction during a continuous drag test. Burnishing most likely begins as the lubricant becomes fully depleted and/or local overcoat asperities have been polished, the head slider begins to make significant sliding contact with disk overcoat asperities, and wear particles accumulate in the head-disk interface. Burnishing therefore begins in the contact start/stop test after the coefficient of static friction has reached (and passed) a maximum value, and in the continuous drag test when the acoustic emission rms shows continuous significant activity.

REFERENCES

(1) Bhushan, B. and Forehand, S., "In-Situ Instrumentation for Localized Wear Studies of Magnetic Thin-Film Disks," *Proc. Instn. Mech. Engrs. Part J: J. Eng. Trib.,* 1997 (in press).

(2) Forehand, S.M and Bhushan, B., In-Situ Studies of Wear Mechanisms In Magnetic Thin-Film Disks, *Tribol. Trans.,* 1997 (in press).

AN INVESTIGATION OF THE DYNAMICS OF PROXIMITY RECORDING SLIDERS USING ACOUSTIC EMISSION AND PHASE DEMODULATED LASER INTERFEROMETRY

Thomas C. McMillan and Frank E. Talke
Center for Magnetic Recording Research, University of California, San Diego, La Jolla, CA 92093, USA

ABSTRACT

In order to reduce magnetic spacing, proximity recording sliders and smoother disks with thinner protective overcoats have been employed in current hard disk drives. Proximity recording sliders are designed to remain in lightly loaded contact with the disk during operating conditions in order to keep the head-disk spacing low and minimize the amount of disk wear. In general, the flying height of magnetic recording sliders is investigated by using multi-chromatic interferometry, whereby a glass disk is substituted for the real disk. Since the surface roughness of a glass disk is substantially smaller than that of real magnetic disks, interferometric flying height data are not sufficient to describe the dynamics and flying behavior of proximity recording sliders, and alternative techniques for characterizing slider-disk interactions are highly desirable.

Traditionally, acoustic emission (AE) sensors have been used to monitor slider-disk interactions. AE sensors detect the energy associated with slider-disk contacts, and since slider-disk contacts excite slider body resonant frequencies, the AE signal contains information about the details of slider-disk interactions (1).

Another technique for studying the dynamic behavior of sliders on real magnetic disks is phase demodulated laser interferometry, also called heterodyne interferometry(2-4). In this technique, phase information from relative spacing changes between slider and disk is obtained by phase demodulation of a carrier frequency, in our case 2 MHz. One beam, modulated at 38 MHz, is reflected from the backside of the slider and the other beam, modulated at 40 MHz, is reflected from the disk. The two beams are then interfered generating a carrier frequency of 2 MHz. The 2 MHz slider-disk signal is then compared to a 2 MHz reference signal as the slider changes its spacing or relative position on the disk. The difference between the two signals is the relative phase shift. The resulting change in spacing is determined from the relative phase change and the wavelength of light used in the interferometer.

In order for this type of interferometer to be used to measure spacing changes between slider and disk, the slider has to be visible from the backside. Current suspensions allow only limited access to the backside of the slider, since they cover most of the slider area. To overcome this shortcoming, we have built a focusing and positioning system, which enables the measurement beams to be directed through small slots or holes in the suspension. The focusing and positioning system can vary the separation between the two beams from 0.5 mm to 2.0 mm and position the beams either side-by-side or one in front of the other. The variable beam separation is controlled by an axicon pair, and the variable beam alignment is controlled by a rotation stage centered along the optical axis of the focusing system. Previous heterodyne interferometers did not have this capability (2-4).

In this paper, we have analyzed the dynamics of proximity recording sliders during contact start/stop using the newly developed heterodyne interferometer, as described above, with improved focusing and positioning optics. Tri-pad sliders and negative pressure tri-pad sliders are investigated and the relative spacing change during start-up is correlated with the acoustic emission signal.

REFERENCES

(1) McMillan, T. C, Swain, R. C., Talke, F. E., An investigation of slider take-off velocity using the acoustic emission frequency spectrum", *IEEE Trans. Mag.*, Nov. 1995, vol. 31, no. 6 pt. 1, pp. 2973-5.

(2) Henze, D., Mui, P., Clifford, G., Davidson, R. J., A multi-channel interferometric measurements of slider flying height and pitch", *IEEE Trans. Mag.*, Sept. 1989, vol. 25, no. 5, pp. 3710-12.

(3) Zhu, L. Y., Hallamasek, K. F., Bogy, D. B., A measurement of head/disk spacing with a laser interferometer", *IEEE Trans. Mag.*, Nov. 1988, vol. 24, no. 6, pp. 2739 - 41.

(4) McMillan, T. C., Talke, F. E., A ultra low flying height measurements using monochromatic and phase demodulated laser interferometry", *IEEE Trans. Mag.*, Nov. 1994, vol. 30, no. 6 pt. 1, pp. 4173-5.

TRIBOELECTROMAGNETIC PHENOMNEA OF HYDROGENATED CARBONFILMS LUBRICATED WITH PERFLUOROPOLYETHER FLUID

KEIJI NAKAYAMA
Mechanical Engineering Laboratory, Namiki 1-2, Tsukuba, Ibaraki 305, Japan
HIROSHI IKEDA
Tsukuba Research Center, Technical Research Laboratory, Nippon Sheet Glass Co., Ltd., 5-4, Tokodai, Tsukuba, Ibaraki, 300-26, Japan
STÉPHANIE NGUYEN
Laboratoire de Mechanics des Contacts, Institute National des Sciences Appliquees de Lyon, Batiment, Avenue Albert Einstein, 69621 Villeurbanne Cedex France

Abstract

Triboelectromagnetic phenomena refers here the phenomena embracing tribo-emission of electrons, ions, and photons, tribo-charging and formation of microplasma state at and near sliding contacts. In this study, triboelectromagnetic phenomena has been investigated in a frictional system with a diamond pin sliding on a hydrogenated carbon film lubricated with a PFPE oil. The frictional system simulates the head/disk sliding contacts with a head coated by a diamond like carbon film and a magnetic disk coated with a hydrogenated carbon film. Triboemission of negatively and positively charged particles, tribo-charging and friction coefficient were measured in air, using a specially developed triboelectromagnetism measuring apparatus (Fig. 1).

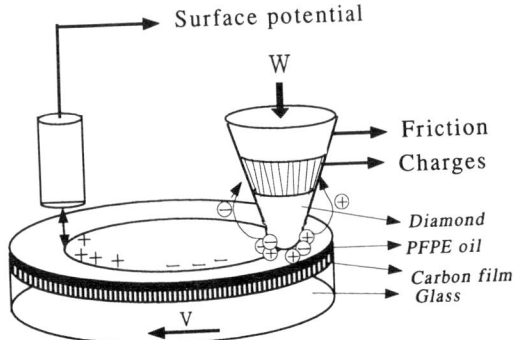

Fig. 1 Triboelectromagnetism measuring apparatus

The diamond pin had an included angle of 90 degrees and a tip radius of 300μm. The hydrogenated carbon films were deposited on a K^+ ion exchanged soda lime glass substrate using DC magnetron sputtering method. The thickness of the carbon films was 20nm and the hydrogen content in the film, C_H, varied from 0 to 43 at.%. Negatively and positively charged particles were detected with a charge detector, surface potential generated by tribo-charging was measured with a electrostatic voltmeter and friction coefficient was detected by strain gauges. The signals of the charged particles, surface potentials and friction coefficients were sampled simultaneously using a common trigger unit, amplified and then computer processed. Details of the experimental procedures are described elsewhere (1-3).

Figure 2 shows the dependence of the surface potential on time with three kinds of hydrogen level in the carbon film. As seen in Fig. 1, Tribo-charging increased with the hydrogen content, i.e. with increase of the resistivity, ρ, of the carbon film. Intense triboemission of the negatively and positively charged particles were observed also. The emission intensity depended on the hydrogen content and the negatively charged particles were detected more intensely than positively charged particles.

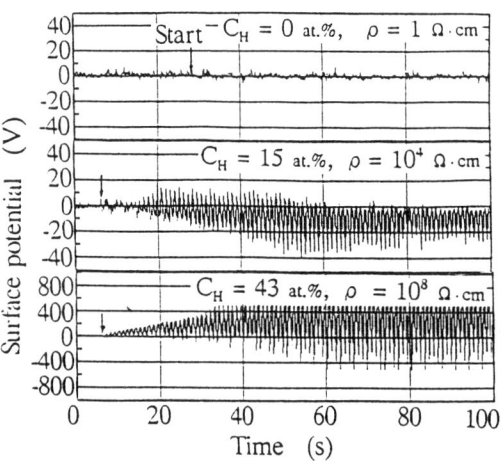

Fig. 2 Surface potential with time

REFERENCES

[1] K.Nakayama, H. Ikeda, *Wear*, No.198, pp.71-76 (1996).
[2] K. Nakayama, H. Ikeda, H. Sato and K. Yamanaka, *Tribology Trans.*, in press.
[3] K. Nakayama, B. Bou-Said, and H. Ikeda, *ASME J. Tribology* in press.

WEAR CHARACTERISATION OF MAGNETIC DISK MEDIA USING ATOMIC FORCE MICROSCOPY

Y MAN and A S CHEKANOV
Data Storage Institute, 10 Kent Ridge Crescent, Singapore 119260
K F LOW and S C LIM
Department of Mechanical and Production Engineering, National University of Singapore,
10 Kent Ridge Crescent, Singapore 119260

ABSTRACT

In the last decades or so, magnetic recording has become the predominant technology for the storage of digital information in modern computer systems. In order to increase the recording density in computer disk drives, it is necessary to minimise the spacing between the read/write head and the magnetic disk medium. In the more recent rigid disk drives, the head-to-medium separation can be as small as 25 to 30nm. With the head flying at such a close proximity to the disk medium, occasional contacts between head and medium are inevitable; such contacts invariably lead to wear damages on the head as well as on the disk media. Continued improvements made to disk media performance has led to the development of substrates made of glass rather than the more-conventional aluminium alloy. Such disk media offer improved performance over the aluminium-substrate media in areas such as rigidity, surface roughness, and impact resistance. It would therefore be of interest to investigate the wear behaviour of this new group of glass-substrate disk media.

The wear characteristics of both the 1.8-inch and 2.5-inch glass-substrate magnetic disk media reported in this paper were investigated using the atomic force microscope (AFM). The AFM, with its ability to provide high resolution information at the micro- and nano-level, enables the systematic investigations of materials response to wear at such a scale. Diamond tips of two different tip radii were specially prepared using mechanical means for the series of nanowear tests carried out. The extent of wear is described by the depth of wear scar on the scanned area. The results obtained show that at a constant load, media wear (in terms of wear depth) generally increases with increasing number of scanning (wear) cycle but not always in a uniform manner; there are cycles during which very little wear (in terms of wear depth) is observed. Similar observations were also made when increasing load was used: the wear depth increases in general with increasing load, but the increase in wear depth can be sudden; very little increase in wear depth is observed over a range of load, and then suddenly a large increase is seen. The behaviour with changes in load is more dramatic than when different scanning cycles were used. As the different layers, such as the lubricant and overcoat layers, of the disk are worn away by the diamond tip, different AFM images of the worn areas showing different features are obtained.

A more time-efficient method for micro/nanowear test using the AFM is proposed for the first time. We have called it the "zoom" method. The basic principle is to make the wear marks (for different testing conditions, for example) inside one another instead of at different parts on the sample's surface. This method is most useful when the wear marks are made first using one type of tip and later imaged by another type (usually the normal AFM tip).

The procedure is as follows: A square wear mark, say 80μm by 80μm, is first made on the surface of interest at a known load. Next, another square wear mark, now smaller with a size of say 60μm by 60μm, is made (at the same load) within the first, larger square, but with one corner of the two squares coinciding. In so doing, the smaller area would have experienced a larger number of scanning (wear) cycle. A third smaller square, say 40μm by 40μm, is then made with one of its corner coinciding with the corners of the two larger ones, and so on. The AFM used for the present investigation only allows the making of four square marks in this manner because any larger number of squares would make measurements of wear depth difficult.

This method has been able to produce reasonably useful results when used on glass-substrate magnetic disk media. Comparing with the normal method where one tip is used to make separate wear marks for different scanning cycles and returning to these wear marks to image the wear features using another tip, the "zoom" method has the convenience of not needing to locate the wear marks after the diamond tip (used for wear tests) has been replaced by the normal tip for AFM scanning. It is able to save a considerable amount of time and effort for the operator.

Submitted to *Wear*

TRIBOLOGICAL CONSEQUENCE OF MOLECULAR ORIENTATION OF POLYMER LUBRICANT FILMS

CHAO GAO, TAM VO and JOEL R. WEISS
Akashic Memories Corporation, 304 Turquoise Street, San Jose, California 95134, USA

ABSTRACT

The objective of this paper is to show how molecular orientation of functional end groups of perfluoropolyether (PFPE) lubricants play an important role in the tribological performance of thin film magnetic disks. These disks typically have an amorphous carbon overcoat upon which a thin lubricant layer is deposited. It was found that the wear durability of these polymer molecule films (~2 nm) with planar end groups is a few times better when the planar end group is oriented more perpendicular to the carbon surface [molecular orientation index (MOI) ~ 1.5] as compared with those which are oriented more parallel to the carbon surface [MOI ~ 0.5 to ~ 1.0]. No significant difference in lubricant bonding (as quantified by the residual film thickness after stripping with a good solvent) was detected over the range of MOIs studied. Nor was there a detectable relationship with hydrophobicity, as measured by water contact angle. It was inferred from an evaluation of thermal effects and storage times that a smaller MOI value corresponds to a lower free energy state of the lubricant film. Recording head burnishing of the surface also had a minimal effect on MOI. A mechanism of locally flexible but globally immobile polymer molecules appeared to account for the observed better tribology performance of lubricant films of MOI ~ 1.5. We expect that this tribological effect of molecular orientation is a general phenomenon for *thin polymer* lubricant films (< ~ 5 nm) and not confined only to magnetic head-disk tribology.

Experimental:

The tribological performance of the lubricant films on magnetic disk surfaces was evaluated using drag-mode contact start-stop testing. In the testing, a disk was clamped to a spindle and rotated at a relatively low speed of 300 RPM (so that an air bearing could not be established by the recording head). The head load was 3.5 grams. The disk rotated for only a few seconds before being brought to a complete stop. The cycle was repeated until a build up of friction / stiction occurred to a set level or a time limit was achieved. During the starting and spinning of the disk, static and kinetic frictional forces at the interface were measured using strain gauges. The lubricant fluid was coated onto the carbon surface of the disk using a dip-coating technique. The thicknesses of lubricant films were measured using FTIR (Fourier Transform Infra-red Spectrometry) and XPS (X-ray Photo-electron Spectroscopy). The MOI values were obtained from glancing-angle FTIR absorbency of the functional end group and mathematical derivation.

Definitions:

A molecular orientation index (MOI) was defined as 1 for randomly oriented functional end groups. The maximum MOI can be mathematically calculated as high as 3 for lubricant molecules oriented with their functional end groups perpendicular to the surface, while the minimum MOI is 0 if lubricant molecules oriented with their functional end groups parallel to the surface. Figure 1 schematically illustrates the meanings of limiting MOI values with respect to molecular orientation of functional end groups denoted as circles. The uncertainty (3 standard deviations) on MOI values in our analysis is ± 0.12 nm.

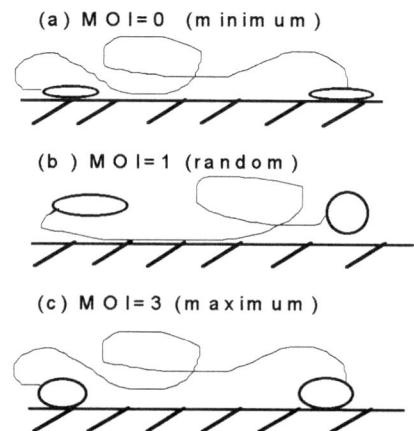

Fig. 1. Schematic for molecular orientation with respect to disk surface.

SURFACE CHARACTERIZATION AND MICROMECHANICS CONTACT SIMULATIONS FOR MAGNETIC HEAD-DISK INTERFACES

E. R. KRAL[1] and K. KOMVOPOULOS[2]
[1]IBM Almaden Research Center, 650 Harry Road, San Jose, CA 95120, USA
[2]Department of Mechanical Engineering, University of California, Berkeley, CA 94720, USA

ABSTRACT

In view of the significance of asperity-level micromechanics to the head-disk interface in magnetic storage devices, finite element indentation simulations were performed to obtain an understanding of the initiation and evolution of plastic deformation and the development and distribution of tensile stresses in carbon thin films, which can lead to cracking and spalling of the surface. Due to varying length scales and sizes of asperities occurring under typical contact conditions, simulations using surfaces based on the actual contact interface topography may yield more accurate information about the stress and strain fields arising under contact conditions.

Characterization of the contacting surfaces with fractal geometry was used to provide a scale-independent description of the surface that can be used for contact simulations on any length scale. It has been common in the disk drive industry to quote only the root-mean-square (rms) roughness, σ, in order to describe a surface. However, additional statistical parameters, such as the skewness, S, and kurtosis, K, are required to fully embody the contact mechanics characteristics of the interacting surfaces. Figure 1 shows two surfaces with identical rms roughness, but different fractal surface parameters and higher-order statistical descriptions. Both surfaces were generated using the algorithm developed by Wang and Komvopoulos (1994). Figure 1(a) is a composite surface generated with fractal parameters calculated from surface scans of an ultra-smooth NiP disk substrate and an Al_2O_3/TiC slider, and Fig. 1(b) is a fictitious surface used for illustrative purposes only. It can be seen that the two surfaces, though of the same rms roughness, possess different contact mechanics characteristics. Indentation simulations were performed with both surfaces until first yield on a finite element mesh corresponding to a carbon-coated thin-film disk. The load at which plastic deformation was initiated is noted in Fig. 1 as the parameter F_y. The surface in Fig. 1(a) produced a yield load of almost twice the surface shown in Fig. 1(b), indicating that the former contact interface can withstand higher loads without undergoing plastic deformation. Thus, though the rms roughness is the same in both cases, the performance of the contact interface will be significantly different

Table 1 summarizes results from 2-D and 3-D finite element contact simulations on carbon films 5-20 nm thick in terms of the indentation displacement, Δ, the plastic zone size, $(h_p/h)_{max}$, the maximum plastic strain, ε_{eq},

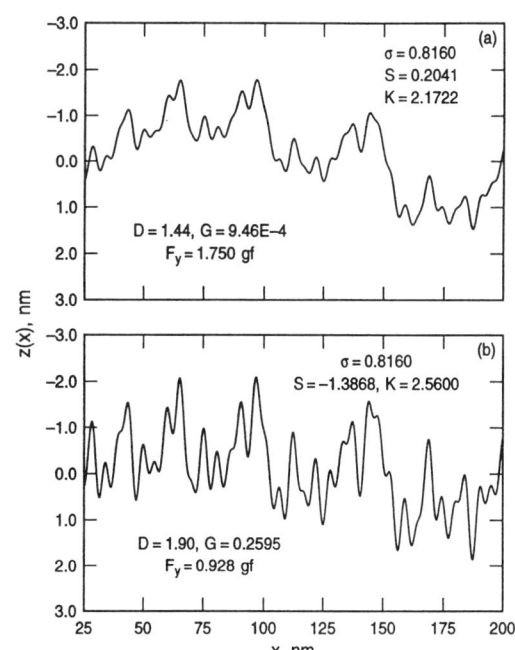

Fig. 1 Fractal surfaces with same rms roughness

Table 1. Finite element results

	C (nm)	Δ (nm)	F (gf)	σ_1^{max} (GPa)	ε_{eq}^{max} (%)	$(h_p/h)_{max}$
2D plane-strain μ=0.0	20	0.50	3.8 (3.2)	0.79	1.50	0.125
	20	0.75	7.0 (5.9)	0.89	2.35	0.188
	10	0.25	2.0 (1.7)	0.35	0.29	0.125
	10	0.50	5.0 (4.2)	0.54	1.43	0.250
	5	0.25	2.9 (1.7)	0.36	0.31	0.25
	5	0.50	5.0 (4.2)	0.56	1.47	0.50
3D μ=0.1	10	0.25	N/A	0.59	1.37	0.25
	10	0.50	N/A	0.71	2.07	0.25

and the maximum tensile principal stress, σ_1. In addition, equivalent loads for the given contact conditions, F, are given for both a "center-pad" slider and a "guppy" slider. Results indicate that for a 10-nm carbon film, the maximum allowable load for the center-pad slider is ~2 g and for the guppy slider is 1.7 g. Results from 3D simulations consistent with the 2D results are also given in Table 1.

REFERENCES

Wang, S., and Komvopoulos, K., 1994, "A Fractal Theory of the Interfacial Temperature Distribution in the Slow Sliding Regime: Part I-Elastic Contact and Heat Transfer Analysis, Part II-Multiple Domains, Elastoplastic Contacts and Applications," *Journal of Tribology*, Vol. 116, pp. 812-823 and 824-832.

EFFECTS OF LASER TEXTURED DISK SURFACES ON A SLIDER'S FLYING CHARACTERISTICS

YONG HU
Iomega Corporation, 800 Tasman Drive, Milpitas, CA95035, USA
DAVID B. BOGY
Department of Mechanical Engineering, University of California, Berkeley, CA94720, USA

ABSTRACT

Recently, laser texturing has captured the attention of head/media interface engineers of magnetic hard disk drive industry because it provides precision in the landing zone placement while eliminating the transition zone of a mechanically textured landing zone. It also offers excellent tribological performance in terms of low CSS stiction and good durability(1). This paper models the effects of laser bumps and laser textured disk surfaces on the flying characteristics of a sub-ambient pressure slider.

To simulate the slider's dynamic response to the moving laser bumps and textures, the generalized Reynolds equation modified by the FK model for the slip-flow correction and the equation of motion of the slider are simultaneously solved. We directly generate the three-dimensional laser bump and texture in the numerical simulator. The mesh size and time step are refined such that they are comparable to the bump dimension (2).

The Headway AAB slider is used in this study. Table 1 summarizes the simulated flying characteristics for two outer rail (OR) fly conditions. Inner rail (IR) flies over the laser textured landing zone. The fly height (FH) increase and the non-decaying modulation can be explained, if the moving laser textured disk surface is visualized as a series of bumps, each providing additional excitation to the slider. These continuous bump excitations are the driving force that increases the fly height and maintains the constant fly height modulation. Compared to the case of both rails flying over the textured landing zone, the slider with the OR flying over the smooth data zone has a much smaller fly height gain/modulation and a substantial decrease of roll. The former is due to the reduced continuous bump excitation to the slider in the vertical direction, while the latter is associated with increased unbalancing excitation in roll direction between the two side rails. Figure 1 plots the non-dimensional 2D air bearing pressure contour for the case of OR flying over the smooth data zone. The stippled pressure peaks embedded on the smooth pressure profile are generated due to the air compression by the moving laser bumps on the disk surface. After entering from the slider's leading edge, these stippled pressure peaks move gradually to the trailing edge portion where they exert the largest perturbation to the slider's motion due to the smallest slider-disk separation there.

As a summary, the moving laser texture generates moving stippled pressure peaks embedded on a smooth pressure profile. These moving pressure peaks increase the slider's trailing edge FH and maintain a constant magnitude FH modulation. The effect increases as the bump spacing decreases and the central dome height and rim diameter increase. Flying the OR over the smooth data zone, while keeping the IR over the textured landing zone, decreases the FH gain and FH modulation, but increases the roll loss and roll modulation.

Outer Rail Fly Condition	Over Data Zone	Over Landing Zone
FH Increase (nm)	0.0154	0.7439
FH Modulation (nm)	0.0576	0.2231
Roll Decrease (μrad)	0.5484	0.0207
Roll Modulation (μrad)	0.1204	0.045

Table 1: Flying characteristics versus OR fly condition. "Sombrero" type, bump spacing = 100×50 mm, rim height = 20 nm and center height = 30 nm

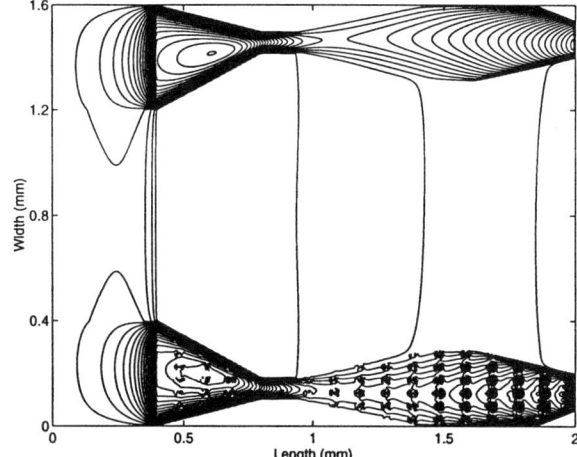

Fig. 1: 2D air bearing pressure contour with OR flying over the smooth data zone. "Sombrero" type, bump spacing=100×50 μm, center height =30 nm, rim height=20 nm and rim diameter=40 μm

REFERENCES

(1) Baumgart, et al., IEEE Transactions on Magnetics, Vol. 31,1995, pp.2946-2951.
(2) Hu, Y., "Head-Disk-Suspension Dynamics," Ph.D. Dissertation, University of California at Berkeley, 1996.

SIMULATION OF LUBRICATION PROCESSES AT MAGNETIC STORAGE HEAD-MEDIA INTERFACES

A P HORSFIELD, R ZUGIC and A P SUTTON
University of Oxford, Department of Materials, Parks Road, Oxford OX1 3PH
C G HARKINS
Hewlett-Packard Laboratories, Palo Alto, CA

ABSTRACT

During the sliding of a carbon coated recording head over a carbon coated magnetic recording medium, the perfluorinated polyether (PFPE) lubricants used to reduce wear and lower friction are found to degrade. There are believed to be two general classes of mechanism leading to degradation: electronic and mechanical.

Possible electronic mechanisms include dissociative attachment and dipolar dissociation. In the former mechanism an electron attaches to a molecule, leading to a change in the bonding that favours the dissociation of the molecule. In the latter an electron scatters inelastically from a molecule, promoting the molecule to an excited state that can relax by dissociation.

Possible mechanical mechanisms include: local heating leading to thermal degradation; local high stresses leading to bond scissioning.

We have performed density functional calculations for a series of hydrocarbon molecules and their fluorocarbon analogues (which are simple prototypes of the perfluorinated polyethers used as lubricants). We looked at the neutral ground state, the state with one additional electron, and the lowest energy excited state. We find that dissociation always occurs once the molecule is placed into an excited state, but only occasionally once an electron has been attached. Both mechanisms appear to contribute little to the total degradation: promotion to the excited state requires a large energy (about 10eV) so is highly improbable; the average life time for electron attachment is too short to lead to dissociation.

Even though most electrons that attach to a molecule remain attached for only a very short time compared with the period of vibration of the molecule, it is possible that the short period of attachment might have a significant effect on the dynamics of a molecule, and thus promote dissociation, at least at a finite temperature. To study this, molecular dynamics has been carried out using an efficient *ab initio* total energy electronic structure method, with the impulse treated by the temporary attachment of an electron introduced in a simple manner. We find that for hydrocarbons the temporary attachment has very little effect on the dynamics of the molecule, and so cannot contribute to degradation.

The mechanical degradation mechanisms were studied using molecular dynamics with a simple molecular mechanics model for the inter-atomic interactions. To gain a proper understanding of the system we first studied the properties of: an isolated droplet of PFPE; a PFPE film on a carbon surface (with active groups on it, such as OH); a PFPE film between static carbon surfaces. For the isolated droplet we observe that hydrogen atoms prefer to be at the surface rather than in the bulk, in agreement with experiment. For the film on the surface we looked for bonding between the film and surface groups, and for layering of the lubricant. Similarly for the film between the surfaces we look for layering as a function of pressure and temperature.

After these preliminary simulations we performed sliding simulations. Here we measured the temperature and coefficient of friction during sliding, and looked for breaking of bonds which mark the degradation of the lubricants. A novel thermostat is used to provide an accurate representation of the thermal properties of a thick substrate.

TRIBOLOGICAL PROPERTIES OF UV-BONDED LUBRICANT: A STUDY USING THE SPM.

A.S. CHEKANOV and T.S. LOW
Data Storage Institute, National University of Singapore, 10 Kent Ridge Crescent, Singapore 119260.

O. KOLOSOV
Department of Materials, University of Oxford, Parks Road, Oxford, OX1 3PH, England.

ABSTRACT.

Effects of UV-irradiation on tribological characterestics of the PFPE lubricating film were investigated using the AFM. Force-distance curve measurements, frequency shift and wear test were used to analyze the changes in meniscus force, lubricant transfer and wear resistance of the lubricated disk surface.

INTRODUCTION.

Lubricating film on the hard-disk surface plays a major role in prevention of early disk wear during start-stop process of the disk drive. It was found that decreasing of the lubricant film thickness reduces the stiction between the head slider and the disk during the head take-off. However, reduced lubricant film thickness also leads to faster lubricant film depletion at the head's landing zone on the disk due to a squeezing action of the head slider. This leads to an earlier onset of the wear of the disk coating. Usual way of preventing of unwanted lubricant migration on the disk surface is to anchor, graft the chemically inert lubricant chains (typically PFPE) to the disk protective coating using some chemically active groups and physical/chemical bonding procedures (thermal bonding, UV- and electron bonding). UV-irradiation of the lubricant film may result in significant enhancement of the lubricant tribological performance. It was shown [1] that optimum amount of UV-bonded lubricant (about 15-30%) provides excellent lubricant performance and significantly decreases the stiction force without sacrificing of the wear durability of the thin-film disk. Analysis of the bonding action of the UV-irradiation shows [2] that lubricant chain attaches itself to the disk surface via dissociative electron attachment which results in formation of a radical and a F ion. This may cause degradation of the lubricant performance and disk drive failure. In fact, lubricant transfer onto the slider rail, which is detrimental to the disk drive performance is often attributed to the lubricant degradation through lubricant molecular chain scission by the head slider impact. Lubricant resistance to chain scission may deteriorate after UV irradiation thus causing excessive transfer of the degraded lubricant on to the slider rails.

EXPERIMENTAL.

In this work we analyze the effects of UV-irradiation on the PFPE lubricant film (Z-dol) using AFM technique. PFPE film was applied to the smooth metallic (Ni-P) and glass substrate using hand wiping and dip coating techniques. Lubricant thickness was controlled by AFM [3]. Variation of the meniscus and adhesive force with thickness and bonded/free lube ratio were observed using the Force vs. distance curve measurements [4]; transfer of the lubricant on the AFM tip was detected and measured using the resonant frequency shift method ("Accurex", TopoMetrix). Wear mechanism of the pristine and UV-irradiated disk surface was analyzed using the nano-wear test with diamond ("Triboscope", Hysitron Inc.) and silicon tips. It was found that selective UV-bonding of the lubricant using special masks further improves its tribological performance in terms of reliability and wear-resistance. Free lubricant at the unexposed areas serves as a reservoir for the replacement of the lubricant at the exposed, UV-bonded areas.

REFERENCES.

1. H.J. Lee, R. Zubeck, D. Hollars, J.K. Lee, A. Chao, M. Smallen ,"Enhanced tribological performance of rigid disk by using chemically bonded lubricant", J. Vac. Sci. Technol. A 11(3), 1993, pp. 711-714.
2. G.H. Vurens, C.S. Gudeman,l.J. Lin, J.S. Foster, " Mechanism of Ultraviolet and Electron Bonding of perfluoroethers', Langmuir 8, 1992, pp. 1165-1169
3. C.M. Mate, M.R. Lorentz, V.J. Novotny, "Atomic force microscopy of polymeric liquid films", J. Chem. Phys. 90(12), 1989, pp. 7550-7555.
4. A. Chekanov, T.S. Low, S. Alli, S. Agarwal, S. Smith, "An AFM study of Environmental Contaminants and Lubricant Morphology on the Magnetic Hard-Disk Surface", IEEE Trans.Mag. 32(5), 1996, pp.3726-3728

HELICAL SCAN HEAD AND TAPE CONTACT BEHAVIOUR : OPTIMIZATION OF TRIBOLOGICAL AND MAGNETIC ASPECTS

F. SOUCHON and P. RENAUX
DMITEC, LETI / CEA, 38054 Grenoble Cedex 9, France
Y. BERTHIER
Laboratoire de Mécanique des Contact, INSA de Lyon, 20 Avenue Albert Einstein, F-69621 Villeurbanne, France

A new silicon integrated head has been developed to achieve high density tape recording. Head to tape spacing and gap height must be especially as small as possible to limit losses and to increase the efficiency of the head. Tribologicaly, these magnetic conditions require to ensure the contact between head and tape while controlling the degradations to decrease the head wear.

The study of these tribological functions is tackled by a phenomena approach in which statistical experimental design are used rather than by a completely parametric approach. In tribology, a parametric approach limits often the range of extrapolations and the research of solutions. Indeed, three sciences (mechanics, materials and physico-chemistry) take part in the contact behaviour through elements of the tribological triplet successively. This triplet is composed by a *mechanism* which houses the contact (helical system), by the bodies in contact or *first bodies* (head and tape), and by particles which are located between the first bodies or *third body* (1)(2). Consequently, the dynamic of head/tape contact behaviour is reconstructed according to the contribution of each element of the triplet by means of head and tape observations (optical profiler and atomic force microscope), tape deformation measurement (laser vibrometer), surface analysis (XPS and AES characterisations).

The results show that the contact conditions are imposed by : global solicitations on a large scale by the helical system, and local solicitations on a small scale by the head and its housing. Physically, these solicitations are materialised by a tent effect shown in Fig. 1 which determines the local conditions of contact. This work allows to distinguish for each degradation the contributions due to bodies in contact from those due to contact conditions (3).

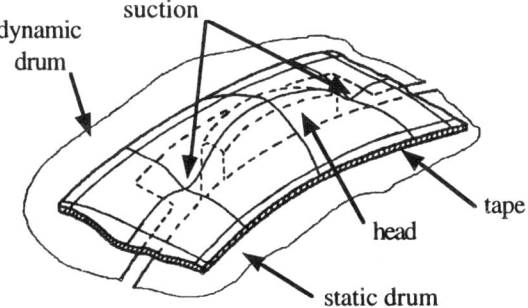

Fig. 1: Tent effect

The degradations are the result of particles detachment caused by velocity accommodation mechanisms by shearing. These degradations are described for the case : silicon head / metal particle tape.

For the tape, the accommdation leads to a plastic deformation of magnetic particles shown in Fig. 2, and so, allows to feed a origin flow of third body composed for the most part by the binder (3).

 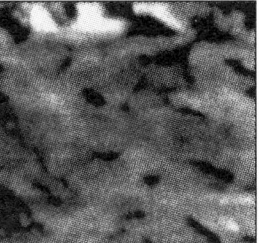

before wear after wear
Fig. 2: Tape degradation (AFM)

For the head, the shearing leads to a superficial tribological transformation shown in Fig. 3 which the expansion is controlled by the third body stemed from the tape. The superficial tribological transformation, initiated during the lapping, ends on a detachment of particles which is the cause of the head wear (3).

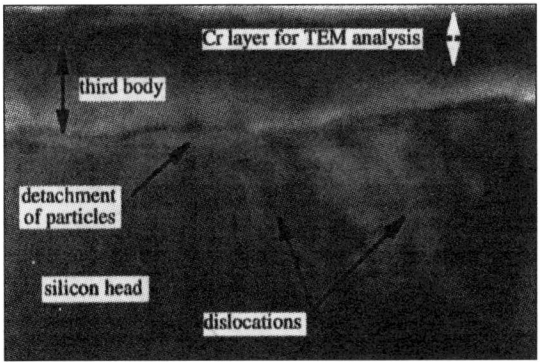

Fig. 3: Head degradation (TEM)

Once the phenomena are understood, the convergence of tribological and magnetic conditions is achieved to optimize the head performances and to limit the head wear.

REFERENCES
(1) Y. Berthier, Elsevier Tribology Series, Vol. 21, page 161, 1992.
(2) M. Godet, Wear, Vol. 100, page 437, 1984.
(3) F. Souchon, Thèse : Tribologie d'un enregistreur magnétique hélicoïdal, Insa de Lyon, 252 pages, 1997.

A STUDY ON HEAD SLIDER FLYING HEIGHT CONTROL UTILIZING ELECTROSTATIC FORCE (METHOD AND EXPERIMENTAL RESULTS)

SHIGERU TAKEKADO and TETSUO INOUE
Energy and Mechanical Research Laboratories, Toshiba R&D Center 1 Komukai Toshiba-cho, Saiwai-ku, Kawasaki 210 Japan

INTRODUCTION

Head slider flying height becomes even lower than 100 nm to achieve high recording density of HDD. One of the key issues to efficiently pack the large capacity data on a disk surface is keeping the constant flying height throughout the data zone. Previous works have emphasized the pressure control by the air bearing slider design(1)-(5). This report presents a new method for controlling head slider flying height, in which the electric potential is applied between magnetic thin film on a disk and a Winchester type head slider. Experimental results showed that head slider flying height can be controlled by the impressed voltage.

EXPERIMENTS

Two sizes of head sliders (3.97X3.24 and 2.58X2.0 mm) were used in the experiments, that are taper flat positive pressure sliders. Bigger one is contactually running and smaller one is flying on a disk. Disks were fabricated by glass and overcoated by SiO2. In contact experiments, AC voltage on DC voltage was impressed between the slider and the magnetic thin film on a disk. Displacements at a leading edge was measured by Laser Doppler Vibrometer. Displacements of trailing edge could not be measured. In flying experiments (Fig. 1), DC voltage was impressed between the slider and the magnetic thin film on a disk. Flying heights were estimated by the space loss equation of reproducing level.

RESULTS

In the contact experiments, The displacements were proportion to AC impressed voltage (Fig.2). The sensitivity of displacement/voltage were reduced as rotational speed of a disk increased (Fig.3). In the flying experiments, flying height were reduced as impressed voltage increased(Fig.4).

CONCLUSIONS

A new method of controlling head slider flying height and attitudes of a contact head slider is expressed, in which the electric potential is applied between the magnetic thin film on a disk and a taper flat type head slider. Experimental results showed that the attitudes and clearances of a contact head slider can be controlled by the impressed voltage.

REFERENCES

[1] White,J.W.; ASME Press Series AISS, Vol.3,(1991),P.1
[2] S.Yoneoka,et al; IEEE Trans. MAG,Vol.27,No.6,(1991),P.5085
[3] C. Hardie,et al; IEEE MAG. Rec.conference,(1993-9),P F3
[4] F.Hendriks; International Tribology Conference Yokohama,(1995),P. 19
[5] K.Higashi,et al, International Tribology Conference Yokohama,(1995), P.21

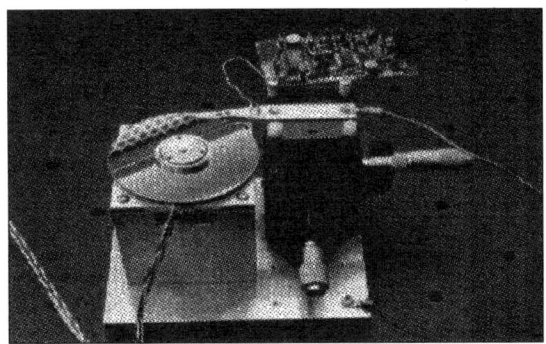

Fig.1 experimental devices

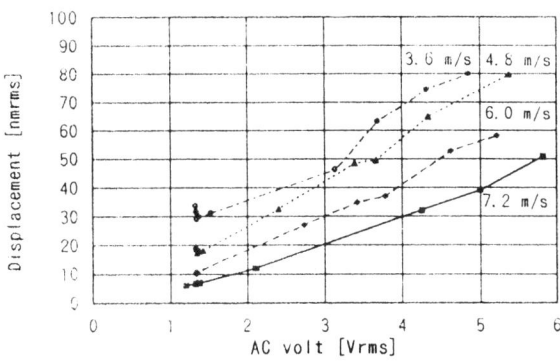

Fig.2 Head Slider-Disk volt vs Leading Edge Disp. AC on DC volt(7.07V)

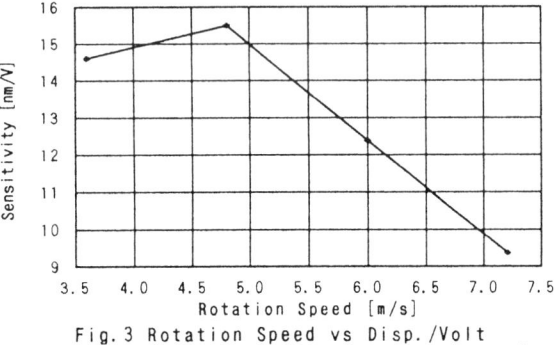

Fig.3 Rotation Speed vs Disp./Volt Sensitivity AC on DC volt(7.07V)

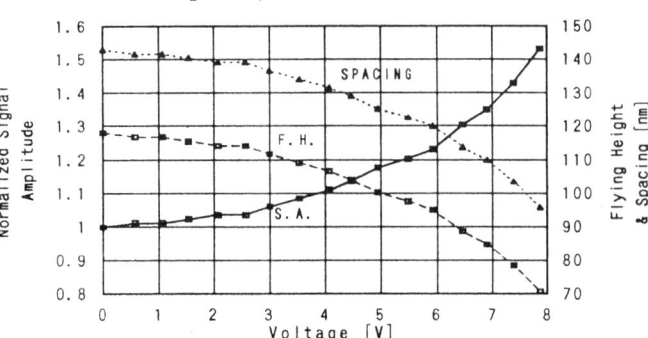

Fig.4 Head Slider Flying Height Control by Voltage. Voltage vs signal amplitude, flying height & spacing

Wear Measurements for Proximity Recording Heads

PAUL W. SMITH

Applied Magnetics Corporation, 75 Robin Hill Road, Goleta, California 93117

ABSTRACT

A focused ion beam is used to create wear fiduciary marks on subambient pressure air bearing sliders, and both atomic force microscopy and optical profilometry are evaluated for localized wear measurement. A group of test heads is oriented such that wear is expected to occur on an alumina surface at nominal operating speed. Data from both measurement methods are used to examine wear behavior as a function of contact start/stop cycling. Wear results, evaluated by both AFM and optical profilometry, show patterns consistent with the flying attitude of the heads, and excellent agreement between the two measurement techniques on the worn parts.

Previous work by Varanasi, et al. (1) made use of a sharp pointed indentor to create wear references. These were measured both before and after wearing of the interface using the AFM. Hsia, et al. (2) offered an improved technique using a focused ion beam to create accurately dimensioned wear gages in the trailing region of the slider

The current research was conducted using a relatively stiff, sub-ambient pressure air bearing. The air bearing used is similar to the "Type 3" omega bearing analyzed by Hsia, et al. (3). In this previous study, the tripad air bearing was found to have a compliance, at the contact point, of approximately 4.4 nm/mN, compared with 1.7 nm/mN for the omega design.

Typical AFM and optical profiler data for 4 longitudinal slots are shown for comparison in Figure 1.

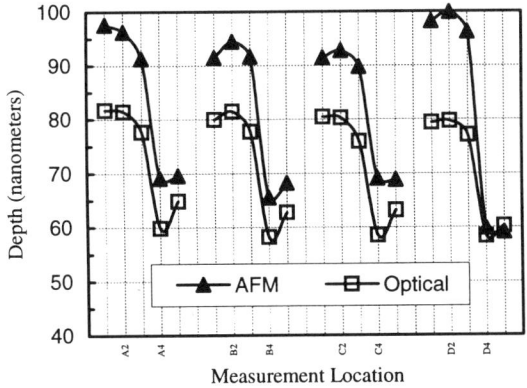

Figure 1: Comparison of AFM & Optical Measurement Methods

Because the slot length extends through both alumina and AlTiC materials, the variation in material properties (and hence milling rates) becomes apparent. In general, the slot depths are well-controlled and predictable. Data from both AFM and optical profilometer show similar trends, but there is a consistent, material dependent offset in depth values. Since the wear measurement will be a differential one, it is important only that a repeatable reference datum be established at the bottom of the slot. Statistics for repeated measurements are summarized in Figure 2.

Measure	AFM	Optical Profilometer
σ_{rpt} (repeatability)	5.89 nm	0.33 nm
σ_{rpd} (reproducibility)	8.02 nm	0.03 nm
σ_{ε} (standard error)	9.95 nm	0.33 nm

Figure 2: Comparative Measurement Errors

Previous work (2,3) concluded that, even with a relatively compliant air bearing in contact on the alumina material, significant wear could be obtained. To examine the case of wear in an operating head-disk interface, 4 sliders were selected with high nominal flying pitch angles, and minimal recession of the alumina surface from the reference plane of the adjacent air bearing. These parts were first measured for slot depth, using both AFM and optical profilometry. After 30,000 contact stop-start cycles, the measurements were repeated.

This small-sample evaluation of the technique for a relatively stiff, sub-ambient pressure air bearing showed good agreement between AFM and optical profilometer measurements, both for low wear cases, and for an anomalous case exhibiting high wear. Good correlation between slider flying attitude and wear location was obtained.

REFERENCES

(1) A. Varanasi, F. E. Talke, and M. Azarian, poster session at APMRC Conference, Singapore, November 30, 1995.

(2) Y.T. Hsia, B. Rottmayer, and M.J. Donovan, IEEE Trans. Magn., Vol. 32, No. 5, September, 1996, pp. 3750-3752.

(3) Y. T. Hsia and M. J. Donovan, Tribology of Contact/Near-Contact Recording for Ultra High Density Magnetic Storage, TRIB-Vol. 6, ASME, 1996.

EVALUATION TECHNIQUE OF HEAD/TAPE CONTACT USING ACOUSTIC EMISSION

K MATSUOKA, K TANIGUCHI and Y UENO

AVC Products Development Lab., Matsushita Electric Industrial Co., Ltd., Matsuba-cho, Kadoma, Osaka, 571 Japan

ABSTRACT

A new methodology has been developed for both predicting magnetic head wear on-line, without direct measurement of wear volume, and for analyzing head/tape contact phenomena using acoustic emission (AE) sensing. It uses the direct relationships between AE signal power and wear, and between AE signal power and the normal force acting on the magnetic head. These relationships, examined by Matsuoka et al. (1), can be expressed as:

$$V_{rms} = \alpha\sqrt{kNv} + \beta \quad (1)$$

where V_{rms} is the AE r.m.s. voltage, α is the square root of the measurement system impedance when the AE signal propagates, k is the wear coefficient of the head, N is the applied normal force, v is the relative sliding speed, and β is the noise in the system. According to equation (1), the wear coefficient, k, can be calculated by measuring AE r.m.s. voltage when the applied normal force, N, acting on the head, and the relative sliding speed, v, are known. Similarly, N can be calculated by measuring AE r.m.s. voltage when k and v are known.

The AE transducer is directly mounted onto the head assembly of a VHS format VCR mechanism in order to measure head/tape contact behavior. This technique has been used to (a) monitor the effect of relative humidity on head wear, as shown in Fig. 1, (b) investigate the wear characteristics of Mn-Zn ferrite as a function of crystallographic orientation, without direct measurement of the wear volume, and (c) evaluate tape abrasiveness.

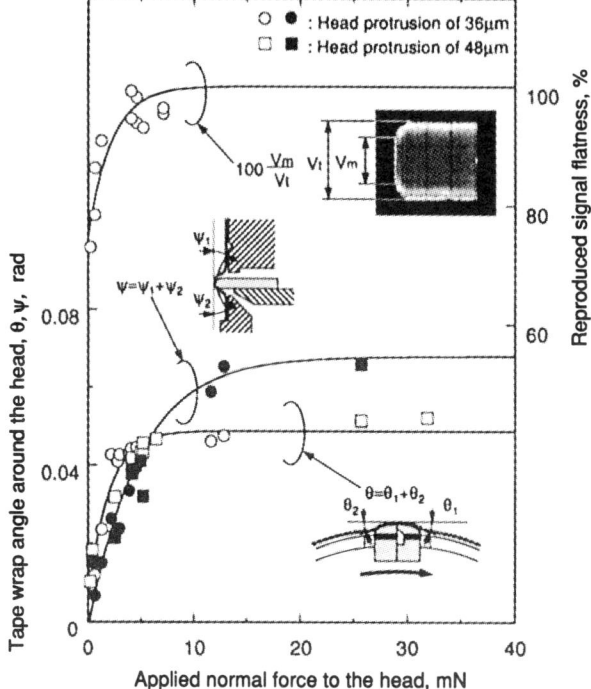

Fig. 2: Tape wrap angle around head and reproduced signal flatness as a function of applied normal force to head

The results obtained in these experiments demonstrate that the evaluation method using AE is useful for head wear prediction.

Explicit correlations were also obtained for the normal force acting on the head, the tape wrap angle around the head, and the reproduced signal flatness, as shown in Fig. 2. The flatness depends greatly on the tape wrap angle in the longitudinal direction, rather than the transverse direction. Both reproduced signal flatness and tape wrap angle can be predicted by measuring the applied normal force using the AE signals.

Moreover, the power of longitudinal tape vibration at the head exit region in the low frequency range correlates with the normal force caused by head impact, as observed by AE. The power of longitudinal tape vibration can be predicted by analyzing AE signals.

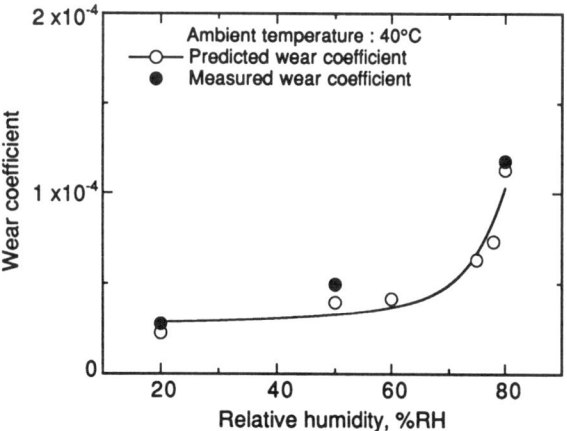

Fig. 1: Predicted wear coefficient of Mn-Zn ferrite using acoustic emission technique and measured values by wear test (2), as a function of relative humidity

REFERENCES

(1) K. Matsuoka, D. Forrest, and M. K. Tse, Wear, Vol. 162-164, pp. 605-610, 1993.
(2) D. Forrest, M. Matsuoka, M. K. Tse, and E. Rabinowicz, Wear, Vol. 162-164, pp. 126-131, 1993.

FRICTION MEASUREMENTS ON RIGID DISKS, THE DIFFERENT LUBRICATION REGIMES

H. VISSCHER and D.J. SCHIPPER
University of Twente, Faculty of Mechanical Engineering, Tribology Group P.O. Box 217, 7500 AE Enschede, NL
P.J. GRUNDY
University of Salford, Department of Physics, Joule Laboratory The Crescent, Salford M5 4WT, UK

INTRODUCTION

In order to further increase the storage density of magnetic recording media, the separation between head and medium should be as small as possible, see e.g. (1). For a maximum storage density the head should be in nominal contact with the magnetic layer of the medium. This is called contact recording, (2). Contact recording has been achieved in principle in tape drives where the relative velocity between head and medium is small. In commercial disk drives and video tape drives where the velocity is high, the head and medium are still separated by a very thin lubricating air film. However, the air film thickness has been reduced considerably in recent years
and experimental rigid disk drives have been build which utilise contact recording. It is predicted that at the end of this decade the head and medium of rigid disk drives will be in nominal contact. Furthermore, the dimensions of the head-disk interface will be reduced in future (that is, smaller diameters and substrate thickness) and the load will be much lower (smaller than 50 mN), such that the nominal pressure in the contact will be very low (below 30 kPa). In addition, the magnetic layers and protective layers will become thinner (below 50 nm) and smoother (< 5 nm CLA roughness) while the speed of the disk will be increased (> 3600 rpm). In order to achieve this it becomes important to understand the friction, wear and lubrication processes at head-medium interfaces.

EXPERIMENTS

This paper describes friction measurements carried out on rigid disks in order to obtain information about the lubrication transitions of a head-disk interface. A nano-indentation hardness tester has been made suitable for this type of measurements. The instrument can be used as a very sensitive pin-on-disk instrument in order to simulate a head-disk interface. The friction has been measured as a function of velocity under the same conditions as at which head-medium interfaces usually operate. The measurements give valuable information about the friction process and the transitions between the different lubrication modes (see for example Fig 1). The transitions were characterised by means of the lubracation number, L, given by,

$$L = \frac{\eta v}{pR}$$

in which η is the viscosity of the lubricant between head and disk, v is the sum velocity of slider and disk, p is the nominal pressure in the contact of the head-disk interface, and R the combined roughness of slider and disk, given by,

$$R = \sqrt{R^2_{slider} + R^2_{disk}}$$

where R_{slider} and R_{disk} are the CLA roughness of the slider and disk respectively.

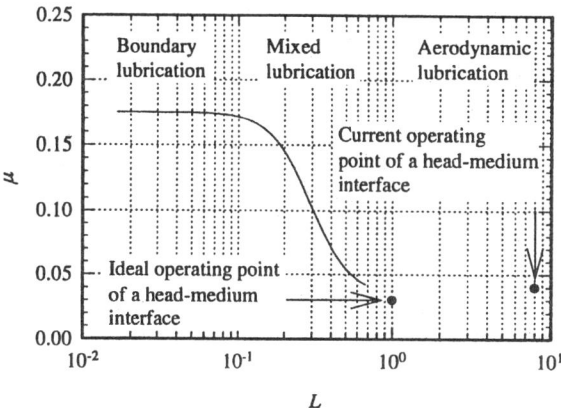

Figure 1 Coefficient of friction, m, of a head-disk interface as a function of the lubrication number L.

REFERENCES

(1) Bertram, H.N. and R. Niedermeyer, "The effect of spacing on demagnetization in magnetic recording", IEEE Trans. Magn., Vol. MAG-18, No. 6, Nov. (1982), pp. 1206-1208.

(2) Concepts in contact recording. 1991 STLE/ASME Tribology Conference, October 13-16, 1991, St. Louis, Missouri, (Adams, G.G. Ed.), The American Society of Mechanical Engineers (ASME), New York, 1992.

ASSOCIATIVE MEMORY APPROACH FOR DETERMINING OPTICAL PARAMETERS VIA ELLIPSOMETRY--APPLICATION TO MAGNETIC HARD DISKS

E R PHELTS[1], Y-H CHEN and J L STREATOR
G. W. Woodruff School of Mechanical Engineering
Georgia Institute of Technology
Atlanta, GA 30332-0405

ABSTRACT

The magnetic layer of a conventional magnetic hard disk is protected by the combination of a carbon overcoat and a liquid lubricant layer. From both quality control and testing points of view, it is of interest to monitor the thicknesses of these layers. When the optical properties (i.e., index of refraction and extinction coefficient) of the substrate and film are known initially or can be measured prior to film deposition, the thickness of the film can be determined analytically from an ellipsometer measurement at a single angle of incidence. However, mathematical complexities arise when multiple films are present, particularly when the layers' optical constants are not known *a priori* and cannot be determine prior to film deposition. In this case, the ellipsometry equations cannot be inverted analytically and must be solved using a numerical scheme.

Depending on the number of unknowns, measurements are generally required at multiple angles of incidence. To determine these parameters one generally constructs an error function based on the difference between the measured values of delta and psi and those computed from ellipsometry theory for a given set of optical constants and layer thickness values. One then iterates on values of these input parameters until the error function approaches zero. Conventional approaches to this minimization process often include gradient searches but, to achieve accuracy, these techniques require the user to implement relatively small "reasonable ranges" with which to bracket the parameters to be determined. For highly nonlinear, multivariate functions such as those of ellipsometry, these methods often prove inadequate. Gradient methods respond to local behavior of the error function, which can be misleading when there is cyclical dependence on the parameters.

In the present work, an associative memory approach (AMA) is investigated as an alternative technique for ellipsometer parameter extraction. The AMA method is based on the development of a "memory matrix" that relates system inputs to system outputs. This relationship is expressed as (1)

$$\{p\} = [M]\{r\}$$

where $\{p\}$ is the vector of system input parameters, $\{r\}$ is the vector of system output parameters, and $[M]$ is the memory matrix. The above equation represents a linearized inversion of a non-linear input-output relationship. In the present study, $\{r\}$ denotes the vector containing the values of delta and psi at up to three angles of incidence, while $\{p\}$ is the vector of values for the unknown parameters, including thicknesses, indexes of refraction and extinction coefficients for the layers in question. The memory matrix $[M]$ is constructed from numerous input-output pairs via the equations of ellipsometer theory (2). The iteration proceeds by continually updating $[M]$ and $\{p\}$ until a sufficiently small error is obtained.

In the present work ellipsometer measurements are conducted on lubricated and unlubricated magnetic hard disks with and without carbon overcoat. The AMA method is compared with a conventional gradient search for its ability to determine desired input parameters. It is found to have a relatively fast, robust convergence even when large "reasonable ranges" are used in bracketing unknown parameters. The effectiveness of the AMA method is attributed to two primary features. First, only the initial memory matrix $[M]$ requires many function evaluations. During the iteration itself, $[M]$ is updated at each step using only a single input-output relationship. Secondly, because the AMA method does not rely on a local derivative of the error function, it is better suited to find overall trends. This feature is useful when the initial guess is relatively far from the desired solution and when the function has cyclical dependency on its arguments, as in the present case.

REFERENCES

(1) R. E. Kalaba, Journal of Optimization Theory and Applications, Vol. 76, No. 2, pp. 207-223, 1993.
(2) R. M. A. Azzam, N. M. Bashara, Ellipsometry and Polarized Light, Elsevier Science Publishers, pp. 283-289, 1989.

---

[1] Now a graduate student in the Department of Mechanical Engineering at Stanford University.

WEDNESDAY 10 SEPTEMBER

Page number

W1	EHL and boundary lubrication – II	179
W2	Automotive and other applications	195
W3	Wear by hard particles – I	211
W4	Wear in ceramic systems	227
W5	Bio-tribology	243

THE INFLUENCE OF ELLIPTICITY RATIO UPON THE ELASTOHYDRODYNAMIC LUBRICATION OF SOFT-LAYERED POINT CONTACTS

D. DOWSON* and D. WANG**
Department of Mechanical Engineering, The University of Leeds, Leeds, LS2 9JT, UK
T. & N. Technology Ltd., Cawston House, Cawston, Rugby, Warwickshire, CV22 7SA, UK.

ABSTRACT

Elastic-isoviscous elasto-hydrodynamic lubrication problems are encountered in a number of physical situations, including water lubricated elastomeric bearings and seals and animal joints lubricated by synovial fluid.

Formulations of the problem for line contact geometries have been considered by a number of authors over the past thirty years. The line contact study of Jin et al (1) in 1993 led to limiting film thickness equations for either a compressible fluid and a Poisson's ratio of 0.4 or an incompressible fluid and a Poisson's ratio of 0.5. A generalised solution for intermediate conditions was been reported recently by the same authors (2).

Dowson and Yao (3) provided numerical solutions for elliptical contacts in which entraining motion was directed along the minor axis of the equivalent dry contact zone. They demonstrated that the relatively simple constrained column model could be used in the elasticity calculation to yield accurate solutions to the elastohydrodynamic lubrication problem if the thickness of the layers of soft bearing material was less than the width of the contact zone and if Poisson's ratio was less than about 0.4 to 0.45.

We now consider an extension to the range of solutions to the elastic-isoviscous problem for elliptical contacts in which entrainment can occur along either the minor or major semi-axes of the equivalent dry contact patch. The assumptions are consistent with earlier analyses, namely that the solids are smooth, that the lubricant is an isoviscous Newtonian fluid and that the total elastic deformation of the equivalent soft layer of bearing material can be assessed with adequate accuracy from the constrained column model.

Solutions have been produced by the recently developed *Effective Influence Newton* (EIN) method outlined by Dowson and Wang (4,5). The general features of the pressure distributions and film shapes are outlined and numerical solutions for the central and minimum film thicknesses are presented for ellipticity ratios ranging from 0.025 (entrainment along the major axis) through unity (circular contact) to 32 entrainment along the minor axis). A typical film shape in which the film thickness is almost constant in the transverse, side-leakage, direction, while demonstrating a near linear decrease in the entraining direction is shown in Figure 1. The complete range of solutions has enabled the central and minimum film thicknesses to be calculated for entrainment along either of the principal axes of the nominal dry contact zone and for a very wide range of geometrical configurations in elliptical, elastohydrodynamic conjunctions.

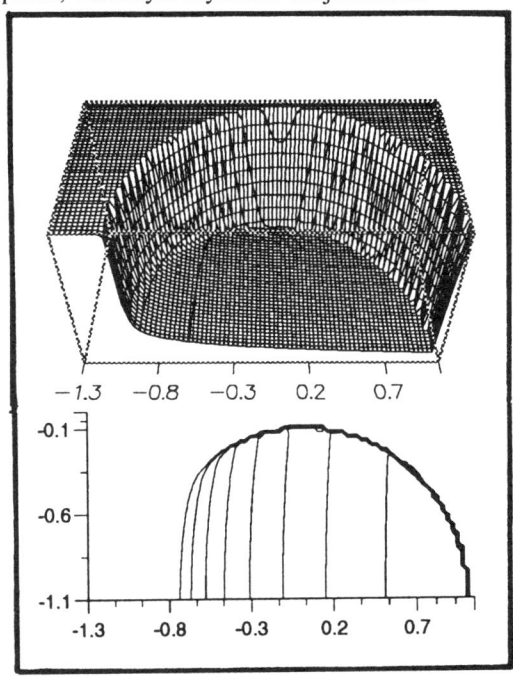

Figure 1. Film Shape for Elliptical Conjunction.
($k=0.6$, $U= 0.7875 \times 10^{-6}$, $W = 0.5185 \times 10^{-4}$, $Ht = 0.002$; $Hmin = 1.68 \times 10^{-6}$, $Hcen = 2.09 \times 10^{-6}$)

REFERENCES

(1) Z.M. Jin, D.Dowson and J. Fisher, WEAR, **170**, 281-284, 1993.
(2) Z.M.Jin, D. Dowson and J. Fisher, 19th. Leeds-Lyon Symposium on Tribology, Elsevier, 545-555, 1993.
(3) D. Dowson, and J.Yao, Proc. I. Mech. E., Pt. 'H', **208**, 43-52.
(4) D. Dowson,and D. Wang, WEAR, **179**, 23-97, 1994.
(5) D. Dowson and D. Wang, 21st. Leeds-Lyon Symposium on Tribology, Elsevier, 565-582, 1995.

VISCOELASTIC ROLLERS ELASTOHYDRODYNAMICALLY LUBRICATED WITH A PSEUDO-PLASTIC FLUID

ABDALLAH A. ELSHARKAWY
Mechanical and Industrial Engineering Department, College of Engineering & Petroleum,
Kuwait University, P. O. Box 5969, Safat 13060, Kuwait.

ABSTRACT

In some lubricated contacts, one or both of the bounding solids may be made from low elastic modulus material such as rubber and polymers. Using such materials provides better damping properties, less rolling noise generation, and less weight. These materials exhibit time-dependent behavior in their relationships between stress and strain and thus they are considered as viscoelastic materials. Typical examples are the lubrication between the plate cylinder and the blanket cylinder of the lithographic printing press (MacPhee et al., 1991) and lubrication of plastic gears (Adams, 1993).

In this study the analyses of Lin and Lin (1990) and Elsharkawy (1996) are extended to incorporate both the viscoelastic behavior of the rotating rollers and pseudo-plastic behavior of the lubricating oil into the elastohydrodynamic lubrication analysis. The surface displacements of the bounding solids are calculated from linear-viscoelastic half-space theory. A standard linear model is used to characterize the viscoelastic behavior of the rotating cylinders. A cubic equation obtained from the empirical curves for pseudo-plastic fluids is used as the relation between shear stress and shear strain rate. The modified Reynolds equation and the viscoelasticity equations of the bounding solids are solved simultaneously within the lubricated conjunction by means of the iterative Newton-Raphson method. The effects of the non-Newtonian behavior of the lubricant and the viscoelastic behavior of the bounding solids on the pressure profiles and film shapes are presented and discussed.

Both the viscoelastic behavior of the bounding solids and the pseudo-plastic behavior of the lubricating oil have been incorporated into elastohydrodynamic lubrication analysis of line contacts. The results showed that:

1. The viscoelastic behavior of the bounding solids has a significant influence on the dry contact pressure and lubricated contact
pressure distributions. As the viscoelastic behavior of the bounding solids increases, the pressure profiles shift to the exit side of the conjunction and this will be the main cause for rolling friction in these applications.
2. The minimum film thickness predicted from viscoelastic the analysis is greater than that of elastic analysis.
3. The minimum film thickness predicted from non-Newtonian analysis is less than that obtained from a Newtonian analysis.

REFERENCES

Adams, C. E. (1993), "Lubricants and Lubrication of Plastics Gears," Journal of Gear Technology, September/October 1993, pp. 42-44.

Elsharkawy, A. A. (1996), "Visco-Elastohydrodynamic Lubrication of Line Contacts," Wear, Vol. 199, No. 1, pp. 45-53.

Lin, T. R., and Lin, J. F. (1990), "The Elastohydrodynamic Lubrication of Line Contacts with Pseudoplastic Fluids," Wear, Vol. 140, pp. 235-249.

MacPhee, J., Shieh J., and Hamrock, B. J. (1991), "The Application of Elastohydrodynamic Lubrication Theory to the Prediction of Conditions Existing in Lithographic Printing Press Roller Nips," Advances in Printing Science and Technology, pp. 242-276.

A PARTIAL ELASTOHYDRODYNAMIC LUBRICATION POINT CONTACT MODEL FOR HOMOGENEOUS AND LAYERED SURFACES

DUCAI WANG
SI C. LEE
The Ohio State University

ABSTRACT

Many concentrated machine elements including gears, rolling element bearings, cams, and tappets operate in the regime of partial elastohydrodynamic lubrication. In partial-ehl, the applied load is shared between the asperity contacts and the lubricant film. Failing to maintain proper level of partial-ehl condition may lead to irreversible damages precipitated by scuffing, pitting, or an unacceptable rate of wear. Two important factors which must be considered in partial-ehl modelling are the surface roughness and the surface topography. These factors are known to have a significant effect on the lubrication performance parameters which include the average film thickness, friction, subsurface stress, and the asperity flash temperature. The ability to accurately predict these parameters is vital for the successful understanding of scuffing, pitting, and wear behaviour. A partial-ehl point contact model for homogeneous and coated surfaces has been developed. The present model employs the effective influence Newton method and the multilevel integration technique. The surface topographies used in the partial-ehl calculations were real digitised surfaces and numerically generated rough surfaces. A large computational grid size of 513 x 513 was used to accurately represent the surface roughness features. The effects of two asperity aspect ratios were investigated: the pure transverse roughness and the pure longitudinal surface roughness.

According the present results, when the average partial-ehl film thickness to composite root-mean-square roughness ratio ($\Lambda = h/\sigma$) was greater than 3.0, the transverse surface yielded a slightly higher film thickness than the longitudinal surface, in agreement with Patir and Cheng's average flow model. However, for smaller values of Λ, the transverse surface resulted in a thinner lubricant film than that of the longitudinal surface. This effect was primary due to increased side leakage along the transverse grooves. Furthermore, it was found that at even smaller values of Λ, the partial-ehl pressure distribution approached that of the dry contact case. The lubricant pressure rose sharply at the regions of asperity interaction and decreased to nearly zero at non-asperity-interaction regions.

Couette flow dominance was also observed during the present investigations. This was especially significant in the longitudinal surface case where the asperity roughness pattern parallels the sliding motion. During contact, the surface roughness decreased and the valleys of the asperities became enlarged, giving greater spaces for lubricant to flow. These asperity level deformations enhanced the lubrication process resulting in a favourable film thickness condition. In the transverse surface case where the asperity pattern is perpendicular to the sliding motion, the roughness increased the flow resistance across the contact. Since the lubricant flows in the direction of least resistance, a significant amount of fluid loss occurred through the sides of the contact. Nevertheless the strong Couette action still dominated the fluid flow by opening up several large longitudinal grooves across the transverse roughness along the sliding direction.

EFFECT OF 3-DIMENSIONAL ROUGH SURFACE TOPOGRAPHY ON THE ELASTOHYDRODYNAMIC LUBRICATION IN POINT CONTACTS

DONG ZHU (Member of the ASME)

Eaton Corporation, Corporate R & D Center, 26201 Northwestern Highway, Southfield, MI 48037, U.S.A.

ABSTRACT

In this paper a full numerical analysis using a multi-grid scheme and a 3-dimensional deterministic model for the rough surface EHL in point contacts is presented. Since both contacting surfaces can be rough and moving at different velocities, the solution is strongly time-dependent. For comparison purpose, both optically measured real machined surfaces (with irregular asperities and random roughness heights) and 3-dimensional sinusoidal waves are used. The computer program developed appears to be a useful tool that can handle different types of rough surface topography and different Hertzian contact ellipticity values in a wide range of operating conditions. For each surface, its roughness, orientation angle and asperity wavelengths in both x-(rolling) and y-directions can be changed. In reality, the rough surface EHL is very complicated, and its behavior has not been satisfactorily understood. This paper presents a number of calculation cases focusing on the study of effects of surface roughness, orientation and relative sliding speed on the EHL characteristics, such as the average film thickness around the center of Hertzian contact, the minimum film thickness and the maximum pressure peak height.

Dimensionless EHL parameters for the analyzed circular contact cases are $U=2.298 \times 10^{-10}$, $G=4000$ and $W=5.04 \times 10^{-6}$, and the maximum Hertzian pressure $P_h = 1.358$ GPa. The composite RMS roughness varies from zero to 0.8 µm, and the slide-to-roll ratio, S, from zero up to 60%. Three surface orientation angles, 0°(longitudinal), 45°, and 90°(transverse), are chosen in the present study. It is found for typical ground surfaces that (i) the effect of surface orientation on the EHL film thickness does not appear to be significant for the cases analyzed; (ii) as the roughness increases, the average film thickness is only slightly increased, but the raise of maximum pressure peak seems to be quite significant; (iii) when the sliding speed goes up, the film thickness increases slightly while the pressure peaks are raised drastically.

In order to avoid any unpredictable influence due to randomness of real machined rough surfaces, cases for 3-dimensional sinusoidal waves are also analyzed under the same operating conditions. For longitudinal wavy surfaces, the wavelength in x-direction is 1.5A (A is the Hertzian radius), and that in y-direction is 0.25A. For transverse wavy surfaces it is the other way around. For isotropic both wavelengths are 0.25A. Similar behavior is observed as described above for ground surfaces, except that (i) the film thickness caused by the transverse waves is slightly greater than those by isotropic and longitudinal ones, but the difference still appears to be quite insignificant; (ii) the maximum pressure peak for ground surfaces appears to be higher than that for sinusoidal waves, if all other conditions are the same or similar. The higher pressure peaks might be caused by the secondary roughness of machined surfaces, as described in (1).

A few cases for an elliptical Hertzian contact are also presented in the paper. The ellipticity is 4.0, quite close to those in some ball bearings. Although the results are still preliminary, observed behavior appears to be similar to that for the cases of circular contact.

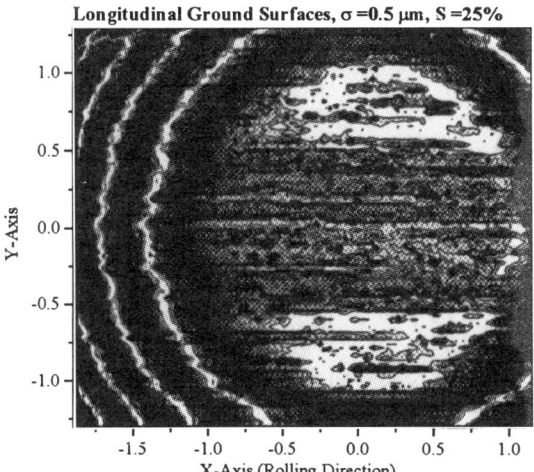
Longitudinal Ground Surfaces, σ =0.5 µm, S =25%

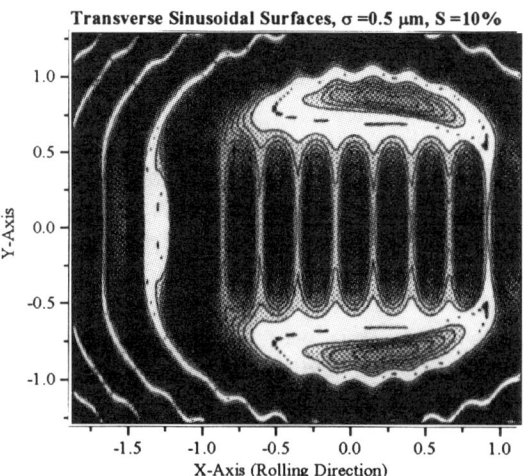
Transverse Sinusoidal Surfaces, σ =0.5 µm, S =10%

REFERENCES

(1) D. Zhu, and X. Ai, ASME Preprint 96-TRIB-24.

Submitted to *ASME Journal of Tribology*

A SIMPLIFIED DIAGRAM OF SEIZURE CRITERIA FOR LUBRICATED SLIDING ROUGH SURFACES

TAKERO MAKINO

Nagasaki R&D Centre, Mitsubishi Heavy Industries, Ltd., 1-1 Akunoura, Nagasaki 850-91, JAPAN

ABSTRACT

The demand for low energy consumption and compactness in industrial machines requires machine elements to have higher load capacities. In order to meet this requirement, designers have to careful choices regarding the materials and dimensions of machine elements. The Pressure-Velocity(P-V) map provides useful information and solutions for such problems. Most designers usually prepare their own P-V maps and use them in deciding the specifications of machine elements. As the demands for high efficient products have become even greater in the recent market, it has become common to specify the characteristics of surface roughness not only in the detailed design process but also in the conceptual or primary design process. The aim of this paper is to present an idea which takes account of roughness effects in the P-V map for a lubricated sliding system and to provide a simple design diagram for the primary design process.

The P-V map basically represents the energy dissipation of the sliding system. Hence the rearrangement of the map as a temperature map proposed by Lim and Ashby(1) for dry friction is a first step. This idea can then easily be extended to lubricated surfaces by taking account of reduction of friction coefficients except for the effects of the oiliness and EP agents.

In this paper we assume that the seizure mechanism involves two different elements : one is plasticity dominated seizure and the other is temperature dominated seizure. The first criterion is defined as the limiting value of a normal load where the plastically deformed asperities of the surface start interacting each other. Generally this criterion does not depend upon velocity because it is simply given by a static contact calculation. In our formulation, however, the velocity effect appears as a reduction of asperity contact load due to mixed lubrication. The second criterion is given as a constant surface temperature line calculated by Archard's formula(2). Although this line inherently includes the velocity effect, the additional effect of sliding velocity with the reduction of effective load is also taken into account.

The velocity effect, i.e. the load capacity due to fluid film formation, is considered as follows. The surface is assumed to have isotropic or longitudinal roughness because most of the surfaces of machine elements have such orientation. It has an important characteristic in the capability of fluid film formation. Namely the roughness effects are less effective for the long bearing whose roughness orientation is isotropic or longitudinal(3). Since the fluid film pressure is likely to be boosted up at the inlet wedge of the contact, a simple long bearing approximation can be used to evaluate the fluid film effects.

Figure 1 shows a typical seizure diagram of steel/steel sliding system, which was calculated with the above assumptions. According to the results of experiments with the steel pad/disk sliding rig lubricated by a light mineral oil (no additives) at a very low speed, it was found that a friction coefficient of 0.15 is the velocity-independent boundary friction coefficient of this system. Critical temperature was given as 150°C(4). The lines on this diagram represent the plasticity and temperature dominated criteria calculated theoretically. The symbols indicate the experimental results. The roughness parameters in the calculation were not identical to those in the experiment but these were given as the average values of several lapped specimens.

Despite a lot of simplifications, the diagram has been able to describe the experimental results. It is also interesting that the plasticity dominated criteria are close to the mild to severe wear transition zone which was empirically given by Lim and Ashby(1).

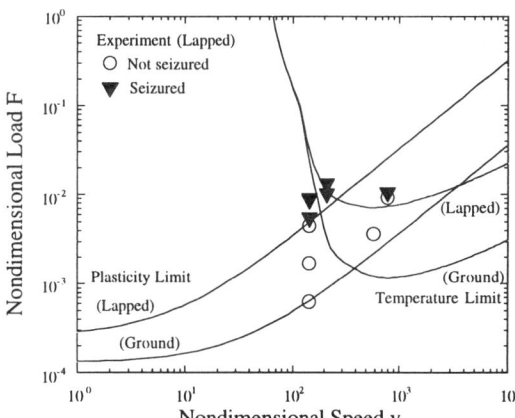

Fig.1 Seizure Diagram

REFERENCES

(1) S.C.Lim and M.F.Ashby, Acta metallurgica, Vol.35, p.1, 1987
(2) J.F.Archard, Journal of Applied Physics, Vol.24, p.981, 1953
(3) N.Patir and H.S.Cheng, Trans. ASME, Journal of Lubrication Technology, vol.101, p.220, 1979
(4) K.Saki, S.Watanabe, K.Sudo and S.Ishibashi, Mitsubishi Juko Giho, Vol.24, p.115, 1987

SOME PROPERTIES OF A CONTINUOUSLY ADJUSTABLE HYDRODYNAMIC FLUID FILM BEARING

DR J K MARTIN
Faculty of Technology, Open University, Milton Keynes, MK7 6AA, UK

DR D W PARKINS
School of Industrial and Manufacturing Science, Cranfield University, Cranfield, MK43 0AL, UK

ABSTRACT

A novel form of adjustable fluid bearing has been devised whereby the hydrodynamic conditions can be changed during operation in an externally controlled manner. The principle can be applied to conventionally orientated journal bearings, i.e. a shaft rotating within the bearing housing; to inverse orientations, i.e. a rotor on a fixed shaft; and to thrust bearings. The embodiments are included in international patent applications filed by British Technology Group Ltd. (1)

Investigations have been carried out on both theoretical and practical models for journal bearings. The theoretical governing Reynold's equation has been derived and expanded to include non-uniform variations in film shape profiles in 2 dimensions, and allow viscosity to vary with temperature and pressure. A computer model has been developed to produce converged pressure field solutions also taking account of the influence of pressure forces on the shape of the fluid film profile.

The computer model, once initiated, automatically operates iteratively combining temperature, viscosity and pressure field finite difference solutions with fluid film shape profiles generated by finite element models of the adjustable segments. Comprehensive studies for the inverse orientation rotor arrangement have predicted improvements over current bearing designs in terms of stiffness, damping and rotational accuracy. Also predicted is the ability to sustain a given rotor eccentricity – including zero – for particular loads and changes in load.

Test rigs have been designed and built for both orientations of journal bearings and test results obtained which closely parallel results from the theoretical models. For the inverse orientation arrangement a means of imparting loads to the rotor by non-contacting electro-magnets was used. The capability to maintain or adjust a given centre of rotation position was repeatedly demonstrated. Displacement coefficients were obtained by a method adapted from a selected orbit technique extended by Parkins. (2)

Figs. 1 and 2 show the effect on direct displacement coefficients of increasing adjustments for the adjustable segments or fingers. All results refer to the inverse orientation version and were obtained within a few minutes for a constant rotor load and speed, and with oil inlet temperatures controlled. The finger adjustments are designated by zero, small, medium and large, where the large setting represents approximately 75% of the zero case radial clearance. Similar results have been obtained for the conventional orientation version of the bearing. It is hoped these will be published separately.

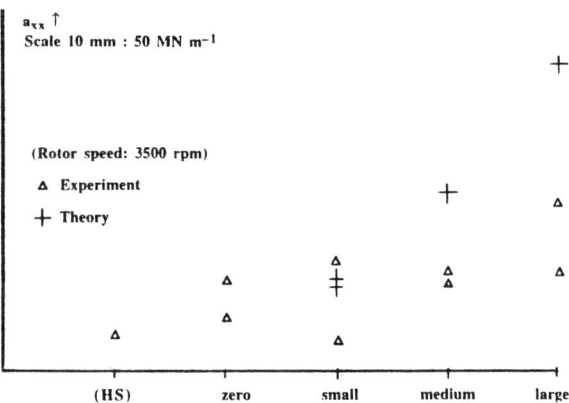

Fig. 1: Displacement coefficient, a_{xx} effect of finger adjustments

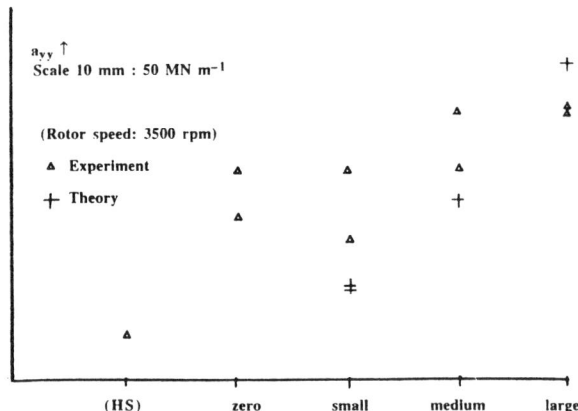

Fig. 2: Displacement coefficient, a_{yy} effect of finger adjustments

REFERENCES

(1) J K Martin, D W Parkins, *Fluid Film Bearings*, International Patent Application, World Intellectual Property Organisation. WO–95/29346, November 1995.

(2) D W Parkins, *Measurement of Oil Film Journal Bearing Damping Coefficients – An Extension of the Selected Orbit Technique*. Trans ASME Jnl. Trib. Vol. 117, October 1996.

EHL CHARACTERISTICS OF ROUGHENED ELASTOMER SURFACES

MASANORI OGATA and ATSUNOBU MORI

Faculty of Engineering, Kansai University, 3-35, Yamate-Cho 3, Suita, Osaka 564, Japan

ABSTRACT

In the run-in process of lip-type oil seals, the sealing surface of the lip is generally roughened to have micro-undulations aligned perpendicularly to the sliding direction. This favorably leads the sealing surface to be well lubricated, and also to have the pumping ability against the pressure increment across it.

In recent publications on lip-type oil seals, some researchers have attempted to apply the EHL theory to such problems of lubrication and sealing mechanisms.

In order to understand the reason why a lip-type oil seal can be well lubricated whereas it effectively seals an oil even in stationary conditions, in the previous paper (1), the authors discussed the difference in deformation characteristics between stationary contact and hydrodynamically lubricated contact of an undulated elastomer surface pressed on a smooth rigid surface.

It was found that the micro-undulation of an elastomer surface can easily be flattened out if it is under stationary conditions without hydrodynamic pressurization in the lubricant. It was also found that the undulation completely flattened out in such a manner can be formed again when it is situated under EHL sliding conditions.

REFERENCE

(1) E. Miyazaki, M. Ogata, A. Mori, and Y. Shimotsuma, Proceedings of the International Tribology Conference Yokohama '95, Vol.II, pp.1039-1042, 1996.

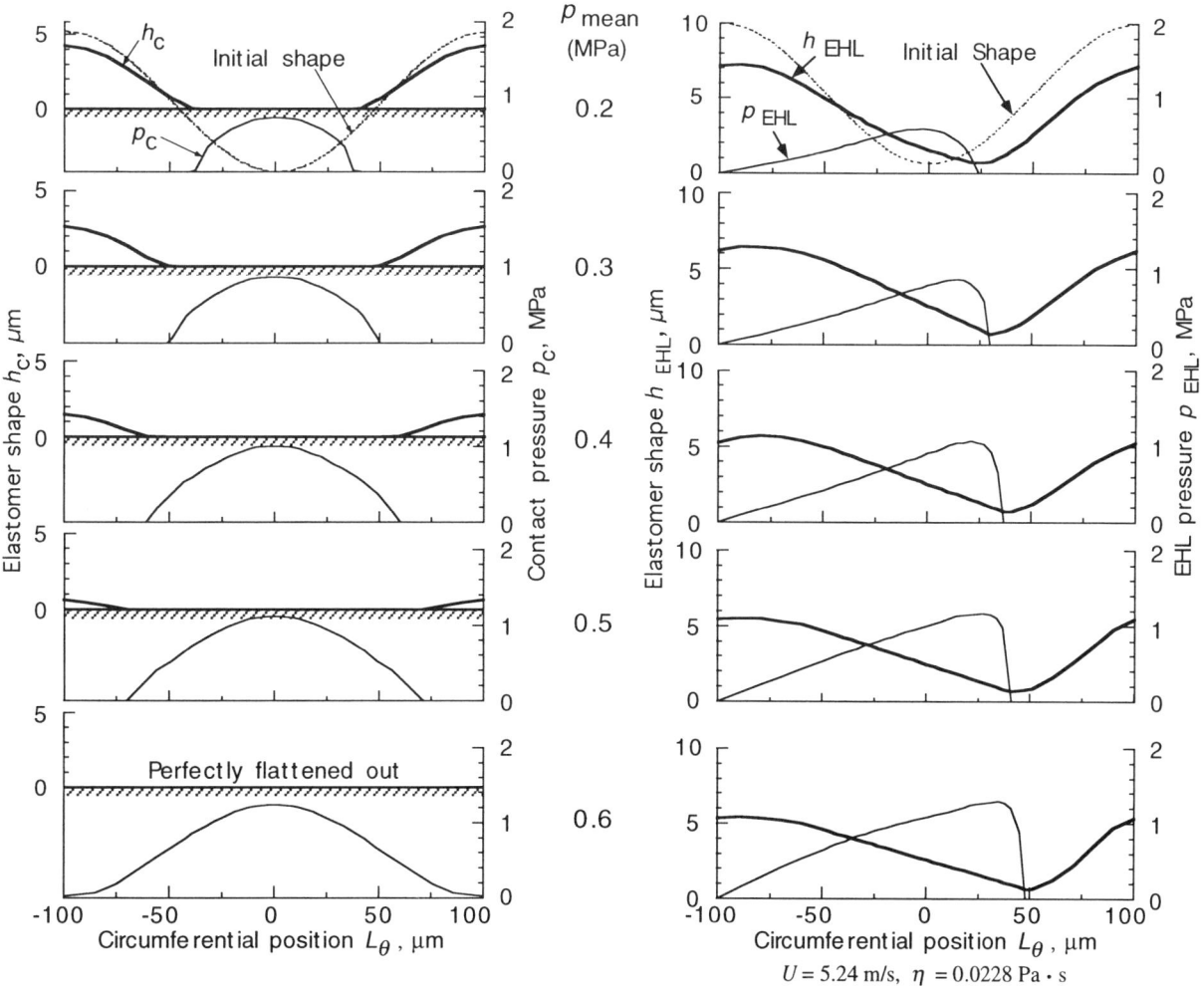

Fig.1 : Stationary

Fig.2 : Sliding

DYNAMIC CHANGES OF SURFACE FILMS DURING BOUNDARY LUBRICATION

JOSE CASTILLO, DONG W. BAI and KENNETH C. LUDEMA
Dept. of Mechanical Engineering and Applied Mechanics, University of Michigan, Ann Arbor, MI 48109-2125
TODD WIELAND and FRED NERZ
Cummins Engine Company, Inc. Columbus, Indiana 47202-3005

ABSTRACT

Wear and friction behavior in boundary lubrication are strongly influenced by surface films on the sliding parts. These surface films change dynamically while sliding takes place. In spite of this importance few studies have focused on the progression of change of those films.

This study focused on simultaneous monitoring of dynamic changes on the surface and friction forces between the sliding parts, with the objective of finding cause-effect relations that could lead to a better understanding of the process of friction and wear.

The monitoring of dynamic changes in surface films was performed by a Division–of–Amplitude–Photopolarimeter Mueller–Matrix Ellipsometer. The wear test rig used is a Cylinder(Pin)-on-Disk type specially designed to be coupled with the Ellipsometer. The friction force was measured using a cantilever beam instrumented with strain gages. The specimens were made of 1080 steel and M2 steel. The disc roughness and hardness were varied to study their influence. Bearing rollers (52100 Steel) were used as the second sliding body. The initial Hertzian contact pressure (maximum) was 900 MPa. Specimens were fully submerge lubricated in low viscosity hydrocarbons (Diesel fuel type) with various additives.

Besides monitoring friction force and surface changes, analysis of wear scars on the roller, and evaluation of the final roughness on the disks were performed on the disk specimens. Microscopic analysis of the surface, including Nomarski Differential Interference Contrast Microscopy was performed on several of the tests as another way of analyzing the formed films.

The main conclusions of the study are:

(1) Films are formed mainly as "patches" corresponding to the localized contact areas during the sliding of the surfaces. The surface roughness is not significantly changed by these films.

(2) No connection was found between the thickness and the protective properties of formed surface film. This implies that mechanical and chemical characteristics of these films are more important than their thickness or covered area.

(3) Significant changes on the surface are registered with the Ellipsometer when scuffing occurs, indicating the capacity of the technique to monitor dynamic changes taking place on the surface under boundary lubrication.

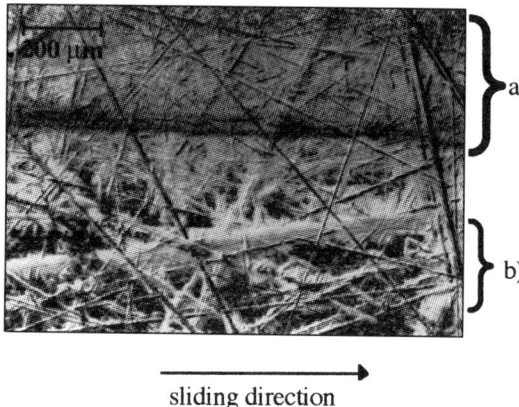

Fig. 1: Nomarski Differential Interference Contrast Microscopy of a specimen rubbed under boundary lubrication. The two zones shown correspond to:
(a) the non-rubbed area (lighter color) and (b) the rubbed area (dark patches).

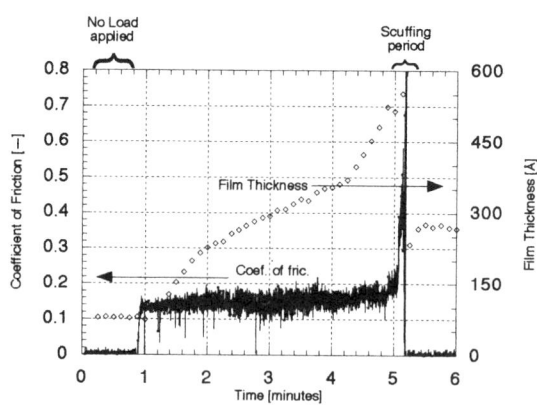

Fig. 2: Simultaneous Coefficient of Friction and Ellipsometer Film Thickness Analysis for one specific case where scuffing (sudden increase on friction) occurred at around 5 minutes.

FRICTION OF SLIDING SURFACES CARRYING BOUNDARY FILMS

K. A. BLENCOE, G W ROPER* and J. A. WILLIAMS
Cambridge University Engineering Department, Trumpington Street, Cambridge, CB2 1PZ, UK
*Shell Research Ltd, Thornton Research Centre, PO Box 1, Chester, CH1 3SH, UK

ABSTRACT

As the severity of loading on a lubricated engineering contact increases so its macroscopic behaviour becomes less dependent on the bulk properties of the intervening fluid film and more strongly influenced by the near-surface lubricant rheology and the direct interaction of asperities on the opposing load bearing solid surfaces. Commercial lubricants invariably contain a chemically complex additive package one of whose roles is to produce protective boundary layers on the opposing solids. The lubricant technologist aims to produce films which are physically robust enough to survive in extreme contact conditions, so preventing damaging metal to metal contact, and yet slippery enough to maintain low coefficients of friction. These 'third bodies' which separate the solid surfaces may have rheological or mechanical properties which are very different from those observed in the bulk.

Classical elasto-hydrodynamic theory considers the entrapped lubricant to exhibit very marked piezo-viscosity while the conventional picture of boundary lubricant layers is one of more solid films whose shear strength τ is linearly dependent on local pressure p. This relation is consistent with Amonton's laws irrespective of the geometric details of the surface topography. However, the properties of adsorbed or deposited surface films produced under real operating conditions are likely to be more complex than this. We have looked quantitatively at the influence of the pressure dependence of the shear strength of such surface layers on the overall friction coefficient of contacts exhibiting both model and measured surface profiles (1). The analysis results in plots of coefficient of friction versus a service or load parameter such as that illustrated in Fig. 1: P is the nominal pressure, H_s the hardness of the softer surface and σ and R and N the standard deviation, peak radius and asperity density of the surface topography. Curve (A) is for the case in which $\tau=0.1p$ while curve (B) represents a layer with a pressure dependence enhanced by a small parabolic term as suggested for multiple layers of stearic acid by ref. (2).

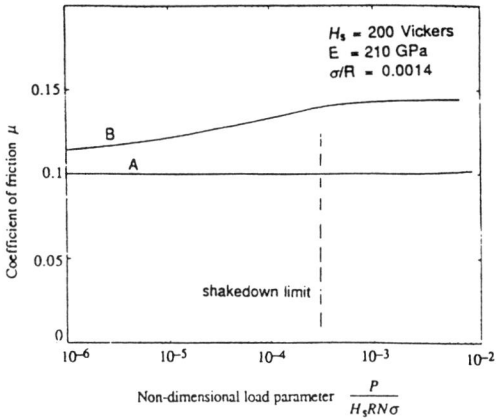

Fig. 1: Plot of overall coefficient of friction vs normalised load parameter.

In principle any variation or dependence of layer shear strength with pressure can be incorporated and this information might come, for example, from either physical experiments or computed molecular dynamic simulations. The results of the analysis indicate that variations in the macroscopic coefficient of friction which may occur as the load is varied have their roots both in the statistical nature of the surface and the mechanical response of the lubricating boundary film. The effect running-in has on friction and safe working loads through topographical changes can also be highlighted. The value of this analysis is that it attempts to combine the behaviour of films at molecular dimensions with the topography of surfaces measured at an engineering scale and so gives an indication of the full-size effects that can be achieved by chemical or molecular surface engineering.

REFERENCES

(1) K. A. Blencoe and J. A. Williams, Wear Vol 203-204, pp 722-729 (1997)
(2) R. S. Tinsit and C. V. Pelow, J. Tribol. Vol 114 pp 150-158 (1992)

SCUFFING BEHAVIOR OF AL-SI ALLOY IN LUBRICATED RECIPROCATING CONTACTS

Z. K. YE and H. S. CHENG
Center for Surface Engineering and Tribology, Northwestern University, Evanston, IL 60208

ABSTRACT

Scuffing of aluminum-silicon alloy pistons at times still occurs in engines and degrades their performance. To prevent and predict scuffing it is necessary to understand the scuffing mechanisms of aluminum-silicon alloy against cast iron under severe engine operating conditions. In this paper the scuffing behavior and mechanisms of the Al-Si (11-13%) alloy against cast iron in lubricated reciprocating contacts are investigated, and the influence of surface roughness, surface profile, lubricating condition and lubricant properties on scuffing is investigated.

To simulate the reciprocating motion in engines a newly designed reciprocating tribo-tester with a stroke of 82.55 mm was used. Three lubricating methods, L1, L2, L3 are used. In L1, lubricant is injected directly to the contact of the specimens. In L2 and L3, lubricant is fed at a distance away from the contact to simulate the normal and starved splash lubrication in engines. The Al-Si alloy piston and skirt specimens were cut from the production pistons and liners.

The scuffing results of nine specimens with different skirt roughness heights and textures using lubrication method L2 are shown in Fig. 1. The rougher surface gave lower scuffing pressures; however, the scuffing temperature were almost same, around 194°C, suggesting the existence of a critical temperature. In another set of tests with a comparable base oil without any additives, the scuffing temperatures were much lower, around 136°C.

Thus, the additives in 10W30 oils appear to improve the scuffing resistance.

The effect of lubrication method on scuffing is shown in Fig. 2. In L1, no scuffing occurred for average pressure as high as 34.5 MPa, since copious lubricant was spurting directly to the contact. In L2 and L3, the lubricant supply was much less copious, scuffing occurred readily. The scuffing pressure for L3 was lower because of the lower lubricant supply. However, the scuffing temperature were same for both tests, around 196°C.

SEM micrographs taken from the scuffing area of the skirt specimen with 10W30 lubricant show that large scale of fracture, severe scratching, and material pull-out occurred. Some dark color region were also found on the scuffing area. EDX analysis indicate the existence of sulfur and phosphorus perhaps within a protective film on the contact area. The oxygen and iron peaks in the EDX spectrum imply that surface oxidation and iron transfer from the cast iron liner also occurred during scuffing.

From the above observations, it appears that scuffing is a result of successive breakdowns of several protective films including adsorbed, chemically reacted film, oxidation film, and the surface layer of the substrate itself. The onset of scuffing is related to a critical temperature for Al-Si alloy/cast iron reciprocating contacts lubricated with 10W30 engine oils.

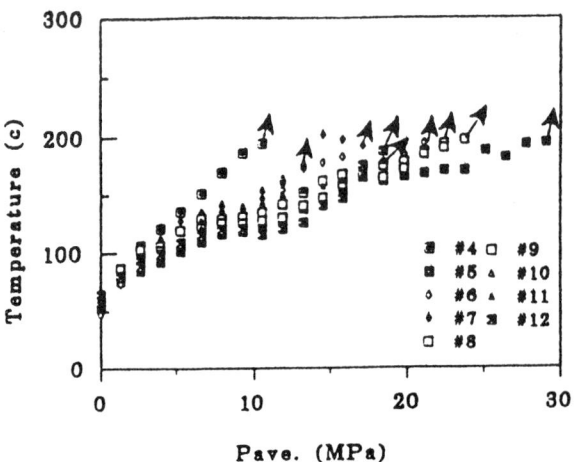

Fig. 1 Temperature rise in the scuffing with 10W30/L2

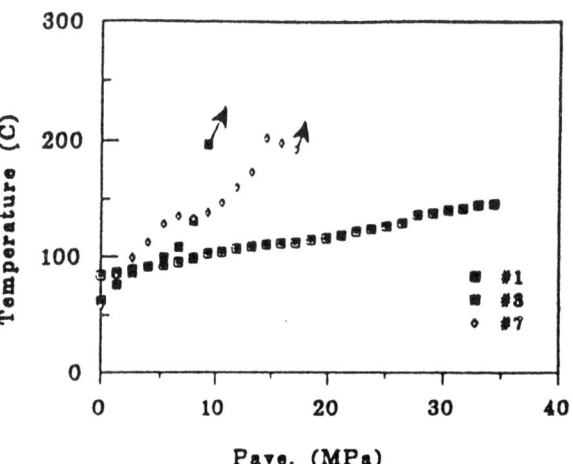

Fig. 2 Temperature rise in L1(#1), L2(#7), and L3(#3) with 10W30

THE BOUNDARY LUBRICATION OF GLASS FIBRES

SA JOHNSON and M J ADAMS
Unilever Research Port Sunlight Laboratory, Quarry Road East, Bebington, Wirral, Merseyside, L63 3JW, UK
S BISWAS, D R WILLIAMS and B J BRISCOE
Dept. of Chem. Eng. and Chem. Tech., Imperial College, Prince Consort Road, London, SW7 2BY, UK

ABSTRACT

During the past twenty years, it has been established that the measurement of the friction between pairs of crossed fibres is a particularly effective way to investigate the fundamental mechanisms associated with boundary lubrication. Often the fibres are sufficiently fine that significant flexure can be observed during relative sliding. The quantification of this flexure, together with a knowledge of the fibres' mechanical properties, allows the friction developed at the contact to be calculated. In addition, the normal load and the actual contact area can be estimated with some certainty due to the well-defined geometry. This allows the calculation of the interfacial shear stress as a function of contact pressure and has formed the basis of several fibre tribometers (1-4), and also the present equipment.

Since the above fibre couple is sub-critically damped and involves highly compliant fibres, intermittent motion is often observed if the static friction is greater than the dynamic value. Boundary lubricants affect the frictional force and the resulting intermittent motion. Several other factors are also important in influencing the frequency and amplitude of the intermittent motion; these include the geometry, topography and dynamics of the two contacting fibres as well as the imposed sliding velocity. When the intermittent motion is of a stick-slip nature, the data obtained in these fibre experiments include the amplitudes of each stick and slip event, together with estimates of the amount of micro-slip that occurs during each stick phase (2). The amplitudes of the stick phases provide information about statistical distributions of static frictional force values and these can be interpreted in terms of established tribological models.

This paper describes the use of a novel PC-based vision system, which is based on a video microscope (Infinivar, Infinity Photo-Optical Company, Colorado, USA) and a frame grabber (CFG, Imaging Technology Inc., Massachusetts, USA), to automatically and conveniently record the stick events between fine fibres of various diameters. An orthogonal configuration (2) is employed with a cantilever fibre being pressed and slid against a fully-incarcerated fibre. Two micro-stepper motor driven translation stages are used to initially position the cantilever fibre, and then to move the fully-incarcerated fibre at a constant velocity, V.

The applied normal load, W, is determined by the imposed flexure of the cantilever fibre (generally in the range from 5 to 200 µm) and the cantilever length, L. The frictional force is calculated in a similar manner from the cantilever flexure that occurs in the plane of the fully-incarcerated fibre. The applied normal load is augmented by autoadhesion. This additional force can be estimated from the friction-axis intercept of a fitted curve through the average static friction versus applied load data.

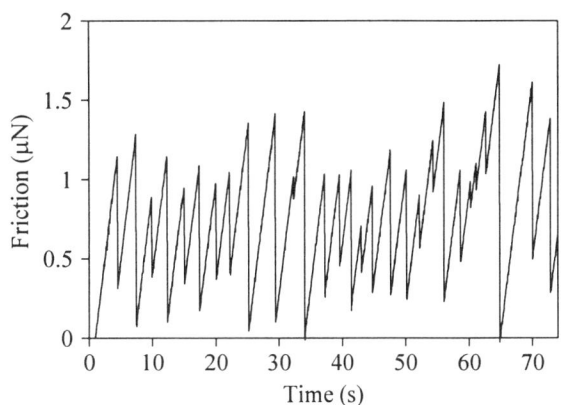

Fig. 1: The friction between two clean glass fibres

The software (system integration done by SMS Ltd., London, UK using Visual C++, Microsoft Corporation, Washington, USA and LabWindows, National Instruments Corporation, Texas, USA) allows flexural information to be acquired at a rate of about 25 Hz, and displayed in real time. The typical range of microscope magnifications used was such that a cantilever flexure of one pixel corresponded to a flexural displacement of between one and four µm. The noise on this data was generally about ± 1.5 pixels.

Figure 1 shows some typical data. These were recorded for a 9 µm diameter cantilever fibre sliding against another fibre of a similar diameter for $L = 1.8$ mm, $W = 1.7$ µN and $V = 38$ µm/s. The pixel noise level results in a force resolution of ± 20 nN for the particular cantilever length used.

Frictional force distributions and interfacial rheological parameters, obtained using this new equipment, will be presented for both clean and boundary-lubricated glass fibres.

REFERENCES

(1) I.C. Roselman and D. Tabor, J. Phys. D **9**, 2517 (1976).
(2) B.J. Briscoe, A. Winkler and M.J. Adams, J. Phys. D **18**, 2143 (1985).
(3) I. Lee, J. Mat. Sci. **29**, 2102 (1994).
(4) J.Y.C. Law, PhD thesis, Univ. of London (1994).

Replenishment of Grease in a Ball and Disc Apparatus

P-O LARSSON
Luleå University of Technology, Division of Machine Elements, SE-971 87 Luleå, SWEDEN

ABSTRACT

In grease lubricated applications, the replenishment of grease into the contact area between moving surfaces is very important to ensure long life and maintenance free operation of a machine element. The equipment used in this investigation was a ball and a disc, similar to those used in optical interferometry rigs to measure lubricant film thickness. The lubricant geometry, trapped between the ball and the disc around the Hertzian contact has a characteristic butterfly wing shape. This geometry can be simplified to two ellipses and an inlet distance m* in front of the Hertzian contact. To evaluate the importance of base oil type, thickener type, base oil viscosity, entrainment velocity, relative slip of the ball and the disc, side slip motion and temperature, an investigation of grease geometry around an elastohydrodynamic contact was carried out using image processing and statistical methods. Surface velocity in the direction of motion of the ball was held constant at 0.15 m/s in the centre of the contact. Slip, a difference in relative velocity between the ball and disc, was varied from -20 % up to +20 %. Side slip motion was accomplished by rotating the axis of rotation of the ball with respect to the disc. For this apparatus the maximum side slip angle was 26.5°, this angle gives a slip 'vector' in the direction of disc rotation of -11 %. The temperature of the lubricant was varied from 20°C to 40°C. The lubricant formation around the contact area was studied using a microscope connected to a conventional VHS video camera. Eight specially manufactured lubricating greases were investigated. Two different base oil types were used, a naphthenic mineral oil and a mixture of polyalphaolefin and diester, each with two different viscosities. The principal factors affecting this geometry were found to be side slip, slip with side slip, base oil type, viscosity and temperature. Increasing side slip increases the inlet distance. Negative slip with side slip gave a larger inlet distance than that with positive slip and side slip. Pure rolling and no side slip resulted in starvation and the inlet distance, m*, became zero for the lubricating greases.
The elliptic shape of the side wings became more circular when changing from naphthenic to PAO base oil. Introducing slip will increase the semi-minor axis ratio, i.e. the semi-minor axis of the inner ellipse divided by the semi-minor axis of the outer ellipse. Semi-major axis ratio was not affected by the investigated parameters. Inner ellipse area increased when changing the base oil type from naphthenic to PAO. Introducing slip also increased the ellipse area ratio. Increasing the viscosity or decreasing the temperature moved the inner ellipse centre towards the contact centre.

REFERENCES

(1) Pemberton, J., Cameron, A., "*A Mechanism of Fluid Replenishment in EHL Contacts*", Wear, Vol. 37, No. 1, pp. 185-190, 1976.
(2) Gohar, R., *Elastohydrodynamics*, pp. 290-294, ISBN 0-85312-820-0, 1988.
(3) Chiu, Y.P. "*An analysis and prediction of lubrication film starvation in rolling contact systems*", ASLE transactions, Vol. 17, No. 1, pp. 22-35, 1974.
(4) Patel, H., "*Distribution of oil and bearing surfaces*", M.Sc. Thesis, London, 1976.
(5) Åström, H., Östensen, J.O., Höglund, E. "*Lubricating grease replenishment in an elastohydrodynamic point contact*", Journal of Tribology, Vol. 115, No. 3, pp. 501-506, 1993.
(6) Larsson, P.O., Jacobson, B., Höglund, E., "*Oil Drops Leaving an EHD Contact*", WEAR, Vol. 179, pp. 23-28, (1994).
(7) Larsson, P.O., Jacobson, B., Höglund, E., "*Oil Drop Formation at the Outlet of an Elastohydrodynamic Lubricated Point Contact*", Journal of Tribology, Vol. 117, No. 1, pp. 74-79, 1995.
(8) Larsson, P.O., Jacobson, B., "*Grease Drop Formation at the Outlet of an EHD Contact*", International Tribology Conference in Yokohama, October 29- November 2, 1995.
(9) Wen, S.Z., Ying, T.N., "*A Theoretical and Experimental Study of EHL Lubricated With Grease*", Journal of Tribology, Vol. 110, pp. 38-43, 1988.
(10) Parkins, D. W., May-Miller, R., "*Cavitation in an Oscillatory Oil Squeeze Film*", Trans. ASME, Journal of Tribology, Vol. 106, pp. 360-365, 1984.
(11) Jacobson, B.O., Hamrock, B.J., "*High-Speed Motion Picture Camera Experiments of Cavitation in Dynamically Loaded Journal Bearings*", Journal of Tribology, Vol. 105, pp. 446-452, 1983.
(12) Myers, R. R., Miller, J. C., Zettlemoyer, A. C., "*The splitting of thin Liquid Films. Kinematics*", J. Colloid Science, Vol. 14, pp. 287-299, 1959.
(13) Coyle, D.J., Macoska, C.W., Scriven, L.E. ,"*Reverse Roll Coating of non-Newtonian Liquids*", Journal of Rheology, 34(5), pp. 615-636, 1990.
(14) Bousfield, D.W., Keunings, R., Marrucci, G. , Denn, M.M.," *Non-linear Analysis of the Surface Tension Driven Break-up of Viscoelastic Filaments.*", J.Non-New. Fluid Mech., Vol. 21, pp. 79-97, 1986.
(14) Leonov, A.I., "*The effect of surface tension on Stretching of Very Thin Highly Elastic Filaments*", Journal of Rheology, 34(2), pp. 155-167, 1990.
(15) Box, G.E.P., Hunter W.G., Hunter J.S. *Statistics for experimenters*, Wiley Interscience, pp. 306-351, pp. 374-433, pp. 453-497, 1978.
(16) Larsson, P.O., "*Grease Filament Formation with Circular Contact Geometry*", Submitted for publication, 1996.

NUMERICAL ANALYSIS FOR THE INFLUENCE OF WATER FILM ON ADHESION BETWEEN RAIL AND WHEEL

HUA CHEN and AKIYOSHI YOSHIMURA
Railway Technical Research Institute, 2-8-38 Hikari-cho, Kokubunji-shi, Tokyo 185, JAPAN
TADAO OHYAMA
Koyo Seiko Co., LTD, 11-15 Ginza 7-chome, Chuo-ku, Tokyo 104, JAPAN

ABSTRACT

Adhesion force is known to decrease with an increasing rolling speed when there exists water or snow between rail and wheel. Up to date, only Ohyama and Ohya[1] tried to theoretically clarify the behavior of adhesion between rail and wheel under water lubrication by applying EHL theory. They have taken Barus's pressure-viscosity coefficient in their analysis, which is usually applied for oil lubrication. However, as reported by Bett and Cappi[2], the viscosity of water has a peculiar dependence on temperature and pressure that is quite different from the viscosity of oil.

In this paper, by taking Bett and Cappi's viscosity values of water, we determined the influence of important factors such as velocity, pressure, temperature on the water film thickness for smooth surface and developed an equation relating these factors with the film thickness. We also calculated the case of rough surface using a modified Reynolds' equation based on average flow model[3] and compared it with smooth surface solution.

For numerical analysis, we used Newton-Raphson method under various parameters such as velocity 100 ~500 km/h, pressure 500~1000 MPa, and temperature at 2.2, 10, 20, 30, and 50°C. In the analysis taking the roughness effect into account, we assumed that both wheel and rail surfaces have the same roughness structure and a Gaussian distribution of surface irregularities. The contact pressure of asperities was calculated from the relationship developed by Greenwood and Tripp[4].

Fig.1 shows calculated results at 20°C for smooth surface. It is observed that the film thickness increases as velocity increases or pressure decreases. Although not presented, we could also find that the film thickness increases as temperature decreases.

The dimessionless film thickness of water H^* at the center has the following relationship with EHL parameters, which was derived from the solutions of numerical analysis by using linear regression method.

$$H^* = 2.349\, U^{0.617}\, W^{-0.274}\, G^{0.02}$$

In order to show the difference in numerical solutions between smooth and rough surface under the same condition, one of calculated results is displayed in Fig.2. Here, hydrodynamic roughness parameter Λ of 2.5 and isotropic roughness pattern ($\gamma = 1$) were used. It is shown that the film thickness of rough surface is larger than that of smooth surface.

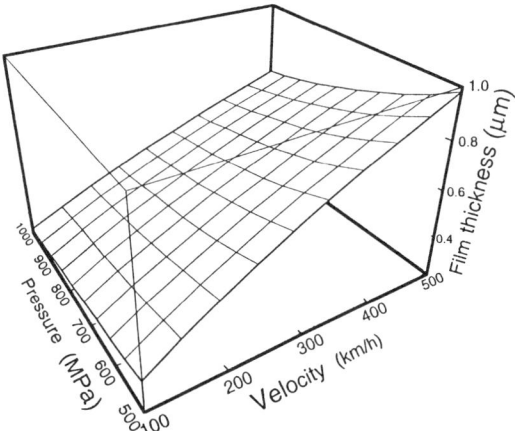

Fig.1: Influence of velocity and pressure on the film thickness at 20°C

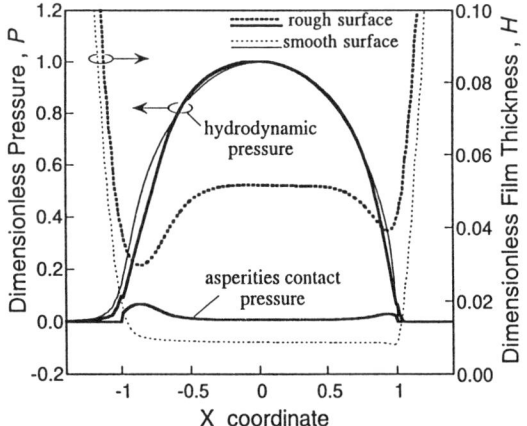

Fig.2: Hydrodynamic pressure and film thickness for smooth and rough surfaces

REFERENCES

(1) T.Ohyama, M.Ohya: J. Jpn. Soc. Mech. Engng., Part C 52-475, 1986, 1037-1046pp.
(2) S.Seki, R.Matuo(translators): Physical properties and Structure of Water, Misuzu Co., page 228, (1984).
(3) N.Patir, H.S.Cheng: Trans. ASME, Vol.101, 1979, 220-230pp.
(4) J.A.Greenwood, T.H.Tripp: Proc. I. Mech. Engng., Trib. Vo.185, 1971, 623-633pp.

A MICRO MODEL FOR MIXED ELASTOHYDRODYNAMIC LUBRICATION WITH CONSIDERATION OF ASPERITY CONTACT

D. Y. HUA and H. S. CHENG
Center For Surface Engineering And Tribology, Northwestern University, Evanston, IL 60208

ABSTRACT

For most of the machine elements with conformal contacts such as gears, rolling bearings and cams, the film thickness is usually in the same order of the surface roughness. In this low λ (film thickness to roughness ratio) region, the lubricant pressure is usually not sufficient to separate the asperities of the two mating surfaces. The total load is carried not only by lubricant but also by the contact asperities. In this mode, lubrication is interrupted by the asperity contact and lubricant is forced to go around the asperity contact area. Asperity contact can finally cause surface failure. Many efforts have been made to understand the effects of surface roughness on lubrication performance. As a deterministic method, micro-EHL has become an effective tool and made considerable advancements in rough contact lubrication (1-3). Nevertheless, the numerical methods proposed in the literature are limited to a certain degree of λ ratio, where solid-to-solid contact is not allowed. For the purpose of understanding and predicting tribological failure, a more accurate model for lubricated rough contact is needed to investigate the transition from full film to boundary film lubrication.

A general model for mixed elastohydrodynamic lubrication with consideration of interactions between lubricated contact and rough solid-to-solid contact is given in this paper. In the case of mixed lubrication, the lubricant film can be penetrated by asperity contact. The total load is carried partially by lubricant and partially by the contact asperities. Reynolds equation is only valid in the lubricated area, Ω_f, where there is a lubricant film to separate the mating surfaces. Chemical or physical boundary film may protect the surface in the asperity contact area, Ω_c. Nevertheless, in concerning about the contact pressure between the asperities, it is reasonable to neglect the boundary film effect by using the elastic/plastic dry contact theory.

A two dimensional time dependent Reynolds equation is considered for elliptical contact problem. Assumption of iso-thermal condition and Newtonian rheology is used. The ambient pressure at inlet, outlet and side boundary is used as boundary condition. At the boundary of Ω_c, no flow condition is automatically satisfied as the lubricant film is considered to be zero. It is appropriate to assume that the lubricant pressure equals to the micro contact pressure at the boundary of Ω_c as the pressure boundary condition. Both lubricant pressure and asperity contact pressure contribute to the elastic/plastic deformation of the surfaces. Asperity contact pressure at Ω_c is determined by the separation of two surfaces with elastic/plastic deformation theory. The total load is the summation of lubricant pressure at Ω_f and asperity contact pressure at Ω_c.

Numerical procedure for solving mixed elastohydrodynamic lubrication coupled with asperity to asperity contact needs an iteration routine to balance the pressure and film thickness between lubrication and asperity contact. During the iteration, asperity contact area Ω_c is determined when the film thickness is equal to zero. Reynolds equation, deformation of surfaces, viscosity-pressure relation of lubricant and density-pressure relation are solved in lubricated area Ω_f to obtain lubricant pressure and film thickness. Asperity contact pressure is determined by the remaining deformation (abstracting the deformation contributed by lubricant pressure from total deformation) needed to keep the asperity in contact. The load carried by the lubricant is updated once the pressure is convergent. The iteration between lubrication and asperity contact is carried out until the total load is balanced by lubricant pressure and asperity contact pressure.

Numerical results have shown the characteristics of lubrication performance when asperity contact presents. A typical result of film shape and pressure distribution under slide-roll mode with one spherical asperity instantly at the center of contact is shown in Fig. 1. Load ratio and contact area ratio are introduced to evaluate the tribological performance in mixed lubrication region.

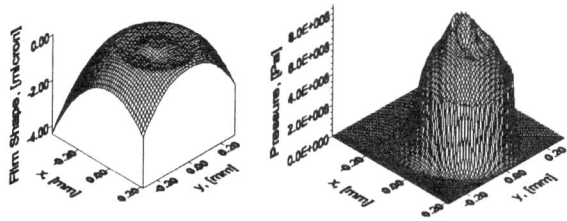

Fig. 1 Film Shape and Pressure Distribution

REFERENCES
(1) C.H. Venner, A.A. Lubrecht, Journal of Tribology, Trans. ASME, V.116, page 751, 1994.
(2) L. Chang, A. Jackson, M.N. Webster, STLE Tribology Trans., V.37, page 435, 1994.
(3) X. Ai, H.S. Cheng, STLE Tribology Transactions, V.37, page 323, 1994.

STATIC FRICTION OF CONTACTING REAL SURFACES IN THE PRESENCE OF SUB-BOUNDARY LUBRICATION

ANDREAS A. POLYCARPOU and IZHAK ETSION
Faculty of Mechanical Engineering, Technion-Israel Institute of Technology, Technion City, Haifa 32000, ISRAEL

ABSTRACT

A model for calculating the static friction coefficient of contacting real (rough) surfaces in the presence of very thin liquid films (sub-boundary lubrication) is developed. The liquid has a very high affinity for the surfaces and its thickness is of the order of the surface roughness average. An extension of the Greenwood and Williamson (GW) asperity model and an improved Derjaguin, Muller and Toporov (DMT) adhesion model are utilized for calculating the contact and adhesion forces, respectively. The effects of the liquid film thickness and the surface topography on the static friction coefficient are investigated. A critical film thickness is found above which the friction coefficient increases sharply. The critical thickness depends on the surface roughness and the external normal load. This phenomenon is more profound for very smooth surfaces and small normal loads, in agreement with published experimental work on magnetic hard disk interfaces.

INTRODUCTION

The static friction coefficient between two real (rough) surfaces in the presence of extremely thin liquid films has received considerable attention in the computer disk industry. Many experiments were performed, e.g., (1), to study the effects of both lubricant thickness and surface roughness on stiction. Also, a number of analytical models have been suggested, sharing the common assumption that stiction results from a significant increase in adhesion due to the meniscus forces of the adsorbed film at the interface.

This paper presents a static friction coefficient model, based on the sub-boundary lubrication mechanism suggested in (2). This mechanism is based on both the strong lubricant-solid bond and on the extremely thin lubricant thickness. The model offers an alternative mechanism to the meniscus model, that can cause high adhesion forces when isolated liquid bridges are unlikely to be formed.

The static friction coefficient after sufficiently long rest time is defined in the form

$$\mu = \frac{Q}{F} = \frac{Q}{P - F_s} \qquad [1]$$

where μ is the static friction coefficient; Q is the tangential force necessary to shear the junctions between the contacting asperities; F is the external normal force, which is equal to the actual contact load, P, minus the adhesion force, F_s, acting between the surfaces in contact. The contacting surfaces are modeled using an elastic-plastic model (3) which is based on the Greenwood and Williamson model (4). The details in calculating the individual forces appearing in Eq. [1] are given in the full version of the paper.

RESULTS AND DISCUSSION

Fig. 1 shows the results of the static friction coefficient versus dimensionless external load, F^*, for four plasticity indices ranging from $\psi = 0.25$ to $\psi = 2.5$ for very smooth to rough surfaces, respectively, and two dimensionless film thicknesses, $t_o^* = 1.0$ (—) and $t_o^* = 0.01$ (- -). Also, shown in the figure by the solid thick lines are the results for the unlubricated case.

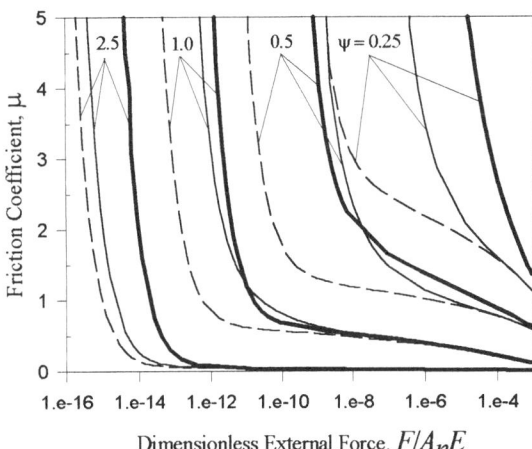

Fig. 1 μ versus F^*

For all cases considered, dry or lubricated, there exists a critical external force below which the static friction coefficient increases sharply. The critical force phenomenon can be understood from Eq. [1]. At high external loads, the adhesion force, is negligible compared to the contact load, and the ratio of Q/P is almost constant. As F decreases, F_s becomes more and more significant and approaches the value of P, causing the sharp increase in μ.

The physical explanation to the critical film thickness phenomenon is that as the lubricant layer thickness increases, it causes the adhesion force to increase accompanied by a decrease in the surfaces separation. As a result of this decreased separation more asperities are brought into contact and the tangential force required to shear them increases, resulting in the increase in the static friction coefficient.

REFERENCES

(1) C. Gao, B. Bhushan, *Wear*, Vol 190, pp 60-75, 1995.
(2) H. M. Stanley, I. Etsion, D. B. Bogy, *ASME Journal of Tribology*, Vol 112, pp 98-104, 1990.
(3) W. R. Chang, I. Etsion, D. B. Bogy, *ASME Journal of Tribology*, Vol 109, pp 257-263, 1987.
(4) J. A. Greenwood, J. P. B. Williamson, *Proc. of the R. Soc. of London*, Vol A295, pp 300-319, 1966.

COMPARISON OF THE STATIC FRICTION SUB-BOUNDARY LUBRICATION MODEL WITH EXPERIMENTAL MEASUREMENTS ON THIN FILM DISKS

ANDREAS A. POLYCARPOU and IZHAK ETSION
Faculty of Mechanical Engineering, Technion-Israel Institute of Technology, Technion City, Haifa 32000, ISRAEL

ABSTRACT

Results from a sub-boundary lubrication (SBL) model, for calculating the static friction coefficient of contacting real surfaces in the presence of very thin liquid films, are compared with published static friction measurements, performed on different magnetic storage hard disks. Four levels of surface roughness represented by the standard deviation of surface heights σ and by the GW plasticity index ψ are studied, ranging from $\sigma = 43.1$ nm ($\psi = 1.15$), for the roughest surfaces to $\sigma = 2.95$ nm ($\psi = 0.39$), for the smoothest surfaces. In all four cases good correlation is obtained between the model and the experimental results, suggesting that the SBL model is a reliable model that accounts for the main parameters that influence stiction in thin film disks. A critical film thickness of the lubricant, above which the static friction coefficient increases sharply (stiction), is predicted by the model in agreement with the experiments. A physical explanation of the stiction phenomenon, in relation to the SBL model is offered.

INTRODUCTION

In this paper, the SBL static friction model of (1) is compared with experimental results from three different references, (2) - (4) that were performed independently, and represent two types of tests: (a) Experiments performed on thin film disks with different surface roughness conditions, variable lubricant and adsorbed water vapor film thickness, and constant normal loads, (2), (3); and (b) experiments that were performed at a constant water film thickness and variable small normal loads (4). These experiments contain several uncertainties related to difficulties in accurate measurements of various properties, such as, micro-hardness of very thin coatings, surface roughness properties and surface tension of very thin lubricant films, especially in the presence of relative humidity. These are pointed out and accounted for in the model in the form of confidence limits.

RESULTS AND DISCUSSION

Fig. 1 presents the comparison of the SBL friction model with the experimental results from (2). The experimental values (solid circles) were obtained from Fig. 5 of (2). In this particular case, the energy of adhesion, $\Delta \gamma_1$, contains some uncertainties since the experiments were performed in the presence of both lubricant and humidity. The two dotted lines present the limit envelopes for $\Delta \gamma_1$ values of 0.05 N/m and 0.14 N/m, corresponding to either pure lubricant or pure water, respectively. The actual $\Delta \gamma_1$ values is somewhere in between these limits. When this is properly accounted for in the model, good correlation is

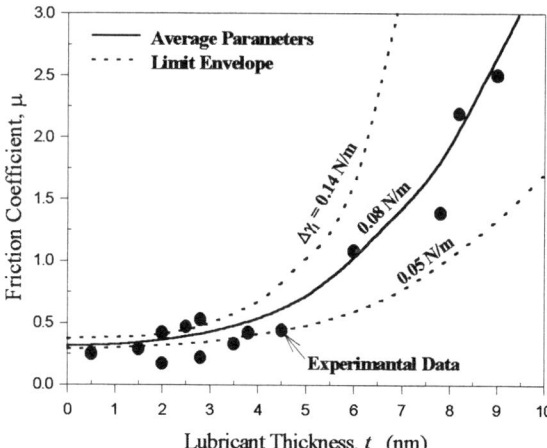

Fig. 1 Comparison of the SBL friction model with experimental data from (2).

obtained with the experimental results (see the solid line).

To an engineer, it may seem counter intuitive that the addition of a molecularly thin lubricant layer, causes the static friction to increase rather than to decrease. The explanation of this phenomenon is that by adding a lubricant layer at a solid interface, two events may happen. First, the surface free energy is reduced which presumably should decrease friction, but most importantly, the lubricant thickness bridges the gaps between initially non-contacting asperities of the mating surfaces, thereby increasing substantially the surface area over which adhesion forces are active. As a result, the adhesion force increases substantially, the separation between the surfaces decreases, more junctions are formed between an increasing number of contacting asperities, and hence, the friction force increases.

In summary, the sub-boundary lubrication static friction model, which is derived from first principles, contrary to semi-empirical models, can be considered a potential reliable tool for predicting friction coefficients in typical thin film disks. Its reliability depends, however, to a great extent on the level of accuracy in measuring the values of some key parameters.

REFERENCES

(1) A. A. Polycarpou, I. Etsion, *ASME Journal of Tribology*, in press, 1997.
(2) H. Tian, T. Matsudaira, *ASME Journal of Tribology*, Vol 115, pp 28-35, 1993.
(3) C. Gao, B. Bhushan, *Wear*, Vol 190, pp 60-75, 1995.
(4) I. Etsion, M. Amit, *ASME Journal of Tribology*, Vol 115, pp 406-410, 1993.

LOSSES IN HIGH SPEED ENGINE VALVE GEAR

R H SLEE

University College, Dublin, Eire, and T&N Technology Ltd., Rugby, CV22 7SA, UK

ABSTRACT

High speed 4-stroke gasoline engines are limited in engine speed by both allowable mean piston velocity (~ 21m/sec), and loss of valve trajectory control arising from the progressing inability of valve springs to resist exponentially rising inertia forces, and spring resonance effects (1). The high loads thereby demanded may invoke substantial friction throughout the valve control system.

Valve gear losses are rarely computed due to supposed unpredictable fluctuations in the coefficient of friction between cam and follower, and uncertainty in the level of reverse torque imparted by the closing flank of the cam. The spring force used in practice commonly exceeds the necessary theoretical values by 15 percent or more, thus friction loss procedures cited in the literature may involve significant errors (2).

For all types of valve gear, the peak corrected valve spring force is obtained by adjusting the normal spring force by a compensation C_k which varies depending upon valve gear stiffness, as set out in the following Table:

Cam-Finger-Valve	Cam-Tappet-Valve	Cam-Rocker-Valve	Hi Cam-Pushrod-Rocker-Valve	Cam-Tappet-Pushrod-Rocker-Valve
C_k = 1.15	=1.2	=1.3	=1.4	=1.5

Using a C.E.C. Cam and Tappet testing machine, work was undertaken by the Author to determine the level of cyclic energy recovery in representative spring controlled cam/tappet systems. A typical torque/angular displacement diagram is shown in Fig. 1, for a 300° period cam; the left plot in this polar diagram demonstrates a positive input torque (valve opening); the right sector shows a negative input torque (valve closing) as energy is recovered from the compressed valve spring. The difference in area of the respective plots is about 60 percent, repeatable within around 2 percent for most cams. (3)

A similar exercise undertaken for inverted tappet thimbles, and valve tips exposed to oblique rocker forces, shows that lateral forces play a significant role in losses, as components tilt under cam or rocker action.

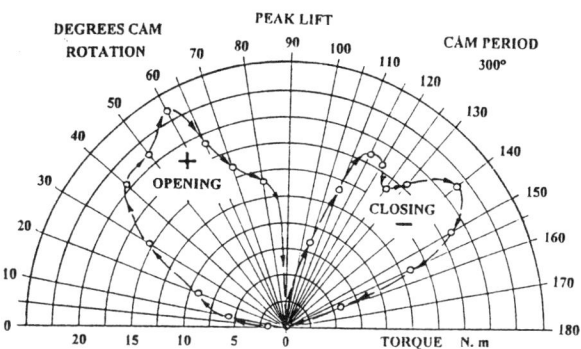

Fig. 1

Torque fluctuations during camshaft rotation

The paper presents a generic expression for power loss at each valve, assuming an average coefficient of friction between cam and tappet in sliding systems. With finger followers, account is taken of the transverse travel across the valve stem tip. Cam-rocker layouts require inclusion of rocker mass and geometry, and because of curved contacting surfaces often exhibit higher values of friction. Where a roller follower is employed, there is a demonstrable reduction in friction at low speed; as cam speed rises, the friction increases due to skidding, depending upon the relative radii of the roller, cam base circle, and the camshaft speed.

In a sample calculation against measured values on a 124cm^3 engine (4), computed losses at 10000 r/min were 0.32 kW; measured losses totalled 0.336 kW.

REFERENCES

1. SLEE R H, High Output Engine Design with Performance and Loss Approximations. PhD Thesis, UCD, (1996)
2. HEYWOOD, J B, Internal Combustion Engine Fundamentals, McGraw-Hill, N.Y. (1988)
3. VAN HELDEN, A K et al, Dynamic Friction in Cam/Tappet Lubrication, SAE 850441 (1985)
4. FROEDE, W, Mechanical Losses in Racing Motorcycle Engines, ATZ, Vol. 55 (1963).

THE DEVELOPMENT OF SCUFFING FAILURE IN AN AUTOMOTIVE VALVE TRAIN SYSTEM

J C BELL and P J WILLEMSE*

Shell Research Ltd., Shell Research and Technology Centre - Thornton, Chester, CH1 3SH, U K

ABSTRACT

The vast majority of experimental investigations of scuffing are of a short-term nature, focusing on failures that occur during the running-in process, in tests of systematically increasing severity. However, in many practically important engineering systems, such as automotive valve trains and hydraulic pumps, scuffing can occur after an extended period of steady wear, without an increase in the severity of the operating conditions. A common characteristic of failure in such systems is that, after successfully running-in the contacting surfaces, their roughness is observed to increase gradually with time. As steady wear progresses, the topography of the surfaces tends to develop a ridged appearance, with long asperities and valleys oriented in the sliding direction, which often exhibit a degree of correlation between the contacting surfaces, i.e. valleys on one surface matching peaks on the other (1).

Recent theories of scuffing suggest that local plastic deformations in asperity contacts lead to failure because of the resulting disruption of protective surface layers and possibly also catalytic degradation of the lubricant. Thus, one possible explanation for long-term scuffing is that the gradual increase in the roughness of the contacting surfaces eventually causes the severity of asperity interactions to generate sufficient plastic deformation for these processes to occur. To test this hypothesis, the development of cam follower surface topography and wear was monitored at regular intervals during a series of valve train wear tests in a fired engine. A representative set of results is shown in Fig. 1. Early

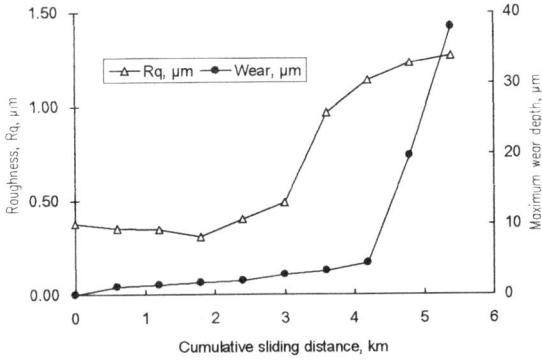

Fig. 1: Evolution of follower roughness and wear

in the wear process the roughness diminished with sliding distance and the wear depth increased at a steady, acceptable rate. After about 2 km of sliding the roughness began to increase to values greater than the original roughness, but with only a small increase in wear rate. Scuffing occurred between 4.2 and 4.8 km, causing a dramatic increase in the rate of wear.

The evolution of the severity of contact between the cam and follower surfaces was analysed using a statistical surface roughness model. After a significant amount of wear had occurred, the topographies of the cam and follower surfaces could be described to a good approximation, within the limits of the macroscopic Hertzian contact dimensions, as an array of cylindrical asperities with their axes oriented in the direction of sliding. The model for the steady state sliding of rough surfaces developed by Kapoor, Williams and Johnson (2) was adapted for the contact conditions described above. The results are shown in Fig. 2, in which the

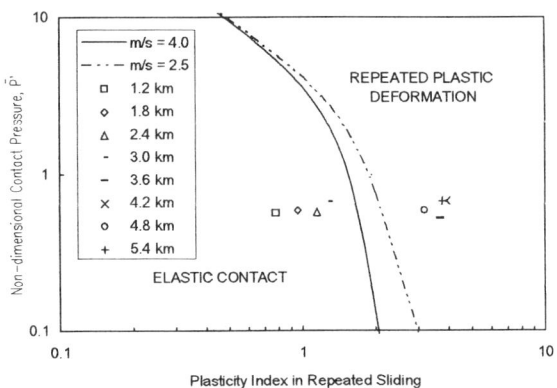

Fig. 2: Shakedown map

solid and broken lines indicate the elastic shakedown limits for two values of the asperity height cut-off ratio "m/s". Contacts progressively approached the elastic shakedown limit until a sliding distance of 3 km, beyond which a large step into the repeated plastic deformation regime was observed, most likely because of the onset of ratchetting wear (3). A similar development of contact severity preceded each observation of scuffing.

REFERENCES

(1) G. W. Roper and J. C. Bell, SAE Paper 952473.
(2) A. Kapoor, J. A. Williams and K. L. Johnson, *Wear*, Vol. 175, pp. 81-92, 1994.
(3) A. Kapoor, K. L. Johnson and J. A. Williams, *Wear*, Vol. 200, pp. 38-44, 1996.

*Present address: Koyo Seiko Company, Almere, The Netherlands.

THE INFLUENCE OF CRANKSHAFT GEOMETRY ON THE SECURITY OF PLAIN BEARING SYSTEMS

C S CROOKS, D D PARKER, I T GRAHAM
Glacier Vandervell Ltd, Kilmarnock, Ayrshire, Scotland, KA1 3NA, UK

ABSTRACT

The security of an engine crankshaft bearing system depends on many factors. Selection of the appropriate bearing material might appear the most important consideration. However the increasing demands on the bearing system place greater emphasis on all the components, e.g. the oil, the bearing housing and journal. In addition to the development of new materials, understanding of the interaction of the system components is fundamental to optimising performance and reliability.

Development of tests simulating the crankshaft bearing operating conditions has permitted investigation of some of these interactions. Arising directly from field experience a study of the effect of journal surface finish revealed smoother surfaces to give profound improvements in system compatibility[1].

Investigations are now covering second order texture effects, such as circumferential journal lobing. Deviations in roundness profile are often encountered on automotive crankshafts. Depending on amplitude and frequency the result may range from severe bearing wear to fatigue cracking and seizure. Normally the deviations are irregular and composed of small and large amplitude features of various lengths or frequencies. Some however are quite regular and single lobe features have also been seen. (Fig. 1)

DeHart and Smiley[2] found that for steadily loaded bearings regular lobing patterns had a strong adverse effect on wear. Experience from gasoline and high speed diesel engines indicates that severe bearing damage can result. This includes wear, scuffing, overheating, and fatigue. The single lobed shaft (Fig. 1) resulted in extensive fatigue of a con rod bearing in a passenger car after 1000 miles. A multi-lobed shaft in a diesel engine generated rapid wear, fatigue cracking and seizure in less than 100 miles.

The authors have devised a method for producing lobed shafts for the dynamically loaded Sapphire[1] bearing fatigue test machine. This involved synchronised deflection of the shaft during grinding using a piezo electric transducer. The desired shape and height of the lobes was obtained by controlling the transducer excitation voltage and ramp rate.

This paper is concerned with the fatigue life of plain bearings operated against regular lobed shafts. Nodular cast iron test shafts were produced with a range of lobe frequencies from 0-25 and amplitudes (peak to valley height) of up to 6 microns. Test bearings 54mm internal diameter were made from an aluminium tin silicon alloy. The oil inlet temperature was 140C and the peak specific load 92 MPa. Testing was continuous until the onset of fatigue was detected by a temperature rise measured on the bearing back.

The results (Fig. 2) show a marked reduction in fatigue life when the amplitude and frequency are high. However maximum fatigue life is not apparent where lobing is virtually zero. Contrary to the findings of DeHart and Smiley the data appears to suggest the existence of a lobe condition that may prolong bearing life. It is possible with 3 lobes that the effect is sensitive to the position of lobe peak relative to peak load. A solution is conceived where a flatter portion of the journal is carrying the load and the extent of the load carrying region is increased[3].

Work is continuing to investigate the effects at lower peak loads and hence fatigue lives more representative of crankshaft applications. Whilst fatigue is a most important characteristic it is also hoped that other properties such as wear and conformability can be evaluated.

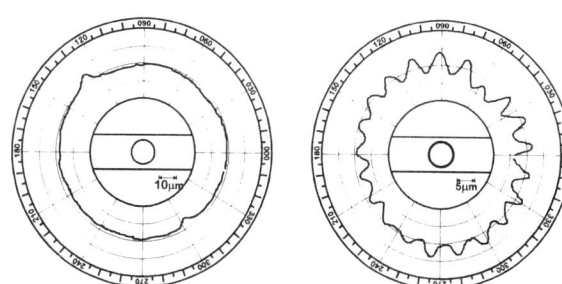

Fig 1. Crankshaft Journal Profiles

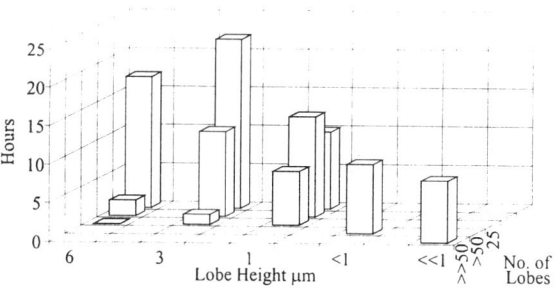

Fig 2. Fatigue Life Results

REFERENCES

(1) C S Crooks, D D Parker, SAE Paper 960983, 1996.
(2) A O DeHart, J O Smiley, SAE Paper 813C, 1964.
(3) D S Mehenny, C M Taylor, G J Jones, H Xu, SAE Paper 970216, 1997.

A NOVEL AUTOMOTIVE BEARING DESIGNED FOR INCREASED DURABILITY AND DYNAMIC STABILITY

M.J. BRAUN, and F.K. CHOY
Dept. of Mechanical Eng., University of Akron, Akron, Ohio, USA
F. DIMOFTE
NASA Lewis Research, Cleveland, Ohio, USA
J. SORAB
Ford Research Laboratory, Ford Motor Company, Dearborn, Michigan, USA

ABSTRACT

Designing bearings for increased durability in adverse environments has been a goal of industries spanning the spectrum from aerospace to automotive. For the automotive industry, this application concerns the design of bearings that can function with low viscosity engine oils and still maintain load carrying capacity comparable to that of conventional bearings, while exhibiting improved dynamic characteristics that will ensure enhanced durability, reduced maintenance and superior performance. The authors have designed and numerically tested a novel bearing geometry that imprints a wave pattern on the stationary surface of the bearing casing.

GEOMETRY

The three lobe wave journal bearing presented here is different from the classical journal bearing in that that it has a slight, but precise variation in geometry, that is a wave form superimposed on a circular profile with a mean diameter that crosses the wave at half of its total amplitude, Figure 1. The wave amplitude is equal to a fraction of the bearing clearance. The number of waves and the wave amplitude configuration to be chosen for a specific application can be computationally optimized

METHOD OF SOLUTION

The steady-state and dynamic bearing performance will be predicted based on a numerical code which uses a perturbation solution for the complex form of the Reynolds equation[1].

$$\frac{\partial}{\partial \theta}(\frac{h^3}{\nu}\frac{\partial p}{\partial \theta}) + \frac{\partial}{\partial z}(\frac{h^3}{\nu}\frac{\partial p}{\partial z}) = 2\Lambda\frac{\partial(h)}{\partial \theta} + i(4f\Lambda)\frac{\partial(h)}{\partial \theta}$$

where $\Lambda = \frac{6\mu\Omega}{p_a}(\frac{R}{C})^2$.

For the solution one has to define a starting geometry of the fluid film. When the load W is applied, the shaft must find an equilibrium position at an eccentricity, e, such that the load capacity of the bearing, F, balances the applied load, W. The pressure generated in the fluid can be calculated by integrating the Reynolds equation. Under dynamic conditions the shaft center rotates in an

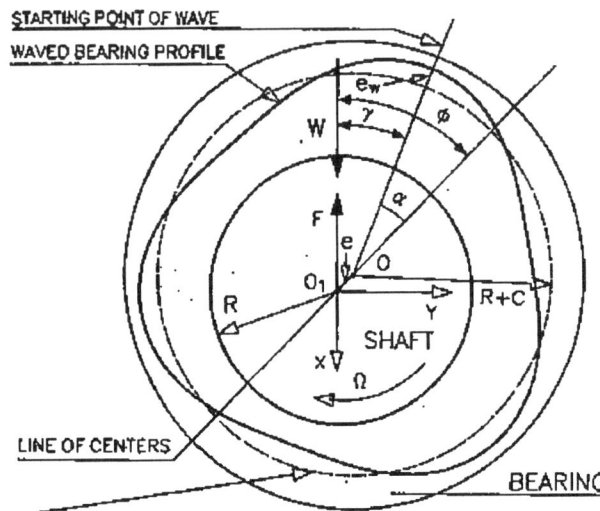

Figure 1. Bearing Geometry

orbit around its static equilibrium position. The resulting bearing dynamic forces are nonlinear functions of the whirl amplitude, especially if the amplitude is relatively large. The bearing's dynamic force components, Fx, and Fy, are functions of the bearing dynamic coefficients.

RESULTS

In numerical experiments, this bearing has demonstrated superior stiffness and damping characteristics relative to plain bearings, even with fluid viscosities as low as 2 mpa.s. From the parametric runs we have identified that the optimum combinations for the wave bearings geometries studied here, when compared to the true circular bearing, are the ones with a wave amplitude ratio WAR=0.2, and a wave position angle $\gamma = 45$, and WAR=0.4, $\gamma = 20$ respectively.

REFERENCES

1. Dimofte, F., "A Waved Journal Bearing Concept with Improved Stead-State and Dynamic Performance", NASA CP 1036, pp. 419-429, 1993

THE MECHANISMS OF DIESEL AND GASOLINE VALVE TRAIN WEAR AND THEIR CONTROL BY THE LUBRICANT

J A MC GEEHAN and P R RYASON
Chevron Products Company, Global Lubricants, Richmond, California 94802, U.S.A.
E S YAMAGUCHI
Oronite Global Technology, Chevron Chemical Company, Richmond, California 94802, U.S.A.

ABSTRACT

This paper describes our research on the mechanism of valve train wear occurring in automotive diesel and gasoline engines, and how this wear can be prevented by the proper engine oil formulations.

In gasoline engines, the mechanisms have been defined in the Ford 2.3-liter overhead cam engine and the GM 3.8-liter V-8. It was found that the mechanisms of wear included: scuffing, fatigue, abrasion, and corrosion. We were able to identify engine blow-by as a primary factor affecting camshaft wear. Nitric acid is the culprit in the blow-by and is derived from nitrogen oxides reacting with water. Even in the presence of blow-by, camshaft wear can be controlled by the proper selection of zinc dithiophosphate (ZnDTP) and detergent. Electron Spectroscopy of Chemical Analysis (ESCA) indicated that the sulfur, zinc, and phosphorus from the ZnDTP are adsorbed on to the surface. Since surface chemistry depends in part on surface temperatures, we measured the camshaft surface temperatures in fired engines. We found that the maximum temperature in the Ford at Sequence V-D conditions and in the GM at Sequence IIID conditions were similar (200°C), explaining why both engines required similar ZnDTPs for wear control. (1)(2)(3)

In diesel engines, the wear mechanisms vary according to the valve train system, but in all cases a protective film from the ZnDTP must be formed and the soot adequately dispersed to prevent its removal.

In the GM 6.2-liter roller follower system and Cummins M-11 slider follower system, soot polishing wear followed by contact fatigue or adhesive and abrasive wear can be eliminated by the proper use of secondary ZnDTP in formulations with adequate soot dispersancy. ESCA indicated that high surface concentrations of phosphorus, zinc, and sulfur are adsorbed on low wear slider follower surfaces, when the soot is well dispersed and the ZnDTP is at sufficient levels.

In regard to roller follower systems, which are predominantly used in U.S. emission-controlled engines, it is important to protect the bronze pin in the steel roller from corrosion by the proper selection of ZnDTP and detergent. Under marginal lubrication conditions that occur when the engine is frequently stopped and started, use of a steel pin as the axle of the silicon nitride roller minimizes roller startup friction and results in the formation of a protective film from the ZnDTP on the steel pin. In addition, cam and roller surface finishes, hardness, and geometry are critical to preventing roller follower roller sliding and contact fatigue of the cams.

These findings underscore the importance of using an effective additive package in the presence of factors, such as engine blow-by and soot, which are antagonistic toward ZnDTP antiwear film formation and retention.

This work is supported by EHD film measurements, X-ray diffraction determination of the residual cam stress and retained austenite, Scanning Electron Microscope analysis, and Electron Spectroscopy for Chemical Analysis of cam and rollers surfaces.

REFERENCES

(1)   J A Mc Geehan, E S Yamaguchi, and J Q Adams, SAE Paper 852133, 1985.
(2)   J A Mc Geehan and E S Yamaguchi, SAE Paper 892112, 1989.
(3)   J A Mc Geehan, J P Graham, and E S Yamaguchi, SAE Paper 902162, 1990.

ENGINE FRICTION : THE INFLUENCE OF LUBRICANT RHEOLOGY

R.I. TAYLOR

Shell Research Limited, Shell Research & Technology Centre, Thornton, P.O. BOX 1, CHESTER, CH1 3SH, UK

ABSTRACT

The sensitivity of engine friction to lubricant rheology has been determined for a modern fuel efficient engine, the Mercedes Benz M111 2.0 litre gasoline engine, for both cold-starting and warmed up engine conditions. The variation of lubricant viscosity with both temperature and shear rate was taken into account. Results are reported for the variation of engine friction for different monograde lubricants, and the distribution of frictional losses between the piston assembly, valve train and bearings has been estimated.

INTRODUCTION

The automotive industry is striving to produce engines with improved fuel economy, to conserve natural resources and to limit CO_2 emissions. One way to alter an engine's fuel consumption is to lubricate it with an oil of differing rheology, and in recent years lower viscosity oils containing friction modifiers have become more commonplace. An engine's fuel consumption is related to friction in the three major engine components, (i) the piston assembly, (ii) the valve train, and (iii) the bearings. This work has shown that it is the valve train which is the critical component since friction losses there increase with decreasing lubricant viscosity, whereas in the other components, the friction loss decreases with decreasing viscosity.

Validated engine friction models for the piston assembly[1], the valve train[2], and the bearings[3], that take account of lubricant viscosity variations with both temperature and shear rate have been used to study engine friction for the Mercedes Benz engine under fully warmed up and cold starting engine conditions.

FULLY WARMED UP ENGINE SIMULATIONS

Simulations were carried out assuming an engine speed of 2500 rpm, and medium load conditions (the peak combustion chamber pressure was 32 bars). Table One shows results for the predicted total engine friction power loss, and the distribution between the various engine components, for monograde lubricants for a boundary friction coefficient, f, of 0.12. Table Two shows a second simulation using $f=0.08$ (more typical of oils containing friction modifier additives).

	Total loss (W)	% P	% VT	% B
SAE-10W	1643.6	29	38	33
SAE-30	1618.3	39	26	35
SAE-50	1631.3	52	8	40

Table One : Engine friction simulations with $f=0.12$

	Total loss (W)	% P	% VT	% B
SAE-10W	1420.7	33	29	38
SAE-30	1472.7	42	19	39
SAE-50	1584.4	54	6	40

Table Two : Engine friction simulations with $f=0.08$

The above work showed that virtually all the boundary friction occurs in the valve train, and that the total friction loss does not always monotonically decrease with decreasing viscosity.

COLD-START ENGINE SIMULATIONS

Further engine simulations were carried out assuming cold-start conditions. The individual engine components were assumed to warm up with a representative time constant t_0. Table Three shows results of the simulations for different times, t, after engine start up for an SAE-10W monograde oil, assuming $f=0.08$, together with the percentage distribution between engine components.

t/t_0	Total loss (W)	% P	% VT	% B
0.0	6659.8	37	0.1	63
1.0	1499.3	47	13	40
3.0	1407.8	34	28	38
5.0	1412.1	33	29	38
∞	1420.7	33	29	38

Table Three : Cold start engine friction simulations

The above results showed that friction modifier effects would be less important under cold start conditions.

CONCLUSIONS

Realistic engine friction models have been used to predict engine friction losses for monograde oils under fully warmed up and cold starting engine conditions for a Mercedes Benz M111 gasoline engine. The distribution of losses amongst the engine components was also predicted. The work enabled conclusions to be reached on the likely impact of friction modifiers to reduce friction in this engine. In the full paper[4], multigrade oils have been considered too.

REFERENCES

(1) R.I. Taylor, T. Kitahara, T. Saito & R.C. Coy, Proc. Intl. Trib. Conf., Yokohama, pp 1423-1428, 1995
(2) T. Colgan & J.C. Bell, SAE892145
(3) A.R. Davies & X.K. Li, J. Non-Newtonian Fluid Mech., **54**, pp 331-350, 1994
(4) R.I. Taylor, submitted to I.Mech.E. J. Eng. Trib.

WEAR IN VEHICLE ENGINES - AN OVERVIEW

L. ROZEANU
Dept. of Materials Entineering, Techion - Israel Institute of Technology, Haifa 32000, Israel
F.E. KENNEDY
Thayer School of Engineering, Dartmouth College, Hanover, NH 03755 USA

ABSTRACT

A vehicle engine can be treated as an assembly of interacting friction systems. Although there is a huge amount of literature regarding wear of individual vehicle systems, some important bits of information somehow do not reach the designer, producer and user, owing to poor communication links. This lack of application of tribological principles harms the user and damages the prestige of the field of tribology as well. The aim of the present paper is to provide an overview of the subject, focusing on some specific topics.

The path of wear of a vehicle engine contains four time-interdependent segments: design, production, nominal service, and decay. The 'running-in' step is not independent; it really belongs to the 'production' section, discreetly transferred to the customer. Nevertheless, it will be included in the discussion.

At the design level, there are four important items which are insufficiently understood: tolerances, surface roughness, materials selection, and lubricants. Some good specifications at the design stage end up being futile because the designer has neglected to consider the changes which occur during production and service. Regarding tolerances, it must be remembered that parts of a friction couple which are made of different materials expand differently when heated, substantially changing the dimensional fit of the couple. Surface roughness is another quantity that changes during service. In addition to the initial surface roughness resulting from the production process, each friction surface develops its own roughness; this change in roughness occurs faster if sliding conditions are more difficult or the surface was rougher to begin with.

Regarding materials selection, there are conflicting requirements. For example, engine bearings are often made of nonferrous materials. These materials must have good metallurgical compatibility with the shaft material and good embedding properties in the sector where the minimum film thickness is found. In that sector, the strength of the softest material must satisfy the *critical load* requirement; the mating component should have sufficient wear resistance, and the part undergoing symmetric wear should be made out of the harder material.

Although an engine incorporates many individual friction systems, the selection of an engine lubricant is usually dictated by one set of components, the hydrodynamically lubricated bearings. Multigrade lubricants with an additive package are usually specified, with the viscosity of the lubricant being selected based on initial clearance of the bearings and expected operating temperatures. Yet the bearing clearance will change gradually as bearing wear occurs, and the viscosity will vary locally owing to different temperatures and rates of shear. Even more importantly, the viscosity of the multigrade lubricants cannot usually be determined accurately by the most commonly used viscosity-temperature relationships, particularly as the oil degrades during operation.

The production process influences the wear of engine components in two ways: unfavorable combination of parts with different tolerance deviations, creating either 'too loose' or 'too tight' fits, and lax assembly procedures.

The running-in period for engine components should be mild and rapid. For the first 500-1000 km of driving, a chemically active lubricant should be used, containing no graphite or other solid lubricant. Then the filter should be changed, the oil should be thoroughly drained, and a regular engine lubricant of appropriate viscosity should be used.

To limit wear of engine components during operation, the engine should be run with few stops, mild acceleration, hot but stable engine temperature, and assisted lubrication. It should be noted that a single oil grade cannot be used for the entire engine life, since the film thickness is so dependent on lubricant viscosity, which is drastically affected by temperature.

The variation of wear with time is usually shown as being initially rapid, then almost horizontal during service life, and growing again toward the end. The oversimplified equations usually suggested for the calculation of wear are insufficiently defined. They do not contain essential terms which control the strength of materials, such as temperature and velocity. Actually, the wear vs. time seems to be a variant of a family of natural decay curves, such as Figure 1. This model treats the engine as a naturally perishable system, characterized by the same evolutionary path as food cans, electric batteries, life. This analogy broadens the meaning of wear of complex friction systems, which is different from that of a single friction couple.

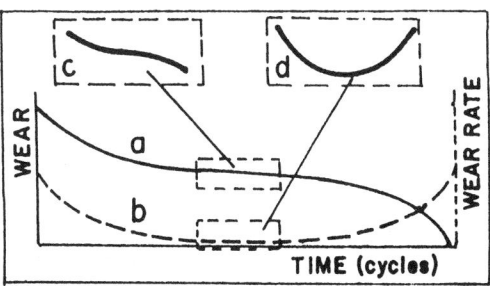

Figure 1: A model of engine wear as a decay function. a) wear (decay) path. b) rate of wear. c) and d) inflection points.

BENCH TESTS USED TO STUDY SOOT RELATED WEAR IN THE PISTON RING - CYLINDER LINER SYSTEM

MIKAEL ÖDFALK
Daros Piston Rings, Box 604, S-433 51 Partille, Sweden
CECILIA TÄRNFJORD
Scania, Structural Strength and Materials Technology, S-151 87 Södertälje, Sweden
STURE HOGMARK and STAFFAN JACOBSSON
Ångström Laboratory, Department of Materials Science, Box 534, S-751 21 Uppsala, Sweden

ABSTRACT

High levels of soot in diesel engine oil often lead to increased wear of components operating under boundary lubrication conditions. Such conditions exist, for example, in the valve train and in the piston ring - cylinder liner system. The wear mechanisms for soot related wear are, however, not fully understood. In an attempt to throw more light on this situation a series of bench tests have been carried out. The objective with the tests was to sort out the main factors influencing the wear rate when the oil is soot contaminated. Accordingly, the influence of normal load, temperature and soot content has been investigated.

Test specimens made from actual engine components (piston ring and cylinder liner) were used. Hence the contact geometry was similar to the real case. The tests were performed with a reciprocating pin-on-plate rig. A mineral base oil with dispergent, detergent and ZDTP additives was used. The soot content of the oil was varied by the addition of different amounts of carbon black.

It was seen that soot had an effect on the wear rate of both piston rings and cylinder liners under running in as well as under steady state conditions. The wear increased quite dramatically even for low soot concentrations compared to clean oil, see figure 1. The steady state wear rate is typically increased by an order of magnitude or more when the oil is soot contaminated.

A statistically designed test series was run with the following parameter combinations.

Parameter	Low	High
Soot concentration (%)	0,18	1,53
Temperature (°C)	80	140
Normal load (N)	70	280

As can be seen from figure 2, the interaction of soot content and temperature has a strong effect on the wear rate. The main explanation for this is that the combination of lower temperature and high soot concentration lead to a viscosity increase in the oil thus changing the lubrication regime from boundary to mixed or full film lubrication. Similarly, low normal load, combined with high soot concentration will promote hydrodynamic film formation.

SEM investigations of worn surfaces on both rings and liners revealed that they were remarkably smooth and free from signs of plastic deformation. Thin reaction films containing mainly S, P, and Zn were found on some specimens but most specimens were free from reaction films. A mild mechanical or chemo-mechanical polishing seems to be the governing wear mechanism for soot related wear.

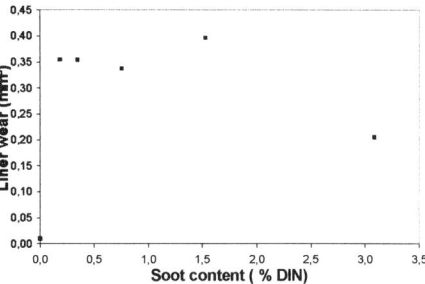

a)

b)

Figure 1, Wear versus soot content, (a) cylinder liner, (b) piston ring. Normal load 280 N, temperature 80°C, test duration 20 hours.

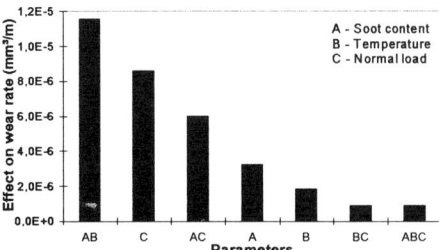

Figure 2, Pareto diagram over the influence of soot content, temperature and normal load on the steady-state wear rate of the cylinder liner specimen

IDENTIFICATION OF TRIBO-MUTATION EFFECTS IN ROAD VEHICLE BRAKES

ČEDOMIR DUBOKA, ŽIVAN ARSENIĆ and JOVAN TODOROVIĆ
University of Belgrade, Faculty of Mechanical Engineering, YU-11000 Belgrade, 27. Marta 80., Yugoslavia
EGON - CHRISTIAN VON GLASNER
Mercedes-Benz A.G., Commercial Vehicles Division, Head of Advanced Engineering, Stuttgart, Germany

ABSTRACT

Friction and wear in tribo-systems, and specially in automotive brakes are subject to significant variations with respect to the main influencing factors - slidding speed, friction surface specific pressure and interface temperature. Different empirical and theoretical models relate friction and/or wear to the dominant factors, but comparison of predicted behaviour to that performed under service conditions is possible only experimentally. Numerous tests were realized at the University of Belgrade, Faculty of Mechanical Enginee-ring, enabling identification of effects of the main influencing factors on friction and wear. Many of them show that there also must be some "residual" or side effects, caused by inter-relations of dominating factors between each-other, generating additional mutations of friction and wear, and these are "tribo-mutations"[1, 2]. In this paper, an investigation of the brakes for 16t truck, 3.5t off-road vehicle, and 1.35t car is presented. Test results are also processed in such a way to "eliminate" partial influence on friction and wear of one dominating factor, or simultaneous effect of two of them. However, such an elimination was not effective, and there still remain some residual effects. Friction may be acquired "in real time", and variations of that kind are obvious, but wear follows a "delayed" process, where such variations are not so obvious. However, based on these experiments, similar conclusions may be derived, providing evidence for experimental identifi-cation of tribo-mutations effects in road vehicle brakes.

THE EXPERIMENT

Friction and wear in automotive brakes depend on the design, physical, chemical and tribological properties of applied friction material and metal counterpart element, but also on operation conditions or service loads - sliding speed, interface temperature and pressure. 3D "matrix-type" inertia dynamometer tests were used to study tribo-mutations in automotive brakes. Single full stop brake applications for the given combination of initial speed, pressure and temperature were followed by consecutive step-by-step variation of initial load conditions. Results show large variation of friction and wear. Further data processing for "elimination" from friction torque values, for example, of partial influence of one or simultaneous influence of two of dominating factors, for example, was undertaken. Fig. 1 shows new results for "relative braking torque", i.e. torque values devided by corresponding values for two (of three) dominating factors, representing these factors only implicitly. The "convergence" of test results may be seen, withun zones that may be easily identified. There is a significant difference both in shape and in scale of the specific braking torque values, that may be interpreted both by stochastic and systematic influence. Further analysis are needed, but we may suggest that variations of braking torque with respect to both, the shape and the scale, are not proportional to the applied pressure, or initial brake speed, or initial interface temperature. Similar may be shown for wear test results, and therefore we assume that both in the case of friction and wear there must be side effects, imputed to tribo-mutations. The detailed elaboration of this shall be presented orally and in the final manuscript.

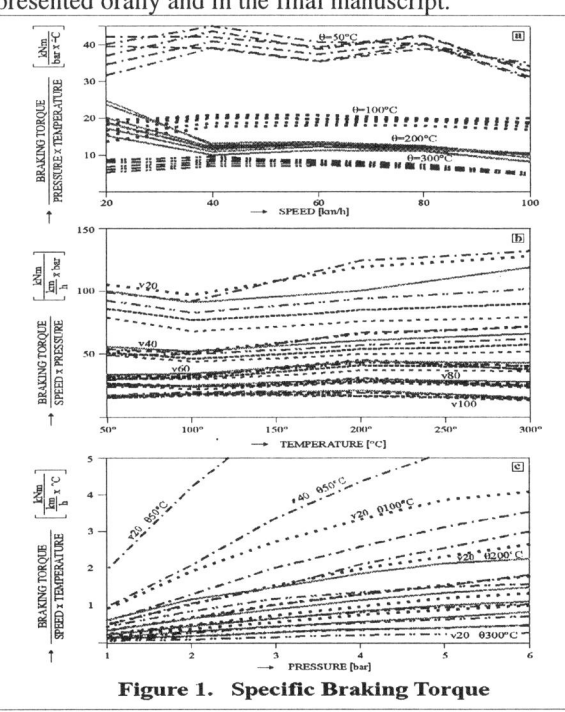

Figure 1. Specific Braking Torque

REFERENCES

[1] Milosavljević M., Mariotti G. V., Duboka Č.: Tribo-mutation effects on wear in friction mechanisms, Proc. "Science and motor vehicles 95", Paper YU-95354, pp. 188-191, Yugoslav Society of Automotive Engineers, Belgrade, YU, 1995

[2] Duboka Č., Arsenić Ž., Milosavljević M.: Tribo-mutations in tribo-mechanical systems, Proc. II International Conference on Tribology BALKAN-TRIB 96, pp. 703-710, Thessaloniki, GR, 1996.

Submitted to *Tribology International*

THERMAL ASPECTS OF DISC BRAKES FOR HIGH SPEED TRAINS

M G EL-SHERBINY and M ABD-RABOU
Mechanical design and production department, Faculty of Engineering, Cairo University, 12361 Giza, EGYPT.
O KRETTEK
Analytishe Verfahren in Fahrzeug -und Transporttecknik, RWTH-Aachen, Germany.

ABSTRACT

With the 512 Km/h speed world record by TGV-Atlantique of the SNCF on May 1990, the braking system of the high speed trains have to satisfy the new demands of the high rates of energy dissipation. An analytical analysis model have been developed to study the behaviour of both the ventilated and solid disc brakes during the different braking modes. The temperature history of different points through the disk thickness are obtained and plotted. Also, stress history of the same points is calculated during stop braking.

INTRODUCTION

The analytical developed code simulate the braking process of the High Speed Train Disc Brake (HSTDB) conditions as semi-finite plate with definite thickness. The developed code HSTDB calculates the temperature history of different points through the disk thickness. A plot routine is developed to demonstrate automatically the temperature and tangential stress of the previously chosen points. Also the temperature history of the contact surface is simulated through both drag and multiple barking.

ANALYTICAL ANALYSIS

The temperature distribution along the brake disc can be obtained by a complete formulation and solution employing the mathematical techniques. A great number of investigators had contributed in the thermal analysis of disc brakes. Several investigations were directed towards the determination of the expected temperature rise during single stop braking (1), or during continuous braking (2), or during multiple repeated braking (3). These analysis indicates that at single brake stopping, the area of the friction surface should be as maximum as possible to minimize the resulting rise in temperature. However, for continuous drag braking and multiple braking, convective capacity is essential.

STOP BRAKING

Assuming that the contact area forms a continuous band on the rubbing path and the heat is generated uniformly over the band source providing that the retardation is constant, the errors introduced by approximating the cylindrical portion of the disc into a flat rectangular plate and the heat flow to be one dimensional are small enough to be neglected. Using the solution of the general heat conduction equation for transient temperatures in finite hollow cylinders gives results in a good agreement with the finite element results.

DRAG BRAKING

It is generally accepted during braking to consider the rate of heat generation to be constant and the temperature through the disc is constant. Both the heat transfer coefficient and the disc convection must have a great consideration in the design process to obtain minimum surface temperature. Increasing convection area is achieved in the recent designs by using ventilated discs. Ventilated disc brakes at High Speed Trains have also shown a drawback attributed to the loss of about 3 kW/disc with a traveling speed of 270 km/h (4).The using of solid disc brake represents a good replacement of the ventilated disc brakes as recommended in the design of TGV to avoid the power loss.

REPEATED STOP BRAKING

Repeated stop braking may lead to a high peak temperature and therefore it must be studied, specially for the case of suburban trains. The variation of the train speed versus time under the conditions of constant maximum running speed, constant retardation and acceleration is taken into consideration. Also the time interval between any successive braking is constant. The cooling during resting may be neglected. This results is overestimating the temperature and hence provide a more conservative result.

REFERNCES

(1) M Fermer, R Lunden"Transient Brake Temperatures found by use of analytical solutions for finite hollow cylinders" J. of Rail and Rapid Transit, ImechE, Vol 205, Part F, 1992.

(2) Y Takuti, , N Noda "Thermal stress problems in industry 2:Transient thermal stresses in a disc brake" Journal of thermal stresses, 1979.

(3) E Saumweber "Temperaturberechnung in Bremsscheiben fuer ein belibebiges Fahrprogram" Leichtbau der verkehrsfahrzeuge, H. 3, 1979.

(4) D Russell, A Williams " The design and development of a brake disc for high speed trains" J. of rail and rapid transit, IMechE, Vol 204, Part F, 1990

FRICTIONAL CONDITIONS IN ROLLER FREE-WHEELS

K DÜRKOPP and W JORDEN

Laboratorium für Konstruktionslehre (LKL), University of Paderborn, Pohlweg 47-49, D-33098 Paderborn, Germany

ABSTRACT

Roller clutches are one of the most important types of frictionally engaged free-wheels. These self-instructing machine parts index depending on their direction of rotation. The functioning of the grip roller clutch depends on the specific tribological conditions at the clamping zones, which are different from those of other machine elements.

The most influential parameter in the design of the grip zone in a roller free-wheel is the clamping angle. Different tribological conditions (lubrication, material, stress level, motional conditions, etc.) determine wear and working life of the clutch, the frictional forces in the clamping zones and thus the best value for the clamping angle. The designer of frictional free-wheels needs to know the right value of the coefficient of friction for optimising the behaviour of the clutch with regard to transmittable torque and durability during different using conditions.

The specific tribological system at the critical places of contact in a roller clutch cannot simply be compared with other machine elements (1). Thus, models of calculation for fatigue life have not yet been verified enough and are not of sufficient scope (2). A complete description of the specific relations in the whole system of the clamping-roller free-wheel is neccessary but has not been determined yet.

Former research work was mainly concerned with wear, now frictional conditions are recieving more attention. To get more insight in this special tribological system a single clamping gap has been considered (fig. 1). A specific test stand was developed to measure the frictional forces during the clamping process (3). The examinations give an insight into what happens during the clamping process in detail. The specific interrelationships between rolling and sliding friction have been analysed. In observing the whole indexing period, rolling friction can not be neglected. Both contact places of the roller have different conditions (see fig 1,2) during the indexing operation.

Furthermore, the coefficient of friction can be derived from the measured forces. The angle of inclination of the clamping ramp can be widened. At the point when the roller starts popping out of the clamping gap, the quotient of the clamping forces of the last secure indexing can be derived.

$$\frac{F_t}{F_n} = \tan\alpha < \mu$$

Thus, the functioning of the free-wheel is dependent on the geometrical conditions (tan α) and on the frictional conditions (μ) in the clamping gap. Different parameter combinations (motional conditions, different lubricants, stress level, etc.) have been tested.

Recent research work seeks to describe the relations in the contact zones with numerical approaches (FEM) to underline the empirical test results.

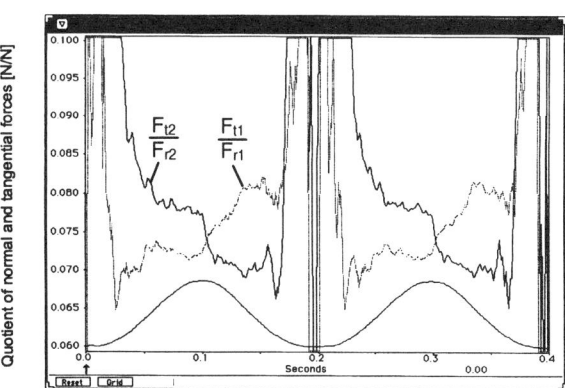

Fig. 2: Print screen of on-line determined frictional conditions at a roller clamping gap

REFERENCES

(1) Dürkopp, K.; Böhnke, H.-J. and Jorden, W., Specific friction and wear mechanisms in clamping-roller free-wheel clutches, Wear, 162-164 (1993) 985-989

(2) Welter, R., Die Lebensdauer von Klemmkörperfreiläufen im Schaltbetrieb, RWTH Aachen, Fakultät für Maschinenwesen, Diss. 1990.

(3) Dürkopp, K.; Jorden, W.: Friction and wear in Roller Clutches. In: Proceedings of the International Tribology Conference, Proceedings III, Seite 1561-1566, Yokohama, Japan, 1995.

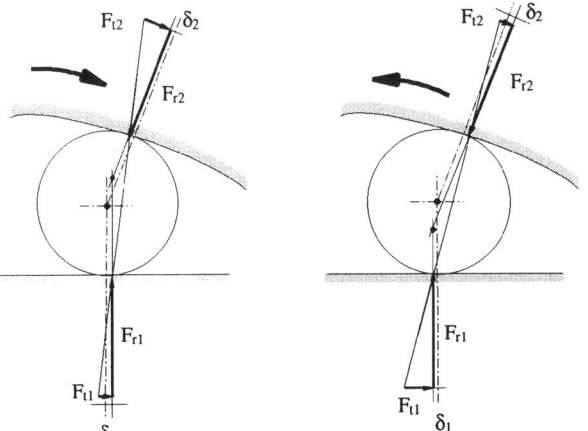

Fig. 1: Clamping forces with rolling friction

Submitted to *ASME Journal of Tribology*

INVESTIGATION OF FAILED SLEW RING BEARING.

Dr J Beard
AEA Technology plc, National Centre of Tribology, Risley Warrington. WA3 6AT.
Mr G Wright
British Steel Engineering Steels, PO Box 50, Aldwarke Lane, Rotherham S60 1DW.

Large slowly-rotating bearings are used in many difficult and arduous situations. They are often exposed to severe environments in applications like the north sea where they are found on single point mooring buoys and cranes. Other particularly testing conditions are experienced on wind turbines, excavating machinery and radio telescopes. The National Centre of Tribology has over many years been involved with the investigation of bearing failures from all of these applications.

One recent failure was on an unusual application, which was on the roof bearing of a large electric arc furnace. The bearing allows the roof lid to slew open for fast loading. Although steel making is a difficult environment, an acceptable bearing life should be achievable. Yet the bearing which was intended to last the 30 year life of the plant, failed within two years of installation.

The price of failure is high. Replacement of a 2 m diameter three row bearing is in excess of £100 000, but the potential loss in production can dwarf this figure. Bearings in such applications must therefore be reliable and carry a very low risk of failure over the planned lifetime.

The bearing failure was investigated by the owner of furnace British Steel with the assistance of the National Centre of Tribology. This paper describes the results of a metallographic investigation combined with an engineering analysis to establish the cause of the failure and determine the remedial action necessary to prevent the failure from reoccurring.

The investigation uncovered problems familiar to those involved with more conventional sized industrial rolling element bearings. The quality of lubrication and the ingress of contamination are well recognised problems to the bearing specialist. However, they are problems that still have to be fully recognised by many OEM's.

Efficient sealing of large rolling element bearings is undoubtedly difficult, but best practice is rarely seen. Designers often assume that proprietary seals, which are usually chosen on price, will last indefinitely and be fully efficient at keeping out of dirt, debris and liquid.

Grease provides an important element in the barrier system for keeping out water and debris. Frequent grease injection is necessary to flushed through the bearing and maintain as far as possible a full fill. Lubrication of large bearings is often by multipoint grease injection. For a large bearing, gravity determines the flow path of the grease. It is important that lubricant is directed at each row of rolling elements. In the present case a primary cause of the bearing failure was due to the low volume of grease injection and to poorly positioned greaseway drillings to the top row.

Poor lubrication causes an increase in frictional traction which results in higher subsurface shear stress thus reducing bearing fatigue life. A bearing life reduction factor of at least four was estimated to have occurred. However, the absence of a grease barrier allows the ingress of water, and internal corrosion of the bearing reduced the life fatigue by at least a further factor of 4. Thus, overall the bearing life was reduced by more than a factor of 16.

The load transmitted through the rolling elements is a key factor in determining the fatigue life of the bearing. The bearing must be capable of carrying the static loading as well as the dynamic load.

A large slew bearing is unusual in that it has to carry a large moment. For bearing sizing this moment has to be translated into an equivalent axial load. However, a larger bearing relies on the structure for its stiffness and any distortion can increase the load on the elements.

Finite element analysis of the structure revealed that although out-of-flatness distortion of the structure was not a problem, the load on the rollers was greater than predicted analytically.

Slew bearing are bolted to the structure. Inadequate ring bolting can lead to bearing distortion and reduced life. Problems with the bolting were identified due to embedment beneath the bolt head, and joint separation due to prying. Both analytical and FE analysis predicted joint separation under the specified bolt preload. This was considered a further factor in reducing bearing life.

THE MEASURED OIL FLOW CHARACTERISTICS OF A CONNECTING ROD LARGE-END BEARING

G J JONES

T&N Bearings Group, Argyle House, Joel Street, Northwood Hills, Middlesex HA6 1LN, UK

ABSTRACT

Operating temperature is one of the most important influences on the performance of an engine crankshaft bearing, since this determines the viscosity of the lubricant and thus the performance of the bearing. This temperature is in turn strongly dependent on the oil flow through the bearing since this carries away most of the heat generated by frictional losses. At the design stage, an understanding of the oil flow characteristics of a bearing is vital for predicting bearing performance.

An experimental program has been carried out to measure both the oil flow and the temperatures in a connecting rod large-end bearing of an automobile engine. The influence of a number of bearing parameters has been investigated. These included the operating conditions, i.e. the oil supply temperature and pressure, and the geometry of the bearing, i.e. the clearance, the bearing width and the oil grooving arrangement.

The experiments were carried out on a 2.0 litre, 4-cylinder, fuel injected, naturally aspirated gasoline engine. The crankshaft was modified to allow direct measurement of the oil flow through the large-end bearing of one of the connecting rods. Temperatures were also measured at a number of locations on the rod, including four close to the surface of the large-end bearing. Signals from the thermocouples were carried back from the rod via a two-bar telemetry linkage

The bulk temperature of the oil in the engine sump was capable of being controlled, independently of the engine operating condition, by means of external cooling or heating. The oil supply to the instrumented large-end bearing was delivered via an external pump circuit which allowed control of the oil feed pressure to that bearing.

Measurements of oil flow and temperature were made at steady state conditions for a range of load and speed conditions, with the bulk temperature of the oil in the sump being controlled to a specified value. At each speed condition, the oil feed pressure to the large-end bearing was set to a number of different values, and readings taken when the temperatures had stabilised

Results from a test using the standard geometry, i.e. a plain large-end bearing fed via a single oil hole in the journal, showed that the predominating influence on oil flow rate was engine speed (Fig. 1). The effect of oil feed pressure on the oil flow rate was significantly less.

These results suggest that the flow rate through the bearing is governed more by the hydrodynamic pumping action within the bearing than the pressure at which oil is fed to the bearing.

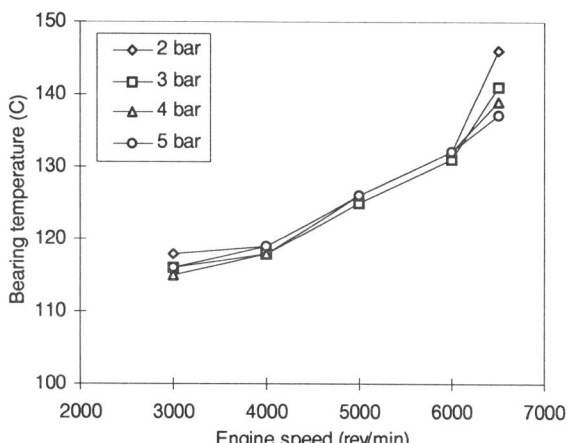

Fig. 2. Bearing temperature for 100°C oil feed

Fig. 2 shows the measured bearing temperature with the sump and oil feed temperature controlled to 100°C. This increased with speed, reflecting the increase in the shearing losses within the bearing. However, it appeared to be little effected by the oil feed pressure, except at the very high speed condition, where the bearing may have been starved of oil at low values of feed pressure.

This study complements some earlier work on the effects of bearing geometry on bearing temperatures (1). Results from these series of tests are being used to refine techniques for predicting the performance of crankshaft bearing systems.

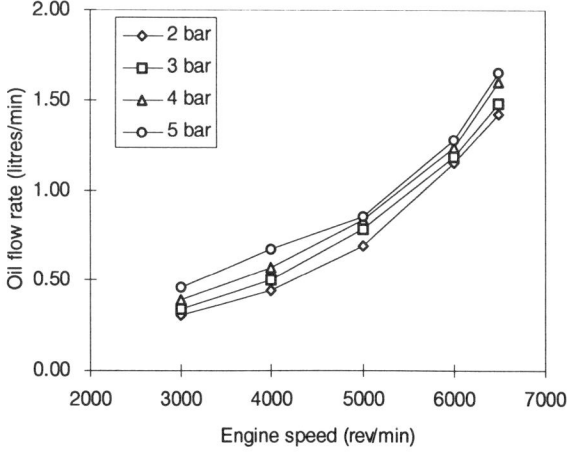

Fig. 1. Oil flow rate for 100°C oil feed temperature

REFERENCES

(1) G.J. Jones, SAE 960989, 1996

PTFE FACED THRUST BEARINGS: STATE OF THE ART REVIEW AND HYDRO-GENERATOR APPLICATION IN THE UK.

J E L SIMMONS
Department of Mechanical and Chemical Engineering, Heriot-Watt University, Edinburgh, EH14 4AS, UK
R T KNOX
Michell Bearings, Newcastle upon Tyne, NE15 6LL, UK
W O MOSS
First Hydro Company, Dinorwig, Llanberis, LL55 4TY, UK

ABSTRACT

Although other materials have been examined, whitemetal, with its well known strengths and weaknesses, remains the world's preferred choice as the facing material in most heavy duty, hydrodynamic bearing applications. An important exception is in the countries of the former Soviet Union (1) and China where, prompted by a series of persistent bearing failures in the 1960's and 1970's, PTFE lined thrust pads have been developed for major thrust bearing applications. By 1990, PTFE faced thrust bearings had been installed in the majority of hydro-electric power stations of the Soviet Union (2). Much higher operating loads (up to 6.5 MPa) than usual for whitemetal bearings are claimed. Start up loads are also higher and high pressure forced lubrication systems at start up and shut down are eliminated.

PTFE is suggested (3) as having a number of benefits as a bearing surface (very low coefficient of friction when paired with steel (0.04 - 0.09), chemical stability, excellent dielectric properties) and an important negative feature in that its mechanical properties are highly temperature dependent. Thus, a temperature rise from 20 °C to 80 °C leads to a decrease in compressive strength by a factor of three. Even more significant, PTFE is subject to excessive creep, that is to say deformation under load, even at low temperatures. These latter properties prevent PTFE being used as a bearing facing material in an unmodified form. The design approach adopted (1, 4) involves the application of sheet PTFE to a liner made from wire coils. The PTFE is applied under pressure which causes it to flow into gaps in the wire coil to a depth of 1 mm to 1.5 mm. In this way a firm bond is formed between the PTFE facing material and the wire coil under-layer.

Based on guidance contained in the literature together with much previous experience in the design of whitemetal bearings, a PTFE faced thrust pad design has been developed for applications in the UK and elsewhere. As recommended, the pads comprise a steel backing, a wire mesh intermediate layer and the PTFE bearing surface. An extensive factory-based experimental programme has been carried out to prove the design. Four offset pivot pads with a total working surface of 60940 mm² were fitted in a standard bearing assembly and subjected to a wide range of duties with specific loads up to 10.1 MPa and rotational speeds of 1500 rev/min, equivalent to 35.5 m/sec at the mean pressure diameter of the bearing.

The pilot PTFE pads performed well in practice even though loads were applied considerably in excess of the usual maximum service loading for whitemetal bearings of 3.5 to 4 MPa. On completion of the experiments, the PTFE pads were closely inspected and found to be in excellent condition. The results of the experimental work were sufficiently promising to encourage the decision to proceed with an initial industrial application at a major UK facility.

A set of PTFE faced pads were designed for a unit at the Ffestiniog Power Station, North Wales, a pumped storage scheme completed in the 1960's. The 12 pads, with a total bearing surface area of 1.07 m², were installed in September 1996 and removed for inspection on 4 May 1997. In this period the bearing experienced over 2000 hours of operation including almost 900 separate times when the unit was started up under load. Full examination of the pads showed them to be in excellent condition with no signs of wear. The pads will next be re-examined in two years time.

The results of this work point encouragingly to the potential for the wide application of PTFE faced thrust bearings in many other hydro-generator applications. The team involved in the current work expect to commence further trials at Dinorwig, one of Europe's largest power plants, later in 1997.

The full version of this paper has been submitted to the IMechE, Journal of Engineering Tribology.

REFERENCES

(1) A.M. Soifer et al., Izvestiya Vysshikh Uchebnykh Zavedeniy, Mashinostroenie, No 7, 1966.
(2) A.E. Aleksandrov, N.G. Plaxtonov, Gidroteknicheskoe Stroitelstvo, No 11, pp20-24, 1990.
(3) A.E. Aleksandrov, Gidroteknicheskoe Stroitelstvo, No 9, pp12-14, 1981.
(4) Y.I. Baiborodov et al., Gidroteknicheskoe Stroitelstvo, No 10, 1977.

MODELING ISSUES FOR PAINTED AUTOMOTIVE FASCIA UNDERGOING 3D FRICTION INDUCED DAMAGE

D. J. Mihora
FM Analysts, Inc., Santa Barbara, CA 93111
A. C. Ramamurthy
FORD MOTOR COMPANY, Scientific Research Lab, Dearborn MI 48121

ABSTRACT

Friction Induced Damage (FID) with automotive fascia is a very rapid and dramatic event. It can take but a single pass of the counterface to introduce catastrophic damage. (Multiple wear cycles are often unnecessary.) This single cycle FID is a very common occurrence with the many types of automotive decorated plastic fascias.

There has been progress in the development of more wear-resistant FID materials using 2D testing. The SLIDO instrument in the constant force tests (mode 1) can readily distinguish localized material failure as either paint delamination and/or plastic cohesive failure. These tests are usually performed with a rigid foundation under the decorated plastic specimens. It should be noted that automotive fascia never see such an inelastic boundary constraints in real life. The best approaches to relate 2D and 3D tests are underway.

Analysis and tests are being performed to understand the complicated stresses under the cylindrical counterface when rubbing highly deformable 3D fascia. It is the different stress states during the large fascia deformations that strongly influence tribological response. The contact stresses between 2D and 3D specimens has been shown to be quite different (1) in finite element analysis. In 2D tests, the boundary constraints have a major influence on the contact stresses under the counterface. The 2D contact stresses can be most unconventional. The classic Hertzian stresses become almost indistinguishable from the contact stresses when foam materials are present. Fig 1 illustrates the large distortions

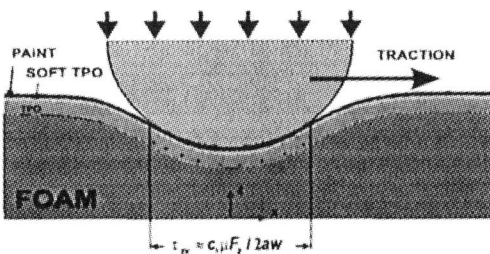

Figure 1 Constant force testing of 2D specimens on very soft foundations.

of the plastic (TPO). There is compression and a lack of tension under the counterface. The shear stresses dominate the damage onset.

Differences in load applications between 2D and 3D specimens also occur. Most of the real world user profiles on complex fascia are specific deflection histories. The counterface traverses a time vs. deflection trace rather than a time vs. loading trace. This difference is very important as compliant decorated fascia behave differently between force and deflection constraints.

The 3D observations with SLIDO 1 and analyses indicate the shear stresses under the counterface are primarily conformable bending induced. Fig. 2 shows the configuration proposed for SLIDO 1 tests that have been analyzed with NIKE3D (2) for

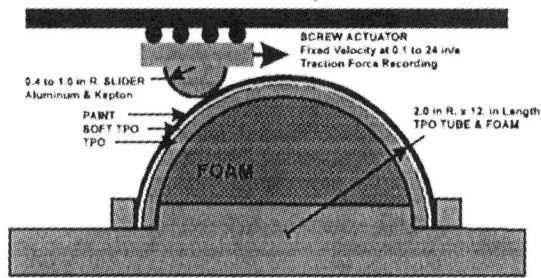

Fig. 2. Deflection constraints applied to deforming Fascia during tribology tests.

these tribology problems.

In general, the 3D geometric deformations around the counterface will reduce FID under the counterface. This occurs because there are now minimal tension stresses under the counterface. However, there may eventually be tension stresses far removed from the counterface centerline when the deflections become extreme. Tests results showing very large deformations do indicate some tension damage far removed from the counterface centerline.

REFERENCES
(1) D. Mihora, A. Ramamurthy, WEAR, 203-204 (1997), 362-374.
(2) B. Maker, NIKE3D: a Nonlinear, Implicit, 3D, Finite element code for solid and structural mechanics, UCRL-MA-105268, Lawrence Livermore Nat'l Labs.

DIRECT OBSERVATION OF DEFORMATION AND FRACTURE ABOUT SCRATCHES IN BRITTLE SOLIDS

V. H. BULSARA, S. CHANDRASEKAR, and T. N. FARRIS
Schools of Engineering, Purdue University, 1287 Grissom Hall, West Lafayette, IN 47907-1287, USA

ABSTRACT

The contact between a hard particle and the surface of a brittle solid is of fundamental interest for understanding material removal and wear in ceramics and glasses by abrasion, erosion and machining (1). It also forms the basis of a method for estimating indentation hardness and fracture toughness of brittle solids (2). One approach of analyzing such contacts involves a study of the contact of a hard, rigid indenter, having a well defined geometry shape, against the surface of a model brittle material. The advantage of this approach is that many details of the contact zone such as its extent, load, and stress fields are sufficiently well defined to interpret the observations. The loading of the indenter may be done quasi-statically as in a hardness test, or in a configuration where the indenter is loaded against and slid with respect to the specimen. In the latter case, the contact involves a combination of normal and tangential (friction) forces. In this research, the deformation and fracture phenomena occurring about scratches in brittle solids due to the action of a sharp, pyramidal diamond indenter are studied. A unique aspect of our scratch experiments is the use of in-situ observations to directly image the deformation and fracture processes occurring in the contact zone. This is done by scratching an optically transparent solid and imaging the contact region through the thickness of the specimen. A variety of optically transparent materials have been studied such as soda-lime glass, sapphire and spinel.

Scratch experiments were done by translating a Vickers indenter against the specimen surface at velocities of a few millimeters per minute with the normal load being kept in the range of 10g to 700g. The contact region was observed and photographed in-situ using a high resolution, optical microscope and video imaging system while forces were measured with a piezoelectric force transducer attached to the indenter. Several unique observations have emerged from the in-situ scratch studies.

First, a mode of fracture was observed in the load range of 75g to 160g in soda-lime glass where the scratch generated was devoid of any visible cracking throughout the scratching process: however, median and lateral cracks were observed to pop up and propagate back along the length of the scratch when the indenter was unloaded from the specimen surface. Figure 1 shows a micrograph of this type of fracture. In a survey of the literature to date, there is no mention of this mode of fracture around scratches. Instead, these cracks were thought to have formed along with the scratch during sliding.

The second major observation pertains to the formation of plastic, micro-cutting type of chips when scratches are made in soda-lime glass at very light loads. While such chips have been observed in polishing and machining debris, the present observations have shown for the first time their actual formation in scratch experiments with sharp diamond indenters. Their appearance suggests that they are formed by shear deformation, similar to chips formed in the cutting of ductile materials. Furthermore, the observations indicate that this is the characteristic mechanism of material removal at light loads.

Lastly, we have observed that, in both soda-lime glass and poly-crystalline spinel, chevron cracks arise due to median-type cracks forming ahead of the sliding indenter and deviating to either side of the scratch track.

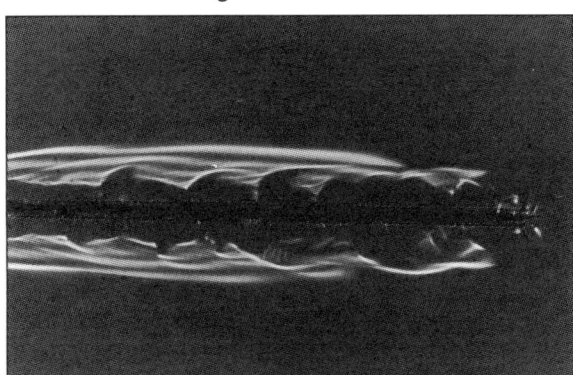

Fig. 1: Optical micrograph of a scratch track generated at a load of 120g in soda-lime glass, showing lateral and median cracks which formed during unloading of the indenter

REFERENCES
(1) M. V. Swain, "Micro-fracture about Scratches in Brittle Solids", Proc. R. Soc. Lond. A 366, pp.575-597, 1979.
(2) B. R. Lawn, Fracture of Brittle Solids, Cambridge University Press, 1994.

VALIDATION OF THE EROSION MAP FOR SPHERICAL IMPACTS ON GLASS

M.A. VERSPUI*, P.J. SLIKKERVEER+, G.J.E. SKERKA+, I. OOMEN*, G. DE WITH*
*Eindhoven University of Technology, P.O. Box 513, 5600 MB, Eindhoven, The Netherlands
+Philips Research Laboratories, Prof. Holstlaan 4, 5656 AA, Eindhoven, The Netherlands

ABSTRACT

The substrate behaviour under impact and erosion is assumed to be identical to quasi-static indentations. Depending on the loading conditions, the material behaviour of surfaces under spherical indentations will show various transitions. The relations describing the various transitions (1-7) can be represented in an erosion map, which is a quantitative representation of the expected surface behaviour under impact from spherical particles.

In the literature erosion maps can be found for the plastic and elastic regimes and their characteristic fracture patterns. Thorough investigation lead to a combined erosion map of both the elastic and plastic regimes. Figure 1 gives an example of such an erosion map.

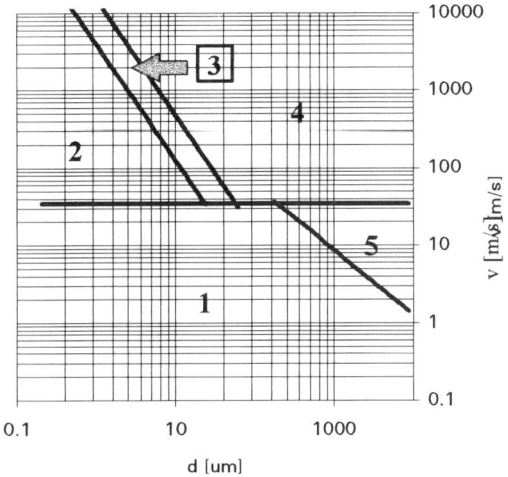

Figure 1: Theoretical erosion map for AF45 borosilicate glass. The horizontal line represents the transition between elastic and plastic behaviour. Area 1: elastic impact, area 2: plastic deformation without cracks, area 3: plasticity + radial cracks, area 4: plasticity + radial and lateral cracks, area 5: ring and cone cracks.

To verify whether this theoretically derived erosion map predict the actual transitions in material behaviour well, impact experiments with spheres of various sizes were performed on a borosilicate glass with a hardness of 5.1 GPa. The materials used are listed in table 1.

Erosion experiments using a normal flux were performed under various velocities with an abrasive-jet machine. In contrast with the glass beads, no particle degradation could be observed for the Zirblast beads. The highest erosion rates were found at normal impact. A transition between elastic and plastic behaviour was found, conform the prediction, around 40 m/s for the Zirblast beads. The velocity exponent in the equation for the erosion efficiency $E = \text{constant}[\text{velocity}]^x$, for the spheres (x=7.0) is much higher than the theoretically derived value of x=2.3.

Table 1: Particles used for the erosion experiments.

Material	Size [μm]	H [GPa]	ρ [kg/m^3]
Zirblast (SEPR)	0-70 70-125	7-9	3850
Glass beads	4-44 44-88	5.6	2500

More erosion experiments with a very low particle flux and a high scan rate were performed to study the single impact sites. These single impact experiments show the existence of ring and cone cracks in the plastic regime. The results indicate a slight translation of the transition lines for lateral and radial cracking. It can be concluded that it seems as if the regimes in the theoretical erosion map are predicted pretty well in a qualitative manner. However, the transition lines are not as sharp as indicated in the figure. More detailed information will be present after finishing the experimental work, which is still in progress.

REFERENCES

(1) S.M. Wiederhorn, B.R. Lawn, J. Am. Cer. Soc., 60 (9-10) 1977, pp. 451-458.
(2) D.B. Marshall, B.R. Lawn, A.G. Evans, J. Am. Cer. Soc., 65 (11) 1982, pp. 561-566.
(3) D.B. Marshall, J. Am. Cer. Soc., 67 (1) 1984, pp. 57-60.
(4) B.R. Lawn, A.G. Evans, D.B. Marshall, J. Am. Cer. Soc., 63 (9-10) 1980, pp. 574-581.
(5) R. Hill, Mathematical theory of plasticity, 1950, Oxford, University Press.
(6) I.M. Hutchings, J. Phys. D.: Appl. Phys. 25, 1992, pp. A212-A221.
(7) I.M. Hutchings, Key Eng. Mat. 71, 1992, pp. 75-92, ed. by J.E. Ritter.

THE EFFECT OF PARTICLE SIZE ON THE SPECIFIC ENERGY OF MATERIAL REMOVAL IN SLURRY EROSION

RYAN B HARTWICH* and HECTOR McI. CLARK**

* Butler Manufacturing, R&D
13500 Botts Road
Grandview, MO 64030-2897, USA

** Consulting Metallurgist
3001 Westdale Road
Lawrence, KS 66049-4411, USA

ABSTRACT

Measurements of erosion rates and specific energies for material removal have been made in a slurry pot erosion tester using a wide range of particle sizes. In general, specific energies increase with decreasing particle size until particle rebound ceases as particle entrapment occurs at the target surface, i.e., when $d_p < \sim 100$ μm.

INTRODUCTION

The calculation of specific energies (in Jmm^{-3}) for material removal in erosion by particle impact at 90°, termed 'deformation wear' by Bitter (1), designated ε, and at glancing angles, termed 'cutting wear', φ, has afforded a means of comparing material behaviour in erosion. Ductile materials typically have approximately equal values of ε and φ while brittle materials have low values of ε and very high values of φ. These values of ε and φ give rise to the characteristic 'ductile' or 'brittle' forms of the plot of erosive wear vs particle impact angle.

APPARATUS

Erosion tests were conducted on 6061-T6 aluminium, acetal and Pyrex glass in a slurry pot in which a pair of vertical cylindrical specimens, 5 mm diameter, were rotated on a radius of 51 mm about a central axis to give a nominal velocity relative to the slurry of 18.7 ms^{-1} using 40 g of carefully sized SiC, d_p between 20 and 770 μm, as erodent in 5 litres of diesel oil. Surface profiles were measured before and after erosion using a computer-controlled LVDT to determine wear as a function of angular location on the specimens.

DATA PROCESSING

The raw wear data yielded plots of wear rate as a function of erodent particle impact angle α through an analysis of particle motion in slurry flow incorporating the effect of the squeeze film on particle impact, (2), which also gave values of the normal and tangential velocities, allowing calculation of the kinetic energy of impact. Specific energies ε and φ were calculated from:

$$W_T = \frac{M}{2}\left\{\frac{(V_I^2 - K^2)}{\varepsilon} + \frac{V_T^2 \sin 2\alpha}{\varphi}\right\}$$

where W_T is the total wear, M the mass of particles impacting, V_I and V_T the normal and tangential particle velocities at impact, and K the minimum normal impact velocity to produce material loss (positive for glass, ceramics and some polymers, zero for ductile metals), (3). The value of ε is calculated from the wear at normal impact and φ deduced by subtraction.

RESULTS AND DISCUSSION

All specific energy results for Al, acetal and Pyrex glass are shown in Fig. 1. The aluminium showed erosive wear at all particles sizes down to 20 μm, while wear in acetal was undetectable below 80 μm and in Pyrex glass below 120 μm. This is ascribed to the resilience of acetal and to the absence of sufficient impact energy to initiate fracture in glass. Below $d_p \sim 100$ μm specific energy for removal of aluminium decreases, i.e., the wear rate is greater than that attributable to the dissipation of particle impact energy. This is believed to be due to a change in erosion mechanism as impacting particles fail to rebound but become trapped by the liquid squeeze film at the target surface, then removing material by being dragged across the surface, (2).

CONCLUSIONS

(1) Specific energy values for material removal from all materials by both deformation and cutting wear increase with decreasing particle size.
(2) Specific energy values cannot be taken as characteristic of particular materials.

REFERENCES

(1) J. G. A. Bitter, Wear **6** (1963) 5 - 21.
(2) K. K. Wong and H. McI. Clark, Wear **160** (1993) 95 - 104.
(3) H. McI. Clark and K. K. Wong, Wear **186-187** (1995) 454 - 464.

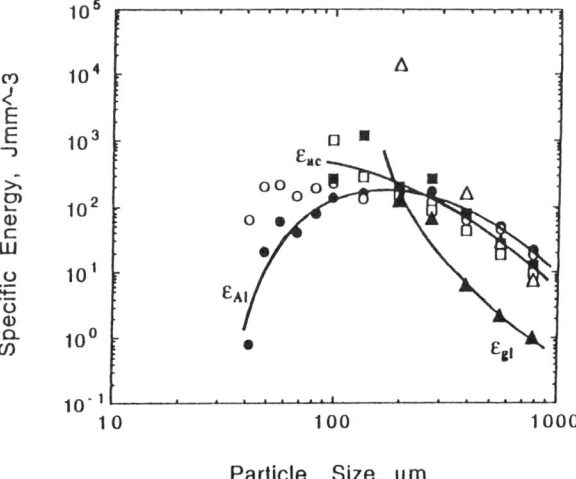

Fig. 1: Plot of specific energies ε and φ for all particle sizes. Closed symbols ε, open symbols φ, circles Al, squares acetal, triangles Pyrex glass.

Submitted to *Wear*

MEASUREMENTS OF SPECIFIC ENERGIES FOR EROSIVE WEAR USING A CORIOLIS EROSION TESTER

HECTOR McI. CLARK and JOHN TUZSON
John Tuzson and Associates, 1220 Maple Avenue, Evanston, IL 60202, USA

ABSTRACT

Measurements of erosion rates and specific energies for material removal have been made under sliding bed conditions using the Coriolis slurry erosion tester.

INTRODUCTION

The present Coriolis tester has been developed from the machine, described by Tuzson (1), with a view to providing a test machine, simple of operation, to simulate wear under sliding bed conditions, i.e., by erodent particles moving over, but not impacting upon, the target surface, as found in slurry pumps and cyclones.

APPARATUS

The Coriolis erosion tester consists of a 152 mm diameter steel rotor containing a 12.5 mm diametral channel through which slurry, supplied to a central chamber, is constrained to flow. As the rotor turns, at speeds up to 7000 rpm, slurry is flung outwards along a 1 mm wide channel in each of two nitrided specimen holders and bears down on the flat plate test specimens which form the bottom of the two channels. The erodent used was 220 mesh (62 µm) crushed alumina suspended in water. For mild steel run at 5000 rpm using 0.5 kg of erodent in 4 litres of water the test requires about 100 sec, and gives a wear scar approx. 50 µm deep. The profile of the scar is measured at any location using a computer-controlled LVDT yielding a set of traces for the conditions specified here shown in Figure 1.

Rotation speed, erodent particle size and slurry concentration may be varied independently. The machine has also been used effectively with dry erodent.

The unworn target surface is used as a datum to allow calculation of the scar cross-sectional areas at each location (representing the severity of erosion) which can then be plotted as a function of distance from the rotor center. Such a set of plots for erosion at different rotation speeds is shown in Figure 2.

SPECIFIC ENERGIES

The slurry flow speed has been measured using a stroboscope and has been shown to be approximately 0.5 of the rotation speed $R\omega$, where R is the rotor radius and ω the angular velocity. The specific energy for material removal from the present work has been calculated as:

Material	Sp. En. (J.mm^{-3})
Aluminium (6061-T6)	10
1020 HR steel	100*
M2 tool steel	160
Rubber	11

*corresponds very closely with that for cutting wear by Clark and Wong (2).

The specific energy was calculated as:

$$\text{Sp. En.} = \frac{0.5\,\omega^2 R\,\text{(Weight of erodent in water)}}{g\,\text{(Cross-sectional area of wear scar)}}$$

CONCLUSIONS

1) The Coriolis erosion tester gives a simple and rapid way of assessing both the erosion resistance of materials and their specific energies under sliding bed conditions.

2) The test uses a wear scar profile measuring device that gives more information than wear measurement by mass loss.

3) Comparison of different materials is easily made by testing under standard conditions.

REFERENCES

(1) J. Tuzson, J. Fluids Eng., **106** (1984) 135-140.
(2) H. McI. Clark and K. K. Wong, Wear, **186-187** (1995) 454-464.

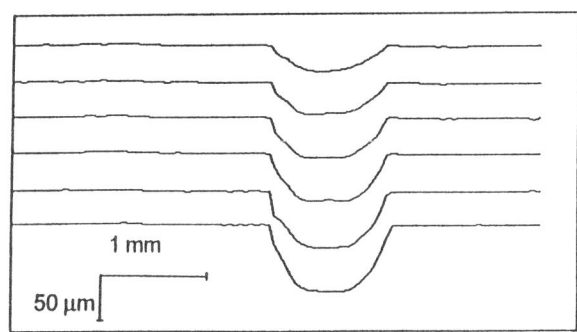

Fig. 1: Raw wear data, 1020 steel. Each traverse consists of 100 steps, each 50 µm. First trace at 7 mm, spacing 4 mm.

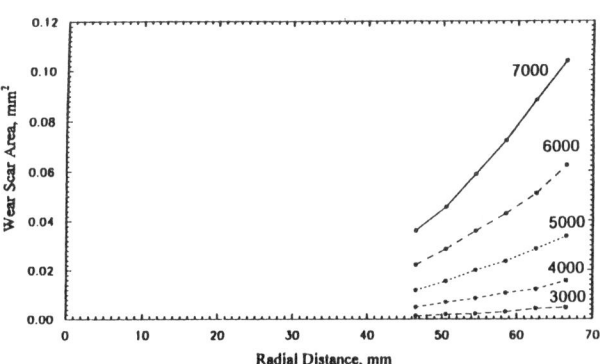

Fig. 2: Wear scar area vs. distance from machine center for 1020 steel, 0.5 kg Al$_2$O$_3$ in 3.6 litres of water at speeds indicated in rpm.

Effects of velocity and applied potential on the erosion of TiN based PVD coatings in aqueous slurries

H.W.WANG AND M.M.STACK

Corrosion and Protection Centre, UMIST, PO Box 88, Manchester M60 1QD, U.K.

ABSTRACT

The present study addresses the erosion-corrosion of TiN coated mild steel (prepared by magnetron sputtering ion plating) in aqueous alkaline slurries containing alumina erodent. This formed a preliminary investigation into the erosion-corrosion behaviour of various PVD coatings of TiN in aqueous slurries for a range of pH values, and for various particle sizes, concentrations and impact velocities. The apparatus used was a roating cylinder electrode.

Potentiodynamic polarisation results, obtained in a carbonate/bicarbonate buffer solution containing 150 μm alumina particles, showed that the TiN coatings exhibited higher corrosion resistance than the mild steel, over a wider potential range. Polarisation curves for TiN from -1.0 to 0.83 V, for erosion at a range of velocities, showed that the dissolution of TiN was negligibly smaller (<2 μA cm^{-2}) than that of the uncoated mild steel. Increases in velocity shifted the E_{corr} value from -0.55 to -0.26 V SCE and slightly enhanced the TiN dissolution. However, the overall corrosion current was much lower. For the uncoated mild steel, the erosion-corrosion behaviour was characterised by strong anodic dissolution (>400 μA cm^{-2}), following an initial cathodic reaction and subsequent passivation. The passivation region of mild steel was reduced with increases of impact velocity and completely disappeared above 4 m s^{-1}. Thus, the overall corrosion of uncoated mild steel was significantly assisted by erosion.

The total weight loss due to erosion-corrosion was very low for TiN by comparison to that for the uncoated mild steel. The morphology of the surfaces following erosion-corrosion, showed exceedingly large unattacked areas for TiN, in sharp contrast to the uniform, severely abraded surface for mild steel, especially at speeds above 3-4 m s^{-1}.

In recent years, there have been significant advances in the development of erosion-corrosion mapping approaches for erosion in aqueous slurries(1). The maps have been used to indicate regimes of degradation and magnitudes of the wastage rate, and mathematical models have been developed to construct the transition boundaries(2). The erosion-corrosion map for the uncoated mild steel, Fig. 1(a), showed that the "low" wastage regime reduced at intermediate applied potentials, at the transition to dissolution affected conditions. However, with further increases in the applied potential to passive corrosion conditions, this regime shifted to higher velocities. For the TiN coated material, Fig. 1(b), the "low" wastage region increased to high velocities, i.e. 8 m s^{-1}, at potentials up to 0.1 V SCE. At potentials greater than 0.1 V SCE, the rapid reduction in the "low" wastage region on the map indicated that, in a very narrow range of conditions, the performance of the coating was inferior to that of the base material. This suggests that selection of coatings for erosion-corrosion resistance in aqueous conditions should be approached with caution.

This paper summarizes the results to date on the erosion-corrosion of PVD coatings in aqueous conditions. Criteria for defining regimes of damage for such coatings are suggested. The generation of erosion-corrosion mechanism maps, showing the possible degradation regimes for the coatings, is also addressed.

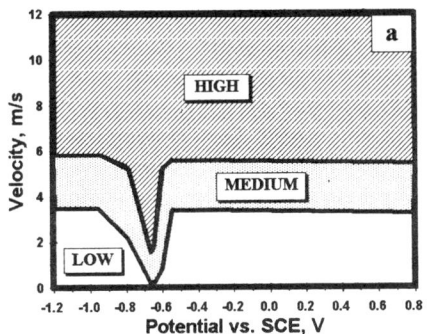

(a) mild steel substrate

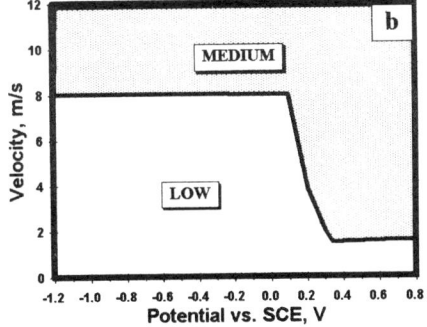

(b) TiN coating

Fig. 1: Erosion-corrosion "wastage" maps for erosion in the carbonate bicarbonate slurry.

REFERENCES

1. M.M. Stack, S. Zhou and R.C. Newman, Mat. Sci. and Tech. 12, 3, 261-268, 1996.
2. M.M. Stack, N. Corlett and S. Zhou, Mat. Sci. and Tech., 12, 8, 662-672, 1996.

THE EROSION BEHAVIOUR OF CVD DIAMOND AND BORON CARBIDE COATINGS

DW WHEELER, RJK WOOD, DC LEJEAU*, BG MELLOR#
Department of Mechanical Engineering,
*Department of Aeronautics and Astronautics,
#Department of Engineering Materials,
University of Southampton, Highfield, Southampton, SO17 1BJ, UK

ABSTRACT

The erosion performance of two coatings produced by chemical vapour deposition (CVD) is reported. It was thought these coatings, when applied to choke valve trims, would significantly reduce erosion damage from sand which, along with oil and gas, is often produced in hydrocarbon fluids extracted from offshore oil wells. Samples were tested in water-sand slurry and air-sand erosion facilities which enabled the erosion behaviour of coatings to be studied at particle velocities in the range 10-340m/s.

Results for the erosion performance of CVD diamond and boron carbide coatings deposited on tungsten and sintered tungsten carbide-cobalt substrates are presented. Their hardnesses are 3 to 8 times that of the erodent and, therefore, are likely to be resistant to sand erosion. The thickness of the diamond samples ranged from 10μm to 250μm in both lapped (R_a 0.2μm) and as-grown (R_a 1-7.5μm) form. This enabled the effect of lapping and residual stress on the erosion rate to be evaluated. Previous work[1] suggests a surface R_a of less than 0.5μm enhances the erosion resistance of coatings. The boron carbide coatings were 15μm thick and were grown on a titanium carbide diffusion barrier of approximately 1μm thick.

All samples were tested at a nominal impingement angle of 90°, chosen because maximum erosion occurs at this angle for brittle materials. Initial tests in the water-sand slurry rig utilised a 2.1% wt. mixture of angular silica sand in water. Two different erosion conditions were employed with mean sand sizes of 135μm and 235μm at velocities of 16.5m/s and 28.5m/s respectively. The samples were also tested in the air-sand rig[2] using sand with 194μm mean diameter at a particle velocity of 270m/s and flux rate of 0.5kg/m²/sec. This blend of sand was formulated to replicate the size distribution found in some North Sea oil fields. The erosion rates, expressed as μm³/impact, were calculated and compared with the erosion results of uncoated sintered tungsten carbide-6% cobalt currently used in valves. Characterisation of microstructure, surface topography, composition and the level of residual stresses in the coatings were undertaken using Talysurf profilometry, scanning electron microscopy for both pre-test and post-test, as well as Energy Dispersive Spectroscopy (EDS), X-Ray Diffraction (XRD) and Raman Spectroscopy. This enabled wear to be assessed and probable erosion mechanisms to be determined. The experimental results of water-sand and air-sand were unified onto a single graph of Erosion Rate against Particle Kinetic Energy.

The results showed that CVD diamond was the more erosion-resistant coating at all energies, and its erosion rate was an order of magnitude lower than sintered WC-6%Co at a velocity of 270m/s. The erosion mechanism was found to proceed by a gradual chipping mechanism in the early stages, later leading to the formation of holes of 100-200μm diameter. As the erosion continues these increase in size and number until spalling of the coating occurs. It was also found that coatings of <15μm, cannot sustain the stresses generated by the impact of the particles with the coating surface and spall rapidly. The boron carbide samples were found to be only marginally more erosion resistant than sintered WC-Co. XRD analysis revealed that this could be due to the presence of undesirable B_xC phases other than $B_{13}C_2$ in the coating as well as its insufficient thickness. Boron carbide erodes by a nano-chipping mechanism leading to rounding of the original sub-micron grains. Later work on these samples will focus on thicker (>20μm) coatings in an attempt to ascertain the optimum thickness for use on valves.

Based on these results, the optimum thickness of CVD Diamond appears to be in the region of 60-120μm. Extrapolation of the erosion rates indicates that the operational performance of a choke valve trim with a lapped coating 60-120μm thick, could be significantly better than the currently-used sintered WC-6%Co.

ACKNOWLEDGEMENTS

The authors thank De Beers Industrial Diamond Division, MPA, and Boart Longyear for the supply of samples, the LINK programme, and the industrial partners for their support.

REFERENCES

1. D.W. Wheeler, R.J.K. Wood, Proc. 8th Int. Energy Week, Houston, Jan.1997, Bk 5, 33-40.
2. R.J.K. Wood, D.W. Wheeler, "Design of a high velocity air-sand jet impingement erosion facility", These proceedings, C491-584.

EXPERIMENTAL INVESTIGATION OF CAVITATION EROSION IN JOURNAL BEARINGS

Dr. MATTHÄUS WOLLFARTH
Now: Freudenberg Dichtungs- und Schwingungstechnik, Technisches Entwicklungszentrum, D-69465 Weinheim, Germany
Prof. Dr. RUDOLF HALLER
Institut für Maschinenkonstruktionslehre und Kraftfahrzeugbau, Universität Karlsruhe, Kaiserstr. 12, D-76128 Karlsruhe, Germany

ABSTRACT:

For a long time cavitation erosion in journal bearings seemed to be a harmless type of damage of these elements. But recently it occurs more and more in combustion engines due to more arduous operating conditions. Therefore we investigated cavitation erosion in journal bearings at the Institute of Machine Design and Automotive Engineering in Karlsruhe using a test rig.

To realize a sufficient oil supply of the connecting rod bearings in combustion engines the main bearings of the crankshaft are partially grooved and there is a rotating cross drilling in the crankshaft to connect the different bearings. Due to this design cavities can be formed when the oil flow is accelerated or suddenly interrupted at the end of the groove. There are four main causes of cavitation erosion in journal bearings:

Flow cavitation:
At the end of oil grooves and at the edges of bore holes cavitation erosion can be caused by increased velocity of flow.

Impact cavitation:
In partially grooved journal bearings, cavity formation can occur behind the suddenly stopped oil flow:

Suction cavitation:
The present oil supply is not sufficient to fill the suddenly expanded clearance between the bearing and the shaft during a radial acceleration of the shaft.

Discharge cavitation:
This mechanism of cavitation erosion is also caused by radial acceleration of the shaft in annular grooved bearings.

Our investigations concentrated upon cavitation erosion by flow and impact cavitation because these types are also most interesting in practice.

A comparison of simulated flow cavitation erosion (left) with some from practice is given in figure 1. Both damaged areas show similar appearances. The electro-deposited overlay is partially removed at the edge of the shell. Also a very good conformity can be observed in cases of impact cavitation erosion (simulated damage left), figure 2. Both photographs show the characteristic shape of this kind of cavitation.

The test rig proved its ability to simulate cavitation erosion in journal bearings, reproducible and comparable with such from practice. Therefore we were able to investigate the influence of different parameters on the appearance of this type of bearing damage. We investigated the influence of film thickness, the design of the oil groove, the bearing material system and the contents of air in the oil. We obtained the following conclusions:

- There is a relation between the position of minimal film thickness and impact cavitation erosion.
- Flow cavitation erosion can be reduced with a smoothly designed end of the oil groove but there is no alleviation of impact cavitation erosion.
- Bearing shells with aluminium-tin alloys are more resistant against cavitation erosion than with lead alloy overlays.
- Flow and impact cavitation erosion will be reduced by increased contents of free air in the oil.

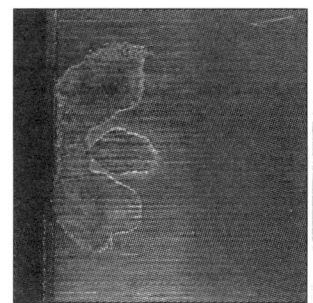 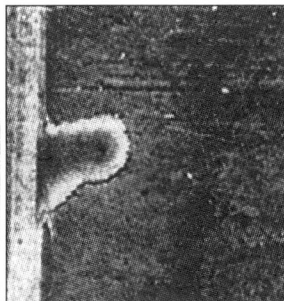

Fig. 1: Comparison of flow cavitation erosion

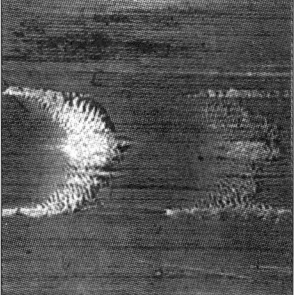 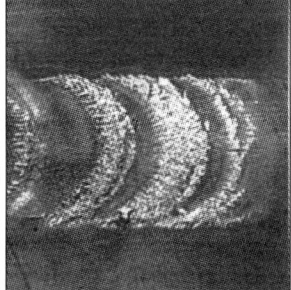

Fig. 2: Comparison of impact cavitation erosion

REFERENCES

M. Wollfarth, Experimentelle Untersuchungen der Kavitationserosion im Gleitlager, Dissertation Universität Karlsruhe, 1996

THE EROSION OF INTERMETALLIC ALLOYS

R.L. HOWARD

Department of Materials Engineering, University of Cape Town, Rondebosch, 7700 South Africa.

Present address: Corrosion & Protection Centre, UMIST, PO Box 88, Manchester M60 1QD, UK

ABSTRACT

Intermetallic compounds consist of well-defined (stoichiometric) proportions of two or more metals in an ordered atomic configuration, with a predominance of unlike nearest neighbours and stronger bonds between atoms than usually exist in disordered materials. As a result of the strong bonds and ordered atomic structure, intermetallics generally possess high work hardening rates, limited room temperature ductility and toughness and good mechanical property retention to high temperature.

Intermetallics for engineering applications are alloyed to optimise mechanical properties, particularly to improve the ductility and toughness of these materials. As a consequence of these mechanical properties and their low density, intermetallic alloys are under consideration for aircraft components, automotive valves and turbocharger rotors, amongst other applications.

The cavitation erosion and particle erosion performance of intermetallic alloys is relevant to their future use and this paper provides an overview of some of the work which has been conducted in this area.

Cavitation erosion can occur in a fluid system in which there are cyclic pressure fluctuations which produce cavities in the fluid. The collapse of these cavities causes pressure pulses which can impose large stresses (100-1000 MPa) at high strain rates (10^4 - $10^6 s^{-1}$) on adjacent surfaces, causing rapid erosion of material.

The cavitation erosion rates of intermetallic alloys are lower than many engineering materials and are comparable to commercial erosion-resistant hardmetals (1). This cannot simply be ascribed to a high bulk hardness, since these intermetallics have only moderate hardness as shown in Fig. 1 (2). Rather, the primary reason for the superior cavitation erosion performance of intermetallic alloys appears to be their high work hardening rates, as measured in conventional tensile tests and from measurements of surface hardening due to erosion (2)(3).

In particle erosion conducted with air-blast apparatus, material removal from the surface of intermetallic alloys occurs by a ductile mechanism, despite the limited ductility of intermetallic alloys (4). Plastic deformation is noticeable on eroded surfaces and greater erosion occurs at oblique impact than at normal impact, in common with ductile metals (5)(6).

Consequently, the particle erosion rates of intermetallic alloys are similar to austenitic stainless steels.

Particle erosion involves a multiaxial stress distribution. The material immediately below an impacting particle is plastically constrained by the surrounding material, which also inhibits fracture. The inhibition of fracture may contribute to the observed ductility of intermetallics in particle erosion, in contrast to their limited ductility in conventional tensile tests.

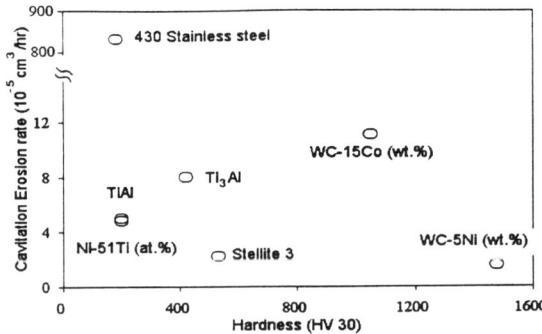

Fig. 1: Cavitation erosion rate versus hardness.

The superior cavitation erosion resistance and reasonable particle erosion resistance of intermetallic alloys means that they are suitable for components in hydrodynamic systems such as valves and impeller vanes.

REFERENCES

(1) R.N. Wright and D.E. Mikkola, Mater. Sci. Eng., Vol. 26, pp 263-268, 1976.

(2) R.L. Howard and A. Ball, Acta Mater., Vol. 44, No. 8, pp 3157-3168, 1996.

(3) T. Okada and S. Hattori, Proc. 41st Ann. Meeting Soc. Mater. Sci.- Japan, pp179-181, 1992.

(4) P.J. Blau, in ASM Handbook, Vol. 18, pp 772-777, 1992.

(5) B.J. Marquardt, D.M. Baker and J.J. Wert, in Proc. Int. Conf. Wear of Materials, ASME, pp 693-698, 1985.

(6) R.L. Howard and A. Ball, Wear, Vol. 186-187, pp 123-128, 1995.

AMBIENT AND ELEVATED TEMPERATURE EROSION MECHANISMS IN THERMAL SPRAY COATINGS

SAIFI USMANI, SANJAY SAMPATH and JONATHAN GUTLEBER
Center for Thermal Spray Research, State University of New York, Stony Brook, NY 11794-2275

ABSTRACT

Thermal spray methods, because of their versatility and economy in processing a diverse range of materials on a variety of substrates, are extensively used to deposit overlay coatings in the energy generating industry. In addition, thermal spray methods are being explored for the synthesis of Functionally Graded Materials (FGMs), which are comprised of continuously or discontinuously varying compositions and/or microstructures [1,2].

This work investigates the ambient temperature (25°C) erosion behavior of NiCrAlY and partially stabilized zirconia (PSZ) FGMs and the ceramic overlayer, and, examines the elevated temperature (500°C) erosion behavior of the ceramic overlayer. Additionally, the effect of starting particle size and thermal cycling on the porosity, strength and the ambient temperature erosion resistance of ceramic coatings was examined. Some results of this work are presented in the following figures.

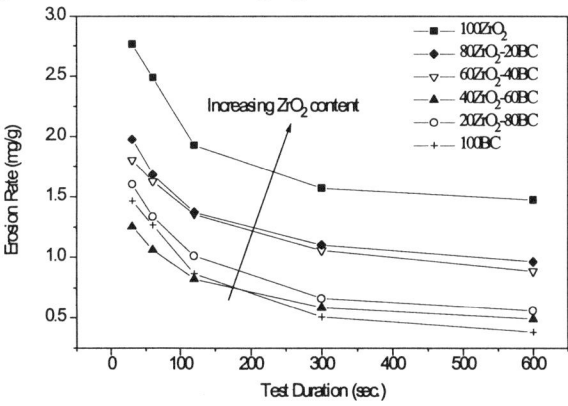

Fig. 1: Erosion rates of the PSZ and alumina coatings and the FGM layers at 25°C.

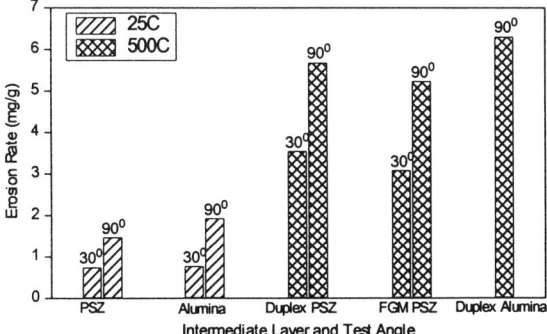

Fig. 2: The comparative steady-state erosion rates of the PSZ and alumina coatings deposited either on a NiCrAlY bond coat (Duplex) or on the FGM inter-layers at two erodent impact angles

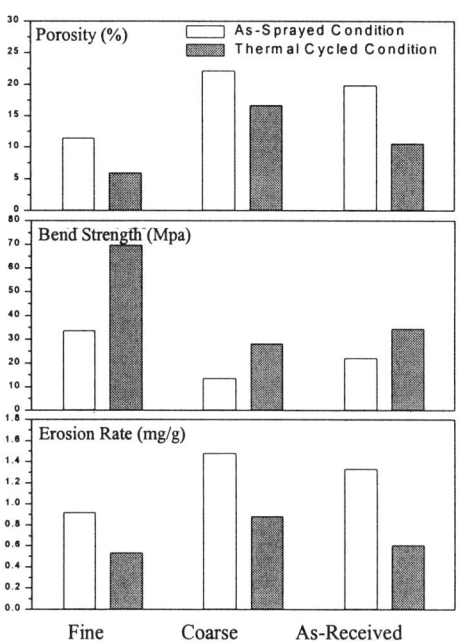

Figure 3: Porosity, bend-strength and erosion rates before and after thermal cycling

The results in Figs. 1-3 were analyzed on the basis of the microstructure and erosion mechanisms unique to thermal spray coatings. These results and analyses lead to the following conclusions:

1. The erosion rate of the FGM coatings decreases with an increase in the metallic content of the exposed FGM layer (Fig. 1). The addition of only 20% NiCrAlY to a PSZ coating can substantially increase its erosion resistance.

2. The incorporation of an FGM interlayer rather than a NiCrAlY interlayer helped to marginally improve the erosion performance of the PSZ coatings. This is attributed to reduced CTE differential and, hence, reduced thermal stresses in the specimen with the FGM interlayers (Fig. 2). The erosion behavior of the ceramic coatings at two impact angles is similar to that of sintered ceramics.

3. A greater fraction of coarse particle sizes in the starting powders is related to insufficient melting of particles in the plasma flame, and hence to higher porosity and lower bend-strength and erosion resistance of the coatings (Fig. 3). Thermal cycling improves the erosion behavior of the coatings, possibly due to sintering effects.

REFERENCES:

(1) S. Sampath, H. Herman, N. Shimoda, T. Saito, MRS Bulletin, 20, 1, (1995), p. 27.
(2) S. Usmani, S. Sampath, JOM, Nov. 1996, p. 51.

AN INVESTIGATION ON EROSION OF CERMET MATERIALS

G E D'ERRICO, S BUGLIOSI and M TOLOMELLI
Istituto Lavorazione Metalli, Consiglio Nazionale delle Ricerche, via Frejus 127, 10043 Orbassano-TO, ITALY

ABSTRACT

This study is focused on erosion behaviour of diverse cermet grades subject to solid particle impingement under diverse impacting conditions. These grades are commonly used with application to cutting tool production (1-3).

These cermets are material compositions whose typical structure is formed by a core composed by TiCN (the primary ceramic phase) and a shell composed by W, Ta, Nb, Ti (C, N) (the secondary ceramic phase which is around 25÷40% of the primary one), with a binder phase composed of Ni and/or Co.

Erosion events are simulated by use of an air jet eroder machine under oblique and normal impact and diverse erodent flows (4-6). The abrasive medium is an alumina based powder with typical percentage composition: Al_2O_3(95.8%), TiO_2(2.6%), SiO_2(1%), Fe_2O_3(0.2%), MgO(0.2%), ZrO_2(0.1%), the remaining 0.1% being composed of alkalis. The mean grain size of the particle is ≈70μm, and the Knoop hardness is ≈21.6 GPa.

Results of erosion tests are presented and discussed in terms of evolution of erosion rate with time, taking into account the material composition of the diverse cermets, their hardness, density, and porosity. Erosion rate is expressed as the ratio between the volume loss of target materials and the weight of impacting erodent.

An attempt is made to propose a model for estimating the erosion performance by observing the material behaviour in approaching their steady-state erosion rates. Steady-state erosion rate is defined as the mean of the erosion rate values obtained after the initial transient is vanished.

Evolution of the target material damage is tracked by means of SEM observations performed during the erosion events. This is aimed at studying the scar appearance in order to support an interpretation of the erosion test results from a morphological point of view.

Modelling is based on an analysis of erosion rates *versus* erodent flow variation, and on a description of erosion events as a time-dependent process.

REFERENCES

(1) J.D.Destefani, *Tooling and Production*, Vol. 59, No. 10, 1994, pp. 59-62.
(2) H.Doi, *Proceedings of 1st International Conference on the Science of Hard Materials*, (Almond, Brookes and Warren eds.), Rhodes, 23-28 Sept. 1984, Institute of Physics Conference, Series No. 75, Adam Hilger Ltd, Bristol and Boston, pp. 489-523.
(3) G.E. D'Errico, S. Bugliosi, D. Cuppini, E. Guglielmi, *Proceedings of 11th International Conference on Wear of Materials*, San Diego-CA, April 20-23, 1997.
(4) H.C. Meng and K.C. Ludema, *Wear*, Vols.181-183, 1995, pp. 443-457.
(5) I. Finnie, *Wear*, Vols. 186-187, 1995, pp. 1-10.
(6) American Society for Testing and Materials, "Standard Practice for Conducting Erosion Tests by Solid Particle Impingement Using Gas Jets", ASTM G76-83.

Approaches to construction of materials selection maps for exposure to elevated temperature erosion.

M.M. STACK AND D. PEÑA
Corrosion and Protection Centre, UMIST, P.O. Box 88, Manchester, M60 1QD, U.K.

ABSTRACT

Erosive wear by solid particles at elevated temperatures is a major issue in energy conversion processes such as coal combustion(i.e. fluidized bed combustion) and in the minerals processing industries such as alumina production. In recent years, there has been extensive interest in characterizing the degradation in terms of "regimes", ranging from "corrosion-dominated" to "erosion-dominated" behaviour. Various mapping approaches have been developed to establish the transitions between such regimes as a function of the main erosion and corrosion parameters(1).

Despite such work, there has been little interest in devising approaches towards selection of the most appropriate material for the exposure conditions, a very critical issue for engineers attempting to optimize the performance of materials for the exposure conditions. The initial laboratory results for wastage of Fe and Ni based materials in laboratory simulated fluidized bed combustion conditions(2-4) detail the effect of alloying additions such as Cr content on the wastage rate. For MMC based materials, the effects of reinforcement additions on the degradation rate also demonstrate clear trends which may be used as a basis for selection of MMCs for erosion resistance at elevated temperatures(3-4).

However, erosion-oxidation resistance is only one of three important factors which affect materials selection decisions at elevated temperatures. High temperature processes such as creep and oxidation may limit the application of the material and therefore such considerations must be incorporated into the material selection process. Hence, the approach to construction of *materials selection* maps for elevated temperature erosion resistance is to combine these factors into an overall map, in the following steps:

(i) Generation of materials performance maps for resistance to (a) oxidation (b) erosion-oxidation and (c) creep, based on standard limitations imposed by the requirements of the process(i.e. heat exchanger tubes in fluidized beds)
(ii) Generation of a materials performance map for combined resistance to (a), (b) and (c).
(iv) Refinement of the map based on various *selection*(5) criteria (ease of fabrication, cost versus predicted lifetime etc) to generate a *materials selection* map.

A schematic diagram of a proposed *materials selection* map for elevated temperature erosion resistance is given in Fig. 1.(3,4). The boundaries reflect the trend for oxidation to provide protection against erosion at elevated temperatures. They also demonstrate the temperature limitations of low alloy steels (due to poor oxidation and creep resistance) despite their "acceptable" erosion-corrosion resistance in low velocity conditions(2). For erosion at low temperatures, the general trend is for the higher strength materials to provide better erosion resistance, with the exception being ceramic materials, where low fracture toughness may limit their application at high velocities, Fig. 1.

This paper outlines the various steps in the construction of materials performance maps for combined erosion-oxidation, oxidation and creep resistance at elevated temperatures. The limitations of the current approach are outlined. Future issues, such as the generation of erosion-oxidation maps from theoretical methods, are also addressed in this talk.

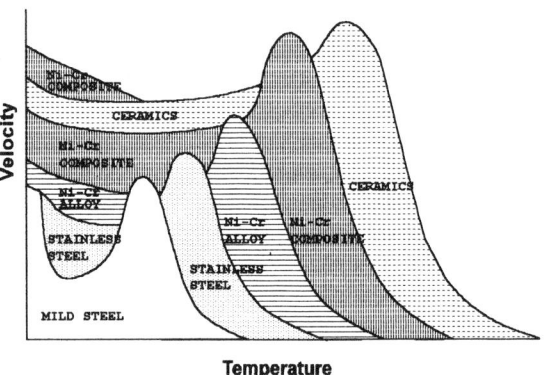

Fig. 1: Schematic diagram of a materials selection map for elevated temperature erosion resistance(3-4).

REFERENCES

1. G. Sundararajan, Wear, 145, 251-282, 1991.
2. A.J. Ninham, I.M. Hutchings and J.A. Little, Corrosion, 46, 4, 296-301, 1990.
3. M.M. Stack, Trib. Int., 1997 (in press)
4. M.M. Stack, Mat. at High Temp, 1997 (in press)
5. M.F. Ashby, Materials Selection in Mechanical Design, Pergamon, 1992.

THE INFLUENCE OF DYNAMIC STRAIN-AGEING ON THE EROSION BEHAVIOUR OF MILD STEEL

M S BINGLEY, A BERHE
The University of Greenwich, Wellington Street, Woolwich, London SE18 6PF, UK
A J BURNETT
The Wolfson Centre for Bulk Solids Handling Technology, The University of Greenwich, Wellington Street, Woolwich, London SE18 6PF, UK

ABSTRACT

The erosive properties of three compositionally similar mild steels : a BS 4360 pipe steel, a black mild steel and an EN3A steel, were investigated. A "rotating disc accelerator" type of erosion tester was used with olivine sand as the abrasive. The erosion resistance of the steels increased in the order given above with the pipe steel behaving particularly poorly. The reason for the contrasting behaviour of the three steels was examined.

The black mild steel had the highest ambient temperature hardness/strength (a consequence of it's smaller grain size) whilst the other two steels had similar values to each other. Ductilities of all the materials were virtually identical. Charpy impact tests revealed that the steels had different ductile-brittle impact transition temperatures although all the materials exhibited virtually identical impact energies between room temperature and 100°C. Initial results, therefore, whilst highlighting some differences between the materials failed to explain the trends in erosive properties. For instance, it might have been expected that the black mild steel having a superior combination of strength/hardness and ductility would have the best erosion resistance.

It has been suggested that high temperatures may occur in the plastically deformed zone adjacent to an impacting particle.(1) Tensile and impact tests were therefore carried out at temperatures of up to 600°C in order to determine material behaviour at elevated temperatures. The tensile tests revealed startling differences between the three steels. The two steels with the poorest impact resistance (the pipe steel and black mild steel) displayed serrated stress-strain curves throughout the 200-300°C temperature range, with a corresponding dramatic decrease in ductility (see Fig 1) and sometimes increase in strength. This phenomena can only be explained by dynamic strain ageing. EN3A, the steel with the highest erosion resistance, did not appear susceptible to it. Charpy impact tests on the black mild and pipe steels indicated an identical embrittlement effect but perhaps due to the higher strain rate this was observed at slightly higher temperatures (300-400°C). The effect was again absent in the EN3A steel.

It is proposed that the dynamic strain-ageing phenomena observed during the tensile and impact tests on the black mild and pipe steels (but not the EN3A steel) also occurs during erosion. This would explain the apparent anomalies in erosion behaviour. The temperature required for the onset of dynamic strain-ageing has been shown to be strain-rate sensitive. In erosion, local strain-rates during particle impact may be much greater than in a Charpy impact test suggesting that the local temperature rise is at least 300°C but may be even higher. The general implication of this work is that it is inappropriate to relate erosion test results to room temperature mechanical properties.

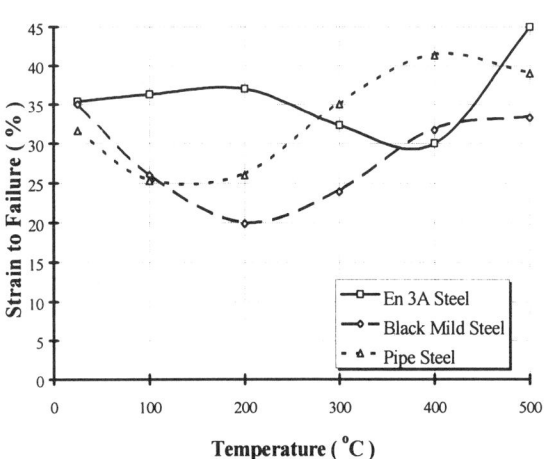

Fig 1. Strain to failure versus test temperature.

REFERENCES

(1) IM Hutchings and AV Levy, Thermal effects in the erosion of ductile metals, Wear, 131(1989) 100-121pp.

EROSIVE WEAR OF POLYURETHANE ELASTOMERS

J I MARDEL
Centre for Advanced Materials Technology, Monash University, Clayton, Victoria, 3168, Australia
M FORSYTH
Department of Materials Engineering, Monash University, Clayton, Victoria, 3168, Australia
A J HILL
CSIRO Division of Materials Science and Technology, Clayton, Victoria, 3168, Australia

ABSTRACT

Thermoplastic polyurethane elastomers (TPUs) are widely utilised as wear resistant materials in the Australian mining industry. They have been used to replace metals in the manufacture of ore screens as they have a high level of resistance to wear, cutting and tearing damage in highly abrasive environments. This work examines the relationship between the erosive wear performance of a TPU and its morphology.

TPUs are segmented block copolymers of the form $-(H_A-S_B)_x-$ where H_A refers to the hard segment and S_B the soft segment material. At room temperature, the soft segments are flexible whilst the hard segments are rigid. Due to thermodynamic incompatibilities between the hard and soft segments, phase separation can occur leading to the formation of a two phase system, or morphology, within the TPU (1)(2). The thermal history of TPUs can directly affect the mechanical and physical properties (3)(4) due to alterations in the morphology.

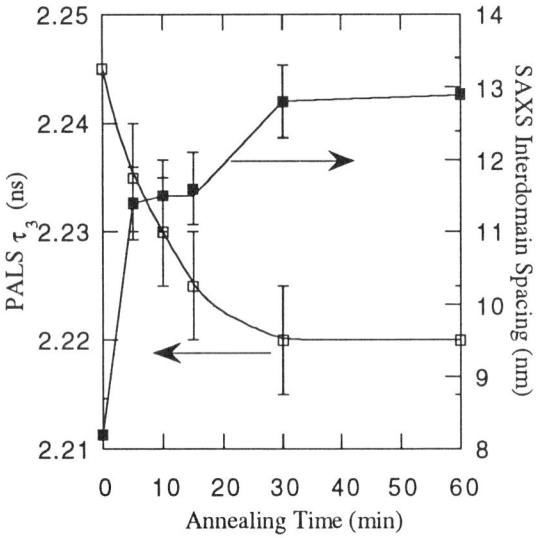

Fig 1. Relationship between change in free volume (from PALS) and interdomain spacing

Samples of the TPU were annealed under various conditions and the morphological development within the material was measured by small angle x-ray scattering (SAXS) and positron annihilation lifetime spectroscopy (PALS). The relationship between the interdomain spacing (as measured by SAXS) and the PALS τ_3 para–meter (which is related to free volume size within the TPU) is presented in Fig. 1. The effect of annealing at 155°C on the erosive wear performance (angle of incidence = 30°) of the TPU is shown in Fig. 2.

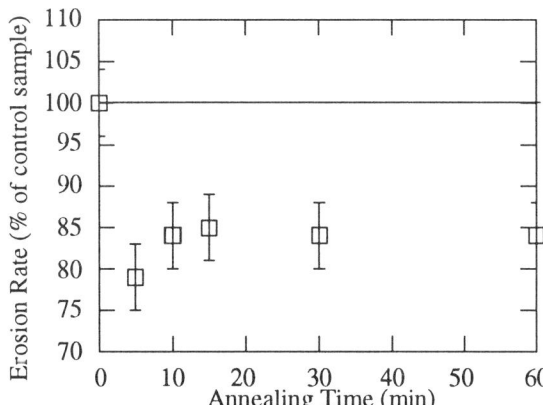

Fig 2. Effect of annealing TPU at 155°C on relative erosive wear performance

The wear mechanisms were observed using SEM and were similar to those reported previously(5)(6). Annealing did not appear to lead to a change in the wear mechanisms, indicating that the improvement in erosive wear performance is directly related to the change in morphology caused by the annealing process.

REFERENCES

(1) J T Koberstein, T P Russell, Macromolecules, Vol. 19, 1992, 714-720pp.
(2) Y Li, Z Ren, H Yang, B Chu, Macromolecules, Vol. 26, 1993, 612-622pp.
(3) S N Lawandy, C Hepburn, Elastomerics, Vol. 112, 1980, June, 24-26pp, October, 43-48pp.
(4) G L Wilkes, R Wildenauer, J. Appl. Phys, Vol. 44, 1975, 4148-4152pp.
(5) J Li, I M Hutchings, Wear, Vol. 135, 293-303pp.
(6) J I Mardel, K R Chynoweth, A J Hill, Materials Forum, Vol. 19, 1995, 117-128.

DESIGN OF A HIGH VELOCITY AIR-SAND JET IMPINGEMENT EROSION FACILITY

RJK WOOD, DW WHEELER
Department of Mechanical Engineering, University of Southampton, Highfield, Southampton, SO17 1BJ, UK.

ABSTRACT

The design and capabilities of a new high velocity air-sand jet erosion test facility are described. It can be used to study the erosion behaviour of materials and coatings which may be used in components which may operate in highly erosive environments such as offshore control valves. These valves suffer from erosion caused by the impingement of sand-containing hydrocarbon fluids on the valve interior. Impingement velocities as high as 400m/s can occur, especially if high pressure dry gas or low levels of liquid hydrocarbons are being produced. The high cost of replacing damaged subsea valves means that ultra-hard wear-resistant coatings for internal parts of these valves are highly desirable. Candidate coatings can be evaluated in this rig where erosion conditions similar to service ones are simulated. It allows coating behaviour to be studied at impact velocities of up to 340m/s.

In the rig, air is produced by a 30kW rotary screw compressor and dried to a dew-point of -70°C before passing through a receiver, pressure regulator and flow meter. Sand is then injected into the dry air-stream in an injection chamber. This air-sand mixture is then accelerated down a 16mm diameter steel tube, 1 metre in length, into the erosion chamber where it strikes the test specimen. Using Computational Fluid Dynamics (CFD), the compressible flow characteristics of the rig have been investigated, allowing optimisation of the nozzle diameters and particle acceleration length. Particle volume fractions were also computed, allowing guidelines to be constructed for minimising particle-particle interactions at the target surface and nozzle exit. CFD was also used to predict the degree of jet spreading as well as sand particle trajectories. The erosion chamber has the facility to vary the sample to air-sand jet impingement angle from 90 to 20° as well as the sample to nozzle stand-off distance.

Full details of all aspects of the rig calibration are discussed, in particular sand feed rates as well as pressures and particle velocities against air flow rate. The air flow rate can be varied between 40 and 380m^3/hr, at 2 barg and 15°C, whilst the sand injection system has a turn down ratio of 100:1 to enable feed rates of between 1 and 100g/min, or flux rates of 0.05 to 5.3 kg/m^2/sec to be obtained. A sub-angular quartz sand blend was used with a similar size distribution to that found in North Sea oil fields. It had a broad size distribution in the range 90-355μm with an average size of 194μm. Minimal particle-particle interactions within the nozzle are predicted as the inter-particle distances are between 7 and 46 times the mean particle diameter. The highly energetic nature of the test means that sand degradation is a significant factor. The change in size distribution over the course of a test is also investigated.

The particle velocities at different air flow rates were measured using an Imacon 790 high speed camera. By using speeds of up to 10^6 fr/sec individual particles can be tracked then image processed and the velocities calculated. A calibration graph is presented which indicates that sonic velocities are attainable for smaller sand sizes.

The initial results obtained using this rig are also discussed. These were performed using AISI1020 carbon steel, sintered tungsten carbide-6% Cobalt, and two types (D-gun and HVOF) of thermally sprayed WC-Co-Cr coatings, 130μm and 200μm respectively, on carbon steel.

The carbon steel and sintered tungsten carbide samples were tested at velocities of 148m/s and 270m/s. Employing a sand flux of 0.5kg/m^2/sec and a 28mm stand-off distance they were tested at an impingement angle of 90°. The tests were periodically interrupted to weigh the samples to obtain cumulative volume loss against time. This also enabled the onset of steady state erosion conditions to be determined. The effect of varying the sand flux on the erosion rate was also investigated and found that a 17% reduction in erosion rate with an increase in flux over the range 0.5 to 7.5kg/m^2/sec. This finding, which is independent of velocity, is thought to be due to particle-particle interaction at the target surface. Increasing the flux rate had no significant effect on the diameter of the erosion scar, at a given velocity and sand loading, and thus the effect of airborne particle-particle interactions can be considered negligible.

The thermally sprayed coatings were tested at a particle velocity of 148m/s at a flux rate of 0.5kg/m^2/sec for 60 minutes. The steady state erosion rates of the D-gun and HVOF samples were 824μm^3/impact and 251μm^3/impact respectively. These compare to 470μm^3/impact for carbon steel and 23 μm^3/impact for the sintered tungsten carbide.

ACKNOWLEDGEMENTS

The authors wish to thank Alister Forder for CFD predictions, DTI/EPSRC for funding, and Rutherford-Appleton laboratory for the high speed camera loan.

EFFECT OF GRINDING ON SURFACE ROUGHNESS AND WEAR OF SILICON NITRIDE

V S NAGARAJAN, L K IVES and S JAHANMIR
National Institute of Standards and Technology, Gaithersburg, MD 20899, U. S. A.

ABSTRACT

Although a number of investigations have been conducted regarding the effects of speed, load, environment and lubricants on the wear of silicon nitride (1), a systematic study on the possible effects of machining induced damage on wear has received little attention. Grinding can result in a considerable amount of surface and subsurface damage in ceramics. The severity of the damage depends on such grinding conditions as feed rate, depth of cut, grinding wheel grit size, etc. Previous studies have indicated that there exists a strong correlation between the microstructure and mechanisms of material removal and damage formation during grinding in a variety of ceramics (2). In this paper, we discuss the effect of grinding induced subsurface damage on the wear of a commercially available sintered reaction-bonded silicon nitride (<1% porosity, 700 MPa fracture strength) sliding against steel.

The silicon nitride specimens were ground under a range of conditions using horizontal spindle surface grinders with resin-bond diamond-grit wheels. The grinding conditions were varied by selecting a number of different combinations of table speed, downfeed, and wheel grit size. The surface roughnesses of the ground specimens were measured using a stylus profilometer. Roughness increased as the grinding conditions became more severe. Specimens ground under the least-severe and most-severe conditions (specific removal rates of 1.25 mm^2/s and 15.6 mm^2/s respectively) were selected for the wear study. A polished silicon nitride sample was also studied for comparison. The wear tests were performed using a block-on-ring tribometer. The silicon nitride specimens were used as the blocks. A carburized AISI 4620 steel (average roughness= 0.18 μm) was used as the ring. A commercial 10W/30 engine oil was used as the lubricant. The wear tests were conducted with two different orientations of the blocks with respect to the grinding direction. In one case, the sliding direction was parallel to the grinding direction. In the other case, the sliding direction was perpendicular to the grinding direction. The wear volumes of the blocks and the rings were calculated from the cross-sectional areas obtained from surface profile traces across the wear scars.

The coefficient of friction remained constant independent of surface preparation, i.e., polished or ground. The wear of the ground silicon nitride specimens was significantly higher compared to wear of the polished specimens. The direction of sliding did not have a significant effect on wear of silicon nitride ground under the least-severe condition, nor was there an effect on the wear of the steel ring. Under the most-severe grinding condition, however, the wear of silicon nitride was higher when sliding was parallel to the grinding direction; wear was also higher on the steel ring. The wear of both the silicon nitride and steel increased as the severity of grinding was increased. The wear of steel was two to three times higher than that of the silicon nitride under all conditions. The wear scar of the polished specimens appeared smooth without any evidence for fracture; whereas, for the ground specimens, the wear surfaces were rough and fractured in appearance. There was, however, little difference between the least and the most severely ground specimens. Wear of the polished specimens was dominated by a tribochemical mechanism. But, the material removal process of the ground surfaces was dominated by a microfracture process.

Damage induced during grinding usually consists of near-surface intergranular and intragranular microcracks, subsurface median type cracks (3), and near-surface compressive residual stresses (4). The median type cracks produced by grinding did not play a dominant role in these wear tests, although these cracks decreased the fracture strength by about 30%. The initial wear rate increases with the severity of grinding condition in connection with the increase in the thickness of grinding induced damage layer. The lower wear rate while sliding perpendicular to the grinding direction is attributed to the larger compressive residual stresses in that direction.

REFERENCES

(1) Jahanmir, S., *Friction and Wear of Ceramics*, Marcel Dekker, New York (1994).
(2) Xu, H. H. K., and Jahanmir, S., *Ceram. Eng. Sci.*, 16, pp. 295-314 (1995).
(3) Rice, R. W., *Machining of Advanced Materials*, S. Jahanmir (Ed.), NIST SP-847, U. S. Government Printing Office, Washington DC, pp. 185-204 (1993).
(4) Marshall, D. B., Evans, A. G., Khuri Yakub, B. T., Tien, J. W., and Kino, G. S., 1983, *Proc R. Soc. Lond., A 385*, pp. 461-475 (1983).

Accelerated Wear of Ceramics by Tribo-chemical Effects

R. G. KAUR and T. A. STOLARSKI
Department of Mechanical Engineering, Brunel University, Uxbridge, Middlesex, UB8 3PH
D. A. COATES and A. GELDER
Castrol International, Pangbourne, Reading, Berkshire, RG8 7QR

ABSTRACT

Ceramics are hard and brittle, therefore, machining such materials is time consuming, difficult and expensive. A low cost, efficient-machining process that can remove material rapidly while maintaining a good surface fininsh is needed. The rate of removal during grinding of ceramic balls is very small when conventional methods are used. Grinding is usually carried out under relatively low loads and speeds as cracks are introduced by vigorous attack with abrasives (1).

The general aim of the studies presented is to find a correlation between the wear mechanisms and surface modifications induced by interactions of tribochemical nature and to identify the most effective combination of parameters in producing high removal rates. These objectives were achieved by studying the friction and wear of silicon nitride balls using a ball-on-plate tribotester and a modified four ball machine. Also the contact area was analysed using high Resolution Scanning Electron Microscope (SEM).

The sliding friction coefficients of tool steel in contact with silicon nitride were measured under boundary lubrication conditions at temperatures ranging from 30-150°C using the TE70 Micro Friction Machine.

The following test conditions were used:
Test Sample   Ceramic ball sliding on steel plate
Lubricants   T80853, T80854, T80855, T80856
               T80857, T80884, Kemet
Contact Stress-304MNm^{-2}, Stroke Length-0.3mm
Frequency-45Hz, Max. Velocity-0.6ms^{-1}

The test results are presented in Fig.1 and Fig 2 shows the SEM image of the wear scar on ceramic ball.

Commercial grinding slurries are normally used in industry for grinding ceramics. An average content of abrasive particles in these slurries is 0.1g per 50ml of liquid. The same amount of 15µm diamond abrasive particles was added to 50ml of each of the test lubricants. Silicon nitride balls of 6.5mm in diameter were used. All the tests were carried out in a configuration typical for the four ball machine (2). The load on a set of nine balls was 400N and the speed was 3000rpm. Test duration was 1 hour.

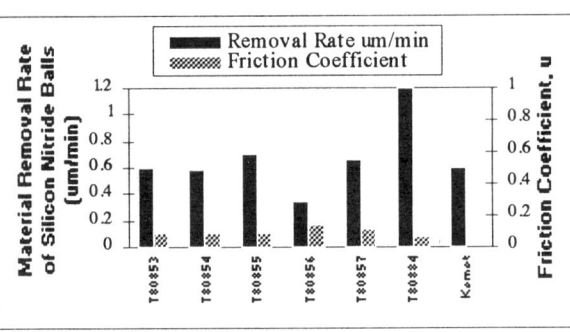

Fig.1: The Material Removal Rates in different lubricants with 15µm diamond particles and the Friction Coefficient values for a ceramic ball sliding on a steel plate using those lubricants

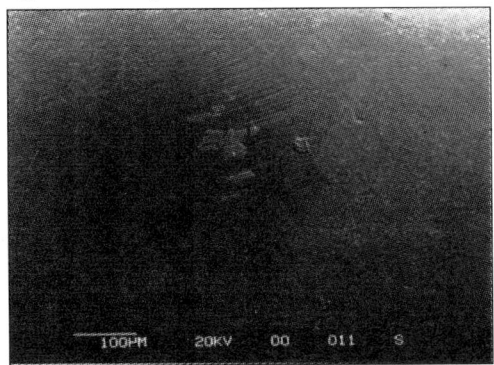

Fig. 2: Wear Scar on Ceramic Ball after Friction Test with T80884 Liquid

The results showed that the highest removal rate and lowest friction coefficient was obtained using T80884 liquid. The lowest removal rate and highest friction coefficient was obtained using T80856 liquid. In Fig. 2, groove marks can be seen on the contact area of the ceramic ball which appear to be similar to the machining marks on the steel plate. This could indicate that a tribochemical reaction was modifying the surface of the ball.

All the result of the research are presented and comprehensively explained in the full length paper of the same title.

REFERENCES

(1) T. A. Stolarski, Ceramics International, 1992, Vol.18, pp.379-384.
(2) R. Ahmed and M. Hadfield, Tribology International, 1997, pp.129-137.

Sliding Wear of Some Si$_3$N$_4$/SiC-Based Ceramic Composites for Face Seal Applications

RICHARD WESTERGÅRD, ALEXANDRA ÅHLIN, NIKLAS AXÉN and STURE HOGMARK
Uppsala University, The Ångström laboratory, Dept. of Materials Science, Materials Science Div
S-751 21 Uppsala, Sweden

ABSTRACT

A number of new Si$_3$N$_4$/SiC-based ceramic composites, intended for face seal applications, have been investigated with the emphasis on the sliding wear and friction behaviour. In particular, the influence of water, vapour, or oil on the friction, wear and tribofilm formation has been studied. The ceramics consisted of Si$_3$N$_4$ and SiC in the proportions 7:3. TiN, 30 %, and BN (hex.) in different amounts up to 8% was also added. A pure Si$_3$N$_4$ was investigated for reference purposes. All specimens were produced by hot isostatic pressing.

The TiN added to the composites was thought to form a friction and wear reducing tribofilm consisting of titania. Previous investigations of the Si$_3$N$_4$/TiN tribosystem has been performed by Kustas et al (1) and Skopp et al (2,3). Hexagonal BN has a layered lattice structure similar to graphitic carbon and a low friction is expected, as has been shown by Rowe (4).

X-ray diffraction was used to reveal the phase composition. Scanning electron microscopy (SEM) was used to study the microstructure on polished and etched specimens. The hardness was assessed using Vickers indentation. A comparative investigation of the thermal shock resistance was performed by studying crack growth during successive heating and cooling. A pin-on-disc tribometer was used to assess the friction coefficient and the sliding wear rate. The wear was quantified as mass loss or with surface profilometry. The chemical composition of the films was studied using ESCA, GDOES and AES.

All materials showed a very dense micro-structure without any signs of pores. The Si$_3$N$_4$ and SiC grain sizes was around 1 µm. The BN was present as elongated structures often considerably less than one µm wide. X-ray diffraction revealed no significant amounts of mixed phases. It was also observed that the Si$_3$N$_4$ in the specimens containing BN had transformed completely to the preferred β-phase, thus indicating that BN acts as a sintering aid.

Crack propagation studies of hardness indentation corners indicated that BN grains and grain boundaries were the mechanically weakest links. The thermal shock resistance of all composites was found to be lower than that of pure Si$_3$N$_4$. Of the investigated composites the specimens containing BN displayed the largest thermal shock resistance.

In the sliding tests all materials formed a ca. 1 µm thick oxide rich tribofilm in dry tests. Water lubrication strongly reduced the wear (two orders of magnitude) and the tribofilm formation became much less pronounced. Water also reduced the friction from between 0.8-1.2 to 0.2-0.6. Lubrication with oil led to insignificant wear and the friction was reduced to around 0.07. In dry wear no significant differences between the materials could be measured. With water, however, the composite with BN displayed a much lower wear rate than the other composites.

It was found that tribochemical reactions occurred on most surfaces resulting in tribofilm formation. ESCA and GDOES depth profiling showed that the tribofilm was much richer on oxides and poorer on nitrides relative to the bulk. There was a tendency towards enrichment of silicon in the tribofilm. Very low amounts of boron and slightly lower titanium concentrations in the top layer relative to silicon could be detected. Low angle X-ray diffraction revealed the tribofilm to be partly amorphous.

The strong difference in wear rate between dry and lubricated wear in general, and the clear separation in material performance in the water lubricated tests in particular, indicate that the formation and maintenance of a lubricating film should be crucial of the material's performance in a face seal application.

Wear and friction behaviour as well as tribofilm formation mechanisms are discussed on the basis of the present results. It is concluded that water lubrication significantly reduces friction and wear of the composites. Of the present specimens, those containing BN appear to be the most promising ceramic composites for face seal applications.

REFERENCES

(1) F. M. Kustas, B. W. Buchholtz, Trib. Trans., 39, pp 43-50, 1996
(2) A. Skopp, M. Woydt, K. H. Habig, Wear, 181-183, pp 571-580, 1995
(3) A. Skopp, M. Woydt, Trib. Trans., 38, pp 233-242, 1995
(4) G. W. Rowe. Wear, 3, 274, 1960

OPTIMUM SURFACE LAYER OF TRANSFORMATION TOUGHENED CERAMICS FOR IMPROVED WEAR PERFORMANCE[1]

N B THOMSEN
Danfoss A/S, DK-6430 Nordborg, Denmark/
Risø National Laboratory, P.O. Box 49, DK-4000 Roskilde, Denmark

ABSTRACT

The poor rolling and sliding wear performance of transformation toughened ceramics (TTC) such as partially stabilized zirconia (PSZ) is believed to be a result of the phase transformation of precipitates located immediately beneath the contacting surface. These precipitates, which on the one hand enhance the toughness of the ceramic by preventing the growth of surface cracks, induce a surface uplift due to their transformation which alters the rolling/sliding conditions and acts as a source of cracks (for details see (1), (2) and (3)).

There is thus a need to alter the microstructure in such a way as to provide the maximum possible crack tip shielding without exceeding a tolerable surface uplift. In order to find a compromise solution, an optimization problem is formulated for an idealized model consisting of an edge crack normal to the contacting surface and an infinite, periodic distribution of transformable grains in the layer immediately beneath this surface.

The stress and displacement fields due to the infinite arrays of transformable inclusions needed to determine surface uplift and the toughening effect when an edge crack is present in the surface layer are obtained through the use of complex potential functions. The expressions for the complex potentials and further details on their derivation may be found in (4) and (5). The shielding effect on an edge crack positioned on the symmetry plane of the inclusions is found by equating the stress on the crack due to inclusions to that due to the interaction of dislocations used to model the crack (4)(5).

The objective is to maximize the crack tip shielding by varying the volume fraction of transformed material and size of the layer without exceeding a prescribed allowable surface uplift.

It is shown that
1. the volume fraction of the transformable phase in the near-surface region should be close to zero in order to avoid excessive surface uplift. This value is considerably lower than volume fractions pertaining in peak-aged PSZ (50-75%) and may explain why this material has exhibited poor performance in a variety of tests and applications.
2. below the near-surface region, but where contact stresses still may be expected to cause transformation, the volume fraction of transformable phase reaches values less than 25%.

For more details on results see (4) and (5).

The optimum solution would suggest that the region immediately beneath the wear surface in TTC should contain only a small amount of transformable material compared to the bulk of TTC. This might be achieved in practice by subjecting the surface to heat treatment, for instance by lasers. The size of the near-surface region and the distribution of transformable phase within and beyond it is determined by the load and crack configuration, as well as the maximum allowable surface uplift. There is however a need for further exploration of this problem. The optimization techniques are eminently suitable for such an exploration, thus putting what has been hitherto a trial and error procedure on a sound physical basis.

ACKNOWLEDGMENTS

Support for this research from the Australian Government through the Overseas Postgraduate Research Scholarship Scheme is gratefully acknowledged. Current financial support from the Danish Academy of Technical Sciences (ATV) is greatly appreciated.

REFERENCES

(1) B. Budiansky, J.W. Hutchinson, J.C. Lambropoulos, International Journal of Solids and Structures, Vol. 19, No. 4, page 337, 1983.
(2) J.F. Braza, H.S. Cheng, M.E. Fine, Scripta Metallurgica, Vol. 21, page 1705, 1987.
(3) H.G. Scott, Wear of Materials (K.C. Ludema, ed.), page 8, 1985.
(4) N.B. Thomsen, B.L. Karihaloo, ASME Journal of Tribology, Vol. 118, page 740, 1997.
(5) N.B. Thomsen, Advances in Fracture Research (B.L. Karihaloo, Y.-W. Mai, M.I. Ripley, R.O. Ritchie, eds.), Proceedings of the Ninth International Conference on Fracture (ICF9), Vol. 2, page 1105, 1997.

---

[1] The work reported was initiated and partly performed at the School of Civil and Mining Engineering, University of Sydney, NSW 2006, Australia

MODELS FOR CERAMIC LUBRICATION BY TRIBOPOLYMERIZATION AT HIGH LOADS AND SPEEDS

M J FUREY and B R TRITT
Virginia Polytechnic Institute and State University, Blacksburg, VA 24061-0238, USA

C KAJDAS and R KEMPINSKI
Warsaw University of Technology, Plock, Poland 09-400

ABSTRACT

An experimental study of ceramic lubrication by tribopolymerization at high loads and high speeds is presented. The information obtained is used to develop new models of anti-wear action of selected monomers on ceramics in tribological processes—an approach that has been demonstrated to be effective in both liquid and vapor phase studies (1-3).

In the first phase of this research, the effects of load and speed on ceramic wear were determined for six different monomers of widely-varying structure. The monomers—used at concentrations of 1% in a hydrocarbon carrier fluid (hexadecane)—consisted of: (a) one condensation-type monomer, a partial glycol ester of a long-chain dimer acid; (b) four vinyl-type addition monomers varying in structure; and (c) an unusual nitrogen-containing monomer found in previous studies to be effective in reducing ceramic wear even at very low concentrations, e.g., 0.01%.

The tests were carried out in a pin-on-disk machine with alumina-on-alumina. A two-factor, two level designed experiment was conducted for the hexadecane carrier and each of the six monomer-containing fluids at loads of 40 and 160N and sliding speeds of 0.25 and 1.00 m/s. Thus, the range in frictional heat generation was 16 to 1.

The results of the wear tests in this phase of the study were rather surprising and changed our thinking on the mechanism(s) by which monomers can act to reduce ceramic wear. For example, at low speeds—regardless of load—the monomers used were very effective in reducing wear, with reductions ranging from 37 to 98 percent depending on the monomer and load. However, at high speeds, the monomers were generally ineffective; in some cases, increases in wear were observed. This was unexpected.

Possible reasons for this behavior are discussed, including thermal degradation of organic/polymeric films at high surface temperatures, effects of triboemission, and tribochemical reactions producing aluminum soaps or complexes. The worn ceramic specimens were also examined using scanning electron microscopy and FTIRM (Fourier Transform Infrared Microspectrometry). There is evidence of tribopolymerization on the ceramic wear regions but not in all cases of anti-wear action. Other tribochemical reactions also occur. It also appears that the alumina wear debris can, in combination with the surface products formed, provide some benefit.

In an attempt to model and understand the behavior of monomers used in our research on ceramic lubrication, a second phase of this study was conducted in three areas, i.e., (a) the use of a chemical modeling computer program to examine the possible orientation of monomers on ceramic surfaces prior to tribochemical reactions with and on the surfaces, (b) the calculation of surface temperatures produced by friction using a more sophisticated approach developed by Vick and (c) the testing of hypotheses designed to explain the results of the first phase of this study. Indeed, there appears to be a relationship between wear volume and calculated surface temperatures assuming contact area governed by either elastic or plastic deformation. There is an effective intermediate temperature range for the monomers; above this (e.g., above ca 400-500°C), the compounds used in this first study do not reduce ceramic wear. But the results of recent additional wear tests show that indeed it is possible to affect significant reductions in ceramic wear at high loads and speeds, i.e., at high levels of frictional heat generation, by the use of new monomers. With compounds designed to polymerize on rubbing surfaces at higher temperatures and/or form more thermally stable polymeric films, we have been able to demonstrate ceramic wear reductions ranging from 95 to 99% at high levels of frictional heat generation. This information is used to develop new models of ceramic lubrication by tribopolymerization—a technique shown to be effective for metals as well.

ACKNOWLEDGMENTS

The authors wish to acknowledge the National Science Foundation as well as the Energy-Related Inventions Program of the U.S. Department of Energy for their support of this research.

REFERENCES

(1) M. J. Furey and C. Kajdas, Proc. 4th International Symposium on Ceramic Materials and Components for Engines, Goteborg, Sweden, Elsevier, pp. 1211-1218, 1992.
(2) M. J. Furey and C. Kajdas, U.S. Patent 5,407,601, Compositions for Reducing Wear on Ceramic Surfaces, issued April 18, 1995.
(3) J. C. Smith, M. J. Furey and C. Kajdas, Wear, Vol. 181-183, pp. 581-593, 1995.

MICROSTRUCTURE AND TRIBOLOGICAL PROPERTIES OF ALUMINA CERAMIC WITH LASERDISPERSED METALLIC ADDITIONS

KLAUS PRZEMECK and KARL-HEINZ ZUM GAHR

University of Karlsruhe, Institute of Materials Science II and Karlsruhe Research Center, Institute of Materials Research I, P.O. Box 3640, D-76021 Karlsruhe, Germany

ABSTRACT

Advanced alumina ceramics have a lot of applications as materials for components with high tribological, mechanical and thermal requirements. Favourable properties of alumina such as high stiffness and hardness, low density, temperature stability or corrosion resistance are used for instance in seal-rings, draw-cones, guides or bearing parts.

However, monolithic alumina can suffer severe problems under high tribological and/or mechanical loads owing to inherent brittleness and the resulting lack of defect tolerance. Self-mated monolithic alumina, or alumina mated to metallic materials, shows relatively high friction coefficients and low wear resistance in unlubricated sliding contact (1). Tribological behavior of ceramics is mainly determined by plastic deformation, microfracture, abrasion, and formation of tribochemical reaction layers or interfacial layers (2). An improved performance can be expected by multiphase ceramic materials. Surface modification of a monolithic alumina can be sufficient if considering that tribologically induced interactions take place only in a relatively thin surface zone. Recent studies showed (3) that laser treatment of alumina ceramic offers a high potential for producing multiphase surface zones of thicknesses between about 100 to 300 µm by embedding oxides or hard particles which results in substantially improved tribological properties. Ductile metallic phases embedded in brittle ceramic materials can lead to materials with increased fracture toughness (4).

The aim of the present paper was to study the tribological behaviour of surface modified alumina ceramic with laserdispersed metallic additions. It was assumed that a relatively soft and ductile metallic phase may improve the tribological properties by avoiding intercrystalline microcracking in the contact area.

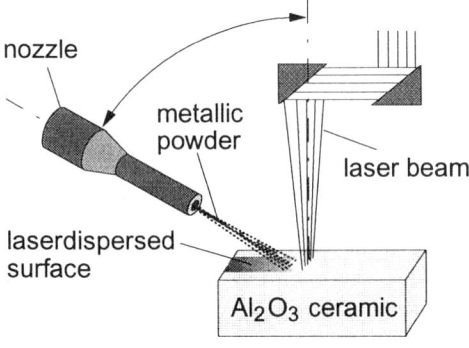

Fig. 1: Laser surface modification of Al_2O_3 ceramic

Metallic particles (W, Nb) were dispersed in slightly porous, commercially available alumina ceramic using infrared CO_2 laser radiation (Fig. 1). The thickness of the modified surface zone ranged between 100 to 500 µm depending on the metallic powders and process parameters used. The volume fraction of the metallic dispersoids was varied in the alumina matrix between 5 to 30 vol-%. Fig. 2 shows the microstructure of the monolithic alumina and the tungsten-laserdispersed alumina ceramic, respectively.

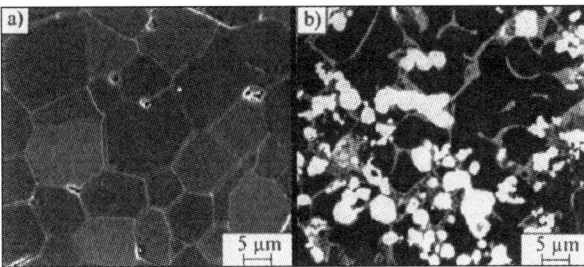

Fig. 2: Microstructure of (a) monolithic and (b) tungsten-laserdispersed Al_2O_3 ceramic (20 vol-% W)

Tribological behavior of the modified ceramics was analysed in unlubricated oscillating sliding contact against balls of Al_2O_3, cemented carbides (WC6Co) and steel (100Cr6), respectively. The tests were carried out with different normal loads and relative humidities in laboratory air at room temperature. In addition, the influence of the surface modification on the microstructural and mechanical properties such as hardness, fracture toughness, and Young's modulus was investigated.

It became obvious that the microstructure of the modified ceramics varied substantially depending on the type of the metallic additions and the process parameters used. The multiphase alumina showed improved fracture toughness but reduced hardness and Young's modulus. In comparison to monolithic alumina lower friction coefficients and higher wear resistance were measured on the multiphase alumina.

REFERENCES

(1) K. Kato, Wear, 136 (1990) 117 - 133.
(2) K.-H. Zum Gahr, W. Bundschuh, B. Zimmerlin, Wear, 162 - 164 (1993) 269 - 279.
(3) K.-H. Zum Gahr, C. Bogdanow, J. Schneider, Wear, 181 - 183 (1995) 118 - 128.
(4) X. Sun, J. Yeomans, J. Am. Ceram. Soc., 79 (1996) 2705 - 2717.

WATER-LUBRICATED AND DRY SLIDING WEAR OF YTTRIA STABILISED TETRAGONAL ZIRCONIA POLYCRYSTAL AND SILICON CARBIDE COUPLES

A M RICHES, J A YEOMANS AND P A SMITH
Department of Materials Science and Engineering, University of Surrey, Guildford, Surrey, GU2 5XH, UK.

ABSTRACT

The aim of this work is to investigate the sliding wear behaviour of zirconia and silicon carbide for use in a pump as part of a sub-sea gas well control system. Both water-lubricated and dry sliding wear are studied and wear rates and mechanisms compared.

Yttria stabilised tetragonal zirconia polycrystal (YTZP) pins were subjected to sliding wear against sintered silicon carbide discs using a high speed pin-on-disc machine. For each test, a single flat-ended pin was held, via a spring mechanism which was in turn attached to the tool post of a lathe, such that a contact pressure of 14 MPa was applied. The pins were polished prior to each test. The silicon carbide disc was held in the chuck of the lathe and relative sliding speeds up to 6 m s^{-1} were achieved.

Two series of experiments were conducted under identical conditions except that one was water-lubricated and the other was without lubrication. Wear rates of the pin were measured by mass loss and revealed that under water-lubricated conditions the wear rate was extremely geometry dependent. The alignment of the polished pin end governed whether or not there was a progressive wear scar across the face of the pin and this caused scatter in the results of nominally identical tests, see figure 1. Under water-lubricated conditions no wear was measured for pins that were polished flat in-situ on the rig. It was thought that the surfaces of the pin and disc were separated by a layer of water, i.e. that hydrodynamic lubrication was achieved. Higher pressures were then employed by polishing a taper on the end of the pin to reduce its area and investigate the breakdown of the water layer and cause wear of the pin. This breakdown began to occur at a pressure of 70 MPa.

An order of magnitude higher wear rate of the pin was observed under the dry conditions. A considerable temperature increase accompanied the unlubricated tests and a fluctuating white glow was observed at the sliding interface.

Disc wear was more difficult to quantify, due to transfer of YTZP from the pin, but was less severe than for the YTZP.

Scanning electron microscopy was used to examine the sliding surfaces and elucidate likely wear mechanisms. As expected from the wear data, this technique revealed a milder YTZP wear mechanism for water lubricated sliding. The surface of the water lubricated worn pins exhibited plastic deformation in the form of grooves parallel to the sliding direction. Areas of surface smoothing were observed, adjacent to which were rougher areas showing intergranular fracture and grain removal. Superimposed over some areas was a network of fine cracks tentatively assigned as thermal shock damage. Fracture on a much larger scale was observed on the surface for the pins from the series of dry experiments. Large flakes were formed on the surface and were subsequently removed by brittle fracture.

Wear of the disc was much milder in both cases, leading to a general smoothing of the surface. Grooves were observed in the wear track, as a result of loose debris. Transfer of YTZP to the disc was seen and was much more evident for unlubricated sliding. These results indicate that loss of lubrication could lead to severe wear in conditions which would normally produce mild wear. Further work is required to investigate the breakdown of the hydrodynamic lubrication layer under conditions of increased load or surface roughness and also to investigate the role of the tetragonal to monoclinic transformation on the wear mechanism using X-ray diffraction.

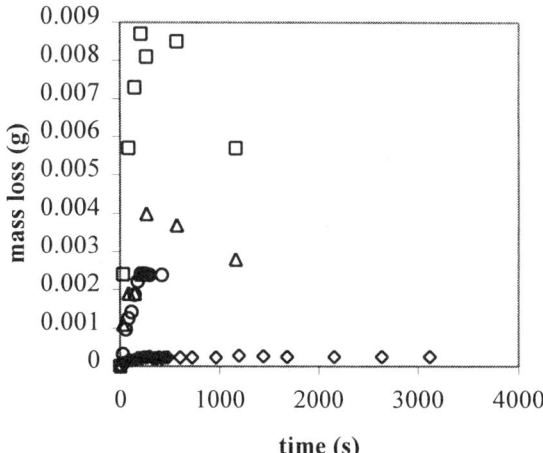

Figure 1: Scatter in mass loss results for four nominally identical water-lubricated tests at a contact pressure of 14 MPa.

ACKNOWLEDGEMENTS

The authors would like to acknowledge the support of EPSRC and BG plc in the form of a CASE award for AMR.

FUNDAMENTAL FRICTION AND WEAR PROPERTIES OF NEW POROUS CARBON MATERIALS : WOODCERAMICS

KAZUO HOKKIRIGAWA

Department of Mechanical Systems Engineering, Faculty of Engineering, Yamagata University, Yonezawa 992, Japan

TOSHIHIRO OKABE and KOUJI SAITO

Industrial Research Institute of Aomori Prefecture, Hirosaki 036, Japan

ABSTRACT

Woodceramics are new porous carbon materials obtained from wood or woody materials impregnated with phenol resin, and carbonized in a vacuum furnace at high temperature (1). (Fig. 1) During the carbonizing process, the wood or woody material changes to soft amorphous carbon and the impregnated phenol resin changes to hard glassy carbon. Woodceramics have superiour tribological characteristics. Because of their porous structure, they can be impregnated with various lubricants. They can be used for friction materials in high-temperature or corrosive environments because of their superiour corrosion resistance. They are sutable for friction materials for high-speed rotational elements, because of their very low density (0.6 - 1.1 g/cm^3).

The purpose of this investigation is to analyse the fundamental friction and wear properties of woodceramics in sliding contact with several materials. It then discusses the possibility of practical use of woodceramics as friction materials.

Ball on plate type friction apparatus was used. Plate specimens were made of woodceramics, which were made of medium density fiberboard (MDF) impregnated with phenol resin and carbonized in a vacuum furnace at 400°C - 2000°C. Ball specimens (R = 1.5mm, 4.0mm) were made of alumina, silicon nitride and bearing steel. Experiments were carried out unlubricated in air, impregnated with base oil and in water, at several normal loads and sliding velocities. The following principal results were obtained.

(1) Vickers hardness of woodceramics carbonized at 800°C - 1500°C takes high value. (Fig. 2)

(2) Woodceramics carbonized at 800 - 2000°C have consistently low friction coefficient (μ = 0.13-0.15) when unlubricated in air, impregnated with base oil, and in water. (Fig. 3)

(3) The specific wear rate of woodceramics carbonized at 800°C - 2000°C is less than 10^{-8} mm^2/N, which is low enough to allow woodceramics to be used in practical applications such as dry rubbing bearings, wet friction materials in automobile clutches, etc.

REFERENCES

(1) T. Okabe, K. Saito & K. Hokkirigawa, J. of Porous Materials, Vol. 2, No. 3, 1996, 207-214pp.

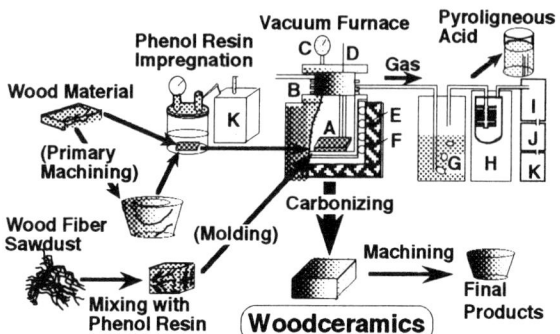

Fig. 1: A schematic diagram of making process of woodceramics

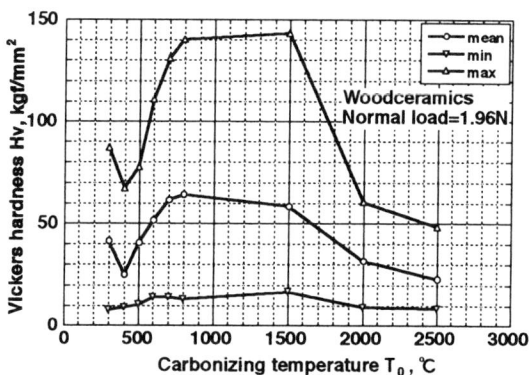

Fig. 2: The effect of carbonizing temperature on Vickers hardness

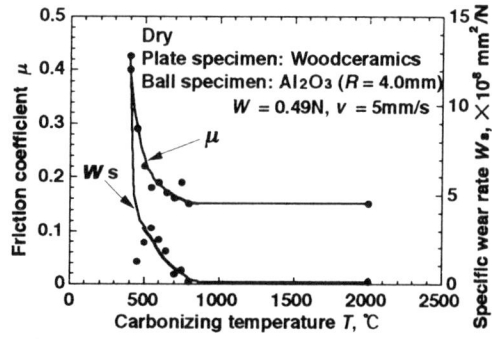

Fig. 3: The effect of carbonizing temperature on the friction coefficient and specific wear rate

THE FRICTION AND WEAR MECHANISMS OF CERAMICS AND AE CHARACTERISTICS

T. HISAKADO and H. SUDA
Department of Mechanical Engineering, Gunma University, Kiryu, Gunma, 376, Japan
M. SEKINE
ShinMeywa, Ltd., Sano, Tochigi, 327, Japan

INTRODUCTION

The use of acoustic emission(AE) event generated by the interacting surfaces has been studied by many investigators (1)(2). But it has hardly reported that the relationships between the friction and wear characteristics of ceramics and AE event signals were discussed considering the surface damages estimated from the topographies of worn surfaces.

The aim of this investigation was to study the use of AE event count rate for estimating the wear rate and surface damage due to sliding contact between ceramics under boundary lubrication, and the relationship between the wear rate and the various topographical parameters of worn surfaces.

EXPERIMENTAL APPARATUS AND PROCEDURE

The pin-disk wear test apparatus was used in the experiment. The wear tests were undertaken for combinations of a Si_3N_4 pin (d=12.7mm in diameter) against five kinds of ceramic disks under the conditions lubricated with oleic acid always supplied on a disk surface at the rate of 0.08g/h, under applied loads of W=3.92N and 5.88N and at four different sliding velocities of v=4, 6, 8, 10m/s, in air at $25 \pm 1°C$ with R.H. (relative humidity) $45 \pm 5\%$. The total sliding distance for all experiments was L= 60 km. AE signal was detected by an AE transducer contacting the AE transmission bar in contact with a pin and AE event signals from the counter was recorded at the rate of each 10 sec during the experiments.

EXPERIMENTAL RESULTS AND DISCUSSION

Fig.1 shows that except for vs. Si_3N_4 (W=3.92 N), the greater the AE event count rate, the greater the wear rate of a pin becomes, where the AE event count rates were averaged over a sliding distance. This suggests that the wear rates of pins can be estimated from AE event count rates for each combination. On the other hand, the AE event count rate for vs. Si_3N_4 disk (W=3.92 N and v=4 m/s) are high in spite of the low wear rate of the pin. This means that the oxidized oleic acid layers increase the AE event count rate because the AE event signal results from generating and wearing off a part of the layer, and back-transfer of the wear particles. But for vs. Si_3N_4 (W=3.92 N and v=10 m/s), when the oxide layers (3) such as SiO_2 and/or $Si(OH)_4$ are formed due to the temperature rise at the interface, the wear rate of the pin increases without increasing AE event signals.

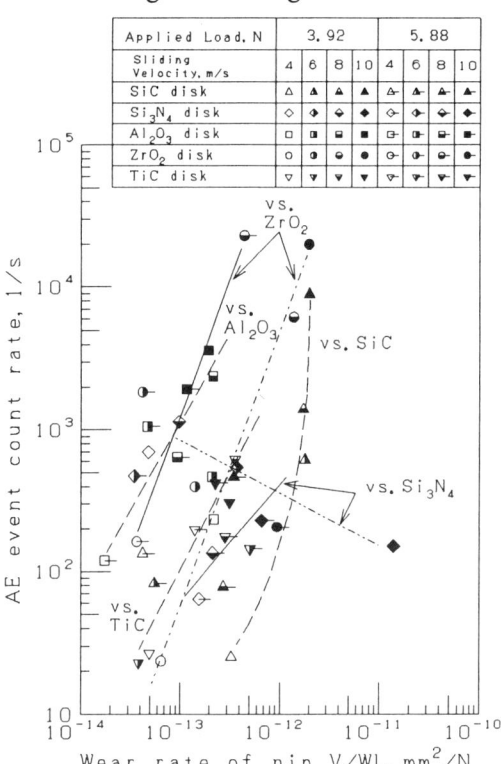

Fig.1 : AE event count rate and wear rate of pin

CONCLUSIONS

1. AE event count rates except for vs. Si_3N_4 increased with an increase in the wear rates of pins and disks.
2. For a pin vs. Si_3N_4 disk, when the surface layers such as the oxidized oleic acid layers were formed, the more those, the less the wear rates and the greater the AE event count rates become, but when the oxide layers (SiO_2 and/or $Si(OH)_4$) were formed due to the temperature rise at the interface, the wear rates of pins increased without increasing AE event signals.
3. The wear rates of pins and AE event count rates depended on the parameters of the worn surface topographies such as the mean depth of micro-grooves, the depth of mean line on pins and so on.

REFERENCES

(1) B.E. Klamecki and J. Hanchi, Trans. ASME, J. Trib., Vol.112, 1990, 469-476pp.
(2) K. Matsuoka, D. Forrest and M. K. Tse, Wear, Vol.162-164, 1993, 605-610pp.
(3) H. Tomizawa and T.E. Fischer, ASLE Trans., Vol.30, No.1, 1987, 41-46pp.

TRIBOLOGICAL CHARACTERISTICS OF PSZ AND TZP CERAMICS IN FRETTING AND RECIPROCATING SLIDING

G.B. STACHOWIAK and G.W. STACHOWIAK
Tribology Laboratory, Department of Mechanical and Materials Engineering, University of Western Australia, Perth, Australia

ABSTRACT

The unlubricated friction and wear characteristics of transformation toughened zirconia ceramics, Mg-PSZ and Y-TZP, under fretting and reciprocating sliding conditions were investigated. Ceramic-ceramic, ceramic-steel and steel-steel combinations were tested and the results compared based on the levels of friction, wear scar sizes and wear scar morphology. Wear and friction tests were carried out on a high frequency reciprocating wear test rig. In the fretting tests, two cylindrical specimens positioned at 90° to each other were employed, while in reciprocating sliding a pin-on-plate configuration was used.

The average steady-state coefficients of friction with standard deviations for fretting and reciprocating sliding are compared in Table 1. Overall, self-mated ceramics exhibited higher friction than self-mated steel or steel-ceramic pairs. It was found that the steady-state coefficients of friction of the steel-zirconia and steel-steel couples were about 0.6-0.7 and those of the self-mated zirconia about 0.8-0.9. The reciprocating friction tests were accompanied by high noise while the fretting tests were quiet.

Material	Fretting	Reciprocating sliding
PSZ-PSZ	0.87±0.03	0.81±0.07
TZP-TZP	0.91±0.06	0.82±0.07
steel-PSZ	0.59±0.05	0.68±0.06
steel-steel	0.63±0.03	0.57±0.03

Table 1. Comparison of the coefficients of friction in fretting and reciprocating sliding.

Wear scar profiles from PSZ-PSZ, TZP-TZP, steel-steel and steel-PSZ contacts after 10^6 fretting cycles at 5N and 50μm stroke are shown in Fig. 1. It was found that due to the absence of oxidative wear the self-mated PSZ and TZP ceramics were more resistant than self-mated steel to fretting wear in air of medium humidity. Microfracture, tribolayer formation and its subsequent delamination were the dominating wear modes occurring on zirconia ceramics, in contrast to oxidative wear observed on steel samples. It was found that the wear resistance of TZP ceramics was lower than that of PSZ ceramics under fretting conditions when the stroke length greatly exceeded the grain size of TZP ceramics. It appears that the ceramic resistance to fretting depends on the ratio of stroke to grain size, i.e. the lower the ratio the higher the resistance to fretting wear.

When steel samples were fretted against zirconia ceramics, a fragmentary metallic film transfer onto the ceramic surface occurred and surface polishing was observed on the ceramic wear scars. Since the combined wear of zirconia and steel was small, this combination of materials can be considered as a good choice to control and reduce fretting damage in air with medium humidity.

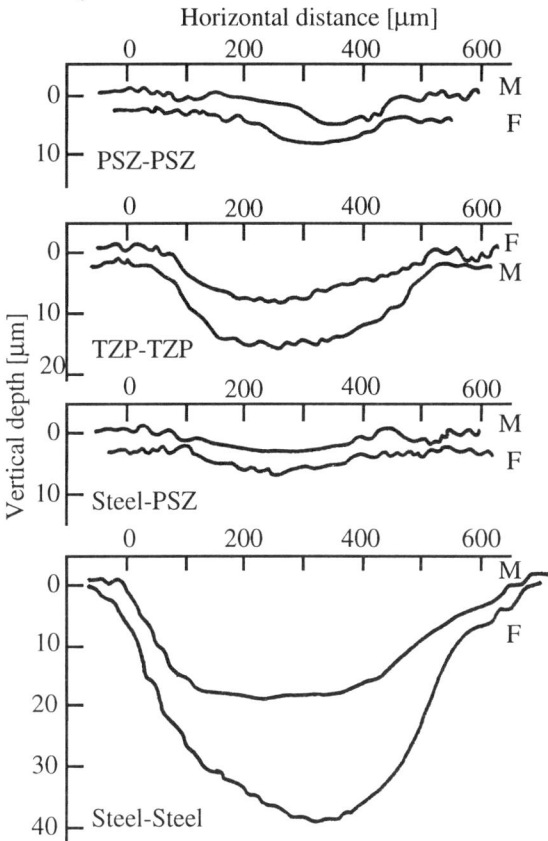

Fig. 1: Wear scar profiles from PSZ-PSZ, TZP-TZP, steel-steel and steel-PSZ fretting contacts; M-moving and F-fixed samples.

In reciprocating sliding in air of medium humidity both self-mated PSZ and TZP ceramics exhibited high wear coefficients of about 10^{-5} mm^3 N^{-1}m^{-1}. The wear modes were similar to those found in fretting.

The introduction of water into the ceramic-ceramic contacts resulted in the decrease of friction coefficient to about 0.5 and two-fold reduction in wear rates under certain conditions. Surface polishing was the dominating wear mechanism with occasional shallow pitting. It was found that TZP ceramics were more susceptible to cracking in water than PSZ ceramics.

THE INFLUENCE OF MICROSTRUCTURE, MECHANICAL PROPERTIES AND TOUGHENING MECHANISM TO THE ABRASIVE WEAR RESISTANCE OF TZP AND ZTA CERAMICS

JARI KNUUTTILA and TAPIO MÄNTYLÄ
Institute of Materials Science, Tampere University of Technology, P.O.Box 589, FIN-33101 Tampere, Finland

ABSTRACT

Zirconia ceramics are known as the toughest commercial oxide ceramics and are widely used in applications where high strength, toughness and wear resistance are required. The highest toughness values are achieved with tetragonal zirconia polycrystals (TZP) in which the unique toughening mechanisms of zirconia can be fully exploited. These mechanisms have been identified to be microcrack toughening, transformation toughening and also recently ferroelastic domain switching has been suggested to contribute to the enhanced toughness (1). Grinding has been observed to cause tetragonal-to-monoclinic transformation and consequently higher bending strengths due to surface compressing stresses.

In this work the influence of ferroelastic domain switching (FDS) on the improved wear resistance of tetragonal zirconia is studied and compared to other toughening mechanisms. Several, both laboratory produced and commercial tetragonal zirconias with varying yttria content were investigated for microstructure, mechanical properties and wear resistance. Also ceria stabilized zirconia, zirconia toughened aluminas (ZTA) and two aluminas with varying purity were studied for comparison. Their grain size, porosity and phase structure were determined. From mechanical properties the 4-point bending strength, hardness and fracture toughness were measured. The fracture toughness was determined by indentation and SENB -methods.

The abrasive wear resistance of the materials was measured by a rubber wheel abrasion equipment and quartz sand was used as the abrasive. Inorder to reveal the response of zirconia to high surface stresses rough grinding experiments were also performed.

The results show the importance of processing conditions and the resulting microstructure to the wear resistance of ZTA. Small grain size achieved by proprietary attrition milling procedure during suspension preparation is essential for improved wear resistance. Ball milling was not able to break agglomerates and large grain size resulted during sintering. SinterHIP process further improved the wear resistance over conventional sintering in air.

With TZP materials the ferroelastic domain switching was verified to be main operating toughening mechanism in low stress abrasion. In Figure 1 the ferroelastic domain switching of the 3YTZP surface due to wear is presented. Due to FDS the orientation ratio of tetragonal (002)/(200) peaks has changed. In high stress abrasion also transformation toughening contributed to the toughening effect. Thus monoclinic phase together with romboedric phase was formed in surface layers.

Ball milled ZTA and commercial ZTA had large zirconia grain size and thus showed a reduced stability against transformation from tetragonal to monoclinic. This caused some increase in monoclinic phase at the surface and reduced wear resistance. Very high wear resistance was obtained with alumina strengthened zirconia containing 29 vol% of Al_2O_3 in tetragonal zirconia matrix. The best wear resistance from pure tetragonal zirconias was achieved with 12mol% CeO_2 stabilized TZP. It also showed the highest orientation change due to FDS.

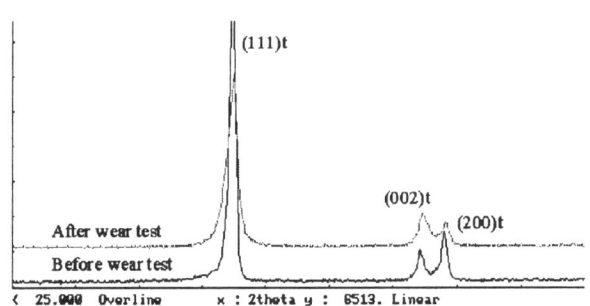

Fig. 1. Ferroelastic domain switching in TZP

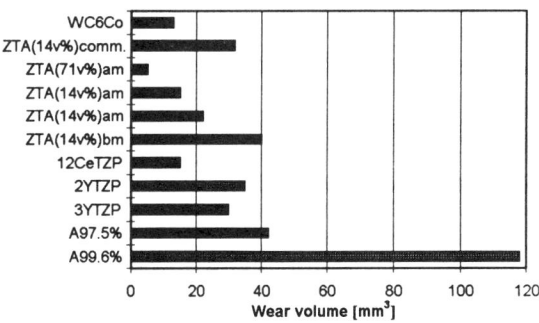

Fig.2 Wear results as volume loss (am=attrition milled, bm=ball milled).

REFERENCES

(1) A.V.Virkar, R.L.K. Matsumoto, J.Am.Ceram.Soc., Vol. 69, [10], C-224-C226, 1986

ALL - CERAMIC ROLLER BEARINGS - ADVANTAGES, LIMITS AND PROSPECTS

M. ROMBACH, W. PFEIFFER and T. HOLLSTEIN
Fraunhofer-Institut für Werkstoffmechanik (IWM), Wöhlerstr. 11, 79108 Freiburg, Germany

ABSTRACT

Ceramic materials have been used for more than 15 years in hybrid roller bearings for the rollers or spheres. In all-ceramic roller bearings, rings and rollers are made of ceramic materials. Some of the advantages of all ceramic roller bearings (1) are that they can

– operate under dry rolling conditions, e.g. in vacuum,
– be lubricated by low viscous fluids like hot water,
– resist aggressive environmental conditions like acids, soils or hot gases and temperatures up to 600°C,
– be used in nutrition industry without capsulation due to their non-toxic wear debris.

Silicon nitride is commonly used for all-ceramic roller bearings because of its high strength and fracture toughness. Due to the totally different material properties of ceramic materials compared to steel, the failure and wear mechanisms in steel and ceramic roller bearings are different, especially under poor lubrication conditions. As a result, an appropriate design, the determination of load capacity, lifetime and further improvements to all-ceramic roller bearings can only be achieved by ceramic orientated experimental and theoretical investigations.

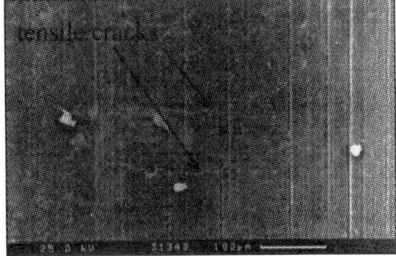

Fig. 1: Raceway with tensile cracks.

Investigations under dry rolling and water lubrication have shown that tensile cracks initiate the wear process. The load capacity under static and rolling loading conditions is therefore limited by the formation of tensile cracks. When cracks can be avoided, reliability and lifetimes are very long. Figure 1 shows a SEM micrograph of a raceway with tensile cracks that will lead to material removal in the later stage of the wear process. On the basis of these results, load capacity can be determined by calculating the load which is causing tensile cracks. The steps in the modell are (2):

– Calculation of normal and tangential contact stresses in a sphere-raceway system and calculation of the tensile stresses resulting from the contact stresses.
– Simulation of the fracture causing natural defects in ceramics by consideration of a fracture mechanical half-penny shaped surface crack.

Figure 2 shows the results for an all-ceramic roller bearing in the rolling loading condition for a fracture toughness of 4,7 MPa√m and a crack radius of 40 micrometers.

R_g is the radius of the raceway groove, R_s is the radius of the sphere. The maximum allowable normal force on

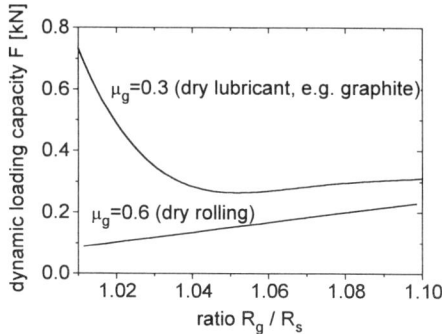

Fig. 2: Load capacity

one sphere of the bearing is shown as function of the ratio between raceway radius and sphere radius for two different sliding friction coefficients typically for dry conditions (0.6) and lubrication with a solid lubricant like graphite (0.3). Two details of the results have to be emphasized: The load capacity strongly depends on the sliding friction coefficient and is quite low for dry rolling. For a given sliding friction coefficient, the load capacity depends on the ratio S between raceway radius and sphere radius. For dry rolling S should be large, for good lubrication conditions it should be small.

Scientific investigations and practical experience show, that there are some issues on which further developments should be focused:

– Silicon nitride is used for bearings because of its high wear resistance and strength. It is in many cases resistant against corrosive attack but new results show, that the corrosion resistance depends strongly on the composition of the glassy phase. At present research is focused on finding additive systems which can improve the corrosion resistance.

– The load capacity in dry rolling is quite low. Solid lubrication like graphite and especially self-lubrication materials like Silicon nitride / Boron nitride or Silicon nitride / Titanium nitride might be able to reduce the sliding friction in contact.

REFERENCES

(1) Popp M., Keramikwälzlager: Prozeßsicherheit für Pumpen und Rührwerke, in Process, 10, 1994.
(2) E. Sommer, M. Rombach, Load capacity and design optimisation of full ceramic roller bearings, Mechanics in Design, Toronto, Canada, Editor.: S. A. Meguid, University of Toronto, S. 437-445, 1996.

RECIPROCATING WEAR OF ION IMPLANTED CERAMICS AND COATINGS

S.J. BULL, S.V. HAINSWORTH AND T.F. PAGE
Materials Division, Department of Mechanical, Materials and Manufacturing Engineering, University of Newcastle, Newcastle-upon-Tyne, NE1 7RU

ABSTRACT

Ion implantation has been used to modify the surface and near-surface properties of materials for some time now and its use to improve the wear resistance of metallic components is well documented. The technique involves the injection of high energy ions into the surface where they slow down and come to rest forming surface structures such as solid solutions and precipitates as well as defects such as dislocations. The effect of these surface modifications is to increase surface hardness and abrasive wear resistance of the implanted layer. In addition the implantation process can modify surface chemistry (e.g. by enhanced oxidation) and alter the formation of transfer layers leading to reductions in friction and adhesive wear. There is generally a compressive stress generated by the implantation process which can lead to reductions in surface-nucleated fatigue failure. For this reason ion implantation has been widely used to improve the wear performance of components such as injection moulding screws, barrels and moulds or high performance bearings where small amounts of wear cannot be tolerated.

The effects of ion implantation on the wear performance of ceramics and coatings is much less well understood. Although it has been shown that ceramics can be hardened by ion implantation and crack suppression and changes in friction are also possible there is little or no long-term data available to determine if these changes in mechanical properties have a beneficial effect on wear performance. Similarly, although it has been demonstrated that implantation of hard chromium plate can be beneficial, the mechanisms for improvement are not well understood.

In this study we have investigated the effect of nitrogen ion implantation on the wear performance of single crystal silicon and hard chromium plate. In both cases it can be demonstrated that ion implantation improves wear performance and the treatment may be optimised to maximise wear lifetime. However, the reasons for the improvements in performance are different in each case.

Steel coupons coated with 12μm hard chromium plate were implanted with 75keV nitrogen in the Blue Tank facility at Harwell to doses in the range 10^{17} to 10^{18} ions cm^{-2}. In addition {100} and {111} oriented single crystal silicon wafers were implanted with 50keV nitrogen in the same facility to doses in the range 5×10^{13} to 6×10^{17} ions cm^{-2}. The samples were characterised by scanning electron microscopy, nanoindentation and scratch testing prior to reciprocating wear testing against either a steel sphere or a diamond cone.

For the hard chromium plated samples the wear rate measured after 1km total sliding distance in the reciprocating wear test against a 3mm diameter 52100 steel sphere at an initial Hertzian contact pressure of 1GPa showed a minimum at a dose of 4×10^{17} ions cm^{-2}. This corresponds to the dose at which the hardness of the sample, determined from nanoindentation measurements, is maximum. The hardening is due to the presence of a fine array of CrN precipitates - the softening at high doses is due to overaging of these precipitates so that dislocations are able to bow around them more effectively. The best wear performance is thus associated with an optimum size and distribution of precipitates produced by ion implantation which will depend on both dose and also dose rate and beam heating.

For the silicon samples there was no change in hardness after ion implantation except for the very highest dose sample when nitrogen bubble formation occurred turning the surface layer into a honeycomb which is softer than the original material. However, there is a considerable reduction in the wear of the silicon in reciprocating wear tests against a Rockwell C diamond slider at 10N normal load due to the suppression of fracture in and around the wear track. This arises due to the presence of compressive stresses induced by the implantation treatment, but also is due to reductions in surface tractions produced by changes in friction after implantation.

Implantation has thus successfully reduced wear in each of the cases investigated, but the mechanisms of wear reduction are considerably different.

ANALYSIS OF THE FORCES FLUCTUATIONS DURING A SCRATCH TEST ON CERAMIC MATERIALS

V. JARDRET*, H. ZAHOUANI, J.L. LOUBET, T.G. MATHIA
Laboratoire de Tribologie et Dynamique des Systèmes, Ecole Centrale de Lyon, 36 Avenue Guy de Collongue,
69 131 ECULLY FRANCE
R. TRABELSI
Céramiques Techniques Desmarquest, Direction des Céramiques Spéciales, 63 rue Beaumarchais,
93100 MONTREUIL FRANCE
B. BOUALI
CPE Laboratoire de Chimiométrie et de Synthèse Organique Appliquée, 43 Boulevard du 11 Novembre 1918,
69100 VILLEURBANNE, FRANCE
* Metals and Ceramics Division, Oak Ridge National Laboratory, P.O. Box 2008, MS 6116,
OAK RIDGE, TN 37831, USA

ABSTRACT

Ceramic materials are used in a large range of mechanical and electronic devices. But these materials are still limited in their application by their brittle behavior. In particular, surface damage of ceramic components may considerably decrease their lifetime. An understanding of abrasion resistance and the associated deformation mechanisms is then of primary importance in the use and machining of ceramic materials. Instrumented scratch testing has been shown to be a useful tool to model machining process as well as to characterize the abrasion resistance of materials. In this work, the analysis of both the normal and tangential force fluctuations during a scratch test was done to describe the wear behavior and fracture processes of ceramic materials.

The measurement of both normal and tangential scratch forces was done at the acquisition rate of 20,000 Hz.

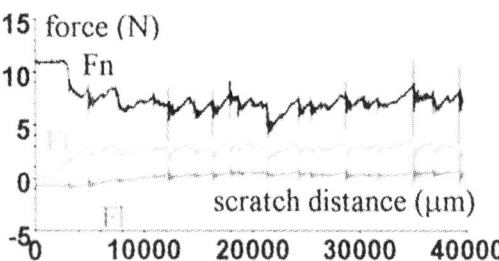

Fig.1: Forces measured during a scratch experiment on alumina ; the force variations are related to the fracture mechanism

The fluctuations of the normal force was related to the worn volume estimated with the help of the three-dimensional topographic relief of the scratched surface.

The abrasion energy is equal to the total scratch energy over the worn volume. It was studied as a function of penetration depth and scratch speed.

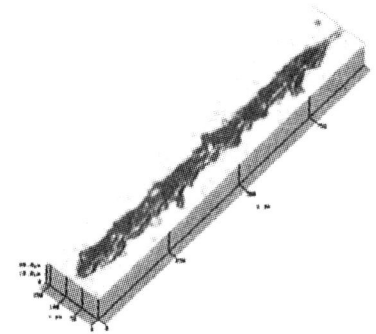

Fig. 2: Three dimensional topographic measurement of a scratch on alumina

The characterization method developed in this work is then used to investigate the effect of the fracture toughness on the wear behavior of two different sintered alumina. These alumina were sintered with different chemical additives which leads to a 30 % difference of their fracture toughness.

The physico-chemical action of a large serie of fluids on the abrasion energy of alumina was clearly emphasized and related to their wetting properties.

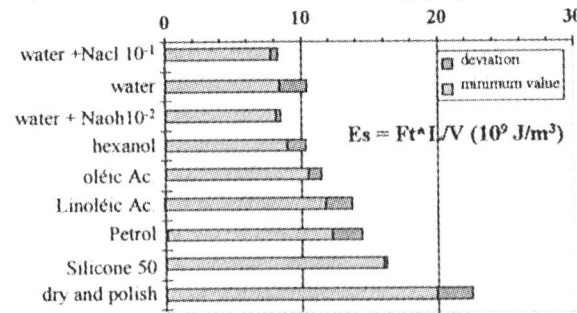

Fig. 3: Action of the environment of the surface on the specific abrasion energy Es of alumina

This research was partially sponsored by the Division of Materials Sciences, US Department of Energy, under contract DE-AC05-96OR22464 with Lockheed Martin Energy Research Corp.

WEAR MAPS OF Si₃N₄ CERAMIC CUTTING TOOL

LIU You-Rong, LIU Jia-Jun, ZHU Bao-Liang
(Tsinghua University, 100084 Beijing, P.R. China)
ZHOU Zhong-Rong
(Southwest Jiaotong University, 610031 Chengdu, P.R. China)
Leo VINCENT, Philippe KAPSA
(Ecole Centrale De Lyon, 69131 Ecully Cedex, France)

ABSTRACT

Ceramic cutting tools have been applied widely for high speed finishing and high removal rate due to their unique mechanical properties. However, it doesn't mean that ceramic tools can machine effectively any kinds of workpiece at any cutting conditions (1). Tool wear and fracture usually occur when applying the ceramic tool in an incorrect way. In this paper the wear maps of Si_3N_4 ceramic cutting tool for cutting 1045 plain carbon and 302 stainless steels were investigated in order to provide the basis for its reasonable utilization and to reveal the wear mechanisms at different cutting conditions.

The ceramic cutting tool used in this research was Si_3N_4 with dimensions of 13 × 13 × 8mm and a rake angle of -6--7°. Cutting tests were carried out without coolant on a C620 lathe. The cutting parameters were selected as follows: cutting speed 50 ~ 260m/min, cutting depth 0.25mm, feed rate 0.1 ~ 0.5mm/rev and cutting distance 300m.

The mean flank wear VB value was measured using a tool microscope. According to the measurement results the wear maps with different regions can be obtained on a 2-dimensional diagram. The wear morphologies and mechanisms in various regions were investigated by the SEM analysis. The temperature distribution on the flank face of Si_3N_4 tool at the conditions of different cutting speeds and 0.1mm/rev feed rate was measured by a thermal vides system (2).

Fig.1 shows the wear map of Si_3N_4 tool for cutting 1045 plain carbon steel. Fig.2 shows the wear map for cutting 302 stainless steel.

Based on these experimental results, the relationships between cutting conditions, cutting temperatures and wear mechanisms were discussed in detail. The main conclusions of this research can be drawn as follows:

(1) The flank wear of Si_3N_4 ceramic tool when cutting 302 stainless steel was always higher than that for cutting 1045 steel due to the higher cutting temperature of the former.

(2) The main wear mechanisms of Si_3N_4 tool were microfracture and abrasive wear in the conditon of lower cutting speed and adhesion-induced fracture and delamination at higher cutting speed. The optimum region was in the conditon of medium cutting speed.

(3) The temperature distribution on the tool surface when cutting stainless steel was about 20% in average higher than that for cutting 1045 steel. The high cutting temperature can easily promote adhesion and adhesion-induced failures of the ceramic tool.

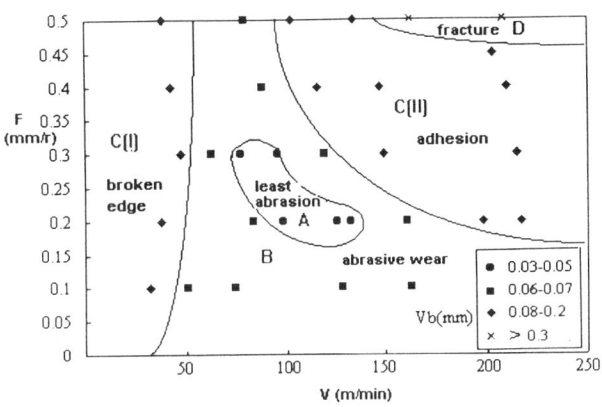

Fig.1 Wear map of Si_3N_4 tool for cutting 1045 steel

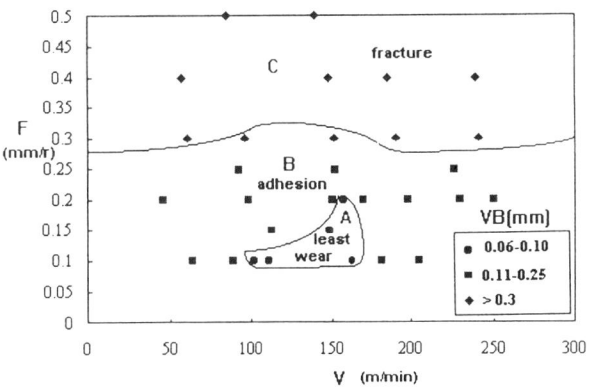

Fig.2 Wear map of Si_3N_4 tool for cutting 302 stainless steel.

REFERENCES

(1) S.F. Wayne Tribology Transactions 33(1990)619-626

(2) Liu Yourong, Liu Jiajun. "The temperature distribution near cutting edge of ceramic cutting tools measured by the thermal video system" to be published in J. of Progress in Natural Science 1997.

Submitted to *Jornal of Materials and Engineering Performance*

Influence of the combined effects of counterface roughness and ageing following gamma in air on the wear of UHMWPE

[1]Besong AA, [1]Hailey JL, [2]Ingham E, [3]Stone M, [4]Wroblewski BM, [1]Fisher J.

1 Department of Mechanical Engineering, University of Leeds, LS2 9JT, England.
2 Department of Microbiology, University of Leeds, LS2 9JT, England.
3 Department of Orthopaedics, Leeds General Infirmary, England.
4 Centre for Hip Surgery, Wrightington Hospital, Wigan, England.

Introduction
The wear of UHMWPE bearing surfaces and wear debris induced osteolysis is a major cause of long-term failure in prosthetic hip joints. Damage to polished femoral heads *in vivo* can cause increased wear of the UHMWPE acetabular cups. Oxidation and ageing, after sterilisation by gamma irradiation in air, can affect the mechanical properties and hence the wear resistance of UHMWPE. The reduction in mechanical properties, such as impact strength and fatigue resistance is expected to lead to a reduced wear resistance of UHMWPE. This study investigated the combined effects of ageing after sterilisation by gamma irradiation and counterface roughness on the wear of UHMWPE.

Materials and Methods
Wear rates were studied on a uni-directional tri-pin-on-disc tribometer. UHMWPE wear pins, manufactured from polyethylene acetabular cups of different ages (12 months and 120 months) after sterilisation by gamma irradiation in air, were tested at a sliding speed of 80mm/s on a stainless steel counterface in a protein-containing lubricating medium. The load on each pin was 80N, giving a contact stress of 12MPa at the contact surfaces. The wear of the aged specimens were compared to that of control specimens that were manufactured from cups that had not been sterilised. Three different counterface roughnesses (R_a = 0.01, 0.07 and 0.11µm) were used. The wear surfaces were tested 1mm below the initial articulating surfaces of the cup, a position of high oxidative degradation.

Results
The wear rate of all materials increased markedly as counterface roughness increased, but to different extents depending on the age of material. The wear rate of UHMWPE which had been sterilised by gamma irradiation in air was shown to increase significantly with ageing time for all the counterface conditions. The combined effect of ageing and increase in counterface roughness had a dramatic effect (up to 2000 fold) on the wear rate. The results of wear factors are shown in table 1. Both an increased roughness of the counterface and degradation of material properties due to ageing are likely to change the dominant wear processes and type of debris produced, as well as the volumetric wear rate. Quantification of the combined effect of these two important tribological variables provides further understanding of the causes of high wear rates of UHMWPE in vivo.

Table1. Wear factor ± standard error (x 10^{-9} mm^3/Nm)

Roughness (R_a)	Non Irradiated UHMWPE (NI)	12 Months Aged Irradiated UHMWPE AI(12)	120 Months Aged Irradiated UHMWPE AI(120)
0.01µm	7 ± 1	20 ± 2	310 ± 110
0.07µm	270 ± 30	360 ± 80	1000 ± 270
0.11µm	1700 ± 300	1100 ± 100	14300 ± 1400

Reference
[1] Fisher J, Hailey JL, Chan KL, Shaw D, Stone M.
   J of Arthroplasty. Vol.10 No5 (1995), 689-692
[2] Sun DC et al.
   21st Annual meeting, Society for Biomaterials, 1995.
[3] Furman B, Li S.
   21st Annual meeting, Society for Biomaterials, 1995.
[4] Trieu HH, Avent RT, Paxson RD.
   21st Annual meeting, Society for Biomaterials, 1995.

Acknowledgement
This work was supported by the Brite Euram Project 7928 and ARC, UK.

THE EFFECTS OF GAMMA IRRADIATION ON THE FATIGUE WEAR RESISTANCE OF ULTRA HIGH MOLECULAR WEIGHT POLYETHYLENE

M. CHOUDHURY AND I.M. HUTCHINGS

Department of Materials Science and Metallurgy, University of Cambridge, Pembroke Street, Cambridge, CB2 3QZ, UK

ABSTRACT:

The wear of Ultra High Molecular Weight Polyethylene (UHMWPE) in total joint replacement components is considered to be the primary problem in orthopaedics today. Analysis of retrieved knee tibial components shows that fatigue is the major factor contributing to failure. This is associated with pitting, cracking, and delamination (1). Virgin UHMWPE is effectively chemically inert and stable. However, gamma sterilization in air initiates chain scission, free radical formation and oxidative degradation, weakening the polymer and increasing its brittleness. A certain degree of cross-linking also occurs. The normal sterilization dose is 2.5 Mrad from a Co-60 gamma source. This paper investigates the effect of irradiation dose on the fatigue wear mechanisms. Since motion at the knee joint involves a combination of rolling and sliding, a new experimental method which cyclically loads and unloads the polymer under conditions of rolling contact has been used.

Plane samples of ram-extruded GUR-415 UHMWPE (70 x 70 x 8 mm) were subjected to 2.5 and 5.0 Mrad gamma doses in air. Non-irradiated material was also studied. Mechanical loading involved the rolling of rigid spheres (12.7 mm diameter) over the polymer surface under distilled water. Cyclic loading and unloading was achieved by using a bearing cage containing the spheres, which rotates under an applied axial load. This produces a circular wear track with each ball exerting a compressive contact stress of ca 70 MPa. A peristaltic pump circulates the fluid and ensures a constant temperature environment. Tests were run up to a maximum of 5 million loading cycles, simulating ca 5 years of clinical usage. Scanning electron microscopy (SEM) was used to examine the resulting wear tracks. The effects of irradiation dose on the morphological properties of UHMWPE have been detailed elsewhere (2-4).

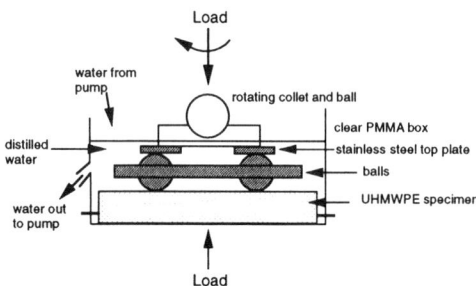

Figure 1 Schematic representation of the experimental set-up

SEM examination show no cracking or pitting on non-irradiated UHMWPE after 5 million cycles. Irradiation to 5 Mrad results in cracking and pitting similar to that observed on a retrieved tibial implant (PCA design, 4 years *in vivo*), which failed by severe delamination and oxidative degradation. Figure 2 shows a thin section microtomed perpendicular to the articulating surface (A = articulating surface). The cracks observed were similar to those formed *in vitro* (5 Mrad PE after 5 million cycles).

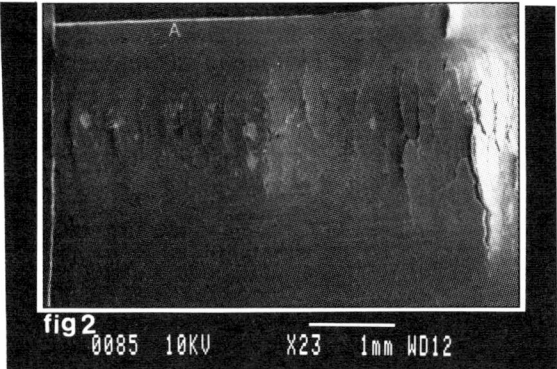

These results are consistent with the view that irradiation in air increases the crystallinity of PE due to chain scission and structural re-arrangement of lower molecular weight fragments. This reduces the ductility of the UHMWPE and increases its susceptibility to abrasive (2-4) and fatigue wear. Similar fatigue-related wear mechanisms on gamma-air sterilized samples have been reported under repetitive high stresses and *sliding* motion (5). It is concluded that non-irradiated PE has a high resistance to rolling contact fatigue wear. Irradiation to 2.5 Mad (current orthopaedic practice) produces little adverse effects, but doubling this dose results in cracking and pitting after only 5 million cycles. This may be of particular concern for components sterilized to 5 Mrad. Finally, this test method is capable of producing cracking and pitting wear typical of that seen in retrieved tibial components. The work suggests that both irradiation dose and cyclic loading contribute to these mechanisms.

References: (1) R.W. Hood et al, J. Biomed. Mater. Res., 17, 1983, pp 829-842. (2) M. Choudhury and I.M. Hutchings, Fifth World Biomaterials Congress, 1996, Toronto, Canada. (3) M. Choudhury and I.M. Hutchings, *Wear* 203-204, 1997, pp 335-340. (4) M. Choudhury and I.M. Hutchings, 43rd Annual Meeting of the Orthopaedic Research Society, February 1997, San Francisco, California. (5). J.V. Hamilton et al, 67th Meeting, AAOS, 1997, San Francisco, California.

IN VITRO SIMULATION OF CONTACT FATIGUE DAMAGE FOUND IN UHMWPE COMPONENTS OF KNEE PROSTHESES

JOHN H. CURRIER, JESSICA L. DUDA, JOHN P. COLLIER, DANIEL K. SPERLING, BARBARA H. CURRIER and FRANCIS E. KENNEDY

Thayer School of Engineering, Dartmouth College, Hanover, NH 03755 USA

ABSTRACT

Studies of knee prostheses retrieved from patients after in vivo service have reported fatigue-related cracking and delamination as the primary modes of failure of tibial bearing inserts made from ultra-high molecular weight polyethylene (UHMWPE). However, the cracking and delamination failure modes have not previously been produced under laboratory (in vitro) test conditions. Recent work has shown that the mechanical properties of UHMWPE degrade over time after having been sterilized with gamma radiation in air (1). The purpose of this study is to develop in vitro tests which reproduce the failure modes found in vivo and to establish a link between radiation-induced material property degradation and contact fatigue failure of UHMWPE tibial components.

In vitro simulation of fatigue loading is carried out on a single station knee simulator, and on a rolling and sliding wear tester. Tibial bearings for the knee simulator are gamma sterilized-in-air, implantable UHMWPE (GUR415) bearings taken from manufacturing inventory. The rolling/sliding disks are machined from GUR415 bar stock and either gamma sterilized in air and accelerated aged, or left unirradiated.

Cracking and delamination of samples gamma sterilized in air and aged are observed in both types of tests. The rolling/sliding samples show signs of damage in as few as 130,000 cycles with an estimated Hertzian stress of 15 MPa and 25% sliding. However, damage is limited to the surface and no subsurface fatigue cracks are seen. In the knee simulator, fatigue damage similar to that found in vivo is visible in gamma irradiated and aged components after as few as 150,000 cycles at a load of 1223 N. Control tests on sterilized, but unaged knee components showed no fatigue cracking out to 2.5 million cycles.

The knee simulator tests enable a comparison of wear modes for a series of components of the same design, with increasing severity of subsurface oxidation that has developed with increasing time after irradiation. The only components that show wear modes similar to those seen in retrieved tibial components are those components with significant subsurface oxidation, and the fatigue cracking initiates subsurface (Figure 1).

Recently published results show significantly reduced strength and ductility in the subsurface layer (depth > 0.7 mm) of UHMWPE bearings that have been gamma irradiated in air and aged (1). Other research indicates that Young's modulus of the oxidized polyethylene in the subsurface zone could be double that of the material at the surface of the component (2). In contrast, both never sterilized and recently sterilized specimens have relatively uniform properties, while accelerated-aged specimens show reduced ductility and increased modulus in the surface, as opposed to subsurface, layer.

Articulation of a knee prosthesis results in normal contact stresses which are maximum at the surface and shear stresses that are maximum subsurface (3,4). Increased stiffness of an oxidized subsurface layer within the UHMWPE increases the contract stress and the von Mises stress (5). The increased subsurface von Mises stress increases the driving force for initiation of fatigue cracks in the oxidized subsurface layer. Reduced toughness of the oxidized subsurface layer compounds the problem of elevated shear stress, and leads to much easier crack propagation within the subsurface layer.

UHMWPE that is gamma sterilized in air and aged shows subsurface fatigue damage during in vitro testing that is similar to damage found in vivo. The presence of the oxidized subsurface layer is shown to be a critical factor in the development of subsurface fatigue cracks. These results are important to the interpretation of in vitro knee simulations used to assess performance of tibial bearings.

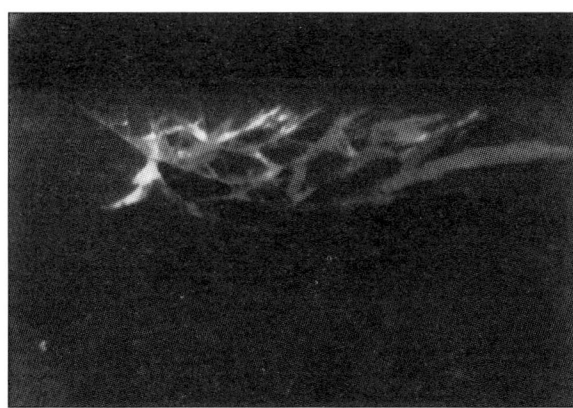

Figure 1. Thin cross section of the component tested in the knee simulator for 150,000 cycles at 1223 N, showing fatigue cracks initiating subsurface.

REFERENCES

(1) Collier JP, Sperling DK, Currier JH et al., J Arthroplasty 11:377, 1996.
(2) Kurtz SM, Rimnac CM, Li S et al., Trans ORS 19:289 1994.
(3) Bartel DL, Bicknell MS, Wright TM, J Bone Joint Surg 68A:1041, 1986.
(4) Collier JP, Mayor MB, McNamara JL et al., Clin Orthop 273:232, 1991.
(5) Plumet S, Duborg, MC, INSA Lyon, Internal Report 1997.

PREDICTION OF CYCLIC STRAIN ACCUMULATION OF UHMWPE IN KNEE REPLACEMENTS

E ALEXANDER REEVES, DAVID C BARTON and JOHN FISHER
Department of Mechanical Engineering, University of Leeds, Woodhouse Lane, Leeds, LS2 9JT, UK
DAVID P FITZPATRICK
DePuy International Ltd., St Anthony's Road, Leeds, LS11 8DT, UK

INTRODUCTION

Replacements of diseased joints do not usually last for a patient's lifetime. The main mode of failure of knee prostheses that have been sterilised by gamma irradiation in air is delamination of the ultra-high molecular weight polyethylene (UHMWPE) component. New methods of sterilisation mean that delamination may become less frequent and that cyclic plastic strain accumulation (or ratchetting) may become more important.

This study advances finite element analysis of knee replacements from the time independent to time dependent regime in order that cyclic strain accumulation can be modelled.

MATERIALS AND METHODS

The analysis was performed using the ABAQUS finite element code. This allowed contacts to be modelled along with the elastic, visco-plastic material properties of the UHMWPE.

The material model used in the finite element analysis was determined from the intra-cycle deformations from the cyclic loading tests for the elastic properties and from the permanent set of the specimens from the constant stress creep tests for the time dependent plastic properties. All tests were carried out on GUR-415, a type of UHMWPE at 37°C (body temperature). The rate of plastic strain accumulation in the material model was linked to the amount that the stress was above the yield stress of the material.

Three models were developed: in the first the surfaces were conforming in both directions and the geometry was regarded as axisymmetric; in the other two models the surfaces were unconforming allowing plane strain conditions to apply. One unconforming model included the anatomical anterior-posterior motion known as roll-back. The knee joint force and rollback motion during normal gait were specified in the analysis as simplified wave-forms. This allowed the walking cycle to be repeated over a large number of cycles.

To date, the analysis has been completed for seven thousand, two hundred walking cycles running on a SUN Sparc workstation, for each of the two unconforming cases. This number of gait cycles would be completed in a few days by a patient with a knee replacement.

RESULTS

The results showed the characteristic maximum subsurface stresses at the contact. The stresses were lower in the conforming prosthesis than unconforming prosthesis, which lead to lower plastic strain accumulation. This result agreed with other studies using time independent finite element analyses.

The time dependent analysis in this study showed that the joint kinematics alter the stress pattern in the material. The stress history at a particular part of the polymer was similar to the load history for the unconforming case without rollback, whereas it was more complex for the case with rollback. This lead to a difference in initial plastic strain accumulation rate. It also lead to a streak of plastically deformed material below the surface.

Initially the strain rate was lower in the model with the rollback, but after a number of cycles the strain rate reduced in the static model (without rollback) due to the increased conformity. Extrapolations of the results showed that the prosthesis with rollback may well have accumulated higher plastic strains than the static prosthesis after a larger number of gait cycles.

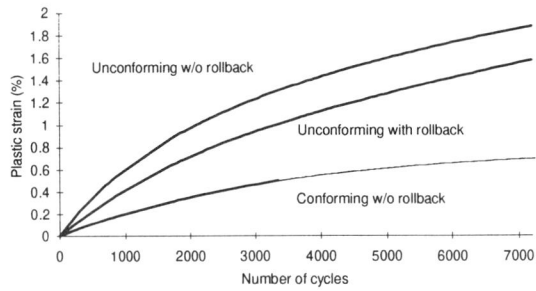

Fig. 1: Plastic strain accumulation in three types of knee replacement

DISCUSSION

This study has provided an advanced analysis tool which can be used in the design process of future knee replacements with various types of polyethylene whose time dependent properties are known.

The result of our research is fully and comprehensively explained in the paper of the title "A two dimensional model of cyclic strain accumulation in ultra-high molecular weight polyethylene in knee replacements" which has been submitted to the IMechE Journal of Engineering in Medicine.

Submitted to *IMechE Journal of Engineering in Medicine*

PLASTIC STRAINS IN UHMWPE DUE TO INTERACTIONS WITH MICROSCOPIC STAINLESS STEEL ASPERITIES IN TJR

C.M. MCNIE, D.C. BARTON, J. FISHER.
Department of Mechanical Engineering, University of Leeds, LS2 9JT, UK.
M.H. STONE
Leeds General Infirmary. Leeds, UK.

INTRODUCTION

It is now recognised that Ultra High Molecular Weight Polyethylene (UHMWPE) particles generated at the articulating surfaces of total hip replacements produce adverse cellular reactions leading to osteolysis and loosening of the artificial joint (1). Recent studies at Leeds University have detected many sub-micron size UHMWPE wear particles in both laboratory tests and retrieved tissue samples (2,3).

The surface roughness and the presence of microscopic defects or scratches on the polished surface of the metal head are important factors influencing the generation of micron size particles (4). In particular, the height of the asperity lip, formed either side of a scratch, has been found experimentally to be a critical parameter controlling the wear rate. Finite element stress analysis has therefore been used to investigate the relationship between the geometry of an asperity on the surface of the stainless steel femoral head and the plastic strain accumulation in the UHMWPE acetabular cup during a single pass of the asperity.

METHOD

A two-dimensional plane strain finite element model consisting of a UHMWPE semi-infinite block and a stainless steel surface containing a single asperity has been analysed using ABAQUS. The UHMWPE was assumed to follow a simplified elastic-plastic material model obtained from material testing at *in vivo* strain rates. The geometry of the asperity was developed from surface characterisation of a series of explanted Charnley stainless steel femoral heads containing scratches. In the retrieved specimens, scratch asperity heights were found to vary from 0.02µm to 8.9µm, with a mean of 0.96µm. The geometry was made non-dimensional by defining an aspect ratio as the asperity height divided by the half width; the measured ratios varied from 0.01 to 0.44. In the finite element model, a parabolic asperity of height 1µm and varying width was defined to give aspect ratios ranging from 0.05 to 0.2. The contact between the two surfaces was simulated by compressing the asperity into the UHMWPE by its full height (1µm) and then allowing the two surfaces to slide at a relative velocity of 22mms^{-1}.

RESULTS

It was found that the maximum Von Mises equivalent stress predicted on the surface of the UHMWPE during a single asperity pass increased linearly with the asperity aspect ratio up to a maximum of 52MPa for an aspect ratio of 0.2. The equivalent surface plastic strain was found to increase non-linearly with increasing aspect ratio up to 49% plastic strain, shown in fig 1. This compares with zero plastic strain produced on a smooth surface contact.

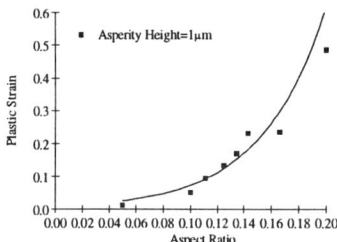

Fig. 1: Variation of plastic strain with aspect ratio.

DISCUSSION

Although the plastic strain to cause failure is as yet unknown, these results demonstrate that plastic strain accumulation rates increase dramatically with increasing asperity aspect ratio. Previous experimental studies have indicated that a 1µm high scratch lip can cause a 20 to 70 fold increase in wear rate compared to a smooth surface. The high plastic strains predicted in this study offer an explanation for the highly elevated wear rates, but also indicate that the aspect ratio of the scratch lip will be a critical determinant of wear. Both the height and width of the lip should be considered in future studies of femoral head damage and the resulting wear in the UHMWPE.

REFERENCES

[1] M. Jasty et al., Curr Op Rheumatology, **4**: 204-209, 1992
[2] J. Hailey et al., J of Eng Med, **210**: 3-10, 1996
[3] A.S. Shanbag et al., JBJS, **76-B**: 60-67, 1994
[4] H. Amstutz et al., Clin Ortho & Rel Res, **276**: 7-18, 1992

ACKNOWLEDGEMENTS

This work was supported by The Wellcome Trust, London, UK.

ELECTRON BEAM CROSS-LINKING OF UHMWPE AT ROOM TEMPERATURE, A CANDIDATE BEARING MATERIAL FOR TOTAL JOINT ARTHROPLASTY

O. K. Muratoğlu, C. R. Bragdon, D. O. O'Connor, E. W. *Merrill, M. Jasty, and W. H. Harris
Orthopaedic Biomechanics Laboratory, Massachusetts General Hospital, Boston, MA 02114

INTRODUCTION

Wear of ultra-high-molecular weight polyethylene (UHMWPE) components and wear debris induced osteolysis are major causes of failure in total joint arthroplasty. A method of improving the wear properties of UHMWPE is cross-linking which could be achieved by chemical modifications (such as peroxides, silanes, etc...) or irradiation [1-3]. Peroxide cross-linking has been shown to improve the abrasion resistance of UHMWPE in sand-slurry experiments [1]. However, the by-products of this chemistry and residual un-dissociated peroxides could jeopardize the long-term stability of the cross-linked material. Irradiation, on the other hand, should be used with caution for cross-linking of UHMWPE, in that, incomplete reaction of radiation-induced free-radicals could remain active for extended periods of time. Consequently, in the long term, oxidation reactions could lead to the embrittlement of the component and suppress the benefits of radiation cross-linking. Therefore, in radiation cross-linking of UHMWPE, to ensure the long-term chemical stability and preserve the high wear resistance of the polymer, the reaction of the free radicals with each other should be carried to completion. The present study describes a method of radiation cross-linking with complete elimination of free radicals. The thermal properties and oxidation levels along with the wear behavior of this cross-linked material are also presented.

MATERIALS AND METHODS

The specimens (n=60) used in this study were cylindrical shaped 'hockey pucks' (height = 4cm; diameter = 9cm) cut from GUR 4150 bar stock material. The pucks were irradiated to 25, 40, 50, 100, and 200kGy (e-beam) at a dose rate of 3kGy/min in air, followed by heating to 150°C under vacuum and cool down to room temperature at a rate of -10°C/min. This process will be referred to as cold irradiation with subsequent melting (CISM). The e-beam source used in this study was a 10MeV linear accelerator (linac) of AECL (Pinawa, Canada) operated at 1kW. Following the CISM process, the specimens were analyzed using a Perkin Elmer differential scanning calorimeter (DSC-7) at a heating rate of 10°C/min, a BioRad UMA 500 infra-red microscope with an aperture size of 200X20μm^2, and electron spin resonance (ESR) to determine the free radical concentration. The variations in the mechanical properties at different absorbed radiation doses were determined using ASTM-D638 at a strain rate of 0.02sec^{-1}. The DSC specimens were 2mm thin sections cut parallel to the top surface of the pucks while the ir spectra were obtained from 50μm thin films cut parallel to the axes of the pucks. The wear properties were determined using a bi-axial pin-on-disk wear tester. Cylindrical pins (height = 9 mm, diameter = 13 mm) were machined from e-beam CISM specimens that received varying radiation doses and unirradiated bar stock. The first 0.5mm on the surface of cross-linked pins was machined away to remove the slightly oxidized surface layer. The pin-on-disk wear experiments were carried out to 2 million cycles with a weight measurement at every 0.5 million cycles.

RESULTS AND DISCUSSION

The ESR measurements indicated that after melting there are no trapped free radicals left in the irradiated polymer at any of the applied radiation doses. The infra-red analysis showed that a 0.5 mm thin surface layer oxidizes during the cross-linking reactions and subsequent melting. Within the oxidized region, the DSC measurements showed that the crystallinity was 27%. Beneath the oxidized skin layer, the cross-linked polymer did not show any traces of a carbonyl peak. The overall crystallinity and melting temperature in the bulk of the CISM material did not vary significantly with radiation dose (48.5±1.0% and 139±1°C). Figure 1 shows the variation in the mechanical properties as a function of absorbed dose. The POD wear rates as a function of absorbed dose are depicted in Figure 2.

The interrelation of wear rate with absorbed radiation dose showed a sigmoidal dependence. The wear rate dropped significantly up to 100-150kGy, above which it reached a plateau. This ~35 fold decrease in the wear rate with no significant change in thermal and physical properties and no trapped free radicals strongly suggests that CISM is a feasible modification route for the improvement of the wear resistance of UHMWPE.

REFERENCES

1. Himont Tech. Info. Bull. HPE-116 (1987). 2. McKellop, H., Harv. Med. Schl. Hip Cse. (1996). 3. Jasty, M. et al., Harv. Med. Schl. Hip Cse. (1996). 4. Jasty, M. et al., ORS, 42:21 (1996).

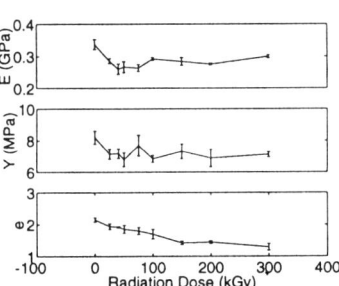

Figure 1. E=secant modulus, Y=yield stress, e=true strain at failure.

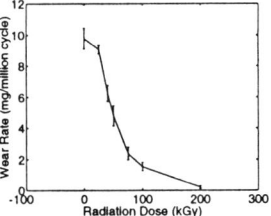

Figure 2.

*Massachusetts Institute of Technology, Cambridge, MA, 02139

DEVELOPMENT OF A NEW PIN-ON-DISK TESTING MACHINE FOR EVALUATING POLYETHYLENE WEAR

D O O'CONNOR, C R BRAGDON, J D LOWENSTEIN, V PREMNATH, M JASTY, and W H HARRIS
Orthopaedic Biomechanics Lab, Mass. General Hospital, GRJ 1126, 55 Fruit St., Boston, MA 02114 USA

INTRODUCTION

A standard screening test for evaluating bearing materials for total joint replacements has been the reciprocating pin on disk or flat on flat wear test which has been used by the automotive industry. This test is attractive because it is relatively easy and inexpensive to perform, the test parts are easy to fabricate, and results can be generated in a short period of time. However, this test, when applied to ultra high molecular weight polyethylene on metal bearing samples, results in abnormally low wear rates, an order of magnitude lower than the rate of fluid absorption, making comparative studies with this test difficult.

Pooley et al showed that the rate of wear of HMWPE is dependent on the amount of cross shear imparted to the polyethylene surface. Work in our laboratory suggests that this is may also be true for UHMWPE. The purpose of this study was to fabricate a pin on disk wear tester capable of imparting various amounts of cross shear to a polyethylene pin under physiologic loads in order to evaluate its effect on the wear rate.

MATERIALS AND METHODS

An X-Y table was modified and fitted with two large stepper motors rated for 350 oz-in of torque in order to control the two-dimensional motion of the table. The two motors were controlled independently by a 386PC with two signal generator cards capable of four channels of output. The table was mounted on an MTS Bionix servohydraulic testing machine.

A cylindrical plexiglass chamber was constructed which firmly held a highly polished 2" diameter cobalt chrome disc. The 7mm diameter polyethylene pin was machined from ram extruded bar stock, sonically cleaned, weighed, and mounted in a collet. A Paul type load curve was used with a peak load of 60 lbs which resulted in a peak contact stress of 6MPa. The stepper motors were driven with two square waves which were out of phase by 90o, resulting in the pin tracing a 0.5cm X 1cm rectangular path against the CoCr disc. The peak load occurred around one half of the rectangle and the preload, representing the swing phase of gait, around the other half. This combination of motion and load approximates the pattern of motion, distance traveled, and load application experienced by a point on the femoral head against a polyethylene cup during normal gait. The test was run at 0.5hz and polyethylene pin was cleaned and weighed every 300,000 cycles. At the end of the test, the wear surface was studied using a scanning electron microscope.

RESULTS

At the end of the test, the appearance of the surface of the polyethylene pin had been substantially altered by the wear process. The concentric machine marks were no longer present and the wear surface had been polished to a highly reflective, mirror-like finish. Weight loss measurements of the polyethylene pin averaged 2.6mg every 300,000 cycles (8.8mg/million cycles).

Scanning electron microscopy of the worn polyethylene surface show fine fibular structures Pdrawn from the substrate. The fibrils measured up to 5µ in length and less than 1µ in width. Underlying the fibril structures, the surface had a rippled appearance of 20µ in width which appear to be intersected by perpendicularly oriented ripples. This is in contrast to the appearance of the pin under unidirectional motion where the ripples appear to run only perpendicular to the direction of motion.

Particles retrieved from the serum were of variable size and shape ranging from less than a micron to several microns. The particles were generally round or slightly elongated, while these were spindle shaped.

DISCUSSION

The wear results from this unique pin on disk testing machine suggest it may be a very useful screening tool for evaluating bearing materials for use in joint replacements. Contemporary, unidirectional pin on disk wear rates of UHMWPE are reported to be on the order of 0.10mg/million cycles, half of the weight loss being accounted for by serum soak controls. We have shown that when bidirectional motion is applied to the test, the mechanism of wear and the rate of polyethylene wear become consistent with the observations made from clinical and retrieval studies of total hip arthroplasties.

REFERENCES

Pooley and Tabor, Proc. R. Soc. Lond. A, 329, 251-274, 1972.

THE ROLE OF INTERSTITIAL FLUID PRESSURIZATION AND EQUILIBRIUM FRICTION COEFFICIENT ON THE BOUNDARY FRICTION OF ARTICULAR CARTILAGE

GERARD A. ATESHIAN, LAURA H. WANG and W. MICHAEL LAI
Department of Mechanical Engineering, Columbia University, Mail Code 4703, New York, NY 10027, USA

ABSTRACT

Articular cartilage has excellent tribological properties which have been attributed, over the years, to various lubrication modes. McCutchen (1) argued that cartilage interstitial water pressurizes under loading, such that a greater proportion of the applied load would be supported by the fluid rather than the cartilage skeleton, leading to a reduction in the friction coefficient. Recent expriments by Forster & Fisher (2) duplicated McCutchen's findings; these authors (2), and ourselves (3,4), proposed a formula for relating the cartilage friction coefficient to the load-sharing between the solid and fluid phases of cartilage.

Despite the variety of lubrication modes proposed for articular cartilage, there are no studies which have directly correlated a theoretical model for cartilage friction against experimental data. In the present study, we propose a refined theoretical model for boundary friction in articular cartilage where the friction occurs predominantly as a result of solid-to-solid interactions. Experimental verifications are provided as well.

The challenge in this formulation is to precisely define the solid-to-solid load at the contact interface of soft porous-hydrated materials. Using the framework of the biphasic theory for cartilage (5), we have derived the following expressions for the solid-to-solid normal load, W_n^{ss}, and the total normal load, W_n, at the interface of two cartilage layers, denoted by 0, 1:

$$W_n^{ss} = \int_\Gamma \mathbf{n} \cdot \left(-\varphi^{s0}\varphi^{s1} p\mathbf{I} + \sigma^e\right)\mathbf{n} d\Gamma$$

$$W_n = \int_\Gamma \mathbf{n} \cdot \left(-p\mathbf{I} + \sigma^e\right)\mathbf{n} d\Gamma$$

where p is the interstitial fluid pressure and σ^e is the elastic stress of the solid cartilage matrix in either layer; $\mathbf{n}$ is the unit normal to the contact interface Γ, and φ^s is the solid fraction of cartilage at its contacting surface. The effective friction coefficient of articular cartilage is given by $\mu_{eff} = \mu_{eq} W_n^{ss}/W_n$, where μ_{eq} is the friction coefficient achieved under equilibrium conditions when interstitial fluid pressurization has subsided.

This formulation was employed to predict frictional measurements of bovine cartilage plugs performed in physiological saline, using a confined compression apparatus. A total of ten specimens were used (radius=3.175 mm, thickness =2.36±0.28 mm). A stress-relaxation test was performed on the cartilage while the sample was simultaneously rotated under continuous motion against a flat, smooth, cylindrical indenter (plexiglas or stainless steel). The normal load and frictional torque were measured as a function of time. The normal load was curvefitted with the biphasic theory to predict the material properties of the tissue. Using these properties, theoretical predictions of μ_{eff} were obtained using the above theoretical model (Fig. 1); curvefitting yielded values of μ_{eq}=0.125±0.025. For all specimens, the nonlinear correlation coefficient between theory and experiment averaged r^2=0.92±0.12. By plotting the predicted fluid pressure at the contact interface, it could be directly observed that μ_{eff} achieved its lowest value when the pressure was greatest.

This study presents an original mathematical formulation of a theoretical model for cartilage friction and provides the first direct verification of any theoretical cartilage friction model against experimental results. It confirms that interstitial fluid pressurization can reduce the friction coefficient of cartilage, as proposed by McCutchen (1), though our model does not require a "weeping" mechanism (1) to explain this effect. Further experiments are planned to verify the overall validity of the proposed model.

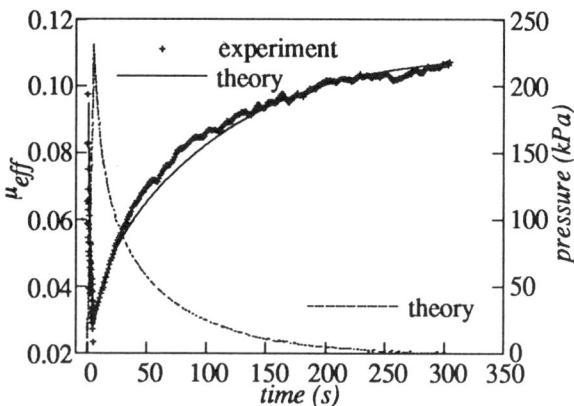

Fig. 1 μ_{eff} (theory vs experiment) and p (theory)

REFERENCES

(1) CW McCutchen, Wear, Vol. 5, 1-17pp, 1962.
(2) H Forster, and J Fisher, Proceedings of the IMechE, Part H, Vol. 210, 109-119pp, 1996.
(3) GA Ateshian, ASME Summer Bioengineering Conference, Vol. BED 29, 147-148pp.
(4) GA Ateshian, ASME Journal of Biomechanical Engineering, Vol. 119, 81-86pp, 1997.
(5) VC Mow, SC Kuei, WM Lai and CG Armstrong, ASME Journal of Biomechanical Engineering, Vol 102, 73-84pp, 1980.

THE ADAPTIVE MULTIMODE LUBRICATION IN NATURAL SYNOVIAL JOINTS AND ARTIFICIAL JOINTS

TERUO MURAKAMI, HIDEHIKO HIGAKI and YOSHINORI SAWAE
Department of Intelligent Machinery & Systems, Faculty of Engineering, Kyushu University,
Hakozaki, Higashi-ku, Fukuoka 812-81 Japan
NOBUO OHTSUKI
Graphic Science, Faculty of Engineering, Kyushu University, Ropponmatsu, Chuo-ku, Fukuoka 810 Japan
SHIGEAKI MORIYAMA
Department of Mechanical Engineeering, Faculty of Engineering, Fukuoka University, Jonan-ku, Fukuoka 814-01, Japan
YOSHITAKA NAKANISHI
Department of Intelligent Machinery & Systems, Faculty of Engineering, Kyushu University, Fukuoka 812-81 Japan

ABSTRACT

The natural synovial joints with very low friction and long durability under various daily activities appear to operate not only under fluid film lubrication but also under other various lubrication modes (1)-(4). In natural synovial joints such as hip and knee joints during walking, the elastohydrodynamic lubrication (EHL) mechanism including micro-EHL is likely to play the main lubricating role. Under severer thin film conditions with local contacts, however, various supplemental lubrication mechanisms such as weeping, boundary and gel-film lubrication modes seem to synergistically operate to protect articular cartilages. In this paper, the lubrication mechanisms in both natural joints and artificial joints with artificial cartilages are examined by pendulum and simulator tests.

First, it was shown in pendulum tests of pig shoulder joints that both concentration of hyaluronic acid or viscosity and adsorbed film formation of proteins and phospholipids exerted significant effect on frictional behaviour in swinging motion immediately after loading of 100N. Under high load of 1 kN, both had slight influence on friction, and low friction was observed under wide-ranging viscosity conditions, since high load similar to body weight probably enhanced the squeeze film effect due to improved congruity. Natural synovial joints appear to operate in various appropriate lubrication modes corresponding to the severity.

Next, lubrication modes in sliding pairs as knee joint models of stainless steel spherical surface and pig tibial cartilage specimen or polyvinylalcohol (PVA) hydrogel specimen during walking were examined in simulator tests. In these tests, the influence of lubricant viscosity and addition of protein on frictional behaviour was evaluated. To evaluate lubrication mode in similar severity to natural knee joint, half load condition was applied as shown in Fig.1. In this study, the effect of proteins on friction was investigated.

Figure 1 shows the results of simulator tests for natural articular cartilage and artificial cartilage (PVA hydrogel). For both compliant materials, the addition of γ-globulin to sodium hyaluronate (HA) solution maintained low friction and protected rubbing surfaces under thin film conditions. These phenomena are discussed from the viewpoint of adaptive multimode lubrication (3)(4).

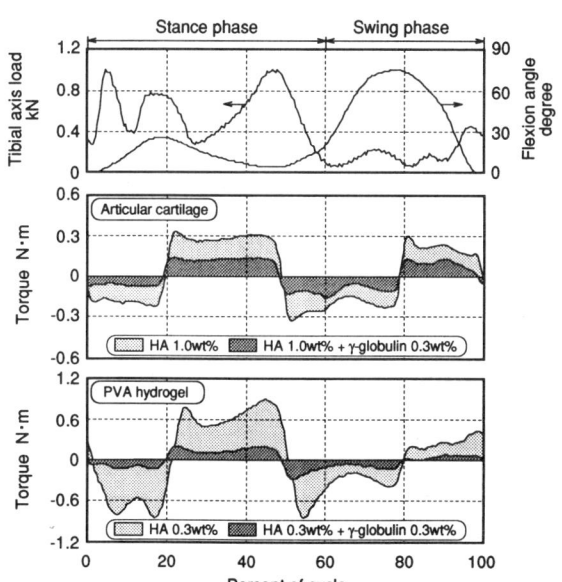

Fig.1: Effect of addition of protein on frictional behaviour of articular cartilage or PVA hydrogel against stainless steel

REFERENCES

(1) D.Dowson, Proc. IMechE, Pt3J, Vol.181, 1966-67, 45-54pp.
(2) A. Unsworth, D. Dowson and V. Wright, Ann. Rheum. Dis., Vol.34, 1975, 277-285.
(3) T. Murakami, JSME International Journal, Ser. III, Vol.33, No.4, 1990, 465-474pp.
(4) T. Murakami, N. Ohtsuki and H. Higaki, Proc. Int. Tribol. Conf. Yokohama 1995, 1996, 1981-1986pp.

NONDESTRUCTIVE AND NONINVASIVE OBSERVATION OF FRICTION AND WEAR OF HUMAN JOINTS BY ACOUSTIC EMISSION

HANS-JOACHIM SCHWALBE, MONIKA TITZE, THOMAS KREIS,
Fachhochschule Gießen-Friedberg, FB MF, Germany
RALF PETER FRANKE
Universität Ulm, Zentralinstitut für Biomedizinische Technik, Abt. Biomaterialien, Germany

ABSTRACT

Quality control in the orthopaedical diagnostics according to DIN EN ISO 9000ff requires methods of nondestructive process control, which do not harm the patient neither by radiation nor by invasive examinations. To gain an improvement of health economy quality-controlled and nondestructive measurements have to be introduced in the diagnostics and therapy of human joints and bones. A non invasive evaluation of the state of wear of human joints and of cracking tendency of bones is, as of todays´s point of knowledge, not established.

Friction and wear of human joints cause acoustic emission by joint action especially under load. The acoustic emission (AE) analysis is based on the phenomenon that sound is emitted in a characteristical and well differentiable way by cracking processes in the bone and by friction in human joints.

Hence an apparatus for testing wear and friction in the knee joint was developed which makes it possible to simulate a combined angle-dependent roll-glide friction. The first results allow a clear differentiation of damaged and undamaged joints. The acoustic emission signals of different artificially set defects were stored electronically and compared with acoustic emissions of known joint defects of patients during a clinical test. Fig 1 shows a typical amplitude course of a friction acoustic signal. The diagnosis of this knee joint was gonarthrosis. This defect could be detected at the same angle of the knee by several successive tests. An analysis of acoustic emissions allows an evaluation of the mechanism of natural sound generation. Characteristical events so far are the elevated level of continuous friction in cases of extended cartilage damage, stick-slip effects of local damage and differential settlement of the cartilage with increasing load.

Acoustic emission analysis allows to inform about the extent and angle of the defect. The state of abrasion or erosion can be evaluated.

The initiation of fracture correlates with a sudden release of stored elastic energy. The fracture in a human bone often starts in a microscopic area of the layer between compacta and spongiosa by reason of different compliances of the materials. The energy release rate of fracture causes a burst type of acoustic emission.

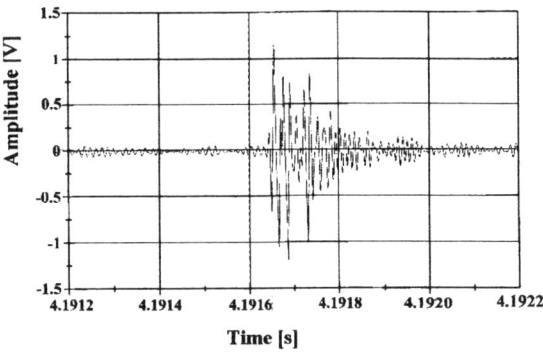

Fig.2: AE of first microfracture in the femur

Fig 2 shows a burst signal of first microfractures in the femur caused by loading due to knee-bending. This microfracture is necessary according to a working hypothesis for the physiological remodelling of the bone. There fore the test can be looked upon as nondestructive testing. By registrating the local load of the bone the cracking resistance can be defined.

Existing fractures can be clearly detected. Under load friction at the fracture surface of the bone causes continuous acoustic emission.

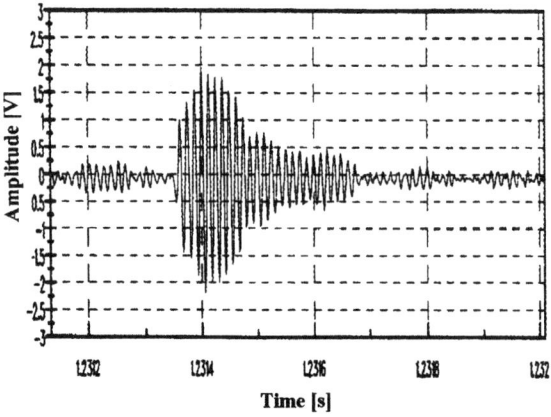

Fig 1: AE of friction in a knee joint

FRICTION IN "CUSHION FORM BEARING" FOR TOTAL KNEE JOINT REPLACEMENTS UNDER ADVERSE LUBRICATION CONDITIONS

T STEWART AND J FISHER
Department of Mechanical Engineering, University of Leeds, Leeds LS2 9JT, UK

Z M JIN
Department of Mechanical & Manufacturing Engineering, University of Bradford, Bradford, BD7 1DP, UK

ABSTRACT

"Cushion form bearing" utilising a compliant layered surface has been recently proposed as an alternative total joint replacement to the current polyethylene one (1, 2). Both experimental and theoretical studies have shown that under normal steady-state walking conditions, full fluid film lubrication can be generated to separate the two bearing surfaces in "cushion form bearing" for both hip and knee joint replacements (1, 3, 4). However, under other conditions such as standing still or moving slowly, the lubricant films may break down. Therefore, the purpose of this study is to investigate experimentally the friction in "cushion form bearing" for knee joint replacements under these adverse lubrication conditions.

The friction was measured in a pendulum joint simulator between a flat composite cushion layer (made from medical grade polyurethanes with elastic moduli of 20 and 1000 MPa respectively) and a model metallic femoral component. The applied load, sliding velocity and angular rotation of the femoral component were varied along with different lubricants (water, serum and high viscosity oil), in order to create a wide range of tribological conditions such as :

- high cyclic loading - in particular, increased load in the swing phase to reduce squeeze-film action
- variable stroke length to investigate lubricant starvation effect
- start-up under constant load to investigate high start-up friction
- constant loading - constant loading with variable velocity and angular rotation

Friction coefficients with both water and bovine serum have been found to rise from 0.02 during normal walking to 0.15 under high cyclic loading as shown in Figure 1. Decreasing the angle of rotation increased the friction due to the stroke length effect and the lubricant starvation. Constant loading had a similar effect on the friction, preventing lubricant entrainment that had previously occurred during the swing phase of walking. Start-up friction was found to peak after 60 s of stationary loading. The effect of increasing the lubricating film thickness using high viscosity oil resulted in low friction coefficients of 0.01 to 0.005 respectively under the high cyclic loading and decreased sliding velocities.

Harsher conditions of increased loading, decreased sliding velocity and decreased stroke length reduced the fluid entrainment into the contact compared to that was allowed during normal walking conditions. Consequently, damage to the polyurethane bearing surfaces such as scratches has been observed. However, contacts starting with thick elastohydrodynamic fluid films were undamaged, and even smoothed due to localised creep in the contact. Therefore it is very important to maximise the lubricating film thickness in the design of the "cushion form bearing" for total knee joint replacements in order to delay the breakdown of fluid film lubrication which can occur under certain loading conditions.

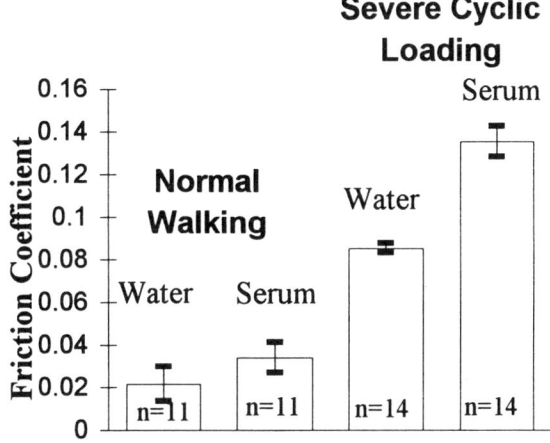

Fig. 1 : Peak friction coefficients

REFERENCES

(1) A Unsworth et al, Eng in Medicine, 1988, Vol. 17, 101-104pp.
(2) D. Dowson, Proc of IMechE Conf on the Changing Role of Engineering in Orthopaedics, 1-5, 1989.
(3) D D Auger et al, Proc of IMechE, Part H, Vol. 209, 73-81pp.
(4) G McClure et al, Proc of IMechE, Part H, Vol. 210, 89-93pp.

ADSORPTION-CONTROL OF SYNOVIA CONSTITUENTS ONTO ARTIFICIAL JOINT MATERIALS BY MEANS OF ELECTRIC FIELD : EVALUATION OF TRIBOLOGICAL CHARACTERISTICS

YOSHITAKA NAKANISHI, TERUO MURAKAMI and HIDEHIKO HIGAKI

Department of Intelligent Machinery and Systems, Faculty of Engineering, Kyushu University,
Hakozaki, Higashi-ku, Fukuoka, 812-81, Japan

ABSTRACT

The adsorption of synovia constituents on sliding materials has an influence on tribological characteristics in boundary and mixed lubrication regime. Consequently, the useful life of artificial joints may be decided by the lubricating ability of the adsorbed film, as far as the full fluid lubrication (1)(2)(3) are not perfectly maintained in daily activities. The purpose of this study is to investigate the possibility of adsorption / desorption-control of synovia constituents onto artificial materials by means of an electric field, in which tribological characteristics are improved.

Mixed lubrication property are evaluated by using a reciprocating and a roller on flat testers (4)(5). In friction tests of a conductive silicone rubber or a UHMWPE sliding on a stainless steel, the existence of protein in lubricant causes the deterioration of friction and wear characteristics. This result seems to be derived from the adhesion of protein on the hydrophobic and low elastic surface. In a stainless steel sliding on a stainless steel, on the other hand, adsorption of protein on sliding materials seems to promote both stability of friction and low wear rate. These results suggest that there are alternative methods of improving the tribological characteristics; promoting adsorption or desorption of protein onto sliding materials.

As the notable examples of adsorption / desorption control by means of an electric field, we show the results of a conductive silicone rubber cylindrical surface sliding on a stainless steel plate: Remarked decrease in friction is observed by means of the electric field (3V, sinusoidal wave, f=1 - 10Hz), in which lubricants are water solution of sodium hyaluronate (HA) with or without physiological concentration of γ-globulin. In HA solution with physiological concentration of albumin, the frictional characteristics are improved in an direct current in which the stainless steel is anode in the mixed lubrication in which hydrodynamic effect is dominative. Judging from the above, potential and instability of electric double layer on rubbing surfaces produced by an electric field seem to change not only adsorption / desorption behaviour of synovia constituents but tribological characteristics.

The instability of electric double layer may produce the weak-bond between adsorbates and sliding materials. To elucidate this effect, friction tests applying the burst waveform-electric filed shown in Fig.1 were undertaken (Fig.2). Burst waveform give improvement

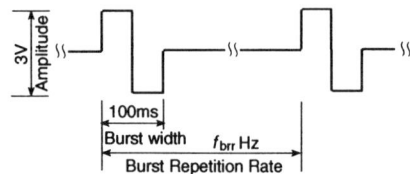

Fig.1: Waveform of electric field applied to rubbing surfaces

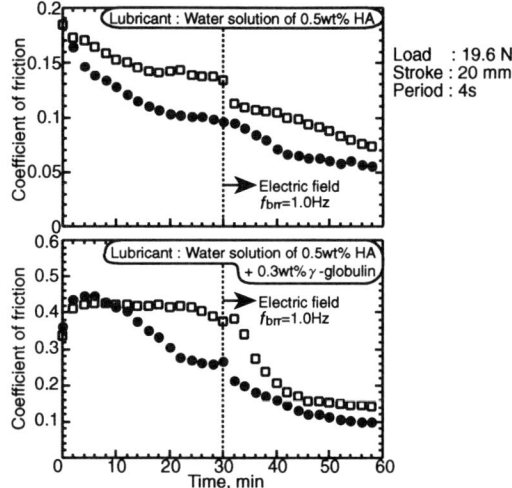

Fig.2: Effect of electric field on friction of conductive silicone rubber cylindrical surface on stainless steel plate

in frictional characteristics in HA solution with γ-globulin lubrication. This type of waveform may be more suitable than sinusoidal wave and direct current in respect of reducing the surface damage of sliding materials and the energy consumption.

REFERENCES

(1) D.D.Auger, D.Dowson, J.Fisher, Proc. IMechE, Part H, Vol.209, page 73, 1995.
(2) T.Murakami, Design of Amenity, Kyushu University Press, page 319, 1995.
(3) K.Ikeuchi, M.Ohashi, N.Tomita, M.Oka, Journal of Japanese Soc. for Clinical Biomech. and Related Res., Vol.15, page 381, 1995 (in Japanese).
(4) Y.Nakanishi, T.Murakami, H.Higaki, Elastohydrodynamics'96-Fundamentals and Applications, Elsevier Science, 1997, in press.
(5) Y.Nakanishi, T.Murakami, H.Higaki, Proc.Int.Conf. New Frontiers in Biomech.Eng., 1997, in press.

LUBRICATION AND WEAR PROCESS IN A Si_3N_4-ON-Si_3N_4 HIP JOINT

KEN IKEUCHI and MINAKO OHASHI
Research Center for Biomedical Engineering, Kyoto University, Kyoto 606-01, Japan

INTRODUCTION

The polyethylene particles from hip prostheses cause loosening due to bone loss (1). A hip prosthesis consisting of a ceramic ball and a ceramic socket seems to be promising for permanent replacement, because a ceramic is more durable and wear resistant than polyethylene. While alumina is the only material for hip prostheses of ceramic-ceramic combination at present, silicon nitride seems to be another candidate ceramic because it is more tough than alumina. According to the sphere-on-flat experiment (2), the mating faces were polished to ultra-smooth by tribo-chemical reaction with water. At the same time, friction decreased drastically because quasi-hydrodynamic lubrication condition was reached. The purpose of this study was to examine the tribological characteristics of a flat surface of silicon nitride ceramic sliding on itself with a thrust collar apparatus.

APPARATUS AND METHOD

Thrust collar apparatus (Fig.1) was used to investigate friction and wear during sliding in distilled water. The contact area of the specimens (Fig.2) is 300 mm^2 and the equivalent radius is 9.7 mm. The initial CLA roughness was 0.03 μm. Sliding velocity was 40 mm/s. During the experiment, the motor was stopped, the specimens were removed and cleaned with distilled water and acetone. Then they were dried using vacuum pump to measure weight loss and observe the surface with AFM.

RESULT AND DISCUSSION

Test was performed at 20 °C while temperature rose to 32 °C. Coefficient of friction (Fig. 3) was kept about 0.2 until sliding distance reached 10 km. Then, the specimens were examined with a scanning probe microscope (SPM/AFM) and SEM. As friction became very high after restarts, contact pressure was decreased to 1.5 MPa because available torque of the driving motor was limited. During the second stage, coefficient of friction fluctuated with time. The maximum value was 0.5 and the minimum value was 0.002. Fluctuation increased further during the third stage where the sliding distance was longer than 19 km. The authors suppose that the reason of such high and unstable friction after each restart is the loss of wear particles, products of tribo-chemical reaction and contaminants from the sliding faces.

REFERENCES

(1) B.M. Wroblewski, Orthop. Clin. North. Am., Vol. 19, 1988, 627-630pp.
(2) Y.S. Zhou, K. Ikeuchi and M. Ohashi, Accepted for publication by WEAR.

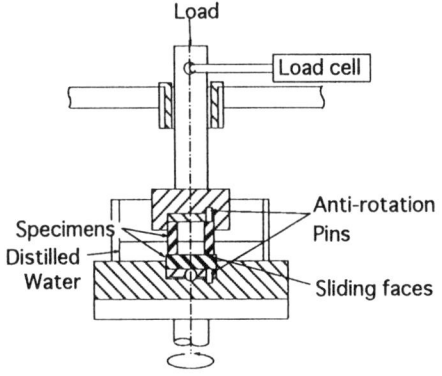

Fig. 1: Cross sectional view of apparatus

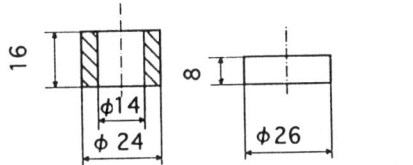

Upper specimen   Lower specimen

Fig. 2: Dimensions of ceramic specimens

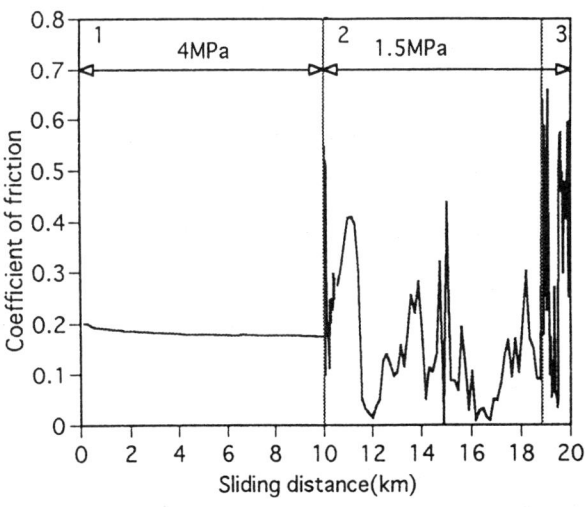

Fig. 3: Transition of friction coefficient

POLYMER TRANSFER FILMS ON CEMENTED TOTAL HIP REPLACEMENT FEMORAL STEMS

J E COOK, A J C LEE and R M HOOPER
School of Engineering, University of Exeter, North Park Road, Exeter, EX4 4QF, UK

INTRODUCTION

Total replacement of the hip joint is a well-established technique, with an estimated one million operations being performed world-wide each year. The generation of particulate debris at the artificial implant interfaces may be a major factor affecting the long term survival of the prostheses (1).

Investigations have been performed into fretting between the femoral stem and polymethylmethacrylate bone cement interface as a source of particulate debris. Parallel studies of femoral implants retrieved from patients and laboratory tests have been undertaken

CLINICAL STUDIES

Optical and electron microscopy were used to study retrieved hip stems. A pattern of wear, characteristic of fretting damage, was observed on the medial-posterior and anterior-lateral surfaces. These regions correspond to the presence of high stresses due to torsion of the implant (2). Localised polymer transfer films were also observed, which has not been previously reported.

LABORATORY STUDIES

To identify positively the wear mechanism observed on the retrieved femoral implants, three different laboratory investigations were performed in parallel. A range of orthopaedic alloys, with both matt and polished surface finishes were tested using each technique for two million cycles, in Ringer's solution at 37°C.

A soft impresser technique, shown in Figure 1, replicated cyclic fatigue loading of a bone cement asperity against a metal surface.

A micro-sliding apparatus was used for small scale *in vitro* fretting experiments between orthopaedic metal samples and polymethylmethacrylate specimens. Relative micromotion of an amplitude ranging between 10μm and 100μm was produced by a electro-vibrator and tests were conducted at varying contact pressures,

Fretting tests of full size femoral implants within a cement mantle were performed. A hydraulically driven apparatus was utilised to apply cyclic axial and torsional loads to the hip stems, with the objective of reproducing the wear phenomenon observed on retrieved specimens.

RESULTS

A localised polymer transfer film was observed on the metal alloy surfaces of each specimen tested using each of the three methods. There was no damage to the metal surfaces tested using the soft impresser technique. However, fretting damage and metallic debris was observed on specimens tested in the micro-sliding apparatus and the full size femoral stem testing machine. Examination of the laboratory specimens has shown that adhesion of the polymer occurs adjacent to wear damaged areas, where relative slip of the interface has occurred. The full size femoral stem testing machine replicated the wear phenomenon observed on explanted stems, allowing direct comparisons of different designs.

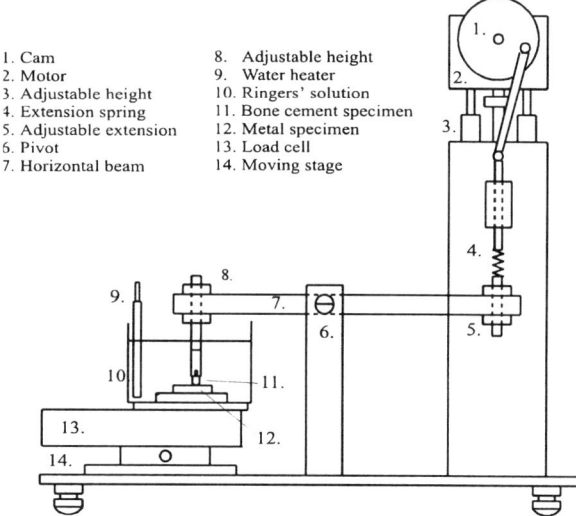

1. Cam
2. Motor
3. Adjustable height
4. Extension spring
5. Adjustable extension
6. Pivot
7. Horizontal beam
8. Adjustable height
9. Water heater
10. Ringers' solution
11. Bone cement specimen
12. Metal specimen
13. Load cell
14. Moving stage

Fig 1: Soft impresser technique apparatus

DISCUSSION

The adhesion of polymer may be explained in terms of chemical bonding and / or mechanical interlock between the polymethylmethacrylate and the passive oxide layer present on the surface of the orthopaedic alloy. The implications of a transfer film will be discussed with reference to the Exeter hip implant, which employs subsidence of the stem taper within the cement mantle to facilitate the transfer of force. If the cohesive strength of the polymer is exceeded by the adhesive forces between the metal and polymer, the long term fixation of the implant may be affected. It is believed that the transfer film phenomenon may be an important factor in the propagation and discharge of particulate debris in joint replacements.

REFERENCES

(1) P.P. Anthony, G.A. Gie, R.S.M. Ling, C.R. Howie, The Journal of Bone and Joint Surgery, Vol. 72-B, No. 6, Pages 971-979, 1990.

(2) N. Berme, J.P. Paul, Journal of Biomedical Engineering, Vol. 1, Pages 268-272, 1979.

Submitted to *IMechE Journal of Engineering in Medicine*

BIOTRIBOLOGY, SYNOVIAL JOINT LUBRICATION, AND OSTEOARTHRITIS

M J FUREY, M O SCHROEDER, H L HUGHES, M C OWELLEN, L J BERRIEN,
H VEIT, E M GREGORY and E T KORNEGAY
Virginia Polytechnic Institute and State University, Blacksburg, VA 24061-0238, USA

ABSTRACT

This study is part of an ongoing research effort to explore possible connections between tribology, synovial joint lubrication, and osteoarthritis. As defined by Sokoloff, osteoarthritis is "an extremely common, noninflammatory, progressive disorder of movable joints, particularly weight-bearing joints, characterized pathologically by deterioration of articular cartilage and formation of new bone in the subchondral areas and at the margins of the joint." Although a surprisingly large number of models and theories of synovial joint lubrication have been proposed, practically all of them focus on friction—which of course is not wear—and most do not consider the biochemistry of the system. The emphasis of this study is on cartilage wear and damage and in particular, how they are influenced by changes in the biochemistry of the fluid environment. The project involves researchers in Mechanical Engineering, Animal Science, Biochemistry, and Veterinary Medicine.

The following paper describes results obtained from *in vitro* tests of bovine articular cartilage using a tribological device described previously (1). The test device is capable of measuring deformation, friction and normal load under controlled conditions of loading and reciprocating velocity. The studies performed include two test configurations: (a) cartilage sliding on stainless steel and (b) cartilage sliding on cartilage. In both configurations, a 6.35 mm diameter upper sample of bovine cartilage was placed in sliding contact with a 25.4 mm diameter lower sample of highly polished stainless steel or bovine cartilage. Loads ranging from 50-70 N were applied to the upper sample yielding an average pressure of 1.6-2.2 MPa in the contact area. Post-test analysis included biochemical assay of test lubricants to determine the amount of wear from hydroxyproline analysis, as well as scanning electron microscopy and histological sectioning to determine the extent of surface and subsurface damage.

In **cartilage-on-stainless steel** tests, scanning electron microscopy and histological sectioning showed distinct differences in surface and subsurface damage created using different lubricants. Tests with a buffered saline lubricant resulted in the most severe damage. Large wear tracks were visible on the surface of the cartilage plug, as well as subsurface voids and cracks. When hyaluronic acid, a constituent of the natural synovial joint lubricant, was added to the saline reference fluid, less severe damage was observed. Little to no cartilage damage was evident in tests in which the natural synovial joint fluid was used as the lubricant. Another interesting result was that a thin film of transferred material was observed on the stainless steel disks after tests with the buffered saline lubricant. Examination of the film with Fourier Transform Infrared Microspectrometry shows distinctive bio-organic spectra which differ from that of the original bovine cartilage.

In **cartilage-on-cartilage** tests, the most severe wear and damage occurred during tests with buffered saline as the lubricant. The damage was less severe than in the stainless steel tests, but some visible wear tracks were detectable with scanning electron microscopy. Histological sectioning and staining of both the upper and lower cartilage samples show evidence of elongated lacunae and coalesced voids that could lead to wear by delamination. An example is shown in Fig. 1. The proteoglycan content of the subsurface cartilage under the region of contact was also reduced. When synovial fluid was used as the lubricant, no visible wear or damage was detected.

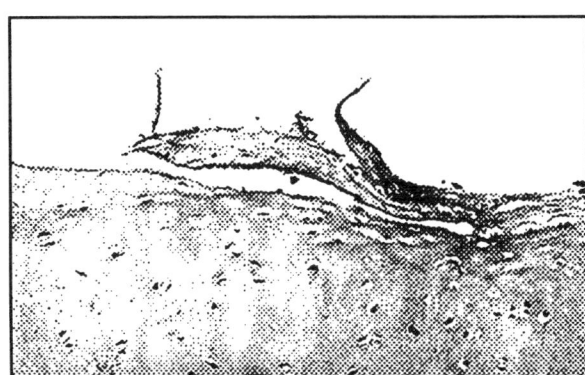

Fig. 1 Cartilage damage produced by sliding contact

It has been demonstrated in these *in vitro* tests with bovine articular cartilage that the biochemical nature of the fluid environment has significant effects on the severity of wear and subsurface damage. Furthermore, scanning electron microscopy, histological sectioning, and biochemical analysis provide effective means to quantify the degree of cartilage wear and damage. The study to examine possible connections between tribology and osteoarthritis is continuing.

REFERENCES

(1) M. J. Furey, "Joint Lubrication," Chap. 23 in *The Biomedical Engineering Handbook*, editor Joseph D. Bronzino, CRC Press (1995), 333-351.

COMPUTER BASED METHOD FOR THE ANALYSIS OF WEAR PARTICLES FROM SYNOVIAL JOINTS AND JOINT REPLACEMENTS

G.W. STACHOWIAK and G.B. STACHOWIAK
Tribology Laboratory, Department of Mechanical and Materials Engineering
University of Western Australia, Perth, Australia
P. CAMPBELL
Joint Replacement Institute, Orthopaedic Hospital, Los Angels, California, USA

ABSTRACT

Wear particles commonly occur in both synovial and artificial joints. The particles can be extracted from the synovial fluid or periprosthetic tissue and their morphology assessed. It seems that particle morphology can provide vital information about the condition of a joint and this can be used in diagnosis and prognosis of joint diseases and also in the assessment of wear processes occurring in the artificial implants. Populations of wear particles were collected and examined in a scanning electron microscope (SEM). The SEM images were digitized and the shape of the particles was analysed by specially developed computer software.

Metallic wear particles from hip and UHMWPE particles from knee replacements were obtained from the periprosthetic tissues. The particles were separated from the soft tissue by chemical methods. The chemical treatment involved tissue digestion in papain or NaOH followed by ultracentrifuging over a variable density gradient. Light UHMWPE particles were collected from the top of the column, while heavier metallic particles were deposited at the bottom in the form of a pellet.

Wear particles from natural joints were extracted from synovial fluid aspirated from healthy and osteoarthritic knee joints. Particles collected from the natural joints and metallic particles collected from the hip prosthesis were deposited on glass slides using ferrography technique while UHMWPE particles from the knee prosthesis were dried and deposited on a carbon tape. All particles collected were examined in a scanning electron microscope. Digitized SEM images of the particles were obtained and analysed on a Power Mac 6100 computer. A specially developed software was used to numerically characterize particles boundary. The following numerical descriptors were used to characterize the particle shape: particle size, boundary fractal dimension and shape parameters such as form factor, roundness, convexity and aspect ratio. Elemental composition of the particles was determined by energy dispersive X-ray spectroscopy.

In this study Ti - based wear particles from a failed hip prosthesis and UHMWPE particles from a failed knee prosthesis and wear particles from healthy and osteoarthritic natural joints were analysed and compared. The examples of particles obtained from artificial and natural joints together with their boundary descriptors are shown in Figure 1.

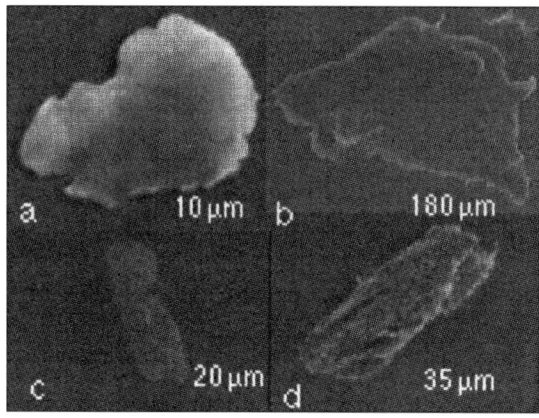

Fig. 1: Examples of particles obtained from artificial and natural joints; a) Ti alloy wear particle from a failed hip prosthesis D=1.0443, b) UHMWPE particle from a failed knee prosthesis D=1.0519, c) wear particle from a healthy knee joint D=1.0446, d) wear particle from an osteoarthritic knee joint D=1.0826.

Populations of particles from artificial and natural joints have been analysed. The results obtained indicate that changes occurring in wear particle shape during joint's wear are reflected in changes of their numerical parameters.

The findings of this work demonstrate that computer image analysis of the particle shape is a useful method that could be employed in the assessment of the joint condition and characterization of wear processes occurring. It appears that the particle shape descriptors can be used in the assessment of the synovial joint condition, i.e. in the diagnosis and prognosis of joint diseases. It was found that the shape and size of particles generated in joint replacement prostheses is characteristic for a particular type of prosthesis and is related to wear processes which contributed to the failure of the prostheses.

SUMMARY:

• Image analysis of wear particles found in artificial and natural joints appears to be a useful method for the characterisation of wear processes occurring in these joints,

• Shape and size of particles generated in joint replacement prosthesis can be related to a joint condition while the shape of wear particles found in synovial joints is related to the degree of osteoarthritic changes occurring in joints.

THREE DIMENSIONAL COMPUTER IMAGE ANALYSIS OF WEAR PARTICLES FROM NORMAL AND ARTHRITIC SYNOVIAL JOINTS USING NUMERICAL MEASURES OF WEAR PARTICLE MORPHOLOGY

D PANZERA, T B KIRK, R V ANAMALAY and M KUSTER
Department of Mechanical and Materials Engineering, The University of Western Australia, Nedlands, Western Australia, 6907, Australia

ABSTRACT

Cells, soft tissue and cartilaginous particles generated by wear of the articular cartilage are all present in synovial joints. Such debris allows the assessment of the wear and lubrication processes operating in a particular joint at a specific time. The large variety of debris within a joint can, however, make distinction between populations of particles from different joints difficult. Pathological joints are characterised by wear anomalies, where severe wear of the articular cartilage is symptomatic of osteoarthritis and anomalous wear debris can cause inflammatory arthritis (1). Since little is known of the wear processes occurring in either normal and pathological synovial joints, wear particle analysis provides a useful tool for the study of wear in human joints because the size, shape and composition of wear debris elucidates the wear process by which the particles were generated.

Computer wear particle analysis requires the debris to be imaged and subsequently, from that image, certain measurements are extracted, and the particle is characterised according to these measurements. Fractal based particle analysis of synovial wear debris, imaged in two dimensions, has previously been used to assess the particle morphology but this technique, however, is not the most appropriate method of characterisation for all particle types.

To overcome this limitation, software comprising other parameters, besides fractal descriptors, have been developed to extract numerical descriptors from three dimensional wear particle images, obtained via a Laser Scanning Confocal Microscopy (LSCM) (2), thus comprising an image analysis system used for characterising synovial wear particles. The LSCM used in this study includes a transmission sensor to simultaneously acquire a transmission image (two dimensional boundary information), shown in Fig. 1, and reflected image (three dimensional surface information). From a series of reflected images scanned at regular depths of the particle, a Height Encoded Image (HEI) can be composed (3), where grey level corresponds to surface contours, as in Fig. 2.

The transmission sensor can provide better images for boundary analysis when compared to the reflected image. This type of analysis of debris from synovial joints has already been shown to be less invasive and more sensitive than arthroscopy for the detection of joint damage and the developments described promise to improve the ability of objective analysis to distinguish between different joint injuries and diseases.

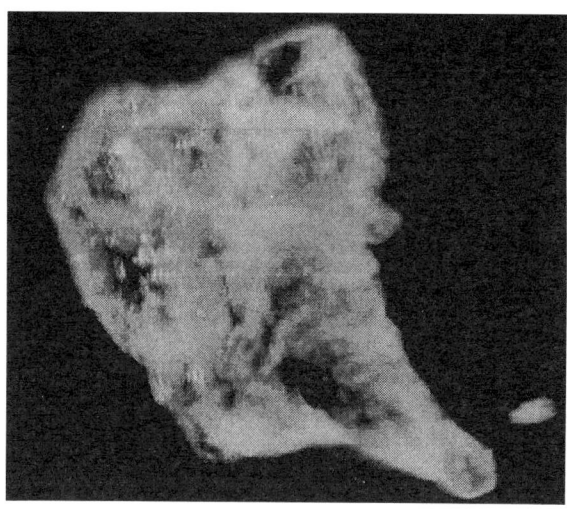

Fig. 1: Transmission image (field of view 75 μm).

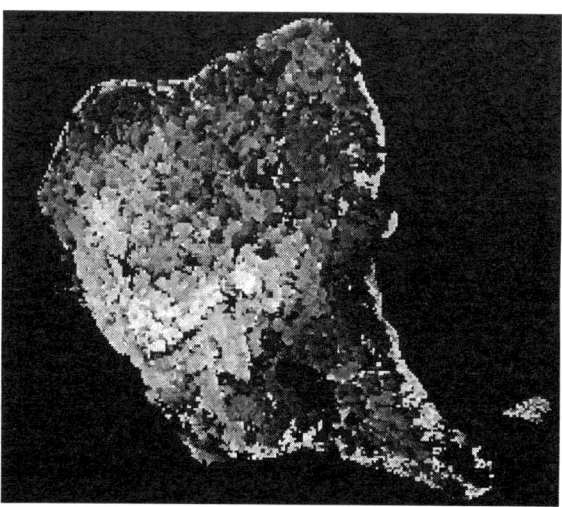

Fig. 2: HEI image (field of view 75 μm).

REFERENCES
(1) C. H. Evans, R. A. Mazzochi, D. D. Nelson, H. A. Rubash, Arth Rheum, 27, p200, 1984.
(2) P.M. Delaney, M. R. Harris, R. G. King, Applied Optics, Vol 33, No 4, p573-577, 1994.
(3) R.V. Anamalay, T.B. Kirk, D. Panzera, Wear 181-183, p771-776, 1995.

THURSDAY 11 SEPTEMBER

		Page number
TH1	Rolling bearings	263
TH2	Industrial problems and test methods	281
TH3	Wear by hard particles – II	297
TH4	Wear in polymer systems	313
TH5	Manufacturing and maintenance	327

Performance Prediction for Aircraft Gas Turbine Mainshaft Ball Bearings

TEDRIC A. HARRIS
Mechanical Engineering Department, Pennsylvania State University, University Park, PA 16802, USA
ROGER M. BARNSBY
Mechanical Components Chief, Pratt & Whitney, United Technologies, East Hartford, CT 06108, USA

ABSTRACT

The internal thrust loads in aircraft gas turbines are carried by angular-contact ball bearings, one on each of the compressor-turbine driven shafts. There are typically two or three concentric shafts, rotating at individual speeds to generate the required aerodynamic loading with a resultant thrust load on each of the ball bearings. The designs of these bearings are based primarily on the required life or fatigue endurance. Estimation of this parameter is currently accomplished using empirically developed, factor-based calculation methods; for example, material-life and lubrication-life factors. These life calculation methods are based on the 1947 published work of Lundberg and Palmgren **LP** (1) as modified by the indicated empirical factors. Moreover, the **LP** life prediction methods consider only the Hertz stress loading of the ball/raceway contacts and only the subsurface failure mode of the raceways.

Experience has demonstrated in many instances that bearing fatigue lives predicted by this method are less than those achieved in practice. Experience has also shown the potential for ball fatigue as well as for bearing raceway fatigue. A more accurate and complete performance prediction system would allow improved optimization of bearing designs, with associated savings in engine performance, weight, and cost, as well as improved reliability.

As thrust-to-weight ratio increases with new engine designs, smaller bearing compartments are necessitated. To extrapolate bearing design to these more demanding applications requires a stress-based rather than a factor-based bearing life prediction method. This means that in addition to the Hertz stresses, ball/raceway contact frictional stresses and ring hoop and residual stresses need to be included in the performance prediction system. The Ioannides-Harris **IH** (2) fatigue life model allows the consideration of these stresses in the life calculation in addition to the effect of a fatigue limit stress.

To correctly employ the **IH** model, it is necessary to accurately determine the contact surface and subsurface stresses. The calculation of the normal stresses acting on the elliptical contact areas is according to the methods of Hertz. Calculation of surface frictional stresses requires a detailed thermal analysis of each contact. Lubricant temperatures must be calculated at the contact inlet and within the contact. The former influences the lubricant film thickness via the viscous properties of the lubricant entering the contact; the latter influences the frictional stresses via the viscosity of lubricant in the contact.

The friction in the contact is the result of the degree of sliding and the lubrication regime in the contact. Harris (3) shows a means to determine contact loads, sliding velocities, lubricant film thicknesses, and frictional forces in the contacts. Iteration is used in the resulting computer program to establish a compatible friction-temperature solution for each contact.

Subsurface stresses are calculated according to the methods of Ahmadi et al (4). In the life analysis, octahedral shear stress is considered as the failure-initiating stress. The life is calculated for each contact using the equation:

$$L = A_{IH} \left(\frac{Q_c}{Q} \right)^3$$

where A_{IH} is a factor comparing **IH** life to **LP** life of the subsurface stress distribution, Q_c is the basic dynamic capacity of the contact and Q is the normal load on the contact. Bearing life is determined from the statistical combination of the contact lives including the consideration of ball failures.

This method is shown to give reasonable agreement between predicted and actual application lives.

REFERENCES

(1) G.Lundberg and A.Palmgren, "Dynamic Capacity of Rolling Bearings", *Acta Polytechnica, Mech.Eng.Ser. 1*, RSAEE, No.3, 7, 1947.
(2) E.Ioannides and T.Harris, "A New Fatigue Life Model for Rolling Bearings", *ASME Trans, J.Trib*, 107, pp367-378, 1985.
(3) T.Harris, *Rolling Bearing Analysis, 3rd Ed*, Wiley (1991).
(4) N.Ahmadi et al, "The Interior Stress Field Caused by Tangential Loading of a Rectangular patch on an Elastic Half Space", ASME 86-TRIB-15, 1986

IMPROVEMENTS IN THE PERFORMANCE OF GAS TURBINE ENGINE BEARINGS BY SURFACE MODIFICATION

A. DODD

NSK-RHP European Technology Centre, Mere Way, Ruddington, Nottingham, NG11 6JZ, U.K.

ABSTRACT

The manufacturers of aircraft gas turbine engines are continually improving the efficiency and reliability of their engines and thereby reducing the life-cycle cost of the unit. This invariably results in greater demands being placed on the engine components particularly the mainshaft bearings requiring them to operate at higher speeds and loads with increased durability. For the bearings to achieve this bearing materials are required with improved rolling contact fatigue (r.c.f.) and wear resistance together with greater tolerance to poor lubrication conditions.

The development of new materials to meet these requirement is costly and slow due to the safety critical nature of the application. Alternatively surface treatments can be used to enhance the properties of existing bearing materials, particularly thermochemical treatments such as nitriding and nitrocarburising. Dezzani et al(1) has shown that nitrocarburised bearing raceways made in M50 and M50 NiL materials performed better than untreated raceways in conditions of contamination, surface flaws, marginal lubrication and traction. The nitrocarburising treatment, however, involved high temperatures (>500°C) and long process times (up to 32 hours) and a final grinding operation is required to remove the compound layer which performs unfavourably in r.c.f.(2).

Outlined in this paper is a nitriding process developed by the NSK-RHP European Technology Centre for M50 and M50 NiL bearing materials. The process is performed at significantly lower temperatures and reduced process times compared to other nitriding/nitrocarburising treatments for bearing materials. Furthermore, the process conditions are such that the formation of a compound layer is prevented and distortion is maintained within acceptable levels thereby negating the need for a final grinding operation. Fatigue testing of M50 and M50 NiL materials under high Hertzian contact stresses and marginal lubrication has shown that nitriding significantly improves the r.c.f. resistance of both materials, see Figure 1.

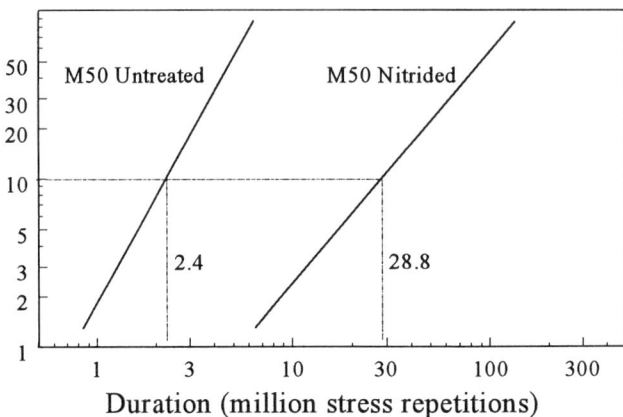

Figure 1. Fatigue test results for M50 material.

REFERENCES

(1) M. M. Dezzani et al, "Hybrid Ceramic Bearings For Difficult Applications", Presented at the Int. Gas Turbine and Aeroengine Congress & Exposition, Houston, Texas, June 5-8, 1995.

(2) J. F. Braza, "Rolling Contact Fatigue Properties of Ferritic Nitrocarburised M50 Steel", Tribology Transactions, Vol. 35, 1, 1992, pp 89-97.

The Experimental Measurement of Cage Drag in High Speed Rolling Element Bearings

R J CHITTENDEN
Department of Mechanical Engineering, The University of Leeds, Leeds LS2 9JT
CN MARCH
Tribologic Ltd, c/o Department of Mechanical Engineering, The University of Leeds, Leeds LS2 9JT

ABSTRACT

The development of rolling element bearings has taken place over many tens of years. The associated numerical models have reached very advanced levels, Harris (1) or Gupta(2), but consideration of their constituent analyses, carried out in the initial phase of the project from which the work reported here forms but part, indicated that some areas had received only limited attention. (See Dunker(3) for a general overview of the activities associated with this work.) In particular the prediction of cage drag / windage, important in bearings typical of aerospace applications, was found to have an elementary basis and an experimental programme was therefore undertaken to determine-

(a) whether the form of the existing, classical, model was satisfactory and

(b) what could be done to take account of the two phase fluid (air and oil) existing in the bearing chamber.

To provide the most direct measurement of cage drag/ churning a test apparatus was designed in which a model representing the cage and rolling elements was driven by means of a variable speed motor and step up belting. The outer raceway was fixed and the inner raceway was belt driven at the required bearing speed by a second variable speed motor. This configuration allowed the cage drive torque, and hence that attributable to cage drag/ churning, to be measured by a transducer whose only parasitic loss was that from the air bearing providing support for the cage model.

Results were obtained for three sizes of roller and ball bearing (43mm, 60mm and 140mm PCD), all being tested with both jet and under raceway oil supply. The cage drive torque was measured at cage speeds of half the inner raceway speed and at speeds +/-20% of this value. The maximum inner raceway speed attained was 40,000rpm. To illustrate the results obtained Figures 1 and 2 show the steady state torque values required to drive the cage on the 60mm pcd bearing at speeds up to approximately 14,000rpm.

Though ready comparison of the results obtained with those derived from the classical approach was difficult, due to the problem of dealing with the air / oil mixture in the bearing chamber, the indications were that there was significant disagreement between the new experimental values and those produced using common assumptions for high speed bearings of this type.

REFERENCES

(1) TA Harris 'Rolling Bearing Analysis' 3rd edition, J Wiley & Sons, New York, 1984

(2) PK Gupta, 'Advanced Dynamics of Rolling Elements', Springer-Verlag, New York, 1984

(3) 'Advances in Techniques for Engine Applications', ed R Dunker, EC Aeronautics Research Series, J Wiley & Sons, Chichester, 1995

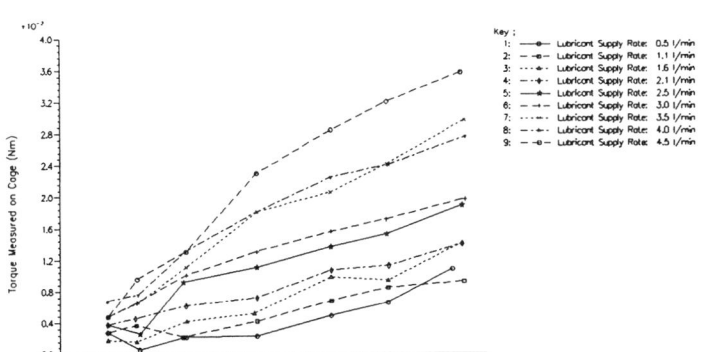

Figure 1 - Measured torque for 60mm pcd ball bearing model with under race oil feed (Inner race speed = 1.8 x Cage speed)

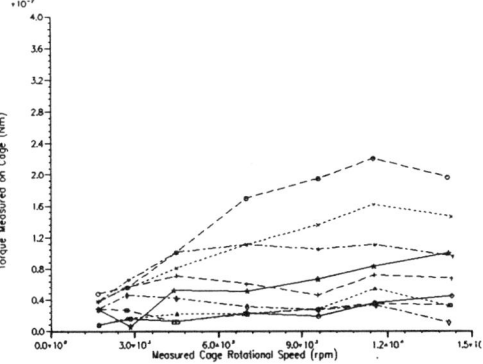

Figure 2 - Measured torque for 60mm pcd ball bearing model with under race oil feed (Inner race speed = 2.4 x Cage speed)

Experimental and Theoretical Investigation on Rolling Contact Fatigue of 52100 and M50 Steels Under EHL or Micro-EHL Conditions

D. NELIAS[1], M.-L. DUMONT[1], F. COUHIER[1], G. DUDRAGNE[2] and L. FLAMAND[1]

[1] L.M.C. Bât. 113, CNRS UMR 5514, INSA de Lyon, 20 Av. A. Einstein, 69621 Villeurbanne Cedex, France
[2] SNR Roulements, BP 2017, 74010 Annecy Cedex, France

ABSTRACT

The purpose of this investigation is to clarify the role of roughness on rolling contact fatigue. We adopt the nomenclature proposed by Tallian (1), where two main types of rolling contact fatigue are identified: i) spalling, which is failure by the formation of macroscopic craters in the contact surface as a result of fatigue crack propagation in the Hertzian stress field, ii) surface distress, which is defined as an asperity-scale spalling fatigue. A link is made between our experimental observations and the initiation mechanisms proposed by Cheng et al. (2,3).

Tests have been carried out on a high-speed twin-disk machine. The operating conditions have been selected to encompass typical jet engine applications. More precisely, the twin-disk machine reproduces the operating conditions of high-speed rolling bearings at the raceway-rolling element contact, i.e. rolling and sliding speeds, contact pressure, lubricant, temperature, material and surface finish. The lubricant used is qualified under the MIL-L-23699 specifications.

Tests have been carried out for two rolling bearing steels (52100 and M50), two surface roughnesses (grinding with and without polishing) corresponding to EHL and micro-EHL conditions, three normal loadings (1.5, 2.5 and 3.5 GPa) and under pure rolling or rolling plus sliding conditions. The mean contact velocity is set equal to 40 m/s for each test. The slide-to-roll ratio has been chosen after the measurement of traction curves and the determination of the scuffing limit at 1.5, 2.5 and 3.5 GPa.

No surface damage has been observed up to 50.10^6 cycles for tests with smooth specimens, independently of material, normal pressure, slide-to-roll ratio, and for both driver or follower disks.

Tests with rough specimens have produced a typical surface damage, called here surface distress, made of a large population of asperity-scale micro-cracks and micro-spalls. No difference in the behavior of M50 versus 52100 steels was demonstrated. It should be noted that surface distress is slightly different for tests under pure rolling conditions (see Fig. 1) and tests under rolling plus sliding conditions (see Fig. 2). Thus, the role of the friction direction is underlined.

From these experimental observations, it can be assumed that former micro-spalls themselves induce formation of further surface damages. To verify this assumption, calculations have been carried out using a transient EHL model. The influence of an indent in a line contact, standing for a micro-spall, is studied. Surface pressure and associated sub-surface stress field are analysed versus the sliding direction.

Numerical results which are presented are useful to understand the surface damage evolution for tests with and without sliding. They show that existing micro-spalls tend to propagate on the contacting surfaces in the friction direction, due to pressure peaks and associated high local stress concentration. That contributes to explain why rough surfaces after rolling plus sliding tests are more damaged than those observed after pure rolling tests.

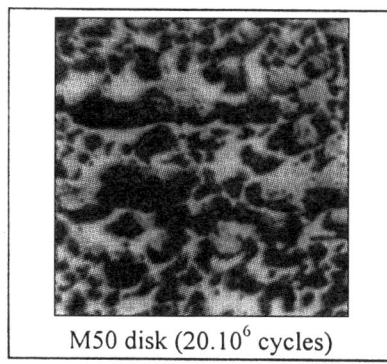

Fig. 1: Surface distress under pure rolling

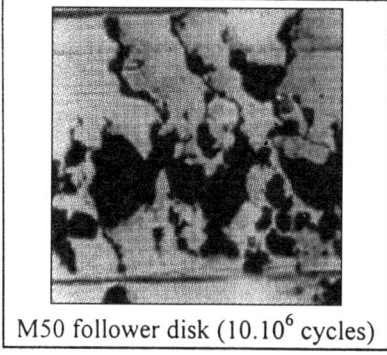

Fig. 2: Surface distress under rolling plus sliding

REFERENCES

(1) Tallian, T.E., 1992, *Failure Atlas for Hertz Contact Machine Elements*. ASME Press, New-York.
(2) Cheng, W. et al., 1994, "Longitudinal Crack Initiation Under Pure Rolling Contact Fatigue," *Trib. Trans.*, 37, 1, pp. 51-58.
(3) Cheng, W. et al., 1994, "Experimental Investigation on Rolling/Sliding Contact Fatigue Crack Initiation with Artificial Defects," *Trib. Trans.*, 37, 1, pp. 62-73.

EFFECT OF THREE DIMENSIONAL RANDOM SURFACE ROUGHNESS ON FATIGUE LIFE OF A LUBRICATED CONTACT

XIAOLAN AI
The Timken Company, Canton OH 44706

ABSTRACT

The increasing demand for power density in modern machinery requires bearings to operate under severe lubrication conditions. The lubricant film thickness is often the same order of the magnitude as surface roughness. Surface roughness plays a significant role in altering contact pressure and lubricant film thickness. Consequently, surface roughness related contact failure becomes an important issue. To completely understand the failure mechanism, precise knowledge of contact pressure and the associated interior stress is often a must.

The recent advances in computing power and numerical technique provide us an opportunity of examining the rough EHL contact with greater detail by using a deterministic approach. A fast multigrid EHL solver developed by Ai and Cheng (1) has been extended in conjunction with the 3-dimensional stress solver to study the EHL point contact problem with measured real surface profiles. The intent of this study is to explore the impact of surface roughness on contact pressure, the interior stress field and, ultimately, on contact fatigue life for surfaces produced by different manufacturing processes

Three types of surface roughness profiles produced by different manufacturing processes have been examined. Table 1 tabulates roughness parameters of these surfaces.

Table 1

Surface	R_a (μm)	σ (μm)	Avg. slope
Process A	0.104	0.131	1.144×10^{-2}
Process B	0.037	0.047	0.604×10^{-2}
Process C	0.035	0.044	0.423×10^{-2}

The EHL contact model was solved numerically by using the multigrid method [see (1)] for contact pressure distributions and surface tangential traction stresses. As an approximation, the perfect visco-plastic model was used to describe the non-Newtonian behavior in lubricant shear stress calculation. The interior stresses were subsequently computed based on Boussinesq and Cerruti patch solutions. For detailed descriptions on numerical algorithms the reader is referred to Ai and Lee (2).

For contact with random surface roughness, interior stresses at any location within the solid experience random variations. To account for stress variations due to random surface roughness, an effective stress concept based on damage accumulation theory has been employed. The basic postulate is that operation at any given cyclic stress amplitude will cause fatigue damage. The seriousness of the damage depends upon the number of cycles of the operation at that stress level and the total number of cycles before failure at that stress level. The damage has an unrecoverable and cumulative nature. Using the power relationship to approximate the material's S-N-P curve, the contact fatigue life can be expressed as,

$$\frac{1}{N} = C_0 \left\{ \int_v \tilde{S}_{\it eff}^{\bar{\beta}} \phi^z dv \right\}^{1/\beta}$$

where $\tilde{S}_{\it eff}$ is the effective stress within an infinitesimal volume of dv and can be evaluated from the time-history record of the equivalent stress S.

Results show that surface roughness can cause significant fluctuations in contact pressure. Pressure fluctuations in turn produce noticeable variations in interior stress field, particularly in the near-surface layer. As a result, the effective stress factor at the near-surface is increased. The maximum effective stress factor for surface produced by Process A is noticeably higher than these produced by Process B and Process C.

Relative life calculations, based on the effective stress, suggest a significant reduction for surfaces produced by Process A, and noticeable improvements for Process B and Process C as compared with Process A. The fatigue life for Process B is slightly longer than Process C but significantly longer than Process A under studied conditions.

REFERENCES

(1) X. Ai and H. S. Cheng, Tribology Transactions, Vol. 37, No. 2, pp 323-335, 1994.
(2) X. Ai and S. C. Lee, 1996, STLE, Tribology Transactions, Preprint No. 96-AM-4J-1.

EXPERIMENTAL STUDY OF ROLLING CONTACT FATIGUE : INITIATION OF A SURFACE BREAKING CRACK AND ANALYSIS OF ITS GROWTH

V.BORDI-BOUSSOUAR, B.VILLECHAISE
Laboratoire de Mécanique des Solides de Poitiers, SP2MI Bd 3 Téléport 2 BP179 86960 Futuroscope Cedex
D.NELIAS
Laboratoire de Mécanique des Solides, INSA Bât 113, 20, avenue A.Einstein, 69621 Villeurbanne Cedex
Ch.DORIER
EDF Direction des Etudes et Recherches, Service EP Dpt Machines, 6, quai Watier, 78401 Chatou

ABSTRACT

Experiments are conducted on a two-disc machine in the aim to understand the mechanisms of rolling contact fatigue crack growth. An original process is used to analyse the growth of surface cracks under repeated rolling contacts.

In the first stage rolling contact fatigue cracks are initiated at the bottom of notches created in two test discs by electric-erosion. The two discs run during the same number of cycles; therefore, assuming the mechanism is reproducible the cracks initiated are identical in the two discs. A destructive metallographic inspection of the first disc gives their features (length, angle of inclination).

On the other hand, the surface of the second disc is machined to eliminate the notches so that only surface breaking cracks remain in the disc.

The second stage consists in the propagation of these cracks in lubricated rolling contact fatigue conditions. The fractographs of the specimen gives information concerning crack path, growth rate and the role of lubricant in the propagation of such a crack.

INFLUENCE OF LOAD AND LUBRICANTS (MINERAL AND SYNTHETIC) ON EHD FILMTHICKNESS AND ROLLING CONTACT FATIGUE LIVES OF STEEL AISI 52100 BALLS.

Fernández Rico, E.; García Cuervo, D.; Tucho Navarro, R.; Montes Coto H.J.
Oviedo University. Construction and Fabrication Engineering Department. Mechanical Engineering Area
Campus of Viesques s/n, 33203 Gijón, Asturias. Spain. Phone: 34-8-5182061. Fax: 34-8-5182160

ABSTRACT.

This work has evaluated the influence of the load and of the type of lubricant in the thickness of EHD film and the rolling-contact fatigue lives of AISI 52100 steel balls. The studied lubricants include two mineral oils of several viscosities and five synthetic oils of three families in several viscosities. It was determined the "h" thickness in order to predict the lubrication regime. On the other hand the S-N fatigue curves were calculated by means of Weibull plots and the fatigue mechanism was evaluated.

The used test machine is a British-made Seta-Shell 1980 Four-Ball E.P. Lubricant Tester. The 12.7 mm diameter test balls used in this study were made from a single heat of carbon-vacuum-deoxidised AISI 52100 steel with hardness RC65.

The typical rolling-contact fatigue results of the tested mineral and synthetic oils under a max. Hertz stress of 8709 MPa are illustrated on Weibull coordinates in Fig.1.

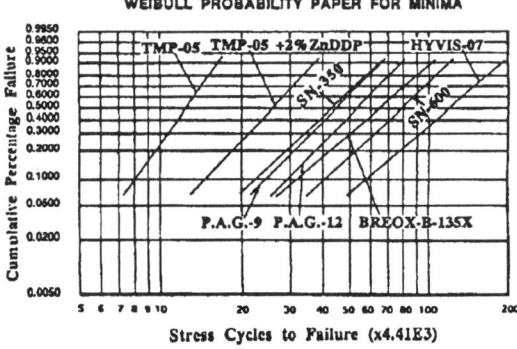

Fig.1 Weibull plots of tested rolling-contact fatigue.

A typical pitting is demonstrated in Fig. 2. The pitting happened a 16th minute of the rolling fatigue test with S.N.-350 mineral oil under maximum Hertz stress 8709 Mpa.

It is clear from the results above that for lubricants of different oil families with similar viscosity at atmosphere pressure, they can give very different rolling contact fatigue live to steel balls. One possible explanation is that they have different values of pressure-viscosity coefficients and this adds to different film thickness and results in different rolling fatigue lives. To verify this, it is necesary to measure the pressure-viscosity coefficients of the lubricants used and predict the elastohydrodynamic (EHD) film thicknesses. Estimated results of pressure-viscosity coefficients and film thickness are included in Table 1.

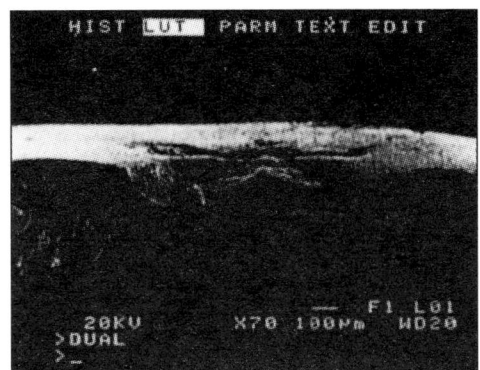

Fig. 2 A typical pitting appeareance.

Table 1. Estimated results of pressure-viscosity coefficients and film thickness.

Lubricant		Viscosity at atm. (Pa·s)	Press.-visc. coeff. α ($*10^8$ Pa^{-1})	Predict. film thickness ($*10^{-6}$ m)
HYVIS-07	23°C	0.3450	2.30	0.387
	80°C	0.0213	1.45	0.047
S.N. -600	23°C	0.295	1.60	0.296
	80°C	0.0204	1.10	0.040
P.A.G. - 12	23°C	0.1439	1.55	0.175
	80°C	0.0190	0.90	0.037
TMP-05	23°C	0.0859	1.10	0.104
	80°C	0.0139	0.70	0.024

CONCLUSIONS

The following conclusions can be drawn from the results presented above:
1. Synthetic polybutene HYVIS-07 gives the longest rolling-contact fatigue life. This may be attibuted to its higuest estimated pressure-viscosity coefficient.
2. The rolling fatigue performance of a synthetic oil could be better or worse than that of a mineral oil, depending on the nature of the synthetic oil family.
3. A higher viscosity at atmosferic pressure in synthetic oil will lead to a longer rolling fatigue life. But the influence of viscosity at atmosphere in synthetic oil on the rolling fatigue life of steels balls will be less than that of changing synthetic oil family.
4. Synthetic oil with E.P. antiwear additive has a longer rolling fatigue life compared with the case without additive.

EFFECT OF PARTICULATE CONTAMINATION IN GREASE-LUBRICATED HYBRID ROLLING BEARINGS

LARS KAHLMAN* and IAN M. HUTCHINGS

Department of Materials Science and Metallurgy, University of Cambridge, Pembroke Street, Cambridge CB2 3QZ, UK
*present address: SKF Nova AB, Chalmers Teknikpark, S-412 88 Göteborg, Sweden

ABSTRACT

The presence of hard contaminants in lubricated rolling bearings is a major cause of premature failure, for example by fatigue or by abrasive wear of rolling elements, races or cages (1). However, the introduction of silicon nitride balls or rollers running against steel races, in the hybrid bearing concept, is likely to show favourable changes in the interactions of hard particulate contaminants with the bearing components.

Generally, hybrid bearings show much better performance than all-steel bearings, including increased fatigue life in high speed tests. They also show advantages over conventional steel bearings due to lower overall thermal expansion, higher stiffness, greater tolerance to mechanical damage and lubricant starvation, greater scope for the use of grease as a lubricant, and good electrical insulation between inner and outer races which can be beneficial in electrical machinery. These factors can be exploited in the desgn of new products with superior performance or lower cost.

Wear in rolling bearing systems in the presence of contaminant particles can be generated by polishing and abrasive action between sliding surfaces. Polishing wear due to micrometre-sized abrasive particles is today regarded as a dominant factor in the lifetime of lubricated steel machine components (2).

In this investigation the performance of all-steel and hybrid rolling bearings was compared in the presence of two well-defined particulate contaminants: titania (TiO_2; anatase) and silica (quartz) particles. The experiments were carried out with grease-lubricated thrust bearings at 2060 rpm. The cages, for both the all-steel and hybrid bearings, were of steel.

The anatase contaminant was a relatively soft oxide with a small particle size (<1 μm). Its hardness (450-500 HV), melting point and particle size are similar to those of typical particles generated within all-steel rolling bearings, such as iron carbides, iron oxides or manganese sulphide.

The silica (SiO_2; α-quartz) contaminant was a harder material (900-1200 HV) with a larger particle size (75-103μm). It is the most prevalent natural abrasive, with a hardness similar to that of hardened bearing steel, but still significantly below that of the silicon nitride balls.

A significant beneficial effect was observed in the hybrid bearings in the presence of fine anatase particles, as lower cage wear and running temperature were recorded than in an uncontaminated, grease-lubricated bearing. This improvement was associated with the formation of thin and relatively soft titania films on the silicon nitride ball surfaces. In the all-steel bearing, however, the presence of fine titania particles increased the wear of both the cage and the balls. The titania particles apparently acted as much more effective abrasives in the all-steel bearing and thus caused greater wear, associated with less plastic deformation of the particles.

In the all-steel bearings, silica particles caused severe wear of the balls and cage, whereas in the hybrid bearing no wear could be detected on the balls, and cage wear was significantly lower than for the all-steel bearing, as seen in Fig. 1. The reduced wear of the steel cage was associated with the formation of composite surface layers on it.

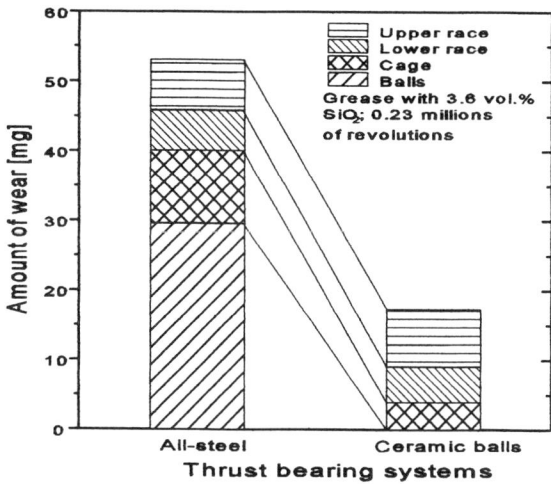

Fig. 1. Mass loss in bearings tested with quartz contamination.

It was concluded that hybrid bearings are considerably more tolerant of lubricant contamination than all-steel bearings, if the contaminant particles are softer than the silicon nitride balls. In some cases the addition of small soft micrometre-sized particles can even result in improved bearing performance, and may offer a promising new supplementary method of lubrication for hybrid bearings (3).

REFERENCES

(1) R.S. Dwyer-Joyce, R.S. Sayles and E. Ioannides, Wear, Vol. 175, pp 133-142, 1994.
(2) P.R. Ryason, I.Y. Chan and J.T. Gilmore, Wear, Vol. 137, pp 15-24, 1990.
(3) L. Kahlman, Swedish Pat. Appl. 9600881-8, 1996.

EFFECTS OF DEBRIS PARTICLES IN SLIDING/ROLLING EHD CONTACTS

G K NIKAS, R S SAYLES and E IOANNIDES*

Department of Mechanical Engineering, Tribology section, Imperial College of Science Technology and Medicine,
Exhibition Road, London, SW7 2BX, UK

*Also in SKF Engineering and Research Centre B.V., Postbus 2350, 3430 DT Nieuwegein, The Netherlands

ABSTRACT

The influence of debris particles in concentrated contacts has been studied in the past, mainly experimentally. Most theoretical studies were focused on purely rolling, isothermal contacts. The present work is a summary of the theoretical work done in the frame of a research project dealing with the thermo-elasto-plastic effects of relatively soft particles in EHD contacts. The theoretical formulation is based on the theory of thermoelasticity and covers several areas of study. The mathematical modeling and computer simulation involve the following.

- The motion and deformation of soft debris particles in the inlet and Hertz zone of line EHD contacts for mixed boundary conditions (friction and lubricant flow related).

- The 3-d thermoelastic distortion of the counterfaces surrounding debris particles and the construction of a wear map showing safe and unsafe regions of operation of the contact. More specifically, the 3-d thermoelastic stress-strain-displacement fields due to the presence of a debris particle in the contact are found at every position during the passage of the particle trough the EHD gap. The damage effects of the particle are clearly demonstrated through 3-d diagrams of flash temperatures, elastic and thermal stresses-strains-displacements around the contact.

- The motion and agglomeration of particles in the inlet zone of point EHD contacts. The analysis is based on 3-d solution of the steady state Navier-Stokes equations for viscous fluids. Debris particles are randomly put in the fluid flow in front of a ball rolling/sliding on a flat surface and their trajectories in the fluid are simulated. The study reveals the behavior of each particle in the flow in an attempt to correlate fluid starvation and scuffing with various operational parameters of the contaminated contact.

The present work covers sliding/rolling (and not only purely rolling) contacts, it incorporates a 3-d thermal stress analysis, and all 3-d calculations are based on as few as possible simplifying assumptions. Material thermal properties are allowed to vary with temperature and thermal anisotropy is incorporated into the theoretical model (assuming orthotropic materials). Realistic boundary conditions are used throughout the analysis.

Some of the most important results are the following:
- Ductile debris particles are responsible for high flash temperatures (due to frictional heating) in sliding EHD contacts (Fig. 1). Thermal stresses may often play an important role in the yielding and failure of such contacts. The results show that even small and soft debris particles can be quite destructive. A somehow unknown scuffing like failure mode is revealed.

- Particles tend to agglomerate in the inlet zone of EHD contacts causing fluid starvation and probably scuffing. The effects of the various operational parameters like debris size, slide/roll ratio, lubricant film thickness on particle entrainment are thoroughly demonstrated. The correlation with relevant experiments is quite supportive, which means that costly and time consuming experiments can be avoided and be replaced by an effective computer simulation.

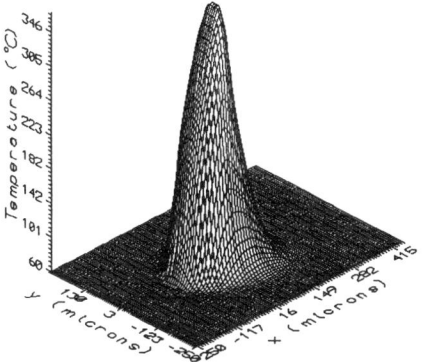

Fig. 1: Temperature distribution on a surface caused by frictional heating when a 30 μm particle is trapped in a sliding EHD contact.

Submitted to *Wear*

STUDY ON FRETTING OF ROLLING BEARINGS IN OSCILLATORY MOTION

K ICHIMARU, N IZUMI, N MIMURO and T MORITA
Department of Mechanical Science and Engineering, Kyushu University, Hakozaki, Higashi-ku, Fukuoka, 812-81, Japan
T SHIOTA
Factory Automation Division, Kyushu Matsushita Electric Co., Ltd., Tateishi-cho, Tosu, 841, Saga, Japan

ABSTRACT

In the raceways of rolling bearings subjected to external vibration while stationary quite heavy wear, which is known as false brinelling, may occur. This type of damage is identified as fretting and can occur similarly in rolling bearings in small angle amplitude of oscillatory motion. In this study endurance tests of a deep groove ball bearing 6805 in oscillatory motion (oscillation amplitude 1 to 14.5 degree) has been conducted to investigate the mechanism of false brinelling and the effect of operating and environmental conditions using a test rig in which eight bearings can be tested simultaneously. The progress of wear was monitored by the relative displacement of the inner ring to the outer ring and the fretting life was defined with total repetition number until the relative displacement reached to a threshold.

In each experimental condition the fretting lives of test bearings follow the Weibull distribution. For unlubricated condition, statistical fretting life L_{10} decreases linearly in a logarithmic scale with increasing in load, oscillation angle (Fig. 1) and frequency, and the following equation estimating fretting life was obtained.

$$L_{10} = K(Q/C_0)^{-a} \beta^{-b} N^{-c}$$

where Q, C_0, β and N are radial load, static load rating, oscillation angle and frequency respectively, and K, a, b and c are positive constant.

Grease extends the fretting life L_{10}, but the effect is not significant in low environmental temperature. And the effects of oscillation angle and frequency on the fretting life in grease lubrication differ from that in the unlubricated condition; L_{10} has the minimum at a certain oscillation angle as shown in Fig. 1 and the effect of frequency is larger than in unlubricated condition. These findings suggests that the behavior of grease, namely the oil-film formation between balls and raceways, has significant effect.

The state of oil-film formation was measured with the electric resistance method and a new parameter, an averaged cumulative insulating voltage E_{ac} defined by the following equation is introduced to correlate the relative displacement and the insulating voltage $E_{AB}(t)$.

$$E_{ac}(T) = \frac{1}{T}\int_0^T E_{AB}(t)dt$$

As shown in Fig. 2 the fretting life depends on the state of oil-film formation evaluated with E_{ac} and the difference in L_{10} between oscillation angle of 5 degree and 8 degree shown in Fig. 1 can be attributed to the state of oil-film formation.

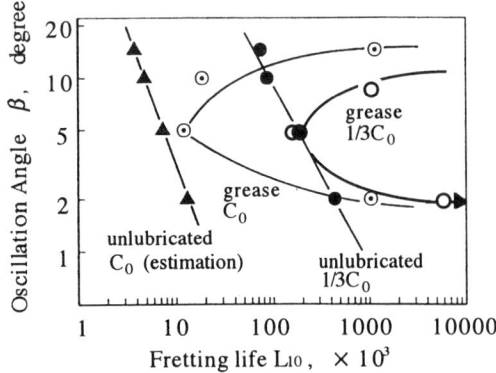

Fig. 1: Effect of oscillation angle on fretting life (400 cpm)

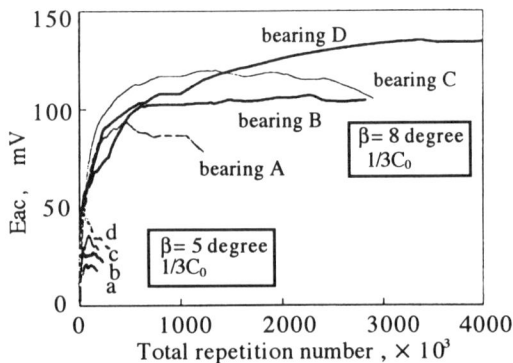

Fig. 2: Changes in averaged cumulative insulating voltage E_{ac} (400 cpm)

FATIGUE LIFE PREDICTION OF THIN HARD PVD COATINGS ON THE RACES OF HYBRID ANGULAR CONTACT BEARINGS WITH CERAMIC BALLS AND STEEL RINGS

K.-D. BOUZAKIS AND N. VIDAKIS

Laboratory for Machine Tools and Machine Dynamics Aristoteles University of Thessaloniki, GR-54006, Greece

ABSTRACT

In the present paper, the fatigue adequacy of thin hard PVD coatings, used in hybrid bearings, is investigated using a new coating fatigue lifetime prediction model (1). The project research aim was to increase the rotational speed and the load carrying capacity of angular contact ball bearings of the machine tool spindles, by means of ceramic rolling elements and PVD coated steel races (2). On the other hand, the minimisation of lubricant requirements was a major assignment.

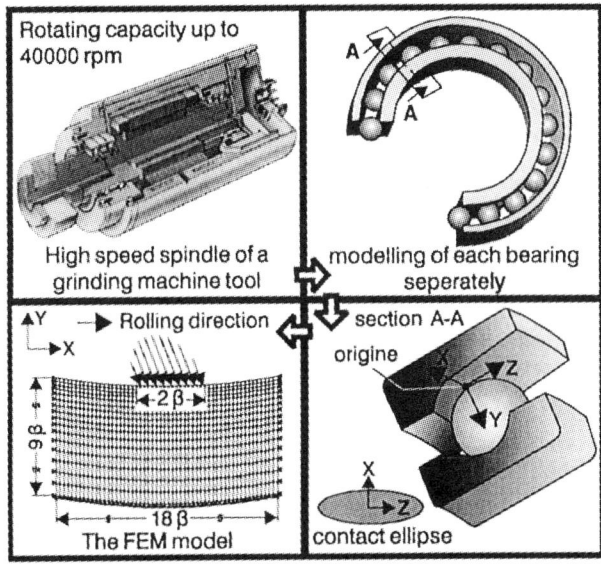

Fig. 1: Modelling of coated hybrid bearings

The fatigue behaviour of coatings, a critical parameter for the performance of the whole bearing, was investigated with the aid of a Finite Elements Method simulation of the most stressed contact regions. The simulation considers the entire geometric specification of the contact region, the contact load as well as the effect of friction (see figure 1). The calculations are conducted for various bearing types, preloads and rotational speeds, as well as for different coatings. The fatigue attitude of the hardest TiAlN and the softest CrN of the examined coatings is presented here.

The maximum von Mises stress values, which are obtained by means of the previously described procedure, changing each influencing factor separately, are compared with the safe fatigue bounds derived by the corresponding Smith diagram of each coating, for stressing in the elastic region (1).

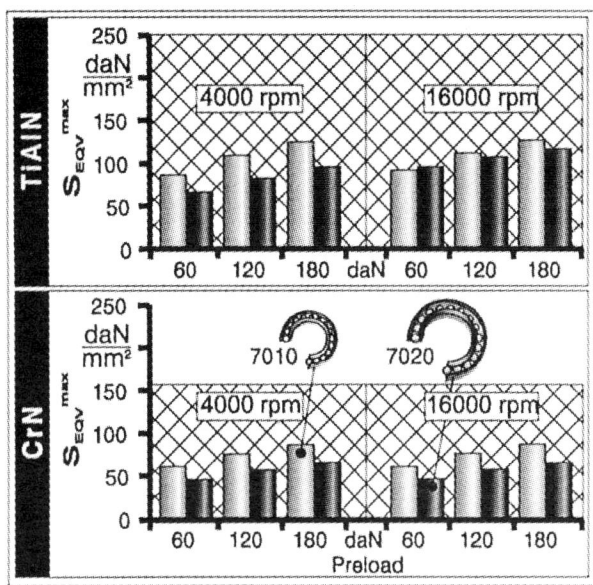

Fig. 2: The effect of preload and rotational speed on coatings fatigue risk.

Figure 2 illustrates fatigue prospects for both examined coatings, as a function of axial preloads and spindle rotational speeds, in two different cases i.e. that of the 7010 bearing with mounting diameter 50 mm and of the 7020 one with 100 mm. The hatched area in each diagram corresponds to stresses, associated with coatings continuous endurance. The enlargement of rotational speed increases the fatigue risk, due to the effect of the centrifugal forces. This effect is more evident in the case of the 7020 bearing, due to the larger diameter and herewith increased mass of the rolling elements. For the same level of axial preload, even if at lower speeds the bigger bearing develops lower stresses, due to its dimensions and number of rolling elements, at higher speeds this situation inverts. The preload growth leads to the enlargement of the stresses developed and this is more obvious in the case of the small bearing. The smaller radii of curvature of the contact bodies and the lower number of rolling elements that characterise the smaller bearing, are associated with higher contact stresses.

REFERENCES

(1) Bouzakis K.-D., et.al., Surface and Coatings Technology 86-87 (1996):549-556.
(2) Weck M, Hanrath G., Production Engineering Vol. IV/1 (1997): 79-82.

MECHANISM OF OCCURRENCE OF MICROCRACKS PRIOR TO ROLLING CONTACT FATIGUE

HISASHI MIYOSHI and TAKEJI TSUBUKU
Mechanical Engineering Laboratory, Namiki 1-2, Tsukuba, Ibaraki 305 JAPAN

ABSTRACT

Rolling contact fatigue arises in cylindrical steel surfaces over many stress cycles. The phenomena can be generally classified as either 'pitting' or 'severe flaking failure' (or *spalling failure*). Pitting failure is a surface deterioration in which fan-shape pits develop after the appearance of several hairline cracks on the surface of the rolling contact. Spalling failure is a destructive phenomenon of the surface, whereby relatively thicker flakes are ejected [1]. The latter occurs at higher Hertzian than those associated with pitting. The well-known reduced stress τ_R is a useful criterion in analyzing the onset of plastic deformation of material, and can be written as

$$\tau_R = \left\{ \frac{1}{6}\left[(\sigma_x - \sigma_y)^2 + (\sigma_y - \sigma_z)^2 + (\sigma_z - \sigma_x)^2\right] + \tau_{xy}^2 + \tau_{yz}^2 + \tau_{zx}^2 \right\}^{\frac{1}{2}}. \quad (1)$$

In (1), the x-, y-, and z-axes are respectively chosen in the direction opposite to the relative movement of the normal load on the peripheral surface of the roller, parallel to the roller axis, and in the radial direction of the roller. Decrease of the rollers' diameters during rolling contact gradually increases Hertzian pressure, expressed by

$$P_{H,0} < P_{H,1} < P_{H,2} < \ldots < P_{H,n-1} < P_{H,n} < P_{H,n+1} < \ldots, \quad (2)$$

where $P_{H,0}$ and $P_{H,n}$ represent the Hertzian pressures at initial static contact and after n stress cycles, respectively. This suggests that when once a region under the contacting surface is deformed plastically and experiences plastic compressive strain, the region will suffer a net plastic deformation in the subsequent stress cycles.

Provided P_H is maintained from $P_{H,C} = 3.11\sigma_Y/\sqrt{3}$ (σ_Y is yield stress) to $P_{H,S,C} = 4\sigma_Y/\sqrt{3}$, the above idea can explain the development of paired hairline cracks prior to pit formation (Fig.1). The angle α between the direction vertical to the direction of hairline cracks and the x-axis, namely

$$\alpha = \tan^{-1}\left[2\tau_{xy}/(\sigma_y - \sigma_x)\right]/2, \quad (3)$$

is essential on the side of the driven roller (follower), where $\sigma_x = \sigma_{x,0} + \hat{\sigma}_x + \sigma_{x,r}$, $\sigma_y = \sigma_{y,0} + \sigma_{y,r}$, $\sigma_z = \sigma_{z,0} + \sigma_{z,r}$, $\tau_{xy} = \hat{\tau}_{xy}$, $\tau_{yz} = \tau_{zx} = 0$. The stresses $\sigma_{x,0}$ and $\sigma_{z,0}$ are obtained theoretically [2]. Furthermore, $\sigma_{x,r}$, $\sigma_{y,r}$, and $\sigma_{z,r}$ are experimentally obtained as residual stresses, and $\sigma_{y,0}$ is given by $\upsilon(\sigma_x+\sigma_z)$, where υ is Poisson's ratio. The magnitude of τ_R along any part of the rim is primarily determined by $\sigma_{x,0}$, $\sigma_{y,0}$, and $\sigma_{z,0}$. In the contacting surface, however, it is mainly influenced by $\hat{\sigma}_x$ and $\hat{\tau}_{xy}$ emerging in the displacement of the contacting area. The components

$$\hat{\sigma}_x = \hat{A}(1+x), \quad \hat{\tau}_{xy} = -\hat{A}y, \quad (4)$$

are derived from a stress function $\phi = \hat{A}(1+x)y^2/2$.

Pitting occurs after the residual compressive stress $\sigma_{x,r}$ in the outer layer levels-off during operation. Therefore, according to (4), increases in the magnitude of $\hat{A}$ leads to increases in $\hat{\tau}_{xy}$, proportional to $\sigma_{x,r}$. This is thought to ultimately produce a pair of initial microcracks.

By contrast, when P_H is greater than $P_{H,S,C}$, the residual compressive stress $\sigma_{x,r}$ in the outer layer of the contacting surface gradually decreases from its peak value of the early stages of operation [3]. Meanwhile, spalling failure takes place. The decrease in $\sigma_{x,r}$ can be explained as follows. The tube's diameter increases as a result of severe plastic deformation during operation, and stretches the periphery of the roller. Note that diameter includes the layer in which the maximum reduced stress exists, namely the region between line a and b (Fig.2). A pair of microcracks arise at A, some distance from the z-axis. The shear tears the material along a line just below which the plastic flow of material takes place (y-axis).

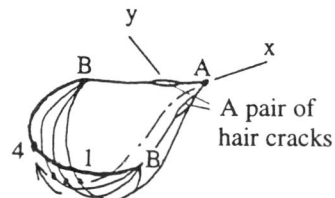

Fig.1 Typical formation of a pit.

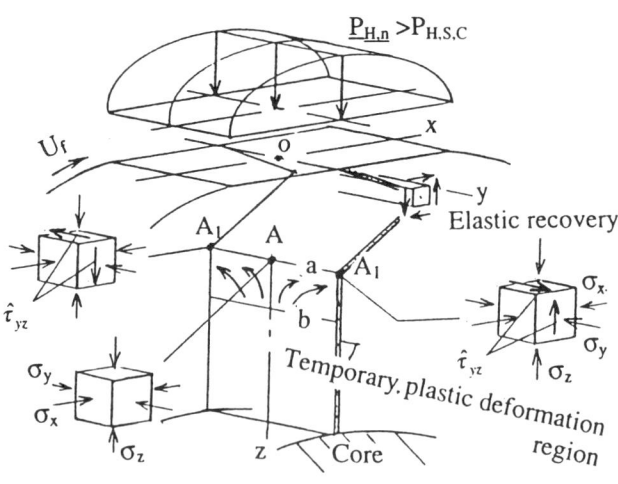

Fig.2 Formation of temporary, plastic deformation region.

REFERENCES

[1] A. P. Voskamp, G.E. Hollox, ASTM STP 987 (1988) 102.

[2] J.D. Smith, C. K. Liu, TRANS. ASME (1953) 157.

[3] K. Furumura, S. Jouta, A. Fujii, NSK Bearing Journal No. 643 (1982) 1.

Propagation of Rolling Contact Fatigue Cracks by the Fluid Entrapment Mechanism

K MATSUDA, A V OLVER and E IOANNIDES
Department of Mechanical Engineering, Imperial College, Exhibition Road, London SW7 2BX, UK

ABSTRACT

A theoretical study has been carried out of the effect of a rolling contact load on a cracked surface using a two dimensional finite element formulation. The model incorporates the following features:

- a frictionless crack inclined at an angle to the surface
- a moving Hertzian contact with sliding friction
- the presence of an inviscid, incompressible fluid
- calculation of the stress intensity factor using the method of Murakami (1)

The situation is similar to that modelled by Bower (2) except for his assumption that the crack internal pressure could not exceed that at the crack mouth. The present use of the finite element method makes this arguably unrealistic assumption unnecessary.

The model used in this analysis is illustrated in Fig.1. A semi-elliptical Hertzian contact pressure $p(x)$ with a surface traction $q(x)$ was assumed to move on a semi-infinite half space from left to right. The normal and tangential contact stresses were defined as follows:

$$p(x) = p_0 \sqrt{1-(x-e)^2/a^2}$$
$$q(x) = f \cdot p(x)$$

Where $f > 0$ means that the surface traction acts in the direction of motion of the load.

Figure 2 show the normalized pressure of fluid trapped inside the crack for the crack inclination angle $\beta = 25°$ and normalized crack length $c/a = 0.5$. The abscissa shows the normalized distance e/c from the crack mouth to the center of Hertzian contact pressure. The fluid pressure was determined by the condition that the volume of fluid trapped when the crack mouth closed remained constant and that uniform pressure acted at the crack surface. When the contact pressure acting at the closed crack face shifted from compression to tension, the crack face was judged to reopen. In the case of $f=0$ and -0.1, the crack is pulled open by the load just reaches the crack mouth, as reported by Bower (2) under the condition of $f=-0.05$. When the load moves a small distance further, the crack mouth closes. The volume of trapped fluid decreases with increasing coefficient of friction f, and becomes almost zero when f exceeds a critical value (about 0.1). As the load moves over the surface, the fluid is forced towards the tip of the crack. Simultaneously, the fluid pressure increases even when it exceeds the contact pressure at the mouth of the crack. The maximum fluid pressure occurs after the center of Hertzian contact pressure has passed over the crack mouth, and increases with increasing negative friction coefficient. The crack mouth reopens just after the load passes over the crack mouth, and the fluid escapes from the crack, which causes a sharp drop in the fluid pressure.

The results show that entrapment and subsequent pressurization of the fluid can occur leading to a significant effect on the stress intensity factor. Entrapment was found to be highly dependent upon the direction of the friction force with respect to the contact motion and upon the angle between the crack and the surface. The results provide possible explanations for both the observed crack angles in rolling contact fatigue and for the dependence of crack propagation on the direction of friction.

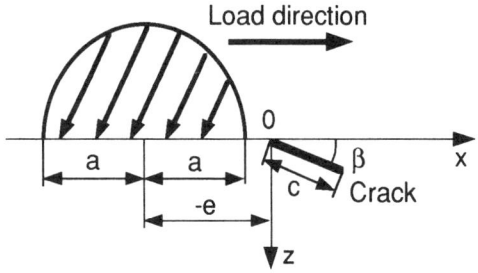

Fig.1: Analytical model.

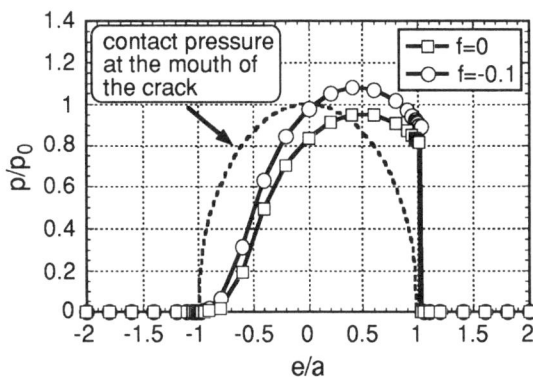

Fig.2: Variation of dimensionless fluid pressure ($\beta = 25°$, $c/a = 0.5$).

REFERENCES

(1) Y. Murakami, Engng Fracture Mech, Vol.8, 1976, pp.643-655.
(2) A.F.Bower, ASME Journal of Tribology, Vol.110, 1988, pp.704-711.

SUCCESSIVE OBSERVATION OF PITTING CRACKS IN ROLLING CONTACT FATIGUE

KAZUNORI ICHIMARU and TAKEHIRO MORITA
Faculty of Engineering, Kyushu University, Hakozaki, Fukuoka, 812-81, JAPAN

INTRODUCTION

In order to investigate the mechanism of pitting type of rolling contact fatigue, experimental studies have been conducted using two-roller machine. In this study, the surfaces of test rollers were recorded in video tape at every adequate loading cycles and evident pitting cracks observed after a long running were followed from an early stage of running by reversing the time. This technique makes the observation of initiation and propagation of cracks possible. Behaviors of surface cracks investigated using rollers having three different roughness orientations are reported.

EXPERIMENTAL CONDITIONS

The test specimens were rollers of diameter 60 mm and contact length 10 mm. Materials of test rollers were Cr-Mo steel SCM440(HV $\cong$ 440) and C steel S45C (HV $\cong$ 220). The lower hardness rollers (F roller) were circumferentially ground. On the other hand, the higher hardness rollers (D roller) were finished in three different roughness orientations by a tool grinding machine: circumferential, axial and oblique ones. The surface roughness of D and F rollers was 0.6 - 0.8 μmRa and about 0.1 μmRa, respectively. The rotational speed of D roller was 1190 rpm and the speed of F roller was reduced at the ratio 28/29 by gears. The applied load was 400 kN/m which corresponds to theoretical Hertzian maximum pressure 0.98 GPa. Gear oil without EP additives (viscosity: 40C 230.9 cSt, 100C 19.22 cSt, specific weight: 15/4C 0.895) was supplied and its inlet temperature was 50C.

RESULTS

Relation between axial width of pitting cracks and load cycles is shown in figure 1. Pitting cracks were most easily caused at oblique roughness compared with at longitudinal roughness or at traverse one. This coincided with increase in surface hardness by running, and therefore corresponded to the severity of asperity contact. The process of pitting formation is supposed to be following.

At an early stage of running, numerous fine shallow cracks may be observed, but most of them disappear by wear in further running.

The first appearance of cracks which should be developed to pitting (initial crack of pitting) is caused after occurrence of plastic flow and strain hardening induced by

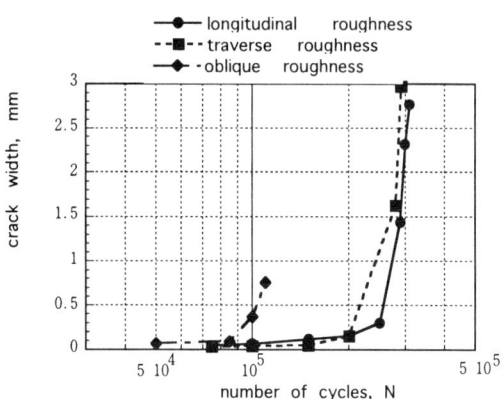

Fig. 1: Increase in width of crack on the surface

severe asperity contacts. In this stage, it is observed that initial surface cracks which became evident pitting were oriented to normal to rolling-sliding direction in spite of roughness orientations. It is supposed that the tensile stress on the surface which should be caused by friction and asperity contact may have significant role on the initiation of pitting.

In the following stage ($8 \times 10^4 \sim 2 \times 10^5$ cycles in longitudinal or traverse roughness), the lengths of surface cracks increased very slowly, however the cracks were supposed to be developing in internal direction. If the crack could not be propagated to sufficient depth, the rolling contact fatigue would become the form of micro-pitting or peeling. The driving force of crack propagation is supposed to be repetitive contacts of asperities which was changed by the contact geometry at every loading cycle. It is confirmed by fracture mechanics analysis (1) that different asperity contact conditions during several passages of crack through contact zone have a significant role of crack propagation and therefore, crack propagation velocity is slow.

In the final stage where crack propagates fast, penetration of oil into crack will play a part of crack propagation.

Similar results which were gotten in the experiments of very low speed (214 rpm) was omitted in this abstract.

REFERENCES

(1) K. Ichimaru, T. Morita, Y. Murakami, C. Sakae, Proceedings of International Tribology Conference, Yokohama 1995, vol.III, 1339-1344pp.

DESIGN AND OPERATING CHARACTERISTICS OF HIGH-SPEED, SMALL-BORE, ANGULAR-CONTACT BALL BEARINGS*

STANLEY I. PINEL
Pinel Engineering, Placentia, California U.S.A.
HANS R. SIGNER
Signer Technical Services, Fullerton, California U.S.A.
ERWIN V. ZARETSKY
NASA Lewis Research Center, Cleveland, Ohio U.S.A.

ABSTRACT

Small aircraft turbine engines (total airflow, 0.45 to 4.5 kg/sec (1 to 10 lb/sec)) require rolling-element bearings to operate at temperatures to 218 °C (425 °F) and speeds above 2.5 million DN. (DN is defined as the shaft speed in revolutions per minute multiplied by the bearing bore in millimeters.) To achieve speeds of 3 million DN with angular-contact ball and cylindrical roller bearings, Zaretsky et al. (1) and Signer et al. (2) proposed the concept of bearing thermal management in which underring-lubricated bearings are also provided with outer-ring cooling. The concept recognizes that total and flexible thermal control of all bearing components is essential to achieving a reliable high-speed, highly loaded bearing. Research reported by Zaretsky et al. (1), Signer et al. (2), and Bamberger et al. (3,4) showed that bearings could be operated reliably for long periods at speeds to 3 million DN with lubrication through annular passages extending radially through the bearing split inner ring and lands (underring lubrication) and with outer-ring cooling.

The computer program SHABERTH was used to analyze a 35-mm-bore, angular-contact ball bearing designed and manufactured for high-speed turbomachinery applications. Parametric tests of the bearing were also conducted in a high-speed, high-temperature bearing tester. Four bearing and separator designs were studied (Fig. 1). These were (1) a split-inner-ring, angular-contact ball bearing (design 1) with lubrication through its inner ring, a 24° unmounted (nominal) contact angle, and an inner-ring-land-guided separator (Fig. 1(a)); (2) the same as design 1 but with a 30° unmounted contact angle (design 2); (3) the same as design 1 but with a double outer-ring-land-guided separator (design 3, Fig. 1(b)); (4) a jet-lubricated, angular-contact ball bearing with a relieved inner ring, a 24° unmounted contact angle, and a single outer-ring-land-guided separator (design 4, Fig. 1(c)). Thrust load was 667 N (150 lb), combined radial and thrust load was 222 and 667 N (50 and 150 lb), shaft speed ranged from 28 000 to 72 000 rpm, (1 to 2.5×10^6 DN) and oil-inlet temperature was 394 K (250 °F). Oil was fed to the bearing at 0.76 to 1.89 liters/min (0.2 to 0.5 gal/min). Outer-ring cooling was provided in some tests. For jet lubrication the oil jet velocity was 20 m/sec (66 ft/sec). The lubricant was

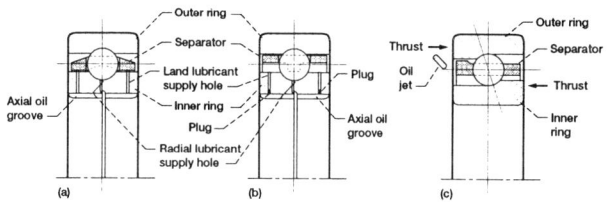

Fig. 1: 35-mm-bore, angular-contact ball bearing.

a neopentylpolyol (tetra) ester that met the MIL–L–23699 specification.

The predicted bearing life decreased with speed because of increased operating contact stresses due to changes in contact angle and centrifugal load. For thrust loads only, the difference in life between the nominal 24° and 30° contact angle bearings was insignificant. However, for combined loading the 24° contact angle bearing gave longer life. Optimal operating conditions were obtained for design 1 by using outer-ring cooling in conjunction with low lubricant flow rates. Lower temperatures and power losses were obtained for design 4. Inner-ring temperatures were independent of lubrication mode and cage design. SHABERTH gave a reasonably good engineering correlation between predicted and actual bearing power losses and predicted and actual inner- and outer-ring temperatures.

REFERENCES

(1) E.V. Zaretsky, E.N. Bamberger, H. Signer, Operating Characteristics of 120 mm Bore Ball Bearings at 3×10^6 DN, NASA TN D–7837, 1974.

(2) H. Signer, E.N. Bamberger, E.V. Zaretsky, Parametric Study of the Lubrication of Thrust Loaded 120 mm Bore Ball Bearings to 3 Million DN, J. Lubr. Technol., Trans. ASME, Series F, Vol. 96, No. 3, 1974, pp. 515–525.

(3) E.N. Bamberger, E.V. Zaretsky, H. Signer, Effect of Speed and Load on Ultra-High-Speed Ball Bearings, NASA TN D–7870, 1975.

(4) E.N. Bamberger, E.V., Zaretsky, and H. Signer, Endurance and Failure Characteristics of Main-Shaft Jet Engine Bearings at 3×10^6 DN, J. Lubr. Technol., Trans. ASME, Series F, Vol. 98, No. 4, 1976, pp. 580–585.

---

*Submitted for publication to Tribology Transactions, STLE, 840 Busse Highway, Park Ridge, IL 60668–2376, U.S.A.

THE EFFECT OF RETAINER ON THREE DIMENSIONAL MOTION OF A BALL IN THRUST BALL BEARING

AKIRA URA, TSUYOSHI KAWAZOE and AKIRA NAKASHIMA
Department of Mechanical Engineering, Nagasaki University, Bunkyomachi 1-14, Nagasaki-shi, 852, Japan

Abstract

The effect of a retainer on the ball motion in a thrust ball bearing is not negligible. Present paper reports the experimental results obtained from the measuring a ball motion in a remodeled retainer hole of thrust ball bearing using Hall effect sensors to be fitted with at three direction(radial direction X, circular direction Y and thrust direction Z). In case of ball bearings, balls have the freedom of three dimensional motion and observation of the motion becomes more difficult. The paper by Hirano(1) reported first previously that a magnetized ball could successfully be used for tracing the motion of the ball. In order to observe the motion of a ball in a thrust ball bearing without disturbance, change in flux due to the rolling motion of the magnetized ball has been observed through Hall effect sensor(2)(3). In measuring the ball motion which each ball collides repeatedly without a retainer, it is not easy to measure the three dimensional motion through Hall effect sensor due to difficulty to be fitted with especially in a thrust ball bearing. Accordingly a long hole along circumferential direction, in which three balls can be loaded, was proveded so as to simulate a condition without retainer. As a result the behaviour of the three dimensional ball motion in thrust ball bearing could be analized through the device using Hall effect sensor. A ball used for test is bearing steel(SUJ) ball with a diameter of 25/32 inch and the total numbers of balls are 8 including three balls loaded in long hole for measurement. Test loads are static loads of 50N, 100N, 150N and 200N. Tested revolution per minute are 550rpm to 2500rpm and the relation between a ratio of centrifugal force to thrust load (indicated by Z·Fc/Pa) and a ratio of ball rotation to a retainer revolution (indicated by Wb/Wc) and also Lissajous figures were obtained. Lubricating conditions are a grease lubrication and a dry condition without any lubricants. Figure 1 shows one example of the behaviour of each ball motion which is loaded in a remodeled retainer hole. It is shown clearly that each ball is interfered with and deviated from the rigid line indicating the theoretical values of the ball motions and also shown that the ball behaviours are different in each motion.

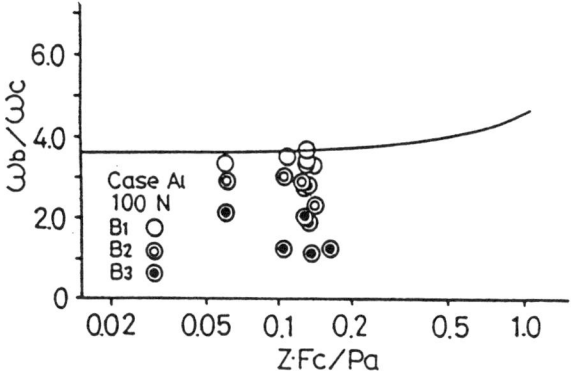

Fig.1 Variation of the Ratio of Ball Rotation to Retainer Rotation

In consequence a few points resulted in as follows;
[1]: In case of three balls loaded in a long hole, each ball rotation is easier to be slipped at the end ball than at the top and second balls due to colliding with retainer hole edge. [2]: In case of three balls a motion of each ball is generally disturbed much more than in case of a ball in a retainer hole. [3]: A ball motion is easy to be slipped much more under grease lubricated condition than under dry condition and the effect of retainer on an orbital revolution of a ball is remarkable especially under a light load such as 50N. [4]: In dry condition, a ball rotation become unstable over 0.5 of a ratio of a centrifugal force to thrust load (Z·Fc/Pa) independently of the weight of a load.

References

(1) F.Hirano, H.Tanoue, Wear vol.131, No.3, 1961
(2) K.Kawakita, Jour. JSLE, 32, 5, 1987
(3) A.Ura, A.Nakashima, H.Kisu, Trans. JSME 53-490, 1987.

FRICTIONAL TORQUE INSTABILITIES OF SOLID-LUBRICATED BALL BEARINGS

SHINGO OBARA[1] and KOJI MATSUMOTO
National Aerospace Laboratory, 7-44-1 Jindaijihigashimachi, Chofu, Tokyo 182 Japan

INTRODUCTION

Solid lubricated ball bearings exhibit relatively large variation of frictional torque compared with liquid lubricated bearings (1). There are cases in which the frictional torque suddenly rises. These rises are much larger than the torque variation in a steady state and appear periodically as shown in Fig. 1.

The purpose of this study was to clarify characteristics of the torque-rises through a survey of about 140 test-results obtained in our laboratory for more than ten years.

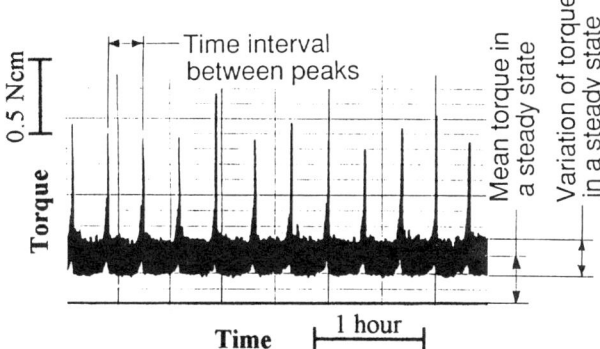

Fig. 1: **Frictional torque of a ball bearing lubricated with a PTFE based composite retainer.**

BEARING-TESTS

About 90 screening-tests in air and 50 tests in vacuum (2) had been conducted using 26 types of polymer-based composite retainers. Balls and races were pre-coated with a sprayed PTFE film or a sputtered MoS_2 film. Full-complement ball bearings were also tested. Most of the tests were carried out using #7204 type angular contact ball bearings under a thrust load of 50 N at a rotational speed of 2000 rpm.

The survey was made by reading values from charts. When torque increased at intervals of over 3 minuets and the peak-values were more than 1.5 times as high as the mean torque in a steady state, we judged that the torque-rises appeared.

RESULTS

For the bearings with a retainer, the torque-rises appeared in about 50% of the tests in air and 20% of those in vacuum. On the other hand, the full-complement ball bearings did not show periodic torque-rises. These results suggest that the motion of a retainer was responsible for the torque-rises.

On the whole, the torque-rises tended to appear for bearings showing low frictional torque with small variation in a steady state. For composites containing polyimide, periodic torque-rises were not recognized in the charts, but high torque with large fluctuation was seen both in air and in vacuum.

In vacuum, the torque-rises were often observed for PTFE composites containing Ag or MoS_2, but not for that containing Mo. It has been confirmed that wear-life of transfer films from the former composites is shorter than that for the latter composite in vacuum (3). Wear of transfer films seems to affect the motion of a retainer.

In air, most of the bearings having PTFE based composite retainers exhibited torque-rises. From the fact that the wear of balls and raceways in air was higher than that in vacuum, it is considered that the transfer film in air was easily worn away and this led to torque-instabilities.

It was found from numerical simulations for a PTFE composite retainer that when some balls of a bearing are assumed to rotate without transfer films, a difference in orbital velocities between balls with and without a transfer film occurs, and this causes unstable motion of the retainer and large torque-variation of the bearing. This phenomenon may be a cause for the torque instability. It was also confirmed by the calculation that the large torque-fluctuation of the polyimide based composite was caused by the severely unstable motion of the retainer due to a high friction coefficient of this material.

CONCLUSIONS

The torque-rises tended to appear for bearings showing low frictional torque with small variation in a steady state and short wear-life of a transfer film. To prevent torque-instability, materials which provide good adhesion and long life of the transfer film should be selected as a retainer.

REFERENCES

(1) S. G. Gould, E.W. Roberts, NASA CP-3032, 1989, 319-333pp.
(2) M. Nishimura, K. Seki, Y. Miyakawa, NAL TR-1019, 1989 (in Japanese).
(3) M. Minami, M. Suzuki, M. Nishimura, Trib. Trans., Vol. 36, 1993, 95-103pp.

---

[1]Present Address: National Space Development Agency of Japan, 2-1-1 Sengen, Tsukuba, Ibaraki 305, Japan

PRESENT LIMITS OF OPERATION OF PRODUCT LUBRICATED BEARINGS IN PUMPS

JONATHAN WATKINS

The Glacier Metal Company Limited, Argyle House, Joel Street, Northwood Hills, Middlesex, HA6 1LN, UK

ABSTRACT

Interest in product lubricated bearings is growing because of a number of benefits derived from these bearings: improved pump performance and reliability, enhanced pump maintainability, elimination of the cost of maintenance of separate systems for lubrication oil and shaft seals, reduction in pump system and weight, reduced power consumption, elimination of oil disposal costs, elimination of potential for product contamination, and elimination of leakage of a hazardous fluid to the environment.

This trend has allowed two types of tilting pad product lubricated bearings to grow in favour: ceramic and polymer coated bearings. Ceramic bearings, formed from both sintered and reaction-bonded silicon carbide have extremely high resistance to impurities, have operated between -200°C and 380°C, have high load capacities and can be lubricated by almost any liquid such as crude oil, acids or liquefied gas. Polymer coated bearings utilise a PEEK based bearing layer bonded onto metal backing via a bronze sinter layer. Made to oil lubricated tilting pad bearing designs, PEEK based bearings are able to withstand temperatures over 250°C and have found a significant market in high temperature bore hole pump applications.

For a given size of tilting pad bearing under given lubricant conditions, a limiting envelope can be produced for safe operating load *vs* sliding speed (1). This limiting envelope is defined by three different conditions shown in Fig. 1:

- Film thickness limit (low speed operation)
- Mechanical limit (medium speed operation)
- Temperature limit (high speed operation)

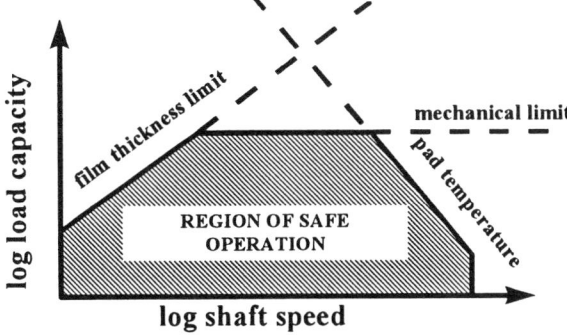

Fig. 1: Limiting Operating Envelope for Tilting Pad Bearings

The basic film thickness limit is not usually the limiting constraint for oil lubrication, but becomes a real restriction with low viscosity fluids as used in product lubrication. Applications of both bearing materials have run successfully in pumps with calculated film thickness of under 2µm.

The peak of the operating envelope formed by the film thickness limit and temperature limit is truncated by the mechanical limit. The design features that limit the pad design to a specific load are pad thickness and pivot area. Silicon carbide bearings, owing to their high compressive strength have been tested up to specific loads exceeding 32MPa; equivalent PEEK based bearings have been tested beyond 10MPa. Excessive overloading of bearings can result in the compressive failure of the silicon carbide or a wiping of the surface PEEK based material.

The pad temperature limit applies mainly to the PEEK based bearings. When PEEK is used with high viscosity lubricants, shear action can cause large temperature rises at high sliding speeds because of the poor surface thermal conductivity of the material. The attendant reduction in film thickness results in a fall in load capacity. For silicon carbide, on the other hand, its high thermal conductivity and temperature capability ensure that sliding speed is not usually a limitation.

Apart from the aforementioned operating envelope, there are other factors to consider: lubricant corrosiveness, cleanliness and dry running. Both silicon carbide and PEEK are inert to almost all process liquids; however, silicon carbide alone, owing to its high hardness, can tolerate fine abrasive particles (for example over 1000 ppm by weight of 25µm silica) without any effect on bearing operation. PEEK based bearings require filtration similar to those that are oil lubricated (10-20µm). Dry running must be avoided at all times for both materials, although diamond like coatings (DLC's) may extend the dry running capability of lightly-loaded silicon carbide bearings (2).

Research continues to extend the limits of operation of product lubricated bearings by the use of new monolithic ceramics, hard coatings and novel tilting pad bearing designs.

REFERENCES

(1) A.J. Leopard, ASLE 30th Annual Meeting, 1975.
(2) M. Findus, H. Knoch, Proceedings of 14th International Pump Users Symposium, 1997, 93-98 pp.

EXPERIENCES OF A TRIBOTESTER FOR THE EVALUATION OF JOURNAL BEARINGS

PETER ANDERSSON
VTT Manufacturing Technology, P.O.Box 1702, FIN-02044 VTT (Espoo), Finland

ABSTRACT

Sliding bearings and similar well-defined sliding surfaces form an unevitable part of numerous machines and mechanical devices. With certain exceptions, sliding surfaces are usually produced onto the respective components during their fabrication, instead of being aquired as mass-produced standard items added to the machine components. In order to end up with sliding surfaces having optimum properties, the producer of a machine component therefore often demands more and specified knowledge in the materials selection for sliding surfaces as well as their production processes and run-in properties. Simple tribological tests, such as the pin-on-disc test, provide information on the performance of material combinations, but for learning more about the behaviour of real components full-scale experiments with real bearings under relevant loads and speeds are much more valuable.

The present paper describes a tribological test equipment originally designed in 1989 for the experimental evaluation of water-lubricated ceramic journal bearings, particularly under initial conditions of boundary or mixed lubrication. The operating requirements taken into consideration were as follows:

- An equipment suitable for ceramic shaft sleeves and journal bearings, but also for other types of materials.
- Maximal test bearing dimensions for SEM analysis, with certain degree of freedom in the choice of specimen design.
- Testing at low, medium or high speeds and loads, at room tamperature.
- Convenient lubricant change and exchange, with unlubricated tests as an alternative.
- Rapid exchange of specimen and good testing economy.
- Easy to build, use, maintain and understand.

The design aspects and the properties of the equipment are described in the paper in greated detail. Apart from the basic construction of the test equipment, which has remained unchanged during the years, testing demands encountered have made it necessary to add various new features to the equipment. Friction torque and temperature measurements are constantly included in the testing practice, while some test programs have required supplementary transducers, *e.g.* for the measuring of vibrations, normal load, oil film thickness and hydrodynamic oil pressure. A cross-sectional view of the equipment is presented in Fig. 1.

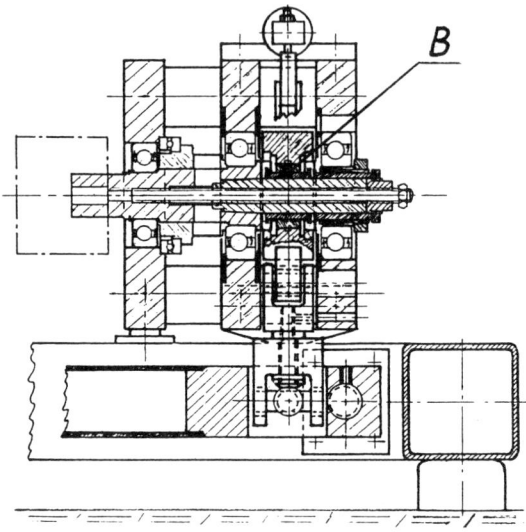

Fig. 1. Cross-sectional view of the VTT journal bearing tester. The test bearing is located at B.

As an example of the use of the tribotester and results obtained with it, a brief description of experimental investigations with water-lubricated ceramic journal bearings is included in the paper. The ceramics studied are different grades of alumina, zirconia, silicon nitride and silicon carbide based materials (1). In the work, particular attention is given to water-lubricated tests with silicon carbide shaft sleeves and journal bearings (2)(3). Except for the presentation of the studies on the water-lubricated ceramic bearings, reference is made to work on unlubricated journal bearing assemblies made from alumina and silicon carbide (4)(5), respectively, and to work on oil-lubricated tests with bearings made from metallic materials.

REFERENCES

(1) P. Andersson, P. Lintula, Tribology International, 27(1994)5, pp. 315-321.
(2) P. Andersson, J. Juhanko, A-P. Nikkilä, P. Lintula, Wear, 201(1996)1, pp. 1-9.
(3) P. Andersson, A-P. Nikkilä, P. Lintula, Wear 179(1994)1, pp. 57-62.
(4) P. Andersson, A. Blomberg, Wear 170(1993)2, pp. 191-198.
(5) P. Andersson, A. Blomberg, Wear, 174(1994)1, pp. 1-7.

BEARING FAILURES DUE TO THERMAL TRANSIENTS: DIAGNOSIS, ANALYSIS, AND SOLUTIONS

SANDY POLAK
Neale Consulting Engineers Ltd, 43 Downing St, Farnham, Surrey GU9 7PH, UK

INTRODUCTION

Failures of rolling element bearings caused by thermal transients used to be very uncommon, but have become more common in recent years. Such failures can be very difficult to diagnose, as usually all that remains of the bearing is blackened and molten. Analysis of the design can confirm the diagnosis, and suggest appropriate design or operational changes.

PRINCIPLE

This type of failure typically occurs in the first few minutes of the warm-up period when a machine is started from cold, and once triggered, the failure can progress to seizure in a few tens of seconds.

The reason for the failure is that the shaft and inner race of the bearing are warming up faster than the housing and outer race, and therefore thermal expansion of the inner relative to the outer causes a loss of the radial internal clearance. Once this is lost, heat generation may rise dramatically, leading to lubrication breakdown and rapid seizure.

The above is clearly a simplified description, as not all bearings which lose internal clearance will necessarily fail by this mechanism.

OPERATING CONDITIONS AND DESIGN FEATURES WHICH INCREASE THE PROBABILITY OF THERMAL RUNAWAY

There are a number of factors which increase the risk of thermal transient induced failures. Some of these are listed below, and a fuller discussion appears in the paper.

>High running speeds.
>Rapid acceleration.
>Low ambient temperatures.
>External heat inputs to shaft, or cooling of housing.
>Hollow shafts (wall thickness < 20% dia.).
>Heavy, thick section housings.
>Insulated bearing races.
>Small bearing internal clearance.
>Axial constraints (pair of locating bearings).

ANALYSIS METHOD FOR ASSESSING RISK OF THERMAL RUANAWAY

When analysing individual failures, it is appropriate to carry out a detailed study of the heat flows, temperatures, and thermal expansion effects occurring during the transient warm-up phase. Such analysis is inevitably limited by uncertainties, particularly regarding the rate of heat generation, and the amount of heat which is removed by lubricant flow rather than conduction. For a new design, these uncertainties will be greater, as there will be no measured temperatures. Therefore, for analysing a new design, it is appropriate to use a simplified method. Such a method is described in the paper.

The results of a typical calculation are shown in figure 1, for a design with a thin wall hollow shaft of 50mm diameter running at 6000 rpm. As can be seen in this example, the temperature differential peaks at about 37 deg C after about 400 seconds.

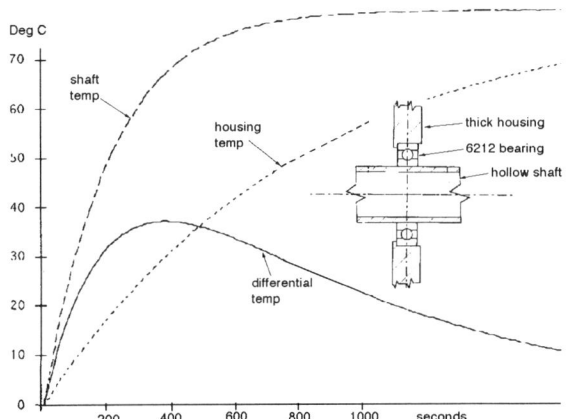

Figure 1. Typical transient temperatures

The relative dimensional change, ie the reduction in clearance, is simply calculated, and in this case the change is about 15 µm radial. This can exceed the as fitted internal clearance for a normal clearance bearing.

This is an extreme case, and the peak differential temperature for a similar design with a solid shaft is substantially less, approximately 15 deg C. Therefore it can clearly be seen that the design of shaft and housing has a substantial effect on the transient differential temperature, and therefore on the probability of thermal runaway.

In practice there are many other effects which may make things worse. For example, if there is another source of heat such as a gear on the same shaft, the temperature difference will be greater. Therefore the above analysis method can only give a rough indication of whether there is likely to be a problem. If the transient temperature differential is calculated to be more than about 25 to 30 deg C, then some precautions such as increased bearing clearances or design changes to the housing are appropriate.

SOLUTIONS

Changes to design or to operating conditions are discussed in the paper, which can be useful when trying to avoid or cure thermal runaway problems. Also the general design and operational trends which are contributing to the increasing number of occurrences of this problem are discussed.

THE DEVELOPMENT OF AN FZG TEST FOR OPEN-GEAR LUBRICANTS: THE SOUTH-AFRICAN EXPERIENCE

PHILIP L DE VAAL
Department of Chemical Engineering, University of Pretoria, Pretoria 0001, South-Africa

ABSTRACT

The standard FZG-test (1) was developed for the performance evaluation of gear oils. Present-day adaptations of this test method to evaluate the performance of open-gear lubricants include reduction of circumferential speed and extended operation at a specified load stage.

Open-gear lubricants are traditionally greases with a consistency resembling that of a high-viscosity oil. In addition to this, lubricants using a variety of technologies are in use, including butiminous products that need to be thinned down with a solvent to enable distribution to the point of application. More modern products are mineral-oil based with a variety of additives and which are applied intermittently and in small quantities, mostly in the absence of a solvent.

In the South-African mining and power industries, large quantities of these products are consumed during milling operations. Most open-gear mill-drives are lubricated on a once-through basis. Due to increasing concern about disposal of used lubricants, lubricant manufacturers are continuously improving products with the aim of reducing the amount of lubricant used.

In order to obtain a reliable laboratory test to determine the performance properties of such lubricants, the standard FZG test was adapted by decreasing the circumferential speed and changing the method of lubrication of the gear surfaces from dip-lubrication to dripping the lubricant directly into the meshing gears (2).

In this paper experience to date using this modified test method is discussed. In-situ performance of various products in use on mill gears are compared with test results obtained on the FZG-apparatus.

An alternative approach to defining a failure is also suggested. In the FZG test for gear oils, failure is determined either by means of a specified increase in mass loss per load stage compared with the average mass loss over the preceding load stages, or by visual evaluation of the damage on the gear tooth faces. Neither of these methods are applicable in the case of open-gear lubricants, since many formulations contain solid additives that are in some cases abrasive. A moderate extent of polishing of gear surfaces is acceptable and in the case of new gears, desirable.

As shown in Figure 1, lubricants with moderate abrasive tendency remove more material from the gear surfaces than is acceptable in the case of a gear oil. This results in a relatively large, gradual mass loss, although no sudden mass loss due to film breakthrough can be observed. When the gear teeth are inspected, the tooth surfaces are polished, the machining marks having been removed through abrasive action of solid additives present in the lubricant.

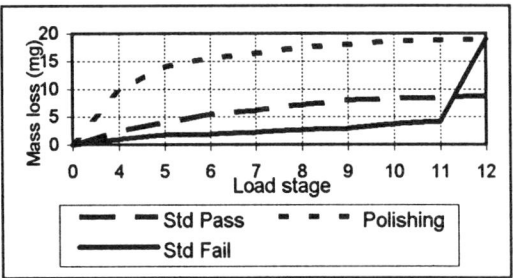

Fig.1: Typical FZG test results

It is suggested that the following criteria be used when defining the load and wear performance characteristics of an open-gear lubricant:

- number of load stages performed.
- total mass loss during test.
- mass loss per unit energy transferred
- mass loss after a 30 hour run at load stage 10.

Failure criteria have to be set taking the expected lifetime of the gears in operation into consideration - correlating the amount of wear during the FZG-test with the measured wear under typical operating conditions.

Accurate determination of wear on open gear teeth during operation is highly dependent on the measuring technique used. The traditional method of taking tooth imprints at regular intervals need to be improved in order to relate measurements to a required wear rate of approximately 20 μm/year.

REFERENCES
(1) DIN 51354, 1970, Testing of lubricants; Mechanical testing of gear oils in the FZG gear rig test machine, January 1970.
(2) P L de Vaal, H J Le Roux, Lubrication Engineering, Vol. 51, No.3 (1995), pp 198-202

OPTIMAL SURFACE ROUGHNESS WITH RESPECT TO NOISE GENERATION IN AUTOMOTIVE GEARBOXES

NASER AMINI and B.G. ROSÉN
Chalmers University of Technology, Göteborg, Sweden, naam@pe.chalmers.se

ABSTRACT

Unwanted noise induced in gearboxes has been one of the most discussed subjects during their evolution. Unpleasant components in the noise from gearboxes have many different sources. The macrogeometry as well as the microgeometry of the tooth play significant roles for the noise characteristics. This paper takes into account solely the effect of the tooth microgeometry on the noise.

During the investigations, the existence of an important micro-waviness, defined as "undulations", on surfaces has been established. Also, the effect of undulations on the noise characteristics has been experimentally studied.

INTRODUCTION

It is well known that the tooth shape contributes strongly to sound activation in gears. Therefore, all parameters concerning macroform have been studied by many researchers and their effects are predictable. In most cases tolerances are well defined. Finishing processes, such as grinding and honing, are used to achieve the predicted macroform. As a consequence of applying such operations, the microform of the tooth is also accomplished. New parameters representing the microform become important subjects for optimization.

UNDULATIONS AND NOISE

In earlier studies by the authors (1), finishing processes used to produce surfaces have been reviewed. They showed that most of these processes yield surfaces dominated by a harmonic component called "undulations". The micro-behavior of the process is the origin of these surface components. The most typical wavelength of such a systematic variation on surfaces is about 0.5 mm. A peak to valley of 4-5 μm does not surprise.

Figure 1 illustrates a gear surface, typically dominated by undulations. Studies concerned with the processes involved for production of this kind of surface showed that the gear grinding operation does not act as it is expected to behave, as common plane grinding operations do. In plane grinding, the tool works on the surface by its single abrasive particles, while redressing itself; however in gear surface grinding, the tool has to be redressed quite often, since the macrogeometry of the tooth has to lie within the demanded tolerances. Therefore, the tool is redressed using a master gear. The master gear in turn maps its mean diamond grain size on the grinding tool. Hence, the grinding tool's actual topography comprises high hills and deep valleys which makes the tool peel the work surface by a group of abrasives, not single particles. These strong formations on the tool surface then easily cause a harmonic waviness on the work surface.

The effect of all systematic variations (waviness - undulations) on system dynamics is a well-known phenomenon. The noise assessment of surfaces, like that shown in figure 1, showed that as long as undulations existed on gear surfaces, the pure tones in the noise measured from the gearbox had a significantly high level, and these pure tones usually

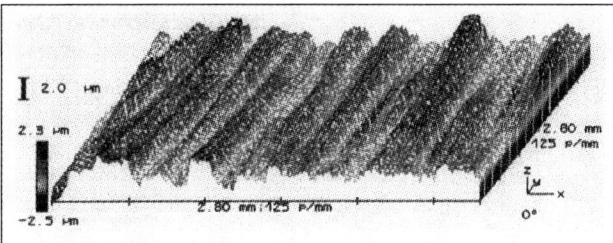

Figure 1: Ground gear surface. Note the harmonic variation on the surface. Measurement performed by a stylus equipment, sampling 0.8 μm.

are perceived as unpleasant to the human ear. This result is based on the noise assessment from a real automotive gearbox (2). The test was a comparison performed on a sample with undulations and the same sample after the undulations had been removed.

CONCLUSION

The amplitude, wavelength, and direction of undulations are critical properties on gear surfaces. An optimal gear surface is one that does not include a regular pattern of variation. A stochastic surface is favorable since it can generate a white-noise floor-level which might be able to overshadow the mesh frequency and its harmonics (the pure tones).

REFERENCES

(1) N. Amini et al. Transactions of the 7th Int. Conf. of Metrology and Properties of Engineering Surfaces, 1997, Part I, 6-16 pp.

(2) N. Amini, et al. Proceeding of the ASME 16th Binomial Conf. On Mechanical Vibration and Noise. 14-17 Sept. 1997, Sacramento, California.

EFFECTS OF THE LATEST RESULTS ABOUT TRIBOMUTATION FOR PROCESSING TRIBOLOGICAL PROBLEMS OF COMBUSTION ENGINES

ANDREAS GERVÉ

IAVF Institut für Angewandte Verschleißforschung GmbH, Im Schlehert 32, D-76189 Karlsruhe

ABSTRACT

Every year the industry make great efforts to optimize tribosystems in engines by very expensive tests. In spite of many investigations on friction and wear it is still not possible to calculate wear. Predictions can only be made from experience, and are poorly defined.

A lot of investigations on tribometers and real engines using the online radionuclide wear measuring technique show us more and more, that we are not yet able to understand the complicated physical and chemical reasons for the tribological properties. The model of tribomutation shows us a new way into a usefull theory of friction and wear:

* The energy input caused by friction which is a function of the stress and other parameters of the tribosystem takes place within short time appearing energy islands, which are statistically distributed over the surfaces. The density of the energy within these tiny energy islands is very high.

* As one result of these high energy densities, the material's composition and structure will be mutated within a very thin layer of less than 100 nm underneath the surface. As another result wear occurs.

* Together with the wearing surfaces, the mutating compositions and structures migrate into the depth of the material. The tribomutation influences significantly the wear behaviour in technical products.

The friction energy E_F of a tribosystem can be represented by three terms:

$$E_F = E_H + E_{TM} + E_W$$

E_H = energy causing the heat
E_{TM} = energy causing the tribomutations of the materials such as the lattice structure and the composition and of the lubricants
E_W = binding energy of the wear particles causing the wear

In a process of internal feedback the friction energy E_F mutates the materials M, hence these tribomutated materials change the tribological properties, the wear W and the friction F. The changed friction F again alters the tribomutation and so on.

The influence of tribomutation has to be seen in a chain of changes of the bulk materials M_B by friction during the manufacturing to M_O and by friction in the tribosystem in an engine or tribometer to M^*.

The mutation of the materials by friction, the tribomutation, is the fundamental idea to explain the decreasing wear rates during the running-in period of a tribosystem at constant running conditions as well as the nonlinear connections between wear and stress of a tribosystem.

Meanwhile we use the ideas of tribomutation for example to optimize the break-in procedure for truck engines, to explain the change from low wear to very high wear in spite of constant operating conditions, and to improve the transfer of results from tribometers into the industrial practice.

Up to now it was not possible to refute the model of tribomutation. Thus there is the urgent need to increase the knowledges about the triboinduced mutations of materials and their influence to the tribological properties in practise. For that reason surface analyses have to be carried out to understand the composition and structure of the materials and their mutation caused by friction and wear. The complementary sensitive wear analysis by means of the radionuclide technique will then enable the invention and the proof of new models and theories about and wear friction.

This is a difficult but interesting challenge to modern physicists, chemists and engineers.

REFERENCES

(1) A. Gervé, Radioisotopes in Mechanical Engineering, Fourth United Nations Conference on the Peaceful USES of Atomic Energy, AED-Conf. 71-100-55, May 1971

(2) A. Gervé, B. Kehrwald, L. Wiesner, T. W. Conlon, G. Dearnaly, Continuous Determination of the Wear-reducing, Effect of Ion Implantation on Gears by Doublelabelling Radionuclide Technique, Material Science and Engineering 69, 221-225 (1985)

(3) K. H. Müller, Oberflächenanalytische Verfahren zur Charakterisierung von Festkörperoberflächen in dünnen Schichten, Nachrichten aus Chemie, Technik und Laboratorium, Band 37/1989

(4) A. Gervé, Zur Strategie der Verschleißforschung, 8. Internationales Kolloquim "Tribologie 2000", Technische Akademie Esslingen, 14-16 Januar 1992

(5) A. Gervé, Tribomutation - Eine Herausforderung für die Tribologieforschung, GfT Tribologie-Fachtagung 1994, Göttingen, 22. und 23. November 1994

METALLURGICAL AND METROLOGICAL EXAMINATIONS OF THE CYLINDER LINER-PISTON RING SURFACES AFTER HEAVY DUTY DIESEL-ENGINE TESTING

MARK SHUSTER, FRED MAHLER
Technology Resource Park, Dana Corporation, Ottawa Lake, Michigan, USA
DAN CRYSLER
Perfect Circle Products Division, Dana Corporation, Richmond, Indiana, USA
DAVE DAMANCHUK
Sealed Power Division, Dana Corporation, Muskegon, Michigan, USA

ABSTRACT

The tribological system Cylinder Liner - Piston - Piston Rings plays a very important role in providing performance and endurance of the engine. To increase the reliability of this system we must identify the nature and mechanism of abnormal and/or severe wear processes and provide conditions to transform this into mild, non-severe types of wear.

This paper will discuss the results of metallurgical and metrological examinations of cylinder liner and piston ring surfaces from 6-cylinder 4-stroke 12-liter Class 8 on-highway diesel engines. Analyses were performed before and after a variety of standard bench tests ranging from 150 to 4,000 hours. Ring coatings evaluated included chrome plating, molybdenun-based plasma spray, and HVOF chrome carbide-nichrome. Hardened and non-hardened cylinder liners were included in the evaluation, as well.

Metrological analysis of the cylinder liners and piston rings included surface profile and the roughness parameters R_A, R_Z, R_K, R_{VK}, R_{VK*}, Skewness, and Oil Retention Volume.

The working surfaces of piston rings and cylinder liners from 400 and 1000 hour tests were free of any type of severe wear mechanism. However, detailed metrological and metallurgical analyses revealed important differences in the nature of the mild wear mechanisms of the different coating systems studied. For example, chrome plated rings had unfavorable surface roughness values related to oil retention, as compared to the plasma and HVOF sprayed coatings. New chrome plated ring surfaces exhibit a positive Skewness (S_K) value compared to the negative S_K values for the other coatings. Furthermore, as these surfaces accumulated test time, the differences in these characteristics became larger. Comparison of surface roughness parameters associated with lubrication retention predicts the chrome plated piston rings to have poorer lubrication than the other two coatings evaluated.

Metallurgical analysis of the tested piston rings revealed a wear particle attached to the chrome plated surface, which was identified as iron by EDX analysis. It was confirmed to have originated from the hardened cylinder liner. A similar wear particle was identified on the working surface of the molybdenum-based plasma sprayed ring, but none was found on the HVOF chrome carbide-nichrome coated ring. Wear mechanisms of the three different piston ring coatings, as well as those of the hardened and non-hardened cylinder liners relative to ring coating and test operating conditions, will be discussed.

Severe wear mechanisms were observed on a hardened cylinder liner run with chrome plated piston rings following a 4,000 hour test, and on an HVOF chromium carbide-nichrome sprayed ring after a 150 hour accelerated test. Detailed metallurgical and metrological comparisons point to a significant difference in the nature of these.

The catastrophic damage to chrome plated piston rings after nearly 4,000 hours of testing with a hardened gray iron cylinder liner has the character of fatigue due to local overheating, probably associated with the insufficient lubricating property of chrome plated surfaces.

The cracking of chrome carbide-nichrome sprayed coatings is associated with splat delamination, a common mechanism in thermal spray coatings. It appears that most cracks initiate at the surface, but may initiate below the surface, as well. They can be the result of localized thermo-mechanical processes and easily propagate along boundaries between hard chrome carbides.

This paper will show the correlation between metallurgical and metrological parameters and the efficiency of the cylinder liner-piston rings tribological system. Combining metallurgical and metrological analyses provides a new perspective and enables us to better understand the root cause of damage to the piston ring-cylinder liner system.

Submitted to *Tribology International*

THE FRICTION FORCE DURING STICK-SLIP WITH VELOCITY REVERSAL

FREDERIK VAN DE VELDE and PATRICK DE BAETS
Department of Mechanical and Thermal Engineering, University of Gent,
Sint-Pietersnieuwstraat 41, B9000 Gent, Belgium

ABSTRACT

Stick-slip is the phenomenon of intermittent motion caused by a velocity dependent friction force in combination with mechanical elasticity of the system of which the friction interface is part. Mostly stick-slip is explained refering to the simplified mechanical model shown in figure 1, consisting of a mass m which is connected by a linear spring with stiffness k to the frame and against which a plate with constant velocity v slides. During the stick phases the mass moves together with the bottom plate to the right due to the (static) friction force between them. The spring is stretched out during this motion resulting in a linearly increasing spring force to the left. When the spring force equals the static friction force, the mass m is accelerated to the left if the friction force decreases with increasing relative sliding velocity. The mass accelerates to the left until the spring force equals the instantaneous friction force; thereafter it continues to move to the left due to its kinetic energy, decelerating because the friction force (to the right) is larger than the spring force (to the left). After it has reached its extreme position, the mass is accelerated to the right again by the resultant of friction and spring force. When it reaches the velocity v of the bottom plate, two situations can occur. Usually, the mass sticks to the bottom plate because the instantaneous spring force is acting to the left in most stick-slip situations. For severe stick-slip however, the spring force can act to the right (the spring is compressed) at this moment of relative velocity equal to zero and can be sufficiently large to overcome the instantaneous (kinetic) friction force between mass and plate. In this case, the mass can continue to accelerate to the right and a reversal of the relative sliding velocity and friction force can take place.

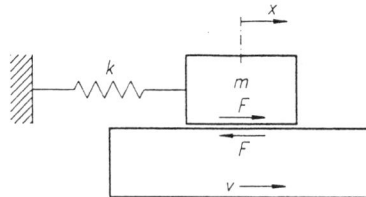

Fig. 1: Simple mechanical model explaining stick-slip

This phenomenon is observed during laboratory experiments for oil lubricated cast iron center-pivoted pads sliding against a cast iron ring in combination with a very flexible tangential spring. Figure 2 shows the typical shape of the experimentally obtained relations between friction force and relative sliding velocity during stick-slip with velocity reversal. Clockwise loops (with increasing time) of the friction force versus the relative sliding velocity are obtained before the velocity reversal, while counter-clockwise loops are found after the (first) velocity reversal (in some cases, different velocity reversals are observed). The influence of lubricant viscosity and additivation on the friction coefficient-relative sliding velocity relation is investigated and the evolution of the friction force at the fast changing sliding velocity, which is typical for stick-slip behaviour, is discussed qualitatively refering to the Stribeck curve (also shown in fig. 2), which corresponds to steady state conditions. The decrease of the friction before the (first) velocity reversal follows from the reduction of the real contact area due to the increase of the normal distance between the sliding specimens by hydrodynamic pressure generation. The increase of the friction coefficient after the velocity reversal is caused by the growth of metal-metal contact resulting from squeezing of the lubricant from the contact zone. It is remarkable that the friction coefficient increases after the velocity reversal even though the relative sliding velocity exceeds its minimum value necessary for hydrodynamic lubrication (see fig. 2). This is caused by the fact that the center-pivoted pads do not tilt quickly enough into the right sense to maintain hydrodynamic pressure generation after the velocity reversal. The delay of the tilting of the pads into the right sense after each velocity reversal is also indicated by the gradual increase of the friction measured during subsequent, manually excited (with initial spring deflection larger than during stick-slip), oscillations with average relative sliding velocity larger than the minimum for hydrodynamic lubrication.

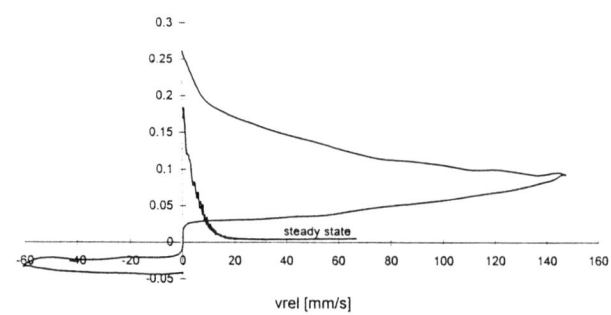

Fig. 2: Friction coefficient versus relative sliding velocity during stick-slip with velocity reversal

EVALUATION OF BOUNDARY FRICTION CONDITIONS IN PASTE PROCESSING USING A RING COMPRESSION TEST

M. J. ADAMS, B. J. BRISCOE and S. K. SINHA

Department of Chemical Engineering & Chemical Technology, Imperial College, London SW7 2BY, UK.

ABSTRACT

One of the important challenges in the field of paste processing is the determination of the boundary conditions or the wall friction between rigid equipment walls and the deforming paste. The wall friction is an important variable during processing as the interface controls the energy transmission from the wall to the paste and hence the efficiency of the whole processing operation. Ring compression tests, where specimens of specified geometry in a ring shape are compressed between flat parallel platens, has been used for characterising hot metal processing operations in order to obtain the friction coefficient between the platen and the ring specimen [1]. The analyses for the ring compression are available for purely plastic materials and assume that the material undergoes no strain or strain rate hardening during the bulk deformation. The final solutions of these analyses are presented as the changes in the geometry of the ring as a function of wall friction. These approaches do not incorporate any bulk rheological parameters for the evaluation of the boundary friction coefficient.

In the present work the Coulombic wall friction coefficient between a model paste (visco-plastic) and flat rigid plate has been estimated by performing ring compression tests. The calibration curve which is obtained from the plastic analysis and relates the changes in the geometry of the ring as a function of the friction coefficient, is used to estimate the wall friction. The interface characteristics are changed by varying the boundary conditions and the data are compared with similar results obtained in other tests such as wedge indentation and upsetting (compression of cylinders between platens). A model paste material commercially known as plasticine was used for the experiments. The paste was moulded and cut as rings of dimensions 60:30:20 mm (outer radius : inner radius : thickness). The ring specimens were compressed between flat parallel platens under three different wall boundary conditions; acetate paper, talcum powder and silicon grease. The test was carried out on an Instron machine and compressive force was recorded as a function of the change in the dimensions of the ring specimens.

Figure 1 shows % decrease in inner radius as a function of the decrease in the thickness of the ring specimen under three wall boundary conditions. The solid curves shown in the figure are obtained from the plasticity analysis using a lower upper bound approach. The data predicts the values of friction factor, m, for silicon grease, talcum powder and acetate paper as 0.8, 0.6 and 0.05 respectively. These values correlate very well with the boundary conditions obtained other tests such as parallel plate compression [2] and wedge indentation [3]. In order to convert the friction coefficient, m, into friction factor, m, the following relationship proposed by Kudo was used; $\mu = m / \sqrt{3}$. For the acetate paper case (m=0.8) the data tends to deviate from the theoretical predictions for higher reductions in the thickness of the specimen. This is beleived to be due to the inhomogeneous flow field generated in the bulk of the ring specimen. Figure 2 shows the flow visualisation picture in the bulk of the ring specimen. This figure clearly shows the different flow regimes viz. wedge formation in the top and bottom parts and divergent flow in the central part of the specimen. Summarizing, the results obtained in this study shows that ring compression test provides an easy and accurate means for characterising the wall boundary conditions for paste materials.

REFERENCES

1. Avitzur, B. in Metal Forming, Mc Graw Hill, 1965.
2. Adams, M J, Briscoe, B J and Sinha, S K; Proc. 20 Leeds-Lyon Symposium 1994, Elsevier Science, p. 223.
3. Adams, M J, Briscoe, B J and Sinha, S K; Phil Mag.A, Vol. 74, 5, (1996), p.1225.

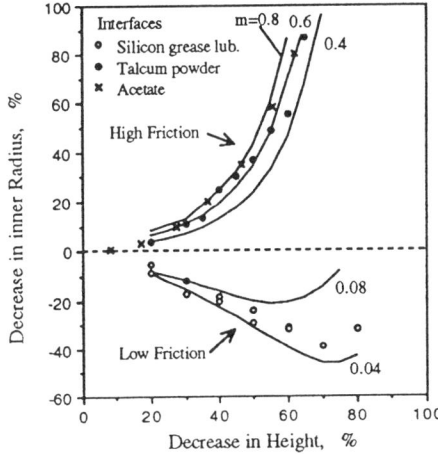

Figure 1 % decrease in inner radius as a function of % decrease in height of plasticine rings (6:3:2).

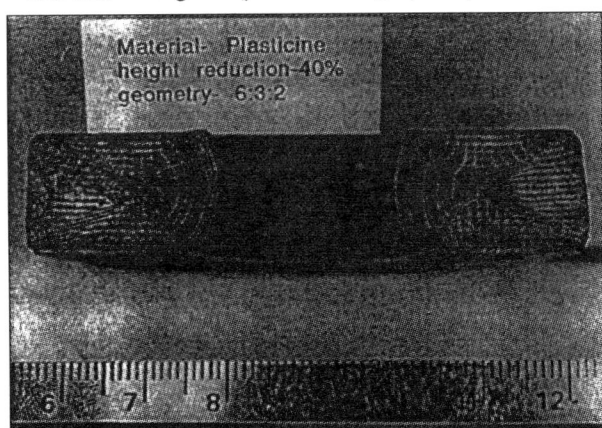

Figure 2 Grid distortion in the bulk of a plasticine ring specimen. The picture shows highly non-homogeneous flow field leading to the formation of wedges on the top and bottom parts of the specimen due to wall friction.

MEASUREMENT OF WEB FRICTION

K G BUDINSKI

Materials Engineering Laboratory, Eastman Kodak Company, Rochester, New York 14652-4347, USA

ABSTRACT

This paper concerns the measurement of the friction of plastic and similar flexible webs in contact (some angle of wrap) with cylindrical surfaces. There are numerous ways for measuring friction coefficients of materials in relative contact (1–4), but to be meaningful, the testing technique should simulate the tribosystem of interest. The subject of this paper is the capstan friction test. It uses the principle (and calculations) that apply to capstans that have been used for centuries to employ the friction force to produce mechanical advantage. The capstan test rig is illustrated in Fig. 1.

It is the purpose of this paper to propose the use of the ASTM G 143 standard which covers this test as the preferred technique to study web transport systems that involve a web in contact with rollers. This tribosystem is used to transport plastic sheets, rubbers, paper, fabrics, wrought metals and many industrial products. If the roller/product friction is too low, there may be transport tracking problems. If the friction is too high, it may lead to breaking of the web. Transport friction needs to be controlled. This paper presents the equations governing the test, describes the ASTM test procedure, and presents data on the effect of variables on the test.

The measured parameters in the capstan test are the tensions in a flexible web in contact with a cylinder and the angle of contact:

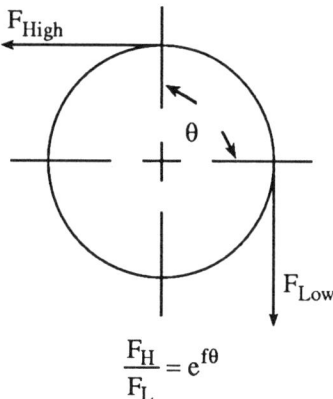

$$\frac{F_H}{F_L} = e^{f\theta}$$

where f is the friction coefficient.

Fig. 1: Capstan friction formula

The ASTM standard test can be done on a common tensile tester as shown in Fig. 2, or it can be done on a special test rig. The procedure does not limit the speeds and loads that can be used, only the specimen configuration—a flexible material in contact with a cylinder.

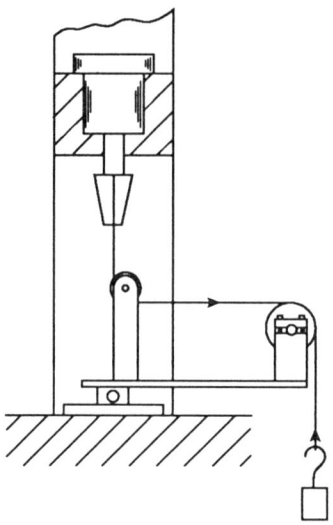

Fig. 2: Use of a tensile tester for capstan test

The repeatability of the test was determined to be acceptable in interlaboratory tests; the coefficient of variation was less than 10 percent. The effects of temperature, humidity, velocity, surface cleanliness and web tension depended on the particular tribosystem. The friction of plastic webs is affected by most of these factors when sliding against themselves or metal rollers. If the tensions in this test are high, the web surface may be damaged. If this happens, the measured friction coefficient becomes quite low (approaching 0.1); the test is no longer measuring friction of surface *a* sliding on surface *b*. It is measuring the marking of surface *a* by surface *b*.

As with all tribotests, care must be used in interpreting results, but it has been determined that this test correlates with manufacturing web/roller transport systems and its use is recommended for these kinds of applications.

REFERENCES

(1) D. Dawson, History of Tribology (London: Longman Gray, Ltd.) 1979.

(2) ASTM G 115. Standard Guide for Measuring and Reporting Friction Coefficients (W. Conshohocken PA: American Society for Testing and Materials) 1996.

(3) K. G. Budinski, "Wear Testing" in Tribology in the USA and the Former Soviet Union: Studies and Applications, V. N. Belyi et al, N. Miskin, Ed. (New York: Allerton Press) 1994.

(4) P. J. Blau, Friction Science and Technology (New York: Marcel-Dekker Inc.) 1995.

THE FRICTIONAL PROPERTIES OF PAPER: COMPARISON OF THREE TESTING METHODS

G. de SILVEIRA and I. M. HUTCHINGS

University of Cambridge, Department of Materials Science & Metallurgy, Pembroke Street, Cambridge, CB2 3QZ, UK

ABSTRACT

Paper production, printing and processing into artefacts requires knowledge of the many parameters influencing paper friction. However, the interaction of paper surfaces during sliding friction is not well understood due to the numerous chemical and physical factors involved.

Paper-paper friction is strongly influenced by the presence and composition of the wood extractives, papermaking additives and ambient conditions. The physical circumstances in which friction is important are also diverse: as in the friction of flat paper surfaces in stacks or winding around rolls and capstans during which the paper strip is under tension.

Friction testing methods recommended in various standards use either an inclined or horizontal plane. Both methods involve a plane paper sample attached to a weighted sled moving over a fixed sheet. However, each standard method specifies different critical details for the determination of results which are both operator- and machine-dependant (1). In the horizontal plane method, the tangential force needed to drag the sled along the plane at a fixed speed is measured with a calibrated load cell and both the static and kinetic coefficient of friction (μ) are determined. In the inclined plane test only static friction can be calculated from the angle (θ) at which the sled starts to move by the expression:

$$\mu_s = \tan \theta$$

A new strip-on-drum apparatus has been devised to provide a better simulation of the contact conditions in papermaking and converting operations, in which paper webs are held under tension during sliding contact around mandrels or rolls,. In this method the tension is measured by a load cell at the fixed end of a paper strip in weighted contact with a rotating drum covered by the second sample. Detailed consideration of the mechanics of the test shows that the tension in the strip and the mean pressure between the strips, during slow rotation, vary around the contact arc (2). Continuous measurement of the tensions T_1 and T_2, and the arc of wrap (in our case 90°) allows estimation of both static and kinetic friction from the expression:

$$\mu = (2/\pi)\ln(T_2/T_1)$$

Paper products included in this investigation were non-recycled and recycled Kraft board, newsprint, mineral-filled writing papers and cotton linter samples. Before friction testing by all three methods, samples were characterised in terms of pulp composition, degree of interfibre bonding and presence of contaminant as these are believed to have considerable effects on frictional behaviour.

The results indicate that not only are various parameters involved in determining paper friction but that their significance is different depending on the testing method employed. The coefficient of friction (COF) varies over a wide range (Fig.1) confirming that it is not an intrinsic property of the material and that the presence of wood extractives, as in the newsprint sample, can have a major influence on friction.

The three test methods used gave reproducible values for the coefficient of friction, but there were consistent differences between the methods. Several factors can be identified which underlie these differences; the most important are load dependence of friction which arises from the elastic nature of the contact deformation and modification of adhesive forces due to the presence of surface wood extractives. The strip-on-drum test differs significantly from the other two in that one of the paper strips is under substantial tension during the test, which may modify both the sheet structure and its response to compressive loading. Repeated sliding over the same area produces significant changes in static friction associated with morphological changes in the paper, as detected by SEM and AFM.

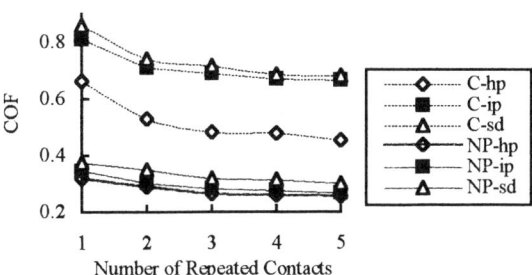

Fig.1: Coefficient of friction for cotton linter (C) and newsprint (NP) samples determined by the three test methods: inclined-plane (ip), horizontal-plane (hp) and strip-on-drum (sd).

REFERENCES

(1) A.Johansson, C.Fellers, D.Gunderson & U.Haugen, Int. Paper Physics Conf., TAPPI Press, p.5-15, 1995.
(2) J. Sato, G. de Silveira & I.M. Hutchings, Tribology International (in press), 1997.

UNDERSTANDING AND QUANTIFICATION OF ELASTIC AND PLASTIC DEFORMATION DURING A SCRATCH TEST

V. JARDRET*,**, H. ZAHOUANI*, J.L. LOUBET*, T.G. MATHIA*.

*Laboratoire de Tribologie et Dynamique des Systèmes, Ecole Centrale de Lyon, 36 Avenue Guy de Collongue
69 131 ECULLY (FRANCE)
** Oak Ridge National Laboratory, Metals and Ceramics Division, MS 6116, P.O. Box 2008
Oak Ridge, TN 37831, USA

ABSTRACT

An understanding of abrasion resistance and the associated surface deformation mechanisms is of primary importance in the materials engineering and design of many important industrial components undergoing wear and abrasion. Instrumented scratch testing has been shown to be a useful tool for characterization of the abrasion resistance of materials. Although most studies on scratch resistance have been limited to the theoretical case of purely plastic materials, experiments on metals and polymers have shown that the contact mechanics and indentation behavior are strongly influenced by the elastic behavior.

In this work, the measurement of the normal and tangential scratch forces, the penetration depth relative to the initial surface and the three-dimensional topographic relief of the scratched surface are done.
They have allowed us to acurately calculate the actual contact area between the indenter and the material, taking into account both elastic deformation and pile up phenomena. This contact surface was used to estimate the real mean contact pressure during scratch testing. This pressure was compared to the static hardness of the studied materials, as well as to the classical definitions of the scratch hardness.

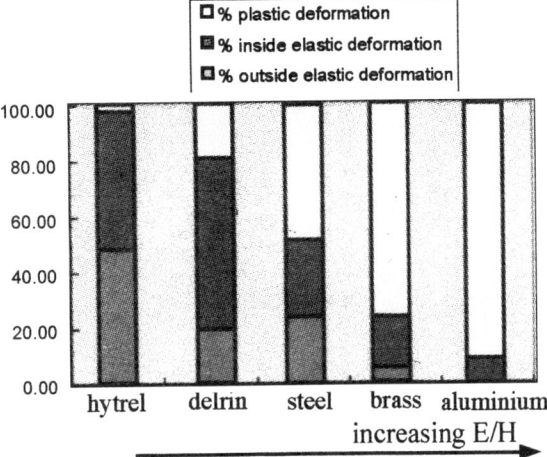

Fig.1: Evolution of the elasto-plastic behavior of different materials with the ratio of the elastic modulus over the hardness

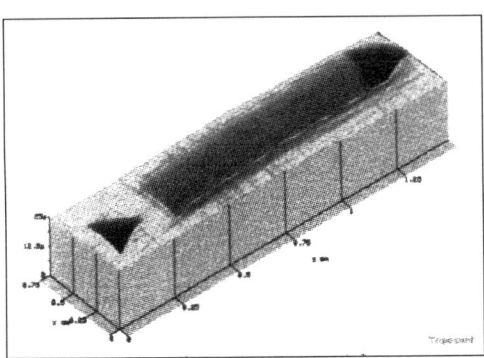

Fig.2: Three dimensional topographic measurement of a scratch on acetate polymer

The ratio between the plastic and elastic deformation during a scratch test with a Berkovich indenter was then related to the ratio of the elastic modulus over the hardness for the tested bulk materials.

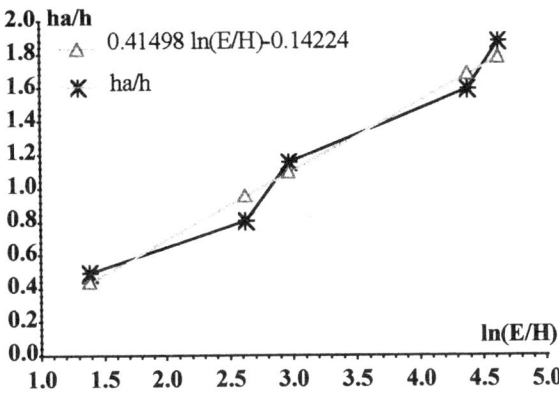

Fig.3: Evolution of the ratio of the contact depth over the penetration depth versus the ratio of the elastic modulus over the hardness

This study was performed on a wide range of materials from polymers to metals and demonstrates the importance of taking both elastic and pile-up into account in order to accurately understand and characterize the scratch resistance of materials.

This research was partially sponsored by the Division of Materials Sciences, U.S. Department of Energy, under contract DE-AC05-96OR22464 with Lockheed Martin Energy Research Corp.

A ROBUST METHOD FOR THE ON-LINE MEASUREMENT OF WEAR AND FRICTION DURING SLIDING TESTS

M G Gee
Centre for Materials Measurement and Technology
National Physical Laboratory, Queens Road, Teddington, Middlesex TW11 0LW, UK

ABSTRACT

The measurement of the wear and friction that are generated by the sliding contact of samples are normally carried out using pin-on-disc or reciprocating geometries. Measurement of wear is normally carried out by several techniques after the test has completed including the measurement of loss of volume through the mass change of test samples, through the measurement in the change in sample dimensions, and through the measurement of the size and shape of wear scars by profilometry. Monitoring the relative movement of the two samples towards one another also gives useful information on the total wear that is taking place as the test proceeds.

The measurement of friction is carried out more simply with a load cell which is used to restrain the movement of the sample holder giving a measure of the frictional force that is generated at the test contact.

This paper describes two developments which improve the acquisition and analysis of these on-line measurements of wear and friction.

The first of these is the use of non-contact optical probe technology to give an accurate measure of the wear to both samples in a wear couple simultaneously. Two different types of probe are available. These are triangulation focus probes, and fibre-optic probes. In pin-on-disc testing, up to three of these probes can be used to measure the relative displacement of the pin sample, the depth of the wear track on the disc, and the displacement of a reference surface which is normally the surface of the test disc away from the wear track. The particular advantages of these probes are that they are non-contact, and that they can measure displacement at rates up to 200 kHz. Initial results from tests carried out on TiN coated samples are given.

The second development is the adoption of techniques which allow for the near real time measurement of the distribution of friction and wear signals as the test proceeds. Modern data acquisition systems allow the capture of large quantities of data which can then be processed to obtain information that was not previously available. In the scheme presented here, large (about 10,000 samples) sets of data are acquired every few seconds. The data sets can be analysed in several ways.

Thus the statistical distribution of the wear and friction signals is calculated for each sample, and used to provide information such as the mean and standard deviation which can be stored and used to provide the trend of wear and friction mean and range with time as the test proceeds (Fig. 1). A similar method can also be used in reciprocating tests, but here more care is needed in interpretation due to the alternating nature of the motion. Each graph shows three trend lines. The middle line is the mean value, and the outer values show the spread calculated on a single standard deviation basis. The graphs also show a large jump in wear displacement after about 4000 s, presumably due to sudden removal of materials from the wear surface.

Fig.1: Wear displacement and friction results

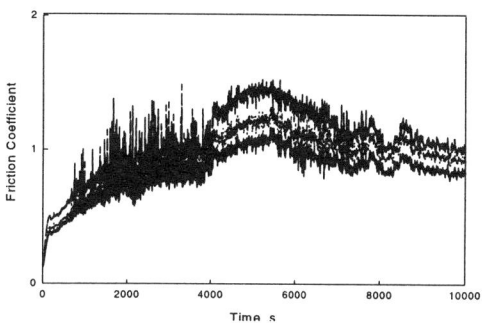

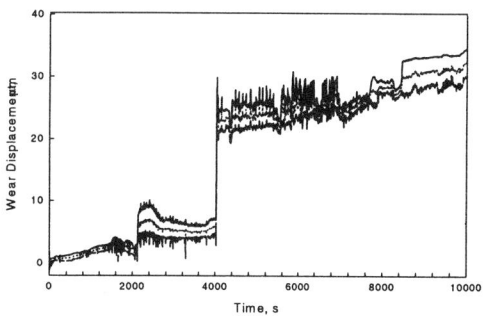

for TiN coated disc worn against alumina ball in reciprocating test system at 5 N applied load, 2 Hz and 5 mm stroke.

In addition, the data samples can be subjected to transient analysis such as frequency analysis to provide information on the dynamics of the test.

These measurement and analysis methods are applied to the real-time evaluation of the wear and friction of an TiN coated sample.

THE DEVELOPMENT OF TESTING OF POLYMER-MATRIX COMPOSITES FOR POLYMER-MATRIX COMPOSITES RUNNING UNDER HIGH-SPEED RECIPROCATING CONDITIONS.

R.W. BAYLISS
Morgan Materials Technology Limited, Stourport-on-Severn, Worcestershire, DY13 8QR, UK
C.A. STIRLING
Morganite Special Carbons Limited, Gosport, Hampshire, PO12 4LJ, UK
G.A. PLINT and A. ALLISTON-GREINER
Plint & Partners Limited, Wokingham, Berkshire, RG41 2FD, UK

ABSTRACT

The tribology of the polymer-matrix composites used in industrial gas compressor applications is not widely understood. This is due in part to the difficulty in emulating the conditions experienced by the materials operating as pistons rings. To improve the understanding and to aid in materials development, a high-speed reciprocating test rig with an environmental enclosure was developed.

The rig is based on a standard PLINT TE77 reciprocating machine fitted with an on-line wear monitor. This test rig may be operated under full computer control and allows load, speed and test temperature to be controlled independently. Data, in the form of apparent wear (as a "wear gap" from a capacitance transducer), sample and counterface temperatures, and friction (from a piezoelectric transducer), to give coefficient of friction values, are recorded continuously. However, to be capable of emulating the operating conditions of compressor applications the standard machine required extensive modification.

High speeds were achieved with a relatively long stroke at high frequency. Over a 50 mm stroke the rig was required to run at up 40 Hz, giving a mean speed of 4 m.s^{-1}. Modifications to the drive linkage and balancing design were necessary to minimise machine vibration with such operating conditions.

Particularly problematic was ensuring meaningful measurements at high speeds and low loads (applied pressures of c. 5 p.s.i. / 3.5 x 10^4 Pa) in a system where a moving specimen of elastic material is constrained only by the applied load. At an early stage in development it was found that at high speeds / low loads the moments exerted on the sample during acceleration/deceleration caused loss of contact of the test surface with the counterface. This was manifest in the departure of the friction trace from a near-ideal square wave. The solution was modification of the test sample size and shape, and the sample carrier design.

The design and materials selection for the reciprocating loading arrangement was critical. The moving mass was minimised to obviate resonances in the system. However, in the process, stiffness was compromised, and this had consequences for maintaining linear motion of the sample. Lateral constraint of the loading arrangement was necessary. Furthermore, weld fatigue failures required redesign of the loading forks. With the various modifications the rig would run hemispherical samples (currently 18 mm diameter) at 4 m.s^{-1} at applied pressures of 5 to 25 p.s.i. (4 to 18 x 10^4 Pa).

Initial tests were carried out at ambient temperature with PTFE to demonstrate the effectiveness of the friction and wear measurements. When similar conditions were applied to filled PTFEs significant frictional heating was found. Differences in friction coefficient between materials resulted, at higher loads and speeds, in a wide variation in measured counterface temperature. Furthermore, during frictional transients, the superimposition of thermal expansion on the wear trace made measurement of the underlying wear more difficult. It was therefore necessary to carry out tests at elevated temperature (>130°C).

As frictional heating could be a significant factor in testing, it was critical that sample temperature was known. Even at moderate operating frequencies failures occurred in the RTD sensor and connections. A robust system for leading out the signal was designed which incorporated a selected cable and guide used in the robotics industry.

Further testing showed that it was essential to ensure stability was reached in terms of isothermal conditions (for the test sample), and that the wear rate and coefficient of friction values had stabilised. This meant that some long run-in times were necessary, in some cases these extended to 72 hours. However, once stabilised, subsequent alterations in terms of load and speed could give meaningful data after much shorter step times. These shorter step times also varied from one sample type to another. To determine how well the actual rig performs a series of matrix tests was designed to measure rig variability. The results show that once stability has been reached then typically wear rate and friction level values can be obtained within a tolerance band of about 5%, which is believed to be exceptional for this type of testing.

SCUFFING TESTING AT HIGH SPEEDS

D M NICOLSON and R S SAYLES

Tribology Section, Department of Mechanical Engineering, Imperial College of Science Technology and Medicine, South Kensington, London SW7 2BX

ABSTRACT

The paper presents a comprehensive review of scuffing and scuffing testing and addresses some of the restrictions in measuring transmission lubricant load carrying capacity, notably repeatability and correlation of test results with field experience. Existing test methods are reviewed, and reasons for poor correlation of results with field experience are explored.

On the basis of this study, the design and development of a mini-disc machine test rig using relatively small discs supported in hybrid hydrostatic/hydrodynamic half-bearings was undertaken (Figure 1).

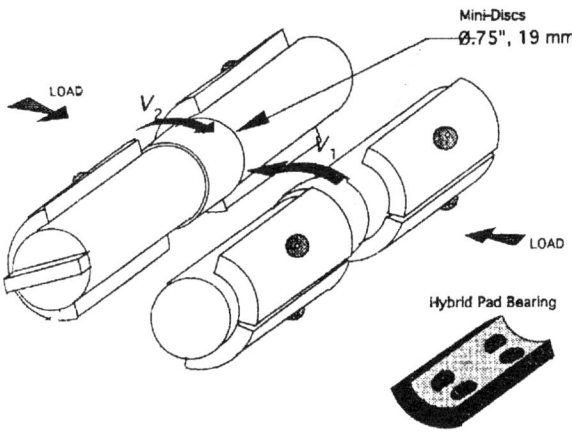

Figure 1 *Mini-Disc and Hybrid Bearing Support*

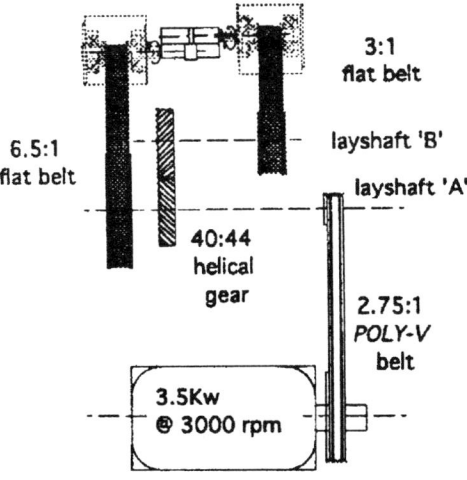

Figure 2 *Schematic of Drive Train*

The drive system is shown schematically in Figure 2 and was designed to run up to disc speeds of 125000 rpm at slide-roll ratio's defined by the layshaft gearing (37% for the set-up depicted in Figure 2).

Results are presented which extend present knowledge to higher speeds, and which are analysed for their own significance, and compared with results from other methods in a critical manner, for example Figure 3 which shows a frictional power intensity comparison.

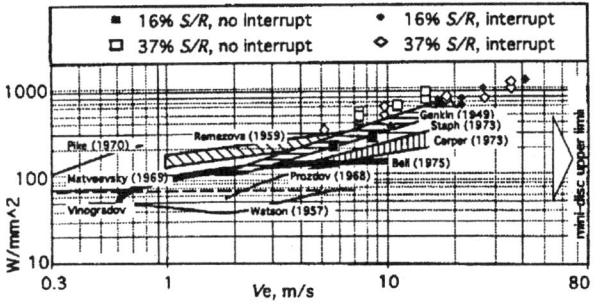

Figure 3 *Comparison of Some of the Paper's Experimental Results with Dyson's (1975) Review Data*

Results show that at higher speeds, following the onset of scuffing, significantly more adhesive transfer of material takes place than is seen with lower speed scuffing, and this situation occurs relatively quickly. Transfer of material is predominantly from the slower to the faster moving surface, as has been observed by others in the past, in disc-machines, and other mechanically similar configurations. However in the tests presented here anomalies to this are also evident where circumferential bands of material transfer can be seen on the slower moving surface. These and other wear and surface modification effects are described and analysed in terms of the engineering science principles involved.

The paper concludes by discussing how such effects might influence the performance of high speed gearing and other tribological elements subject to operating conditions in the range explored.

Dyson, A. (1975) "Review of Scuffing", *Trib. Int. V8*, pp77-87 & 117-121.

THE PARTICLE SIZE EFFECT IN ABRASION STUDIED BY CONTROLLED ABRASIVE SURFACES

RICKARD GÅHLIN and STAFFAN JACOBSON

Uppsala University, Ångström Laboratory, Department of Materials Science
Box 534, S-751 21 Uppsala, Sweden

ABSTRACT

The effect of abrasive particle size on the wear rate is well known. For small particles (corresponding to fine grade abrasive papers) the wear rate increases with increasing particle size. Above some critical size (coarse grade) the wear rate becomes almost independent of further size increases.

Several theories have been presented to explain the size effect. We have examined a theory based on the particle shape, specifically the bluntness of the abrading tips. Blunt shapes produce less wear than do sharp. The outermost tips of abrasive particles and surfaces normally exhibit a bluntness, due to wear or fracture.

Thus, in situations of shallow particle penetration, *only the blunt part is engaged* and the resulting wear rate is low. The average penetration is shallow if the load is distributed over numerous tips, as is the case in abrasion against fine grade abrasive papers.

Accordingly, no shape effect should be expected if the abrasives were ideally sharp.

To test this theory, the abrasive wear produced by ideally sharp tips was compared with that of well-defined blunt tips.

Extremely well defined abrasive surfaces of pyramidically shaped tips were produced by micro mechanical etching of silicon wafers (Fig. 1). One "ideally sharp" shape (tip radius in the nanometer scale) and two blunt shapes (radii about 3 and 9 μm) were produced. The tip blunting was produced by polishing the etched wafers with diamond particles.

The different particle sizes (paper grades) were simulated by varying the packing density of the tips, rather than their individual size. All tips were of the same size (height of about 70 μm, a base diameter of 100 μm and a cone angle of 100°) while the packing densities were chosen to be: 88 tips/mm^2 (fine grade), 24 tips/mm^2 (medium grade) and 6 tips/mm^2 (coarse grade). In this way the packing density, in accordance with an abrasive particle size, corresponds to a typical abrasive groove size.

Wear rates were obtained by wearing cylindrical specimens against the silicon abrasive disc in a pin-on-disc configuration, while continuously measuring the specimen shortening. The worn pin material was tin (HV 3) with a cylinder diameter of 2 mm. The load was 1,2 N, the speed 18 rpm and the sliding distance about 10 m.

The results showed that the sharp tips exhibited a negligible size dependence, while the wear rate due to blunted tips was significantly reduced for the more closely packed! (Fig. 2).

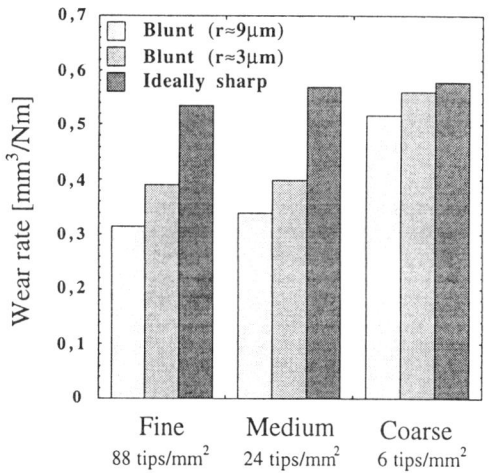

Fig. 2: The tip shape effect

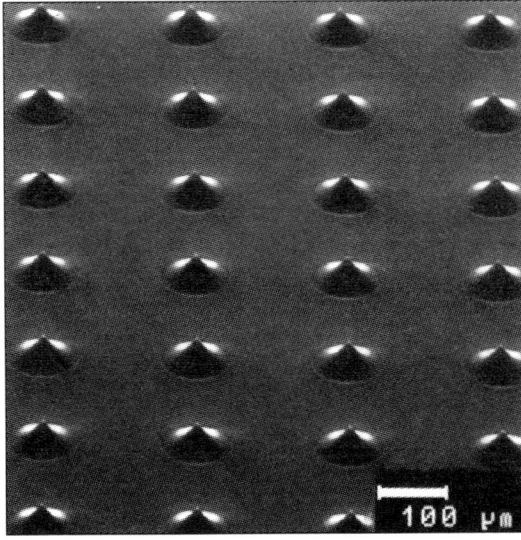

Fig. 1: Abrasive surface produced by micro mechanical etching of silicon wafers (medium grade, i.e. 24 tips/mm^2).

The main conclusions from this work are:
• The shape theory for the particle size effect has been confirmed – ideally sharp tips do not show any size effect!
• Clogging of the abrasive surface also results in a particle size effect.
• Other small shape differences (such as cone angle) also have strong impact in the wear rate.
• The micro mechanical etching techniques applied have proven a very strong tool for making fundamental research on abrasion.

STRUCTURAL FEATURES OF ABRASIVE WEAR OF STEELS

I.I. GARBAR

Department of Mechanical Engineering, Ben-Gurion University of the Negev, P.O.Box 653, Beer-Sheva, 84105, Israel

ABSTRACT

The process of wear in metals is determined by structural changes in its surface layers, which are affected by plastic deformation of the material. These structural changes are responsible for the metal hardening, negative hardening and failure under friction, and therefore can be used for substantiation of structural indicators of material wear resistance (1).

It is well-known that the abrasive wear resistance of pure metals and annealed alloys is proportional to their hardness. However, wear resistance of steels after heat treatment is lower than those for pure metals with the same hardness (2). Therefore, the relationship between wear resistance and hardness for annealed and heat treated steels under abrasive wear vary, and initial hardness cannot be used as a unique indicator of abrasive wear resistance.

To study the reasons for this difference, a structural investigation of steels after various heat treatments was conducted. Samples of AISI 1020, 1040 and 1080 steels were used. The initial hardness of the samples ranged from HV221 – for annealed steel AISI 1020, to HV868 – for water-quenched steel AISI 1080. Two-body abrasive tests on silicon carbide abrasive paper of grit size 1200–240 were carried out in a friction machine under identical conditions for all samples. X-ray studies of the samples were conducted before and after these tests. Both the integral width of diffraction lines and the interplane distances of the crystal lattice were determined.

The structural changes were shown to have different values (Fig. 1). These values varied inversely with initial hardness of the samples for both annealed and heat treated steels.

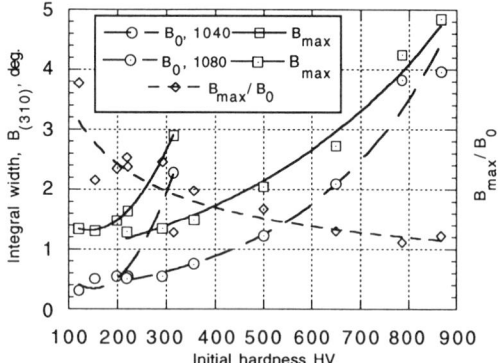

Fig. 1: Maximal changes in integral width of diffraction line (310) α-Fe after abrasive wear for steels of varying hardness.

Fig. 1 shows the changes in integral width of the diffraction line (310) α-Fe after abrasive wear for steels of varying hardness. As can be seen from the results, these changes became smaller as the hardness of the steels (due to heat treatment) increased.

Thus, it can be assumed that structural changes in surface layers of steels after abrasive wear can be used as a unique indicator of abrasive wear resistance, for both annealed and heat-treated steels.

The residual stresses in the surface layers of the steels depend on abrasive grit size and heat treatment regimes (Fig. 2). These results show that the residual stresses observed are compression stresses in the most cases. For steels with the greatest hardness, compression stresses are quite large. The largest is evident after abrasive wear with grit sizes 320 and 240. It should be pointed out that after surface grinding, the compression residual stresses in surface layers are reduced, and in some specimens this changes into the tension residual stresses.

From these results, we can assume that abrasive wear resistance depends on the sign and the magnitude of residual stresses in the surface layers of steels, as well as on the structural state of the surface layers of the steels.

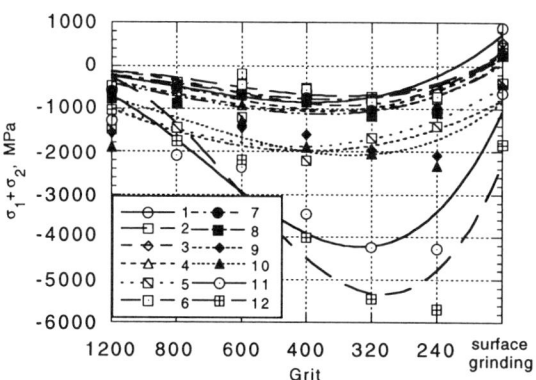

Fig. 2: Residual stresses after abrasive wear as a function of grit size for steels with different Vickers hardness: 1–121; 2–152; 3–198; 4–221; 5–315; 6–219; 7–292; 8–357; 9–500; 10–650; 11–785; 12–868; 1–steel 1020; 2-5–steel 1040; 6-12–steel 1080.

REFERENCES

1. I.I. Garbar, Wear, Vol. 181-183, 1985, 50-55pp.
2. M.M. Kruschov and M.A. Babichev, Investigations into the Wear of Metals, USSR Academy of Sciences, 1960.

EROSION AND ABRASION OF SOME COMMON FACE SEAL MATERIALS

H. ENGQVIST, N. AXÉN and S. HOGMARK
Uppsala University, The Ångströmlaboratory, Department of Materials Science
Box 534, S-751 21 Uppsala, Sweden

ABSTRACT

Commonly used face seal materials are silicon carbide, various cemented carbides and for some applications alumina. The ceramics have a high hardness and corrosion resistance but are much more brittle than cemented carbides. The most striking disadvantage with cemented carbides in seal applications is their sensitivity to corrosive environments, where the Co binder phase is attacked. Substitution of Co with Ni improves the corrosive resistance. Another route to avoid the sensitivity to corrosion is to reduce the amount of metallic binder phase to nearly zero. Such binderless cemented carbides have shown higher corrosive and oxidative wear resistances than the conventional cemented carbides (1). The general tribological performance of binderless cemented carbides have, however, not yet been thoroughly studied.

In this investigation five common face seal materials have been evaluated in abrasion and erosion tests; silicon carbide, alumina, two cemented carbides with 6 % Co but different WC grain sizes (1 and 7 μm) and a binderless cemented carbide consisting of tungsten carbide with 3% TiC and 2% TaC/NbC by weight. The tungsten carbide grains size of the binderless material was about 1 μm.

In the abrasion tests the materials were ground with silica, silicon carbide or diamond particles in the size range of 5-15 μm. The erosion tests were performed with 80, 200 and 600 μm silicon carbide erodents. The angle of impingement was 45° and the erodent velocity 70 m/s. The materials were also evaluated in scratch tests using a pyramidal Vickers- and a sperical diamond tip. The scratches as well as the worn surfaces were investigated with Scanning Electron Microscopy (SEM) and Atomic Force Microscopy (AFM).

In all tests the cemented carbides showed the highest, the binderless cemented carbide an intermediate and the ceramics the lowest wear resistance. In abrasion, the coarse grained cemented carbide showed a higher wear resistance than the cemented carbide with the smaller tungsten carbide grains. In the abrasion tests, the binderless cemented carbide was worn by a preferential removal of the TiC-phase, as revealed from topo- and compo-mode SEM studies. This is believed to be the main cause of the lower abrasion resistance of this material compared to the other cemented carbides. The wear resistance of the ceramics were much lower than the tungsten carbide based materials in all abrasion tests.

The abraded surfaces, as investigated by AFM microscopy, of both the Co-bonded and the binderless cemented carbides were similar and showed much larger amounts of ductile grooving than the ceramics.

Also in erosion the ceramics showed the lowest wear resistances. In contrast to the abrasion results, the binderless cemented carbide performed about equally well as the cemented carbides in erosion. The ranking of the tungsten carbide based materials was, however, dependent on the size of the damage caused by the individual erodent impacts. The coarse grained cemented carbide performed relatively better in erosion with small erodents.

In erosion the wear mechanisms are largely plastic for the binderless cemented carbide, whereas the ceramics were worn by micro-fracture. For the cemented carbides a dependence of the size of the individual erodent damages to the micro-structure of the materials was observed (2). While the large erodents caused damage involving large numbers of grains also for the coarse grained material, the smaller erodents caused damage the size of individual grains. For small erodents, it was observed that impacts close to grain boundaries were more likely to cause micro-fracture than hits in the grain centre.

CONLCUSIONS

In both abrasion and erosion, the binderless cemented carbide showed a higher wear resistance than the tested ceramics. In abrasion, however, it showed a slightly lower wear resistance than the Co- bonded cemented carbides.

In abrasion, the binderless cemented carbide was worn by a preferential removal of TiC grains.

In erosion, the binderless cemented carbide behaved in a much more ductile manner than the ceramics, thus reminding more of the other cemented carbides.

REFERENCES

(1) S. Imasato, K. Tokumoto, T. Kitada, S. Sakaguchi, Int. J. of Refractory Metals and Hard metals, 13, 1995, 305-312.
(2) K. Anand, H. Conrad, Materials Sience and Engineering, A 105/106, 1988, 411-421.

Submitted to *Tribology Letters*

INFLUENCE OF HEAT TREATMENT ON THE ABRASIVE WEAR BEHAVIOUR OF HVOF SPRAYED WC - Co COATINGS

D. A. STEWART, P. H. SHIPWAY and D. G. McCARTNEY
Department of Materials Engineering and Materials Design, University of Nottingham,
University Park, Nottingham, NG7 2RD.

ABSTRACT

In order to produce high quality wear resistant thermally sprayed WC-Co coatings it is necessary to limit the amount of decomposition of the powder that occurs during deposition. The high temperatures experienced in the flame result in melting of the cobalt and subsequent dissolution of the carbide, and the rapid solidification as the molten splats impact on the substrate results in the formation of an amorphous binder phase of cobalt containing tungsten and carbon. The spray process also leads to decarburization of WC into W_2C and in extreme cases tungsten.

Post spraying heat treatment has been shown to have a beneficial effect on the wear resistance of WC-Co thermally sprayed coatings through the re-crystallization of the amorphous binder phase (1). However, the temperatures required for this were high (~853°C).

In this work a sintered and crushed WC-17wt%Co powder was used as the feedstock for HVOF spraying of coatings onto mild steel substrates in which decomposition was limited through the use of an oxygen-lean flame. Coatings 200μm thick were produced and consisted of WC and W_2C in an amorphous cobalt based binder. The as-sprayed coatings were heat treated for 50 minutes in an inert atmosphere at temperatures between 250°C and 1100°C. All the coatings were characterized through the use of XRD, SEM, TEM, microhardness and profilometry. At temperatures above 700°C re-crystallization of the amorphous phase into the eta carbide Co_6W_6C was observed; the grain size of the eta carbide increased with heat treatment temperature. Residual stress state calculations were performed based on change in curvature measurements to determine the stresses in the coatings.

As-sprayed coatings were in tension but at the heat treatment temperatures due to differential thermal expansion coefficients of the coating and steel substrate, the tensile stresses were significantly increased. The net result was that microcracking occurred and the room temperature residual tensile stresses were generally diminished or became compressive.

Three-body abrasive wear tests using a modified dry sand rubber wheel test with 425-600μm alumina abrasive were performed on the as-sprayed and heat treated coatings under two loads of 50N and 75N.

The wear rate was found to decrease with heat treatment temperature up to 600°C after which it increased (Figure 1). This trend was mirrored by the microhardness data which had a maximum at 700°C. For comparison, abrasive wear tests under the same conditions were performed on two conventional sintered tungsten carbides with 11%Co and 24%Co; data for these materials are also shown in Figure 1.

The trends in the wear behaviour of the coatings are attributed to a number of factors, namely the re-crystallization of the amorphous phase into the eta carbide, the residual stress states of the coatings and microcracking within the coating caused by the heat treatment. The latter two effects are believed to have had the greatest effect on the wear resistance as significant changes in wear rate were observed for heat treatment temperatures below that at which the amorphous phase re-crystallized.

This work demonstrates that heat treatment of WC-Co HVOF sprayed coatings can lead to substantial improvements in abrasive wear resistance. Whilst optimum improvements are seen for heat treatments at 600°C, significant improvements are seen (21% and 36% respectively for the two loads examined) after treatment at 250°C; this lower temperature is obviously significantly more attractive for industrial practice.

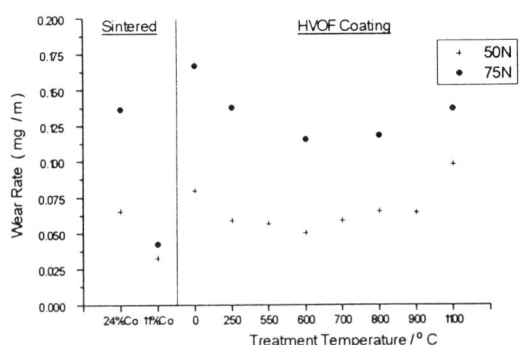

Fig. 1: Wear rate of sintered and HVOF sprayed WC-Co

REFERENCE

(1) J. Nerz, B. Kushner & A. Rotolico, Microstructural Evaluation of Tungsten Carbide-Cobalt Coatings, Journal of Thermal Spray Technology, Vol. 1, No. 2, 1992, 147-152pp.

CRITICALLY EVALUATED ABRASION WEAR TESTS ON WC/CO HARDMETALS

M G Gee, B Roebuck and W P Byrne
Centre for Materials Measurement and Technology
National Physical Laboratory, Queens Road, Teddington, Middlesex TW11 0LW, UK

ABSTRACT

WC/Co hardmetals are extensively used in applications where abrasion resistance is critically important. A number of previous studies have examined the abrasion resistance of hardmetals, but normally only a single test method has been used [1,2]. By contrast, this paper examines the results of abrasive wear tests carried out on a range of these materials. The measurement methods that were used were the ASTM G65 dry sand rubber wheel test, the ASTM B611 steel wet slurry test, the miniature abrasion (ball cratering) test, and scratch testing. The results of the abrasion tests were correlated with the mechanical properties of the materials and also the results of indentation testing.

In contradiction to conventional wisdom about wear testing, in the ASTM B611 test it was found that the scatter in results for this test was much less than in other mechanical tests (Fig. 1). This means that this wear test can be used to discriminate between hardmetals with a similar composition and structure. Measurements were also carried out with slurries based on water with different pH values so that the effect of possible synergism between wear and corrosion could be determined.

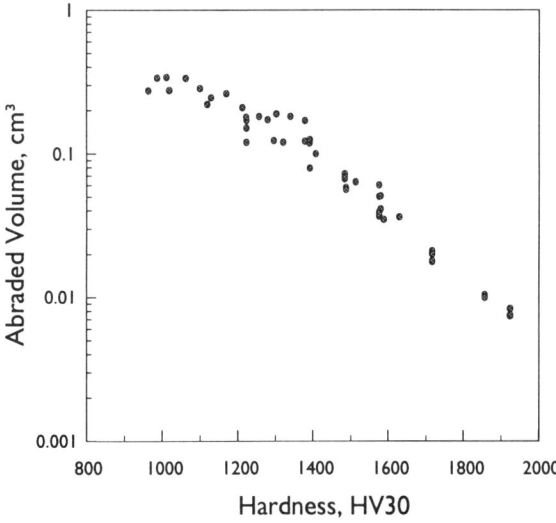

Fig. 1: ASTM B611 abrasion test results

Scratch tests were made using the commonly accepted 200 mm radius Rockwell indenter under loads of 50N. The tests were carried out in a computer controlled test system which enabled multi-pass as well as single pass experiments to be carried out whilst the frictional and normal load, and sample and indenter displacements were continuously monitored.

In the miniature abrasion tests it was found, unusually, that increasing wear was obtained as the hardness of the samples increased (Fig. 2).

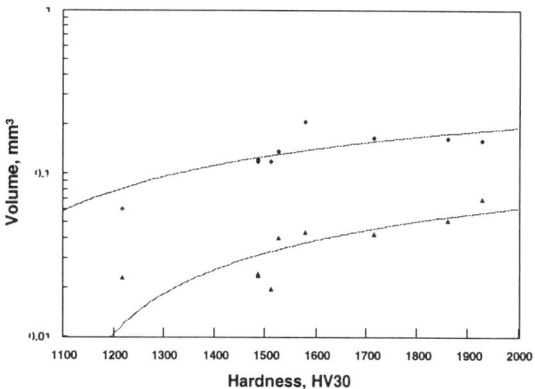

Fig. 2: Results of miniature abrasion test

The results of the abrasion tests are also correlated with an examination of the worn surfaces and an analysis of the mechanisms of wear that occurred (Fig. 3).

Fig. 3: Surface of WC/Co sample abraded in ASTM B611 test.

REFERENCES

(1) J Larsen-Basse and N Devnani, Science of Hard Materials, Institute of Physics Conference Series Number 75, pp 883-895, 1986
(2) D G F Quigley, S Luyckz and M N James, Int J Refractory & Hard Materials, Vol 15, pp73-79, 1997

WEAR OF ABRASIVES: EFFECTS ON MEASURED FRICTION AND WEAR

JORN LARSEN-BASSE
National Science Foundation, 4201 Wilson Blvd., Arlington, VA 22230, USA

ABSTRACT

Friction and wear data were evaluated for a WC-12w/o Co composite abraded under two-body conditions by SiC abrasives in a range of grit sizes. This was done as a follow-up on previous work (1,2) which had demonstrated that friction in two-body abrasion of metals is remarkably repeatable and very closely related to bulk hardness. The reason is that the hardness controls the nature of the deformation near and around a contacting abrasive grit, as its depth of indentation affects the mechanism and depth of deformation and thereby the corresponding local friction force. Qualitative comparison of the friction results with Khruschov's classic abrasion resistance-hardness graph (3) suggested that abrasion might be one wear mode in which friction and wear possibly are closely related. The results further indicated that for hard materials the measured friction and wear data may be strongly affected by a synergistic wear effect: not only do the abrasives wear the sample, but a hard test surface can cause considerable wear of the contacting abrasives and, depending on the extent of the contact, change their ability to cause wear while they still carry some of the load.

A number of samples were cut from a WC-12w/o Co plate with a Vickers hardness of 1215. The samples had rectangular cross sections, 2 mm on one side and ranging between 1.25 and 6.25 mm on the other. They were tested by abrading in slow sliding in the two prime directions for each individual sample under constant load on fresh tracks of SiC abrasive papers. The results showed that both wear rate and friction depend strongly on specimen size in the direction of sliding, having an initially strong drop and then a leveling out for longer samples. The reduction is more pronounced the finer the grit size is. For example, a 3 mm length in the sliding direction reduces the removal capability of grit 600 paper by 80% under the conditions used here; the corresponding number for grit 120 is about 43%. The friction data support the earlier conclusions about the nature of the contact deformation: that it covers a continuum from plastic chip formation to plowing to elastic-plastic surface deformation to purely elastic contact. The lowest friction coefficient measured was about 0.25 which was obtained for very fine grit and long samples; it most probably indicates that the grit-sample interaction is primarily elastic.

The results also show an unexpected dependence of friction and wear on specimen width, for constant length in the direction of sliding. Both increase as the specimen width is reduced below a certain size, which is proportional to the grit size; the relationship indicates that there is a "repeatability of contact conditions" scale of about 80 times the grit size. The increase in friction and wear below the transition specimen width is explained as due to an increase in depth of grit indentation as the specimen size decreases below this "repeatability" value. (Above it, increased load serves primarily to bring more grits into contact). By assuming that the corresponding increase in friction force is responsible directly for the simultaneous increase in wear rate one can calculate the horizontal stress which causes this particular incremental wear. The result is 5 x H, where H is the bulk hardness. This would appear to be a very reasonable result, as the corresponding deformation is expected to be primarily plastic. This approach may have promise in attempts to link friction and wear. On the other hand, it was concluded that making a direct link between full friction and wear is not readily possible, at least not for these hard materials. The reason for that is that much of the load is carried by elastic contacts and they are responsible for a very large, but not very predictable fraction of the friction force, while they cause little or no wear.

REFERENCES

(1) J. Larsen-Basse, Proc. Int. Tribology Conf., Yokohama, JST (1996), 91-96.
(2) J. Larsen-Basse, WEAR (in press).
(3) M.M. Khruschov and M.A. Babichev, Investigations Into the Wear of Metals, USSR Academy of Sciences, Moscow, 1960 (in Russian).

THE EFFECT OF ABRADENT PARTICLES COMMINUTION ON THE INTENSITY OF THREE - BODY ABRASION

S.F. SCIESZKA and A.S.M. JADI
Technical University of Silesia, Akademicka 2, 44-100 Gliwice, POLAND

ABSTRACT

An attempt was undertaken to present the most important parameters effecting the intensity of abrasive wear and to look for any hitherto neglected factors which make an undesirable gap in the understanding of wear processes and in the knowledge of tribological modelling. It was revealed that the comminution or size reduction process of abradent (third-body particles) in the three-body abrasive conditions is the omitted and quite significant factor. Prasad and Kosel highlighted (1) that the abrasive particles undergo some degradation during the course of tests carried on the rubber wheel arasion tester (RWAT), however, they did not take it into account because they considered that these crushed particles represented a very small portion of the total number of abradents that a specimen may encounter during the process of wear. Scieszka (2) studied the abrasive wear of materials in friction contact with mineral particles and he presumed that the abrasion intensity could be attributed to individual particle collapse and the sudden release of elastic strain energy and the formation of new sharp cutting edges which are an additional source of damage.

The relationship between the comminution of abradent and the intensity of wear has not been investigated so far. This deficiency led to extensive experiments with a wide range of variables i.e. four different mineral abradents (coal, alumina, quartz and bauxite) with fixed particles size range (between 300 to500μm, six different cast iron alloys (Table 1) and six different loads.

Table 1. Chemical composition of cast irons

Cast iron grade	C	Si	Mn	Cr	Ni	Mo	Cu
CI 1	1,30	0,76	0,70	13,0	0,17	0,30	-
CI 2	1,90	0,50	0,35	19,2	0,30	0,06	-
CI 3	3,20	1,20	0,55	24,4	0,18	-	0,09
CI 4	3,24	3,80	0,25	0,25	0,01	-	0,12
CI 5	3,30	2,10	0,65	0,65	0,10	-	0,30
CI 6	3,60	1,96	0,63	0,63	0,85	0,15	0,09

Abrasion wear tests were carried out on RWAT according to ASTM G-65 specification. The volumetric wear of specimens, W_V, was calculated after each test. Sieving analysis were conducted after each run and a comminution factor, C_f was calculated from the equation:

$$C_f = \frac{m_{s1}}{m_s}$$

where: m_{s1} is the weight of the abradent after test which came through sieve ϕ 0,3mm,
m_s is the total weight of the abradent.

The investigated relationship was presented in a form of equation (Table 2) and graphically (Fig.1).

Table 2. Values of contant a and exponent b

$W_v = a (1+C_f)^b$	
Rubber wheel	Steel wheel
a = 0,403	a = 0,613
b = 0,450	b = 5,360

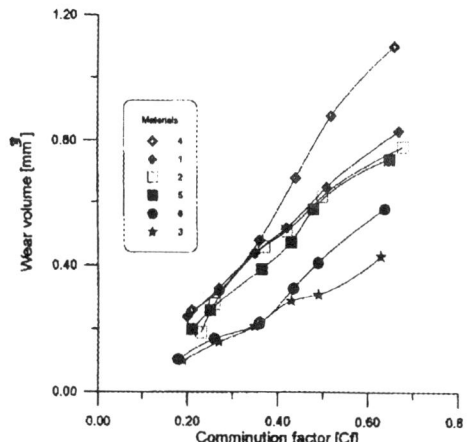

Fig.1. The effect of comminution factor on the wear (rubber wheel and coal)

Careful examination of the abradent particle after each test revealed that even for the lowest load the abradent particles did not remain unbroken. The results obtained revealed that at least part of variability between tests carried out under nominally identical conditions can be ascribed to negligence of the size reduction process monitoring, hence that the recommendation for some changes in standard procedure should be considered.

REFERENCES

(1) S.V. Prasad, T.H. Kosel, Wear, Vol.95, 1984, pp 87÷102.

(2) S.F. Scieszka, Laboratory method for combined testing of abrasiveness, grindability and wear in mineral processing systems, Preprint No 1991-AM-5F-1, STLE, pp 1÷10.

Submitted to *Wear*

A NUMERICAL INVESTIGATION OF SOLID PARTICLE EROSION EXPERIENCED WITHIN OILFIELD CONTROL VALVES

A F FORDER and M T THEW
Department of Mechanical Engineering, University of Southampton, Highfield, Southampton, SO17 1BJ, UK
D HARRISON
BP International Ltd, Research and Engineering Centre, Cherstey Road, Sunbury-on-Thames, TW16 7LN, UK

ABSTRACT

The petroleum industry relies upon regulation of its reservoirs, allowing time specified drainage. This enables production revenues to be maximised through strict reservoir management. This control is provided through the installation of valves or 'chokes' on individual risers from the reservoir. Such chokes act to dissipate energy, hence provide a means to obtain the correct process conditions.

Extracted petroleum fluids are commonly accompanied by solids, usually sand, due to the geological composition of the reservoir rock. The size, shape and concentration of the sand in the process stream may vary; however, one universal feature is its erosive potential.

The dissipative nature of the choke, renders it susceptible to sand erosion, through the high flow velocities induced. At the 'vena contracta', the flow velocity can approach sonic when the process stream is wholly or partially gaseous. As an indication of the severity of this problem, extreme cases have seen loss of pressure containment in as little as 8-14 days. In general, a choke would be expected to experience loss of flow control in approximately 12-24 months, when operating under 'normal' conditions. Normal denotes sand concentrations of 10-50 ppm by wt, particle size 100 - 350 μm.

Although such erosion may not be eliminated, design evolution and careful material selection can reduce its rate and effect on the flow. Design evolution, traditionally of the trial and error form, may be accelerated through the application of complex numerical techniques. This paper details developments based upon a commercial CFD (Computational Fluid Dynamics) code, namely CFX. Such developments have allowed solid particle erosion predictions, in terms of both intensity and location, to be made for the complicated geometries experienced in chokes.

Predictions are made for the turbulent flow field, using the established turbulence closure models of CFX, and hence the trajectories of the suspended particles via the momentum coupling. A data feed from the particle trajectory calculation, provides the dual equation erosion model with the necessary impact variables. The model incorporates contributions from two constituents of erosion generated upon the particle impact angle, i.e. low angle cutting and high angle deformation, a philosophy first generated by Finnie (1).

In this instance, the Hashish (2) model provides the cutting contribution, whilst Bitter (3) presents the deformation requirements.

A complication experienced within the choke is the use of differing materials. Here the body may be an AISI 4130 steel, whilst the control surfaces are commonly tungsten carbide. The problem presented is two fold; first the erosion equations must have the versatility to be applicable to a wide range of materials. Secondly, the material types must be identified to the erosion model. In the first instance, a solution is obtained by the generation of a database. This database not only holds the relevant material constants for each material type, but subtle re-formulations of the erosion equations to suit each material type. Such alterations have been derived by the present author, AFF.

The material type is declared through a unique interaction between the CFD code and the programmed erosion model. This interaction, following a simple naming scheme, allows individual materials to be identified throughout the computational domain used to describe the physical coordinates of the choke.

A further benefit of holding knowledge regarding material types, is an ability to describe the coefficient of restitution. The coefficient is used to denote rebound behaviour at each particle impact. One may readily appreciate that a soft material, such as AISI 4130 steel, will have a differing rebound signature than that of a tungsten carbide, so the correct rebound signature must be specified for each material.

The results produced by the investigation are displayed as coloured three-dimensional surface plots of the chokes internals. Providing information on the distribution of particle impact velocity, angle of impact, particle kinetic energy at impact and erosional intensity. Currently, the methods outlined above are being refined by the comparison of predicted and experimental data obtained from full valve tests. The computer code will become a valuable tool to the choke valve engineer.

REFERENCES

(1) Finnie I; *Erosion of Surfaces by Solid Particles*; Wear; Vol. 3; 1960; pp 87-103.
(2) Hashish M; *An Improved Model of Erosion by Solid Particles*; Proc. 7th Int. Conf. on Erosion by Liquid and Solid Impact; 1988; Paper 66.
(3) Bitter J.G.A; *A study of the Erosion Phenomena, Parts 1 and 2*; Wear; Vol. 6; 1963; pp 5-21, 169-190.

THE DESIGN AND USE OF A LARGE SCALE EROSION TEST RIG

RICHARD WELLMAN and FELIX VAN BORMANN
Eskom, Technology Research and Investigations, P B 40175 Cleveland, 2022, South Africa

HAROLD JAWUREK and JOHN SHEER
School of Mechanical Engineering University of the Witwatersrand, Johannesburg, South Africa

ABSTRACT

This study made use of a large scale erosion test facility to determine the relative erosion resistance of various metal spray coatings and compares the results with those obtained using a standard pneumatic lab size test rig. The preliminary results from the testing of regenerative airheater packs are presented The design and operation of the large scale test rig is discussed in detail.

The test rig was initially designed to examine the erosion characteristics of heat transfer elements of rotary air heaters under local ash conditions. This information is to be used to optimise the design of air heater elements which should result in an increased life span of the airheater elements. The test rig has already been used successfully on a number of different air heater packs under various conditions. This section of the project is still underway and while preliminary results are presented here more comprehensive results will be published in the future.

Little work in the field of regenerative air heater element erosion has been done for the European and North American markets (1). This is due to the low ash and high sulphur coals that are used. Most of the failures in these continents are due to acid corrosion, which is the main focus of their research. However, some work has been done in the USSR on the erosion of air heaters (2).

The main test chamber of the erosion rig can accommodate heat transfer elements with a face area of 300 x 300 mm and up to 500 mm long. However, there are also other points in the rig at which test pieces can be inserted into the flow, for the erosion testing of pipes or flat plates. The rig operates at velocities between 7 and 64 m/s with an erodent feed rate of 0.25 to 4 kg/s. The system operates on a once-through philosophy, taking fly ash from the conveyor belts at the bottom of the electrostatic precipitators and then returning the ash into the precipitators. The rig is the largest known erosion test facility and its potential use extends beyond air heater applications.

A test rig of this size fills the gap between time consuming in situ testing and small scale lab erosion testing. Furthermore it is possible under the accelerated conditions to test materials to destruction, which is not always feasible with lab size equipment which normally measures mass loss in milligrams.

The use of metal spray coatings for the protection of boiler tubes has been steadily increasing over the last few years in the South African power industry. It has become necessary to evaluate new coatings before using them extensively in the stations. Since in-situ testing is too time consuming and the results from previous laboratory tests conflict with what is experienced in the boilers, it was decided to investigate using a large scale test rig. This resulted in the airheater erosion test rig being used for the erosion testing of metal sprayed coatings.

The difference in the results from the two different test rigs has been attributed to the large incubation period experienced in testing the coatings. Subsequent to testing the coatings in the large scale erosion rig it became evident that a steady state condition had not been reached in the lab test rig. This was further supported by the difference in the surface roughness of the coatings after testing the coatings in the two different rigs. The samples tested in the large scale rig had a significantly lower surface roughness than those tested in the lab scale rig.

REFERENCES
(1) D E Gorski, Ljungstrom Technical Conference 1992
(2) A A Vasilev V I Dombrovsky, Thermal Eng. Vol 25 No. 1 July 1978

ABRASION AND EROSION OF MATERIALS FOR WEAR PROTECTION IN OIL SANDS MINING AND PROCESSING EQUIPMENT

R J LLEWELLYN and H M HAWTHORNE
National Research Council Canada, Vancouver. B.C. Canada, V6T 1W5
J. OXENFORD
Syncrude Canada Ltd., Edmonton, Alberta, Canada, T6N 1H4

ABSTRACT

Wear costs the Canadian mining, oil and gas producing industries over C$1billion per year(1). The burgeoning oil sands sector which currently supplies over 20% of Canada's oil requirements, is particularly severely affected. It's two commercial plants alone, spend over C$ 75 million per year on replacement parts and associated labour charges. This paper reviews the approach taken to address the problem at Syncrude Canada Ltd., which is the larger operation with a 1996 annual output of 73.5 million barrels of oil.

INTRODUCTION

Mining and processing of the vast oil sands deposit in Northern Alberta, Canada, result in a multiplicity of wear attack situations. These range from impact and high stress abrasion in mining at very low ambient temperatures, to fine particle erosion/corrosion in elevated temperature slurry streams during bitumen conversion to lighter hydrocarbon products.

Such diverse conditions necessitate the use of a wide variety of protection systems, encompassing relatively soft elastomers to extremely hard cermets and ceramics.

WEAR MATERIALS SELECTION PROCEDURE

A paucity of relevant performance data and documented experience was available at the start of production in 1978.

To improve this situation, the dry sand rubber wheel wear test method (ASTM G65) was chosen for qualification assessment of candidate materials. It was considered to be a reasonable simulation of the low stress sliding abrasion mechanism responsible for most of the severe damage encountered in oil sands mining.

Data generated has enabled superior materials to be introduced into mining service. Particular successes have been achieved through the more widespread use of bulk weld overlay and laminated white iron wear plates, hardfacing alloy improvements and bimetallic casting developments(2,3).

Typical benefits from the program are exemplified in Figure 1. This illustrates the 80% reduction in costs for bucketwheel reclaimer digger teeth and the 10% increase in throughput, which has been achieved since 1990-91, by the use of extended coverage of novel mixed carbide hardfacing on a 495 HB hardness steel base.

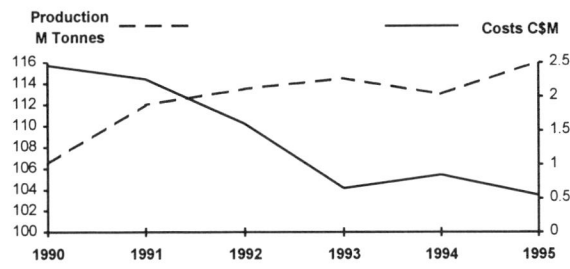

Fig.1 BWR production and tooth costs

ASTM G65 test data have also been used effectively in materials selection for applications involving slurry processing and pipelining. Notable improvements have been achieved in the application of thermal spray coatings, ceramics and cermets in separation equipment, pumps and valves(2,4).

This work has contributed significantly to a decrease in unit production costs from over C$25 per barrel in 1978-9, to its current level of less than C$14.

To help maintain this trend, the program has recently been expanded to include evaluation methods which simulate specific wear mechanisms which still cause high losses or which will present increased challenges in the future. These include high stress abrasion and slurry erosion testing.

REFERENCES

(1) National Research Council Canada. Report No 26556
(2) R.J.Llewellyn,J.C.Tuite, Welding Journal, March 1995, pp55-60
(3) A. Khan,D.Reid,K.Obaia,D.Adamic, Mine Planning and Equipment Selection, Balkema,Rotterdam, 1995.
(4) R.J. Llewellyn, G. Hurtubise, Int. Symp.in Advanced Ceramics,Vancouver, Canada, 1995.

THE WEAR PROPERTIES OF ULTRA-FINE GRAINED WC-Co ALLOYS

C ALLEN and V A PUGSLEY
Department of Materials Engineering, University of Cape Town, Rondebosch 7701, South Africa

I T NORTHROP
Boart Longyear Research Centre, PO Box 1242, Krugersdorp 1740, South Africa

ABSTRACT

Recent interest in the production and properties of ultra-fine grained WC-Co parallels the increasing attention being given to nanocrystalline materials in general. Because of the high proportion of constituent atoms lying at interfaces or grain boundaries, nanocrystalline materials often display new and unique combinations of properties.

Ultra-fine grained WC-Co is defined as having a grain size of between 0.1 and 0.5µm and cannot therefore strictly be termed nanocrystalline. However, it too has been shown to give the high combination of strength and hardness characteristic of these materials[1]. Furthermore, the development of alternative powder production methods, notably the spray conversion process[2], raises the possibility of further refinements in grain size.

It is generally accepted that the wear resistance of WC-Co alloys increases with decreasing grain size[3,4] and hence one would expect superior properties from ultra-fine grained alloys. Indeed the abrasion resistance of these alloys has been found to be approximately double that of the most resistant conventional material, far outstripping the increase in hardness[5]. This leads one to suspect the action of significantly different material removal mechanisms.

However, a full characterisation of the wear properties of these materials and investigation into the modes of material removal occurring has yet to be published. To rectify this situation, this paper presents the current results of an on-going project.

A series of ultra-fine grained WC-Co alloys of varying binder content but with a constant grain size of approximately 0.3µm has been sintered from powder produced through the spray conversion process. These have been subjected to cavitation erosion, and slurry erosion in carrier fluids of varying corrosivity. Material removal mechanisms have been elucidated using the scanning electron microscope.

The results are compared to those obtained for conventional hardmetals with the same range of cobalt contents and grain sizes of up to 4µm. Differences in performance are related to variations in material properties such as strength and ductility, and the consequent changes in deformation characteristics and material removal mechanisms.

---

[1] L.E. McCandlish, V. Kevorkian, K. Jia and T.E. Fischer, Adv. Pow. Met. and Part. Mat, Toronto 1994, pp 1-9

[2] L.E. McCandlish, B.H. Kear, and J. Bhatia, Spray conversion process for the production of nanophase composite powders (U.S. Pat. App. S.N. 433 742)

[3] S.F. Wayne, J.G. Baldoni and S.T. Buljan, Tribology Transactions, Vol. 33, pp 611-617, 1990

[4] M.K. Keshavan and N. Lee, Met. Powder Rep., Vol. 42, No. 12, pp 866-869, 1987

[5] K. Jia and T.E. Fischer, Wear, Vol. 200, pp 206-214, 1996

OBSERVATIONS ON, AND THE MODELLING OF, THE EROSIVE WEAR OF A LONG RADIUS PNEUMATIC CONVEYOR BEND

A J BURNETT, A N PITTMAN and M S A BRADLEY
The Wolfson Centre for Bulk Solids Handling Technology, University of Greenwich,
Wellington St., Woolwich, London, SE 18 6PF, UK

ABSTRACT

Results of a programme of work designed to derive a more efficient way of predicting the life of pneumatic conveyor bends are presented in this paper.

Detailed measurements were taken from tests carried out on a single bend in an industrial scale pneumatic conveyor for a controlled set of conveying conditions. Results obtained for the wear rate of the steel used to construct the bend were found for a range of particle impact conditions using a rotating disc accelerator type erosion tester. Analysis of these two groups of results enabled a model capable of predicting the life of the bend to be developed.

Several previously unreported trends on the effect of the variables involved in such a tribological system on the life of the bend were observed. These trends will be discussed in this paper which has been submitted to the IMechE Journal of Engineering Tribology for publication.

INTRODUCTION

Puncture of components due to erosive wear is frequently encountered during the operation of pneumatic conveyors. Predominately wear occurs where changes in direction of flow occur, i.e. in bends (1,2). An increase in the particle velocity causes the puncture rate to increase and particle concentration within the pipe bore also effects the bend wall penetration rate as does the geometry of the bend (3).

TEST METHODS

All the tests that were undertaken used mild steel for the bend wall material and olivine sand was used as the abrasive.

Pneumatic Conveying Tests

Tests were carried out using a bend of 53mm bore and 750mm radius. Measurements of the rate of bend wall puncture, superficial air velocity before the test bend and particle concentration were undertaken. The conveying conditions were accurately controlled to give a mean particle velocity of 25.6 m/s and a particle flow rate of 0.259 kg/s.

Laboratory Erosion Tests

A rotating disc accelerator erosion tester was used to determine the wear behaviour of the mild steel. Tests were carried out at three particle concentrations, 1, 4 and 13 kg/m^3, each at three particle velocities, 15, 25 and 35 m/s (4).

RESULTS

Pneumatic Conveying Test Results

Two observations were made:-

- Bend wall puncture occurs in the region where the particles strike the bend wall for the second time. The number of particle impacts per unit area of bend wall increases at the puncture point because of the effects of bend geometry on the particle trajectories.
- The wear patterns observed were similar to the shape of the curve of aberration for a concave cylindrical mirror (1).

Laboratory Erosion Test Results

The following two observations were made:-

- Erosion rate increases with an increase in particle velocity, but decreases with an increase in particle concentration.
- Erosion rate does not exhibit a peak when plotted against the angle of particle impact (4).

A power law model was developed to predict the erosion rate with changes in angle of particle impact, the magnitude of its velocity and the particle concentration (1,4).

THE PREDICTIVE MODEL

Pivotal to the model used to predict the life of the pneumatic conveyor bend was the power law erosion rate model derived from the laboratory erosion tests. Results obtained from using this model for the mean conveying test conditions were modified by using a multiplication factor. The multiplication factor was derived from the geometry of the curve of aberration for a cylindrical concave mirror. This factor accounted for the intensification in the concentration of particle impacts at the bend wall puncture point.

REFERENCES

(1) A.J. Burnett, PhD Thesis, University of Greenwich, London, UK, 1996.
(2) G.J. Wright, PhD Thesis, University of Witwatersand, Johannesburg, S.A., 1994.
(3) D. Mills, Thames Polytechnic, Workshop Notes for exhibition, SOLIDEX '86, Harrogate, Yorkshire, UK, 1986 (copy available from A.J. Burnett).
(4) A.J. Burnett, M.S. Bingley and M.S.A. Bradley, Pneumatic and Hydraulic Conveying Conference, Palm Coast, Florida, USA, April 1996, (Engineering Foundation, New York, USA), publication expected in Powder Technology in 1997.

EFFECT OF CARBIDE VOLUME FRACTION AND MATRIX MICROSTRUCTURE ON THE WEAR RESISTANCE OF HIGH CHROMIUM CAST IRON BALLS TESTED IN A LABORATORY BALL MILL

E ALBERTIN
Foundry Laboratory, Technological Research Institute - IPT, São Paulo, Brazil
A SINATORA and D K TANAKA
Department of Mechanical Engineering, Polytechnic School, University of Sao Paulo, Brazil

ABSTRACT

The carbide volume fraction effect on the wear resistance of high chromium cast irons was evaluated in ball mill wear testing of six martensitic materials with carbide volume fractions ranging from 0 to 41%. The matrix microstructure effect was studied on 30% carbide volume fraction cast irons, with martensitic, pearlitic or austenitic matrices.

50 mm diameter cast balls were used as test specimens. The carbon contents of cast alloys ranged from 1.65 to 3.54% and the chromium contents were adjusted to keep the Cr/C ratio around 6.5. Air-quenching from 980°C followed by sub-zero treatment (liquid N_2) and tempering at 250°C presented tempered martensitic matrices. For the 30% carbide alloy, fully pearlitic or austenitic matrices were also obtained. In addition, a "matrix-steel" alloy, with 8%Cr and 0.8%C, was cast, providing a 0% carbide condition.

144 balls (groups of 12 to 24 balls of each material) were tested simultaneously in a 40 cm diameter laboratory ball mill. Hematite iron ore, phosphate rock or quartz foundry sand (AFS 100) were wet ground. The tests were continuously carried out for 200 hours.

Quartz sand presented highest wear rates, from 6.5 to 8.6 μm/h for the martensitic balls, while the least wear rates were observed for the phosphate rock, from 1.4 to 2.9 μm/h. Wear rate for hematite ranged from 2.5 to 3.6 μm/h. A summary is shown in Fig. 1.

Increase in carbide volume fractions resulted in decrease on wear rates when the abrasive was hematite or phosphate rock, but the opposite effect was observed for the quartz sand. These results suggest that while carbides are effective barrier to the wear by relatively soft hematite or phosphate rock, quartz particles can break exposed carbide ramifications. Since these abrasives are harder than the martensite, the matrix is rapidly worn out, continuously exposing new carbide branches which can be removed by breakage.

The alloy with highest carbon content presented poor wear resistance due to spalling of coarse carbides formed in hypereutectic solidification. On the other hand, eutectic microstructures presented good performance, even against the quartz abrasive, which can be attributed to almost complete protection of the matrix by hard carbides present in this case.

Pearlitic balls presented high wear rates, followed by the austenitic ones, while the balls with martensitic matrix presented the best performance. The differences between performances were more noticeable for softer abrasives. In phosphate rock tests, as the matrix is replaced from martensitic to austenitic and to pearlitic, the wear rate increased from 1.5 to 2.8 and to 6.4μm/h respectively. For quartz, the wear rates ranged from 7.8 to 8.3 and to 11.1μm/h for the same matrix change.

A global analysis of these results showed that the hardness ratio between abrasive and matrix was a determinant parameter. In addition, balls wear profiles showed that non-martensitic balls presented deep subsurface carbide cracking, due to matrix deformation. When these pre-cracked carbides rise out to the surface they are readily removed, increasing wear rate.

Pin-on-disc tests, using SiC or alumina sandpapers, performed on trepanned test coupons from the balls, the carbide volume fraction effect on martensitic materials presented same trends observed in the ball mill test. On the other hand, when different matrices effect are compared, austenitic samples performed better than the martensitic ones, which results are conflicting to those observed in milling tests. These results show that pin tests used in material selection for balls can be misleading in presence of retained austenite.

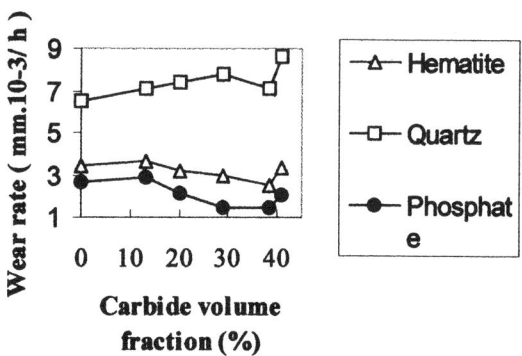

Fig. 1: Effects of carbides and abrasives

GROOVING ENERGY MEASUREMENTS WITH A PENDULUM TECHNIQUE

O. VINGSBO[†], F. ATTIA[††]
[†]Dept of Mech. Engr., [††]Dept of Technology,
University of Houston, Houston, TX 77204, USA,

ABSTRACT

One pass grooving experiments are performed with a so called Uppsala Pendulum (1), (2). It consists of a modified Sharpy pendulum, in which a radially protruding cemented carbide tip is used as a grooving tool. A horizontal specimen is placed in a holder at the bottom position of the pendulum swing, in a way to let the tip make an arcuate groove of adjustable size during the downswing. The grooving energy E corresponds to the difference between the downswing and the upswing angles, and can be obtained by taking readings from the standard gauge of the pendulum. The specific grooving energy e is defined as e = E/W, where W is the weight of the removed material. Since the groove is arcuate, it is possible to calculate its volume $V(\lambda)$, and weight $W(\lambda)=\rho \cdot V(\lambda)$ for a given groove length λ, (ρ=density) and an e(W) curve can be plotted with the aid of a number of grooves of different sizes λ, each corresponding to a point on the curve. Such e(W) curves are used for ranking different materials with respect to abrasion resistance, taking the severeness of the abrasive attack (the groove size) into account.

The pendulum is equipped with a triaxial force tranducer, built into the shaft immediately above the tip. The tangential force f(x) is recorded as a function of the tip position x during a grooving event. The integral

$$\int_0^x f(y)dy$$

is recorded, and corresponds to the work done for each position x of the tip in the groove being cut, i.e. E(x). E(x) increases with x until it reaches the final grooving energy

$$E(\lambda) = \int_0^l f(y)dy$$

for the complete groove. An e(W) curve can now be constructed in two ways:
1) making a series of grooves of different λ_i and plot the $e(W_i)$ curve
2) making one groove of length $\lambda_{i,max}$, and construct an $e(W_i)$ plot based on a series of integrated energy values for each $x=\lambda_i$.

Obviously, the two techniques are not exactly equivalent. The one-groove e(W) curve corresponds to the successive growth of one and the same groove, whereas each point of the multi-groove curve corresponds to one of a series of successively larger complete grooves. Particularly, the W_i values will not be the same in the two cases. The one-groove alternative is experimentally less tedious.

A third alternative is to take an f(x) reading, and calculate the e value for a groove of constant depth d, corresponding to that particular x value. For a constant d the energy will be

$E(x)=f(x) \cdot x$, and the volume will be

$V(x)=A \cdot x$, where the groove cross section area A is also a constant. Consequently, the specific grooving energy e is independent of x, and a function of d only, according to

$e \cdot \rho = E/V = f(x) \cdot x/A \cdot x = f(x)(/A$.

The grooving element is a square pyramid with 90° apex angle, truncated to a 1x1 mm square flat tip, and A is a function of d

$A=d(d+1)$ (in mm^2)

rendering a specific grooving energy

$e=f(d)/d(d+1)$

e can now be plotted versus the depth for grooves of constant depth d, using the same experimental data as for the previous two alternatives.

The paper contains a comparative study of these three different specific grooving energy concepts.

REFERENCES

(1) U. Bryggman, S. Hogmark, O. Vingsbo, Wear 112 (1986) 145-162.
(2) O. Vingsbo, S. Hogmark, Wear, 100 (1984) 489-502.

CASE STUDIES OF HIGH RATE TRIBOLOGICAL EVENT: OBLIQUE IMPACT OF STONES INTO MULTI-LAYER PAINTED AUTOMOTIVE SURFACES

D. J. Mihora
FM Analysts, Inc., Santa Barbara, CA 93111
A. C. Ramamurthy
FORD MOTOR COMPANY, Scientific Research Lab, Dearborn MI 48121

ABSTRACT

A major consumer complaint is the chipping of exterior automotive paint. It is well known from field observations and laboratory tests that various sizes and shapes of roadway stones lofted into the painted panels is the cause of paint chipping. The size and shape of roadway stones is most diverse. Trial and error experiments to understand and improve the quality of multi-layer paints have not been totally successful – partially because the physics of the paint "chipping-event" underneath the stone are not totally understood. Finite element analysis of dynamic conditions associated with paint chipping has been initiated for both smooth and rounded stones with masses of 0.1 to 5 grams. These initial insights about stresses within the thin paint layers are aiding in the design of forthcoming new laboratory test equipment. These results also point toward materials research of specific paint layers now known to be deficient in specific physical properties.

This program is a first attempt to understand the tribology under the stone impact site when different shaped rocks impact painted exterior automotive surfaces at oblique impact angles and different velocities. The matrix of possible CAE cases was condensed to the minimal number that could provide meaningful parametric insights.

For expediency, a number of finite element simplifications were made to accomplish the goals of completing the 25 case runs. The numerous finite element simplifications can be revisited at a future date. Several of the major simplifications include limiting the number of finite elements to approximately 20,000. Also, the paint layers did not incorporate an adhesion failure criterion. Relatively simple failure conditions were employed to approximate the locus of failure. These oblique impact cases utilize only 5 to 10 percent of the intrinsic modeling capabilities of DYNA3D and TrueGrid for CAE. These results are not the definitive presentation of CAE capabilities. However, they clearly provide parametric insights and demonstrate the ability to identify locations and modes of failure within the complex new paint systems. To provide better insights, the response with aluminum and steel substrates will be provided.

Significant work was accomplished to acquire realistic plastic and viscoplastic properties of materials to failure. Tables 1 and 2 show no fail (NF) conditions as well as times (μs) to paint failure onset. Generally, paint failure is initiated in the colored base coat (blue) by the extreme shear stresses under both the rounded and sharp stones.

Table 1 Steel sphere and painted steel

V (m/s)	$\gamma = 90°$	$\gamma = 45°$	$\gamma = 15°$
18	N.F.	N.F.	N.F.
36	N.F.	10	15.

Table 2 Sharp edge stone and painted steel.

V (m/s)	$\gamma = 90°$	$\gamma = 45°$	$\gamma = 15°$
40	28.	28.	N.F.
80	N.F.	11.	22.

The most damaging impact for the stones is the 45 deg. trajectory for both steel and aluminum substrates. The 90 deg impacts do indeed produce the largest dynamic compression stresses (10 to 13 KBar) but compressive stresses are not involved in the paint chip initiation. Compression aids in the stabilization of the paint layers during the friction induced shearing event which occurs in the first 30 μs after the initial impact.

Clearly, the sharp stones initiate damage much earlier than the steel sphere. The aluminum plate also introduces chipping earlier (easier) than a stiffer steel foundation. Wave propagation and ringing and spallation are not factors in the initiation of paint chip failure but contact stresses and tribology are significant.

london1.doc

The Effect of Counterface Surface Roughness and its Evolution on the Wear and Friction of PEEK and PEEK-Bonded Carbon Fibre Composites on Stainless Steel

D M ELLIOTT and J FISHER.
Department of Mechanical Engineering, University of Leeds, Leeds, LS2 9JT, UK
D T CLARK
Research Unit for Surfaces, Transforms and Interfaces, Daresbury laboratory, Warrington, WA4 4AD, UK

ABSTRACT

Carbon-carbon composites, with their low density, high strength and high melting point, are being used in an ever increasing number of applications. It is, therefore, important to gain a thorough understanding of the range of frictional forces and wear mechanisms that these composites exhibit. As a basis for comparison of the tribological behaviour of these carbon-carbon composites some aspects of the tribology of PEEK and the composite pre-cursor PEEK-bonded carbon fibre (APC2) have been investigated.

This present work investigates the time-dependent variations in dry sliding wear rate and friction of PEEK (450G and 100P) and APC2 on stainless steel (316S16) and measures the evolution of wear scratches with optical profilometry. A three-pin-on-disk tester was used, operating at 0.18 m s^{-1} with a pressure of 1.0 MPa on each 5 mm diameter pin. Pin wear factors were calculated from mass lost after successive 15 km runs and friction force was monitored at intervals by a strain-gauged torque beam. Two series of 90 km tests were performed for each pin/disk combination. The stainless steel disks had either a fine-ground or polished surface. The roughness of the ground stainless steel surface was chosen to coincide with the optimum value for low wear rate of the PEEK pins as reported by Ovaert and Cheng (1). Ra values for this surface were between 0.1 and 0.2 μm measured by contacting stylus (Rank Taylor Hobson), corresponding to values of 0.29 to 0.39 μm measured by a non-contacting laser profilometer (UBM). The polished stainless steel surfaces had Ra values of 0.02 to 0.03 μm (UBM) or 0.0065 to 0.0080 μm (RTH). To maximise pin/disk contact area, the PEEK and APC2 pin surfaces were run-in on 1000-grit silicon carbide paper and ultrasonically cleaned in acetone before the first wear test. There was no significant difference in steady-state wear between these pins and those tested with as-moulded surfaces. The disks were analysed at 15 km intervals with the UBM profilometer, providing 3D surface scans and 2D profiles for roughness parameters without damaging the wear track.

The 450G had a higher molecular weight than the 100P which was made at least two years before the 450G and contained black specks of degraded PEEK (up to 500 μm dispersed at random) which effected the wear rate. The manufacturing process has now been improved and the 450G contained a negligible amount of specks.

The 450G PEEK wear factors were similar for both counterface surfaces. However, the friction forces measured against the polished disk were up to three times higher than those against the ground disk, (0.35+/-0.01), decreasing to similar values as the polished surface became scratched. With 100P PEEK the friction was seen to increase by about 30% during interaction between any large black speck and the disk. A marked difference in wear factor was observed between the two types of PEEK. The 450G had the lowest wear factor of 6 x 10^{-6} mm^3 N^{-1}m^{-1} on both the polished and ground counterface, whereas the 100P had wear factors of 1 x 10^{-4} mm^3 N^{-1}m^{-1} on ground and 2 x 10^{-5} mm^3 N^{-1}m^{-1} on polished stainless steel.

The APC2 pins showed very little difference in wear factor between the ground and polished disks. However, a steady decrease from 1 x 10^{-6} to 1 x 10^{-7} mm^3 N^{-1}m^{-1} over a distance of 90 km was measured as more fibres were ground into a graphite film lubricant and the friction dropped from 0.35 to 0.18.

The wear factors and friction coefficients of the APC2 samples were lower than those for PEEK; but the Ra of the steel counterface changed from 0.022 to 0.089 μm after 15 km against PEEK and from 0.022 to 0.576 μm against APC2.

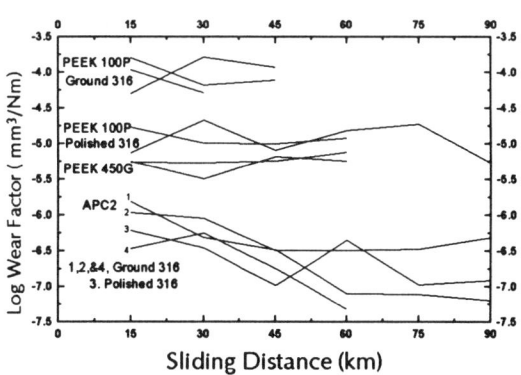

Fig. 1: Wear Factors of PEEK and APC2

REFERENCES

(1) T C Ovaert and H S Cheng; *Wear*, vol. 150, 1991.

POLYMER WEAR IN PARTICULATE CONTAMINATED WATER.

Dr P J TWEEDALE
AEA Technology plc, National Centre of Tribology, Risley Warrington. WA3 6AT.

ABSTRACT

The importance of hard particulate contaminants on determining the in-service life obtained from components in water powered machinery is not fully understood, and has not been adequately investigated before for polymeric materials. The existing state of knowledge on the performance of materials or the interactions between debris, polymer, counterface and fluid is poor.

The wear of polymers in water has received treatment in the past(1-3), but the major interest has been in the evaluation polyethylene (especially UHMWPE) for joint prosthesis applications(4-6). The wear of polymers, especially those capable of forming seals in water powered machinery for use in particulate contaminated media, has however received much less attention.

Contaminant influence in the operation of tribological contacts has been addressed in many studies, from the solid particle in a dry(7) or wet(8) contacts to the effect of abrasive particle shape on abrasive wear(9). The influence of particle entrapment in the life of oil lubricated rolling elements bearings is a topic of much research interest (10).

A test procedure capable of determining the wear rate of polymers in particle contaminated water has been developed using a pin-on-disc configuration. A range of currently available engineering polymers (UHMWPE, PEEK, PPS, thermoformable PI, aromatic liquid crystalline polymer) capable of being formed into seal components were evaluated. The specific wear rate of these polymers in both clean and abrasive particle contaminated water environments over a range of counterface speeds were determined.

Table 1. Dry and clean water specific wear rates.

SPECIFIC WEAR RATE (m^3/Nm) x 10^{-16}	DRY 0.1m/s	WET 0.1m/s	1m/s	10m/s
UHMWPE	2.7	58.7	151	205
UHMWPE + 20% PTFE	6.5	589	853	93
PEEK	61.5	137	40.5	62
PEEK + 20% PTFE	11.9	53.6	21.9	25.8
ACETAL	2.96	-	1160	-

A number of conclusions were drawn from this work:

UHMWPE is still probably the best material for seal materials because of the balance of surface to bulk properties that gives it good performance in dry contacts, also seems to be effective in dealing with particulate contaminated water.

The engineering polymers seem to offer no useful improvement over UHMWPE, other than the ability to be directly formed and filled into seal components.

Although UHMWPE is relatively weak materials in comparison it wears at the same or a lesser rate than tougher and stronger engineering polymers under these particulate contaminated conditions.

The engineering polymers do however provide a degree of insurance in water lubricated contacts that may not be optimally designed for loading and frictional work dissipation i.e. is being over stressed.

Table 2. Particulate contaminated wear rates.

SPECIFIC WEAR RATE (m^3/Nm) x 10^{-16}	SPEED		
	0.1 m/s	1 m/s	5 m/s
UHMWPE	200	110	28.2
UHMWPE + 20% PTFE	667	225	36.4
PEEK	653	225	183
PEEK + 20% PTFE	256	260	116
ACETAL	-	722	-
VECTRA	-	100	57
Torlon	-	115	206
PPS + 20% PTFE	-	776	716

All measurements made at 1.78 MPa contact pressure and against 316 stainless steel of initial counterface roughness, Ra = 0.2 μm (lay in the direction of motion).

REFERENCES

(1) M. Watanabe. 'Wear of Materials, '79.' p 573-580.
(2) A.I.G. Lloyd, R.E.J. Noel. Tribology International, 21(2), 83-88, (1988).
(3) K. Tanaka. 'Wear of Materials, '79.' p 563-572.
(4) D. Dowson, M.M. El-Hady Diab, B.J. Gillis, J.R. Atkinson. in 'Polymer Wear and its Control', (1974). Ed. L.H. Lee. ASM. Che. Soc., New York.
(5) D.A. Miller, R. Ainsworth, J. Dumbleton, D. Page, E.H. Miller, C. Shen. Wear, 28, 207, (1974).
(6) D.M. Tetreault, F.E. Kennedy. Wear, (1989).
(7) G. Huard, J. Masounave, M. Fiset, Y. Cote, D. Noel. 'Wear of Materials, '87
(8) B.J. Briscoe, P.J. Tweedale. 'Polymers in Offshore Engineering.' p 3/1-12, 1989.
(9) M.A. Moore, P.A. Swanson. 'Wear of Materials - '83', p 1-11.
(10) R.S. Sayles, J.C. Hamer, E. Ioannides. Proc. Inst. Mech. Engrs. Conf. 'Aerospace Bearings.'

NANOMETER SCALE WEAR STUDIES ON POLYSTYRENE

D. D. WOODLAND and W. N. UNERTL

Laboratory for Surface Science and Technology, University of Maine, Orono, ME 04469-5764, USA

ABSTRACT

The processes that result in the onset of wear are not well understood. In this paper we describe the effects of molecular weight (MW) and thermal processing on the initial wear caused by a single asperity sliding on the polystyrene surface. The asperity is the silicon nitride tip of a scanning force microscope (SFM). Applied loads were ≤ 200 nN with contact radii in the range 5 - 25 nm resulting in contact pressures of 0.1 - 2.5 GPa. Sliding speeds were up to 0.3 mm/s. Preliminary results have been published elsewhere (1).

Earlier studies of polystyrene showed that characteristic equally spaced ridges form perpendicular to the sliding direction but disagreed on the dependence on molecular weight (2) (3). Tanaka et al. showed that the polystyrene surface is more viscoelastic than the bulk (4).

The polystyrene films in this work were cast from toluene solution onto clean glass slides. Two different MWs were investigated, 24k and 210k, since the bulk tensile strength changes by an order of magnitude in this range. Some samples were annealed at 130 °C for 1 hour. The abrasion tests were carried out using Si_3N_4 tips with nominal tip radii of 20 nm and force constants of 0.03 N m^{-1} and 0.10 N m^{-1}. Wear was induced by repeated abrasion cycles over 4 μm x 4 μm areas, each cycle consisting of a single raster scan over the area. Within each set of experiments the same tip was used and all tests were carried out at room temperature (≈ 25 °C) and approximately 40% relative humidity.

All four surfaces investigated (24k as prepared, 24k annealed, 210k as prepared, and 210k annealed) were initially very smooth with rms roughness values at or below 0.5 nm. The abrasion pattern formed consists of well oriented ridges running perpendicular to the scan direction (see Figure 1).

The major results are: [1] Root-mean-square roughness of the region increases with number of abrasion cycles, which is in qualitative agreement with Ref. 3. [2] The spacing of the parallel ridges is independent of applied load, F, which rules out a power law dependence, $F \propto \lambda^m$, suggested in Ref. 3 to support a Schallamach mechanism for the nanometer scale abrasion. [3] The ridge spacing increases linearly with sliding speed v; i.e., v=Aλ where the slope A is about 130 Hz for 24k MW and 270 Hz for 210k MW. This suggests that characteristic surface vibrations such as suggested by Fukahori and Yamazaki (5) might be involved in the onset of wear. [4] Heat treatment prior to abrasion significantly reduces the rate of wear by a single asperity. [5] Heat treatment also results in slight increase in ridge spacing, λ, by about 10%.

The qualitative features of the observed wear correlate with the MW dependence of the tensile, flexure, and impact strengths of bulk polystyrene. JKR contact mechanics analysis (6) using the parameters of bulk polystyrene show that adhesion between the asperity and substrate can account for a large part of the MW dependence of the initial wear. Recent work by Russell (7) suggests near surface rearrangements of the polymer chains are responsible for the increased wear resistance of the heated samples rather than the removal of trapped solvent

The mechanism of parallel ridge formation is not yet clear but stick-clip phenomena are likely to be an important factor. Preliminary results, some obtained with a new nanotribometer (8), on stick-slip during the formation of single grooves will be presented.

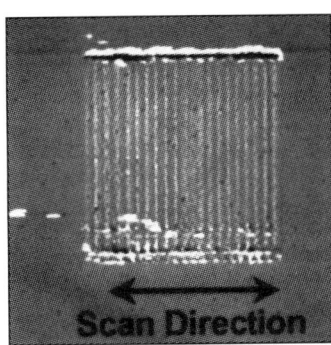

Fig. 1: Abrasion pattern on 24k polystyrene, image is 6.5 μm x 6.5 μm, 40 nm grayscale.

REFERENCES

(1) D. D. Woodland and W. N. Unertl, Wear, Vol. 203-204, page 685, 1997.
(2) O.M. Leung and M.C. Goh, Science, Vol. 255, page 64, 1992.
(3) G.F. Meyers, B. M. DeKoven and J.T. Seitz, Langmuir, Vol. 8, page 2330, 1992.
(4) K. Tanaka, A. Taura, S.R. Ge, A. Takanhara, and T. Kajiyama, Macromoleclues Vol. 29, page 3040, 1996.
(5) Y. Fukahori and H. Yamazaki, Wear, Vol. 178, page 109, 1994.
(6) K.L Johnson, Contact Mechanics, Cambridge University Press, Cambridge, 1985.
(7) T. P. Russell, Univ. of Massachusetts, private communication.
(8) See paper (LO7/C491-920) to be presented at this meeting.

WEAR CHARACTERISTICS OF PTFE COMPOUND CRANKSHAFT SEALS

H SUI, H POHL and U SCHOMBURG
University of the Federal Armed Forces Hamburg, D-22039 Germany

ABSTRACT

Radial lip seals made from PTFE-compounds are utilised more and more frequently for the sealing of crank shafts in the automotive industry as a results of their excellent temperature properties and chemical resistence. PTFE seals have a very long service life compared to elastomeric seals. They finally fail due to wear. Wear is negligible under ideal conditions e.g. if the oil is clean. But in field operation the oil is polluted with dust. Service life of a good for purpose designed seal is mainly determined by its resistance against wear.

Due to wear the seal loses material. This leads to changes in the contact width and stress. Moreover the depth of the grooves decreases so that the dynamic pumping capacity of the seal is reduced. The design engineer wants to estimate how the cross section geometry and stress distribution are continuously changed by loss of mass due to wear. Unfortunately experiments are highly time and cost consuming, so that it is desirable to simulate the increasing wear by progressively reducing the volume of the material.

Previous research efforts on wear of PTFE seals concentrated on the experimental aspect (1). The numerical analysis with finite elements is extremely difficult, as not only wear but also contact, large strains and strong nonlinear material properties have to be taken into account. Because of the nonlinear elasto-visco-plastic material properties of the sealing material the mounting process of the seal has to be followed in the finite element procedure to get a proper initial contact stress distribution.

In this paper a newly developed finite element procedure is presented (2). The reduction of the lip thickness at each contact node is calculated under the assumption that for a certain time period (a small time step) the lip wears out according to the magnitude of the contact stress (maximum wear at maximum contact stress). Thus the finite element mesh has to be updated at all contact nodes in each incremental step. Strains and stresses from the previous run are interpolated onto the new mesh. A new contact solution can then be performed. This process will be repeated as many times as needed until the accumulated time reaches the desired service life.

Numerical results [Fig.1] are in good agreement with long term tests. The contact pressure distribution gets continuously smoother as the contact peaks of the manufactured thread disappear gradually. Moreover the whole contact pressure distribution widens and the lip gets thinner in the contact region.

The numerical and experimental results indicate also under which radial load the seal can withstand insufficient lubrication without much wear.

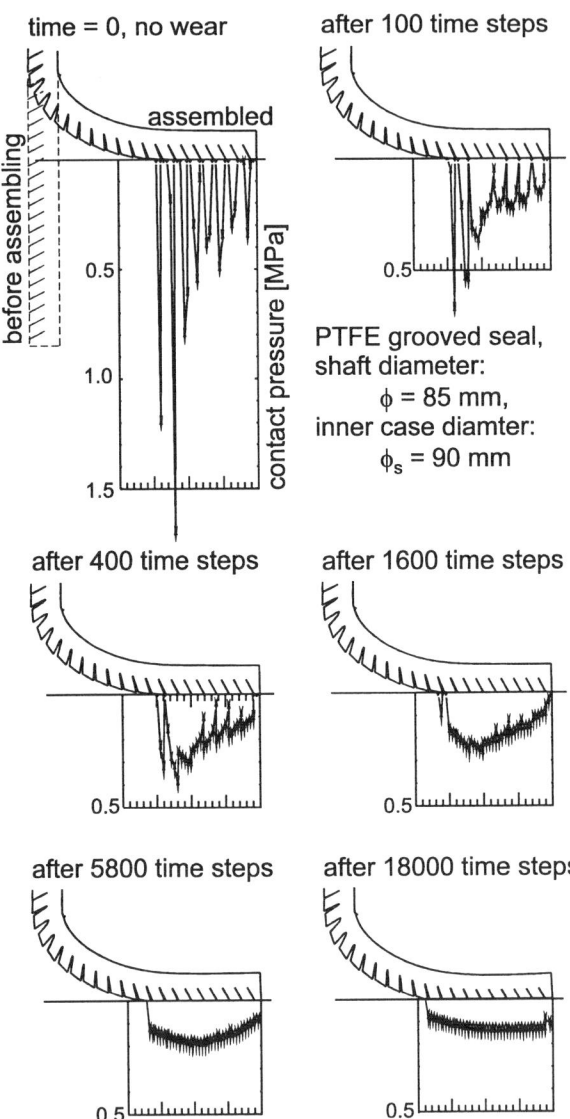

Fig. 1: Contact pressure distribution during wear

REFERENCES

(1) Hoffmann, Chr., Müller, H.K., Haas, W.: „Reibung von PTFE-Wellendichtringen", Konstruktion 48, Springer Verlag, 1996, Germany

(2) Sui, Hai: „Tribologische Untersuchungen der PTFE-Dichtung", unpublished report, Universität der Bundeswehr Hamburg, Sept.1996

FRICTION AND WEAR PROPERTIES OF PTFE-BASED COMPOSITE IN LUBRICATED SLIDING FRICTION AT HIGH SPEED

T.AKAGAKI
Hachinohe National College of Technology, Hachinohe, Aomori 039-11, JAPAN
K.KATO
Tohoku University, Sendai, Miyagi 980-77, JAPAN
M.KAWABATA
Tribotex Incorporation, Obu, Aichi 474, JAPAN

ABSTRACT

Although various kinds of PTFE (polytetrafluoroethylene)-based composites have been developed, there have been only a few works reported on friction and wear behavior of the composite in oil lubricated friction at high sliding velocity(1-2). In this study, the friction and wear properties of the composite rubbed against steel at high sliding velocity were studied.

Experiments were conducted with a block-on-ring wear tester. The block material was PTFE-based composite (1.5 μ mRa, 66HD). It contained the glass fiber (15wt.%) and the molybdenum disulfide (5wt.%). Pure PTFE (0.34 μ mRa, 55HD) and white metal (WJ2, 0.33 μ mRa, 26HV) were also tested for the comparison. The ring material was a forging steel (SF55, 0.18 μ mRa, 189HV). The sliding velocity was varied in the range of 1.0 to 19.0 m/s. The applied load was 294N. The lubricant was a non-additive turbine oil (ISO VG46). The oil was supplied at a flow rate of 23cc/min with a micro-tube pump.

Fig.1 and Fig.2 show the friction and wear behavior for three kinds of block materials, respectively. The block material has no obvious effects on the friction behavior until ~10m/s and low friction is maintained. As the sliding velocity increases over ~10m/s, the coefficient of friction tends to increase, depending on the block material. In WJ2, it increases suddenly and reaches up to ~0.1 at 19m/s. It also fluctuates largely (0.06~0.14). Corresponding to that, the specific wear rate of the block increases from ~10^{-7} to ~10^{-5} (mm^3/Nm). Thus, seizure occurs at the sliding velocity above ~15m/s. In contrast, the coefficient of friction for the composite is 0.02~0.05 at 19m/s. Although the friction increases appreciably at high sliding velocity, the value is relatively small. The specific wear rate is almost constant at (3~4)x10^{-7}, not depending on the sliding velocity. Although the friction behavior of the pure PTFE is almost the same as that of the composite, the specific wear rate is higher than that of the composite and its value is ~2x10^{-6} at 19m/s.

When the steel rubbed against the composite, the ring temperature, measured at 1mm below the ring surface, became above ~100 ℃ at 15 and 19m/s. Therefore, it is expected that the oil film is not useful for preventing the contact between surfaces. Under such a severe condition, the composite maintains relatively low friction and low wear loss. Based on the stribeck curve, the lubrication mode was the mixed lubrication. Thus it is concluded that the composite has the high capacity to maintain the mixed lubrication.

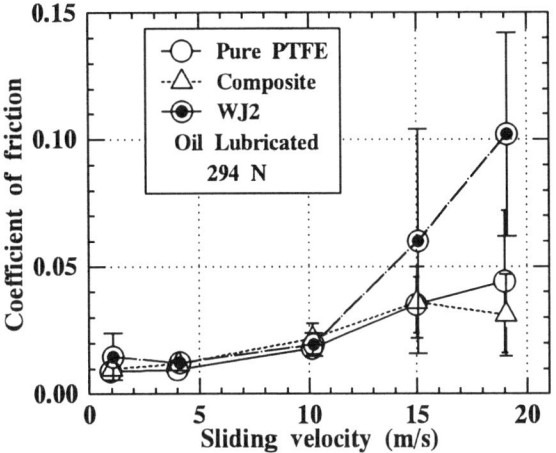

Fig.1: The relationship between the coefficient of friction and the sliding velocity.

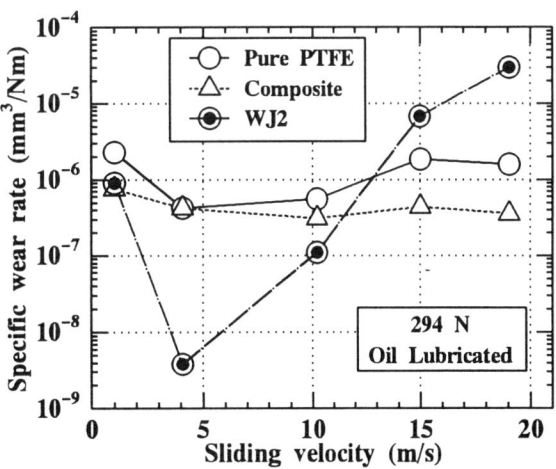

Fig.2: The relationship between the specific wear rate of the block and the sliding velocity.

REFERENCES
(1) S.Uno et al, Proceedings of JAST Tribology Conference, Kanazawa, October 1994, 297-300pp.
(2) T.Akagaki et al, ibid.,Kitakyusyu,1996, 523-525pp.

SLIDING CHARACTERISTICS OF THERMOPLASTIC POLYIMIDE BASED COMPOSITES

MASAKI EGAMI
Bearing Engineering R&D Center, NTN Corporation, 3066, Oyumida, Kuwana, Mie, 511, Japan

ABSTRACT

Polyimide (PI) has excellent heat resistance and self-lubricity. However, its poor moldability prevents applications from spreading. Recently, a thermoplastic polyimide (TPI), which is injection moldable, was developed featuring a flexible segment in its molecular chain (1). In this research, TPI with different additives was studied to see the effect of additives on friction and wear properties when run against aluminum at high temperatures. Also, the wear mechanism of the composite TPI was investigated at different temperatures.

Five kinds of specimens were tested as shown in Table 1. The sliding tests were performed on a journal type test rig. The test specimen is a ring shape, with an inner diameter of 20mm and a width of 5mm. The tests were carried out at a sliding speed of 9m/min, radial load of 34.3N, and temperature of 250℃ for 50h. Aluminum JIS A5052 was used as a mating material. The sliding characteristics of one compound, TPI-4, were studied at different temperatures.

From the test results at 250℃ shown in Fig.1, it was determined that the wear resistance of the composite TPI / PEEK alloy with PTFE was greatly improved by adding oxybenzoyl polyester (OBP) and that the coefficient of friction decreased with the addition of graphite. The IR spectra of the transfer film on the mating aluminum indicated that OBP may have the ability to make a thin homogeneous and lubricious transfer film when mixed with TPI and PTFE. The composite with OBP, graphite and aramid fiber, specimen TPI-4, showed superior wear resistance and a low stable coefficient of friction.

The sliding characteristics of specimen TPI-4 at various temperatures are shown in Fig.2. The coefficient of friction of TPI-4 tended to decrease with increasing temperature. The specific wear rates were considered low throughout the temperature range tested. On the sliding surface of the specimens tested below 200℃, the sliding surface of the specimens had flat areas with worn grooves corresponding to the top of asperities on the mating shaft. The wear debris was small. On the 250℃ test specimen, however, the sliding surface was very rough and contained a crack. The debris from this test was large and flat. From these results, it was assumed that the wear mechanism changed between the temperatures of 200℃ and 250℃. The wear mechanism was simple adhesive wear below 200℃ and at 250℃ the wear mechanism changed to adhesive wear, dominated by fatigue fracture. This change was due to the reduction of the composite's modulus at about 230℃ caused by the glass transition temperature of TPI.

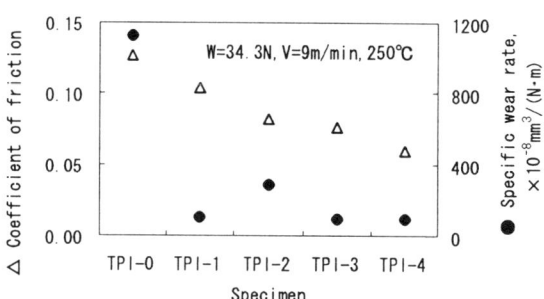

Fig.1 Sliding characteristics of TPI specimens

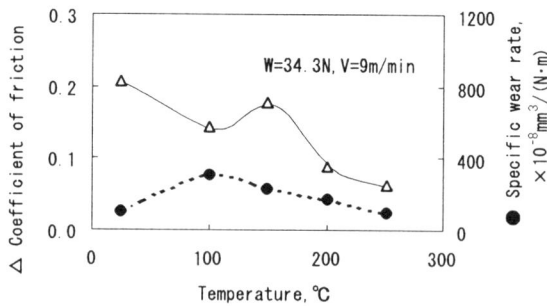

Fig.2 Sliding characteristics of TPI-4

REFERENCES

(1) K. Ito, Machine Design, Vol.10, No.10, page 107, 1994.

Table 1 Composition of the TPI specimens

Unit ; vol%

Specimen No	TPI-0	TPI-1	TPI-2	TPI-3	TPI-4
TPI/PEEK alloy	80	60	65	60	55
PTFE	20	25	25	25	25
Aramid fiber	-	5	5	-	5
Oxybenzoyl polyester	-	10	-	10	10
Graphite	-	-	5	5	5

HYSTERESIS AND TRIBOLOGICAL PROPERTIES OF STYRENE-BUTADIENE RUBBER

DARIUSZ M BIELIŃSKI and LUDOMIR ŚLUSARSKI
Institute of Polymers, Technical University of Łódź, Żeromskiego 116, 90-924 Łódź, POLAND

INTRODUCTION

Friction of rubber is a very complex process, involving both the surface and the bulk phenomena. Two main components of the friction force (F) can be distinguished and namely adhesional (F_A) and hysteretical (F_H) respectively:
$F = F_A + F_H$
Elasticity of the material, on the one hand creating its unique properties, on the other hand limits engineering application of rubber, owing to its poor tribological performance. There is a lot of studies on tribological properties of elastomers, mainly dealing with effects of their modifications. However, data concerning composition of the friction force is still not available. Knowledge of the relationship between hysteretical and adhesional components, together with information on structure of the material, create possibility of modelling and prediction of tribological behaviour of rubber.

EXPERIMENTAL

Influence of the crosslink density and structure on the coefficient of friction and hysteresis loss of styrene-butadiene rubber (SBR) was studied. Dicumyl peroxide (DCP), tetramethylthiuram disulphide (T), sulphur with mercaptobenzothiazole (S_8+MBT), or sulphur with diphenylguanidine (S_8+DPG) systems were used as curing agents. Despite the network structure, they influence the surface energy of vulcanizates. Rubber specimens of various thickness were tribologically tested against a stainless steel counterface. Friction force values, extrapolated to the hypothetical sample of zero thickness, let the adhesional component of friction to be estimated.

RESULTS & DISCUSSION

Tribological performance of the vulcanizates strongly depends on the sliding speed. Higher crosslink density lowers the coefficient of friction of the rubber only if goes together with the high sliding speed. In the case of the "high speed" friction, the hysteresis component is not able to be fully developed due to limited growth of the adhesion forces. Smaller dynamic deformations result in the smaller hysteresis loss leading finally to lower, in comparison to the "low speed" experiments, values of the coefficient of friction. For the "low speed" process the coefficient of friction increases with an increase of the crosslink density. Experiments performed point out the hysteresis loss being responsible for this. The surface energy of the vulcanizates, increasing slightly with an increase of the crosslink density is likely to make the adhesion forces stronger. Together with a longer lasted frictional contact it leads to stronger dynamic deformations accompanying friction. These in turn, combined with better mechanical properties of the cured to higher extent material, results in the higher hysteresis loss and an increase of the coefficient of friction.

The sliding speed value also changes tribological performance of the vulcanizates of different crosslinks structure. The curing systems applied were chosen to study the influence of the sulfidity, changing from C-C (DCP), mono- (T) to polysulfide (DPG) crosslinks, on the coefficient of friction. "Sulphur" crosslinks are longer and more labile than "peroxide" ones, what provides better fatigue properties, hot tear and tensile strength. Their disadvantages in comparison to the short C-C, covalent crosslinks concern poorer resistance to permanent set, creep and stress relaxation properties. It should be also noticed that the influence of the sulfidity of crosslinks on the surface energy of the rubber vulcanizates is far more pronounced than the crosslink density. At the "low speed" conditions the higher the sulfidity of crosslinks the lower the friction. The reversed tendency was obtained in the case of the "high speed" experiments. It can be concluded that the ability to accumulation of energy cannot be fully developed by polysulfide crosslinks, subjected to high frequency (sliding speed) deformations. The lowest coefficient of friction obtained at "semi-equilibrium" conditions for the polysulfide crosslinks (DPG) points on the more important role of the hysteretical than adhesive component in friction of elastomers, as this particular system brings the lowest hysteresis loss and the highest surface energy (adhesion) of the vulcanizates studied.

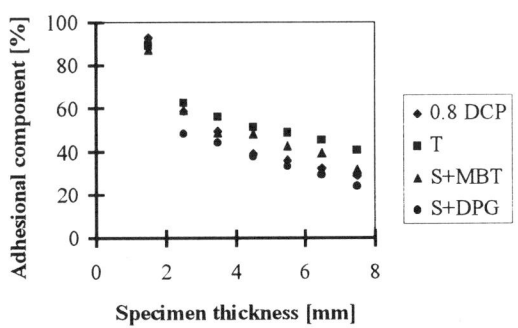

Fig. 1: Example of the friction force composition for the "low speed" experiment (v=0.02 m/s, N=4N)

ACKNOWLEDGEMENT

Authors would like to thank Shell Int. Co. (Branch in Warsaw) for the help in donation of SBR samples.

REGULARITIES OBSERVED AT METALS AND POLYMERS FRICTIONAL HEATING

P N BOGDANOVICH
Metal-Polymer Research Institute named after V.A. Belyi of Belarussian Academy of Science
32a Kirov Str., 246652 Gomel, Republic of Belarus

Theoretical approach to heat processes at actual contact spots has been well developed (1-4). Experimental investigations deal mainly with maximum temperature measurement in the friction zone (4-6). The aim of the work was to obtain experimental data about temperature fields at sapphire-metal and sapphire-polymer friction contact. Test geometry was the flat surface of stationary sapphire disc against side surface of a metal or polymer rotating disc. Sliding velocity varied in the 5-90 m/s range. Heat processes in the friction zone were examined using the optical scanning electron microscopy.

Spots elongated in sliding direction were found to be the sources of heat irradiation. Their size, contour and maximum temperature T_m are dependent on the mating surface mechanical properties, as well as on friction surface microgeometry and P,V-condition. With increasing sliding velocity V and pressure P the spots become more elongated in sliding direction and T_m monotonously rises. When approaching one of the pair members melting temperature the dependences $T_m(P, V)$ obtain a more flattened curve. Such a behaviour is characteristic of all the studied materials except for polyethylene, for which an active growth of T_m is observed under high loads and sliding velocity. The maximum experimental temperatures and those calculated using the Block and Kuhlmann-Wilsdorf equations (1,4) were compared. The comparison have shown that the dependences of P,V-condition are governed by the same law. It should be noted, that the experimental temperature is higher the calculated one and the divergence grows with increasing normal load and activity of the pair materials.

It is demonstrated that during sapphire friction over the silica glass and copper the surface layer of the deformed material can transmit into the local melting state.

The temperature peak of the extremely loaded spots could significantly exceed melting temperature T_{ml} of aluminum and polyethylene. The main causes of inequality $T_m > T_{ml}$ are the thin layer structural transformations and heat induced tribochemical reactions in the actual contact spots, which are accompanied by a considerable heat generation. As far as heat generation is localized on small contact areas tens micrometers in size, temperature gradient across the spot radius is so high (reaching 10^4 degree/mm) that the spot boundaries experience pulse effect of the thermal stress able to exceed the contact stress.

Proceeding from the analysis of heat and mechanical processes in the friction contact, wear can be treated as the result of two simultaneous processes different in their time and spatial level. The first one is the recurrent breakage of contacting asperity peaks due to the low-cyclic fatigue, plastic displacement or momentum thermal decomposition of the material intermittent with prolonged repose. As some model experiments have shown, recurrence of the process can be attributed to the thermoelastic instability of the friction contact. The second process is the material fatigue failure in the vicinity of contact spots. Investigation of damage kinetics have demonstrated that the fatigue crack propagation is the periodic process, where the pulse growth of fatigue cracks is accompanied with prolonged stabilization of their dimensions. The intensity and propagation depth of fatigue processes are predetermined by the contact thermal stress.

REFERENCES

(1) H.Block, JME Proc,Vol.2,1937, 225-235 pp.
(2) F.F.Ling, Surface Mechanics, 1973
(3) J.F. Archard, R.A.Rowntree, Wear, Vol. 128, 1988, 1-17pp.
(4) D.Kuhlmann-Wilsdorf, Materials Science and Engineering, Vol.93,1987, 107-133 pp.
(5) T.F.I.Quinn, W.O. Winer, ASME Journal of Tribology, Vol.109, No 62, 1987, 290-295pp.
(6) S.Chandrasekar, T.N.Farris, B.Bhushan, ASME Journal of Tribology, Vol. 112, No3, 1990, 535-540 pp.

SIMULATION OF THE TIME-DEPENDENT WEAR AND SURFACE ACCUMULATION BEHAVIOR OF PARTICLE-FILLED POLYMER COMPOSITES

THIERRY A. BLANCHET and SUNG WON HAN
Department of Mechanical Engineering, Aeronautical Engineering, and Mechanics
Rensselaer Polytechnic Institute, Troy, NY 12180, United States

ABSTRACT

With regards to composite materials with wear resistance provided by hard particulate fillers it has been demonstrated that an inverse rule-of mixtures can be used to describe composite steady-state wear rate $K_{c_{ss}}$

$$\frac{1}{K_{c_{ss}}} = \frac{x_{fb}}{K_f} + \frac{x_{mb}}{K_m}$$

where x_{fb} and x_{mb} represent bulk volume fractions, while K_f and K_m represent the specific wear rates, of filler and matrix materials respectively. Considerably higher wear rates, however, may be experienced during the initial stages of sliding, before steady-state conditions are attained. This can result from the preferential removal of less wear-resistant matrix from the sliding surface, with remnant filler receding into the underlying matrix, particularly in the case of polymer matrices experiencing cold flow viscoplastic deformation. The resulting development of a subsurface concentration profile, with filler volume fraction x_f varying from x_{fb} in the bulk to x_{fs} at the sliding surface, eventually attains a steady-state. Steady-state surface filler volume fraction $x_{f_{sss}}$ is shown to be

$$x_{f_{sss}} = \frac{1}{(1+(\frac{1}{x_{fb}}-1)K_f^*)\sigma_f^*}$$

where σ_f represents the filler contact pressure. Contact pressures may be non-dimensionalized (denoted by *) by dividing by the nominal composite contact pressure, while wear rates K may be non-dimensionalized by dividing by the specific wear rate of the matrix K_m.

In many cases the initial run-in wear behavior of a composite is as important, if not more important, than the steady-state behavior. In such cases, an appropriate description of time-dependent wear behavior would be particularly useful. The approach of surface filler volume fraction x_{fs} with increasing sliding distance S towards its steady-state value can be approximated as an exponential decay

$$x_{fs}(S^*) = x_{fb} + (x_{f_{sss}} - x_{fb})(1-\exp\{-CS^*\})$$

where sliding distance is non-dimensionalized by multiplying by the product of matrix specific wear rate and nominal contact pressure divided by the filler particle size. Correspondingly composite wear rate would asymptotically approach its steady-state value.

$$K_c^*(S^*) = 1-(1-K_f^*)\sigma_f^*(x_{f_{sss}} - (x_{f_{sss}} - x_{fb})\exp\{-CS^*\})$$

Composite wear volume Q_c as a function of sliding distance that accounts for initial transient run-in contributions in addition to steady-state wear is therefore

$$Q_c^*(S^*) = (1 - (1-K_f^*)\sigma_f^* x_{f_{sss}})S^* + \frac{(1-K_f^*)\sigma_f^*(x_{f_{sss}} - x_{fb})}{C}(1-\exp\{-CS^*\})$$

The parameter C is the decay constant describing the asymptotic approach towards steady-state conditions.

A simulation has been developed to investigate the transient wear behavior of particle-filled composites with randomly distributed filler particles. Program inputs include non-dimensional filler specific wear rate, contact pressure, and bulk volume fraction. Program outputs include time-dependent surface filler volume fraction, composite wear rate, and wear volume. The simulation was run for 100 combinations of input filler specific wear rate, contact pressure, and bulk volume fraction. In each case, resulting steady-state composite wear rate and surface filler volume fraction were in agreement with that predicted theoretically. Also the functionality of the asymptotic approaches to steady-state surface filler volume fraction and composite wear rate predicted by the simulation were indeed approximated by the exponential decay functions previously offered. Curve-fitting output from the numerous simulations of surface filler volume fraction as a function of sliding distance, the dependence of the exponential decay parameter C on filler specific wear rate and contact pressure, and bulk filler volume fraction was approximated.

$$C = -4.3K_f^* x_{fb}\sigma_f^* + 1.15K_f^* x_{fb} + 2.2K_f^* \sigma_f^*$$
$$+ 1.0x_{fb}\sigma_f^* - 0.39K_f^* - 0.575x_{fb} + 0.14\sigma_f^* - 0.0126$$

In the development of this model, it is presumed that time-dependent run-in wear behavior of particle-filled composites is due to the development of a steady-state volume fraction profile. Composites which possess such a volume fraction profile prior to sliding should be resistant to run-in wear effects, instead adopting the lower steady-state rate of wear at the onset of sliding. This effect is demonstrated experimentally. As simulation output also includes the development of volume fraction profile with increasing sliding distance, it may serve as a guide to the engineering of graded concentration profiles in composites that resist run-in wear.

Submitted to ASME Journal of Tribology

PV DIAGRAMS FOR HIGH TEMPERATURE THERMOPLASTIC BEARINGS

S. MARX
Sachsenring Entwicklungsgesellschaft mbH, 08058 Zwickau, Germany
R. JUNGHANS
Lehrstuhl Tribologie, Technical University of Chemnitz-Zwickau, 09107 Chemnitz, Germany

ABSTRACT

The friction and wear behaviour of sliding bearings made from high temperature thermoplastics was investigated to assess their suitability for dry sliding applications and to generate data for engineering calculations. The influence of load, sliding velocity and temperature on friction and wear of plain bearings from modified polyaryletherketone (PEEK 10CF/10PTFE/10Gr; PEEKK 15CF/15PTFE; PEKEKK 30CF) and other materials was investigated. Tests were conducted on a plain bearing test apparatus (1, 2).

PV diagrams are mainly used for representation of operating limits of friction materials or sliding bearings (3). In addition, the diagrams offer the possibility of a simple graphical representation of testing results (wear rates, friction coefficient, temperatures) as data base for engineering calculations and for comparison of testing results with theory of effects of speed and contact pressure to friction coefficient and wear rate.

In general, the operating performance is significantly influenced by the operating conditions (P, V, T) and the precise construction of the tribological system (e.g. bearing width, diameter, wall thickness, housing). The general problems of calculating thermoplastic sliding bearings are the very complex and quantitatively mostly unknown interactions in the sliding contact.

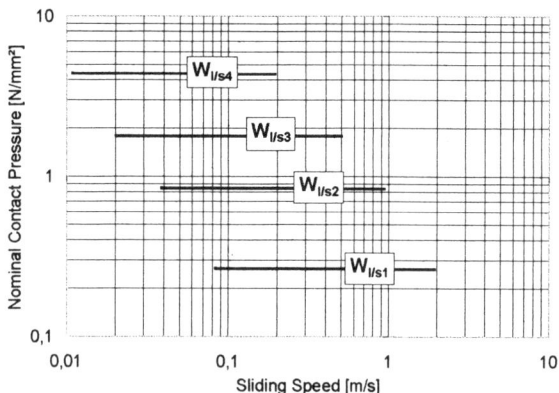

Fig. 1: Graphs of linear wear rate $W_{l/s} = f(P, V)$ (scheme of ideal notion; $W_{l/s1} < W_{l/s2} < W_{l/s3} < W_{l/s4}$)

The comparatively high friction coefficient under dry sliding conditions and the low heat conductivity of polymers can lead to a remarkable generation of heat in the contact area and to very critical temperatures. The materials tested provide reliable operation of dry sliding bearings to temperatures over 250 °C.

The specific wear rate (unit: mm³/Nm) of the investicated bearings cannot be seen to be constant, as widely-used simplified theories imply (3). In some cases wear rate even drops with increasing contact pressure, speed or temperature, which is connected with a change in the dominant wear mechanism. For the graphs of linear wear rate (unit: µm/km) there is a significant difference in shape and orientation between ideal notion (Fig. 1; no influence of speed) and the testing results (Fig. 2).

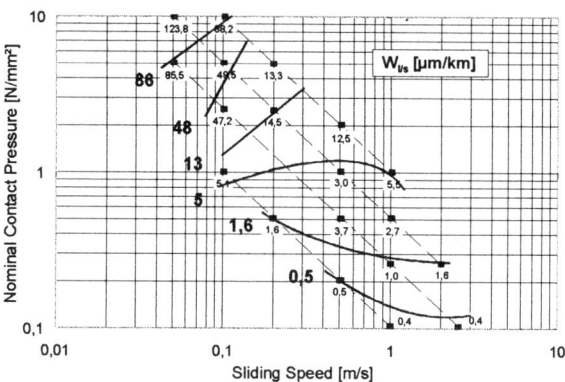

Fig. 2: Graphs of linear wear rate $W_{l/s} = f(P, V)$ for bearings of PEEK 10CF/10PTFE/10Gr (counterpart 100Cr6, hardened, polished; R_a = 0.18 µm; bearing B = D = 30 mm, wall thickness 5 mm; C* = 16 °/oo)

Higher temperatures support the generation of smooth transfer layers on the shaft surface. For bearings of PEEK 10CF/10PTFE/10Gr, for example, that leads to higher friction coefficients as a result of increased adhesion, to higher wear rates at contact pressures below 1 MPa but to lower wear rates at contact pressures above. Thus, a step-like wear rise as noticed for unheated tests in the range of these pressure is not detectable.

The representation of testing results in PV diagrams offers a good clearness and the advantage of a combined deference to the main influence factors. It corresponds to the complex character of the system in question, which does not allow direct analytic calculation of friction and wear from material properties and working parameters.

REFERENCES

(1) S. Marx, R. Junghans, Wear 193 (1996) 253-260 pp.
(2) S. Marx, Doctoral Thesis, 1997 (in German)
(3) J. K. Lancaster, Tribology International 6 (1973), 219-251 pp.

SURFACE FATIGUE OF ENGINEERING POLYMERS IN ROLLING CONTACT

T. A. STOLARSKI AND S. M. HOSSEINI
Department of Mechanical Engineering, Brunel University, Uxbridge, Middlesex, UB8 3PH, UK
SHOGO TOBE
Department of Mechanical Engineering, Ashikaga Institute of Technology, 268 Ohmaecho, Ashikagashi, Japan

ABSTRACT

Due to their ease of manufacture, reduced weight and low cost of processing, engineering polymers are increasingly used in rolling contact applications where the predominant mode of loading leads to surface fatigue. Although polymer rolling contacts are used in over 1000 applications in more than 100 industrial fields, fundamental understanding of their failure mechanism is insufficient and currently available information about their performance does not allow for their rational design (1). The study summarised here is a part of a wider programme devoted to systematic investigations leading to a better understanding of surface fatigue of polymers and factors governing their performance in rolling contact (2).

Three polymers, namely polymethylmethacrylate (PMMA), acetal (A) and polycarbonate (PC), were selected for studies of fatigue crack initiation and propagation due to surface cyclic loading. Two lubricating liquids were used; a base oil (Talpa-20) and a brake fluid. Contact stresses resulting from the contact configuration of a ball-on-plate ranged from 1.5 GPa to 2.5 GPa.

Test apparatus used was based on a well know four ball machine. In this apparatus, three steel balls with diameter of 12.7 mm were in loaded contact with a flat polymer disc of 30 mm in diameter and 5 mm thick. The balls, driven by the upper ball attached to a spindle of the apparatus, were housed in a stainless steel cup and were free to roll over the face of the disc located at the bottom of the cup. During a typical test, the cup was filled with one of the lubricating liquids used. Tests were interrupted at regular intervals for the inspection of contact path produced by the motion of the lower balls. Depending on the load on the contact, the inspections were carried out every 5 min at the highest loads applied and every 20 min at the lighter loads. Such frequent inspections were necessary in order to capture early signs of surface distress and the beginning of surface damage initiation. At this stage the inspections of the contact region were only cursory and the thorough post-test examinations using scanning electron microscopy (SEM) and atomic force microscope (AFM) were performed later.

The time to failure as a function of applied load (surface fatigue strength) strongly depends on the type of lubricating liquid and is different for each polymer studied. In case of brittle polymers (PMMA and PC) the process leading to failure can be divided into three distinct stages: crack initiation, stable growth of cracks and final fast agglomeration of cracks leading to damage on a macro-scale. On the other hand, acetal discs did not show any signs of surface fatigue and their performance was excellent.

Figure 1 is an example of performance of PMMA discs given in the form of a S-N curve. Both PMMA and PC were found to fail in a manner similar to surface fatigue of ferrous materials. A typical surface damage of a brittle polymer is shown in Fig.2.

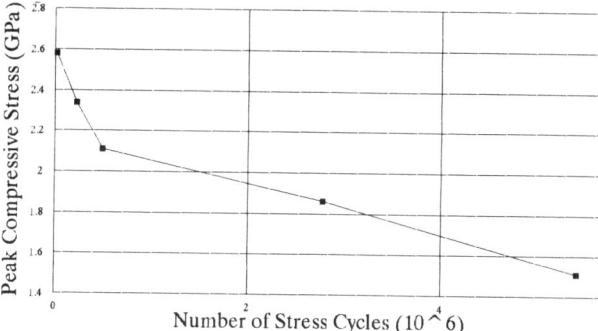

Fig.1: S-N curve for PMMA tested in base oil

Thermoplastic polymers, represented in the study reported here by acetal, do not normally fail through creation of cracks and their subsequent agglomeration.

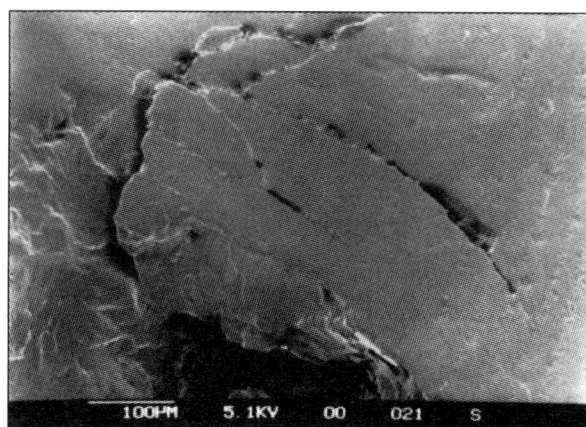

Fig.2: SEM image of PMMA disc damage

REFERENCES

(1) C.C.Lawrence, T.A.Stolarski, Wear, vol.132, 183-191, 1989.
(2) S.M.Hosseini, T.A.Stolarski, J.Appl.Polymer Sci., vol.56, 311-316, 1995.

Submitted to *Wear*

FRETTING BEHAVIOUR OF PMMA UNDER LINEAR AND SPIN MOTIONS

B.J. BRISCOE and T.C. LINDLEY
Imperial College of Sciences, Technology and Medicine, Prince Consort Road, London, SW7 2BJ, UK
A. CHATEAUMINOIS
Département Matériaux - Mécanique Physique, UMR IFoS 5621, Ecole Centrale de Lyon, 69131 Ecully, France

ABSTRACT

Small amplitude vibrational contacts referred to as 'fretting' are known to induce complex contact zone kinematics. At low displacement amplitudes, the contact area can generally be divided into two domains : a central stick domain without any sliding and an external corona where micro-slip takes place. For higher displacement amplitudes, a gross slip condition is reached and micro-slip prevails over the whole contact zone. This latter condition is often associated with particle detachment and wear. According to the third body concept [1], the wear processes are strongly dependent on the contact zone kinematics which control the ability of detached particles to be detrapped from the contact. These effects have mostly been investigated under linear fretting motions. On the other hand, few attention has been paid to more complex contact zone kinematics combining for example linear motions and load axis spin.

The objective of this paper was to investigate the wear processes occurring under such loading conditions in the case of a fretting contact between a PMMA flat and Steel ball. A specific fretting device has been designed which allows to impose an oscillating rotation of the ball around its load axis. By varying the angle between the axis of rotation and the surface of the PMMA specimen, it was possible to achieve contact conditions ranging from pure torsion to stationary rolling (Fig.1). In addition, some tests were carried out under linear sliding conditions using a modified tension-compression hydraulic machine described elsewhere [2]. The fretting wear behaviour of the PMMA specimens was analysed by means of SEM and laser profilometry. No significant degradation of the Steel ball was found to occur during the tests.

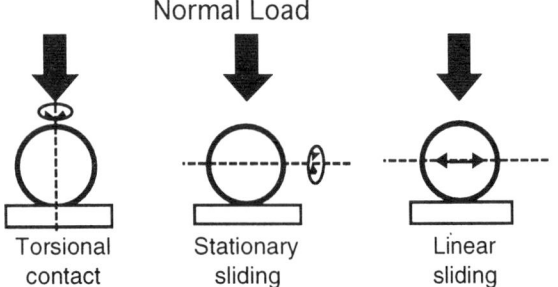

Fig.1 Contact loading conditions

The test conditions were selected in order to achieve gross sliding conditions. The displacement amplitude (linear sliding conditions) or the twist angle (torsion and rolling conditions) ensured that the same integrated displacement was achieved in the contact whatever the test configuration.

Strong differences were noted in the wear scars as a function of the contact zone kinematics. Under stationary rolling, the progressive accumulation and compaction of wear debris in the middle of the contact area led to the formation of a roll perpendicular to the sliding directions. The formation of this roll was attributed to the reduced mobility of detached particles in the centre of the contact area, as a result of the high values of the normal pressure. Wear particles were detrapped mostly from the ends of the sheared roll. A similar mechanism of roll formation was noted under linear sliding conditions. The number and the size of the rolls were correlated to the displacement amplitude. In addition to particle detrappment from the ends of the rolls, a significant amount of wear debris was found to escape from the moving edges of the contact area, by virtue of the decreased overlap ratio of the counter-bodies. On the other hand, torsional contacts resulted in a radial arrangement of the rolls with only a limited detrappment of detached particles.

These results emphasised the strong influence of the contact zone kinematics on the rheology of third body and the associated wear processes.

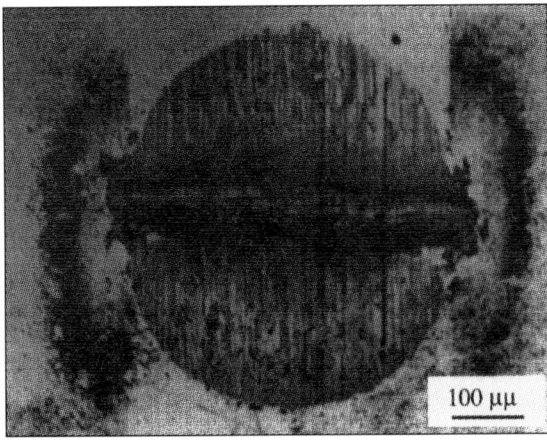

fig.2 : Wear scar under stationary rolling conditions

REFERENCES
(1) Godet, M., Wear, Vol. 100, 1984, pp. 437-452
(2) Krichen, A., Kharrat. M, Chateauminois, A., Tribology Int, Vol. 29, N°7, 1996, pp. 615-624

TRIBOCHEMICAL PROCESSES IN POLYMERS.

A.KRASNOV, I.GRIBOVA, V.MIT,
A.N. Nesmeyanov Inst. of Organo-Element Compounds of Rus. Ac. of Sciences, Vavilova 28, Moscow,Rus.
M.BRONOVETS,
Interdisciplinary Scientific Tribology Council of Russian Academy of Sciences, Moscow, Russia.
V.ADERIKHA
Institute of Mechanics of Metall-Polymeric Systems, Kirova str. 32 "A", Gomel, Belarus.

Imaginations about primary tribochemical processes mainly have been obtained with the help of the special friction machine, placed in the mass-spectrometer camera. The developed design of the direction unit allows to use the mass-spektrometer system of the direct imput of a sample and to measure the counterbody's temperature. The developed setting makes it possible to place the sample as far as 5 mm from ionisation box,that distinctly shorten the distance to the ionisation zone for the escaping products.

The given paper considers processes, taking place during friction of amorphous aromatic polymers of similar molecular weight : polycarbonate(2), polyphenilenoxide(1), new amorphous polyetherketone(3),[1], polyarilate(4) and thermostable polyhetero-arylenes(5-16): polyimides(14-16), polyphenylquinoxalines(5-11), polyoxydiazoles(12-13). Were also examined aliphatic polymers such as polyolephynes, polyamides and polyacetales. During polymer friction of both aromatic and alyphatic types it has been found that tribodestruction has two main directions. One of them is the direction of side substitutes (diphenylation and dehydrogenation reactions). The following transformation results in formation of branched and cross-linked structure. Another one - in main polymer chain (with opening of heterocycle and C-C bonds) with formation of the low molecular weight products of destruction.

While carrying on the examination of polymer friction in the mass-spectrometer camera, it has been shown that non-cyclizied and thermically weak fragments of polymer chain link suffer most active degradation.

Under the condition of friction it was obserwed that some synthetic processes were not complited. They are: cyclization of polyimides, forming of azometine cross links in the copolymers of polyphenylquinoxaline and reaction at the edge groups.

Tribochemical processes easily carrying out at not thermostable "defect" groups leads to sharp increasing of not stable friction character and polymer wear. In that this effect considerably differs from thermostability, where smallest quantities of "defect" fragments do not considerably effect the destruction temperature.

The presence of gaseous and low molecular weight products in the zone of interaction with the steel counterbody conferms the fact that the examined polymers do not undergo the real "dry" friction.

It was shown by the number of investigations that low molecular weight products of polymers tribodestruction can play a role of "tribochemical" lubricant, lowering friction coefficient. As those ones can be used both liquid olygomere products, presence of which in the friction zone have been shown not once, and gaseous ones.

It was also discovered that the tribodestruction products composition gets change with the increasing of thermostability and rigidity of polymers. That means predomination of the "small" products of sidechain substituents splitting off. Their number is much bigger than that of fragments which characterize the main polymer chain decomposition. This leads to the cross-linked structure formation at surface layers wich is thermostable and containg deeply destroyed portion as well. As a result thermostable cyclochained polymers exibit lower wear than less thermostable heterochain analogs. The obtained results make it possible to link wear, in general, with 2 characteristics of investigated polymers such as thermostability (T_d, K) and cohesion energy densites (δ^2, kal/cm^3) .

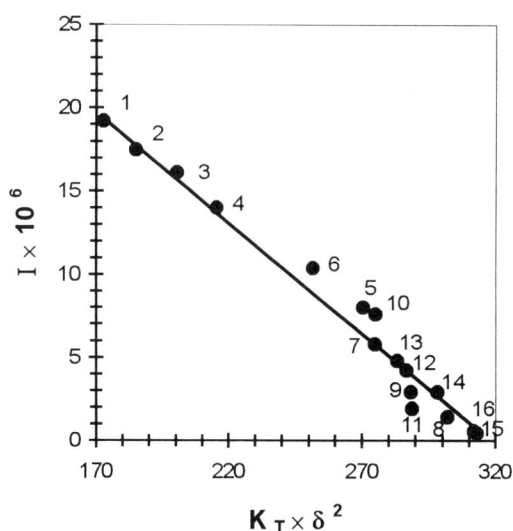

REFERENCES [1] A.P.Krasnov, B.S. Liosnow ,G.I. Gureewa, S.N. Salaskin, W.W.Shaposhnikowa, Polymer Science 1996,A,v.38,N12,1956-1961pp.

Interrelation between abrasion and scratch hardness for PMMA

B. J. Briscoe, E. Pelillo and S. K. Sinha
Department of Chemical Engineering & Chemical Technology,
Imperial College, London SW7 2BY, UK.

Abstract

The modelling of the abrasion of polymers had been a preoccupation of many tribologists. Material removal in polymers, in contrast to that of metals, during abrasive contact presents a number of problems due to many factors and the way in which these factors affect the process of wear. Some of the phenomena occuring during the abrasive wear process are visco-elastic flow, ductile-brittle transition, fatigue etc. which are influenced by the asperity geometry, attack angle, severity of deformation, temperature and so on. Hence, while modelling or simulating abrasive wear of polymers, these factors must be considered. The scratch hardness experiment is one example where some of these conditions can be simulated in part for individual asperity contacts.

In the present paper an attempt is made to model the abrasive wear of PMMA using scratch hardness data. A number of studies have been carried out in the past to relate scratching with abrasive wear for metals however such data are not widely available for polymers [1]. In order to simulate the abrasive wear, scratches are produced both on the same track (unidirectional multiple pass) and adjacent to each other such that they interact with each other during scratching as may be experienced in an actual wear process. Abrasive wear data are also produced by sliding a polymer pin on abrasive papers of various grades to obtain results for different mean asperity angles. The results show that the scratch hardness data can be correlated with the wear data if the contact conditions in the two cases are simulated.

The material chosen for this study was PMMA (ICI, UK). The scratch test (single pass) was performed on a simple lever type scratch machine using different cone angles as indenters. The wear data were collected on a pin-ondisc machine using counterface discs of different roughnesses.

Initial results are plotted in Figure 1 and Figure 2. Figure 1 shows wear rate for scratching using single pass for different attack angles (90-α; where α is the included semi cone angle). The wear rate increases linearly as the attack angle is increased. This is expected as an increase in the attack angle changes the material removal process from plastic deformation to cutting mode. In contrast to the magnitude of wear rates for single pass scratching the wear rate shown in Figure 2 for abrasive wear on pin-on-disc machine is low by an order of magnitude. There are many reasons for this. First, the single pass scratching process does not represent the actual wear process in a pin-on-disc machine where sliding is carried out repeatedly on the same track. The other important points are the contact pressure influence and the distinction between single point abrasion and abrasive wear. A higher load in a single point scratching test increases the depth of indentation which will influence the deformation mechanism. On the other hand an increase in the normal load in surface contact could increase the contact area withough affecting the depth of penetration of individual asperities into the polymer surface. Currently these points are under investigation and will be reported in the congress.

References

1. Evans, P. D. The hardness and abrasion of polymers, PhD Thesis

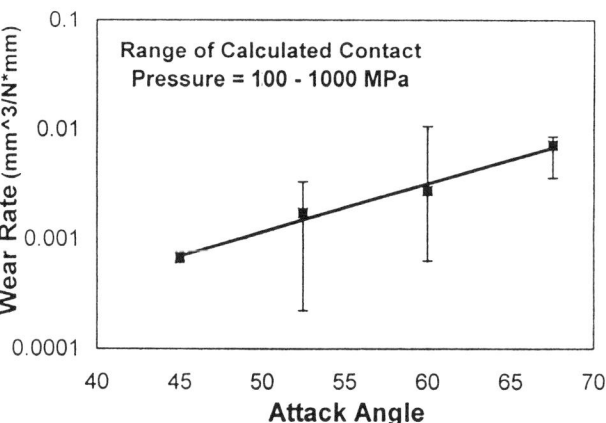

Figure 1 Wear rate as a function of attack angle for single pass scratching of PMMA surface by steel conical indenters.

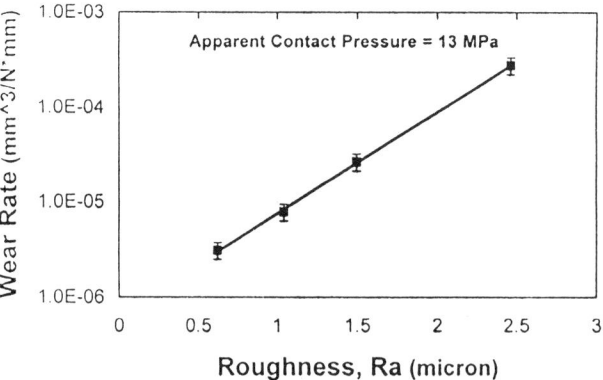

Figure 2 Wear rate for PMMA measured as a function of surface roughness of brass counterface on a pin-on-disc machine at a mean contact pressure of 13 MPa.

THE INFLUENCE OF CUTTING EDGE ROUGHNESS ON FLANK WEAR OF SINTERED DIAMOND TOOLS

KATSUHIKO OZAKI
Process Technology Research Laboratory, Kobe steel, LTD, Nishi-ku, Kobe 651-22, JAPAN
KOJI KATO
School of Mechanical Engineering, Tohoku University, Aramaki-Aza-Aoba, Sendai 980, JAPAN

ABSTRACT

Sintered diamond has large wear resistance, and it has been used for cutting tools(1),(2). The cutting process is influenced by the roughness of the cutting edge. However, no results show how roughness of the cutting edge of sintered diamond tools affects the cutting performance of that. Therefore, this paper focuses on the influence of cutting edge roughness of sintered diamond tools on flank wear. Sintered diamond tools of grain sizes 1 ~2 um were used to cut Al-16%Si-Alloy. The surface roughness of the workpiece, the cutting force and the flank wear were measured. Cutting tests were carried out under conditions of continuous dry cutting.

The flank wear of the roughest cutting edge tool was the lowest, and the wear of the smoothest edge tool was greater than that of any other tool. Therefore, the flank wear was in inverse proportion to the cutting edge roughness as in Fig 1. Flank wear is caused by contact and the slip process between the flank face and the finished surface of the workpiece. The amount of flank wear depends on the sliding distance, the load and the coefficient of wear(3). In this case, the sliding distance was the same for each tool. The pressure at the flank face of the roughest tool is bigger than that of the smoothest tool. In this case, the smoothest edged tool had the broadest contact area, because the surface roughness of the workpiece cut by the smoothest edged tool was smooth, while the plastic work of the finished surface at the flank face of the smoothest edgee tool was bigger than any other as in Fig 2. The thermal energy of the smoothest edged tool was largest. Therefore, the wear rate of the smoothest edged tool was the highest.

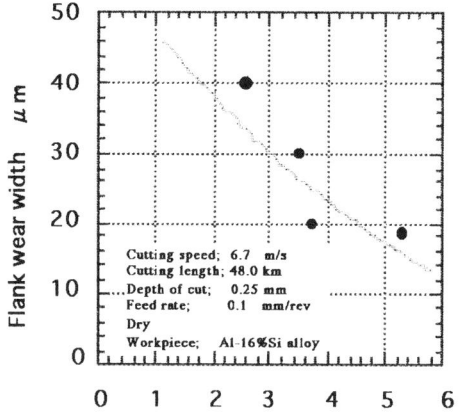

Fig 1 Relation between roughness of cutting edge and flank wear

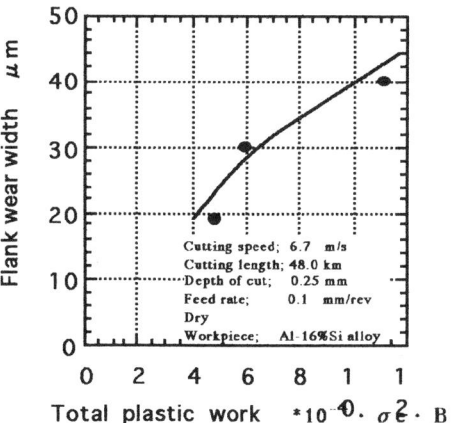

Fig 2. Relation between plastic work and flank wear

REFERENCES

(1) Ramulu, M., Machining of Graphit/Epoxy Composite Materials With PCD Tools, Trans. of ASME, 113, Oct. (1991) 430-437.
(2) Watanabe, M. and Toshikawa, M., Cutting of Hypereutectic Al-24%Si alloy by Diamond Coated Cutting Tools, JSPE, 56, 12 (1990) 99-104.
(3) Crompton, D., Hirst, W., Howse, M.G., The wear of diamond, Proc. R. Soc.Lond. A. 333 (1973) 435-454.

RULER™ AND USED ENGINE OIL ANALYSIS PROGRAM

ADRIAN JEFFERIES
Castrol International, Technology Centre, Whitchurch Hill, Pangbourne, Reading,
RG8 7QR, UK
JO AMEYE
Socomer Scientific Instrument Dpt, Nieuwbrugstraat 73, B-1830 Machelen, Belgium

ABSTRACT

This paper presents the results of research to evaluate the Remaining Useful Life Evaluation Routine (RULER™) for engine oils. The RULER™ studied in this paper is based on a voltammetric method (1). The RULER™ instrument uses voltammetric techniques to apply a controlled voltage ramp through the electrode inserted into the diluted oil sample. As the voltage potential increases, the antioxidant additives become chemically excited This causes a current to flow which peaks at the oxidation potential of the antioxidant. The height of the peak is related to the concentration of the additive in the solution.

A lubricant has many functions such as protecting engines from wear, maintaining cleanliness and acting as a coolant. The components of a lubricant are predominantly base fluid(s) containing specifically designed additives. It is these additives which provide the lubricants' functions and also protect the base fluid from degradation, as the fluid transports the additives around the engine. The majority of base fluids used in modern diesel and gasoline engines are hydrocarbon based and are prone to oxidative degradation.(2)

The Remaining Useful Life (RUL) is the length of engine/equipment operating time from the time a lubricant is sampled, until large changes in the lubricant properties occur, at which point the lubricant cannot function properly.

In a first step the precision of the voltammetric analysis was tested. From the standard deviation of this mean the precision of the measurement was found to be 2.5% RSD. Many formulated oils were tested, showing a characteristic RULER™ Voltammogram. Typically two additive peaks are detected (for diesel engine oils) usually one from the ZDDP, and one from a phenol/aminic antioxidant. The data allows the user to monitor the depletion of both additives. If the total effective antioxidant capacity depleted rapidly to a low level this could suggest the oil is inappropriate for the engines particular duty cycle. For this study the total RULER™ area was used to compare with other standard used oil analytical techniques.

A series of samples from different equipment types/manufacturers, both engine (Mercedes Benz EURO 2 test, BMW Turbo Diesel test) (fig 1) and field trials, were analysed using the RULER™ instrument. The antioxidant capacity of each used engine oils was determined and compared to the new oil.

The results showed that the antioxidant capacity depleted with time, but that the rate of depletion varied depending on operating conditions. The "RULER™" results were compared to other standard oxidation and physical tests, Differential Scanning Calorimetry (D.S.C)., Fourier Transform Infra Red (FTIR), Total Acid Number (TAN), Total Base Number (TBN) and Viscosity to determine any correlation between the techniques.

It was also found that different oils gave characteristic RULER™ traces, which could be used to identify erroneous top-ups, or mislabelling of samples. It was concluded that the RULER™ instrument can determine rapidly the effective antioxidant concentration in a used oil. When compared to the original oil, a plot of additive depletion versus oil age can be determined, allowing an estimate of the oil's condition to be made.

The RULER™ instrument is a useful monitoring system for engine oils, due to its rapid measurement, accuracy, low sample volume, and it's freedom from interferences such as soot.

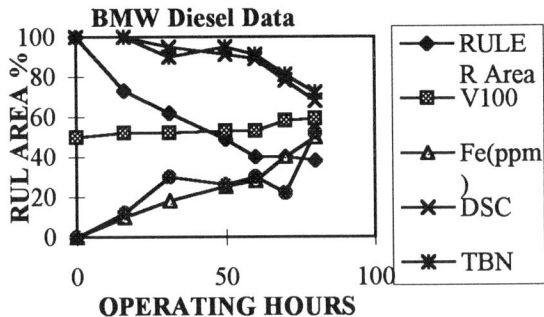

Fig. 1 : RULER analysis on BMW Diesel Test

REFERENCES

(1). Kauffman, R.E., STLE CRC Handbook of Lubrication and Tribology, (Vol III), Booser, E.R. ed., CRC Press, Boca Raton, FL, pp 89.(1994)
(2). Mortier and Orszulik: "Chemistry and Technology of Lubricants", Blackie and Son, 1992, pp 94.

Three Dimensional Characterization of Surfaces for Sheet Metal Forming

M. Pfestorf and U. Engel and M. Geiger
Institute of Manufacturing Technology, Egerlandstr. 11, 91058 Erlangen, Germany

ABSTRACT

In the recent years a precise characterization of the surface' topography especially in sheet metal forming became more and more important. One of the reasons is the continously growing sophistication of the forming process facilitated not only by specific and closely tolerated properties of the sheet but by specific surface properties as well, given by the topography. Today quite different technologies for surface texturing are available yielding a broad diversity of topographies in practice. For that view, it is evident that the roughness characterization of technical surfaces based on a simple 2d description cannot be sufficient anymore. Additionally, some functional properties of the topography like the tribological behaviour in forming processes are almost impossible to be described using conventional 2d parameters. In response to the great importance of friction in forming processes the development of new and more intelligent 3d surface parameters is thus essential.

New functional parameters can be derived from a mechanical rheological model which is about to be developed in order to understand and to describe the complex interaction between tool and workpiece taking place at the interface. Within this model, the load on a surface is transmitted by three totally different kinds of bearing ratios. These are the solid contact area as well as the static and dynamic lubricant pockets. The ratio of solid contact corresponds to the relative amount of the real contact area. The dynamic lubricant pockets represent those regions of lubrication where during the forming process the lubricant can be squeezed out of the loaded area. In these regions the load can only be transmitted by a hydrodynamic pressure. In contrast to the dynamic lubricant pockets, the static lubricant pockets have no connection to the boundary of the loaded area. Thus the lubricant is trapped in these pockets and a hydrostatic pressure can be built up. To characterize the topography following the idea of the model, 3d surface parameters have to be defined. The relative amount of solid contact corresponds to the *material area ratio*. The 3d surface parameters for the static and dynamic lubricant pockets are the closed and the open void area ratio respectively. Refering to the model, the *closed void area* are those regions, which have no connection to the boundary of the evaluation area. In contrast, the *open void areas* are those regions which have a connection to the boundary of the evaluation area.

Fig. 1 shows the results of a calculation of the surface parameters for a single crater from a laser-textured surface with separated lasertex craters. The material area ratio corresponds to the Abbott curve which is well known from the 2d-surface analysis. The open and closed void area ratio also result as a function of the penetration of the surface. The area ratios are calculated within planes parallel to the mean plane. The first plane at a penetration of 0 % touches the highest asperity of the topography e.g. the last plane touches the deepest valley. On the first plane, at a penetration of 0%, only open void areas are present. With further penetration of the surface, the open void area ratio decreases as the material area ratio increases. In the third surface plot a section of the crater can clearly be seen. As there is no connection to the boundary of the evalation area, the crater is characterized by the closed void area. With further penetration, the open void areas disappear i.e. there are only closed void area ratio and material area ratio. At the end, 100 % material area ratio remains. At least, two significant surface parameters can be derived from this diagram. First of all, this is the maximum of the closed void area ratio α_{clm}. The other one is the closed void volume V_{cl} which can be calculated by intergration of the closed void area curve. It can be proved, that the parameters are within very small variations at an evaluation area of 9 mm^2 [1].

As it can be shown by the results of a strip drawing test as well as by the results of a ring compression test combined with torsion, both parameters are very suited to characterize the tribological properties. Additionally the superiority of the 3d surface parameters to the 2d parameters can be shown by the investigation of special effects as e.g. the mechanism of microfilm lubrication.

ACKNOWLEDGEMENTS

Special thanks are given to the Deutsche Forschungsgemeinschaft (DFG) for financial support of the research project.

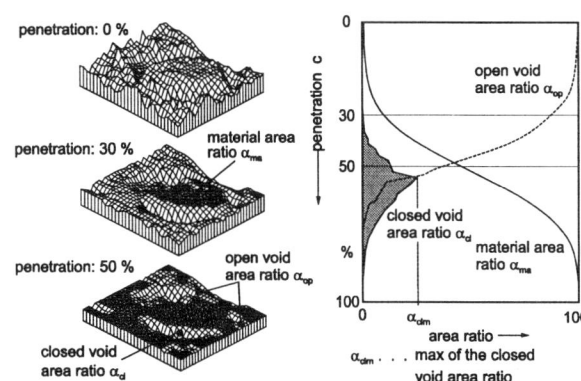

Fig. 1: 3d surface parameters at a lasertex crater

REFERENCES

(1) M. Pfestorf and U. Engel and M. Geiger, Proceedings of the 7th International Conference on Properties and Metrologies of Engineering Surfaces, 2-4April, Göteborg, 1997, to be published in: Journ. Engineer. Manufact. Tech

Submitted to *Wear*

DETERMINATION OF THE MINIMUM QUANTITY OF LUBRICANT FOR SHEET METAL FORMING

D SCHMOECKEL and J STAEVES
Institute for Production Technology and Forming Machines Technical University Darmstadt,
Petersenstraße 30, D-64287 Darmstadt, Germany

ABSTRACT

Component quality and process safety in sheet metal forming are decisively influenced by tribological fringe conditions. Besides sheet material and tool material, the type and quantity of lubricant applied to the sheet are of major importance. In cases of doubt, excessive quantities of lubricant are frequently used. Both for economic and ecological reasons, a method is therefore needed which provides simple means for assessment of the optimum quantity of lubricant.

A focal point of the studies is the topography of the sheet material, as this may be anticipated to have a major impact on the lubricant quantity required. Since two-dimensional surface parameters can describe the sheet surface inadequately only, a corporate research project for development of 3D parameters adapted to sheet metal forming was started (1,2).

As is the case in two-dimensional measuring, also the values of 3D indexes are decisively influenced by filtering. Sheet materials have different waviness depending on the manufacturing processes. For assessment of the amount of lubricant, waviness must be filtered out so far as to make the measuring area correspond to the material surface in tribological contact. The filters used in 2D-metrology have the disadvantage of being oriented at the mean profile, but not at the tribological contact surface. Comparison of the filtering methods was made on the basis of the void volume which was defined to be 100 % for an artificial surface without waviness.

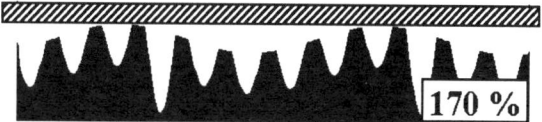

Fig. 1: Model surface with waviness

The void volume calculated for the surface including waviness is too high by 70 %.

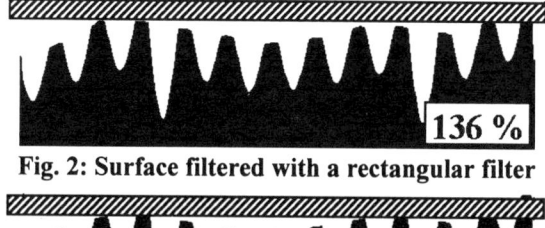

Fig. 2: Surface filtered with a rectangular filter

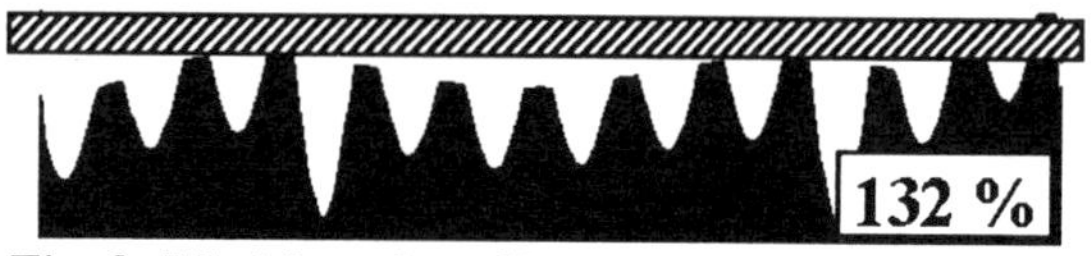

Fig. 3: Model surface filtered with a double filter

Both the rectangular filter and the double filter with striation suppression (DIN4776) just calculate part of the waviness out of the profile. In contrast with this, using the ball filter results in 103 % and thus approximately the actual void volume.

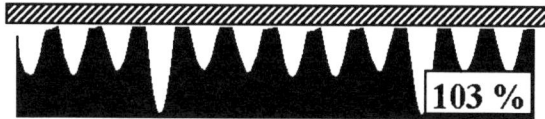

Fig. 4: Model surface filtered with a ball filter

In order to adapt the ball filter to the requirements of sheet metal forming, a modified ball filter was developed by PtU which permits a defined number of points to penetrate into the ball.

Fig. 5: Principle of the modified ball filter

Apart from suppressing isolated peaks, this permits to achieve that filtering is oriented at the elastically and plastically deformed contact surface.

In order to assess the influence of the filtering methods, a representative cross section of the sheet materials currently used by the automobile industry for producing car body parts were evaluated on the friction testing equipment. The results clearly show the advantage of the modified ball filter.

References

(1) Wagner, S., 1996, 3D-Beschreibung der Oberflächenstrukturen von Feinblechen, Dr.-Ing. Dissertation, Technical University Stuttgart, Germany
(2) Geiger, M., Engel, U., Pfestorf, M., 1997, New developments for the qualification of technical surfaces in forming processes, Annals of the CIRP, 46/1

Further Information: www.ptu.th-darmstadt.de

INFLUENCE OF DEFORMATION ON THE LUBRICATION REGIMES IN SHEET METAL FORMING

D.J. SCHIPPER, R. ter HAAR, and E.G. de VRIES
University of Twente, P.O.Box 217, 7500 AE Enschede, NL
H. VEGTER
Hoogovens IJmuiden B.V., P.O.Box 10.000, 1970 CA IJmuiden, NL
N.L.J.M. BROEKHOF
Quaker Chemical Europe B.V., P.O.Box 39, 1420 AA Uithoorn, NL

INTRODUCTION

In sheet metal forming friction plays an important role with respect to deformability and product quality. During sheet metal forming, for instance deep drawing, the material is subjected to large deformations. The influence of the deformation mode and the degree of deformation of the sheet on the lubrication behaviour as present in sheet metal forming is investigated.

FRICTION TESTER

For this purpose a new friction tester is developed (1) consisting out of a deformation unit (tensile tester) and a friction measurement unit, schematically shown in Fig. 1. With the deformation unit sheet material can be deformed (mode and degree) in a controlled way as present in sheet metal forming. During an friction experiment the friction measurement unit moves along the deforming strip measuring simultaneously friction and normal force using piezo-electric force transducers. The specimen holders are connected to the supports by using elastic joints to avoid static friction components in the overall friction signal. With this friction tester it is possible to measure friction separately from the deformation forces.

EXPERIMENTS

The measured friction is represented in generalized Stribeck curves, see Fig. 2. The lubrication number L is defined by $\eta v/(pR)$ in which η is the viscosity, v is the velocity, p is the nominal pressure and R the combined surface roughness. On the basis of these curves, A) the influence of the deformation mode (elastic or plastic) and degree of deformation on the transitions between the different lubrication modes and B) the influence of the deformation on the friction level is shown.

REFERENCES

1) Haar, R. ter, 1996, "Friction in Sheet Metal Forming", PhD. Thesis University of Twente, Enschede, The Netherlands.

2) Schipper, D.J., 1988, "Transitions in the Lubrication of Concentrated Contacts", PhD. Thesis University of Twente, Enschede, The Netherlands.

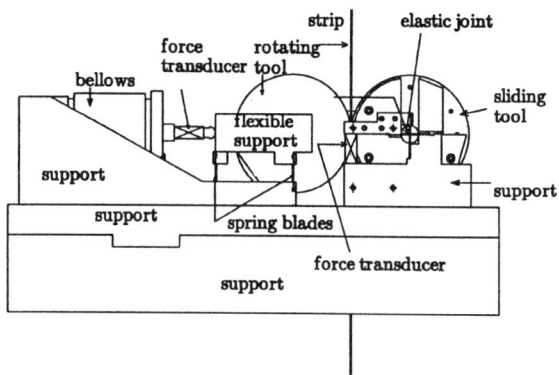

Figure 1 Friction measuring device.

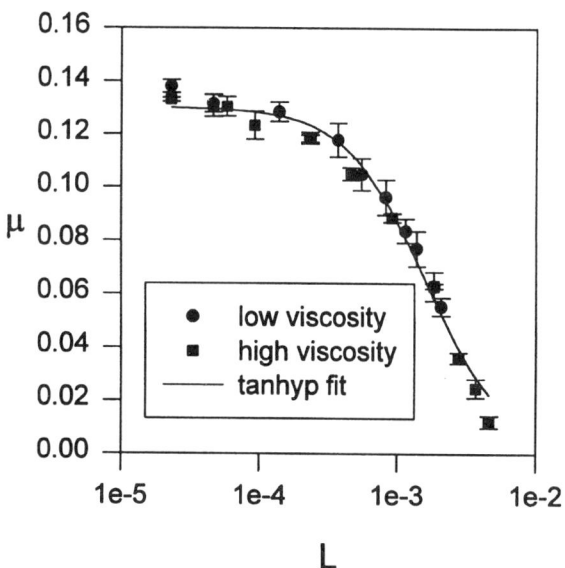

Figure 2 Friction versus lubrication number.

TRIBOLOGICAL SIMULATION OF ALUMINIUM HOT EXTRUSION

THOMAS BJÖRK and JENS BERGSTRÖM
Department of Technology, materials engineering, University of Karlstad, Sweden

STURE HOGMARK
Department of materials engineering, University of Uppsala, Sweden

INTRODUCTION

Hot extrusion is a very cost efficient method of aluminium forming. An Al-billet at typically 550°C is pushed through a profile opening in a die. The extrusion is then halted for 5-15 s in order to change billets, whereafter a new billet is extruded. This is called the extrusion stop and start cycle.

A main problem is wear on the bearing surface which may deteriorate the surface shape of the profile.

Nitrided hot work tool steels are commonly used while new surface treatments are being introduced.

Full scale tests with extrusion dies [1] are expensive, thus there is a need for an inexpensive and accurate evaluation method. Therefore a block-on-disc test machine has been designed and built. This paper describes the test machine, its major features, fields of application and some interesting outputs.

EXPERIMENTAL

A specimen is pressed upon a rotating Al-cylinder. Normal and friction forces along with sliding speed and temperature are varied and continuously recorded. The temperature is at maximum 620°C and is measured on the test specimen. To imitate the oxygen-free metal contact in extrusion the tests are carried out in a tight chamber with inert argon atmosphere.

Studies of the stop and start cycle are enabled, also including high speed data acquisition, up to 5 kHz. The tests are evaluated by friction measurements, wear quantification and surface characterisation.

RESULTS AND DISCUSSION

Comparison was performed of the wear mechanisms on used extrusion dies and worn test specimens of H13 hot work tool steel in two surface treated conditions, CVD/TiC+TiN and nitrocarburizing

In both cases the CVD coating showed initially mild chemical and in the long run adhesive wear (fig 1a). Once the coating was locally worn through, the aluminium flow digged deep craters in the steel substrate.

On the other hand, on both nitrocarburized dies and test specimens, the nitrogen rich surface zone was worn through in furrows. Those are created by an initial pick-up of coating fragments (fig 1b) and a subsequent ploughing by aluminium, creating the furrows.

It has been suggested that most of the wear has its origin in the stop and start cycle [2], when aluminium and tool material may stick to each other chemically. Thus strong adhesive forces are in action at the restart.

In wear tests, the sliding friction was in the range $1.0<\mu<1.4$, with only slight differences between CVD-TiC+TiN and nitrocarburized specimens. The start friction was in general high compared to the sliding friction (fig 2). The level and the extension of the start peak is believed to originate from the bonding between the tool surface and aluminium in combination with a velocity dependent friction behaviour.

The amount of wear was measured by weight loss after the test. On a nitrocarburized specimen a 20 µm deep crater was worn in the substrate while a TiC+TiN specimen only showed a slight polishing of the coating.

CONCLUSIONS

- Since the laboratory test reproduces the wear mechanisms of actual extrusion, this is an accurate tribological testing method, which should be useful in the development of new surface treatments for the application.
- High coefficients of friction, almost independent of surface treatment indicate that the contact is mostly aluminium-aluminium.
- The wear resistance is much better for a CVD-TiC+TiN coated specimen than for a nitrocarburized specimen.

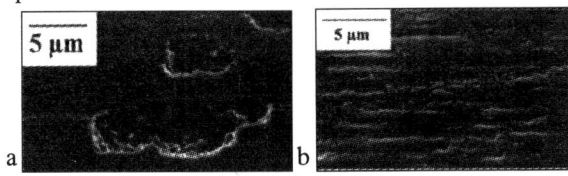

Figure 1. a) Nitrocarburized and b) CVD-TiC+TiN

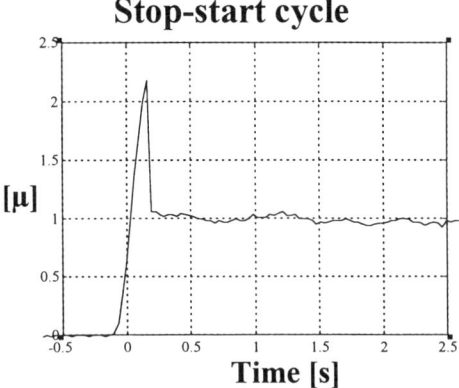

Figure 2. Start friction peak with CVD-TiC+TiN.

REFERENCES

1. Extrusion Technology Seminar II. 1992 p.233
2. Extrusion Technology Seminar II. 1996 p.1

EFFECT OF PRE-WELDED BLANKS ON THE WEAR OF DEEP DRAWING TOOLS

M VERMEULEN, J SCHEERS and J-M VAN DER HOEVEN
OCAS, Research Centre of SIDMAR, the Flat Product Division of the ARBED Group
J. F. Kennedylaan, 3, B-9031 Zelzate, Belgium

INTRODUCTION

Pre-welding of blanks is demonstrating a rapid increase in use in the automotive industry, because of its potential to reduce car body weight and its flexibility in design by combining different materials. It enables an optimisation of sheet material with respect to mechanical strength and corrosion resistance.

Die design however might need to be modified because of the presence of a heat affected zone (HAZ) near the weld, showing a hardness increase (about 2.5 X) that can cause severe problems of wear or galling in high loaded die areas, such as radii. Galling is a dedicated term in deepdrawing. The incipience of galling is an adherence problem with very small material pick up from the soft sheet onto the die. Subsequent work-hardening and further accumulating up to macro-asperities occur, scratching the following blanks and resulting in unacceptable galls on the finished parts.

In this work, the effect is studied of the laser welded seem in the tribological behaviour of some different die materials and hard coatings.

EXPERIMENTAL TECHNIQUE

Test method

Experiments were carried out on a draw bead simulator (Fig. 1), similar to this reported in (1).

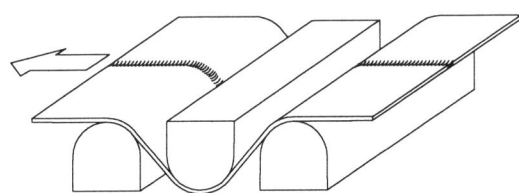

Fig. 1: Draw bead simulator test rig

The test procedure consists of drawing a first oiled strip through fully cleaned dies, followed by subsequent dry (cleaned) strips until visible galling on the strip occurs. Due to the decreasing amount of lubricant available for the next strip draw, this procedure is fairly severe. When no lubrication failure occurs after 10 strips, the test is stopped as this reflects excellent anti-galling behaviour. Further test conditions are reported in (2). Dies were selected from commercially available materials, already in use in different automotive stamping plants.

Test analysis and results

Test materials have been analysed by visual inspection (#: number of strips until galling), extensive profile measurements on the dies, and by comparing friction forces. See Fig. 2 and table below.

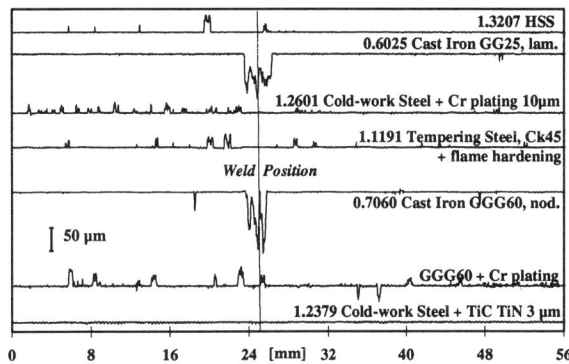

Fig. 2: Profile measurements on dies

tool / Ra [µm]	galling #	wear	friction
1.3207 HSS / 0.05	mild (6)	none	↑
0.6025 GG25 / 0.50	none	large	stable
1.2601+Cr(10µm)/0.50	severe (3)	none	↑↑↑
1.1191+harden. / 0.40	severe (3)	none	↑↑
0.7060 GGG60 / 0.25	mild (5)	large	stable
same + Cr(10µm) / 0.6	severe (2)	less	↑↑
1.2379+TiC TiN / 0.75	none	≈ none	↑

CONCLUSIONS

In spite of the absence of thickness overlap of the butt laserweld, even the small increase of hardness of the HAZ leads to deviating sliding behaviour compared to non welded blanks. Popular cast irons suffer from severe wear at the weld seem, as they are quite soft. Hard(ened) tool steels on their own do not prevent galling. The use of hard coatings is therefore recommended. Cr-plating does not seem to avoid the galling problem, while TiC-TiN is performing well without galling nor wear, in spite of its higher roughness.

REFERENCES

(1) J. Shey, Lubricant effects in drawing coated sheets over nitrided die surfaces, Lubrication Engineers, Vol.52, No.8, pp. 630-636, 1996.

(2) M. Vermeulen, J. Scheers, The influence of pre-welding in the galling and wear behaviour of deepdrawing tools, to be published.

WEAR OF SUPERABRASIVE CUTTING TOOLS

S. A. KLIMENKO, Yu. A. MUKOVOZ, G. P. KUDRYAKOV, and M. Yu. KOPEIKINA
V.Bakul Institute for Superhard Materials of the Nat. Ac. Sci. of Ukraine,
2 Avtozavodskaya St., Kiev, 254074, Ukraine

ABSTRACT

Wear behavior of cutting tools made of monocrystalline diamond and diamond- and CBN-based polycrystalline composites exhibits specific features when machining varions materials.

Cutting tools of diamond and diamond-based composites are used in machining copper, brass, magnesium, precions metals, bronze, aluminium, polymeric composite materials. When cutting materials which have no abrasive inclusions, the tools fail because of both prolonged adhesive wear and accidental reasons (dynamic, action of vibration, impacts, accidental inclusions in the material, etc.).

Cutting copper and aluminum alloys, as well as polymeric composite materials, which contain hard and abrasive particles, is accompanied by a low tool wear due to the abrasive and adhesive interactions of materials being in contact. In this case, cutting tool wear is almost independent of cutting temperature. Its effect shows up most vividly when cutting titanium, zirconium, their alloys, tungsten, and hard alloys, the cutting temperature may be as high as 2000 K (1). In this case, tool wear is characterized by an intensive adhesive interaction with a workpiece material as well as by carbon diffusion into the workpiece material.

Iron, nickel, cobalt and alloys based on them as well as other metals which interact actively with carbon, and metals, in which carbon dissolves, make a separate group. Diamond tool failure when turning the above metals at high speeds is characterized by a specific mechanism of wear, i.e. eutectic.

Tools of CBN-based composite materials are more efficient in machining the above materials.

Depending both on the nature of the materials being in contact and contacting conditions, different wear mechanisms of the above tools show up, and in most cases, they define the extreme behaviour of the tool life - cutting speed relationship (2), (3). Supported are data on abrasive- mechanical and abrasive- chemical mechanisms of CBN tool failure, which are related with phase transformations in boron nitride, its oxidation and the effect of got free particles of the tool material, as well as on adhesion and diffusion wear mechanisms.

Chemical interaction between a CBN-based tool material and a workpiece material to form interaction products in a liquid phase is shown by experiment and verified by the theory.

One of the ways to control the interaction between elements of the materials in contact and environment is the usage of technological media, gaseous in particular (Fig.1).

Overall, depending both on the workpiece material and machining conditions, superabrasive cutting tools, in machining different materials, are subject to a veriety of wear mechanisms: adhesive, abrasive, diffusion, eutectic, oxidizing, or to a combined wear mechanism (a combinetions of adhesive, abrasive and mechanical mechanisms, adhesive and diffusion mechanisms, abrasive and oxidizing mechanisms, etc).

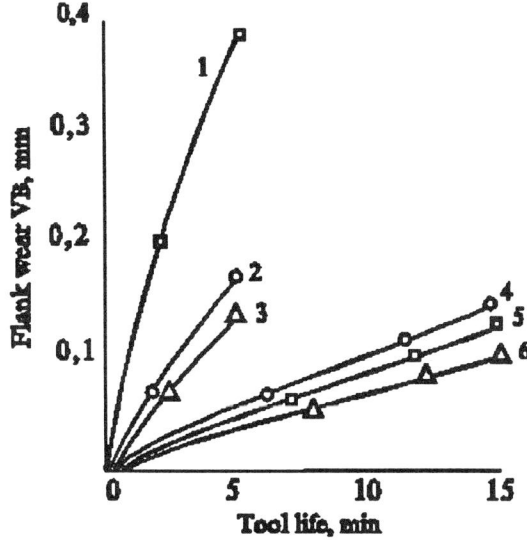

Fig. 1: Effect of gaseous media on cBN tool wear when turning a Ni-Cr-B-Si coating: 1, 5 - air; 2, 4 - argon; 3, 6 - nitrogen; 1, 2, 3 - v=3.01 m/s; 4, 5, 6 - v=0.76 m/s.

REFERENCES

(1). Müller-Hummel P., Lahres M., IDR, 2, 1995, 78.
(2). Karyuk, G.G.(Ed.): Technological Features of Machining Using Polycrystalline Superhard Material Tools, Naukova Dumka, Kiev, 1991 (in Russian)
(3). Advanced Ceramic Tools for Machinihg Application-II /I.M.Low and X.S.Li (Ed.), Trans Tech Publications Ltd, Switzerland, 1996.

NOTCH WEAR OF TOOLS AND THE ROLE OF WORK HARDENING AND OXYGEN POTENTIAL

H. CHANDRASEKARAN and JAN OLOF JOHANSSON
Swedish Institute for Metals Research,
Drottning Kristinas väg 48, S-114 28 Stockholm, Sweden

INTRODUCTION

Localised removal of tool material from rake/flank surface of cutting tools at the depth of cut line is termed notch wear. While the role of austenitic steel composition on the evolution of tool wear including notch wear for un-coated tools has been reported recently (1), work hardening as well as chemical interaction between the cutting medium and the tool/work materials have been identified as critical to notch wear evolution quite early (2). However, little information is available on the role of work hardening distinct from oxygen potential on notch wear mechanism during the turning of austenitic stainless steels. The present work is an attempt to rectify this situation.

TEST METHODOLOGY

In our studies 2-level experimental design approach was used to investigate the influence of parameters such as work hardenability, oxygen potential, feed, tool edge radius as well as nose radius on the resulting tool wear in general and notch wear in particular. Cutting speed (V = 100 m/min) and depth of cut (t = 1,5 mm) were kept constant. Two comparable austenitic stainless steels, namely SS 2353 and SS 2375, differing in their propensity to strain harden mainly due to difference in their N content (respectively 0,08% and 0,18%), were used as work materials. Coated as well as plain hardmetal inserts without chip breakers were used as tools. Novel features of the present investigation are
* the use of controlled turning operation for inducing work hardening resulting in different work piece hardness and
* the use of gas mixtures ($N_2 + O_2$) and ($N_2 + Ar$) with comparable thermal properties to distinguish the predominantly chemical effects of oxygen on wear.

RESULTS AND CONCLUSIONS

Turning high N content steel (SS 2375) with a large cutting edge radius (~ 70 µm) at large feeds induced severe work hardening, resulting in hardness almost twice as the one resulting from a low N content steel machined with a sharp tool insert (edge radius ~ 20 µm). When these two work pieces were evaluated, the notch wear was affected adversely by edge radius, oxygen potential, material (hardness) and feed in that order (Fig. 1). Analysis of the role of multiple factors showed that the sensitivity of notch wear to oxygen potential to be greater at higher feeds, probably from enhanced thermal effects.

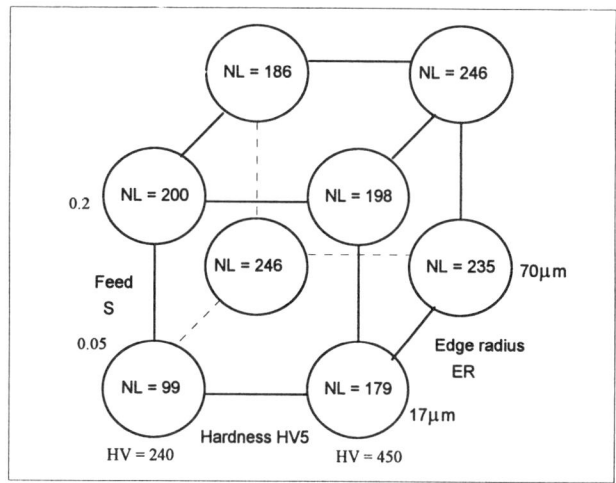

Fig. 1 Notch wear (NL in µm) results from 2-level turning tests.

SEM studies of the notch regions of the tool enabled us to distinguish the mechanical and metallurgical aspects of micro-wear mechanisms. Although success in reaching the original objectives was partial due to material limitations, the influence of work hardenability (through hardness) and chemistry of the work material have been collectively identified.

Further, the present study has also shown that the role of strain hardenability and oxygen potential to be complimentary. When a threshold level of this combination is reached, notch wear is initiated through the removal of the tool coating. Subsequent propagation of notch wear is related to the depletion of exposed binder phase from the tool substrate, followed by interrupted chip adhesion.

REFRENCES

1. H. Chandrasekaran and Jan Olof Johansson, Annals of CIRP, 42/1, 1994, pp. 101-105
2. M.C. Shaw, A.L. Thurman and H.J. Ahlgren, Trans. of ASME, J. Engr. for Industry, 1966, pp.142-146

AN EXAMINATION OF THE TRIBOLOGICAL CONTACT CONDITIONS IN MACHINING USING PVD TiN COATED CUTTING TOOLS

E D DOYLE and Q T PHAN
School of Mechanical and Manufacturing Engineering, Swinburne University of Technology,
Johns St, Hawthorn, Victoria 3122, Australia

ABSTRACT

Understanding the frictional interaction between the chip and the rake face of the cutting tool is a particularly complex and challenging problem. Unlike "normal" friction in conventional sliding, the initial contact near the cutting edge between the chip and the tool involves intimate contact so that the distinction between real and apparent area of contact is not relevant. It is important, therefore, to understand the contact condition between the chip and the tool in order that one might better design surface coatings to reduce wear and promote longer tool life.

Over the last fifty years, there has been considerable debate over the nature of the contact conditions between the chip and the rake face of uncoated cutting tools. There have been many contributors to this debate, for example, Zorev (1) concluded, on the basis of split tool experiments, that the chip sticks to the rake face initially and then gives way to sliding contact. Trent (2) proposed that the nature of the chip/rake face tool contact is predominantly one of seizure contact. The seizure conditions usually occur in an area close to the cutting edge. The concept is one in which there is no relative movement at the chip/tool interface, consequently chip movement takes place by secondary shear within the chip material immediately adjacent to the chip/tool interface. In contrast, Doyle et al (3) showed relative movement at the chip/rake face interface using a transparent sapphire tool albeit at low cutting speeds on ductile workpiece metals. They concluded that the contact region could be divided into zones in which the severity of sliding contact varied. In something of a synthesis of these ideas, Wright and Thangaraj (4) concluded that conditions of sliding and/or seizure would prevail depending on the conditions at the chip/tool interface and the machining parameters. In the present study simple turning experiments were carried out using powder metallurgy high speed steel (HSS) inserts (PM-M41) on a medium carbon steel workpiece (K1050, 240 ± 30 HVN). The machining conditions were selected as follows: cutting speed 51m/min, feed rate 0.224mm/rev and depth of cut 1.25mm. Under these conditions, the HSS inserts gave a tool life of the order of 20-25 minutes compared with less than 2 minutes in the uncoated condition. Machining was interrupted at certain time intervals and the inserts examined in a scanning electron microscope. Examination of the TiN coated rake face of the PM-M41 insert after 30 seconds of cutting, revealed a very clear pattern of three zone contact. The first zone extending 200µm from the cutting edge showed no material transfer. The second zone extending 200-375µm from the cutting edge showed a distinct line of deposit of material transfer. The third zone extending 375-750µm from the cutting edge showed evidence of rubbing contact without any significant material transfer. It should be noted that the location of maximum temperature generation was in the third contact zone. After 2 minutes of cutting, the first zone remained the same. The second zone changed in that some of the deposit has been removed exposing a layer of non-conducting material being a mixture of silicate (type 2) and sulphide (type 1) inclusions). In the third zone, a deposit of material transfer from the chip had formed. After 4 minutes, the latter deposit gave way to crater formation, with a significant deposit of transfer material in the crater. The deposit of non-conducting material was evident in a region around the periphery of the crater. After 11 minutes of cutting, the contact conditions were much the same. After 23 minutes of cutting, the insert was close to failure. The crater had grown in width and depth. Significantly, the region adjacent to the cutting edge was free of material transfer. This region extended 125µm up the rake face and the TiN coating was still intact.

In conclusion, results from this study of machining with continuous chip formation, using a TiN coated HSS tool, suggest that the chip/rake face contact length can be usefully broken down into three zones, each involving different interfacial contact conditions.

REFERENCES

1. N N Zorev, Metal Cutting Mechanics, Pergamon Press, Oxford, 1966.
2. E M Trent, Metal Cutting and the Tribology of Seizure, Part II, Movement of Work Material over the Tool in Metal Cutting, Wear, Vol. 128, 1988, 47-64pp.
3. E D Doyle, J G Horne and D Tabor, Frictional Interactions between Chip and Rake Face in continuous Chip Formation, Proc. R. Soc. London, A. 336, 1979, 173-183pp.
4. Wright and A Thangaraj, Correlation of Tool Wear Mechanism with New Slip Line Fields for Cutting, Wear, Vol. 75, 1982, 105-122pp.

FRICTION AND LUBRICATION EFFECTS IN THE MACHINING OF ALUMINIUM ALLOYS

W Y H LIEW, I M HUTCHINGS* and J A WILLIAMS

Department of Engineering, Cambridge University, Trumpington Street, Cambridge, CB2 1PZ, UK
*Department of Materials Science & Metallurgy, Cambridge University, Pembroke Street, Cambridge, CB2 3QZ, UK

ABSTRACT

Friction between the rake face of a cutting tool and the freshly formed chip surface plays a vital role in influencing both the ease of cutting and the quality of the resultant machined surface. The existence of clean surfaces together with the high local hydrostatic stresses favour the formation of strong adhesion between the cutting tool or insert and the machined component. These adhesive bonds can lead to poor surface integrity although their extent can be limited by the provision of a suitable machining lubricant.

In an effort to identify the essential lubricating aspects of fluid activity, as opposed to any role as a coolant, experiments involving the orthogonal machining of precipitation hardened aluminium alloys, principally 2014, have been carried out in controlled low pressure environments in which cutting speed and temperature have been varied while using a variety of tool materials and lubricating species. The results, some of which are shown in Figs. 1 and 2, illustrate the effect of cutting speed on measured rake face friction coefficient for a variety of conditions: in Fig. 1 the material was in condition T4 (naturally aged) while in Fig. 2 it had been artificially aged to condition T6.

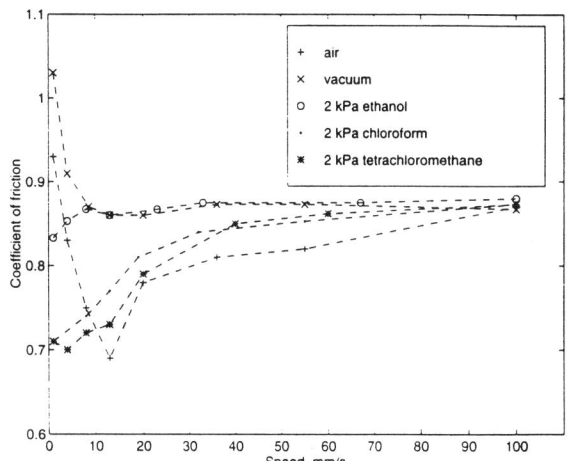

Fig. 2: Rake face coefficient of friction vs. cutting speed for aluminium 2014-T6 in various atmospheres: HSS tools.

The observed variations in cutting forces and chip forms indicate that there can be unexpectedly subtle, but significant, interactions between the metallurgy of the workpiece, the nature of the surface of the tool and the surrounding environment. These are not wholly consistent with conventional theories of vapour phase lubrication in which transport of the lubricant has been assumed to control the effectiveness of the lubricating agent (1-3). The implications of these observations for the complex tribological system constituted by the combination of workpiece, tool surface and local environment will be discussed.

REFERENCES

(1) T. Wakabayashi, J. A. Williams and I. M. Hutchings, Proceedings of the IMechE, Part J, Vol 209, pp 131-136 (1995)
(2) T. Wakabayashi, J. A. Williams and I. M. Hutchings, in Tribology in Metal Cutting and Grinding, MEP, London (1992)
(3) T. Smith, Y. Naerheim and M. S. Lan, Tribology International, Vol 21, pp 239-247 (1988)

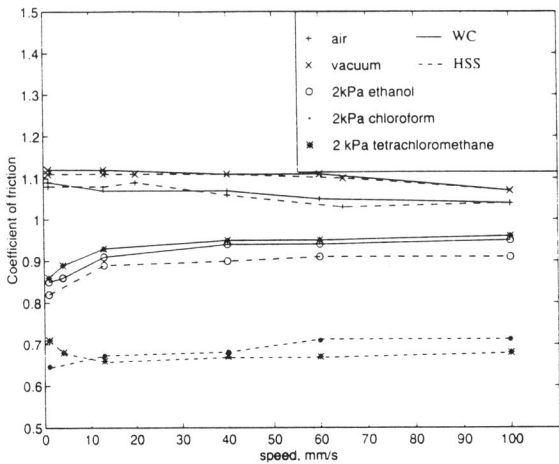

Fig. 1: Rake face coefficient of friction vs. cutting speed for aluminium 2014-T4 in various atmospheres: HSS and tungsten carbide tool material.

Submitted to *Tribology Letters*

HEAT PARTITION AND TEMPERATURES IN GRINDING

Y. JU, T. N. FARRIS, and S. CHANDRASEKAR
Schools of Engineering, Purdue University, 1282 Grissom Hall, West Lafayette, IN 47907-1282

ABSTRACT

A theoretical analysis of heat partition and surface temperatures for the grinding of steel with both aluminum oxide and CBN wheels is presented. The model considers interaction of both the wheel and grain contact length scales. On the wheel contact length scale, the heat partition between the fluid and workpiece is considered. On the local scale, the partition of local grinding heat to the abrasive grain, workpiece, and chip is considered. The model includes the transient process occurring between the grain and workpiece. Thus the model is accurate for all speeds and feeds as well as both up and down grinding. With the aid of partition functions, the problem is solved using a direct numerical approach for continuity of temperature everywhere in the contact zone.

The numerical predictions of the model are shown to agree with experimental results available in the literature (1). It is found that heat partition is a strong function of position inside the grinding zone, varying over a wide range depending on grinding conditions. Eight numerical examples show that CBN grinding and creep-feed grinding can reduce the heat partition to the workpiece and the workpiece temperature significantly.

Based on the full numerical results, a simplified model is developed to help understand the effects of grinding conditions on heat partition and workpiece temperature. The simplified model assumes constant partition functions and matches average temperatures. The simplified model is shown to be very accurate for down grinding and fairly accurate for up grinding. Note in up grinding the sliding surfaces move in opposite directions which is not as amenable to constant partition as the case of velocities in the same direction. The effects of grinding conditions on the energy partition temperature are studied systematically. It is found that heat partition is determined by grinding conditions including the material removal rate, specific energy, real contact area ratio, thermal properties of the materials, fluid, speed combination, and type of cut(down or up grinding).

The presence of the fluid inside the grinding zone can reduce the heat flux into the workpiece and the workpiece temperature significantly. However, it is very difficult to keep the fluid active for conventional grinding of steel with aluminum oxide due to fluid boiling. Figure 1 illustrates both the full model (symbols) and the simplified model (lines). The material removal rate is kept constant so that depth of cut decreases as workpiece velocity increases. Note that as the workpiece speed increases the fluid boils as illustrated by the simple model for wet grinding being more appropriate for low velocity and the simple model for dry grinding being more appropriate for high velocity. For conventional grinding of steel with CBN, or creep-feed grinding of steel with aluminum oxide or CBN, it is possible to keep the fluid active and therefore to reduce thermal damage. It is also found that a moderate ratio of the workpiece velocity to wheel velocity gives high temperatures and therefore should be avoided. That is, in form finish grinding high workpiece speed leads to low temperature whereas for creep-feed grinding, low workpiece speed leads to low temperature.

Arguments are made by Ju (2) that temperature controls many aspects of ground surface integrity. In particular, global temperature controls residual stress and local temperature controls metallurgy.

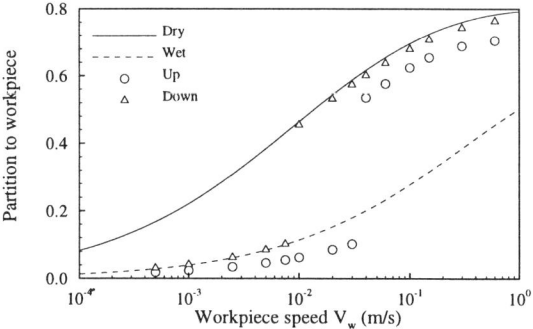

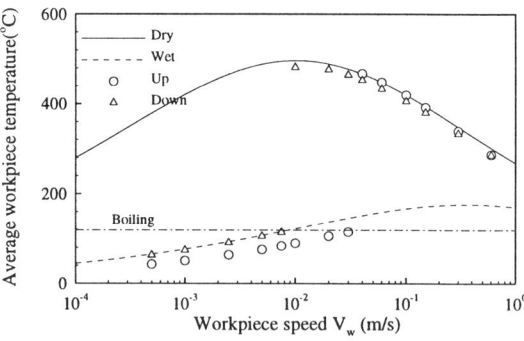

Fig. 1: The effects of the workpiece speed on the partition and workpiece temperature for the grinding of steel with an aluminum oxide wheel

REFERENCES

(1) S. Kohli, C. Guo, S. Malkin, ASME Journal of Engineering for Industry, Vol. 117, 1995.
(2) Y. Ju, Ph.D. dissertation, Purdue University, 1997.

MECHANISMS OF CHEMICAL-MECHANICAL POLISHING OF SiO2 DIELECTRIC ON INTEGRATED CIRCUITS

J. LEVERT, F. MESS, R. SALANT AND S. DANYLUK
George W. Woodruff School of Mechanical Engineering, Georgia Institute of Technology, Atlanta, GA 30332-0405, USA
R. BAKER
Rodel,Inc., 451 Bellevue Ave., Newark, DE 19713, USA

ABSTRACT

Chemical-mechanical polishing (CMP) refers to the rubbing of soft (conformal) pads over surfaces in the presence of chemically-active slurries that contain abrasives (1). The pad is typically 'conditioned' (roughened) to produce asperities of known predictable size. CMP is used to create planar surfaces of dielectric (silicon dioxide), and metal and polymeric films that are deposited on single crystal silicon substrates of integrated circuits. Dielectric CMP is typically performed on a slurry-flooded polyurethane pad mounted on a rotating platen. The slurry is an aqueous fluid (~ 11 pH) containing a suspension of silica particles (~ 200 nm diameter). Silicon wafers are polished with the circuit side down on pads that rotate. The contact of the pad and slurry with the surface layers on the silicon produces optically-smooth surfaces for further processing such as photolithography. CMP is a complex tribological problem that involves chemical reactivity of the slurry, interaction of the abrasive, and the deformation and contact of pad asperities with the wafer. This paper describes experiments to address the mechanical contact of the pad asperities with a silicon wafer surface.

The CMP experiments were carried out by polishing 100 mm diameter silicon wafers on a commercial polishing machine (Logitech, Inc.) with a 300 mm diameter platen using a commercially-available polyurethane pad and slurry (Rodel, Inc.). The wafers were wax mounted on a fixture and positioned in an auxiliary apparatus that held the wafers stationary relative to the platen. The normal load was applied through linear bearings, and vertical wafer displacement measurements were obtained in situ by three capacitance probes (2).

The principal result of this research can be seen in Figure 1 (3). This figure shows the vertical displacement of a silicon wafer (μm) versus the speed of the platen (m/s) for low (0.35 kPa) normal loads, typical (48.3 kPa) normal loads with a smooth pad, and typical normal loads with a conditioned (roughened) pad. These data show that full film hydrodynamic lubrication can occur if the pad is unconditioned, and this hydrodynamic condition can be reduced if the normal load is increased and the pad is glazed (worn smooth), and that negative displacement results when the pad is conditioned. Negative displacement implies that a vacuum force is associated with the roughness of the pad. Additional experiments revealed that the interfacial slurry pressure was, on average, negative during typical CMP. Measurements of the coefficient of friction were also consistent with the change from the hydrodynamic lubrication condition to the negative displacement.

These results suggest that the pad asperities mechanically contact the wafer surface during typical CMP. Also, the surface roughness of the pad determines the transition from a hydrodynamic slurry condition to a vacuum condition. This transition can greatly increase the wafer-to-pad mechanical contact load and consequently affect the polishing rate.

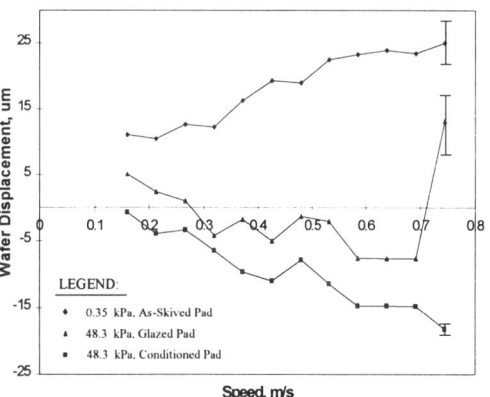

Fig. 1: Wafer displacement (μm) versus platen speed (m/s) for three different normal loads and pad conditions.

REFERENCES

(1) L. M. Cook, Abstracts of 1995 SSDM - of the Japan Society of Applied Physics, page 581.
(2) J. A. Levert, et al., Transactions of International Tribology Conference - Yokohama, October, 1995.
(3) J. A. Levert, et al., *Mechanisms of Chemical-Mechanical Polishing of SiO2 Dielectric on Integrated Circuits*, Tribology Transactions (submitted 1996).

CHARACTERIZATION OF WEAR PARTICLES AND THEIR RELATIONS WITH SLIDING CONDITIONS

AKIHIKO UMEDA, JOICHI SUGIMURA and YUJI YAMAMOTO
Department of Energy and Mechanical Engineering, Kyushu University, Fukuoka, 812-81, Japan

ABSTRACT

A multi-layer neural network and a self-organizing neural network were applied to recognition and classification of wear particles, worn surfaces and relations between them.

Lubricated sliding experiments were conducted in which a bearing steel ball was slid against a carbon steel disk under two different loads and with paraffinic oils containing different additives. The test conditions are shown in Table 1. Microscopic images of both worn surfaces and wear particles were analyzed by image processing.

Characteristics of wear particles were described by four parameters, namely representative diameter, elongation, roundness and reflectivity. The particles had different features according to sliding conditions. A multi-layer neural network with the back propagation algorithm learned the differences in particle parameters, and recognized similar features for particles produced under similar lubrication conditions.

A self-organizing neural network was applied to classify the wear particles on a feature map. The competitive learning allowed the network to learn data such that wear particles from similar sliding conditions appeared at units close to each other on the feature map, as shown in Fig. 1. In addition, weight vectors of the feature map were analyzed to obtain relations between the particle parameters and sliding conditions.

Relations between wear particles and worn surfaces were studied using two textural parameters, namely the angular second moment and the contrast. It was found that the wear particles had relatively lower reflectivity than worn surfaces, and textural parameters of wear particles did not always coincide with those of worn surfaces, suggesting that the particles were generated at particular area of the worn surfaces that had oxide films, or heavy plastic deformation occurred during removal of the particles.

Another application of the self-organizing neural network was made to the microscopic images. After learning a number of square image data on worn surfaces of all the experiments, the feature map organized itself so that brighter image data were located near the bottom-left corner and darker image top-right corner, while images with higher contrast appear at top-left and bottom-right corners. The number of data located at each of the unit is shown in Fig. 2 for 400 data from a worn surface and 20 data of large particles in A2 and D2. The figures demonstrate that the surface of A2 has more image data than D2 in bottom-left half on the map, i.e. the images are brighter, and that the particles have data only near the top-right of the map and distributions are different from the surface particularly in A2.

Table 1: Experimental conditions

Lubricant	Load 58.8N	176N
H60	A1	A2
H60 with stearic acid 0.1wt%	B1	B2
H60 with TCP 1.0wt%	C1	C2
H60 with DBDS 0.5wt%	D1	D2
H60 with ZnDTP 1.0wt%	E1	E2

Sliding speed: 6.28mm/s, Sliding distance: 203.4m

D2	C2			A2
			A1	
C1	B2			
				E1
B1		D1		E2

Fig. 1: A feature map of wear particles by a self organizing neural network

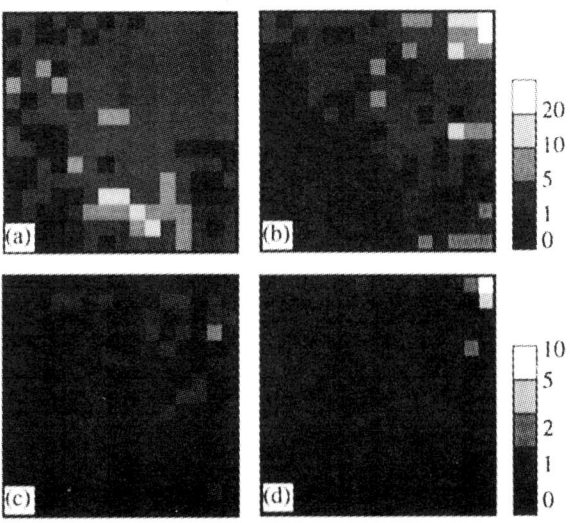

Fig. 2: Feature maps of image data; (a) worn surface of A2, (b) worn surface of D2, (c) wear particles of A2, (d) wear particles of D2

WEAR PARTICLE ATLAS - A COMPUTERISED WINDOW ON WEAR AND ITS UNDERLYING CAUSES

B.J.ROYLANCE
Department of Mechanical Engineering, University of Wales, Swansea, SA2 8PP, U.K.

ABSTRACT

A new wear particle atlas is described which has recently been developed to assist in wear research. It also has application in the field where morphological analysis of wear debris is undertaken as part of machinery health monitoring procedures used in conjunction with condition-based maintenance activities. It is based on Hyper-Text Mark Up Language (HTML) methods of data presentation and manipulation. The aim in exploiting these modern computer-assisted procedures for use in particle analysis is to provide more effective and rapid means of analysing particles viewed in the microscope through direct comparison with a library of definitively identified and labelled particles.

INTRODUCTION

Morphological analysis of wear particles based on the ferrography method has been carried out for the past 25 years, (1). The principal morphological attributes of the particles are: size, outline shape, edge detail, surface texture, colour and thickness. Numerous manual forms of wear particle atlas have come into use around the world since the first atlas was introduced in 1976, (2). Only a very few atlases have been linked to computer-assisted processing and analysis procedures. The principal advantage of introducing HTML methods is the exploitation of the mouse operated 'click-on' facility for instantly displaying highlighted features and coupling them to immediate cross-referencing with other sections of the document.

ANALYSIS STRATEGY

Reference to Figure 1 shows that the wear particle atlas is a crucial element in the overall analysis strategy when undertaking a wear research investigation or, alternatively, when participating in a condition monitoring programme. Methods are presently being developed to automate much of the analysis through the use of modern artificial intelligence methods, such as neural net. Identifying the particle type through morphological analysis in terms of its associated wear mode represents an important part of establishing the primary active wear mode, its location and severity.

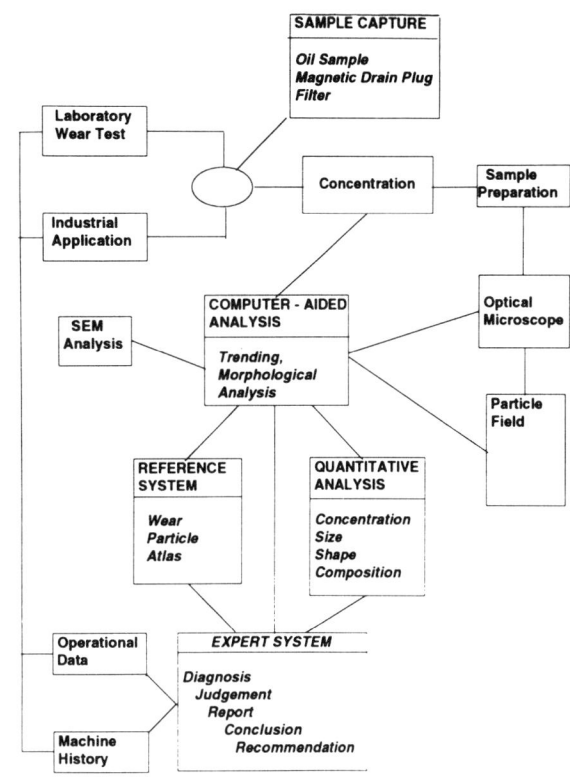

Fig. 1: Sampling, processing and analysing wear particles

CLOSURE

In connection with fundamental wear studies aimed at establishing the underlying causes of wear, the wear particle atlas is a powerful aid to identification of particles captured non-intrusively from the wear site.

REFERENCES

(1) W.W.Seifert and V.C.Westcott, WEAR, Vol.21,1972, 27-42pp.
(2) E.R.Bowen and V.C.Westcott, Final Report to Naval Air Engineering Center, Lakehurst, N.J., Contract No. N00156-74-C-1682,1976.

A Novel Approach to the Assessment of Friction loss in Viscoelastic Materials

M O A MOKHTAR
Faculty of Engineering, Cairo University, Cairo, Egypt
S A R NAGA and A M EL-BUTCH
Faculty of Engineering, Mataria, Helwan University, Cairo, Egypt

Abstract

This paper presents a novel experimental technique to assess the frictional behavior of a wide range of viscoelastic materials.

A pendulum impact test rig has been designed and constructed to perform single touching impact.
In order to measure the coefficient of friction, two pizo electric force transducers have been used, one fitted under the specimen to measure the normal force and another one fitted directly in front of the specimen to measure the tangential force. To reduce the effect of the frictional force between the specimen and lower transducer on the tangential force value, a piece of Teflon has been inserted between them.

The energy consumed in both friction and elastic deformation of the specimen can be read from the reduced upswing of the pendulum as shown in Fig. 1. The energy consumed during the touching impact has been found to correlate strongly with the coefficient of friction.

Tests were carried out on reinforced polyamide 66 specimens. A comparison between the previously published work on the frictional behavior of reinforced polyamide 66 using traditional pin and disc machine (1) and the present recorded friction results showed a good agreement Fig. 2. Due to the test conditions, the attained experimental results lie between the static and kinetic coefficient of friction and a same trend of increase of the coefficient of friction with fiber content up tilll a limiting content has been recorded Fig. 2.

Of interest to note that there is a direct relation between the energy loss and the recorded coefficient of friction. the ratio between the energy loss to the original potential energy which is equal to Cos θ is presented in Fig. 2 to confirm the linear relation between the lost energy and the coefficient of friction.

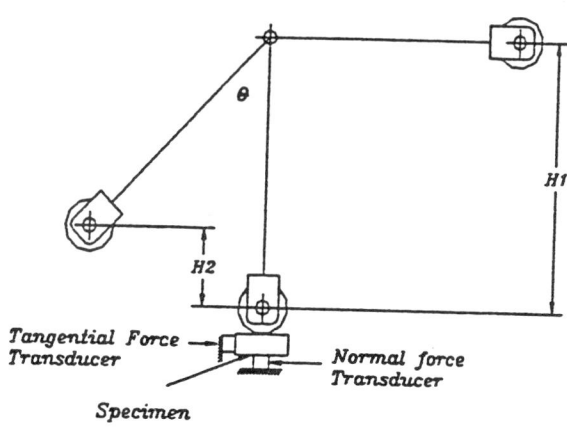

Fig. 1 Schematic drawing of the test

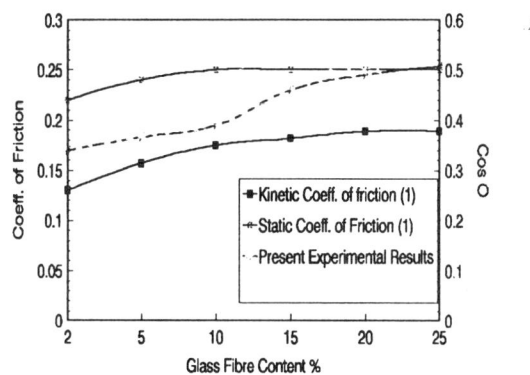

Fig. 2 Effect of glass fiber content on the coefficient of friction

References

- Soheir A. Naga, "Effect of Glass Fiber Content on the Frictional Behaviour of Glass Fiber Reinforced Polyamide 66", 8th. International Colloquium, Tribology 2000, 14-16 Jan., 1992

FRIDAY 12 SEPTEMBER

		Page number
F1	Environmental issues	345
	Seals	349
F2	Tribology in extreme environments	353
F3	Micro- and nano-tribology	361
F4	Thermal effects of tribo-chemistry	369
F5	Measurement and simulation – II	377

BIODEGRADABLE LUBRICANTS

REBECCA GOYAN, DR. ROGER MELLEY, WILLIAM ONG, PETER WISSNER
Maryn Research Ltd.

ABSTRACT

The purpose of this paper is to introduce some of the chemistry behind biodegradability and some of the environmental regulations from a Canadian perspective. Four specific applications are presented as examples of eco-compatible products currently in testing or use in North America.

Lubrication is a vital component of both domestic and industrial processes. Recently, environmental behavior of lubricants, such as emissions, safe handling, toxicity, and biodegradability have come into consideration.

Simple, biodegradable lubricants were in widespread use, up until the discovery of petroleum oil in the late 1800's. With the development of the petrochemical industry, large volumes of base stocks became readily available, displacing natural products for reasons of lubricity, stability, and economics. Along with advancement of analysis and awareness of the environment have come increasing concern over the effects of petroleum products being released into the environment, leading to a reinvestigation of vegetable oils and readily biodegradable products.

Biodegradability is the ability of a substance to be decomposed by microorganisms. Although there are yet no standards to determine biodegradability, several test methods exist to determine the extent to which materials are biodegradable. The three most common methods include the CEC L-33-T82, EPA 560/6-82-003 "Shake Flask" test, and the OECD 301 series of methods.

Fluids which are considered environmentally friendly must not only be biodegradable, but also relatively non-toxic, in both their initial form and degradation products. Their effects on flora and fauna must be minimal. There are two common tests to evaluate toxicity: the Microtox and Rainbow trout bioassay.

The most common mineral oil substitutes consist of vegetable oils (natural esters) and synthetic esters. Rapeseed and canola oil are the most common base stocks for vegetable based-lubricants. Additives for biodegradable lubricants must also be biodegradable and non-toxic, or at least not interfere with the biodegredation of the base fluid. Additives which are considered appropriate for biodegradable lubricants must contain no chlorine or heavy metals and should not be controlled under any occupational health and safety regulations.

In North America, biodegradable fluids have found widespread use in hydraulic and forestry applications, although performance of other biodegradable lubricants is being tested and evaluated in a wide variety of equipment. These applications include saw guide oils, valve oils, turbine oils, and rail curve greases.

The forestry industry currently uses environmentally friendly chain bar oils and has expanded to using biodegradable lubricants for the blades and saw guides in the mills. A biodegradable, nontoxic natural and synthetic ester based fluid was developed specifically for saw guides. This product has not only decreased operational costs, but has solved the mineral oil related health problems.

The North American pipeline network spans a wide variety of regions. Gas transmission valves are controlled remotely using the pressure of the gas in the lines. However, significant amounts of petroleum oil is leaked when these valves are used. A biodegradable valve actuator oil was developed based on biodegradable synthetic esters which would be both biodegradable and non-toxic, in both the new and used forms.

Alberta's power generation is closely linked to the river system, both in the coal fired thermal plants and the hydro-electric dams. The potential for adverse environmental effects from accidental release has become a concern to the point where increased scrutiny has caused many facilities to pro-actively investigate the use of environmentally friendly fluids. After 6000 hours of testing, a synthetic ester based fluid seems to be comparable to the mineral oil, with the added benefit of being 99% biodegradable in 21 days.

Curves on rail tracks can cause excessive wear of the rail and wheel flanges. These tracks are often found in environmentally sensitive areas, such as Canadian National Parks. A biodegradable grease based on a low pour point synthetic ester was developed and is currently being evaluated by major Canadian railroads. The biodegradable rail curve grease has cut the rate of wear in half as compared with the mineral based grease. Although the biodegradable grease is higher in cost, the initial layout will soon be recovered in terms of clean-up costs, rail and wheel wear savings and man-hours in maintaining frozen lubricators.

Although the potential for use of eco-compatible lubricants is great, the supply of suitable base stocks is limited. The use of eco-compatible lubricants needs to be limited to those areas of highest need and greatest chance for success.

Submitted to *Lubrication Engineering*

WEAR BEHAVIOUR OF THE PISTON/GUDGEON PIN IN A HERMETIC COMPRESSOR WITH REPLACEMENT CFC REFRIGERANTS

S. SAFARI and M. HADFIELD
Department of Mechanical Engineering, Brunel University, Uxbridge, Middlesex, UB8 3PH

ABSTRACT

The use of man-made chloroflurocarbon (CFC) is the principal reason leading to depletion of the ozone layer. International agreements, such as the 1987 Montreal protocol and its later revisions aim to reduce both the production and consumption of many CFC compounds (1). Refrigeration and air conditioning systems within buildings are a major application for the CFC's and hence the need to replace them by new environmentally friendly refrigerants. This paper is concerned with a specific component design that has been affected by this recent environmental legislation. Refrigeration compressors using CFC refrigerants are being replaced with hydrochlorofluorocarbon (HCFC) and hydrofluorocarbon (HFC) refrigerants. Compressors in refrigeration systems are lubricated by oils which dissolve refrigerants. The traditional mineral oil lubricants are also being replaced with synthetics to maintain the oil's compatibility with the new refrigerants. The implications of replacing the refrigerant and lubricant on the tribological performance of the compressor piston are not fully understood and the gudgeon-pin surface of a hermetically sealed domestic refrigeration system is of particular interest.

The novelty of this research is the investigation of new refrigerant/lubricant combinations on wear behaviour of the compressor gudgeon pin system. This is achieved by using realistic experimental modelling of the sliding contact and combining a refrigeration system with a bench wear testing machine. This research investigates the performance of a sample material under the condition as closely related to a real situation as possible. A small domestic compressor rig is used to evaluate the lubrication and wear behaviour of the piston, particularly at the gudgeon pin interface. The refrigeration rig is tested according to the British Standard specification for the compressor and the running time is 500 hours. The refrigerant used for these particular experiments is KLEA 134a and the lubricating media is refrigerant soluble synthetic based oil with different viscosity/density.

The early indications suggest that there is a great deal of plastic deformation in the region near the groove of the gudgeon pin (figure 1). The severe sliding friction of the pin inside the piston could have a crucial factor in deterioration of the pin surface. Surface chemical analysis of the gudgeon pin also reveals the deposition of materials such as Al, Si and Zn from the small end of the conrod (2). This is mostly due to the lower hardness of the conrod to that of the gudgeon pin. The use of a micro-friction machine is of great importance to evaluate and predict the true behaviour of Aluminium alloy components of different compositions in sliding contact with that of high Chromium mild steel (3). The bench test rig results (micro-friction machine) and compressor results are compared and correlated. This approach allows for practical improvements to the design of piston/gudgeon-pin system lubricated with new refrigerant/lubricant combinations.

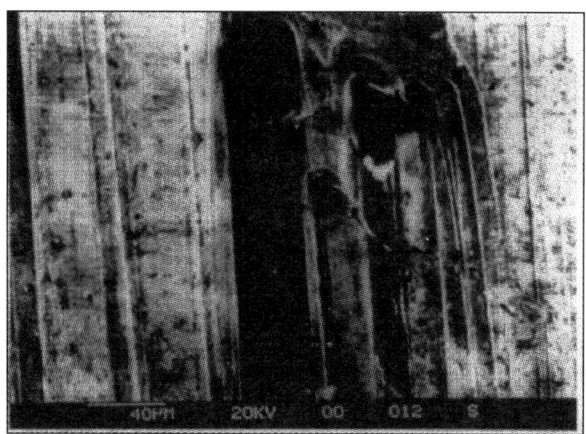

Fig. 1: Plastic Deformation near the groove of the gudgeon pin resulting from sliding contact with the small end of the conrod.

This area of research is of international concern as the pace imposed to change the refrigerants has resulted in unknown effects on compressor components. The full details of the results and analysis will be included in the final paper.

REFERENCES

(1) C. E. James, M Sc Building Services Engineering, Brunel University, 1996.
(2) T Yoshimura, H Akashi, A Yagi and T Nagao, Proceedings of the 1996 International Refrigeration Conference at Purdue, West Lafayette, Indiana, USA, 115-120pp.
(3) H Yoon, T Sheiretov, C Cusano, Proceedings of the 1996 International Refrigeration Conference at Purdue, West Lafayette, Indiana, USA, 139-144pp.

GEO-TRIBOLOGICAL ASPECT OF THE NOJIMA FAULT AT M=7.2 KOBE EARTHQUAKE

Y. Enomoto* and T. Asuke*
Mecanical Engineering Laboratory, Namiki 1-2, Tsukuba, Ibaraki 305 Japan
Z. Zheng
Geoscience Co. Ltd., Higashi Ueno 6-1-1, Taito-ku, Tokyo 110 Japan,
Y. Mizuta
JEOL Ltd., 1-2 Musahino 3-chome, Akishima, Tokyo 196 Japan

Abstract

Shear rupture of the crust, that is earthquake, is a largest tribosystem, where stick-slip motion occurs almost periodically at an interval of some 100-1000 years. Seismic physico-chemical processes at precursor stage of the final faulting have been focused on special attention because of potential applicability of imminent earthquake prediction.

A mag 7.2 earthquake at a depth of 14.3 km hit Kobe and Awaji island (so called Kobe E.Q.) on 17 January 1995. The earthquake was accompanied by the surface rupture of Nojima fault on the northern side of Awaji island as shown in an insert of Fig. 1. The lightning and luminosity spreading from the ground near the epicenter of the earthquake was witnessed by many inhabitants.

In order to find any geological evidences of EQL, core drillings were conducted at the Nojima fault where earthquake lightning (EQL) was witnessed. Typical cross secti1nal view of the fault drilled is shown in Fig.2. The fault gouge of about several centimeter thick existed between mudstone zone and weathered granite zone. The detailed study of the fault gouge showed anomalous features in terms of tribological response and remanet magnetization properties as described in the following:

1) the fault gaouge is harder than mating mudstone and weathered granite, nevertheless lamellar pattern, possibly formed due to the shear action of the fault movement was realized, and

2) natural remanet magnetization of the harder fault gouge is about 100-1000 times as high as that of the mudstone and weathered granite near to the fault and weak fault gouge at 14 m deep.

The fact 1) is unusual as a tribological resonse, in which shear deformation always occurs in a softer material, but not in a harder material, when they slide with each other.

It is known that frictional heating during a fault movment causes melting of the fault gouge at depth. So one possible explanation of the anomalous tribological features of the fault gouge is that the gouge melted by frictional heating at depth was squeezed out along the fault plane and then solidified near the ground surface. However, the model is mprobable in the present case, because the higher magnetization of the harder gouge could not be explained. As a result of detailed magnetization studies, we confirmed that the harder gouge is a strong geological evidence, where an EL current of as high as 1,000 amperes possible passed through the fault plane, causing spark plasma sintering of the weak clayish gouge and generating a strong magnetic field that induced anomalous magnetization (1).

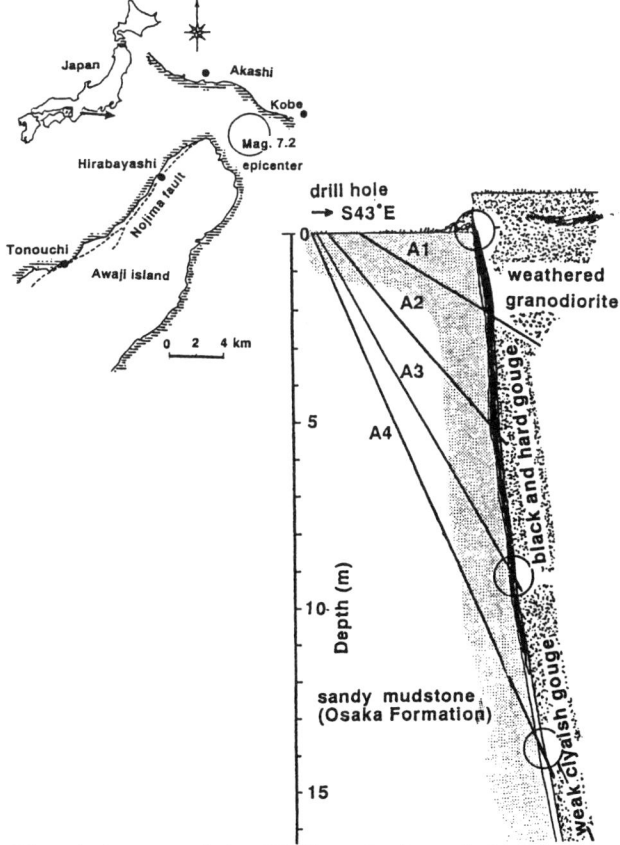

Fig. 1 A map of the Nojima fault and the cross sectional view accompaied by the Kobe E.Q.

(1) Y. Enomoto, Z. Zheng, submitted to Science

MECHANICAL SEAL BEHAVIOR DURING TRANSIENT OPERATIONS

SUSAN R. HARP and RICHARD F. SALANT
School of Mechanical Engineering, Georgia Institute of Technology, Atlanta, GA 30332-0405, USA

ABSTRACT

While most mechanical seals spend a large proportion of their life in steady-state operation, they frequently experience transients in operating conditions. The latter include startup, shutdown, and a variety of speed and pressure histories. It is necessary to assure satisfactory seal performance during these transients. One must avoid excessive leakage, high contact forces, and high heat generation rates, and also maintain stability and tracking of the seal faces.

Most practical operating transients are slow compared to characteristic dynamic times. Transient times of the order of one second or larger are typical. Provided the seal is stable and tracks, the seal behavior during the transient will be dominated by the face deformation history, which determines the coning history. As the shaft speed and sealed pressure vary, the changing mechanical and thermal loads will alter the mechanical and thermal deformation of the faces, and hence, the coning. This will alter the lubricating film thickness distribution, which affects the hydrodynamics of the film. The film pressure distribution will change, as well as the contact pressure distribution, the contact area, the heat generation rate, and the leakage rate.

It has been the objective of the present work to develop a model and computational scheme to predict the performance history of a mechanical seal during transient operation, as described above, assuming the seal's stability and tracking is assured (1). Such a model could be used to obtain an initial estimate of a seal's transient performance, including a history of the coning. Then, dynamic analyses could be used to check the stability and tracking at various times, using the previously computed coning as input.

In the present work, the model is applied to a particular seal, under particular operating conditions, and subjected to a particular transient, to illustrate its use and to display some of the types of seal behavior one can expect during transient operations. However, the model is intended to be applicable to a wide variety of seals, operating conditions, and transients.

The model consists of four primary elements: force balance, fluid mechanics analysis, contact mechanics analysis, and deformation analysis. It is assumed that the seal is axisymmetric, perfectly aligned, stable, and experiences only axial translation.

The force balance is justified by considering the equation of axial motion of the floating face. Order of magnitude estimates for practical operating transients indicate that the inertial term (the product of the mass and acceleration) is at least six orders of magnitude smaller than the individual forces acting on the face. It is therefore reasonable to neglect the inertial term, and use a quasi-static analysis in which the forces on the floating seal face are in balance at each instant of time. In such an analysis, the seal history is taken into account only through the changing operating conditions of the transient and the squeeze film term in the Reynolds Equation.

The fluid mechanics analysis consists of a finite difference solution to Reynolds Equation for the film pressure.

The contact mechanics analysis uses a plastic contact model, and assumes a Gaussian distribution of asperity heights, to yield contact pressures.

Both the film pressure distribution and the contact pressure distribution are strongly dependent on the film thickness distribution. The latter, in turn, is strongly dependent on the mechanical and thermal deformation of the seal faces. Such deformation is produced by the pressure distribution between and around the faces, the contact forces, centrifugal forces, temperature differences across the seal faces, and heat generation in the sealing gap. Since the deformation computation must be imbedded in an iteration loop, conventional finite element analysis for each iteration would be computationally expensive. Instead, an influence coefficient method is used, in which it is assumed that the deformation is linearly related to the applied forces and heat fluxes. Separate influence coefficients are defined for the sealed pressure, the centrifugal force, the film and contact pressure force, the heat generation rate, the clamping force, and the temperature difference between the sealed and ambient fluid. These influence coefficients (which are generally matrices) are computed off-line, using a finite element analysis.

To determine the seal conditions at any instant of time, an iterative procedure is necessary due to the close coupling of the fluid mechanics, contact mechanics, and deformation processes, as described above. The analysis requires that the deformations produced by the forces reciprocally produce the same forces. At each instant in time, the solution process further seeks the conditions for which the forces on the floating seal face are in balance.

REFERENCES

(1) S.R.Harp, <u>A Mathematical Model of a Mechanical Seal Under Transient Operating Conditions</u>, M.S. Thesis, Georgia Institute of Technology, 1996.

PHYSICALLY BASED MODELING OF SLIDING LIP SEAL FRICTION

DIRK A. WASSINK and KENNETH C. LUDEMA
Dept. of Mechanical Engineering and Applied Mechanics, University of Michigan, Ann Arbor, MI 48109-2125
VIESTURS G. LENSS and JOEL A. LEVITT
Ford Motor Company, P.O. Box 2053, Dearborn, MI 48121-2053

Sliding lip seals, applied in hydraulic actuators and other hydraulic systems, develop friction forces which vary strongly with sliding conditions; thus, they may contribute significantly and variably to actuator or system dynamics. Estimation and control of seal friction requires a simple model, based as much as possible on physical processes and parameters.

A model for lip seal friction sliding with constant speed is developed. Three energy dissipation mechanisms form the basis for the model: 1) viscous shear loss in the lubricant; 2) hysteretic losses due to roughness-imposed deformation of the seal material, and 3) hysteretic losses due to near-surface seal deformation caused by varying intermolecular forces at the sliding interface. The contribution of each of these mechanisms is calculated for a given set of sliding conditions, and the three components are summed. The number of model constants has been kept less than ten; many of these have physical bases as well, but are difficult to determine by measurement or analysis. The constants remain fixed over all the simulations.

An important aspect of the friction model is the role of the lubricant in modifying deformations in the seal material during sliding. At low speeds, the losses due to viscous shear of the lubricant should be small, unless the molecularly thin lubricant film begins to behave as a solid. However, seal deformations caused by shaft roughness and varying intermolecular forces at the surface should be large at small speeds and increasingly attenuated by the lubricant as the sliding speed (and hydrodynamic film thickness) increases. The second and third model components account for attenuation due to the hydrodynamic film.

Model simulations of sliding lip seal friction track experimentally observed trends in friction with temperature, hydraulic pressure, shaft roughness, oil viscosity and seal visco-elastic properties, which were also measured in this study. In particular, the model accounts well for conditions under which a peak in the friction-sliding speed curve appears. Simulation results are presented in Figure 1 for a seal with a glass transition at about -25°C, a sealed pressure of 1.4MPa, a surface roughness of R_a=0.25μm, using a light paraffin oil lubricant at temperatures of 24°, 41°, 54°, 71° and 88°C. As Figure 1 suggests, the model component attributed to near-surface deformations in the seal caused by ordering-disordering cycles (2) in the thin lubricant film contributes most significantly to friction at low speeds. This component shows a friction peak which shifts toward higher speed and lower friction values as temperature increases. The friction peak occurs at speeds about three orders of ten lower than would be expected for the same temperatures in dry sliding, following the observations of Grosch (1) for a rubber pad sliding on steel without lubricant.

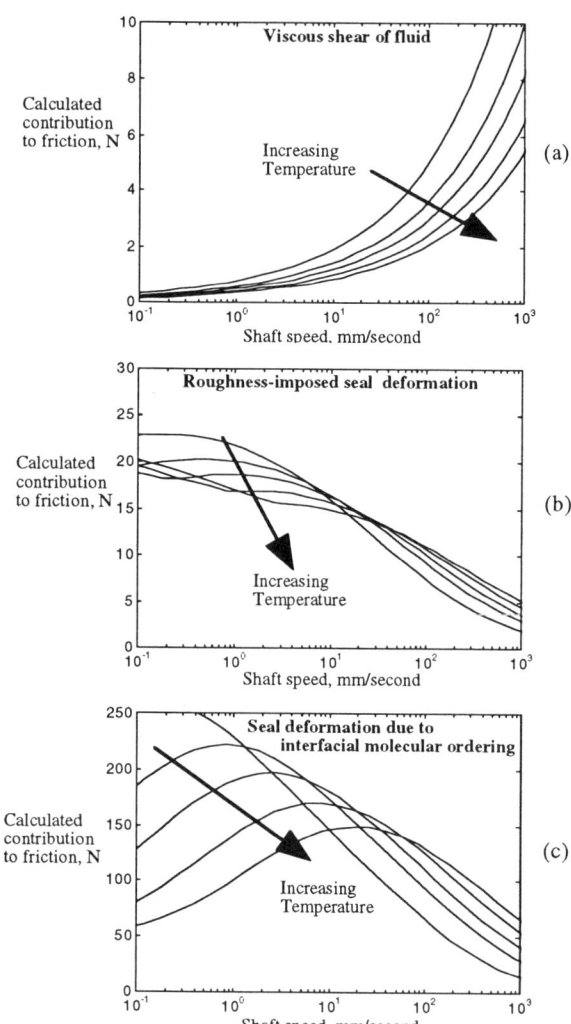

Fig 1: Seal friction contributions due to (a) viscous shear of lubricant, (b) roughness-imposed seal visco-elastic deformation, and (c) visco-elastic seal deformation due to changing intermolecular forces at the interface as predicted by the model.

REFERENCES

(1) Grosch, K.A., Proc. Roy. Soc., A274, pp 21-39.
(2) Israelachvili, J.A. in *Fundamentals of Friction: Macroscopic and Microscopic Processes*, I.L. Singer and H.M. Pollock eds., Kluwer, 1996, pp 374-376.

A NOVEL HIGH PRESSURE (UP TO 5000 PSI / 340 BARS) POLYMERIC ROTARY SHAFT SEAL

M. S. KALSI
Kalsi Engineering, Inc., 745 Park Two Drive, Sugar Land, TX 77478 USA

Even though the conventional elastomeric and high performance plastic seals (e.g., chevron packings, spring energized PTFE seals, V-seals, and U-cup seals) are widely used in rotary shaft sealing applications, their performance is limited to relatively low pressure (P) and velocity (V) combinations. These seal designs suffer from the same basic drawback, i.e., the seal material directly rubs against the rotating shaft surface, thus resulting in relatively high friction and excessive heat generation at the sealing interface. Brute force approaches primarily focusing on material improvements have extended the PV values to 200,000 psi x ft/min (69 bars x m/s), which is the current limit for conventional elastomeric/plastic seals. In general, operation at pressures above 500 psi (34 bars) has been found to be unacceptable when surface speeds are 400 ft/min (2 m/s) or more. Operaton above the PV limit of 200,000 psi x ft/min leads to failure due to blistering, scorching, or melting of the seal material and grooving of the shaft.

A novel polymeric rotary shaft seal has been developed that overcomes these limitations and dramatically extends the PV range by a factor of 15 to 20 and pressures up to 5,000 psi (340 bars). This is accomplished by successfully employing the hydrodynamic lubrication principle in conjunction with a multiple modulus material construction (Fig. 1). The seal is installed with a radial interference in the gland with the same ease as an O-ring and statically performs in the same manner. However, the seal's unique geometry creates a wavy contact footprint on the lubricant side which, during relative rotation of the shaft, develops a velocity component (V_n) that wedges a film of lubricant at the shaft-to-seal interface causing the seal to hydrodynamically *lift* away from the shaft surface and ride on the film. During rotation, the direct contact between the seal and shaft is eliminated, the friction coefficients are very low (typical hydrodynamic range), heat generation is minimized, and wear of both seal and shaft surfaces is virtually eliminated. The sharp edge of the seal on the environmental side is designed to exclude environmental abrasives from entering the seal interface. A high modulus material is employed near the dynamic interface to bridge the extrusion gap under higher pressures. Laboratory tests have confirmed that the seal is capable of operating at pressures up to 5,000 psi and PV values up to 3.7 million psi x ft/min (1,275 bars x m/s) with very low friction coefficients in the range of 0.01 to 0.02 (Fig. 2). The design is covered by U.S. Patents 4610319, 5230520, and other pending U.S. and foreign patents.

The seal design is based upon elastohydrodynamic analyses and optical interferometric studies to predict film thicknesses and friction coefficients over a wide range of pressures and speeds. The seal is being used successfully in several critical high pressure applications in drilling described in Reference 1.

Reference
1. M. S. Kalsi, et al. *A Novel High-Pressure Rotary Shaft Seal Facilitates Innovations in Drilling and Production Equipment*, SPE/IADC 37627, 1997.

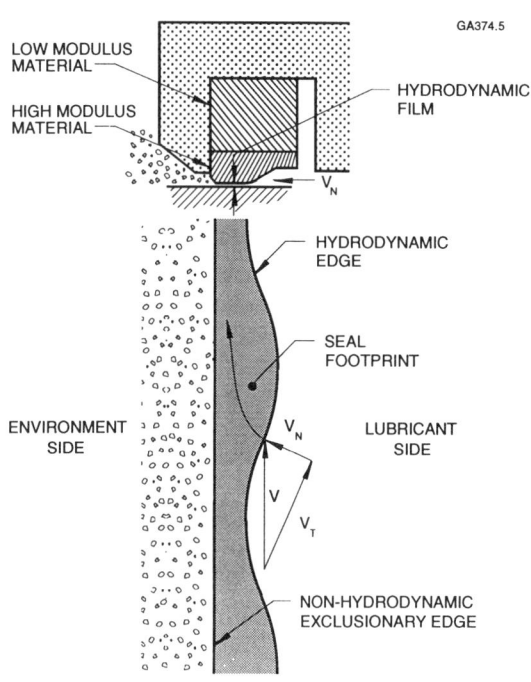

Fig. 1: Principle of Patented Rotary Seal

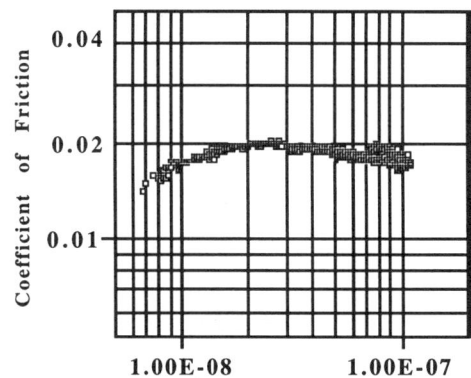

Fig. 2: Low Friction Coefficients Exhibited by the New Seal Design Tested at 5,000 psi and Speeds Up to 700 ft/min

Submitted to *ASME Journal of Tribology*

NEW MODEL OF INTERACTION OF FRICTION ZONES IN DEVICES FOR SEALING ROTATING SHAFTS

A.V.CHICHINADZE
Assoc. of Engineer-Tribologists, Kursovoi per.,17,AIT, Moscow, 117049, Russia
O.G.CHEKINA
Inst. for Problems in Mechanics, pr.Vernadskogo,101, Moscow, 117526, Russia
A.YU. KOJAEV
Gubkin State Academy of Oil and Gas, Lenin pr.,65, Moscow 117917, Russia

ABSTRACT

Most contemporary models of rotating seals are based on an assumption that a thin fluid film exists between a shaft and a seal. The methods of EHL theory are used for prediction of operation parameters [1]. Modern numerical techniques allow to take into account various effects, including roughness influence in such models. However, there are some issues that call for the development of alternative models.

1. All models based on hydrodynamics predict that leakage is strongly dependent on the viscosity. However, our experiments with rubber-metal seals revealed no pronounced influence of viscosity on the leakage for such different fluids as water, kerosene, petrol, and heavy mineral oil (Fig.1).

2. It is recognized that leakage rate (and hence the lubrication and wear of seal and shaft surfaces) is strongly dependent on roughness. Seal operation is poor for surfaces that are too smooth [2].

These facts indicate that for some types of seals the pattern of the fluid motion and hence the amount of leakage is determined mainly by geometrical parameters. The notion of composite roughness is not sufficient for the process description; the relative motion of two rough surfaces is likely to be an important mechanism of liquid transportation.

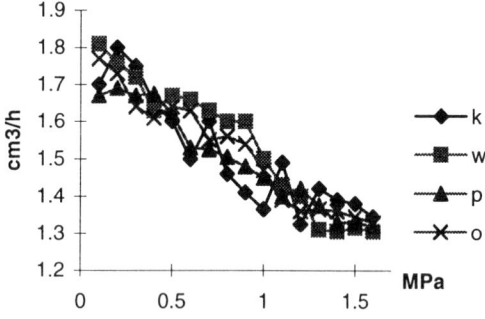

Fig.1 Leakage vs.pressure

The following model is proposed. A contact of two rough surfaces is considered. The ratio of pressure to elastic modulus of deformed material is assumed to be large enough to provide large area of contact. Then, the areas of no contact form a system of microvolumes (voids) moving randomly inside the area of nominal contact during mutual seal/shaft sliding. The voids generated at the seal edge that is connected with the sealed fluid, capture some amount of the fluid. The wandering of filled voids in contact zone provides seal-shaft contact lubrication. If a filled void reaches the opposite edge of the contact zone, an elementary leakage event takes place.

Based on the above assumptions, a computer model is developed that allows one to simulate the influence of pressure, seal macrogeometry, roughness, presence of micro features (such as grooves) on leakage. It predicts some incubation period during which no leakage occurs, and a steady-state leakage regime. Such integral parameters as leakage rate and fluid distribution along the leakage direction inside the area of nominal contact ($l(x)$) can be calculated. The latter parameter determines the quality of lubrication and, hence, the wear of the seal. Figure 2 illustrates $l(x)$ for the initial and the steady-state stages of seal operation for a symmetric seal with parabolic cross-section. At the initial stage, only a part of the contact zone is lubricated. Steady-state lubrication is low in some regions. A more uniform lubrication can be achieved by optimizing the seal cross-section shape.

The proposed model needs further development and experimental verification. However, it does predict some qualitative features of rubber-metal seal operation.

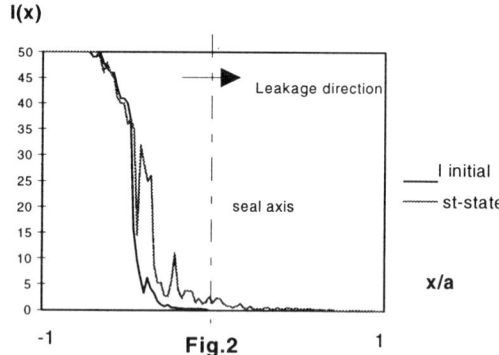

Fig.2

REFERENCES

(1) R.F. Salant, Trans. of the ASME, Ser.F, Journal of Tribology, Vol.118, No.2, April, 1996,292 - 296 pp.
(2) R.L. Dega, Trans. of the ASME, Ser.F, Journal of Lubrication Technology, Vol.90, No.2, April, 1968, 382 - 394 pp.

DEVELOPMENTS IN FLUID LUBRICANTS FOR SPACECRAFT APPLICATIONS

S GILL

ESTL, AEA Technology plc, Risley, Warrington, WA3 6AT, UK

ABSTRACT

Since its formation in 1972, ESTL (European Space Tribology Laboratory) has maintained a continual research programme into fluid lubricants for spacecraft applications.

This paper will draw together the data as presented over the years and discuss the limitations that currently exist to oil/grease performance in the vacuum of space. The routes ESTL research is following to enhance this performance will be discussed.

Early spacecraft were short-lived with short tribological life requirements. Early usage of silicone oils was quickly abandoned as they tended to degrade and creep away from the contact zones.

The arrival of the perfluoropolyalkylether (PFPE) oils and greases, with their very low vapour pressure and high molecular weights (~10,000) promised great advances. It was soon realised however that there were drawbacks. In vacuum they polymerise when in tribological contact with steels, a phenomenon first reported by ESTL (1), and they also do not readily accept performance enhancing additives. Despite these drawbacks they have proved the mainstay oils/greases for use in space for many years when operated in open bearing systems.

Much work has been performed with a view to delaying the onset of this polymerisation. The most widely used technique has been via using TiC-coated balls which has been shown to delay the degradation by a significant margin (2). From work performed on boundary lubrication simulated testers it is also clear that thin film lubricant coatings under the PFPE also assist in this process, although this technique has not yet found widespread acceptance. Recent developments have shown that Braycote 601 PFPE grease operating above a coating of ion-plated lead allows linear motion screw devices to outperform either lubrication technique in isolation by a significant margin. This approach has been baselined for a number of programmes.

A further approach, in which ESTL has not been active, is the inclusion of additives in the PFPEs. Early attempts at this were unsuccessful, however recent results are more encouraging, and indeed there are other papers about this area of study at this conference.

Other sealed mechanisms in Europe like momentum and reaction wheels, where low torque performance is paramount and subsequently oil quantities are at a minimum, have continued to utilise highly refined mineral oils inherited from the gyro industry, like SRG-60 and KG80. The original supplies of these oils have now been used up and since the oil manufacturers have ceased their production, the wheel manufacturers are now looking for alternatives.

The most recent oil to enter the spacecraft mechanism arena is a synthetic oil formulated by Pennzoil, called Pennzane SHF X-2000 (marketed by W F Nye Inc). This oil is a multiply alkylated cyclopentane, and its chemical constituents of carbon and hydrogen preclude any tendency to polymerise. A further benefit is that this oil readily dissolves additives, and terrestrial applications often use an anti-oxidant to inhibit breakdown during running. Its molecular weight is nominally 910, and so claims that the base oil vapour pressure is as good as that of the PFPEs are difficult to substantiate, however it is gaining widespread acceptance in both sealed and open systems. ESTL has performed long term tests which failed after a couple of years in vacuum due to oil starvation (3) and preload loss due to wear of the races. The interesting effect in these tests was that the bearings which failed were unsealed to the vacuum chamber, whereas a labyrinth delayed the loss of oil from other bearings significantly.

Further tests are now underway at ESTL with new batches of oil, both with and without additives. For comparison purposes tests are being performed both in air and vacuum under a soft preload at a slow speed which infers boundary lubrication conditions. The torque results initially were comparable, however with extended test durations the two samples in vacuum have shown a decrease in torque implying that oil is probably being lost.

The majority of work performed at ESTL on liquid lubricants has involved oils, however the most recent test campaign has compared the performance of a range of vacuum greases against temperature when operated in vacuum. Future work is planned to look in more detail at the endurance performance limits for these greases.

REFERENCES

(1) K T Stevens, Proc 1st European Space Mechanisms and Tribology Symposium, ESA SP-196, December 1983, pp 109-117.

(2) S Gill et al, Proc 5th ESMATS, ESA SP-334, April 1993, pp 165-170.

(3) S Gill & R A Rowntree, Proc 6th ESMATS, ESA SP-374, August 1995, pp 279-284.

TWO-PHASE FLOW IN FLOATING-RING SEALS FOR CRYOGENIC TURBOPUMPS

MAMORU OIKE
Tohoku University, Institute of Fluid Science, Sendai, Miyagi 980-77, Japan
MASATAKA NOSAKA, MASATAKA KIKUCHI and SATOSHI HASEGAWA
National Aerospace Laboratory, Kakuda Research Center, Kakuda, Miyagi 981-15, Japan

ABSTRACT

Reusable rocket engines are required for future space transportation systems to reduce costs. The success of reusable rocket engines depends on the durability of their elements, especially bearings and shaft seals for turbopumps. The floating-ring (FR) seal shows promise for application in reusable rocket engines because it does not entail rubbing contact under steady operations. Since a typical radial sealing clearance (h) of the floating-ring seal is 30 to 40 μm, the leakage rate of the floating-ring seal is much higher than that of the face contact seal. Therefore, a reduction of the leakage rate is required for the floating-ring seal.

Beatty and Hughes (1) introduced a mathematical model for the prediction of leakage rates in annular shaft seals. Their parametric studies indicated that vapor production through pressure drop or viscous heat generation in the seal is an important mechanism for limiting the leakage rate. Based on these studies, we considered a two-phase flow in the sealing clearance to reduce the leakage rate because of the low mass flux. However, experimental results (2) are insufficient to explain the effect of the two-phase flow in the floating-ring seal on the leakage rate.

In this paper, a flow visualization study on the two-phase flow in floating-ring seals for cryogenic fluid was carried out to identify the two-phase flow area inside the sealing clearance induced by viscous frictional heating and the pressure drop. Four types of the polycarbonate floating-ring were tested at a sealed pressure range of 1.0 to 1.4 MPa and a rotational speed range of up to 40000 rpm, LN_2 being used as the sealed fluid. The effect of the two-phase flow area (A_2) on leakage was investigated by comparing leakage measurements (Q) and the values of $A_2=L_2/L$ obtained from observational results, as shown in Fig. 1. A_2 increased with increasing rotational speed; however, the increase of A_2 did not always bring about a decrease in the mass leakage rate, as shown in Fig. 2. Moreover, the effect of the two-phase flow on the reduction of leakage was discussed by comparing the measurements of Q with the computational results (Q_{liquid}) obtained from the incompressible flow equation. It was confirmed that the effect of the two-phase state on the reduction of leakage can be broadly classified into the following two effects: effect of the all-liquid choked flow (1) and effect of the two-phase flow area inside the sealing clearance. In a high degree of subcooling conditions, the increase in A_2 tends to depress the reduction effect on the leakage rate.

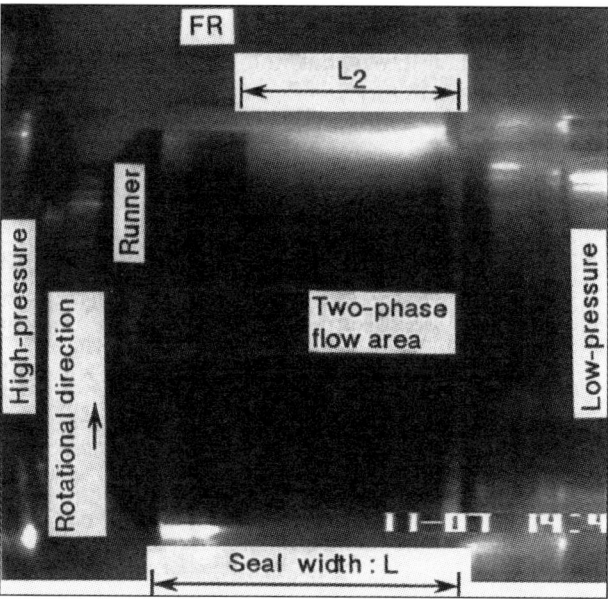

Fig. 1: Observation of two-phase flow in seal FR:A
(N=24000 rpm, ΔP_s=1.25 MPa, T_0=96.6 K, L=30mm)

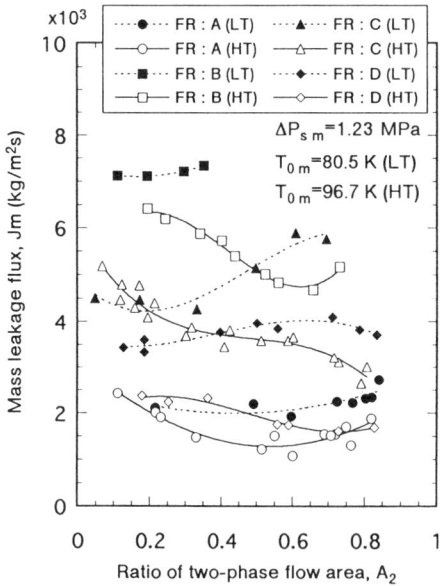

Fig. 2: Effect of tow-phase flow area on leakage

REFERENCES

(1) Beatty, P.A. and Hughes, W.F., ASME Journal of Tribology, 112, 1, pp 372-381. (1990).
(2) Oike, M., Nosaka, M., Kikuchi, M. and Hasegawa, S., Proceedings of the International Tribology Conference, Yokohama 1995, pp 1871-1876. (1995).

TRIBOLOGICAL BEHAVIOUR OF MATERIALS AT CRYOGENIC TEMPERATURES

W HÜBNER, TH GRADT, TH SCHNEIDER, H BÖRNER
Federal Institute for Materials Research and Testing (BAM), D-12200 Berlin, Germany

INTRODUCTION

Tribologically stressed cryotechnical systems are characterized by the low temperatures and the special environmental demands. Low temperatures affect the mechanical properties of the materials. Multicomponent systems are affected by thermal stresses caused by different coefficients of thermal expansion. The environmental conditions in inert gases are similar to vacuum. In hydrogen or oxygen the materials can react with the environment even at low temperatures, because of the activation of the surfaces.

The technical solutions for cryo-tribological systems acquired up to now do not meet the all requirements of new cryotechnical systems. E.g., the tribosystems developed for the launchers for the space shuttle or the Ariane rockets, powered by LH_2 and LOX, can be used for magnets with superconducting coils or for hydrogen-fuelled airplanes only with reservation (1). Often informations are missing about the low temperature suitability of materials.

On this background at BAM a project was started for investigations of materials at low temperature stressing and, in future, in hydrogen to obtain materials characteristics for design in cryogenics.

EXPERIMENTAL

For low temperature investigations the tribometers were designed and constructed. CT1 can be operated as bath cryostat or as continuous flow cryostat. In flow cryostat operation the sample is surrounded by gaseous helium and the temperature can be adjusted between 4,2 K and room temperature. In bath cryostat operation the chamber is filled with the cryogen and the sample is submerged in the liquid. The temperature corresponds to the boiling point of the medium used (LN_2: 77 K; LHe: 4,2 K). The boiling liquids cause effective cooling of the sliding contact so that measurements with high friction power are possible.

CT1 is designed for easy handing and rapid cooling. It is appropriate for a fast survey of a great number of materials, but it is limited to low cooling powers, low loads and inert environments.

To overcome these limitations further tribometers were constructed. Tribometer CT 2 was designed for investigations in LHe and LH_2. It works with a bath cryostat for higher frictional powers. Tribometer CT 3 is similar to CT 1 with respect to the cooling operation, however, the device is designed for pressure up to 20 bar. This enables investigations in hydrogen beyond the critical point what is of importance e.g. for the design of LH_2 pumps. CT 3 is suitable also for tests of sliding and rolling bearings with loads up to 10 kN.

Experiments were carried out with coatings and polymers under the standard stress parameter of a normal force of 5 N, a sliding speed of 0,2 m/s and a sliding distance of 1,800m.

RESULTS

TiN-coated steel samples showed friction and wear results similar to room temperature behaviour: The main influences observed were caused by the environmental conditions. For ADLC-coatings on steel substrates internal stresses created by mismatch of the thermal expansion coefficient led to premature failure of the samples; best adhesion is achieved with single layered and thin coatings.

Polymers showed strong dependence on the temperature: all materials exhibit dramatically decreased friction and wear. However, differences could be observed in the topography of the sliding surface (2). From the morphology of worn surfaces of polyimide it can be deduced that the adhesion is changed in the low temperature regime. At room temperature the high friction owing to the adherence to the metal counterface caused intense crack formation. At 77 K in helium only some small cracks, and in LN_2 only deformation marks at the surface are detected.

The influence of the molecular structure can be discussed with the PA 6 and PTFE results. Both show drops in friction and wear and exhibit similar topographies of the sliding planes. However, whereas PTFE transfer to the steel surface takes place down to the lowest temperature caused by its straight molecular structure, PA 6 causes abrasive wear of the counterbody. The molecular structure of PA 6 is characterised by hydrogen bonds which become very strong at low temperatures. At 77 K in He-gas the transfer of frictional heat is low, leading to higher temperatures in the contact area which enables material transfer. Improved cooling in LN_2 results in stronger bonds of the PA 6, strong enough to scratch the steel counterbody.

REFERENCES

(1) T Gradt, T Schneider, W Hübner, Materialprüf., Vol. 39, No. 1-2, 1997, 20-24pp.
(2) W. Hübner, T. Gradt, H. Börner, Tribol. u. Schmierungstechn. vol. 32, No.4, 1993, 150-177pp.

ANALYSIS OF TWO-PHASE FLOW IN CRYOGENIC DAMPER SEALS. THEORETICAL MODEL, MODEL VALIDATION AND PREDICTIONS *

GRIGORY ARAUZ and LUIS SAN ANDRÉS

Department of Mechanical Engineering, Texas A&M University, College Station, Texas 77843-3123

ABSTRACT

Cryogenic fluid damper seals operating close to the liquid-vapor region (near the critical point or slightly sub-cooled) are likely to develop a two-phase flow region which affects the seal performance and reliability. An all-liquid, liquid-vapor, and all-vapor, i.e. a "continuous vaporization" adiabatic bulk-flow model for prediction of damper seal dynamic forced response is presented. The two-phase region is regarded as a homogeneous saturated mixture in thermodynamic equilibrium. Temperature in the single-phase regions, or mixture composition (quality) in the two-phase region are determined from the fluid enthalpy transport equation. The large axial flow velocities typical of annular seals determine negligible heat conduction through the seal walls as compared to the heat carried by fluid advection.

Static and dynamic force performance seal characteristics are obtained from a perturbation analysis of the governing bulk-flow equations. The solution, expressed in terms of zeroth and first-order fields, provides the static parameters (leakage, torque, velocity, pressure, temperature and mixture quality [λ]), and dynamic force coefficients for specified excitation frequencies. The solution method of the flow equations is a CFD control volume algorithm on staggered grids.

Computed predictions for static seal characteristics, leakage and entrance pressure drop as depicted in Figure 1, correlate well with existing measurements for a liquid nitrogen seal with two-phase at the seal exit plane (1). Other calculations also agree well with test data for a gaseous nitrogen seal. Flow rates and mixture quality predictions reproduce other numerical results for a liquid oxygen seal case operating over a full range of two-phase flow conditions (2).

The effects of two-phase flow on the dynamic force coefficients and stability of damper seal operating with oxygen are discussed, see Figure 2 for results on direct stiffness coefficients vs. supply temperature for various ratios of whirl frequency (ω) to rotational speed (Ω). Here fluid compressibility effects, particularly for mixtures with low mass content of vapor, are of importance and determine an increase on seal direct stiffness and reduction of the whirl frequency ratio. The variations of mixture sound speed along the seal (Figure 3) are paramount in determining the seal dynamic forced response.

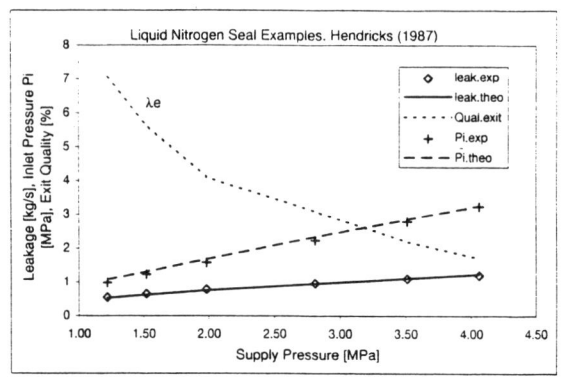

Fig. 1: Comparisons of numerical predictions with experimental results for LN$_2$ seal (1).

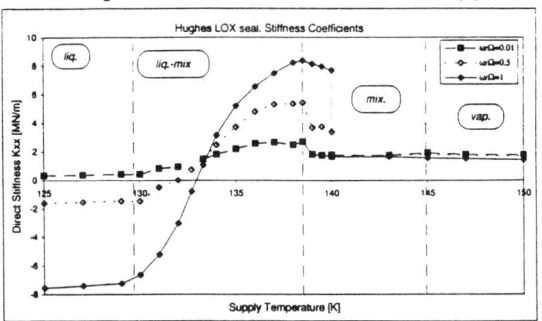

Fig. 2: Variation of dynamic direct stiffness vs. supply temperature in a LO$_2$ damper seal.

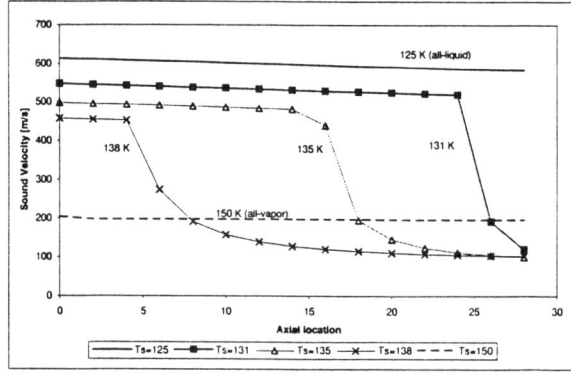

Fig 3. Variation of sound velocity along damper seal.

REFERENCES

(1) Hendricks, R.C., 1987, "Straight Cylindrical Seals for High Performance Turbomachinery," NASA TP-1850.

(2) Beatty, P.A., and Hughes, W.F., 1987, "Turbulent Two-Phase Flow in Annular Seals," *ASLE Transactions*, Vol. 30, pp. 11-18.

*Accepted for publication in the ASME *J. of Tribology*.

THE PROBLEMS OF THE SERVICE LIFE PREDICTION IN SPACE TRIBOLOGY

Yu. N. DROZDOV

Department of Friction, Wear and Lubrication, Mechanical Engineering Research Institute of Russian Academy of Sciences, Moscow, Russia

ABSTRACT

Space tribology has been growing for more than 40 years, the experience in this field accumulates together with the creation of artificial Earth satellites, orbital stations, planetary rovers, devices and equipment for the space, the Moon, Mars, Venus, and the other planets study.

Further development of space technique and tribology will be induced by the works directions in the fields: observation of the Earth for the resource reconnaissance, for the meteorology and movement management; the maintenance of global communication and navigation in "the Earth—the Moon" system; the creation and maintenance of multimission large-scale stations and industrial complexes performance; research and assimilation of long-distant space; creation space energetic complexes; derivation to the space area harmful waste; creation of protection systems from asteroids and available terrorism; for the space tourism organization etc. If now there are 400-500 ton/year of useful loads derive to orbit, than after 10 years it will be 8-10 times as much. The mass of objects, their value, resource of active performance are grow up. For example, international space station "Alfa" will be gathered have 454 tons and cost more than 100 billion of dollars, with the period of predicted performance 15 years. The tribological reliability of mechanical systems in this conditions is on importance.

The methods of assurance of normal operation in space of gear trains, rolling and sliding bearings, cam mechanisms, guideways, sealing elements, docking assemblies, separate connectors, hinges, latches, and complex mechanisms and devices have the "space" particularity. Above all it is necessary to have in view the absence of an oxidative medium, increased vaporization of the lubricants, high thermal stressing, high gradient of temperatures, the presence of radiation, the weightlessness and other factors.

It is great value of the physical simulation of tribological processes in application to space in laboratory conditions. However, experimental and theoretical researches are necessary on principle, it is impossible to form the mathematical model and to be steady in the success of real constructions performance without them.

The great class of complicated tribological problems were needed to be solved during the creation of "Lunokhods" transmission, of station complex "Luna", during the creation of the system "Energiya - Buran", and other space systems(1). In Russia, during the creation of the system "Energiya -Buran" there were developed 600 new technologies, 1000 specimens of industrial equipment, 30 kinds of construction materials, that have a significant meaning for the general engineering.

We worked out the methodology, that allows to evaluate the service life of joints of friction from self-lubricated materials, with sold lubrication coatings and with application of plastic lubricant materials. Space tribology inspired the creation of foundations of mechanical systems resource designing (by wear), necessary on principle for general engineering. In the produced methods physical, chemical and mechanical processes are considered, that accompany the wear process, determined and probabilistic approaches are used.

In connection with the relatively limited energy capacity of autonomous systems, with aspiration to the minimization of the mass and size characteristics, maintenance of the high reliability of the friction assemblies of space vehicles, and the difficulty involved with the condition of repair operations, it is advisable to determine in the design stage the expected energy losses and the reliability of tribological systems. The traditional methods of design of the friction assemblies, usually functioning in a medium of liquid lubricant materials, for the conditions of the work in open space are not applicable because of the significant difference in the nature of the surface damage (wear) of rubbing bodies in vacuum. International association of scientists and engineers, who carry on experimental and theoretical researches is necessary for the managing and successful solution of the strategic problems of space tribology.

REFERENCES

(1) Yu. N. Drozdov, Development of Tribology for Extremal Conditions, Tribology in the USA and Former Soviet Union: Studies and Applications, 1994, 305—319 pp.

WEAR RESISTANCE OF SUPER ALLOYS AT ELEVATED TEMPERATURES

JAMES L. LAWEN, JR. and SAIM DINC
GE Corporate Research & Development Center, Research Circle, Schenectady, NY, 12309, USA

SALVADOR J. CALABRESE
Rensselaer Polytechnic Institute, 15th Street, Troy, NY, 12180, USA
Troy, NY

ABSTRACT

Although tribological properties have been provided in many past studies, their applicability is often limited. Typically, the reported data is isolated to a single operating temperature or involves only a particular alloy wearing against itself. However, mechanical designs often involve different materials contacting and wearing against each other. Operating temperatures can also widely vary.

In response to this, an extensive sliding wear testing program was established to evaluate the wear resistance of several material combinations currently used in high temperature applications such as ground based gas turbines and aircraft engines. This type of data will provide a tool for selecting material combinations for minimizing wear and optimizing the life expectancy of component parts. Nickel and cobalt base superalloys, referred to as NBA and CBA, respectively, and iron base stainless steels were tested in different combinations, and their wear rates compared to determine optimal wear resistance.

The alloys are typically operated between 290°C and 480°C and were, therefore, tested within or near this range. Since all conditions except temperature were maintained, the results are presented in the form of wear rate (or weight loss per unit time) for relative comparison between material combinations. For reference, wear coefficients, hardness, and friction coefficients are also provided for some of the tests. Many tests other than reported within the paper were conducted. However, the results included do reflect the general trends within the material combination sets.

Typical test results are shown in Figures 1 and 2. Figure 1 shows a bar graph of the wear rates of CBA2 against itself and several other cobalt base alloys at 425°C. CBA2 has a much higher wear rate against itself than against the other cobalt base alloys. For this combination, a large quantity of loose debris was present on the sample surface near the wear track. Examination of the oxides and wear tracks showed that the wear track for this couple was continuously worn during sliding removing most of the protective oxide as well as some CBA2 metal from the sample surface.

Figure 2 shows the wear rates of NBA2 against various nickel base alloys at 425°C. The wear rates range from very high at 14.1 mg/hr to very low at 0.1 mg/hr. An important observation in this test set is the results of NBA2 against NBA4 as compared to NBA2 against NBA5. Although the mechanical properties of the three materials are similar, their wear rates are entirely different even though the tests were run at the same temperature and provided almost equal friction levels (0.38 and 0.39).

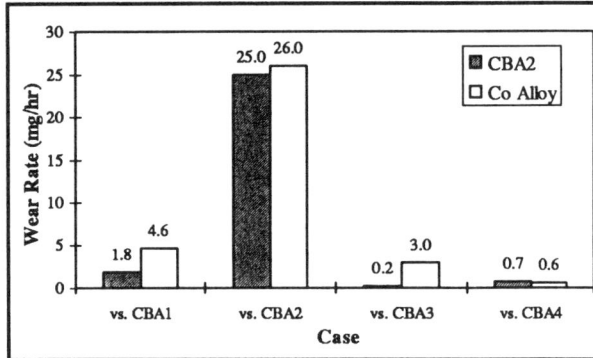

Fig. 1: Wear rates of CBA2 vs. various cobalt base alloys at 425 °C

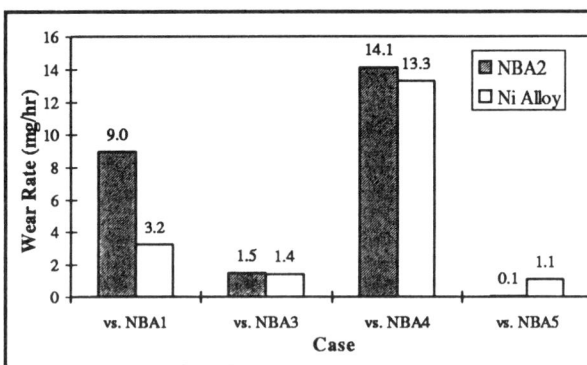

Fig. 2: Wear rates of NBA2 vs. various nickel base alloys at 425 °C

The complete test program results show that an alloy's wear resistance is highly dependent on operating temperature and its coupling with another material. The influences of friction, hardness, and oxide formation on the alloy's wear resistance are also presented and discussed in the paper of this title, which has been submitted to the Journal of Tribology for publication.

Advanced Material for High Temperature and High Speed Cylindrical Roller Bearings

Kenji Fujii, M. Itayama, K. Ito, Y. Fujii and S. Yokoi
Research Institute of Advanced Material Gas Generator, 4-2-6 Kohinata, Bunkyo-ku, 112 Tokyo, Japan

ABSTRACT

The Advanced Material Gas-Generator(AMG) R&D Project was initiated in 1993 as a ten year program with a joint investment by the Japan Key Technology Center and fourteen participating domestic companies. The program objective is to establish basic key technologies for next generation gas-generators using advanced materials. The gas-generator should have features such as a significantly lower fuel consumption with reduced weight and size. It should be environmentally acceptable, relative to future industrial, marine and aerospace gas-turbine requirements.

The investigation of a cylindrical roller bearing that is capable of operating at higher speeds(3-4 million DN) and temperatures (300-400°C) than current limits(2.3 million DN and 210°C) is a basic theme of the advanced components technology study for the project (1).

If the bearing operates at severe conditions, then advanced materials for the bearing become necessary. The first stage of investigation on this high speed roller bearing project was to conduct studies to identify suitable bearing component materials necessary to complete the remainder of the plan. The following results were obtained:

1. Race material

The higher speed will cause a centrifugally induced, very higher tensile hoop stress in the rotating bearing inner ring. A bearing using conventional material M50 is limited to approximately 2.3 million DN. The higher temperature will affect the hardness and reduce bearing life. A new carburized material which was modified to contain more chromium than M50, was identified with properties superior to that of M50 material. As Figure 1 shows, the new material has 5 times higher fracture toughness than M50 material. It has a higher hardness at high temperatures than M50 as well (2).

2. Roller material

At high speed, centrifugal forces at the roller/outer raceway contact produce a load which reduces bearing life. Low mass Silicon Nitride rollers are an excellent countermeasure for reducing the centrifugal load. In addition, under high speed operation, Silicon Nitride rollers strongly affect bearing operating temperature favourably (3).

3. Cage material

At higher speeds, a lower mass and a low friction material is necessary to reduce contact loads and increase sliding speed. Silicon Carbide particle reinforced aluminum matrix composites have a mass that is 60% lower than a steel cage and as Figure 2 shows, they have a lower friction coefficient than conventional cage materials as well.

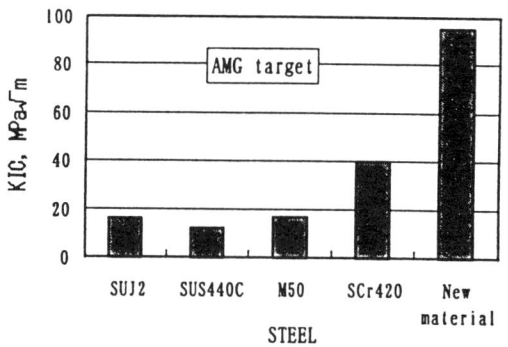

Fig. 1: Fracture toughness of various steels

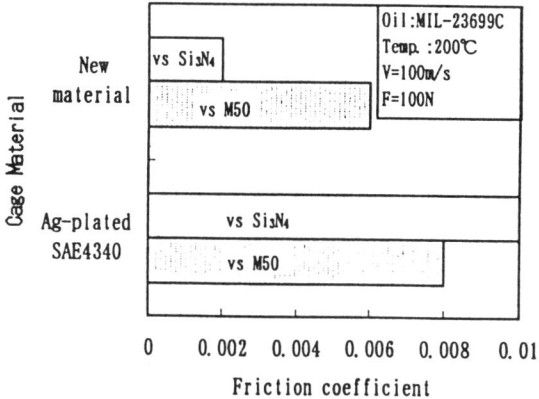

Fig. 2: Friction coefficient of cage materials

REFERENCES

(1) M.Itayama, Proceedings of the 1995 Yokohama Int. Gas Turbine Congress, vol. III, page 231, 1995.
(2) K.Ito, Proceedings of JAST Tribology Conference Kitakyushu, page 440, 1996.
(3) M.Itayama, Proceedings of the 74th JSME Fall Annual Meeting, vol. IV, page 13, 1996.

FRICTION AND PULL-OFF FORCES ON SUBMICRON-SIZE ASPERITY

YASUHISA ANDO
Mechanical Engineering Laboratory, Namiki 1-2, Tsukuba, Ibaraki, Japan
TAKASHI HORIGUCHI
Chuo University, Kasuga 1-13-27, Bunkyo-ku, Tokyo, Japan

ABSTRACT

The friction and pull-off forces between asperity with various radii of curvature and a scanning probe of an atomic force microscope (AFM) were measured. The asperity was created by using a focused ion beam (FIB) to mill patterns on a single crystal plate of silicon through a sputtering process. Fig. 1 shows an example of AFM image of asperity on a silicon plate. By changing the milling conditions, we obtained asperity with the radius of curvature of about 70 to 300 nm. The probe tip of the AFM was a square flat, 0.7 x 0.7 μm^2. The contact area calculated from the JKR theory (1) was in the range of 50 to 500 μm^2.

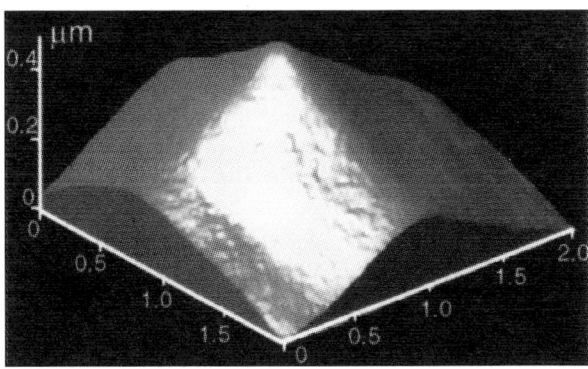

Fig. 1: AFM image of the silicon asperity processed by FIB

When measuring the pull-off force, we used the force-curve mode that shows the force required to pull the scanning probe tip off the specimen. The friction force was measured from the torsion of cantilever. Fig. 2 shows relation between the friction and pull-off forces measured on silicon asperity with various radii of curvature. The friction force was proportional to the pull-off force, thus agreeing with previous results for forces measured with extremely low normal load using a reciprocating friction tester (2). That report states that the friction force is proportional to the sum of the normal load and the pull-off force. In the experiments here involving an AFM, the normal load was less than 5 nN, which is much less than the pull-off force, resulting in the friction force being proportional to only the pull-off force.

Fig. 3 shows relation between the friction and pull-off forces measured on an asperity with radius of curvature of 110 nm. When the relative humidity was changed from 2 to 64%, the pull-off force on a spherical asperity was independent with the relative humidity. If the Laplace pressure is considered, the attraction force caused by water bridge between a spherical asperity peak and a flat probe surface is theoretically proportional to the asperity curvature and does not change by the relative humidity.

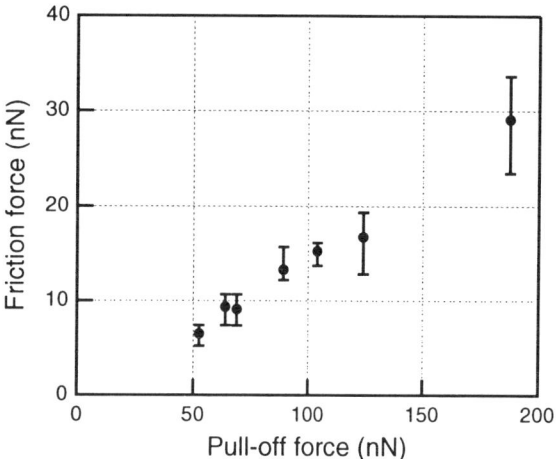

Fig. 2: Relation between friction and pull-off forces measured on silicon asperity

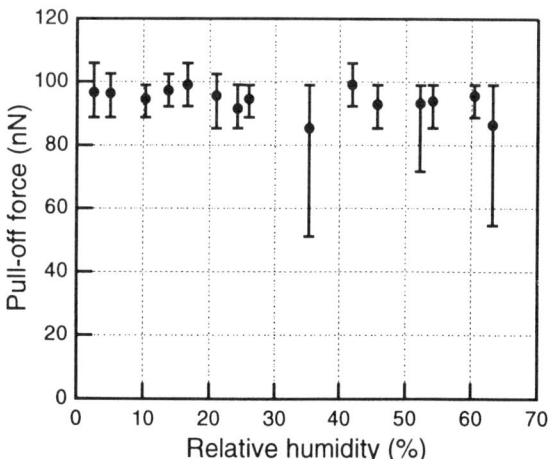

Fig. 3: Relation between pull-off force and relative humidity

REFERENCES

(1) K L Johnson, K Kendall, A D Roberts, Proceedings of the Roy Soc. Lond A, Vol. 324, 1971.
(2) Y Ando, Y Ishikawa, T Kitahara, Transactions of ASME Journal of Tribology, Vol. 117, No. 4, 1995.

Submitted to *Wear*

FRICTION DURING ATOMIC FORCE MICROSCOPE SCANNING OF IONIC SURFACES

ALEXANDER L. SHLUGER
Department of Physics, University College London, Gower Street, London WC1E 6BT, UK
ALEXANDER I. LIVSHITS
The Royal Institution of Great Britain, 21 Albemarle Street, London W1X 4BS, UK

ABSTRACT

Models of contact between Atomic Force Microscope (AFM) tips and surfaces are often considered as a prototype for the interaction between two rough surfaces and from this perspective are useful for understanding of adhesion, friction, and tribocharging. Torsional deflection of the AFM cantilevers produced by lateral(friction) forces can be measured with high accuracy using various experimental techniques. Monitoring these deflections, one can construct surface images which in many cases have lattice periodicity and are complementary to those obtained by monitoring the vertical cantilever deflections. The nature of atomic periodicity in friction force variations observed in many such experiments even at high tip loads is the topic of current debate.

As the AFM tip and the surface in contact make a common system with particle and energy exchange, there are at least two sources of instabilities which may cause friction. One can be called "internal" and appears when there is dynamic exchange of ions between the tip and the surface, or ions are displaced inside the surface and cannot return back to their original sites during the scanning cycle. This depends on the tip structure and scanning conditions and is determined by the character and speed of the tip and surface relaxation. Another, "external", source of instabilities is much better studied and is due to softness of macroscopic cantilever spring. This induces stick-slip tip behaviour even when the adiabatic potential of the tip-surface system is periodic. This mechanism explains observed friction force surface images (1). However periodicity of the tip-surface adiabatic potential even at high tip loads remains a mystery due to strong surface disruption during scanning. One possible answer to this question could be that the tip contamination by the surface material plays some role in interaction between two surfaces. Mechanisms of sliding involving tip modification seem to be more general than those based entirely on the rigid tip structure and geometry.

To study these mechanisms, we developed a theoretical model of SFM using a molecular dynamics method for the calculation of the interaction between a crystalline sample and a tip nanoasperity, combined with a semiempirical treatment of the mesoscopic van der Waals attraction between tip and surface, and the macroscopic parameter of cantilever deflection. The main features of the AFM experiment were modelled, including force vs. distance curves at various tip positions on the surface, and scanning of a perfect LiF surface in contact regime. As a model for a hard oxide tip we used MgO (2).

To have a representative AFM model, we first considered the contact formation which consists of surface indentation and tip retraction. The surface indentation is accompanied by large displacements of the surface ions inside the surface and their adsorption onto the tip. The 'jump-off point' determines stability of AFM operation in attractive regime. It depends on the depth of surface indentation and afterwards changes during scanning if tip is modified. After determining the force vs. distance parameters of our AFM model, we modelled the surface scanning with the tip velocity of 1 m/s, and at repulsive force of 1 - 2 nN.

Initially, the tip-surface interaction exhibits irregular stick-slip behaviour due to the transient adsorption of Li^+ and F^- ions onto the tip and strong surface and tip distortions. Our simulations demonstrate several processes of atomic instabilities and charge transfer responsible for "internal" friction. However, after some period of transformation the structure of the cluster adsorbed on the tip stabilises and the interaction becomes periodic. This demonstrates that tip contamination due to adhesion to the surface atoms may promote periodic AFM imaging, if the adsorbed surface material makes stable structures on the tip. Further calculations showed that the adsorbed cluster can adjust its structure to direction of scanning by exchanging atoms with the surface and changing its structure. We believe that this dynamic transformation of the surface material on the tip during scanning could be a general effect. This tip modification makes the tip-surface interaction less destructive which we call "self-lubrication".

REFERENCES

(1) T. Gyalog, M. Bammerlin, R. Luthi, E. Meyer, and H. Tomas, Europhys. Lett. Vol. 31, pp. 269-274, 1995.
(2) A. I. Livshits and A. L. Shluger, Faraday Discussion No. 106 (at press).

Superlubricity Effect In Contact Mode AFM Due To Ultrasonic Vibration Of The Sample

F. DINELLI, O. KOLOSOV, G. A. D. BRIGGS
Department of Materials, University of Oxford, Parks Road, Oxford, OX1 3PH
S. K. BISWAS
Indian Institute of Science, Bangalore, India

ABSTRACT

The atomic force microscope (AFM) provides a useful means of investigating friction at the nanometre level (1). In recent years, many studies have been undertaken to understand the nature of friction and wear in these conditions.

A modified AFM has been recently developed to sense the dynamic mechanical properties of a surface. This microscope, called the ultrasonic force microscope (UFM), is a contact mode AFM in which the sample is vibrated at an ultrasonic frequency. The free resonance frequency of the cantilever is more than two orders of magnitude lower than the excitation frequency and so, although there is some cantilever vibration at that frequency, its vibration amplitude is small compared with the excitation amplitude. The tip-sample distance is thus modulated. If the ultrasonic amplitude is high enough to reach the strong non-linearity of the force vs distance curve, an additional 'ultrasonic' force acts on the tip (2).

In this paper we present a study of dynamic friction in an ambient environment in a AFM when ultrasonic out-of-plane vibration is applied to the sample. The normal load is kept constant through a feedback circuit while the friction force is measured using a lock-in amplifier (3). Three different materials have been used as samples: glass, silicon and mica. We have used cantilevers of various elastic constants (from 0.05 to 0.24 N/m) with tips of different materials (SiO_2 and Si_3N_4) and various radii (nominally from 10 to 50 nm).

Measurements without the ultrasound show that the friction force vs. load curve is almost linear for all the cantilever-sample combinations used. This suggests that a multi-asperity contact is taking place.

In the figure the representative dependencies of dynamic friction and ultrasonic force on the ultrasonic amplitude are presented (Si sample, Si_3N_4 microlever). The graph shows measurements for two different values of the normal load F_i (i = 1, 2). As the ultrasonic amplitude is increased from zero the friction force decreases at first slowly. This decrease always commences at very low ultrasonic amplitudes irrespective of the normal load. When the amplitude reaches a critical value a_{ci}, a cantilever deflection shift onsets due to the strong non-linearity of the force curve. At amplitudes above a_{ci} the contact begins to break for part of the ultrasonic cycle and the friction force rapidly goes to zero. The same results were obtained for various cantilever-sample combinations.

The vanishing of friction at amplitudes higher than a_{ci} can be explained by the fact that the contact is broken for part of the ultrasonic cycle. The decrease at low amplitudes where the contact is not broken is more puzzling. In fact at low amplitudes the average normal load is equal to the initial set value and according to the Amontons law the friction force should not vary. The pull-off force is not significantly modified by ultrasound and it varies by less than 5% over the whole range of applied amplitudes. A mechanism based on slip at the minimum contact area in each cycle may also be discounted. In fact in a set of static friction measurements, we also observed friction reduction at a small ultrasonic amplitude which was not dependent on the maximum applied shear force.

Therefore we suggest that such a behaviour indicates the presence of a liquid layer formed by water (possibly with organic contaminants) organised in a solid-like structure between the tip and the sample sustaining the load. Such a layer should exhibit a viscoelastic behaviour. At ultrasonic frequencies, even at low ultrasonic amplitudes, the viscosity does not allow the soft layer to relax leading to a momentarily reduction of the normal load and, therefore, of the friction force.

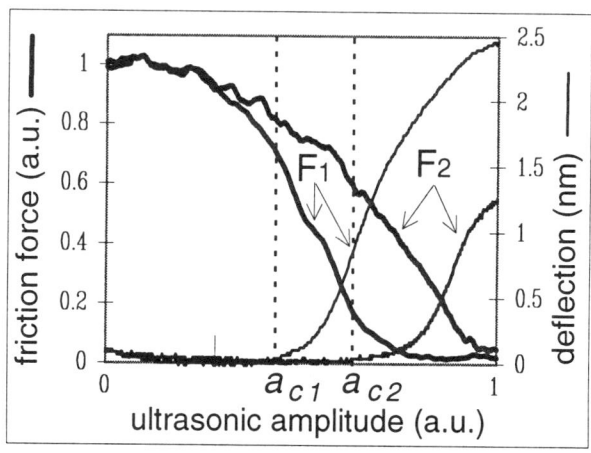

REFERENCES

(1) C. M. Mate, G. M. McClelland, R. Erlandsson and S. Chiang, Phys. Rev. Lett. **59**, 1942 (1987)
(2) O. Kolosov and K. Yamanaka, Jpn. J. Appl. Phys., Vol. 32, p. L1095, 1993
(3) J. Colchero, M. Luna and A. M. Baró, Appl. Phys. Lett., Vol 68, p. 2896, 1996

A NANO-TRIBOMETER FOR SINGLE ASPERITY SLIDING CONTACTS

W. N. UNERTL and S. D. DVORAK
Laboratory for Surface Science and Technology, University of Maine, Orono, ME 04469-5764, USA

ABSTRACT

We describe a new nanotribometer specifically designed to study sliding contacts with contact diameters in the range from a few nanometers to a few micrometers. This instrument currently operates in ambient air atmosphere, and is optimized to operate in the gap between the ultra-low loads of scanned probe microscopes and the mN loads of more traditional microtribometers. Based on the design and operation of a standard atomic force microscope, this nanotribometer is patterned after a prototype metrology instrument developed at NIST by Schneir, et al. (1). Like the NIST instrument, the nanotribometer consists of the three stacked stages shown below in Fig. 1: the flexure stage, the metrology stage, and the force head.

Several innovative features of the nanotribometer make it especially suitable for quantitative friction and wear studies at intermediate force and length scales. A piezo-actuated flexure stage allows horizontal sample displacements up to 45 µm at speeds up to 1 mm/s. Samples held in this stage can be heated to 120 °C in situ. A Zygo interferometer measurement and feedback system is mounted on the metrology stage, and gives accurate and repeatable horizontal positioning the sample stage and reduces thermal drift. This feedback system allows us to perform cyclic wear experiments, precision scratch tests and nanolithography experiments.

A Queensgate piezostack with integral capacitance gage controls the vertical sample position. A feedback system eliminates uncertainty on the vertical axis due to piezo non-linearities. This capacitance gage yields topography information accurate to within one nanometer

The nanotribometer force head uses an optical balance technique for measuring normal and lateral forces. Bending and twisting modes of a cantilever force sensor are used to simultaneously measure the normal (applied) and lateral forces exerted on the cantilever probe tip. A continuous PIN diode position-sensitive-detector (PSD) is used to monitor the bending and twisting of the cantilever. This detector allows the measurement of much larger forces than the standard four-element quad detector used on similar scanned probe instruments. The nanotribometer force head is designed to provide clearance for a variety of home-built cantilever force sensors, as well as accommodate in situ calibration optics and filters.

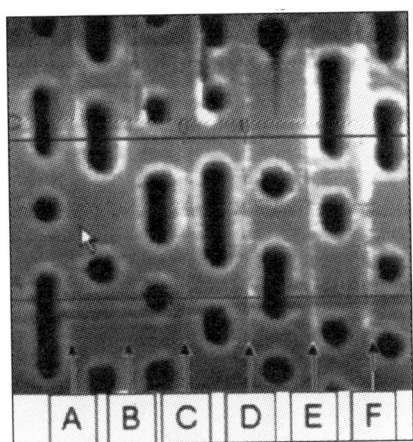

Fig. 2: A cyclic wear test on polycarbonate

This instrument has been used to perform single-pass scratch and cyclic wear experiments under controlled load and speed to investigate the onset of wear in polymers. The image shown in Fig. 2 was made by the nanotribometer after a typical wear test, in this case, on an uncoated compact disk. The image shows six parallel tracks, 10 µm long, which are vertical in this image. Each track represents 50 cycles at 6 µm/s, at a constant normal force, ranging from 400 nN (track A) to 1.4 µN (track F).

Fig. 1: The nanotribometer, showing the
A - flexure stage, B - metrology stage,
and C - force head

REFERENCE
(1) J. Schneir, T.H. McWaid, J. Alexander, and B.P. Wilfley, J. Vac. Sci. Technol. B 12 (1994) 3561.

THREE-DIMENSIONAL MOLECULAR DYNAMICS ANALYSIS OF ATOMIC-SCALE INDENTATION

W. YAN and K. KOMVOPOULOS
Department of Mechanical Engineering, University of California, Berkeley, CA 94720, USA

ABSTRACT

In this paper, the deformation of a dynamic solid substrate indented by a hard tip is investigated, and the effects of the substrate temperature and indentation speed on the interfacial force, elastic-plastic material behaviors, and energy dissipation are elucidated.

Valuable insight into nano-scale surface properties of materials has been obtained from molecular dynamics (MD) simulations. Landman et al. (1) considered the indentation of a gold substrate by a nickel tip (or vise versa), and examined the evolution of adhesion, indentation, and necking phenomena at the tip/substrate interface. Harrison et al. (2) studied the effects of surface crystallographic direction, temperature, normal load, and scanning velocity on the friction coefficient of two relatively sliding hydrogen-terminated diamond surfaces, and Belak and Stowers (3) presented results for the stresses in a metallic substrate resulting from indentation by a hard ball or scraping by a diamond tool.

The indentation of an fcc solid substrate by either a single atom or an fcc rigid tip is analyzed here using a recently developed MD simulator. The properties of either argon or copper are used for the intra-substrate and interface interatomic potentials. The atomic motion at the side and bottom boundaries of the substrate is fully constrained, whereas the atoms of the surface and interior of the substrate exhibit a dynamic behavior which satisfies the equations of motion. After the substrate reaches an equilibrium state corresponding to a desired temperature (T_d), indentation is simulated by moving the tip (or single atom) toward and into the substrate up to a certain depth (loading) and, subsequently, retracting it back to its original position at the same speed (unloading).

The generation of a single dislocation in a Lennard-Jones (L-J) solid is revealed in the single-atom indentation. Significant hysteresis loops are observed in the normal force versus tip-to-substrate distance curves for a rigid tip indenting a dynamic substrate, such as shown in Fig. 1. The rising portions of the loading curve are due to the elastic compression of the substrate, whereas the sharp drops of the force correspond to dislocation movement. In the unloading process, a sawtooth-like response is obtained due to the formation and elongation of a neck between the tip and the substrate.

It is found that both the compressive yield strength and the elastic unloading stiffness decrease with increasing substrate temperature and decreasing indentation speed, as shown in Fig. 2. In addition, the energies dissipated due to plastic deformation and thermal heating within an indentation cycle are found to decrease significantly with increasing substrate initial temperature, and the energy dissipated due to plastic deformation to be higher than that due to thermal heating.

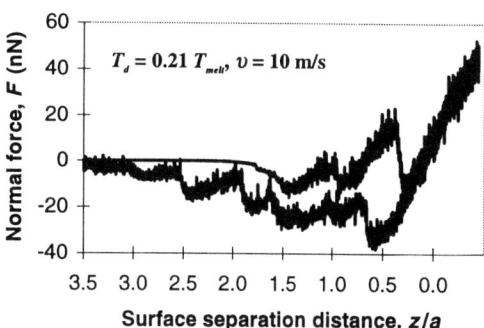

Fig. 1: Force vs. distance for copper

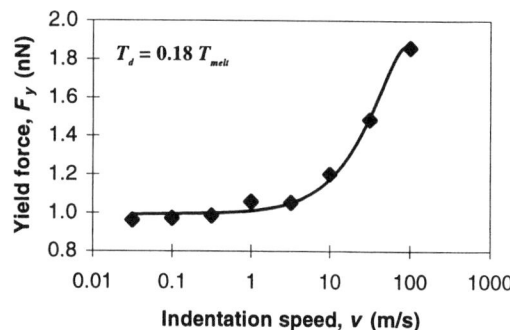

Fig. 2: Yield force vs. indentation speed for argon

In conclusion, the results of this study have elucidated the atomic-scale deformation processes occurring in L-J and metal substrates during indentation. The dependence of different mechanical properties on the substrate temperature and indentation speed was also analyzed.

REFERENCES

(1) Landman, U., Luedtke, W. D., and Ribarsky, M. W., *MRS Symposium: New Materials Approaches to Tribology: Theory and Applications*, pp. 101-117, 1989.
(2) Harrison, J. A., White, C. T., Colton, R. J., and Brenner, D. W., *Phy. Rev. B*, Vol. 46, pp. 9700-9708, 1992.
(3) Belak, J., and Stowers, I. F., *Fundamentals of Friction: Macroscopic and Microscopic Processes*, pp. 511-520, 1992.

Submitted to ASME Journal of Tribology

MICROSCOPIC FRICTION INVESTIGATIONS OF THIOLIPID LB FILMS ADSORBED ON MICA BY MEANS OF A MODIFIED AFM

D. GOURDON, N. BURNHAM, A. KULIK, E. DUPAS, F. OULEVEY and G. GREMAUD
Institut de Génie Atomique, Ecole Polytechnique Fédérale de Lausanne, 1015 Lausanne, Switzerland

ABSTRACT

We performed lateral force microscopy on thiolipid amphiphilic Langmuir-Blodgett (LB) films physisorbed on mica substrates with a silicon tip of an atomic force microscope. At ambient temperature, these films incorporate two phases analogous to a solid (crystalline phase) and a fluid (amorphous phase); six-fold star-shaped solid-like domains are surrounded by a fluid-like area (fig.1). As seen in the friction (lateral force) image, each arm of the star corresponds to a different value of the measured friction, most likely due to the orientation (packing and tilt) of the molecules within each domain (1).

Lateral-force and normal-force measurements were performed on a commercially available AFM (M5, Park Scientific Instruments) under ambient conditions. V-shaped cantilevers of 0.6 µm thickness (PSI) with theoretical normal and lateral spring constants of ≈ 0.03 N/m and ≈ 3 N/m, and sharpened conical pure silicon tips (10 nm nominal radius of curvature) were used in this study.

Quasi-static measurements were performed by using "friction loops". We observed the torsion of the cantilever during a forward and a reverse scan of the same sample zone (line). The lateral (friction) forces were first measured as a function of (normal) applied load and sliding velocity (scan amplitude and frequency). Then, to study the dependence of the friction on scan direction relative to the principal axis of an arm of a star, we turned the sample, which was mounted on a rotating holder. In such a way, we investigated the influence of the solid-domain orientation respective to the scan direction on
-the maximum friction force and
-the friction-loop shape (asymmetry between left-to-right and right-to-left scans) (2).

We found that at a fixed velocity and in a positive range of loads (cantilever in compression), the lateral force increases with applied load in a linear fashion whatever domain is studied.

We observed that friction was not directly dependent on *scan frequency* while it was dependent on the product *scan size* x *scan frequency* (= *relative velocity*). Within the velocity range of 0.01 to ≈ 50µm/s, the lateral force signal initially increases monotonically and quickly with velocity (static regime) and then stabilizes when the tip begins sliding.

The friction force and the asymmetry in the quasi-static "friction-loops" were found to be dependent on the domain orientation respective to the scan direction. A maximum in friction is observed when the arm is oriented perpendicularly to the scan direction while a minimum is observed when it is oriented parallely to the scan direction, whatever arm is sudied. "Opposite" results were found regarding the asymmetry. Together, friction and asymmetry enhance and map molecular packing and (radial) tilt.

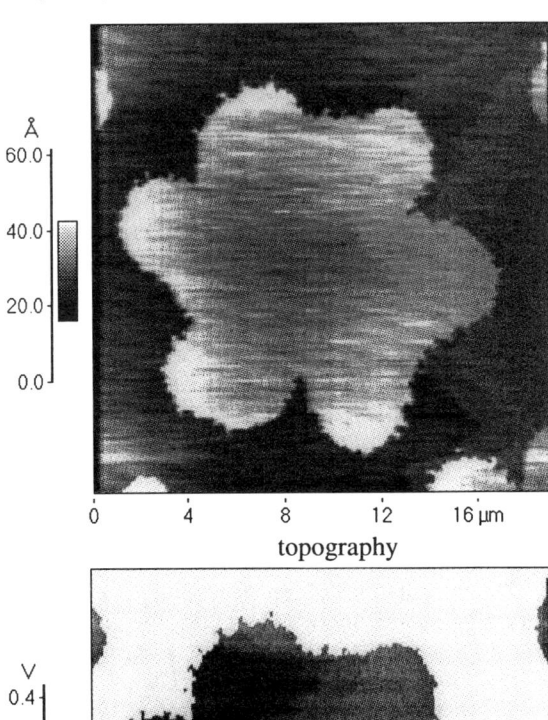

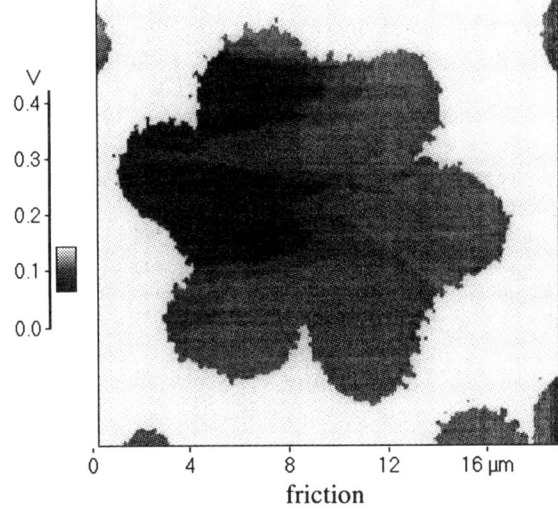

Fig.1. Topographic and lateral-force images of the thiolipid film on mica. The stars are formed from the solid-like phase, and the matrix from the fluid-like phase.

REFERENCES
(1) L. Santesson et al., J. Phys. Chem., Vol. 99, No. 3, 1995.
(2) U. D. Schwarz et al., Physical Review B, Vol. 52, No 20, 1995.

SURFACE MECHANICAL PROPERTIES MEASUREMENTS BY COUPLING NANOINDENTATION EXPERIMENTS AND IMAGING PROCEDURES

S BEC and A TONCK

Laboratoire de Tribologie et Dynamique des Systèmes, UMR 5513, Ecole Centrale de Lyon, B.P. 163, 69131 Ecully Cedex, France

Surface mechanical properties measurements of solids and thin films are of interest to tribologists as they are involved in frictional behaviour and wear performances of surfaces. Load and depth-sensing indentation is a well adapted method for the determination of near surface mechanical properties and is therefore extensively used (1)(2). At the nanometer scale, the actual indenter/material contact area must be carefully determined to obtain reliable values of mechanical properties from an indentation test. At this scale, the contact area is affected by the local surface roughness, by the geometrical tip defect and by the possible formation of plastic pile-up (or sink-in) around the indent. Parameters such as heterogeneity in surface and in film thickness also make it difficult to conduct and to interpret nanoindentation tests. Classical load versus displacement nanoindentation measurements without imaging have no protection against these sources of errors.

A new method, which couples nanoindentation experiments and imaging procedures has been developed (3). Both are performed with the same original three-axial instrument. It derives from the Surface Force Apparatus developed in our laboratory and is also designed for friction experiments at the molecular scale. This instrument can operate either in controlled displacement mode or in normal force controlled mode. This feature and the sensitivity of the three displacement transducers (measuring displacements in three directions X, Y and Z) and the two force transducers (measuring normal and tangential forces) allow us to perform accurate quasi-static and dynamic measurements of load and displacements as well as topographic images, without having to move the sample. The experimental procedure includes three main stages :

a) <u>local choice of the indented area</u> with a preliminary image of the surface topography,

b) <u>indentation</u> with continuous quasi-static measurement of the normal load and simultaneous dynamic measurements of the normal and tangential contact stiffnesses versus penetration depth,

c) <u>visualization of the residual indent</u> with a second topographic image of the indented area. The actual contact area is determined from this topographic image (see figure 1). This allows us to quantify piling-up effects in terms of increase of contact area.

Fig 1. : Topographic image of a residual indent on a gold layer (maximum load : 1009 µN). Determination of the actual contact area including plastic pile-up.

In this paper, the apparatus and the method are described. In the case of layered materials, we determine the local plastic properties. On the other hand, the elastic measurements give a global information from which we obtain the elastic properties of the layer with a simple modelling. On rough and heterogeneous materials such as friction films (4), a specific supplementary development of the method makes it possible to determine the mechanical properties of the film with a good accuracy.

REFERENCES

(1) J.B. Pethica, R. Hutchinson, W.C. Oliver, Phil. Mag. A, 48, 593, 1983.

(2) M.F. Doerner, W.D. Nix, J. Mater. Res., 1, 601, 1986.

(3) S. Bec, A. Tonck, J.M. Georges, E. Georges, J.L. Loubet, Phil. Mag. A, 74, n°5, 1061, 1996.

(4) S. Bec, A. Tonck, Tribology Series, 31 : The Third Body Concept : Interpretation of Tribological Phenomena, D. Dowson et al. (Editors), Elsevier Science, 1996, pp 173-184.

Cost effective tribo-corrosion resistant coatings for aggressive marine environments

Y PUGET and R J K WOOD
Department of Mechanical Engineering, University of Southampton, Highfield, Southampton, SO17 1BJ, UK
K TRETHEWEY
Department of Engineering Materials, University of Southampton, Highfield, Southampton, SO17 1BJ, UK

Onboard every ship and off-shore oil platform, there are typically several kilometres of pipework handling sea water for cooling and fire main purposes. These pipe networks are currently constructed from cupronickel which is relatively expensive and has a lifetime of only 7 years. Cupronickel also limits the sea water flow rate in order to minimise erosion-corrosion attack which can perforate the pipework if sea water velocities exceed 3 m/s. Modern designers require more compact, cheaper and lighter pipe networks which can handle higher sea water flow rates. This pipework must be resistant to erosion-corrosion induced by the impact of solid particles and cavitation as well as high velocity (10-20 m/s) enhanced corrosion. One solution under consideration is that of coated mild steel pipework but little relevant tribo-corrosion performance information is available for suitable candidate coating systems which currently undermines the selection of such coatings. This paper covers some initial slurry erosion tests on a wide range of coating compositions to establish suitable coating systems for future erosion-corrosion testing.

Slurry erosion tests have been performed on plastic, ceramic, metallic and composite coatings which have been applied by a variety of deposition techniques such as thermal and plasma spraying, HVOF and D-Gun, Electro-plating, brushing and Physical Vapour Deposition (PVD) on mild steel coupons. Sand and fresh water slurry jet erosion tests at 30° jet impingement angle and at two different jet velocities of 10 m/s and 20 m/s are presented.

PVD coatings were not considered to be appropriate for immersion in sea water, since traces of "rust" Fe_2O_3 spread over the specimen surfaces after a few minutes of immersion in fresh water. Electroplated coatings did not provide adequate protection to the steel substrate since after being subjected for 90 minutes to a slurry at 20 m/s, the coatings were stripped and the underlying substrates were exposed to the slurry. Both plasma sprayed ceramic coatings and brushed polymeric coatings provided satisfactory resistance against sand particle erosion and corrosion by fresh water, since after erosion of the coated specimens for 90 minutes, the underlying steel was not exposed, and no trace of corrosion was visible on the coupon surfaces.

The ceramic thermal sprayed coatings were much more resistant against hard particle erosion than polymeric coatings. After being eroded for 90 minutes at a jet velocity of 20 m/s, the volume loss per particle impact of the least resistant thermal sprayed coatings was 2.48 times lower than that of the most resistant polymeric coating, namely Polyurethane (PU). However, the cost of polymeric coatings were substantially lower than that of thermal sprayed coatings and therefore, under the conditions of erosion selected, the brushed PU coating was considered to offer the best combination of erosion-corrosion resistance for the lowest price. Polyurethane has also been recognised previously as a good coating system for combating immersed conditions where corrosion and abrasion are considered a major cause of substrate breakdown (1).

The resistance of PU coatings against erosion by air borne particles (2) or by fresh water slurry has already been studied, and the degradation of PU coated metal substrate immersed in static sea water has also been investigated by electrochemical techniques (3). However, no work has been done to study the resistance of this coating system when subjected simultaneously to corrosion by flowing sea water and erosion by the sand particles carried by the flow.

In order to further investigate the erosion-corrosion resistance of polyurethane coatings, other slurry erosion tests will be performed on PU coated steel coupons. A new rig entirely made of plastic will allow circulation of a sea water slurry without contamination by the rig components. Thus, the synergism between erosive and corrosive attack will be determined by comparing the erosion rates obtained using fresh water and sea water slurries. The corrosion resistance of coated coupons will be assessed by AC Impedance and in-situ electrochemical noise techniques.

Acknowledgements

Financial support provided by the DRA and the University of Southampton is gratefully acknowledged. Thanks are also due to Dr Guy Denuault for his advice and assistance.

References

(1) J.G. Tucker, 1986, Anti-Corrosion, Sept., pp 10-13
(2) J. Li, I.M. Hutchings, 1990, Wear, **135**, pp 293-303
(3) C.T. Chen, B.S. Skerry, 1991, Corrosion, **47**, no 8, pp 598-611

APPLICATION OF THE THERMOMECHANICAL WEAR TRANSITION MODEL TO TRIBOLOGICAL DESIGN

RICHARD S. COWAN and WARD O. WINER
Woodruff School of Mechanical Engineering, Georgia Institute of Technology, Atlanta, GA 30332-0405 USA

INTRODUCTION

As thermal and mechanical demands on machine components increase, the ability to identify the transition between mild and severe wear from sliding contact becomes increasingly important to a designer concerned about component cost, quality and reliability. Through a parametric study, equations developed from a thermomechanical wear transition model have been analyzed, given changes in the magnitude of a series of variables controlled by the designer. Parameters that significantly have an impact on the potential for thermomechanical wear of ceramics, metals and polymers are identified and discussed with reference to frictional heating, contact area and failure strength.

BACKGROUND

In the thermomechanical wear transition model, a circular contact of radius, a, and a maximum Hertzian pressure, P_o, occur when a total normal load, F_N, is applied between a hemisphere, representative of an asperity, and a semi-infinite flat surface. Should the bodies be moving at a relative velocity, V, thermal effects are anticipated from the frictional heat that is generated.

The total stress field of the sliding contact, σ_{ij}, can be decomposed into two parts, comprised of isothermal and thermal stresses. The former may be calculated from the analyses of Hamilton and Goodman [1]. The latter may be obtained via the Navier's equation with a thermoelastic displacement potential as demonstrated in [2]. Expressed as

$$\sigma_{ij} = P_o \{ \bar{\sigma}_{ij}^i(\underline{x}, \nu, f) + G_t \cdot \bar{\sigma}_{ij}^t(\underline{x}, \nu, F_o) \}, \quad (1)$$

the total stress field is a function of contact locale, $\underline{x}$, Poisson's ration, ν, friction coefficient, f, and Fourier number, F_o. The functions $\bar{\sigma}_{ij}^i$ and $\bar{\sigma}_{ij}^t$ represent non-dimensional isothermal and thermal stresses. The variable G_t, signifying the relative importance of the thermal stress contribution, can be used as a thermomechanical wear-control parameter. It is defined as

$$G_t = \frac{E\alpha_t}{k(1-\nu)} \gamma f V a, \quad (2)$$

where E, α_t and k are the asperity material's elastic modulus, thermal expansion coefficient, and thermal conductivity. The fraction of total frictional heat entering this hemisphere is the heat-partition factor, γ. Under the assumption that plastic deformation at the surface is a wear mechanism, the stress field and frictional heat source may induce severe wear through material yielding or fracture.

THERMOMECHANICAL WEAR ANALYSIS

To determine the potential of thermomechanical wear, the real area of contact is needed for computing the Hertzian pressure. The presence of lubrication influences this calculation. For instance, hydrodynamic lubrication (thick film) infers the total separation of asperities, such that the load is primarily carried by fluid pressure. In mixed lubrication, part of the load is carried by the asperities, separated by a molecularly thin lubricant film. In its most severe form, considerable asperity interaction occurs, resulting in boundary lubrication. Depending on the load, speed, lubricant viscosity, contact geometry and surface roughness of both surfaces, the contact may shift between any one of these regimes.

The region of contact also experiences a temperature rise when components in relative motion are mechanically engaged. At the micro level, this increase can be substantial as evidenced by hot-spots. A localized change in material properties, an increase in chemical reactivity, and ultimately, failure of the mechanical system may result. Toth [3] indicates that a transition temperature exists in which the response of a material (e.g., ceramic) to an applied stress may change from brittle to ductile behavior. Hence, the influence of frictional heating based on equations governing the flow of heat must be considered in order to minimize the uncertainties associated with material behavior and material property values.

In this presentation, the aforementioned concepts are used to demonstrate how the thermomechanical wear transition model can be applied to avoid or promote material failure. Wear maps are generated, and used to verify predictions with experimental results.

REFERENCES

[1] G. M. Hamilton and L. E. Goodman, *Journal of Applied Mechanics*, **33**, 1966, p. 371.
[2] B. Y. Ting and W. O. Winer, *Journal of Tribology*, **111**, 1989, p. 315.
[3] L. E. Toth, *Transition Metal Carbides and Nitrides*, Academic Press, 1971, p. 169.

THERMAL EFFECTS IN LARGE HIGH SPEED HYBRID JOURNAL BEARINGS

D. IVES and W.B. ROWE

Liverpool John Moores University, School of Engineering and Technology Management, Byrom Street, Liverpool L3 3AF, UK

INTRODUCTION

Slot entry bearings have been the subject of several investigations since their original conception as hydrostatic bearings for use with either a gas or liquid lubricant film (1). They have demonstrated the ability to operate in the hybrid (hydrostatic/hydrodynamic) mode and optimisation procedures based on load/total power have been developed (2).

In recent years considerable attention has been given to the application of hybrid journal bearings to large high speed industrial systems such as turbines and generators (3). Such applications are characterised by high power dissipation acompanied by a corresponding rate of heat generation. It is usual for hybrid bearings to be considered as low temperature rise machine elements, enabling design and analysis to be performed using a uniform effective viscosity for the bearing. However, in a series of tests on large diameter high speed slot entry hybrid bearings (4) it was found that the peak temperature in the bearing exceeded that which was considered safe for bearings with white metal linings.

The source of the temperature problems was associated with the physical geometry of the bearings. The two rows of pressurised slots one near each axial extremity of the bearing (a/L=0.1) resulted in little axial flow in the central region of the bearing at low eccentricity ratios. Thus the majority of cool fluid pumped into the bearing flowed directly out over the axial lands and the fluid in the centre of the bearing was subjected to constant recirculation. Heat generation in this central region resulted in higher than normal fluid temperatures and conduction became the principal mode of heat transfer.

THEORETICAL ANALYSIS

The theoretical analysis took the form of a finite difference solution of the Reynolds equation with variable viscosity, coupled with a heat balance for the fluid film. The heat balance accounted for heat generation and heat convection in the fluid film, dissipation of friction power and pumping power, heat conduction to the bearing and journal surfaces and heat conduction in the bearing structure. Due to the complex structure of a hybrid slot entry bearing, which incorporates oil supply grooves and channels, conduction was assumed to be radial only and the temperature at the outer surface of the bearing was set at a constant for the whole surface.

TEST RIGS AND BEARINGS

Two test rigs were used for the work, a 40mm diameter rig at Liverpool John Moores University and a 350mm rig at G.E.C. Alsthom, Stafford. Initially standard slot entry bearings with two rows of 12 equally spaced slots were used. As the extent of the temperature problems became apparent an additional asymmetric configuration was designed for the 350mm rig. This configuration consisted of five slots per row in the lower half of the bearing with a low pressure axial groove in the upper half.

DISCUSSION OF RESULTS

For the 40mm bearings the variable viscosity model provided close agreement with the experimental values and previous isoviscous models. The significance of this is that provided a reasonable estimate of effective viscosity can be obtained, isothermal techniques are still useful for predictions involving small diameter bearings.

Except for the asymmetric configuration the 350mm bearings initial performance predictions showed a relatively poor correlation with experimental results, but were an improvement on predictions using isoviscous theory. Futher, the variable viscosity model failed to predict the excessive surface temperatures observed during testing.

As a result of futher investigations, it was realised that the flow conditions in the bearing structure could lead to stagnation or even backflow of hot fluid from the bearing into the structure. This rendered the initial constant temperature boundary condition for the bearing structure inadequate. Modification of the boundary condition greatly improved correlation with experimental results and indicated that effective cooling could be provided by judicious placing of the inlet fluid supply channels in the bearing structure.

REFERENCES

(1) G.L. Shires and C.W. Dee, Southampton University, Gas Bearing Symposium, Paper 7, 1967.
(2) W.B. Rowe and D. Koshal, Wear, Vol. 64, No. 1, 1980, 115-131pp.
(3) D. Ives and W.B. Rowe, STLE Tribology Transactions, Vol. 35, No. 4, 1992, 627-634pp.
(4) J.E. Brown, G.S. Khera, N.C. Lees and G.D. Whale, Report 5/953u1065, G.E.C. Alsthom, Stafford, 1985.

CHARACTERISATION OF THERMAL EFFECTS IN COMPRESSION AND WEDGE INDENTATION OF "SOFT SOLIDS"

M. J. ADAMS
Unilever Research, Port Sunlight Laboratory, Bebington, Wirral, L63 3JW, UK.
B. J. BRISCOE, D. C. KOTHARI and C. J. LAWRENCE
Department of Chemical Engineering & Chemical Technology,
Imperial College of Science, Technology & Medicine, London, SW7 2BY, UK.

ABSTRACT

Wedge indentation and compression experiments have been performed to study the effect of heated walls on the friction of a model soft-solid material. The experiments were interpreted by comparison with solutions of modified slip-line field and lubrication theories respectively. The friction relationships deduced were further analysed in terms of Arrhenius activation energy.

INTRODUCTION

Soft solids represent a large and growing fraction of the materials processed in industry. Many products such as foods and household products may be classified as soft solids, while other materials such as ceramics are processed via soft-solid intermediate states. These materials are generally elasto-viscoplastic, and undergo large plastic strains during processing. Considerable effort has been invested in the design of suitable experiments and the determination of sufficient models for the intrinsic or bulk constitutive behaviour. A number of simple experimental techniques have been identified (1), but the difficulty remains of accurately accounting for the contribution of wall friction to measured loads and deformations. Indeed the characterisation of wall friction, or interface rheology, has itself become a preoccupation (2).

An attempt has been made to influence the frictional boundary conditions at solid surfaces by heating the surfaces. As long as the bulk of the solid remains at ambient temperature, the contrast between the heated and isothermal experiment gives a fair measure of the influence of wall friction.

EXPERIMENTS

The experiments were conducted on the model paste Plasticine™ (Peter Pan Playthings Ltd.) This is a concentrated dispersion of small particles of clay and other minerals in a mineral oil. The properties seem to be reproducible within a given batch and the material is useful for experiments, being stable and resistant to drying. Specimens were left for two hours at room temperature to allow any residual stresses to relax before conducting an experiment.

A set of stainless steel wedges was used for the first series of experiments. These had included angles of 30°, 60°, 90°, 120° and 150°, and breadth of 180mm. The material was pressed into large rectangular blocks 90mm deep by 100mm wide by 90mm broad. The greatest dimension was perpendicular to the wedge indenter so as to approximate a semi-infinite solid. The wedges were mounted to the cross head of an Instron testing machine and were positioned carefully just at the surface of the soft solid. The machine was set to run at a constant speed up to a depth of 10mm and then remain in that position. The load was recorded at suitable intervals automatically, using a sensitive load cell on the cross head. Ten cross head speeds were used, ranging from 0.5mm/min to 500mm/min on a logarithmic scale between. The wedges were maintained at temperatures of 20°C to 80°C in increments of 5°C. Thus a matrix of 650 experiments was performed with approximately 50 load-displacement measurements per experiment.

In the second series of experiments, thick disks of Plasticine of height 20mm and diameter 80mm were squeezed between parallel platens to a height of 10mm. The platens were made of smooth stainless steel, but various materials could be interposed between the platens and the specimen to modify the interface friction. The platens were heated and maintained at the same set of temperatures as the wedges. A reduced range of velocities was used to accommodate the maximum load on the Instron testing machine. The load was again recorded automatically.

Flow visualisations were conducted for some of the compression and indentation experiments, to assess the degree of homogeneity of the deformation and identify the occurrence of wall slip or shear bands within the material.

RESULTS

The indentation experiments were interpreted using a simple power-law model relating average indentation pressure and nominal strain rate for the bulk material. Constraint factors were calculated on this basis, and were compared with the values predicted by slip-line field theory. Three different friction models were used and the best model was selected for determination of the coefficient. The compression experiments were interpreted with the aid of a lubrication theory modified to allow for wall-slip. In both cases, the temperature dependence of the wall friction was further interpreted using the Arrhenius relationship.

REFERENCES

(1) S. Sinha, "The Rheology of Soft Solids Pastes", PhD Thesis, Imperial College (University of London, 1994).
(2) G. Corfield, "The Constrained Flow of Pastes", PhD Thesis, Imperial College (University of London, 1996).

CONTACT MECHANICS WITH REGARD TO WEAR AND GENERATION OF HEAT DUE TO FRICTION

V M ALEXANDROV, M A BRONOVETS and E V KOVALENKO
Institute for Problems in Mechanics RAS, 101, Prospect Vernadskogo, Moscow, 117526, Russia
S P ZARITSKI
Technical and Engineering Center "Orgtechdiagnostika", 13/1 Karamzina str., Moscow 117463, Russia

ABSTRACT

Classical schemes of static local contact, which are the basis for calculation and designing a great many of modern machines' parts, are too simplified and do not always meet the requirements of practice.

In these classical schemes the quality of the surfaces of bodies in contact is idealized, the effect of generation of heat is not taken into account, the evolution of surfaces in contact, caused by their wear, is neglected. These and some other questions pertaining to the modern problems of contact mechanics are touched upon in this paper (1-13).

Consideration is given to quasistationary plane and spatial contact problems of disconnected thermoelasticity for bodies with coatings (linear guides, cylinder and spherical sliding supports and so on) with regard to wear and generation of heat due to friction. Proposed are highly general relations of friction coefficient, contact thermal resistance, the wear law and the law of roughness smoothing resulting from contact pressure and temperature.

Between the bodies in contact there exist conditions of both ideal and nonideal thermal contact in the presence of a thin interstitial third body with variable in depth heat conduction containing distributed heat sources.

As a rule, the problems reduce to the solution of nonlinear integral equations with respect to the main characteristics of contact, for the solution of which asymptotic methods are developed.

It is shown that these exist such critical speeds of relative motion of bodies at which the loss of regimes of the quasistationary heat conduction and the thermal force stability of junction work, respectively, occurs. The wear caused by melting of one of the bodies in contact, when the melted material is pressed out from the region under the punch with its resulting settling, is studied.

The critical speed of relative motion of bodies, at which the melting process begins, has been found.

REFERENCES

(1) V.M. Alexandrov, E.V. Kovalenko, J. Appl. Mechs Tech. Phys., No.3, 1985, 129-131pp.
(2) V.M. Alexandrov, J. Phys-Chem. Mechs. Materials, No.1, 1986, 116-124pp.
(3) V.M. Alexandrov, G.K. Annakulova, Friction and Wear, Vol. 11, No.1, 1990, 24-28pp.
(4) V.M. Alexandrov, Mechs. of Solids, No.6, 1990, 36-42pp.
(5) V.M. Alexandrov, G.K. Annakulova, Friction and Wear, Vol. 13, No.1, 1992, 154-160pp.
(6) V.M. Alexandrov, Mechs. of Solids, No.5, 1992, 73-80pp.
(7) A.A. Yevtushenko, E.V. Kovalenko, J. Appl. Maths Mechs., Vol. 57, No.1, 1993, 148-156pp.
(8) E.V. Kovalenko, A.A. Yevtushenko, Friction and Wear, Vol. 14, No.2, 1993, 259-269pp.
(9) E.V. Kovalenko, Friction and Wear, Vol. 15, No.4, 1994, 549-557pp.
(10) V.M. Alexandrov, J. Probl. Mach. Build. and. Mach. Reliab., No.5, 1995, 70-75pp.
(11) A.A. Yevtushenko, E.V. Kovalenko, J. Appl. Maths Mechs., Vol. 59, No.3, 1995, 485-492pp.
(12) A.A. Yevtushenko, E.V. Kovalenko, Mechs. of Solids, No.4, 1995, 56-62pp.
(13) E.V. Kovalenko, A.A. Yevtushenko, E.G. Ivanik, Friction and Wear, Vol. 17, No.3, 1996.

Submitted to *ASME Journal of Tribology*

APPLICATION OF FRICTIONAL INTERACTION AND CHEMICAL REACTION FOR MICRO-PATTERNING OF SILICON

DAE-EUN KIM
Department of Mechanical Engineering, Yonsei University, Shin-chon Dong 134, Seoul, Korea
JAE-JOON YI
Hyundai Motor Co., Mabook-ri, Korea

ABSTRACT

The application of Micro-Electro-Mechanical-Systems (MEMS) to key technologies has been receiving much attention in recent years. One of the critical issues regarding MEMS lies in the efficiency of its fabrication process. Since the dimensions are of micro-scale and the commonly used material is silicon based, the fabrication process used to process semiconductors are employed to make MEMS parts. However, such process is only suitable for mass production, and therefore, would not be viable for producing a limited number of parts. There is a need to develop new processing techniques for fabricating micro-scale components cost effectively with flexibility.

In this work a novel and economical method of generating three dimensional micro-patterns on single crystal silicon without the need for a mask is presented (patent pending). Though this technique cannot produce all complex shapes for MEMS application, it is capable of fabricating patterns that are of micrometer scale with flexibility. Unlike the laser writing method, the process does not require extensive capital investment.

The technique introduced in this work is based on fundamental understanding of frictional interaction between two solids in relative motion while in contact under a light load (1). Micro-patterning is done through a two-step process that involves mechanical scribing and chemical etching. When a brittle solid such as silicon is slid against a hard solid to produce a pattern on the surface, the brittle material tends to fail by fracture. However, if the applied load is properly controlled, gross generation of cracks can be avoided. In fact, under controlled conditions, the silicon can be slid against a hard solid without undergoing fracture. However, under such a condition the surface is hardly damaged, and therefore, no appreciable amount of material is removed. During this interaction micro-machining does not occur. However, the fact that appreciable amount of force is needed to scribe the silicon surface under such a condition suggests that some amount of damage is being incurred in the surface region of the silicon.

The basic idea behind the process introduced in this work is to impart energy along the prescribed track through frictional interaction between the tool and the workpiece. This must be done without detectable surface damage or wear. Then, by exposing the surface to a chemical etchant under controlled condition preferential chemical reaction is induced along the track to form a protrusion or a groove. The desired dimensions of the pattern can be controlled by varying the normal force during scribing and by the etching condition. The normal force is selected based on the Hertzian contact model as well as a modified Vickers hardness test model for a desired width.

A precision machine was built to verify the feasibility of the new micro-machining concept described above. The machine is equipped with an x-y table which is controlled by a computer and has a resolution of 5 μm. A sharp diamond tool was used as the tool. Also, semiconductor strain gage was used to measure the frictional force during the scribing process. Micro-patterning experiments were performed using a (100) silicon wafer. It is shown that this process is capable of generating micro-patterns with dimensions of about 5 μm width and 1 μm height (Fig. 1). This method of micro-machining may be used for making patterns in MEMS applications cost effectively. Also, this work demonstrates the applicability of frictional interactions in micro-fabrication process.

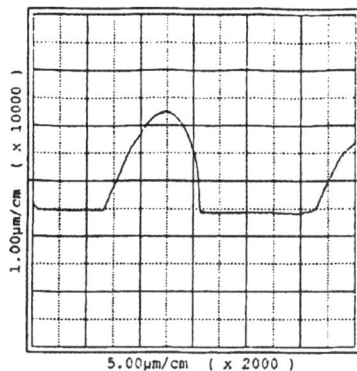

Fig. 1 Profile of micro-pattern fabricated on Si

ACKNOWLEDGMENT

This work has been supported by Korea Ministry of Education through Mechanical Engineering Research Fund (ME94-E-09).

REFERENCE

(1) D. E. Kim, N. P. Suh, Wear, Vol 149, 1991, pp. 199-208.

EVIDENCING SUPERLUBRICITY FROM MICRO TO MACRO SCALE

MOTOHISA HIRANO

Information Hardware Systems Laboratories, Nippon Telegraph and Telephone Corporation, Musashino, Tokyo 180, Japan

JEAN MICHEL MARTIN

Laboratoire de Tribologie et Dynamique des Systems, UMR 55 13, Ecole Centrale de Lyon, F-69131, Ecully, France

ABSTRACT

A new regime of friction - superlubricity - where friction completely vanishes has recently been discovered (1)(2). It has been theoretically shown that superlubricity occurs even at the strongly interacting interfaces where realistic inter-atomic potentials, such as metallic bonds, operate (1). Other researchers, however, have shown that it occurs only at weakly interacting interfaces (3). The purpose of this study is to present experimental evidence for superlubricity by measuring friction in systems ranging in size from micro to macro scale.

The experiment with mica (4) showed the possibility of the existence of superlubricity. The friction of cleaved single-crystal surfaces was measured in the elastic contact regime as a function of the lattice misfit between surfaces. The experimental results were consistent with our prediction that friction vanishes when, for example, surfaces contact incommensurately. The measured friction forces were anisotropic with respect to the lattice misfit, i.e., they decreased (increased) when the contacting surfaces approached being incommensurate (commensurate).

The direct observation of superlubricity is presented by examining atomically-clean surfaces using ultra-high-vacuum scanning tunneling microscopy. Sliding with atomic-scale elastic contact is achieved by using tunneling between a tip and a surface (Fig.1). In this experiment, Si(001) is scanned against W(011) on a polycrystalline W tip and the friction was measured as a function of the lattice misfit between the surfaces. Friction of magnitude 8×10^{-8} N, which is comparable to the calculated value, is observed when the contact is commensurate. However, when the contact is incommensurate, friction is not observed in this measurement which can resolve friction forces of 3×10^{-9} N.

Superlubricity in a macro-scale experiment is evidenced by using a ultra-high vacuum tribotester coupled with a preparation chamber. Pure nanocrystallized MoS_2 was deposited on different substrates and friction was measured using a reciprocating pin-on-flat friction machine. The maximum contact pressure was about 1 GPa for a contact area of diameter 100 microns. The vanishing of the friction force is effectively observed in UHV (friction coefficient below 10^{-3}) and this corresponds to two crystal orientation mechanisms: orientation of MoS_2 basal planes (0001) parallel to the sliding direction and frictional anisotropy due to the rotation of stacks of crystallites around the c-axis. In this more practical case, it is remarkable to observe that there is natural tendency of MoS_2 to reorient in the interface to give the expected phenomenon.

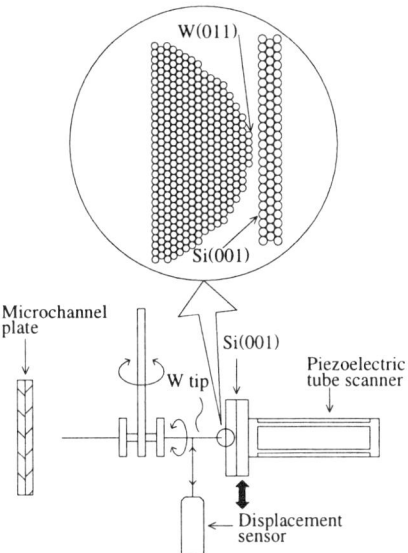

Fig. 1: UHV-STM friction measurement system

We thus conclude that the observed frictional anisotropy stemming from the differences in the commensurability of the contacting surfaces implies the existence of superlubricity as a new regime of friction.

REFERENCES

(1) M. Hirano and K. Shinjo, Phys. Rev. B47, 1990, 11837-11851pp.

(2) K. Shinjo and M. Hirano, Surf. Sci. 283, 1993, 473-478pp.

(3) For example: J. B. Sokoloff, Surf. Sci. 144, 1984, 267-272pp.

(4) M. Hirano, K. Shinjo, R. Kaneko, and Y. Murata, Phys. Rev. Lett. 67, 1991, 2642-2645pp.

(5) M. Hirano, K. Shinjo, R. Kaneko, and Y. Murata, Phys. Rev. Lett. 78, 1997, 1448-1451pp.

(6) J. M. Martin, C. Donnet, and T. Le Mogne, Phys. Rev. B48, 1993, 10583-10586pp.

Submitted to *Tribology Letters*

MEASUREMENT AND CHARACTERIZATION OF WEAR PARTICLE SURFACE TOPOGRAPHY

P. PODSIADLO and G.W. STACHOWIAK
Tribology Laboratory, Department of Mechanical and Materials Engineering
University of Western Australia, Perth, Australia

ABSTRACT

The fractal nature of wear particle shapes and surfaces, reflecting the chaotic nature of wear processes involved in their formation, has been observed. Recently fractal methods have been employed in the numerical characterization of particle shapes but, as yet, little quantitative consideration has been given to the characterization of 3D topography of wear particle surfaces. The main difficulties are associated with the acquisition of accurate 3D data of a particle surface, assessment of a particle surface anisotropy and lengthy computation time required to obtain the surface fractal dimension. These problems are addressed in this paper.

3D surface elevation maps of wear particles were obtained by SEM stereoscopy and a non-contact optical profiler (NOP). The stereoscopy technique involves the acquisition of two images of the same particle taken under slightly different angles followed by noise reduction, edge detection, thresholding, stereoscopic matching, calculation of elevation points and interpolation. An automated stereoscopic matching algorithm developed earlier (1) has been used in this work. The NOP is based on the principle of fringe-modulation interferometry (2). An example of 3D stereoscopic and interferometric images of a brass wear particle is shown in Fig. 1. The particle was generated during sliding of a brass pin against a ceramic plate (3).

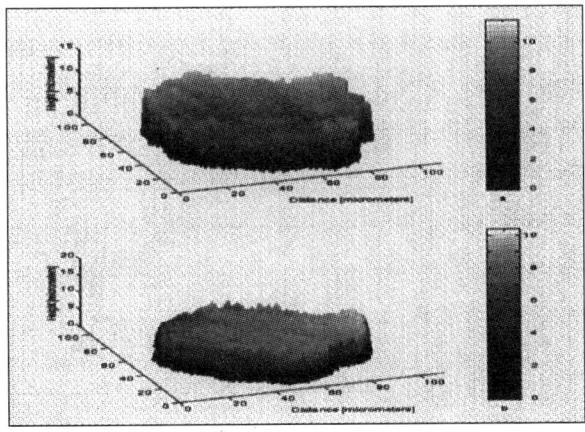

Fig. 1: Example of 3D stereoscopic (a) and interferometric (b) images of wear particle.

A specially modified Hurst Orientation Transform (HOT) (4) to suit wear particle surface data has been used to characterize the surface topography of particles. The calculation of HOT involves searching all pairs of pixels in a large neighbourhood in order to build a table of maximum differences. From this table, the Hurst coefficients are calculated in all directions and plotted as a function of orientation to reveal the surface anisotropy. Three new surface texture parameters, $i.e.$ texture minor axis (S_{ta}), texture aspect ratio (S_{tr}) and texture direction (S_{td}), have been developed based on the rose plots of the Hurst coefficients. S_{ta} is a parameter used to describe the wavelength of a surface's significant roughness components, S_{tr} is used to identify the anisotropy or isotropy of a particle surface while S_{td} indicates the dominating direction of surface texture.

The modified HOT method was applied to a number of wear particles found in tribological systems. The resulting plots obtained after the application of the HOT method to images of the particle displayed in Fig. 1 are shown in Fig. 2. The overall shape of these plots is elliptical with the largest values centred around a left direction. This implies that the particle surface exhibits a lower fractal dimension in this particular direction than in any other direction, $i.e.$ the surface is anisotropic. This anisotropy is confirmed by visual examination of the particle surface topography. It was observed that the particle exhibits dominating patterns, $i.e.$ grooves, corresponding to this direction. S_{ta}, S_{tr} and S_{td} were also calculated and are shown in Fig. 2.

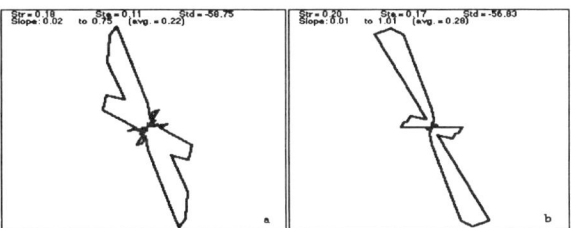

Fig. 2: Rose plots of Hurst coefficients.

SUMMARY:

• 3D wear particle surface topography can successfully be assessed by means of SEM stereoscopy and NOP,

• rose plots of the Hurst coefficients and surface texture parameters developed are suitable for visual interpretation and quantitative measurement of particle surface roughness and anisotropy.

REFERENCES

(1) P. Podsiadlo, G. W. Stachowiak, in press, Wear, 1997.
(2) P. J. Caber, J. Martinek and R. J. Niemann, Proceedings SPIE 2088, Laser Dimensional Metrology, Photonex '93.
(3) G. W. Stachowiak, G. B. Stachowiak, A. W. Batchelor, Wear, Vol. 132, 1989, 361-381pp.
(4) P. Podsiadlo, G. W. Stachowiak, Journal of Computer-Assisted Microscopy, submitted, 1997.

Submitted to *Wear*

THREE DIMENSIONAL IMAGING AND ANALYSIS OF MACHINE COMPONENTS

R.V. ANAMALAY, T.B. KIRK and D. PANZERA
Department of Mechanical and Materials Engineering, University of Western Australia, Nedlands, Western Australia.

ABSTRACT

It is known that surface topography or surface texture governs the nature of the interaction between surfaces. Tribological phenomena such as friction and wear depend primarily on the nature of the real area of contact between surfaces which is in turn dependent upon the distributions, sizes and shapes of the surface asperities (1). In order to understand this interaction, it is important that the topography be properly characterized. But before a surface can be characterized, it has to be recorded. Traditional surface recording techniques, such as profilometry, deal mainly with one or two dimensions only (2). This is inadequate to properly characterize surfaces, especially since most engineering surfaces are anisotropic by nature.

Three dimensional surface recording techniques such as specialized profilometers and Laser Scanning Confocal Microscopy or LSCM offer the ability to record the three dimensional topography of surfaces. However, three dimensional information gathering stylus profilometers have typically low scanning rates (2). Up to now however, the analysis methods have not matched the recording methods. Currently, the analysis methods involved are still mainly in one and two dimensions.

In this study, machine components like :
[a] honed engine cylinder liners and
[b] roller bearing surfaces
were imaged and analyzed . Images were acquired using a Laser Scanning Confocal Microscope or LSCM. The LSCM has certain advantages over stylus methods in that it :
[a] is a non-contact and non-destructive method of analysis,
[b] requires no special sample preparation is needed
[c] offers much better lateral resolution than stylus profilometers (which are limited by stylus size). The lateral resolution of the LSCM depends largely on the numerical aperture (NA) of the lens used
[d] it offers the ability to record data in three dimensions by taking successive optical sections at different focal planes. These optical sections can then be used to build a three dimensional model of the surface (3).

In this study, images were acquired and two types of images were generated : a maximum brightness projection or MBP and a height encoded image or HEI.

A MBP in an image in which every part is in focus whereas a HEI is an image where the colour of the pixel determines the height of the feature recorded. The HEI is particularly useful because from it can be extracted surface parameters. Figures 1 and 2 are example of a MBP and HEI respectively of a lapped metal surface.

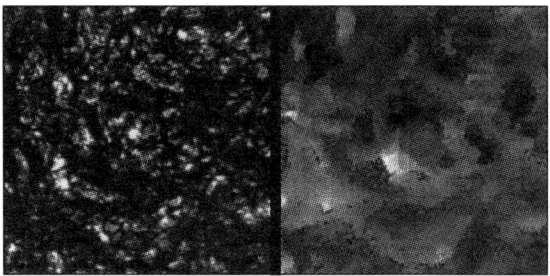

Figure 1: MBP   Figure 2: HEI

In this paper, two dimensional parameters such as the surface fractal dimension, autocorrelation function and the bearing ratio are used to characterize the experimental surfaces used. The paper also examines the need for three dimensional parameters and how existing two dimensional parameters may be extended to deal with three dimensional data for the topographic assessment of surfaces. The proposed paper will also provide examples of the assessment of these surfaces and outline the capabilities and deficiencies in current surface analysis techniques for these applications.

REFERENCES

(1) J. Halling and K. A. Nuri, Principles of Tribology, pages 18-19, Macmillan Press, 1975.
(2) M. G. Gee and N. J. McCormick, The Application of Confocal Scanning Microscopy to the Examination of Ceramic Wear Surfaces, A230-A235, J. Phys D: Appl. Phys. 25 (1992).
(3) R. V. Anamalay and T. B. Kirk, The Advantages of Laser Scanning Microscopy (LSCM) over Conventional Stylus Profilometry for the Assessment of Surfaces, pages 1-8, Condition Monitoring in Perspective, 1994 Centre for Machine Condition Monitoring Forum.

AN AFM ANALYSIS OF SURFACE TEXTURES OF METAL SHEETS CAUSED BY SLIDING WITH BULK PLASTIC DEFORMATION

HIROSHI IKE

(The Institute of Physical and Chemical Research [RIKEN], 2-1 Hirosawa, Wako, Saitama 351-01, Japan)

ABSTRACT

Surface textures of metal sheets before and after metal forming have called for much attention because of their functional properties, *e.g.*, lubricating ability, adhesion of painting layers to the metal matrix after forming, and image clarity after painting.

The sheet surface texture is governed both by the initial surface asperities and by lubrication and deformation. In metal forming, the asperities of the sheet are deformed by surface asperities of the tool under the specific lubrication mechanism. In this sense, lubrication is one of the decisive factors of the surface textures of formed products. However, the difficulty of the evaluation of the actual lubrication mechanism have been preventing the entire understanding of formation of surface textures.

On the other hand as a basis of metal forming in an extreme precision, the ability of faithful negative transfer of surface microgeometry of the tool to workpiece under the boundary lubrication should also be made clear in the nanoscopic sense. The recent approach of EHD to boundary lubrication (1) seems to apply to metal forming as well.

Before 1990's surface textures of deformed and finished sheets were evaluated mainly by either stylus profiler or specular gloss meter. However these instruments have various limitations. Recent developments in Atomic Force Microscope (AFM) and Scanning Laser Microscope (SLM) provide many possibilities to obtain profiles or three-dimensional images of a product surface,.

In the present paper the above instruments are complementarily used to investigate the surface microtexture of the metalworked products and to evaluate actual lubricant film thickness.

STRIP DRAW TEST

A strip draw test of aluminium sheet with specially roughened initial surface was carried out to observe a typical mixed lubrication regime. The commercially-pure aluminium half hard strips were preliminarily roughened by rolling. These strips were drawn at a low thickness reduction (below 1% without bulk plastic deformation) and at a high reduction (10%) using a cylindrical dies of 50 mm radius at a low speed (0.86 mm/s). A machine oil was used as a lubricant.

By a SLM the area ratio of boundary contact region was checked. The surface roughness measured by an AFM at a boundary contact region was larger at low reduction without bulk plasticity where the micro-valleys prevail than at high reduction where the boundary contact region prevails.

SHEET COMPRESSION TEST

A newly devised sheet compression test illustrated in **Fig. 1** was conducted. A gauge block, with one of the most smooth surface commercially available is used as a flattening tool. A backup cylindrical tool is designed to give variation of contact regime (mixed to boundary) as a function of tool diameter, sheet surface roughness, lubricant and tool speed. Two paraffinic base oils (P8 and P500) were used as lubricants. The kinematic viscosity were 7.95 and 91.8 mm^2/s at 40°C, respectively. The surface roughness of the tested sheet is summarized in **Table 1**. The surface roughness is assumed to be independent of scratches caused by sliding because of lack of directionality. The roughness ranges within the length of the average lubricant molecule or several times of the width.

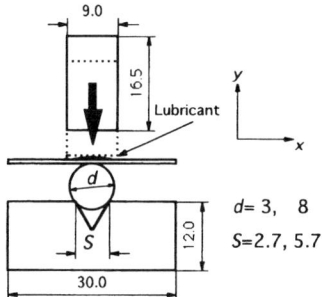

Fig. 1 Schematic of sheet compression tester.

	Regime	Ra/nm	Ry/nm
P8	Mixed lub.	0.94	7.0
P8	Boundary lub.	0.45	2.5
P500	Mixed lub.	1.2	7.5
P500	Boundary lub.	0.37	2.4
Gauge block	Smooth	0.25	1.6
Gauge block	Rough	0.85	3.5

Table 1 Surface roughness of boundary lubricated portion. (Backup tool diameter: 3mm, Contact force: 10kN, Contact pressure: *ca.* 170MPa)

CONCLUSION

From the surface roughness measurement by AFM over the boundary contact region, it is concluded that the surface roughness is comparable with the length of lubricant molecule or several times of the molecule width. This seems to suggest the reduced lubricant film thickness in the metalworking processes and the limit of currently obtainable smoothness and traceability.

REFERENCE

(1) H. A. Spikes, *Thin Films in Tribology*, ed. D. Dowson *et al.*, 1993, 331-346 pp, Elsevier.

LABORATORY SIMULATION OF WORK ROLL WEAR IN HOT MILLS

J GOODCHILD and J H BEYNON
Department of Mechanical Engineering, University of Sheffield, Mappin Street, Sheffield, S1 3JD.

ABSTRACT

Since the introduction of high speed steel (HSS) rolls in the 1980s to the finishing stands of hot strip mills, improved wear resistance has been observed over traditional roll materials. In most cases findings such as these have relied on data taken from hot mills. The hot roll wear test facility currently being developed at the University of Sheffield will enable simulation of rolling conditions to take place. Off-line laboratory testing can be conducted at much lower cost than full scale mill trials and are a useful precursor to trials on the mill.

This paper discusses the design, construction, and commissioning of a hot roll wear and thermal fatigue rig to simulate conditions which are comparable with those encountered in a modern hot strip mill.

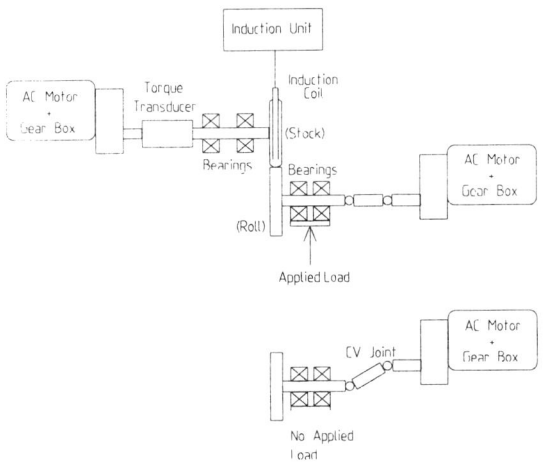

Figure 1. Schematic View of Test Rig Showing Roll Disc and Shaft Arrangement in the Loaded and Unloaded Positions.

The test rig (Figure 1) is in essence a twin disc machine where the discs are in rolling sliding contact. Each disc is independently driven through an in-line gear box by two 7.5kW A.C. motors, such that the unidirectional slip between discs can be controlled accurately. A shaft cooling system is necessary in order to run the equipment at elevated temperatures for long periods. The rig is designed to accommodate roll discs of nominally 230mm diameter and width of 25mm. Stock discs are machined from mild steel and are 230mm diameter and 15mm wide. The disc representing the stock material is heated by a 15 kW induction heating unit via a two piece coil to typical operating temperatures. These temperatures are monitored using a laser focusing two colour infrared pyrometer. Mill data taken from British Steel plc Hot Strip Rolling plant at Port Talbot South Wales were used to calculate the thermal cycle, contact times and contact patch length which become critical when focusing on the thermal fatigue aspects of roll damage. These data were then scaled to suit the test sample dimensions. The roll disc may be cooled between contacts to simulate the fluctuations in thermal stress at the roll surface seen in the mill. There are two sources of oxide within the roll gap; that which is grown on the roll surface, and that which is transferred to the roll surface from the strip. In order to look at the wear of work rolls it is necessary to grow a representative scale on the stock disc before commencing a test. To aid this, an environment chamber is to be constructed such that the stock disc may be brought to working temperatures in inert surroundings before exposing it to the atmosphere and creating a scale layer of known composition. Contact pressures between roll and stock discs representative of practical rolling loads are achieved by the use of a servo hydraulic ram.

Initial tests encompassing high chrome roll materials will be used as a validation process for the rig leading to characterisation of roll wear and fatigue. These mechanisms will then be investigated to ensure that the initiation and propagation of roll surface damage recreated on the rig is comparable with that on the mill. The rig can also be used to rank different roll materials. However the ultimate goal of this work is roll life prediction and the production of guidelines for the design of new roll materials.

ACKNOWLEDGEMENTS

The authors wish to thank British Steel Strip Products for their contributions financial or otherwise. Research work carried out with financial aid of the European Coal and Steel Community.

MEASUREMENT OF FRICTION UNDER SIMULATED METALWORKING CONDITIONS IN MINIATURISED TEST SYSTEMS

M G Gee, M Loveday and M R Brookes
Centre for Materials Measurement and Technology
National Physical Laboratory, Queens Road, Teddington, Middlesex TW11 0LW, UK

ABSTRACT

In metalworking, there is an increasing use of sophisticated modelling procedures which are used to predict the way that metal workpieces will deform during processing. These models are being used to optimise the process conditions, so that the metalworking can be carried out more reliably, giving improved yields and improved final product performance through the reduction or elimination of defects in the finished product.

The success of these modelling techniques, which are often based on a finite element approach are dependent on the availability of reliable data on materials behaviour, in particular the friction that is generated between the deforming metal workpiece and the tool.

Whilst the values of friction at the contact interface can be estimated from measurements made on instrumented metal-working and other simulation test systems, a large number of simplifying assumptions have to be made in the analysis which lead to large errors in the friction values obtained (1). Moreover, since the relative speed and contact pressures are altering across the contact the workpiece and tool in metalworking, the value of friction that is estimated can only be an average value for the whole process and will not be specific to particular contact conditions.

As an alternative approach, the metalworking contact conditions were simulated in two miniaturised test systems which enable values of friction under well controlled contact conditions to be measured.

The most important consideration is that in many metalworking processes the tool is cold and the workpiece is hot.

The first test system that was used in this study was based on a reciprocating test rig modified so that a single stroke experiment could be carried out. The tool sample was kept cold by water cooling, whilst the workpiece sample was heated to temperatures up to 1000 °C by a SiC furnace. Experiments were carried out by bring the samples into contact at the required test load whilst the samples were moving relative to each other.

In the second test system, the workpiece sample was a thin strip of the metal under test which was heated by a DC current in a specially designed sample holder (Figure 1). The DC self-heated workpiece sample replaced the pin in a conventional pin-on-disc test system. In both test systems the frictional and normal forces, and the relative displacement in the friction and normal directions were all recorded continuously by computer, with the DC heating for the self-heated pin also controlled by computer.

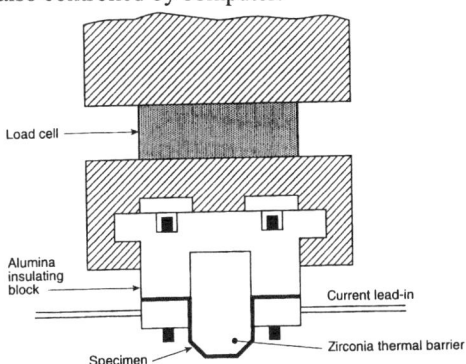

Fig. 2: DC heated pin

In preliminary experiments using an AISI 52100 bearing steel as a simulated tool material, and mild steel as the hot workpiece material, friction coefficients in the range 0.30-0.45 were observed in tests where the initial workpiece temperature was up to 1000 °C (Figure 2).

Fig. 2: Results of friction measurement at initial

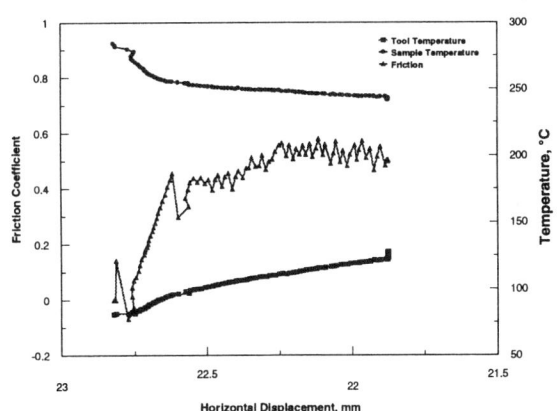

workpiece temperature of 300 °C

REFERENCES

(1) J. A. Schey, Tribology in Metalworking: Friction, Lubrication & Wear, American Society of Metals, 1983.

ANALYSIS OF SLIDING WEAR MECHANISM OF OXIDE FILMS ON HOT ROLL SURFACES BASED ON IN-SITU OBSERVATION OF WEAR PROCESS BY THE CCD MICROSCOPE TRIBOSYSTEM

Kazuo HOKKIRIGAWA, Teruko KATO, Tsutomu FUKUDA and
Masanori SHINOOKA

Department of Mechanical Systems Engineering, Faculty of Engineering,
Yamagata University, Yonezawa 992, Japan

ABSTRACT

The purpose of this investigation is to analyze microscopic sliding wear mechanism of oxide films on hot roll surfaces based on in-situ observation of wear process. Dynamic wear processes of oxide films on hot roll surfaces under unlubricated and water lubricated conditions were observed successively by using a CCD microscope tribosystem shown in Fig.1. Upper ball specimen is made of bearing steel (SUJ-2, R=1mm, H_v=7.51GPa). Lower plate specimens are made of representative hot roll materials such as adamite(AD), nickel grain cast iron(GH), high chromium cast iron(HCR) and high-speed tool steel(SKH). The surfaces of plate specimens are covered with oxide films with the thickness of about 5 μm.

Wear modes of oxide films on four different hot roll surfaces were classified into three types, such as "ploughing or powder formation", "flake formation" and "spalling". Wear modes observed in all wear tests are summarized as the wear mode map shown in Fig.2, where horizontal axis is a dimensionless contact pressure P_{max}/H_s (maximum contact pressure / Vickers hardness of substrate of hot roll) and vertical axis is a friction coefficient μ. The wear map shown in Fig.2 is useful in order to suppress severe wear mode of "flake formation" or "spalling". In the mild wear mode of "ploughing or powder formation" of oxide film, specific wear rate w_s increases with increasing of dimensionless parameter S_w^* as shown in Fig.3, where S_w^* is defined as follows [1];

$$S_w^* = (1+10\mu) H_v R_{max}^{1/2} / 5K_{Ic} \qquad (1)$$

where μ is friction coefficient, H_v is Vickers hardness of oxide film, R_{max} is maximum surface roughness and K_{Ic} is fracture toughness of oxide film.

It can be concluded that Fig.2 and Fig.3 give useful information for improving wear resistance of oxide films on hot roll surfaces.

REFERENCES

(1) K. Hokkirigawa: Wear maps of ceramics, Bulletin of the Ceramic Society of Japan, **32**(1)(1997)19.

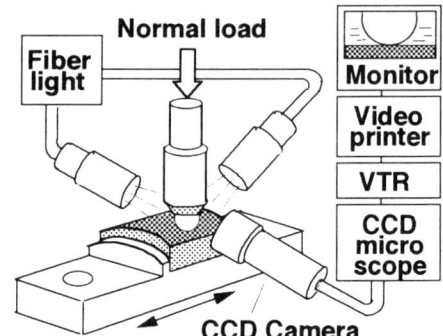

Fig.1 CCD microscope tribosystem

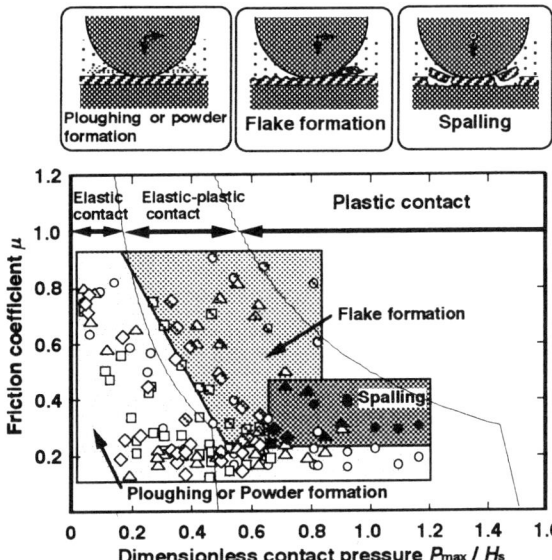

Fig.2 A wear mode map of oxide films

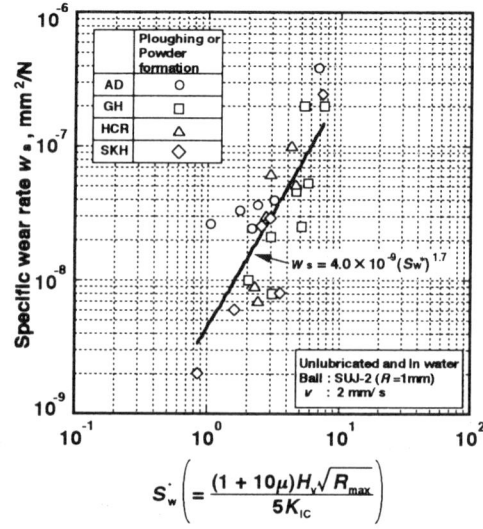

Fig.3 Relationship between dimensionless parameter S_w^* and specific wear rate w_s

RESEARCH ON INTELLIGENT SIMULATION SYSTEM OF CYLINDER-PISTON-RING TRIBOSYSTEM FOR MARINE DIESEL

YAN LI and XU JIUJUN
Institute of Metal & Technology, Dalian Maritime University, Dalian (116026), P.R.CHINA

ABSTRACT

Bad quality fuel, in which the sulfur and abrasive content are very high, is used for marine diesel engine. So the working environment for the cylinder-piston-ring tribosystem is worst than that of any other diesel engine. The wear rate of cylinder and piston-ring is often very high.

Although many simulation research works which could predict the wear rate have been done, it is difficult to apply these results to the real system because of the lack of the simulating rules and theory model for wear problem.

This paper introduces the intelligent simulation method and intelligent simulating system for wear test (WTIS) which is developed for the first time. The cylinder-piston-ring tribosystem of marine diesel engine is chosen as the study object. The intelligent simulating model is built on the basis of wear test and system analysis of the chosen tribosystem. The fuzzy logical theory and artificial intelligence technique are used for modeling and result verification. This simulating result, which describing the state of tribosystem with a creditability, can represent the characteristics of the system. So the research method introduced in this paper is more reasonable.

The following key problems are solved in this paper.

1. SYSTEM ANALYSIS

With the help of the systematology, the relationship among the elements in the selected tribosystem and between the system and the environment can be easily found. So the basic rules and facts that control the wear state of cylinder and piston-ring can be concluded. These rules and facts are the main parts of the knowledge base of the simulation system. The logical relationships that described by the rules compose the intelligent structure of the simulation model.

2. MODELING

The simulation model that is built on the basis of the system analysis consist of the two main parts, fuzzy inference and integration calculation. The WTIS system synthesizes the input parameters and the knowledge on the base under controll of the inference machine. At the same time, the integration method is used to calculate the lubrication state between the cylinder and piston-ring.

3. FUZZY INFERENCE MECHANISM

The knowledge base, which is the basis of the fuzzy inference machine, is mainly composed of fuzzy facts and fuzzy rules. During the inference procedure, the creditability of the rules should be changed automatically according to the cited freqency and the successive rate. Those which can make the rules correctly reflect the wear characteristic should be cited prior to any others. This would be established as the selfstudy mold of the WTIS system.

4. KNOWLEDGE EXPRESSION

According to characteristic of the knowledge and the needs of the inference, the knowledge should be expressed in many forms. The most important one should be the rules expression that used to show the logical relationship among the facts. These rules is composed of the following parts: conditions, coefficient, condition creditability, relation, result, result creditability and rule creditability. These rules can describe the logical structure and intelligent relations of the facts.

5. THE DEVELOPMENT OF WTIS SYSTEM AND EXAMPLE

The intelligent inference system is developed in the FoxProw for Windows environment in this paper. Using this language for programming the WTIS system, a very beautiful and friendly interaction interface can be created in shorter period than do using any other special intelligent program language.

The WTIS system is applied to RLB48 marine diesel engine (owned by TianJin department of COSCO) and the simulation result is found satisfaction in comparing with the measurement value.

REFERENCE

[1] Yan Li, Wear and reliability technique of diesel engine, Jiao Tong Publication House, 1992.
[2] Xie Youbai, Lubrication and seal, 1986,6,P1 ~ 10.
[3] Xu Jiujun, Yan Li, Transportation and computer, 1997,1,p1-4,9
[4] Xu Jiujun, yan Li, Proceedings of the second academic annual meeting of youths of liaoning province(Da Lian) 1995,10: 163~166.
[5] Xu Jiujun, doctorate degree paper, Dalian Maritime University, 1996, 4.

DIRECT OBSERVATION OF FRICTIONAL SEIZURE DURING SLIDING OF STEEL ON AL 6061 DISK

MARGAM CHANDRASEAKARAN, XING HUTING & ANDREW WILLIAM BATCHELOR,
School of Mechanical and Production Engineering, Nanyang Technological University, Singapore 639798

ABSTRACT

Seizure of sliding contacts have always posed a challenge to tribologists over the past decades as it often occurs without prior warning. Various researchers have proposed theories and models for the frictional seizure of tribological contacts. These theories however predict the conditions of seizure well, but lack the experimental evidence to explain the mechanism operative during seizure which is vital in the design of new improved type of materials for tribological contacts. In-situ observation of seizure were carried out by Spikes et al using a sapphire disk[1-2]. The present investigation focuses on the in-situ observation of frictional seizure using a X-ray microscope to test the accuracy of current models as well as find the actual mechanism of seizure. The seizure experiments were carrried out using a model pin on disc apparatus custom built in house shown in Figure 1. A steel pin sliding on aluminum disk was selected as a model contact and the sliding tests were carried out at different speeds with varying loads. Cyclic wear of aluminum disk was observed during sliding. It was observed that at low loads and sliding speeds, the scuffing or seizure occurred predominantly due to plowing and adhesive bonding in case of mild steel specimens. At higher sliding speeds, the seizure was predominantly due to rolling of wear particles to form filaments which was was pressed and bonding occured between the wear sheets and the two nascent surfaces. Stainless steel specimens seized at lower sliding speeds possibly due to atomic transfer and chemical bonding and while at high speeds plowing was predominant. The time taken for the seizure in case of mild steel samples were comparatively higher than those of stainless steel samples inspite of the higher hardness levels and strength of the later. This was possibly due to the sliding of metals with similar crystal structure.

REFERENCES

1. Enthoven, J.; Spikes, H.A, Infrared and visual study of the mechanisms of scuffing, Tribology Transactions v 39 n 2 Apr 1996. p 441-447.
2. Enthoven, J.C.; Cann, P.M.; Spikes, H.A.,Temperature and scuffing, S T L E Tribology Transactions v 36 n 2 Apr 1993. p 258-266.

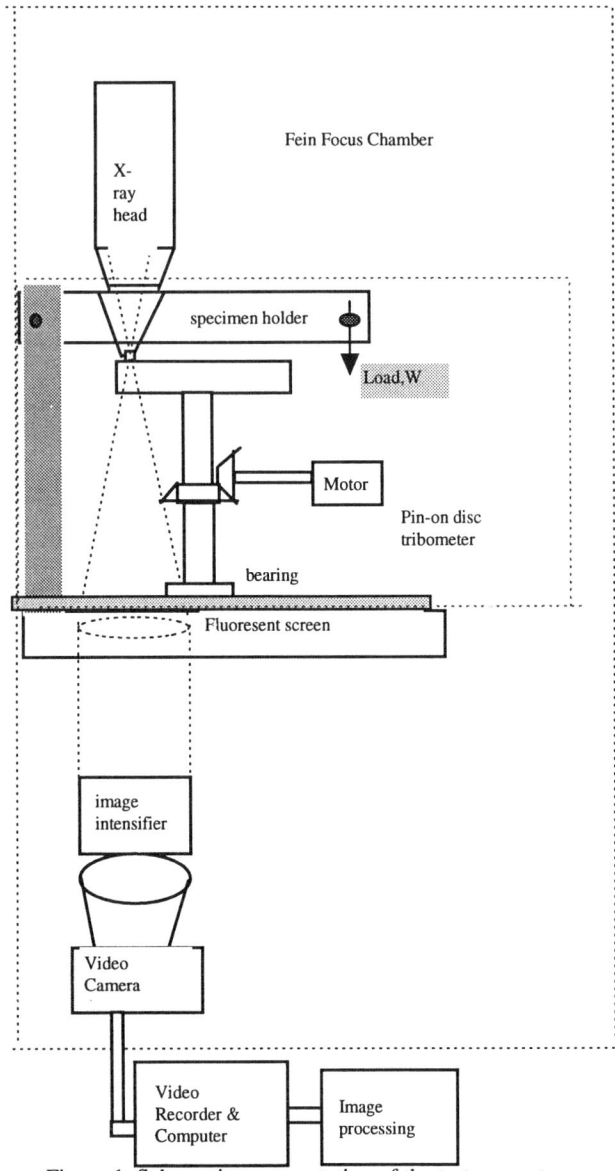

Figure 1. Schematic representation of the test apparatus

MONDAY
8
SEPTEMBER

**Corresponding
oral session**

                                                  Page number

M1/TU1   Thick film bearings 387

M2/TU2   Lubricant chemistry and rheology.. 443

M5   Solid lubrication 473

M4/TU4   Surface engineering.. 479

HIGH SPEED SUPPORTS USING AEROSTATIC AERODYNAMIC BEARING

B.B. AHUJA, P. P. CHIKATE and S. K. BASU
Department of Production Engineering
Government College of Engineering, Pune - 411 005, India

ABSTRACT

The spindle units developed by the authors are based on conical bearings with aerostatic - cum - aerodynamic support, incorporating a hybrid design. Such spindle support working on aerostatic - cum - aerodynamic conditions operate upto 70,000 rpm.

The present work is confined mainly to the consideration of hybrid conical air bearing set up of the feed hole (jet feed) type. Conical bearings take the form of combined journal and thrust bearings, thus providing for radial and axial loads. The speed of the spindle can be very high with adequate stability but on lighter load.

Air is admitted to the bearing clearance maintained at $\pm$ 44 microns through 8 number of equally spaced supply orifices in each plane located around the circumference. The conical bearing bush is designed for two plane admission, with a semicone angle of 10°. Assumptions are based on the fact that the viscosity is kept constant, common supply pressure is getting divided into large and small ends of the bearing and flow continuity equations are dealt separately for large and smaller ends. Down stream pressure at an orifice is assumed to be equal to supply pressure.

Experimental results are found in close agreement with the theoretical work in regard to no-rotation cases. This validates the available mathematical models(1). In the case of static response of the bearing, the interactions between the three major parameters, the radial load, supply pressure and the bearing design parameter were projected on 3D response surface curve. The trend shows that the optimum field of working requires 18 N as radial load to get the benefit of optimally minimum eccentricity ratio.

DELTA, the bearing design parameter theoretically estimated, is dependent on the radial clearance and supply pressure, and also varies inversely with cube of radial clearance. eccentricity ratio is dependent on supply pressure P_s and radial load W_r, connected by generalized equation showing exponential relationship, with a fairly high regression coefficient.

$$\varepsilon = 0.6 * W_r^{1.09} * P_s^{-0.67}$$

The authors found out the rigidity from the empirical relationship obtained through experimental observations connecting the parameters - dynamic deflection, and supply pressure for no rotation condition and relating the parameters dynamic deflection, supply pressure and conical bearing number in rotational cases. The patterns obtained for rigidity for air hybrid bearings, as seen from the response curves, support the observations by Fuller and Powell in their respective publications (2).

The rigidity in the case of air bearing, rises with increase in dynamic deflection for increasing conical bearing number rotation case with constant supply pressure. In case of no-rotation it increases with dynamic deflection and constant supply pressure.

For consideration of bearing design and operational parameters on radial and axial load capabilities, stiffness and flow rate, the theoretical predictions of the radial as well as axial load coefficients, axial flow for various values of supply pressure ratios, eccentricity ratios and design parameter (DELTA) have been tabulated. The estimated values and magnitude compare reasonably with the experimental results obtained.

During no rotation of air bearing, at higher value of supply pressure, eccentricity ratio remains low, particularly in the region of 0.4 - 0.5, for maximum radial load bearing capacity. Though there is no provision for reducing ambient pressure but studies made with different values of supply pressure, response surface curves, when analysed indicate that for a given conical bearing number, reduction in supply pressure from 6.078 x 10 5 N/sq.m to 4.052 x 10 5 N/sq.m does not make any significant change in radial load carrying capacity of conical bearing.

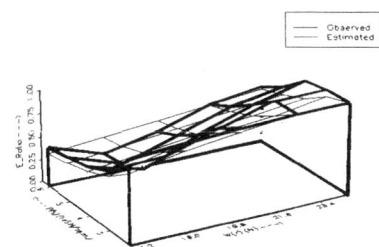

Fig 1. W_r Vs P_s for observed and estimated eccentricity ratios under no rotation

REFERENCES
(1) K. Srinivasan and B.S. Prabhu, "Analysis of Externally Pressurized Gas-Lubricated Conical Bearing" Wear, 86, (1983) pp 202-210.

(2) DD Fuller "Theory & Practice of Lubrication for Engineers, (2 nd Edition) 1984, John Wiley & Sons, NewYork.

AXIAL PROFILE EFFECTS OF JOURNAL BEARINGS

MAIDO AJAOTS and MART TAMRE

Department of Precision Engineering, Tallinn Technical University, Ehitajate tee 5, Tallinn, EE 0026, ESTONIA

ABSTRACT

The paper deals with friction problems of boundary lubricated journal bearings. The aim of the paper is to analyse the misalignment and axial shape effects of rubbing surfaces encountered with the bearings of precise devices.

The study is based on the joint analysis of the isothermal contact problem of a bearing and journal equilibrium conditions on the one hand (1), and on the deformation-adhesion friction model (2) and experimental results' analysis on the other hand.

The positive effect of surface macro geometry serves as sum of several phenomena: contact pressures reduction; friction coefficient decrease due to friction components change; formation of opportune conditions for removing wear debris from friction region under the affect of transversal gradient of contact pressure and formation of favourable conditions for leading lubricant into the friction region (3). An essential role of the contact pressure transverse gradient of a journal bearing became evident.

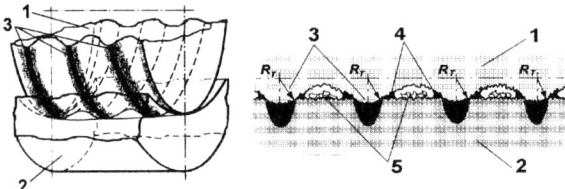

Fig. 1: Axially profiled bearing, 1 - journal, 2 - bearing bush, 3 - arched segments, 4 - lubrication zones, 5 - grooves.

The guided wear debris removal from the rubbing zone prevents breaking the boundary layer between the rubbing bodies and increases the friction region by accumulating the products leading to a characteristic selforganizing process decreasing total friction and wear as result. A bearing arched segment curvature radius $R_{T1,2}$ leading to friction losses minimum can be predicted by the formula

$$K_{G4} = \frac{(9.0n-5.0)\times 10^7 \psi^{(5.8-15.1n)}}{C_0^{(2.7-12.8n)} \overline{F}_k^{2.0}} \left[\frac{K_x \Delta^{0.67} E_2}{\tau_0 (1-\mu_2^2)^{0.3}} \right]^{(4.3-22.7n)},$$

where $\quad K_{G4,5} = \left(5\times 10^{-5} + \overline{R}_{T1,2}^{-1}\right).$

An axial segment curvature radius less than critical leads to accelerated formation of wear debris able to break the boundary layer.

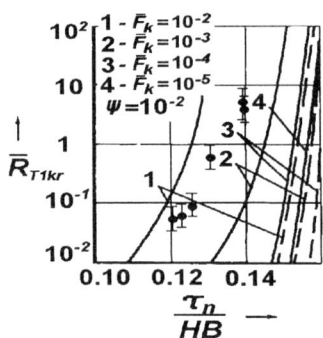

Fig. 2: Critical radius of journal arched segment vs. relative shear stress, —— steel - bronze, - - - steel - polyamide PA6.

The results demonstrate that a minimum of friction dissipation energy accompanies with a characteristic axial shape of the journal bearing. The effect can be used at bearing design.

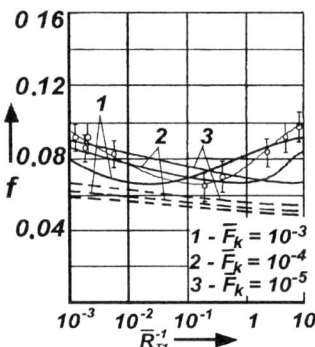

Fig. 3: Friction coefficient vs. radius of curvature of journal arched segment (r=4 mm, surface micro-geometry parameter Δ=0.001, ψ=0.01): —— steel - bronze, - - - steel - polyamide PA6.

Friction torque of bearings with arched segments is up to 50 % and variation of friction up to 30 % smaller compared with cylindrical bearing characterised with load asymmetry.

REFERENCES

(1) M.Ajaots, M.Tamre, Calculation of contact areas and pressures at misaligned journal bearings. Soviet Journal Friction and Wear, V.14, 1993, 334-340pp.
(2) I.V.Kragelsky, M.N.Dobychin, V.S.Kombalov, Friction and Wear, Pergamon, 1982.
(3) M.Tamre, Misaligned Journal Bearings. Proc.of the Estonian Acad. of Sciences, Eng., N°1, 1995, 87-97pp.

ON THE SHAPE OF THE LUBRICANT FILM FOR THE OPTIMUM PERFORMANCE OF A ROUGH SLIDER BEARING

P I ANDHARIA and G M DEHERI
Department of Mathematics, Sardar Patel University, Vallabh Vidyanagar, Gujarat, 388120, INDIA
J L GUPTA
Department of Mathematics, B V M Engineering College, Vallabh Vidyanagar, Gujarat, 388120, INDIA

ABSTRACT

This investigation concerns the analysis of the lubricant film profile for a rough slider bearing working with an incompressible lubricant for optimum load carrying capacity. The bearing surfaces are assumed to be stochastically rough. The roughness of the bearing surface is modelled by stochastic random variable with non-zero mean, variance and skewness. An attempt has been made to find the shape of the lubricant film profile for the bearing such that the load carrying capacity of the bearing is optimum. The governing differential equation for the lubricant pressure is the Reynolds equation which is stochastically averaged with respect to the random roughness parameter. The optimum film profile is found to be a step function with the step location and step height ratio depending on the parameters : mean (α), standard deviation (σ) and measure of symmetry (ε) of the stochastic random variable characterizing the surface roughness of the bearing. Results for the step location, step height ratio and the bearing performance characteristics such as load carrying capacity of the bearing, centre of pressure, frictional force and coefficient of friction for different values of α, σ and ε are numerically computed and tabulated. It shows that step height ratio H_0 and coefficient of friction **f** increase with increasing α, σ and ε while the load carrying capacity **w** and frictional force **F** decrease with increasing α, σ and ε. Further it is observed that step location and centre of pressure shift towards the outlet edge with increasing values of α while they shift towards the inlet edge with increasing values of σ and ε.

HYDRODYNAMICAL ANALYSIS OF CIRCULAR CONTACT BETWEEN ROUGH SURFACES AT IMPACT LOADING

GABRIEL ANDREI and MIHAI JASCANU
Department of Mechanical Engineering, "Dunarea de Jos" University of Galati
47 Domneasca Street, Galati, 6200, RO

ABSTRACT

The initial stage of the lubricated circular contact at impact loading is characterised by a low level of pressure and the absence of any elastic deformation of the contacting bodies (1). That is why the lubricated impact in its early stage is hydrodynamically approached. The case of a body with spherical contacting part bouncing on a flat surface covered by a thin lubricant layer is analysed theoretically. The surface topography has a significant effect on the passing to the EHD regime. The model is based on the Reynolds equation for isothermal conditions, assuming an isoviscous and incompressible lubricant. Pressure distribution within the lubricating film can be obtained from:

$$\frac{\partial}{\partial x}\left(h^3 \frac{\partial p}{\partial x}\right) + \frac{\partial}{\partial y}\left(h^3 \frac{\partial p}{\partial y}\right) = 12\eta \frac{\partial h}{\partial t} \quad (1)$$

The equation of motion for the impacting ball is:

$$a_0(t) = \frac{w(t)}{m} \quad (2)$$

The variation of pressure during the normal approach can be determined by solving the equation (1) and (2) simultaneously. The film thickness, in the case of transverse roughness can be calculated from:

$$h(x, y, t) = h_0(t) + \frac{x^2 + y^2}{2R} - a \cdot \cos\left(\frac{2\pi x}{\lambda}\right) \quad (3)$$

The force is found by integrating the pressure on the current contact area. The velocity and the rigid separation are obtained from:

$$v_k = v_{k-1} + a_{0k-1}\Delta t \quad (4)$$

$$h_{0k} = h_{0k-1} + v_{k-1}\Delta t + a_{0k-1}\frac{(\Delta t)^2}{2} \quad (5)$$

This theoretical model was used to analyse the initial stage of the lubricated impact. It is also studied the influence of the ball mass, radius of curvature, impact velocity, initial rigid separation and viscosity on the pressure distribution and film thickness profile. The effect

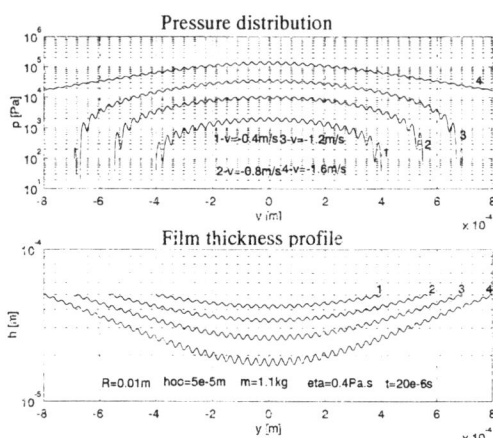

Fig.1: Influence of impact velocity

of impact velocity is shown in Fig.1.

An interesting result is that the pressure in the case of rough surface is lower than those obtained in the smooth surface case (2). This is in agreement with the results of Majumdar and Hamrock (3). In the early stage of the collision, the influence of the mass is negligible. The ball radius has an insignificant effect on the pressure and film thickness. The increase in the impact velocity has a decisive influence on the pressure values and film thickness profile. It was possible to determine, for given conditions, the transition to the EHD regime. The decrease in the initial rigid separation brings obvious changes after 20 μs, when the pressure starts to increase rapidly. The presence of the surface roughness has no effect on the influence of the viscosity, related to smooth surface case. The analysis of inertia effect in the early stage of the impact, based on Reynolds' number, Re, shows that there is no significant change in the nature of the fluid flow.

REFERENCES

(1) R.Larsson, E.Höglund, ASME Journal of Tribology, 1994, Vol.116, 770-776 pp.
(2) G.Andrei, Proceedings of the 7th International Conference on Tribology Rotrib '96, Bucharest, Vol. 3, pages 52-60, 1996.
(3) B.C. Majumdar, B.J. Hamrock, ASME, Journal of Lubrication Technology, Vol. 104, 1982, 401- 409pp.

EFFECTS OF ELASTIC DEFORMATIONS OF A BEARING LINER ON THE PERFORMANCE OF ORTHOGONALLY-DISPLACED PRESSURE DAM BEARINGS WITH ROTOR FLEXIBILITY EFFECTS

S.K. ANGRA, N.P. MEHTA and S S. RATTAN
Department of Mechanical Engineering, Regional Engineering College, Kurukshetra-136119, India

ABSTRACT

It is well known that elastic deformations, whether permanent or transient, have significant effect on the hydrodynamics of bearings. Thus it is necessary to couple the elastostatic (solid) and hydrodynamic (fluid) aspects of the bearing for proper evaluation of its performance.

Elastohydrodynamic lubrication is a form of fluid film lubrication where elastic deformations of the lubricated surfaces become significant. Theory of elastohydrodynamic (EHD) lubrication has developed significantly during the last four decades. In the early stages of its development it remained confined mainly to the problems of heavily loaded contacts as in rolling contact bearings and gears. However, in the recent past the use of EHD lubrication theory is enlarged to include the study of hydrodynamic bearings as significant distortion of the bearing elements were observed under the action of hydrodynamic pressures. With this view in mind the elastohydrodynamic study of orthogonally-displaced pressure dam bearing which is considered to be the most stable among the various two lobe bearings has been undertaken.

Solution of bearing problem considering the flexibility effects of the liner involves the simultaneous solution of the Reynolds equation of the lubricant and the elasticity equation in the bearing liner. While finding the solution, it is assumed that the liner is enclosed in a rigid housing and the thermal effects are neglected. To consider the flexibility of the bearing liner, a dimensionless deformation coefficient Ψ as a function of the journal speed, geometry of the bearing, viscocity of the lubricant, thickness of the bearing liner and the modulus of elasticity of the liner material is defined.

The deformation of the bearing liner is obtained by using a three dimensional elasticity model using hexahedral elements. Finite element method has been used to calculate the deformation of the liner induced by the pressure of the oil film, which is in turn influenced by this deformation causing an increment of the oil film thickness.

The results show that peak pressures decrease slightly in the upper lobe, whereas, there is redistribution of pressures in the lower lobe with the peak pressure decreasing and shifting towards the direction of rotation as the elasticity of the liner is increased. Attitude angle and oil flow decrease, whereas, eccentricity ratio and minimum film thickness increase as the flexibility of the liner is increased. There is marginal increase in the friction coefficient as the liner flexibility is increased.

For rigid rotor, stability increases in terms of zone of infinite stability and minimum threshold speed as the flexibility of the liner is increased. In case of flexible rotors, the liner flexibility increases the zone of infinite stability, however, the minimum threshold speed first decreases and increases afterwards

ON VARIATIONAL PROBLEMS IN THE GAS LUBRICATION THEORY

YURY BOLDYREV
Physical & Mechanical Faculty, St Petersburg State Technical University,
Politechnicheskaya 29, St Petersburg, 195251, Russia.

ABSTRACT

The investigation of variational problems of the lubrication theory was begun by lord Rayleigh (1). His work was far ahead of needs of his time. The researches of Rayleigh were continued by American scientists Maday and Rohde since the begining of the sixtieth years. In Russia this work was carried on, basically, in Leningrad Politechnical Institute (nowadays Saint Petersburg State Technical University).

The results of investigations in area of two- and one-dimensional variational problems in the theory of compressible and incompressible lubrication are presented in this report. The report covers the questions of correct statement of variational problems in the lubrication theory, and questions of tackling the difficulties arising in the process of its solving, and application of obtained outcomes.

The variational problems of the theory of lubrication are studied within the framework of mathematical model, based on the Reynolds equation, which has the following dimensionless form in the case of gas lubrication

$$div(h^3 p \nabla p - \Lambda p h \underline{v}) = \sigma \partial(ph)/\partial t ,$$

where p and h are dimensionless pressure and clearance, $\underline{v}$ is velocity of sliding, and Λ and σ are physical criteria of similarity. Profile function h here plays the role of the control and belongs to the close domain: $h_{min} \leq h \leq h_{max}$, where h_{min} and h_{max} are given. Initial and boundary conditions to Reynolds equation are determined by the construction of a considered unit. As a functional of the variational problem we can take any stationary performance of the lubricant layer, for example, the load capacity or force of friction and etc. In optimization of dynamic characteristics it can be "a critical mass" of the rotor or elements of dynamic rigidity matrix. Such criterion as time of "surfacing" of lubricated rotor is of special interest in non-stationary problems. Thus, in general case, for selected criterion we have the Lagrange problem of the calculus of variations for equation with partial derivatives. The first example of solution of this problem in two-dimensional case was given in work (2). If necessary the statement of a problem can be supplemented with other restrictions, for example, by isoparametric conditions. The distinctive feature of the Reynolds equation, from viewpoint of the modern approaches of the calculus of variations is circumstance, that the control function h enters into main part of differential operator of this equation. It generates serious difficulties – the variational problem becomes nonclassical. The approaches developed on study of the first variation of functional ascending to Euler and Lagrange are inapplicable to this problem. In some cases solution does not exist in the form of piecewise smooth profile h (3). The solution takes on the form of shuttering control which is realized, for example, as indefinitely alternated areas with $h=h_{min}$ and $h=h_{max}$. It is obvious, that in this case the Reynolds equation does not hold and requires the generalisation of the initial problem. Generalized Reynolds equation is the form of the asymptotic equation for pressure field, that is known as Whipple equation. In most general case the coefficients of this equation are matrices. The problem of optimization becomes the problem of definition of eigenvalues and eigenvectors of these matrices, which characterize some limit periodic microstructure of the profile geometry. In last years the gas lubrication has found important application in new generations of seals for turbocompressor machines. New interesting class of variational problems appears in this connection (4). These problems are given in the report. The examples of the solutions of variational problems for various criteria and different types of bearings and seals are presented. Also we consider the problem of developing the optimal "technological profiles".

REFERENCES

(1) Lord Rayleigh . Philosophical Magazine. Vol. 35, No.1, 1918 1-12pp.
(2) Y. Boldyrev ,V. Troickii. Izv. Akad. Nauk SSSR. Mech. Zhidc. Gaza. No.5, 1975, 34-39pp.
(3) Y. Boldyrev. Izv. Akad. Nauk (Russia). Mech. Zhidc. Gaza. No.2, 1992, 3-10pp.
(4) Y.Boldyrev, S. Lupuleac, J. Shinder. ECMI-96, Book of Abstracts, 1996, 523-524pp.

VISCOUS FLOW BETWEEN ROTATING DISKS

N. M. BUJURKE
Department of Mathematics, Karnatak University, Dharwad - 580 003, INDIA
N. P. Pai
Department of Mathematics, M. I. T. Manipal - 576 119, INDIA

ABSTRACT :

Viscous flow between rotating disks is reinvestigated using (i) computer extended series and (ii) formal power series. (1) The problem admits similarity transformation which enables in transforming governing equations to coupled non linear o.d.e. Follwing cases, leading to specific type of flows, are studied for calculating lift.
(a) Both disks stationary and the flow is due to injuction only.
(b) One of the disks is rotating and another one is stationary.
(c) Both disks corotate with same speed.
(d) Both disks counterrotate with same speed.

In each case it is possible to generate large number ot terms of series solution using recurrence relation. These series have finite radius of convergence. For recasting the series into new useful form the location and nature of nearest singularity, restricting the convergence, are estimated using Domb-Sykes plot.

In case (a) Domb-Sykes (2) plot shows the square root singualarity restricting the convergence of the series. Reversion of the series followed by Pade' approximants (5) of recasted series results in analytic continuation beyond the region of convegence of the series representing lift.

In case (b) the sign pattern of the co-efficients of the series is random but they decrease in magnitude. Pade' approximants provide useful information about lift. The analysis of cases (c) and (d) resulted in using Euler transformation for recasting the series into new form whose region of validity increased considerably for the calulation of lift. In all the cases the formal power series solution is also obtained. The two point b.v.p. is converted into unconstrained optimization problem which is solved using Brown's method (3). As the Reynolds number increases the terms of the power series required for convgence is quite large and this restricts the use of formal power series. Comparsion of power series solution with computer extended regular perturbution series (in R, the Reynolds number) is presented. Above analysis confirms with available pure numerical solutions and is valid for much larger values of Reynolds number compared with numerical findings (4).

REFERENCES

(1) N. M. Bujurke, N. P. Pai and P. K. Achar, Proc. Indian Acad. Sci. Vol. 105, No. 3, pp 353-369, Aug 1995.
(2) M. Van Dyke, Quart Jl. Mech. and Appl. Math. Vol. 27, p 423, 1974.
(3) G. Byrne and C. Hall (eds), Numerical solution of system of nonlinear Algebraic Equations Academic press, New York, 1973.
(4) C. Y. Wang and L. T. Watson, ZAMP Vol-20 p 773, 1979.
(5) G. A. Baker, Essentials of Pade' Approximants, Academic press, New York , 1975.

NON-LINEAR ANALYSIS OF FLEXIBLY SUPPORTED FINITE TURBULENT FLOW OIL JOURNAL BEARINGS

A K CHATTOPADHYAY
Department of Mechanical Engineering, B.E. College(D.U.), Howrah-711103, West Bengal, India
S KARMAKAR
R&D Department, Warthington Pump India Ltd., 22 Ferry-Fund Road, Panihati-743176, West Bengal, India

ABSTRACT

Bearing support structure plays a very important role in the successful operation of high speed machinery. Although, the bearing pedestal frequently acts as a flexible mounting, there is a practical advantage in having the bearings flexibly supported, to facilitate ease of alignment.

Lund (1) made a theoretical study on the threshold speed of a flexible rotor with damped supports and concluded that damped flexible supports considerably increased the threshold speed. Boffey (2,3) examined the stability characteristics of a gas bearing on flexible damped supports using the linearized ph solution. He showed two distinct whirl zones, a lower one associated with an instability of the rotor and the upper one with the bearing. Kirk and Gunter (4,5) studied the influence of flexible supports on the synchronous unbalanced response of a single mass flexible rotor and optimised the support system characteristics to minimise the rotor amplitude and forces transmitted over a given a speed range. However, all the analyses were confined to laminar flow regime. Hashimoto et al. (6) studied the dynamics of turbulent journal bearings. They used the short bearing assumptions in their model of study. They concluded that turbulence played a significant role on the dynamic behaviour of the rotor bearing system. They, however, did not consider the flexibility of the support system in the model of their analysis.

The progressive increase in size or speed of the rotating machinery and also the use of fluid with low kinematic viscosity, the oil-film flow in the bearing may become turbulent. It is, therefore, felt necessary to study the stability characteristics of a flexibly supported rotor bearing system with turbulent oil-film flow.

The aim of this paper is to present a theoretical analysis to obtain the transient response of a flexibly mounted rotor-bearing system for different support conditions (stiffness and damping) and different oil-film Reynolds number. The dynamics of the system is studied by calculating the components of fluid-film force by solving the generalised Reynolds equation modified to include the effect of turbulence and using these in the non-linear equations of motion of the journal and bearing. The modified Reynolds equation used in the present analysis is that of Hashimoto et al. (6).

The computational scheme consists in solving the modified Reynolds equation for pressure distribution by using finite difference method with S.O.R. The pressure distribution thus obtained is used to calculate the oil-film forces in two mutually perpendicular directions. The equations of motion of the rotor and the bearing are solved by space technique using the fourth order Runge-Kutta method as the tool for solving the resulting eight first order differential equations of motion of the rotor bearing system in two mutually perpendicular directions in which the components of the fluid-film forces were calculated. The resulting solution is then used to solve the modified Reynolds equation to further calculate the components of the fluid-film force under the changed operating condition. The process is continued stepwise to give the transient solution of the non-linear system.

The results indicate that for particular fixed values of aspect ratio, steady state eccentricity ratio, mass parameter, stiffness and damping parameters of support system, an increase in turbulence leads towards more stable orbit.

REFERENCES

(1) J.W. Lund, ASME Transactions, J. Applied Mech., Vol. 87, 911-920pp.
(2) D.A. Boffey, Gas Bearing Symposium, University of Southampton, paper 12, 1969.
(3) D.A. Boffey, Proceedings of the first World Conference in Industrial Tribology, New Delhi 1972.
(4) R.G. Kirk and E.J. Gunter, Res. rep. No. ME-4040-105-710, University of Virginia, 1971.
(5) R.G. Kirk and E.J. Gunter, ASME Transactions, J. Engg. for Industry, 1972, 221-232 pp.
(6) H. Hashimoto, S. Wada and J. Ito, ASME Transactions, J. Tribology, Vol. 109, No.2, 307-314 pp.

A SOLUTION OF PROBLEM ABOUT LONG POROUS CYLINDRICAL BEARING

A.V. GREKOVA and A.V. CHIGAREV
Department of Theoretical Mechanics, Belarussian State Polytechnic Academy,
Scoriny Ave., Minsk, Belarus

ABSTRACT

Using the porous bearing we can realized duty liquid friction, which can be provided by the liquid supply into the clearance. The structure of these bearing allows supply liquid trough porous body.

The problem about the lubrication of the long porous cylindrical bearing having eccentricity is considered. The lubricant supply is realized under pressure through the fixed porous bearing into clearance between the bearing and the uniformly rotation solid shaft. The lubricant is viscous incompressible liquid.

The problem have been reduced to joint solution of Stokes equations (equations of viscous liquid movement), which describe a movement in lubrication layer, and Laplace equation for pressure in porous body under the following boundary conditions:

-the pressure function in the lubrication layer is equal the pressure function in the porous medium on the boundary of the media;

-the liquid velocity dependence on the pressure on the boundary of the porous body is subordinated to Darcy low;

-the pressure on the external bearing surface is the liquid supply function

$$P = P_0 (1+\delta \sin \omega t)(1+ \sum_{n=1}^{\infty} \varepsilon^n (a_n \cos n\theta + b_n \sin n\theta)),$$

here θ is polar angle, ε is relative eccentricity.

-the liquid do not slide on the surface, i.e. the liquid velocity on the shaft is equal the velocity of corresponding shaft points.

Two small parameters ε and δ are in the problem. This made possible to carry out the problem analysis.

In limit case (when $\delta=0$) under given boundary conditions the solutions can be written in the form of series in power of ε (bearing eccentricity). The analytical expressions of the first approximations for liquid velocity and pressure have been obtained.

The calculation was carried out for $k=0.1$, $\varepsilon=0.1$, $a_1=1$, $b_1=1$ and $P_0=0$. The pressure distribution along the bearing circumference and its dependence on permeability k of porous layer are presented in Figure 1.

It was determined that loading value on the surface of shaft decrease with increase of permeability.

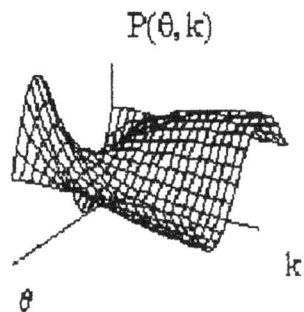

Fig.1: Pressure in the bearing for $\delta=0$

If we neglect the eccentricity that is if we suppose $\varepsilon=0$ ($\delta \neq 0$), we obtain the solution for pressure and velocity.

Considering ε and δ as a small parameters we obtain the problem solution to within the second power of small parameter. The pressure distribution along the bearing circumference and its dependence on time are shown in Figure 2.

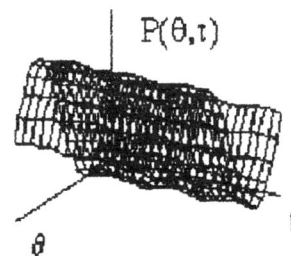

Fig.2: Pressure in the bearing for $\delta=0.5$

Showing in the figures Functions were calculated for such bearing, that the ratio of the external and internal diameters of this bearing is 1.1, the ratio of diameters of the shaft and the bearing is 0.9.

AN IMPROVED THD ANALYTICAL MODEL FOR A NONCONTACTING, ALIGNED MECHANICAL FACE SEAL

MIRCEA D. PASCOVICI and TRAIAN CICONE
Polytechnic University of Bucharest, Spl.Independentei 313, Bucharest 77206, ROMANIA

ABSTRACT

Temperature distribution in the fluid film, and in the seal rings is important for the prediction of the operating instabilities due to thermal distortions or due to phase change. Also, the temperature of the stator is frequently used to control the operating conditions of a mechanical face seal (MFS).

Only few analytical approaches for predicting temperature distribution in a MFS were published. Buck (1) presents a series development solution based on a constant heat flux entering the rotor. The model proposed by Pascovici and Etsion (2) is based on the assumption that the heat flow path in the rotor can be approximated by a set of straight lines, inclined at a fixed angle, φ, to the seal face. The solution overestimates the face temperature variation in the radial direction -see Etsion and Groper (3). In both THD models, the stator is assumed thermally insulated.

In the present paper, the heat rejected by the stator continues to be neglected, but the heat recirculated inside the stator, due to radial temperature gradient in the interface, is taken into account. A simplified axisymmetric THD model of a noncontacting seal is developed to obtain an analytical solution. The heat flow path in the rotor is approximated by the same set of straight, parallel lines (2). The heat flow path in the stator is approximated by a set of concentric circles, centered in the middle of the seal face. Therefore, in the inner half of the seal gap the heat produced by viscous friction, q^*, is partitioned between the rotor, $q_R = \lambda q^*$, and the stator $q_S = (1-\lambda)q^*$. In the outer half of the seal gap, the heat recirculated by the stator is added to the heat produced locally by viscous friction and it is entirely rejected via the rotor, $q_R = q + (1-\lambda)q^*$. All the other assumptions are the same with those presented in (2).

The solution consists of 3 linear equations which must be solved consecutively to find the coefficient of heat partition at inner radius, λ_i, and the temperatures at both ends of seal gap. The temperature variation along the gap can be found by linear interpolation. The solution allows easy parametrical study on thermal behaviour of MFS, including the effect of the stator.

For negligible stator conductivity, the heat transfer analysis can be improved considering the recirculation of the heat in the rotor; thereby, the recirculated and the straight heat flows are superposed. In this case, the accuracy of the analytical solution can be evaluated by comparing the results of the fluid mean (seal face) temperature with the results obtained by an "exact" numerical solution (3). As can be seen from Fig. 2, the improvement of the present model is important especially for the inner face temperature.

REFERENCES

(1) Buck, G.S., Proceedings of the 6th Int. Pump Users Symp., Houston, Texas, pp 9-15, 1989.
(2) M.D.Pascovici. and I. Etsion, Trans. ASME - J. of Tribology, Vol. 114, pp 639-645, 1992.
(3) I. Etsion and M. Groper, 14-th Int. Conf. on Fluid Sealing (BHRA), Firenze, Italy, 1994.

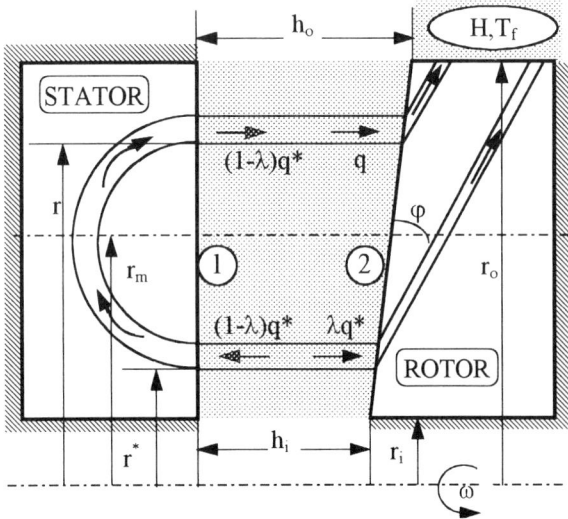

Fig. 1: Seal model

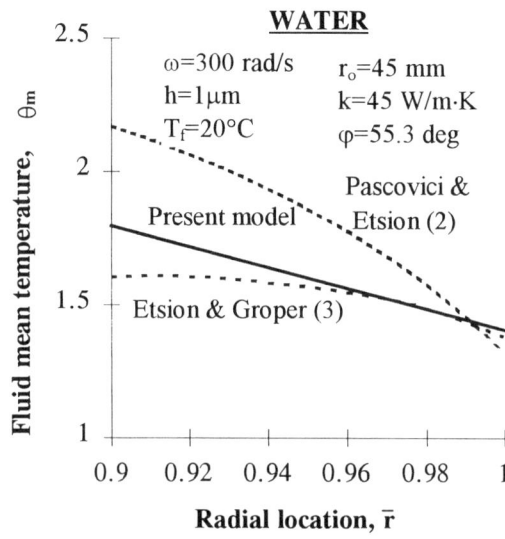

Fig. 2 Comparison with other solutions

EFFECTS OF OPTIMIZATION OF OIL GAP PROFILE IN A HYDRODYNAMIC THRUST BEARING

LESZEK DABROWSKI and MICHAL WASILCZUK

Faculty of Mechanical Engineering, Technical University of Gdansk, Narutowicza 11/12, 80-952 Gdansk, POLAND

ABSTRACT

Oil film profile and its influence on hydrodynamic bearing performance has often been an object of investigation. At first the investigations have been done for one-dimension bearings, later in two dimensions but, due to complexity of calculations, usually for isothermal model of the oil film. Typically obtaining maximum possible load-carrying capacity at a given minimum film thickness has been the objective of the optimization process. Although methods of optimization have varied the optimum shape has been similar. As a rule in an optimum shape bearing the pad should form a pocket with enlarged inlet area and raised sides and outlet area in order to restrict side leakage, e.g. (1). The results of theoretical investigations are difficult to be put into practice because:

a) complex shape of the bearing surface is difficult to be machined with desired accuracy, and

b) profiled bearing surface deforms during operation changing the favourable profile of the oil gap.

A bearing in which initially flat, elastic thrust plate deflects during operation to form a favourable shape of the oil gap can be the solution to the above mentioned problems.

The task of designing an optimum-shape bearing has been divided into two steps:

(1) determining the optimum shape of the oil gap for a given load/speed conditions,
(2) designing a bearing which deflects during operation (due to pressure and temperature) to form an optimum shape of the oil gap.

A favourable shape of the bearing surface has been estimated using a thermo-hydrodynamic model of the oil film which takes into account 2 D viscosity variation (2). Shape of the bearing surface has been assumed to be the result of deflection of an elastic plate, therefore it was described with spline functions. In the course of calculations the shape was changed to obtain maximum load-carrying capacity. The shape of the bearing surface which was obtained from calculations is similar to the results of the others - enlarged inlet area and raised sides- the main difference is that the surface is smooth and therefore it is more likely to be obtained in a real bearing as a result of thermoelastic deformations.

An effort of designing an optimum shape bearing must be preceded by estimation of potential improvement of performance of an optimum shape bearing in comparison with traditional designs. Calculations of potential bearing performance carried out for various shapes of the oil gap, including flat and deformed shape of a tilting-pad bearing were performed for the following shapes of the oil gap:

- optimum shape, as estimated by Rohde (1)
- favourable spline shape, as described above
- flat pad of 1/2.25 taper
- thermally deformed pad

Calculations were performed for the following data: outer diameter 172 mm, inner diameter 112 mm, speed 2000 rpm (sliding speed 15 m/s), ISO VG-32 oil supplied at 40° C. Minimum film thickness for all cases has been set to 15.4 μm and for that thickness other important parameters have been calculated. The results are presented in the table:

Shape	Parameter		
	Specific load	Friction loss	Maximum temperature
	MPa	kW	deg. C
Rohde	4.09	1.50	70.1
Spline	4.13	1.34	66.9
Flat pad	2.59	1.16	64.4
Deformed pad	2.10	1.13	58.5

CONCLUSIONS

The results show that optimum shape bearings offer almost two fold increase in load-carrying capacity over a tilting-pad bearing. On the other hand in optimum shape bearings maximum temperatures and friction losses are somewhat higher.

The potential improvement in performance is an incentive to take up the next step of the programme of designing a bearing in which shape of the oil film will be close to the optimum due to deflections of an elastic thrust plate.

REFERENCES

(1) Rohde S. M., McAllister: On the optimization of fluid film bearings. Proc. R. Soc. Lond. A. 351, 481-497 (1976).

(2) Dabrowski L., Wasilczuk M.: On the accuracy of theoretical models of hydrodynamic thrust bearings. Proceedings of the conference BALKANTRIB 96, Thessaloniki 1996

INCREASING EFFICIENCY OF TURBINE-GENERATORS BY APPLYING HYDRODYNAMIC, LEADING EDGE GROOVE, BEARING TECHNOLOGY

SCAN M. DECAMILLO
Kingsbury, Inc., 10385 Drummond Road, Philadelphia, PA, 19154 U.S.A.

ABSTRACT

Results of a study are presented which show a significant improvement in efficiency is obtainable in large turbine-generator applications by applying leading edge groove technology to the system's hydrodynamic thrust and journal bearings. Leading Edge Groove (LEG) is a method of lubrication that delivers cool oil directly into the oil film to lower bearing power loss and pad temperatures. The study is in response to an increasing demand for efficient machines in a growing, international energy market. The demand is also driven by Deregulation and Privatization where resulting competition now more strongly dictates advantages of efficient, cost effective equipment.

BACKGROUND

While direct lubrication of thrust bearings has been successfully implemented for more than fifteen years, tests on journal bearings have only recently been documented for small (100 mm) high speed journals with favorable results (1). Literature has also extrapolated this data to project benefits of LEG lubrication in large (546 mm) turbine pivoted shoe journal bearings (2). Most recently, data has been acquired from actual tests of 460 mm (3) and 510 mm LEG pivoted shoe journal bearings. Where each prior report and test focused on a particular bearing, this study assesses all large bearings in a typical turbine-generator.

STUDY

Fig. 1 tabulates flooded and LEG thrust bearing field data from large turbines (2). There is a noticeable reduction in thrust bearing power loss and oil flow. Note how large the oil flow and loss of the two flooded journal bearings are compared to thrust bearing.

Fig. 2 is test data from generator tests of flooded and LEG journal bearing designs (3). Significant reductions in flow and loss were obtained for the two large bearings.

Fig. 3 lists bearings typical in size for a typical 300 MW turbine-generator train, developed for purposes of this study. Smaller HP-IP and exciter journals are not included in the study. Flows and losses from the above test data are used where appropriate, and other bearings are calculated in proportion to the test data. In comparison of bearing losses for such a T-G set, flooded bearing oil flows are on the order of 4360 liters per minute, and power losses approach 2000 kW. Using LEG test data, it would be possible to reduce oil flows 50 percent, and to reduce losses from 1976 down to 1338 kW which represents a reduction on the order of 32 percent.

CONCLUSIONS

Test data supports that leading edge lubrication technology can be used to increase the operating efficiency of large, turbine-generator sets. The study indicates that power loss from the four large pivoted shoe journal bearings in such a turbine-generator train are great in comparison to the thrust bearing losses, and that significant reductions in power loss are obtainable from the journal bearings in addition to the thrust bearing. More detail is presented in the full report along with discussions of additional benefits in regard to reducing lubrication system size, and the implications of the increased machinery efficiency in terms of cost savings.

762 mm (12x12) thrust bearing.   Two 546 mm 4-pad journal bearings.

	TURBINE				
	THRUST	PJ1	PJ2		
BRG TYPE	STD	STD	STD		TOTAL
Oil Flow (lpm)	1726	682	682		3090
Loss (kW)	646	377	377		1400

BRG TYPE	LEG	STD	STD		TOTAL
Oil Flow (lpm)	960	682	682		2324
Loss (kW)	500	377	377		1254

Fig. 1 Turbine Thrust Bearing Field Data.

				GENERATOR		
				PJ3	PJ4	
BRG TYPE				STD	STD	TOTAL
Oil Flow (lpm)				635	635	1270
Loss (kW)				288	288	576

BRG TYPE				LEG	LEG	TOTAL
Oil Flow (lpm)				318	318	635
Loss (kW)				181	181	363

Fig. 2 Generator Journal Bearing Test Data.

Thrust Bearing	762 mm (12x12) thrust bearing.
LP Turbine Bearings	Two 546 mm 4-pad journal bearings.
HP-IP & Exciter Bearings	Not included in the study.
Generator Bearings	Two 457 mm 4-pad journal bearings.

	TURBINE			GENERATOR		
STD bearing	THRUST	PJ1	PJ2	PJ3	PJ4	TOTAL
Oil Flow (lpm)	1726	682	682	635	635	4360
Loss (kW)	646	377	377	288	288	1976

LEG bearing	THRUST	PJ1	PJ2	PJ3	PJ4	TOTAL
Oil Flow (lpm)	960	341	341	318	318	2277
Loss (kW)	500	238	238	181	181	1338

Fig. 3 Typical T-G Bearings - Estimated Data.

REFERENCES

(1) Edney, S., "Profiled LEG Tilting Pad Journal Bearing for Light Load Operation," Texas A&M Proceedings, 1996
(2) Brockwell, K., "Analysis and Testing of the LEG Tilting Pad Journal Bearing," Texas A&M Proceedings, 1994
(3) DeCamillo, "Performance Tests of an 18-Inch Diameter, LEG Pivoted Shoe Journal Bearing," IC-HBRSD Proceedings, China, 1997

INERTIA EFFECT AT CONVERGENT AND DIVERGENT STEP WITH POSSIBLE INTEREST IN SEALS AND HYBRID BEARINGS

S. GALETUSE
Airspace Engineering Faculty, "Politehnica" University of Bucharest, RO

ABSTRACT

The idea of using grooved surface in non-contact seals, for to benefit by the pressure loss, is very old. In recent years, the utilization in turbo-pumps for liquefied gases, multistage centrifugal pumps, high pressure fuel turbo-pumps, etc. has been on large scale. Many papers have studied various flow aspects of the above applications or flow for plane surfaces with inlet pressure drop.

In this paper, we wish to obtain some analytical relations of practical interest. For this purpose, we have to evaluate the influence of inertia forces in the vicinity of the convergent and divergent steps. Note that, in both pocket and land area, the film thickness is constant so we can consider that the flow is approximately inertialess for the entire length.

A method to find out the pressure drop at the convergent step is to apply the linear momentum balance. Burton (1) and Pan (2) have obtained useful results for some similar problem using this method. Using an approximate solution for the pressure integral over the vertical wall, it was obtained pressure drop as

$$\Delta p = 1.2 \frac{\rho Q^2}{h_2} \left(\frac{1}{h_2} - \frac{1}{h_1} \right) \quad laminar$$

$$\Delta p = (0.16\xi + 0.84) \frac{\rho Q^2}{h_2} \left(\frac{1}{h_2} - \frac{1}{h_1} \right) \quad transition$$

$$\Delta p = \left(1 + \frac{9.6}{\mathrm{Re}^{0.6}} \right) \frac{\rho Q^2}{h_2} \left(\frac{1}{h_2} - \frac{1}{h_1} \right) \quad turbulent$$

where h_1, h_2 are pocket and land thickness, and coefficient ξ can be calculate by the formulae (3)

$$\xi = 1 + \frac{60}{\mathrm{Re}^{0.6}},$$

with Re effective Reynolds number.
For the divergent step it was obtained similar relations but the product between density and squared flux is divided by h_1 instead of h_2.

The method was applied also to an annular seal. In order to check the obtained analytical relations, some experimental data and numerical data where used, and an acceptable agreement was obtained. In order to point out the influence of the parameter $K = h_1 / h_2$ on the dimensionless flux the results were used for N pockets and total length l. It can be seen in the Fig. 1 that it is possible to obtain a smaller flux if the parameter $(\mathrm{Re} h_2 / l)N$ has a convenient value. Note that, it is possible to obtain a negative effect if the parameter K or N is not correct chosen.

The results given in this paper are valid for pocket length much bigger than the film thickness. If the pocket length is small, so that the recirculation appears, another model should be considered.

Finally, one can add that the formulae for pressure drop and pressure rise at the step could be useful for the pressure distribution calculation in the lateral direction for hybrid bearings (hydrodynamic and hydrostatic). Indeed, when the lubricant has small kinematics viscosity, or journal speed is big the operating Reynolds number rises and the inertia effect have to be taken into account.

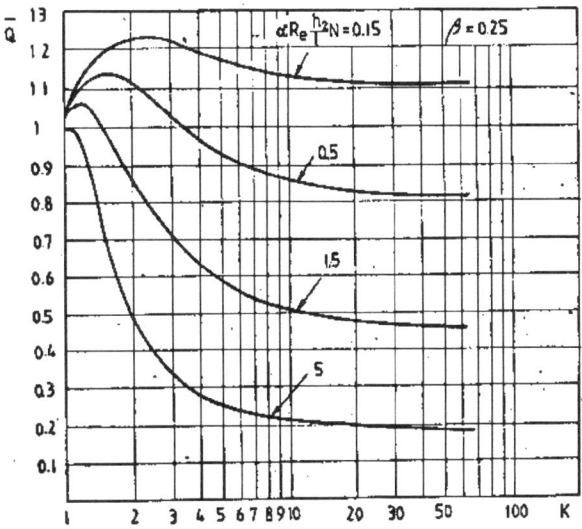

Fig. 1: Dimensionless flow $\overline{Q}$ versus K ratio

REFERENCES

(1) R.A.Burton, ASME Journal of Lubrication Technology, Vol..90, No. 2, 1968,443-449pp.
(2) C.H.T.Pan, ASME Journal of Lubrication Technology, Vol. 96, No. 1, 1974, 80-94pp.
(3) S.Galetuse, Revue Roumaine des Sciences Techniques, Vol. 20, No. 4, 1975, 485-506pp.

Submitted to *ASME Journal of Tribology*

THERMO-ELASTIC DEFLECTIONS OF BIG END BEARING

HEINZ GLÄSER
Westsächsische Technische Hochschule, Dr. Friedrichs-Ring 2A, 08001 Zwickau Deutschland
TOMASZ KUBIAK and ANDRZEJ MŁOTKOWSKI
Department of Materials Strength, Technical University of Łódź, Stefanowski Street 1/15, 90-924 Łódź, Poland
STANISŁAW STRZELECKI
Institute of Machine Design, Technical University of Łódź, Stefanowski Street 1/15, 90-924 Łódź, Poland

ABSTRACT

Performances of the engine bearings are affected by thermal and elastic deflections of bearing liner and housing. These deflections affect the entire structure of bearing, change the oil film geometry, affect the static and dynamic characteristics (1)(2). For the failure-free operation of bearing, more information regarding thermo-elastic deflections should be available at the early design stage of the bearing. Thermo-elastic effects are important in the evaluation of oil film thickness, maximum pressure and temperature, i.e. bearing characteristics deciding about reliable operation of single bearing and the bearing system.

The state of investigations in the thermoelastohydrodynamic lubrication, allowing to receive enough data on the journal bearing performance is not regarded as satisfactory. Some researchers suggested, that elastic deflections might be the cause of the observed increase in cavitation angle and decrease in peak pressure. In the case of assumption of soft shell bearing supported in a rigid housing the investigation shows that: the ratio of the peak pressure to the average pressure is reduced, the bearing stiffness is reduced, the minimum oil film thickness for a given load is reduced by elastic deflection at eccentricity ratios less than unity, and the friction is also limited.

For investigation an effect of thermo-elastic deflections on the bearing performance, the geometry, Reynolds, energy, viscosity equations coupled with the thermoelastic equations should be solved to allow the determination of pressure distribution in oil film and the load capacity, that is the ground for computation of journal centre trajectory (3). The position of maximum pressure load and the corresponding pressure distribution in the oil film can be found from the journal centre trajectory. The calculated pressure and temperature distributions are the input data for computation of bearing liner and housing deflections and stresses. The calculations assume the full bearing solution. Negative values of oil film pressures are neglected. Variations of viscosity with the pressure are ignored for the considered values of pressure. The procedure does not consider the journal deformations as very small in comparison to the bearing deformations. The developed computer program (3)(4) applied in calculation solves the basic thermohydrodynamic equations for the different sets of bearing geometric and exploitation parameters.

The task of the investigation is to determine the thermo-elastic deflections of big end bearing liner and housing operating in the condition of adiabatic oil film and with aligned and misaligned axis of journal and sleeve. The calculation of minimum oil film thickness, maximum pressure and temperature for the different parameters of the bearing have been carried out too. As an example of calculation, the deflections of the cylindrical journal bearing with aspect ratio L/D=1.0, clearance ratio ψ=1.5‰ and aligned axis of journal and sleeve are shown in Fig 1.; the deflections were determined for one of the points on the journal centre trajectory.

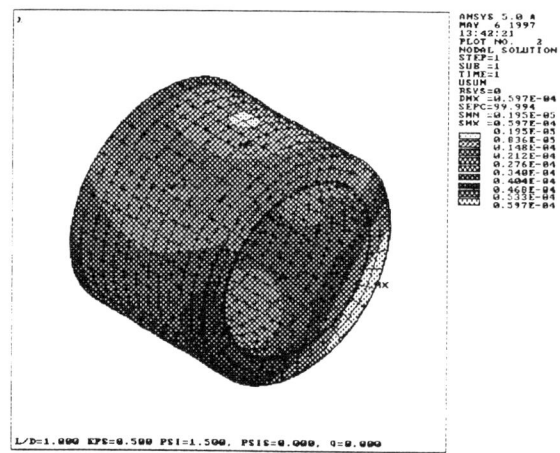

Fig.1 Deflections of the bearing caused by the pressure and temperature of oil film

REFERENCES

(1) P.K.Brighton, C.J.Hooke, J.P.O'Donoghue, Proceedings of the Institution of Mechanical Engineers. 182.3N, 1967, 183-191pp.
(2) H. Gläser, T. Kubiak, A. Młotkowski, S.Strzelecki,. Journal of KONES 1995, Vol.2, No.1 ,142-147pp.
(3) H. Gläser, A. Młotkowski, S.Strzelecki, Proceedings of the 4th Conference of the Egyptian Society of Tribology, Cairo 1995, 95-103pp.
(4) H. Gläser, A. Młotkowski, S.Strzelecki Proceedings of the International Tribology Conference. Vol 3 Yokohama 1996,1273-1278pp.

THE METHOD OF ADJOINT EQUATIONS FOR SOLVING PROBLEMS OF DYNAMICS FOR GAS LUBRICATED BEARINGS

B S GRIGORIEV and G K IZMAILOV

Physical & Mechanical Faculty, Technical University, Politechnicheskaya 29, St. Petersburg, 195251, RU

ABSTRACT

The objective of this paper is to present a new method for solving problems mentioned at the title. This method permits separately to solve the problem describing the gas film and equations of motion for mobile bearing elements. As a result, the entire problem is essentially simplified.

Let us consider a self-acting gas bearing in the steady-state position. Denote $p(x,t)$ the gas film pressure, $h(x,t)$ the gas film thickness in a point $x = (x_1, x_2)$ at a moment t. Describe the state of bearing mobile elements by the number of parameters which form a vector $\underline{\varepsilon}$ of generalised displacements $\underline{\varepsilon} = (\varepsilon_1, \varepsilon_2, \ldots, \varepsilon_n)$. The gas film thickness can be written for many types of bearings as $h(x,t) = g(x) + \underline{\varepsilon}(t) \cdot \underline{\eta}(x)$, were a function $g(x)$ and a vector-function $\underline{\eta} = (\eta_1, \eta_2, \ldots, \eta_n)$ are given, a dot means the inner product of vectors.

Suppose at a moment $t = 0$ the bearing leaves its steady-state position in consequence of some external action. Then it moves in a small vicinity of that state. Thus the pressure and the displacements can be written in the form $p(x,t) = p_0(x) + \delta p(x,t)$, $\underline{\varepsilon}(t) = \underline{\varepsilon}_0 + \delta\underline{\varepsilon}(t)$, where dynamic perturbations $\delta p, \delta\underline{\varepsilon}$ are small with respect to corresponding steady-state values $p_0, \underline{\varepsilon}_0$. The gas film restoring force $\delta\underline{W}$ is given by the formula

$$\delta\underline{W}(t) = \int_\Omega \delta p(x,t)\underline{\psi}(x)d\Omega, \tag{1}$$

where the integration is made over the lubricated area Ω and $\underline{\psi}$ is the known vector-function. The equation for displacement $\delta\underline{\varepsilon}$ contains $\delta\underline{W}$ where the value δp can not be explicitly expressed by $\delta\underline{\varepsilon}$ and its derivatives. For this reason the equations for δp and $\delta\underline{\varepsilon}$ have to be solved jointly that is the severe problem.

The method of adjoint equations permits to derive another formula for $\delta\underline{W}$, namely

$$\delta\underline{W}(t) = -\int_0^t \left[\underline{\underline{D}}(\tau)\delta\dot{\underline{\varepsilon}}(t-\tau) + \underline{\underline{G}}(\tau)\delta\underline{\varepsilon}(t-\tau)\right]d\tau - \underline{\underline{D}}(t)\delta\underline{\varepsilon}(0) \tag{2}$$

Here a dot above a symbol refers to the derivative d/dt and coefficients $\underline{\underline{D}}$ and $\underline{\underline{G}}$ are given by:

$$\underline{\underline{D}} = \sigma\int_\Omega p_0 \underline{v}\underline{\eta}\,d\Omega, \quad \underline{\underline{G}} = \int_\Omega (\underline{\underline{R}}_h^* \underline{v})\underline{\eta}\,d\Omega$$

In these formulas two neighbour vectors form a dyad. The vector-function $\underline{v} = (v_1, v_2, \ldots, v_n)$ is so called adjoint function with respect to δp. It is the solution of the problem

$$\sigma h_0 \frac{\partial \underline{v}}{\partial t} + \underline{\underline{R}}_p^* \underline{v} = 0, \quad \underline{v}\big|_{t=0} = \frac{\underline{\psi}}{\sigma h_0} \tag{3}$$

The boundary conditions for $\underline{v}$ are the same as for δp. The matrix operators $\underline{\underline{R}}_h^*$ and $\underline{\underline{R}}_p^*$ are

$$\underline{\underline{R}}_h^* \underline{v} = p_0(3h_0^2 \nabla p_0 - \Lambda \underline{U})\nabla\underline{v}$$
$$\underline{\underline{R}}_p^* \underline{v} = -p_0\,div(h_0^3 \nabla\underline{v}) - \Lambda h_0 \underline{U}\nabla\underline{v}$$

where Λ and σ denote the compressibility number and the squeeze number respectively, $\underline{U}$ is the sliding velocity vector of bearing surface.

The problem (3) is named as adjoint one with respect to the problem for δp in connection with the functional (1). The formula (2) for $\delta\underline{W}$ on the contrary of (1) includes $\delta\underline{\varepsilon}$ and $\delta\dot{\underline{\varepsilon}}$ explicitly. The adjoint function $\underline{v}$ as it follows from (3) does not depend on these values. Finally, the equation of the motion for $\delta\underline{\varepsilon}$ using the formula (2) neither forms a system with the equation for δp (it is not contained in (2) yet) nor with the equation for $\underline{v}$. That is why the adjoint function was introduced.

It can be shown that the adjoint function from a physical point of view is the measure of the influence on $\delta\underline{W}(t)$ of perturbations arising in the lubricated film at previous moments of time $(0 < \tau < t)$.

As an example of the theory results about the stability of the thrust spiral grooves bearing under any small external actions are presented.

USE OF COMPUTATIONAL FLUID DYNAMICS IN HYDRODYNAMIC LUBRICATION

P Y P CHEN and E J HAHN

School of Mechanical and Manufacturing Engineering, University of New South Wales, Sydney NSW 2052, Australia

ABSTRACT

The extent to which the assumptions made in developing the Reynolds equation for hydrodynamic lubrication, particularly the neglect of inertia terms, are justified, has been the subject of numerous investigations (1). More recently, using approximate velocity solutions, it has been shown that inertia terms could contribute significantly to the dynamic behaviour of squeeze film damped rotor bearing systems, this contribution being proportional to the so-called squeeze Reynolds number $\omega c^2/\nu$ (2)(3)(4). All of these investigations utilise a simplification of the Navier-Stokes equations, with certain numerical constants obtained by approximating assumptions and averaging methods.

As a result of recent developments in computational fluid dynamics (CFD) and the availability of large computer software packages, it should now be possible to analyse hydrodynamic lubrication problems in non-simple geometric domains, where neglect of inertia terms is unlikely to be justified, eg dynamically loaded journal bearings or squeeze film dampers with shallow circumferential grooving.

In this paper, the applicability of CFD to such problems is initially established by reproducing the analytical pressure and velocity solutions to the Reynolds equation for the two-dimensional slider step and journal bearing subjected to constant unidirectional loading, assuming constant fluid properties. This is achieved by solving numerically the momentum equations:

$$\frac{\partial p}{\partial x} = \mu\left(\frac{\partial^2 u}{\partial x^2} + \frac{\partial^2 u}{\partial y^2}\right) - \rho\left(\frac{\partial u}{\partial t} + u\frac{\partial u}{\partial x} + v\frac{\partial u}{\partial y}\right)$$

$$\frac{\partial p}{\partial y} = \mu\left(\frac{\partial^2 v}{\partial x^2} + \frac{\partial^2 v}{\partial y^2}\right) - \rho\left(\frac{\partial v}{\partial t} + u\frac{\partial v}{\partial x} + v\frac{\partial v}{\partial y}\right)$$

and the continuity equation:

$$\frac{\partial u}{\partial x} + \frac{\partial v}{\partial y} = 0$$

where, in order to equate to the Reynolds equation solutions, all terms in heavy type are neglected. Analytical short bearing solutions are reproduced by expanding the above equations to three-dimensions, adding the momentum equation in the z-direction, again neglecting all inertia terms and retaining only the viscous terms $\mu\frac{\partial^2 u}{\partial y^2}$ and $\mu\frac{\partial^2 w}{\partial y^2}$. Finally, applicability of CFD for dynamically loaded journal bearings and squeeze film dampers is established by reproducing both long and short bearing solutions for journal bearings whose centres execute synchronous circular orbits around the bearing centre.

Conditions under which the temporal inertia terms, the convective inertia terms and the viscous terms may or may not be neglected are investigated. In particular, short journal bearings and squeeze film dampers are studied over a wide range of operating conditions. It is shown that in most situations the approximate velocity solutions in references (2)(3)(4) predict reasonably well the effect of the convective inertia terms. The temporal inertia terms were found to have significant contributions, mainly under dynamic loading conditions. In finite width (ie three-dimensional) applications, the hitherto neglected viscous terms were also found to be significant, with their inclusion resulting in differences in the stiffness coefficients of up to 20% in certain cases.

Thus, under operating conditions when inertia and viscous terms need to be considered, CFD has been shown to be a most useful tool in analysing slider and journal bearings, including squeeze film dampers.

REFERENCES

(1) O. Pinkus, B. Sternlicht, "Theory of Hydrodynamic Lubrication", McGraw-Hill, 1961
(2) L.A. San Andres, J. M. Vance, ASLE Transactions, Vol. 30, 1987, 63-68pp.
(3) A. El-Shafei, S. H. Crandall, Rotating Machinery and Vehicle Dynamics, ASME Publication DE-Vol. 35, 1991, 219-228pp.
(4) J. Zhang, J. Ellis, J. B. Roberts, ASME Transactions, Vol. 115, 1993, 692-698pp.

PERFORMANCE OF MICROGROOVED JOURNAL BEARING UNDER STEADY AND DYNAMIC LOADING

D J HARGREAVES and D ARMATYS
School of Mechanical, Manufacturing and Medical Engineering, Queensland University of Technology, PO Box 2434, Brisbane 4001, Australia

INTRODUCTION

Conventional hydrodynamic lubrication is based on the assumption of perfectly smooth surfaces. The possibility of improving bearing performance by modifying bearing surface geometry has attracted the attention of researchers in recent years.

An experimental investigation was undertaken to determine the performance of microgrooved journal bearings operating under both steady and dynamic loading conditions.

EXPERIMENTAL APPARATUS

A shaft supported on two rolling element bearings and is arranged so that the bearing to be tested is mounted on the cantilevered section of the shaft. Three bearing were tested; one without any microgrooving, referred to as a plain journal bearing, a bearing with longitudinal microgrooving, that is with the ridges and valleys running in the direction of rotation, and a bearing with transverse micrgrooving, that is with the ridges and valleys running perpendicular to the direction of rotation. Bearing bushes were 39.76mm diameter and 25mm long. All microgrooves were 0.006mm deep with a pitch of 0.02 mm. An ISO VG 46 oil was used and each bearing was 25 mm wide.

For the steady load cases, the load was applied by dead weights hanging from the cantilevered bearing. A cam arrangement was used to activate a pneumatic cylinder which in turn loaded the test bearing once per revolution for dynamic loadings.

The friction force was measured by using a strain gauged bar to resist the frictional torque between the rotating shaft and the floating bearing bush.

TEST RESULTS

Each bearing was tested with loads of 116, 194, 291 and 388 N at speeds from 400 to 2000 rpm. A typical plot of frictional torque versus rotational speed for a steady load of 291 N load is shown in Figure 1. The corresponding plots for the plain and longitudinal grooved bearings for other loads were similar to those shown here. The shape of these lines implies that the bearings operate in the fluid film regime of lubrication. The calculated film thickness for the lowest speed and highest load was 8 microns. For all loads, the frictional torques for the plain and the longitudinal grooved bearings were very similar although the difference becomes more significant as the speed increases.

However the transverse grooved bearing

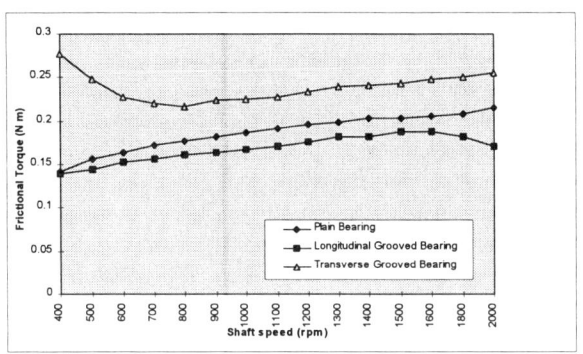

Figure 1 Frictional torque of plain and microgrooved journal bearings (load of 291N)

operates in the mixed lubrication regime for speeds less than about 800 rpm as shown by the change in the slope of the frictional torque-speed curve. Similar results were found for loads of 194 N and 388 N. For the 116 N load, this bearing seemed to operate with a full fluid film.

These results are consistent with the results of (1) who indicates that the change in frictional forces are insignificant for longitudinal grooving but are significant for transverse grooving.

A load of 498 N was applied once per revolution whilst the shaft rotated at 400 rpm for the dynamic loading test. The maximum friction torque for longitudinal and transverse micrgrooved bearings respectively exhibited about 90% and 120% of the friction torque of the plain bearing.

CONCLUSIONS

Preliminary testing of microgrooved journal bearings has shown that for a steadily applied load, there is an insignificant difference in frictional torque between the plain and the longitudinal grooved bearing. There is however a significant difference between the plain and the transverse grooved bearing.

For the case of a dynamically applied load, the transverse grooved bearing again exhibited a significantly higher frictional torque than both the plain and the longitudinal grooved bearing. Longitudinal grooving should help in retaining oil in the bearing when is subjected to dynamic loads, that is help reduce the side flow from the bearing ends and so make the squeeze mechanism more powerful.

REFERENCE

H G Elrod, Leeds-Lyons Symposium, 11-26, 1977

Submitted to *Tribotest*

PRESSURE DISTRIBUTION DUE TO SQUEEZE EFFECTS IN PLAIN SLIDER BEARINGS

D J HARGREAVES and W SCOTT

School of Mechanical, Manufacturing and Medical Engineering, Queensland University of Technology, PO Box 2434, Brisbane 4001, Australia

ABSTRACT

Investigation into the failure of a coal crusher bearings where the plain journal bearings rotated about a stationary shaft at the same speed as the rotating load led to some doubt as to the effectiveness of the oil feed arrangements. Since a plane inclined thrust bearing in the same machine also failed, and because the inclined surfaces were moving in relation to a fixed collar, the question arose as to whether the oil feed requirements for this arrangement were the same as that for a moving collar and stationary fixed inclined surfaces. An inclined surface moving across a fixed flat surface is the linear equivalent of the synchronously loaded journal bearing. Under conditions of rotating bearing bushes and a stationary shaft, the flow requirements could differ from that of a steady unidirectional loaded bearing due to the squeeze action of the hydrodynamic film in comparison to the wedge film formation. However, it would appear that most authors consider the flow requirements for steady, unidirectional loaded bearings to be the same as that for dynamically loaded bearings, for example (1,2).

As a precursor to studying the actual journal bearing problem, an experimental investigation using a flat thrust pad was undertaken. A test apparatus was built which could have either the flat surface moving and the inclined surface fixed or the inclined surface moving and the flat runner stationary. Pressures were measured in the oil film at 14 locations for various operating conditions, for example angle of inclination, relative speed, minimum film thickness and oil feed arrangements. These pressure tapping points are arranged along the two centrelines of the 80mm square pad. The lubricant was supplied to the pad-disc interface by a positive displacement pump via a flow control valve and a flowmeter was used to measure the actual flow rate. A detailed description of the apparatus can be found in (3).

Lubricant (ISO VG 68) was supplied to the leading edge of the pad from a tube which rotates with the pad when the pad is moving or is stationary if the pad does not move. It was found that consistent results could only be achieved when the lubricant was supplied close to the leading edge of the pad and close to the disc. This observation is consistent with the belief that the details of the lubricant supply arrangements can have a significant effect on the load carrying capacity of the bearing.

Initially, tests were performed by setting the film shape, that is the inlet and outlet film thicknesses. The next set of results used a pad which could pivot about either the centre of the pad or 5mm behind the pad centre.

Typical results are shown in Fig 1.

REFERENCES

(1) FA Martin, Tribology International, Vol 16, No 3, pp 147-164, 1983

(2) FA Martin, Trans ASLE, Vol 26, No 3, pp 381-392, 1983

(3) Hargreaves, DJ; Scott, W and Beach, J (1995): Feed pressure flow for synchronously loaded plain bearings, Int'l Tribology Conf, Japan, Oct 1995

(4) B Jakobsson and L Floberg, Trans Chalmers University of Technology, Report No 203, Sweden, 1958

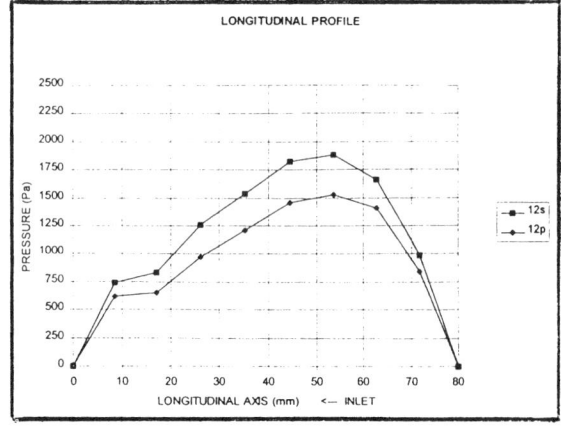
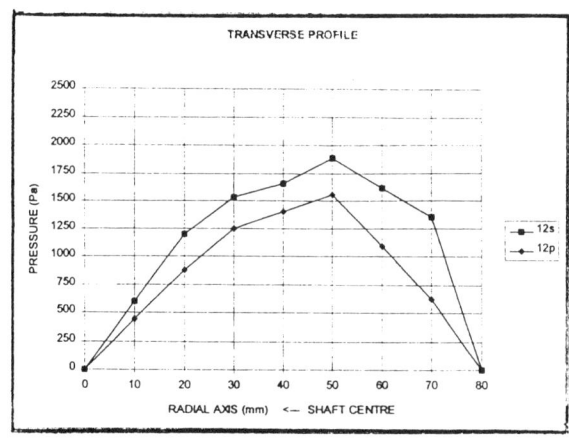

Figure 1 Metered flow sample chart - Case 12s and 12p - N = 30 rpm, h_i = 0.75 mm, h_o = 0.5 mm, Q = 1.21 lpm

RAPID AND GLOBALLY CONVERGENT METHOD FOR DYNAMICALLY LOADED JOURNAL BEARING DESIGN

H HIRANI and K ATHRE
Department of Mechanical Engineering, Indian Institute of Technology Delhi, New Delhi -16, INDIA
S BISWAS
Industrial Tribology, Machine Dynamics and Maintenance Engineering Centre, Indian Institute of Technology Delhi, New Delhi - 16, INDIA

ABSTRACT

The design of dynamically loaded journal bearing is normally carried out, using the graphical Mobility method or by using analytical curve fits for journal bearing solution. Since such methods do not provide procedural insight, it often leads to bearing size which is not optimum, and the designer might experience difficulties in adopting these methods to the design of bearing under consideration. A fast globally convergent design methodology is presented in this paper which is accurate, fast, and converges to the solution from almost any starting points. This method combines the rapid convergence of the Newton's method and a globally convergent strategy alongwith Reason & Narang[1] solution technique for rapid design and performance evaluation of dynamically loaded bearing.

Reason and Narang developed an analytical pressure expression, based on short and long bearing approximations for steadily loaded bearings. On the same principle, an analytical pressure expression for dynamically loaded journal bearing is proposed. The starting and ending angular positions of the pressure curve are predicted by simple algebraic equations. The journal orbit is determined by treating equations of motion as an "inverse problem" and evaluated by using the Newton-Raphson root finding method. The Newton's method for solving these equations, often gives erroneous solution, if the initial guess is not sufficiently close to the roots. The Newton method requires a correction at each step for the set of equations of motion. In this paper a globally convergent technique is used for this purpose. The technique is: collapse all dimensions into one, always first try the full Newton step and check at each iteration that the proposed step reduces the load function and if not so, backtrack along the Newton direction until an acceptable step is determined. The Newton step is a descent direction, hence finding an acceptable step by backtracking is guaranteed.

To illustrate the validity of the present study, the Ruston and Hornsby 6 VEB MK-III Marine Diesel Engine big-end connecting rod bearing is analysed. The comparison of the minimum film thickness and maximum film pressure obtained by the proposed methodology and those obtained by other methods is illustrated in Table 1.

TABLE 1: Results of the Ruston-Hornsby Diesel Engine big-end con. rod bearing

Various Methods	Maximum Pressure MPa	Min. Film Thickness μm
Booker's Short Bearing	35.95	5.01
Goenka's New Curve Fits [2]	34.57	3.52
Finite Element Method [2]	34.40	3.47
Proposed Scheme	34.27	3.69

The results indicate that the solution accuracy of the present method matches well with the elaborate and time consuming finite element analysis, but at much less computational cost. The analytical pressure expression used in the present technique reduces the number of iterations. The globally convergent Newton-Raphson method makes solution rapid and highly convergent. The proposed design methodology is rapid, accurate for wide range of L/D ratios, easy to understand and use.

REFERENCES

1. Reason, B. R., and Narang, I.P., "Rapid Design and Performance Evaluation of Steady State Journal Bearings - A Technique Amenable to Programmable Hand calculator", Trans. ASLE, 1982, Vol. 25, pp. 429-444.
2. Goenka, P.K., "Analytical Curve Fits for Solution Parameters of Dynamically Loaded Journal Bearing", Trans. ASME, Journal of Tribology, 1984, Vol. 106, pp. 421-428.

SOLUTION OF THE RAREFIED GAS LUBRICATION EQUATION USING AN ADDITIVE CORRECTION BASED MULTIGRID CONTROL VOLUME METHOD

YONG HU

Iomega Corporation, 800 Tasman Drive, Milpitas, CA95035, USA

DAVID B. BOGY

Department of Mechanical Engineering, University of California, Berkeley, CA94720, USA

ABSTRACT

The ever shortening product cycle for magnetic recording disk drives demands a fast and accurate numerical prediction of the slider's flying characteristics during the design stage. A computationally efficient additive correction based multigrid control volume method is developed for the solution of the very high bearing number and shaped rail air bearing problems.

The finite difference control volume schemes for discretizing the Reynolds lubrication equation are based on various convection-diffusion formulations. These schemes use the original nonlinear governing equation. The discretization equation is derived by representing the total flux at the control volume interfaces using the different convection-diffusion formulations which include the central difference, upwind, hybrid, power-law and exponential schemes. By doing so, the four rules (consistency at control-volume faces, positive coefficients, negative-slope linearization of the source term and sum of the neighboring coefficients), which constitute the underlying guiding principles for a physically realistic solution, are guaranteed to be satisfied (except when using the central difference).

One of the factors affecting the convergence of a scheme is the ability to smooth out errors of all frequencies in the solution. It was showed that the conventional iterative methods such as the line-by-line method are only efficient in smoothing out those error components whose wavelengths are comparable to the mesh size; error components with longer wavelengths are removed at progressively slower rates. The motivation for the multigrid methods is to solve the equations on a hierarchy of grids so that all frequency components of the error are reduced at comparable rates. Inexpensive iteration on the coarse grid rapidly diminishes exactly those components of the error that are so difficult and expensive to reduce by fine grid iteration alone. An additive correction based multigrid method is implemented for the solution of the resulting discretization equations. The method is based on the principle of deriving the coarser grid discretization equations from the fine grid discretization equations. An adaptive-cycling version of the multigrid method is described and applied to the 50% tripad and Headway AAB sliders (1).

Grid	Disk Speed (rpm)	Ratio of CPU Time (Single-grid/Multigrid)
98×98	5400	15.2
194×194	5400	39.7
194×194	2700	38.5

Table 1: Ratio of CPU time between the multigrid and single-grid methods for the 50% Headway AAB slider

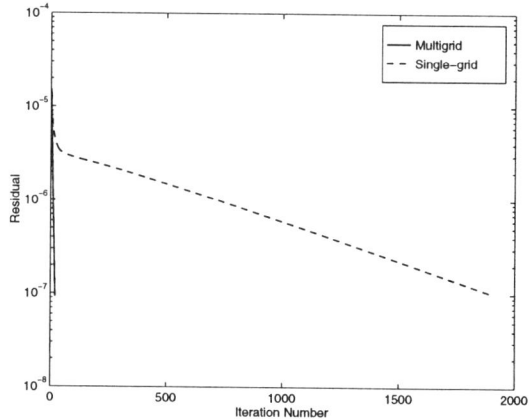

Fig. 1: Comparison of the convergence rates between the multigrid and single-grid methods for the 50% Headway AAB slider with the finest grids of 194×194 and the rotation speed of 5400 rpm

Table 1 and Figure 1 show the comparison of the convergence rates between the present calculations and those of the single-grid calculations. The study demonstrates the current method to be rapidly convergent with savings in CPU requirements by a factor ranging from 3.9 to 39.7 depending on the slider type, grid number and bearing number as compared to the single-grid method. The performance gets better as the increases of the grid number and the bearing number. The efficiency of the multigrid method over the single-grid method is further dramatically improved for today's shaped rail sub-ambient pressure sliders.

REFERENCES

(1) Hu, Y., "Head-Disk-Suspension Dynamics," Ph.D. Dissertation, University of California at Berkeley, 1996.

EXPERIMENTAL INVESTIGATION ON THE DYNAMIC BEHAVIOR OF AN AIR LUBRICATED MULTI-LEAF BEARING

PYUNG HWANG
School of Mechanical Engineering, Yeungnam University, Gyongsan, Gyongbuk, Korea
SUNG-IN KWON
School of Mechanical Engineering, Graduate School, Yeungnam University, Gyongsan, Gyongbuk, Korea
DONG-CHUL HAN
Department of Mechanical Design and Production Engineering, Seoul National University, Seoul, Korea

ABSTRACT

Aerodynamic bearings, in comparison with hydrodynamic bearing, some advantages such as low friction, no contamination and high temperature running. Because of these advantages, aerodynamic bearings are applied in many fields such as small turbo machinery. But sometimes aerodynamic bearings are unsuitable for those fields because of instability at high speed rotation, deformation of journal at high temperature and requirements of extremely fine finish and so on. To eliminate these problems, aerodynamic-elastic bearings were developed. There were many investigations in these bearings, but those investigations did not carry out exactly on experimental method about dynamic characteristics.

In this study, measurement of the vibration amplitudes of rotor supported by air lubricated multi-leaf bearing were carried out by the experiment of shaft vibration in various speed, and then dynamic characteristics and the stability of the system are analyzed by using these results. Especially, by using the transient data, we found the startup & shutdown characteristics that have been as a problem in air lubricated multi-leaf bearing system.

We use the air lubricated multi-leaf bearings, one of the aerodynamic-elastic bearings, in this experiments. The experimental units consist of two multi-leaf bearings and rigid rotor supported by these bearings. In order to remove axial movement of rotor and reduce rotational resistance caused by friction, we used the externally pressurized thrust bearing at either end of the rotor. To obtain rotating power, we setup 24 turbine blade in the middle of the rotor, and supply compressed air through 3 nozzles. To obtain the data of the shaft vibration amplitude, we use two non-contact inductive displacement and keyphasor sensor. We analyzed the data with A/D converters, personal computer, FFT and so forth.

In the results of this experimental investigation, relatively severe unstable state appears at startup. The starting instability is being disappeared as rotational speed is increased, but severe instability takes place over 20,000 rpm suddenly. In the shutdown process, the system repeats the same as startup process. Startup and shutdown instability results from resonance and friction between rotor and bearing leaves, as it has been considered to design air lubricated multi-leaf bearing.

The severe instability takes place over 20,000rpm is come from oil whirl effect, thus more system design consideration is needed. In the view of synchronous vibration component(1X), the system is stable over 20,000rpm. We know the factors that effect severe instability takes place over 20,000rpm is non-synchronous components.

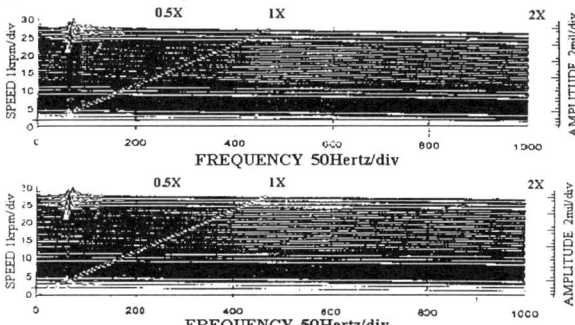

Fig. 1: Spectrum cascade plot of the rotor supported by air lubricated multi-leaf bearing

By experimental dynamic analysis of this rigid rotor and air lubricated multi-leaf bearing system, we can estimate stability of this system as follows. (1)Using air lubricated multi-leaf bearing which has stable region between 13,000 rpm to 21,000 rpm. At the lower part of stable region, instability phenomena occurs because of its initial operating friction and system resonance. Above this instability range, we can see whirl frequency of this system at startup condition. (2)The stability of air lubricated bearing is effected by external air pressure when using turbine rotational system at shutdown operation. And when using a multi-leaf bearing, it must be considered that there are instability problem caused by initial friction at low speed. (3)For the accurate dynamic analysis, transient analysis is more valuable in predicting system stability from this experiment.

REFERENCES

(1) W. A. Gross, Fluid Film Lubrication, John Wiley & Sons, 1980
(2) P. Hwang, Ph. D. Thesis, Seoul National University, 1989
(3) D. C. Han, J I Kim, Journal of KSME, 1986, 399-407pp., 1986
(4) L. Licht, ASME Journal of Lubrication Technology, 1969, 477-493pp

Submitted to ASME Journal of Tribology

MEASUREMENTS OF THE STATIC LOAD (ON PAD) PERFORMANCE AND PAD TEMPERATURES IN A FLEXURE-PIVOT TILTING PAD BEARING*

CHRISMA JACKSON and LUIS SAN ANDRÉS
Department of Mechanical Engineering, Texas A&M University, College Station, Texas 77843-3123

ABSTRACT

Advances in high speed turbomachinery demand reliable bearing supports for vibration attenuation and rotordynamic stability. Tilting pad bearings offer low cross-coupling forces and the suppression of oil whip. Flexure-pivot, tilting-pad (*FTPB*) bearings are constructed using an EDM, and with the pads attached to the bearing by thin webs giving rotational stiffness. These bearings offer no pad flutter or pivot wear, no stack-up of tolerances, allow control of cross-coupled stiffness, viscous damping between pads and casing, and tolerance to rotor axial misalignment (1).

Measurements of bearing displacements and pad edge temperatures in a four shoe *FTPB* under static loading towards a pad are presented. The test journal is mounted on a rigid shaft supported on precision ball bearings and driven by a variable speed motor. The housing is suspended by cables and holds the test bearing composed of four 80° arc pads, 50% offset and null preload. The bearing length, diameter and radial clearance are 46, 127 and 0.178 mm, respectively (see Figure 1). The pad rotational stiffness equals 1,125 N.m/rad. Steady state tests with an ISO VG 22 oil are conducted at rotational speeds of 1.8, 3.0 and 4.5 krpm, with loads applied to the bearing through a jack and load cell mechanism.

Measurements include bearing X&Y displacements, oil flow rate, and temperatures at the inlet and exit, and pads' leading and trailing edges. The tests show a bearing attitude angle of approximately 30° due to the large rotational stiffness. Figure 2 depicts the Sommerfeld number $[\mu \Omega LD(R/C)^2/W]$ vs. the measured journal eccentricity and with comparisons to an effective viscosity FD model (2). The bearing does not show subsynchronous whirl even at the null load condition. At the largest test speed (4.5 krpm), a drop in the bearing load capacity occurs due to a rise in the lubricant and pad temperatures and a reduction of the lubricant viscosity.

The pad edge temperatures $\{T_i\}_{i=1,8}$ follow well known trends with the largest magnitudes at the trailing edge of the loaded pad as shown in Figure 3 for tests at 3.0 krpm. The pad temperatures are found to be more sensitive to changes in the journal speed than to variations in the applied load.

*Accepted for publication in the *STLE Tribology Tansactions*

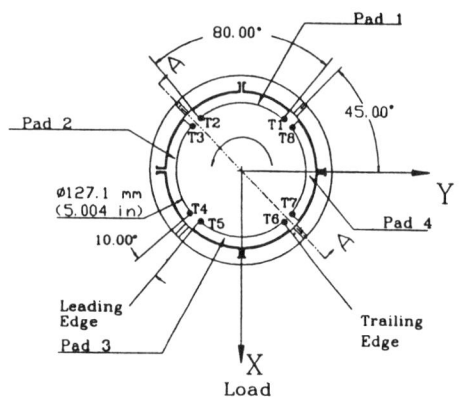

Fig. 1: Test flexure-pivot tilting pad bearing and placement of thermocouples.

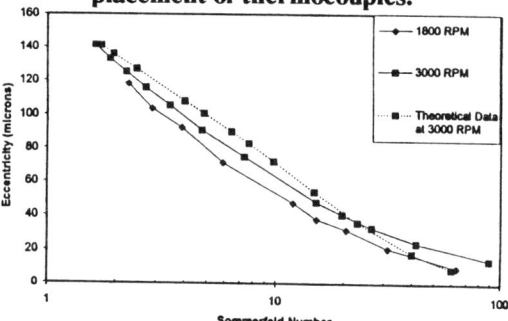

Fig. 2: Measured bearing eccentricity versus Sommerfeld numbers at two journal speeds.

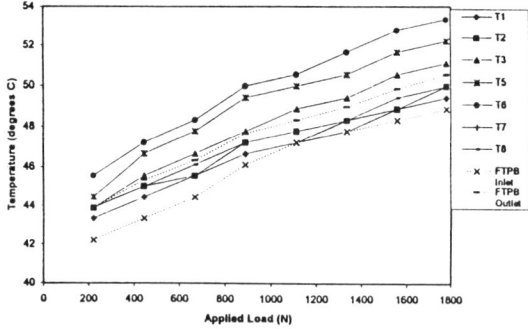

Fig. 3: Measured pad edge temperatures versus applied load at 3.0 krpm.

REFERENCES

(1) Zeidan, F., "Developments in Fluid Film Technology," *Turbomachinery International*, Vol. 9, pp. 24-31, 1992.

(2) San Andrés, L. A., " Turbulent Flow, Flexure-Pivot Hybrid Bearings for Cryogenic Applications," *ASME Journal of Tribology*, Vol. 188, pp. 190-200, 1996.

ANALYSIS OF LUBRICATION IN POROUS METAL BEARINGS BASED ON EXPERIMENTS

MOMCILO R. JANKOVIC and ALEKSANDAR B. MARINKOVIC
University of Belgrade, Department of Mechanical Engineering, 27. mart 80, 11000 Belgrade, YU

ABSTRACT

In literature of porous metal bearings is known hydrodynamic lubrication theory based on same theory of massive sliding bearings, only with addition oil flow through porous material of sleeve. The calculations of porous metal bearings, as a specific sort of bearings includes non dimensional constructional parameter (ψ), which depends of permeability (Φ), sleeve bearing wall thickness (H) and clearance (c) (2).

$$\Psi = \frac{\Phi H}{c^3}$$

Somerfelds number (S) is load characteristics of Porous Metal Bearing:

$$S = \frac{W}{U\eta b}\left(\frac{c}{r}\right)^2 \quad \text{here is}$$

W - bearing load, U - peripheral speed
η - oil viscosity, b - bearing length
r - bearing shaft radius.

Numerical solution of equation for oil film pressure gives diagrams of the load characteristics (Ocvirks load parameter), the friction parameter and the load angle in dependence of the construction parameter (ψ) for different values for relative eccentricity (e). For analysis of lubrication in porous metal bearings very important is diagram which gives dependence of reciprocal Somerfelds number value from the construction parameter. This theoretical facts are known, still in paper: Morgan V.T. and Cameron A.: Mechanism of lubrication in porous metal bearings (London 1957), (1).

Many experimental investigations and experience from exploitation showed that porous metal bearings works most of time in area of boundary and mixed lubrication. But, on higher speed and lower radial load, lubrication can be very near to hydrodynamics.

Here, authors are trying to proof this, using the results of experimental investigation of porous metal bearings produced in Yugoslavia (3). On "homemade" machine USL1 were testing journal porous metal bearings made of bronze (CuSn10) and bronze with graphite addition (CuSn10+1%C) which dimensions are $\varnothing 20 / \varnothing 30 \times 20$mm. Bearings worked on five different speeds in range of v=(1...5)m/s, loaded by force W=(750...150)N, and values of temperature and coefficient of friction were noticed when they become constant. Values of Somerfelds number (S) and constructional parameter (ψ) were calculated and placed as points in diagram which separate areas of hydrodynamics and boundary lubrication (4).

As you can see at the picture, on higher speeds lubrication is at least very near hydrodynamic form. But for lower speed values and higher loads, that is sure boundary lubrication. This can be indicated through value of relative oil film thickness by relation:

$$\lambda = \frac{h_{min}}{(R_{q_1}^2 + R_{q_2}^2)^{1/2}} \quad \text{where are}$$

h_{min} - minimum of oil film thickness
$R^2_{q_1}, R^2_{q_2}$ - average square deviation contact surfaces

Also coefficient of friction that has been calculated, based on hydrodynamic lubrication theory of finite porous journal bearings, can be compared with values from experiments. This comparation can show actually lubrication form.

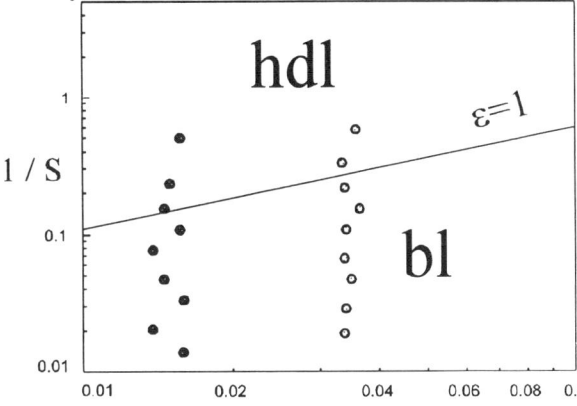

Fig.1: Working regime points in experiments

REFERENCES

(1) Morgan V.T., Cameron A., Mechanism of lubrication in porous metal bearings, Conferenceof Lubrication and Wear, London 1957.

(2) Jankovic M., Stevanovic M., Isledovanie vlijania nagruzki i skorosti skolzenia na koficient trenja, Masinostroiteljstvo, Moscow 1995.

(3) Marinkovic A., The operational performances investigation of Self-lubricating journal Porous Metal bearings produced in Yugoslavia, Msc. Thesis, University of Belgrade, Department of Mechanical Engineering, 1994.

(4) Jankovic M., Marinkovic A., Lubrication Frequency Analysis for Porous Metal Bearings, International Paper: Tribology in Industry, Vol. 2, Kragujevac, 1996.

DYNAMIC CHARACTERISTICS OF THREE-LOBE JOURNAL BEARING WITH THE DIFFERENT SLEEVE SHAPE

WIESŁAW KANIEWSKI and STANISŁAW STRZELECKI
Institute of Machine Design, Technical University of Łódź, Stefanowski Street 1/15, 90-924 Łódź, Poland

ABSTRACT

The analysis of vibrations and stability of the systems of journal bearings demands the knowledge of spring and damping coefficients of oil film. In journal bearings the directions of external load and respective movement of journal centre do not generally coincide and that is the reason for arising of the unstable state of equilibrium of a rotating shaft supported by journal bearings, and occurring a self-excited whirling of the shaft. When the journal vibrates, squeeze film pressure is also generated, in addition to the wedge oil film pressure. These two kinds of pressure give rise to spring and damping forces of oil film and therefore affect the onset speed of the above mentioned whirling, the critical speeds and amplitude of the rotor system. The spring and damping coefficients of oil film may be obtained by solving Reynolds equation. They are determined for the positions of journal in the line of static equilibrium allowing the dynamic estimation and stability analysis of the rotor operating in the journal bearings.

The multilobe bearings, mostly used in the slightly loaded, high speed machines are characterised by good damping of vibrations and the stable operation. Geometry of the multilobe bearings is introduced in the literature but there is a lack of information on the static and dynamic characteristics of three-lobe bearing with the different profile shape used, e.g. in the bearing system of the grinders spindles or in the high compressors. Typical three-lobe journal bearing is composed of single circular sections whose centres of curvature are not in the geometric centre of the bearing. The geometric configuration of the bearing as a whole is discontinuous and not circular. Pericycloidal three-lobe journal bearing (1) as another kind of multilobe bearing is characterised by three hydrodynamic oil films on the journal perimeter. Pericycloid is a continuous curve that is a trajectory of plane point of circle undergoing pure rolling with internal curvature on a fixed circle. Continuous curvature of the operating surface is an important feature of the pericycloidal sleeve. Such a configuration allows simultaneous machining of the whole surface by simple workshop techniques and hence precise shape, dimensional accuracy. Good surface finish can be obtained irrespective of the number of lubricating grooves.

The paper introduces the dynamic characteristics expressed by the spring and damping coefficients of oil film for the three-lobe journal bearing with the cylindrical and pericycloidal shape of sleeve. The laminar and adiabatic flow of oil in the lubricating gap of the bearing of finite length and pressurised oil supply has been assumed. Two different relative lengths of the bearing, clearance ratios and preload values were assumed. The Reynolds, energy and viscosity equations were solved numerically in the conditions of static equilibrium position of the journal (2). The results of calculation in form of the static and dynamic characteristics can be used in the design and optimisation process of three lobe bearings.

Some results of calculation of the spring and damping coefficients for the pericycloidal journal bearing with the relative length $L/D=1.0$, pericycloid factor $\lambda_{per}=0.5$ and three lubricating surfaces are introduced in Table 1 and Table 2 where: $g_{11} \div g_{22}$ are the spring and $b_{11} \div b_{22}$ are damping coefficients., ε- relative ecentricity, α_{eq} - static equilibrium position angle. So - load capacity of the bearing.

ε	α_{eq}	g_{11}	g_{12}	g_{21}	g_{22}
0.2	315.2	0.2730	-0.4276	0.5613	0.4365
0.4	318.6	0.3585	-0.4437	0.5965	0.6926
0.6	320.6	0.4918	-0.5239	0.6712	1.1212
0.8	321.8	0.6899	-0.7076	0.8088	3.9084

Table 1 The spring coefficients of pericycloidal bearing

ε	So	b_{11}	b_{12}	b_{21}	b_{22}
0.2	0.125	1.0231	0.1387	0.1454	1.1380
0.4	0.274	1.0324	0.5743	0.5965	1.28296
0.6	0.468	1.0575	0.7621	0.7964	1.5803
0.8	0.747	1.1465	0.8708	0.9320	2.0997

Table 2 The damping coefficients of pericycloidal bearing

The program developed for calculation of the spring and damping coefficients can be used in the theoretical investigation of dynamic characteristics of cylindrical and multilobe journal bearings under assumptions of adiabatic or isothermal oil film.

REFERENCES

(1) S. Strzelecki, Ph. D. Thesis, Technical University of Łódź, 1976.
(2) S. Strzelecki, Proceedings of the Conference „Engineering of Bearing Systems", Gdańsk, 1996,453-460pp.

EXPERIMENTAL INVESTIGATION INTO THE LOAD CAPACITIES AND FRICTION FORCES OF GAS-DYNAMIC JOURNAL BEARINGS

TIMO KERÄNEN and TATU LEINONEN
Department of Mechanical Engineering, University of Oulu, PB 444, FIN-90571 Oulu, Finland

ABSTRACT

Most of the theories enabling designers to calculate the load capacities and friction forces of gas bearings are based on the well-known Reynolds equation. Some theories include additional factors such as surface roughness. The objective of this work was to show experimentally which factors are needed in theoretical calculations. The air is used as the lubricant and there is always a certain roughness to the surfaces of the bearings. The roughness structures are oriented in many directions relative to the direction of rotation, in addition to which there are shape inaccuracies in the bearing surfaces and the bearings may become misaligned during rotation.

The fact that the load capacities of gas bearings, especially with small working clearances, increase with surface roughness is of significance in practical applications, e.g. in gyroscopes and cryogenic expanders. The earlier after starting up that a bearing begins to rotate gas-dynamically, the shorter is the metallic friction time and the less wear occurs. Likewise, the lower the rotational speed at which a bearing is still carrying as it stops, the less wear there will be.

The aim here was to study experimentally the load capacities and friction forces of gas-dynamic journal bearings. Journal bearings (D = 5, 10 or 15 mm, L/D = 1, 1.5, 2) were designed and manufactured for this investigation. They could be loaded with different weights. At first the bearing clearances, which were used when calculating the theoretical load capacities and friction forces, were determined. The diameters of the bushes and the shafts were measured. The radial clearance between the shape inaccuracies and roughness asperities of the bush and the shaft is a half of the difference between the bush and the shaft diameters. This is known as the free clearance.

Since the roughnesses and shape inaccuracies increase the air volume between the bushes and shafts, it is also essential to take into consideration the distance between the roughness and shape inaccuracy peaks and valleys for both the bushes and the shafts when calculating the total clearances. In the first case the halves of the circular tolerances, and in the second case also the halves of the arithmetical mean deviations of the profiles for both the bushes and shafts as well, are added to the free clearances to obtain the total clearances.

The theoretical threshold speeds (the speed at which a bearing begins to rotate gas-dynamically), and the theoretical friction forces of smooth surface bearings were calculated with help of the Raimondi (1) curves. The load capacities and friction forces of the bearings were measured in a test rig designed for this purpose. When comparing the measured and the theorical load capacities and friction forces it appeared that the measured results were higher than the theoretical results for smooth bearings. In the second case the test bearings began to rotate gas-dynamically at a speed which was 46 % lower than that predicted by the smooth bearing theory. The measured friction forces were 52 % higher than the theoretical values.

Although the measured load capacity and friction force values were much higher than the theoretical values according to Raimondi´s (1) smooth bearing theory, the theory can still be considered valid, because the surface roughnesses increase the load capacity and the friction force according to Arakere´s, Nelson´s and Rankin´s (2) theory. According to Czyzewski´s and Titus´s (3) theory the bearing misalignment increases the load capacity.

The results showed that it is essential to take the additions given by shape inaccuracies and surface roughnesses with when determining the bearing clearance. It was concluded that the load capacity and the friction force of a gas-dynamic journal bearing increase with the surface shape inaccuracies, in addition to the surface roughnesses and, for the part of load capacity, to the misalignment of the bearing.

REFERENCES

(1) A. A. Raimondi, Trans. Amer. Soc. Lub. Engrs.,Vol. 4, No. 1, 1961, 131 pp.
(2) N. K. Arakere, H. D. Nelson, R. L. Rankin, STLE Tribology Transactions, Vol. 33, No. 2, 1990, 201-208 pp.
(3) T. Czyzewski, P. Titus, Wear, Vol. 114, 1987, 367-379 pp.

THEORETICAL AND EXPERIMENTAL DYNAMIC CHARACTERISTICS OF A HIGHLY PRELOADED THREE-LOBE JOURNAL BEARING

GREGORY J. KOSTRZEWSKY
Rockwell Automation, Dodge, 1225 Seventh Street, Columbus, Indiana, 47201, USA
DAVID V. TAYLOR
De La Rue Faraday, Chantilly, Virginia, 20171, USA
RONALD D. FLACK and LLOYD E. BARRETT
Department of Mechanical and Aerospace Engineering, University of Virginia, Charlottesville, Virginia, 22903, USA

ABSTRACT

Fluid film bearings exert a strong influence on the performance of rotor-bearing systems. The bearing's dynamic properties have a direct impact on machine stability, unbalance response, and load capacity. Consequently, the determination of bearing coefficients of known reliability is of practical importance.

The purpose of this paper is to extend previous comparison studies of a plain 2-axial groove journal bearing (1) by examining the measured and predicted performance of a fixed geometry bearing having a bore profile significantly different from a cylinder. A noteworthy factor is the explicit calculation of uncertainties for all of the experimental results, which provides a good estimate of both the confidence in the experimental results and the degree to which theory and experiment should match.

The predicted performance of a three-lobe journal bearing with a preload factor of 0.75 is compared with the measured performance. Operating eccentricity and dynamic coefficients versus Sommerfeld number are compared for three shaft speeds and various steady loads.

An average magnitude and phase method is used to derive the experimental linearized bearing coefficients using synchronous, sinusoidal excitations (2). Numerical results are based on a model which solves Reynolds equation and allows for a variety of thermal effects including circumferential and cross-film viscosity and temperature variation. A numerical perturbation of position and velocity about the steady-state operating position is used to calculate the theoretical coefficients.

Agreement between measured and predicted eccentricity ratios is very good, within 9%, for Sommerfeld numbers below about 0.7. Agreement in attitude angle is very good, within 6°, for Sommerfeld numbers less than about 0.4. At high Sommerfeld numbers, disagreement between theory and experiment increases to 30% or more. In qualitative terms, agreement in journal position improves as load increases and as shaft speed increases.

At Sommerfeld numbers less than 0.4 to 0.7, agreement between the numerical and experimental dynamic coefficients is very good, typically within the uncertainty of the measured data. The magnitude and the slope of the coefficients vs. Sommerfeld number match well, as represented by the vertical principal stiffness coefficient shown in Fig. 1. Disagreement in coefficient magnitude by as much as 100% is seen at high Sommerfeld numbers, although there is general agreement in trend for the predicted coefficients vs. Sommerfeld number.

The level of agreement found provides the designer a reasonable level of confidence in modeling the dynamics of a highly non-cylindrical bearing geometry for Sommerfeld numbers less than about 1.0. The need for additional research is indicated to account for the discrepancies in predictions at small steady loads, or high Sommerfeld numbers.

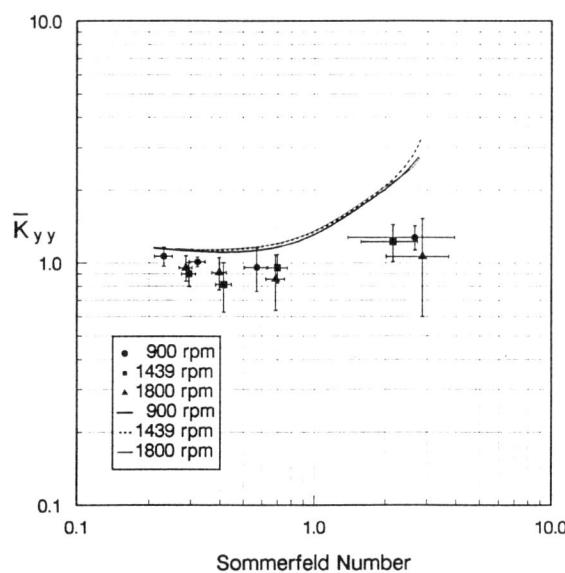

Fig. 1: Comparison of $\overline{K}_{yy}$ vs. S for three speeds

REFERENCES

(1) G Kostrzewsky, R Flack, and L Barrett, Tribology Transactions, Vol. 39, No. 3, pp. 571-578, 1996.
(2) R Flack, G Kostrzewsky, and D Taylor, Tribology Transactions, Vol. 36, No. 4, pp. 497-512, 1993.

INFLUENCE OF BEARING PAD DESIGN ON THRUST SLIDER BEARING CHARACTERISTICS

LESZEK KUŚMIERZ

Department of Mechanical Engineering, Technical University of Lublin, Nadbystrzycka Street 36, Lublin, Poland

ABSTRACT

The oil film shape distinctly influences the thrust slider bearing characteristics (1), (2). This shape depends mainly on the type of the bearing pads: stationary or tilting pad, pad flexibility, the manner of pad support, the pad dimension proportions. In the case of tilting pad thrust bearing the oil film geometry results both from the pad inclination and from the pad elastic and thermoelastic deflections.

This paper presents some results of computations of stationary loaded thrust bearing with elastically deformed tilting pads. The four types of the pad design were considered. The mathematical model of the bearing is described by the set of equations:
- oil film geometry equation
- temperature-viscosity relation
- oil film distribution equation
- energy equation
- temperature of pad friction surface equation.

The adopted theoretical model is based on the following assumptions:
- bearing is operating in steady-state conditions
- oil flow in the bearing is laminar
- oil viscosity changes with temperature only
- pressure and temperature across the oil film thickness are constant
- heat transfer within the oil film takes place in the way of convection
- bearing pad deformations resulted from the pressure distribution and temperature gradients across the pad.

The set of mentioned above equations is solved using the finite difference method and iterative procedure. Results of these computations : values of oil film pressure and temperature in mesh points are transferred to the FEA system (3), in which the pad deflections are computed.

In Fig. 1 are shown the four discrete models of the examined bearing tilting pads respectively : P1, P2, P3 - flexible pads, P4 - conventional pad. The exemplary results of computations resulted from the following assumptions: inner and outer radius of the pad : R_i=128 mm, R_a=238mm, the relative width of the pad B^*=1.2, the pad thickness H=5 mm (for P1,P2,P3 types of the bearing pad), H=20 mm (for P4 type), oil viscosity η=0.082 Pas (in 30°C), tangential velocity v=18.3 m/s, the leading edge temperature T_p=50°C, the bearing pad load F= 15 kN.

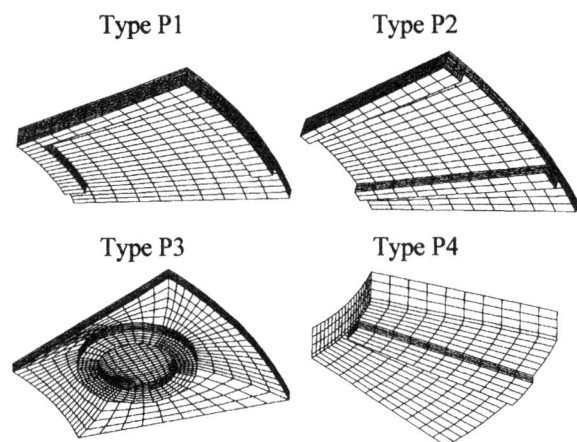

Fig.1: Discrete models of bearing pad (support side view)

The results of calculations for adopted values of input data : the minimum film thickness h_{min}, the moment of friction M_t, the inlet oil flow Q_{in} and maximum pressure in the oil film p_{max} are presented in the Table 1.

Table 1: Bearing characteristics

Type of pad	P1	P2	P3	P4
h_{min} [mm]	0.0361	0.0364	0.0314	0.0314
M_t [Nm]	118.92	124.57	112.74	109.20
Q_{in} [m³/s]x10⁻³	0.289	0.216	0.298	0.320
p_{max} [MPa]	3.8578	4.1856	3.9281	3.7902

REFERENCES

(1) L.Kuśmierz, G.Ponieważ, Methodological Foundations of Computer Aided and Computer Automated Design, I International Seminar and Workshop, Wrocław, Poland, 1993, 107-114 pp.
(2) L. Kuśmierz, G. Ponieważ, Proceedings of the International Conference „Computer Integrated Manufacturing", Zakopane, Poland, 1996, 253-259 pp.
(3) FEA System ALGOR.

A COMPARATIVE THERMAL ANALYSIS OF THE STATIC PERFORMANCE OF DIFFERENT BORE PLAIN BEARINGS

MING-TANG MA
Department of Engineering and Product Design, University of Central Lancashire, Preston, PR1 2HE, UK
C M TAYLOR
Department of Mechanical Engineering, University of Leeds, Leeds, LS2 9JT, UK

ABSTRACT

Fixed profile bore bearings are extensively used in turbo-machinery due to their high load, high speed capability and their ability to dissipate significant energy in a small volume whilst offering outstanding reliability. The thermal performance of these bearings has, however, received little attention. The objective of this paper is to present and compare the calculated steady-state performance characteristics of five commonly used types of journal bearings with the consideration of thermal effects. These bearings are two-lobe circular (CL), elliptical (EL), offset-half (OH), three-lobe (3L) and four-lobe (4L) types. The results have been obtained by employing a detailed thermohydrodynamic (THD) analysis developed by the authors (1)(2).

The THD model incorporates the effect of temperature variation across the film thickness. The film pressures were obtained by solving the generalised Reynolds equation. The two-dimensional energy and heat conduction equations were solved separately for the film and bush temperatures on the bearing centre-plane, and the results were applied to the entire bearing width. In addition, the temperature fade in the exit region of bearing lobes and the oil mixing phenomenon in grooves were modelled in order to establish satisfactorily the inlet temperature to a bearing lobe.

The minimum film thickness, maximum film temperature, power loss and flow rate are the four important performance characteristics of oil lubricated journal bearings under steady-state operating conditions. In this paper, the calculated results of these parameters together with the maximum film pressure and journal temperature are presented and compared for the five bearing types in various conditions. In addition, the effect of groove angle and loading direction on the performance of a sample four-lobe bearing is examined.

Fig. 1 compares the results obtained for the five bearing types over a range of specific loads at a fixed shaft speed of 4000 rpm and a preload (offset) ratio of 0.5. The journal diameter, aspect ratio and clearance ratio used were 110 mm, 0.7 and 0.0015 respectively. From this figure, the performance of the bearings can be summarised by Table 1.

Table 1: Summary of bearing static performance

Brg. Code	Max. film temperature	Power loss	Lubricant flow-rate	Load capacity
CL	Moderate	Low	Low	Very good
EL	Low	Moderate	High	Good
OH	Low	Low	Very High	Good
3L	Moderate	Moderate	Low	Moderate
4L	High	High	Low	Poor

In conclusion, generally the static performance of the non-circular bearing types is inferior to that of the circular bearing. Of the non-circular types, the offset-half and elliptical bearings have a better overall performance; the four-lobe bearing has the poorest behaviour. In addition, an increased groove angle has a favourable effect on the bearing performance.

REFERENCES

(1) M-T Ma and C M Taylor, in *Plain Bearings—Energy Efficiency and Design*, Mechanical Engineering Publications, London, 1992, pp. 31–44.
(2) M-T Ma and C M Taylor, in *Proc. of the 20th Leeds-Lyon Symposium on Tribology*, ELSEVIER, 1994, pp. 431–444.

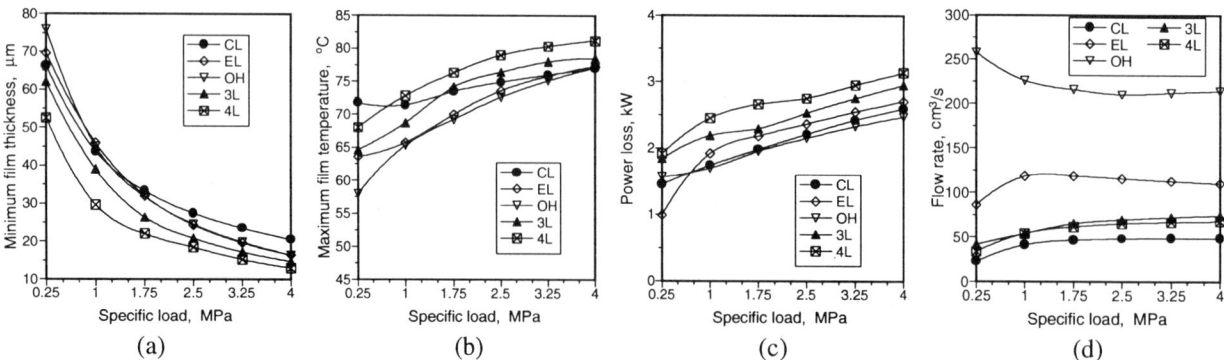

Fig. 1: Calculated performance characteristics of five bearing types at different specific loads

OIL FLOW IN PLAIN STEADILY LOADED JOURNAL BEARINGS: REALISTIC PREDICTIONS USING RAPID TECHNIQUES

F. A. MARTIN
Consultant to The Glacier Metal Co Ltd, Argyle House, Northwood Hills, Middlesex HA6 1LN, UK

ABSTRACT

The aim of the paper is to produce a rapid method for predicting lubricant flow in plain journal bearings, suitable for desktop computing. Flow data, already available from rigorous solutions considering the influence of film reformation, are used together with experimental evidence to develop unique design charts. These together with the relevant equations associated with the chart give the designer an aid for predicting flow 'instantly' which maintains the same flow characteristics as that from the more time consuming rigorous solutions.

These design charts, linking the actual flow Q with those from rapid solutions, give a normalised actual flow Q/Q_p as a function of the normalised hydrodynamic flow Q_h/Q_p. The main input data consists of two flow terms, a hydrodynamic flow Q_h and a feed pressure flow Q_p, which have been in common use in rapid design procedures over many decades. Simplified methods for calculating these flow terms are given in the appendix to the paper.

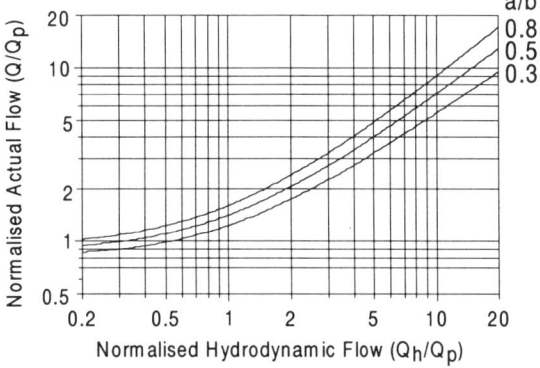

Fig.1. Predicted flow Q based on data considering reformation boundary conditions

The flow prediction chart Fig. 1, was developed for a bearing with a *feed groove at the maximum film thickness position*, using flow data (including the film reformation effects) from Dowson, Miranda & Taylor (1). Predicted flows from Fig. 1 are considered acceptable, generally being within 5% of the flows from Dowson et al as shown in Fig. 2.

Without any modification, the same flow prediction chart can be applied to a bearing *with two axial grooves*, although not quite so precisely. Fig. 3 shows predicted flows derived from Fig. 1 compared with new data from Claro (2).

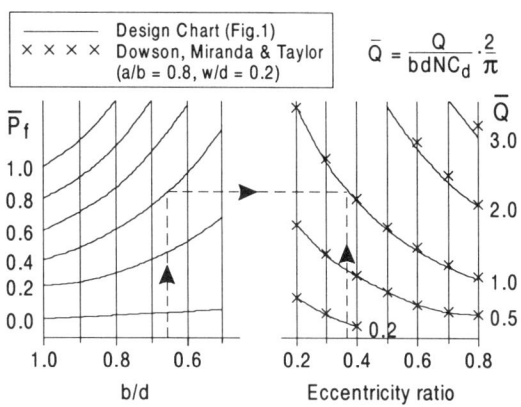

Fig. 2 Comparison of predicted flow for bearing with a groove at maximum film position

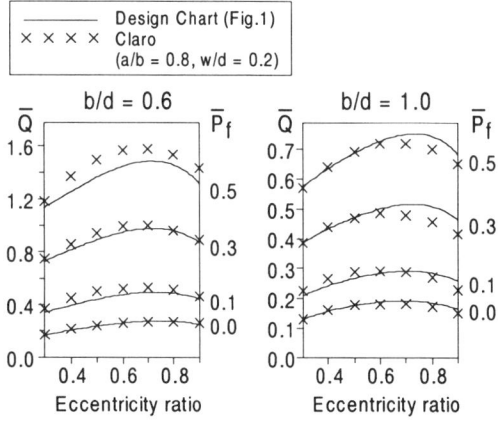

Fig. 3 Comparison of predicted flow for a bearing with two axial grooves

The designer now has a method for flow prediction to use in interactive programs, which gives the opportunity to assess the effect of design changes instantly. The design chart in Fig. 1 and associated equations give realistic flow predictions for a bearing with an axial groove at the maximum film thickness position and also for a two axial groove bearing, one of the more commonly used types of bearing. Other lubricant feed arrangements are considered in the full paper.

REFERENCES

(1) D. Dowson, A.A.S. Miranda, C.M. Taylor, Proc. Inst. Mech. Engrs. Vol. 199, C2, pp 95-102, 1985
(2) J.C.P. Claro, PhD Thesis, Dept. of Mech. Eng., Minho University, Portugal, 1994

CONSTITUTIVE LAWS AND THE CORRESPONDENCE PRINCIPLE FOR THE DYNAMICS OF GAS LUBRICATED TRIBOELEMENTS

B. MILLER and I. GREEN

George W. Woodruff School of Mechanical Engineering, Georgia Institute of Technology, Atlanta, Georgia, 30332-0455

ABSTRACT

A new method for characterizing the dynamic behavior of gas films in triboelements is developed that is based on an expansion of the step jump method. In the new method, the dynamic character of the gas film is preserved in the form of its force response to a step jump stimulus. Transforming this step response into the frequency domain yields the frequency dependent stiffness and damping properties of the gas film. By curve-fitting the step response with a thermodynamically admissible analytic function and using the elastic-gas film correspondence principle, it is possible to determine the system characteristic equation in analytic form and to find closed form solutions for stability and forced responses. The new method offers a time savings compared to direct numerical solution methods, and it is much more conducive to a parametric study.

INTRODUCTION

The number of analysis techniques used in the study of gas lubricated triboelements is limited because the gas film governing equation, the unsteady Reynolds equation, is nonlinear in pressure and a closed form solution is not possible. Consequently, only two methods are commonly used. The first, and most widely used, is a direct numerical method (1), in which the equations of motion and the Reynolds equation are solved simultaneously using some numerical procedures. Another technique is a linear analysis by the perturbation method (2), in which the Reynolds equation is linearized about an equilibrium flying height and pressure distribution. A new method of analyzing the dynamics of gas lubricated triboelements is developed here that yields a closed-form solution for the motion, gives significant insight into the character of the gas film, and is conducive to a parametric study.

CHARACTERIZING GAS FILM PROPERTIES

The new method is based on an expansion of a technique called the step jump method that was originally developed by Elrod et al. (3). The primary assumption is that the change in force generated by the gas film is linearly related to the displacement, and therefore, the total gas film force response can be found by superposing a series of step responses using Duhamel's integral. A distinct advantage of this technique is that the stiffness and damping properties of the gas film are completely characterized by its response to a step jump. Therefore, a constitutive law is formed when the step response is approximated in closed form by an analytic function using a curve-fitting algorithm. Originally, Elrod et al. (3) employed Laguerre polynomials to approximate the step response. However, Miller and Green (4) proved that this function violates the Second Law of Thermodynamics. Hence, they propose three other functions that unconditionally satisfy the law: a Prony series, a series of complementary error functions, and a series of Bessel functions of the first kind of order zero. Once the constitutive law is obtained in closed form, it acts as a kernel of solution for the Reynolds equation. As a result, it is no longer necessary to generate repeatedly a solution for the Reynolds equation in the process of finding a solution to the equations of motion.

RESULTS AND CONCLUSIONS

For simplicity, all the theory and examples presented in the paper relate to a generic gas lubricated, thrust slider bearing of infinite width. The bearing pad has only one degree of freedom, which is translation in the vertical direction.

The solution procedure first requires generation of the step response, which is calculated directly by numerical solution of the unsteady Reynolds equation only once. Then constitutive laws are formed by approximating the step response with the three analytic functions stated earlier. These three functions were chosen because they unconditionally comply with the Second Law of Thermodynamics and have both time and Laplace domain representations. Next, the contribution from the gas film is incorporated directly into the dynamic system model using the elastic gas film correspondence principle. The result is an analytic expression for the equations of motion in the Laplace domain. Transforming the appropriate expression into the time domain either directly or numerically yields, then, the time response. Results from this method, which include closed-form expressions for the characteristic equation and the forced response, compare well with conventional numerical methods.

REFERENCES

(1) V. Castelli, J. T. McCabe, ASME Journal of Lubrication Technology, Vol. 89, pp. 499-509, 1967.
(2) K. Ono, ASME Journal of Lubrication Technology, Vol. 97, pp. 250-260, 1975.
(3) H. G. Elrod Jr., J. T. McCabe, T. Y. Chu, ASME Journal of Lubrication Technology, Vol. 89, pp. 493-498, 1967.
(4) B. Miller, I. Green, ASME Journal of Tribology, Vol. 119, No. 1, pp. 193-199, 1997.

FORCED COOLING OF A SLIDER THRUST BEARING

EDWARD MURDZIA

Institute of Machine Design, Technical University of Łódź, Stefanowski Street 1/15, 90-924 Łódź, Poland

ABSTRACT

The problem of carrying-away the heat generated in the slider thrust bearings with tilted pads is very important in these types of bearings (1)(2)(3)(4)(5). The decrease of oil film temperature and the segments is especially necessary in the bearings operating at the high slide speeds, e.g. 20 m/s and more.

The paper concerns the thrust slide bearing with the segment of special design allowing for the flow of oil inside the segment. The oil flow through the channels and recesses reaching the oil lubricating gap what means that assure the cooling and more uniform temperature distribution. The suggested design permits for the flow of oil to the zone of inlet under pressure. In this way the generating of oil film is easier and shortens the period of mixed friction conditions during the start and run-off.

For the considered bearing and under assumption and simplification of theory of hydrodynamic lubrication, the equations which describe the mathematical model were derived. These equations concern the Reynolds, energy, viscosity and oil film geometry with the boundary conditions of pressure and temperature fields.

All equations were solved numerically by the method of finite differences. The program solve equations giving the load capacity (W), friction torque Mt), angular coordinate of resultant force point (φw) and the increase of temperature in oil film (Tmax-Tp).

Theoretical investigations were verified on the test rig, designed for thrust bearing with tilting pad. All theoretical results were compared with the experimental ones. The effect of cooling, by the mean of special design of segments, on the characteristics of bearing was stated. Significant effect is at the circumferential speeds over 20 m/s.

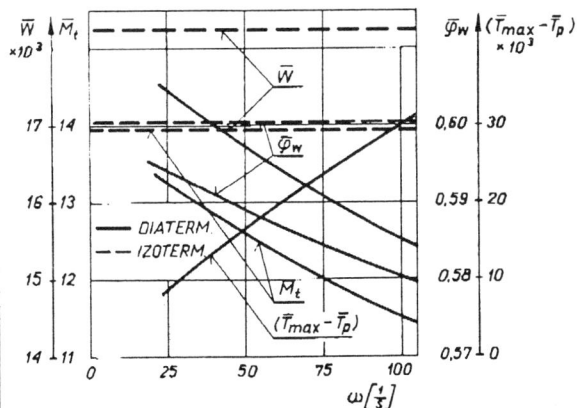

Fig.1 Effect of angular speed on bearing characteristics.

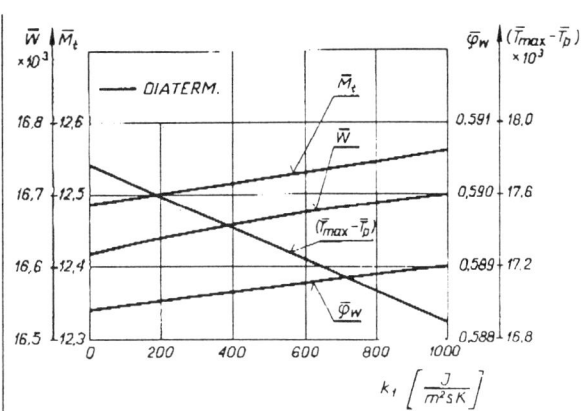

Fig.2 Effect of heat transfer coefficient on bearing characteristics.

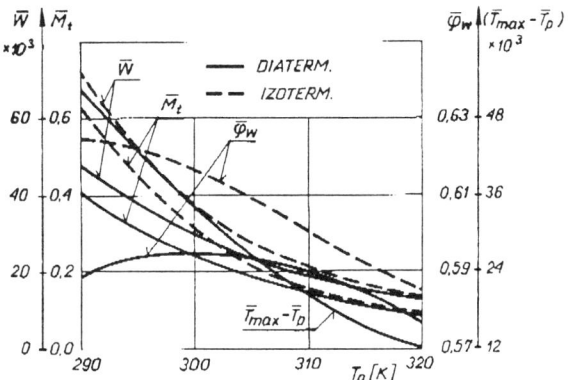

Fig.3 Effect of inlet oil temperature on bearing characteristics.

REFERENCES

(1) T.Tokar, I. Sajcuk, Neizotermiceskaja zadaca smazki upornych podsipnikov s ucetom teplootvoda v telo poduski, Masinovedenie, 1/1973.

(2) F.Detinko, M.Zicharevic, Izmenenie temperatury poperek masljanoj plenki i otvod tepla v podusku podpjatnika. Razvitie gidrodinam. teorii smazki, Moskva, 1970.

(3) C.Rodkiewicz, The Thermally Boosted Oil Lubricated Sliding Thrust Bearing, Trans. ASME, vol. 96, ser.F, 2/1974.

(4) H.Tahara, Forced Cooling of a Slider Bearing with Wedge Film, Trans. ASME, vol.90, ser.F, 1/1968.

(5) E.Murdzia, Termohydrodynamiczne charakterystyki lozyska wzdluznego, Doctor Thesis, Lodz, 1981.

COMPRESSIBLE NARROW GROOVE ANALYSIS-PART 1: DERIVATION

CODA H. T. PAN
Engineering Consultant, Millbury, Massachusetts 01527, USA

ABSTRACT

Compressible Narrow Groove Analysis (**CNGA**) makes available the versatility of narrow groove analysis (1) for the study of gas films with no restriction on the pressure level as may be related to the compressibility number and the Burgdorfer-Knudsen number. The technique of High Resolution Modeling of Thin Films (**TF-HRM**) (2) is combined with a two-scale analysis to formulate **CNGA**. To allow for a state of arbitrary rarefaction of the gas film, the data bank of Fukui and Kaneko (3) is utilized through an empirical formula, which covers the full range of the data bank with a single analytical expression. A graphical comparison of the numerical computed result with the original data values is shown in Figure 1 for two values of the accommodation coefficient $(1.0, 0.8)$.

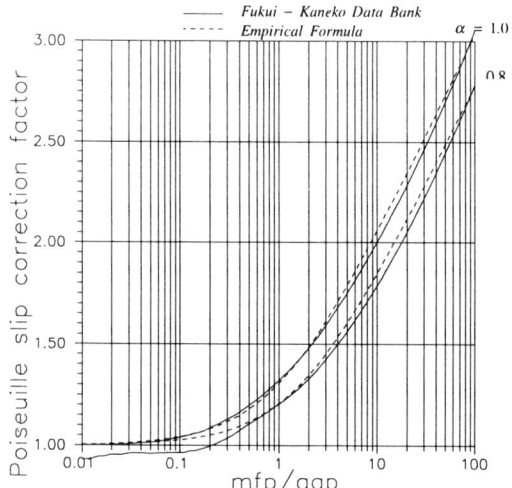

Figure 1   Rarefaction parameter

TF-HRM for the construction of pressure profile in a gas film has been further refined
- to deal with a Gas Film at Arbitrary Rarefaction (**GFAR**), and
- to ensure a single-valued high resolution pressure profile along a surface step via a Flux Shear Rule (**FSR**).

Simplifications based on Narrow Groove Analysis is retained by imposing normal flux continuity across steps, and is rendered applicable to **GFAR** by employing a closed form formula adapted from the classical work of Harrison (4) for the azimuthal pressure profiles between the steps. The robust computation procedure for such pressure profiles (2) thus becomes applicable for any combination of compressibility and Knudsen numbers.

Groove-wise continuity retains the quasi-one-dimensional character in that through flux is computed by azimuthal integration of the transverse flux. As cross-groove pressure variation is recognized, a linear variation of the transverse flux between steps must be postulated, in contrast to the assumption of a constant value according to the Locally Incompressible Narrow Groove Analysis (1). Groove-wise pressure profile is represented by that along one of the two steps of the pattern period, that along the other step being fixed by the azimuthal pressure profile. **FSR** is explicitly observed in the construction of the groove-wise pressure profile.

REFERENCES

(1) S. Whitley, L. G. Williams, UKAEA-IG Report 28 RD/CA, 1959.
(2) Diego Arturo Jäger, Dissertation for the degree of Doctor of Engineering Science, Columbia University, 1987.
(3) S. Fukui, R. Kaneko, ASME Journal of Tribology, 1990, 78-83 pp.
(4) W. J. Harrison, Trans. Cambridge Phil. Soc., XXii, 1913, 6-54 pp.

COMPRESSIBLE NARROW GROOVE ANALYSIS- PART 2: COMPUTATION OF PRESSURE FIELD IN A SPHERICAL DEVICE ROTATING IN EITHER DIRECTION

CODA H. T. PAN
Engineering Consultant, Millbury, Massachusetts 01527, USA

ABSTRACT

Compressible Narrow Groove Analysis (**CNGA**), as derived in a companion paper (1), is a model implementation of **T**hin **F**ilm **H**igh **R**esolution **M**odeling (**TF-HRM**) for gas films. This paper describes the numerical procedure to compute the pressure field in a centered spherical device, as sketched in Figure 1, which has general design features originally intended for a high performance gas bearing gyroscope (2). In the present study, such a device would operate either in the pressurizing mode by rotating with the sense indicated in the sketch, or in the evacuating mode by rotating in the reverse direction. This example demonstrates the prowess of High Resolution Modeling for gas films in dealing with the combination of a large compressibility number and a high rarefaction state.

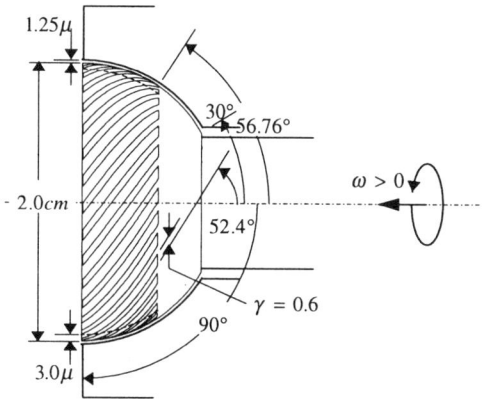

Figure 1   Spiral grooved spherical device

In the pressurizing mode, the number of groove patterns is varied to bring out the significance of the local compressibility number. As illustrated in Figure 2, increased local compressibility, associated with reduced number of groove patterns, causes successive degradation of the pressurization capacity. Eventually, excessive local compressibility causes "choking" and suppresses further pressure rise as marked by the open circles.

With reversed rotation the device operates in the evacuating mode. Pressure reduces rapidly toward the interior and creates a high vacuum state in the gas film concurrent with a large compressibility number. The evacuation operation is relatively insensitive to the number of groove patterns as shown in Figure 3. Evacuation profiles are virtually independent of the number of groove patterns. In all cases, evacuation "choking", again marked by open circles, occurs very close the inner terminal circle, where the pressure level is less than 10^{-10} atm.

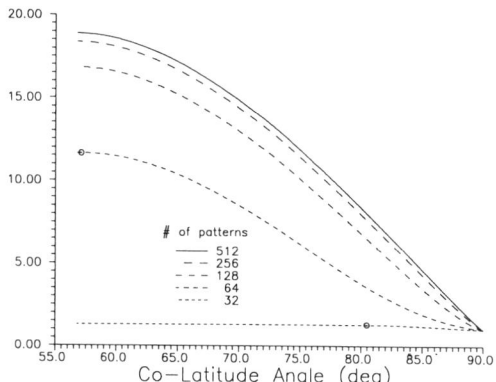

Figure 2   Groove-wise pressure profile for pressurization

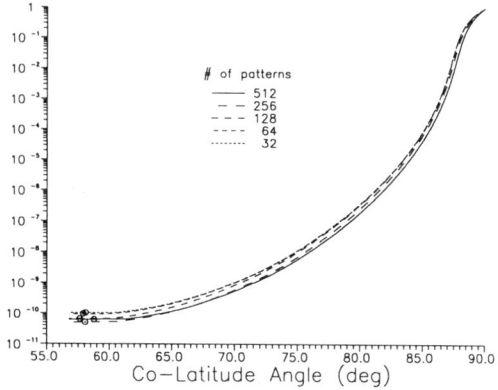

Figure 3   Groove-wise pressure profile for evacuation

REFERENCES

(1) C. H. T. Pan, to be published in Journal of Tribology.
(2) W. H. Keating, C. H. T. Pan, ASME Journal of Lubrication Technology, 1968, 753-760 pp.

A COMPUTATIONAL VERIFICATION TO THE LUBRICATION THEORIES OF FILM JOURNAL BEARINGS

J Y PENG and A K TIEU
Department of Mechanical Engineering, University of Wollongong, Wollongong, NSW 2522, Australia

ABSTRACT

The Reynolds equations for laminar or superlaminar flows have contributed a great deal to film lubrication theories. The intention of this contribution is to adopt a computational approach to verify the turbulent lubrication theories. Based on the concept of continuity, the lubrication theories and the velocity distributions obtained by LDA experiments are investigated.

Refer to Fig. 1, Eq. (1) is established to describe the flow continuity on which the Reynolds equations are based. A control volume with the axial length dz at the center of the bearing is selected, and one section of this volume is located at the position of maximum pressure with film gap H, i.e. $\left.\frac{\partial p}{\partial x}\right|_{p=p_{max}} = 0$, and the other is at the location of minimum film thickness. On the basis of Eq. (1), we can obtain Eq. (2).

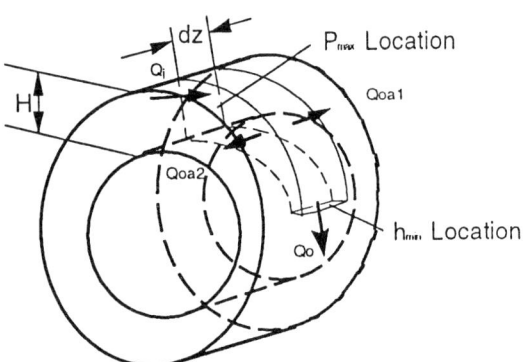

Fig. 1: Diagram of a journal bearing

$$Q_i - Q_o - Q_{oa1} - Q_{oa2} = 0 \qquad (1)$$

$$\frac{UH}{2}dz - Q_o - 2\sum_{i=0}^{m} h_i \int_0^{h_i} u_z dy = 0 \qquad (2)$$

Then the solutions to the Reynolds equations by different theoretical models at different Reynolds numbers can be generated and the angles at which the maximum pressure happens can be found. The models we considered were (1) the k-ε model; (2) Constantinescu theory; (3) Tieu & Kosasih theory; (4) The law of the wall. From these results, we find that the angles generated from different theory models under the same working conditions are the same, and the value of $\frac{h}{H} = \frac{1-\varepsilon}{1+\varepsilon \cos \theta_{p_{max}}}$ changes little. For the selected

volume, Eq. 1 can be defined as $Q_{oa} = \frac{UH}{2}dz - Q_o$ or Eq. (3) nondimensionally.

$$\overline{Q}_{oa} = 0.5 - \overline{Q}_o \qquad (3)$$

Consider the following two cases: (1) the laminar flows and (2) the superlaminar flows. In the former case, $\overline{Q}_o = \frac{Bx}{12}\frac{h}{H} + 0.5\frac{h}{H}$, if the calculated values are assigned to the parameters of this equation, we can obtain the $\overline{Q}_o$ and $\overline{Q}_{oa}$; Employing theoretical models mentioned above, We can obtain the values of $\overline{Q}_o = \frac{h}{H}\int_0^1 \overline{u}_x d\overline{y}$. Based on the above considerations, the difference among the different superlaminar theories is identified. Finally, the velocity distributions generated from LDA experiments are compared with the theoretical models.

REFERENCES

(1) Burton R.A., and Carper, H.C., 1967, "An Experimental Study of Annular Flows with Applications in Turbulent Lubrication", ASME Journal of Lubrication Technology, 89, 38.
(2) Constantinescu, V.N., 1959, "On Turbulent Lubrication", Proceedings of The Institution of Mechanical Engineering, London, 173, 38, 881.
(3) Elrod, H.G., and Ng, C.W., 1967, "A Theory for Turbulent Fluid Films and Its Application to Bearings", ASME Journal of Lubrication Technology, 89, 381.
(4) Galetuse, S., 1974, "Experimental Study on The Interference of Inertia and Friction Forces in Turbulent Lubrication", ASME Journal of Lubrication Technology, 96, 164.
(5) Granville, P.S., 1989, "A Modified Van Driest Formula for the Mixing Length of Turbulent Boundary Layers in Pressure Gradients", ASME Journal of Fluids Engineering, 111 94.
(6) Harada, M., and Aoki, H., 1988, "Analysis of Thrust bearings Operating in Turbulent Regime", ASME Journal of Tribology, 110, 555.
(7) NG, C.W., and Pan, C.H.T., 1965, "A Linearized Turbulent Lubrication Theory", ASME Journal of Lubrication Technology, 87, 675.
(8) Tieu, A.K., Kosasih, 1994, "A Study of Fluid Velocities in Tribological Fluid Film", Transactions of the ASME, 116, 133.

ANALYSIS OF PIVOTED PAD THRUST BEARINGS ON REPEATED STARTS AND STOPS OF A MACHINE OPERATING UNDER THE INFLUENCE OF SHAFT VOLTAGES

HAR PRASHAD
BHEL, R&D Division, Vikasnagar, Hyderabad - 500 093 (India)

ABSTRACT

Impedance, capacitance and charge accumulation on the surfaces of hydrodynamic journal and thrust bearings operating under the influence of electrical current have been theoretically evaluated (1) (2). Various authors have reported that the passage of current through hydrodynamic journal and thrust bearings in a zone of load carrying oil film causes craters to develop on the liner surface over time, and damages the bearing. Also, it causes loss of load carrying capacity of the bearing over time. Theoretical analysis leading to the reduction in life of hydrodynamic journal as well as thrust bearings under the influence of electric current has been reported (3) (4).

In a hydrodynamic pivoted pad thrust bearing, the variation in oil film thickness between the pads and thrust collar forms capacitors of varying capacitance from leading to trailing edge. In this paper, a study is reported on the capacitive effect and life estimation of hydrodynamic pivoted pad thrust bearings on repeated starts and stops of a machine operating under the influence of shaft voltages to determine the increase in charge accumulation on pad liners of a bearing with time as soon as the machine is started and the gradual leakage of the accumulated charges on pad liners as the shaft voltage falls as soon as the power supply to the machine is switched-off. Under these conditions the variation of shaft revolutions to accumulate charges and discharge of the accumulated charges on the liner surface of pads of a hydrodynamic pivoted pad thrust bearing at various levels of bearing to shaft voltage is analysed. Also, the variation of safe limits of starts and stops with the ratio of bearing to shaft voltage is studied. The diagnosis has a potential to study the transient effect of the shaft voltages of a tilting pad thrust bearing during start and stop cycle of a machine.

REFERENCES

1. H, Prashad., Theoretical Evaluation of Capacitance, Capacitive Reactance, Resistance and Their Effects on Performance of Hydrodynamic Journal Bearings, ASME, Journal of Tribology, Vol.113, pp 762-767, October 1991.
2. H, Prashad., "An Approach to Evaluate Capacitance, Capacitive Reactance and Resistance of Pivoted Pads of a Thrust Bearing" STLE, Tribology Transactions, Vol. 35, 1992. 3, pp 435-440.
3. H, Prashad., Theoretical Evaluation of Reduction in Life of Hydrodynamic Journal Bearings Operating Under the Influence of Different Levels of Shaft Voltages ,STLE, Tribology Transactions, Vol.34, 4, pp 623-627 1991.
4. H, Prashad., Analysis of the Effects of Shaft Voltages on Life Span of Pivoted Pad Thrust Bearings", BHEL Journal, issue 1, 1993.

THERMOELASTOHYDRODYNAMIC ANALYSIS OF TILTING PAD JOURNAL BEARING - THEORY AND EXPERIMENTS

D. SUDHEER KUMAR REDDY, S. SWARNAMANI and B. S. PRABHU
Machine Dynamics Laboratory, Department of Applied Mechanics, Indian Institute of Technology,
Madras - 600 036, India

ABSTRACT

Thermal and elastic effects play a very important role in the lubrication of tilting pad journal bearings, especially when the rotor-bearing system operates at heavier loads and higher speeds. When a high speed bearing is subjected to heavy load, pressure and viscosity of the lubricant film change significantly. The pads deform due to pressure developed in the fluid film and therefore, the film profile gets modified. The viscosity of the lubricant also changes with the change in temperature, all these factors affect the hydrodynamic pressure distribution in the fluid film on which the performance characteristics of the whole bearing depends. So to get more realistic design of tilting pad journal bearing, the effect of elastic deformation of pad and variation of viscosity of the lubricant with temperature should be taken into account. To include all these effects in the theoretical analysis of a tilting pad journal bearing, the simultaneous solutions of Reynolds equation, energy equation, equation and deformation equation are needed.

In this paper, a theoretical and experimental investigation has been carried out on the performance of tilting pad journal bearings. The analysis code has been developed based on the simultaneous solution of Reynolds equation, energy equation and deformation equation. The formulations have been done using finite element method. Theoretical results in terms of steady state and dynamic performance characteristics have been obtained for different geometrical parameters, loading configurations of a tilting pad journal bearing.

In the present work an experimental investigation has been carried out on the steady state and dynamic performance characteristics of a four pad tilting pad journal bearing. Pad temperature, bearing outlet oil temperature measurements have been made using thermocouple measuring system. Coast down time experiments were conducted on a rotor-bearing test rig and the coefficient of friction of a tilting pad journal bearing has been obtained experimentally.

Effect of oil hole and groove position on the operating reliability of journal bearings

GERHARD REISCHKE

Fac. of Engineering Sciences, University of Rostock, Rostock, D-18051, Germany

ABSTRACT

Heretofore, only approximate estimates were made of the operating reliability of dynamically stressed bearings with interrupted circumferential grooves.

Because of the time-dependent position of the shaft in the bearing clearance space, the expansive dimensioning must be carried out by a computer (1), (2).

Important decisions during bearing dimensioning procedure are the geometry and the position of oil supply and oil distribution. For the oil supply from the bearing liner a single oil hole or a groove is used in the most cases. The groove is performed as a circumferential (360°-) groove or as an interrupted circumferential groove in the less loaded half of the bearing liner. The oil hole and the groove are important to secure oil flow for lubrication and cooling. But both, hole and groove, alter the hydrodynamic pressure distribution and diminish the load-carrying capacity of the journal bearing. For a journal bearing with a single oil hole and a circumferential (360°-) groove the bearing characteristics are known (1). In practice methods, which the engineer uses, allow the consideration of an oil supply hole in maximum film thickness position or a circumferential (360°-) groove in the middle position of the bearing liner. But for internal combustion engine slide bearings an interrupted circumferential groove (extent about 180°) in the lower loaded half of the bearing liner is very important. This case has so far been assessed approximately. Also the presumption that the oil supply is located in the maximum film thickness position can't keep up for dynamically stressed bearings.

To ascertain the load-carrying capacity and other hydrodynamic characteristics for journal bearings with interrupted circumferential grooves, the Reynolds differential equation is solved numerically for the basic cases of pure sliding and pure squeeze (1). The feed pressure flow was ascertained additionally.

It follows from the calculations that the load-carrying capacity decreases in the groove area. The decrease of load-carrying capacity is smaller for a constant length-to- diameter ratio b/d, if eccentricity ratio ϵ increases. As well the load-carrying capacity depends on ratio b/d for a constant eccentricity value ϵ. It decreases considerably in the groove area. For b/d = 0,2 and ϵ = 0,97 it has only an amount of 20% of the value for a journal bearing without groove. Therefore a precise consideration is necessary of this mode of oil supply, which finds wide use in internal combustion engines; with small b/d ratios.

The solution of the Reynolds differential equation shows normally, that as a result of a single oil hole a decrease of the load-carrying capacity occurs. But the extent of a hole in circumferential direction is small and the load-carrying capacity is affected only in a small part of circumference in comparison with the interrupted circumferential groove.

The calculated discrete values of the bearing characteristics are the basis of approximation equations, which are included in a widely spread computer program (2) for the designing of dynamically stressed plain journal bearings. Now the improved program allows the consideration of any interrupted circumferential grooves. It is applicable to any interrupted grooves (up to 180°) and allows the calculation of characteristics, which describe the bearing behavior and design quality, i.e. absolute minimum film thickness h_{0min} and peak pressure p_{max}.

The effect of a circumferential groove on the operating reliability of internal combustion engine slide bearings was studied in (3). Among other things it was established, that for suitable groove positions and groove dimensions x_N, b_N the bearing behavior is the same as for a bearing with an oil supply hole. An unfavorable groove position should be avoided also as a groove extent $x_N > 180°$ or a groove width $b_N > 0,2 \cdot b$.

Besides the present calculation method results in sufficient exact values for the case of an oil supply hole.

The developed engineering method leads into expressive values for the expected estimate values of bearing calculation. The computer program is applicable to any groove extent (up to 180°), groove positon and groove length. It is now possible to optimize the position of the oil supply and oil distribution.

REFERENCES

(1) Lang, O. R.; Steinhilper, W.: Gleitlager - Berechnung und Konstruktion von Gleitlagern mit konstanter und zeitlich veränderlicher Belastung. Berlin: Springer-Verlag 1978

(2) Affenzeller, J.; Gläser, H.: Lagerung und Schmierung von Verbrennungsmotoren. In: Die Verbrennungs- kraftmaschine. Neue Folge Bd. 8. Wien, New York: Springer-Verlag 1996

(3) Reischke, G.: Hydrodynamic characteristics of journal bearings with interrupted circumferential grooves and the calculation of dynamically stressed bearings. Doc. thesis, University of Rostock, 1994

TENSOR PROPERTIES OF BEARING AND SEAL ROTOR DYNAMIC CHARACTERISTICS

J. T. SAWICKI
Associate Professor, Department of Mechanical Engineering
Cleveland State University, Cleveland, OH 44115, U.S.A.

M. L. ADAMS
Professor, Dept. of Mechanical & Aerospace Engineering
Case Western Reserve University, Cleveland, OH 44106, U.S.A.

ABSTRACT

A seldom mentioned property of journal bearing and seal rotordynamic coefficients is that they are components of single-point second rank tensors, just like stress and mass moment of inertia components. Therefore, some potentially improved physical insights and other improvements in dealing with computational and experimental treatments are not commonly utilized by specialists in the rotordynamics field. For example, when experimentally determining bearing or seal rotordynamic coefficients, if one makes the critical measurements (force and displacement vector components) simultaneously in a multitude of cartesian coordinate systems, then the tensor transformation property of the stiffness, damping and virtual mass arrays can be employed to extract the coefficients that are most consistent with the general linearized mathematical model. When decomposing any of these arrays into their respective symmetric and skew-symmetric portions, the separate portions are also tensors and thus additional physical insights can be derived from the type of tensor each yields. For example, the skew-symmetric portions are of course invariant to orthogonal transformation (coordinate system rotation) i.e., they are isotropic tensors. Similarly, the symmetric portions have real eigenvalues, i.e., principal values and principal directions, just like the stress and moment-of-inertia tensors. The conceptualization and design of more informative rotordynamical experiments is also a logical consequence of full utilization of the tensor transformation properties of bearing/seal rotordynamic arrays.

Since the bearing and seals rotordynamic coefficients are in fact derivatives of forces, i.e.,

$$K_{ij} \equiv -\frac{\partial F_i}{\partial x_j}, \quad C_{ij} \equiv -\frac{\partial F_i}{\partial \dot{x}_j}$$

therefore experimentally extracted coefficients are quite sensitive to the measurement errors. Adams and Rashidi (1) proposed the potential improvement of measurement accuracy by deducing the damping coefficients from the forward whirl mode at the threshold of instability, and the stiffness coefficients from an accurate steady-state locus curve.

For the bearing and seal coefficients the following standard cartesian tensor transformation is satisfied,

$$K_{ij} = T_{ki}T_{\ell j}K'_{k\ell}, \quad C_{ij} = T_{ki}T_{\ell j}C'_{k\ell}$$

where T is an orthonormal coordinate transformation matrix formed by the direction cosines of the ith measuring coordinate system located at γ_i with respect to the reference coordinates x-y. The above mentioned tensor property of K_{ij} and C_{ij} creates the opportunity to extract the rotordynamic coefficients based on the simultaneous measurement of vibration in a multitude of coordinate systems, as shown in Fig. 1.

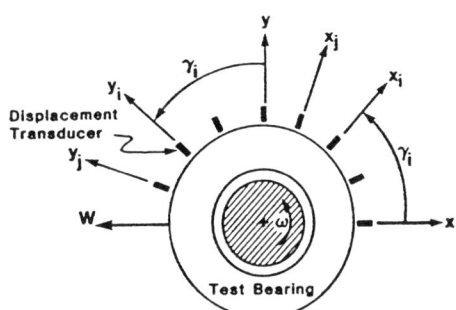

Fig. 1: Multiple displacement measurement coordinate systems

Such approach is justified by actual nonlinear dynamical behavior of bearings and seals, which imposes a coordinate-dependent fit of experimental measurements. Thus, for any operating condition of the bearing there exist coordinate system orientations which are the least sensitive to the measurement errors and best accommodate non-linearities. Using statistical methods for *tensor averaging* one can quite accurately extract K_{ij} and C_{ij} using an experimental approach based on either impedance method or instability threshold (1).

REFERENCES

(1) Adams, M.L., and Rashidi, M., ASME *J. of Vibration, Acoustics, Stress, and Reliability in Design*, Vol. 107, pp. 404-409, 1985.

THE EFFECT OF AIR-ENTRAINED LUBRICATING OIL ON THE VIBRATION OF UNI-DIRECTIONALLY LOADED PLAIN JOURNAL BEARINGS

W SCOTT
School of Mechanical, Manufacturing and Medical Engineering, Queensland University of Technology,
PO Box 2434, Brisbane 4001, Australia
S KNIGHT
Fuchs Australia Pty Ltd, 744 Hunter Street, Newcastle West NSW 2302, Australia

ABSTRACT

The theoretical analysis of fluid film plain journal bearings is attributed to Reynolds [1] whose resulting equation is derived from the Navier-Stokes and continuity equations. Solution of the Reynolds' equation yields the pressure distribution around the circumference of the bearing. In solving Reynolds' equation for a liquid lubricant, the general approach, for simplicity, is to assume that the viscosity and density of the lubricant are constant throughout the film, ie isothermal and incompressible conditions. In many cases these assumptions are clearly invalid.

Mineral oil, which is the most commonly used liquid lubricant, can dissolve a certain percentage of air. The amount of air dissolved depends upon the pressure on the liquid and at atmospheric pressure is about 8-12%. Whereas air in solution has little effect on most systems, air in entrainment, ie in the form of bubbles, can be very damaging in terms of system performance and wear of components.

In a plain journal bearing, operating under full fluid film conditions, the pressure increases from near atmospheric at entry to hydrodynamic film zone to a maximum at the minimum film thickness. Air entrained as bubbles at entry may be forced into solution at the higher pressure and then released again at the divergence immediately beyond the minimum film thickness (the cavitation boundary). This will cause instability.

Although theoretical investigations have been carried out to predict the performance of hydrodynamic journal bearings with bubbly oil [2] and experiments have been carried out to determine the effect of gas entrainment on squeeze film damper performance [3], relatively little has been done on the effect of bubbly oil on the performance of plain journal bearings which includes the aeration/air release properties of the lubricant and, of course, less on the effect of air entrainment on the vibrational characteristics so that this form of condition monitoring may be employed to detect aeration problems. The likelihood of the situation arising in practice is great as agitation, and hence air entrainment, of the lubricant in its reservoir is common.

Preliminary tests were carried out to investigate the effect of air entrained lubricant on the vibration of a plain hydrodynamic journal bearing. The results showed a significant increase in vibration, especially at a frequency corresponding to the rotational speed of the shaft (Figure 1). Investigations are continuing with varying conditions and amounts of air entrainment in the lubricant.

REFERENCES

1 Reynolds, O (1886) - On the theory of lubrication and its application to Mr Beauchamp Tower's experiments, including an experimental determination of the viscosity of olive oil, Phil Trans Roy Soc, Vol 177, pp 157-234

2 Chamniprasart, K; Alsharif, A; Ratagopal, KR and Szeri, AZ (1993) Lubrication with Binary Mixtures: Bubbly Oil, Trans ASME Jnl of Tribology Vol 115, pp 253-260

3 Feng, NS and Hahn, EJ Effects of Gas Entrainment on Squeeze Film Damper Performance, Trans ASME, Jnl of Tribology, Vol 109, pp 149-154 (1987)

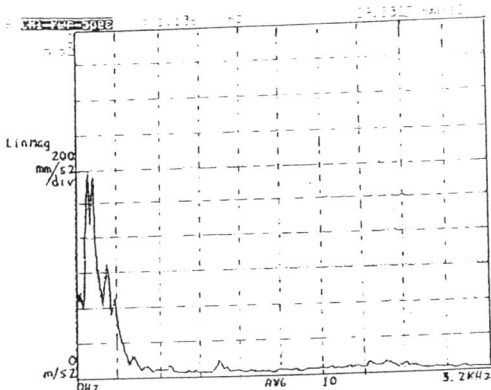

Figure 1 a - Journal Bearing (0-3.2 kHz) - Plain Oil

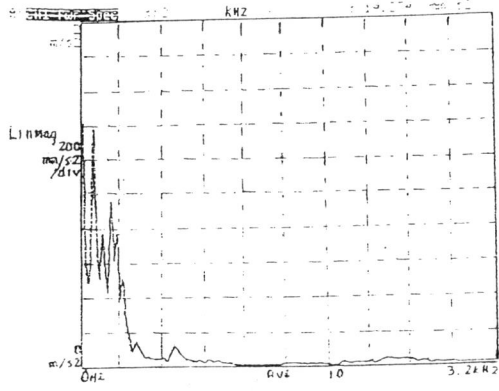

b - Journal Bearing (0-3.2 kHz) - Air Bubbled Oil

Figure 1 Power Spectrum of Vibration Measured at Journal Bearing

A STUDY ON CHARACTERSTICS OF SLOT-ENTRY HYDROSTATIC /HYBRID JOURNAL BEARINGS USING F E M

SATISH C SHARMA, VIJAY KUMAR, S.C. JAIN, (Late) R.SINAHASAN and M.SUBRAMANIAN
Department of Mechanical and Industrial Engineering
University of Roorkee, Roorkee-247 667, INDIA

ABSTRACT

The slot-entry journal bearing is a simple configuration of a non-recessed externally pressurised bearings in which the fluid is fed to the bearing gap through a number of narrow slots, which have the same order of the width as the bearing clearance. The slot-entry journal bearings have a number of advantages including an improved load carrying a capacity over recessed bearings, accurately predictable performance, reduced sensitivity to variation of viscosity and improved tolerance of manufacturing errors. Their main area of application is in grinding wheel spindles.

The available studies [1-5] concerning the slot-entry journal bearing (as far as the authors are aware) are mainly limited to their static performance, however, owing to their suitability for high speed applications, their dynamic characteristics i.e. rotordynamic coefficients, stability threshold speed etc. are also important. The present work is aimed to study comprehensively the static and dynamic performance of slot-entry hydrostatic/hybrid journal bearings. The theoretical study has been carried out using Finite Element Method Technique. To obtain the fluid film pressure distribution in the bearing clearance space, the Reynold's equation has been solved together with the slot restrictor flow equation. In order to find the equilibrium position of the journal for a specified external load (Wo), an iterative scheme has been used. After establishing the steady state solution, the static and dynamic bearing performance characteristics were obtained. The non-dimensional performance characteristics of a double row slot-entry journal bearings of symmetric and asymmetric configurations in terms of maximum pressure (p_{max}), minimum film thickness (h_{min}), bearing flow(Q), stiffness and damping coefficients ($S_{ij}, C_{ij}, i,j=1,2$) and stability threshold speed bearing operating and geometric parameters : Aspect ratio (L/D) = 1.0, concentric pressure ratio (β^*) = 0.5, speed parameters (Ω) = 0.0 and 1.0, slot width ratio (SWR) = 0.25, 0.5 and 0.75, external load (W_o) = 0.8 to 1.6, number of slots per row = 12 and 6 respectively for symmetric and asymmetric configurations. In order to have a better understanding of their performance vis-a-vis other non-recessed bearings, the performance of slot-entry journal bearings have been compared to that of similar hole-entry compensated journal bearings and using other flow control devices i.e. capillary, orifice and constant flow valve restrictors for the same bearing operating and geometric parameters

The threshold speed for slot-entry journal bearings along with the hole-entry journal bearings compensated by capillary, orifice and constant flow valve restrictors has been shown in Fig.1, for Ω =1.0, β^*=0.5 and land width ratio=0.25. The curve indicates that for a given external load (W_o), the symmetric slot-entry configuration provides a larger stability margin than asymmetric hole-entry journal bearing configuration compensated by capillary, orifice and constant flow valve restrictors. The results computed in this study are fully and comprehensively explained in the paper of this title. The authors have written this paper in anticipation of its being read widely. The authors hope that it will be published in full.

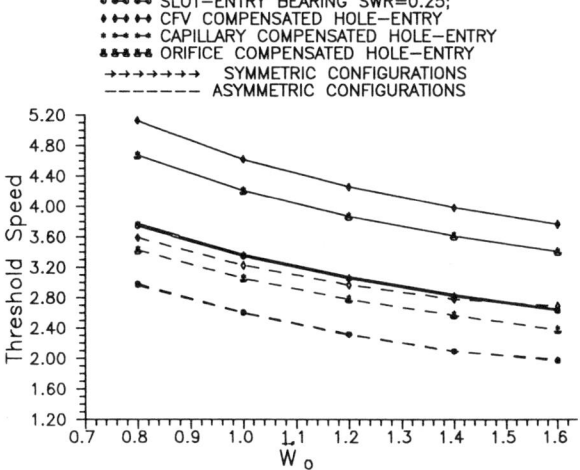

FIG. 1 COMPARISON OF THRESHOLD SPEED

REFERENCES

(1) C.W.Dee, G.I.Shires,ASME Transactions, Jr.Lub.Tech., Vol.93, No.2,1971,441-450pp.

(2) W.B.Rowe, S.X.Xu, F.S.Chong and W.Weston Tribo.,Int., Vol.15, No.6,1982,339-348pp.

(3) D.Ives, W.B.Rowe,Tribo., Int., Vol.35,No.4, 1992, 627-634pp .

(4) S.Xu,STLE, Tribo.Trans., Vol.37, No.2,1994,285-292pp.

(5) K.Cheng,W.B.Rowe,Tribo. Int., Vol.28, No.7, 1995,465-474pp.

DEFLECTION AND STRESSES IN DYNAMICALLY LOADED JOURNAL BEARING WITH PERIPHERAL OIL GROOVE

TSUNEO SOMEYA
Musashi Institute of Technology, Tamazatsumi 1-28-1, Setagayaku, 158 Tokyo, Japan
STANISŁAW STRZELECKI
Institute of Machine Design, Technical University of Łódź, Stefanowski Street 1/15, 90-924 Łódź, Poland

ABSTRACT

Internal combustion engine journal bearings are dynamically loaded elements of an engine. One of the important problems of assuring the bearing reliability arises during dimensioning of the geometry and the position of oil supply point. The single hole or the peripheral 360° grooves supply the oil to the bearing. Both, hole and groove change the oil film pressure distribution decreasing the value of resulting force of the bearing and causing the increase of stresses and deflection of bearing material.

To obtain the data on the performance and stresses of dynamically loaded journal bearings manufactured with the peripheral oil groove, the Reynolds differential equation is solved numerically giving the pressure distribution (1)(2). Integrating the pressure distribution around positive values, give the resultant force. Oil film resulting force allows computation of journal trajectory and finding the position of maximum pressure load with corresponding oil film pressure distribution, which is used, as input data, for determining, e.g. the elastic deflections and stresses of bearing liner and housing .

The journal centre trajectory is obtained by equating the oil film forces generated by the position and velocity of the journal centre with respect to the bearing centre, to the external load at any instant (3).

Computation of stresses in the bearing liner caused by oil film pressure was performed by the finite element's method. The dynamically loaded finite length cylindrical journal bearing with the peripheral oil groove was modelled as two-layer bearing with elastic bearing material and steel housing as outer layer (3). The soft bearing material and housing are represented by the mesh of hexahedral elements. Solid element with eight nodes and three degrees of freedom in the node has been used in the calculation (3).

The developed program input data were as follows: length and diameter of the bearing, dimension of the peripheral oil groove, diametrical clearance, viscosity of the oil, speed of the journal, external load applied to the bearing in the function of the crankshaft angle.

The paper presents the procedure of defining the stresses and deflections in the bearing material and housing of cylindrical journal bearing with the peripheral oil groove.

For assumed width of peripheral (360°) oil groove, situated in the middle plain of the bearing, relative clearance and length of the bearing, the journal centre trajectory, minimum oil film thickness, maximum oil film pressures together with the deflections and stresses in the bearing structure were determined.

Some results of theoretical investigation of bearing static characteristics for the bearing with aspect ration L/D=0.52, relative eccentricity ψ=1.5‰ are introduced in the Table 1 where γ- attitude nagle, H_{min} - minimum oil film thickness, p_{max} maximum oil film pressure. The stresse caused by the pressure and temperature are introduced in Fig.1

Width of peripheral groove[m]	ε -	γ [°]	H_{min} [μm]	p_{max} [MPa]
0.004	0.943	347.7	2.75	59.82
0.006	0.950	347.8	2.42	68.23
0.008	0.956	347.9	2.12	77.58

Table 1

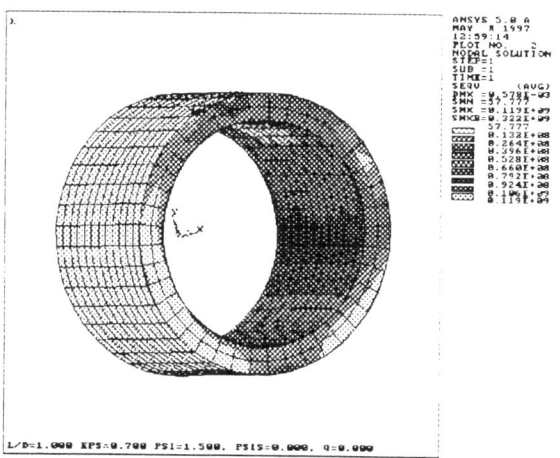

Fig.1 Stresses in the bearing with the peripheral oil groove

REFERENCES

(1) H. Gläser, T. Kubiak, A. Młotkowski, S.Strzelecki,. Journal of KONES 1995, Vol.2, No.1 ,142-147pp.
(2) H. Gläser, A. Młotkowski, S.Strzelecki Proceedings of the International Tribology Conference. Vol 3 Yokohama 1996,1273-1278pp.
(3) T.Someya, S.Strzelecki, Proceedings of the VIth International Symposium INTERTRIBO'96. The High Tatras, Bratislava, 1996,224-227pp.

Submitted to *Wear*

EFFECT OF CAVITATION ON THE DYNAMIC CHARACTERISTICS OF CYLINDRICAL JOURNAL BEARING

JERZY MACIEJ STASIAK and STANISŁAW STRZELECKI
Institute of Machine Design, Technical University of Łódź, Stefanowski Street 1/15, 90-924 Łódź, Poland

ABSTRACT

Operating parameters of radial journal bearings such as: the load capacity, friction force, position of journal with respect to the sleeve or spring and damping properties of oil film are the functions of oil flow in the bearing gap (1)(2)(3). Generally this flow is considered as non isothermal with creating the air bubbles at the decrease of pressure below the oil tension strength. There are two phenomena connected with the discontinuities of oil contained in the bearing gap: cavitation and suction of the oil from the outside space of bearing. In the calculation methods applied up to now these phenomena are not considered and the assumption is that in the "cavitated" part of oil lubricating gap the pressure is equal to the ambient pressure and the calculation problem concerns only the determination of the boundaries of this part of bearing gap where the over pressure exists. The rest of the lubricating gap is ignored during calculation.

The oil film pressure distribution in so defined oil film describes the Reynolds equation under assumption of the respective boundary conditions determining the place of oil film disruption-φ_k. Mostly the Gümbel ($p_{(\varphi=180°)} = 0$) boundary condition or Reynolds ($p_{(dp/d\varphi=0)} = 0$) are assumed:

The first condition gives the solution of problem in the situation of film disruption at the point where the bearing gap has the minimum value (h_{min}) at $\varphi_k = 180°$ and the second one is applied in the case when the oil film ends in not so large distance behind h_{min}, i.e. $\varphi_k \geq 180°$. The above mentioned boundary conditions can be applied for the calculation of the heavy loaded bearings where the relative eccentricity is little less than one, and φ_k slightly larger than 180°. The result of the authors' investigations (2)(3) and literature point out, that:

1) At small loads of the bearing the full oil gap is filled with oil. The oil film pressure distribution on the bearing peripheral shows the presence of positive and negative values.
2) Overcoming of some, critical value of load causes the rapid rise of cavitation zone i.e. in the oil gap appear the longitudinal stabile gas bubbles which causes the shortening of the length of oil film.
3) At a very large load of the bearing, the eccentricity closes the value one and the cavitation zone fully fills the divergent part of oil lubricating gap.

On the basis of the introduced oil flow in the hydrodynamic journal bearing the boundary condition allowed for the determination of the oil film end has been defined as $dp/d\varphi = A/H^2$ with A as the function of bearing operating conditions and the properties of oil.

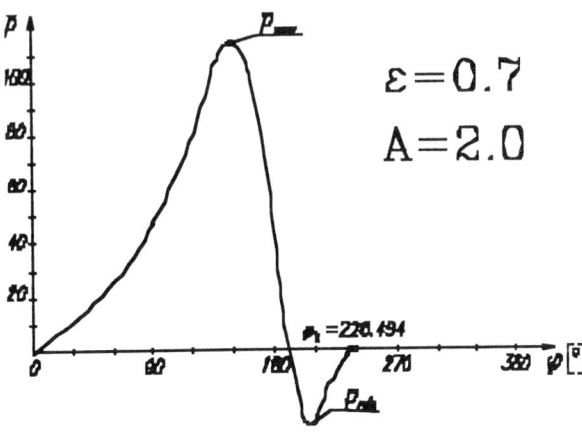

Fig.1 Oil film pressure distribution in the cavitated oil film of cylindrical journal bearing for defined value of the relative ecentricity ε

Solving the geometry, Reynoldes, energy and viscosity equations, the oil film pressure distribution in the full oil lubricating gap, load capacity and spring and damping coefficients (4) of oil film in the presence of cavitation can be calculated. In the case of slightly loaded bearings the oil film is characterised by a larger volume of air bubbles and this oil - air mixture has better damping properties that affect the spring and damping coefficients.

The paper introduces the results of calculation of oil film pressure as well as the spring and damping coefficients of cylindrical journal bearing under assumption of cavitated oil film.

REFERENCES

(1) S. Strzelecki, Ph. D. Thesis, Technical University of Łódź, 1976.
(2) T. Merc, J.M. Stasiak, Cavitation in statically loaded slide bearings. Mechnical Overview. No. 20/1983. 5-9 pp.
(3) T. Merc, Ph.D. Thesis, Technical University of Łódź, 1981.
(4) S.Strzelecki, Proceedings of the VIth International Symposium INTERTRIBO'96, The High Tatras, Bratislava, 1996. 212-215 pp.

EFFECT OF SLEEVE SHAPE ON THE STRESSES DISTRIBUTION IN THE BEARING MATERIAL

STANISŁAW STRZELECKI

Institute of Machine Design, Technical University of Łódź, Stefanowski Street 1/15, 90-924 Łódź, Poland

ABSTRACT

In practice fulfilling the condition for journal and sleeve axis parallelism as well as coaxiality of sleeves bores is problematic. The increase of precise positioning of these elements causes a significant increase of machining costs.

The useful load, e.g. rotor is generally supported between two bearings and it causes a slight bending of shaft axis in relation to the bearing. Misaligned shaft affects the bearing, significantly diminishing the range of safe bearing operation. Misalignment of journal bearing can be caused by position errors too.

Journal bearings operating in the conditions of misaligned axis show the stresses concentration on the edges and in the central part of the sleeve. As the result fluid film lubrication is disturbed causing mixed friction conditions and transient conditions of unstable operation.

Application of journal bearings with convex sleeve e.g. hyperboloidal bearing allows to extend bearing operation range without the stress concentration on the edges of sleeve. These bearings successfully carry the external load in conditions of misaligned axis of journal and the sleeve and eliminate the necessity of using self--aligning bearings.

In the literature there is a lack of data on the operating conditions and edge stresses of misaligned hyperboloidal journal bearings running in the conditions of adiabatic oil film and static equilibrium position of the journal. Comparison of the edge stresses distribution between the cylindrical and hyperboloidal bearing can give information about the use of both bearings in different applications. Assumption of static equilibrium position allows calculations of the spring and damping coefficients of oil film that are necessary for analysis of rotor-bearing system stability.

For the considered hyperboloidal bearing, pressure, temperature and viscosity fields have been received by iterative solution of the Reynolds', energy and viscosity equations. Adiabatic oil film, laminar flow in the bearing gap as well as aligned and misaligned orientations of journal in the sleeve were considered. All calculations have been performed in the conditions of static equilibrium position of the journal. Calculated pressure and temperature fields are the input data for receiving the edge stresses in both types of bearings.

The resulting force, attitude angle, maximum oil film pressure, temperature, minimum oil film thickness and edge stresses have been computed for two different values of aspect ratio, assumed shape and inclination ratio coefficients.

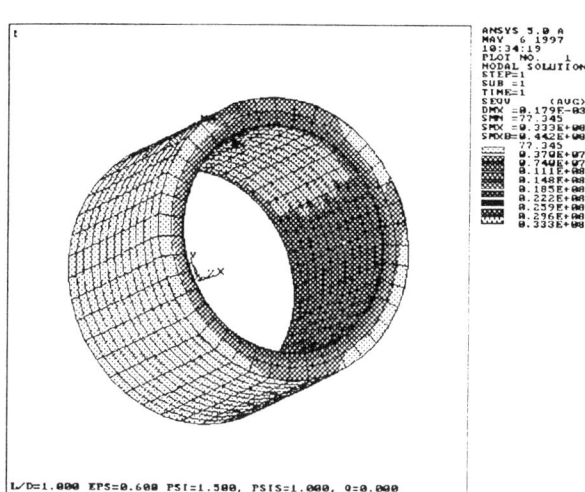

Fig 1 Stress distribution in the hiperboloidal bearing structure

The distribution of stresses in the axial cross-section of hyperboloidal bearing operating in the conditions of non-isothermal oil film is characterised by larger values of stresses in the middle plain of the bearing as compare to the cylindrical bearing; the maximum value of stress equals 33 MPa for the bearing with aspect ratio L/D=1, relative eccentricity $\varepsilon=0.6$, clearance ratio $\psi=1.5‰$ and aligned position of shaft and sleeve axis.

REFERENCES

(1) Someya T., Strzelecki S., Proceedings of the VIth International Symposium INTERTRIBO'96, Stara Lesna-Tatranska Lomnica, April 23-26, 1996. 224-227pp.

(2) Strzelecki S., Proceedings of the 4th International Tribology Conference AUSTRIB'94, Vol.II, Perth, 1994, 579-585pp.

OPERATING CHARACTERISTICS OF DYNAMICALLY LOADED TWO-LOBE JOURNAL BEARING

STANISŁAW STRZELECKI

Institute of Machine Design, Technical University of Łódź, Stefanowski Street 1/15, 90-924 Łódź, Poland

ABSTRACT

Two-lobe bearings are used not only in the medium speed and loaded turbomachines but also in the design of another dynamically loaded bearing systems. The shape of the separate lobes is machined as the cylindrical, pericycloidal or typical two-lobe bearing. The symmetrical displacement of the lobes in the peripheral direction assures good stability conditions at the medium loads and the higher speeds of the journal. Asymmetric two-lobe bearing with the wider lobe situated in the bottom of the sleeve and at the vertical load, make it possible to apply larger loads at the higher speed of the journal. There is almost no information in the literature on the results of investigation and the use of dynamically loaded symmetric and asymmetric two-lobe journal bearings. The evaluation the possibilities and performance of these bearings can be achieved by the calculation of the journal centre trajectory under dynamic load. Investigation of the centre trajectory of journal operating in the two-lobe bearing allows determination of the extreme operating characteristics of the bearing, expressed by the minimum oil film thickness, maximum oil film pressure and temperature this also gives an answer on the reliable operation of the bearing (1)(2). The correct design of bearing is assured when the minimum oil film thickness, maximum pressure and temperature for the assumed set of bearing parameters are known and these extreme values do not affect the bearing material and oil properties.

Comparing the oil film resultant force to the outer dynamic load gives at any moment the values of eccentricity and attitude angle determining the journal centre trajectory. Partial differential equation of oil film pressure distribution was replaced by an approximating finite difference equation. Developed computer program (2)(3) with energy and viscosity equations enables to incorporate the adiabatic or mean temperature oil film and together with the assumed external dynamic load make it possible to compute the eccentricity, attitude angle determining the journal centre trajectory, as function of crankshaft rotation angle.

The paper introduces the trajectory of the journal centre computed for the different length and relative clearance ratios of 2-lobe bearing. The symmetrical and asymmetrical two-lobe bearings operating in the conditions of the dynamic load were considered. Two different dynamic load characteristics were applied in the investigation. The acquaintance of the extreme parameters for different geometric and operation parameters of two-lobe bearing allows to choose the best design and operating parameters of the bearing fulfilling the conditions of long, reliable operation.

The oil film pressure, temperature and viscosity distribution have been determined on the base of Reynolds, energy and viscosity equations (3)(4). These equations were derived under the assumption that in the bearing gap exists the laminar flow of non compressible Newtonian fluid characterised by the constant heat conductivity and specific heat. For the simplicity of considerations the pressure and viscosity were assumed as constant on the thickness of oil film layer.

The caculatins were carried out with three different aspect ratio **L/D** of the bearing, clearance ratio $\psi=1.5‰$, two different segment clearance ratio for the upper and lower lobe of the bearing as well as oil supplied under pressure to the bearing. Some results of caculations for the different bearing aspect ratio are introduced in the Table 1.

L/D	H_{min} [μm]		p_{max} [MPa]	
	Symm.	Unsymm	Symm.	Unsymm
0.52	6,14	6.73	64.32	61.54
0.8	10.85	12.63	30.63	26.94
1.0	13.65	18.14	21.21	18.02

Table 1: Minimum oil film thickness H_{min} and maximum oil film pressure p_{max} for symmetrical and unsymmetrical two-lobe bearings with different aspect ratio.

Asymmetry of the lobes has an effect on the journal centre trajectory and causes the increase of minimum oil film thickness as well as the decrease of maximum oil film pressure.

REFERENCES

(1) Hahn W.: Dissertation T.U. Karlsruhe, 1957.
(2) Someya T., Strzelecki S.: Proceedings of the VIth International Symposium INTERTRIBO'96, Stara Lesna-Tatranska Lomnica, April 23-26, 1996. pp.224-227.
(3) Strzelecki S.: Proceedings of the Vth International Symposium INTERTRIBO '93, Bratislava, 1993, pp.163-167.
(4) Strzelecki S., Szkurłat J.: Proceedings of the VIth Tribological Conference, 6-7 June, Budapest, 1996

A THERMOHYDRODYNAMIC LUBRICATION ANALYSIS TO DESIGN LARGE TWO-PAD JOURNAL BEARING WITH COOLING DITCHES

S.TANIGUCHI
Hiroshima R&D Center, MHI., Mihara, JAPAN
Y.OZAWA
Takasago R&D Center, MHI., Takasago, JAPAN

T.MAKINO
Nagasaki R&D Center, MHI., Nagasaki, JAPAN
T.ICHIMURA
Takasago Machinery Works, MHI., Takasago, JAPAN

ABSTRACT

In recent years, steam turbines and gas turbines for electric utilities have increased in capacity to meet tremendously increasing electric power demand, and these shaft diameters have to be large enough not only to transmit large torque but also to survive sudden torque variation due to short circuit fault conditions. As the journal bearings of these turbines have been operating under the high speed condition near 100m/s of sliding velocity, it causes considerable increase of bearing surface temperature and friction loss, and decrease of stability margin against self-excited vibration of rotors supported in the bearings. Consequently it has become an important problem to operate these bearings safely with less friction loss.

From these view points, a new style of bearing called two-pad tilting pad bearing was developed as shown in Fig.1. It is a hybrid type whose lower half consists of two tilting shoes with some cooling ditches penetrating them, and whose upper half is a semicircular sleeve with a viscous pump to sluice oil into the cooling ditches and a wide scoop to reduce friction loss.

In order to predict the bearing performance and the viscous pump performance accurately, the turbulent thermohydrodynamic lubrication analysis was done in connection with the flow analysis of oil circulating system including the cooling ditches and viscous pump. For the tilting shoes, the previously published(Taniguchi etal, 1990)turbulent Reynolds equation taking account of three-dimensional temperature variation in the oil film and also heat transfer condition on the lubricating surface was used. Ten simultaneous equations concerned with the oil flow paths were introduced. For instance, the pressure rise ΔP along the viscous pump, the pressure drop $\Delta P'$ along the cooling ditches and the pressure equilibrium between both pressures are described respectively as below.

$$\Delta P = \frac{\beta \mu_p l_p}{G_{xp} b_p t_p^3} \left(\frac{U b_p t_p}{2} - Q_p \right) \quad (1)$$

$$\Delta P = \frac{\rho l_c}{2 d_c} \left(\lambda_{cl} V_{cl}^2 + \lambda_{cl} V_{cl}^2 \right) \quad (2)$$

$$P_s + \Delta P = P_d + \Delta P' \quad (3)$$

Another seven flow equations are omitted at here. These complicated equations were solved simultaneously and used to design a bearing of 535mm diameter which was selected looking for ahead into the future.

Full size bearing tests were carried out to check the static and dynamic performances of the newly developed bearing and to confirm the accuracy of this calculation method. This bearing was certified to operate safely with large minimum film thickness and low metal surface temperature, namely the former was 229μm and the latter was 68.0℃ at the normal condition of 3000rpm with the load of 290kN. Furthermore good agreement was obtained with the theoretical and experimental results of the bearing performance and viscous pump performance, consequently the newly proposed design method for two-pad journal bearing with cooling ditches was proved to be of practical use.

REFERENCE

Taniguchi,S.,Makino,T.,Takeshita,K. and Ichimura, T., ASME Journal of Tribology,Vol.113,1990,542-549pp.

NOMENCLATURE

b_p, l_p, t_p = width, length and depth of viscous pump
d_c, l_c = hydraulic diameter and length of cooling ditch
G_{xp} = turbuleut viscosity correction coefficient
P_d, P_s = discharge pressure and supply pressure
Q_p = flow rate at viscous pump
U = sliding velocity
V_c = flow velocity through cooling ditch
β = efficiency of viscous pump
ρ, μ = density and viscosity of lubricant
λ_c = coefficient of fluid friction at cooling ditch

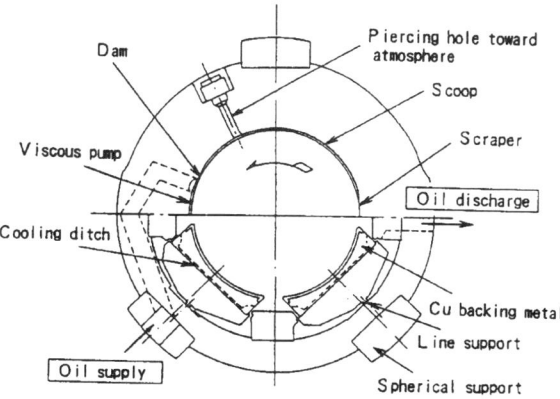

Fig.1 Two-pad bearing assembly

AN EXPERMIMENTAL DETERMINATION OF REYNOLDS STRESSES IN SUPERLAMINAR HYDRODYNAMIC JOURNAL BEARINGS

A.KIET TIEU, ENBANG LI and JINJYI. PENG
Department of Mechanical Engineering, University of Wollongong, NSW 2522, Australia

ABSTRACT

Several theories on turbulent flows in bearings applied various concepts such as Boussinesq eddy viscosity, bulk flow characteristics, Kolmogorov-Prandtl energy model with one equation and two equation k-ε transport model. Basically, the various theories apply curve fitting techniques to experimental data for Couette and Poiseuille flows in large parallel flow gaps, and then develop empirical constants in the expression for turbulent stress in thin film flow well beyond the range of parameters of the original data.

The work in this paper describes the application of a recently developed two-dimensional LDA system that can determine the Reynolds stresses in a journal bearing through the measurements of circumferential and axial velocity distributions.

A journal bearing test rig was designed for the experimental studies of turbulent flow. The bearing housing dimensions were 90×90×140 mm. In this paper the authors have obtained for the first time, experimental velocity profiles in the circumferential and axial directions in the above bearing operating with gaps from 0.37mm to 0.70mm.

The LDA system consists of all solid state components. To provide two dimensional velocity measurements, two orthogonal set of fringes were produced. The fringes were labelled using two different optical frequency shifts. This allows the Doppler signals from each set of fringes to be separated. The measuring volume (40μm long x 20μm dia) was focused in the bearing gap of 370μm at Reynolds number 300 to 3500. Simultaneouus measurements of circumferential and axial velocities at a gap of 0.37mm were obtained.

From the mixing length equation :

$$-\overline{u'v'} = \varepsilon \frac{d\overline{u}}{dy} = l^2 \left| \frac{d\overline{u}}{dy} \right| \frac{d\overline{u}}{dy} \quad (1)$$

the mixing lengh can be calculated from the measured velocities and then compared against the modified Van Driest mixing length expression:

$$l^+ = k\, y^+ \sqrt{\tau^*}\, (1 - e^{-y/\lambda^*}) \quad (2)$$

where λ^* = Van Driest factor

$\tau^* = 1 + p^+ y^+$

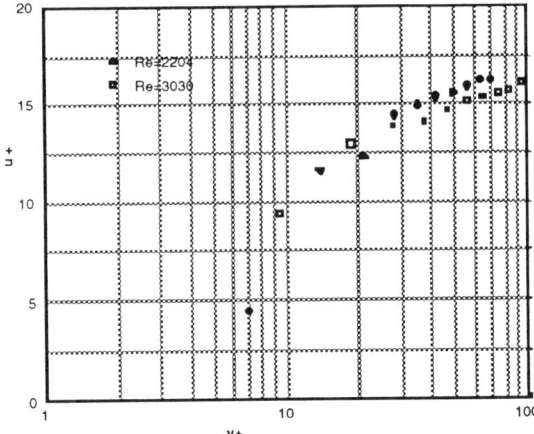

Figure 1 Circumferential profiles at the gap of 370 microns

The circumferential velocity profiles for a gap of 370 microns were shown in Figure 1 and Reynolds stresses u'v' in Fig.2.

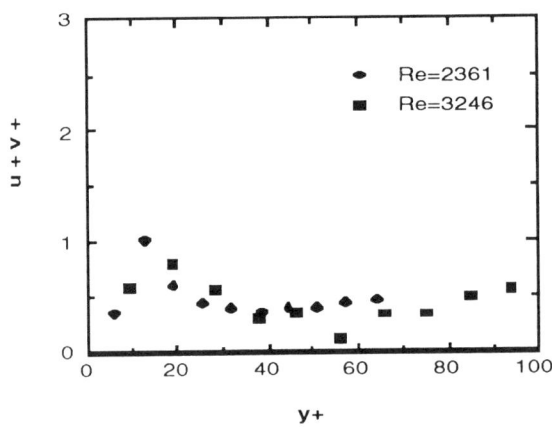

Fig. 2 Measured Reynolds stresses at a gap of 370 microns

The Reynolds stresses show significant changes in the range $y^+ < 20$, and remains relatively steady at higher values of y^+.

STABILITY ANALYSIS OF A RIGID ROTOR SUPPORTED ON HYDRODYNAMIC JOURNAL BEARINGS WITH ROUGH SURFACES USING STOCHASTIC FEM

RAM TURAGA, A.S. SEKHAR and B.C. MAJUMDAR
Department of Mechanical Engineering, Indian Institute of Technology, Kharagpur-721 302, INDIA

ABSTRACT

The effect of surface roughness on the stability characteristics of hydrodynamic journal bearings has been studied. Roughness has been considered to be a stochastic variable, which is stationary, ergodic with mean zero as given by Christensen (1) and Christensen and Tonder (2). Both one-dimensional roughness (longitudinal and transverse) and two-dimensional roughness (isotropic) models have been considered. Linear stability analysis has been performed following the method given by Majumdar, Brewe and Khonsari (3). However the perturbed equations here are solved using the variational formultion in the similar line of Kilt and Lund (4) and by the stochastic finite element method (5). Using an approximation to Gaussian distribution (2) and C=0.149, where C=c/Δr, the stability of finite journal bearings were obtained. This implies that the eccentricity ratio (ϵ_0) can take a maximum value slightly over 0.85. From the obtained pressure distribution, four stiffness and four damping coefficients are evaluated which are non-dimensionalised.

The results indicate that there is a significant influence of surface roughness on the non-dimensional stiffness and damping coefficients. For Isotropic roughness at ϵ_0 =0.81 and b/d=0.5 the stiffness coefficient K_{rr} decreases by 15.7%, $K_{\Phi r}$ decreases by 11.5%, $K_{r\Phi}$ decreases by 1.5%, $K_{\Phi\Phi}$ decreases by 2.8%. similarly in case of longitudinal roughness at ϵ_0 =0.81 and b/d=0.5, K_{rr} increases by 4%, $K_{\Phi r}$ increases by 4%, $K_{r\Phi}$ increases by 47.1%, $K_{\Phi\Phi}$ increases by 1.1%. In case of Transverse roughness at ϵ_0 = 0.81 and b/d=0.5, K_{rr} increases by 26%, $K_{\Phi r}$ increases by 15.6%, $K_{r\Phi}$ increases by 137.8%, $K_{\Phi\Phi}$ increases by 21.6% over smooth case. In case of transverse roughness there is a significant change in the values of one of the cross coupled stiffness coefficients at high eccentricity ratios. Damping coefficients for isotropic roughness at ϵ_0=0.81 and b/d=0.5, D_{rr} decreases by 11.2%, $D_{\Phi r}$ decreases by 8.8%, $D_{r\Phi}$ decreases by 8.4%, $D_{\Phi\Phi}$ decreases by 4.4%. Similarly in case of longitudinal roughness D_{rr} increases by 14.8%, $D_{\Phi r}$ increases by 15%, $D_{r\Phi}$ increases by 4.9%, $D_{\Phi\Phi}$ increases by 4.4%. In case of Transverse roughness at the same ϵ_0 and b/d, D_{rr} increases by 0.9%, $D_{\Phi r}$ decreases by 2.5%, $D_{r\Phi}$ increases by 2.7%, $D_{\Phi\Phi}$ increases by 7.3%.

Using the stiffness and damping coefficients calculated above, non-dimensional mass parameter M and whirl ratio (λ), measure of stability have been found. Isotropic roughness reduces the M by about 21% at ϵ_0 = 0.81 for b/d=0.5 and by 10.42% at ϵ_0 = 0.81 for b/d=1.0. Correspondingly there is an increase in the case of whirl ratio. Longitudinal roughness has a marginal effect on M and λ. The reduction in M in this case is by about 4% at ϵ_0 = 0.81 for b/d=0.5 and by 2.68% at ϵ_0 = 0.81 for b/d=1.0. Transverse roughness has maximum influence on M and λ. It is observed that M increases by about 315% for b/d=0.5 at ϵ_0 = 0.81 and by about 89.06% for b/d=1.0 at ϵ_0 = 0.81 over a smooth bearing. The whirl ratio shows a decrease of lesser magnitude (about 100% at b/d=0.5 and 34.12% at b/d=1.0).

The change in b/d ratio has a significant effect on the stability characteristics of journal bearings. At ϵ_0 = 0.81, for smooth bearing M is higher by 21.5% for b/d=0.5 than b/d=1. Similarly in case of isotropic roughness M is higher by 10.9% for b/d=0.5 compared to b/d=1. In case of longitudinal roughness M is higher by 20.1% for b/d=0.5 compared to b/d=1. In case of transverse roughness M is higher by 167.1% for b/d=0.5 over b/d=1. From this it can be seen that surface roughness has maximum influence on bearings with lower b/d ratio.

REFERENCES

(1) H. Christensen, Proceedings of the IMechE, Vol. 184, 1, 55, 1013-1026pp, 1969.
(2) H. Christensen, K. Tonder, ASME Journal of Lubrication Technology, 1973, Vol. 95, 166-172pp.
(3) B.C. Majumdar, D.E. Brewe, M.M. Khonsari, ASME Journal of Tribology, 1988, Vol. 110, 181-187pp.
(4) P. Kilt, J.W. Lund, ASME Journal of Tribology, 1986, Vol. 108, 421-425pp.
(5) Ram Turaga, A.S. Sekhar, B.C. Majumdar, STLE Tribology Transactions, (in press), 1997.

INERTIA EFFECTS OF VISCOPLASTIC LUBRICANT IN CURVED SQUEEZE FILM

E WALICKI, A WALICKA and D RUPIŃSKI
Department of Mechanics, Technical University of Zielona Góra, ul. Szafrana 2, Zielona Góra, POLAND

ABSTRACT

Many fluids of engineering interest appeaar to exhibit yield behaviour, where flow occurs only when imposed stress exceeds a critical yield stress. To describe the rheological behaviour of such viscoplastic fluids the Ostwald - de Waele and Bingham models are used. Recently, the non-linear model of Shulman (1) has been succesfully applied. The constitutive equation for this model is the following:

$$\tau = \left[\tau_0^{1/M} + (\mu\dot{\gamma})^{1/N}\right]^M$$

where τ is the shear stress, τ_0 is the yield shear stress, μ is the coefficient of plastic viscosity, N and M are the non-linearity indices. By reducing of the coefficients in the Shulman equation one can obtain simpler models describing the flow of viscoplastic fluid.

The paper deals with the laminar squeezing flow of a viscoplastic fluid in the clearance of small thickness between two curvilinear surfaces of revolution, having a common axis of symmetry, as shown in Fig. 1.

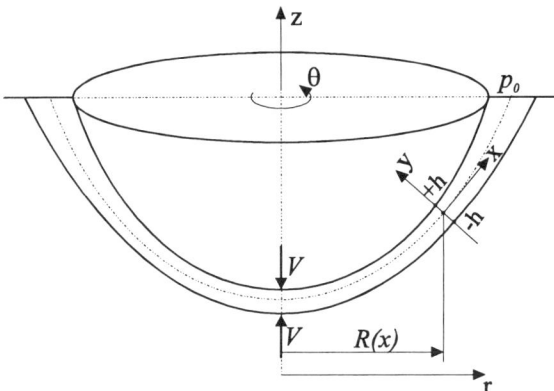

Fig. 1: Bearing clearance between curvilinear surfaces

Using the method of integral approaches one obtains the following equation for pressure distribution:

$$\frac{\rho}{k+1}\left(\frac{1}{R}\frac{dR}{dx}+\frac{\partial}{\partial x}\right)\int_{-h}^{+h} v_x^{k+2}dy =$$

$$= -\frac{dp}{dx}\int_{-h}^{+h} v_x^k dy + \int_{-h}^{+h} v_x^k \frac{\partial \tau_{yx}}{\partial y}dy$$

where τ_{yx} is the non-zero component of stress tensor. Its solution has a form (2) (3):

$$p(x,t) = p_R(x,t) + p_I(x,t)$$

where $p_R(x,t)$ is the solution in the Reynolds approximation (without inertia, $k=0$) and $p_I(x,t)$ is the correction term connected with the inertia.

The study has been performed for the following viscoplastic fluids: Herschel-Bulkley ($M=1$), Bingham ($M=N=1$), Ostwald - de Waele ($M.=1$, $\tau_0 = 0$) and for Newtonian fluid ($M=N=1$, $\tau_0 = 0$).

The dimensionless pressure distribution $\tilde{p}$ depends on the Reynolds number R_λ of squeeze flow and the viscoplastic indices.

Consider for example the squeeze film of the Ostwald - de Waele fluid between two disks. The dimensionless pressure distribution is shown in Fig. 2. The value of $R_\lambda = 0$ indicates the case of flow without inertia.

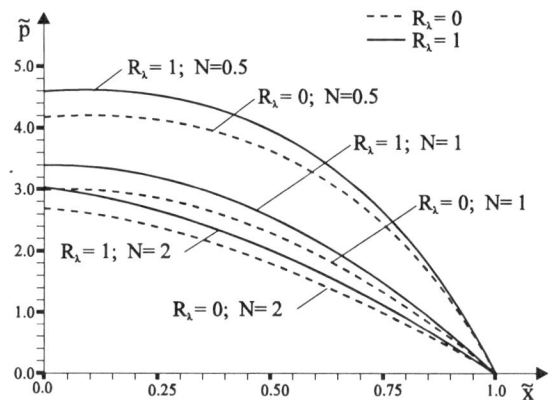

Fig. 2: Dimensionless pressure distribution for $k=1$ (energy integral approach)

REFERENCES

(1) Z. P. Shulman, Convective heat transfer of rheologicaly complex fluids [in Russian], Energy, Moscow, 1975.
(2) G. H. Cowey, B. R. Stanmore, J. Non-Newt. Fluid Mech. Vol. 8, 1981, 249-260 pp.
(3) E. Walicki, A. Walicka, D. Rupiński, Atti IV Conv. AIMETA Trib. 1996, 83-90 pp.

PERFORMANCE OF A CURVILINEAR THRUST BEARING WITH MICROPOLAR LUBRICANT

E WALICKI and A WALICKA
Department of Mechanics, Technical University of Zielona Góra, ul. Szafrana 2, Zielona Góra, POLAND

ABSTRACT

Presently there exist several new developments in fluid mechanics that are concerned with structures within the fluid. Certain fluids, e. g., emulsions, solution of polymers, polymer melts and suspensions are known to fall beyond the domain of applications of the classical theory of Newtonian fluids. Many theories incorporating particle microstructure have been postulated. Eringen's theory (1) of micropolar fluids is characterized by the presence of suspended rigid particles with microstructure. This theory can be used to model lubricants containing suspended additive particles.

The problem considered is that of a laminar flow of an incompressible micropolar fluid in a thrust bearing shown in Fig. 1.

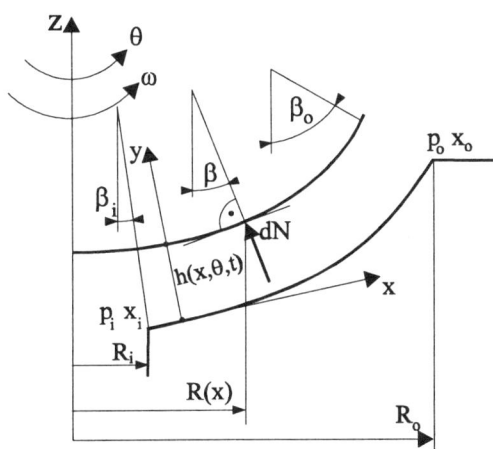

Fig. 1: Coordinate system in the bearing clearance

Considering the micropolar fluid flow in the bearing clearance and taking into account the results of (2), (3) and (4) one obtains the modified Reynolds equations:

$$\frac{1}{R}\frac{\partial}{\partial x}\left[Rh^3\left(G\frac{\partial p}{\partial x}-\frac{3\rho\omega^2 RI}{10}\frac{dR}{dx}\right)\right]+\frac{1}{R^2}\frac{\partial}{\partial\theta}\left(h^3 G\frac{\partial P}{\partial\theta}\right)=$$

$$=3\mu\omega(2+\delta)\frac{\partial h}{\partial\theta}+6\mu(2+\delta)\frac{\partial h}{\partial\theta}+6\mu(2+\delta)\frac{\partial h}{\partial t}$$

where G, I, δ are given functions of micropolar parameters. The dimensionless solution of this equation (or dimensionless pressure $\tilde{p}$) in the case of symmetry depends on the parameter:

$$\pi_p = \frac{3\rho\omega^2 R_o^2}{20 p_o}\frac{I}{G}.$$

Having the pressure distribution one can find the load capacity which also depends on $\varepsilon = R_i / R_o$.

Considering for example a thrust bearing modelled by two disks one obtains the dimensionless pressure distribution and the load capacity shown in Figs 2 and 3. The negative values of π_p are for micropolar fluids but the positive ones - for Newtonian fluids.

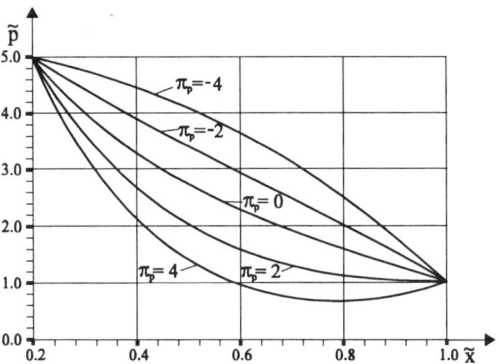

Fig. 2: Dimensionless pressure distribution $\tilde{p}$

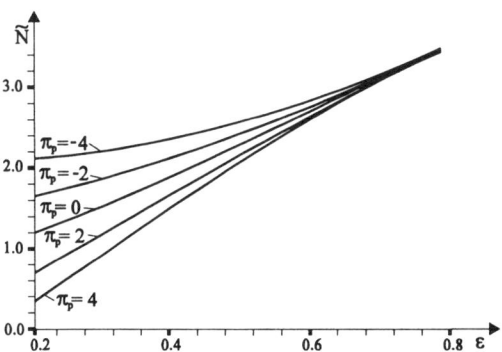

Fig. 3: Dimensionless load capacity $\tilde{N}$

REFERENCES

(1) A. C. Eringen, JMM, Vol. 16, No. 1, 1996, 1-18 pp.
(2) A. J. Willson, Appl. Sci. Res., Vol. 20, 1969, 338-355 pp.
(3) A. Walicka, Micropolar flow in a slot between rotating surfaces of revolution, WSI, Zielona Gora, 1994.
(4) E. Walicki, A. Walicka, AME, Vol.1, No. 1,1995, 25-55 pp.

PERFORMANCE CHARACTERISTICS OF AN LEG, NON-EQUALIZING, TILTING PAD, HYDRODYNAMIC THRUST BEARING

JOSEPH J. WILKES and MATT MARCHIONE
Kingsbury, Inc., 10385 Drummond Road, Philadelphia, PA, 19154 U.S.A.

ABSTRACT

Presented are data from tests of a newly designed hydrodynamic tilting pad thrust bearing. The accomplished goal of this thrust bearing is to perform at high speeds and high loads with reduced film temperatures, lower oil flow requirements and reduced frictional losses as compared to conventional flooded designs. The intention was to optimize the design for a single direction of rotation in order to achieve peak performance. This thrust bearing incorporates leading edge groove (LEG) lubrication technology and offset spherical pivots in a thin, non-equalizing size frame.

The authors also compare, examine and analyze the different aspects of the new bearing, such as pivot location, shape of pivot, and method of lubricant supply. Comparisons are made against an 8 pad, flooded, line-contact, center pivot thrust bearing of approximately the same size. The new bearing is an 8 pad, 273 mm O.D. x 148 mm I.D., with 31,666 square mm of bearing area. The data in this paper are obtained from tests run on a test rig capable of speeds between 4,000 and 14,000 rpm and bearing specific loads between .7 MPa and 4.14 MPa. Data from over 100 sensors were recorded and analyzed. A range of operating conditions was tested including variations in load, speed, oil supply flow rate, and pad material.

The LEG design is a directed lube method, incorporating an oil feed groove at the leading edge of the pads which supplies cool oil directly into the hydrodynamic film. Since oil is effectively supplied directly to where it is needed, less oil flow is required. This serves two benefits: a smaller lubrication system and reduced frictional losses in the bearing. Although LEG technology has been utilized for 15 years in equalizing bearings, results have not been presented using it in the thin, non-equalizing style.

Designing a bearing for one direction of rotation greatly improves the bearing as compared to the one size fits all condition. Offset pivots are an example of this rational. It has been shown by many authors over the years that offset pivots significantly improve the thrust bearing's performance with respect to film temperatures and load carrying capacities. Also since most bearings today are instrumented with temperature sensors at the 75/75 position, bi-directional bearings are inevitablyl used primarily in one direction.

Another feature of this new bearing which has been found to improve performance is its spherical pivot which offers three advantages. First, it allows the pads to find the proper tilt for balanced forces within the oil film. Second, it allows for misalignment and cannot be edge loaded. Third, it provides some elastic compliance at the pivot which helps equalizes the load between the pads.

Line contact pivots have been thought to prevent pad distortion due to the pad being in contact across the width on the pad. However, since most of the pad distortion in high speed applications is thermal, the line contact strip is ineffective against this type of crowning.

The following graphs compare the new LEG bearing to the flooded thrust bearings operating at 2.76 MPa.

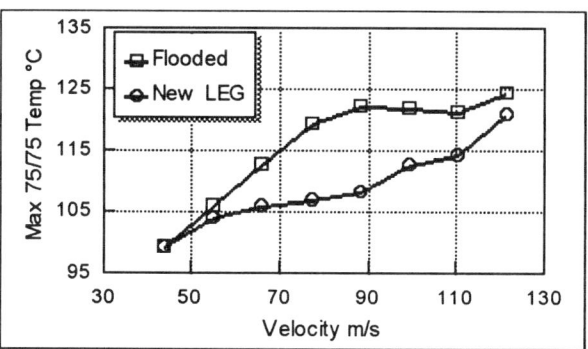

Fig. 1 Maximum measured 75/75 pad temperature versus speed at 2.76 MPa unit load.

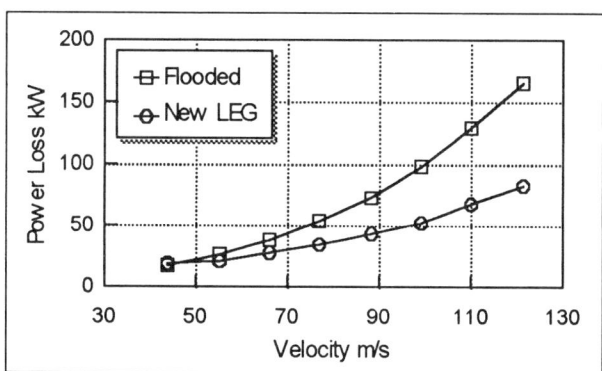

Fig. 2 Frictional power loss versus speed for double element bearings at 2.76 MPa unit load.

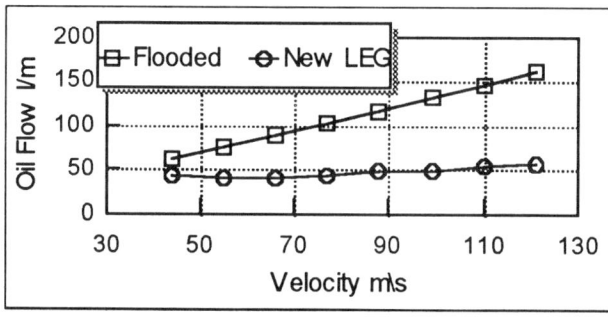

Fig. 3 Oil flow versus speed for loaded side bearing at 2.76 MPa.

A FRAMEWORK FOR THERMOHYDRODYNAMIC LUBRICATION

W. R. D WILSON.
Center for Surface Engineering and Tribology, Northwestern University, Evanston, IL 60208-3111, USA.

ABSTRACT

Thermohydrodynamic lubrication (THL) models treat the temperature rise due to viscous dissipation within the lubricant film which generally reduces the lubricant viscosity and consequently the hydrodynamic load capacity of the lubrication system. The problem is complicated by the multiplicity of heat transfer paths from the site of dissipation to the surroundings. This has impeded a general understanding of THL phenomena.

The method of lumped parameters is applied to a simple slider bearing (as an example) to explore THL. This approach leads to a better understanding of the interplay between the different modes of heat transfer and the influence of the associated temperature rises on performance and how this can be captured by the use of appropriate non-dimensional measures.

In a slider bearing, heat generated by viscous dissipation is carried away by the lubricant (by convection) or passes through the lubricant film (mainly by conduction) into the pad and runner surfaces. Heat entering the pad surface is mainly conducted to the pad backing. Heat enters the runner by conduction and is carried away by convection. Each heat transfer path can have a thermal resistance associated with it. Thus, the complete "lumped" system can be represented by a network of thermal "resistances" carrying a total "current" equal to the viscous dissipation. The relative importance of different heat transfer paths can be estimated by comparing their resistances. Typical temperatures, (or "voltages"), can be calculated using the methods of circuit analysis.

Appropriate non-dimensional measures for the THL problem can be constructed by multiplying the various thermal resistances by the viscous power dissipation and the temperature coefficient of viscosity. Two sets of such measures are useful; one using the film thickness and the other using the load support as parameters. These measures must be combined with non-dimensional measures of the various "sink" temperatures (inlet lubricant temperature, inlet runner temperature and pad backing temperature) to provide a complete set of non-dimensional measures for the problem.

Solutions to the THL problem for a particular geometry can be presented as thermal correction factors given as a function of the measures described above. Two types of thermal correction factor are useful in design, since thermal corrections may be applied to either the load or film thickness.

Figure 1 is shows the results of the lumped variable solution for the infinitely wide plane slider bearing neglecting heat transfer in the solid surfaces. The results are presented as contours of non-dimensional film thickness (or thermal correction factor) H* defined by

$$H^* = h / h_{iso}$$

where h is a typical film thickness and h_{iso} is the isothermal film thickness under the same operating conditions. The contours are plotted on the plane of two thermal loading parameters $M_{\ell v}$ and $M_{\ell d}$. $M_{\ell v}$ is a related to convection in the lubricant while $M_{\ell d}$ is related to conduction. They are defined by

$$M_{\ell v} = \frac{\alpha \bar{p}}{\rho_\ell c_\ell}$$

and

$$M_{\ell d} = \frac{\alpha \mu_0 U^2}{k_\ell}$$

respectively, where $\bar{p}$ is the average bearing pressure, U is the sliding speed and $\alpha, \rho_\ell, c_\ell, \mu_0$ and k_ℓ are the lubricant temperature coefficient of viscosity, base viscosity, density, specific heat and conductivity respectively. The $M_{\ell v}, M_{\ell d}$ plane in Figure has been partitioned to show the different regimes of THL.

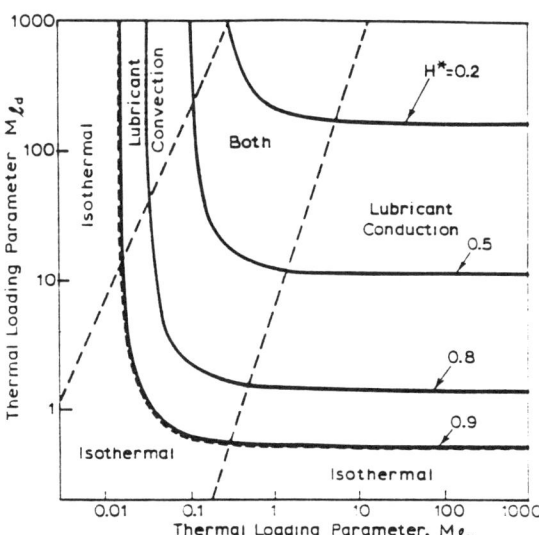

Fig. 1: Regimes of THL for slider bearing

The influence of heat transfer in the solid surfaces and the various sink temperatures can be represented by adding additional thermal loading parameters. The extension of the methodology to other cases is also discussed.

NUMERICAL CALCULATIONS OF RIVLIN ERICKSEN OIL FLOW AND OPERATING PARAMETERS FOR SHORT JOURNAL BEARING

D. WISSUSSEK and K. WIERZCHOLSKI
Institute of Design Engineering University of Essen, Germany
Institute of Applied Mechanics University of Szczecin, Poland

ABSTRACT

The paper shows the method of determination of the operating parameters such as friction forces, load capacities and friction coefficients in the short slide journal bearings for non-Newtonian Rivlin Ericksen lubricant.

We are considering the laminary, stationary, isotermic and incompressible lubricant flow without inertia forces i.e. convection terms are neglected.
The bearing operating parameters of short journal bearings for Newtonian classical oil are well known(1). Now we have to determine the correction values of friction forces, Sommerfeld Numbers, and friction coefficients caused by the Rivlin Ericksen oil properties in short bearings (2),(3).

In this paper an attempt is made to find the method of numerical analysis which is performed for analytical solutions of determination of the differences between bearing operating parameters (capacity values, friction forces, friction coefficients) for short journal bearings and non-Newtonian oil with impurities and corresponding operating parameters for Newtonian oil without impurities.

Values subjected to the numerical simulations are as follows (2),(3):
- Friction forces:

$$F_{R\Sigma} = \frac{bR^2 \omega \eta_f}{\varepsilon}\left(F_{R1} + A_\alpha^* \Delta F_{R1}\right)$$

- friction coefficient:

$$\left(\frac{\mu}{\psi}\right)_\Sigma = \frac{\mu}{\psi} + A_\alpha^* \Delta\left(\frac{\mu}{\psi}\right)$$

- Sommerfeld numbers:

$$S_{o\Sigma} = S_o + A_\alpha^* \Delta S_o$$

We are using following notations:
ω - angular journal velocity, R-radius of the journal, ε - radial clearance, ψ - relative radial clearance=ε/R, η-oil dynamic viscosity, 2b - bearing length, A_α^* -dimensionless small parameter, μ - dimensionless friction coefficients. Moreover we have: F_{R1}, μ/ψ, S_o - classical values for Newtonian oil and ΔF_{R1}, $\Delta(\mu/\psi)$, ΔS_o - correction values caused by the Rivlin Ericksen non-Newtonian oil properties. Some calculated friction coefficient and its corrections versus dimensionless bearing lengths L_1=b/R and relative eccentricities λ are presented in Fig.1. and Fig.2.
Main conclusions obtained in this paper are as follows:
➢ If bearing dimensionless length L_1=b/R decreases from L_1=1/4 to L_1=1/16 then increases (decreases) friction coefficient (friction force) and its changes or corrections for constant relative eccentricity.

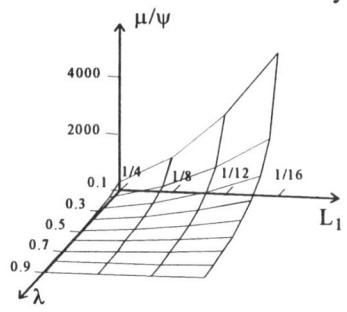

Fig.1: Three dimensional illustration of classical dimensionless friction coefficient

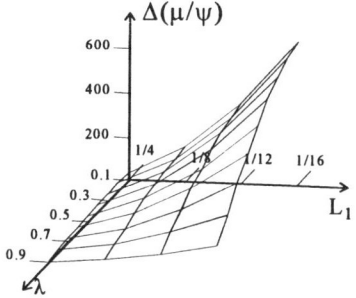

Fig.2: Non-Newtonian dimensionless corrections of friction coefficient

➢ If eccentricity ratio increases from 0.1 to 0.9, then decreases (increases) friction coefficient (friction force) and its changes for constant bearing length.

➢ If dimensionless length of short journal bearing decreases form L_1=1/4 to L_1=1/16, then dimensionless positive Sommerfeld S_o numbers for Newtonian oil and negative Sommerfeld Number corrections ΔS_o are decreasing for constant relative eccentricity λ.

➢ If dynamic viscosity increases (decreases), then total friction coefficient increases (decreases) for fixed bearing length value in range from L_1=1/4 to L_1=1/16.

➢ If dynamic viscosity increases (decreases), then total friction force increases (decreases) for fixed length L_1.

REFERENCES
(1) W. Steinhilper, R. Röper, Springer Verlag, Berlin N.Y, 1994.
(2) K. Wierzcholski, D. Wissussek, A. Miszczak, System Modelling Control, 8, Vol.2, page 394-399, 1995
(3) K. Wierzcholski, D. Wissussek, Tribologia 1, (145) page 11-24, 1996.

Step Response Characteristics of Hydrostatic Journal Bearings with a Self-Controlled Restrictor Employing a Floating Disk

S.YOSHIMOTO
Dept. of Mech. Eng., Science University of Tokyo, 1-3 Kagurazaka Shinjuku-ku Tokyo 162 JAPAN
K.KIKUCHI
Olympus Co. Ltd. 2951 Ishikawa-cho Hachioji-Ciity Tokyo 192 JAPAN

ABSTRACT

This paper describes the step response characteristics of hydrostatic journal bearings with a self-controlled restrictor employing a floating disk. This type of bearing can achieve very high static stiffness by controlling the mass flow rate of the fluid entering the bearing clearance using a floating disk. Many design parameters affect the step response characteristics of the proposed bearing. Therefore, influences of each design parameter on the step response characteristics are theoretically investigated in this paper. Furthermore, the theoretical results are compared with the experimental results in order to verify the theoretical predictions. It is consequently found that the proposed bearing consistently shows a stable step response irrespective of the step-load directions.

Figure 1 shows the schematic configuration of the proposed journal bearing. The proposed bearing is divided into four pads. Two of them have a capillary restrictor and the self-controlled restrictor is installed in another two pads. In Fig.2, the detailed drawing of a self-controlled restrictor is illustrated and the oil flow is also described.

Figure 3 shows the static characteristics of the proposed bearing when the static load is imposed in the y_1 direction. It is clearly seen that the proposed bearing can achieve very high static stiffness. In Fig.4, the step response characteristics are shown when the direction of the step load is varied. Three curves are almost the same irrespective of the step load direction though the proposed bearing is asymmetric with respect to the bearing center.

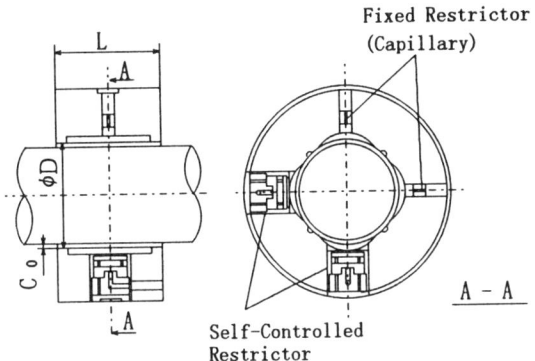

Fig. 1: The proposed journal bearing with a self-controoed restrictor

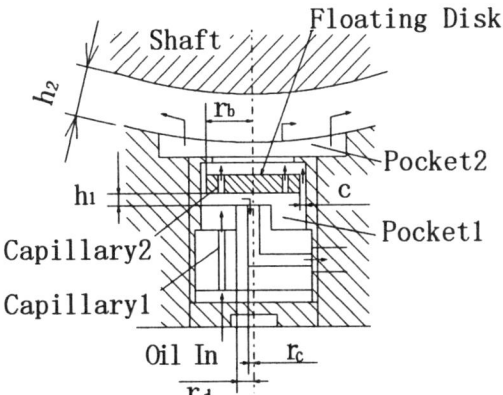

Fig. 2 : Detailed drawing of a self-controlled restrictor and flow pattern of operating fluid

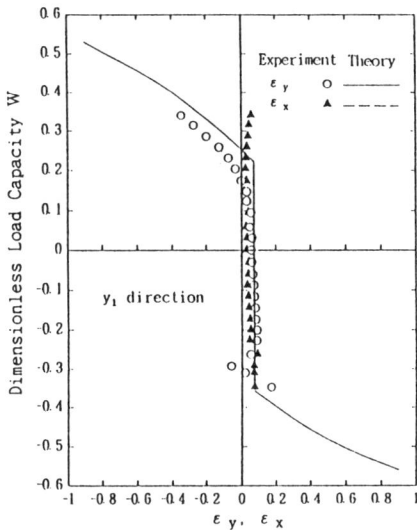

Fig. 3 : Static characteristics of the proposed bearing

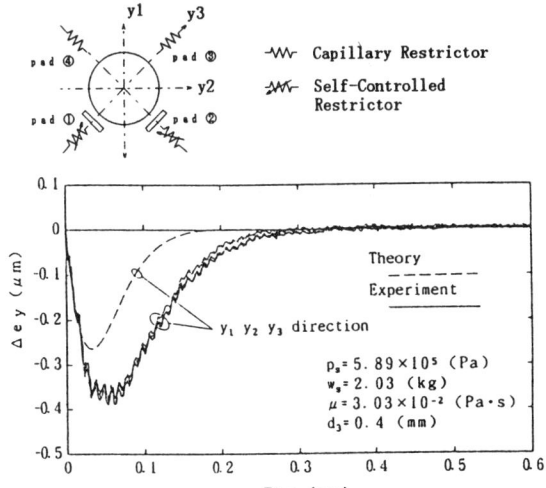

Fig. 4 : Step response characteristics (effect of the step load direction)

DYNAMIC PERFORMANCE OF SELF-COMPENSATED AEROSTATIC BEARING WITH TIME DELAY

BI ZHANG, YUNQUAN SUN and ZHENQI ZHU
Precision Manufacturing Center, University of Connecticut, Storrs, CT 06269, USA

ABSTRACT

Although time delay always exists in self-compensated aerostatic bearings, its effect has been ignored or under estimated in bearing design and performance evaluation. In this paper, time delay is considered in a newly established dynamic model for a self-compensated aerostatic bearing to achieve infinite stiffness, and verified by experiments. The model can be used to predict the static and dynamic responses of self-compensated aerostatic bearings. Sensors and actuators with a high resolution and acurracy are used in testing the dynamic responses of the bearing and the results show a good repeatability. The method used and results obtained in this study are beneficial to the understanding of the operation principles of various self-compensated aerostatic bearings.

The total time delay incurred in the whole process is associated with moving masses and transmitting forces through pressurized air films, and be considered as a single delay in the system model for simplicity. Such a consideration is often used to simplify a complicated system. In this study, a single pure time delay is considered in the feedback loop of the bearing system. The static and dynamic responses of the bearing can be obtained with a fixed time delay constant using Nyquist plots as shown in Fig. 1.

$$\frac{X_1(s)}{F(s)} = \frac{G(s)}{1 + G(s)H(s)e^{-sh}}$$

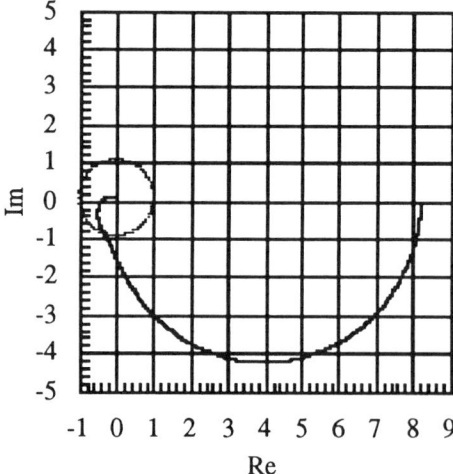

Fig. 1 Nyquist plot of the open loop transfer function

The static and dynamic responses of a self-compensated aerostatic bearing are investigated both analytically and experimentally in terms of step, impulse, and frequency responses. An impulse, sinusoidal and step inputs were applied to the thrust plate to investigate the static and dynamic responses of the self-compensated aerostatic bearing. The impulse input was used to obtain power spectrum, magnitude frequency response and phase frequency response of the bearing. The power spectrum was used to decide the natural frequency, while the phase frequency response provided more information on the effect of time delay, the time delay constant, and the phase shift of the bearing. The step responses were used to test the conditions under which positive, infinite and negative stiffnesses were observed. The sinusoidal responses were used to confirm the natural frequency of the bearing and to observe the effects of self-compensation closely.

An analytical model is established for a self-compensated aerostatic bearing with a time delay. Experimental results demonstrated that time delay exists in the bearing although it is as small as in milliseconds. The results presented should particularly be useful in understanding the underlying mechanisms of self-compensated aerostatic bearings to achieve an infinite or negative stiffness, and in comprehending the dynamics of these bearings under different operating conditions as well as in designing such bearings. As a result of this study, the following points can be made:

a. Time delay is identified in a self-compensated aerostatic bearing through the impulse response of the bearing system. The time delay constant is estimated to be in the range of milliseconds, which agrees to that predicted by the analytical model.

b. Positive, infinite, and negative stiffnesses are predicted by the model and verified by the experiments through step and sinusoidal responses. The effect of time delay on the bearing responses at a low frequency is not as strong as at a high frequency.

c. The phase difference between the thrust plate and regulating ring determines the bearing responses to an external load. Self-compensation reaches the maximum effect when phase difference is 180°, and the minimum effect when 0° at which the bearing becomes resonant.

REFERENCES

1. Mizumoto, H., T. Matsubara, N. Hata, and M. Usui. "Zero-Compliance Aerostatic Bearing for an Ultra-Precision Machine." Precision Engineering Vol. 12 (No. 2 April 1990): 75-80.
2. Tully, N. "Static and Dynamic Performance of an Infinite Stiffness Hydrostatic Thrust Bearing." Journal of Lubrication Technology, Transactions of the ASME (1 1977): 106-112.
3. Zhang, B. and Y. Q. Sun. "Principle of Self-Compensation for Infinite Stiffness." submitted to Journal of Tribology, Transactions of the ASME, 1996.

FINITE ELEMENT ANALYSIS OF HERRINGBONE GROOVE JOURNAL BEARINGS: A PARAMETRIC STUDY*

NICOLE ZIRKELBACK and LUIS SAN ANDRÉS
Department of Mechanical Engineering, Texas A&M University, College Station, Texas 77843-3123

ABSTRACT

Presently, the herringbone groove journal bearing (HGJB) has important applications in miniature rotating machines such as those found in the computer information storage industry. Grooves scribed on either the rotating or stationary member of the bearing pump the lubricating fluid inward thus generating support stiffness and improving its dynamic stability when operating concentrically. The narrow groove theory (NGT), traditionally adopted to model the concentric operation of these bearings, is limited to bearings with a large number of grooves. A finite element analysis is introduced for prediction of the static and rotordynamic forced response in HGJBs with finite numbers of grooves.

Results from this FEM analysis are compared to available experimental data as well as to estimates from the NGT (1) in Figure 1. As shown, the path of the journal center, illustrated by the journal attitude angle (ϕ), is orthogonal to the applied load direction at low journal eccentricities indicating cross-coupling due to hydrodynamic effects; while at higher eccentricities, the journal center moves toward the direction of the load. The NGT predicts well at low eccentricities, with differences between the predicted theory and the test data increasing with eccentricity. The present method (cavitated case) accurately predicts the attitude angle throughout the range of journal eccentricities.

Many rotating machines using HGJBs have vertical shafts and/or light loads, and hence, rotordynamic stability concerns emphasize the importance of predictions at the concentric position. To determine the optimal design characteristics, a bearing geometry parametric study is conducted to determine optimum rotordynamic force coefficients for a bearing running concentrically with 20 grooves. The results conducted confirm the optimum geometry given by NGT studies.

The temporal variation of the bearing reaction forces and force coefficients are of importance in HGJBs with a small number of grooves. As the grooved journal rotates at a fixed eccentricity, the film thickness changes repeatedly from the ridge clearance to the groove clearance. These periodic force variations, most prominent for bearings operating at large journal eccentricities, can possibly induce rotordynamic parametric excitations. Figure 2 shows the HGJB radial ($-F_X$) and tangential (F_Y) forces for increasing journal centricities (ε_X). The HGJB with a small number of grooves has a lower load capacity than the bearing with a large number of grooves at small journal eccentricities (ε). This trend is reversed for large journal eccentricities, and where the bearing reaction forces show large periodical variations with a frequency equal to the number of grooves times the journal rotational speed.

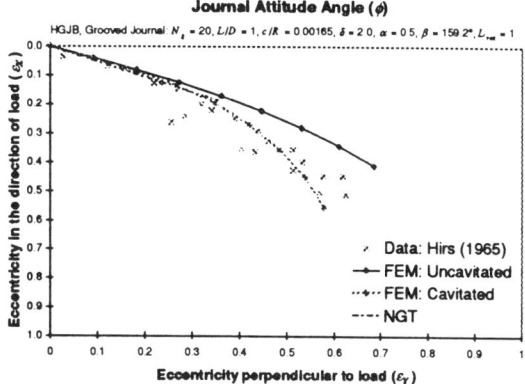

Fig. 1: Comparisons of the present FEM with experiments and the NGT (1).

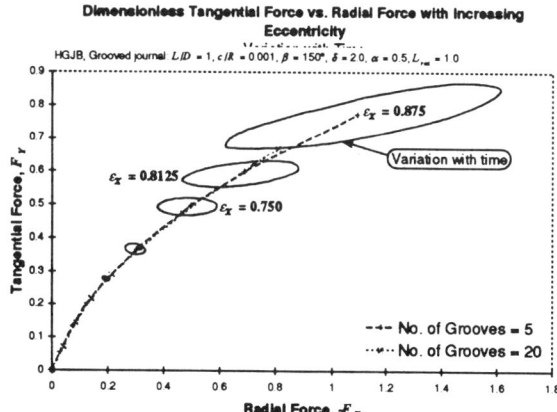

Fig. 2: Time variations of the dimensionless tangential and radial forces with increasing eccentricities.

REFERENCES

(1) Hirs, G. G., 1965, "The Load Capacity and Stability Characteristics of Hydrodynamic Grooved Journal Bearings," *ASLE Transactions*, Vol. 8, pp. 296 - 305.

*Accepted for publication in the ASME *J. of Tribology*.

INFLUENCE OF MICROBIAL CONTAMINATION ON WATER SOLUBLE METALWORKING FLUIDS PERFORMANCES

MARGARETA BALULESCU and JEAN-MICHEL HERDAN
ICERP S.A.-LUBRICANTS & ADDITIVES - Ploiesti, Romania

INTRODUCTION

Metalworking processes require different types of fluids depending on work condition and final product characteristics. Water based metalworking fluids are used in turning, milling, drilling, grinding. Considering the mineral oil content, there are three classes of soluble metalworking fluids: conventional (>40% mineral oil), semi-synthetic (<40% mineral oil) and synthetic fluids. In the formulation of a MWF there are additives that ensure special properties, such as extreme-pressure additives, corrosion inhibitors (amine, amine derivatives and boron compounds), emulsifying agents (ethoxylated products, petroleum or synthetic sulfonates), sometimes biocides.

BIODEGRADATION OF METALWORKING FLUIDS

Microbial contamination is a natural occurring process. Microorganisms come in the emulsion with water, contaminated concentrate or from the air. As the emulsion is usually 95% water, the temperature 35°C, and the fluid is a good nutrient for them, bacteria and fungi thrive in these media and degradation of the fluid occurs. There are many problems associated with microbial growth in emulsions: corrosion, emulsion instability, health and environment risks, high maintenance costs. A special issue with the behavior of emulsions in use is the nitrosamine formation. Nitrosamines are of high risk for human health being cancer inducers. They can appear in a bacterial contaminated emulsion in the presence of nitrites and secondary amines. Hydrogen sulfide is another product, released from sulfate-reducing bacteria. This compound gives so called 'Monday morning smell' in workshop, a smell of rotten eggs and is corrosive, irritating and highly toxic for human.

OXYMETRY

We studied different new formulations and additives alone behavior under microbial attack. Biodegradability of additives and fluids was tested by oxygen uptake measurement. In these tests we used bacterial inoculum isolated from contaminated fluids. Among the additives we tested, amine based corrosion inhibitors (CI) had the highest biodegradation rate, followed by emulsifying agents (EA), EP-additives (EP) and mineral oil (MO).

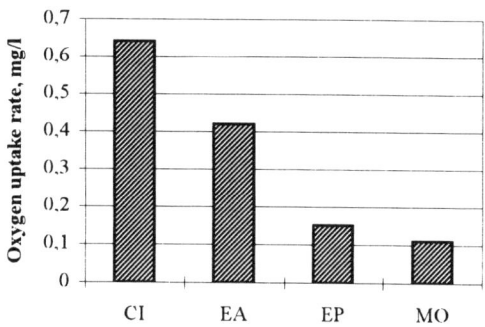

Fig.1: Oximetry of mwf components

LABORATORY TESTS

These preliminary results confirmed the behavior of fluids in microbial challenge tests and field trials. In a long term test, emulsions were subjected to frequent inoculum addition, and the properties and performances were monitored. Conventional fluids were biodegraded in the absence of biocides and lost their anti-corrosion and emulsion stability properties.

Semi-synthetic and synthetic fluids were more stable to microbial attack. Vegetable oil based fluids were easily biodegraded in a very short time and needed biocide as tank side addition in order to be used in an workshop.

FIELD TRIAL

Field trials with different types of fluids were consistent with laboratory results. When biocide was not used for a mineral oil containing fluid, biodegradation occurred in two weeks. The main problems related to this process were drop in pH, corrosion, oil separation.

Semi-synthetic and synthetic fluids were functional at high microbial levels, but major problems as worker safety and production stop due to pipes and filter blockage with molds could not be avoided.

Vegetable oil based fluids were tested only with biocide addition and microbial growth was kept under control.

In conclusion we consider that microbial contamination is a potential risk for fluid performances and a good behavior in use is strongly related to microbial control in the emulsion.

Submitted to *Industrial Lubrication and Tribology*

BREAKDOWN CHARACTERISTICS OF LUBRICATING OIL UNDER THE INFLUENCE OF VOLTAGE

S.BISWAS and S.K.BHAVE
Failure Analysis Group, Corp, R&D Divn., BHEL,Hyderabad - India
S.K.BISWAS
Dept of Mech Engg, Indian Institute of Science, Bangalore, India

ABSTRACT

Rolling element bearings employed in AC motors are prone damage by shaft current. (1,2) The damage occurs due to metal removal by pitting of the rolling elements. Subsequently, indentation of the balls/rollers on races produces flutes or corrugations. Rolling element bearings used in AC induction motors are known to fail in the above manner when subjected to certain environment of electric fields and dynamic forces. Detailed understanding of the breakdown of the EHD films under electric stresses is not available in literature. An attempt has been in this paper in that direction.

In order to understand the breakdown characteristics of oil film in a roller bearing, in rotating condition under the application of AC and DC voltages, experiments were conducted using a test roller bearing on a test rig. The wave shapes of bearing voltages as observed in oscilloscope were studied under different test conditions. Relationship between the applied voltage and current flowing through the bearing were also observed under different loads and speed of rotation. The experiments were designed to simulate conditions of AC motor roller bearings in service.

It was observed that although oil film could withstand higher DC voltages than AC voltages, the similarity in behaviour under both types of voltages were unmistakable.

At low impressed voltages, there was no significant flow of current. The disturbances in wave shapes was found to be linked to the flow of current. The plots of current against impressed voltage established that the initiation and the magnitude of flow were dependent on the film thickness and also on the type of voltage. The lubricating film in the roller bearing under running condition behaved in a complex combination of capacitive and resistive nature when subjected to an electric field. A limiting voltage could be detected through current increased in an approximately exponential fashion. This paper describes in detail the above characteristics and their interpretation which lead to better understanding of the problem of electrical pitting in rolling element bearings.

REFERENCES

1. F.F.Simpson et al, Paper 27, Lubrication and Wear Conf. 1993, IMechE 299-304
2. S.Andreason. Passage of current through rolling bearings,Ball Bearing Journal,153, 6-12, 1968.

EFFECT OF LUBRICANT ADDITIVES ON OPERATION OF MACHINE FRICTION PAIRS

JAN BURCAN

Institute of Machine Design, Technical University of Lodz, Poland

ABSTRACT

The effective way of solving the problem of lubricating a small-size bearing is the use of unconventional lubricants or special additives, such as boron nitride added to typical oils increasing lubricity. In the research a 3 per cent mixture of the "Merck" boron nitride with various degree of granulation (BN1, BN2, BN3) and base oil SAE 10/90 was used as well as special gear-hydraulic oils (49/97, and Tszp8) containing additives increasing their lubricating abilities. The results were then compared with the characteristics obtained from research on a bearing working under spinning friction and lubricated with oil mixed with magnetically active substances - Fig. 1.

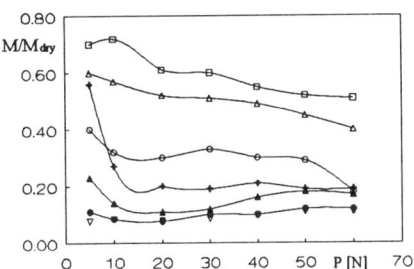

Fig. 1. The change of relative moment of friction in relation to the moment of friction without lubricating as function of load

□ ŁT4S3 greas	○ BN-1	▲ BN-2
△ mag. active greas	+ BN-3	● 49/97
	▽ Tszp8	

The moment of friction was recorded on a stand continuously recording instantaneous values of the moment of friction and two others measuring the average values of the moment of friction. While the first one makes it possible to measure values of the moment of friction from 0.01 to 0.3 Nm as precisely as 0.0002 Nm, the latter measure the values ranging from 0.05 to 0.3 Nm, with accuracy to 0.01 Nm. The research was conducted on plates made of steel, both raw and hardened, and of bronze, sinter and plastic. They mate with bearing ball made of steel ŁH 15. In such frictional pairs, under the burden of the shaft, plastic strains, accompanied by smearing of a softer material, local melting and intensive wear occur. These also occur if there is enough oil with no additives in a bearing. However, adding substances increasing lubricity, like boron nitride, considerably decreases the disadvantages. Motion resistance and wear are reduced, motion becomes more stabile, i.e. oscillations of relative moment of friction are reduced.

Not only do the quantitative changes occur but so do qualitative ones, which is very well illustrated by the change of relative moment of friction in relation to the moment of friction without lubricating (M/M_{dry}) as function of load - figure 1.

Figure 2 shows the viscosity characteristics of the lubricants used in the research.

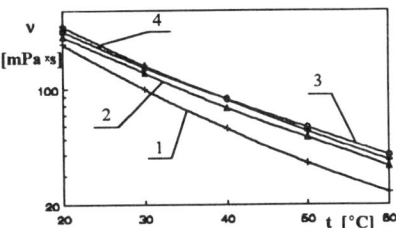

Fig. 2. The viscosity characteristics of the lubricants used in the research: 1 - motor oil 40, 2 - the mixture of motor oil and a magnetically active lubricant, in relation 1:1, before magnetisation, 3 - the same mixture immediately after magnetisation, 4 - the same mixture a week after magnetisation

The analysis leads to the conclusion that it is possible to obtain a magnetically active lubricant whose lubricating capabilities are very similar to those of nodes in living organisms in which synovia has the properties of the non-Newtonian liquid (1). It manifests itself in the change of properties depending on the load. In ordinary conditions the viscosity of the synovial liquid is greater, by the order of magnitude, than the viscosity of water and it changes with the shearing velocity from 0,01 up to 100 pois (with the decrease of shearing velocity the viscosity of synovia increases and the lower the velocity the more distinct the effect described). The increase in the viscosity of the synovial liquid occurs also during jogging, jumping, overload or overweight. Besides, the viscosity of synovia drops with the increase in the temperature of the body.

The analysis of the results obtained for oil with boron nitride additives and the one without additives, magnetically active lubricant and oil with other than boron nitride additives show a significant reduction of the friction coefficient and contact temperature.

REFERENCES

(1). Jan Burcan, Magnetically active lubricant - capabilities, prospects, Scientific Bulletins of Łódź Technical University, No 765 vol. 85, 1996, pp 23-30.

EFFECT OF TEMPERATURE, SUBSTRATE TYPE, ADDITIVE AND HUMIDITY ON THE BOUNDARY LUBRICATION IN A LINEAR PERFLUOROPOLYALKYLETHER FLUID

BÜLENT ÇAVDAR
Advanced Technology Center, Borg Warner Automotive, Lombard, Illinois, USA

ABSTRACT

The molecular structures of fluorinated lubricants and hydrocarbon lubricants are similar except that the hydrogen-carbon bonds are replaced by the much stronger fluorine-carbon bonds in fluorinated lubricants. The strong fluorine-carbon bonds give higher thermal and oxidative stability, chemical inertness and dielectric strength to the fluorinated lubricants over hydrocarbon lubricants. Furthermore, the high molecular weight fluorinated copolymers with ether linkages have low volatility, wide liquid temperature ranges, and good viscosity-temperature characteristics as liquid lubricants. Due to the above and other desirable properties, per-fluorinated polyalkylether lubricants (PFPAEs) find applications in spacecrafts, precision instruments, electronic, nuclear and magnetic media industries in spite of their very high production costs (1).

In boundary lubrication the chemical interactions of lubricant with the rubbing materials and surrounding atmosphere affect interface development at the contact zone. Earlier, it was shown that the organic or inorganic nature of interface films formed in hydrocarbon lubricants affects the friction and wear performance of tribological systems operating under boundary lubrication regimes (2). Formation of an inorganic film of FeF_3 in the boundary lubrication of steel surfaces in PFPAE fluid was reported by Carré (3). More recently, the dual nature of interface films, i.e. the presence of a fluorinated organic film above the inorganic FeF_3, was also shown for a linear PFPAE lubricant (4). There is a need to better understand the behavior of PFPAE fluids under boundary lubrication so that their interaction with metal surfaces may be modified by additives or by other means.

In this study, the effects of temperature, substrate steel type, an antiwear additive, and atmospheric humidity on the protective surface film formation in a linear PFPAE fluid were studied. The measured performance characteristics were coefficient of friction, area of wear scar, final surface roughness and rate of surface film formation. The high temperature (150°C) caused higher friction and wear as compared to lower temperatures (50°C and 100°C). The high friction and wear at 150°C was explained by the lack of organic adsorbate film formation over the sliding surfaces at 150°C. It was proposed that humidity reduces friction and wear by increasing the rate of formation of perfluorocarboxylate species over inorganic surface layers. A fluorinated tertiary alcohol additive reduced the friction at the first hour of sliding at 150°C. However, the friction was increased towards the end of the five hour sliding tests due to the loss of entire additive by evaporation. The rate of surface film formation and wear protection was higher on M50 steel than on M50-NIL steel because M50 steel surface reacts more with the PFPAE degradation products.

REFERENCES

(1) T. W. Del Pesco, Perfluoroalkylpolyethers, in R. L. Shubkin (ed.), Synthetic Lubricants and High-Performance Functional Fluids, M. Dekker, New York, 1993.
(2) B. Çavdar, K.C. Ludema, Wear 148, (1991) 305-361.
(3) Carré, D.J., ASLE Trans., 29, 121-125, (1986).
(4) B. Çavdar, J. Liang, P.J. John, Trib. Trans., 39(4), (1996), 779-786.

(Note: This paper has been accepted for a publication in Wear Journal in 1997).

THERMAL DECOMPOSITION OF ZINC DIALKYLDITHIOCARBAMATES.

G. CELICHOWSKI and S. PŁAZA
Department of Chemical Engineering and Environmental Protection, University of Łódź,
Pomorska 163, 90-236 Łódź, Poland

L. COMELLAS RIERA
Institut Quimic de Sarria, Via Augusta 390, 08017 Barcelona Spain

ABSTRACT

Zinc, antimony and oxothiomolybdate dialkyldithiocarbamate complexes have antiwear and antioxidation properties, antimony complexes are used also as extreme pressure agents while molybdenum ones as friction modifiers' additives.

Phosphor presents in zinc dialkyldithiophosphates which are widely used multifunctional additives disactivated platinum catalyser in engine's exhaust system. Using dithiocarbamate complexes alone or in combination with others lubricating oil additives gives hope to resolve this problem.

Our investigations are focused on zinc dialkyldithiocarbamates (ZnDTC) because this additives are the most used in lubricants and they have also lower toxic properties.

In our opinion understanding mechanism of ZnDTCs tribochemical reactions in friction contact is very important in multifunctional action of these additives. We have investigated the thermal decomposition of ZnDTCs because the heat is one of many forms of energy that we can find during friction but it is the major one, in our opinion, and the most influences on tribochemical reactions. Our investigations are focused on ZnDTC's decomposition in different conditions. Four types of experiments were made:

1. Analyse of ZnDTCs decomposition (alone and in the presence of Fe and Fe_2O_3 powders) using thermogravimetrical method.
2. Flash Vacuum Pirolysis (FVP) to investigate ZnDTC's decomposition in very short time (10^{-5}s) and in high temperature (300-1000°C). This decomposition was done in pirolyser that contained inert quartz. These temperatures, in our opinion, fulfil conditions close to that we can find in real friction in asperite top with very short time and high temperature (flash temperature). Collection in low temperatures trap products of FVP were analysed using Gas Chromatography with Mass Spectrometry detection (GC-MS). In products of FVP we found among others: alkylthiocyanate, alkenes, secondary amines, thiols and others compounds.
3. Decomposition of ZnDTCs in pirolyser directly connected with GC-MS analysing system. These experiments were done at 250, 300 and 500°C with using additive alone, additive mixed with iron and additive mixed with iron oxide powder. In these experiments we have analysed decomposition product "in situ".
4. Tribological tests with using Amsler and four ball machine with GC-MS analyse of products formed from ZnDTCs during decomposition in real friction conditions.

Comparing the results of 1-3 tests with friction test 4 give us information about way of ZnDTCs tribochemical transformation during acting of this additives.

Based on our experiments results obtained and literature data (1-3) probably mechanism of ZnDTCs action in frictional condition is proposed.

REFERENCES

(1) S. K. Sengupta, A. S. Kumar; Thermochimica Acta, Vol. 72, 349-361. 1984
(2) A. S. Kumar; Thermochimica Acta, Vol. 104, 339-372. 1986.
(3) J. O. Hill, J. P. Murray, K. C. Patij; Reviews in Inorganic Chemistry Vol 14, 363-387. 1994.

Submitted to *Tribology International*

RARE EARTH COMPOUNDS: A BRANCH OF PROMISING LUBE ADDITIVES OF 1990'S

CHEN BOSHUI, DONG JUNXIU, YE YEE and JIANG SONG

Department of Petrochemistry, Logistical Engineering College, Chongqing 630042, China

ABSTRACT

Rare earth compounds, because of their peculiar physical and chemical behaviors, have been widely employed in metallurgy, electric industry, petrochemical industry, space industry etal, even in medicine and agriculture with increasing improvment of the preparation, separation and purification techniques of rare earth elements and compounds since 1960's. Although many characteristic natures are still unknown, the unique performances of rare earth compounds have attracted scientists of various fields. More extensive investigations and applications of rare earth compounds have under the circumstances spured since 1980's. In recent years, the interests in tribological applications of rare earth compounds, especially lanthanides and lanthanas, in the fields of wear-resistant metals, ceramics and macromolecular materials, seems to be increasing with the increase of determination. But unfortunately, the use of rare earth compounds, especially oil soluble ones, as lube additives is still in its infancy. Jost[1] denoted that LaF_3 increases the service life of bonded coatings by 2-4 times and the load carrying capacity of lubricating greases and pastes by 10-100%. Segaud[2-3] reported that some lanthanide holide dispersions enhance lubricity of lubricating oils and aqueous solutions significantly. Some other publications [4-5] also indicate that many rare earth compounds provide excellent antiwear and friction-reducing abilities. Rare earth compounds are powerful and potential lube additives and have become one of the most interested research topics for tribochemists.

Rare earth compounds as antiwear and friction-reducing additives, in the fields of wear-resistant metals, ceramics and lubricating oils and greases are reviewed in the present paper. A series of oil soluble rare earth compounds, e.g. rare earth dialkyldithiophosphates (REDDP), were prepared. The structures of REDDP were characterized and their basic chemical compositions are shown below:

$$RE[S(S)P(OR)_2]_3$$

where RE=La, Pr, Sm, Eu and Gd;
R=iso-octyl

Tribological performances of REDDP were evaluated and compared with ZDDP. REDDP and ZDDP as lubricating oil additives was blended into a base oil in the proportion of 1.0% respectively. The tribological results are shown in table 1. Each datum in this table was repeated three times and the average values were reported.

Table. Tribological results of REDDP and ZDDP

	LaDDP	PrDDP	SmDDP	EuDDP	GdDDP	ZDDP
P_B, N	1117.2	980.0	1185.8	1048.6	1048.6	823.2
D, mm	0.684	0.714	0.763	0.857	0.758	0.865
μ	0.065	0.054	0.051	0.067	0.057	0.078

Notes: P_B--maximum nonseizure load; D--wear scar diameter; μ -- average friction coefficient. D and μ were tested under 588N for 30 minutes.

Table 1 indicates that REDDP provides excellent antiwear and friction-reducing capacities, better than ZDDP. The excellent tribological performances of REDDP have been testified in the industrial applications preliminarily. The industrial test results also show that REDDP exhibits good anti-oxidation ability. REDDP is indeed a promising and attractive multifunctional lube additive.

REFERENCES
(1) H. P. Jost, Industrial Lubrication and Tribology, Vol.44, No.2, 1992.
(2) C. Segaud, US Patent 4946607, 1990.
(3) C. Segaud, US Patent 4946608, 1990.
(4) B. S. Chen, Wear, Vol.196, No.4, 1996.
(5) B. S. Chen, Chinese Journal of Tribology, Vol.22, No.4, 1994.

QUANTITATIVE ESTIMATION AND PREDICTION OF TRIBOLOGICAL PERFORMANCE OF PURE ADDITIVE COMPOUNDS THROUGH COMPUTER MODELLING

G. S. CHOLAKOV, K. G. STANULOV, P. DEVENSKI and H. A. IONTCHEV,
University of Chemical Technology and Metallurgy, boul. Kl. Ohridsky N 8, Sofia 1756, Bulgaria

ABSTRACT

Recently new opportunities for numerical description of chemical structure and molecular design have emerged (1). The present work describes an algorithm for computer modelling of tribological performance of lubricant additives. It is illustrated with a practical example - mathematical description and modelling of published experimental data for 78 four ball machine weld loads (IP 239) of aliphatic and aromatic organic sulphides at different concentrations in 6 different base oils.

The weld load, **WL**, achieved with lubricants, containing organic sulphides, is assumed to be contributed by the weld load of the base oil, **WLO**, the properties of the film, formed by the additive, which depends on its chemical structure, presented by molecular descriptors, Md_f, and - the rate of a chemical reaction, **r**, controlling the film formation:

$$WL = K1*WLO + K2*Md_f + K3*r$$

The kinetic equation for modelling of **r** is derived from a simplified presentation of the action of sulphide additives in the determination of the weld load, as controlled by the adsorption and decomposition of the sulphide to release sulphur (2):

```
R-CH2  H2C-R              2 R-C°-H2 +
  |      |
  S------S          →
  :      :                   2 S⁻ ↓
------------------       ------------------
    Metal                      Metal
   Sorption                   Reaction
```

The reaction rate, **r**, in the above case is given by:

$$r = k_r * b_A * C_A / (1 + b_A * C_A)$$

where C_A is the bulk concentration of the additive, k_r - the rate constant ($k_r = k_0 * \exp(-E_a/RT)$), b_A is the adsorption coefficient ($b_A = b_0 * \exp(Q_A/RT)$), **Ea** is the activation energy, Q_A - the heat of adsorption of the additive on the particular metal, **R** - the universal gas constant, **T** - the temperature. It is further assumed that the energy terms in the kinetic equation are proportional to relevant descriptors of the molecular structure of the particular sulphide - MD_{Ea}, and MD_{Qa}. A phenomenological model with the weld load as the dependent variable, and **WLO**, C_A, and the molecular descriptors - as independent variables is thus obtained.

The values of the descriptors of the additive structures, **MD**, are estimated with molecular mechanics computer simulation. The most suitable descriptors are selected from parameters, characterizing the energy distribution in the minimized energy molecular models of the additives through multiple regression, applying a stepwise variable selection procedure. The constants and descriptors of the phenomenological model, which predict the experimental weld loads with a standard mean relative deviation less than 15 % of the average between experimental and calculated values, are presented in the full text of the paper. The prediction of all weld loads is better than the experimental reproducibility of the parameter (52 % according to IP 239). The selection of molecular descriptors - total molecular energy, sulphur release and torsional energies, dipole moment, dipole - charge interaction energy are also relevant to the phenomena, being described.

The full text illustrates also an opportunity for computer modelling of tribochemical performance with statistical molecular design procedures (3). A predictive model is derived from a representative sample of only 12 data sets, selected by statistical design from the matrix of the principle components of the independent variables, to represent the whole chemical class. The predictions of this model are also better than the IP 239 reproducibility of the weld load.

Finally, approaches for future development of computer modelling of additive performance for scientific and industrial purposes are discussed. The necessity for compilation of data banks with tribological properties of pure additive compounds, the role of kinetic and adsorption studies, etc. are emphasized. Connection to systematic formulation and design of additive products and modern refinery modelling concepts is also discussed.

REFERENCES

(1) A. Horwath, Molecular design, Elsevier, 1992.
(2) E. Forbes, Wear, Vol.15, Page 87, 1970.
(3) P. Geladi, M-L. Tosato. in Practical Applications of QSAR in Environmental Chemistry and Toxicology, Kluwer Acad. Publ., Dodrecht, Page 171, 1990.

Submitted to *Wear*

SYNTHESIS OF NEW ADDITIVES FOR LUBRICANT OILS AND INDUSTRIAL GREASES

V. DINOIU, D. FLORESCU and CHIRIACA STANESCU
Lubricants and Additives -ICERP, 291A Republicii bvd., RO-2000-Ploiesti, ROMANIA

ABSTRACT

The multifunctional additives play a major role in today lubricant oils and greases and many studies describe in details chemistry, synthesis and application of new compounds with antioxidant, antiwear, antirust, anticcorosive and extreme- pressure activity (1). The objective of this paper is to present the results concerning the synthesis of some additives based on hydrocarbon soluble metalic complexes of thio-bis phenols and their utility as additives for oils and greases. The organic metalic complexes are the reaction product of a hydrocarbon substituted thio-bis-phenol and a source of metal. Thio-bis phenols are produced from the reaction of phenols having nonyl or dodecyl group in para-position and a source of sulfur ,e.g.chemical sulfur or sulfur halides (2). In order to study only the influence of the nature of the metal on the EP-AW and antioxidant properties of the metalic complexes of thio-bis-phenols (MeTBPh), the metal studied were: Me = Zn, Cd, Cu, Mo, Bi .

The performance as additives was studied in a parafinic oil at a concentration of 1.5 %.

A four-ball machine was used for studying antiwear and extreme-pressure of MeTBPh. The criterion used to asses the antiwear (AW) performance of additives is the mean wear diameter (d*), i.e. the arithmetric average of the wear diameters of succesive loads of 60-80-100 and 150 kgf. For ZnTBPh, the wear scar diameter d*=2,67. The extreme- pressure properties were determined by measuring the weld load on a four ball machine according to the DIN 5130 test method. The results of Four-ball AW and EP test are presented in Table 1:

MeTBPh	Weld Load (kgf)	Wear Scar Diameter (150 kgf,1minute) (mm)
ZnTBPh	140	2,9
CdTBPh	150	2,5
CuTBPh	140	2,75
MoTBPh	150	2,56
BiTBPh	200	3,2

It may be observed that all synthesised compounds present a good antiwear performance in parafinic oil and a low extreme - pressure activity. Only BiTBPh shows a good extreme pressure performance, having a higher value of weld load (200kgf). The paper presents, also, the influence of additives concentration on AW and EP properties of MeTBPh Antioxidant Performance was determined by Test for oxidation of oils in slight thin and by Panel Coking Test Metod. The Test for oxidation in slight thin is used for determining the detergency and antioxidant tendency of finished oils where are in contact with steel surfaces at elevated temperature in a slight thin for six hour at 315° C and determining the weight of deposites. The Panel Coking Test is used for determining the tendency of finished oils to form solid decomposition products in contact with surfaces at elevated temperatures for six hours. The results of both tests show a good antioxidant performance in parafinic oil. ZnDTPh with nonyl group in para-position, has better activity than ZnDTPh with dodecyl group in para-position. It is also clear that at 1.5 % concentration in oil, ZnDTPh showed the best antioxidant performance than the others METBPh (3) (thermo-oxidation stability = 28.2 mg).

The copper-corrosion behaviour of the synthesised products was determined by the ASTM D 130 ; this test shows that only compound with Bi is highly corrosive and the others are least corrosive towards copper.

The antioxidant performance of MeDTPh in industrial greases was determined by the Oxygen bomb method (ASTM D 942). The degree of oxidation after a given period of time is determined by the corresponding decrease in oxygen pressure. For all synthesised products, the pressure drop is less than 0.33 barr (33kpa) in Lithium and Non Soap greases. ZnDTPh shows good antioxidant propertie, comparative with conventional diphenil amine which is usually used in greases.ZnDTPh blended with sulfurised olefines in 1:1 ratio, presents good extreme-pressure activity(230 Kgf) at 3% concentration in Lithium grease.

REFERENCES

(1) E.S.Forbes, Antiwear and extreme pressure additives for lubricants. Tribology 1970, 145
(2) K. Coupland et al, U.S. Patent 4.248.720 (C 10M 1/54) 1981
(3) V.Dinoiu, D.Florescu and E.Kiss, Rom. Patent 1995

THE DESCRIPTION OF LUBRICATION ACTION OF CUTTING FLUIDS TRIBO-ACTIVE COMPONENTS

V.A., GODLEVSKI, A.V. VOLKOV, V.N, LATYSHEV L.N. MAURIN
Phys. Dept., State University Ivanovo, Ermak st, 39, 153025, Ivanovo, Russia

The chemical activity of cutting fluids (CF) now is extremely undesirable from ecological reasons. It forces to reduce the chemical-active additives concentration. Therefore it is reasonable to use the surfactants (SA), which even in small concentration are able to decrease tool-chip adhesion. It can be explained only by the phenomenon of boundary lubrication.

Literature analysis (1) and our investigations (2) show, that, as a rule, the boundary lubrication layer formation on tool-chip interface goes in some stages, which of them may be characterized by time periods: 1. liquid penetration into contact zone (τ_l); 2. liquid evaporation with (or without) thermal decomposition of components CF (τ_g); 3. chemical adsorption layer forming (τ_c); 4. physical adsorption with forming of monolayer lubrication films (τ_p); 5. forming of supramolecular (multi-layer) lubrication films (τ_m).

We shall consider the stage of physical adsorbed layer formation by SA in contact zone. The physical adsorption flows much slower than chemical one and consequently it is more limiting in general chain of lubrication effects. The time limiting these processes is the period of tool–chip contact. On Figure the model of single cylindrical capillary is represented (see also (1-3)). The capillary has one open end, faced to environment, other end is closed and moves along the tool with chip's velocity – u. The last determines the capillary lifetime τ_c.

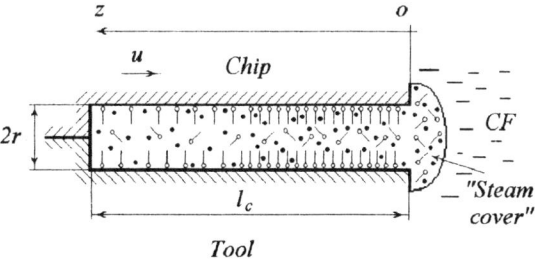

Figure: Single capillary located in tool-chip interface surrounded by evaporating SA solution: • – solvent molecules; o— – molecules of SA; r, l_c – capillary dimensions; oz – axis

Let's consider the process after finish of "microdroplet evaporation" when the gas pressure has come in equilibrium state (2). It can be shown that kinetic of SA adsorption is determined only by adsorption activity of capillary's walls. Using the law of conservation of mass and principles of non-equilibrium thermodynamics we shall receive two equations:

$$\left.\begin{array}{l}\dfrac{dn}{dt_1}=-\dfrac{d\beta}{dt_1}\\[6pt]\dfrac{d\beta}{dt_1}=-\dfrac{\tau_c}{\tau_p}(\beta-n\nu)\end{array}\right\}\quad\begin{array}{l}t_1=0:\ n=1,\\ \beta=0,\end{array} \quad (1)$$

where n – non-adsorbed SA dimensionless concentration; β – adsorbed SA dimensionless concentration; $t_1 = t/\tau_p$ – dimensionless time; ν – proportionality factor (tangent of inclination angle of the isotherm).

Solving the equation (1) for β, and updating β into SA concentration a per unit of capillary surface for the moment τ_c, we receive:

$$a(\tau_c)=\frac{rc_l l_l}{2l_c}\frac{\nu}{1+\nu}\left[1-\exp\left(\frac{(1+\nu)\tau_c}{\tau_p}\right)\right], \quad (2)$$

where c_l – SA concentration in liquid phase; l_l – liquid state penetration length.

Equation (2) shows that for increasing of concentration of SA on capillary walls (i. e. improvement of lubricant action) it is necessary to increase values: c_l, l_l/l_c, τ_c/τ_p, ν, connected with technological parameters of cutting.

REFERENCES
(1) Williams J.A., Bulletin du Cercle D'etude des Metaux, Vol. 14, 1980, 211-241 pp.
(2) Godlevski V.A., Volkov A.V., Latyshev V.N., Maurin L.N., Trenije i Iznos, Vol. 16, 1995, 938-949 pp. (In Russian).
(3) Latyshev V.N., Increasing of Cutting Fluid Efficiency. Moscow: Mashinostrojenije, 1985. (In Russian).

SELECTIVE TRANSFER OF MoDTP FILMS DURING FRICTION IN VACUUM

C. GROSSIORD, J.-M. MARTIN and Th. LE MOGNE
LTDS, UMR 5513, Ecole Centrale de Lyon, BP 163, 69131 Ecully Cedex, France
Th. PALERMO
Institut Français du Pétrole, 1-4 avenue de Bois-Préau, 92506 Rueil-Malmaison Cedex, France

INTRODUCTION

For many years, metal dithiophosphates have been used as antiwear additives in motor oils. Their action during friction is to create a tribochemical film known to be the formation of a solid transition metal phosphate glass on steel surfaces (1). Among them, molybdenum dithiophosphate (MoDTP) is recognised to have antifriction properties.

We are studying the formation mechanisms of the films and the chemical reactions induced by friction in vacuum.

EXPERIMENTAL

The MoDTP films are generated in a Cameron-Plint friction machine. At 60°C, a steel cylinder rubs on a steel flat (ASI52100) under 350N and for 1 hour in base oil containing 1% weight of MoDTP.

The tribochemical films are studied in a ultrahigh vacuum (UHV) reciprocating pin-on-flat tribometer (2). Before friction, the chemical composition of the wear surfaces are characterised by *in situ* surface analysis tools as Auger Electron Spectroscopy (AES) and X-ray Photoelectron Spectroscopy (XPS). After friction tests in vacuum, the wear scars on the steel pin and the wear tracks on the flat are analysed by AES.

AES depth profiles on transfer films are recorded to determine their thickness and composition.

RESULTS

The friction coefficient curve recorded in UHV is presented in fig.1. We can observe three stages. At the beginning and for few cycles, the friction coefficient is near 0.3 (A). Then it decreases to a very low value near 0.01 (B). Finally the film is broken and the friction coefficient rises to 0.8 (C). AES spectra are recorded at each point (A, B and C) on the wear scar of the pin and the wear track of the flat. Very low friction seems to be correlated to the presence of sulphur, molybdenum, carbon and iron (fig.2) whereas in point A and C, the transfer film contains also phosphorous and oxygen.

Depth profiles of the transfer films on the pin reveal different chemical compositions from the top surface to steel. We will discuss the results which indicate chemical transformations of initial films during friction in vacuum.

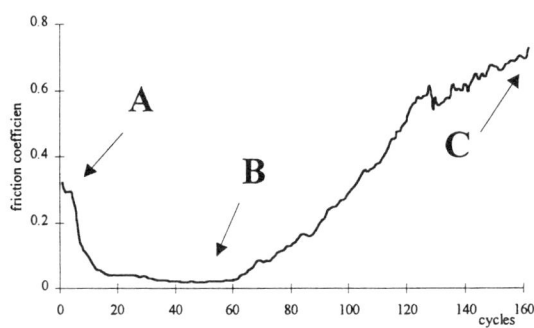

Fig.1 : Three characteristic periods in the friction coefficient in UHV.

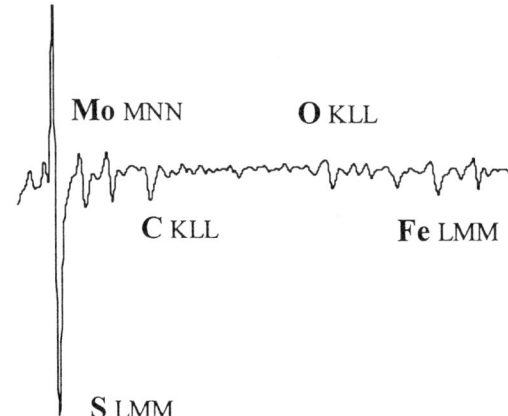

Fig.2 : AES analysis in scar B on the pin.

REFERENCES

(1) M. Belin, J.-M. Martin, J.-L. Mansot, STLE Trans, 32(1989), 410.
(2) J.-M. Martin, Th. Le Mogne, C. Grossiord and Th. Palermo, Tribology Letters, 3(1997), 87-94.

TRIBOLOGY OF METALWORKING FLUIDS - BENCH TEST AND FIELD TRIAL

J.M. HERDAN
ICERP S.A. Lubricants & Additives, 291A Republicii Bd., 2000 Ploiesti Romania
A. PALUSAN
RULMENTUL S.A. Brasov Romania

INTRODUCTION

From a strictly tribological point of view metalworking fluids (MWF) have to reduce the friction and wear between the tool and metal and to facilitate the penetration of the tool edge in the superficial metallic layer. Beside this function they have other important tasks like removal of heat produced by external and internal friction, maintaining of cleanliness of the tooling area by fast transporting of swarf, and protection of metal, tools and machines against corrosion. Two classes of MWF are commonly used in metal cutting and forming: neat oils and soluble oils. Both neat and soluble oils contain antiwear and extreme-pressure additives (E.P.), and some of them also contain friction modifiers. These additives make MWF able to accomplish their tribological function mentioned above.

ANTIWEAR AND E.P. ADDITIVES

Many investigations have been made to better understand the action mechanism of antiwear and E.P. additives. It is generally accepted that antiwear additives for adsorbed layers on the metal surface. These layers reduce friction and wear between the metallic parts in reciprocal movement. E.P. additives act in conditions of boundary lubrication, when the lubricant film is formed and destroyed continuously, and metal-to-metal contacts arise. In this situation the additive is decomposed due to the local high pressure and temperature, the decomposition products react with metallic surface and form chemically bonded layers, acting as solid lubricants. The efficacy of antiwear and E.P. additives depends therefore on the reactivity of the chemicals towards the metal. Three different AW/EP additives where considered in our investigations: chlorinated paraffin containing 55% chlorine (additive A), a sulphur containing additive (additive B) and a phosphorus compound developed in our laboratories (additive C). These additives are specific for the formulation of metalworking straight oils and soluble oils. Our target was to replace chlorinated paraffin, and to find correlation between the lab and bench tests and the behaviour of the fluids in field trials.

LABORATORY TESTS

Differential thermal analysis (DTA) of the above mentioned additives was performed. DTA curves show endothermic peaks between 250 and 290 °C attributed to the decomposition of the additives. DTA curves for mixtures of the additives with iron powder show a first endothermic peak at 170-195°C, and an exothermic peak at 210-245°C. The first peak was attributed to the decomposition and the second one to the reaction of the decomposition products with the metal. The lower is the temperature of these two peaks, the higher is the reactivity of the additive to the metal. The reaction of the decomposition products of the additive with the metal leading to the formation of the superficial layer is responsible for the E.P. activity. The more reactive additive was the sulphur containing product B, hence we expected the same behaviour in E.P. bench tests.

BENCH TESTS

Both metalworking neat oils and soluble oils where formulated using additives A, B, and C. Antiwear and E.P. characteristics of the neat oils and 1:20 dilution of the soluble oils where determined on Four Ball, Falex and Timken machines. In all these tests formulations containing additive B showed the best performances, that confirms the action mechanism and the supposition mentioned above that more reactive is the additive, better is its E.P. behaviour.

FIELD TRIAL

In the real life the things are nevertheless more complicated. Not only the E.P. properties are responsible for the performances of the fluids, but also the detergency and anti-corrosive properties. In the formulation of MWF an equilibrium must be kept between all these demands. Bench tests allow only a preliminary screening of the fluids. Carefully driven field trials are necessary for the final decision, especially in systems with big tanks. In the field trials the behaviour of the fluids was traced for long time, by collecting and analysing fluid samples, and by monitoring the tooling performances, tools life and the surface quality of the machined workpieces. The field trials where realised on individual machines. The fluids where tested in the most important operations in ball bearing manufacture as: turning, grinding and fine grinding. The trials emphasised that lab and bench tests have only a limited reliability. Chlorinated paraffin can be successfully replaced with either additive B or C.

PREPARATION OF MAGNETIC FLUIDS HAVING ACTIVE GASSES RESISTANCE AND ULTRA LOW VAPOR PRESSURE FOR MAGNETIC FLUID VACUUM SEALS

T KANNO, Y KODA, Y TAKEISHI, T MINAGAWA and Y YAMAMOTO
Research Department, Research & Development Division, NOK CORPORATION,
25 Wadai, Tukuba-shi, Ibaraki-ken, 300-42, JAPAN

ABSTRACT

In widely known, it is not too much to say that properties of magnetic fluid vacuum seals are dependent on properties of magnetic fluids that is composed of the magnetic fluid vacuum seals. There are very few reports(1) on perfluorinated oil based magnetic fluids in order to give lower vapor pressure and better active-gas resistant property against gasses using for manufacturing semiconductors. At present, the pressure of vacuum obtained by the vacuum seals using such magnetic fluid is limited to around 10^{-7}Pa. In this work, the preparation method of magnetic fluids having a ultra low vapor pressure, 7×10^{-10}Pa, and a resistance of active gasses is developed.

A base oil was used perfluoropolyether oil, that has superior chemical stability and low vapor pressure. For ultra fine magnetic particles being dispersed in the perfluoropolyether base oil, perfluoropolyether surfactants were newly designed and synthesized two perfluoropolyether surfactants, having same molecular structure unit as the base oil and a preferable chain length. One of which was hexafluoropropylene oxide oligomer acid sodium salt (HFPO-COONa) and the other was hexafluoro propylene oxide polymer amino dodecylamido (HFPO-DADo). HFPO-COONa was synthesized by the hydrolysis of hexafluoropropylene oxide oligomer methyl ester(MW=1500) and HFPO-DADo was synthesized by reacting the methyl ester (MW=6700) and 1,12-diaminododecane(DADo). And the magnetic ultra fine particles, prepared by co-precipitation method, was dispersed in the perfluoropolyether base oil by using the synthesized perfluoropolyether surfactants. The preparation flow chart of the magnetic fluid is shown in Fig. 1. The prepared magnetic fluid had the saturation magnetization of 35.0mT and the viscosity of 6000mPa·s.

A vapor pressure of the magnetic fluid was measured by Vapor Pressure Measurement (ULVAC). And a resistance test for active gasses is that the magnetic fluid was subjected in Chlorine gas (1.3×10^4Pa,24h) or Tetrafluoromethane (CF4) plasma (4Pa,300W,5h). The result was confirmed that the obtained magnetic fluid has the vapor pressure of 7.0×10^{-10}Pa (at 293K)(Fig.2), calculated with Cox chart(2), and had the good resistance of the both active gasses.

Furthermore, the ultimate pressure and the outgassing and leak rate, by a build-up method(3), of magnetic fluid vacuum seals, in which the prepared magnetic fluid has been used, was evaluated. The ultimate pressure was 8.0×10^{-9}Pa, and the outgassing and leak rate was 1.5×10^{-10}Pa·m^3/s of similar level to the amount of 1.2×10^{-10}Pa·m^3/s when the piping between the seal unit and the chamber was closed.

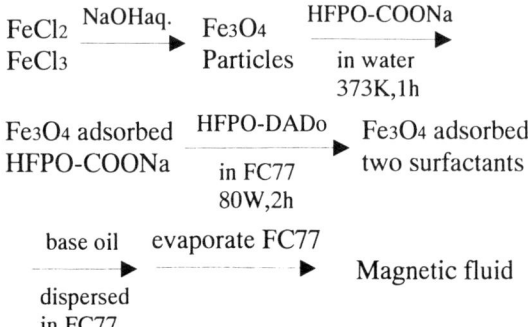

Fig. 1: Magnetic fluid preparation flow chart

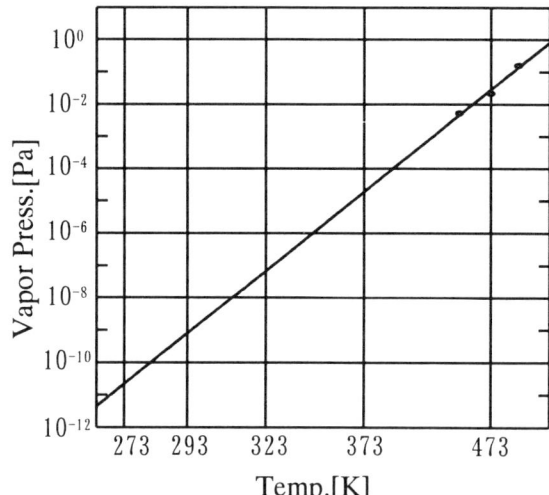

Fig. 2: Vapor pressure chart of magnetic fluid

REFFERENCE
(1) R.Kaiser,USPatent,3784471,1974
(2) R.A.Kumar,S.R.Reddy,A.Satyanarayanan, Chem.Eng.World,Vol.8,No.11,page85,1973
(3) M.Kikuchi,Nucl.Fusion,Vol.26,No.2,page 223,1986

EFFECTS OF OVERBASED METAL SULFONATES ON THE FRICTION AND WEAR IN HOT METALWORKING OF STAINLESS STEELS

KUNIO GOTO

Corporate Research and Development Laboratories, Sumitomo Metal Industries Ltd.,
1-8 Fuso-cho, Amagasaki, Hyogo, JAPAN

ABSTRACT

Hot rolling of stainless steels often causes seizure of the steel on the work rolls, resulting in a roughening of the surfaces of the work rolls, in turn, leads to the formation of surface flaws on the hot rolled product. The purpose of this paper is to present a lubricating composition for hot rolling which can effectively prevent surface flaws during hot rolling of stainless steels.

A new method to improve the lubricity was proposed: both " the Principles of Hard and Soft Acids and Bases (1)" for improving chemisorption force and " the super fine particles(2) " for preventing metal/metal contacts.

Experiments have been conducted with a lubricated disc-block type friction test(3) at elevated temperature. Block specimen can be heated up to 1100℃ during the test in a high frequency induction heater. Disc specimens were made of high chromium iron(2.8wt%C-18wt%Cr) and high carbon high speed steel (1.9wt%C-3.5wt%Cr-2.6wt%Mo-5.4wt%V)as roll materials. Block specimens were made of a ferritic stainless steel. The lubricants used in this study were mineral oil based metal sulfonates(30wt%), Lithium grease based graphite(10wt%), mineral oil based powder of Fe_2O_3 (10wt%), Conventional oil(synthetic ester). Mineral oil is a viscosity of 90 cSt at 40℃. Lubricant was supplied to frictional surface of disc specimen as neat oil. After friction test, frictional surface flaws of disc specimen and friction coefficient were estimated.

Fig.1 shows that a lubricating composition based on overbased metal sulfonates were effective for preventing seizure in hot metalworking. Fig.2 shows that a calcium sulfonate has the highest lubricity among the alkaline earth salts of the sulfonic acid having the same base number(TBN) and the lubricity of the lubricating compositions containing the metal sulfonate having a higher base number is better.

I guess that overbased metal sulfonates easily react with oxide layer of disc specimens, as overbased metal sulfonates belong to hard bases and oxide layer of disc specimens belong to hard acids.

The overbased calcium sulfonate contains fine particles(less than about 100 angstrom) of a calcium carbonate. When the overbased calcium sulfonate is mixed with a base lubricating oil such as mineral oils, synthetic lubricating oils, the fine particles form a colloidal dispersion in the oil, which liberates the corresponding oxide such as CaO in hot rolling temperature range.

As a result, the fine particles are introduced into the working interface in a uniform and stable manner, and prevent metal/metal contacts.

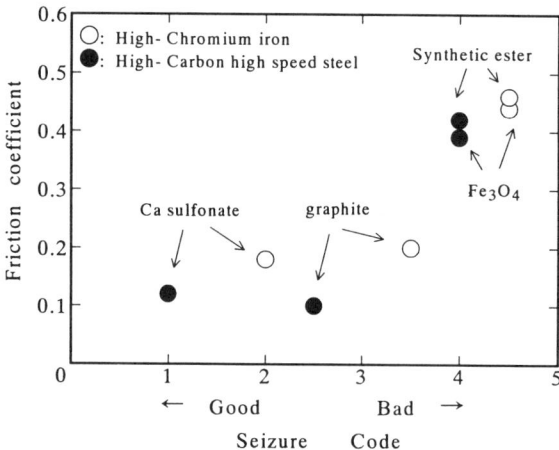

Fig.1: Friction test results

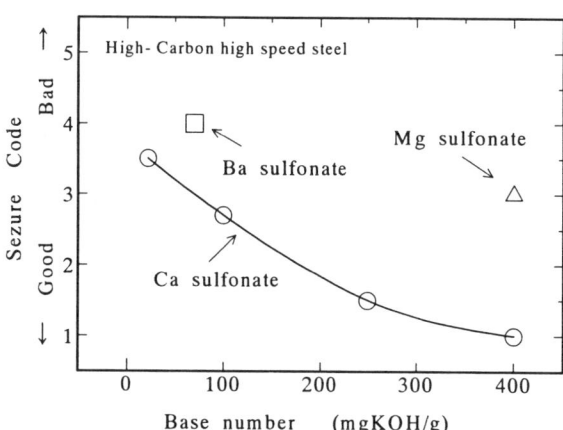

Fig.2: Effects of alkaline earth metal sulfonate and base number on surface flaw

REFERENCES

(1)R. G. Pearson (ed.), Hard and Soft Acids and Bases, Dowden, Hutchinson & Ross, Inc., (1973)
(2)K. Goto, T. Shibahara, Proc. of JAST Tribol. Conf., Tokyo, (1996) 427-428
(3)K. Goto, T. Shibahara, CAMP-ISIJ, vol.7(1994)1365

"BI-GAUSSIAN" REPRESENTATION OF WORN SURFACE TOPOGRAPHY IN ELASTIC CONTACT PROBLEMS IN ROTARY SEALS

S E LEEFE
BHR Group Limited, Cranfield, Bedford, MK43 0AJ, UK

ABSTRACT

Rotary seals commonly operate in the mixed friction regime, where leakage and friction depend on surface topography. In order to predict and optimise performance, mathematical models are employed, which require characterisation of surface roughness and specification of operating conditions. One of two methods is typically employed. In the 'conventional' approach leakage is estimated from the Reynolds equation modified by flow factors, whilst friction is based on a calculation of the real area of contact. In an alternative ('numerical') approach, a detailed map of surface topography is fed directly into the Reynolds equation which is solved by multigrid techniques.

Each approach has problems. The numerical approach is computationally expensive; implies 3-D topography measurement; and requires the specification of 3-D texture parameters suitable for quality control. The conventional approach, as usually practised, relies on assumptions about the statistical distribution of surface roughness heights. This is because both the published flow factors and the most widely understood computation of real contact area assume that this distribution is Gaussian, whereas in most worn surfaces it is not.

There is, however, no inherent necessity to use a Gaussian roughness distribution in the conventional method. Three elements are required: 1) a suitable alternative topography representation; 2) incorporation of this roughness representation into an appropriate model of contact; and 3) computation of flow factors through such roughness. This paper describes the first two elements of such a model.

If the bearing area curve is plotted for a surface with a Gaussian distribution of roughness, with the cumulative probability axis on a normal probability scale, the result is, by definition, a straight line whose slope is R_q. Investigation of many worn tribological components has shown that typical measured roughness, when plotted in this way, is characterised by two straight line segments, as shown in Fig. 1 for a carbon-graphite mechanical seal face. This form is typical of surface roughness whose topography consists of plateau regions, separated by valleys.

The bilinear or "bi-Gaussian" form is strongly suggestive of the removal of peaks from an underlying roughness scale. This roughness representation is characterised by three physically meaningful parameters: the rms values of the two scales of roughness and the location of the knee-point on the bilinear curve, which represents the depth of peak removal.

Mechanical contact between surfaces is dominated by the behaviour of roughness asperities. In the present context two questions follow. Firstly: does the statistical distribution of asperity peak heights follow the same bi-Gaussian form? Measurements on a preliminary sample of seal faces from a variety of services suggest that it does. Secondly: is contact elastic in nature? One argument suggests that any initial inelastic contact results in the wear of asperities so that asperity tip curvatures in-service are characteristic of elastic contact. Measurements on the preliminary sample of used seal faces supports this view. It is therefore possible to construct a model of elastic contact based on measured asperity height distribution, tip curvature and area density assuming that only plateau top roughness contributes to the real area of contact. This is explained in the full text of the paper (1).

Given that the same bi-Gaussian surface topography is exhibited in components after wear in excess of 1 mm depth, surface topography must be self-replicating and not simply the result of the truncation of initial roughness resulting from the finishing process. This suggests that attention to material microstructure may yield a particular combination of bi-Gaussian texture parameters which represent optimum tribological performance.

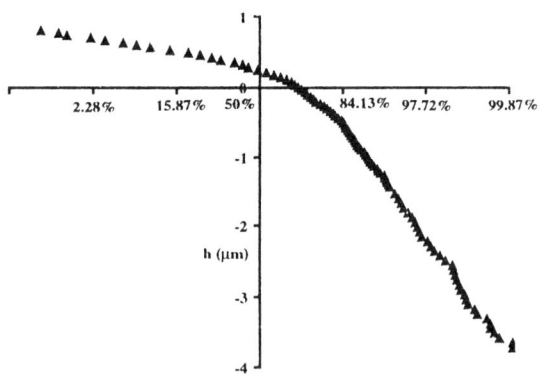

Fig. 1: Bearing area curve for worn carbon graphite

REFERENCES

(1) S.E. Leefe, "Bi-Gaussian" representation of worn surface topography in elastic contact problems. Accepted for publication in the Proceedings of the 24th Leeds-Lyon Symposium on Tribology, Sept 1997.

Submitted to *Leeds-Lyon Symposium on Tribology Proceedings*

THE STUDY OF ANTIWEAR SYNERGISM BETWEEN AN OIL-SOLUBLE CERIUM LIGAND AND BORATE

LIGONG CHEN and JUNXIU DONG

Dept. of Petrochemistry. Logistical Engineering College, Chongqing630042, China

ABSTRACT

Borate dispersion system and organic borate have been found use in engine and gear oils owing to its extraordinary oxidative stability and load-carrying capacity. On the other hand, rare earth metals are often utilized to modify the mechanical and metallurgical properties of steel or as catalysts to boronize steel in chemical heat treatment. In recent years, some progress of rare earth compounds as antiwear and/or extreme pressure additives for lubricating oils and greases have also been made(1-2).

This presentation aims at describing the antiwear synergism between a borate ester and an oil-soluble cerium ligand, cerium dioctyl dithiocarbamate (CeDTC), and postulating the mechanism of tribocatalytical boronization of the ligand to rubbing surface based on surface analysis.

Both the cerium ligand and borate ester in this study were prepared by the authors. Each of them showed good friction-reducing and antiwear abilities. However, antiwear synergism was achieved even half of each additive was added toghther into the ISO VG32 base oil. The initial seizure load evaluated with a HQ-1 block-on-ring test rig and MQ-800 four-ball machine confirmed this conclusion, as shown in Table 1.

Table 1. Initial Seizure Load Evaluated with Block-on-ring Test Rig and Four-ball Machine(N)

lubricant	block-on-ring	four-ball
ISO VG32	800	392
VG32+5.0% Borate	1600	862
VG32+5.0% CeDTC	1300	803.6
VG32+2.5%Borate +2.5%CeDTC	2100	960

More evidence for antiwear synergism between them was added by the smallest wear scar diameters after four-ball tests for 1800, 2700 and 3600 seconds under the load of 294N and 588N as compared with those of each individual additive.

In order to reveal the mechanism of antiwear synergism, Auger electron spectroscopy with argon ion sputtering was applied to examine the rubbing surfaces of steel balls. Depth profile showed that the atomic concentration of boron in the wear scar tested in the oil with 2.5wt.% borate and 2.5wt.% CeDTC was 2-3 times higher than that tested in the oil with 5.0wt.% borate. This indicated that cerium could stimulate boron atoms decomposed from borate to diffuse into the rubbing surface.

X-ray photoelectron spectroscopy disclosed that there were at least two boron-containing substances generated on the wear scars after tested in borate-containing oil: boron oxide and ferric boride, and the content of boride was much higher in the presence of CeDTC. It was of note that boride was the boronizing product of steel in chemical heat treatment. As a result, the antiwear synergism between CeDTC and borate was ascribed to that CeDTC played a tribocatalytical role in the boronization of borate to the rubbing metallic surface. Besides, theoretical elucidation was supplied to explain the synergism according to thermodynamic theory.

REFERENCES

(1) B. Chen, J. Dong and G. Chen, Wear, Vol.196, No. 1, pp.16-20, 1996.
(2) Y. Lian, L. Yu and Q. Xue, Wear, Vol.181-183, pp.436-441, 1995.

TRIBOCHEMICAL CHARACTERISTICS OF BISMUTH DIOCTYL DITHIOCARBAMATE

LIGONG CHEN and JUNXIU DONG

Dept. of Petrochemistry. Logistical Engineering College, Chongqing630042, China

ABSTRACT

Organic bismuth compound seems to be one of the most promising antiwear and extreme pressure additive for lubricants(1-3). In this study, another oil-soluble bismuth compound, bismuth dioctyl dithiocarbamate (abbreviated as BiDTC) was prepared. Its chemical structure was detected by infra-red spectroscopy(IR) and C^{13} nuclear magnetic resonance spectroscopy(NMR), and its elemental composition was measured by inductively coupled plasma spectroscopy (ICP). Its tribological behaviours were evaluated with a block-on-ring test rig and a four-ball machine. Futhermore, surface analysis was carried out by means of Auger electron spectroscopy(AES) and X-ray photoelectron spectroscopy(XPS) to characterize the surface film generated by the additive under boundary lubrication conditions.

Combined with the elemental composition of the compound measured by ICP, the analytical results of IR and C^{13}NMR confirmed the following formula of BiDTC:

$$\left[\begin{array}{c} C_8H_{17} \\ C_8H_{17} \end{array} \!\!\! N\!-\!C \!\!\! \begin{array}{c} S \\ S \end{array} \right]_3 Bi$$

The friction-reducing ability of BiDTC was evaluated with a HQ-1 block-on-ring test rig after the compound was added into ISO VG32 mineral paraffinic base oil in the concentration of 0.5wt.%, 1.0wt.% and 2.0wt%. 16.7%, 24.2% and 27.5% reduction of friction coefficient was respectively obtained compared to that of the base oil.

The load-carrying capacity of BiDTC was tested with a MQ-800 four-ball machine. Test results showed the initial seizure load increased from 400N of the base oil to 840N with the increase of BiDTC content from 0 to 5.0wt.%. Whereas the welding load of BiDTC increased from 1500N of the base oil to 8000N.

In addition, antiwear performance of the additive was also investigated by measuring the mean wear scar diameters of lower balls after respectively tested in 2.0wt.% BiDTC oil under the load of 294N and 588N for 1800, 2700 and 3600 seconds. The results showed that BiDTC could effectively control the wear of the ball.

AES survey disclosed the existance of bismuth, carbon, oxygen, sulfur, phosphorus and nitrogen in the surface film formed by the additive on the ball under boundary lubrication conditions. Depth profile with argon ion sputtering gave the distribution condition of the elements in atomic concentration and showed that the concentration of bismuth exceeded that of sulfur 3 times. However, the molar concentration of bismuth the the molecule of the compound was only one sixths of that of sulfur.

Analytical results of XPS identified the chemical composition of the surface film on the rubbing surface. Among the organic and inorganic species, bismuth sulfide and oxide were found in the outmost layer. In addition to the above two bismuth substances, metallic bismuth was found in the inner layers after argon ion sputtering for 10, 20 and 30 minutes, whereas sulfur was mainly present in sulfate and sulfide in outer layer and only sulfide in inner layer. these species produced by tribochemical reaction were very conducive to the reduction of friction and wear. An action mechanism of BiDTC was elucidated based on the analysis.

REFERENCES

(1) R. Otto, NLGI Spokesman, Vol.57, No.2, pp.6-13, 1994.
(2) D.K. Tuli, R.Sarin, A.K.Gupta and A.H. Kumar, Lubrication Engineering, Vol.51, No.4, pp.298-303.
(3) L. Chen and J. Dong, Proceedings of 10th Intl. Collq. Trib., pp.1223-1230, 1996

THE INFLUENCE OF CHEMICAL COMPOSITION ON THE POUR POINT DEPRESSANT PROPERTIES OF METHACRYLATE COPOLYMERS USED AS ADDITIVES FOR LUBRICATING OILS

PAULA LUCA, M. FLOREA, S. BALLIU and DOINA CATRINOIU
Additives and Lubricants - ICERP, 291A Republicii bvd., RO-2000 Ploiesti, Romania

ABSTRACT

Pour point depressant (PPD) efficiency of polymethacrylates (PMA) on different lubricating oils is known to be related to the average length of the side chains and the nature of the base oil (1,2). It is however not very well understood how chemical structure influences their PPD properties. Regularly, mixtures of high and middle cut methacrylates are used in the synthesis of these compounds. PPDs and VI improvers based on PMAs including short alkyl methacrylates or polar methacrylate monomer and also methacrylate - styrene copolymers were reported as additives that keep the depressant function (3-5).

The aim of our work was to establish the effect of short alkyl methacrylates (C1-C4) and styrene content on the PPD effectiveness of methacrylate copolymers as additives for paraffinic base oils. Copolymers with different average side chain length were obtained, based on mixtures of high and middle length alkyl methacrylates and including various amounts of the above mentioned monomers. The examination of the efficiency of these products was carried out at low (below 0.3 wt. % polymer) and high concentrations (about 4-5 wt. % polymer), in two base oils with close viscosities but different original pour points.

Figure 1 and 2 illustrate the depressant effect of copolymers, with different alkyl group length and different styrene or C1 content, blended in a paraffinic base oil with initial pour point of -3 °C, and kinematic viscosity of 6.2 $mm^2.s^{-1}$. A higher styrene content requires a longer average side chain of the methacrylate mixture to give an effective PPD on the same oil. Short alkyl chain methacrylates shift the range of the effective average chain length toward higher values, and extend this range, especially at high concentrations (VI Improvers) where otherwise this is severely restricted, as it can be seen in Fig. 2. The

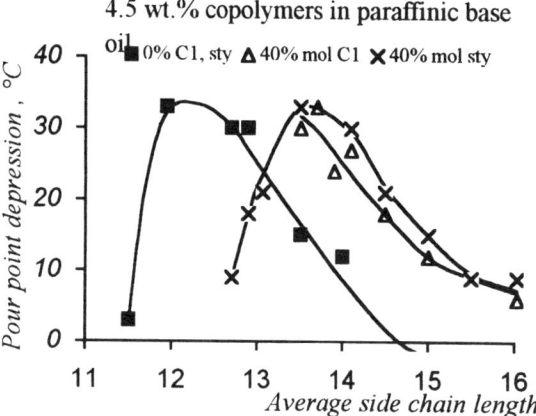

Fig. 2: The influence of styrene and C1 content on the pour point depressant properties of VI Improvers

averages of the alkyl group lengths were calculated taking into account only the methacrylate esters with long and middle alkyl groups.

The results can be explained in terms of the interactions that occur, in the conditions of wax crystallisation, between high paraffins and the alkyl groups of methacrylate copolymers in competition with intra and inter-molecular interactions of the macromolecules. Modifications in the architecture of the macromolecules affects the equilibrium of these complex interactions. Also, the extent of the intermolecular interactions between polymer molecules increase with concentration.

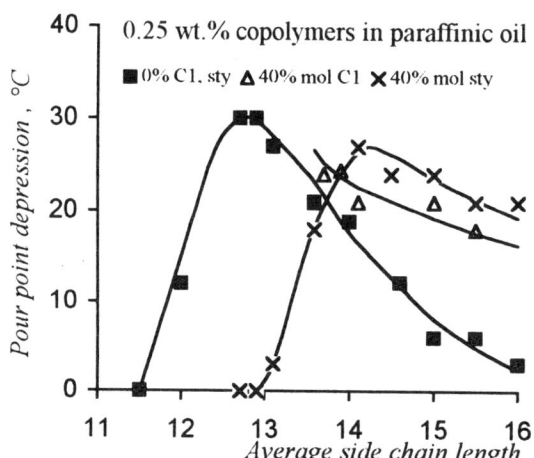

Fig. 1: Influence of the styrene and C1 content on the PPDs effectiveness

REFERENCES

(1) J. Denis and J.P. Durand, Rev. IFP, **46**, 637-49 (1991)
(2) J. Denis, Rev. IFP, **42**, 385-98 (1987)
(3) B. E. Wilburn, W. J. Heilman, PCT Int. Appl. WO 89/01507 (1989)
(4) Pennewis, S. et al., U.S. Patent No. 4,290,925
(5) Ahmedov, A. I., Khim. Tekhnol. Topl. Masel., **10**, 26-8

PYROMELLITIC ESTERS WITH ENHANCED AROMATIC CONTENT AND A MIXED STRUCTURE CONSIDERED AS TRIBOLOGIC FLUIDS.

LIVIU E. MIRCI
Department of Chemical Engineering, University "Politehnica" Timisoara, Pta. Russel No.4,1900 Timisoara, ROMANIA

JEAN M. HERDAN
ICERP, Research Institute of Oil Distilery and Petrochemistry Ploiesti, Bd.Republicii No.291A,2000 Ploiesti, ROMANIA

ABSTRACT

As it is well known, the aromatic nucleus presents a particularly high thermal resistance and this feature is implicitly attributed to the compounds that possess this function. The pyromellitic structure preserves this characteristic and pyromellitic esters are known for their inherent stability. On the basis of these premisses one may anticipate that an increase in the aromatic content might adequately increase the thermal behaviour of the presumptive compounds obtained. However, the aryl radical automatically raises the freezing points of the substance involved. In order to use such materials in the tribological field one may modify the structure so as to hold this parameter within the suitable range. This purpose is supposed to be achieved by realizing a mixed structure in which a significant amount of aliphatic chain should be present.

The previous results recorded, taking into account the trimellitic anhydride, were delivered in Japan (1) and Germany (2); they were also extensively described in the patent awarded (3). The objective of this paper is to present the results concerning the synthesis and characterisation of some tetraesters of pyromellitic anhydride with superior normal or branched aliphatic alcohols as well as with an alkyl-aryl alcohol, respectively. By varying the molar relationship between these alcohols three alternatives have been considered and, as a result, three series of products have been produced. These series are illustrated in the following general structures shown by the formulae I, II and III:

R_2OOC—⟨⟩—$COOR_1$ R_1OOC—⟨⟩—$COOR_2$
R_2OOC—⟨⟩—$COOR_2$ R_1OOC—⟨⟩—$COOR_1$
         I   III

$(R_1)R_2OOC$—⟨⟩—$COOR_1(R_2)$
$(R_1)R_2OOC$—⟨⟩—$COOR_1(R_2)$
         II

R_1 = alkyl-aryl chain, like -CH_2-CH_2-O-C_6H_5
R_2 = normal or branched aliphatic chain like C4, C2-6, C8 and C10

On the basis of this programme one may investigate the contribution of the aromatic ring (content) to all properties through a relevant comparison between the three series. It becomes also possible to study the influence of the aliphatic chain (both length and structure) on physical-chemical and tribological properties. The non-symmetrical structure of these tetraesters is expected to determine a diminution of the assembling capacity, thus preventing, to a certain extent, the crystallization at low temperatures.

At the same time, this structure keeps its valuable features at high temperatures. The compounds mentioned above have been performed in a solution esterification process within one or two stages (4).

These tetraesters with a mixed structure were first characterized within the classical field for organic compounds, as follows: (a) their saponification values were quasi-theoretical; this fact confirms the purity and the validity of the admitted structures; (b) their densities present a linear variation depending on the increase in number of carbon atoms of the aliphatic alcohol; series I is inferior to series II which in its turn is inferior to series III; (c) their refractive index showed a similar dependence, thus increasing with the increase in aromatic content; (d) their dynamic viscosities showed a pseudoplastic behaviour, the absolute values range between 1.6 Pa.s and 122 Pa.s.

The analysis of the values of the tribological parameters permits to state the following considerations: (a) the flash points recorded were high, the main values ranging between 280°C and 295°C; (b) the flow points were not as low as presumed, the best values occurring at about -27°C to -24°C for the series I (high aliphatic content) while series II and III showed values around 0°C or even positive; (c)the kinematic viscosity and the viscosity index presented also a better response for series I as compared to series II and III, the values of the viscosity index ranging between 50 and 100; (d) the thermogravimetric analysis showed a very good thermal resistance even at 300-350 °C; the loss of weight did not exceed 15-20% for most of the terms of all series: (e) the wear spot diameter (four ball test) for non-additivated specimens shows values of 0.55-1.15 mm; when additivated with 1.5% Zn-dithiophosphate, this parameter showed better values such as 0.30-0.44 mm.

All these properties combined, along with high and versatile viscosity, suggest that these products may be considered for use as base oils, thermal-resistant fluids, or in any application where stability is of primary importance.

REFERENCES

(1) L.Mirci, A.Pruncu, Intern. Tribology Conf., Yokohama, Japan, 29 Oct.-2 Nov. 1995, Vol.II, p.863-867
(2) L.Mirci, M.Maties, 10th. Intern. Colloq. Tribology, Ostfildern, Germany, 9-11 Jan. 1996, Vol.I p.373-377
(3) L.Mirci, Rom.Pat. 109644, 31 March 1995
(4) L.Mirci, Rom.Pat. 111760, 31 December 1996

VISCOMETRIC STUDIES OF OIL-IN-WATER LUBRICANTS

M B NABHAN
Department of Mechanical Engineering, University of Bahrain, P.O.Box 32038, Bahrain

ABSTRACT

An interesting property of oil-in-water (o/w) mixtures which has attracted the attention of tribologists is that they can form thicker hydrodynamic films than those expected from their apparent rheological properties. The oil phase is believed to separate out on the metal surface in the concentrated conjunctions lubricated with o/w emulsions (1-3).

To study this phenomenon, experiments are conducted in a specially designed viscometer rig, where different concentrations of o/w emulsions are used as the lubricant. The rig is composed of a metal tube which has a straight and parallel cylindrical ground bore and steel droppers of similar tolerance features. The droppers are allowed to descend in the tube which is filled with the emulsion at a uniform velocity. The radial clearances between the descending droppers and the tube are very small.

It is found that, even at the oil concentrations as low as several percent, higher values of the emulsions viscosity are obtained in the concentrated conjunctions in relation to their bulk values. The results are shown in Fig.1. These results can be quantitatively explained by a theory (4) which assumes trapping of oil particles between metal surfaces which is correlated to the displacement energy at the oil-water-metal boundaries.

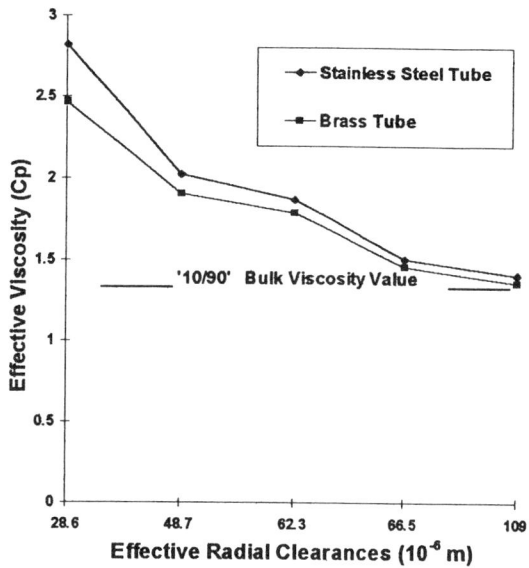

Fig. 1: **Variation of effective viscosity with clearances for a '10/90' emulsion**

REFERENCES

(1) H Hamaguchi, H A Spikes, and A Cameron, Wear, Vol.43, 1977, pp.17-24.
(2) G T Y Wan, P Kenny, and H A Spikes, Tribology International, Vol.17, No.6, 1984, pp.309-315.
(3) Y Kimura, and M Okada, JSLE International Tribology Conference, Tokyo, 1985, pp.937-942.
(4) Y Kimura, and K Okada, Tribology Transactions, Vol.32, No.4. 1989, pp.524-532.

Urea Grease Life Formula for Ball Bearings

H. Nakashima, M. Minami
NTN Corporation, 3066 Higashikata, Kuwana, 511 Mie, Japan

ABSTRACT

Under the operating conditions of high temperature and high rotational speed, grease life is more critical to bearing performance than rolling contact fatigue life. Therefore, it is important for bearing design to accurately estimate grease life. The grease life formula presented by E.R. Booser is currently the most widely accepted. However, several new formulas have recently been reported.

Urea greases are widely used in applications where grease life is critical to bearing performance. Synthetic oil, which has good heat resistance properties, is mainly used as the base oil in urea grease. We have developed a grease life formula for urea grease, based on our test results. These tests were conducted at different temperatures, rotational speeds and loads. This formula is compared to other grease life formulas which have been presented.

Grease life tests of various urea greases were performed on 6204-size sealed ball bearings under various test conditions, as shown in Table 1.

Table 1 : Test conditions

	Influence of temperature	Influence of rotational speed	Influence of load
Bearing	6204ZZC3	←	←
Temperature, °C (End surface of outer ring)	120, 150, 180	150	150
Rotational speed, rpm (Inner ring rotation)	10000	10000, 15000, 20000	10000
Load, N	Fa = Fr = 67	←	Fa = Fr = 67, Fa = 294, Fr = 67
Number of grease	20	3	5

From these test results, the influences of temperature, rotational speed and load upon grease life were evaluated. Based on these evaluations, the following grease life formula has been developed.

$$\log L = -0.0295T - 2.02 \times 10^{-6} V - 9.14 P/C + 8.51 + K$$

where L : 50% Grease life T : temperature (°C)
  V : rotational speed(dmN) P/Cr : bearing load
  K : correcting factor for the kind of urea greases

Figures 1 and 2 show the relationship between the test results and calculated lives. In the case of E.R. Booser's and Ito's formulas, the difference between the test results and the calculated grease lives were substantial. However, the new grease life formula which we have developed was superior to the others, even when comparing test results conducted by other researchers. To further improve the new grease life formula, test data for various bearing sizes and various test conditions, such as outer ring rotation, will be needed.

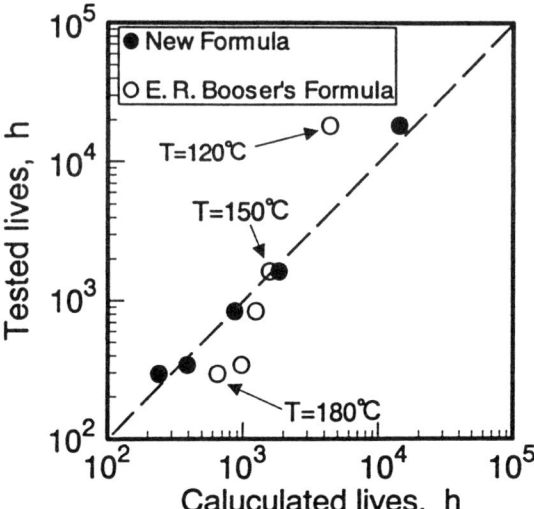

Fig.1 : Comparison the calculated lives with tested results

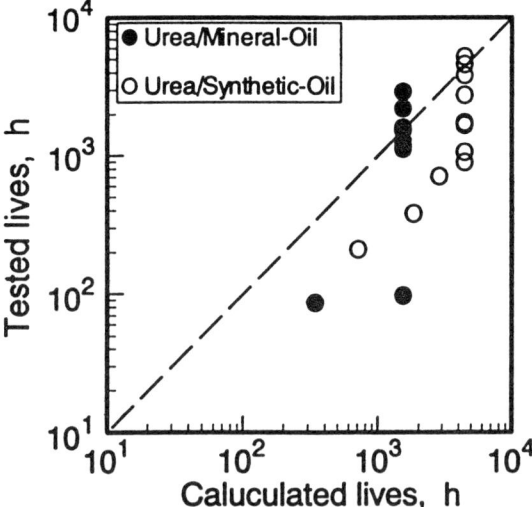

Fig.2 : Comparison the calculated lives using Ito's formula with tested results

REFERENCES

(1) E. R. Booser: J. ASLE Lubrication Engineering, 30, 1974, 536pp
(2) Y. Ito et al.: NSK Technical Journal No.660, 1995, 8pp

REACTIVITY OF ANTIWEAR ADDITIVES INVESTIGATED BY MEANS OF ELECTROCHEMICAL METHOD IN TERMS OF THEIR TRIBOLOGICAL EFFECT

Dariusz OZIMINA
Department of Technical Chemistry, Technical University of Kielce, 25-314 Kielce, Poland

ABSTRACT

The paper presents the electrochemical method of evaluation of both antiwear additives reactivity and the formation of surface layers on the basis of some selected organometallic compounds.

The fundamental assumption has been made that electric energy connected with electron transfer through the metal/lubricant interface
is observed in friction processes in addition to heat and mechanical energy (1) (2) (3). The electric energy enables fast diagnosing of the processes and changes taking place on metal surfaces.

The results have been obtained by means of electrochemical simulation method in the initial stage of the method application and by testing correlated with the results of tribological experiments.(4).

The method enables the evaluation of antiwear additives reactivity for a selected metal in non-aqueous media. The reference system used in property evaluation of subsequently tested substances is determining their effect upon non-active gold surface. Fig. 1 presents cyclic chronovoltammetric curves which enable determining the effect
of non-active gold surfaces upon Zn DTC (an additive commonly used in lubricants). While analysing the curves it is easily observed that the run of polarisation curve connected with formation of mono-layer is completely different from those of cycle 2 and subsequent cyclic curves. The bigger the number of cycles and the thickness of surface layer the smaller the reaction currents. The process is clearly shown on separate charts. Testing of other additives from Me DTC and Me DTP groups may be connected with UDP- underpotential deposition irrespective of reactions taking place on that surface. However, the further cycles point at formation of levelling layers of smaller specific area which reduce the reaction current in subsequent cycles. In the actual friction knot the interface metal/lubricant occurs. As a result, the metal surface functions as a working electrode on which heterogenic electrochemical processes accompanying tribological processes take place. The observation of mono and poli - layers formation with the simultaneous recording of the formation conditions by means of electrochemical scanning tunnel microscopy ESTM confirm the processes.

On the basis of the conducted experiments the attempt has been made to determine the mechanism of surface layer formation on metals as a result of reactivity of model lubricant of high tribological activity. The diagnostic element of the method comprises: quantity of passing electric charge, initiating conditions and conditions of surface layer formation.

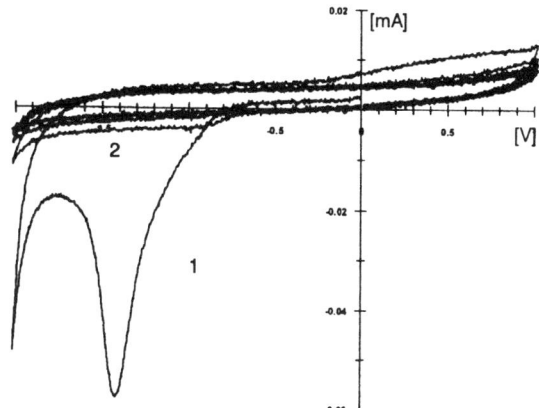

Fig. 1. Cyclic chronovoltommetric for Au in non-aqueous media solution ZnDTC

REFERENCES

(1) Heinicke, G.: Tribochemistry. Akad. - Verlag Berlin 1984
(2) Suh, N.P.: Tribophysics - Prentice - Hall, Inc. New Jersey 1986
(3) Nakajama, K.: Wear, 1994, 178, 61 - 67 pp.
(4) Ozimina, D.: ZEM (Exploitation Problems of Machines), 1995, Vol. 30, 521 - 536 pp.

LABORATORY EVALUATION OF GEAR OILS WORKING UNDER EXTREME CONDITIONS

W PIEKOSZEWSKI, M SZCZEREK, W TUSZYŃSKI and M WIŚNIEWSKI
Institute for Terotechnology, ul. Pułaskiego 6/10, PL - 26-600 Radom, Poland

ABSTRACT

The four-ball extreme-pressure tester is one of the most widely spread machines used for investigation of lubricants. According to the known method for testing anti-seizure properties of lubricants on the four-ball machine the contact load is stepwise increased in a number of the test runs until the boundary layer is broken and heavy scuffing begins. In many cases, however, the lubricants which have similar four-ball characteristics show very different anti-seizure behaviour during machine operation. In this work an attempt has been made to develop new test procedure which enable more precise evaluations of anti-scuffing properties of lubricants.

The essential of the proposed method (1) is that the four-ball test is to be performed with the continuously increasing contact load and the instantaneous values of this load and the moment of friction are to be recorded. Performing of such a test enables a modified four-ball apparatus (Figure 1) equipped with the motorised loading (2). The operation of the motor 9 driving the guide screw 6 which moves the dead weight 16 and all measuring functions are controlled by specialised PC software.

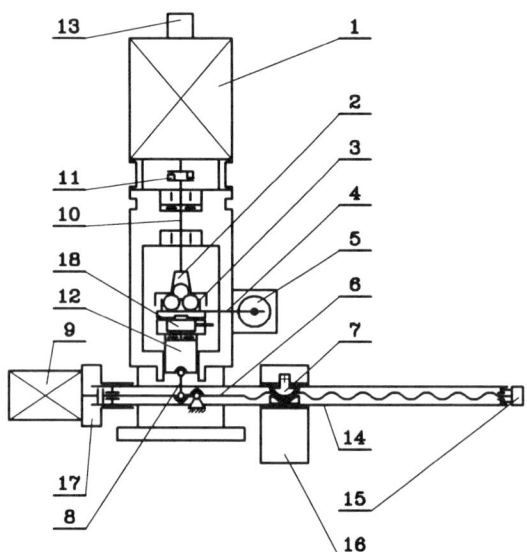

Fig 1: Scheme of the modified four-ball machine

The rotational speed during the new test is constant (500 ± 20 rpm) and the load increases from 0 to 7848 N (800 kgf) with the constant speed of 408.75 N/s. Tests is terminated when the upper ball stops which indicates that the extreme-pressure level of the lubricant is exceeded.

The ambient temperature is chosen as 20 ± 2 °C.

For verification of the new method 5 gear-hydraulic oils (C4 grade) were examined. These oils have shown in the common four-ball test the same last non-seizure load P_n=785 N which means that their anti-seizure properties according to existing standards are the same. However, in tests with linear increasing contact load these oils have very different profiles of the moment of friction. Figure 2 shows an example of the test results for two oils. Heavy scuffing occurred for oil A earlier as for oil B and the moment of friction for oil A has a rapid increase while the curve for oil B is more flat.

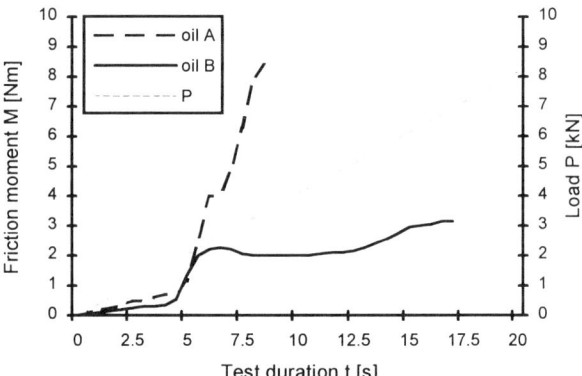

Fig 2: Experimental results for two C4 grade gear-hydraulic oils

As quantitative parameters describing the resistance of lubricants to seizure under extreme-pressure conditions following factors are proposed: the increase of the moment of friction ΔM, limiting pressure of seizure p_{gz} and wear scar d. The results of the laboratory evaluation of anti-seizure properties have been confirmed in machine operation. Oil B has parameters ΔM, p_{gz} and d significantly better than oil A and high-power planetary gears lubricated with oil B work successfully while for oil A severe failures occurred. Therefore, the proposed method enables the identification of the anti-seizure properties of the high-performance oils which is impossible by means of standardised four-ball tests.

REFERENCES
(1) Patent Application P 311066, Patent Office of the Republic of Poland, 1995.
(1) Patent Application P 309531, Patent Office of the Republic of Poland, 1995.

MODELING OF REACTION FILM FAILURE IN GEAR LUBRICATION

QINYU JIANG

Department of Mechanical Engineering, Dalian Railway Institute, Dalian 116028, P.R. China

ABSTRACT

The failure mechanism of reaction films generated on the gear flanks lubricated by oils, in which anti-scuffing additives(e.g. ZDDP) are contained, has been a controversy and a complicated problem in the field of chemi-tribology. Some researchers regarded that reaction films cease to be effective due to mechanical wear, whilst some others disputed the failure mechanism on that the loss of reaction films is subjected to a certain chemical process of decomposition. The different viewpoints lead to the impossibility of a uniform criterion for evaluating the scuffing load in a wide range of running speed, as a result, the experimental phenomenon that the scuffing load decreases in low speed range and increases in high speed range cannot be explained by a single failure mechanism. This paper attempts to model the reaction film failure according to the different failure mechanisms.

In low speed range, wear of the reaction film occurs because the EHL film is difficult to form, and cannot protect the surfaces from scuffing. With the increase of speed, the EHL film increases in thickness and is unlikely to be broken. In this case, due to high pressure and high temperature, decomposition of the reaction film occurs instead of wear. Variation of the failure mechanism with the running speed is the key to the problem that the scuffing load decreases first, next reaches a valley, and then increases as reported in many literature, such as in (1).

The growth of a reaction film can be modelled physically by a diffusion process, and a modified equation for the growth of the reaction film reads(2)

$$X = KC_r \exp(-E_r/RT) K_d t^{1/2} = K_0 t^{1/2} \qquad 1$$

where X is the thickness of the reaction film, and t the reaction time; the denotation of the other parameters see (2), and all the parameters can be represented by only one overall parameter K_0.

The loss of a reaction film, according to wear mechanism, can be described by Archard's wear model(3)

$$Y = a \cdot P \cdot V \qquad 2$$

where Y is the thickness of worn reaction film, a the wear coefficient peculiar to a reaction film; P is the contact pressure and V the pitch speed of gears.

The loss of reaction film, according to the decomposition mechanism, can be described by a chemical process(3)

$FeB \rightarrow Fe + B$
$X - Z \qquad Z$

$$Z = X'\left\{1 - \exp\left[-tA \exp\left(-E_d/RT\right)\right]\right\}$$
$$= X'[1 - \exp(-bt)] \qquad 3$$

where Z is the thickness of the decomposed element; X' is a part of the formed film participating in the decomposition reaction, and $X'/X = C_0 < 1$.

If the critical state is approached under which scuffing is likely to occur, the relations between the growth and loss of the reaction film are as following

$Y/X = c_1$ (wear mechanism, WM) 4

$Z/X = c_2$ (decomposition mechanism, DM) 5

where c_1 and c_2 are the critical value of the occurrence of scuffing failure.

For gear transmission

$$t = K' P^{1/2}/V \qquad 6$$

where K' is a overall parameter for considering geometric and material factors as well as the conversion the tangential speed into pitch line speed.

Considering all the above equations, we get

$$P^{3/4}V^{3/2} = c_1 K_0 K'^{1/2}/a = C_1 \quad \text{(WM)} \qquad 7$$

$$P^{1/2}V^{-1} = -\ln(1 - c_2/C_0)/bK' = C_2 \quad \text{(DM)} \qquad 8$$

This model accounting for different failure mechanism fits well most experimental results, as reported by Winter in Fig 1.(1).

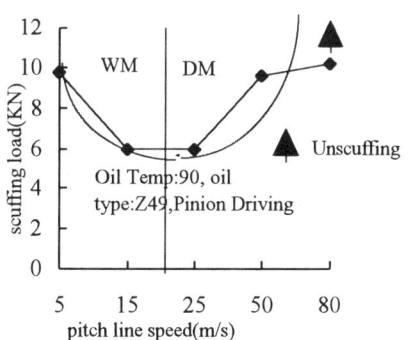

Fig.1 Variation of scuffing load with speed of gears

REFERENCES

(1) H.Winter,K.Michaelis,,AGMA,90FTM8.
(2) H. So, Y.C. Lin, Wear177(1994)105-115.
(3) J.Halling, Principles of Tribology. The Macmillan Press LTD.1975.
(4) Q.Y. Jiang, Doctoral Dissertation, Harbin Institute of Technology, Harbin, China,1995

ELECTRIFICATION OF OIL AND FILTER ELEMENT - PART II
PRODUCTION OF FREE RADICALS AND OIL AUTOXIDATION

AKIRA SASAKI and SHINJI UCHIYAMA
KLEENTEK Industrial Co., Ltd. 2-7-7 Higashi-ohi, Shinagawa-ku, Tokyo 140, Japan
TAKASHI YAMAMOTO
Dept. of Mech. Eng., Tokyo University of Agri. & Tech., 2-24-16 Nakamachi, Koganei, Tokyo 184, Japan

ABSTRACT

The first paper demonstrated that static electricity would be generated in the process of filtration by friction of oil with a filter element which are dielectric and that spark discharges of the accumulated electric charge on the filter element were strong enough to crack the oil molecules (1).

This paper is intended to investigate if the filtered oil has free radicals which accelerate oil oxidation. Two kinds of oil samples were investigated; one was the oil in which static electricity accumulated on the filter element was discharged with sparks and the other the circulating oil which passed through a filter element for 30 days.

The spark discharge test device which was explained in the first paper was used for the test. The static electricity generated on a filter element by friction with oil was taken out of the filter element by induction through the insulated filter housing. The test device with a pair of electrodes was placed in the 150 ml of inhibited turbine oil. Whenever spark discharges happened between the pair of electrodes, gas bubbles and a trace of carbon blacks were ejected in the oil. The oils in which spark discharges happened were analyzed by Gas Chromatography. The analyses showed that the main components of the gases in the oil samples were hydrogen and the low molecular weight hydrocarbons which were produced by cracking of oil molecules.

Oil samples were taken after 500, 2000 and 3000 spark discharges. Immediately their TANs were measured. The TANs of them were 0.08 mgKOH/g as much as that of the new oil. The new oil and the oils having had 500, 2000 and 3000 spark discharges were put in the glass bottles and kept in a dark locker at room temperature for 6 and 9 months. The TANs of the oils having had 500, 2000 and 3000 spark discharges and having been stored for 6 months were 0.09 mgKOH/g, 0.36 mgKOH/g and 0.56 mgKOH/g respectively and those having had 2000 and 3000 spark discharges and having been stored for 9 months were 0.40 mgKOH/g and 0.74 mgKOH/g respectively, although the TAN of the new oil remained unchanged after 6 and 9 month storage. The IR Spectra of the oils which were stored for 6 months after 500, 2000 and 3000 spark discharges showed substantial absorption at 1720 cm^{-1}, 1690 cm^{-1} and 1650 cm^{-1} which indicate C=O stretching and oil oxidation (2). It is known that autoxidation commences by the production of free radicals in oil (3). The fact that oil oxidation has progressed in the oil which has been kept in a dark place at room temperature suggests that free radicals are produced in oil by cracking of oil molecules.

When the 1500 ml of test oil was cleaned with the grounded depth filter, spark discharges were not seen. After having been cleaned for 300 hours, the oil was cooled down to room temperature for 24 hours and 150 ml of oil was taken in a plastic bottle with 350 ml volume and the sealed bottle was stored in a dark place at room temperature. The plastic bottle was found deformed and dented after having been stored for 4 months. The TANs of the oil were 0.33 mgKOH/g and 0.77 mgKOH/g respectively after 6 and 9 month storage, although that of the new oil was 0.02 mgKOH/g and remained unchanged by storage. The oil sample after 4 month storage was analyzed by IR Spectroscopy. A sharp absorption was found at 1750 cm^{-1} which indicated oil oxidation. The facts that TANs of both the oil having been exposed to spark discharges and the oil having been cleaned with depth filter became substantially high after having been stored in a dark place at room temperature for several months and that IR spectra of them showed sharp absorption indicating C=O stretching demonstrate that filtration produces free radicals and promotes oil oxidation.

Oil oxidation is not favorable to hydraulic and lubricating oils. This study demonstrates that there is some possibility that conventional mechanical filters crack oil molecules by spark discharges of the accumulated static electricity and accelerate deterioration of oils which must be protected, although they can remove particulate contaminants with micron sizes.

REFERENCE

(1) Sasaki, A., et al, L07/C491-274, WTC, Sept., 1997.
(2) Horiguchi, H., IR Chart Book (Japanese), Sankyo Shuppan, 1977, Chapter 17.
(3) Water, W. A., Mechanisms of Oxidation of Organic Compounds, Mrthuen & Co. Ltd., 1964, Chapter 2.

TRIBOCHEMISTRY AND EP ACTIVITY ASSESSMENT OF Mo-S COMPLEXES IN LITHIUM BASE GREASE

T SINGH* AND M F SAIT
Product Development Centre, Hindustan Petroleum Corporation Ltd,
Haybunder Road, Bombay - 400 033, India

ABSTRACT

The use of molybdenum disulphide as an excellent solid lubricant is well known, however, its insolubility in oil prevents its use in liquid lubricants. Oil-soluble organomolybdenum compounds, e.g., molybdenum dialkyldithiocarbamates (MoDTC) and molybdenum dialkyldithiophosphates (MoDTP), have been used as excellent anti-friction, anti-wear and extreme pressure additives in lubricating oils and greases. These additives are known to increase the load-carrying capacity of lubricants and reduce fuel consumption and power loss by reducing friction (1-3).

Lubricants containing oil-soluble organo-molybdenum compounds, sometimes referred as third generation lubricants, have been known to show outstanding advantages over conventional lubricants in enhancing the component life, reducing operating temperatures and greatly extending the lubrication intervals.

It is believed that the molybdenum and sulphur present in the additive form a low-friction surface film during operation of a machine under high loads. In view of these observations and our interest (4) in search for better extreme pressure additives, in this paper we are reporting few potential molybdeno-sulphur complexes as a potential extreme pressure additives for lithium base lubricating grease and their tribochemistry by AES and SEM techniques.

Lithium base greases are widely used in the industrial machinery and automobile (5). These greases are formulated with various compounds with a view to optimising the end use. In steel mill applications, lithium-12-hydroxystearate greases, formulated with extreme pressure additives, have performed well and it varies from inter plant to coke oven. These greases, have shown better performance, e.g., high drop point, better anti-wear and extreme pressure properties, better oxidation stability, pumpability as well as better wheel bearing performance as compared to conventional calcium base grease. The use of certain S-P and Pb-S system in the lithium base grease has been on record (6).

The blends of bis(1,5-diaryl-2,4-dithiomalonamido)dioxomolybdenum (VI) complexes in lithium base grease are evaluated for their extreme pressure activity in a 'four-ball test' using 12.7 mm diameter alloy steel ball specimen. The additives, bis(1,5-di-p-methoxyphenyl-2,4-dithiomalo-namido)dioxo-molybdenum (VI) and bis(1,5-di-p-chlorophenyl-2,4dithiomalonamido)di-oxo-molybdenum (VI) exhibited lower values of wear scar diameter at higher load and higher values of weld load, flash temperature parameter and pressure wear index as compared with lithium base grease without additives.

The topography and tribochemistry carried out by means of Scanning Electron Microscopy and Auger Electron Spectroscopy techniques, respectively.

REFERENCES
(1) P C H Mitchell, Wear, Vol.100, p 281, 1994.
(2) Zhen En-cai Qian Xiang-lin, Wear, Vol. 130, p 233, 1989.
(3) Y Yamamoto, S Gondo, T Kamakura, N Tanaka, Wear, Vol. 112, p 79, 1986.
(4) T Singh, V K Verma, Indian J. Tech., Vol.28, p 649, 1990.
(5) C J Boner, NLGI, USA, 1983.
(6) A G Izcue, NLGI Spokesman, Vol. 44, No. 11, p 280, 1980.

RAPIDLY BIODEGRADABLE HYDRAULIC FLUIDS ON THE BASIS OF RAPE SEED OIL

R. ŠRAJ and J. VIŽINTIN

University of Ljubljana, Faculty of Mechanical Engineering, Centre of Tribology and Technical Diagnostics, Bogišičeva ul. 8, 1000 Ljubljana, Slovenia

ABSTRACT

Environmental awareness and the protection of natural resources are two of the most important problems today. Their part to pollution contributes also toxic and non biodegradable hydraulic fluids. The reasons for pollution are mostly leakage and accidents. These initiatives have prompted several lubricant companies to develop environmentally acceptable formulations of hydraulic fluids on the basis of rape seed oil.

The properties of refined rapeseed oils in comparison to mineral base oils are widely discussed in the literature (1-5). The rapeseed hydraulic oils can offer significant environmental and other advantages with respect to resource renewability, biodegradability and toxicity when compared to mineral oil. However, they cannot be freely substituted for mineral oils in hydraulic system. On one hand the rapeseed hydraulic oils offer a good lubricity, on the other hand they are oxidative and hydrolytic instable because of the chemical nature of triglycerides.

The development, laboratory and field testing of the rapeseed based hydraulic fluids are described in the paper. The first step at testing rapeseed based hydraulic fluids that we have done is testing of their physical, chemical and mechanical properties with the standardizational methods. The results of these tests had to be in the limit that are regulated by the standards for hydraulic fluids (6). When the standard tests were finished, the testing was proceeded on the hydraulic system in laboratory and on hydraulic system in praxis. In the framework of the project of testing hydraulic fluids we also test hydraulic components. Up to date hydraulic components are optimised for the use of hydraulic fluids on the basis of mineral oils. Some additional conditions before the use of biodegradable hydraulic fluids have to be fulfilled for the sureness of working and attaining optimal life time of hydraulic components and hydraulic fluids. For this purpose we carry out tests on two identical testable hydraulic systems in the laboratory in the Center of Tribology and Technical Diagnostics. These two hydraulic systems are projected on the basis of the working analysis of more hydraulic systems in practice that operate with hydraulic fluids on the basis of mineral oils. The tests go on simultaneously on both testable hydraulic systems with two different hydraulic fluids on the basis of rapeseed oil and on the basis of mineral oil. Hydraulic fluids and hydraulic components are tested in various working conditions, the tests are permanent and cyclic. The working parameters of cyclic tests are based on the statistical analysis of working conditions of hydraulic systems in practice.

Through simulated working conditions on a laboratory hydraulic system, the behaviour of rapeseed hydraulic oils with respect to their oxidative and thermal stability, anti-wear performance, hygroscopic characteristics and particle build-up has been studied. When the laboratory tests were finished, the testing was proceeded on hydraulic system on the dredger in praxis.

Results of the tests have shown that the hydraulic systems satisfactorily operate with the developed hydraulic fluid based on rapeseed oil with some restrictions of working parameters: the working temperatures have to be within a certain interval and it is necessary to prevent pollution of fluids with the substances that accelerate their disintegration. Biodegradable hydraulic fluids are specially sensitive to temperature higher than 80°C and lower than -10°C and to the presence of water, that should be below 0,2%. Such development will in future surely exceed the application of refined rapeseed oils as hydraulic fluids.

REFERENCES

(1) Pelzer E.: Normung und Vergaberlichtlinien des Umweltzeichens von umweltschonenden Hydraulikflüssigkeiten, Kontakt&Studium, Band 402, Expert Verlag 1993

(2) Kempermann C., Remmelmann A., Backe W.: Biologisch schnell abbaubare Hydraulikflüssigkeiten, Technische Akademie Esslingen, Lehrgang 19151/68.356, Februar 1995

(3) Anželj M., Arnšek A., Vižintin J.: The use of rapidly biodegradable oils in Slovenia, Proceedings of the Int. Conference on Tribology SLOTRIB '94, Gozd Martuljek, Slovenia, November 10-11, 1994

(4) Vižintin J., Šraj R.: Rapeseed based oils tested for hydraulic systems, Proceedings of the VI. Tribological Conference, Budapest, Hungary, 6.-7. June 1996

(5) Šraj R., Vižintin J.: Environmentally acceptable hydraulic fluids based on rapeseed oil, Proceedings of the Int.Conference on Tribology SLOTRIB '96, Gozd Martuljek, Slovenia, November 13-14, 1996

(6) VDMA - Einheitsblatt 24 568 - Biologisch schnell abbaubare Hydraulikflüssigkeiten - Technische Mindestanforderungen

INFLUENCE OF THE RATIOS AMONG ADDITIVES, CONTAINING BORON, SULPHUR AND NITROGEN ON ANTIWEAR PROPERTIES OF EP PACKAGES

K. G. STANULOV, H. HARHARA, G. S. CHOLAKOV, and H. A. IONTCHEV,
University of Chemical Technology and Metallurgy, boul. Kl. Ohridsky N 8, Sofia 1756, Bulgaria

ABSTRACT

The use of boric compounds in combinations with conventional additives is a subject of theoretical and practical interest for the modern lubricant technology. The polyelement character of such packages may have different influence on the properties of the lubricant (1)(2)(3). Compatibility among additives may lead to synergism in some and antagonism in other functional effects, depending on chemical structure, balance of active elements, and other factors.

The present work investigates antiwear properties of mixtures of sulphur, nitrogen and boron containing additives in up to 4.5 % overall additive concentration in a base mineral oil with kinematic viscosity 218 mm^2 s^{-1} at 40 °C. The antiwear properties of the doped oils are characterized by the average wear scar diameter (d_{av}), estimated on a standard four ball tribometer in a 60 min test at 75 °C, under 400 N normal load. The optimization is performed with a designed experiment, following a D-optimal design (4) of 20 points and 3 factors: x1 = concentration of borated nonyl phenol with 1.13 % boron content; x2 = concentration of sulphuruzed esters of plant fatty acids with sulphur content of 11 %, and x3 = concentration of "pure" grade oleyl amine. Each of the factors is varied within a 0 - 1.5 % concentration range, and the regression equation is of the type:

$$Y = b_0 + \Sigma b_i x_i + \Sigma b_{ij} x_{ij} + \Sigma b_{ii} x_i^2 + \Sigma b_{ijk} x_i x_j x_k + \Sigma b_{iii} x_i^3,$$

where b_0, b_i and b_j, etc. are the coefficients, x_i, x_j, x_k, etc. - the independent variables, and Y - the dependent variable (d_{av}) of the regression model.

The constants of the model, derived from the experimental results and its statistical parameters are reported in the full text of the paper. The model was used for the creation of different diagrams by varying the concentrations of the additives.

Fig. 1 shows one of these diagrams. It describes the influence of the boric and the sulphur containing additives on the wear scar diameter at constant concentration of the oleylamine, x3 = 0.37 %.

The increase of the concentration of the boric additive decreases the antiwear properties of the doped oils. However, various specifications of doped industrial oils, require that the wear scar diameter diameter should be less than 0.5 mm. Such diameters can be realized with the compositions included in the hatched

area, shown on Fig.1., if the boric concentration is below 0.7 %. Alternatively, if the concentration of the boric additive in the oil is above the critical 0.7 %, wear scar diameters will be higher than 0.5 mm, regardless of the concentration of sulphuruzed esters.

If the concentration of the boric additive is less than 0.5 %, a wear scar diameter of 0.3 mm can be mantained, while varying the concentrations of the sulphur additive between 0.35 and 1.20 %, depending on the desired level of the other functional properties - antiscuffing, antioxidant, etc.

The optimization approach, illustrated in detail in the full text of the paper, provides an appropriate tool for the study of polyelement additive packages, and their influence on the wide variety of functional properties of different lubricants. It allows for the formulation of the most economically feasible composition with a guaranteed property balance, while also saving time and money for research.

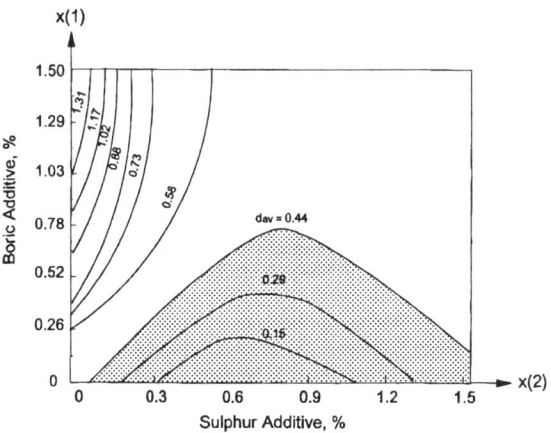

Fig.1. Areas of constant values of the wear scar dia. at concentration of the oleyl amine, x3 = 0.37 %.

REFERENCES

(1) W. Liu, Z. Jin, Q. Xue, Lubr. Sci., Vol. 7, No. 1, Page 49, 1994.
(2) K. Stanulov, G. Cholakov, I. Cheriisky, Oxid. Comm., Vol. 17, No. 3, 220, 1994.
(3) US Pat. 3 666 6662, 1972.

TRBO-ECONOMIC ASPECT OF CUTTING FLUID APPLICATION

BOGDAN VASILJEVIC and MIROSLAV BABIC
University of Kragujevac, Faculty of Mechanical Engineering, Kragujevac, Yugoslavia

ABSTRACT

Optimal application of cutting fluids (CF) in metal working industry assumes multidisciplinary approach to this problem. Relaying only on CF manufacturers' recommendations, which is the most frequent case in practice, does not guarantee minimal costs. The multidisciplinarity of the optimization problem of working with CF assumes engagement of the group of experts of different profiles. The areas that should be covered are: tribology, economy, management, and ecology.

Roughly, this problem can be divided into three basic parts:

- Determination of tribological characteristics of cutting fluids that are used in the manufacturing process, what assumes determination of correlation between the cutting tool life and concentration of the applied cutting fluids, and corresponding tool life and machined surface quality;

- Analysis of CF application costs, starting from CF purchasing, preparation of emulsions and solutions, pouring, replacing the worn CF, maintenance, pouring out, and reservoir cleaning.

- Analysis of costs of decomposition or storage of worn CF, as well as costs caused by hazardous effects of cutting fluids on workers and environment.

Creation of tribological data base with CF tribological characteristics in the concrete manufacturing conditions represents the most complex point of optimization of working with cutting fluids. Determination of cutting fluids tribological characteristics, expressed through the tool life and the machined surface quality in the longer time interval, on the chosen machines, and experimentally in laboratory conditions through the wear resistance, represents the basis of the method. Through the statistical data processing of investigations' results the optimal cutting fluid concentrations and the best kinds of cutting fluid are determined, from the tribological aspect.

Fig. 1 show the difference between the best concentration of cutting fluids based on tribological investigations and the optimal concentration of cutting fluids based on total costs analysis.

Tribological characteristics of cutting fluids determined in this way are then brought into correlation with manufacturing operation costs, namely the tool costs. In this way is possible to model the influence of deviation from optimal concentration from tribological aspect of costs increase. The curve that represents this type of costs has the minimum in the area of optimal concentration.

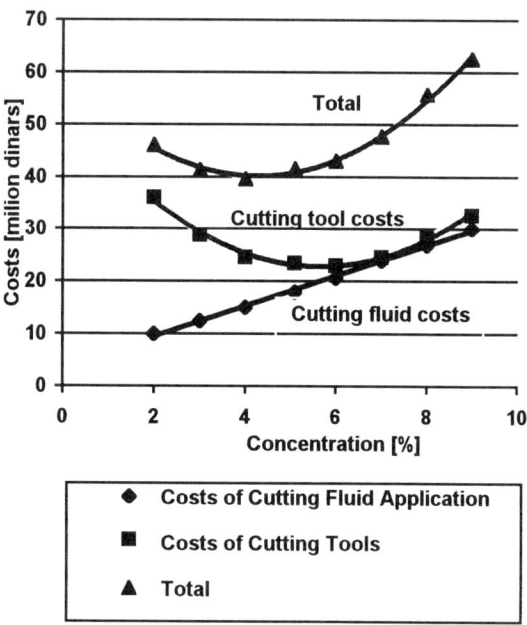

Fig. 1

Based on the proposed model are formed matrices of costs of working with CF, as well as matrices of related costs, like the cutting tool costs and the rejects caused by CF. Thus formed costs matrices are, by computer, summarized according to types of applied CF, applied concentrations and machining types. Results of investigations show that the increase of the CF concentration for only 1 % from the optimal one, causes significant increase of total costs.

REFERENCES
(1) T.Mang, A.Tweebeeke, Economical evaluation of the application of watermiscible metalworking fluids, Lubrication Engineering, 6, 1983.
(2) B. Vasiljevic, Costs of Application of cutting fluids, Tribology in industry, 2, 1983.

TRIBOLOGICAL PERFORMANCE AND ACTION MECHANISM OF CERTAIN S, N HETEROCYCLIC COMPOUNDS AS THE POTENTIAL LUBRICATING OIL ADDITIVES

Y WAN, W YAO, X YE, L CAO, G SHEN
Department of Chemistry and National Tribology Laboratory, Tsinghua University, Beijing 100084, P. R. China
Q YUE, T SUN
R & D Center, Tianrun Chemical Co. LTD, Beijing 100085, P. R. China

ABSTRACT

Organic compounds containing active elements such as S, P, Cl, N, B as well as organometallic compounds have been widely used as antiwear and extreme pressure additives in lubricating oil. Recently, the friction and wear properties of heterocyclic compounds are currently attracted a considerable mount of research effects (1-3). In view of these observation and our continued effects to design and develop benzothiazole derivatives as potential additives (3), it was considered worthwhile to prepare new additives and to evaluate tribological performances.

The additive, 2-(N-octylcarbamoylmethyl)thio-benzothiazole(BB-2), was prepared in-house. The compound was characterized by the TLC method, element analysis, IR and ^{1}HNMR spectrophotometeric techniques.

The antiwear properties of the novel additive were evaluated by four-ball machine under the following conditions: rotating rate: 1480 rpm, testing duration: 15 min, load: 20, 30, 40, 50 kg, temperature: 20 °C. The load-carrying of additive was obtained by GB 3142-82, similar to ASTM D2783. For comparison, the lubricating performance of zinc butyloctyldithiophosphate, a commercial product, was also evaluated.

The EP performance of additives in base stock was evaluated as maximum non-seizure load, Pb value. The results summarized in Table 1 indicate that the novel compound improved the EP property of base stock. The load-carrying capacity is equal to that of ZDDP.

Table 1 Comparison of load-carrying capacity of novel additive and ZDDP

Additives	Concentration (wt%)	Maximum non-seizure load (kg)
BB-2	0.5	78
ZDDP	0.5	78
Base stock	-	40

The antiwear properties of additives in base stock as the function of load are shown in Figure 1. It is clear that the novel additive has better lubricating performance than base stock. Furthermore, the performance is similar to that of ZDDP, especially at higher load.

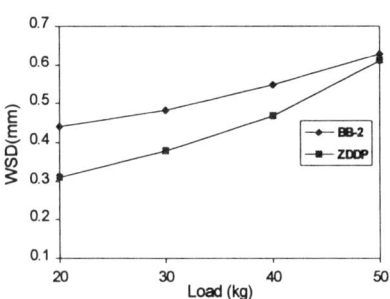

Figure 1 Effects of applied load on the wear properties of novel additive at concentration of 0.5wt%

Tribochemical reaction and acting mechanism of the novel compound using X-ray photoelectron spectroscopy(XPS) and scanning Auger microprobe (SAM). It can be seen that the surface film is consist of carbon, sulfur, oxygen and nitrogen. Furthermore, there is a sulfur-rich layer on the surface. The sulfate layer only exist in the outer surface. FeS_2 are the main components in the inner layer which is effective at reducing the wear and increasing load-capacity.

With respect to the standpoint of coordination chemistry theory, the molecular structure of the novel compounds described here is an excellent polydentate ligand with strong coordinate capacity. Therefore, the novel compounds can interact with the surface to form a stable protective layer which is effective at reducing the wear and inhibiting the corrosion of metal. In addition, the sulfur element in the compound reacts with the surface to produce a chemical reaction film on the metal surface which can further improve the extreme pressure characteristic and load-carrying capacity.

REFERENCES

[1] T. Singh and C. V. Chandrasekharan, Tribl. Int., 26(1993)245-250
[2] T. Singh, A. Bhattacharya and V. K. Verma, Lubr. Eng., 46(1989)681-685
[3] Y. Wan, Q. Pu, Q. Xue and Z. Su, Wear, 192(1996)74-77

ESTIMATION OF THE LIFE OF A LUBRICANT USING THERMOGRAVIMETRIC ANALYSIS

FUMIHIKO YOKOYAMA and TAKAHISA KIMIJIMA
Ishikawajima-Harima Heavy Industries Co., Ltd. Japan

ABSTRACT

It is one of major concern to determine the stability of oils, in particular the oil life, when lubricants are used at high temperature. In general, the stability of oils have been estimated using isothermal bulk oxidation tests such as corrosion & oxidation test (C&O test) conforms to Federal-Std-Method-791. In these tests, changes in properties, such as total acid number and viscosity through oxidation, are measured. However, these tests generally take long time and need large amount of sample. On the other hand, thermogravimetric analysis (TG) enable the test duration and the sample amount to minimize and is easy and convenient to estimate the stability of oils (1). In addition, it is easy to calculate the thermal factor based on TG-curve obtained from TG. Therefore, TG was adopted in this study to estimate the stability of oils, in particular the oil life.

The objectives of this study are to investigate the method of estimation of the life of a lubricant using TG and to evaluate the correlation with C&O test.

TG experiment involves heating a thin film of sample (10mg) under flowing synthetic air (Ar: 40ml/min, O$_2$: 10ml/min) and measuring the weight loss as temperature is raised from room temperature to 500℃, at 10℃/min.

Applying the Arrhenius rate law, reaction rate quotation was supposed to be as followed([a]). Thermal factors, such as activation energy, power of reaction and frequency factor, were calculated from TG-curve basically using Freeman-Carroll method (2).

$$-dX/dt = k X^n = A e^{-E/RT} X^n \quad \cdots\cdots \text{[a]}$$

Where k is the rate constant, X is the apparent residual weight fraction at time t, n is the power of reaction, A is the frequency factor, E is the activation energy, R is the gas constant, and T is the absolute temperature.

Putting these thermal factor into the integral form of [a], and the apparent residual weight fraction with time at variable temperature were obtained. In addition, it is possible to predict the oil life with temperature using transformation of this integral form of [a]. The curve of Figure 1 illustrates the oil life with temperature in the case of X=0.01, 0.001. This is the results of oils meet MIL-L-23699.

In addition, C&O test were also conducted regards these oils. In this isothermal oxidation bulk test, the oil life was determined as the time to be reached the changes in TAN of 3mgKOH/g or the ratio changed in viscosity of 25% at each temperatures. The dot of Figure 1 shows the oil life obtained from them.

Considering the correlation between the oil life obtained from the estimation using TG and the oil life obtained from C&O test, they have good correlation regards the slope. Putting 0.001-0.01 into X in the transformation of this integral form of [a], it is available to estimate the oil life without conducting the C&O test. This method using TG can be effectively used in estimating the oil life. In addition, it thought to be suitable to screen high temperature liquid lubricant candidates.

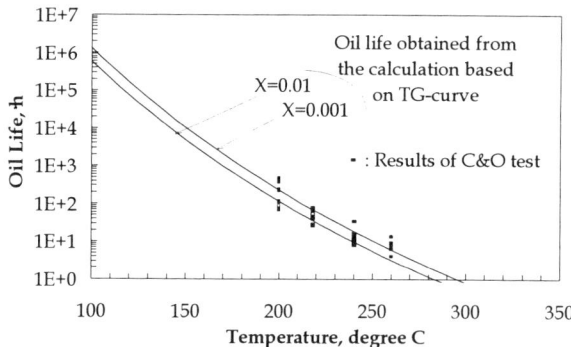

fig. 1: Oil life with temperature
(Lubricants meet MIL-L-23699)

REFERENCES
(1) F. Yokoyama, Proceedings of the International Tribology Conference, Yokohama, 1995, 857-861
(2) E. S. Freeman and B. Carroll, Journal of Physical Chemistry, 62, 1958, 394

TRANSFER AND OXIDATION BEHAVIOUR OF MoS_2 IN TRIBOLOGICAL CONTACTS

S. DEBAUD, S. MISCHLER and D. LANDOLT
Ecole Polytechnique Fédérale de Lausanne
EPFL-DMX-LMCH, CH-1015 Lausanne Switzerland
Phone (+41 21) 693 29 56, Fax (+41 21) 693 39 46

INTRODUCTION

Molybdenum disulphide MoS_2 is widely used in tribological applications because of its solid lubricating properties. For example sputtered MoS_2 coatings can be applied on cutting tools for dry machining. The performance of MoS_2 is often limited by several mechanisms such as film transfer and chemical interactions with oxygen and with the mating material.

The present work was initiated with the aim to characterise the transfer and oxidation behaviour of MoS_2 coatings as a function of the nature of substrate and mating material occurring in sliding contacts. For this frictional test are carried out using a reciprocating wear test rig in inert (nitrogen) or oxidative (oxygen) controlled atmospheres. Worn surfaces are characterised using AES (Auger Electron Spectroscopy) and XPS (X-Ray Photoelectron Spectroscopy).

EXPERIMENTAL SETUP

A Rod-on-Plate reciprocating tribometer was used in a controlled atmosphere chamber. (Fig 1).

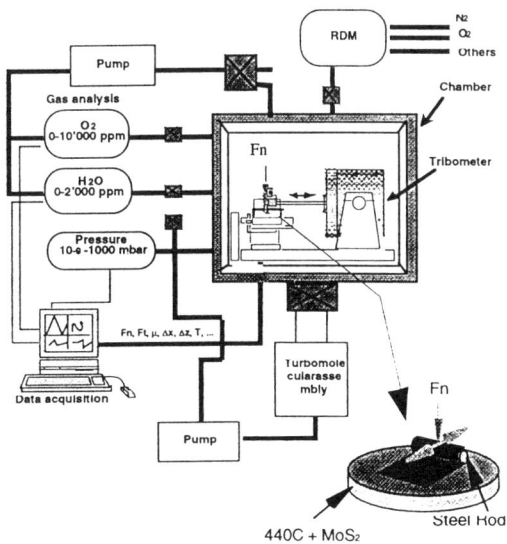

Fig. 1 : Experimental setup

FRICTIONAL BEHAVIOUR

The presence of oxygen in the test atmosphere leads to higher values of the wear rate and of the frictional coefficient (Fig. 2) than in nitrogen.

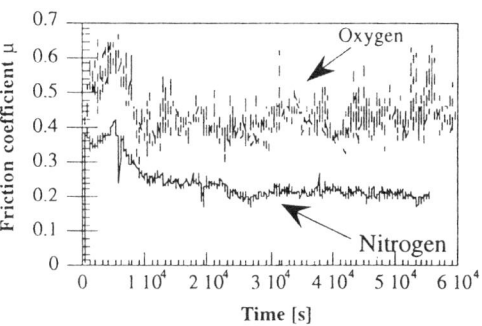

Fig. 2 : Friction coefficient vs Time for Oxygen and Nitrogen environmental conditions.

AES ANALYSIS

Auger analysis (Fig. 3) show that the chemical nature of transfer film observed on the rods depends on the environment: in oxygen atmosphere no sulphur is found on the transfer film which consists mainly of Fe-Mo oxide whilst in the nitrogen atmosphere transfer of molybdenum disulfide is found.

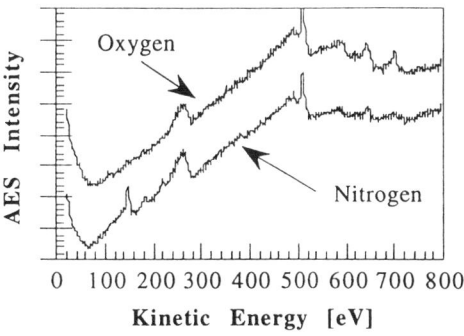

Fig. 3 : AES survey of transfer film for Oxygen and Nitrogen environmental conditions.

CONCLUSIONS

These preliminary results show the important role of transfer film formation and of surface oxidation for the tribological behaviour of MoS_2 coatings.

THE INFLUENCE OF SOLID LUBRICANTS ON PEARLITIC RAIL STEEL FATIGUE LIFE

D I FLETCHER and J H BEYNON

University of Sheffield, Department of Mechanical Engineering, Mappin Street, Sheffield, S1 3JD, UK

ABSTRACT

The use of lubricants on railway systems to minimise rail and wheel wear has recently progressed from the use of grease and liquid lubricants, to the use of solids such as graphite and molybdenum disulphide. These solids are most often applied from train based applicators, as opposed to track based application of previous lubricant types. The implications of the change in lubrication method for the rolling contact fatigue life of the rail have been investigated.

An existing twin disc test machine was used to produce a contact experiencing a rolling sliding motion comparable with the rail-wheel contact, and known to produce valid results for dry and water lubricated conditions. The machine was based around a Colchester Mascot 1600 lathe used to supply up to a 5.6kW drive to the rail test disc, and a 4kW DC motor used to drive the wheel test disc. Speed variation of the DC motor was used to maintain the combined rolling and sliding motion between the test discs. The machine was modified to apply the new types of lubricant to the contact in a realistic manner. Pearlitic rail steel was used in all tests, and all test discs were cut from parent rail and wheel material.

To monitor crack growth a non-contact eddy current probe scanned the rail disc throughout testing, detecting cracks early in their development. Tests were typically stopped with cracks of length 400 μm, allowing their early growth stages to be examined. Fatigue life was defined by the detection of a crack giving an eddy current output trace of height equal to that given by a spark eroded calibration crack.

Trials were run with three proprietary solid lubricant types based on different lubricant combinations, to establish validity of the test method when compared with field trials of the same lubricants. A good correlation was found with preliminary field experience. Results of the initial trials indicated that solid lubricants did not have the detrimental effect on rail fatigue life that has been found with water lubrication. It was identified that different proprietary lubricants could produce up to an order of magnitude difference in rail rolling contact fatigue life. The extent of penetration by the lubricants into microcracks on the rail surface was thought to be central to the rolling contact fatigue life variation seen in the preliminary trials. The preliminary test work also highlighted the need for a comprehensive overhaul of the test machine.

The machine overhaul included fitting a more powerful computer controlled 7.5kW AC motor in place of the 4kW DC motor, and interfacing all instrumentation to a computer to provide comprehensive data logging. A Pentium 75 computer running National Instruments LabView software was used to perform the tasks of machine control, data logging, and providing a user control panel.

Following the overhaul the new machine was capable of speeds from 100RPM to 1600RPM, and could maintain the sliding to rolling velocity ratio at between 0% and 20%. Over the length of the test slip could be maintained to within ±0.02 of the set percentage value, neglecting disc diameter changes during the test. The most demanding tests on dry contacts were restricted to lower speed and slip ranges due to vibration problems, but testing of lubricants was possible over a wide range of slip, load and speed combinations. Water lubrication was found problematic at speeds above 400RPM due to rapid ejection of the water from the contact, which was not found at lower speeds. Normal loading of the discs was applied hydraulically, as before, with new instrumentation recording the contact load showing Hertzian contact pressure in the range 900MPa to 1800MPa being maintained to an accuracy of ±1%. Throughout tests the load, torque due to the tangential force transmitted through the contact, speed of discs, and total numbers of disc revolutions were recorded, giving a full record of the experiment.

The most recent data available from the new machine includes a series of tests to establish the capabilities of the re-built machine, and a series to examine the repeatability of results under particular test conditions. Following this, further work will move away from testing of commercial lubricants, and will examine the development of rolling contact fatigue under variable load and speed situations, as are experienced by rails in service. This work will take place with both water and oil based lubricants containing dispersed solid lubricant particles.

INVESTIGATION OF RUBBING BEARINGS WITHOUT EXTERNAL LUBRICATION

KOZMA MIHÁLY

Institute of Machine Design, Technical University of Budapest, H-1111 Budapest, Műegyetem rkp. 3.

ABSTRACT

Rubbing bearings running without external lubrication are favourable in many applications because of their low maintenance costs and other benefits. A lot of research works have been devoted to develop bearing materials of which plain bearings running without maintenance and lubrication can be manufactured. Polymers and metals with self-lubrication properties are able to fulfil the demands on running without external lubrication

Suitable designed plain bearings made of polymers or polymer linings can carry heavy load at low sliding speed. Some polymers have unbeneficial tribology properties limiting their application. To improve the load carrying capacity and friction of polymers reinforcements and solid lubricants are added into them. Our experiments and also other's one proved, that the tribology properties of filled polymers are not always better than the unfilled one's, or even they can be worse. At low load the contain of silicon oil or MoS_2 do not decreased the wear rate of polyamide (Fig. 1).

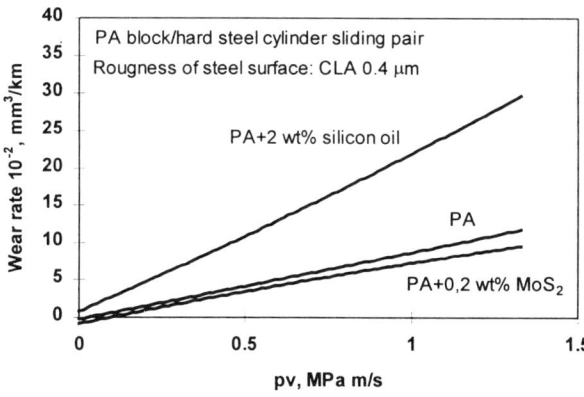

Fig. 1 Wear rate as function of *pv* product

In the case of beneficial effect of fillers the working temperature limits the applicability of polymer composites decreasing their strength and causing softening or decomposition of running surfaces.

Therefore in heavy duty applications and at higher operating temperatures it seems to be reasonable to use metals made suitable to run without external lubrication as bearing materials. There are some metal-matrix composites (e.g. Copper-based, nickel-based, iron-based metal matrixes) being able to carry heavy loads at moderate speed without external lubrication even at higher temperatures (200-700 °C).

Aluminium alloys are advantageous structural material, but have disadvantageous tribology properties: the pure aluminium is soft, its sliding surface can be easily damaged. However, long since sliding bearings are made of aluminium alloys: solid bushes as well as thin wall bushes with aluminium linings which are able to run reliably with proper lubrication but not without it.

According to the data published in technical literature the friction and wear properties and also the scuffing strength of aluminium can be improved by fillers making possible to slide aluminium surfaces without external lubrication. The effectiveness of fillers in aluminium depends on the properties of the metal matrix and the sort and amounts of fillers. We have found some contradictions between the results in literature and the results of our laboratory experiments made on aluminium matrix composites/steel sliding pairs. Investigations carried out on rubbing bearings made of porous aluminium filled with reinforcements (B_4C powder) and different solid lubricants (graphite, MoS_2) showed, that in dry conditions the friction and wear are high, and oil impregnation of the bearings was necessary to reach beneficial friction conditions. The aluminium composite bearings impregnated with oils can run without external lubrication at low friction coefficient and at high load carrying capacity (Fig. 2).

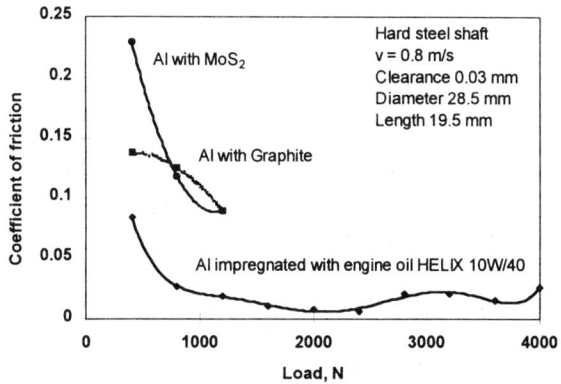

Fig. 2 Coefficient of friction of aluminium composites as function of load

TRIBO BEHAVIOUR OF SOME METALS AND FILLED OR UNFILLED PLASTIC USING SOLID LUBRICANTS

K.R.SATYANARAYAN
Army Institute of Technology, Dighi, Pune - 411 015, India
S.K.BASU
Department of Production Engineering, Government College of Engineering, Pune - 411 005, India

ABSTRACT

The abstract deals with the tribological behaviour of some ferrous, non-ferrous and polymeric materials used for various applications of machine tools in dry and using solid lubricants. Solid lubricants used are MoS2 and Griphite powder mixed with industrial oil in varying concentrations. The experiments were done to find the friction and wear characteristics of materials like EN-8, EN-31, Brass, Copper, PTFE, Nylon, Filled PTFE and Filled Nylon mated with steel using grease based solid lubricants like graphite and MoS2 powder mixed with base oil in varying proportion of 10 to 30 %.

From the results obtained, 'Iso-wear' curves were drawn based on the equations generated through statistical design of experiments. Iso-wear curves are useful to designers and users to determine the treatment combinations of various parameters, so that the wear rate remains uniform and within threshold values. One set of such curves in dry condition is shown in Fig 1. The generalised equation for wear, developed in each case was subjected to sensitivity analysis and the optimum value of the concentration of solid lubricants mixed with industrial oil for minimum wear, found out in each case based on experimental observations. Some representative curves have been plotted to know the optimum values of MoS2 concentration in industrial oil that will cause minimum wear of the mating parts from all such characteristic curves, we confirm that an optimum concentration of 20 % MoS2 in industrial oil in all cases gives the minimum wear of the part. Based on the results of statistically designed experiments, generalised relationship was found out in the form

$$W = C * p^{\alpha} * V^{\beta} * t^{\gamma}$$

Where C - constant, p - pressure on the mating surface in kg /sq.cm., V - velocity of sliding in cm/sec and t - hours. Similar experiments were done with MoS2 filled PTFE and Nylon.

This finding is of primary importance to designers specifying such solid lubricants to reduce the friction and wear. Reduction of wear which is substantial can be seen by using optimum concentration of MoS2 in oil. Graphs from Pin & Disc recording show also the average coefficient of friction in case of M.S., Brass, Copper, PTFE, Nylon while mated with EN 31 under dry as well as lubricated conditions using 20 % MoS2 in oil.

Table 1 : Values of the constant and experimental Wear Equation(2)

Material Pair (Mated with EN31)	Condition	C	α	β	γ
PTFE	Dry	3.08	0.33	0.19	0.23
Nylon	"	0.73	0.06	0.06	0.09
Copper	"	0.32	0.53	0.34	0.21
Brass	"	18	0.32	0.31	0.29
EN-8	"	15.6	0.49	0.22	0.42
Silver-Steel	"	2.8	0.66	0.25	0.31

Brass	Lubricated MoS2	0.44	0.0049	0.10	0.20
Copper	20% Conc.	0.70	0.02	0.02	0.02
EN-8	Indus. Oil	0.67	0.01	0.05	0.02
Silver -Steel		0.10	0.01	0.10	0.14

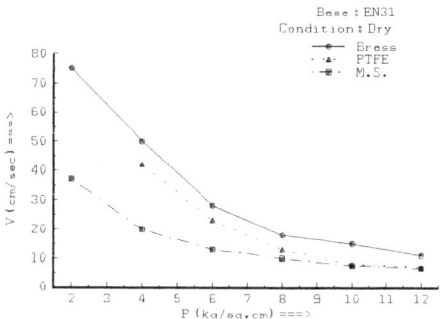

Fig 1 : Iso-wear Curve

REFERENCES :

(1). S. K. Basu, : 'Improvement of Surface Integrity & Frictional Behaviour of Ferrous and Non-ferrous metals & Alloys using Solid Lubricants' C.S.I.R. Project Report, Council of Scientific & Industrial Research, New Delhi, India, 1995

(2) S.C. Goswami and Bandwe S. B. : 'Tribological Behaviour of some metals and Polymers in Dry and Lubricated Condition', M.E. thesis, Dept. of Metallurgy, Govt. College of Engineering Pune - 411 005,India

DEVELOPMENT OF SUPER FINE POWDER MOS₂ + GRAPHITE + PTFE BONDED COMPOSITE COATING

X.Y. SHENG, J.B. LUO, S.Z. WEN

National Tribology Laboratory, Department of Precision Instrument, Tsinghua University, Beijing, 100084, China

ABSTRACT

Solid lubrication is very important in space machines. The reduction of the expenditure of energy caused by friction can save the limited energy of space machines and simplify their structure. Bonded coating is used widely in space machines especially when a stable low static friction coefficient is needed(1). A bonded composite coating is developed in this paper with low static friction coefficient of 0.03. The average diameter of MoS$_2$ used in this paper is 0.8 μm. PTFE used in the experiment is acid PTFE coating material Teflon®. Anti-high temperature bonding agent polymer is used and mixed with MoS$_2$ and graphite in acetone. The mixture is sprayed on the cleaned substrate under the pressure of 5 × 10^5 Pa and then heat at 108°C for more than thirty minutes. The cooperation of solid lubrication MoS$_2$, graphite and PTFE is considered in the development of the coating. The effect of the size of MoS$_2$ on the friction coefficient is discussed. The effect of temperature on the static friction coefficient is also investigated. At last, a bonded composite coating is made which is excellent in friction reduction, anti-high temperature and cohere to the substrate.

We can conclude:
1. The cooperation of MoS$_2$, graphite and PTFE contributes to friction reduction of the coating;
2. The use of super fine powder is also a reason that the coating has a low friction coefficient;
3. The friction-induces basal plane orientation occurs very easy for super fine MoS$_2$ powder.

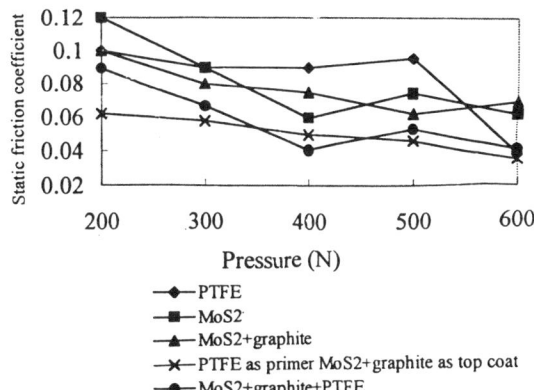

Fig.1: Static friction coefficient of the coatings

REFERENCES

(1) X.Y. Sheng, J.B. Luo, S.Z. Wen, Lubrication Engineering, Vol.2, 1997.

WEAR OF BRONZE-IRON-GRAPHITE COMPOSITES

GUOWEI ZHANG
Laboratory of Solid Lubrication, Chinese Academy of Sciences
WILL SCOTT
School of Mechanical, Manufacturing and Medical Engineering, Queensland University of Technology, PO Box 2434, Brisbane 4001, Australia

ABSTRACT

This study focuses on bronze-iron-graphite self lubricating composite materials. Although a great deal of effort has been devoted to material combinations and/or composite fabrications to obtain low friction and wear there seems to have been little detailed research on bronze-iron-graphite composites.

Seven materials, with the compositions given in Table I were fabricated by pressing the powders (with sieve analysis given in Table 2) to 687 MPa to form "coins" of 25mm diameter and 6mm thick. These were introduced into a tube furnace under a reducing atmosphere (3% hydrogen and 97% nitrogen) flowing at 1 l/min and heated at a rate of 20° C/min to a designated temperature of either 800°C or 875°C. After sintering for 60 minutes the compacts were drawn slowly to the end of the tube and cooled to 60°C before being removed from the reducing atmosphere. Specimens 5mm square were machined from each coin shaped compact. Their surfaces were ground to an average roughness, Ra, of between 0.4 and 0.8 microns. Some were soaked in a bath of ISO VG 22 mineral oil for 60 mins at a temperature of 120°C.

Table 1 Serial numbers, compositions and theoretical densities of the materials

Serial number	Fe	Cu	Sn	Graphite	Theoretical density
010	0	88.5	10	1.50	8.35
460	39	53.99	6.1	0.92	8.16
550	49	45.14	5.1	0.77	8.11
640	59	36.29	4.1	0.62	8.06
552	48.02	44.23	5.0	2.75	7.70
554	47.04	43.33	4.9	4.73	7.34
556	46.06	42.43	4.79	6.72	7.01

Table 2 Sieve analysis of metal powders (total %)

Size (micron)	ASC 100.29	H-jet
+180	1.2	0.1
150-180	8.7	1.0
106-150	17.5	1-9
75-106	24.3	8-24
45-75	24.9	22-42
-45	23.4	30-64

The specimens were subjected to density, hardness and impact strength measurements as well as friction and wear tests on a Timken Extreme Pressure Lubricant Testing machine.

Results of the effects of the different variables were obtained as follows:
- The effect of iron content on the density, relative density, hardness and impact strength
- The effects iron content on wear and friction
- The effect of graphite content on the density, hardness and impact strength
- The effect of graphite content on wear and friction
- Wear and friction of bronze-iron-graphite composites in contact with copper rings (for evaluation as an electrical sliding contacts material)

It was found that the hardness and impact strength of the composites sintered at 875°C were higher than those sintered at 800°C. The hardness increased with increase in iron content for both sintering temperatures. However, although the impact strength increased with iron content for the 875°C sintering temperature, it reduced with increased iron content for the 800°C sintering temperature.

Iron affects the wear performance of iron-bronze composites by two different processes ie (i) increasing the mechanical strength, which may reduce the wear of the composites but may increase the ploughing wear of the matings, and (ii) changing each other's solubility and/or reactivity of the sliding surfaces, which may affect the adhesive wear.

Graphite has opposing effects on the wear performance of metal-graphite composites ie (i) reducing wear by forming a graphite film on the sliding surfaces and (ii) increasing wear by weakening the metal matrix. The overall effect depends on which of the two opposing effects is dominant under the conditions appertaining. For the experimental conditions of this study the increase in iron and graphite content did not significantly affect the coefficient of friction or the wear rate of the mating steel rings. However, it had significant effects when sliding against copper rings.

It was found that the wear of iron-bronze-graphite composites is always superior to that of the sintered bronze, the latter of which is widely used as a self-lubricating material. Iron-bronze-graphite composite is a potential candidate for various sliding contact applications which currently utilise sintered bronze materials.

APPLICATION OF COMPLEX TECHNOLOGICAL METHODS FOR MATERIALS PROCESSING TO GIVE INCREASED DURABILITY OF FRICTION PAIRS UNDER CONDITIONS OF LOW-CYCLE LOADING

L M ABRAMOV, A S ASTAKHIN and I L ABRAMOV
Kovrov State Technological Academy, 601910, Russia, Vladimir Region, Kovrov, Mayakovsky st. 19.

ABSTRACT

Under conditions of low-cycle operation, under high contact stresses exceeding the plasticity limit, the separation of the surfaces can lead to breakdown of the lubricating film. One approach to the problem of increasing the durability of friction pairs, working under conditions of high specific loading, is to increase the plasticity limit of the surfaces and produce material structures better able to resist the extreme contact conditions.

Our approach to solving this problem is based upon the following algorithm:

- manufacture and preparations of friction pairs by conventional methods
- using methods of processing to obtain required levels of physical-mechanical properties in products
- producing surface layers having desired properties (wear resistance, etc.) including composite coatings, produced by laser processing

Current technologies, including laser technology, provide opportunities for solving problems connected with the development of improved surface properties of machine components, irrespective of their design features, by determining the level and distribution of micro-hardness and residual stresses in the component surfaces. Experience shows that the best results in practice are obtained by joint (duplex) processing, such as combined thermal pulsing and pulsing deformation.

The influence of pulsing temperature–mechanical processing on micro-hardness distribution as a function of depth in machine components manufactured from 40X and P6M5 steels has been investigated. Measurements were made on cylindrical and flat specimens before and after processing by controlled explosions on the surfaces. It was established that explosive loading produced surface hardening.

The micro-hardness of 40X increased from 4340 to 6750 MPa to a processed depth of 0.2 ±0.02mm, and from 6670 to 10,420 MPa to a depth of 0.13 ± 0.01 for P6M5. It was established that a given process produces a characteristic level of internal pressure: residual compressive stresses were observed in surface layers, which were virtually zero in central regions.

Comparative tests on industrial components (cutting tools, gear wheels) showed increase service lives of 30–40% on average It should be noted that this method of processing is considerably cheaper and more productive compared with conventional methods of hardening.

In another practical problem the wear resistance of a cutting tool, made from P6M5 steel was increased by means of a surface layer The essence of the process consisted of applying a layer of special material containing special elements on to cutting blanks of the tool, and subsequent explosive processing. This produced a modified surface layer of depth 0.4±0.06 mm. Micro-hardness was increased on average by 1.35 to 1.4 times, and wear resistance by 1.64 to 1.9 times.

The reason for the increase in micro-hardness (and wear resistance) is based upon the special metallurgical structures, not only on their initial nature but also the high-speed nature of their formation and elimination under conditions of significant thermal energy density at processing.

Thus, pulsing temperature–mechanical processing can be applied as one of the most universal methods for improving the quality of surface layers of machine components.

REFERENCES

(1) V.S. Ivanova, et al. Sinergetika and factors in materials science. Moscow. Acad. science, 1994.

(2) L.M. Abramov, A S Astakhin. Proceedings of II ISTC "Wear resistance of machines", Part 2, page 88, 1996.

POSSIBILITIES FOR IMPROVEMENT OF TOOLS FOR METAL POWDER PRESSING

MIROSLAV BABIC and STJEPAN PANIC
University of Kragujevac, Faculty of Mechanical Engineering, Kragujevac, Yugoslavia
MIODRAG ZLATANOVIC
Iniversity of Belgrade, Faculty of Electrical Engineering, Belgrade, Yugoslavia

ABSTRACT

Application of the metallurgy based products is today practically unlimited, and in many cases the only possible one as well. In that, the most present are products based on the sintered metals and their alloys. Fulfillment of strict requirements with regards to dimensions, shapes and density of the sintered parts, as well as the economic indicators of the machining process, is directly caused by decrease of numerous, mutually superimposed negative consequences of friction and wear in the contact of the working elements of tool and powder, during the sintering process and pushing out the pressed parts.

Results that are presented and analyzed in this paper, are related to tribological effects of application of different tool materials (tool steel S4150, with a hardness 61 HRC, high speed steel S7680 with a hardness 62 HRC, tool steel obtained by powder metallurgy ASP-23 with a hardness 62 HRC, hard metal with a hardness 70 HRC, tool ceramics Al_2O_3 with a hardness 60 HRC), and to different procedures of contact surfaces modifications. In the paper are given results of tribological investigations of three types of modification: a) by changing of the base material properties in the surface layer by the plasma nitriding procedure, b) by forming the TiN coating by application of the reactive magnetron sputter ion plating procedure, and c) by duplex treatment - plasma nitriding and TiN coating.

Tribological investigations are of the model type and they were conducted on the tribometer with the pin on disc contact pair geometry. As pins were used the samples made of the tested tool materials with the unmodified and modified front (contact) surfaces. Disks were the pressed pieces of the Fe powder with precisely defined characteristics realized in the stable sintering conditions.

The contact conditions were modeled based on data obtained by measurements of radial and residual sintering pressure in real sintering conditions. As the simulation criteria were used the dominant mechanisms of the abrasive wear and process temperatures in the real system and on the model.

The obtained results point to very significant differences in tribological behavior of tested base materials. Especially superior characteristics, both in regard with friction and wear, correspond to hard metal samples. In all the tested cases the contact surfaces modification procedures have exhibit great potentials with respect to improvement of tribological properties of base materials, and especially from the aspect of wear resistance (Fig. 1.). In that, different materials are characterized by different degrees of improvement. Thus, for instance, the application of the TiN coating and the duplex treatment have especially positive effects on samples made of high speed cutting steel ASP-23.

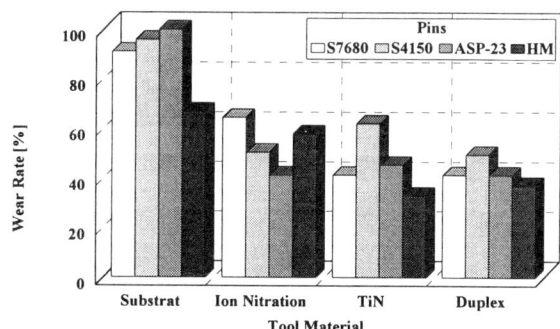

Fig. 1: **Wear rate for different tool materials**

Established effects of the contact surfaces modification, besides in view of contribution to savings of high quality materials, through increase of tool working life, are especially important from the aspect of their substitution with materials of lower tribological quality and price.

REFERENCES

(1) M. Babic, Tribo-economic Aspect of Contact Surfaces Modification, Tribology in industry, 2, 1996, 45-50

(2) F. R. Mallender, D. S. Coleman, Tool Materials, their Friction Coefficients and Estimated Wear Rates in Compation and Ejection of Iron Powder and Zinc Stearate Mixtures, Pow. Met. Int. v. 23, No 5, 1991, p. 121.

STUDY OF ADHESION AND WEAR RESISTANCE OF SEVERAL HARD COATINGS

H. HOUMID BENNANI, J.Y. RAUCH, J. TAKADOUM
Laboratoire de Microanalyse des Surfaces, ENSMM, 26, chemin de l'Epitaphe,
25030 Besançon cedex, France
C. ROUSSELOT
Laboratoire de Métrologie des Interfaces Techniques, 4 place Tharradin,
25020 Montbéliard, France

ABSTRACT

In the present paper, we have studied adhesion and tribological properties of the following materials :
- TiN and TiCN obtained by reactive ion plating
- Diamond-like carbone films (DLC) prepared using a radio frequency plasma system (PECVD)
- $Ti_x Al_{1-x} N$ films with various composition produced by rf reactive magnetron. The composition of the alloy varied between TiN and AlN, x ranging from 0 to 1.

higher friction coefficient (Table 1, columns 2 and 4) while the harder film (DLC) gives the lowest wear volume for the steel ball due to its low friction coefficient. The TiCN coating shows an intermediate behaviour between TiN and DLC films.

Coating	Hardness (Hv)	Lc (N) (± 3)	Friction coefficient (steel ball)	Friction coefficient (alumina ball)
TiN	2 100	25	0.65	0.20
TiCN	2 600	39	0.55	0.15
DLC	3 800	13	0.15	0.15
$Ti_{0.25} Al_{0.75} N$	1 775	22	0.80	0.90
$Ti_{0.33} Al_{0.67} N$	1 930	26	0.80	0.80
$Ti_{0.50} Al_{0.50} N$	2 170	23	0.60	0.80
$Ti_{0.75} Al_{0.25} N$	2 350	23	0.60	0.75

Table 1 : Column 2 : hardness of the coatings, column 3 : critical scratch test load, columns 4 and 5 : friction coefficients.

The coatings were deposited on Z85WDCV6542 steel with a hardness of 880 Hv.

The coating thicknesses were determined from metallographically prepared cross-sections and varied between 3 and 4 µm.

Hardness of the coatings was measured using a Vickers micro-hardness tester using a light load of 15 g. The hardness values are reported in table 1 (column 2).

The scratch tests were performed using the CSEM-Revetest automatic fitted with a Rockwell C diamond stylus. The critical scratch-test loads (Lc) obtained for all the films are presented in table 1 (column 3). Sliding tests were carried out using a pin-on-disk tribometer. The experiments were conducted in air at room temperature. The riders were spheres of polycrystalline alumina (applied load 25N) or 100C6 steel (applied load 10N) with a diameter of 5 mm. The motion was reciprocal and the length of the wear track was 15 mm. The total sliding distance was 8 m.

The values of friction coefficient are presented in columns 4 and 5 (table 1).

Fig. 1 shows the volume wear values of steel after sliding against TiN, TiCN and DLC. It appears that the wear of the steel ball does not increase with the hardness of the coating but depends on the coefficient of friction between steel and the film. A higher wear volume was found after sliding against TiN (the softer coating) which presents the

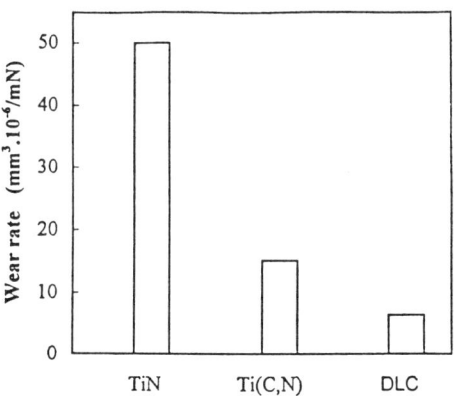

Fig.1 : Wear volume of the 100C6 steel ball after sliding against TiN, TiCN and DLC.

REFERENCES

(1) J. Takadoum, H. Houmid Bennani, M. Allouard, Surf. Coat. Technol., Vol 88, page 232, 1996
(2) J. Takadoum, H. Houmid Bennani, D. Mairey, Z. Zsiga, Accepted for publication in Journal of the European Ceramic Society
(3) J.Y. Rauch, C. Rousselot, N. Martin, J. Takadoum, V. Schmitter, Presented at 10th International Conference on Thin Films, Salamanca, 23-27 September 1996, Spain

MANUFACTURING AND REBUILDING OF ROLLING MILL ROLLS BY SURFACING

PAVEL BLAŠKOVIČ

Department of Flux Covered Arc Welding and Surfacing

Welding Research Institute, Račianska 71, 832 59 Bratislava, Slovakia

ABSTRACT

Production of rolls for rolling mills and their renovation represents a phenomenon which is economically very significant for metallurgists. It can be simply proved that the roll as a tool significantly affects the quality of rolled products and also continuity of production related to its life.

In the world we ever more often encounter the metallurgical rolls with two, even more surface layers, where the quality of working surface can be separated from that of base body of the roll. Such is also the principle of rolls where the working surface is obtained by welding technologies.

Moreover, the new types of steel rolls /1, 2/ (HSS) have shown great possibilities of their renovation but also their production by welding technologies.

We shall indicate what is now available for the mentioned purposes.

The main advantage of surfacing technology application is a good control over the process micrometallurgy and consequently improved quality and purity of the deposited layer and most recently also fabrication of composite weld overlays.

The mostly used surfacing technology is submerged arc process. It is used for deposition of block rolls, biller rolls, prefinishing rolls and for the rolls of continual casting line.

The base metal of surfaced rolls is selected from the following steel types:

1. 0.6C, 1Cr, 0.3Mo - steel in accordance with STN 42 2866.5

2. 0.8C, 1.4Cr, 0.3Mo - steel in accordance with STN 42 2865.9

3. 0.5C, 1Cr - forged steel in accordance with STN 14 161

4. 0.6C - forged steel in accordance with STN 12 060.9

5. 0.5C, 0.8Cr - steel in accordance with STN 42 2739.5.

The choice of base metal depends on overall loading of roll and its dimensions.

For submerged arc surfacing the rolling mill rolls and rolls for continual casting, more than ten types of tubular wire and flux cored strip electrodes, including the flux combination from the production of Welding Research Institute Bratislava, types are used.

In addition to submerged arc surfacing, for surfacing the continual casting rolls also open arc self-shielding process is applied.

The third technology, which is concerned in relation to to production of renovation of rolls is the electroslag surfacing /3/. Up to now we have surfaced and service tested two types of materials. It has been shown that this technology can be applied mainly by the switch from production of rolls of high-chromium alloy to the high-speed steel materials (HSS).

REFERENCES

/1/ Bryant, J.M: Rolls 2000 - The Challenge. In.: Rolls 2000, Birmingham UK, March 1996

/2/ Hashimoto, M. - Shibao, S.: Recent Technical Trends of Hot Strip Mill Rolls at Nippon Steel Corporation.

In.: Rolls 2000, Birmingham, UK, March 1996

/3/ Blaškovič, P.: Present State and Future Perspectives of Surfacing Rolling mill Rolls in Slovakia

Doc. IIW: SC-S-48/94.

THE NEW IRON-BASE COMPOSITES - HARDFACING MATERIALS FOR ABRASIVE CONDITIONS

PAVEL BLAŠKOVIČ

Department of Flux Covered Arc Welding and Surfacing,

Welding Research Institute, Račianska 71, 832 59 Bratislava, Slovakia

NINEL A. GRINBERG

All-Russian Pipeline Construction Research Institute, Okhruznoj Prospect 19, 105 058 Moscow, E-58, Russia

ABSTRACT

Abrasive wear represents the most common mechanism of wear in the nature. The machine parts exhibited to abrasive wear must be made of high-alloyed alloys resistant against such a type of wear or they must be provided by a protective layer of similar materials. In this case high-chromium materials are widely applied, mainly high-chromium alloys, where the main role in high resistance is played by chromium carbides /1/.

From the viewpoint of welding technologies the application of high-chromium electrodes is demanding from the aspect of protection of welder and environment due to carcinogenic effects of chromium.

In this respect one of possibilities to solve this problem was an attempt to substitute the high Cr content in the weld deposit by other elements which would guarantee the same life of weld deposit.

The development concept followed from a composite material, where composite material represented the hard particles of titanium and chromium diborides which cooperated with abrasive particles and the basic matrix was selected in dependence on intensity of impacts, so that the structure consisted of a tough austenitic matrix containing very brittle martensitic phase /2/.

The diborides of titanium and chromium have a high melting point (2200 - 2600 °C). The are prepared by melting as synthetic phases characterized with high wear resistance owing to minimum amount of structural defects /2/.

Microhardness of titanium and chromium diborides obtained by the SHS method (self hardening synthesis) is comparable to the level of microhardnes of SiC carbides /2/.

Beside the manual arc surfacing we deal also with mechanization of this process and its application in industrial practice. In that case, with the aim to improve the micrometallurgy of weld deposit, the physical fields have been applied (ultrasound, magnetic field) to refine the structure and assure higher uniformity of composite distribution /3/.

REFERENCES

/1/ Vocel, M. - Dufek, V.: Tření a opotřebení strojních součastí (Friction and wear of machine parts). SNTL Prague, 1976

/2/ Grinberg, N.A.: New hardfacing materials for abrasive conditions. International Tribology Conference, Yokohama Japan, Oct. - Nov. 1995

/3/ Blaškovič, P.: Program COST 516 Tribology.SK1.

NITRIDING OF HARD IRON ELECTRODEPOSITION AND ITS EFFECTS ON WEAR RESISTANCE

GAO CHENGHUI and ZHOU BAIYANG
Department of Mechanical Engineering, Fuzhou University, Fuzhou, 350002, P.R.C.

ABSTRACT

The iron coating which is hard, brittle, with fine crystal grain and high residual stress(1) is usually applied to restore the parts due to wear, corrosion or failure in working. Using the traditional heat treatment can not strengthen the coating because of recrystallization. Nitriding at temperatures lower than the recrystallizating temperature has been employed for hard iron coating to gain a hard-facing and antiwear coating.

After nitriding the structure changed from the black network into white one. Some white plates were discovered in the cross section. The black plate in the coating as-deposited had not been observed until now. The Auger spectroscopy analysis showed that C, O contents of network were higher than the matrix and N contents of the white network were higher than that of the black network or the matrix in the nitrided iron coating. The X-ray phase analysis showed that there were four phases of α-Fe, Fe_3N, Fe_4N and Fe_3O_4 in the nitrided coating. The dispersed particles and white network which formed a frame like a beehive in the nitrided coating were all nitrides.

The microhardness, crystal size, microstrain, macrostress and binding strength of the coating with substrate are shown in Tabel 1. Heated at 450℃ the microhardness, microstrain decreased and the grain size increased. The tendency became obvious when heating temperature was over 500 ℃ (2). Nitrided at 450 ℃ the macrostress of the coating changed from tension stress to compression stress, which was in favour of improving fatigue strength and wear resistance of the coatings. Nitriding impeded the growth of grain and the decrease of microstrain, and raised the microhardness because of the dispersion strengthening of the nitride. Heated and nitrided at 450℃ the binding strength increased due to the release of H and the diffusion of C, N atoms near the binding boundary.

There existed many grain boundaries, dislocations and the network in the hard iron coating. It is the loose network and these defects that make nitriding of hard iron coating perform at a lower temperature in shorter times. During nitriding the solute N atoms and dispersed nitride particles would hinder the migration of dislocations and boundaries, and hence impeded the recrystallizating process.

The relationship of the wear losses with sliding times appeared linear. The wear rates of the coatings as-deposited, heated and nitrided at 450℃ were 0.844, 0.813 and 0.196 mg/h and the friction coefficients were 0.016, 0.014 and 0.010 respectively. Nitriding can raise the wear resistance of hard iron coating evidently and decrease the friction coefficient. The observation of wear tracks by SEM showed that the main worn forms of the iron coating as-deposited and heated at 450 ℃ were brittle peeling and groove respectively. There were no pits, pile-up and scuffing on the worn surface of the coating nitrided at 450℃, and the grooves were the most shallow.

The comparison of wear resistance of the iron coating as-deposited and nitrided with that of Cr coating showed that the nitrided iron coating had highest wear resistance. Its wear resistance was 8.4 times higher than that of the iron coating as-deposited and 3.6 times higher than that of Cr coating. The wear resistance of the spheroidal cast iron matching with the nitrided iron coating was 2.8 times as high as that matching with the iron coating as-deposited and 1.7 times as high as that matching with Cr coating.

REFERENCES

(1) C H Gao, S X Huang, Journal of Fuzhou University, 1984, 12(4):87 - 95
(2) S X Huang, C H Gao, Journal of Fuzhou University, 1986, 14(4):53 - 60

Table 1 Mechanical properties of the coatings

state	Hm MPa	r nm	$|\Delta d/d|$ $\times 10^{-3}$	σ_{ma} $\times 10^2$ Mpa	σ_{bi} $\times 10^2$ MPa
As-deposited	6320	42.4	2.23	5.0	1.83 ± 0.22
Nitrided	7870	48.0	1.80	-4.1	2.76 ± 0.11
Heated	5100	102.1	1.08		2.43 ± 0.13

"TRIBOLOGICAL BEHAVIOUR OF PLASMA SPRAYED Al_2O_3 COATINGS IN SLIDING AGAINST BEARING MATERIALS".

Montes Coto, H.J.; Fernández Rico, J.E.; Cuetos Megido, J.M.; Cadenas Fernández, M.

Oviedo University. Construction and Fabrication Engineering Department. Area: Mechanical Engineering.

Departamento de Construcción e Ingeniería de Fabricación. Universidad de Oviedo. Campus of Viesques S/ N, 33204 Gijón, Asturias. Phone: 98-5182344. Fax: 98-5182360

ABSTRACT.

In this work has studied the tribological behaviour of Al_2O_3 plasma sprayed coatings versus two important bearing materials, tin base antifriction LgSn80 (White Metal) and elasthomer Thordon XL (Fernández, 1995) (Fernández, 1996).

Friction and wear experiments were performed using a conformal contact block-on-ring friction and wear tester with contact area 1 cm^2. The ring specimen, coated with Al_2O_3 by plasma spray, is driven to rotate against the block, which is made of either Thordon XL or LgSn80.

In dry sliding three normal loads (45,3 - 61,3 and 88.5 N - Pressure 0.453 MPa, 0.613 MPa and 0,885 MPa) was used. Only one (680 N - 6,8 MPa) in lubricated conditions, in this cases, the ring specimen is partly covered with a lubricating oil bath. Commercial mineral oils SAE 30 and SAE 140 without additives were used. To monitoring the temperature rise at the contact surface, one thermocouple was used.

Table 1: Friction Coefficients of the tested pairs

Materials Combination	Friction coefficient µ		
	45.3 N	61.3 N	88.5 N
Al_2O_3 / Th XL	0.44	0.57	0.81
Al_2O_3 / LgSn80	0.55	0.60	0.57

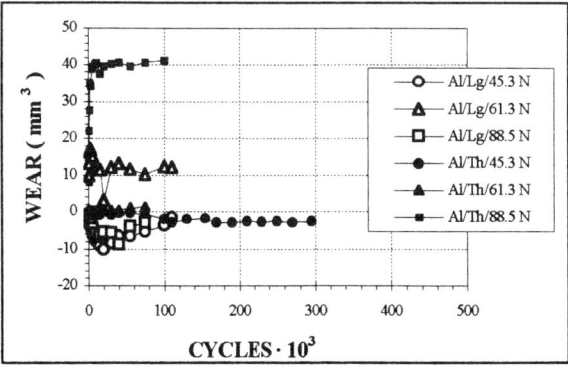

Fig. 1: Wear of Al_2O_3 coating Vs. Lgsn80 and Thordon XL

In dry sliding, the LgSn80 coating comes to disappear completely in the sliding surface centre, finding LgSn80 adherence on the Al_2O_3 surface in the three studied cases. Against Thordon XL the lowest Al_2O_3 wear occurs for the 45.3 N load. In the lowest loads, a light tendency to the Thordon XL adhesion on the Al_2O_3 surface is observed (Fig 1).

Negative wear in lubricated tests is produced (see Fig. 2), for this reason the Al_2O_3 wear could not evaluate by loss weight. It was due to oil absorption in the ceramic pores and particles adhesion block's wear proceeding. Plasma Emission Spectrometry analysis (PES) was carried out in order to determine the wear metal elements contents in the oil.

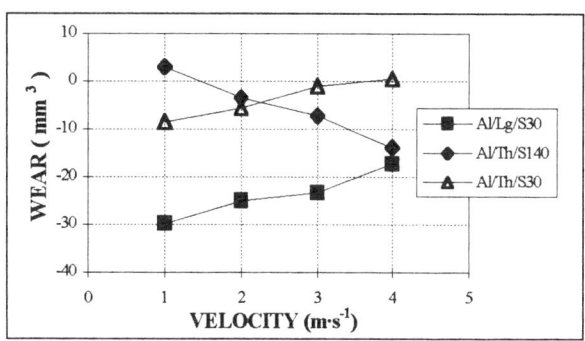

Fig. 2: Velocity and oil viscosity influence in the Al_2O_3 wear Vs. LgSn80 y Thordon XL.

CONCLUSIONS.

PES, SEM, EDS and roughness profiles analysis allows us to know that in both dry and lubricated sliding the coating Al_2O_3 was not obviously worn when is confronted to Thordon XL, while against LgSn80 suggests a adhesive wear mechanism with a little polishing to Al_2O_3 coating. In all studied cases the Al_2O_3 coating wear is negligible in comparison with the blocks to them opposed.

REFERENCES.

Fernández, J. E. et Al.: "Friction and Wear behaviour of Thordon XL and LgSn80 against Plasma Sprayed Al_2O_3 Coatings". XIII Congr. Brasileiro e II Congr. Ibero Americ. de Eng, Mec. Brazil. 1995

Fernández, J. E.; et Al.: "Friction and Wear behaviour of Thordon XL and LgSn80 in sliding against Plasma-Sprayed Cr_2O_3 Coatings". Tribology International. Vol 29 No. 4. 1996. pp. 323-331.

BUCKLING FORMATION OF CERAMIC COATING DURING WEAR

DONGFENG DIAO

Department of Mechanical Engineering, Shizuoka University, Hamamatsu 432, JAPAN

ABSTRACT

Ceramic coatings are widely used in tribological elements to give better tribological performance than those of bulk materials. Difficulties arise because friction and wear of ceramic coatings will cause sudden buckling to occur. It has been recognized that this phenomenon is related to the residual compressive stresses in the coating, mechanical properties of the coating, contact state in friction and wear, etc., but due to ultra thin nature of the coating, measurement of the critical thickness for the buckling has been very difficult. As a result, the mechanism of the buckling is not fully understood. In this paper, six ceramic coatings were deposited on WC-Co substrate by chemical vapor deposition. The tribological properties under no buckling condition were measured with the tester of a diamond pin(cone of 200μm radius) on the ceramic coating disks. The buckling in coatings of Al2O3/TiC, Al2O3/TiC/TiN, TiN/Al2O3/TiC were observed by using the scanning electron microscope. In order to understand the mechanism of the buckling of ceramic coating during wear, a analysis based on the buckling theory was made and a simple formula for expecting the critical thickness for buckling was given. The main results will shown as following.

Figure 1 shows representative variation of friction coefficient of Al2O3/TiC against diamond pin under no buckling condition(arrow A or C) and buckling condition(arrow B) and the SEM examinations of wear track at the positions A, B and C. There are three stages in the wear. The plastic flow occurs at high asperities in the beginning frictional cycles(arrow A) and on whole contact interface after a critical cycles(first stage). The buckling formation at the position B which correlates well with the sudden increase in friction coefficient(second stage). After the buckling of coating, the buckled film will be broken by the following pass of the pin, when the coating is completely lost, the friction conffcient will be stable(third stage, arrow C).

Figure 2 shows a wear-induced buckling model for understanding the mechanisms of buckling formation. There are four steps in the buckling process, (1) Initiation of micro-cracks at the interface between a coating and a substrate, which is usually caused by maximum von Mises stresses at the interface; (2) Propagation of micro-cracks along the interface, which is caused by maximum shearing stresses; (3) Formation of a long delamination at the interface, which is caused by linking the micro-cracks; (4) Buckling formation of coating from the interface, which is caused by the compressed residual stress or maximum contact compressed stress.

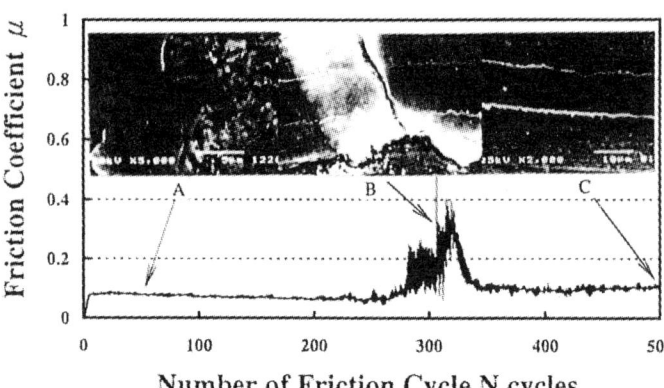

Fig.1: Representative variation of friction coefficient with frictional cycles and SEM examinations wear track at the positions A, B and C for Al2O3/TiC/TiN at sliding speed of 1mm/s

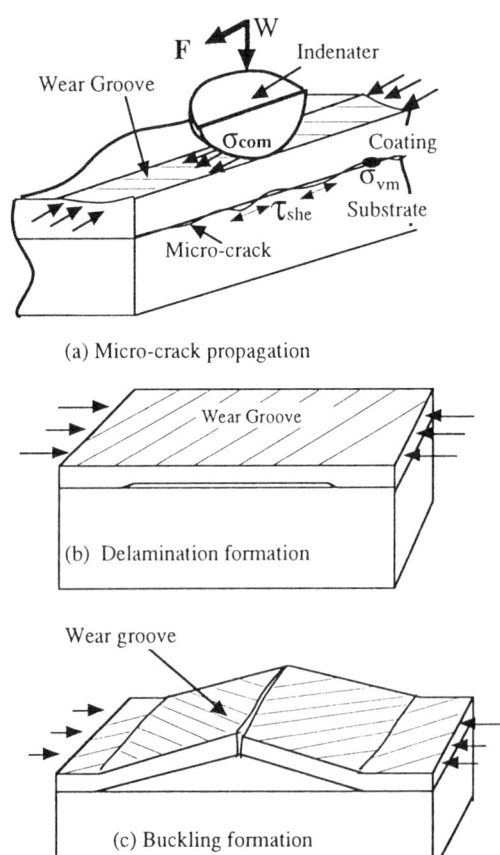

Fig.2: Wear-induced buckling model

SURFACE MODIFICATION BY TRIBOCHEMICAL TREATMENT

JUNXIU DONG, GUOXU CHEN, LIGONG CHEN, SONG JIANG

Department of Petrochemistry, Logistical Engineering College, Chongqing 630042, China

ABSTRACT

The phenomenon of Tribochemical Treatment of Metal Surfaces was discovered in the study of interaction between antiwear lube additives and frictional surfaces. The extremepressure antiwear action of a mew type nonactive antiwear additives, which do not corrode metals, is clarified that permeating layers rather than chemical reaction films are formed in the frictional subsurfaces, modifying metal surfaces tribochemically. This is in many ways similar to that in chemical heat treatment and is therefore named Tribochemical Treatment of Metal Surfaces[1]. Some rare earth coordination compounds not only possess excellent antiwear and friction-reducing capacities, but also tribo-catalyze boronization during friction[2]. Antiwear additives with soft metals such as copper and tin tribodiffuse and metallize on friction surfaces[3]. These phenomena are also responsible for Tribochemical Treatment of Metal Surfaces.

The technique of tribochemical treatment can be employed to improve the surfacial performances of machine parts. The surfaces of machine parts need not hardening before running-in, they can be modified by formulating additives containing boron, rare earth elements, etc. into lubricating oils during running-in period. Therefore, the two processes-surface treatment and running-in can be combined into one. For example, a Lanthanum-Boron additive was used as the medium for running-in of gear. After running-in, the anti-seizure load stage of the gear, which made of 20CrMnTi steel raised from 7th stage to 9th stage in FZG gear test.

The theory of Tribochemical Treatment of Metal Surfaces not only explains the action mechanisms of nonactive antiwear additives, but also unifies the action mechanisms of traditional antiwear additives with sulfur and phosphorus. Iron sulphate and phosphate are formed by traditional sulfur and phosphorus additives on friction surfaces, the essence of which is tribochemical treatment to form reaction films, similar to sulphation and phosphorizaton in chemical heat treatment of metal surfaces. Like thiocyanization in chemical heat treatment, lube additives containing thiocyanicate also provide good antiwear and friction-reducing abilities.

As a result, Tribochemical Treatment of Metal Surfaces is a common feature that lube additives modify the friction surfaces in the friction processes. The theory of Tribochemical Treatment of Metal Surfaces satisfactorily explains the extreme-pressure antiwear mechanisms of lube additives. It also plays dominant theoretical roles in the preparation of many new lube additives which can strength, modify and heal frictional metal surfaces. The discovery and applications of Tribochemical Treatment of Metal Surfaces enrich the new branch of Tribology------Tribochemistry.

Some new antiwear additives have been used in industry. In Dongfeng Boilers Works of Sichuan province for example, the reject rate of U-shape stainless tubes for 300MW electricity generating boilers is reduced from 50% to 5% by incorporating 5wt.% of an organo-borate containing nitrogen additive into the cold extruding oils. The annual savings amount to more than 1 million yuan, while the use of organo-borate containing nitrogen additive costs only 5000 yuan..

REFERENCES

(1) J.X.Dong, etal, Lubrication Engineering, vol.50, N0.1, 1994
(2) B.S.Chen, Wear, vol.196 No.4,1996
(3) J.B.Yao, J.X.Dong, Lubrication Engineer, vol.50, No.9, 1994

SLIDING FRICTION AND WEAR MECHANISMS OF DIAMONDLIKE CARBON FILMS ON CERAMIC SUBSTRATES*

A. ERDEMIR and G.R. FENSKE
Argonne National Laboratory, Energy Technology Division, Argonne, IL 60439, USA

M. HALTER
Prude University, Department of Materials Science and Engineering, West Lafayette, IN 47906, USA

ABSTRACT

In this study, we present the results of a mechanistic study on hydrogenated and hydrogen-free diamondlike carbon films. Specifically, we describe the friction and wear mechanisms of hydrogen-free and hydrogenated films produced on silicon carbide (SiC) substrate by ion-beam deposition. Sliding tests were performed with pairs of alumina (Al_2O_3) and magnesia-partially-stabilized zirconia (MgO-PSZ) balls and DLC coated SiC disks in a pin-on-disk machine at velocities of 0.1 to 2 m/s and at contact loads of 1 to 10 N in open air of 30-50% relative humidity and dry N_2. The sliding distance was 10 km.

The results showed that both films provided friction coefficients of 0.05 to 0.15 to sliding surfaces in open air. The initial friction coefficients were in the range of 0.08 to 0.2 but decreased steadily as sliding continued and reached steady-state values of 0.05 (for hydrogenated films) and to 0.15 (for hydrogen-free films). The friction coefficients remained constant at these values until the end. The friction coefficients were generally lower under heavier loads and reached low steady-state friction coefficients faster at higher sliding velocities than at lower velocities. Wear rates of the ceramic balls were $\approx 10^{-8}$ $mm^3/N \cdot m$ during sliding tests in air, while the wear rates of the hydrogen-free and hydrogenated films were so low that they were difficult to measure even after 10 km of sliding distance.

The friction coefficients of hydrogen-free films were rather high (0.3-0.4) in dry N_2, but those of the hydrogenated films were very low; i.e., ≈ 0.02 (especially during sliding under higher loads and at higher velocities). The wear rates of ceramic balls were in the range of $5 \times 10^{-7} - 10^{-7}$ $mm^3/N \cdot m$ against hydrogen-free films in dry N_2, while the wear rates of those balls slid against the hydrogenated films were much lower, i.e., $\approx 10^{-9}$ $mm^3/N \cdot m$. There were visible wear tracks on hydrogen-free films after test in dry N_2.

The wear-debris particles found in and around the wear scars and tracks were analyzed by Raman spectroscopy and by electron microscopy. The debris particles were black and made of a carbonaceous material. Electron microscopy revealed that these particles were extremely fine (10-50 nm in size). Raman spectra of the transfer layers on the sliding ball surfaces, as well as of the debris particles in and around the wear tracks, revealed Raman bands indicative of micrographitization. Specifically, as shown in Fig. 1, the micro-laser Raman spectra of black transfer layers and debris particles indicate that their structural chemistry is quite different from that of the original DLC film, but has Raman features similar to those of the crystalline graphite (provided as a standard). Two broad Raman bands at 1354 and 1594 cm^{-1} match D and G lines of crystalline graphite. The Raman line shapes of the carbonaceous transfer layers are not as sharp as those of the crystalline graphite. This may be due to the very fine nature of the debris particles and to phonon damping. It also indicates that the degree of structural disorder in carbonaceous transfer layer is very high. Therefore, the carbonaceous transfer layers or debris particles found at sliding interfaces are not necessarily in the form of a crystalline graphite, but in a disordered state. Mechanistically, this suggests that during sliding, DLC films transform to a graphitelike precursor (mainly because of high mechanical and thermal loading) and are removed from the surface as very fine debris particles. The low friction coefficients (i.e., 0.02-0.05) observed during tests under high loads and at high sliding speeds can be attributed to the much faster formation of these graphitelike precursors at sliding interfaces.

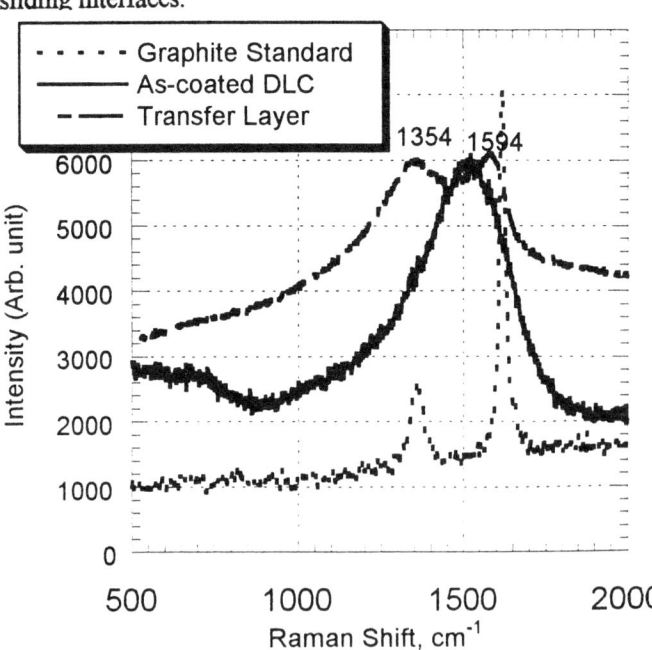

Fig. 1. Raman spectroscopy of transfer layer, graphite and DLC film.

*Work supported by U.S. Department of Energy under Contract W-31-109-Eng-38.

TRIBOLOGICAL PROPERTIES OF DEPOSITED DIAMOND-LIKE FILMS: MOLECULAR DYNAMIC SIMULATION

S.A. FEDOTOV, A.A. EFIMCHIK, A.V. BYELI,
Physical-Technical Institute, 4 Zhodinskaya str., Minsk, Republic of Belarus

ABSTRACT

Ion beam assistant deposition (IBAD) consists of ion bombardment of a coating during its growth and seems to be a promising technique for deposition of multicomponent wear resistant coatings. One of the main advantages of this technique is that parameters of atom flux depositing the coating and ion flux assisting the deposition are maintained independently and in the case of carbon coatings, containing both graphite and diamond-like phases can be obtained. At least nine parameters of the process effectively influence relationships between of sp, sp^2, and sp^3 hybridizations, and development of efficient technique for IBAD simulation is essentially important for proper selection of deposition parameters.

In this particular research ion beam sputtering of a carbon target was used for coating deposition, and argon ions were used for ion assistance. The simulation of the coating growth was performed on the basis of potential by Tersoff to describe the interactions between atoms and structural relaxation of atomic ensemble. The TRIM code was used to calculate the flux density and energy distributions of the deposited atoms. The simulation of IBAD process revealed the influence of ions energy, and the ratio of depositing and assisting beam intensities on the microstructure of the coatings.

Transmission electron microscopy, electron diffraction and tribological tests confirmed that IBAD with parameters, predicted by simulations, results in deposition of coatings consisting of fine diamond-like inclusions in amorphous matrix and revealing optimum tribological properties.

STUDY ON THE WEAR RESISTANCE OF COMBINED BORIDE COATINGS ON STEELS

N. GIDIKOVA
Institute of Metal Science, Bulg. Acad. Sci., Sofia 1574, 67, Shiptchenski prohod St, Bulgaria

YU. SIMEONOVA
Space Research Institute, Bulg. Acad. Sci., Sofia 1000, 6, Moskovska St, Bulgaria

E. ASSENOVA and KR. DANNEV
Tribology Centre, Technical University - Sofia 1756, Bulgaria

ABSTRACT

On the basis of transient metal vanadium, using thermodiffusive treatment, combined boride coatings on steels of medium carbon content have been produced. The phase content, concentration of the transient metal in depth of the coating, as well as microstructure and microhardness of both vanaded and combined coating have been studied.

The obtained coating (V+B) on steel at temperatures higher than 1200° and duration of the process more than 11 hours exhibits a zone structure. In its surface zone the composite VB_2 with microhardness 23.10^3 MPa is being formed.

The subsurface layers of this coating are significantly different as regards of the phase content, depending on the boration conditions. Normally, the subsurface layer is composed of α hard solution of vanadium in iron, which ensures conditions for better plastic bondage with the basic material. Ferrous borides have not been detected. This circumstance makes the studied combined coating significantly different from the coatings of the type /$Me_{transient}$ + B/, e. g. coatings of zirconium-boron, titanium-boron, chromium-boron on the same kinds of steel, obtained by the authors in some earlier investigations (1) (2) (3).

Above characteristics of the newly obtained coatings have led to improvement of the wear resistance of the friction surface, which has been proved for dry friction at vacuum (10^{-4} Pa) conditions. As a result of the preliminary thermo-diffusive treatment of steel with vanadium, a decrease of the coefficient of friction from 0,322 (for mono-boride steel) to 0,145 for the combined coating has been observed. The intensity of wear also decreases accordingly from 1,94 to 1,03 $g/m^2.km$.

It has been found that the decrease of wear is more strongly stated for middle-carbon steels. For a high-carbon steels, an intense formation of ferrous carbides in the zone, which lies closest to the basic material, has been observed during the process of high-temperature diffusive treatment of the material directed to the coating creation. The carbides lower the plastic properties of the coating and worsen its wear resistance.

REFERENCES

(1) N. Gidikova, Yu. Simeonova, D. Petkova, Material Science and Engineering, A184, 1994, L1-L4.

(2) N. Gidikova, R. Kovacheva, Yu. Simeonova, Praktische Metallographie, Band 33, 1996, 154-160pp.

(3) N. Gidikova, Yu. Simeonova, D. Petkova, Journal of the Balkan Tribological Association, Vol. 2, No. 3, 1996, 161-164pp.

Tribological Properties of TiN Thin Film Coated on Alloy Tool Steels

Y. HARUYAMA and N. YOKOI
Department of Mechanical Systems Engineering, Toyama Prefectural University, Kosugi-machi, Toyama 939-03, Japan
S. KAWAMURA
YKK Corporation, 200 Yoshida, Kurobe 938, Japan
K. SHIOZAWA
Department of Mechanical and Intellectual Systems Engineering, Toyama University, 3190 Gofuku, Toyama 930, Japan
Y. KIMURA
Founding Office of Faculty of Engineering, Kagawa University, 1-1 Saiwai-cho, Takamatsu 760, Japan

ABSTRACT

Titanium nitride (TiN) has been considered as one of the promising coating materials on steel substrates, particularly suitable for cutting and metalworking tools because of its hardness, low coefficient of friction and chemical inertness. The present paper describes a series of experiments on friction, wear and life of TiN thin film coated on tool steels when slid against aluminum alloy, aiming at application to dies for hot extrusion of aluminum alloys.

Sliding experiments were made on a ring-on-disk friction tester, the sliding system of which was placed in a chamber conventionally sealed to control the atmosphere. Coating of TiN had been made by a thermal CVD process on the disk surface of an alloy tool steel JIS SKD61 (ATS) and a high-speed tool steel JIS SKH51 (HSTS). To ensure adherence to the substrate, intermediate layers of Ti(C,N) and TiC were formed between the TiN film and the substrate. The total film thickness was about 8 μm. Their surface was either coated (Ra 0.034 − 0.038 μm) or finished by lapping (Ra 0.017 − 0.018 μm), and had Vickers hardness of 1700 − 1880. The disk specimen was slid against an end surface of a ring, 20 and 25.6 mm in the internal and external diameter, respectively, made of an aluminum alloy JIS A6063. Unlubricated sliding was made under a load, 1 − 70 N, at a sliding speed, 0.01 − 0.20 m/s, in air or in nitrogen atmosphere. Friction was continuously monitored, while wear was determined with the disk by weighing or by profilometry of the track after a run.

The coefficient of friction is rather high, particularly in the nitrogen atmosphere, Fig 1. However, its values are almost constant from the outset of sliding till failure of the film which in indicated by the termination of the data. No significant difference in the coefficient of friction due to the different substrates is found. Figure 2 shows increase in the wear amount of the disk specimens with sliding distance. It is clear that wear is much larger when slid in air, which is caused by the abrasion with fine hard debris of aluminum oxide. In nitrogen, little oxidized debris are formed, but gross transfer of metallic aluminum to the disk surface takes place, which lowers wear amount while causes higher friction with sliding between like metals. Nevertheless, the film life is generally longer in air. Localized detachment of the thin film takes place after considerable wear occured in air, while higher friction in nitrogen seems to have caused incipient detachment of the thin film. The difference in the substrate little affects the wear amount, but the film life is much prolonged with the HSTS substrate.

Extrusion in industry is conducted in air. However, it is likely that the particular geometry of the process prevents oxygen to reach the sliding interface to form oxidized debris, and therefore the tribological behavior may become similar to that in nitrogen in the present experiments. More detailed comparison between the real process and the experiments is necessary.

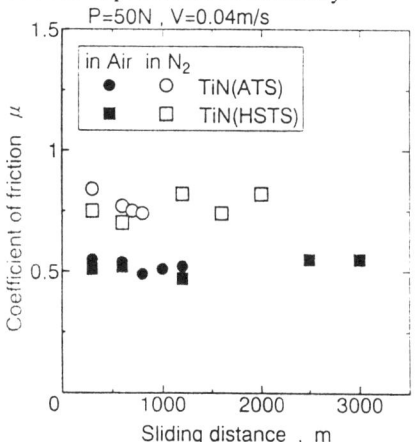

Fig.1: Variations in the coefficient of friction with sliding distance

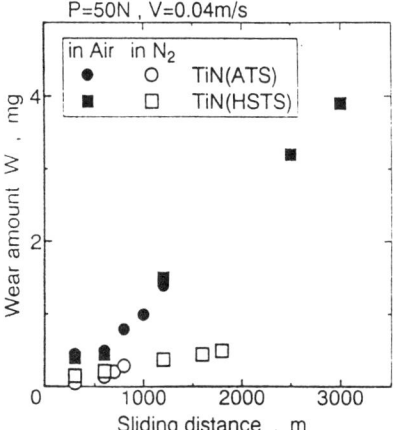

Fig.2: Variations in the wear amount with sliding distance

STUDY ON THE INDENTATION CHARACTERISTICS OF THE BASE METAL WITH HARDER FILM

AKIYOSHI KOBAYASHI and YASUHIKO DEGUCHI
Department of Mechanical Engineering, Meijo University, Shiogamaguchi 1-501,Tenpakuku, Nagoya, JAPAN

ABSTRACT

Coated hard thin films have been widely used to improve the tribological properties (wear and friction characteristics) of primarily many machine parts and cutting tools. However, the mechanisms that contribute to the improvement of the tribological properties are not still clarified. In order to improve the characteristics of friction and wear, it is very important to clarify the mechanical properties of the coated thin films.

In this paper, as a basic step to examine the tribological characteristics of hard thin films, the indentation characteristics of a base metal with harder thin film are examined theoretically and experimentally. The theoretical analysis of the indentation characteristics of a base metal with harder thin film in the elastic region was done based on Hertzian theory. The theoretical analysis of the elastic-plastic regions were done based on the extended Hertzian theory. Here, the base metal is under plastic deformation, while the coated thin film is under elastic deformation in elastic-plastic region. Figure 1 shows the displacement of the base metal.

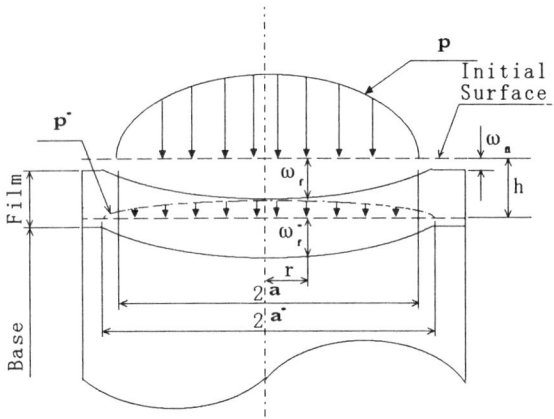

Fig.1: Displacement of base metal

In this experiment, the base metal is brass and two kinds of thin films, Ni and Cr, were used. Ni film is harder than the base metal, the film is ductile. Cr film is much harder than the base metal, but the film is brittle.

Figure 2 shows the quantitative comparisons of the theoretical and experimental values of the depth of indentation in the elastic, the elastic-plastic and the plastic regions. The both values of the elastic and the plastic regions show good agreement. While in the elastic-plastic region, the experimental values are higher than the theoretical ones.

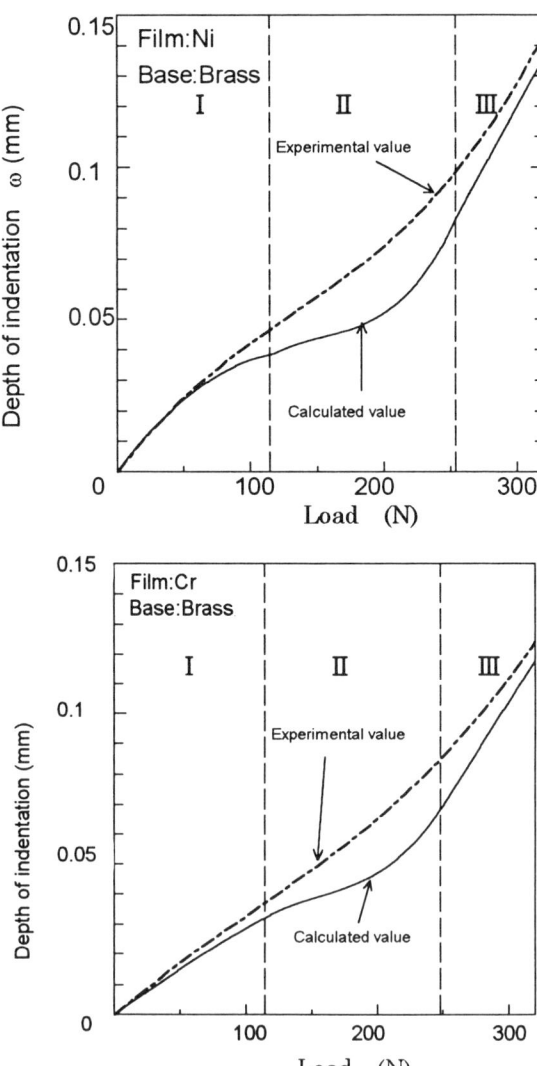

Fig.2: Comparison theoretical value and experimental one

THE INFLUENCE OF THE VACUUM CONDITIONS ON THE ION BOMBARDMENT INDUCED HARD WEAR RESISTANT LAYERS FORMATION

F F KOMAROV, V V PILKO, A V FRANTSKEVITCH
The Institute of Applied Physics Problems, Kurchatova street, 7, 220064, Minsk, Belarus

ABSTRACT

The suitability of rapid, high-current density ion implantation at elevated temperatures under various vacuum conditions as a process to engineer solid surfaces for tribological applications is discussed. Two effects that can lead to a change in wear resistance are considered: the condensation of the carbon containing film and the sample heating.

The inhomogenity of the data published in the field of the surface sheets wear resistance changes under reactive ions bombardment is caused mostly by two main factors: unstandartized test methods and the different implantation conditions.

Previous investigations have shown that large amounts of implanted nitrogen tend to increase the wear resistance of the steel surface layers by up to order of magnitude (1).

One of the main processes that may occur during ion implantation is the condensation of the residual gases components on the target surface. These disadvantages are presumed to disappear when using ion implantation at elevated temperatures.

The high-throughput ion implantation systems works usually in the target chamber pressures range from 10^{-5} to 10^{-6} Torr. In the case of targets held at room temperature the C/N ratio decreased in this pressures range from 4 to 0.1. By ion implantation at temperature elevated from 20°C (the standard temperature of the diffusion pumps cooling) to 350-400°C (the standard temperature of the diffusion pumps heaters) the diffusion oil condensation rate decreased up to two orders of magnitude and gives the possibility to achieve the C/N ratio near to 3/4.

The present work was undertaken to investigate the composition and wear resistance of carbon-nitrogen films formed by vacuum condensation of the carbon containing residual gases under nitrogen ion bombardment.

The samples were hardened and polished steel disks with the diameter of 25 mm. The nitrogen ion implantation was performed at various thermal contacts with the 20°C cooled sample holders in a commercial vacuum system evacuated by a rotary pump and an oil diffusion pump and in high-voltage implanter evacuated by a turbomolecular pump. The pressure during the implantation was maintained at 8×10^{-6} Torr and 2×10^{-6} Torr, respectively. The ion beam current and acceleration voltage were in ranges $2\text{-}20\mu A/cm^2$ and 50-200 keV, respectively. The total implantation dose was $2\times10^{17}\text{-}2\times10^{18}$ at/cm^2.

The composition of the formed surface sheets was analyzed by the Rutherford Backscattering method.

The wear curves were obtained using the pin-on-disk method at load of 0.1 N. These curves represent the functional relationship between the friction coefficient and the sliding distance. In all the tests the pins were made of polished GCr6 with Hv=280 kg/mm^2 and have the radii of 2 mm. Ethanol was employed as a lubricant.

It was found that for good cooled samples the thickness of ion beam assisted carbon containing films have linear dependence on the implantation time and on the residual gas pressure. For the obtained in this case C/N ratio near to 4 the wear resistant layers are formed only at the implantation doses higher than 5×10^{17} at/cm^2. It corresponds to the middle ion penetration ranges close to condensed film depth. At lower implantation doses the adhesion of the condensed layers to the surface and its wear resistance was found to be low.

By the implantation at elevated temperatures the C/N ratio was obtained to be less than 1 in the wide depth range. The wear resistance increased by up to order of the magnitude.

It may be concluded from the results that

(a) The wear-resistant films with the best properties are formed at C/N ratios less than 1 in the middle ion penetration range.

(b) The implantation technique results in forming of CN layers with an extremely good adhesion to the substrate. The improved wear resistance persists to a depth of two projected ranges of the nitrogen ions.

(c) It can be summarized that phase formation after nitrogen ion implantation in the condensed films can be described by a model of implantation-induced phase transformations.

REFERENCES

(1) F.F.Komar , V.V.Pilko, V.A.Yakushev, V.S.Tishkov, NIM B 94, 1994, 237-239 p.p.

Submitted to *Surface and Coatings Technology*

MULTILAYER COATINGS CONTAINING MoS₂ FOR TRIBOLOGICAL APPLICATIONS

K. J. Ma and A. Bloyce
School of Metallurgy and Materials,
The University of Birmingham, Birmingham, B15 2TT, U. K.
J. Hampshire and D. G. Teer
Teer Coatings Ltd., Hartlebury Trading Estate, Worcestershire, U.K.

ABSTRACT

MoS$_2$ coatings deposited by closed field magnetron sputtering are dense and adherent and have superior tribological properties compared to MoS$_2$ coatings deposited by other techniques. The tribological properties have been further improved by the co-deposition of small amounts of metal with the MoS$_2$ (1). This study is devoted to the enhancement of endurance by the addition of a thin layer of Au adding to the metal containing MoS$_2$ coatings. A series of coating system including MoS$_2$, TiN/MoS$_2$, MoS$_2$/Au and TiN/MoS$_2$/Au coatings were deposited on M2 high speed steel substrates using unbalanced magnetron sputtering. The pin on disc wear test was performed at an applied load of 40 N (using 5 mm WC/Co as a ball) and high relative humidity (~ 45%). High resolution microstructural examination was used to interpret the wear mechanisms occurring in the these tests.

Experimental results show that the addition of a Au film increases the endurances of MoS$_2$/Au and TiN/MoS$_2$/Au multilayer systems over equivalent coatings without Au, as shown in Fig. 1.

In the case of TiN/MoS$_2$/Au multilayer systems, a relatively high coefficient of friction (0.15) was measured in the initial sliding, which indicates that intrafilm flow occurs within thin layer of Au to accommodate relative movement of the two surfaces. Repeated sliding allows the Au to pile up at the edge of wear tracks, thinning the Au layer. The MoS$_2$ layer gradually becomes involved in the plastic flow process if the shear stress developed is above the shear strength of the MoS$_2$ coatings. The friction coefficient rapidly decreases to a stable value ($\mu = 0.045$) as MoS$_2$ shears. Continuing sliding allows the Au to combine with the orientated MoS$_2$ to form a very thin Au-MoS$_2$ composite layer. The transfer and detachment alternately of layered Au-MoS$_2$ wear debris occur at this stage. Both intrafilm plastic flow within orientated MoS$_2$ and Au-MoS$_2$ composite layers and interfilm sliding between the Au-MoS$_2$ composite wear debris are thought to dominate the friction process. After more than 15000 cycles sliding, the MoS$_2$ layer totally combines to form a Au-MoS$_2$ composite layer, the plastic flow and interfilm sliding will occur in Au-MoS$_2$ composite layer to accommodate shear and normal stresses. The friction coefficient gradually increased to a second stable value ($\mu = 0.15$). An endurance of over 50000 cycles was measured in this case. It is believed that Au or Au-MoS$_2$ composite layer can effectively prevent oxygen or moisture reaction with MoS$_2$ and hence significantly increase the wear life. The transfer mechanisms and detailed tribochemical reaction still require further study.

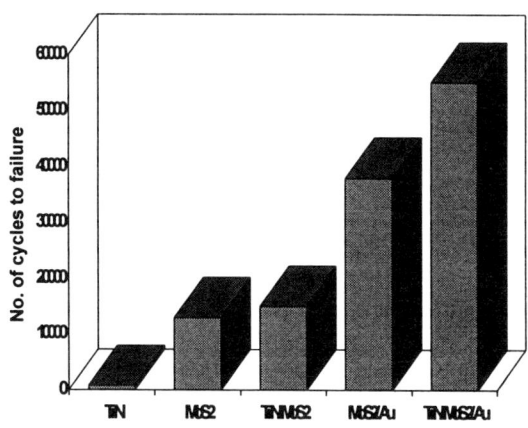

Fig. 1: Comparison of the endurance of the coatings during the pin-on-disc wear test; using a 5 mm dia. WC/Co ball as the pin, and the sliding speed was 150 rpm.

REFERENCE

(1) D. G. Teer, J. Hampshire, V. Fox and V. Bellido-Gonzalez, International Conference on Metallurgical Coatings and Thin Films, San Diego, California, April 21-25, 1997.

THE SCRATCH TEST: ATLAS OF FAILURE MODES

J. MENEVE and D. HAVERMANS
Materials Technology Centre, Vlaamse Instelling voor Technologisch Onderzoek, Boeretang 200, 2400 Mol, Belgium
J. von STEBUT
Laboratoire de Science et de Génie des Surfaces, Ecole des Mines, Parc de Saurupt, 54042 Nancy Cedex, France
N.M. JENNETT, J.P. BANKS and S.R.J. SAUNDERS
Centre for Materials Measurement and Technology, National Physical Laboratory, Teddington, TW11 0LW, UK
D. CAMINO and D.G. TEER
Teer Coatings Ltd., 290 Hartlebury Trading Estate, Worcs. DY10 4JB, United Kingdom
P. ANDERSSON and S. VARJUS
VTT Manufacturing Technology, P.O. Box 1702, FIN-02044 VTT, Finland

ABSTRACT

Of the various engineering test methods for measuring coating adhesion, the scratch test is most widely used because it is a quick and simple test and has been successful in the investigation of the effects of deposition variables, ageing, etc. (1-5). The test consists of drawing a diamond stylus across the specimen surface under increasing normal load until some well defined failure event is observed in a regular way along the scratch channel at a load which is called the critical normal load (L_c). L_c can be determined by in situ techniques such as acoustic emission detection, lateral force monitoring and measurement of the scratch depth, but these methods cannot discriminate between interfacial and cohesive failures nor can they always detect the very first failure events. Therefore, examination by microscopy remains the only means of associating a failure event with a measured critical normal load.

Not all of the observed failure modes are relevant as a measure of adhesion. Failure events such as cohesive failure within the coating or substrate may occur but are not related to the coating/substrate interfacial bond strength. These cohesive failure events, however, can be equally important to determine the behaviour of a coated component in a particular application.

In spite of its widespread use, the results obtained from scratch testing may reflect a variety of failure processes. Since these processes are not always clearly defined when explaining results, the scratch test is often held in low esteem. Much of this could be avoided if test procedures were standardised. The value of the scratch test would then be greatly enhanced. Within this scope, there is an absolute need for a catalogue of scratch test failure modes that can be referred to by scratch test users when reporting critical normal loads.

The target of the present study was to prepare an atlas of scratch test failure modes. To that end, eight different coating-substrate composites were produced, based on four distinct coating types (TiN, SiC, Al_2O_3 and a-C:H) on ductile (hardened steel) and brittle (cemented carbide) substrates. Surprisingly, however, only few of the failure events reported in literature were obtained. Therefore, it was decided to scratch a greater series of coated specimens available in the participating laboratories, even though some of the failure modes reported in literature may be irrelevant from a practical point of view, e.g. originating from brittle coatings on unrealistically soft substrates. A wide range of failure events could be inventoried, which were classified into cracking events, spallation events, plastic deformation, etc.

Under a reflected light microscope, micrographs were taken from each observed failure mode. A concise description was given to each catalogued failure event, with the emphasis on discriminating between interfacial and cohesive failure events.

Although the present study does not claim to be exhaustive, it should provide a solid basis for the more standardised reporting of results obtained with the generally recognised scratch test method.

The results of this study will be used to produce an improved European Standard on the scratch test, based on earlier work within the European Standards Committee CEN TC 184/WG 5 'Test Methods for Ceramic Coatings' (6).

REFERENCES

(1) K.L. Mittal, Electrocomponent Sci. Technol., 3 (1976) 21.
(2) A.J. Perry, Thin Solid Films, 107 (1983) 167.
(3) J. Valli, J. Vac. Sci. Technol., A4 (6) (1986) 3007.
(4) P.A. Steinmann and H.E. Hintermann, J. Vac. Sci. Technol., A7 (3) (1989) 2267.
(5) D.S. Rickerby, Surf. Coat. Technol., 36 (1988) 541.
(6) ENV 1071-3:1994 E: Determination of Adhesion by a Scratch Test, CEN Central Secretariat, Stassartstraat 36, 1000 Brussels, Belgium.

This work is supported by the European Commission under Standards, Measurements and Testing contract MAT1-CT94/0045.

USE OF THIN FILM, LASER AND PLASMA IMMERSION ION IMPLANTATION TECHNOLOGY TO REDUCE WEAR AND FRICTION IN AUTOMOTIVE APPLICATIONS

BARRY LESLIE MORDIKE
Institut für Werkstoffkunde und Werkstoffkunde, Technische Universität Clausthal,
Agricolastr. 6, 38678 Clausthal-Zellerfeld, Germany

DETLEV REPENNING
o.m.t. Oberflächen- und Materialtechnologie GmbH, Seelandstr. 65, 23569 Lübeck, Germany

ABSTRACT

The increasing demands made on automobile components has resulted in the development of new technologies and coating materials to improve the surface properties. The paper will discuss some of these innovations with examples in automobile applications.

The first application discusses the surface remelting of camshafts using lasers. Previously the camshafts, made in cast iron, were either chill cast or TIG melted. Laser surface remelting produces a thin ledeburitic layer with better properties than the alternative methods. The advantages are more economical production and improved performance. These will be detailed in the poster.

There are many applications where thin coatings of diamond like c-Me-H films produce a significant improvement in wear resistance and reduction in friction. C-Me-H films have been applied to tappets, valve stems, turbo-supercharger parts, shock absorber parts, draw keys, rocker arms, gear teeths, petrol pump parts, piston pins and rings. The main advantage is the very low coefficient of friction against most engineering metals. It can also be used in nonlubricated or dry conditions such as in petrol pumps. The effect of coating the tappets on oil temperature and effective friction will be shown as well as photographs of the other applications.

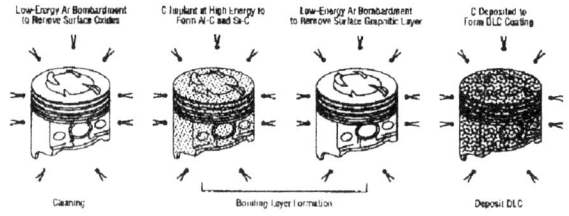

Fig. 1: Schematic diagram showing the sequence for producing highly adherent diamondlike-carbon films on aluminium 390 alloy by PI^3 (1)

Plasma immersion ion implantation (PI^3) is a new hybrid technology using elements of ion implantation as well as plasma nitriding. In addition and due to the high energetic ion bombardment thermal diffusion can be used to obtain thicker layers than in conventional beam line ion implantation. Furthermore the process can be used in an Ion Beam Enchanced Deposition (IBED) mode producing various kinds of metal and ceramic coatings. This enables the treatment to be adapted to the material structure and state and also to the intended service conditions. Furthermore economic and processing aspects make PI^3 an interesting and potential technique for surface modification of automotive components (Fig. 1). Unfortunately the number of tested components is still small because this new technology has to compete with conventional ion implantation as well as with plasma nitriding. The latter is a well known and established technology for surface protection of many automotive parts.

Nevertheless PI^3 offers some important advantages in comparison to nitriding.
- lower treatment temperatures, reducing the degree of distortion, surface roughening and over-tempering associated with higher temperatures
- separation of plasma generation from target bias, reducing the likelihood of arcing and allowing better process control

The advantages over conventional ion implantation techniques can be summarised as follows:
- more uniform coverage
- fast and treatment time is independent from the surface size
- thicker treated layers if thermally induced diffusion is exploited, while the surrounding plasma easily prevents outward diffusion of the implanted species and may also promote the uptake during pulse pauses
- lower treatment costs
- the easing of line-of-sight restrictions
- the ability to treat complex shapes;
- the ability to scale to large targets (or batches), offering the prospect of a technologically simple implanter design

The introduction of metal ions, the possibility of IBAD (or IBED) in addition to conventional implantation at room or elevated temperature makes PI^3 to a universal tool for surface modification of automotive components. Possible applications and reasons for the reluctant uptake of this new and promising technology will be discussed.

REFERENCES

(1) J.V. Mantese, I.G. Brown, N.W. Cheung, G.A. Collins, MRS Bulletin, August 1996, 52-56

SOME WAYS OF CREATION OF MATERIALS DESIGNATED FOR TRIBOTECHNICAL PROCESSES

F I PANTELEENKO
Department of Construction Materials Technology, Polotsk State University, Belarus

ABSTRACT

Generalization and sistematization of results of our investigations, focused on elaboration and creation of materials, surfaces and coatings, designated for different fields of industry, have been carried out for the last 15 consequent years. The classificational scheme of the types of structures of wear- resistant materials is offered. Materials can be relatively divided into homogeneous and heterogeneous ones according to the morphology of their structure.

Heterogeneous metal-, polymer-, non-organic-base materials constitute the second group. Their heterogeneous structure, combining softer and more solid phases and interpreting the Sharpy's principle, is the one which envisages future development to the utmost degree in the field of wear-resistant and anti-friction materials obtaining. The simplest examples of them are compositional coatings including graphite, MoS_4, B_4N, SiC, Al_2O_3, SiO_2, AlN.

Different technological methods of creations of materials, surfaces and coatings have been used: thermal processing, including highly concentrated energy sources; chemico- thermal processing; fusion (plasma, lazer, gaso- flaming); methods of powder metallurgy.

The perspective of these methods, which allow to make the structure smaller and to find optimal correlation in the structure of solid and soft constituents, to create materials of eutectic type with set properties, has been discovered.

Concrete examples of the above listed technologies realization have been given.

Thus, elaborated technology of lazer tempering (lazers "Cyprus-1", "Quantum-16") of small- sized parts allowed us to solve the problem of strengthening [of the saw chains axles (made of steel 65G) parts that are regularly being worn, at the expense of the high speed of the heating process and the process of heat elimination that follows (the axles that come in contact with the material that has a heat-conducting index (factor) not lower than 150 Vt/m·K in the function range of temperatures)].

Chemico- thermal processing (chroming, chromo- titaning) carbonaceous steel types containing 0.45, 0.8, 1.2 per cent of carbon and alloyed chromo-tungsten-manganese tool-steel containing 0.9 per cent of carbon with preliminary applications of diffusion zinc or chemical nickel - phosphorus surfaces and coatings allowed to intensify the carbide coatings formation.

Chemico-thermal processing was held in powder-like alumino-thermal compounds at the temperatures of 850-1050 °C. As a rule, after chemico-thermal processing, they get carbide coatings of a homogeneous type with the layer thickness of about 0.02 mm, and with preliminary coatings, heterogeneous coatings with the thickness of about 0.1 mm, have been obtained. With that, the greatest quantity of the carbide phase is situated in the middle part of the coating, which allows to hold fine. Final mechanical processing operations, and provides better adjustment and reduces fragility.

The technology allows to regulate the correlation of solid and soft constituents in the coating.

Higher degree of wear-resistance and greater functioning resource of such coatings with dry sliding friction as compared to traditional carbide coatings (with the contents of 2-20 per cent of α-solid solution insertion) have heen revealed.

A class of new cheap self-fluxing, ferrum-based powders has been elaborated.

Fluxing and consolidating alloying elements were inserted by the means of diffusion. It excludes their waste and allows to get from them the required features of the coatings that can be regulated in future.

A theoretical basis of the diffusion alloying of microobjects, with a number of peculiarities, has been worked out.

Self-fluxing powders are being used to get lazer, plasma, gaso- flaming fused in ordinary atmosphere, wear- resistant surfaces and coatings for the restorations of different parts and mechanisms in thermo- energetics, internal combustion engine.

A complex of properties of these coatings was examined and found out that the best wear- resistance is the property of overeuthectic boron- containing compositional coatings, containing 48 ... 65 per cent of superfluous boride phases.

The second important direction of applying of boron- containing fusions or compositional materials, obtained by the methods of powder metallurgy is obtaining of the cutting instrument on the base of powder steel R6M5, that has 40- 60 °C higher indexes of heat- firmness. The results of investigations of higher heat- firmness together with some conclusions have been drawn up thereafter in the work.

THE EFFECT OF GRANULOMETRIC COMPOSITION OF POWDER COPPER-GRAPHITE SYSTEM ON TRIBOLOGICAL PROPERTIES OF THIN FILMS FORMED BY ELECTROCONTACT SINTERING

V.A.KOVTUN and Y.M.PLESKACHEVSKY

V.A.Belyi Metal-Polymer Research Institute of the Belarussian Academy of Sciences, 32a Kirov Str., Gomel, Belarus

ABSTRACT

Copper-based composite materials are widely utilized due to their high antifrictional properties, corrosion resistance and thermal conductivity (1)(2). To assist their antifrictional characteristics, durability and self-lubricity solid additives are used as fillers, i.e. graphite, sulfides and metal selenide. Still, graphite is most often used owing to its low friction coefficient comparable with that of a liquid oil film.

One of the promising ways of improving friction joints life is application of thin film powder copper-graphite coatings by electrocontact sintering (3). To form a thin-film coating by electrocontact sintering a metallic substrate is used. The substrate is made as a strip with a loose copper-graphite powder layer placed between two electrodes. The materials are mechanically activated by simultaneous compression and transmission of an electric current (20-30 kA, 3-4 V). Electrocontact sintering utilizes the heat energy generated by the electric current in the contact points between the powder particles themselves and at the interface. The affected powder particles bond with each other and with the metallic substrate. The processes of coalescence and diffusion-based sintering proceed during such treatment. This method promotes rational introduction of the maximum permissible lubricant components into the metal matrix and their efficient use in the range of the coating limiting wear. The processes effecting structural formation and tribological properties of the thin film graphite-containing coatings are dependent on the filler and matrix contact behaviour during electrocontact sintering. The powder system interaction character and that of binary blends, which components are devoid of mutual solubility and only one of them takes part in sintering (for e.g. copper-graphite powder system), are predetermined, to a great degree, by the filler dispersity and the main component particle size.

The present work has studied the effect of granulometric composition and particle size of the copper-graphite and copper-coppered graphite powder systems on tribological and physico-mechanical properties of thin film (90...100 μm) high-filled (up to 30 mass %) coatings formed by electrocontact sintering.

Granules of coppered graphite were used as the copper matrix filler in the copper-coppered graphite systems. The granules were graphite particles clad with a 3-7 μm copper film of dendrite structure.

As investigations have shown, the higher friction coefficient is observed in samples with clad graphite granule size below 100 μm within the whole range of copper particle size variation from 20 to 250 μm. Under such circumstance the coatings show strong wearing. At copper particle size less 50 μm and that of coppered graphite above 100 μm practically all samples display 0.15-0.2 friction coefficient under stabilized regime.

The best triboenginering characteristics (0.1-0.11 friction coefficient, 0.12 μm/km wear rate) possesses composite coating with 100-160 μm copper particle size and 100-200 μm coppered graphite granule size. Nevertheless, enlargement of granule size above 200 μm leads to their uneven distribution in the coating bulk, mechanical failure and spalling of coarse particles at friction. Tensile testing evidences that higher breaking stress (80-90 MPa) is shown by copper-graphite coatings which copper particle size is about 50 μm in the whole studied range of varying coppered graphite granule size from 20 to 250 μm. The dependence here is of extreme character with a minimum.

Microhardness investigations of copper-graphite coating structural components have shown that upon friction tests the copper matrix microhardness increases from 1000-1100 to 1300-1400 MPa. This is, probably, assisted by the coating boundary layer strengthening during friction.

Calculation of copper-coppered graphite particle ratio has demonstrated that a considerable change takes place in the componental particle number proportion which difference reaches 10 and more times. This results in variation of packing and number of metal copper-copper contacts which influences the coating formation process and, as a consequence, its tribological and physico-mechanical properties.

REFERENCES

(1) V.A. Belyi, Sov. J. Friction and Wear, Vol.3, No.3, 1982, 389-395pp.

(2) K.C.Owen, M.J.Wang, C.Persad and Z.Eliezer, Wear, Vol.120, 1987, 117-121pp.

(3) V.A.Kovtun, V.B.Shuvalov and V.V.Yashin, Proc. Moscow Int. Composites Conf., Elsevier, Oxford, 1991, 969-972pp.

INFLUENCE OF THE CURRENT DENSITY ON THE TRIBOLOGICAL PERFORMANCE OF HARD COATINGS

ANDREIA ANA POPESCU, IOAN TUDOR and MARIA BERTALAN,
"Petroleum-Gas" University of Ploiesti, 39 Bucuresti Blvd. P. O. Box 10, 2000 Ploiesti, Romania
MIHAI BALACEANU and EUGEN GRIGORE,
National Institute of Physics for Lasers, Plasma and Radiation Physics, P. O. Box MG 36, Bucharest- Magurele
Romania

ABSTRACT

As it is well known (e.g. (1)), thin film properties depend significantly on the energetic particle bombardment (ions and neutrals) on the substrate. In this work tribological properties of some hard coatings (TiN, TiC, Ti(C,N) and (Ti,Al)N) deposited on substrates differently bombarded by energetic particles during the deposition process are presented. Various substrate current densities for the same bias voltage were obtained by mounting the substrates in different positions within the deposition chamber.

TiN, TiC, Ti(C,N) and (Ti,Al)N hard coatings were deposited by hollow cathode discharge process (2). The samples were mounted on special supports (Fig. 1) in the positions indicated by 1-7.

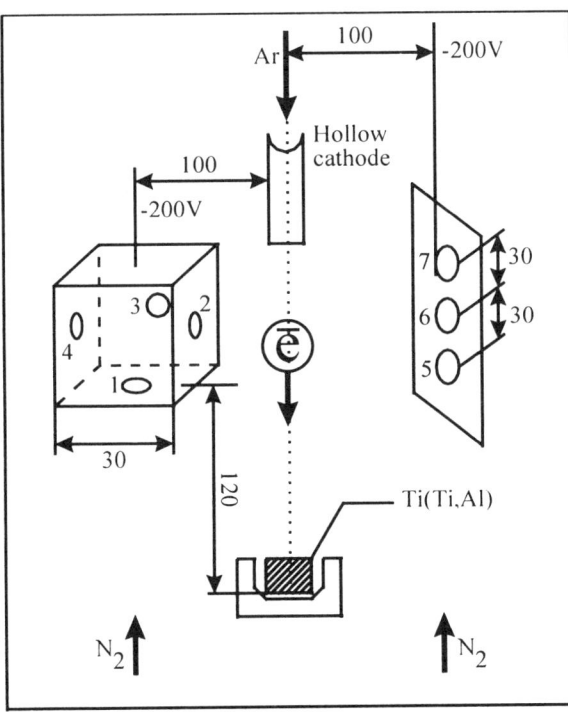

Fig. 1. Schematic of the deposition apparatus

The substrate current densities, determined by plane probe measurements under standard deposition conditions, are shown in Table 1 (bias voltage is -200V).

Table 1 Substrate current density (mA/cm^2)

Substrate position						
1	2	3	4	5	6	7
6.8	4.1	1.5	1.1	4.4	3.7	2.5

The tribological characteristics of the coatings (wear resistance and friction coefficient) under lubricated conditions were investigated by using an Amsler machine (main testing conditions: sliding speed = 0.38 m/s; sliding distance = 342 m; load = 20 kgf.). The wear resistance was determined by measuring the wear depth (d) of the coatings. The wear behaviour of some investigated coatings is shown in Fig. 2.

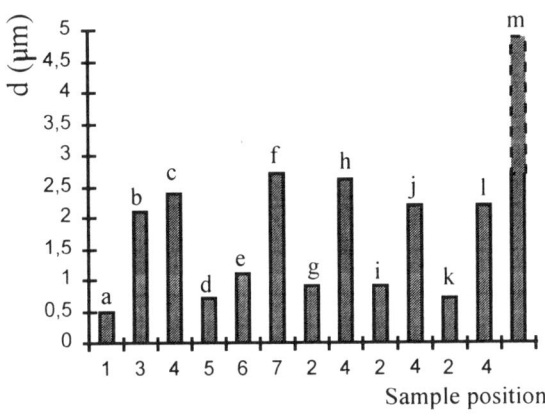

Fig. 2. Wear resistance of various coatings (a-f: TiN; g-h: TiC; i-j: Ti(C,N); k-l: (Ti,Al)N; m: steel).

It should be noted a much lower wear of the coated samples in comparison with the uncoated one and an increase of the wear resistance with the increasing current density for all the coatings. The measured friction coefficient values were ranging from 0.07 to 0.15 depending mainly on the coating type.

REFERENCES

(1) K. S. Fancey, C. A. Porter, A. Matthews, J. Vac. Sci. Technol. A Vol. 13, No. 2, p. 428, 1994
(2) M. Balaceanu, E. Grigore, G. Pavelescu, I. Tudor, A. Popescu, Proc. 2nd Intern. Conf. Tribology Balkantrib '96, Thessaloniki, p. 419, 1996.

THE TECHNOLOGICAL DEVELOPMENT OF A TRIBOLOGICALLY USEFUL SURFACE LAYER ON ALUMINIUM ALLOYS

A. Posmyk and W. Skoneczny
Institute of Engineering Problems, University of Silesia, Katowice, Poland

ABSTRACT

To obtain tribologically useful surface layers (SL) the electrochemical treatment called hard anodic treatment (AHC or ANOX-Layer) is applied. Formation of the hard oxide on the aluminium alloys increases the possibilities of its application in different fields of industry. By means of the right choice of anodizing technology, the chemical construction of the electrolyte and parameters of the technological process one can programme properties of the obtained coatings. With AHC used for sliding couplings during the process of formation of the surface layers. Anodic of aluminium alloys makes it possible to obtain the layer by many important properties from the tribological point of view.

The possibility of dripping some lubricating agent during preparation of the coating makes it possible to apply AHC for sliding couplings which do not have to be lubricated during operating e.g.: in non-lubricated compressors used in power engineering for suppression, in pneumatic drives of robots and manipulators used in medicine, the food industry and pharmaceutical industry.

One can „programe" useful properties of the anodic hard coatings according to the demands at the stage of fabrication. They have been applied hitherto to perform sliding kinematic pairs of the technical objects operating under conditions of dry friction (1), limited lubrication (2) and conventional lubrication.

In the first case AHC mate with plastics containing film forming materials (materials which assist sliding film formation during operating) e.g. PTFE, graphite, MoS_2. Under the conditions of both limited and conventional lubrication these coatings, after right modification, mate with cast iron piston rings or steel cylinder bearing surfaces of the fuel engines.

It is necessary to form anodic hard coating with properties appropriate to each particular application. For unlubricated couplings the oxides have small porosity as well as small surface roughness. Sometimes sealing of the coatings before co-acting is used. For lubricated couplings the high porosity and modification by chosen metal is required.

Porosity of AHC is of great importance taking into consideration the increase of its wear resistance, sorption of the lubricating agents, possibility of the sliding film formation from the PTFE containing material as well as determining flexibility of AHC to be modified further. Porosity of the oxide film depends on pH of the electrolyte, homogeneity of the oxidated metal, surface roughness, duration of the process, anodic current density and temperature of the bath. Dependence of the porosity on the current density for the coatings obtained in the SAS electrolyte is shown in Fig.1.

Under the conditions of technically dry friction the sliding film forms on AHC during operating. The sliding film changes character of the co-operation from plastic/AHC to plastic/plastic. Mating with AHC, the plastic shows different values of the friction factor which depend on the technological parameters of the anodic treatment. The friction factor is described by the following equation:

$\mu = 0.11+0.022j+0.025T+0.022t+0.02j^2-0.003jt$

Formation functions of the composite coating mating with cast iron are described by the equations:

$\mu = 0.1+0.01T-0.01T^2-0.01t^2+0.01Tt$
$W_l = 17.6+13.6T$

where: W_l - linear wear of the coating, j - anodic current density T - temperature of the electrolyte during modification, t - time of modification.

Summing up, it is necessary to point out that it is possible to form tribological properties of the surface layer of the aluminium alloys within a very wide range by the technological method. The ceramic AHC is the basis to form the tribologically useful surface layer. By means of the right choice of the chemical constitution of the electrolyte, as well as parameters of technological process, one can form coatings which mate with both plastics under conditions of the technically dry friction and cast iron or steel under limited lubrication. Mating of AHC with cast iron and steel is possible after the correct modification process which results in formation of the AHC + M composite coating.

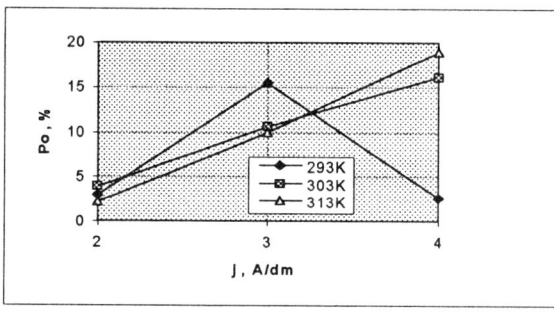

Fig. 1. Dependence of porosity of the AHC deposited in the SAS electrolyte.

REFERENCES

(1) W. Skoneczny, A. Tokarz, Mechanism of destructive changes in oxide-covered surfaces caused by friction. Wear, Nr 169 P. 209-214, 1993

(2) A. Posmyk, Z. Legierski, Abribfeste Kompositüberzüge auf ANOX - Schichten und ihr tribologisches Verhalten. Tribologie und Schmierungstechnik Nr 6 P. 324-328, 1995

TRIBOLOGICAL TESTING OF TiN-COATINGS IN DRY SLIDING CONTACTS - EVALUATION OF AN INTERNATIONAL MULTLABORATORY PROJECT

ERICH SANTNER and NORBERT KOEHLER
Bundesanstalt fuer Materialforschung und -pruefung (BAM), D-12200 Berlin

ABSTRACT

In the frame of Versailles Advanced Materials and Standards (VAMAS) Programme a group of 23 laboratories from 11 countries did more than 200 friction and wear test runs on TiN-coated samples under identical conditions with the final aim to develop agreed wear test methodologies for inorganic coatings.

All participants received „identical" test samples and had to perform the tests following a consistent sample handling and test procedure:
- 1) Steel ball (M50) / TiN-coated steel(M50) disc;
- 2) TiN-coated M50 ball/ TiN-coated M50 disc;
- 3) Si_3N_4 ball / TiN-coated M50 disc;
- coating thickness 4.5 µm on discs, 7 µm on balls;
- 10N load; 0.1 m/s sliding velocity; room temperature; 50% rel. humidity; 1000 m sliding distance (200 m for the test couples with TiN-coated balls);

RESULTS

The spread, means and standard deviations of reported friction and wear values are shown in Fig. 1.

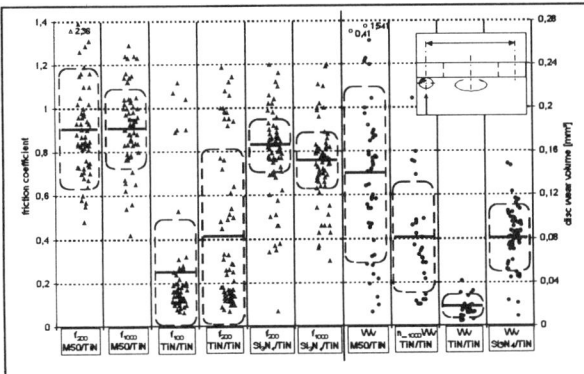

Fig. 1: Reported friction and wear values

M50 /TiN: Friction coefficient (sliding distance /m/),
$\mu(1000)_m$ = 0.91 ± 0.18 (mean ± std.dev.).
M50 /TiN: Wear volume $W_{v\,disc}$ (1000 m sliding)
$W_{v\,disc}$ = 0.14 mm³ ± 0.08 mm³ (mean±std.d).

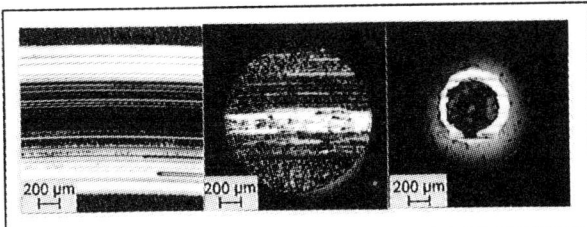

Fig. 2: Wear scars on TiN-disc, M50-ball, TiN-ball

TiN/TiN: Friction coefficient (sliding distance /m/),
$\mu(200)_m$ = 0.41 ± 0.40 (mean of 64 values).
After cancelling results from experiments with worn through TiN-coatings on the balls (Fig. 2, right) 27 from 64 reported results remained despite the reduced sliding distance.
$\mu(200)_m$ = 0.19 ± 0.07 (mean of 27 values).

TiN/TiN: Wear volumes $W_{v\,disc}$ (200m sliding)
Only 27 values have been reported, presumably due to low wear in consequence of the short sliding distance, and only 8 for the case of intact ball coatings.
$W_{v\,disc}$ = (0.016 ± 0.01) mm³ (mean of 27)
$W_{v\,disc}$ = (0.008 ± 0.005) mm³ (mean of 8)
Extrapolation to 1000 m sliding distance would give:
$W_{v\,disc}$ ~ (0.04 ± 0.025) mm³.

TiN/TiN: Wear volumes $W_{v\,ball}$ (200 m sliding):
27 of 68 reported values were from intact coatings.
$W_{v\,ball}$ = (0.001 ± 0.0009) mm³ (mean of 68)
$W_{v\,ball}$ = (0.0004 ± 0.0002) mm³ (mean of 27)
Extrapolation to 1000 m sliding distance would give:
$W_{v\,ball}$ ~ (0.002 ± 0.001) mm³ (mean of 27)

Si_3N_4 / TiN: Friction coefficient (sliding distance /m/),
$\mu(1000)_m$ = 0.76 ± 0.13 (mean of 64)

Si_3N_4 / TiN: Wear volume $W_{v\,disc}$ (1000 m sliding):
$W_{v\,disc}$ = (0.08 ± 0.03) mm³ (mean of 57).

Besides the rejection of results with worn out TiN-coatings on the balls (Fig. 2, right) also the friction values from three laboratories have been rejected which had been far below the mean and from one laboratory which had been far above the mean for all three different friction couples.

CONCLUSION

- Steel ball/TiN test couple showed unreproducible TiN wear particle transfer and embeding to the balls (Fig. 2, centre) with unhomogenious wear of TiN-coatings (Fig. 2, left), steel transfer and high scatter of wear measurements (Fig. 1).
Is not suited for standardization!

- TiN/TiN test couple has the disadvantage of possible wear down of ball coatings and very low wear values for short distance runs.
Can not be recommended for standardization!

- Si_3N_4/TiN reveals the highest reproducibility of friction and wear measurements.
Recommended for standardization of tribological coating tests!

LASER BEAM IRRADIATION EFFECTS ON TRIBOLOGICAL PROPERTIES OF LOW PRESSURE PLASMA SPRAY FILMS

SHINYA SASAKI and HIROFUMI SHIMURA

Mechanical Engineering Laboratory, 1-2 Namiki, Tsukuba-shi, Ibaraki 305 JAPAN

ABSTRACT

We have developed a new surface modification technique which couples a high-power CO_2 laser processing with a low pressure plasma spray coating method[1] as shown Fig.1. In order to clarify a laser beam irradiation effect on tribological property of coatings, metal alloy films, Triballoy, Ti/Ni, Mo/Cu, Cr_3C_2-Ni-Cr, were synthesized under a variety of laser irradiating conditions

Tribological properties of each spray films were evaluated by sliding testers under lubrication with oil and/or without oil at high temperature. The films, which were synthesized by simultaneous laser irradiation, exhibited excellent tribological properties with stronger adhesiveness on substrates, lower percentage of micro-porosity and higher hardness compared with low pressure plasma coating films.

XRD analytical results suggest that a metastable state, which is formed by the laser irradiation during the spraying process, is closely related to an anti-wear and a low friction mechanism. The formation of metastable state is considered to increase the surface hardness, to accelerate chemical activity for lubricant and to adsorb friction energy, which causes wear damage, by a phase transformation of it's crystal structure.

REFERENCES

[1] S.SASAKI et al. : "Laser assisted plasma spray coating for carbon matrix composites", Proceeding of the 4th International Symposium on Ceramic Materials and Components for Engines, Elsevier (1992) 409.

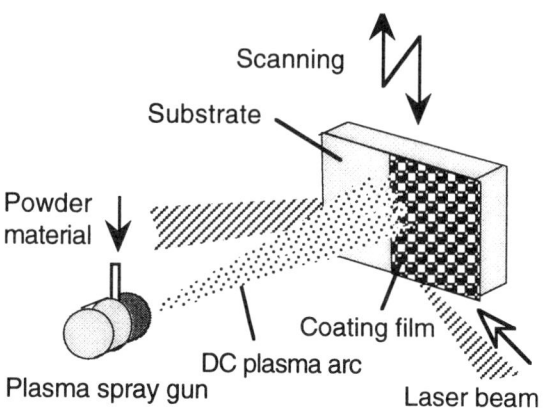

Fig. 1 Laser and plasma hybrid spray method

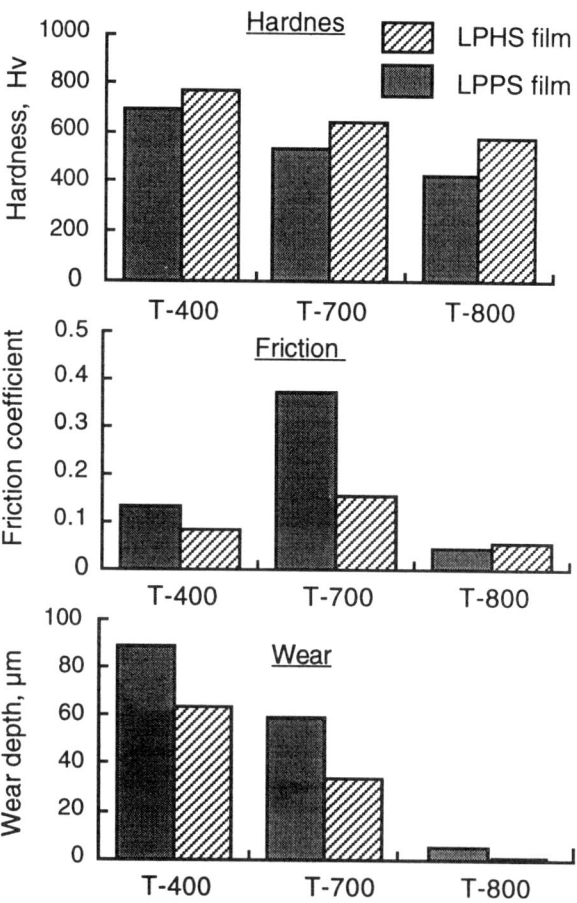

Fig.2 Tribological properties of Triballoy films
Ball on plate sliding tester, Load 98N,
Velocity 10 Hz, Amp. 50mm, Oil lubrication

NON-CONVENTIONAL METHODS FOR MULTICOMPONENT, ANTIWEAR SURFACE LAYERS PRODUCING

JAROSLAW SEP

Faculty of Mechanical Enginnering and Aeronautics, Rzeszow University of Technology, Al. Powstancow Warszawy 8, 35-959 Rzeszow, Poland

ANDRZEJ POSMYK

Silesian University, ul. Sniezna 2, 41-200 Sosnowiec, Poland

ABSTRACT

The paper presents two methods of producing surface layers consisting of two different materials placed side by side. Results of the tests confirming advantageous tribological properties of such layers are also presented.

The first method, applicable to cylindrical patrs, makes possible to obtain the surface layer that is composed of base material and modifying material, in the shape of wire, placed in helical groove, that had been machined in the part; after that the surface is burnished in order to obtain stable joint of base and modifying materials (1).

Tribological tests of such layers were realized in the system roller (specimen) - fixed block (counter-specimen); tribological load was 2400 N, and friction velocity was 0.45 m/s. Specimens were made from 45 steel (0.45% C), counter-specimens were made from 55 steel (0.55% C), lubricating agent was Selektol Specjal SAE SD 20W/40 oil. Modifying material covered about 25% of the surface.

The best results have been achieved using silver wire as the modifying material (elemination of seizure, decrease in wear by 90% and in friction coefficient by 35% in relation to the specimens without such a layer).

Advantageous results have been also achieved using copper wire as the modifying material (increase in friction distance to seizure by 520%, decrease in wear by 83.5% and in friction coefficient by 17% in relation to specimens without such a layer) and using brass wire as the modifying material (increase in friction distance to seizure by 80%, decrease in wear by 50% in relation to specimens without such a layer).

The results, that have been achieved using aluminium and lead wire as the modifying material, were not advantageous.

It has been also stated that for considered variants of processing there are essential differences, both in variations of friction force during tests and in chemical constitution of the surface

The second method takes advantages of waves formed on the surface of the base material in result of explosive cladding of soft materials (for example aluminium and its alloys). Oxide coatings, obtained in result of hard anodic oxidation of aluminium alloys, form the perfect surface layer in sliding contact with plastics and cast iron. From tribology point of view, it is very interesting to produce such layers on high-strength materials. It is possible to extend application of hard anodic coatings by applying oxidizable alloy on the one, that is difficult to oxidize.

The paper presents results of the tests concerning usability of the sliding joint, consists of Al (oxidizable) - AlCu4MSi (difficult to oxidize) materials, after explosive cladding, followed by grinding of the plated surface. As a control mechanism a ratio „k", between initial thickness of the layer that has been cladded on the substrate and the average wave height was taken. For k>1 aluminium coating was continous, for k = 0.5 aluminium covered about 60% of the surface (the overlaid material - aluminium - remains only in cavities of the waved material). Then anodic oxidation has been carried out, in order to obtain Al_2O_3 coating on aluminium (2).

Tribological tests of such specimens were carried out in sliding connection with cast iron. Tests conditions: unit pressure 8 MPa, friction velocity 1 m/s.

It has been stated that decrease in ratio „k" results in decrease in tribological wear. For k=2 average tribological wear of the specimens was 2 µm, for k=1 average tribological wear was 1.7 µm (decrease by 15%) and for k=0.5 average tribological wear was 1.3 µm (decrease by 35%). Tribological wear of the specimens with discontinous Al_2O_3 coating was smaller than tribological wear of the specimens with continous Al_2O_3 coating.

It has been also stated that decrease with ratio „k" (for 2 to 0.5) did not result in variations of friction coefficient.

REFERENCES

(1) A. Posmyk, J. Sep: Verschleßmindederne antiadhäsive kompositüberzüge an ausgewählten Werkstoffen. 10 th International Colloquium, Esslingen, 1996, 1929-1934 pp.

(2) A. Posmyk: Programming of tribological properties of the explosion plated joints. Tribologia vol. 132, No.6, 1993, 101-111 pp (in polish, summary in english).

EFFECT OF ION-IMPLANTING ON THE WEAR OF TOOLS

HAN ZHANG SHOU
Shanghai Research Institute of Materials, Shanghai, 200437, China
JIANG ZHONG SHI
Kun-Shan Special Coating Factory, Da-Shi, Kun-Shan, Jiang-Su, 215323, China

ABSTRACT

Ion-implantation is a technique included ionizing an implanted element, accelerating ionized atom to a high level of energy and bombarding the surfaces of metal with accelerated ions, and an established method for modifying the surface properties of a wide range of materials (1). Since a metal vapour vacuum arc (MEVVA) source was developed by Brown and coworkers (4) in 1985, the implantation cost has been cut down, the implantation of heavy metal ions has become available and the progress of ion implantation industrialization has been accelerated. The purpose of this paper is to find out the effect of cobalt ion implantation with heavy current density on the tribological performance of high speed steel (HSS) and the working life of HSS tools.

Specimen were made of HSS, type $W_6Mo_5Cr_4V_2$. Its heat treatment was oil quenching at 1050 C, tempering for 1h at 500 C and then cooling with furnace.

The Vickers-hardnesses, the friction coefficient and the wear resistance of specimens were measured and the working life testing of drills and tape was done as well.

Improved factor	Implanting fluence, × $10^{17} cm^{-2}$			
	0	1.5	2.3	3.5
Drill	1	1.14	1.2	1.32
Tap	1	~1	1.06	1.14

Properties		Implanting fluence, × $10^{17} cm^{-2}$			
		0	2	4	6
Microhardness	Value	813	1035	931	1113
	Factor	1	1.27	1.15	1.37
Friction coefficient		–	0.13	0.15	0.14
Scratch Section (μm^2)	Value	0.561	0.485	0.301	0.144
	Factor	1	1.16	1.86	3.89

The results show that the variation of the coefficients of friction were not obvious in the range of Co-implanting fluences $2\sim6 \times 10^{17} cm^{-2}$ and there were the decrease of scratch wear and the extension of working life with the increase of Co-implanting fluences.

REFERENCES

(1) P. J. Evans and F. J. Paoloni, Surface and Coating Technology, 65(1994)175-178.
(2) G. Deanaley and N. E. W. Hartley, Thin Solid Films, 54(1987)215.
(3) Christen A. Straede et al., Proceeding of SMMIB'95, Spain, 1995.
(4) I. G. Brown et al., Appl. Phys. Lett., 47(1985)358.

INFLUENCE OF RESIDUAL STRESSES ON TRIBOLOGICAL BEHAVIOUR OF HARD COATINGS

B.ŠKORIĆ and D.KAKAŠ
Institut for production engineering, University of Novi Sad, Trg D. Obradovića 6
21000 Novi Sad, Yugoslavia
T.GREDIĆ
Institut for Nuclear Research, IBK Vinča, Belgrade, Yugoslavia

ABSTRACT

The generation of residual stress in coating depends upon different factors. In this paper is present the measurement of residual stress of TiN and (Ti,Al)N coatings on substrates and in combination with plasma nitrided steel. The variation in the properties of the coatings is caused by different growth condiditions. Film growth at the atomic level can be important for the development wear resistant coatings that the examination cover the full spectrum of pre-coating substrate and control sputter parameters.

In this present study, unlubricated wear tests were conductet. With specially hardware and software it was possible "on-line" monitoring the develop of wear process.

Use of hard coatings to improve the wear behaviour of machine parts and equipment is now common industrial practice (1)(2).

Hardened and polished samples in the form of pin ⌀5x50 mm were used for the investigation of fundamental tribological properties. Prior the depositio the substrates of high speed steels (DIN S 6-5-2) were plasma nitrided. The surface roughness, Ra, was measuring using a stylus type (Talysurf Taylor Hobson) instruments. The coating process did not significantly change the surface roughness (Ra=0.322 μm).

The phase composition, inter planar distances and preferred orientation in polycrystalline hard coatings were analyzed by X-ray diffraction (XRD). The spectra were recorded at an angle step of 0.02° and the line intensities were measured during 1s per step. The X-ray source was Co Kα radiation.

The hardness values are also shown to demonstrate the increase in surface hardness due to nitriding at low pressure and coating (TiN-$HV_{0.03}$=2700, PNlp/TiN-$HV_{0.03}$=3500). The microhardness values indicate the composite coating has the potential to offer some good tribological properties.

The scratch test was chosen for coating adhesion and cohesion investigation. A CSEM scratch tester with a Rockwell C diamond that was drown over the surface of the coated sample with increasing normal force up to 100N (PNlp/TiN or TiAlN), without the coating detached or cracked.

The present coating method can produce dance structures, high hardness and the high critical load values can be achieved. Tribological tests confirm that these composite coatings are wear resistant and provide very low friction coefficient (Fig.1) and contact temperature.

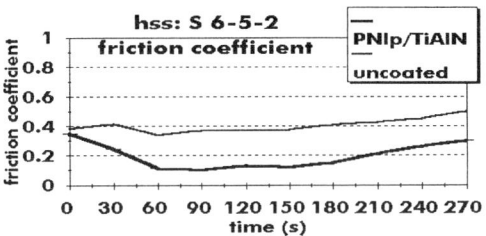

Fig.1 Friction coefficient vs sliding time

The coating show (111), (200) and (220) TiN diffraction peaks. The interplanar distances are d (2.4642 Å, 2.1300 Å, 2.0456 Å).

The residual stress was evalueted using values for E, the Young's modulus, ν the Poisson's ratio, and a_o, the equilibrium lattice parameter.

The compressive stresses of 2170 Mpa was measured in the coating.

The data analysis is ilustrated in Cohen-Wagner plots (Fig.2), where the values of lattice parameter of planes parallel to the sample surface against a trigonometric function.

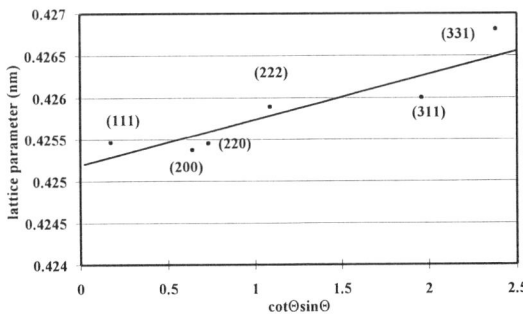

Fig.2 Cohen-Wagner plot for PNlp/TiN

We can see an expansion of the lattice in normal direction, caused by a compressive stress in the film.

REFERENCES

(1) J.A.Sue and H.H True, Surface and Coating Technology,1990, 44/45,709.
(2) T.P. Chang, M.E. Graham, H.S. Sproul, Surface and Coating Technology, 1992, 54/55, 495.

EFFECTS OF ENVIRONMENTS ON FRICTION AND WEAR OF DIAMOND-LIKE CARBON FILMS PRODUCED BY PLASMA ASSISTED CVD TECHNIQUE

AKIHIRO TANAKA, KAZUYUKI MIZUHARA, AND KAZUNORI UMEDA
Mechanical Engineering Laboratory, Agency of Industrial Science and Technology, MITI, Namiki, Tsukuba, Ibaraki, 305 Japan
MYOUG-WAN KO, SEONG-YOUNG KIM, and SANG-HYUN LEE
Korea Academy of Industrial Technology, Sihwa Industrial Complex, Jungwang-Dong, Siheung, Kyunggi-Do, 429-450 Korea

ABSTRACT

The diamond-like carbon (DLC) films were deposited on a silicon substrate by a RF plasma assisted CVD apparatus. The friction and wear of DLC films were investigated in various environments such as dry gases, dry air, humid air, and so on. A ball-on-disk type friction tester housed in an airtight vessel was used for friction and wear experiments; the friction counterpart was a SiC ball.

The friction coefficient in dry gases and air is shown in Fig. 1. The friction coefficient in both nitrogen and argon is fairly lower than that in dry air.

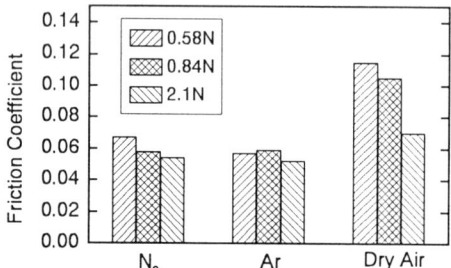

Fig. 1. Friction coefficient of DLC films in variuous gas environments

Figure 2 shows the relationship between friction coefficient and relative humidity. When the relative humidity is less than about 50%, the friction coefficient is scarcely affected by the humidity. On the other hand, when the relative humidity is more than about 50%, the friction coefficient increases with increasing the humidity. Figures 1 and 2 suggest that the friction coefficient of DLC films increases when some substances produced by tribo-chemical reaction exist on friction surface.

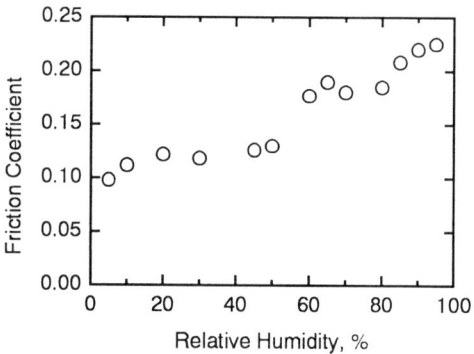

Fig. 2. Relationship between friction coefficient of DLC films and relative humidity

The relationship between wear rate and relative humidity is shown in Fig. 3. The wear rate in dry air is a little higher than that in high humid air: the dependency of wear on relative humidity is reversed to that of friction. From the observation of wear fragments, it was found that the wear fragments in high humid air adhere more strongly on friction surface than those in dry air; this difference of adhesion is probably related to the difference of wear in dry air and in high humid air.

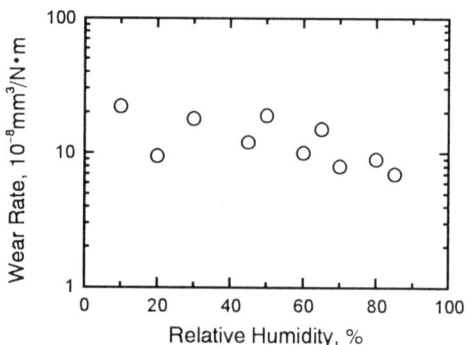

Fig. 3. Relationship between wear rate of DLC films and relative humidity

THE TRIBOLOGICAL PROPERTIES OF MoS$_2$ / METAL COMPOSITE COATINGS DEPOSITED BY CLOSED FIELD MAGNETRON SPUTTERING.

V C FOX, J HAMPSHIRE, D G TEER

Teer Coatings Ltd, 290 Hartlebury Industrial Estate, Hartlebury, Kidderminster, Worcs, DY10 4JB

ABSTRACT

MoS$_2$ is the most widely used lamellar compound solid lubricant material for space applications, (1) and is used in release mechanisms, precision bearing applications, main weather sensor bearings and gimbal bearings. In this paper a new use of MoS$_2$ coatings is presented, as a solid lubricant coating for use on cutting and forming tools.

In the past, MoS$_2$ coatings deposited by sputtering have consisted of a first few layers of MoS$_2$ with a dense coherent structure (2) followed by an open columnar structure with only the first few layers of coating providing the wear resistance.

In the present work the MoS$_2$ coatings have been deposited in the Closed Field Unbalanced Magnetron Sputter Ion Plating system as developed by Teer Coatings. The coatings deposited by this method are adherent and have a dense coherent structure producing superior tribological properties compared to coatings deposited by other techniques. It is thought that the high ion current densities at low substrate bias voltages which are characteristic of the Closed Field Unbalanced Magnetron Sputter Ion Plating system are responsible for the dense, coherent and adherent coatings.

The tribological properties of these coatings have been tested at a humidity of between 40 and 50% using reciprocating and pin on disc friction wear machines and the results are presented. The properties of the coatings were improved by optimising the deposition parameters, in particular by reducing the amount of water vapour present in the atmosphere of the coating system.

The equipment and deposition methods are described in some detail.

The properties can be further improved by co-deposition of small amounts of metal with the MoS$_2$.

The results of scratch adhesion, pin on disc and reciprocating wear tests are reported and compared to those from earlier coatings.

The MoS$_2$ / metal composite coatings are harder, much more wear resistant and also less sensitive to water vapour in the testing atmosphere. They give excellent wear resistance at loads as high as 140N on the reciprocating wear tester rubbing against a 5mm diameter WC pin. The coefficient of friction is as low as 0.02 at a humidity of 40% and the coatings and adhesion scratch tests indicate critical loads of >120N.

A number of practical applications are given and in particular large improvements in the performance of cutting tools are reported.

Figure 1 shows the results of 100N reciprocating wear tests on progressive MoS$_2$ and MoST coatings developed during this study. It can be seen that the MoST coating shows greater wear resistance than the MoS$_2$ coating developed in 1996. It should be noted that the initial coating produced in 1994 had an endurance of 100 cycles in this test, and this coating was a significant improvement compared to coatings produced by other techniques at this time.

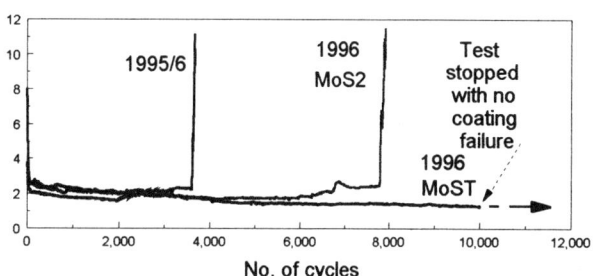

Figure 1. 100N reciprocating wear test on progressive MoS$_2$ coatings.

REFERENCES.

(1). M R Hilton, P D Fleischauer, Surface and coatings technology, 54/55, 1992, 435-441

(2). P D Fleischauer, Thin solid films, 154, 1987, 309-322

ACCELERATED ELECTROSPARK DEPOSITION AND THE COATING'S WEAR BEHAVIORS

P Z WANG, G S PAN, Y ZHOU, J X QU and H S SHAO
Beijing Graduate School, China University of Mining & Technology

ABSTRACT:

Electrospark deposition (ESD) is a pulse-arc microwelding coating process used mainly in two kinds applications. One is to enhance the performance of electrical contact points (1), while the more common application is to increase service life of many parts subjected to wear(2-6).

ESD possesses some unique advantages. The principal one is that the coatings are metallurgically bonded to a metal substrate with such allow total heat input that the bulk substrate material remains at or near ambient temperature(2). There are also some limitations, such as the low coating efficiency. Low charge-discharge frequency is the most important factor that limits coating efficiency. In our research, new switch components together with controlling and driving circuits have been applied and developed. The new design allows higher charging voltage in order to reduce the period and increase the spark frequency.

Commercial WC-8%Co is used as electrode, substrate specimens are prepared with quenched AISI 1045 steel, 500HV0.05. All depositions were performed in air. Wear experiments were finished on MM200 type wear tester.

Table 1 Comparison between convertional and new ESD euipment

	D9130	NEW
maximum coating thickness, μm	50-60	100
roughness, μm	1.6-6.3	2.6-6.3
coating rate*, cm^2/min	0.3-0.5	1-2

* for highest coating thickness.

Comparison between the newly designed accelerated ESD equipment and a conventional D9130 type is shown in Table 1, it can be seen that coating rate has been increased by 3-4 times, while the lowest value of surface roughness increases a little.

The influences of process parameters on coating thickness are shown in figure 1. The important fact revealed in figure1 is that the coating thickness has a highest value at a special depositing time, that is about 3min in the experiments of present paper, thus excessive depositing time is useless or even harmful. Surface roughness increases with the increase of pulse energy, pulse frequency and depositing time.

X-ray diffraction reveals that the main phases are M_6C type carbides such as Fe_3W_3C, Co_3W_3C, while W_2C, WC were also present. Underneath the ESDed coating there is the thermally affected substrate. The hardness of the ESDed coating is 1400-1600HV0.5.

Comparative wear experimental results between quenched and tempered AISI 1045 steel, with and without ESD coating, show that wear resistance is increased by 5-8 times.

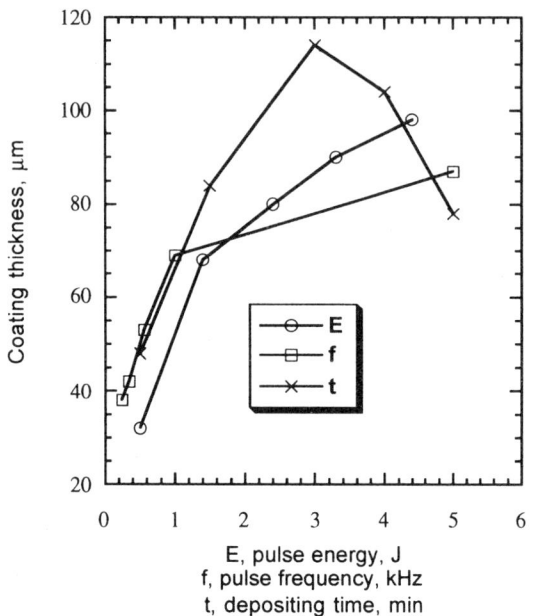

E, pulse energy, J
f, pulse frequency, kHz
t, depositing time, min

Figure 4 The relationship between coating thickness and process parameters.

SEM observation on the worn surfaces of the two groups specimens revealed the wear mechanisms: For steel-block/steel-ring pair, severe adhesive wear is the dominating mechanism, while for coating-block/steel-ring pair just mild wear occurred, including oxidization and melting. Wear debris also reflects the different wear characteristics. So, it is the difference in composition, microstructure and hardness brought about by ESD coating that prevents adhesion and increases the wear resistance.

REFERENCES
(1) G L Sheldon, et al, Surface and Coatins Technology, Vol. 36 (1988), No 1-2, 445-454
(2) R N Johnson, et al, Journal of Vacuum Science.and Technology. A4(6), Nov/Dec 1986, 2740-2745
(3) P H Thornfon, et al, Metals Technology, Feb.,1979, 69-74
(4) G L Sheldon, et al, Wear of Materials (ASME), 1985, 388-396
(5) R N Johnson, Thin Solid Films, Vol.118(1984), No.1, 31-47
(6) E A Brown, Wear, Vol.138(1990), No.1-2, 137-151

SLIDING WEAR OF PVD MULTILAYER COATINGS FOR WEAR RESISTANCE APPLICATIONS

U. WIKLUND, O. WÄNSTRAND, M. LARSSON, S. HOGMARK
Materials Science, Uppsala University, Box 534, S-751 21 Uppsala, Sweden

ABSTRACT

Much of the current development in surface coatings technology is focused on multilayered coatings. These coatings consist of layered combinations of two or more materials, often nitrides or carbides. This concept has proven to offer both a high hardness (1) and an improved toughness (2) as compared to single layer coatings. It is also indicated that this type of coatings has the potential to improve the tribological properties (3).

Friction and wear properties of five hard PVD multilayer coatings were studied experimentally using a ball-on-disc tester. TiN/CrN, TiN/MoN, TiN/NbN, TiN/TaN multilayer coatings with lamellae thicknesses close to 5 nm were evaluated together with a commercial TiN/TiAlN multilayer coating (Balinit® Futura). In addition, a TiN single layer coating was included as a reference.

Coatings were deposited both on speed steel discs (ASP2030), and on ball bearing balls. TiN was deposited by reactive electron beam evaporation and CrN, MoN, NbN and TaN was deposited by reactive magnetron sputtering. Arc evaporation was utilised in the case of TiN/TiAlN.

All coatings were approximately 4 μm thick and, except for the TiN/TiAlN, they were smooth with R_a-values around 0.2 μm. The arc evaporation process yielded a rougher surface ($R_a \approx 1$ μm) for the TiN/TiAlN. The composite hardness varied from 2500 HV for the TiN coating to 3600 HV for the TiN/TaN coating, as measured with a load of 25 g.

The sliding wear properties of the coatings were evaluated in a ball-on-disc equipment. The ball is exposed to continuous contact with the coated disc while each point in the wear track on the disc only experiences intermittent contact. Self-mated coated disc/ball pairs were evaluated together with coated discs against uncoated balls of steel and alumina, respectively. The tests were performed in ambient atmosphere with a normal load of 5 N, a sliding speed of 50 mm/s and a sliding distance of 30 m. A separate test with 250 m sliding distance was run with alumina balls to obtain significant wear of the disc coatings.

Coated ball. During the first few meters of sliding, several coatings, and especially the TiN/NbN multilayer coating, showed extremely low friction (0.14) and little wear. After this initial sliding, however, the friction leveled off at 0.5-1.0 with the lowest value and also least wear for the TiN coating. The oxidation properties of the coatings showed to be important. Relatively high wear rates were found for the TiN/CrN and the TiN/MoN, see Fig. 1.

Steel ball. Relatively high friction coefficients (0.6-1.0) were recorded in this test and the wear of the steel ball varied largely between the coatings. The TiN/TaN multilayer coating wore the steel ball least, despite displaying the highest friction. The largest amount of wear of the steel ball was found for the TiN/TiAlN, as a result of its high hardness and surface roughness.

Alumina ball. In contrast to the above tests, it was possible to obtain measurable wear of all disc coatings with the extended test utilising uncoated alumina balls. This test also revealed high coefficients of friction (0.8-1.0) for all coatings, except for TiN/MoN ($\mu=0.4$). This coating also performed superior regarding the wear of the coated disc. The highest wear was found for the TiN/TaN coating.

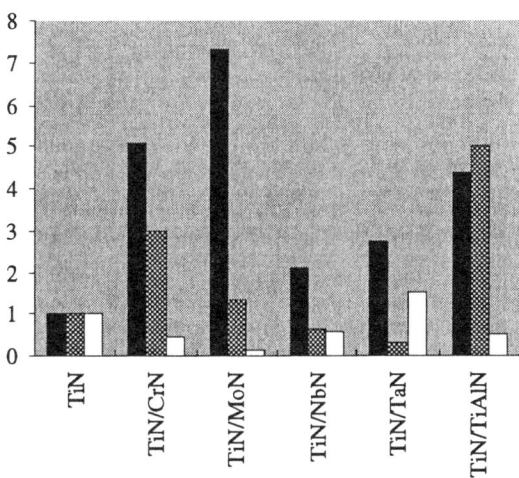

Figure 1. Normalised wear. Black bars correspond to wear of the coatings applied to the ball in the self-mated test. Grey bars show wear of the uncoated steel ball. White bars represent wear of the disc coatings when tested against the alumina ball.

REFERENCES

(1) H. Holleck, J. Vac. Sci. Technol. A4 (6), Nov/Dec 1986
(2) U. Wiklund, P. Hedenqvist, S. Hogmark, Surf. Coat. Technol., in print
(3) M. Nordin, U. Wiklund, M. Larsson, P. Hedenqvist, S. Hogmark, NORDTRIB '96, Vol. I, 16-19 June 1996, Bergen, Norway

THE ELECTRO DISCHARGE TEXTURING AND HARD CHROMIUM PLATING OF WORK ROLLS IN METAL ROLLING

J P THOMAS
Texturing Technology Ltd, P O Box 22, Port Talbot, West Glamorgan, SA13 2YJ, UK

INTRODUCTION

The maintenance of roll surface texture is generally accepted as a pre-requisite to good strip surface quality and is recognised as a major cost factor in the rolling process.

The surface texture transmitted by the work roll to the strip is of great importance. It is recognised as being an essential characteristic in the behaviour of the sheet during pressing, painting and coating. This surface requirement is defined by the customer and is continuously developing as more is understood of the technological considerations.

ELECTRO DISCHARGE TEXTURING

Electro Discharge Texturing (EDT) is a method of texturing surfaces between 0.5µm and 10+µm. It uses controlled electrical pulses discharging between the electrode and the component surface which are separated by a dielectric fluid.

The system which was developed from Electro Discharge Machining (EDM) is now in use within the metals industry where, due to its consistency and reproducibility it has largely replaced Shot Blasting as a method of producing textures on production process rolls.

The machine considered is of fixed electrode design and is capable of operating with either positive or negative electrode polarity. Additional flexibility is provided by operating in either Impulse Mode, where the energy pulses are controlled by the very rapid switching on and off of the power circuitry, or Capacitor Mode, where pulses are provided by setting the discharge of previously loaded capacitors. Combined, these advantages permit flexibility and control of the relationship between Peak Count (Pc) and Roughness (µm) within predetermined limits and allows tolerances of plus or minus 4% of that specified, regardless of material hardness.

Figure 1 shows the results of this on steel strip surface and the much greater consistency obtainable when this method is contrasted with Shot Blasting is clearly indicated (1).

All the above mentioned factors result in a rolled surface with a controlled relationship between Ra and Pc. This results in greatly improved formability and when combined with the higher level of consistency of the engineered surface a lower waviness (Wa) and thus a better distinctness of image which translates into better paintability.

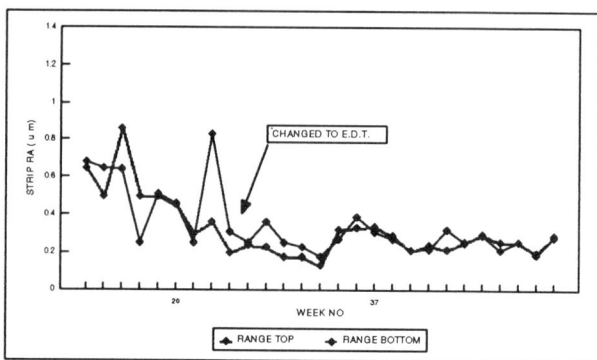

Fig. 1: Strip Surface Roughness Consistency

CHROMIUM

Hard chromium coating of work rolls for both the aluminium and tinplate rolling sector has been in use for over 10 years with quantitative improvements, both in quality and roll life due to improved wear characteristics being experienced (2).

Operational advantages achieved by the use of chromium coated work rolls include:
- Extension of roll life by improved wear characteristics, with rolled tonnage increasing by a factor of 2.2 times to 2.6 times, depending on mill stand.
- Improved reflectivity (cleanliness) of the product by an average 16 points.
- Reduced costs.

CONCLUSION

The results of combining both these techniques clearly show the superiority of the Electro Discharge Process over traditional Shot Blast textures especially when combined with accurately controlled Chrome Deposit technology. Thus ensuring that compliance to ever tightening customer surface specification is more readily accomplished.

REFERENCES

1. J P Thomas and B W Morgan, Institute of Materials Conference Rolls 2000 1996, p 176 - 185.
2. R R P Court, Court Holdings Ltd. private correspondence.

IMPORTANCE OF MACHINING ON TRIBOLOGY OF LUBRICATED SLIP-ROLLING CONTACTS OF Si_3N_4, SIC, Si_3N_4-TIN AND ZrO_2

UTE EFFNER and MATHIAS WOYDT
Federal Institute for Materials Research and Testing (BAM), Unter den Eichen 44-46, 12200 Berlin, Germany

ABSTRACT
Introduction

Ceramic ball bearings, particularly Si_3N_4 hybrid bearings, are commercially available products in all branches of industry. All-ceramic bearings have proven to be effective in dry high-temperature conditions, while hybrid bearings (with ceramic rolling elements and metallic rings) have been used in low-temperature, high-speed applications and have now a market share of about 2%. Promising commercial applications include use in high-speed machine tool spindles, instrument bearings, turbomolecular pumps, and other applications where high rotational speeds and/ or deficient lubrication may occur. Additionally, Si_3N_4 bearing materials have proven a superior performance also in some corrosive environments.

Today high manufacturing costs hinder a wider application of ceramics in tribocontacts involving high slip-rolling stresses.

Experimental

The slip-rolling friction and wear tests were performed in a twin disc tribometer of the Amsler-type at 3 GPa over 2 million revolutions. Unadditivated paraffinic oil and water were used as lubricants. The surfaces of the ceramics were machined with different processes, resulting in different surface roughnesses (i.e. rough and fine honed, rough and fine grounded, rough and fine lapped and rough and fine polished). Ceramic materials like HIP-Si_3N_4 (NBD 200), SSiC (EKasic D), HIP-ZrO_2 (htc-PSZA) and GPS-Si_3N_4-TiN (EDM) were investigated as self-mated couples. This paper summarises the achieved results.

Results

Si_3N_4, Si_3N_4-TiN and ZrO_2 exhibits in paraffinic oil generally a small wear coefficient in the range of 10^{-9} mm^3/Nm and their wear coefficient correlates with the initial surface roughness and the material removal rate. The lowest wear coefficients exhibited ZrO_2. With a reduction of the hertzian pressure to 1,5 GPa, SSiC exhibit the same tribological behaviour as the other ceramics. The wear coefficient in water is for HIP-Si_3N_4 two orders of magnitude higher. SiC , Si_3N_4-TiN and ZrO_2 surfaces show under water lubrication pitting after the tribological tests and can not operate under water lubrication.

The results presented here does not consequently support the machining philosophy „ best as possible".

The slip-rolling wear coefficient depends at 3 GPa on the surface geometry, the ambient media and the surface roughness and consequently on the machining process as well as on the material removal rate and lie for SiC in the range of 10^{-7} mm^3/Nm and for HIP-Si_3N_4, Si_3N_4-TiN and ZrO_2 in the range of 10^{-9} mm^3/Nm.

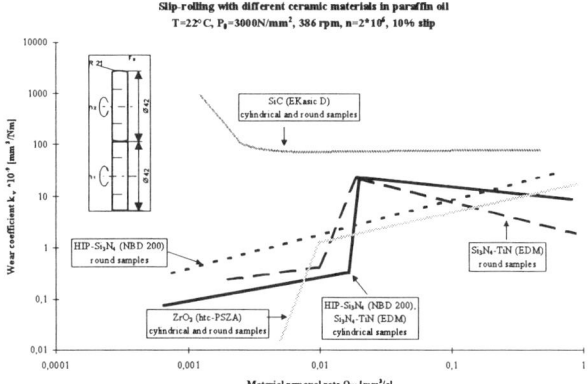

Fig. 1: Wear coefficient depends on the material removal rate

If it is acceptable to allow a linear running-in wear in the range of 0.5µm to 3µm, polishing with long machining times can be exchanged for ZrO_2, Si_3N_4 and Si_3N_4-TiN by fine grinding, honing or lapping. These three materials exhibits their best slip-rolling wear resistance in unadditivated paraffinic oil for spherical surfaces for a roughness in the range of 0.5µm≤R_{vk}≤0.9µm.

The slip-rolling wear coefficient of the SSiC at 1,5 GPa is smaller than 10^{-7} mm^3/Nm. Under these conditions, the wear coefficient decreases with decreasing material removal rate.

Water increases the slip-rolling wear of all four ceramics tested and enhances the formation of pitting and conical cracks.

The achieved results have clearly demonstrated, that through the appropriate machining process and parameters the machining costs can be further reduced without a loss in tribological performance.

Acknowledgements

This work was supported from the German Research Foundation (DFG), D-53170 Bonn.

INVESTIGATION OF THE HARD LUBRICATION COATINGS IN OPEN SPACE AROUND MOON

V M YAROSH and A A MOISHEEV
Lavochkin Association, 24 Leningradskaya street, Khimki -2, Moscow region 141400, Russia
M A BRONOVETS
Interdisciplinary Scientific Tribology Council, 101 Prospect Vernadskogo, Moscow 117526, Russia
A S LOPATIN
JSC "ROS", 65/4 Leninsky Prospect, Moscow 117917, Russia

ABSTRACT

In this paper presented the results of measurement on the autonomous instrument "Friction simulator" (FS) of the friction coefficient of the hard lubrication coatings in open space around moon. Lavochkin Association developed and produced the (FS), fig 1.

Fig.1: Friction simulator: external view

The set of the FS includes: the FS autonomous instrument (dimensions: 315x180x230 mm, mass 4,0 kg); two five-channel strain-measuring converters (converter dimensions: 225x120x170mm, mass 3,0 kg) and control unit for the tested assembly (dimensions: 240x60x180 mm, mass 1,5 kg). The FS consumes electric current of 0,4 A with constant voltage of 27 V.

The FS allows one to test materials and coatings for friction and wear simultaneously in nine friction pairs (Fig. 2): three pairs by the "shaft-bush"(2,3) scheme, six pairs by "pin-on-disk"(4,5) rotary motion.

In this paper presented investigations of the hard lubrication coatings. A load-speed mode of the friction-assembly operation was selected such, that it correspondent to the real-assembly modes of the spacecraft mechanism and provided a testing resource no more than 120-150 hours. The sliding speed is 0,8 cm/sec, pressure 9,8 MPa for the "shaft-bush" scheme and sliding speed is 1cm/sec, pressure 9,8 MPa for "pin-on-disk" scheme.

Total operation time of the FS was 128 hours during 15 months. There were conducted 18

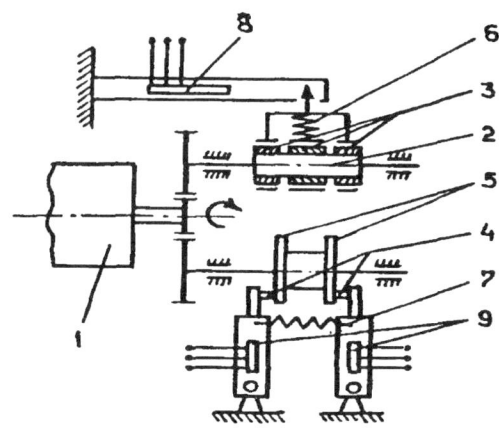

1- drive; 2,3- shaft-bush; 4,5 - pin-disk; 6,7 - spring; 8,9- tensobeams

Fig.2: Scheme of the friction simulator

communication contacts. The first turn-on of the FS took place on the 30th day of the flight. The non-operating state of the FS during a long period of time was selected to determine an influence of the space flight factors upon the structure and measuring systems of the FS instruments, upon indicating the process of sublimation of the adsorbed gases, water vapours and cleanness of the friction surfaces.

The mean of *the initial friction coefficient* was 0.24-0.27 for the "shaft-bush" scheme and 0.16 -0.18 for "pin-on-disk" scheme, *mean rotation-friction coefficient* was 0.12-0.14 for the "shaft-bush" scheme and 0.09-0.10 for "pin-on-disk" scheme, approximately in 10 hours of the FS operations in a near-Moon orbit *the friction coefficient* was 0.10-0.12 for the "shaft-bush" scheme and 0.07 -0.08 for "pin-on-disk" scheme, by the end of 15th month of operating the FS in a near-Moon orbit *the friction coefficient* decrease to 0.03-0.04 for the "shaft-bush" scheme and for "pin-on-disk" scheme 0.02-0.03 accordingly.

Comparing the results of laboratory tests of the FS with the result on circumlunar orbit, it might be noted that *the friction coefficient* in technological vacuum of laboratory installation was some higher: 0.05-0.06 for the "shaft-bush" scheme, and 0.03 - 0.04 for "pin-on-disk" scheme.

THE INVESTIGATION OF RELATIONS BETWEEN STRUCTURAL AND MECHANICAL PROPERTIES OF HARD COATING USING SCRATCH- AND CALO-TEST METHODS

MIODRAG ZLATANOVIĆ, RADOMIR BELOŠEVAC, AMIR KUNOSIĆ
Faculty of Electrical Engineering, Bulevar Revolucije 73, 11120 Belgrade
NADA POPOVIĆ
Institute of Nuclear Sciences "Vinča", P.O. Box 522, 11000 Belgrade

ABSTRACT

The relations between structural and tribological characteristics of the TiN coatings were investigated by using several characterization techniques. Samples made of steel grades DIN S 6-5-2 and DIN X210Cr12 in the form of disc 3 mm thick and 30 mm in diameter with a channel for the fractional cross section analysis, were plasma nitrided at low pressure and subsequently coated with the TiN coating using reactive magnetron sputter ion plating technique. The deposition system consists of single magnetron with permanent magnets mounted around the deposition space to form closed field configuration and additional anode parallel to the target surface (1). The same processing equipment was used for plasma nitriding at low pressure and for the deposition of hard coatings on nitrided substrates. The strong influence of plasma nitriding pretreatment at low pressure on the properties of hard coatings was found.

Successful application of duplex treatment requires high compatibility between the structure and properties of two layers. The structural properties of the composite plasma nitrided/TiN coated layer were investigated using optical microscopy, SEM analysis and X-ray diffraction. The coating microhardness was measured as well as the microhardness distribution over cross-section of the composite structure using Vickers method. Microhardness distribution of composite layers is relevant for the efficacy of duplex technology. The adhesion at coating/plasma nitrided layer interface is of critical importance for the application of combined technology. The critical load for adhesion was evaluated by the scratch-test method. Calo-test is a simple method which was used mainly for determination of the coating thickness. A modified calo-testing apparatus was applied to the track produced during scratch test and gave interesting information about failure mechanism (2). The scratch test equipment was also used for testing the abrasive wear properties of variously treated samples. Coefficient of friction and tangential force were measured using constant normal load of 10 N on the diamond stylus tip, and adhesive failure of coating was not found by optical microscope inspection. Scratch energy density as a measure of abrasive wear properties may be calculated from tangential force F_t, removed volume V_s and scratch test length l_s according to the relationship (3) : $\omega = F_t \cdot l_s / V_s$. The values of scratch energy density for various substrate treatments are represented in Fig. 1.

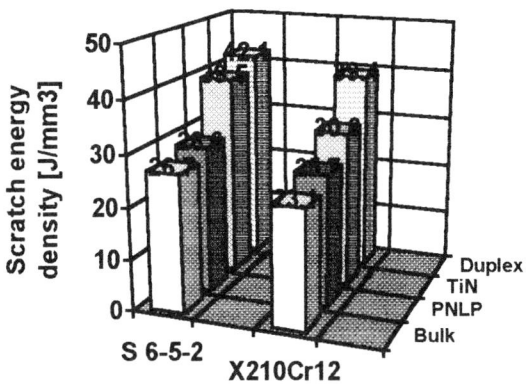

Fig.1 Scratch energy density for differently plasma treated substrates

The abrasive wear resistance is found to increase with nitrogen diffusion pre-treatment of all substrates investigated. The values of corresponding frictional coefficients indicate beneficial influence of plasma nitriding in all cases. The results of XRD analysis and tribological tests are in agreement with conclusions from Ref. 4 about supreme wear characteristics of coatings with (220) preferred growth orientation.

The structural and wear investigations showed superior characteristics of the composite structure compared to the single hard coating.

REFERENCES

(1) M. Zlatanović, R. Beloševac, A. Kunosić, N. Popović, Ž. Bogdanov, Surface & Coatings Technol., 74-75 (1995) 844
(2) B. Pierret, V. Bellido - Gonzales, N. Stefanopoulos, J. Witts, J. Hampshire, D. G. Teer, New Uses of Ball Cratering in Coating Property Analysis, ICMCTF'96, San Diego, 1996.
(3) K. Hock, G. Leonhardt, B. Bucken, H.-J. Spies, B. Larisch, Surface & Coatings Technol., 74-75 (1995) 339
(4) A. Matthews, H.A. Sundquist, Proc. Internat. Ion Engineering Congress ISIAT&IPAT 83, Vol. II, Ed by T. Takagi, Kyoto 1983, p.1325

TUESDAY 9 SEPTEMBER

Corresponding oral session

		Page number
M3/TU3	Wear and friction in sliding systems	517
TU5	Magnetic storage systems	579
F1	Rotordynamics	593
TH5	Tribology in manufacturing and maintenance	597
TU1	Tribology education and training	633

LOW FRICTION IN UNLUBRICATED SLIDING OF CAST IRON AGAINST SILICON NITRIDE+

S. K. SINHA*, K. ADACHI** and K. KATO**
*Department of Chemical Engineering, Imperial College, London SW7 2BY, UK.
** School of Mechanical Engineering, Tohoku University, Sendai, 980, Japan.
+ submitted to IMechE Journal of Engineering Tribology

ABSTRACT

The sliding pair of ceramics against metals has a certain advantage over other pairs in tribological applications as the large amount of heat energy produced during surface interaction can be easily dissipated through the metal part[1]. This is not possible in a ceramic/ceramic pair as the accumulation of interfacial heat leads to rise in temperature (ceramics are usually bad conductors of heat) causing thermally induced changes such as softening, sintering and chemical reactions leading to high wear rates. In this study, sliding test was carried out on a ball-on-disc apparatus. Silicon nitride balls ($R_{max.}$=0.02 μm; $R_{max.}$ of a surface is the mean distance between peaks and valleys of asperities on the surface) were slid against cast iron discs of different roughnesses under varying conditions of normal load and sliding velocity. Friction coefficient (μ) was measured as the ratio of friction force (F) to normal load (L).

Figure 1 shows friction coefficient measured as a function of number of cycles of revolution of silicon nitride balls against cast iron discs. The data are presented for different initial roughnesses ($R_{max.}$) of the disc under a fixed normal load of 15N and sliding velocity of 0.025 m/s. The plots show that for a highly polished ($R_{max.}$=0.05 μm) cast iron surface, the coefficient of friction maintains a very low value of 0.1 for up to 2.5×10^4 cycles of revolution. From the literature data this is the lowest friction coefficient measured for ceramic/metal pair so far. The friction coefficient is a strong function of the initial surface roughness of the disc. The number of cycles for low friction condition varies with the disc roughness and also with the normal load. In order to study the effect of normal load on friction coefficient for the smoothest ($R_{max.}$=0.05 μm) disc we carried out tests at different normal loads under the same sliding velocity and keeping the roughness of the disc as $R_{max.}$=0.05 μm. These tests showed that for a 15 N normal load the number of cycles providing friction coefficient of 0.1 was greatest. This indicates that this load (15 N) is optimum for the present set of experimental variables. The wear rates measured for the cast iron disc and silicon nitride ball for the condition used in this study are 10^{-6} mm^{-3}/Nm and 10^{-7}-10^{-8} mm^{-3}/Nm respectively.

A study of the mechanism of low friction showed that the characteristics of the evolving microstructure of the cast iron disc was extremely important during the sliding process. A scanning electron microscopic study of the initial cast iron surface showed that the roughness of the disc greatly influences the presence of *in-situ* graphite on the surface of the disc. SEM pictures show remarkable changes in the presence of graphite on the cast iron surface. Smoother cast iron surface show greater presence of graphite on the surface. This *in-situ* supply of graphite to the interacting surface is a very effective lubricant and hence the friction coefficient is found to be low for smooth discs. When the measured friction coefficient is high the cast iron surface either shows very minimal presence of graphite or almost no presence at all. For such cases, there is invariably a metallic film on the surface of cast iron.

In most of the previous studies[2-4] involving ceramics and metals under dry sliding condition the friction coefficient was recorded as greater than 0.2. The reason for this observation can be explained from the results obtained in the present study. The low friction mechanism operates only under the optimum conditions of the surface roughness, normal load and sliding velocity. In the absence of these conditions the benefits of graphite at the interface is eluded due to the formation of a metallic layer. For such dry sliding conditions, the in-situ presence of a solid lubricant is also very essential.

REFERENCES

1. Aronow, V. and Mesyet, T., J. Tribology, **108**, 16-21 (1986).
2. Zhou, YiMin G. L., Zhou, Jing En and Zhou, Qing De, Wear, **176**, 39-48 (1994).
3. Gautier, P. and Kato, K., Wear, **162-164**, 305-313 (1993).
4. Ravikiran, A. and Pramila Bai, B. N., Wear, **181-183**, 544-550 (1995).

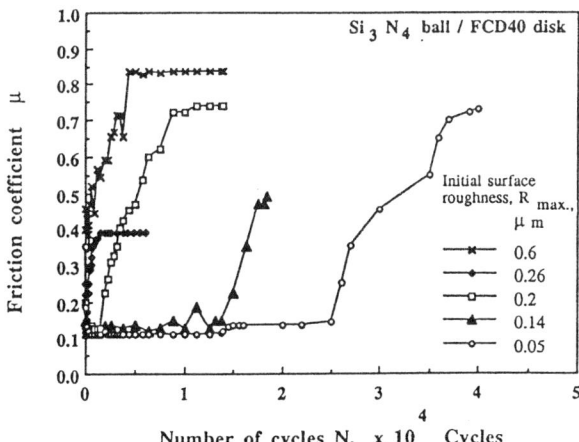

Figure 1 Friction coefficient as a function of the number of cycles of rotation for different initial disc surface roughnesses. Normal load = 15 N, Sliding velocity=0.025 m/s

THE EFFECTS OF GAS COMPOSITION ON FRETTING WEAR.

Dr J Beard and Mr N A Strong
AEA Technology plc, National Centre of Tribology, Risley Warrington. WA3 6AT.

It is well-established that gaseous environment can have large effects on the sliding wear rates of metal couples. This paper reports a series of factorial experiments carried out on a number of gas-cooled nuclear reactor materials to investigate the effect of changes in coolant composition on fretting wear. The investigation centred on evaluating the change in wear brought about by the addition of carbon monoxide (CO), moisture (H_2O), hydrogen (H_2) and methane (CH_4) relative to a reference high purity carbon dioxide gas environment. Tests were carried out on three material combinations: mild steel against itself, AISI 316 stainless steel against itself and Nimonic PE16 against 20/25/Nb stainless steel.

All tests were carried out under an applied normal load of 100N for a test duration of 100 hours, and at a constant test frequency of 25Hz.

Despite differences in detail the wear behaviour of the three material combinations overall was very similar. At low temperatures carbon monoxide had the largest effect on all the materials reducing wear rates by up to a factor of 5.

With the exception of 316 stainless steel, at low temperatures moisture had a detrimental effect and resulted in an increase in specific wear rate by a factor of 2. At higher temperature the presence of moisture was beneficial or had no detectable effect. In some instances, moisture appeared to produced an interaction with carbon monoxide. However, at higher temperatures all significant effects were small and in practical engineering terms generally negligible.

Extensive surface examination and metallography showed at low temperature that the addition of CO or moisture to the gas environment markedly reduced the amount of oxide present on the contact surface of mild steel.

The presence of carbon monoxide appears to increase the ratio of metal to oxide present on the surface. At low temperatures this has a beneficial in reducing wear, whereas at higher temperatures CO tended to impede the formation of a protective oxide callous.

As expected temperature had the largest influence on the fretting process. At the higher temperatures a wear resisting oxide callous (sometimes termed a 'glazed oxide') was formed over the surface on both contacting specimens. This resulted in a net volume gain (ie negative wear) at the higher test temperatures.

Metallography shows the callous to be around 60 microns thick. It is postulated that the callous is produced by direct oxidation of the surface material. Mechanical working of the surface increases its chemical reactivity resulting in a much higher oxidation rate than might be expected under static oxidation conditions.

The main test programme was carried out at 100 microns slip amplitude. The effect of increasing the slip amplitude was studied with the mild steel material but only at a temperature of 250°C. The tests carried out at 200 microns amplitude appeared to negate or reduce the effect of the gas additions.

The effect of amplitude may be linked to the increase in mechanical working of the surface. The greater shear deformation produced at higher slip may accelerate the formation the oxide callous. This is analogous to static oxidation, where mechanical damage to a surface can enhance formation rates. Establishing an oxide callous more quickly may reduce the period over which the gas additions can affect the wear process.

One of the most unexpected findings occurred with all material combinations and was not associated with an effect of gas composition. Despite the identical geometry of the moving and static specimens, at a temperature that was different for each of the materials, the specimens were subject to a completely different wear behaviour which has been termed asymmetrical wear. Further work is required to understand this process, but it is suspected that it may be caused by a small temperature gradient across the specimens. Asymmetrical wear is an important effect to understand, it opens the possibility of selectively controlling which surface will wear even with like-on-like material combinations.

Hybrid Model of Tribological Interactions Between Soil and Tillage Tool.

BALLA, J. - BROZMAN, D. - KUBIK, L.
University of Agriculture, Tr.A.Hlinku 2, 949 76 Nitra, Slovak Republic

ABSTRACT

Recently, investigators and engineers in experimental mechanics have developed a hybrid method [1] based on concepts of the standards ISO 9001. In this paper the methodology has been applied to problems of intensive abrasive wear of tillage machine active tools. Initial measurement and computer modelling indicate that the knowledge gained in first two-dimensional approach can be used to predict a three-dimensional ploughshare wear.

An original geometry of ploughshare cutting edge is changing very quickly as a result of its interaction with soil. Important factor of blade behavior is its asymmetrical position in ralation to direction of motion. It results into increase of draft force and energy input in ploughing process. Up to 20% increase of draft force in case of using blunted blade was fund in total comparing to the original shape of blade [2]. The approximation effort to exact solution of cutting edge changes during interaction with soil leads to the mathematical theory of systems what requires measuring and modelling of blade wear in various soil and operational conditions.

The measurement of wear on physical model were realized by optical moire method. Wear distribution is illustrated as a contour map resulting from moire patterns (fig.1) which overlap on the tool surface. The results of real wear distribution measured differentially on in operation worn ploughshares are compared with the results of mathematical finite element model of pressure distribution at the cutting edge of tool.

In first stage the mathematical model based on the 2-D Laplace's equation of potential incompressible flow

$$\frac{\partial^2 \Phi}{\partial x^2} + \frac{\partial^2 \Phi}{\partial y^2} = 0$$

applied to the tool cutting edge of various radius was used. Considering continuation of wear proportionally to the pressure distribution, the progress of cutting edge shape changes was modelled (fig.2).

Fig.1. Moire interferogram of partially worn ploughshare cutting part.

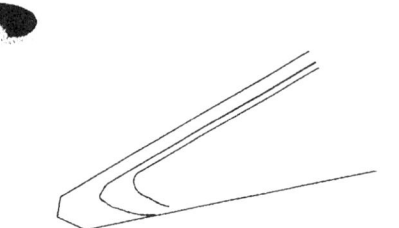

Original and worn ploughshare profiles

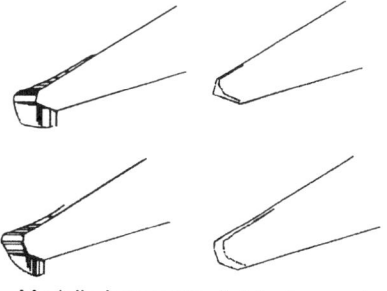

Modelled pressure distribution and corresponding wear

Fig.2

For second stage and approximation to reality the mathematical model will be corrected for using of compressible continuum.

REFERENCES

[1] D.Brozman, Agricultural Engineering, Vol.42, No.3, 1996, 81-86pp

[2] J.Balla, V.Gubka, In: Reliability of Agricultural Machines, Banska Bystrica, 1990

RUNNING-IN;
the effect of ZDDP film formation.

T J BENHAM
Physics Department, Chalmers University, Gothenburg, Sweden.
G WIRMARK
Applied Physics Department, Volvo Technological Development, AB Volvo, Gothenburg, Sweden

ABSTRACT

It is well known that the highest wear rates and highest frequency of tribologically related failures in machinery occur at the beginning of service life. A well designed running-in procedure, where the two mating surfaces adapt to each other, can often substantially decrease the number of these early failures.

During the running-in of a cylinder and piston ring the honing pattern plays an important role in enabling the surfaces to conform to each other by wearing rapidly and sustaining the high local stresses involved, without suffering from severe damage (1). Oil additives assist with this surface re-formation and enable the surfaces to run under more severe conditions than otherwise possible.

In this study, tests have been carried out using a new high speed test rig which re-produces different conditions in the piston ring - cylinder liner contact system and has been used with a variety of oils, ring and liner materials.

Using well run-in specimens, this rig has been used to create a set of Stribeck curves at 4 different loads. This has been done by changing the temperature, hence oil viscosity, during the test. With base oil the usual 3 lubrication conditions (boundary, mixed and hydrodynamic) were identified, with the Boundary to Mixed lubrication regimes showing a load dependence of $W^{-1/2}$. The Mixed to Hydrodynamic transition was found to be largely load insensitive.

ZDDP addition showed a modification of the boundary region in which the tribo-film exhibited viscosity, resulting in the increase of friction coefficient with reducing temperature until oil film thickness effects dominated. The ZDDP delayed the transition from Boundary to Mixed lubrication with respect to the viscosity, which is believed to be due to formations of tribo-film on the surface (2). A further effect was that the friction coefficient increased with a decrease in load.

By comparing these Stribeck curves with previous tests a number of important factors in the running-in have been identified. They often act in combination and have been found to act differently under the different conditions.

- The ZDDP acts detrimentally to wear when not acting in an anti-wear role in severe contacts.

- Severe conditions in the high wear boundary regime are very quickly mitigated. Long term wear is then much lower than conditions predict.

- Capacitance measurement has shown extremely rapid tribo-film generation with ZDDP irrespective of conditions. It also showed gradual oil film generation with base oil in un-disturbed conditions, due to surface conformity. This conformity is quickly lost if the surfaces are disturbed for any reason, i.e. a temperature change.

- The load dependence and ZDDP effects on friction prevent its use solely, as a criterion of the regime in which a contact is operating.

REFERENCES.
(1) Barber, G.C. and K.C. Ludema, *The break-in stage of cylinder - ring wear: A correlation between fired engines and a laboratory simulator.* Wear, 1987. **118**: p. 57 - 75.

(2) Tripaldi, G., A. Vettor, and H. Spikes, *Friction Behaviour of ZDDP Films in the Mixed, Boundary/EHD Regime.* SAE, 1996. : p. 73-83.

COMPATIBILITY OF TRIBOSYSTEMS

NICOLAY A. BUSHE
All-Russian Research Institute of Railway Transport,
10, 3-d Mytishchinskaya, 129851, Moscow, Russia

ABSTRACT

Current understanding of friction involves concepts of compatibility. Compatibility of tribosystems is understood as the ability of tribosystem to provide an optimal functioning according to selected criteria in a given range of operating parameters (1). Number of work on compatibility considerably increased lately resulted in significant increase of the compatibility criteria. This enables to obtain more realistic description of tribosystems which vary in character and environmental factors.

To estimate tribosystem's response to the intensification of friction regime an approach has been worked out. For instance, the considerable difference in the tribotechnical characteristics between two antifrictional materials has been obtained when they are tested under conditions of step loading (Fig. 1). Favourable response is recorded when thin, as a rule soft mobil films are formed in the contact area The same reaction is noted with hard, seizure resistant oxide films, or with coated protective hard or soft films (2).

Compatibility of tribosystem is exposed during running-in process at mixed lubrication conditions. When materials are compatible to achieve good results it is essential to provide non-rigid mixed lubrication regime when share of liquid lubrication film is more than boundary film.

Basing on the concepts of compatibility, the selection of antifrictional and journal materials as well as rational regimes of running-in for various types of diesel engines has been done.

The physical concepts of compatibility of tribosystems are suggested. They consider specific state of surface layers which, regardless of friction regimes and type of material, endure plastic deformation and partake in physico-chemical reaction.

During friction active surface layer behaves according to the patterns of the profound plastic deformations characterized by the rotation effects and nonstability.

State of active surface layer is defined by his energy parameters studied by many scientists. The physical concepts of compatibility have been studied by S. V. Fedorov from the ergodynamic attitude, by L. I. Bershadsky, B. I. Kostetsky from the structural-energy adaptation attitude (3).

Utilization of the scientific concepts of tribosystem's compatibility enables to solve some problems of the rational selection of materials, design, surface coatings and tribosystem operating conditions (4).

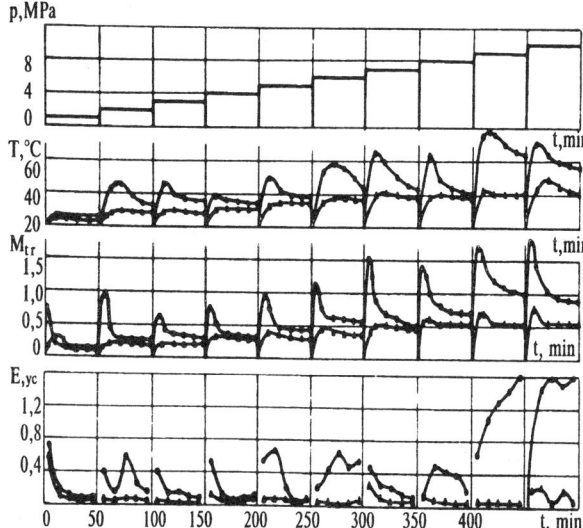

Fig. 1. Temperature (T,°C), friction moment (M_{tr}), wear rate ($E_{,yc}$) of bearing alloys under step loading (P, MPa) testing conditions. (Continions line - brass with 22%Pb, dotted line - aluminium alloy with 20%Sn).

REFERENCES

(1) N. A. Bushe, V. V. Kopytko. Compatibility of rubbing surfaces, Moscow, Nauka, 1981.
(2) N. A. Byshe. Evaluation of the role of metallic materials in tribosystem compatibility, Tribology in the USA and the former Soviet Union Studies and Application Allerton Press, Inc. New York 10011, 1994.
(3) Fundamentals of tribology. Textbook, Edited by A. V. Chichinadze, Moscow, 1995.
(4) V. S. Avduevsky, M. A. Bronobetz, N. A. Bushe, V. M. Shkolnikov. Theoretical and applied aspects of modern technology, Proceedings of International conference "Energodiagnostics", Vol. 1, p. 31-61.

FRICTION AND WEAR BEHAVIOUR OF C^4 Al_2O_3/Al COMPOSITES UNDER DRY SLIDING CONDITIONS

LORELLA CESCHINI, GIAN LUCA GARAGNANI and CARLA MARTINI
Institute of Metallurgy, University of Bologna, viale Risorgimento 4, 40136 Bologna Italy
GLENN S. DAEHN
Dept. of Mat. Science and Eng., The Ohio State University, 116W 19th Ave Columbus OH 43210 U.S.A.

ABSTRACT

The wear resistance of MMCs can be improved by increasing the volume fraction of the reinforcing ceramic phase (HRC) by as much as 70%. Among the various types of HRC composites, a new Al_2O_3-Al *co-continuous ceramic/metal composite* (referred to as C^4 material) has been recently produced at The Ohio State University (1). A new method of preparation, based on a displacement reaction between a silica precursor (whose frame is faithfully reproduced in the composite product) and molten Al, allows the production of a composite material where both the Al_2O_3 (~70vol.%) and Al (~30vol.%) phases are continuous and interpenetrating (2). The aim of the present work was to investigate the tribological behaviour of C^4 Al_2O_3/Al composites under dry sliding conditions. Friction and wear tests were carried out using a computer-controlled slider-on-cylinder tribometer. The stationary sliders were constituted by the composite under investigation, in the form of prismatic blocks (5x5 mm cross section) surface finished to Ra≅0.3 μm roughness. The rotating cylinder (40 mm in diameter) was an AISI 1040 carbon steel surface hardened to 64 HRC hardness and finished to Ra≅0.07 μm. The tests were carried out at room temperature in laboratory air, at applied loads in the range 5 to 30 N and sliding speeds in the range 0.3 to 1.8 m/s, for sliding distances up to 10 km. Both friction resistance and total wear were continuously measured and recorded as a function of the sliding distance. The maximum wear depths on the C^4 samples were also measured at the end of each test using a stylus profilometer. Wear scars and debris were characterized by means of scanning electron microscopy (SEM), electron probe microanalysis (EPMA), with an EDS analyzer and X-ray diffraction (XRD) analysis.

The coefficients of friction were always quite high, ranging from about 0.6 to 1.1, with a tendency to decreases (down to 0.4) at the highest sliding speed. The C^4 material exhibited, under the adopted testing conditions, a high wear resistance related both to the high hardness and the good interfacial bonding between the interpenetrating phases.

The maximum wear depth, evaluated by stylus profilometer, was always lower than about 80 μm even under the more severe sliding conditions. Under similar testing conditions, conventional aluminium alloys (AA 6061 and AA 2014) reinforced by about 20 vol% of Al_2O_3 particles showed a worse wear behaviour with maximum wear depths up to about 800 μm (3).

The third-body (4) produced by the abrasive action of the hard ceramic phase on the counterfacing steel, is mainly constituted by Fe_2O_3 as demonstrated by SEM and XRD analysis. Iron oxides appear both on the wear scars (as protective plateaus) and in the debris.

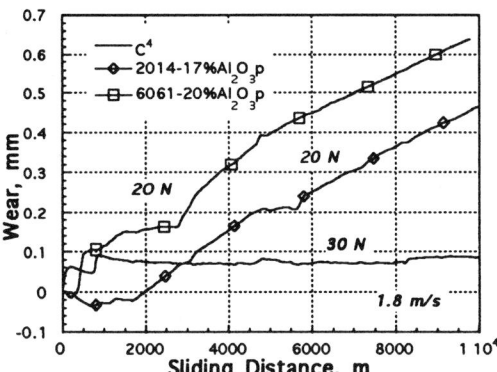

Fig.1: Comparison of the wear behaviour of the C^4 with conventional Al_2O_3 particles reinforced Al alloy matrix composites.

REFERENCES

[1] M.C.Breslin, *US Patent No. 5,214,011* Patented May 25, 1993.

[2] Breslin M.C., Ringnalda J., Xu L., Fuller M., Seeger J., Daehn G.S., Otani T., Fraser H.L., *Mat.Sci.Eng. A*, 1995 V195, N1-2 (Jun 1) 113.

[3] Ceschini L., Garagnani G.L., Palombarini G., Poli, *Proc. 4th European Conference on Advanced Materials and Processes (EUROMAT '95)*, A.I.M. Ed.- Milano (1995) 381.

[4] Godet M., *Wear*, 136 (1990) 29.

TO THE PROBLEM OF NECESSITY OF SCREENING ACTION OF THE THIN LUBRICAION FILMS ACCOUNT WHEN VALUING CRITICAL TEMPERATURES UNDER BOUNDARY LUBRICATION CONDITIONS

A. CHICHINADZE, N. POLYAKOV

Russian Engineering Academy, Association of Tribology Engineers, Kursovoi Str. 12, Moscow, Russa

ABSTRACT

Up to the present in the tribological science the conception has been formed according to which one of the determining factors is the maximum temperature on the friction suface. The methods of the heat calculations for different friction units are well- known (1)(2)(3)(4).

At present in the heat calculation of friction mechanisms calorific characterictics of materials taken from the tables are used (2). In some cases such approach is not quite corect. In theory and practice it is determined that subsbance film on the material surface changes its calorific characteristics. The films of oxides, oils, other organic and non-organic combinations differ essentially from materials of friction. When the friction is nonstationary heat penetrates into the body at some depth.

The film substances make worse penetration conditions of the heat into bodies depth. During friction this leads to the increase of temperature of friction surfaces. This phenomenon depends upon the time, calorific impulse, the width of substance film, materials calorific characteristics and other parameters. This effect (screen effect) is displayed on microcontact more essentially.

The created calculation method allows to determine the calorific characteristics and temperature in the zone of lubricated friction contact. For calculation the formulas of heat transmission theory for flaky structures are used. Results can be essentially different from the traditional ones. For example consider the structure metal + mineral oil film. Maximum temperature on the friction surface was determined according to the formulas (1)(3)(4) considering the hypothesis of summing up the temperatures and conditional heat parameters. The dependence of the temperatures on the depth of the oil film is shown in Fig.1.

Our researches have the shown, that heat calculation has to take into consideration the depth of the film, the duration of the temperature flash and other parameters affecting the screening action.

Such approach allows to obtain tolerances of modes of friction at the maximum temperature for other mechanisms, working at the boundary lubrication conditions. Using checked calculation will allow to choose the materials, the lubricant and the friction modes more precisely. All this will improve the projection quality of the new friction mechanisms.

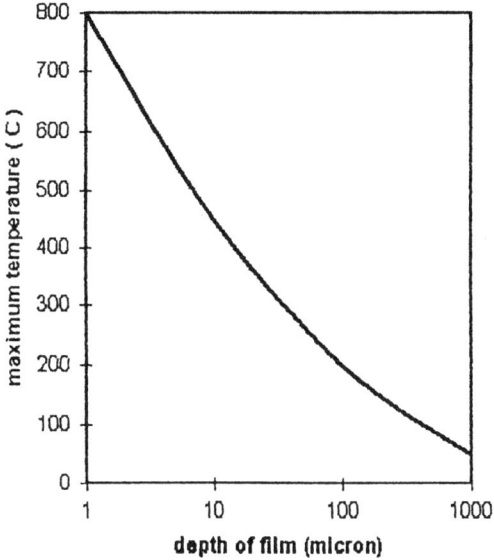

Fig.1. The maximum temperature depends on the depth of the film

REFERENCES

(1) M. Khebda, M. : Machinebuilding, vol 3, page 730, 1992

(2) H. Blok, Proc. 7 Round Table Disc on Marine Reducation Gear, p.p 3- 25, 1969

(3) A. Chicinadze M.; Science and Technica, page 778, 1995

(4) A. Chicinadze M.: Science and Technica, vol 1,2, 1994.

QUANTITATIVE CORRELATION OF WEAR DEBRIS

UNCHUNG CHO and JOHN A. TICHY
Department of Mechanical Engineering, Aeronautical Engineering and Mechanics
Rensselaer Polytechnic Institute, Troy, New York 12180-3590, USA
chou@rpi.edu and tichyj@rpi.edu

ABSTRACT

The size, number and shape of microscopic wear particles in a lubricant or working fluid are strong indicators of the machine's health -- and consequently its reliability. Considerable interest has been expressed in the analysis of wear debris for prediction of machine reliability. Most present methods, e.g. Ferrography, involve subjective identification of wear modes based on the visual assessment of wear debris (abrasive wear, fatigue wear, etc.) as an intermediate step to decision-making as to maintenance actions [1]. The accuracy of these methods depends on human expertise and the results may not be precisely reproduced. Therefore, there has been a long-standing need to speed up the process and reduce its subjectivity. In this work, image acquisition equipment, image processing and statistical software are used to permit systematic analysis of wear debris as shown in Fig. 1.

Tests are performed to generate wear debris of differing categories. Five wear variables are chosen for wear tests: load, contact geometry, material, surface finish and lubricating oil. The used oil is collected for further quantitative studies.

Images of wear debris are captured with image acquisition equipment consisting of a microscope, a CCD camera, a frame grabber and a computer with image processing software. After the images of wear debris are captured, binary images are generated by thresholding. Then, morphological attributes of 2-dimensional binary image -- size, curvature and shape -- of wear debris are quantified with morphological parameters [2, 3].

Among the broad spectrum of statistical data analysis, analysis of variance (ANOVA) is applied to find which morphological parameters are affected by the difference of wear conditions. The morphological parameters are affected by the difference of wear conditions as follows:

- Size parameters are the most strongly affected.
- Shape parameters are moderately affected.
- Curvature parameters are slightly less affected than shape parameters.

It is shown that size parameters are the most strongly affected by the different wear conditions. The result indicates that the change of wear conditions can be detected by observing the proper parameters.

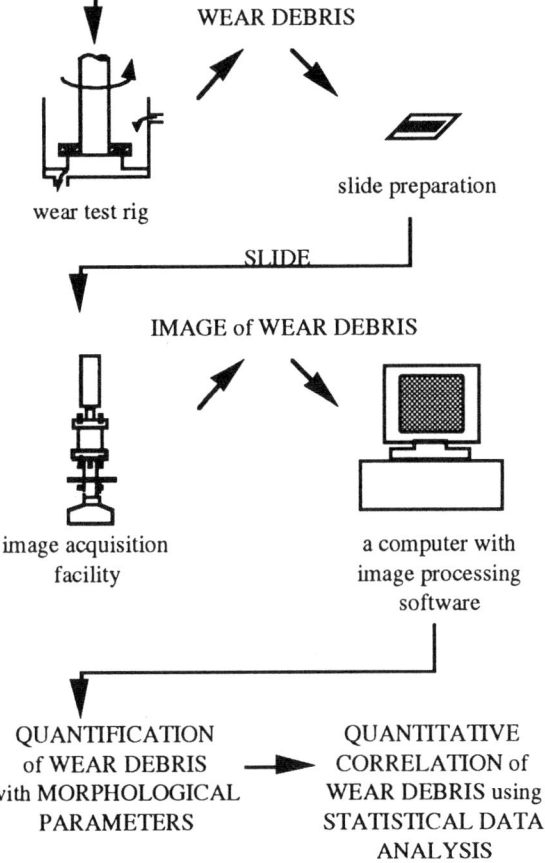

Fig. 1: Procedure for quantitative correlation of wear debris

REFERENCES

(1) D. W. Anderson, Wear Particle Atlas (Revised), Naval Air Engineering Center, No. NAEC-92-163 (1982).
(2) B. J. Roylance and S. Raadnui, The morphological attributes of wear particles their role in identifying wear mechanisms, Wear, Vol. 175, 1994, pp. 115-121.
(3) T. B. Kirk, D. Panzera, R. V. Anamalay and Z. L. Xu, Computer image analysis of wear debris for machine condition monitoring and fault diagnosis, Wear, Vol. 181-183, 1995, pp. 717-722.

FRACTALS IN WEAR PARTICLES

JIA CHUNDE, SHI YINONG, ZHANG ZHIJUN

Shanyang Institute of Technology, No.81, Wenhua Road, Shenyang, Liaoning 110015 P. R. China

ABSTRACT

The fractal theory, which was founded twenty years ago, has found its application in many science fields. However, you can find very little use of fractal views in the study of friction and wear, needless to say the fractal study of wear particles that is growing. As we know, it is difficult for conventional wear test methods to reveal the relationship between the appearance of wear particles and their forming mechanism. So what we have employed in our study is the dynamic direct observation technique that makes it possible to describe the real phenomena of friction and wear more reliably by means of fractal study. The objective of this paper is to present the result of fractal dimensions concerning with different process and forms of wear particle formation and growth.

The wear tests were done with a special wear test apparatus in which the wear behavior of metal pin versus glass could be seen directly and the relations between wear particle forms and their growth, and also the formation mechanism could be distinguished clearly.

The direct observation wear tests found: In single pass wear under dry friction, there were two typical types of models for wear particle formation and growth according to the wear conditions. When using copper pin against glass in oxygen, the wear particle grew up through the stage of materials mutual adhesive transfer and plowing. When using Aluminum pin against glass in oxygen, the rolling bar-like wear particle grew up in rolling adhesive manner.

The fractal measuring is based on the following principal:

$$N = F \cdot r^{-D} \quad (1)$$

Where N is the number of squares that cover the boundary of wear particle, F is a constant that depends upon the shape of wear particle, r is the magnitude of the square, while D is the fractal dimension.

The fractal study result of different wear particle growth stage of copper pin versus glass in one single pass is shown in Fig.1, showing that in the growing process of wear particle with mutual adhesion transfer mechanism, the fractal dimensions increase with the distance from the "leading edge" (2) and assumes the same tendency with the average diameter and the area fraction of wear particles. The fractal dimensions could be described as the state function of particle growing.

The N(r)-r curves (fig. 2) show that various wear particles with different growth mechanism have a very different shape. The wear particle with the mutual adhesive growth mechanism has a more zigzag boundary than that with rolling adhesive mechanism due to their different growth process.

It can be concluded that different types of wear particles controlled by different formation mechanisms and the different growth stages for one type of wear particle could be described by their fractal dimensions mathematically, and the fractal representations were reliable as long as the relationships between wear particle shape and their growth mechanism are determined precisely.

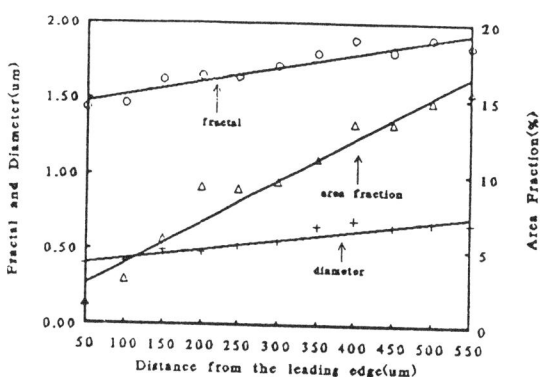

Fig.1 Variations in fractal, diameter and area fraction

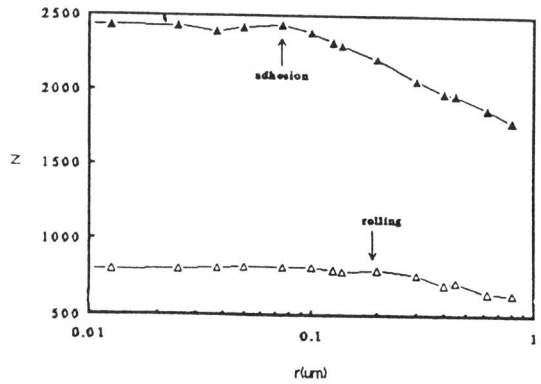

Fig.2 N(r)-r curves of various wear particles

REFERENCES

(1) Kleiser T., Bocek M. Z. Metalk, vol.77, Page 582, 1986.

(2) C. Jia, Tribology 2000, 8th International colloquium, 14-16 January 1992, Page 2.4-2, 1992.

NEW MODEL OF FRICTIONAL CONTACT

N B DEMKIN

Tver State University of Technology, 22, A.Nikitin emb., Tver 170026, Russia

The computer model of frictional contact has been developed on the following basis. The variation of real contact pressure during the contact deformation is described by unified function for elastic, elastoplastic and plastic deformation. Thus, the difference in deformation mechanisms of individual asperities does not influence on calculation of contact parameters. At the same time, the percentage of elestic, elastoplastic and plastic asperities may be estimated, as well as the corresponding parts of total contact load.

The above contact model takes into account the probability of interaction of individual asperiries as a function of their heights, radii and spacing. A single asperity is believed to be a spherical segment. Asperity height obeis beta-distribution function, and it is believed, that asperiries are uniformly distributed on contact surface. In case where one of the contacting surfaces is more rough than another, the latter is considered as a smooth one. The criterium is tenfold difference of average asperity heights. Otherwise, both surfaces are considered as rough ones.

The above computer contact model has been used for the calculation of equivalent roughness parameters as a function of the number of contacting asperiries and their height distribution (1). The equivalent roughness permits to replace the contact of two rough surfaces with the contact of rough surface with smooth one. In the latter case the calculation of contact parameters is easier. The computer modelling shows, that equivalent roughness height strongly depends on asperity height distribution and the number of contacting asperities.

Let us introduce the coefficient K_r as

$$K_r = \frac{R_p}{R_{p1} + R_{p2}}$$

where R_p is the equivalent maximal height of asperiries above the mean line, R_{p1} and R_{p2} are the maximal height of asperiries above the mean line for the first and second surfaces respectively. The maximal value of K_r obviously equals to unity. Practically it is less than unity. Our results show, that the value of K_r strongly depends on asperities height distribution and on the number of asperiries. For the surface with asperiries of equal height $K_r=1$ when $N>20$ (N is the number of contacting asperities). For the surface with normal distribution of asperities heights $K_r=0.67$ when $N=20$, $K_r=0.77$ when $N=100$ and $K_r=0.84$ when $N=1000$. Thus, K_r tends to unity when the number of contacting asperities increases. In other words, the equivalent asperities height depends on the contact area and the contact load.

For the calculation of equivalent roughness height above the mean line the following approximative formula has been proposed:

$$R_p = (R_{p1} + R_{p2})[1 - e^{(-\frac{10}{v_1+v_2})}]$$

where v_1 and v_2 are the coefficients of height distribution function (1).

When the elastoplastic solids are in frictional contact, the surface asperities deforme in different way, thus the friction force is different on each single contact spot. According to the molecular-mechanical theory of friction (2), the friction force consists of two parts: adhesive one and deformative one. On the basis of the above theory the following formula have been obtained for the friction coefficient:

$$f = (\tau_0/GH_m)(Y_c/Y)^\omega + \beta + 0.55\sqrt{Y/r} + \\ +0.19\alpha\sqrt{Y_u/r}$$

where τ_0 and β are the parameters of specific friction force: $\tau=\tau_0+\beta q_r$, q_r is the real contact pressure, G and ω are coefficients, Y is contact deformation, Y_c is critical deformation, r is mean asperity top radius, α is hysteresis coefficient, Y_u is elastic recovery of a single asperity.

The above computer model permits to calculate the total friction force as a sum of frictional forses on single contact spots. The computer modelling show, that the dependence of friction force on contact load is not monotone, i.e. there is a minimum value of friction force at some load value.

REFERENCES
(1) N. Demkin, Journal of Friction and Wear, Vol.16, No.6, 1995.
(2) I.V. Kragelskii, Friction and Wear, Butterworths, London, 1965.

A NEW SCUFFING MODEL FOR THE EP-LUBRICATED ADDITIVE CONDITION

ENRIQUE F. ESCOBAR
Jaramillo Universidad de la Serena, Chile

ABSTRACT

The present study is concerned with the development of a new 3D-Model for the scuffing mode of failure. The main object is referred to the role of additives on the improvement of the scuffing capacity by gearing, disks and other machine elements by rolling-sliding. The observed mechanism giving rise to the scuffing limit, correspond to a linear viscoelastic response of the friction site, as a process, depending on the sliding speed as the dominant variable and two response coefficients containing all the tribology of the model. Thereafter, the mechanism of scuffing, could be explained as a tribo energetical instability of the contact, represented by a combined variable $[\bar{\vartheta}/\mu_{\xi(t)}]$ where $\bar{\vartheta}$ is the temperature and $\mu_{\xi(t)}$ a relaxant friction coefficient due to the action of the additives. This procedure permits to account for the effects of the EP-additive oils, in order to develop a calculating criterion which agrees well with experiment. The proposed method provides a simple means to predict the scuffing limit, because for each additive condition, only three input parameters are required, which can be obtained from a few experimental scuffing tests. Furthermore, the proposed short and economical experimental test procedure to evaluate the scuffing limit, emerges as a useful and novel tool for researcher and engineers avoked to the scuffing subject. The whole method leads to important insights into the rheology of contacts and its process thermodynamic, clarifying the concept of the lubricant as machine element.

FRICTION AND WEAR IN HIGHLY LOADED SLIDING CONTACTS UNDER MIXED LUBRICATION CONDITIONS

L. DETERS, G. FLEISCHER and D. WEINHAUER

Machine Elements and Tribology, University of Magdeburg, Universitaetsplatz 2, D-39106 Magdeburg, Germany

ABSTRACT

Highly loaded contacts with high sliding portions can be found e.g. at cam/follower-systems, gearings and rolling element systems with high creep ratios. Under standard lubrication conditions the well-known equations of the EHL theory are available for the calculation of these contacts (1), (2). However, at low entrainment velocities and/or very high specific loads on the contact the lubricant film thickness can be remained under the critical one, i.e. the transition film thickness. Then mixed lubrication will be met. With this two states of frictions will occur at the same time, and that are the friction of the hydrodynamic lubrication and the solid friction. The external load is supported partly by the EHL lubricant film and partly by the asperities, which are deformed elastically and plastically then.

The friction force of mixed lubrication at highly loaded contacts can be computed by adding up the friction forces of EHL and solid friction. The friction force of solid friction is mainly caused by deformation of asperities, which are located on the macro-surfaces being in contact (Fig.1). The adhesion between the asperities can be neglected, as the surfaces are covered with a thin lubricant film.

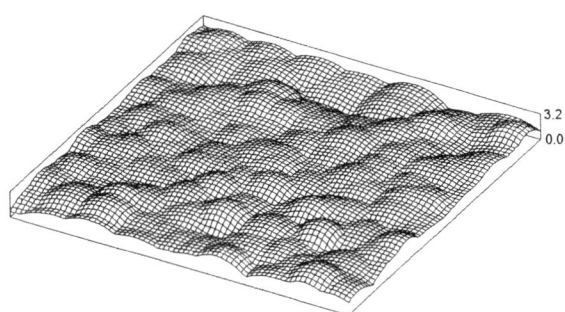

Fig. 1: Modelled surface (200µm x 200µm)

For the calculation of the part of solid friction within the friction of mixed lubrication first of all the work of deformation at the representative asperity is determined, which can be received in dependence on the surface topographies, the external load and the material properties (3). After that with the aid of the distance, on which the friction force operates at the representative asperity, the friction force itself and from this the coefficient of friction can be calculated. In the following the overall friction force can be received with the number of asperities being in contact. The finding out of the part of friction of EHL within the friction of mixed lubrication results from the known solutions of the EHL theory and the assumption, that the shear stress in the lubricant film is proportional to the local velocity gradient and the lubricant viscosity.

The calculation of wear is closely coupled with that of the friction. It is based on the hypothesis, that the friction energy, which is brought into a tribological stressed surface, is partly irreversibly accumulated (4). The stored energy is continously increased such a time until a critical threshold is reached, and then a wear particle is separated. The height of the wear is directly proportional to the number of stress cycles, to the ratio of the friction force stressed distances on the counterbody to that on the observed body and to the solid friction force and inversely proportional to the friction work used for the wear volume and to the width of the contact. When empirical determined data about the overall wear of the two bodies being in contact are available, the wear profiles on the contact surfaces can be calculated approximately.

A comparison of the theoretical results with test data demonstrates good accordance. The tests were carried out on a test machine, working like a cam/follower-system, at which the load shows sinusoidal fluctuations, what can be seen e.g. by the curves of the friction force in Fig.2.

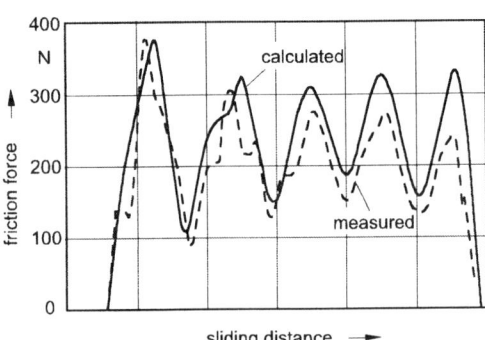

Fig. 2: Calculated and measured friction force

REFERENCES

(1) G. Kümpel, Tribologie und Schmierungstechnik, Vol.33, No.4, p.299-304, 1986.
(2) M. Wallinger, Dissertation, Techn. Universität Clausthal, 1983.
(3) D. Weinhauer, Dissertation, Otto-von-Guericke-Universität Magdeburg, 1996.
(4) G. Fleischer, Wiss. Z. d. TU Otto von Guericke Magdeburg, Vol.34, No.8, p.55-66, 1990.

DETERMINED AND PROBABILISTIC APPROACHES TO THE CALCULATION OF THE RESOURCE UNDER WEARING CRITERION.

Yu. N. DROZDOV
Department of Friction, Wear and Lubrication, Mechanical Engineering Research Institute of Russian Academy of Sciences, Moscow, Russia
V.J. MUDREAC, S.Y.DINTU
Department of Technological Industrial Equipment, Technical University of Moldova, Chisinau, Moldova

ABSTRACT

The authors have an idea to join the determined and probabilistic approaches. Moreover in the structure of rated equation for the wear intensity the components of different nature are put into, principally characterized the wear process (1), (2), (3).

In application to the sliding bearings, lubricant liquids of which can contain and a certain percent of mechanical admixtures, such relation has a form:

$$I = a_0 \left(\frac{p}{HB_{1,2}}\right)^{a_1} \left(\frac{\lambda}{h}\right)^{a_2} \left(\frac{E_{1,2}}{\sigma_0}\right)^{a_3} (1+\alpha K)^{b_1} \left(1+\beta\frac{HB_a}{HB_{1,2}}\right)^{b_2} \left(1+\gamma\frac{Sd_a}{V}\right)^{b_3}$$

where $p/HB_{1,2}$ — dimensionless complex, that characterize the stress state in the contact; p — specific load, MPa; $HB_{1,2}$ - hardness of the working surfaces of shaft and bearing, MPa; λ/h — complex, that determine the contact conditions, lubrication mechanism. Under the absence of mechanical admixtures in the lubricant liquid $\lambda = (R_{a1}^2 + R_{a2}^2)^{1/2}$, where $R_{a1,2}$ - average standart deviation of the surface roughness, mkm. Under the presence of mechanical admixtures $\lambda = d_a$, if $d_a > (R_{a1}^2 + R_{a2}^2)^{1/2}$ and $\lambda = (R_{a1}^2 + R_{a2}^2)^{1/2}$, if $d_a < (R_{a1}^2 + R_{a2}^2)^{1/2}$; d_a - reduced diameter of the particles, mkm; h - thickness of the lubricant film, mkm; $E_{1,2}/\sigma_{01,2}$ — complex, taking into account plasticity and fatigue durability of materials; $E_{1,2}$ — modulus of elasticity of the first kind, MPa; $\sigma_{01,2}$ — the limit of fatigue of materials, MPa; $1+\alpha K$ — the criterion of the concentration of mechanical admixtures; α — the coefficient, taking into account the degree of the influence of particle concentration on the wear; K — concentration, kg abr./kg of liquid; $1+\beta HB_a/HB_{1,2}$ — the criterion of admixture hardness; β — the coefficient, taking into account the influence degree of coefficient of particle hardness on the wear; HB_a — hardness of the mechanical admixtures, MPa; $1+(\gamma S \cdot d_a)/V$ — criterion of the particles form; γ — the coefficient, taking into account the influence degree of the particle geometry on the conjunction wear; $(S \cdot d_a)/V = K_\phi$ — the shape coefficient; S, d_a, V — correspondingly the surface, reduced diameter and the particle volume; a_0 — the coefficient, taking into account physical-chemical contact characteristics; a_i, b_i ($i = 1,3$)— the exponents, appointed on the base of the existing information about the wear of the materials or testing of the analogous specimens.

The probability of the sliding bearings fail-safe work

$$P(T) = 0,5 + \Phi\left(\frac{X_{max} - X_0 - M(\gamma)T}{\sqrt{\sigma_{x_0}^2 + D(\gamma)T^2}}\right)$$

where T— the bearing resource, h; X_0, σ_{x0} — an average range and average standart deviation of the initial meaning of the output parameter (radial tolerance), mkm; X_{max} — the limit meaning of the output parameter, mkm;

The bearing resource under the different meanings of probability of the fail-safe work is determined by the formula

$$T = \frac{-(X_{max} - X_0)M(\gamma)}{[D(\gamma)Z_p^2 - M(\gamma)^2]} + \frac{\sqrt{(X_{max} - X_0)^2 M(\gamma)^2 + [D(\gamma)Z_p^2 - M(\gamma)^2][(X_{max} - X_0)^2 - \sigma_x^2 Z_p^2]}}{[D(\gamma)Z_p^2 - M(\gamma)^2]}$$

where Z_p - quantile of the normal distribution.

To the basis of the calculation of the indexes of sliding bearings operating reliability put the information, obtained from the operate organizations.

As a result of the data processing about the buildup 154 electropumps ЭЦВ6, 45 - ЭЦВ10 - 63-110 и 75 - ЭЦВ10 - 63 - 150 obtained the reliability indexes of the bearings in the operating conditions.

The comparison shows, that an average relative deviation of the calculated meanings of the probability of the sliding bearings fail-safe work from the operating equals~11%. The relative deviation of the average resource meaning (T_{av}=11140h) from the operating ($T_{av. op.}$=12900h) equals~14%, that proves the usefulness of the proposed method.

REFERENCES

(1) Yu.N. Drozdov, Mashinovedenie, №3, 1980, 93-99 pp.
(2) V. P. Koqaev, Yu. N. Drozdov, Прочность и износостойкость деталей машин, page 319, 1991.
(3) Yu. N. Drozdov, V.G. Pavlov, V. N. Puchkov, Friction and Wear in Extremal Conditions, Handbook [in Russian], Mashinostrienie, Moscow, page 224, 1986.

WEAR BEHAVIOUR OF COPPER-GRAPHITE COMPOSITE

S.A. EL-BADRY
Azhar Univ.,Faculty of Science, physics dept., Cairo, Egypt.
S.F.MOUSTAFA
Central Metallurgical R&D Institute, P.O.Box 87 Helwan, Cairo, Egypt.

ABSTRACT

Copper matrix alloy, copper reinforced with either Cu-coated or uncoated graphite particulate composite materials were fabricated by powder metallurgy technique. The wear and friction properties of the investigated materials has been assessed in air using a pin-on-ring dry wear test. Pins of 7.9±1 mm diameters and 12 mm in length, were rubbed against rotating steel ring (SAE 1045), at a constant sliding velocity of 0.2 ms^{-1} and within a load range of 40 to 450 N.

At low loads, the wear rates of graphite-coated composite were lower compared with the matrix and graphite uncoated composite. With increasing the applied load, sever wear rate of the matrix specimen took place at load of about 120 N, while graphite composites could withstand loads up to 450 N in case of coated graphite composite. After that, severe wear rate took place and the specimen fracture in large chips (microgrooving mechanism). The uncoated graphite composites could bear loads up to 300 N. In pure copper, however, the mechanism of wear was of three regimes of wear: oxidational wear, seizure, and fusion or melt wear(1)(2). However, the mechanism of wear in composites was completely different.

Both cases of coated and uncoated graphite composites were behaved in almost the same manner, namely, strain induced delamination, and sub-surface delamination (microgrooving). The cause of better wear resistance of graphite composites could be explained as follows: the graphite particles were smeared on the copper surface by extrusion from the matrix (3). The smeared graphite film decreases the coefficient of friction and also lowers the wear of the composites appreciably.

REFERENCES

(1) S.C. Lim and M.F. Ashby, Acta Metall., 35, 1987.
(2) R. Antoniou and D.W.Borland, Mater.Sci. Eng., 93, 1987.
(3) S.F.Moustafa, Canadian Metall. Quart., Vol 33, No3, 1994.

WEAR SIMULATION OF SPUR GEARS

ANDERS FLODIN and SÖREN ANDERSSON
Machine Elements, Department of Machine Design, Royal Institute of Technology, S-100 44 Stockholm, Sweden

ABSTRACT

Predicting wear in spur gears, which is regarded in this paper, has been calculated using an approach where the wear depth on a tooth surface is being integrated over time using the Euler integration method. Variations in pressure distribution over the surface is regarded as well as changes in curvature and surface profile of the teeth as they wear. The paper focuses on the use of an appropriate wear model for simulating the conditions present between high performing lubricated gears.

SIMULATION FOUNDATION

In a study of mild wear of spur gears, Andersson and Eriksson (1) introduced the concept of "single point observation". Flodin and Andersson also used this concept in a previous paper (2) since it complies well with numerical methods and computer simulation. In this paper the authors have used existing software developed by themselves, which includes the single point observation method. For a more detailed discussion on the software and the single point observation method see (1),(2).

The determination of the contact pressure at each pressure cell or point and at each time step, is simplified by modelling the surfaces as *elastic foundation models* or *Winkler surface or mattress models*, see figure 1.

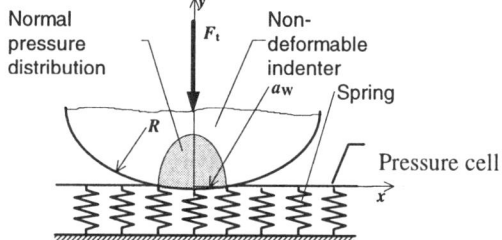

Fig. 1: Principle of mattress model of surface

Three different wear models are presented and used in simulations. They are:

1. A modified Archard's wear equation were the wear depth, h, is proportional to pressure, velocity and time in contact.

$$\frac{dh}{dt} = k \cdot p \cdot v$$

2. An oxidational model has been derived (3) were the wear depth, h, is proportional to factors such as frictional heat, flash temperature and contact width.

$$\frac{dh}{dt} = \frac{J}{(C^2 \cdot a - 2 \cdot C \cdot E \cdot a^2 + E^2 \cdot a^3) \cdot b \cdot l_p} \cdot \exp\left(-\frac{Q_p}{M \cdot a - S \cdot a^2 + V}\right) \cdot v_p$$

3. A physisorption model (4) were the wear depth, h, is proportional to factors such as flash temperature, sliding velocity and contact pressure is also used in simulations.

$$\frac{dh}{dt} = k_m \cdot \left[1 - \exp\left[-\frac{X}{v_P \cdot t_o} \cdot \exp\left(-\frac{E}{R \cdot T_P}\right)\right]\right] \cdot \frac{p_P \cdot v_P}{p_m}$$

The Flodin-Andersson approach give the following result, see figure 2 were the initial wear is plotted.

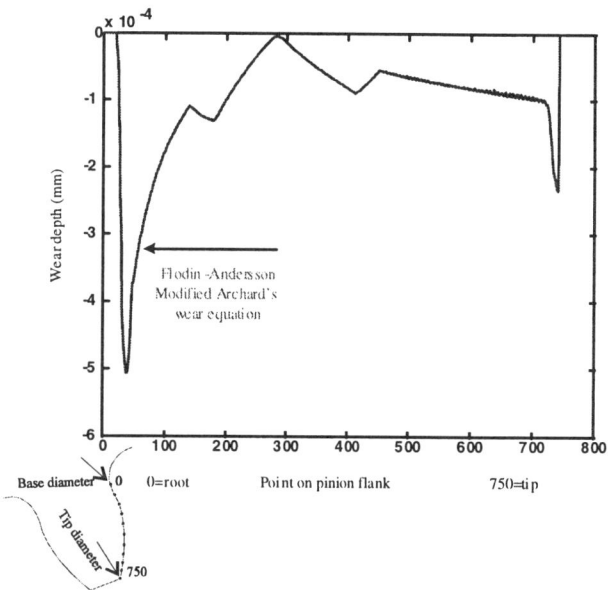

Fig. 2: Initial wear on pinion

REFERENCES

(1) Andersson, S. and Eriksson, B., (1990), Prediction of the Sliding Wear of Spur Gears, Proceedings of NORDTRIB'90, ISBN 87 983538 0 2.

(2) Flodin, A. and Andersson, S., (1996), Simulation of Mild Wear in Spur Gears, Proceedings of NORDTRIB'96. To be published in Wear.

(3) Quinn, T.F.J. Rowson, D.M. and Sullivan, J.L., (1965), Application of the oxidational theory of mild wear to the sliding wear of low alloy steel, Wear, vol. 65, pp 1-20.

(4) Wu, S and Cheng, H.S., (1991), *A sliding wear model for partial EHL contacts,* Transactions of the ASME, vol. 113, January 1991, pp. 134 141.

CORRELATION BETWEEN FRETTING CRACKING PROCESS AND CLASSICAL FATIGUE VARIABLES

S. FOUVRY*, Ph. KAPSA*, L. VINCENT**
Ecole Centrale de Lyon, CNRS, BP 163, 69131 ECULLY CEDEX, France
* LTDS (UMR 5513), ** IFoS (UMR 5621)

ABSTRACT

Cracks in contacts induced by fretting are uneasy to detect and can lead to catastrophic damages (1). However, recent developments have shown that crack appearance in fretting contacts can be predicted by applying multiaxial fatigue criteria (2). It implies to determine the stress evolution imposed on each point of the contact during the fretting cycle. The two controlling crack nucleation parameters, which are the local shear stress amplitude $\hat{\tau}$ and the tensile state p_H, are calculated and combined with the classical fatigue variables such as the alternated shear and bending fatigue limits of the material (τ_d and σ_d respectively). To quantify the cracking risk, the Dang Van formulation consider a maximisation which permits to extract the scalar variable "d" (3):

$$d = \max_{t,\vec{n}} \left[\frac{\|\hat{\tau}(\vec{n},t)\|}{\tau_d - \alpha \cdot p_H(t)} \right] \text{ with } \alpha = \frac{\tau_d - \sigma_d/2}{\sigma_d/3}$$

When d is superior to 1 then there is a risk of cracking.

Performed below as well as on the surface this analysis permits one to localise and estimate the intensity of the maximum risk. This maximum is observed at the contact borders along the sliding axis on a sphere/plane contact when the friction coefficient is superior to 0.3.

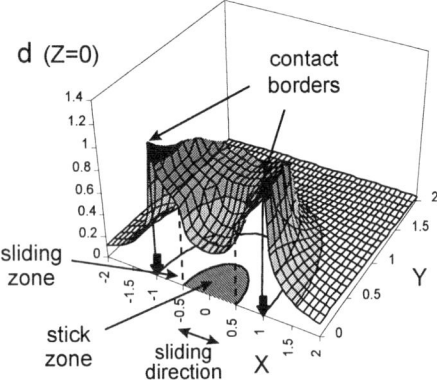

Fig. 1 : Surface distribution of the Dang Van cracking risk for a sphere/plane contact.

This high cycle fatigue is supposed to be transposed for any situation since the loading path is known (i.e. by F.E.M analysis for complex geometry). In fact the quantitative prediction is barely altered by the sharp gradient of stress which characterises the contact loading. Indeed a very small volume of matter is stressed in the contact compared to the classical fatigue situations.

To estimate the contact failure from the macroscopic fatigue approach it is then fundamental to take into account the size effect. The contact stress point analysis is then replaced by a mean stress approach $\bar{\Sigma}(V(l))$ where the loading state is averaged on an elementary cubic volume V(l). Fig.2 shows that depending on this volume, the cracking risk distribution is strongly modified.

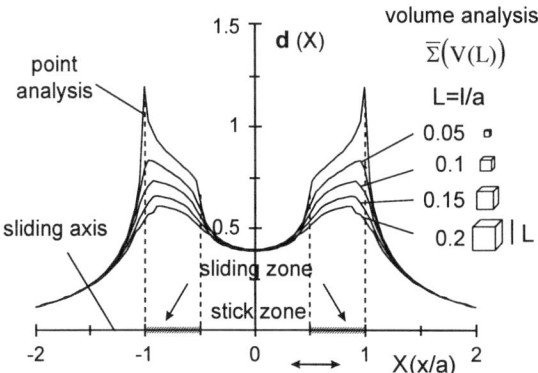

Fig. 2 : Evolution of the Dang Van cracking risk as a function of the elementary volume V(l).

To estimate the critical cubic volume, an accurate experimental identification of the crack nucleation at 10^6 cycle has been performed. The analysis was performed on a well characterised low alloy steel (τ_d = 428 MPa, σ_d =690 MPa) verifying the Dang Van model whereas, optimised partial slip contact have been obtained thanks to the use of a large smooth 52100 steel ball (Ra<0.1 µm, R=50 mm). A direct correlation between the contact crack nucleation and the prediction given by the model is obtained if the contact loading is averaged on an elementary volume of about 5 to 6 µm edge. Such a critical elementary volume is analysed in terms of the probability to activate a micro-defect. Taking into account the contact size effect various predictive fretting cracking diagrams are proposed as a function of the contact loading and the fatigue properties of the materials.

REFERENCES

(1) R.B. Waterhouse, Fretting Fatigue, 1981.
(2) S. Fouvry, Ph. Kapsa, L. Vincent, K. Dang Van, Wear 195, 1996, 21-34 pp.
(3) K. Dang Van, ASTM STP 1191,1993, 120-130pp.

DYNAMIC PROCESSES IN MACHINES WITH DRY FRICTIONAL UNITS

F.R.GEKKER

Bauman Moscow State Technical University, Bauman str., 5, 107005, Moscow, Russia

ABSTRACT

The report being presented scales with the results of more than 30 years research connected with the solution of dynamical problems of mechanical systems with dry friction the development of the methods of calculation of the machines with dry friction units (1). The report includes the following parts.

Development of a generalized dynamic model of friction units (Fig. 1) taking into account thermal processes, geometric, rheologic and frictional characteristics of a friction pair, as well as possible relative shutdowns of frictions surfaces. This model included two without inertia rough surfaces which connected with mass bodies by means spring - dissipate units.

Mathematical methods of investigating dynamic processes in machines with dry friction units (1). Accurate and approximate methods. Fields of their application. Methods of mathematical modelling.

Forced oscillations of dynamic systems with dry frictional units. Systems with one or two degrees of freedom (1).

Damping the oscillations in machines by dry friction elements. Optimization of parameters of the system of absorbing oscillations in machines with dry friction elements. Optimization of parameters of elastic-frictional dampers built-in clutch discs for reducing torsional vibrations in car and tractor transmissions caused by non-uniform operation of piston engines (2).

Dynamic and thermal processes modelling in car and tractor transmiisions with the clutch engagement. Calculation of the engagement life (2).

Especial attention was spare phisics of friction automatic oscillations in the machines. In the investigation of this oscillations the dynamic model of a friction unit was assumed, taking into account friction surfaces movement in normal and tangential directions to the friction surfaces (3) - (5). On the basis of this model the design procedure of multi-disc aviation brakes was developed. The problems of stability of the braking systems movement excluding the appearance of frictional automatic oscillations were also considered.

In order to out friction units calculathion it is necessary to have the frictional and wear characteristics of friction pair and a series of real parameters of the systems being calculated (3). Hence, in his papers the author devotes great attention to the development of the methods of obtaining multifactor models, describing the change of friction coefficient and value of the so-called release of friction surface depending on the slip velocity, pressure and temperature on the friction surfaces (6). The method of determination of rheological characteristics of the surface layers of the bodies in friction were developed.

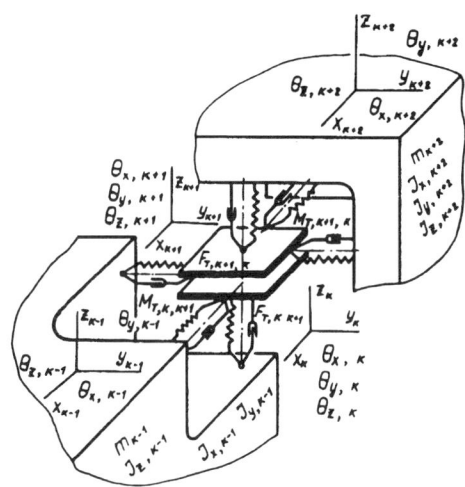

Fig. 1: The generalized dynamic model of dry friction units

REFERENCES

(1) F.R.Gekker. Dynamics of machines operating without lubrication of friction units. Moscow, Machine Building (Mashinostroenie), 1983, 168 p.

(2) F.R.Gekker ... Friction Clutches of transport vehicles and tractors. - M.: Mashinostroenie, 1989, 344 p.

(3) F.R.Gekker, S.I.Khairaliev. Effect of rougness and rheology properties of contacting bodies on steady sliding modes. High School news. Machine Building (Mashinostroenie), 1986, N 5, 23 - 27 pp.

(4) F.R.Gekker, S.I.Khairaliev. About the stability of body sliding over a moving bass. Friction and Wear, 1992, Vol. 13, N 4, 581 - 587 pp.

(5) F.R.Gekker, S.I.Khairaliev. On self-excited friction vibrations in brakes, DE - Vol. 84 - 1, 1995 Design Engeneering Techical Conferences, ASME, 1995, Vol. 3, Part A, 1229 - 1231 pp.

(6) P.V.Kuzhelev, F.R.Gekker, V.A.Sopkin. Mathematical modelling of friction propeties and wear resistance of friction couples. Friction and Wear, 1989, Vol. 9, N 3, 528 - 533 pp.

ESTIMATION OF WEAR INTENSITY OF FRICTION SURFACES

G R GEVORKIAN, Y S ARUSTAMIAN, V R GEVORKIAN
Department of Mechanics and Machine Sciences, State Engineering University of Armenia
105 Terian St., 375009, Yerevan, Republic of Armenia

In the normal running mode the machines and mechanisms friction parts usually operate in conditions of steady-state wear. It is accepted that given invariable friction parameters this, mode is characterized by a constant value of wear intensity, representing the ratio of wear value differentials (linear, in respect to mass and volume) to the friction path. However, standard investigations on a four-ball friction machine indicate that the dependence of volume wear (ΔV) on friction path (l) is usually non-linear, resulting in wear intensity being variable.

The lubricants were tested on the four-ball friction machine in a heavy mode of boundary friction. The machine was completed by a series of high-sensitivity electro-tensometric dynamometers, providing variable rigidity and constant sensitivity.

Industrial oil I-40A, cable oil C-220, and vacuum oil VM-1 as well as solutions in some of the most common industrial additives DF-11 (dialkyldithiophosphate of zinc), LZ-23K (ethylene-bis opropile xanthate) and LZ-318 (trichloromethyl ether 2-mercaptobenzthiazole) in different ratios (5%) were used as lubricants. Steel balls with diameter 12.7×10^{-3} m were used as friction bodies. The hardness was RC 62-65, and the roughness was 0.08-0.16μm.

The experiments were carried out at a loading of 73.6 N, which was subsequently increased by 73.6 N. Each experiment was carried out at a constant load, new surfaces and fresh lubricants. The experiments were completed when reaching the binding load. The maximum duration of the experiment was 60 sec. After each experiment the diameters of the friction spots on the lower balls were measured in two mutually perpendicular directions. The final results was based on the mean value of six subsequent measurements.

The dependence of volume wear ΔV on friction path l in ordinary coordinates are non-linear, and semilogarithmic coordinates they are linear with angular coefficient, with analitical expression:

$$\alpha = (1/l) \ln \Delta V / \Delta V_0 \qquad (1)$$

where ΔV_0 is the volume of the elastically deformed contact spot, α is the angular factor with dimension m^{-1} conventionally named as wear intensity factor.

The above mentioned experimental results suppose that the well known concept of wear intensity needs a correction. It is to be replaced by a parameter, remaining constant in invariable friction conditions. With this in view, it is of considerable interest to investigate the nature of α variations in variable wear conditions. A data processing was carried out for a series of experiments where along with varying the axial load, slip velocity and system rigidity one could vary the lubricants (through variously purposed additives). Besides a consequent juxtaposition of lubricants antiwear characteristics with those calculated for the factor α was implemented.

The experiments show, that any change in one of friction parameters conditionin wear, leads to the corresponding change in α, thus suggesting that the factor α may be characteristic for evaluating the wear intensity. Base on it one can judge the wear processes taking place. Hence, while considering the evaluation problems of antiwear properties of the lubricant, one can interpret the experimental results through the factor α.

At the same time, a certain regularity follows, i.e. for addition-containing lubricants regardless to the wear conditions, low values of seizure always correspond to high values of α, and vise versa. These observations enable making an assumption that the processes of wear and seize are based on certain phenomena of general nature related to the chemical nature of adhesive layers, being formed on the surfaces of friction bodies in the contact on one hand, and the interaction character between two conjugated surfaces, on the other hand. For the above discussed experiments, this type of correlation may be expressed in terms of the dependence of a conventional parameter $(P_k-P)/P$, characterizing the anti-scoring properties of the lubricant, on the reciprocal wear factor $1/\alpha$.

$$(P_k-P)/P = -8.56 \cdot 10^{-6} (1/\alpha)^2 + 9.1 \cdot 10^{-3} (1/\alpha) - 0.062 \qquad (2)$$

Thus, the factor α is a correlating parameter for wear processes taking place on the contact of friction surfaces in conditions of boundary lubrication. The dependence represents the principle possibility of forecasting and quantitative evaluation of the critical seize load by means of α values calculated from the results of a single experiment for an arbitrary chosen load P. These loads are essentially characteristical for friction nodes of numerous machines and mechanisms.

ACKNOWLEDGMENTS

The authors are thankful to the Alexander von Humboldt Foundation, Bonn, Germany for their financial support.

MAGNETIC AND ELECTRON MICROSCOPIC ANALYSIS OF WEAR PRODUCTS FOR SLIDING TRIBOCONTACTS

E.S.GORKUNOV, V.V.KHARLAMOV, S.V.SYTNIK

Institute of Engineering Science, Russian Academy of Sciences (Urals Branch), .91 Pervomaiskaya st., GSP-207, 620219, Ekaterinburg, Russia

ABSTRACT

Deciphering and analysis of information stored in the frictionally peeled wear products are of a great importance not only for determination of the residual service life of the tribocontact, but also for watching (monitoring) its in-service performance.

The wear products have an effect on capability of the oils to conserve their lubricant properties in the course of time and on the tribocontact penetrability under mixed conditions of the friction. In present work the method has been developed to receive information through the relationship between the geometric, physical and mass properties of the wear products, laws of their distribution and their magnetic properties. Then the electromagnetic characteristics of wear particles could be coded and converted into electric signal for processing by specially developed computer programs. To realize this method the waste lubricant (with the wear products) has been used. The magnetic properties of the set of wear particles as a ferrodielectric material with different values of density were studied in d.c. and a.c. magnetic fields. Using the quality method, the permeability and total loss as depended on the frequency were measured in a.c.fields to find an upper bound for the frequency range of measurements. To obtain additional information about variations of physical properties of the materials the averaged d.c.magnetic properties over the given set of particles separated from the lubricant were measured: saturation magnetization M_S, coercivity with respect to magnetization H_{cM} and magnetic flux density $B(H)$. The value of M_S measured in maximum magnetic field strength of 1360 kA/m was equal to 527 kA/m, and coercivity H_{cM} was equal to 47.1 A/cm. Due to internal stress in the wear particles the values of maximum magnetic permeability, magnetic flux density and saturation magnetization decreased, whereas the coercivity increased as compared with the initial state of the material of friction surfaces. This fact allows to apply the magnetic methods for evaluation not only internal stress in the wear perticles, but also chemical transformations in the surface layers on the areas of the physical contact. As a result of the experiments on the elements of the friction pair after several thousand of cycles of the friction interaction, it has been shown that there is a relationship between accumulation of the flaws in the surface layers of tribocontacts and variations of the magnetic permeability. Due to this fact, one can monitor the accumulation of damages in surface tribocontact without analysis of the wear products and determine the position of the second critical point on the wear curve. However, this circumstance does not eliminate the necessity of analysis for wear particles that contain additional information about variations of the wear intensity, structural accomodation of the tribocontact, surface temperatures and other friction characteristics.

The study of the shape and grading content for separated particles after their washing was carried out with the help of the electron microscope DS-340 Tesla. The picture from the microscope was recorded on the diskette for subsequent processing by means of the personal computer. The electron-graphics method allows to analyze the shape and sizes of the particules, and it is possible to evaluate the chemical composition of wear products by means of the special adapter for the scanning electron microscope. Using the computer image processing programs, that were developed to this end, it is possible to draw curves for the distribution of the particles according to their size. In present work, under heavy-loaded conditions of rolling friction with a slip (an automobile transmission) the sizes of the wear particles were from 10 to 200 mm. The appearance of prevalent amount of coarse particles (with size about 150 ÷ 200 mm) in the lubricant made possible to predict a damage of the investigated friction subassembly. The effect of the wear particles on lubricant ageing (deterioration of lubricant properties) was evaluated with the help of the infrared spectrometer JR-470, according to the absorption intensity at the wave numbers between 3400÷2700 cm^{-1} and 1740 ÷ 1550 cm^{-1}.

The infrared spectrum measurements on the waste transmission oil touch upon the vibration range of saturated hydrocarbons. The wave number is displaced from 2850 to 2830 cm^{-1}, and insignifically intensive absorption is observed at the wave number 1160 cm^{-1}. These effects may result from the appearance of new organic compounds such as $R-SO_2-R$.

Complex computer data processing of the mentioned parameters permits to monitor the development of the processes on the friction surfaces during the different operational periods of the investigated tribocontact.The method has been implemented to evaluate the residual service life of the friction subassemblies in automobiles. By now on the base of the completed investigations the model of the instrument has been developed and made. It provides to get a quantitive assesment for the wear intensity of the tribocontact service life during the operational period of the machine or mechanism.

DYNAMICS OF MOTION OF THE SLIDING BODY WITH DUE REGARD TO NON-STATIONARY PROCESSES IN FRICTION CONTACT ZONE

G.S.GOURA
Engineering and Ecology Department, Sochi State University for Tourism and Resort Studies.
Sovetskaya street, 26a, Sochi, 354000, Russia

P.Painleve paid attention to the discrepancy and the duality of the differential equation solutions in the classical mechanics problems with dry friction according to Coulomb and with identical initial conditions. The discussion (1) revealed the opportunity of the paradox elimination, if the bodies are considered as deformed in the friction contact zone.

It was experimentally shown that when a tangential force is applied to one of the metallic bodies of a frictional pair, the elastoplastic deformation takes place in the zone of preliminary displacement and near it with no sliding at the interface. The interacting junctions and sublayers of the counterface are translated due to considerable adhesion and normal force at multiple contact points. When under deformation by tangential forces the base material resistance reaches the maximum defined by the preliminary displacement value (δ) the rigid body will move together with joined mass of sticking base material on the certain slip line. Now the body is in "fresh" not yet deformed condition and the process starts again in the same way. Similar phenomena are mentioned by some authors in connection with friction of metallic bodies and polymers and also in soils and other media (2). The mass which we called an apparent mass depends on the velocity:

$$m_0(v) = m_0 e^{-kv/(1-qv)} \quad (1)$$

The author has made an attempt to deduce differential equation of variable mass system in the motion velocity range characteristic for non-stationary processes in the friction contact zone under sliding rigid body on deformed foundation.

After modelling foundation as Kelvin-Fright rheological model and introducing function cx (where c is elastic constant) as Fourier series the equation may be written:

$$\{m+m_0 e^{-kv/(1-qv)}[1-kv/(1-qv)^2]\}dv/dt+$$
$$+\mu v+c[\delta/2-\delta/\pi \sum_{n=1}^{b}(1/n)\sin(2\pi nx/\delta)]=T-F \quad (2)$$

where m is rigid body mass; m_0 is apparent mass when motion velocity is zero; k and q are parameters responsible for rheological body properties; e is a base of natural logarithm; t is time; v is moving body velocity; μ is viscosity coefficient; x is displacement; T is thrust; F is friction force on interface.

When the rigid body velocity is $v > 1/q$ the above mentioned non-stationary processes of the apparent mass displacement in the zone of frictional contact have no time to develop, and friction processes take place on the interface, and the differential equation of motion according to Newton will be as follows:

$$mdv/dt = T - F. \quad (3)$$

The equation (2) and (3) well explain the differences between static and dynamic friction.

In the analysis of the equation (2) it is evident that a required thrust can be decreased by apparent mass reduction. This can be done by several methods. Physical and chemical methods are aimed at adhesion reduction on the interface. Mechanical methods use kinematics parameters of the system by excitation of bearing surface oscillations of certain amplitude and frequency. Amplitude value A must exceed the preliminary displacement value δ, the motion velocity on frictional contact surface must exceed $A\omega$, where ω is angular frequency. In technological methods soft coats on the hard support are used.

The main features of motion of the system are the following: 1) rigid body motion on deformed foundation must be divided into two stages: a) non-stationary regime, b) quasistationary regime; 2) non-stationary motion regime from the starting moment till a certain velocity at which an apparent mass equals zero is described by non-linear differential equation of variable mass motion; Apparent mass value is determined by physical, mechanical and chemical properties and conditions on interface especially by adhesion, loads, stationary contact time and motion velocity; 3) the problem of rigid body sliding on deformed foundation refers to the problems of non-holonomic system with constraints dependent on mass change process; 4) the principal vector of resistance forces to sliding, in general case is not situated in the interface; 5) forced oscillations acting on frictional contact reduce friction force. Amplitude-frequency range depends considerably, as it is seen from the equation, on the apparent mass value and preliminary displacement which are function of physical and mechanical properties of material - the modules of elasticity and shear.

REFERENCES
(1) P.Painleve, The Lectures about Friction (in rus), Moscow, 1954.
(2) G.Goura, Priroda (Nature), № 7, 1973, 82-84pp, (in rus).

MODELLING AND WEAR CALCULATION ON FRICTION

D.G. GROMAKOVSKY, A.N. MALYAROV and Y.P. SAMARIN

Research Institute for Mechanical System Reliability Problems, Samara State Technical University, Galaktionovskaya Street, 141, Samara, Russia

ABSTRACT

Kinetic interpretation of wearing undertaken in the present work is founded on the termofluctuation conception of material's strength which is developed under influence of Russian Scientific Academy member S.N.Gzurkov's ideas (1).

Periodical destruction and removal of some totality most high surface roughness' ledges in the form of wear particles stand out as important circumstance in the microscopic picture of wearing. Then next ledges totality come into operation and it is register on trials in the form of "steps" on the wear graph curve.

According this effect structural state is rebuild periodically (iterative) from strengthening to unstrengthening and destruction during a working in the localised volume.

Deformed material's microvolume - V_d, which will have critical destruction accumulation is defined by real contact area - ΔA_r having the depth - h is separated from under-situated material by particular layer with high damages concentration formed under surface during friction process.

$$V_d = \Delta A_r \cdot n_r \cdot h, [m^3]$$

where n_r - number of real contact spots.

Proceeding from statistical analysis of wear particles' size and form, damaged surface fractography and some schematization of the process in worked out model the total quantity of bonds - λ subjected to destruction in etch totality of localized material's microvolumes V_d for one kinetic cycle is estimated.

$$\lambda = \lambda' \cdot \frac{S_{cr}}{S_{Va}} \cdot \frac{V_d}{V_{cr}},$$

where λ' - number of athom bonds in one activation volume V_a; S_{cr} and S_{Va} - surfase areas of statistical mean particle, [mm^2]; V_{cr} - statistical mean of wear particles' volume, [mm^3].

The time necessary for destroying single bond (by S.N.Zhurcov) is defined as

$$\tau = \tau_0 \exp\left(\frac{U_0 - \gamma\sigma}{RT}\right), [s]$$

The activation parameters of the kinetic equation are defined by a new experimental method which permits to take account for environment influence effect and tout state in surface layers arising from normal and tangent stresses on friction.

Taking into consideration the size requirements, correlations (1),(2),(3), presence of relaxsation of damaged bonds - ϑ, initial damageness of material - ξ, dissipation properties' characteristics of contact (absorption coeffitient - ψ) and other, base calculation model of materials' wear rate have got following expression

$$J_V = \frac{\Delta A_r \cdot n_r \cdot h}{\xi \cdot \lambda \cdot \vartheta \cdot \tau_0 \exp\left(\frac{U_0 - \gamma\sigma}{RT}\right)}, [m^3/h]$$

where U_0 - activization energy of fatique destruction of surface, [kJ/mol]; γ - structure sensitive coefficient; τ_0 - constant time (period of athoms oscillation), $\tau_0 \approx 10^{-13}$ [s]; R - gas constant; T - temperature, [K]; σ - normal stress, [MPa]; ψ - absorption coefficient.

Methods and programming maintenance are worked out for wear calculation of the typical friction units combinations: "shaft-bush", "ball-raceway", "surface-surface", etc.

The calculation proposes two basic systematic versions - prognosis for a present constructions and calculation of tribo-units' parameter optimisation with intention of ensuring given recourse.

REFERENCES

(1) D.G.Gromakovsky. Friction and Wear. Vol.18, No.1, 1997.

Signal-Analytic Modeling of Hydraulic-Actuator Sliding Seal Friction Dynamics

GEE-SERN HSU, DIRK A. WASSINK, KENNETH C. LUDEMA AND ANDREW E. YAGLE,
The University of Michigan, Ann Arbor, Michigan 48109, USA,

JOEL A. LEVITT,
Ford Research Laboratories - MD 3182, The Ford Motor Company,
Dearborn, Michigan 48121-2053, USA.

ABSTRACT

Measurements reveal that the friction produced by the reciprocating sliding of the shaft of a double-ended and balanced hydraulic actuator through the actuator elastomeric lip-seals is a rapidly varying nonlinear function of: piston velocity; hydraulic pressure, and temperature. This function also involves slowly changing: elastomer relaxation spectra; sliding surface fracture; metal sliding-surface texture; fluid composition and viscosity; elastomer/fluid/metal chemistry, and debris formation. Lacking comprehensive understand-ing, a highly simplified physical model was combined with system identification methods (ARMAX modeling) to predict friction during the first tens of reciprocating cycles. An extended Kalman filter was then used over the following 6,000 cycles to adapt the ARMAX model to the slowly-varying actuator properties.

MEASUREMENTS

Data were obtained with a laboratory apparatus similar to the actual power steering actuator, but differing so as to keep the net hydraulic force on the piston rod zero [1]. The reciprocating piston rod was driven using an eccentric cam or a precision screw. Fluid temperature was kept constant, while the fluid pressures in the actuator were varied to explore the joint effect of time varying pressure and rod velocity on friction dynamics.

CALCULATIONS

In generating the simplified physical model, it was assumed that: the lip seals were made of a linear elastomer; the hydraulic fluid was Newtonian, and there were only two significant energy dissipating mechanisms; deformation of seals, and shearing of the fluid between the seals and the rod. The simplified physical model was nonlinear in velocity and pressure, and lead to a linear difference equation with unknown coefficients and of unknown order. These were determined by minimizing the mean square difference between the model and the measurements. This simplified difference equation captured many of the features of the early measurements, but not at low reciprocating frequencies.

The simplified model was incorporated as the exogenous input of an **A**uto**R**egressive **M**oving **A**verage with e**X**ogenous input or ARMAX model [2]. This adds filtered zero-mean white noise to the previous difference equation. The values of the additional coefficients are found by using several cycles of data and minimizing and whitening the one-step-ahead prediction error. Prediction of the early measurements was excellent.

An extended Kalman filter [3], initialized by the ARMAX model, was used to deal with the evolution of friction dynamics as the actuator wore during the next 6000 reciprocation cycles. This filter adapts the ARMAX coefficients, as new data is received, and prepares a new set of coefficients from the old after each new measurement. The results were excellent.

The extended Kalman filter has a further desirable property, that not all prior data need be saved, and, therefore, it can be used for on-line friction compensation or control.

REFERENCES

1: Wassink,D.A., V.G. Lenss J.A. Levitt, K.C. Ludema and M.A. Samus, "Friction Dynamics in Sliding Lip Seals", *Proc. of the 47th National Conf. on Fluid Power*, 23-25 April 1996, Chicago, 205-211.
2: Ljung, L., *System Identification: Theory for the User*, Prentice-Hall, New Jersey, 1987.
3: Jazwinski, A.H., *Stochastic Processes and Filtering Theory*, Academic Press, Boston, 1970.

THE EFFECT OF STROKE ON THE FRETTING WEAR BEHAVIOUR OF A FINE PARTICULATE REINFORCED ALUMINIUM ALLOY MATRIX COMPOSITE

Q HU, I R McCOLL and S J HARRIS

Materials Engineering and Materials Design, University of Nottingham, University Park, Nottingham NG7 2RD, UK

ABSTRACT

Particulate reinforced aluminium alloy composites can offer considerable improvements in tensile and fatigue behaviour, when manufactured by a powder metallurgical route involving mechanical alloying. For example, AMC217 (Aerospace Metal Composites), which is an Al-Cu-Mg (2124) alloy reinforced with 17 volume% of 3 μm silicon carbide particles, achieves a 50% improvement in both proof and fatigue strength (R = 0.1), over the equivalent monolithic alloy (2024), both materials in the T4 condition (1). However, joining of these materials will normally necessitate the use of mechanical fasteners or adhesives so that their fretting wear behaviour is an issue

It has been shown previously (2) that for fretting distances up to 80 m, AMC217-T4 exhibits an enhanced fretting wear performance over the equivalent monolithic alloy, tested at a stroke of 40 μm against a medium carbon steel. At longer fretting distances, wear of the composite accelerates and exceeds that of the monolithic alloy. However, an anodizing surface treatment has been shown to improve the performance of the composite by a factor of five over these longer fretting distances, at least up to 800 m, at a nominal contact pressure of 5 MPa (3).

In this paper the effect of stroke on the fretting wear behaviour of this composite, at a nominal contact pressure of 10 MPa, with and without the anodizing surface treatment, is reported. Anodizing was carried out in sulphuric acid, producing a layer 8 to 10 μm in thickness. Fretting tests were again carried out against a medium carbon steel counterface, using a crossed flat-on-flat specimen arrangement to give a relatively large nominal contact area of 120 mm^2. Fretting strokes from 40 to 120 μm were investigated.

Wear of the composite (Fig. 1) and counterface both increase with fretting stroke, as well as with fretting distance. At the shorter strokes cracking of the composite is observed, both parallel to its surface and inclined to it, the former resulting in a degree of delamination. Transmission electron microscopy indicates the presence of a spinel, $FeAl_2O_4$, in the debris ejected from the fretting interface, suggesting that a degree of mechanical alloying takes place in the interface during fretting. At the longer strokes, debris transfer to the counterface is less pronounced, with a greater fraction of debris being ejected from the interface. Furthermore, this ejected debris exhibits a lower degree of oxidation, and, transmission electron microscopy/X-ray diffraction suggests that the debris particles are themselves agglomerations of sub-micron particles, the size of which increases with fretting stroke.

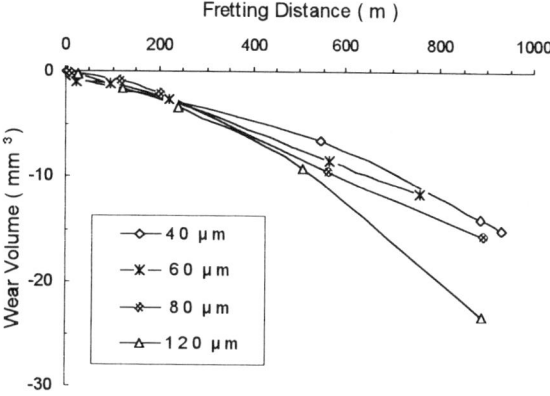

Fig. 1: The effect of fretting stroke on composite wear volume.

Anodizing of the composite has a marked effect on its fretting wear behaviour, particularly at the longer fretting distances, when wear is reduced by a factor of three, at the lowest value of stroke. This behaviour is interpreted in terms of the anodized layer, which rapidly fragments, initially preventing adhesion/welding and debris transfer, and then combining with material abraded from the counterface to produce beds of loose spherical debris. At the longer values of stroke these beds are less well developed, although spherical debris particles are distributed across the whole fretting interface.

REFERENCES

(1) Z W Huang, I R McColl and S J Harris, *Mat. Sci. Eng. A* **215** 1-2 (1996) 67-72.
(2) I R McColl, S J Harris and G J Spurr, *Wear* **197** (1996) 179-191.
(3) I R McColl, S J Harris, Q Hu, G J Spurr and P A Wood, *Wear* **203-204** (1997) 507-515.

TRANSIENT ANALYSIS OF SLIDING CONTACT SYSTEMS

H N ILIEV
Department of Mechanical Engineering, University of Zimbabwe, P.O. Box MP167, Harare, Zimbabwe

ABSTRACT

The performance and life-time of sliding contact systems such as bearings, brakes, clutches, slide-ways, electric contacts, etc., are characterized not only by the amount of wear, friction and contact pressure, but also by their distribution over the contact area and their variation with time. Unlike wear of materials theory, in analysing wear of sliding contact systems it is necessary to consider contact surface configuration and sliding motion kinematics. Consideration of sliding contact systems as contact pairs with distributed parameters provides an opportunity for optimization by means of parameters distribution.

Because of wear contact surface configuration changes with time which conditions contact pressure redistribution. The latter causes change in wear intensity. In this way mutual influence of surface profile and contact pressure distribution brings about transitional process. Under certain conditions transitional process leads to stationary regime, when contact surface profile reproduces itself. Transition to steady-state wear regime leads to reduction in wear rate and friction losses to their minimum. Stationary regime is optimum regime with respect to wear and friction. Transitional processes in sliding contact systems are characterized by interdependent variation of contact pressure, friction forces and wear rate.

Premeditative contact surface shaping can be used to shorten transitional process, reduce wear rate and in this way increase life-time of sliding contact systems (1).

Steady-state wear analysis of sliding contact joints is presented in (2)-(4).

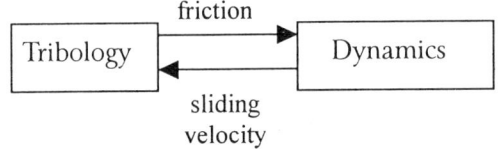

Fig. 1: Tribology and dynamics connection

In this paper transitional processes in sliding contact systems are discussed from the point of view of mechanics. Analysis is limited to unlubricated systems and lubricated systems in boundary regime of lubrication.

Applying macroscopic approach to modelling of sliding contact systems continuous model of contact interaction has been used and normal wear regime is proposed. Rough surface layer is considered to be smooth continuous layer whose properties differ from bulk body properties. Bulk body is assumed rigid or elastic body. Surface layer properties are determined from experiment. Materials combination of sliding contact system is tested on small samples. Test is designed to provide even distribution of contact pressure and sliding velocity. Elementary contact interaction is described in terms of contact normal pressure p, tangential pressure τ, contact deformation δ and linear wear J as follows:

$$\delta = kp, \quad \tau = \mu p, \quad J = \int (mvp)dt,$$

where k, μ, m are experimentally determined coefficients of contact deformation, friction and wear respectively, v - sliding velocity, t - time.

For analysis of transitional processes in sliding contact systems knowledge of sliding velocity is required. It comes from dynamics consideration of mechanical system, where sliding contact system is joined in. On the other hand knowledge of friction, required in dynamics analysis, comes from tribology (Fig.1).

Mathematical model of transitional processes formulated reflects the interdependency of friction and wear on the one hand and sliding dynamics on the other hand. Algorithm for numerical solution of dynamic equations is developed.

Sliding systems with translational and rotational sliding like disc clutch, rotating cylinder - friction block, elastic slider are considered in detail. Some of the theoretical results obtained are verified experimentally.

Transient analysis presented is of interest to design of sliding contact systems.

REFERENCES

(1) H. Iliev, Design for wear resistance through macroshape of sliding contact surfaces, Proc. 10th international colloquium tribology, 9-11 January 1996, Esslingen, Germany, vol. 1, p. 1027-1031.
(2) A. Pronikov, Reliability of machines, Mashinostroenie, Moscow, 1973, 592p.
(3) R. McLeish, J. Scorecki, Analysis of wear in a spherical joint, Proceedings of the IMechE, vol. 183, 1968.
(4) R. Kilburn, Reducing wear in an electromagnetic clutch, Trans. of ASME, Ser. B, 2, 1968.

CONTACT INTERACTIONS IN ELECTRICAL CONTACTS

V V IZMAILOV
Tver State University of Technology, 22 A.Nikitin emb., Tver, 170026, Russia

Electricsl contacts are a kind of tribological joints. Naturally, they have some specific features. They are the following: specific range of contact materials, specific range of lubricants (if they are used), specific geometry of contact parts and so on.

Tribological approach to electrical contacts has been activated recently, but there still remain some problems, such as calculation of contact resistance and other contact parameters, making of materials and lubricants with desired properties, prediction of contact lifetime and others.

CONTACT RESISTANCE

The theory of rough surface contact (1) enables to calculate the contact resistance through the roughness parameters and physical and mechanical properties of contact materials. Here we present aproximative formulae for contact resistance estimation. More precise formulae are presented elsewhere (2). The following approximative formulae provide upper and lower estimations of contact resistance value R_c:
for plastic contact:

$$\frac{\rho s_m}{3}\sqrt{\frac{H}{NA_n}} \leq R_c \leq \frac{\rho s_m H}{5N}$$

where ρ is the mean specific resistance of contact materials, s_m is the mean spacing of the profile irregularities, A_n is a nominal contact area, H is a microhardness of the softer contact material, N is a normal contact load;
for elastic contact:

$$\frac{\rho}{2}\sqrt{\frac{s_m R_a E}{NA_n}} \leq R_c \leq (0.6)\frac{\rho R_a E}{N}$$

where E is an equivalent Young's modulus, R_a is c.l.a. roughness.

It should be remembered that the above formulae are valid for contacts without tarnish films. If the contact surface is covered by thin tarnish film, its resistance has to be included in total contact resistance (2).

ELECTRICAL CONTACT NOISE

Fluctuation of contact resistance (electrical contact noise) is the undesirable feature of any electrical contact - static or sliding. Our experiments on contact noise reveal the following results. Spectral intensity of contact noise in static contacts is approximately inverse proportional to frequence (so-called 1/f- noise). The most probable nature of contact noise in this case is formation and disappearance of fritting junctions. Sliding electrical contact is also a source of electrical noise. The principal causes are contact discreteness, wear debries, variation of surface film thickness. Theoretical analysis predicts that spectral intensity of contact noise in sliding contact has to be inverse proportional to f^α, where α is about 2.

MATERIALS FOR SLIDING ELECTRICAL CONTACTS

Materials for electrical contacts have to meet conflicting demands: they have to be good conductors and good antifrictional materials. The way out is in division of the conductive and antifrictional functions between components of contact material. The example is composite material. Our experience shows that fusible metals (such as tin, lead, indium and so on) are effective as antifrictional additives. The optimal compositions have been established for high-current and low-current contact materials.

Surface coatings also may be considered as a realisation of the above way (i.e. electrical and frictional functions division). For instance, tin and tin-alloy coatings have low contact resistance. Sometimes it is lower than that of gold plated contacts (because of low hardness of tin). But low corrosion resistance of tin results in contact resistance increasing in long time operation, especially in sliding contacts. Insulating wear debries provoke high level of contact resistance and its significant fluctuations. But if wear debries are moved off from contact zone, tin coating is not worse than noble coatings.

A new perspective kind of contact materials are liquid metals, such as gallium and gallium alloys. Liquid metal as an intercontact medium permits to increase contact area and decrease contact resistance. In fact, a contact becomes soldered joint, but at the same time it remains separable. Our experiments show that in liquid metal contact resistance is reduced by a factor of 100 - 1000 in comparison with dry contact. Wear rate decreases twofold.

REFERENCES

(1) N. Demkin, V. Izmailov, Tribology International, Vol.24, No.1, 1991.
(2) V. Izmailov, Proceedings of the Int. Symposium on Electrical Contacts, Theory and Applications (ISECTA), Almaty, Kazakhstan, 1993, 179 - 187 pp.

ESTIMATION OF LINEAR WEAR VALUE OF SURFACES BASED ON THEIR TECHNOLOGICAL INHERITANCE

J JABŁOŃSKI
Rzeszów University of Technology, 35-959 Rzeszów, ul. W. Pola 2, Poland

ABSTRACT

Estimation of the functional properties (FP) of machine elements during formation of their surface layer (SL) is difficult; the most probably reason of it is the great level of complication of the function:

$$FP = f(TP) \qquad [1]$$

where TP - technological parameters of mechanical working. This problem can be easier when equation [1] can be change on two equation:

$$SLP = f(TP) \qquad [2a]$$

$$FP = f(SLP) \qquad [2b]$$

where SLP - surface layer parameters. This paper presents the possibilities of estimation of the amount of linear wear of cylindrical surface during running-in process. The model derived by J.M. GOLUBEV and W.J. NEBOLSIN was used. This model of linear wear takes into consideration parameters of roughness obtained after manufacturing process. Execution of subject was done basing on the following assumptions:

a) counter-surface is ideal plane,
b) wear resistance of the machined surface is caused by:
1 - wear resistance of the material,
2 - properties of surface layer as a result of mechanical working (they are dependent of machining conditions,
3 - the following function is the analysed model of linear wear during running-in process (1):

$$Z_l = cR_pR_m/(0.5+0.066 k)(R_p+R_m) \qquad [3]$$

where:
Z_l - coefficient characterising the linear wear in the running-in process,
$c = 0.6$ - the coefficient dependent on the sliding conditions,
R_p, R_m ISO roughness parameters,
k - parameter indicating the symmetry of the roughness amplitude distribution.

In the experiment oscillate burnishing was selected (OB). The variable parameters are: rotational speed of workpiece -n, the force of the ball pressure - P, frequency - f, feed - p and amplitude of the oscillation -A.

The complex machining process OB, the great amount of variable parameters and the lack of precise dependencies among them were the reasons of the fact that the multidimensional object of control of the one exit was selected. Static model based on the active experiment, called factorial analysis was selected. It is therefore necessary to plan experiment, which aim is to find extreme of function of five variables:

$$FP = f(TP) = f(P,p,n,f,A) \qquad [4]$$

Basing on the initial testing it was established that function [4] is continuos. It was assumed that it is possible to approximate unknown extreme characteristics [2], in the surrounding of the starting point (P_0, p_0, n_0, f_0, A_0), by hipersurface of second degree. According to procedure of Box-Wilson (2) the regression coefficients were computed : b_0, b_1, ...b_k changing the variables of the OB process on the three levels:

$$Z_l = b_o + b_1P +...+ b_5A + b_{11}P^2 +...+ b_{55}A^2 + b_{12}Pp + ...+ b_{45}fA \qquad [5]$$

The analysis of the model was done using Powell's method. In the second stage, similar model (finding extreme) was analysed.

$$FP = f(TP) = f(P,p,n,f,A) \qquad [6]$$

Then the hypotheses about simultaneous finding of extreme of models [4], [5] for the same surface after OB was verified. The possibility of using substantiality coefficients in order to find extreme values of these models was proved (3). The assessment of substantiality was done using t-Student test.

REFERENCES

(1) J. M. Golubev, W. J. Nebolsin, Wlijane serochovatosti poverchnosti na nacalnyj iznos. Riga 1972, page 53 - 63 (in Russian).
(2) G. E. P. Box, K. B. Wilson, On the eksperimental attainment of optimum conditions, Journal of the Royal Statistical Society, Series B, No 1, 1951.
(3) J. Jablonski, The Possibility of Using the Substantiality Coefficient of Factorial Analysis in Manufacturing Proces Optimisation, The 7th International Conference on Metrology and Prop. of Eng. Surfaces, Goteborg, Sweden 1997 page 500 - 504.

ANALYSIS OF THE INTERACTION BETWEEN RADIAL LIP SEAL AND SHAFT COVERED WITH CERAMIC LAYER (Al_2O_3)

WITOLD JORDAN, ANDRZEJ MRUK, BOLESLAW STOLARSKI and JAN UNARSKI
Cracow University of Technology, Al.Jana Pawla II 37, 31-864 Krakow, PL

ABSTRACT

Research on action mechanism of rotary shafts radial lip seals reveal that the rotation of the shaft and the resulting surface slide towards ring seal edge produce series of phenomena in the seal zone. The most important is a progressive loss of the contact between the lip and the shaft surface as consequence of emerging a lubricant layer of a hydrodynamic character. Immobilisation of the shaft results in hydrodynamic pressure failure and the occurrence of an important radial force causes the oil squeezing out of the seal gap and consequently, the steel shaft surface being smooth, some time later a dry contact effect appears.

As a result of these phenomena, the character of friction and wear between the ring lip and the shaft changes; from technical dry friction and mixed friction during the start-up to fluid friction while operating. It is notable that for the ring rubber lip wear and, consequently, for the seal life vital factors are technical dry and mixed friction (1).

Thus the life of the seal can be increased by a reduction of technical dry and mixed friction phase. It is particularly important for intermittent running mechanisms.

To achieve this, the authors decided to take advantage of some specific qualities of ceramic coatings plasma sprayed on the shaft steel surface, e.i. of such coatings open porosity.

This porosity should ensure advantageous friction and wear effects also during the start-up, making impossible total squeezing out of the oil from the sealing gap.

The investigations carried out intended to determine the influence of shaft ceramic coating characteristics on the ring lip seal wear while starting up as compared with the steel shaft.

The experiments were carried out on a laboratory stand allowing a simulated operation of the sealing pair.

Steel shafts interacting with the ring lip were covered with plasma sprayed Al_2O_3 ceramic layer. Deposited layers have been submitted to an abrasive machining (grinding), having as result a thickness of 0.8 (mm) and porosity - $R_a= 0.9$ μm.

The porosity of examined coatings (total sample porosity) was 6-8.5 %, and the average pore radius - 180 nm.

The rings were examined in repeatable working cycles of the sealing pair, one cycle run being as follows:
- start-up - operation of the pair during 1 min. at the velocity n = 3000 t/min
- stop - 9 min.

Total number of cycles - 600.

As measurable wear characteristic, the width of ring lip seal contact area has been assumed.

The results of studies on the wear of the lips of the rings interacting with ceramic coated shafts have been compared with these obtained for steel shafts without ceramics. They are presented in Fig. 1.

As it appears, the wear of the lips of the seal rings interacting with ceramic coated shafts is much lower than in the case of steel shaft without ceramics, which confirms our preliminary assumed thesis concerning the role of the porosity in tribological phenomena occuring in lip and shaft contact zone.

On the basis of test results it can be stated that the pores, characteristic for ceramic plasma sprayed coatings, are filled with oil which eliminates in sealing pair at start-up moment dry friction determining the wear of the lip.

The observations resulting of the investigations induce the authors to consider the effect of ceramic coatings deposited on a steel shaft on the sealing lip dynamics. The problem is only signalled by the authors of this paper.

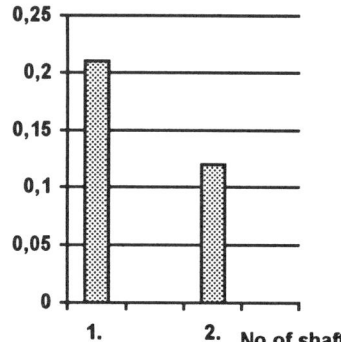

Fig.1. Wear of the lips of the rings interacting with:
1 - steel shaft without ceramic
2 - steel szaft covered with ceramic layer

REFERENCES

(1) W.Jordan, Dissertation, Cracow University of Technology, 1989.

TRIBOLOGICAL CHARACTERISTICS OF Cu-BASED METALLIC FRICTION MATERIAL OF HIGH PERFORMANCE

HYO-JUNE AHN
Samsung Motors Inc., Shinho-dong 25 17, Kangseo-gu, Pusan, Korea
CHAE-HO KIM and SEOCK-SAM KIM
Department of Mechanical Engineering, Kyungpook National University, Taegu, Korea

ABSTRACT

Friction and wear characteristics of a friction brake is primarily determined by the nature of interactions between two contact surfaces. It depends on the properties of materials. Rhee reported that the wear of organic linings decrease with increasing thermal conductivity(1). The function of a friction brake are to generate a retarding force by converting mechanical energy to thermal energy and to act as a storage and dissipation element for this thermal energy. When friction brake disc is in sliding contact under severe loading conditions, local high temperature may occur as a result of excessive frictional heating near the contacting surface. The brake disc is subjected to severe thermal loading by high frictional heating on the braking surfaces. Because of a combination of thermal heating and the mechanical loading, numerous cracks are frequently found on the surfaces of brake drum(2) and they made the wear of friction material. Recently, Limpert(3) investigated thermal conditions leading to surface rupture of cast iron rotors.

This paper dealt with the friction and wear mechanism of Cu-based metallic friction materials. Wear test between gray cast iron and friction material was carried out to investigate macroscopic and microscopic wear characteristics of friction brake under various sliding conditions. In every experimental condition, friction coefficient and wear volume were measured and compared with each other. The worn surfaces were observed by SEM. The experimental was performed under room temperature and room pressure. The temperature of surface of disc was taken continuously by the non-contacting thermometer. The wear volumes of disc were gained by the surface profilometer.

Fig. 1 shows the friction coefficient and surface temperature as a function of sliding distance under the load of 70kgf and sliding speed of 2m/s. The surface micrograph of friction material after the test under the condition of Fig. 1 is shown in Fig. 2.

The main results obtained from this experiment are as follows;
1. Friction coefficient tends to exhibit larger values for smaller applied load and sliding speed.
2. Increasing applied load and sliding speed, the temperature of contact surface increased gradually and reached to a stable value.
3. Wear loss of friction material against gray cast iron was larger for larger applied load and slower sliding speed due to thermal effects.

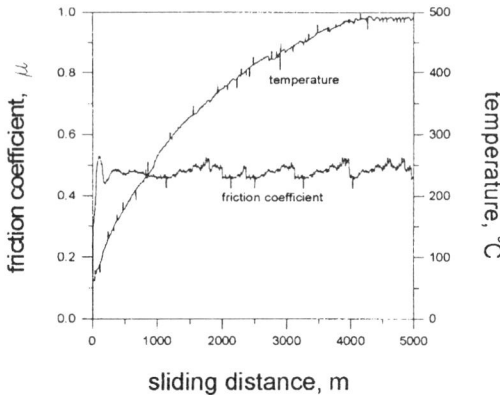

Fig. 1 : Friction coefficient and surface temperature as a function of sliding distance (load : 70kgf, sliding speed : 2m/s)

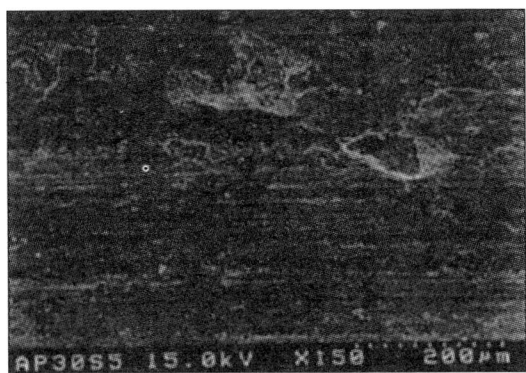

Fig. 2 : SEM photograph of the surface of friction material after sliding 5000m with condition of Fig. 1.

REFERENCES

(1) Rhee, S.K., DuCharme, R.T and Spurgeun, W.M., SAE 720056, 1972, 1-7pp
(2) Kennedy, F.E. and Ling, F.F., Trans.ASME, J, of Lub, Tech, Vol, 96, 974, 497-507pp.
(3) Limpert, R., SAE 720447, 1972, 1507-1520pp.

FRICTIONAL NON-CONDUCTIVITY OF LOW-ENERGY SLIDING CONTACTS

V. G. KURANOV
Saratov State Technical University, Saratov, Russia

ABSTRACT

This paper reports the result of investigations of the phenomenon of the Frictional Non-conductivity (FN) of low energy sliding contacts and effects connected with this phenomenon. The possibility of using these effects in friction assemblies are showed, also at the realisation of selective transfer effect. The contradiction is permitting in the frictional conductivity explanation at the joint action of oxide and frictional polymer films.

Key words : frictional non-conductivity, selective transfer, wear-durability, contacting, adhesion, oxidizing, polymer formation, activation, diffusion.

CONCLUSIONS

1. Friction non-conductivity of low-energy sliding contacts arises in the Pt contacts on the score of Pt oxidizing by the internal oxygen. The absence of oxygen in alloys excludes the oxidising and leads to the adhesion.
2. There are manifest the possibility of materials to the self-organization and the possibility to rebuild the surface structure for better wear resistance in FN mechanism. There are also manifest catalytic Pt properties such as the possibility to accumulate and to return the oxygen for the catalytic reactions on Pt surface. In friction the metal surface activation by the plastic deformation is proceeding accompanied by the surprising effect - the oxidising of itself surface (the friction catalytic self-oxidizing).
3. Jump effect in the FN mechanism do not permit to limit the oxide film thickness to the limits of tunnel conductivity. It is the reason why there is not succeed to combine the conduct and wear-resistance properties by traditional methods. It is possible only particular solving of the task by the application of materials oxidising selectively (Au and Ag alloys).The other way is to keep the "pause" state when the film formation do not accompany by the jump, for example, at the polymer formation in organic vapours, where the jump does not arise and the invisible film formation is possible. More full solving is possible with selective transfer mechanism application, where the lubrication role are played by the metal-plated film.

THE SYSTEM OF INTERACTION IN SLIDING PAIRS AND TEST MRTHODS
IVAN KRYVENKO
Kyiv International University Of Civil Aviation
Avenue Kosmonawta Kamarova, Kyiv, 252058, Ukraine

ABSTRACT

The new theory of friction gives possibility to create test methods of friction sliding pairs in boundary and unlubricated friction on prupose of instantaneous rate of wear quantity contribution of separate sorts of interaction in wearing.

The results of our investigations detecor effect (effect of Fiveg) in pairs of friction, conceptions and ideas another scientists and olso conceptions of basic and overlapping sciences have been used for the new structure interaction.

The fundamental tribology level is interaction of microcontacts: gas and hydroldinamic; resilient lowadhesion; tribochemical impacts (impulses of triboactivation); bridges of microwelding. Gas and hydrodinamic interaction depends on availabilyty of gas and fluid in the microholes of friction surfaces. Resilient lowadhesion interactions are realized in dinamic contact of surfaces microrughnesses. This surfaces are covered with he metastable, cellular secondary structure and with the boundary lubricating layers. Transference within the lowadhesion interractions and tribochemical impacts depends on such phenomenas as: the delay of yield of material in very rapid increase of stress till the value that exceeds fluid yield of material in very rapid increase of stress till the value that exceeds fluid yeild limit; the relaxation of metastable secondary and bondary lubrisating lauers; autocanalic creation of strong adhesion connection and so on.

When bucking resilient lowadhesion interaction, it spasmodielly devellopes in tribochemical impact and triboplasma formates. In triboplasma, on a level with seperation of wearing, grain dewelopes restoration of lowadhesion due to co-operative process and interaction with an environment secondary structure. Cells, that has had their day are changed with the new ones.

On the part of surfase which size oscillates in course of time with low frequency the bridges of microwelding are developed. Bridges of microwelding are more large breach of resilient interaction than tribochmical impact. Bridges of microwelding make a great contribution in wearing at the expense of material turbulization in dislocation-vacant interactions .

Resilient lowadhesion interactions are acccmpanied by the convertibe transformation of external mechanical work into secondary work of resilient stress waves in the volume of solid body where this secondary work changes into warmth at the expense of imperfect resilience (of internal friction } . In all the other microcontacts interactions external work unconvertibly within the microcontact in the warmth that extends round the details with thermal condactivity

Microparticles in all sort of microcontact interaction within changing their guantitative relation in extremely broad bounds describe vast spectrum of f riction conditions and tipes of wearing that take piace in engineering practice .

Surface interaction of macrolevel descript on : the microcontacts interaction destribution ; medium thicness of secondary structures ; medium duration of their living . Kinetic (dynamic} description of friction pair depends on all level interaction . The result of these interactions is wearing . Instantaneous rate of wear is the most important description , but one can't measure it with traditional test methods Interconnection between rate of wear and cinetic descriptions are established not only on the basis mathematical correlation but also on the level of physical functional connection with interractions of microcontacts . When one interactions the averoge contact electrical resistence ction (detector effect is stimulaed by triboplasma) is increased when one increases share of tribocemical impacts in general guantity of the most intensive microcontact interaction , that shows the decrease of rate of wear . Rectificant current flows from friction pair element with increased rate wear. Thermodinamical quality factor of inner friction system is increased at the expense of microcontact revesible interactions share. The description of secondary structures (thickness and lifetime) is connected with the factors of kinetic description oscilation in course of timeThese connections were used for creation for friction pair instanteneous rate of wear methods (1)(2)(3).These methods give possibility to revel the content of interactions in friction system (so called "blak box" in traditional test methods). The value of information that was obtained in such control is increased repeatedly in comparison with information that was obtained in traditional test methods.

REFEREHCES

(1) Pat. USSR No **888017, GO1N19/02.**

(2)**Pat. USSR N0 1285347, GO1N3/56.**

(3) И И Кривенко, заводская лаборатория, N0 9, 1996.

TOOL STEELS WITH ADDED CERAMIC PARTICLES

A J LEONARD and W M RAINFORTH
The University of Sheffield, Department of Engineering Materials, Sir Robert Hadfield Building, Mappin Street, Sheffield, S1 3JD, UK

ABSTRACT

Extrusion dies are typically manufactured from the hot work die steel H21 (0.35%C, 3.5%Cr, 9.0%W, 0.5%V, 0.3%Mn). Despite the wear resistant nature of this material dies must be frequently repaired by welding to ensure accurate size and shape tolerances on the manufactured article. The wear resistance of such materials has been attributed to the presence of a large volume fraction of hard carbides which precipitate on solidification and subsequent heat treatments (1). The volume fraction, type and size of carbides is largely limited by traditional casting and hot working routes.

The current work investigated a novel approach to increasing carbide volume fraction by the addition of a FeWTiC™ master alloy which yields TiC particles known to improve wear resistance (2). Additions of up to 7 vol.% TiC were made to melts of H21, under vacuum, with TiC recovery of >70% being achieved. Homogeneity in the TiC distribution was achieved in casts containing up to 5 vol.%, above which agglomeration of the added carbide and incomplete dispersion of the powder were observed. TEM studies and mechanical property tests (4-point bending) revealed an excellent bond between the matrix and the added carbides. The principle carbide type, in addition to M_6C normally present in H21 (3), was identified as TiC, with the only evidence of W found in the precipitation of the non-equilibrium η-carbide $M_{12}C$ (4) at the interface between TiC and matrix. The addition of the TiC did not alter the maximum hardness significantly (604H_v for the H21, 594H_v for the H21+TiC), developed after a double temper at 550°C, but did change the time to maximum hardness from 1h in the H21 to 1.5h in the H21+TiC.

Wear testing was undertaken using a block-on-ring configuration at a sliding speed of $1ms^{-1}$ and loads in the range 54-254N. Figure 1 shows the results of wear tests on samples containing nominally 0, 3 & 5 vol.% TiC. At loads of <154N all casts showed similar wear behaviour. Although some fragmentation of TiC was observed, the particles remained within the surface and the predominant wear mechanism was one of oxidation. However at the highest load, additions of TiC showed an improvement in wear compared to unreinforced H21, with a 3 vol.% addition yielding the best wear performance. At this load the wear mechanism was found to be predominantly oxidation and delamination. Subsurface cracking was found to nucleate at M_6C particles in the matrix rather than at the TiC interface. Considerable flow of the matrix martensite around the TiC was observed without signs of debonding or cracking of the TiC.

Cross-sections through worn surfaces showed that equivalent tensile strains in excess of 14 were generated during wear tests on H21. The microstructural changes at the extreme surface of the steel as a result of wear were investigated by TEM back-thinning preparation techniques. A region of fine grained oxide and metallic iron was observed, the mean crystallite size was 15-30 nm. Little evidence was found of fracture of the TiC-matrix interface. Despite the abundance of carbides generated in the matrix material during casting and heat treatment, very few were found in the surface region. This, and direct bulk temperature measurements made during wear testing, indicated substantial heating at the surface.

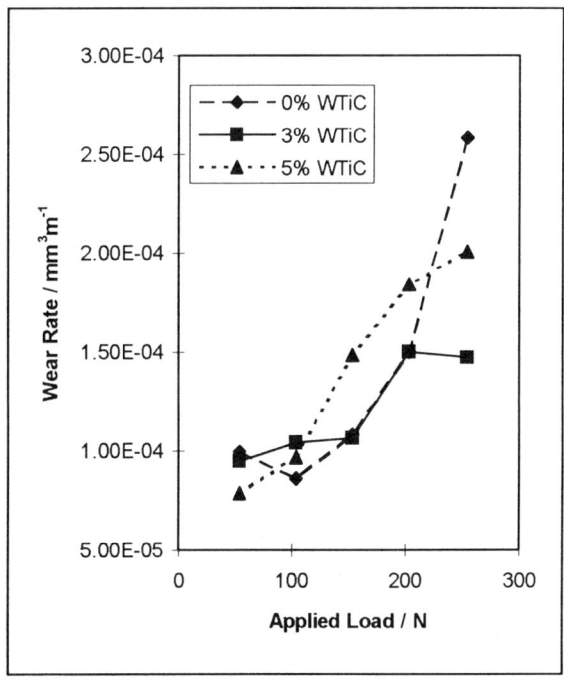

Fig. 1: The wear rate of H21 containing different volume fractions of TiC with applied load

REFERENCES
(1) P.L. Hurricks, Wear **26** (1973) 285-304.
(2) J.V. Wood, K. Dinsdale, P. Davies & J.L.F Kellie, Materials Science & Technology **11** (1995) 1315-1320.
(3) K. Kuo, Journal of the Iron and Steel Institute **173** (1953) 363-375.
(4) E. Lugscheider, H. Reimann & R. Pankert, Zeitschrift fur Metallkunde **73** (1983) 321-324.

AN EXPERIMENTAL STUDY OF FRICTION INDUCED VIBRATION IN A CERAMIC CONTACT

JACEK IGOR LUBINSKI, KRZYSZTOF DRUET, TADEUSZ LUBINSKI
Technical University of Gdansk, Faculty of Mechanical Engineering, ul. Narutowicza 11/12, 80-952 Gdansk, Poland

ABSTRACT

The study was performed using a tribometer with a changeable dynamic properties of moving members (1). Friction couple comprised two alumina specimens, 1 and 2 (Fig. 1). The upper specimen, 2, was attached at the end of a spindle having a pulley at the other end. The lower specimen, 1, was supported in load transmitting elastic membrane with glued strain gauges, 4, to measure real load and was constrained in a friction force measuring system with two bent beams, 3. Driven member of the friction couple comprised transmission pulley and a spindle with the upper specimen, 2, attached at the end, 6, was rotated by a dc electric motor, 8, and belt transmission. The tests were run under variable friction parameters.

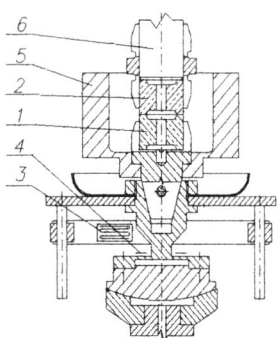

Fig. 1. Specimen set-up

The course of friction was identified on the base of measurement of:
(a) lower specimen displacement in two directions: friction surface and its normal directions with strain gauge tribometer measuring systems (1),
(b) instantaneous rotational velocity of upper specimen cinematic member with a tachometer,
(c) acceleration with accelerometers fixed to deliberately chosen places,
(d) noise spectrum with an acoustic analyser.

The measurements and following calculations enabled to describe produced oscillations. Stable oscillations of tick-slip type for very small velocities or quasi-harmonic for higher velocities were obtained. Examples of the results are presented in Fig. 2.

Main conclusions are as follows. Oscillations of both kinematic members appeared simultaneously but they were different in amplitude and frequency. Oscillation forms of both friction members depend on their individual dynamic properties but are mutually interfered by counter partner movement via forces acting at the interface. Both normal and friction forces change. There is a dependence between coefficient of friction sliding velocity and the velocity changes.

The paper with details will be submitted to Wear.

a)
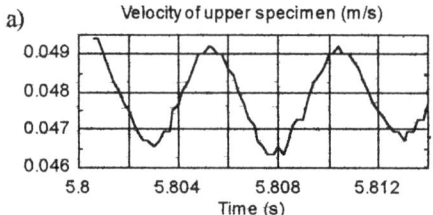

b)
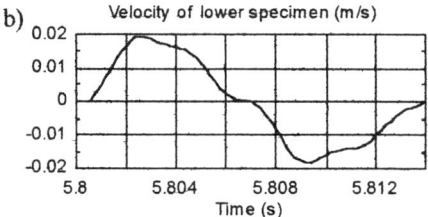

c)
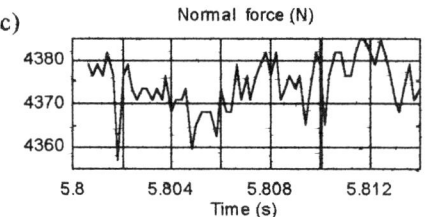

d)
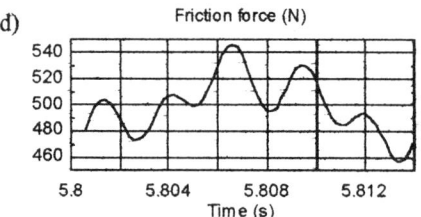

e)
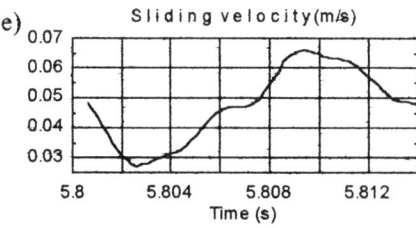

Fig. 2. Computed data in the same time domain

REFERENCES

(1) Lubinski T, Testing of friction dynamics on a tribometer PT-3/96, Proc. of the Congress.
(2) Lubinski T, Druet K, Lubinski J I, Dynamic features of a tribometer affect friction and wear test results, Proc. of the Congress.

PHYSICAL-CHEMICAL FUNDAMENTALS OF FRICTION AND WEAR OF FRICTION SURFACES OF VEHICLES WHEELS AND RAILS

Ju.M. LUSHNOV, A.V.CHICHINADZE, S.S. PETRAKOVSKY
Mechanical Engineering Recearch Institute of the Russian Academy of Sciences

ABSTRACT

The effectiveness of using and many economic criteria of transport vehicles service are determined by a level and reability of cohesion of their wheels with hard road coating. For practice it is important to choose of a possibly higher rated coefficient of cohesion providing relatively not so large losses due to wheels and rails wear and additional losses of energy connected with a surplus sliding of wheels. In all these factors the important role belongs to the primary friction state of rubbing bodies in much depending upon the weather conditions. Because of convenience of organization of movement of rail-way transport, independing of sufficient difference in climatic conditions of sections of rail-way network the rated coefficients of cohesion of vehicles are chosen practicaly the same. Sometimes it leads in separate regions of rail road network and during different seasons to significant super-normative wears of rubbing elements and to different damage of equipment as well as to rather large non-productive energy losses resulted in a significant raising of vehicles power. Here, as a rule, no due attention is paid to a search of ways of a more effective using of friction units, effective applying of materials, and decreasing resistance to a movement of rail-way transport.

A step solving these problems can become a deeper understanding of the process of interaction between wheels and rails and mechanism of their primary friction properties changes, (1,2)

Summing up factors which determine friction properties of wheels and rails, friction regularities of a way and particularities of cohesion of locomotives with rails gives a base for plotting a nomogram (2) permitting to judge of the most probable realizable coefficient of cohesion of a locomotive connected with changing weather conditions. One can estimate a probability of realization of locomotive cohesion for given condition in general case using the formula:

$$P(\Psi) = P\{\mu \, /(P/P_s ; a)\} P(\Psi/\mu) P(\Psi/\upsilon) P(\Psi/\Psi')$$

The nomogram includes dependencies establisting a correlation between a level of humidifying wear tracks on rails and a possible for realization friction coefficient μ, water vapour pressure P/P_s, and on the quantity of moisture on the path of friction have been taken into account, and a tie between the friction coefficient μ and the locomotive cohesion coefficient in the moment of friction (Ψ - with sand, Ψ' - without sand in contact), between a change coefficients of cohesion Ψ' and whilst changing a locomotive speed υ.

Dependencies are given for positive temperatures T, and - for negative temperatures.

All these factors enable using meteorological data to get a notion of the most probable level of realization of this or that with a locomotive coefficient of cohesion in some definite moment of a day or night, a month or a year, for this or that region of rail-ways.

Analysis of the data shows that the most unforvourable for locomotive work (like cars) are I, II, III and XII months. On the rest months of a year frictional possibilities of this friction unit are evidently under-used. These regulations with some versions are seen under an analysis of conditions of work of locomotives in various rail-road regions. It is possible to state that analoguos regulations for the first time established for locomotives can be seen during cars, trams and other vehicles service. Worked out by us an approach to estimation of operating conditions of transport can become a base for some special maps of divisions into districts of friction possibilities of transport vehicles designing. These data may be useful both for planning measures aimed at more effective using existing transport vehicles and for planning operating conditions in the new geographical regions of the world.

The increase of speeds and loads gives the high increase of lateral wear of wheels and rails.

The increase of μ in the zone of a contact between a wheel and rail can give surface temperatures about 100-600 ºC. That is why we have very high speeds of phisical and chemical processes here. These processes change molecular and deformation components of the friction processes and the wear of wheels and rails.

The dependence $\mu(T)$ can reach meanings from 0,5 till 0,2÷0,3. An atomic composition will be changed. It depends on loads of wheels and rails. Experiments showed it is possible to receive a solid and unsolid layer of the material. Molecular and corrosive - mechanical process influence hard on the level of the wear of wheels and rails. These process will be more activated with the increase of trains speeds.

REFERENCES

(1) Физико-химическая механика сцепления., Труды МИИТа, вып. 445, М., 1973 г.
(2) Исаев И.П., Лужнов Ю.М. Проблемы сцепления колес локомотива с рельсами., М.: Машиностроение, 1985 г.

DEFORMATION-AND-ELECTRODYNAMICAL THEORY OF FRICTION - A WAY TO IMPROVE RELIABILITY OF TECHNOLOGICAL SYSTEMS

V. A. LYASHKO and M. M. POTEMKIN
V. Bakul Institute for Superhard Materials of the Nat. Ac. Sci. of Ukraine,
2 Avtozavodskaya St., Kiev, 254074, Ukraine

ABSTRACT

All technological systems currently available are based on the use of a great number of friction couples, 80-90% of which fail because of wear. Clearly the reliability of the systems essentially depends on performance characteristics of friction parts and the main potentiality for increasing their reliability is in the methodology of friction units designing. However, the designer's decisions should have a comprehensive theoretical substantiation, otherwise the development of technological systems of a reasonable reliability is associated with an added costs on experimental studies.

Unfortunately it should be noted that the existing theories of friction do not reveal the physical nature of the friction process, and particularly concerning the adhesion component.

To elucidate this and other phenomena occurring in friction, we propose to use a new line in the science of friction and wear, i.e. the deformation-and-electrodynamical theory of friction which is based on the consideration of two processes: deformation process and electrodynamical process. In this case, the deformation component of the friction process is considered on the principals of thermodynamics of open systems, while the electrodynamical component is based on the hypothesis that the friction is of a field nature, and whereby the interaction of electromagnetic fields which surround bodies being in contact, is the reason for initiation of friction proper (i.e. of that what is meant by the molecular component).

According to the nature of the interaction with the magnetic field, all existing materials are divided into three main groups: ferromagnetics, paramagnetics and diamagnetics. Ferrum, nickel, cobalt, etc. as well as a number of chromium and manganese compounds, which are in most common use in mechanical engineering be long to ferromagnetics, as they are of a domain structure, i.e. they have local regions of spontaneous magnetization.

Materials which are widely used in friction couples and as additives to oils to ensure antifrictional conditions (phosphorus, sulfur, copper, zinc, tin, lead, etc.) are diamagnetics and possess diametrically opposite properties: they have no domain structure and magnetic charge carriers are oriented in the direction opposite to that the action of the external magnetic field. For this reason, this problem requires further analysis.

As an example of a simplest model object we consider a friction couple which consists of two ferromagnetic materials (with juvenile surfaces) with no lubricant and gaseous media. This corresponds, e.g. to the contact of two steel specimens in vacuum.

In this case, by virtue of the presence of the domain structure, as two parts are coming into contact with each other there occurs interaction between their magnetic fields. As vectors of the spontaneous magnetization of the domains being in contact are of mismatched orientation in space, to ensure a minimum of free energy of a new system, the precession of the magnetic charge carriers will occur to release excess energy as heat.

If the released heat suffices to initiate the development of molecular bonds, the system energy will decrease because of their formation.

The relative movement between two surfaces of parts being in contact initiates two moving electromagnetic fields. The fields are alternating, and the frequency of their modulation is defined by the sizes of domain structures, the distance between magnetic charge carriers and the rate of relative movement.. According to the principles of electrodynamics we can consider the superposition of these fields, and if their frequencies do not coincide, the mutual effect of the parts can be considered separately. When the modulation frequencies coincide, the resonance may occur, which causes an inerease in the heat liberation.

Cleary the amount of the as-liberated heat, with the relative movement between the parts, essentially increases (as the value of induced currents is proportional to the rate of magnetic flux change). This in turn causes the intensive seizing of the parts a and higher wear rate.

The analysis performed is in full agreement (in a qualitative sense) with experimental results. This points to the fact that the approach being developed holds much promise.

SLIDING WEAR OF HEAT TREATED PLAIN CARBON STEEL AGAINST ALLOY STEEL UNDER BIO-FUEL CONTAMINATED LUBE OIL

M. A. Maleque, H. H Masjuki and M. Ishak
Department of Mechanical Engineering, University of Malaya, Malaysia.

ABSTRACT

The effect of hardness and vegetable oil based lube oil additive on wear of materials or tribological components have been reported in the literature (1, 2). However, little attention has only been paid by the present authors (3) to the use of palm oil diesel (POD) as an additive. An investigation has been made on sliding wear and friction of heat treated plain carbon steel against alloy steel using a tri-pin-on-disk type of wear and friction apparatus under POD contaminated lube oil. A systematic study has been made with 0%, 5% and 100% POD lube oil. Results from this study will be useful in material selection for tribological components in diesel engine.

Tests were performed at a constant speed of 500 rpm, giving a sliding speed of 1.83 m/s, normal load of 100 N and at ambient temperature. Heat treated plain carbon steel for pin and alloy steel for rotating disc were used in this study.

The wear and friction properties of pin materials were investigeted in terms of specific wear rate, friction coefficient and shown in Figs. 1 and 2 respectively. Results indicate that better wear resistance and friction properties were found from 5% bio-fuel (POD) blended lube oil. This may be attributed to the fact that POD contains fatty acids which forms soap film only when blended in small amount.

TABLE - I

Make up lubricant	FT-IR Analysis			
	Soot (A/0.1 mm)	Oxidation (A/0.1 mm)	Nitration (A/0.1mm)	Anti-wear (A/0.1 mm)
Plain lube oil	0.08	0.56	0.00	-0.02
5% POD blended lube oil	0.09	0.80	0.00	-0.02
Plain POD	0.14	4.60	0.04	-0.04

From FT-IR spectroscopy analysis (Table I) a quantitative data for the degradation of the lube oil additives was found. The results showed that plain POD increases the oxidation products viz. oxygen, nitrogen and sulphate whereas, 5% POD blended lube oil decrease its value.

Kinematic viscosity (numerical value) of different lube oils viz. plain lube oil, bio-fuel blended lube oil and plain POD were 104 cSt, 99.6 cSt and 4.5 cSt respectively which means that bio-fuel contaminated lube oil exibits better physical properties. The degradation of the lube oil additives, expressed as negative sign under the heading of anti-wear products was also found to be higher for pure POD. Corrosive wear and pits on the specimen surface (photographs has not shown here) were found when pure bio-fuel was used as lubricating oil.

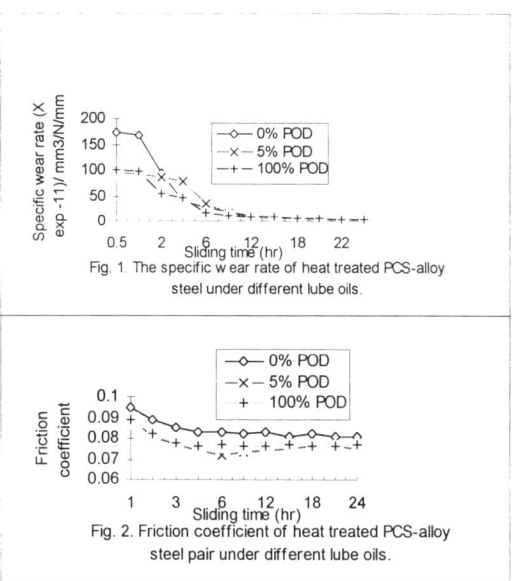

Fig. 1. The specific wear rate of heat treated PCS-alloy steel under different lube oils.

Fig. 2. Friction coefficient of heat treated PCS-alloy steel pair under different lube oils.

Sliding wear and friction tests performed under bio-fuel contaminated lube oil, show that better tribological performance could be obtained from 5% POD blended lube oil.

REFERENCES

1. K. Hokkirigawa and K. Kato, **Wear**, 123 (1988), 241.
2. P. Studt, **Tribology International**, Vol-22 (1989), 111.
3. H.H. Masjuki, M.A. Maleque, **Wear**, 198 (1996), 293.

THE ANALYSIS OF ASSEMBLY: CAM-FOLLOWER WEAR IN RELATION TO MATERIAL PROPERTIES AND HEAT TREATMENT

JACEK MICHALSKI and PAWEŁ PAWLUS
Rzeszów University of Technology, 35-959 Rzeszów, ul. W. Pola 2, Poland

INTRODUCTION

The assembly: cam-follower is one of the most important tribological systems of internal combustion engine. Many types of wear-resistant materials have been adopted for automotive engine cams and followers, to meet the requirement of various engine builders. These materials include high chromium cast irons, tungsten cemented carbides, silicon nitride ceramics and ferro-based powder metals with a high chromium content (1)(2). A few studies have been reported in which investigation were made of the relation between wear resistance and individual material properties as well as heat and chemical treatment (3)(4)(5). An attempt to rectify this problem is made in this paper.

THE SCOPE OF THE INVESTIGATIONS

The analysed assemblies: cam-follower from gasoline, four stroke, water cooled internal combustion engines were tested. The camshaft was mounted in engine block. These engines were assembled in automotive vehicle. The cams have synthetic profile: the pressure on the top was 425 MPa (force 600 N) and on cam nose 1150 N (rotational speed was 4000 rpm).

The same experimental procedure was adopted to these engines. Various pairs: cam-followers were measured after running-in of the engines. After 2,000 traction km engines were disassembled and the same elements were measured. Also, the qualitative analysis of cam- follower wear was done.

EXPERIMENTAL RESULTS

Firstly, the analysis of wear of eight various camshafts collaborated with the hydraulic followers made from the same materials was done. The material of follower was grey cast iron of the frontal surface hardened during casting process (chilled cast iron). Metallographic structure contained changed ledeburite (hardness 61-62.5 HRC).

Camshafts No.No. 1 and 2 were done from spheroidal fine pearlite cast iron, their cams were quenched using induction method; their core matrix structure contained small grains of free ferrite (smaller than 5%). However the structure of hardened layer had fine martensite and nodular graphite. Various thickness of hardened layer was the main difference between these cams. After the additional tests it was found that cam No 1 of bigger thickness contained bigger amount of retained austenite and tensile residual stresses (in the other case compression stresses were found). The linear wear was bigger for cam No 1. For two cams the abrasive wear (severe on top and mild on nose) and fatigue wear (cracks and delaminated particles) were found.

On the front of follower collaborated with cam No 1 the mild abrasive wear was found; however the follower collaborated with cam No 2 showed abrasive wear and delaminated particles.

The same cams after hardening were additionally subjected to ion nitriding (No 3 and 4). In both cases only small abrasive wear was found.

The cams No.No 1, 2 were also (after quenching) was sulfonitrided in the soaking pit in the atmosphere of ammonia and sulphur (No 5 and 6). The big fatigue and abrasive wear was found.

Additional cams were not quenched; they were subjected only to ion nitriding (No 7) and sufonitriding (No 8). In these cases, also abrasive and fatigue wear (cracks, delaminated particles) were found.

Two kinds of wear can be explained by the possibility of the tensile stresses existing in the sliding contact (Hertz theory). It will be particularly described in the paper.

It was found that cams No 3 and 4 were the best.

Other two assemblies cam - follower were analysed. Camshaft was done from different chilled pearlitic grey cast iron. First cam collaborated with follower made from chilled grey cast iron (after hardening) - it was pair No 9, but the second cam collaborated with 40H steel after toughening and superficial hardening, No 10. The observed abrasive wear was very mild.

CONCLUSION

It was found that assembly (10) were the best from all the analysed. It is recommended by the authors of this paper. In full paper version the attempt is made in this paper to explain these experimental results.

REFERENCES

(1) M. Kano and I. Tanimoto: Wear mechanism of high wear-resistant materials for automotive valve trains. Wear, 151 (1991), 229-235.
(2) M. Kano and I. Tanimoto: Wear resistance of ceramic rocker arm pads, Wear, 145 (1991), 153-160.
(3) D.F. Diao, K. Kato, K. Hayashi: The maximum tensile stress on a hard coating under sliding friction. Tribology International, 27, No. 4, 1994, 267-272.
(4) N.P. Suh: An overview of the delamination theory of wear. Wear, 44, 1977, 1-16
(5) R. B. Waterhose (ed.): Fretting fatigue. London: Applied Science Publishers 1981.

ESTIMATING THE INFLUENCE OF TEXTURE ON THE RESISTANCE AND WEAR OF BIMETALS

A MILOSAVLJEVIĆ, R PROKIĆ-CVETKOVIĆ and Z RADAKOVIĆ
Faculty of Mechanical Engineering, University of Belgrade, 27. marta 80, Belgrade, Yugoslavia
O ĐORĐEVIĆ
SARTID 1913, Smederevo, Yugoslavia
Ž BLEČIĆ
Faculty of Metallurgy, Podgorica, Yugoslavia

ABSTRACT

The final goal in developing and applying all multi-layered materials, and also bimetals, in reflection to monometallic materials, is in getting better mechanical and economic features (e.g. improved wear and corrosion resistance; less specific weight; higher resistance to crack initiation; saving expensive material, etc). High possibilities for producing bimetals as a combination of different monometals goes in favour of producing new material qualities, and at the same time widening its field of application.

Wide area of bimetal application requires constant development of mechanical and structural characteristics. This can be achieved by applying various regimes of cold plastic deformation, whereas the material structure gains a crystallographic texture orientation that predominantly effects material behaviour. (1,2)

Influence of texture components and their characteristics to the condition of surface layers in the tested materials in cases of a given regime of cold plastic deformation is considered. Efforts were made to apply a treatment regime and produce effective texture that would enhance resistance and wear features.

Hot rolled sheets of low-carbon steel were joined to brass Cu76Zn24, Cu77Zn23 and Cu78Zn22, separately. All brass specimens were deformed in rolling regimes RI and RII. Larger degrees of deformation were in case of regime RI ($l_d/h_m \geq 5$), with a smaller number of passes, while in regime RII ($l_d/h_m \approx 0.7$), the number of passes was larger, with a smaller deforming degree (l_d -deformation zone length, h_m -mean thickness of rolled sheets).

A special Horta method was used to evaluate the texture on three brass alloys - surface bimetal layers (3, 4,5,6).

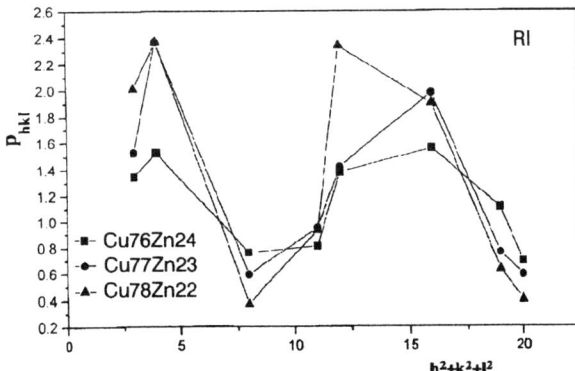

Figure 1 - Crystal plane type dependence on p_{hkl} for regime RI

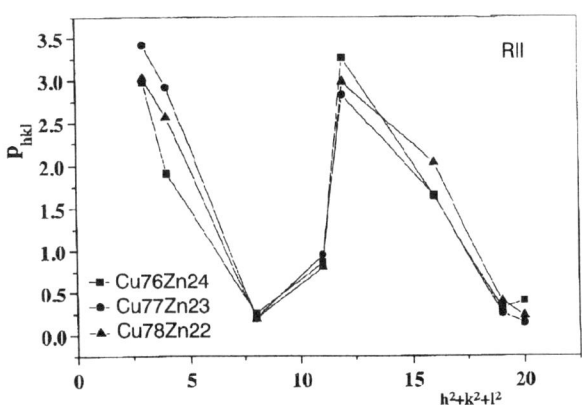

Figure 2 - The crystal plane type dependence on p_{hkl} for regime RII

Figures 1 and 2 show the distribution of texture components expressed by the p_{hkl} parameter.

The change in structure showed a finer grain for all alloys in the case of regime RII. The rise in material resistance, decrease in wear, and also uniform distribution of texture components (Fig. 2), all go in favour of regime RII as the optimal.

REFERENCES

(1) J.C. Wright, New materials for sheet metal working - Part 3, *Sheet Metal Industries* (1974) p.126-129
(2) Р.М. Кључников, Получение биметаллов для електротехнических устроицтв холодного плакирования, *Черная металургия* (1971) с.101-105
(3) P.J. Horta, W.T. Roberts, R.O. Wilson, *TMS-AIME*, **245** (1969) 2525
(4) V. Šijački-Žeravčić, A. Milosavljević, M. Rogulić, The change of strengthening mechanism in Cu-Cr alloy after different thermomechanical treatment, *Residual Stresses in Science and Technology*, Macherauch and Hauk, Eds., DGM Information Verlag ICRS 1, Vol.2 (1986) p.671
(5) V. Šijački-Žeravčić, M. Rogulić, Determination of texture and macro residual stresses in the copper alloy with 0.4% chromium after different thermomechanical treatment, *J. Serb. Chem. Soc.* **60** (7) 601-609 (1995)
(6) A. Milosavljević, V. Šijački-Žeravčić, M. Rogulić, V. Milenković, The influence of the mechanical schedule on deformation of AlMgSiCu alloy texture, *Proc. 2nd Intern. Cong. on Heat Treat. of Mat.*, Florence (1982) p.189-194

WEAR RESISTANCE OF ALUMINA FIBRE δ-TYPE REINFORCED Al 4% CU MATRIX COMPOSITE

S.F. MOUSTAFA and F.A. SOLIMAN *
Center Metallurgical R & D institute, Cairo, Egypt
Faculty of Eng. Cairo Egypt.

ABSTRACT

Many investigations reported that the incorporation of hard reinforcements (1-4) into an aluminium alloy improved the wear resistance

Effect of Particle sizes on wear behaviour was studied (5,6). They reported that the wear resistance increased with volume fractions and sizes of reinforcements and they found that a major wear mechanism was the abrasion wear in AL/SIC and Al/A203 particulate composites. But there are some other studies which suggested that addition of reinforcements slightly increase the wear resistance as the volume fraction of reinforcements increases up to a points at which the wear resistance becomes worse. They attributed this results to the third body abrasion. In previous work (7) the wear properties of Al/Al2O3 fibre of δ. type was considered up till now the effect of types of reinforcements sliding speed and hardness on wear behaviour have not been understood clearly

In this investigation the wear behaviour of Al-4%Cu Alloy and its composite containing alumina fibres of δ- type has been studied at two condition at room temperature and at elevated temperature of 150°C using a standard pin-on-ring tribometer The results show that for both two conditions the composite showed higher wear resistance and lower coefficient of friction as compared with those of matrix alloy.

The wear mechanisms of matrix alloy tested at room temperature were of three mechanisms namely oxidational induced dilamination and seizure followed by melt wear. At elevated temperature of 150 °C however the matrix wear mechanism is seizure and melt wear. In case of composite materials the wear mechanisms were different at low loads. Oxidational induced dilamination was predominant increasing the applied normal load up to 80 N. high strain induced dilamination mechanism took place at higher load (beyond 120 N) sub-surface dilamination was the activated mechanism

REFERENCES

1   Z.C. Feng and K.N Tandon Scripta Metal V32 No 4,. P 523, 1995
2- F. Rana and D.M. Stefanescu Metal Trans V 20A, P.1564, 1989
3- A.T. Alpas and J Zhang Metal Trans V25A: P 969, 1994
4- S, K. Srivastava, S. Mohan V Agarwala and R.C. Agrawala Metal Trans V.25A P851 1994.
5   S Skolianos and T. Zkattamis Mater Sci. And Eng. V. 163A P 107 1993
6   F.M. Hosking F.F. Portillo R. Wunderlin, and R. Mehrabian J. Master Sci , V 171 P477 1982.
7   S.F. Moustafa Wear V185, P 189, 1995

ANALYSIS OF THE INTERACTION OF PLASMA SPRAYED CERAMIC (ZrO_2) COATING - CAST IRON SLIDING PAIR

WITOLD JORDAN, EDWARD KOLODZIEJ and ANDRZEJ MRUK
Cracow University of Technology, Al.Jana Pawla II 37, 31-864 Krakow, PL

ABSTRACT

Research on the application of ceramic heat-insulating materials used as coatings on the surfaces of internal combustion engine elements like piston rings, cylinder barrels, pistons, is one of the tendency in internal combustion engines development. Rings, cylinder barrel and piston constitute a friction pair and therefore their upper layers must meet also, apart from the others, tribology requirements.

So far the research has proved that from the standpoint of coatings resistance to thermal loading changes the application of ceramic coatings made of solid solution ZrO_2 partially stabilised with oxides Y_2O_3, CaO, MgO or rare earths mixture (Ln_2O_3) (1,2) is the most advantageous.

For the tests, a plasma method of applying two-layers coatings: sub-layer and principal ceramic layer, has been admitted. Coatings were deposited on the AlSi12 base. Ni-Cr was applied on the sub-layer, and powder composed of 9% wt Ln_2O_3 + 91% wt ZrO_2 on the main ceramic layer. The sprayed layers thickness was: sub-layer 0.1 - 0.3 mm, main layer 0.8-1 mm and their porosity was 10-12 %.

The following sliding pair has been used for tribological tests: the ceramic coating as the sample and cast iron used for internal combustion cylinder barrels as the counter-sample.

Tribology tests included the determination of the friction coefficient (static and kinematic) as well as the wear of friction pair working under hard conditions, e.g. in technical dry friction, when the oil lubrication was mean and the counter-sample was covered with graphite powder.

The main tribology tests were carried out on a hydraulic drive stand performing a to-and-fro motion at a low friction velocity in the range 0.01 to 0.1 m/s and a stroke of 0.2 m (3).

The coefficient of friction value and the sample wear value after an assumed worktime have been accepted as main assessment criteria of tribological properties. Furthermore, porosity changes in the sample and counter-sample have been analysed.

Changes in friction coefficient value in function of the pair worktime, in dry friction, the lubricating substances being oil SD 20W/40 and graphite, are shown in. Fig. 1.

Friction force was measured continuously, while the value of friction force along the counter-sample stroke (i.e. friction path) was registered periodically.

As far as the character of friction coefficient changes is concerned, it is similar in all the cases, however the differences of values occuring between them seem to be worthy of notice (Fig.1), since it appears that there is no possibility to produce a convergent lubricating film. Nevertheless, there is a significant difference between using oil or graphite as lubricant.

Analysing the changes of porosity parameters in sample as well as in counter-sample, it can be proved that these changes are closely connected with friction conditions.

The results of tribology research reveal advantages of sliding pair under examination, since even in dry friction conditions there was no seizing of the operating pair. Though working conditions of the pair were not too hard, still they were unfavourable enough to affirm that for another kind of pair the wear would be much more intensive.

On the basis of these preliminary research results the following can be stated:

- properties of the examined sliding pair promise some practical possibilities to use it in internal combustion engines for cylinder barrel parts in the area of interacting with the piston.

- in case of using graphite as lubricant for the examined sliding pair, an important decrease of friction coefficient and its stabilisation along the friction path are achieved.

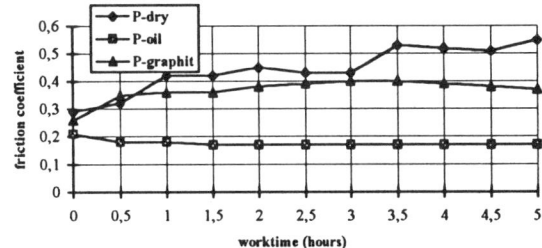

Fig.1: Changes of friction coefficient during the test; test conditions: p = 0.92 (Mpa), v=0.05 (m/s).

REFERENCES

(1) P.Vincenzini, Industrial Ceramics, Vol.10, No.10, 1990, 113-126pp.
(2) A.Mruk, Monograph No.198, Cracow University of Technology, 1995.
(3) W.Jordan, E.Kolodziej, A.Mruk, B.Stolarski, Monograph No.210, Cracow University of Technology, 1996, 53-61pp.

WEAR MECHANISM OF COPPER ALLOY WIRE SLIDING AGAINST IRON-BASE STRIP UNDER ELECTRIC CURRENT

HIROKI NAGASAWA
Railway Technical Research Institute, Kokubunji, Tokyo,185, JAPAN
KOJI KATO
School of Mechanical Engineering, Tohoku University, Sendai, 980-77, JAPAN

INTRODUCTION

In electric railways, wear rate of trolley wire decides the life of the wire. In order to search for methods of decreasing wear, wear rate of wire is expressed with two parameters from results of laboratory wear test under electric current. Three types of wear particles are observed.

EXPERIMENTAL PROCEDURE

Material of trolley wire are Cu-Cr-Zr alloy, Cu-Sn alloy and pure copper which were reported before (1). They are 4 mm in diameter and round in section. Contact strip is an Fe-base sintered alloy 25mm long in sliding direction.

The wire is wound around a rotary disk of 492mm diameter in two rows. Strip is pressed against wire. DC electric current flows from wire to strip. Sliding velocity is 1.4 to 5.6 m/s, normal load 10 to 40 N, and current 10 to 30 A (100V) or 1 mA (1.5V).

Wear amounts of wire and strip are calculated from their weight loss. Friction force is measured from strain of support box of strip. Contact loss of strip is detected by measuring electric potential between wire and strip.

RESULTS

Wear rate of wire: Wear rate of wire is related to contact pressure and temperature rise on sliding surface (2). Under electric current flow condition, heat generated on wear surface of unit length of wire q is

$$q = q_f + q_r + q_a \quad [J/m]$$

where q_f is frictional heat, q_r is joule heat and q_a is arc heat when contact loss occurs (3).

So wear rate of wire is expressed with two parameters as shown in Fig. 1. One is qt which is the product of heat q by sliding time t per unit length of wire. The other is normalized pressure p defined as follows;

$$p = F/(AH)$$

where F is normal load [N], A is contact area [m^2], H is Vickers hardness [Mpa] (2).

In Fig.1, wear rate of wire is expressed in wear volume per unit length at one pass of strip [mm^3/m].

Wear mechanism: The shape of wear particle is classified into three types which are shown in Fig. 1. Flake type is observed in Reg. 1, where heat of current can be negligible. Roll type is observed in Reg. 2, where current flows without remarkable arc. Block type is observed in Reg. 3, where contact loss occurs with arc.

CONCLUSIONS

Wear rate of trolley wire is expressed with two parameters of p and qt as a wear map. Three types of wear particle shape are recognized and they are shown in three regions in the map.

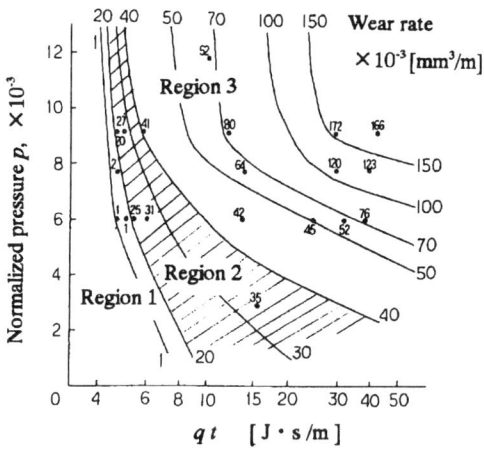

Fig.1 : Wear map of wire under electric current

REFERENCES

(1) H.Nagasawa and K.Kato: Proc. Int. Trib. Conf., Yokohama 1995, Vol.1, 343-348pp.

(2) S.C.Lim and M.F.Ashby: Acta Metallurgica, Vol.35, No.1, 1987, 1-24pp.

(3) O.Oda and Y.Fujii: Journal of JSLE, Vol.29, No.1, 1984, 66-71pp.

WEAR RESISTANCE OF Fe-Cr-C STEELS UNDER SLIDING FRICTION

V.G. NOVITSKII and V.I.TIKHONOVICH
Physico-Technological Institute of Metals and Alloys of the National Academy of Sciences of Ukraine, 34/1 Vernadsky Avenue, 252580 Kiev 142, Ukraine

ABSTRACT

The work is devoted to the optimizing composition of Fe-Cr-C steels, destined for parts of circulation component of power pumps which are influenced by erosion, corrosion and sliding friction.

Many years' tests gave possibility to establish that a base cast steel containing 1.2% C and 15% Cr has a good erosion and corrosion resistance. The increase of wear resistance of this steel under condition of sliding friction can be achieved by additional alloying which will lead to the desirable changes of the initial steels structure and to the transformation of near-surface layers' structure at friction.

Wear resistance is an integral characteristic of alloys. Such an approach takes into consideration numeral interrelation between structure, properties and factors of external influence. The ability of alloys to resist to wear is considered as a structure-sensitive characteristic and one of ways of synthesis of wear resistant alloys is the creation of materials with the structure capable to absorb energy introduced from outside and to scatter it by means of reverse phase and structure transformations. Wear resistance of structure in this case is defined by ability of quick reconstruction of initial structure into the state advantages for these conditions. (1), (2).

For defining characteristics of such state, which is formed at sliding friction there were implemented studies of a base steel, additionally alloyed by the elements expending area of γ-phase existence (Mn, Ni, Cu) as well as the elements arrowing this area (Al, W, V, Nb).

Heat treatment conditions were selected with the help of mathematics planning. They consist of hardening from 1060-1080 °C into an oil and tampering at 560-600 °C

Friction and wear tests were implemented in accordance with the following scheme: a sample (insert) - counterbody steel containing 0.2% C, 13% Cr (shaft) in water; sliding speed - from 1 to 5 m /sec, unit load- 1 and 5 Mpa.

Characteristics of structure and phase composition were studied by X-Ray structural and micro- X-Ray spectral methods (3), (4). The results received showed that additional alloying exerts considerable influence on wear resistance of steels.

The greatest effect under conditions of normal wear display elements the content of which in alloys does not exceed 1.0 wt. % .

On the basis of studies implemented it was established the following:

-within a range of alloying the studied steel by Mn (0.7-7.5%), Ni (0.2-4.2%), Cu (0.2-4.7%), Al (0.2-1.9%), W (0.1-3.2%), V (0.3-4.5%), Nb (0.02-0.4%) the change of the sliding friction wear curve has an extreme character;

-the greatest influence on steel wear resistance has alloying by elements in quantities which not lead to the formation of own carbides but are dissolved in a matrix and/or in complex carbides;

-the chromium content in near- surface layers at additional alloying the studied steel bears an extreme character; at maximum chromium content the steel has maximum wear resistance;

-at minimum wear in near -surface layers - an intensive $\alpha \Leftrightarrow \gamma$ transformation takes place; a syncretic 50:50 ratio between α- and γ- phases and dislocation density equality are established;

-at minimum steel wear the maximum construction of structure- stressed state in near-surface friction layers is observed what corres-ponds to the in crease of the third kind distortion in γ-phase.

REFERENCES

(1) I.M.Lyubarskii, L.S.Palatnik, Metallophizika treniya, 1976, 176 p. (in Russian).
(2) V.I. Tikhonovich,Litye iznosostojkie materialy, 1978, 3-18 pp. (in Russian).
(3) V.V. Nemoshkalenko, V.V. Gorsky, V.I. Tikhonovich ets., Acta metallurgica, 1978, Vol. 26, 705-707 pp.
(4) V.G. Novitskii, V.I. Tikhonovich, N.A. Kaltchuk, Friction and Wear, 1996, Vol. 17, N5, 690-693 pp. (in Russian).

MEASUREMENTS OF PLASTIC STRAIN AROUND AN INDENTATION CAUSED BY THE IMPACT OF A WC BALL

Y I OKA, M MATSUMURA, H OHNOGI and K NAGAHASHI
Department of Chemical Engineering, Hiroshima University, Higashi-Hiroshima, 739 Japan

ABSTRACT

The aim of this paper was to consider the damage mechanisms caused by solid particle impact. The effect of particle shape and impact angle were investigated by examining the plastic strain distributions, caused by a round WC particle and an angular SiC particle on two metallic materials. Regular square grids with 300 or 1000 lines per inch were made on the surfaces using a photoengraving technique. The strain and distortion of the grids was determined by measuring the distance between the grids (1).

The principal shearing strain distribution on the cross-sectional surface, by the WC ball was found to be dependent upon the type of material and impact angle. In the case of iron, the position of the maximum shearing strain was observed below the rim of the indentation. For aluminium, the maximum occurred below the center of the indentation, irrespective of the impact angle.

Figure 1, shows an indentation on the surface of an aluminium specimen caused by the impact of a 3 mm WC ball. Well-strained grids and spalled areas can be seen around the indentation. Although the value of the plastic strain in the intruded area and on the apex of the lip could not be obtained, because of broken grids, it was considered to be relatively high. The location of the lip formed from the extruded material by the impact of the WC ball, was unaffected by the type of the material and impact angle. The maximum lip height at shallow impact angles (20, 30 and 40 degrees), was observed at the side, rather than at the front of the impact crater. The crater configuration traced with a profilometer revealed the highest lip coincided with the maximum shearing strain. Material removal did not occur with the impact of a single spherical projectile at any impact angle. The volume ratio of the lip to the depressed section was under 1.0.

Figure 2 shows a typical appearance of an indentation caused by the impact of a 3 mm SiC angular particle under the same impact conditions as in Fig. 1. Well-strained grids and spalled areas were again observed, but material removal was now possible with the single impact. It was found that the maximum lip height was observed at the side of crater.

In conclusion, the plastic strain on the impacted surface was generally larger than on the cross sectional surface of the specimen under any impact conditions. The value reached a maximum with an impact angle of 20 degrees for the aluminium specimen, and 30 degrees for the iron specimen for impact with an angular particle. This result was consistent with the mechanisms of erosion damage discussed in the previous paper (2). It was found that particle shape had a greater effect on damage to the soft material, aluminium, that than the harder material, iron.

REFERENCES
(1) Y I Oka, M Matsumura, H Funaki, Wear, Vol. 186-187, p 50, 1995.
(2) Y I Oka, H Ohnogi, T Hosokawa, M Matsumura, Wear, Vol. 203-204, p 573, 1997.

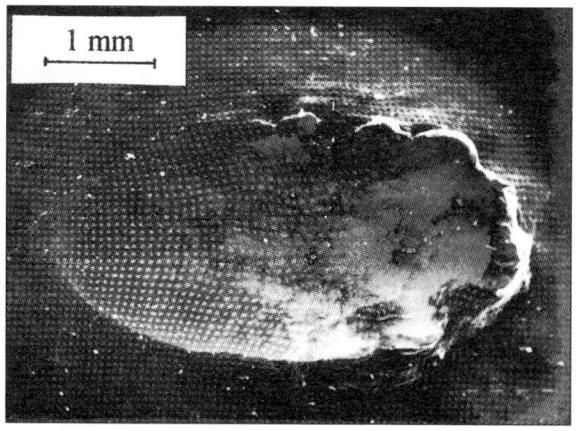

Fig.1: An impact crater by a WC ball at 100 m/s, 20 deg

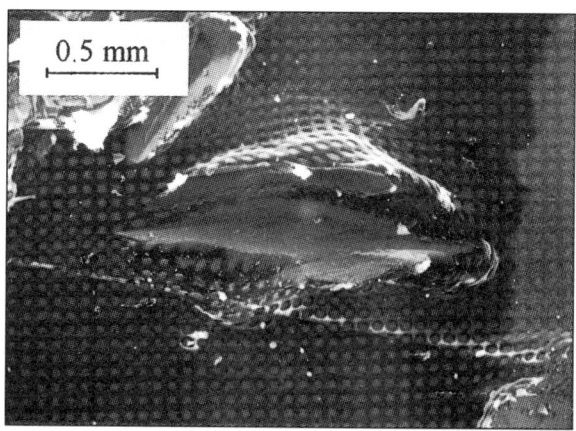

Fig. 2: A typical impact crater by a SiC particle at 100 m/s, 20 deg.

Computer Image Analysis of Wear Particles in a Combination of Space and Frequency Domain Descriptions in Three Dimensions

Z. Peng and T.B. Kirk
Department of Mechanical and Materials Engineering,
The University of Western Australia, Nedlands, WA6907, Australia.

ABSTRACT

Wear particles are produced in running machinery. The wear particle analysis (1) is now an well accepted method for machine condition monitoring, and it can provide useful information on fault diagnosis. With the development of image acquiring and analysis techniques (2) in recent years, automatic computer analysis of wear particles by both their boundary morphologies and real surface topographies is possible, and is being developed rapidly. Most of earlier attempts, however, are based on traditional shape and surface roughness parameters. It has been found that those descriptors have difficulties to distinguish some types of wear particles such as laminar, fatigue chunk and severe sliding wear particles, due to their similar boundary morphologies and complex surface topographies.

In order to identify the above three types of wear particles, another very different approach to describe shape and surface roughness is available, which is often known as spectral analysis (3)(4). In spectral analysis methods, the fast Fourier transform is a useful technique to study shapes and surface textures of wear particles in frequency domain description. However, the two-dimensional spectral technique has not been widely applied in studying surfaces of wear particles. The main reason is because the characterisations of the two-dimensional FFT series analyses techniques are not familiar to many engineers and researchers. In addition, the complexity of wear particle morphologies is another explanation. In fact, two-dimensional fast Fourier transform and the power spectral analyses of wear particle surfaces can reveal valuable information about surface textures (5). In this paper, three-dimensional wear particles have been studied by spectral analysis together with the analysis in the space domain. Three types of wear particles are investigated in this paper and the analysis results both in space and frequency domain have been discussed and applied to distinguish them.

The study has shown that the spectral technique has a potential advantage in indicating the texture direction of surfaces, while the space domain parameters can reveal morphology characteristics of wear particles. Based on above characteristics of these two methods, severe sliding particle can be easily distinguished from the others by its strong anisotropic surface texture. Laminar debris is different from fatigue chunk in that it has greater height aspect ratio (H.A.R.) and relative small R_q (see table 1). As two methods represent different means to obtain quantifiable values that can be used to describe and separate different types of wear particles, a combination of frequency and space domain descriptions appears to be the best way to study wear particles.

Table 1. Analysis of three types of wear particles

Particle Types	Rq (µm)	H.A.R	Texture Direction Index	Variations of Spectral Intensities
Severe Sliding	8.7	97.8	0.75	22.05
Laminar	10.9	102.4	0.43	2.97
Fatigue Chunk	24.5	64.6	0.51	4.01

Note:
1. R_q: Root-Mean-Square roughness of R_a;
2. H.A.R.: major dimension-to-thickness of particles;
3. Texture direction index: is the average amplitude sum divided by the amplitude sum of the texture direction;
4. Relative variation of spectral intensities: measure spectral intensities in five directions on a surface and calculate the ratio of the maximum arithmetic difference among them to the average spectral intensity.

References

(1) Naval Air Engineering Center, Wear Particle Atlas, Report NAEC-92-163 (revised), 1982.
(2) R.V. Anamalay, T.B. Kirk, D. Panzera, Numerical Descriptors for the Analysis of Wear Surface Using Laser Scanning Confocal Microscopy, Wear, 181-183 (1995), pp771-776.
(3) W.P. Dong and K.J. Stout, Two-Dimensional Fast Fourier Transform and Power Spectrum for Surface Roughness in Three Dimensions, Proc Instn Mech Engrs, Vol. 209, 1995, pp381-391.
(4) K.J. Stout, P.J. Sullivan, W.P. Dong, E. Mainsah, N. Luo, T. Mathia and H. Zahouani, The Development of Methods for the Characterisation of Roughness in 3 Dimensions, 1994 (Commission of the European Communities).
(5) Z. Peng and T.B. Kirk, Two-Dimensional Fast Fourier Transform and Power Spectrum Analysis of Wear Particles, accepted by Tribology International, 1997.

TRIBOLOGICAL INVESTIGATION OF ANTIFRICTION MATERIALS

PETKOVA D. D.
Technical University - Gabrovo, 4 H. Dimitur Street, Gabrovo 5300, Bulgaria

ABSTRACT

The contemporary level of technology development is characterized by higher demands on surface strength, which appears to the chef factor that determines reliability and durability of machine assemblies and parts.

The common relationship between friction and wear found by (2) and its concomitant phenomenon of structural adaptability, allows better use of the strength capacity of available structural materials as well as better use of operating environment components and promotes engineering of anti friction and wear resistant friction pairs.

The "friction system" pertains to overt/open thermodynamic systems which interact with external medium through exchange energy of matter. In this case the oxidizing agents will be the oxygen contained both in air and the lubricant used as well as the chemically active additives found in the lubricant (S, F, Cl, etc.) (2).

Tribological experiments were performed using a computer aided measuring unit, "Stribo"(1). It incorporates two test rigs: STI-1 for reciprocating motion and STI-2 for rotary motion. Experiments have been carried out in different modes of operation for slide bearings of various types: sliding speed v up to 2.7m/s and pressure p up to 15MPa.

Investigations of the constructive materials (steel with different C and Cr contents and bronze with Sn an Al, Fe, Ni defining different Cu contents) are presented by normal conditions of mechanical-chemical wear with and without lubrication. Investigations of cermet materials in critical wearing mode are presented by the same conditions with mineral oil M10D (cinematic viscosity by 100° C 10-18 cSt and SAE viscosity class - 30). The cermet samples are on Fe base with different contents of C=1.5%; 2.5% and Cu=2.5% and different porosity 15 and 20%. For these samples additional carburizing in vacuum was made.

All tested samples are cylindrical in shape with diameter of 11.2 mm (friction area 1 cm^2) and length of 20 mm. The type of counter body used with STI-1 is a plate with dimensions 120x60x10 mm and the counter body used with STI-2 is a disk with dimensions ⌀180x10 mm. The counter body for both rigs are made of identical materials, i.e. steel having chemical composition as follows: (0.36-0.44)% C; (0.17-0.37)% Si; (0.5-0.8)% Mn; (0.8-1.1)% Cr; 0.25% N; microhardness HRC 50 and original roughness R_a=0.63 μm.

These experimental results reveal the quantitative aspect of both friction and wear processes with regard to tribological performance parameters, such as friction coefficient, temperature values on close proximity with the friction zone and rate of wear. Studied is the structure of surface layer prior to and friction which determines the qualitative aspect of these processes.

Secondary structure of I type obtained in different operating modes, enhance wear resistance of surface layer.

This work presents the results of experimental investigation done with materials such as carbon-steel of different carbon and chrome content; bronze with tin and aluminium iron and nickel, aluminium allows, etc. Tests have been carried out in conditions of usual mechanical and chemical wear which and without lubricant. Experiments with Fe-based materials of metal ceramics, having various content of carbon and copper as well as different porosity, are quoted in friction of boundary lubrication mode and identical conditions of wear with available lubricant - mineral oil M10D.

Cermet materials feature a porous structure, about 15-35 %. Pores absorb oil and in heating they lubricate the bearing, i.e. they have a self lubricating effect. The Fe-based cermet could entirely replace a number of the available bearings of steel, bronze and aluminium alloys that are now in use.

REFERENCES

(1) Petkova D.D., "Wear Measuring by Reciprocation Sliding", 4th International Tribological Symposium INSYCONT'94, Krakov, Poland, vol. XXV, №4'94 (136), p.p. 485-491 (English)

(2) Kostetzky B.I. and team, "Surface Strength of Materials by Friction", Technica, Kiev, Ukraine, 1978, 292 p. (Russian).

OXIDATIONAL WEAR MODELLING : PART III- THE EFFECTS OF SPEED AND ELEVATED TEMPERATURES

T.F.J. QUINN

Wear Consultant, 5 Hunningham Grove, Solihull, B91 3UR

ABSTRACT

This paper is the final of a series in which complex Oxidational Wear Models are applied to a tribosystem involving the unlubricated sliding wear of high-chromium ferritic steel pins against austenitic stainless steel disks at speeds (U) between 0.23 and 3.3 m/s (a) without any external heat injection and (b) at 2 m/s with externally-induced disk temperatures (TDS) in the range 200 to 500 deg C. The applied loads (W) were also varied between 12.5 and 100N. These are the materials, speeds, temperatures and loads relevant to the exhaust valve systems of deisel engines, systems that often exhibit valve seat wear problems.

In Part I of the series (1), the speed of sliding (U) and the applied load (W) were the independent variables. The main thrust of that paper was to show that, by assuming that the Original Theory of Oxidational Wear (2) was relevant, and by measuring the oxide thicknesses on the pin and the disk in the scanning electron microscope, it was possible to compute tentative tribological oxidation constants (AP and QP) for these particular steels from associated measurements of the wear rate coefficient and the division of heat at the sliding interface. Although the computed values of the Arrhenius Constant for tribo-oxidational (AP) were consistent with the values published (3) for the static oxidation of iron, those values computed for the Tribo-oxidational Activation Energies (QP) ware all lower than the published values.

In Part II(4), some further experiments were analysed in which the disk was externally heated at 400 degC, the speed being maintained at 2 m/s and the applied loads being varied between 12.5 and 100 N. Under these conditions, one can no longer ignore the "out of contact" oxidation at the general temperature of the surface (TPS). The General Theory of Oxidational Wear (5) takes into account the oxidation of the real areas of contact (a) while they are in contact at the contact temperature (TF) and (b) while they are out-of-contact at the general surface temperature of the pin (TPS). Part II(4) gives all the details of how the General Theory was used to deduce values of tribological Activation Energies, varying from 51.459 kJ/(mol-K) at 12.5 N down to 44.491 kJ/(mol-K) at 100 N.

Table1: TPS, N, TF and QP for various Loads and Speeds

U= 0.23 m/s

W (N)	TPS (deg.C)	N	TF (deg.C)	QP (kJ/(mol-K)
12.5	35.4426	0.32723	185.855	22.962
18.75	39.7577	0.48346	193.554	22.29
25	46.7579	0.70744	209.922	21.95
37.5	46.5873	0.97868	180.361	20.952
43.75	52.0896	1.1363	176.52	20.895
50	46.9858	1.21629	126.327	20.225

U= 1.00 m/s

12.5	49.6115	6.0317	188.723	24.13
25	71.5171	12.9015	223.435	23.78
37.5	80.5251	17.254	173.256	23.28
62	116.361	28.7204	213.76	23.8
68.75	116.445	31.0659	196.285	23.49
75	156.627	35.0161	256.829	25.53

U= 2 m/s

12.5	56.2505	24.7205	255.45	24.67
31.25	114.792	65.5665	370.706	26.21
37.5	97.8856	72.0058	271.227	24.49
50	146.306	98.1426	610.932	30.16
75	301.775	208.546	714.589	34.88
81.25	308.891	210.963	698.266	35.03
87.5	338.169	250.898	776.359	36.13

U= 3.3 m/s

7	40.3503	37.5073	248.757	25.538
25	97.0031	137.702	338.599	26.474
31.25	106.95	207.503	306.529	26.05
37.5	133.477	202.519	399.228	27.672
43.75	160.088	237.436	433.06	29.031
56.25	185.51	300.07	440.399	30.04
62.5	198.455	351.154	440.416	29.05

In this paper, the wear experiments are analysed in terms of the General Theory of Oxidational Wear in the way described in Part II of the series (4). Table 1 gives a summary of the analyses obtained for the unheated experiments which formed the subject of Part I. From this table, one can see that the General Theory gives consistent values for the Tribological QP's, varying only slightly from about 21 kJ/(mol-K) at 0.23 m/s up to about 29 kJ/(mol-K) at 3.3 m/s. The magnitudes of N (the instantaneous number of contacts, TF (the temperature at the real areas of contact) and TPS (the general temperature of the pin surface) are all discussed in detail in the poster, as also are their variation with both load (W) and speed of sliding (U)

Table 2 summarises the experiments carried out at a constant speed of sliding of 2 m/s, the loads being varied from 12.5 to about 100N, while the disk temperatures were maintained, by external heating, at values of 200, 300, 400 and 500 degC. One can see that these elevated temperature experiments give consistently higher values for the Tribo-oxidational Activation Energies than those for the unheated disk experiments. Also. they do not change much with either load or disk temperature. In fact, one can deduce an average value of QP = 44.07 kj/(mol-K) with a linear spread of 4.48 kJ/(mol-K).

Table 2: TPS, N and QP for various Loads and Disk temperatures

W (N)	TPS (deg.C)	N	TF (deg.C)	QP (kJ/(mol-K)	|AveQP-QP (kJ/(mol-K)

TDS=200 deg.C

12.5	264.352	32.9383	701.855	39.858	4.20842
31.25	280.084	75.8147	659.953	37.7	6.36642
43.75	304.357	93.2197	553.903	38.298	5.76842
50	407.605	112.619	576.192	53.4	9.33358
56.25	411.347	127.437	586.005	53.1	9.03358
62.5	418.724	140.518	570.032	53.1	9.03358
75	441.185	181.58	578.621	53.69	9.62358

TDS=300 deg.C

12.5	323.528	21.8643	443.171	44.22	0.15358
25	340.875	44.777	502.221	42.11	1.95642
37.5	371.328	71.6547	638.874	42.65	1.41642
50	390.81	98.1914	634.175	41.75	2.31642
50	407.605	114.3	747.943	42.9	1.16642
56.25	390.227	110.779	674.329	41.225	2.84142
62.5	373.467	115.161	567.991	39.68	4.38642
62.5	418.724	148.222	764.073	42.45	1.61642
68.75	397.703	130.986	636.189	40.6	3.46642
75	378.206	140.809	589.6	38.9	5.16642
81.25	425.939	203.059	740.85	41.25	2.81642
93.75	419.104	208.724	640.759	40.12	3.94642

TDS=400 deg.C

12.5	429.246	28.2601	700.183	51.445	7.37858
25	439.492	50.5733	480.728	48.27	4.20358
37.5	464.057	92.8238	684.044	48.1	4.03358
50	456.535	114.916	608.55	45.378	1.31158
56.25	455.077	121.647	519.016	44.58	0.51358
68.75	470.451	165.054	597.574	44.31	0.24358
100	516.782	320.301	700.276	44.57	0.50358

TDS=500 deg.C

25	553.186	105.395	927.016	53.4	9.33358
37.5	592.189	155.384	692.222	53.815	9.74858
50	639.047	251.425	655.994	54.19	10.1236
75	637.994	403.961	694.26	51	6.93358

REFERENCES

(1) T.F.J. Quinn, Wear, 153, 179-200, 1992
(2) T.F.J. Quinn, Physical Analysis for Tribology, Cambridge University Press, Cambridge, 1991
(3) D. Caplan and M. Cohen, Corrosion Science, 6, 321, 1966
(4) T.F.J. Quinn, Wear, 175, 199-208, 1994
(5) T.F.J. Quinn, Proc. 3rd Tribology Congress (Eurotrib'81) Warsaw, September 21-22, Elsevier, Amsterdam, p.198. 1982

THE EFFECT OF PROCESSING ROUTE, CARBIDE SIZE DISTRIBUTION AND HARDNESS ON THE WEAR RESPONSE OF A CHROMIUM BEARING STEEL IN DRY SLIDING WEAR

M.R. RAMADAN
Sirte Oil Company, PO Box 385, Tripoli, Libya
D.N. HANLON, W.M. RAINFORTH
Department of Engineering Materials, The University of Sheffield, Mappin St. S1 3JD, UK.

ABSTRACT

The commercial attractiveness of rolling mill work rolls with greater resistance to wear and fatigue damage is substantial, providing reduced rolling mill down times, increased roll life, reduced product surface defects and improved stock gauge tolerances. Relatively little is known about the wear mechanisms which limit the useful life of rolling mill work rolls[1,2].

In this work, a steel of composition 0.8%C/5%Cr, representative of industry standard cold mill work roll materials, has been manufactured by conventional processing (casting and forging) and by spray forming. Conventionally processed material was subject to both heavy and light forging reductions. After heat treatment the microstructure of all materials consisted of a martensitic matrix containing a dispersion of $(Fe,Cr)_7C_3$ carbides[3].

Spray forming provided a microstructure similar to that of the heavily forged material but with a finer average carbide size (table 1). Furthermore, the spray formed material did not contain the occasional coarse carbides (~2μm) found in the forged material (Table 1)

	Carbide Size (μm)	
Material	Mean	Maximum
Spray Cast	0.128	0.5
Heavily Forged	0.208	0.9
Lightly Forged	0.444	2.0

Table 1: Mean and maximum carbide sizes for materials in all processing conditions.

Comparison of the spray formed and conventionally processed materials thus enabled the effect of the size and distribution of $(FeCr)_7C_3$ carbides on wear to be investigated. Tempering of the quenched steel was undertaken in the temperature range 200-500°C, which yielded a large range of hardness from 470-780H_{V30}. Wear testing was undertaken in pure sliding using a block on ring configuration, against high speed steel (M2) counterfaces at 1m/s and loads of 154N and 304N.

The wear coefficient was approximately constant as a function of initial hardness for both loads tested (figure 1). However, the average wear coefficient was lower for the spray formed material (k=1.38x10^{-5} mm^3/Nm) compared to the heavily forged material (k=1.72x10^{-5} mm^3/Nm) and the lightly forged material (k=1.97x10^{-5} mm^3/Nm) when tested at 304N. This demonstrated that a finer carbide size imparts superior wear resistance. The friction coefficient was 0.49-0.63 irrespective of load or material temper.

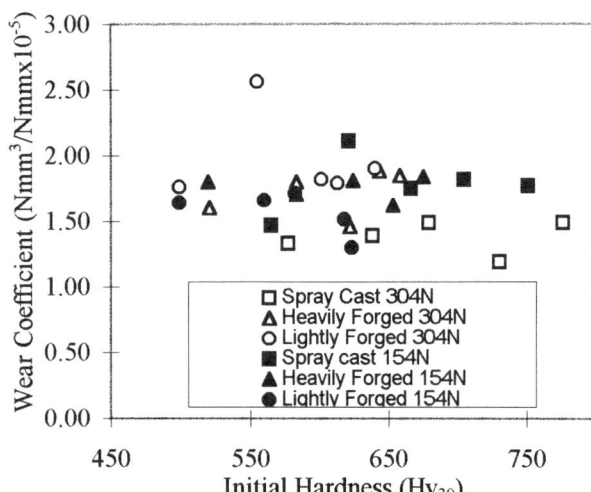

Figure 1: The variation of wear coefficient with initial specimen hardness for materials in all processing conditions.

Observation of the wear debris indicated that wear was predominantly by delamination and oxidation (wear debris consisted entirely of oxide and metallic plates). Characterisation of worn surfaces indicated that surface strain was limited to a depth of ~5-20μm. The absence of a difference in wear behaviour as a function of hardness was shown to be a result of heating effects, which tempered the martensitic matrix and led to re-solution of carbide particles in the near surface region.

REFERENCES

(1) S. Spuzic, K.N. Strafford, C. Subramanian and G. Savage, *Wear*, (1994), **176**, p261
(2) R. Price, T. Gisborne, D.N. Hanlon, W.M. Rainforth and C.M. Sellars, "*Conf. Proc. Rolls 2000*" The Institute of Materials London, (1996), p250
(3) D.N. Hanlon and W.M. Rainforth, *Wear*, (1997), in press.

TRIBOCHEMICAL REACTIONS OF Fe AND Cu ALLOYS + GLICERINE SYSTEM

J.SENATORSKI and R.MARCZAK
Institute of Precision Mechanics, Duchnicka Str. 3, 00 967 Warsaw, Poland
A.KUZHAROV
Don State Technical University, Gagarin Sq.1, 344700 Rostov-on-Don, Russia

ABSTRACT

In the work of A.Kuzharov and co-authors (1), it was indicated that the appearance of self-organisation in a tribological system (thermodynamically open) is bounded with the oscillatory character of the course of tribochemical reactions in the friction area. Their external symptoms are oscillatory changes of values of different physico-chemical characteristics of tribological systems, and observed - in these conditions - anomalously low values of the frictional resistance and wear intensity. Representation of there analysis is based on simple, linear models of reaction courses of the type:

$$\xrightarrow{\alpha} A \underset{k_2}{\overset{k_1}{\rightleftarrows}} B \underset{k_4}{\overset{k_3}{\rightleftarrows}} C \xrightarrow{\gamma}$$

where:
- k_1, k_2, k_3, k_4 - coefficients,
- α - intensity coefficient of an entrance of the substrate A to the system,
- γ - intensity coefficient of an exit of the product C from the system.

It was proved that parameters k_i of the above-mentioned tribological reactions, and also a level of „openness" of the tribological system (expressed by proportion of α and γ coefficients informing on the flow, in a contact zone, of mass flux with an intensity that depends on the value of external inputs p, v, T), lead to theoretical solution indicating the possibility of occurrences of oscillatory changes of the A, B and C component concentration in a tribological reaction versus time.

One should expect a similar course of characteristics of other friction - coefficient of friction and wear intensity.

In the fifties, D.N. Garkunov observed that friction parts of refrigerating compressors were worn out at an anomalously low level. Analysis of a mechanism of that effect, carried out together with I.V. Kragelsky, showed that copper atoms were a thin leyer of particular properties was formed. These scientists obtained a similar, low-frictional layer in a frictional pair steel-copper alloys lubricated with glycerine (2). This effect is known in the literature as:
- Garkunov's phenomenon,
- selective transfer phenomenon,
- no-wear friction effect.

That last term is connected with the fact that, in some cases, the wear intensity can diminish even 1000 times! This effect can occur in specific conditions, when fluid friction does not exist in the frictional pair, however the value of friction coefficient is comparable with fluid friction coefficient, and the wear intensity diminishes by 2÷3 orders.

An exploitational surface layer is formed as a result of the running-in process. It is a comlex process but it is possible to distinguish two simple processes which occur simultaneously, interact mutually and overlap. These are:
- the process of mutual fitting of collaborating surfaces through plastic strains and abrasion,
- the formation process of a layer of oxides or other metallic compounds and a monomolecular layer of chemisorbed compounds on the working surfaces of frictional bodies.

The transformation of a technological surface layer into an exploitational surface layer occurs in friction zone under the influence of external forces with the participation of active components of a lubricating material.

The essence of this transformations is a stable bonding of organic compounds with metal (directly or, for example, throughout oxides).

Some concluding remarks summarize the presented results:
- An appearance of self-organisation in the state of no-wear friction can be stated on the ground of transition kinetics of the investigated system. An oscillatory character of the changes of wear intensity in the functions of time is related with a mechanism of the mass transfer process on conditions of selforganisation.
- On the ground of the obtained results, one should conclude that in the stage of no-wear friction chemical or electrochemical processes dominate.

REFERENCES

(1) M. Grzywaczewski, A. Kuzharov, W. Mirosznikov, ZEM (in Polish), No 3-4, 1995.
(2) D.N. Garkunov, I.V. Kragelsky, A. Poljakov, Selective transfer in friction pairs (in Russian), Moscow, 1969.

FRACTAL TRANSITION MODEL IN PREDITING STATIC FRICTION COEFFICIENT

X.Y. SHENG, J.B. LUO, S.Z. WEN

National Tribology Laboratory, Department of Precision Instrument, Tsinghua University, Beijing, 100084,China

ABSTRACT

It has been demonstrated that there is a fractal characteristic in most engineering surfaces. Considering Weierstrass-Mandelbrot (W-M) function as the expression of surface profile, Majumdar and Bhushan have proposed a factal model of elastic-plastic contact. In present work, the relationships between static friction force and fractal parameter, material properties have been gotten by using a simplified stress field on the surface at the back edge of the contact boundary and M-B fractal contact model. The static friction coefficient is calculated. The predicted static friction coefficient increases with the increase of normal load. This means that static friction coefficient is low under small normal load condition and coincides with the experimental results on Scanning Probe Microscope (SPM) and Lateral Force Microscope (LFM), Atom Force Microscope(AFM). Effects of the surface fractal parameter D, G, and the material parameter φ on the static friction coefficient are discussed. When the fractal dimension D is between 1.4-1.6, the static friction coefficient increases with the increase of D. When the fractal dimension D is between 1.6-1.9, the static friction coefficient decreases with the increase of D. The static friction coefficient decreases with the increase of G and Φ.

A fractal transition model is proposed which takes into account the change of fractal dimension D. In the calculation of contact load or real contact area, what we really concern is the real contact part, not the whole surface. But the topography of the real contact part is changed during the two surfaces are in contact condition. The changed parameter should be used in the calculation. With the increase of load, the factual dimension of the real contact part will become smaller with a limit of 1 for the surface profile and 2 for the surface. The predicted load-contact area relationship based on fractal transition model is not simply an exponential function as that of proposed by Majumdar-Bhushan[1][2] and Greenwood-Williamson[3]. The prediction correlates better with the experimental results than the G-W and M-B contact model. The predicted static friction coefficient based on fractal transition model increases with the increase of load, and then decreases with the increase of load. The transition point is near D=1.5. This can explain that the static friction coefficient is very small under the small load and the decreases with the increase of load under higher load condition.

REFERENCES

(1) A. Majumdar and B. Bhushan, ASME Journal of Tribology, 112(1990)205-216.
(2) A. Majumdar and B. Bhushan, Journal of Tribology, 113(1991)1-11.
(3) J.A. Greenwood and J.B.P. Williamson, Proc. R. Soc. London, Ser. A, 295(1966)300-319.

WELDING WAVES ON THE SURFACE CONTACT SPOTS OF SOLIDS

A.A. SHTERTSER
Design & Technology Institute of High-Rate Hydrodynamics,
Tereshkovoi 29, Novosibirsk, 630090, Russia

ABSTRACT

The seizure phenomenon is caused by welding of solid bodies over contacting area during friction or joint deformation. To avoid seizure in most cases lubricants are used in tribology systems. At the same time seizure phenomenon is used in technologies of metal joining by friction, roll and pressure welding. Thus it's very important to understand the mechanisms controlling the process of welding.

Perhaps everybody in childhood watched on the process of mercury balls coalescence. The moving force of the process is surface tension of material caused by existence of surface free energy. Coalescence occurs very quickly as mercury has small shear strength and viscosity. For all solids which have positive surface free energy bonding over the contact zone is thermodynamically justified. And if there are not surface layers (lubricants, oxides, hydroxides, absorbed moleculars) preventing direct contact between surface atoms their adhesion is quite possible. But in presence of material strength (low mobility of atoms) the bonding process is usually suppressed. This obstacle (strength) is overcome by action of high compressive and shear stresses appearing at contact zones during deformation or friction.

It is shown in considerations below that growth of welded area can result from propagation of welding wave (WW). The driving force of this process is surface tension (existence of positive surface free energy).

Let us consider two-dimensional situation and suppose that two pieces (parts 1 and 2) of the same material are subjected to direct contact along their plane surfaces (see fig.). Part 1 and part 2 lay at Y>0 and at Y<0 correspondingly. The plane XZ is the contact area of materials. According to modern welding theories first acts of bonding occur at "active" points (1). Let the bonding appears at "active" point X=0. In further considerations we suppose that material flow can be described by hydrodynamic equations for viscous liquid. This approach is often used in mechanics of high loading. And it permitted for example to describe behaviour of thin surface layers (lubricants, absorbed moleculars etc.) under compressive stresses (2). The usage of hydrodynamic equations including surface free energy shows that the event of bonding at X=0 initiates the propagation of two WW moving on opposite directions (points A and B in the fig.).

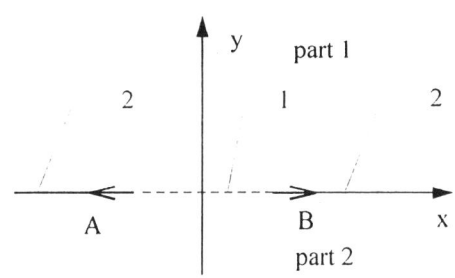

Fig. Broadening of bonded area by WW propagation.
1-bonded area (A≤ x ≤B); 2-initial unbonded area (x<A and x>B); Points A and B are moving at velocity U_{ww}.

For isothermal process the WW velocity is:
$$U_{ww} = -(\eta/4d\rho) + [(\eta/4d\rho)^2 + E/\rho]^{1/2}$$
where η- viscosity, d- interatomic distance, ρ- density, E- Young's modulus of material. When $\eta \rightarrow 0$ then U_{ww} tends to it's maximal value which is the sonic velocity of material:
$$U_{ww} = [E/\rho]^{1/2}$$
Actually in solid state all metals have very high values of η. In this case we have:
$$U_{ww} = 4\theta\gamma/\eta = 2Ed/\eta$$
where γ is surface free energy for which there exist theoretical formula $\gamma = Ed/20$. The data from (3) permits to estimate η for different deformation velocities ϵ'. For aluminium it gives $\eta \sim 8.9 \cdot 10^6$ Pa·s at $\epsilon' \sim 1$ s^{-1} and $\eta \sim 7.8 \cdot 10^5$ Pa·s at $\epsilon' \sim 10$ s^{-1}. This yields $U_{ww} \sim 4.3$ μm/s and $U_{ww} \sim 48.7$ μm/s for these two values of ϵ' respectively. The seizure time depends on U_{ww} and on surface density of active points.

REFERENCES

(1) Yu. L. Krasulin. Vzsaimodeistvije Metalla s Poluprovodnokom v Tverdoi Faze, Moscow, Nauka, 1971.
(2) A.A. Shtertser. Combustion, Explosion, and Shock Waves, Plenum Publ. Corp., vol.31, No.6 (1995),P.711-714.
(3) H.J.Frost, M.F.Ashby. Deformation-mechanism Maps, Cheljabinsk, Metallurgija, 1989

The effect of Wear Debris as a Third Body in the Wear Behavior of Al_2O_3-Al-SiC (DIMOX) CMC against SiC

R. ARVIND SINGH*, V. JAYARAM** AND S.K. BISWAS*
* - Dept. of Mechanical Engg., **- Dept. of Metallurgy, Indian Institute of Science, Bangalore - 560 012, INDIA.

ABSTRACT

We report here the remarkable abrasion wear resistance (normal pressure: 2.92 to 5.42 MPa, sliding speed: 0.25 to 3 m/s) of a soft (relative to other structural ceramics) melt oxidized ceramic matrix composite Al_2O_3-SiC-(Al,Si) **(1)**. The wear studies were conducted with and without removing the wear debris, against a glass bonded SiC abrasive disc to simulate the two body and three body wear situations. The wear behavior of this composite has also been compared with that of Zirconia toughened Alumina (ZTA, Al_2O_3-4wt.% $ZrO2$) a common cutting tool material, under identical conditions.

The observed wear mechanisms in ZTA were microabrasion and surface fracture. The trends in wear rate with sliding velocity and normal load were similar to that observed by others **(2)**. In the case of composite, the wear is by microabrasion and delamination. At sliding speeds greater than 2 m/s its performance is comparable to that of ZTA (Fig 1). The following explanations are suggested for this behaviour.

(1) All worn surfaces of the composite were found covered by a tribofilm rich in metal (Al). The presence of this metal phase in the film increases slightly with increase in velocity. With increased sliding velocity, the co-efficient of friction is reduced resulting in a lowering of the surface tractions and subsurface stresses that are responsible for delamination.

(2) The thermal shock resistance of a material is given by, $R = \sigma(1-\mu)\lambda/\alpha E$.

σ: Fracture stress, μ: Poisson's ratio, λ: Thermal conductivity, α: Thermal expansion coefficient and E: Young's modulus. (Volume averaged rule of mixtures was used to calculate R).

The composite having an estimated R of 5450W/mK as compared to 2220 W/mK for ZTA, is expected to be more resistant to surface fracture and thermal shock than ZTA.

(3) When abrading the composite pins, the SiC counterface is smeared with the debris which is more prominent in the two body case than in the three body interaction. In the case of two body abrasion, the absence of debris brings the pin surface in contact with SiC grit (counterface), resulting in increased wear rate (Fig. 1). In the wear process, the SiC grits get blunted and faceted. This phenomenon is prominent in the two body case resulting in the reduction of friction coefficient.

The wear rate of the composite becomes comparable to that of ZTA in the test speed range greater than 2m/s. The ability of the composite to generate a metal rich tribofilm when rubbed and its high thermal shock resistance are responsible for its unique performance.

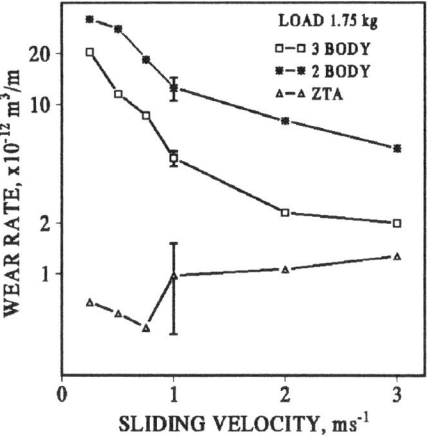

Fig.1 Wear Rate vs. Sliding Velocity

REFERENCES

1. Vikram Jayaram, Rampada Manna, Manjunath G. Kshetrapal, Jaydeep Sarkar and Sanjay. K. Biswas, J. Am. Ceram. Soc., 79(3) 770-72, 1996.
2. J. Denape and J. Lamon, J. Mater. Sci., 25, 3592 - 3604, 1990.

ON LINE TRIBOSCOPIC FRICTION AND WEAR MONITORING OF AL-BASED, QUASICRYSTALLINE MATERIAL IN BALL ON FLAT CONTACT GEOMETRY

J.M. SORO AND J. VON STEBUT
Laboratoire de Science et Génie des Surfaces, URA CNRS 1402, INPL-École des Mines 54042, Nancy, France

ABSTRACT

Friction force is often, erroneously, taken as an indicator of absolute contact severity, sometimes even without any final wear volume measurement. Because of continuously changing contact conditions during a real wear experiment (real contact area modification, material transfer, debris generation etc.), the subject of more reliable on-line wear assessment has been addressed in various studies (1, 2, 3).

Specimens (pure AlSiCuFe quasicrystalline and 1/1 approximant AlSiCuFe), 30mm in diameter are prepared by powder metallurgy. They are ground flat and polished to $R_A = 0.1 \mu m$. Vickes surface hardness is respectively 8.5 GPa and 10 GPa.

Friction and wear are studied by means of our triboscopic friction rig (2) under 3 GPa of contact pressure. Friction and system "topography" (pin/flat distance) are monitored on-line as a function of dry sliding distance and pass number up to 400 unidirectional cycles (sliding speed : 1mm/s). $\varnothing$1.58mm cemented carbide and AISI 52100 ball bearing steel balls were chosen as pins. "Topography" modifications are tracked by a stylus profilometer pick-up in horseback position on top of the rider. Periodically the rider is placed at a reference position outside of the wear track, for rider wear and/or material pickup from the flat.

Virgin friction (extending over roughly 10 cycles) is low ($\mu \sim 0.1$) for all of the four ball/flat combinations. It is not clear whether this is specimen specific or just an initial surface contamination artefact. Only for the combination steel ball/AlSiCuFe flat did this transition regime show an incubation phase lasting over 50 passes of dry sliding under controlled laboratory conditions (22°C, 50% RH).

For all combinations the *permanent friction regime* is very unstable. For both types of specimens it is around $\mu = 0.2$ for the WC ball riders and roughly $\mu = 0.3$ for the steel ball.

Analysis of the on-line, triboscopic data gives the following information :
a.) *ball modifications* :
-For the WC rider there is indication of transfer layer formation of fluctuating thickness (1 - 2μm).
-For the steel ball rider there is considerable dispersion with an average of zero apparent wear..
b.) *system topography changes (system wear)* :
- For the cemented carbide ball no perceptible modifications are observed for either specimen.
- For the steel ball rider on the pure quasicrystal there is a decrease in system wear distance (final value: $\Delta Z=-4\mu m$) building up continuously over the entire experiment. For the steel rider sliding on the approximant flat, topography remains unchanged during the incubation phase of friction, thus showing the beneficial effect of this situation.

Off line surface damage analysis
This final "inspection" by means standard metallurgical microscopy and scanning stylus profilometry (SSP) both for the wear track and for the WC ball slider. SSP is indispensable for final validation of the on-line information.

Ball surface modification :
-For the cemented carbide riders. SSP shows the absence of wear. A slight, flat transfer layer is observed for the approximant specimen.
-For the steel ball riders we observe pure truncation when sliding on the pure quasicrystal while both truncation and transfer coexist for the approximant.

Wear tracks :
Wear depths assessed by SSP are on the order of the on-line information (~2μm). The instantaneous contact pressure, is in general reduced as ball wear and material pick-up reduce the effective contact pressure.

Specimen brittleness is evidenced on micrographs of the wear tracks after 400 cycles. WC riders produce brittle chipping for both specimens. The severity is attenuated for the approximant phase. This should be due to contact pressure reduction owing to build up of a roughly flat transfer layer. The mechanism of reduced contact pressure is active for the steel riders, with brittle damage reduced to tensile type, hertzian cracks.

In conclusion : on-line triboscopic information, while giving useful information, well beyond non specific friction shoud always be validated by off-line wear volume and damage analysis if useful information for the tribologist and the surface engineer is to be gained.

REFERENCES

(1) M. Ouadou, T.G. Mathia, B.Clavaud, R.Longeray, P.Lanteri in "Mechanical Identification of Composites", H. Vautrin & H. Sol Eds. Elsevier Applied Science, p303
(2) A. Darbeïda, J. von Stebut, M. Assoul, J. Mignot, J. Mech. Tools Manufact. 35, (1995), 177
(3) M.G. Gee, this conference, session TH2
(4) J. von Stebut, J.M. Soro, Ph. Plaindoux, J.M Dubois in "New Horizons in Quasicrystals, A.I. Goldman, D.J Sordelet, P.A. Thiel, and J.M. Dubois, World Scientific 1997, p.248

Submitted to *Surface Engineering*

KINETICS OF WEAR-FATIGUE DAMAGE AND ITS PREDICTION

L A SOSNOVSKIY
Scientific and Industrial Group "TRIBOFATIGUE", P O Box 24, Gomel, 246050, Republic Belarus
A V BOGDANOVICH
Scientific Center of Machines Mechanics Problems of the Belorussian Academy of Sciences,
Skoriny Avenue, 12, 220072, Minsk, Republic Belarus

ABSTRACT

For elaborating a theory of wear-fatigue damage accumulation the following main principles are accepted (1)(2)(3):

1. The emergence and growth of wear-fatigue damage is determined, generally, by four phenomena: mechanical fatigue, friction and wear, thermal and corrosion processes.

2. All these processes are interconnected and a limiting state of material in a general case is determined, consequencely, due to not only one phenomenon but their joint action with each other.

3. The kinetics of wear-fatigue damage is conditioned not by the whole energy U given but only its effective part

$$U_* = \lambda U, \quad \lambda \ll 1.$$

4. The criterion of fracture of material is a condition when effective energy can reach the limiting value U_o - energy of breaking of interatomic bonds:

$$U_* = U_o.$$

5. The energy U_o is considered a fundamental characteristics for a given substance. It does not depend on test conditions and damage mechanizms.

6. The effective energy is a function of four variables: frictional U_p, mechanical U_σ, thermal U_T and electrochemical U_e energies:

$$U_* = F(U_p, U_\sigma, U_T, U_e).$$

7. The Limiting State of one of the elements of a cycle load-bearing system appears when in its dangerous volume an effective energy can reach the critical value.

8. The dangerous volume is determined by the conditions of contact interaction of elements of a cycle load-bearing system, physico-mechanical properties of material, loading regime, a number of constructional, technological and operation factors.

9. Wear-fatigue damage accumulation is a non-linear function of time in a general case.

10. If

$$\omega_T = f_1(U_T/U_o), \quad \omega_\sigma = f_2(U_\sigma/U_o),$$
$$\omega_p = f_3(U_p/U_o), \quad \omega_e = f_4(U_e/U_o)$$

are relative damage measures determined by thermal processes (index "T"), mechanical fatigue (index "σ"), friction and wear processes (index "p") and electrochemical phenomena (index "e"), the hypothesis of limiting state may be submitted by equation:

$$\varphi(\omega_T, \omega_\sigma, \omega_p, \omega_e, t) = 1,$$

were t is a test time.

The distribution function $F(X)$ of $X=N_{cal}/N_{exp}$ value is plotted in Fig. 1, where N_{cal} and N_{exp} are calculated and experimental life under fretting-fatigue.

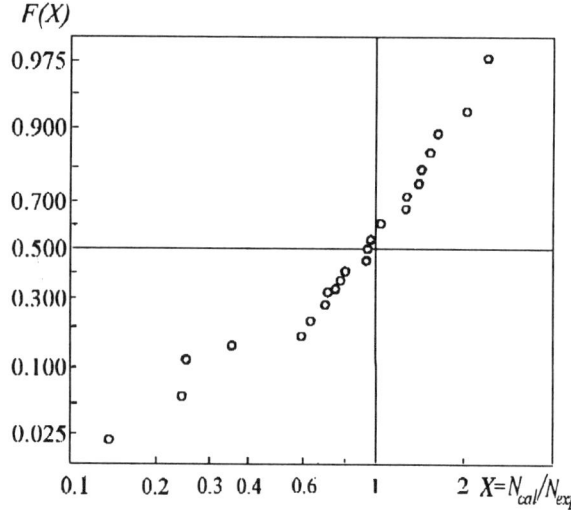

Fig. 1: Distribution function $F(X)$ of $X=N_{cal}/N_{exp}$ value

The paper presents a cumulative wear-fatigue damage model based on the principles mentioned above. The recommendations on the parameters estimation of this model are given. The experimental results of its testing for some loading conditions are discussed.

REFERENCES

(1) L A Sosnovskiy, Statistic Mechanics of Fatigue Fracture, Minsk, 1987, 288p.
(2) L A Sosnovskiy, News of Belorussian Academy of Sciences. Series of the Phys.-Tech. Sci., No 4, 1991, 87-92pp.
(3) L A Sosnovskiy, N A Makhutov, Strength of Materials No 1, 1993, 11-23pp., No 3, 1993, 17-29pp.

SLIDING-MECHANICAL FATIGUE: DIRECT AND BACK EFFECTS

L A SOSNOVSKIY
Scientific and Industrial Group "TRIBOFATIGUE", P O Box 24, Gomel, 246050, Republic Belarus
A V MARCHENKO
Belorussian State University of Transport, Kirova Street, 34, 246022, Gomel, Republic Belarus

ABSTRACT

The influence of friction and wear processes on fatigue resistance of materials and cycle load-bearing systems elements is called direct effect (1)(2).

The results of wear-fatigue tests of Carbon 45 steel (specimen) – to-glass-filled polyamide ПА66КС (counterspecimen) are shown in Fig. 1. The tests were carried out on SI type machine. The perculiarity of these experiments is that during sliding friction of polymer against steel physical wear of a steel specimen is not detected. Nevertheless, the life and limiting stress σ_{-1} for steel specimen decrease with a rise of contact presure p.

Direct effect can be described by equation:

$$\sigma_{-1p} = \sigma_{-1}\left[1 - K(p,T,S)\right]^{1/m_v},$$

where $K(p,T,S)$ is function of influence of friction and wear processes and m_v is its parameter.

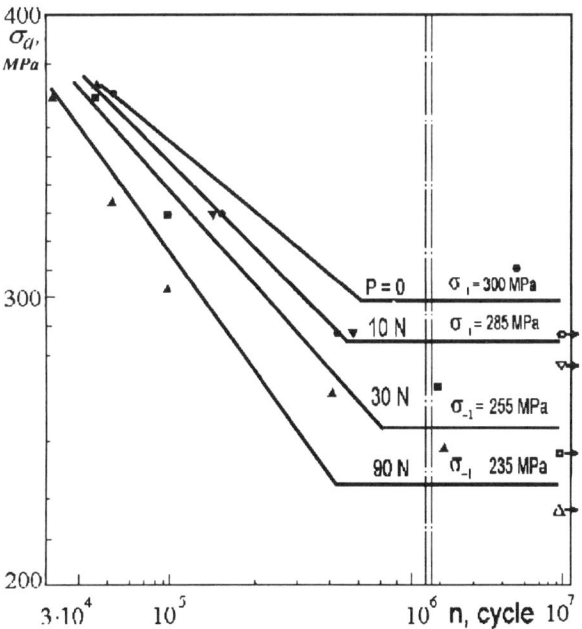

Fig. 1: Fatigue curve of steel specimens drawn on contact pressure. Parameter p under sliding friction

The influence of cyclic stresses on the characterictics change during friction and wear processes is called back effect in tribo-fatigue (1)(2).

Fig. 2 shows the influence of stress amplitude in a steel specimen on wear intensity of polimeric counterspecimen. It appears to be marked degree. If cyclic stresses increase from 170 to 300 MPa than resulted wear intensity increments from 110 to 180 per cent (comparing to one in a simple friction unit when $\sigma_a = 0$) (3).

Back effect can be described by equation:

$$I_\sigma = \frac{I}{\left[1 - K(\sigma,T,V)\right]},$$

where $K(\sigma,T,V)$ is function of influence of mechanical fatigue processes, I and I_σ are wear intensity for $\sigma_a = 0$ and $\sigma_a = const$ correspondingly.

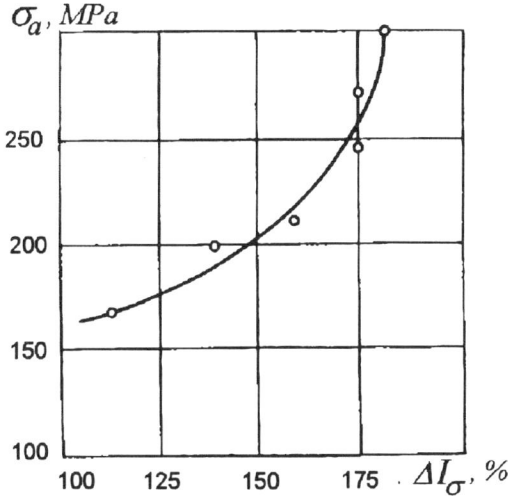

Fig. 2: Increment of wear intensity of the polimeric counterspecimen from stress amplitude in the metallic specimen (Carbon 45 steel) at $p = 5.7$ MPa

The paper gives detailed information on the conditions and tests results and the analysis of the main regularities of direct and back effects during sliding-mechanical fatigue.

REFERENCES

(1) Belorussian СТБ 994-95 Standard "Tribo-fatigue. Terms and Definitions", Minsk, 1995, 98p.
(2) L A Sosnovskiy, N A Makhutov, V A Shurinov, Plant Laboratory, No 9, 1992, 49-66pp.
(3) L A Sosnovskiy, Reliability and Lifetime of Machines and Constructions, No 9, 1986, 93-102pp.

A DISCUSSION ON THE WEAR MECHANISM OF DIAMOND TOOL

MASAO UEMURA

Department of Mechanical Engineering, Toyohashi University of Technology Toyohashi, 441, Japan

ABSTRACT

The diamond which is in contact with iron at high temperature transforms into graphite (1) and carbon atoms of the diamond diffuse into iron (2). Therefore, the graphtization of diamond would play an important role on the wear of the diamond. The surface of a diamond is quite stable because of the terminated hydrogen and the desorption of hydrogen is known as a cause of graphtization (3). We apply these knowledge of the diamond surface to understanding of the wear of diamond tool cutting Fe, Ni, Co, Cu, Ti and Al.

It is possible that the oxidation of a clean surface of work materials was the first to occur and then the formed oxide reduces the C-H bonds of the diamond surface to carbon. Because the standard free energy of chemical adsorption of hydrogen on a diamond surface was not clear, we estimated the reduction of the C-H bond for the oxidation of methane by the oxides. In this case, the relation between the equilibrium partial pressure of H_2O and CH_4 can be showed as follow.

$$\{P(H_2O)^2 / P(CH_4)^2\} = 10^{-\Delta G_T^0 / 19.14}$$

where, $P(H_2O)$ and $P(CH_4)$ are the equilibrium partial pressure of H_2O and CH_4. ΔG_T^0 is the standard free energy change of the reaction. The calculated equilibrium pressure of H_2O stood in the order of Cu > Ni > Co > Fe ≫ Ti > Al and it was 1×10^3 Pa for FeO and 1×10^{-8} Pa for TiO_2 at 700°C for a pressure of 10^5 Pa of CH_4. Because the pressure of H_2O of 1×10^3 Pa was real, the surface layer of diamond which was cutting Cu, Ni, Co or Fe would transform into graphite. Copper, however, has no solubility of carbon and no ability to make carbides (4). The wear rate of the diamond cutting pure copper or soft copper alloy would be extremely low, because the graphite film acts as a solid lubricant.

In case of Fe, Ni or Co, the diamonds would wear by the diffusion of the carbon into the work. To make sure this wear mechanism, we calculate the wear rate of the diamond using the following equation.

$$Z = \frac{L}{S}\left(\frac{A_{dD}}{A_{dM}}\right)^3 \int_0^\infty C_0\left(1 - \frac{2}{\sqrt{\pi}}\int_0^{\frac{x}{2\sqrt{Dt}}} e^{-y^2} dy\right) dx$$

where, Z is the wear depth, L is the cutting distance, S is the contact length, t is the diffusion time (t=S/V), V is the cutting speed, C_0 is the solubility limit of carbon, D is the diffusion coefficient, A_{dD} and A_{dM} are the neighboring atomic distances of the diamond and the metal. The calculated wear rate agrees with the experimental results (2) for the single diamond grain grinding tests, if the temperatures at cutting interface were 560°C for Fe and 600°C for Ni.

REFERENCES

(1) C. Phaal, Ind. Dia. Rev., Vol. 25, 486-489pp, 1965.
(2) N. Ikawa and T. Tanaka, J. Japan Soc. of Precision Eng., Vol. 37, 824-828pp, 1971.
(3) B. B. Pate, Surface Science, Vol. 165, 83-142, 1986.
(4) C. J. Smithells, Metals Reference Book, Butterworths, 1996, pages 204 and 511.

CONTACT AND THERMAL ANALYSIS OF THE WEAR PROCESS IN LINEAR BEARING

KÁROLY VÁRADI - ZOLTÁN NÉDER - TIBOR BERCSEY
Institute of Machine Design, Technical University of Budapest, H-1521 Budapest, Hungary
WALDEMAR STEINHILPER
Institute of Machine Element Design and Gear Technics, University of Kaiserslautern, Germany

ABSTRACT

Linear bearings are frequently used to transfer load under linear alternating motion. The sliding motion and the load transfer produce frictional energy loss and in this way heat generation. To reduce friction and wear lubricants are used to create boundary lubrication.

The aim of this analysis is to study the wear process, to evaluate the contact and thermal parameters for the original and also for the worn surfaces, based on the measured surface roughness data for different cases.

In the present analysis due to the slow sliding a "near dry" lubrication is assumed. In this case Hertz type contact may be expected and the EHL effect may be ignored.

The developed contact algorithm (1), (2) can simulate the elastic and in an approximate way, the elastic-plastic sliding contact behaviour in the vicinity of the asperities by ignoring the effect of the tangential forces on the vertical displacement.

During the thermal analysis total sliding is assumed over the real contact areas. The frictional energy may be calculated if the contact pressure distribution, the sliding speed and the coefficient of friction are known. The friction energy is converted into heat sources, that is partitioned between the two contacting bodies.

At first the contact parameters (real contact areas and contact pressure distributions) between the measured surfaces of the bronze linear bearing and a steel counterpart having smooth surface were evaluated. There are basic differences in the ratios of the real contact areas over nominal contact area representing the benefit of the worn (run-in) surfaces.

At second both elements, being in sliding contact, have their real surfaces. The contact problem was solved at several sliding positions while the upper surface was slid over the bottom one. In Figs 1a and 1b the locations of the real contact areas are shown for the original and the worn surface pairs at a certain sliding position.

For the worn surfaces the real contact areas are "more circular" than for the original sliding surfaces. Regardless the finer surface of the linear bearing during the wear process the real contact areas follows the surface pattern of the steel counterpart, because the surface of the counterpart practically did not changed. If the counterpart had finer surface the contact pressure distribution would dominantly be elastic one and it would produce less wear and lower local heat generation.

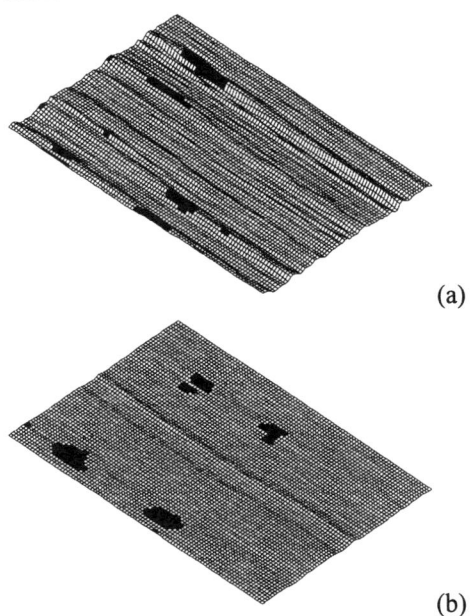

Fig. 1: The real contact areas between the original (a) and worn (b) sliding surfaces at a certain sliding position

ACKNOWLEDGEMENT

The authors are indebted to Dipl.-Ing. U. May at Institute of Machine Element Design and Gear Technics, University of Kaiserslautern for the surface roughness measurements.

The presented research task was sponsored by the Hungarian National Scientific Research Foundation (OTKA T023351) and was part of the German-Hungarian scientific research co-operation (TET 57/96).

REFERENCES

(1) K. Váradi, Z. Néder, Tribologia, 3-1996, 237-261
(2) K. Váradi, Z. Néder, Third Biennial Joint Conference on Engineering Systems Design & Analysis, Montpellier, France, July 1-4 1996.

FULL PUBLICATION

K. Váradi, Z. Néder, T. Bercsey, W. Steinhilper, Tribology Transactions, (preprint at the 1997 World Tribology Congress)

UNSTATIONARY TRIBOMECHANISMS; A SURVEY ON MATERIAL DETERIORATION MECHANISMS OCCURRING ALTERNATIVELY, SIMULTANEOUSLY AND IN PARALLEL

TOPI VOLKOV, MATTI SÄYNÄTJOKI and PETER ANDERSSON

Technical Research Centre of Finland (VTT), VTT Manufacturing Technology, P.O. Box 1702, 02044 VTT, Finland

ABSTRACT

A modern piece of moving mechanical assembly basically consists of a magnitude of subsystems, each designed for a specific purpose. Included in the function or mission of a component is almost always a requirement to withstand various kinds of physical loads or chemical attacks. In ideal cases, when the requirements due to function and environment are clearly identifiable, classical rules can be applied in the design of a machine part. In frequent cases, typically related to machinery optimized for high performance, the true spectrum of physical and chemical requirements acting on the respective components and subsystems is complex, and the response of the material of the components is more or less problematic to predict.

The problem of prognostication of a structural material under a spectrum of physical and chemical requirements arises from a lack of knowledge in the combined effect of what in our present understanding are related to as specific material deterioration mechanisms. Such specific and fairly well documented mechanisms are, for instance, known as wear mechanisms, fatique crack formation mechanisms, corrosion modes, creep and brittle fracture. Certain material deterioration mechanisms that represent the borders between the main groups of mechanisms, have been as well widely documentated, with tribochemical wear and stress corrosion as representative examples.

The most unpredictable situation, however, arises when more than one material deterioration mechanism is able to act on the same surface or volume of material. In such cases the mechanisms may alternate, each one leaving their specific traces and causing a new starting point for the following mechanism to act. An example of this is wear and corrosion occurring alternatively on a steel surface.

Two or more material deterioration mechanisms, phenomenologically treated as separate ones, may act simultaneously causing a combined damage to the surface or bulk of the component. An example of material deterioration mechanisms with possibility to act simultaneously is sliding wear and mechanically generated fatique focused onto the same surface. Predictions of the response of a material to such combined requirements are succesful only if statistical support information is available.

The third aspect on combined requirements on a mechanical component, with respect to its ability to withstand material deterioration mechanisms, is that of different requirement spectra at different location of the same piece of material. This delicate problem arises from a traditional desire to manufacture a singular component from a singular workmaterial. The choice of material may be optimized for the most relevant function of the component, both regarding its production and its use, but problems can arise at locations on the component that are of secondary or tertiary importance for the holistic performance of the particular component. An example of a component subjected to requirements in parallel is a shaft transmitting a torque pivoted by a journal bearing in which wear of the shaft surface occurs.

In the present paper, a survey on the principal material deterioration mechanisms and their submechanisms are presented. Secondly, examples of material deterioration mechanisms occurring alternatively, simultaneously and in parallel are presented and analysed. Finally, a model (Fig. 1) for the analysis of unstationary tribological situations in general is presented in a generalized and simplified manner.

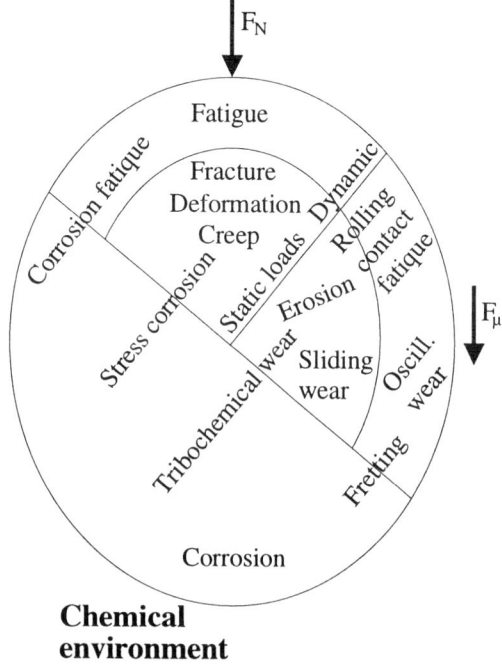

Fig. 1: Material deterioration mechanisms

FRETTING WEAR OF MACHINE PARTS

VYACHESLAV F. BEZYAZICHNY and ALEXANDER N. SEMYONOV
Rybinsk academy of aviation technology, Pushkin st., 53, 152934, Russia

One of specific and the most dangerous types of material destruction in contact interaction zones is fretting wear - corrosive-mechanical wear of contacting bodies during short oscillating relative motions. In areas damaged by fretting wear the processes of setting, abrasive failure fatigue-corrosive failure are proceeding. These processes are observed as a rule in the point of contact of tightly contracted components if as a result of vibration between surfaces microscopic shear displacement appears. Bolted and riveted joints of frames, bearing fits, hinged joints, clutches, pipe couplings and others are subjected to wearing during fretting wear (corrosion).

Fretting wear control carried out in the process of friction showed that at the initial stage of the first cycles the samples parting - "negative wear" and a sudden increase of friction coefficient occur. Then wear rate stabilizes and the curve takes the usual form (fig. 1).

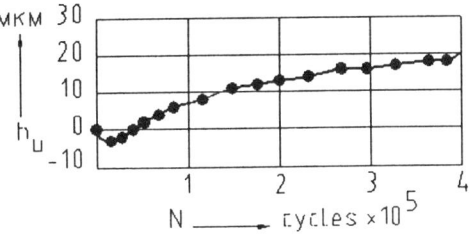

Fig.1 Dependence of wear on the number of cycles

The absolute load value affects the wear rate by external friction of any kind because it determines the area of actual contact. The data obtained (fig. 2) showed that wear rate grew with the increase of contact pressure, linear variation of wear rate being observed in the investigated range after established wear had begun.

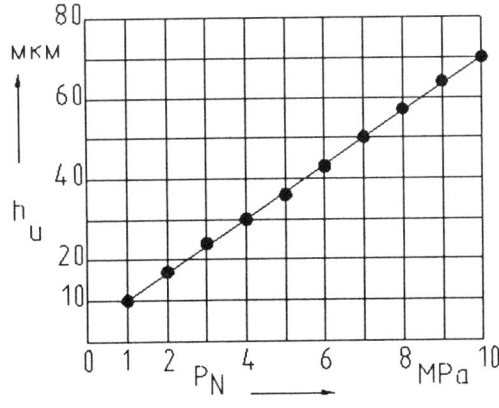

Fig.2 Dependence of fretting wear on the unit load

Dependence of fretting wear on vibromotion amplitude is represented in fig.3 and is good approximaxed by linear function. Under the conditions of the small amplitude fretting the surfaces have a more uniform character and are covered with a dense layer of wear products. With the increase of the vibromotion amplitude the non-uniformity of surfaces increases, the processes of setting and plastic deformation are intensified. With the increase of slipping amplitude the number and value of the applied microvolumes of a surface layer increases, the removal of wear products from the friction area is facilitated, their damping ability decreases.

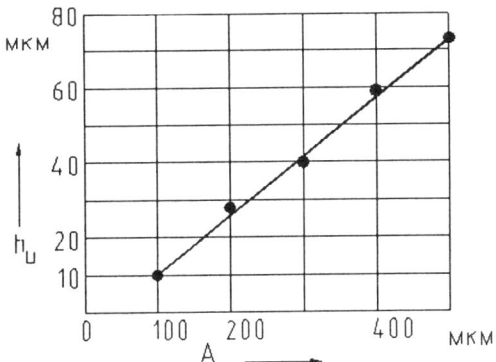

Fig.3 Dependence of fretting wear on slipping amplitude

The study of fretting wear and friction surface dependences by means of an electronic microscope allowed to distinguish three characteristic stages for the part of driving wear mechanism and the proceeding regularities.

The first stage is the stage of adhesive interaction which is developed in intensive plastic deformation setting, depth tearing out, mutual metal transfer on contacting surfaces.

The second transitional stage of fretting wear, the stage of running-in consists in wearing adhesive layers of transferred and redeformed material and the formation of nominal contact area.

The formation of a nominal contact area, the stabilization of temperature-force processes determined by the set up level of force and the coefficient of friction characterizes the transition to the third stage, the stage of a set up wear. Accordingly the processes of a friction area destruction are stabilized: the fatigue mechanism becomes dominating.

The suggested fretting wear model describes general regularities of all stages proceeding on the base of studying driving mechanisms of a surface layer destriction.

Submitted to *Surface Engineering*

DYNAMICS OF STICK - SLIP MOTION EXPERIMENTAL AND THEORETICAL STUDY

MARIAN WIERCIGROCH, V W T SIN and Z F K LIEW
Department of Engineering, Kings College
University of Aberdeen, Aberdeen AB9 2UE, UK

ABSTRACT

The stick-slip is a phenomenon combining a stop and a portion of periodic motion, and it occurs in mechanical systems with dry friction characteristics. This terminology is often used in problems related to the machine tools, where unfavourable velocity regions of sliding parts may produce so-called self-sustained vibrations.

The study of stick-slip caused by a variable friction characteristics has attracted a considerable interest for many years, but in the same time it has suffered from the lack of a generalised model. Only recently the models can capture complex dynamic behaviour, where a connection between a dry friction force and chaotic vibrations has been made. Although a formal and rigorous description of the phenomena associated with the self-sustained vibration has been tried by many investigators, the majority of them assumed a difference between static and dynamic friction is acting on the dynamic system. This approach is only correct when the relative velocity between sliding parts has the same sign.

The above mention limitation and the desirability of providing a more general approach to the description of a friction force generated on sliding surfaces were the main motivations for this study. A novel approach of viewing systems with a dry friction through the dynamics of systems with discontinuities (1)(2) will be shown by analysing a single degree of freedom based excited oscillator with a Coulomb damper (3). A set of compression springs, mass, viscous damper and pneumatically controlled Coulomb damper comprises the oscillator driven by a shaker, which is depicted in Fig. 1.

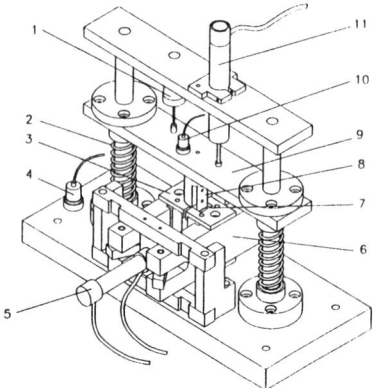

Fig. 1: Experimental oscillator

The friction force is generated by a clamping mechanism utilising the lever principle and pneumatic cylinder. Various friction pads including steel, brass, aluminium and Teflon are used to simulate different friction materials.

The oscillator mounted onto a shaker was excited with a chosen waveform and measurements of the absolute mass acceleration, friction force, and relative displacement between the base and mass were taken. The time signals were capture using the data acquisition systems and then processed to Force-Velocity and Force-Displacement characteristics, which are useful to analyse stick-slip motion. Fig. 2 shows typical above mentioned characteristics obtained for the following data set
- steel-steel combination,
- clamp force of 35N,
- base acceleration of 34.3 m/s^2 and
- forcing frequency of 18Hz.

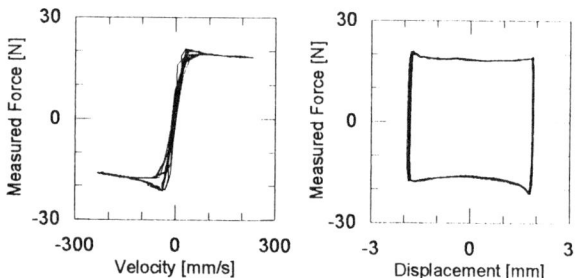

Fig. 2: Force-Velocity and Force-Displacement graphs for stick-slip motion

The nonreversibility of the dry friction characteristics to be incorporated in the oscillator has been modelled by the exponential function (2) and then solved numerically using the algorithms outlined in the monograph on the discontinuous systems (1), which are using switch functions to detect the time when stick-slip occurs or relative velocity is equal to zero.

REFERENCES

(1) M. Wiercigroch, M. Scientific Press of Silesian Technical University, Gliwice. Dynamics of Discrete Mechanical Systems with Discontinuities, 1994.
(2) M. Wiercigroch Journal of Sound and Vibration Vol. 175, No. 5, 700-704 pp.
(3) Z.F.K. Liew. University of Aberdeen, BEng Thesis. Theoretical and Experimental Study of Based Excited Linear Oscillator with Coulomb damper, 1995.

INNER AND OUTER OXIDATION PROCESSES AND THEIR EFFECT ON METAL-POLYMER FRICTIONAL INTERACTION

A. L. ZAITSEV and Yu. M. PLESKACHEVSKY

V.A. Belyi Metal-Polymer Research Institute, BAS, Kirov St. 32a., 246652 Gomel, Belarus

ABSTRACT

Despite a great number of publications concerning oxidizing wear process investigation, the influence of inner and outer sources of oxidation on physico-chemical processes of metal and polymer friction remains obscure (1). This question is of strong significance for studying such important phenomenon as frictional transfer. Any acceptor of electrons that saturates the environment is the source of outer oxidation. Inner oxidation results from desorption of adsorbate, decomposition of metal oxide films, chemisorbed substances, products of oxidation on the polymer friction surfaces, in the bulk and in boundary layers, as well as due to chemical reactions of active structural fragments of the polymer.

The aim of the present paper is to summarize the results of oxidative wear studies of metal-polymer joints in view of outer and inner oxidation sources influencing frictional physico-chemical processes.

on frictional properties is rendered by sources of inner oxidation. Metal oxides and oxygen-containing structural fragments of polymer macromolecules govern physico-chemical processes, in particular, catalytic reactions of intermediate polymer oxidation, formation of metal carbonates and oxyhydroxides. Certain concentration of metal oxides promote wear-resistance increase by means of metalorganic substances formation.

The role of outer oxidation sources is connected with decreasing ability of polymer tribodestruction products in reactions with metal and its oxides that reduces adhesional interaction. For nonpolar polymers outer oxidation leads to formation of oxygen-containing groups and polymer transfer increase. As a result, seizure in metal - polymer contact and wear intensity of materials grow.

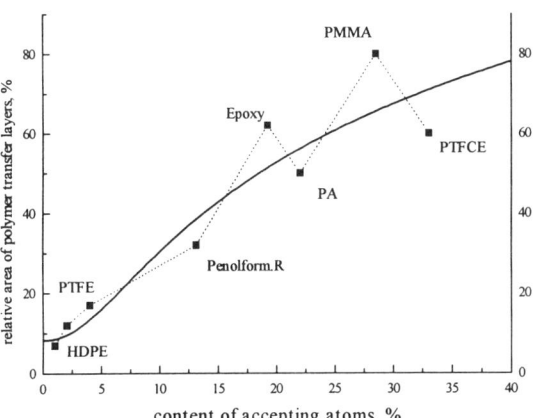

Fig. 1 Dependence of frictional transfer on electron accepting atom content in the polymer

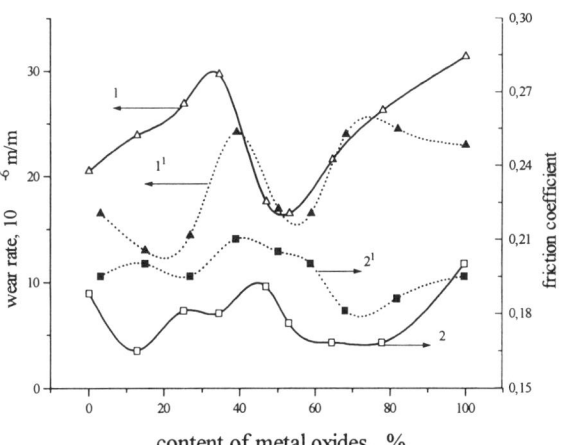

Fig. 2. The dependence of HDPE wear rate (1) and friction coefficient on oxide film content on metal at friction in inert gas (1, 2) and air (1^I, 2^I)

The obtained experimental results (2), presented in Figs. 1, 2, allows an assumption that physico-chemical processes of friction causing polymeric material transfer on the metal surface are predetermined by the content of electron accepting atoms in the friction zone and degree of metal surface oxidation as well as by polymer chemical structure and oxidizing activity of environment.

Using the example of thermoplastic and thermosetting polymers friction on steel, hard alloy and titanium it was established, that the greatest influence

REFERENCES

(1) D.H. Backley, Surface Effects in Adhesion, Friction, Wear and Lubrication. Elsevier Publ. Comp. 1981.

(2) A.L. Zaitsev Influence of hard alloy surface oxidation on frictional properties of polyethylene. Sov. J. Friction and wear, 1989, v. 10, № 5, pp. 886-890.

COMPLEX TRIBOLOGY SYSTEMS: MODELING APPROACH.

SERGEY M. ZAKHAROV and ILYA A. ZHAROV
All-Russian Research Institute of Railway Transport,
10, 3-d Mytishchinskaya, 129851, Moscow, Russia

ABSTRACT

Tribology objects are parts of machines. The reliable design of machines and tribology elements can be achieved when they are considered as inseparable parts of entire mechanical systems. Computer aided modeling is suggested as a way of studying and optimization these systems. As a complex tribology system may consists of many elements and the processes to be modeled are of different origins, the methodology has been worked out. The methodology is based on energy consideration approach and system analysis. The model of complex tribology system is presented in the form of an oriented graph, thus allows the evaluation of the level of complexity and optimization the efficiency of calculation (1). This methodology has been applied to three complex tribology systems. A short description of these systems is given bellow.

1. Model of engine crankshaft, its bearings and supports. In this model it is considered that the elastic crankshaft rotates in the main bearings which are placed in the elasticity deformed cylinder block supports. The linear and the angular elasticity parameters of the supports and the crankshaft are defined basing on the finite elements technique. The model of dynamically loaded bearing considers the misalignment and the angular displacements of the crankshaft journals as well as mixed lubrication conditions (2,3,4,5). A new method to obtain hydrodynamic forces and moments resulted from the angular displacements of journals and supports was suggested (3,5). The results of modeling are: load, moments, stress and strength safety factor in crankshaft elements, bearing journal paths, oil film thickness at the center of bearings and at its edges, oil flow, temperature change and friction loses. In addition to the existing bearing performance evaluation factors the journal deviation criteria, the probability of bearing scoring and the wear rate prognoses have been introduced (6). The model and its modifications has been on-going basis for study and improvement of several types of diesel locomotive engines.

2. Wheel/rail interaction model. The model of this complex tribology system considers non-linear rail car dynamics, determination of the contact points for real geometry of rail head and wheel set taking into account the angle of attack (7), calculation of vertical and lateral forces, components of wheel set slip and corresponding friction forces. Simulation of a freight car movement in curves and along straight sections with the given distribution of track irregularities in the plane and profile gives, among other parameters, friction power of wheel flange and rolling surfaces. Using the results of study of wheel/rail interaction surfaces (8) this enables to calculate the volume wear rate of interacting elements. This model was applied to evaluate the influence of track and car parameters on wheel flange/thread and rail head wear.

3. Roller bearing - railcar model (9). The model of this complex tribology system considers the dynamics of bearing elements, the tribological relations between them, the influence of unbalanced masses of the car and the action of track elements. The interaction between rollers and bearing cage is described using the hydrodynamic, or the mixed lubrication, or the Herzian contact theories. Some results of modeling obtained for the railcar roller bearing were compared with the existing experimental data.

REFERENCES

(1) S.M.Zakharov, I.A.Zharov. Methodology of modeling of complex tribology systems, Journal of Friction and Wear ("Allerton Press"), Vol.9,N 5,1988.
(2) S.M. Zakharov, I.V. Sirotenko, I.A.Zharov. Simulation of tribosystem "crankshaft, its bearings and supports" of engines, Journal of Friction and Wear, Vol.16,N 1,1995.
(3) S.M.Zakharov, I.A.Zharov. Prediction, approximation and area of application of plain journal bearing hydrodynamic characteristics considering journal axis angular displacement. Journal of Friction and Wear, Vol 16, No 1, 1995.
(4) S.M.Zakharov, I.A.Zharov. Calculation of dynamically loaded plain journal bearings considering journal axis angular displacement and mixed lubrication. Journal of Friction and Wear , Vol.17, No 4, 1996.
(5) S.M.Zakharov, I.A.Zharov. Modeling of the engine mechanical system: crankshaft, its bearings and supports,SAE International Congress, 1997, Paper 970511
(6) S.M.Zakharov, I.A.Zharov. Tribological criteria of the crankshaft bearings performance evaluation. Journal of Friction and Wear, Vol.17.No5,1996.
(7) V.M.Bogdanov etc. Simulation of wheel/rail interaction ,wear and damage accumulation. Journal of Friction and Wear, Vol.17,No 1, 1996.
(8) S.M. Zakharov, I.V.Komarovsky, I.A.Zharov. Study of wheel/rail interaction surfaces. Proceedings of 7th International Conference, Metrology and Properties of Engineering Surfaces, Goteborg.1997
(9) S.M.Zakharov, V.I. Tzurkan. Simulation of large railway roller bearings, Journal of Friction and Wear, Vol.15, N 4, 1994.

GREASE LUBRICATION IN FRETTING

Z. R. ZHOU

Tribology Research Institute, Southwest Jiaotong University, 610031 Chengdu, P. R. China

Ph. KAPSA and L. VINCENT

LTDS, UMR 5513, Ecole Centrale de Lyon, 69131 Ecully, France

ABSTRACT

Crack nucleation and propagation and contact wear have been well identified in both fretting fatigue and fretting wear (1,2). Facing the high risk of service failure induced by fretting, various palliatives have been tried. Grease, one of the classic palliatives for wear, has been widely used in mechanical system on various contact mode components. This paper concerns fretting behaviour using grease lubrication and under dry conditions.

All fretting tests were carried out on fretting devices developed for tension-compression hydraulic machines. A sphere-flat contact was used. The fretting tests were performed under laboratory controlled conditions. The principal mechanical parameters ranged in the following intervals: slip amplitude, from 10 to 150 µm; normal load from 100 to 600N; frequency from 0.5 to 5Hz; number of cycles from 0 to 3×10^5 cycles. The sphere was a specimen used for ball bearing. The flat piece was made of a low-alloyed steel. The grease used is a classic mineral oil grease thickened by a polymer. The grease viscosity is 100 cst at 40°C.

Variations in the tangential force versus the displacement amplitude as a function of the fretting cycles were recorded. The contact surface and its cross-section were subsequently examined by OM, SEM, and tridimensional profile analysis. The influence of displacement amplitude, frequency, normal load, etc on fretting behaviour was investigated.

The benefits of grease as a palliative was sligth at low slip amplitudes, grease even accelerated the wear depth at larger amplitude; High frequency was not favourable for grease as a palliative for fretting (3); The benefit of grease for fretting diminished with an increase in the sphere radius.

In order to visualise the grease evolution at the interface during fretting tests, a glass/flat contact was used: From the friction log four stages were often observed. This data was combined with image analysis: (a) For the first cycles were completed, the two surfaces were separated by grease; The coefficient of friction was about 0.05; (b) From cycle 100 to cycle 1000, the grease was gradually eliminated from the contact and the coefficient of friction increased up to 0.5. Some bubbles were expelled from the contact edges in the fretting direction and the number of bubbles increased as a function of the number of cycles; (c) A vraiation in the coefficient of friction was observed between cycles 1000 and 5000. More bubbles and some metallic particles were expelled; (d) A relatively stable stage then established itself: the bubbles almost disappeared and the coefficient of friction remained stable.

Three lubrication conditions has been determined: Effective lubrication condition was characterized by effective protection of the surface during fretting; Mixed lubrication condition in most cases appeared particularly at the transition stage: Non-lubrication condition could be observed for slip conditions with a high normal load or a high frequency, the lubricant was completely expelled from the contact.

The investigations have proven that self-repair (penetration of the grease by fretting action) is proposed to have taken place during the tests. Given these results, a ratio was proposed of the Hertzian contact area radius to the displacement amplitude. This seems to be one of the most important parametes for the occurrence of self-repair. A great number fretting tests demonstrated that the effectiveness of grease as a palliative for fretting increased as a function of the ratio. Effectiveness was slight for a low ratio value (<10%); a strong variation in the coefficient of friction during testing was observed, and the self-repair action occurred when this ratio exceeded 20%.

REFERENCES:

(1) R. B. Waterhouse, Fretting-Fatigue, Applied Science, London, 1981.
(2) Z. R. Zhou and L. Vincent, J. of tribology, ASME, Vol. 119, pp.36-42, 1997.
(3) Y. Berthier, L. Vincent and M. Godet., Wear 125, pp.25-38, 1988.

MICROSCOPIC DEWETTING OF THIN LIQUID FILMS

CHAO GAO, PETER DAI, ANDY HOMOLA and JOEL WEISS
Akashic Memories Corporation, 304 Turquoise Street, San Jose, California 95134, USA

ABSTRACT

Experimental observations on droplet formation as well as a mechanism for micro dewetting are presented. It is widely observed that droplet formation (dewetting) occurs on magnetic disk surfaces (coated with a thin layer of perfluropolyether liquid) due to either lubricant additives in the lubricant, or organic contamination from improper handling/storage/drive operation. Regardless of the details on contamination/additives, the droplets are typically 0.5 to 5 μm in diameter (not a sphere, see Fig. 1) with heights ranging from 10 nm to 100 nm, as measured using AFM. Those droplets can be detrimental in file operation. The flying height of a magnetic head over a disk surface is typically less than 50 nm, hence the magnetic head may pick up liquid from these microscopic droplets. The effect of the excess liquid on the magnetic head is two folds. During file operation, it may affect the flyability of the magnetic head causing magnetic signal degradation (or head crash in a worse situation). During parking cycle (power off, for example) it may induce unacceptable high stiction (static friction force) due to excess meniscus formation at the interface.

It is further observed that for the same conditions (the same amount of additive or the same amount of hydrocarbon contamination), droplet formation occurs earlier for thicker lubricant films with larger size of droplets. Such dewetting phenomenon cannot be derived from macroscopic theory, in which dewetting is based on contact angle between the liquid and the solid. The contact angle is related to surface tension of the liquid, surface tension of the solid and interfacial tension of the liquid-solid interface. This contact angle concept has difficulty in dealing with two or more liquids on a solid. This is so because any liquid profile satisfying the contact angle between two liquids cannot satisfy Pascal's law, i.e. no pressure gradient can exist in static fluids. Furthermore, contact angle is not a well defined quantity at micrometer scale, nor surface tension for films with thickness in the nm range.

The proposed mechanism for microscopic dewetting is based on Lifshitz theory of electromagnetic fluctuation for van der Waals interaction. For simplicity, the Hamaker constant for a three component system (see Fig. 1) is

$$A \approx \frac{3}{4}kT\left(\frac{\varepsilon_1 - \varepsilon_2}{\varepsilon_1 + \varepsilon_2}\right)\left(\frac{\varepsilon_3 - \varepsilon_2}{\varepsilon_3 + \varepsilon_2}\right)$$
$$+ \frac{3h}{4\pi}\int_{v_1}^{\infty}dv\left(\frac{\varepsilon_1(iv) - \varepsilon_2(iv)}{\varepsilon_1(iv) + \varepsilon_2(iv)}\right)\left(\frac{\varepsilon_3(iv) - \varepsilon_2(iv)}{\varepsilon_3(iv) + \varepsilon_2(iv)}\right)$$

where h is the Planck's constant, k Boltzmann constant, T the temperature, $v_1 = (2\pi kT/h) = 4\times10^{-13}$ s^{-1}, and ε_1, ε_2, and ε_3 are dielectric constants for the solid (Medium 1), the lubricant (Medium 2), and the contaminant or additive (Medium 3), respectively. The first term is none-dispersion (zero-frequency or entropic energy) whereas the second term is dispersion and suffers retardation effect due to finite speed of electromagnetic wave of the interactions. Depending on details of lubricant thickness, contaminant/additive amount, and ε_1, ε_2, and ε_3, not only the strength ($A = A_{v=0} + A_{v>0}$) of the interactions on Medium 3 is changing, but also the sign of A may change. When A is negative, the van der Waals force on the contaminant/additive is repulsive, favoring spread of the contaminant/additive. However, when A is positive for most cases, the van der Waals force on the contaminant/additive is attractive, favoring the building up of the contaminant/additive, hence dewetting occurs. The humidity is believed to modify the situation to a four component system and to accelerate dewetting by enhancing materials migration/transportation.

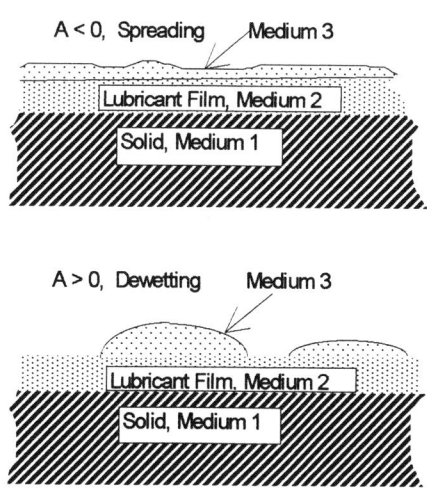

Fig. 1. Wetting (A<0) and dewetting (A>0, droplets).

RUNNING-IN EFFECTS ON THIN-FILM MAGNETIC DISKS

YOUICHI KAWAKUBO and MIEKO ISHII
Mechanical Engineering Research Lab., Hitachi Ltd., Kandastu, Tsuchiura, Ibaraki 300 Japan
TETSUJI HIGASHIYA* and SADANORI NAGAIKE
Data Storage and Information Retrieval Div., Hitachi Ltd., Kouzu, Odawara, Kanagawa 256 Japan

ABSTRACT

Since the sliding area of magnetic heads is much smaller than that of disks, and head wear is therefore faster, it is more important to predict head wear than disk wear when designing near-contact/in-contact magnetic recording rigid-disk storage devices.

It has been common to predict the amount of wear by using Archard's equation, which assumes that wear increases in proportion to sliding distance. However, our previous study showed that there were running-in effects on thin-film magnetic disks(1)(2). Therefore, we analyzed the progress of head wear on thin-film magnetic disks, in single-track pin-on-disk wear tests (STWT), using transparent quartz pins. We also performed multi-track-seeking wear tests(MTWT) with the same testing period(24 hr) and varied seek-stroke lengths to determine the effects of seek operations in actual hard-disk drive-usage conditions.

It was found that pin wear increased in proportion to the square-root of the sliding distance in STWT(Fig. 1). The pin wear also increased in direct proportion to the applied load. From these results, we derived a new experimental wear equation to include the running-in effect.

$$V = k_s \cdot W \cdot (L/L_s)^{0.5} \quad \cdots \cdots \cdots \cdots (1)$$

Here, V is the wear volume, W is the applied load, L is the sliding distance, and k_s is a standard specific wear rate at a fixed sliding distance L_s. Wear volume calculated using this equation is plotted on Fig. 1 with L_s equal to 1 km. It is clear that the new experimental wear equation can explain measured results very well.

Next, we extended our new equation (1) to include the effects of seek operations. Considering the seek stroke length B and the head width b, the number of the slide-over tracks becomes B/b. We then obtain the following equation (2).

$$V = k_s \cdot W \cdot ((L/L_s)\cdot(B/b))^{0.5} \quad \cdots \cdot (2)$$

The results of MTWT using Al_2O_3-TiC pins is shown in Fig. 2(a). This pin wear also increased in proportion to the square-root of the seek-stroke length. Measured and calculated results are compared in Fig. 2(b). It is clear that the test results are well accounted for by this extended equation. We consider this to confirm the accuracy of our extended wear equation.

The decrease in roughness of the disk surface after a MTWT is shown in Fig. 3. We consider this to be a likely cause of the running-in effect.

This type of exponential wear rate decrease was reported before and the reason was considered to be the roughness decrease of tapes(3). We consider this type of running-in due to the roughness decrease of recording media to be a common characteristics of magnetic recording systems.

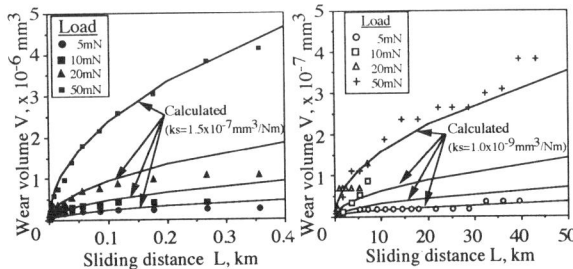

(a) on a textured disk (b) on a plane disk
Fig. 1: Change in wear volume during STWT

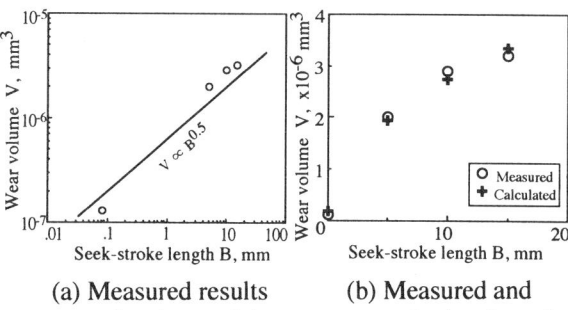

(a) Measured results (log-log scale) (b) Measured and calculated results
Fig. 2: Change in pin wear volume with respect to seek stroke length in MTWT

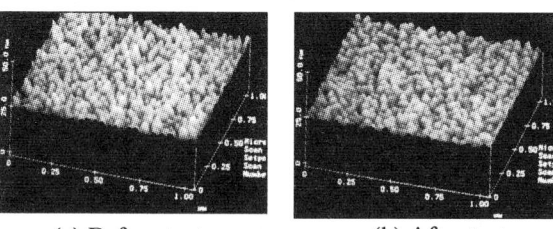

(a) Before test (b) After test
Fig. 3: Change in disk surface profiles over course of MTWT

REFERENCES

(1) Y. Kawakubo, Y. Yahisa, Trans. ASME, J. of Tribology, Vol. 117, No.2, 1995, 297-301pp
(2) Y. Kawakubo, M. Ishii, Y. Kokaku, Y. Yahisa, T. Yamamoto, Proc. of International Tribology Conf. `95 Yokohama, JAST, 1996, 1805-1810pp
(3) K. Matsuoka, D. Forrest, M.-K. Tse, E. Rabinowicz, JSME Trans., Ser. C, Vol. 60, No. 576, 1994, 2609-2614pp (in Japanese)

ESTIMATION OF MENISCUS FORCE CONSIDERING THE ELASTIC SURFACE DEFORMATION

YOSHIKAZU KOBAYASHI and KENJI SHIRAI
Department of Computer Science, College of Engineering, Nihon University, Tamura-machi, Koriyama, 963, Japan
KAZUHISA YANAGI
Center for Surface Metrology, Nagaoka University of Technology, Nagaoka, Niigata 940-21, Japan

ABSTRACT

A decrease in the meniscus force caused by contact between a magnetic disk and the head slider is considered a favorable condition, because durability and reliability of the magnetic disk drive are thereby improved. So far, the surface topography has been modeled geometrically to calculate the meniscus force, and many models, assuming various approximate conditions, have been used(1)(2). However, there are differences between these models, and they are not required to agree with actual experiment results. So, in order to accurately calculate meniscus force, a model, which takes into consideration the elastic deformation caused by the contact of two surfaces, is proposed in this study. The elastic deformation in this study is a macroscopic one, caused by the interaction between contact points.

The isotropic surfaces of four magnetic disks (DATA1-DATA4) with glass substrate were measured with an atomic force microscope, and the resulting data were used for calculation. Fig.1 shows the examples of 3-D topography plots. Table 1 shows the basic 3-D surface roughness parameters, SRq, SRt (3) and the topography parameters which are needed for meniscus calculation in this study, where, μ and σ show the average value and the standard deviation of the height of all peaks respectively. C and λ are constants and standard deviation in exponential function. The exponential distribution was applied to peaks greater than a height of $(\sigma + \mu)$. $SDtotal$ and $SD\sigma$ show the peak densities for all peaks and selected peaks which were greater than a height of $(\sigma + \mu)$, respectively. $Rall$ and Rs are average values of all peaks' and selected peaks' radius of curvature.

As a result of this calculation, it is understood that meniscus force grows rapidly, compared with an increase of the liquid film thickness L at a random surface. The calculated values, using the proposed model, are estimated more roughly than the Greenwood and Williamson model(4) (Fig. 2). The peak distribution of the surface of the disk has a strong relation to these calculated values. Moreover, the values influence the size of the real contact areas between two surfaces (Fig. 3), where Ar is the real contact area and Ao is the apparent contact area.

REFERENCES

(1) Gao, C., Tian, X., and Bhushan, B., STLE tribology Transactions, Vol.38, pp.201-212,1995.
(2) Tian, H., and Matsudaira, T., Jour. of Trib., Vol. 115, pp. 28-35, 1993.
(3) Stout, K. J., "Three-Dimensional Surface Topography ; Measurement, Interpretation, and Applications," Penton Press, pp. 363, 1994.
(4) Greenwood, J. A., and Williamson, J. B. P., Proc. Roy. Soc (London), Series A., Vol.295, pp. 300-319, 1966.

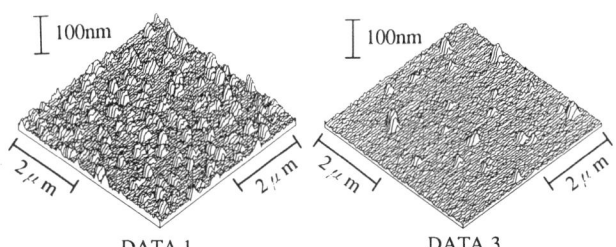

Fig. 1 : 3-D topography plots of the samples

Table 1 : Surface topography parameters

	SRq (nm)	SRt (nm)	μ (nm)	σ (nm)	SDtotal (/μm²)	λ (nm)	SDσ (/μm²)	Rs (nm)	Rall (nm)
DATA 1	8.4	72.5	11.5	10.7	18.3	5.1	4.2	63	73
DATA 2	6.2	81.3	9.1	7.9	23.1	8.6	3.4	59	76
DATA 3	3.7	49.9	4.8	5.0	24.6	16.7	1.6	68	130
DATA 4	2.8	38.3	4.9	4.7	24.4	7.4	2.7	66	136

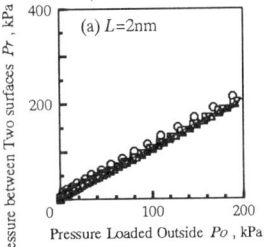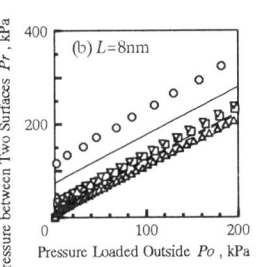

Sign	Model	Sample	Sign	Model	Sample
○	Proposed	DATA 1		GW	DATA 1
□	Proposed	DATA 2	------	GW	DATA 2
△	Proposed	DATA 3	— — —	GW	DATA 3
▽	Proposed	DATA 4	— — —	GW	DATA 4

Fig. 2 : Relation between pressure Po and pressure Pr

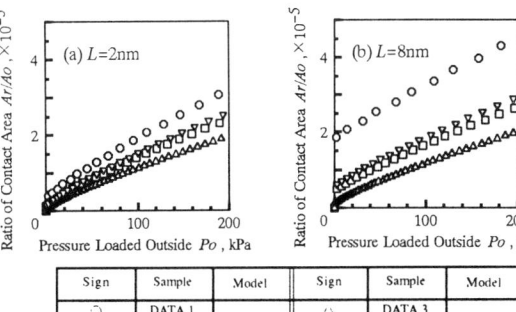

Sign	Sample	Model	Sign	Sample	Model
○	DATA 1	Proposed	△	DATA 3	Proposed
□	DATA 2		▽	DATA 4	

Fig. 3 : Ratio of contact area Ar/Ao for pressure Po

MOLECULAR ORBITAL SIMULATIONS ON DEGRADATION OF LUBRICANT FOR MAGNETIC DISK

SATOSHI MATSUNUMA
Hitachi Ltd. Central Research Laboratory, 292 Yoshida-cho, Totsuka-ku, Yokohama 244, Japan

ABSTRACT

The molecular orbital method seems to give a deep understanding of tribochemistry of perfluoropolyether (PFPE) by discussing reaction mechanism, interaction, bonding character, and electron property (1)(2). In this study semi-empirical molecular orbital simulations on degradation of perfluoropolyether (PFPE) which is used as a lubricant for a magnetic disk are presented.

Two processes of degradation such as a low energy electron-induced decomposition and a reaction initiated by interaction with ionic species on the surface of disk were simulated by PM3 calculations (3) on energetic and structural parameters including heat of formation, bond order, and charge density, in a similar manner as reported previously (1)(4)(5).

Calculated heat of formation of anion radicals of PFPEs were lower than that of neutral species. These results postulated that a low-energy electron attachment to PFPE easily occurs on the electron-rich surface. The formed anion radical had weakened CO bonds, and cleavages of these bonds were simulated by a stepwise calculation of heat of formation with a geometrical optimization except two atoms of the stretched bond as the reaction coordinate. Calculated final structure and charge density suggested that products were an anion and a neutral radical. The results from this simulation showed close agreement with that from the spectroscopic study on the electron decomposition of PFPEs (6).

The anion formed from anion radical of perfluoro methylene oxide was simulated to successively degrade to produce carbonyl fluoride and a shortened anion. This successive degradation to form carbonyl fluoride was considered as a model for experimental results well known as unzipping mechanism (7).

The cation radical which was formed through the interaction with a high-energy electron was expected to degrade to a cation and a neutral radical. The formation of carbonyl fluoride was simulated in the case of the decomposition of perfluoro linear propylene oxide cation radical.

Neutral radicals of PFPE were also simulated to decompose to carbonyl fluoride and a shortened radical. But degradation of neutral radicals needed larger activation energy than that in the case of ionic species.

Meanwhile an approach of proton to the oxygen site of PFPE, as the initial step of acidic cleavage on aluminum surface, formed a proton-coordinate intermediate with decreasing total energy by 496.0 kJ/mol. The cleavage of the weakest C-O bond of the intermediate gave an alcohol and a carbonium cation. The formed carbonium cation was expected to degrade to shorter carbonium cation and carbonyl fluoride successively considering of the estimated energy.

The interaction between PFPE and hydroxyl group which has a little chance to exist in hydrate smoothly gave alkoxy anion and alcohol with energy-stabilization by 470.0 kJ/mol just as reported on perfluorodimethylether (2).

Some cations such as AlF_2^+ and $Al(OH)_2^+$ of active site models on material surface were simulated to interact with PFPE to be followed by the subtraction of the F atom from PFPE as described bellow.

$$CF_3OC_2F_4OCF_2OCF_3 + M^+$$
$$\rightarrow CF_3OC_2F_4OCFOCF_3^+ + M\text{-}F$$

M^+ and M-F indicate AlF_2^+ or $Al(OH)_2^+$ and AlF_3 or $AlF(OH)_2$, respectively. The formed PFPE cation was postulated to degrade to a PFPE with a CFO group which has been detected in the wear test of a magnetic disk. The energy diagram in the interaction between PFPE and $Al(OH)_2^+$ is shown in Fig. 1.

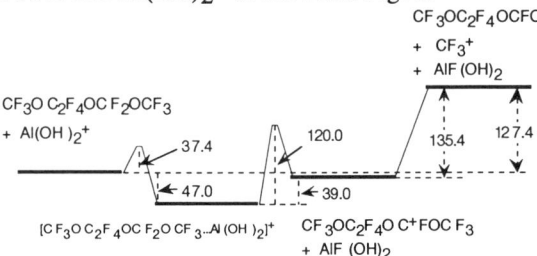

Fig.1: Energy diagram (Energy in kJ/mol).

In conclusion molecular orbital simulations showed that the interaction of PFPE with a low-energy electron or ionic species weakened some CO bonds of PFPE and led to the degradation to ion and other products. The formed ions of PFPE were postulated to decompose successively to carbonyl fluoride and shortened ions. These results suggested that the ionic process and the ionic intermediate play an important role in degradation of PFPE on a magnetic disk.

REFERENCES

(1) S. Matsunuma, T. Miura, H. Kataoka, Tribology Transactions, Vol. 39, 1996, 380-385pp.
(2) W. Morales, STLE Preprint NO. 95-TC-5A-4.
(3) J. J. P. Stewart, MOPAC Ver. 6.0, QCPE #455; T. Hirano, Ver. 6.01.
(4) S. Matsunuma, Proceedings of the International Tribology Conference, Yokohama 1995, 1129-1134pp.
(5) S. Matsunuma, Y. Hosoe, Tribology International, Vol. 30, 1997, 121-128pp.
(6) G. Vurens, R. Zehringer, D. Saperstein, ACS Symposium Series, Vol. 485, 1992, 169-180pp.
(7) S. Mori, W. Morales, NASA TP-2910, 1989.

TRIBOLOGICAL EVALUATION OF THE STREAMING MODE PERFORMANCE OF CARBON-COATED METAL EVAPORATED TAPES

STEVEN T. PATTON and BHARAT BHUSHAN
Computer Microtribology and Contamination Laboratory
Department of Mechanical Engineering
The Ohio State University
Columbus, OH 43210-1107

ABSTRACT

Metal evaporated (ME) magnetic tape is a leading candidate for ultra high density magnetic tape recording applications. However, durability of ME tape is still in question. Previous studies have shown that commercial ME tape is less durable than commercial metal particle (MP) and barium ferrite particulate tapes (1)-(4). Other studies investigated the effects of temperature and relative humidity on the performance of ME and MP tapes (5), (6). Waviness (of about 60 nm peak-to-valley and about 250 µm wavelength) of the ME tape surface used in the previous studies led to poor high density recording performance due to spacing loss of signal and accelerated and localized tape wear at high contact points. In play/rewind cycling experiments with wavy ME tape, head-to-tape spacing was reduced by about 10 nm over about 850 play/rewind cycles (where tape failure occurred), and damage areas initiated at high points or bumps on the tape surface were connected by lateral cracks (driven by longitudinal tape tension) across the tape width at tape failure. In the previous studies, high asperities on the virgin tapes were severed off during the record pass, and those that remained on the tape surface increased friction force and tape wear by three body abrasion early on in play/rewind cycling tests. Lower friction and virtually no wear were observed later in cycling tests when loose debris were no longer on the tape surface and wear mechanism was adhesive.

In this study, the three kinds of ME tape tested for friction, wear and magnetic performance are: (a) diamondlike carbon (DLC) coated tape without waviness, (b) non-DLC tape without waviness and (c) non-DLC tape with waviness. Using a commercial video cassette recorder as a magnetometer, and the Wallace equation, changes in rms head output level were correlated to changes in head-to-tape spacing as tape wear occurred during play/rewind cycling tests. Interface stability and recording performance at a 0.6 µm recording wavelength were measured to bit level resolution using a dropout counter. Development of the experimental apparatus with nanometer vertical and sub-µs temporal resolutions has enabled unprecedented understanding of the interplay of friction, wear and surface topography in a sliding contact.

Comparing non-DLC tapes (b) and (c), it was found that tape (b) without waviness (relatively flat) had superior magnetic performance of higher head output and lower dropout frequency due to less spacing loss of signal. However, durability of tape (b) which failed after 350 play/rewind cycles was worse than that of wavy tape (c) which failed after about 850 play/rewind cycles. Premature failure of tape (b) was due to a catastrophic three body abrasive wear mechanism, and this kind of failure was averted by wavy tape (c) due to abrasive particles settling down into non-contacting valleys on the tape surface. Comparing DLC coated tape (a) and non-DLC tape (b), both without waviness, improvement presented by DLC coating was marginal at best. Durability of DLC coated tape (a) which failed after 450 play/rewind cycles was only slightly better than non-DLC tape (b) which failed after 350 play/rewind cycles. DLC coated tape (a) had inferior magnetic performance of lower head output and higher dropout frequency as compared to non-DLC tape (b). The 10 nm thick DLC coating reduced head output by spacing loss of signal and increased dropout frequency by acting as a source of debris (flaking of the coating).

ME tapes used in this study all exhibited poor durability as compared to the known durability of particulate tapes. Ultimate failure for all of the ME tapes always involved formation of lateral cracks across the tape surface, and cracks usually propagated through wear scars or coating defects. Further work is required to attain adequate durability of ME tape.

REFERENCES

(1) S. T. Patton and B. Bhushan, Tribol. Trans., Vol. 38, 1995, pp. 801-810.
(2) S. T. Patton and B. Bhushan, ASME J. Trib., Vol. 118, 1996, pp. 21-32.
(3) S. T. Patton and B. Bhushan, IEEE Trans. Magn., Vol. 32, 1996, pp. 3684-3686.
(4) S. T. Patton and B. Bhushan, Proc. Instn. Mech. Engrs., Part J: J. Eng. Trib., submitted for publication, 1997.
(5) S. T. Patton and B. Bhushan, J. Appl. Phys., Vol. 79, 1996, pp. 5802-5804.
(6) S. T. Patton and B. Bhushan, IEEE Trans. Magn., in press, 1997.

MECHANISM OF TAPE DEBRIS FORMATION AND POLE TIP WEAR IN COMPUTER TAPE DRIVES

WILLIAM W. SCOTT and BHARAT BHUSHAN
Computer Microtribology and Contamination Laboratory
Department of Mechanical Engineering, The Ohio State University, Columbus, OH 43210-1107, U.S.A.

FRANK SHELLEDY and SUBRATA DEY
Storage Technology Corporation, South 88th Street, Louisville, CO 80028-8110, U.S.A.

ABSTRACT

The increase in spacing between magnetic heads and magnetic tapes results in signal loss in magnetic tape drives. The wear of tapes and read/write heads can cause an increase in spacing. This wear has become worse with the move toward the use of harder Al_2O_3-TiC substrates for heads in linear tape drives. Observations indicate that the use of these hard materials results in more staining and debris on the read/write element and in more pole tip wear. The more extensive use of a less abrasive, metal particle (MP) tape, as opposed to CrO_2 tape, makes the problem worse yet.

In this study, we analyze the causes of, and possible remedies for the generation of stains or loose tape debris at the head/tape interface. We also consider the wear of pole tips (also known as pole tip recession or PTR) in inductive write heads.

Functional tests are conducted using thin-film Al_2O_3-TiC inductive heads with Co-Zr-Ta poles and thin-film Ni-Zn ferrite inductive heads with Ni-Fe poles. Each of the three drive operating conditions (wrap angle, sliding speed, and tape tension) that affect the interaction between a certain head/tape pair, is varied with respect to the others. The friction force, measured with a strain gage load cell, is monitored throughout the tests. In this way, we are able to find the relative influences of these three conditions and of the friction force on tape debris formation.

Optical microscopy and computerized image analysis are used to measure the amount of tape debris (measured as the surface area of the head covered by debris) and its distribution on the head. It is shown that tape debris consists of three types: magnetic particle rich debris, binder rich debris, and adherent debris (or stain). The generation of magnetic particle rich debris is shown to be approximately proportional to friction force for a given head substrate material. This statement is true, independent of the change in the operating condition that accounts for the change in friction force. So, the change friction force is sufficient to explain the change in magnetic particle rich debris generation. The generation of binder rich debris does not follow such a simple relationship. Adherent debris appears only on the Al_2O_3-TiC head. Since it collects near the pole tip and is difficult to remove, it, compared to the other tape debris types, probably has the potential for causing the most spacing loss problems.

Possible solutions for the problem of debris formation, such as adding a CrO_2 tape leader to each reel or cleaning after every one-hundred hours with CrO_2 tape, are also investigated.

Atomic force microscopy is used to measure PTR. It has been shown elsewhere that PTR can grow 20 to 35 nm over short sliding distances (500 - 1000 km) (1) (2). In this paper we compare the PTR for the heads with different substrate materials.

REFERENCES

(1) B. Bhushan, S.T. Patton, R. Sundaram, S. Dey, "Pole Tip Recession Studies of Hard Carbon-Coated Thin-Film Tape Heads", Journal of Applied Physics, Vol. 79, 1996, pp. 5916-5918.
(2) S. Patton and B. Bhushan, "Micromechanical and Tribological Characterization of Alternate Pole Tip Materials for Magnetic Recording Heads", Wear, Vol. 202, 1996, pp. 99-109.

EFFECT OF SLIDER BURNISH ON DISK DAMAGE DURING DYNAMIC LOAD/UNLOAD

M. SUK and D. GILLIS
IBM Corporation, Storage Systems Division, San Jose, CA 95193

ABSTRACT

In today's disk drives the read/write transducer carrying slider rests on a disk surface while the disk drive is not in operation. During the power-up and power-down phases of the drive, the slider slides across the disk surface in contact below speeds where sufficient air-bearing is generated to lift the slider off the disk surface. This process of starting and stopping is called contact-start-stop and usually takes place in a dedicated start-stop zone on the disk. With increases in storage densities, disk surface roughness of both data and start-stop zones have been decreasing rapidly. However, controlling the roughness of the start-stop zone while ensuring acceptable level of head/disk interface stiction and durability performances has become a formidable task. To overcome some of the difficulties, the disk drive industry has been investigating new lubricants, tougher carbon overcoats, and non-mechanical texturing processes such as the laser texturing process (1).

An alternative technology commonly called load/unload (L/UL) has been implemented since early disk drives with 14" removal media. In the middle of 1980's, L/UL was reintroduced into large and small form factor files to avoid head/disk stiction and durability problems all together. In this type of a system, the slider carrying suspension sits on top of a ramp which lifts the slider completely off the disk surface when the drive is not in operation. During the drive start phase, the suspension slides off the ramp at a controlled speed and the slider loads onto the disk surface while the disk is spinning at some designed speed.

One potential problem with this type of system is head/disk contacts at high horizontal velocities leading to disk damage. Previously published studies have focused on variations in the parameters, such as the loading velocity, slider static attitude, that can lead to head/disk contacts (2). None of the earlier work, however, has addressed the relationship between disk damage and slider geometry. Jeong and Bogy (3) did study disk damage due to impacts, but their study focused on impact induced demagnetization that may result in possible unrecoverable data loss.

In this study, we show that an edge of the air-bearing surface contacting the disk during the L/UL process can induce disk damage. We observe that at 7200 rpm, disk damage occurs after running for 20K L/UL cycles almost independent of slider's roll-static-attitude and pitch-static attitude.

From the resistance methods, however, we observe that the number of contacts decreases exponentially with L/UL cycles and that the likelihood of disk damage can be minimized by 3-dimensionally rounding the slider. Figure 1 shows the cumulative number of hits as a function of L/UL cycles as detected by the resistance circuit. The solid line is the best fit line. The total number of recorded contacts was 408. It is clear the rate of head/disk contacts decreases exponentially as a function of L/UL cycles. In fact, after 20K L/UL cycles, the rate of contact decreases from 10 hits per 100 L/UL cycles to fewer than 0.2 hit per 100 L/UL cycles. This type behavior is generally observed when contacting surfaces wear with time. In fact, a well burnished slider rarely causes any disk damage thus leading to an interface with much improved reliability.

Finally, a simple Hertzian contact stress analysis indicates that the contact stress at the head/disk interface can be significantly decreased by increasing the radius of curvature of the contacting edge of the slider.

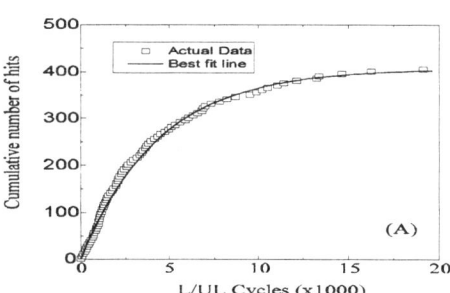

Fig. 1: Cumulative number of contacts as a function of L/UL cycles.

REFERENCES

(1) Baumgart, P, Krajnovich, D.J., Nguyen, T.A., Tam, A.C., 1995, "New laser texturing technique for high performance magnetic drives," *IEEE Trans Magn.*, Vol. 31, No.6, pp 2946-2951, Nov., 1995.

(2) Jeong, T.G. and Bogy, D.B., 1992, "An Experimental Study of the Parameters that Determine Slider-Disk Contacts During Dynamic Load-Unload," ASME Journal of Tribology, Vol. 114, pp 507-514, July, 1992.

(3) Jeong, T.G. and Bogy, D.B., 1993, "Dynamic Loading Impact Induced Demagnetization in Thin Film Media," *IEEE Trans. Magn.* Vol. 29, Vol. 6, pp 3903-3905, Nov., 1993

INVESTIGATION OF THE FLYING BEHAVIOUR OF PROXIMITY SLIDERS AND COMPARISON WITH SIMULATION RESULTS

STEFAN WEISSNER, TOM MCMILLAN, ERIC BAUGH and FRANK E. TALKE
Center for Magnetic Recording Research; University of California, San Diego
9500 Gilman Drive; La Jolla, CA 92093-0401

ABSTRACT

As the flying heights of modern sliders decrease and proximity recording becomes more common, the need for accurate numerical predictions of the flying behavior of the slider grows. This is necessary in order to generate an appropriate slider design that provides low contact forces, and thus low wear, by supporting the load of the suspension almost entirely with air pressure.

A recent slider design that is used in magnetic hard drives is the subambient tri-pad slider. This slider combines a third pad and a recessed area, the so-called cavity. This cavity generates a subambient pressure that acts against the positive pressure beneath the rails and the third pad. Since the pressures on the rails and in the cavity oppose each other, the flying height is only weakly affected by velocity and can be kept nearly constant for all radius/skew combinations.

A typical subambient pressure tri-pad design was chosen to compare measured flying height data with numerically predicted results. The numerical calculations were performed with a simulation program using finite element analysis. A mesh with approximately 6500 degrees of freedom (Fig. 1) was generated to closely model the slider air bearing surface which was measured using an optical profilometer.

To obtain close agreement between experimental and numerical results, it was found that it is necessary to include the recessed area around the third pad in the numerical calculations since this region generates a subambient pressure due to the diverging channel created by the etch step at the rear of the primary air bearing surface.

The influence of various design parameters such as crown and cross-crown, pivot point location and third pad geometry was investigated. In addition, a Monte Carlo analysis was performed to study the effect of manufacturing tolerances on the flying behaviour (Fig. 2).

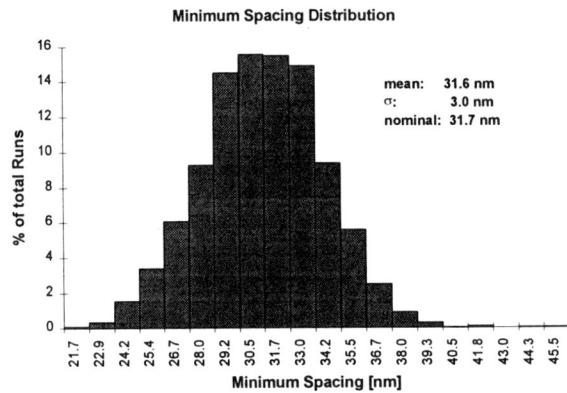

Fig. 2: Minimum spacing distribution

For the experimental flying height measurements, a smooth glass disk was used. In actual operating conditions, the slider is in partial contact with the surface asperities of a carbon-coated magnetic disk. To model this situation, a modified Greenwood-Williamson contact model (1) was implemented, resulting in excellent agreement between experimental results and numerical predictions.

REFERENCES

(1) M. Wahl, H. Kwon, F. Talke, "Simulation of asperity contacts at the head/disk interface of tri-pad sliders during steady-state flying", Tribology Transactions, Vol. 40, pp 75-80, Jan 1997

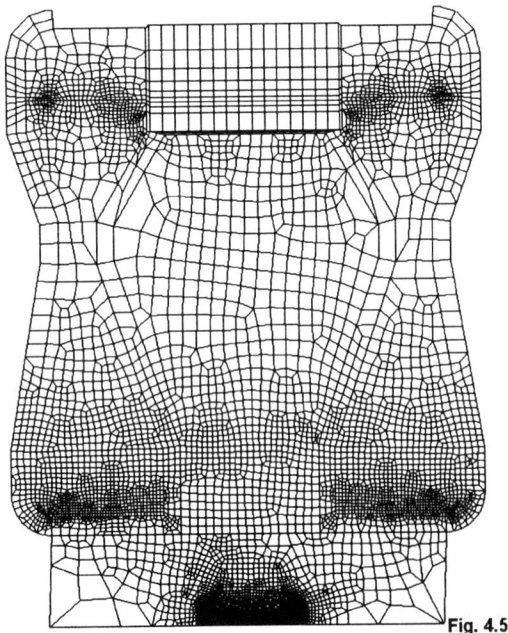

Fig. 1: Mesh for flying behaviour simulations

EFFECT OF SLIDER MATERIAL ON FRICTION AND DURABILITY OF LUBRICATED THIN-FILM DISKS

JUNGUO XU and BHARAT BHUSHAN
Computer Microtribology and Contamination Laboratory
Department of Mechanical Engineering
The Ohio State University
Columbus, OH 43210-1107

ABSTRACT

Ever increasing magnetic recording density requires a head slider to fly close to the disk surface which may lead to an intermitted contact condition resulting in wear on both slider and rigid disk surface (1). In order to maintain or increase the durability of head-disk interface (HDI), alternate slider materials are being explored.

Recently, there has been growing interest in silicon carbide (SiC) as a substrate material for picosliders (35%) and femtosliders (30%) with magnetoresistive (MR) type read heads. Silicon carbide has high thermal conductivity which provides better cooling during powering of the MR heads. Silicon carbide is also a preferable substrate for small sliders because it machines with less edge rounding, less pullouts, and high smoothness. In addition, its high thermal conductivity leads to lower interface temperatures which generally results in less degradation of the interface (2) (3).

Objective of this research is conduct drag and CSS tests using various ceramic sliders and conduct detailed material analyses of the interface component to elucidate the failure mechanisms, particularly to address the question on why does silicon carbide exhibits better friction and wear properties. Results of this study are the subject of this paper.

Mn-Zn ferrite, calcium titanate, Al_2O_3-TiC and SiC sliders were used in drag tests, stiction tests and contact-start-stop (CSS) tests. SiC slider exhibits the best tribological performance at all humidities as compared to other sliders. Relevant mechanisms are revealed with the measurement of contact angle of perfluoropolyether (PFPE) lubricants on slider materials, lubricant transfer tests, scanning electron microscopy (SEM), X-ray photoelectron spectroscopy (XPS) and Fourier transform infrared (FTIR) analyses on the slider surfaces before and after tests.

SiC slider is found to have the best tribological performance among the sliders tested in drag and CSS tests. The reasons can be attributed to the lubricant transfer which leads to the formation of a lubricant layer on the slider surface and tribochemical reaction which leads to the formation of a gel-like hydroxide layer. Both of them protect the HDI from direct solid contact and are responsible for the good tribological performance of SiC slider. Following are detailed conclusions:

1) SiC slider has the lowest coefficient of static and kinetic friction and the longest durability followed by Mn-Zn ferrite, Al_2O_3-TiC and calcium titanate sliders.

2) High relative humidity (85%) makes the durability of Mn-Zn ferrite, Al_2O_3-TiC and calcium titanate sliders significantly shorter. However, SiC slider still exhibits a long durability at high humidity.

3) Local lubricant stain was found on Al_2O_3-TiC and calcium titanate slider surface after sliding while it was not found on Mn-Zn ferrite and SiC slider surface.

4) PFPE lubricants (Z-Dol and Z-15) have the lowest contact angles against SiC compared with those against Mn-Zn ferrite, Al_2O_3-TiC and calcium titanate. It indicates that SiC is easily wetted and transferred by the lubricants.

5) The thickness of transferred lubricant on SiC surface reaches to 4 nm in static contact tests. Mn-Zn ferrite shows a thinner thickness of transferred lubricant layer compared with SiC while Al_2O_3-TiC and calcium titanate only has locally transferred lubricant layer. After lubricant transfer, the SiC surface becomes more hydrophobic.

6) Oxide was found on the virgin surface of SiC and Al_2O_3-TiC slider. The amount of oxide on SiC slider surface increases after sliding and high relative humidity leads to a larger amount of oxide formation. No changes was found on Al_2O_3-TiC slider surface before and after sliding. The oxide thickness for both SiC and Al_2O_3-TiC is less than 2 nm.

7) Hydroxides were found on the SiC surface after sliding at high relative humidity (RH = 85%) while no hydroxide was found on Al_2O_3-TiC surface.

REFERENCES

(1) Bhushan, B., 1996, *Tribology and Mechanics of Magnetic Storage Devices*, Second edition, Springer-Verlag, New York.
(2) Bair, S., Green, I., and Bhushan, B., 1991, *ASME Journal of Tribology*, Vol. 113, pp 547-554.
(3) Bhushan, B., 1992, *ASME Journal of Tribology*, Vol. 114, pp. 420-430.

LUBRICANT FLOW ON MAGNETIC HARD DISKS
- EFFECT OF AIR FLOW IN THE SPACE BETWEEN CO-ROTATING DISKS OR BETWEEN DISK AND WALL -

M YANAGISAWA

Functional Devices Research Laboratories, NEC Corporation, Miyamae-ku, Kawasaki 216, Japan

ABSTRACT

Lubricant thinning on rotating magnetic disks (spin-off) is an important problem to be solved, because a thick lubricant layer effectively suppresses bouncing and reduces wear, particularly, for contact recording systems (1).

Lubricant flow on magnetic hard disks was theoretically and experimentally studied for A) a single rotating disk, B) co-rotating disks, C) a single rotating disk and a static wall (Fig.1). Lubricants flow outward because of the shearing force at the air-lubricant interface in addition to centrifugal force. Different thinning behaviors of PFPE lubricants are observed in each of the three cases, because the air shear depends on air flow distribution in the space between co-rotating disks, or between the rotating disk and the static wall. The measured thinning behavior showed good agreement with calculated values, integrated from the momentum equation (2) (Eq.1), when air shearing forces τ (Eq.2) at the lubricant-air interfaces was introduced through the air-flow equation 3. As a result, the effect of the air shear on lubricant thinning was minimized when the distance between two disks, or between the disk and wall, was decreased to below a critical spacing. The critical spacing was 1.0 mm for case B (Fig.2), and 0.1 mm for case C. The critical spacing for the case C is smaller than for case B because the air shear over the co-rotating disk surfaces is larger than that over the surface of the disk rotating against the static wall.

REFERENCES

(1) Yanagisawa, M., Sato, A, and Ajiki, K., ASME, New York, TRIB-Vol.6, 1996, 25-32pp.

(2) Emslie, A.G., Bonner, F.T., and Peck, L.G., J.Appl.Phys., Vol.29, 1958, 858-862pp.

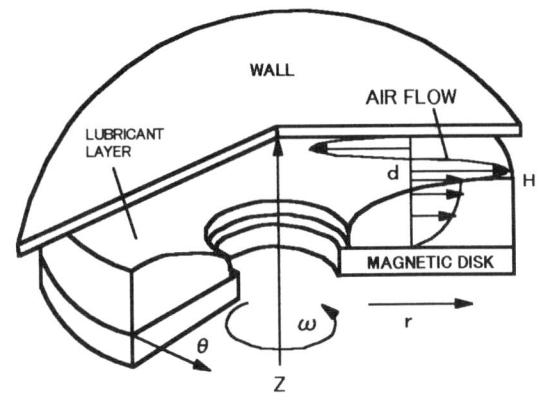

Fig.1: Lubricant flow model for case C.

$$-\eta \frac{\partial^2 v(r,z,t)}{\partial z^2} = \rho \omega^2 r \quad [1]$$

$$\tau = \eta_a \frac{du}{dz}\bigg|_{z=H_L} = \frac{\eta_a r\omega}{d} F'(0) \quad [2]$$

$$\frac{d^2F}{d\xi^2} - R^{1/2}J\frac{dF}{d\xi} - R(F^2 - G^2) = 0$$

$$\frac{d^2G}{d\xi^2} - R^{1/2}G\frac{dG}{d\xi} - 2FG = 0 \quad [3]$$

$$\frac{dJ}{d\xi} + 2RF = 0$$

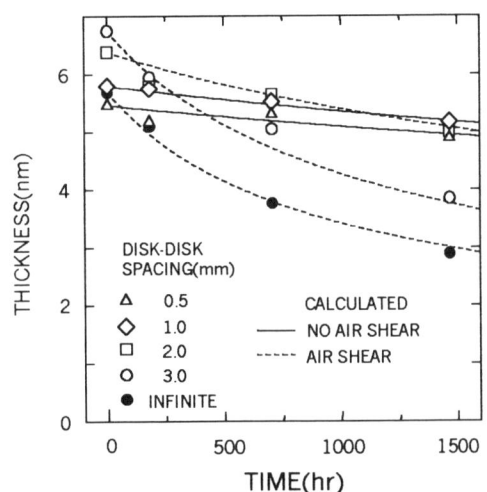

Fig.2: Thinning behavior for a variety of spacings between co-rotating disks.

LUBRICANT DESIGN FOR CONTACT RECORDING SYSTEMS

M YANAGISAWA, A SATO, and K AJIKI

Functional Devices Research Laboratories, NEC Corporation, Miyamae-ku, Kawasaki 216, Japan

ABSTRACT

Contact recording systems have been studied for future magnetic recording disks with a high recording density (1) (Fig.1). Key tribological technologies for ultra-low spacing and low wear are required for contact systems (2). Liquid lubrication systems (wet system) (Fig.2) particularly play an important roll in reducing mechanical spacing and improving wear performance. However, a lubrication design concept for contact recording systems has not been established.

In this study, new contact recording systems were studied for future magnetic recording disks with a high recording density. A molecular design of lubricants for contact systems was discussed from the viewpoint of bouncing and wear for 30% contact sliders. The following conclusions were reached.

1) The liquid lubrication system plays an important roll in reducing the bouncing height of contact sliders. Particularly, surface energy is effective for the bouncing suppression because of the meniscus force of lubricants (Fig.3). MPBT is a candidate for an effective lubricant. A minimum bouncing height of 3 nm and low wear were obtained for ion-etched 30% contact sliders by the optimization of pad design and lubricant material.

2) MPBT lubricant showed high wear performance. The result shows that the combination of light load force and MPBT lubricant is an effective for approaching a zero wear system for contact recording systems.

REFERENCES

(1) H Hamilton, Concepts in Contact Recording, ed. Adams. G.G., TRIB-Vol. 3, ASME, 1992, 13-23pp.

(2) M Yanagisawa, A Sato, and K Ajiki, Tribology of Contact/Near-Contact Recording for Ultra High Density Magnetic Storage, TRIB-Vol.6, ASME, 1996, 25-32pp.

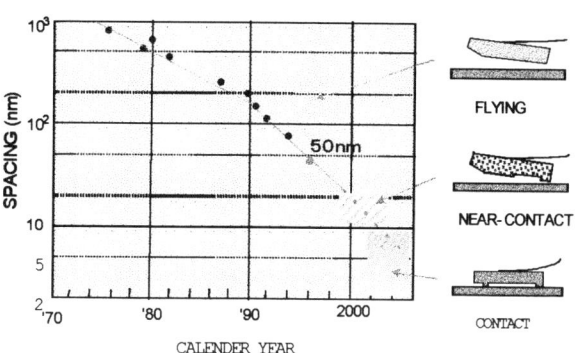

Fig.1: Spacing trend and contact system.

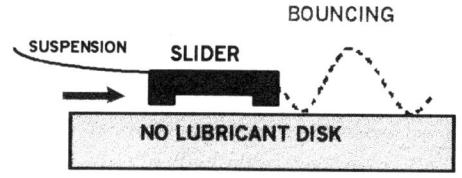

Fig.2: Wet system for bouncing suppression.

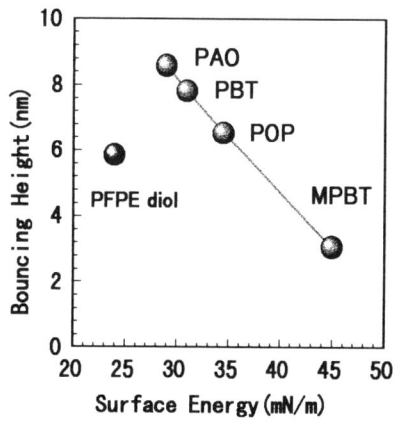

Fig.3: Bouncing height as a function of surface energy.

CONFIGURATION OPTIMIZATION OF A NEGATIVE PRESSURE MICROHEAD SLIDER FOR MEETING THE DESIRED HEAD-DISK INTERFACE PERFORMANCES

TAE-SIK KANG and DONG-HOON CHOI
School of Mechanical Engineering, Hanyang University, Sungdong-Gu, Seoul, Korea, 133-791
SANG-JOON YOON
Technical Center of DAEWOO MOTOR CO. LTD, Chongchon-dong, Puk-Gu, Inchon, Korea, 403-714

ABSTRACT

This study presents a design approach to determining the configuration of a negative pressure (NP) slider by using an optimization technique in order to meet the desired head-disk interface (HDI) performances. The desired HDI performances considered in this study are to minimize the variation in flying height from a target value, to maintain the pitch angle as large as possible, to keep the roll angle as small as possible, and to keep the outside rail to fly lower than the inside rail, over the entire recording band. Since the variations of flying height h, pitch angle α, and roll angle β as a function of a radial position are known to be smooth and continuous over the recording band, we concentrate on the smallest and/or the largest values of these flying attitude parameters. Denoting the smallest value and the largest value over the recording band by subscripts min and max respectively, the above performance requirements can be mathematically expressed as:

- to minimize $\left|h^* - h_{\min}\right| + \left|h^* - h_{\max}\right|$,
- to maximize $\alpha_{\min}$,
- to minimize $\beta_{\max}$,
- to satisfy $\beta_{\min} \geq 0$,

where h^* denotes the target value of the flying height specified by the designer.

As design variables, we choose several slider configuration parameters which seem the most influential on air-bearing characteristics. Denoting a vector of the design variables as **x**, we can formulate a multi-criteria optimization problem as follows: find the design variable vector **x** to

minimize $\Psi = w_1\left[\left(1 - H_{\min}\right)^2 + \left(1 - H_{\max}\right)^2\right]$
$- w_2 A_{\min}^2 + w_3 B_{\max}^2$,

satisfying $1 - A_{\min} \leq 0$,
$B_{\min} \geq 0$,
$x_i^L \leq x_i \leq x_i^U$, $i = 1, 2, \cdots$,

where $H = \dfrac{h}{h^*}, A = \dfrac{\alpha}{\alpha^L}, B = \dfrac{\beta}{\beta^U}$.

Notice that, three flying attitude parameters h, α and β are normalized by the target value h^*, the lower limit value α^L, and the upper limit value β^U, respectively. This normalization is important for successful convergence to the optimum design since it makes each performance index have the same order of magnitude regardless of units used.

To determine flying attitude parameters, we use the static analysis method proposed by Choi and Yoon (1). The pressure calculation for the static analysis is carried out by solving a generalized lubrication equation based on the linearized Boltzmann equation (2). For rapid pressure calculation, a Poiseuille flow rate database (3) is also used. For optimization, we utilize the sequential quadratic programming (SQP) method in Automated Design Synthesis (ADS) developed by Vanderplaats (4).

To validate the proposed design approach, a computer program is developed and applied to the configuration design of the Guppy slider positioned by a rotary actuator. A variety of optimum configurations are automatically obtained without any difficulty for various decisive parameters (determined prior to slider design) such as target flying heights, disk sizes, skew angle ranges, preloads, and disk rotation speeds. This excellent adaptation to various design situations in an efficient manner demonstrates the effectiveness of the suggested approach

REFERENCES

(1) Choi, D.-H. and Yoon, S.-J., ASME Journal of Tribology, Vol. 116, pp. 90-94, 1994
(2) Fukui, S. and Kaneko, R., ASME Journal of Tribology, Vol. 110, pp. 253-262, 1988
(3) Fukui, S. and Kaneko, R., ASME Journal of Tribology, Vol. 112, pp. 78-83, 1990
(4) Vanderplaats, G.N., Engineering Design Optimization, Inc., 1985

Effect of Environment on the Friction/Stiction and Durability of Lubricated Magnetic Thin-Film Disks

Zheming Zhao and Bharat Bhushan
Computer Microtribology and Contamination Laboratory, Department of Mechanical Engineering
The Ohio State University, Columbus, OH 43210-1107, USA

ABSTRACT

Environmental humidity plays a significant role in the tribological performance of head-disk interface (HDI) (1)(2). High stiction and friction arises as a result of the development of menisci at the asperity contacts. Recent studies by authors (3)(4)(5)(6) have been conducted to systematically understand the effects of the degree of chemical bonding of the lubricant and humidity on friction/stiction and durability in drag tests, and methodology for roughness measurement and contact analysis. No studies have been reported to date which examine the effects of degree of bonded fraction on tribological performance at various levels of humidity in contact start and stop (CSS) tests. The purpose of this research is to study these effects in CSS tests and compare the tribological performance with that in drag tests.

The single-disk tester was set-up in a humidity-controlled chamber, in a class-100 clean room environment at a temperature of 22±1°C. Static, kinetic friction and durability measurements were performed on unlubricated and lubricated disks. The disks with untreated, partially bonded and fully bonded films of a polar perfluoropolyether lubricant were tested at different levels of humidity. Thin-film Al-Mg disks used in this study had a diamond-like carbon overcoat with an rms roughness of 10.4 nm. The chemical structure of the PFPE lubricant (Z-DOL) is $HO-CH_2-CF_2-O-(CF_2-CF_2-O)_m-(CF_2-O)_n-CF_2-CH_2-OH$. The two rail taper-flat microsliders fabricated from Al_2O_3-TiC (70-30 wt%) were chosen for tests with an rms roughness of 1.3 nm.

Static friction is low for unlubricated disks and disks with a fully bonded lubricant film as compared to disks with untreated and partially bonded lubricant films. Static friction gradually increases with an increase in the relative humidity and the effect of humidity at high humidity values, is more pronounced in the case of disks with untreated and partially bonded lubricant films.

At 80% RH, the unlubricated disks show some effect of the humidity on the coefficient of kinetic friction whereas the fully bonded disks do not.

For fully bonded disks, the durability in the drag tests is insensitive to humidities and at high humidity it is equivalent of the unlubricated disk. In the CSS tests, durability decreases with an increase of the relative humidity and levels off at a lower level which is equivalent to that of the unlubricated disk at moderate to high humidities. For partially bonded and untreated disks, durability generally decreases from low to moderate humidities and drops in all cases at high humidity.

For partially bonded and untreated disks, at 5% and 50% RH in drag tests, higher durability is achieved by the partially bonded disk followed by the untreated disk. At high humidities, the durability in drag tests is comparable to that of the fully bonded and unlubricated disks. In the CSS tests, the durability of partially bonded and untreated disks is comparable at all levels of humidities. Thus, the partially bonded disk is the best in drag tests and is as good as the untreated disk in CSS tests. The mobile fraction of lubricant film allows mobility to protect exposed areas and the immobile fraction can prevent from being disturbed by the adsorbed water molecules. Besides, the immobile fraction will not form meniscus forces in a long-term stiction test. It is clear that degree of bonded fraction is key to friction/stiction and durability of the disks in different environments.

A further examination of the slider surfaces after durability tests provides more information on the failure mechanisms at the head-disk interface. Most of the debris is deposited in the leading and trailing areas. There is no direct effect of the humidity on the debris deposition on slider surfaces after durability tests. The deposited debris can be removed by a dry cotton swab. XPS spectrum shows that the debris on the slider surface most probably consists of the lubricant and the disk overcoat since there are no visible worn patches or scratches on the cleaned slider surface under an optical microscope. It appears that the coexistence of mobile and immobile fractions of the lubricant film is most efficient in collecting the debris, acting as cement to collect the debris on the interface, particularly the slider surface.

In general, partially bonded lubricant films are desirable for low friction/stiction and high durability and lubricated disks perform best at low to moderate humidities. Operation at high humidities must be avoided.

REFERENCES

(1) B. Bhushan, *Tribology and Mechanics of Magnetic Storage Devices*, second edition, Springer-Verlag, New York, 1996.
(2) B. Bhushan, M.T. Dugger, *ASME J. Tribol.*, Vol. 112, 1990, 217-223pp.
(3) C. Gao, B. Bhushan, *Wear*, Vol. 190, 1995, 60-75pp.
(4) Z. Zhao, B. Bhushan, *Wear*, Vol. 202, 1996, 50-59pp.
(5) B. Bhushan, Z. Zhao, *IEEE Trans. Magn.*, Vol. 33, 1997, 918-925pp.
(6) B. Bhushan, *IEEE Trans. Magn.*, Vol. 32, 1996, 1819-1825pp.

PASSIVE MAGNETIC BEARINGS: CONIC BEARING

ROBERTO BASSANI and SEBASTIANO VILLANI
Dipartimento di Costruzioni Meccaniche e Nucleari - Università di Pisa - Via Diotisalvi, 2 - 56126 Pisa - Italy

ABSTRACT

Magnetic passive bearing is a device that uses magnetic forces in order to preserve a gap between two surfaces with relative motion. They are interesting for a drastic reduction, on the whole, of friction, wear and energy loss phenomena. The present possibility of producing Rare-Earth permanent magnets (Samarium-cobalt and Neodymium-iron-boron) is making possible the realisation of maximum energy product (BH_{max}) and high-coercitivity permanent magnets, for magnetic levitation passive support systems. In Fig. 1 the property achievement is shown. Permanent magnets have traditionally been limited to low wattage devices but now there is a definite trend to apply permanent magnets in larger equipment and for a new motion control industry. The instability of magnetic passive bearings doesn't compromise their utilisation in hybrid levitation systems, constituted by magnets and nonmagnetic bearings. The possibility to develop stable systems of magnetic bearings is investigated (1).

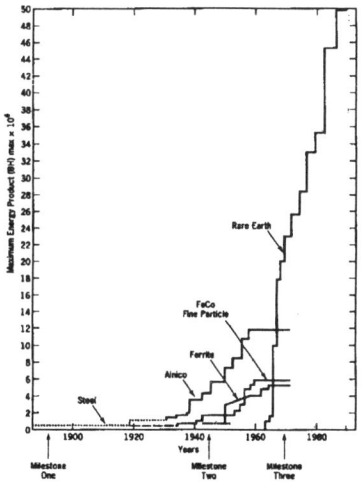

Fig. 1 - Progress in property development.

The increasing attention toward passive magnetic bearing can be explained also with stiffnesses growing values (10^3-10^4 N/m), against the (10^6 N/m) of active magnetic bearings stiffnesses. We can say that ring-shaped permanent magnets are the most used in passive magnetic bearings construction, either because they are the most suitable in replacement of conventional mechanical radial and axial bearings, or their easy realisation (2). An analytical and numerical method, based on equivalent surface currents densities theory for magnetic fields modelling, and Ampére's theory for calculating forces and stiffnesses versus geometric and magnetic characteristics of rings (3)(4), has been tested and then used for characterising and synthesising the behaviour of some basic configuration of axially and radially magnetised ring-shaped magnetic bearings (5).

The equivalence has been verified at the same geometrical and physical conditions. In order to complete the description and analogy between magnetic and mechanical bearings, we have studied the behaviour of magnetic bearings with permanent magnets or with rings impiled in a conic shape (Fig. 2) conic-shaped. It is clear that angular bearings with rolling elements (ball or cylinder) constitute the mechanical equivalent of these magnetic conic-shaped bearings. Also in the case of Meissner repulsive levitation a bowl-shaped object is required to achieve an energy potential well and to make the stability achievable.

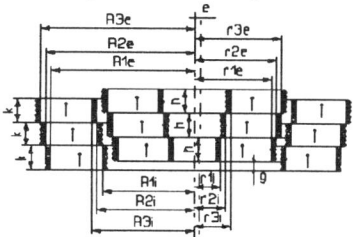

Fig. 2 - Equivalent surface-current representation.

The modified and adapted analytical and numerical method has been used in order to study the behaviour of conic-shaped with axial, radial and then angular magnetization magnetic bearings. The graphic results (Fig. 3) of the research are been obtained using the general equation of forces that is synthetically:

$$F = -\frac{\mu_0 j_1 j_2}{4\pi} \oint\oint \frac{r(ds \cdot ds')}{r^3}$$

In Fig. 3 are presented the axial and radial components for one couple of magnetic bearing.

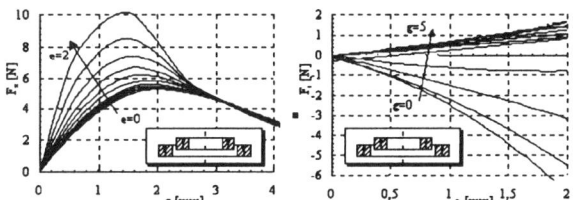

Fig. 3 - Axial and radial component versus axial (g) and radial (e) displacement.

REFERENCES

(1) R. Bassani, Atti del Dip. di Costruz. Mecc. e Nucleari, DCMN 004(94).
(2) R. Bassani, S. Villani, Atti del Dip. di Costruz. Mecc. e Nucleari, DCMN 005(94) and 006(94).
(3) R. Bassani, S. Villani, Atti III Convegno AIMETA di Tribologia, 3-4 Sept. 1994, 55-63 pp.
(4) R. Bassani, S. Villani, Applied Mechanics and Engineering, Vol. 1, No.1, 77-96.
(5) R. Bassani, S. Villani, Atti IV Convegno AIMETA di Tribologia, 3-4 Oct. 1996, 213-220 pp.

CHAOS IN THE UNBALANCE RESPONSE OF A RIGID ROTOR IN JOURNAL BEARINGS

R D BROWN and G DRUMMOND
Heriot-Watt University, Edinburgh, EH 14 4AS
P S ADDISON
Napier University, Edinburgh, EH14 1DG

ABSTRACT

Hydrodynamic journal bearings operating at high eccentricity generate oil film forces which can be extremely nonlinear. For some conditions the calculated response due to high levels of unbalance is aperiodic as shown by Ettles et al in 1978 (1). The operating conditions for these cases are mostly indicated as stable using a linear model.

Recently calculations have been carried out with a rigid rotor supported on an oil film modelled by a short bearing theory have shown that the aperiodic behaviour predicted by Ettles is probably chaotic. Various techniques have been used to demonstrate the nature of the nonlinear response. These include bifurcation diagrams, spectral analysis and estimates of fractal dimension.

The aperiodic boundary of Ettles was shown to be very similar to a chaotic boundary Brown et al (2) when compared on a force ratio basis. This ratio was unbalance force divided by the gravitational load. When this ratio exceeded a value a little greater than one chaotic behaviour was obtained. For force levels of this nature it is clear that the overall bearing load is momentarily very low and hence intermittently unstable. This is a classic situation for chaotic behaviour.

Figures 1 and 2 show a typical bifurcation diagram and a Poincare plot. Both of these examples were obtained for high eccentricity and high unbalance loads. The Poincare or return mapping is often used to examine the cyclical behaviour of nonlinear systems. A limit cycle syncronous response mapping will produce a single point. For chaotic systems this mapping produces a fractal structure as can be seen in Figure 3.

Recently some experimental evidence has been published by Adiletta et al (3). They point out that when chaotic motion is present the oil film thickness can achieve very low values that may affect bearing operation.

The required levels of unbalance to establish chaos are only an order of magnitude greater than acceptable levels for rotating machinery and thus could be experienced with in-service erosion or minor damage. The subsequent non-synchronous response could result in fatigue and potential shaft failure.

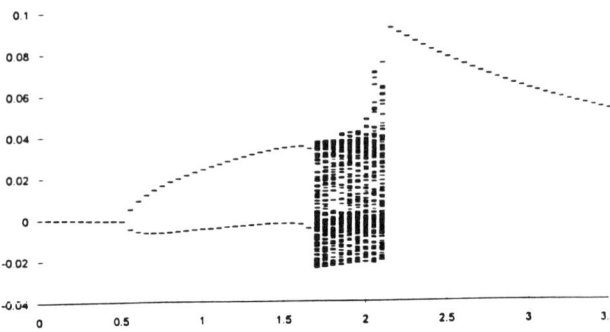

Fig. 1: Typical bifurcation e = 0.8, $\omega\sqrt{C/g} = 2.5$

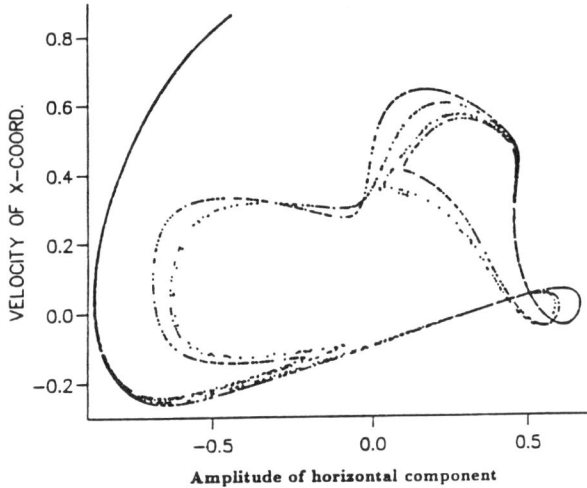

Fig. 2: Typical Poincare map e = 0.778

REFERENCES

(1) A G Holmes, C M M Ettles and I W Mayes, International Journal for Numerical Methods in Engineering, Vol.12, 1978.

(2) R D Brown, P S Addison and A H C Chan, Nonlinear Dynamics (Kluwer) Vol.5, pp 421-432, 1994.

(3) G Adiletta, A R Guido and C Rossi, Nonlinear Dynamics (Kluwer), Vol.10, pp 251-269.

INFLUENCE OF FLUID INERTIA FORCES ON THE SUDDEN UNBALANCE RESPONSES OF SQUEEZE FILM DAMPER SUPPORTED ROTORS

GUO YIN-CHAO and MENG GUANG
Northwestern Polytechnical University, Xian, China
E J HAHN
School of Mechanical and Manufacturing Engineering, University of New South Wales, Sydney NSW 2052, Australia

ABSTRACT

The unbalance value of a rotor may increase suddenly in service, eg due to blade loss, thereby jeopardising the safety of the rotor system. Earlier analyses show that a well-designed squeeze film damper (SFD) support system has the ability to suppress the harmful effects of sudden unbalance response (1)(2), but the influence of fluid inertia on SFD oil film forces have generally been ignored. However, recent theoretical and experimental research has shown that the influence of fluid inertia forces is sometimes quite large and should be considered. In their experiments, Tecza et al (3), showed that fluid inertia may be a significant factor in determining the dynamic characteristics of SFD's. Tichy (4) provided an explanation for the importance of fluid inertia forces in SFD's as distinct from journal bearings, while more recently, San Andres and Vance (5) using momentum methods, obtained the oil force coefficients for short cylindrical SFD's.

Hence, this paper investigates the influences of fluid inertia forces on the sudden unbalance response of a flexible (Jeffcott) rotor supported on SFD's. The system's equations of motion are solved for both centralised and uncentralised dampers, using a simple differentiable function to better express the sudden unbalance process. It is shown that for sudden unbalance response at constant rotating speed, fluid inertia can shorten the transient process and decrease the transient vibration amplitude. For sudden unbalance response while accelerating through a bistable region, if the sudden unbalance occurs before the bistable region, the response will follow the large orbit solution of the bistable response and will jump down when the bistable region ends; whereas if the sudden response occurs within the bistable region, the response will follow the small orbit solution and the bistable jump will not occur. In all cases, fluid inertia forces help shorten the transient time and reduce the transient amplitude after sudden unbalance.

REFERENCES

(1) G. Meng, E. J. Hahn, Sixth International Conference on Vibration in Rotating Machinery, IMechE Conference Transactions, 1996, 651-659pp.

(2) M. Sakata et al, ASME Transactions, Journal of Engineering for Power, 1983, 480-486pp.
(3) Tecza et al, ASME Paper 83-GT-177.
(4) J. A. Tichy, ASLE Transactions, Vol. 25, 1982, 125-132pp.
(5) L. A. San Andres, J. M. Vance, ASME Transactions, Journal of Engineering for Gas Turbines and Power, Vol. 108, No. 2, 1986, 332-339pp.

A NEW AND COST EFFECTIVE APPROACH TO CONDITION MONITORING FOR THE BLENDING OF NEW LUBRICANTS

EMANUEL AKOCHI-KOBLÉ, MICHEL MURPHY (Member STLE) and DAVID PINCHUCK
(Member, STLE)
Thermal Lube Inc. 255, Avenue Labrosse, Pointe-Clare, Québec, Canada H9R 1A3

ASHRAF A. ISMAIL*, FREDERICK R. VAN DER VOORT AND JUN DONG
McGill IR Group, Department of Food Science and Agricultural Chemistry
Macdonnald Campus of McGil University, Ste Anne de Bellevue, Québec, Canada H9X 3V9

TAEHYUK CHOI
Yeongcheol Park, Jonghoon Lee, Woo H Shin Yukong Limited, Korea

ABSTRACT

This paper describes the application of an FTIR based Continuous Oil Analysis and Treatment system (COAT) for the condition monitoring of new lubricants during formulation process. The manufacturing of lubricants involves blending, on weight or volume basis, of specific additives into a selection of base oils. The choice and amount of additives will directly influence the grade and viscosity characteristics of the blended lubricant. Batch to batch variability due to under- or over-dosage of one or more additives can have adverse effects on the performance of the finished product.

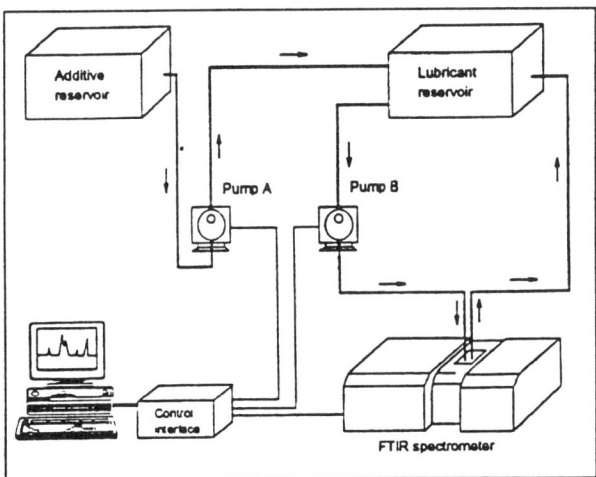

Figure 1: Schematic diagram of the COAT system

Further more, most additives are not sold in pure state, but rather in a carrier fluid at a given concentration due to their polymeric nature and high viscosity; thus, a change of supplier or an error in the formulation of the concentrate can adversely influence or compromise the quality of finished products. The above errors can be greatly reduced when each additive concentrate is pre-analyzed and the amount closely monitored during production. FTIR spectroscopy offers the flexibility to accomplish the above without any sample preparation, and in a rapid and cost effective manner.

The system (Fig.1) makes use of a proprietary dedicated software interface (COAT Scan) to manage the entire analysis process, from sampling to data analysis, and to subsequent corrective actions. What makes this system unique is that COAT Scan allows to take full advantage of a powerful infrared analytical approach, through automation of routine operations such as data collection, data processing and analysis. The continuous monitoring along with the extensive data collection and real time analysis capabilities of this technology makes it applicable to used oil condition monitoring, thus resulting in extending the service life of machinery components and lubricants therefore, reducing down time, disposal and oil changes. This is certainly a cost-effective approach as lubricant disposal costs will soon exceed new lubricants costs. In a recent study, the service lives of various lubricants were increased 3 to 4 folds by continuous monitoring key additives. Should readings for any particular additive fall below the pre-established limits, the system will initiate a replenishment process.

The first phase of this work was to monitor the entire process of manufacturing automotive and hydraulic synthetic lubrication fluids. As a first step, a comprehensive infrared data base was compiled. Subsequent to that, the infrared spectra of all targeted additives i.e. antiwear, viscosity improver and anti-oxidant etc.. were recorded at their pre-established levels for each type of product. This data base was used as a reference to monitor the actual lubricant state during production.

The second objective, a on going collaboration between Thermal-Lube Inc. Canada and Yukong Limited, Korea, was to develop various calibration models for condition monitoring of automotive (gasoline and diesel) and industrial mineral lubricants produced by Yukong Limited. Calibration algorithms relating peak height to absorbance and multivariate approaches, depending on spectra characteristics, were successfully employed. These calibrations were to be used for lubricant condition monitoring in production facilities.

TRIBOLOGICAL PHENOMENA AT EXTRUSION OF HIGH FILLED COMPOSITES

V.G.BARSUKOV and J.F.SVEKLO
Research Center on Resource Savings of Belarus Academy of Sciences, Tysengauz sq. 7, Grodno, 230023, Belarus, CIS
V.Ja.PRUSHAK
Research Institute on Resource Savings, Kozlov st. 69, Soligorsk, 223710, Belarus, CIS

Methods of screw and plunger extrusion are very extended at manufacture of profile articles made from different materials, but processing of high-filled composites by these methods is kept back due to unsatisfactory known factors like the effect of friction of the filler with itself as well as with working units of equipment on forming processes.

In high filled composites contact interactions among filling particles and between the particles and walls of forming equipment are characterized by high local pressure; and at the same time there is dry, boundary, and semi-liquid friction. An application of Coulomb's friction law for an analysis of shear in such systems is restricted.

The choosing calculations' scheme is founded on the fact that Coulomb's friction law is valid only for the loads low in comparison with contacting solids hardness, i.e. when actual contact area is smaller than nominal one. At high pressure as it has been market in number of papers (1), (2) the non-linear asymptotical dependency of friction force on pressure is achieved (Fig.1.)

The influence of tribological factors on rheological features of high filled composites at extrusion processing is studied using the piecewise linear approximation of friction force on applied pressure.

It was shown that in general case there are two rheological zones in the extrusion channel: viscous-elastic and viscous-plastic ones. The first zone is located near the outlet of the channel and the second - at the inlet. A position of the boundary between these zones is determined by a ratio of external and internal friction forces and depends on a cross-section shape of the extrusion channel. For the main cross-section shapes of the extrusion channels (ring, circle, rhomb, and rectangle) the extreme friction forces are estimated. An exceeding of those values results to disappearance of the viscous-elastic zone and spreading of the viscous-plastic one on the hole length of the extrusion channel.

Fulfilled investigations showed that in processing material a region of tensile stresses appears near the outlet of the extrusion channel under influence of friction forces. The maximum values of these stresses equal to the value of tangential tensions at the material bulk owing to friction forces. Comparing trajectories of the main tensile stresses calculated using numerical Runge-Kutt method with trajectories of inclined cracks in the material shows that an origin and development of such cracks take place when the viscous-plastic zone of the composite is within the region of the tensile stresses. The recommendations on calculation of the extreme friction forces to prevent cracking are proposed (3).

An appearance near the inlet to the extrusion channel of the zone of deformation absence, known as 'dead zone', where irreversible flow of the composites practically absent, may be considered as rheological phenomena caused by friction forces.

In the space between flights of screw the friction forces cause the energy dissipation of the drive on the sliding surface. It follows by increasing of the composite temperature on 5-30 °C and more versus initial values. It may cause thermodestruction of the nonthermostable material such as PVC-compositions and decreasing of productivity of equipment manufacturing.

REFERENCES

(1) B.A.Druyanov, Applied theory of porous solids plasticity, Moscow, Mashinostroenie, 1989.
(2) E.M.Makushok, T.V.Kalinovskaya, A.V.Bely, Mass-transfer in friction processes, Minsk, Nauka i technika, 1978.
(3) V.G.Barsukov, B.I.Kupchinov, E.M.Lapshina, Stretch strain in extrusion profiles from high-filled moulding materials, Vesti Akademii Nauk Belarusi, Phys.-techn. part. J., Vol.1, p.11-14, 1994.

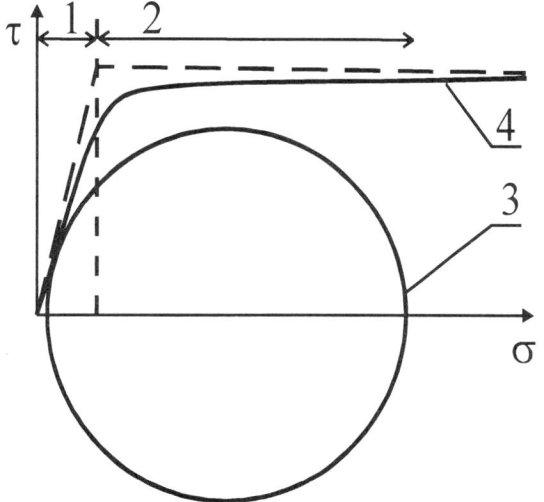

Fig.1 The dependence of specific friction force τ on pressure σ. 1 - zone of Coulomb's friction law validity; 2 - zone of Prandtl's friction law validity; 3- Mohr circle; 4- real dependence of specific friction forces on pressure (the Mohr circle envelope).

THE INFLUENCE OF CUTTING FLUID COMPOSITION ON THE WEAR OF HIGH SPEED STEEL TOOLS IN INTERMITTENT CUTTING

E BERMINGHAM
Bermingham Lubetec Services, Beaconsfield, Bucks, HP9 1RY, UK
J L HENSHALL
Mechanical Engineering Department, Brunel University, Uxbridge, Middlesex, UB8 3PH, UK
R M HOOPER
School of Engineering, Exeter University, Exeter, Devon, EX4 4QF, UK

ABSTRACT

High speed steel (HSS) tools are being replaced for many continuous cutting operations, but the size and complexity of some tools, e.g. as used in gear hobbing, mean that HSS will continue to be used for the foreseeable future. In these situations, cutting fluids and additives are used extensively in industry to prolong the working life of the tool by reducing the rate of wear during cutting. Shaw (1) has concluded that the selection and application of metal cutting fluids is much more of an art than a science and Trent (2) states that the influence of active lubricants on the rate of tool wear is very complex and that too little attention has been given to the mechanisms by which lubricants affect tool wear. There is little science in the method of choosing which is the optimum in a particular operation. Most recommendations and choices are still based on general guidelines, as contained in several handbooks, previous experience or very time-consuming, and hence expensive, in-service trials.

Thus the aim of the present work was to develop an appropriate tool wear assessment procedure, and to use this to evaluate the performance of seven mineral oil based cutting fluids. The gear hobbing operation was simulated by using a single point high-speed steel tool (composition {wt%} C-1.1: Co-8.0: Cr-3.7: Mo-9.5: V-1.2: W-1.5) intermittently cutting a low alloy steel (composition {wt%} C-0.13: Cr-1.01: Mn-0.54: Mo-0.49: N-0.25: P-0.018: S-0.034: Si-0.2: V-0.005).

Single point intermittent cutting was achieved by mounting the tools, which were shaped in-house from standard HSS rods of 9.5 mm diameter, in a flycutter attached to the head spindle of a vertical milling machine. The cutting speed varied between 0.3 m/s and 1.1 m/s, thickness of cut, t_i, between 0.09 and 0.37 mm and the depth of cut was constant at 1.6 mm.

The cutting fluids used covered a range of values of viscosity (5-39 cSt at 40°C), elemental sulphur (0-0.8%), sulphurised fat (0-10%), chlorinated paraffin (0-23%), and free fatty acid 0-0.2%). The amount of cutting with a particular fluid was limited to ensure little depletion of the additives.

The undeformed chip thickness was used as the main control parameter regarding the severity of the machining operation. The worn tool from each test, and chips from some of the tests, were examined by optical microscopy and scanning electron microscopy, with attached energy dispersive X-ray analysis.

Conventionally, the rate of tool wear is assessed in terms of one particular measure of tool wear, e.g. length of flank wear scar. A different approach to determining effectively the rate of tool wear was developed in this work. The problem with the conventional approach is that the trials have to be conducted until 'failure' occurs, which may require prolonged machining, and it is not necessarily *a priori* possible to determine the critical wear mechanism when this may change as a result of the composition of the cutting fluid. The methodology adopted required an assessment of the relative extent and smoothness of the worn regions on the flank and rake faces, the degree of 'built-up edge' on the rake face and the degree and intensity of 'tempering' colours in the vicinity of the cutting zone.

It was found that the variation in viscosity had no discernible effect, and also the limited amount of free fatty acid had no effect. These observations would be as generally expected. The effects of the sulphur and chlorine containing levels on the wear processes can be summarised as follows:

1. <u>Low t_i (0.09 & 0.1 mm)</u>

   Increasing the chlorinated paraffin content decreased rake face temperature and decreased the rate of flank wear or built-up edge formation. Increasing the elemental sulphur increased the rate of flank wear.

2. <u>Medium t_i (0.14 & 0.19 mm)</u>

   Increasing the chlorinated paraffin decreased the flank face temperature. Increasing the elemental sulphur or sulphurised fat decreased the rate of rake face wear. Increasing the sulphurised fat resulted in an increase in the flank face wear rate.

3. <u>High t_i (0.29 & 0.37 mm)</u>

   Increasing the chlorinated paraffin decreased the flank face temperature. Increasing the sulphurised fat tended to increase the rates of both rake and flank face wear.

REFERENCES

(1) M. C. Shaw, Metal Cutting Principles, Clarendon Press, Oxford, 1984.
(2) E. M. Trent, Metal Cutting, Butterworth's, London, 2nd Edition, 1984.

COMPUTATIONAL DETERMINATION OF MANUFACTURE CONDITION RESULTING IN A PRE-SET WEAR RESISTANCE.

VYATHESLAV F. BEZJAZICHNY
Rybinsk academy of aviation technology, Pushkin st.53, Rybinsk, 152934, Russia

A functional relationship is established between the cutting conditions, the tool geometry on the one hand and the surface parameters and accuracy of machining on the other hand. The investigations were made taking into account the work and tool material properties, the rigidity of the machine-fixture-tool-workpiece complex, dimensions of the workpiece machined and size of the cutting tool.

It thus follows that cutting condition and the toll geometry are functions of the surface characteristics, machining accuracy, work and tool material properties, workpiece, and cutting tool dimensions and the rigidity of the machine-fixture-tool-workpiece complex.

It is advisable to select cutting conditions which give optimum temperature providing minimum relative change in dimensional size and, therefore, maximum tool life.

Operation at optimum speed of cutting provides the most favourable parameters of the quality of the surface layer. Thus, at optimum speed of cutting minimum asperity height of the surface machined and minimum approach of the contacting surfaces under load are observed. Cutting at optimum speed gives minimum wear of the surface machined.

When the optimum speed of cutting is known the wear rate value for the known machining conditions may be determined from the theoretical formula.

Thus, the wear rate is a function of the physical and mechanical properties of the work and tool material, of the cutting conditions and of the geometrical parameters of tool cutting edge.

It may be assumed that the qualitative characteristic of the wear rate relating it to the machining conditions is correct. Increasing feed leads to increases both in the asperities height and spacing, and, as a result, to the subsequent increase of the wear rate of the surface machined. Increase of the cutting tool nose radius will reduce the asperities height and spacing, producing a reduction of the wear rate of the surface machined. Increasing the depth of cut will reduce the wear rate, if the ratio of depth of cut is more than nought point two. With further increases in the depth of cut the wear rate practically will not change because of analogous changes in surface roughness parameters when the depth of cutting is increased.

As a rule, quantative results of the calculations are found to be somewhat different from actual values of the wear rate obtaind in the calculations.

In order to obtain better of the results of the calculations with the actual values of wear resistance given above, the following dependencies have been established.

$$\frac{R_{max}}{b^{\frac{1}{v}} \cdot r} = \frac{C_0 Б^{C_1} Γ^{-C_2} Д^{2.2} M^{0.8} (1-\sin\alpha)^{0.17}(1-\sin\gamma)^{0.1}}{E^{1.2} Э^{C_3} \left(\frac{\delta}{\rho_1}\right)^{0.19}}$$

where:

$$Б = \frac{Va_1}{a}; Γ = \frac{\lambda_p}{\lambda}βε; Д = \frac{a_1}{b_1}; M = \frac{b}{b_1}; E = \frac{\rho_1}{a_1}; Э = \frac{s}{r}$$

R_{max} is the maximum asperities of the profile; b and v are parameters of power approximation; C_0, C_1, C_2, C_3 are coefficients, dependent on work material properties.

	Heat-resistant allous	Titanium alloys	Stainless and heat-resistant steels	Carbon and alloy steels
C_0	0.0195	0.0334	0.0438	0.0915
C_1	0.22	0.10	0.10	0.06
C_2	0.33	0.33	0.32	0.26
C_3	2.74	2.62	2.62	2.58

S is the feed; V is the speed of cutting; r is the tool nose radius; a is the coefficient of temperature conductivity of the work material; λ and λ_p are the coefficients of the heat conductivity of the work and tool material, respectively; ρ_1 is the cutting edge radius; β and ε are the wedge and tool nose angle, respectively; a_1 and b_1 are underformed chip thicknees and with of cut; b is total length of the engaged cutting edge.

Thus, it appears possible to calculate tribological characteristics of the surfaces machined and machining conditions providing specified operation characteristics (by theoretical methods).

Wear intensity of surface machined:

$$J_h = 0.078 \frac{\alpha^{0.5} \varepsilon_1^{2.1} m_1^{0.5}}{n} R_z^{0.5x+0.364}$$

where: α - coefficient, dependent on contact type ; ε_1 - relative rapprochement of surfaces ; n - number of cycles of influence, coming to material destroy, m_1 and x - coefficients, dependent on Rz.

PREDICTION OF THE TOOL WEAR IN GEAR HOBBING

K. –D. BOUZAKIS, A. ANTONIADIS AND S. KOMPOGIANNIS

Laboratory for Machine Tools and Machine Dynamics Aristoteles University of Thessaloniki, GR – 54006, Greece

ABSTRACT

Gear hobbing is an efficient method of gear manufacturing. Due to the fact that during the cutting process every hob tooth always cuts in the same generating position, while in the various generating positions the formed chip has different geometry, the resulting tool wear is not uniform on any particular hob tooth. In order to overcome this problem, the hob is shifted tangentially after a certain number of cuts. Mathematical models to calculate the progress of hob wear in the individual generating positions, considering the existing process parameters, were presented (1). In this publication, in order to calculate flank wear regarding the complicated chip geometry, equivalent chip dimensions, such as the cutting length l, the chip thickness h_s and the characteristic chip form (chip group) were introduced. Based on these calculations, a computer algorithm for the determination of the hob flank wear, which depends on the shifting conditions, was presented.

Simulating the hobbing process with the aid of a computer program, it is possible to determine the length, thickness and group of every chip in the various cutting and generating positions. With the aid of these parameters, the progress of the flank wear on a hob tooth during cutting in the same generating position in all successive cutting positions along the gear width can be determined, as demonstrated in figure 1. This procedure is repeated for all generating positions.

To optimize the shift displacement and amount, the course of the flank wear versus the number of hobbed gears is calculated in every individual generating position as well as the wear distribution at the hob teeth. The calculated number of hobbed gears and the occurring width of the flank wear, using various shift conditions, is shown in figure 2. The shift displacement is expressed as a multiple of the hob axial pitch ε. Using such diagrams the shift displacement and amount can be determined with respect to a prescribed maximum value for the flank wear.

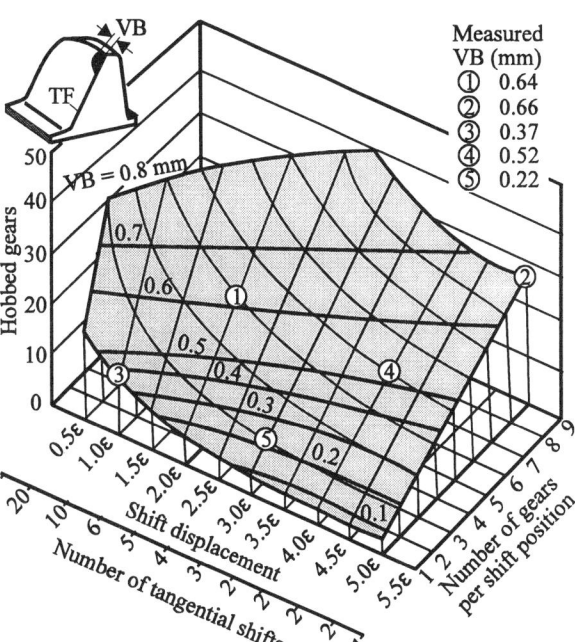

42CrMo4 V, R_m = 920 N/mm^2, S6-5-2-5, up-cut hobbing
m=4 mm, α=20°, β=0°, n_i=9, ε=1.39 mm,
z_1/z_2=1/23, d_{a1}=80 mm, b=25 mm, s_a=2 mm/rev, v=20 m/min,

Fig. 2: Number of hobbed gears for various shift conditions and occurring width of flank wear.

REFERENCES

(1) Bouzakis, K. -D, Konzept und technologische Grundlangen zur automatisierten Erstellung optimaler Bearbeitungsdaten beim Waelzfraezen. Habilitationsschrift, TH Aachen, VDI-Verlag, Fortschr. Br. 02, Nr. 42/81, (1980).

(2) Bouzakis, K. -D, Antoniadis, A., Annals of the CIRP, Vol. 44/1, (1995): 75-78.

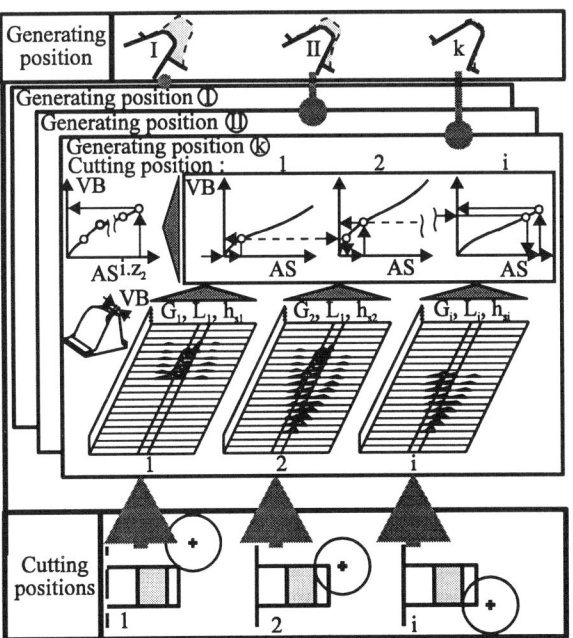

Fig. 1: Determination of the hob flank wear in the successive cutting positions of a generating position.

METAL CUTTING AND MODEL OF COMPLEX OPTIMISATION OF CUTTING CONDITIONS

FRANC CUS and JOZE BALIC
University of Maribor, Faculty for Mechanical Engineering, Institute for Production Systems,
Smetanova 17, 2000 Maribor, Slovenia,

ABSTRACT

The rationality and economy of manufacturing which are a result of material and energy saving and shorter machining times, depend to a large extent on the right choice of selected cutting conditions and required product quality. This paper deals with the development of a model for complex optimisation of cutting conditions showing that the right way from optimal cutting conditions to product quality is via process quality.

The increase in productivity in industrial production requires not only organisational measures but also engagement of all production for a complete activation potentials of all available manufacturing facilities (1). The following measures are possible for activation of technological reserves:
- optimal technological procedure,
- optimal selection of tool,
- optimal combination of machining material and cutting material,
- determination of optimal cutting values.

With introduction of more and more demanding machining system the need for more reliable technological information has increased. This requires a thorough analysis of conditions in the cutting zone (the tool- workpieces- chip system). In this zone the stresses, friction, high temperatures and deformations occur.

As the process of mechanics of the cut in this area is very complicated, since the tribological rules continuously intermingle, it is difficult to give absolute statements on their mutual influences.

All values influencing the cutting process which last a very short time, are transformed into an equilibrium model. On this model there must be equilibrium between the energy (power) fed into the cutting process and the distributed and/or transformed energy (power).

During face milling, the first contact between the cutter and machined material is decisive because of high mechanical and thermal loadings; this is connected with impact loadings which primarily depend on the geometrical shape of the cutter and position of milling cutter - axis with respect to input plane.

For functional dependence of mutual influences a computer program has been developed which ensures computer-aided simulation of all types of contacts during face milling and variation of geometrical values.

The FEM method also ensures the stress analysis in the workpiece for plane or space state. A model has been made for simulation of the face milling process, where the stresses and displacements are checked by calculations in case of very expensive workpieces before their batch production starts.

If they are not within the permissible limits, it is necessary to take the required actions. It is necessary to change the cutting values or the geometric values of the tool. Frequently it is only necessary to change the workpiece clamping method (mechanical parts with thin walls).

As the investigation of cutting processes required much experimental and theoretical work and since a great number of data are applied. The Faculty of Mechanical engineering, University of Maribor has organised an information centre for cutting conditions INCERP (1, 2). Its purposes are:
- automatic collection of data on cutting,
- analysing of cutting processes,
- computer-aided optimisation of cutting values.

The output data from the process described are guidelines for cutting conditions collected in tables. In designing the information centre for cutting conditions it was not possible to ignore the demand for the technological data bank which must be actively included in the computer-aided integrated manufacture - CIM.

REFERENCES

(1) F. Cus, Automatisches Datenerfassungssystem für Zerspanungen, Werkstatt und Betrieb, München, 120 (1987) 11, p. 923-926.
(2) F. Cus and J. Balic, Face Milling on the Influence of Various Geometrical Shapes of Cutter on Cutting Conditions, 3rd International Conference on Advanced Manufacturing Systems and Technology, Proceedings, p. 279-285; Udine, Italy, 1993.
(3) J. Kopac, M. Sokovic and A. Smolej, Strategy of Machinability of Aluminium Alloys for Free Cutting, 30th Int. MATADOR Conference, Proceedings, p. 156-161, UMIST, Manchester, 1993.

EFFECTS OF OIL TRACTION VELOCITY AND RELATIVE SLIDING VELOCITY ON FRICTION IN MILD STEEL FORMING

KUNIAKI DOHDA and ZHRGANG WANG

Department of Mechanical System Engineering, Gifu University, 1-1 Yanagido, Gifu, Japan

ABSTRACT

Friction in the lubricated forming process derives from viscous shear of the hydrodynamic film of the lubricant and from shearing of the real contact with or without boundary films (1). The real contact ratio is mainly determined by the amount of the lubricant at the contact interface. The thickness of the lubricant film is proportional to the oil traction velocity at the contact zone inlet (2). The viscous shear stress of the lubricant film is determined by the lubricant properties, the film thickness and the relative sliding velocity. The relative sliding velocity especially affects the lubricant properties and the lubricant film thickness (3). This suggests that the oil traction velocity at the contact zone inlet and the relative sliding velocity at the contact interface are the main factors determining the lubricated forming process friction behavior.

The purpose of this research is to discuss the independent effects of the oil traction velocity at the contact zone inlet and the relative sliding velocity on the friction behavior.

The principle of the tribometer used is shown in **Fig. 1**. A workpiece is drawn into the gap between rolls by the friction force on the traction roll. The surface of the traction roll is roughened to 40 μ mRy to increase the friction force. The normal force and friction force can be measured during the process. The rotation speed of the friction roll can be set to 1/1, 1/4, 1/15 of that of the traction roll by changing gears.

A workpiece is cut off from a rolled mild steel sheet 0.9 mm in thickness with a dull surface. The friction roll was made of D2 die steel with a 0.15 μ mRy polished surface finish. Two paraffinic mineral oils with different viscosity were used.

Fig. 2 shows the friction coefficient μ against the relative sliding velocity ΔV. Under the present experimental conditions, μ decreases linearly with increasing ΔV, and the gradient is almost independent of the viscosity of the lubricant, but becomes smaller with increasing V. With increasing V, μ changes little for high-viscous lubricant P460 but increases for low-viscous lubricant P26.

Rolled surfaces show that the rubbed portions become interrupted shortly and the slip band becomes more marked with increasing ΔV. With an increase of V, the surface roughness becomes somewhat larger. These results mean that the friction coefficient is determined by the lubricant film performance.

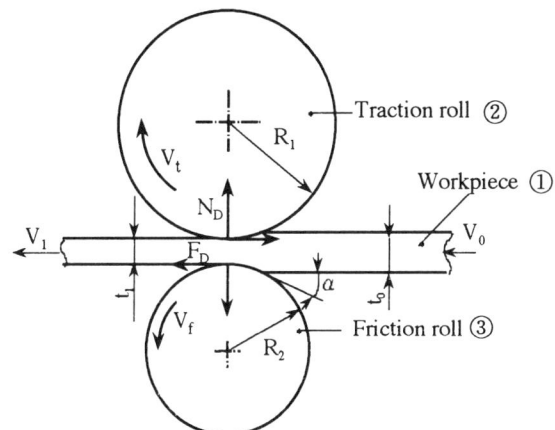

Fig. 1: Illustration of the rolling-type tribometer

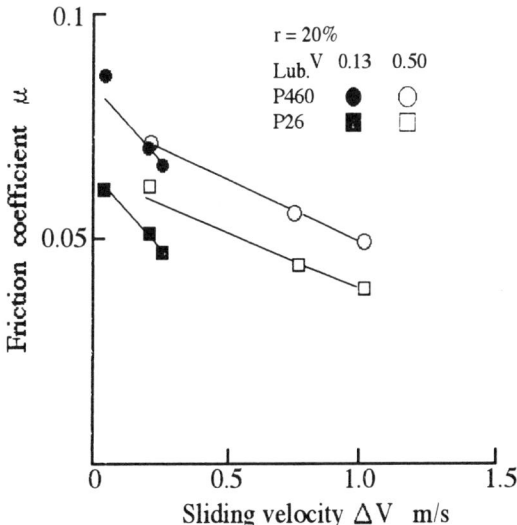

Fig. 2: Relation between friction coefficient and sliding velocity

REFERENCE

(1) Kasuga, Y. and Yamaguchi, K., Bull. of JSME, 11-44, pp. 344-365, 1968.
(2) Mizuno, T., Journal of JSTP (in Japanese), Vol. 7-66, pp. 383-389, 1966.
(3) Wang, Z., Kondo, K. and Mori, T., ASME Journal of Engineering for Industry, Vol. 117, pp. 351-356, 1995.

TEROTECHNOLOGY & POSSIBLY ROLE EXPERT SYSTEMS IN THIS PROCESS

ERIC M. DRAGAN
Company Sloboda, Ratka Mitrovica b.b., 32000 Cacak, Yugoslavia
DROBNJAK DJ. VELIMIR and MILOSEVIC V. MILOVAN
Fabrika reznog alata, Hajduk Veljkova 37, 32000 Cacak, Yugoslavia

ABSTRACT

Terotehnology-maintance technology of tehnical sistems-demands the computer application with optimisation of this process as the aim. A demonstration of aplication artificial intelligence and expert systems in following of keeping in order technical systems is given in this paper. Many problems connected with maintance system reliability, diagnosis can be presented in the knowledge basis in the appropriate way.

In our case, the shell-system Kappa-PC which belongs to software tools for expert systems construction, is used for knowledge presentation. The importance of this expert system in technological processes is given on the figure 1.

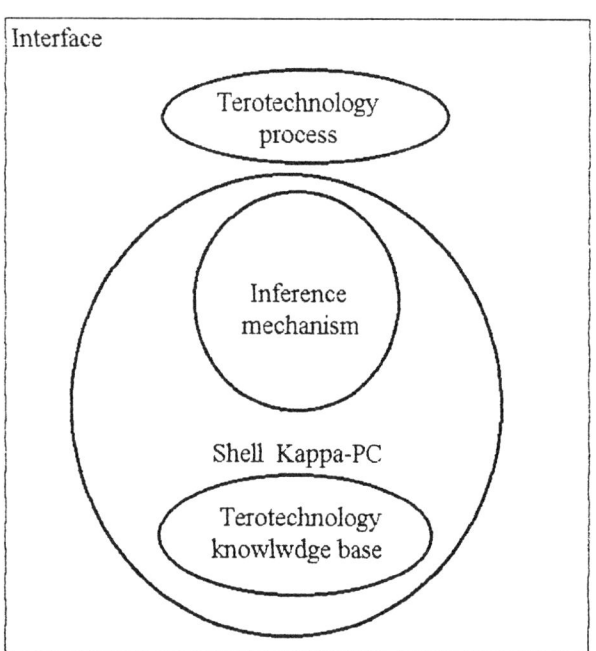

Fig. 1: Expert systems in terotechnology process

KAPPA-PC uses objects as the main form of knowledge representation. Physical entities or abstract concepts are grouped into a hierarchical structure of classes, subclasses and instances. Characteristics, or slots, describe each class and are inherited by any object below in the object hierarchy (1)(2).

These redefined slot options describe the type of information held in the slot and the number of values stored there. A slot can be of type text, numerical, Boolean or the name of another class of object held in the system. The system stores values in a multiple-value slot as a list and can perform list operations on them, including finding the position of a particular element in the list or ascertaining if an item is a member of that list.

The program can obtain new information from established information throughout object-oriented programming, rule-based programming, and functional programming. It reasons about objects using rule-based and functional programming. Rules, functions, and methods are itten using KAL (the Kappa Application Language). The C interface provides additional functionality. Here, we'll concentrate on inference mechanisms available in object-oriented and rule-based programming.

KAPPA-PC rules can be used in both forward and backward inferencing. Pattern-matching rules are written in KAL using an IF-THEN format. Patterns make rule execution dependent on the existence of slot values in objects. Once the system activates a rule, it seeks all descendants of that class that fulfill the IF premises, and enables the Then action to be executed (3).

CIM (include tribology) (4) is one base goals in past of research in technological processes. Today is certain what artificial intelligence would be one of the most important tools in realization of CIM structures. This paper is attempt of implementation some elements of artificial intelligence in classical tribology technologies (5).

REFERENCES

(1) T. Helton, Object-oriented expert system tool Kappa-PC 1.1., AI Expert-software Review, March 1991., 65-67.

(2) T. J. Lydiard, Kappa-PC, Expert product reviews, October 1990., 71-77.

(3) F. Arlabose, Knowledge-based systems for manufacturing, 3rd CIM Europe Conference, 19-21 May 1987, Knutsford, U.K.

(4) V. Devedzic, Expert systems for work in real time, Institute "Mihajlo Pupin", 1994.,Beograd, 19-22.

(5) D. Eric, V. Drobnjak, M. Milosevic, D. Radovic, Artificial Intelligence and Knowledge Base as Support in CAM-Systems, IMEC'96, August 7-9, 1996, Storrs, U.S.A.

SOME MECHANICS AND SIMULATIONS OF ROTARY CUTTING

L.A. GICK
Kaliningrad State Technical University, 1 Sovietsky Avenue, Kaliningrad 236000, Russian Federation
A.A. MINEVICH
Physical Technical Institute, 4 Zhodinskaya Street, Minsk 220141, Belarus

ABSTRACT

Despite a variety of the methods, generally process of metal cutting is based on application of the same fundamental principle. A tool blade normally done as a hard motionless wedge is sliding over a workpiece and removing a chip. A work is usually being done by a small area of cutting edge, which is operating in permanent contact with a workpiece material. This results in permanent warming up, wearing off, and comparatively quick going a tool out of operation.

The situation can be significantly changed, if the tool blade obtains additional movement in direction tangential to the cutting edge (1-2). In this case the permanent change of the working blade sectors results in improvement of contact conditions, temperature fall in the zone of cutting, increase of the working cutting edge length, and, respectively, decrease of the period of real operation of each cutting edge point. Finally, it provides higher productivity of machining, tool life and smaller surface roughness (3-4). This idea is being most easily performed in constructions of round rotating cutting tools. In this case cutting process combines sliding, which provides the main working process of chip removal, and rolling of the cutting edge over the cutting surface. Except the former Soviet Union, this variant of machining was studied by research groups in USA (M.C. Shaw, P.A. Smith, N.H. Cook), Italy (S. Amari), UK (W. Chen, D.K. Aspinwall), etc. It is notable, that in rotary cutting a significant improvement of cutting performance is reached without application of better tool materials. Higher performance can be provided by the same HSS and cemented carbide inserts owing to original kinematic solution, changing the conditions of frictional interaction in tribosystem.

Two principle schemes of rotary cutting known in technical literature are as follows:
1) when the face of the round insert is used as the rake face and the peripheral surface as the flank face], and 2) when the peripheral surface is used as the rake face.

In present study scheme 2 was taken to be the basic as more general one. On our opinion this scheme possesses a number of advantages as compared with the first scheme. They involve the following features: 1) stability of tangential movement (rotation) in a wide range of crossing angles between a tool and a workpiece axes; 2) possibility of making more rigid, strong and efficient bearing in comparison with the scheme 1 (in the scheme 1 this is not possible because of limitation of its dimensions by a diameter of cutting insert); 3) provision of a low surface roughness ($Ra<1.25$ μm) after machining even in cutting at a high cutting feed, and therefore higher productivity of machining.

In spite of attractive properties the process of rotary cutting is still not completely understood, because in the most of studies, the approach based on natural tests of rotary tools was undertaken. Really it only helps to accumulate the data on case studies.

Meantime, the authors think that it is possible to find out general features of the concerned process using some simulations by means of free cutting with additional tangential movement. Creating a model of rotary cutting, we have revealed the need in some new distinctive parameters, including cut-off angle, chip orientation angle, chip shrinkage in direction of chip formation. In tribotests the components of cutting force, contact length, temperature in the zone of cutting, friction coefficients, chip characteristics, specific work of deformation and heat generation were determined as a function of cut-off angle in both orthogonal and oblique cutting.

Based on the results of these simulations and natural tests an explanation of rotary cutting mechanics and a physical picture of friction interaction will be given. The simulations carried out in the work are applicable to creation of new machining operations, using rotary tools.

REFERENCES

(1) E.G. Konovalov, L.A. Gick, Cutting by Round Rotary Tools, Minsk, 1969. (In Russian).
(2) L.A. Gick, Rotary Cutting of Metals, Kaliningrad, 1990, 254 p. (In Russian).
(3) L.A. Gick, Proc. Conf. "Wear of Materials", Orlando, FL, Apr. 7-11, 1991, v. 2, p. 679-682.
(4) L.A. Gick, A.A. Minevich, Proc. 4th Int. Tribology Conference AUSTRIB'94, Dec. 5-8, 1994, Perth, Western Australia, v.II, p. 489-493.

Submitted to *ASME Journal of Tribology*

TRIBOLOGICAL INVESTIGATION OF FUNCTIONAL EFFICIENCY NEW TYPE MULTIFUNCTIONAL ADDITIVE FOR CUTTING FLUIDS

M.GOLOGLAVIC-KOLB
NIS-Refinery Beograd
M.B.POPOVIC
Laboratory for fuel and lubricants-Car factory Zastava Kragujevac

ABSTRACT

Tribological investigation on efficiency a new multifunctional additive type for cutting fluids was realised on the tribometar "Pin on Disk" .Contact geometry is line. Chemical investigation performed by instrumental analytic methods.

Tested additive is blend of multifunctional synthetic bases that contain natural products based on vegetable oils.

The experimental program included examination of some deferent cutting oils. In this paper will be show results for three conventional straight cutting oils and for three cutting oils with new multifunctional additive type.

The contact pairs on the tribometar made of carbon steel (Disk) and High Speed Steel (Pin). The results of the investigation are to present anti-wear properties and coefficient of friction concerning the determination on the these tribometer. Physical property will be present trough viscosity of the different cutting oil's types.

For example, the friction coefficient of the different cutting oil's type (for same condition of the examination) showed in figure 1.

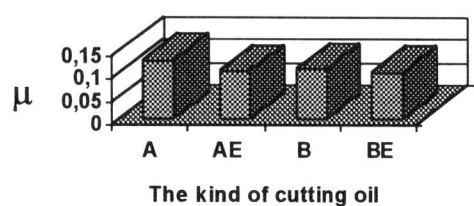

Fig.1: The friction coefficient of the different kind of cutting oils

Anti-wear properties of the different kind cutting oil's types showed in figure 2. Viscosity for different cutting oil's types (oils with and without new multifunctional additive) is given in figure 3.

Fig. 2: Anti-wear properties of the different kind of cutting oils

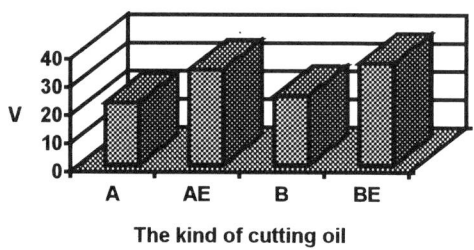

Fig.3: Viscosity of the different kind of cutting oil

Our research is are confirmed that new formulations give substantial benefits:
- powerful lubricate in boundary regime
- effective protection against wear
- lower viscosity variation with temperature
- chemical stability at high temperature

These properties assure good surfaces finish, reduce power consumption and prolong tool life.

REFERENCES

(1) W.J.Bartz, EMW '94 Study of Metal Cutting and Forming Processes, Effectiveness of coolants considering technical and ecological aspects, page 041,1994.
(2) B.Ivkovic, B Tadic,Proceedings of Balkantrib'96, Tribometry and Tribodiagnostic, page 834,1996.

ON THE AUTOMATED LABELING OF INDIVIDUAL SHAPE DISTINCTION OF WEAR PARTICLE IMAGES

A Y GRIGORIEV
Metal-Polymer Research Institute, Kirova St., 32-A, 246652 Gomel, BELARUS
V A KOVALEV
Institute of Mathematics, Kirova St., 32-A, 246652 Gomel, BELARUS
H-S AHN
Korea Institute of Science and Technology, 39-1 Hawolgok-dong, Songbuk-Gu, Seoul, 136-791, KOREA

ABSTRACT

The morphological analysis of wear particles is one of widespread diagnostic method in the tribology. It is well known that the morphology of wear particles indicates a type of wear process and current condition of a machinery. The shape of the particles is usually described in terms of their area, perimeter, size of the bounding rectangle, form factor, etc. There are reasonable simple interpretations of such features. However, they do not always characterize objects in a unique fashion. It is possible to show that there are debris, while distinct in shape, with an identical area, perimeter, and form. The selection of the representative set of morphological features is a complicated task because of insufficient knowledge of a correlation between the shape peculiarities and wear process. In addition, there are certain difficulties in formulating of an expert's perception for shape differences in the quantitative categories. Thus, labeling of shape distinctions is one of the ways to formulate the differences mathematically and to construct a feature set for debris classification and wear diagnostics.

The basic idea of the method is to represent the particle shape by the co-occurrence matrix of chord length distribution. For this purpose the contour of a particle is first extracted. The corresponding planar closed curve is specified by a finite set of samples, i.e., image pixels. Next, the chord lengths for all possible pairs of pixels are measured and the matrix is calculated. We used the following two-dimensional co-occurence matrix:

$$M = \| m(r,a) \|,$$

where m is the number of pairs of chords with length ratio r and angle a between them. The columns of the matrix are normalized:

$$p = m(r_i, a_j) / \sum_j m(r_i, a_j).$$

In this form the matrix represents the conditional probability of the measurements. Each row of the matrix represents the chords ratio distribution in angle a. It should be pointed out that the values of matrix elements (frequency of co-occurrence) are independent of various particle image transformations such as shift, rotation and rescaling that is characterize the particle shape in a unique fashion. To avoid of ambiguity in angle determination (obtuse or not) $sin(a)$ instead of angle a. To reduce the computational complexity both dimensions of the matrix were quantized by 10 levels.

The matching and labeling of individual shape distinctions of wear particles were conducted by the corresponding procedure of matrix element comparison. For this aim we used coefficient of similarity defined as follows:

$$k = 1 - |p^0 - p| / p,$$

where p^0 ($p^0 \neq 0$), p are the corresponding elements of primary standard and tested matrix. The labeling of pixels on the particle contour which more/less then predefined threshold of k allows us to visualize its shape similarity/dissimilarity. The result of such operation for two particles is shown in Fig. 1.

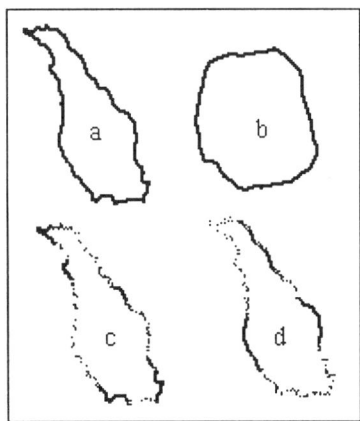

Fig. 1: The result of wear particles matching: a, b – initial images; c – dissimilar and, d – similar shape features of the particle 'a' comparing 'b'.

It is possible to find typical/atypical elements of matrices for some sets of particles obtained at different wear stages. As a result, the shape features corresponding different wear stage can be extracted. The experiment performed with the wear particles of a steel-steel tribosystem shows good agreement with results of experts' analysis of debris differences.

Submitted to *Tribology International*

APPLICATION OF MOLECULAR-MECHANICAL THEORY OF FRICTION TO TRIBOLOGY OF THE CUTTING PROCESS

W. GRZESIK

Department of Manufacturing Engineering and Automation, Technical University of Opole, PO Box 321, 45-233 Opole, Poland

ABSTRACT

The action of cutting creates exceptional tribological conditions at the chip-tool interface including plastic contact (seizure) between the jointed materials and intensive heating of the contact area by the energy generated in plastic deformation and friction. In consequence, it is evident that under cutting conditions strong adhesion, which is a thermally activated process, can not be avoided (1). It is generally accepted (2) that sliding and sticking friction can exist simultaneously in two distinct regions at the tool rake face but seizure commonly occurs in the region adjacent to the cutting edge.

Iwata et al. (3) showed that the adhesion force between C45 carbon steel and P10 uncoated carbide reached the maximum value at a specific temperature of 1050 °C. Moreover, the coefficient of friction was about 0.8 and the ratio of the absolute temperature initiating strong adhesion to the melting point of the softer metal (θ_a/θ_m) was between 0.49 and 0.52.

In the present investigation the molecular-mechanical theory of dry sliding friction was used to estimate the shear strength of the adhesive junction and to seperate the mechanical and adhesive components of the total coefficient of friction. The basic assumption was that the minimum undeformed chip thickness corresponded to the transition from ploughing to micro cutting (4).

The shear strength of the adhesive junction was expressed as (4) (5)

$$\tau_a = 0.34 L \rho \ln(\Theta_m / \Theta)$$

where L is the heat of melting, ρ is the density of the material, Θ_m is the melting point of the metal, Θ is the average interface temperature and Θ_m/Θ is the homologous temperature.

The total coefficient of friction was determined by summing the adhesive and mechanical components, i.e. $\mu = \mu_a + \mu_m$.

Experiments were conducted with a carbon steel and an austenitic stainless steel (equivalent to AISI 1045 and AISI 304, respectively) using P10 and P35 carbide grades. Workpieces in the form of cylindrical tubes were turned under orthogonal conditions with cutting speeds ranging from 42 m/min to 240 m/min and feed rates ranging from 0.08 mm/rev to 0.27 m/min. During each test the cutting temperature was measured using thermo-electromotive force signals and the chip thickness ratio of each produced chip was calculated.

In order to compare the conditions under which adhesive bonds are formed both the ratio of the shear strength τ_a to the yield stress of the work material at room temperature and the ratio of the interface temperature to the melting point of a given material were analysed.

It was established that the interface temperature ranges from 410 °C to 560 °C for carbon steel and from 510 °C to 650 °C for stainless steel. This means that the ratio of θ_a/θ_m varies from 0.27 to 0.36 for carbon steel and from 0.31 to 0.41 for stainless steel. The values of τ_a obtained seem to be consistent with the shear stresses acting on the shear plane.

In general, the values of μ produced when turning a C45 carbon steel are higher than those for an austenitic stainless steel. As can be seen from Fig. 1 μ ranges between 0.8 and 1.1 for a carbon steel, and between 0.61 and 0.82 for a stainless steel. As a result, the variations of μ_a are between 0.7 and 0.58 for carbon steel, and between 0.42 and 0.35 for stainless steel.

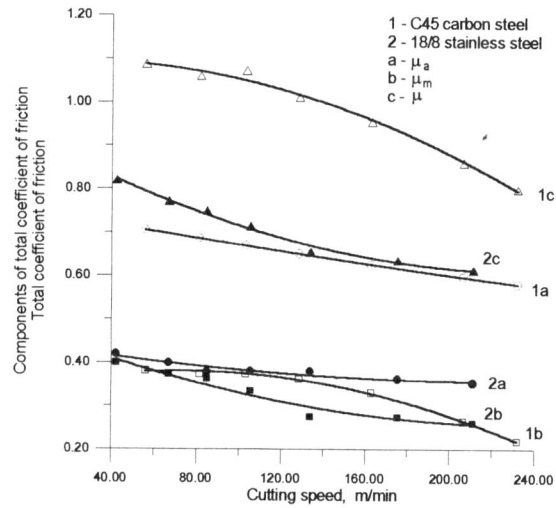

Fig. 1: The cutting speed effect on the coefficient of friction and its components

REFERENCES

(1) E.M. Trent, Metal Cutting, Butterworths, London, 1989.
(2) J.A. Bailey, Wear, Vol. 31, 1975, 243-275pp.
(3) K. Iwata et al., Wear, Vol. 18, 1971, 153-163pp.
(4) I.V. Kragelsky et al., Friction and Wear. Calculations Methods, Pergamon, Oxford, 1975.
(5) W. Grzesik, Wear, Vol. 194, 1996, 143-148pp.

FRICTION MEASUREMENT APPARATUS AND PROCESS MONITORING TRANSDUCER FOR SHEET METAL FORMING AND OTHER TRIBOLOGICAL PROCESSES

S. HAO
Seagate Technology, Advanced Products Development, Minneapolis, MN, 55435, USA
B. E. KLAMECKI, S. RAMALINGAM
Department of Mechanical Engineering, University of Minnesota, Minneapolis, Minnesota, 55455, USA

ABSTRACT

Two types of experimental apparatus have been developed. One type is test apparatus for determining friction between a test sheet metal strip and a cylindrical surface. The apparatus are extensions of the strip drawing test pioneered by Duncan (1) in which the capstan model of friction is used to calculate strip-cylinder coefficient of friction. Rather than measuring tension force in one leg of the strip specimen and deformation in another specimen leg as in the original test configuration, strip tensions are measured directly in the sections of interest. Three types of apparatus were built. One of the three is shown in Figure 1. Each can be mounted in a tensile test machine and the forces acting on the strip specimen measured directly by the test machine load cell and load cells built into the apparatus.

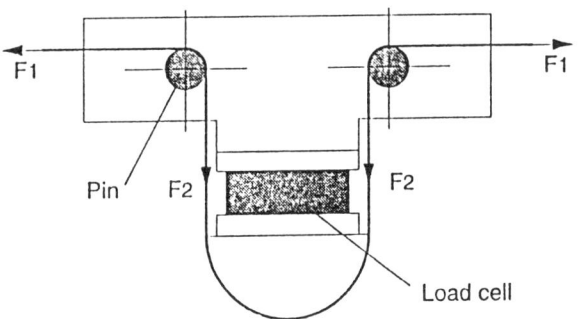

Fig. 1: One of the test apparatus

The three test apparatus were used to determine the effects of pin radius, pin material, strip material, wrap angle, strain, strain rate and lubrication on pin-strip coefficient of friction.

In addition to laboratory apparatus for measuring sheet metal coefficient of friction, a transducer for measuring friction force acting on a surface was developed. The transducer can be built into a surface and used for in-process monitoring of friction force. For example, the test surface can be a die or punch surface in a sheet metal forming process. A schematic representation of the transducer is shown in Figure 2. The active part of the transducer body rests on piezoelectric elements. Transducers were built in which only normal force sensitive and both normal force sensitive elements and shear force sensitive elements supported the test surface.

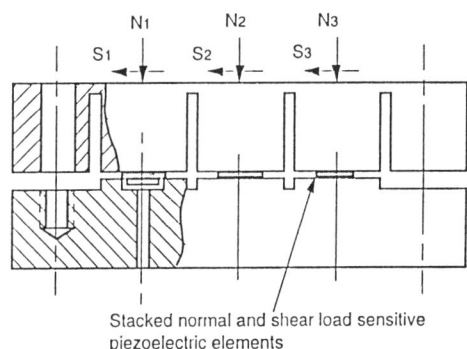

Fig. 2: Surface forces tranducer

The full sensitivity matrices relating sensor outputs to applied loads were obtained from calibration tests. These tests showed little cross sensitivity between normal and shear force measurements and between normal forces acting over different parts of the transducer surface. The sensors were demonstrated in sliding tests and accurately measured normal load distributions and friction forces.

Laboratory test apparatus for measuring sheet metal coefficient of friction and transducers useful for in-process monitoring of normal and friction forces have been developed, characterized and tested. Details are available in the work of Hao (2). A paper describing the work has been submitted to the American Society of Mechanical Engineers Journal of Tribology.

ACKNOWLEDGMENT

This work was supported by the National Science Foundation under Award Number DDM - 9022550.

REFERENCES

(1) J. L. Duncan, B. S. Shabel and J. G. Filho, "A Tensile Strip Test for Evaluating Friction in Sheet Metal Forming," SAE Technical Paper, No. 780391 8p., 1978.
(2) S. Hao, "Optimal and Intelligeny Computer Controlled Sheet Metal Forming," PhD Theses, University of Minnesota, 1994.

MACHINABILITY OF STEELS AND AUSTEMPERED DUCTILE IRONS - RUNKING TEST ON TRIBOMETER PIN ON DISK

B.IVKOVIC
Masinski Fakultet, Univerzitet u Kragujevcu, S.Janjica 6, Kragujevac, YU
D.Jesic
TRIBO TEHNIK, Titov trg 6/4, Rijeka, CR

ABSTRACT

Machinability is defined as the ease with which a metal can be machined. It is one of the principle factors affecting a products utility, quality and production cost. Besides a large number of machinability test in the past, the ability to quantity machinability properties of materials remain a big problem.

The large number of machinability tests can be classified as being either absolute or ranking tests. The results of absolute machinability test give possibility to form analytical expression of machinability using material properties. The runking test, therefore, compare the performance of two or more materials in the same cutting condition. On the base runking test machinability index is created for a group materials.

In the base, the runking tests of machinability are created on the judgement of tool wear process during machining. The magnitude of wear land on relief face of cutting tool is common wear parameter.

Two friction pairs exist in the cutting zone. In the both, the sliding contact between machining material and tool are realised during cutting process. The critical tool wear parameter is wear land on the clearance face in tribo-mechanical system "relief face and machining surface".

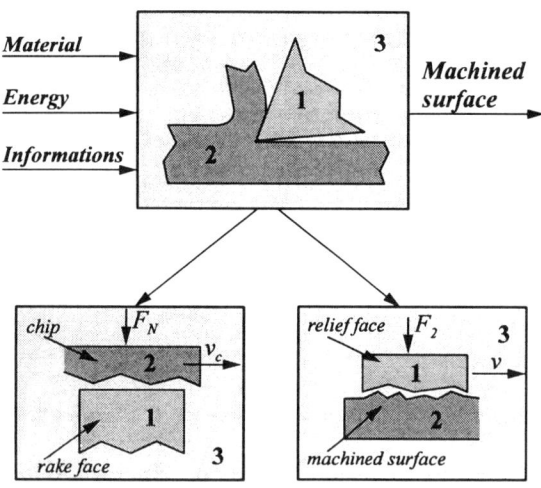

ASTM G 83 gives proposal for tribological properties of tool steels determination by the point contact on Tribometer "Crossed-Cylinder Apparatus". Runking tests of tribological properties of tool steels can be make by measure wear track on the tool steel cylinder.

In the milling process the line contact is realise between tool clearance face on machining surface on the workpeace. This is sliding contact with sliding speed that is equal cutting speed.

The normal load in this tribo-mechanical system is radial component of the cutting force.

The same contact condition, it is possible realise on Tribometer Pin (Block) on Disk between Block is made of tool materials (HSS or TC) and Disk is made of workpiece material (different kind of steels, for instance).

$b_1 = 0.444$ mm
$b_2 = 0.438$ mm
$b_3 = 0.445$ mm
$b_4 = 0.455$ mm

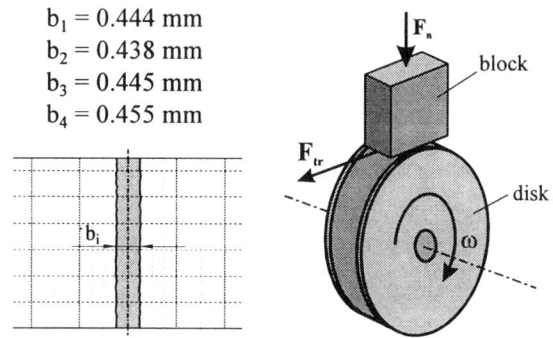

In the one investigate programme, it is realised many experimental operations in these wear parameter (width and debt of scar) on HSS Block and PQ index in cutting fluid were measured after three hour contact.

Scar width on HSS Block

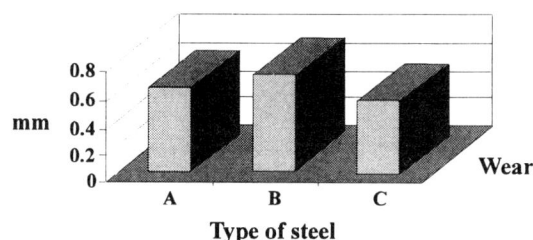

Machinability index for three kind of steels can be form by magnitude of scars on HSS Block and the known connection between tool life, tool wear and cutting speed.
A part of these results is presented in this paper.

REFERENCES
(1) B.Ivkovic,B.Tadic, Proceedings of Balkantrib'96, Tribometry and Tribodiagnostic, page 834, 1996
(2) B.Ivkovic,B.Tadic, D.Jesic, J.BTA, Vol.2, No3, page 136-140,1996

DIAGNOSTICS OF TRIBO-MECHANICAL SYSTEMS BY COEFFICIENT OF DYNAMIC BEHAVIOR

BRANISLAV JEREMIC, NENAD MILIC, PETAR TODOROVIC and IVAN MACUZIC
Department for Manufacturing Engineering, Faculty of Mechanical Engineering, University of Kragujevac, S. Janjic 6, 34000 Kragujevac, YU, e-mail: ceter@uis0.uis.kg.ac.yu

ABSTRACT

Diagnostics of technical systems, namely their tribo-mechanical systems (TMS), includes theoretical analysis, methods and measuring equipment for identifying the system's state in conditions of limited number of information. In the process of diagnostics, the diagnosis is set, and it is related to making conclusions, namely to determination of the system's state. Identification of the technical system state is connected with choice and monitoring the changes of diagnostic parameters that belong to the TMS.

In the process of exploitation, due to development of tribological processes, the changes occur in conditions of contact realization, and accordingly, changes of the nature of the friction force. The friction force always includes the dynamic component. The dynamic component of the friction force varies with changes of topography of the contact surfaces, coating destruction rate, lubricating conditions, etc.

For the TMS dynamics, the diagnostic parameter is introduced which is called the coefficient of dynamic behavior. This parameter, by its nature, represents the ratio of the effective value of the friction force dynamic component and the average value of the friction force. Investigations related to evaluation of the sensitivity of the mentioned parameter's changes, were conducted on "the disk-on-disk" tribometer with contact surfaces made of steel alone and of steel coated with the TiN and ZrN coatings. Investigations were performed without lubricants, and with solid lubricants based on PTFE and MoS$_2$.

The measurements setup and the used software enabled the A/D conversion (Fig. 1), data acquisition and analysis of the friction process dynamics, through the power spectrum in the frequency domain, and during different phases of the lubricant's layer destruction.

For quantification of dynamic processes in the contact zone we introduced the coefficient of dynamic behavior K_d (1) which represents the ratio of effective value of the friction force dynamic component $\Delta F_{\mu eff}$ to the average value of the friction force $\overline{F}_\mu$ (Fig. 2):

$$K_d = \frac{\Delta F_{\mu eff}}{\overline{F}_\mu}$$

All the obtained results point to the conclusion that the coefficient of dynamic behavior can be used as the diagnostics parameter of the tribo-mechanical systems from the aspect of the friction process stability.

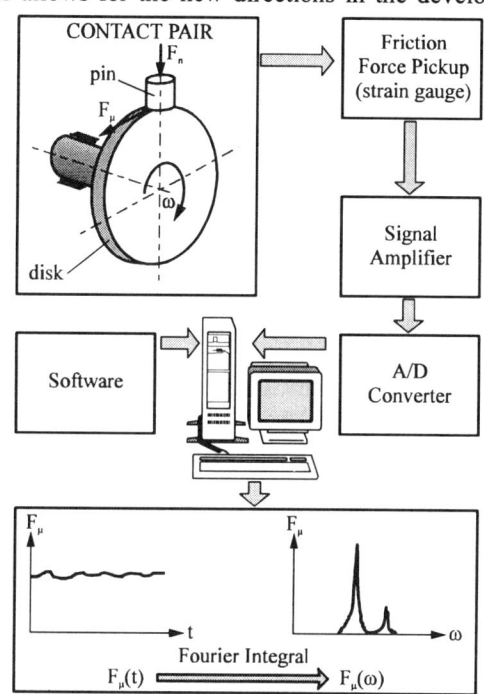

Fig. 1: Measurements setup

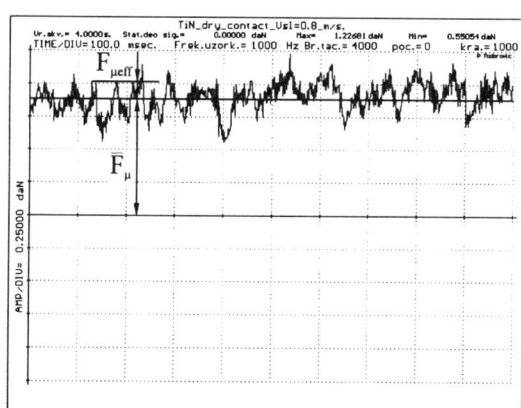

Fig. 2: Typical form of the friction force in the time domain

This allows for the new directions in the development of the diagnostics equipment and systems of automatic control of the TMS lubrication, where the stable friction process is required, namely the minimum dynamic component of the friction force.

REFERENCES

(1) B. Jeremic, M. Babic, M Meyer, P. Todorovic, N. Milic, Tribomechanic System Diagnostics Throught the Friction Process Stability, Balcantrib '96., 398-405pp.

A FERROUS DEBRIS MONITOR TO MEASURE TOTAL AND SEVERITY OF WEAR

M.H. JONES
Department of Mechanical Engineering, University of Wales Swansea

ABSTRACT

A Time Dependent Particle Quantifier is described which is capable of measuring the concentration of ferrous wear debris suspended in a lubricant and the severity of wear associated with the particle size of this suspended debris. A coding system is proposed: PQ index (total wear): initial TDPQ slope (large particles): final TDPQ slope (small particles). Correlation with existing measurements is detailed.

Users of spectrographic oil analysis have, for some considerable time, been aware of the limitation of this technique in relation to the accuracy of this method when analysing samples which contain particles larger than 5 um. Emission, absorption and inductivity coupled plasma spectrometers are each limited in their ability to analyse these large particles.

The direct-read ferrograph, the Wear Particle Analyser (Tribometrics), the ferrous wear debris monitor (PQ90) and the use of magnetic plugs have addressed this problem. These techniques are restricted to measuring ferrous wear debris. Most equipment however, which generate large particles, gearboxes for example, produce ferrous wear debris and it is this equipment and material which is monitored.

The ferrous debris monitor (PQ90), which is a sensitive magnetometer, measures the quantity of ferrous wear debris in a sample without the disadvantages of time and dilution requirements associated with the other techniques. The unbalanced condition is measured and displayed as the PQ Index. This index is measured in arbitrary units which may be correlated with the D1 and Ds ferrographic measurements. These measurement units, like all concentration measurements obtained by spectroscopy, should not be considered absolute measurements due to the particle size sensitivity of each technique.

The Time Dependent Particle Quantifier (TDPQ) technology offers both portable and laboratory based instruments the capability of measuring the total amount and severity of wear occurring in mechanical equipment. This instrument, in its portable mode, would also provide a simple site screening technique capable of selecting fluid samples which should be sent to a central laboratory for more detailed tests.

EXAMPLE: LOCOMOTIVE GEARBOX

ppm ICP Acid Digested	ppm ICP Diluted Only	PQ Index
200	23	101
340		176
330		201
450		226
650	45	351
720		401
830		451
990	65	601
1000		701
1900	100	1276

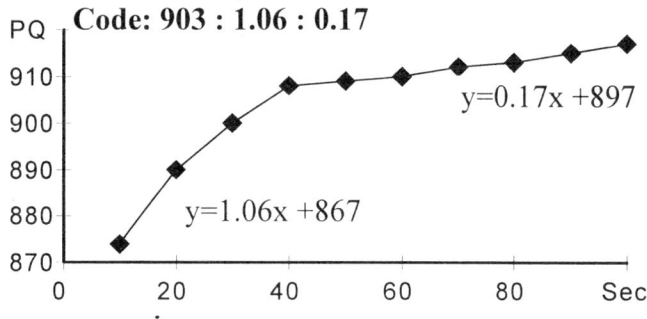

SAMPLE 27 Example of severe wear
Code: 903 : 1.06 : 0.17

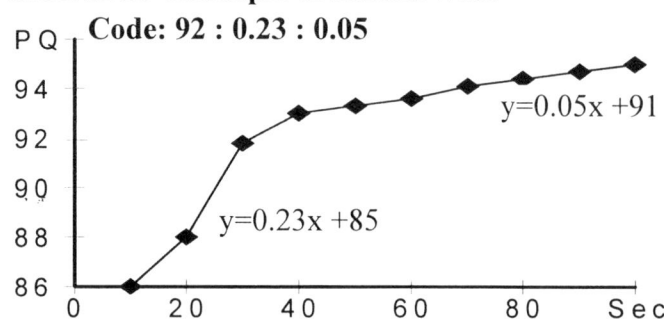

SAMPLE 35 Example of normal wear
Code: 92 : 0.23 : 0.05

SOME RESULTS IN APPLYING BASIC WEAR EQUATIONS IN MINING EQUIPMENT MAINTENANCE

F. KORONKA, I. ANDRAS and D. JULA
Department of Mining Equipment, University of PETROSANI, Universitatii str. 20, 2675 Petrosani, ROMANIA

ABSTRACT

The results of friction process are energy losses and wear by material detachment, which means a change in the contact surface's initial state. The necessity of knowing and limiting wear represents the main challenge in tribology. During the force transmitting process from an element of friction linkage to another, in the presence of a relative motion, the real contact areas are destroyed at one or more loading cycles, and wear particles appear. Friction linkage's elements discreetly reduce their dimensions in the normal direction at the frictioning surface.

The wearing analytical calculus presumes the assessment of the specific intensities of wearing in correlation with physical - mechanical properties of the material and micro-geometric characteristics of contact surfaces for the micro-roughnesses elastic and plastic contact. From the calculus, results that both for plastic contact and elastic contact, the external load represented by nominal pressure has definitive influence over wear intensity.

By many expert's opinion, the wear process depends and can be classified by the main occurring phenomena which are: thermo-physical, mechanical and chemical. Other experts believe that mechanical phenomena represent the main factor in the surfaces wear and the characteristic type of wear would be the abrasive wear .(1)

Regarding the wear dependence upon other parameters it had been accepted the following two conclusions:

1. wear increases with friction length or with process duration, increase which generally is not linear;

2. wear decreases with the increasing of surfaces hardness (Though, we have some exceptions.)

On the wear dependence upon the burden, contact area and other factors, there are different opinions. Concerning wear development, at the beginning, at the new friction coupling, the intensity or the rate of wear quickly increase because of the initial surface conditions, which gradually tend to accommodate. After that, comes a longer period of normal functioning with a stable wear, and finally, the wear becomes destructive, with irreversible effects, situation when the friction linkage is out of order.(2)

The study of the wear intensity, by deriving appropriate formulas, allows taking the necessary measures for life increasing in some concrete situations, decreasing wear intensity or, allows the charge and maximum friction coefficient assessment for an imposed and pretended duration.

The paper is related to the basic mathematical laws of those processes in order to offer a theoretical basis to the assessment of the effects of wearing in order to avoid by appropriate design, manufacturing and maintenance the shoot-downs due to excessive wearing. On this background, equations were derived for the assessment of wear intensity in case of elastic and plastic surfaces.

From calculus results that both for plastic contact and elastic contact, the external load represented by the nominal pressure p has definitive influence over wear intensity. Also, wear intensity gets minimized when friction coefficient is minimum, other words spoken when micro-geometry complex parameter has an optimum value characteristic to normal functioning stage of the friction linkage. The study of derived equations allows taking the necessary measures for life increasing in some concrete situations, decreasing wear intensity or allows the change and maximum friction coefficient determination for an imposed duration.

This becomes very important in the case of mining machines which operate in very aggressive environment, being loaded by important and variable forces. The maintenance and lubrication of wearing elements can accidentally being weak, the rate of shoot-downs caused by excessive wearing is higher. For this reasons, the study of basic wearing phenomena occurring in the most sensitive elements is very important. The knowledge of the basic mathematical laws of those processes can offer a theoretical basis to the assessment of the effects of wearing in order to avoid by appropriate design, manufacturing and maintenance the shoot-downs due to excessive wearing.

REFERENCES

(1) D. Pavelescu, Tribotehnica. Editura Tehnica, Bucureşti, 1983.
(2) A. Tudor, Contactul real al suprafetelor de frecare, Editura Academiei, Bucureşti, 1990.
(3) I.W. Kraghelski, Friction, Wear, Lubrication (Tribollogy Handbook) Mir Publishers Moscow,1986.
(4) R.D. Arnell Tribology, Mac Millan, London, 1991

RELATIONSHIPS BETWEEN TOOL WEAR, CUTTING TEMPERATURE AND TOOL-WORK TEMPERATURE

PAVEL KOVAČ

University of Novi Sad, FTN, Institute for Production Engineering
21000 Novi Sad, Trg D. Obradovića 6, Yugoslavia

ABSTRACT

Tool life testing generally requires time and material spending and therefore is relatively expensive procedure. With knowledge that heat load determines tool wear rate, was made an attempt for developing a method for predicting tool wear parameters trough thermodynamic cutting functions (1). Statistical experimental design is used for empirical tool wear relationships prediction

Tool-work thermocouple thermo-voltage was measured during milling. This method was an acceptable way to evaluate average interface cutting temperature. Cutting insert from cemented carbide was insulated from tool holder as well, to avoid parasitic thermo-voltage. The cold junction on tool was moved away from cutting zone by added cemented carbide inserts parts of the same kind hard metal to avoid parasitic thermo-voltage. The measured values were stored and processed by the PC by developed software. Calibration for determination of average temperature was not done and only values of average thermo-voltage are presented, what is for analyse goal irrelevant.

Work material was bar of 100 x 120 x 600 mm of steel AISI 1060 (C. 1730). A single tooth face milling cutter of 125 mm diameter, with cemented carbide P 25 insert SPAN 12 03 ER was used. Width of flank wear land VB, depth of crater KT and width of crater KB was measured on a tool microscope, when cutting was stooped. Adopted tool life criterion was VB = 0,3 mm. The experiment was carried out for different cutting speed (v), and for feed per tooth (s) and depth of cut (a) constant, according to a planing of experiment. The ranges of variation was: v=2,95 - 4,63 m per s. The width of cut was 100 mm with central position of tool.

Cutting temperature measurement was performed by standard inserted CrNi-Ni thermocouple, 0,1 mm diameter, prepared and built in a cemented carbide insert seat of the milling cutter, nearest as possible to the point with maximal cutting temperature (2). The measuring point was on the same place during whole experiment for all cutting conditions.

Width of flank wear land VB, width of crater KB and depth of crater KT are determined for three values of cutting speed. Average thermo-voltage U, and cutting temperature θ in the point of the insert support surface measured by inserted thermocouple is measured versus cutting time for three values of cutting speed.

Predicted are tool life values for tool wear criterion VB = 0,3 mm.

The progress of cutting temperature versus time is approximately the same regardless like tool wear curves versus time. Cutting temperature in a point change due to varying cutting speed and going on wear of cutting tool. The time progress of cutting temperature can be derived into three distinctive stages. The initial stage during which a very rapid increase occurs. The second normal stage occurs generally vary at a constant tool wear rate. The final stage of progress often happens rapidly with great possibility of tool failure.

On the basis of the time progress of cutting temperature and tool wear it is possible to obtain functional relationships among them.

In the investigated wide interval of cutting speed the average thermo-voltage remains constant until tool wear parameters have three distinctive stages.

This type of time progress of average thermo-voltage has certain advantage for tool life determination. If tool life is determined according tool wear curves, cutting must be continued until third stage when tool is worn. When correlation between average thermo-voltage and tool life is determined sufficient is to determine average thermo-voltage with short cutting tests in the first tool wear stage.

Based on experimental results are determined constants in relationships cutting temperature versus speed and parameters of tool wear VB (KT, KB)

$$\theta = 114,51 \cdot v^{0,492} \cdot VB^{0,223}$$

Relationships for tool wear parameters VB, KB and KT versus cutting speed and cutting temperature are determined as well:

$$VB = 5,855 \cdot 10^{-10} \cdot v^{-2,206} \cdot \theta^{4,484}$$

With this relationships it is possible to indirect determine tool wear parameters by measuring cutting temperature for different cutting speeds.

REFERENCES

(1) P. Kovač, Proc. of Internat. Conf. Prod. res. ICPR, Hefei, 1992, pp 2055-2058
(2) P. Kovač, D. Milikić, Proc. of CSME Mech. Eng. Forum, Vol. 3, Toronto, 1990; pp 283-288.

PREDICTION OF SERVICEABILITY OF STANDARD MATING PARTS.

VLADISLAV LOZOVSKII, MICHAEL IAMPOLSKII and GENNADY SHELIKHOV
Scientific & Technical Center, 140003 Liubertsy-3, a/b 8, Moscow Region, RUSSIA

ABSTRACT

The operational reliability and lifetime of aeronautical and other modern technical equipment essentially depends on unfailing operation of such standard Contact-Loading Matings (CLM) as rolling-contact bearings (turbo-compressor rotor and blade base units), gearings (gearboxes, distributing mechanisms), and splined joints (transmission units).

CLM failures are caused by material contact fatigue, its flacking, the formation of cavities at parts' working surfaces and the further seizure or breach of a mating.

We suggest the instrumental way of CLM serviceability prediction at the initial stage of machines' operation or repair. This way allows one to reveal some objective indications showing that CLM parts' actual loading exceeds the allowable level which guarantees the reliable operation without failures caused by material contact fatigue. The actual loading level depends on contact pressure, temperature and relative travel speed at the condition of using lubricant.

The mentioned method is based on the results of investigating numerous failures of the matings of this kind. The experiments were carried out on different types of machines. It was established that loading conditions which are accompanied by material contact fatigue at the initial stage of CLM operation naturally lead to typical micro-injuries of parts' surfaces due to seizure. These micro-injuries are accompanied by changes in physical and mechanical properties and may be reliably detected by using the methods of metallophysical analysis.

Primary indications are the local tears and metal transfer from one mated part to another accompanied by the formation in these sites metal microvolumes featuring changed (as compared to the initial state) magnetic properties.

These indications can be revealed by analyzing under the microscope the condition of parts' surfaces using the films with a domain structure (Fig 1).

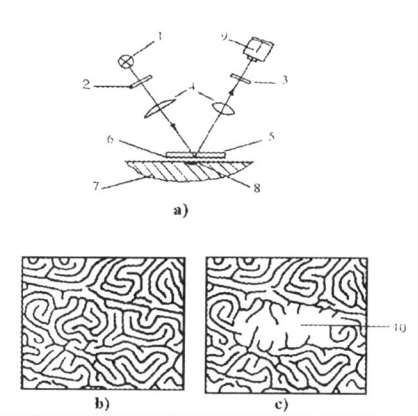

Fig1: Scheme of detecting a section of the test object damaged due to seizure.

a) Scheme of analyzing the damage under the microscope
1 - light source, 2 - polarizer, 3 - analyzer, 4 - optical system, 5 - backing, 6 - ferromagnetic film, 7 - test object, 8 - test object damage caused by seizure, 9 - photo camera, 10 - domain structure of the ferromagnetic film positioned over the damaged section.

b) Domain structure of the ferromagnetic film with no damage on the test object

c) Domain structure of the ferromagnetic film superimposed onto the damaged section of the test object

HIGH SPEED PHOTOGRAPHY OF THE CHIP-TOOL INTERFACE

V. MADHAVAN, S. CHANDRASEKAR, and T. N. FARRIS
Schools of Engineering, Purdue University, 1287 Grissom Hall, West Lafayette, IN 47907-1287

ABSTRACT

The atomically clean contact surfaces, extremely high contact pressure, high temperatures, temperature gradients and the nature of the tangential forces applied to the chip make the tribology of the chip-tool interface in machining a truly unique one.

In the machining of ductile metals a freshly generated chip, in a state of incipient plastic flow, moves over the rake face of the cutting tool under high contact pressure. Due to the shear deformation suffered by the chip material and due to frictional heating because of the relative motion between the chip and the tool under the high contact pressure, the temperature at this interface can be very high. The continuous flow of the clean chip surface progressively removes contaminants from the tool rake surface resulting in two atomically clean surfaces in contact with each other. Under these circumstances it will normally be expected that a layer of the softer material, namely the chip, will get stuck to the tool and relative motion between the chip and the tool will occur by shear within the chip (1).

But it has actually been observed, in studies using specially designed transparent tools to directly observe the interface conditions while cutting is in progress, that this need not be the case. Optical microscopy and high-speed photography at framing rates up to 4000 frames/sec have been used to record and analyze the evolution and the steady state distribution of sticking and sliding zones in the contact region. The results have shown that under some conditions there is intimate sliding contact between the chip and the tool at and near the cutting edge. This is the region where the contact pressure is highest. Further away from the cutting edge and close to the end of the chip-tool contact, sticking or transfer of chip material to the tool rake surface is observed. Deposition of the chip material on the tool rake surface initially occurs near the end of contact and progressively extends outward and away from the cutting edge as the length of contact between the chip and the tool increases. Metal transfer is greatly inhibited when cutting in the presence of lubricants, leading to lower cutting forces.

The apparently anomalous behavior of sliding under high contact pressure and sticking under low contact pressure could be due to one of two causes. (i) The constant flow of incoming workpiece material to become part of the chip could result in the tangential stress for movement of the chip over the tool being applied right till the interface (2). This is different from the norm in most tribological situations. (ii) The action of the environment on the fresh chip surface could lead to chemical reactions which enhance the adhesion between the chip and tool surfaces leading to sticking at the edge of the contact region (3).

A finite element analysis of machining, which simulates material separation close to the cutting edge without using artificial separation criteria is being used to understand the influence of the peculiar nature of the stress distribution on the tribology at the chip-tool interface. Experimental studies of the temperature field at this interface are also currently in progress using infrared focal plane array cameras and dual wavelength imaging for emissivity compensation.

REFERENCES

(1) E. M. Trent, Metal Cutting, Butterworths, London, 1984.
(2) V. Madhavan, S. Chandrasekar, and T. N. Farris, Direct observations of the chip-tool interface in machining, Submitted to the symposium on predictable modeling of machining as a means for bridging the gap between theory and practice, ASME Winter Annual Meeting, Dallas, TX, Nov. 1997.
(3) E. D. Doyle, J. G. Horne, and D. Tabor, Frictional interactions between chip and rake face in continuous chip formation, Proc. R. Soc. Lond. A 366, pp.173-187, 1979.

CONDITION MONITORING OF THE MACHINING CENTER

NENAD MILIC, BRANISLAV JEREMIC, MIROSLAV BABIC
University of Kragujevac, Faculty of Mechanical Engineering, Department for Manufacturing Engineering
S. Janjic 6, 34000 Kragujevac, YUGOSLAVIA, E-mail: milic@uis0.uis.kg.ac.yu

ABSTRACT

Results of the exploitation investigations of the machining center's diagnostic parameters, which had the aim to develope the diagnostic system for condition based maintenance of the group of the same machining centers working in batch production conditions and high time efficiency ratio are presented in the paper.

Conducted investiogations included state monitoring of the tested technical systems through dynamic behaviour, temperature and wear particles as a diagnostic parameters. From the aspect of realizing high quality and machining precision, it was necessary to dedicate a special attention to the main spindle and their bearing system. Dynamic behaviour of the main spindles was monitored by amplitude-frequency characteristics of the main spindle housing motion acceleration and temerature measurements were conducted by thermocouple placed in the main spindle housing of the front couple of the inclined ball bearings. By the ferrography method (through PQ index), contents of wear particles in the lubrication oil of the main transmission gearbox, having closed lubricated systems, were monitored. Obtained results were numerous, and some of the characteristic examples are given in figures 1,2 and 3.

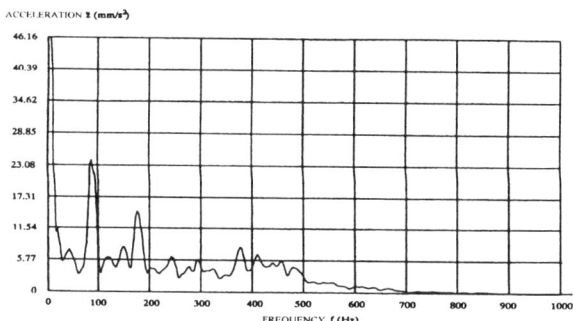

Fig. 1: Amplitude-frequency characteristic of the motion acceleration along the Z-axis at 5500 rpm for the machining center 58349 after 12564 working hours

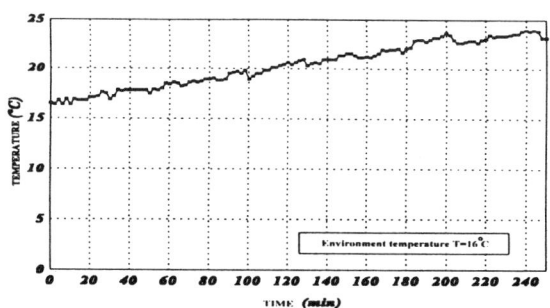

Fig. 2: Temperature in the spindle bearings housing for the machining center 58350 at the normal regime of exploitation after 10489 working hours

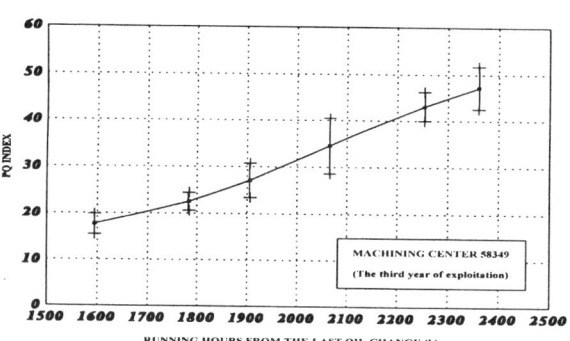

Fig. 3: The PQ index as a function of exploitation time for the machining center 58349

The basic harmonic of 92 Hz at 5500 rpm (Fig. 1) originated from the spindle's unbalance, and the second harmonic are the consequences of the spindle being out of plane and inclined.

Obtained results of investigations of the main spindle dynamic behaviour show good repeatability and give possibility of definition of variation trend of chosen parameter - acceleration amplitude of oscillatory motion in time. In consideration of adequote sensitivity of the previous, it could be reliably used as a diagnostic parameter of tested machining centers state.

Monitoring of the thermal state shows small sensitivity of the temperature changes, which was expected on account of small speed of tribological processes unfolding, which caused their variation. This parameter can be used, by all means, as a diagnostic one for long term monitoring of technical systems state, when temperature change presents some unregularity of the processes that are unfolding in the system.

Obtained values of PQ index are relatively small, which pointed to the low contents of wear particles in the tested samples, so the monitoring time should be increased. For monitoring of a tribomechanic system state, as main transmission is, PQ index can be quite reliably used as a general macroindicator of the same system, and it follows quite well the integral wear processes which take place inside the system.

REFERENCES

(1) V. Wowk, Machinery vibrations: measurements and analysis, McGraw Hill, Inc., New York, 1991.
(2) N. Milic, Development of the diagnostic system for the machining centers maintenance, M.Sc. Thesis, Faculty of Mechanical Engineering, Kragujevac, 1994.
(3) W. Hernandez, Using High-Frequency Vibration Analysis, Maintenance Technology, April 1995, pp. 34-36.

Submitted to *Tribology International*

CONTROL RESARCH OF PARALLEL MOVEMENT OF MACHINES ON THE BASE OF CHANGEABLE TRIBOLOGICAL EVENTS

MIOMIR JOVANOVIĆ, RADIĆ MIJAJLOVIĆ,
Mechanical Faculty University of Niš, 18000 Niš, 14 Beogradska St, YU,
MIODRAG ARSIĆ, DRAGAN DENIĆ
Faculty of Electronic Engineering University of Niš, 18000 Niš, 14 Beogradska St, YU,

ABSTRACT

The cranes displacement at their traveling is the result of the tribological random events caused by their own geometrical deformities, irregularities of the crane paths, the dynamical influences, the damage of the crane and the path caused by the exploitation. The displacement effect disturbs regular work, increases the damages of the system which leads to early interruption of exploitation.

The solving of this important task requires the replacement of classic solutions by electronical control of the parallel crane position in moving. The reliability of this solution is based on the information of absolute position of the left and the right side of the crane. By applying of the new methods (1) pseudorandom positioning encoder (with 12-bit output resolution) which eliminates previous defects, is developed (2).

The identification of the absolute position is based on "window feature" of pseudorandom binary sequences (PSBS), $\{S(p)/p=0,1,...,(2^n-2)\}$. According to this, any n-tie binary word $\{S(p+n-k)/k=n,...,1\}$, obtained by scanning PSBS, by window width n, $\{S(k)/k=n,...,1\}$, is unique and can completely identify the absolute window positron **p** towards the beginning of the sequences. According to the carried out simulations on encoder (2), applying two such measuring position transducers, can give reliable information on both crane sides. Applying this information on crane driving control, the crane displacement would be prevented. The solution with one microprocessor has been developed, Fig. 1.

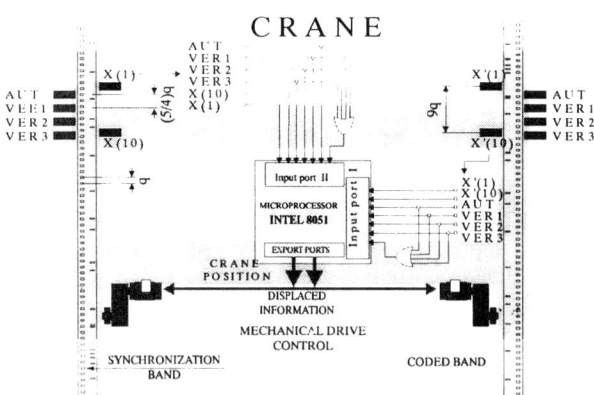

Fig. 1. The measurement - transducer system for positioning and prevention of the crane displacement

The serial reading of 10-bite pseudorandom code is carried out by two sensor heads on both sides of the crane (x(10) and x(1), that is x'(10) and x'(1). The fixed space of the sensor heads on (n-1)·q distance is a reference for the detection of possible error at code reading. Depending on the crane traveling direction, one head is used for forming the basic pseudorandom code word, and the second one for the control code. Comparing these two code words, the accuracy of code reading is checked. The head reading AUT provides the synchronizing impulses and the heads VER provide the information on crane traveling direction. The sensor heads VER are arranged according to vernier law. The vernier law uses the relation:

$$S = (1 + \frac{1}{m+1}) \cdot q = (1 + \frac{1}{3+1}) \cdot q = \frac{5}{4} \cdot q$$

The obtained total resolution is now q/(m+1)=q/4, which corresponds to the resolution obtained by pseudorandom code with width of 12 bites. Due to width reducing of pseudorandom code word, the time of conversion reduces 4 times. In comparison with the classic encoder, 12-bite pseudorandom encoder provides the multiplicity of the maximum crane speed approximately 16 times (2). The conversion of pseudorandom/natural code is written in machine language applying **Intel** microprocessor developing system **SDK-51** at 12 MHz. Applying the position transducer with 12-bit output resolution, permits the distribution of crane path range to $2^{12}-1=4095$ quantisation steps (3). For example, this provides that on crane path length 400 m, getting measure resolution of **q**=10 cm. This speed satisfies the maximum speed of bridge and reloading crane defined by present standards. The processor system is linked with motor-driving groups by which the crane displacement is corrected according to proportional law at coarse.

REFERENCES

(1) Arsić M., Denić D., "Digital transducer for length and position measurement", <u>Automation '92 Conference,</u> Scientific society of measurement and automation, pp. 1-11, Budapest, 1992.

(2) Arsić M., Denić D.,Pešić M., "The measurement position transducers of the transporting machines with 12-bit output resolution", <u>Proceeding of ETAN,</u> Vol.II, pp.235-242, Jugoslavija,1992,

(3) Arsić M.,Jovanović M., Denić D: "The measurement transducer systems for positioning and anti-displaced control at portal and bridge cranes", First International Conference TM'93, Vol.2, pp.190-195, Vrnjačka Banja, Jugoslavija,1993.

FEM SIMULATION OF FRICTION TESTING METHOD BASED ON COMBINED FORWARD CONICAL CAN – BACKWARD STRAIGHT CAN EXTRUSION

TAMOTSU NAKAMURA and ZHI-LIANG ZHANG
Department of Mechanical Engineering, Shizuoka University, 3-5-1, Johoku, Hamamatsu, Shizuoka 432 JAPAN
NIELS BAY
Institute of Manufacturing Engineering, Bldg.425, Technical University of Denmark, DK-2800 Lyngby, DENMARK

ABSTRACT

A new friction testing method based on combined forward conical can-backward straight can extrusion, is proposed in order to evaluate friction characteristics in server metal forming operations. Fig.1 shows the design of the testing method, which enables measurement of the friction characteristics on punch while piercing into a billet. A cylindrical workpiece is deformed simultaneously into a straight can and conical can. The friction coefficients is estimated on the conical punch in the forward conical can extrusion using theoretical calibration curves representing the relationship between the forward can height H_L or the backward can height H_U, the die friction characteristics and the conical punch friction characteristics without requirements of measuring the forming load and the flow stress of the workpiece.

Before determining the conical punch friction characteristics the die friction characteristics should be estimated by a friction test based on combined forward straight can-backward straight can extrusion as shown in Fig.2.

Fig.3 shows the relationship between the normalized can height H_u/D_0 and the punch stroke S_P/D_0. The calibration curves are obtained by the FEM simulation of a combined forward straight can-backward straight can extrusion, The friction coefficient μ_D along the die wall should be estimated when some experimental points are plotted on the $H_u/D_0 - S_P/D_0$ diagram of Fig.3.

Fig.4 presents curves of constant friction coefficient μ_{LP} in a diagram plotting the upper can height H_u/D_0 versus the die friction coefficient μ_D. The calibration curves are obtained by the FEM simulation of a combined forward conical can-backward straight can extrusion. Plotting measurements of corresponding values of H_u/D_0 and μ_D, the friction coefficient μ_{LP} can be estimated.

Some preliminary experimental friction tests have been carried out with aluminum alloy A6061 and varying lubricants. Fig.5 shows the friction coefficient μ_{LP} on the conical punch determined from Fig.4. The conical punch friction coefficient μ_{LP} for every lubricant was higher than the die friction coefficient μ_D for the lubricant.

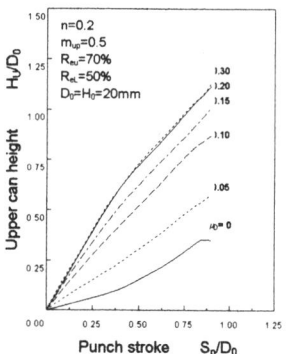
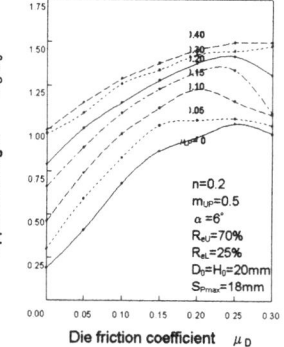

Fig.3 Calibration curves for μ_D on H_u/D_0 and S_P/D_0 diagram in combined straight cans extrusion test.

Fig.4 Calibration curves for μ_{LP} on H_u/D_0 and μ_D in combined forward conical can-backward straight can extrusion test.

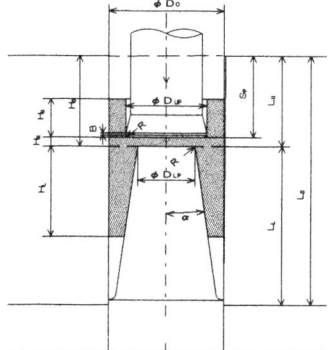
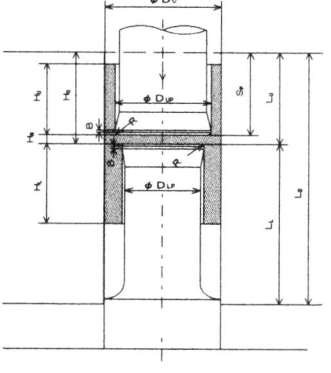
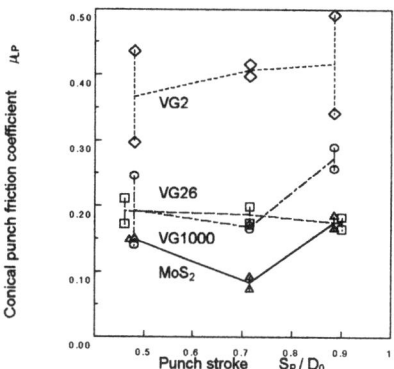

Fig.1 Design of combined forward conical can-backward straight can extrusion test

Fig.2 Design of combined straight can extrusion test

Fig.5 Conical punch friction coefficient μ_{LP} for A6061 billets with various lubricants.

ANALYSIS OF DYNAMIC BEHAVIOR OF TOOLS IN METAL TURNING

MSc. NEDIĆ BOGDAN

Mechanical Engineering Faculty, University of Kragujevac, S. Janjić 6, 34000 Kragujevac, Yugoslavia

ABSTRACT

All the processes and phenomena that are happening in the localized zone of cutting have, not static, but prominent dynamic character. The dynamic character of processes are caused by dynamic character of shaving forming process, variable cutting depth, nonhomogeneity of the workpiece material, forming of coating on the cutting wedge tip, instability of the system tool - workpiece - machine, workpiece geometry, etc.

Development of the mathematical model of tool dynamic behavior has an objective to establish the influential parameters and their effects at output values of the cutting process.: the quality of the machined surface, tool wear, accuracy, etc.

The voluminous experimental investigations were conducted before the mathematical modeling approach. By them the frequency of free damped vibrations and the non-dimensional damping factor of the turning knives of different dimensions of cross-section and different overhanging lengths. Those investigations were done by the usage of accelerometer and appropriate hardware and software. Fig. 1 show the obtained experimental data of the change of free damped vibrations frequency due to dimensions of tool cross-sections and overhanging length.

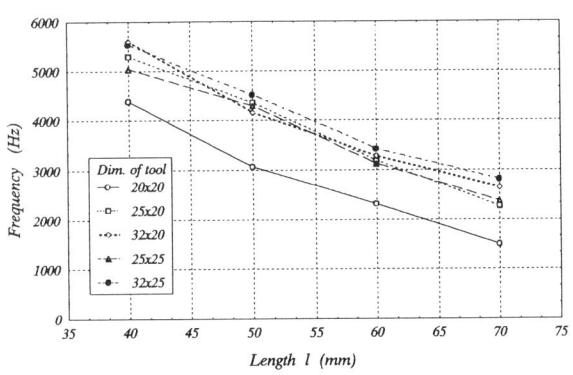

Fig. 1.

Another step at experimental procedure was the measurement the cutting force and acceleration of the tool tip at various cutting conditions and with various tool wear. In order to measure the cutting force and acceleration and to do the signal analysis at real time the monitoring system was formed.

Fig. 2 shows the amplitude-frequency spectrum for the signal of the cutting force F_2 at the accessory movement direction, with the tool wear of $h=0,3$ mm.

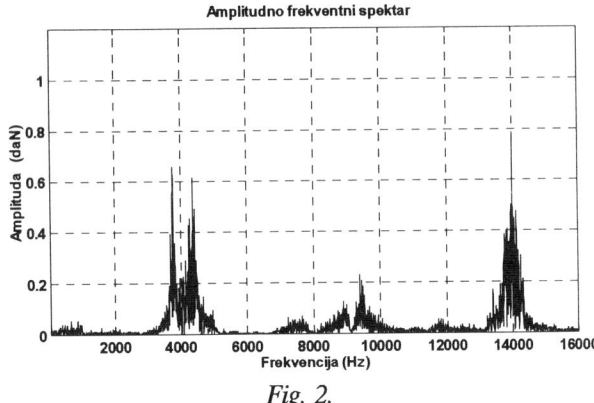

Fig. 2.

With in these investigations the adequate conclusions were drawn about the possibilities of the developed monitoring system for monitoring the cutting process and prediction of the tool failure.

Mathematical modeling and experiments researches have shown that cutting process is tabbing place at very complicated dynamic conditions which are represented by dynamic character of the sheaves forming process, corresponding tool strength and machine elements, changeable strength of the workpiece, wave-shaped surface of the workpiece, various cutting depth and change of the analyzed values with time, e.g. with tool wear. With some specific values of the cutting conditions parameters the phenomenon of resonance at the system tool - workpiece - machine, which effects very badly at machining surface quality, wear and tool failure. There are also the increase of the cutting force dynamic components and of the amplitudes at power spectrum of the acceleration signal with the increase of the tool wear.

The developed mathematical model of the tool dynamic behavior and experimental investigations enabled the development of the monitoring system and cutting process management based on monitoring and analysis of the cutting-force dynamic components and analyzing and processing the acceleration signal.

REFERENCES:

[1] B. Nedić, Development of the management method for the metal cutting process on the basis the cutting parameters' dynamic characteristics, doctor dissertation in manuscript, Faculty of mechanical engineering, Kragujevac, 1997.

SOME ASPECTS OF MAINTAINING THE MACHINE TOOLS GEOMETRICAL PRECISION PARAMETERS

V. OGNJANOVIĆ, M. DJURDJANOVIĆ and P. MILOSAVLJEVIĆ
University of Niš, Faculty of Mechanical Engineering, Beogradska 14, 18000 Niš, FR Yugoslavia
D. SOLDATOVIĆ
Pump Factory "Jastrebac", Bulevar 12 februar 82, 18000 Niš, FR Yugoslavia

ABSTRACT

In the machine tools process of work the geometrical accuracy parameters greatly influence the quality of machining. Machine tools geometrical accuracy is tested in two ways: by direct and indirect methods.

The direct method is mainly used by the machine producers. It covers measuring and testing of the geometrical and kinematics deviations. The measured values should be within the standard prescribed values. The indirect method testing is not prescribed by the standard, because it contains only the recommendations. The way of indirect testing is regulated by the machine producer's norms.

This work gives the case of testing the geometrical accuracy parameters by the direct method, with the machine tools already in use, namely:

- a universal lathe (stand horizontal position, lathe centres axis matching, internal cone centricity and operating spindle parallelism, etc.)
- a vertical lathe (transversal support horizontal position, parallelism and orthogonal position, etc.)
- a vertical mill (horizontal, parallel and perpendicular positions, operating spindle internal cone centricity, etc.) (1).

The values of these parameters have been measured in various time intervals. By methods of discrete medium square approximation and by using the programme package "ORIGIN" mathematical functions have been obtained, representing the pattern of parameters change within the time for the universal lathe, vertical lathe and vertical mill respectively:

$$y = 0,07136 + 0,00086\, x + 0,00009\, x^2$$

$$y = 0,29446 - 0,059\, x + 0,00286\, x^2$$

$$y = 0,16804 + 0,00112\, x + 0,00042\, x^2$$

Mathematically set patterns of important parameters elementary and integral changes, regardless of the approximations made, give the basis for the system analysis and diagnose (indicate) the state of the machine.

For more detailed analysis of the geometrical accuracy parameters pattern change, it is suitable to graphically represent the equations stated, like in Figure 1, for the universal lathe (the graphs of other machines can be represented in a similar way).

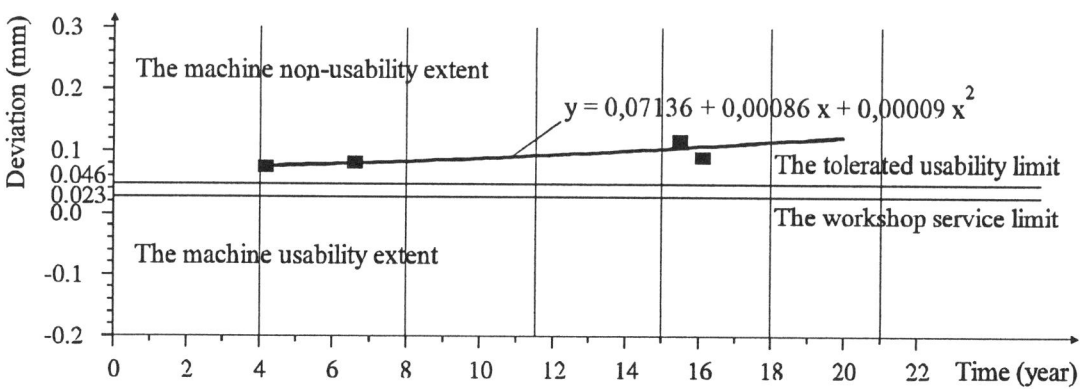

Fig. 1: The universal lathe geometrical accuracy parameters medium values

Based on the analysis of the graphical representation of the parameters medium values pattern change and the machine tools producer's recommendations, the maintenance cycles have been defined together with the description of the process activities structures in the course of maintenance. The information on the character of parameters values changes, make it possible for us to foresee and plan the activities in the process of machines maintenance.

In that way, the measures taken are preventive both in view of keeping the machine operation accuracy and in view of extending its life-time (1).

REFERENCES

(1) P. M. Milosavljević, Technical Life-time Models of Machins and their Influence to the Maintenance Cycle Determination, MS Thesis, University of Niš, Faculty of Mechanical Engineering, Niš, 1997.

THE DEFORMATION DEGREE IN THE TECHNOLOGY OF PLASTICITY

Predrag V. Popović, Dušanka M. Vukićević and Dragan I. Temeljkovski
Department of Mechanical Engineering University of Niš, 18000 Niš, Serbia, Yugoslavia.
Božidar Jovanović
Institute for the Quality of life and Work Environment "1MAJ", 18000 Niš, Okt. rev.1/II, Yugoslavia

ABSTRACT

It is known that the friction forces acting on the tool and workpiece surfaces bring about the material non-uniform deforming at plastic deforming. Starting from the known definition of the deformation non-uniformity degree, this paper deals with the effects of the two influential factors - the deformation degree and roughness of the tool operating surface - upon the non-uniformity of cylindrical workpieces deforming and the expression is derived for exact calculation of the deformation non-uniformity degree.

The analysis is base on the assumption of the cutting as the technique for tool operating surface heathen which causes differences in horizontal and vertical roughness and that is reflected in the friction intensity deflection as the function of the change of the radial direction of the material sliding upon the tool contacting surface which is reflected in the fact that the cylindrical workpiece contacting surfaces are not transformed into the circles of greater diameters but into ellipses. The corrected expression given in the paper for deforming the deformation non-uniformity degree includes the given phenomenon as that the values obtained by its use in calculating the deformation on non-uniformity degree more realistically express the deformation non-uniformity with respect to the familiar expression for calculating the given degree.

This paper gives the results of the experimental research of the tool operating surface and work piece material properties roughness effect on the cylindrical elements deformation non-uniformity. The obtained results show that the correction factors should be introduced in the deformation non-uniformity degree calculation that take into consideration deflections in the friction forces intensity on the tool and workpiece contact surfaces depending on the material sliding direction along the tool operating surface.

Since the contacting surface of the workpiece obtained from the cylindrical prepared model can be freely described by ellipse in this case while remaining at the same time within the limits of the allowed deviations, and since starting from the general form of the ellipse equations (according to the Figure 1.), we obtained the expression for calculating the degree of the radial non-uniformity of the deformation in the form:

$$n_{dr} = 1 - \frac{\pi}{4 \cdot A_o} \cdot (1-\varepsilon) \cdot \frac{d_p^2 \cdot d_u^2}{d_p^2 \cdot \cos^2\alpha + d_u^2 \cdot \sin^2\alpha}$$

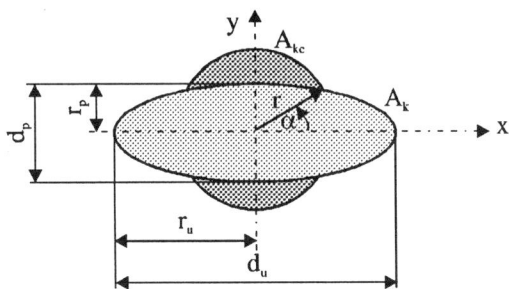

Fig. 1: Contacting surface of the workpiece

On the basis of everything said in this paper the following conclusions can be drawn. they are:

- The obtained expression which expresses a degree of the radial deformation non-uniformity can be used the intensity of the non-uniformity of the material plastic deforming as well as for obtaining the law of change of this very intensity according to the quantity of work required.
- The law of change of a degree of the deformation non-uniformity according to the quantity of the work required can be obtained for every tool used in practice by compressing one cylindrical test tube in it and by using the expression.
- Knowledge of the law of change of a degree of the deformation non-uniformity according to the quantity of the work required is without any doubts important for practice since it can prevent undesirable consequences in the manufacturing process.
- The experiment performed verifies the theoretical hypothesis that the quality of the working surface of the tool and the features of the processing item material influence both the known "vertical" uneven deforming, and the radial uneven material deforming.
- The results obtained are extremely important in defining the processing mode of the working surfaces of the tools and their quality, as well as, regarding the process, the quality of the processing product and the kind of material features.

REFERENCES

/1/ P. Popović, Lj. Bogdanov, V. Stoiljković, Journal Technique, Vol. 9, Beograd, 1974, 1516-1819pp.
/2/ P. Popović, D. Vukićević, D. Temeljkovski, Proc. 30 year of the machinery - Niš,Niš, 1991,133-137pp.
/3/ P.Popović, D. Vukićević, D.Temeljkovski, Proc. 7th Yug.Symp. of the plasticity, Pula 1991,134-138pp.

INFLUENCE OF TRIBOLOGICAL PROCESSES ON MECHANISM OF FORMATION OF COPPER-LAYERS DURING STEELS CUTTING

STANISŁAW PYTKO
Faculty of Mechanical Engineering and Robotics, Academy of Mining and Metallurgy, Al. Mickiewicza 30, 30-059 Kraków, Poland

STANISŁAW MARZEC
Faculty of Technology, Silesian University, ul. Żeromskiego 3, 41-200 Sosnowiec, Poland

ABSTRACT

In many technological processes copper layers on surfaces of machine elements are layed as one of production operations. Between others such coatings are applied before electrolytic chromium plating and prior to carbonizing processes on the surfaces that shall not be carbonized. Preferably it is a separate operation preceding the main process.

The authors have developed a novel proocess that makes possible constitution of a coating of different metals (the copper coating is the subject of this work) on surfaces of steel elements at the machining process. In this process the water solutions containing metal complexes are used instead of typical emulsions or oils. The coating is constituted from metals of a/m complexes. There were developed liquids containing s Cu complexes but also Cu+Zn, Cu+Cr, Cu+Ni, Cu+Mo. Composition of cooling-lubricating liquid used in this process is patent pended (1). The coating produced in this technology may serve as a decorative one, anticorrosive, protective against carbonizing and as precoat before electrolytic chromium plating. The authors have developed this method to eliminate etching before electrolytic chromium plating as this process is very hazardous as regards to ecology. This cooling-lubricating liquid has good lubricating characteristics (four-ball machine tests gave results up to 800dN seizure force when modification additives were used), as well as good antibacterial and antifoam properties.

Such coating of a metal contained as a metal complex in cooling-lubricating liquid is formed on the surface as a result of tribological phenomenae that ocurr in the area of cutting edge. It is known from investigations of the process that temperature in the edge area may achieve 1000K, normal stress above 1000MPa and also very high are tangential stresses evolved by friction (2).

At machining process the surface is uncovered, so it is very clean, without any iron oxides, or other adsorbed compounds. Such surface being highly activated chemically and mechanically allows for penetration of cooling-lubricating liquid components. Machining process could be realized in such a way that prior to a tool edge a crack appears resulting in vacuum where easily penetrates the liquid used in machining process. In the contact with hot surface of machined metal of temperature about 1000K, water evaporates rapidly and metal content does cover an active surface uncovered some microsconds earlier. It is a shock.

In the paper is presented mechanism of constitution of the coating consisting of three zones i.e. 1-Cu, 2-transient Cu+ Fe and 3-deformed bulk material. Some characteristics of the coating are also described like roughness, wear resistance, chemical composition (Auger electrons spectroscopy method), initial stress.

The result of this technology is coating of machined steel element by another metal but also better surface smoothness is achieved, the fact of importance in finishing processes and higher is tool edge lifetime in comparison with processes where typical emulsions have been used. The layers of copper on the edge is deteriorated by chips however then it is continuosly regenerated.

The proposed liquids do not create hazard for man and environment and have obtained positive opinion from the Labour Medicine Institution. Our technology when accompanied with electrolytic chromium plating process (Cr III) will be an ecologically clean technology. Besides this technology when applied in practice gives good economical efects.

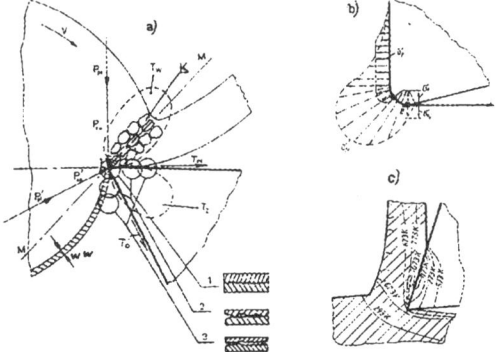

Fig.1: Schematic diagram of forces and phenomena during metal cutting
a) Distribution of forces, b) Distribution of stresses
c) Distribution of temperature

REFERENCES

(1) S. Marzec, S.Pytko, A.Zochniak, Patent No.167431, Warszawa Poland, 1995.
(2) S.Pytko, S.Marzec, International Tribology Conference.Yokohama '95,1697-1702 pp.

CONDITION-BASED MAINTENANCE - THE BENEFITS OF COUNTING THE COST

B.S.RAJAN,
Glaxo Wellcome Operations, Ulverston, Cumbria, LA12 9DR, U.K.
B.J.ROYLANCE
Department of Mechanical Engineering, University of Wales, Swansea, SA2 8PP,U.K.

ABSTRACT

Large potential cost benefits exist for high capital cost - high consequential loss plant, but the extent to which similar pro-rata benefits are realisable in the case of lower value, batch process plant is more difficult to quantify. Some results are presented which were obtained from a detailed investigation of batch process operations and maintenance in the pharmaceutical industry. Of particular interest is the effect of consequential costs when machinery fails which is critical to the maintenance of production, such as compressors and fans, and for which there is no provision for standby equipment.

INTRODUCTION

A mathematical model has been devised for critical component operation on a plant-wide basis (1). It was derived using actual data obtained over a five year period in which the maintenance cost was predicted and compared with the actual costs over the period. The advisability of utilising a condition-based strategy was assessed in relation to alternative maintenance strategies.

The equation developed for determining the direct costs of machine breakdown, C_d in relation to the initial capital cost is given as:

$$C_d = C_i * I_p * I_c * I_{pr} * K_d$$

where C_i = initial cost of machine corrected to present day value

I_p = Power index
I_c = Criticality index
I_{pr} = Process index
K_d = Direct costs factor

RESULTS

To determine K_d, data was obtained for a large number of components in the same category.

Figure 1 shows results obtained for pumps operating in the power range 1 to 250kW. A total of 329 pumps were surveyed covering centrifugal and positive pressure, single and multistage, and vacuum. Most of the pumps operated on a standby basis, and there were therefore, no consequential costs incurred when breakdown occurred. Other results will be presented for fans and compressors in which consequential costs were incurred. The implications are considered in relation to establishing the correct basis for determining whether the use of a condition-based maintenance policy is justified in terms of cost benefits.

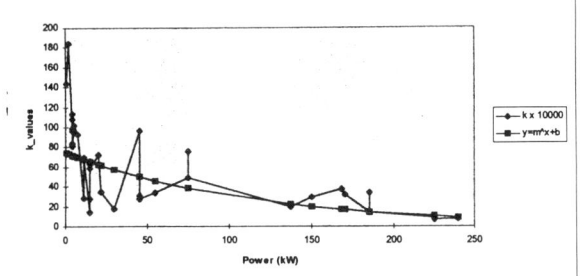

Fig. 1: Curve fit for Kd values

CONCLUSION

A suitable mathematical model for predicting maintenance costs in batch process operations has been developed which permits the costs of maintenance to be predicted and related to the type of maintenance strategy adopted.

REFERENCES

(1) B.S.Rajan and B.J.Roylance, Proc. Int. Conf. Integrated Monitoring, Diagnostics and Failure Prevention, Mobile, AL, USA April 1996, 725-736pp.

Submitted to Leeds-Lyon Symposium on Tribology Proceedings

INFLUENCE OF SURFACE ROUGHNESS PROFILE ON THE FRICTIONAL BEHAVIOR IN THE SHEET METAL FORMING PROCESSES

SOO-SIK HAN and KEE-CHEUL PARK
POSCO Technical Research Laboratory. Pohang-shi, Kyungbuk, 790-785, KOREA

ABSTRACT

The current trend in sheet metal forming industry is the growing importance of the formability of sheet metal. For this, the sheet metal that has good frictional characteristic is required. As is known from literature, many parameters are involved in the frictional behavior of a given sheet metal forming processes, like workpiece material, tool material, tool geometry, surface roughness, etc. Especially, the surface roughness is of major importance in frictional behavior of sheet metal. Therefore, a better control of the surface roughness of sheet metal is necessary(1)(2).

The influence of surface roughness on the frictional behavior of sheet metal was investigated in this study by newly developed stretch bending type friction measuring system that can evaluate the frictional behavior of sheet metals at the actual situation in sheet metal forming process as close as possible. The developed friction test method shows good reliability and repeatability in the fields of analyzing the frictional behavior.

To isolate the friction test results from influence of mechanical properties of material, cold rolled steel sheets with same mechanical properties were used in this study. Wide ranges of surface roughness were produced on the sheet surface by temper-rolling with bright rolls, surface corrosion with acid, surface grinding with diamond powder. Friction tests were performed with test specimens of 40mm in width and 400mm in length. To investigate the effect of lubricant, three types of lubrication conditions, nonlubrication, lubrication with low viscous oil (20.2 Cps), lubrication with high viscous oil (907.2 Cps), were adopted. The tool surface was polished with #2000 abrasive paper at right angles to the sliding direction before each friction test. Average contact pressure was 2kg/mm^2 and sliding speed was 20mm/sec.

For a determination of the influence of surface roughness, it is obvious that the surface roughness should be measured carefully. This has been done in the following way. On the each test specimen surface, scratchfree part of the surface was selected and cleaned with acetone. On this area, roughness measurements were carried out by using noncontact type precision 3D surface roughness measuring instrument GMB600. Each measuring area was 5.2 x 5.2 mm. From all specimens, the cut edges were deburred, the strips were cleaned with acetone and wiped dustfree with a clean cloth. Lubricant was applied to each specimen generously immediately before testing.

The effect of the surface roughness on the friction coefficient with different lubrication condition is shown in Fig.1 and Fig.2. As surface roughness is smoother, the friction coefficient tends to decrease. In the low viscous lubrication, the influence of surface roughness on the friction behavior is very small but in the high viscous lubrication, surface roughness has great influence on the friction behavior of sheet steel. In the nonlubrication condition, it is easy to occur stick-slip phenomena as surface roughness of specimen is smoother. The shape of roughness peak also has influence on the frictional behavior. Sharp peak causes higher friction coefficient than dull peak.

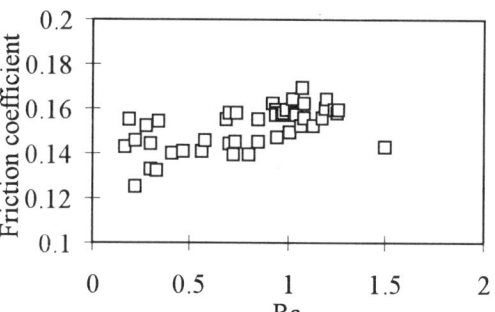

Fig. 1 : Friction coefficient with Ra (low viscosity lubricant, 20.2Cps, 25°C)

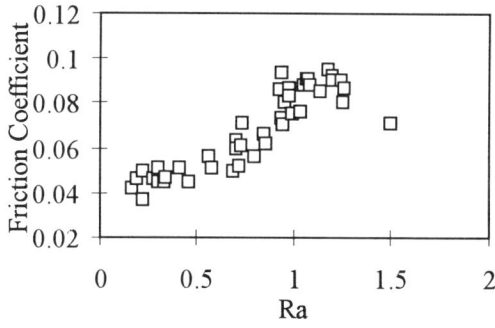

Fig. 2 : Friction coefficient with Ra (high viscosity lubricant , 907.2Cps, 25°C)

REFERENCES

(1) Kudo H., A. Azushima, Proceedings of the 2nd ICTP, page 347,1987
(2) Emmens, W. C., Proceedings of 15th IDDRG, page 63, 1988

IMPORTANCE OF STRAIN ANALYSIS IN TRIBO-MODELING IN DEEP DRAWING

M STEFANOVICH and S ALEKSANDROVICH
Faculty of Mechanical Engineering, S. Janjic 6, 34000 Kragujevac, YUGOSLAVIA

ABSTRACT

In deep drawing of parts of irregular geometry made of thin metal sheets, as for instance are the parts of car body, in certain zones of the part exist significant differences in stress-strain schemes. Also, significant are differences in the way of realization of the contact between the metal sheet and tool, namely in the contact macro geometry, values of pressure, speed and temperature, friction is useful in certain zones, and in others it is not, etc. In that sense several physical tribo-models were developed, elementary and complex ones, that are related to characteristic zones of the piece that is being drawn: sliding between the flat surfaces of the die and the holder, sliding over the draw bead, bending with tension over the die edge, two-way tension-stretching under the front of the drawing tool, and pure deep drawing in the zone of tangential compression on the flange (1).

Results of tribological investigations can be divided into three groups, according to the nature of parameters that represent the measured variables in realization of corresponding modeling.

The first group consists of physical variables, that are presented as function of pressure, speed, temperature, etc. (friction force and coefficient, roughness parameters, sliding length in galling tests, weighting amount of wear, etc.).

The second group consists of the so called macro-indicators for the whole piece, and their relations, for example the drawing force, the blank holding force, limiting drawing ratio, average friction coefficient, limiting drawing depth, etc.

The third group of parameters is directly related to realized strain fields on the drawn pieces and for their determination one assumes knowledge of strain analysis. In experiments are used the measuring grids, that are being applied to sheet metal surfaces by the special procedure before drawing (2). Strain distribution is being determined for the characteristic section area, or the complete piece that is being drawn.

By its structure and nature, parameters of this group belong to the "interior" indicators and in direct way describe effects of tribo-effects in deep drawing. In experiments were used different materials for deep drawing, several lubricants, different geometries of blank and tools, strain rates and specific pressures in contact. For characteristic models- stretching and pure deep drawing - are enumerated results of the principal strain distributions as function of location, with indicators of the distribution homogeneity, distribution gradient, etc., for different contact and other conditions (Fig. 1). Strain distributions in forming limit diagram enable determination of the plastic reserves, strain paths etc. Also, given are the results of influence of contact conditions for non proportional forming, that exists in multi-phase deep drawing (Fig. 2).

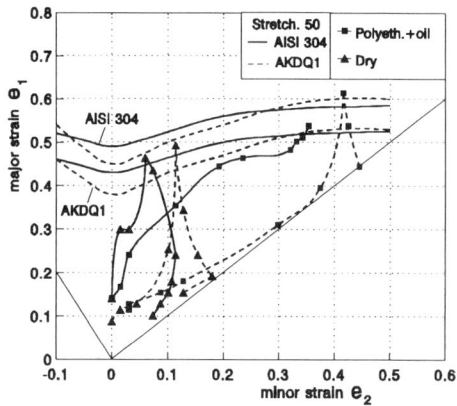

Fig. 1: Strain distribution in FLD for AISI 304 and AKDQ1 steel sheets

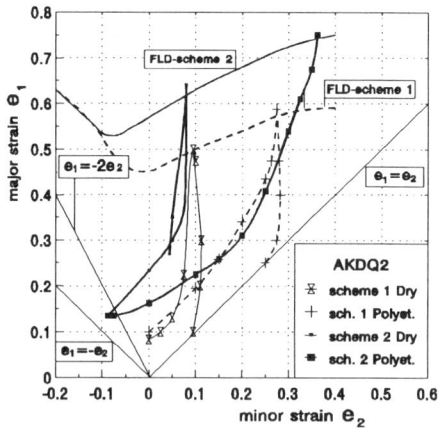

Fig. 2: Strain distribution in FLD for different strain paths (AKDQ2 steel sheet)

REFERENCES

(1) M. Stefanovich, S. Aleksandrovich, Proceedings of the Balkantrib '96, Thessaloniki 1996, 214-221 pp.
(2) Dinda S. at all., How to Use Circle Grid Analysis for Die Tryout, ASM, Metals Park, Ohio, 1985.

TRIBOLOGICAL AND MECHANICAL PROPERTIES OF TOOLS AND WORKPIECE'S MATERIALS IN THE TECHNOLOGY OF PLASTICITY

Dragan I. Temeljkovski, Dušanka M. Vukićević and Predrag V. Popović
Department of Mechanical Engineering University of Niš, 18000 Niš, Serbia, Yugoslavia.
Božidar Jovanović
Institute for the Quality of life and Work Environment "1MAJ", 18000 Niš, Okt. rev.1/II, Yugoslavia

ABSTRACT

It is known that in the process of material's plastic reshaping there are internal and extend friction forces. It is also known that between the friction coefficient at gliding in engineering structures and between the friction coefficient in the processes of material plastic forming there are considerable differences manifested as the differences of their intensities.

Starting from identification and analysis of the differences in sliding of a friction couple in engineering constructions and metal forming processes necessary mechanical properties of tools and workpieces attributes are defined, namely the relational parameters which an serve as the basis for choosing appropriate materials for tool's manufacture.

The theoretical background for a more systematic, detailed and all inclusive experimental researches and testing are established in order to calculate the optimum values of correction factor (f_k). It has been done by determining of the partial meritorious correction factors (f_{ki}) in order to prolog the tool life and to reduce uneven wearing and deformation. Having obtained the correction factor value, the unreliable selection of material is eliminated on the grounds of experience.

To conclude, having in mind the theory of plastic compression which implies that tangent and normal contacting stresses along the contacting surface do not have the same intensity and since the tool must meet the domain of elastic deformities, and the piece must remain within the range of plastic deformities, it is more correct to determine ($\Delta\sigma$) on the grounds of expression:

$$\Delta\sigma \geq f_k \cdot (K_{gr} - R_{Eo})$$

where:
$\Delta\sigma$ - the difference between the material elasticity limit of the tool and the material of the piece
$(\Delta\sigma = R_{Ea} - R_{Eo})$;
R_{Ea} - material elasticity limit of tool;
R_{Eo} - material elasticity limit of the piece;
K_{gr} - the limiting specific resistance of the piece material to deforming and
f_k - correction factor, which can be expressed as the result of partial correction factors:

$$f_k = \prod_{i=1}^{i=6} f_{pki}$$

where each correction factor $f_{pki} > 1$.

Partial correction factors include:
- variation in mechanical features of the piece and tool material;
- variation in temperature on contacting surfaces of the tool and piece, reflecting both on the mechanical features of the piece and tool and the friction coefficient;
- variation in quality of the working surface of the tool, i.e., its shabbiness and
- the presence of other attachments and solid dirts in the material of the piece.

Preliminary research into this area show that the value of the correction factor remains with the following limits:

$f_k = 1,4 \div 1,5$ during processing in the cold state;

$f_k = 1,6 \div 1,7$ during processing in the warm state.

On the basis of everything said in this paper the following conclusions can be drawn. they are:

* This postulates a theoretical basis for systematic, larger and more encompassing experimental research and examination in order to calculate the optimum values of the correction factor f_k, by determining the values of merritory partial correction factors (f_{ki}), in order to prolong the age of the tool and to diminish uneven deformities.

* Obtaining the optimum value of the correction factor eliminates the unreliable method of tool material choice, based on experience.

REFERENCES

/1/ H. Djukić, P. Popović, Metal Forming, University of Mostar, Mostar, 1988, 425pp.
/2/ B. Musafija: Metal forming, University of Sarajevo, 611pp.
/3/ D. Vukićević, D. Temeljkovski, P. Popović, Journal TRIBOLOGY IN INDUSTRY, Universityu of Kragujevac, Kragujevac, Vol.XV, No 4, 1993, 181-184pp.

THEORETICAL AND EXPERIMENTAL INVESTIGATIONS OF THE EFFECT OF THE RADIAL CLEARANCE BETWEEN BARREL AND SCREW ON THE EXTRUDER FLOW RATE

NIKOLA VOJKOVIC and ZORAN CANIC
Holding Company Kablovi, DD Institute, 35000 Jagodina, Yugoslavia
SLOBODAN MITROVIC
Faculty of Mechanical Engineering, Department for Manufacturing Engineering, 34000 Kragujevac, Yugoslavia

ABSTRACT

The aim of this paper is an experimental and theoretical investigations of the effect of the radial screw clearance, which is a part of the trybomechanical system barrel - screw - polymer, on the extruder flow rate. The rise of the functional radial clearance which occurs thanks to the wearing process during the exploitation of an extruder can greatly reduce the extruder output which, in turn, can greatly reduce the production rate of an extrusion line.

The experimental work was carried out on the single screw, industrial, 90 mm extruder using LDPE as a polymer being processed. This extruder is exclusively used in everyday industrial practice as the extruder for outer samy-conducting layer during the production of power cables insulated by XLPE. The theoretical approach applied in this work is the "Metering section theory" based on flat plate model of an extruder (1, 2, 3).

The experimental investigations has been organized in order to obtain the experimental dates of the extruder Net flow rate, Drag low rate and pressure and temperature dates of the melt in the extruder head. In this case, the pressure of the polymer in the extruder head is assumed to be the pressure drop of the polymer in the metering section of the extruder (1, 2).

The measurements were taken for the screw speeds of 5 (min^{-1}), 10 (min^{-1}) and 20 (min^{-1}) and each experimental point was determined by measuring 3 samples of the extruder output per one minute with time gap of two minute between each measurement. The stabilization times between each two screw speeds were 15 minutes. The average value of the three measurements was taken to be the representative one..

The variation of the pressure in the extruder head was accomplished through the replacement of the dies (nozzles) in the way that the screw is stopped for a while, the nozzle (die) is changed with new one and, finally, the screw speed is set to the minimum chosen value. When the measurements are being carried out for each screw speed, the same procedure is repeated until every given nozzle is used.

Leakage part of Drag flow and Leakage part of Pressure flow which are actually the subject of this paper are caused by the existence of the functional radial clearance, greatly depend on the wearing process development (radial clearance is increasing). They were calculated mostly on the bases of the experimental results but when it was necessary the calculations were performed on the bases of the theoretical approach..

Calculations and graphical presentations of the performed experimental and theoretical investigations show that the leakage flow parts can not be neglected and that the radial clearance is an important parameter which should be permanently controlled during the exploitation period of the single screw extruder in order to control and to prevent the production rate drop of the extrusion line. The fractions of the leakage flow parts obtained in this way are presented on the Fig. 1.

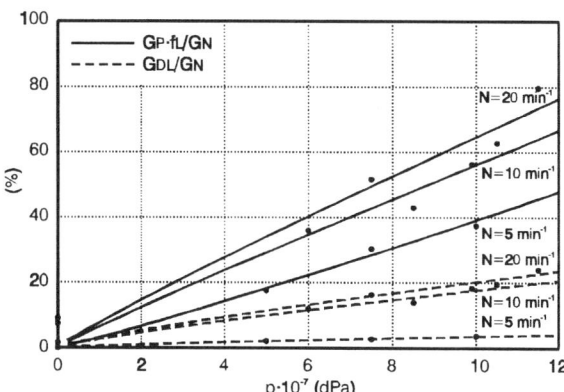

Fig. 1: The fractions of the leakage flow parts

REFERENCES

(1) Fisher, E. G., "Extrusion of plastics", London, Iliffe books Ltd., 1964.

(2) I. Klein and D. I. Marshall, "Computer Programs for Plastics Engineers", Reinhold Book Corporation, 1968

(3) Tadmor, I. Klein, "Engineering principles of plasticating extrusion", R. E. Krieser publishing Z. Co., 1970.

A STUDY ON COMPREHENSIVE CONDITION MONITORING OF WORM GEAR

TONG WANG, XIE GUAN GAO, LING HENQIO, MING LAI WONG
Jiao Tong University, Shanghai, 200030, China
YU JAN LEE and XIN XIANG WANG
Shindler Co. Shanghai, China

ABSTRACT

The operating condition monitoring of machinery-electrical installment is an important means for increasing the reliability, decreasing maintenance cost, predicting damage and preventing failure. Worm drive is one of the most typical devices which are widely used. It is reported that full life monitoring to the important power transmission device has been paid to the research of the separate condition monitoring. So the research work of comprehensive monitoring for power transmission is very necessary and very urgent.

Temperature monitoring: The temperature on the friction surface can be measured by the following methods: imbedded thermal couple method, digital temperature measure meter, dynamic thermal couple method and Kr temperature method. The oil temperature can be monitored by general temperature meter or temperature meter of XWDI type. For example, on the roller test machine M-10 type, IR temperature measure meter of HD-400 type, digital temperature gauge and imbedded thermal couple--XY recorder, etc. are used to measure the surface temperature. The results obtained with three methods above mentioned are almost same.

Noise monitoring: The noise gauge of type ND6 is used to monitor the noise of the worm gear reducer continuously in this test research. The gauge is one meter apart from the case. The noise along 5 different directions have been measured.

Vibration monitoring: In order to monitor its vibration, an accelerator sensor is placed on the case side in the directions of axis X, Y and Z. A tape recorder of TEAC-R81 type is used to record the vibrating signal and original analyses are carried out with the spectrum analysis meter of B&K 2034 type.

Efficiency monitoring: Here efficiency monitoring is in fact rotational speed and torque monitoring of the input and output shaft of the case. The input speed is kept steadily. Measure the rotational speed and torque, then convert them into the transmission efficiency of the worm reducer.

IR Spectrum monitoring: In this research, the specimen materials are steel #45(worm material) and bronze(worm gear). Several kinds of anti-wear oil additives are used in the test. Before test, the IR spectrum of the specimen surface layer are made out with an IR gauge of PE-983 type, which will be taken as the basis of comparison.

Ferrography monitoring: During the monitoring process different oil sample are collected from several oils and under different load class by a certain regulation. Then it is measured with a ferrography gauge of spinning type XTP-1. The results also show that the different kinds of oil additives have great effect on the anti-wear capability of the worm and gear couple.

CONCLUSION: Not only in analysis study but also in practical application, the above mentioned monitoring methods have a great value in dynamics research, in the study of surface failure mechanism, and in the developments of comprehensive monitoring to mechanical installments. Especially the wide application of IR spectrum, frequency spectrum, and temperature monitoring will greatly push forward the development of comprehensive condition monitoring technology.

REFERENCES

[1] D.W. Duddley "Gear Technology- Past, Present and Future", Proceedings of International Conference on Gearing, Nov. 1988, China.

[2] T. Wang, L.H. Qiu, X.G. Gao etc. "Comprehensive Condition Monitoring of Gear Operation", Proceedings of the Eighth Conference on Power Transmission, Nov.1993, China.

Evaluation of A Self-propelled Rotary Tool in The Machining of Aerospace Materials

Z. M. WANG, E. O. EZUGWU and A. GUPTA
School of Engineering Systems & Design, South Bank University
Borough Road, London SE1 0AA, England

ABSTRACT

Aerospace materials, such as titanium and nickel based alloys, generally cause problems (such as short tool life and severe surface abuse of machined workpiece) during machining owing to several inherent properties of the materials. In this study, a self-propelled rotary tool (SPRT) was developed and used for machining two aerospace superalloys, IMI 318 (Ti-6Al-4V) and Inconel 718.

The SPRT incorporating cemented carbide inserts exhibited superior wear-resistance and up to 60 times improvement in tool life when machining IMI 318 compared with those used for conventional turning due to the reduction in relative cutting speed, the use of the entire cutting edge and the lower cutting temperature as a result of the rotation of the tools during machining (Figure 1) [1, 2]. Excessive chipping at the cutting edge was the dominant tool rejection criterion when machining IMI 318 with the SPRT. This can be attributed to thermal and mechanical shock induced by continuous shifting of the tool edge during machining. Welding of work materials onto the cutting edge and subsequent pullout also encourages the chipping of the rotary tool.

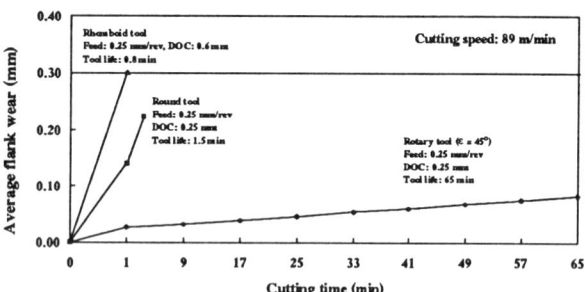

Figure 1: Comparison of wear characteristics when machining IMI 318 with various cutting tools

When machining Inconel 718 with the SPRT, tool life increased by three to four times more than when machining with round and rhomboid inserts respectively. This improvement is not as remarkable as when machining the titanium alloy due to severe attrition and abrasion wear mechanisms during rotary cutting as a result of vibration caused by the low stiffness of the rotary cutting system, rapid work hardening of the machined surface as well as the presence of hard carbides in the Inconel 718 superalloy.

Machining IMI 318 with the SPRT gave lower cutting (15-20%) and radial (25-35%) forces than when machining with round inserts under conventional turning due to reduced amount of work done in chip formation and lower friction on the rake face of the tool under rotary cutting. An increase in the inclination angle led to a reduction in the cutting force due to increased rotary speed and higher effective rake angle of the SPRT, while the feed force increased due to increased feed resistance at high inclination angles.

Surface finish produced with the SPRT was greatly affected by the stability of rotary cutting system. Increase in feed rate slightly deteriorated the surface finish due to increased smearing action between the tool and the workpiece. Surface finish improved by reducing inclination angle due to a corresponding increase in effective nose radius of the round cutting tool [3, 4].

REFERENCES

[1] Chen, W.Y., PhD Thesis, University of Birmingham, UK, 1993.

[2] Chen, P., JSME Int. J., Seroes III, 35(1),1992, P.180-185.

[3] Armarego, E.J.A., Annals of the CIRP, 42(1), 1993, P.49-54.

[4] Chang, X., Chen, W.Y., Pang, M. and Zhong, G.J., Journal of Engineering Manufacture, 209, 1995, P.63-66.

THE CUTTING ZONE MODELLING WITH THE AID OF BOUNDARY ELEMENT METHOD

WOJCIECH ZĘBALA
Production Engineering Institute, Cracow University of Technology, Al. Jana Pawła II 37, 31864 Kraków, PL

ABSTRACT

An analytical model of the cutting zone during precision machining was worked out (2)(3). The purpose of modelling was the marking out of stresses and movements distributions in a cutting layer and a tool nose. A geometrical model was divided into two separated zones, the part of the workpiece and the tool nose with a cutting edge roundness. Material constants were assigned to both zones (invariable in any point of the zone).

The cutting edge (**T**) and the chip (**Ch**) are two deformable bodies in the contact - fig.1. The region of the tool body Ω^T and the region of the chip Ω^{Ch} are bounded by surfaces: Γ^T and Γ^{Ch}. The working medium $\mathbf{f}^0$ causes the tool and the chip approach. The result of this approach are considered in the moment of a real contact of the chip with the tool face. The symbol Γ_C means the contact surface. At the beginning, the localisation of the Γ_C surface is unknown. Parts of the surfaces Γ^T and Γ^{Ch} (lying on the bodies boundary) which are in contact in the next moment can be marked out. The surfaces Γ^T and Γ^{Ch} do not have to be identical as surfaces after deformation of the surface Γ_C. $\underline{\Gamma}_C^T$ and $\underline{\Gamma}_C^{Ch}$ are the symbols of surfaces which are transformed into surface Γ_C after deformation. The following relations should proceed:

$$\Gamma^T \supset \Gamma_C^T \supset \underline{\Gamma}_C^T \qquad \Gamma^{Ch} \supset \Gamma_C^{Ch} \supset \underline{\Gamma}_C^{Ch}$$

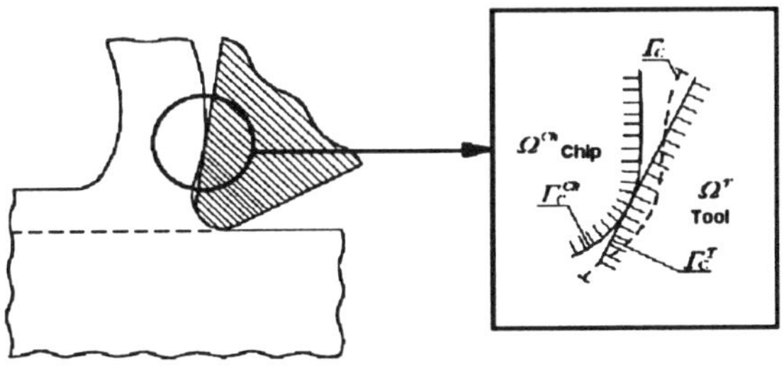

Fig. 1: *A contact surface of the tool face and the chip*

For the every bodies being in the contact we can write the boundary integral equation in the form (1):

$$c(x)u(x) + \int_\Gamma P(x,y)u(y)d\Gamma = \int_\Gamma U(x)p(y)d\Gamma + \int_\Omega U(x,y)b(y)$$

with the boundary conditions: movements field (**u**), surface loads (**p**) and volume forces (**b**) on the boundary Γ:

$$u_i(x) = u_i^0(x), \qquad x \in \Gamma_1$$
$$p_i(x) = \sigma_{ij}n_j = p_i^0(x), \qquad x \in \Gamma_2$$
$$\Gamma_1 \cup \Gamma_2 = \Gamma \text{ and } \Gamma_1 \cap \Gamma_2 = 0$$

A special experiment was carried out in order to mark out the load value on the cutting zones boundaries. Three components of cutting forces F_P, F_f and F_C were taken by a piezo-electric dynamometer (Kistler type 9257B) installed under a tool holder on a lathe slide connected to a computer system (4). Cutting force signals were passed through a charge amplifier (Kistler type 5019A) and recorded on a data recorder. The measured components of the cutting forces were resultant of the forces acting towards three surfaces: face, shearing and cutting edge fillet. The knowledge of the surface forces distribution on the body boundary made possible to mark out the movement in any point $x \in \Omega$ (model volume). Fig.2 shows the symbolic form of the shear stress distribution in: work, chip and tool.

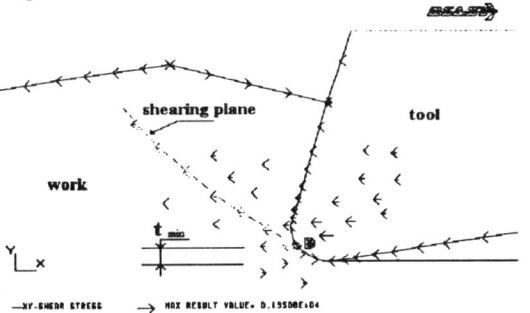

Fig.2: *The distribution of shear stress - symbolic form*

The biggest marks (with arrow) allow to find the localisation of the theoretical shearing line (plane). This line starts at the point **D** which is called the „death point". The minimal thickness of cutting layer (t_{min}) can be identification.

REFERENCES

(1) T. Burczyński, Boundary Element Method in the Mechanics, WNT, W-wa 1995.
(2) W. Zębala, J. Gawlik, Proceedings of the University of Rzeszów, Vol.20, Nr 3, page 81, 1996.
(3) W. Zębala, J. Gawlik, Proceedings of the Symposium Intertribo'96 in Slovakia, page 119, 1996.
(4) W. Zębala, Proceedings of the Inter. Conference on Production Research in Israel, page 98, 1995.

KELVIN TECHNIQUE FOR ON-LINE RUBBING SURFACE MONITORING

A. L. ZHARIN

Tribology Laboratory, Belarussian State Powder Metallurgy Concern, 41 Platonov St., Minsk, 220600, Belarus

ABSTRACT

A method for continuous non-destructive in-process and on-line monitoring of changes in the electron work function of a rubbing surface has been developed. The method is based on the Kelvin technique for measuring contact potential differences. Kelvin technique is a really two dimensional method in the case of metal surfaces. Surface resolution is within macroscale (typically, some mm^2). But depth of layer under measurement is about Debye screening length, which depends on the density of free electric charge carriers and in the case of metallic surfaces approximately equal to between atomic distance.

The work function possesses high sensitivity to practically all events accompanying friction in the contact zone, namely, plastic deformation, creation of new surface material, adsorption, oxidation, phase changes, redistribution of alloy components, etc. The technique can be used to investigate tribological materials for a wide range of conditions, including load, sliding speed, and environment, with or without lubrication. It can be used with practically any friction testing machine.

At present, this is the only method which is sensitive to both surface and near-surface defects and permits study of one of the two interacting surfaces during sliding. Using this method one may speak of monitoring a friction surface state in the literal sense of the word.

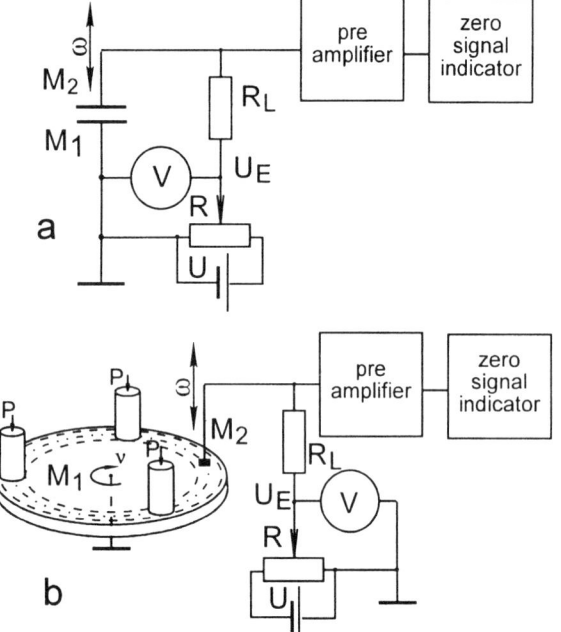

Fig. 1. Functional diagram of the generalized Kelvin technique (a) and its modification for tribology (b)

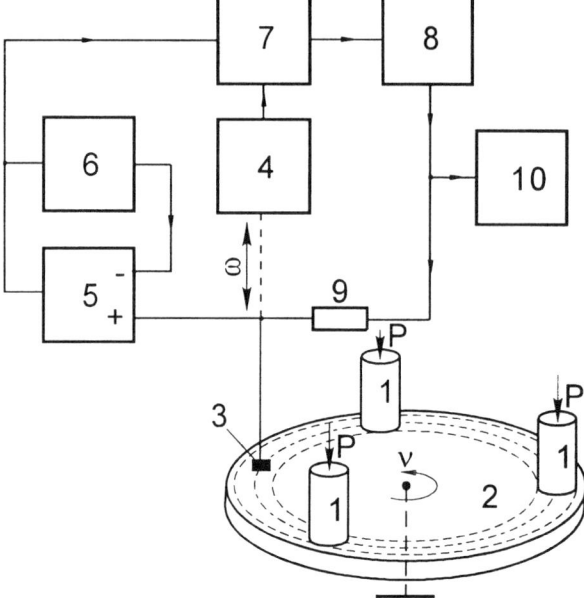

Fig. 5. Schematic diagram of the developed system. 1 - pins; 2 - rotating grounded disk; 3 - reference electrode (nickel or gold); 4 - electromechanical modulator; 5 - differential amplifier with high-Ohmic inputs; 6 - feedback circuit; 7 - phase-sensitive detector; 8 - integrator; 9 - high-Ohmic resistor; 10 - recorder.

Experiments have demonstrated that the technique is applicable for investigation and analysis of a wide range of processes on rubbing surfaces, e.g., running-in processes, formation and depletion of lubricating layers, responses of a surface to changes in friction conditions, etc. Currently the technique is being applied in two major directions:
- determination of critical points with respect to changes in normal load, with relevance to materials optimization and selection;
- studies of the kinetics of friction processes, including periodic changes which may be related to those in fatigue.

The current paper will describe problems of practical implementation of the method (theoretical consideration of the measurement sell, error determination, electronic design etc.). It will also describe some typical results.

REFERENCES

(1) A.Zharin, Contact Potential Difference Technique And Its Application In Tribology, Minsk, Bestprint, 1996. (in Russian).

INTEGRATION OF HYPERMEDIA AND ARTIFICIAL INTELLIGENCE TO TRIBOLOGICAL DECISION MAKINGS

K CHENG
Department of Engineering, Glasgow Caledonian University, Cowcaddens Road, Glasgow G4 0BA, UK
W B ROWE
School of Engineering and Technology Management, Liverpool John Moores University, Liverpool L3 3AF, UK

ABSTRACT

Tribology is an inter-disciplined subject concerned with engineering, physics, chemistry, and materials science. Tribological knowledge exists in a wide range of locations. Access to many sources of information is often inconvenient for practising engineers. This situation results in suboptimal utilisation of tribological knowledge in many engineering decisions. Therefore, it is important to create interactive intelligent tribological information systems which are readily available for a broad community of mechanical engineers, who need apply tribological knowledge to specific, practical design or operating problems (1).

The integration of advanced technologies such as hypermedia and AI with conventional tribology techniques is seen as a potentially fruitful avenue to make increase the accessibility of tribological knowledge. More importantly, the approach offers a high level aid for engineering decision makings with intelligent interactive and multimedia features (2).

This paper describes using hypermedia as a tool with application to tribological decision making. The association attributes of hypermedia and tribological decision making are explored. The attributes are related to the associative linkage in hypermedia and the association actions within tribological analysis and synthesis. A developer based approach is presented to developing a tribological hypermedia based system. Application exemplars are provided for the following categories of tribological decision makings:

- tribo-component selection such as the selection of bearing type or configuration, component materials, and lubricants.
- tribological design such as the design of bearings, gears, cams, chain drives, belt drives, clutches, brakes, seals, bushings, and fasteners.
- tribological failure diagnosis such as the diagnosis of failure modes including wear, galling, lubrication failure, and leakage.
- tribology documentation with multimedia features.

The decision expertise and knowledge has to be acquired from experienced tribologist and authoritative publishing sources. The knowledge acquisition is the bottle-neck in developing a hypermedia based engineering decision support system.

Figure 1 shows the menu card of a hypermedia based design support system for externally pressurised journal bearing (3). The system includes two user support modules and six functional modules which can support a full design procedure of an aerostatic or hydrostatic bearing.

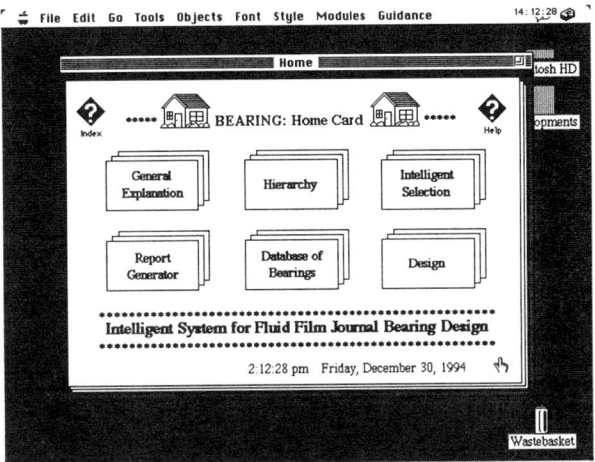

Fig.1: A hypermedia based bearings design system

A number of application exemplars are provided and illustrate that, through the integration of raster, vector, text, video, and audio data, hypermedia can add new dimensions to tribological decision making. The paper concludes with a discussion on the potential benefits, and the future applications of hypermedia in tribology research and development.

REFERENCES

(1) T.E. Tallian, Transactions of the ASME, Journal of Tribology, Vol.109, July 1987, pp.381-387.
(2) K. Cheng, Journal of Materials Processing Technology, Vol. 61 No.1-2, 1996, pp.143-147.
(3) W.B. Rowe and K. Cheng, Annals of the CIRP, Vol. 41/1/1992, pp.209-212.

PHILOSOPHER/MECHANIC MEDIATION - THE PEDAGOGICAL BRIDGE AND PRACTICAL EXPLOITATION OF RESEARCH INFORMATION

Eur Ing Roland R Gibson-Carqueijeiro
Engineering Consultant, Charnwood House, 29A Lingfield Road, Wimbledon Common, London SW19 4PU

ABSTRACT

To review the state-of-the-art, define current research, explore future opportunities and consolidate progress in tribology at a world level, is the declared aim of this congress and one might ask what can an ordinary engineer contribute to such a congress. The array of speakers and the quality of their work is impressive, but its use to the practitioner is questionable. The doctoral pursuits nudging technology forward are laudable and necessary, but inaccessible to the ordinary engineer. Perhaps justifiably our institution stopped providing members with free papers some years ago. So why am I talking here today?

I am talking because tribology has been my life. My involvement began as with many practitioners before my industrial life: it started with being involved in the dismantling, setting up and maintaining the running parts of my bicycle. We have seen the growth of specialisations in science and maybe we could show on the typical phase space diagram a state of chaos (1). As with music, where we define certain frequencies to represent a framework within which we can express ourselves, we are able to draw back from the state of chaos in science by defining so called rules of behaviour, so that the ordinary engineer can understand, explain and put things right in the world we build.

It was impossible to know when I started, that I would spend most of my life concerned with tribology, or lubrication as it was then called. My training and academic work followed the typical Mechanical Engineering route. This started with an apprenticeship and evening classes and only later involved academic work. I am pleased to say that it was only after a considerable time in industry and full familiarisation with the needs of machinery, that I returned to university to study Tribology. Tribology can afflict equipment during all stages of the life of their component mechanisms and I had been involved in the design, development, operation, maintenance and decommissioning of machinery in the real world before I began discussions with Dr Jost and Professor Dowson and went to Leeds to enter the Tribology masters course.

I am grateful for my practical background because it developed an intuitive understanding of phenomena. Amidst the conscious conception of how mechanisms work, the practitioner develops an appreciation of the subtle interaction of the components within mechanisms. It is these conceptions which need to be explored and this means defining the physical interaction of bodies, analysing the traditional theoretical definitions, explaining their behaviour and interpreting the current research thinking appropriately for the practitioner.

I took comfort from the concept of the engineer acting as the interpreter, standing between the philosopher and the mechanic (2) and the ingeneer (3) or ingenieur (4) more rightly being an exponent of ingenuity rather than engines (3). This paper will employ actual examples of tribological problems solved during 40 years experience to illustrate some real needs of the practitioner and our industry.

REFERENCES

(1) J.G.Gleick,. 'Chaos Making a New Science'. *ABACUS*, 1987.
(2) D.Dowson, 'History of Tribology', *Longman, London*, 1979.
(3) Sir H Ford. Proceedings of the IMechE, Vol. 191, Pages 36/77, 1997.
(4) FEANI registration classification.

MERSEYSIDE TRIBOLOGY CENTRE - A TECHNOLOGY TRANSFER INITIATIVE

D. IVES, J.W. HADLEY and B. MILLS
Liverpool John Moores University, School of Engineering and Technology Management, Byrom Street, Liverpool
L3 3AF, UK

INTRODUCTION

In December 1995 a new technology transfer initiative, the Merseyside Tribology Centre, was launched during a visit by Dr. H. Peter Jost to Liverpool John Moores University. The aim of the Centre is to develop a partnership with industry to improve and expand manufacture and to enhance the industrial training of graduate engineers through the correct application of the science of friction, lubrication and wear.

The idea of tribology centres is not novel, indeed the initial DES report of 1966 (1) resulted in the founding of three such centres, UKAEA Risley (now NCT), Leeds University and Swansea University College. Since then many universities and colleges have taken up tribology to differing extents and offer a range of research, consultancy and courses. Liverpool John Moores University is a Designated Centre of Tribology and has offered tribology from undergraduate level to Ph.D. research since 1965. However, one of the findings of a later DTI report (2) indicated the difficulties involved in operating a tribology centre on a commercial basis in a university, especially with small companies that have limited resources. Yet often it is these companies that would benefit most from the transfer of tribological skills and knowledge.

The circumstances on Merseyside have been compounded by the loss of traditional industries and this has been a factor in the region receiving Objective One status under the European Community European Regional Development Fund (ERDF). Thus the Merseyside Tribology Centre was conceived primarily to assist small to medium enterprises (SME's) on Merseyside by using graduate engineers supported by academic staff and laboratory facilities. This will aid the development of existing and new industries and help to train and retain graduates on Merseyside.

FUNDING

The Merseyside Tribology Centre could not have been initiated without the unique opportunities available within the framework of ERDF. For the first three years of operational life the Centre is funded jointly by Liverpool John Moores University, ERDF and Merseyside industry. To date industry has contributed in excess of £160,000 by way of equipment donated. This funding arrangement allows industry access to expertise and equipment in the Centre at significantly reduced costs. The intention is that the Centre will gradually develop into a self financing unit within the University.

OPERATION

There were some initial problems setting up the Centre and some refinement of the envisaged mode of operation was required. This has resulted in the Centre's ability to offer a wide range of assistance packages from simple telephone advice through to research and consultancy. However the principal vehicles for project work are final year BEng. undergraduate projects and graduate trainee projects.

The undergraduate projects work in a similar fashion to many other academic institutions. They are supervised by staff from the Centre, but due to their nature they are limited in time and scope. In contrast, the graduate trainee projects are targeted specifically at the requirements of the participating company and are run over an appropriate time scale, typically three months. At the end of a project the company has the solution to a problem and the possible benefits of increased operational efficiency. The graduate gains the benefit of additional industrial training and thus enhanced employment prospects. The participating company may opt to retain the graduate engineer.

PROGRESS TO DATE

The Centre has now been operating in excess of one year and despite a relatively slow start, many projects and activities are now taking place. At present some 20 companies are involved with the Centre and 20 projects have been successfully completed. These include wear testing of hard materials and coatings, PTFE wear in aviation fuels, wire drawing lubricants, grease development and testing and analysis of bearing and gear failures. The vast majority of these projects have been conducted by undergraduates or graduate trainees. Three of these projects have developed into research programmes or major technology transfer projects, and have drawn in additional funding from other sources.

REFERENCES

(1) Lubrication education and research. Department of Education and Science, HMSO 1966
(2) The introduction of a new technology. Department of Trade and Industry, Committee on Tribology Report, 1966-1972, HMSO 1973.

THE TRIBOBRACHYSTOCHRONE

N. MANOLOV, V. DIAMANDIEV
Society of the Bulgarian Tribologists, 1000 Sofia, 5A Slavyanska St., Bulgaria

ABSTRACT

The problem about the brachystochrone is formulated by Iohann Bernoulli in the 17. century. It considers the obtaining of equation of a curve going through an initial and an end points from a homogeneous gravitational field which should satisfy the requirement: minimum for the time necessary for a slider moving on this curve with a zero initial velocity to pass the curve length to the second point. The solution of this problem performed the foundation of the variation calculation in mechanics, mathematics and physics (2). In the absence of friction, it has been found that this curve coincides with a trajectory of a point on the surface of rotating disc without sliding, a curve known in kinematics as a cycloid.

Attempts have been done the problem on the brachystochrone to be generalised also for the case of friction, but the results did not satisfy the requirement of tribology because of their fragmentary and situation character (1).

The present paper completes a system tribological study on the problem of the brachystochrone under the complex name of tribobrachystochrone. The cases of influence of gravitational, viscous and inertial friction are considered. Every particular case is examined in relation both to the existent in literature results and to the findings in this paper. This approach approves the reliability and validity of the obtained theoretical results and their interpretation.

In the case of dry gravitational friction of the slider, the cycloid of the classical Bernoulli's solution keeps in principle, but the following changes are established:

The opening of the cycloid's ark is increased proportionally to the term $\sqrt{(1+\mu^2)}$, where μ is the coefficient of friction.

The tribocycloid is turned clockwise relatively to the cycloid at the angle α around a horizontal axe perpendicular to the cycloid's plane, where tag $\alpha = \mu$.

Depending on the working conditions of the slider, each succeeding component of the total friction increases α, and changes the form and size of the basic cycloid.

The common feature in all case is the fact the obtained curves are tribological and brachystochronic.

The practical interest towards this work increases with the verification of the final part of the paper: it appears that brachystochronee are simultaneously the profiles of surfaces, on which sliders moves realising minimal energy and material losses.

For tribology, the above analysis has a theoretical-methodological character (3) (4).

REFERENCES

(1) D. Dimitrovski, M. Miyatovich, M. Lekic, The problem of minimum work of friction forces. Annual of Physics of the Faculty of natural and mathematical sciences, Skopje, 40, pp. 33-43.

(2) L. Pars. An introduction to the calculus of variations, London, 1962.

(3) N. Manolov, V. Diamandiev, THE MODELS OF TRIBOLOGY, Society of the Bulgarian Tribologists, Sofia, 1995 (in Bulgarian).

(4) N. Manolov, THE CONTACT, Society of the Bulgarian Tribologists, Sofia, 1995 (in Bulgarian).

TRIBOLOGY IN BELARUS REPUBLIC

YU.M. PLESKACHEVSKII, V.V. KONCHITS AND N.K. MYSHKIN
The Metal-Polymer Research Institute of Belarus Academy of Sciences, 32a Kirov Street, Gomel,
Republic of Belarus

ABSTRACT

The paper reviews the evolution of tribology in Belarus Republic. The appearance of tribology directly relates to Professor Vladimir. A. Belyi (1922-1995). He created the first tribological laboratory at the Belarus Institute of Railway Engineers (at present Belarus Transport University), Gomel in 1954. This laboratory grew into the world-known Metal-Polymer Research Institute of the Belarus Academy of Science. Professor Belyi has established the Society of Tribologists in Belarus in 1992 and was its first Chairman.

Tribology evolved together with the expansion of industries in Belarus, machine tools and instruments building, microelectronics, chemical and light industries, leading to the appearance of new research tribological centers and groups in the Academy of Science (Institute of Reliability and Durability of Machines and Physical-Engineering Institute), Powder Metallurgy Research Institute, Minsk, Gomel and Brest Polytechnic Institutes, Gomel and Novopolotsk Universities, Technological Institute in Minsk, at a number of plants. At present 22 tribological centers and some groups are quite active in Belarus. Many of them were included into the review of I. Zinger devoted to tribology in the former USSR (1). More than 150 Belorus researchers published their theoretical and applied tribological results during the recent years (1990-1996).

The first serious assessment of the achievements of Belarus tribologists was accomplished during the International Symposium "Nature of Friction of Solids" (Gomel, 1969). A large National Conference "Friction and Wear of Materials" was held in Gomel in 1982. The First Soviet-American Conference "New Materials and Technologies in Tribology" in Minsk in October 1992 attracted many researchers. Second Conference has been held this September, 1-6. Soviet "Journal of Friction and Wear" has been published in Belarus since 1980.

The major research trends in Belarus tribology was outlined by Prof. A. Sviridonok (2). In the sphere of **triboanalysis** the studies of microstructural modifications occurring in metals and alloys at friction are interesting. Great attention is paid to the solution of thermal problems at high-speed friction. A search is continued to find analytical and experimental methods of description of real surfaces with the account of roughness at the nanometer level, frictional transfer pattern.

The Belarus scientists are very active in the **contact mechanics**. They have accomplished extensive studies in **tribomaterials**. It is the development of the materials based on metallic powders and alloys, composites based on polymers and wood, lubricants for machining of metals, materials with electromagnetic properties. The main trend in **tribotechnology** is application of coatings with improved performance. The great attention in **triboengineering** is paid to various friction parts: tooth gears, bearings, and brakes. Investigations are underway in the domain of **tribodiagnosis** and **tribotesting**. Databases on structural and tribological materials are developed.

There was no special tribology program in the USSR. The **national program "Triboengineering"** is at present being performed in Belarus (3). An analysis of carrying out the national programs in the field of material science, mechanical engineering, resources and power saving and also proposals of enterprises and research institutes allowed to form the program. It contains 6 sections: Fundamentals of Tribology, Devices and Methods for Tribosystem, Diagnosis, Materials for Friction Units, Lubricants, Friction Surface Modification and Triboengineering Units and Designs. In the framework of the program 86 projects are developed by 78 organizations. More than 30 industrial enterprises use the results of the program, including the largest in Belarus: Minsk Tractor Plant, Belarus Automobile Plant (Zhodino), Belarus Metallurgy Plant (Zhlobin), "Planar" (Minsk) and others.

The new **national program "Surface Protection"** is being formed in 1997. The program will contain such tribological sections, as Wear-Resistant Materials; Oils, Lubricants and Additives; Tribotechnical Units, Surface Engineering.

In spite economical difficulties, Belarus has a good capacity to supply industries with the recent achievements in triboengineering. There are many examples of the industry efficiency and machinery competitiveness increase on the base of tribology.

REFERENCES

(1) I. Zinger, Information Bulletine ESN 93-03
(2) A.I. Sviridenok, Journal of Friction and Wear, Vol.16, No. 3, 1995
(3) Yu.M. Pleskachevskii, V.N. Savitskii, V.V. Konchits, and V.A. Barabas, Journal of Friction and Wear, Vol. 16, No. 3, 1995

TRIBOLOGY SCIENCE, TECHNOLOGY AND EDUCATION IN ARMENIA

A K POGOSIAN

Graduate School, State Engineering University, 105 Terian Street, Yerevan, 375009, Armenia

ABSTRACT

As far back as in 50's in Armenia the investigations in the field of providing wear resistance and durability of cutting tools were set up. Tribological problems are actual in machine-tool industry, in agricultural machine building industry, automotive industry, instrument engineering, electrical machine industry, as well as in the field of composite tribomaterial science, material cutting etc.

One of the most prosperous directions of tribological studies is the calculation and design of machine parts and friction nodes on the basis of composite polymer materials instead of conventional metal parts. Scientific principles on creating of new polymer- based self-lubricating composites with the minerals as fillers, calculating methods on the selection of materials and estimation of wear resistance of friction nodes, as well as physical simulation of friction and wear processes on the small-size laboratory samples and forecasting of serviceability of friction nodes according to the results of accelerated tests are worked out (1). In the base of composite design the principle of self-lubricating mechanism by using the properties of friction transfer film is put (2). A new class of self-lubricating materials, called SIPAN, has been developed.

One of the most urgent problems of Armenian economy is the creation and utilization of lubricants based on local materials and industrial waste, as well as restoration of used oils, taking into account the market demands for wasteless technologies, environmental protection and urgent needs of the Republic. New composite lubricants have been developed based on wastes of oleic acid industry and I-40A type industrial oil and the additives with high antiwear and antiscoring properties have been proposed. The generalized dependence of lubricants' antiwear and antiscoring properties is revealed and this correlation permits to forecast the antigrip properties of lubricants by means of few simple tribological tests.

In recent years in Armenia significant development has got the theory of providing motion stability at friction (Stick-Slip process). Complex methods of reducing friction oscillations in the guide-ways of machine tools, robots and other machines are worked out (3). For the first time the task of motion stability conformable to the sliding guide-ways is solved on the basis of space model of oscillation process of friction nodes of slider-guider type. A new class of composite materials NASPAN promoting to decrease the scale of vibrodisplacements are created.

The dynamics of the breaking process, as well as the action of vibration loading on the wear and friction properties of brake materials are investigated.

The problem of creation of the asbestos - free friction (brake) composite materials with mineral fillers of local origin is of greatest importance and crucial for environment protection. A new class of asbestos - free composite friction materials BASTENIT is created along with its varieties (4). A classification of the obtained materials is carried out in accordance to the operational characteristic properties of the vehicles. The second environment - protection advantage of BASTENIT from the viewpoint of squeak background decrease has been revealed by means of the squeak (shaking) oscillations' spectrum.

Tribological studies are actual in the other fields of Armenian machine building. Prosperous are the developments of antifriction composite materials by the methods of powder metallurgy (5) and the study of wear process of cutting instruments and tools at the treatment of various materials (6).

Tribology education in Armenia is mostly carried out at the State Engineering University within the whole scale of educational system: in BE, ME and PhD programms (7). Obviously the school and new generation of the specialists in the field of tribology in Armenia have already formed. In the development of Tribology in the Republic a significant role plays, organized in 1974, Armenian Tribology Committee (ATC).

REFERENCES

(1) A.K.Pogosian, Friction and Wear of Filled Polymer Materials. Moscow, 1977, 139 p.
(2) A.K.Pogosian, Tribology in the USA and the former Soviet Union, Allerton Press, New York, 1994, 271-283pp.
(3) P.V.Sysoev, M.M.Bliznetc, A.K.Pogosian, Antifriction Epoxside Composites in Machine Tool Building. Minsk, 1990, 231p.
(4) A.K.Pogosian, P.V.Sysoev, N.G.Meliksetian, Polymer-Based Friction Composites. Minsk, 1992, 218p.
(5) Composite Materials and their Treatment, Ed. by N.V. Manoukian. Yerevan, 1985.
(6) The Quality of Machine Parts Surface, Ed. by M.V. Kasian. Yerevan, 1985.
(7) A.K.Pogosian, Principles of Tribology (Textbook). Yerevan-Athens, 1994-95, 296p.

TRIBO-FATIGUE IN ENGINEERING EDUCATION

L A SOSNOVSKIY
Scientific and Industrial Group "TRIBOFATIGUE", P O Box 24, Gomel, 246050, Republic Belarus
A S SHAGINYAN
Gomel Polytechnical Institute named after P.O.Sukhoy, Prospect Oktyabrya, 48, 246746, Gomel, Republic Belarus

ABSTRACT

Before the emergence of Tribo-fatigue as a new scientific direction of Tribology and Fatigue Strength the notions of such mechanical phenomena as friction and wear, fatigue of materials and assembly elements, rolling-mechanical fatigue, fretting-fatigue, sliding-mechanical fatigue have been developed separately. Therefore, in the curriculum of many higher technical schools the information on the mechanical properties, mentioned above, is delivered in different disciplines and disconnected both from the point of its content and the period of learning.

The tribo-fatigue has formed a scientific and engineering concept about the complex of mechanical phenomena which are realized simultaneously and closely in real systems and can limit their durability as well (1). The tribo-fatigue course in higher technical schools is to reflect this important process in the engineering education.

The first introductory course on tribo-fatigue was delivered in 1986/87 for the students of the mechanical faculty of the Belorussian Institute for Railway Engineers (Gomel, Republic Belarus) (2).

At present the "Foundations of Tribo-fatigue" discipline is introduced at the Gomel Polytechnical Institute named after P.O.Sukhoy for mechanical engineers.

The lecture course (18 hours) includes the following topics (2)(3):
– classification of assembly units of machines and equipment operated under friction along with simultaneous cyclic pulsating load (such units are called cycle load-bearing systems);
– main information on types and mechanizms of wear-fatigue damage and methods of its prediction and prevention;
– main principles of direct and back effects under sliding-mechanical fatigue, rolling-mechanical fatigue and fretting-fatigue;
– methods and machines for wear-fatigue tests of materials and cycle load-bearing systems models.

The students are to master methods of the analysis and calculation of cycle load-bearing systems during the practical hours (18 hours).

Analogous course on tribo-fatigue is delivered for two faculties in the Belorussian State University of Transport.

Both higher institutes use the literature (1)(2)(3)(4)(5)(6) in their curriculums.

The 2nd International Symposium on Tribo-fatigue was held in october 1996 in Moscow (7). There were following host organizations of the symposium: the Russian Academy of Sciences, the Belorussian Academy of Sciences, the National Academy of Sciences of Ukraine and the Scientific and Industrial Group "TRIBOFATIGUE". It recommended to introduce the course of tribo-fatigue in curriculum of different higher institutes of Russia, Belarus and Ukraine.

The SI machines for wear-fatigue tests (designed in the Scientific and Industrial Group "TRIBO-FATIGUE") are used for carrying out laboratory works (4) (18 hours). Students study the construction of SI-01, SI-02 and SI-03 machines and their peculiarities. They take possession of wear-fatigue tests methods (on fretting-fatigue, sliding-mechanical fatigue and rolling-mechanical fatigue), and they also acquire knowledge of measuring methods of wear, friction moment and others. Learning the principles of infomation-controlling system based on personal computer takes main attention.

The detailed analysis of the course "Foundations of Tribo-fatigue" is given. In this paper which deals with the contents of the first text-book on tribo-fatigue the authors of which are famous specialists in tribology and fatigue strength. The teaching experience of the tribo-fatigue course is given.

REFERENCES

(1) Belorussian СТБ 994-95 Standard "Tribo-fatigue. Terms and Definitions", Minsk, 1995, 98p.
(2) L A Sosnovskiy, Complex reliability on fatigue and wear resistance criteria (Foundations of Tribo-fatigue), Gomel, 1988, 56p.
(3) L A Sosnovskiy, Complex wear-fatigue damage and its prediction, Gomel, 1991, 188p.
(4) Tribo-fatigue–95. SI-series Machines for wear-fatigue tests, Ed.: L A Sosnovskiy and M S Vysotsky, Gomel, 1996, 88p.
(5) Трибофатика. Трыбафатыка. Tribo-fatigue. Tribo-ermüdung, Terminologic Dictionary, Ed.: L A Sosnovskiy, Minsk - Gomel, 1996, 135p.
(6) Some Words on Tribo-fatigue, Ed. and Comp.: A V Bogdanovich, Minsk - Moscow - Gomel - Kiev, 1996, 136p.
(7) 2nd International Symposium on Tribo-fatigue, Abstracts of Papers, Moscow, 1996, 104p.

DEVELOPMENT TENDENCY OF TRIBOLOGY IN THE COUNTRIES OF THE FORMER USSR (1990-1996)

A.I. SVIRIDENOK

Research Center on Resource Savings of Belarus Academy of Sciences, Tysengauz sq. 7, Grodno, 230023, Belarus, CIS

Practically all the main tribological directions were actively developing in the former USSR countries in the post-war decades urged by rapid growth of engineering, military, space and other branches of industry. Russian tribological science occupied foremost international positions; above 2,000 tribologists .productively worked and published their investigation results till the beginning of 90-s (1, 2).

Collapse of the USSR state in 1991 and abrupt aggravation of economic situation have affected Soviet tribology. At first (1991-1993), the period of swift disintegration was observed followed by cut down of tribological staff. Many of talented specialists moved to non-scientific spheres, some were hired or immigrated to USA, Israel, Poland and other countries of the world.

Then, the period of relative stabilization began, when publication of monographs and papers has somewhat revived.

The volume and scope of tribological investigations and developments in the former USSR can be most filly presented via analyzing publications in the Soviet Friction and Wear Journal which is translated and published in English by Allerton Press Inc. (New York). In the period between 1990-1996 more than 978 papers of 1130 authors have been published (Table 1.). Among them about 100 scientists who have published 3 to 7 works. As before rather fruitful are Russian tribologists conducting their research in 64 scientific and educational establishment, those of Ukraine in 32 institutions and Belarus in 25 organizations. Unfortunately, there are few actively publishing tribologists in Estonia, Turkmenistan, Armenia, Uzbekistan and Azerbaidjan.

Lately, the subject matter of manuscripts has noticeably changed. More and more attention is paid to practical problems and there appeared publications reflecting the results of earlier secret works. There are also perceptible changes for scientists in CIS countries in what concerns their international contacts, joint projects, participation in conferences. Significant joint projects have been fulfilled (3). The scope of micromechanics and contact thermodynamics, nanotribology and tribodiagnostics, biotribology, triboecology and triboinformatics has shifted too.

The release of new fundamental handbooks on tribology should be noted (4, 5) used in training engineers-tribologists in the Universities of Moscow, Kiev, Gomel, Grodno and etc.

REFERENCES

(1). V.A. Belyi, A.I.Sviridenok, Soviet J. of Friction and Wear, vol. 8, no. 1, 1987.

(2). P.Jost. Report of the 1989 tribology group mission to the Soviet Union / Mech. E. 1989.

(3). Tribology, Investigation and Applications. Experience of the USA and CIS countries, Ed. by V. Belyi, K.Ludema and N. Myshkin, Mashinostroenie (1993), Allerton Press (1994).

(4). Fundamentals of Tribology, Ed. by A.V. Chichinadze, Textbook, Nauka i Tekhnika, 1995.

(5). I.B.Fux, I.A.Buyanovsky, Introduction into Tribology, Textbook, Moscow, Neft i Gaz. 1995.

Table 1.

Main tribology divisions	Publications in years							Total in 1999-1996	
	1990	1991	1992	1993	1994	1995	1996	items	%
Triboanalysis	43	27	45	31	35	34	33	248	25.4
Tribomaterial science	47	45	39	35	32	29	39	266	27.2
Tribotechnology	27	24	20	28	24	18	9	150	15.3
Tribotechnics	31	30	26	15	24	38	25	189	19.3
Tribomeasurements and tribomonitoring	10	15	9	22	10	1	11	78	8.0
Triboinformatics	3	3	1	0	3	1	0	11	1.1
Triboecology	0	0	0	0	1	3	0	4	0.4
Reviews	6	5	4	2	5	7	3	32	3.3
Total	167	149	144	133	134	131	120	978	100.0

DEVELOPMENT AND APPLICATION OF A NUMERICAL, TRIBOLOGICAL DATA COLLECTION "TRIBOCOLLECT"

M. WOYDT, E. SANTNER and D. PEDZINSKI
Federal Institute for Materials Research and Testing (BAM), D-12200 Berlin

Each year thousands of tribological tests were performed and published, but the access to the results for engineers and researchers is time-consuming. Inconsistent tribological specifications hinders the validation of these data and the comparison between them.

To overcome this, BAM has developed the basis of a format and identified fields for organising sliding, fretting and rolling friction and wear data coming out from screening tests under unlubricated and boundary or mixed lubricated conditions and has compiled an extensive data collection with 12,100 data records for a variety of examined materials couples.

Up to 140 attributes (version 1.2) can now describe comprehensive and complete a tribological test. The terminology was checked, with existing standards DIN 50320, 50323 and 50324 and ASTM G118-93 and running EC research projects, like EUFRETTING or FASTE.

"Tribocollect" contains only critically evaluated own numerical data and is focussed on the following area application:

- Comparative analysis of the tribological behaviour of materials
- Assistance for the selection of candidate materials for a specific tribological operation
- Influence of operating conditions on tribological quantities
- Input, documentation and management of tribological results

"Tribocollect", version 1.2, is now an operational, tribological data collection with 12,100 data records available for industrial customers (Daimler Benz AG, Robert Bosch GmbH, BMW AG). For the version 1.2 a software is evaliable. His structure and the precise defined formats may guide in the future scientists and engineers to organise and publish their tribological tests and results in a complete and comprehensive way. This data collection enables resaerchers, engineers, developers and consultants to retrieve the appropriate material couples in a fast and convenient way, to get reliable data about the tribological behaviour as well as they have now a tool to come up with innovative tribological solutions on short notice.

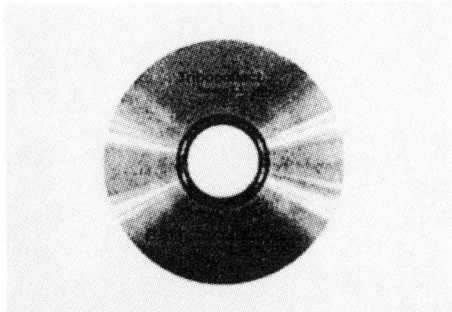

The numerical, tribological data collection "Tribocollect" on CD-ROM

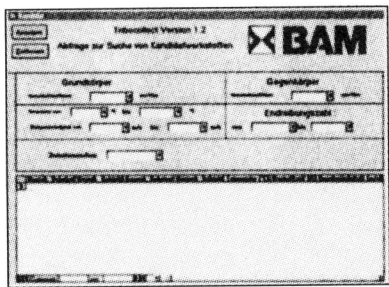

Example of a menu to retrieve candidat materials in Tribocollect

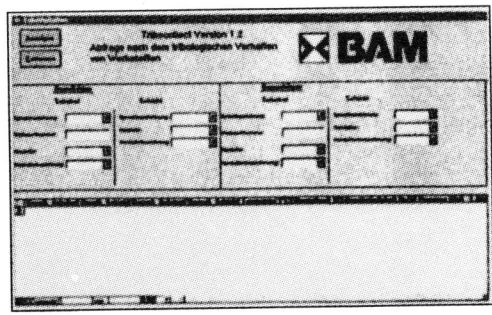

Example of a menu to search for the tribological behaviour of materials in Tribocollect

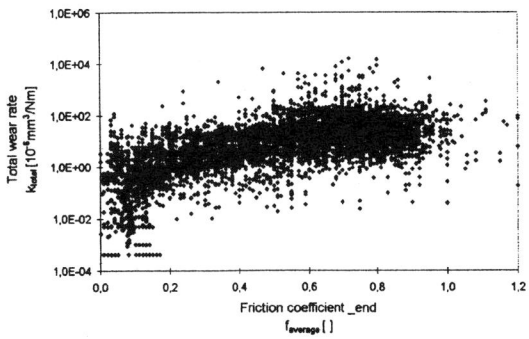

Plot of the hits from the filter condition "total wear rate" versus "coefficient of friction at test end"

EXPERIENCE OF ORGANIZATION AND CARRYING OUT THE LABORATORY-RESEARCH WORKS ON TRIBOLOGY IN HIGHER INSTITUTES OF LEARNING

J P ZAMYATIN and A J ZAMYATIN

Rybinsk Academy of Aviation Technology, Pushkin st. 53, Rybinsk, 152934, Russia

ABSTRACT

Tribology as general technical discipline was introduced into the programs of many Russian higher educational establishments for enough long time. Recently the tribological specialities were organised in the number of them and here the tribology had become the special discipline. Moreover, the questions of contacting, friction, lubrication and wear are examined during the faculty raising the level of engineers skill. That is why it can be considered that in Russia a great experience on tribology teaching both in the student and engineering audience has been accumulated.

The questions of methodology and technical provision of tribological laboratory-research works in higher institutes of learning of the aviation bias are discussed in the paper.

Usually the subjects of the work are divided on two directions: fundamental (general tribological) and applied.

When experimentally examine the fundamental problems executors acquire the skill of rational use of the tribotest and tribomeasuring equipment, tests planning and carrying out, functional-statistical processing of obtained experimental information. The main attention was given to: technology of the appraisal of geometric, physics-mechanical and friction fatigue properties of friction surfaces, and also to the qualities of base oils and additives for them; methods of determination the characteristics of contacting, friction, lubrication and wear of materials and solids by statistic information; organisation of the subject and knowledge data bases. The model of rolling, sliding and rolling with slippage friction pairs are adopted as the objects of investigation. The examples of main such objects are shown on Fig. 1.

The laboratory-research works, tied with complex research of the tribocharacteristics of aircraft and their engines main movable conjugations, as well as with the increase of the durability of aircraft construction wear surfaces were attributed to the applied. The works of this cycles were carried out on special benches. The objects of investigation are: high-rotation rolling bearings; heavily loaded gears; sliding bearings and hinges; piston groups of heavily-duty engines for small aviation; contact and groove sealing units; precision friction pairs of fuel-distribution apparatus; friction units and pairs under the influence of lower, abnormally high and high temperatures, aggressive gas mediums and also vacuum and radiation; nominally motionless ("turbine blade bandage shelf - next blade bandage shelf" and so on), subjected to intensive fretting wear; constructions with erosive wear of surfaces, including contacting with fuel combustion products.

The concrete subjects of laboratory-research work on both directions are adduced in the paper. The offered test methods, equipment and devices, as well as the technology of functional-statistical interpretation and use of obtained results are described. Requirements to the drawn up and defence of the account on fulfilled works are examined. The data of the interrogations of graduates about directions and profit of laboratory-research works in their engineering practice are cited.

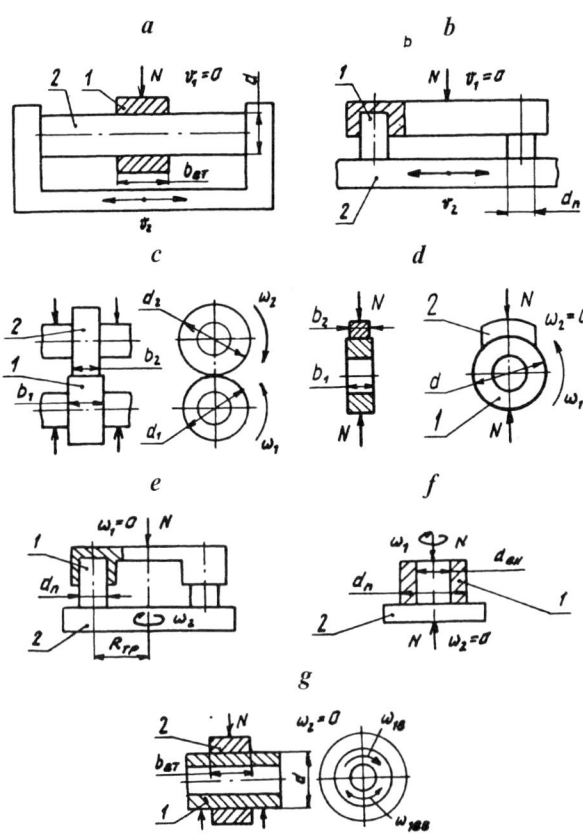

Fig.1: Laboratory-research models of friction pairs

WEDNESDAY 10 SEPTEMBER

Corresponding oral session		Page number
TU1/W1	EHL and boundary lubrication	645
W3/TH3	Wear by hard particles	665
W4	Tribology of ceramic materials	675
M5/F5	Measurement and simulation	691
W5	Bio-tribology	731
W2	Practical applications of friction and wear	749

SIMPLIFIED ANALYSIS OF NON-NEWTONIAN EFFECTS IN A CIRCULAR ELASTOHYDRODYNAMIC CONTACT AND COMPARISON WITH EXPERIMENT

K J H SHARIF, C A HOLT, H P EVANS AND R W SNIDLE
School of Engineering, University of Wales, Cardiff CF2 3TA, UK

ABSTRACT

The paper describes a simplified non-Newtonian numerical model of an elastohydrodynamic point contact. It is based on the control volume flow continuity technique described by Holt et al (1) Observation of the shear stress values produced in the solutions to the problem obtained by Holt has led to the consideration of a simplifying approximation in the non Newtonian formulation for calculation of flow rates in which the resultant shear stress is approximated by the shear stress in the predominant direction of sliding. This assumption gives easily calculated expressions for the flow rates, and the simplification leads to direct evaluations of the surface shear stress components. Results of the simplified model are found to agree well with the more exact treatment of Holt et al under realistic operating conditions. A thermal treatment of the oil film for sliding contacts is also included. Experimental measurements of film thickness in contacts with varying degrees of sliding have been carried out which show good agreement with the simplified theory.

In the control volume approach the key calculation is that for the mass flow rate in each of the axis directions. These flow rates are functions of the pressure, film thickness, pressure gradient and viscosity as follows.

$$\phi_x = \phi_x\left(h, p, \frac{\partial p}{\partial x}, \frac{\partial p}{\partial y}, \eta\right) \quad \text{and} \quad \phi_y = \phi_y\left(h, p, \frac{\partial p}{\partial x}, \frac{\partial p}{\partial y}, \eta\right)$$

The flow rates are calculated from the Eyring model so shear stress components are related to shear strain rates according to

$$\frac{\partial u}{\partial z} = \frac{\tau_o \tau_x}{\eta\sqrt{\tau_x^2 + \tau_y^2}} \sinh\left(\frac{\sqrt{\tau_x^2 + \tau_y^2}}{\tau_o}\right)$$

with a similar form for the y-direction. The coupling of the shear stresses in the two directions into these shear strain rate expressions leads to a pair of transcendental equations in the surface shear stress components that must be solved by a numerical method to give the flow rates ϕ_x and ϕ_y. This can be achieved by a Newton method, which although quick and reliable, leads to long run times because of the number of times the calculation must be repeated. Observation of shear stress values in solutions obtained in this way led to the approximation of the resultant shear stress $\sqrt{\tau_x^2 + \tau_y^2}$ by the shear stress in the predominant direction of sliding, τ_x, so that the shear strain rates are related to the shear stress components by

$$\frac{\partial u}{\partial z} = \frac{\tau_o}{\eta} \sinh\left(\frac{\tau_x}{\tau_o}\right); \quad \frac{\partial v}{\partial z} = \frac{\tau_o \tau_y}{\eta \tau_x} \sinh\left(\frac{\tau_x}{\tau_o}\right)$$

This assumption gives easily calculated expressions for the flow rates, and the simplification leads to direct evaluations of the surface shear stress components without recourse to an iterative solution of simultaneous transcendental equations. Full details are given in the paper. This simplification has the virtue that it quantifies the essential reduction in effective viscosity brought about by sliding, but avoids the complication of the fully coupled formulation. It may be regarded as a means of rapidly progressing to a solution which can then be refined with a few iterations of the full formulation. Alternatively, based on comparison with fully coupled results, it can be seen to give a satisfactory solution for most practical conditions where there is a predominant fixed sliding direction. In order to compare predictions of the model with experimental measurements of film thickness a thermal analysis was also introduced: this is fully described in the paper. Comparison of the theoretical predictions of the full and simplified treatments under typical engineering conditions at slide/roll ratios of up to 1.5 gave very satisfactory agreement.

Experimental measurements were made in an optical EHL rig under conditions from pure rolling up to a slide/roll ratio of 1.5 using steel/sapphire surfaces. Good agreement was found between the film thickness measurements and the predictions of the non-Newtonian/thermal model.

REFERENCE

(1) Holt, C.A. Non-Newtonian effects in point contact elastohydrodynamic lubrication. *PhD Thesis, University of Wales* (1994).

NEW APPROACHES IN TRIBOLOGY: USING SELF-ORGANISATION PHENOMENA IN FRICTION JOINTS

V S AVDUEVSKII and M A BRONOVETS
Interdisciplinary Scientific Tribology Council, 101 Prospect Vernadskogo, Moscow 117526, Russia
V D BABEL' and D N GARKUNOV
Russia's Public Association on Selective Transfer and Self-organising Systems under Friction, 16 Poklonnaya str., Moscow 121170, Russia
V F PICHUGIN
Gubkin State Academy of Oil and Gas, 65 Leninsky Prospect, Moscow 117917, Russia

ABSTRACT

New approaches in create boundary lubrication moving joints and friction units of the machines, devices and equipment are consider. Its are based on the using the principles of the contact phenomena self-organisation to choice and creation materials and lubricants for the friction pairs.

The theoretical foundations of the self-organising dissipative structures under friction are described. It is shown that the acceleration of the sublayer deformation diffusive flows occurs by means of the selective dissolution of the friction pair structural materials and the maintaining of the dynamical dissipative structures is possible due to an exchange by the matter with the environment.

The role of defects and dislocation in the generation and maintaining of the selective transfer under friction is investigated. It has been established that the surface layers generated by the selective transfer have an excess of vacancies, but low dislocation density. The low friction coefficients in the selective transfer are stipulated by the diffusive-vacancyonal mechanism of the rubbing surface deformation.

Particularity of chemical processes in · the friction zone when using chemical active elements in composition additives to lubricants, forms of the substance transfer from one to another friction surfaces and back without carry wear products from friction zone, achievement the conditions with very low resistance to relative motion of friction surfaces are presented.

The theoretical studies are complemented by the experiments using, in particular slip X-ray beam, photo-electronic spectroscopy, and electronic-probe microanalysis as well as the tribological testers and the field tests of machine components lubricated by the special composition. In particular, the unusually low friction coefficients in the boundary lubrication mode are comparable with thouse in the hydrodynamical lubrication mode. The application on the developed lubricant composition containing the cooper and tin salts result to the reduction of the internal combustion engine running-in time by several times and improve, at the same time, antiwear, antifrictional, and load capacity characteristic.

Essential degrease of the oil bulk temperature in gear transmission under influence special additives, also increasing lifetime and decreasing of friction loss of the internal-combustion engine supplying continuously modification oil devices during work of the engine are presented.

Technical applications developments of the new approaches to the study and development tribosystems, which open the new ways in tribology are demonstrate. Receiving effects may be more impressive, if are used complex approach to developments of the new technologies to creation machines and its exploitation.

REFERENCES

(1) D.N. Garkunov, Tribotekhnika, Mashinostroenie, M., 424 pp., 1985 (in russian).

(2) A.A. Polyakov, Some Aspects of the No-wear Effect Developments, No-wear Effect and Tribotechnologies, №1, 1994, 3-18 pp. (in russian).

(3) M. A. Bronovets., S.P. Zaritski., A.S. Lopatin, Basic Trends of Frictional Interaction and Tribodiagnostics, Technology Showcase: Integrated Monitoring, Diagnostics and Failure Prevention, Proceedings of a Joint Conference, USA, Mobile, Alabama, April 22-26, 1996, 111-118 pp.

A NON-SYMMETRIC DISCRETIZATION FORMULA FOR THE NUMERICAL SOLUTION OF E.H.L.

JAN ČERMÁK
Department of Thermomechanics, Technical University of Brno, Technická 2, Brno, 616 69, The Czech Republic

ABSTRACT

The numerical simulation of the elasto-hydrodynamic lubrication process, especially the fluid film thickness computation, is based on the numerical solution of the Reynolds equation coupled with the film shape equation. It is necessary to replace the non-linear integro-differential Reynolds equation by a non-linear system of algebraic equations. Such a step, so called discretization, is performed using a discretization scheme. Till now, there has been used both finite difference and finite element schemes.

In this paper, the control volume method (1) is employed to obtain an appropriate discretization scheme for the numerical solution of the Reynolds equation. A new, non-symmetric discretizetion formula has been developed and a benchmark computation has been performed. The isothermal point (circular) contact case will be used to demonstrate the core of the problem. Note, that the line contact behavior is very similar.

Let us consider the two-dimensional, steady state Reynolds equation. Integrating the equation over a rectangular control volume results in (formally):

$$\Gamma_e \left(\frac{\partial p}{\partial x}\right)_e + \Gamma_s \left(\frac{\partial p}{\partial y}\right)_s - \Gamma_w \left(\frac{\partial p}{\partial x}\right)_w - \Gamma_n \left(\frac{\partial p}{\partial y}\right)_n =$$

$$= 12 u (\rho h)_e - 12 u (\rho h)_w, \text{ where } \Gamma = \frac{\rho h^3}{\eta}.$$

Let us use such control volume, as on fig. 1. The pressure terms can be expressed via the weighted arithmetic average and the two point approximation of the derivatives. To ensure, that the velocity terms will not cause an oscillation (2), they should be expressed via up-wind. The non-symmetric control volume (fig. 1) has been designed just to reduce the first order up-wind scheme error. So, the discretization formula reads:

$$\left(\frac{\partial p}{\partial x}\right)_e = \frac{p_E - p_P}{x_E - x_P}; \quad \Gamma_e = \frac{3}{4}\Gamma_P + \frac{1}{4}\Gamma_E;$$

$$\left(\frac{\partial p}{\partial y}\right)_s = \frac{p_S - p_P}{y_S - y_P}; \quad \Gamma_s = \frac{1}{2}\Gamma_P + \frac{1}{2}\Gamma_S;$$

$$(\rho h)_e = (\rho h)_P. \text{ (} w \text{ and } n \text{ wall terms are similar.)}$$

A benchmark computation has been performed for the following Dowson & Higginson's dimensionless parameters: $U=3.043*10^{-12}$, $W=9.763*10^{-7}$, $G=3560$. Three important dimensionless film thickness values has been observed with increasing number of grid nodes: the global minimum H_{min}, the minimum at the center line ($y=0$) $H_{min,0}$ and the central value ($x=0, y=0$) $h_{0,0}$. Results are introduced in table 1.

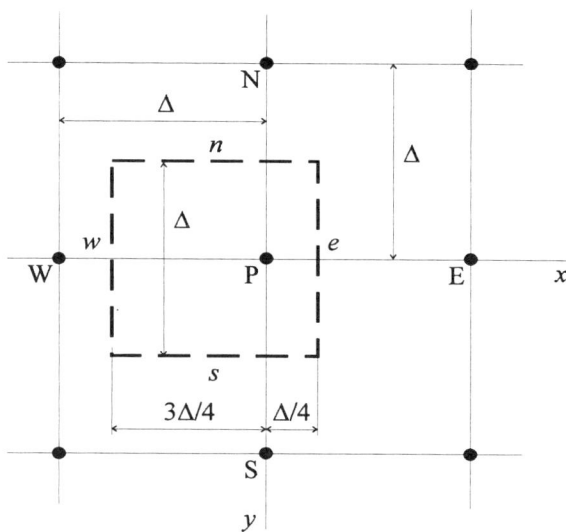

Fig. 1: Non-symmetric control volume
(Walls are dashed and denoted by corresponding small letters. Grid nodes are denoted by capital letters.)

number of grid nodes	H_{min} * 10^{-6}	$H_{min,0}$ * 10^{-6}	$H_{0,0}$ * 10^{-6}
79*79	2.7189	5.6610	6.4880
118*118	2.8969	5.7572	6.6415
157*157	2.9038	5.7987	6.6931
196*196	2.9220	5.7635	6.7146
235*235	2.9182	5.7390	6.7246
274*274	2.9151	5.7270	6.7296

Table 1: Resulting dimensionless film thickness for different number of grid nodes

In comparison with the classical finite-difference formula, used for example in (3), the introduced scheme gives smaller discretization error for a given grid density. Moreover, the central film thickness value is underestimated by the discretization error. It is safe from the design point of view.

REFERENCES

(1) S. V. Patankar, Numerical heat transfer and fluid flow, Hemisphere, Washington, 1980.
(2) H. Okamura, Proc. 9th Leeds - Lyon Symposium on Tribology, 1982, 313-320 pp.
(3) A. A. Lubrecht et al, ASME JOT, 1988, 503-507 pp.

Submitted to *Tribology International*

A WAY TO REDUCE END EFFECTS IN FINITE LENGTH LINE CONTACTS

E. N. DIACONESCU and M. L. GLOVNEA
Mechanical Engineering Department, University of Suceava, 1 University Street, Suceava, 5800, Romania

ABSTRACT

A line contact of finite length, as encountered very often in roller bearings, is subjected to an important pressure concentration near the front edges. Although various longitudinal profiles of contacting bodies have been proposed to reduce these end effects, the only theoretical derivation of an optimum profile belongs to Lundberg (1). The improvement reported later by Reusner (2) remains at a statement level because no details of the profile or of its actual derivation are given.

In spite of the fact that in principle Lundberg proposed a correct inverse solution to the problem, due to the lack of computation techniques, the involved integrals were solved analytically under certain simplifying assumptions. A numerical dimensionless analysis of the problem, performed by Diaconescu (3), indicates that these assumptions introduce significant errors in the longitudinal profile, especially towards the contact ends. The derivatives of the integral equation of the actual profile that stems from a Lundberg pressure distribution show that the ordinate, the slope and the curvature of this profile possess discontinuities at contact edges. Consequently, Lundberg pressure and its associated profile do not satisfy compatibility equations of displacements and the restriction formulated for contact pressure by Boussinesq (4) and later on by Johnson (5).

In order to comply with the requirements of basic equations of elasticity and of contact pressure, a modified Lundberg pressure is proposed. This is generated by adding frontally two quarters of pressure ellipsoids to the Lundberg pressure, as shown in Fig. 1, thus removing the pressure discontinuity.

The longitudinal profile that generates such a pressure distribution is derived numerically. This is continuous and possesses continuous derivatives in the lateral ends of the contact, as required by the fundamental equations of elasticity.

The small axial extension of contact pressure is then determined numerically such that this pressure to lead to a nearly uniform distribution of maximum Huber-Mises-Hencky (HMH) stress along the contact, with no stress risers in or near contact edges.

For instance, Fig. 3 shows profiles of maximum HMH stress along the contact taken at various for a practical case, when $a = 2b$ and $c = b/5 = a/10$. The variation of equivalent stress along the contact depends on the depth below the surface, but the maximum values, reached at different axial positions, are practically the same.

It can be concluded that, at the expense of incomplete use of one tenth of the contact length, a good stress distribution is achieved.

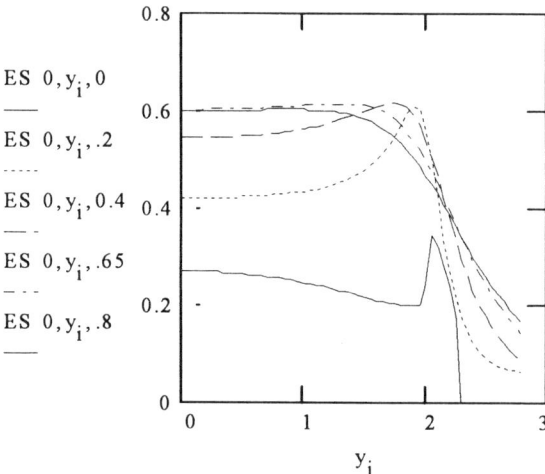

Fig. 3: Equivalent HMH stress along the contact

REFERENCES

(1) G. Lundberg,, Forschung Gebiete Ingenieurwesens, Vol. 10, 5, 1939, 201-211pp.
(2) H. Reusner, Ball Bearing Journal, Vol. 230, SKF, June, 1987, 2-10pp.
(3) E. Diaconescu, Acta Tribologica, Vol.3, No.1-2, 1995, 7-12pp.
(4) J. Boussinesq, Application des potentiels á l'etude de l'equilibre et du mouvement des solides élastiques, A. Blanchard, Paris, 1969.
(5) K.L. Johnson, Contact Mechanics, Cambridge University Press, London, 1985.

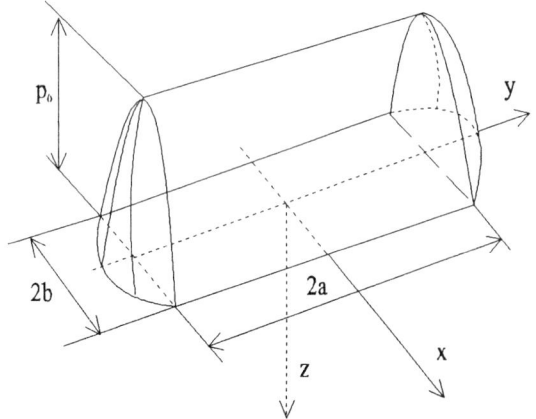

Fig. 1 Corrected pressure distribution

ELASTOHYDRODYNAMIC FILM THICKNESS IN MIXED LUBRICATION

G. GUANGTENG and H. A. SPIKES
Department of Mechanical Engineering, Imperial College, London SW7 2BX, UK

ABSTRACT

In mixed lubrication regime, both surface roughness and entrainment speed influence the film separation, and the lambda ratio, the ratio of film thickness to the undeformed surface roughness, has been used in predicting lubrication state and machine component lifetime.

This work is aimed at investigating directly the influence of roughness of realistic, engineering surfaces upon film thickness formed by lubricant base fluids. The ultra thin film interferometry is used to measure mean film thickness in a rolling EHD contact between a ball bearing and a coated glass plate. The steel balls have a range of isotropic surface roughness characteristics. The two test surfaces are designated as 'Rough1' (60 nm rms.) and "Rough2" (100 nm rms.) respectively.

Fig. 1 shows the mean film thickness results versus rolling speed for the two roughnesses with SHF41 as the lubricant. The mean film thicknesses are nondimensionalised with the rms. roughness of the corresponding surface. The film thickness results for a 'smooth' ball is also included for comparison. It is indicated that a full film lubrication is only realised when the lambda ratio (λ) is above 2. As the lambda ratio further reduces ($\lambda < 1$) the film thickness decreases more rapidly. The mixed lubrication regime seems to span a large range in terms of the lambda ratio ($0.1 < \lambda < 2$).

It is important to realise that the above mentioned film thickness means the lubricant film due to hydrodynamic effect, whilst the total separation between two surfaces is the sum of this film and the effective roughness. With a recently developed technique, it is now possible to measure film profiles so that the effective roughness in static contact can be estimated.

Fig. 2 shows the mean separation in static contact versus contact load. This may serve as an indicator of the order of the effective roughness in the EHD contact. It can be seen that the mean separation asymptotes to 10 nm (for Rough1 surface) at the load used in the film thickness measurements. This demonstrates that the total separation is the measured film thickness plus a value close to 10 nm for Rough1 surface.

From the current work, the results can be summarised as follows.

(1) A full film lubrication is realised when $\lambda > 2$.

(2) The mixed regime spans a large lambda ratio range, $0.1 < \lambda < 2$.

(3) It is important to appreciate the precise definition of lubricant film thickness in rough surface conditions.

Experiment details and some typical film profiles are shown in the poster.

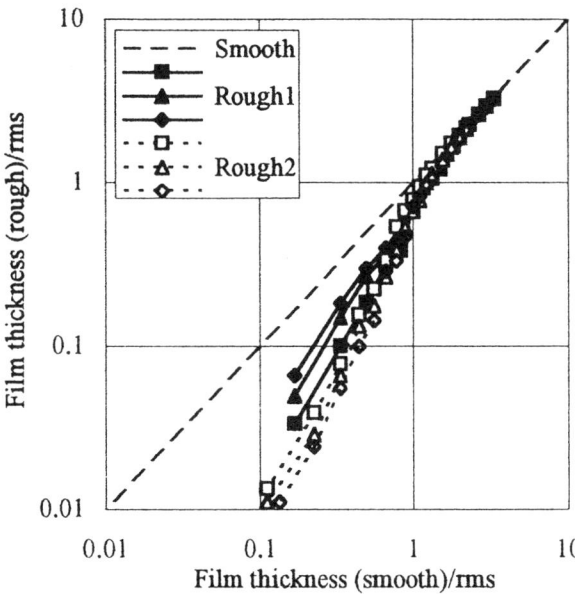

Fig. 1: Comparison of lambda ratios

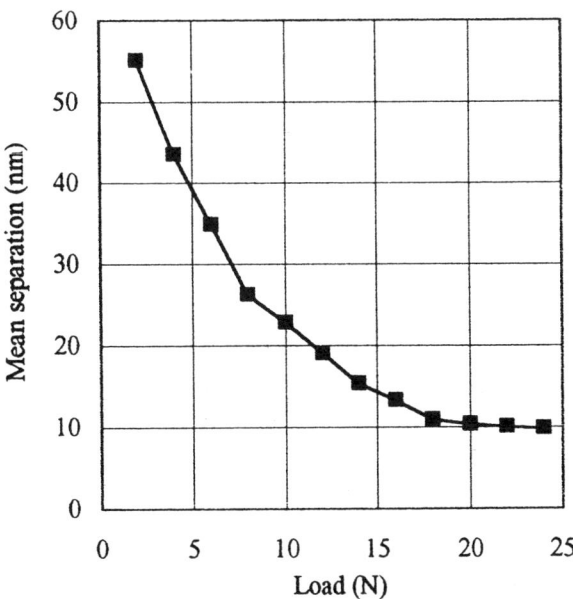

Fig. 2: Mean separation in static contact

ELASTOHYDRODYNAMIC FILM THICKNESS MAPPING BY COMPUTER DIFFERENTIAL COLORIMETRY

M HARTL, I KŘUPKA and M LIŠKA
Faculty of Mechanical Engineering, Technical University of Brno, Technická 2, 616 69 Brno, The Czech Republic

ABSTRACT

This paper concerns the development and application of an experimental technique that enables accurate, quick and automatic evaluation of EHD film thickness distribution with high resolution from chromatic interferograms obtained from a conventional optical test rig. This technique overcomes some limitations of conventional optical interferometry and is intended for the study of transient and quasistatic phenomena.

The technique is based on the idea of replacing human eye by simple differential colorimeter for interference colour evaluation. The lubricant film thickness is determined by comparing colour coordinates between digitized EHD interferogram and digital colour chart with the help of a colour difference formula. This comparing is realized by newly developed computer software system and gets over the inability of human eye to perceive colours precisely and accurately. The digital colour chart has been obtained from Newton rings for static contact. Its colours are described quantitatively and their change with growing film thickness can be plotted as a curve in the CIE-chromaticity diagram (Fig. 1).

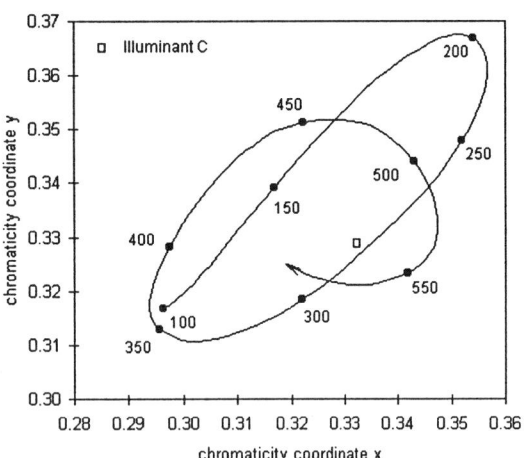

Fig. 1: Interference colours in CIE-chromaticity diagram (figures are optical thickness in nm)

In this study interferograms were recorded through the industrial microscope with photomicrographic equipment on 35 mm reversal colour film Kodak and digitized by the Kodak Photo CD System. The CIELAB colour difference equation (1), which is simply the Euclidean distance between two colours specified in CIELAB coordinates, was used for comparing EHD interferograms with the digital colour chart. This colour chart contains L*, a*, b* values with appropriate film thicknesses from 60 to 800 nm with the difference of 1 nm.

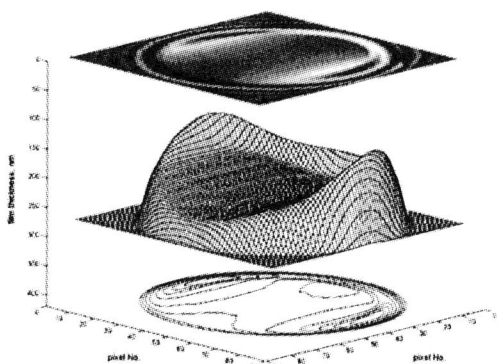

Fig. 2: Gray scale interferogram, corresponding mesh surface and contour plots of film thickness

Fig. 2 shows the reconstruction of the lubrication film thickness distribution from the chromatic interferogram of 80 pixels x 80 lines for the surface velocity of 0.0541 m/s, max. Hertz pressure 0.425 GPa and naphthenic base oil. It consists of the digitized interferogram converted into gray scale form, corresponding mesh surface and contour plots of the film thickness.

To validate this technique its accuracy has been checked and a comparison with conventional monochromatic interferometry has been done. Technique resolution has been checked through the observation of the local film thickening just before the elastohydrodynamic exit constriction for both pure rolling and sliding conditions (2) and was found to be better than 3 nm within the range 60 - 800 nm.

REFERENCES

(1) F. W. Billmeyer, Jr., M. Saltzman, Principles of Color Technology, New York, 1981, 25-66 pp.
(2) I. Křupka, M. Hartl, M. Liška, The Study of Elastohydrodynamic Lubrication of Rolling/Sliding Point Contacts by Computer Differential Colorimetry, World Tribology Congress '97.

This paper was submitted to Tribology Transactions for publication.

ELASTOHYDRODYNAMIC LUBRICATION AND TRIBOLOGY DESIGN FOR FINITE ROLLER CONTACTS

MA JIAJU and XU WEN

Department of Mechanical Engineering, Zhejiang University, Hangzhou 310027, P.R.China

CHEN XIAOYANG

Bearing Institute, Shanghai University, Shanghai 200072, P.R.China

ABSTRACT

In this paper, a numerical analysis of isothermal elastohydrodynamic lubrication (EHL) is presented for a finite roller, rolling over an elastic half-space. In the analysis, the wedge term of the discrete Reynolds equation is taken the form of one-order complete backward difference. The Reynolds viscosity-pressure and the Dowson-Higginson density-pressure dependences are employed for a mineral oil. To improve calculation accuracy, the elliptic-paraboloid method(1) is used to calculate surface deformation. The Jacobi iteration procedure is constructed for improving convergence.

To investigate the effects of roller geometry on EHL characteristics, several different profiles have been considered. For a lubricating finite cylindrical roller with direct generator, the edge effects are obviously displayed in the ends, where the film pressure is much higher than the average. But, the degree of pressure concentration is weakened as compared with that in dry contacts and the minimum of the film thickness is taken place in the ends. When the roller with such generator is used in engineering, the influence of the edge effects and the side leakage is critical. Under low velocity and heavy load, the lubricant film in the ends could not be established. The asperity contact or dry friction will be taken place in the interface. The immaturity failure of the material will be initiated. The service life will be reduced.

From the analysis of the roller with logarithmic profile, the edge effects would be disappeared. Under low velocity, the well-distributed film pressure along axial direction are displayed, and under high velocity, the film closure contractions in the ends cause closed effects to the advantage of establishing film with enough thickness. Meanwhile, the roller contacts have outstanding load characteristics under EHL condition. A comparison of film thichness with load is shown in Fig.1.

In order to increase the load carrying capacity and the service life of a roller contacts, the roller must be appropriately profiled. Generally, it is believed that the roller with logarithmic profile is a advanced design(2). But, Lundberg's profile(3) is not an optimal one because it was obtained in approximate condition, in

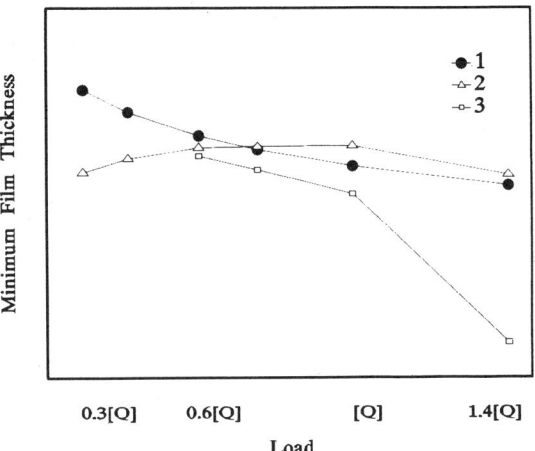

(1) infinite line contact (2) the middle of roller
(3) the end of roller

Figure.1: Comparison of the variation of film thichness with load for the roller with logarithmic profile (u'=1.06e-10, G=2387.7)

which the width/length ratio of the contact area should approach to zero, and in the ends, his profile is discontinuous. For practial engineering application of the logarithmic profile, a new tribology design of roller profile is given as.

$$T(y) = f\frac{[Q]}{\pi E'b}\ln\frac{1}{1-(1-0.3\frac{a}{b})(\frac{y}{b})^2}, \quad |y| \leq b$$

When the roller with this new profile is rolling over a lubricating elastic half-space, some marked EHL performance are shown in comparing with the contacts with other profiles, such as crowned or dub-off profiles. The logarithmic profile might be an optimal selection in tribological design of finite line contact elements.

REFERENCES

(1) Liu Shuang-Biao, Ma Jia-Ju, Chen Xiao-Yang. Tribology international. 1993. 26: 443 – 448
(2) Reusner H. The Ball Bearing Journal, 1987, 203: 2 – 10
(3) Lundburg G. Forchung auf dem Gebiete des Ingenieurwesns. Sept./okt 1939.10 (5): 201- 211

FRICTION AND WEAR MODEL OF A HEAVY-LOADED PLAIN BEARING

V.L.KOLMOGOROV, V.V.KHARLAMOV, S.V.PAVLISHKO

Institute of Engineering Science, Russian Academy of Sciences (Urals Branch), .91 Pervomaiskaya st., GSP-207, 620219, Ekaterinburg, Russia

ABSTRACT

The mode of mixed friction in a lubricated radial sliding bearing is discussed, with the working surfaces of the journaland the bush being incompletely separated by the lubricant. In this case, a part of the radial load on the bearing is taken by the areas of actual contact (aac), the other part being taken by the lubricant placed between the areas. The journal is modelled as an absolutely rigid cylinder Laving regular roughness in the form of triangular prisms located along the generatrix; the bush is modelled as a rigid-plastic body. The mechanical and thermal characteristics of the bush and the lubricant are specified. Only those cases of the mixed mode of friction is discussed when mechanical interactions prevail over tribochemical, triboelectrical and other ones. The boundary layer of the friction zone is modelled as a porous medium, and the lubricant flow between the aac obeys the laws of liquid flow through porous media. The wear of the bush is assessed on the basis of the phenomenological model [1], [2] of damage cumulation in cyclic plastic deformation on the aac and defect healing by heating. The paper deals with the stationary and unsteady modes of friction and the transition from liquid to mixed friction [3]. The criterion of transition has been found, and the transition has proved to be smooth. Solving the boundary value problems for the motion of the rigid triangularwedge in the rigid-plastic semispace results in defining the characteristics of the stress-strain state of the bush, the strain component of the friction force and friction moment, the rate of damage cumulation and healing in the plastically deformable layerof the bush and the intensity of wear on the aac. Solving the boundary value problems of lubricant flow betweenthe aac, heat conductivity and plasticity in common has enabled us to deformine the sharesof the aac and the lubricant layer in the response to extrnal loading and the shares of the moments of fluid and dry friction in the mixed mode.

Solving nonstationary boundary value problems has offered the criteria and conditions of transition from mixed to fluid friction and vice versa.

The mathematical model comprises:
• Equation of plane flow of perfectly plastic incompnessible material describing the fields of stresses and strain rates in the vicinity of the plastic wave (stress and strain rate tensor components), the field of velocities of material particles within the plastic wave.

• Equation of heat conductivity

$$c\rho \frac{dT}{dt} = \lambda \nabla^2 T + q.$$

• Differential equation of damage cumulation and healing in the boundary layer of contact

$$\frac{d\Psi_i}{dt} = \frac{H(t)}{\Lambda_p} - \Psi_{i-1} \exp(-\beta t).$$

• Condition the failure of plastically deformable layer

$$\Psi(t_n) = \sum_{i=1}^{n} \Psi_i^a \geq 1.$$

• Equation of liquid flow in a porous media with movable walls

$$V_i = -\frac{K(i)}{\mu} \frac{\partial p}{\partial i} - U_i.$$

• Equation of continuity of liquid flowing in a porous media

$$\Pi \frac{\partial \rho}{\partial t} + \mathrm{div}\, V = 0.$$

The model enables one to predict friction and wear parameters for newly designed heavy-loaded tribocouples and to select material proferties of bering minimum wear with specified external mechanical and thermal effects on the sliding bearing.

REFERENCES

(1) V.L.Kolmogorov, Friction and wear model of a heavy-loaded sliding pair, Part I, Metal damage and fracture model, Wear 194(1996), p. 71-79.

(2) V.L.Kolmogorov, V.V.Kharlamov, A.M.Kurilov, S.V.Pavlishko, Friction and wear model of a heavy-loaded sliding pair, Part II, Application to an unlubricated journal bearing, Wear 197(1996), p. 9-16.

(3) V.V. Kharlamov, Mathematical Model of Transition Process from Liquid Friction Mode to Mixed Mode on an Example of Heavy-Loaded Sliding Contacts, Trans of Tribology Conference, The institution of Engineers, Australia, Brisbone, 1990, p. 192-195.

THE ACTIVATION ENERGY OF REACTION LAYER FORMING UNDER BOUNDARY LUBRICATION AS THE CRITERIA OF EP PROPERTIES OF LUBRICANT

I.A.BUYANOVSKY
Mechanical Engineering Research Institute of Russian Academy of Scienses, Griboedova, 4, Moscow, 101830, Russia
I.S. PANIDI, M.PINDEL, A.V.KOZEKIN and V.A.TROFIMOV
State Academy of Oil and Gas, Leninsky, 65, 117343, Moscow, Russia

ABSTRACT

Formerly the kinetic model of seizure was developed by one of the autores for connection of reaction layer forming on friction surfaces under boundary lubrication conditions and the temperatures of boundary EP layer failure under friction (1). Contact stresses, speed of solids under friction, properties of rubbing materials, reactant content and its activity, metal-metal and metal-lubricant interaction also are took into account in the kinetic model. At the base of this theory the energies of activation of reaction layer forming for a lot of sulfurcontained EP additives dissolved in white oil were estimated by the way of "friction-temperature" diagrams treatment. These diagrams were obtained using the results of the friction tests of tested lubrication compositions by the Matveevsky method (2) at slowly rotating four-ball KT-2 maschine with gradually heating friction unit from 20 to 300"C. Temperatures of chemical modification of rubbing surfaces under friction for boundary lubrication conditions (effective temperatures of low friction obtaining due to reaction layer appearance on the friction surfaces) and temperatures of begining of friction decreasing due to destruction of investigating addtives and metal-reactant interaction are measured.

Investigating media were produced as solutions of a number of sulfurized oligomeres in inactive white oil as EP additives. These additives were sintezed and investigated by the some of autors (3).The temperatures of decomposition of said additives were measured and compared with temperatures of chemical modification of rubbing surfaces under boundary lubrication conditions. Welding loads of investigating compositions were determined using usual four-ball machine tests. Activation energies calculated were compared with welding loads of corresponding lubrication compositions (fig.1). The results presented are discussed and kinetic theory of reaction layer forming and failure of them under friction (1) is supposed to be suppoted by the results presented. According to this it is useful to utilizate the values of the energy of activation of boundary layers forming under friction as criteria of EP properties of lubrication materials due to their invariableness in a wide range of variing of operative factors vaiu-es. The possibility of using of the values of these activation energy values as AW properties criteria of lubricants also discussed.

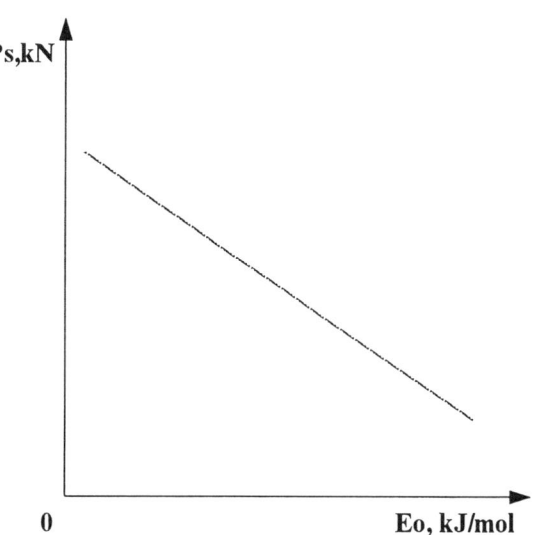

Fig.1: The effect of activation energy of chemical modification (Eo) on the welting load (Ps) for media investigated

REFERENCES

(1) I.A.Buyanovsky. Friction and wear (russian journal 'Trenie i isnos'). Vol.14, No1, 1993, 129-142pp.

(2) R.M.Matveevsky. Tribology, Vol.1, No2, 1968, 115-117pp.

(3) M.Pindel, A.V.Kozekin, I.S.Panidi, V.A. Trofimov, Chem. and Techn. of fuel and oil, 1997, №2,16-18pp.

Transition Between Elastohydrodynamic Lubrication and Boundary Lubrication

Jianbin Luo, Shizhu Wen
National Tribology Laboratory, Tsinghua University, Beijing 100084, CHINA

1. Experiments

A technique of relative optical interference intensity is used to measure the film thickness. The resolution of the instrument could reach 0.5 nm in the vertical direction and 1.5 μm in the horizontal direction. The viscosity and the reflective index of lubricants used in experiments are given in Table 1.

Table 1—Lubricants

Lubricants	Viscosity (mPa•s/20°C)	Refractive index
Mineral oil A	190	1.4871
Mineral oil B	84	1.4836
Mineral oil C	36	1.4771
Lubricant 13604	17.4	1.4733
Lubricant 13602	4.21	1.4616
Paraffin liquid	30	1.4612

Note: Lubricant 13602 and 13604 are of pure alkane.

2 Results
2.1 Transition from EHL to thin film lubrication

In the high speed region in Fig.1 (above 5 mm/s), the film thickness changes signific-antly and the speed index is about 0.69 (Fig.1, curve b) which is very close to that in EHL. The film becomes thinner as the speed decreases. When the film thickness is less than a critical thickness which is related to the atmospheric viscosity of lubricants, the speed index is only about half of the EHL one and the relation between the film thickness and the speed does not follow the EHL theory any more. This point is believed as the transition point from EHL to TFL. In addition, the film thickness measured is much higher than that calculated from Hamrock-Dowson formula (Fig.1, curve c). This indicates that the viscosity is much higher than that of bulk oil when the film is sufficiently thin.

3.2 Effect of time

In the TFL regime, the film thickness of some lubricants varies with the running time under certain conditions. Factors, such as load, velocity, and viscosity of lubricant, influence significantly on the relationship of film thickness and time.

3.3 Effect of substrate surface energy

The larger the surface tension, the thicker the total film and the critical film will be (Fig.2). This indicates that the surface tension adsorbs the lubricant molecules nearby to adhere to the interface and orients them.

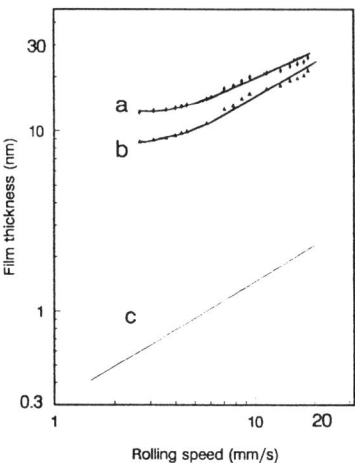

Fig.1 Film thickness in the contact region
Load: 4 N, T=25°C, Ball: φ23.5 mm, Lubr. 13604, a-measuring data, b-film thickness without static film, c-calculated from Hamrock-Dowson formula

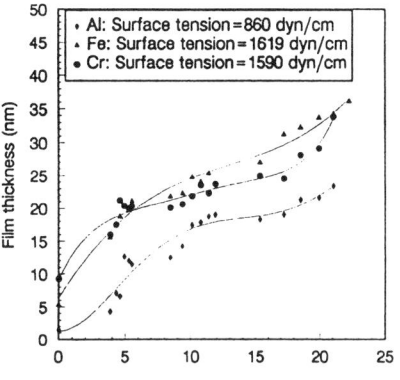

Fig.2 Film thickness with different substrate
Lubr. 13604, Load: 4 N, T=18.5°C, Ball: φ23.5 mm

4. Conclusion

In the thin film lubrication, The EHL rules is no longer suitable. The critical film thickness, at which the transition from EHL to TFL happens, is related to the initial viscosity of lubricant. The film is also related to the running time and the surface energy of substrates.

EXPERIMENTAL STUDY OF ULTRATHIN LIQUID LUBRICATION FILM THICKNESS AT THE MOLECULAR SCALE

H. MATSUOKA and T. KATO

Department of Mechanical Engineering, The University of Tokyo, 7-3-1 Hongo, Bunkyo-ku, Tokyo 113 Japan

ABSTRACT

Ultrathin liquid lubrication film thickness is measured by a setup in which mica is used as solid surfaces and sliding with low velocity is carried out. Discretization of the film thickness is observed when the liquid film is of several molecules thick. It was found that the force acting on the surfaces did not depend on the sliding speed of the surfaces. Analysis of the experimental data shows that the discretization is due to the structure of the liquid, namely the layering of the liquid molecules.

INTRODUCTION

Many researches have been published recently on the characteristics of the liquid film in the vicinity of a solid surface (1,2) and it has been found that the thin film shows quite different properties from the bulk. These are thought to be by the intermolecular force. It is considered, thus, the predictions based on the continuum theories must be modified when intermolecular forces cannot be neglected such as in the confined liquid film of the several nanometers thick. We observed the discretized and larger film thickness than that predicted by conventional EHL theory (3). The results are analyzed and discussed in the present work and it will be shown that the results are due to the layering of the liquid molecules.

EXPERIMENTS AND DISCUSSIONS

Film thickness of the order of several nanometres was measured by using an experimental setup in which two molecularly smooth mica surfaces in crossed cylinder configuration were slid past each other with liquid (n-hexadecane, cyclohexane or OMCTS (octamethylcyclotetrasiloxane)) between the surfaces. Examples of the results using OMCTS are shown in Fig. 1 where film discretization is observed and measured thickness is larger than the hydrodynamic theory (solid line) or conventional EHL theory (dotted lines). The discretized film step is about 1 nm which is nearly equal to the molecular diameter of OMCTS. Measured film force is then divided into two parts; Fh (hydrodynamic force) and Fs (additional force) as shown in Fig. 1, and Fs is plotted against film thickness (Fig. 2). It is seen from the figure that Fs decreases exponentially (dotted line in the figure) with the film thickness. Further experiments showed that Fs did not depend on the sliding velocity, thus the force is due to the static force like intermolecular one.

CONCLUSIONS

From these observations and analysis, we concluded that the additional force is the structural force of the liquid (or solvation force) which is due to the molecular layering and exhibits oscillation (attractive and repulsive force) with a period of molecular diameter but decays exponentially with separation. Fig. 3 shows an example of simulation results (solid line) that the structural force is modeled by $Fs = A\exp(-h/B)\cos(2\pi h/B)$ where A and B are constants and h the film thickness.

REFERENCES

(1) Van Alsten & S. Granick, Phys. Rev. Lett., 61, p. 2570, 1988.
(2) G. Guangteng & H. A. Spikes, STLE Trans., 39, 2, p.448, 1996.
(3) H. Matsuoka & T. Kato, ASME J. of Tribology, 118, 4, p.832, 1996.

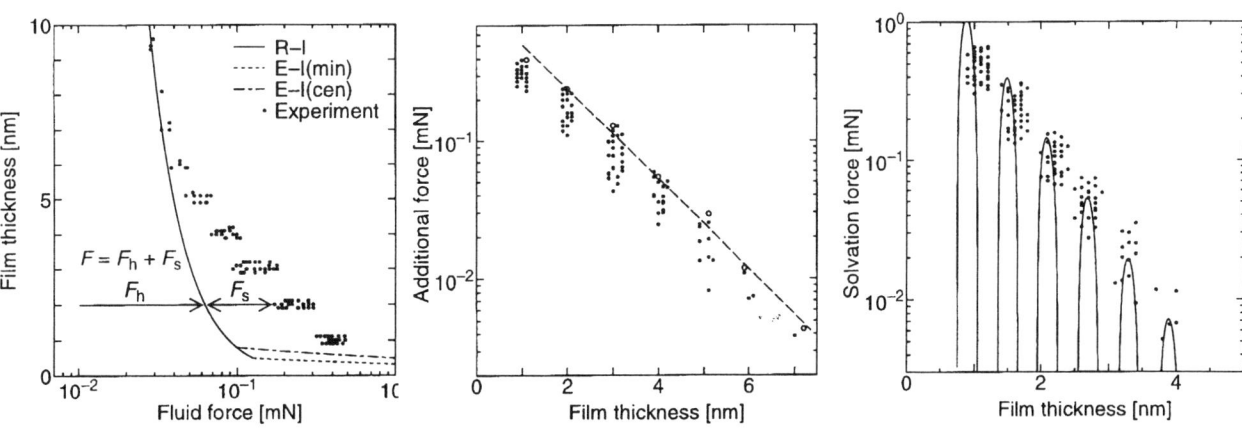

Fig. 1 Experiments (OMCTS, u:200 μ m/s)

Fig. 2 Additiona force Fs (OMCTS, u:200 μ m/s)

Fig. 3 Modeling by exp-cos curve (cyclohexane u:200 μ m/s)

ELASTOTHERMOHYDRODYNAMIC EFFECTS IN A THREE LOBE JOURNAL BEARING

K.PRABHAKARAN NAIR and SHIVLAL YADAV
Department of Mechanical Engg., Regional Engineering College, Calicut, India

ABSTRACT

In this paper, the effects of elastic deformation of bearing and variation of viscosity with both pressure and temperature on the static and dynamic performance characteristics are presented.

Though the bearings are generally designed using the data developed with the assumptions that their surfaces are rigid and the viscosity of the lubricant is constant, the bearings operating at high speed and carrying heavy load, need analysis and design, which take into account their elastic deformation and also the viscosity variation of the lubricant with both pressure and temperature. The bearing deformation may quite often have magnitudes of the order of film thickness, thus affecting the clearance geometry to an extent such that the actual performance characteristics are significantly different from those computed with rigid bearing assumption. The viscosity variation with both pressure and temperature may also affect the performance characteristics appreciably. Generally, the non-circular bearings show better dynamic stability than plain cylindrical bearings. The available literature indicates that a rigorous elastothermohydrodynamic analysis of a three lobe bearing is scant. So, it is felt that there is a need to recompute the performance characteristics of a three lobe bearing considering the effects of bearing deformation and variation of viscosity with both pressure and temperature on the performance characteristics.

In the present analysis three dimensional momentum and continuity, elasticity, energy and heat conduction equations are used to compute pressure distribution in the flow field, the deformation of bearing liner, temperature distribution in the flow field and the bearing body. The solutions of lubricant flow, elasticity and thermal equations are obtained using the finite element method and a direct iteration scheme.

The static characteristics in terms of the load capacity, attitude angle and the end leakage and the dynamic performance characteristics in terms of the stiffness coefficients, damping coefficients, Threshold speed and the damped frequency of whirl are obtained for the following three cases.

(1) Isoviscous
(2) Piezoviscous
(3) Thermopiezoviscous

On all these three cases the static and dynamic performance characteristics are computed for a wide range of values of deformation coefficient and several values of eccentricity ratio (e) and the computer results shows that the bearing deformation and variation of viscosity with both pressure and temperature affects the bearing performance characteristics appreciably.

Fig.1 shows the variation of load capacity with deformation coefficient for the three cases studied and the results show that the effects of bearing flexibility and variation of viscosity with both pressure and temperature are quite significant.

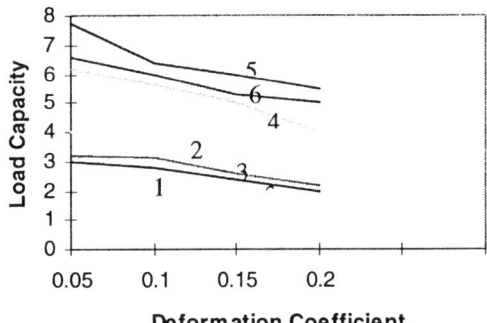

Fig.1 : Load Capacity Vs Deformation Coefficient

1,4. Isoviscous (e=0.2, e= 0.3)

2,5 .Piezoviscous (e=0.2, e=0.3)

3,6.Thermopiezoviscous(e=0.2, e= 0.3)

REFERENCES

(1) D.T. Gethin, J.D. Medwell, Tribology international, vol-18,1985
(2) W.A. Croshy, Wear, Vol- 143, 1991
(3) Khonsari and S.E. Wang, Trans. ASME, Journal of Tribology, Vol- 113, 1991.

AN ACCURATE THERMAL ANALYSIS OF TRACTION IN ELASTO-HYDRODYNAMIC ROLLING/SLIDING LINE CONTACTS

RAJ K. PANDEY
Research & Engg. Division, Kirloskar Brothers Ltd., Chintan Bldg., Mukund Nagar, Pune-411 037, INDIA
MIHIR K. GHOSH
Dept. of Mech. Engg., Institute of Technology, Banaras Hindu University, Varanasi-221 005, INDIA

ABSTRACT

In Elastohydrodynamic lubricated (EHL) contacts viz. ball and roller bearings, gears and traction drives etc., an accurate estimation of traction force (traction coefficient) and temperature rise due to sliding is essential from the view point of power loss and failure of lubrication in the contacts. The prediction of the EHL traction analytically have been done under certain assumptions and considerations by several researchers e.g. Cheng(1), Kannel & Walowit (2), Crook(3) and Gupta et al (4). However, the upto date results reported in the literature are usually for either high loads, low rolling speeds & low slips or for low loads, high rolling speeds and high slips. Moreover, surface temperature variation of the disks are neglected in the most of the theoretical analyses. The viscosity-pressure-temperature relationship adopted by the most of the researchers is of Barus' type which usually predicts higher viscosity in the contact zone and leads to prediction of higher traction coefficients in the contact region. The present work is concerned with an accurate contact zone analysis of thermal traction for high rolling speeds upto 35 m/s, maximum Hertzian pressure upto 1.52 GPa, and for slip varying between 0.5 - 35 % for EHL line contacts using the numerical method developed by Elrod and Brewe(5). The convective heat transfer in the film along with the variation of density with pressure and temperature have been taken into the account. A more realistic viscosity model i.e. Roelands' viscosity relationship have been adopted. Also, the surface temperature variation of the disks due to conduction of heat has been taken into account.

Significant reduction in traction coefficient at high rolling speeds due to thermal effects have been found. Large increments in traction coefficient have been determined at high contact pressures e.g. at 1.52 GPa as compared to 0.4 GPa at a particular rolling speed. The results of the present work have been compared with the Cheng's theoretical work & Crook's experimental results of traction in fig.1 for rolling speeds of 4 m/s and 6 m/s. It is observed that large differences exist at low slip between present work & Cheng's work. But at the same range of slip, there is good matching between Crook's experimental work and the present work. The deviation between present theoretical work & that of Cheng's is due to the fact that in Cheng's theory the temperature-pressure-viscosity relationship used may not be adequate to approximate the viscosity. It seems over-estimation of the viscosity causes high traction coefficient in his approach. But the predictions obtained by present approach are consistent with the available experimental data, which justifies accuracy of Roeland's viscosity model used for this work. From the present work, it seems there is good correlation with Crook's experimental work upto 30% slip as observed in fig.1. Therefore, in order to estimate traction correctly in heavily loaded, high speed contacts, an accurate thermal analysis is necessary with appropriate Newtonian viscosity model viz. Roeland's viscosity relationship is adequate for correct estimation of traction coefficient in the contact.

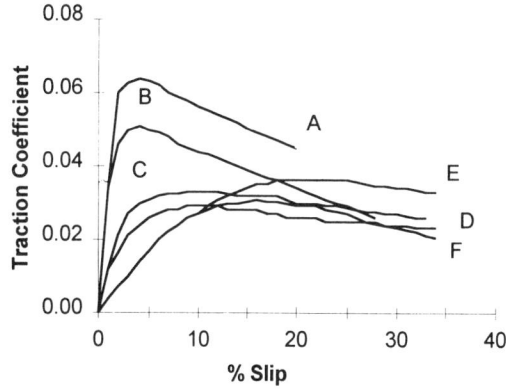

Fig. 1 : Comparison of authors' traction data with Cheng's theoretical & Crook' experimental data, P_H=0.49 GPa

REFERENCES

(1) H.S. Cheng, Trans. ASLE, Vol.8, No.4, 1965, pp.397-410.
(2) J.W. Kannel, J.A. Walowit, Trans. ASME, 1971, pp.39-46.
(3) A.W. Crook, Roy. Soc.London, Ser. A, 1961, pp.237-258.
(4) P.K.Gupta, L. Flammand, D. Berth and M. Godet, Trans. ASME, 1981, pp. 55-64.
(5) H.G.Elrod, D.E. Brewe, NASA Tech. Memo. 88845, 1986.

Submitted to *Wear*

IDENTIFICATION OF THE LUBRICATION CONDITIONS IN EHD CONTACT

W PIEKOSZEWSKI, M SZCZEREK, M WIŚNIEWSKI and J WULCZYŃSKI
Institute for Terotechnology, ul. Puławskiego 6/10, PL - 26-600 Radom, Poland

ABSTRACT

In the recent years a number of theoretical solutions have been elaborated in order to study the effect of the surface roughness on the EHD film thickness. These theories, operating with averaged or local Reynolds equation, delivered some interesting results (1). The purpose of this work is therefore to verify experimentally these results.

The experimental determination of the lubrication conditions has been performed on the block-on-ring test rig. Block and ring were electrically insulated from each other in order to identify the micro-short-circuits caused by metallic contact of asperities. All measured values of the number and time of the micro-short-circuits were recorded on the hard disc (2).

The values of the relative metallic contact time were approximated with regression lines. The slope of these lines is characterised by means of the critical contact load P_{cr}, ie the force that, for sliding velocity of 1 m/s, would cause a full metallic contact. Some results are summarised in Figure 1.

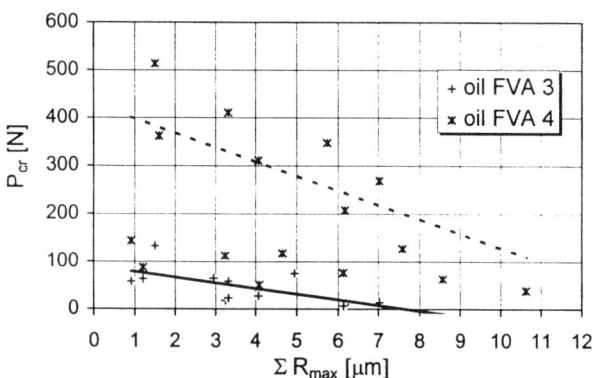

Fig 1: Critical load *vs* the sum of the maximum roughness depths

The trend lines for critical load as a function of the sum of roughness height ΣR_{max} reveal small slope that means that the surface texture has moderate effect on EHD film thickness. The critical load for oil with higher viscosity (FVA4) is greater than for the less viscous one.

Comparison of pairs smooth roller/rough block and *vice versa* is shown in Fig. 2 in which nearly smooth surface of specimen is signed with symbol "0", anisotropic one with cross machining traces has symbol "X" and isotropic texture with longitudinal traces - "||".

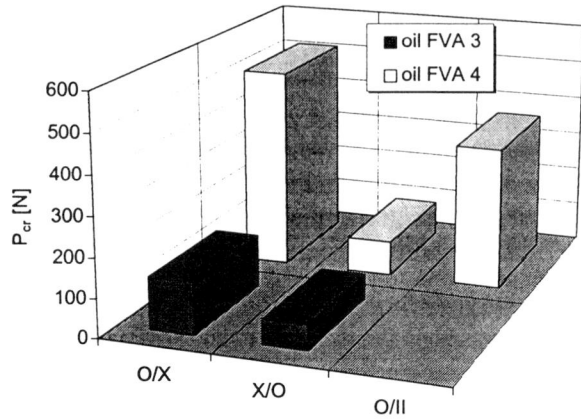

Fig 2: Critical loads for the contact of rough specimens with a nearly smooth ones

First symbol concerns the block and second - the roller. The roughness values are: O/X ΣR_{max}=1.51 μm; X/O ΣR_{max}=1.21 μm; O/II ΣR_{max}=1.61 μm. It can be seen that the critical contact load is higher if the rough surface is slower and lower if the smooth one is slower.

The experimental figures for critical contact load are in good agreement with numerical results for film thickness; the main parameter of the surface micro-geometry influencing the EHD film thickness is the sum of the peak-to-valley heights of co-operating elements. This effect is, however, moderate and elements with smooth surfaces do not have significantly greater load capability.

In additional analysis a mulitple regression model for the EHD film thickness parameter at 10% metal-to-metal contact has been obtained (3). These results are presented in the full version of this paper.

REFERENCES

(1) M Wiśniewski, M Szczerek, EHD lubrication in severe loaded contacts of rough elements, Problemy Eksploatacji, No 11, 1994, 288-291pp.

(2) W Piekoszewski, M Szczerek, M Wiśniewski, Conditions of the breaking of the EHD film between rough bodies (in Russian), 3 Int Symp SLAVYANTRIBO-3, 1995, Rybinsk, Russia.

(3) W Piekoszewski, PhD Thesis (in Polish), Poznań University of Technology, 1997.

THE THERMAL INFLUENCE FUNCTION METHOD AND ITS APPLICATIONS IN TEHD STUDIES

FANGHUI SHI and QIAN (JANE) WANG
Department of Mechanical Engineering Florida International University Miami, FL 33199 U.S.A.

ABSTRACT

Thermal analyses for a tribological problem include the temperature and thermal-deformation calculations. Jaeger's heat-source method for flash temperature and Boussinesq's integral method for elastic deformation are commonly used for contacts of surfaces whose structures can be simplified into half spaces. For tribological contacts that cannot be modeled by half-space problems or boundary conditions cannot be simplified, numerical procedures, such as FDM (finite difference method) and FEM (finite element method), are utilized for the temperature analysis. Because the temperature and thermal deformation calculations are intermediate steps in a lubrication analysis, combining FEM procedures into an iteration process may be considerably costly or even impractical. It is necessary to simplify the calculations process without losing the accuracy. The idea of the influence-function method for calculating the surface elastic deformations (Woodward and Paul, 1976) for finite structures is extended to develop an influence function method for the steady-state temperature analysis and thermal deformation calculation. Once numerically calculated, the thermal-influence-functions can be treated as input files in the process of computation. They will function in the same way in the temperature and thermal deformation analyses as the elastic influence function does in elastic deformation calculations.

INFLUENCE FUNCTION FOR TEMPERATURE, τ_t

For a tribological element subject to frictional heating over the contacting surface, the boundary conditions may involve the convection with environment, isolation, and prescribed temperatures. The method of variable separation is used by assuming $T(x,y,z)$ as a summation of T_c due to the non-homogeneous convection and T_q due to the heat flux and the homogeneous convections. Temperature T_c can be solved by a FEM formulation. Similarly, T_q can be solved in another variational formulation involving the heat flux on the contacting surface region. Considering a unit heat flux, q_0, acting over a differential area, ΔS_0, in the vicinity of a surface point, (x_a, y_b, z_c), the corresponding temperature represented by $\tau_t(x_i, y_j, z_k, x_a, y_b, z_c)$ may be expressed as follows:

$$\int_\Omega \left\{ k_x \frac{\partial \tau_t}{\partial x}\frac{\partial v}{\partial x} + k_y \frac{\partial \tau_t}{\partial y}\frac{\partial v}{\partial y} + k_z \frac{\partial \tau_t}{\partial z}\frac{\partial v}{\partial z} \right\} d\Omega + \sum_{i=0}^{m} \left\{ \int_{S_i} \beta_i \tau_t v ds_i \right\} = q_0(x_a, y_b, z_c)\Delta S_0$$

Therefore, temperature T_q due to frictional heat flux q can be obtained through the following summation:

$$T_q(x_i, y_j, z_k) = \sum_{a,b,c \subset S_0} \tau_t(x_i, y_j, z_k, x_a, y_b, z_c) q(x_a, y_b, z_c)\Delta S_0$$

where $\tau_t(x_i, y_j, z_k, x_a, y_b, z_c)$ is defined as the influence function for temperature. Thus the total temperature can be expressed in the following form:

$$T(x_i, y_j, z_k) = T_c(x_i, y_j, z_k) + T_q(x_i, y_j, z_k) =$$
$$= T_c(x_i, y_j, z_k) + \sum_{a,b,c \subset S_0} \tau_t(x_i, y_j, z_k, x_a, y_b, z_c) q(x_a, y_b, z_c)\Delta S_0$$

INFLUENCE FUNCTION FOR THERMAL DEFORMATION, $\vec{G}_T$

The thermal deformation caused by the temperature rise in a tribological element may be solved with the same procedure as mentioned in the previous section. Assuming a unit temperature rise, Δt_0, in a differential volume, $\Delta \Omega$, in the vicinity of (x_l, y_m, z_n), the corresponding deformations, $\{d_t = d_{tx}, d_{ty}, d_{tz}\}$ of a point, (x_i, y_j, z_k), can be solved by a FEM formulation. Because the major concern in a tribological analysis is on the surface, the thermal deformation of a surface point, (x_a, y_b, z_c), can be expressed as:

$$\{u_T\} = \begin{Bmatrix} u_{Tx}(x_a, y_b, z_c) \\ u_{Ty}(x_a, y_b, z_c) \\ u_{Tz}(x_a, y_b, z_c) \end{Bmatrix} = \begin{Bmatrix} \sum_\Omega d_{tx}(x_a, y_b, z_c, x_l, y_m, z_n)\Delta T(x_l, y_m, z_n) \\ \sum_\Omega d_{ty}(x_a, y_b, z_c, x_l, y_m, z_n)\Delta T(x_l, y_m, z_n) \\ \sum_\Omega d_{tz}(x_a, y_b, z_c, x_l, y_m, z_n)\Delta T(x_l, y_m, z_n) \end{Bmatrix}$$

The vector form of the influence function for thermal deformation can be defined as $\vec{G}_T$:

$$\{\vec{G}_T(x_i, y_j, z_k, x_l, y_m, z_n)\} = \begin{Bmatrix} d_{xt}(x_a, y_b, z_c, x_l, y_m, z_n) \\ d_{yt}(x_a, y_b, z_c, x_l, y_m, z_n) \\ d_{zt}(x_a, y_b, z_c, x_l, y_m, z_n) \end{Bmatrix}$$

Replacing the FEM procedures by summation operations considerably saves the computation time. A sample problem that could take about two hours for the full-TEHD process to finish on a Pentium computer used only a few minutes with the influence-function method on the same computer.

A MIXED-TEHD MODEL FOR JOURNAL-BEARING CONFORMAL CONTACTS, PART I: MODEL FORMULATION AND APPROXIMATION OF HEAT TRANSFER CONSIDERING ASPERITY CONTACT

Fanghui Shi and Qian (Jane) Wang
Department of Mechanical Engineering
Florida International University
Miami, FL 33199 U.S.A.
(305) 348-1973

ABSTRACT

Modeling of mixed lubrication is an important step that leads to the study of failure transitions for journal-bearing conformal contacts working under adverse conditions. Mixed lubrication is complicated in nature, and it is controlled by the structural deformation, lubricant and lubrication, and asperity contact. These controlling factors are all functions of temperature. The friction as a result of asperity contact in mixed lubrication is expected to produce a large amount of heat, which may considerably increase the temperature of the journal, the bearing, and the lubricant. The structure distortion due to the thermal-elastic deformation may interact with the change of the lubricant film, altering the characteristics of the asperity contact. A mixed-TEHD (thermal elastohydrodynamic) study is needed for a better understanding of the mixed lubrication and failure transitions.

Figure 1 presents a typical bearing-lubricant-journal system working at a large eccentricity ratio and subject to various thermal and elastic boundary conditions. The bearing surface is rough and the journal surface is assumed to be smooth, both of them are subject to thermal-elastic deformations.

A Mixed-TEHD model was developed for journal bearings working at large eccentricity ratios. It consists of a mixed-lubrication process which considers the roughness effect and asperity contact, a thermal process for temperature analysis, and a thermal-elastic process for the deformation calculation. The interactive journal, lubricant, and bearing were treated as an integrated system. The temperature distribution in this system was numerically analyzed and the comparison with the published experimental data suggested that the integrated system can yield a satisfactory accuracy in thermal analyses. The influence of the changes of lubricant flows as a result of asperity contact on the system heat transfer was numerically investigated. The results (Fig. 2) revealed that the roughness effects on both lubrication and heat transfer should be properly considered in order to obtain accurate thermal-tribological analyses for mixed-TEHD problems.

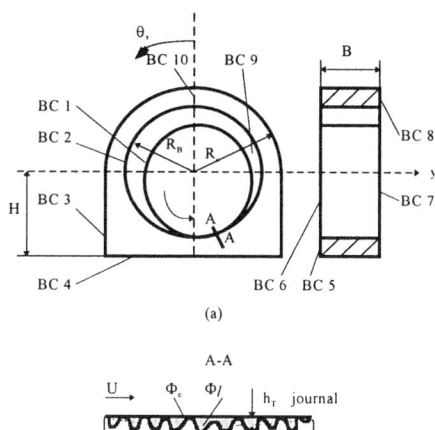
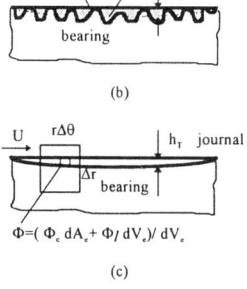

Figure 1: Description of the mixed-TEHD model. BC 1 through BC 10 are boundaries of the system.
(a) The bearing-lubricant-journal system.
(b) Asperity contact and heat generations.
(c) An equivalent flow model for heat transfer analysis.

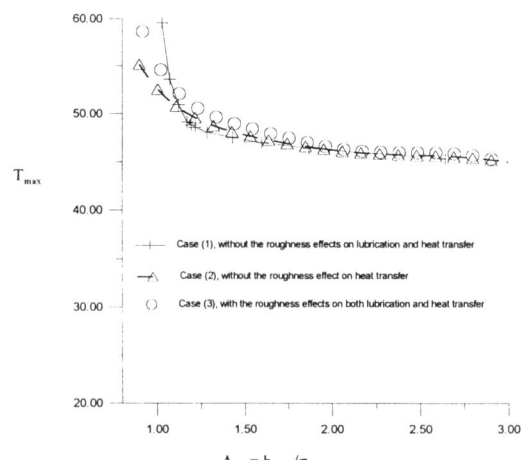

Figure 2: The maximum temperatures calculated with different considerations.

MULTILEVEL SOLUTION OF ELASTOHYDRODYNAMIC LUBRICATION OF HELICAL GEARS WITH TOOTH PROFILE MODIFICATION

LI WEI and MENG HUIRONG

Beijing Graduate School, China University of Mining and Technology, Beijing 100083, P.R.China

CHANG SHAN and CHEN CHENWEN

Department of Mechanical Engineering, Harbin Institute of Technology, Harbin 150001, P.R.China

ABSTRACT

The contact line on the surface of a helical gear tooth profile is inclined, when meshing in (or meshing out), the contact line is in (or out) at one end of tooth top or tooth root. The ideal modification method is the method which can make modification depth vary with contact line marching and the modification depth on a contact line is basically same, the meshing in end or meshing out end is the maximum modification depth, Reference (1) set up this modification method (shown in Fig.1) which received the effect of reducing vibration and noise.

Whichever modification method is all lack of EHL research. Specially as to helical gear, because the geometric parameters and kinematic parameters at various points on several instant contact lines after modification are different, it is difficult to give a convergent EHL nummerical calculation programme at every point which can calculate the lubrication parameters of load distribution, pressure distribution and oil film thickness.

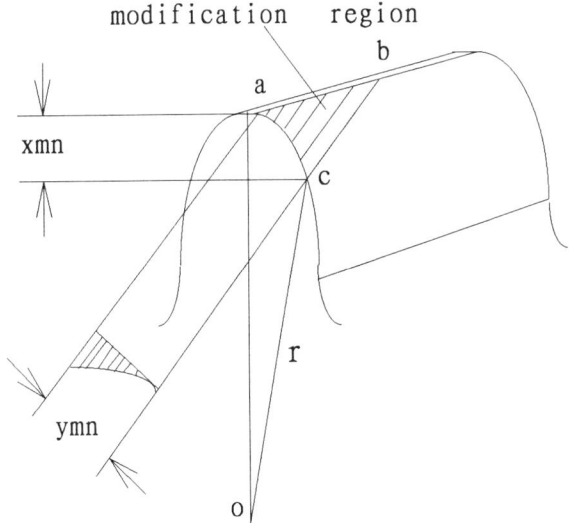

Fig.1: Geometric relation and modification region of helical gear

In the course of common EHL calculation (2), it is always that load information is known while pressure distribution and oil film shape is to be calculated. But as to helical gear transmission,. Once oil film appears, when the error is not taken into account, the oil film thickness at every point of a line is equal, that is to say:

$$H_0^i = C \qquad i = 1, 2, \cdots, n$$

In which, i is the number of dispersed point on the contact line of helical gear. The formula can be seen as the boundary condition of helical gear Reynolds equation.

Seen from above, in the course of solving helical gear EHL, the oil film thickness is known while the load information is unknown. So using common method, it is impossible to directly calculate all the pressure distribution and load distribution of helical gear on the contact line at a meshing instant.

This paper presents a new method for analyzing the lubrication performance of modification helical gears. The multilevel technique (3) is used to solve the simultaneous Reynolds, elasticity, energy, thermal interface temperature and load balance equations with real tooth geometry and kinematics conditions of helical gears. The pressure distribution, minimum film thickness and maximum temperature rise on every engagement point along instantaneous contact line before and after the modification have been investigated. The computed results show that adding appropriate modification does good to the lubrication of helical gears. The calculation results have also been compared with some experimental results with well conformity. The present analysis provided a method and program for estimating performance of EHD lubrication of modification helical gears. In addition, some special approaches are adopted to increase the computation speed and some new interesting conclusions have been obtained.

REFERENCES
(1) M..weck, G.Mauer, ASME Journal of Mechnaical Design, Vol.112, 1990
(2) D.Dowson and G.R.Higgson, Elasto-hydrodynamic Lubrication, Porgamon Press, 1966
(3) A.A.Lubrecht, Journal of Tribology, Trans ASME, Vol.108, 1986

ELASTOHYDRODYNAMIC ANALYSIS OF CONNECTING ROD DESIGNS FOR IMPROVED BEARING PERFORMANCE

H. Xu

Engineering Analysis Department, T&N Technology Ltd, Cawston House, Rugby, Warwickshire, CV22 7SA, UK.

ABSTRACT

Modern engines have been designed to operate at much higher speed and power output rate. Through an engine cycle, a connecting-rod bearing is expected to go through a complex sequence of structural distortion. In order to effectively understand its operational behaviour, it is essential that the elastohydrodynamic lubrication (EHL) theory is adopted. A tool based on such a theory enables us to achieve much more realistic predictions of the performance of engine bearings, and interactions between engine connecting rods and bearings[1,2]. The development and adoption of such a tool has greatly enhanced bearing manufacturers' knowledge of engine bearings.

A comprehensive study was carried out to investigate the influence of the big end housing designs upon the performance of connecting rod bearing. The purpose of the study was to improve bearing performance through modifying the structure of the bearing housing, and to derive an optimal design of the connecting rod big end structure. The study was carried out by employing a predictive too for the analysis of the elastohydrodynamic lubrication of dynamically loaded bearings. The engine bearing was considered as a part of the cylinder system, instead of an isolated independent part.

The study was based upon the connecting rod of a modern gasoline engine. The study involved a detailed study of the essential bearing performance parameters and the characteristics of the EHL solutions. Efforts were then made to identify the appropriate design changes to the structure of the connecting rod big end which then lead to improvements in these critical bearing performance parameters.

The study has shown that the performance of the connecting rod bearing was significantly affected by the stiffness of the supporting housing structure. The relative stiffness profile of the structure was also shown to be influential, which including the presence of the bolt holes, and the ways the clamping bolts were fitted.

It was observed that the minimum oil film thickness could be substantially changed by modifying the bearing housing structure. The changes caused to other performance parameters more modest. Nevertheless, a change in the overall bearing frictional power loss of between 5-10% was achieved, which is still considered to be significant.

Figure 1 shows the through minimum oil film thickness of the bearing fitted into a troubled connecting rod design. The dotted line represents the location of the minimum film measured in degree of bearing angles. There is a clear indication of the bearing suffers from the problem of surface touch down at the location between 100 to 120 degree bearing angle, as measured from the crown position of the rod half bearing. A connecting rod with the optimised big end structure can greatly improve the lubrication condition of the bearing, in which the through cycle minimum oil film thickness of the bearing is significantly increased.

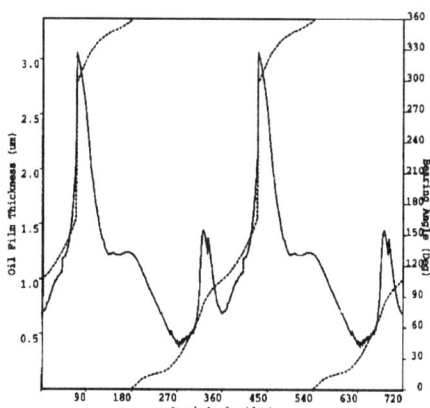

Figure 2. through cycle variation of the magnitude (μm) and location (crank angle degree) of the minimum oil film thickness as predicted with an optimal rod design

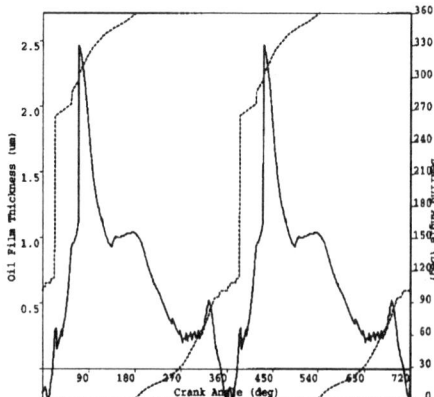

Figure 1. through cycle variation of the magnitude (μm) and location (crank angle degree) of the minimum oil film thickness as predicted with the original rod design

REFERENCES:

1. Goenka, P K and Oh, K P, `An Optimum Connecting Rod Design Study - A Lubrication Viewpoint', Trans ASME, J. Tribology, Vol 108, 1986, p487.
2. Xu, H, `Effects of EHD Contacts Upon the Bearing and Housing Behaviour', SAE Technical Paper 960987. 1996.

INFLUENCE OF SURFACE ROUGHNESS AND TEST METHOD ON SCUFFING

AKIRA YOSHIDA and MASAHIRO FUJII
Faculty of Engineering, Okayama University, 3-1-1, Tsushima-naka, Okayama, 700, Japan

ABSTRACT

The evaluation of scuffing has been performed by two cylinder test or four ball test. In these tests two kinds of test methods have been used. One is the step-load method, in which the load is increased stepwise at a constant velocity ratio, and the other is the step-speed method, in which the velocity ratio is increased stepwise under a constant load. The correlation between these two test methods are not clarified and the method to evaluate the scuffing are not defined. In this study the influence of surface roughness and test method was examined with test cylinders, which had a wide range surface roughnesses, by both the step-load method and the step-speed method. Two cylinder tests were performed under #68 turbine oil lubrication. The lubricating oil was controlled at 313K and was supplied at a rate of $100ml/min$. The test cylinders were made of bearing steel SUJ2. The dimensions of test cylinders were $40mm$ in diameter, $10mm$ in width and $60mm$ in crowning radius. The surface was ground and the surface roughnesses were about 0.05, 0.5 and $1 \mu m Ra$. The hardness was about $HV 800$. In case of the step-load method, the mean rolling speed of two cylinders was $1000 rpm$, the specific sliding σ_h of the faster cylinder was 50 to 90%, and the Hertzian pressure p_{max} was increased from $800\ MPa$ at a load step of $100MPa$. In case of the step-speed method, the Hertzian pressures were 1000, 1500, 2000, 2500 and $3000MPa$, the rotating speed n_1 of the slower cylinder was $200rpm$ or $600rpm$, and the rotating speed of the faster cylinder was increased at a speed step of $100rpm$.

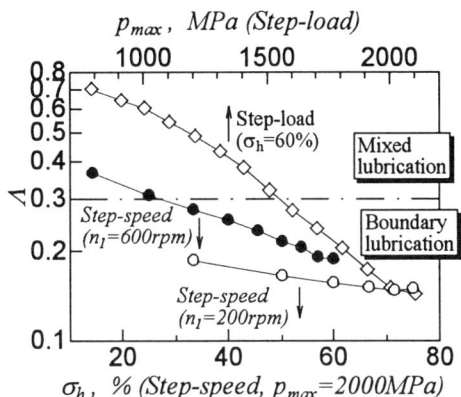

Fig. 2: Example of the variation of film parameter Λ ($\Sigma Ra = 1\mu m$)

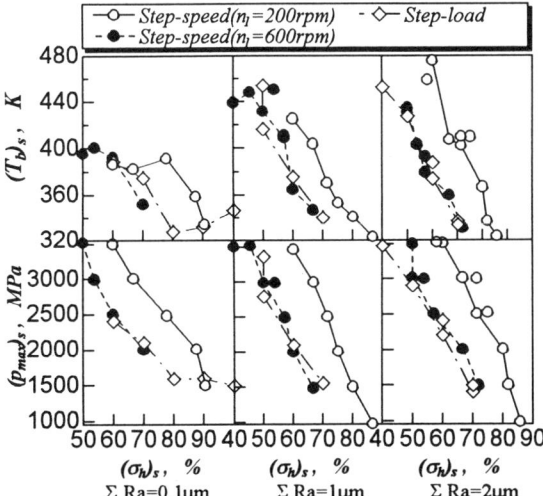

Fig. 1: Relation between $(\sigma_h)_S$, and $(p_{max})_S$ and bulk temperature $(T_b)_S$ on the outer surface of cylinder at scuffing occurrence

Figure 1 shows the relation between $(\sigma_h)_S$, $(p_{max})_S$ and the bulk temperature $(T_b)_S$ on the outer surface of cylinder at scuffing occurrence. ΣRa was the sum of the center line average transverse roughnesses of two cylinders before test. The smaller the surface roughness was, the higher the scuffing load and the specific sliding were. On the other hand, $(T_b)_S$ was higher in case of the larger surface roughness. The result by the step-load method was in good agreement with that by the step-speed method ($n_1=600rpm$), while the values of $(p_{max})_S$ and $(T_b)_S$ in the step-speed method ($n_1=200rpm$) were greater than those in the step-load method. The degrees of the formation of oxidative film and of the reduction of surface roughness depended on the test method and the initial surface roughness. Figure 2 shows an example of the variation of film parameter Λ. The frictional surface was in boundary lubrication state when $\Lambda<0.3$ (1). In case of the step-speed method ($n_1=200rpm$) the lubrication condition at the early stage of the test was the boundary lubrication, where the frequency of the asperity interaction was high, and then the surface roughness was reduced and the oxidative film was formed on the frictional surface. Thus the scuffing capacity might be overestimated by the step-speed method ($n_1=200rpm$). Similarly, in case of $\Sigma Ra=0.1\mu m$, T_b was low and Λ was large at the early stage, then the frequency of the asperity interaction was low and the oxidative film was formed little. Thus $(T_b)_S$ was lower than that in case of $\Sigma Ra=1$ and $2\mu m$ as shown in Fig.1.

REFERENCE
(1) K.Fujita and F.Obata, Bulletin of the JSME, Vol.25, No.205 (1982), p.1156.

A NUMERICAL ANALYSIS OF CRANKSHAFT BEARINGS IN MIXED LUBRICATION

CHAO ZHANG and ZHIMING ZHANG
Research Institute of Bearings, Shanghai University, Shanghai, 200072, P.R. China

ABSTRACT

The nominal minimum film thickness h_{min} in engine bearings is of the same order of magnitude as the surface roughness and it often becomes such thin that the bearings run in the mixed lubrication. The objective of this paper is to study the effects of two sided purely longitudinal, transverse and isotropic roughness on the engine bearings in the mixed lubrication, using Christensen's stochastic model of hydrodynamic lubrication of rough surfaces (1) and Greewood and Tripp's method for calculating the nominal contact pressure (2), and considering the running-in effect. Reynolds equation is solved by FDM with the Reynolds boundary conditions. The motion equations are integrated simultaneously using a fourth order Runge-Kutta numerical scheme.

The same roughness structure $\sigma_1 = \sigma_2$ (σ_1 and σ_2 denote standard deviations of roughness height distribution of bush and journal) has bigger h_{min} than the different structure and the structure effect is the biggest for the isotropic case and the smallest in the longitudinal case. The longitudinal and isotropic roughness give the biggest and smallest h_{min} respectively as shown in Fig. 1. As compared to the smooth bearing, when $\sigma_1 = \sigma_2 = 0.4\mu m$, h_{min} of three roughness textures are bigger; when $\sigma_1 = 0.7\mu m$ and $\sigma_2 = 0.1\mu m$, h_{min} is bigger only in the longitudinal case.

Roughness increases the cycle maximum film pressure and induces osillations of maximum film pressure in the contact zones as shown in Fig. 2. The degrees of the osillations are stronger for the same roughness structure than for the different structure and are strongest in the transverse case.

The total contact load is much smaller than the hydrodynamic load and the maximum real area of contact fraction is only 0.008%. The effects of contact load can be neglected. The roughness distribution of the running-in surface deviates from the Gaussian distribution and makes its asperity peak lower than the latter. This deviation affects the degree of contact.

The effects of roughness are closely tied up with the roughness textures and structures, the features of nominal geometry, journal mass, and operation conditions (3)(4).

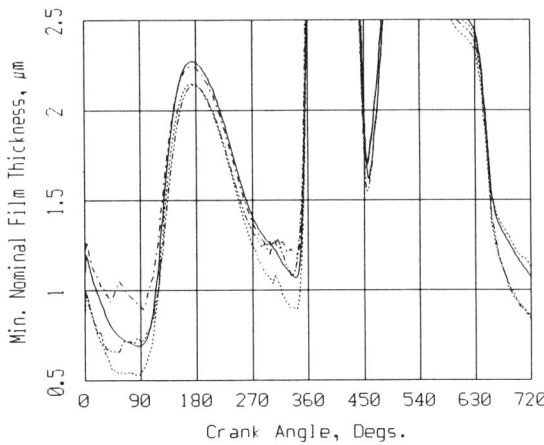

Fig. 1: Min. Film Thickness ($\sigma_1 = 0.7\mu m$, $\sigma_2 = 0.1\mu m$
———— Smooth, ------- Longitudinal,
············ Isotropic, —·—·— Transverse,)

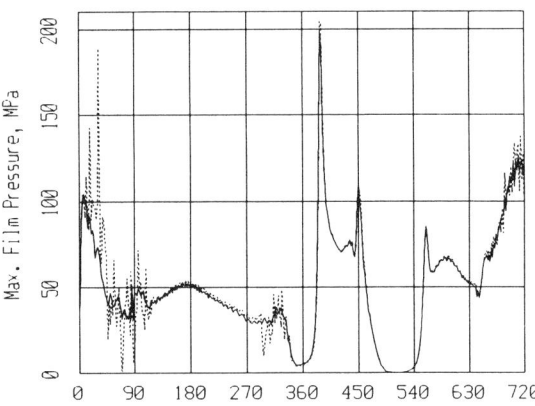

Fig. 2: Max. Film Pressure ($\sigma_1 = 0.7\mu m$, $\sigma_2 = 0.1\mu m$
———— Smooth, ············ Transverse, ,)

REFERENCES

(1) Christensen, H., Proc. Instn. Mech. Engrs., Vol. 184, Part 1, No. 55, pp. 1013-1026. (1969-1970).
(2) Greewood, J. A. and Tripp, J. H., Proc. Inst. Mech. Eng. (London), Vol. 185, pp625-633(1970-71).
(3) Chao Zhang and Zhiming Zhang, Proc. of the 2nd International Conference on Hydrodynamic Bearing-Rotor System Dynamics, Xi'an, China, 1997, pp.18-24.
(4) Chao Zhang and Zugan Qiu, to appear in STLE Tribology Transactions.

ABRASIVE ACTION OF OXIDED WEAR DEBRIS IN PIN ON DISK SLIDING FRICTION ON CARBURIZED STEEL

T. Balon
EdF, CNEN les Reacteurs, 2 Av. du General de Gaulle F-92141 Clamart, France
G. Zacharie
EdF, RNE/EMA, C.R des Renardieres, F-77250 Moret sur Loing, France
H. Darbeïda
Sedis R&D, 35 r. des Bas Trevois, F-10000 Troyes, France
J. von Stebut,
LSGS, URA CNRS 1402, INPL-Ecole des Mines F-54042 Nancy Cedex, France

ABSTRACT

Carburizing is commonly retained to enhance surface hardness of steel for friction and wear applications, in particular for gear boxes where the thickness of the treated layer may be more than 1mm.

The present study is focused on 17CrNiM06 gear steels with approximately 1mm thick carburized surface layers, some of which were oxidised in a final phase This conventional treatment is followed by the standard quenching + annealing procedure and final grinding, leading to specimens with varying surface hardness (6.6 GPa < Hv < 8.4 GPa) and toughness.

Micromechanical testing, proven very successful on nitrided 316L stainless steel (1), is retained to elucidate a possible contribution of brittleness to surface damage.

Surface topography modifications are monitored by means of scanning stylus profilometry (SSP) off-line after quasi static testing or on-line with a triboscopic technique (2), tracking friction and wear as a function of sliding distance and cycle number.

In "quasi-static loading" surface damage is studied by means of high load Vickers indentation and scratch testing. For the latter well as all friction experiments a $\varnothing$ 1.58 cemented carbide ball is selected as indenter/pin. Standard progressive loading and single pass, sclerometric operation under 10N normal load reveal the surface damage to be *ductile ploughing*, as opposed to nitrided layers where brittle, tensile type cracking had been observed (1).

In low cycle, friction fatigue, multipass scratch operation in standard laboratory environment (21°C, 50% RH.) after 4000 cycles of 4 mm in unidirectional dry sliding at 1 mm/s, under 10N normal load, the essentially ductile damage is confirmed; except for formation of negligible redeposition transfer layers, no brittle damage is observed in the wear track of the carburized surface.

In pin on disk operation, (except for sliding speed of roughly 5m/s) all contact parameters were chosen identical. This leads to substantial material transfer to the cemented carbide ball/slider with consecutive, preferential redeposition within and on either side of the gouged centre part of the wear track. Owing to energy dispersive X ray analysis (EDX) these redeposition transfer layers are found to be highly oxidised as expected to occur at the high sliding speeds retained in the present study.

The striking information in pin on disk operation comes from reflected light microscope inspection and quantitative SSP analysis of the cemented carbide wear pins with a hardness of roughly 1700 GPa :

Instead of an expected, essentially circular truncation of the cemented carbide ball, we find formation of an arch morphology, with practically no wear in the centre (constantly running in the hollow part of the wear track), while the side parts (in intermittant contact with the redeposition layers on both sides of the hollow part) are worn flat.

Detailed analysis leads to the following conclusions :

During virgin contact, the contact pressure is essentially hertzian. Over the prevailing contact width of the ball a primary transfer layer forms continuously, protecting this part of the ball from wear. As a consequence, owing to adhesive wear in the real contact area, highly work hardened debris are evacuated from the disk material towards both sides, thus giving rise to redeposition on the disk and 3 body abrasive wear of the cemented carbide ball. It can be conjectured that the carbide wear debris, once generated, will then join in and continue to feed the above wear mechanism.

ACKNOWLEDGEMENTS

This study is part of T. Balon's work for the engineering degree of the Conservatoire National des Arts et Métiers in Paris.

REFERENCES

(1) T. Roux, A. Darbeida, J. von Stebut, J.P. Lebrun, D. Hertz, H. Michel
34° Journées du Cercle d'Étude des Métaux, École des Mines de St. Étienne, 1995, paper 24,
Surf. and Coatings Technology **68/69** (1994) 582-591

(2) A. Darbeida, J. von Stebut, M. Assoul, and J. Mignot
J. Mech Tools Manufact. **35** (1995), 177

EFFECTS OF WALL MATERIAL AND BULK SOLID PROPERTIES ON CHOICE OF WALL MATERIAL FOR DESIGN OF HOPPERS AND SILOS

BRADLEY MSA, PICKERING J, BIRKS AH, PITTMAN AN
The Wolfson Centre for Bulk Solids Handling Technology
University of Greenwich
Wellington Street
London SE18 6PF

INTRODUCTION AND OBJECTIVE

Knowledge of the frictional characteristics between bulk solid and constraining wall is essential when designing equipment for handling bulk materials. Minimising this friction is economically desirable. The friction is known to be heavily dependent upon the finish on the steel; this research programme was intended to evaluate the less well-known effects of the mechanical properties of the underlying steel itself.

METHOD

Steels of different hardnesses, finished in the same way, were subjected to friction tests with the same bulk solid. Measurements were made with the bulk solid moving both parallel to and normal to the surface finishing scratches. Hardness and surface roughness were also measured on the steel samples.

MATERIALS AND FINISHES CHOSEN

A bulk solid of angular particles significantly harder than the steels was used, to isolate the effects of steel hardness. Different size ranges of particles were used in order to evaluate the effect of this variable (coarse, fine and wide particle size range). Five different steels were chosen, covering a 2:1 range of Vickers Hardness Number. All were finished in the same way, to a specification commonly used in the industry.

FINDINGS

* Surface roughness varied, Ra values (both along and across the finish marks) increasing with hardness.

* Friction correlated with hardness, reducing as hardness increased in spite of the increased roughness of the harder steel samples.

* Friction was higher with the fine bulk solid, lower with the coarse one, and in-between with the wide size range.

* The increase in friction when the bulk solid was moving across, rather than along, the finish marks, was interesting. With fine bulk solid, the increase showed a clear tendency to reduce with reducing hardness, whereas with the coarse bulk solid it increased with reducing hardness; for the wide size distribution, it was lower with intermediate hardnesses and higher at the extremes of hardness.

CONCLUSIONS

The different surface finishes were all sufficiently fine, with respect to the particle sizes involved, that variation in this did not of itself have an influence. The concept of a "threshold" surface finish, beyond which friction is no longer reduced, is supported.

Lower friction with increased steel hardness confirms that reduced depth of "ploughing" at the particle contact points more than offsets the increased resistance of the harder steel to the process of ploughing as the contact points move. Modelling has been undertaken to support this.

Higher friction with the smaller particle sizes shows that the greater number of smaller contact points result in a greater sum of ploughing forces, in spite of the smaller load at each point. Various possible reasons for this have been considered.

Increased friction in sliding across the surface finishing marks, compared with parallel to them, is hardly surprising, but the differences in the increase with different combinations of bulk solid and wall material is extremely interesting. This has given strong clues about the relative significance of the processes which contribute to the friction.

Harder steels are worth considering economically purely for the reduction in friction. The finishing process to be used, and the Ra value achieved, needs to be considered carefully as clearly the harder steels take a rougher surface when finished in the same way. The concept of the "threshold" surface finish should be borne in mind. Higher friction with the small particles shows that friction should be measured using the smaller fraction of a sample where the bulk solid could be variable or subject to segregation in handling.

MODELLING THE PARTICLE DYNAMICS IN A ROTATING DISC ACCELERATOR EROSION TESTER

A J BURNETT, R J FARNISH and A R REED
The Wolfson Centre for Bulk Solids Handling Technology, University of Greenwich,
Wellington St., Woolwich, London, SE 18 6PF, UK

ABSTRACT

This paper presents the result of a programme of work designed to derive a more efficient model for predicting the dynamics of the abrasive particles within a rotating disc accelerator erosion tester.

Detailed measurements were taken of particle velocity, particle concentration and particle jet divergence for the erosion tester in use at The University of Greenwich (1,2,3). From these results an improved model was developed to model the particle dynamics in such testers. All known models are compared with some results recently obtained from the use of this model and erosion tester.

Some unexpected results obtained from the rotating disc type of tester will be explained in the context of a detailed analysis of the particle dynamics in such a device. Methods for improving our understanding of the operation of such testers will be discussed.

This paper has been submitted to Wear.

INTRODUCTION

Rotating disc accelerator erosion testers have been used over many years to determine the wear properties of materials in laboratories (4,5,6). However, only a small amount of detailed work has been carried out into observing the detailed operation of such testers (1,2,3,5). An overview of the operation of such testers is overdue especially considering the previously poorly observed importance of the intensity of particle impacts per unit area of target on the erosion rate that occurs (2,3,7,8).

TEST EQUIPMENT AND MEASUREMENTS

The rotating disc accelerator used to provide the results described in this paper is similar to that originally proposed by Kleis and Söderberg (4,5). A rotating disc is used to accelerate particles to the desired velocity. Ten target holders with edge protection are equi-spaced around the disc. Particles are fed into the rotating disc using a vibratory feeder.

Measurement of the extent of particle jet divergence and particle velocity were carried out using proven techniques.

THE PARTICLE DYNAMICS MODEL

The new model for the particle dynamics for the rotating disc accelerator erosion tester is presented. Initially the model provides a result for the particle velocity vector at exit from the rotating disc. This is achieved by using a finite differencing model based on the forces acting on the particles in the channels in the disc. Particle impact angles on the targets are determined using methods proposed by Söderberg (5). Particle jet divergence is accounted for based upon the use of an empirical model determined from test results.

This new model is compared with earlier models.

DISCUSSION

It has been found that this form of tester does not yield results that show the peak in erosion rate versus angle of particle impact that is commonly expected. This fact is concluded to be due to the relative absence of inter-particulate collisions in the region of the wear scar for low angles of impact when compared to the situation commonly seen in other non-rotating forms of erosion tester. This is due to the dynamic nature of the jet of particles in the rotating disc tester.

The variation of the intensity of particle impacts per unit area and its effect on the erosion of the targets in a rotating type test rig are also discussed.

Improvements in the new particle dynamics model are discussed with detailed reference to results obtained from the erosion tester for the erosion of a series of mild steel targets.

REFERENCES

(1) A.J. Burnett, S.R. De Silva, A.R. Reed, Wear, Vol. 186-187, 1995, 168-178pp.
(2) A.J. Burnett, PhD Thesis, The University of Greenwich, London, UK, 1996.
(3) A.J. Burnett, M.S. Bingley and M.S.A. Bradley, Pneumatic and Hydraulic Conveying Conference, Palm Coast, Florida, USA, April 1996, (Engineering Foundation, New York, USA), publication expected in Powder Technology in 1997.
(4) I.R. Kleis, H.H Uuemois, L.A. Uksti, T.A. Pappel, Int. Conf. on Wear of Materials, Dearborn, Michigan, April 1979, ASTM, 212-218pp.
(5) S. Söderberg, S. Hogmark, U. Engman, H. Swahn, Tribology International, Vol. 14, No. 6, 1981, 333-343pp.
(6) A. Shimizu, Y. Yagi, H. Yoshida, T. Yokomine, Journal of Nuclear Science and Technology, Vol. 30, No. 9, 1993, 881-889pp.
(7) P.H. Shipway, I.M. Hutchings, Wear, Vol. 174, 1994, 169-175pp.
(8) P. Chevallier, A.B. Vannes, Wear, Vol. 184, 1995, 87-91pp.

CATHODE WEAR IN ALUMINIUM ELECTROLYSIS

XIAN-AN LIAO and HARALD A. ØYE

Institute of Inorganic Chemistry, The Norwegian University
of Science and Technology, N-7034 Trondheim, Norway

ABSTRACT

Carbon cathode wear is caused by a combined process of physical wear and chemical corrosion. A room temperature test (1)(2) has been developed for the study of the physical wear. The main results are:

1. The wear increases with alumina content, velocity and pressure. Their effects can be described with power equations:

$$W_R = kx^n$$

The exponent n depends on the experimental conditions. Alumina content has the largest effect ($n \geq 7$), velocity intermediate ($n=2-4$) and pressure lowest ($n=0.5-1.3$).

2. Relative abrasion resistance

CS graphite	100
Ordinary graphite	110–140
Semigraphitized carbon	90–110
Semigraphitic carbon	60–90
Anthracitic carbon	10–30

The chemical corrosion is caused by aluminium carbide formation and dissolution or by direct electrochemical dissolution of carbon.

The reaction between carbon and aluminium

$$4Al(l) + 3C(s) = Al_4C_3(s)$$

is thermodynamically favoured at all temperatures. The standard Gibbs' energy change for the reaction is -147 kJ at 970 °C. Aluminium carbide is found dissolved in the cryolitic melt and aluminium, but the amount formed is much less than the quantities suggested by the favourable Gibbs' energy of formation for the reaction. Usually, contact between liquid aluminium and carbon does not give appreciable carbide formation below 1000 °C. The reaction is enhanced by the presence of cryolitic melts, but the amount is still very small in a real aluminium electrolysis cell considering such a large negative Gibbs' energy change.

A test equipment shown in Fig.1 was designed to study the wear of carbon cathode materials in aluminium electrolysis at 960-1035 °C. The main results are:

1. Slight wear occurs in molten aluminium.
2. Due to chemical corrosion the wear in a mixture of cryolitic melt-aluminium is much larger than in aluminium only. The wear increases linearly with velocity. The velocity dependence of the chemical corrosion is much less than that of the physical abrasion. The typical wear rate of graphite at $V=0.55$ m/s is about 26 cm/year.
3. Addition of alumina to a cryolitic melt-aluminium mixture strongly suppresses the chemical corrosion. The typical wear rate of graphite at $V=0.55$ m/s, excess alumina content=0.70 g/ml is about 7.2 cm/year.
4. Polarization increases the wear rate. The typical wear rate of graphite in an alumina-saturated electrolyte at CCD=0.21 A/cm^2, $V=0.55$ m/s and T=985 °C is about 52 cm/year. The wear is decreased to about 3.5 cm/year when excess alumina content is increased from zero to 0.90 g/ml.
5. Graphite and anthracitic carbon have approximately the same resistance to the chemical corrosion due to carbide formation and dissolution.
6. In a cryolitic melt-alumina slurry without aluminium and without polarization, the wear of amorphous carbon is about one fourth to one fifth of CS graphite which is similar to the results for the room temperature test.

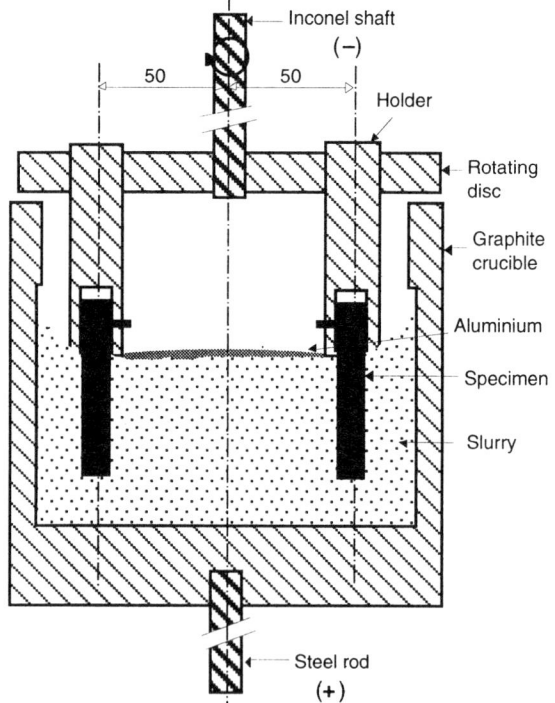

Fig. 1: Set-up for carbon cathode wear study at high temperatures

REFERENCES

(1) X. Liao, H. Øye, Carbon, Vol.34, No.7, pp649-661, 1996.
(2) X. Liao, H. Øye, Tribologia, Vol.15, No.3, pp3-34, 1996.

NEW MANGANESE AUSTENITIC WEAR RESISTANT STEELS

A. MAGNEE,
University of Liège; Dept Metallurgy and Materials Science, 2 rue A. Stévart, 4000 - LIEGE (B)

ABSTRACT

A little more than a century ago, R.A. Hadfield's research resulted in a very special type of manganese steel, whose usual composition as well as thermal treatment have remained unchanged since (Carbon content between 1 and 1.45 %, an Mn/C ratio of 10 and cooling by water). (1)

The low production cost and simple preparation of manganese steel still ensure a wide range of uses for Hadfield steel, especially in erosion-resistant moulded parts (crusher plates, hammers).

Numerous attempts at modifying the original composition were made or claimed in patents that have remained largely unexploited, due to the limited improvements in performance obtained. Current compositions have remained virtually unchanged since the implementation of the ASTM-128 norm in 1964.

A systematic study of the Fe-C-Mn-X system has been undertaken and research has been conducted in two areas : a percentage increase in volume of carbides and a modification of the stability characteristics of the matrix. Resistance to erosion has been studied in some 90 experimental alloys by adding elements (X) acting either on the austenitic matrix (Ni, Cu, Co, Mn) or on the forming of carbides with different stoichiometries (Cr, Mo, W, Ti, Nb, V).

The range of chemical composition is as follows: 0,4 < C < 1,8 %; 1 < Mn < 16 %; 0 < X < 10 %.

In terms of theoretical analysis, the modelization of the hardening effect of the new alloys has been developed (2).

The kinetics of the martensitic transformation of metastable austenites is described by an auto-catalytic asymptotic model which does not take into account the transformed martensitic volume fraction V but rather the transformed volume fraction compared to volume fraction at saturation V_S. The relationship is given by :

$$V = \frac{V_S}{1+\frac{\varepsilon^B}{A}}$$

A is a parameter defined by temperature and B is a characteristic parameter of the auto-catalytic reaction related to the chemical composition. The greater the value of A, the more rapidly saturation is reached, and for the same value of A, the lower the value of B, the more unstable the material. ε is the true deformation of the steel under impact.

The increased resistance of the new alloys to erosion damage results from an optimalized combination associating classical strain hardening with a hardening reinforced by austenite-martensite phase transformation under shock.

In terms of industrial applications, several alloys are now being used successfully (figure 1) : they combine real toughness with an erosion resistance is twice as high as that of Hadfield steel (3, 4).

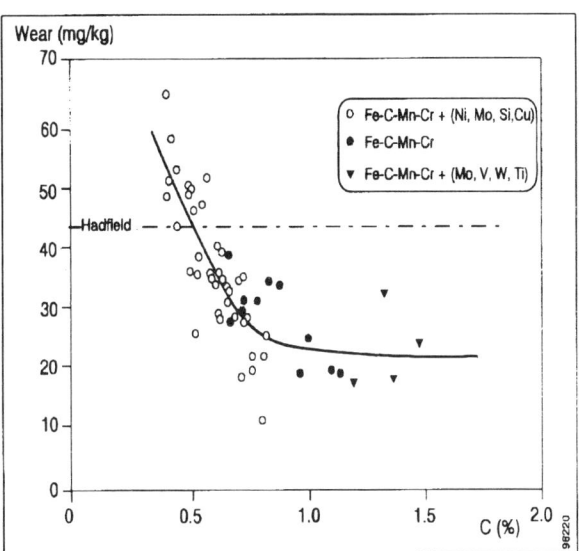

Fig.1: Erosion of Fe-C-Mn-X steels as a function of their carbon content (impact angle = 30° ; velocity = 75 m/s ; abrasive : MgO).

REFERENCES

(1) K.J. IRVINE, F.B. PICKERING, Austenitic manganese steel : effect of heat treatment. J.I.S.I. 1956, pp. 135-139.

(2) INTERNATIONAL MANGANESE INSTITUTE, High carbon manganese austenitic steels. Monography, 1995, p. 26.

(3) PRATIQUE DES MATERIAUX INDUSTRIELS, chap. 9, Matériaux résistant à l'érosion, Dunod-Paris, ISBN2- 100201689, Déc. 1996.

(4) A. MAGNEE, Generalized law of erosion : application to various alloys and intermetallics. Wear, 1995, 181-183, pp. 500-510.

THE INFLUENCE OF BULK PROPERTIES AND MICROSTRUCTURE ON THE SOLID PARTICLE EROSION OF STEEL

D J O'FLYNN AND A J BURNETT
The Wolfson Centre for Bulk Solids Handling Technology, The University of Greenwich, London SE18 6PF
M S BINGLEY
The University of Greenwich, Wellington St., Woolwich, London SE18 6PF

ABSTRACT

The effect of the bulk properties on the erosion resistance of steel was determined for three impact angles at three impact velocities. Two steels were chosen for testing due to their suitability for heat treatment. These were treated into four different microstructures, selected to provide a range of different mechanical properties.

The microstructures tested were Pearlite and Spherodised Iron Carbide, from a 0.8 % carbon steel, EN42, and Pearlite and Upper Bainite, from an alloy steel, En24.

Erosion testing was conducted using a 'rotating disc accelerator type' erosion tester, based on the design by Söderburg et al. (1). Results were typical of ductile metals, with maximum erosion occurring at low angle impact, shown in Fig. 1. With safety in mind, the erodent selected for this test programme was a fine olivine sand, traded under the name 'Renova fine'.

Hardness, tensile, fatigue and impact tests were conducted to measure the bulk properties of each steel. These were conducted at a range of temperatures from room temperature to 400°C, as temperature increases may be of significance to erosive wear.

Scanning electron microscopy, (SEM) was used to examine eroded surfaces and micro-hardness readings were taken through cross section of the wear scar to determine any changes in sub-surface properties due to erosion.

These SEM photographs of the eroded surfaces showed a marked difference between wear at low angle impact to wear at normal impact, implying a different mechanism of material removal for these extreme conditions. Intermediate angles of impact resulted in wear scars exhibiting both mechanisms of erosion. At a low angle of impact, a considerable amount of deformation was observed with significant 'smearing' of the surface. At normal impact, evidence of a 'platelet' mechanism of erosion existed. Between different microstructures there was little to suggest different mechanisms, only varying amounts of wear.

From Vickers micro-hardness readings taken within the wear scars it was found that each material displayed a degree of strain hardening at the worn surface, with the greatest increase in hardness being observed for those specimens eroded at normal impact, shown in fig. 2. The influence of velocity on this increase in hardness was seen to be so small that it was deemed to have no influence, though the depth of this hardened layer was dependent on the impact velocity. It is considered that the material properties of this work hardened layer might be of more significance to the erosion resistance of steel than the properties of the bulk material. A test programme to measure the properties of these layers is currently in progress.

Relationships between material properties and erosion rates are discussed in this paper, although no simple relationship between individual bulk properties and erosion resistance was found.

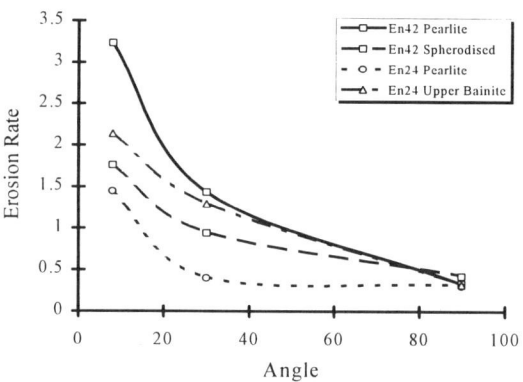

Fig 1: Erosion rate with angle for 15 m/s impact velocity

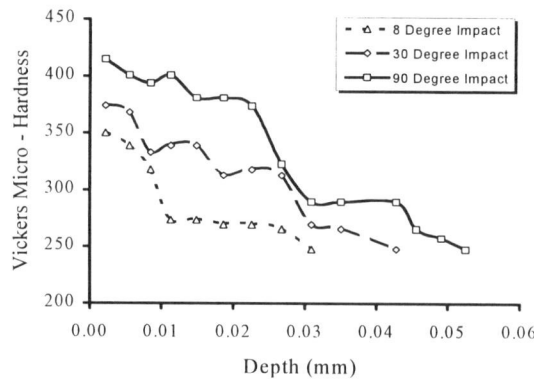

Fig. 2: Hardness with depth in worn surface, for upper bainite.

REFERENCES

(1) S. Söderburg, S. Hogmark, U. Engman, H. Swahn, Tribology International, Dec. 1981, 333-343pp.

CAVITATION EROSION OF METAL SURFACES OF COOLING SYSTEM OF DIESELS AND ROLLING BEARING FATIGUE LIFE IN DIFFERENT WATERBASE FLUIDS

DR. LJUDMILA SHABALINSKAYA
Podmoskovny Branch of Scientific Research Tractor Institute, Chekhov of Moscow Region, Russia

ABSTRACT

Mineral oils provide the overwhelming majority of hydraulic fluids. However, there is a need for effective fire-resistant fluids in working environment where fire is a risk. Underground mining and hydraulic equipment of foundries are ones of the areas where fire-resistant fluids are often used.

For demanding hydraulic applications glycol fluids and HFC-fluids are used. They have the advantage of being suitable for use at lower temperatures than others (-30°C to -50°C). With correctly selected performance additives such fluids can be used as antifreezes for a diesel. That is why there are two aspects, on which to focus — mechanical performance and system compatibility. Such fluids examples are Shell Irus Fluid C (1) and VLP-Fluids (2). The important aspect of mechanical performance includes rolling bearing fatigue life, steel-corrosion-fatigue durability and cavitation erosia.

The Unisteel rig is used to evaluate Shell Irus Fluid C performance in this respect and statistical analysis of the results enable the L10 and L50 lives of rolling bearing to be estimated (Method IP305).

The Corrosion-fatigue rig is used to evaluate VLP-20 Fluid's performance in the same respect compared to mineral oils MGE-46V and MGE-25T on Steel lamella specimens, immersed in the environment of the fluid (or mineral oil) and have been under a life circle load as shown in Table 1.

Condition of tests		Life in environment, circle, 10^{-4}		
Specimen's metal	Frequency, circle, 10^{-4}/hour	VLP-20	MGE-46V	MGE-25T
Steel 08KP	3	24	30,8	22,8
Steel 08HGT	8,2	3,75	4,12	3,48

Table 1.

The fractures of specimens after the test were estimated by electron microscope scanning (SEM). These positive results finally served as an allowance for an actual pump test.

The final test of the VLP-20 Fluid compared to mineral oil MGE-46V was carried out using hydraulic system with axial Piston pumps of Zauer type GST-90 during 500 hours, increasing the load to 2000 s^{-1}, temperature to 100°C, pressure to 35 MPa. Results of this test served as the allowance into operation.

If the user wants to employ a waterbase fluids as an antifreeze, it is especially necessary to focus on the problem of the cavitation fracture, because it is working under vibration and in the aggressive environment at the same time.

This problem can be solved ultrasonically at magnetostrictive vibrator (3). The degree of the protection from the cavitation erosia (C,%) of metal surfaces with this method can be determined through the two final formulas:

$$V = \frac{G}{\eta \cdot \rho \cdot S \cdot t} \quad \text{and} \quad C = \frac{V_{water} - V_{antifreeze}}{V_{water}} \cdot 100\%$$

where V - speed of the cavitation erosia, $mm/hour$; S - square of irradiator of the vibrator, mm^2; ρ - density of specimen, mg/mm^3; t - time of test, $hour$; η - coefficient (0,10); G - mass loss, mg.

Using this method for the estimation of the influence of waterbase fluids on the cavitation erosia of different metal surfaces it is possible to clarify conditions of the operation of the equipment. For example, with the results, submitted in Table 2, the user of Euclid R 170-dumper gains the possibility of selecting an antifreeze providing the best protection from the cavitation erosia.

Specimen's metal	V [C] in environment					
	VLP-1 [1]	VLP-10 [1]	Lena-65 [2]	Lena-65+DCA4 [3]	Lena 65+Lub-recool [3]	Water
Aluminum (AL9)	75 [67]	106 [53]	224 [0]	45 [80]	90 [60]	224 [1]
Steel 38HMU	12,8 [67]	18,1 [53]	15,5 [60]	7,7 [80]	15,5 [60]	39 [1]
Iron Cummins-KTA-50C-bush)	14,4 [70]	20,3 [58]	17,6 [60]	8,7 [65]	17,6 [65]	48,7 [1]

[1] - Propylenglycol fluids; [2] - Ethylenglycol fluids; [3] - Coolant additives.

Table 2.

In other cases in order to reduce the possibility of cavitation in hydraulic equipment, inlet system should avoid negative pressures by having adequate and unrestricted pipelines and by siting the pump to give a full fluid head.

REFERENCES

(1) S.Hirst, Waterbase Fluids, Seminar by Shell about lubricants, Moscow, 1990.
(2) L.Shabalinskaya, V.Kelbas, Construction And Road Building Machinery, No.10, page 35, 1996.
(3) G. Stativkin, V.Yanchelenko, Patent SU 1538100 A1 [51] 5 G 01 N 3/56.

Submitted to *Synthetic Lubrication*

EROSIVE WEAR OF DREDGE PUMPS

JAN SUCHÁNEK
Deptartment of Heat Treatment and Tribology, SVÚM a.s., Areál VÚ, P.O.Box 17, 190 11 Praha 9, CZ
MILAN VOCEL
TRIBO-S Praha, Na Kuthence 16, 160 00 Praha 6, CZ

ABSTRAKT

Machine service life and their service efficiency are affected by their wear caused most often by treated medium, impurities of environment or only by interaction of functional surfaces in relative motion, remarkably. In accordance with our long-term knowledge, wear of machines and hence their service life are influenced by 3 main groups of reason:
- machine concept and design including a material choice
- production quality
- treated medium and machine service conditions.

Every of the above mentioned groups can be estimated to participate in machine wear of 1/3, roughly, even when significant deviations affected by a given medium, service or other reason occur in individual cases. Typical case in which acted all these reasons was extreme erosive wear of dredge pumps used in hydraulic transport.

One of the limiting factors of hydraulic transport is low service life of some parts of whole system, dredge pumps, especially. Low service life of dredge pumps has an influence on whole system performance and is insufficient from economical point of view (1). At Oslavany coal-fired plant, service life of impellers (50-100 service hours) and further exposed pump parts was very low during service. Pumps ran at the 50 MW block dredge station where granulated slag was hydraulic transported on a slag dump at a distance of 600 m from a slag tap unit burning black coal with about 50% ash content. The angular granulated slag was very brittle and able to be cleaved into smaller particles with very sharp edges. These slag particles caused very intensive erosive wear of the leading edge and end impeller blades, packing rings and dredge pump casing. The main reason of local erosive wear on packing rings was an inaccurate assembly in pump repairs.

The impellers were originally cast of austenite manganese steel which is used under impact abrasive wear conditions, conventionally. However, particles of transported material slip along the surface or impact into the surface of exposed parts at small angle. Therefore, the dominant mechanism of material damage is microcutting. The surface layers of a part aren't dynamically affected, intensively, so that plastic deformation and phase transformation of manganese austenite to martensitic-type ε-phase associated with a remarkable increase of hardness and resistance against wear by particles don't occur. Erosive resistance of an austenite manganese steel is, therefore, small, at the level of medium carbon hardened and low-tempered construction steels, roughly (2,3). Laboratory tests of the group of selected wear resistant cast steels and white cast irons in a water-sand mixture by the use of an apparatus with a abrasive vessel made to select more suitable materials for exposed dredge pump parts possible It is true, Oslavany power station was succeeded in an increse of impeller service life by the order of 4-5 by the use of Cr12Mo1-type white cast irons, however, economy of hydraulic transport was still unacceptable for high repair costs of dredge pumps.

The final solution of the problem was found by a change of the fly-ash handling system concept. Hydraulic transport of ashes from combustion chamber space was replaced by redler conveyers and further by removing on a slag dump by means of track haulage. Repair costs were reduced, significantly, and designed parameters of power station block were gained, successfully. Analysis showed the designer of energetic block hadn't taken into account specific conditions of burning process and the character of black coal with a high portion of ballast matters. Design of dredge pump and materials used for its functional parts were also unsuitable. During their maintenance, some pump part incorrect assembly participated in their low service life,too. Solution of mentioned cases of low service life of dredge pumps using in ash matter hydraulic transport at power stations showed the principal requirement for ensuring long-term and reliability function of transport of abrasive particles and was whole project processing with regard to the character of transported material, required system output parameters and necessary service life of critical parts.

REFERENCES

(1) M Vocel,J Polách, Slévárenství, Vol.XIV, No.8, p. 326 1966.
(2) Suchánek, J. et all: Strojírenství, Vol. 30, No.5, p. 294, 1980.
(3) M Vocel, A Šustek, Strojírenství, Vol. 29, No. 7, p. 429,1979.

ABRASIVE WEAR BEHAVIOUR OF SOME LOW ALLOY STEELS

B.D.TRIPATHI, M.L.NARULA, V.S.DWIVEDI & S. JHA

Research & Development Centre For Iron & Steel
Steel Authority of India Limited
RANCHI-834002 (INDIA)

ABSTRACT

The most effective way of enhancing the wear resistance of a component is by improving its hardness[1] in combination of other mechanical properties. The industrial use of stainless steel as cold sinter screen in sintering plant is due to low hardness and good punchability but this is a poor wear resistant steel and very expensive too. This led to investigate a substitute[2]. The resistance to dry sliding wear behavior was evaluated for two high strength low alloy steels i.e.1Cr-0.3Mo steel and 0.5Cr-0.2Mo-0.7Ni steel both containing Cu,V,& B as alloying elements, using a pin-on-disk machine at different sliding distances, loads and speeds. The wear pins were made of above said alloy steels plates while the disk was made of CW-4 material which had hardness of Rc63. The wear pins were rubbed against the disk and the weight changes were measured as a function of sliding distances. Before evaluating the sliding wear properties, plates were austenitised at 1133^0 K for 45 min. and quenched in oil followed by tempering at different temperatures. The tempering behavior of the alloys revealed that there was no appreciable amount of change in hardness upto 573^0K but above this, softening was observed due to microstructural changes and the hardness decreased.

In present experiment, the data were collected for 1255m travelled and at the interval of 125.5m. The plots between weight loss and sliding distances revealed a linear relationship. It was also observed that the slopes between weight loss and sliding distances for pins were affected by the heat treatment conditions. The wear rate was minimum for pins in as-quenched condition. A relative wear and hardness for mild steel, stainless steel alongwith low alloy steels are shown in fig-1. It was also found that due to frictional heating the wear pins ,in as-quenched condition, are tempered after certain number of sliding distances and thus the wear rates are changed. The study, made on the effect of loads on wear behavior, revealed that 1Cr-0.3 Mo steel was more resistance to sliding wear than 0.5Cr-0.2Mo-0.7Ni steel. The increase in temperatures was found to be linear with loads at constant speed. It was also observed that as the speed was increased the wear loss first deceased and then increased. The abraded surface and debris were observed in optical microscope as well as in scanning electron microscope. The debris was found to be consisting of shining iron particles and iron oxides. The oxide was analyzed with help of EPMA.

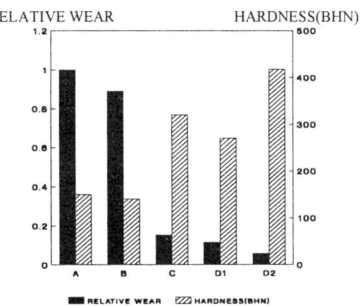

A-MILD STEEL   B-STAINLESS STEEL-304
C-.5Cr-.2Mo-.7Ni STEEL   D-1Cr-.3Mo STEEL
FIG-1 RELATIVE WEAR VsHARDNESS OF STEELS

REFERENCES

1) D.A.Rigney, Wear, 175(1994)63
2) B.D.Tripathi, Development of wear resistant steel for cold sinter screen, Report NO.R&D:22:02:2424:02:95 1996, RDCIS, SAIL, Ranchi

MIRROR WEAR SURFACE STRUCTURE OF ALUMINA CERAMICS IN UNLUBRICATED SLIDING AT HIGH TEMPERATURE

KOSHI ADACHI and KOJI KATO
School of Mechanical Engineering, Tohoku university, Sendai 980-77, Japan

ABSTRACT

It is well known that limit of roughness of smooth surface of ceramics is determined by pores appeared on the surface (1).

In this paper, possibility of surface finishing method by using sliding contact under high temperature for pore-free mirror surface are shown.

In unlubricated sliding of alumina against itself, relatively smooth surface compared with grain size are obtained in the region of relatively low load, low temperature (<300°C) and relatively high temperature (>800°C). Especially, the smooth surface obtained after sliding at temperature of 800-900°C and contact pressure of 2-3GPa can be called "mirror surface" whose maximum surface roughness is below 100nm of limit value defined for mirror surface. The condition for these two types of surfaces are summarized in Fig. 1 as a function of normal load and temperature.

From detailed observations of smooth surfaces, it can be clear that they are formed by two processes such as micro-wear of asperity peaks (head of original grains) and filling in surface hollows (like valleys, pore etc.) by fine wear particles (Fig. 2). Furthermore, it becomes clear that mirror surface can be obtained when fine particles are sintered at the surface hollows (pores) during friction as shown in Fig. 3.

In order to prove the agglomeration and sintering of fine particles under applying friction, behavior of fine particles under friction and/or high static contact pressure at 900°C are observed with two model experiments.

On the basis of the model experiments, it can be confirmed that fine particles are sintered under high pressure and high temperature of more than 2GPa and 900°C, respectively. And the surface hardness of agglomerated particles under 3Gpa at 900°C show high value of 15GPa as a results of sintering.

Finally, frictional conditions which can give mirror surface are proposed by analyses from both view points of fracture mechanics(2) and sintering of fine particles, and agrees well with experimental results.

REFERENCES

(1) J. Watanabe, Precision Machinings for Ceramic Materials, Journal of the Japan Society of Precision Engineering, 54, 7, 1988, 1227-1230.
(2) K. Adachi, K. Kato and N. Chen, Wear Map of Ceramics, Wear, 203-204, 1997, 291-301.

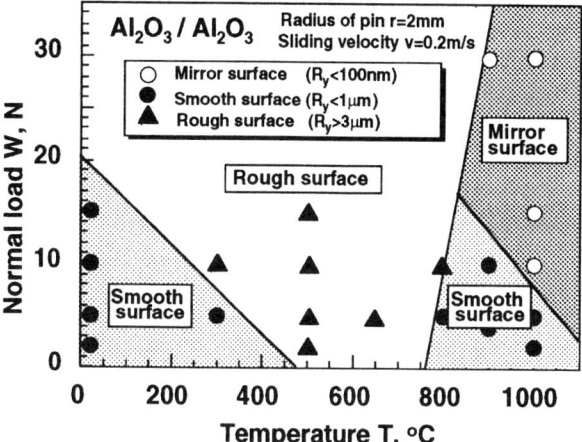

Fig. 1 : Region of smooth surface and mirror surface obtained after sliding contact.

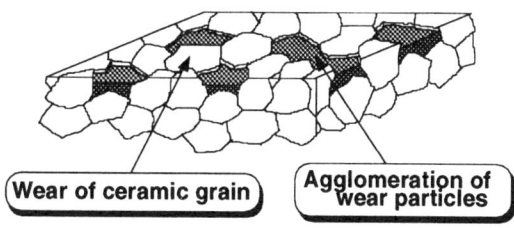

Fig. 2 : Schematic diagram of general appearance of smooth surface.

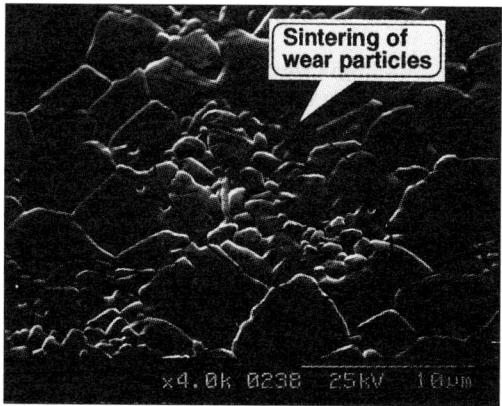

Fig. 3 : SEM image of thermally etched mirror surface obtained after sliding under 30N of normal load, 0.2m/s of sliding velocity and 900°C of temperature.

SURFACE TEMPERATURE AND WEAR MODE OF SILICON NITRIDE IN DRY ROLLING-SLIDING CONTACT

MAKOTO AKAZAWA

Department of Mechanical Engineering, Miyagi National College of Technology, Natori-shi, 981-12, Japan

ABSTRACT

In the dry rolling or rolling-sliding contact of silicon nitride, there are four types of wear mode with the magnitude of the contact pressure and the slip ratio introduced in the rolling. The first mode is the mild oxidative wear mode with filmy silicon oxide wear particles and this mode appears at the low contact pressure or at the low slip ratio. The second is the severe oxidative wear mode with the initiation of the micro cracks and wear particles are flaky silicon oxide. The third is the mechanical wear mode with fine crystalline silicon nitride wear particles. And the fourth is the severe mechanical wear mode with large silicon nitride wear particles. Fig. 1 shows the wear mode diagram at 1000 rolling cycles(4).

It is known that silicon nitride would oxidize under normal sliding conditions by tribochemical reaction with H2O (1). In dry rolling and rolling-sliding contact, the oxidative wear particles were observed as mentioned above(2)(3).

In the wear process of silicon nitride, surface oxidation and initiation of cracks are important.

The objective of this study is the measurement of the contact surface temperature which is closely related to tribochemical reaction.

Wear tests were carried out a ring-on ring wear tester at room temperature and in air at 50 % relative humidity. The rotational speed of the lower specimen was 800 rev/min and the rotational speed of the upper specimen was reduced from 800 to 560 rev/min (slip ratio was from 0% to 30%). The maximum hertzian pressure was varied from 0.25 Gpa to 1.5 Gpa. Ring specimens of silicon nitride (including 10 wt% additives; SiO2, MgO and Al2O3). The diameter of the specimen was 32mm and the thickness were 8mm for the lower specimen and 6mm for the upper specimen. The surface temperature was measured by using an infrared image thermograph.

Fig. 2 shows the surface temperature after 3200 rolling cycles. The surface temperature increased 1to 127.7° C at the low contact pressure and at the high slip ratio but at the low slip ratio temperature rise was only 13° C at the highest pressure. From these results the main cause of the temperature rise was the slip and this may accelerate the wear of silicon nitride .

The temperature of the region of oxidized surface (or oxidized wear particles attached on the wear track) was higher than the substrate.

The temperature of the lower specimen was always higher than the upper specimen. The higher temperature region was observed on the wear tracks caused by the geometry of the specimen.

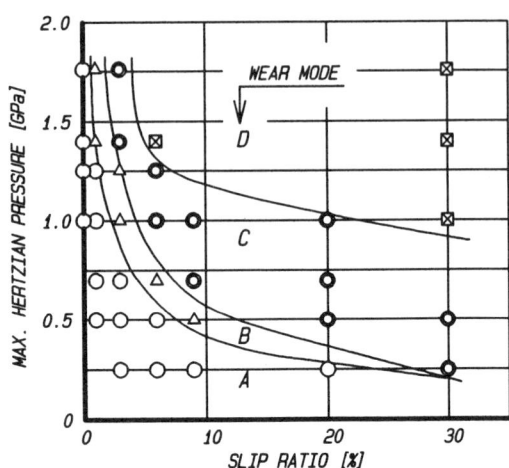

Fig.1: Wear mode diagram of silicon nitride in dry rolling - sliding contact

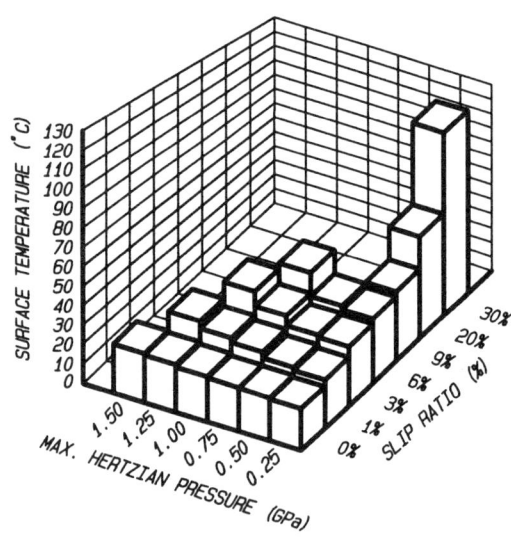

Fig. 2: Surface temperature after 3200 rolling cycles

REFFERENCES

(1) H. Tomizawa, T.E.Fischer, Wear, 105, 1985, page 21.
(2) M.Akazawa, K.Kato,K.Umeya, Wear,110,1986, page 285.
(3) M.Akazawa, K.Kato,Wear,124,1988,page123.
(4) M.Akazawa, Proc. of the 5th EUROTRIB, Vol.3,1989,page 126.

LUBRICATION OF NON-METALLIC MATERIALS

HYUN-SOO HONG
The Lubrizol Corporation, Wickliffe, OH 44092, U.S.A.

ABSTRACT

The use of ceramics in tribological systems has increased in recent years due to their high temperature strength and stability, light weight, chemical inertness, and high rigidity (1-4). Silicon nitride (Si_3N_4), silicon carbide (SiC), and Zirconia (ZrO_2) have been evaluated for tribological applications. Among these ceramics, Si_3N_4 is considered as a leading candidate for the tribological components which require lubrication.

A recent study (5) showed that molybdenum dithiocarbamate (MoDTC) performed better than ZDP in metal/ceramic contact. Oil containing MoDTC provided lower specific wear rates and friction coefficients for a Si_3N_4/cast iron pair than for a 52100 steel/cast iron pair, while, oil containing ZDP showed increased specific wear rates for the Si_3N_4/cast iron pair than for the 52100 steel/cast iron pair. The formation of a thick MoS_2 film on the contact area was observed on the cast iron specimen tested against Si_3N_4. Friction coefficients also decreased when tested with MoDTC compared to ZDP.

In this study, the effect of calcium overbased sulfonate detergent and ashless succinimide dispersant on the antiwear and antifriction performance of MoDTC and ZDP was investigated in Si_3N_4/cast iron and cast iron/cast iron contacts.

Hot-pressed Si_3N_4 (3.25 g/cc, 2000 kg/mm2) and cast iron were used as a plate specimen and cast iron was used as a pin specimen. A mineral oil containing 0.3 %w sulfur was used as a base oil and 1.0 %w of secondary zinc dialkyldithiophosphate (ZDP) or molybdenum dithiocarbamate (MoDTC) were added as an antiwear additive. 1.0 %w of calcium overbased sulfonate and 4.0 %w of ashless succinimide were added as detergent and dispersant.

The addition of calcium overbased sulfonate or ashless succinimide to MoDTC or ZDP resulted in higher specific wear rates for the cast iron/cast iron pair compared to cast iron/cast iron tested with MoDTC and ZDP alone. The addition of calcium overbased sulfonate and ashless succinimide to MoDTC or ZDP showed lower specific wear rates than those of the cast iron/cast iron pair tested with MoDTC or ZDP containing calcium overbased sulfonate and ashless succinimide.

For cast iron/Si_3N_4 pairs, the addition of calcium overbased sulfonate and/or ashless succinimide to ZDP increased specific wear rates compared to those tested with ZDP alone. However, the addition of calcium overbased sulfonate to MoDTC did not increase the specific wear rates, and the addition of calcium overbased sulfonate and ashless succinimide to MoDTC in cast iron/Si_3N_4 pairs increased specific wear rates slightly.

The performance of MoDTC with detergent and/or dispersant was better than ZDP with detergent and/or dispersant. However, it was observed that there is a definite time delay to reach steady state friction coefficients was observed for cast iron tested against Si_3N_4 with oils containing MoDTC and calcium overbased sulfonate and ashless succinimide.

To understand the time dealy to reach steady state friction coefficients, additional tests were done with an oil containing MoDTC and calcium overbased sulfonate. The tests were stopped before and after the transition and Auger Electron Spectroscopy (AES) analysis was done on the tested specimens. The AES analysis showed that the surface is covered with high concentrations of oxygen, calcium, and sulfur before the transition. Molybdenum was not observed before the transition. However, after the transition, the presence of a layer with a thickness of about $250°A$ containin the high concentration of sulfur and molybdenum was observed. This is followed by layers containing calcium, oxygen, and sulfur which was observed before the transition. The AES study clearly shows that the calcium overbased sulfonate goes to the surface first and delays the formation of MoS_2 film.

REFERENCES

(1) Advanced Ceramic Materials - Technological and Economical Assessment, Noyes Publication, Park Ridge, New Jersey, 1985.
(2) A Review of the State of the Art and Projected Technology of Low Hear Rejection Engines, National Academy Press, Washington, D.C., 1987.
(3) Wayne, S.F. and Buljan, S.T., SAE paper #880676, 1988.
(4) Habeeb, J.J., Blahey, A.G., and Rogers, W.N., I. Mech.E., 1987, pp. 555.
(5) H. Hong,, Lub. Eng., 50, 1994, pp. 616 - 622.

TRANSFER OF A THERMAL FLUX BY SURFACES COVERED WITH THIN Al₂O₃ LAYERS

WITOLD JORDAN, ANDRZEJ MRUK, JAN TALER and BOHDAN WEGLOWSKI
Cracow University of Technology, Al.Jana Pawla II 37, 31-864 Krakow, PL

ABSTRACT

The objective of this paper is an experimental analysis of heat exchange through surfaces covered with thin Al₂O₃ ceramic layers.

Ceramic layers 0.65 - 0.85 mm thick were plasma coated on the samples made of AlSi alloy. Examined coatings porosity was 9-12 % and it was an open one.

The evaluation of heat exchange through surfaces covered with thin ceramic layers has been done by determining the heat transfer coefficient equivalent α_{eq} and the heat resistance of the coating R_p. To determine these values, the tested sample was heated up from the head surface by constant heat flux q. The side surface of the sample was thermally insulated and its back coated surface was cooled. Heat exchange on this surface proceeded by convection.

During examination on a test stand temperatures were measured by a data acquisition system, the frequency of the sampling being 3 s (1, 2). The heat flux density, the heat transfer coefficient equivalent and the ceramic layer heat resistance were calculated on-line, by formula:

$$\alpha_{eq} = \frac{q}{T_{sc} - T_p} = \left(\frac{1}{\alpha_k} + R_p\right)^{-1}$$

The values of the heat transfer coefficient equivalent for 3 samples with ceramic layer (α_1, α_2, α_3) and without layer (α_4) are drawn up in Fig.1. As it can be seen, these values for samples 1, 2, 3 with ceramic layer are higher as compared with the sample clean 4. These results are surprising.

It can be explained by the intensifying effect of the layer surface roughness on the convection heat exchange. Additionally, layer open porosity influences surface extension like fins on the flat surfaces of heat exchangers. Moreover, the increase of the heat transfer coefficient equivalent value is affected by heat exchange from radiation. In the case of the sample with a smooth, clean surface (aluminium alloy) its emissivity is very low as compared with the rough surface of ceramic layer.

On the basis of test results it can be stated that, in the case of thin-walled ceramic coatings deposited on metal surfaces, thermal conductivity resistance through such layers is not the only factor determining heat exchange through such complex systems. Other important factors affecting heat exchange are: geometrical microstructure characteristics of the surface (roughness) and of the layer (porosity), since they determine both heat transfer coefficient and ceramic coating emissivity. They can considerably intensify heat transfer. It can prove that, in spite of the low thermal conductivity of such layers, heat transfer coefficient equivalent taking into account convection heat exchange coefficient and conventional thermal resistance of the layer, can be much higher than on a surface clean.

Taking into consideration the above, it is not possible to define heat flux taken through surfaces covered with such layers, when knowing only their thickness and heat conductivity value. In such conditions heat exchange can be estimated only experimentally.

Heat transfer coefficient equivalent [W/m2K]

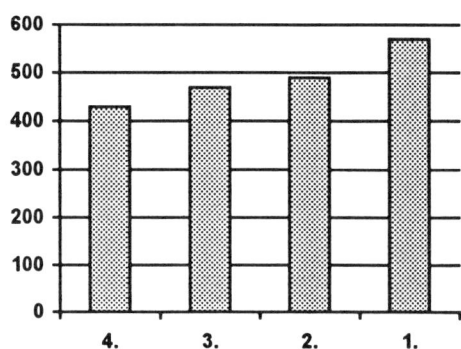

Fig.1. Calculated values heat transfer coefficient equivalent for the sample with a ceramic layer (1, 2, 3) and without layer (4).
q = 4800 [W/m²], T_p = 294 K

NOMENCLATURE

T_{sc} - temperature of wall surface at the place of contact with ceramic layer
T_p - cooling air temperature
α_k - convection heat transfer coefficient

REFERENCES

(1) A.Mruk, W.Jordan, J.Taler, St.Lopata, B.Weglowski, SAE Technical Paper No. 941779.
(2) W.Jordan, St.Lopata, A.Mruk, J.Taler, B.Weglowski, ZEM, No.2, 1994, 213 - 226 pp.

TRIBOLOGICAL CHRACTERISTICS OF Cr_3C_2 BASED CERAMICS IN HIGH-TEMPERATURE AND HIGH-PRESSURE WATER

Satoshi KITAOKA, Kazumi KASHIWAGI and Yoshimi YAMAGUCHI
Japan Fine Ceramics Center, 2-4-1 Mutsuno, Atsuta-ku, Nagoya, 456, Japan
Yasuo IMAMURA and Yutaka HIRASHIMA
Denki Kagaku Kogyo Co.,Ltd., 1 Shinkai-cho, Omuta, 836, Japan

ABSTRACT

Advanced ceramics have been attracting great attention for engineering applications in nuclear power plants under conditions of high-temperature and high-pressure water. Cr_3C_2 based materials, particularly, are candidates for tribological materials because of their excellent corrosion resistance at high temperature. This study, as part of a series of investigations on the tribology of advanced ceramics, reports the effects of temperature and sliding speed on the tribological characteristics of monolithic Cr_3C_2 (CRC), Cr_3C_2-TiC (CRT) and Cr_3C_2-Ni (CRNF) composites in self-mated sliding in deoxygenated water at room temperature, 120° and 300°C, and discusses the friction and wear mechanisms.

At the testing temperatures up to 120°C, the total wear amounts of both plate and disk specimens for CRC and CRT increased with rising the testing temperature and decreased with increasing the sliding speed. At 300°C, however, the total wear amounts of CRC and CRT at the lower sliding speed were smaller than that at 120°C and increased with increasing the sliding speed. The total wear amounts of CRNF significantly increased with increasing the testing temperature and sliding speed, especially much larger than those of CRC and CRT at the higher sliding speeds.

Fig.1 shows each wear component contribution to the wear volume of CRC relating to the wear mechanisms. The wear behavior of CRC was mainly controlled by tribological oxidation to produce H_2, CO_2 and various hydrocarbon gases etc. in similar to that of CRT. The wear depression of CRC and CRT at the lower sliding speed in water at 300°C are probably the reason why the partially hydrated oxide particles, which fully covered the worn surfaces, prevent direct contact between the plate and disk, resulting in reducing the magnitude of the local contact stresses. The variation in the sliding speed dependence of the wear amounts of CRC and CRT at the testing temperatures between below 120° and 300°C may be related to the reduction of the hydrated particles, which acts as a lubricant, by ready dehydration due to rising frictional surface temperature with increasing the sliding speed in water at 300°C compared with below 120°C. The wear of CRNF is governed by microfracture, where adhesive wear is considered to be accelerated by decreasing the plastic flow pressure of nickel region due to rising the frictional surface temperature and abrasive wear of this region also seems to be progressed by the Cr_3C_2 wear debris.

This R & D project was performed under the sponsorship of the Advanced Nuclear Equipment Research Institute (ANERI) who made the research contract with the Agency of National Resources and Energy, Ministry of International Trade and Industry.

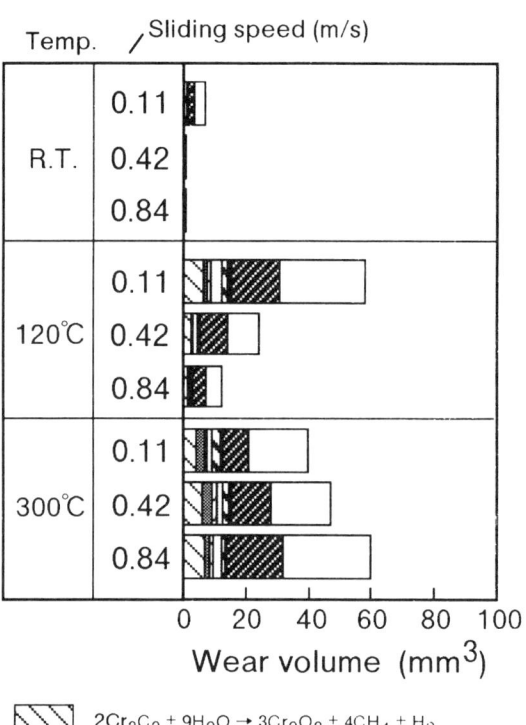

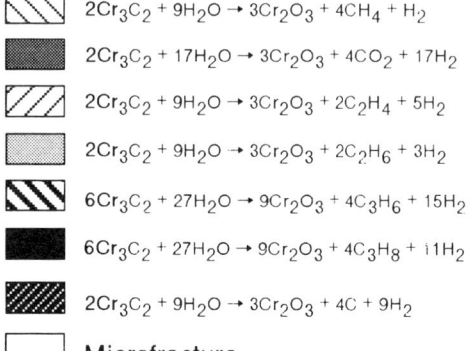

$2Cr_3C_2 + 9H_2O \rightarrow 3Cr_2O_3 + 4CH_4 + H_2$

$2Cr_3C_2 + 17H_2O \rightarrow 3Cr_2O_3 + 4CO_2 + 17H_2$

$2Cr_3C_2 + 9H_2O \rightarrow 3Cr_2O_3 + 2C_2H_4 + 5H_2$

$2Cr_3C_2 + 9H_2O \rightarrow 3Cr_2O_3 + 2C_2H_6 + 3H_2$

$6Cr_3C_2 + 27H_2O \rightarrow 9Cr_2O_3 + 4C_3H_6 + 15H_2$

$6Cr_3C_2 + 27H_2O \rightarrow 9Cr_2O_3 + 4C_3H_8 + 11H_2$

$2Cr_3C_2 + 9H_2O \rightarrow 3Cr_2O_3 + 4C + 9H_2$

Microfracture

Fig.1 : Total wear volume and each wear components as a function of the sliding speed for CRC.

WEAR MODES OF VC-NiBSi SINTERS ASSESSED BY X-RAY RESIDUAL STRESS ANALYSIS

B. LAVELLE
URA-CNRS-445, INP-ENSCT, 118 Route de Narbonne, 31077 Toulouse Cedex, France
I. C. GRIGORESCU
Department of Materials Technology, INEVEP, S. A., P.O. Box 76343, Caracas 1070A, Venezuela
J. LIRA
Department of Materials Science, Universidad Simón Bolivar, Sartenejas de Baruta, Caracas, Venezuela.

ABSTRACT

Wear induced residual stresses (RS) were analysed on the worn surface of VC-NiBSi composites by X ray diffraction. Tensile and compressive RS states were revealed and related to wear modes.

INTRODUCTION

X ray RS analyse by the standard "$\sin^2 \Psi$" method allows the detection of tribologically induced compressive RS. Tensile RS are also expected when wear occurs by surface micro-fracture, however this kind of stress require more complex mathematical treatments to be revealed (1). In the present work, a simple experimental comparison between bi-axial bending stresses and wear induced RS made both tensile and compressive RS states evident.

EXPERIMENTAL

Composites of 15 vol % VC in a 2.0Si, 1.5B, 0.2Fe, 0.01Mn, 0.03Cr, bal. Ni matrix were prepared by mechanically mixing the precursor powders, uniaxial pressing at 10 ton cm^{-2} and heating at 1050 °C during 5 min. under argon flow. Pin-on-disk tests were performed by using a low contact pressure set-up, where the pin had a square contact face of 4 x 4 mm and it was made of VC-NiBSi composite. Two materials were employed for the disk samples: quenched and tempered 100C6 steel (62 HRC) and alumina-zirconia in as-sintered condition. Wear tests were performed under the following conditions: air and Mobilgear 626 oil as environment, 10 cm/s sliding speed, 1000 to 20000 m sliding distance, 5 to 50 N normal load. A Cu source and a four circle goniometer in a Ψ set-up were used for X ray measurements. RS evaluation procedure was described elsewhere (2). It took into account the dispersion modes of $\Delta\Theta$ values for four selected observation Ψ angles: $\pm 30°$: $\pm 45°$, where Θ is the diffraction angle on the [331] plane of Ni grains. The bi-axial applied load (tensile or compressive) that brought RS close to zero was identified as representative for the RS state induced by wear.

RESULTS

Tensile RS state were systematically associated with material transfer and back-transfer between mating surfaces; this wear mode developed during the un-lubricated contact between VC-NiBSi pins and disks made of either 100C6 steel or alumina-zirconia and it could be detected by scanning electron microscopy (SEM) and energy dispersive X ray spectroscopy (EDS). The equivalent tensile bi-axial load was in the range of 44 - 62 MPa. Compressive RS states reached equivalent loads of about 200 MPa and they were found on wear scars produced under lubrication or by un-lubricated running-in condition. These surfaces showed mild morphological modifications and any possible changes in their chemistry were inferior to EDS detection limit.

REFERENCES

(1) J.W. Ho, C. Noyan, J. B. Cohen, *Wear*, 84 (1983), 183-202 pp.

(2) B. Lavelle, I. C. Grigorescu, J. Lira, *Proc. Conf. EUROMAT '95*, Vol. G-H 379-382 pp.

The Influence of Toughness on the Wear of Y-TZP Couples

E. LILLEY*, C. VROMAN†, and M. R. PASCUCCI†
* Saint-Gobain/Norton Industrial Ceramics
NRDC, Goddard Rd., Northboro, MA 01532 USA
† Worcester Polytechnic Institute
100 Institute Rd., Worcester, MA 01609 USA

Yttria stabilized zirconia is an important engineering ceramic. It is used in applications which range from artificial hip joints to dies used in metal forming. In wear applications the question arises, how transformable should the zirconia be? If the material is highly transformable, then the K_{IC} will be high and it will strongly resist the propagation of cracks. However, the friction forces can cause a surface transformation from monoclinic to tetragonal. If this surface transformation is extensive, e.g., 20% monoclinic, then the transformational stresses will cause cracking. Alternatively, if the Y-TZP ceramic is only slightly transformable with a low K_{IC}, then it will be much less effective in resisting the propagation of cracks, but it will be much less likely to transform to monoclinic on the wear surface.

The toughness of Y-TZP materials can be most strongly varied by changing the yttria concentration. At 2 mole% K_{IC} values as high as 20 MPa m$^{1/2}$ have been found, however, such materials are not used in commercial applications, because these are considered to be too unstable. At 3 mole% yttria, the K_{IC} is about 5 MPa m$^{1/2}$. Almost all the engineering zirconias have this composition. The toughness can also be varied slightly by changing the grain size and more significantly by phase partitioning which alters the compositions of the material grains producing low yttria grains with about 2 mole % yttria and high yttria grains of about 8 mole% yttria. In this study we have made both 2 and 3 mole% yttria ceramics. We have also phase partitioned the 3 mole% material to produce a range of K_{IC} values. All the ceramics were HIP'ped.

A pin-on-disk study of fine grain zirconia balls on a zirconia disk was performed under dry conditions in air. The test conditions were as follows: Speed: 5mm/sec, Loads: 9, 18, and 27N, Distance: up to 100m, Ball Diameter: 3/8", Ball Surface Roughness: (R_A): less than 10 nm, Track Diameter: ≈40mm.

Materials: K_{IC} (MPa m$^{1/2}$)

Yttria Mole %	Sinter + HIP 1350°C	Additional 2 Hours, 1450°C	Additional 15 Hours, 1450°C
2	15.1	-	-
3	5.2	5.4	6.6

The wear tests were interrupted at intervals, the surfaces cleaned off and the wear scar on the ball measured with optical microscopy. The test was then continued using the same ball in the same position. The wear on the plate was very slight, at the nanometer level.

The wear volume in mm^3 was measured and plotted as a function of sliding distance for different loads and different K_{IC} Y-TZPs. An example is shown for 3 mole % yttria having a K_{IC} of 5.4 MPa m$^{1/2}$. The slopes of the linear, steady state portions of these graphs were taken to determine the wear coefficient k which is widely used in the literature for comparing wear data and is defined as:

$$k = \frac{Vol\ (mm^3)}{Load\ (N)\ Distance\ (m)}$$

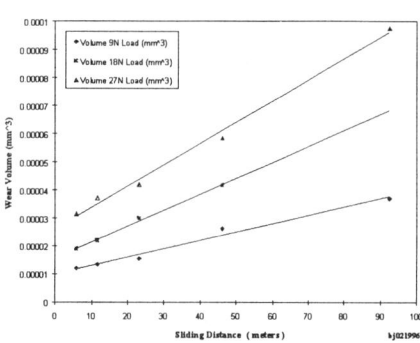

The wear coefficient results for different materials are presented in the table below tested with a 27N load.

Wear Coefficient as a Function of Toughness

K_{IC} (MPa m$^{1/2}$)			
5.2	5.4	6.6	15.1
7 x 10^{-8}	3 x 10^{-8}	2 x 10^{-8}	4.8 x 10^{-8}

They range from k = 2 to 7 mm10^{-8}/ Nm.

We conclude that under these wear conditions the toughness has no strong effect on the wear rate. This is opposite to the earlier study by Fischer et al.(1)

References

(1) Fischer, T.E., Anderson, M.P., and Jahanmir, S., *J. Amer. Ceram. Soc.*, 72, 252-257, (1989).

IMPROVEMENT OF TRIBOLOGICAL PROPERTIES OF C/C COMPOSITES AT HIGH TEMPERATURES

AKIHIKO MATSUI and AKINORI YASUTAKE

Nagasaki R&D Center, Mitsubishi Heavy Industries Ltd., 5-717-1 Fukahori-machi, Nagasaki 851-03, JAPAN

ABSTRACT

Carbon/Carbon (C/C) composites offer a high specific strength and an excellent heat resistance and are attractive materials for applications involving a high temperature and motion as disk brakes of aircraft and high speed train. Effects of sliding conditions on the friction and wear properties of C/C composites have been reported(1),(2). However, the effects of sliding conditions as atmospheric elements and temperature on the friction and wear properties remains unclarified. It has been reported that the frictional stabilities of carbons were largely associated with the adsorption of water vapor(3), the transition temperature to high friction and wear increased with partial pressure of the vapor and, for a series of n-paraffins, also increased with molecular size(4). However, the transition temperature is at most 400 ℃ and the control of atmosphere is difficult from the viewpoint of their applications. Therefore, the frictional stabilities until high temperatures through improvement of the materials are required.

In this study, dry friction tests of C/C composites up to near 1000℃ were performed to evaluate the effects of the temperature relating to evaporation of adsorbed water vapor and oxidation of carbon on the friction and wear properties. And observations of friction surfaces, cross sections and the state of formed wear particles were made, which allowed us to compare their friction and wear properties. Additionally, the tests of C/C composites incorporated with additives, hydrocarbons of variable carbon-number and water soluble inorganic compounds of variable melting point, were performed. Friction tests were performed under surface contact dry sliding conditions and run under constant conditions where the normal pressure was 0.4MPa and the speed of sliding was 0.7m/s in all cases.

Firstly, the tests were run under air of variable the atmospheric humidity and vacuum conditions at room temperature (RT.). The friction and wear increased with decreasing the humidity in air, and much more increased under vacuum conditions. Then, the tests with increasing temperature at the rate of 10℃/min. and those at constant temperatures by infrared lamp-heating were performed. As shown in Fig.1, the friction coefficient of C/C composites suddenly increased at about 200℃. Although the sliding surface at RT. was flat and smooth, that of 200℃ was rough and covered with wear particles. In order to increase the transition temperature to high friction and wear, the friction tests of C/C composites made from chopped fibers incorporated additives, that were hydrocarbons of variable carbon-number and water soluble inorganic compounds of variable melting point, were performed. C/C composites were impregnated with liquid hydrocarbons or aqueous solution of inorganic compounds, then were dried under vacuum condition. Fig.2 shows that the transition temperature to high friction increased with increasing the heat-resistant temperature of hydrocarbons and inorganic compounds, which were plotted as boiling point and melting point, respectively. From the results of Mass Spectrometry, it is considered that the friction and wear properties of C/C composites are improved by the increasing of the adsorbing capacity of water vapor and the decreasing of the oxidation temperature.

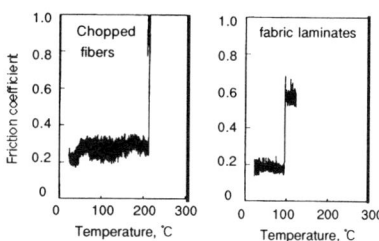

Fig.1: Friction coefficient with temperature

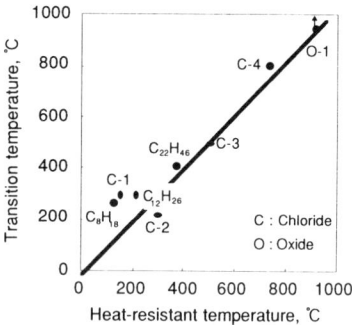

Fig.2: Transition temp. to high friction vs. heat-resistant temp. of additives

REFERENCES

(1) Kimura, S., Yasuda, H., Narita, N., Journal of JSLE, 28, 3, pp.185-191, (1981).
(2) Awasthi, S., Wood, J. L., Ceram. Eng. Sci. Proc., 9, 7-8, pp.553-559, (1988).
(3) Savage, R.H., Journal of Applied Physics, 19, 1, pp.1-10, (1948).
(4) Lancaster, J.K., Pritchard, J.R., J.Phys.D: Appl. Phys.,14, pp.747-762, (1981).

AN INTEGRATED TESTING METHOD FOR CERMETS ABRASION RESISTANCE AND FRACTURE TOUGHNESS EVALUATION

S.F. SCIESZKA and K. FILIPOWICZ
Technical University of Silesia, Akademicka 2, 44-100 Gliwice, POLAND

ABSTRACT

Both main sources of distress, the abrasive wear and the brittle fracture lead to massive losses of function in the mining equipment, involving high maintenance costs as well as replacement costs and extended down time for worn or mal-functioning equipment and tools (1). In the industrial processes involving e.g. direct impact and rubbing contact between the mining tools and rock, the main question remains unanswered-namely what is the optimal balance between wear resistance and fracture toughness of the carbide rock bit materials. The methodology and apparatus presented in the paper might be able to fill the gap.

During the course of this investigation the abrasive wear resistance and the fracture toughness of eight selected cermets (Table 1) were evaluated using a novel integrated testing procedure. The apparatus used simulated the tribo-conditions between the mining tools and rock including the rock comminution process.

Table 1 Mechanical properties of cermets

No	Cermetal (grade)	Composition WC%	Co%	E GPa	HV$_{30}$ GPa	K$_{IC}$ MPa m$^{1/2}$
1	CM1	94	6	630	15,5	8,6
2	CM2	94	6	630	14,3	10,5
3	CM3	91	9	590	12,5	11,2
4	CM4	89	11	580	11,5	11,7
5	CM5	89	11	570	11,,0	12,8
6	CM6	85	15	540	10,5	16,1
7	CM7	94	6	630	12,5	14,9
8	CM8	90,5	9,5	580	10,5	15,9

The apparatus consisted of a cylindrical grinding chamber (ϕ 30 mm) in which a drive shaft was mounted, to one end of which a disc was attached. The bottom of the chamber held the charge of granular rock (coal or sandstone) and the specimen (cermet cutting element) was attached to the underside of the disc. The apparatus was described elsewhere in full detail (2). The machine was operated for a fixed number of revolutions. During the tests, torque, normal load, temperature and shaft displacement were recorded. After each test the specimen was removed then ultrasonically cleaned and then dried and weighed to the nearest 0,0001 g. The tests were then repeated using a new sample of coal or sandstone but with cutting cermet edge the same as before (Fig.1).

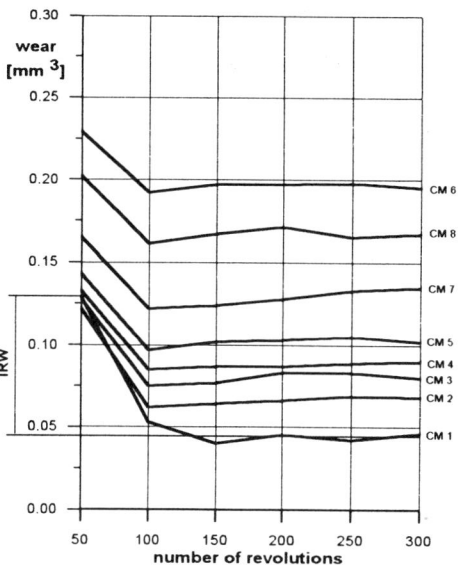

Fig.1: The effect of a number of revolutions on volumetric wear of the triangular specimens in contact with granulated sandstone

In this way the initial rise of wear (IRW) due to brittle fragmentation and chipping of the edges during tribo-contact with rock particles was evaluated. From IRW a new surface (S) produced by the fracture of the edges was computed. Based on the results from the carefully controlled repeated abrasion tests the stress intensity factor K$_{CRW}$ was calculated from the equation:

$$K_{CRW} = 0,20419 \left(\frac{E}{HV_{30}}\right)\left(\frac{1}{S}\right) EI^{1/2} Q^{1/2} \; [MPa\,m^{1/2}]$$

where: E is Young modulus [GPa]
HV$_{30}$ is Vickers hardness [MPa]
S is new crack surface [mm^2]
EI is energy input for each test [Nm]
Q is normal load acting on the shaft [N]

The equation is a final outcome of the study on the empirical correlation between the fracture and wear indicators of the selected cermets obtained from the novel integrated testing procedure.

REFERENCES

(1) W.E. Jamison, Tools for drilling, in W. Winer and M.B. Peterson (ed), Wear Control Handbook, ASME, New York, 1980, pp 859÷887.

(2) S.F. Scieszka and R.K. Dutkiewicz, Testing abrasive wear in mineral comminution, Int. Journal of Mineral Processing, 32, 1991, pp 81÷109.

CERAMICS UNDER HIGH SPEED EHD SLIDING CONTACT

ECKHARD SONNTAG

CeramTec AG, Fabrikstrasse 23-27, D-73207 Plochingen, F.R.G.

ABSTRACT

Advanced ceramics as partially stabilized circonia oxide ZrO_2, alumina oxide Al_2O_3, and silicium carbide SiC and SiSiC were used as fixed rings being in sliding contact with a rotating steel hook. Sliding velocity was up to 45 m/s. Lubricant was mineral oil, synthetic oil, synth fat, bee-wax, PA6 and PES under boundary conditions (1).

High speed temperature measurements and recalculations showed contact temperatures between 100 and 500 °C depending on friction power, contact area and heat conductivity of the used ceramic.

FEM simulations for certain conditions and materials confirmed the measured temperatures and revealed the mechanism for thermodynamic wear of those ceramics (Fig.1).

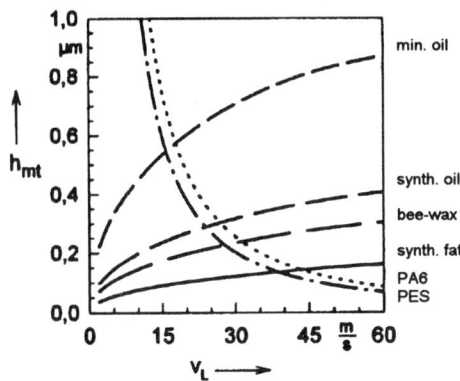

Fig. 2: Effective theoeretical film thickness h_{mt} as function of sliding velocity and different lubricants

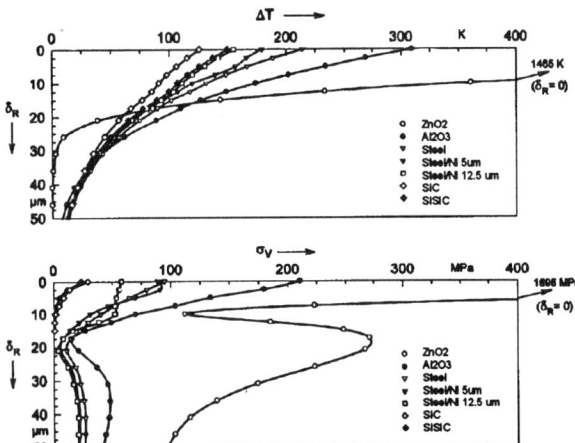

Fig. 1: Simulated temperatures and corresponding van Mises stress for different materials but same friction power

Using the well known equation for isothermal film thickness (2)(3):

$$h_{min,isoth.} = 2,65 \cdot E'^{-0,03} \cdot R^{0,43} \cdot L_S^{0,13} \cdot F_N^{-0,13} \cdot u^{0,7} \cdot \eta_T^{0,7} \cdot (\alpha_{p200})^{0,54}$$

with a thermical corrective (4)

$$C_{th} = \frac{3,94}{3,94 + L_{th}^{0,62}} \quad \text{and} \quad L_{th} = \eta_T \cdot \frac{\alpha_T \cdot u^2}{\lambda_s}$$

the effective theoretical film thickness was calculated to

$$h_{mt} = C_{th} \cdot h_{min,isoth.}$$

These formulas gave the following results for the film thickness in respect to rotational velocity (Fig.2).

The lubricating film not only helps to decrease friction, the temperature measurements also revealed that this film supports a more homogenious temperature distribution along the contact area. The minimum film thickness developing on the ring gliding track must be higher than the surface rouhgness to reach minimum wear.

RESULTS

It was found, that the surface roughness changed dramatically in both ways for the different ceramics used: from smooth to rough and vice versa to a specific equilibrium roughness depending on the load.

For increasing velocity of the rotating steel hook the ceramic surface roughness has to be appropriate to the used lubricant to reach fully EHD-contact. Mechanism on friction and wear can be explained as a function of h_{mt}.

REFERENCES

(1) E. Sonntag, Fortschr.-Ber. VDI Reihe 1, Nr.252, Düsseldorf: VDI-Verlag 1995.

(2) R.R. Mikic, R.T. Roca, Trans ASME, J. Appl. Mech., 96E (1974), 801 ff.

(3) F. Bowden, D. Tabor, The Friction and Lubrication of Solids, Part II, Oxford Univ. Press: Clarendon Press, Oxford, 1964.

(4) U. Schmidt, H. Bodschwinna, U. Schneider, Antriebstechnik 11/1987, 55-60; 12/1987, 55-60.

TRIBOLOGICAL BEHAVIOUR OF CARBON-CARBON COMPOSITES

B VENKATARAMAN and G SUNDARARAJAN

Defence Metallurgical Research Laboratory, Kanchanbagh P.O., Hyderabad-500 058, India

ABSTRACT

Carbon fibre reinforced carbon matrix composites known as carbon-carbon composites (CCC) are increasingly used as brake pad material in aircraft owing to their light weight, increased strength especially at high temperature and higher specific heat capacity (1). Although a high and optimum friction is required for the brake materials, the carbon is known to undergo transition in friction coefficient once contacting interface attains a critical interface temperature (2). The critical temperature depends on the trade off between the quantity of heat generated at the interface and heat sinking ability of the mating materials. For like material contact, when the configurations of the mating surfaces are similar, the heat sinking ability of both the mating materials will be identical. Then the interface temperature largely depends on the heat generated at the interface which in turn depends on the normal pressure (P), linear sliding speed (V) and the friction coefficient. Therefore, the influence of P and V on friction behaviour is important to understand in order to identify the safe limit (normally expressed as the product of P and V) of the brake pad. In brake pads the configuration of the rubbing surfaces may be similar or dissimilar. In similar contact it is normally a disc rubbing on another identical disc. The dissimilar contact represents a case where one of the rubbing surfaces is a slab or shoe when the other is a disc. The PV will be further influenced by such variation in the relative patterns of the mating surfaces. The present work aims at evaluating the influence of PV on the friction behaviour of CCC composites with disc on disc and slab on disc configurations.

Three types of aircraft CCC brake pad materials were used as the test specimens. A homebuilt Friction dynamometer was used for the friction tests. One of the mating surfaces was stationary while the other was rotating. Unlike conventional brake dynamometer in which the tests are conducted in a deceleration mode, in the present dynamometer, constant sliding speed was maintained for each test by a DC motor. The normal load measured by a load cell was applied by means of a hydraulic loading system. A torque transducer was used to measure the friction force from which the friction coefficient was estimated as a function of sliding distance. A wearing thermocouple was used to measure the interface temperature. Two types of specimen configurations namely disc on disc and sample on disc were tested. The tests were carried out with a wide range of P and V. The μ map with axes of P and V was obtained. The transition of μ from a low value to a high value was associated with the interface temperature. The calculated and measured interface temperature were found match in the present investigation. Fig.1 represents the friction map over a range of normal pressure P and sliding speed V for two types of composites coded as CC1. This material was tested in disc on disc configuration. It can

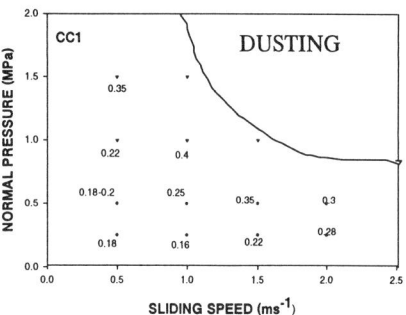

Fig 1. Friction map on Normal pressure (P)-Sliding speed (V) axes for CC1.

be noticed that the transition to dusting wear occurred beyond a critical PV value of 1.5 MPa m/s. The transition PV (or critical PV) is same for CC1 and CC2. Chen & Ju (3) also observed that in disc on disc configuration the PV limit was 2.5 MPa m/s. However, the present study indicated that it varied with the sample configuration. For example, for sample on disc the transition PV value was 27 MPa m/s. Such variation has been rationalized based on the interface temperature. The analysis carried out in the present investigation indicates that the heat partition factor in sample on disc configuration is only 0.025 while it is 0.5 in disc on disc configuration. The friction undergoes transition when the interface temperature reaches a critical value of about 400 °C. In the case of sample on disc configuration, the critical temperature is reached at a higher PV value because of the lower heat partition factor. In order to have the safe operation at higher PV values, the preferred configuration is a sample on disc configuration in which one of the mating element is a shoe and the other is a disc. The present work highlights the importance of P-V map in designing a carbon-carbon brake pad.

REFERENCES

(1) J.V.Weaver, Aero. Journal, 76, 695, 1972.
(2) J.K.Lancaster, Treatise on Materials Science and Technology, vol 13, 141, 1979
(3) J.D.Chen and C.P.Ju, Wear, 174, page 129, 1994

TRIBOLOGICAL PROPERTIES OF SELF-LUBRICATING COMPOSITE CERAMICS CONTAINING BARIUM CHROMATE

KAZUNORI UMEDA, SOKICHI TAKATSU and AKIHIRO TANAKA

Mechanical Engineering Laboratory, MITI, Namiki 1-2, Tsukuba-shi, Ibaraki, 305 JAPAN

ABSTRACT

Tribological properties of self-lubricating ceramics, $BaCrO_4$, $BaCr_2O_4$, Al_2O_3-$BaCrO_4$, Al_2O_3-$BaCrO_4$-Ag and Al_2O_3-$BaCrO_4$-$BaCr_2O_4$, were studied in air at temperatures from room temperature (RT) to 1000 °C. These ceramics were sintered with spark plasma sintering (SPS) method. The sintering was carried out in a graphite mold under conditions of 800 °C to 1300 °C of temperature, 50 Mpa of pressure and 3 to 5 min of time.

With a pin-on-block reciprocal friction tester, friction coefficient and specific wear rate of ceramics were measured by friction with alumina ball. The temperature was set by step heating and cooling between room temperature and 1000 °C with 100 °C interval. At each step, the temperature was kept for 10 min.

Friction coefficient of $BaCrO_4$ is 0.7 at RT and bellow 0.4 above 800 °C. Al_2O_3-$BaCrO_4$ and Al_2O_3-$BaCrO_4$-Ag show slightly lower friction coefficient than simple $BaCrO_4$ through the test temperatures. Friction coefficient of Al_2O_3-$BaCrO_4$-$BaCr_2O_4$ is nearly the same revel with other specimens at 800 °C. Friction coefficient is remarkably decreased with increase in content of solid lubricant up to 30 % and then it shows almost the same value in the higher contents.

Specific wear rate was measured after 60 min friction at 800 °C and the total load of 9.8 N. Wear of simple $BaCrO_4$ is 200 times larger than that of Al_2O_3-50%$BaCrO_4$. It is decreased with increasing alumina content. Specific wear rare of tested specimens ranges from the order of 10^{-4} to 10^{-6} mm^3/Nm and Al_2O_3-50%($BaCrO_4$-$BaCr_2O_4$) shows the best wear resistance. Table 1 shows the friction coefficient and the specific wear rate of self-lubricating ceramics at 800 °C

Soft layer rich in lubricative components is formed on the wear scar of the ceramics. In the case of (Al_2O_3-50%$BaCrO_4$)-30%Ag, hardness on the wear scar is Hv 80 to 150, while that on the outside of the wear scar is Hv 250 to 470 (Fig. 1). By ESCA analysis, peak intensity of Ba, Cr and Ag on the wear scar are 1.5 to 2.5 times higher comparing to those on the outside, while the peak intensity of Al and O is decreased on the wear scar. Results of EPMA show that O, Ba and Ag are uniformly distributed on the wear scar. Intensity of Ba and Ag is stronger on the wear scar, while that of O is weaker.

Table 1 Tribological properties of self-lubricating ceramics at 800 °C

Specimen	Specific wear rate, mm^3/N	Friction coefficient
$BaCrO_4$	8.6x10^{-3} *	0.39*
Al_2O_3-50%$BaCrO_4$	4.1x10^{-5}	0.36
(Al_2O_3-50%$BaCrO_4$)-30%Ag	1.6x10^{-5}	0.36
(Al_2O_3-90%$BaCrO_4$)-30%Ag	1.9x10^{-4}	0.36
Al_2O_3-30%($BaCrO_4$-$BaCr_2O_4$)	6.0x10^{-6}	0.36
Al_2O_3-50%($BaCrO_4$-$BaCr_2O_4$)	<1.0x10^{-6}	0.36

Laod=9.8N(*1.96N), Sliding speed=1.2m/min

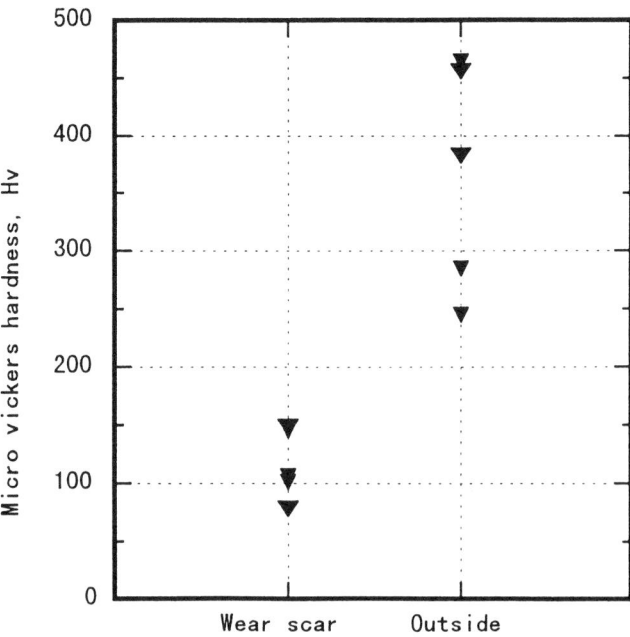

Fig. 1 Micro vickers hardness of (Al_2O_3-50%$BaCrO_4$)-30%Ag

IN-SITU FORMATION OF TRIBOLOGICALLY EFFECTIVE OXIDE INTERFACES IN SIC-BASED TRIBOCOUPLES DURING DRY SLIDING

ROLF WÄSCHE and DIETER KLAFFKE
Federal Institute of Materials Research and Testing (BAM)
Unter den Eichen 44 - 46, D 12203 Berlin, Germany

ABSTRACT

Based on silicon carbide binary and ternary ceramic particulate composite materials in the system SiC - TiC - TiB$_2$ have been prepared by a conventional ceramic manufacturing process. The tribological properties of the ceramic composites were characterised with the oscillating sliding method with two different sliding strokes (0.2 mm and 0.8 mm). Measurements were carried out at room temperature in air of different relative humidities with counterbodies of SiC and alumina. Table 1 summarises the experimental parameters.

Table 1: Parameters of Oscillating sliding experiments

Ball	SiC, Al$_2$O$_3$
Disk	SiC SiC-TiC SiC-TiC-TiB$_2$
Stroke Frequency Normal Force Number of cycles	$\Delta x = 0.2, 0.8$ mm $\nu = 20, 10$ Hz $F_n = 10$ N $n = 100\,000, 600\,000$
Temperature	25°C
Rel. Humidity	5 %, 50 %, 100 %

The system wear is strongly dependant on the phase composition of the composite material. Especially at low relative humidities the wear of the Alumina/SiC-TiC-TiB$_2$ system is drastically reduced as compared with the SiC/SiC and the Alumina/SiC system. The normalised coefficients of system wear for SiC/SiC was found to be at $35*10^{-6}$ mm^3/Nm and for SiC/SiC-TiC-TiB$_2$ at $6.1*10^{-6}$ mm^3/Nm which is reduced further for alumina/SiC-TiC-TiB$_2$ below $1.0*10^{-6}$ mm^3/Nm.

The Figure summarises the results of the volumetric wear for both bodies, ball and disk. The comparison of the wear results shows that in the case of the composite materials a tendency is observed for reduced wear of the ball and a reduction of the system wear in general. As main wear mechanisms fatigue, grain pullout at high contact stresses and the tribologically induced formation of oxide layers have been indicated. The wear debris and the wear scars were investigated by different microanalytical methods. The investigation by optical and scanning electron microscopy revealed the formation of thin wear reducing films during the tribological process within the contact interface. However, the films do not cover the whole interface area but contain distinct and pro-bably isolated parts. Therefore, deduced from the experimentally obtained facts, the film formation seems to be balanced by growth and destruction processes, where tribooxidation plays a key role in the formation kinetics of the film. The composition of the in-situ formed films is of oxidic character but contains also non-oxidic wear debris of the bulk ceramics.

Both SiC / SiC and SiC / Al$_2$O$_3$ tribo couples showed the formation of unstable tribochemically formed reaction layers. In the case of the binary and ternary composite materials as disk materials stable reaction layers were formed causing obviously the observed drastically reduced system wear.

Although the exact composition and structure of these thin layers which are formed in the interface are not yet understood in the very detail a model is proposed for the explanation of the wear behaviour of the different tribocouples. This model comprises 2 different stages of wear and film formation, which are different for the tribocouples according to the composition of the disk material. The film formation itself is mainly steered by the presence of Ti-phases:

- In the running in period intense surface contacts in single asperities of the bulk material take place, accompanied by high contact stresses. This stage is characterised by both fatigue and grain pullout leading to pits in the surface. Pulled out grains are crushed, partly oxidised, mixed with unoxidised debris and filled in the pits.
- In the steady state phase of reduced wear the film is formed by spreading the oxidic wear debris over parts of the interface, thus reducing the high contact stresses and fatigue and grain pullout processes also. The pits in the surface act as reservoirs for the wear debris.

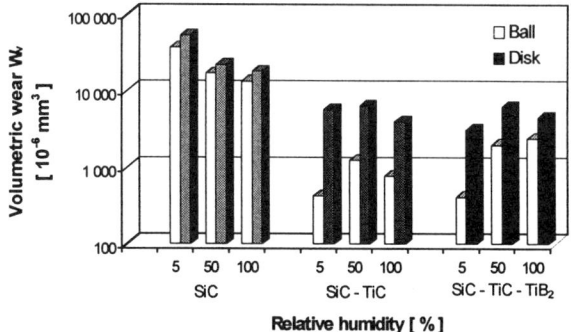

Volumetric wear of ball and of disk after tests with Al$_2$O$_3$ ball, different R.H.. T = 24 °C, n = 600 000

REFERENCES

R.Wäsche and D.Klaffke, 10 Int.Coll.Trib., TA Esslingen, Germany, ed. W.J.Bartz, 09-11.1.96 Vol.3, p 2371-2380

Submitted to *Tribology Letters*

INFLUENCE OF INTERPOSED PARTICLES ON FRICTION AND WEAR PERFORMANCE OF SILICON CARBIDE

YUJI YAMAMOTO, LIN YANG and JOICHI SUGIMURA
Department of Energy and Mechanical Engineering, Kyushu University, Fukuoka 812-81, Japan

ABSTRACT

Wear debris generated during runs considerably influenced the friction and wear characteristics of silicon carbide in sliding contact (1)(2). To study the influence of wear debris, sliding contact tests in which silicon carbide or silicon dioxide particles with different sizes were intentionally introduced between the rubbing surfaces, were carried out in different dry atmospheres using a ring-on-disc sliding contact apparatus.

The effects of size and the kind of the supplied particles and their changes during runs were investigated. The results are shown in Fig.1 and Fig.2. Interposing silicon carbide particles had little effect on frictional characteristics in argon and oxygen atmospheres. The wear characteristics were to some extent improved by interposing silicon carbide.

With silicon dioxide particles, however, the friction decreased considerably compared with silicon carbide particles. Since silicon dioxide particles were lower in hardness and fracture toughness than silicon carbide, they were rapidly broken to smaller ones during a run leading to mitigating local stress concentration and suppressing brittle fracture of the rubbing surfaces. In addition, silicon dioxide of ionic material easily formed a surface thin film compared with silicon carbide of covalent materials (3). The wear characteristic was significantly improved by interposing silicon dioxide particles as shown in Fig.1 and Fig.2.

REFERENCES
(1) M.Godet, Wear, Vol.100, page 437, 1984.
(2) Y.Yamamoto, A.Ura, Wear, Vol.154, page 141, 1992.
(3) O.O.Ajayi, K.C.Ludema, Wear, Vol.140, page 191, 1990.

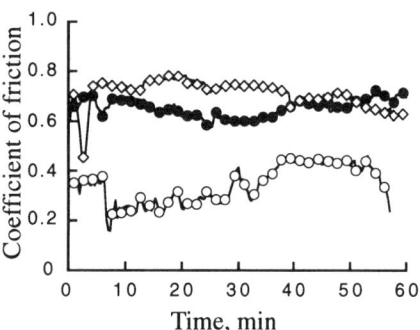

(a) Change in coefficient of friction

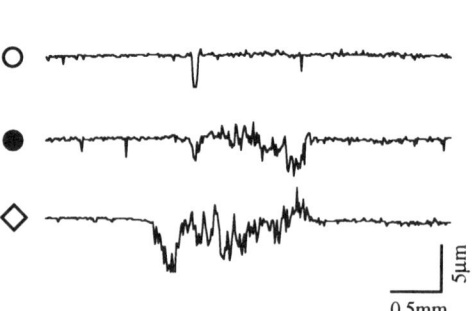

(b) Surface roughness after a run

Fig. 1: Effect of interposed particles in dry argon: contact pressure 130kPa, sliding speed 9mm/s, ○: SiO_2, 1.7μm, ●: SiC, 3μm, ◇: no addition

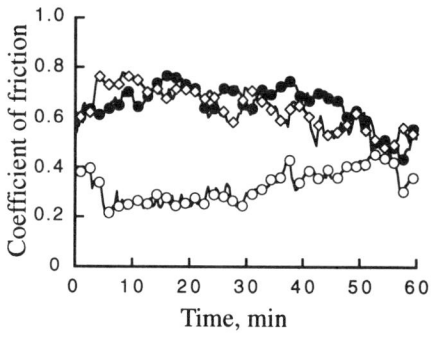

(a) Change in coefficient of friction

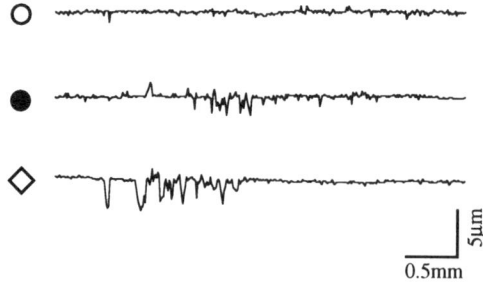

(b) Surface roughness after a run

Fig. 2: Effect of interposed particles in dry oxygen: contact pressure 130kPa, sliding speed 9mm/s, ○: SiO_2, 1.7μm, ●: SiC, 0.7μm, ◇: no addition

Dependence of Sliding Wear Behaviour of High Performance SiC- and Si₃N₄-Ceramics on Different Machining Procedures

ROLF ZELLER, WULF PFEIFFER and THOMAS HOLLSTEIN

Fraunhofer Institut für Werkstoffmechanik, Wöhlerstraße 11, 79108 Freiburg, Germany

Abstract

Because of their good mechanical properties and corrosion resistance high performance ceramics like SiC and Si_3N_4 are becoming much more commonly used in applications with enhanced tribological requirements. For long term action under these conditions and for life time prediction one has to know the way in which wear depends on machining. Therefore the surface states produced by different machining procedures were detected and correlated to the wear behaviour.

For the investigations a high performance sintered SiC (EKASIC D) and a gas pressure sintered Si_3N_4 (SN-N 3208) were used. The specimens were machined by the Werkzeuglabor der RWTH Aachen (WZL), the Fraunhofer-Institut für Produktionstechnologie Aachen (IPT) and by the Institut für Werkzeugmaschinen und Fertigungstechnik der TU Berlin (IWF) within the scope of a collaborative project.

The sliding wear tests performed in this project are a model for face sealings. To avoid fatigue of the surface layers like in pin on disk arrangement, the plane functional faces of the specimen rings are pressed to each other in the sliding wear testing equipment in such a way that the normal contact pressure is constant in the whole contact zone.

The mass loss due to wear was detected by weighing the specimens with an extremely precise balance. The mean sliding velocity of the wear tests was of the amount of 0.3 ms^{-1} (180 rpm). There was a difference of about 15 % in sliding velocity between the outer and the inner margin of the contact face. Purified water cooled and lubricated the specimens; it poured over the outer shape of the specimens as well as into the void between the pairs. The temperature of the specimens did not rise over 45°C during the tests. The mean normal contact pressure was of the amount of 2 MPa.

Several SiC and Si_3N_4 specimens with laser caved, electro discharge machined, and ultrasonically machined surface states were examined and compared with polished and lapped ones regarding their tribological behaviour.

For a quantification of machining induced as well as wear induced surface damage, the depth distributions of microplastic deformation were measured by X-rays. X-ray residual stress measurements of various machining states after different sliding distances were performed, too.

The most important results are:

- Laser caving enhances the wear resistance of polished surfaces of SiC specimens up to the order of the lapped ones.

- Only laser caved Si3N4-specimens reveal an acceptable wear behaviour, which is comparable to that of the polished SiC-specimens.

- Electro-discharge machined SiSiC-specimens, as well as ultrasonically machined SiC-specimens, show an excellent wear behaviour.

- The amount and the depth of the damage, measured by x-rays, are the higher, the coarser the specimens are machined.

- Different machining procedures leave neglectible low surface residual stresses, which become superposed by wear induced residual stress rather soon.

One can classify the influence of the different abrasive machining procedures into three categories: Machining procedures that produce a »smooth« surface of low damage reveal a bad wear behaviour because of considerable adhesive wear processes. Machining procedures that create a »rough« and strongly damaged surface show a disadvantageous wear behaviour due to deficient surface integrity. The relatively best wear behaviour is caused by machining procedures that produce medium roughness and moderate surface damage.

Thus one can expect high improvement of wear resistance only by such machining procedures, which produce a hydrodynamical bearing topography without decreasing mechanical capability. The results of the laser caved, the electro discharge machined, and the ultrasonically machined specimens show that this way is of great promise.

MONITORING THE BEHAVIOUR OF TRIBO-SYSTEMS WITH BOUNDARY LUBRICATION

G ABRAHAM, P ARTEMCZUK, F FRANEK, A MATZNER

Institute of Precision Engineering - Department of Tribology, Technical University Vienna, Floragasse 7, 1040 Vienna, Austria

ABSTRACT

Many efforts have been made to understand the nature and the formation of boundary layers within tribo contacts (1). Electrical measurement methods, especially for measuring the electric conductivity between metallic friction elements have already been used for describing the behavior of boundary layers (2)(3).

But most of the standard test equipments for the performance of lubricants do not allow in-situ-studies of the formation of oil-films and boundary layers. Usually the surface of the test specimen is analised after a test to detect layers and to quantify wear effects.

An electric measurement method has been developed to monitor the reactions of the lubricant.

Comparing measurements have been carried out on a oscillating friction wear test machine and on an adapted pin-disc-machine.

The measuring device consists of a HF-oscillator developed at the Institute of Precision Engineering, Technical University Vienna, allowing to monitor the electrical impedance between metallic elements of a lubricated tribo-system.

The measuring principle has the advantage that both capacitive and resistive components of the impedance of the lubrication layers determine the output signal. So all possible states of lubrication can be monitored including full metallic contact between the sliding elements and hydrodynamic lubrication with a separating oil film.

The measurement results shown below are from a test with the oscillating friction and wear test machine under the following test conditions:
- Test configuration: Ball on disc
- Load: 50 N for 5 min; 100 N increase every
  4 min; max 1200 N - depending on the lubricant performance.

The test runs are characterized by three typical phases. At low loads a separating oil-film ist formed. The impedance depends on the load and the type of lubricant as well as on the temperature. Increasing the load reduces the average oil film thickness. The capacity rises while the resistance is decreasing until a limit is reached that is determined by the formation of a boundary layer. In this case the impedance of the lubrication layer is no longer dependent on the load. It remains constant until the boundary layers break through at increasing loads. Then the impedance is mainly an ohmic resistance that is defined by the resistance between the two metallic surfaces.

The characteristic value of impedance indicating the formation of boundary layers depends on the geometry of the tribo system and on the calibration of the measuring device but seems to be constant for different test conditions. Whenever this value had been reached during a test run the formation of surface layers could be proved by EDX-analysis as well as reduced wear effects, documented by scanning electron microscope photographs of the probe surface after the test.

The effects described that can be monitored by the HF-measurement method can not be detected by measurement of the friction force that is used in most cases.

The figure below shows the results of a typical test run.

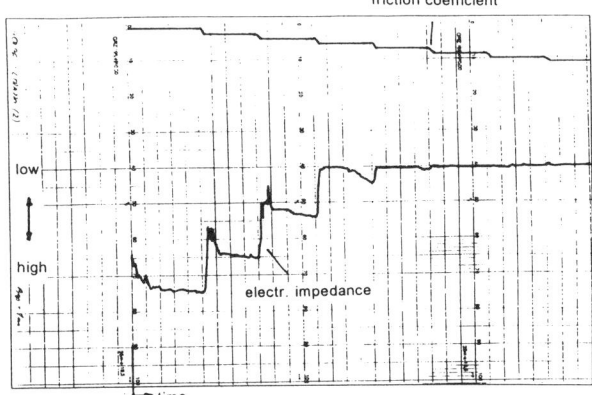

REFERENCES

(1) LUO, J.; SHI, B.; WANG, R.; CAO, N.; WEN, S.: Study on the characteristics of lubricant film at nanometer scale. Proceedings of the International Tribology Conference, Yokohama (1995), Page 1083 ff.

(2) SCHRÖDER, H.; JANTZEN, E.: Untersuchungen zur Mischreibung und Reaktionsschichtbildung. Tribologie und Schmierungstechnik, 2 (1989), 62-66

(3) MYSHKIN, N.; KONCHITS, V.: Evaluation of the interface at boundary lubrication using the measurement of electric conductivity, Wear 172 (1994), 29-40

THE DEVELOPMENT OF ENDOSCOPES FOR THE THREE DIMENSIONAL IMAGING OF SURFACES

R.V. ANAMALAY, T.B. KIRK and D. PANZERA
Department of Mechanical and Materials Engineering, University of Western Australia, Nedlands, Western Australia.

ABSTRACT

The morphologies of engineering surfaces are critical to the way in which they interact and the wear processes that operate. The recording of and subsequent analysis of the microstructure of surfaces can therefore be a valuable tool for quality control and condition monitoring of machine components. Existing techniques, such as profilometers, generally provide limited two dimensional recording and analysis of surfaces in a plane normal to the surface. These instruments tend to be laboratory based instruments and cannot generally be used on machine components *in-situ*. Three dimensional stylus instruments generally have low scanning rates and can damage the specimen by the movement of the stylus across its surface(1).

Inspection of components can therefore be impossible due to the necessity for dismantling a machine to inspect the components. The studies to be outlined in this paper focus on two deficiencies in the current image recording and analysis procedures:

[a] Real surfaces have a three dimensional topography while analysis is generally only performed on a 2D section.

[b] Generally, only laboratory analysis is available, whereas components must be inspected *in-situ* if surface analysis is to be a useful condition monitoring tool.

The first of these problems is countered by the use of laser scanning confocal microscopy or LSCM techniques to provide three dimensional data for analysis. By taking successive two dimensional optical sections of a surface at different focal planes, a three dimensional model of the surface can be built up from adjacent two dimensional optical sections(2).

A further advantage of fibre optic laser scanning confocal microscopy is its adaptability to endoscopic examination of surfaces due to its compactness and flexibility(3), thus eliminating some of the difficulties in examining machine components *in-situ*. The paper will report on the development of such an endoscope and its use on engineering surfaces. The paper will also report on further developments which are currently being implemented to make internal examination of machines practical in industry. Currently, coating of the surfaces with fluorescent dye is required to use the endoscope. Figure 1 below is a fracture surface imaged using a fluorescent dye on the LSCM and figure 2 is a height encoded image or HEI of the same surface. The different shades of gray correspond to different heights of the surface features. As can be seen, fluorescent methods lend themselves well to the LSCM.

 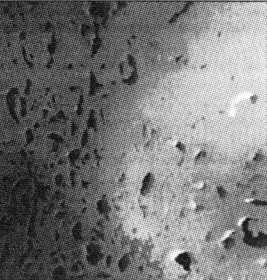

Fig 1: Fracture Surface   Fig 2: HEI of figure 1

The developments currently being proposed are intended to facilitate imaging of the surfaces using reflected light, thus eliminating the need for fluorescent dyes. This would further simplify the condition monitoring of machine components, and also the inspection of manufactured items for quality control purposes.

REFERENCES

(1) M. G. Gee and N. J. McCormick, The Application of Confocal Scanning Microscopy to the Examination of Ceramic Wear Surfaces, A230-A235, J. Phys D: Appl. Phys. 25 (1992).
(2) R. G. King and P. M. Delaney, Confocal Microscopy, Materials Forum 18, pages 21-29 (1994).
(3) L. Giniunas, R. Juskaitis and S. V. Shatalin, Endoscope with Optical Sectioning Capability, Applied Optics, Vol 32, No 16, 1 June 1993.

IN SITU MEASUREMENT OF ROLL SURFACE 3-D TOPOGRAPHY IN SLIDING AGAINST STEEL SHEET

KATSUMI ANDO
Process Technology Research Laboratories, Nippon Steel Corporation, Futtsu-city Chiba 293, Japan
KOJI KATO
School of Mechanical Engineering, Tohoku University, Sendai 980-77, Japan

ABSTRACT

Many rolls for handling steel sheets are used in steel rolling and processing plants. It is well known that the friction and wear properties of rolls for handling steel sheets are much influenced by the roll surface topography, so the roughened rolls with wear-resistant coating are widely used. But a few researches on friction and wear behavior of such rolls against steel sheet (1) have been done.

In friction and wear experiments of rolls against steel sheets in sliding, friction and wear is changed with the surface topographical changes of rolls by wear. The quantitative measurement of the roll surface topography during experiments is considered to be more effective to examine friction and wear mechanism, but only a few researches was carried out.

To examine friction and wear behavior of rolls against steel sheets in sliding, the 3-D surface topography measurement - friction and wear experiment system was developed for *in situ* measurement of 3-D topography of roll surface without removing roll specimen. Fig. 1 shows the outline of this system. The auto-focusing laser displacement sensor is used as the non-contact measurement device in this system; therefore the high-speed measurement of roll surface profile with high resolution is possible.

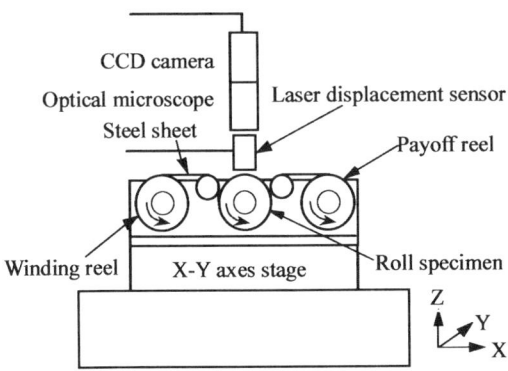

Fig.1: The outline of 3-D surface topography measurement - friction and wear experiment system

Two kinds of the roughened carbon steel rolls were prepared for experiments. One was a shot blasted roll which had the random surface roughness (Ra=3.3μm), and the other was a columnar embossed roll which had the regular surface (maximum height=36.5μm). The sliding friction and wear experiments of two kinds of rolls against stainless steel sheets and *in situ* measurement of 3-D topography of the roll surface were performed on this system. The 3-D contour maps and the surface roughness parameter of the roll surface topography at the same point on the roll were obtained by *in situ* measurement every 101 meters at sliding distance.

Fig. 2 shows the coefficient of friction against stainless steel sheets and the 3-D contour maps of each roll surface against sliding distance. The coefficient of friction of each roll is increased from the beginning and became almost constant with sliding distance. It is realized that the friction and wear behavior of each roll is different by the roll surface topography.

The analytical results of these experimental data show the effectiveness of this system to examine friction and wear mechanism of rolls in sliding against steel sheets.

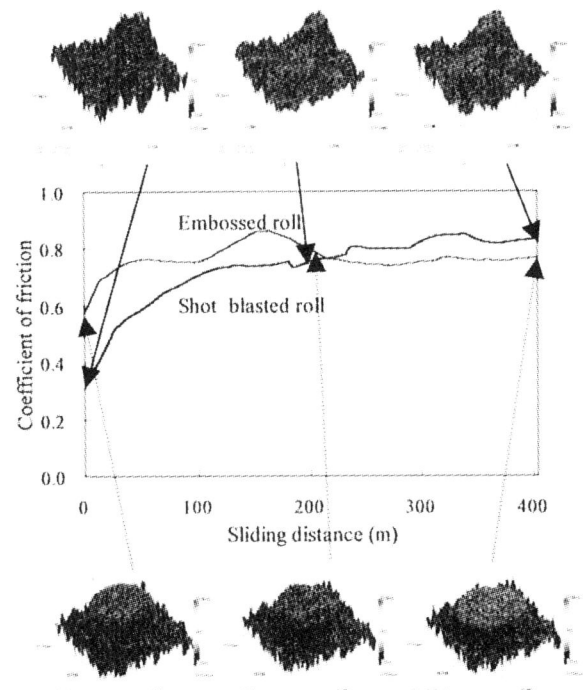

Fig. 2: The coefficient of friction against stainless steel sheet and the 3-D surface topography of rolls versus sliding distance

REFERENCES

(1) K. Ando, S. Shibamoto, K. Kato, Proceedings of the International Tribology Conference, Yokohama 1995, 145-150pp

INVESTIGATION ON A THRUST BEARING TILTING PAD

R BASSANI, E CIULLI and P FORTE
Dipartimento di Costruzioni Meccaniche e Nucleari, Università degli Studi di Pisa, Via Diotisalvi 2, 56126 Pisa, Italy

ABSTRACT

The tilting pad thrust bearings are often used in the centre pivoted configuration, preferred for rotation in either direction, foolproof assembly and minimum stocks. The present work is a contribution to the understanding of the involved phenomena through a study on tilting pads of same shape and dimensions with different pivot position.

An experimental investigation is carried out on the pads using a particularly versatile test rig. Simultaneous measurements of film thickness and friction are made. The tilt angle of the pads is measured by optical interferometry with monochromatic light; in order to make the measurements possible, a glass disk simulates the shaft in contact with the pad (Fig. 1).

 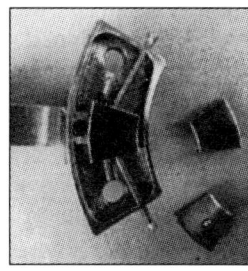

Fig. 1: Experimental arrangement (left) and pads (right) with pad-case

The tilt angle (α) is evaluated using the formula:

$$\alpha = \operatorname{atan}(\Delta h/\Delta x) = \operatorname{atan}((\lambda/4n)/\Delta x)$$

where Δh is the difference of film thickness corresponding at the distance (Δx) between a dark fringe and the nearest light one, λ the wavelength of the used light and n the refractive index of the lubricant. Typical interference images are shown in Fig. 2.

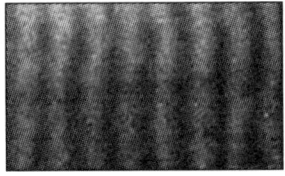 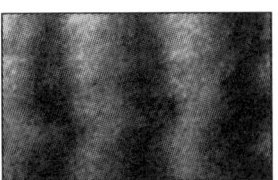

Fig. 2: Interference images obtained with 0.2° (left) and 0.08° (right) tilt angle

The friction coefficient is estimated by force measurements thanks to an aerostatic bearing for the supporting pad structure.

Experimental results can be compared with numerical ones. A FEM code is used to calculate the pressure field of the finite thrust pad (Fig. 3). Plane triangular finite elements are employed assuming pressure linearly dependent from the nodal values. An iterative procedure yields the tilt angle using the pressure boundary conditions and the pivot zero torque condition.

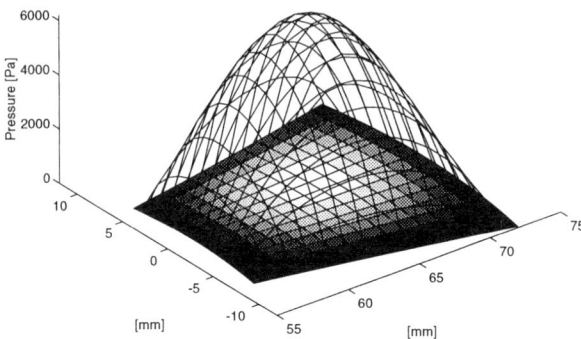

Fig. 3: Numerical pressure field

Static pad characteristics are calculated for different values of geometric parameters and operating conditions. An example is shown in Fig. 4.

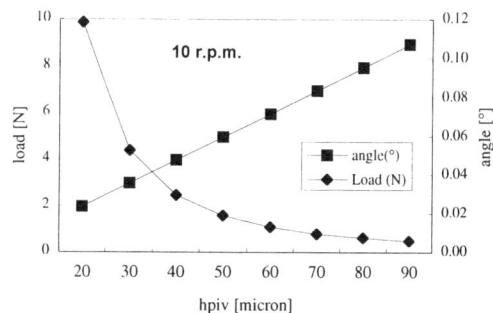

Fig. 4: Load and tilting angle as function of the pivot film thickness

First results show the feasibility of the optical approach but in the meantime its complexity. The use of a video camera allows the investigation of transient conditions such as start up.

REFERENCES

(1) Y K Wang, C D Jr. Mote, Journal of Tribology, Vol.116, 1994, 521-527pp.
(2) R Bassani, E Ciulli, Proceedings of the 23rd Leeds-Lyon Symposium on Tribology, Leeds, UK, 10-13 September 1996.
(3) P Forte, Proceedings of the 4th AIMETA Tribology Congress, S.Margherita Ligure, Italy, 3-4 October 1996, page 115 (in Italian).

DIRECT OBSERVATION OF FRICTIONAL SEIZURE DURING SLIDING OF STEEL ON AL 6061 DISK

MARGAM CHANDRASEAKARAN, XING HUTING & ANDREW WILLIAM BATCHELOR,
School of Mechanical and Production Engineering, Nanyang Technological University, Singapore 639798

ABSTRACT

Seizure of sliding contacts have always posed a challenge to tribologists over the past decades as it often occurs without prior warning. Various researchers have proposed theories and models for the frictional seizure of tribological contacts. These theories however predict the conditions of seizure well, but lack the experimental evidence to explain the mechanism operative during seizure which is vital in the design of new improved type of materials for tribological contacts. In-situ observation of seizure were carried out by Spikes et al using a sapphire disk[1-2]. The present investigation focuses on the in-situ observation of frictional seizure using a X-ray microscope to test the accuracy of current models as well as find the actual mechanism of seizure. The seizure experimements were carrried out using a model pin on disc apparatus custom built in house shown in Figure 1. A steel pin sliding on aluminum disk was selected as a model contact and the sliding tests were carried out at different speeds with varying loads. Cyclic wear of aluminum disk was observed during sliding. It was observed that at low loads and sliding speeds, the scuffing or seizure occurred predominantly due to plowing and adhesive bonding in case of mild steel specimens. At higher sliding speeds, the seizure was predominantly due to rolling of wear particles to form filaments which was was pressed and bonding occured between the wear sheets and the two nascent surfaces. Stainless steel specimens seized at lower sliding speeds possibly due to atomic transfer and chemical bonding and while at high speeds plowing was predominant. The time taken for the seizure in case of mild steel samples were comparatively higher than those of stainless steel samples inspite of the higher hardness levels and strength of the later. This was possibly due to the sliding of metals with similar crystal structure.

REFERENCES

1. Enthoven, J.; Spikes, H.A, Infrared and visual study of the mechanisms of scuffing, Tribology Transactions v 39 n 2 Apr 1996. p 441-447.
2. Enthoven, J.C.; Cann, P.M.; Spikes, H.A., Temperature and scuffing, S T L E Tribology Transactions v 36 n 2 Apr 1993. p 258-266.

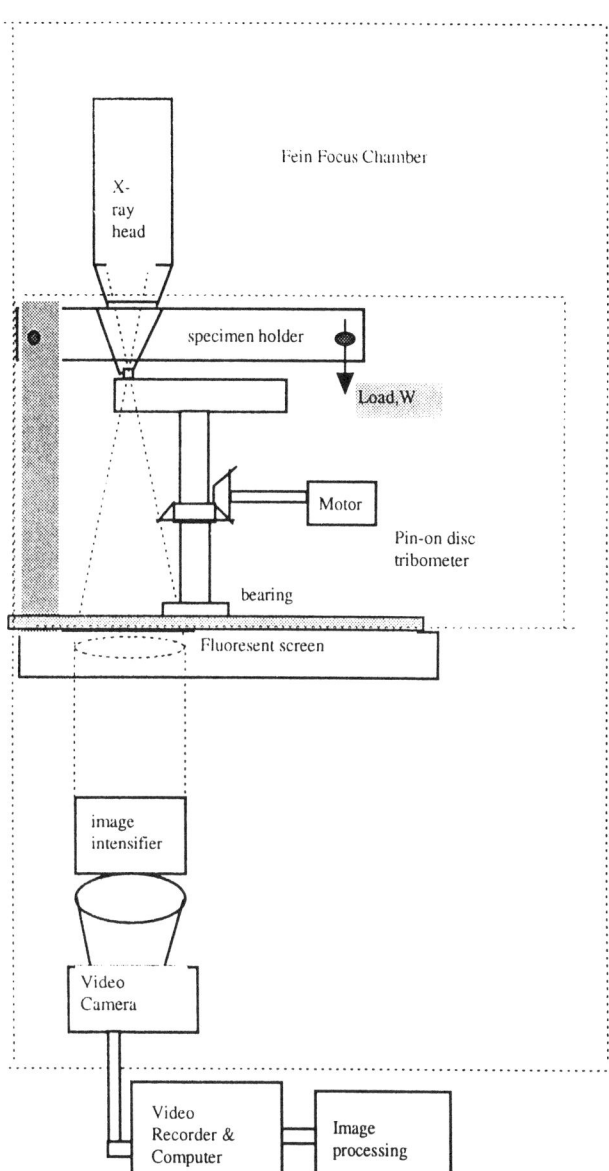

Figure 1. Schematic representation of the test apparatus

The development of a high speed reciprocating test for studying the running-in of the cylinder liner - piston ring system.

T J BENHAM
Physics Department, Chalmers University, Gothenburg, Sweden.
G WIRMARK
Applied Physics Department, Volvo Technological Development, AB Volvo, Gothenburg, Sweden

ABSTRACT

In the testing of cylinder liner surfaces it is normal to use sections cut from engines produced in the production line of an engine manufacturer. This however limits the studies to those surfaces used in production, and piston rings made to match the dimensions. With the curved specimens there are also problems with alignment and particularly radius matching as the piston rings are made larger than the bore diameter. To avoid these problems a flat honing technique has been developed to produce specimens for a new high speed reciprocating sliding rig. Capacitance measurement of the oil film thickness has been used in addition to monitoring friction in order to give more information of the prevailing conditions as the test proceeds.

The liner specimens were cut as blocks from lorry engine liners then honed as required with the same stones as used in production to reproduce the roughness parameters required in a lorry engine. These were then tested against cylindrically ground ring specimens for which the material or the coating could be chosen. This gave a well defined cylinder on flat geometry. Both specimens were held for testing in re-locatable clamps so that the tests could be interrupted for profilometer measurement of the wear, then the test could be continued. 3D profilometry was used and evaluated by a Volvo in-house program which is able to allow for specimen distortions.

The high sliding speed was desired in order to approach the speed of a piston in an engine. Tests have been run to study the running-in behaviour at typical loading conditions for a lorry diesel, and a car petrol engine using oils with and without anti-wear and detergent additives. SiC particles were also added to some oils to represent contamination left by a production honing process. Carbon black was also used to represent soot contamination of the oil.

Although the tests have been performed at 0.9 m/s (full design speed: average 2.4 m/s over cycle) a full range of lubrication conditions has been generated. This was made possible by the wide range of conditions available. Temperatures between 30 and 140 ∞C were possible at this sliding speed and 500 N load. The temperature being measured on the liner specimen.

This versatility in specimen manufacture and handling has enabled different materials and conditions to be studied over short and intermediate running-in periods. Even when using pure base oil, capacitance measurement across the oil film shows that there is running-in of the surface within a very short time and a continuous oil film, even at dead centres, for the remainder of the test.

There is quite a thick build up of tribo-film on the liners when ZDDP is used in the oil with its thickness apparently increased by higher bulk specimen temperature. The honing plateaus appear burnished with or without the use of ZDDP with few very small scratches in the sliding direction, indicating the existence of mild conditions generated by the high sliding speed.

When abrasive particles are added to the oil the anti-wear additive is not seen to have a significant effect on the wear. Even a very small concentration of hard third body causes a marked increase in the wear; the 9.3µm SiC was added at the rate of 1:4600 wt/wt. The concentration of third body was found to be more important than the size; the 27 nm carbon black was used at the rate of 1%.

METHODS AND INSTALLATIONS DEVELOPED FOR FRICTION FATIGUE TESTS OF MATERIALS

V F BEZYAZICHNY and V J ZAMYATIN
Rybinsk Academy of Aviation Technology, Pushkin st. 53, Rybinsk, 152934, Russia

ABSTRACT

In accordance with the fatigue theory of wearability of I.V. Kragelsky and his collaborators the number of cycles of interaction of the individual projections of the rough layer till the separation of the wearing particles and therefore the durability of the friction surface can be calculated according to the formulas:

$$n_e = \left[\frac{2r\theta\sigma_0}{K_{f(e)} l_e v^{0,5}} \right]^{t_e} C_e,$$

$$n_p = \left[\frac{2^{2/3} r e_0}{K_{f(p)} l_p v^{0,5}} \right]^{t_p} C_p$$

where e and p are the factors corresponding to the elastic and plastic contact of the microroughnesses; r is the average radius of the projections tops curvature; v is the parameter of the degree approximation of the starting section of the supporting curve; θ is the elastic constant of Kirchgoff; l is the average distance between the spots of the real contact; $K_{f(e)}$, $K_{f(p)}$, is the dimensionless parameter which takes into account the effect of the fatigue damages summation in the condition of the real loading of the microroughnesses; σ_0, t_e, e_0, t_p are the parameters of the friction fatigue (1,2). Determination of the values r, v, θ, l, K_f, C at present is not a big problem (1,2). The main problem is to determine parameters of the friction fatigue σ_0, t_e, e_0, t_p.

At the present stage of the tribology development the only method of evaluation of σ_0, t_e, e_0, t_p is the experimental - calculated way which is based on the abrasion of the tested sample by the standard abrasing sample - indentor of the right geometric shape (1) However, on one hand the testing cyclometers are not made in big series and on the other hand the only types of cyclometers known today don't meet modern requirements of productivity, results reproduction, accuracy of estimation, kinematics of the relative travel of the contacting surfaces. We develop multiindentoric cyclometers with the spheric and cylindrcal abrasing indentors as well as the methodology of interpretation of data received with their help to improve the technology of parameters determination of the friction fatigue of the materials.

In this report the constructions of the robot - type cyclometers are being presented. The cylindrical indentors are arranged uniformaly along the rotor periphery in them. Kinematically the system of rotation of the rotor and the cylindrical rollers is based on the principle of the rotations addition.

Structurally it is made in the form of the planetary and differential tooth - lever mechanism, the carrier of which is in the rotor. Engagement of the central gear with the intermediate wheels can be external and internal.

The total movement velocity of the operating points A of the indentors in relation to the tested sample is determined in the following way:

$$v_A = w_H (r_1 \pm r_2) + \left[w_H \left(1 \pm \frac{r_1}{r_2} \right) \mp w_1 \frac{r_1}{r_2} \right] r_A,$$

where the upper sign of addition or subtraction refers to the outer gearing and the lower - to the inner; r_1, r_2 are the radiuses of the original circles of the central and intermediate gears; w_H, w_1 - are the angle velocities of the carrier (in our case it's the rotor) and the central gear.

In the regime of testing there is one or two abrased samples made according to the required technology. If two samples are used they are mounted on the rotating platform. At every moment of the one sample can be treated by the indentors - instruments while the wearability of another is being measured along the whole surface of friction simultaneously. The probing of the surface is done with a special probe. The probe sensors are connected with the decoder of signals, amplifier and processor. During the work of this device not only the wearability of the samples is controled but also the arising forces of friction, angle velocities and the total number of the rotor rotations. The actual pressure can be variable along the abrased sample because of it's configuration (triangular, rhombic etc.). The parameters values of the friction fatigue are determined according to the wearability epure, number of the cycles of the sample and indentor interaction, pressure and other condition of contacting.

REFERENCES

(1) Kragelsky I.V., Dobyichin M.N.,Kombalov V.S. Fundamentals of Friction and Wearability Calculations. - M.: Machinebuilding, 1977. - 642 p.
(2) Methods of Calculating Evaluation of the Friction Surfaces of the Machine Parts. - M.: Standarts Publishing House, 1979. - 100 p.

THE QUANTITATIVE ESTIMATION OF THE WEAR PROCESS AND INHOMOGENEITY OF THE SURFACE PROPERTIES OF THE FRICTION PAIRS

CRIETININ O. V., CVARTALOV A.R., KOUDRIAVTSEV S.A., LAHONIN A.N.
State Technical University, Minin street, 24, Nizhny Novgorod, 603600, Russia

ABSTRACT

The processes, occurring in the area of the contact between the work surfaces (for example between the tool and the blank in cutting) are extraordinary sensitive to the changes in the conditions. The inconstancy of the cutting tool, the inhomogeneity of the blank material influence much upon these processes. For a quantitative evaluation of the phenomena taking place in the contact area it is proposed the technique based upon the measuring and analysing the signals of the variable part of the thermo-electromotive force tool- blank (E) and of the acoustic emission (A).

The zone of the contact may be presented as a kind of set of active microareas, inside of which the drawing together, formation of the communication, elastic and plastic deformation and at last destruction of that communication are executing. These processes simultaneously but with different modes are displayed in arising signals: the pulses of the mechanical pressure cause the signal A and the thermal burst, which accompany the plastic deformation, cause the signal E (1).

The automated system for the registration of the E-A signals is created. This system permits to register the signals with the sampling rate 1 MHz.

The characteristics of the contact process occurring at the given moment are defined by a lot of conditions from which the local properties of the blank material are the most fast varying. The changes in microstructures, zones of the local hardening, congestions of the inhomogeneity areas create the powerful disturbing effect on the interaction process. For extracting the information about the wear process rate from signals E and A it is necessary to resolve the problem of the information filtering.

On some level of the consideration detail it is possible to allocate the areas, inside of which the properties of the blank material change slightly. The size of such areas is 50 - 1000 μm in dependence on external conditions. In the space of the E and A signals these areas are corresponded the stationarity intervals with definite spectral characteristics (2).

Let us assume, that within the limits of these zones of the inhomogeneity the type of the contact process does not change, then the opportunity of using them for a valuation of the influence on the interaction process of the various external conditions arises.

For an automatic detection of the intervals borders the criterion of the relative change of the signals spectrum form is used. The frequency range at analyses is 15 - 100 kHz. Each found interval of the uniformity is identified, i.e. related to one of the probable classes. This process is based on the analysis of the correlation matrix of the E and A signals characteristics. As sharing attribute the parameters, applied in the recognition of the images theory, may be used, for example the significance of the excess and asymmetry of the E and A signals distribution, the mutual spectral density of these signals and other.

The snap of the E, A signals to the place of their generation on the work surface was provided, it has allowed to receive the important conclusion that each local condition is corresponded the definite type of the contact interaction. The valuation of the wear rate **I** within the interval belonging to class **k** may be based on the equation:

$$I = \alpha_k \exp\frac{\gamma_k P(A) - \beta_k}{P(E)};$$

where: **P(E), P(A)** - the signals E and A powers in limits of the found class; $\alpha_k, \beta_k, \gamma_k$ - the constants, describing the influence of the class **k** on the wear rate. The integrated valuation **Is** is calculated as sum of the separate classes contribution. For deriving the significance of the $\alpha_k, \beta_k, \gamma_k$ on preliminary stage the set up of the technique is produced.

The positive feature of the offered express techniques consists in the extraordinary small costs of the tested materials and time in combination with good statistical reproducibility: for tests it is necessary to reproduce the procedure of interaction with the required set of conditions during a few seconds.

REFERENCE
(1) Crietinin O.V., Cvartalov A.R., Lahonin A.N., Golubev N.J. Eurometalworking 94, Udine, Italy, 081-1...081-5 pp.
(2) Crietinin O.V., Cvartalov A.R., Koudriavtsev S.A., Lahonin A.N. Balkantrib'96 Proceedings of the 2nd international conference on tribology.- Thessaloniki-Greece, 1996, 776-780 pp.

A DYNAMIC PROBE OF TRIBOLOGICAL PROCESSES: TRANSIENT CURRENT GENERATION*

J. T. DICKINSON, S. C. LANGFORD, W. FAULTERSACK, AND H. YOSHIZAKI
Department of Physics and Materials Science Program,
Washington State University, Pullman, WA 99164-2814 USA

ABSTRACT

An important component of friction during rubbing of two surfaces arises in the rapid, transient making and breaking of adhesive bonds between asperities. When conductors are drawn across polymers and inorganic crystalline materials (i.e., insulators), continuous detachment between the two surfaces generates charge separation due to contact electrification. We have devised sensitive circuits for detecting instantaneous transient currents generated by this process while simultaneously measuring the normal and lateral forces as a conducting stylus is moved across insulating surfaces. These measurements are extensions of experiments on dynamic transient currents generated by propagating cracks at interfaces.[1,2] The experiments are performed in high vacuum as well as in controlled atmosphere.

Using hard conducting tips (tungsten carbide-colbalt) on softer substrates (polymers such as PMMA, polyethylene, polycarbonate, and polystrene, and single crystal inorganics such as MgO, Al_2O_3, and SiO_2) we relate these currents to the extent of damage to the substrate, the contribution of adhesion to the frictional force, and the physics of contact charging. Present tip radii used are a few μm in dimensions with efforts to extend down to smaller sizes. Time resolved measurements reflect temporal statistics of make-break interactions - presently, time resolutions of μs have been achieved. The analysis of both the magnitude and fluctuations in the current are informative regarding the complex micromechanics of tip asperity/substrate interactions.

Using transient currents as a sensitive probe of electrical activity during a tribology experiment we have found evidence for electrochemically-enhanced corrosion and galvanic corrosive wear of Al-6061 alloy in the presence of a perfluoropolyether (PFPE) lubricant, Fomblin Z-DOL. Galvanic current and potential measurements were carried out in three experimental configurations: (1) an Al stylus was translated across a Z-DOL-coated Al substrate in a typical wear geometry, (2) an Al electrode was brought down perpendicularly onto an identical, Z-DOL-coated Al electrode, and (3) in a simple electrolytic cell. All configurations yielded readily measurable currents with corresponding potential differences of ~300 mV. Corrosion during long-term Z-DOL exposure in galvanic cells was confirmed by the detection of significant Al concentrations in the oil. Analysis of oil employed in non-galvanic cells (same geometry, but no external electrical connection) showed essentially background Al concentrations. Scanning force microscopy of Al surfaces exposed to Z-DOL in galvanic cells were actually smoother than as-polished surfaces due to an electropolishing effect. The consequences of galvanic corrosion regarding lubricant lifetime and wear in applications of PFPE lubricants, including magnetic disks, is discussed. These measurements along with simultaneous detection of electron emission and decomposition products emitted directly into vacuum support a previously proposed lubricant degradation mechanism involving chemical reactions promoted by electron-molecule interactions.[3,4]

*This work supported by the U. S. National Science Foundation, Surface Eng. and Tribology Program.

REFERENCES

(1) K. A. Zimmermann, S. C. Langford, and J. T. Dickinson, J. Appl. Phys. 70, 4808 (1991).
(2) Sunkyo Lee, L. C. Jensen, S. C. Langford, and J. T. Dickinson, J. Adhesion Sci. Technol. 9, 1 (1995); 9, 27 (1995).
(3) J.-L. Lin, C. Singh Bhatia, and J. T. Yates, Jr., J. Vac. Sci. Technol. A 13, 163 (1995).
(4) J. T. Dickinson, S. C. Langford, W. Faultersack, and H. Yoshizaki, "Application of transient current measurements: evidence for galvanic corrosive wear of aluminum by a polyperfluoroether lubricant," submitted to *Wear*.

QUANTITATIVE DEPTH PROFILE ANALYSIS OF BOUNDARY LAYERS BY GD-OES

SENAD DIZDAR
Machine Elements, Department for Machine Design, Royal Institute of Technology, KTH, 100 44 Stockholm, Sweden

ARNE BENGTSSON
Swedish Institute for Metals Research, Drotning Kristinas väg 48, 114 28 Stockholm, Sweden

ABSTRACT

Favourable lubricant properties in boundary lubrication are normally obtained by alloying the base oil with tribophysically and tribochemically active substances (additives). These substances physically and chemically adsorb on or react chemically with the metal contact surface and form friction and wear reduction boundary layers (see Godfrey (1) and Sakurai (2)).

Under high contact loading only (tribo)chemically reacted boundary layers can allow friction and wear reduction (2). These layers are products of reactions between oxidised metal surface and chlorine, sulphur, phosphorus-containing additives as well as ZDDP. Recently, numerous research programmes considering tribochemical reactions in boundary layers have been carried out and some contribution to a better understanding of their nature made (3). However, since the boundary layers formation is complex and transient, localised in a nanoscale depth and extremely requested to material analysis methods, the structure and its implications on the friction and wear reduction properties are still poorly understood.

The elemental depth profiles were obtained by glow discharge optical emission spectroscopy (GD-OES). This technique is based on cathodic sputtering of the sample surface in a small, low-power argon plasma. The sample surface layer is continuously eroded by the plasma at a rate of approximately 20 nm/s. The sputtered material is atomised and electronically excited in the plasma, and detected by means of element-specific optical emission. This gives an elemental depth profile in the form of emission intensities as a function of sputtering time. This depth profile is then quantified into elemental concentrations as a function of depth by means of calibrating with reference materials of well-known composition and sputtering rate (4). GD-OES is characterised by short analysis times, high sensitivity and the ability to measure a large number (>20) of elements simultaneously. In comparison with well-known surface analytical techniques based on electron spectrometric methods (AES, XPS), GD-OES is characterised by considerably higher sensitivity for most elements, 10-100 times higher sputtering rate and a better multi-element capability. On the other hand, GD-OES lacks lateral resolution, the sputtered spot is typically 2 - 4 mm in diameter.

Olofsson and Dizdar suggest a way to describe structure implications on the friction and wear properties of the boundary layers (5). They studied surface topography, wear and the elemental depth profile of the boundary layers of boundary lubricated spherical roller thrust bearing. The bearings were lubricated with two different commercially available lubricating oils. In quantified depth profile obtained by the GD-OES measurements tribochemically active elements are easily detected, even in minute amounts. Olofsson and Dizdar observed that measured total enrichment depths of oxygen, D_O, and tribochemically active elements, D_E, (TCHAE, in this case P, S, Zn) correlate with changes in surface topography and commercial lubricating oils (Fig.1).

REFERENCES

(1) D. Godfrey in: P.M. Ku (ed.), Interdisciplinary approach to friction and wear: Proc. NASA Symp., NASA, SP-181, Washington 1968.
(2) T. Sakurai, Journal of Lubrication Technology, 103 (1981) 473-485.
(3) Q. Xue, W. Liu, Lubrication Science, 7 (1994) 81-92.
(4) A. Bengtsson, Spectrochimica Acta, 49B (1994) 411-429
(5) U. Olofsson, S. Dizdar, Submitted for publication, 1997.

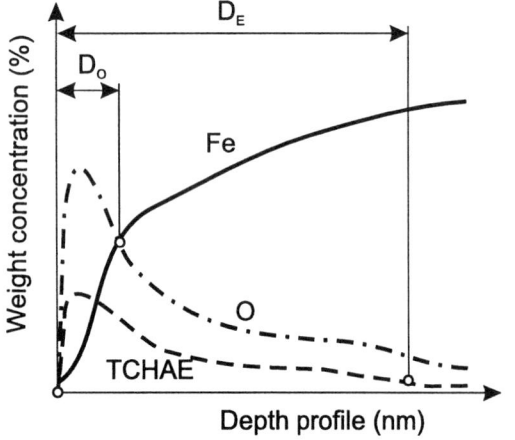

Fig. 1: A schematic view to a depth profile of tribochemically reacted boundary layers (5).

AUTOMATIC TRIBOTECHNICAL COMPLEX

N. F. Dmytrychenko, P. V. Nazarenko, R. G. Mnatsakanov, L. S. Bratitsa.

Departament of Mechanical Engineering. The Kyiv International University of Civil Avition, 252058, Kyiv -58 Komarova, 1, Ukraine.

ABSTRACT

Results of research of lubricant action and wear process under frequently repeated startup and stop condition with the help of Automatic Tribotechnical Complex (ATC) used for estimation of antiwear and antifriction properties of lubricant materials have been submitted. The quantative characteristic, establishing interrelation between an elastic-plastic deformable condition of structural-sensitive properties of friction-pair material and lubricant layers.

The serviceability of friction units depends on a series of factors. The necessity of friction-pair tribotechnical characteristic improvement is primarily important for avoidance of dangerous deterioration of the friction process under condition of extreme modes of operation development- decreasing or termination of lubricant material delivery, short-time local excess of allowable dynamic and temperature modes (1,2). That is why the actual question is to develop technically substantiated and experimentally confirmed methods of the quantative estimation of lubricating properties and durability depending on conditions of operation mode of friction pairs.

The realization of expressed above considerations can be possible if to make research using Automatic Tribometrical Complex (ATC) which includes the machine of friction (СМЦ-1), microprocessor equipment and tele-video devices.

The Automated System of friction surface Technical Condition (ASTC) has been developed for diagnostic purposes.

The system is based on the application of tele-video equipment, microprocessor devices and it consist of two independent subsystem, one of which provides television picture of the friction path on the monitor screen, while the other makes autamed processing of video information in order to detect possible defects, to determine their quantative parameters, to gather information and to make decision. Mathematical and software programs have been created to obtain the express processing of digital pictures supplied by the machine of friction. The results of the express processing are collected and directed to computer printer in the format graphic dependence and tabulated data.

The mechanism of physico-chemical modifying of surface layers has been investigated by methods of scanning electron microscopy (SEM), augar electron spectroscopy (AES) and x-ray analysis.

The estimation of tribotechnical parameters has been carried out the basic of analysis of the processes occurring at the beginning of friction surfaces motion. In this case physico-chemical properties of thin layers of lubricants are fully displayed.

The possibility and the intensity of action development at the beginning of motion characterise in many aspects the ability of the tribotechnical complex to avoid dangerous deterioration of friction process.

Thickness of a lubricant layers, moment of friction and temperature have been recorded during tests. At some stages of experiments roughness / Ra/, linear wear and specific work of friction have been also determined.

REFERENCES

(1) Dmytrychenko N. F. , Mnatsakanov R. G. Raiko M. V. Grabchak V. G. Mechanism of oil and lubricant action under transient condition of bearing operation. J. Phys D: Apll 24 (1991).

(2) Грошев Н. Б., Дмитриченко Н. Ф., Рыбак Т. И. Надежность сельскохозяйственной техники, Киев Урожай, 1990.-192с.

ESTIMATION OF REAL CONTACT AREA OF PAPER-BASED FRICTION MATERIALS FOR WET CLUTCHES BY INTERFERENCE METHOD USING SCANNING LASER MICROSCOPE

Masao EGUCHI, Tomoo NIIKURA and Takashi YAMAMOTO
Faculty of Technology, Tokyo Univ. of Agriculture and Technology, Koganei, Tokyo 184, JAPAN

ABSTRACT

Paper-based friction materials for wet clutches are desired to have good friction - velocity characteristics for smooth engaging and torque-transmitting. In order to understand the characteristics, it is necessary to consider the real contact area contributing friction generation. Since paper-based friction materials are composite materials that have elasticity and porosity, the contact mechanism is not clear for the complexity.

Optical techniques by total internal reflection and interference allow direct observation of real contact areas although with an inherent disadvantage that one of the contacting surfaces must be transparent to the light (1). Total internal reflection method using polarized light allows to obtain a real contact image of rough surface like as paper-based friction materials (2), but its adjustment in setting up is complicated and the obtained image has normally aspect ratio. The optical interference method can offer a correct real contact image with high accuracy corresponding to light wavelength, and also can observe non-contacting surrounding area. However, the interference method does not succeed in observation of real contact area of rubber seal that has rough surface and lower surface reflectivity (3).

Scanning laser microscope has the confocal optical system and the special memory system that stores vertical location of focal points, then these systems provide images with high contrast and deep focus under an atmospheric circumstance. We attempt to measure the real contact area by interference method with the laser-scanning function and the confocal optical function in this study. An apparatus for pressing a specimen into a glass plate is made and the measuring system is constructed. Contacting area of a paper-based friction material pressed against a glass plate is observed with the scanning laser microscope. Clear optical interference images are conveniently obtained in spite of low reflectivity and high roughness of material. This may be attributed to function of the confocal optical microscope system that only detects faint interference images. From the observation of Hertzian contact area using steel balls, the validity of this method is confirmed with interference fringes based on the two beam optical interference theory.

The real contact area of a paper-based friction material is determined by a computer-processing for monochromatic image in which threshold value is set at the intensity of zeroth order fringe. In this method, higher order dark fringes as shown in Fig. 1(a) appear when a rough surface as a paper-based friction material is observed. This makes estimation of real contact area too complex. Figure 1(b) shows a distribution of intensity in 256-tone grey scale image on the x-x' line as shown in Fig. 1(a). This figure indicates that the brightest zones attain maximum level of 255 and the zones of zeroth order dark fringe recognized as the real contact area coincide to the level of less than 125. This treatment of the interference image makes distinction of real contact area clear. Surface pressure dependence of real contact area of paper-based friction materials is determined from these processed data of interference images. Real contact area of paper-based materials is proportional to surface pressure within 3MPa. According to this method, measurement of real contact area and gap between a glass plate and a low refractive rough surface is successfully carried out.

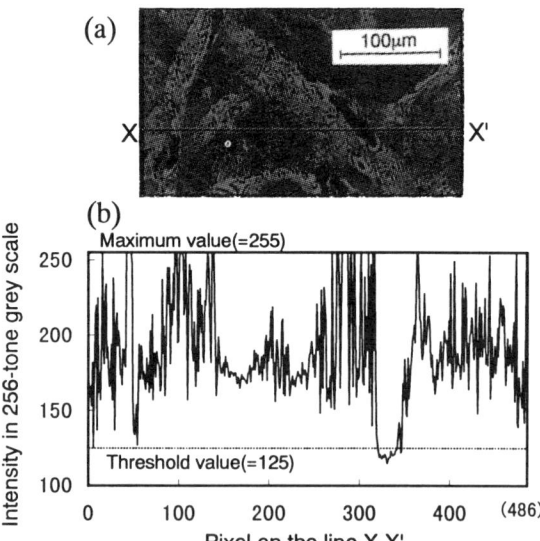

Fig. 1: Intensity distribution of interference

REFERENCES

(1) B. Bhushan, ASLE Trans, Vol. 28, No.1, page 75, 1985.
(2) C. Otani, Y. Kimura, J. of JAST, Vol. 39, No. 12, page 1042, 1994.
(3) C.R. McClune, D. Tabor, Tribology International, Vol. 11, page 219, 1978.

THE APPLICATION OF A RELIABILITY ANALYSIS APPROACH TO THE ASSESSMENT AND PREDICTION OF WEAR

J L HENSHALL
Department of Mechanical Engineering, Brunel University, Uxbridge, Middlesex, UB8 3PH, UK
J KNEZEVIC
School of Engineering, Exeter University, Exeter, Devon, EX4 4QF, UK
M G GEE
CMTS, National Physical Laboratory, Teddington, Middlesex, TW11 0LW, UK

ABSTRACT

There have been many improvements in the wear resistance of components, either as a result of the development of new materials and/or surface coatings, or improved design and manufacture. Nevertheless the problem of wear is still present in many engineering applications.

The causes and descriptions of wear have been well elucidated for many years, and there is a very large database of wear rates derived from laboratory tests. It is however generally only qualitatively, or at best semi-quantitatively, possible to correlate existing laboratory-derived data with in-service performance requirements. The reasons for this include inter alia: (a) material differences, e.g. there may well be differences in surface finish (1), composition and/or microstructure between nominally identical materials: (b) operational conditions, e.g. speed, vibration (2), unidirectional vs reciprocating: and (c) environment, e.g. humidity, temperature. These factors are, of course, in addition to the scatter that may occur even in laboratory tests run under constant operating conditions (3), whereas the actual in-service conditions will vary substantially throughout a component's lifetime. This leads to the observed significant differences in the lifetimes of nominally identical components, where the predominant failure mode is wear related.

The purpose of this study is to start from the premise that there will be variability in wear processes and rates in the vast majority of applications, and to consider the use of a statistical approach to determining component lifetimes. The methodology is based on the reliability analysis developed by Knezevic and co-workers, using the concept of a Relevant Condition Indicator, RCI (4-6). The requirements of the RCI are: (a) that it provides a full description of the relevant condition of the item/component/system, (b) that it changes continuously monotonically during the operating time, and (c) that it provides a numerical measure of the condition of the 'item'. The reliability of a system can then be described in terms of the probability of the RCI being less than some proscribed limiting value.

Wear results on AISI 316 stainless steel vs itself, and alumina, again vs itself, have been used as examples to describe the use of this approach. The data have been analysed in terms of the volume of wear as a function of both [load x distance, i.e. Pd] and [load x distance/ hardness, i.e. Pd/H]. The probability density function and cumulative probability distribution functions at particular values of Pd and Pd/H are determined as the best fit of a normal, exponential, log-normal or Weibull distribution.

Using the RCI approach the reliability of a system as a function of time, R(t), can be described by:

$$R(t) = \int_{RCI_{in}}^{RCI_{lim}} f(RCI,t) \, dRCI$$

where $f(RCI,t)$ is the probability density function of RCI at time t, RCI_{in} is the initial value and RCI_{lim} is the limit value of the relevant condition indicator. In this work RCI is the volume of wear, and time is effectively replaced by 'equivalent time', i.e. Pd, or Pd/H. The measure of wear used as the RCI could be either volume of wear, mass of wear, or wear scar depth. The choice of parameter would depend upon the particular application, and may even involve an indirect estimator of wear volume, e.g. remnant particle weight in a lubricating fluid.

Since the expected operating life of the system/component is equal to the area beneath the reliability function, it is possible to estimate either the operating life at the design stage using laboratory data or the remnant life using inspection derived wear data with this approach.

REFERENCES

(1) M G Gee and E A Almond, Materials Science and Technology, Vol. 4, 877-884, 1988.
(2) M G Gee and E A Almond, Journal of Materials Science, Vol. 25, 296-310, 1990.
(3) M G Gee, ASTM Special Technical Publication No. 1167, 24-44, 1992.
(4) J Knezevic, Reliability Engineering, Vol. 19, 29-39, 1987.
(5) J Knezevic, Reliability, Maintainability and Supportability, McGraw-Hill, 1993.
(6) J Knezevic and J L Henshall, Materials 88, F1-F2, 1988.

A TRIBOLOGICAL STUDY OF A ROTARY DIESEL FUEL PUMP

DAVID JILBERT, CHRIS D RADCLIFFE and CHRIS M TAYLOR
Department of Mechanical Engineering, The University of Leeds, Woodhouse Lane, Leeds, LS2 9JT, UK

ABSTRACT

The environmental impact of road transportation relating in particular to diesel engined vehicle emissions has caused wholesale changes to the demands placed on diesel fuel injection equipment. (1) Vehicle emissions have been improved greatly by the use of highly refined fuels (typically 0.05% sulphur) coupled with increased injection pressures in the range from 10 to 100 MPa (100 to 1000 bar).

A research study has been undertaken in association with a manufacturer of diesel fuel injection equipment to assess the effect of surface finish and clearance on fuel distribution rotor scuffing performance.

The fuel distribution rotor is an important integral part of the rotary diesel fuel injection pump; its primary function being to deliver an exact quantity of high pressure diesel fuel in sequence to each fuel injector. The 19 mm diameter rotor operates in a highly toleranced bore having a typical diametral clearance of only 2μm over the entire 45 mm rotor length. Each rotor is manufactured to a precise specification being round to within 0.4 μm with a typical surface roughness of 0.04 μm Ra. Diesel fuel provides the only means of lubrication in the contact where it experiences shear rates of the order of $10^6 s^{-1}$ and pressures in excess of 20 MPa (200 bar).

The work outlined in the paper has been conducted on a dedicated test apparatus using real fuel pump rotor and barrel components. The apparatus has been fitted with rotor and barrel pairs of differing surface finishes and clearances tested over a range of operating speeds and pressures.

The tribological behaviour of the rotor has been studied by measuring the drive torque, pressure and rotor orbit along with associated determination of changes to the surface topography and overall geometry.

In the absence of a conventional lubricant, the rheological properties of the diesel fuel in such a demanding environment are of obvious importance and can influence the overall rotor performance. Fuel viscosity at both high and low shear rates have been determined prior to, and after testing over a range of temperatures. High shear rate testing of the fuel has been performed using the Ravenfield BS/C high shear rate viscometer.

The topic of lubricity (2) has been addressed during the project and fuel samples have been tested on the High Frequency Reciprocating Rig (HFRR) (3) to establish a lubricity ranking. All current test work has been undertaken at room temperature and pressure.

In association with the experimental work, which has been the main emphasis of the study, it has been possible to compare and contrast the results with data generated from two analytical sources. One system generates results using a commercially available finite element analysis package whilst the other is computer program based working from first principles.

These two very different approaches to the predictive modelling of the rotor orbit have been invaluable in verification of both the experimental results and the analytical software.

The use of capacitance probes to measure rotor orbit has been undertaken with considerable success. It has been possible to detect with certainacy rotor motion to within 0.1 μm using the calibrated transducers positioned at stratecgic points in the outer rotor barrel. Figure 1 below depicts a typical response of a probe to radial rotor motion caused by the injection of high pressure fuel into the rotor contact.

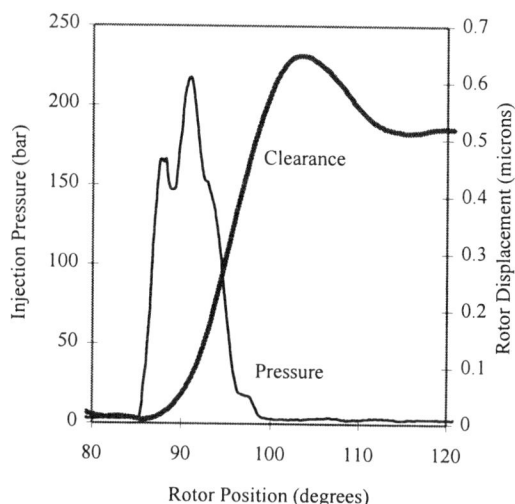

Fig. 1: Rotor Displacement due to Pressure

REFERENCES

(1) J.C.Wang, D.J. Reynolds, SAE Technical Paper Series, 942015, 1994.

(2) H.A.Spikes, D. Wei, Wear, Vol. 111, 1986, 217-235pp.

(3) J.W. Hadley, B. Mills, SAE Technical Paper Series, 932692, 1993.

DEVELOPMENT OF THE TESTING METHOD FOR FRICTION VARIABLE DEVICES

NATASA JOVIC

Department of Mechanical Constructions, Mechanical Engineering Faculty, University of Kragujevac, Yugoslavia

ABSTRACT

The subject of friction is a very ancient one, but it is still a highly controversial one. A very significant part of total energy consumption is expended in overcoming frictional losses during relative motion which makes the reduction of friction an extremely important problem of modern technology. However, it must not be overlooked that very many processes of everyday life are dependent for their effectiveness on the presence of friction in large enough amounts. Hence, the provision, where required, of sufficiently large friction is also a task of grate importance. Friction mechanical transmissions belong to the group of devices where the maintenance of high friction is required.

The paper gives the elements of the experiment done at the Laboratory for Mechanical Transmissions, Faculty for Mechanical Engineering, Kragujevac.

INTRODUCTION

The power and motion transmission by friction is specific in many ways. It is carried out on the basis of friction forces at the contact zone. Relative motion between the friction surfaces of the friction variable device is rolling with sliding. The following miscellaneous processes may occur during rolling and use up energy: sliding elastic hysteresis, plastic deformations and adhesion process. This shows that power transmission by friction is quite a complex one and it is difficult to set down its quantitative laws.

The investigation described here belongs to the group of simulation ones. It was carried out at the test rig designed and made at the Laboratory for Mechanical Transmissions, Faculty for Mechanical Engineering, Kragujevac.

TEST METHODOLOGY

The test rig has the open power circuit with electrical brake and consumer. Driver aggregate type KR-11/2C (37-180 o/min) is the production of "Prva petoletka" FUD Brus Corporation. Experimental variables are the load of the investigated transmission (M_o, n) and elements of the tribomechanical couple.

The principle of experimental research is as follows: the work regime of the driver conical disc is chosen and electrical brake gives the load torque. The contact geometry and friction coefficient of the observed friction couple are known. Those define the work conditions. Observed trybomechanical system is conical disc-friction wheel. Contact materials are metal-textolite (polymer), resp. Nominal contact is linear. Relative motion is rolling with sliding.

Function of the research system is provided by continuous generation of torque and rotation speed signals (monitoring) and periodical analysis of the contact area condition. Change at torque signal indicates some of progressive wear types and therefore it is important to generate it. After established work conditions the demontage of transmission had been done at preliminary chosen working periods. The roughness parameters, topography of contact area and wear are measured. The Talysurf 6 is used to obtain some of the data and as part of data acquisition system, altogether with connected PC.

Obtained files enabled evaluation of some statistical parameters, as autocorrelation and spectrum density function. These parameters bring out the contact area condition more clearly.

CONCLUSIONS

Based on experimental research, some essential conclusions may be noticed:

♦ The friction variable transmission is very specific kind of transmission. Numerous appearances take place at contact zone and none of them is quite known.

♦ Experimental research described in this paper is specially created for purpose of investigation power and motion transmission by friction. This includes purposely-constructed test rig, measurement system and developed software. The test rig is capable of adaptation for various trains.

♦ The comparative diagrams show the great accordance with theoretical examinations. However, some of properties are discovered, such as principle of "autoadjustment". This principle is closely attached to ratio unstability, which is characteristic for this kind of transmissions. Further more, it is interesting to observe the dimensional change of contact area due to appearance described above.

♦ Obtained data have good repeatability and great accordance with theoretical examination. Therefore the prediction of using this very method for testing frictional variable devices is proved.

REFERENCES

(1) Jovic, N., Investigation of the conical friction transmission's tribological process, master thesis, MFK, July 1994.,
(2) Brcic, V., Cukic, R., Experimental methods at construction designs, IRO "Gradjevinska knjiga", Beograd, 1988.,
(3) Pantelic, I., Introduction of the engineering experiment theory, "Prosveta", Novi Sad, 1976.

MEASUREMENT OF TRIBOLOGICAL CHARACTERISTICS IN RESEARCH OF JOURNAL BEARINGS MATERIALS

WIESŁAW KANIEWSKI and STANISŁAW STRZELECKI and RYSZARD WÓJCICKI
Institute of Machine Design, Technical University of Łódź, Stefanowski Street 1/15, 90-924 Łódź, Poland

ABSTRACT

The bearing material in conjunction with lubricant and shaft materials has to display good sliding and dry-running properties, adequate behaviour in respect to wear, running in and bedding in. The pressure distribution in oil film causes within the bearing material a three-dimensional stress state that in the case of variable loading can lead to the fatigue damage. In addition, the temperature gradients produce thermal stresses on the plain bearing material. The basic body of the engine bearing is the sleeve containing the bearing material on the inner surface of the housing. The counterpart is the journal made from hardened steel of higher quality. The design of this tribosystem and further operation demands the knowledge of the phenomena occurring during the operation of bearing. The structure of tribosystem consists the characteristics of elements in which the material characteristics, especially in the case of journal bearings, are of great importance.

The friction varies in relation to the specific load, slide surface roughness and bearing's temperature. It is very important to receive very small value of the friction coefficient and minimum wear through full period of bearing operation. The tribological characteristics of tribosystem, i.e. the friction torque, temperature of the bearing, conditions of lubrication alter according to the developing wear. The properties of the bearing material decide mainly about operating conditions in the range of dry or mixed friction.

The paper introduces the experimental investigations of the journal bearings consisting of three different bearing materials with the journal made from hardened steel as the counterpart. The effect of wear on the tribological characteristics of the journal bearing operation was investigated. Wear was measured by the mean of weight method.

The test rig, equipped in the computerised data transmitting system, allows the full investigation of plain journal bearings and different frictional materials operating in the conditions of mixed, fluid and dry friction under room and higher temperatures. Fast variation of the quantities influencing the phenomena occurring during operation of the bearing or the probe contacting the steel anti probe and their complicity has caused the necessity for applying the computer in the process of experimental investigations. The design of test rig and prepared algorithm of the control and measurements allow the input of the quantities such as the load of the probe, rotational velocity, temperature and pressure of the supplied lubricant. The measurements of input and output values are done by the electronic-mechanical systems.

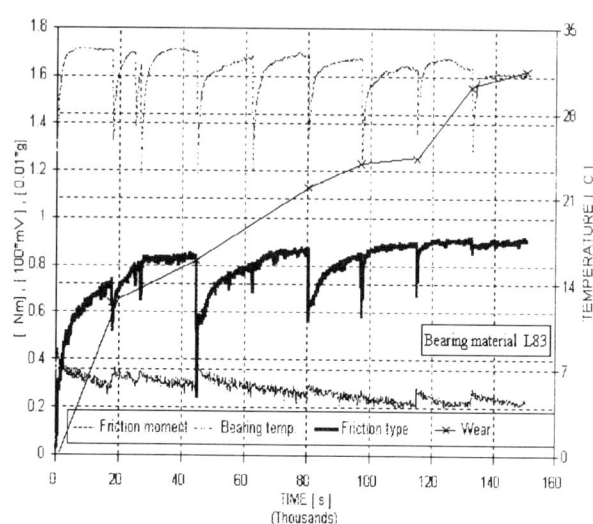

Fig.1 Identification of friction type [mV], moment of friction [Nm], temperature [°C] in the bearing

The results of experimental investigation of the bearing materials have allowed for the following conclusions:
1. There is a good agreement of the character of relative changes of friction type and bearing temperature as well as friction torque, the wear of sleeve and friction type.
2. The characteristics have the similar course for both investigated bearing materials.
3. The value of wear is larger in the case of bearing material Ł16.

The developed test rig allows for the complex frictional investigations and analysis of the different tribological pairs of materials.

REFERENCES

(1) Wójcicki R.: X Conference "Methods and Means of CAD", Warszawa 1995,225-232pp.
(2) Wójcicki R.: XVII Symposium of Machine Design. Proceedings, Vol. II, Lublin, 1995, 1024-1030pp.
(3) Wójcicki R.: Journal of Kones, Internal Combustion Engines, Proc. of the Conference, Vol.2, No.1, Warsaw-Poznań, 1995,268-274pp.
(4) Strzelecki S., Wójcicki R. Proc. of the VI Tribological Conference,6-7 June 1996, Technical University of Budapest, Budapest 1996, 79-84pp.

SURFACE FAILURE IN THRUST CONES AND THE INFLUENCE OF ROLLING AND SLIDING SPEED IN CONCENTRATED CONTACTS

D A KELLY and C G BARNES
Department of Engineering, University of Leicester, University Road, Leicester, LE1 7RH, UK
L M RUDD
Defence Research Agency, Pyestock, Hants, GU14 0LS, UK

ABSTRACT

The work presented (1,2) elucidates mechanisms of surface failure of relevance to thrust cone bearings in particular and concentrated contacts in general. Thrust cones are conical rims mounted near the periphery of single helical gears arranged so that the axial force generated is conveniently reacted.

In the case of a particular full scale thrust cone design, prediction of minimum oil film thicknesses of the order of the surface roughness raised concern that there was a possibility of scuffing and wear. That such bearings might operate at quite diverse rolling and sliding speeds also raised the possibility that failure by plastic deformation might intervene. A lack of adequate criteria to define the effects of rolling and sliding speed on mechanisms of failure focused attention on their crucial influence.

A recent micro-tribological model incorporating salient features of scuffing in mixed lubrication conditions is introduced. It is shown to give the form of two boundaries to a regime in the sliding/rolling speed domain in which running-in may be expected to precede scuffing. The predicted form of boundary may be written as:

$$U = V/2 - P/\ln(1 - Q/V),$$

where U is the rolling or mean surface speed, V is the sliding or difference in surface speed, and P and Q are constants. Above the bounded regime scuffing without running-in may be expected. Below it failure by plastic deformation may be expected.

Results of tests on circumferentially ground discs, at sliding and rolling speeds in the ranges 0 to 14 and 3 to 22 m/s, respectively, are reported for EN36/EN36 disc pairs and for EN36/EN24 disc pairs. They confirm that, depending on speed, failure occurs either by scuffing, with or without prior running-in, or by plastic deformation. The model is shown to rationalise the variations in failure mechanism observed and the results are used to determine values for P and Q that define the general locations, Figures 1 and 2, of boundaries to the regime in which running-in precedes failure by scuffing.

Observations made during small-scale thrust cone simulation tests are also reported and interpreted.

The full scale helical thrust cone design is assessed in the light of the findings and general implications for engineering practice are briefly discussed.

REFERENCES

(1) D. A. Kelly, C. G. Barnes, and L. M. Rudd, "Aspects of Thrust Cone Tribology: Part 1, Effects of Slide to Roll Ratio on Surface Failure Mechanisms in Twin-Disc Tests", submitted to Proc. I. Mech. E., Journal of Engineering Tribology, Dec, 1996.

(2) L. M. Rudd, C. G. Barnes, and D. A. Kelly, "Aspects of Thrust Cone Tribology: Part 2, Surface Failure in Thrust Cones and the Influence of Rolling and Sliding Speeds in Concentrated Contacts", submitted to Proc. I. Mech. E., JET, Dec, 1996.

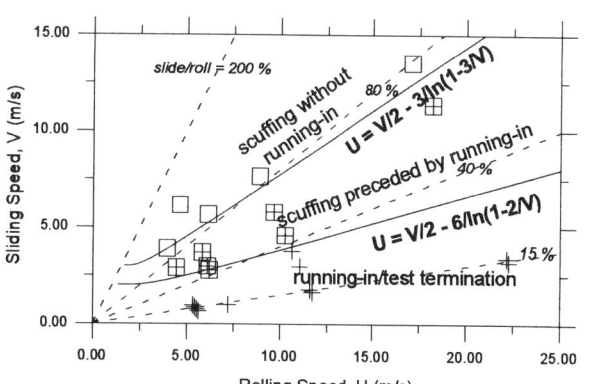

Fig. 1. Sliding Speed v Rolling Speed for EN36A/EN36A cylindrical disc tests

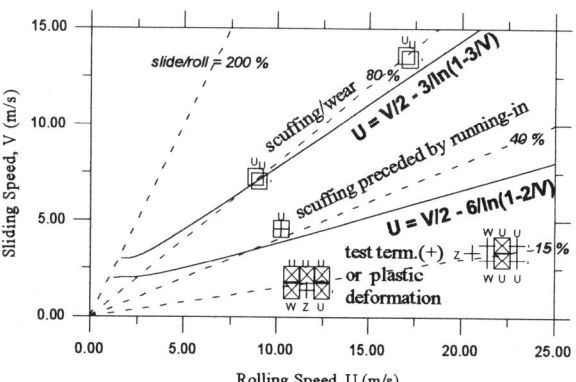

Fig. 2. Sliding Speed v Rolling Speed for EN36A/EN24 U, W, and Z, cylindrical disc tests

DYNAMIC FEATURES OF A TRIBOMETER AFFECT FRICTION AND WEAR TEST RESULTS

TADEUSZ LUBINSKI, KRZYSZTOF DRUET, JACEK IGOR LUBINSKI
Technical University of Gdansk, Faculty of Mechanical Engineering, ul. Narutowicza 11/12, 80-952 Gdansk, Poland,

ABSTRACT

The paper explains mutual dependence of dynamic features of a machine and friction produced between its moving members. The problem is of great importance to improvement of practical usefulness of friction experiments. Large spread of friction and wear test results is still not elucidated. There is a great number of various friction tribometers in world laboratories. Their dynamics is ignored even though it may be one of the most influential factors to friction produced.

It is obvious that for a specific tribo-system the course and effects of friction are functions of instantaneous friction parameters e.g. load, velocity and temperature actually existing in the friction zone.

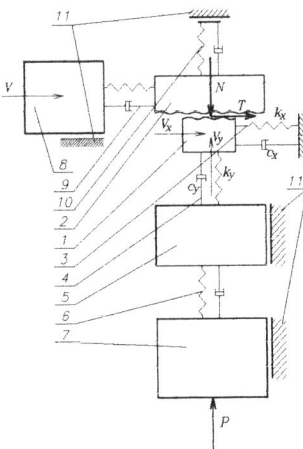

Fig. 1. Dynamic model of tribometer PT-3/96 (

In tribometers friction parameters measured are applied to members remote from friction zone. Friction measurement systems installed in most tribometers comprise an elastic element to provide large strain necessary for a proper accuracy. This determines system dynamics. Even identical types of tribometers, for example four ball apparatuses, are being equipped with different measurement systems. Literature reports evident influence of measuring systems on friction course and wear rate. So intermediate members and various measuring systems can influence the test parameters actually applied to friction surfaces examined. This very situation is shown in Fig. 1 which depicts the dynamic model of tribometer PT-3/96. The tribometer is described in an accompanying paper (1). Numbering in the above drawing corresponds to the one used in Fig. 1 in ref. (1) where a detailed legend is given. Tribometer dynamics should be understood as stiffness and inertia of members transmitting motion and load to specimen contact.

Neither the friction nor the applied parameters are constant in time. Varying velocity and load will evoke dynamic response from the system thus actual friction parameters will be different from those intentionally applied. Additionally any mechanical system in motion introduces its own dynamic input resulting from form technological imperfections such as rotating mass eccentricity etc .

Friction may be accompanied by vibration. An example of friction induced oscillations generated to examine their influence experimentally is presented in an accompanying paper (2). Vibration may be self-induced or penetrate to friction contact in many test situations. Literature data considering the dependence between system stiffness, friction induced vibration, and wear are not consistent. In one case friction induced vibration caused transition from mild to severe wear increasing wear rate even to few ranges of value. Another reports inform about beneficial affects of vibration on friction, wear rate, load carrying capacity and temperature rise. Vibrating cutting tools improved quality of machining. A commutator with slight waviness performed better than an exactly cylindrical one. Very heavy objects are easily moved about vibrating tables. Technical practice serves with many more similar examples. Despite the inconsistency strong mutual dependence between friction and system dynamics exists with no doubt, and may be the source of abnormal spread of many friction test results.

The authors are carrying out the research program using deliberately developed and equipped tribometer PT-3/96, already mentioned, of own design. Details will be given in a paper submitted to Wear.

The complex research goal is not only how to improve test results repeatability but how to use the results in engineering with an acceptable confidence. If the dependence being under consideration really exists in tribometers then a similar dependence must be true in any machine. Equations to predict wear rates seem to be easier to obtain experimentally than in any theoretical way. They are, similarly to stress analysis equations, of desired in computer aided design applications.

REFERENCES

(1) Łubiński T, Testing of sliding friction dynamics on a tribometer PT-3/96, Proc. of the Congress.
(2) Lubinski J I, Druet K, Lubinski T, An experimental study of friction induced vibration in a ceramic contact, Proc. of the Congress.

TESTING OF FRICTION DYNAMICS ON A TRIBOMETER PT-3/96

TADEUSZ LUBINSKI

Technical University of Gdansk, Faculty of Mechanical Engineering, ul. Narutowicza 11/12, 80-952 Gdansk, Poland,

ABSTRACT

A tribometer has been developed to investigate mutual relations between dynamic characteristics of a tribometer and a friction produced itself at a test model. An exemplary test model consists of two specimens, 1 and 2 (Fig. 1). Load is applied by a hydraulic plunger actuator, 7, via a servo-valve. Moving member of the friction couple comprises transmission pulley and a spindle with the upper specimen, 2, attached at the end is rotated by a dc electric motor, 8, and a corrugated or poly-V belt transmission, 9. All moving functional members are supported by hydrostatic bearings, e.g. 6 and 10. Fluid films in these bearings guarantee high precision of measurement and insulation of friction zone from external mechanical interference. The self-aligning bearing 6 allows test samples to align for proper contact without the necessity of running-in.

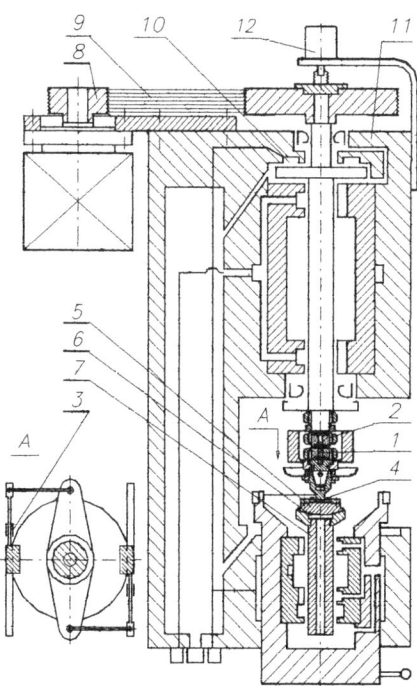

Fig. 1: Scheme of tribometer PT-3/96 head

A computer system controls test run, digitises and stores output signals representing (a) lower specimen displacement in sliding direction, (b) the specimen normal displacement and (c) rotational velocity of upper specimen that are produced accordingly by: (a) strain gauges, 3, glued on two bent beams, (b) attached to membrane, 4, and (c) tachometer, 12, linked to the spindle pulley. In addition to these measurements local accelerations of various components can be measured by accelerometers fixed to them. Calculation is possible of forces at friction contact, sliding velocity, accelerations and power spectra of friction induced oscillation considering inertia and possibly damping forces (1).

The most typical sets of specimens (test models) that may be tested on the rig are shown below:

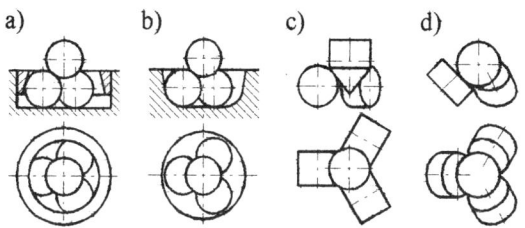

Fig. 2. Point contact area test sets

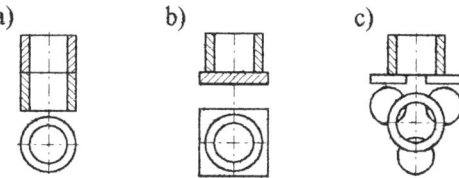

Fig. 3. Annulus contact area test sets

Methodology of experimental friction examination is a very important domain of tribology but not developed properly to enable professional verification of theoretical mathematical models or create equivalent experimental ones. In all situations we deal with a model, a procedure and an apparatus. Standards describe only the first two; apparatus details are out of their description. Standardisation of not only test models and procedures, but test apparatus as well seems to be justified. Design details should be internationally agreed to allow for the use of identical tribometers on a world-wide scale. Fulfilment of the task needs research to establish mutual influence of friction and tribo-mechanical system dynamic (1).

To perform the defined examination certain moving parts of the tribometer PT-3/96 are replaceable to obtain various dynamic system parameters, e.g. mass, moment of inertia or constraint stiffness of moving test-tribo-dynamic-system members.

Paper with details involved in the vital and persistent goal in engineering will be submitted to Wear.

REFERENCES

(1) Lubinski T, Druet K, Lubinski J I, Dynamic features of a tribometer affect friction and wear test results, Proc. of the Congress.

Submitted to *Wear*

SEIZURE OF RAIL AND WHEEL STEELS

D. MARKOV

Department of Metals, Railway Research Institute, 3 Mytishchinskaya 10, 129851, Moscow, Russia

ABSTRACT

Severe form of wear often occurring on the side surface of rails and wheel flanges in a small radius curves is associated with formation of extensive grooves and scratches which according to the ASTM standard G 40-92 and Russian standard GOST 27674-88 can be qualified as scoring. Provisionally, until wear terminology is formed, term *seizure* refers to the severe form of wear which is characterised by intense plastic deformation (plowing) of one or both surfaces within a small real contact area - seizure nucleus. Terms *scoring* and *galling* refers to types or mechanisms of seizure.

The dependence of seizing load vs. hardness of movable (H_b) and motionless (H_t) rollers in air is shown in Fig. 1. Area of scoring existence is diapason of hardness within 700 HV and slow sliding speeds. Area of galling existence is hardness of motionless roller higher than 700 HV or high sliding speeds. When hardness of movable roller is higher than 700 HV and sliding speed is low, neither scoring nor galling has place even under extreme load.

Scoring. Main feature of scoring is plastic deformation of *both surfaces*. During scoring two wedges form, one on the top and other on the bottom roller, Fig. 2. As a result of wedges interaction the rider - hard movable protruding nucleus - arises ahead the groove and semigouge. Role of the adhesion in formation of scoring is insignificant. There is only mechanical bonding between wedges. Mean value of wedge's hardness is 1.6 time higher than bulky hardness of roller: $H_{wedge} / H_{bulk} > 1.6$ - that is the condition of rider's development. Wedges never develop on the top and bottom rollers simultaneously. They form one by one and develop by turns very quickly but not instantaneously. It needs some sliding distance for plastic deformation, hardening of wedges, development of rider, groove and gouge.

Galling. Basic features of galling is plastic deformation of only one roller surface and transfer of material. Role of adhesion in galling is very big. Adhesive wear always accompanies the contact of two bodies even if there is a slight touch, but only the progressive transfer of material leads to galling. Every time when asperities are interacting, only small particles of metal transfer even if operation pressure is extreme and surfaces are pure. Galling protrusion gradually arises on the friction surface. Final stage of galling can be qualified as a welding in adhesive wear process. Gross energy generates mainly during plastic deformation and so soft metal always welds on the hard surface. In contradiction to the rider, the galling protrusion strongly attaches to both surfaces and breaks brittle when rollers move apart.

Some conclusions follow from scoring mechanism of seizure: Slippage in rolling-sliding motion should be periodic changed in order that scoring might develop. Period of slippage alteration must include pure sliding phase with sufficient duration. Therefore there is slippage threshold of seizure for longitudinal slippage, which does not depend on pressure and which differs for tribosystems with different rigidity. There is critical slope of wheel flanges and critical rolling speed of wheels which prevent seizure due to elimination of pure sliding phase.

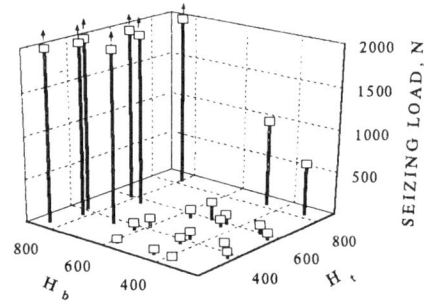

Fig. 1: Seizing load vs. hardness of Amsler's rollers

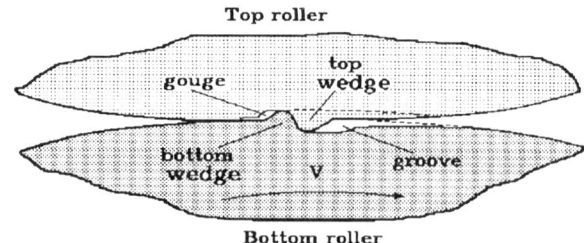

Fig. 2: Scheme of scoring formation

The results of this research will be exhaustively explained in paper '*Laboratory tests for seizure of rail and wheel steels*' in 1997 in Wear.

BOUNDARY LUBRICATION DETERMINATION OF FRICTION MATERIAL

G MIKOLASCH, A PAUSCHITZ, F FRANEK, G ABRAHAM

Institute of Precision Engineering - Department of Tribology,
Technical University Vienna, Floragasse 7, 1040 Vienna, Austria

ABSTRACT

The primary task of friction materials is to guarantee a high and constant friction moment and avoid high wear rates.

Friction materials are generally operating dry, but in some clutches and in automatic vehicle transmissions they run lubricated. Lubricants remove heat from the contacting surfaces and influence the friction and wear characteristics by forming reaction layers.

The formation and destruction of reaction layers in wet friction materials depends on friction material, interfacial medium and operating variables.

A new measuring device - called High-Frequency-Impedance-Measuring-System (HIFIMS) - developed at the Institute for Precision Engineering, Department of Tribology, Technical University Vienna, has been applicated to a standard disc-friction-and-wear-apparatus to determine the contact intensity and the generation of reaction layers.

HIFIMS consists of a HF-oscillator, which oscillates as a function of the electrical impedance connected to the oscillator input.
The HF-oscillator drives two external capacitors. One of the capacitors is formed by the metal mating samples and the intermediate liquid. The oscillator frequency depends on the distance of two electrical conducting metal surfaces, lubricant performance, reaction layers, series capacitor and additional parallel capacitors (1).

Using an additional series capacitor a precise adjustment is possible. The measuring voltage on the series capacitor decreases to very low values and thus avoids critical electric field intensities. Thus a detraction of intermediate liquid is avoided and a falsification of the measured value is excluded, and also if there is only a small clearance between the two plates and the gap is filled with a dielectric.

The test rig is based on the twin disc principle. The basic body is fixed on a clamping unit which is connected with the friction measuring device. The counteracting disc is driven by a motor unit.

The test body is a steel disc with a porous friction layer, which is made of a blend of different metal powders. A steel disc is used as counteracting body and different oil as intermediate liquid.

The center of the fixed basic body is connected with a separate lubrication tank so that the gap between the friction plates is supplied with a lubricant which also works as a cooling agent.

The fixing devices of the discs are electrically isolated from each other and from the bench, and connected to the measuring device. A voltage signal for the HIFIMS is transmitted to the disc in motion by means of a coupling capacitor.

Physical and chemical variations of the intermediate body or of the friction layers lead to a change in the HIFIMS signal - in some cases before the coefficient of friction shows a remarkable change in the tribological condition.

Boundary layers, which are formed by physical and/or chemical reactions, are destroyed by solid body friction process and wear particles. This can be dedected by the HF-oscillator from the decreasing output frequency. The generation of a boundary layer leads to an increasing HIFIMS output.

A lot of friction and wear effects cannot be detected by simply measuring the friction moment or analysing the surface after the tribological process. The monitoring by HIFIMS gives additional information of formation and destruction of boundary layers.

REFERENCES

(1) ABRAHAM, G.; FRANEK, F.; EBRECHT, J.: Evaluation on lubricant quality by monitoring the electrical impedance of lubricating films. Proceedings, 6th International Congress of Tribology, Budapest, EUROTRIB (1993), Vol.4, 99-105pp.

INFLUENCE OF SURFACTANTS ON THE STATE OF SURFACE MICROVOLUMES OF THE METAL AT FRICTION

OLEG A. MISHCHUK and YURI L. ISHCHUK
Ukr.NDINP "MASMA", 46 Palladin Avenue, Kyiv 252680, Ukraine

ABSTRACT

As known the surfactants are the organic materials which have the high adsorption activity. However this terminology is acquiring the new sense at friction. It is its cause that physical and chemical processes in the friction contact zone are highly energetic. There are a plastic deformation of a metal, its surface doping by the elements of lubricant environment, a formation of the protective solid films with the ultracrystalline structure, the metal wear etc. Even at the antiwear conditions the scale of these processes is exceeding on a few orders the scale of the tribochemical transformation adsorption layer and the surface energetic influence. Thus at friction, the surfactants must have such surface activity which has the effect on the physical and chemical state of the relatively thick surface microvolumes of metal. For example, it had been found that the soap thickener of greases with various cation furthers the formation of the friction metal surfaces with various chemical composition and tribotechnical characteristics (1). Now, regularities of similar surfactants` action are not practically studied.

In this work the search and the study of main factors of the surfactants` effect on the state of thick metal surface layers at friction are the purpose. The model base greases with various soap thickeners were the objects of this investigation. Their properties is fully described in (2). We were implying that the greases must not have sulphur, phosphorus, chlorine and other electronegativity elements, with the exception of oxygen from this row, in their molecule composition.

At the same time with tribology of friction pairs, the states of the contact friction zone and friction surfaces were investigated by the methods of current-voltage characteristics, scanning electron microscopy and Auger spectrometry (Auger microprobe JAMP-10S). New possibillities of analysis of Auger spectrum and depth-profiles of elemental compositions are used also (1).

It is showed that physical and chemical states of metal surface layers are changing at friction as a result of their deformation, oxidation and carbonization. The carbonization of metal surfaces at friction has restrained their oxidation. It had been found that the source of carbon for theirs is the organic environment. The study of the structure and the chemical state of friction surfaces had witnessed the existence of the correlation and the interaction between various surface layers, where the adsorption layer is one of them.

The results of our investigations were generalized in the following way.

The certain self-organization of diffusive zone of surfaces takes place as the result of competitive diffusion of oxygen and carbon in the surface layers of metal. It leads up to the division of the diffusive zone on the clear formed surface layers with various phase composition.

The thin surface oxygen layer which is one of theirs has properties of ionic membrane. It regulates the diffusion of elements (carbon and oxygen) from the adsorption layer to the metal at friction.

Various properties of the surface oxygen layer are formed as a result of its various doping by atomic carbon at friction. A few correlations between the properties of adsorption and carbon-doped oxygen layers were studied.

Consequently, the new action mechanism of surfactants at friction, according to which their influence on the state of the metal surface layers and on the values of friction and wear is connected with selective doping these layers by oxygen and carbon, is discovered. The existence of the carbon-doped nanoscaling oxygen layer, diffusion transparency of which is regulated by surfactants, is one of main factors of this mechanism. Such mechanism, for example, may be significant for the soap greases.

All results of our research is fully stated in the paper of this title.

REFERENCES

(1) O.A.Mishchuk, B.A.Godun, *Soviet J. Friction and Wear*, Allerton Press (USA), 1993, vol. 14, No. 2, pp. 141-144.

(2) Yu.L.Ishchuk, *Composition, Structure and Properties of Plastic Greases*, Naukova Dumka (Kyiv, Ukraine), 1997, 516 p.

ON THE EFFECT OF MISALIGNMENT ON THE PERFORMANCE OF U TYPE LIP SEAL

M O A MOKHTAR, M A A MOHAMED AND M E ELGEDDAWY
Mech. Design and Prod. Dept., Faculty of Engineering, Cairo University, Cairo EGYPT
S A Y YASSEN
El-Nasr Pharma Chemicals Co., Cairo EGYPT

ABSTRACT

In the literature on the radial lip seal performance, great attention has been paid to the sealing lubrication mechanism and factors related thereto. (1). The Present experimental work is, however, an endeavor towards correlating the operating conditions to the rate of fluid leakage under different misalignment conditions.

The experimental results have been recorded under three values of operating pressure ,(7,10 and 13 bar), two values of shaft running speed,(377 and 655 rpm) and using two types of hydraulic oils, (37C and 100C having kinematics viscosity at 40°C of 37 and 100 cst respectively).

The Effect of angular misalignment: The results, figure 1,show that under test conditions, the leakage rate is reduced with slight increase of angular misalignment.

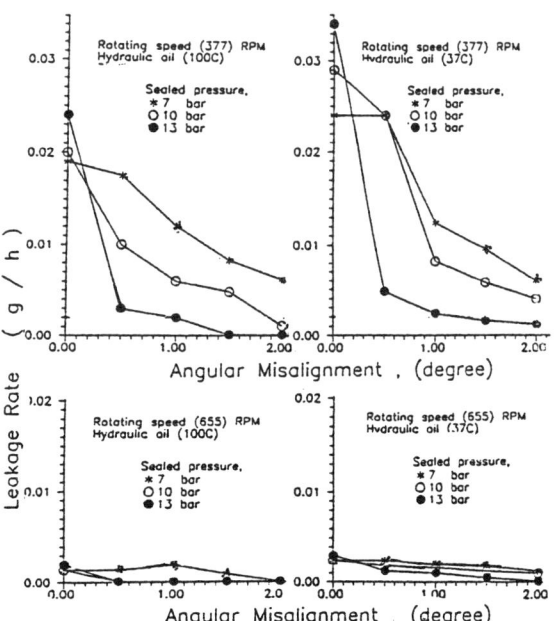

Fig.1 : Effect of angular misalignment on leakage

Effect of bore eccentricity : Results as shown in figure 2 show that the leakage rate increases, in general, with the increase in bore eccentricity with running shaft. Higher fluid pressure render higher values of leakage rates.

DISCUSSION OF RESULTS

It was earlier postulated that surface tension is responsible for retaining the lubricant film at the seal interface. This hyposis could not be supported by later work (2). As a result of small contact width at lip interface broad temperature rise occurs and the mechanism of sealing becomes complicated. Based on seal pumping action, it could be concluded (3) that the pumping action of the seal is counter balanced by the capillary forces of the oil/air interface on the air side. Recently, on the bases of coupling between the sealing mechanism and the lip microgeometry, the sealing mechanism is viewed in terms of physical reverse pumping process.

The effect of misalignment can be discussed in light of the increase in the seal/shaft contact pressure with the increase in the degree of misalignment. Though this would result reduced leak, ashown figure 1, an expected higher contact temperature and possibly would result a reduced sealing life.

Under eccentric seal mounting with the shaft (bore eccentricity) the contact pressure distribution will vary along the seal circumference giving an almost loose contacts at some regions with increased possibility of leakage as shown in figure 2.

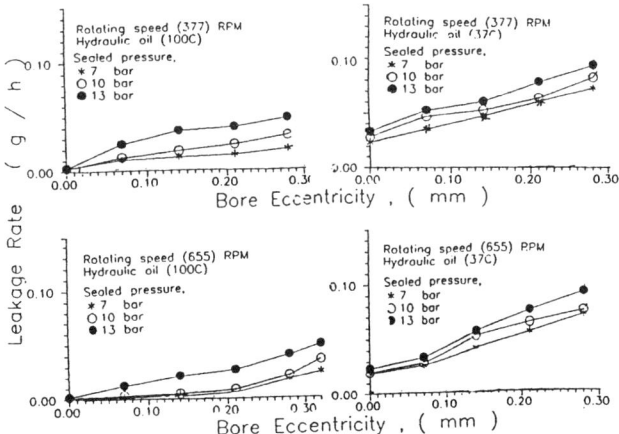

Fig.2: Effect of bore eccentricity on leakage

REFERENCES

1. Mokhtar,M.O.A. et al, "Effect of oil viscosity on the seal behaviour of u-type seal", to be published
2. Gabelli,A. and Poll,G.,"Formation of lubricant film in rotary sealing contact I",ASME JoT,1990,pp.1-8
3. Stakenborg,M.J.L.,"On the sealing mechanism of radial lip seals"Tribology Int.21,1988.pp.335-40.

Submitted to *Wear*

A HIGH-SPEED FRICTION STUDY OF ELASTOMERS

D F MOORE and K KELLY
Department of Mechanical Engineering, University College, Dublin 4, Ireland

ABSTRACT

This paper describes results some results and conclusions from an ongoing study of hysteresis friction on elastomers at University College Dublin. The work has been carried out within the Tribology Design Centre using natural and synthetic rubber material. The approach has been to regard rubber as a viscoelastic material that can be represented using separate spring and damping elements. The complexity required to simulate actual properties is introduced by making these elements both frequency and temperature - dependent. This representation has been described as a complex Voigt model (1), and its great advantage is minimum mathematical complexity with realistic property modelling.

The mechanism of adhesion dominates in elastomeric friction at creep sliding speeds [0 →0.1 m/s]. Here, the well-known Williams-Landel-Ferry (WLF) transformation (2) permits a useful interchange of frequency and temperature effects. The relevant temperature of interest can be considered constant, having a nominal or ambient value. At slightly higher sliding speeds [0.1→5 m/s], the hysteresis mechanism appears, taking the form of delayed recovery of the elastomer after indentation by discrete surface asperities in a sliding situation. The interface temperature in this situation can still be considered reasonably constant if a lubricant is used to suppress adhesion. At medium to high sliding speeds [5 →50 m/s], frictional heating dominates to such an extent that ambient temperature is irrelevant. Experiments described in this paper have established in this sliding speed range a thermo-visco-elastohydrodynamic balance, wherein the following conditions apply:

(a) The hysteresis mechanism now dominates in the form of interactive frequency-stiffening and temperature-softening effects,

(b) The adhesional mechanism has been suppressed using a lubricant and smooth ball-bearing and cylindrical surface elements, and

(c) Interfacial temperature is largely determined by the speed of sliding through the mechanism of frictional heating.

Since each sliding speed (or frequency of indentation) selected has its own self-determined interface temperature rather than a constant pre-selected value, the concept of a WLF transformation as originally proposed has no meaning or benefit at high sliding speeds. In these experiments, coefficient-of-friction vs. frequency plots show no pattern. However, when temperature effects (measure with an insertion probe) are normalised to one value, the same data transforms to a single curve with a distinct viscoelastic peak. Simple mathematical analysis, using a combination of relaxation times and complex Voigt modelling equations, determines the method of normalisation used. The important conclusion is that whereas such normalisation at high sliding speeds has no physical relevance, it appears to validate the basic assumption of viscoelastic behaviour (3) for the rubber-like materials used in this study.

The paper concludes with a discussion of how the relative magnitude of viscoelastic frequency-stiffening and temperature-softening effects in high speed hysteresis friction can be varied by selecting the type of surface asperities.

REFERENCES

(1) D.F.Moore, Wear, Vol.158, 1992, pp.185-192
(2) M.L.Williams, R.F.Landel, J.D.Ferry, Jour.Amer.Chem.Soc., Vol.77, 1955, p.3107
(3) D.F.Moore, Viscoelastic Machine Elements, Butterworth-Heinemann Ltd., Oxford, 1993, p.52.

Submitted to *Leeds-Lyon Symposium on Tribology Proceedings*

DETERMINATION OF TRIBOTECHNICAL PROPERTIES OF NEW COMPOSITE MATERIALS MADE OF IMMISCIBLE COMPONENTS USING THE DIAGRAMS OF CONTINUOUS INDENTATION

V. P. ALEKHIN, S. I.BULYTCHEV, V.V. POROSHIN, A.D.SHLIAPIN
Moscow State Industrial University, Avtozavodskaya St.,16, 109280, Russia

ABSTRACT

One of the main trends of tribology consists in revealng, classification and detailed study of structural levels of deformation and tribofracture and using this as a base for adequate models of friction processes [1]. Wear resistance depends on geometric parameters of contacting surfaces of materials and correlates to mechanical properties of surface layers. Due to the roughness and waviness of the surface thr contact of two solids is always discrete , while wear occurs as a result of the action of the local stresses and strain in the region of local contact . Therefore testing materials using the microindentation is rather advanced technique to study the features of materials' wear. The possibility itself to gain the information about physico-mechanical properties using the microhardness tests [2] stimulated the progress of indentation fracture mechanics as applied to the wear problem [3]. However, it has been experimentally shown that microhardness test with continuous control of indentor parameters delivers new possibilities for the solution problems dealing with friction and wear, fatigue and brittle fracture. This is achieved due to simultaneous determination of a number of physico-mechanical properties of material and its structural characteristics.

The microhardness tests accompanied by continuos registration of indentation parameters possess high resolution ability. E.g. the equipment of "Nanoindentor Co" [4] has the resolution with respect to the displacement of about 10^{-9} m. On microhardness testing at small loading when indentation sizes are less than those characteristic of the phase inclusions, the registered values of microhardness will be distributed inside the groups, the number of which equals to that of the phases. The scattering inside each group will be determined by the anisotropy of microhardness of each phase and number of tests.

Hence the experimental microhardness histogram may be used to determine the microhardness values of separate phases, while the height of histogram peaks will give the volume fractions of the phases.

Increase of loading results in the indentation sizes significantly exceeding that of the phases inclusions. In this case the hardness becomes independent of the indentor positioning and the hardness of macroscopically uniform material is measured. Its value is expressed by one fixed . For intermediate values of the indentation sizes, when they are comparable with those of the phase inclusions the view of histogram will be also intermediate. Thereby the dependence of microhardness variation coefficient upon tjr indentation diameter determines the scale of structural inhomogeneity of a material. Using the approaches resembling those of stereometric metallography one can estimate the change of the shape and distribution of the particles over their dimension in the run of wear simultaneously with the determination of the materials' physico-chemical properties

Experimentally the theory was verified using the materials made of immiscible components, produced with the help of technology described in [5].

The group of materials based upon the equilibrium diagrams having miscibility gap is rather important. According to NASA estimation in case of proper microstructure of such alloys a variety of new composite materials may be obtained having a number of different special properties such as low friction coefficient, high damping and wear capacity, high thermal stability, etc. Preliminary study of such alloys as Al-Sn-Pb, Fe-Cu-Pb, Fe-Cu-Pb-Sn, Fe-Pb, Fe-C-Pb, Co-Pb etc obtained in this laboratory confirmed the prediction. Apart from extremely high properties the above alloys possess an outstanding model structure, suitable for carrying out gauge experiments for new testing instruments and methods. This is also the case for continuous indentation technique.

REFERENCES

(1) Tribology. Investigations and applications: Experience of USA and republics of CIS Collection ed. By V.A. Bely, K. Ludema, N.K.Myshkin. Moscow, Mashinoctroeniye, 1993, 452 pp.
(2) Bulytchev S.I., Alekhin V.P. Testing Materials by Continuous Indentation, M., Mashinistroenie, 1990, 224 pp.
(3) Kolesnikov Yu.V., Morozov E.M.. Mechanics of contact fracture.. Moscow Nauka.:, 1989, 221pp..
(4) Pharr, W.C. Oliver, D.R.Clarke. The Mechanical Behavior of Silicon Small-Scale Indentation. J. of Electronic Materials, Vol.19, ¹ 9, 1990, pp. 881-887.
(5) Avraamov Yu.S., Shliapin A.D. et.al. " Structure and Properties of Al-Sn-Pb Alloys with High Lead Content", Design, Technology and Economics in Motor Car Building, Moscow, 1987, 3pp.

Submitted to *ASME Journal of Tribology*

APPLICATION OF EDDY CURRENT NDT FOR ESTIMATION OF WEAR LOSS

L. RAPOPORT and Y. BILIK

Department of Mechanics and Control, Center for Technological. Education., P.O.B. 305, Holon 58102, Israel

ABSTRACT

Material loss through wear is one of the important parameters that determine the behavior of contact pairs. Weighing, measurement of geometrical parameters, profilography are mainly applied in evaluation of wear loss.

More recently NDT become widely applied in tracking friction and wear behavior, one of these method being that based on acoustic emission. A definite relation has been found between the signals of acoustic emission, the friction coefficient and the character of fracture, but no such connection could be established between the signals of acoustic emission and wear loss. In this relation, eddy current method (ECM) can be applied in wear measurement. Because of the ECM possibility to inspect the clearance between a coil and a wear surface, wear losses can be measured. This may be of great interest in tribology due to the rapid action and the comparatively low cost of this measurement system.

The main problem in the application of ECM to wear measurement is that of selectivity viz. the distinction between the "useful" signal indicating the change in wear depth and the signals caused by such undesirable surface effects as distinctions in roughness, strain hardening, and possible phase transformations (1, 2).

The research presented here is one of the first works on the application of ECM to the determination of wear loss.

Eddy current device (ECD, UH-Locator, Hocking, UK) with local eddy current probe (ECP, type 310 P24) of $d = 3\ mm$ and working frequency of 2 MHz was used in order to measure the wear loss causing the clearance between the probe and the surface concerned. The problem in wear loss evaluation is the measurement of the signal indicative of the probe average clearance increment $(h_{av.})$ across the wear track (3). As a result, the dependence of $(h_{av.})$ vs. position ECP within wear track named " clearance profile" was obtained. ECP was fixed in a three-coordinate mechanism in order to calibrate the device and to measure the change in the clearance.

Dry friction tests were carried out on a pin-on-disk machine with a sliding velocity, V, of 0.45 m/s and loads from 2 N to 13 N. Ferromagnetic (steel) and non-magnetic materials (bronze) were used in this study. The effects of load and time were studied in this experiment. The wear losses as represented by the area of wear tracks ("clearance profiles") and by the weight loss were compared. Based on a preliminary theoretical and experimental study, the optimal original clearance between the surface tested and the probe was determined to be of 0.2 mm.

It was established that the sensitivity to the change in clearance between a probe coil and a standard virgin surface is ~ 0.4 μm (gain = 800) for bronze and ~ 0.8 μm (gain = 400) for steel. The results of a comparison between the wear loss as measured by ECM and that determined by the weight method showed that the maximum difference between the values of wear loss as measured with ECM and that determined by weight method, is less than 15-20 % a result that satisfies engineering practice. The measurement of "clearance profiles" at different sections of the specimen showed that the non-uniformity of the wear in the wear track due to the wobble of sample may be also measured. . "Clearance profiles" of wear tracks can be described as integral profiles as distinct from the roughness profiles. A good correlation between the wear volume calculated by ECM and that determined by the weight method testifies the integral "clearance profiles" reflecting mainly the geometrical form of the wear track.

The important peculiarity of the wear process is the deformation and structure change of the surface layers. It is shown the possibility to take into account the change in the noise signals with change of roughness, and also small constant varies in structure and deformation state that allowed to enhance the accuracy of ECM.

On the basis of this non-contact method inexpensive simple apparatus can be developed for the measurement of wear or clearance in different contact conditions. The finish goal of this work is a dynamic inspection of the wear tracks when the velocity of the specimen is quite high.

REFERENCES

(1) F. Forster, Z. Metallkunde,, Vol. 43, 1952, 163-171pp .
(2) H.L. Libby, Introduction to electromagnetic nondestructive test methods, John Willey & Sons, 1971.
(3) Yu. Bilik, A.L. Dorofeev, Soviet Non-destructive testing, in USA, Vol. 17, 1981, 447-452pp.

Presented on the First World Tribology Congress, 8-12 September 1997, London

PRESENTATION OF A HIGH FREQUENCY TRIBOMETER FOR MICROSCOPIC FRETTING

P. REHBEIN
Robert Bosch GmbH, Department FV/FLM, Postfach 106050, 70049 Stuttgart, Germany
J. WALLASCHEK
Heinz Nixdorf Institut, Universität GH Paderborn, Fürstenallee 11, 33102 Paderborn, Germany

ABSTRACT

A test rig was developed for high frequency fretting conditions (20 kHz, 9-48 µm amplitudes) with superimposed macroscopic sliding (about 1 Hz, 1-10 mm stroke length). It is suitable for different sample geometries, mainly bloc-on-cylinder and normal loads range from about 10 N to 100 N. Using this tribometer the high frequency frictional behaviour as well as the wear characteristics of polymer/metal and alumina/alumina couples were studied with regard to engineering applications. In this contribution the test rig and some typical results of the friction characteristic are presented. The results are useful used for solving tribological vibration problems in sliding contacts, especially frictional drive applications as ultrasonic piezo motors and other automotive components.

INTRODUCTION

Presently, there are no high frequency tribometers commercially available for mechanical testing at high frequencies and micrometer-amplitudes. Also there are no tribometers available which are able to allow the investigation of oscillations superimposed by macroscopic sliding. High frequency vibrations (in the kHz-regime) of sliding contacts cannot be investigated by conventional oscillatory test rigs working below 100 Hz. Because of high frequency fretting problems in technical applications there is need for a special test rig realising such tribological working conditions. Additionally there still exist many fundamental questions with regard to the high frequency friction and wear processes that can be systematically investigated using such a test rig.

CONSTRUCTION OF THE TRIBOMETER

A resonant high-power piezo actuator system consisting of a converter, booster and a sonotrode is used to excite the high frequent oscillations of the test specimen which is of ring form. The cylindrical contact surface is oriented perpendicular to the motion. The test specimen is mounted at the end of the sonotrode (fig. 1). Working in resonance (at 20 kHz) 9-48 µm axial amplitude corresponding to an average relative velocity of more than 1.5 m/s can be reached.

Simultaneously, the whole exciting system (converter, booster and sonotrode) is moved by a reciprocating slider in order to avoid local thermal heating and to allow wear particles to be transported away. The wear conditions that can be reproduced on the test are representative for a wide variety of engineering applications like e.g. ultrasonic motors or nozzles or electrical contacts under fretting.

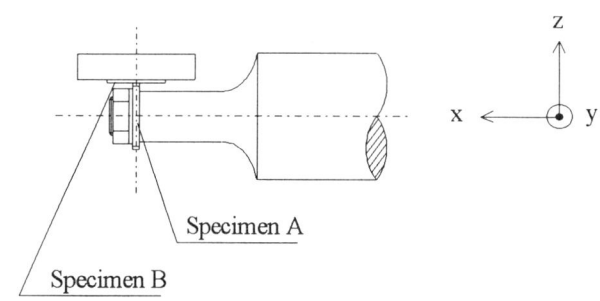

Fig. 1: Detail view: sonotrode and friction couple

A three axis dynamic force sensor is used to measure both the dynamic friction force $F_{T(t)}$ and the dynamic normal force $F_N + F_{N(t)}$. The high frequency friction coefficient can be calculated for each cycle as

$\mu_{HF} = F_{T(t)} / (F_N + F_{N(t)})$.

Friction and wear appearance of polymer/bearing steel couples (PA6, PTFE, PTFE + carbon fibers, PI, PI + PTFE) were investigated leading to the following conclusions for the friction behaviour:
The high frequency friction coefficient is always lower than the corresponding sliding or static coefficient:

$\mu_{HF} < \mu_S$.

The high frequency friction depends on the operating conditions and shows temporary fluctuations.

The wear was analysed quantitatively by profilometry and the wear appearance was studied by SEM to obtain information about the wear processes. A strong lubricating influence of the PTFE containing materials could be validated.

Submitted to *Wear*

NANO METROLOGY OF CYLINDER BORE WEAR

B.-G. ROSÉN, R OHLSSON & T. R. THOMAS
Department of Production Engineering, Chalmers University, S-41296 Gothenburg, Sweden

ABSTRACT

Cylinder bores are multi-process surfaces whose roughness is difficult to characterise for tribological purposes by conventional methods. Statistical approaches may be used to compute asperity densities, summit curvatures and so on, but suffer from the usual disadvantage of tending to infinite values in the absence of a short-wavelength cut-off. A useful advance in tribological roughness assessment would be to find a means of establishing an appropriate scale of measurement.

Using a form of the plasticity index corrected for anisotropy, a short-wavelength limit λ_p is derived below which asperities will not take part in long-term tribological interactions. A general relationship is obtained between three dimensionless numbers, the short-wavelength limit λ_p normalised by the topothesy Λ (1), the fractal dimension D and the material ratio (the ratio of the Hertzian elastic modulus E' to the hardness H)

$$\left(\frac{\lambda_p}{\Lambda}\right)^{2D-1} = \frac{-1175}{2D-1}(2\pi)^{2D-4}\Gamma(4-2D)*\cos\left(\frac{2-D}{2}\right)\left(\frac{E'}{H}\right)^2$$

From this relationship (Fig. 1) the appropriate scale of roughness measurement for any tribological investigation of a fractal surface may be determined.

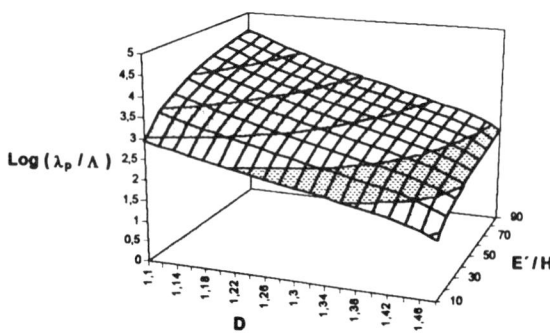

Fig. 1: Dimensionless critical wavelength as a function of material ratio and fractal dimension.

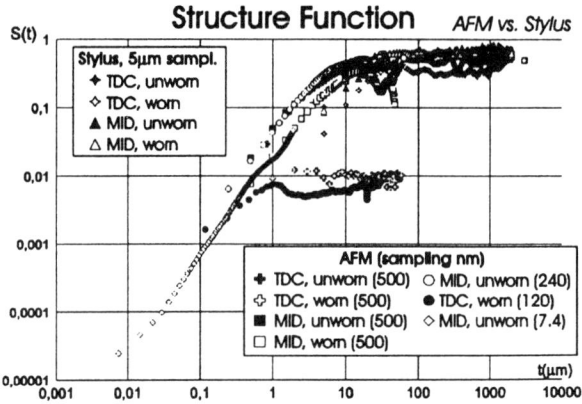

Fig. 2: Structure functions of regions of the same cylinder liner in worn and unworn conditions, measured with AFM and stylus instrument.

With a stylus instrument and an atomic force microscope, a number of cylinder bores were measured at locations of both high and medium wear before and after running in (2). By inspection of an ensemble of structure functions (3), it appears that cylinder bore surfaces are multifractal (4), with a corner frequency (5) at about 20 µm, corresponding to the average size of a honing grit. Below this length the surfaces are self-similar fractals down to the limits of AFM resolution. The longest fractals have a dimension of about 1.2 and a topothesy of 0.13 nm. The short wavelength limit using the above relation turns out to be about 40 nm, well below the range of instruments usually employed to measure tribological surface roughness.

REFERENCES

(1) Sayles, R. S. and T. R. Thomas, "Surface topography as a non-stationary random process", *Nature* **271**, 431-434 (1978)

(2) Rosen, B.-G., R. Ohlsson and T. R. Thomas, "Wear of cylinder bore microtopography", *Wear* **198**, 271-279 (1996).

(3) Sayles, R. S. and T. R. Thomas, "The spatial representation of surface roughness by means of the structure function: a practical alternative to correlation", *Wear* **42**, 263-276 (1977).

(4) Russ, J. C., *Fractal Surfaces*, Plenum Press, New York (1994).

(5) Majumdar, A., and C. L. Tien, "Fractal characterisation and simulation of rough surfaces", *Wear* **136**, 313-327 (1990)

LASER INTERFEROMETER SYSTEM FOR MEASUREMENT AND DIAGNOSIS OF SHAPE AND POSITION OF LARGE-SIZE ELEMENTS OF FLAT AND CYLINDRICAL SLIDING BEARINGS

ANDRZEJ RYNIEWICZ
Cracow University of Technology, Production Engineering Institute, Al. Jana Pawła II 37, 31-864 Kraków, POLAND

ABSTRACT

Development of new manufacturing techniques of machine parts aims at improving their performance also through minimising shape and position errors of kinematics pairs - sliding bearings to the technologically justified level. The paper present computer aided system for certain shape and position bearing surfaces errors analysis for quality assurance. The described system allows to assess large- size sliding bearings elements basing on accepted criteria. Analysing system co-operates with the laser interferometer measuring system which performs translations and rotational displacement measurements.

General version enables to evaluate the errors of straightness and flatness of bearing surfaces, their distance variation, movement velocity, its fluctuation of orientation of the travelling element.

Nominal conditions and assessment criteria are determined by the design and manufacturing requirements, including the shape of the sliding surface. Straightness and flatness errors are calculated in accordance with the relevant standard, parallelness errors is directly connected with the reference surface in the measuring sub-area.

The main functions of the system are as follows:
- communication with the measuring laser interferometer system for automatic data collection,
- data analysis basing on accepted geometrical criteria for sliding elements of bearings,
- graphical representation of the obtained results,
- sliding bearings correction data verification,
- setting the task parameters (kind of shape errors to inspect, way of determining positioning errors of the sliding surfaces).

The block diagram of the measurement and diagnosis system is shown on fig. 1.
Procedure „analysis" comprises two main pairs.

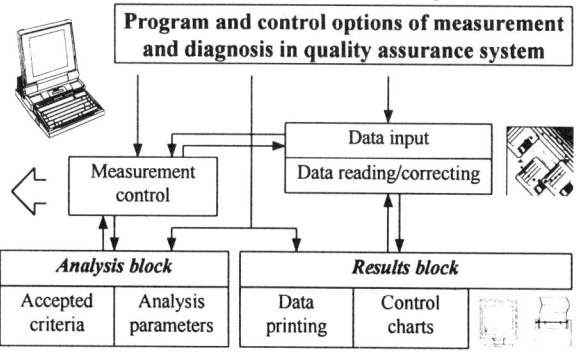

Fig. 1: The measurement and diagnosis system

The first one concerns shape and position errors, where as the second one includes positioning and movement parameters of the elements.

Shape and position errors are represented by: flatness, straightness, parallelness, orthogonality and relative orientation of element. The above are the subject of the analysis. Nominal conditions and assessment criteria are determined by the design and manufacturing requirements, including the shape of the element. Straightness and flatness errors are calculated in accordance with the relevant standard, parallelness errors is directly connected with the reference surface in the measuring sub-area.

Results of analysis are presented in the graphic and analytical form as the synthetic factors of assessment:

- Shape Deviations - the „Straight Synthetic Index" (SSI) which mathematical description is shown in the equation:

$$SSI = \begin{bmatrix} DIST \\ FLAT \end{bmatrix}$$

$$DIST = \frac{F_{DIST}(x,y)}{DIST_0}$$

$$FLAT = \frac{F_{FLAT}(x,y)}{FLAT_0}$$

$F_{DIST}(x,y)$ - function of the real distance changes between researched surfaces, $DIST_0$ - the basis distance value, $F_{FLAT}(x,y)$ - function of the surface's flatness in the defined region, $FLAT_0$ - the basis flatness value.

- Supporting Flatness Errors Function,
- Harmonic Characteristics, Elements Errors Analysis.

Additionally, the program generates a 3D picture of the analysed guiding elements, taking into account their actual mutual position and orientation in space.

The system of assessment may be applied for various type of processing flat and cylindrical sliding bearings with free configuration of sliding surfaces and its task is to improve operational properties of the equipment at the stage of final technological inspection in quality ensuring systems.

REFERENCES

(1) Ryniewicz A. Application and adaptation laser interferometer systems for measurements of certain travelling units parameters - rapport CPBP 0220, Cracow University of Technology, Cracow 1991

(2) Ryniewicz A. Introduction to the project of computer aided accuracy measurements of the processing machines parts - rapport Cracow University of Technology, Cracow 1994

SPECIAL FEATURES OF WEAR OF COPPER AND ITS ALLOYS WITH SUBMICROCRYSTALLINE STRUCTURE UNDER CONDITIONS OF DRY AND LIQUID FRICTION

F.A. SADYKOV, N.P. BARYKIN and I.R. ASLANYAN

Institute for Metals Superplasticity Problems, Russian Academy of Sciences, Khalturina 39, Ufa (450001), Russia

ABSTRACT

Submicro- and nanocrystalline materials are known to posses a unique combination of physical and mechanical properties such as strength, hardness, fracture toughness and others (1). It is urgent to study the operating characteristics of these materials, wear resistance in particular (2).

The work presents the results of the investigation of the influence of a microcrystalline (MC) and submicrocrystalline (SMC) structures on wear resistance in copper (99.9 wt%), bronze CuAlFe (Al-9.0 wt%, Fe-2.5 wt%) and brass CuZnPb (Zn-39.0 wt%, Pb-0.9 wt%) on wear resistance. The MC and SMC structures in these materials was obtained by deformation-heat treatment methods. Wear investigations were carried out at dry and liquid friction using metalcluding lubricants under the following conditions: sliding velocity, $0.78 \text{ m} \cdot \text{s}^{-1}$; pressure, 0.2 to $10 \text{ N} \cdot \text{mm}^{-2}$ and sliding distance, 0.5 to 100 km. The optical and scanning electron microscopies were used for study of the structure and morphology of the worn surfaces.

The some results of the investigation, presented in Table 1, indicate an essential effect of some structural parametrs (grain size, phase composition, texture) on dry wear resistance of these materials. This data are concern to pressures: $0.2 \text{ N} \cdot \text{mm}^{-2}$, $0.6 \text{ N} \cdot \text{mm}^{-2}$, $4 \text{ N} \cdot \text{mm}^{-2}$ for copper, brass, and bronze, respectively.

As one can see, the wear intensity of materials (copper, brass, and bronze) with MC and SMC structures is less than for coarse-grained ones.

Moreover, it was established, that deformation texture significially influences of the wear resistance. The specimens, in which the direction of sliding is along the axis of extruding possess less wear intensity than in the perpendicular state. After the annealing at 200 °C the influence of texture on wear of copper considerably decreases.

The wear of brass and bronze depends on phase composition too.

The microhardness distribution at surface layer after the friction showed that the thickness of a hardened surface layer for material with SMC structure is less than for coarse-grained one.

The analysis of the results allowed us to reveal a special feature of wear of materials with the SMC structure under hydrodynamic friction.

The investigation of the morphology of the worn surfaces revealed the main mechanisms of "mild" wear.

Material	Treatment	Wear intensity $(\text{mg} \cdot \text{mm}^{-3})$
Copper [a]	Initial state	6.7×10^{-8}
Copper [b]	Cold extrusion ($\parallel$)	5.5×10^{-8}
	Cold extrusion ($\perp$)	10.5×10^{-8}
	Cold extrusion + annealing at 200 °C ($\parallel$)	10.4×10^{-8}
	Cold extrusion + annealing at 200 °C ($\perp$)	12.0×10^{-8}
Brass [a]	Initial state	2.8×10^{-7}
	Tempering at 840 °C	5.1×10^{-7}
Brass [c]	Cold rolling + annealing at 300 °C	2.2×10^{-7}
Bronze [a]	Initial state	6.2×10^{-7}
Bronze [c]	Cold rolling + annealing at 600 °C	5.0×10^{-7}
Bronze [b]	Torsion strain + annealing at 600 °C	1.8×10^{-7}

$\parallel, \perp$ - Orientation of the sliding direction regard to the extrusion axis.

[a], [b], [c] Coarse-grained, SMC and MC structures, respectively.

Table 1: The data of wear of materials

REFERENCES

(1) H. Gleiter, Nanostruct. Mater., Vol. 1, No.1, 1992.
(2) F.A. Sadykov, J. of Mater. Eng. and Perf., Vol. 4, No.1, 1995.

POLYCOMPONENT LOADING SYSTEMS IN TRIBOPHATICS

A.S.SHAGINYAN
Gomel Polytechnic named after P.O.Sukhoi, pr. Oktyabrya, 48, 246746, Gomel, Belarus

ABSTRACT

Actual machines are normally subjected to combined loads changing randomly or according to definite laws.

Mechanical loading of machines is accompanied by a number of factors to be taken into account during standard specimens tests [1,2].

Full-scale tests to create actual loads of machines and units are the most correct methods to evaluate their strength and service life.

Special polycomponent loading systems with electrohydraulic servosystems creating all loading components of transport facilities (cars) axle loads: P_z — loaded car weight; P_y — braking effort; P_x — side drift offsetting effort on the hairspin road ($P_x = 0$ on the straight road) were developed in Gomel Polytechnic named after Sukhoi P.O.

Fig.1 schematically shows the axle of a car wheel with P_z, P_y, P_x forces and the polycomponent system of car wheel loading.

It goes without saying that P_z and P_y loads correspond to a road profile and P_z, P_y component values depend on car speed and weight. P_x value will change according to low-frequency loading programme taking into account the road hairspin, speed and weight of a transport facility.

For programming of the system presented a computer or magnetograph can be used. They send time input signal corresponding to specified service life of a specimen and reflecting actual operational loadings of the specimen tested, their intensity, mode of changes and frequency spectrum.

Bench test system provides for direct loading through car wheel, there also being investigated axle fatigue strenght and carring capacity together with the analysis of all bearing support triboparameters.

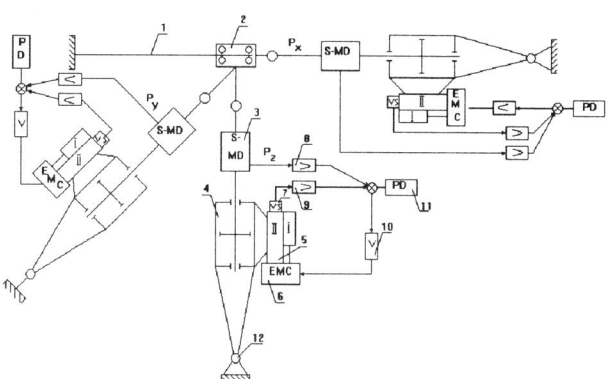

Fig.1. Electrohydraulic polycomponent loading servosystem: 1—car wheel axle; 2—wheel axle support; 3—strain-measuring device; 4—hydrocylinder; 5—double-stage hydraulic booster; 6—electromechanical converter; 7—control valve sensor; 8,9—strain-measuring device feedback amplifier and spool-valve amplifier; 10—error signal amplifier; 11—programming device; 12—joint.

REFERENCES

(1) L.A.Sosnovsky, Wear test methods of power systems and their models. Friction and wear, vol.14, issue № 5, Minsk, 1993.

(2) A.S.Shaginyan, Dynamics of general-purpose test machines with electrohydraulic drive, International Symposium "Tribophatics - 93", Gomel,1993.

IMPROVEMENT OF LUBRICATION FOR CAM AND FOLLOWER

M.SOEJIMA and Y.EJIMA
Faculty of Engineering, Kyushu Sangyo University, Matsukadai, Higashiku, Fukuoka, 813 Japan
Y.WAKURI
Faculty of Engineering, Fukuoka University, Nanakuma, Jonanku, Fukuoka, 814-01 Japan
T.KITAHARA
Faculty of Engineering, Kyushu University, Hakozaki, Higashiku, Fukuoka, 812 Japan

ABSTRACT

In order to improve the lubrication of cam and follower for the valve train of internal combustion engines, the influences of materials and lubricating oil properties on the friction and scuffing characteristics have been examined with a test rig by increasing the valve spring pre-load to make the contact condition severe. The friction between cam and follower is directly measured and the change of friction coefficient diagram is recorded to detect the occurrence of abnormal friction and scuffing failure (1)-(4).

Test specimen materials of the cam are hardened ductile cast iron and hardened S48C steel, and those of the follower are chilled cast iron, sintered metals (Ni-base / Co-base) and silicon nitride ceramics. Test lubricants are base oils of SAE30, SAE10W, SAE5W and 80N, and oil additives of 1ry/2ry ZnDTP, TCP, MoDTP and MoDTC were used in the test.

As an example of test results the comparison of changes of average friction coefficient over contact duration with running time on oil additives obtained under the constant camshaft revolution speed and oil temperature is shown in Fig.1. From the experiment the followings have been made obvious. Firstly, for the cam material, the hardened ductile cast iron is superior in scuffing resistance to the hardened steel though it is higher in friction, and for the follower material, the hard sintered metal and the sillicon nitride ceramics are superior in scuffing resistance to the chilled cast iron and are lower in friction when mating the hardened steel cam. Secondly, as the viscosity of the base oil becomes low, the friction decreases due to some oil additives but the scuffing resistance always becomes small. The mono-additive of MoDTP remarkably decreases the friction and the one of ZnDTP is much effective against the scuffing. Also, in the case of mixed additive, (MoDTP · 1ry ZnDTP) is less in scuffing resistance though the friction is lower, but (MoDTP · 2ry ZnDTP) is superior in anti-scuffing though higher in friction, and specifically (MoDTC · 1ry ZnDTP) improves the lubrication in both respects of the reduction of friction and the anti-scuffing property, but (MoDTC · 2ry ZnDTP) is higher in friction.

Further the effects of lubricating oil supply through the oil-hole of camshaft on the reduction of friction and the prevention of scuffing have been examined. The oil-hole was located at each of the seven positions on the width center of the base circle, two flanks, two shoulders and nose along the cam profile. The changes of maximum contact load for scuffing occurrence with the oil-hole position are shown in Fig.2. As other conclusions of present studies the followings have been clarified. Then, lastly, the oil supply from cam surface oil-hole is effective means to improve the lubrication and the effect becomes the largest when the oil-hole is arranged between the flank and the shoulder of cam corresponding to the valve opening.

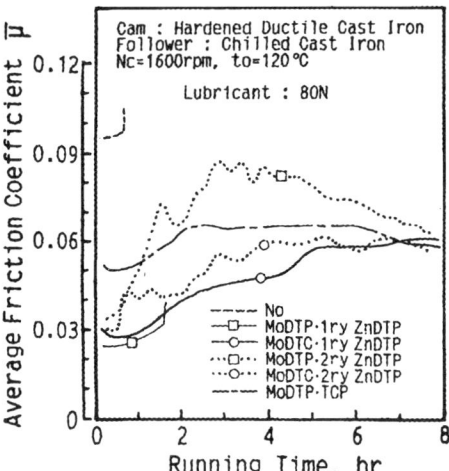

Fig. 1 : Comparison of average friction coefficient on oil additives (Base oil : 80N, 1600rpm)

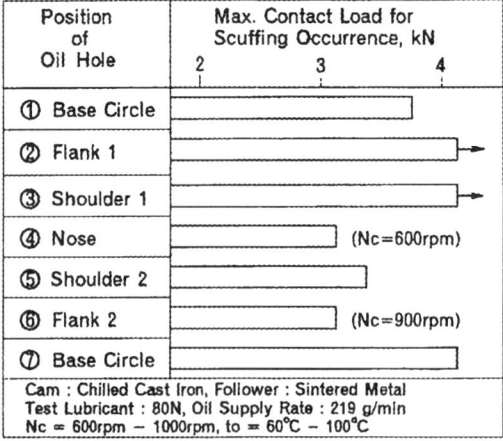

Fig. 2 : Changes of max. contact load for scuffing occurrence with oil-hole position

REFERENCES

(1) M. Soejima, et al., Proceedings of the 6th International Congress on Tribology, Budapest, 1993, 329-334pp.
(2) Y. Wakuri, et al., SAE Paper No.952471, 1-15pp.
(3) M. Soejima, et al., Japanese Journal of Tribology, Allerton Press, Vol.40, No.8, 1995, 761-775pp.
(4) M. Soejima, et al., Proceedings of International Tribology Conference, Yokohama, 1996, 1483-1488pp.

METHODS AND MACHINES FOR WEAR-FATIGUE TESTS OF MATERIALS AND THEIR STANDARDAZATION

L A SOSNOVSKIY
Scientific and Industrial Group "TRIBOFATIGUE", P O Box 24, Gomel, 246050, Republic Belarus

V N KORESHKOV
State Committee on Standardazation, Metrology and Sertification, Starovilensky Trakt 93, 220053, Minsk, Republic Belarus

O M YELOVOY
Scientific Center of Machines Mechanics Problems of the Belorussian Academy of Sciences,
Skoriny Avenue, 12, 220072, Minsk, Republic Belarus

ABSTRACT

The development of tribology resulted in creating a special class of test equipment — machines for friction and wear tests. The development of fatigue fracture mechanics brought about the creation of a special class of test equipment as well — machines for fatigue tests. In connection with this numerous and generally recognized tests methods, including standard ones were elaborated.

The development of tribo-fatigue, in its turn, results in building up a new class of test equipment — machines for complex wear-fatigue tests (1). The designer and manufacturer of such machines is the Scientific and Industrial Group "TRIBOFATIGUE" (Republic Belarus). Methods and machines for sliding-mechanical fatigue tests (SI-01 machine) and rolling-mechanical fatigue tests (SI-02 machine) in main rotary motion (Fig. 1 and 2) were elaborated first. A full-complete desk-top module SI-03 machine for wear-fatigue tests of materials is being elaborated at present.

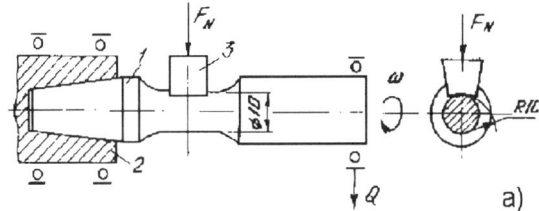
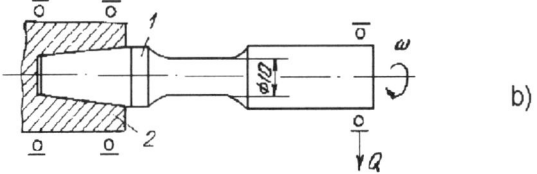
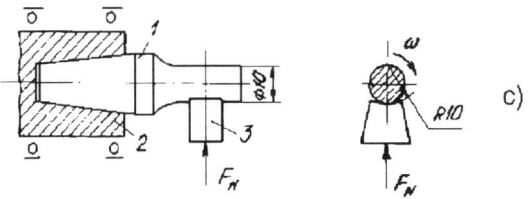

Fig. 1: Tests scheme (1 - Specimen, 2 - Spindel, 3 - Counterspecimen); a) Sliding-mechanical fatigue; b) Mechanical fatigue; c) Sliding friction

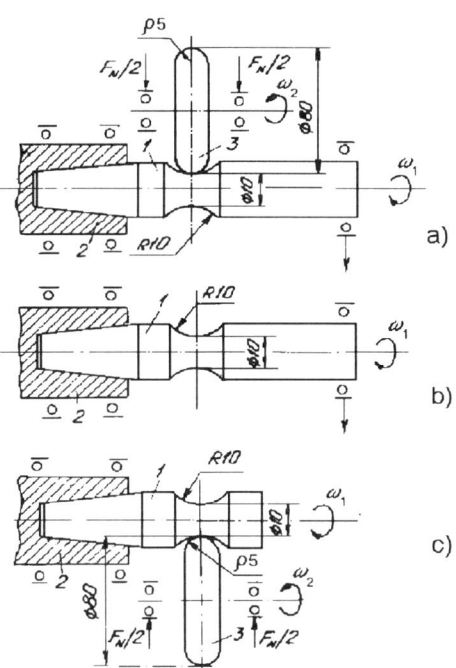

Fig. 2: Tests scheme (1 - Specimen, 2 - Spindel, 3 - Roller); a) Rolling-mechanical fatigue; b) Mechanical fatigue; c) Rolling friction

The SI-01 and SI-02 machines are designed for large-scale tests. With this aim the elaboration of the Standard of Belarus on the wear-fatigue tests methods and the main technical requirements for wear-fatigue test machines have been working out as well (2).

The methods of wear-fatigue tests are discussed, constructional features of machines and their technical characteristics are described, typical results of such tests are given. The general principles of information-controlling system and the experimental results acquisition under specialized software are presented. Some information on the notions of standards elaborated for wear-fatigue tests methods are given.

REFERENCES

(1) L A Sosnovskiy, Friction and Wear, Vol. 14, 1993, No 3, 937-952pp.

(2) Tribo-fatigue–95. Problems of Standardazations, Ed.: L A Sosnovskiy, V N Koreshkov, Gomel, 1996, 87p.

COMPUTER-AIDED OPTICAL SENSING IN THE STUDY OF WORN SURFACES

JOICHI SUGIMURA and YUJI YAMAMOTO
Department of Energy and Mechanical Engineering, Kyushu University, Fukuoka 812-81, Japan

ABSTRACT

Optical microscopes have long been used in tribology researches to observe sliding surfaces because of their readiness to provide preliminary information and their relatively easy operation. While it is recognised that information from microscopic observation is only qualitative, experts are able to deduce, not only just rough pictures of worn surfaces, but also detailed features of damages on the surfaces, or even predict causes of the damages from their observation. This has an implication that, if such processes of detection, recognition and comprehension of features are performed by a computer, a classical microscope can be turned into a powerful intelligent measuring device.

This paper describes a microscope image analysis system currently developed for quantitative determination of geometrical and optical characteristics of worn surfaces. Figure 1 is a schematic diagram of the system. A sample stage which is controlled by a computer to move in three directions allows the system to, like the remote sensing by an artificial satellite, scan a wear track automatically.

Three dimensional shape measurement of worn surfaces is made by taking a number of images while moving the stage vertically, and finding focal heights for each point in the images by image processing. Figure 2 shows a photograph of a worn surface and a 3-D representation of its geometry viewed from the bottom-left corner of the photograph.

Geometrical patterns produced by wear particle removal and plastic deformation are also described by texture descriptors. Surface films formed by chemical reaction of the surface with lubricants are charecterised by reflectivity, colour and texture descriptors (1). These are compared and related with data determined by other devices such as profilometers and XMA.

The technique is useful in studying worn surfaces without special devices for obtaining particular information, as well as in classifying surface modifications and damages in a quantitative way. In addition, relationships found between the surface features and wear debris features can effectively be used in condition monitoring.

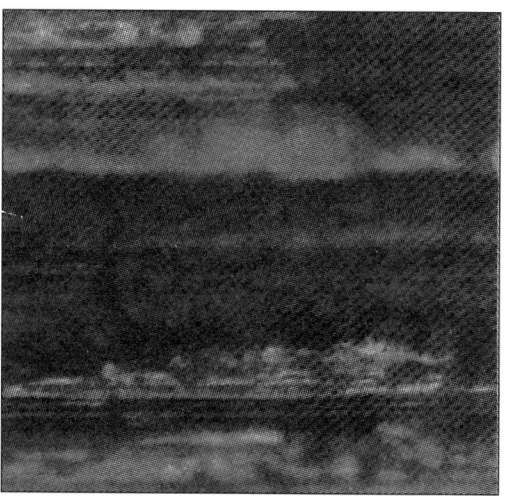

Fig. 2 Worn surface and its 3-D geometry obtained with the present system; 0.45%C steel disk surface slid against a bearing steel ball; sulphur-containing oil, 58.8N, 2.09mm/s

REFERENCE

(1) J. Sugimura et al., Proc. JAST Trib. Conf., Morioka, pp.295-298, 1992 (in Japanese).

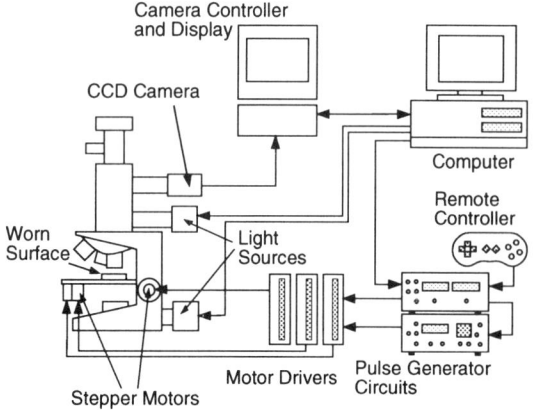

Fig. 1 Microscope sensing system

EXPERIMENTAL ANALYSIS OF AIR FLOW IN A MAGNETIC BEARING

M.R. MACKENZIE, A.K. TIEU, AND E. LI
DEPARTMENT OF MECHANICAL ENGINEERING, UNIVERSITY OF WOLLONGONG, N.S.W. AUSTRALIA

This paper reports on the design of a test rig for the measurement of air flow in a magnetic bearing using a laser Doppler anemometer (LDA) (figure 1).

The bearing consists of a steel tube 500mm long supported at each end by four horse-shoe shaped electromagnets wound with 300 turns of copper wire. The shaft has a mass (m) of 0.67kg and has an inside and outside diameter of 38mm and 41mm respectively.

The position of the shaft at each end of the shaft was measured by two laser diode/photodiode sensors and these provide feedback for control of the magnets. The linear signal from the photodiode sensors was amplified to provide an output of $k_p = 1567$V/m. The noise level at the output was measured to be less than ± 5mV corresponding to a sensitivity of $\pm 2\mu$m in position.

Magnetic bearings are inherently unstable. The bearing was stabilized with a proportional derivative (PD) controller $G_{pd}(s) = k_{pd}(s + z_{pd})$ which artificially adds damping to the system. With position feedback from the photodiode the control function of the magnetic bearing is

$$\frac{x(s)}{v(s)} = \frac{G(s)}{1 + k_p G_{pd}(s) G(s)}$$

$$G(s) = \frac{k_i}{R\left(s^2 - \dfrac{k_x}{m}\right)}$$

where x is the gap between the magnet and shaft, v is the magnet coil voltage, k_i and k_x were obtained experimentally to be 4.57 N/A and 5390 N/m and R is the coil resistance (3.5Ω). The system is stable for PD gains $k_{pd} > k_x R / m z_{pd} k_p k_i$. With a PD zero ($z_{pd}$) located at 45Hz this corresponds to $k_{pd} > 0.17$. In actual fact the control function is more complicated than that given above due to the inductance of the coil and to the necessity of including a low pass filter for the PD controller.

The airflow is measured with a LDA designed to have a very high spatial resolution (17μm measurement volume). It has previously been demonstrated capable of measuring at distances as close as 30μm from the flow wall. For reasons of economy the LDA uses solid-state components but these also allow for the instrument to be of very small dimensions providing for convenient operation.

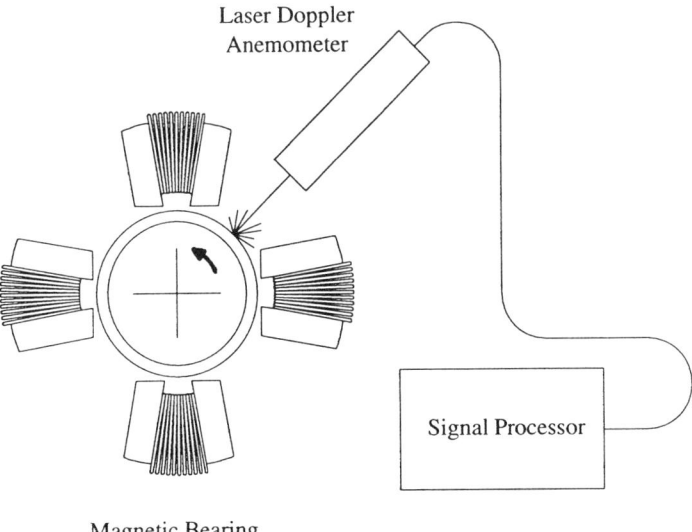

Figure 1. Schematic diagram of magnetic bearing test rig.

DEVELOPMENT OF WEAR SENSORS BASED ON LASER BEAMS REFLECTION, LASER PROFILING AND CUTTING FORCE MEASUREMENT

BOGDAN VASILJEVIC

University of Kragujevac, Faculty of Mechanical Engineering, Kragujevac, Yugoslavia

ABSTRACT

Development of reliable wear sensors presents the basic prerequisite for qualitative optimization of manufacturing processes. In literature from this area we are meeting the whole series of algorithms for the ACO control of machine tools. The great number of algorithms, unfortunately, do not give the answer to the basic question: how to reliably and precisely on-line determine the cutting tools wear degree.

In the paper are presented results of complex, indirect on-line determination of tool wear, by comparative monitoring of the laser beam reflection Sr[%], and laser profile measuring of the freshly machined surface Rmax, together with measurement of the cutting force Fz[daN] and medium cutting temperature. It was also possible to monitor vibrations in the tool - working piece system, as well as periodically measure the wear belt width on the cutting tool back surface hsr[mm], on the universal microscope (Figure 1.).

During investigation, the special attention was focused on possibilities of laser applications as the wear sensor, what required construction of special equipment for positioning and bringing in the compressed air into the cutting zone. Beside determination of the known correlation between the cutting tool wear degree, we tried to form the correlation between the cutting tool wear hsr[mm] and degree of the laser beam reflection Sr[%], and laser machined roughness parameters Rmax[μm].

For experimental investigation we chose the metal machining by milling, as the most convenient for laser application, and the complete investigation was done without application of cutting fluids, with medium cutting speeds. Investigation results show the high degree of correlation between the measured degree of tool wear and the degree of laser beam reflection, and the laser determined roughness parameters, as well as the cutting forces.

Correlation dependencies presented in this paper can represent the starting basis for further investigations in the field, with the task to finally obtain the reliable on-line wear sensor as the basis for broader application of the ACO algorithms for machine tools control.

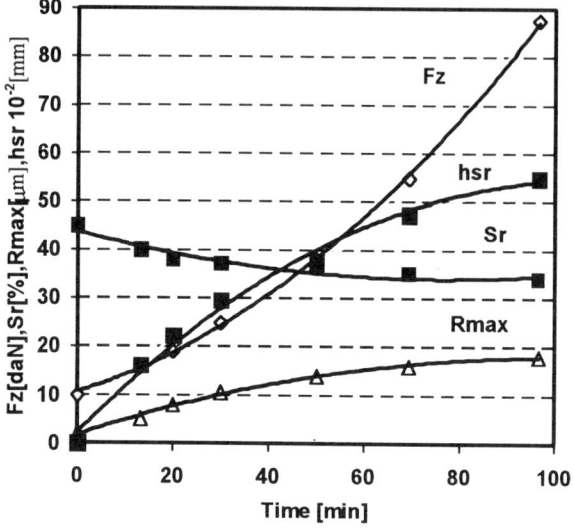

Fig.1

The basic problems that appeared during investigations were elimination of influence of vibrations on the measurements' accuracy, as well as efficient removal of shavings from the freshly machined surface, for the purpose of efficient application of the laser method. The application of the multi-channel fast A/D and D/A converters, connected to the PC, as well as for this investigation specially developed computer program, enabled obtaining of the whole series of useful information that are presented in tables and graphs in this paper.

REFERENCES

(1) Beckmann P., Spizzichino A., The scattering of electromagnetic wawes from rough surfaces, The Macmillan Co., new York, 1963.
(2.) Shiraishi M., In-Process Measurement of Surface Roughness in Turning by Laser Beams, ASME Conference, Chikago, 1980.
(3) Vasiljevic B., Jeremic B., Babic M., Laser Method Sensitivity as a Function of Contact Surface Microgeometry Parameters, Journal of the Balkan tribological association, Vol. 1, No. 4, 319-325 pp, 1995.

TRIBOLOGY TESTING WITH SMART MACHINES AND METHODOLOGIES

L.D. WEDEVEN

Wedeven Associates, Inc., 5072 West Chester Pike, Edgmont, Pennsylvania 19028-0646, USA

ABSTRACT

New technology in motion control and instrumentation allows highly flexible test machines to simulate a multitude of lubrication and failure mechanisms of engine and drivetrain systems. A schematic of the Wedeven Associates, Inc. Machine (WAM3 version) is shown in Figure 1. Various test specimen geometries can be used, which give point or elliptical contacts. The mechanical control and data systems are designed to provide a highly flexible tribo-testing platform for lubricant and material evaluation, along with detailed hardware simulation. Test temperatures range from cryogenic to 800°C.

If one sub-divides a lubricated contact into structural elements, a rational approach to testing can be developed. A new approach provides fundamental understanding and performance prediction. Entraining velocity and sliding velocity are key parameters, along with replication of material properties, surface texture and thermal environment. A wide range of operating conditions can be obtained by controlling surface velocity vectors in both amplitude and direction. Precise control of oil film thickness (h) relative to surface roughness (σ) allows multiple lubrication and failure mechanisms to be invoked. The independent control of h/σ and the shear strain within the contact provides testing methodologies for wear, scuffing and micro-pitting. From testing experience, tribology performance can be mapped with respect to specific operating hardware. A map of interactive and competitive failure modes is shown in Figure 2.

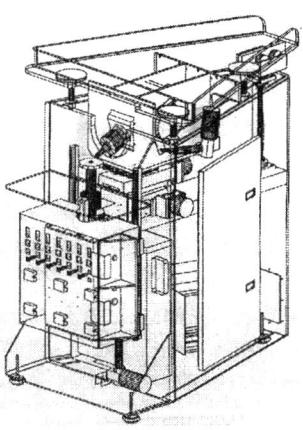

Fig. 1: Schematic of WAM3 test machine

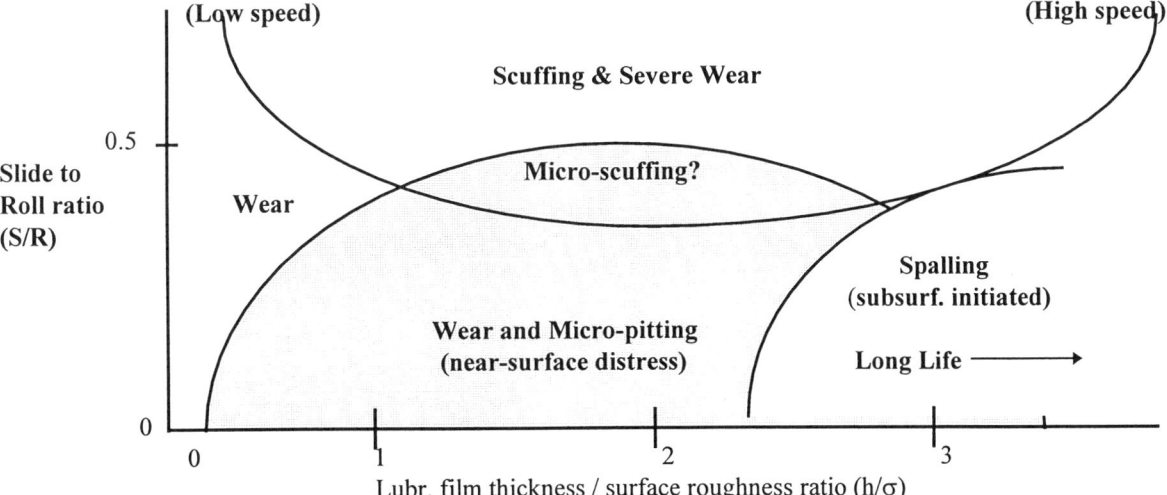

Figure 2: Schematic of life-limiting surface failure modes (boundaries are postulated)

MESUREMENT OF OIL FILM THICKNESS BETWEEN HELICAL GEAR TOOTH PROFILE

LI WEI   TANG QUNGUO   and   YANG SHAOKUI

Beijing Graduate School, China University of Mining and Technology, Beijing 100083, P.R.China

ABSTRACT

The minimum oil film thickness between two gear surfaces reflects the gear lubrication condition, and it is one of the important factors which affect gear transmission characteristics and lifetime. So searching for new method to precisely monitor gear lubrication condition is always one of the main scientific directions for EHL searcher about gear.

So far, there are many method of measuring EHL film thickness. Generally speaking, they are devided into several kinds: electric measuring method, light measuring method, magnetic resistance method and so on. All these method have sucessfully been used on disc test set and have received many useful results. But due to these methods' own characters and demands, under existing condition, their direct application to measuring gear surface lubrication state is still difficult and it still has some limits.

For example, light interference method (1) demands that one of the measured elements must be made of transparent material,laser transmission method (2) is an indirect measuring method, and due to too many middle link, axial torsional, bending deflection and the install error of bearing and gear, its measuring result is not precise, in 1952, Lane and Hughes (3) using resistance method, measured the lubrication state of gear meshing. They insulated gear body and axle, explained why only the average lubrication state in the course of gear meshing can be measured but the instant lubrication state on the contact lines in the gear contact areas (one tooth or two teeth) can not be.

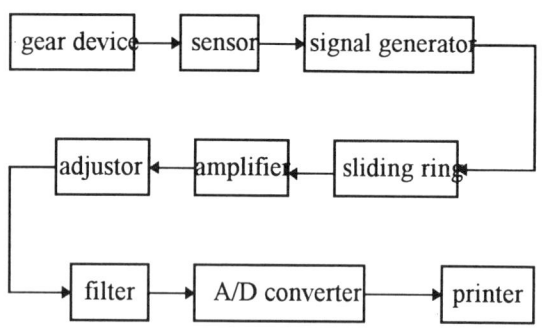

Fig.1: Measuring system block diagram

By now, the measuring study of helical gear lubrication is still little seen. In this paper, according to the meshing characters of helical gear (in the course of meshing, the every instant contact line is straight line), adopt light   pervious to decide the location of instant contact line, use high precise line cutting machine to cut slot along contact line, make sensor by means of ambushing copper wire, measure the tooth surface lubrication state of helical gear under many working conditions. Its measuring system block diagram is shown in Fig.1.

Something has to be explained: to better dynamically measure the helical gear oil film capacity under full film lubrication state, a analogue circuit is designed. First, obtain the part voltage U_X of gear surface oil film capacity to alternate voltage input signal U by way of RC integral circuit, then obtain direct voltage signal U_{X0} by way of projectile pole tracer, amplyfier, semiwave adjustor and filter, at last (by data collecting system) input the measured signals into a computer by way of data collecting system and use the computer to record and analytical handle.

The output voltage between two oil film end is:

$$U_X = \frac{U_0}{\sqrt{1+(2\pi f c_X R)^2}} \sin(2\pi f t + \phi - 90°)$$

in which, C_X is the oil film capacitance to be measured, U_0 is the amplitude of source voltage, φ is the phase difference between the source voltage and electric current.

Seen from the formula once the U_X is measured, the oil film capacitance C_X can be calculated, and then the film thickness value to be measured can be known from the mark curve.

According to the measuring theory in this paper, the gear lubrication state under various working conditions are dynamically traced and measured on GZ150 high speed gear experimental rig,and the effects of the rotating speed and load to lubrication states are analyzed. Experiment results show that: this method is a valuable way to trace helical gear lubrication state. Meanwhile this experiment method can also be used in the film thickness measuring of other high speed lubrication element, such as cam or roller bearing etc.

REFERENCES

(1) A.Cameron, R.Gohar, Proc. R. Soc. London. Ser. A. , 1966
(2) Y.Jibin, Q.Yulin, ASME Journal of Tribology, Vol.112, 1990
(3) T.B.Lane, J.R.Hughes, British Journal of Applied Physics, Vol.3, Oct., 1952

A STUDY ON THE TRIBOLOGICAL CONDITIONS OF INTERFACE FILM FORMATION UNDER BOUNDARY LUBRICATION BY THE DIRECT OBSERVATION TECHNIQUE

SHI YINONG and JIA CHUNDE

Shenyang Institute of Technology No,81,Wenhua Road,Shenyang,Liaoning 110015, P.R.China

ABSTRACT

The most important property of a Lubricant is its ability to generate thick、elastohydrodynamic (EHD) film in concentrated contacts. Many methods have been developed to study the thickness、shape、pressure distribution, friction coefficient and the chemical properties of the film, as long as the film thickness measuring was concerned. The capacitance-type, X ray-type, optical interferometer or Laser interferometer method seemed to be more practical (1),while studying of the interface film from EHD to boundary lubrication state, most previous studies were focused on the friction coefficient variation. In order to determine the relation between the tribological conditions and the interface film variation which will indicate the change of lubrication type ,a new way developed in this study was to use direct dynamic observation technique which enable a viewer to see the interface film and its variation combining with measuring friction coefficient and the film thickness.

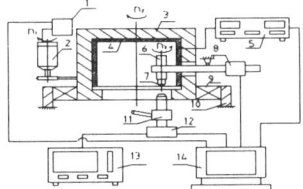

Fig.1:Schematic diagram of test device

1.Frequency trans 2.High speed spindle mortor(up to 12000r/min) 3.Chamber 4.Thermoelectricity ceramic materials 5.Temperature controller 6.Air motor(1000-300000r/min) 7.Specimen 8.Torque and pressure sensor 9.Bearing 10.Frame 11.Optical microscope 12.Optical and digital camera 13.Monitor 14.Computer

Figure 1 shows a schematic diagram of the test apparatus. A optical microscope with the amplification of 400x ,through which the interface film variation could be seen directly when using metal ball or pin against glass or Cr. coated glass disc was employed. The identification was based on the changes of lubricant film color, friction coefficient as well as wear behaviors when changing the pressure, temperature, slide speed and the lubricants. With the help of the thermoelectric-

ity ceramic materials coated on the chamber internal surface,the chamber temperature could be regulated from room temperature to 400°C.The glass disc was driven by a variable-frequency high speed spindle motor,and specimen involving cylinder plane,ball or pin was driven by a variable velocity super high speed pneumatic motor or just kept fixing, through this way, very wide relative speed range from 0.1m/s to 40m/s was obtained. The above technique made it possible that the test on oil-gas lubrication for super high speed bearing could be carried out. The normal pressure was adjusted to keep Hertzian contact stress from 3000 Mpa to 4500 Mpa for different specimen. Higher resolution dynamometer could distinguish the small change of friction coefficient. Connected with a camera, a monitor was used to show the friction interface,and a computer image and data processing system was also employed.

As a example, the interface film variation from three different lubricants is shown in Fig.2. The direct

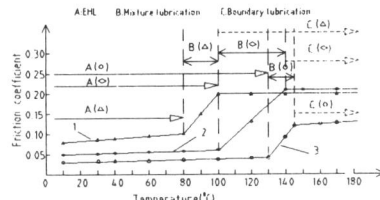

Fig.2. Effects of temperature and lubricants on lubrication type

1:200 SN base oil 2:Two-stroke engine oil 3:200 SN base oil with ultramicro-powder additive of PTFE; P=4000Mpa; V=0.5m/s; Test time:10 min for each new metal ball

observation shows:(1) There was a conversion section for friction coefficient of each lubricant under special tribological condition ;(2) The oil film color was unchanging on the left section of the turn point ,while the oil film color turned to dark on right section, which suggested that the lubrication type had changed from the EHL state to boundary lubrication even to lubrication failure;(3) The higher quality lubricant, the higher temperature at which one lubrication modle took place was corresponded;(4) The good correspondence between the results from direct dynamic observation on interface film variation and the results from friction coefficient record indicated that the technique used in this study was reliable.

REFERENCE:

(1)J.Jakobsen,P.C.Larsen, " Interferometric Deformation Measurement of Elastohydrodynamic Loaded Surface " Transactions of ASME,Journal of Lubrication Technology, Vol. 100,Oct,1987,S. 508-509.

SURFACE AND SUBSURFACE MATERIAL STATE ON THE STAGE OF SLIDING FATIGUE PREFRACTURE

A. L. ZHARIN and N. A. SHIPITSA

Tribology Laboratory, Belarussian State Powder Metallurgy Concern, 41 Platonov str., Minsk, 220600, Belarus

ABSTACT

Using originally developed technique for continuous non-destructive monitoring of changes in the electron work function of a rubbing surface we have found that the kinetics of steady state friction are characterized by regular periodic changes of the friction surface electron work function integral value (Fig. 1). The period, amplitude and harmonic contents of such changes are determined by the properties of materials and testing conditions.

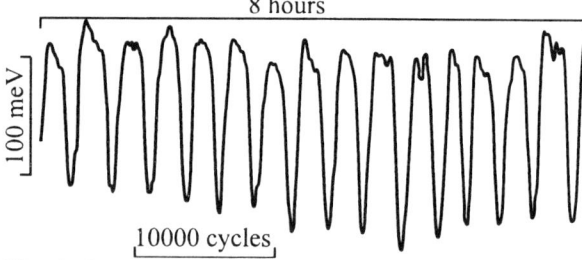

Fig. 1. Record of the brass rubbing surface electron work function periodic changes during 8 hours.

Additional investigations have shown that such periodic changes are related to sliding surface fatigue. Of particular interest is the subsurface material state in the prefracture stage. In this case the distribution of microhardness in the subsurface region is not uniform. There is a redistribution of alloy components, e.g., an increase in the zinc and decrease in the copper concentrations (for the case of brass), and a super-long (millimeters) subsurface damaged zone parallel to the surface. The main feature of the damaged zone is that it looks like an extended system of voids rather than a crack. Also it can be noted that the metallographic structure of the material band between the surface and the damaged zone is not clearly revealed by metallographic etching, but the bulk material under the band shows a clear metallographic structure. The unresolved region can be explained by the small subgrain structure in this band. This was confirmed by microhardness data as well.

A generally similar structural arrangement was reported earlier (2). In this work the structures found in a finely abraded surface of brass are summarized diagramatically and their schematic representation includes the subgrain band at the surface. From the results described above it is reasonable to extend this schematic representation by adding voids under the subgrain band (fig. 2). In such a model stresses at the rubbing surface are transmitted to the deeper material layers through the subgrain band without changes in dislocation density in the subgrain. The main processes governing subsurface fatigue in such a model are void

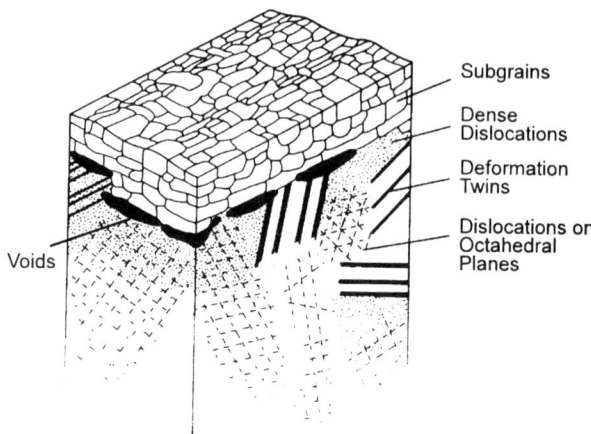

Fig. 2. Schematic representation of the worn surface for the case of the prefracture stage of subsurface fatigue.

nucleation and growth, while crack propagation processes play a relatively small role. Two reasons may be considered responsible for void nucleation and growth. The first one is the existence of significant shear stresses at the boundary between the subgrain band and the bulk material, as, in general, rearranging the near surface dislocation density into subgrain boundaries will be associated with a volume change. The second one is void nucleation and growth due to processes of nucleation, migration and annihilation of the dislocations. The second case seems more likely as it is indirectly confirmed by the observed alloy component redistribution. Large numbers of dislocations could move together with their atmospheres of impurity atoms to the accumulation zone.

Based on the fatigue-like nature of the rubbing surface work function periodic changes integral value, one expects a correlation with known characteristics of bulk fatigue fracture. Curves similar to bulk fatigue curves (Weller's curves) were plotted using the work function periodicity data. Since it is possible to plot curves similar to those of Weller using data on the rubbing surface periodic work function changes, approaches used for conventional fatigue processes can be applied to evaluate fatigue fracture parameters for rubbing surfaces.

REFERENCES

(1) A.Zharin, Contact Potential Difference Technique And Its Application In Tribology, Minsk, Bestprint, 1996. (in Russian).

(2) L. E. Samuels, E. D. Doyle and D. M. Turley, in: Fundamentals of Friction and Wear of Materials, ASM Materials Science Seminar, 1980, 13.

A STUDY OF THE WEAR PROPERTIES OF TWO GRADES OF UHMWPE AS USED FOR ORTHOPAEDIC PROSTHESES.

P S M BARBOUR and J FISHER
Dept. of Mechanical Engineering, University of Leeds, Leeds, LS2 9JT, UK.
M H STONE
Leeds General Infirmary, Great George Street, Leeds, LS1 3EX, UK.

INTRODUCTION

The use of Ultra High Molecular Weight Polyethylene as a bearing material in total replacement joint prostheses has been common practice for over 30 years. This material was primarily chosen due to its high bio-compatibility, low friction and high wear resistance. However, the mixed lubrication mechanism that operates within this bearing couple does lead to the wear of the UHMWPE. Over a number of years the release of polymer wear debris, in the form of micron sized particles, eventually leads to a biological reaction and the failure of the bone to implant interface. This leads to the loosening and failure of the prosthesis.

This study investigates the wear resistance of two grades of UHMWPE, which are currently used for joint prostheses, under conditions similar to those found in vivo and compares their wear properties.

MATERIALS AND METHODS

The two grades of UHMWPE used in prosthesis are GUR 1120 and GUR 4150HP. The former material is produced in Europe by Hoechst AG and is compression molded into slab form by Poly Hi Solidur. The latter material is manufactured in the USA by Hoechst Celanese and is ram extruded into bars by Poly Hi Solidur Menasha Corporation. Apart from the differences in processing method the GUR 4150HP has a molecular weight of 7.3 million as compared to 4.4 million g/mol for the GUR 1120 (1).

Specimens from a single batch of each material were tested on a six station pin on plate style apparatus with bovine serum lubrication. The load was 240 N/pin with a contact area of 20 mm^3 to give a contact pressure of 12 MPa. A number of tests were conducted with varying values for the counterface roughness of the cobalt chrome plates. These conditions are given in table 1.

Test	Ra, mean ± SD (µm)	Total sliding distance (km)
Rough 1	0.068 ± 0.010	175
Rough 2	0.064 ± 0.004	175
Smooth	0.008 ± 0.002	125

Table 1 : Wear test conditions.

These test conditions and the counterface roughness values were designed to simulate the wear processes occurring on new (smooth) and damaged (rough) femoral heads.

RESULTS

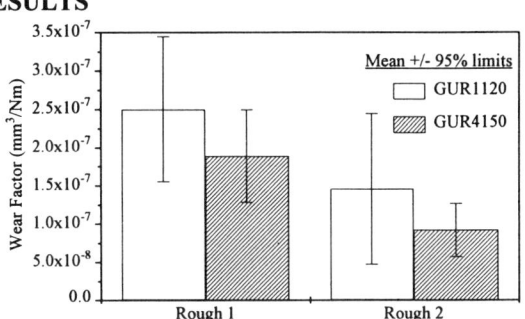

Figure 1 : Wear factor results for rough tests.

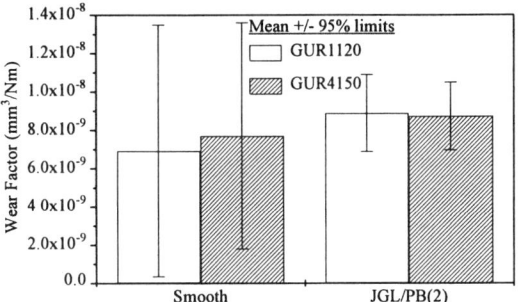

Figure 2 : Wear factor results for smooth tests.

DISCUSSION

The results shown in figures 1 and 2 indicate that there is no statistically significant difference in the wear resistance of these materials at 95% confidence limits. However, for the rough counterfaces the GUR 4150HP is producing a consistently lower wear factor than the GUR 1120.

The slight difference in wear resistance may be due to the higher molecular weight or the different processing methods which result in a material with fewer fusion defects and hence fewer potential fracture sites. The implication that a difference in wear resistance was only observed under rough counterface conditions implies that different wear processes highlight different material property requirements. On the smooth counterfaces an adhesive/fatigue mechanism will operate whereas under rough conditions an abrasive wear process will predominate.

ACKNOWLEDGMENT

This work was supported by Brite Euram project 7928.

REFERENCES

(1) Hostalen GUR Technical Data, Hoechst AG, Frankfurt am Main, Germany.
(2) Lancaster J, MSc Thesis, University of Leeds, 1995.

Effect of head size and loading regime on the wear of UHMWPE acetabular cups in a hip joint simulator.

A A Besong, R J A Bigsby, P S M Barbour, J Fisher.
Department of Mechanical Engineering University of Leeds, Leeds, LS2 9JT, England.

Introduction
The current challenge to attenuate the effect of polyethylene wear and wear debris in replacement joints has led to a large increase in the development and manufacture of hip joint simulators. The hip joint simulator has seen various designs of differing complexities. The majority of these simulators apply simple non-physiological loading and motion patterns that produce a wear path that is not representative of that found in-vivo. In an attempt to replicate the physiological hip joint conditions, more complex and expensive hip simulators have also been developed. The Leeds I physiological anatomical hip simulators have three axes of loading and three independent motions that are applied to precisely replicate the walking cycles defined by Paul. These motion and load patterns produce a wear path and wear scar that is identical to that found in vivo during normal walking.

This study compared the volumetric wear rate of a 32mm UHMWPE acetabular cup, using a full physiological hip simulator with all the load and motion components applied, to the that of a 28mm UHMWPE acetabular cup on the same simulator but with a simplified loading configuration, using one axis of time dependent vertical loading.

Method.
Two tests were conducted on the Leeds I physiological anatomical hip joint simulator. This machine was equipped with two articulating stations and one non-articulation station that was used as a creep control station. In the first test, the full physiological (3 axes) load and motion components were applied and the volume changes of the 32mm GUR 1120 UHMWPE acetabular cup articulating on zirconia ceramic were measured by a co-ordinate measurement machine (CMM). A second test was conducted, using the same machine, on a 28mm UHMWPE and zirconia ceramic joint combination. The motions in both tests were identical but in the latter test only the femoral axis load was applied. In both tests bovine serum was used as the lubricant. The non-articulating station was used to differentiate wear and creep.

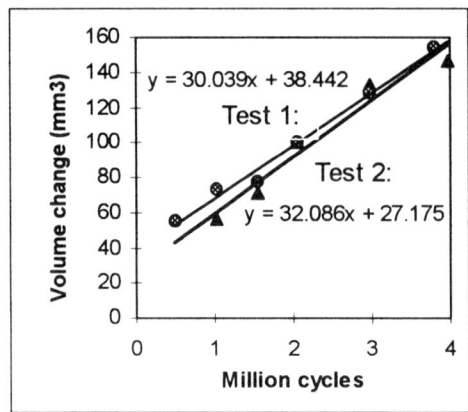

Figure 1: Linear regression analysis of wear volume for test 1 (full loading configuration) and test 2 (femoral axial load only).

Results
The wear scar in the cups in both tests was localised to the superior quadrant as found on clinically retrieved specimens. The surface roughness of the zirconia ceramic heads remained the same throughout both tests. The slope of the linear regression lines (figure 1) of the two tests, excluding the initial volume change that was mostly due to creep, were 30.03 ± 3.24 mm^3/ 10^6 cycles for test 1 and 32.09 ± 3.4 mm^3/10^6 cycles for test 2. In both tests the regression line of the creep station was horizontal and had a similar volume intercept as the regression line of the wear data. This means that one could associate the initial volume change (intercept on the volume axis) with creep.

The difference in wear rate between the two tests was not statistically significant ($P \gg 0.05$). The tests were not sufficiently sensitive to differentiate the small difference predicted from sliding distances due to the different head sizes. From these results, it appears that eliminating the medial/lateral and anterior/posterior loads did not have a significant effect on the wear rates in this hip joint simulator. The volumetric wear rates at 30 - 32 mm^3 per million cycles are lower than average clinical wear rates. This is to be expected as the UHMWPE, although irradiated was less than two years old, and the femoral heads remained undamaged throughout the test. These volumetric wear rates can be extrapolated to predict less than 1mm linear penetration at 10 years in vivo.

INFLUENCE OF SPECIMEN MISALIGNMENT ON WEAR BEHAVIOUR USING THE MTS HIP WEAR SIMULATOR

S BLATCHER, K E TANNER and W BONFIELD
IRC in Biomedical Materials, Queen Mary and Westfield College, Mile End Road, London E1 4NS, U.K.

INTRODUCTION

In the UK, 18% of total hip replacements performed in 1995 were revisions (1). There is strong evidence linking the generation of particulate wear debris with aseptic loosening, a major cause of prosthesis failure (2). Hip wear simulators can be used to predict the performance of implant materials by subjecting them to the loads and movements of gait.

For wear tests of 5 or more million cycles, test methodology is critical, as minor parameters can have gross long term effects. Acetabular test specimens are often mounted in low modulus fixtures to facilitate removal of the cups for gravimetric analysis. These fixtures may be aligned initially, but long term loading may produce creep resulting in misalignment. Although the MTS hip wear simulator demonstrates excellent levels of load uniformity and reproducibility (3) the influence of specimen alignment has not been analysed. In this study, the relationship between specimen alignment and frictional torque was examined for a range of load conditions.

MATERIALS AND METHODS

A 28 mm alumina ceramic femoral head was mounted on the upper support column of one test station. The corresponding ceramic acetabular cup was mounted in a specially designed stainless steel alignment rig and attached to the base plate of the simulator. The base plate is situated directly beneath the head support column at an angle of 23°.

To achieve alignment between the rotational axes of the acetabular and femoral components, the cup is mounted at a specific height above the base plate. The base plate is then raised by a hydraulic actuator until the cup engages the head. Mounting the cup at an incorrect height produces eccentric alignment. A bearing mechanism compensates for eccentric alignment by allowing horizontal motion of the femoral head.

Using the alignment rig it was possible to mount the acetabular cup -3 mm, -2 mm, -1 mm, 0 mm, 1 mm and 2 mm from the theoretical alignment position. A rotational velocity of 0.5 Hz and loads from 100 N to 4 kN were applied at each position. The horizontal translations of the femoral head column were monitored using a dial gauge and frictional torques were measured using the built in torsion cells.

RESULTS

The translational data (Fig. 1) shows that both load and cup position affect femoral head movement. At positions -3 and -2 mm, increasing load leads to reduced translations. At 0, 1 and 2 mm the opposite effect is observed. Position -1 mm appears to be on a threshold as the translations decrease to a minimum at 1 kN but then increase with load.

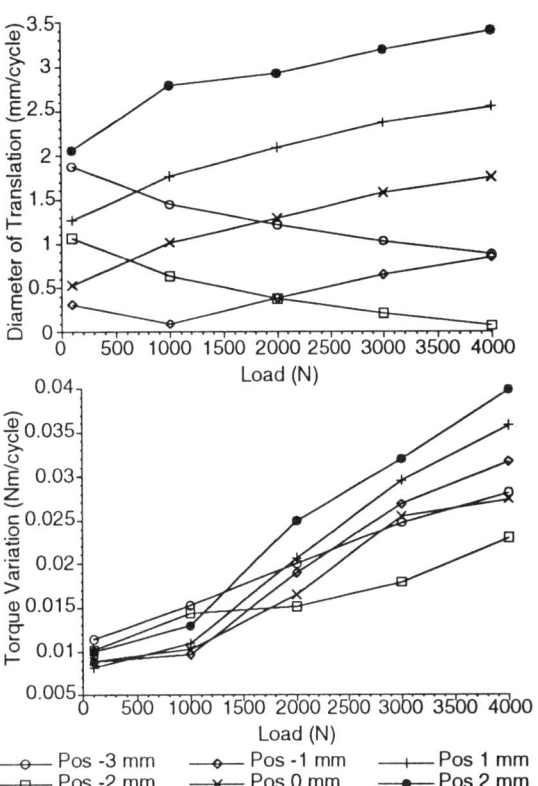

Figure 1: Variation in Femoral Head Translation and Frictional Torque with Cup Position

A constant load and rotational velocity should produce uniform frictional torque. The torque variation in Figure 1 represents the difference between the maximum and minimum torque per cycle. It can be seen that frictional torques were not constant and that increasing load leads to increased torque variation. No clear relationship was apparent for cup position.

DISCUSSION

For loads up to 2 kN an alignment position of -1 mm produces minimum head translation and torque variation. At loads from 2 kN to 4 kN, a position of -2 mm produces minimum translation and torque variation. Thus, for the dynamic loads of 0 to 4 kN used in wear testing, a low modulus fixture must be designed to minimise the variation in head translation and consequently frictional torque.

REFERENCES

1 Department of Health, R & D Priorities for Biomaterials and Implants, 1996, p. 7.
2 Kobayashi *et al*. Proc.I.Mech.E., 1997, **211H**, 11-16.
3 Mejia & Brierley, Bio-Med. Mats. & Eng., 1994, **4**, 259-271.

MARKED IMPROVEMENT IN THE WEAR RESISTANCE OF A NEW FORM OF UHMWPE IN A PHYSIOLOGIC HIP SIMULATOR

M. JASTY, C.R . BRAGDON, D.O.O'CONNOR, * O.K.MURATOGLU, *V. PREMNATH,.* E. MERRILL
AMTI Inc., 176 Waltham St., Watertown, MA 02172. *MIT, 77 Mass. Ave., Cambridge, MA 02139

INTRODUCTION

Efforts to develop new forms of ultrahigh molecular weight polyethylene (UHMWPE) with higher wear resistance have been hampered by the lack of information on the mechanisms of wear of this material in the hip. Recent studies from retrieved acetabular components have shown that the UHMWPE at the surface undergoes large strain plastic deformation and subsequently breaks up due to multi-directional motions in the hip. We postulated that highly cross linked UHMWPE would be resistant to this large strain plastic deformation and thus would have substantially improved wear resistance. We created a highly cross linked UHMWPE by physical methods rather than chemical cross linking.

The purpose of this study is to compare the wear resistance of this new material to conventional UHMWPE in a hip simulator capable of generating physiological conditions of motion in the human hip joint after THR.

METHODS

UHMWPE ram extruded bar stock (GUR 415) was extensively cross linked while molten, using high energy irradiation resulting in a material having a lower modulus and lower crystallinity than conventional bar stock. Acetabular liner components were machined from this material as well as from conventional UHMWPE to a 62 mm outer diameter and a 32 mm inner diameter.

The wear testing was done using the twelve station hip simulator capable of reproducing closely the physiologic conditions of gait and associated loads in calf serum maintained at 37°C. The motions used were $\pm 23°$ of flexion/extension, $\pm 10°$ of external/internal rotation and $\pm 8.5°$ abduction/adduction. The exact wave form of each arc of motion and their relationships to each other were applied to the motion actuators to reproduce the physiologic pattern.[1]

Five conventional UHMWPE and five melt cross linked UHMWPE components were wear tested for 2.7 million cycles. One additional component from each bank of 6 stations on the hip simulator served as a load soak control (to account for fluid absorption during the test) in which only the load was applied, but the motion was not.

The amount of wear was measured by the gravimetric method. The components were cleaned and dried, and weighed before and after the wear testing. The weight loss of the components was subtracted from the weight gain of the load soaked control and the result was multiplied by the density of the polyethylene to yield the wear volume. All tests were carried out to 2.7 million cycles.

RESULTS

The mean wear rate per million cycles in the components made of conventional UHMWPE was 29 ± 3.9 mg comparable to the wear rates found in specimens retrieved at autopsy from patients. The mean wear rate per million cycles in the components made of melt irradiated UHMWPE was - 2.8 ± 2.3 mg, meaning that the low wear cups gained slightly more weight than the load soak control specimen. This is a highly significant and substantial difference. The wear rates of the components made of this new material are, in essence, zero at nearly three million cycles. CMM data confirmed that no detectable wear had occurred.

Visual examination of the articular surfaces of the components also confirmed the wear data. The components made of conventional UHMWPE all showed the typical highly worn area with a glassy finish separated from the unworn area by a visible ridge. The components made of melt irradiated UHMWPE showed no wear and, in fact, the machining marks from the manufacturing process were still visible after nearly 3 million cycles.

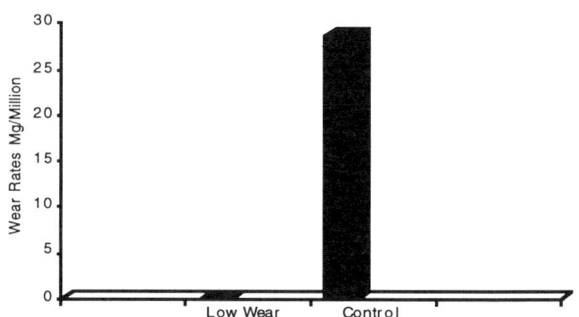

CONCLUSIONS

Recent studies have considerably expanded our understanding of the mechanisms of wear of UHMWPE in the hip. Reorientation of the polyethylene under the predominantly unidirectional wear in the hip is a key feature of the wear. With that understanding, it is now possible to improve the wear resistance of UHMWPE. A new form of melt irradiated, highly cross linked UHMWPE was created without introducing chemicals to produce the cross links. This new material has markedly improved wear resistance in a sophisticated hip simulator up to nearly 3 million cycles, showing no detectable wear at nearly three million cycles under physiologic conditions.

REFERENCES

1. Bragdon et al. Proc Instn Mech Engrs., V.210, 157-65, 1996.

THE IMPORTANCE OF MULTIDIRECTIONAL MOTION FOR THE WEAR OF POLYETHYLENE IN THE HIP

C.R. BRAGDON, D.O. O'CONNOR, J.D.LOWENSTEIN, JR., and W.D.SYNIUTA *
Advanced Medical Technology, Inc., 176 Waltham Street, Watertown, MA 02172

INTRODUCTION

Efforts to improve the wear performance of ultra high molecular weight polyethylene used in total hip replacements have been hampered due to the lack of information on the mechanisms of wear of this material in the hip. Studies from retrieved acetabular components suggest that unidirectional motion leads to orientation of the surface layer of the polymer and fibril formation in the direction of motion, and that multi-directional motions lead to rupture of these fibrils and the liberation of submicron wear particles. In order to simulate these mechanisms of polyethylene wear in the hip, a sophisticated hip wear simulator was built which could independently simulate the three different motions of the hip, (flexion/extension, internal/external rotation, and abduction/adduction), along with the loads, lubrication conditions and temperature during normal gait.

METHODS

Metal backed acetabular components with 62 millimeter outer diameter and 32 millimeter inner diameter were used in the study. The cups were mounted in the anatomical position above the femoral head within a sealed test chamber and were completely submerged in calf serum maintained at 37 degrees centigrade. The calf serum was continually recirculated through the test chamber by the use of a peristolic pump. Joint load profiles were derived from the instrumented prosthesis data of Bergmann et al. and were applied through the heads using a hydraulic actuator. Flexion/extension motion was applied to the acetabular component using a separate hydraulic actuator. Internal/external rotation was applied hydraulically to a shaft holding the femoral head, and the abduction/adduction motion was applied to the acetabular component using a servoelectric device.

For the first set of experiments, a simplified gait cycle similar to that used in other wear testers was used. This consisted of +/- 23o of flexion/extension and +/-10o of internal/external rotation applied in an sinusoidal manner at a rate of 2 Hz. The two motions were in phase and resulted in curvilinear tracking of the femoral head against the polyethylene component.

For the second set of experiments, physiological motion of normal gait were applied to the components. These data were obtained by Johnston et al from direct measurements using a three-dimensional goniometer on patients during normal gait. The motions used were +/- 23o of flexion/extension, +/-10o of external/internal rotation and +/-10o abduction/adduction. The exact wave forms and their relationships to each other were applied to the motion actuators which resulted in the femoral head tracing near rectangular paths against the polyethylene component, the exact size and shape of the path varying across the surface of the cup. This complex pattern of motion is similar to that predicted for normal gait using Johnston's motion data. Wear volume was measured by determining the weight loss of the cups after they were cleaned and dried.

RESULTS

Surprisingly, the simplified gait cycle resulted in no measurable wear of the polyethylene component out to 7 million cycles. This is in stark contrast to the results obtained with the physiological motion. The average wear rate was 14 mg/million cycles with the physiologic wear patterns. No wear particles could be retrieved from the serum with the simplified gait cycle, whereas abundant micron and submicron wear particles were recoverable from the serum with the physiologic motion.

On examination of the articulating surface using the electron microscope, the appearance of the worn surfaces with the physiologic motion was similar in appearance to surfaces of retrieved polyethylene components. The wear surface had a fine fibular appearance, with the size of the fibrils ranging from submicrons to a few microns. The surface appearance resulting from the simplified gait cycle had a rippled appearance at high magnification, the ripples being perpendicular to the direction of motion.

CONCLUSIONS

This study points to the importance of multidirectional motion for the wear of polyethylene in total hip replacements. With unidirectional reciprocating motion, the polyethylene surface appears to become oriented and strain hardened and does not wear, whereas multidirectional motion leads to drawing and shearing of the polyethylene into particles from the surface, producing wear.

REFERENCES

1. Bragdon et al. Proc Instn Mech Engrs., V.210, 157-65, 1996.

ANALYSIS OF POLYETHYLENE WEAR DEBRIS IN TOTAL ARTIFICIAL JOINTS USING MASS AND FREQUENCY DISTRIBUTIONS AS A FUNCTION OF PARTICLE SIZE

J FISHER, AA BESONG, PSM BARBOUR, MJ KING, BM WROBLEWSKI, JL TIPPER, E INGHAM, M STONE
Biomedical Engineering Research Group, University of Leeds, Leeds LS2 9JT

INTRODUCTION

The identification of micron and sub micron size ultrahigh molecular weight polyethylene (UHMWPE) wear particles as a cause of wear debris induced osteolysis has prompted the development of several different methods for debris isolation and quantification of particle size distributions. Both automated image analysis and Coulter particle counting have been used to quantify the frequency distribution of the particle sizes in the range 0.1 to 10 µm. These studies have shown the mode of the particle size distribution to be in the range 0.3 to 0.6 µm. There is recognition that much larger debris outside the size range measured is present in clinical samples, and that while it may not be particularly biologically active, its mass may substantially contribute to the total wear volume. In this study we present an analysis of the mass distribution as a function of particle size as an effective way of quantifying wear particles from *in vivo* and *in vitro* samples.

MATERIALS AND METHODS

Ten different particle distributions were analysed. Four samples of UHMWPE were isolated from simple configuration laboratory wear tests. Four samples were isolated from tissues retrieved from revision hip prostheses. Two theoretical particle size distributions were analysed with a mode of 0.1 µm. The particle size distribution for the eight experimental and clinical samples were determined by two dimensional particle sizing, and the mode of the particle size distribution was in the range 0.1 to 0.5 µm. The distribution of the mass of the wear debris as a function of particle size was determined for all the samples in two size ranges, 0.1 to 10µm and greater than 10µm. The mass distribution was determined by a method of sequential filtering. The wear particles were isolated from the tissues and laboratory lubricants using a multistage digestion process, involving potassium hydroxide, nitric acid, chloroform/methanol extraction and centrifugation. This method digested the tissue and the serum lubricant such that it could be passed sequentially through 10 µm and 0.1 µm filters, to collect the polyethylene debris in the sample. The filters were then dried and weighed to obtain the mass of the debris in each size range. The particle size distribution was obtained using SEM images from each of the filters. The mass and frequency distributions, as a function of size, allowed the total number of particles to be estimated.

RESULTS

The table shows the mode of the particle size distribution, the percentage mass for the two size ranges and the estimated number of particles per mg of wear debris for each of the ten samples. The theoretical distributions T1, T2 were selected to demonstrate the extreme cases that can be found when two particle distributions have the same mode, and different mass distributions. The total number of particles produced varied from 200×10^6 to 10^{12} particles per mg of debris. The patients P1 to P4 and laboratory samples L1 to L4 all had a mode of the size distribution in the range 0.1 to 0.5 µm. However, the mass distributions varied considerably, with between 5 and 82% of the mass of the debris being less than 10 µm, and the estimated total number of particles per mg of debris varying from 500 to 8,200 million. The increased mass in the large size range dramatically reduced the total number of particles per unit mass.

Sample	Mode of Particle Size Distribution µm	Percentage Mass in size range		Estimated No. of Particles per mg
		0.1 to 10 µm	> 10 µm	
T1	0.1	100	0	1000 10^9
T2	0.1	2	98	0.2 10^9
P1	0.1 - 0.5	82	18	8.2 10^9
P2	0.1 - 0.5	80	20	8.0 10^9
P3	0.1 - 0.5	38	72	3.8 10^9
P4	0.1 - 0.5	5	95	0.5 10^9
L1	0.1 - 0.5	71	29	7.1 10^9
L2	0.1 - 0.5	36	64	3.6 10^9
L3	0.1 - 0.5	10	90	1.0 10^9
L4	0.1 - 0.5	6	94	0.6 10^9

DISCUSSION

This method for analysing UHMWPE debris, using the mass distribution as a function of size, has the ability to differentiate samples which have similar particle size distributions and demonstrated a large difference in the resulting total number of particles. The power of the technique arises from the quantification of debris over a size range which takes account of the contribution made by the larger debris. This method has the potential to illicit further understanding of the relationship between tribological designs, materials, clinical variables and wear debris induced osteolysis.

TRIBOLOGICAL ASPECTS OF APPLICATION OF DENTAL MATERIALS FOR FIXED PROSTHODONTICS FACINGS

WOJCIECH RYNIEWICZ, ANDRZEJ GALA

Jagiellonian University, Collegium Medicum, Dep. Of Prosthetic Dentistry, 30-102 Kraków, ul. Syrokomli 10, Poland

ABSTRACT

Materials for preparing fixed prosthodontics must not be harmful for surrounding tissue and for the whole body. In addition, they have to meet specific biological, technical, mechanical and tribological requirements. Most of contemporary dental materials fulfill such needs.

Due to high standard set by prevention in applying fixed prosthodontics, very important is a selection of material and accuracy of machining during manufacturing process.

The paper presents results of researches of materials commonly used for esthetic facing of fixed prosthodontics of the following types: crown, bridges, inlays, onlays. The researches comprised porcelain (Vita, Ivoclar Excelco Shofu, Synspar), composite and acrylic materials (Biodent /De Trey/, Duropont /Nowodent/, Chromasit /Ivoclar/, Artglass /Kultzer/, Stellon /De Trey/, Palapont /Spofa Dental/). Furthermore experiment included silver-palladium alloy (Spall)

The aim of these researches was to determine durability of these materials and to analyse microgeometry of the samples which underwent precise polishing and samples after wear tests.

Tribological tests were performed on Four-Ball Wear Tester Machine / GE Modification manufactured by Roxana, Illinois, USA. The friction pair in the system ball - three discs consists of three discs made of a given dental material fixed in the special base with seats pressed to the rotating ball.

The measure of antiwear properties of dental materials is scar diameter on tested discs. The indexes for a material tribologic properties are also: contact pressure, friction coefficient recorded continuously in tests.

In the fig. 1 is presented diagram of antiwear properties of tested dental materials.

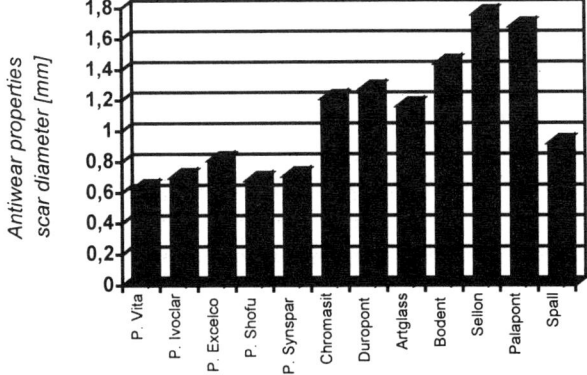

Fig.1: Results of tests of tribological properties

Tests of surface microgeometry were made on profilometer Hommel Tester.

The objective of tests was estimation of samples surface before and after wear tests. Results of surface microgeometry tests were determined by:
- roughness parameters (R_a, R_z),
- graphmetric analysis

In the fig. 2 are presented exemplary profilograms of tested materials.

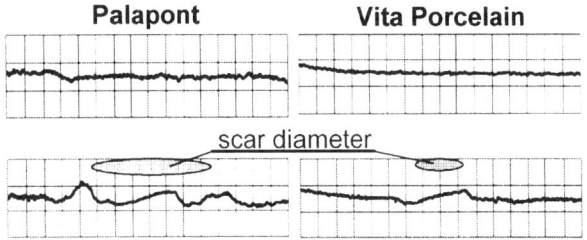

Fig. 2: Exemplary profilograms before and after wear tests

Tests confirmed usability of the applied method because when parallel estimation of tribologic properties is made it is possible prediction of behaviour of those material in clinic practice.

Irregardless of the material used a very important factor is a roughness of its surface which results from both a material structure and possibility of its being machined.

Surface asperities and their specific features (specially in the outer area of the scar) is connected with a kind of applied material and makes it possible for the dental plaque to deposit and at the end gingival trauma and periodontitis.

Geometry disturbances of material for facing of fixed prosthodontics resulting from the type of material, accuracy of polishing and biological wear can cause mechanical irritation, may enable food particles to deposit and can prevent self cleaning, thus making it more difficult to preserve hygiene.

REFERENCES

(1) L. Hupfauf, Teilprothesen, Urban & Partner, Wrocław 1997.

(2) M. Kleinrok, Zasady wykonania protez stałych - korona lana i licowana pocelaną, Lublin 1995.

(3) S. Majewski, Zastosowanie protez stałych w leczeniu braków uzębienia, Wydawnictwo AM, Kraków 1991.

(4) A. Ryniewicz, Sposób badania materiałów konstrukcyjnych i smarowych kulistą przeciwpróbką, Technika Smarownicza, z 5-6, 1978.

POLYETHYLENE PARTICLES OF A "CRITICAL" SIZE ARE NECESSARY FOR THE INDUCTION OF IL-6 BY MACROPHAGES *IN VITRO*

TIM GREEN AND EILEEN INGHAM
Department of Microbiology, University of Leeds, UK.
JOHN FISHER
Department of Mechanical Engineering, University of Leeds, UK.

INTRODUCTION

UHMWPE wear particles play a critical role in wear debris induced osteolysis leading to the aseptic loosening of total artificial joints. Wear particles stimulate macrophages to produce mediators of bone resorption *in vitro*. Recent *in vitro* cell studies have utilised commercially available particles, differing greatly in their shape and size distributions from clinically generated debris. In addition these studies have mainly used macrophage cell lines which may not respond in the same way as primary macrophages. The pseudomembrane from patients with failed prostheses has shown that 100% had levels of IL-6 whereas 73% and 53% demonstrated levels of IL-1β and TNF-α respectively. Moreover the presence of IL-6 receptors on osteoclasts suggests IL-6 plays a crucial role in osteolysis. The purpose of this study was to challenge primary macrophages with polyethylene particles that had been separated into definitive size ranges in order to determine the "critical" size range of polyethylene particles for macrophage IL-6 secretion.

MATERIALS AND METHODS

Macrophages were collected by peritoneal lavage of C3H mice and were resuspended in macrophage culture medium RPMI 1640 medium. Polyethylene ceridust 361 was suspended in RPMI 1640 and was sequentially filtered through 10, 1, 0.6, 0.4, 0.2 and 0.1μm cyclopore polycarbonate membranes and the weight of debris calculated. SEM and image analysis were used to determine the particle size distributions of 0.21 ± 0.069μm, 0.49 ± 0.11 μm, 4.3 ± 1.89μm and 7.2 ± 3.15μm. GUR1120 UHMWPE resin with a mean size of 88μm ± SD 29μm was used to resemble the large shards of debris generated in clinical samples. The particles were washed from the filters and resuspended in RPMI 1640 in the separate size ranges. A 1% agarose solution was prepared in RPMI 1640, of which one volume was mixed with two volumes of each particle suspension. The mixtures were then added to 48 well plates in 200μl volumes in triplicate and centrifuged at 800g which resulted in a superficial layer of polyethylene particles. Peritoneal macrophages were cultured at 5×10^5 cells.well^{-1} with the particles held in the agarose at particle volume (μm^3) : cell ratios of 10:1 and 100:1 for 24 hours in triplicate. Macrophage viability was measured after the 24 hours by MTT conversion. IL-6 in the culture supernatants was assayed by ELISA.

RESULTS

Macrophages cultured with polyethylene particles with a mean size of 0.49μm and 4.3μm at volume:macrophage ratios of 100:1 elicited the release of significantly higher levels of IL-6 after 24 hours relative to macrophages that were cultured without any particles. IL-6 production was not significantly raised in cultures of macrophages with 0.2μm, 7.2μm or 88 μm particles at 10:1 or 100:1 (figure 1). Cell viability was not affected by any of the treatments.

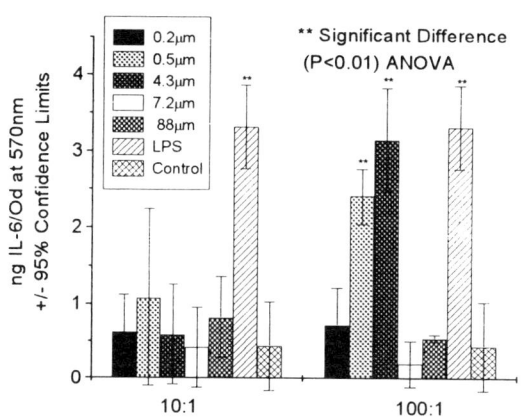

Figure 1: IL-6 Production by Peritoneal Macrophages Stimulated by Different Size/Volume of Polyethylene Particles

DISCUSSION

These experiments have shown that polyethylene particles can cause primary macrophages to produce elevated levels of IL-6. The findings suggest that the volume of particles and their size is a critical factor in macrophage activation. Smaller particles (<7.2μm) activate macrophages to produce higher levels of cytokine than larger particles, providing the volume is great enough. However particles in the 0.21μm size range did not stimulate IL-6 production at either of the volumes tested. This study indicates that the most biologically active particles will be in the phagocytosible size range, approximately 0.5-10μm.

THE LUBRICATING ABILITY OF BIOMEMBRANE MODELS WITH DIPALMITOYL PHOSPHATIDYLCHOLINE AND γ-GLOBULIN

HIDEHIKO HIGAKI, TERUO MURAKAMI and YOSHITAKA NAKANISHI

Department of Intelligent Machinery & Systems, Faculty of Engineering, Kyushu University, Fukuoka, 812-81, JAPAN

INTRODUCTION

Single continuous biomembrane-structure and osmiophilic stratums were successfully detected on articular surfaces of human and animals by using a transmission electron microscope. The superficial membrane-structure is supposed to affect the contact behavior as boundary friction(1)(2)(3)(4). Joint lubrication is theoretically and experimentally considered that fluid film plays a main role in most of physiological motion and some supplemental mechanisms as the boundary lubrication function to protect the articular surfaces, depending on the severity of rubbing condition(5)(6). The lubricating ability of biomembrane models of superficial bilayers on articular surface were examined through two kinds of friction tests.

MATERIAL AND METHODS

The effect of addition of riposomes of Lα-dipalmitoyl phosphatidylcholine (Lα-DPPC) and human serum-derived γ-globulin in a saline solution of sodium hyaluronate was evaluated through pendulum friction tests. The reciprocating frictional behaviour in sliding pairs of pig articular cartilages and glass plates was also studied to elucidate the tribological role of those constituents in Langmuir-Blodgett films as models of boundary lubricating films. Since it was reported that palmitic acid was main fatty acid in synovial fluids(7), dipalmitoyl phosphatidylcholine was used as a phospholipid in the membrane-structure. The γ-globulin is one of hydrophobic globular proteins which ratio of non-polar residues was 74.1 % and similar to those of LGP-1(8)(9) and PSLF(10). Non-polar residues of the globular proteins and glycoproteins are considered to assosiate with phospholipids in biomembrane structure which hydrophobic groups meet inside.

RESULTS

In pendulum friction tests, the frictional characteristics of pig shoulder joints were confirmed to depend on the viscosity of lubricants only in the physiologically low load condition and immediately after the loading. Detergent (polyoxyethylene p-t-octylphenyl ether, Triton X-100) was successfully used to remove adsorbed molecules from the articular surfaces. It was confirmed from the change of wettability of the dried articular surfaces. By the surface treatment with the detergent, the friction coefficient of natural synovial joints was significantly increased in mixed lubrication mode with a saline solution of sodium hyaluronate of 0.2 g/dl. The addition of riposomes of Lα-DPPC or γ-globulin significantly improved the boundary lubricating ability of the articular surfaces treated with the detergent, depending on the quantity of those additives. It appears that the riposomes of Lα-DPPC and γ-globulin can form protective films on the articular surfaces and keep the superior frictional characteristics.

In reciprocating friction tests, pig synovial fluid and a water solution of hyaluronic acid were used as lubricants. The synovial fluid significantly had superior lubricating ability under physiologically high load condition to the sodium hyaluronate solution of equivalent viscosity. It seems to be responsible for the boundary lubricating ability of constituents other than hyaluronic acid. Langmuir-Blodgett films of Lα-DPPC on the glass plate kept up low and stable friction coefficient, depending on the accumulating layer number. In conditions of mixed films with Lα-DPPC and γ-globulin, the frictional behavior was improved with increasing the quantity of γ-globulin.

CONCLUSION

We propose a model of boundary lubricating film on the articular surfaces, which effective adsorbed films are composed of proteins, phospholipids and other conjugated constituents. The excellent lubricasting ability was demonstrated through two kinds of friction tests for the models. The superficial membrane-structure appeared to be controlled by hydrophobic groups in those amphiphilies.

REFERENCES

(1) D. Guerra, et al., J. Submicrosc. Cytol. Pathol., 28(3), 385-393, 1996.
(2) P. F. Williams, et al., Proc. Instn. Mech. Engrs., 207, 59-66, 1993.
(3) T. B. Kirk, et al., J. Orthop. Rheum., 6, 21-28, 1993.
(4) B. A. Hills, J. Rheumatol., 16(1), 82-91, 1989.
(5) D. Dowson, Proc. Instn. Mech. Engrs., 181, Pt.3J, 45-54, 1966-67.
(6) K. Ikeuchi, Lubricants and lubrication(Ed. D. Dowson, Elsevier), 65-71, 1995.
(7) J. L. Rabinowitz, et al., Clin. Orthop., 190, 292-298, 1983.
(8) D. A. Swann, et al., Biochem. J., 225, 195-201, 1985.
(9) D. A. Swann, et al., J. Biol. Chem., 256, 5921-5925, 1981.
(10) G. D. Jay, Connective Tissue Res., 28, 71-88, 1992.

Submitted to *IMechE Journal of Engineering in Medicine*

IN VITRO GENERATION OF CLINICALLY RELEVANT STERILE UHMWPE WEAR DEBRIS SUITABLE FOR USE IN CELL CULTURE STUDIES

J. BRIDGET MATTHEWS, ALFRED A. BESONG*, JOHN FISHER* and EILEEN INGHAM
Department's of Microbiology and Mechanical Engineering*, University of Leeds, Leeds LS2 9JT, UK
MARTIN STONE
Department of Orthopaedic Surgery, Leeds General Infirmary, Leeds LS21 3EX, UK
B. MIKE WROBLEWSKI
Centre for Hip Surgery, Wrightington Hospital, Wigan WN6 9EP, UK

INTRODUCTION

Wear particles are critical in the periprosthetic osteolysis which results in the long-term failure of total joint replacements. Their composition, number, size and shape and the volume of debris, are all thought to influence this macrophage-mediated process. In order to identify the mechanism of prosthetic loosening, cell-particle interactions have been studied *in vitro* utilising commercial particles, however, these differ widely from clinical debris in terms of their uniform shape and size distributions. Furthermore, they may be contaminated with microbes or endotoxin, both of which will also activate macrophages. Indeed endotoxin i.e. lipopolysaccharide (LPS), is used routinely to stimulate cytokine synthesis by macrophages *in vitro*. This study aimed to generate characterised, sterile, UHMWPE wear particles *in vitro*, comparable to those found *in vivo*, and suitable for use in cell culture studies.

MATERIALS AND METHODS

UHMWPE (GUR 412) debris was generated using a tri-pin-on-disc tribometer (1). Stainless steel counterfaces and the wear test rig were sterilised and all procedures were carried out aseptically within a Class I Laminar Flow Cabinet. Tests were performed on rough counterfaces ($R_a = 0.07 \pm 0.01$) lubricated with sterile pyrogen-free water, with a unidirectional motion of $0.12 m.s^{-1}$ and a constant load (stress = 12MPa). Lubricants were sequentially filtered through 10μm and 0.1μm membranes and the weight of debris calculated. Mass and particle size distributions were determined by SEM and Image Analysis. Samples of lubricant, from the start and finish of tests, was plated onto Sabourauds dextrose, nutrient and heated blood agar to detect any microbial contamination. Debris, resuspended in sterile culture medium at $10 \mu g.ml^{-1}$, was analysed for endotoxin using a quantitative LAL test.

RESULTS

Wear debris, produced using conditions comparable to those which are currently used in the majority of bio-engineering laboratories, is invariably contaminated. All debris analysed in the current study was, therefore, generated aseptically. All results are expressed as the mean ± 95% confidence limits. Following filtration and mass determination, the mean Wear Factor was calculated to be $2.1 \times 10^{-6} \pm 1.3 \times 10^{-6}$ $mm^3.Nm^{-1}$ with 80% (+8.7 & -10.4%) and 20% (+10.5 & -8.6%) of the total mass of debris on the 10μm and 0.1μm filter respectively. All the particle types seen clinically i.e. shards, platelets, ribbons and sub-micron particles were observed. Image analysis of debris on the 0.1μm filter determined a mean particle thickness of $0.88 \pm 0.13 \mu m$ with 1.88×10^{10} particles per mg of debris. Whilst the majority of particles were within the 0.1-0.5μm size range (75.2%), they did not account for the majority of mass which was concentrated above 10.0μm (85.7%) - Table 1. No microbial growth was observed for either pre- or post-test lubricant plated out on any of the media, therefore, sterility was maintained throughout the procedure. Low endotoxin levels (1.2-1.3 $EU.ml^{-1}$) were also achieved in wear debris samples. Measurement of the comparative activity of LPS used in cell culture determined that this equated to approximately 100pg LPS per ml or per 10μg of debris.

	Length of particles (μm)				
	0.1-0.5	0.5-1.0	1.0-5.0	5.0-10	>10
No. of particles (%)	75.2 +12.7 -15.4	14.2 +8.5 -6.7	9.9 +9.0 -6.4	0.5 +0.9 -0.5	-
Mass of particles (%)	1.9 +5.7 -1.9	2.0 +4.5 -1.9	7.1 +11.3 -6.1	2.6 +5.4 -2.5	85.7 +12.9 -22.8

Table 1: % Distribution of UHMWPE wear particles determined by particle length

DISCUSSION

This study has, for the first time, produced fully characterised, sterile UHMWPE wear debris *in vitro*, comparable, in both morphology and size distribution, to debris generated clinically *in vivo* (2). Importantly, no microbial contamination and only low levels of endotoxin (≡100pg LPS/10μg debris) were detected. As LPS is used experimentally in ng or μg concentrations, the levels present within this debris was 10-10,000 fold less than would activate macrophages. This will, therefore, enable the macrophage response to clinically relevant particles to be accurately studied *in vitro*.

REFERENCES: 1) Randall, J.E. & Dowson, D. Engng in Medicine, 1984, 13, 55-66. 2) Hailey, J.L; Ingham, E; Stone, M; Wroblewski, B.M. & Fisher, J. Proc. Instn. Mech. Engrs; 1996, 210, 3-10.

QUANTIFICATION AND SIMULATION OF THIRD BODY DAMAGE, AND ITS EFFECT ON POLYETHYLENE WEAR IN ARTIFICIAL HIP JOINTS WITH DIFFERENT TYPES OF FEMORAL HEAD MATERIALS

H MINAKAWA and J FISHER
Department of Mechanical Engineering, University of Leeds, Leeds LS2 9JT, UK
M STONE
Department of Orthopaedic Surgery, Leeds General Infirmary, Leeds LS1 3EX, UK
E INGHAM
Department of Microbiology, University of Leeds, Leeds LS2 9JT, UK
B M WROBLEWSKI
Centre for Hip Surgery, Wrightington Hospital, Wigan, UK

INTRODUCTION

The wear of ultra-high molecular weight polyethylene (UHMWPE) is recognized as one of the major problems in total hip joint arthroplasty. The roughening of the femoral head can increase both the wear volume and the number of wear particles generated. Damage to the femoral head can be cased by third bodies from bone, bone cement, metallic particles and hydroxyapatite. Stainless steel, cobalt chrome, titanium, alumina and zirconia ceramic femoral heads retrieved at revision surgery were examined and the wear of UHMWPE against femoral head materials which had been simulated third body damage was studied.

MATERIALS AND METHODS

1. Analysis of explanted femoral heads

Ten stainless steel, Ten cobalt chrome, seven titanium, ten alumina ceramic and three zirconia ceramic explanted femoral heads were analyzed using 2D contacting and 3D non-contacting profilometry. The heads were measured in four fixed positions which were two articular positions and two non articular positions and also in areas of macroscopic observed damage. The scratches were characterized by R_{pm}.

2. Simulation of third body damage and wear of UHMWPE

Wear plates of stainless steel 316, cobalt chrome alloy, alumina and zirconia ceramics were highly polished and simulated third body scratches. Third body scratches were simulated on the plates using diamond stylus with a load of 0.4N. Each plate had eight equal scratches 5 mm apart. The scratches were analyzed using 2D and 3D profilometry.

Studies of the wear of UHMWPE were carried out against scratched counterfaces in a six pin-on-plate reciprocating machine. Wear pins were UHMWPE (GUR 415) which were non-irradiated and had 5 mm diameter of contact surfaces. Bovine serum was used as a lubricant. Tests were carried out for 24 hours giving a sliding distance of 15km for metallic plates and 48 hours giving a sliding distance of 30km for ceramic plates.

RESULTS

1. Analysis of explanted femoral heads

Table 1 shows the mean R_{pm} for explanted femoral heads.

Femoral head materials	Rpm (µm)		
	Articular	Non-articular	Damaged
Stainless Steel	0.068	0.059	0.400
Cobalt Crome alloy	0.095	0.056	0.446
Titanium alloy	0.241	0.130	0.556
Alumina ceramic	0.018	0.016	0.023
Zirconia ceramic	0.025	0.022	0.025

Table 1

2. Simulation of third body damage and wear of UHMWPE

Table 2 shows the mean of Rp for simulated scratches on the femoral head material plates.

Femoral head materials	Mean of Rp (µm)
Stainless Steel	1.01
Cobalt Crome alloy	0.39
Alumina ceramic	0.05
Zirconia ceramic	0.06

Table 2

Table 3 shows the wear factor of UHMWPE against the scratched counterfaces.

Femoral head materials	Wear factor ($x10^9 mm^3/Nm$)
Stainless Steel	45.0
Cobalt Crome alloy	16.8
Alumina ceramic	8.6
Zirconia ceramic	9.9

Table 3

CONCLUSIONS

1) Ceramic heads were more resistance third body damage and remain smoother compared to metallic heads *in vivo*.
2) Ceramics were also more resistant to be simulated third body damage than metallic alloys and produced less polyethylene wear in the laboratory tests.
3) The wear factor of UHMWPE against the damaged femoral head materials were dependent on Rp.
4) This study demonstrated the benefit using the hardest ceramic materials for the femoral head of hip prosthesis.

A 12 STATION, UPRIGHT HIP SIMULATOR WEAR TESTING MACHINE EMPLOYING OSCILLATING MOTION REPLICATING THE HUMAN GAIT CYCLE

D O O'CONNOR, C R BRAGDON, J D LOWENSTEIN, B RAMAMURTI, D W BURKE, W H HARRIS Orthopaedic Biomechanics Lab., Mass. General Hospital, GrJ 1126 55 Fruit St. Boston, MA 02114

INTRODUCTION: Wear of UHMWPE used in joint replacement surgery is the number one problem of total joint replacement. Proper evaluation of factors governing the wear needs to be performed under conditions as close to a physiologic situation as possible. The goal of this project was to develop a hip simulator that incorporated as many physiologic characteristics as possible into a testing apparatus which also had unique monitoring capabilities in conjunction with great flexibility in the characteristics of the loading cycle and wear pattern.

DESCRIPTION OF MACHINE: A new hip simulator has been developed in conjunction with A.M.T.I. (Advanced Mechanical Technology Inc., Watertown, Ma.). The machine incorporates 12 identical stations, all of which have built in six-component load cells. The kinematics of the machine closely match that of the human gait cycle (1). Simulation includes + 23 deg. of flexion, up to + 10 deg. of abduction and up to + 10 deg. of head rotation. The acetabular cup is mounted in an upright "physiologic" position. The femoral head applies a joint reaction force through a non-constrained post which is attached to a hydraulic cylinder. The "BERGMAN" curve of load across the hip joint (2) is used in this 12 station machine. Each station has an individual sealed chamber in which the articulating components are protected. The polyethylene wear rates generated by this machine using physiologic loads (2) and wear pathways (1) are identical to those measured in specimens retrieved at autopsy of successful THR after years of in vivo service.

DISCUSSION: This hip simulator represents several major advances in the investigations in total hip joint articulations and especially in the study of polyethylene wear. Moreover, we have defined the physiologic path that any given point on the femoral head takes in articulating with the polyethylene and have reproduced that unique and specific wear path in the simulator. The upright position of the articulation allows particles to escape out of the articulation, as in the human experience. This wear tester, in contrast to some prior machines, produces wear in the polyethylene that is eccentric in direction but cylindrical in contour, and thus creates the type of wear path that occurs in humans.

Ref:1. Ramamurti et al, Soc. Biomaterials. #347, 1995, 2. Bergman et al., J. Biomechanics Vol.26. No.8 pg. 969,1993.

The effect of embedding UHMWPE on the bearing surface of femoral component of artificial joint

B WANG, K OKADA and Q OYANG
Department of Mechanical System Engineering, Faculty of Engineering, Yamanashi University, Kofu, JAPAN
T IDE
Department of Orthopedic surgery, Yamanashi University, Medical college, Kofu, JAPAN

ABSTRACT

The main reason for the poor tribological properties of artificial hip joint has been demonstrated to be the hard solid to solid contact between bearing surfaces(1,2). In order to improve this weakness, we designed a new type of Co-Cr alloy head with ultra high molecular weight polyethylene(UHMWPE) pieces embedded on the surface. In this paper, we report the experimental results on the new artificial hip joint.

On the top of ordinary Co-Cr alloy head seven small holes symmetrical to the central point were made and small pieces of UHMWPE were inserted into these holes. A plan view of the machined head is shown in Fig.1.The common UHMWPE cup was used in the test to slide against the new head. The cup is fitted and the head is reciprocated in horizontal direction similar to the movement of human hip joint. The applied load was chosen as 500N and frequency of the movement as 2/sec corresponding to a sliding velocity of 0.04m/sec. Joints were first tested under conditions of lubrication using distilled water and then 0.9wt% isotonic sodium chloride solution containing 0.04wt% hyaluronic acid.

The wear depth-time curve is shown in Fig.1. In all cases the wear of embedded joint is always less than that of ordinary one. At the end of 25th day, the total wear depth for embedded joint is 0.05mm, it is about a half of that of ordinary one 0.1mm. Comparing the wear in the two different solutions, it can be seen that in the poorer lubricant (distilled water), the wear improvement of embedded joint is more distinct. In terms of weight loss, embedding UHMWPE on head surface is much more effective. The weight loss of the cup for embedded joint is reduced by about 30% of that of ordinary joint. The weight of embedded head showed an increase instead of decrease after test. This demonstrates that either the weight loss was very less or large amount of UHMWPE transferred from the cup to the head forming a layer adheres strongly and thus is difficult to remove from the head surface.

BSEM observations indicated that the wear surface of ordinary cup is composed of cleavage areas, which are usually caused by fatigue(2) and sliding areas, but for the cup of embedded joint, only sliding area was found. Conpared with this, wear debris of ordinary joint is large and displays the shape of flakes. Thus we get the conclusion that the fatigue wear seems to be prevented in the wear of embedded joint.

Two effects are thought to be responsible to the wear improvement of UHMWPE embedded joint. First is the thin well distributed UHMWPE transfer film formed on the head. The thin flat film plays the action of a solid lubricant and prevented the direct contact between soft cup and hard alloy head. Second, we consider the uneven elastic deformation of the UHMWPE pieces. Because the great difference of elastic modules of these two materials, the exposed section of the UHMWPE column is deformed into a concave shape. The hollow formed by the concave can be regarded as a container for lubricating liquid.

REFERENCES
(1)J.Fisher and D.Dowson, Proc.Instn.Mech.Engrs., Vol.1.No.205, 73-79pp.
(2)S.Bahadur and V.K.Jain, Wear Mater,1981,707-713pp.

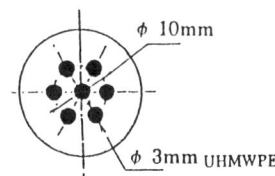

Fig.1. UHMWPE embedded Co-Cr alloy head component.

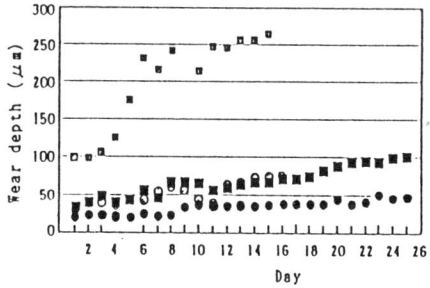

● - Embedded head in isotonic sodium chloride solution
■ - Ordinary head in isotonic sodium chloride solution
◐ - Embedded head in distilled water
▣ - Ordinary head in distilled water

Fig.2 Wear depth as a function of time.

INVESTIGATION INTO THE EFFECT OF PROTEOGLYCAN MOLECULES ON THE TRIBOLOGICAL PROPERTIES OF JOINT TISSUES

JENNY PICKARD, EILEEN INGHAM and JOHN FISHER
Biomedical Engineering Research Group, Mechanical Engineering Department, University of Leeds, Leeds, LS2 9JT.
JOHN EGAN
3M Health Care Limited, Rotherham, S66 8RY.

INTRODUCTION

Human synovial joints have to withstand complex and harsh loading regimes, being subjected to dynamic and impact loading cycles under sliding and rolling conditions. These conditions often occur after considerable periods of stationary loading.

Periods of constant loading and no sliding will breakdown the fluid film regime resulting in asperity contact known as the boundary lubrication regime.

The aim of this study was to investigate the role of the water bonding properties of the proteoglycan molecules within the cartilage matrix in the mixed and boundary lubrication regime.

MATERIALS AND METHODS

Three types of tissue were tested in this study; healthy bovine cartilage, bovine meniscus and bovine cartilage that had been degraded with chondroitinase AC to remove the chondroitin sulphate from the proteoglycan molecules in the cartilage.

The coefficient of friction of these samples was measured using a sliding friction rig. Both 30N and a 3.5N loads were applied to a 3mm diameter test sample against a smooth cast cobalt chrome counterface. A 25% serum, 75% Ringer's solution was used as the lubricant. The coefficient of friction was measured for each specimen after various lengths of stationary loading. The 4mm/s sliding speed of the counterface and the stroke length of 40mm ensured boundary lubrication of the cartilage.

The indentation tests were carried out on both tissue types under a constant load of 5.6N. The specimens were immersed in the serum/Ringer's solution and the displacement was measured with respect to time over a 45 minute loading period.

RESULTS

Figure 1 shows the results from the friction test under a 30N load. All three materials were tested at this load revealing that the meniscus has a much more rapid increase in friction to its equilibrium value than both the cartilage and the degraded cartilage. All three materials reach the same maximum value of friction.

Figure 2 also shows friction test results but under the lower load of 3.5N. Initially the cartilage and the degraded cartilage follow the same friction curve but they begin to separate at the 20 mins point with the degraded cartilage being slightly lower, and reaching a lower maximum friction value than the control cartilage.

The indentation tests are shown in figure 3. In this test the bovine cartilage and the meniscus follow almost identical paths but the degraded cartilage increases much more quickly to its terminal value.

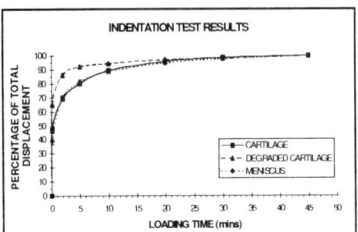

Fig. 1: Indentation Results

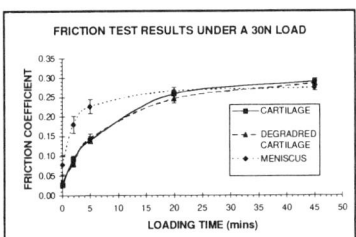

Fig. 2: Friction Results at 30N

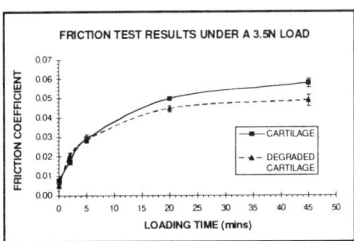

Fig. 3: Friction Results at 3.5N

DISCUSSION

The non-linear rise in displacement of the articular cartilage is thought to be associated with the flow of water, due to a pressure gradient within the cartilage. This restriction may be associated with the proteoglycan content of the tissue. The chondroitin sulphate is thought to play a major role in water retention. The cartilage has more proteoglycan molecules than meniscus, but the degraded cartilage has none. The proteoglycan content does not appear to have the same effect on the coefficient of friction. In particular, modification of the biphasic properties of articular cartilage by removing the chondroitin sulphate did not modify the friction properties. Further work is required to understand the changes in the coefficient of friction with time under mixed and boundary lubrication conditions.

Submitted to *IMechE Journal of Engineering in Medicine*

POLYMER ENDOPROSTHESES FRICTION MATERIAL WITH CARTILAGE-SIMULATING STRUCTURE

LEONID PINCHUK, YELENA TSVETKOVA, VICTOR GOLDADE
Metal-Polymer Research Institute named after V A Belyi
32a, Kirov St., 246652 Gomel, Republic of Belarus

ABSTRACT

Metal-polymer endoprostheses consisting of metallic rod with ball head and adjoint polymer cup are widespread in medical practice. Their main deficiency is low wear resistance that does not allow to consider tribological problem of endoprostheses solved at present time. It is stipulated first at all by the fact that endoprostheses friction pair is very far from a complex biomedical system - a human joint.

The aim of the present paper consisted in modeling endoprostheses polymer friction material with cartilage-simulating structure.

The phase transformations of polymer-solvent systems leading to formation of high molecular compound jellies are the scientific basis for solution of this problem proposed by us. Jellies on the basis of polyethylene and mineral oils widely used at prosthesis are applied for more than ten years in triboengineering for suppression of corrosion-mechanical wear of metallic friction parts.

Ultra-high molecular weight polyethylene (UHMWPE) Hostalen 4120 (Germany) was used in our experiments. Its modification was done by means of medical vaseline oil (VO). The friction properties were also studied with the help of standard methods and pendulum friction machine containing only one (investigated) friction pair which imitate cartilage friction conditions. Supermolecular structure of UHMWPE was studied by a scanning microscope.

The porous polymer matrix containing fluid is formed at modifying UHMWPE VO. By means of technological methods one can regulate sizes of pores within 1-20 μm that correspond to parameters of a human cartilage porous system. The comparison of porous structures on Fig. 1 with photographs of cartilage surface at the same magnification proves that in fact they are full analogous.

The examined structures are not in equilibrium and undergo syneresis at VO content more than 5% mass. Maximum separation of fluid phase takes place after 100-200 hours of sample formation and reaches 1% of a sample mass. Then the mass of extracted fluid is exponentially decreases approaching zero in 40-45 days.

The modification of UHMWPE by VO results in changing deformation and strength characteristics of samples. In order not to reduce by plasticization the strength of material that does not participate in friction, the possibility of local modification of UHMWPE sample friction surface by thermal treatment in VO was investigated. Size increase of samples at VO treatment in isothermal conditions takes place according to exponential law, the numerical parameters of which depend on temperature of thermal treatment. The maximum sizes of samples and time within which they approach constant value grow at temperature increase.

The strength UHMWPE cylindrical samples, faces of which undergo thermal treatment in VO, monotonously decreases depending on thermal treatment time. The higher the thermal treatment temperature, the higher the strength reduction speed. Apparently it is the consequence of intermolecular plasticizing of UHMWPE that is spread on a considerable part of the volume. At the same time the insignificant change in strength of samples after thermal treatment at 125°C is a evidence of a principles possibility of forming friction layer with porous structure without decreasing the strength of the whole sample.

Materials	Friction coefficient		Wear rate, mg/h
	P=3.5MPa V=0.5m/s	P=7MPa V=0.5m/s	P=6.3Mpa V=0.25m/s
UHMWPE	0.09	0.06	0.52
UHMWPE after treatment in VO at 125°C, 3h	0.05	0.03	0.31
UHMWPE+ VO(10%)	0.04	0.03	0.15

The analysis of data in the table shows that tribotechnical characteristics of UHMWPE are improved at VO introduction. It is obviously connected to participation in friction of a material that has a structure of polymeric matrix, in pores which is contained a lubricating liquid. The wear rate of UHMWPE-samples, subjected heat treatments in VO, a some higher, than ones produced by compacting of a components mixture. As appear, it is induced by the grows of wear rate of the porous layer during the friction. Essential result is achievement of reasonably low of friction coefficient (f=0,03-0,050) without additional submission of a lubricating liquid in a zone of friction. Tribotechnical tests were done in continuous slip conditions when restoration of a deformed porous matrix practically did not take place. This allows to hope that at work of such pair in a human body in repeated loading conditions and at synovial fluid lubrication the friction coefficient will be reduced to the level characteristics of natural joints.

EFFECTS OF SYNOVIA CONSTITUENTS ON FRICTION AND WEAR OF UHMWPE SLIDING AGAINST PROSTHETIC JOINT MATERIALS

YOSHINORI SAWAE, TERUO MURAKAMI and JIAN CHEN

Department of Intelligent Machinery and Systems, Faculty of Engineering, Kyushu University, Hakozaki, Higashi-ku, Fukuoka, 812-81, Japan

ABSTRACT

In many studies about the friction and wear of UHMWPE, the pin-on-disk or pin-on-plate test method has been used. Although these screening test methods were simple and inexpensive, results are likely to vary with test conditions, especially with the composition of the lubricant used in these tests (1,2). Synovia, the lubricant for total joint prostheses after implantation, contains the various constituents such as several kinds of serum proteins, hyaluronic acid and also several kinds of lipids. Each of these compositions presumably has a different effect on the friction and wear of UHMWPE *in vivo*. Therefore, the effect of each constituent must be identified to understand the *in vivo* friction and wear mechanism of UHMWPE, and to simulate them in the laboratory test to get the useful test results. The purpose of this study is to examine the effect of major synovia constituents, such as albumin and hyaluronic acid, on the friction and wear of UHMWPE in the pin-on-disk test.

In this study, bovine serum albumin and sodium hyaluronate, a kind of hyaluronic acid, were dissolved in saline respectively and their solution were used as lubricants in pin-on-disk tests to evaluate their effects. Sodium hyaluronate used in this study has a molecular weight of 9.8×10^5. Saline and diluted bovine serum were also used as control lubricants.

Pin specimens were cut from UHMWPE (GUR415) block, and 316 stainless steel and almina ceramics were used as disk specimens. The average surface roughness and vickers hardness are 0.018μm (R_{rms}) and 180 for stainless steel disk and 0.014μm (R_{rms}) and 2300 for alumina disk. In sliding tests, pin specimens were pressed against disk specimens to give rise to mean contact pressure of 3MPa, and articulated at the constant sliding speed of 20mm/s. Each test was run for the sliding distance of 34.6km and temperature of atmosphere inside the test section was controlled to keep around 37°C.

The friction between the pin and the disk specimen was measured continuously, and the wear rate of UHMWPE was determined by the weight loss of the pin specimen mesured at intervals of three days. The liquid bath was washed and then filled with a fresh lubricant every 3 days. In addition, distilled water was supplied to the lubricant every 24 hours to adjust the concentration.

The minimum and maximum values of friction coefficient and the wear rate for each combination of a disk and a lubricant are summarized in Tables 1 and 2. The albumin solution gave the similar wear rate of the diluted bovine serum and prevent the formation of the UHMWPE transfer layer on stainless steel disk. Although the static adsorption of serum proteins from lubricant was not clearly observed by AFM, albumin and other proteins might prevent the adhesion of UHMWPE to the stainless steel and affected the condition of the sliding interface. However, friction coefficient in the albumin solution was as high as in saline, and the wear mechanism in the albumin solution derived from a fine structure of the worn surface observed by SEM was significantly different from that in the bovine serum. Therefore, further studies with other constituents of serum are required to elucidate the wear mechanism.

On the other hand, sodium hyalunate could reduce the friction and wear between UHMWPE pin and almina disk. It was presumably responsible for the increased viscosity of the sodium hyaluronate solution.

Table 1 Friction and wear of UHMWPE sliding against stainless steel

Lubricant	Saline	Albumin solution	Bovine serum
Friction coefficient Min. - Max.	0.02 - 0.2	0.11 - 0.16	0.04 - 0.06
Wear rate mm^3/(N•m)	1.4×10^{-8}	7.1×10^{-8}	5.8×10^{-8}

Table 2 Friction and wear of UHMWPE sliding against Alumina

Lubricant	Saline	Albumin solution	Bovine serum	Sodium hyaluronate solution
Friction coefficient Min. - Max.	0.05 - 0.15	0.1 - 0.14	0.05 - 0.1	0.02 - 0.05
Wear rate mm^3/(N•m)	1.5×10^{-8}	2.2×10^{-8}	3.4×10^{-8}	1.2×10^{-8}

REFERENCES

(1) H. McKellop, I.C. Clarke, K.L. Markolf and H.C. Amstutz, J. Biomedical Mater. Res., 12 (1978) 895-927.
(2) P. Kumar, M. Oka, K. Ikeuchi, K. Shimizu, T. Yamamuro, H. Okumura and Y. Kotoura, J. Biomedical Mater. Res., 25 (1991) 813-828.

EVALUATION OF LUBRICATION AND FRICTION IN ELBOW JOINT ATTACKED BY RHEUMATOID ARTHRITIS AFTER OsO_4 TREATMENT

RUMEN STOILOV, IORDAN SHEITANOV
Clinics of Rheumatology, University of Medicine, Urvich str., 13, 1606 Sofia, Bulgaria.
ZHETCHO KALITCHIN
IG "Bulgaria", Ltd., D. Gruev str., 42, 1606 Sofia, Bulgaria.

ABSTRACT

Synovial joints act as mechanical bearings that facilitate the work of the muscoloskeletal machine. As such, normal joints are remarkably effective with coefficient of friction lower than those obtainable with manufactured journal bearings. The lubrication in the joints is mainly determined by the chialinic cartilage which covers the joints' surfaces and the specific viscosity of the synovial fluid [1]. The authors [2] proved that osmic tetroxide is not dipterous to normal human cartilage and appears to adhere to and affect only inflamed cartilage. The osmic tetroxide treatment of knee joint is known in Bulgaria [3] and it turned out to be very effective towards Rheumatoid Arthritis patients for local therapy of knee joint.

The objective of the present paper is to present the lubrication in the elbow joint (EJ), estimated mainly by its effusion and the range of motion, using the clinical results after intra-articular injections of 1 % osmium tetroxide water solution, combined with Lidocain and Cortisone.

Sixteen rheumatoid arthritis patients with elbow joint synovitis and effusion were selected for the study. In the group we studied 5 men and 11 women. All the patient have been treated with NSAID and SAARD, continuously.

The observed results show that the chemical synovectomy is a very effective additional method of treatment of inflammatory joint diseases. The patients are influenced very well and are in full remission in the whole period of observation from 12 to 24 months. The lubrication in the elbow joint is restored best in the I and II radiographic stage of the disease. Allergy and other side effects are not registered in the group of patients studied.

The decrease of pain is observed in all patients. Its average value is 3.2 before the treatment and at the 6-th month the value was 0.7. It is significant that in 56 % of the patients is observed sharp decrease in the pain at the end of the first month. The average values of the joint effusion are obtained by measuring the circumference in centimeters. Before the treatment its average value is 22.9 cm, while at the end of the sixth month it is 19 cm. In Figures 1, is given the range of motion - parameter giving us direct information about the lubrication and friction in the joint.

The EJ function, i.e. its range of motion is measured in degrees. This parameter increased from an average value of 68 degrees before the treatment to 113 degrees. The improvement of this parameter, giving information about the tribological properties inside EJ are with high statistical probability - $p < 0.001$. Further increase of the volume of motion after the sixth month are not observed. Obviously, the lubrication and friction in EJ are deteriorated as a result of synovitis, due to the decrease of the viscosity of the synovial fluid. It is caused by the increased production of lyposomic enzymes [1]. The improvement in lubrication and friction are proved also by measuring the flexion contracture of EJ.

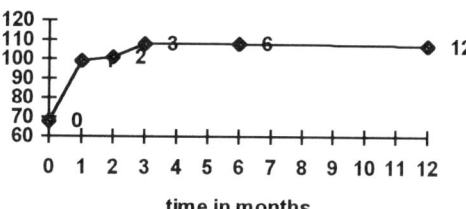

Figure 1. Dependence of the range of motion of EJ in degrees versus the time in months.

The chemical synovectomy with OsO_4 is very effective method for treatment of the arthritis of EJ. It gives no side effects. The parameters showing that the lubrication and friction is restored in EJ are joint function, range of motion, flexion contrcture and pain. They decrease more rapidly after the treatment than the joint edema. The improvement of all the studied by us parameters are observed until the sixth month. The best results are obtained after the local treatment of patients which are in the I and II X-ray stage of RA.

REFERENCES

1. Schumacher, H. Ralph, 1993, Primer on the Rheumatic Diseases, (Tenth Edition),.
2. Swann, D.A., Silver, F.H., Slayter H.S. et al. 1985, *Biochem. J.*, 225, 195-210.
3. Stoilov, R., Kalitchin, Zh.D., Sheitanov, J., 1995, *Journal of the Balkan Tribological Association*, 1, No 2, p. 201 - 206.

QUANTITATIVE COMPARISON OF POLYETHYLENE WEAR DEBRIS, WEAR RATE AND HEAD DAMAGE IN RETRIEVED HIP PROSTHESES

J.L. TIPPER, E. INGHAM, J.L HAILEY, A.A. BESONG, AND J. FISHER
Departments of Microbiology and Mechanical Engineering, University of Leeds, LS2 9JT
M. STONE
Department of Orthopaedic Surgery, Leeds General Infirmary, Leeds, LS1 3EX
B.M. WROBLEWSKI
Centre for Hip Surgery, Wrightington Hospital, Wigan.

A number of methods have been reported for the isolation and characterisation of micron and sub-micron sized UHMWPE wear debris which is a major cause of osteolysis and failure of total artificial joints. These methods generally ignore the contribution of the larger (>10μm) particles. We present a digestion procedure which allows isolation and characterisation of all sizes of UHMWPE wear debris generated in 19 explanted Charnley hip prostheses. The total mass of wear debris, the size and mass distributions were determined and compared with femoral head damage and linear wear rates of the acetabular cups.

Nineteen explanted Charnley prostheses with a mean implant life of 13 years (range 10 to 20 years) were retrieved at revision with acetabular tissue. Acetabular cups were measured to determine total linear wear. Damage to the femoral heads was measured and compared with a new Charnley prosthesis. The wear debris was isolated from 1g of randomly selected acetabular tissue by digestion with potassium hydroxide. UHMWPE wear debris was collected by sequential filtration through 10μm and 0.1μm filters, which were weighed to determine the total mass of debris. Particle size distribution was determined by 2-D particle sizing using SEM images. The distribution of the mass of the wear debris as a function of particle size was determined. The mass and frequency distributions allowed the total number of particles per gram of tissue digested to be estimated. In addition, the total number of particles generated over the lifetime of the prostheses were calculated.

The mode of the frequency distribution of the particle sizes was 0.1-0.5μm for all patients. Table 1 shows the masses of wear debris isolated from 1g of tissue. There was no correlation between the total mass of wear debris isolated and the wear rate. The estimated total number of particles varied from patient to patient (range 1.5×10^8-1.3×10^{10} particles per gram of tissue) as did the number of particles generated over the lifetime of the prosthesis. The mass distributions varied considerably, with 18-97% of the debris being <10μm. There was an association between the proportion of UHMWPE wear debris that was <10μm and the penetration rate and head damage R_{pm}. At high penetration rate or R_{pm}, a large proportion (>60%) of the debris was <10μm.

Previous studies have found limited differences in particle size distributions for polyethylene wear debris. Analysis of the mass distribution as a function of size has allowed differentiation between the UHMWPE wear debris from different patients. The lack of correlation between wear and mass of debris retrieved may be explained by variations in the clearance and/or retention of the UHMWPE particles. There is an indication that femoral head damage R_{pm} and increased linear wear rate may affect the size of the UHMWPE wear debris, but other factors are clearly influencing the total number of particles retrieved. These may include oxidative degradation of the UHMWPE or patient factors such as activity and tissue response.

Implant life (yrs)	Total mass of debris (μg)	% mass of UHMWPE debris >10μm	% mass of UHMWPE debris 0.1-1μm	Est. total no. of particles (x 10^9)	Total no. particles/ lifetime (x 10^9)
19	618	20	80	4.9	14,736
18	779	8	92	7.1	7,360
18	280	15	85	2.8	10,880
17	168	20	80	1.3	10,384
15	423	41	59	2.5	7,104
15	113	82	18	0.2	760
14	698	38	62	4.3	645
13	194	65	35	0.65	1,281
12	786	13	87	6.8	1,409
12	893	11	89	7.9	13,278
12	59	54	46	0.27	3,979
11	1350	3	97	13	3,918
11	57	35	65	0.37	8,164
11	308	51	49	1.5	3,537
11	433	10	90	3.9	10,422
10	202	8.5	91.5	1.8	1.8
10	331	4	96	3.2	3,878
10	21	29	71	0.15	6,489
10	331	34	66	2.2	1,438

Table 1 Mass of UHMWPE wear debris and number of particles isolated from 1g acetabular tissue from 19 patients.

THE NEW SOLUTION OF THERMO-TRIBOLOGICAL PROBLEMS FOR INVOLUTE GEARS OF REAL SIZES

Dr. BRONISLAV E. GURSKY
Moscow State Technical University by name N.E.Bauman, Elements of Machines' Chair,
2-nd Baumanskaja 5, 107005 Moscow, Russia

ABSTRACT

Intensification of the use of machinery stimulates the development of the calculation, design and technological solutions aimed at improving the reliability of gear transmissions as a construent part of machines. From this viewpoint the priority direction is the rise of the active tooth surfaces' endurance. The insufficient endurance of these surfaces manifests under scuffing, scoring, pitting etc. thermal damage conditions. The manifestation of these damage kinds is began from narrow zones - the zones of minimum endurance (ME); for pitting - the zones of minimum contact endurance (MCE). The study of these zones' manifestation has varied importance, both scientific and technical. For example, the roller (ball) tests are used often for the imitation of the involute tooth meshing operation. However these tests simulate the only involute profile point during constant test conditi ons. The simulate of such profile point does not always correspond to real involute meshing durability. Therefore the necessity in additional tests for purpose the determination of critical regimes for real mesh is possible. For the purpose of money and time economy it is proposed to use the roller (ball) tests which correspond to the only involute profile point with the minimum endurance.

The analitical methods are worked in order to determine the zones of ME (MCE) for new spur and helical gears. The zone of ME is showed in consequence of the uneven distribution of the load and friction's warmth along the active tooth surface. The progressive destruction of the whole active tooth surface is began from the destruction of the ME's zone. It follows to expect that the zones of ME will be manifested on the surface zone with maximum temperature value. For the purpose of the exact determination of the ME's zone the presented solutions are based on the Blok's model, but our solutions are used for involute gears of real sizes. Therefore the presented solutions take into consideration: the tooth meshing geometrical and kinematical parameteres of real sizes, the contact duration of each surface point, the contact speed inequality, the contact heat flow sharing etc.

The determination of ME's zones allows to develop improved radial thermal hardening surface technology. This technology is managed for purpose to rise the durability of the ME's zones.
The presented technology management allows to obtain the risen hardness and thickness of surface and subsurface hardening layers and thus to receive the risen durability of the whole gear spare part banks of new machines. And thus at the same time we have the decrease of the spare part bank and the increase of the significant durability of new machines.

The instruction on the ME's zones is the next step in the instruction on the "weakest" machine link. The instruction on the ME's zones is of special importamce for large gears (diameter more than 0,5 m, tooth height more than 20 mm and weight from some tens kilogrammes to some tones and more) because it is not obligatore grinding after radial thermal hardening.

The essence of the such model may be shown for cylindrical involute gears. The special feature of solution of the problem consists of the fact that the physical contact itself is not considered. Instead we consider the field of engagement as the region of interaction of two contacting half-spaces imitating the mating of a pair of gears. The general contact line, i.e., a line heat source, moves at a constant velocity over the engagement field and, in various sections of this field, merge with the contact lines of each of the mating teeth for various lengths of time. The heat source moves so fast that the heat has no time to propagate in front of it, and is absorbed by the bodies of the mating teeth along the normal to the surface only under the contact line. The large heat capacity of gears, the closed state of the contact zone and the short time span of the process allows us to consider such a process as adiabatic. The temperature field is considered as quasistationary. On the basic of such an approach, mating of a pair of gears is considered as a contact between two half-spaces bounded by plane rectangular surfaces equal in dimensions to the engagement field and loaded by the set of all loads directed along the normal to the surface and applied at each point of the active tooth surface - acting simultaneously, with the duration of their contact taken into account. Under the contact duration we understand the time of stay of a given point of the active tooth surface within the limits of the contact spot.

It is possible that the determination of the zones of minimum endurance will be the perspective for the both science and technical directions.

THE HYDRAULIC MODELS OF THE SEAL AND METHODS OF CALCULATION OF LEAKAGE

POROSHIN V., SHEIPAK A.
Moscow State Industrial University, Avtozavodskaya St., 16, 109280, Russia

In present work the two-dimensional flow through a channel with a real microgeometry is considered[1]. The creation of adequate hydromechanical models of hydraulic devices is an important problem. It considers about the bottom from individual problems: flow of a viscous uncompressible liquid through normally unhermetic stationary flat seal. Distance between surfaces is accepted random and is described by the function with the normal distribution or any other.

The solution of a hydrodynamical problem is found in approximation of the theory of a lubricating layer in the form obtained by Reynold's, thereby the similarity criteria have the following order of magnitude: Euler number and Reynold's number do not exceed the ratio of characteristic length of the seal to the ratio menu gap, while the Struchala number doesn't exceed 1. The area of flow is set as follows: on the x axis - [0, a], on an y axis - [0, b], on an z axis - [0, h (x, y)]. The mathematical model is reduced to a differential partial equation of an elliptical type for a field of pressure:

$$\frac{d}{dx}\left(h^3 \frac{dp}{dx}\right) + \frac{d}{dy}\left(h^3 \frac{dp}{dy}\right) = 0;$$

$$v_x = \frac{1}{2\mu}\frac{dp}{dy}(z^2 - zh);$$

$$v_y = \frac{1}{2\mu}\frac{dp}{dy}(z^2 - zh).$$

On finding the solution one firstly generates the randomly distributed values and then using them - normally distributed ones, with the parameters as follows: max h(max) and min value of a the gap; is accepted equal to arithmetic mean while mean square value is accepted equal to 1/6 of the amplitude.

The solution of an equation for pressure is found out by a sweep method. Then the partial derivatives of pressure with respect to x and y are calculated, the values of derivatives allow to receive value of the component of speeds. The consumption of fluid was calculated by integration of components of speed over y axis at. Area of an integration represents a rectangle with the sides [0, a] and [0, h (x, b)].

$$Q = \int_0^h \int_0^b v_x \, dy \, dz.$$

The program is realised on the language C++ and works with an operational system DOS of any versions.

The value of leakage through seal depends on geometric distribution parameters: hmax = 14 μm and hmin = 2 μm. A factor of a variation of outflow is 24 %, asymmetry - 0.114 (its standard deviation - 0.405), kurtosis - 0.149 (its standard deviation - 0.128).

The created database can be used further for the improvement of the various programs of account of outflow, flows etc. On the bas's of the simulation one can find the optimum technological solution for designing and fabrication of seal.

Phenomenon of obliteration is considered as an adsorption of molecules of polar - active substances (PAS) on internal walls of seal, through which liquid, containing PAS penetrates. The PAS molecule has a static dipole moment; in the account Lennard-Johns potential is accepted. The molecule is accepted spherical, we neglect inertial forces. The condition of preservation of PAS molecules number permits to close the system from algebraical equation of the sixth degree and from ordinary differential equation of the first order.

Numerical evaluations have shown, that PAS molecules move basically along the direction of a gradient of pressure, slowly coming nearer to walls. All molecules, located in initial moment in determined area, adhere to a wall.

References

1. Sullivan, P.J., Poroshin, V.V., and Hooke, «Application of a three-dimensional surface analysis system to the prediction of asperity interaction in metallic contacts.», Int.J.Mach.Tools Manufact Vol.32. No. 1/2, pp. 157-169, 1992.

EXPERIMENTAL STUDY ON STATIC AND DYNAMIC CHARACTERISTICS OF ANNULAR PLAIN SEALS WITH POROUS MATERIALS

S KANEKO, H KAMEI and Y YANAGISAWA
Department of Mechanical Engineering, Nagaoka University of Technology, Kamitomioka, Nagaoka, 940-21, JAPAN
H KAWAHARA
Miura Industries Co., Ltd., 7 Horie, Matsuyama, 799-26, JAPAN

ABSTRACT

It is now well known that high-pressure annular seals, such as impeller seals and balance piston seals, have a significant influence on the rotordynamic response of high-performance turbomachinery. As a result, extensive efforts have been made to theoretically predict and to experimentally measure the dynamic force response of these seals. Black et al. (1)(2) first explained the influence of seal forces on the rotordynamic behavior of pumps. Childs (3)(4) provided a comprehensive program for the analysis of turbulent annular seals based on Hirs' turbulence bulk-flow model (5) and included inlet swirl velocity effects.

For improving the efficiency and the stability margin of pumps, it is generally required that annular seals reduce the leakage flow rate and cross-coupled stiffness coefficients and increase the main stiffness and main damping coefficients. Recent research efforts have therefore been focused on designing annular seals satisfying the above specifications, and various seal configurations have been proposed.

The present study is concerned with the experimental investigation of the static and dynamic characteristics of the annular plain seals with porous materials. Based on our theoretical predictions (6)(7), porous materials were applied to the seal surface by insertion into the middle of the seal in order to yield a high stable operation speed by a stiffer rotor support.

The experimental apparatus employed in the present study was similar to that used by Kanemori and Iwatsubo (8), i.e., test seals fixed in an outer cylinder were not shaken relative to a rotor and the rotor was independently driven by two motors to yield both a spinning and a whirling motion. In this experiment, the fluid-pressure distributions in the seal clearance were measured for various combinations of the spinning and whirling velocity with two pressure transducers mounted in the rotor. The circumferential film-pressure distributions were able to be measured, since the transducers rotated at a relative velocity of the spinning to the whirling in a fixed coordinate system located on the seal clearance, which rotated at the rotor whirling velocity. Axial pressure distributions were also measured by moving the laterally fixed outer cylinder in the axial direction at short intervals. Dynamic forces were therefore obtained by integrating the measured film-pressure distributions along the axial and circumferential directions.

Experimental results show that annular plain seals with porous materials have a higher leakage flow rate, larger main stiffness coefficients, and smaller cross-coupled stiffness coefficients and main damping coefficients than conventional annular plain seals with solid surfaces. In the porous seals, an increase of approximately 30 percent in the leakage flow rate and reduction of the same amount in the main damping coefficients are obtained, whereas the main stiffness coefficients for the porous seals are four to six times as much as those for the solid seals due to the increase in the hydrostatic force induced by a function of the hydrostatic porous bearing. This suggests that the quantitative effects of the porous materials on the main stiffness coefficients are much more significant than the effects on the leakage flow rate and the other dynamic coefficients. The larger main stiffness coefficients for the porous seals yield larger radial reaction force for a small concentric whirling motion, which would contribute to rotor stability from the viewpoint of increasing speed limits due to a stiffer rotor support.

REFERENCES

(1) H F Black, 1969, J. Mech. Eng. Sci., Vol.11, pp.206-213.
(2) H F Black, 1971, ASME Paper 71-WA/FF-38.
(3) D W Childs, 1983, Trans. ASME J. Lubr. Technol., Vol.105, pp.429-436.
(4) D W Childs, 1983, Trans. ASME J. Lubr. Technol., Vol.105, pp.437-444.
(5) G G Hirs, 1973, Trans. ASME J. Lubr. Technol., Vol.95, pp.137-146.
(6) S Kaneko, 1989, Trans. ASME J. Tribol., Vol.111, pp.655-660.
(7) S Kaneko, 1990, Trans. ASME J. Tribol., Vol.112, pp.624-630.
(8) Y Kanemori, T Iwatsubo, 1992, Trans. ASME J. Tribol., Vol.114, pp.773-778.

LUBRICANT FOR DISK DRIVE SPINDLE MOTOR WITH HYDRODYNAMIC FLUID BEARING

RAQUIB U KHAN and GREGORY I RUDD
Motor Design and Development Center, Seagate Technology, Scotts Valley, CA 95067

ABSTRACT

Numerous papers and articles in books and journals have been devoted to the development of latest generation lubricants for application in everything from cameras to spacecraft mechanisms to disk lube in the computer disk drive. But very few articles have discussed lubricants specifically for high speed miniature self pumping liquid hydrodynamic bearings for spindle motors in disk drive applications. Self pumping hydrodynamic bearings (HDB) were originally used in precision gyroscopes and motors in magnetic tape drives and are only very recently finding application in computer disk drive spindles. Because of the very low non-repetitive run-out and high damping, hydrodynamic fluid bearings are getting more and more attention as the areal density and speed of the next generation of computer disk drives increases

HDB's generate an internal pressure in the fluid film that separates the bearing surfaces. When grooves are placed on the bearing surface or surfaces with a pattern other than axial, there will be a component of the rotational circumferential velocity along the grooves. A pumping action will be created along the grooves and lubricant will consequently flow in that direction. The retardation or stopping of the lubricant flow at the discharge end of the grooves generates the hydrodynamic pressure by continuous pumping action. The shear stress resulting from the relative motion occurs entirely within the fluid film; no contact occurs. For HDB's, boundary lubrication occurs only during start-stop of the spindle. This situation is unlike typical boundary or elastohydrodynamic lubrication applications (e.g. roller, ball and sliding bearings) where the bearing surfaces are often in contact and shear occurs at the interface of solid surfaces with only a very thin oil film.

From these considerations, two broad requirements for hydrofluids are identified and described in this paper. The *mechanical performance* is a paramount concern in the hydrodynamic regime; power consumption, stiffness and damping- are directly related to the bulk lubricant viscosity at any particular temperature and for a particular geometry. On the other hand, *reliability* depends mainly on the hydrofluid's chemical properties.

These requirements are critical as the disk drive spindle motor has very stringent constraints regarding power, stiffness, damping, long life performance, and low contamination. The objective of this paper is to discuss the types of lubricants studied and the required important lubricant properties. The paper will also highlight some of their physical properties such as viscosity, viscosity-temperature behavior, vapor pressure, surface tension, and evaporation.

SLIDING FRICTION AND HEAT GENERATION IN PRESSURIZED U-CUP SEALS UNDER LOW-SPEED RECIPROCATING CONDITIONS

T. C. OVAERT and J. A. BADGER
Department of Mechanical Engineering, The Pennsylvania State University, University Park, PA 16802 USA

ABSTRACT

A test apparatus was designed and constructed to investigate the friction and heat generation characteristics of sliding polymeric u-cup seals used in high-pressure nitrogen springs. Tests were run at three speeds; and friction and temperature data were obtained. Seal specimens were examined under a scanning electron microscope before and after each test. Evidence of degradation and thermal softening of the seal material was found for the highest relative speed tests where elevated temperatures were measured. In addition, a transient finite-difference thermal model of the high-pressure chamber in the test apparatus was also developed. Numerical simulations of the three tests were performed and temperature predictions from the model were compared to those measured at different locations in the test apparatus. Predicted values exhibited reasonable agreement with the measured values, particularly under steady-state conditions. The variations observed for certain transient conditions appeared to be due to overestimated convective heat transfer coefficients at lower speeds, and due to the transient nature of thermal resistances between the seal cartridge and the high-pressure chamber, which may vary with a change in temperature. Surface roughness profiles of the rod and of the new and worn seals were taken, and film thicknesses were estimated for each test using an approximate asperity-scale elastohydrodynamic analysis. A discussion was made regarding the potential lubricating film support at the seal-rod interface for each test.

INTRODUCTION

Friction in a lubricated reciprocating seal depends on many factors such as the presence or absence of a lubricant film under the sealing surface, the surface roughness of the seal, the surface roughness of the shaft, the velocity of the shaft, the seal material, the shaft material, and the lubricant properties. Perhaps the most important of these is the presence or absence of a lubricant film. Lewis (1) discussed the three classic regimes (boundary, mixed, and hydrodynamic) of lubrication based on the film thickness. He noted that seal friction is strongly dependent on the extent of surface separation and the properties of the lubricating fluid. In addition, Belforte, Raparelli, and Velardocchia (2) investigated lip seal friction as a function of velocity. It was found experimentally that for non-lubricated contact, friction is largely independent of velocity; for lubricated contact, friction is a function of velocity. The frictional characteristics of seals change as the seal becomes worn due to repeated cycling.

RESULTS

The following figures show the measured friction force (at the beginning and end of the test) and the test and model (rod center-line and chamber) temperatures at the most severe test condition. The friction force increased during the test causing the seal material to soften after approximately 120 minutes, causing the divergence between the test and model predicted temperatures.

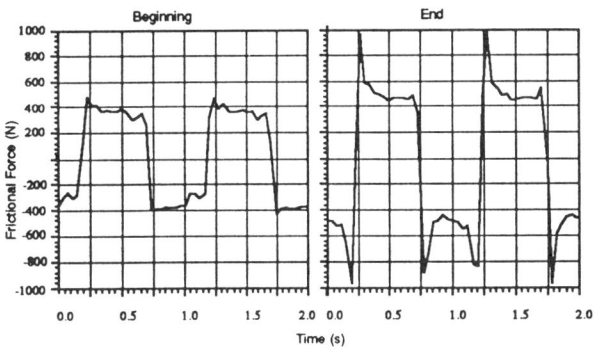

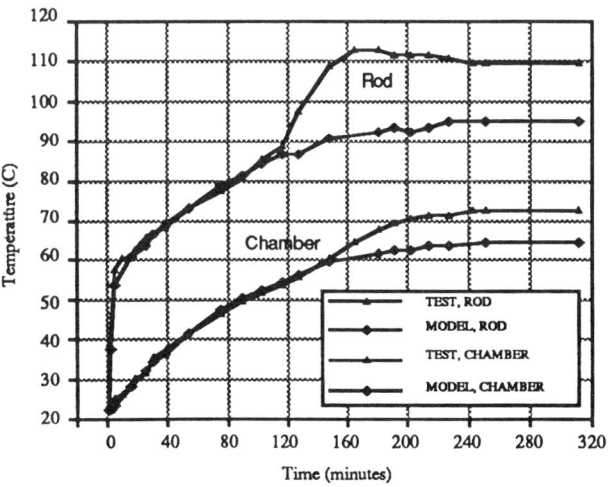

REFERENCES

(1) M.W.J. Lewis, Lubrication Engineering, Vol. 42, No. 3, 1986.
(2) G. Belforte, T. Raparelli, and M. Velardocchia, Lubrication Engineering, Vol. 49, No. 10, 1993.

Submitted to *ASME Journal of Tribology*

EFFECT OF VISCO-ELASTIC PROPERTY ON FRICTION CHARACTERISTICS OF PAPER-BASED FRICTION MATERIALS FOR OIL IMMERSED CLUTCHES

TOMOYUKI MIYAZAKI and TAKAYUKI MATSUMOTO
R&D Department, NSK Warner K.K., 2345 Aino, Fukuroi, Shizuoka 437, Japan
TAKASHI YAMAMOTO
Department of Mechanical Systems Engineering, Tokyo University of Agriculture and Technology,
2-24-16 Naka-cho, Koganei, Tokyo 184, Japan

ABSTRACT

Paper-based friction materials for oil immersed clutches in automobile automatic transmissions are made of fibers, fillers and resin. The structural properties of the materials include porosity and visco-elasticity. The relationship between porosity and friction characteristics were studied by one of the present authors (1)(2). The same author studied the carbonization characteristics influencing the compressive deformation and the friction in a durability test (3).

In this study, the relationship between visco-elastic deformation and friction characteristics is discussed under severer conditions compared to the previous study. Three kinds of materials (A, B and C) with different properties were used. In order to vary the visco-elasticity while retaining the ingredients almost constant in the material, the formulation of the sample materials required changing the compression rate during the clutch plate bonding process and changing the volume percentage of resin.

Visco-elastic deformation properties of these materials were measured, in wet conditions, by using a tensile and compressive strength measuring apparatus. Figure 1 shows the variation in the mean value of longitudinal modulli of elasticity as a function of compressive force. From the compressive deformation characteristics and material properties, it is found that porosity percentage or density of the material is an effective means to increase compressive deformation in the friction materials. The longitudinal modulus of elasticity of the material becomes higher with the increasing density. It was inversely proportional to the porosity percentage of the material. As the result of shearing deformation test, it is found that difference in shearing deformation among the materials is much smaller than that of the compressive deformation. Shearing modulli are slightly influenced by the density of the material or by the volume percentage of resin.

The friction test was conducted by a process combining breaking-away and continuous slip, with a newly designed friction testing machine. A friction and steel plate underwent mutual sliding before the test as a running-in process. Friction test results at the contact pressure of 0.5 MPa are shown in Fig. 2. From these results, it is found that the longitudinal modulus of elasticity imparts a marked effect on the friction coefficient. Also, the effect of elasticity on the friction coefficient becomes bigger when the normal load increases. For a sliding range larger than 100 mm/sec, the friction coefficient of each material observed at 0.3 MPa was almost the same as that at 0.5 MPa. Therefore, the effect of elasticity on friction coefficient appears primary in the low sliding velocity range.

In conclusion, from the measurements of compressive and shearing deformation, the density of the material is more effective at increasing the compressive deformation than the shearing deformation. Even if the longitudinal modulus of elasticity of each material is different, the shear modulus of each material is almost the same. The friction coefficient of the material that has the lower longitudinal modulus of elasticity is higher, while that of the material having higher longitudinal modulus of elasticity is lower.

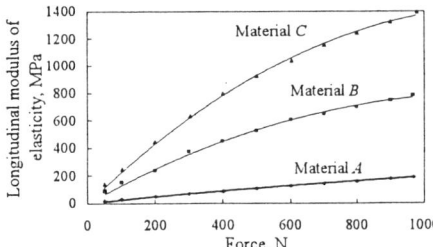

Fig. 1: Variation of longitudinal modulus of elasticity with compressive force

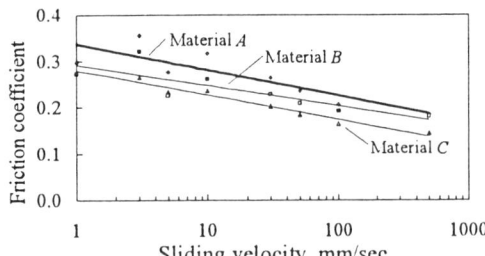

Fig. 2: Variation of dynamic friction coefficient (Contact pressure: 0.5 MPa)

REFERENCES

(1) T. Matsumoto, SAE Transactions, Section 6, Vol. 102, pp.2417 - 2424, 1993.
(2) T. Matsumoto, SAE Tech. Pap., 941032, 1994.
(3) T. Matsumoto, ASME Transactions, Vol. 117, pp. 272 - 278, 1995.

EFFECTS OF FILM TEMPERATURE ON PISTON-RING LUBRICATION FOR REFRIGERATION COMPRESSORS CONSIDERING SURFACE ROUGHNESS

HIRONORI NAKAI
Advanced Tech. Lab.,Mayekawa Mfg. Co., Ltd. (MYCOM), 2000 Tatusawa , Moriya, kitasohma,Ibaragi, Japan
NOBUMI. INO
Mayekawa Mfg. Co., Ltd. (MYCOM),2-13-1 botan, Koto-ku, Tokyo, Japan
HIROMU HASHIMOTO
Dept. of Mechanical Engineering, Tokai university, 1117 Kitakaname, Hiratsuka, Kanagawa, Japan

ABSTRACT

This paper describes the theoretical modeling for piston-ring lubrication considering the combined effects of surface roughness and oil film temperature variation for refrigeration compressors. In the modeling, the piston-ring is treated as one-dimensional dynamically loaded bearing with combined sliding and squeezing motion.

Assuming that the circumferential flow of oil can be neglected, the pressure generated in the lubrication film is estimated by the following one-dimensional modified Reynolds equation considering the surface roughness. (1)

$$\frac{\partial}{\partial x}(\phi_x \frac{h^3}{\mu_m} \frac{\partial p}{\partial x}) = \frac{U_2}{2}\frac{\partial}{\partial x}(h_T - \sigma\phi_s) + \frac{\partial h_T}{\partial t}$$

where μ_m is the viscosity corresponding to the mean film temperature $Tm.(T_m(x) = \frac{1}{h}\int_0^h T\,dy)$. ϕ_x and ϕ_s are the flow factors by Patier and Cheng.(2) Assuming that the flow in the oil film is the laminar and the circumferential oil flow can be neglected, the energy equation is obtained. Then, integrating the energy equation from 0 to h in y direction and considering the heat produced due to the contact of asperities at $y=h$, the averaged type of energy equation is obtained as follows :

$$\rho C_v \int_0^h u\frac{\partial T}{\partial x}dy = k\int_0^h \frac{\partial^2 T}{\partial y^2}dy + \int_0^h \tau_{xy}\frac{\partial u}{\partial y}dy + S$$

where S is the heat produced per unit time and unit real area.(3)

Based on the above mentioned theoretical model, the cyclic variation of the minimum oil film thickness, the friction force and oil film temperature are calculated. Fig.1 shows the cyclic variation of minimum oil film thickness. The minimum film thickness increases with an increase of composite roughness near the top dead center. Fig.2 shows the oil film temperature variation in the case of supply oil temperature of T_0=70℃, in which ΔT_m is defined as

$$\Delta T_m \equiv \int_{x_i}^{x_o} T_m(x)\,dx / b - T_0$$

,The effects of surface roughness on the oil film temperature variation is significant near the top dead center. The temperature variation for larger composite roughness of σ =2.12 μ m is much smaller than the variation for smaller composite roughness of σ =1.06 μ m. This means that the temperature variation is mainly led by the film thickness rather than the composite roughness.

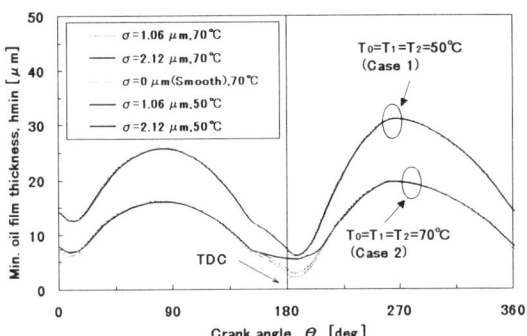

Fig. 1 Relationship between minimum oil film thickness and crank angle

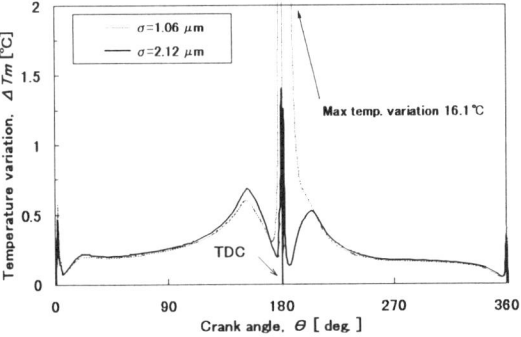

Fig.2 Relationship between oil film temperature variation and crank angle, T_0=70℃ (Case 2)

REFERENCES

(1) H. Nakai, N. Ino and H. Hashimoto, Trans. ASME, Journal of Tribology, Vol.118,1996, 286-291pp.
(2) N. Patir and H. S. Cheng, ASME Journal of Lubrication Technology, Vol. 101,1979,220-230pp.
(3) B. Bhushan, "Tribology and Mechanics of Magnetic Storage Devices, Springer-Verlag, 1990, 367pp.

A POSSIBLE MECHANISM FOR RAIL DARK SPOT DEFECTS

M KANETA, K MATSUDA, K MURAKAMI, and H NISHIKAWA
Department of Mechanical Engineering, Kyushu Institute of Technology,
1 - 1, Sensui-cho, Tobata, Kitakyushu, 804, Japan

ABSTRACT

Rail dark spot defect, also termed squat failure or shelling, which is a kind of rolling contact fatigue failure and occurs frequently on running surfaces of railway rails carrying high speed traffic, is one of the most dangerous rail failures. The dark spot defect is characterized by a principal crack propagating at an acute angle to the rail surface in the direction of traffic, and a second crack growing in the opposite direction to traffic.

The purpose of this investigation is to reproduce the dark spot defects using a newly developed two-disk rolling/sliding contact machine and to show how and why the dark spot defects occur.

The experiments were mainly carried out under following six running modes. The total number of contact cycles was mainly 10^7.
(A) Dry running.
(B) Water running.
(C) Dry running followed by wet running.
(D) Wet running followed by dry running.
(E) Dry running followed by wet and dry runnings.
(F) Repetitions of dry and wet runnings.

Dark spot defects, which were very similar to those appeared in actual rails, were produced only under frequent repetitions of dry and wet runnings. Common features of the dark spot defect observed in this study are summarized as follows:
(1) A principal crack composing the dark spot defect initiates from a micropit produced a posteriori, and propagates at an acute angle to the surface in the direction of motion of the load. Only some of large numbers of micropits produced become dark spot defects.
(2) When the principal crack extends and attains a certain depth, a second crack branches from the principal crack and grows at a slightly less shallow angle as compared with the principal crack to the surface in the opposite direction to motion of the load. The cracks also extend perpendicular to the direction of motion of the load.
(3) Although the extended principal crack has many branched cracks directing toward the surface, the second crack has very few branched cracks.
(4) The periphery of crack at the contact surface is depressed and becomes black. The depth of depressed area increases with the growth of the dark spot.
(5) The dark spot cracks may propagate extensively and down deeply into the subsurface even when the depressed or blackened area is small.
(6) The number of the dark spot defect tended to increase with an increase in the number of the repetition.
(7) The traction force plays an important role for the occurrence of dark spot defects.

The features of dark spot defect obtained in the present study are almost the same as those of actual dark spot defects. As a result, we can conclude that a dark spot or squat defect is caused by frequent repetitions of dry and wet runnings. Even in actual railroad rails, the same running mode would be realized on rainy days. When a high speed traffic passes through on railway rails wetting with rain water, the surfaces of the rails become dry, because the rain water on the rails is removed gradually by wheels of high speed traffic.

On the basis of fracture mechanics approach, Kaneta and Murakami (1)(2) have confirmed that the Way's hypothesis (3) should be accepted as a crucial mechanism, and arrived at the following conclusions. A surface initiated tiny fatigue crack under rolling/sliding contact extends along the original crack plane by shear mode crack growth. As the crack extends, growth might be altered to tensile mode. This transition is caused by hydraulic pressure resulting from fluid penetration into the crack. The crack faces existing at the slower moving surface may receive two kinds of oil hydraulic pressure action. One is the direct action of fluid pressure induced by contact pressure through an opening crack mouth and the crack propagates in the direction of motion of the load. The other is oil blockade action caused by a closing crack mouth and the crack propagates in the opposite direction of motion of the load. Although it seems to be very difficult to seal the mouth of the crack under conditions where the surfaces are separated with water and the crack is filled with water, dry running after wet running would easily bring about the state. That is, the second crack, which characterizes the dark spot defect in railway rails, seems to be produced by the latter hydraulic pressure mechanism. The cracks, which branch from the principal crack and direct towards the surface, seem to be produced by the former mechanism.

REFERENCES

(1) Kaneta, M., and Murakami, Y., Tribology Int., Vol.20, 1987, pp.210-217.
(2) Kaneta, M., and Murakami, Y., J. Tribology, Trans. ASME, Vol.113, 1991, pp.270-275.
(3) Way, S., J. Appl. Mech., Trans. ASME, Vol.2, 1935, pp.A49-A58.

A POSSIBLE MECHANISM FOR SURFACE CRACKING OF CEMENTED CARBIDE RINGS IN MECHANICAL SEALS

KENJI MATSUDA and MOTOHIRO KANETA
Department of Mechanical Engineering, Kyushu Institute of Technology, Tobata, Kitakyushu, 804, Japan
SHINGO MATSUI
Eagle Industry Co., Ltd., Sakado, Saitama, 350-02, Japan

ABSTRACT

Cracking, which occurs on the sealing surface as radial cracks, is one of the most serious failures in mechanical face seals, because it brings about excessive leakage. The cracking can be classified into two types; one is "through cracks" that run radially across the surface of the seal ring and the other is numerous shallow cracks that occur in the contact zone. The authors (1) have proved that the former cracks, which generally occur in rings made of brittle materials such as ceramics, are caused by thermally-induced tensile circumferential stresses. However, they could not apply their conclusions to the latter cracks, which are often observed on the surface of cemented carbide rings having much higher thermal conductivity as compared with alumina ceramic rings.

The purpose of this study is to clarify factors which induce cracking characterized by shallow numerous cracks confined within the rubbing surface, and to propose a possible mechanism for cracking.

The experiments were carried out by paring two kinds of cemented carbide rings with an alumina ceramic ring or a silicon carbide ring fabricated by reaction bonding under conditions with and without distilled water to be sealed. Residual stresses in ring surface before and after the experiments were measured by the X-ray diffraction method. The experimental results show that numerous shallow cracks occur when the mean friction torque exceeds a critical value without distinction of running time and acceleration time. They occurred even under dry conditions; the lower the bending strength, the higher the frequency of occurrence of cracks. It has also been found that the residual surface stresses are tension in the direction of sliding and compression in the direction perpendicular to the sliding direction; the shift of residual surface stresses from compression to tension never depends on the shape of the specimen.

The thermo-electromotive force was also measured using the ring made of the silicon carbide as the rotational specimen. Its value suggests that such a high temperature as about 3000°C is generated at the contact surface. This result also suggests that there are certainly hot spots or contact patches and that the actual contact area is very small and consequently, the contact pressure is very high.

Through these experimental results, the authors have arrived at a conclusion that surface cracking is caused by dynamical action due to high contact and tangential stresses resulting from the concentrated contact patches that move around the contact surface rather than thermal stresses produced as a result of frictional heating. To elucidate this conclusion, thermal and elastic-plastic analyses taking into account the thermal stresses of the subsurface layer of the seal ring were conducted using the two dimensional finite-element technique. In this analysis, a semi-elliptical Hertzian contact pressure with a surface traction was assumed to move on a semi-infinite half space. The frictional heat was assumed to be generated at the nodes within the contact zone. The yield strength and tangent modulus for cemented carbide materials used in present experiments were estimated using the three dimensional finite element method for analyzing the Vickers hardness test developed by one of the authors(2).

Even if the contribution to the stress distribution of the normal and tangential contact forces is ignored, the release of the compressive thermal stress caused by movement of the contact patch results in a residual tensile stress on the surface, as reported by Kennedy et al. (3). However, in this case, the residual tensile stress produced in the direction perpendicular to the direction of sliding is higher than that produced in the direction of sliding. Therefore, it seems to be difficult to explain by their thermocracking mechanism why cracks run perpendicular to the direction of sliding. Analytical results show that the dynamical action of a high contact pressure together with a high tangential force plays a very important role in the occurrence of cracking and generation of tensile residual surface stress in the direction of sliding. Furthermore, it has been found that under conditions of high contact pressure and low sliding speed, the maximum tensile stress in the direction of sliding occurs not after the release of the contact force but near the trailing edge of the active Hertzian contact pressure.

REFERENCES

(1) M. Komiya, K. Matsuda M. and Kaneta, Tribology Transactions, Vol.37, 1994, pp.245-252.
(2) Y. Murakami and K. Matsuda, ASME Journal of Applied Mechanics, Vol.61,1994, pp.822-828.
(3) F.E. Kennedy and S.A. Karpe, Wear, Vol.79, 1982, pp.21-36.

MODELING THE TRANSIENT BEHAVIOR OF ROTARY LIP SEALS

RICHARD F. SALANT
School of Mechanical Engineering, Georgia Institute of Technology, Atlanta, GA 30332-0405, USA

ABSTRACT

Despite its widespread use and economic importance, the operation of the rotary lip seal has been poorly understood until recently. In the absence of analytical design tools, the design and development of such seals is presently done empirically, with expensive and time-consuming test programs. There is therefore a strong need for analytical design tools.

One such design tool has recently been developed. A comprehensive steady-state model of lip seal operation has been constructed (1), based on the accumulated experimental results of many investigators. While this model has substantially increased our understanding of the lip seal, it is not applicable during startup, shutdown, and other transient operations.

During transient operations the lubricating film between the lip and the shaft, which protects the lip during steady-state operation, breaks down or is not fully established. The resulting contact between the lip and the shaft produces both mechanical and thermal damage to the lip. The accumulated damage over many transient operations can eventually lead to seal failure. Thus, it is clear that the transient behavior plays an important role in determining the life of the seal. It would therefore be extremely useful to the designer to understand and be able to predict the seal behavior during transients. This would allow the development of seal designs that would minimize lip damage and extend seal life. For example, the ideal seal design would be such that during shutdown the lubricating film would not abruptly collapse, but would be maintained as long as possible to allow for a "soft landing" at the end of the transient. During startup, the ideal design would promote rapid development of the film.

In the present transient model, the seal lip is modeled as an elastic body, whose surface consists of a uniform distribution of asperities with initially circular cross-sections. The lip surface is separated from the perfectly smooth shaft surface by a lubricating film. The asperities on the lip can deform in the circumferential direction along with the bulk lip material, due to the action of shear forces induced by shaft rotation and transmitted through the film. On the air-side of the seal, a meniscus separates the sealed fluid from the atmosphere.

The model uses a quasi-steady elastohydrodynamic analysis to compute the seal behavior at each instant of time. The analysis is comprised of a fluid mechanics analysis of the lubricating film, an elastic structural analysis of the lip, an analysis of the meniscus separating the sealed fluid from the atmosphere, and an iterative computation procedure.

The fluid mechanics of the film is governed by the Reynolds equation. Since cavitation occurs, the Elrod formulation is used. It is discretized, satisfying the appropriate internal cavitation boundary conditions. The resulting difference equations are solved for the pressure distribution and density in the cavitation zones, using a five level V-cycle multigrid technique, and the ADI (alternating direction, implicit) relaxation scheme.

To find the average film thickness and the circumferential displacement of the lip surface (and the distortion of the asperities), it is necessary to compute the deformation of the lip. This is done using an influence coefficient method, which assumes the displacements from the static configuration vary linearly with the applied normal and shear stresses. The influence coefficients, as well as the static configuration (and static contact pressure distribution) are computed off-line with a finite element analysis.

To find the conditions at the liquid-air interface, the meniscus is treated as two-dimensional, with a simplified analysis. The location of the meniscus relative to the edge of the lip, at each instant of time, is computed using a mass balance and taking account of the leakage rate through the seal.

The overall computation scheme is iterative. At each instant of time an initial film thickness distribution is assumed. This would generally correspond to the film thickness distribution computed at the previous time step. Then the fluid mechanics analysis is performed, yielding fluid pressure and shear stress distributions. These are then used in the structural analysis of the lip to compute radial and circumferential displacements, which then yield a new film thickness and asperity shape distribution. The latter are used to re-compute the fluid pressure and shear stress distributions. This process is repeated until convergence is obtained. The meniscus location is adjusted depending on the computed leakage rate.

This model allows prediction, for a given transient, of the histories of the average film thickness distribution, pressure distribution, cavitation region locations, torque, and circumferential displacement of the lip surface.

REFERENCES

(1) R. F. Salant, ASME Journal of Tribology, Vol. 118, pp. 292-296, 1996.

AN EXPERIMENTAL INVESTIGATION OF A FLOW EFFICIENT VISCOUS PUMP

ITZHAK GREEN AND MARLENE MAINLAND
George W. Woodruff School of Mechanical Engineering, Georgia Institute of Technology, Atlanta, Georgia 30332 USA

INTRODUCTION

A semicircular lobe (C-lobe) design had been devised as an idea for a viscous pump (1). Then it was successfully tested and proven physically feasible (2). However, significant deviations between theoretical and experimental flow capacities were found, particularly at high-pressure and low-flow-rate conditions. Subsequent work (3) has revealed that a nonsymmetric V-lobe design is analytically superior to the C-lobe design, by being capable of producing higher flow rates under the same operating conditions and geometrical constraints. This work was dedicated to the experimental validation of the C- and V-lobe viscous pumps. It was found in the experiments that the V-lobe design is indeed significantly superior to the C-lobe design. However, both pumps suffered from profound deviations between theoretical predictions and experimental results, again at high-pressure and low-flow-rate conditions. These deviations make either pump less than optimal. This inconsistency was investigated through the use of the finite element method. It was found that a conceptual overlook created the enigma. Hence, further design modification would be necessary to impose best conditions for utmost pumping capacity.

NONSYMMETRIC V-LOBE DESIGN

The design features and the optimization of the geometrical parameters for C- and V-lobe (Fig. 1) designs were thoroughly discussed in (3). The lobes protrude from a stator disk forming, respectively, large and small gaps, C, and, c from a rotor that rotates in proximity to the lobe legs. Viscous drag forces fluid from the low pressure (p=0) zone to the high pressure (p=P) zone. Net pumping occurs when the viscous induced flow is greater than pressure induced flow loss.

ANALYSIS AND RESULTS

All previous analyses assumed ideally that the pressure induced flow passes the lobe leg widths transversely (enabling meanwhile closed-form solutions). That is, no flow losses may occur at the leg ends, mathematically expressed by an ideal boundary condition, $\partial p/\partial s$ = constant (see Fig. 1). Since the inconsistency between theoretical and experimental results occurred at high pressure pumping it was suspected that ideal boundary condition did not truly apply in the tested pumps. A different, less than ideal, boundary condition, p=0, was tried. Since this no longer lends itself to a closed-form solution the finite element method was used to solve the Reynolds equation.

The analysis revealed that the experimental results and the numerical predictions match very closely when a more realistic boundary condition was used (see Fig. 2).

CONCLUSIONS

Both C- and V-lobe designs provide viable viscous pumping mechanisms. The nonsymmetric V-lobe was found also experimentally far superior, capable of producing higher pumping capacity. Very good agreement was obtained between experimental and numerical results when a realistic (yet not ideal) boundary condition was applied in the latter. Some simple improvements in the lobe design will impose the best (close to ideal) boundary condition for utmost pumping.

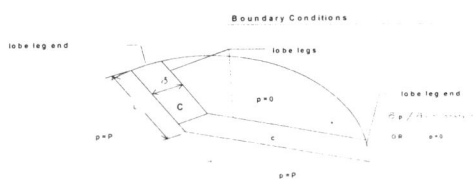

Figure 1 Nonsymmetric V-lobe design and boundary conditions

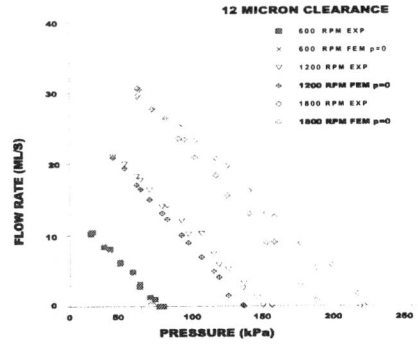

Figure 2 FEM $_{p=0}$ model compared to V-lobe experimental results at c-12 μm and various speeds.

REFERENCES

(1) Etsion, I., and Yaier, R., *ASME J. of Trib.*, **110,1** (1988) 93-99.
(2) Green, I. et al., *ASME J. of Trib.*, **113,4** (1991) 725-728.
(3) Mainland, M., and Green, I., *ASME J. of Trib.*, **114,3** (1992) 515-523.

STUDY OF LUBRICATING ADDITIVES AND PROFILE MODIFICATION FOR IMPROVING THE CAPABILITY OF GEAR TEETH TO RESIST SCORING

SONG JIANG, JUNXIU DONG, DACHANG DU

Department of Petrochemistry, Logistical Engineering College, Chongqing 630042, China

ABSTRACT

The gear teeth scoring is one of general failures, so it is very important to improve the capability of gear teeth to resist scoring[1]. Profile modification can improve the load-carrying capacity of gears, either lubricating additives can. Do they have coordination effect? This effect of mechanics and tribochemistry is seldom studied before, because mechanics scholar and tribochemistry scholar studied it respectively. Therefore, it isn't known that there is the coordination effect, it hasn't been brought into play. In this paper, authors plan to demonstrate existence of the coordination effect and discuss it simply.

The experiment was conducted on the gear test machine CL-100. Organo-boron additive, in short for B agent, was selected. Because it not only has excellent wear resistance and friction-reducing ability, but also good oxidation stability and compatibility with sealing materials, volatility, toxicity and unpleasant odor[2] The lubricant used for this test was ISO VG68 oil. The gear material is 20CrMnTi. Top of tooth was modified. The profile modification curve is straight line, the profile modification value is 50mm and the profile modification highness is 1.421mm[3]. To demonstrate the existence of coordination effect, comparative experiments were conducted at 1450rpm.

The comparative experimental results are listed in Table. They shows: Under the same running condition, the scoring capability of gears with profile modification is 2 load level higher than that of gears without profile modification, the scoring capability of gears with additives is 2 load level higher than that of gears without additives, the scoring capability of gears with profile modification and additives is 5 load higher than that of gears without profile modification and additives. Therefore, profile modification and lubricating additives have coordination effect.

Table The scoring capability (load level)

Lubricant \ Gear	ISO VG 68 oil	ISO VG68 oil with 5% B agent
No profile modification	7	9
profile modification	9	12

Organo-boron additive and profile modification were used to resist scoring, the coordination effect was discovered. The load-carrying capacity of gears was improved from 7 load level to 12 load level in this paper. X-ray photoelectron spectroscopy(XPS) and scanning auger electron microprobe(SAM) are used to analysis the surface of gear teeth. The study showed that boron atoms had permeated into the surface and formed a protective layer containing FeB,Fe2B etc., which enhanced the antiwear ability of the gear teeth surfaces.

REFERENCES

(1) Yoshio Terauchi, ASLE Lubrication Engineering, vol.40, No.1,1984

(2) Adams, J.H.,Lubrication Engineering, vol.35, No.5, 1977

(3) Jiang Song, Journal of Shanghai Jiao Tong University, vol.34, No.3,1989

(4) Dong Junxiu, et al , Lubrication Engineering, vol.50, No.1, 1994

THE EFFECT ON HONED CYLINDER SURFACE FINISHES DURING A SIMULATED RUNNING-IN

Mr MARK HENDEL
Sunnen Products Limited, Maxted Road, Hemel Hempstead, HP2 7EF
Dr BRIAN SNAITH
University of Central England, Perry Barr, Birmingham, B42 2SU

ABSTRACT

Major developments have been made in respect to the type of texture, imparted onto cylinder liners/bores in the internal combustion engine, via the process of Honing. This is due to the view that the topography of the cylinder surface is influential on the engines performance. Preliminary research has indicated that the 'ideal' surface topography will improve engine performance, prolong the life of the engine, reduce oil consumption and finally diminish the amount of harmful emissions.

This paper summaries a project on the implementation of an economical means to simulate mechanical wear using a simple air compressor. The aim being to produce an alternative to engine test cells or track testing to study mechanical wear and performance of different honing specifications.

TEST DETAILS

The honing specifications required were produced in the compressor bore using a hydraulic single vertical spindle honing machine. Three operations being required to create the final topography. These included *Rough Hone* - primary for stock removal followed by *Base Hone* -leaving scratches for oil retention and *Plateau Hone* - producing the final plateau (bearing area) on the surface. The actual specifications, achieved employed in the tests consisted of those currently used by Petrol and Diesel engine manufacturers.

The mechanical wear test rig consisted of a two-stage air compressor, the principle of which is very similar to that of the reciprocating engine i.e. a more dominant amount of wear on the thrust side to that on anti-thrust. It was run with a constant back pressure of 6 bar on the compression side of the piston face thus simulating the idealised conditions to that of a combustion engine. After obtaining surface measurements on the newly machined surface, subsequent measurements were taken at 30 minute intervals, up to a maximum of 240. Further testing not producing any significant changes to the bores.

A comprehensive surface topographical analyses of the bore was performed using a Somicron Surfscan stylus type surface mapping system. Standard surface texture parameters, conforming to ISO and DIN standards were computed.

RESULTS

One of the recognised means of assessing bore topographies is via the DIN4776 standard. This characterises the surface profile into three regions i.e. extreme peaks, core and extreme valleys. The associated parameters being R_{pk}(reduced peak height), R_k(core roughness) and R_{vk}(reduced valley depth). An 'ideal' surface will be free of large peaks and have adequate valleys for oil retention. It should therefore have a very low R_{pk}, large R_{vk} and low R_k. Figures 1 and 2 show typical test results for Diesel and Petrol specification respectively. Included on the graphs, for comparison, are the amounts of apparent wear after running an engine for 1000 miles. It is clearly evident that the diesel specification reaches a similar stage of being run-in much sooner than the petrol one.

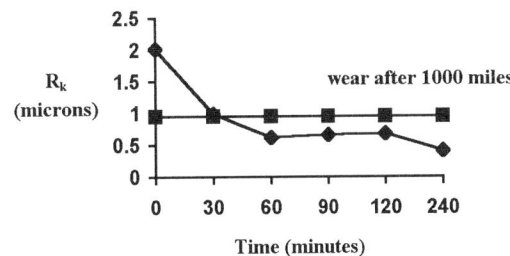

Figure 1 Diesel Specification, R_k

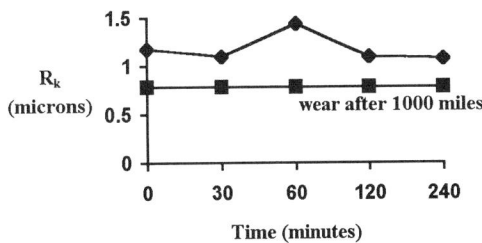

Figure 2 Petrol Specification, R_k

CONCLUSION

These initial tests have highlighted the use of this simple test compressor as means of simulating engine wear. It provides a fast and economical means of analysing honed surface specifications.

Submitted to *ASME Journal of Tribology*

ANALYSIS OF FRICTION FORCES IN TUBE ROCKET LAUNCHERS

MOMČILO MILINOVIĆ, D. ŽIVANIĆ, P. AŠKOVIĆ

Faculty of Mechanical Engineering, University of Belgrade, 27. mart 80, 11000 Belgrade, Yugoslavia

ABSTRACT

This paper considers theoretical and experimental aspect of unguided rocket projectile motion through the launcher of tube type. Major performances that influences shoot precision are initial rocket velocity and initial rotation (rpm) of projectile at launching tube muzzle. Both parameters are achieved by rocket engine forced motion.

Initial muzzle velocity is realized by total impulse of reactive thrust force, diminished by friction forces in launching tube. Initial rotation at tube muzzle could be realized by forced rocket motion through the screw gutter formed in the launching tube (one or more) or by space inclined nozzles driven by axial rocket thrust force.

In first case friction force is parasite resistance for both rotational and axial motions influencing on diminishing of muzzle initial velocity and projectile angle velocity at the launching tube muzzle (fig. 1).

In second case friction resistance is a part of directional force that form rotational motion using screw principle by the action of normal forces on sides of screw channel. That force acts axial and tangential on the launching tube (fig 2).

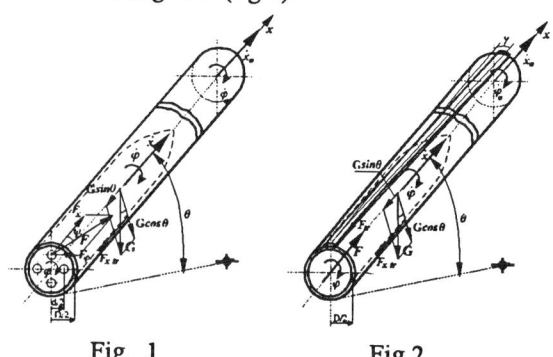

Fig. 1   Fig 2

Paper considers both cases and analyzes forces that act on missile and launcher inside the launching tube. By Lagrange equations of second order this paper evaluates theoretical aspects of action of forces mentioned above. Taking in account calculations and experimental evaluations of the model, paper analyzes total energy losses. All losses generated by friction forces are related on projectile muzzle velocity and projectile initial rotation velocity losses.

Analyzed results with and without influence of angular velocity on friction coefficient gave significant information on relation between angular velocity $\dot{\varphi}$ and initial axial velocity of projectile $\dot{x}$.

Paper considers theoretical and stochastic experimental research of friction forces in static conditions. Boundary parameters that have influence on friction forces are determined. Average value and distribution of friction force is experimentally determined. Total influence of mentioned parameters is evaluated on the basis of experimental results.

Experimental results, and statistical description of obtained data, shows normal law of distribution and possible tolerance field that contains 95.5% of all friction forces that appeared during static test. That field is about ± 18 % up and down of average value of force measured in all tests. Error made in mathematical treatment is not of important significance in that case.

Experimental results also gives data about necessary design parameters specially for tolerance field between dimensions of rocket to launcher integration. This tolerance field can be proved by measurement of friction forces tolerances (it's dispersion) and given as a regulations for final control.

REFERENCES:

(1) D. Živanić, 1990, Influence of self-propelled multitube rocket launcher oscillations on dispersion and cadence of launching, master thesis, MF Belgrade

(2) M. Milinović, 1993, Principles of rocket and launcher design, lectures, MF Belgrade

(3) R.R. Slaymaker, 1966, Calculation of slide bearings, University of Belgrade, Scientific book, Belgrade

(4) D.J. Vitas, M.D. Trbojević, 1981, Machine members, Scientific book, Belgrade

THE INFLUENCE OF GEARING PARAMETERS ON DAMAGING OF GEAR FLANKS

MILETA RISTIVOJEVIC, BOZIDAR ROSIC, VESNA LACARAC
Faculty of Mechanical Engineering, 27. marta 80, 11000 Belgrade, Yugoslavia

The wearing intensity of active gear flanks is dependent on the value of contact pressure, sliding velocity and the thickness of oil film. Surface damaging due to fatigue is the consequence of the large contact pressures, oil presence, rolling and sliding of contact surfaces and could be observed as small dimples -pitting on active surfaces of the gear flunks. Occurrence of pitting causes the increase in load concentration along undamaged parts of the gear flunks and concurrently gives rise to conditions for the development of other forms of damaging of gear flunks. Since reliable method for determination of damaging of the surfaces of gear flunks is not available, than the intensity of contact stress presents good characteristic for determination of the surface strength of gear flunks.

In the majority of cases, gear pairs defined by the pressure angle of the tool of 20^0 are being used (1),(2),(3),(4). However, more stringent requirements in terms of specific mass, load carrying capacity, gear dimensions are imposing the application of gear pairs defined by the pressure angle of the tool different from 20^0. In addition, mentioned requirements are influenced by addendum modification coefficient. Therefore, the influence of the pressure angle of the tool and addendum modification coefficient on the surface strength of gear flunks will be addressed separately, as well as concurrently.

Considering the influence of the pressure angle of the tool and addendum modification coefficient, the following equation is given:

$$\frac{k}{k_o} = \frac{tg\alpha'_w \cos^2 20^o}{tg\alpha_w \cos^2 \alpha}$$ (1) where:

k_o - contact pressure on the gear flunks, while $\alpha=20^0$
k - contact pressure on the gear flunks, while $\alpha \neq 20^0$
α - the pressure angle of the tool
α'_w - the pressure angle when $\alpha=20^0$
α_w - the pressure angle when $\alpha \neq 20^0$

The equation (1) for the value of $\alpha=25^0$ is shown in fig. 1. Based on this equation it could be concluded:
- the influence of addendum modification coefficient on contact stress of gear flunks is strongly pronounced for small number of conjugated teeth and negative value of addendum modification coefficient, while positive values of addendum modification coefficient have no influence on load carrying capacity of gear flunks.
- the influence of the pressure angle of the tool on the contact stress of gear flunks is more visible for the lower sum of the addendum modification coefficient of mating gears. For negative values of the sum, by decreasing the number of teeth of mating gears, the influence of the pressure angle of the tool on contact stress is becoming more marked.

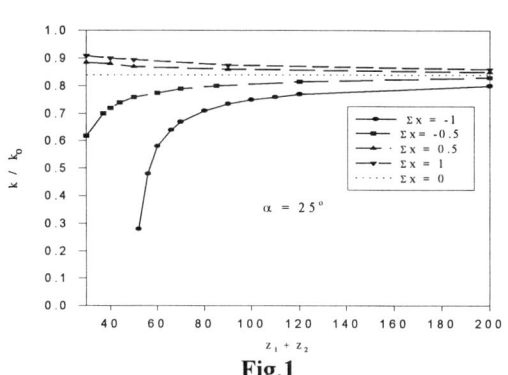

Fig.1

The influence of the pressure angle of the tool and addendum modification coefficient on wearing of gear flunks is addressed in terms of specific sliding. Assuming uniform wearing of the mating gear flunks at the ends of the active line of action the following equation is formed:

$$\rho_{E2}\,\rho_{A2} = u^2\,\rho_{E1}\,\rho_{A1}$$ (2) where:

u - cinematic ratio
ρ_{A1}, ρ_{A2} - radius of curvature of involute profiles of the tooth at the addendum circle of mating gears
ρ_{E1}, ρ_{E2} - radius of curvature of involute profiles at the points corresponding to root diameter of mating gears

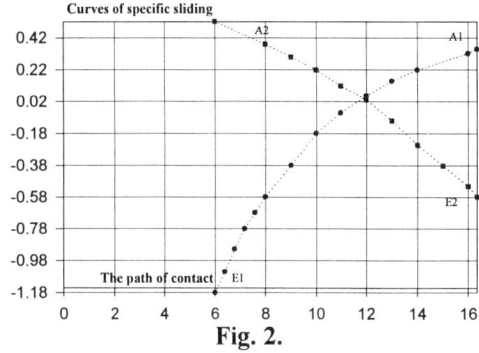

Fig. 2.

Over the contact period, along the path of contact, the absolute difference of specific sliding of teeth profiles of pinion and spur wheel is decreasing with increasing the pressure angle of the tool, as indicated in Figure 2.

REFERENCES
(1) Buckingham, E., Analytical Mechanic of gears, New York 1963
(2) Veriga, S., Mechanical elements III, Faculty of Mechanical Engineering, Belgrade, 1984
(3) ISO TC-60 6336-2
(4) Kudrajcev, V.N., Zubcatije peredaci, Masgiz, Moskva, St Petersburg, 1957

ON INFLUENCE OF FRICTION AND WEAR ON STRESS STATE OF LUBRICATED MOVABLE COUPLINGS OF AXIAL - PISTON PUMP

V. V. MESHKOV and T. A. KOROCHKINA
Tribomaterials Science Department, Metal-Polymer Research Institute, 246652, Gomel, Belarus

ABSTRACT

The performance peculiarity of bearing disc—axial bearing and distributing disc—rotor couplings is rather rigid conditions of deformation and friction interaction, and the action of the local repeated impact loading as well.

The estimation of the stress state of the parts responsible for the tribocouplings resource from the point of view of dynamics is considered to be urgent. For the estimation the finite element method is used.

The calculated data analysis for the bearing disc shows that for the case of the common operation conditions (the minimum friction force) the highest are the axial normal stresses σ_Z. The maximum σ_Z is reached in the centre of the axial bearing sliding track in the area of applied impact loading. This is the stress concentrator as closed isolines of high stresses appear here, Fig. 1.

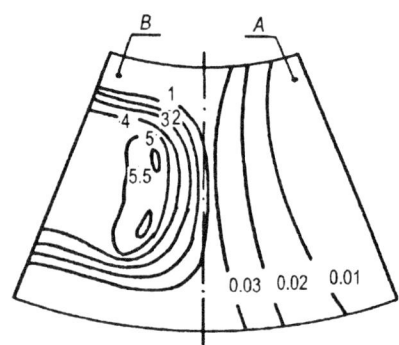

Fig. 1: Distribution of stresses σ_Z (MPa) in the bearing disc at pressure of the working liquid of $P = 20$ MPa and friction coefficient $f = 0.02$: A — intake, B — forcing zones

Under the emergency conditions the friction coefficient f may go up from 0.02 to 0.5. In the case of the stable operation conditions of the bearing disc — axial bearing coupling, the axial normal stresses σ_Z are the highest and tend to monotonous decreasing from 5.5 MPa to 2.8 MPa with the friction force growth. In contrast, the tangential stresses τ_{XY} being the minimum considerably rise from 0.5 MPa to 16.2 MPa for the case of the maximum friction force. Even at friction coefficient of 0.15 stresses τ_{XY} get higher than stresses σ_Z.

In contrast to the bearing disc, stresses σ_Z in the distributing disc are considerably higher than stresses τ_{XY} under the stable operation conditions of the distributing disc—rotor coupling. At $f > 0.45$ stresses τ_{XY} become higher than stresses σ_Z, Fig. 2.

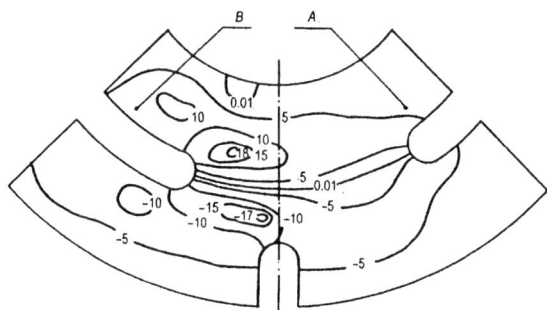

Fig. 2: Distribution of stresses τ_{XY} (MPa) in the distributing disc at $P = 32$ MPa and $f = 0.5$

This is probably caused by the greater contribution of the inertia forces due to the action of the locally applied repeated impacting load.

Thus, we may come to the conclusion that under the common couplings operation conditions the area of the applied impact loading being the stress concentrator is the most dangerous. Probably, crack generation occurs exactly here and leads to the discs fatigue failure. With the friction force growth the tangential stresses τ_{XY} predominate in this area of the applied impact loading. This results in the acceleration of the process of crack generation at some depth under the surface and, thus, in the increase of rubbing surfaces wear.

To improve the bearing and distributing discs wear resistance the following recommendations have been elaborated based on the calculated data analysis: 1) to reduce stresses in the area of the applied impact loading spherical cavities were made to enlarge the contact area and increase the amount of lubricant; 2) to increase the resistance to dynamic loading high-chromium steel was replaced by low-alloy steel which possesses a greater contact fatigue limit.

The tests of the axial-piston pumps with the couplings made in accordance with the above recommendations show their considerably increased durability due to the decrease of the precision couplings wear.

NUMERICAL MODEL OF WEAR IN PISTON - CYLINDER JOINT

V V MESHKOV
Tribomaterials Department, Metal-Polymer Research Institute, 246652, Gomel, Belarus
V I TSVETKOV and L A LIPSKY
Department of Programming, Branch of Institute of Mathematics, 246000, Gomel, Belarus
V N ABRASHIN
Department of Numerical Methods, Institute of Mathematics, 220604, Minsk, Belarus

ABSTRACT

The rational way of increasing the reliability and durability of machines is estimating wear of sliding joints, including a piston - cylinder joint.

The aim of the given work is developing a model to estimate stress state, temperature and wear of the surface layer of a piston made of a composite material at non-stationary sliding in a cylinder along a "fresh" surface.

The elastic behaviour of the piston material is described with a hyperbolic system of differential equations of the first order taking into account temperature. To calculate plasticity Mises condition and stress deviator correction are applied. To solve a quasilinear equation of thermal conductivity and a set of differential equations the difference scheme of multicomponent alternating-direction method is used by Abrashin et al (1).

The piston wear criterion is as follows. If

$$J_2 \geq \frac{1}{3}\sigma^2\left(\overline{J_2}, T\right),$$

where $J_2, \overline{J_2}$ are second invariants of stress deviator and strain rates tensors, σ is ultimate stress of a piston material, T is temperature, wear occurs, otherwise no wear occurs. Like Kennedy in (2), we accept that as a result of wear, the ring-like band of the material is removed from the piston sliding surface.

The calculation is carried out for acceleration conditions of a piston made of a fibre-reinforced thermoplastic sliding along a "fresh" surface of a metallic cylinder up to maximum velocities $U_1=100$ m/s and $U_2=500$ m/s.

It is revealed that during the whole time of sliding in most of the piston surface layer elements normal tensile stresses develop, while in the subsurface layer compressive stresses prevail (3).

For U_1 and U_2 during the whole time of sliding the values of tangential stresses in the subsurface layer are approximately two times larger than the corresponding stresses in the surface layer.

For U_1 the piston begins to wear when the sliding time approaches the time corresponding to the maximum piston acceleration and continues till the end of sliding. For U_2 wear takes place during the whole time of sliding. Before reaching the maximum acceleration, the wear was observed to be nonuniform when at different times different elements of the piston surface layer were subject to wear. Starting from that moment till the end of sliding elements of the surface layer adjoining the working flat surface of the piston are mainly subject to wear.

When the volume contents V of fibres in the composite grows from 0.1 to 0.3, wear decreases, Fig.1. There is probably optimum contents of fibres in the composite at which the minimum wear is observed.

The simulation gives an opportunity to select piston material components and increase its wear resistance. The comparison of the stand test results with the numerical calculation shows the model effectiveness.

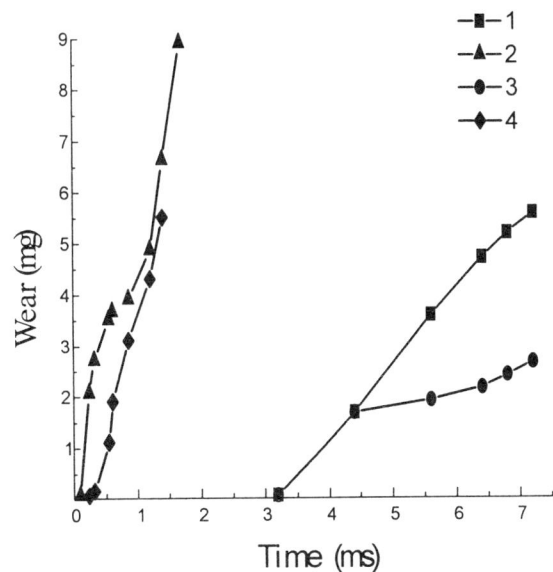

Fig. 1: Piston wear at $U_1=100$ m/s (1,3) and $U_2=500$ m/s (2,4); V=0.1 (1,2) and V=0.3 (3,4)

REFERENCES

(1) V N Abrashin et al, J. Differential Equations, Vol.28, No.2, 1992, 249-260 pp.
(2) F E Kennedy, F F Ling, J. Lubrication Technol., Vol.96, No.3, 1974, 497-508 pp.
(3) V V Meshkov et al, J. of Friction and Wear, Vol. 16, No.3, 1995, 34-39 pp.

WORLD TRIBOLOGY CONGRESS - LONDON 1997

TRIBOLOGY CRITERION OF OPTIMIZATION OF CONSTRUCTION MACHINERY MANIPULATORS

JANOSEVIC DRAGOSLAV
IMK "14 OKTOBAR" Institute of Construction Machinery, Krusevac, Yugoslavia
Profesor D. Sc. VINKO JEVTIC
Faculty of Mechanical Engineering, University of Nis, Yugoslavia

INTRODUCTION

Manipulators of construction machinery, represent generally kinematic chain consisting of many members. The members of the chain are levers connected by joints of the fifth cluss. The elements of joints are sliding bushes and shafts. Driving mechanisms of manipulators consist of actuators - hydraulic cylinders of double actions connected, directly or indirectly, to the members of kinematic chain of manipulator.

Manipulators drive mechanisms are defined by their tranformative and transferred parameters. The transformative parameters are the piston diameter (d_{i1}) and hidraulic cylinders connecting rod diameter (d_{i2}), (Fig 1).The transferred parameters represent co-ordinates (a_i, b_i) of the joints where the hydraulic cylinders are connected to the members of manipulator kinematic chain.

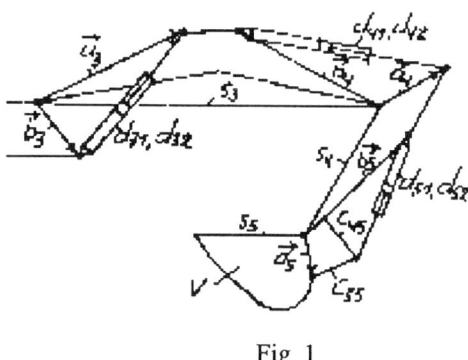

Fig. 1

Jt is obvions that the same function of drive mechanisms can be achieved with less transformative and higher transferred parameters and vice versa.

Possible variational solutions at synthesis of the mechanisms are the question of the choice of the optimal parameters. For that reasons it is set the system of the function (1) where tribology criterion of mechanisms optimization was defined

CRITERION

The function of the manipulators, besides other things, has the tribology occurances-friction and wearing between the elements of the mechanisms joints of the manipulators. The consequences of tribology occurances are the loss of the effective power of the drive mechanisms as well as reduced life-time of the joints elements.

At synthesis of the manipulators mehanisms it is set tribology criterion of the optimization aiming that the loss of the power due to friction resistance in the mechanisms joints becomes minimal: $K = \min (N_{ti})$.

On the base of the hydraulic excavator, mass 110000 kg, with manipulator of back-hoe, it is given the loss of the power (Fig 2) due to friction in the joints for two (A11,A12) different variant of drive mechanisms.The variant A11 is,with less transformative and higher transferred mehchanisms parameters.The variant A12 is vice versa.The loss of the power is defined by means of computer, and dynamic simulation of the variational solutions of the excavator with the same manipulation task.

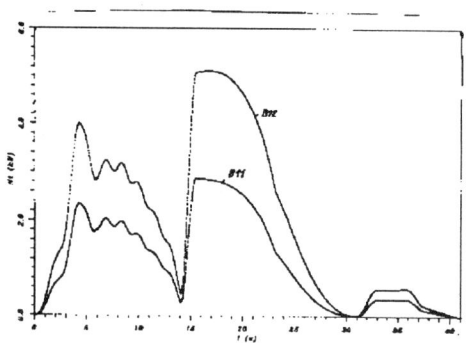

Fig.2

Tribology analysis that was performed shows that at the synthesis of the manipulators drive mechanisms it is necessary to find that kind of mechanisms that have less transformative and higher transferred parameters till the limit that allows flexible stability of hydraulic actuators and the space for their assembling.

LITERATURE

(1) Metods for the optimal hidraulic transmission system synthesis working equipment of hydraulic excavator, series Mechanical engineering, vol 1 No 1, University of Nis, 1994.

TRIBOLOGICAL PROPERTIES OF ELECTROLYTICALLY POLISHED SURFACES OF CARBON STEEL

TAKESHI NAKATSUJI
Department of Mechanical Engineering, Kobe City College of Technology, 8-3, Gakuen Higashimachi, Nishiku, Kobe, Hyogo, Japan
ATSUNOBU MORI
Department of Mechanical Systems Engineering, Kansai University, 3-3-35, Yamatecho, Suita, Osaka, Japan

ABSTRACT

Previously, it was shown that electrolytic polishing of tooth faces had successfully elongated pitting durability of medium carbon steel gearing under severe loading and poor lubrication conditions, irrespective of reducing surface hardness (1).

This paper investigates tribological properties such as friction, surface temperature rise, wear and scuffing of electrolytically polished surfaces in lubricated contact under severe loading using a ball on disk machine. The results are compared with those of mechanically finished surfaces.

The electrolytic solution was a mixture of 70 % - sulfuric acid and 85 % - phosphoric acid of weight ratio of 1 to 2. The bath temperature was controlled at 50 ± 5 °C. Applied voltage was 10 V, and the current density was 15 A/dm^2. Application time was 30 min. Such a condition yields lusterless surfaces suitable for severe contact. The electrolytic polishing under the above mentioned condition removed the ground marks, and formed many micro-pores over the surface, as shown in Fig.1. Moreover, in the electrolytically polished surfaces, the oxygen and phosphate films were thickened.

In consequence, as shown in Fig.2, in the mixed lubrication regime, the electrolytic polishing promoted oil film creation by yielding many micro-pores on the steel surfaces. In the boundary lubrication regime, it effectively lowered the friction coefficient and surface temperature rise by yielding a soft boundary layer composed of oxidized and/or phosphoric compounds on the steel surfaces.

Owing to such properties, electrolytically polished steel surfaces exhibited higher endurance limit to scuff and longer durability for pitting.

ACKNOWLEDGMENT

This research was financially supported by the Kansai University Special Research Fund, 1995.

REFERENCE

(1) T. Nakatsuji, A. Mori, Y. Shimotsuma, STLE Trib. Transactions, Vol.38, 2, 1995, 223-232pp.

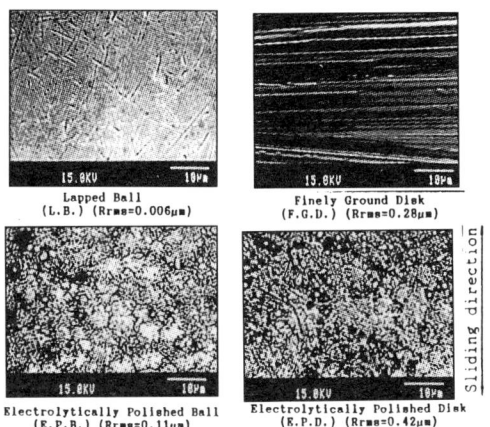

Fig.1: Scanning electron micrographs of surfaces of balls and disks

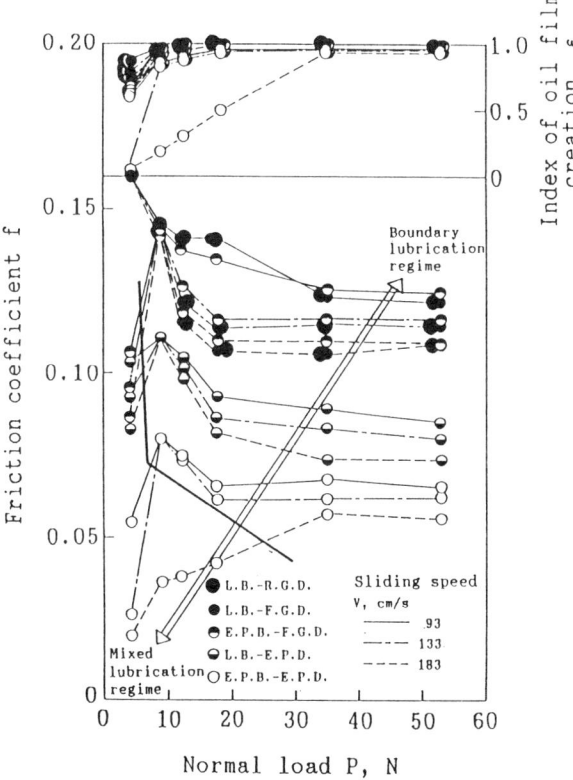

Fig.2: Steady state friction coefficient, f, and oil film creation index, ε, vs. normal load, P

RESEARCH ON THE INFLUENCE OF COMPOSITE MATERIALS AND SiO₂ PARTICLE COATS ON TRIBOLOGICAL PROPERTIES OF THE FRICTION PAIR SEAL - SHAFT

J PADGURSKAS and R RUKUIZA
Department of Machinery Production and Service, Lithuanian University of Agriculture,
4324 Kaunas-Akademija, Lithuania
M VÖTTER and V WOLLESEN
Konstruktionstechnik 2, Hamburg-Harburg Technical University, Denickestr. 17, 21071 Hamburg, Germany

ABSTRACT

Friction pair shaft - radial lip seal (RLS) limit the reliability of internal combustion engines. Therefore each attempt to increase the reliability and to improve the work conditions for this friction pair is very important.

The work was based on common research project (supported by German Government) and previous investigations of German scientists which show that the oil convey and sealing mechanism of RLS could be explained by non-Newtonian behavior of oil (1). Aim of tests was to research the possibility of decreasing friction losses and increasing the durability of the friction pair shaft - RLS, using metallpolymeric oil additives (SURM and MKF), polymeric (Foleox), ceramic (Al_2O_3 -TiC), diamond-like-carbide (DLC), NABA-technology coats on the shaft and into the RLS vulcanized SiO_2 particles. The use of special oil additives and NABA-technology coat was oriented into creation the selective transfer effect (STE) in friction pairs (2).

Experiments have been accomplished on the special stands for seals created in Hamburg-Harburg TU. Different combinations of tribological materials have been tested using oil Pentosin 15W-40CE/SG and fluorine rubber material RLS. At first the pairs was tested for 9 different speeds on the friction torque stand. Later after 500 hours (at 3000 rev/min) run in special wear stand was registered the depth of the wear pit on the shaft using roughness measurement device.

The most significant decrease of the friction torque in friction pair „shaft - radial lip seal" we have using tribomaterials which are creating STE, such a NABA-technology coat or oil additives MKF and SURM (fig.1). The use of with the SiO_2 particles vulcanised seals increase general quantity of friction torque but these seals have 15...20% higher radial load and real friction torque could be accordingly lower as control version. The friction torque for all tribomaterials in low speed depends on the work time, and for the speed over 2 000 rev/min the friction torque was virtually constant for all time of the tests. The reduction rate of friction torque for different tribomaterials comparing with the control version is approximately equal ~ 5...20 %.

The wear resistance of shaft in sealing friction pairs significant increase with hard coats hard coats (DLC and ceramic) but more important should be the use of special oil additives and cheap and simple technologies coats like a Foleox, ceramic and especially NABA-technology which are also improving wear resistance these friction pairs.

Our experiments show that in solving the problem of increasing the reliability of these friction pairs could be modified all components of tribological system: surfaces of both friction materials, lubricant and friction environment. Most efficient way for sealing friction pairs was the modification of lubricant or friction surface of the shaft with materials which create new friction conditions making the servovitic film between friction surfaces according STE. This film decreases the friction factor and protects the surfaces of main materials undertaking the friction loads.

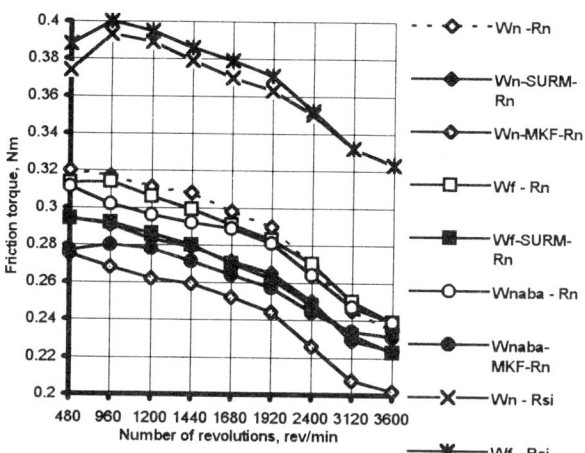

Fig.1: Friction torque as a function of rotation speed for various test combinations.

REFERENCES

(1) de Oliveira, V.Wollesen, M.Vötter, Schmierungs- und Dichtvorgänge bei Radialwellendichtringen // Tribologie und Schmierungstechnik. - 43. Jahrgang, 1/1996. - 32 - 36 pp.

(2) D.N.Garkunov, Tribotechnology [in Russian], Moscow, Mashinostroenie, 1989, 328p.

METHOD FOR CALCULATING MULTI-SUPPORT CRANK-SHAFT TRIBO-SYSTEM OF INTERNAL COMBUSTION ENGINES

V.N. PROCOPIEV, J.V. ROJDESTVENSKIY, S.R. SIVRIKOVA
Tribotechnics laboratory, Chelyabinsk State Technical University, Lenin avenue 76, Chelyabinsk, 454080, Russia

ABSTRACT

For the reliable working of internal combustion engines, tribocontact is necessary to supply a lubricant layer between surfaces of friction over the course of the engine's running cycle, and it in turn depends on the pressure of the lubricant supply and the complex processes of the lubricant flow in the whole multielement tribosystem. The main lubricant consumers are the complexly loaded crankshaft main bearings. The main journals are functionally connected with each other not only by hydraulic, but also elastic relations through bend deformations of the multisupport crankshaft and its bearings, rendering significant mutual influence to hydrodynamic parameters of main bearings describing its loading. Thus the main bearings of the crankshaft and the system of its oil supply have to be considered as one tribosystem taking into account hydraulic and elastic relations.

Modelling of the multisupport crankshaft tribosystem presents severe mathematical and computing problems. The authors gained experience in the solution of specific problems of crankshaft main bearings' calculation (1), calculation of its elasticity (2), modelling of the oil supply system of the internal combustion engine tribocontacts (3). In the report is shown the possibility of the calculations of the multisupport crankshaft in view of the elastic and hydraulic relations of its main bearings as an integral multielement tribosystem.

The calculations are based on the equation system describing a great number of conditions such as the distribution of hydrodynamic pressure in a lubricant layer of main bearings, movement of friction surfaces, stress-deformable condition of crankshaft and bearings, thermal balance of main bearings and also on component and topological equations of the lubricant flow in the tribosystem channels. The equations are solved numerically.

The multisupport crankshaft of the internal combustion engine is simulated as a spatial elastic beam, leaning on nonlinear-elastic support. Static uncertainty is revealed by a method of forces.

For the analysis of the oil supply system a library of the basic components is developed. The components of the main bearings with complex loading are presented in the form of macromodels resulting from thermal and hydrodynamic calculation and determination of movement trajectories of the shaft journals.

The algorithm of calculation consists of three stages. The first one determines the basic hydrodynamic and thermal parameters of a main bearing as a function of time (crankshaft turning angle). For the solution of the Reynold's equation a method of finite differences' approximations on a sequence of grids was used. Availability and arrangement of lubricant sources on the surfaces of a journal and a bearing (pockets, flutes, apertures) and also deviation from the round-cylindrical form of a bearing are taken into account. The equation system of a method of forces is considered in combination with the system, describing a movement of the bearing journal. The solution was found by a method of consecutive approximation. The equation of the thermal balance was solved in parallel with the calculation of the movement trajectory of the main journals at each time interval.

At the second stage the results of the hydrodynamic calculation of tribocontacts are used for formation of their nonlinear macromodels and for the calculation of hydraulic resistance during lubrication flow in the clearances between the surfaces of friction. Method of power circuits is taken as the basis of algorithm for the calculation of the hydrosystem. The mathematical model of the devised system is built by joint transformation of component and topological equations. Fields of hydrodynamic pressures and consumptions in all central points of the oil supply system are the result of the static problem solution.

To specify values of loading parameters for the main bearings at the third stage, a hydrodynamic calculation is again carried out with known values of temperatures and pressure at the input (received at the second stage of calculation).

As an example, a crankshaft tribosystem of the automobile diesel engine is considered. A complete picture is received of the distribution of pressure and charges in all central points of a tribosystem with the account of the rigidity characteristics of the engine crankcase and crankshaft, nonaxialities of main bearings and journals, arrangement of lubricant flutes and apertures in the bearings and journals, oil supply channels and sources.

REFERENCES

(1) В.Н. Прокопьев, Вестник машиностроения, №5, 1979, С.26-30.
(2) Ю.В. Рождественский, Н.А. Хозенюк, Техническая эксплуатация, надежность и совершенствование автомобилей: Сб. научн. тр., Челябинск: ЧГТУ, 1996, С.11-24.
(3) V.N. Procopiev, J.V. Rojdestvenskiy, S.R. Sivrikova, Proceedings Aristoteles University, 2nd International conference on trybology «Balcantrib-96», Thessaloniki, Greece, 1996, 478-484pp.

THE PREDICTION OF MECHANICAL SEAL PERFORMANCE AND EXPERIMENTAL VALIDATION OF CSTEDY^{SM} SOFTWARE

GUANGRUI ZHU

Technology Department, John Crane EMA, 40 Liverpool Road, Slough, SL1 4QX, UK.

ABSTRACT

Correct prediction of mechanical seal performance is essential to reduce the time and cost of new product development. For many years the knowledge gained from Tribology research has laid a sound foundation for seal performance prediction. As a results of long term commitment of John Crane unique R&D team, CSTEDY^{SM}, a PC based software, has been developed and has been continuously refined and validated since 1992. Now it has been implemented to John Crane subsidiaries all over the world for daily seal performance analysis and design activities. This paper offers an overview of CSTEDY^{SM} development and its experimental validation of the software.

The CSTEDY^{SM} analysis assumes that seals operate under axisymmetric condition. Finite element method is adopted for elasticity and heat transfer analysis. The balance of interface pressures including hydrodynamic, hydrostatic and asperity contact pressure to the external load is achieved using Newton-Raphson iteration. The viscosity, density, and phase condition of the fluid within the deformed interface are allowed to vary with pressure and temperature. A user friendly interface provides a platform for user to define one's problem, retrieve material and fluid information from databases and to check the definition of the problem. CSTEDY^{SM} predicts the film thickness; pressure, and temperature distributions across the seal interface; stress distribution; elastic distortions; seal power consumption; leakage rate and asperity contact load etc. The output can be presented in either digital or graphic form. Figure 1 shows typical distortion and temperature distribution of a high performance Type 48 seal for petrochemical and oil refinery industry.

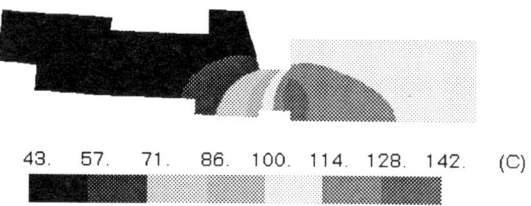

Figure 1. Isotherms and distortion (x100) of a type 48LP seal using kerosene at a speed of 1750rpm and a sealed pressure of 30bar.

In order to investigate seal basic mechanisms and to assist developing theoretical model a test apparatus has been constructed and commissioned to a very high standard since 1991. Two seals are assembled back to back in a test cell. The seals can be tested at a PV value up to 140 MPa-m/s. The seal friction torque, leakage rate, cell temperature, temperatures near seal inner (ID) and outer diameter (OD) can be constantly monitored and recorded using a PC based data logging system. The surface roughness of the face and the seat is accurately measured before and after each test using digitised Talystep, Talysurf, and Talyrond. Scan Electronics Microscope is also available for further inspection.

One hundred and nine test cases of 3 different geometry and material seals resulting in more than 5200 hours have been conducted to validate CSTEDY^{SM} predictions. These cases cover a wide range of operating conditions with load parameter (G) varying of 10^{-9} to 10^{-6}. Each test case is correlated with CSTEDY^{SM} predictions. Excellent agreement is achieved on the seal friction torque with an averaged difference being about 9%. The predicted temperatures in the seat near face ID and OD correlate with experimental measurements quite well. The averaged difference is within 5%. The comparison of leakage rate is also quite satisfactory. CSTEDY^{SM} predictions are well within the upper and lower limit of measured leakage rate. The averaged difference is only 0.14 g/h. Figure 2 is an example of a comparison of measured friction torque, seal leakage rate, temperatures near ID and OD of seal using water at a shaft speed of 3000rpm.

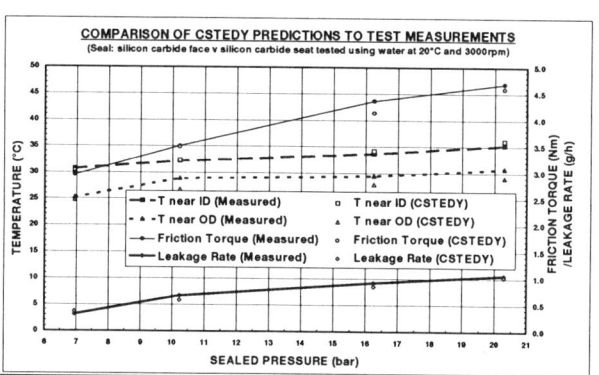

Figure 2. Comparison of test results with CSTEDY^{SM} predictions (tested at 20°C & 3000rpm).

To summarise, a sophisticated computer software has been developed to predict mechanical seal performances. The core of the model has been refined against test results. Extensive test results have confirmed the correlation and accuracy of the model. It gives a great confidence in application of the software to mechanical seal development, trouble shooting, special product designs and applications. It is estimated that the new product development period is cut by 50% due to the implementation of CSTEDY^{SM}.

THURSDAY/FRIDAY 11 SEPTEMBER

Corresponding oral session		Page number
F3	Micro- and nano-scale tribology	773
TH1	Rolling bearings	795
F4	Thermal effects in tribosystems	815
F4	Tribochemistry and corrosion wear	823
TH4	Tribology of polymers	831
TH2/TH5	Industrial problems and solutions	843
F2	Tribology in extreme environments	861
W2	Practical applications of friction and wear	869
F1	Environmental issues	901

SELF-ORGANISATION IN CONTACT: SURFACE GROWTH AND DESTRUCTION ALGORITHMICAL MODELS

E. ASSENOVA and K. DANNEV
Tribology Centre, Technical University - Sofia 1756, Bulgaria

ABSTRACT

The idea in this paper is in combining computational experimental methods (fractal structure generation, cellular automata method) with some fundamental methods in physics and methodology in order to observe the evolution of the interface or the contact through direct simulation, revealing laws through algorithms.

The crucial role of surface geometry and substance in third body formation is incontestable. Surface geometry is of interest not only in physics and chemistry, but indeed in various tribological situations It is known that surface topography is a non stationary random process, in which the variation in height correlates with sample length. Thus, devices with different resolution and range display different values of these statistical parameters for one and the same surface. Conventional methods seemingly fail. Their main inconvenience is in the following: although surface roughness subsists concurrently on different scale levels, its representative parameters depend only on some specific scale levels, among them device resolution and sample length. It appears to be reasonable to use self-affine or scale invariant parameters in the process of surface roughness and contact identification, i.e. to use fractal structures (1). Of the manifold methods to acquire a fractal surface, the most suitable for surface roughness modelling appears to be the method of diffusion-limited-aggregation (DLA). With DLA simulation, the structure grows automatically; it is not given in advance. This can be seen as a type of self-organisation of the concerned object (3) (5).

Tribological systems are naturally complex, and being effected by numerous contact interconnections at different scale levels, triboprocesses are dynamic and developing in time and space, and consequently are computation irreducible (4) (5). For systems of that kind, no simple formulae have been found. More abstract models, of the kind of evolution-algorithmical models using direct imitation and presenting physical laws through algorithms, are found to suit better to contact properties modelling.

The benefit in the above model seems to be the idea of simulation of some physical processes in the generation of real surfaces using an algorithmic approach. In general, fractal, dendritic, columnar or single crystal surface growth and destruction can be described not only as a kinetic process by various continuum models (with a set of differential equations) or by discrete models, but also with an algorithmic approach.

The algorithmic approach is related to direct simulation, presenting physical laws not through formulae, but through algorithms, that is a set of rules and procedures to determine the behaviour of the system.

A general way in this direct simulation of systems imposed lately in many different domains are *cellular automata* (2).

The cellular automaton provides some important advantages which are favourable for tribological modelling (5):

* The CA gives the possibility a system to be modelled as a whole. The universally used, but in fact absolutely arbitrary and unnecessary division into explicit discrete pieces, like division of a tribosystem in body and counterbody, is thus being avoided.

* By CA-models suits the principle: the whole determines the properties of the parts. Every parts possesses characteristics depending on its relations with the remainder, i.e. what is crucial, is the **interaction** with the environment. If modelled with accentuation on the interaction, a system reveals as decisively contact system.

* In the CA the states that a cell can occupy represent all those "integral" properties we recognise in the object. However, the particular state of the cell depends besides on its relations with the whole system, also on the fluctuations in the system, i.e. the element of randomness is included in the CA-model too.

The attempts of the authors in this sense, develop above concept for tribological demands in different situations: simulation of contact deformation, computing pneumatic-hydraulic contact conductance, interaction between rough surfaces and fluids, etc. A crucial point is the development of growth and destruction algorithmic models for contact surfaces in different wear conditions.

REFERENCES

(1) F. F. Ling, EUROTRIB'89, Vol.2, Helsinki.
(2) T. Toffoli, N. Margolus.Physica D, 45(1990)230.
(3) E. Assenova, K. Dannev, 7th Trib. Coll. Esslingen'90.
(4) E. Assenova, K. Dannev, EUROTRIB'93, Budapest.
(5) E.Assenova, K.Dannev,EUROTRIB'96, Thessaloniki.

Effect of Surface Roughness on Indentation Measurements.

M. S. BOBJI and S. K. BISWAS

Department of Mechanical Engineering, Indian Institute of Science, Bangalore, 560 012. India.

ABSTRACT

Indentation methods are being increasingly used to evaluate the mechanical properties of materials. With the advent of the depth sensing indentation methods and better imaging techniques like Atomic force Microscopy, the size of the indents made has gone down to few tens of nanometers.

The effect of the roughness of the test surface on the hardness estimation is negligible if the indentation depths are much greater than the surface roughness. The roughness wavelengths which affect a physical process are determined by the length scale of the process. In the case of indentation it will be the size of the indent made.

For an elastic-plastic indentation, the indentation pressure depends on the imposed strain. As the indenter penetrates the rough surface, the radius of curvature of the asperities encountered increases. This means that the effective strain changes with penetration and hence the hardness varies with penetration.

The hardness (H) as measured by indenting a single spherical asperity with a spherical indenter can be expressed as a function of the asperity radius (R_a), indenter radius (R_i) and the angle (θ) between the line joining the centers and the measurement axis as, (1)

$$H = H_s \left(\frac{R_a \cos^2 \theta}{R_i + R_a} \right)$$

This relation is verified experimentally by carrying out depth sensing indentation experiments with a spherical indenter at the macroscopic level in an universal testing machine (10 ton capacity).

As θ is small, the above relation can be approximated for $R_i \ll R_a$ as

$$\frac{H}{H_s} = 1 - \frac{R_i}{R_a}$$

The exact relation between R_a and the penetration depth δ depends on the nature of the rough surface. A general form of such relation could be written as

$$R_a = K_1 (\delta/d_r)^m$$

where d_r is a roughness parameter such as root mean square roughness. Thus

$$\frac{H}{H_s} = 1 - \frac{K_2}{(\delta/d_r)^m} \qquad \ldots (1)$$

Weiss (2) pointed out that the effect of the roughness can be accounted for by adding an error term in the displacement. This would give,

$$\frac{H_r}{H_0} = \frac{Af(\delta \pm \delta_e)}{Af(\delta)} = \left(1 \pm \frac{\delta_e}{\delta}\right)^n$$

where H_0 is the bulk hardness and Af is the area function of the indenter. 'n' is 1 for spherical indeter and is 2 for conical indenter. Comparing this equation with equation 1 for a spherical indenter, it is clear that δ_e is related to the roughness parameter and the indenter geometry. Thus,

$$\frac{H_r}{H_0} = \left(1 - \frac{k}{\delta/\delta_r}\right)^n \qquad \ldots (2)$$

Numerical simulations were carried out to test the applicability of this expression. The rough surface is simulated using Weierstrass-Mandelbrot function. The statistical properties of such a self-affine fractal surface remain the same at the different magnifications. A conical indenter with a spherical tip is brought into contact with this surface. At first the contact is established with a single asperity which deforms with increasing penetration and the neighboring asperities will come into contact with the indenter. The contact area consists of many tiny islands. The load supported by each of these island is calculated using Johnson's spherical cavity model for elastic-plastic indentation.

It is assumed that the asperities are conical with a base area equal to the area of the contact island. Summing up the load due to all the asperities in contact, the total load required to produce a given penetration is estimated. Hardness is obtained, as in the depth sensing experiments, by dividing this total load with the area obtained from the area function of the indenter.

The results match with the equation 2 remarkably well and the value of k is a function of indenter geometry and depends on whether the hardness measurement is carried out by imaging or depth sensing experiments.

REFERENCES

(1) M.S.Bobji, M.Fahim and S.K.Biswas, Tribology letters, Vol 2, page 381, 1996.

(2) H.J.Weiss, Physics status solidi A Vol 129, page 167, 1992.

LUBRICATING PROPERTIES OF ORDERED MOLECULAR LAYERS STUDIED WITH MOLECULAR TRIBOMETRY

K BOSCHKOVA, J M BERG and B KRONBERG
Institute for Surface Chemistry, P.O. Box 5607, SE-114 86 Stockholm
P M CLAESSON
Laboratory for Chemical Surface Science, Department of Chemistry, Physical Chemistry, Royal Institute of Technology, SE-100 44 Stockholm, Sweden

ABSTRACT

In this study the lubricating properties of systems exhibiting molecular layering are investigated. The systems are studied with a new type of Surface Forces Apparatus (1), capable of shearing the two surfaces with respect to each other, measuring the lateral (friction) force. The shearing speed and normal load can be controlled and varied during an experimental run. The Surface Forces Apparatus makes it possible to determine the absolute distance between the surfaces, the dimensions of the contact radius, and the surface deformation during an experimental run. This makes it possible to measure the friction between two surfaces separated by a layered medium as a function of the number of layers.

Molecules that are linear or pseudo-spherical, such as straight chain alkanes or OMCTS (octamethyl-cyclotetrasiloxane), exhibit short range ordering of the molecules in the liquid state. Since the lubricating properties of a liquid can be expected to be dependent on the degree of order in the liquid, there are reasons to believe that the lubricating properties of alkanes are very sensitive to additives.

The ordered layer structure in OMCTS will give rise to oscillations in force-distance measurements, with a period of about 8 Å. Measurements of normal forces in dry OMCTS between two mica sheets were performed, and were found to be consistent with previous investigations (2). A shearing motion is applied and the friction force is measured. In figure 1 the friction force during shear is shown, at a point corresponding to a repulsive oscillation peak at D=56 Å. Between each plateau in the figure the normal force is decreased. The shearing is continued for about 80 s for each value of the normal force.

In figure 2 the friction force is plotted as a function of the normal load. The friction coefficient was found to be around 0.26.

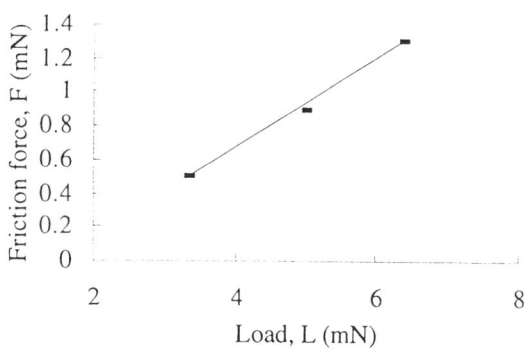

Fig. 2: Measured friction force plotted against load for sliding of two mica surfaces separated by a film of OMCTS at D=56 Å.

REFERENCES

(1) J. N. Israelachvili, P. M. McGuiggan, J. Mater.. Res. **5** (1990)
(2) H. K. Christenson, R. G. Horn, J. N. Israelachvili J. Colloid Interface Sci. **88**, 2223 (1983)

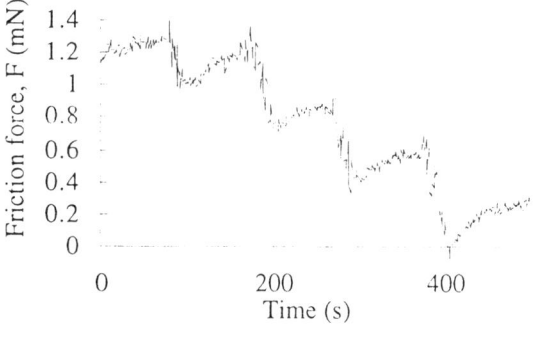

Fig. 1: Measured friction force versus time for a stepwise decreasing normal load. Mica surfaces separated by a film of spherical molecules of OMCTS at D=56 Å.

ESTIMATION OF MOLECULAR GAS SQUEEZE FORCES IN VIBRATION-TYPE SPM MEASUREMENTS

S. FUKUI, Department of Applied Mathematics and Physics, Tottori University, Tottori 680, Japan
S. UMEMURA, NTT Laboratory, Musashino-shi, Tokyo 180, Japan   (*email:fukui@damp.tottori-u.ac.jp*)
R. MATSUDA, School of Administration and Informatics, University of Shizuoka, Shizuoka 422, Japan
R. KANEKO, Department of Opto-mechatronics, Wakayama University, Wakayama 640, Japan

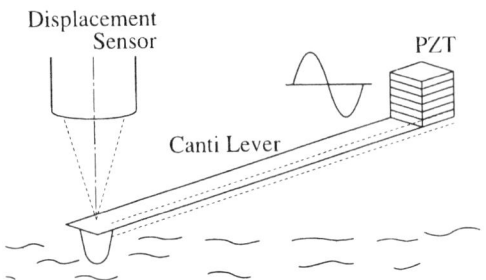

Fig. 1 Vibration-type SPM

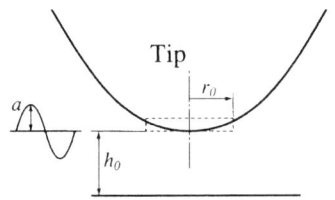

Fig. 2 Squeeze motion between the tip and the test surface

ABSTRACT

In micromechanical systems such as a vibration-type scanning probe microscope (SPM) (1), as shown in Fig. 1, or capacitive accelerometer (2), molecular gas film lubrication effects, especially squeeze gas film effects, are not negligible between the tip of the vibrating cantilever probe and the test surface (see Fig. 2). To analyze the gas film lubrication effect with ultra-small mean spacings, $\bar{h}$, of less than 0.1 μm, one should use the molecular gas film lubrication (MGL) equation (3), which is the Reynolds-type gas bearing equation for arbitrarily small spacings which was derived from the linearized Boltzmann equation. The MGL equation has an additional local flow rate coefficient, $\widetilde{Q}_p = Q_p/Q_{pcon}$, in pressure flow terms, which denotes the ratio of the flow rate of plane Poiseuille flow in arbitrarily small spacings to that in the continuum.

To analyze the squeeze gas film characteristics, we applied the perturbation method to the MGL equation (4) with respect to a sinusoidal squeeze motion with an infinitesimal amplitude, and obtained the analytical form of the nondimensional reacting forces, W, which consists of an in-phase component that corresponds to a stiffness force, W_1, and an out-of-phase component which corresponds to a damping force, W_2.

Assuming the tip shape of the SPM probe facing the test surface as a circular plate, we calculated W, W_1 and W_2. The time dependent W with respect to the squeeze motions reveals that as the mean spacing, $\bar{h}$, decreases, the peak values of W decrease and the phase delays increase. The relationships between W, W_1, W_2 and the modified squeeze number, σ', defined as

$$\sigma' = \sigma/\widetilde{Q}_p(\bar{h}) = 12\mu\omega r_0^2/p_a\bar{h}^2\widetilde{Q}_p(\bar{h}),$$

the amplitude, W, and the stiffness, W_1, monotonically decrease as σ' decreases. The damping force, W_2, however, has its maximum value at around 10 as shown in Fig. 3.

When the radius of a circular plate in the vibration-type SPM probe is 1 μm and the mean distance, $\bar{h}$, and the vibration amplitude are 100 nm and 20 nm, respectively, the air film stiffness force is found to be on the order of 10^{-6} N/m, and moreover, the squeeze damping force overwhelms the stiffness force.

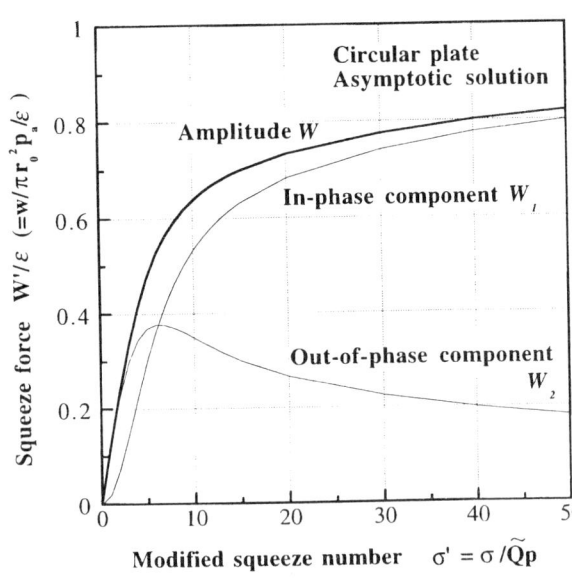

Fig. 3 Relationships between W, W_1, W_2 and σ'

REFERENCES

(1) U. Hartmen, *Advances in Electronics and Electron Physics*, Vol. 87 (1994) pp. 49-200
(2) T. Veijola, H. Kuisma, J. Lahdenpera and T. Ryhanen, *Sensors and Actuators* A, Vol. 48 (1995) pp. 239-248
(3) S. Fukui and R. Kaneko, *Handbook of Micro/ Nanotribology*, ed. Bhushan, B., CRC Press (1995) Ch. 13
(4) S. Fukui, R. Matsuda and R. Kaneko, Trans. ASME, *J. Tribology*, Vol. 118 (1996) pp. 364-369

THE BOUNDARY ELEMENT METHOD APPLIED TO FRICTIONAL CONTACT PROBLEMS UNDER NORMAL AND TANGENTIAL LOADING

R.S. HACK and A.A. BECKER
Department of Mechanical Engineering, University of Nottingham, University Park, Nottingham, NG7 2RD, UK

ABSTRACT

The Boundary Element (BE) method is now well established as an accurate numerical tool in stress analysis. Its surface-only modelling capability and high resolution of stresses makes it very suitable for contact problems, in particular those requiring a high degree of accuracy. Unlike the Finite Element (FE) method, the BE approach is capable of directly incorporating the contact relationships of equilibrium and compatibility into the equation solver, without the need for special contact elements, such as gap or interface elements used in some FE programs.

Although both the FE and BE methods have been successfully applied to the study of contact problems involving unidirectional loading cases, where the normal contact stresses are of primary interest, there have been problems when multidirectional loading is applied. This will result in a contact zone, where stick and partial-slip conditions will coexist within the contact zone. It is common to find numerical instabilities at the edge of the contact zone and within the partial-slip contact zone.

The BE and FE methods are used to evaluate the normal and tangential contact stresses and the relative tangential displacements of the surfaces of the classic Hertzian contact problem of a cylindrical punch on a foundation, where normal and tangential loads are applied, as shown in Fig. 1, where; $R_p=15$, $H_p/R_p=0.133$, $R_p/W_p=1.5$, $R_p/H_f=1.5$, $R_p/W_f=2$, $E_p=2.1 \times 10^4$, $\nu_p=0.29$, $E_f=2.1 \times 10^3$, $\nu_f=0.45$, $\mu=0.5$

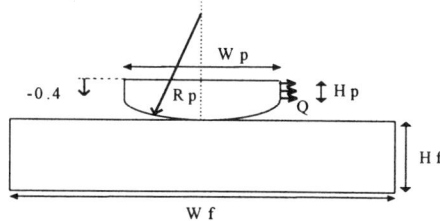

Fig. 1, Cylinder on a Foundation

In this case the normal load is applied as a vertical displacement of the cylinder, and the tangential load is applied as a horizontal thrust Q in steps of 100, 200, 400 and 800. The dimensions are chosen to coincide with Kikuchi & Oden (1988).

The BE solutions were obtained by using the BEACON software, Becker (1989), and the FE solutions were obtained by using the "contact pairs" function of ABAQUS (HKS, 1996).

From Fig. 2, both the FE and BE methods, generated a classic Hertzian normal contact stress distribution curve for a cylinder in contact with a flat surface. Fig. 3 shows the shear stress distribution at the contact interface, where it can be seen that the BE data do not match precisely the FE data, but there is agreement as to the extent of the stick and partial-slip zones within the contact region. Fig. 4 shows a close agreement in the trend of the predicted slip of the two surfaces within the contact region.

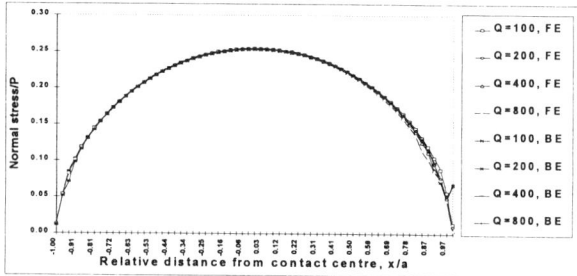

Fig. 2 Normal contact stress distribution

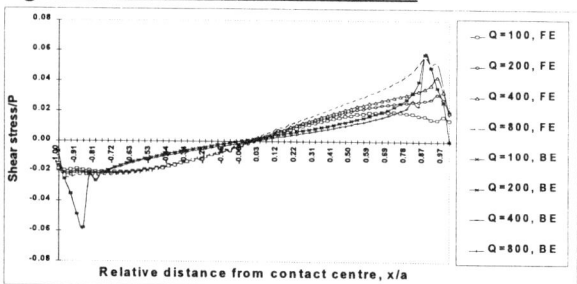

Fig. 3 Shear contact stress distribution

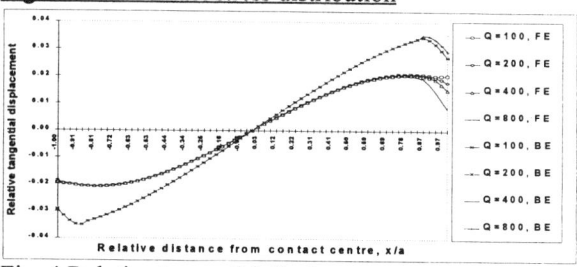

Fig. 4 Relative tangential displacement

REFERENCES

ABAQUS version 5.5, User's Manual, (HKS Inc., Rhode Island, U.S.A.)

Becker A.A., 1989, "A boundary element computer program for practical contact problems", Modern practise in Stress and Vibrational Analysis, (ed. J.E. Mottershead), Pergamon Press, pp 313-321

Kikucki N. & Oden J.T., 1988, "Contact problems in elasticity: A study of variational inequalities and finite element methods, SIAM, Philadelphia.

FRICTION AND TRIBOCHEMISTRY BETWEEN DIAMOND SURFACES IN SLIDING CONTACT

JUDITH A. HARRISON, MARTIN D. PERRY, and STEVEN J. STUART
United States Naval Academy, Chemistry Department, Annapolis, MD, 21402-5026, USA.

ABSTRACT

Because the development of new technological applications involving diamond coatings involves the motion of diamond on diamond, understanding the tribological properties of diamond is paramount. Knowledge of the atomic-scale mechanisms that give rise to friction and wear might ultimately aid in the design of coatings with specific tribological characteristics. To this end, molecular dynamics (MD) simulations have been used to examine the atomic-scale phenomena governing the tribology of diamond surfaces.

The forces governing the interatomic interactions were derived from a reactive empirical bond-order (REBO) hydrocarbon potential (1). This potential energy function is unique in its ability to model chemical reactions and the associated changes in hybridization. The tribology of diamond surfaces in sliding contact has been previously investigated as a function of crystal face, temperature, sliding velocity, crystallographic sliding direction, surface state, and in the presence of third-body or debris molecules (2-4).

Microscopic friction is usually accompanied by wear, and energy is dissipated into the bulk material by the movement of dislocations. In contrast, atomic-scale friction may occur with or without wear. Simulations show that two hydrogen-terminated diamond surfaces in sliding contact exhibit no wear even at fairly large loads (2,3). The sliding causes vibrational excitation of the diamond surfaces. This excitation is transferred to the bulk and eventually dissipated as heat. This process is the essence of "wearless" atomic-scale friction (3). This excitation increases as a function of load, and therefore, the friction increases (Fig. 1).

Replacing some of the hydrogen atoms on one of the diamond surfaces with hydrocarbon groups dramatically affects the friction (Fig. 1). The presence of the longer chain alkyl groups, e.g., $-C_2H_5$, reduces the friction dramatically at moderate to high loads. The size and flexibility of these larger hydrocarbon chains, in conjunction with the conformations they adopt while sliding, are responsible for the observed reduction (2).

At fairly large loads, wear of the diamond surfaces is initiated during sliding when $-C_2H_5$ groups are present one surface (4). These simulations were the first simulations to observe tribochemical reactions, that is, reactions that are initiated by, and occur during, sliding. Complex, nonequilibrium radical chemistry is initiated by the loss of hydrogen atoms from the chemisorbed $-C_2H_5$ groups. Subsequent reactions include abstraction of hydrogen from both diamond surfaces, radical recombination, transient adhesion of the two surfaces, and the formation of debris.

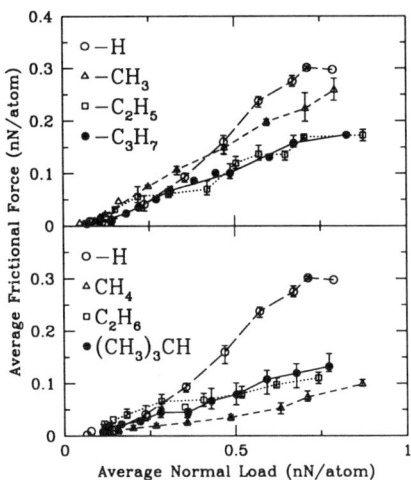

Fig. 1: Friction as a function of load for two hydrogen-terminated diamond (111) surfaces with chemisorbed groups and in the presence of third-body molecules in the upper and lower panels, respectively.

Small third-body molecules trapped between the hydrogen-terminated diamond surfaces also dramatically affect the friction. In general, the third-body molecules act as a boundary layer between the two diamond surfaces reducing the amount of vibrational excitation the surfaces can impart to one another. Therefore, any observed friction arises from the interaction of the third-body molecules with the diamond surfaces. The small methane molecule interacts the least with the diamond surface during sliding, and therefore, induces the smallest amount of friction. Tribochemical reactions were observed when ethane and isobutane molecules were trapped between the diamond surfaces.

ACKNOWLEDEMENTS

This work was supported by The Office of Naval Research under Contract No. N00014-97-WR20019.

REFERENCES

(1) D.W. Brenner, Phys. Rev. B., Vol. 42, page 9458, 1990.
(2) M.D. Perry and J.A. Harrison, J. Phys. Chem., Vol. 101, page 1364, 1997.
(3) J.A. Harrison *et al.*, Thin Solid Films, Vol. 260, page 205, 1995.
(4) J.A. Harrison and D.W. Brenner, J. Am. Chem. Soc., Vol. 116, page 10399, 1994.

TRIBOLOGY OF SILICON UNDER THE LOAD OF MILLINEWTON

K. HIRATSUKA and K. TAKACHI
Department of Precision Engineering, Chiba Institute of Technology,
2-17-1, Tsudanuma, Narashino-shi, Chiba 275 JAPAN

ABSTRACT

In micro-machine technology, silicon plate is often used as the substrate on which the small machines are assembled. In some cases, it is in a sliding contact with various metals or ceramics. So it is important to understand the fundamental tribological properties of silicon under the light loads. From this background in this study, we have carried out the friction experiments of silicon with SUJ2 and silicon nitride.

Single crystal silicon (111) plate specimen is rubbed against ball specimen of SUJ2 steel and silicon nitride under the loads from 1 to 10 mN using ball-on-plate friction rig. The ball diameters are 0.8, 1.2, 1.6, 2.0, 4.0, 6.0mm. The sliding velocity is 20mm/s. All tests are carried out in laboratory air condition without lubricants.

Fig.1 shows the effect of load on the coefficient of friction between silicon plate and SUJ2 ball of various diameters. The regression curves are also drawn in the graph with the regression equations. It is clear that the coefficient of friction is proportional to about -1/3rd power of the load. In the rubbing between silicon nitride and silicon, the coefficient of friction against load shows the same behavior and the regression indices of the load are ranged from -0.287 to -0.345.

Fig.2 shows the effect of ball diameter (d) of SUJ2 specimen on the coefficient of friction. The coefficient of friction increases with $d^{0.181}$ and $d^{0.137}$ in the loading conditions of 0.98mN and 2.94mN, respectively. In heavier loads, it is indifferent to the ball diameter. In the rubbing between silicon nitride and silicon, the diameter does not have the effect on the friction, either.

When the ball with radius r_1, Young's modulus E_1, Poisson's Ratio v_1, contacts the plate elastically with Young's modulus E_2 and Poisson's Ratio v_2, the diameter of the contact circle a is given by

$$a = \left\{ \frac{3}{4} WR \left(\frac{1-v_1^2}{E_1^2} + \frac{1-v_2^2}{E_2^2} \right) \right\}^{\frac{1}{3}}$$

If the friction appears at this area, the coefficient of friction μ is calculated by

$$\mu = \frac{s_i A}{W} = \pi s_i \left\{ \frac{3}{4} \left(\frac{1-v_1^2}{E_1^2} + \frac{1-v_2^2}{E_2^2} \right) \right\}^{\frac{2}{3}} W^{-\frac{1}{3}} R^{\frac{2}{3}}$$

, where s_i is the shear strength per unit area.

If the friction force arises at the elastic contact zone, the coefficient of friction would be proportional to the -1/3rd power of the load and 2/3rds power of the radius of the spherical specimen. The results that the coefficient of friction is proportional to the -1/3rd power of the load means that the friction force appears at the elastic contact area. On the other hand, the results that the coefficient of friction is almost indifferent to the diameter of the ball means that the friction force appears at the plastically deformed area. This inconsistent results are not clarified in detail. But it is considered that when the diameter of ball decreases, the contact pressure at the silicon surface increases, so that the micro-fracture inside the silicon occurs, which results in the increase in the resistance to shear fracture.

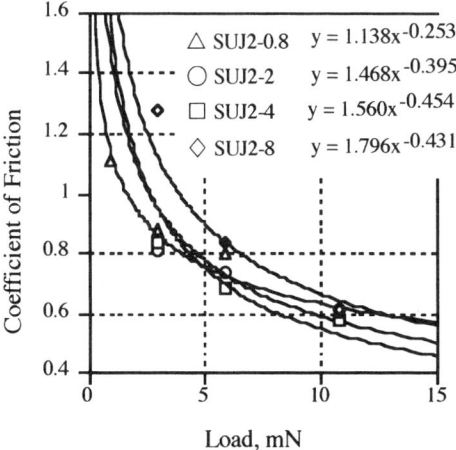

Fig.1: Effect of load on the coefficient of friction

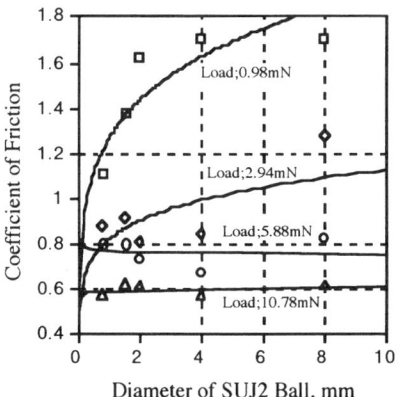

Fig.2: Effect of ball diameter on the coefficient of friction

Substrate Surface Energy Effects on Liquid Lubricant Film at Nanometer Scale

Jianbin Luo and Shizhu Wen
National Tribology Laboratory, Tsinghua University, Beijing 100084, CHINA
Lawrence K.Y. Li
Department of Manufacture Engineering, City University of Hong Kong, Hong Kong

1. Experiments

The film thickness is measured by means of the relative optical interference intensity (ROII) technique with a resolution of 0.5 nm in the vertical direction and 1.5 μm in the horizontal direction.

In order to probe the relationship between the surface energy of the substrate and the thickness of the lubricant film, different coatings made by the plasma assistance sedimentation (PAS) technique are used to check the variation of the oil film thickness with the quantities of other parameters, such as the surface energy of substrate, rolling speed, running time, etc. The thickness of coating can be determined by the adsorbed energy of the x-ray.

3 Results and Discussion

3.1 Combined surface roughness in thin film lubrication (TFL) regime

The measured combined surface roughness is closely related to the film thickness when it is in the order of nanometers. This can be explained by the fact that when film thickness is smaller than the height of the combined asperities, the lubricant film between the high asperities of the combined surface becomes very small. The lubricant in these thin film region is found to be able to sustain much higher pressure than the bulk lubricant. Hence, deformations of the high asperities in this region will take place.

3.2 The effect of surface energy on film thickness in TFL Regime

In order to check the effect of surface energy on the film thickness, a Chromium coating which is close to the steel in the surface energy and an Aluminum coating which is far from the steel in that are chosen to be formed on the surfaces of the balls. The larger the surface tension of the substrate, the thicker the total film and the critical transition film will be in the contact region.

Fig.1 shows that the lubricant film thickness is closely related to running time with different substrate surface energy, that is, the film in the contact region will become thicker with time during the running process. This is because that the larger the surface energy especially the dispersive energy, the more the molecules move to the interface from the bulk and the stronger the orienting force filed will be, and then the thicker the ordered film will be.

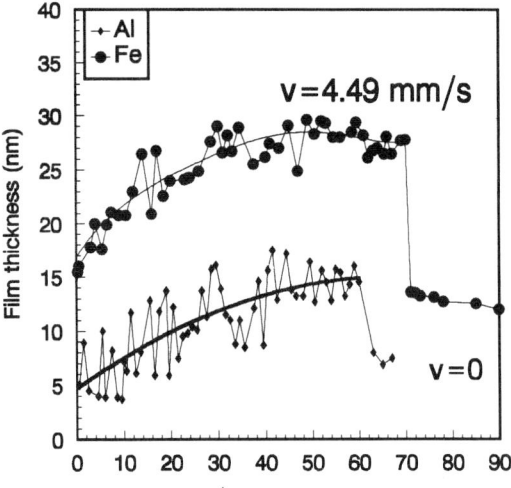

Fig.1 Film thickness changing with coatings and running time
Temperature: 18°C, Load: 7 N
Rolling speed:4.49 mm/s, Ball φ23.5 mm

4. Conclusions

The combined surface roughness of the contact surface changes significantly with the liquid lubricant film thickness and the contact pressure when the operation is in the thin film lubrication regime. The film thickness is larger in TFL when the substrate has higher surface energy. The TFL film grows and reached a steady value with time and the amount of increase in film thickness depends on the substrate surface energy.

CONTACT OF DETAILS FROM THE POROUS AGGLOMERATING MATERIALS

VLADISLAV MAKSAK and TATIANA MAKSAK
Khakas Technical Institute of Krasnoyarsk State Technical University, Shetinkin Street 27, Abakan, Russia
NIK KUPRIYANOV
Tomsk Politechnical University, Lenin Street 30, Tomsk, Russia

ABSTRACT

The practice of operating machines and mechanisms shows that their precision, vibrostability to a considerable extent are determined by processes occuring injointing of interfacing parts.

A model of porous sintered material has been developed to describe processes of contacting.

The model of powder material is presented with assemblage of coalescing spherical particles in the form of cubic packing. When the material is pressed particles of the radius "r" are deformed. Interaction among the particles is done in the circuis contact's site of the radius a_0. Under pressing two particles will approach at the value of $2\delta_0$ which is approachment of two plastically compressed spheres.

The values a_0 and δ_0 are obtained in the result of solving the contact problem on plastic compression of spherical segments with the help of the method of variable parametres of elasticity. Variable parametres of elasticity E^* and μ^* depend on the intensity of stresses in the critical point of spherical segment and found out by the method of successive approachment.

The elastic restorability of particles occurs after removing of compacting pressure. Elastic restorability δ' presets linear off - loading and is defined according to Herts' problem.

The radius of the contact's site in the result of elastic restorability changes and becomes equal to a.

The model of the material supposes that after sintering the contact boundary among the particles is preserved. The contact surface of the porous powder body is presented by a set of spherical segments comparable with the size of powder particles. Let's examine deforming of a single microprojection which is formed by the surface particle of powder material, smooth, rigid plane. Approachment of the particle with the peane will be defined by deformation of microprojection Δ_{m_i} and its shearing relative to neighbouring particles Δ_{r_i}. Common approachment of the contact microasperity with the smooth plane

$$\Delta_i = \Delta_{m_i} + \Delta_{r_i}$$

Numerical analysis of the equation for shearing allows to approximate it in the form of

$$\Delta_i = 1.65 \cdot K_n \cdot (1-\mu^2)^{\frac{2}{3}} \Big/ E^{\frac{2}{3}} \cdot R^{\frac{1}{3}} \cdot N_i^{\frac{2}{3}},$$

where K_n - voids ratio.

Transition from the contact of single asperity to the contact of rough surface is done according to the methods by Demkin. Approachment of rough surface of a porous body with smooth plane will be written in the form of

$$\Delta = \left(\frac{1.5\pi \cdot K_n \cdot I \cdot R \cdot H_{\max}^V \cdot N}{b \cdot A_c} \right)^{\frac{2}{2 \cdot v+1}},$$

where b and v -parametres of the curve of base surface; $H_{\max}$ - maximum heightof the projection; A_c -contour area of the contact.

Experimental investigations have been conducted on the original installation on the optic-mechanical system of measuring microshears. Scale factor for registration of approachment is $0.02*10^{-6}$ m.

The technological process of making blanks from the powder consists in cold pressing and sintering according to the industrial technological conditions. After sintering compactings are subjected to mechanical treatment when the surfaces taking part in contacting are not treated. The samples from the powder of titanium, iron and copper have been investigated.

The general character of influence of compressive force on the value of approaching bodies from porous materials in the investigated range of changing porosity (8 ÷ 41%) remains the same as for compacted ones. That is rigitidy of the contact is small at the initial stage of the its loading and increases and in the increased loads the contact between compressive force and approximation tends to the linear one.

With the increase of porosity resistance of the surfaces decreases and the conditions of sealing single microprojections in the regulations of their contact approachment begin to play a more significant part.

A NEW DEVICE FOR TESTING MICROMOTORS AND ROTATING MICROSYSTEMS

A MATZNER, G ABRAHAM, P ARTEMCZUK, W BRENNER, F FRANEK, C HERZA

Institute of Precision Engineering - Department of Tribology, Technical University Vienna, Floragasse 7, 1040 Vienna, Austria

ABSTRACT

New manufacturing techniques offer completely new possibilities for miniaturization of mechanical components like motors, gearboxes etc.

Applications of microsystems can be found in sensors and actuators for example in medical equipment.

The parts within microsystems have dimensions of some 100 microns or less. As soon as there are relative motions within the system tribologic problems occur because the influence of friction and wear is much higher than in systems of „usual" size.

For describing the behavour of motors usually the torque and rotational speed are of interest. Micromotors offer only low torques at sometimes very high rotational speeds. Not much load is allowed to be applied to such motors and for this reason usual methods of measuring the motor torque can not be used in the case of micro-motors (1). Measuring the electrical supply current of the motor is a method already used for small motors but in case of micromotors it fails because the low efficiency of the motors causes large errors.

A new device for testing micromotors using new methods for application of load and measurement of torque and rotating speed has been developped at the Institute of Precision Engineering, Technical University Vienna.

The application of load is realized by mounting a very lightweight specially shaped „brake body" onto the shaft. Increasing the load is done by dipping the rotating „brake body" into a liquid of defined viscosity.

The diving depth is controlled by a stepping motor drive allowing steps down to 2 microns.

The brake moment is depending on the diving depth of the „brake body". The „brake vessel containing the liquid is mounted on a torsion element allowing the vessel to turn within a small angle depending on the torque. This angle is measured by detecting the position of a laser beam that is reflected by a mirror mounted on the „brake vessel". The measuring range ist set by chosing the dimensions of the torsion element.

The rotating speed ist also measured by an optoelectronic equipment. The „brake body" ist equipped with reflective sides allowing to reflect a laser beam onto an array of optoelectronic sensors. During the rotation of the motorshaft the reflected laser beam hits the sensors in a defined sequence allowing to calculate the rotational speed as well as irregularities of the rotation.

The device offers also the possibility of testing other rotating micro-systems, for example gearboxes and bearing with micromechanical dimensions.

S c h e m a t i c

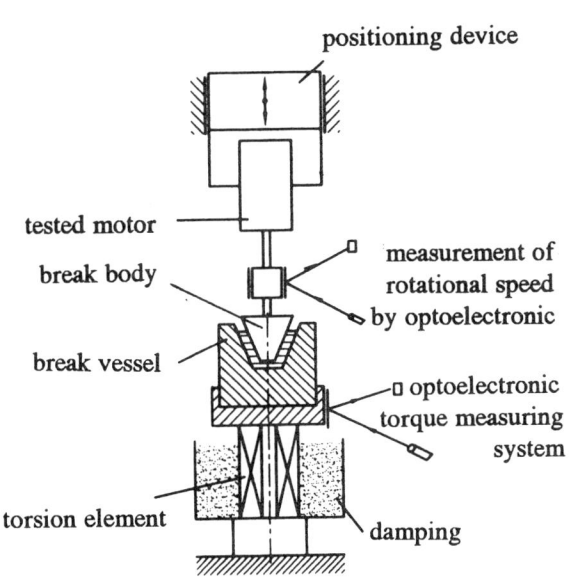

Technical data - measuring range
rotational speed   0 bis 400 000 min-1
torque   0,5 nNm bis 10 mNm
(1nNm = 0,000 000 001 Nm)

REFERENCES

(1) FRANEK, F.; MATZNER, A.: Messung kleiner Motordrehmomente. Feinwerktechnik und Messtechnik 96 (1988) 12, 549-552pp.

(2) BRENNER, W.; HADDAD, G.; POPOVIC, G.; VUJANIC, A.; ABRAHAM, G.; MATZNER, A.: The measurement of minimotors and micromotors torque characteristic. Proceedings of the 21st International Conference on Microelectronics, Nis, Serbia (1997)

REPULSIVE AND ATTRACTIVE VAN DER WAALS FORCES: ADHESION AND FRICTION MEASUREMENTS

A MEURK and L BERGSTRÖM
Institute for Surface Chemistry, PO Box 5607, SE-114 86 Stockholm, Sweden

ABSTRACT

Among the many contributions to the interaction between surfaces there is one type of interaction which is always present, the van der Waals (vdW) interaction. Symmetric material combinations always display an attractive vdW force regardless of the intervening medium. Asymmetric combinations can show either an attractive or repulsive vdW force depending on the dielectric properties of the materials and medium. The magnitude and sign of the interaction is determined by the Hamaker constant for a fixed geometry.

We have designed the systems Si_3N_4-Diiodomethane-Si_3N_4 and Si_3N_4-Diiodomethane-SiO_2, from calculation of Hamaker constants where attractive and repulsive vdW forces, respectively, are expected (1). These inorganic material combinations in diiodomethane show an interaction where the vdW force is dominating and other surface forces kept to a minimum.

The resulting interactions have been probed by performing force measurements with an atomic force microscope (AFM). Both the sign and magnitude of the force-distance curves corresponded well with theoretical calculations. The symmetric system showed an attraction on approach and a huge adhesion on retraction, whereas the asymmetric system yielded repulsion on both approach and retraction with no sign of adhesion (Fig. 1).

Although the interaction results were expected, it was surprising to notice the absence of adhesion even when a high force was applied in the repulsive system. Two surfaces of relatively high surface energies are expected to adhere when separated. However, it may be argued that the surfaces never really experience direct contact in AFM force measurements when strong repulsive forces dominate. According to elastic contact mechanics models (2) a negative load is required to separate two adhesive surfaces in contact, thus at zero applied load a finite friction force is present at contact.

We have measured friction forces as a function of contact force between Si_3N_4 surfaces in air and diiodomethane, and Si_3N_4 and SiO_2 in diiodomethane. As expected, measurements in air resulted in a strong adhesion force and a frictional component at zero applied load. The friction force increased with load and could be described by JKR theory, indicating single-asperity contact at low applied loads (Fig. 2). This behaviour is due to the presence of a thin film between the materials which fills up pores and cracks thus creating smooth surfaces.

In diiodomethane the friction force was no longer proportional to the area of contact which could be seen in the linear dependence on load. The absence of adhesion made a JKR fit difficult to assess and the contact was entirely assumed to be of multiple-asperity origin.

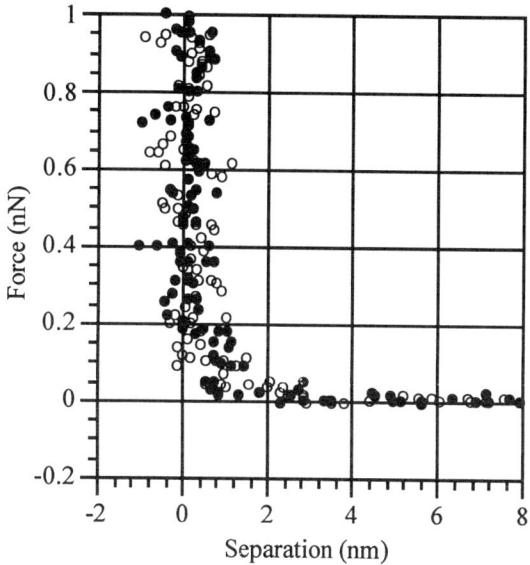

Fig. 1: vdW-force for Si_3N_4 - diiodomethane - SiO_2

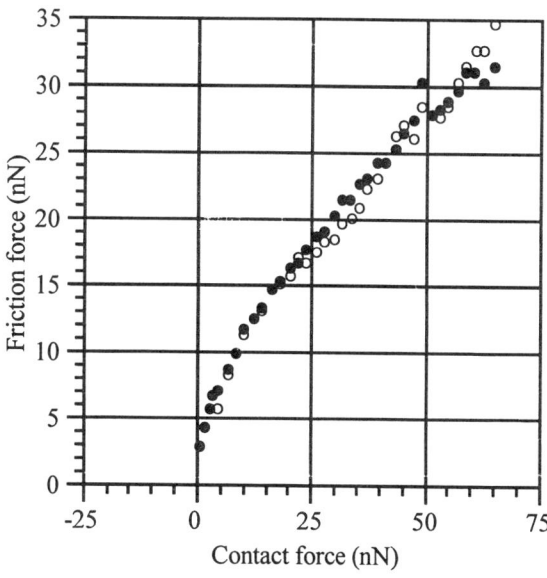

Fig. 2: Friction vs. load for Si_3N_4-surfaces in air

REFERENCES

(1) A. Meurk, P. F. Luckham L. Bergström, Langmuir, in press.
(2) K. L. Johnson, K. Kendall, A. D. Roberts, Proc. R. Soc. Lond. A., Vol. 324, 301-313pp, 1971.

NANOMETER DEEP WEAR AND MECHANICAL PROCESSING OF MUSCOVITE MICA BY ATOMIC FORCE MICROSCOPY

S. MIYAKE and T. OTAKE
Department of System Engineering, Nippon Institute of Technology
4-1 Gakuendai, Miyashiro-machi, Saitama 345 Japan

ABSTRACT

The atomic force microscope (AFM) is a powerful tool that was developed by Binnig et al (1) to investigate the topography of a sample surface on an atomic scale. Several attempts have also been made to use the AFM to examine the local modification of surfaces(2).

In my previous study, I evaluated the atomic wear of layered crystal structure material based on topographic changes caused in the atomic surface by sliding. Layered crystal structure materials, such as muscovite mica, have basal planes with weak interaction which makes their surface to cleave and to observe on an atomic scale. Wear to a depth of about one nm in the mica surface corresponds to the distance from the surface of one cleavage plane to the surface of the cleavage plane immediately below it(3). The processing of the muscovite mica was therefore performed by silicon tip sliding, to approximately this depth(4).

In this report, images of mica using atomic-scale topography and atomic-scale friction force microscopy (FFM) were observed simultaneously. Then, the atomic wear phenomenon was evaluated from the topographic changes caused in the atomic surface by sliding. Micromechanical processing was performed by applying these wear mechanisms.

Figure 1 shows the dependence of the profile and depth of wear grooves on load after two sliding cycles. A wear groove with a width of nearly 100 nm was formed. The depth of the wear grooves changed discretely with the amount of load. Wear began with a 500 nN load. The depth of this groove was nearly 0.8 nm. With 500, 1000 and 1500 nN loads, grooves onenmdeep were formed. The wear depths were mainly 0.8 and 1.0 nm, which corresponded to the distance from the top of the SiO4 layer to the cleavage plane of potassium below it, and from the top surface of SiO4 to the top surface of the next SiO4 layer below it, respectively. The interface between K-SiO4 and SiO4-K was weakly bonded, therefore, removed depths of 0.8 and 1.0 nm were predominant. Heavier loads caused damage to layers deeper than one nm: with 2500 and 3000 nN loads, the wear depths increased to 2 nm; with 3500 and 4000 nN loads, about 3 nm-deep grooves were formed. From these results, it was concluded that one-nm-deep mechanical processing from the surface of one cleavage plane to the next can be achieved by sliding the tip several times with a slightly heavier load than the critical load at which the wear begins.

With a load slightly heavier than the critical load, repeated sliding of the tip generated a one-nm-deep groove, and removed residual potassium from the surface. For example, a groove with four steps, each one nm deep, was processed by step-by-step mechanical sliding. The alphabetic and Chinese characters (2000 nm X 600 nmX 1 nm deep) as shown in Fig. 2 were formed by mechanical scanning computer numerical control.

REFERENCES

(1) G.Binnig, C.F.Quate and C.Gerber: Phys. Rev. Lett. Vol.56, 1986, 930-933pp.
(2) E.J. van Loenen, D.Dijkkamp, A.J.Hoeven, J.M. Lenssinck and J.Dieleman: Appl. Phys. Lett. Vol.55, No.25, 1989, 1312-1313pp.
(3) S.Miyake: Appl. Phys. Lett., Vol.65, No.8, 1994, 980-983pp.
(4) S.Miyake: Appl. Phys. Lett., Vol.67, No.13, 1995, 2925-2927pp.

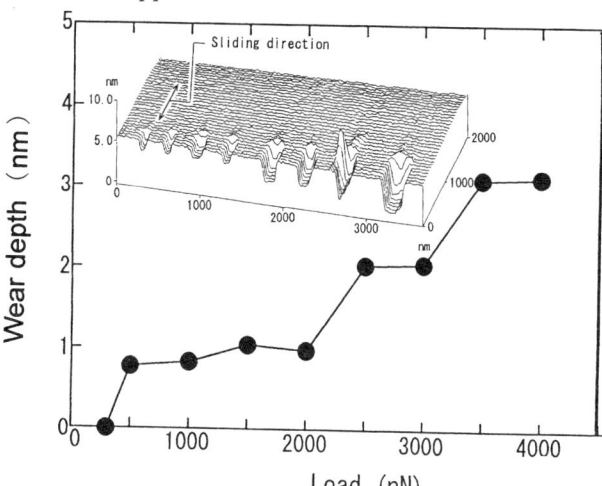

Figure 1 Wear depth dependence on load.

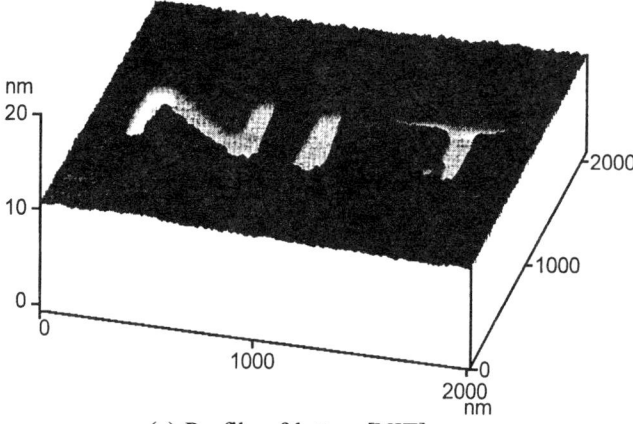

(a) Profile of letters [NIT]

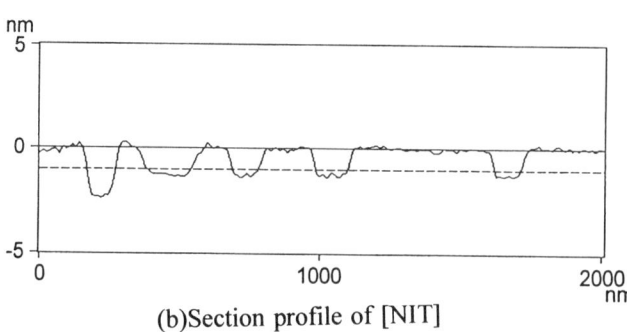

(b) Section profile of [NIT]

Figure 2 Processed letters [NIT]

ASPECTS OF DAMAGES EVOLUTION IN THE SUPERFICIAL LAYERS IN WEAR PROCESS

L. PALAGHIAN
Department of Machine Design, Dunărea de Jos University, Domnească Street 47,
6200 Galaţi, ROMANIA

ABSTRACT

This paper presents compared results obtained on studying the damaging phenomena in the superficial layers in severe wear and, separately, in fatigue process. The investigations were done for the superficial layer of samples bearing plane fatigue bending and of a tribomodel roll - half-bearing.

The damage was estimated by the help of the parameters for fine structure: the dislocation density, the internal micro-stresses of second order. Their evolution and accumulation in time produce the initiation and development of cracks and other kind damages (1).

The experimental results pointed out that for both types of loading (severe wear and fatigue bending) the damages of the superficial layer have two levels of evolution: a) the microstructure level: the modification are induced by dislocation mobility; b) the macro-structure level characterized by initiation of the micro-cracks in the layer being adjacent to the superficial one and followed by the development of fatigue cracks or wear micro-probes, respectively.

The authors' researches pointed out that for the level a) the dislocation density (Fig.1) and the micro-stresses of second order (Fig.2) have a similar evolution for both types of damaging processes. This process has several stages with modifications of different frequencies, intensities and time intervals. These structural modifications have as effects plastic micro-deformations and local hardening processes at

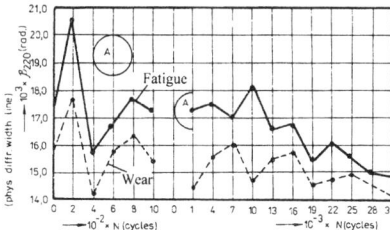

Fig. 1: The evolution of the density dislocation

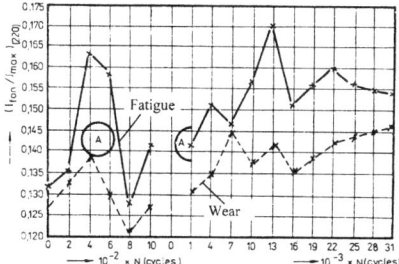

Fig. 2: The evolution of the internal micro-stresses of second order

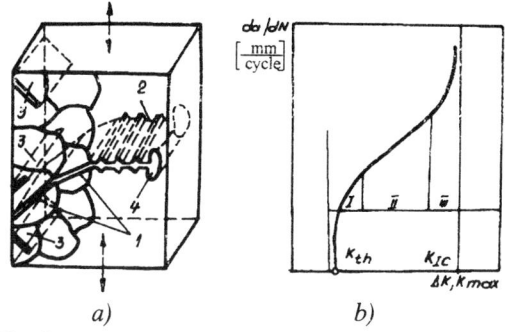

Fig. 3: The stages of fatigue damaging process

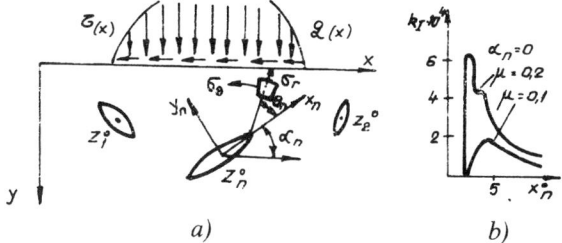

Fig. 4: The evolution of development rate for the crack

the level of crystalline grains.

The level b) of damaging is characterized by the growth of the main fatigue crack (Fig. 3, a) and, for the severe wear (Fig. 4, a) the micro-pores meet and unit producing a specific wear profile. Both damaging processes may be pointed out by the help of the stress intensity coefficient defined in the mechanics of cracks. As for fatigue, the evolution of fatigue cracks may be described with a relation type Paris $da/dN = f(\Delta k)$ (Fig.3, b) (2). As for wear process, will be characterized by the coefficients of stress intensity specific of types I and II for fracture. These will determine a complex coefficient for the stress intensity (fig. 4, b) $k_I + k_{II}$ (3).

The conclusion of this paper underlines that at the microstructure level, the damages caused by wearing processes have a fatigue character and at the macrostructure level, the deterioration may be described by the help of relationships resembling to those used for fatigue.

REFERENCES

(1) M. Klesnil, P. Lukas, Journal Iron Steel Institute, vol. 105, July, 1967

(2) P. Paris, I. Erdogan, Journal of Basic Engineering, ser.D 85, 1963

(3) I. Kudish, Friction and Wear, vol. 7, 1986.

FRICTIONALLY HEATED MICROCONTACTS IN HYDRAULIC COMPONENTS

J. PEZDIRNIK and J. VIŽINTIN
Faculty of Mechanical Engineering, University of Ljubljana, Aškerčeva 6, 1000 Ljubljana, Slovenia

F. VODOPIVEC
Institute of Materials and Technologies, Lepi pot 11, 1000 Ljubljana, Slovenia

ABSTRACT

In hydraulic components two elements are frequently in oscillatory sliding contact. When a transverse force occurs it is transmitted from the oscillating piston to the stationary bore wall. The surfaces of contacting elements are not ideally shaped. Due to incorrect cylindrical shapes of the spool and/or its bore both sliding surfaces are not completely in contact. The transverse force is therefore transmitted from the spool to the bore wall only over a few microcontact areas resulting in the asperities which protrude from one or both surfaces.

We are dealing with plastic asperity deformation and we calculate the contact radius using Bowden - Tabor's equation stated also at Czichos (1). The heat flow is treated to be conducted into material of both contacting elements through differentially thin layers with their areas changing with depth. The contact area of oscillating element's asperity is considered to be of circular shape, according to Archard's theory from 1959 (2). Further the sliding surface of the stationary element **A** is considered as an oblong sliding surface taking into account the stroke **s** of the oscillating element **B** (Fig. 1).

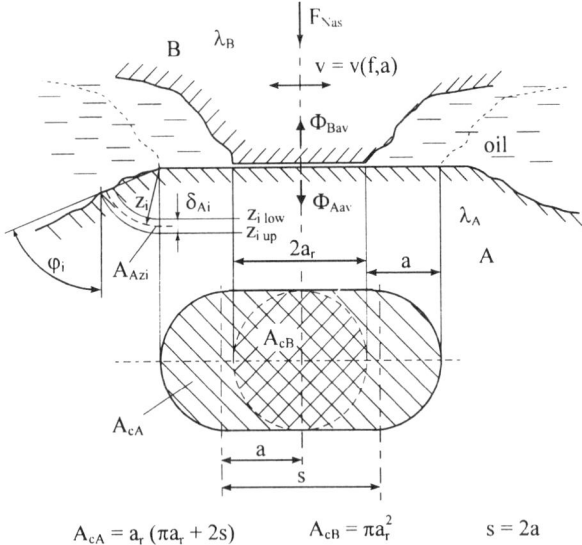

Fig. 1: Circular and oblong sliding areas of the single contact spot at an oscillating asperity.

At the contacting asperities the frictional work is transformed into heat. Considering further that surface temperature at microcontact of both elements **A** and **B** is the same, the heat flow is divided between both elements according to the ratio influenced upon thermal conductivities of the contacting materials and the areas of the sliding surfaces of this contact. Microcontact temperature rise and time of temperature rise depend on heat flow Φ, contact radius a_r, oscillation frequency f and amplitude a, angle φ of the contour's inclination of the asperity and material density ρ, thermal conductivity λ and specific heat c.

Paper presents shortly a way how to calculate microcontact interface and subsurface temperatures of an asperity which is exposed to oscillatory sliding movement and loaded with a normal force F_{Nas}. Further this paper gives a simple mathematical model for calculation the time of microcontact temperature rise from the starting bulk temperature up to the maximum temperature.

Laboratory experiments were carried out with pairs of specimens with concave/convex sliding surfaces. The oscillating specimen was a small piston and the stationary one was a half bushing. We have tested at frequences from 10 Hz up to 500 Hz and amplitudes of linear oscillations from 5 μm up to 300 μm. The oscillating piston was loaded with various normal forces from 5 N to 400 N. The bulk temperature of both specimens was 50 °C. Sliding surfaces at most experiments were lubricated with hydraulic oil according to ISO VG 46. When enough great force was transmitted to the bushing surface only over some microspots of deformed asperities, the damage occured in these surface regions. The analysis of the material at these damaged spots showed microstructural changes to the depths up to some ten micrometers under microcontact interface.

Due to the operating limits of the laboratory testing machine at higher frequencies lower amplitudes were used. Regarding furthermore the properties of the materials used at the testings, Peclet numbers were in the range up to about 1.0; rarely they overcame the value 1.5 and only exceptionally they reached the values up to about 2.5.

REFERENCES

(1) H. Czichos, Tribology, Elsevier Scientific Publishing Company, Vol. 1, 1978.
(2) J. F. Archard, R. A. Rowntree, Wear, 128 (1988), 1-17 pp.

COMPUTER SIMULATION OF ATOMISTIC MECHANISMS OF ADHESIVE FRICTION, WEAR and SEIZURE

VLADIMIR POKROPIVNY, VALERY SKOROKHOD and ALEKSY POKROPIVNY
Institute for Problems of Materials Science, National Academy of Science, 252680, Kiev - 142, Ukraine

ABSTRACT

In the last decade a development of atomic-force microscopy has given impetus in the study of many interface phenomena including the old and challenging problem of adhesive friction and wear at nanoscopic level. This results to the development of a frictional-force microscopy, nanotribology and nanofabrication technology. To simulate the interface phenomena and the adhesive friction on a basis of molecular-dynamic method we elaborated the program SIDEM/AFMF. This program involved three main routines, namely, the construction of the tip-surface interface, relaxation part and friction scenario. Atomistic mechanisms of adhesive friction, wear and seizure have been computer simulated on atomic scale in Fe-Fe and W-Fe (fig.1) tribo-couples (1)(2)(3).

The set of interface characteristics has been calculated during stick-slip friction, namely, an internal and external friction force, a viscosity, a total displacements of all atoms, a displacements of mass center, an adhesive force, an adhesive energy, a number of adhesive bonds, a frequency spectrum et al. Structure transformations in conjunction with variation of friction force and other friction numerical characteristics were studied and described in details. It was shown that a quasi-chaotic dynamics of friction force and viscosity in process of adhesive friction and wear appear to be strictly deterministic. The small peaks (spectral noise) are caused by deformation of adhesion bonds. The creating and breaking of this bonds is responcible for the large peaks. The huge peaks are caused by atomic ordering of friction contact. This peaks are illustrated in fig.2. The mechanism of adhesive friction was shown to consist of elementary acts, such as a creation, compression, stretch and break of a single cohesive interfacial interatomic bond. In constant height mode the mechanism of the boundary adhesive wear such as a stick of a single asperity atom on a surface was found to involve a friction epitaxy, contact boundary migration, viscous dragging and asperity amorphisation, finished by a dropping of W-atom and sticking it on a surface. The loaded sharp asperity penetrates into substrate. In constant force mode the seizure mechanism was examined, exhibiting a plastic deformation of slider under great load. Also the isoforce imaging of surface roughness like in an atomic-force microscope was obtained under small load. Force criteria for the wear, seizure and fracture of asperity during friction have been suggested.

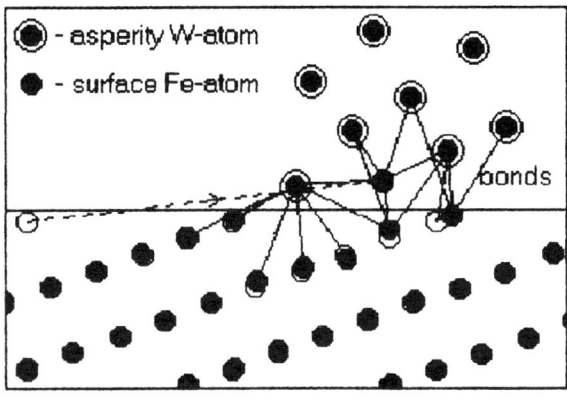

Fig.1: Atomic-sharp W-asperity over (114)-α-Fe-surface in the elementary act of adhesive wear

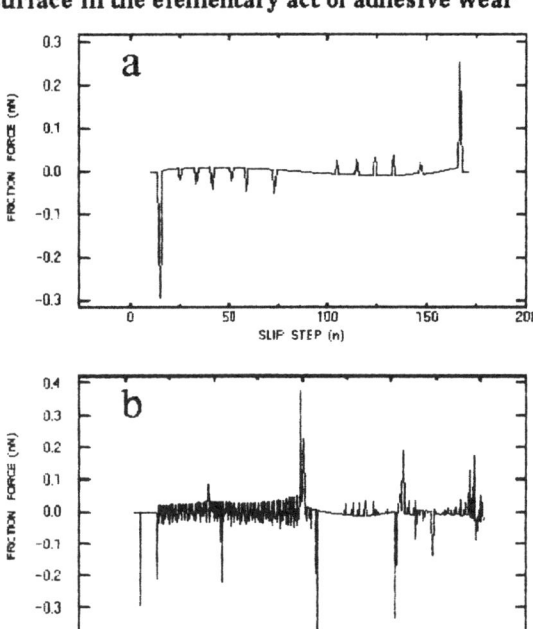

Fig.2: Friction force at n^{th} slip step in the elementary acts of adhesive friction (a) and wear (b)

REFERENCES

(1) A.Pokropivny, V.Pokropivny, V.Skorokhod, Sov. Tech. Phys. Lett., Vol. 22, No.2, p. 1-7, 1996.
(2) V.Pokropivny, V.Skorokhod, Sov. Tech. Phys. Lett.,Vol. 22, No.9, p. 70-77, 1996.
(3) V.Pokropivny, V.Skorokhod, A.Pokropivny, Friction and Wear, No.5, 1996; Materials Letters, No.4, 1997.

Submitted to *Tribologia*

CONTACT OF ROUGH CURVED SURFACES UNDER NORMAL LOAD

P. K. RAJENDRAKUMAR and S. K. BISWAS
Department of Mechanical Engineering
Indian Institute of Science
Bangalore-560 012, India.

ABSTRACT

Wear and fatigue characteristics of engineering surfaces are dependent on the contact pressures and the resulting subsurface stress distributions in the bodies which are in contact. Real surfaces are rough and hence the contact between two real surfaces can be considered in general as a contact between two curved rough bodies. The present paper analyses the above problem in its simplest form as a plane strain problem of a rigid cylinder of radius r, indenting an elastic half space bounded by a plane of uniformly spaced cylindrical asperities. This is now a Class-A punch indentation problem (1) with incomplete contact at the base of the punch profiles so that the contact pressures are bounded (and equal to zero) at the ends of the regions of contact. Using the method of complex variables, the above problem has been reformulated in terms of a complex potential $\Phi(z)$, where $z = x+iy$; $i=\sqrt{-1}$, to yield $2N$ number of non-linear simultaneous equations in $2N$ variables (which are the end points of the regions of contact of N punches) which can be solved by Newton's iterative scheme. The contact pressure distribution on each contact region can now be obtained easily (see (2) for details).

Known the contact pressure distribution, the subsurface stress components and the corresponding equivalent (principal) shear stresses at different points inside the elastic half space are evaluated (3).

The present analysis based on complex variable method shows that the contact except at the central axis, occurs not at the asperity summits, but at the inward flank of the asperities. As the load increases, the asperities are seen to rotate in conformity with the global displacement of the flat surface, and the contact positions move back towards the original summit position. The asperity contacts are dynamic and non-Hertzian in nature due to the mutual interaction of tractions on them.

The subsurface stress analysis shows that the equivalent shear stresses at the subsurface show a local maximum at the near-surface subsurface of every contacting asperity and a global maximum at the subsurface. The comparative values of these local and global maxima are seen to depend on the asperity tip radius ρ, asperity pitch λ and the applied load P. Fig.1 shows a threshold surface in these variables such that if the contact configuration is above this surface (Response I) the global maximum is larger than the maximum of the local maxima and vice versa (Response II). This result implies that for certain geometries the asperities can persist to be elastic even when the subsurface starts to yield.

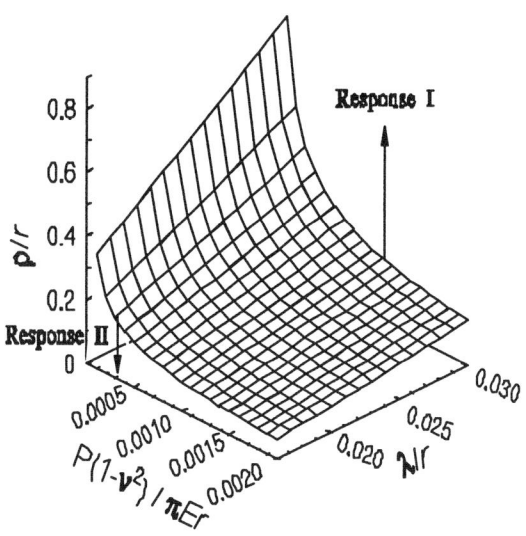

Fig. 1: Variation of threshold contact configuration.

REFERENCES

(1) Muskhelishvili, N. I., Some Basic Problems of the Mathematical Theory of Elasticity, Groningen: Noordhoff, 1953.

(2) Rajendrakumar, P. K. and Biswas, S. K., Mechanics Research Communications, Vol. 23, 1996, 367-380pp.

(3) Johnson, K. L., Contact Mechanics, Cambridge University Press, 1985.

NANOSCALE SCRATCHING TO CHARACTERISE MAGNETIC HARD DISC OVERCOATS

Nick Randall

CSEM Instruments, Jaquet-Droz 1, CH-2007 Neuchâtel, Switzerland

ABSTRACT

With the continuing search for higher and higher recording densities of magnetic hard discs and thus greater recording capacities, the typical thickness of such a disc's surface carbon coating has had to be reduced accordingly down to the 10-20nm range. The characterisation of such thin films has led to much activity in the fields of nanoindentation, nanoscale scratching and nanowear so that the mechanical properties of a film at this scale can be accurately measured independently of the substrate material.

The Scanning Force Microscope (SFM) has already proved its worth in this domain for evaluating wear properties, lubrication effects and elasticity/adhesion to name but a few. By using micromachined silicon cantilevers that have been coated with a layer of diamond, it is possible to perform nanoscale scratch tests on even the hardest of thin films, such as diamond-like carbon (DLC) which is commonly used in the hard disc industry. The same coated cantilever tips can subsequently be used to image the scratched area and thus gain vital information concerning the response of the sample material. By monitoring the lateral force signal of the scratching tip, in addition to the normal force, it is possible to obtain frictional data related to the wear of the thin film. The SFM thus provides a very versatile tool for the study of tribology at a nanometric scale and, especially in the magnetic recording industry, has great potential for further development. In this paper we demonstrate the use of a commercial SFM instrument for a comparative study of several different hard disc overcoats in terms of scratch resistance, hardness and frictional properties.

SLIDING FRICTION BEHAVIOR OF Si LUBRICATED BY H_2S GAS

I L SINGER
Code 6176, Naval Research Laboratory, Washington DC, USA.
T LeMOGNE, Ch DONNET AND J M MARTIN
Laboratoire de Tribologie et Dynamique des Systèmes Ecole Centrale de Lyon, Ecully, France.

ABSTRACT

Reciprocating sliding tests were performed on Si(100) surfaces exposed to H_2S gas (1 x 10^{-2} to 1 Torr) in a multi-analytical, UHV tribometer. Tests were conducted at a fixed gas pressure with a spherical pin (radius 1.6 mm) of either SiC or Al_2O_3. Each test was begun with a 10-20 cycle run at a fixed load (from 0.5 to 2N) and fixed speed (from 0.2 to 4 mm/sec), then subsequent runs were made (without lifting pin from track) at different speeds and at different loads.

Friction coefficients with both pins increased dramatically with increasing speed. With SiC, the friction coefficient at lowest speed (v=0.2 mm/sec) consistently reached the superlow value of µ=0.001 to 0.002 and, at highest speeds (>2 mm/sec), leveled off at values from 0.1 to 0.2. Al_2O_3 pins, however, gave friction coefficients at all speeds an order of magnitude larger than those of SiC. With Al_2O_3, the friction coefficient at lowest speed was µ=0.01 to 0.04 and, at highest speeds, reached values close to 1. Friction coefficients displayed some pressure dependence at highest speeds: the friction coefficients at 1 Torr were somewhat lower than those at 1 x 10^{-2} Torr. No load dependence of the friction coefficient could be seen.

In situ Auger analysis of wear tracks indicated Si sulfide and, possibly, sulfur formed inside the Si track. A Si sulfide layer was also found on fine (< 1 micron) debris particles surrounding the wear track and on the SiC pin. *Ex situ* analytical studies (EDX, TEM and Auger) will also be reported. Discussion will focus on the relationship between tribochemical films and friction (1)(2), and the anomalous velocity-dependent friction behavior.

REFERENCES

(1) I.L. Singer, T. LeMogne, Ch. Donnet, and J.M. Martin, J. Vac. Sci. Technol., Vol. 14, page 38, 1996.
(2) I.L. Singer, T. LeMogne, Ch. Donnet and J.M. Martin, Tribol. Trans. Vol. 39, page 950, 1996.

CORRELATION BETWEEN MAXIMAL ROUGHNESS HEIGHT AND THE MEAN ARITHMETIC DEVIATION OF THE PROFILE FROM THE MEAN LINE OF MACHINED SURFACE

S St SEKULIC
Institute of Industrial System Engineering, Faculty of Technical Science
University of Novi Sad, 21000 Novi Sad, D. Obradovic Sq 6, YU

ABSTRACT

The maximum roughness height and the mean arthmetic deviation of profile from the mean line, for statistically valid sample, in exponential

$$R_{max} = B \ R_a^a$$

and linear form

$$R_{max} = a \ R_a + b$$

have been requested, in this papar.

On the basis of experimental dates given by turning, milling and grinding data processing was provided (from large sample, the 150, for each of the individual machining process and for collective dates, independent of kind of the machining processes, for very large representative sample N=450>50).

For the kind of machining processes, the next mathematical models in liner form for:

- turning

$$R_{max} = 5{,}22 \ R_a + 2{,}02 \ ; \ (r = 0{,}88)$$

- milling

$$R_{max} = 9{,}24 \ R_a - 5{,}44 \ ; \ (r = 0{,}87)$$

- grinding

$$R_{max} = 6{,}84 \ R_a + 0{,}92 \ ; \ (r = 0{,}96)$$

and for collective dates, independent of kind of the machining processes

$$R_{max} = 5{,}58 \ R_a + 1{,}99 \ ; \ (r = 0{,}88)$$

and for corresponding exponential models for:

- turning

$$R_{max} = 5{,}99 \ R_a^{0{,}95} \ ; \ (r = 0{,}94)$$

- milling

$$R_{max} = 6{,}03 \ R_a^{1{,}11} \ ; \ (r = 0{,}83)$$

- grinding

$$R_{max} = 7{,}92 \ R_a^{0{,}85} \ ; \ (r = 0{,}98)$$

and for collective dates

$$R_{max} = 7{,}62 \ R_a^{0{,}95} \ ; \ (r = 0{,}97)$$

are given.

After experimental data processing, convenient mathematical models are evaluated by the correlation coefficient value and on the basis of the magnitude of exponents in the exponentional relationship (if the exponent in exponential relationship is near one, the linear relationship between variables exist).

The magnitude of exponent, in exponentional relationship is a=0,85 point deviation from linear relationship. Since in liner relationship, the correlation coefficient is hight, the both relationships can be used.

Referring to the above montioned, we conclude:

- for mathematical models of correlation which connected the maximal roughness height and the arithmetic deviation of the profile from the mean line of the machined surface, the exponential and linear relationships can be used and

- in proposed mathematical models, there are a very strong correlations between observed parameters.

REFERENCES

(1) S. Sekulić, P. Kovač, Tribology in Industry, Vol.XV, No.4, 1993 (In Serbian), 38p.
(2) S. Sekulić, P. Kovač, Proceedings of the INTERTRIBO 93, Bratislava, 87-92pp, 1993.
(3) S. Sekulić, P. Kovač, Journal of the Balcan Tribological Assotiation,Vol.2, No.1,1996, 24-31pp.
(4) S.Sekulić, P.Kovač, Proceedingc of the INTERTRIBO 90, High - Tatras, 1990.
(5) S. Sekulić, P.Kovač, Proceedings of the INTERGRIND 88, Budapest, 1988.
(6) S. Sekulic, P. Kovac, Publication of the School of Engineering sciences, Novi Sad, Vol. 19 & 12, Novi Sad, 1988 & 1989.

Submitted to *Surface Engineering*

MOTIFS AND SPECTRAL CHARACTERISATION OF ANISOTROPIC MORPHOLOGY OF ENGINEERED SURFACES. INCIDENCE IN TRIBOLOGY

H. ZAHOUANI, R. VARGIOLU, M. DURSAPT, T.G MATHIA

Laboratoire de Tribology et Dynamique des Systèmes, UMR CNRS 5513.
Ecole Centrale de Lyon, 36 Avenue Guy de Collongue, 69131 Ecully cedex France
& Ecole Nationale d'Ingénieurs de St Etienne 42000 France.
Institut Européen de Tribologie

ABSTRACT

The fluid flow phenomena between contacting rough surfaces, are interesting for the lubrification mechanisms, and the sealing problems. For example the metal-to-metl sealing parts are sometimes used as separable fluid connectors. It is well known that the surface roughness of contacting parts governs the leakage across the metal seals, essentially the directional properties of roughness plays an important role on the leakage rates [1].

The roughness of a surface prepared by mechanical machining usually has a directional property, which can be represented as a function depending on three fundamental parameters:

$$z(x,y) = f(\rho, \lambda_s, \theta) \qquad (1)$$

where ρ represents the height of roughness, λ_s the wavelength morphology and θ the roughness direction of the different components.

The relationship between surface topography and its functional performance has been partially investigated using two dimensional surface analysis. But even this limited analysis is not yet fully understood. However, it is becoming increasingly obvious that full understanding of the connection between surface topography and functional performance can only be achieved if a three-dimensional approach is used.

In this paper we show a new development in the three-dimensional characterisation of rough surfaces. The basic idea is to consider roughness as a combination of components defined by the roughness amplitude, wave length, local and overall direction of various components.

The directional parameter is used both to identify the anisotropic components by an appropriate decomposition of the topography into local motifs, with a representation similar to the two-dimensional Fourier transform, where the direction represents the topographic phase in the form of a morphological rose [2].

The morphological rose is generated by computing the density of motifs oriented in a direction θ, identified with respect to the x axis in the original sample. The geometrical form of the morphological rose represents the local and global motif anisotropy and allows quantification of the phase of surface topography. The example in figure (1) shows the morphological rose of random and oriented texture.

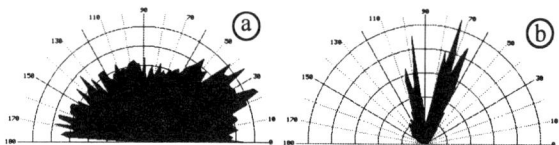

FIG. 1 Morphological rose of random (a) and oriented texture (b).

When the local heights of each motif ρ are known, the morphological rose versus the motifs level can be represented:

$$R_M(\theta) = f(\Delta\rho) \qquad (2)$$

This original representation of the anisotropy versus the level of motifs family can be used to study the volumetric aspect of anisotropic surface topography and as a multi-scale analysis of the roughness. With this new approach we can take further steps in the contribution of directional properties of rough surfaces in the leakage rate, lubrication mechanisms, friction and wear.

[1] N. PATIR, H.S. CHENG, "An average flow model for determining effects of three-dimensional roughness on partial hydrodynamic lubrication transaction of the ASME vol. 100, 1978, pp 12-18.

[2] H. ZAHOUANI, V. JARDRET, T.G. MATHIA, "Mophological characterisation of rough material". Surface modification technologies VIII, Edited by T.S Sudarshan and M. Jeandin. The Institute of Materials 1995, pp 135-147.

THREE-DIMENSIONAL FUNCTIONAL ANALYSIS OF SURFACE MORPHOLOGY

E MAINSAH
IBM UK Ltd., Hursley Park, Winchester, Hants., SO21 2JN
H ZAHOUANI
Laboratoire de Technologie des Surfaces, l'Ecole Centrale de Lyon, Ecully, CECEX, France

ABSTRACT

A number of numeric and visual tools have been developed over the years to assist the experienced engineer in the analysis of surfaces. This is because it is accepted that none of the existing techniques is a panacea but that they are mostly complementary rather than competitive and can all play a role in the characterisation of engineered surfaces.
This Paper attempts to draw together different characterisation techniques including visual, statistical, fractal and functional. It demonstrates the usefulness of each of these techniques in the characterisation of a wide range of engineering and engineered surfaces.

INTRODUCTION

Three dimensional surface analysis is used increasingly in industry and has been the focus of intense research in the last few years. This has been encouraged partly by the increasing realisation that surfaces interact in 3-D and that the study of tribology would be limited if all The statistical technique is perhaps, the oldest and over the years a number of parameters have been developed although not all authors have welcomed this[1]. Recently, Stout et al.[2] have attempted to limit the proliferation of parameters by defining a limited set of 14 parameters for general use. This work was intended to form the basis of an international standard in 3-D topography and although the long term goal is some way away, there is evidence to suggest that many of the parameters are being used by both researchers and instrument manufacturers and indeed has won the support of respected experts within the topography community[3]. A list of the 14 parameters is shown on Figure 1; their detailed definitions can be found in [2]. The set includes statistical, functional and visual parameters.

The fractal technique was developed some years ago and was thought to be of only limited use in an industrial environment; it has been slowly gaining in popularity and has now been applied to a number of interesting real life problems, for example, bearing problems, the simulation of anisotropic surfaces and correlation studies between surface fractal dimensions and process parameters[4]. It seems clear from the literature, that the fractal technique, with its greater parameter stability - because of the scale-independent nature of the approach - offers scope for the control of processes. It is at the conceptual level that the approach is likely to meet the greatest resistance, as a number of technicians and quality control engineers still find the concepts hard to grasp.

An increasing trend is the use of artificial neural networks for process control. Recently, Mainsah and Ndumu[5] have investigated the use of back-propagation networks as well as adaptive resonant theory (ART2a) networks to classify and characterise engineering surfaces. They have successfully classified ground, honed and electro-discharge textured surfaces with greater than 90% consistency; in another work they were able to split a large number of honed surfaces from an automobile cylinder bore into one of two categories - worn and unworn. This is seen as the preliminary stages of an automatic data characterisation tool that can be used on the shop floor for surface quality control - it is expected to extend the work to cover data capture from a vision system. This should speed up the system considerably.

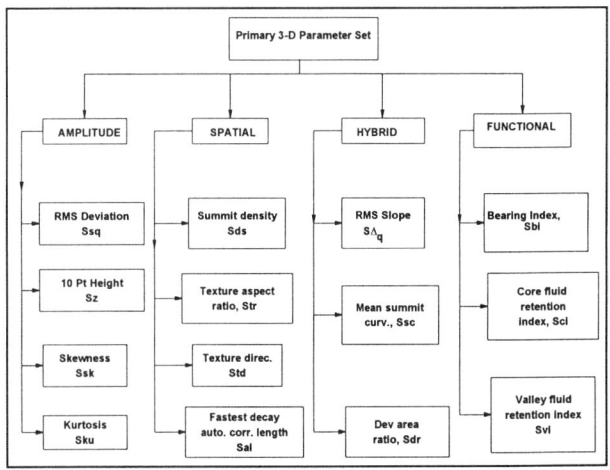

Fig. 1: Schematic showing Stout Parameter set.

REFERENCES

[1] Whitehouse, D J (1982); "Parameter rash - is there a cure?" *2 Int Conf on Met. Prop. Eng. Surfaces,* pp1 -16.
[2] Stout, K J et al. (1993), *Dev of Methods for the Characterisation of Roughness,* CEC, Brussels.
[3] Thomas, T R (1997); Trends in surface roughness", *7 Int Conf on Met.Prop. Eng. Surfaces.*
[4] see *7 Int Conf on Met. and Prop. Eng. Surfaces.*
[5] Mainsah E and Ndumu, D T, *ibid,* pp 191 - 198.

AN INVESTIGATION INTO THE POSSIBILITY OF THE EXPERIMENTAL DETERMINATION OF THE CONTACT DEFORMATION MODE

A. L. ZHARIN, E. I. FISHBEIN and N. A. SHIPITSA

Tribology Laboratory, Belarussian State Powder Metallurgy Concern, 41 Platonov str., Minsk, 220600, Belarus

ABSTACT

The possibility of the experimental determination of the contact deformation mode by the method based on recording the relative changes of the electron work function of the surface before and after loading is considered. The choice of this parameter is explained by the fact that the electron work function is highly sensitive to the surface defects density and mainly to the density of the dislocation spots on the surfaces. Experiments have been conducted to determine how deformation influences the work function for the cases of the simple uniaxial tension, contact loading and sliding contact loading.

Specimens were tested by tension with simultaneous registration of the work function (Fig. 1). Experiments showed that the work function increases slightly in the elastic region and decreases sharply when the yield point is reached. Based on the above results it is possible to determine the type of contact deformation for real surface layers which are as thick as the Debye screening length and approximately equal to the interatomic distance for the cases of metal and alloy surfaces.

The experiments with contact deformation were carried out with a cubic samples. After each subsequent step of loading the work function of the contact and lateral sample surfaces were measurement and its dependencies vs. load were plotted. In the case of the contact surface the inflection of the curve is appeared at the considerable lower loading than for the case of the lateral surface and this can be explained by the plastic processes beginning at the real contact spots.

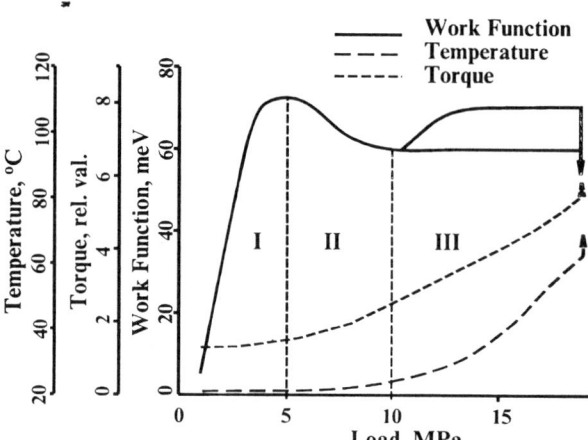

Fig. 10. Bronze rubbing surface electron work function, torque and surface temperature vs. normal load curves.

Studies of a wide range of materials have shown similar qualitative dependence of the rubbing surface electron work function on normal load (Fig. 2). In general there are three specific zones. First, there is an increase of the work function with load in zone I. In zone II the friction surface work function decreases. With further loading in zone III there is very little change until the beginning of scoring, which is detected by increases of the friction force and the bulk temperature of the sample. During scoring the value of the work function decreases sharply. It has been found that during long-run trials the wear rate is very low at loads corresponding to zones I and II. For zone III damaged spots on the surface and high wear rate are observed.

Additional investigations of specimens via independent methods as well as studies concerning changes of the work function during simple uniaxial deformation (Fig. 1) allow us to interpret the results obtained on the basis of dislocation interactions. We suggest that the first zone corresponds to mainly elastic and early stages of plastic deformation without a significant increase of dislocation concentration near the surface. In the second zone plastic deformation dominates, but these processes are not saturated yet from the view-point of the density of dislocations (in this zone the work function decreases with increasing load, as in Fig. 1). In the zone III plastic processes are also important, but the dislocation concentration and substructure saturate; i.e., there is a dynamic process involving generation and annihilation of dislocations and the creation of micropores and microcracks.

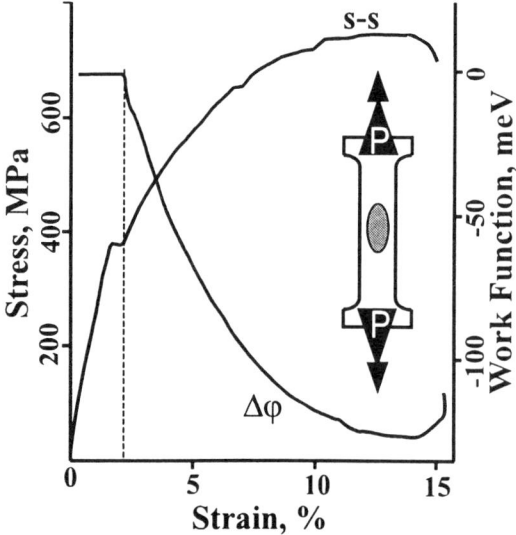

Fig. 5. Results of simultaneous measurements of the regular stress - strain diagram and the work function for medium carbon steel.

REFERENCES

(1) A.Zharin, Contact Potential Difference Technique And Its Application In Tribology, Minsk, Bestprint, 1996. (in Russian).

SURFACE ROUGHNESS ANALYSIS AFTER HIGH SPEED HYBRID BALL BEARING TESTING

J. AMPUERO, U. MÜLLER AND D. DELFOSSE
Swiss Federal Laboratories for Materials Testing and Research, Überlandstr. 129, 8600 Dübendorf, Switzerland
P. PAHUD, I. MALLABIABARRENA AND O. GOEPFERT
Swiss Federal Institute of Technology, Mechanics Department, Ecublens, 1015 Lausanne, Switzerland
H. BOVING
Centre Suisse d'Electronique et de Microtechnique SA, Jacquet-Droz 1, 2007 Neuchâtel, Switzerland

ABSTRACT

The potential advantages for the application of ceramic materials and ceramic coatings to rolling element bearings were demonstrated several years ago (1) (2). Since then, hybrid (ceramic balls and steel races) or pseudo-hybrid (TiC coated steel balls and steel races) ball bearings have become particularly attractive for high speed spindle-bearing systems.

Although most of the developments with ceramic components have been carried out on hot isostatic pressed silicon nitride, other materials show competitive characteristics. A serious evaluation of the different hybrid ball bearing materials for spindles requires carefully performed experiments on high speed testing machines together with meticulous characterization of surface topographies. This contribution shows possibilities and limitations of modern, high resolution, three dimensional assessment techniques for the surface roughness analysis as a methodology to evaluate tribological interactions between hybrid ball bearing components.

In this study, five different ball materials (100Cr6, 100Cr6+MOVIC® - MoS_2 -, AISI440C+TiC, Si_3N_4 and ZrO_2) have been investigated under controlled high speed test conditions (Table 1). The surface topography of balls and races was evaluated after testing using Scanning Electron Microscopy (SEM), Atomic Force Microscopy (AFM) and Laser Scanning Profilometry (LSP).

During operation the surface roughness of the bearing races increases as a consequence of ball-ring interaction. This interaction is strongly influenced by the ball material. The LSP roughness comparison in Figure 1 shows that the race surface alteration is highest when steel balls are used and very low when ceramic balls are used.

Bearing type	SNFA VEX25
Rotational speed	50'000 rpm
Test duration	50 hours
Preload	240 N
Lubricant	Grease Lubcon182

Table 1. Test conditions

Figure 2 shows an AFM ball roughness analysis using a variable measurement-field-size technique (3). The obtained roughness profile fully characterizes the actual topography of this very finely polished surface. The comparison between a new and a used ball surface shows clearly the scales of local ball-race interaction.

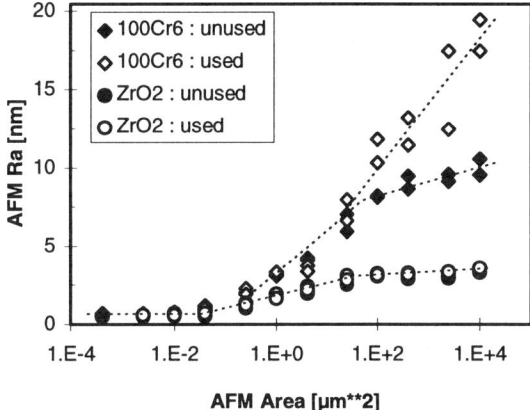

Figure 2. AFM roughness analysis of ball surfaces

The comparative evaluation of both ball and the raceway enables the wear process and the interactions occuring during hybrid (metal-ceramic) rolling contact testing to be better characterized and understood. The advantage of using ceramic balls can also be quantified in terms of raceway roughness alteration.

REFERENCES

(1) C. F Bersch, *"Overview of Ceramic Bearing Technology"*, in *Ceramics for High Performance Applications*, Vol. 2, J. J. Burke, A. E. Gorum and R. N. Katz eds., Brook Hill Publishing Co., Chesnut Hill, MA, 1978, p 397.

(2) H. J. Boving and H. E. Hintermann, in Thin Solid Films, 153, 1987, 253-266pp.

(3) U. Müller and J. Ampuero, *"Rasterkraft-Untersuchungen an Kugeln aus Kugellagern"*, in Oberflächentechnik '96, DGO/SGO/ÖGO-Meeting, Friedrichshafen, 1996.

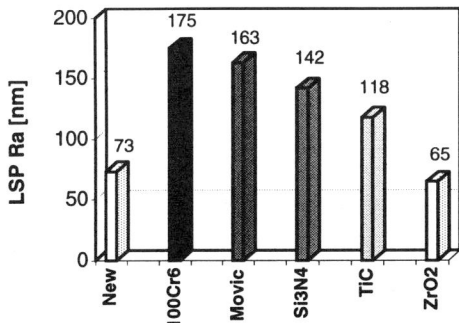

Figure 1. LSP roughness of outer ring raceways (measurement length = 3.5 mm)

THE PROFILED CYLINDRICAL ROLLER DESIGN IN CONSIDERATION OF GRINDING UNDERCUT EFFECT

X Y CHEN
Research Institute of Bearings, Shanghai University, 149 Yangchang Road, Shanghai, 200072, P. R. China
J J MA
Department of Mechanical Engineering, Zhejiang University, 20 Yugu Road, Hangzhou, 310027, P. R. China

ABSTRACT

Roller bearings are used to support heavy loads. If the generatrices of roller and raceways are absolutely straight and if the roller length does not coincide with the raceway width, then notwithstanding relatively light loading, edge effects will arise at the ends of the contact surfaces. So, the rollers of a cylindrical roller bearing are radiused slightly toward their both ends to relieve high edge stress concentration caused by their finite length and by misalignment.

Until now, the theoretical basis of the roller axial profile designing methods are mainly elastostatic analysis which assuming a roller contacts with a boundary plane of half infinite body, the crowning value is ideally designed for only one condition of loading (1). However, it is well known that most of raceways of the cylindrical roller bearing have one or two grinding undercuts on their ends which are needed for refining manufacture (2).

Based on the profiled roller geometrical equations and numerical calculation of three-dimensional elastic contact problems of finite length (3), a new numerical method for the crown drop evaluation considered the grinding undercut and the profiled roller symmetrization effects was developed in this paper. Which makes it possible to profile the cylindrical rollers axially until the longitudinal pressure distribution is neither any edge effects nor high pressure concentration at the center.

Fig. 1 shows the pressure distributions of a 3 μm crowning value, which is a radial drop evaluation at the point shown in Fig. 2, logarithmic profile roller contacted with two kinds structure bearing raceway. It is clearly appeared in Fig. 1b that under the same loading condition, the edge effects arises again at the raceway ends. Because of the refining manufacture of crowned raceways is more difficult than the crowned rollers, and it is not easily to machine the radial form designed by means of the raceway width along the whole length of the longer roller. That is the main reasons why this new method be put forward. According to this method, the edge effects at the raceway ends could be eliminated when the crowning value is 25 μm at the same measure point for 10020 N load, as shown in Fig. 2, or reduced

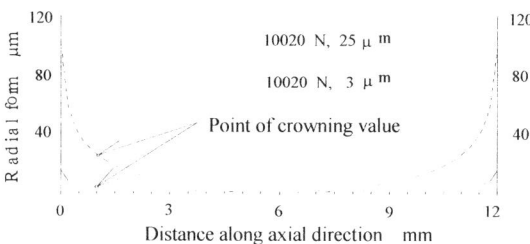

Fig. 2: Roller profiles for different contacts

the applied load. And some other key problems in profiled cylindrical roller design are,

1. Except edge effect should be avoided, contact stress concentration level at contact surface center should be also reduced.

2. The optimum crowning design depends on the maximum roller load and the bearings structure.

3. The design load of roller crowning value should take the maximum practical load at least and the low limit of roller tolerance should be taken as controlling the design crown in the maximum roller load.

This method could also be used for profiled taper rollers or needle rollers design which used in the taper roller bearings or needle roller bearings with the same kind of structure.

REFERENCES

(1) H. Reusner, Ball Bearing J., No. 230, pp.2, 1987.
(2) T. A. Harris, Rolling Bearing Analysis (3rd Edition), John Wiley & Sons, Inc. 1991.
(3) X.-Y. Chen and J.-J. Ma, Mocaxue Xuebao, Vol. 13, No. 1, pp.25, 1993. (in Chinese)

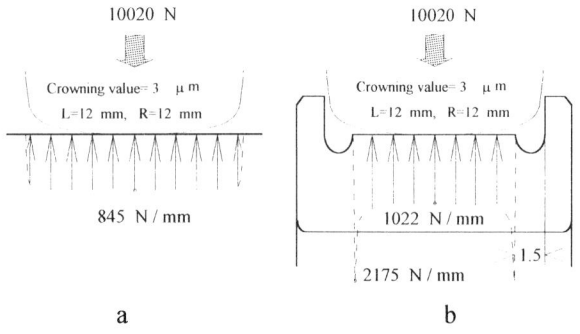

Fig. 1: Pressure distribution at different contacts

THE INFLUENCE OF INITIAL STRESSES UPON ROLLING CONTACT FATIGUE

G. FRUNZA and E. N. DIACONESCU

Mechanical Engineering Department, University of Suceava, 1 University Street, Suceava, 5800, Romania

ABSTRACT

Many solid bodies contain initial stresses in their natural state of mechanical equilibrium in the absence of external loading, as a result of their manufacture or history. The initial state stress tensor is symmetric, satisfies the equilibrium equation without body forces and the boundary conditions on the surface, Hoger (1).

Initial stresses often have a detrimental effect on the service of the object through the occurrence of failures such as yield, fatigue and fracture under the design limit loads. The presence of a particular residual stress field in a component may also have a positive effect in special situations.

Although important progress has been achieved to the date in research of contact fatigue, the phenomenon in not entirely understood yet. In particular, the stress state into an element subjected to fatigue results from superposition of three-dimensional contact and interface stresses upon the initial stress state.

The contact stresses depend on the geometry of contacting elements, their elastic-plastic properties and the applied load. The interface stresses are produced by shear of oil film or by friction. Therefore, they are influenced by oil film thickness, by viscosity and solidlike proprieties of the lubricant and by applied shear rate or shear strain.. The initial stresses depend on mechanical, thermal and chemical factors and they may increase or diminish the operating stresses. Finally, they influence the nucleation and propagation of fatigue cracks, affecting the fatigue life.

Many hypothesis have been formulated to date to explain the later influences. Most of the developments are based on the theories of fracture mechanics and of three-axial fatigue. The number of experimental results to validate these theoretical hypotheses is quite small. Consequently, this paper reports new results concerning the rolling contact fatigue life influenced by macroscopic initial stresses induced in the specimen by torsion and bending, as indicated in Fig. 1, (2, 3).

Experimental investigations indicate cleraly that initial shear stresses induced by torsion and normal positive stresses produced by bending decrease the fatigue life in a smaller extent, Fig. 2. The initial compressive stresses induced by bending have a beneficial effect upon fatigue life, Fig. 3.

Various theoretical hypotheses concerning the criteria used to designate the components of the stress state responsible for fatigue fracture are analysed and compared with newly obtained experimental results. This tends to indicate that Dang Van criterion for multi-axial fatigue adapted by Diaconescu (4) for the case of rolling contact failure is able to predict better the observed fatigue lives than other criteria.

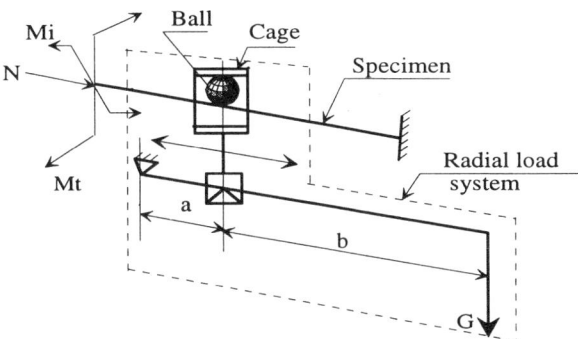

Fig. 1: Schematic showing layout of fatigue tester

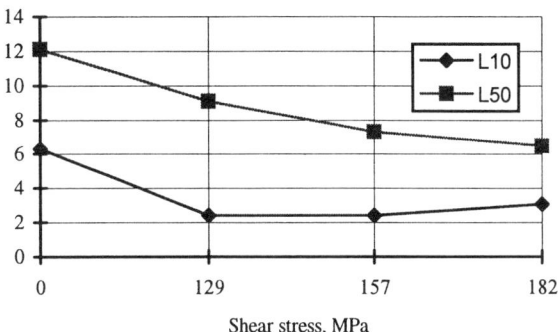

Fig. 2: Effect of share stresses upon fatigue life

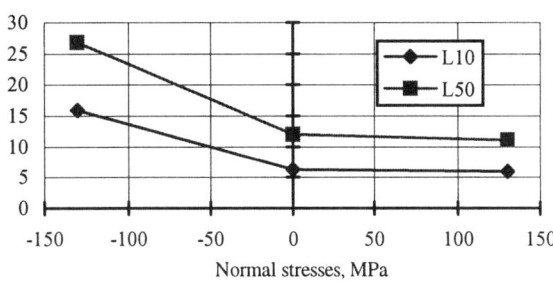

Fig. 3: Effect of normal stresses upon fatigue life

REFERENCES

(1) A. Hoger, Jnl. of Elasticity 31, 1993, 219-237pp.
(2) G. Frunza, Acta Tribologica, Vol. 3, No.1-2, 1995, 53-61pp.
(3) G.Frunza, Ph.D. Thesis, University of Suceasva, 1996.
(4) E.N. Diaconescu, Decisive Stresses for Rolling Contact Fatigue, Scientific Report, ERB-CIPA-3510-PL-92-4085, INSA Lyon, 1993.

RACEWAY MICRO-GEOMETRY AND DYNAMIC EFFECTS IN FRICTION TORQUE OF ROLLING BEARINGS

M. D.GAFITANU, C. RACOCEA and G.D. HAGIU
Department of Machine Design & Vibrations, "Gh. Asachi" Technical University of Iasi
22 Copou Bulevard, 6600 Iasi, ROMANIA
V. CIOAREC
Rolling Bearings Factory of Alexandria
1 Tr. Magurele Str., 0700, Alexandria, ROMANIA

ABSTRACT

There are presented the results of geometrical measurements concerning the deep groove ball bearings raceways symbole **625**.

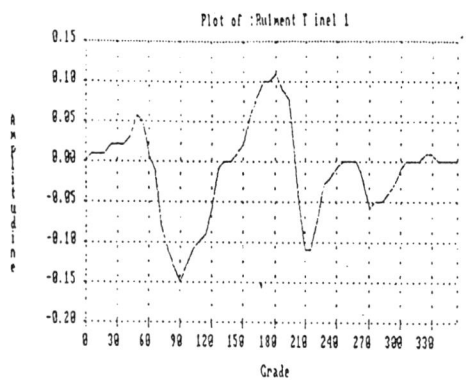

Fig. 1: Profilogramme

There measurements were oriented especially to the roughness and the out of roundness, but the profilogrammes obtained offered valuable data on the waviness, and its processing, with immediate consequences on the dynamic friction torque variation.

The rolling bearings measured and analyzed have the following main caracteristics: deep groove radial ball bearings **625**, quality class **Po**, bore diameter **5 mm**, outher diameter **16 mm**.

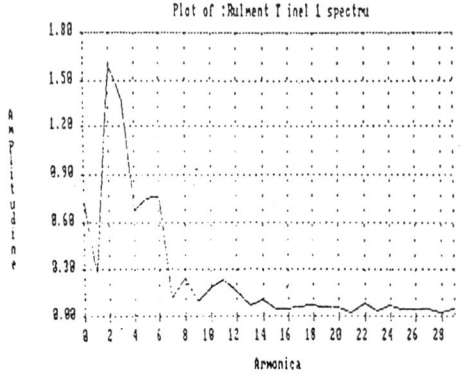

Fig. 2: Spectrum analysis

A Talysurf and Talyround instrumentation was used. The statistical analysis of the results offered preliminary data: medium roughness inner ring **0,059 μm**, outer ring **0,0759 μm**, out of roughness inner ring **2,17 μm**, outer ring **1,5 μm**. There is no significant corelation between these caracteristics and the vibration level of the mounted bearing.

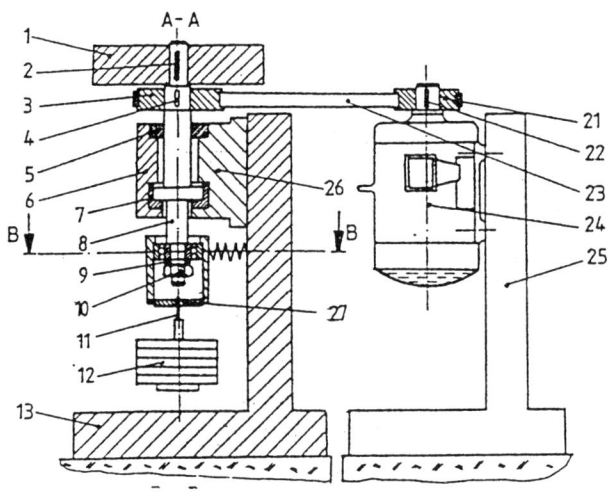

Fig. 3: Original test rig

The raceway profile was analysed via **FFT** and the median modulus of the first 30 harmonics was put into evidence together with the observed dispersion inner ring $\bar{x} = 30.87 \cdot N^{-1.0588}$, outer ring $\bar{x} = 11.72 \cdot N^{-1.363}$ in accordance with various other experimental results.

Bibliography

1. VERGHILES, A.M., Matematiceskaia Mode Vozmusciaevo Momenta Malonagrujenah sarikovah opor kacenia, Treniei Iznos, t.3, nr.3, 1982, p. 545-552.
2. GAFITANU, M.D. et al. Rolling Bearings, t.II, Ed. Tehnica, Bucuresti, 1985.

NUMERICAL MODEL FOR LIFETIME PREDICTION OF ROLLING MECHANICAL ELEMENTS

SREČKO GLODEŽ, ZORAN REN and JOŽE FLAŠKER
University of Maribor, Faculty of Mechanical Engineering, Smetanova ul. 17, 2000 Maribor, Slovenia

ABSTRACT

A new numerical model for lifetime prediction of rolling mechanical elements is presented. The computational model comprises contact conditions and simulates a complete fatigue process including the conditions required for the fatigue crack initiation and simulation of the fatigue crack growth. The service life of contacting mechanical elements can be easily estimated from the numerical results and with consideration of some particular material parameters.

INTRODUCTION

In modern industry, the designer is commonly restricted by requirement that mechanical elements, like wheels, gears, bearings, cams, *etc.*, should carry high loads at high speed with both size and weight kept to a minimum. For such applications, predicting component operational failure becomes crucial to ensuring an adequate design. Mechanical elements subjected to rolling contact conditions fail by several mechanisms and the most prominent among these is surface pitting (1). The process of surface pitting can be visualised as the formation of small, surface-breaking or subsurface initial cracks that grow under repeated contact loading. The pitting process can be divided into two stages- the fatigue crack initiation and the fatigue crack propagation until final failure. Therefore it follows that the service life can be estimated from a number of stress cycles N_o required for crack initiation and a number of stress cycles N_f required for a crack to propagate from the initial to the critical crack length:

$$N = N_o + N_f.$$

FATIGUE CRACK INITIATION

The theory of dislocation motion on persistent slip bands is used to describe the process of fatigue crack initiation. Following this assumption, Mura and Nakasone (2) proposed an analytical model for determining the number of stress cycles required for fatigue crack initiation:

$$N_o = \frac{\gamma}{h\left[\log\frac{8l}{h} - \frac{3}{2}\right](\Delta\tau - 2\tau_f)} \cdot \frac{2-f}{f},$$

where $\Delta\tau$ is the applied stress amplitude on the slip layer, γ is the surface energy of the crack, τ_f is the frictional stress of the material, h is the width of a slip band, $2l$ is the length of dislocations pileup and f is the irreversibility factor of dislocations pileup.

FATIGUE CRACK PROPAGATION

The developed model takes into account the short crack growth theory, where the process of crack propagation is characterised with successive blocking of the plastic zone by slip barriers (e.g. grain boundaries) and the subsequent initiation of the slip in the next grain when enough strain energy is generated in a grain. The crack growth rate da/dN is assumed to be proportional to the plastic displacement at the crack tip $\Delta\delta_{pl}$ (3)

$$\frac{da}{dN} = C_o\left(\Delta\delta_{pl}\right)^{m_o},$$

where C_o and m_o are material fatigue constants, that can be determined experimentally.

COMPUTATIONAL MODEL

To study the process of crack initiation and crack propagation in the contact area of rolling mechanical elements (bearings, gears, *etc.*) an equivalent model of two cylinders is used (4). The equivalent cylinders have the same radii as are the curvature radii of treated mechanical elements. The stress field in the contact area of rolling cylinders is determined with the finite element method. It is assumed that the crack is initiated in a single grain at the position of the maximum equivalent stress when the criterion for the crack initiation is fulfilled. The maximum stresses due to contact loading of mechanical elements with smooth surfaces and good lubrication always appears at some depth under the contacting surfaces. After the fatigue crack is initiated its growth under repeated fatigue loading is simulated with the virtual crack extension method (VCE) in the framework of the finite element method. From this analysis the relationship between the stress intensity range ΔK and the crack length a ($\Delta K = f(a)$) can be determined. From this relationship the corresponding plastic displacement at the crack tip $\Delta\delta_{pl}$ can be determined and with proper integration of the crack rate equation the service life of the mechanical element can be estimated (1).

REFERENCES

(1) S. Glodež, J. Flašker and Z. Ren, Fatigue Fract. Engng Mater. Struct., Vol. 20, No. 1, pp. 71-83, 1997.
(2) T. Mura and Y. Nakasone, ASME J. Appl. Mech., Vol. 57, pp. 1-6, 1990.
(3) A. Navarro and E.R. de los Rios, Philosophical Magazine A, Vol. 57, pp. 15-36, 1988.
(4) K.L. Johnson, Contact mechanics, Cambridge University Press, Cambridge 1985.

Submitted to *ASME Journal of Tribology*

RIGIDITY AND DAMPING CHARACTERISTICS OF HIGH SPEED BALL BEARINGS

G. D. HAGIU, M. D. GAFITANU, and C. RACOCEA
Department of Machine Design & Vibrations, "Gh. Asachi" Technical University of Iasi
22 Copou Boulevard, 6600 Iasi, ROMANIA

ABSTRACT

The dynamic state of a grinding main spindle is conditioned, mainly, by the dynamic characteristics of the assembly components, i.e. shaft-bearing-housing (1,2). While the dynamic characteristics of the shaft and housing were approached by reliable mathematical models, the bearing assembly is difficult to model. Moreover, if the interfaces shaft/inner ring and housing/outer ring are carefully controlled and considered as "rigid joints", the bearing elements interactions, and consequently, its rigidity and damping characteristics can be considered as having a major influence on the main spindle dynamic state (3,4). The aim of this paper was to develop an analysis concerning the rigidity and damping characteristics of high speed ball bearings mounted on grinding machine main spindles. A theoretical dynamic model was suggested and an adequate validation on a test grinding machine main spindle was achieved.

The main steps of the analysis were:
1. Determination of the dynamic characteristics of ball/raceway contact in EHL conditions.
2. Determinaton of the bearing dynamic characteristics and the elaboration of the dynamic model.
3. Dynamic model validation.

For a ball bearing loaded by a radial force F_r, axial force F_a, and bending moment M the load - displacement correlations are given by:

$$F_r = \delta_r (K_r + iH_r); \quad F_a = \delta_a (K_a + iH_a); \quad M = \varphi (K_m + iH_m),$$

where $\delta_r, \delta_a, \varphi$ are the displacements between bearing rings, K_r, K_a, K_m are the overall rigidities and H_r, H_a, H_m are the overall hysteretic dampings. From these considerations, for a high speed ball bearing the dynamic model presented in Fig. 1 was proposed.

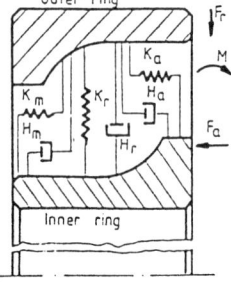

Fig. 1: **Dynamic model for a high speed ball bearing**

The validation of the proposed dynamic model was achieved by a theoretical and experimental analysis of the dynamic state of a test grinding machine main spindle. Theoretical and experimental amplitudes of the transversal vibrations of the test main spindle in controlled speed conditions for various values of bearings preload F_p and test force F_g were determined.
A comparison between theoretical and experimental results obtained is presented in Fig. 2. The experimental results comfirm largely the theoretical ones.

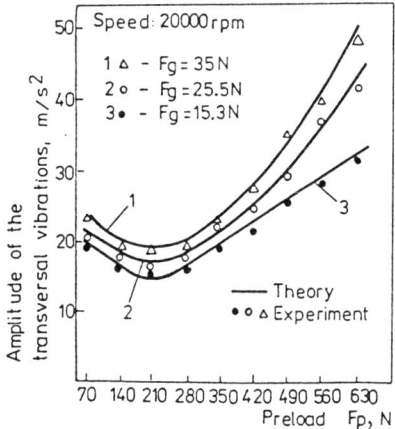

Fig. 2: **Theory - experiment correlation**

From the results obtained some conclusions may be drawn:
1. The agreement between theoretical and experimental results obtained highlighted an appropriate estimation of the dynamic characteristics of high speed ball bearing.
2. Some measured vibrations amplitudes were less than the theoretical ones; in this case can be advanced the assumption that the bearing interfaces, i.e. shaft/inner ring and housing/outer ring could be, in high speed conditions, supplementary sources of vibrations damping.

REFERENCES

(1) D. N. Reshetov, T. Portman, Accuracy of machine tools, Hardcover, 1989.
(2) M. Weck, Ball Bearing Journal, No. 208, 1-3 pp.
(3) T. L. H. Walford, B. J. Stone, J. Mech. E., 197C, 1983, 225-232 pp.
(4) L. I. Mc Lean, E. I. Hahn, ASME Transaction No.107, 1985, 402-409 pp.

FRICTION TORQUE OF NEEDLE ROLLER THRUST BEARINGS

TETSUZO HATAZAWA, JUJIRO KAGAMI and TAKAHISA KAWAGUCHI
Faculty of Engineering, Utsunomiya University, 2753 Ishii, Utsunomiya, Tochigi, JAPAN
KUNIO YAMADA
School of Science and Engineering, Teikyo University [Former], 1-1 Toyosatodai, Utsunomiya, Tochigi, JAPAN

ABSTRACT

Needle roller thrust bearings have generally very high load carrying capacity due to the line contact between the rollers and raceways. However, these bearings also experience a large amount of relative sliding between the rollers and raceways because of bearing geometry. Consequently, it was considered that these bearings are not suitable for high speed operation. The operating performances have been understood qualitatively[1], but it is considered that the friction characteristics under various operating conditions are not sufficiently clear.

In this study, friction torque of the needle roller thrust bearings was investigated theoretically and experimentally. In order to evaluate the friction torque, at the beginning, roller motions of the bearing, the rotating velocity and orbital motion, are clarified under the condition that the energy loss due to the friction generated at the contact between the rollers and raceways become minimum. Secondly, by using these results, the friction torque are obtained. Following the theoretical study, experimental investigation were carried out with a wide range of thrust loads and rotational speeds. Various types of bearing having different roller diameters or different pitch diameters were prepared and tested.

As the needle roller thrust bearings are limited to low speed operation together with high thrust load, there are many asperity contacts and fluid film effects are almost negligible. Therefore it is considered that the friction behavior is similar to that in boundary lubrication. The roller motion and friction torque of the bearings are analyzed under the conditions mentioned above.

The rotating velocity of roller about its own center ω_r, the orbital motion on the bearing axes ω_c and bearing torque T obtained theoretically are as follows.

$$\omega_r = (l_g/2r)\omega_h, \quad \omega_c = \omega_h/2$$

$$T = 0.25 W_t \mu l_p$$

where ω_h is the angular velocity of the washer, l_g the bearing pitch radius, W_t the thrust load, r the roller radius, μ the coefficient of sliding friction, l_p the roller effective length. As can be seen from above equation, torque T is proportional to W_t, μ and l_p, not dependent on the rolling velocity n (rpm), roller radius r, bearing pitch diameter D and number of rollers N.

Fig. 1 shows the relationships between the thrust load W_t and the friction torque T, at the various rotational speeds n, obtained from the experiments. In case of these experiments, the bearing pitch diameter D_m is 40 mm, and this bearing type have 12 number of rollers. As can be seen in Fig. 1, characteristics of the friction torque with the thrust load is nearly proportional to the thrust load W_t. Fig. 1 also shows the effect of rotational speed n on the friction torque T. The torque has a very little tendency to being small as the speed n is higher. It is considered that the fluid film is effective only slightly, but the differences in these speed range become to be small as it can be negligible. In this case, the contact lubrication mechanism may be governed by boundary lubrication. On the other hand, theoretically, the equation of the torque is not including the speed parameter. Consequently, it may be considered that experimental results are agree with the theoretical ones.

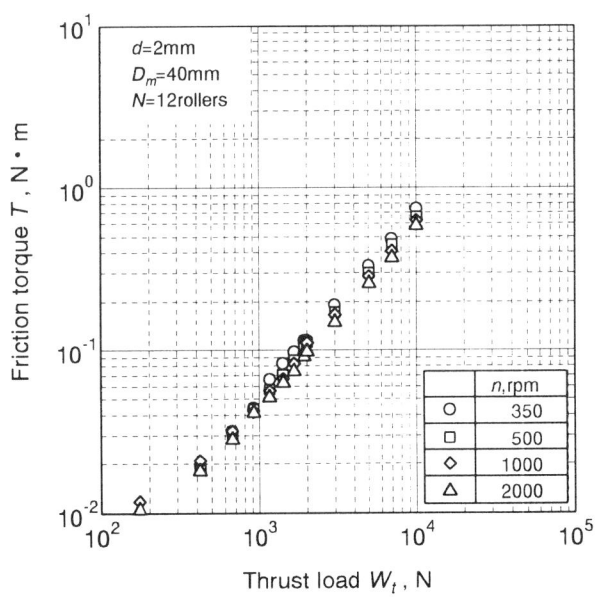

Fig. 1: Friction torque vs. thrust load

REFERENCES

(1) T. A. Harris: Rolling Bearing Analysis, John Wiley & Sons Inc.(1966)32.

INVESTIGATION OF SURFACE GENERATION AND ROLLING RESISTANCE PROPERTIES OF STEEL IN DRY AND LUBRICATED ROLLING CONTACTS

DAE-EUN KIM and KUM-HWAN CHA
Department of Mechanical Engineering, Yonsei University, Shin-chon dong 134, Seoul, Korea

ABSTRACT

The rolling resistance and wear that occur in rolling elements affect their performance as well as life, and therefore, these properties should be properly optimized in bearing applications. It has been well established that resistant force and wear that occur during rolling motion depend on several factors such as material type, hardness, subsurface microstructure, applied load, and speed (1).

The purpose of this work is to investigate the generation of surface topography and rolling resistance properties of steel in dry and lubricated rolling contacts.

Rolling contact experiments were conducted with a plate-on-ball type tribotester that was built using a drill press. SM45C (C: .45%), steel was used as the disk material and for the balls, STB2 chrome steel was used. Sixteen balls, with 6.35mm diameter, were placed continuously along the thrust-bearing race placed on the bottom and the disk was placed above the balls. A load of 48 kgf was applied to the balls. The ball and the thrust-bearing race had a conformity factor of 0.53, and therefore, the contact stress between the two surfaces was intentionally made to be significantly less than the stress between the ball and the disk. The disk was driven by the drill press spindle motor at 900 rpm. To obtain the rolling resistance in real time the thrust-bearing race was mounted on a thrust bearing which in turn was constrained from rotating by a force sensor beam from which the restraining torque could be measured. The signal was picked up by an A/D converter and recorded by a PC based data acquisition system. The generation of surface topography and subsurface microstructure were characterized using an optical microscope, SEM, and a stylus type surface profilometer.

The steel disks were turned under mild (Disk A) and severe conditions (Disk B, C) to attain different levels of initial surface deformation layer. The intention was to investigate the effect of microstructure and the state of deformed layer on the rolling contact characteristics.

The rolling resistance behavior in the early stages of rolling (about 75000 stress cycles) in dry condition varied between the two groups of specimens. As can be seen in Fig. 1, the specimen with less subsurface deformation showed an increase in the rolling resistance which eventually dropped to a steady state value while the specimens with greater deformation increased gradually to the steady state value which was quite similar for the two cases. Such a behavior of rolling resistance is attributed to plastic deformation work done during groove formation on the disk specimen at the initial stage of rolling.

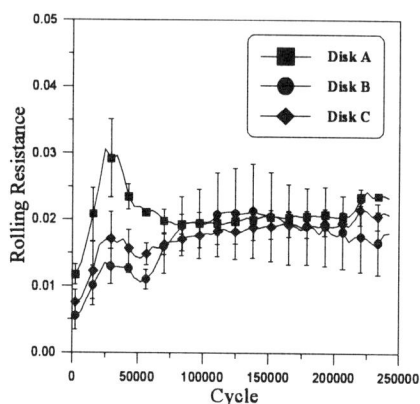

Fig. 1 Rolling resistance of SM45C with respect to cycles in dry condition

The cross-sectional view of the microstructure shows that surface traction has a definite effect on the morphology of the surface region. That is, significant slip seems to have taken place between the ball and the disk during the rolling test.

Rolling tests were also performed in lubricated conditions using DTE24 oil and grease. The rolling resistance values were about 50% lower than those of the dry rolling case. The surface generation effects were significantly less compared to the case of dry rolling contact. An interesting observation was noted in that for some cases the maximum plastic strain occurred below the surface where shear stress was predicted to be the maximum by Hertzian contact analysis. This finding was contrary to the results of the dry rolling contact where maximum material flow consistently occurred at the surface.

The results of this work show that the rolling resistance behavior depends on the state of the deformed layer. Also, lubrication can reduce the plastic flow at the surface but may still have an effect on the subsurface strain.

REFERENCE

(1) M. R. Hoeprich, J. of Tribology, Vol. 114, 1992, pp. 328-333.

AN EVALUATION OF DIAGNOSTIC TECHNIQUES FOR CHARACTERISATION OF BALL BEARING MISBEHAVIOUR

S.D. LEWIS
European Space Tribology Laboratory, AEA Technology plc, Risley Warrington WA3 6AT, U.K.

ABSTRACT

In many industries, an increasing requirement is to diagnose incipient failures in tribological components in order to prevent costly plant downtime and un-planned maintenance. In no application is this requirement more clearly demonstrated than in spacecraft mechanisms where in-service maintenance remains impracticable.

Condition monitoring techniques are now widely used in terrestrial applications, and sophisticated techniques for information extraction from, for example, accelerometer signals are widely documented and discussed in the literature. However, in the opinion of the author, there is a large gap between the capabilities of the signal processing tools, and current understanding of the of signals generated by different kinds of ball bearing misbehaviour (except perhaps for gross effects such as fatigue spalling which is quite easily detectable). If diagnostic techniques are to be more widely and effectively adopted (specifically in the field of condition monitoring in spacecraft), then as a first requirement it is imperative to gain a better insight into the nature of signals generated by both "misbehaving" and "well- behaved" bearings.

This paper discusses experimental work carried out by the European Space Tribology Laboratory, aimed at characterising the behaviour of ball bearings using a number of accepted diagnostic techniques identified in an earlier review (1). The test facility developed specifically for this work is unusual in that it allows precise control of bearing misalignment, and is adjustable during bearing operation. This is achieved by use of tilting housing actuated by micrometer screw for coarse adjustment or piezoactuator with sub-micron resolution for fine adjustment (Figure 1). Experimental measurements of shaft radial runout, bearing torque and structural acceleration are simultaneously acquired and processed to provide an insight into the behaviour of the test bearings.

Historically, low-pass filtered torque data has been used for identification of bearing misalignment and misbehaviour such as "cage hang-up". Therefore initially, data was analysed conventionally in the time domain, and in the wide-band frequency domain using Power Spectrum and Cepstrum analysis as well as statistical measures such as Skewness and Kurtosis. However, with the exception of low-pass filtered torque data, none of these techniques proved capable of characterising and discriminating well behaved (well aligned, lubricated and stable) bearings from their misbehaving counterparts (misaligned, poorly lubricated bearings with tight cage pocket clearances).

However, adoption of narrow-band analysis of signals using the same techniques did permit both qualitative and quantitative assessments of the operating characteristics of the bearings. It was therefore concluded that all of frequency domain techniques investigated offer potential for characterising bearing misbehaviour, but that a narrow band approach, perhaps focused ball group, cage rotational, or ball-pass frequencies is essential. A particular interest is the use of narrow-band time-frequency analysis technique (2), in which the amplitudes of signals in specific frequency ranges is monitored over time. In future work, it is expected that this technique will be investigated further.

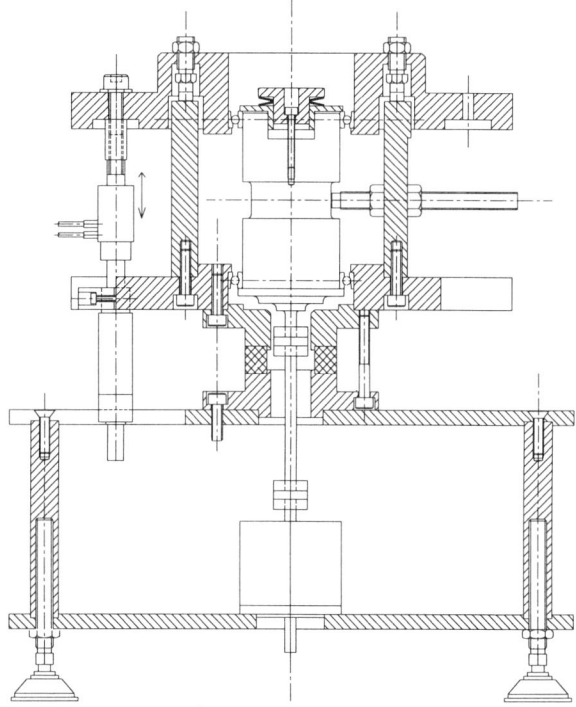

Figure 1 General Assembly of Test Facility

REFERENCES

(1) S. Lewis "Bearing Diagnostics Review", ESTL/TM/144, 1994
(2) L. Nicollet. et. al. "Noise analysis of dry lubricated ball bearings for a scanning earth mechanism", Proc. Sixth European Space Mechanisms & Tribology Symposium, 1995

ACCELERATED ENDURANCE TESTING MACHINE AND METHOD OF ROLLING BEARING FATIGUE LIFE*

XingLin LI and YanLiao ZHANG
Hangzhou Bearing Test & Research Center**, 310022, P.R.China

INTRODUCTION

Ball and roller bearings are used in practically every piece of mechanical equipment worldwide. These bearings are composed of an outer ring, an inner ring and a number of rolling elements. Many times these individual components or their materials are evaluated using element testers. However, because the bearing components act as an integral unit, the total bearing assembly must be evaluated as well.

A principal tool in almost inexhaustible need for bearing evaluation is the endurance test machine. This machine tests the bearings to damage and, by the data generated, allows us to make statistical inferences about the performance and life of the population of bearings.

Because this testing is so important, we have established an exhaustive endurance testing program. This program at its inception began testing bearings on an older design ZS type machine. It was our experience that these machines did not give us repeatability from test to test. With the ZS type machine, the vibration inherent in the machine and test setup cause the load to vary rather than remain constant.

We have developed the new generation endurance test machine--AETM (Accelerated Endurance Testing Machine) of rolling bearing fatigue life.

THE MAIN PARAMETERS OF AETM
*Bearing types: ball & roller bearings
*Bearing dimensions: 10-60 mm (inner diameter)
*Bearing numbers: 2 or 4 pieces
*Max. radial load: 100 ± 1.5% KN
*Max. axial load: 50 ± 1.5% KN
*Testing speeds: 1000+500m (m=0,1,2,..., 8) r/min (± 0.5%)

THE PERFORMANCE OF AETM
*Automatic controlled steady loading
*The outer rings temperature automatic display, absolute error ± 1 ℃.
*Test time automatic display, absolute error < 1 min.
*Automatic alarm & stop
*Testing process intelligent controlled with Pentium computer.

TEST METHOD & CONDITION

In terms of three type test machines--ZS type, B type (imported from U.S.A.), AETM B10-60R type, comparison testing for more than 110,000 hours with four different levels of stress, we put forward an AETM test method remain the consistency of contact fatigue failure mechanism. The AETM test method compared with usual one was time-saving more than 4/5, & cost-saving about 4/5, condensed the development period of new product of rolling bearing greatly, it had obvious benefits either economical or social. The new generation AETM provides the basis for the improvement of test technique of Chinese rolling bearing industry.

REFERENCES
(1) XingLin LI, Accelerated Endurance Testing Machine of Rolling Bearing Fatigue Life, BEARING, 1993, No.5, P47
(2) XingLin LI et al, Proc. Int. Tribol. Conf., Nagoya, JAST(1990)803-808
(3) Tedric A. HARRIS, Rolling Bearing Analysis (Third Edition), John Wiley & Sons, Inc.(1991)
(4) XingLin LI et al, Synopses of Int. Tribol. Conf., Yokohama, JAST(1995)211-2,P205

---

*Supported by
Zhejiang Natural Science Foundation.
**With assistance of UNDP/UNIDO.

DETERMINATION OF TIME SPAN FOR APPEARANCE OF FLUTES ON TRACK SURFACE OF ROLLING ELEMENT BEARINGS UNDER THE INFLUENCE OF ELECTRIC CURRENT

HAR PRASHAD
BHEL, R&D Division, Vikasnagar, Hyderabad - 500 093 (India)

ABSTRACT

Flow of *current* depends on the bearing impedance, which is a function of the lubricant characteristics, resistivity, oil film thickness and voltage across a bearing (1). In general, the effect of electrical currents on contact temperature, slip bands initiation and the life of rolling contact bearings have been discussed (2). In addition, investigations have been carried out on corrugation patterns and the theoretical evaluation of impedance, capacitance and instant charge accumulation on the surfaces of bearings (3). It has been established that above a certain level of cylic stress (the fatigue limit) some crystals on the surface of the specimen develop slip bands as a result of shearing of atomic planes within the crystals. With an increase in the number of contact cycles these slip bands broaden and intensify the points where separation occurs within one of the bands and a crack is formed. No study has been reported regarding appearance of crack of certain width on the track surface of roller bearing after initiation of slip bands for a bearing operated under the influence of electric current.

In this study, a theoretical model has been developed using continuum theory of Griffith to determine energy per unit area required for development of corrugations on the track surfaces after appearance of slip bands under the track surfaces. Also, an expression is deduced for the energy input per unit area on the track surfaces by the amount of current passed through the bearing at the measured shaft voltage in a given span of operation. By the pitch of corrugations derived using bearing dimensional and operational parameters along with the developed model and expressions, the time span for the development of flutes/corrugations on the track surface after appearance of slip bands under the track a has been determined. Similar time span has also been ascertained expermentally and found to match closely with theoretical value.

REFERENCES

1. H, Prashad., "Effects of Operating Parameters on the Threshold Voltage and Impedance Responses on non- Insulated Rolling-Element Bearings under the Action of Electric Current", Wear, 117 1987, pp 223-240.
2. H, Prashad., "Analysis of the Effects of an Electric Current on Contact Temperature, Contact Stresses and Slip Band Initiation on the Roller Tracks of Roller Bearings", Wear 131 1989 pp 1-14.
3. H, Prashad., "Investigations on Corrugated Pattern. on the Surfaces of Roller Bearings Operated under the Influence of Electrical Fields", Lubrication Engg., 44(8), 1988, pp 7l0-718.

ROTATION NON-UNIFORMITY VERSUS RACEWAYS MICROGEOMETRY IN ROLLING BEARINGS: A POSSIBLE GLOBAL CONTROL METHOD

C. RACOCEA, M. D.GAFITANU, and F. TARABOANTA
Department of Machine Design & Vibrations, "Gh. Asachi" Technical University of Iasi
22 Copou Bulevard, 6600 Iasi, ROMANIA
V. CIOAREC
Rolling Bearings Factory of Alexandria

ABSTRACT

The manufacturing errors influence both the size and time variation of the friction torque of rolling bearings. The control of this variation could be considered as a valuable nondestructive test method reflecting the raceways geometrical quality.

A special test rig was designed together with an adequate measurement chain and test programme.

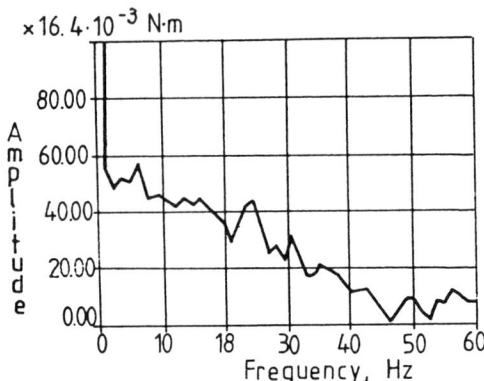

Fig. 1: Experimental results analysed in frequency domain

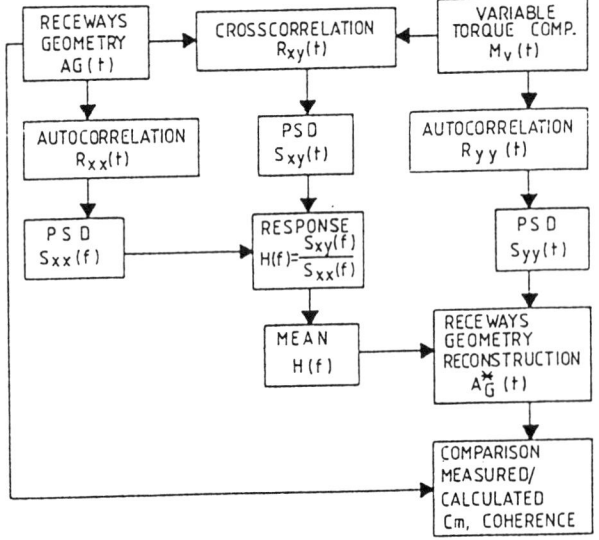

Fig. 2: The coherence function in the frequency domain

The data processing demonstrated a good correlation between the theory rolling body contacts with the raceways and the experimental results.

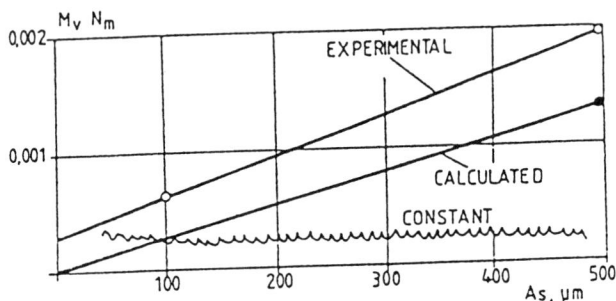

Fig. 3: Theory-experiment correlation

It was possible to reconstruct the geometry of the raceways profiles starting with the friction torque time variation with a sufficient precision.

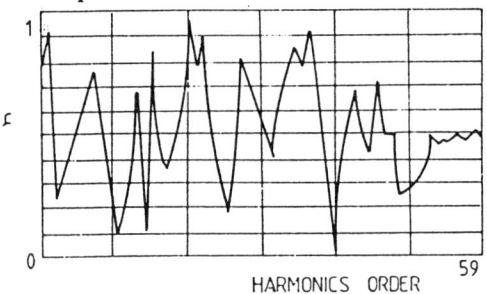

Fig. 4: Coincidence factor versus harmonics order

The correlation was estimated via coincidence parameter (fig.4) and the coherence function, and important consequences regarding the tehnological control of rolling bearings were formulated. A simple and useful control method of rolling bearings manufacturing precision was suggested.

Bibliography

1. BRAUN,S. Trans., ASME, Vibro acoustics & Stress in design, 106, Jan. 1984, 1.
2. STRUM, A. et al., Walzlogerdiagnostik fur Machinen und Anlagen, VEB Verlag Technik, Berlin, 1985.

Submitted to *Wear*

THE INFLUENCE OF LUBRICATING GREASES ON ROLLING BEARINGS FAILURES

RADOSLAV RAKIĆ
NIS-Naftagas promet, Bulevar Oslobodjenja 69, 21000 Novi Sad, Yugoslavia
ZLATA RAKIĆ
Vojvodjanska banka a.d., Miroslava Antića 2, 21000 Novi Sad, Yugoslavia

ABSTRACT

The correct use of lubricants is essential. Significant savings can be made in terms of energy consumption, replacement parts, maintenance costs and the reduction of machinery down time.

The aim of the study presented here is to investigate the influence of lubricating greases on the rolling bearings failures.

The rolling bearings have poor conformity between surfaces, very small contact areas and very high unit loading. In rolling element bearing operation, elastohydrodynamic lubrication films are generated which reduce the interaction of the contacting surfaces. For rolling element bearings, numerous methods have been developed to calculate the thickness of these films, including those by Dowson and Higginson (1) etc.

Selecting a lubricating grease for rolling bearings is somewhat more complex then selecting an oil. Variations in penetration levels, thickeners, additives and so on, result in numerous lubricating greases from which the selection can be made. Consideration include range of operating temperatures, load, presence of moisture or water contamination as well as the method of application. The author has presented the flowchart of grease selection procedure for rolling bearings in function of all relevant influential factors.

The problems appearing at all rolling bearings very often represent the consequences of tribological processes development on their contact surfaces. The consequences of tribological processes development on contact surfaces of rolling bearing elements cause the failure and lack of machine ability to carry out working. The tribological processes on contact surfaces of rolling bearings have been studied by Rakić (2,3) in function of all influential factors.

The experimental investigation of the influence of lubricating greases on rolling bearings failures has been carried out on 120 machine tools in three periods of time, each being 20.000 working hours. The results of the experimental investigation are presented for three NLGI consistency number and three type of lubricating greases:
- lithium multipurpose greases, ISO-L-XCCDA NLGI 1, 2, 3.
- lithium greases for extreme pressure, ISO-L-XCCEB NLGI 1, 2, 3.
- aluminum complex greases, ISO-L-XCFHB NLGI 1, 2, 3

The life time of the rolling bearings up to the failure mostly shows large deviation. By the aid of a probability and statistic method, it was possible to determine the influence of lubricating greases on failures, life and reliability of rolling bearings.

Figure 1 shows the curves of the rolling bearings reliability "R" versus time "T" as a function of three type of lubricating greases NLGI 3.

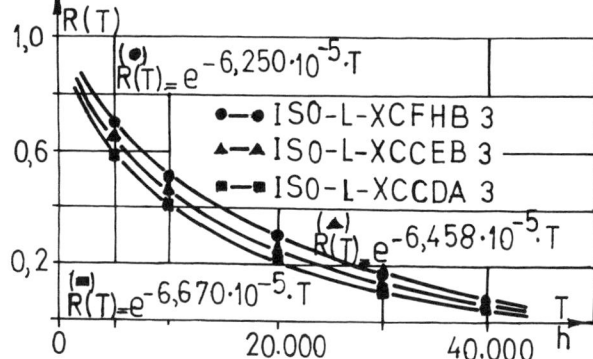

Fig. 1: Reliability of rolling bearings versus time

The following conclusions can be drawn from the results presented above:
- the rolling bearings failures were found to be affected by both NLGI consistency number and type of lubricating grease.
- increasing the NLGI consistency number of these lubricating greases tended to reduce the failure rate of the rolling bearings and led to a longer rolling bearings life.
- the lubricating grease ISO-L-XCFHB NLGI 3 gives the longest rolling bearings life among all the lubricating greases under presented operating conditions.

REFERENCES

(1) D. Dowson and G. R. Higginson, Elastohydrodynamic Lubrication, Pergamon Press, Oxford, 1966.
(2) R. Rakić, Proceedings of the 6th Int. Congress on Tribology, Vol.2, pp.228-233, 1993
(3) R. Rakić, Proceedings of the 13th Int. Conference on Production Research, pp. 274-276, 1995.

THE EFFECT OF DESIGN ON FRETTING WEAR OF A BALL BEARING

M SHIMA[1], Li QIJUN[2], S AIHARA[3], T YAMAMOTO[4], J SATO[5], R B WATERHOUSE[5]

1) Tokyo Univ. of Mercantile Marine, Tokyo, Japan
2) Riken Co. Ltd, Saitama, Japan
3) NSK Ltd, Kanagawa, Japan
4) Tokyo Univ. of Agriculture & Tech., Tokyo, Japan
5) Saitama University, Saitama, Japan
6) Nottingham University, Nottingham, England

ABSTRACT

When a stationary ball bearing is subjected to external vibration, fretting wear often takes place. In general the amount of wear is very small, but when the damaged bearing is operating noise and vibration may be generated.

In this paper the effect of a change in design of a ball bearing is described based on the results of numerical and experimental analysis to reduce fretting wear. Since fretting wear in ball bearings predominantly results from repeated Heathcote slip (1), (2) and as there is good correlation between the area of damage and the product of the repeated slip δ and the tangential traction τ at the contact region (3), (4), it is expected that the change can be profoundly reduced by decreasing $\tau \cdot \delta$.

The analysis of $\tau \cdot \delta$ as well as the fretting tests were made for a deep groove ball bearing with eight balls of radius R_b = 3.969mm. The inner diameter of the inner race was 20mm, the outer minimum diameter R_1 = 25.6mm, and the inner maximum diameter of the outer race R_2 = 41.476mm, both of them being kept constant. The groove radius of the inner race was changed in the range of 4.02mm to 4.21mm.

Increasing the radii of curvature of the inner and outer races by a small amount reduces $\tau \cdot \delta$ as shown in Fig. 1. Although the groove radius of the inner race of 4.02mm is usually used in practice for this type of bearing, if the radius is increased by only 0.08mm (2%) the value of $\tau \cdot \delta$ at both the centre and around the ends of the contacting surface can be reduced by over 30%. The same tendency can be seen for the contact of the outer race and ball. Such marked reduction in the value of $\tau \cdot \delta$ is attributed to the distinctive decrease in the relative slip due to changing the groove radius of the inner race. This result suggests that the fretting wear in ball bearings can be drastically reduced with the increase in the groove radii.

This prediction is confirmed experimentally. Figure 2 shows the percentage increase in damaged area over the Hertzian contact area plotted against the groove radius of the inner race.

REFERENCES

(1) M Shima, Qijum Li, T Yamamoto and J Sato: Fretting wear of ball bearing (part 3), Tribologists, Vol.40, No. 9 (1995) 755-762.

(2) Qijun Li, M Shima, T Yamamoto and J Sato: Fretting wear of ball bearing (part 4), Tribologists, Vol.40, No.12 (1995) 1027-1029.

(3) C Ruiz and KC Chen: Fatigue of Eng. Materials and Structures, Inst. Mech. Eng., London, (1986) 187-194.

(4) M Kuno, RB Waterhouse, D Nowell and DA Hills: Fatigue, Fract. Eng. Mater. Struct., 12 (1989) 387-398.

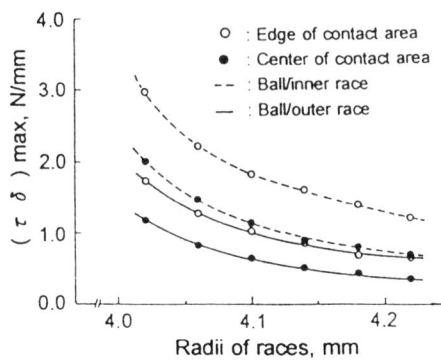

Fig.1. Effect of groove radii on $\tau \cdot \delta$ (coefficient of friction=0.6, Normal load=490N)

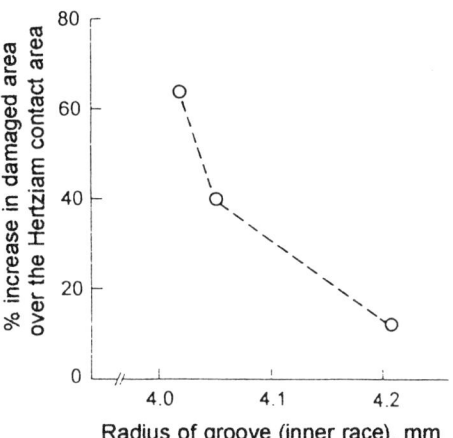

Fig.2. Effect of groove radius on damaged area

LIMITING STATE DIAGRAM OF A CYCLE LOAD-BEARING SYSTEM ON ROLLING-MECHANICAL FATIGUE

L A SOSNOVSKIY
Scientific and Industrial Group "TRIBOFATIGUE", P O Box 24, Gomel, 246050, Republic Belarus
A V BOGDANOVICH
Scientific Center of Machines Mechanics Problems of the Belorussian Academy of Sciences,
Skoriny Avenue, 12, 220072, Minsk, Republic Belarus
S A TYURIN
Belorussian State University of Transport, Kirova Street, 34, 246022, Gomel, Republic Belarus

ABSTRACT

During rolling-mechanical fatigue the limiting state of a cycle load-bearing system can be determined by means of various criteria: fatigue fracture, pitting of critical density, surface plastic waves formation, etc.

The limiting state in such conditions may be generally described by a diagram drawn according to the experimental data on "contact pressure (under rolling friction) — cyclic stress amplitude" coordinates (1)(2).

Fig. 1 presents the limiting state diagram of a Carbon 45 steel (specimen) — 25ХГТ steel (roller) system drawn in the Scientific and Industrial Group "TRIBOFATIGUE" based on the experimental results (3). Wear-fatigue tests were carried out on SI machine.

Point A of the diagram ($\sigma_a = \sigma_{-1}$; $p_o = 0$) is drawn as a result of standard tests of a cylinder specimen 10 mm diameter under bending with rotation (mechanical fatigue) on base 10^7 cycles. The limiting state criterion is fatigue fracture of a specimen into two parts through the main crack propagation in its danger section. Point D of the diagram ($\sigma_a = 0$; $p_o = p_f$) is drawn as a result of standard tests under rolling friction on base $2 \cdot 10^7$ cycles. The limiting state criterion is the initiation of pittings of critical concentration $S_\sigma/S_o = 0.1$, where S_σ - average sum area of pitting and S_o - average rolling friction area.

For drawing AB-part (see Fig. 1) the value of contact pressure p_o has been given and fatigue limit σ_{-1p} on mechanical fatigue criterion has been determined correspondingly. CD-part of the limiting state diagram has been drawn similarly: the limiting pressure $p_{f\sigma}$ on rolling fatigue criterion has been determined for various values of cyclic stress amplitude σ_a. AB and CD curves are described by equations:

$$\frac{\sigma_{-1p}}{\sigma_{-1}} = 1 - \frac{p_o}{p_f}\left(1 - \frac{p_o}{p_f}\right)ln(1-\mu_p)$$

$$\frac{p_{f\sigma}}{p_f} = 1 - \frac{\sigma_a}{\sigma_{-1}}\left(1 - \frac{\sigma_a}{\sigma_{-1}}\right)ln(1-\mu_\sigma)$$

where parameters $\mu_\sigma = 0{,}60$ and $\mu_p = 0{,}92$.

The analysis of a diagram shows in particular:
– the fatigue limit of steel can 1.5 – 1.6 times increase when rolling friction is realized simultaneously under optimum contact pressure (direct effect);
– the limiting pressure during rolling friction can 1.2 – 1.3 times increase if cyclic stresses of optimum level are applied to a specimen simultaneously (back effect).

The paper gives a detailed analysis of the reasons why the system life under wear-fatigue tests is longer than that under purely friction or purely fatigue tests. The experimental results of the investigation of a crack emergence by the AFM-method are discussed.

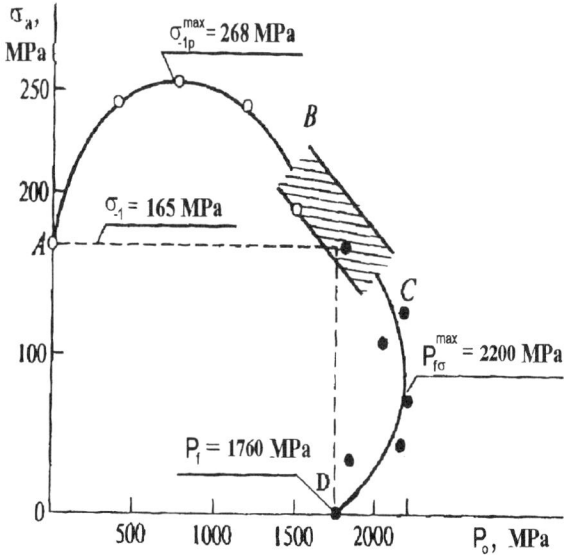

Fig. 1: Limiting state diagram of a cycle load-bearing system carbon 45 steel - 25ХГТ steel

REFERENCES

(1) L A Sosnovskiy, Complex wear-fatigue damage and its prediction, Gomel, 1991, 188p.
(2) L A Sosnovskiy, N A Makhutov V A Shurinov, Plant Laboratory, No 11, 1992, 44-61pp.
(3) L A Sosnovskiy, N A Makhutov, A V Bogdanovich, S A Tyurin, Plant Laboratory, No 2, 1996, 39-42pp.

CALCULATION OF THE FATIGUE LIFE OF REGENERATED LARGE OVERALL DIMENSIONS ROLLING BEARINGS

STANISŁAW STRZELECKI and BOGDAN WARDA
Institute of Machine Design, Technical University of Łódź, Stefanowski Street 1/15, 90-924 Łódź, Poland
STANISŁAW BEDNARCZYK and ANDRZEJ WDOWIAK
Rolling Bearings Plant, Kraśnik S.A., 23-210 Kraśnik, Poland

ABSTRACT

Bearings of the diameter larger than 150 mm are called large overall dimensions bearings. Because of the large mass and dimensions this type of bearings very expensive and their regeneration can give high savings..

Before starting the regeneration it is necessary to calculate the labour demand and the general cost of regeneration and make a good economical analysis. Technical state of each single element and their potential usability must be determined (1). Only bearings in which the wear of the races and rolling elements achieves or exceeds the maximum permissible value are suitable for the regeneration. The qualifying of the bearings to this group should based on the economic calculation motivating the usefulness and profitability of regeneration.

Thus to begin the regeneration the following conditions should be fulfilled:
- the total cost of regeneration should not exceed 60% of new bearing price,
- the quality and life of the bearing can not differ from the parameters of a new bearing,
- time of regeneration must be as short as possible.

Determination of the fatigue life of the regenerated bearing gives an idea about further operation of this bearing. Because in the literature there are no data on the calculation of the life of regenerated bearings it is useful to determine this life on the example of the regenerated deep groove ball bearings operating without problems in industry.

Based on the analysis of the wear and process of regeneration the evaluation of the fatigue life of large overall dimension deep groove ball bearing 618/560 MA applied in the industrial washing machine was considered. The reasons of wear, analysis of wear and calculation of the fatigue life before and after regeneration have been given into consideration.

The fatigue life of investigated bearing 618/560 MA was determined by the method of Lundberg and Palmgren under assumption of restricted life (1)(2). The knowledge of inside geometry of bearing before and after regeneration (3) allowed to determine of the contact load capacity for the stationary external bearing ring and for the internal bearing ring, rotating with respect to the external load. The basic data of the bearing and results of the fatigue life calculation are given in Table 1.

Bearing parameter	Bearing before regeneration	Bearing after regeneration
Ball diameter	36,512 mm	38,1 mm
Race radius	18,826 mm	19,62 mm
Diameter of the outer ring race	656,512 mm	658,1 mm
Diameter of the inner ring race	583,488 mm	581,9 mm
Fatigue life	37,400,000 rev	48,400,000 rev

Table 1: Fatigue life of the 618/560 MA deep groove ball bearing

Analysis of wear, regeneration and calculation of the fatigue life of large overall dimensions ball bearings 618/560 MA have allowed the introducing of the following conclusions:
1) in many cases of failures of large overall dimensions ball bearings there is the possibility of bearing regeneration,
2) the cost of regeneration of a new bearing is about 30% cost of a new one,
3) increase of the diameter of rolling elements and the variations of rings diameters gives the higher chances for correct operation of the bearing after regeneration,
4) the regenerated bearings are characterised by the 30% increase of the theoretical fatigue life with regard to the new bearing.

The last conclusion should not be considered in all cases of regenerated bearings because many different factors have an effect on the fatigue life of the rolling bearings.

REFERENCES

(1) F.T. Barwell, Bearing systems. Principles and Practice. Oxford University Press, 1979.
(2) T. Harris, Rolling Bearing Analysis. New York, Wiley, 1966.
(3) Strzelecki S., Wdowiak A., Bednarczyk S.: Proceedings of the Conference „Engineering of Bearing Systems '96",Gdańsk,1996, 477-484pp

ROLLING FRICTION RESISTANCE IN COMPLEX TRIBOLOGIC PAIR

MICHAŁ STYP-REKOWSKI
Technical and Agricultural University, Mechanical Department, Kordeckiego str.20, 85-225 Bydgoszcz, POLAND

ABSTRACT

The rolling bearings in spite of the competition with other bearing types (sliding, magnetic) further are often applied in many kind of machines as a units affording possibilities for relative motion of these machine elements and they are still an object of many theoretical and experimental investigations for optimization of their constructional features (CF). In tribological meaning the rolling bearing is possible to admit as complex tribologic pair.

In many cases when an application of a typical rolling bearings is not reasonable, the most often of economic cause, the special bearings are applied and then published results of investigations concerning typical bearings not always are possible to use. An example, confirming this thesis is bearing applied in popular one-track vehicles. Their operating conditions: low load and rotational speed and not so high the motion accuracy give the reasons for using the untypical (for rolling bearings) materials and other CF. The general character of tribologic phenomena inside the bicycle bearings are similar as for the typical bearings but they are different in the quantitative meaning.

The some results of the experimental investigations reffering to the relation between constructional features:
- geometrical coefficient $\delta = 2r_2/d_b$ (r_2 - radius of inner raceway curvature, d_b - ball diameter),
- hardness of raceway H,
- load coefficient $k = F_a/F_r$ ($F_{a,r}$ - axial and radial load),

and rolling friction resistance (M_f) of the special bearing are presented in Fig.1. Resistance to motion is one of the most important qualitative features of bicycle bearings because energy for overcoming this resistance is taken from human organism.

In all recorded cases in the ranges of investigated independent variables all functions are monotonic i. e. they have not relative extremum inside the ranges. Therefore it is difficult to select constructional feature values of rolling bearing elements for which bearing will be characterized with the minimum value of resistance to motion. Special optimization procedure was worked out as the solution of this problem.
Procedure consist in solution of following problem:

Min $M_t(\delta, k, H)$ and   Max $L_n(\delta, k, H)$

with boundary values of parameters described by system of inequalities:

$1,01 \leq \delta \leq 1,09$;
$0,04 \leq k \leq 0,4$;
$150 \leq H \leq 500$.

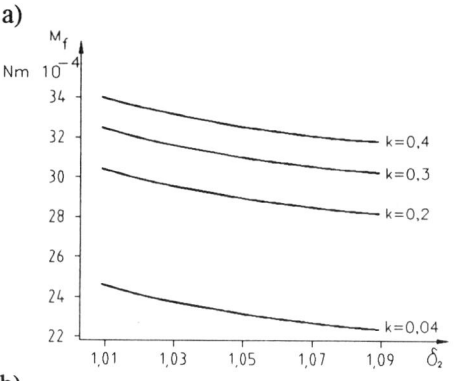
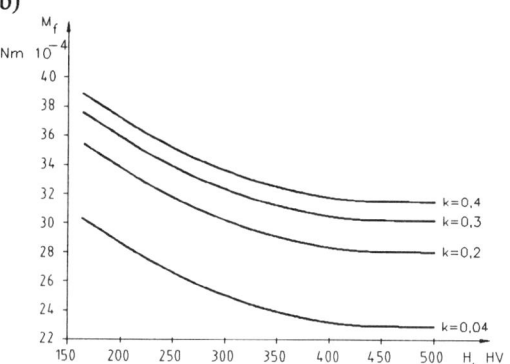
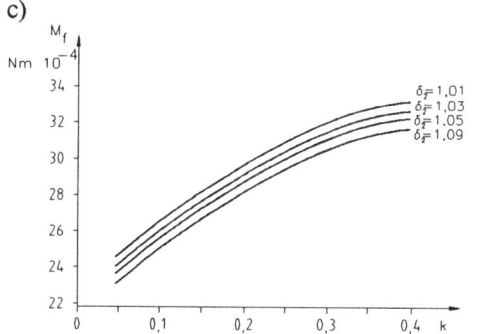

Fig. 1: Examplary relationships between friction moment M_f and: a) coefficient δ, b) raceways hardness H, c) coefficient k

Observed relations between inner resistance to motion and CF are described by following regression equation:

$$M_f = 195,35 - 280,63\delta + 45,00k - 0,07H + 123,50\delta^2 - 46,88k^2 + 8,32 \cdot 10^{-5} H^2$$

The experiments confirm thesis that in the special bearings intensity of tribological phenomena depends on investigated constructional features of bearing elements. They also give informations about phenomena character.

FRICTIONAL PROPERTIES OF VARIOUS KINDS OF PLASTICS AS ROLLING BEARING MATERIAL

NAOHISA TSUKAMOTO
Chiba Institute of Technology, 2-17-1 Tsudanuma, Narashino-shi, Chiba 275, Japan
YOICHI KIMURA
TOK Bearing, Inc., 2-21-4 Azusawa, Itabashi-ku, Tokyo 174, Japan

ABSTARCT

1. Preface

Notwithstanding recently plastic rolling bearings have achieved the important role as the bearings for information processing equipment, their operational characteristics have not yet been elucidated. From this fact, in this research, first, the coefficient of friction of plastic ball bearings was investigated, and this result was evaluated by comparing with that of steel ball bearings which have been used widely.

2. Dimensions of the tested ball bearings

As to the dimensions of the tested ball bearings, inside diameter is 6 mm, outside diameter is 22 mm, width is 7 mm, ball diameter is 3.968 mm, and the number of ball is 6.

3. Experimental conditions

According to the kinds of materials of inner and outer rings, balls and others and with or without lubrication, in this research, the types of the tested ball bearings were divided into 17 kinds from TYPE A to TYPE Q as shown in Table 1.

The experiment was carried out by fixing radial force F at 49N for respective bearing types, and changing number of revolutions to three steps, 500, 1000, and 1400 rpm. This F=49N is the value which was set as the rated force for Type A bearings at the time of normal use. As for the grease lubrication, 0.1g was automatically injected for one ball bearing. The operation was carried out up to 1.5×10^6 in all the experiments.

4. Experimental results and evaulation

The coefficients of friction of respective ball bearing types are shown in Fig.1. The abscissa in the figure represents the peripheral velocity on the circle which connects the centers of ball of a bearing as the indication of rpm. As for the coefficient of friction on the ordinate, the mean value of the coefficient of friction in the total time of experiment was taken. In Fig.1, the coefficient of friction of Type P and Type Q are not shown, but these coefficients of friction are 0.035~0.040.

Table 1 Type of ball bearings and experimental conditions

Type	Materials				Without Lubrication	Lubrication
	Inner race	Outer race	Ball (rolling element)	Retainer		
A	POM	POM	Steel	PP	○	
B	Steel	Steel	Steel	Steel	○	
C	POM	POM	Steel	PP		○
D	Steel	Steel	Steel	Steel		○
E	MM	MM	Steel	PP	○	
F	Steel	POM	Steel	PP	○	
G	Steel	POM	Steel	PP		○
H	PPS	PPS	Steel	PP	○	
I	PPS	PPS	Steel	PP		○
J	PEEK	PEEK	Steel	PP	○	
K	PEEK	PEEK	Steel	PP		○
L	PPS	PPS	POM	PP	○	
M	PPS	PPS	POM	PP		○
N	PEEK	PEEK	POM	PP	○	
O	PEEK	PEEK	POM	PP		○
P	PA	PA	Steel	PP	○	
Q	BP	BP	Steel	BP	○	

POM: Polyacetal (Homopolymer)
PP: Polypropylene
MM: Composite polyacetal filled with fatty acid
PPS: Polyphenylenesulfide
PEEK: Polyetheretherketone
PA: Polyamide (6 nylon)
BP: Biodegradable plastic

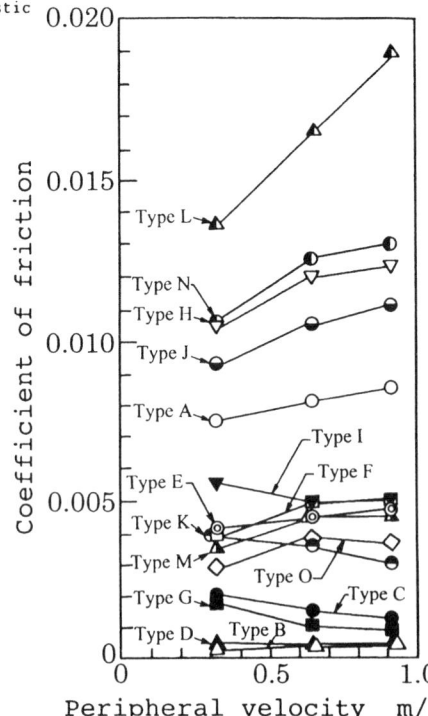

Fig.1 Coefficient of friction

ROLLING CONTACT FATIGUE STRENGTH OF INDUCTION-HARDENED SINTERED POWDER METAL ROLLERS WITH DIFFERENT GREEN DENSITIES AND ITS EVALUATION

AKIRA YOSHIDA and YUJI OHUE

Faculty of Engineering, Okayama University, 1-1 Naka 3-chome, Tsushima, Okayama-shi, Okayama 700 Japan

Abstract

In order to elucidate and evaluate the rolling contact fatigue strength of induction-hardened sintered powder metal rollers with different green densities, rolling with sliding contact fatigue test, FEM analysis of shear stress around the pores and pore distribution analysis were performed.

The fatigue-test results are shown in Fig.1. The green densities of rollers IBL, IBM and IBH are 6.5, 6.9 and 7.3 g/cm^3, respectively. Specimen marks of '30' and '60' indicate the diameter of roller. The rolling contact fatigue strength of the roller increased as the density increased. Especially, roller IHIP60 sintered by HIP had almost the same surface durability as the steel rollers. The failure mode of these sintered rollers was spalling due to subsurface origin cracks. The spalling crack depth corresponded to the depth where the amplitude of the ratio of orthogonal shear stress to Vickers hardness became maximum. Concerning the pore distribution, the pore size tended to decrease, the distance between pores increased and the pore cluster factor D_f by fractal dimension decreased, as the density increased. Any pores could not be observed in roller IHIP60.

For investigating the effects of the pore size and the distance between pores on the stress state of the sintered rollers, the orthogonal shear stress around the pore was calculated, using models with one hole, three holes and five holes in a semi-infinite plane under the Hertzian pressure. The hole size and the distance between these holes were changed, and the stress around those holes were calculated for all models. As the results of these FEM analyses, it could be recognized that the shear stress around hole tended to increase as the hole size increased and the distance between holes decreased. However, the shear stress tended to decrease, when the distance between holes became very close. To calculate the stress concentration factor, the model with three holes arranged parallel to the semi-infinite plane was adopted. Considering these results, the rolling contact fatigue strength was tried to be evaluated using the relationship between the fatigue strength reduction factor β and the modified stress concentration factor $\alpha \cdot (1 + D_f)^{0.8}$. The relationship between β and $\alpha \cdot (1 + D_f)^{0.8}$ is shown in Fig.2. The correlation coefficient γ (= 0.7654) of this relationship was higher than that γ (= 0.7311) of the relationship between β and stress consentration factor α. The rolling contact fatigue strength of the sintered material could be evaluated by that relationship between the fatigue strength reduction factor and the modified stress concentration factor. It could be understood that the rolling contact fatigue strength of sintered materials had to be evaluated not only by the pore diameter and the distance between pores but also the pore cluster factor indicated by fractal dimension.

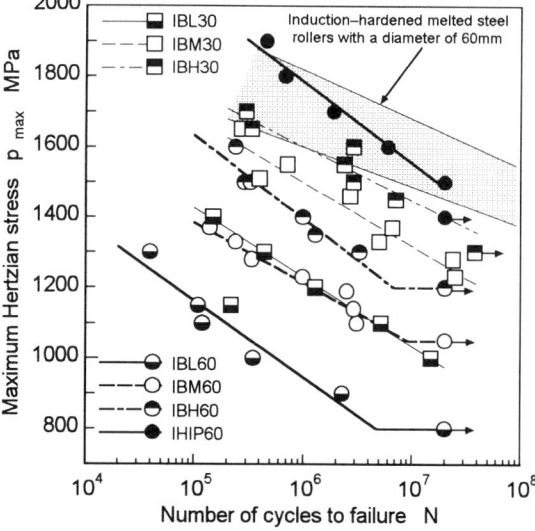

Fig. 1: p_{max} - N_2 curves

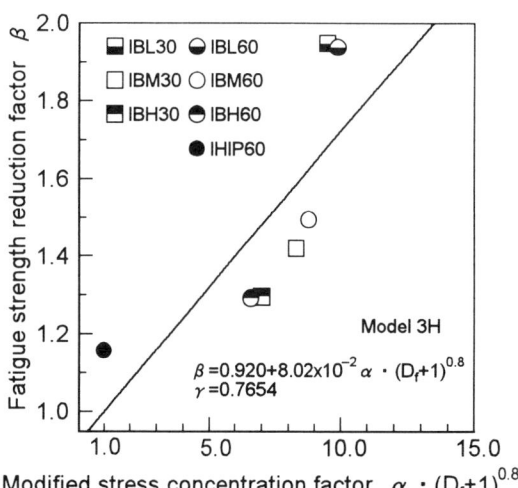

Fig. 2: Relationship between β and $\alpha \cdot (D_f + 1)^{0.8}$ curves

APPLICATION OF MULTI-LEVEL MULTI-INTEGRATION (MLMI) TO CONTACT PROBLEMS: NON-HERTZIAN CONTACTS IN ROLLING BEARINGS

S. NATSUMEDA

Basic Technology Research & Development Center, NSK Ltd.,
1-5-50 Kugenuma-shinmei, Fujisawa, Kanagawa, 251 Japan

ABSTRACT

This paper presents a computer programme for the analysis of non-Hertzian contact in rolling bearings using multi-level multi-integration (MLMI) which was first reported by Brandt and Lubrecht (1), and applied to EHL analysis by Venner (2). The advantage of MLMI is the extreme reduction in computing time compared to the conventional method, i.e. multi-integration (MI), which is expressed in the discretized form as written below.

$$f_{i,j} = \sum_{k=0}^{n_x} \sum_{l=0}^{n_y} D_{i,j,k,l} \cdot p_{k,l}$$

where $f_{i,j}$: object resultant from the contact pressure (displacement or internal stress component), $p_{k,l}$: contact pressure, $D_{i,j,k,l}$: influence coefficient, and n_x, n_y: number of mesh in x and y directions.

As the preparation to the development of the contact analysis programme, the numerical inspection was carried out to find the combination of appropriate values of MLMI parameters, which is depend on the characteristics of influence coefficient and the aspect ratio of mesh rectangle. The results were summarized into tables that are utilized in the contact analysis.

In the programme of contact problem in which the pressure correction is repeated until the condition of contact is satisfied, some techniques were adopted to ensure the fast and stable convergence. For example, the distributive Jacobi's relaxation was used for the pressure correction. This has already been reported by Venner (2), but the improvement was made to ensure the stable convergence for a large value of aspect ratio of mesh rectangle.

Using the developed programme the analyses were performed for the contact problems in roller bearings. Fig.1 shows an example of roller to inner raceway contact in NU208 cylindrical roller bearing. (Notation; p_H: maximum Hertzian pressure, L_e: effective half length of contact in axial direction, b: Hertzian half width of line contact)

Nowadays contact analysis can be performed on a personal computer within a few minutes even if the total number of grid points is more than ten thousand. In the analysis of Fig.1 the 19345 grid points were used but the computing time was within 3 minutes using the PENTIUM 133MHz microprocessor even the analysis included the calculation of internal stress distribution.

REFERENCES

(1) A.Brandt, A.A.Lubrecht, J. Comp. Phys, vol.90, 1990, pp.348-370.
(2) C.H.Venner, Ph.D. Thesis, Univ. Twente, 1991.

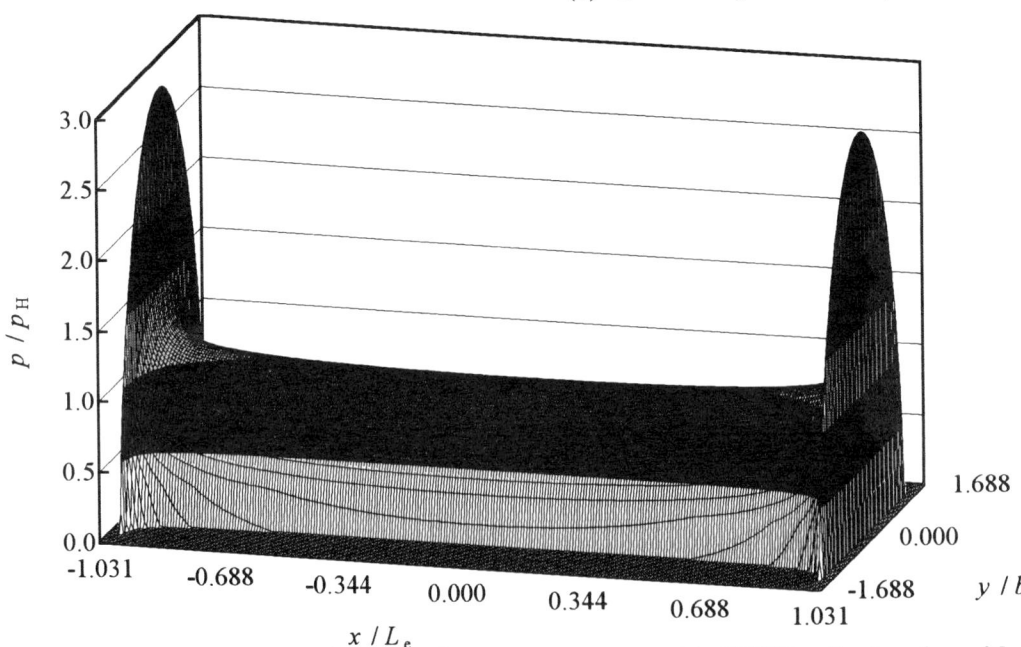

Fig.1: Contact pressure distribution of roller to inner raceway contact in NU208 roller bearing without crowning

ANALYSIS OF THE EFFECT OF THE GEOMETRY OF THE CONFIGURATION OF COOLING DUCTS ON THE HEAT ABSORBTION RATE IN WATER COOLED THRUST BEARINGS

K. K. CHATURVEDI and S. K. ROY
Bharat Heavy Electricals Ltd., Corporate R&D Division, Vikasnagar, Hyderabad-500093, India
K. ATHRE
Deptt. of Mech Engg., Indian Institute of Technology, Hauz Khas, New Delhi-110016, India
S. BISWAS
ITMMEC, Indian Institute of Technology, Hauz Khas, New Delhi-110016, India

ABSTRACT

Large thrust bearings tend to have excessive thermal deformation during operation, which can lead to unstable operation and failure. This can be reduced by circulating cooling water in the channels below the active face. The water-cooled slider bearings were analyzed by Tahara (1). Experimental measurements on these bearings were done by Kuhn (2). Johrde (3) reported development of a large water-cooled bearing. Kawaike et. al (4) described analysis and development of a thrust bearing of double layer construction. The heat transfer in the water-cooled thrust bearings was analyzed by Huffenus and Khaletzky (5) in both two and three dimensions.

In this paper, the heat transfer in a water-cooled thrust bearing has been analyzed by using a local linearization method to model the water ducts of circular shape located in a rectangular grid. A uniform temperature on the active face has been assumed. The effect of varying the geometry of the configuration of the ducts on heat absorption rate is analyzed.

The duct of a circular cross-section is located in a rectangular grid. The conductive heat transfer at the duct normal to its surface, is equated to the convective heat transfer to the water. The thermal gradient normal to the duct surface is expressed in terms of its values along the grid axes. The distance of the node located on the grid from the duct surface is considered. The finite difference expressions for the conduction equation terms are expressed for rectangular grid with varying spacing. Thus, only half the width of the pad between two adjacent nodes is considered, as the temperature distribution on both sides of the node is symmetrical.

The results for the temperature distribution conform to the circular shape of the water duct (Fig.1). Such results cannot be obtained by use of the lumped parameter method, wherein the heat transfer effects due to presence of the water duct are distributed over the nearby nodes, due to the inherent limitations of the method. The results for the effects of the variation of the active face temperature, spacing between the ducts and depth of the duct from the active face are given.

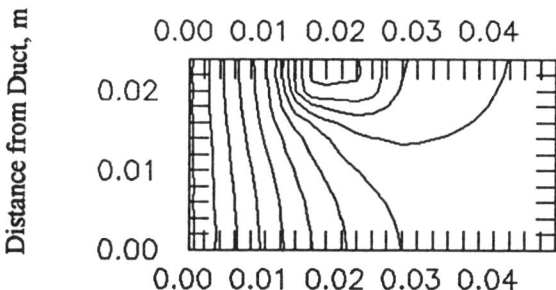

Fig. 1. Temperature Distribution in Water-Cooled Bearing

REFERENCES

(1) H. Tahara, Journal of Lubrication Technology, Trans. of the ASME, Jan., 1968.
(2) E. C. Kuhn, Power, Oct.,1968.
(3) R. S. Johrde, Proc. of CIGRE Conf., Paper No. 11-06, 1976.
(4) K. Kawaike, K. Okano and Y. Furukawa, Trans. of the ASLE, Vol. 22, No. 2, 1979.
(5) J. P. Huffenus and D. Khaletzky, Proc. of the 6th Leeds-Lyons Symposium of Tribology, Sept., 1979.

Submitted to *ASME Journal of Tribology*

TEMPERATURE RISE SIMULATION OF THREE-DIMENSIONAL ROUGH SURFACES IN MIXED LUBRICATED CONTACT

L. QIU and H. S. CHENG
Center for Surface Engineering and Tribology, Northwestern University, Evanston, Illinois 60208

ABSTRACT

The surface temperature at the interface of bodies in sliding contact is one of the most important factors influencing their wear and scuffing behavior. Blok (1) first investigated the temperature rise of contact surfaces under boundary lubrication. Jaeger (2) formulated a method to predict the temperature rise on the surface of a semi-infinite medium for moving uniform heat sources for various shapes. In addition, many papers on surface temperature analysis also have been published. However, more research on surface temperature is still needed for real rough surfaces. In this paper a numerical simulation of the temperature rise of three-dimensional rough surfaces is presented, based on the formulation developed by Lai and Cheng (3). The solution involves a mixed lubricated condition, which includes the effect of solid contact friction and lubricant shearing friction.

The calculation model uses two rough surfaces in sliding with velocities V_1 and V_2 respectively. Under mixed lubricated condition, heat is contributed by the friction of solid contact interface at the contact area and by the friction among viscous oil films at the non-contacting area. The equations of the temperature rise at the interface are given as follows based on Carslaw and Jaeger's theory (4).

Temperature rises on surface 1 and 2 are:

$$\Delta T_i(x,y,t) = \int_0^t \iint_{\Omega_c} f_i \frac{q \cdot (x',y',t')dx'dy'dt'}{4\rho c [\pi \alpha(t-t')]^{3/2}}$$

$$\cdot \exp\left\{-\frac{[(x-x')-V_i(t-t')]^2 + (y-y')^2}{4\alpha(t-t')}\right\}$$

where: $f_1 = 1 - f(x,y,t)$ and $f_2 = f(x,y,t)$
$i = 1$ or 2

The equation relating the two temperature rises in two surfaces is:

$$[T_{2b} + \Delta T_2(x,y,t)] - [T_{1b} + \Delta T_1(x,y,t)]$$
$$= [1 - 2f(x,y,t)] \cdot h(x,y) \cdot q_1 / (2K_f)$$

A numerical simulation has been developed to solve the above equations for the temperature rises of three-dimensional rough surfaces in sliding contact. The heat partition factor $f(x',y',t')$ depends on the properties of their materials and the history of contact surface temperature between them. It changes with the position of the heat sources and time. Equations are discretized to solve for the partition factor $f(x',y',t')$. With the heat partition factor $f(x',y',t')$ known, the temperature rise can be calculated. Moving Grid Method is adopted to calculate the coefficient matrix of the temperature. This greatly reduces the computing requirements. The biquadratic polynomial interpolating functions are used for the construction of the coefficient matrix. An accelerating forward iteration algorithm is used to speed up the process of solving the equations. Fig. 1 shows a temperature rise distribution of longitudinal rough surface

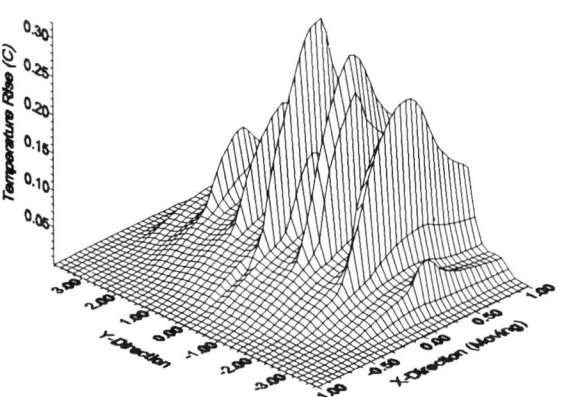

Figure 1. Temperature rise distribution of longitudinal rough surface (x 263.2 C)

By the Critical Temperature Criterion, which assumes that the local frictional coefficient of asperity contact will be changed from the lubricant film to solid asperity contact dependent on the local contact temperature, the program can simulate the surface scuffing propagation. It is useful for failure analysis in practical applications. Results for different contact conditions were verified by comparing them with results from Blok's and Francis' formulae.

REFERENCES

(1) Blok, H., Proc. General Discussion on Lubrication, London, Vol. 2, pp. 222-235, (1937).
(2) Jaeger, J. C., J. Proc. Roy. Soc., N.S.W., Vol. 76, pp. 203-224, (1942).
(3). Lai, W. T. and Cheng, H. S., ASLE Transactions, Vol. 28, 3, pp. 303-312, (1985).
(4). Carslaw, H. S. and Jaeger, J. C., Oxford Press, Second Edition, (1959).

MODERN TESTING TECHNOLOGY OF NEW FRICTION MATERIALS AND FRICTION MATERIALS WORKING CAPACITY PREDICTION FOR MOBILE VEHILES BRAKES

A.V. CHICHINADZE, E.D. BROWN
Mechanical Engineering Research Institute of the Russian Academy of Sciences
4 Maly Kharitonievsky Lane, Moscow 101830, RUSSIA

ABSTRACT

Friction units cause premature failures of motor vehicles and of equipment (in 90% of cases) and lead to wear degradation products harmful impact to ecology. Vehicle testing is rather expensive and it takes much time, information on real longevity of vehicles is obtained too late. By that time a lot of money has been spent on production management, and what is the most important, much time has been spent on manufacture, modification and testing. Testing of a real vehicle is performed, as a rule, in real time scale (not in compressed or accelerated time scale). We consider such a traditional way economically inexpedient, in particular, for large-size friction units.

Decomposition principles of real tribological system and system components analysis including scale factor and materials assessment, are in the focus of friction unit working capacity case study. Special attention is being paid to operational conditions (speed, load, temperature range, including friction heating, stress state of material within a contact zone, medium), structural design, type of material (monometal, bimetal, composite, polymer).

As a rule, physical experiment is longer and more expensive than computer simulation. In this regard, research strategy includes obtaining basic laws of friction materials behaviour, using small-size models in a compressed (accelerated) time scale, the laws, which are necessary and sufficient for decision-making concerning expediency of test materials application, of their structural design in friction unit. The above basic laws of behaviour comprise initial data mathematic modelling through wear and friction thermal dynamics methods. The authors of this paper have been guided by the following main principles:

1. Research margines revealing. Friction pair working capacity margines are being revealed according to loading, temperature, etc., basing on preliminary physical experiment carried out by friction thermal resistivity method. Further tests will present interpolation within these margines.
2. Taking into account scale factor, which makes it possible to obtain the same forces and termal fields, interacting at the model during friction unit operation, that provides identify of processes of physical and chemical mechanics between mathematic model and test sample. For this purpose a complete mathematic description of wear and friction processes is prepared and instead of separate parameters of the process (speed, loading, time, etc.), their simplexes are being used. Simplex is a ratio between similar parameters of a model and a test sample. Such a mathematic description allows to take into account scale factor for each parameter of the process as well as parameter changes while varying control parameters meaning (friction areas, dimensions, time, etc.). Tribological experiments will have accelerated character while compressing time scale in such a way.
3. Mathematic modelling and prediction of friction and wear characteristics for real friction unit with determined design operation conditions.

Solving the system of equations of friction and wear thermal dynamics (the system includes changes in speed, loading, hardness, as well as it also includes changes of wear intensity, friction coefficient, friction heating temperature, taken from preliminary physical experiment. These changes permit to obtain information required by a designer, with 10% error at most.
4. Materials assessment of materials behaviour adequacy for friction pair components, using small-size during physical modelling and further, during operation (X-ray analysis, electrone microscopy, micro- and mucrohardness, unevenness estimation, etc.).
5. Application of this scientific and engineering technique gives positive technological and economic effect for friction units development, design, manufacture, testing and operation. For instance, as for large-size vehicles - positive economic effect thanks to this technique application is up to 50 units per 1 unit of capital investment, that is why equipment cost is rapidly repaid.

REFERENCES

(1) Refernce book on tribological engineering, Vol.1 Chapter 7, and Vol.3, Chapter 7, Mashinostroyenie, 1989-1992.
(2) Tribology Fundamentals, Ed. by A.V. Chichinadze, Chapters 3,4,7,8,9,11,13, Science & Technology, 1995.
(3) Reference book: MACHINERY QUALITY, Ed. by A.G. Souslov, Vol.1,Chapter 2, Vol.2, Chapter 5, Mashinostroyenie, 1996.
(4) A.V.Chichinadze, Mechanical engineering and problems of machines reliability, Issue 5, 71-79pp., Issue 6, 74-83pp., Science, 1996.

MODELLING ASPECTS OF A RATE-CONTROLLED, THERMALLY INDUCED SEIZURE

M. M. KHONSARI
Professor and Chairman, Department of Mechanical Engineering, Southern Illinois University at Carbondale, Carbondale, IL 62901-6603

J. Y. JANG
Department of Mechanical Engineering, University of Pittsburgh, Pittsburgh, PA 15261

M. D. PASCOVICI
Department of Mechanical Engineering, Polytechnic University of Bucharest, Bucharest, 79590, Romania

ABSTRACT

Thermally induced seizure refers to a catastrophic failure of a mechanical component, such as a bearing, that experiences a rapid loss in clearance. The mechanism of failure responsible for this phenomenon is, therefore, thermoelastic expansion of the surfaces resulting from frictional heating.

A bearing is particularly susceptible to a large amount of heat generation during the start up process where the moving part (shaft) is directly in contact with the stationary component (bushing) and the coefficient of friction is large. Under normal circumstances there generally exists a residual layer(s) of lubricant on the surfaces and upon starting from rest, a fresh supply of oil is delivered into the bearing clearance so that seizure is prevented. Yet there are practical situations where the surfaces may experience "dry contact" at start up.

The problem of thermally induced seizure in journal bearings associated with dry contact during the initial start-up period has received much attention in recent years. See for example references (1-4). However, very little information is available regarding thermally-induced lubricated bearing seizure. This phenomenon is known to occur in hydrodynamic journal bearings.

To gain an understanding of the phenomenon of lubricated bearing seizure, a theoretical study is presented that deals with thermally induced seizure in an unloaded journal bearing represented by two concentric cylinders fields with lubricant. The appropriate governing equations and numerical solutions are presented for computing the thermal expansion of the rotating journal and the stationary bush as a function of time.

It is shown that a considerable amount of heat is generated within the clearance space of the cylinder that could indeed lead to a catastrophic seizure defined as a complete loss in the operating clearance. It is shown that prior to the loss of clearance, the friction torque exhibits a rapid rise. Seizure can be avoided, therefore, if the torque exerted by the motor is monitored so that the operation could be shut down prior to the complete loss in clearance. Furthermore, to *extend* the seizure time, at the design stage, one may consider choosing a low-conductivity material for the shaft along with a high-conductivity material for the sleeve. Enlarging the bearing clearance can also be considered at the design stage. Nevertheless, this option is prone to whirl instability problems with is particularly severe in lightly-loaded, vertical rotors. Enhancing the convective heat transfer coefficient can also help to extend the seizure time.

Another parameter that is found to significantly affect the seizure time is the shaft rotational speed which is assumed to remain constant in this study. The variation of speed with time would add to the complexity of the analysis. Further research is needed to quantify the effect of shaft eccentricity on the seizure characteristics of loaded journal bearings.

REFERENCES

(1) Dufrane, K. and Kannel, J. (1989) "Thermally Induced Seizure in Journal Bearings," *ASME* Journal of Tribology, v. 111, pp.288-292.

(2) Khonsari, M. and Kim, H. (1989) "On the Thermally Induced Seizure in Journal Bearings," *ASME Journal of Tribology,* V. 111, No, 4, pp. 661-667.(3) Hazlett, T. and Khonsari, M. (1992a) "Finite Element Model of Journal Bearings Undergoing Rapid Thermally Induced Seizure," *Tribology International*, v. 25, pp. 177-182.

(4) Hazlett, T. and Khonsari, M. (1992b) " Hazlett, T. and Khonsari, M. "Thermoelastic Behavior of Journal Bearings Undergoing Seizure," *Tribology International*, V. 25, pp. 178-183.

A GENERALISED MODEL OF A FRICTIONAL CONTACT: THERMOFRICTIONAL - THERMOELASTIC INSTABILITY OF SLIDING

I.L. MAKSIMOV and V.A. LAZAREV
Department of Theoretical Physics, Nizhny Novgorod University
23 Gagarin ave., Nizhny Novgorod, 603091, Russia

ABSTRACT

A generalized model is proposed of a thermofrictional contact, taking into account thermoelastic effects in the contact zone with due regard to the friction force dependence upon the interface temperature T and sliding velocity v. The criterion is derived for the collective -- thermoelastic - thermofrictional -- sliding instability emergence. Sliding stability diagram is constructed in the parameters range characterising system response with respect to the dissipative, mechanical and thermoelastic disturbances.

INTRODUCTION

It is known that the sliding is accompanied by the intense dissipation localized inside the contact zone. Such a dissipation while affecting the friction coefficient value $\mu(T, v)$ (1) may also cause a supplementary thermoelastic strain, which in its turn appreciably affects sliding dynamics (2). Below new sliding instabilities emerging due to dissipation phenomenon in the frictional contact are reported.

STATIONARY SLIDING STATE

We consider a thermoelastic plate sliding on a plane rigid substrate under action of the tangential force F. Assuming that sliding occurs under fixed vertical displacement one finds the thermoelastic component of the normal force N. The system evolution in time is governed by the Newton equation and the heat balance equation, supplemented by the normal force variation with temperature. The stationary sliding regime $dv/dt = 0$, $dT/dt = 0$ is described by the steady-state temperature $T = T_0 (v_0) = T_0 + F v_0 /(WA)$ (W is the heat transfer constant, A is the mean area of the contact) and sliding velocity v_0, which is determined by the condition: $F = N_0\ \mu(T_0(v_0), v_0)$.

LINEAR STABILITY ANALYSIS

In order to find sliding instability emergence condition one has to linearise the evolution equations with respect to small disturbances of the velocity $\delta v = v - v_0$ and temperature $\delta T = T - T_0$. Searching for the solution in the form $\delta v, \delta T \sim \delta v_0, \delta T_0 \exp(\Gamma t)$ one finds the instability increment $\Gamma = \Gamma_1 + i \Gamma_2$. The condition $\Gamma_1 > 0$ corresponds to the instability emergence. Two types of unstable behaviour are described:

i) an absolute instability ($\Gamma_1 > 0$, $\Gamma_2 = 0$) characterised by the monotonous (exponential) increase of disturbances in time and
ii) an oscillational instability ($\Gamma_1 > 0$, $\Gamma_2 \neq 0$).

The analysis shows that the sliding instability can manifest itself in two ways. First instability regime is characterised by steady plate acceleration accompanied by monotonous temperature increase inside the contact: (sign δv = sign δT). In this case the heat generation rate on the contact surface exceeds the heat removal rate to the cooler. Such a regime we define as a *runaway* instability of sliding. Another scenario of sliding instability emergence is associated with the boundless temperature increase, which takes place during deceleration (sign δv = - sign δT). The latter regime could be specified as a *thermal explosion* instability of sliding.

Sliding stability diagram is constructed in the parameters range representing system response to thermomechanical disturbances of limited magnitude. It is found that thermoelastic response inside the contact zone may provide both stabilising (near the runaway threshold) and destabilising (near the explosion instability border) impact on the dynamics of sliding.

It is shown that a generalized model explains features of the well-known experiment (2) and incorporates the constant-speed-sliding case considered in (3). Specifically, the critical sliding speed defining thermoelastic instability threshold naturally follows from our approach.

NON-LINEAR VIBRATIONS

The non-linear analysis of the increasing perturbations evolution reveals the existence of a stable limit cycle, which represents stationary oscillations of velocity, temperature and strain in the course of sliding. The oscillations frequency and magnitude are found analytically. The conditions for the experimental observation of the predicted sliding features are discussed.

REFERENCES

(1) I L Maksimov, Journal. of Tribology, Vol. 110, 1988, 69-72 pp.
(2) J.R Barber, Proc. Roy. Soc., Vol. A312, 1969, 381-394 pp.
(3) Th. Dow, R A Burton, Wear, Vol. 19, 1972, 315-328 pp.

CONTACT TEMPERATURES AND MICROSTRUCTURAL CHANGES AT FRETTING

B. PODGORNIK, J. VIŽINTIN and J. PEZDIRNIK
University of Ljubljana, Faculty of Mechanical Engineering, Centre of Tribology and Technical Diagnostics,
Bogišičeva ul. 8, 1000 Ljubljana, Slovenia
F. VODOPIVEC
Institute of Materials and Technologies, Lepi pot 11, 1000 Ljubljana, Slovenia

ABSTRACT

The contact temperature is a dependent variable, being a function of size and shape of the real contact area, along with the friction coefficient, normal load, sliding velocity and thermal properties of the contacting bodies. Almost in all sliding situations contact occurs not at a single spot, but at several microscopic contact spots within the nominal contact area. The frictional heating, which is generated only in real contact regions, causes high flash temperatures and relatively steep temperature gradient in the subsurface layer. In any sliding system, the temperature of the contact interface may have a significant effect on the tribological behaviour of the contacting materials. Temperature dependence of the microstructure and the mechanical and physical properties of the contacting solids affect considerably the contact configuration and the wear process. The extent of the temperature rise in the fretting contact zone has been a subject of considerable interest, but the literature reports vary significantly in this matter. Some authors reported very low contact temperatures in fretting conditions, even bellow 10°C. They reported that white phase layers may form in steels without large temperature rise due to the large plastic strains which may be produced under conditions of high hydrostatic pressure. Meanwhile the others reported temperatures in the range from 500 to 1000°C, which also causes formation of white phase layers.

The contact surface temperature model used in most engineering applications of dry and boundary lubricated sliding has been the classical Blok model. This model considers the contact surface temperature problem as a semi-infinite body subject to a single concentrated heat source. The validity of this assumption may be questionable for many practical dry and boundary lubricated sliding conditions. In many sliding situations the size of the contacting bodies is finite and contact occurs not at a single spot, but at several microscopic contact spots within the nominal contact area.

Shearing of asperities in the initial stage of fretting cause detachment of the particles from one of the sliding bodies and because of the small amplitudes only a few particles could appear within the nominal contact area. Therefore the real contact area consists of the limited number of contact spots between asperities of both contact bodies and particles trapped in the interface and only a fraction of nominal contact area may be in actual contact. Frictional heating generated in the contact is therefore generated only at microcontacts and causes large flash temperatures.

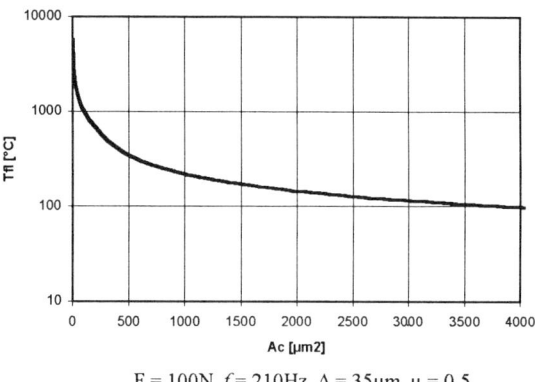

$F = 100N, f = 210Hz, \Delta = 35\mu m, \mu = 0.5$

Fig. 1: Flash temeprature as a function of contact area

The local rapid heating and subsequent quenching caused by the cold bulk of the surrounding material can cause local changes in the surface and subsurface structure.

White phase area starts to form under individual contact spots as a result of high flash temperatures and large plastic strains. With the duration of the contact islands of white phase area grow by coalescence to a single large area.

Several techniques were used to examine the microstructure and microstrucural changes generated by fretting of a AISI 52100 steel at different amplitudes, loads, frequencies and test times. The main purpose of our study is to determine the reason of white phase layer formation, the influence of temperature and pressure and to explain the growth process of the white phase layer at fretting. Results of the experiments, microstructural analysis and comparison between theoretical and experimental results are also presented in the paper.

A MODEL FOR THERMOSLIDING ADHESIVE WEAR

A.TUDOR

Machine Elements & Tribology Department, "Politehnica" University of Bucharest, 313 Independenţei Spl.,Bucharest, RO

ABSTRACT

In recent years it has been noticed some new efforts for systematisation of wear theory (1), (2). The correlation between material, environment and operation parameters defines the wear maps. A localised change in material properties, an enhancement in fatigue and, ultimately, failure of the mechanical pairs can result. Cowan and Winer (3) proposed, based on the thermomecanical wear, a model for the plastic deformation with material field criteria.

In friction contact the tangential stresses are variable and the wear process is accepted as a cumulative fatigue phenomenon (4). In relative sliding process between friction surfaces, the tangential stress (τ) on real contact area determine the fatigue of surface layer and the appearance of wear particles. The tangential stresses are determined by both normal and tangential loads and friction forces. The correlation between normal stresses $\{\sigma\}$ and the friction coefficient f, after the molecular - mechanical friction theory is

$$f = f_0 + \tau_f/\{\sigma\} \qquad (1)$$

where f_0 is the molecular friction component (property of material); τ_f - the shearing strength of surfaces layer (property of both materials in contact defined as property of material pair); $\{\sigma\}$ - the total normal stresses including both isothermal σ^i and thermal σ^t components. The friction fatigue phenomenon is defined by the fatigue law

$$\tau N^c \geq \tau_c = m\sigma_c \qquad (2)$$

where N is the number of cycles for the appearance of wear particle; c - a fatigue material coefficient; τ_c, σ_c, critical tangential and normal strength (brittle or ductile material) for one cycle; m - proportionality constant.

The dimensionless Hertzian fatigue pressure $\bar{p}_{of}$ is obtained as:

$$\bar{p}_{of} = \frac{p_0}{\sigma_{co}G_m} \geq \frac{1}{\Phi(G_t,c_f,F_0) + c_t G_t T^x} \qquad (3)$$

where: p_0 - initial Hertzian maximum pressure; σ_{co} - reference failure strength at reference bulk temperature; G_m - friction fatigue parameter of material; c_t - thermal coefficient; G_t - Cowan and Winer's sliding wear control parameter; c_f - fatigue parameter; F_0 - Fourier number; T^x - dimensionless temperature; $\Phi(G_t, c_f, F_0)$ - normal isothermal and thermal stresses function dependent on the dimensionless isothermal stresses (Hamilton and Goodman's solution (5)) and thermal stresses (Zang and Winer's solution (6)).

The friction fatigue wear maps can be obtained only by numerical solution for the dimensionless Hertzian fatigue pressure, the Fourier number and G_t parameter. In the case when the friction fatigue parameter of material $G_m = 1$ ($\tau_f = 0$ and $N = 1$) and it is accepted the plastic deformation, the friction fatigue wear map has the Zang and Winer's form.

The experiments were conducted on the spherical pin-disc machine. The wear of spherical pin was appreciated by the average diameter of the contact trace. The wear of the disc was evaluated by the average thickness lost of the worn layer. The agreement between experiments and theory, evaluated by the friction fatigue wear map (Fig. 1) is good.

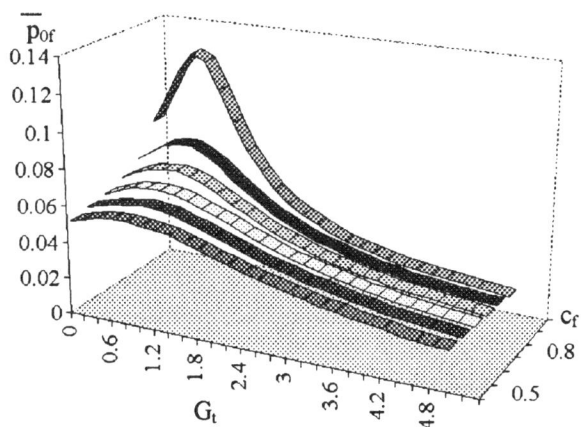

Fig. 1: Map of thermosliding adhesive wear ($F_0=2$)

REFERENCES

(1) K. Hokkirigawa, Eurotrib 93, 5 (1993), 32.
(2) S. C. Lim, M. F. Ashby, Acta Metall., 35, 1 (1987).
(3) R. S. Cowan, W. O. Winer, Tribotest,vol.1, No. 2 (1994).
(4) A. Tudor, Eurotrib 93, 5 (1993), 59.
(5) G. M. Hamilton, L.E. Goodman, Journ. of Applied Mechanics, vol.33, p.p.371 (1966).
(6) J. Yang, W. O. Winer, Journ. Tribology, 113, 262 (1991).

SOME CONTACT PROBLEMS WITH HEAT GENERATION

O.UKHANSKA[1] and V.ONYSHKEVYCH[2]

[1]Departament of Geodesy, University "Lviv Politechnica", Bandera Str. 12, Lviv, 290646, UKRAINE
[2]Departament of Matematics, Ukrainian Forestry University, General Chuprynka Str. 103, Lviv 290057, UKRAINE

ABSTRACT

The work of gaskets, brake blocks, friction strengthens etc. may be simulating by the following problem.

The plane punch of the height H with a plane base is pressed by the force P in an elastic half-plane and moves along creating a line with a constant velocity. The heat exchange between the side surfaces of the punch and the external medium occurs according to New-ton's law with the coefficient of heat exchange γ_a. The upper end of the punch carries out a heat exchange with external medium with the coefficient γ_H. The heat fluxes have been generated due to the action of frictional forces. The unloaded surface of half-plane is heat insulated.

Mathematically this problem may be describe by the equations of thermoelasticity (plane case) with heat boundary conditions

$$\frac{\partial t^{(1)}}{\partial y} = \gamma_H t^{(1)}; \quad |x| \le a, \ y = 0$$

$$\frac{\partial t^{(1)}}{\partial x} = \mp \gamma_a t^{(1)}; \quad 0 \le y \le H, \ x = a$$

$$\lambda^{(1)} \frac{\partial t^{(1)}}{\partial y} - \lambda^{(2)} \frac{\partial t^{(2)}}{\partial y} = -f_T V \sigma_y(x);$$
$$|x| \le a, \ y = H$$

$$\lambda^{(1)} \frac{\partial t^{(1)}}{\partial y} + \lambda^{(2)} \frac{\partial t^{(2)}}{\partial y} = h\left(t^{(2)} - t^{(1)}\right);$$
$$|x| \le a, \ y = H$$

$$\frac{\partial t^{(2)}}{\partial y} = 0; \quad |x| > a, \ y = H$$

and mechanical conditions

$$v(x,H) = f(x) + \delta; \quad |x| \le a; \ y = H$$
$$\sigma_y(x,H) = 0; \quad |x| \ge a; \ y = H$$
$$\tau_{xy}(x,H) = 0; \quad |x| < \infty; \ y = H$$

By using a finite difference approximation of the heat conductivity equation for the punch and corresponding boundary conditions on the coordinate x the solution of problem for the strip is constructed by the method of straight lines. The solution of the thermoelasticity problem for half-plane is obtained by integral Fourier transformation.

Numerical computations were carried out for the case when the material of the strip is steel and the material of the half-space is aluminium. In the result of the computation a change of σ_y sign on the area $[-a,a]$ was revealed. This confirms the existence of half-plane receding zones from the strip. Corresponding figures for the thermal fields, heat fluxes and contact stresses in the interaction bodies are given.

Note, the similar problems are investigated earlier in [1-3].

REFERENCES

(1) J.R.Barber, Q.J.Mech.Appl.Math., 1982, Vol.35, 141-154pp.
(2) V.P.Levytsky, Int. J. Engng.Sci., 1994, Vol.32, 1693-1702pp.
(3) V.P.Levytsky, V.M.Onyshkevych, Int.J.Engng.Sci., 1996, Vol.36, 101-112pp.

REACTIVITY OF NASCENT STEEL SURFACES CREATED BY FRICTION OR ION ETCHING : EFFECT OF OXYGEN AND HEXANE

M. BOEHM* and **, Th. LE MOGNE* and J. M. MARTIN*

* Laboratoire de Tribologie et Dynamique des Systèmes, École Centrale de Lyon, UMR 5513, BP 163 - 69131 Écully Cedex, France.

** Pechiney - Centre de Recherches de Voreppe, Centr'Alp, BP 27, 38340 Voreppe

INTRODUCTION

Highly reactive nascent surfaces may play an important role in the formation of tribochemical films and therefore in boundary lubrication (1). In particular, it is vital for the lubrication processes where large areas of nascent surface are likely to be created as in cold rolling or more generally in metal forming.

The present study proposes to compare the reactivity of different nascent steel surfaces with molecules used to simulate the lubricant components and additives by their chemical function.

EXPERIMENTAL

Chemical nature of nascent surfaces is investigated by X-ray photoelectron spectroscopy (XPS) and by Auger electron spectroscopy (AES) in a ultrahigh vacuum tribometer. Different nascent steel surfaces (Ti-IF steel) are compared. Their are created :

- by repeated cycles of ion sputtering,
- after several friction cycles on an ion etched plane,
- during sliding.

Reactivity of nascent surfaces is studied by static and friction experiments in presence of oxygen and hexane which are known to behave very differently (2). Friction experiments are performed using the pin (AISI52100) on plane tribometer with different conditions of temperature and pressure. The compounds are introduced through a leak valve.

RESULTS

Static experiments show that the nascent surface is highly reactive even in residual vacuum ($P<5.10^{-8}$ Pa) and that this reactivity depends on the cleanliness of the surface. Reactivity is important for oxygen but almost non measurable for hexane.

The friction experiments are in good agreement with the static ones. The high reactivity of oxygen decreases the coefficient of friction by forming a protective oxide layer which prevents strong metal-metal adhesion. This phenomenon occurs even at low exposure of 10^{-3} Pa. This pressure can be linked to the coverage of the whole friction scar by one chemisorbed oxygen monolayer. The rate of oxide formation increases with temperature. Friction experiments are not modified by the introduction of hexane even at a pressure of 1000 Pa. Hexane does not seem to enter into the contact and is likely to be just weakly physisorbed.

This study is part of the 'Contrat de Programme de Recherches : Mise en Forme des Matériaux : contact outil—produit-lubrifiant' between CNRS, l'Irsid, Pechiney CRV, l'ECL (LTDS), l'INSA Lyon (LMC), l'ENSMP (CEMEF), l'INPT (IMF), le Collège de France (PMC), l'Université d'Orsay (LMS) et le CNRS (SCA).

FIGURE

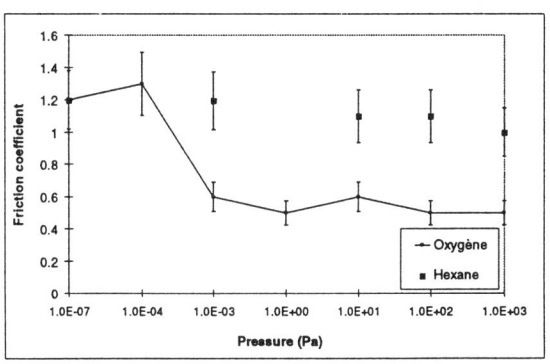

Fig.1 : Evolution of the friction coefficient for different pressures of oxygen and hexane

RÉFÉRENCES

(1) S. Mori and Y. Imaizumi, Adsorption of model compounds of lubricant on nascent surfaces of mild and stainless steels under dynamic conditions, STLE trans., vol. 31, 4 (1988) 449-453

(2) S. Mori, Tribochemical activity of nascent metal surfaces, Proc. International Tribology Conference, Yokohama (1995), 37-42

DRY FRICTION BEHAVIOUR OF SiC BASED CERAMICS AT HIGH SLIDING SPEED

Y.M. CHEN, J.C. PAVY and B. RIGAUT
Tribology laboratory, Technical Center of Mechanical Engineering CETIM, BP 80067
60300 Senlis, France

ABSTRACT

Silicon carbide has been increasingly used in seal applications for its high wear resistance, especially in abrasive, erosive and corrosion environment. But it has often to be used against graphite because high and instable friction was often observed when SiC rubbing on itself. In order to investigate tribological behaviour of self-lubricated SiC, four SiC based ceramics were tested against a sintered SiC under the conditions near to those of sealing applications.

The pin materials were a self-sintered high density SiC(A), a reaction bonded SiC(B), a composite SiC-C produced by chemical vapor reaction of siliconized graphite (C), and a reaction bonded SiC-C composite produced by infiltration of molten silicon in a SiC-C matrix. The two composites have free graphite particles homogenously dispersed in a SiC matrix. The disk material was a sintered monolithic SiC.

The wear tests were carried out on a pin-on-disk machine (flat on flat contact) under dry condition with 50% relative humidity, low pressure (0.25 MPa), and at high sliding speed (10m/s).
The pin diameter was 5 mm and that of disk 160 mm. The sliding distance was 72 km. Each test was repeated three times to analyze reproductibility. Friction coefficient and linear wear of pin were continuously measured during the test (1).

As we can see in figure 1, friction coefficient for the self-lubricated SiC-C (C and D), was just slightly lower than that of the monolithic SiC (A and B). The wear rate of the self-lubricated SiC-C was generally one or two times lower than that of the monolithic SiC (figure 2). If we compare the tests in the same group, we note that tests for the SiC-C were almost 10 times more reproducible than those of the monilithic SiC. In addition, more vibrations and noises were found during the tests of monolithic SiC (2).

So the self-lubricated SiC-C can not only decrease the wear rate by one or two times, but also make the friction behaviour more stable.

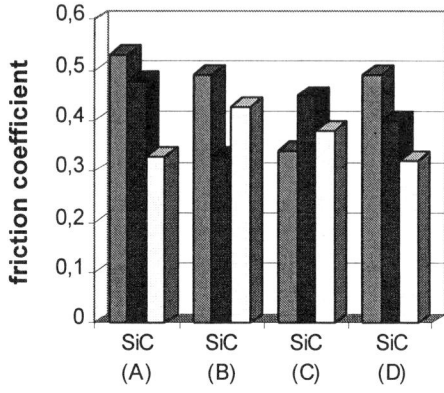

Figure 1 : Friction coefficient

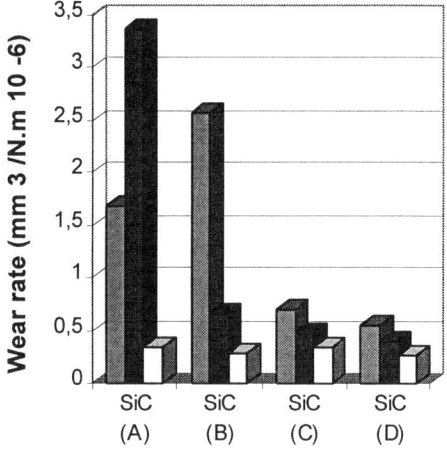

Figure 2 : Wear rate of pin materials

REFERENCES
(1) B. RIGAUT, Y.M. CHEN, J. SAINT-CHELY, Wear Beavior of Al_2O_3, Si_3N_4 and CBN cutting Tool Materials at High Sliding Speed - Journal of STLE, June 1994, 485-489 pp.

(2) Y.M. CHEN, J.C. PAVY, B. RIGAUT, Friction and wear of ceramic composites materials, CETIM report N° 181390, 1995.

A MECHANISM FOR MECHANOCHEMICAL WEAR OF MODEL SYSTEMS: NANO-TRIBOLOGY STUDIES OF CARBONATE AND PHOSPHATE SINGLE CRYSTAL SURFACES IN AQUEOUS MEDIA

J. T. DICKINSON, L. SCUDIERO, N. S. PARK, M. W. KIM, AND S. C. LANGFORD

Department of Physics and Materials Science Program
Washington State University, Pullman, WA 99164-2814 USA

ABSTRACT

In several mechanical wear situations, e.g., those involving biomaterials and applications of mechanochemical polishing,[1] a surface experiences simultaneous tribological loading and corrosive chemical exposure; the combination can greatly increase wear rates. We examine the exposure of single crystal calcite [$CaCO_3$],[2,3] dolomite [$MgCa(CO_3)_2$], and brushite [$CaHPO_4 \cdot 2H_2O$] to buffered aqueous solutions and mechanical stimulation with an Scanning Force Microscope (SFM) tip. Tip radii and cantilever force constants are calibrated. Silicon nitride tips are used with applied normal loads from 0-300 nN, tip radii ~30 nm and tip velocities from 1-200 μm/s. We present the influence of normal force, tip velocity, and solution chemistry on the rates of corrosive wear of calcite and dolomite. Images of the wear of atomic steps can be used to examine the wear rates and propagation of dissolution around the stimulated region. Mechanical stimulation includes small area scans, linear reciprocation, and indentation. Quantitative data on wear rate on the nanometer scale and single atomic layer dimensions is readily obtained.

A very important finding is a highly *non-linear* dependence of the wear rate on the applied normal force, essentially exponential. Our results are interpreted in terms of a mechanically enhanced double kink nucleation mechanism for dissolution. A double kink in an initially straight step could be as small as one or two missing ions. Assuming elastic deformations only, the average compressive stress, σ_{avg}, in a soft material beneath a hard, spherical indentor is given by the Hertz relation:

$$\sigma_{avg} = \frac{2}{3\pi}\left(\frac{12 F_N E^2}{R^2 (1-v^2)^2}\right)^{1/3} \quad (1)$$

where F_N is the contact force, and R is the radius of curvature of the indentor. E and v are the Young's modulus and Poisson ratio of the soft phase, respectively. If we then assume a stress dependent decrease in the activation energy for double kink formation, the growth rate of the wear track, V, is given by a Zhurkov-Arrhenius expression:

$$V = V_o \exp\left(-\frac{E_{act} - v^*\sigma}{kT}\right) = V_o' \exp\left(\frac{v^*\sigma}{kT}\right) \quad (2)$$

where V_o is the appropriate pre-exponential, E_{act} is the zero stress activation energy for double kink nucleation and v^* is an activation volume, and $V_o' = V_o \exp(-E_{act}/kT)$. Combining Eqs. (1) and (2) yields the predicted stress dependence of V. Wear rates on calcite and brushite steps have been acquired over a large range of normal forces. Curve fitting to our experimental measurements on several surfaces yields values of v^* very close to the volume of a *single lattice ion*. Thus, the activation of a single lattice site is sufficient to initiate double kink formation followed by rapid removal of a large number of remaining step ions (until annhilated or "poisoned" by impurities). Studies of planarization of surfaces and formation of desired nanostructures are also presented along with redeposition structures. For example, in brushite we are able to make atomically flat regions of many microns in dimension. Likewise nanoscale structures are also possible.

REFERENCES

(1) M. A. Martinez, Solid State Technology 37, 26 (1994).
(2) N.-S. Park, M.-W. Kim, S. C. Langford, and J. T. Dickinson, Langmuir 12, 4599-4604 (1996).
(3) N.-S. Park, M.-W. Kim, S. C. Langford, and J. T. Dickinson, J. Appl. Phys. 80, 2680-2686 (1996).

This work supported by the U. S. National Science Foundation, Surface Engineering and Tribology Program.

3rd BODY TOPOGRAPHY AND SLIDING SPEED: THE ROLE OF STICK-SLIP AND TRIBOCHEMISTRY

PASCAL DISS and MARCEL BRENDLE
Institut de Chimie des Surfaces et Interfaces (ICSI) (C.N.R.S.) -BP 2488
F 68057 MULHOUSE Cedex (France)

ABSTRACT

Although the determining role of a third body in dry contacts is now generally accepted, it still remains a black box where chemistry and mechanics permanently interact. While tribochemistry is often associated with dry contacts, its contribution to a possible change in surface topography is seldom considered. New asperities may build-up by the reagglomeration of small debris detached from the first bodies. By using a model system where a pin of graphite is made to rub in the pin on disc geometry on thoroughly polished steel, the third body generally consists of isolated transfer particles, directly identifiable to these asperities. Their 2D characteristics are therefore easily quantified by image analysis i.e. by the particle density N and the area fraction X covered by the particle transfer. Previous investigations have shown that these characteristics reflect the influence of most parameters and allowed us to propose various models accounting for the asperities stability in terms of adhesive interactions (1), sliding speed and of stick-slip motions (2). The aim of this work is to complete the previous models by introducing tribochemistry both at the level of the surface interactions and in the cohesive bonding between debris.

The experimental support is provided by the systematic study of the variations of the frictional force and of the particle transfer characteristics for constant sliding distances, as a function of sliding speed **v** and temperature. As illustrated in Fig.1, they all display a linear dependence of the form:

$$y = a \log v + b$$

As long as the slope **a** of these lines remains negative, i.e. until 90°C, these variations may be interpreted in terms of stick-slip as reported in a previous paper (2). In contrast, for temperatures around or above 130°C, the variations display an opposite trend, which appears difficult to be explained by the previous model unless a second phenomenon i.e. the sliding speed dependence of the superficial tribochemistry (including the occurrence of dandling bonds) is taken into account.

The fact that the variations of the area fraction **X** display a slope **a** progressively increasing with temperature from negative values to positive values as illustrated in Fig.2 (a very similar trend is observed for the variations of the particle density) strongly suggests that stick-slip and tribochemistry are always acting simultaneously in some unknown proportion, this proportion varying progressively with temperature.

The temperature dependence of the nature of the superficial oxygenated chemical groups as revealed by XPS analyses is also considered.

The new model accounting for most observations and encompassing the previous models of stick-slip will be described extensively in (3).

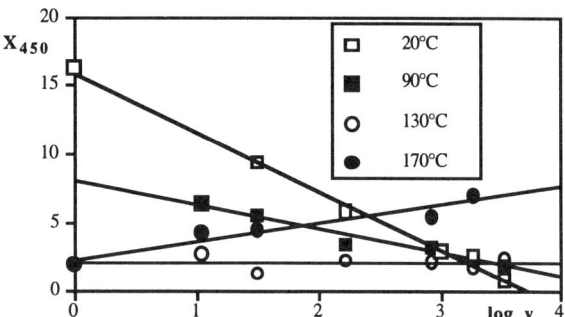

Fig.1 : Variations of the area fraction X_{450} observed after identical sliding distances (here 450 revolutions or 40.5m) under ambient air, as a function of sliding speed v (mm/s). Each line corresponds to a different series of experiments performed at a different temperature and is characterised by its slope **a**.

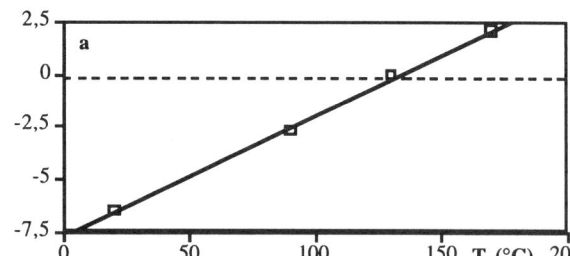

Fig.2 : Variations of the slope a (from X = a log v + b shown in Fig.1) as a function of the corresponding temperatures.

REFERENCES

(1) M. Brendlé, P. Turgis, S. Lamouri, Trib. Trans., (39), 1, page 157, 1996
(2) P. Diss, M. Brendlé, Wear 203-204, page 564, 1997
(3) P. Diss, M. Brendlé, Trib. Trans.(revised form in progress)

THE CORROSIVE WEAR MECHANISM IN SOLUTION

A. IWABUCHI, T. TSUKAMOTO, T. SHIMIZU and H. YASHIRO
Faculty of Engineering, Iwate University, 4-3-5 Ueda, Morioka 020, Japan

ABSTRACT

In the previous papers authors noted that corrosive wear was mainly determined by electrochemical dissolution, where the wear volume was equivalent to the corrosion volume calculated from Faraday's law with the anodic current density obtained by potential pulse (PP) method (1). Then, they proposed the new corrosive wear model based on the concept that the corrosion rate on worn surface was different from that on unworn surface. The mechanical factor was merely to expose the fresh surface, here. This PP method resembles the exposure of fresh surface during sliding. However, the corrosion volume has not actually measured yet for the confirmation. In this paper, wear test was carried out to evaluate the corrosion volume in terms of dissolved iron determined by absorption spectroscopy as well as the calculation by Faraday's law.

An experimental apparatus provided reciprocating motion to a flat specimen of SUS304 (type 304) stainless steel (HV 253), pressed by an Al_2O_3 ball (HV 1439). The solution was 0.1 mol/dm^3 Na_2SO_4 with normal pH of 6.5. The experiment was done in open air without circulation.

The experimental conditions of the wear test were normal loads of 10 N, 30 N and 50 N, a peak-to-peak slip amplitude of 5 mm, a frequency of 8.33 Hz, and the maximum number of repeated cycles of 5×10^4. The potential of the specimen was varied from -1100 mV (vs. SCE) to 1000 mV.

For the calculation of corrosive volume two different polarization curves were obtained: an ordinal one and the PP method. The polarization curves obtained are shown in Fig. 1. From these curves the corrosion, i.e. dissolution of metal from the fresh surface occurs even in the cathodic region for ordinary surface between -199 mV and -1090 mV. In the anodic region metal on the fresh surface dissolves with a rate of four order of magnitude greater than that on the ordinary surface.

The relationship between wear volume and potential at different normal load after 50,000 cycles shows the increase in wear with potential. The wear volume at -1100 mV represents the pure mechanical wear, because every component in the alloy is in the immune region.

Figure 2 shows the three volumes noted above against the potential at 30 N after 50,000 cycles. In the figure three curves are qualitatively and quantitatively quite similar over the potential.

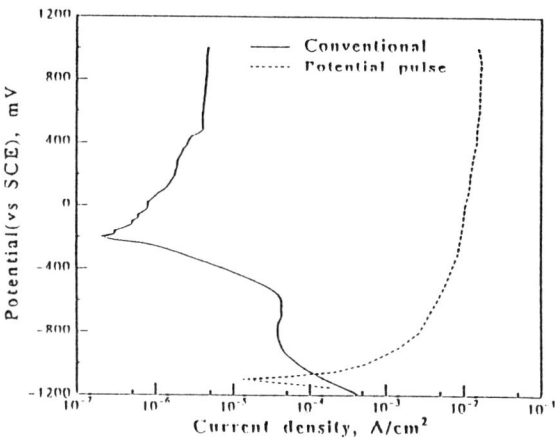

Fig. 1: Polarization curves of the conventional method and the potential pulse method

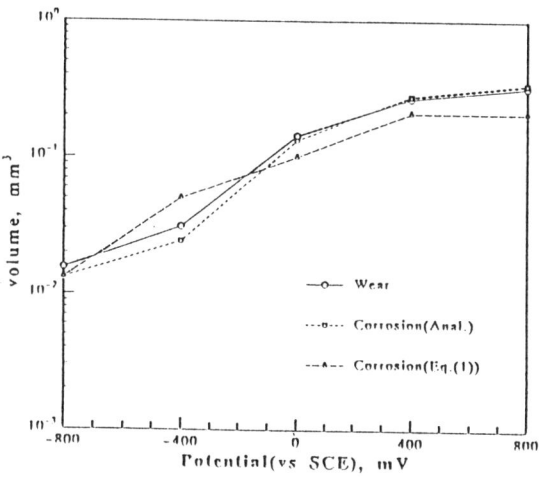

Fig. 6 The comparison between wear volume, calculated volume and analyzed volume against the potential at 30 N

As a result, both corrosion volumes were almost the same as the wear volume. Therefore, it was confirmed that the estimation of the corrosion volume according to Faraday's law with the anodic current density obtained by the potential pulse method was reasonable and the electrochemical factor was significant in the corrosive wear.

REFERENCES

(1) A. Iwabuchi, et al. Wear, 156, pp301-313, 1992.

Submitted to *Lubrication Engineering*

TRIBOCORROSION IN AQUEOUS LUBRICANTS

S. MISCHLER, S. DEBAUD, D. LANDOLT
Ecole Polytechnique Fédérale de Lausanne, Materials Department, LMCH
CH-1015 Lausanne

ABSTRACT

A quantitative model to describe the corrosion behaviour of passive metals subjected to two-body abrasive wear in corrosive environments has been proposed and tested experimentally using a reciprocating pin on plate tribometer equipped with an electrochemical set-up. An alumina pin was used in conjunction with iron, nickel, chromium, stainless steel and titanium alloy plates in sulfuric acid or sodium sulfate media. The model is based on a consideration of multiple asperities contact between the sliding surfaces and it takes into account the passivation behavior of the metal in the corrosive environment. The model predicts the metal removal rate by anodic dissolution as a function of operational parameters (speed, stroke length, stroke frequency, normal force) and of mechanical (hardness) and electrochemical (passivation charge) properties of the metal. A good correlation between the model predictions and the experimental results was found provided no modifications of the pin surface occurred during rubbing and the IR corrected potential in the wear scar was situated well in the passive potential region.

INTRODUCTION

Passive metals such as stainless steels and titanium alloys own their good corrosion resistance to a thin (2-3 nm thick) surface oxide film (passive film) formed by reaction of the metal with the corrosive medium. This film represents a barrier separating the metal from the solution and therefore, it protects the underlying metal against corrosion. Due to their corrosion resistance passive alloys are widely used for tribological applications in corrosive environments (tribocorrosion systems) such as in orthopedic implants, in food processing or in the mining industry, where materials are subject to a combined attack by mechanical (wear) and chemical (corrosion) phenomena. However, in tribocorrosion systems rubbing may lead to a loss of the protective properties of passive films resulting in an increased corrosion rate (wear accelerated corrosion). Wear accelerated corrosion in addition to mechanical wear can lead to rapid degradation of the performance of tribocorrosion systems.

The tribocorrosion behaviour of passive metals has been studied by a number of authors using sliding wear test rigs allowing for electrochemical polarization of the samples. The combination of mechanical and electrochemical methods offers the possibility to carry out friction and wear tests under well defined corrosion conditions by controlling the electrode potential. In addition by measuring the current it is possible to quantify, according to Faraday's law, the metal loss due to corrosion processes which occur during rubbing. A significant increase in anodic current with the onset of rubbing was observed by several authors and interpreted as resulting from a periodic removal of the passive film by abrasion leading to subsequent exposure of bare metal to the corrosive solution followed by repassivation. The extent of wear accelerated corrosion has been found to depend on several parameters such as the nature of the metal and the environment the applied normal force and the electrode potential as well as, for rotating motion devices, from the rotation rate. However no quantitative models describing the extent of wear accelerated corrosion of passive metals in aqueous solutions as a function of these parameters are available at this time.

Previous modelisation attempts were made by Abd-El-Kader and El-Raghy for repeated sliding and Adler and Walters for single scratch tests. Abd-El-Kader and El-Raghy related the materials degradation only to the passivation kinetics by assuming that the degradation proceeds by successive stages of build-up and complete removal of the passive film from the rubbing surface. Their model does not describe however the experimentally observed influence of the normal force. The model proposed by Adler and Walters relates the extent of bare metal surface to the plastic deformation which depends on the normal force. On the other hand the model does not take into account the influence of the electrode potential.

The aim of the present paper is to develop a theoretical model describing quantitatively the corrosion rate of passive metals under tribocorrosion conditions as a function of materials parameter and of mechanical and chemical solicitations. The validity of the model is tested by experimental measurements carried out on a reciprocating pin-on-plate tribometer equipped with an electrochemical set-up using different passive metals and alloys in acid and neutral sulfate solutions.

STUDY OF THE TRIBOCORROSION MECHANISMS UNDER ELECTROCHEMICAL POTENTIAL CONTROL

J. TAKADOUM, H. HOUMID BENNANI
Laboratoire de Microanalyse des Surfaces, ENSMM, 26, chemin de l'Epitaphe,
25030 Besançon Cedex, France
P. ZECCHINI
Laboratoire de Cristallographie et Chimie Minérale, UFR Sciences, Route de Gray,
25030 Besançon Cedex, France
A. BENYAGOUB
Institut de Physique Nucléaire de Lyon, IN$_2$P$_3$-CNRS, 43 bd du 11 novembre 1918, 69622 Villeurbanne Cedex, France

ABSTRACT

Wear-corrosion or tribocorrosion is a joint effect of mechanical and electrochemical processes which contribute to material removal from surfaces in sliding contact.

During a tribocorrosion test, the electrochemical potential of the metal determines the surface properties and affects the friction and wear. Depending on the applied potential, the metal may be protected cathodically or may be in the anodic region. The latter may correspond to active dissolution or passivation of the metal.

The formation of a passivating film with good protective qualities enhances the resistance of the metal to corrosion. During sliding, this film may be broken when the applied load and/or the sliding speed are high enough.

The experiments have been conducted using a new apparatus designed and built for evaluating cojoint action of corrosion and wear. It consists of a teflon celle mounted on a ball-on-disc tribometer. The rider is a ball of polycristalline alumina sliding in a reciprocal motion on the flat specimen. The electrochemical potential of the specimen (the working electrode) was controlled with respect to an Hg$_2$SO$_4$ electrode (ESS). The apparatus enables measurements of friction coefficient under electrochemical potential control (1).

Many authors investigated experimentally friction and wear under conditions of controlled potential and showed that wear process involves combined chemico-mechanical effect (2-5). A synergistic effect between corrosion and wear leads to accelerate material removal.

The total volume of material removal (Vt) consists of three components :

$$Vt = Vw + Vc + Vs$$

where Vw is the pure mechanical wear loss. It is obtained after wear tests conducted on a cathodically protected specimen. Vc represents material removal due to corrosion reaction and Vs represents the synergistic effect when corrosion and abrasion operate simultaneously.

The objective of this paper is to report on a study of tribocorrosion of nickel and 316 steel when sliding against an alumina ball in sulfuric acid.

Using the three-dimensional profilometry, we have measured the volume of material removal for four specimens of nickel tested at different potentials in the anodic region (active dissolution).

The results are presented in figure 1. They show that the synergistic effect between corrosion and wear, which lead to accelerate material removal, increases with potential (corrosion component).

For the specimen passivated (500 mV), the surface of the wear track shows that the metal did not suffer any wear indicating that the protective quality of the passivating film is excellent.

Figure 2 shows the volume of material removal for nickel and 316 steel in the passivating region. It appears clearly that the passive film formed at the surface of nickel protects better the metal against wear.

In the case of the 316 steel, the composition of passivating film was studied by RBS while the iron, nickel, chromium and molybdenum contents dissolved in the electrolyte before and after wear tests were determined using the Inductively Coupled Plasma Spectroscopy (ICP) at various potentials.

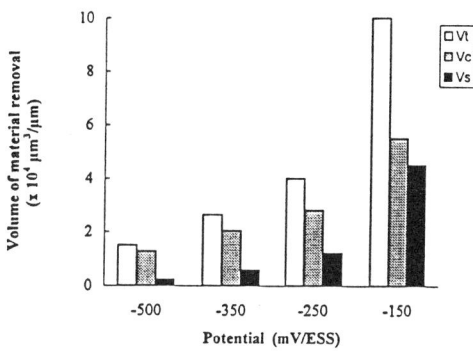

Fig. 1 : Variation of Vt, Vc and Vs with potential. The volume Vw measured in the cathodic region at - 1300 mV is 103 µm^3/µm.

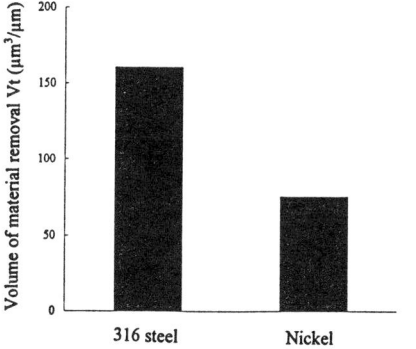

Fig. 2 : Volume of material removal after wear tests in the passivating region.

REFERENCES

(1) J. Takadoum, Corrosion Science, Vol.38, N°4, page 643, 1996

(2) T.C. Zhang, X.X. Ziang, S.Z. Li, X.C. Lu, Corros. Sci., 36 (12), page 1953, 1994

(3) M.H. Hong, S.I. Pyun, Wear, 147, page 59, 1991

(4) K. Miyoshi, Surf. Coat. Technol., 43/44, page 799, 1990

(5) A.L. Grogan, V.H. Desai, F.C. Gary, S.L. Rice, Wear, 152, page 383, 1992

THE INFLUENCE OF ULTRADISPERSED FILLERS BY TRIBOLOGICAL PROPERTIES OF AROMATIC POLYAMIDE

A.I.BURYA and N.T. ARLAMOVA
Department of chemistry, State Agrarian University, Dniepropetrovsk, Ukraine
A.V.VINOGRADOV
Siberian Department Institute of nonmetallic materials, Yakutsk, Russia

ABSTRACT

In order to imrove the tribological characteristics of thermo-resistant polymers materials science aspects of filler choice allow for the ultradispersed inorganic compaunds as a promising filler (1).

We was study the influence of containing (0,2-10 mas.%) ultradispersed fillers the oxinitride of silicon-yttrium (size of particles 0,08-0,12 mcm, spesific surface 42,5 m^2) and β-sialone (hard solution α-Al_2O_3 and AlN in β-Si_3N_4 with size of particles 0,008-0,14 mcm, spesific surface 63 m^2) by tribological properties of composites in based of aromatic polyamide phenilon C-2.

An effect of the use of these fillers is based on formation of spesified supermollecular composite structure.

The filler in polymeric matrix was introduced in revolving electromagnetic field by means of ferromagnetic particles.

The work into experimental spesimens realization the method of compression pressed (the temperature 593K, spesific pressure 55 MPa). The investigation tribological characteristics studied in mashine SMT-1 by scheme "disk-lag" under load 1 MPa, sliding speed 1 m/sec and way of friction 1000 m.

The result of research was showed (tabl. 1), that the using of refractory ultradispersive fillers permit raise the wear-resistant of phenion C-2 in 1,2-2,0 once.

The harshest wear-resistant raise by containing of filler to 5 mas.%. Further increase filler influence by wear-resistant of composites insignificant, in case of oxinitride the wear even increase.

As for friction, that was discovered the presence two minimums by containing of filler 0,2 and 5,0 mas.%, in case dynamic and static friction coefficients. Apparently, the first minimum be conditioned by transformation of structure polymeric binding, in second-begining work frame-work from filler.It is shown an addition ultradispersed fillers lowering the static friction coefficiente, by 3-18 and dynamic by 5-15%. Considerable changes of tribological characteristics are observed at the filler containing 0,2-1,0 mas.%.

The wear-resistant of composites about abrasive particles (pattern-steel 45) testify that only 0,2 and 1,0 mas.% filler increase the wear-resistant.

Results obtained demonstrate good correlation between friction resistant, hardness and viscosity of composites.

The investigation of microstructure the friction surface of composites was discovered, that they wearing on friction mechanism and increasing of percentage containing of filler growth not only quantity longitudinal strips of ploughing, but the degree hoe of surface layers of materials.

Table 1. The tribological characteristics of composites

Properties	The containing of filler, mas.%			
	0	0.2	1.0	5.0
The filler - oxinitride of silicon-yttrium				
Δm, mg	5.70	5.00	3.14	2.80
f	0.45	0.39	0.41	0.37
T,K	387	385	368	372
The filler - β-sialone				
Δm, mg	5.70	4.01	3.02	2.87
f	0.45	0.44	0.46	0.41
T,K	387	380	379	390

Δm - mas. wear; f - static friction coefficient; T - temperature in contact zone

REFERENCES

(1) G.Samsonov, V.Kazakov and other. The mechanical properties of nitride-oxide materials of system`s Al_2O_3-Si_3N_4, Powdery Metallurgy, 1974, No.2, 60-63pp.

INITIAL RESULTS FROM HIGH SPEED RECIPROCATING TESTING OF POLYMER MATRIX COMPOSITES

R W BAYLISS
Morgan Materials Technology Ltd., Bewdley Road, Stourport on Severn, Worcestershire, UK.
C A STIRLING
Morganite Special Carbons Ltd., Quay Lane, Gosport, Hampshire, UK.

ABSTRACT

A reciprocating test rig has been developed to allow the examination of the tribology of polymer-matrix composites in the range of conditions encountered in industrial gas compressor applications. To simulate these conditions a standard proprietary wear testing machine has been extensively modified to allow samples to be run at up to $4ms^{-1}$ over a 50mm stroke. Testing can be undertaken under full computer control, with speed, load and temperature being controlled independently of each other. The sample reciprocates across the counterface surface, which is housed in a nominally sealed atmosphere chamber. A separate moisture control facility has been constructed to feed known gases and gas mixtures into the chamber to blanket the test pairing with a controlled humidity gas. The data from the test are monitored on a continuous basis, viz., friction force, wear gap, speed, load, plate and sample temperature and moisture level. The theoretical background to the development of the test rig is discussed elsewhere (1).

The composites initially of interest generally contain polytetrafluoroethylene (PTFE), sometimes with another binder phase, and with a variety of fillers, including glass fibre, carbon and metal powders. From experience with wear testing of such materials it is known that achieving reproducibility can be difficult, even under less demanding test conditions. Furthermore, materials have to be tested in a range of gases and under known moisture levels. During normal usage the composites may run under very dry conditions and for this reason the separate moisture control facility was constructed, controlling moisture from <1ppmV to 1000ppmV. The gas, typically air or nitrogen, is fed into the wear chamber and exhausted through a bubbler which maintains a slight overpressure in the chamber to prevent laboratory air (moisture) ingress.

The initial results are from a series of matrix experiments where alterations to load and speed were made at fixed temperatures, and where the sample was known to be running under isothermal conditions. The requirement for isothermal conditions is paramount for these materials where frictional heating can cause a rise in temperature and changes in the wear rate process, from low to high or vice-versa.

The initial study was also used to determine the variability in the machine, so some runs under static conditions were also recorded. The results generally on one specific grade against a stainless steel counterface, under ambient air conditions, show that very good repeatability can be achieved with the test rig. A tolerance figure (T), defined from the standard error (SE) and mean (M) as:

$$T = \frac{2 \times SE \times 100}{M}$$

was obtained using two speeds, 20 and 35Hz, and two loads, 10 and 40N. The testing showed that, with repeat testing, a tolerance figure of <10% could be achieved on both the wear rate and the friction coefficient. The data in sets of low load & low speed and high load, low & high speed are clear of each other at the 99% confidence level. The problematic region in terms of instability of friction coefficient was the low load at high speed region. For this region the low load limit was raised. The operational envelope for the machine was determined to be 10 to 50N at 20Hz and 20 to 50N at 35Hz.

The effect of moisture from 1000 down to <1ppmV level was also examined for three grades running under nitrogen. The testing showed:

- Distinct differences in performances between the three grades examined which agreed well with that found with current practice.

- For some grades there appeared to be a transition region at a moisture level of about 100ppmV where a dramatic change in wear rate occurred.

The results from other tests are also presented.

REFERENCES

(1) R W Bayliss, C A Stirling, G A Plint, A F Alliston-Greiner. World Tribology Congress, IMechE, London, 8-12 September 1997.

CORRELATION BETWEEN THE COUNTERFACE ROUGHNESS AND WEAR OF HIGH DENSITY POLYETHYLENE IN SLIDING

B. J. BRISCOE and S. K. SINHA
Department of Chemical Engineering, Imperial College,
London SW7 2BY, UK.

Abstract

Film formation on the counterface during sliding of a polymer against hard surfaces is a common phenomenon in polymer tribology. These films are generally between 2.0 nm and 10.0 nm in thickness. The thickness of the film also depends upon the roughness of the counterface and the chemical nature of the surface. The result of this film formation is that the wear rate of the polymer decreases.

In many cases where a polymer is used as a tribological material, the measurement of the progressive wear of the polymer is difficult due to very low wear rates. In this paper, it is argued that the film thickness and counterface roughness can be a measure of the wear rate of polymer against a hard surface. Experiments were carried out by sliding High Density Polyethylene (HDPE) against steel surfaces and emery papers. The surface roughness of the counterface was measured as a function of the wear rate of the polymer. The results show that the wear rate of the polymer has a linear relationship with the counterface roughness. Initially, the wear rate decreases as the counterface becomes smoother. The wear rate stabilises to a constant value when the counterface attains a certain value of the roughness.

High density polyethylene (ICI, UK) pins were slid against two grades of emery paper with grit sizes of 2 and 6 µm. Wear volume was measured after passing the polymer pin on the same track. The track length was kept at 50 mm. Figure 1 shows data from initial results on the effect of counterface roughness and the number of traversal on the wear volume. The figure shows wear of the polymer per traversal (half cycle) as a function of the number of traversal for two roughnesses (grades of emery paper). The figure shows that the wear rate drops initially with the number of traversal and attains a constant value after about 5 traversals. There is a strong effect of the roughness of the counterface on the wear of high density polyethylene. Further results on this study will be shown in the congress and published elsewhere.

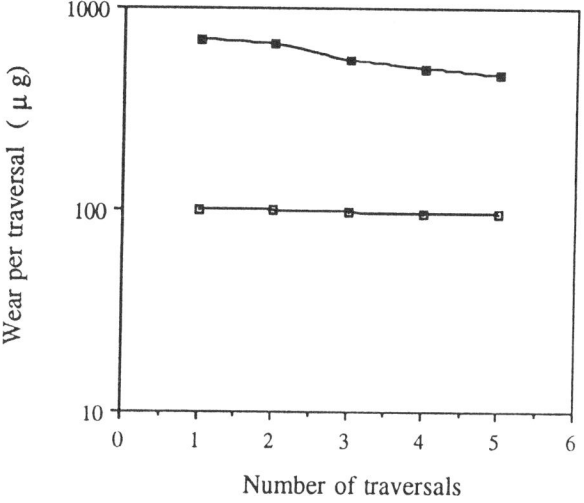

Figure 1: Wear of HDPE for one 50 mm traverse on two grades of abrasive papers (2 and 6 µm) against the number of repeated traversals over the same track. Load=2.5 kg, sliding velocity = 1 mm/s.

THE WEAR AND FRICTION OF GRAPHITE PLASTICS FILLED WITH THERMAL DILATION OF GRAPHITE

A.I. BURYA and A.A. BURYA
Department of chemistry, State Agrarian University, Dniepropetrovsk, Ukraine

ABSTRACT

In order to improve the tribotechnical characteristics of compositive materials on the base of thermo-resistant aromatic polyamide phenilon C-2 new compositive materials-graphitoplasts were recieved. As the filling thermal splitting graphites GL-2 and silver were used. Such graphites are ultrasmalldispersive filling and characterized division of primary graphite crystile on the axle C on thin belts with the foration of folded structure which gives them unique properties and determines prospectiveness of practical usage in different branches of technique. The usage of thermal splitting graphites as the filling in comparison with colloid graphite [1] gives the opportunity to increase graphitoplasts mechanical characteristics and striking viscosity for 10-60 % destructive stress during compression in 1.2 - 1.3 times in particular. On the one hand this explains the advantages of ribbon building of fillers, on the other hand the securing of its equal distribution in polymery matrix (fig.1).

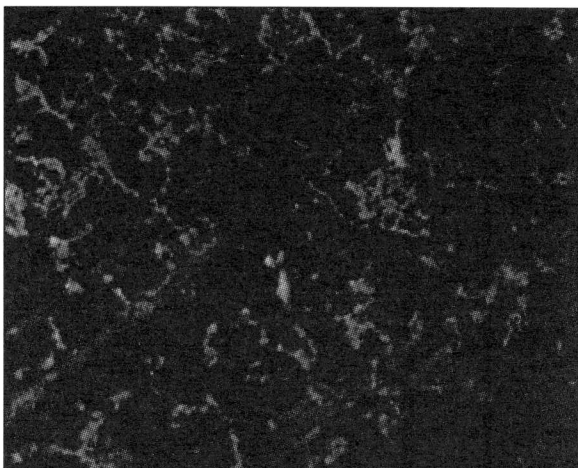

Fig.1: Electronic-microscopic print of the friction surface of graphite plastics with 15 mas % GL-2. Increase 100

The researches of tribological graphitoplasts properties on the disc friction machine were made. The tests were conducted on the steel counterbody (steel 45, thermalworked up to the firmness of 45-48 HRC, R_a = 0.16-0.32 mcm) in the ranks of dry friction during the sliding speed of 1 m/sec, friction ways of 1000 m and various loading.

Characterizing the tribotechnical properties of graphitoplasts it should be mentioned that with the increasing graphite contents, the coefficient of friction and resistance to wear are reduced. The best resistance to wear among all tested compositive materials has the graphitoplast, containing 25 mas. % of silver graphite. Maximum value of PV parameter for this materials is 25 MPa·m/sec (tabl.1).

Tabl.1. The tribological characteristics of graphite plastics

Graphite	The specific pressure, MPa					
	3	5	10	15	20	25
Wear, mg						
GL-2	3.1	3.1	2.6	2.6	4.0	4.9
silver	3.0	3.0	3.0	3.4	4.3	4.3
The friction coefficient						
GL-2	0.17	0.11	0.06	0.05	0.04	0.04
silver	0.18	0.12	0.07	0.05	0.05	0.04
The temperature in contact zone, K						
GL-2	355	358	363	372	386	395
silver	355	358	363	372	386	395

The sliding speed -1m/sec; friction way - 1000 m

Due to the recieved information the graphitoplasts have PV = 10-25 MPa·m/sec in the conditions of dry friction and coefficient (0.04 - 0.06) which can be compared to the friction coefficient given during the work of mobile connections with metal alloys being oiled.

It is shown that graphitoplasts can be used in chemical industry (condensations of pumps), metallurgy (inserts for driving rollers of leading out roller conveyers of pilgrim tube-rolling mills, condensations of compression equipment), agricultural engineering (sliding bearings). The usage of graphite plastics in the rainmaking units DDA and "Fregat" are shown as an example. It is stressed that experimental details from graphitoplasts have the advantages in lasting life in comparison with the fundamental made of bronze more than 2 times.

REFERENCES

(1) A.I.Burya, V.I.Sytar, A.A.Burya et al The Influence of Carbon Fillers Nature of the properties of composites on the Basis of Aromatic Polyamid, Proceedings of the 10th Int. Colloguium "Tribology-Solving Friction and Wear Problems", Vol.2, Esslingen, 1996, 2355-2359 pp.

THE INFLUENCE OF FIBRE ON THE TRIBOLOGICAL CARBON PLASTICS PROPERTIES ON THE BASE OF POLYARYLATE

A.I.BURYA, O.P.CHIGVINTSEVA, N.I.AGAPHONOV
Department of chemistry, State Agrarian University, Dniepropetrovsk, 320027, UKRAINE

ABSTRACT

Nowadays the leading role in the increasing of machines and mechanism, the reducing of their metalocapacity belong to polymer compositive materials. One of the most important advantages of the compositive materials is their ability to create the elements of wares with before hand given properties corresponding to the characteres condition of the work of details and constructions (1).

Complicated aromatic polyefirs including polyarylate on the base dyfenilpropane and aromatic dycarbone acid which are made of hard macromolecules. They are used in many branches of technology where the combination of rather high firmness and ability to workfor a long time in the conditions under the influenece of consicleable mechanical stresses is needed (2).

Carbon fibres having unique physical-mechanical properties and heat-firmness belong to the most perspective filling of thermoplastic connections. In this report the result of tribological researches of polyarylate and composition on its base, filled by the carbon fibres-carbon plastics are given . As a filling the carbon Uglen-9, the content of which was changing from 5 go 35 mas. % was chosen.

Compositions were made accoring to the special method in the help of the nonequal axleed ferromagnet parts due to which it was possible to distribute fibre filling in the connection. Models for the researches were made on the base of the method of compression pressing. Studing of tribological characteristics of initial polymer and carbon plastic on its base was made in the ranks of dry friction along the steel contbody made of steel 45 thermoworked up to the firmness 45 - 48 HRC, having rouhness of surface R_a = 0.16...0.32 mkm while the specific loading is equal to 0.4-0.8 MPa and sliding speed 0.5-2.0 m/s.

The researches work about the influence of exploitation on the tribotechnical characteristics of polyatylate has shown that the increasing of sliding from 0.5 to 1.5 m/s and specific pressure by 0.4-0.7 MPa is accompanied with monotounous increasing of temperature in the contact zone polymer-contbody (on 309 K) and increasing the intensiveness of the line wear for more one stage.

The analysis of the recieved facts gives us change to underline that the given polymer can successfully work at the criterion meaning PV ‹ 1.2 MPa · m/s while the roughing of exploitation is accompanied with catastrophic wear of material. Considerable expansion of polyarylate exploitation range is achieved during armouration it by carbon fibres.

It is settled that the most essential reducing of friction coefficient (in 2.4 times) of carbor fibres come be seen when the contents is 15 mas. % and futher increasing of carbon fibre contents leads to the reduce of this figure only for 33.6 and 48.2 % (fig.1). As for the intensiveness of line wear is concerned

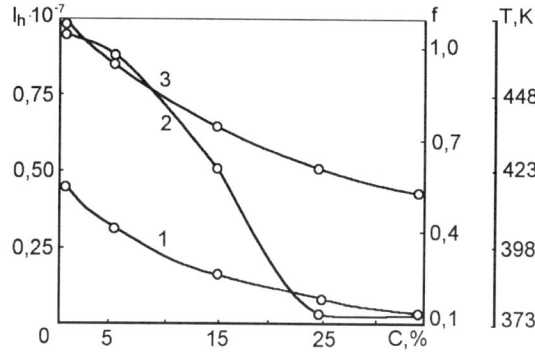

Fig. 1: The influence of fibre contants on the friction coefficient (1), intensiveness of line wear (2), temperature in the zone contacts (3) carbon plastics properties on the base of polyarylate

it is minimum when the fibre contents is 25 mas. %, after that it is insignificantly increasing. It can be noticed good correlation between the fibre contents and temperature in the zone contacts that is apparently explained by the improvement of warmhysical and antifricrional carbon fibres properties with the growth of carbon fibre contents.

REFERENCES

(1) Plastics thermostability of constructional purpose/ Under edition of E. B. Trostyanskaya, Moscow, "Chemistry", 1980, 240 p.
(2) Askadsky A.A. Physical-chemistry of polyarylate, Moscow, " Chemistry", 1968, 211 p.

EXPERIMENTAL INVESTIGATION INTO REPLACEMENT OF BABBIT LINING BY POLYMER LAYERS IN LARGE THRUST BEARING

T. R. CHOUDHARY and Dr. G. VENKAT RAO
Bharat Heavy Electrical Ltd. Research and Development Division, Vikasnagar Hyderabad, India
Dr. A. SETHURAMIAH and Dr. OM PRAKASH
ITMMEC, Indian Institute of Technology Delhi, Hauz Khas, New Delhi, India

ABSTRACT

Thrust bearings with babbit are common in large hydrogenerator. Such bearings present problems during starting due to high friction and need separate oil system to provide hydrostatic support. To avoid such problems low friction polymer bearing layers have been tried as alternative and the papar describes this investigation. Some work was earlier reported in this area by Biber et. al (1).

Study of tribologycal properties for babbit, pure PTFE and three locally developed PTFE composite materials has been done. Friction and wear of these material have been evaluated on a pin-on-disc machine. The PTFE composite with 55% bronze and 5% graphite with low friction and wear has been selected for bearing lining. The evaluation of friction and wear of these materials is shown in table 1.

Two millimeter layer of the selected polymer is then adhered to the steel backing. The bond strength of these layers was carefully compared with babbit lining by shear testing and cyclic stressing test. This lead to the effective selection of suitable adhesive for the purpose. The test results are shown in tables 2 and 3.

Both babbit and polymer lined thrust bearings are being tested with turbine oil lubricant in a simulated thrust bearing test rig. Initial comparisons in the low speed boundary regime demonstrates that the polymer layers have friction which is about 40-45% lower that of babbit layer as shown in table 4. Also, it was observed on the basis of surface roughness changes that wear is lower with polymeric material. Further work with thrust bearing test rig is in progress.

Table 1: Friction and wear of PTFE composites

Material	Pressure Kg/cm²	Sliding distance	Coeff. of friction	Weight loss,g
Babbit	3	8085	0.2155	1.00
Pure PTFE	3	8085	0.135	3.18
PTFE + 35% carbon +5% gra.	3	8085	0.180	0.98
PTFE + 45% bronze + 5% gra.	3	8085	0.168	0.87
PTFE + 55% bronze +5% graphite	3	8085	0.107	0.23

Table 2: Evaluation of bonding strength

Specimen	Frequency of load,Hz	Load,N Max/min	No. of cycle	Observation
Babbit (1)	15	6000/600	51000	Cracking
(2)	15	5000/500	120000	Debonding
PTFE Comp-site (1)	15	6000/600	100000	No sign of debonding or crack
(2)	15	5000/500	2000000	

Table 3: Tensile stresses at surface & interface

Lining Material	Max. load (N)	Tensile stress at mid span (Mpa)	
		on the surface	at the interface
Babbit	6000	160	96
	5000	133	80
Polymer	6000	292	133
	5000	243	111

Table 4: Thrust bearing friction under boundary regime

Load in kgs	speed in rpm	coefficient of friction	
		babbit bearing	polymer bearing
80	116	0.168	0.0924
500	116	0.0876	0.0625
1000	116	0.0611	0.0354

REFERENCES

(1) L.A. Biber et.al, Some aspects of Hydrogenerator Modernization, CIGRE-1986.

STUDY FRICTION AND WEAR POLYMER MATERIALS UNDER DRY SLIDING FRICTION

VLADIMIR I KLOCHIKHIN and YURIY N DROZDOV
Department of Friction, Wear and Lubrication, Blagonravov Mechanical Engineering Research Institute,
Russian Academy of Science, Griboedova Str.4, Center, Moscow, 101830, RUSSIA
ALEXANDER P KRASNOV
Laboratory of Filled Polymeric Systems, Nesmeyanov Institute of Organo-Element Compounds,
Russian Academy of Science, Vavilova Str.28, Moscow, B-334, 117813, RUSSIA

INTRODUCTION

Polymeric compounds have been had high chemical resistance, small friction coefficients and well wear resistance under room temperatures and small operating loads. However under high loads they are deformed besides of pure mechanical strength and under high temperatures the polymers usually are destructed from complex activity of oxygen and heat.

There are number of the ways to enhance their resistance to heat and contact loads. One of the ways to increase the load capacity of them is to apply polymeric coatings on the base of fabric materials with different chemical nuture. The thickness of such coatings are in the interval of 200 -1000 μm.In order to deffend their against thermal destruction under high temperatures the thermostable polymeric fibres and fabrics are involved with high value of chemical connections power. Such polymeric coatings can be used widely under dry sliding friction connections where the usage of thin solid lubricating coatings or grease are impossible or embarrassingly. Another way is to use the geteroheneous materials on the basis of hard thermoplastics reinforced by dispersuos fillers such as thermostable binders and solid lubricants, for example PTFE,graphite and MoS_2.

EXPERIMENTAL

To study the friction and wear polymeric materials under high loads and temperatures they have been separated on four groups:
1. Thermoplastics on the basis of polyamide + graphite/ carbinic fibres, polycarbonats + PTFE/ carbonic powders, polyimides + graphite/MoS_2 ;
2. Thermostable polymers (textolites) from carbonic/ aromatic polyamide/polyoxidiazole fibres;
3. Polymeric coatings on the basis of thermoplastic polymers such as polyethelene + furfuralacetone + epoxy resin with thickness 130-150μm,polyethelene + PTFE + MoS_2 + graphite with thickness 70-80μm;
4. Polymeric fabric coatings on the basis of thermostable polymeric fibres such as polyoxadiazol fibres + phenol-formaldehyde resin with thickness 300-450μm and PTFE + polyimide fibres + phenol-formaldehyde resin with thickness 500-600μm.

The specimens from the first and second groups were made as bushes. The specimens from the third and fourth groups were made as coatings on the steel shaft.The friction pair *bush-shaft* with parameter L/D = 0.6-0.8 for diameter D = 10 mm was tested in air under reciprocating-rotary motion and under following test parameters: load, 10-60MPa, temperature, $24\pm250^{0}C$, speed, 0.0004 - 0.04m/s. Roughness value of counterbody was R_a = 1.25-0.63μm. The clearance in unit was H8/d9 - H8/f9. Countermate material was steel with 0.45% C.

RESULTS

Investigations carried out have been shown that the best results on friction and wear under high loads indicated reinforced materials on the basis of thermostable polymeric fibres and fabrics and polymeric fabric coatings on the basis of thermostable polyoxadiazol fibres with binder from phenol-formaldehyde resin. The best wear resistance under high temperatures had textolite on the basis of polyoxadiazole fibres and polymeric fabric coating on the basis of PTFE and polyimide fibres with phenol-formaldehyde resin as binder. Self-lubricating materials on the basis of thermoplastics textolites were destructed or had plastic deformation onto friction zone under loads more 30MPa.

Research results have been shown that the entering the solid lubricants into antifrictional reinforced materials capable to work on friction pairs under high loads there is not obligatory. It has been shown when the rubbing material had been consisted only two components - polyoxidiazole fibre and phenoformaldehyde resin. Individually each component has not had a good lubricant proporties. But together they design antifrictional material capable to obtain the small value of friction coefficient under rubbing values steel mate.It was got in touch with peculiarity their tribodestruction process under high loads that brought to the formation of the separating soft layer with small shear stresses between friction elements.

Friction surface analysis have been shown that the thermostable compounds on the basis of thermostable polyoxadiazol fibres with binder from phenolformaldehyde resin polyoxadiazol fibres can be used as the base for antifriction materials constrction.

ADHESIVE COUPLING ON INTERFACE AND FRICTION CHARACTERISTIC OF POLYMER COMPOSITES.

V. I. KOLESNIKOV, N. A. MYASNIKOVA, A. V. WOLKOV
Rostov state university of transport communication, sq. Narodnogo Opolcheniya, 2, Rostov-on-Don, Russia, 344017.

High physical-mechanical and tribology characteristic of polymer composites are not only determined with characteristics of components, but also with value and stability of adhesive interaction on interface of matrix-filler. Adhesimetry in polymer material science plays an essential role, as on its basis the common approaches to the choice of components of polymer composites, modification of surfaces of fillers and matrixes can be formulated. It lead to development of materials with the maximum realization of properties included in their structure a component.

The realization of account of adhesive coupling on interface is based on definition of value of van-der-Waals and chemical interaction between components.

The force of molecular interaction between two bodies on the basis of the general theory of van-der-Waals forces is determine by their permittivity functions on a seeming axis of frequencies:

$$\varepsilon(i\xi) = 1 + \sum_{j=1}^{M} \frac{f_j^2 + \xi h_j}{(\omega_j^2 + \xi^2) + \xi g_j},$$

where M - complete number of lines in a spectrum of absorption; ω_j, f_j^2, h_j and g_j are parameters of spectroscopic lines of spectrum.

The energy of chemical interaction can be defined as $E = (E_1 + E_2) - E_{12}$, where E_1, E_2 - energy of models of a polymer matrix and filler, E_{12} - energy of a complex. The chemical connection can arise only then, when the energy of a complex E_{12} appear less, than sum $E_1 + E_2$. To accounts of energy of models we applied an extended Hukel method.

Using the given technique, number of composites on a basis glass-reinforcement ED-20 with various hardener(p-ABA - para-aminobensilanilin, UP-583D - diethylthreeaminomethylphenol, MPhDA - metaphenildiamine) is considered. The results of account of adhesive coupling between components F (N/m^2) at l=0.5 nm and tests of composites for shift durability τ_{sh} (MPa) and specific wear J (mg/km) are shown in tab. 1.

Tab 1.

Component	F	τ_{sh}	J
ED-20	1.89×10^7		
p-ABA	2.02×10^7	58-60	12,4
UP-583D	8.25×10^7	45-50	17,8
MPhDA	1.53×10^8	48-51	18,6

The most similar and stable there will be the interface, if hardener and the polymer matrix have close value of interaction force with filler (there will be no redistribution a component on the interface). The tribology characteristics of such composite should be best. It confirm data in tab.1 for a pair ED-20 - p-ABA.

At creation of glass-reinforcement composites for amplification of adhesive on interface one use such kind of modifying of a surface as primary treating. It is processing of a glass fiber surface by water silicone solutions; therefore there is the chemical connection between appret and filler. Moreover appret is selected so that its molecule contained also groups, capable to chemical action with polymer binder.

In the given work comparative accounts physical (van-der-Waals) and chemical interaction between polymer binder ED-20 and filler - glass fiber (SiO$_2$), and between glass fiber and water silicone solutions

A-174   [CH$_2$-C(CH$_3$)-CO-Si(OCH$_3$)$_3$],
AGM-9   [NH$_2$-(CH$_2$)$_3$-Si(OC$_2$H$_5$)$_3$],
KVM-603   [NH$_2$-(CH$_2$)$_2$-NH-(CH$_2$)$_3$-Si(OCH$_3$)$_3$],
GVS-9   [CH$_2$=CH-Si(OCH$_3$-CH$_2$-O-C$_2$H$_5$)$_3$]

are spent. Results of accounts of force F(kN/m^2) appret-SiO$_2$ and ED-20-appret at l=10nm, the energy E(J/mol) chemical action of appret- SiO2, and wear of composites with specified apprets J (mg/km) are shown in tab.2.

Tab 2.

	A-174	AGM-9	GVS-9	KVM-603
F(app-SiO$_2$)	307.5	275.99	431.03	278.64
F(ED-20-app)	4.81	4.83	4.74	4.89
E(app-SiO$_2$)	373	353	385	352
J	18,7	19,8	17,1	20,3

The greatest physical and chemical interaction with glass is received for GVS-9 (force of interaction of apprets with binder are close on value). Minimum wear is received for a composite with GVS-9.

Especially application of such accounts is effective at realization of comparative estimations, for example, at selection to given filler the binders, hardeners, apprets. Thus, at the stage of designing of a composit material it is possible to predict some strength and friction characteristics and varying component structure of a composite, to achieve increase of its characteristics.

FRICTION PROCESSES IN DEPRESSED ATMOSPHERIC PRESURE CONDITIONS OF SOME SLIDING PAIRS: PTFE COMPOSITES AND RUST - PROF STEL

ZBIGNIEW OLEKSIAK
Fakulty of Mechanical Engineering and Robotics, Uniwersity of Mining and Metallurgy in Cracow,
Al. Mickiewicza 30, 30-059 Krakow Poland

ABSTRACT

The device constructed in University of Mining and Metallurgy is denoted by symbol OSA-02 and patend pended no 164316. It operates on the „ pin on disc „ principle. Schematic of the device is presented in fig.1. Friction tests have confirmed the following research possibilities of of the device:
- simulation of friction and wear processes in the air of 10-2 Tr gauge,
-computer recording of instantaneous values of 16 set parameters of resultaut quantities and particulary: continuous

- measurement of displacements and calculation of instantaneous value of a sample or countersample wear with accurancy up to 0.001 mm,
- continuons obserwation of the countersample surface morphology change at friction process.

Instandaneous image of the countersample operation surface sectors observed and magnified (*100) in optical mikroscope(7) processed by the camvid(8) and elektronic stabilized in transducer(9) are recordet in video recorder(10) and obserwed in monitor(11)

In friction proces of PTFE composite on stainless countersample a slow mouvement of „ film" spread over countersample operation surface has been observed and particulary when the surface was earlier cleaned with spirit.

REFERENCES

(1) Kubiesa R.: Analysis of utilization possibilities of tensometric and induction sensors fo linear wear measurement of samples in friction process. Graduate Thesis in Faculty of Mechanical Engineering and Robotics. Krakow, 1996. Promotor Z. Oleksiak.

(2) Oleksiak Z., Kubiesa R.: The method of contactless measurement of linear wear of friction pairs

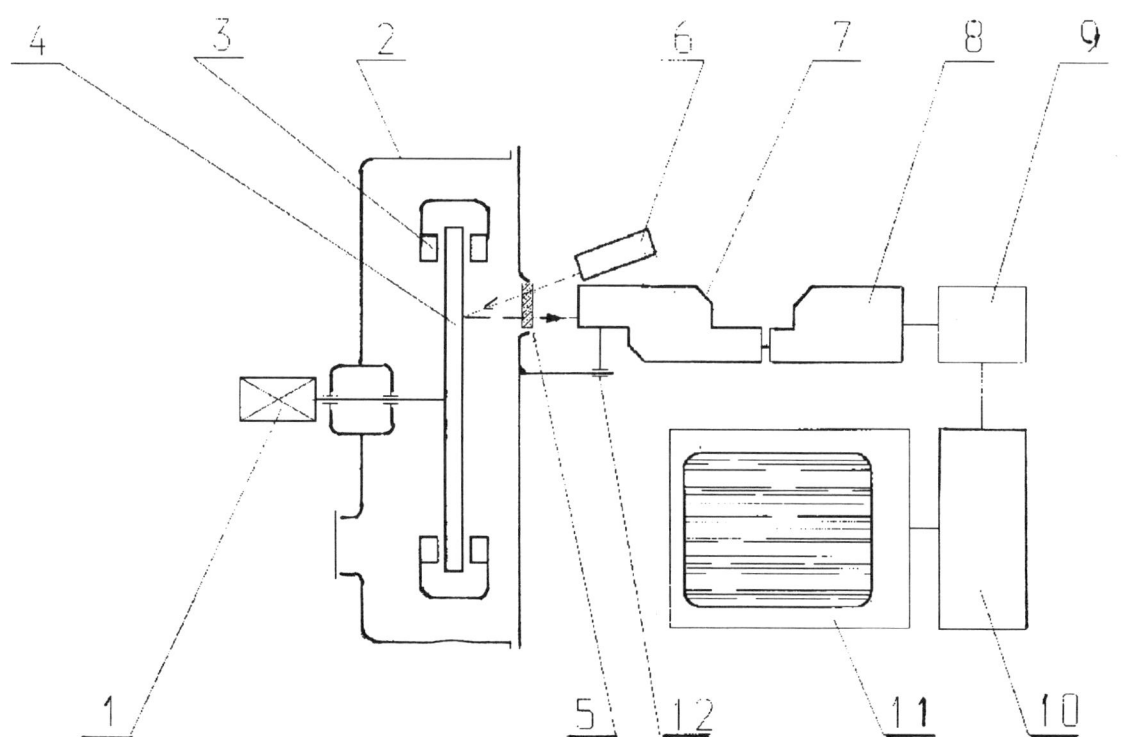

Fig.1. Scheme of device OSA-02 for continuous observation of countersample operationd surface sector in friction process.

Friction-Induced Vibration of Oscillating Multi-Degree of Freedom Polymeric Sliding Systems

RONALD A L RORRER and JOSHUA C BROWN
The Gates Rubber Company, PO Box 5887, Denver, Colorado 80217, USA

ABSTRACT

The purpose of this study is to investigate friction-induced vibration of oscillating systems. Special attention is focused on polymer-on-metal and polymer-on-polymer systems. Experimental and analytical friction results from non-oscillating or unidirectional sliding modes are extended into the oscillating sliding mode. Specifically, this refers to the incorporation of friction(velocity) ($\mu(v)$) relationships estimated from experimental results. The area of primary interest is the conditions under which polymeric based sliding systems exhibit friction-induced vibration.

In general, unidirectional sliding situations are governed by the long term asymptotic behavior of the frictional force function. Friction-induced vibration during oscillatory sliding can be viewed as periodic transient behavior, with each reversal of motion appearing as a transient event. While the imposed frequency of oscillation is always present when oscillatory systems undergo friction-induced vibration, additional self-excited frequencies often are present. In order to fully understand the response of a system possessing either one or both subsystems as polymeric elements, the frictional behavior of the sliding material pair must be coupled with the respective subsystem structural elasticity and hysteresis.

The primary motivation of this study was to extend behavior that can be obtained in unidirectional motion studies of friction-induced vibration to oscillatory situations. The reasons for this are quite simple; linear motion test devices are widespread and typically simpler than oscillatory systems. Friction-induced vibration can also be experimentally determined on a linear test device. With a general understanding of oscillating system dynamics, unidirectional motion information can be extrapolated to qualitatively predict oscillatory behavior. The system properties used in this study are typical of elastomer-on-metal sliding systems.

The system was modeled with a metallic subsystem (m_1, c_1, k_1) sliding on an elastomeric subsystem (m_2, c_2, k_2) as shown in Fig. 1. The input to this frictionally coupled system is a sinusoidal velocity, $V(t) = V_0 \sin \omega t$, and thus a sinusoidal displacement. Masses m_1 and m_2 are coupled by friction as a function of the relative velocity ($V_{rel} = \dot{x}_1 - \dot{x}_2$). The form of friction force is,

$$F_f = F_c \, sgn(V_{rel}) \, (1 - e^{-\beta |V_{rel}|}) \, (1 + (f_r - 1) \, e^{-\alpha |V_{rel}|})(1 + e^{-\gamma |V_{rel}|})$$

where α, β, and γ are constants which determine the velocity scales of the friction force, and, F_c and f_r are constants which scale the initial peak and asymptotic friction values.

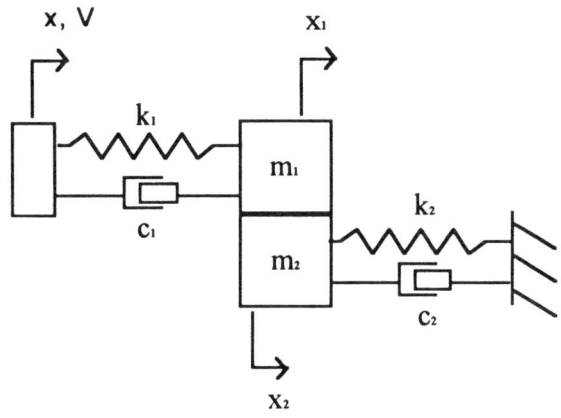

Figure 1. System Model

The resulting system of differential equations were solved numerically. Based upon the amplitude of input oscillation, the following behavior can be exhibited; a single stick-slip followed by oscillatory sliding, a single stick-slip at each motion reversal, multiple stick-slip events during each half-cycle of oscillation, a square wave force(time) behavior, or a higher frequency oscillation superimposed on the driving oscillation frequency.

It is possible to estimate the friction(velocity) relationship from unidirectional sliding experiments for a material pair and then incorporate this relationship into a multi-degree of freedom system model to predict the behavior in oscillatory sliding.

FINITE ELEMENT CONTACT SIMULATION AND EXPERIMENTAL INVESTIGATION OF A DIAMOND INDENTOR COMPRESSED INTO COMPOSITE STRUCTURES

KÁROLY VÁRADI - ZOLTÁN NÉDER
Institute of Machine Design, Technical University of Budapest, H-1521 Budapest, Hungary
KLAUS FRIEDRICH - JOACHIM FLÖCK
Institute for Composite Materials, University of Kaiserslautern, 67663 Kaiserslautern, Germany

ABSTRACT

A non-linear FE contact analysis was prepared to model the effect of a diamond indentor while it is compressed into different composite materials. Two different fibre volume fractions (20% and 50%) and two different fibre orientations (normal and parallel) were modelled.

To solve the non-linear contact problem the COSMOS/M system was used.

In the Finite Element Model the fibre-matrix micro-structure consists of only a few fibres surrounded by a particular polymer matrix material (due to the large number of DOF). The size of the whole FE model is about 100x100x100 micrometers.

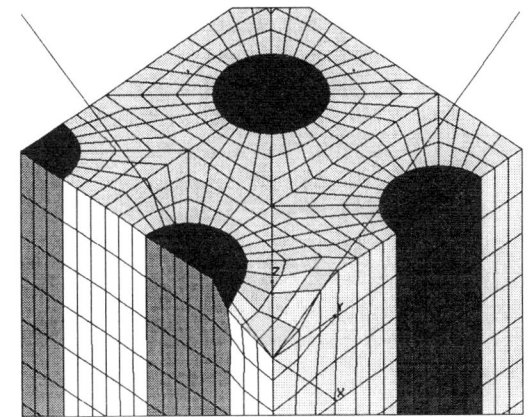

Fig. 1: The deformed shape of the quarter of the micro segment (load step: 0.05)

The FE model (Fig. 1) is fixed along the four sides, and a uniformly distributed load is applied from the bottom of the composite model, which is thus pressed against the rigid and fixed tip of the indentor (tip angle: 115°) located on the top of the specimen. Node-to-surface type contact elements are used. Contact nodes are specified on the top of the specimen, and the indentor is represented by contact surfaces. Fig. 1 shows the deformed shape of the FE model. Due to symmetry reasons only a quarter of the micro-structure is considered. The deformed shape of the same FE model is presented in Fig. 2, representing a higher load. A non-linear contact algorithm is used, material non-linearity is considered and different failure criteria are modelled. Equivalent stress distribution in the fibre is shown in Fig. 3.

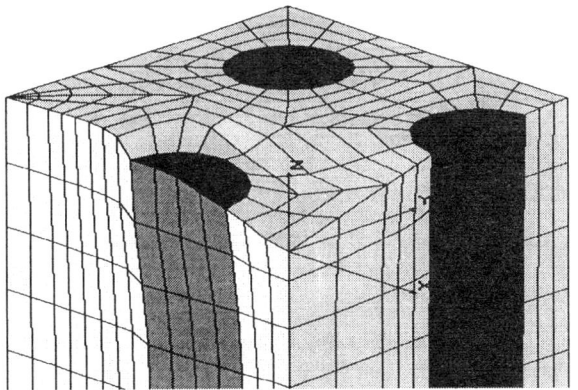

Fig. 2: The deformed shape of the quarter of the micro segment (load step: 0.4)

During indentation the load and/or displacement are measured, furthermore size and exact location of the contact area are checked by optical microscopy. The demaged area is evaluated by scanning electron microscopy as well.

This analysis is aimed to provide preliminary information for failure analysis (like fibre-matrix delamination, debonding and buckling of fibres, etc.) under contact pressure of a continuous fibre/polymer matrix system with hard abrasive particles.

ACKNOWLEDGEMENT

The presented research task was supported by the Deutsche Forschungsgemeinschaft (DFG FR675/19-1) as part of the German-Hungarian research co-operation on contact mechanics of different materials.

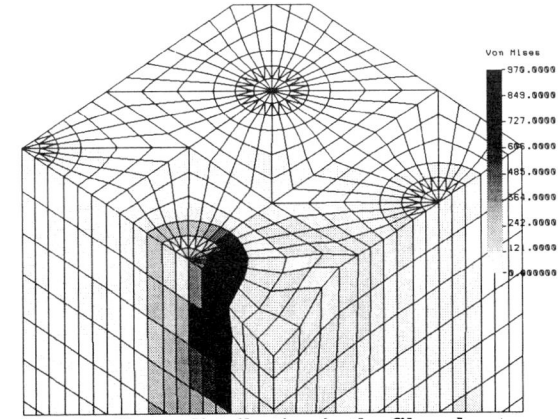

Fig. 3: Stress distribution in the fibre due to diamond indentation

EFFECTS OF FRICTION ON FREE VOLUME HOLES IN PTFE

Q J XUE and F Y YAN
Laboratory of Solid Lubrication, Lanzhou Institute of Chemical Physics, Chinese Academy of Sciences, Lanzhou 730000, P. R. China

ABSTRACT

The correlation between microstructure and macroscopic properties of polymers has been of interest for years (1)(2)(3). Many researchers (4)(5) have used positron annihilation technique (PAT) to study the behaviors of atomic-scale (a few Angstroms) free-volume holes in polymers with respect to their macroscopic properties, such as deformation behaviors, thermal properties, and phase transition processes. Experimental results showed that the randomly distributed holes are directly relevant to the above macroscopic properties. The objective of this paper is to relate the tribological behaviors of PTFE to the nature of its atomic-scale free volume holes.

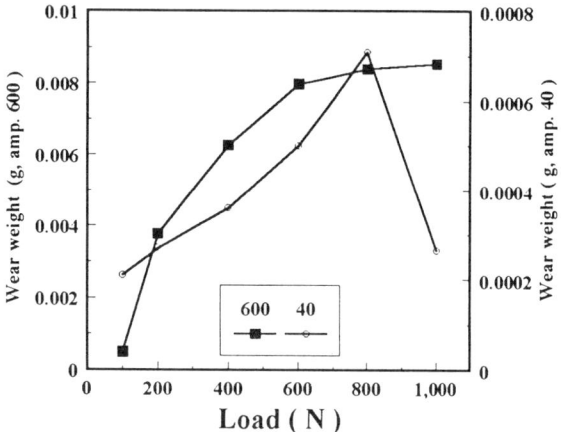

Fig. 1 Wear behaviors of PTFE at amplitude of 600μm and 40μm

The wear behaviors of PTFE were measured on an SRV oscillating and fretting wear tester with the plane contact of a bigger cylinder (SAE52100) end to a smaller cylinder (PTFE) end. Under the oscillating wear conditions, the wear weight of PTFE (Fig.1) increased in parabola with increasing of load, but under the fretting wear conditions, it possessed the biggest wear value at load of 800N.

Positron annihilation lifetime (PAL) method is established as the unique technique that can provide information about the size and concentration of the free volume holes. In this paper, the PAL measurements were carried out by using a fast-slow coincident PAL spectrometer system. Each positron annihilation lifetime (τ) and intensity (I) spectrum of PTFE was resolved into four positron components. For PTFE polymer (5), the third positron component, including τ_3 and I_3, results from the pick-off annihilation of o-Ps (a kind of bound states of positron) trapped at the crystal-amorphous interface. For symmetrical polymer molecules, there exists a linear relationship between τ_3 and free volume size V, expressed as $\tau_3 = 7.80V + 1.29$. And I_3 corresponds to the concentration of free volume holes. The τ_3 and I_3 of PTFE were measured to be: τ_3 1.79 ns (10^{-9} seconds), I_3 11.9%.

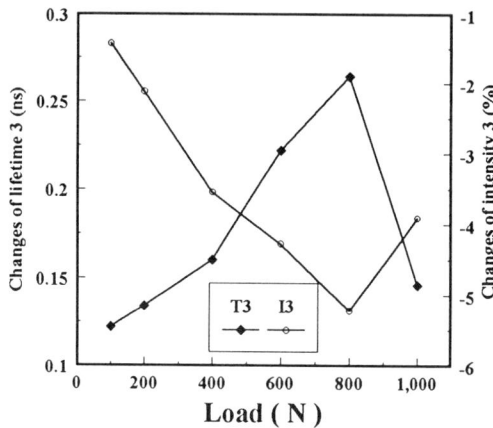

Fig. 2 The changes of the third positron components caused by fretting

It is interesting to notice (Fig. 2), that there exists a correlation between the fretting wear weight of PTFE and the changes of its free volume size at amorphous-crystal interface. The correlation can be formulated to be $\Delta\omega = A \cdot \Delta\tau_3$, where $\Delta\omega$ is wear weight of PTFE under fretting wear conditions, A is a constant, and $\Delta\tau_3$ represents the relative changes of the third lifetime by the fretting. The fact that $\Delta\tau_3$ inversely corresponds to ΔI_3 indicates that the total volume of free volume holes at interfaces are relatively stable in fretting process. The phenomena that the linking and splitting of free volume holes occur at interface, which represent the self-cured function of PTFE, is also found. It can be concluded that the stability of the average size of free volume holes at about interface is one important factor for the fretting wear behaviors of PTFE.

REFERENCES

(1) W. Brandt and I. Spirn, Phys. Rev., Vol. **142**, 231, **1966**.

(2) J. D. Ferry, Viscoelastic Properties of Polymers, Wiley, New York, **1980**.

(3) R. J. Samuels, Structured Polymer Properties, Wiley, New York, **1974**.

(4) Y. C. Iean, Microchem. J., Vol. **42**, 72, **1990**.

(5) C. L. Wang, B. Wang, S. Q. Li, and S. J. Wang, J. Phys.: Condens. Matter, Vol. **5**, 7515, **1993**.

MACHINE WEAR ANALYSIS
A RATIONAL APPROACH TO METHODS INTEGRATION FOR MAXIMUM BENEFITS

PETER G. BALL, F.Inst DGTE; M.STLE; M.MES Aust.
Technical Director, Machine Reliability Services (A'Asia) Pty Ltd.

ABSTRACT

An effective Machine Wear Control programme integrates the use of lubricant, vibration, thermal and performance analyses. This is because it is economically unacceptable to shut down expensive production plant, due to tribological failures. There is considerable concern, that many industries appear to be uncertain, in regard to these issues.

The strategy for integrated Machine Wear Control should be to use only those tests that are relevant. Acquisition of reliable and trendable data, is vital to this concept, and its management. This paper outlines the most significant analysis methods, and provides basic guidelines for their integration. It also explains how this approach can provide maximum benefits, if undertaken correctly.

Data released by the Strategic Industry Foundation indicate annual maintenance costs in Australia are between 30-50% higher than European and Japanese companies and 15% higher than their North American counterparts. This fact could be roughly assessed as being due to a lack of basic understanding in matters concerning machinery wear characteristics, more simply described by the term **Tribology**, the science and technology of interacting surfaces in relative motion. The field of tribology is well positioned to provide the technology for management of maintenance and at the same time co-ordinate the prime requirements of affordable reliability and high equipment availability.

The costs of wear involve not only replacement parts but also lost production and opportunity, through unscheduled downtime. During the mid 1980s these costs to the Australian economy were conservatively estimated at around 6 percent of Gross National Product, and by the year 2000 it is estimated that good tribological practice will save Australian industry close to $AU3 billion per annum. In the current economic climate, Australian Management will, by the use of Condition based Machine Reliability practices, access these potential savings. An improved National fiscal bottom line will reflect this pragmatic approach.

Relevant areas addressed, include the following:
Problem Areas and Failure Modes
Risk, Root cause and Failure Potential Analysis
Measurement Techniques
Vibration Analysis
Performance Monitoring
Thermography
Machine Lubricant Analysis; with three basic methods for consideration under this heading.

BENEFITS

Integrated Machine Wear Control does require an investment in equipment and operational resources before returns from the strategy are realised as 'bottom line savings'. Cost benefits can be considered in terms of cost reductions, the details of which are outlined in the paper.

A strategy of Integrated Machine Wear control can produce the following 'Bottom Line' results:
- 50 - 80% reductions in repair costs
- 30% increase in revenue
- 50 - 80% reduction in maintenance costs
- Spares inventories reduced by more than 30%
- Overall profitability of plants increased by 20 - 60%

In 1990, North Broken Hill Peko Limited Annual Report, Review of Operations, announced that such a strategy, introduced by the Engineering Technical Services Department of Ranger Mine, produced Cost Efficiencies and Savings of $AU1 Million, for Energy Resources of Australia Ltd, Ranger Uranium Mine in the Australian Northern Territory.

CONCLUSIONS

An effective Machine Wear Control programme integrates the use of lubricant, vibration, thermal and performance analyses. Do them on the same machine, at the same time, for maximum benefit.

The strategy for integrated Machine Wear Control should be to use only those tests that are relevant.

A properly integrated Machine Wear Control programme can produce increases in Overall Plant Profitability of between 20 to 60 percent.

REFERENCES

(1). Davidson, J., "The Reliability of Mechanical Systems" IMechE Guides for the Process Industries.
(2). Czichos, H., "Tribology" a systems approach to the science and technology of friction, lubrication and wear.
(3). Ball, P.G., "Improving Profitability Through a Reliability Centred Maintenance Strategy". ICOMS-94 Proceedings, Paper 18, Sydney.
(4). Ball, P.G., "Machine Reliability and the Understanding of Failure Potential", Proceedings: (CM)2 Forum 1995, Centre for Machine Condition Monitoring, Monash University. Melbourne.
(5). Sherwin, D.J., "Improving the Quality of Maintenance", 1992, IES Conferences, Brisbane.
(6). Results of a survey of over 500 plants conducted by Technology for Energy Corporation.

STUDY OF ROLL WEAR PROBLEM IN GRINDING OF HIGH ASH COAL

S.BISWAS, P.B. GODBOLE, B.L.JAISWAL, U.M.CHAUDHARI, S.K.BHAVE
Mechanical Engg Dept, Corp., R&D div., BHEL, Hyderabad, India
S.K. BISWAS
Dept of Mech.Engg. Indian Institute of Science, Bangalore, India

ABSTRACT

Coal containing high ash causes high wear of the grinding elements used in pulverisers in Thermal Power Plants. Reymonds Bowl Mill is the most widely used pulveriser for utility sets in India. Although it offers better wear rate (gm of iron per tonne of coal ground) than that of the other types of mills, the life of the grinding elements is still very low as compared to the life expectation as observed with low ash coal in USA and other countries.

Detailed investigations were carried out at site and laboratory to, (i) understand the cause and mechanism of wear, (ii) identify areas which offered scope for improvement, (iii) formation of possible remedial measures. This paper describes mainly the studies of stress related wear which was one of the key investigation areas.

Study of wear pattern on the rolls showed formations of corrugation or flutes whose severity increased with wear depth. It was also found that the wear profile changed with time. Severe collar formations were noticed at both ends of the rolls. Wear rate was more for higher sized rolls. These observations led to the hypothesis that stresses generated on the grinding interface were unduly high.

The measurement of force on the grinding rolls were carried out by putting force measuring devices on the spring elements which provided the grinding pressure. It was seen that reduction in spring precompression reduced the total forces as well as the dynamic forces. A reduction of spring constant also reduced the dynamic forces as well as wear rate.

Photoelastic model study was carried out using three different contact conditions, i.e. with a bed of coal powder in a flat plate, direct contact with flat plate and contact through a ball. The maximum stress was found in the ball loading case. It was concluded that passage of large size quartz suggests between roll and bull rings segments can cause indentation by plastic deformation. The wear of rolls was thus seen to be caused by the of indentation, in addition to a process of deformation and high stress abrasion.

This paper describes in detail the results obtained from the above studies and the recommendations emerging out of them for improvements of roll life.

A TECHNOLOGICAL LEAP IN THE CORRUGATED BOARD INDUSTRY

J T CONNAL

K.S. Paul Products Limited, Eley Estate, London N18 3DB

ABSTRACT

The paper board industry supplies the global demand for cardboard packaging feedstock, processing paper and starch through corrugated board plants to produce flat corrugated sheets ready for cutting, folding and printing to the contours of the common cardboard box.

Within the corrugated board plant, a single-facer glues a flat sheet of paper to a second sheet which is corrugated by passing through steam heated rolls which a corrugated wave profile. These heated rolls which are up to three metres wide, are supported by large roller bearings.

The corrugated rolls are steam heated to temperatures of up to 210°C. These high temperatures result in mineral and standard synthetic greases carbonising and base fluids evaporating which in turn lead to premature bearing failure. For this reason, corrugator roll bearings have historically been lubricated mainly by total loss or recirculating oil systems.

These systems have made roll changes, from one corrugated profile to another (e.g. heavy duty corrugated to food packaging corrugated sheet) an extremely time consuming and messy task with engineering staff having to dismantle oil lines then clean the resulting spill before resuming production. This procedure was principally responsible for reducing the flexibility of corrugated board plants to produce many different types of corrugated board profile, an ability which has proved essential to the survival of board plants in a commercially competitive environment.

In 1985 (1) an innovative corrugated board manufacturer faced with disposal problems resulting from their total loss oil lubrication system, decided to trial a "fluorinated grease" which consists of a perfluoroalkylpolyether fluid thickened with telomers of tetrafluoroethylene. Unlike conventional greases, this material does not carbonise, the base fluid exhibits very low volatility, even at elevated temperatures and the viscosity does not change greatly, even over a 200°C temperature range.

This particular corrugated board manufacturer soon realised additional benefits in the form of reduced maintenance costs and time since the pumps, filters, coolers and lines associated with the old oil lubrication system could be removed. With all of this redundant equipment stripped from the single facer, roll changes could then be conducted with great speed since there were no oil lines to be disconnected and no oil leaks to clear after switching sets of corrugated rolls.

Within five years, over half of the producers of corrugated board equipment had incorporated this novel "fluorocarbon" grease into their machinery giving greater production flexibility and decreased maintenance costs to the international paper board manufacturers who use their machines.

Within a further five years, almost half of the corrugated board producers in Europe and North America had either purchased new machinery utilising this "fluorinated" grease or had converted their old oil lubricated single facers in-house or through conversion kits which have been introduced by the original equipment manufacturers.

The result of this innovation has been to allow paper board producers to switch between different corrugated board flute profiles with ease and rapidity, which in turn has led to the development of new, highly specialised corrugated board types which would not have been economically produced without an effective grease lubrication system for the corrugator roll bearings.

In addition to facilitating the commercial development of quick change single facer corrugated rolls and new corrugated board types, the use of "fluorinated" grease in the paperboard industry has significantly reduced the amount of waste oil and contamination produced by the international corrugated board industry.

REFERENCES

(1) International Paper Board Industry, June 1992

A TRIBOSYSTEMIC APPROACH TO REFRACTORY LINING DESTRUCTION IN BLAST FURNACES

ION CRUDU
Machines Parts Department, University of Galatzi, 6200, Str. Domneasca no. 47, Fax 040 (036) 461353;
MIRCEA-PETRE IONESCU and VIOREL MUNTEANU
S.C. I.C.P.P.A.M. S.A., Galatzi, 6200, Str. Smardan no. 2, Fax 040 (036) 461238;
ION-FLORENTIN SANDU and PARASCHIV NEDELCU
S.C. SIDEX S.A., Galatzi, 6200, Romania, Str. Smardan no. 2, Fax 040 (036) 461492.

ABSTRACT

In the paper systems theory and tribology principles [1,2,3] are applied to the study of blast furnace brickwork destruction.

Refractory materials are utilised for the construction of blast furnace wall brickwork: graphite, semigraphite, Al_2O_3-SiO_2, SiC, Cr_2O_3, Si_3N_4, $Si_{6-z}Al_zO_zN_{8-z}$ ($0<z<=4,2$). Their characteristics are different as a function of stresses they support in every zone, and two tendences are visible: a) utilisation of refractory materials having great thermal conductivity and simutaneous utilisation of a very effective cooling system of the cold face; b) utilisation of refractory materials having low thermal conductivity but great resistence to simultaneously mechanical, thermal and chemical stresses exerted by blast furnace inner environment.

In the study of factors and processes determining the destruction of refractory brickwork the whole blast furnace may be presumed as a complex tribosystem (Fig. 1) where the constitutive elements are: a) refractory brickwork - stationary triboelement; b) granular solids, liquid iron, liquid slag, blast furnace gas - moving triboelement; c) interface element; d) the tribosystem environment, which may be simultaneously the moving triboelement too represented by liquid iron, liquid slag, or blast furnace gas.

The processes determining the blast furnace refractory brickwork destruction are presented in Fig.2.

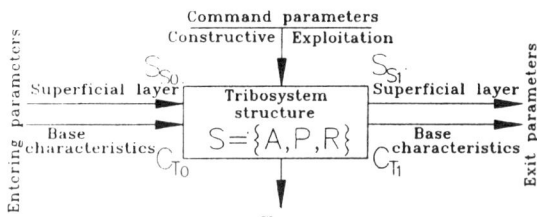

Fig. 1: Complex tribosystem model.

The destruction of blast furnace brickwork refractory materials take place simultaneously at its hot face (superficial layer) and in its volume, and it is a resultant of their interactions with the other parts of the wall (cooling system and armour) and with blast furnace content (raw materials granules, gases, liquid iron, liquid slag) and depends on blast furnace operating

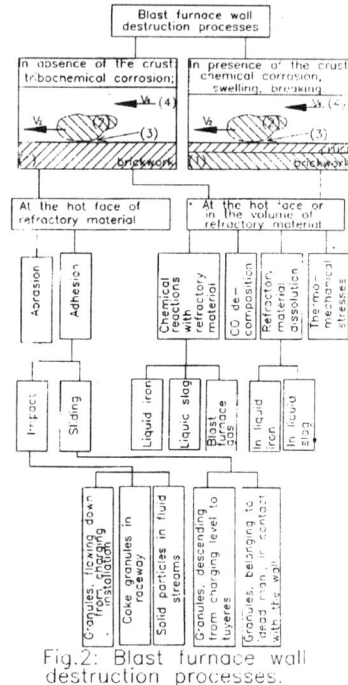

Fig.2: Blast furnace wall destruction processes.

conditions (axial or peripheric gas circulation, descending uniformity of the blast furnace materials column, acid or basic slags).

The destruction is performed by: a) adherent crusts detachment from refractory material hot face and simultaneous rupture of some parts off the lining; b) in the absence of crusts the following wear processes are possible in the superficial layer (at hot face): impact, abrasion, erosion with gases, erosion with liquid iron, erosion with liquid slag; c) in the presence as well as in the absence of crusts the following destruction processes are possible at the hot face and in refractory material pores: chemical reactions between ceramic material and fluids, dissolution of refractory material in liquid iron and in liquid slag, CO decomposition; d) thermomechanical stresses.

REFERENCES

(1) Ion Crudu, ON THE CONCEPT OF TRIBOSYSTEM AND A TRIBOMODELING CRITERION, 4ème Congress Europeen de Tribologie, Ecully, France, 9-12 Septembre, 1985
(2) R. D. Arnell, P. B. Davies, J. Halling, T. L. Whomes, TRIBOLOGY PRINCIPLES AND DESIGN APPLICATIONS, Macmillan London, 1991
(3) Horst Czichos, TRIBOLOGY - A SYSTEMS APPROACH TO THE SCIENCE AND TECHNOLOGY OF FRICTION, LUBRICATION AND WEAR, Elsevier Scientific Publishing Company, 1978

CALCULATION METHOD FOR BLAST FURNACES HEARTH WEAR RATE

MIRCEA-PETRE IONESCU
S.C. I.C.P.P.A.M. S.A., Galatzi, 6200, Romania, Str. Smardan no. 2, Fax 040 (036) 461238;
ION CRUDU
Machines Parts Department, University of Galatzi, 6200, Str. Domneasca no. 47, Fax 040 (036) 461353.

Abstract

Tribological processes have significant importance among the factors influencing the blast furnaces hearth destruction rate (1).

The method consists in integration of Laplace differential equation for steady-state bidirectional heat conductance on selected surface of the hearth axial section, with the help of finite elements method:

$$\frac{\partial}{\partial x}\left(\lambda_x \frac{\partial t}{\partial x}\right) + \frac{\partial}{\partial y}\left(\lambda_y \frac{\partial t}{\partial y}\right) = 0 \quad /1/;$$

Where: t=temperature; λ_x is thermal conductivity in x coordinate axis direction; λ_y is thermal conductivity in y coordinate axis direction.

If thermal conductivity, hot face temperature (equal to liquid iron temperature in evacuation), temperature values in some wall brickwork points, vertical and horizontal thermal flows are known, then remanent brickwork thickness may be determined by calculation.

For calculation the following simplifying hypotheses are considered: a) from the entire hearth refractory lining are taken into consideration only the first, the second, the third, the fourth layers, and the first, the second, and partially the third rings and also the molten iron bath, permanently existing inside the blast furnace hearth, under the tapholes level; b) the entire iron quantity from the metallic bath has the temperature of liquid iron at evacuation owing to free and forced convection currents existing in the molten bath and making the temperature uniform (2).

The calculation stages are: calculus, expandation and totalisation of elemental matrixes, boundary conditions imposing, equations system resolving, convergence verification, measured and calculated temperatures equality verification, isotherms calculation, graphic representation of results (3,4).

Elemental matrix equation is:

$$\begin{vmatrix} +\frac{1}{3}m_x+\frac{1}{3}m_y & -\frac{1}{3}m_x+\frac{1}{6}m_y & -\frac{1}{6}m_x-\frac{1}{6}m_y & +\frac{1}{6}m_x-\frac{1}{3}m_y \\ -\frac{1}{3}m_x+\frac{1}{6}m_y & +\frac{1}{3}m_x+\frac{1}{3}m_y & +\frac{1}{6}m_x-\frac{1}{3}m_y & -\frac{1}{6}m_x-\frac{1}{6}m_y \\ -\frac{1}{6}m_x-\frac{1}{6}m_y & +\frac{1}{6}m_x-\frac{1}{3}m_y & +\frac{1}{3}m_x+\frac{1}{3}m_y & -\frac{1}{3}m_x+\frac{1}{6}m_y \\ +\frac{1}{6}m_x-\frac{1}{3}m_y & -\frac{1}{6}m_x-\frac{1}{6}m_y & -\frac{1}{3}m_x+\frac{1}{6}m_y & +\frac{1}{3}m_x+\frac{1}{3}m_y \end{vmatrix} * \begin{vmatrix} t_1 \\ t_2 \\ t_3 \\ t_4 \end{vmatrix} = \begin{vmatrix} +n_x+n_y \\ -n_x+n_y \\ -n_x-n_y \\ +n_x-n_y \end{vmatrix}$$

/2/

$$m_x = \frac{y_4-y_1}{x_2-x_1}\lambda_x; \quad /3/ \qquad m_y = \frac{x_2-x_1}{y_4-y_1}\lambda_y; \quad /4/$$

$$n_x = \frac{1}{2}q_x(y_4-y_1); \quad /5/ \qquad n_y = \frac{1}{2}q_y(x_2-x_1); \quad /6/$$

Where: x_2-x_1 and y_4-y_1 are the two dimensions of the finite element; q_x and q_y are thermal fluxes in horizontal and vertical directions; t_1, t_2, t_3, t_4 are the temperatures of the four nodes of a square finite element.

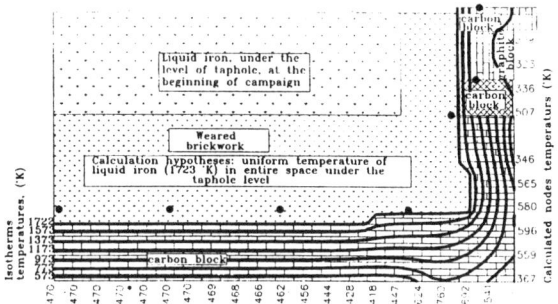

Fig. 1: Axial section through no. 1 taphole of no. 6 blast furnace hearth from S.C. SIDEX S.A. at the end of campaign: a) calculated wear rate: 1722 K isotherm; b) measured wear rate: •

Calculated wear rate and isotherms were represented in Fig. 1. In the same figure were represented for comparison the results of measurements performed during hearth lining demolition at the end of campaign.

In Fig. 1 may be observed: a) near the taphole the hearth wall has a calculated thickness of remanent brickwork of 1000 mm and a measured thickness of 600 mm ; b) the bottom of the hearth has 1000 mm calculated thickness and 1200 mm measured thickness at the axis.

Differences between the calculated and measured values of wall remanent thickness were 200..400 mm.

REFERENCES

(1) Ion Crudu, Mircea-Petre Ionescu, Viorel Munteanu, Ion-Florentin Sandu, Paraschiv Nedelcu, A TRIBOSYSTEMIC APPROACH TO REFRACTORY LINING DESTRUCTION IN BLAST FURNACES, World Tribology Congress, 8..12 Sept. 1997, London, U.K.
(2) F. Yoshikawa, J. Szekely, MECHANISM OF BLAST FURNACE HEARTH EROSION, Ironmaking and Steelmaking, 1981, No. 4
(3) Dan Garbea, ANALIZA CU ELEMENTE FINITE, Editura Tehnica, Bucuresti, 1990.
(4) S. P. Mehrotra, Y. C. Nand, HEAT BALANCE MODEL TO PREDICT SALAMANDER PENETRATION AND TEMPERATURE PROFILES IN THE SUB-HEARTH OF AN IRON BLAST FURNACE, ISIJ International, Vol. 33 (1993), No. 8, pp. 839-846.

METHOD TO DETERMINE RBOT ENDPOINT FOR PAO SYNTHETIC OIL USING FTIR CORRELATION

BRYAN JOHNSON

Arizona Public Service-Palo Verde Nuclear Generating Station, 5801 S. Wintersburg Rd., Tonopah Arizona, USA

Abstract

The cost of oil changes can be broken down into the price of the lubricant and the associated labor costs. A maintenance strategy that reduces the total number of oil changes will save in both labor and material costs providing the oil remains satisfactory for the intended drain interval. If the oil breaks down prematurely, wear to the lubricated machinery will likely result. In some cases, the available window to perform an oil change is limited due to either accessibility or production needs. For machinery of these types, it is extremely important to use an oil which will have enough oxidation reserve to remain chemically stable for an intended operational cycle. Test methods were evaluated to determine an RBOT test endpoint for the PAO oils.

Background/Vendor Recommendations

The used oil test methods typically applied to determine if an oil is oxidized are the change in the oil's acid number and viscosity. These methods can answer the question if the oil is oxidized, however, they are not effective in forecasting remaining oil oxidation reserve. Other test methods are available which can provide information concerning likely remaining oil service life. These test methods have a greater expense and may be more difficult to interpret than the viscosity or acid number tests.

One of the largest applications for oil at our generating station are the cooling tower gearboxes. Each gearbox requires 18 gallons of oil and a total of 144 are in service. . The gearboxes are located approximately 80 feet above the ground which makes changing the oil a labor intensive effort that requires the use of a crane. Production losses are also incurred if any of the gearboxes are out of service during the summer. Any oil change should be performed during the winter or scheduled outage. To lower the cost of an oil change, they are scheduled once every 18 months.

A 220 ISO rust and oxidation mineral oil was initially used in the gearbox application. This oil oxidized, although not severely during the 18 month operating cycle. A decision was made to use a PAO synthetic, in part to improve the lubricating film at the gears and to obtain a 36 month operating cycle. Since superior oxidation stability would be expected from the PAO oil, information concerning its anticipated service life along with a testing strategy to forecast when the end of life would occur were solicited from the oil suppliers. The intent was to test the oil at the 18 month service point and then make a decision if the oil would remain reliable for an additional 18 months. Vendors indicated that this type of oil would be expected to satisfy a 36 month service requirement. The testing strategies proposed to monitor and forecast the service life were based upon changes in acid number (ASTM D 974 Test Method for Acid and Base Number by Color-Indicator Titration) and viscosity changes. Oil samples were recommended every 3 months to monitor the oil condition. This strategy was implemented and the oil required changing after approximately 24 months of service.

Test Methods

The oils did not reach the 36 month service as expected. An PAO oil from a different manufacture was placed into the cooling tower application. The vendor was approached concerning a test strategy. A similar recommendation was made. These oils lasted longer than the initial brand, however, they required changing prior to the 36 month period. Unlike the earlier oil, the second was tested for more than the Total Acid Number and the change in viscosity. Additional test methods included RBOT (ASTM D 2272 Test Method for Oxidation Stability of Steam Turbine Oils by Rotating Bomb) and FTIR (Fourier Transform Infrared) spectroscopy analysis. All testing was performed in house.

The RBOT and FTIR methods are routinely performed on mineral oils tested in the stations laboratory. These methods have provided useful information that has been used to forecast future oil changes. The RBOT test produces a different test curve for PAO synthetic oil than it does for the mineral oil. Laboratory testing was done to evaluate if the mineral oil endpoint (25.4 psi drop) was valid for the PAO oil.

Conclusion

Data from FTIR, Viscosity, and Total Acid Number tests were used to determine the RBOT test endpoint for the PAO synthetic oil used in the cooling tower gearboxes. Oils which were oxidized in an accelerated manner in the lab and used gear oils were tested. It was found that the PAO oil had a reduced test endpoint than the mineral oils.

THE EFFECT OF BREAK-IN STRESS ON THE WEAR BEHAVIOUR DURING LIFETIME

B. KEHRWALD, A. GERVÉ

IAVF, Institut für Angewandte Verschleißforschung GmbH, Im Schlehert 32, D-76189 Karlsruhe

ABSTRACT

Some passenger car manufacturers still want their customers to handle their products with care during the break-in period, giving speed limits or rpm limits for the first hundreds or thousands kilometres in order to achieve a good durability. Race car engines all get different special break-in procedures on the test bench. Truck diesel engines and diesel engines for other industrial purposes get a break-in at the end of the production when adjustment work and quality controls are done on the factory test benches. Although there is a lack of knowledge concerning the infuence of the stress during break-in on the wear-and friction behaviour of the later lifetime Diesel engine manufacturers can explain their needs clearly. They want:

- a short time production test bench occupation per engine,
- a fast development of torque and nominal power output during the first minutes of operation
- a save operation free of pre-damages during break-in,
- a break-in which provides a low lifetime wear as well as low oil consumption and a small scattering of durability.

Studies on the surfaces of engine components like piston rings, cylinder liners, camshafts, tappets, sprocket wheels and other engine parts showed that roughness decrease is not the essential effect during break-in. Sometimes fine production qualities yield increasing roughness during the first operation time. It could be proved in engines as well as on pin- and disc machines that Tribomutation of the sliding surfaces is responsible for the reduction of friction and wear rates during break-in.

Modern methods of surface analysis with mass-and electron-spectroscopy allow to investigate the Tribomutation. Changes of the element distribution, of chemical bindings, and of the structure happen in thin nm-layers by the friction process especially in a dramatic manner during the break-in. Using the knowledge about the Tribomutation and with the help of a continuous observation of the engines wear by using radio nuclide technique it is possible to create engine type specific break-in procedures in order to fulfill the diesel engine manufacturers demands. A feasible high stress which has to be defined in a controlled way produces a good conditioning of the surfaces, but one has to remain under the limit of pre-damages.

For industrial customers optimisations had been successfully carried out concerning break-in procedures at the end of the production line. The developed break-in procedures include adjustment work and quality controls. The best bench occupation usually can be reduced down to times between 6 and 18 minutes depending on the engine type and design. The shortening of the break-in time enabled the manufacturers to lower of about 20 to 30 % of their end of line test benches and save alone by this fact some million German marks. Power output development could be accelerated from more than 50 hours to the range of less than 3 hours. Lifetime wear had been reduced down to 35 %.

FAILURES ANALYSIS OF 2-LOBE JOURNAL BEARINGS OF TURBOGENERATOR

STANISŁAW STRZELECKI

Institute of Machine Design, Technical University of Łódź, Stefanowski Street 1/15, 90-924 Łódź, Poland

WOJCIECH LITWICKI

Power Plant Bełchatów, 97-400 Bełchatów, Poland

ABSTRACT

The operation of the journal bearing in the conditions of the mixed friction is the phenomenon effecting the durability of the bearing and often causing their serious failures. Misalignment of the bearing and rotor axis is one of the reasons of this situation. As result of misalignment there is the significant increase of operating temperature and the concentration of stresses on the bearing edges damaging the bearing material (1)(2)(3) occures.

Journal bearings applied in turbines or turbogenerators have usually 2 operating surfaces. These bearings should be reliable and durable through very long period of time. Any failure of these bearings causes very high losses of time and money.

The 2-lobe journal bearing is one of the journal bearings of turbogenerator TG360 (4). This bearing that is placed between low and mean pressure stages of the turbine should assure very long, reliable and operation without failure's.

The paper introduces some cases of an excessive increase of turbogenerator bearing temperature and as result the shutting-off the turbogenerator with the necessity of bearing repair. For explaining this kind of failure, the geometric parameters of the two-lobe bearing and their effect on the bearing operating characteristics have been considered. The main task of analysis assumed the calculation of the bearing static characteristics in the case of misaligned orientation of journal and sleeve axis. Computed pressures and temperatures' distributions as well as the maximum values of pressure, temperature and minimum oil film thickness allowed for explanation of the failures.

The investigations were carried out at the assumption of adiabatic oil film, the finite length of the bearing, aligned and misaligned orientation of journal and sleeve axis and the vertical load of the bearing. Because for the eccentricities larger than 0.9 there exists the possibility of operation in the mixed friction condition, all necessary characteristics have been determined for the eccentricities smaller than 0.9. The calculations were made under assumption of adiabatic oil film and static equilibrium position of journal. Effect of the misalignment on the operating parameters of the turbogenerator bearing for the relative eccentricity $\varepsilon=0.6$ is introduced in Table 1 where: L/D- aspect ratio, ψ- bearing relative clearance, ψs- segment relative clearance, α eq- angle of static equilibrium position of journal, So - resultant force, pmax, Tmax - maximum oil film pressure and temperature, q - misalignment ratio.

L/D= 0.88, $\psi = 1.5‰$, $\psi s = 1.00$ Table 1

ε -	q -	αeq [°]	So -	pmax -	Tmax [°C]
0.6	0.0	352.4	0.272	0.66	61.9
0.6	0.1	337.4	0.427	1.14	91.3
0.6	0.15	328.6	0.575	1.68	150.3

Table 1: The calculated basic parameters of the two-lobe journal bearing

The results given in Table 1 show the significant increase of maximum oil film pressure and temperature with the increase of misalignment.

As result of the theoretical investigation the conclusions given below were formulated.

1. The program for calculations allows for the theoretical investigation of the multilobe journal bearings in the conditions of aligned and misaligned orientation of journal and sleeve.
2. The misalignment of journal and sleeve axis causes the significant deformations of pressure and temperature distributions in the axial cross-sections of the bearing
3. Maximum values of pressure and temperature increase with the increase of relative length of the bearing and misalignment.
4. Segment clearance ratio has significant effect on the basic parameters of the bearing at both aligned and misaligned .operation.

REFERENCES

(1) Strzelecki S.:. Proceedings .of the 3rd Conference on Tribology, Budapest, 1 (1983), 231-236 pp.

(2) Strzelecki, S.: Proceedings 4th Int. Trib. Conf., AUSTRIB'94, Perth, (1994). 579-585 pp.

(3) Gläser H., Strzelecki S.: Proceedings of the International Tribology Conference, ITC'96, Yokohama 1995.

(4) Strzelecki S., Litwicki W.:. Engineering of Bearings Systems '96. Proceedings of the Scientific-Technical Conference. Gdańsk, 4-5th June 1996.Gdańsk 1996.,461-468pp.

COMPOSITION MATERIALS "MASLIANIT".
THE PROBLEMS OF THE CHEMICAL DISIGN
AND DIRECTION OF THE FRICTION CHARACTERISTICS

V.T. LOGINOV, S.V. ILIASOV, O.M. BASHKIROV, D.V. MINKOV, A.V.GONCHAROV, N.V. LOGINOVA
Special Design Office «Orion», Russia

In 1967 the great sientist-tribologist A.A. Kutkov has expressid an idea of the employment of the plastificator, introduced into the polymer during his convertion, as the liquid grease, transformed in the zone of the tribologue contact, while functioning the friction park. In 1968 in Cambridge University during the period of the probation, he has described with D.Taybor some investigation, corroborated this idea. During 25 years the Special Design Office «Orion», founded by A.A.Kutkov elaborates and serves the customers with the self-lubricated materials with the industrial-commercial name "Maslianit". This idea forms the basis of the technology of their manufacture and utilisation.

The problems of the chemical design and control of the friction descriptions of the "Maslianits" are solved in the SDO «Orion» by several ways, that is:

– synthesis of the composition materials on the basis of the aromatic and greasy polyamides with heightened content of the interblock plastificator. The technologies are elaborated (the innovations of the SDO «Orion»). These materials are used in all directions of the science, technique and all branches of industry;

– synthesis of the new thermostable polymers on basis of the polyimides for the composition antifriction materials, that are used as the matrix of the articles of the vacuum and cosmic technics for the details of the assembly of frition of the apparatus and mechanisms, including small-module, highly precision, touth mechanisms. The new technologies of the two-phasec process of the hardening are;

– direction of chemical and physical modification of the filling mechanisms, that permits to raise sharply the physical-mechanic and tribologue characteristics of the combination of the polyimide and of the polyamide matrices with the filling mechanism – polytetrafluorethylen, polyphosphates and with the classical solid lubricants too – graphite, disulphidum molybdenum, nitridum of titanium are elaborated;

– the utilization of the methods of the structural kinetic modelling of the mobile molecular forms by chemical designing of the composition materials. The peculiarities of antifriction of the solid lubrificants are investigated thanks to the reconstruction of the structure of these flaky combinations by pseudogyration of the cooptation polyhedrons. The polymorphous transitions and the kinematics of the untvist of molecular chains of the polytetrafluorethylen and polyphosphates are studied;

– the utilization of the methods of the supersonic diagnostics of the composition materials and ceramics, that permits to make a prognosis of their mechanical and tribologue characteristics in the course of making. The superconic transformers with the high solved capacity and the heightened everelasting heads are elaborated;

– the utilization of the theories of the biodamages and biocorrosion for a prognosis of the ecological characteristics of the composites and for the biological treatment of the surfaces too, so as to give them specific properties.

Today in different branches of industry the bearings of the belt conveyers are replaced by bearings from composition material "Maslianit" on the basis of the aromatic polyamides, which raises fivefold-tenfold the interrepair periods. "Maslianit" ASMK-12 is utilized practically in all high pressure hydro-electric power station of Russia and countries of SNG. It has found an application for water-gates in the water-ways departments on the Volga, the Kama, the Volga-Baltic way, the Ob, in the locks of Volga-Don shipping canal.

"Maslianit" has found an application in the sea- and rivership-building and ship-repair: in particular, in the bearings and condensations of the working organs of the grab dreaders of the firm Fukushima Ship Yard Matsue, on the buckets of the firm Mitsubichi Nagassaki M.F.G. Ltd (Japan), in the building shipyards and works of Nizhni Novgorod, Tuapse, Murmansk, Severodvinsk, Kaliningrad, Lazarevka, Shigalovsk, Ust-Donetsk.

The thermostable antifriction materials on the polyimides basis of TASM were utilized for the train of dears and tribologue mechanisms of the apparatus of the board sputnik systems in accordance with the programs "Venera", "Mars", "Luna", for the components of the special precision mechanisms of the sewing-machines "AOMZ - Iamato".

"Maslianit" is an indispensable composition material for the equipment of the chemical- and gas industry. It is utilized on a large scale in the compressors of the firms "Manesmer-Mayer" and "Uraka" (German) in the oilrefining works of Russia and the countries of SNG.

One of the advantages of the "Maslianits" (in comparison with another composition materials)is the ecological purity and biodamage stability thanks to the ecological analysis of the components of the composites and thanks to the introduction into the special components on the different stages of the synthesis and analysis.

SIGMA-ANALYSES FOR A MORE EFFICIENT SOLUTION OF TRIBOLOGICAL PROBLEMS OF COMBUSTION ENGINES

W. MACH, A. GERVÉ

IAVF, Institut für Angewandte Verschleissforschung GmbH, Im Schlehert 32, D-76189 Karlsruhe,

ABSTRACT

Caused by different strategies of testing developments of the tribology in combustion engines are performed separately from developments of the mixture preparation and the combustion as well.

Tribological investigations mostly base on experiments on tribometers in order to select materials and on following *durability tests* by test riggs and by engines. Within the durability tests standardised wear coefficients are evaluated discontinually for the new parts as well as for the worn parts. These values represent an integral result which does not express influences of the operational time nor of the operational conditions within the test.

The strategy of durability testings usually leads to the use of more wear resitant but more expensive materials and considers only deficiently that the wear behaviour is a property of a system which is determinated by a number of influencing factors such as operating conditions, micro- and macro-geometry, surface treatments, lubrication etc.

In opposite to this procedure developments of the combustion mainly base on *operational analyses* where the investigated engine properties like exhaust emissions and fuel consumption are measured continuously and can directly be appropriated to the respective operating conditions, time etc. In comparison to results from durability testing results from operational analyses are available within a significantly shorter time and reveal more informations about the system.

By application of continuous wear measurements using the radio nuclide technique operational analyses as described above are brought into use for the solution of tribological problems at IAVF. The engine property 'wear' is recorded continuously in the same way as engine properties from the area of the combustion.

Modifications of engine components as well as at the engine management usually have impacts on more than only one engine property. In order to achieve a more comprehensive knowledge to consequences of modifications therefore in Sigma-projects an as high as possible, reasonable number of engine properties which influence themselves mutually is recorded simultaneously and continuously.

Especially during developments of the mixture preparation and of the combustion process like
> tunings of the exhaust gas recirculation,
> modifications of the injection
> modifications of the ignition timing
> application of new fuels
> installation of new injection systems

consequences of modified parameters of the combustion to the tribological properties as well as to oil consumption and fuel entrainment into the lubricant are evaluated simultaneously.

By the multitude of engine properties which are recorded simultaneously within a Sigma-analysis optimisations for competitive targets are improved and simplified substantially.

As an example of a Sigma-analysis which concerns also the wear behaviour of pistons, piston rings and of cylinder walls it will be presented which influences do have
> the operating conditions,
> parameters like e.g. material, fuel injectors,
>> phasing of the injection, ignition timing
> the operating time

on the engine properties
> wear of piston rings,
> wear of cylinder walls,
> oil consumption / fuel entrainment
> exhaust gas emissions.

Within the example it is demonstrated how modern tribology contributes to shorten engines' development by using the continuous measuring technique and in consideration of mutual influences of the different engine properties.

A FRICTION MODEL OF THE MACHINE TOOLS GUIDES

Prof. Dr. Eng. Georgi Mishev
Department of Mechanical Engineering, Technical University Plovdiv, BG

ABSTRACT

The dynamic behaviour of the machine slides rectilinear movement depends mainly on the friction in the guides. In the present paper the main results of the performed theoretical-experimental investigations of the tangential contact deformations in the machine tool guides are presented as well as their influence on the friction force.

* A theoretical friction model in the machine tool guides is proposed to clarify the mechanism of friction force forming (Fig.1).

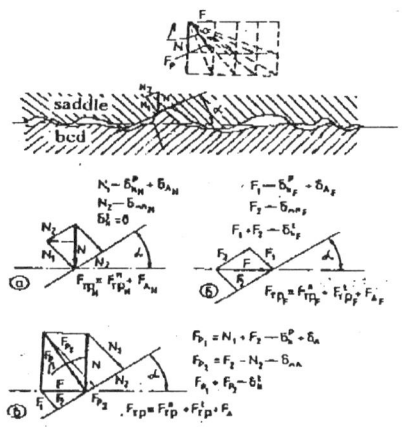

Fig.1 Forces, acting at an average roughness

The contact deformations aries out of the forces acting on the spot contacts. The micro-slidings in tangential and normal direction cause breaking of the micro-weldings in the contact, which leads an adhe-sion component of the friction force to appear. The theoretical model has been proved experimentally (Fig.2);

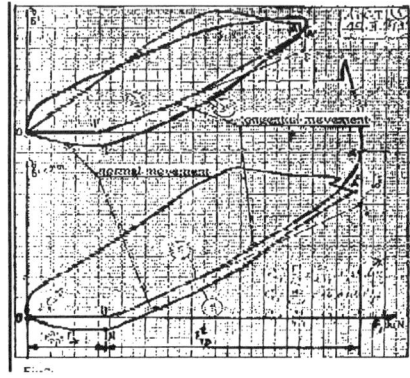

* Due to the experimental prove of the theoretical model an empirical dependence for the tangential contact deformations is obtained;
* By the Buckingham Π-theorem and the Rayleigh method a dimensionless number is obtained X, which connects the tangential contact deformations with the four fundamental parameters of the system: the speci-
Submitted to *Wear*

fic pressure p (normal force N); the slide velocity V; the contact surface quality R_a and the slide mass m:

$$X = \frac{p.R_a^3}{m.V^2} = \frac{X_1}{X_2}$$

where $X_1 = p.R_a^3$ is the stored potential energy of contact and $X_2 = m.V^2$ is the kinetic energy of the moving carriage. This number serves for finding the energetic functions, which define the criteria for the friction formation at the rectilinearly moving machine systems: the friction gives the correlation between the kinetic energy of the micro-slidings at the contact and the accumulated potential energy of the contact pre-deformation. The friction change in the machine tool guides could be achieved by a control of the correlation between the kinetic energy and potential energy at the system contact by change of the four fundamental parameters. In this way, a minimal friction force could be achieved on the machine tool design stage, i.e. a minimal friction coefficient obtained by optimization of the design and operation parameters.

* The experimental investigations confirm that the movement velocity and specific pressure exert the strongest influence on the tangential contact deformations, i.e., on the friction force.
* It is proved experimentally that the slide movement at auto-oscillating regime is a two dimensional function of the tangential contact deformations while the amplitude of the self-excited oscillations is a directly proportional to the same deformations (Fig.3).

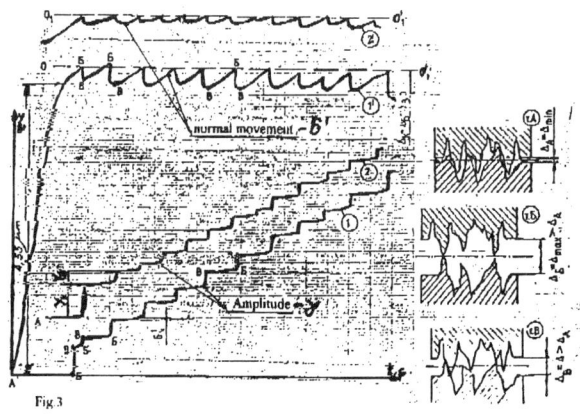

Fig.3

REFERENCES

(1) Filipov D. Seil-Excited oscillations in rectilinear mechanical-systems. The American Society of Mechanical Engineers, Vol.I, Juli 8-13, Canada, 1979
(2) Mishev G. New method for calculating friction force in the guides of machine tools. 2nd international conference on tribology, Thessaloniki, Greece, Proceedings, 547-553, 1996

DEVELOPMENT OF PISTON-RING FOR MARINE DIESEL ENGINE

Kenzo MIURA and Tatsuhiro JIBIKI
Tamano Laboratory, Mitsui Engineering & Shipbuilding Co.,Ltd, Tamahara 3-Chome, Tamano City, Japan
Takao TANAKA and Shinichi MIYAKE
Diesel Engine Factory, Mitsui Engineering & Shipbuilding Co.,Ltd, Tama 3-Chome, Tamano City, Japan

ABSTRACT

Piston-ring for a marine diesel engine is mainly limited to relatively low strength cast iron materials. However, problems such as sticking and breaking of the piston-ring sometimes occur. For developing piston-ring to meet the more severe working condition in the future, suitable and efficient countermeasures are most necessary to improve the reliability of marine diesel engine.

In one direction of improving the material strength of the piston-ring of the marine diesel engine, we MES have so far made efforts to develop a new wear resisting steel in the expectation that it may supersede high phosphorus gray cast iron. In line with the above R & D activities, we carried out wear tests and scuffing tests for the newly developed wear resisting steel. The test results of the new type steel were analyzed in comparison with those of gray cast iron.

Measurements of scuffing characteristics were made using an abrasion tester of pin-on-disk type. The center diameter ϕ of sliding disk was 120mm, and the sliding speed on the test surface was constant at 5 m/sec. The test lubricating oil was a mixture of lubricating oil SAE #30 and white kerosene at 1:1. The test lubricating oil was supplied on the disk which was rotated at a speed 5 m/sec for 30 sec., and then stopped to supply it. After that, the pin was loaded to start the test. The initial load was set to 245N, and after 30 minutes, set to 490N, then further increased by 98N for each 5 minutes. An example of the scuffing test result of MES wear resisting steel, the new type steel with surface modification treatments is shown in Fig. 1.

The following results were confirmed from the experimentally screening tests: The wear amount of the new type steel was smaller than gray cast iron. The scuffing initiation time of the new type steel with sulfurization, the surface modification treatment to improve the initial break in condition of steels, became longer than gray cast iron.

Using the new type steel, we manufactured piston-rings for marine diesel engine. The piston-rings with sulfurization were set to the actual engine, and was driven for more than 160 hours. The surface roughness of piston-top-ring before test and after 29 hours driving test is shown in Fig. 2.

The new type steel possesses the good wear resisting characteristics and the good running in properties. After the initial surface break in condition of the said steel was improved by the surface modification treatment, the piston-ring thus treated has since been offering no problem at all.

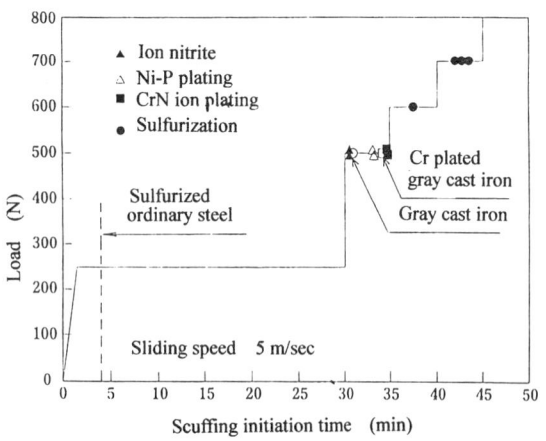

Fig. 1: An example of the scuffing test result of MES wear resisting steel with surface modification treatment

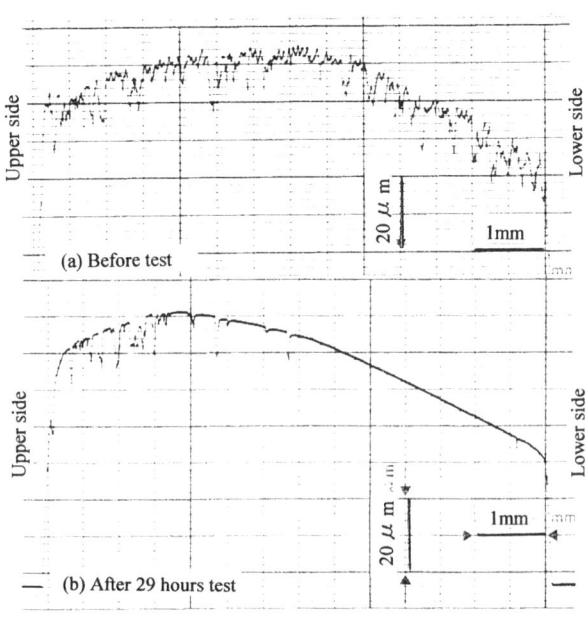

Fig. 2: Surface roughness of piston-top-ring before test and after 29 hours test

THE INFLUENCE OF FRICTION AND DAMPING FORCES ON DYNAMIC LOADING AT CRANES

RADIĆ MIJAJLOVIĆ, ZORAN MARINKOVIĆ, MIOMIR JOVANOVIĆ,
Mechanical faculty University of Niš, 14, Beogradska St., 18000 Niš, YU

ABSTRACT

The influence of the friction forces on dynamic crane forces is analyzed in this paper. Driving engine moment overcomes a resistance when we start the machine. In transitional kinematics regimes, variable friction forces are present and their development is difficult to estimate. These forces influence the dynamic processes and these processes are relevant for crane security calculation. This paper shows mathematical model with there degree of freedom mechanism for lifting driving bridge cranes. By comparison equation movement with and without damping, we can get damping influence on dynamic lifting forces. Damping in elastic joint system provoke dissipation of resistance which influence amplitude and oscillation rules. Torsion elastic-kinetic model with two masses and forces of viscose friction in joint elements are discussed.

THE DYNAMICAL LOADINGS OF THE CRANE MECHANISMS

For the dynamic load analysis of the control crane mechanisms at nonstationar working period, different models are used. The elasto-kinetic model with two rotating masses (1,2) is successfully used today. In these analysis two basic cases are occur. The first one is unloaded elastic link between masses. The second one is loaded link in case of lifting or lowering the load from the air. In the first case, at the acceleration period, the modifications of the dynamically model occur because this period is divided into three motion periods.

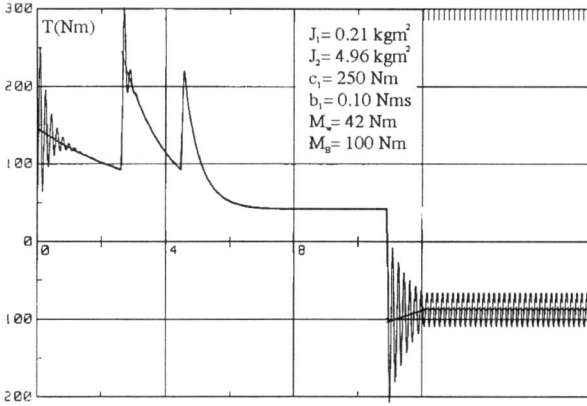

Fig.1 The change of the torsion moment of the driving shaft of bridge crane mechanism

In differential equations, the dumping (caused by viscose friction with constant coefficient), has the dominant influence. The feature of this friction is that the successive amplitudes of the oscillator movement are reduced by geometrical progression. Researches have shown that in this case, the resistance of dry friction (*Coulomb*'s friction) has dominant influence (2). The torsion moment in elastic link of the dynamic model is shown at Fig.1.

WORKING RESISTANCE OF LUFFING DRIVING MECHANISM

Luffing driving mechanism is often realized with a gear accompanied with a power screws. By measuring motor moment during increasing and reducing of crane radius (M_R and M_S) and their comparison with regard to the same crane radius it is possible to reach a conclusion regarding friction in power transmission gear (3). Motor moments during opposite directions of luffing can be defined as:

$$M_R = F_P \frac{d_2}{2} tg(\varphi + \rho_v) \frac{1}{i_R \cdot \eta_R} = k_1 \cdot F_P,$$

$$M_S = F_P \frac{d_2}{2} tg(\rho_v - \varphi) \frac{1}{i_R \cdot \eta_R} = k_2 \cdot F_P.$$

In the above relations F_P - force in leading screw, d_2 - medium diameter of screw pair, φ - angle of the screw thread, ρ_V - reduced screws friction angle, i_R - reducing gear transmission ratio and η_R - utilization factor of reducing gear. k_1 and k_2 are coefficients. The presence of additional resistance in screw pair joint, caused by irregular geometry of a power screws, has been identified. This additional tribological moment, due to irregular geometry of a power screws M_O can be determined. This moment's value is not constant, but does not have great dissipation:

$$M_0 = \frac{k_1 \cdot M_S - k_2 \cdot M_R}{k_1 - k_2}$$

REFERENCES

(1) Marinković Z.: PROBABLE-STATISTIC MODEL OF THE LIFE-TIME CALCULATION OF THE CRANE DRIVING MECHANISM ELEMENTS, Dissertation, Mechanical faculty University of Niš, 1993.

(2) Marinković Z., Mijajlović R., Marković S. : ELASTO-KINETIC MODEL IN DYNAMIC ANALYZIS OF CRANE DRIVING MECHANISMS, 13 International Conference, Mechanical faculty University of Beograd, 1994, pp.239-244.

(3) Jovanović M., Vulić A.: ANALYTIC AND EXPERIMENTAL IDENTIFICATION OF TRIBO-PHENOMENA IN POWER SCREWS OF PORTAL JIB-CRANES, Facta Universitatis, Series Mechanical engineering, No1, Vol.2, 1996. University of Niš,

ECONOMIC AND ECOLOGICAL PROBLEMS OF TRIBOLOGY DECISIONS FOR RAILWAY TRANSPORT

A.T. ROMANOVA, V.A. DMITRIEV, S.S. PETRAKOVSKY
Moscow University of Railway Transport, Obraszova st. 15 Moscow, 101475 Russia
Ju. M. LUSHNOV
MADI - Technical University, Leningradsky prospect, 64, Moscow, 125319 Russia

ABSTRACT

The conception and the model of the industrial and economical system of railway transport are given in the paper. There are the conditions of the research and technical progress and the coefficient of energy resources usage in the model, which show the peculirities of the railway functions. It is opened the time of the crisis beginning in the branch technology development. The crisis has the negative influence on the economic and ecological characters of the branch. It is given the results of analysis which we can see from the high economic and ecological effects from decisions directed on the reduction of friction losses in friction systems of the railway transport. (One of the main way for it is to pay attention to season modifications of the friction coefficient). It is shown results of this problem investigations.

The railway transport is one of the most power-intensive branches of national economy. In a number of power-bearers embraced with the structure of expenditures part of its fuel-energetic balance (FEB) the rail-way transport is one of main consumers in the country. The share of the branch comes about to 50% of the consumption of diesel fuel, 66% of coal, 66% of consumption of fuel oil in the national economy of Russia. Reduction of power-capacity of transport which reached was received mainly for the account of changing of FEB structure of railway transport as a result of electrification of sections of railway. However non-complexity of electrification leads to a insignificant positive change of service indicators: on 1% of investments growth there was often not more than 0,03% of positive change of indicators. And it causes a reduction of economic effect of electrification and decreases saving of power resources in comparison with potentially possible one, (1).

In the report the data are presented showing that by increasing of power-capacity of railway transport one can improve using of other kinds of resources when organizing transportation process and as a result to obtain an increase in the profitableness of the branch. Thus, forming the expenditure part of branch's FEB together with the power-saving must be directed to the increasing an effectiveness of all kinds of resources usage in the production processes of the branch whilst the realization of the chosen strategy of its development. The sharpness of ecological problems of using power resources, limitation on natural resources, transition to market relations dictating a growth of competition between different kinds of transport inside the transport system, also demands a study of possible version of operative and strategic decisions defining the development of FEB of the branch with an account of economic criteria of its functioning, technological and resource limitations of characteristics of its work, (1).

To provide the effective using of power resources on the railway transport one worked out a conception and a metodology of variant prognosis of the fuel-energy balance of the branch with possible resources limitations which allow realization of scientific-technical progress under conditions of the market.

One made a prognosis of specific energetic expenses as to a traction of trains on the perspective up to 2020 in accordance with the proposed method of calculation.

Calculations lead to a conclusion that more 80% of power consumption goes to overcome the work of friction forces, aero-dynamic and omic losses. Now realized on railway networks measures on decreasing forces of resistance of trains movement and, thus, decreasing a dissipative component are not sufficient. Both analysis of boundaries of the dissipative component of power and fuel consumption of trains traction calculated for a perspective up to 2020 and expert estimates allow to single out power consumption connected with a resistance to friction forces. The last ones are even with an estimate on the lower boundary for electric trains traction obtained at a proving ground not less than 42-37% and for an diesel locomotives traction obtained at a proving ground - more than 45%, (1).

Friction losses can be effectively decreased by using already existing home and foreign data on lubricants and optimal design of friction units.

Effectiveness of tribotechnical decisions for overcoming a technological crisis of the branch is confirmed by the estimates given in (1).

It is known that 85% of faults of the rolling-stock occurs because of the wear of friction units. For the last 10 years a number of repairs of cars grew almost in 2 times. However capacities of car-repair plants are not enough. To improve wear resistance is a way to a solution of this problem. Using the resource-saving oils for transport machines can give an economy of oil from 3 up to 6%, decreasing power consumption by 30%, redusing a wear and corrosion by 20-30%.

REFERENCES

(1) Романова А.Т. Экономическое прогнозирование топливно-энергетического баланса железнодорожного транспорта. Деп. в ЦНИЭИуголь 21.09.92, N 5397, 307 с.

Submitted to *Leeds-Lyon Symposium on Tribology Proceedings*

QUALITY ASSESSMENT AND INSPECTION OF SLOW BURNING HYDRAULIC OIL

ANNA RYNIEWICZ
Academy of Mining and Metallurgy, Al Mickiewicza 30, 30-059 Krakow, Poland.

ABSTRACT

In hydraulic system operating in fire hazard conditions there are slow-burning fluids of various kinds used instead of commonly applied mineral oils. There fluids may be applied in devices operating close to fire environment at high temperatures and pressures.

Till now there lack of commonly accepted quality requirements and test methods for hydraulic oils and its test methods reported by hydraulic equipment manufactures. There is lack of an international standard that may serve as a fundamental criterion of such oils quality assessment.

In the paper is presented the system of estimation and control process of slow burning hydraulic fluid based on phosphor esters. That fluid operates in an automatic control system of a 500MW turbine. On the fig. 1 is presented schematic of estimation process and quality control of the hydraulic fluid dented TSO-hs.

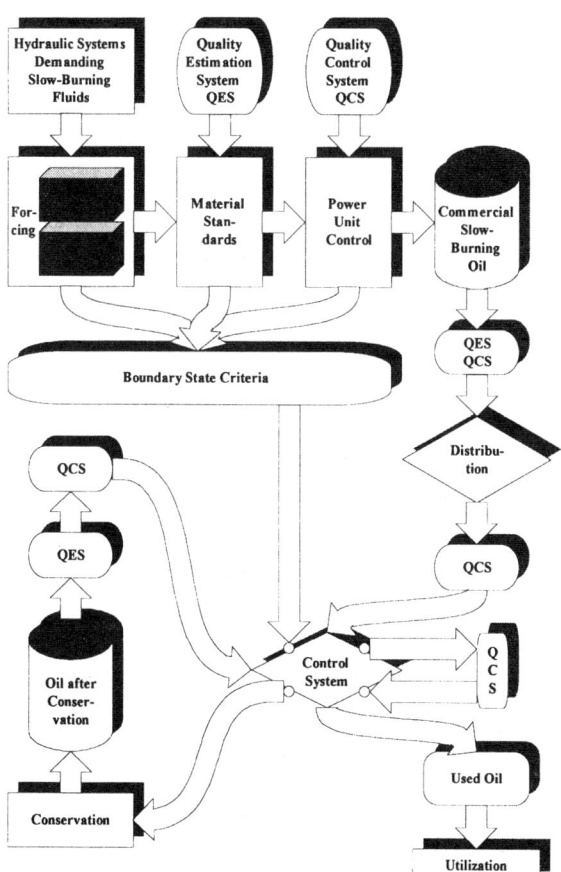

Fig. 1: Scheme of system of estimation and qualify control of slow-burning oil.

To ensure proper operation of power unit automatic system control it is continued the examination of physical-chemical and tribological properties of hydraulic fluid. An immediate examination of the oil quality is performed by the power station maintenance group but once a month a sample is taken for full analysis in spacialized laboratory.

Analized changes of the hydraulic fluid properties are:

- increase of acid number,
- kinematics viscosity diminish,
- hydrolysis resistance lowering,
- change of water content in the system,
- fluid purity,
- lowering of corrosion resistance,
- wear resistance,
- foam resistance.

Slow burning phosphor ester based is a very specific substance when is used in operation. If the fluid ageing processes, catalysed by wear products of the system would go too for and the fluid does achieve boundary state conditions - the process of a fluid regeneration in practically impossible.

Initially the effort was undertaken to apply base fluid with a minimum acid number but then and additional system for the fluid cure has been introduced instead, that operates on replaceable masses principle.

This system is aimed at preservation of a fluid number on a low level and ensuring demanded clority of fluid as well as its dehydration.

On the base of the fluid quality control and tests made in defect states the range and limit values of parameters have been stated that can disqualify the fluid and put it to utilization

The presented system of estimation and quality control of slow burning phosphor esters based fluid allows for effective operation, preservation of defect states and extension by 50% the period of its exploitation in the hydraulic system.

REFERENCES

(1) A. Ryniewicz, Z. Szydło, B. Zachara Slow Burning Hydraulic Oil TSO-hs In Exploitation, 10th International Colloquium Tribology - Solving Friction And Wear Problems, Technishe Akademie Esslingen, 09-11 January 1996 vol 1, (445-454).

(2) A. Ryniewicz Quality Assurance Of Slow Burning Hydraulic Oil TSO-hs In Exploitation, 7th International Daaam Symposium, Technical University Of Vienna, 17-19 October 1996, vol. 1, (343).

STUDY ON THE ANTI-WEAR ABILITY OF SHIPLOCK MATERIAL

X.Y. SHENG, J.B. LUO, S.Z. WEN

National Tribology Laboratory, Department of Precision Instrument, Tsinghua University, Beijing, 100084 China

ABSTRACT

The driving mechanism of shiplock is about 6-12 meter deep in the water, so it is difficulty to repair. Serious wear of shiplock material will bring a series of problems such as hanging of the top part of the gate, difficulty opening of gate and inrush of water(1). A ping on disc tester is used to test the anti-wear ability of 17 kinds of shiplock material. The ping is the tested material and the disc AISI E52100. The tested load is 84.3N. The wear length is 40,000m for every kind of tested material. They are lubricated by water and lithium grease + 3% MoS_2. The relation between wear and hardness, surface roughness and kinds of heat treatment are discussed. Considering Weierstrass-Mandelbrot W-M function as the expression of wear surface profile, the fractal parameters D and G of wear surface are calculated. Wear mechanism is also discussed. The wear under different lubrication are also discussed in the present paper. We can conclude for water lubrication:

1. the nitriding 0.6mm AMS 6470E is the best material;
2. the anti-wear ability has relation with hardness, but not one-to-one correspondence;
3. the higher of anti-wear ability, the smooth of wear surface;
4. the fractal dimension D is higher and fractal roughness parameter G is higher for materials with higher anti-wear ability;
5. the fractal dimension D is about 1.1~1.2 and fractal roughness parameter is lower for materials with lower anti-wear ability;
6. there is three kinds of wear mechanism: plough, spallation and plastic flow.

For lithium grease + 3% MoS_2 lubrication:

1. surface harding AISI 5140 forged steel is the best material;
2. the anti-wear ability has no good relation with hardness;
3. wear rate is 100-1000 times lower than that under water lubrication;
4. there is MoS_2 coating on the tested material, friction is low and the ping can rotate under the load. MoS_2 coating is the main reason for friction reduction and super low wear rate;
5. for most of the tested material, only the boundary contact with disc, so the roughness of the wear surface changes little;
6. the fractal dimension D and fractal roughness parameter G of the wear part has no relation with anti-wear ability;
7. for most of the tested material, wear mechanism is plough. A curious phenomena is observed for surface harding AISI 1045 forged steel: there is spallation like fatigue spallation of thick coating. But there can not be any thick coating. Another curious phenomena is that many particles grow up on the wear surface which contain chlorine.

REFERENCES

(1) X.Y. Sheng, J.B. Luo, Y.Q. Li, S.Z. Wen, Progresses on Tribology, Vol.1, No.2, 1996.

TRIBOLOGICAL STUDIES ON LINEAR MOTION BALL GUIDE SYSTEMS

S. SHIMIZU(Professor), E. SAITO and H. UCHIDA (graduate students)
Department of Precision Engineering, Meiji University, Kawasaki, Japan
C. S. SHARMA and Y. TAKI
Engineering Division, THK Co., Ltd., Meguro-ku, Tokyo, Japan

ABSTRACT

Linear Motion Ball Guide (LMBG) system was introduced in machine tools around 25 years ago, but no work has been reported on the assessment of life of this most critical component, now used for almost all precision machines. In this paper, the accumulated stress cycles on rail and guide blocks, the rolling body load and the equivalent average ball load are calculated for selected LMBG systems comprising a rail with two guide blocks fixed to a common top plate [1]. Considering these calculations, theoretical relationships are derived to determine the basic dynamic load rating, C for varying conformity factor, f and load eccentricity, e_z using 3-parameter life Eq.(1) given below [2],[3].

$$L_{10} - \gamma = \left(\frac{C}{F}\right)^p \quad \ldots\ldots(1)$$

where L_{10} is rated life for 90% reliability and γ is the location parameter or the minimum life.

The effect of change in 'f' on calculated values of basic dynamic and static load rating is illustrated in Fig. 1, where C_0 is the static load rating obtained from permanent deformation of $10^{-4}D_a$. It is observed that C and C_0 for pure radial load, F_y and $C(M_z)$ and $C_0(M_z)$ for pure pitching moment, M_z are larger for closer conformity factor, i.e. smaller 'f' value while they drastically reduce for increasing 'f'. The calculated values of C and C_0 for component representing a rail with one guide block are also included for comparison.

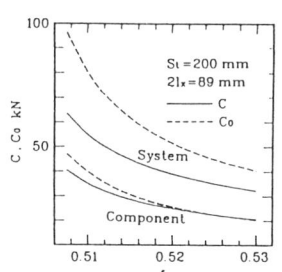
(a) C, C_0 vs. f under pure radial load, F_y

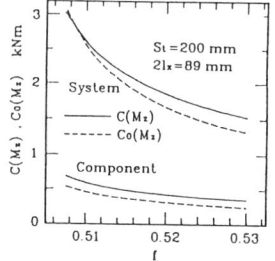
(b) $C(M_z)$, $C_0(M_z)$ vs. f under pure pitching moment

Fig. 1: Effect of conformity factor, 'f' on C and C_0

A similar effect of increasing the value of load eccentricity, e_z which defines the deviation of the point of action of the load from the center of the table is also found in this theoretical analysis.

The theoretical results are further verified by performing experimental life tests for two types of LMBG system using two types of test rigs. Test rig 'A' is driven by a ball screw mechanism, while another test rig 'B' is driven by a crank/slider mechanism. In these cases the number of test specimens varied from 35 to 41. Two sets of specimens with different crowning patterns were used in this experiment in order to compare this result with the theoretical result including the effect of crowning.

The Weibull distribution plots obtained from the failure data in the above mentioned life test for two crowning patterns are shown in Fig. 2.

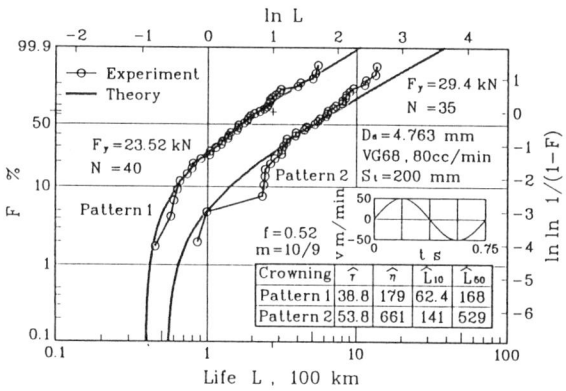

Fig. 2. Effect of crowning patterns on LMBG life

The result indicates an excellent conformity of the test data with the theoretically plotted data. By selecting an appropriate pattern of crowning a higher life may be expected. Accumulation of a large number of stress cycles on the rail due to overlapping stresses caused by two guide blocks, the flaking is initiated mostly on the rail.

REFERENCES

(1) S. Shimizu, H. Takizawa, C. S. Sharma and H Koshiishi, Proceedings of Int'l Tribology Conference (JAST), Japan, Vol III, pp. 1369-1374, 1996.
(2) S. Shimizu and M. Izawa, J of JSLE (Int'l Edition), Vol. 3, pp. 71-76, 1982.
(3) T. Tallian, ASLE Trans., Vol. 5, pp. 183-202, 1962.

SOME FUNCTIONAL PARAMETERS FOR PERFORMANT TRIBOLOGICAL APPLICATIONS - CASE OF SLIDING ON SNOW

T. MATHIA, P. LANTERI*, H. ZAHOUANI, R. LONGERAY*, B. BOUALI*, and A. MIDOL

Laboratoire de Tribologie et Dynamique de Systèmes UMR 5513 CNRS, double site Laboratory
Ecole Centrale de Lyon, F-69280 Ecully & Ecole Nationale d'Ingénieurs de St. Etienne, F-42000 St. Etienne - France
*Laboratoire de Chimiométrie, Université Caude Bernard & Ecole Supérieure de Chimie Physique Electronique, F-69622 Villeurbanne - France

ABSTRACT

As one can easily imagine, sliding on water (boats, canoe - kayaks, surfs. etc....) and on snow (skis, snowboards, sledges,...) can be mastered by nature of sliding surfaces and manufacturing process offering specific topography, end eventually specific coating.

Analysis of friction on snow with premelting and melting phenomena (1) as well as on water involves dynamic wettability characteristics of concerned surfaces (2,3). Sliding involves vector description of movement and therefore directionality's characterisation of surfaces in terms of its non-additive physico-chemical as well as morphological descriptors as shown in Fig 1

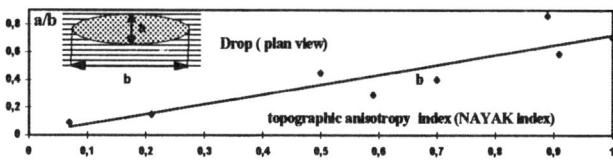

Fig 1: Lateral assymetry of wettability versus morphological anisotropy

Pragmatic contribution of novel approach of taking into account of morphological characterisation will be applied to the determination of simultaneous influence of morphology and of wettabilty on sliding performances as shown in Fig 2.
Particularly the influence of the morphological anisotropy on dynamic wettability in terms of advanced and receding angles and initial sliding conditions is treated and shown in Fig 1. New mechano - chemical treatment in order to master tribological properties will be described (4). This treatment allows simultaneous generation of suitable topography and the chemisorption of specific functional molecules (with hydrophobic groups like - CH_3, $-CF_3$, for example) on the surfaces.

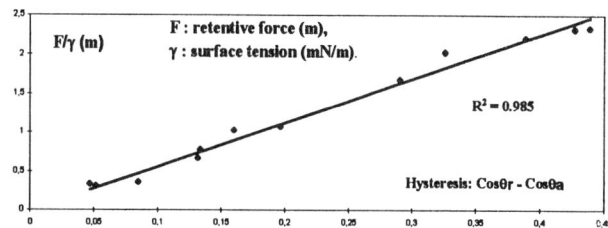

Fig 2: Vertical assymetry of wettability versus mobilyty

Physico chemical properties of surfaces and of the coatings deposited on different morphologies is introduced and modelled thanks to experimental design strategy for modelling and for analysing mixture. At this stage the prediction of optimal composition for requested properties of treated surfaces is possible. (5)
Sliding performance on snow will be correlated with morphological as well as wettability properties. Validation of the model on site will be tested and functional parameters will be extracted.
In case of speed skiing very important gain of performance has been achieved exceeding the speed limit of 230 km/h thanks to presented approach.

REFERENCES

1. L. Wojtczak, J. Rousseau, T. Mathia. « Friction in the Atomic Scale » Applied Mechanics and Engineering, Vol. 1, N° 2, pp 173-191, 1996
2. B.J Briscoe. & K.P. Galvin, Colloids and Surfaces, Vol. 52, pp 219-229, 1991
3. T. Mathia, P. Lanteri, R. Longeray, A. Midol, Cahier de Rhéologie IX pp 85-89, Nov. 1991
4. T.G Mathia P. Lanteri, R. Longray, A. Midol, Proceed. of NORDTRIB'94, Vol. III, pp 621-631, 1994
5. T. Mathia, P. Lanteri, R. Longeray, A. Midol, H. Zahouani, P. Ribot Patent N° FR 94-09 069 (1994)

TRIBOLOGICAL CHARACTERISTICS OF BONDED MoS2 FILMS AT HIGH LOAD CONDITIONS IN VACUUM

KOJI MATSUMOTO and MINEO SUZUKI
Space Technology Research Group, National Aerospace Laboratory, Chofu Tokyo, JAPAN
MASAHIRO KAWAMURA
Kawamura Research Laboratory, Inc., Meguro Tokyo JAPAN
MAKOTO NISHIMURA
Hosei University, Koganei Tokyo JAPAN

ABSTRACT

Tribological characteristics of bonded MoS2 films were evaluated at high load conditions in vacuum, using a Falex type vacuum tribometer. Pin and V-block specimens were based on Falex standard, and made of 440C stainless steel. Tested bonded films were MoS2 + SbO3 + binder and MoS2 + SbO3 + graphite + binder. The binder was polyamideimide for both films. The film was coated on both pin and V-block specimens, and the thickness of the film was 10-15 μm. A diagram of the Falex type vacuum tribometer is shown Fig.1. Applied load was varied by regulating pressure in bellows placed between load-arms. It was increased stepwise from 980 N to a pre-determined value(maximum load) by 490 N every 20 minutes, and then was kept on until film failure. The maximum load tested was from 1960 N to 5880 N. Sliding speed was 0.096 m/s.

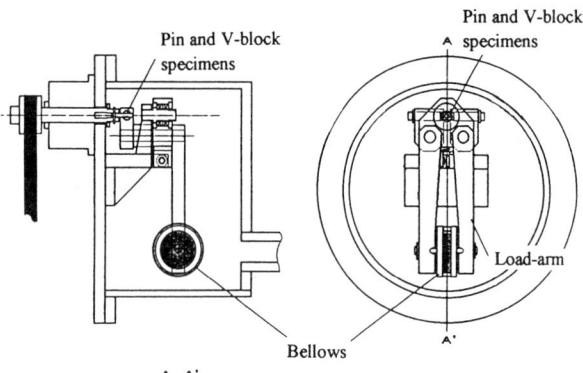

Fig.1 Falex Type Vacuum Tribometer

For both films, torque rose sharply at the beginning of the test and then it showed a little increase despite the load increased. So friction coefficient decreased as the load increased, and the value was as low as 0.04 at a load of 4000 N. Temperature of the V-block specimen rose gradually during testing, and it reached 400-500 ℃ at the film failure. Pressure in the vacuum chamber was 1×10^{-4} Pa before testing, however, it rose to around 10^{-2} Pa at the film failure. No difference in friction characteristics was observed between the two films. However, the addition of graphite was found to have a great effect on wear life. Fig.2 shows wear life of both bonded films against maximum load. Wear life of the film without graphite decreased slightly as the load increased. The effect of the applied load was relatively small for this film. For the film with graphite, wear life was 4-6 times longer than that without graphite at loads less than 4000 N. Even if it dropped drastically at loads above 4000 N, it still remained longer than that without graphite. It has been previously shown that graphite improves tribological characteristics in air and that bonded MoS2 film with graphite operate preferentially under certain load condition[1]. In this study, the beneficial effect of the addition of graphite on the tribological characteristics is also shown under vacuum condition.

Quadrupole mass analyzer show that no difference in outgassing characteristics was observed between the two films. The solvent which was used for spray coating remained in the film. It was clearly detected only at the beginning of the friction test. Furthermore, the film failure can be characterized by an outgassing and pressure increase. By comparison with the spectra at the beginning of the test, new mass peaks of hydrocarbon species appeared when films failed.

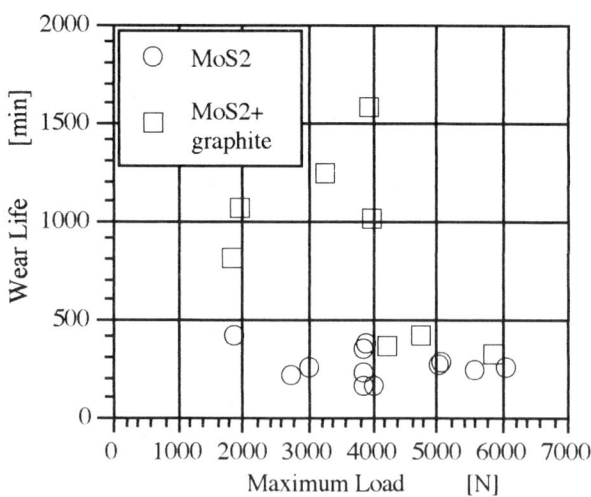

Fig.2 Wear Life against Maximum Load

REFERENCES

(1) M. Kawamura, I. Umeda, Influence of specimen form, contact pressure and sliding speed on tribological characteristics of solid lubricant films, Proceedings of JAST tribology Conference, 611-612, 1991

WEAR MECHANISM OF AN ALUMINUM-CARBON COMPOSITE MATERIAL

MUNEO MIZUMOTO, MASAKI KOYAMA and EIICHI SATO
Mechanical Engineering Research Lab., Hitachi Ltd., Kandastu, Tsuchiura, Ibaraki 300 Japan
HIROAKI HATA
Refrigeration & Air-Conditioning Division, Hitachi Ltd., Ohira, Shimotsuga, Tochigi 329-44 Japan

ABSTRACT

An aluminum-carbon composite, which is a sintered hard carbon material into which aluminum permeates, is a suitable tribological material for compressors using alternate refrigerants.

We did a series of experiments to clarify the tribological characteristics of the aluminum-carbon composite material. We used experimental equipment that could simulate the sliding conditions between a vane and a roller of rotary type compressors. The specimen materials were aluminum-carbon composite for the vane and a cast iron for the roller, and the experiments were carried out in a HFC134a alternate refrigerant environment.

Figure 1 shows a relationship between the amount of wear and the oil supply rate. The amount of wear is in general liable to decrease as the lubricant supply rate increases, with the exception of the cutting type abrasive wear. But the experimental result in Fig. 1 shows the opposite tribological characteristic.

We observed the specimen surfaces to clarify the reason for the above interesting tribological phenomena and got following results.

1) Corners of the carbon particles which protrude from the vane surface were worn down with small supply rates of the lubricant oil, and as a results both of the vane and roller surfaces became very smooth. In contrast with this, the roughness of the both surfaces increased with the large oil supply rates.

2) After the vane and roller surfaces were smoothed with the small oil supply rate, their roughness kept small and did not increase any more with the large oil supply rate.

3) The oil film between the vane and the roller was found to be thicker under the smooth surface conditions by measuring the electrical resistance.

It can be said from the above results that the small oil supply rate makes the vane and roller surfaces smooth, and after being smoothed both surfaces are not worn any more by the help of the thicker lubricant film. Therefore, the wear of the vane and roller decreases as the oil supply rate decreases.

Figure 2 schematically shows the reason why the vane and roller surfaces were smoothed with the small oil supply rates. In the case of large oil supply rates, the lubricant oil located between the two surfaces distributes the load. As a result, the loads on the real contact points of the two surfaces become too small to destroy the brittle carbon particles which protrude from the vane's surface, and they wear the roller surface. In the case of the small oil supply rate, the real contact points of the vane and the roller take almost all of the load, and the brittle carbon particles on the vane surface are destroyed by the concentrated pressures. As a result, the vane surface becomes smooth and does not wear the roller.

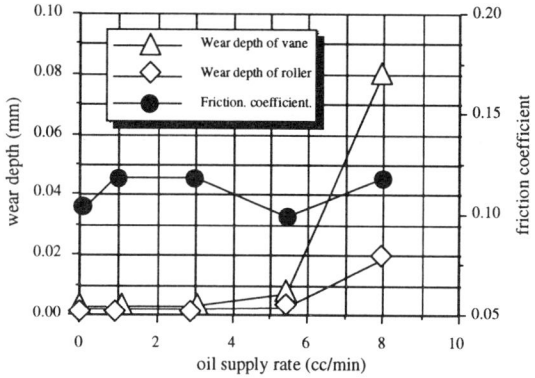

Fig. 1 Relationship between the amount of wear and oil supply rate.

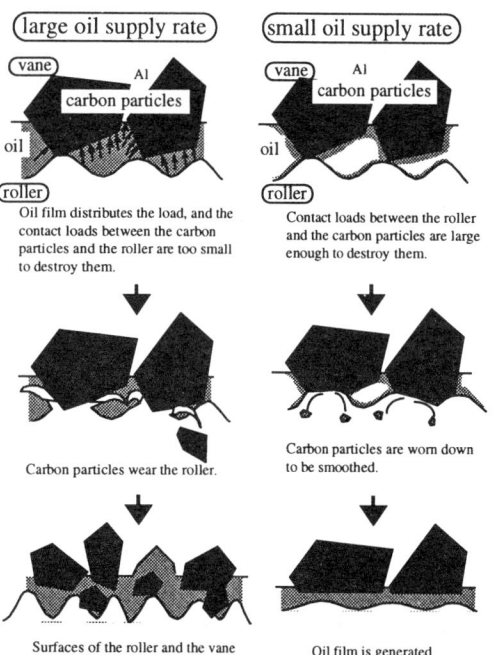

Fig. 2 Wear mechanism of aluminum-carbon composite material

Basic Properties of New Heat Resistant Carburizing Steels for High Temperature and High Speed Cylindrical Roller Bearing

Katsunori Ito, H. Nakashima, K. Fujii and S. Yokoi
Research Institute of Advanced Material Gas Generator, 4-2-6 Kohinata, Bunkyo-ku, 112 Tokyo, Japan

ABSTRACT

As recent interest towards global environmental protection and reduction of petroleum resources are increasing, research and development for innovative gas-generators with improved heat efficiency, lower weight, smaller sized components and reduced discharge of NOx exhaust gases have been studied. The Research Institute of Advanced Material Gas-Generator (AMG) has been developing a next generation gas-generator using advanced materials (1). The target is to develop a cylindrical roller bearing which is capable of operating at 300-400°C and rotational speeds of 3-4MDN (million DN, bore diameter of inner ring times the rotational speed of the inner ring). This bearing requires excellent high temperature properties and high fracture toughness because of the high temperature and high DN conditions.

Several trial materials were produced and tests conducted to develop a new heat resistant steel for bearing races. The new steel was required to have, not only excellent high temperature strength properties, like M50 and M50NiL (current high temperature bearing steels), but also better fracture toughness than M50 and M50NiL.

Figure 1 shows the relationship between the fracture toughness and core hardness of 2 trial materials (A: 5% Cr and B: 6% Cr) and M50NiL, heat treated under the same condition. Both trial materials had higher fracture toughness values than M50NiL, despite their higher core hardness. The fracture toughness of steel B was 120-130MPam$^{1/2}$, which was about 2 times the fracture toughness of M50NiL and also superior to M50 SuperNiL (2). The optimum precipitation of chromium carbides in the core was related to the increase in Cr content.

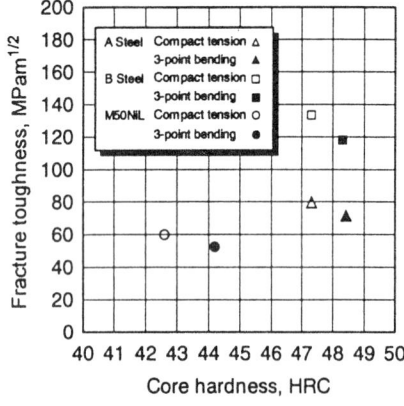

Fig. 1: Fracture toughness and core hardness

Figure 2 shows RCF test results of all of the steels (carburized) in contact with a steel roller (Figure 2a) and with a ceramic roller (Figure 2b). When tested under both roller conditions, the trial materials performed the same or better than M50NiL in rolling contact fatigue.

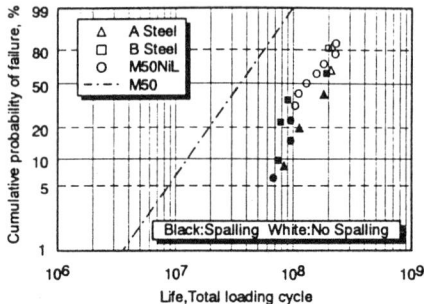

Fig. 2a: RCF test results with Steel roller

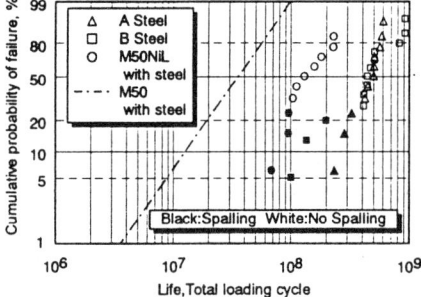

Fig. 2b: RCF test results with ceramic roller

From these results and other experiments, the following was concluded:
1. By increasing the Cr content to 5% the fracture toughness was increased to more than 70MPam$^{1/2}$. Increasing the Cr content to 6% improved the fracture toughness to 120MPam$^{1/2}$.
2. The trial materials were found to have better than or equal hardness and residual stress distributions compared to M50NiL after heat treating.
3. The high temperature hardness and the rolling contact fatigue properties of the trial materials were better than M50 steel and better than or equal to M50NiL steel.
4. The trial materials can be used as a heat resistant steel for cylindrical roller bearing races.

REFERENCES

(1) M.Hiromatsu and S.Seki:Proc.the 1995 Yokohama Int. Gas Turbine Congress, 1, 1995, 203pp.
(2) Harris, T., Ragen, M. and Spitzer, R., Trib.Trans., 35, 1, 1992, 195pp.

AN ATTEMPT TO A PHENOMENOLOGICAL APPROACH OF THE WEAR OF PRESSURIZED WATER NUCLEAR REACTORS COMPONENTS

D CLAIR, B NOEL and Y BERTHIER
Laboratoire de Mécanique des Contacts, INSA Lyon, 20 Av. A. Einstein, 69621 Villeurbanne, FR
M ZBINDEN
Laboratoire Usure et Tribologie, D. E. R. d'Electricité de France (EdF), 77250 Moret sur Loing, FR

ABSTRACT

The wear of control rods against their guides in Pressurized Water Nuclear Reactors (PWR) is mainly due to flow induced vibrations causing tube/guide interactions. The contact kinematics may be impacting, sliding, or a combination of these two elementary motions. Previous studies (1) have laid emphasis on the role of contact kinematics on wear. As the actual kinematics of the components in operating condition are not clearly determined, any modeling or experiment program on wear is faced with a lack of input data. Furthermore, because wear laws do not integrate the evolution of tribological conditions due to third body formation (particle detachment), they are not able to predict the operating life of PWR components.

The experimental program, undertaken by Electricité de France (EdF), aims at clarifying the wear mechanisms associated with pure impact, pure sliding and the combinations between these mechanisms. The lack of rheological data concerning third bodies may be overcomed by an "atlas" relating detached particles morphologies to their cohesion and adhesion (2). It appears that the third body morphology depends on the velocity difference between guides and tubes and range from powder to flakes (Fig. 1) as the kinematics vary from pure impact to pure sliding. The "atlas" is established by performing tests on PEDEBA, an experimental rig which does not exactly reproduce real contact geometry (Fig. 2) but enables a precise control of the samples kinematics (2).

The use of the "atlas" allows the "reconstruction" of the contact life and applied kinematics. The "atlas" validity for kinematics analysis is checked by performing surveys on used components and tested specimens. A simulator reproduces the real contact geometry and conditions (pressurized water, temperature) to create specimens which serve as a reference. Then a comparison of third bodies morphologies obtained with the "atlas" gives an estimation of kinematics applied during the test.

The "atlas" has partially demonstrated its effectiveness as a tool for a kind of "inverse approach" allowing to determinate the contact kinematics of specimens tested at different temperatures from 31°C to 285°C. The variation of contact damage and third body morphology and flow with temperature has been identified. It has been pointed out that temperature acts directly on material properties and indirectly, through its influence on the mechanism (simulator), on contact kinematics. Temperature and strains act simultaneously and lead to subsurface material transformation (Superficial Tribological Transformation, STT). The consecutive change of structure undergone by the skin of the bulk material (304L stainless steel) results in a huge modification of third body birth and life and alter the components life duration.

To be reliable, this approach has to be improved and validated. An experimental rig allowing both geometrical conformity and precise kinematics control is now available. It will permit to sharpen the atlas and to relate degradations and particles morphologies to quantitative data. F.E.M calculation of the impact of tubes against their guides are also under developpement (3). They will strengthen the understanding of particle detachment and wear mechanisms.

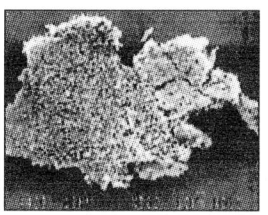 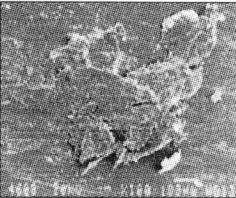

a. Powder/Pure impact   b. Flakes/Pure sliding
Fig. 1 : third body morphologies

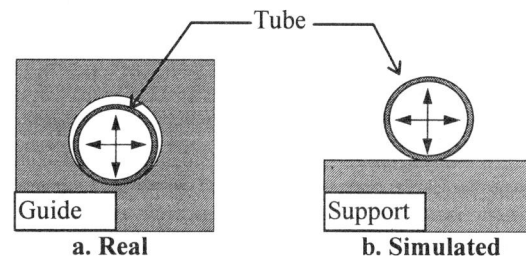

a. Real   b. Simulated
Fig. 2 : contact geometry

REFERENCES

(1) KO, P. L. "The significance of shear and normal force components on Tube wear due to fretting and periodic impacting". *Wear,* Vol. 106, 1985, 261-281 pp.
(2) NOEL, B. "Tribological Degradations in Conformal Contacts due to impact/sliding". *Ph. D. Thesis ISAL 0069*, INSA, 69621 Villeurbanne FR, 1995, 254 p.
(3) CLAIR, D. "Mécanismes et modélisation de l'usure par impacts/glissements". *Rapport N°1, EDF / MTC / LUT*, 77250 Moret sur Loing, FR, 1997, 57 p.

Submitted to *Wear*

TRIBOLOGICAL BEHAVIOUR OF EXPERIMENTALS DEVICE COMPONENTS FOR CONTROLLED NUCLEAR FUSION RESEARCH

A. ORSINI, S. LIBERA, L. VERDINI, E. VISCA

Associazione EURATOM-ENEA sulla Fusione, Centro Ricerche Frascati,
C.P. 65, 00044 Frascati, Rome, Italy.

ABSTRACT

The design of most of the first wall components inside the vacuum chamber of the present experimental TOKAMAK reactors for magnetic confinement controlled fusion requires high thermal loads to be sustained during the operative phases. Among the numerous technical implications there is the need to allow for relevant thermal expansions while retaining the required rigidity and geometrical precision of the components. This results in a rather extensive need for provisions ensuring the relative displacement of components with respect to their supports, by keeping into account that high vacuum and hydrogen atmosphere are the normal operating conditions.

The paper will describe two relevant tribological tests, performed for the qualification of materials and surface treatments, and in particular their behaviour under the design conditions.

The first test refers to the measurement of the tribological behaviour of Carbon Fiber Composites (CFC) on stainless steel sliding support under load. The use of this kind of contact is foreseen in the present design of the ITER TOKAMAK for high heat flux components. In these components CFC acts both as an armour and as a support for the tubes of a high performance heat exchanger that protects the TOKAMAK vacuum chamber and divertor, from the direct plasma interaction. The high temperature reached by CFC under heat flux in the order of 20 MW/m^2 (2000-2500 °C) requires the relative sliding of the armour to be considered with respect to the metallic structure surfaces supporting the components. In particular, the friction coefficients and the erosion rates under short stroke/high pressure conditions were investigated.

The second test refers to the study of the tribological behaviour of coatings and materials to be utilized as sliding pads for large support structures to be installed on rails in the vacuum chamber of the ITER machine. In this case the study aims at selecting the best option for sliding pads avoiding seizing under relevant contact load even after long permanence in high vacuum atmosphere. The selection has been made among a number of stainless steels, coatings on stainless steel substrates and antifriction materials.

The testing procedures and the results are reported in the paper.

LUBRICATIVE PROPERTIES OF THE INFINITE LAYER OXIDES ($Sr_xCa_{1-x}CuO_y$) FOR HIGH TEMPERATURE

MASAMI SASAKI, MASAHIRO SUZUKI and TOSIAKI MURAKAMI
Department of Material Science & Technology, School of High-Technology for Human Welfare,
TOKAI University, Nishino 317, Numazu-city, Shizuoka, 410-03, Japan

INTRODUCTION

Especially, oxide compounds may be expected to be potential solid lubricants under oxidative environments because they will not degrade upon exposure to air at elevated temperature. Recently, various kinds of multi-oxides[1] and mixture of oxides with other compounds[2] have been searched and researched for new solid lubricants which are applicable to high temperature of 600~800°C.

In this study, 'infinite-layer' structure of Ca(Sr)-Cu-O system superconductor was noticed because of the simplest crystal structure build from CuO_2 sheets that sandwich Ca(Sr)[3]. The double oxides were prepared and their lubricative properties were evaluated using a cylinder on plate tribometer at room temperature to 800°C under two testing conditions.

EXPERIMENTAL

Some oxides of $Sr_xCa_{1-x}CuO_y$ with x=0.12, 0.14 and 0.16, (SCCO-1, SCCO-2, SCCO-3, called later, respectively) were prepared from stoichmetric amounts (1/20mol) of $SrCO_3, CaCO_3$ and CuO heated to about 1000°C. The preparation procedure contained griding and mixing of raw material powders, followed by sintering the mixture at 930~970°C, and the procedure was repeated three times to be homogenous double oxide products. The oxide was grinded into powder to be provided as a powder sample (ave.dia.30 μm)Further, the powder was cold-pressed and sintered at 970°C to be provided as a flat plate speciment. It was found in XRD and TG that 'infinite-layer' structure was produced at x=0.14, and that little weight less occurred up to 1000°C.

Except for Ca(Sr)-Cu-O oxide, other oxides and MoS_2 were tested as a reference : $Sr_{0.14}Ca_{0.86}ZnO_y$ (SCZO) in which Cu was replaced by Zn, and $YBa_2Cu_3O_y$ (YBC) which was tested elsewhere[4].

Lubricative properties of the oxide powder and plate specimen were evaluated by a cylinder on plate test rig in which a cylinder (high temp. bearing steel, M50) rotated with 20~26mm dia. track on the plate (34×34), and two different testing procedures.

One is to test a load carring capacity of the sample oil blended with 10mass% oxide powder. Another is to test the dependence of friction on temperature.

RESULTS & DISCUSSION

In the case of sample oil with oxide powders, SCCO-2 with 'infinite-layer' structure gave a lower friction (about 0.1) and higher load carring capacity than the same type oxides, SCCO-1 and SCCO-3, other type of oxides, SCZO and YBC and MoS_2. Also,

Fig.1 shows the variation of friction with temperature in the case of SCCO-2 solid specimen under dry sliding condition. It was seem that SCCO-2 lubricated effectively at high temperature up to 650°C and low friction level (about 0.2) lasted with a little fluctuation over wide temperature. It was noted in SEM and EPMA that the worn surface was smoother than the unworn surface and that thin solid film contained of Ca, Cu, Sr was formed on the rubbing counterface.

Above lubricating test and surface analysis results suggests that SCCO-2 lubricity may be contributed to infinite-layer crystal structure. This laminar structure provides the anisotoropic shear properties with prefered easy shear parallel to the basal planes (CuO_2 sheet and Ca(Sr) sheet). This results in the ability to develop a lubricative transfer film on the counterface.

The critical temperature of 500~650°C was, however, lower than that expected in considering into SCCO-2 thermal stability without decomposition up to 1000°C. Many small cracks on and near wear track were observed clearly after a sudder increase of friction at high temperature above 500°C, but the cause is not cleared. The mechanical strenght reduction or tribochmical reaction on the rubbing surface needs to be investigated further.

CONCLUSION

The double oxide with $Sr_{0.14}Ca_{0.86}CuO_y$ was characterized to be ideal infinite-layer structure of oxygen-deficient pseud-Perovskite crystal structure. The oxide showed lower friction, higher load carring capacity and higher temperature limit (about 650°C) than other composition oxides.

REFERENCES

(1) K.Umeda, Y.Enomoto, A.Tanaka;Proc.Inter.Trib.Conf. (1995) 1181
(2) C.Dellacorte & J.A.Laskowski;Trib.Trans. 40, 1 (1997) 163
(3) M.Azuma, Z.Hiroi, Y.Bandou & Y.Tanaka:NATURE, 356, 30(1992)775
(4) Y.Enomoto, K.Umeda:J.JAST, 34, 2 (1993) 223

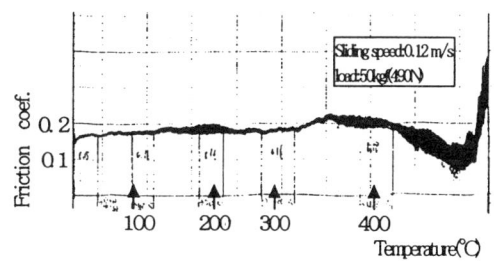

Fig.1 Friction coef. of SCCO-2 vs. temperature

LUBRICATING BEHAVIOUR AND ITS RELATION TO MATHING PAIRS OF IORN RHENATE AT ELEVATED TEMPERATURES

D. S. XIONG and X. B. LI
State Key Lab of P/M, Powder Metallurgy Research Institute, Central South University of Technology, Changsha, 410083, P. R. China
S. Z. LI
Institute of Metal Research, Chinese Academy of Sciences, Shenyang 110015, P. R. China

ABSTRACT

The friction and wear of metallic alloys at high temperatures are controlled by their tribologically generated oxide films. Considering this viewpoint, we have been developing alloys which might be lubricated by their naturally occurring oxide films(1) (2) (3). First, friction tests were run on a variety of oxides which could be formed at high temperatures. In this paper, the iron rhenate was compounded by chemical method, which can be generated during wear process of Fe-Re alloys at high temperatures(4), and its thermal stability was analyzed by TGA(Thermogravimetric Analysis). The friction coefficient of iron rhenate has been determined with a pin-on-dis wear device. The experiments were conducted using five kinds of ball materials (Si_3N_4, Al_2O_3, glass, 440C, HastC) matching seven kinds of disc materials (Al_2O_3, 761 iron base superalloy, glass, stainless steel, bronze, copper and Al-Li). They had a matte finish with different surface roughness. Comparisons of the friction-temperatures behaviors between oxides and alloys were made.

$Fe(OH)_2$ was added into solution of $HReO_4$ and stirred. After the complete dissolution of $Fe(OH)_2$, hydrated iron rhenate ($Fe(ReO_4)_3 \cdot xH_2O$) powder was prepared using the heating evaporation method. The powder of iron rhenate was smeared on frictional orbit of disc. The tests were run at low load (9.8N) and low speed (1m/min).

The TGA result of rhenate shows that except a dehydration weight loss at 80~110°C, there exists an obvious weight loss from 420 to 570°C, which corresponds with the decomposition of $Fe(ReO_4)_3$ as follows:

$$2Fe(ReO_4)_3 \rightarrow 3Re_2O_7 \uparrow + Fe_2O_3$$

Re_2O_7 in decomposition products will sublime and Fe_2O_3 will retain on friction orbit, which accounts for increasing friction coefficient above 600°C.

The research on the lubricating behavior of iron rhenate indicates that:

(1) The iron rhenate has lubricating role from room temperature to 600°C, especially from 20 to 100°C and 400 to 600°C, its friction coefficient is lower than 0.3 and 0.2, respectively. The surface roughness of disc has great effect on the lubricating behavior. The optimum original surface roughness of glass disc is about 3.5μm Ra.

(2) The lubricating behavior of iron rhenate relates to matching pairs. The principles for selecting of matching pairs are: (a) hard ball / relative soft disc (mechanical property) and (b) good compatibility of the oxide with material of disc (chemical property)

(3) The lubricating behaviors of Fe-Re alloys and Cu-Re alloys at elevated temperatures are controlled by iron rhenate and copper rhenate, respectively. The formation of rhenate in the process of friction and wear is base of designing rhenium-bearing alloy with self-lubricating behavior.

REFERENCES

(1) S. Z. LI, X. X. Jiang, F. C. Yin, et al. Materials Science Progress (in Chinese),1989, 3(6): 481.
(2) X. X. Jiang, S. Z. Li, M B Peterson, et al, Materials, Science Progress (in Chinese), 1989, 3(6): 487.
(3) D. S. Xiong, S. Z. Li, X. X. Jiang et al, Trans Nonferrous Met Soc. CHINA, 1995, 5(2): 93
(4) D. S. Xiong, X. B. Li, S. Z. Li, et al, J Cent. South Univ. Technol. (in Chinese), 1995, 26(1): 61

INFLUENCE OF AMORPHYSE COMPOZITE COATING IN FRICTION PAIRS OF DETAILS OF MACHINES ON THEIR DURABILITY IN CONDITIONS OF A LITTLE CYCLES LOADING

L.M.Abramov, A.S.Astakhin

Kovrov technological academy 601910, Russia, Vladimir region, Kovrov, Mayakovsky st., 19.

ABSTRACT

A little cycles of loading is characterized by a high level of maximum pressures, therefore application of materials with increased by the characteristics will allow essentially to increase durability of friction pair in particular if the structure of a material can be optimized on physical-mechanical properties. Such optimization is possible by use of laser processing smelting at final stages of processing of preparations of details of machines (1).

We offer complex technology of reception of compozite coating, consisting from two blocks:

- reception of preparations by onventional methods;
- obtaining to surface layers of products of a necessary level of special properties by modern technological methods.

The essence of technology consists that on working surfaces of details (by use of a substrate from constructional materials (steel)) receive compozite coating, using laser radiation for processing.

Was thus received compozite coating, consisting from two layers (2):
amorphyose layer - thickness 0.1-.15 mm, H = 11 400 MPa;

transitive fine-graine zone of a material - thickness about 0.3 mm, H = 8 300 MPa.

During researches is established, that wearresistance and limit of fluidity of a material of amorphyose coating essentially depend on its thickness. The optimum thickness of coating has made d=0.1-0.15 mm.

With the purpose of increase of hysical-mechanical properties of amorphyose coating processing of a surface the high pressure (3).

Parameters of processing of amorphyse layer was conducted chose from a condition:

1.1 $G_{transit.layer} < P < E_{amorph.layer}$

As a result of processing was obtaing of compozite coating consisting from three layers:
amorphyose layer - thickness 0,1...0,15 mm, H =12 500 MPa;
indenter layer - thickness 0,15 ...0,2 mm, H =10 300 MPa;
transitive layer - thickness 0,17...0,2 mm, H= 8 450 MPa.

In the result a microheterogeneous structure of coating, consisting from firm of monokristal inclusions by the size d= (0,1-0,5) мкм, in regular intervals distributed in a less firm matrix of a transitive layer was received. Thus the fields of ssure from monokristal inclusions were not interference, since the distances between sources made more (3-5) d accordingly.

It is necessary also to note, that a opportunity of reception amorphyose of a layer of a material (thickness up to 50 mkm) on a surface of a product is experimentally established at processing by its high pressure (P>G)

It is possibility, that during contact friction on working surfaces of products will be formed amorphyose layer, which does n ot render essential influence to durability of a product, owing to, its unsufficient thickness.

As real products were processed gear transfers of motorcycles and details of weapon.

The results of operational tests of products have shown a increase of durability in 2-2.5 times, that is stipulated by a structure and properties received compozite material.

REFERENCES

(1) V.S.Ivanova and other. Sinergetika and fractales in material science - Moscow, 1994. - 383 pages,: il.

(2) L.M.Abramov and other. Proceeding of V ISTC " Laser technologies - 95 ", p.328-330, 1995 - 2713 pages.

(3) L.M.Abramov, A.S.Astakhin. . A symposium " Sinergetira, structure and property of materials, self-organizing technologies ", Part 1, page.128, 1996 - 255 pages.

ABOUT THE CUTTER-TUBULAR MATERIAL CONTACT WITH APPLICATION IN DRILLING-PRODUCTION

NICULAE NAPOLEON ANTONESCU, ION NAE and ADRIAN CATALIN DRUMEANU
"Petroleum-Gas" University Ploiesti, 39 Bucharest Blvd., 2000 Ploiesti, Romania

ABSTRACT

The friction phenomenon appears wherever the relative motion exists, respectively where the tendency of relative movement between two bodies. At small sliding speeds, under the mixt or dry friction conditions, the movement can be jetky or intermittent.

The paper presents friction appearance and releasing at the cutter-tubular material contact with applications for the multidimensional tongs used in the drilling-production works. Movement transmission from the tong to the external surface of the tubular material is done through the cutters.

From a tribological print of view the study of the surfaces appropriation and of the phenomena which take place at their direct contact is incomplete. The existent theories are generally based on the statistic results, obtaind by the investidation of a large number of surface pairs characterized by different parameters concerning the material, the surface shape (plane, concave, convex), the manner of mechanical treatment, sligh thermal treatment, normal force size, rugosity parameters of the surfaces in contact, the surface in contact materials compatibility, the medium conditions etc. After the settlement of the contact between the surfaces, the initial surface configuration is remodelled.

This is due to the plastic and elastic deformations which are produced during the surfaces approaching.

Less known is the remodelling process itself, considered as a succesion of sequences in which the microneregularities are plastic and elastic strained, their material is redistributed determining new forms with a superficial rigidity different from that initial, the mechanism stopping when a certain equilibrum state is attained.

Also, less known is the size of the absolute approaching of the surfaces, respectively the moment when the reciprocal displacement of the surfaces to the contact setting stops (1) (2).

Becouse the functional role of the tongs is transmit the construction torque (detachment) of the tubular material, without sliding, the paper presents the criterion of good working of the product, studing the cutter-tubular material contact, taking into account the working conditions.

For the simulation of multidimensional tongs functioning the testing stand presented in figure 1 has been used.

The actual friction coefficient μ_k at the die tubular element contact is calculated by relation:

$$\mu_k = M_t / 3 \cdot R \cdot N$$

in which: M_t represents the tightening (undoring) torque corrensponding to the occurrence of the tubular-material gliding phenomenon; R - external radius of the tubular-material; N - normal force at the die-tubular material contact.

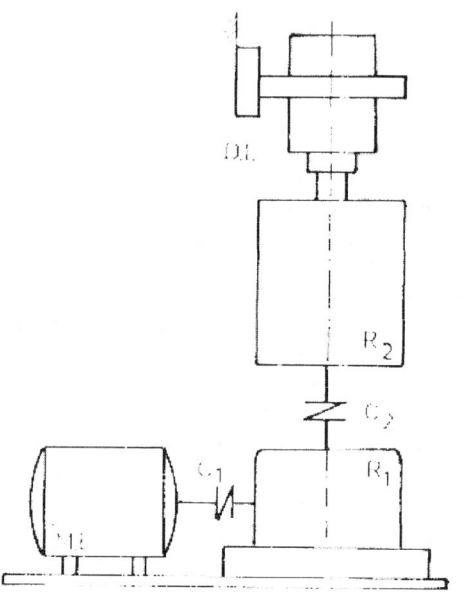

Fig. 1: Stand for determining the friction coefficient
R_1, R_2 - testing reducer; C_1, C_2 - couplind;
D.I. - device; M.E. - direct-current motor.

To execute the experimental determinations it is necessary to marke clear the shape and the dimensions of the samples for the cutters and for the hollow material. The experimental determinations were made in the following working conditions:
- sample cutters;
- sample material-cutters OLC 10, OLC 20;
- sample hollow material;
- material hollow element OLT 65;
- friction system (dry, water, drilling mud, row oil);
- the roughness of the surfaces in contact
$R_z = 25$ μm.

REFERENCES

(1) N.N. Antonescu, and I. Nae, Aspects of How Rotary-Slide Tongs Work, Buletinul Universitatii "Petrol-Gaze", XLV, no. 1-2, 1993, 109-120pp.
(2) I. Costin, Scule pentru foraj si extractie, Editura Tehnica, Bucuresti, 1990.

WEAR BY THERMAL FATIGUE OF THE STEELS USED TO THE MECHANICAL BRAKES CONSTRUCTION

NICULAE NAPOLEON ANTONESCU, ADRIAN CATALIN DRUMEANU and ION NAE
"Petroleum-Gas" University Ploiesti, 39 Bucharest Blvd., 2000 Ploiesti, Romania

ABSTRACT

Metallic structures of the dry friction couplers which are in the componence of the heavy duty mechanical brakes are cyclic thermal stressed due to the heatings that appear as a result of their working. These cyclic stresses have a non-isothermal character, and in time they determine the wear by tribothermal fatigue of the metallic elements from the braking couplers.

The starting point in the experimental investigation of this kind of wear is the cracking degradation of the drums from the mechanical band brakes of the drawworkses that equip the drilling-production rigs. The thermal stresses of these brakes are very hard and they lead to the premature cutting out of action of the metallic drums, much more before their wear by abrasion to reach the limit value. Because the band of these brakes has the blocks made from composed friction material ferrodo like, the biggest part of the quantity of heat (over 90%) which is produced during the braking is taken by the metallic drum. Depending on the load operated by the drawworks, the temperature on the friction surface at the brake block-drum contact reaches values in the range of 500...800°C. In these conditions the thermal stresses which appear hase values which are generally in the range of 100...800MPa, so, much more over the yield limit of the steels used to the drums construction. Due to the cyclic character of the braking processes, the thermal stresses are cyclic too; metallic material degradation is produced by cracking and it has a character similar to that characteristic to low cycle fatigue.

The braking process parameters (the contact pressure, the sliding speed, time) determine the thermal condition on the friction surface of the coupler that determins the durability of the metallic structures of the braking coupler by its effects (thermal strains and stresses) (1)(2).

The experimental determinations concerning the influence of the dry friction process parameters on the durability of the metallic elements of the braking couplers were done on a testing stand equipped with an inferior friction coupler, IVth class, plan-cylinder like. At this coupler the cylinder is just the metallic sample and the plan a friction material disk, ferrodo like. The cylinder (the sample) is pressed on the disk that is in rotary movement with different forces. The main measured parameters are: nominal contact pressure, sliding speed, heating time and the temperature on the friction surface.

The criterion for durability determination is the appearance of the first crack on the friction surface of the sample. On this stand were tested samples made from four types of low and middle alloy steels:

1) 0.35C - 1.4Mn - 0.4Cr - 0.6Ni - 0.6Si;
2) 0.32C - 1Cr - 0.6Ni - 0.2Mo - 0.4Mn;
3) 0.4C - 1.4Mn - 0.25Cr;
4) 0.4C - 0.5Mn - 1Cr - 0.002B.

For example, to emphasize the quantitative influences of nominal contact pressure, sliding speed and time, it was done a multiple regression analysis like:

$$N = K (pv)^a \tau^b$$

where: N is the durability, cycles; p - the nominal contact pressure, MPa; v - the sliding speed, m/s; τ - the time, s; K, a, b - the regression coefficients.

The results of this regression analysis are presented in Table 1.

Table 1: The Regression Analysis Results

Regression analysis characteristics		Steel			
		1	2	3	4
Maximum temperature of the cycle on the friction surface,°C		800			
Variation range for τ, s		12 ... 30			
Variation range for p, MPa		1.388 ... 1.943			
Variation range for v, m/s		2.67 ... 3.12			
Variation range for N, cycles		190 ... 400			
Regression coefficients values	K	260	357	130	63
	a	-0.79	-0.69	-0.54	-0.33
	b	0.43	0.31	0.54	0.68
Multiple correlation coefficient, R		0.98	0.97	0.99	0.98
Partial correlation coefficients	r_{Npv}	-0.96	-0.98	-0.99	-0.97
	$r_{N\tau}$	0.98	0.98	0.99	0.98
Influence weight on the durability, %	pv	52.7	62.1	44.2	27.5
	τ	43.8	35.0	56.0	69.2
Experimental data dispersion around the regression surface for a confidence interval with 95%, cycles		±24	±24	±13	±24

REFERENCES

(1) N.N. Antonescu, and A.C. Drumeanu, Revista Romana de Petrol, vol.3, nr.3,1996, 235 -237pp.
(2) N.N. Antonescu, A.C. Drumeanu and I. Nae, Proceedings of the 10th International Colloquium, Tribology-Solving Friction and Wear Problems, Technische Akademie Esslingen, Ed. W.J.Bartz, 9-11 Jan.1996, vol.2, 899 - 908pp.

TRIBOLOGICAL SIMULATION OF WORM GEAR TEETH

TIBOR BERCSEY and PETER HORAK

Institute of Machine Design, Technical University of Budapest, Műegyetem rkp. 3, Budapest,. H-1111, Hungary

ABSTRACT

The velocity relations and the curvature of the gear tooth surfaces, depending on the meshing position, change along the contact line. The traditional two-roller tribometers are not able to present the relative velocity and radii of curvature changing along the contact line, so it is necessary to create a new test rig for the tribological simulation of spatial gear teeth.

The geometry, the kinematics, the material properties and the load should be considered during the tribological investigations. The aim of this investigation is to connect and compare two of these four factors, the geometry and the kinematics, with the real effects occurring on the gear teeth.

A computer program has been developed to calculate the contact lines, the radii of curvatures and the velocity distribution of the surfaces.

A finite element analysis has been used for the investigation of the displacement and stress distribution of the worm and worm gear. Based on the results, we can say that the maximum stresses are caused by the Hertzian contact of the tooth surfaces on the contact line. The results of the calculations were used to build an approximate model and to set the parameters of the test pieces.

The real radii of curvature can be approximated by linear functions along a section of the contact line. This approximation function can be realized by contact of either a plane and a cylinder or a plane and a cone (Fig. 1).

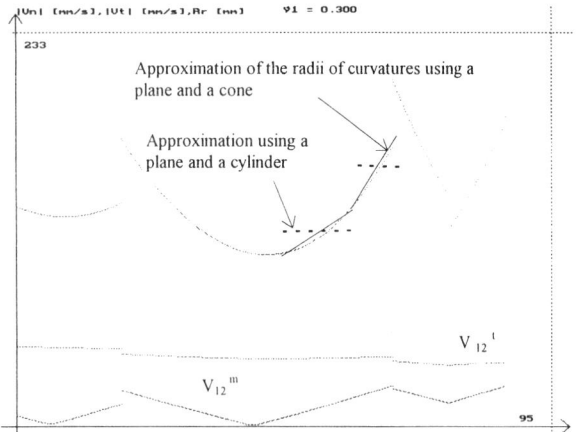

Fig. 1: Approximation of the real radii of curvatures along the contact line

If the plane (disc) and the cylinder (roller) rotate, it is possible to set variable velocity relations along the contact line depending the direction of rotation and the eccentricity of the roller. It is a linear approximation of the real velocity relations of the worm gear tooth surfaces.

The sum of the velocities describes the rolling in the rolling-sliding contact and gives the hydrodynamic pressure in the oil film.

$$v_\Sigma = \frac{1}{2}\left|v_{12}^m + v_{21}^m\right|$$

The sliding of the tooth surfaces causes power loss. The sliding velocity v_s in the tangent plane influences the friction coefficient and determines the direction of the friction force at the contact point.

$$v_s = \left|v_{12}^m - v_{21}^m\right|$$

If the roller is conical then the reduced curvature will change along the contact line. It means that the function of the curvature is approximated by a linear function. The directional change of the velocity v_{12}^m perpendicular to the contact line can be simulated by a rotation around an axis perpendicular to the disc. Figure 2 represents the velocity distribution of this system.

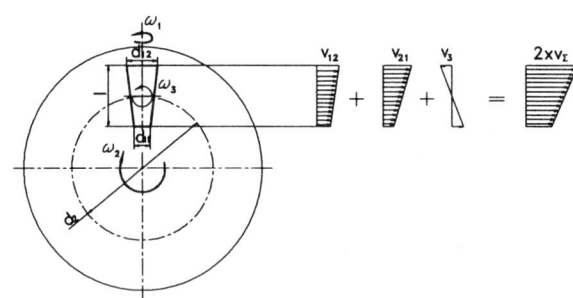

Fig. 2: Velocity distribution between a disc and a conical roller rotating around two axes

For the experimental investigation of worm gear teeth contact, a roller-disc tribometer has been built. It is able to model the rolling-sliding contact of the tooth surfaces both in line and point contacts. The aim of the experimental investigation is to establish the relationship between the coefficient of friction and the various velocity and curvature relations.

INTERACTIONS OF FRICTION MATERIALS WITH A MINERAL-BASED AND A PARTIAL-SYNTHETIC-BASED AUTOMATIC TRANSMISSION FLUID

BÜLENT ÇAVDAR and ROBERT C. LAM
Advanced Technology Center, Borg Warner Automotive, Lombard, Illinois, USA

ABSTRACT

In an earlier study, we investigated the wet clutch performance in a mineral-based fluid (Fluid-M) and in a partial-synthetic-based fluid (1). Since the formulations of the fluids were not readily available, an extensive analytical work was done to identify some basic properties of the fluids. In particular, the viscosity, molecular weight, volatility characteristics and elemental properties were identified. Some of the performance characteristics such as the torque response curve shapes, the compressibility of clutch packs and the thermal degradation of the fluids were discussed (1). In the current study, low speed friction characteristics, the surface interactions between friction materials and ATFs as well as the effects of aging the friction materials in ATFs on the mechanical properties of the friction materials are discussed.

The viscosity characteristics, the additive packages, and the degradation products of Automatic Transmission Fluids (ATFs) affect the alteration of chemical composition and the physical structure of the surfaces of friction materials during a wet clutch operation. The mineral-based fluid (Fluid-M) and the partial-synthetic fluid (Fluid-PS) follow different degradation paths. Fluid-M has a calcium sulfonate material as a detergent additive which helps esterification of degradation products and hence, sludge is formed. On the other hand, Fluid-PS tends to form less sludge but more carboxylic acids. The friction material surfaces were found to be smoother in Fluid-M than in Fluid-PS during sliding due to two main reasons: 1) The pores of the friction material surface were plugged with calcium and sulfur rich sludge in Fluid-M as evidenced by Energy Dispersive X-ray Spectroscopy (SEM/EDX) analysis, 2) Due to the superior viscosity characteristics, Fluid-PS forms thicker fluid film between the friction plate and the separator plate, hence, reducing the coefficient of friction and causing less asperity wear.

The mechanical strength of the cellulosic fiber containing friction materials was considerably reduced upon soaking in ATFs at 160°C for 300 hours. However, the reduction in mechanical strength was more severe in Fluid-PS than in Fluid-M for the following reasons: The partial-synthetic-based ATF contains more phosphorus than the mineral-based-ATF. The chemical structure of the phosphorus additive is also different in Fluid-PS than in Fluid-M as evidenced by Nuclear Magnetic Resonance Spectroscopy (NMR) analysis. Probably, the type of phosphorus additive found in Fluid-PS is more chemically aggressive against the cellulosic fibers than that found in Fluid-M. Furthermore, the carboxylic acids which form in Fluid-PS may also attack the cellulosic fibers. The mechanical strength of the aramid fiber containing friction materials was not reduced by either ATF.

REFERENCES

(1) B. Çavdar, R.C. Lam, "Wet Clutch Performance in a Mineral-Based, and in a Partial-Synthetic-Based Automatic Transmission Fluid" SAE Paper Nr. 970976, (1997).

TRIBOLOGY OF CONSTANT VELOCITY BALL JOINTS

JAMES COLE and Dr GARETH FISH
GKN Technology Ltd., Birmingham New Road, Wolverhampton, WV4 6BW, UK.

ABSTRACT

Constant velocity joints (CVJs), as used to transmit torque from the gearbox to the wheels on front wheel drive cars, are complex tribological systems. This paper explains the tribology of both fixed and plunging ball type CVJs and discusses the lubrication of their concentrated contacts.

Constant velocity fixed ball joints consist of six balls held in the homokinetic plane by a cage, the balls running in toroidal tracks, in the inner and outer raceways. The raceway tracks are very conformal in one plane with the balls, giving highly elongated elliptical contact patches, the lengths of which are of the same order as the ball radius, thus breaking a central assumption of Hertzian contact theory(1). Theoretical ball-track contact pressures are high, typically 2-3GPa. Track surfaces, typically ground or as formed, have roughnesses with Ra values of 0.4 to 0.8μm. Under normal straight ahead driving conditions, the motion of the balls is one of low speed variable sliding and rolling with sinusoidal oscillation of velocity. Due to the differences in radii of the inner and outer raceways and the different loading, the forces controlling the motion of the contact patches cause the amount of sliding to be different within the two contacts. To achieve the high level of refinement required, the ball in the cage window is an interference fit. The cage window contacts act like a brake and further complicate the motion of the ball. The pressure angle on the tracks, typically 40°, also imparts a component of spin into the contact. This is explained in figure 1. This spin is thought to be beneficial to the lubrication of the contacts because the grease which lubricates the joint coats the ball and the spin motion brings that grease into the contact. It also ensures that the point of contact on the balls is constantly changing, which reduces the number of stress cycles experienced by any given point on the ball surface. The cage is controlled by the spherical surfaces on the outer and inner raceways, whose area contacts are under pure sliding in a distorted figure of eight motion with variable loading.

Constant velocity plunging ball joints are similar to fixed ball CVJs except that the balls run in cylindrical straight tracks. In GKN cross-groove (VL type) CVJs the track surfaces are hard-machined (typical Ra values of 0.2 to 0.8μm) or broached (typical Ra values of 1.0 to 4.0μm), and in double offset (DOJ type) CVJs they are either ground or with as formed surfaces with typical Ra values of 0.4 to 0.8 μm. The track conformity with the balls is similar to fixed ball joints, except that contact pressures can be higher typically 2-4GPa. Except when plunging, the motion of the balls is similar to that in the fixed joints. In VL type joints, the role of the cage is too keep the balls in the homokinetic plane, and it is not controlled by the inner and outer raceways. The highest pressures are at the cage window-ball contact. The motion of the balls is therefore governed by the geometry of the joint and by the pre-load applied to the joint. In DOJ type joints, depending on the angle of the joint, the balls are kept in the homokinetic plane either by the cage or by the intersection of the tracks. The tight fits between the cage, balls and inner raceway promote sliding and this is responsible for the higher heat generation in this type of CVJ.

Using steady-state approximations for the maximum velocity of the contact patch, the elastohydrodynamic (EHD) film thicknesses have been determined (2), and very low λ-ratios have been calculated. Fixed ball and DOJ type CVJs are thought to operate in the mixed elastohydrodynamic and boundary lubrication regimes. Contact analysis of VL joints shows that only the tops of the broached track surfaces are involved in the lubrication of the contact, and they can be considered to operate in the μ-EHD regime (3). This gives rise to significant difficulties in developing lubricants for ball type CVJs. From their knowledge of CVJs, GKN has been able to develop enhanced lubricants for its components which has facilitated a downsizing of the components such that the same torque can be transmitted through smaller, lighter driveshafts with no loss of fatigue strength and durability performance.

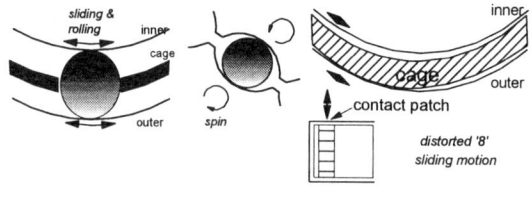

Fig. 1: Fixed ball joint contacts

REFERENCES

(1) K.L. Johnson, "Contact Mechanics" Cambridge University Press, 1987
(2) D. Dowson, B.T. Hamrock, ASME Transactions, F, 1977, Vol. 99, 262-76pp
(3) K.P. Baglin, Private Communication, 1994

STUDY ON THE WEAR OF BASIC ELEMENTS OF CONSTRUCTION AND ROAD BUILDING MACHINERY

D. DANCHEV, V. PANOV, E. ASSENOVA
Technical University - Sofia 1756, Bulgaria

ABSTRACT

INTRODUCTION

The elements, units and assemblies of the construction and road building machinery operate in abrasive and chemically aggressive environment, increasing loads, impact and vibration influence, at low relative velocities (1), (2). Intensity of wear is related to the "service-point" of the worn element requiring fast service action, the lack of which leads to a precocious "limit-state" (2), (3). The solution of tribological problems in above mentioned machinery is essential in regard to the high machinery prices and high running and repair costs.

The paper aims generalisation and analysis of the experienced by the authors theoretical and experimental study on wear in construction and road building machinery widely used in Bulgaria.

CONDITIONS AND OBJECTS OF STUDY

Investigation of hydraulic excavators, loaders and, bulldozers produced in the USA, Germany, Russia, Czech Republic and Bulgaria has been carried out in real operating conditions in the civil engineering, mining industry and irrigation building. Wear has been registered in various basic elements of this machinery, like undercarriage elements (shoes, pins, bushings and links of the track, sprockets, idlers); cutting parts; hydraulic elements, etc. The variation of both geometric parameters of the elements and exploitation parameters (performance, energy consumption, etc.) of the whole machine has been observed. Wear has been studied also in relation with the physical-mechanical properties, the chemical content of the processed soils and the characteristic working conditions, climate, humidity, content of particles and chemically active substances in the air.

RESULTS AND DISCUSSION

Some of the obtained wear relationships for elements of the track links of bulldozers, loaders, excavators: CAT963, D6H and D9N (Caterpillar); ЭО4121 and ЭО5122A (Russia), DH101 (Czech Republic); RH40 (O+K) are exposed in the figures below. Fig. 1 shows wear (in % of the limit amount) in the case when the machines work in two open coal mines. The different wear curves: 1, 2, 3, 4 in Fig. 2 reveal the dependence on the soil type. The figures show that the increasing of wear involves an increase of the working resistance, which leads to increasing (at a different degree depending on the different working conditions and the type of machine) of fuel consumption per 1 m³ dug soil.

CONCLUSION

It appears to be impossible, by means of theoretical and experimental analysis, to determine with sufficient for practice preciseness wear intensity and wear duration of elements of excavating and transport machines, even in the case of machines of equal type and working conditions.

Wear depends weakly on construction differences of the machines produced in one company and more significantly on the working conditions, qualification and experience of the operator.

Study of wear in the different elements of the machine is necessary for any machine and for the concrete operation conditions; measures for the effectiveness of operation, maintenance and management of the machine should be planned.

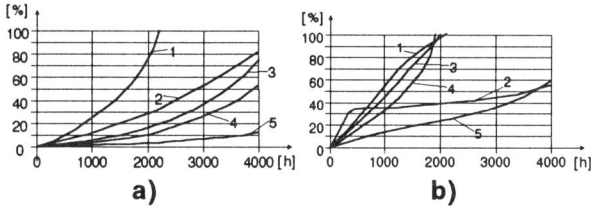

Fig.1: Wear law in a) shoes, b) bushings; 1-D7H(No.1), 2-D9N, 3-D7H(No.2), 4-CAT963, 5-D6H

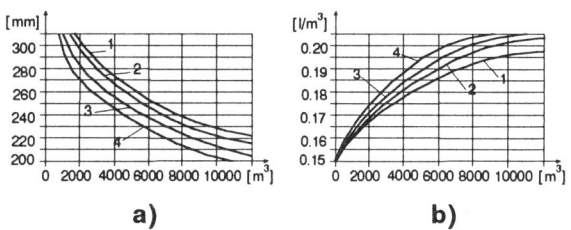

Fig.2: Wear of a) basket tip length, b) fuel consumption, with volume of dug soil.

REFERENCES

(1). Gustov, Yu. I. Improvement of wear resistance of working parts of building machines, Mechanisation of Building, No.5, 1996, pp. 15-16 (in Russian).

(2). Danchev D., V. Panov, Tribological investigations for energy consumption in construction and mine industries, Journal of the Balkan Tribological Association, No 2, 1996, pp 171-176.

(3). Panov, V., D. Danchev, Study on the undercarriage wear of track-type bulldozers and loaders, Journal of the Balkan Tribological Association, No 3, 1996, pp 141-144.

TRIBOLOGY OF CONSTANT VELOCITY PLUNGING TRIPOD JOINTS

Dr GARETH FISH and JAMES COLE

GKN Technology Ltd., Birmingham New Road, Wolverhampton, WV4 6BW, UK.

ABSTRACT

Constant velocity joints (CVJs), as used to transmit torque from the gearbox to the wheels on front wheel drive cars, are complex tribological systems. This paper explains the tribology of plunging tripod type inboard CVJs and discusses the lubrication of their concentrated contacts. GKN manufactures two main types of constant velocity plunging tripode joints: the Glaenzer Interieur (GI) and the Angular Adjusted Roller (AAR).

The GI design of tripod type CVJ, consists of a three-legged inner race (the tripod) on which three rollers, supported on needle bearings, run in cylindrical straight tracks. The roller-track contacts are highly conformal. From Hertzian line contact theory(1), the theoretical contact pressures of the track have been calculated to be typically 1.5-2.5GPa, with an essentially rectangular contact patch. The maximum contact stress on a typical medium sized needle under a high nominal torque was calculated to be 3.8GPa. It is outlined how the trunnion supporting the needles, is of an eccentric profile, which significantly reduces the maximum individual needle load by spreading the total load over a larger number of needles compared with a cylindrical trunnion. This also significantly increases the durability of the needle-trunnion contact against surface initiated rolling contact fatigue.

It is explained how in the straight ahead position, the roller-track contacts are pure rolling, but at angle, an element of sliding is introduced. As the joint angle increases this element of sliding increases, and axial forces are generated from the sliding friction force at the track roller contact. In typical low installation angle applications, this axial force is low and does not affect the refinement of the car on which the shaft is fitted.

To meet the continuing demands for improved driveline refinement, GKN developed the AAR design of tripod type CVJ, an advanced multi-element roller design. The AAR differs from the GI in its roller assembly. The AAR has a spherical trunnion running in a cylindrical inner roller, on the outer surface of which, runs the needles. The outer roller runs on the needles and the outer roller in turn runs in the straight tracks. This is illustrated in Figure 1. The inner roller-trunnion contact is tribologically similar to a spherical plane bearing . This multi-element roller joint eliminates the sliding friction at higher angles present in GI joints, and thus offers the driveline packaging engineer a higher installation angle to improve driveability whilst maintaining the level of refinement. In order to prevent the AAR joint from locking at high angle, which could occur because of the extra degrees of freedom in the needle roller assembly, skew control contact surfaces are included, which introduce further complex tribological contacts, which are explained.

The as-formed surfaces of the outer raceway have typical Ra values of 0.4 to 0.8μm, and the rollers have polished rolling element bearing finishes. The low maximum rolling velocity at normal operating angles, can generate only thin lubrication films (2) and the contacts can be considered to operate in the mixed elastohydrodynamic and boundary lubrication regimes.

GI joints are normally grease lubricated, typically lithium or calcium soap thickened mineral oil blends, whose viscosity is normally ISO VG 68. The sliding friction coefficient of a standard GI grease is between 0.12 and 0.15. Friction modified greases reduce the coefficient of friction from this level to about 0.06, which significantly reduces the axial forces generated, without the loss of durability performance.

Close examination of all the contacts in the AAR joint, give rise to different lubrication requirements of the various contacts. GKN has developed optimised grease solutions to prevent wear and rolling contact fatigue, which give the joint excellent performance in terms of both noise reduction and extended durability.

The different axial force generation of the GI with standard grease, with friction modified grease, and of the AAR joint is illustrated in figure 2.

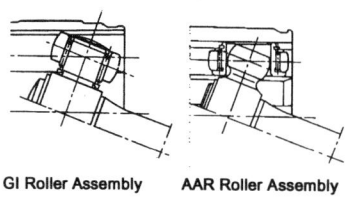

GI Roller Assembly   AAR Roller Assembly

Fig. 1. Tripod joint Assemblies

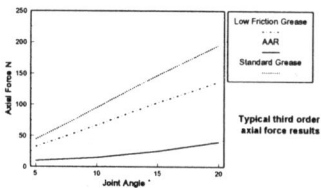

Fig 2. Axial force vs joint angle

REFERENCES

(1) K.L. Johnson, "Contact Mechanics" Cambridge University Press, 1987
(2) D. Dowson, B.T. Hamrock, ASME Transactions, F, 1977, Vol. 99, 262-76pp

TRIBOLOGICAL CHARACTERIZATION OF COMPOSITE POWDER METALLURGY MATERIALS FOR HEAVY DUTY DIESEL ENGINES IN STARVING OR NO LUBRICATION CONDITIONS.

P. GIMONDO
*Centro Sviluppo Materiali, via di Castel Romano 100/102, 00129 Roma, Italy
F. BAUDROCCO
**Iveco T.T.O. Testing, Strada delle Cascinette 424/34, 10156 Torino, Italy
J.A. BAS, J. PENAFIEL and A. BOLARIN
***Ames S.A., C/Tort 15, 08014 Barcelona, Spain

ABSTRACT

In the last few years the idea of developing engines that may work in starving or no lubrication conditions has attracted many researchers and industrial companies.

In the context of the "Clean Car" philosophy adopted by the European Union(EU), an important objective, is the attainement of environmental, economic and political requirements concerning the heavy duty diesel engines.

The realization of a particular valve train system, which could work in absence of lubricants, would give a chance of reducing the engine oil consumption.

Routes to produce such a valve system have been explored, and composites obtained by powder metallurgy(PM) and containing solid lubricants have been tried mostly. The expectation from utilizing these materials, keeping, at least, unchanged the life of engine components, are: a) a reduction of general consumption and fog presence of oil in the cylinder head and, b) a big environmental benefit.

However, further research and development work is needed to fulfill such objectives. Expecially considering that valve zones are one of the critical engine parts, since they undergo severe wear and corrosion-oxidation phenomena during normal operation.

This paper is concerned with the development of various sintered materials(1) utilized for three different engine components: valve cover, valve seat and valve guide. The sintered materials were prepared using solid lubricants such as Cu, C(Graphite) and MnS, dispersed in a metallic matrix, prepared under different vacuum conditions(2). The behaviour of these materials was evaluated by disc-on-disc and fretting tests, in cold and hot conditions, respectively.

The design and manufacturing of the dies for preparing prototypes concerning the three valve components were realized according with the normal production specifications of a Diesel truck. The construction of prototype components was tried, and successfull ones were then evaluated by an engine endurance test. The sintered valve cap and guide, in terms of friction coefficient and reduction of environmental pollution, shown very interesting results. The paper discusses the tribological characterization and the evaluation of these sintered materials.

REFERENCES

1) Matthews, F.L. and Rowlings, R.D.(1987), "Composite Materials Engineering and Science", Chapman & Hall, London.

2) Zum Gahr, K. H. (1990), *Microstructure and wear of materials*, Elsevier, Amsterdam.

A PART FORM CHANGING PROCESS AS A RESULT OF THE WEAR AND THE METHOD OF ITS SOLVING

Vladimir GRIB

Moscow State Academy of Chemical Engineering, Staraj Basmannaj 21/4, 107884, Moscow, Russia.

ABSTRACT

The machine friction assembly wear modeling at the design stage allows the designer to choose optimal combination of structural materials and lubricants, to find optimal designs and operating conditions, and to predict the life and condition of a friction assembly and in this manner, to provide the longevity of a machine.

The machine friction assembles are complex triboengineering systems. Their complexity is caused by the following properties: (a) by the variability of the input (operating condition) and output (function determining) parameters; (b) by the changes of the system structure and properties in time (part form changes, material property changes, movements of mated parts relative to each other, and so on); (c) by the feedback of the input and output parameters; (d) by the distribution of the parameters characterizing the system state (part forms, temperature, stresses, and deformations); (e) by the variety of the processes that are interrelated and running simultaneously (kinematics, mechanical, thermophysical, hydrodynamic, and other processes). In its nature, the triboengineering processes are stochastic.

The friction assembly function and wear prediction problem is actually the friction assembly state change problem. The friction assembly state is described by the combination of the following indicators: part form, mutual location of parts, relative motion law, and temperature, stress, and deformation fields. The friction assembly state varies during the operation from the initial value specified by the design and implemented during the manufacture to the maximum one restricted by the operating instruction specifications.

The part form changing process can be interpreted as a domain that is actually the wearable material space in each point of which the wear rate vector γ is determined. The vector γ is directed along the normal to the friction surface deep into the material being worn. Any point (j) taken on the friction surface will move during the form changing process over the trajectory

$$\rho_j(t) = \rho_j(t_0) + \int_{t_0}^{t} \gamma_j \, dt \quad (1)$$

The absolute value of the parameter γ depends on the tribomating material properties, the environment, the temperature, the stress state, and the rate of sliding in the point at the time moment when it appears on the friction surface. In its turn, the temperature and stress and deformation fields in the parts will vary in compliance with the variations of the part forms, the contact operating conditions, and the part boundary conditions (heat removal conditions, fixing, and others). The form changing process will be described by the movement of a combination of the points located on the friction surface.

The solution of the evolution problems with the distributed parameters and the feedback's in a closed form is possible only in exceptional cases. The most general method of solving such problems consists in discretizing a system being examined in time and space and in implementing the numerical solution methods.

To this aim, the parts and their surfaces are subdivided into finite elements of a simple geometrical form. The parameters characterizing the system state are determined in the finite number of the knotes connecting those elements.

The discretization of the system behavior in time comes to the consideration of a number of the successive system states specified after each random sufficiently-small time period δt.

Accordingly, the equation (1) becomes transformed into the difference expression

$$\rho_j(t_{i+1}) = \rho_j(t_i) + \gamma_j(t_i) \cdot \delta \cdot t$$

This report covers the general methodological approach to the prediction of the machine element form changing during wear. In this case, the friction assemblies and the process running in them are considered as complex multiple-factor systems. The report contains the examples of the wear analyses of some tribomatings (bearing, gear and others).

REFERENCES

Grib V.V. Solving the Triboengineering Problems by Numerical Methods, Moscow, p. 112, 1982.

COMPOSITE ELEMENTS INDICATING STRESS AND TEMPERATURE IN TRIBOLOGICAL PAIRS

TADEUSZ HABDANK-WOJEWÓDZKI
Faculty of Electrical Engineering, Electronics and Automatics,
Academy of Mining and Metallurgy, Al Mickiewicza 30, 30-059 Krakow, Poland.
ANNA RYNIEWICZ, ZBIGNIEW SZYDŁO, BOLESŁAW ZACHARA
Faculty of Mechanical Engineering and Robotics,
Academy of Mining and Metallurgy, Al Mickiewicza 30, 30-059 Krakow, Poland.

ABSTRACT

Algorithms of multiparameter measurement systems of metrologically sensitive tribological pairs that have to signal critical temperatures, based on mathematical models are frequently impossible in realization. Utilization of neuron networks is gainfull but for large, single tribological - sensor systems.

In the case of nodes-sensors where knowledge basis is to be developed it is Fuzzy Logic that is the most advantageous in application. It is an approach resulting in iterative solution of the problem of friction pairs as multiparameter sensors. Its essence is based on nonlinear transformation functions of triangle, "bell" and probabilistic type or on utilization of a certain class of stochastic functions.

When application software is used or mixed software-hardware type means are utilized the problem of friction pair as a sensor may be solved. It is obviously advantageous because of very high resistance sensitivity of semiconducting composites used in such friction pairs in relation to temperature changes. Resistance changes are nonlinear and characteristic for critical concentration span of a composite of percolation characteristics. It is caused by superposition of dilatation stresses constituted in technological process.

In the paper are presented results of electrical properties tests of tribologic pair with the surface covered by semiconductive composites. Tribologic pair consists of a steel ball cooperating with three plates covered by composite layers. Tests were performed with application of typical tribological tool i.e. Four-Ball Wear Test Machine manufactured by Roxana, Illinois, USA.

The composite layers on the plates were made of polyesterimid resin - graphite - metal and have semiconducting properties. Tests of their electric properties have shown that current - resistance vs. temperature characteristics of some compositions (fig. 1) fulfil requirement made for temperature sensors.

Additional tests for compositons in the vicinity of critical concentration have confirmed that that they are piezoresistive composites of high sensitivity when a composites composed of elements with high resistivity after processing been applied. The sets decomposed on temperature and stress components by application of numerical filtration algorithms.

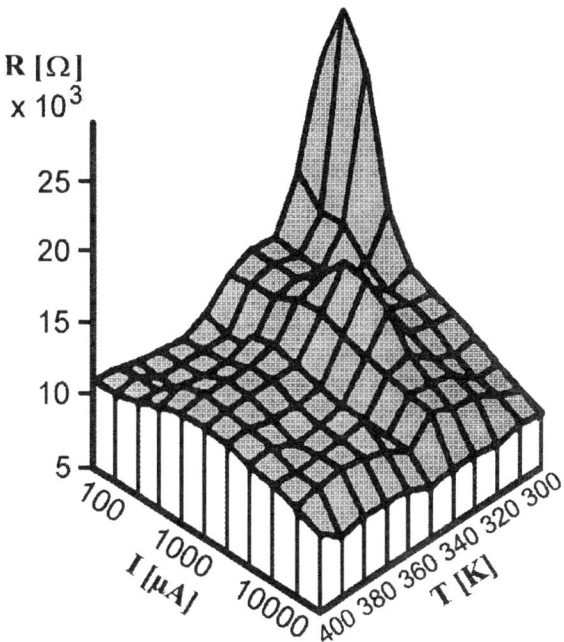

Fig. 1: Variation of resistance vs. temperature with changing supply current for sample compositeresin-poliestroimide-graphite-metal.

From the point of wiev of mechanical design these materials may be formed to conform various properties dependently on application. They could have profitable strenght properties, good lubricity or even both properties may be achieved. Thus dependently on demand they could be applied in lubricated nodes or in design pair operating without lubricant.

REFERENCES

(1) Z. Rojek, T. Habdank-Wojewódzki, Three - Components Compositon for Piezoresistive Pressure Sensor Construction, Proc ISHM'91, Orlando, Fl, USA.

(2) Z. Szydło, A. Ryniewicz, B. Zachara: Ceramic and Basalt Linings Resistivity to Abrasive Slurries, 10th International Colloquium, Tribology - Solving Friction And Wear Problems, Technishe Akademie Esslingen, 09-11 January 1996 vol 3, (2417-2429).

EFFECTS OF KIND OF LUBRICANTS AND SHAPE OF TOOTH PROFILES UPON EFFICIENCY OF GEAR DRIVES

A ISHIBASHI and K SONODA

Department of Mechanical Engineering, Kumamoto Institute of Technology, Kumamoto 860, Japan

H A MOHAMMAD

Department of Mechanical Engineering, Saga University, Saga 840, Japan

ABSTRACT

The effect of the additional torque caused by the friction force at the contacting teeth is neglected in deriving the equations which are widely used for calculating the reference efficiency of a pair of gears (1) (2).

Recently, the authors derived the equations of reference efficiency of gears with a standard tooth profile by considering the effect of the additional torque (3).

In the present investigation, two kinds of the reference efficiency are derived to know the effect of the shape of tooth profiles upon the efficiency. Equation (1) shows the reference efficiency of a gear pair consisting of a driving gear with an all addendum profile and the following gear with the all dedendum profile.

$$\eta_A = 1 - \frac{0.5\mu Z_A (1/Z_A + 1/Z_D) \cdot t_n(\varepsilon_R + 2\varepsilon_o)}{R_{gA} + \mu R_A \sin\alpha + 0.5\mu t_n(\varepsilon_R + 2\varepsilon_o)} \quad (1)$$

where μ is the friction coefficient between meshing teeth, Z_A is the number of teeth of the driving gear with an all addendum profile, Z_D is the number of teeth of the following gear. t_n is the normal pitch. R_{gA} and R_A are the base circle and pitch circle radii of the driving gear, respectively. ε_R is the recessing contact ratio and ε_o is zero when the contact starts at the pitch point. The load sharing ratio between two pairs of teeth is assumed to be 0.5.

The reference efficiency of a gear pair with all dedendum driver and the all addendum follower is expressed by Eq. (2).

$$\eta_D = 1 - \frac{0.5\mu Z_D (1/Z_A + 1/Z_D) \cdot t_n(\varepsilon_A + 2\varepsilon_o)}{R_{gD} - \mu R_D \sin\alpha + 0.5\mu t_n(\varepsilon_A + 2\varepsilon_o)} \quad (2)$$

Z_D is the numbers of teeth of the driving gear with an all dedendum profile, Z_A is the number of teeth of the following gear with an all adendum profile. ε_A is the approaching contact ratio and ε_o is zero when the contact finishes at the pitch point.

When the additional torque produced by the friction force at the tooth surface is neglected, Eqs. (1) and (2) are reduced to Eqs. (3) and (4), respectively.

$$\eta_A = 1 - \mu\pi(1/Z_A + 1/Z_D) \cdot (\varepsilon_R + 2\varepsilon_o) \quad (3)$$

$$\eta_D = 1 - \mu\pi(1/Z_A + 1/Z_D) \cdot (\varepsilon_A + 2\varepsilon_o) \quad (4)$$

Equations (3) and (4) suggest that the reference efficiencies are the same even when an all addendum profile is used for the driving gear or following gear when $(\varepsilon_R = \varepsilon_A)$ and $(\varepsilon_o = 0)$.

It is very difficult to experimentally show the difference in the efficiencies suggested by Eqs. (1) and (2) when a conventional gear drive is used. The authors used an epicyclic differential gear drive for verifying the difference because the power loss in the gear drive is magnified due to power circulation. This means that the small difference in the reference efficiencies is magnified and the difference can be easily detected using a simple measuring apparatus (3).

Figure 1 shows the effects of kind of lubricating oils and the shape of tooth profiles on the power transmission efficiency of a differential gear drive consisting of four external gears. One of the centre gears is fixed and the other is used for the output gear. The two planet gears are engaged with these centre gears and the input power is provided to the planet carrier (planet gear shaft). The gear drive is of the differential type with a reduction ratio of 151. An appreciably greater power, 149 times the input power, circulates in the drive.

It is clearly seen that the efficiency becomes higher when the gears with all addendum tooth profile are used for the centre gears (corresponds to the driving gear). Moreover, a lubricating oil with extreme pressure (EP) additives (hypoid gear oil) brings about higher efficiency in comparison with the oils without EP additives (a cylindrical gear oil and a spindle oil).

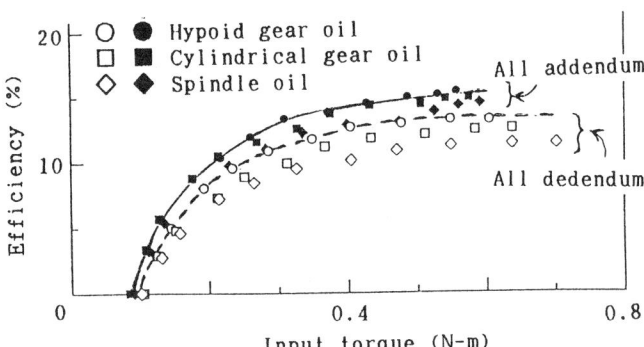

Fig. 1 Effects of tooth profile and lubricating oils on efficiency

REFERENCES

(1) G. Niemann and H. Winter, Maschinen-elemente, Bd. 2, Springer V., 1983, p. 218.

(2) M. Morozumi, Theory and Design Calculation Method for Planetary and Differential Gear Drives, Nikkan-kogyoshinbun, 1989, p. 33.

(3) A. Ishibashi, K. Sonoda and S. Isami, Proc. 7th ASME Inter. Power Trans. and Gearing Conf. 1996, p. 397,

ON-LINE MONITORING OF LUBRICATED WEAR OF BEARING METAL BY A PARTICLE COUNTER

Y.IWAI, S.YOSHINAGA, K.MITAMURA and S.ISHIGURO
Dept. of Mechanical Engineering, Fukui University, Bunkyo 3-9-1, Fukui 910, Japan
M.KAWABATA
Tribotex Co. Ltd., Yamaguchi 45-7, Nagakusa, Obu, Aichi 474, Japan

INTRODUCTION

Oil and wear debris analyses are very useful to monitor lubricated conditions and predict the surface failure of tribo-components. In this study, an on-line measuring system of wear particles using an optical particle counter was developed and applied to the failure detection of sliding wear of tin-based bearing alloy in oil.

EXPERIMENTAL APPARATUS

Our developed particle counter consists of a particle sensor head, an amplifier, A/D converter and a personal computer[1]. The particle sensor was set on the way of the pipe in which the lubricating oil circulated from an external oil tank to the rubbing surfaces. Wear tests were carried out by rubbing a cylindrical tool(0.55% carbon steel) against a plate specimen of tin-based white metal(JIS WJ2, Sn,85.0:Sb,8.5:Cu,5.5:Pb,1.0) in lubricating oil. Paraffin oil(ISO viscosity grade 32) was used. The oil temperature was maintained at $40\pm1^{\circ}C$ in the oil bath.

EXPERIMENTAL RESULTS OF WEAR TEST

Wear tests were conducted at various contact loads and sliding velocity of 0.1, 0.6 and 1.0m/s. **Fig.1** shows the variations in the wear rate as a function of contact load. Three regions of variation in the wear rate with contact load and sliding velocity were determined. In the region 1, wear traces smaller than $1\mu m$ in width were formed regularly on rubbing surface, and the wear particles were very few. In the region 2, as increasing contact load, the area of trace of adhesive wear increased and roughness became larger than $4\mu m$ in depth, and a few plate-like wear particles larger than $40\mu m$ were observed. In the region 3, severe damage caused by detachment of large wear particles was observed on rubbing surface, and a lot of wear particles larger than $40\mu m$ were observed. These appearances suggest that macroscopic metal contacts had occurred and followingly seizure would produce in the region 3. Therefore, it is found that the detection of the changes in wear mode in the region 2 is required to predict the surface failure.

REAL-TIME MEASUREMENT OF WEAR PARTICLES

We measured the size and the number of wear particles in lubricating oil at real-time during the wear test. The number of wear particles increased proportionally with sliding distance for every test under different contact loads. Their increase rates were calculated by dividing the number of wear particles by sliding distance. **Fig.2** shows the increase rates of the number of wear particles of each size range under different contact loads. The increase rates become large with contact load. Especially, at P=200N and 500N in the region 2, the increase rates of large particles more than $40\mu m$ go close to those in the region 3. Therefore, there is good correlations between the distribution of particle size and wear modes described above. From these results, it is concluded that our developed monitoring system using the particle counter can predict severe damage of lubricated wear at real-time.

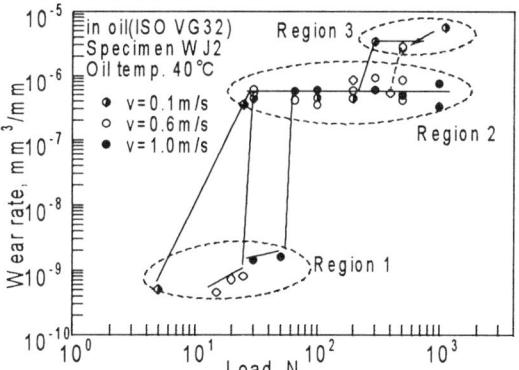

Fig.1 Variations in the wear rate as a function of contact load.

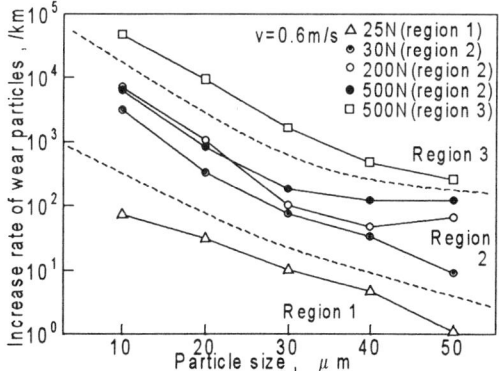

Fig.2 Variations in the increase rate of the number of wear particles of each size range.

REFERENCE

(1) S.Yoshinaga,Y.Iwai et al, Proc. of the Int. Tribology Conference Yokohama 1995, Vol.3 (1996) 1993

RETENTION OF FUEL EFFICIENCY OF ENGINE OILS

MILTON D. JOHNSON, STEFAN KORCEK, RONALD K. JENSEN and ARUP GANGOPADHYAY
Ford Motor Company, Ford Research Laboratory, MD 2629/SRL, Dearborn, MI 48121-2053, USA
KURT SCHRIEWER AND CLARENCE McCOLLUM
Ford Motor Company, Advanced Vehicle Technologies, MD 44/POEE, Dearborn, MI 48121-2053, USA

ABSTRACT

Additive systems containing friction reducing additives that are effective at engine operational temperatures can substantially improve the fuel efficiency of engine oils and, at the same time, provide engine designers with a wider range of design opportunities. However, during engine operation, the friction reducing benefits of these additive systems can be depleted prematurely (1) causing a loss of some fuel efficiency benefits before the end of the service interval. It was concluded that future engine tests designed to evaluate the fuel efficiency of engine oils must involve oil aging prior to the fuel economy determination in order to reflect the effects of customer use. This aging, to be realistic, must be more extensive than that used in the current fuel economy engine test, Sequence VIA, that ages the oil only for the purpose of stabilizing its viscometric properties.

In this work, we are striving to develop an oil aging procedure that simulates the effects of mileage accumulation conducted prior to engine fuel economy certification testing. The effects of additional aging in a modified Sequence VIA test are being investigated. Results reported here were obtained using the same fuel efficient 5W-20 oil, formulated with molybdenum dialkyldithio-carbamate friction modifier, MoDTC, that was evaluated previously (1). The more severe engine aging procedure used in this testing consisted of operating the current ASTM Sequence VIA engine at 3000 rpm, 135 °C oil temperature and 30 kW load. Samples were withdrawn periodically for analysis.

Results from a 96 hour aging test are shown in Table 1 and Figure 1. Oxidation and additive depletion levels corresponding to those observed after 6,672 km

Table 1. Comparison of dynamometer and vehicle aging			
Aging Time, h	Oxidation Abs/cm	Antioxidants %	MoDTC %
0	0	100	100
32	9	72	65
48	12		48
64	14	51	38
80	16	35	18
96	17	26	10
Vehicle, 6672km	16	48	30

of mileage accumulation in a 4.6L engine are reached by 80 hours of aging but the complete loss of friction reducing capability, which was observed after that mileage accumulation, was not apparent until 96 hours in this aging procedure.

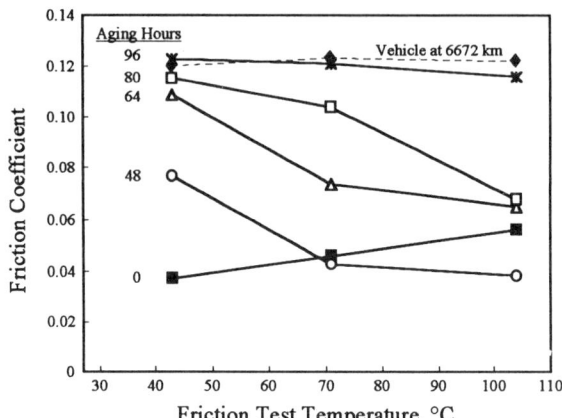

Figure 1. Effect of aging on boundary friction

Testing after the 96 hours of aging, using the fuel economy portion of the current Sequence VIA, showed that the fuel economy improvement of 1.4% in the standard test had been reduced to 0.4% after aging. A portion of the decrease is attributable to viscosity increase; HTHS went from 2.61 to 2.93 mPa·s.

The additive depletion and additive interaction processes that occur during aging of MoDTC containing oils require very careful consideration for prolonging friction reducing effects. MoDTC can be depleted due to antioxidant reactions and also undergoes ligand exchange reactions with zinc dialkyldithiophosphates. Results from studies of factors affecting these interactions and depletion processes (including base oil effects) are being reported separately (2, 3).

The maximum benefits of fuel efficient engine oils will be realized only if changes in friction reducing capabilities and viscometric properties during service are minimized.

REFERENCES

(1) M.D. Johnson, R.K. Jensen, E.M. Clausing, K. Schriewer and S. Korcek, SAE Technical Paper 952532, 1995.
(2) M. D. Johnson, R. K. Jensen and S. Korcek, SAE Technical Paper 971694, 1997.
(3) S. Korcek, R.K. Jensen, M.D. Johnson, A. Gangopadhyay and M.J. Rokosz, World Tribology Congress 1997, ACS Pet. Preprints 1997, planned for Lubrication Science.

DIAGNOSTICS AND REGENERATION OF THE WEAR DAMAGED TEETH GEARS

DANICA JOSIFOVIC
Faculty of Mechanical Engineering University of Kragujevac, Sestre Janjic 6, 34000 Kragujevac, Yugoslavia
SVETISLAV MARKOVIC
Technical High School, Svetog Save 65, 32000 Cacak, Yugoslavia

ABSTRACT

This paper deals with diagnostics and regeneration of teeth gears which are damaged working in normal and special difficult exploitation conditions. Diagnostics and regeneration of teeth gears consist of detection of their damages, cause and character of the same, application of the most rational model of regeneration and method of the working surfaces hardening.

As regeneration of the damaged gears one understands the order of technological operations which as a goal has retrieval of the lost indicators of their working ability. By regeneration one must ensure nominal measures and the required quality of the machined surfaces, as well as the accurate geometrical form and preservation (or even an improvement) of the basic mechanical characteristics of the regenerated tooth gear material.

The basic forms of destruction and damage, depending on the working conditions of the tooth gears, their constructive and technological parameters, can be: tooth fracture, abrasive wear, intrusion or biting, plastic deformation of the materials surface layers, and damages of the tooth front.

In order to establish rationality and profitability of the tooth gears regeneration, it was performed on the example of two tooth gears, shown in Fig.1.

Taking into account the working regimes of these tooth gears, the required quality and hardness which are necessary in order to provide the working life of the tooth gear at the level similar to the newly manufactured ones, we made regeneration by the method of manual metal arc (MMA) welding-on.

Considered gears are of the relatively simple configuration, all the teeth were damaged, they work in difficult exploitation conditions, required is the high level of quality of the machined surfaces, etc. These examples illustrate the magnitude of the economic advantages of the regeneration procedure.

Name	Tooth gear of the I and rear gear	The milling machine synchrone tooth gear
Damaged gears		
Based geometrical parameters	$m=9$, $z=12$, $\alpha_0=20°$, $\beta=0°$, $d_0=108$ mm, $d_k=127,8_{-0,5}$ mm, $d_f=96_{-1,5}$ mm, $b=100_{-0,23}$ mm, $W_2=43,2132$ mm, $xm=+2,7$ mm, C.5421	$m=2,5$, $z=51$, $\alpha_0=20°$, $\beta=0°$, $d_0=127,5$ mm, $d_k=132,5_{-0,5}$ mm, $d_f=122_{-1,0}$ mm, $b=10_{-0,1}$ mm, $W_6=42,3775$ mm, $xm=0$ mm, C. 4320
Kind of damage	Destructive pitting of the working surfaces of the all gear teeth under and near the pitch circle	Wear of the tooth front of the all teeth over 20%
Regeneration method	MMA welding-on by the CASTOLIN 2 and CASTOLIN 680 S electrodes; HRC≥56	MMA welding-on by the CASTOLIN 2 electrode; HRC=55÷58
Regenerated gears	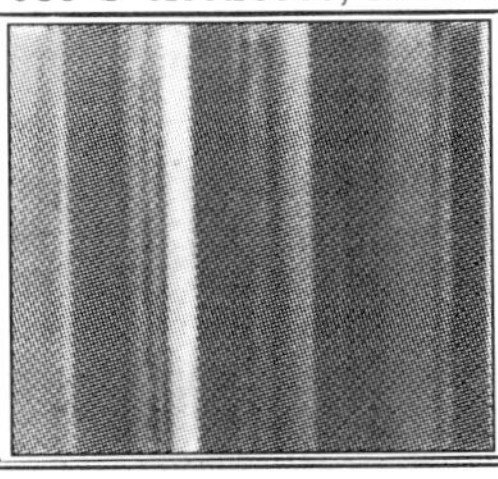	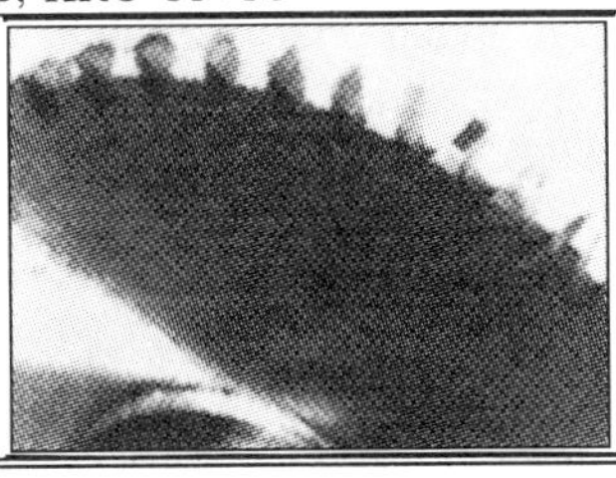

Fig.1: Damaged and regenerated gears

REFERENCES

(1) S. Asheko, P. Klauz, K. Sokolov, Repair of the automobile and heavy transportes machines, (In Russian) Transport, Moscow, 1968, 335 p.

(2) S. Markovic, D. Josifovic, Savings in Material and machining costs of the work tooth gears by the regeneration method, Tribologie in Industry, 1995, vol. 4, p. 120÷126.

DRY SLIDING FRICTION PAIRS WEAR RESISTANCE

VLADIMIR I KLOCHIKHIN

Department of Friction, Wear and Lubrication, Blagonravov Mechanical Engineering Research Institute, Russian Academy of Science, Griboedova Str.4, Center, Moscow, 101830, RUSSIA

INTRODUCTION

Dry sliding friction pairs (SFP) have widespread application into moving units of machines. Most of them have single or reciprocating, or reciprocating-rotary character of motion under sliding. Types of connections such as *shaft-bush, sphere-sphere, plate - plate, pin - on - disk* are representatives of sliding friction pairs.Usually they have friction contact in plate shape.As a rule, they operate under high loads and temperatures and small sliding velocities, therefore oils cannot be used in order to guarantee them high work longevity and wear resistance in long work. They working conditions can be separated on three operating regime that can be seen from the Table.

Table. Working condition ranks for SFPs

Ranks of tribounit loaded	Work duration (cycles)	Operating parameters		
		Load, MPa	Speed, m/s	Temperature, ^{0}C
Normal	>5000	1- 10	to 0.1	-20 - + 80
High	>5000	10- 100	to 0.1	-80 - + 180
Extreme	<5000	> 100	> 0.1	<-80,> +180

An antifrictional materials such as solid lubricating coatings, self-lubricating polymeric materials and grease are used for decreasing of friction loss and are favourable for using in SFP . In each rank exist one or two major parameters that define the ways to guarantee the work duration demanded of friction pair. Existing oppinion that there is a frictional triangle *friction pair - antifrictional material - operating parameters* determining all the sides of tribounit operating is not corrected.It is thought that it does not get in touch with one of the important parameters under dry sliding as constructive-technological factors due to the any dry SFP will be not able to ensure the high wear resistance.To avoid that the some suggestions are offered in given work.

METHODOLOGY

Analysis of SFP operating on food,printing industry branches,railway transport,transporters, trailers and combine machines,cranes,conveyers of heavy industry, sail machines,avia and space branches has been shown that in most degree they work into high-loaded working regime conditions. Therefore it was selected for our propositions.

It is assumed that they can be changed a little for the first and the third ranks,though the common way is the same. *Methodologic way* offered for wear resistance enhancing of dry SFPs permits to take into account more important parameters capable to obtain the replays on the following questions:
1.What antifrictional material type is more serviceable for using into friction pair under given operating conditions ?
2.What kind of constructive and technological parameters in SFP should be taken into account for given operating conditions ?
3.What selected parameterts values are optimum for using in SFP under given operating conditions ?
4.How estimate the work longevity and wear resistance both an antifrictional material selected and SFP itself on the design friction pair stage ?

It is thought that these questions are appeared every time before a lubricating engineer in order to ensure the wear resistance and work longevity of dry SFPs. Sliding bearing (pair *shaft-bush*) with reciprocating-rotary motion, so-called, *cylindrical joint* was used as tested pair for practical illustration our propositions (1).

The kinematics friction pair and constructive-technological factors influence significantly on the wear of antifrictional materials. Each friction pair has a number of such factors that define it wear resistance. If we know the answer on the second question we can find out the answers on the first and third ones with using of operating parameters that only will press their mutual influence on each other. The one of the major tasks of lubricant engineer is to find out the optimal meanings of the constructive and technological factors inside those selected antifrictional material will be had the most wear resistance under given operating conditions.Only after optimal characteristics selected factors have been obtained and the SFP kinematic has been defined, too, you can decide the task about ensuring the most work longevity and wear resistance of SFP with the least expenses on friction loss and exploitation of whole mechanism. It will permit you to construct the tribological recommendations for design of SFP on the draft stage and obtain the replay on the fourth question.

REFERENCES
(1) V I Klochikhin, Int. Res. J. Friction & Wear, Vol. 11, No 3, 1990, 480-489pp.

COMPARATIVE DURABILITY OF MATERIALS IN WEAR

V. A. LYASHKO, M. M. POTEMKIN, and S. A. KLIMENKO
V. Bakul Institute for Superhard Materials of the Nat. Ac. of Sci. of Ukraine,
2 Avtozavodskaya St., Kiev, 254074, Ukraine

ABSTRACT

When analyzing wear conditions, a microvolume of the material under study is considered as an open thermodynamic system, that after overcoming an activation barrier P_1 and accumulating the critical entropy $[S]$ for the τ time, passes over the bifurcation point to non-thermodynamic branch of development. In this case, the equation of life is

$$N = \frac{[S] - S_o}{S_1} / \int_{P_2}^{P_1} g(P)dP,$$

where N is the amount of the loadings, P_1 is the load, which if exceeded initiates a mechanism of the development of the entropy (the value of activation barrier); P_2 is the highest loading; S_o is the entropy value defined by the prehistory; S_1 is a mean value of the entropy developed per unit time; P is the loading; g(P) is the probability of a loading distribution.

Thus, to ensure a high wear resistance, a material should have a sufficiently high activation barrier and a low ability for the development of the entropy.

To assess the activation barrier of material surface layers and their ability to develop the entropy, to use the scratch-hardness test with a cone indenter is advisable as in this case, the linear dimensions of the layer being analyzed are commensurable with those which are subjected to the deformation in the process of exploitation. The ability of the surface under study to develop the entropy with increasing imperfection was assessed from the information of indenter vibrations caused by the tangent force during scanning. In this case, the value of the entropy produced in the material surface layer by scratch-hardness test is proportional to the dispersion of deviation of the sclerometer data unit from its midposition and the dispersion itself characterizes a specific development of the entropy. The time period needed for the system to attain the bifurcation point is found by the equation:

$$\tau = ([S] - S_o)T / KD,$$

where D is the dispersion of deviations of the sclerometer data unit from its mid position; K is the coefficient; T is the temperature. It follows from the above that the length of time before the working surface attains the bifurcation point is defined by the relationship between its initial imperfection and the rate of the increase in the imperfection degree during operation.

Based on the aforesaid, a relative service life of different materials can be assessed by

$$\tau_1 / \tau_2 = ([S_1] - S_{o1})D_1 / ([S_2] - S_{o2} / D_2)$$

In this case, the S_o entropy describes the condition of the material surface layer induced by a certain machining technology, and $[S]$ characterizes the limiting possibilities of a particular structural material. For the simplest case, when working surfaces of a product are made of the same material but machined according to different procedures:

$$\tau_1 / \tau_2 = D_2 / D_1 = F,$$

where F is Fisher's criterion for a dispersion of scretch-hardness testing two surfaces being compared.

Thus, with respect to the service life under the conditions of a bulk accumulation of damages, materials are ranked via comparison between the dispersion of forces of their contact interaction with the indenter in scratch-hardness tests.

Wearing tests of a number of alloyd steels (HB 200-HRC 65) and grey iron under various loading conditions verified the efficiency of the procedure suggested.

The dispersion of deviations of the sclerometer data unit from its midposition is closely connected with the energetic spectral density W of the distribution of the force of a contact interaction between the indenter and the material under study. The comparison between these characteristics (W_1/W_2) of the materials in contact allow us to assess their interrelation in the friction pair.

Based on the W_i parameter, the most efficient material has been chosen for tools to be used in turning hard protective faced and sprayed coatings and the conditions of the tool operation have been optimized.

MENISCUS SHAPE AND SPLASHING OF MAGNETIC FLUID IN MAGNETIC FLUID SEAL

YASUNAGA MITSUYA and KAORU TAKEUCHI
Dpt. of Electronic-Mechanical Engineering Nagoya University, Furo-cho, Chikusa-ku, Nagoya 464-01, Japan
KENJI TOMITA
Data Storage and Retrieval Systems Division Hitachi Ltd., 2880 Kozu, Odawara 256, Japan
SHIGEKI MATSUNAGA
Research & Development Center, NSK Ltd., 1-5-50 Kugenuma-shinmei, Fujisawa 251, Japan

ABSTRACT

Magnetic fluid splashing was experimentally investigated in relation to clearance and magnetic field strength around a gap to be sealed. (1)

Dimensions and tolerance specifications of the experimental seals are presented in Fig. 1. The seal ring and shaft piece are exchangeably mounted on the cylinder and shaft, respectively, for clearance variation. Meniscus shape around the gap was determined from a silhouette formed when the meniscus is illuminated with a laser beam from behind.

The curves in Fig. 2 exemplifies the free surface topography of the magnetic fluid obtained by applying image processing to the meniscus silhouette. The abscissa designates the yoke surface, and the ordinate the shaft piece. From these curves, it is found that rotation-induced surface deformation can be classified into the following three stages: *Stage I*: [0 − 1000 rpm] The meniscus on the shaft side shrinks, while that on the yoke side remains unchanged. *Stage II*: [1000 − 1800 rpm] The meniscus on the shaft side is deformed only slightly but noticeably so on the yoke side. *Stage III*: [> 2000 rpm] There is total meniscus shrinkage.

The meniscus size just before splashing was measured for different amounts of magnetic fluid. The values are designated by rotation height, h, on the yoke surface, and stationary width, w, on the shaft surface. From the experimental results, the splashing initiation speed is found to be roughly in inverse proportion to h and w.

From considerations of magnetic field strength on the yoke surface, it is reasonable to standardize rotation height, h, by using the relationship of $(h - C)/$, where B_{gy} means maximum gap flux density in the gap abutting the yoke surface.

Experimental data are plotted in Fig. 3 after the standardization. Three experiments are found to be superimposed on the same curve, and accordingly, the transformation of $(h - C)/B_{gy}$ is useful as an indicator of the splashing initiation. This indicates the initiation of splashing to be controlled by the magnetic field gradient at the outermost boundary of the meniscus on the yoke surface.

REFERENCE

(1) Mitsuya, Y. et al., 1993, Journal of Magnetism and Magnetic Materials, Vol. 122, pp. 420-423.

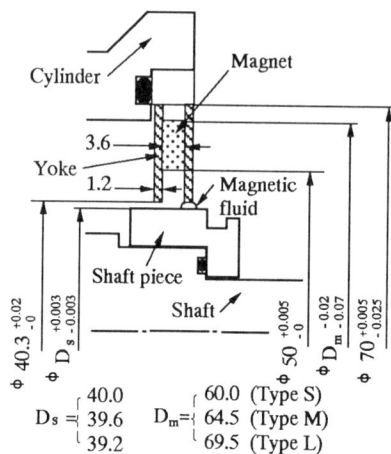

Fig. 1: Magnetic fluid seal structure and dimensions

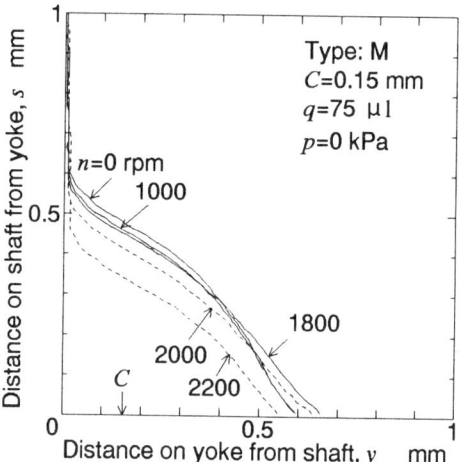

Fig. 2: Deformation of the meniscus shape due to seal rotation

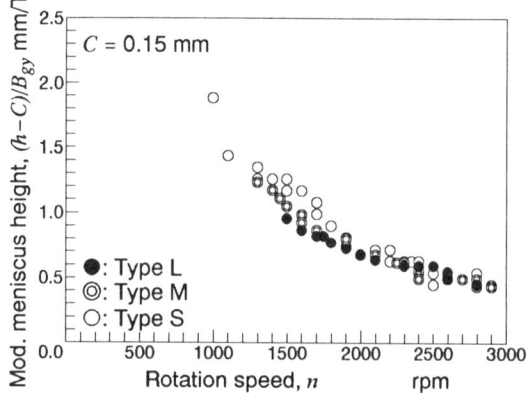

Fig. 3: Splashing initiation speed compared against modified menicus height

Effects of SiC particles on the friction interface between the cast iron shoe and wheel

Toru MIYAUCHI Taro TSUJIMURA
Railway Technical Research Institute, Kokubunji, Tokyo, 185, JAPAN
Jun-ichi NAKAYAMA
Hokkaido Railway Company, Sapporo, Hokkaido, 065, JAPAN
Hiroshi ARAI
East Japan Railway Company, Nagano, 381, JAPAN

INTRODUCTION

'Speed-up' is one of the most important things for a transportation system such as a railway. Especially in Japan, where the emergency braking distance on the existing lines is legally restricted to 600m, development of a high-performance brake shoe has been strongly desired.

The brake performance of cast iron shoe has been improved by adding some alloying elements to precipitate hard phase, such as cementite, carbide and steadite in matrix pearlite phase. We studied to supply ceramics as a hard phase between cast iron shoe and wheel during braking for the purpose of improving brake performance at high speed and confirmed the effect of them by using a full-scale brake tester.

EXPERIMENTAL PROCEDURE

Alloyed cast iron shoe (NHF) and a full-scale brake tester were used in all tests.

First step - we fed SiC, PSZ, Al_2O_3 particles between NHF and wheel during braking. Then the examination condition was as follows : inertia moment about 1200kgm^2, wheel diameter 860mm, shoe load 15kN, initial speed 125km/h.

Second step - we tested NHF including SiC particle blocks in it. Then the examination condition was as follows : inertia moment and wheel diameter were same as above condition ; two shoes used in the test, the shoe load 25kN and initial speeds 35, 95, 135km/h.

RESULTS AND DISCUSSION

The result of first step is indicated in fig.1. Fig.1 shows the relationship between average friction coefficient and kinds of ceramics at 125km/h. SiC was highest average friction coefficient in all ceramics. PSZ was the same as NHF and Al_2O_3 was worse than NHF. Kinds of ceramics are important to improve brake performance. SiC indicated the best brake performance, so we studied how to supply SiC particles and what is the effect of SiC quantity.

We tried to include SiC particle block in NHF, to produce NHF with a changed number of blocks and tested them. Fig.2 shows the relationship between average friction coefficient and number of blocks. The area ratio of one block is about 2% of rubbing area. 4 blocks yields the highest an average friction coefficient and average friction coefficient almost same as to at 135km/h though the number of blocks was increased more than 4 blocks. It indicated the same tendency at other initial speeds.

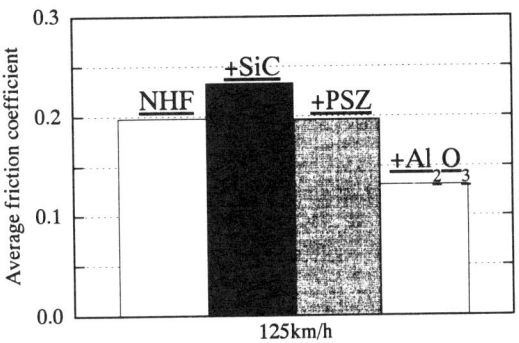

Fig.1 The relationship between average friction coefficient and kinds of ceramics

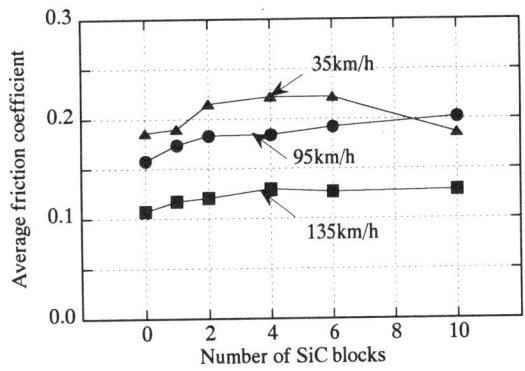

Fig.2 The relationship between average friction coefficient and number of SiC blocks

CONCLUSIONS

Brake performance of alloyed cast iron shoe was improved by supplying SiC particle between the shoe and wheel. Alloyed cast iron shoe with 4 SiC particle blocks proved the most suitable.

ON THE EFFECT OF MISALIGNMENT ON THE PERFORMANCE OF U TYPE LIP SEAL

M O A MOKHTAR, M A A MOHAMED AND M E ELGEDDAWY
Mech. Design and Prod. Dept., Faculty of Engineering, Cairo University, Cairo EGYPT
S A Y YASSEN
El-Nasr Pharma Chemicals Co., Cairo EGYPT

ABSTRACT

In the literature on the radial lip seal performance, great attention has been paid to the sealing lubrication mechanism and factors related thereto. (1). The Present experimental work is, however, an endeavor towards correlating the operating conditions to the rate of fluid leakage under different misalignment conditions.

The experimental results have been recorded under three values of operating pressure, (7,10 and 13 bar), two values of shaft running speed,(377 and 655 rpm) and using two types of hydraulic oils, (37C and 100C having kinematics viscosity at 40°C of 37 and 100 cst respectively).

The Effect of angular misalignment: The results, figure 1, show that under test conditions, the leakage rate is reduced with slight increase of angular misalignment.

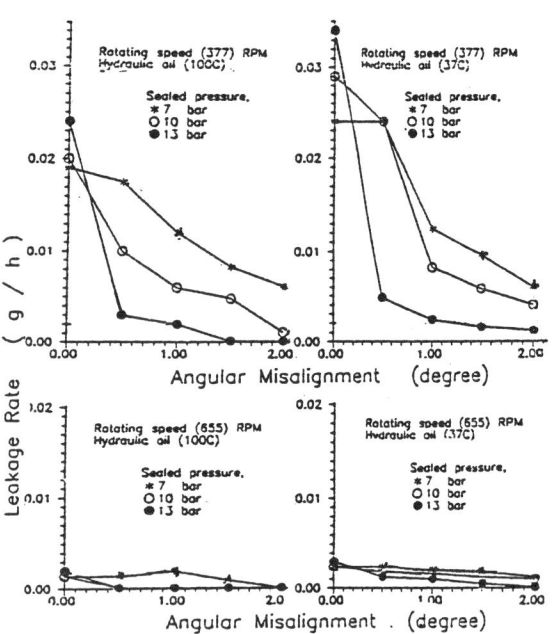

Fig.1 : Effect of angular misalignment on leakage

Effect of bore eccentricity : Results as shown in figure 2 show that the leakage rate increases, in general, with the increase in bore eccentricity with running shaft. Higher fluid pressure render higher values of leakage rates.

DISCUSSION OF RESULTS

It was earlier postulated that surface tension is responsible for retaining the lubricant film at the seal interface. This hyposis could not be supported by later work (2). As a result of small contact width at lip interface broad temperature rise occurs and the mechanism of sealing becomes complicated. Based on seal pumping action, it could be concluded (3) that the pumping action of the seal is counter balanced by the capillary forces of the oil/air interface on the air side. Recently, on the bases of coupling between the sealing mechanism and the lip microgeometry, the sealing mechanism is viewed in terms of physical reverse pumping process.

The effect of misalignment can be discussed in light of the increase in the seal/shaft contact pressure with the increase in the degree of misalignment. Though this would result reduced leak, ashown figure 1, an expected higher contact temperature and possibly would result a reduced sealing life.

Under eccentric seal mounting with the shaft (bore eccentricity) the contact pressure distribution will vary along the seal circumference giving an almost loose contacts at some regions with increased possibility of leakage as shown in figure 2.

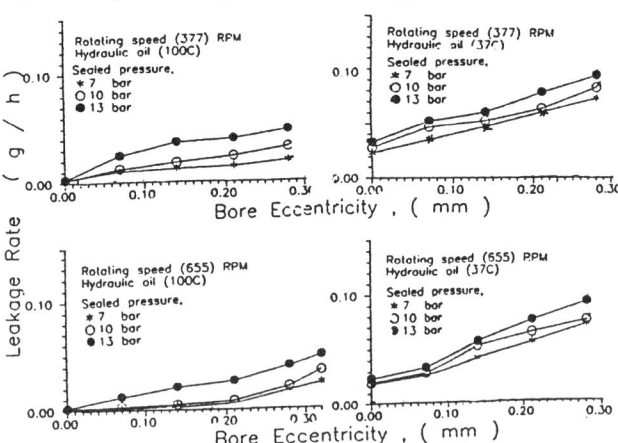

Fig.2: Effect of bore eccentricity on leakage

REFERENCES

1. Mokhtar,M.O.A. et al, "Effect of oil viscosity on the seal behaviour of u-type seal", to be published
2. Gabelli,A. and Poll,G.,"Formation of lubricant film in rotary sealing contact I",ASME JoT,1990,pp.1-8
3. Stakenborg,M.J.L.,"On the sealing mechanism of radial lip seals"Tribology Int.21,1988.pp.335-40.

FRICTIONAL CHARACTERISTICS OF A SOFT PACKED STUFFING BOX SEAL IN ROTARY MOTION

WŁODZIMIERZ OCHOŃSKI, ZBIGNIEW SZYDŁO AND BOLESŁAW ZACHARA
Faculty of Mechanical Engineering and Robotics, Academy of Mining and Metallurgy, Al. Mickiewicza 30, 30-059 Krakow, Poland

INTRODUCTION

For proper choice of power transmission systems for rotary pumps, it is necessary to have known frictional torque, occuring in the soft packed stuffing box seals.

Friction resistance in a soft packed stuffing box seal depend on many factors, like: kind of packing material, gland follower press force, operational medium type together with its pressure and temperature, viscosity, shaft rotational velocity and others [1-4]. Determination of characteristics of the friction torque versus a/m factors is possible but only in experiments.

The objective of this investigations was determination of friction torque in a stuffing box where graphite packing was used, dependently upon the number of sealing rings in the packing and pressure of operational medium at the initial phase of the sealing operation.

EXPERIMENT

Experimental investigations of friction torque in the soft gland packing were carried out on a specially designed test stand. The test stand (Fig.1) consists of the following sub-assemblies: power transmission system, test head, pressurizing system, record-measurement equipment. The following quantities were measured in the experiment: rotational speed, gland force, the pressure of sealed medium, friction torque. In the all tests were used preformed packing rings made of expanded graphite tape.

The following were tests parameters: rotational velocity of the shaft n=1460 rpm (circumferential velocity v= 3.45m/s), gland followers press force P=300N, operational medium: water, operational medium pressure: 1.0, 2.0, 3.0 Mpa, shaft diameter: d=45mm, stuffing box diameter: D=65mm, number of sealing rings in the packing: 1,2,3,4.

Results are formulated as relations between axial press force of the packing, friction torque of a single ring, and the friction torque and friction power losses in the packing upon pressure of the operational medium.

CONCLUSIONS

The following remarks could be proposed on the base of tests:

1. Friction torque in a sealing box with fixed numberof rings does increase with growing pressure of the operational medium.
2. Friction torque in a sealing box with even press of opeational medium does increase with higher number of sealing rings in the packing.
3. The sealing press of the packing is developed as a result of operational medium pressure acting on the sealing packing; the higher pressure of operational medium the stronger press of the packing.

REFERENCES

(1) Austin R.M., Fisher H.J.: BHRA, Report RR no 741, 1962.
(2) Denny D.F., Turnbull D.E.: Proc.Instn.Mech.Engrs, Vol.174, no6, 1960.
(3) Bohner K., Blenke H., Proc. of the 7th International Conference on Fluid Sealing, Paper G4, Nottingham, 1975.
(4) Ochonski W., Zachara B., Proc. of 10th International Colloquium on Tribology, Stuttgart/Ostfildern, 1966.

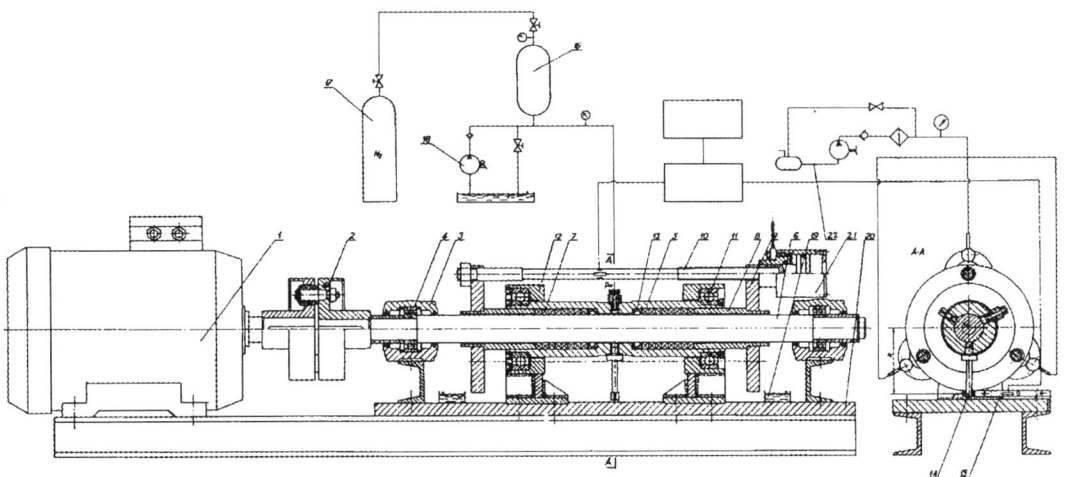

Fig.1.Test stand for measurement of friction torque in soft gland packing.

EROSION AND ABRASION RESISTANCE OF AUTOMOTIVE PAINT COATINGS

MATTHEW J. PICKLES and IAN M. HUTCHINGS
Department of Materials Science & Metallurgy, University of Cambridge, Pembroke Street, Cambridge CB2 3QZ, UK
A.C. RAMAMURTHY
Ford Motor Company, Automotive Components Division, Dearborn, MI 48120, USA

Two novel tribological tests have been developed and applied to study the durability of automobile paint systems. These coatings are complex polymeric systems, typically consisting of 4-6 different layers and may be applied to either steel or polymeric substrates. The paint system must provide resistance to corrosion, chemical attack and UV radiation as well as resistance to particle impacts, scratching and rubbing. Two tests have been developed to investigate the erosion and abrasion resistance of a variety of paint systems and these configurations represent important practical causes of surface damage.

EROSION DURABILITY TEST

This test involves the impact of hard abrasive particles against the coating to produce solid particle erosion. The particles are accelerated along a parallel-sided nozzle and impact the coated specimen which is positioned perpendicular to the outlet of the nozzle (Fig. 1). The resulting erosion scar is circular, with a sharply defined boundary between the areas where the coating is removed down to the substrate and the region where some of the coating is still intact (1).

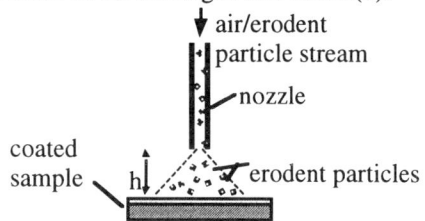

Figure 1: Schematic diagram of the erosion rig

By studying the particle angular flux distribution, it is possible to relate the diameter of the erosion scar to the erosion durability of the coating, Q_c. This is defined as the critical mass per unit area of impacting particles required just to remove the coating down to the substrate. A plot of the scar radius, r, against the logarithm of the erodent mass, $\ln m$, is found to be linear and the value of Q_c is found from the intercept of this line at the abscissa (1). The value of Q_c therefore defines the erosion durability of a coating to solid particle erosion. This technique has recently been extended to study multi-layered coatings and is able to determine the durability of each layer in a multi-layered coating, from a single test (2).

MICRO-SCALE ABRASION TEST

The second tribological test uses a steel sphere rotating against the specimen coating in the presence of a slurry of abrasive particles, to measure the abrasive wear resistance of the coating (Fig. 2). This test is able to examine the uppermost 30 μm of the coating, which for automobile paints is entirely within the 'clearcoat' lacquer coat. The particles produce three-body abrasion within the coating layer and the geometry of the wear crater is imposed by the rotating sphere. The volume of material removed can therefore be calculated from a simple measurement of the scar diameter. By applying the Archard wear equation, the wear coefficient, κ, may be calculated from the slope of a linear plot of the product of the sliding distance and normal load, SN, against wear volume, V (2).

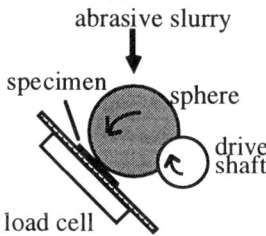

Figure 2: Schematic diagram of the micro-scale abrasion rig

RESULTS

A number of different paint systems have been examined and the effects of the coating cure conditions on the erosion and abrasion processes have been studied (3,4). The tests were able to discriminate well between coatings with different compositions and between different bake conditions. It was found that 'overbaked' coatings (i.e. those baked at a higher than normal temperature) were less durable than coatings cured under normal conditions. 'Underbaked' coatings were found to have intermediate durability and the coatings cured under standard conditions were the most durable.

CONCLUSIONS

The test methods described have both been shown to be highly suited to the study of the tribological responses of automobile paints. Both methods eliminate the need for subjective assessment of the onset of damage, without the need for the measurement of very small mass or volume changes, and are applicable to a wide range of materials.

REFERENCES (1) P.H. Shipway and I.M. Hutchings, *Wear*, **162-164** 148-158 (1993). (2) K.L. Rutherford, R.I. Trezona, I.M. Hutchings and A.C. Ramamurthy, *Wear*, **203-204** 325-334 (1997). (3) M.J. Pickles, I.M. Hutchings and A.C. Ramamurthy, *Proc. ISATA Conf. Paint and Powder Coatings*, Paper 97PA019, Florence, Italy (1997). (4) R.I. Trezona, I.M. Hutchings and A.C. Ramamurthy, *Proc. ISATA Conf. Paint and Powder Coatings*, Paper 97PA018, Florence, Italy (1997).

THE IMPROVED NOISE REDUCTION AND TRIBOLOGICAL CHARACTERISTICS OF COMPONENTS IN TRUCK AXLES USING AUSTEMPERED DUCTILE IRON (ADI)

PEKKA SALONEN
JOT COMPONENTS LTD, Cast Components Div., P.O. Box 40, FIN-03601 Karkkila, Finland

ABSTRACT

The purpose of this investigation was to reduce noise, improve tribological behaviour and increase efficiency in trucks. The chosen truck components for testing were wheel hub and its gear wheel. The excellent damping effect of ductile iron (Fig. 1) compared to steel was utilized in order to achieve the permissible sound levels according to the EEC-Regulation 92/97 EEC.

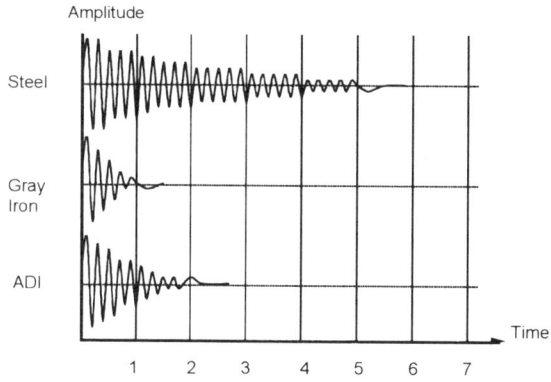

Fig. 1. The damping effect of Steel, Gray Iron and ADI. (5)

A test rig for measuring noise levels in the hub was designed and built in Helsinki University of Technology, Laboratory of Machine Design (1). The test rig consisted of four axles connected to each other and the power was applied to the system with two electric motors. The load was applied by a torque gear situated in the middle of the system.

Three different types of axles with steel and casted components were tested against each other. With several different loads (torque) and rotational speeds the noise levels were measured and the wear characterized from the gear wheels. In order to compare the laboratory tests to the actual applications, two of the casted components were assembled to trucks and the noise levels were measured (1).

The results in laboratory tests showed that using casted component as gear wheel material reduced the peek noise level by 8...10 dB(A). Thank the use of ADI the weight of the total axle could be reduced by 10% which also resulted a 20% drop in the machining costs. The field tests ensured the laboratory tests. The noise level was 5...8 dB(A) smaller with ADI-components compared to steel ones (Fig. 2), (3).

The wear of the gear wheel teeth was determined by assembling the casted components into a truck, which was in normal use and after certain drive distances the axles were opened and the wear determined microscopically. At the same time the oil analysis was made. The wear of ADI components was in running-in moderate, but after the work-hardening no measurable wear could be determined.

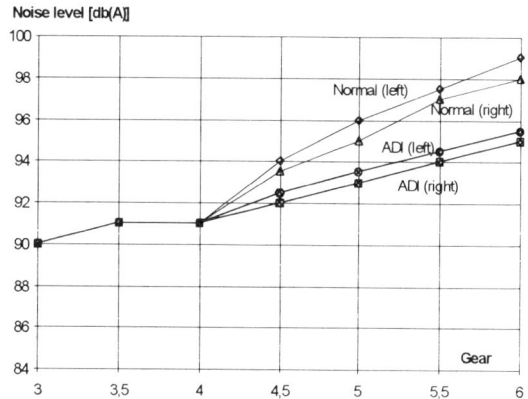

Fig. 2. The noise levels with different gears at field tests. (2)

REFERENCES

(1) Koponen M., "Vetävän akseliston melun pienentäminen", Publ. Nr. C 273, Helsinki Univ. of Tech., Lab. of Machine Design, Otaniemi 1995.

(2) Koponen M., Ylöstalo O., Hautala P., "Akseleiden kenttämittaukset", Publ. Nr. C 275, Helsinki Univ. of Tech., Lab. Of Machine Design, Otaniemi 1995.

(3) Koponen M. et al., "Konstruktiomuutosten vaikutus akselin napameluun", Publ. Nr. C 276, Helsinki Univ. of Tech., Lab. of Machine Des., Otaniemi 1996.

(4) Salonen P., "Bainitoitu pallografiittirauta Kymenite-ADI konstruktiomateriaalina", Koneensuun:n XX kans. Symp:n esitelmät, Publ. Nr. C 270. Helsinki Univ. of Tech., Lab. of Machine Des., Otaniemi 1995, 95-103 pp.

SHOT PEENING TO DEVELOP WEAR RESISTANT SURFACE IN SOIL WORKING TOOLS

S.K. RAUTARAY and M.C. SHARMA

Department of Mechanical Engineering, Maulana Azad College of Technology, Bhopal 462 007, India

ABSTRACT

Abrasive wear of soil engaging components is considered analogous to microfatigue failure (1) which may be reduced by shot peening. Shot peening action causes plastic deformation of work surface and induces surface work hardened layer by virtue of residual compressive stresses. Because of this, shot peening is best applied for surface modification to reduce replacement and maintenance cost of machine parts subjected to fatigue, fretting etc. (2). With this analogy, the present investigation (3) aimed to optimise the shot peening parameters so as to quantify and compare the surface improvements in respect of abrasive wear resistant properties due to peening of a low carbon structural steel being carburized, hardened and tempered to that of a high strength alloy steel normally used for manufacture of rotavator blades. However efforts for minimising abrasive wear of soil engaging components of agricultural equipment are yet under consideration for Indian farmers due to want of appropriate shot peening process package and related cost-effectiveness.

The objective of this study is to present the wear test results of the shot peened carburised structural steel and high strength spring steel, to compare the cost-effectiveness of the treated surfaces and to select an appropriate peening package.

Structural steel (SAE 1022) was carburized, hardened, tempered (CHT) and shot peened at 7 peening intensities. Spring steel (En42) was hardened, tempered (HT) and shot peened under same peening conditions. A wear test set-up was developed to move the blade specimens under simulated conditions. A comparison was made of the results obtained on percentage mass wear and volume wear measured after 5 different durations of test run for all peening conditions including the unpeened state of the material.

The results showed that shot peened surface offered more abrasive wear resistance as compared to unpeened ones (Fig.1). Peening intensity range 0.15-0.25, 0.20-0.30 and 0.30-0.45mm 'A' were considered optimum in respect of wear performance for SAE 1022-virgin, SAE 1022-CHT and En42-HT blade specimens respectively as the variation in per cent mass wear within these ranges of peening intensities for the respective material specimens were significantly not different. However peening intensity range of 0.25-0.30 mm 'A' was considered optimum for SAE 1022-CHT and En42-HT specimens within which both these two material specimens could be equated to each other in respect of their optimum wear performance in terms of service life as the plastically stretched dimple peaks on the surfaces offered higher resistance to abrasion and thereby took longer time to vanish being fatigue off. Higher wear performance of shot peened SAE 1022-CHT specimens coupled with lower unit cost proved to be more cost-effective than En42-HT. Therefore substitution of SAE 1022-CHT (0.25 mm 'A') in place of En42-HT (0.30 mm 'A') was considered to be economically justified and accordingly related peening package was developed. Detailed results of this research has been fully and comprehensively explained in the paper of this title.

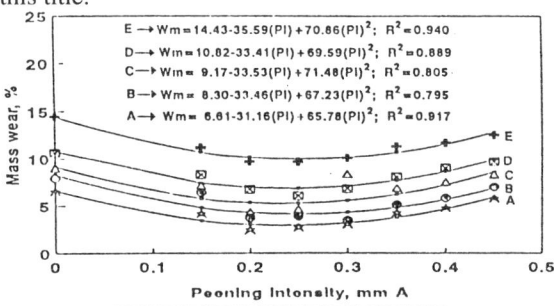

Fig.1:

REFERENCES

(1) A Mubeen, Advanced machine design, Khanna Publishers, India, pp28-45, 1983.
(2) M.C. Sharma, Fatigue and fracture behaviour Ph.D thesis, B.U., India, 1988.
(3) S.K. Rautaray, Fatigue and Wear Characteristics of Shot peened rotavator blade materials, Ph.D Thesis, B.V., India, 1997.

TRIBOLOGY IN VEHICLE TRANSMISSIONS - PRACTICAL METHODS OF IMPROVING FUEL ECONOMY

D SIMNER

Advanced Power Train Technology, Rover Group Ltd, Engineering Building, University of Warwick, CV4 7AL. UK

SYNOPSIS

This paper discusses the contribution of a road vehicle's transmission system to fuel economy and in particular how the tribological aspects can be used to advantage. As background, a brief review is made of the need for improvements in fuel economy and how the whole transmission effects vehicle fuel economy.

Of the different inefficiencies which occur in the transmission, it is possible to improve both the power related and parasitic losses. Test data on efficiencies will be presented, together with the improvements that can be made, and what vehicle fuel economy benefits can result from these changes.

INTRODUCTION

It is proposed that there are four aspects of the transmission system which influence fuel economy:
- The parasitic losses in the transmission units.
- The power related losses at the gear mesh.
- The ratio within the transmission unit.
- The weight and inertia within the transmission.

This work looks at transmission efficiency and some of the experimental results are discussed. The important next step, however, is to be able to use this efficiency data to determine the fuel economy benefit that would result in the vehicle. It is proposed that simulation techniques are important in this. Providing accurate transmission efficiency data is available the resultant effect on fuel economy can be deduced.

IMPROVING TRANSMISSION EFFICIENCY

A number of authors have identified that losses in a transmission can be related to the power transmitted, or to a number of 'parasitic' losses. The latter tend to be dependant on the speed of the transmission and viscosity (hence temperature) of the oil.

The sources of the parasitic losses include the churning of the oil, especially in a dip lubricated gearbox, the drag associated with the bearings and oil seals, and the losses in the oil pump.

The power related losses are usually considered a percentage of the transmitted power of each gear mesh. The work carried out for this study has shown that within the accuracy required for the simulation work this is certainly so. It is proposed that this aspect can be effected by oil additives and by surface treatments on the gears, and while such improvement may not be too significant for fuel consumption it may be important in reducing the operating temperatures when a vehicle is being used at high load.

The methods which can be used to reduce the levels of friction in a transmission and hence the power related losses are:

- Friction modification additives in the lubricant.
- Surface treatments and coatings.
- Modifications to the lubricant base stock.

Parasitic losses can be improved by one or more of the following methods. It is worth making the point that the combined effect of more than one item may not be cumulative.
- Reduction in oil churning; lower oil levels, etc.
- Optimised oil pump design
- Fast warm up at cold start.
- Reduced viscosity lubricants.
- Thermal management to optimise temperatures.

With a number of options available the use of the simulation techniques is valuable to determine the most effective.

PRACTICAL EXAMPLES

To support this work a number of practical tests have been completed on test rig and on development vehicles. Testing, as in fig1, has been used to establish the improvements that can be made to a particular unit.

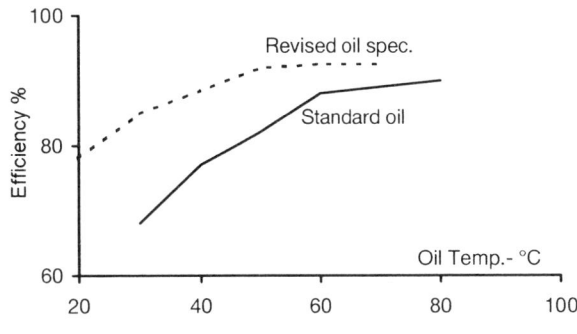

Fig 1: Improved efficiency due to revised oil at constant speed and torque

A major part of the improvement shown here can probably be attributed simply to the change in oil viscosity. Also a low efficiency can be seen at lower operating temperatures due to parasitic losses having a significant effect at moderate operating torques.

Vehicle testing has shown that a lower viscosity oil in the transmission of a vehicle gave an improvement of 1½% over a legislated cycle. In another test a greater improvement was obtained by increasing the speed at which the transmission warmed up from a cold start.

FULL PUBLICATION

It is the intention to submit a paper on this work to the IMechE Journal of Engineering Tribology.

FRICTION REDUCTION POTENTIAL OF ADVANCED ENGINE OILS IN BEARINGS

JAGADISH SORAB and STEFAN KORCEK
Ford Motor Company, Ford Research Laboratory, MD 2629/SRL, Dearborn, MI 48121-2053, USA

ABSTRACT

Advanced low viscosity engine oils containing molybdenum dialkyldithiocarbamate friction modifiers (FM) are gradually finding increasing application in automotive engines for their ability to enhance fuel efficiency. These oils reduce viscous friction in components that operate in predominantly hydrodynamic lubrication and also reduce boundary friction in components operating in boundary and mixed regimes. However, by decreasing oil film thickness, low viscosity oils increase the possibility of mixed-boundary lubrication and metal-to-metal contact. This is particularly true in the case of journal bearings which are currently designed for mostly hydrodynamic lubrication. Design guidelines call for minimum film thickness in the range of ~1 μm. Of course, metal-to-metal contact in journal bearings during engine operation can lead to unacceptably high friction and wear.

Using two complementary experimental techniques [1,2], we have investigated the lubrication regimes in connecting rod bearings lubricated with low viscosity engine oils. Load supporting capacity under conditions ranging from fully flooded to mixed lubrication was measured for several candidate oils using a journal bearing simulator. This instrument simulates the motion of a journal bearing at fixed measurable eccentricities. Friction on both the journal and bearing of a connecting rod under typical engine conditions of speed, load and temperature was measured using a dynamic bearing test rig.

Results indicate that bearings operate in hydrodynamic lubrication under most test conditions, though mixed lubrication conditions were shown to exist at low speeds in heavily loaded journal bearings (Figure 1). Well formulated low viscosity engine oils helped achieve significant reductions in friction in both hydrodynamic and mixed lubrication conditions. Decreases in friction associated with reduced viscosity were observed under most conditions of hydrodynamic bearing operation. In mixed lubrication conditions, associated with low speeds and high bearing loads, an oil with effective friction modification (RO11) produced reduced friction by extending the hydrodynamic lubrication regime. An oil without friction modifier (RO11a) and a very low viscosity oil even with friction modifier (RO12) exhibited sharp increases in friction typical of mixed lubrication and higher bearing wear.

Further tests using a journal bearing simulator confirmed that in mixed lubrication, oils with friction reducing additives exhibit higher load supporting capacity, distinct separation of moving parts (Figure 2),

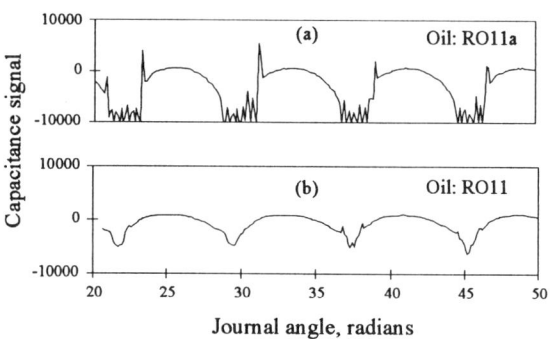

Figure 2: Film thickness measurements for oils (a) without and (b) with friction reducing additives

and reduced friction relative to oils without such additives. Improvement in journal finish appeared to reduce bearing friction at high eccentricities by reducing metal-to-metal contact. Under certain conditions, increase in load supporting capacity was also possible through reduced clearances. These findings suggest that even very low viscosity oils (5W-20) with effective friction reducing additives can extend low friction benefits over the entire range of operation of engine bearings, while protecting durability.

REFERENCES

(1) T. W. Bates, G. W. Roberts, D. R. Oliver, and A. L. Milton, SAE 922352, 1992.
(2) J. Sorab, S. Korcek, C. L. Brower, W.G. Hammer, SAE 962033, 1996.

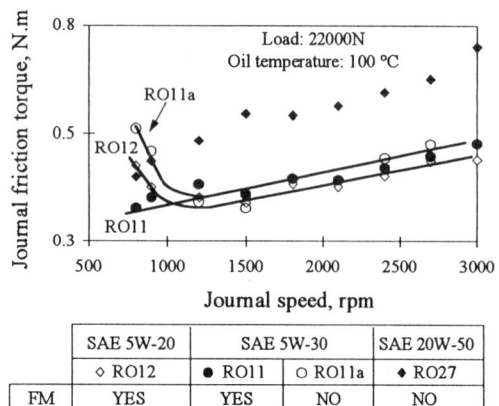

	SAE 5W-20	SAE 5W-30		SAE 20W-50
	◇ RO12	● RO11	○ RO11a	◆ RO27
FM	YES	YES	NO	NO

Figure 1: Friction reduction with low viscosity oils.

COMPARISON OF STATIC AND RUNNING EFFICIENCIES OF PLANETARY GEAR DRIVES

K SONODA and A ISHIBASHI
Department of Mechanical Engineering, Kumamoto Institute of Technology, Kumamoto 860, Japan
S ISAMI
Sumitomo Heavy Industries Ltd., Aichi 474, Japan

ABSTRACT

Efficiency of planetary gear drives can be calculated using reference efficiency (l) (2). However, the calculated efficiency is not so accurate as that obtained by experiments.

The power transmission efficiency of gear drives can be measured conventionally using torque meters consisting of the strain gauges and slip rings. The price of precision torque meters is high and measuring errors become problem when the capacity of the gear drives is large and their rotational speed is very high. In contrast to this the static efficiency can be measured without using the strain gauges and slip rings (3).

The static efficiency of a gear drive can be measured by the following two steps: In the first step, theoretical torque T_1 on the input shaft and the corresponding theoretical output torque on the output shaft, $T_2 = I_o T_1$, are applied statically without making rotation, where I_o is the reduction ratio. In the second step, the torque on the input shaft is increased up to the critical value $(T_1 + \Delta T)$ at which the output shaft with torque T_2 begins to rotate. The static efficiency is calculated from Eq. (1).

$$\eta = \frac{T_2}{I_o(T_1+\Delta T)} \qquad (1)$$

The static efficiency can be measured precisely using two torque arms with dead weight which are used for producing the torques on the input and output shafts. This means that the efficiency of gear drives can be measured without slip rings and strain gauges. However, it is necessary to conduct comparison tests to evaluate whether or not the static efficiency is effectively used instead of the running efficiency of gear drives.

Experiments were conducted using a planetary gear drive of the differential type with a reduction ratio of 31.3, which was designed and made by the authors. As lubricating oil, three representative oils (hypoid gear and cylindrical gear oils and spindle oil) were used for comparison tests. Using strain gauges and slip rings, the running efficiencies were measured by changing the output torque and rotational speed of the input shaft. The static efficiencies were calculated using measured torques and Eq. (1). Figure 1 shows changes in the efficiencies for speed reduction which were obtained using two different methods when the gear drive was lubricated by the cylindrical gear oil. White circles indicate the efficiencies obtained by the conventional method at an input shaft speed of 1800 rpm. Dark circles indicated the static efficiencies calculated using statically measured torques and Eq. (1). From this figure, it is understood that the efficiencies measured by two different method agree well at and near the rated capacity (about 16 N-m). However, under partial loading conditions, the running efficiency is appreciably lower than the static efficiency due to the oil

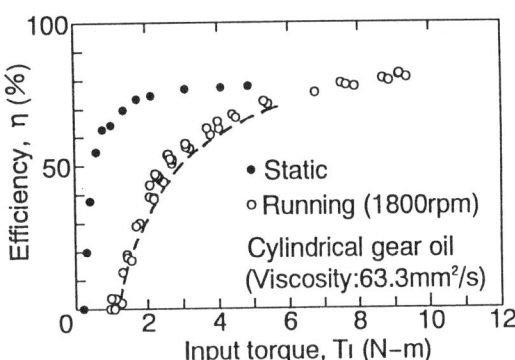

Fig. 1 Static and running efficiencies

churning loss which corresponds to the input torque (about 1.1 N-m) at zero efficiency in the figure. It should be noted that the oil churning loss never included in the static efficiency although the friction coefficient between contacting teeth may be a little higher than that in conventional running tests because no hydrodynamic action hardly occurs due to an extremely low gearing speed.

For the sake of comparison, the oil churning loss (1.1 N-m) was added to the measured torque for the static efficiency, and the modified efficiency was calcuated. The modified efficiency is shown by broken line in Fig. 1, which is nearly equal to the running efficiency measured by conventional method. Similar results were obtained when the kinds of lubricating oils were changed and the gear drive was used for speed increase.

The reason why the modified efficiencies are a little lower than those obtained by running tests may be ascribed to the difference in the friction coefficient between contacting teeth. The difference becomes negligible when a hypoid gear oil is used because the hypoid gear oil prevent metallic contacts even at an extremely low speed.

From this investigation, the static efficiency can be used for evaluating the running efficency under full loading conditions and also partial loading conditions when some modifications are made on the oil churning loss.

REFERENCES

(1) G.Niemann and H. Winter, Maschinen elemente, Bd. 2, Springer V., 1983, p. 218.
(2) M. Morozumi, Theory and Design calculation Method for Planetary and Differential Gear Drives, Nikkan-kogyoshinbun, 1989, p. 33.
(3) A. Ishibashi, K. Sonoda and S. Isami, Proc. 7th ASME Inter. Power Trans. and Gearing Conf. 1996, p. 397.

INDUSTRIAL TRIBOLOGY PROBLEMS CAUSED BY THE SHAFT CURRENT

STANISŁAW STRZELECKI

Institute of Machine Design, Technical University of Łódź, Stefanowski Street 1/15, 90-924 Łódź, Poland

ABSTRACT

Rolling bearing failures in electric machines are due to various reasons from which the vibration during the passage of electric current is one of the most important. This kind of failure significantly decreases the durability of bearing and increases the risk of motor damage as well as the cost of maintenance and overhaul. Passage of electric current in both rolling and journal bearings or so called shaft current is a side-effect in electric machines (1)(2)(3)(4). It is very important to know the source of shaft current, the direction of current flow and the methods of avoiding this inconvenient flow.

The main reasons of the possible difference in the potential on the shaft ends are: the magnetic asymmetry and ring induction. The former one generates the alternating voltage called "shaft voltage" on the shaft ends, which flows through the shaft. The ring induction is the source of homopolar voltage, which can be direct or alternating, and which is a consequence of flow asymmetry in the stator yoke. The shaft voltage results from compensation flow arising in the stator yoke, whose activity is alternate voltage generated on the shafts' ends. As the result the bearing current flows in outer circuit through the shaft, both bearings and supporting frame. Homopolar voltage creates the loop, closing through the shaft, both bearings and supporting frame. The current flows from the shaft ends to the bearings.

Vibrations of rolling bearing are caused by non-linear relationship between, the value of contact deflection and load, the lack of circularity and non uniformity of rolling elements, as well as by technological, assembly and exploitation conditions (5). The level of vibration generated by rolling bearings mounted in machine housing depends on the vibration level of the bearing itself, its kind, i.e. ball, and roller, too. Coupled action of vibrations and electric current passage through the bearing causes the premature failure of bearing.

The paper considers this coupled action in the case of large diameter single deep-groove ball bearing 6326P6C3 of high power electric motor; the second bearing is roller NU326 one (6).

After 1 month of operation an excessive vibration and noise of motor has been stated. Measurements of single amplitudes of displacement have given the resultant value about 48 μm for bearing 6326P6C3 and 40 μm for bearing NU326. Checking the voltage and current values on the side of ball bearing has revealed respectively 0.7V and 5-10A current flowing through the bearing. After shutting down the motor and disassembly, the visual inspection of bearings has shown the wear markings of ladder shape on both inner and outer ring ball races. Uniformly displaced markings of ball contact with outer and inner ring ball race have the shape of elongated ellipse with the length 12 mm and 1 mm width and shifted 2 mm from the symmetry plane of inner ring. Maximum depth of these markings reaches 0.005 mm. Wear of outer ring ball race is characterised by 2 - 3 mm width and 14 mm length markings in form of elongated ellipse, placed symmetrically on the race. The depth of these markings changes from 0.009- 0.014 mm.

It was stated, that the measurements of current value referred to the contact ellipse gives the information about the possibility of durability loss and the circuit consisting of an adequate ohm resistance and capacity allows to avoid the premature failure of the ball bearing. The circuit developed for carrying away the shaft voltage was mounted in two of the units' motor - pump in thermo-electric power plant causing the increase of durability of the bearing.

The described case of premature loss of ball bearing durability and the analysis of the failure reasons allows for the conclusions given below.
1. The vibration during the passage of electric current is the basic factor of fast losses of ball bearing durability.
2. Measurements of current value referred to the contact ellipse gives the information about the possibility of durability loss.
3. Circuit consisting of an adequate ohm resistance and capacity allows to avoid the premature failure of the ball bearing.

The circuit developed for carrying away the shaft voltage was mounted in two of the units motor - pump and the durability of the bearing increased to the catalogue value.

REFERENCES

(1) Allan R.K.: Rolling Bearings. London, Sir Isaac Pitman & Sons, Ltd. 1946
(2) Haus O.: ETZ-A. Bd.85, 1964, H.4, 106-112pp.
(3) Przybysz J.: Turbogenerators. Exploitation and Diagnosis. WNT, Warsaw, 1991 (in polish)
(4) Akagaki T., Kato K., Kawabata M.: International Tribology Conference, Yokohama ,1996.
(5) Krzemiński-Freda H.: Rolling Bearings. PWN, Warsaw, 1985 (in polish).
(6) Pilak E., Strzelecki S., Sliwinski S.: XVII Symposium on Machine Design. Lublin-Nałęczów, Vol.3, 1995, 71-76pp.(in polish)

MATHEMATICAL MODEL OF PISTON IN THE FLUID AND MIXED LUBRICATION

JAN SZKURŁAT
Institute of Machine Design, Technical University of Łódź, Stefanowski Street 1/15, 90-924 Łódź, Poland

ABSTRACT

The resistance of motion generated by the piston and rings assembly has a very important effect on the mechanical efficiency of engines(1)(2). Decrease of piston friction allows for the increase of durability and reliability, decrease of fuel consumption, and environmental pollution. The tendencies for correction of these parameters and concern for the reduction of time and design cost of new solutions in the field of engines need the calculations' models of engines and their elements (3)(4)(5).

The paper concerns the phenomena strictly connected to the lubrication and friction of piston in the cylinder together with the reaction of connecting rod and piston rings.

The aim of investigation is to determine the calculation model of piston lubrication and friction in the steady kinematics, dynamic and heat conditions.

The range of investigation concerns the determination of physical model, statement of mathematical model equations, developing the method of solution of equations system.

The physical model was developed under assumption of the engine piston operating at the constant velocity of crankshaft; on the piston crown the known, uniformly distributed gas pressure is applied which is the function of time. In the ring part there are three rings meshing with the cylinder bearing surface at the mixed friction. Non-cylindrical piston and the cylindrical bearing surface deflect only locally during the contact of their smooth surfaces and undergo the global thermal deflections.

As result of forces and moments acting on the piston and the presence of backlashes, the piston displaces in the plain parallel to the plain of connecting-rod. The lubricating gap is filled up with oil and it is described on the isothermal model of flow. On the basis of the assumed physical model of piston the equations of dynamic equilibrium of piston, connecting-rod as well as the equation of forces acting on the separate piston rings were derived.

The resultant force of oil film occurring in the equilibrium equations of piston was determined on the basis of pressure distribution and geometry of lubricating gap. For solving these equations, the boundary conditions of oil film were assumed. When the minimum oil film thickness is equal zero then comes to the contact of piston skirt with the cylinder bearing surface.

The force and moment of contact pressure, that is in the equations of piston equilibrium are determined on the base of local deflections of piston skirt by mean of dependency received for the cast-iron sleeve and piston made from aluminium-alloys.

The system of equations describing the mathematical model of piston is possible to solve under assumed initial conditions, i.e. the values of displacements and velocities of points of the piston axis and placed in the boundary plains of piston skirt.

On the basis of assumed values, the thickness of oil film can be calculated and checked the possibility of the contact of piston with the cylindrical bearing surface. Next, the equation of oil film pressure distribution is solved and in the case of the contact of piston with the cylinder surface the contact pressure is determined. After that, the rest of forces and moments existing in the equilibrium equation of piston, rings and connecting rod are calculated.

The iteration procedure is applied and the assumed exactness of displacement, velocities on the begin and end of cycle is the condition to stop this procedure. Introduced model together with the method of solution is the basis of computer program allowing the evaluation of characteristic parameters: design of piston - connecting - rod system, macro geometry of piston, load conditions, properties of lubricant on the minimum oil film thickness and the motion resistance of piston engine.

REFERENCES

(1) R.Nozawa, Y.Morita, M.Shimizu, Tribology International, Vol.21, No.1, 31-37pp.
(2) Li, S.M. Rhode, H.A. Ezzat, ASLE Transactions, Vol.26, No.2, 151-160pp.
(3) W.L.Blair, D.P. Hoolt, V.W.Wong, ASME Journal of Engineering for Gas Turbines and Power, Vol.112, Jul. 1990, 287-300pp.
(4) Hoshi, Y.Baba, ASLE Transactions, Vol.30, No.4, 444-451pp.
(5) J.Szkurłat, Journal of KONES, Vol.1, No.1, Warsaw-Lublin, 1994, 657-663pp.

ENERGY-EFFECTIVE CHARACTERISTICS OF LUBRICANT AND WORKING LIQUIDS IN VEHICLE HYDROSYSTEMS

Ivan VERENICH and Nicolai BOGDAN
Belarusian Polytechnical Academy, Chair of Hydrophneumoautomatics,
Prospect Fr.Skoryna, Minsk 220027, Republic of Belarus

Quality indicators, trial methods and criteria of choice of hydraulic liquids and lubricants for vehicles are regulated by international and producers classifications which vary in trial methods as well as in requirements to identical oils. The modern list of requirements to motor oils includes their energy-effective characteristics assessment (1) which is made according to CEC L-54-X-94, Modified Sequence VI-A(2) and other methods.

Fuel saving in vehicle engines is directly connected to loss by friction and, accordingly, to characteristics of lubricants, that is why some researchers suggest forecasting fuel saving by friction coefficient for different lubricant conditions, others - according to change in temperature of cooling liquid, vehicle efficiency or change of viscosity at high temperatures and high speeds of shift.

A quality assessment method by a complex criterion of energy- effectiveness in reference to the characteristics of standard oil is considered. The method is based on running the tested liquid (volume 1 or 1.5 litre) through a closed hydrosystem, which consists of a pamp, a hydromotor, filtering elements and adjusting and recording devices. The filters provide different degree of tested liquid purity.

At trials the following parameters are recorded: temperature, pressure, flow rate, difference in pressure at different parts of a hydrosystem, quality characteristics of hydraulic liquid or lubricant (viscosity, etc.). Energy characteristics of a test bench such as pump power, moment constancy on the hydromotor shaft, number of revolutions are also recorded.

Some kinds of motor and transmission oil, hydraulic liquids of different degree of purity and viscosity within the range of 0-100°C (temperature) and 0.05-10 MPa (pressure) have been trialed. Oil 15W/40, 75W/90 and aviation hydraulic oil were considered as standard.

Energy-effective characteristics assessment was made according to the following formulae:

$$\varepsilon = \alpha_i / \alpha_t;$$

$$\alpha = (1 - \eta) \times (1/\eta - 1) \times \chi ,$$

where α_i and α_t are energy-effective coefficients of tested and standard liquids and η is system efficiency, χ - coefficient depending on the type of hydraulic liquid.

The results of the trials at the start of a vehicle and its frequent stops are obtained. The received dependence of energy-effective cefficient α on the characteristics of tested oils is shown in fig.1 . It is proved that increase in purity at one degree (class) raises energy-effective coefficient by 0.6% for motor oil, by 0.8% for transmission oil and up to 1.3% for hydraulic liquids. A significant decrease of the energy-effective coefficient is noted within the pressure range up to 2 MPa, and an increase - within 6-8 MPa . A rise in lubricant temperature or hydraulic liquid results in viscosity fall that causes the change of lubricant flow rate.

Applying this method one can choose the most appropriate oil or hydraulic liquid for certain operation conditions.

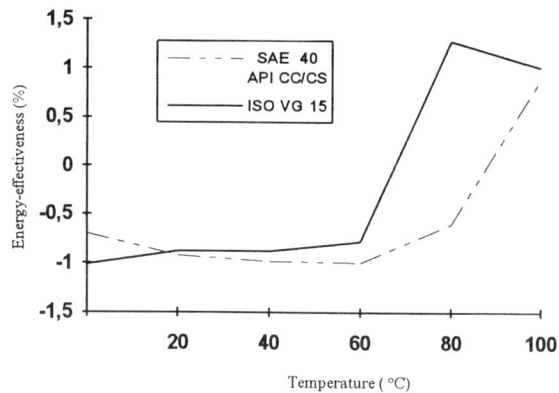

Fig.1: Energy effectiveness with temperature

REFERENCES

(1) M.Gairing - In : 10th Intern. Colloquium "Tribology-Solving Friction and Wear Problems", 9-11 Jan.1996. Esslingen. Supplement, p.123-133.
(2) J.A.M'cGeehan, T.M.Franklin et al - In: 10th Intern. Colloquium "Tribology-Solving Friction and Wear Problems",9-11 Jan.1996. Esslingen, v.3, pp.2249-2263.

DEVELOPMENT OF A LOW FRICTIONAL SHOCK ABSORBER FLUID

SEIICHI WATANABE

Lubricating Oil Research Center, Japan Lubricating Oil Society, 2-16-1, Hinode, Funabashi, Chiba, 273, JAPAN

ABSTRACT

A typical suspension of passenger cars consists of a spring and a shock absorber(SA). A tube-shape damper with a piston and a cylinder contains oil for damping vibration. This oil is called shock absorber fluid(SAF).

The prime quality requirements for SAF are viscometric properties, friction reducing and wear preventing performance, compatibility with seal rubber and other materials, and oxidation stability.

Low frictional property and its durability are especially important for, so called, ride comfort because frictional force occurred in a SA delivers vibration of a tire to a body of cars. A strut type double tube SA for small passenger cars was selected in this new SAF development.

FRICTION ANALYSIS

Frictional resistance in a SA originates in three sliding contacts. They are seal/chromium-plated rod, DU bush bearing/chromium-plated rod and piston/cylinder.

Friction between a oil seal of NBR and a chrome-plated rod results from a radial lip load and its value was observed about 61 N with a base fluid and about 30-38 N with formulated oils, which was relatively unchanged by a load applied transversely to the SA.

Friction between a DU bush bearing and a chrome-plated rod breaks out when a SA receives a bending moment. Its value was about 39 N with a base fluid and about 26-28 N with formulated oils when the SA got a transverse load of 490 N at a normal position.

Friction between a piston of carbon steel and a internal wall of a bearing steel tube also occurs when a SA receives a bending moment. Its value was about 16 N with a base fluid and about 10.8-11.4 N with formulated oils when the SA got a transverse load of 490 N at a normal position.

The frictional force between a chromium surface of a rod and the bush bearing of which surface is coated with PTFE resin was larger than the friction between a piston and a cylinder.

An effect of friction modifiers on these material combinations was also admitted.

DURABILITY TEST

In order to confirm the performance of candidate fluids, SA durability tests were carried out. Test conditions were as follows;

Transverse load: 980~2940 N
Up and down pitch distance ±30 mm
Pitch cycle 5~10 Hz
Oil temperature 70 degree C
Total cycles $2*10^6 \sim 10*10^6$

The durability of low frictional property and wear preventive performance of SAFs were evaluated. Typical results are shown below.

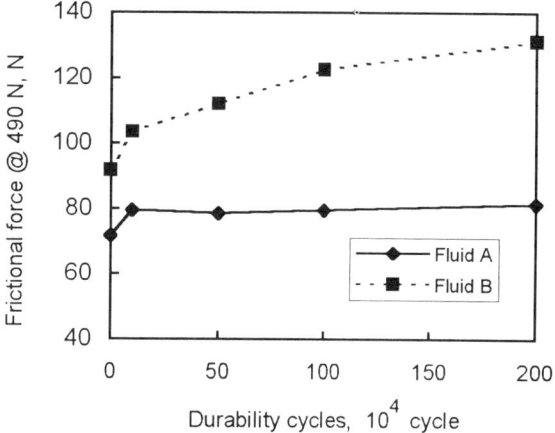

Fig. 1 Durability of low frictional property

COMPATIBILITY

Fluid's compatibility with seal rubber and DU metal was observed with beaker tests. The surface of NBR rubber was roughened by an additive, organic acid phosphate in the fluid because acid phosphate reacted with zinc oxide, an activator in rubber. Therefore, this led to the upper limit for the dosage of acid phosphate.

A compatibility test of fluids with DU metal revealed that fatty acid in some fluids corroded lead in DU bush, so this restricted the addition of fatty acid type oiliness agent.

FORMULATION STUDY

It is one of conclusions that the best friction modifier selection was a combination of organic acid phosphate and dialkyl phosphite. A new SAF containing these phosphorus compounds was developed. The SAF, Fluid A in Fig. 1, reduced friction in SA to the required level.

The SAF, Fluid B, contained zinc dithio-phosphate(Zn DTP) as an anti-wear and anti-oxidation agent, and the Fluid B with an additive combination of Zn DTP and friction modifiers shows high friction.

Submitted to *Industrial Lubrication and Tribology*

THE INVESTIGATION OF THE TRANSITIONAL SEIZURE WITH AN AMSLER TEST SET

ANDRÁS ZALAI
Expert in tribology, H-1085 Budapest, József krt. 51 2/1. Hungary

ABSTRACT

The objective of the investigation is to study a tribologycal system in seizure or seizure close condition. I have detemined the weight changes of the specimens, micro-hardness (HV_{micro}), roughness (R_t) of the worn surfaces.

I have carried out the investigation on test set Amsler A 135 using "small and big discs", at 0,7646 m/s sliding velocity, 500...1500 N load and 0,33...1,33 hour test time. The specimens were made from low strength steel, marked Fe 490. I have used technical grade white oil of low viscosity, highly refined with unfavorable lubrication features to achieve seizure in a short time. R_t was determined by a perthometer S6P, the hardness with a Hanemann micro-hardness tester.

The small disc in the case of surface seized up to 0...100 % resulted in significant loss of weight. The big disc in the case of seized up surface up to 0...75 % indicated gain of weight, that is a part of the wear particles deposited on the surface of the big disc. Less strict test conditions, smaller seized up surface is favorable for the deposit. In case of 75...100 % seized up surface the load is even for the big disc favorable regarding the separation of the wear particles (Fig. 1)

Roughness increases quickly on the seized up surface of the big disc under the R_t increasing effect of the deposit. At 35 % seizure reaches a maximum value and further on decreases until 100 % seizure. On the seized up surface of the small disc the values of R_t describe a curve of similar shape running at lower values. In the case of identical percent of seizure the difference of R_t between the two disc may be 35 μm.

Conclusions of the investigation are the followings:
- In the case of test with relatively small seizure ratio (20...55 %) seized up and not seized up surfaces develop circularly.
- Regarding seizure the change of weight of the disc and the R_t of friction surfaces are authoritative. Change of the micro-hardness of the sliding surfaces effect these factors, since for example the harder surfaces wear each other better and cause greater change in weight.
- Big and small discs behave differently since the load of the small disc, because of the smaller mass of the disc and the more frequent contact of the surface, is greater.
- The adherent wear particles increases R_t greatly and the separating one to a smaller extent.

- Adhesion of the wear particles on the surface may take place at smaller load.
- In the case of initial seizure - that is in the case of 0...30 % seized up surface - the micro-hardness of the discs is low (350...400 HV). At the same time in the case of extensive seizure (75...100 %) micro-hardness increases greatly an reaches 600...730 HV_{micro}. At same time, because of wearing effect, the R_t of both discs decreases significantly.
- On parts seized up to 35...75 % there is a transition between the initial and final sections. The adhesive quantity on the big disc decreases and turns towards decrease of weight. Roughness significantly increases at both discs and following maximum decreases significantly.
- The hardness at both discs first slightly, than greatly, at last to smaller extent increases. Adhesive parts have strong roughening effect.

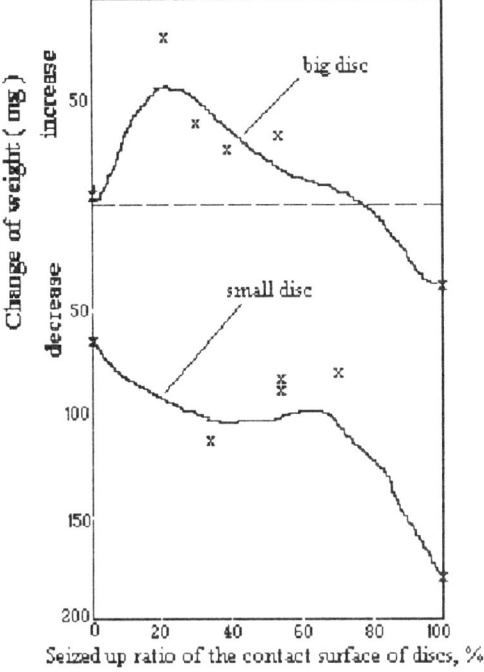

Fig. 1: Change of weight of the disc, mg

Acknowledgment: I express my appreciation for dr. Béla Palásty K. and dr. Gyula Kiss for R_t and the micro-hardness measurements.

AN INVESTIGATION INTO THE TRIBOLOGICAL AND ENVIRONMENTAL ASPECTS OF CUTTING PROCESSES

A KALDOS
School of Engineering and Technology Management,
Liverpool John Moores University, Byrom Street, Liverpool, L3 3AF, UK

ABSTRACT

In component parts manufacturing the dominant finishing manufacturing processes are those which fall into the group of material removal processes, i.e. metal cutting. This is particularly true in the area of discrete product manufacture of component parts in small batches, where significant developments have been experienced over recent years particularly in Computer Integrated Manufacture (CIM) and in the various forms of Flexible Manufacturing Systems.

Cutting processes fall into three distinct groups, processes using single point cutting tools, multi-point cutting tools and the multitude of randomly positioned cutting grits. Whichever the case, the major concerns in metal cutting are the productivity of these processes, the quality of parts, primarily the surface roughness and geometrical accuracy compounded with technical-economical considerations, like optimum tool life, optimum process parameters, material removal rate, specific removal rate and others.

All of these are primarily governed by the extremely complex and interdependent physical-chemical-mechanical, in other word tribological, phenomena in the contact zone of the cutting tool and material causing the tool to wear, the material to be removed from the surface of the blank part and thus generating the required surface geometrical configuration, accuracy and surface quality (1).

Friction between the tool and workpiece depends on a multitude of factors, like process parameters, cutting tool geometry and tool material, surface coating properties, acting forces, pressure between tool and work interface, heat generated during the process, temperature of contact zone and the cutting fluid applied. There are extreme conditions in the contact zone due to the elevated temperature and high pressure which, when cutting steel, may reach the values of over 1,000 °C and 200 MPa respectively.

These tribological phenomena depend on not only the chemical composition and mechanical properties of the material but also depend upon the chemical-mechanical properties of the tool material. In addition, the type and quality of cutting fluid applied are also critical for the cutting operations.

Material removal in the form of swarf requires considerable energy input. The efficiency of energy utilisation can be measured by the chip compression ratio, the specific cutting energy and specific cutting force. With the continuous improvement of both workpiece materials and tool surfaces provided by surface engineering, the rate of energy input has increased dramatically giving rise to environmental considerations (2). In the United Kingdom the machine tool population is around 350,000 units consuming a vast amount of energy generated by power stations causing environmental problems both regionally and in a global scale. Almost 100% of the energy input in metal cutting is converted into heat and dissipated into the environment. The reduction of energy consumption in all industries, including metal cutting industry, is of vital importance.

Compounded to this problem is the extensive use of cutting fluids aiming at reducing the friction in the contact zone and removing the heat generated during the cutting process. At the same time cutting fluids not only represent additional environment problem but generate separation, cleaning, treatment and health hazard problems (2). To address these issues means additional costs for the manufacturing industry.

Attempts have been made to eliminate the use of cutting fluids with limited success, but recent results indicate that dry cutting may result in substantial cost savings. Further to this, the principles of minimum cutting fluid application have emerged (3). Theoretical analysis and practical tests show enormous savings in cutting fluid, cleaning and treatment and separation equipment and labour (3). The application of biodegradable, multi-layer type cutting fluids has been tested with excellent results.

The potentials of tribology in the still unexplored field of cutting processes give rise to further savings in the manufacturing and energy generating industry to meet environmental, health and safety requirements.

REFERENCE

(1) H Czichos, The role of tribology as science and technology - what are the essentials? Tribology International, Vol. 28. No.1. 1 Feb. 1995 pp.15-16.

(2) Anon., Monitoring the performance of metalworking fluids, Machinery and production engineering, 19 July 1996 pp. 24-29

(3) H Popke, Th Emmer, Bohren ohne Emulsion - geht das? Machinenmarkt, April 1997

(4) J Schmidt, R Wassmer, Minimalmengen-Kuhlschmiersysteme - eine interessante Alternative in der Kuhlschmiertechnik, Technische Rundschau, Jg.86 (1994) No.5.

Submitted to *ASME Journal of Tribology*

TRIBOECOLOGY AND TRANSPORT

MARA KANDEVA

Space Research Institute, Bulgarian Academy of Science - 1000 Sofia, 6, Moskovska St.
Bulgaria

ABSTRACT

The deep essence of the ecological crisis lies in the bipolar thinking and behaviour of mankind, in the excluding and destroying of contacts as functional elements of the ecological wholeness "society - environment" and its structures. At the end of the 20. century, the spirit and the revision role of science are determined by the new interdisciplinary sciences.

Triboecology is a modern scientific trend, which is formed on the boundary of the inter-disciplinary sciences tribology and ecology. It can be assessed as an interdisciplinary field embracing the nature, relationships and phenomena of contact formations as autonomous functional elements of the ecological whole and its structures.

The papers deals with some methodological problems of triboecology: methods, principles and values, based on the modern paradigm of tribology concerning the triune character of the functional wholeness (1) (2).

The concrete subject of triboecological analysis in the paper is the system "transport-contact-environment". Contact is being regarded at different structural levels as an element with specific mechanisms of generation, destruction and emission of material, energetic and informational flows in the environment. The influence of contact friction, wear and conductance upon ecological-economic factors of the environment and of the transport means as a complex tribosystem is being studied The basic mechanisms of pollution are the processes of 'combustion" and "friction" and their contact. Friction causes energetic contamination manifested in intensive energy dissipation in the form of heat, noise, vibrations. Cars emit in the environment about 60% of the consumed energy through combustion and friction in the multitudinous contacts. About 30% of the amount of CO_2 in atmosphere has been emitted by the automobiles, which affects the warming of the climate through the "hothouse effect". As a general mechanism of degradation, wear in contacts has a limiting function on the complex parameters of the car: efficiency, power, fuel and lubricant consumption. This change results from the complex contact realisation between friction, wear, lubrication, contact conductance and combustion mechanism and is interpreted as informational contamination of the system 'transport - contact - environment". Informational contamination is of contact nature in regard to the energetic and material contamination and is in strong correlation with the reliability of the system and its structures.

A concrete example is given revealing the possibilities of tribology in the improvement of ecological-economic and functional parameters of automobile, namely those achieved with the use of the unique equipment "TREFA- HART" of the Tribology Centre at the Technical University in Sofia enabling identification, control and regeneration of air automobile filters. The technology lead to improvement of filter resource with 250% and diminishing of fuel consumption and gas emissions to 15%. The project is implemented in the Tribology Centre - Sofia as a service equipment satisfying the needs of local and international transport in the country.

REFERENCES

1. N. Manolov, THE CONTACT, Society of the Bulgarian Tribologists, Sofia, 1995 (in Bulgarian).

2. N. Manolov, V. Diamandiev, THE MODELS IN TRIBOLOGY. Technical University - Sofia, 1995 (in Bulgarian)

3. M. Kandeva, N. Manolov, Device for regeneration of air automobile filters, Patent No. 51388/1993 - Sofia.

4. M. Kandeva, Pl. Markov, Tribotechnology for regeneration of air automobile filters, Tribology, quality, reliability, Bulletin of the Co-ordination Centre of Tribology, Sofia, Vol. 1, 1992 (in Bulgarian).

5. M. Kandeva, Triboecological aspects of automobile filters. Balkan Conference "CONTACT'95", Sofia, Vol. 1, p. 103, Edition of the Society of the Bulgarian Tribologists.

Authors' Index

A

Abd-Rabou M 204
Abe N 64
Abraham G 691, 711, 782
Abramov I L 479
Abramov L M 479, 869
Abrashin V N 765
Adachi K 517, 675
Adams G 46
Adams M J . 138, 189; 289, 372
Adams M L 424
Addison P S 594
Aderika, W 325
Agaphonov N I 835
Åhlin A 229
Ahmed R 156
Ahn B G 120
Ahn H J 544
Ahn H S 155, 607
Ahuja B B 387
Ai X 267
Aihara S 808
Ajaots M 388
Ajiki K 589
Akagaki T 317
Akazawa M 676
Akochi-Koblé 597
Alahelisten A 145
Albertin E 309
Alcock J R 19, 150
Alekhin V P 715
Aleksandrovic S 626
Alexandrov V M 373
Allen C 307
Alliston-Greiner A 294
Almakhlafy A 13
Ameye J 328
Amini N 285
Ampuero J 795
Anamalay R V 259, 378, 692
Anderberg G 145
Andersson P ... 60, 282, 495, 57
Andersson S 53, 531
Andharia P I 389
Ando K 693
Ando Y 361
Andras I 613
Andrei G 390
Andrés L S 408, 441
Angra S 391
Antonescu N N 870, 871
Antoniadis A 601
Arai H 887

Arnšek A 29
Arsenic Z 203
Arsic M 618
Artemczuk P 691, 782
Aruaz G 356
Arustamian Y S 534
Askovic P 762
Aslanyan I R 720
Assenova E 490, 773, 875
Astakhin A S 479, 869
Asuke T 347
Ateshian G A 250
Athre K 9, 405, 815
Attia F 310
Avduevskii V S 646
Axén N 229, 299

B

Babel V D 646
Babic M 470, 480, 617
Badger J A 753
Bai D W 186
Baiyang Z 484
Balaceanu M 499
Balic J 602
Ball P G 843
Balla J 519
Balliu S 459
Balon T 6665
Balulescu M 443
Banks J P 495
Barbour P S M ...731, 732, 736
Barnes C G 707
Barnsby R M 263
Barrett L E 412
Barsukov V G 598
Bartel D 15
Barton D C 246, 247
Barykin N P 720
Bas J A 87
Bashkirov O M 851
Bassani R 593, 694
Basu S K 387, 476
Batchelor W 384, 695
Baudrocco F 877
Baugh E 586
Bay N 619
Bayer R G 67
Bayliss R W 294, 832
Beard J 206, 518
Bec S 367
Becker A A 48, 777
Bednarczyk S 810

Belin M 158
Belin M 73
Bell J C 196
Belosevac R 513
Bengtsson A 700
Benham T J 520, 696
Bennani H H 481, 829
Benyagoub A 829
Bercsey T 571, 872
Berg J M 775
Berger L M 151
Bergström J 332, 783
Berhe A 223
Bermingham E 599
Berrien L S 257
Bertalan M 499
Berthier Y 128, 171, 865
Besong A A 243, 732,
.... 740, 748
Beynon J H 380, 474
Bezjazichny V F 600
Bezyazichny V F 573, 697
Bhave S K 444, 844
Bhushan B 162, 583, 587,
.... 591
Bielinski D M 319
Bigsby R J A 732
Bilik Y 716
Bingley M S 670
Birks A H 666
Biswas S 9, 189, 405, 444, 844
Biswas S K 363, 444, 815,
.... 566, 774, 788, 844
Bjerrum N J 123
Björk T 332
Björkman H 145
Blanchet T A 321
Blaskovic P 482, 483
Blatcher S 733
Blawert C 66
Blecic Z 553
Blencoe K A 187
Bloyce A 56, 494
Bobji M S 774
Boehm M 823
Bogdan N 620, 898
Bogdanovich A V 568, 809
Bogdanovich P N 320
Bogy D B 18, 168, 406
Bohling M 71
Bolarin A 877
Bolyrev Y 392
Bonfield W 733

Borahni G H 129
Bordi-Boussouar V 268
Börner H 355
Boschkova K 775
Boshui C 448
Bouali B 240, 861
Bouzakis K D 157, 273, 601
Boving H 795
Bradley M S A 308, 666
Bragdon C R 248, 249, 734, 735, 742
Bratitsa L S 701
Braun M J 198
Brendle M 826
Brennan N 124
Brenner W 782
Briggs G A D 363
Briscoe B J 189, 289, 324, 326, 372, 833
Brockwell K R 6
Broekhof N L J M 331
Bronovets M A 325, 373, 512, 646
Brookes M R 381
Brown C A 161
Brown E D 817
Brown J C 840
Brown R D 594
Bryant M D 130
Budinski K G 290
Bugliosi S 221
Bujurke N M 393
Bull S J 239
Bulsara V H 211
Bultman J E 76
Bulytchev S I 715
Burcan J 445
Burke B 742
Burnett A J 223, 308, 667, 670
Burnham N A 366
Burya A A 834
Burya A I 831, 834, 835
Bushe N A 521
Buyanovsky I A 653
Byeli A V 70, 489
Byrne W P 301

C

Calabrese S J 358
Camino D 55, 60, 495
Campbell P 258
Canic Z 628
Cao L 471
Castillo J S 186
Catrinoiu D 459
Çavdar B 446, 873
Celichowski G 447
Celis J P 61
Cermak J 647
Ceschini L 522
Cha K H 802
Chandrasekar S ..211, 338, 384, 616
Chandrasekaran H 335
Chandrasekaran M, ... 695
Chang P T 18
Chateauminois A 324
Chattopadhyay A K 394
Chaturvedi K K 815
Chaudhari U M 844
Chekanov A S 165, 170
Chekina O G 352
Chen G 487
Chen H 191
Chen J 746
Chen L 457, 458, 487
Chen P Y P 402
Chen X Y 796
Chen Y H 176, 824
Cheng H S 188, 192, 816
Cheng K 633
Chenghui G 484
Chenwen C 661
Chichinadze A V 352, 523, 549, 817
Chigarev A V 395
Chigvintseva O 835
Chikate P P 387
Chittenden R J 265
Cho K H 141
Cho U 524
Choi D H 120, 590
Choi T 597
Cholakov G S 449, 469
Choudhary M 244, 836
Choy F K 198
Christensen E 123
Chun Y J 120
Chunde J 525, 729
Cicone T 10, 396
Cioarec V 798, 806
Ciulli E 694
Claesson P M 775
Clair D 865
Clark D 313
Clark H M 213, 214
Coates D A 228
Cole J 874, 876
Collier J P 245
Connal J 845
Consiglio R 58
Cook J E 256
Corbett J 19
Coto H J M 269, 485
Couhier F 266
Cowan R S 370
Crietinin O V 698
Crooks C S 197
Crudu I 846, 847
Crysler D 287
Cuervo D G 269
Cunningham J M 81
Currier B H 245
Currier J H 245
Cus F 602
Cvartalov A R 698

D

D'Errico G E 221
Dabrowski L 397
Daehn 522
Dahm K L 68
Dai P 579
Damanchuk D 287
Danchev D 875
Dannev K 490, 773
Danos J C 10
Danyluk S 339
Darbeida H 665
de Baets P 83, 288
de Silveira G 291
de Vaal P L 284
de Vries E G 331
de With G 212
Dearnley P A 68
Debaud S 473, 828
DeCamillo S 6, 398
Deguchi Y 492
Deheri G M 389
Delfosse D 795
Dellacourte C 17
Demkin N B 526
Denic D 618
Deters L 15, 528
Devenski P 449
Dey S 584
Diaconescu E N 23, 648, 797
Diamandiev V 636
Diao D F 486
Dickinson J T 699, 825
Dimofte F 198
Dinc S 358
Dinelli F 363
Dinoiu V 450
Dintu S I 529
Diss P 826
Dizdar S 700
Djurdjanovic M 621
Dmitriev V A 856
Dmochowski W 6
Dmytrychenko N F 701
Dodd A 264
Dohda K 603
Dong H 56

Dong J .... 457, 458, 487, 597, ... 760
Donnelly S G 32
Donnet C 158, 790
Đorđevic O 553
Dorier C 268
Dowson D 82, 179
Doyle E D 138, 336
Dragan D 604
Dragoslav J 766
Drees D 61
Drinkwater B W 43
Drozdov Y N 357, 529, 837
Druet K 548, 708
Drumeanu A C 870, 871
Drummond C 594
Du D 760
Duboka C 203
Duda J 245
Dudragne G 266
Dumont M L 266
Dupas E 366
Dürkopp K 205
Dursapt M 792
Dvorak S D 364
Dwivedi V S 673
Dwyer-Joyce R S 43

E

Ecobar E 527
Effner U 511
Efimchik A A 489
Efstathiou K 157
Egami M 318
Egan J 744
Eguchi M 702
Ejima Y 722
El-Badry S A 530
Eleöd A 44
Elgeddawy M E 713, 888
Elliott D 313
Elsharkawy A A 180
El-Sherbiny M G 204
Engel P A 67, 329
Engqvist H 299
Enomoto Y 347
Erdemir A 488
Erickson L 146
Etsion I 193, 194
Evans H P 645
Ezugwu E O 630

F

Farnish R J 667
Farris T N 211, 338, 616
Faultersack W 699
Fedotov S A 489
Fenske G R 488

Ferguson J H 4
Fernandez M C 485
Filipowicz K 683
Fillon M 8
Fischer A 71
Fish G 874, 876
Fishbein E I 794
Fisher J ...... 243, 246, 247, 253, ...313, 731, 732, 736, 738, 740, ... 741, 744, 748
FitzPatrick D P 246
Flack R D 412
Flamand L 266
Flasker J 799
Fleischer G 15, 528
Fletcher D I 474
Flöck J 50
Flöck J 841
Flodin A 531
Florea M 459
Florescu D 450
Forder A F 304
Forehand S M 162
Forsyth M 224
Forte P 694
Fouvry S 532
Fox V C 507
Franek F 154, 691, 711, 782
Franke R P 252
Frene J 8, 11
Friedrich K 50, 841
Frunza G 797
Ftrantskevitch A V 493
Fujii K 359, 864
Fujii M 663
Fujii Y 359
Fukuda T 382
Fukui S 776
Furey M J 231, 257

G

Gadala M 45
Gafitanu M D 798, 800, 806
Gåhlin R 297
Gahr K-H Z 232
Gál P 154
Gala A 737
Galetuse S 399
Gallagher O J 138
Galun R 65
Gangopadhyay A 882
Gao C 166, 579
Gao X 629
Garagnani G L 522
Garbar I I 298
Gardner W W 3
Garkunov D N 646
Gawne D T 62

Gee M G 293, 301, 381, 703
Geiger M 329
Gekker F R 533
Gelder A 228
Gerve A 286, 849, 852
Gethin D T 13
Gevorkian G R 534
Gevorkian V R 534
Ghosh M K 657
Gibson-Carqueijeiro R 634
Gick L A 605
Gidikova N 490
Gill S 353
Gillis D 585
Gimondo P 877
Gläser H 400
Glodez S 799
Glovnea M L 648
Godbole P B 844
Godlevski V A 451
Goepfert O 795
Goldade V A 745
Gologlavic-Kolb M 606
Goncharov A V 851
Goodchild J 380
Gorkunov E S 535
Goura G S 536
Gourdon D 366
Goyan R 345
Gradt T 355
Graham I T 197
Gredic T 505
Green I 12, 416, 759
Green T 738
Gregory E M 257
Grekova A V 395
Gremaud G 366
Grib V 878
Gribova I 325
Grigore E 499
Grigorescu I C 680
Grigoriev A Y 607
Grigoriev B S 401
Grill A 158
Grinberg N A 483
Gromakovsky D G 537
Grossiord C 452
Grundy P J 175
Grzesik W 608
Guang M 595
Guangteng G 649
Gunsel S 21
Gupta A 630
Gursky E 749
Gutleber J 219

H

Haar R 331
Habdank-Wojewodzki 879
Hachiya K............................. 33
Hack R S 777
Hadfield M 156, 346
Hadley J W 635
Hagiu G D 798, 800
Hagman L............................ 41
Hahn E J 402, 595
Hailey J L 243, 748
Hainsworth S V 239
Haller R 217
Halter M 488
Hampshire J 494, 507
Han D-C 407
Han S W 321
Han S-S 625
Hao S 609
Harada M 5
Hargreaves D J 403, 404
Harhara H N 469
Harkins G 169
Harp S R 349
Harris D W 742
Harris S J 539
Harris W H 248, 249
Harris, T A 263
Harrison D 304
Harrison J A 778
Hartl M 650
Hartwich R B 213
Haruyama Y 491
Hasegawa S 354
Hashimoto H 755
Hata H 863
Hatazawa T 801
Havermans D 495
Hawthorne H M 146, 306
Hayes J A 40
Heidenfelder F 142
Hendel M 761
Henqio L 629
Henshall J L 599, 703
Herdan J M 443, 453, 460
Herrera-Fierro P 35, 37
Herza C F F 782
Higaki H 251, 254, 739
Higashiya T 580
Hill A J 224
Hirani H 405
Hirano M 375
Hirashima Y 679
Hiratsuka K 779
Hiroaka N 74
Hironaka S 137
Hisakado T 235
Hogmark S 145, 202, 229,
.... 299, 332, 509
Hokkirigawa K 234, 382

Holinski R 78, 142
Hollman P 145
Hollstein T 238, 689
Hollway F 138
Holmes A J 31
Holt C A 645
Homola A 579
Honda T 132
Hong H S 121, 677
Hoogendoorn R 28
Hooper R M 256, 599
Horak P 872
Horiguchi T 361
Horsfield A P 169
Hosseini S M 323
Howard R L 218
Hsu G 538
Hsu S M 47
Hu Q 539
Hu Y 18, 168, 406
Hua D Y 192
Hubner W 355
Hughes H L 257
Huirong M 661
Hutchings I M 244, 270, 291,
.... 337, 890
Huting X 384, 695
Hwang D H 131
Hwang P 407

I
Iampolskii M 615
Ichimaru K 272, 276
Ichimura T 431
Ide T 743
Ike H 379
Ikeda H 164
Ikeuchi K 255
Iliasov S V 851
Iliev H 540
Imamura Y 679
Ingham E 243, 736 738, 740,
.... 741, 744, 748
Ino N 75
Inoue T 172
Ioannides E 271, 275
Ionescu M P 846, 847
Iontchev H A 449, 469
Isami S 895
Ishak M 551
Ishchuk Y L 712
Ishibashi A 880, 895
Ishiguro S 881
Ishii M 580
Ismail A A 597
Itayama M 359
Ito H 27
Ito K 864
Itoh K 359

Ives D 227, 371, 635
Ivkovic B 610
Iwabuchi A 827
Iwai T 881
Iwai Y 132
Iwasaki Y 132
Iyer, S S 45
Izmailov V V 401, 541
Izumi N 272

J
Jablonski J 542
Jackson C 408
Jacobson S 127, 202, 297
Jacquemard P 128
Jadi A S M 303
Jahanmir S 227
Jain S C 426
Jaiswal B L 844
Jang J Y 818
Jang S 25
Jankovic M 409
Jardret V 240, 292
Jascanu M 390
Jasty M 248, 249, 734
Jawurek H H 305
Jayaram V 566
Jefferies A 328
Jennett N M 60, 495
Jensen R K 882
Jeremic B 611, 617
Jevtic V 766
Jha S 673
Jiaju M A 651
Jiang J 55, 135
Jiang Q 465
Jiang S 487, 760
Jibiki T 854
Jilbert D 704
Jimbo T 137
Jin Z M 253
Jisheng E 62
Jiujun X 383
Johansson J O 335
Johnson B 848
Johnson M D 882
Johnson S A 189
Jones A H S 55
Jones G J 207
Jones Jr W R 34, 35, 37
Jones M H 612
Jordan W 205, 543, 555, 678
Josifovic D 883
Jovanovic ... 618, 622, 627, 855
Jovic N 705
Ju Y 338
Jula D 613
Junghans R 322

Junxiu D 448

K

Kagami J 801
Kahlman L.......................... 270
Kajdas C 231
Kakas D..................... 149, 505
Kaldos A 901
Kalitchin Z 747
Kalsi M S 351
Kamei H 751
Kamio N 36
Kandeva M......................... 902
Kaneko R........................... 776
Kaneko S 751
Kaneta M................... 756, 757
Kang T S 590
Kaniewski W.............. 410, 706
Kanno T 454
Kapsa P 42, 133, 532, 577
Karmakar S......................... 394
Kashiwagi K 679
Kato K 139, 148, 317, 327,
.... 517, 556, 675, 693
Kato T 39, 382, 655
Kaur R G 228
Kawabata M 317, 881
Kawaguchi T 801
Kawahara H 751
Kawakubo Y...................... 580
Kawamura M 862
Kawamura S 491
Kawazoe T 278
Kehrwald B 849
Kelly D A 707
Kelly K.............................. 714
Kempinski R...................... 231
Kenmoch N 74
Kennedy A R..................... 136
Kennedy F E....... 161, 201, 245
Keränen T.......................... 411
Khan R U 752
Kharlamov V V .. 535, 652, 831
Khonsari M M 818
Kikuchi M 354, 439
Kim C........................ 544, 825
Kim D E 131, 374, 802
Kim S S 141, 544
Kim S Y 506
Kim T H 141
Kimijima T........................ 472
Kimura Y................... 491, 812
King M J........................... 736
Kirk T B 259, 378, 559, 692
Kiss G 154
Kitahara T 722
Kitaoka S........................... 679
Klaffke D................... 147, 687

Klamecki B E.....................609
Klein H 57
Klimenko S A 334, 885
Klochikhin V I............837, 884
Knezevic J..........................703
Knight S.............................425
Knox R T 208
Knuuttila J 237
Ko M W.............................506
Ko P L.................................45
Kobayashi A 492
Kobayashi Y 581
Koda Y..............................454
Koehler N 501
Koizumi H 27
Kojaev A Yu......................352
Kolesnikov V I..................838
Kolmogorov V L................652
Kolodny J..........................161
Kolodziej E 555
Kolosov O...................170, 363
Komarov F F.....................493
Kompogiannis S................601
Komvopoulos K..........167, 365
Konchits V V 637
Kopeikina M Y 334
Korcek S 21, 882, 894
Koreshkov V N..................723
Kornegay E T....................257
Korochkina T V.................764
Koronka F 613
Kostrzewsky G J 412
Kothari D C 372
Koudriavtsev S A...............698
Kovac P 614
Kovalenko E V 373
Kovalev V A......................607
Kovtun V A.......................498
Koyama M.........................863
Kozekin A V 653
Kral E R 167
Krasnov A..................325, 837
Kreis T 252
Krettek O 204
Krupka I............................650
Kryvenko I I......................546
Kubiak T 400
Kubo S..............................139
Kudryakov G P 334
Kukareko V A.....................70
Kulik A J...........................366
Kumar V 426
Kunio G 455
Kunosic A 513
Kupriyanov N A................781
Kuranov V G 545
Kurita M 7
Kuster M 259
Kuzharov A 563

Kuzšmierz L 413
Kwon O K 120
Kwon S I 407

L

Lacarac V 763
Lahonin A N..................... 698
Lai J T................................ 32
Lam R C 873
Landolt D 473, 828
Langford S C 699, 825
Lanteri P........................... 861
Larsen-Basse J................... 302
Larsson B 190
Larsson M......................... 509
Latyshev V N 451
Lavelle B 680
Lawen J R......................... 358
Lawrence C J 372
Lazarev V A 819
Le Mogne T.............. 158, 452,
.... 790, 823
Leahy J G 48
Lee A J C 256
Lee S C............... 52, 181, 660
Lee S H............................. 506
Lee S J.............................. 131
Lee S K............................. 155
Lee Y J 629
Leefe S E 456
Leinonen T 411
Lejeau D C 216
Lenss V G......................... 350
Leonard A J 547
Levert J............................. 339
Levitt J A.................. 350, 538
Lewis S D 803
Li E........................... 432, 725
Li L K Y 780
Li S Z 868
Li X 804, 868
Li Y 383
Liao X 668
Libera S............................ 866
Liew W Y 337, 574
Lilley E............................. 681
Lin H S 14
Lin T Y.............................. 37
Lin W-H............................. 34
Lindley T C 324
Lindsay Smith J R 31
Lipsky L A 765
Lira J 680
Liska M 650
Litwicki W........................ 850
Liu J........................... 143, 241
Liu W 140
Liu Y R............................. 241
Livshits A I....................... 362

907

Llewellyn R J 306
Lobodaeva O V 70
Logan P M 150
Loginov V T 851
Loginova N V 851
Longeray R 861
Lopatin A S 512
Loubet J L 42, 240, 292
Loveday M 381
Low T S 170
Lowenstein J D ... 249, 735, 742
Lozovskii V 615
Lubinski J I 548, 708
Lubinski T 548, 708, 709
Luca P 459
Lucas B N 51
Ludema K C 186, 350, 538
Luo J ... 477, 564, 654, 780, 858
Lushnov J M 549, 856
Lyashko V A 550, 885

M

Ma J J 796
Ma K J 494
Ma M T 414
Mach W 852
Mackenzie M R 725
Macuzic I 611
Madhavan V 616
Magnee A 669
Mahler F 287
Mainland M 759
Mainsah E 793
Majumdar B C 433
Makino T 183, 431
Maksak V I 781
Maksimov I L 819
Maleque M A 551
Mallabiabarrena I 795
Malyarov A N 537
Man Y 165
Manolov N 636
Mäntylä T 151, 237
Mao K 56
March C N 265
Marchenko A V 569
Marchione M 436
Marczak R 563
Mardel J I 224
Marinkovic A 409, 855
Markov D P 710
Markovic S 883
Marsault 14
Martin B 133
Martin F A 415
Martin J K 184
Martin J M .. 375, 452, 790, 823
Martini C 522

Marx S 322
Marzec S 623
Masjuki H H 551
Masuda S R 34
Masuko M 36
Mathews J B 740
Mathey Y 57
Mathia T G 42, 292
Mathia T 240, 792, 861
Mathur N C 121
Matsuda K 275, 756, 757
Matsuda R, ... 776
Matsui A 682
Matsui S 757
Matsumoto K 279, 862
Matsumura M 558
Matsunaga S 886
Matsunuma S 582
Matsuoka H 655
Matsuoka K 174
Matsuoka T 22
Matzner A 691, 782
Maurin L N 451
Maury J 128
Mawatari T 159
Mazibrada L 149
Mc Geehan J A 99
McCartney D G 300
McColl I R 539
McCollum C 882
McDevitt N T 77
McMillan T C 163
McMillan T 586
McNie C M 247
Medley J B 4
Megat Ahmed M M 13
Megido J M C 485
Mehenney D S 16
Mehta N P 391
Melley R 345
Mellor B G 216
Meneve J 60, 495
Merrill E 248, 734
Meshkov V V 764
Mess F 339
Meurisse M H 128
Meurk A 783
Michalski J 552
Midol A 861
Mihaly K 475
Mihora D J 209, 311
Mijajlovic R 618, 855
Mikolasch G 711
Milic N 611, 617
Milinovic M 762
Miller B 416
Mills B 635
Milosavijevic A 553, 621
Milovan M V 604

Mimuro N 272
Minagawa T 454
Minakawa H 741
Minami I 121
Minami M 462
Minevich A A 605
Minkov D V 851
Mirci L E 460
Mischler S 473, 828
Mishchuk O A 712
Mishev G 853
Mistry K N 9
Mit V 325
Mitamura K 881
Mitrovic R 628
Mitsui H 24
Mitsuya Y 886
Miura K 854
Miyake S 784, 854
Miyauchi T 887
Miyazaki T 754
Miyoshi H 274
Miyoshi K 144
Mizuhara K 506
Mizumoto M 863
Mizuta Y 347
Mlotkowski A 400
Mnatsakanov R G 701
Mohamed M A A 713, 888
Mohammad H A 880
Moisheev A A 512
Mokhtar M O A 713, 888
Monmousseau P 8
Moore D F 714
Mordike B L 65, 66, 496
Moreau M 128
Morgan J 124
Mori A 185, 767
Morita T 272, 276
Moriyama S 251
Morohashi A 49
Moss W O 208
Moustafa S F 530, 554
Mruk A 543, 555, 678
Mudreac V I 529
Mukovoz Y A 334
Müller U 795
Munteanu V 846
Murakami K 756
Murakami T 251, 254, 739, 746, 867
Muratoglu O K 734
Murdzia E 417
Murphy M 597
Myasnikova N A 838
Myshkin N 637

N

Nabhan M 461
Nae I 870, 871
Nagahashi K 558
Nagaike S 580
Nagarajan V S 227
Nagasawa H 556
Nagatomi E 31
Nair K P 656
Naka M 27, 33
Nakai H 755
Nakajima A 159
Nakamura T 619
Nakanishi Y 251, 254, 739
Nakashima A 278
Nakashima H 462, 864
Nakatsuji T 767
Nakayama J I 887
Nakayama K 164
Narula M L 673
Natsumeda S 814
Navarro T R 269
Nazarenko P V 701
Nebelung M 151
Nedeljkovic L 846
Neder Z 50, 571, 841
Nelias D 266, 268
Nerz F 186
Nguyen S 164
Nicolson D M 295
Niikura T 702
Nikas G 271
Nishikawa H 756
Nishimura M 64, 862
Nita I 49
Njiwa R K 58
Nodomi Y 74
Noel B 865
Nogi T 39
Northrop I T 307
Nosaka M 354
Novitskii V G 557

O

O'Connor D O... 248, 249, 734,
.... 735, 742
O'Flynn D J 670
Obara S 279
Ochonski W 889
Ödfalk M 202
Ogata M 185
Ognjanovic V 621
Ohashi M 255
Ohnogi H 558
Ohtsuki N 251
Ohue Y 813
Ohyama T 191
Oike M 354
Oka Y I 558

Okabe T 234
Okada K 134, 743
Oleksiak Z 839
Oliver W C 51
Olufsson U L F 41
Olver A V 275
Ong W 345
Onyshkevych V 822
Oomen I 212
Orsini A 866
Otake T 784
Otani C 49
Oulevey F 366
Ouyang Q 134
Ovaert T 753
Owellen M C 257
Oxenford J 306
Oyang Q 743
Oye H A 668
Ozaki K 327
Ozawa Y 431
Ozimina D 463

P

Paciorec J K L 34
Padgurskas J 768
Page T F 239
Pahud P 795
Pai N P 393
Pailharey D 57
Palaghian L 785
Palermo T H 452
Palusan A 453
Pan C H T 418, 419
Pan G 508
Pandey R K 657
Panic S 480
Panidi I S 653
Panov V 875
Panteleenko F I 497
Panzera D 259, 378, 692
Park K C 625
Park N S 825
Parker D D 197
Parkins D W 184
Pascovici M D 396, 818
Pascucci M R 681
Patel V 158
Patton S T 583
Pauschitz A 154, 711
Pavlishko S V 652
Pavy J C 824
Pawlus P 552
Pedzinski 641
Pelillo E 326
Peña D 222
Penafiel J 87
Peng J Y 412, 432, 559

Perry M D 778
Person V 11
Petkova D D 560
Petrakovsky S S 549, 856
Pezdirnik J 786, 820
Pfeiffer W 238, 689
Pfestorf M 329
Phan T 336
Phelts E R 176
Pichugin V F 646
Pickard J 744
Pickering J 666
Pickles M J 890
Piekoszewski W 464, 658
Pilko V V 493
Pinchuck L S 597, 745
Pindel M 653
Pinel S I 277
Pittman A N 308, 666
Plaza S 447
Pleskachevsky Y M ... 498, 575,
.... 637
Podgornik B 820
Podsiadlo P 377
Pogosian A K 638
Pohl H 316
Pokropivny A 787
Polak A 75
Polak S 283
Polyakov N V 523
Polycarpou A A 193, 194
Ponsonnet L 158
Popescu A A 499
Popovic M B 606
Popovic N 513
Popovic P V 622, 627
Poroshin V 715
Posmyk A 500, 503
Potemkin M M 550, 885
Prabhu B S 422
Pramila Bai B N 220
Prasad S V 77
Prashad H 421, 805
Prat P 26
Premnath V 249, 734
Priest M 82
Procopiev V N 769
Prokic-Cvetkovic R 553
Prushak V J 598
Przemeck K 232
Puget Y 369
Pugsley V 307
Pytko S 623

Q

Qijun L 808
Qiu L 816
Qu J X 508

Quinn T F J 561
Qunguo T 728

R

Racocea C 798, 800, 806
Radakovic Z 553
Radcliffe C D 704
Rainforth W M 547, 562
Rajan B S 624
Rajendrakumar P K 788
Rakic R................................. 807
Rakic Z................................. 807
Ramadan M 562
Ramalingam S 153609
Ramamurthy A C 51, 209,
.... 311,890
Ramamurti B 742
Randall N 789
Rapoport L 716
Ratoi M 28
Rattan S S............................ 391
Rauch J Y 481
Rautaray S K 892
Reddy D S K 422
Reed A R............................. 667
Reeves E A 246
Rehbein P 717
Reischke G 423
Ren Z 799
Renaux P 171
Repenning D....................... 496
Reungoat D......................... 10
Riches A M 233
Rico F E 269
Rico J E F............................ 485
Riera L C 447
Rigaut B 824
Risdon T J 78
Ristivojevic M 763
Roach C J 19
Roebuck B 301
Rojdestvenskiy J V 769
Romanova A T 856
Rombach M......................... 238
Roper G W 187
Rorrer R A L 840
Rosén B G 285, 718
Rosic B................................ 763
Ross D 146
Rousselot C 481
Rowe W B 371, 633
Roy S K............................... 815
Roylance B J 13, 341, 624
Rozeanu L 201
Rudd G I............................. 752
Rudd L M 707
Rudolphi A K 127
Rukuiza R........................... 768

Rupinski D..........................434
Rupp M 59
Ryason P R 199
Ryniewicz A 719, 857, 879
Ryniewicz W......................737

S

Sadykov F A 720
Safari S 346
Sait M F...............................467
Saito E 859
Saito K 234
Salant R F 349, 758
Salant R................................339
Salonen P 891
Samajdar I...........................220
Samarin Y P 537
Sampath S 219
Sandu I F.............................846
Santner E...................501, 641
Saravanan A 220
Sasaki A..........30, 74, 466, 867
Sasaki S...................... 79, 502
Sato A 589
Sato E 863
Sato J 808
Sato T 148
Satyanarayan K R 476
Saunders S R J 60, 495
Sawae Y 251
Sawicki J T 424
Sayles R S 271, 295
Saynatjoki M 572
Scheers J 333
Schipper D J 175, 331
Schmoeckel D 59, 330
Schneider T 355
Schomburg U 316
Schriewer K 882
Schroeder M O..................257
Schwalbe H J 252
Scieszka S F.............303, 683
Scott W 404, 425, 478
Scudiero L...........................825
Seitzman L E......................63
Sekhar A S 433
Sekine M............................235
Sekulic 791
Semyonov A N..................573
Senatorski J K 563
Sep J 503
Shabalinskaia L..................671
Shaginyan A S 639,721
Shan C................................661
Shao H S 508
Shaokui Y 728
Sharif K J H 645
Sharma C S 859, 892

Sharma S C......................... 426
Sheer T J 305
Sheipak A 750
Sheitanov I 747
Sheldon B M 161
Shelikhov G....................... 615
Shelledy F 584
Shen G............................... 471
Sheng X............. 477, 564, 858
Shi F 659, 660
Shi J Z 504
Shima M............................ 808
Shimizu S 827, 859
Shimura H 502
Shinada T 125
Shinooka M 382
Shiota T............................. 272
Shiozawa K 491
Shipitsa N A 730, 794
Shipway P H............. 136, 300
Shirai K 581
Shliapin A D..................... 715
Shluger A L 362
Shoda Y.............................. 33
Shogrin B A............... 35, 37
Shou H Z 504
Shtertser A A 565
Shuster M 287
Signer H R....................... 277
Simeonova Y 490
Simmons J E L 208
Simner D 893
Sin V W T....................... 574
Sinahasan R 426
Sinatora A 309
Singer I L.............. 63, 73, 790
Singh R A 566
Singh T........................... 467
Sinha S K.... 289, 326, 517, 833
Sinhasan R 426
Sivrikova S R 769
Skerka............................. 212
Skoneczny 500
Skopp A.......................... 147
Škóric B.................. 149, 505
Skorokhod V 787
Slee R H 195
Slikkerveer P J................ 212
Slusarski L..................... 319
Smeeth M 21
Smith P A 233
Smith P W 173
Snaith B....................... 761
Snidle R W 645
Soejima M................... 722
Soldatovic D................ 621
Soliman F A 554
Someya T 427
Song J......................... 448

Sonntag E 684
Sonoda K................... 880, 895
Sorab J 198, 894
Soro J M 567
Sosnovskiy L A 568, 569,
.... 639,723, 809
Souchon F 171
Sperling D K 245
Spikes H A 649
Spikes H 21, 24, 28
Sraj R 468
Srivastava V K 129
Stachowiak G B.......... 236, 258
Stachowiak G W. 236, 258, 377
Stack M M 135, 215, 222
Staeves J............................ 330
Stals L 61
Stanescu C........................ 450
Stanulov K G.............. 449, 469
Stasiak J M 428
Steenberg T 123
Stefanovic M 626
Steinhilper W 571
Stephenson D J............ 19, 150
Stewart D A....................... 300
Stewart T 253
Stirling C A 294, 832
Stoilov R 747
Stolarski B......................... 543
Stolarski T A 228, 323
Stone M..... 243, 247, 731, 736,
.... 740, 741, 748
Stott F H 135
Streator J L........................ 176
Strijckmans K 83
Strong N A 518
Strzelecki S 400, 410, 427,
... 428,429, 430, 706, 810, 850,
.... 896
Stuart B H.......................... 778
Styp-Rekowski M............... 811
Subramanian M 426
Suchanek J........................ 672
Suda H............................... 235
Sugimori Y.......................... 27
Sugimura J.......... 340, 688, 724
Sui H 316
Suk M 585
Sun D C............................... 67
Sun Y 440
Sundararajan G.................. 685
Sung I H 131
Surappa R A 220
Sutton A P 169
Suzuki M 69, 862, 867
Sveklo J F........................... 598
Sviridenok A I 640
Swarnamani S 422
Swingler J........................... 40

Syniuta W D 735
Systnik S V 535
Szczerek M 464, 658
Szkurlat J 897
Szydlo Z 879, 889

T

Tai H...................................146
Takachi K 779
Takács J 154
Takadoum J.................481, 829
Takahashi S........................125
Takatsu S 686
Takeishi Y..........................454
Takekado S 172
Takeuchi K 886
Taki Y................................859
Taler J 678
Talke F E 163, 586
Tamre M 388
Tanaka A...................506, 686
Tanaka D K........................309
Tanaka M..............................7
Tanaka T...........................854
Taniguchi K 174
Taniguchi S.......................431
Tanner K E 733
Taraboanta F.....................806
Tärnfjord C 202
Taylor C M 16, 82, 414, 704
Taylor D V.........................412
Taylor R I 200
Teer D G.55, 60, 494, 495, 507
Tehnik T............................610
Temeljikovski D I 622, 627
Tewari A 130
Thew M T..........................304
Thomas J P 510
Thomas T R 718
Thomsen N B....................230
Tichy J A 25, 524
Tieu A K 420, 432, 725
Tikhonovich V I.................557
Tipper J L 736, 748
Titze M 252
Tobe S...............................323
Todorovic J 203
Todorovic P 611
Tolomelli M......................221
Tomita K 886
Tonck................................367
Torrance A A 124
Torregrossa F......................57
Tournerie B.................10, 11
Trabelsi R 240
Trethewey K 369
Tripathi B D......................673
Tritt B R............................231

Troczynski T 146
Trofimov V A..................... 653
Tsubuku T 274
Tsuchida Y 859
Tsujimura T....................... 887
Tsukamoto N 812
Tsukamoto T 827
Tsukazaki J........................... 5
Tsvetkov V I 765
Tsvetkova Y A 745
Tudor A 821
Tudor I 499
Turaga R........................... 433
Tuszynski W 464
Tuzson J 214
Tweedale P J 314
Tysoe W T........................ 122
Tyurin S A........................ 809

U

Uchiyama S 30, 466
Uemura M 570
Ueno Y 174
Ukhanska O...................... 822
Umeda K 340, 506, 686
Umehara N 148
Umemura S...................... 776
Unarski............................. 543
Unertl W N................ 315, 364
Ura A................................ 278
Usmani S 219

V

van Bormann F.................. 305
Van De Velde F............ 83, 288
van der Hoeven J M 333
Van der Voort F R 597
Varadi K............... 50, 571, 841
Vargiolu R 42, 133, 792
Varjus S....................... 60, 495
Vasiljevic B 470, 726
Vaughan H 45
Vegter H 331
Veit H............................... 257
Velimir D D J 604
Venkataraman B 685
Vercammen K 60
Verdini L 866
Verenich I A 898
Vergne P........................... 26
Vermeulen M.................... 333
Verspui M A 212
Vidakis N 157, 273
Villani S 593
Villechaise B 268
Vincent L.... 133, 241, 532, 577
Vingsbo O 310
Vinogradov A V 831

Visca E 866
Visscher H 175
Vizintin J 29, 468, 786, 820
Vo T 166
Vocel M 672
Vodopivec F 786, 820
Voevodin A 152
Vojkovic N 628
Volkov A V 451, 572
von Glasner E C 203
von Stebut J .. 58, 60, 495, 567,
.... 665
Votter M 768
Vroman C 681
Vukicevic D M 622
Vukicevic D M 627
Vuoristo P 151

W

Wahl K 73
Wakabayashi T 125
Wakuri Y 722
Walicka A 434, 435
Walicki E 434, 435
Walk S D 76
Wallaschek J 717
Wan G T Y 124
Wan Y 471
Wang B 743
Wang D 179, 181
Wang H 215
Wang L H 250
Wang P Z 508
Wang Q 659, 660
Wang T 629
Wang X X 629
Wang Z M 630
Wang Z 603
Wanstrand O 509
Warda B 810
Wasche R 687
Wasilczuk M 397
Wassink A 350
Wassink D A 538
Watanabe S 64, 899
Waterhouse R B 808
Watkins J 281
Wdowiak A 810
Wedeven L D 727
Weglowski B 678
Wei L 661, 728
Weinhauer D 528
Weisheit A 65
Weiss J 166, 579
Weissner S 586
Wellman R G 305
Wen S 477, 564, 654,
.... 780, 858
Wen X 651

Westergård R 229
Wheeler D W 216, 225
Wieland T 186
Wiercigroch M 574
Wierzcholski K 438
Wiklund U 509
Wileman J 12
Wilkes J J 136, 436
Willemse P J 196
Williams D R 189, 337
Williams J A 187
Wilson W R D 14, 437
Winer W O 370
Wirmark G 520, 696
Wisniewski M 464, 658
Wissner P 345
Wissussek D 438
Wojcicki R 706
Wolkov A V 838
Wollesen V 768
Wollfarth M 217
Wong M L 629
Wood R J K 216, 225, 369
Woodland D D 315
Woydt M 511, 641
Wright G A 68, 206
Wroblewski B M 243, 736,
.... 740, 741, 748
Wu K-R 67
Wu P Q 61
Wulczynski 658

X

Xiaoyang C 651
Xiong D 868
Xu H 662
Xu J 587
Xue Q J 140, 842

Y

Yadav S 656
Yagle A E 538
Yamada K 801
Yamaguchi E S 199
Yamaguchi Y 679
Yamamoto K.K, T 340
Yamamoto T 30, 466, 702,
.... 754, 808
Yamamoto Y 454, 688, 724
Yan F Y 842
Yan W 365
Yanagi K 581
Yanagisawa M 588, 589, 751
Yang L 688
Yao W 471
Yarosh V M 512
Yashiro H 827
Yassen S A 713, 888

Yasumoto K 64
Yasutake A 682
Ye X 471
Ye Z A 188
Yee Y 448
Yelovoy O M 723
Yeomans J A 233
Yi S 374
Yin-chao G 595
Ying T N 47
Yinong S 525, 729
Yokoi N 491, 864
Yokoi S 33, 359
Yokota H 125
Yokoyama F 472
Yoon S J 590
York D 130
Yoshida A 663, 813
Yoshii Y 74
Yoshimoto S 439
Yoshimura A 191
Yoshinaga S 881
Yoshinori S 746
Yoshizaki H 699
Yuan J H 4
Yue Q 471
Yui H 33

Z

Zabinski J S 76, 77, 152
Zachara B 879, 889
Zacharie G 665
Zahouani H .. 42, 133, 240, 292,
.... 792, 793, 861
Zaitsev A L 575
Zakharov S M 576
Zalai A 900
Zamyatin A J 642
Zamyatin J P 642
Zamyatin V J 697
Zaretsky E 277
Zaritski S P 373
Zbinden M 865
Zebala W 631
Zecchini P 829
Zeller R 689
Zhang B 440
Zhang C 664
Zhang G 478
Zhang Y 804
Zhang Z L 619
Zhang Z 664
Zhao Z 591
Zharin A L 632, 730, 794
Zharov I A 576
Zheng L 153
Zhijun Z 347, 525
Zhou Y 508

Zhou Z R 143
Zhou Z R 577
Zhu B L 143, 241
Zhu D 182
Zhu G 814
Zhu Z 440
Zhuang D M 143, 143
Ziege A 71
Zirkelback N 441
Zivanic D 762
Zlatanovic M 480, 513
Zugic R 169